View north over Hut Point Peninsula to Mount Erebus

Scott's *Discovery* and hut at Winter Quarters Bay

Future site of McMurdo Station is above opposite shore, Crater Hill on left, The Gap, and Observation Hill on right.

Emperor penguins and rookery at Cape Crozier

Beyond is the margin of frozen Ross Sea and "The Barrier", the ice-cliff terminus of the Ross Ice Shelf.

These watercolor paintings by Dee Molenaar were originally published in 1985 with his map of the McMurdo Sound area of Antarctica. We are pleased to republish these paintings with the permission of the artist who owns the copyright.

Gunter Faure · Teresa M. Mensing

The Transantarctic Mountains

Rocks, Ice, Meteorites and Water

Springer

Gunter Faure
The Ohio State University
School of Earth Sciences
and Byrd Polar Research Center
275 Mendenhall Laboratory
125 South Oval Mall
Columbus, Ohio 43210
USA
faure.1@osu.edu

Teresa M. Mensing
The Ohio State University
School of Earth Sciences
and Byrd Polar Research Center
1465 Mt. Vernon Ave.
Marion, Ohio 43302
USA
mensing.1@osu.edu

ISBN 978-1-4020-8406-5 e-ISBN 978-90-481-9390-5
DOI 10.1007/978-90-481-9390-5
Springer Dordrecht Heidelberg London New York

Library of Congress Control Number: 2010931610

Cover illustration: A tent camp in the Mesa Range of northern Victoria Land at the foot of Mt. Masley.

Printed on acid-free paper

Springer is part of Springer Science+Business Media (www.springer.com)

We dedicate this book to Lois M. Jones, Eileen McSaveny, Terry Tickhill, and Kay Lindsay who were the first team of women to conduct fieldwork in the Transantarctic Mountains during the 1969/1970 field season.

Prolog

Antarctica! The very word brings to mind images of fierce winds, bone-chilling cold, and utter desolation. Antarctica has the reputation of being a hostile place unfit for human habitation where only foolhardy explorers go to satisfy their craving for adventure. Those of us who have worked there have a very different impression of Antarctica. To us it is a place of unsurpassed beauty where the isolation from the "world" permits time for contemplation and creative thought, where loneliness is transformed into solitude, and where life depends on respectful submission to the weather. Antarctic field-geologists learn to live in harmony with the land because they have come not to test their survival skills, but to discover the geologic history of this place and to learn how geological and biological processes work in this unique environment.

Antarctica is also a continent without borders where scientists from many nations come to work in peace, constrained only by the terms of the Antarctic Treaty. In Antarctica, people of different nationalities and cultures support each other in times of need and enjoy each other's fellowship when they meet on the trail. The spirit of Antarctica has evolved because of the absence of economic competition which has been deliberately excluded by the Treaty. A similar spirit of cooperation is now manifesting itself in the exploration of the solar system because the harsh environment of space and the great effort that is required to survive there demand it. Success in the scientific exploration of Antarctica and of the solar system depends both on international cooperation and on our willingness to respect the natural environmental conditions that exist in these places.

Preface

This book presents an integrated overview of all aspects of the geology of the Transantarctic Mountains in easily readable form. The book can also be used to look up specific information about the geology of this mountain range and it records the names of many of the Earth Scientists who have contributed to the understanding we now have of the origin of the Transantarctic Mountains. In spite of the remote location of Antarctica, tens of thousands of men and women have already been there and many more will visit the continent in the future in order to participate in scientific research or to support the research programs that are undertaken by the nations that have joined the Antarctic Treaty. In addition, hundreds of tourists visit Antarctica annually in order to enjoy its natural beauty and to gain an appreciation of their own good fortune for living in the more hospitable regions of the Earth. Antarctica can teach all of us to respect the natural environment that sustains us on the Earth. This book is therefore intended for the men and women who have already visited Antarctica and for those who may visit this continent in the future in order to work there or to be inspired by its natural beauty.

We confine our attention in this book to the Transantarctic Mountains where geologists of many nations have been working since the International Geophysical Year (1957–1958) and where we have also conducted fieldwork between 1964 and 1995. The Transantarctic Mountains are unusual because, for most of their length, they consist of an uplifted and deeply dissected plateau of flat-lying sedimentary and volcanic rocks that were deposited in Phanerozoic time. These rocks are underlain by a basement consisting of a folded and variably metamorphosed volcano-sedimentary complex that was intruded by granitic rocks in the course of the Ross Orogeny during the early Paleozoic Era. Our objectives in writing this book are to summarize the relevant facts about each of the major rock units in the Transantarctic Mountains, to present the hypotheses that have been proposed to explain their origin, and to make our readers aware of issues that are still unsettled. In this way, we hope to encourage further work on geological problems and to identify areas in the Transantarctic Mountains where additional research may be needed. The information we present is derived primarily from the relevant literature supported, when appropriate, by the results of our own work and that of our students. We assume that our readers have a working knowledge of the technical aspects of Earth science and we encourage them to make up their own minds concerning the hypotheses we present.

Antarctica is important not only because of the rocks that form its crust, but also because of the large ice sheet that covers most of the continent. The glaciation of East Antarctica started during the Neogene and has formed an ice sheet that is more than

3 km thick and contains most of the world's fresh water. The stratigraphy of the ice and its isotopic composition of oxygen and hydrogen record variations of the climate extending upto 800,000 years into the past. The history of the East Antarctic ice sheet is also recorded by the geomorphology of the Transantarctic Mountains and by the deposits of till, gravel, and sand the ice sheet has left behind. The ice of the East Antarctic ice sheet adjacent to the Transantarctic Mountains does not melt, except locally on rare occasions. Instead, it ablates directly into the air. Consequently, terrestrial rock debris and extraterrestrial meteorite fragments that are transported by the ice sheet accumulate on the blue-ice surfaces of its ablation zones. Outlet glaciers, that flow from the polar plateau through valleys in the Transantarctic Mountains to the coast of Victoria Land and into the Ross Ice Shelf, descend to the low elevations of the coast where the ice does melt during the austral summer. In the dry (or ice-free) valleys of southern Victoria Land the resulting meltwater collects in lakes and ponds on the valley floors.

In spite of the harsh climate that characterizes the Transantarctic Mountains, bacteria, algae, lichens, and moss grow in sheltered places in the soil and some plants have adapted by becoming endolithic. Even mites and nematodes have been discovered in the ice-free valleys and algal mats thrive in the warm brines that occur at the bottom of the largest and deepest lakes.

These attributes of the Transantarctic Mountains are reflected by the title of this book because a complete description of this mountain range must address not only the rocks, but must also include the ice, the meteorites, and the water. The study of the Transantarctic Mountains is a multi-disciplinary enterprise including aspects of geology, glaciology, meteoritics, aqueous geochemistry, botany, and zoology.

The relevance of Antarctica to the populated areas of the Earth may become more apparent in case global warming causes the Antarctic ice sheet to start melting, thereby raising sea level and flooding coastal areas worldwide. The resulting loss of living space will require the human population to adjust on an unprecedented scale exceeding the increase of sealevel at the end of the Pleistocene Epoch when the population of the Earth was much smaller than it is today.

Acknowledgments

We are pleased to acknowledge our colleagues, former students, and friends who gave us reprints of their papers, permitted us to use their diagrams and photographs, or otherwise assisted us in the preparation of this book. Their names are recorded here in alphabetical order with our sincere thanks:

Chris Adams, Richard Alley, Ernie Angino, Rose Askin, Loren Babcock, Enriqueta Barrera, Peter Barrett, Charles Bentley, John Behrendt, Paul Berkman, Jane and Phil Boger, Scott Borg, Hal Borns, Terry Boroughs, George Botoman, Margaret Bradshaw, Henry Brecher, David Bromwich, Colin Bull, Richard Cameron, Bill Cassidy, Eric Cherry, Norbert Clauer, Jim Collinson, Charles Corbató, Gary Cotton, Cam Craddock, Bert Crarey, Ghizlaine Crozaz, Ian Dalziel, Mary Davis, Georg Delisle, George Denton, Don DePaolo, Amy Deventer, Marcia Dixon, George Doumani, Gisela Dreschhoff, Rene Eastin, David Elliot, James Elliot, Frank and Carol Faure, Bob Felder, Robert Fleck, Ken Foland, Art Ford, Trey and Sarah Fortner, N.H. Fourcade, Larry Frakes, Ralph von Frese, Harald Furnes, Harry Gair, Bill Gealey, Chris Gero, Dick Goldthwait, John Goodge, Bill Green, Ed Grew, Pieter Grootes, Anne Grunow, Bernie Gunn, John Gunner, Marta Haban, Erik Hagen, Martin Halpern, Warren Hamilton, Bill Hammer, Ralph Harvey, David Harwood, Ann Hawthorne, Knut Heier, Tim Horner, Terry Hughes, John Isbell, Ken Johnson, Ken Jezek, Lois Jones, Barbara Kaelber, Elizabeth Kibler, George Kleinschmidt, Karen Klusmeyer, Chris Koeberl, Jack Kovach, Larry Krissek, Phil Kyle, Leo Laporte and the TWiT Army, Dan Larsen, Brenda and David Lasorsa, Wesley LeMasurier, John Lindsay, Bill Long, Berry Lyons, Ursula Marvin, Paul Mayewski, M.K. McClintock, Bill McIntosh, Barry McKelvey, Garry and Dianne McKenzie, Beverly McMahon, Eileen and Maury McSaveney, John Mercer, Velon (Tex) Minshew, Art Mirsky, Michael Mohlzahn, Dee Molenaar, Joe Montello, Raymond Montigny, Ellen Mosley-Thompson, Sam Mukasa, LeeAnn Munk, John Murtaugh, Dirk Neethling, Kuni Nishiizumi, Robin Oliver, Olav Orheim, Larry Owen, Julie Palais, Robert Pankhurst, Matt Place, Mike Prentice, Doug Pride, Phil Ray, Peggy Rees, Bert Rowell, Peter Rowley, Bob Rutford, Dwight Schmidt, Jim Schopf, Emil Schulthess, Chuck Schultz, Roberta Score, Mary Siders, Andy Sipp, John Splettstoesser, John Sutter, Bernard Stonehouse, Mike Strobel, Ed Stump, Chuck Summerson, Paul Tasch, Karen Taylor, Tom and Edie Taylor, Jim Teller, Franz Tessensohn, Lonnie Thompson, Sam Treves, Fiorenso Ugolini, Chuck Vavra, Bob Walker, Peter Wasilewski, Gerald Webers, Peter Webb, Ian Whillans, J.D.L. White, Shawn Wight, Bob Wilkinson, Terry Wilson, Rebecca Witherow, John Zawiskie, Ed Zeller, and J.H. Zumberge.

Most of all we thank our mountaineering friends who guided us in the field and helped us to work safely in the Transantarctic Mountains and on the polar plateau: David Buchanan, David Reed, John W. Schutt, and Courtney Skinner.

We also owe a debt of gratitude to the administrators of the National Science Foundation of the USA for the financial and logistical support that enabled us to work in Antarctica. We thank them all but mention especially: David Bresnahan, Erick Chiang, Shaun Everett, Helen Gerasimou, Margaret Lanyon, Ann Peoples, Winnifred Reuning, Phil Smith, Mort Turner, Jack Twiss, Jill Vereyken, and Peter Wilkniss.

We also thank the staff of the Byrd Polar Research Center (formerly the Institute of Polar Studies) of The Ohio State University who supported our efforts in many ways: Peter Anderson, Kathleen Doddroe, David Lape, Arthur Mirsky, Rae Mercier, and Lynn Everett.

Our efforts to write this book were greatly facilitated by the assistance we received from Lynn Lay, librarian of the Byrd Polar Research Center, and from Mary Scott and Patty Ditto of the Orton Memorial Library of the School of Earth Sciences at The Ohio State University. We also thank Betty Heath for transforming a very messy manuscript into a neatly typed manuscript.

We gratefully acknowledge the financial support of The Ohio State University at Marion with special thanks to Dr. Greg Rose, Dean and Director of The Ohio State University at Marion. We also thank Petra van Steenbergen, Senior Publishing Editor of Springer, and her assistant, Cynthia de Jonge, for their friendly cooperation during the writing and production of this book.

Last but not least, we freely admit that all errors of omission and commission in this book are entirely our responsibility and do not reflect on the Office of Polar Programs of the National Science Foundation or on The Ohio State University.

Gunter Faure and Teresa M. Mensing

We dedicate this book to Lois M. Jones, Eileen McSaveny, Terry Tickhill, and Kay Lindsay who were the first team of women to conduct fieldwork in the Transantarctic Mountains during the 1969/1970 field season.

Prolog

Antarctica! The very word brings to mind images of fierce winds, bone-chilling cold, and utter desolation. Antarctica has the reputation of being a hostile place unfit for human habitation where only foolhardy explorers go to satisfy their craving for adventure. Those of us who have worked there have a very different impression of Antarctica. To us it is a place of unsurpassed beauty where the isolation from the "world" permits time for contemplation and creative thought, where loneliness is transformed into solitude, and where life depends on respectful submission to the weather. Antarctic field-geologists learn to live in harmony with the land because they have come not to test their survival skills, but to discover the geologic history of this place and to learn how geological and biological processes work in this unique environment.

Antarctica is also a continent without borders where scientists from many nations come to work in peace, constrained only by the terms of the Antarctic Treaty. In Antarctica, people of different nationalities and cultures support each other in times of need and enjoy each other's fellowship when they meet on the trail. The spirit of Antarctica has evolved because of the absence of economic competition which has been deliberately excluded by the Treaty. A similar spirit of cooperation is now manifesting itself in the exploration of the solar system because the harsh environment of space and the great effort that is required to survive there demand it. Success in the scientific exploration of Antarctica and of the solar system depends both on international cooperation and on our willingness to respect the natural environmental conditions that exist in these places.

Preface

This book presents an integrated overview of all aspects of the geology of the Transantarctic Mountains in easily readable form. The book can also be used to look up specific information about the geology of this mountain range and it records the names of many of the Earth Scientists who have contributed to the understanding we now have of the origin of the Transantarctic Mountains. In spite of the remote location of Antarctica, tens of thousands of men and women have already been there and many more will visit the continent in the future in order to participate in scientific research or to support the research programs that are undertaken by the nations that have joined the Antarctic Treaty. In addition, hundreds of tourists visit Antarctica annually in order to enjoy its natural beauty and to gain an appreciation of their own good fortune for living in the more hospitable regions of the Earth. Antarctica can teach all of us to respect the natural environment that sustains us on the Earth. This book is therefore intended for the men and women who have already visited Antarctica and for those who may visit this continent in the future in order to work there or to be inspired by its natural beauty.

We confine our attention in this book to the Transantarctic Mountains where geologists of many nations have been working since the International Geophysical Year (1957–1958) and where we have also conducted fieldwork between 1964 and 1995. The Transantarctic Mountains are unusual because, for most of their length, they consist of an uplifted and deeply dissected plateau of flat-lying sedimentary and volcanic rocks that were deposited in Phanerozoic time. These rocks are underlain by a basement consisting of a folded and variably metamorphosed volcano-sedimentary complex that was intruded by granitic rocks in the course of the Ross Orogeny during the early Paleozoic Era. Our objectives in writing this book are to summarize the relevant facts about each of the major rock units in the Transantarctic Mountains, to present the hypotheses that have been proposed to explain their origin, and to make our readers aware of issues that are still unsettled. In this way, we hope to encourage further work on geological problems and to identify areas in the Transantarctic Mountains where additional research may be needed. The information we present is derived primarily from the relevant literature supported, when appropriate, by the results of our own work and that of our students. We assume that our readers have a working knowledge of the technical aspects of Earth science and we encourage them to make up their own minds concerning the hypotheses we present.

Antarctica is important not only because of the rocks that form its crust, but also because of the large ice sheet that covers most of the continent. The glaciation of East Antarctica started during the Neogene and has formed an ice sheet that is more than

3 km thick and contains most of the world's fresh water. The stratigraphy of the ice and its isotopic composition of oxygen and hydrogen record variations of the climate extending upto 800,000 years into the past. The history of the East Antarctic ice sheet is also recorded by the geomorphology of the Transantarctic Mountains and by the deposits of till, gravel, and sand the ice sheet has left behind. The ice of the East Antarctic ice sheet adjacent to the Transantarctic Mountains does not melt, except locally on rare occasions. Instead, it ablates directly into the air. Consequently, terrestrial rock debris and extraterrestrial meteorite fragments that are transported by the ice sheet accumulate on the blue-ice surfaces of its ablation zones. Outlet glaciers, that flow from the polar plateau through valleys in the Transantarctic Mountains to the coast of Victoria Land and into the Ross Ice Shelf, descend to the low elevations of the coast where the ice does melt during the austral summer. In the dry (or ice-free) valleys of southern Victoria Land the resulting meltwater collects in lakes and ponds on the valley floors.

In spite of the harsh climate that characterizes the Transantarctic Mountains, bacteria, algae, lichens, and moss grow in sheltered places in the soil and some plants have adapted by becoming endolithic. Even mites and nematodes have been discovered in the ice-free valleys and algal mats thrive in the warm brines that occur at the bottom of the largest and deepest lakes.

These attributes of the Transantarctic Mountains are reflected by the title of this book because a complete description of this mountain range must address not only the rocks, but must also include the ice, the meteorites, and the water. The study of the Transantarctic Mountains is a multi-disciplinary enterprise including aspects of geology, glaciology, meteoritics, aqueous geochemistry, botany, and zoology.

The relevance of Antarctica to the populated areas of the Earth may become more apparent in case global warming causes the Antarctic ice sheet to start melting, thereby raising sea level and flooding coastal areas worldwide. The resulting loss of living space will require the human population to adjust on an unprecedented scale exceeding the increase of sealevel at the end of the Pleistocene Epoch when the population of the Earth was much smaller than it is today.

Acknowledgments

We are pleased to acknowledge our colleagues, former students, and friends who gave us reprints of their papers, permitted us to use their diagrams and photographs, or otherwise assisted us in the preparation of this book. Their names are recorded here in alphabetical order with our sincere thanks:

Chris Adams, Richard Alley, Ernie Angino, Rose Askin, Loren Babcock, Enriqueta Barrera, Peter Barrett, Charles Bentley, John Behrendt, Paul Berkman, Jane and Phil Boger, Scott Borg, Hal Borns, Terry Boroughs, George Botoman, Margaret Bradshaw, Henry Brecher, David Bromwich, Colin Bull, Richard Cameron, Bill Cassidy, Eric Cherry, Norbert Clauer, Jim Collinson, Charles Corbató, Gary Cotton, Cam Craddock, Bert Crarey, Ghizlaine Crozaz, Ian Dalziel, Mary Davis, Georg Delisle, George Denton, Don DePaolo, Amy Deventer, Marcia Dixon, George Doumani, Gisela Dreschhoff, Rene Eastin, David Elliot, James Elliot, Frank and Carol Faure, Bob Felder, Robert Fleck, Ken Foland, Art Ford, Trey and Sarah Fortner, N.H. Fourcade, Larry Frakes, Ralph von Frese, Harald Furnes, Harry Gair, Bill Gealey, Chris Gero, Dick Goldthwait, John Goodge, Bill Green, Ed Grew, Pieter Grootes, Anne Grunow, Bernie Gunn, John Gunner, Marta Haban, Erik Hagen, Martin Halpern, Warren Hamilton, Bill Hammer, Ralph Harvey, David Harwood, Ann Hawthorne, Knut Heier, Tim Horner, Terry Hughes, John Isbell, Ken Johnson, Ken Jezek, Lois Jones, Barbara Kaelber, Elizabeth Kibler, George Kleinschmidt, Karen Klusmeyer, Chris Koeberl, Jack Kovach, Larry Krissek, Phil Kyle, Leo Laporte and the TWiT Army, Dan Larsen, Brenda and David Lasorsa, Wesley LeMasurier, John Lindsay, Bill Long, Berry Lyons, Ursula Marvin, Paul Mayewski, M.K. McClintock, Bill McIntosh, Barry McKelvey, Garry and Dianne McKenzie, Beverly McMahon, Eileen and Maury McSaveney, John Mercer, Velon (Tex) Minshew, Art Mirsky, Michael Mohlzahn, Dee Molenaar, Joe Montello, Raymond Montigny, Ellen Mosley-Thompson, Sam Mukasa, LeeAnn Munk, John Murtaugh, Dirk Neethling, Kuni Nishiizumi, Robin Oliver, Olav Orheim, Larry Owen, Julie Palais, Robert Pankhurst, Matt Place, Mike Prentice, Doug Pride, Phil Ray, Peggy Rees, Bert Rowell, Peter Rowley, Bob Rutford, Dwight Schmidt, Jim Schopf, Emil Schulthess, Chuck Schultz, Roberta Score, Mary Siders, Andy Sipp, John Splettstoesser, John Sutter, Bernard Stonehouse, Mike Strobel, Ed Stump, Chuck Summerson, Paul Tasch, Karen Taylor, Tom and Edie Taylor, Jim Teller, Franz Tessensohn, Lonnie Thompson, Sam Treves, Fiorenso Ugolini, Chuck Vavra, Bob Walker, Peter Wasilewski, Gerald Webers, Peter Webb, Ian Whillans, J.D.L. White, Shawn Wight, Bob Wilkinson, Terry Wilson, Rebecca Witherow, John Zawiskie, Ed Zeller, and J.H. Zumberge.

Most of all we thank our mountaineering friends who guided us in the field and helped us to work safely in the Transantarctic Mountains and on the polar plateau: David Buchanan, David Reed, John W. Schutt, and Courtney Skinner.

We also owe a debt of gratitude to the administrators of the National Science Foundation of the USA for the financial and logistical support that enabled us to work in Antarctica. We thank them all but mention especially: David Bresnahan, Erick Chiang, Shaun Everett, Helen Gerasimou, Margaret Lanyon, Ann Peoples, Winnifred Reuning, Phil Smith, Mort Turner, Jack Twiss, Jill Vereyken, and Peter Wilkniss.

We also thank the staff of the Byrd Polar Research Center (formerly the Institute of Polar Studies) of The Ohio State University who supported our efforts in many ways: Peter Anderson, Kathleen Doddroe, David Lape, Arthur Mirsky, Rae Mercier, and Lynn Everett.

Our efforts to write this book were greatly facilitated by the assistance we received from Lynn Lay, librarian of the Byrd Polar Research Center, and from Mary Scott and Patty Ditto of the Orton Memorial Library of the School of Earth Sciences at The Ohio State University. We also thank Betty Heath for transforming a very messy manuscript into a neatly typed manuscript.

We gratefully acknowledge the financial support of The Ohio State University at Marion with special thanks to Dr. Greg Rose, Dean and Director of The Ohio State University at Marion. We also thank Petra van Steenbergen, Senior Publishing Editor of Springer, and her assistant, Cynthia de Jonge, for their friendly cooperation during the writing and production of this book.

Last but not least, we freely admit that all errors of omission and commission in this book are entirely our responsibility and do not reflect on the Office of Polar Programs of the National Science Foundation or on The Ohio State University.

Gunter Faure and Teresa M. Mensing

Contents

Part IV Fire and Ice

Part I
Exploration and Characterization

Chapter 1
The Exploration of Antarctica

In spite of its remote location and its cold and stormy weather, Antarctica has attracted seal hunters, whalers, and scientists all of whom have contributed to the exploration of this continent. The names of these explorers are preserved by many of the topographic and geologic features of Antarctica, as well as by the names of its coastal areas and surrounding oceans. Those who now wish to visit Antarctica can do so in steel-hulled ships or in large transport planes that land on the sea ice or even on the polar plateau. The research stations that are maintained in Antarctica by several nations offer most of the comforts of home. Even scientists who work in the mountains and on the polar plateau are transported into the field by helicopters or fixed-wing aircraft; and they use snowmobiles rather than dogsleds to travel across the polar plateau. In spite of these labor-saving accommodations and devices, the weather in Antarctica has not changed since the epic journeys of Robert Scott, Ernest Shackleteon, and Roald Amundsen. The sign at Williams Field adjacent to Ross Island welcomes travelers with the Maori phrase: "Haere Mai" (Welcome). It could also read: "Enter at your own risk." Antarctica can be a cruel place. Those of us who now work in Antarctica should be aware of the heroic struggle and tragic fate of some of those who preceded us.

1.1 Brave Men in Wooden Sailing Ships

Captain James Cook was the first person to circumnavigate Antarctica between 1772 and 1775 during his three voyages of exploration into the Pacific Ocean (Beaglehole 1955, 1961, 1962, 1967; Bellec 2000). On January 30, 1774, Captain Cook reached a point near the coast of West Antarctica at 71°10'S, 106°54'W. These voyages exposed the crews of his ships to frightful conditions caused by severe cold, violent storms, and dangerous pack ice along the coast. As a result of these hardships and based on what he had seen, Captain Cook reported that any land that may exist close to the geographic south pole was not worth the effort to explore (Stonehouse 2002). In the years that followed Captain Cook's voyages, the ocean around Antarctica was frequented by a multitude of European and American whalers and seal hunters who were primarily interested in making a profit from the sale of whale oil and seal skins (Stonehouse 2002; Victor 1964).

The scientific exploration of Antarctica started by Captain Cook continued 45 years later with the Russian Naval Expedition (1819–1821) led by **Fabian Gottlieb Benjamin von Bellingshausen** (1778–1852) who was a Baltic nobleman of German descent from the island of Saaremaa (Oesel) in Estonia which, at that time, was a province of Tsarist Russia (Debenham 1945). As a midshipman in the Russian Imperial Navy, Bellingshausen had previously participated in the Russian around-the-world expedition (1803–1806) that was commanded by Admiral Johann von Krusenstern (1770–1846) who was also a Baltic nobleman from Estonia. Bellingshausen in Fig. 1.1 circumnavigated Antarctica with two ships named Vostok and Mirnyy and his closest approach to the mainland of Antarctica was 69°S. Bellingshausen acted on orders of the Tsar of Russia who wanted to demonstrate Russia's naval power while at the same time contributing to the exploration Antarctica which had been officially recognized as a landmass at the bottom of the Earth.

During his return voyage, Bellingshausen sighted the ship of the American seal hunter, **Nathaniel Palmer** (1799–1877) off the coast of Deception Island (62°57'S, 060°38'W). The two explorers met on this

G. Faure and T.M. Mensing, *The Transantarctic Mountains: Rocks, Ice, Meteorites and Water*,
DOI 10.1007/978-90-481-9390-5_1,

Fig. 1.1 Fabian Gottlieb von Bellingshausen (1778–1852) was an Estonian nobleman of German descent who became a highly regarded officer in the navy of the Tsar of Russia. For that reason, he was ordered to undertake a voyage of discovery to the Southern Ocean with two wooden sailing ships named Vostok (commanded by Bellingshausen) and Mirnyy (commanded by Lazarev). Both ships successfully circumnavigated Antarctica between 1819 and 1821 and returned safely to their home port. This voyage confirmed that Antarctica was a large continent surrounded by the Southern Ocean (Photo by the Scott Polar Research Institute, University of Cambridge, used here with permission)

Fig. 1.2 James Weddell (1787–1834) and Matthew Brisbane were hunting seals in the South Shetland and South Orkney Islands in 1823 when they turned their ships south and sailed into a large embayment in the land mass of Antarctica now known as the Weddell Sea. They reached a point even farther south than Captain Cook when they decided to turn back. They were fortunate to have penetrated this far into the Weddell Sea and to have escaped unhurt because the pack ice in the Weddell Sea is notorious for trapping ships (Photo by the Scott Polar Research Institute, University of Cambridge, used here with permission)

occasion and exchanged information about their findings. The exploits of the Russian explorer are recorded by the names of the Bellingshausen Sea and Bellingshausen Island (59°25'S, 027°03'W), whereas the memory of the American seal hunter and navigator is preserved in the Palmer Archipelago (64°15'S, 062°50'W) and in Palmer Land (71°30'S, 065°00'W) which is the southern part of the Antarctic Peninsula. In addition, the USA presently maintains a scientific research station called Palmer Station on Anvers Island (64°33'S, 063°035'W) off the west coast of the Antarctic Peninsula.

Even as Bellingshausen was returning from his expedition to Antarctica, a British seal hunter of Scottish ancestry by the name of **James Weddell** (1787–1834) entered upon the stage of Antarctic exploration. Weddell in Fig. 1.2 retired from the British Navy in 1819 and subsequently made a name for himself as a successful seal hunter and expert navigator working in the South Shetland (62°00'S, 058°00'W) and South Orkney islands (60°35'S, 045°30'W) of the southern Atlantic Ocean. In 1823, James Weddell, accompanied by **Captain Matthew Brisbane** in a second sailing ship, turned south from the South Orkney Islands and sailed into a large embayment in the Antarctic mainland now known as the Weddell Sea. On February 20 of 1823 the ships reached their most southern position at 74°15'S, 034°015'45"W, which was 214 nautical miles farther south than the position achieved previously by James Cook. The Weddell Sea is notorious for the slowly rotating pack ice which trapped and, in some cases, destroyed the ships of subsequent expeditions to the area (e.g., Shackleton's ship the Endurance, 1914–1917).

The next actor to appear on the Antarctic stage was the French nobleman and navy officer **Jules Sébastien César Dumont d'Urville** (1790–1842) who was in charge of the French Naval Expedition to Antarctica in 1837–1840 (Bellec 2000). His plan was to follow the course of James Weddell and to attempt to reach a point even farther south than Weddell or anyone else had achieved. He also hoped to find the magnetic pole in Antarctica. After wintering in Indonesia and taking on provisions in Hobart, Tasmania, d'Urville in Fig. 1.3 sailed south-east in January of 1840 and, although he did not succeed in finding the magnetic pole, he did claim for France a segment of the coast of East Antarctica, which he named the Adélie Coast (67°00'S, 139°00'E) in honor of his wife.

On the return trip to France, d'Urville was surprised to encounter an American warship that was part of a fleet commanded by **Lieutenant Charles Wilkes** (1798–1877) who was leading the United States Exploring Expedition from 1838 to 1842 to the southern ocean. On December 26 of 1839 four ships in Wilkes' fleet headed south toward the coast of Antarctica. Wilkes in Fig. 1.4 sailed along the coast of East Antarctica including the part that is now called Wilkes Land (69°00'S, 120°00'E). Only two of the six ships that started this voyage returned to New York in August of 1842.

Fig. 1.3 Jules C. Dumont d'Urville (1790–1842) was a French naval officer and navigator who led the French Naval Expedition to Antarctica from 1837 to 1840. He was an experienced explorer who had worked in the eastern Mediterranean Sea followed by two expeditions to map the coasts of Australia and New Zealand and to map variations of the magnetic field in the South Pacific and the Southern Ocean. He used two ships (Astrolabe and Zéleé) in an attempt to find the magnetic pole in Antarctica. Although he failed to achieve that objective, he named Adélie Land along the coast of East Antarctica after his wife and discovered two large islands off the coast of the Antarctic Peninsula (Photo by the Scott Polar Research Institute, University of Cambridge, used here with permission)

Fig. 1.4 Lieutenant Charles Wilkes (1798–1877) of the US Navy commanded the United States Exploring Expedition from 1838 to 1842. He was an outstanding navigator and hydrographer and was appointed to be the Director of the Navy's Depot of Charts and Instruments. In 1838 he was put in charge of a fleet of six ships that were to sail around the world with side trips into the Southern Ocean for the purposes of geographic exploration and hydrographic research. During the summer of 1840 he sailed close to the coast of East Antarctica where he repeatedly sighted land although the weather was cold and stormy (Photo by the Scott Polar Research Institute, University of Cambridge, used here with permission)

On the day when the sailing ship Astrolabe of Dumont d'Urville encountered the Porpoise of Wilkes' fleet on the high sea off the coast of East Antarctica, the crews of both ships hoisted full complements of sails and passed without saluting each other in contrast to Bellingshausen and Palmer who preceded them as Antarctic explorers.

The claim to Adélie Land which Dumont d'Urville had discovered was not recognized until 1926 because it was not clear whether d'Urville or Wilkes had actually seen it first. Finally, on January 20 of 1950, a group of scientists of the French Polar Expedition landed on the Adélie coast to set up a base which was named Dumont d'Urville in honor of the famous navigator and explorer (Stonehouse 2002).

1.2 Search for the Magnetic Pole in Antarctica

The magnetic field of the Earth and the location of the magnetic poles were of great interest to the seafaring nations of the nineteenth century because the magnetic compass was used to steer ships across the oceans that contained few points of reference. For this reason, several Antarctic explorers of the nineteenth century attempted to map the magnetic field of the Earth and to determine the position of the magnetic pole in Antarctica (e.g., Dumont d'Urville and Charles Wilkes).

The magnetic field of the Earth was also much on the mind of **James Clark Ross** (1800–1862) a British naval officer who had participated in and led several exploratory voyages and dogsled expeditions in the Arctic between 1818 and 1831 searching for the elusive Northwest Passage from the Atlantic to the Pacific Ocean. On May 31, 1831, James Ross in Fig. 1.5 reached the magnetic pole in the Arctic at 70°05'N and 096°46'W. In 1904 Roald Amundsen relocated the pole at 70°30'N and 095°30'W. Still later, in 1947, a Canadian scientific expedition placed the magnetic pole at a spot north of Prince of Wales Island at 73°N and 100°W. The present position of the pole is at 77°18'N and 101°48'W (Emiliani 1992, p. 228).

Evidently, the magnetic pole of the Arctic moves and is not coincident with the geographic North Pole of the Earth. These observations are one of the points of departure for the on-going research to understand how the magnetic field of the Earth is generated, why the magnetic poles move, and why the magnetic field has spontaneously reversed its polarity many times during the course of geologic time.

Fig. 1.5 Captain James C. Ross of the Royal Navy of Great Britain together with Captain F.M. R. Crozier was one of the most successful early Antarctic explorers between 1839 and 1843. The two British Navy Captains discovered the Ross Ice Shelf and sailed along it to the coast of Victoria Land. They also discovered Ross Island and named its two principal volcanoes after their two ships: Erebus and Terror (Photo by the Scott Polar Research Institute, University of Cambridge, used here with permission)

In 1839, Captain James Ross was appointed to lead the British Naval Expedition (1839–1843) to Antarctica. His assignment was to penetrate the pack ice with his two small ships, HMS Erebus and HMS Terror, and to reach the magnetic pole. Ross and his associate, **F.M.R. Crozier** who commanded HMS Terror, set sail on September 30, 1839 and arrived in Hobart, Tasmania in early 1840. After visiting Sir John Franklin, who was the Governor of Tasmania, Ross and Crozier started their voyage to Antarctica on November 13, 1840. Their ships were well stocked with provisions and were staffed with volunteers on

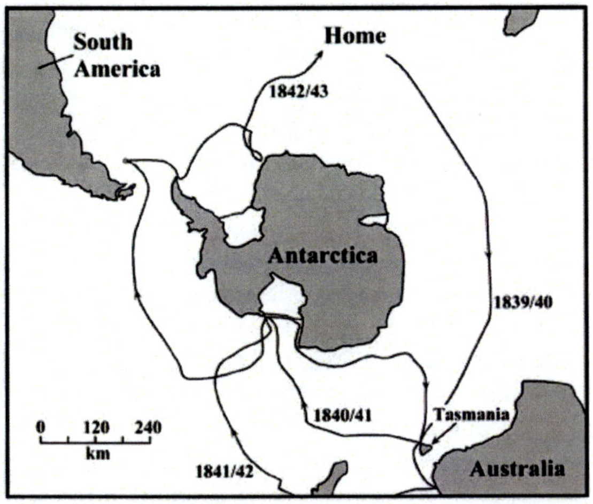

Fig. 1.6 The Antarctic voyage of Ross and Crozier started in 1840 from Hobart, Tasmania. After their return to Hobart, they moved on to Sydney, Australia. From there they set out again to the Ross Sea in 1841 and then headed for home via the Falkland Islands with two more close approaches to the northern tip of the Antarctic Peninsula and to the Princess Martha Coast of Queen Maud Land in East Antarctica (Adapted from Stonehouse 2002, p. 215)

double pay. Ross and Crozier in Fig. 1.6 were able to break through the pack ice which typically surrounds the continent in the early spring of the southern hemisphere and sailed into open water now called the Ross Sea. During this voyage, Ross and Crozier discovered the Ross Ice Shelf and sailed along its high cliff until they encountered the coast of Victoria Land, which Captain Ross claimed for England and named in honor of Queen Victoria.

On January 24, 1841, these explorers also discovered an island at 78°S and 168°E located off the coast of Victoria Land. The latitude of this island exceeded the most southern position reached by James Weddell, which Ross and Crozier celebrated by issuing double rations of rum to their crews. Ross named the two volcanoes on the island after his ships Erebus and Terror and the island itself is known to us as Ross Island. Captain Crozier's name identifies a prominent peninsula on Ross Island where the famous rookery of Emperor Penguins is located.

Ross and Crozier attempted to find the magnetic pole in Antarctica based on a continuous series of measurements they had made of the magnetic field of the Earth ever since they left London. Ross was determined to plant the same flag on the magnetic pole of Antarctica that he had hoisted over the magnetic pole in the Arctic. He was unable to achieve that goal only because the magnetic pole at that time was located behind the mountains of Victoria Land and was not reachable with sailing ships.

After wintering at Hobart, Ross and Crozier sailed south again in December of 1841 and discovered the Bay of Whales along the edge of the Ross Ice Shelf. In 1910 Roald Amundsen set up his base camp called Framheim at this site from where he reached the geographic South Pole in 1911. Ross and Crozier nearly perished when their ships collided in a violent storm among the icebergs of the Ross Sea. They managed to untangle their ships and sailed north to the Falkland Islands. From there they crossed the Weddell Sea with two close approaches to the Antarctic Peninsula and the coast of East Antarctica (Fig. 1.6).

Ross returned to England in 1843 and was knighted by Queen Victoria. Although he had not reached the magnetic pole of Antarctica, his discoveries opened the way for the subsequent exploration of the interior of Antarctica by Robert F. Scott, Ernest Shackleton, Roald Amundsen, and Richard E. Byrd. After publishing a report of his Antarctic expedition (Ross 1847), Ross undertook a final expedition into the Canadian Arctic in 1848–1849 to search in vain for survivors of Sir John Franklin's expedition who had used his two Antarctic ships Erebus and Terror in a failed attempt to find the Northwest Passage. All 129 officers and enlisted men under Franklin's command were later found to have died primarily of lead poisoning (Faure 1998).

The magnetic pole in Antarctica was eventually discovered in 1909 by T.W. Edgeworth David at 72°025'S and 115°16'E in the interior of the continent behind the mountains of Victoria Land. David was a member of the so-called Nimrod Expedition of 1907–1909 that was led by Ernest Shackleton (1874–1922). Since its discovery in 1909, the magnetic pole in Antarctica has moved 900 km to its present position at 65°S and 139°E in the Indian Ocean off the Adélie Coast of East Antarctica (Stonehouse 2002, p. 169). However, nearly 50 years elapsed before the map of the magnetic field in Antarctica was completed during the International Geophysical Year that lasted from 1957–1958 (Victor 1964).

In retrospect, we see that Bellingshausen's circumnavigation of Antarctica between 1819 and 1821 on behalf of the Tsar of Russia may have motivated a series of national expeditions by several countries, including France (1838–1842), USA (1838–1842), Great Britain (1839–1843), Belgium (1897–1899), Germany (1902–1903), and Sweden (1902–1904). All of these early expeditions attempted to explore Antarctica and to claim parts of its territory from the decks of wooden sailing ships. Apparently, the exploration of Antarctica that was initiated by Russia with Bellingshausen's voyage became a priority for other nations.

1.3 Surviving the Antarctic Winter

All of the scientific expeditions prior to 1898 returned to Australia or New Zealand for the winter because the wooden sailing ships they used could not withstand the harsh winter weather and the pressure of the sea ice that forms annually around the coast of Antarctica. The first group of scientists to spend the winter in Antarctica was the Belgian Antarctic Expedition of 1897–1899 in the ship Belgica commanded by **Lieutenant Adrien de Gerlache de Gomery** (1866–1934) of the Belgian Navy. The first mate of the predominantly Norwegian crew was Roald Amundsen and the ship's physician was Dr. Frederick Cook who later claimed to have reached the geographic North Pole. On March 3, 1898, the Belgica became trapped in the pack ice of the Ross Sea west of Alexander Island at 71°30'S near the base of the Antarctica Peninsula and remained stuck for 13 months while it drifted more than 1,000 km with the ice. The scientists on board continued to make observations throughout the long Antarctic winter. The ship finally broke free of the ice on March 14, 1899 and returned home. Amundsen and Cook, both of whom had extensive prior experience in the Arctic, contributed significantly to the ultimate success of the Belgica expedition (Cook 1909).

The second group to winter in Antarctica was led by **Carsten Borchgrevink** (1864–1934) who organized the privately financed British Antarctic Expedition of 1898–1900. Borchgrevink was born in Oslo, Norway, but emigrated to Australia in 1888 and became a school teacher. In 1894 he joined the Norwegian Whaling Expedition of 1893–1895 when it took on supplies in Melbourne. The ship was unable to penetrate the pack ice in the Ross Sea and landed at Cape Adare in northern Victoria Land. Borchgrevink who was neither a sailor nor an explorer decided to return to Cape Adare and eventually obtained financing from a British newspaper man, which explains why this Norwegian-Australian adventurer became the leader of a British Antarctic expedition. He and his team returned to Cape Adare on February 17, 1899, aboard the sailing ship Southern Cross. Borchgrevink intended to penetrate into the interior with dog teams, but the terrain did not permit that. The group survived the Antarctic winter in two prefabricated wooden huts they assembled on the beach at Cape Adare. When their ship returned in January of the following year, Borchgrevink sailed south along the coast of Victoria Land to Ross Island and then followed the edge of the Ross Ice Shelf, which had receded about 48 km since James Ross had been there in 1840. Consequently, Borchgrevink and his men set a new record by reaching the southern latitude of 78°21'S and subsequently improved on that record when they traveled to 78°50'S during a short excursion on the Ross Ice Shelf. The coast of Northern Victoria Land where Borchgrevink and his men wintered over is now called the Borchgrevink Coast.

The fate of the Belgica Expedition was later shared by the German South Polar expedition of 1901–1903 led by **Erich von Drygalski** (1865–1949). His ship, the Gauss, was also trapped in the ice for 1 year from February 1902 to February 1903 about 95 km from the Kaiser Wilhelm II Coast (67°00'S, 090°00'E) of East Antarctica. The scientists on board had only limited opportunities to pursue their scientific objectives except the geologists Philippi (1912a, b) and Rheinisch (1912) who described the alkali-rich lavas of the Gaussberg (66°48'S, 089°12'E), an isolated extinct volcano on the coast of East Antarctica (Drygalski 1912; Sheraton and Cundari 1980; Duncan 1981; Collerson and McCulloch 1983; Tingey et al. 1983; Faure 2001; Lűdecke 2006). The members of the expedition also deployed a captive balloon in order to get a wider view of their surroundings and sent sledging parties to explore the Kaiser Wilhelm II Coast. After Drygalski returned to Germany, his ship was sold to Canada.

The Swedish South Polar Expedition of 1901–1904 led by **Otto von Nordenskjőld** (1869–1928) was forced to spend two winters on Snow Hill Island identified in Fig. 1.7 at 64°28'S and 057°12'W off the east coast of the Antarctic Peninsula. The expedition tried to reach the King Oscar II Coast (65°45'S, 062°30'W) during the austral summer of 1902. However, when dense pack ice prevented the ship from entering the Weddell Sea, Nordenskjőld chose to winter over in a small comfortable hut they erected on Snow Hill Island while his ship, the Antarctic, commanded by Captain C.A. Larsen, headed north to the Falkland Islands. In October of 1902, Nordenskjőld and two of his five companions used dogs to travel south on the Larsen Ice Shelf (discovered by C.A. Larsen in 1893) to a small nunatak at 66°03'S, 062°30'W, which Nordenskjőld named Borchgrevink Nunatak after the Norwegian-Australian-British explorer. In December of 1902 Nordenskjőld and his team explored Seymour Island (64°17'S, 056°45'W) in Fig. 1.7 where they found fossil plants and the bones of large penguins both of which indicated that the climate of this area was warmer in the past than it is at present. However, when their ship tried to pick them up from Snow Hill Island in November of 1902, it was forced to turn back by heavy pack ice. During a second try to reach Snow Hill Island, the Antarctic was trapped by the pack ice and was crushed about 40 km east of Paulet Island which is located off the east coast of Dundee Island (63°30'S, 055°55'W) identified in Fig. 1.7. The crew of 22 men abandoned the ship on February 14, 1903, and walked to Paulet Island where they spent a crowded and uncomfortable winter in a small stone hut they constructed.

Three men, who had previously gone ashore at Hope Bay before the ship sank, tried to walk to Snow Hill Island located about 100 km southeast of their position "as the crow flies." When they realized that they could not reach their destination, they returned to Hope Bay and expected the ship to pick them up. When the ship did not return (because it had sunk), they fortified their tent with a stone wall in Fig. 1.8 and survived the austral winter of 1903 eating penguins and seals. On September 29 of 1903 the three men at Hope Bay tried again to walk to Snow Hill Island and on October 9 accidentally met Nordenskjőld who was exploring Vega Island with two companions and a dog team. After the unexpected reunion, Nordenskjőld guided the castaways to his winter camp on Snow Hill Island. On October 31, 1903, when the ice had begun to break

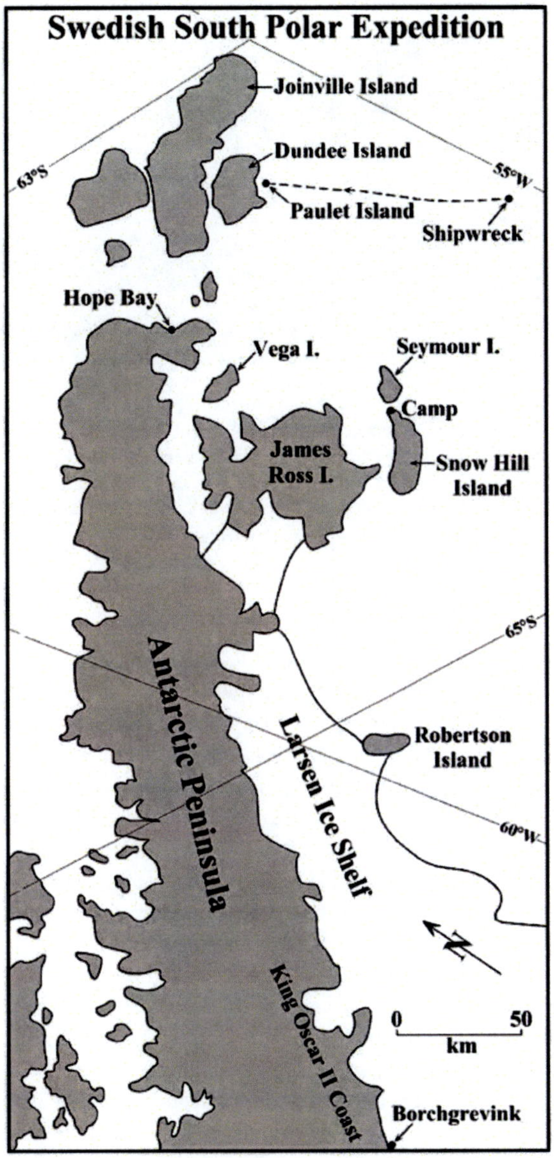

Fig. 1.7 The Swedish South Polar Expedition (1901–1904) was led by Otto Nordenskjőld who set up a winter-over hut on the northern tip of Snow Hill Island close to Seymour Island where Captain C.A. Larsen had previously found interesting fossils. During October of 1902, Nordenskjőld and two companions used dog teams to travel south across the Larsen Ice Shelf to a nunatak on the King Oscar II Coast which they named Borchgrevink. When their ship attempted to pick them up at Snow Hill Island, it was crushed by the pack ice and sank. Captain Larsen and the crew escaped to Paulet Island and eventually rejoined Nordenskjőld. In the end, everyone was rescued by the Argentine navy (Adapted from Stonehouse 2002, p. 259)

Fig. 1.8 This is the stone hut at Hope Bay on the northern tip of the Antarctic Peninsula in which three members of Otto Nordenskjőld's Swedish South Polar Expedition (1901–1904) survived the winter of 1903 after their ship "Antarctic" was crushed by pack ice and sank on February 14 of that same year (Photo by Bernard Stonehouse, published with his permission)

up, Captain C.A. Larsen and five seamen rowed an open boat from Paulet Island to Hope Bay in order to rescue the three men he had put ashore there. However, when they arrived at Hope Bay on November 4, they found a note indicating that the men were walking south to Snow Hill Island. Therefore, Larsen and his men continued rowing south for three more days until they reached the fast ice around Snow Hill Island. They left the boat at the edge of the fast ice and walked 24 km to Nordenskjőld's camp where they were later picked up with Nordenskjőld's party by the Uruguay, a ship of the Argentine Navy commanded by Captain J. Irizar. The Uruguay went on to rescue the remaining survivors on Paulet Island and returned the entire Nordenskjőld expedition to Buenos Aires. In spite of the horrendous ordeal endured by Nordenskjőld and his crew, the expedition lost only one man who died of natural causes on Paulet Island (Nordenskjőld et al. 1905).

When it became apparent that several European nations were actively exploring Antarctica, a French medical doctor by the name of **Jean-Baptiste Etiene Auguste Charcot** (1867–1936) decided that France should not be excluded from this enterprise. Therefore, he raised funds for the construction of a ship he named Le Francais. In August 1903 he set sail for the Antarctic Peninsula and proceeded to map its west coast and off-shore islands. Contrary to most of his predecessors, Charcot decided to winter over in a sheltered bay on Booth Island (65°05'S, 064°00'W), where he and his group of scientists made observations and collected specimens. The ship broke free from the ice in December of 1904, which allowed Charcot to continue to explore the west coast of the Antarctic Peninsula. On February 15 of 1905 his ship struck a submerged rock and sustained significant damage. After making temporary repairs, Charcot returned safely to Buenos Aires where he sold his ship.

The limited success of the Francais Expedition caused Charcot to commission a larger and better equipped ship he named the Pourquoi Pas? ("Why Not?"). He left France on August 15 of 1908 in his new ship and stopped on Deception Island where he was pleased to note that the captains of whaling ships and seal hunters were using the maps he had made during his first expedition. He continued mapping the west coast of the Antarctic Peninsula and wintered at Port Circumcision on Petermann Island (65°10'S, 064°10'W). During the following year, Charcot continued his exploration by mapping the coast even though his new ship was also damaged by running aground. At the end of a productive voyage of exploration during the summer of 1910, Charcot returned to France where he received a warm welcome as a successful and unusually sensitive polar explorer.

Victor (1964, p. 183) reported that Charcot disapproved of the French habit of drinking wine with every meal, partly because the wine barrels took up too much space on his ship. Therefore, he compensated his fellow scientists on board by stocking the ship's library with classical books by Virgil, Dante, Cervantes, Swift, Saint-Simon, Hugo, and Michelet. Charcot later noted with pleasure that his colleagues learned to enjoy these works during the long voyage. Charcot and most of his crew died in 1936 when his ship foundered off the coast of Iceland (Stonehouse 2002).

1.4 The Race to the Geographic South Pole

At the start of the twentieth century the exploration of Antarctica had progressed from observations of the coast to efforts to penetrate into the interior of the continent. More specifically, Robert F. Scott, Ernest H. Shackleton, and Roald Amundsen competed to become the first humans to reach the geographic South Pole.

1.4.1 *Scott's First (Discovery) Expedition, 1901–1904*

The exploration of the interior of Antarctica and the race to the geographic pole started with the British National Antarctic Expedition of 1901–1904. It was jointly sponsored by the British government and by the major British scientific societies of the time and was led by **Commander Robert F. Scott** (1868–1912) of the British Navy (Huntford 1986a). The objective of this expedition was to explore the Ross Sea sector of Antarctica. The expedition was given a new ship christened the RRS Discovery which was designed and built for exploration in Antarctic waters. The third officer of the crew was **Ernest Henry Shackleton** (1874–1922) who was entrusted with the job of preparing the Discovery for a 3-year voyage to Antarctica (Fig. 1.9a and b).

The ship was equipped with two steam engines and carried 335 t of coal. The expedition left England in August of 1901 and reached Lyttelton Harbor on the South Island of New Zealand at the end of November of the same year. After restocking the ship with provisions, the expedition headed south to Cape Adare and followed the coast of Victoria Land to Ross Island. From there, Scott sailed east along the edge of the Ross Ice Shelf to the coast of West Antarctica which he named King Edward VII Land.

Scott had brought along a captive helium balloon in order to view the terrain in Antarctica from a high elevation. He deployed this balloon on February 4, 1902 at the Bay of Whales and took turns with Shackleton to observe and photograph the surface of the Ross Ice Shelf from a height of 790 ft or 241 m (Victor 1964, p. 178). Scott's ascent in a tethered balloon came only

Fig. 1.9(a) Robert Falcon Scott (1868–1912) led two expeditions to Antarctica in 1901–1904 and 1910–1913. Although he reached the geographic South Pole in 1912, he and his men perished on the return trip to his base on Ross Island (Photo by the Scott Polar Research Institute, University of Cambridge, used here with permission) **(b)** Ernest Henry Shackleton (1874–1922) also led two expeditions to Antarctica in 1907–1909 and 1914–1917. Although he did not achieve all of his objectives, all of the men under his direct command survived great hardships. Shackleton died in 1921 on South Georgia Island while on his third Antarctic expedition and was buried in the whalers' cemetery at Grytviken (Photo by the Scott Polar Research Institute, University of Cambridge, used here with permission)

Fig. 1.10 The hut that Robert Scott and his men built in 1902 still stands at Hut Point where their ship, Discovery, was moored for two winters. The hut was not well-enough insulated to withstand the cold which forced the crew to live on the Discovery. Observation Hill looms in the background. The hut is on the list of Historic Sites and Monuments (HSMs) and is being curated by New Zealanders stationed at nearby Scott Base. It still contains some of the provisions and utensils used by Scott and his men as well as frozen carcasses of Weddell seals that were fed to the sled-dogs (Photo by G. Faure)

53 days after a similar ascent by Drygalski along Kaiser Wilhelm II Coast of East Antarctica. The ship then returned to Ross Island where Scott and his men erected a wooden hut that had been built for use on a ranch in the outback of Australia.

This hut in Fig. 1.10 still stands at the tip of Hut Point Peninsula about 1 km from McMurdo Station of the USA at 77°51'S, 166°40'E. This station was initially established in 1955 by Rear Admiral George J. Dufek of the US Navy and was named after Lieutenant A. McMurdo who served with Captain Crozier on the HMS Terror.

Scott's expedition spent the austral winter on the Discovery, which was frozen into the ice next to the hut, because the hut was not sufficiently insulated to be habitable during the Antarctic winter. During the following spring, Robert Scott, Edward Wilson, and Ernest Shackleton made a long trip across the Ross Ice Shelf and came close to the mouth of the Beardmore Glacier in Fig. 1.11. All of the sled dogs died during this trip or were killed and the men had to resort to man-hauling in spite of snow blindness and scurvy. Shackleton became so exhausted that he was unable to pull the sled and could barely walk. The group finally returned to the Discovery Hut on February 3, 1903 in a state of complete exhaustion resulting from inexperience in polar travel, insufficient food, and bad planning (Stonehouse 2002, p. 232). Shackleton was sick for a month after this trip and was sent home on the relief ship on orders from Scott.

While Scott and his companions were exploring a route to the geographic South Pole, a second group of Scott's men, led by Albert Armitage, crossed McMurdo Sound on November 29, 1902 and explored the mountains of Victoria Land in Fig. 1.12. They discovered a large glacier which they named after the geologist Hartley Ferrar and used it to reach the polar plateau before returning to their base camp on Ross Island on January 19, 1903.

About 1 week later, the relief ship Morning arrived, but was unable to free the Discovery. Therefore, the

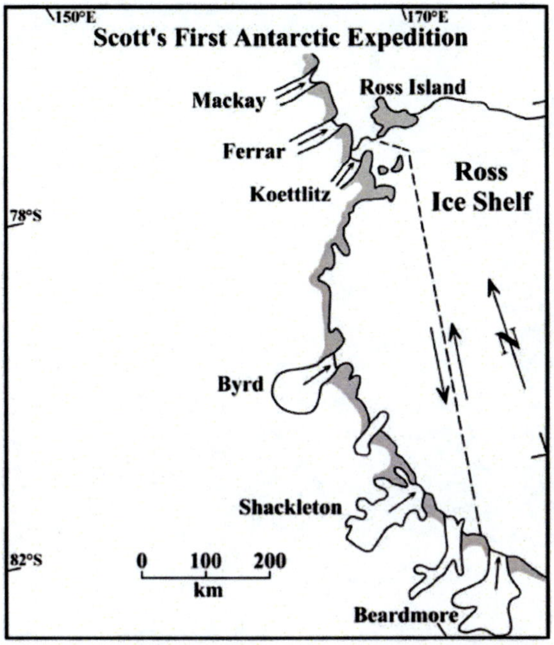

Fig. 1.11 Robert Scott, Edward Wilson, and Ernest Shackleton left their base at Hut Point on November 2 of 1902 and traveled across the Ross Ice Shelf to a point near the mouth of the Beardmore Glacier at 82°16'S. They attempted to use dogs but ended up man-hauling their sleds because their dogs died or were killed. They returned on February 3 of 1903 after 93 days on the trail during which they traveled 1,320 km for a daily average of about 14 km per day. All three suffered from snow blindness, scurvy, and exhaustion. This trip opened the way for Scott's quest to reach the geographic South Pole during his second expedition from 1910 to 1913 (Adapted from Stonehouse 2002, p. 46)

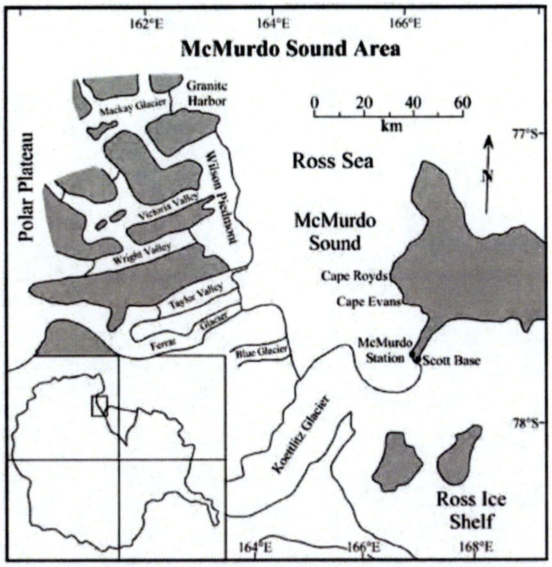

Fig. 1.12 While Scott and his companions were trekking across the Ross Ice Shelf to the mouth of the Beardmore Glacier, a second group led by the navigator and surveyor Albert Armitage crossed McMurdo Sound on November 29 of 1902 and explored the mountains of southern Victoria Land. They were the first humans to enter the ice-free valleys and discovered the Ferrar Glacier which they named after the geologist Hartley Ferrar. They surveyed the glacier and ascended it up to an elevation of 2,750 m above sea level thereby pioneering the route that Scott used in the following year (1903) to reach the polar plateau across which he traveled for about 450 km before turning back (Adapted from Allen 1962)

relief supplies were transferred from the Morning to the Discovery before the Morning departed in early March with several members of the expedition on board, including Ernest Shackleton. The remainder of Scott's group spent a second winter on the ice-bound Discovery at Hut Point.

Scott later led a nine-man party west across McMurdo Sound to continue the exploration of Victoria Land. They reached the polar plateau and traveled across its empty vastness for 450 km before returning to Hut Point in time for Christmas of 1903. About 2 weeks later, two relief ships, the Morning and the Terra Nova, appeared at the ice edge some 32 km from Hut Point. Fortunately, the sea ice in McMurdo Sound began to break up in early February and all three ships were able to sail north together.

1.4.2 Shackleton's First (Nimrod) Expedition, 1907–1909

After Shackleton had returned from the British National Antarctic Expedition (1901–1904) led by Robert Scott, he decided to organize his own expedition that would attempt to reach both the geographic and the magnetic poles in Antarctica and to explore King Edward VII Land in West Antarctica. The resulting British Antarctic (Nimrod) Expedition (1907–1909) left London at the end of July of 1907. The Nimrod, an old wooden sealing ship, was heavily loaded with parts for a small hut, a motor-driven car, nine dogs, ten ponies, and a crew and staff of 39 persons (Huntford 1986b). The race for the South Pole had begun!

On January 23 of 1908 the Nimrod reached the edge of the Ross Ice Shelf where Shackleton intended to set up his base camp in a deep embayment he had discovered

Fig. 1.13 Shackleton's hut at Cape Royds on Ross Island accommodated the 15-man team of the British (Nimrod) Expedition during the winter of 1908 (May to September). The group departed on March 25, 1909, with all hands on board. The interior of Shackleton's hut remains as its occupants left it with canned goods stacked neatly on shelves and in boxes on the floor. In addition, canned goods in wooden boxes are still piled against the outside wall of the hut nearly 100 years after Shackleton and his men departed. The huts on Cape Royds and Cape Evans are now on the list of Historic Sites and Monuments (HSMs) (Photo by T.M. Mensing)

in 1903 (i.e., the Bay of Whales). However, the embayment had disappeared and Shackleton decided to move to Ross Island and landed at Cape Royds where he and his men set up the hut in Fig. 1.13 and unloaded their supplies. The Nimrod departed on February 22 leaving Shackleton and 15 men to winter over. In early March, a group led by William Edgeworth David climbed Mt. Erebus for the first time.

Shackleton was personally interested in reaching the geographic South Pole and therefore began laying depots for the trek south as soon as the weather permitted in September of 1908. On October 28 a group of four men led by Shackleton (i.e., Jameson Adams, Eric Marshall, and Frank Wild) headed south across the Ross Ice Shelf. They reached the mouth of the Beardmore Glacier and followed it up to the polar plateau. The team continued toward the South Pole by man-hauling on limited rations. On January 9 of 1909, at 88°23'S and 162°E, they were only 97 miles from the pole (Stonehouse 2002, p. 39). On this day Ernest Shackleton decided to turn back because they did not have enough food and fuel to get back safely, although they probably could have reached the South Pole. Therefore, the men retraced their steps and returned to Cape Royds on March 1 where the Nimrod was waiting to take them home.

By deciding to turn back when it was not safe to continue, Shackleton failed to reach his goal, but he achieved an even more important goal of bringing his companions back alive. Shackleton is widely admired for taking care of his men at the expense of his personal quest to reach the South Pole.

Before Shackleton and his men had left for the pole, another group consisting of the geologists W. Edgeworth David and Douglas Mawson accompanied by the physician Alister Mackay set out to find the magnetic pole of Antarctica (Mawson 1915). They drove the motor car across McMurdo Sound, but man-hauled sleds while traversing the mountains of Victoria Land. After they reached the polar plateau, they used a dip needle to find the magnetic pole. Finally, on January 15, 1909, they came to a place at 72°25'S, 155°16'E where the dip of the magnetic field differed from the vertical by only 15' of arc. They calculated that the pole would move to their position in 24 h. Therefore, they raised the British flag on the following day and claimed the area for the British Empire (Stonehouse 2002).

A third party consisting of Raymond Priestley, Bertram Armytage, and Sir Philip Brocklehurst explored the Ferrar Glacier and the ice-free valleys of southern Victoria Land west of Ross Island. Both parties were picked up by the Nimrod and, after Shackleton and his men had returned from their trek, the Nimrod headed north and landed safely in Lyttelton Harbor on the South Island of New Zealand. Ernest Shackleton was knighted by the King of England for his achievements in Antarctica, but his career of geographic exploration did not end there.

1.4.3 Scott's Second (Terra Nova) Expedition of 1910–1913

Robert Scott bided his time until Shackleton had returned from his expedition. When it became apparent that his former friend and recent rival did not reach the South Pole, Scott started preparations for another assault on the Pole. The resulting British Antarctic (Terra Nova) Expedition of 1910–1913 got underway on June 1 of 1910. The Terra Nova was a converted Scottish whaling ship commanded by H.L.L. Pennell of the British Navy. The ship had an auxiliary engine and was crowded with 19 ponies, 30 dogs, and three tractors. When Scott reached Melbourne, Australia, on October 12, he received an ominous telegram from Roald Amundsen of Norway:

Beg leave to inform you proceeding Antarctica

Scott knew then that he had a formidable rival in his race to the South Pole.

The Terra Nova reached Ross Island without incident and unloaded a disassembled wooden hut and the supplies of the expedition at Cape Evans where Scott had decided to set up his base camp. By mid-January of 1911, the hut in Fig. 1.14 had been erected and all of the supplies had been stowed.

Next, the Terra Nova took a four-man team led by Griffith Taylor across McMurdo Sound in order to explore the ice-free valleys of southern Victoria Land (Strange and Bashford 2008). The members of this team included Charles Wright, Frank Debenham, and Edgar Evans.

The Terra Nova then sailed along the edge of the Ross Ice Shelf where Scott wanted to establish a secondary base camp at the Bay of Whales. However, the bay was already occupied by the Fram of Roald Amundsen's Norwegian South Polar Expedition. Consequently, Victor Campbell who was to set up Scott's secondary base, withdrew and returned to Cape Evans. The final assignment of the Terra Nova was to sail north to Cape Adare in order to put ashore a group of six men led by Victor Campbell and including Raymond Priestley. They set up a base camp near the huts of Borchgrevink and spent a busy year engaged in biological, meteorological, and geological research. When the Terra Nova returned in January of 1912, Campbell asked to have his group moved to Terra Nova Bay (74°50'S, 164°30'E) located about 450 km south of Cape Adare on the coast of Victoria Land. He intended to spend 6 weeks there doing mostly geological research. However, when the ship later returned to pick up Campbell and his team, Terra Nova Bay was blocked by pack ice. Consequently, Campbell's party was forced to spend a miserable winter in an igloo and an ice cave on Inexpressible Island (74°54'S, 163°39'E). On September 30 of 1913 Campbell and his men started to walk back to Cape Evans on Ross Island and finally reached it on November 7 after 36 days. They survived in part by using supplies left behind by David, Mawson, and Mackay of Shackleton's expedition in 1907/1909. This legendary ordeal which was later described by Priestley (1914), took place at a site that is now reachable in about 2 h by helicopter from McMurdo Station.

Another legendary sledging trip was undertaken on June 27 of 1911 by Edward Wilson, Apsley Cherry-Garrard, and Henry (Birdie) Bowers who man-hauled sleds from Cape Evans to Cape Crozier in order to recover some embryos of Emperor penguin chicks before they hatched. This trip took place in the dark in extremely cold and windy weather. The team did reach Cape Crozier, where they collected three embryos, and returned to Cape Evans on August 2 after a trip that also lasted 36 days.

Fig. 1.14 Scott's hut at Cape Evans on Ross Island from which he started his trek to the South Pole during the British (Terra Nova) Expedition of 1910–1913. Cape Evans in Figure 1.11 is located only 24 km from McMurdo Station compared to 35 km for Cape Royds. Scott's hut at Cape Evans was later occupied by the men of Shackleton's Imperial Trans-Antarctic Expedition of 1914–1917 who established a string of food caches across the Ross Ice Shelf all the way to the mouth of the Beardmore Glacier to be used by Shackleton during his crossing of Antarctica (Photo by G. Faure)

Today, Cape Crozier is reachable by helicopter from McMurdo Station in less than 1 h, but it is off limits to visitors because it is a Specially Protected Area (SPA) that can only be entered with a permit. Cape Crozier is also a Historic Site and Monument (HSM) because of a message post and a stone hut erected by members of Scott's expeditions of 1901–1904 and 1910–1913, respectively.

After the Terra Nova left Cape Evans on January 25 of 1911, Scott started moving supplies to the Ross Ice Shelf to be used on the way to the Pole. On November 3, 1911, after spending the winter at Cape Evans, Scott and a group of 16 men, two tractors, nine ponies, and one dog team started the journey to the Pole traced in Fig. 1.15. The tractors broke down in less than 3 days and had to be abandoned. The ponies were shot as needed to provide food for the dogs and the men. When the group reached the mouth of the Beardmore Glacier on December 9 (36 days after leaving Cape Evans) all of the remaining ponies were shot. The only remaining dog team turned back 2 days later. Another group of four men turned back halfway up the Beardmore Glacier (Atkinson, Cherry-Garrard, Keohane, and Wright). When Scott reached the edge of the polar plateau on January 4 of 1912, four more men were to leave (Bowers, Crean, Evans, and Lashley). However, Scott decided on the spur of the moment to add Bowers to his group that was scheduled to continue to the South Pole.

This group of five men (Robert F. Scott, Edward Wilson, Lawrence "Titus" Oates, Edgar Evans, and Henry "Birdie" Bowers) reached the South Pole on January 18 of 1912. They were profoundly disappointed and demoralized when they discovered that Amundsen and his men had reached the Pole about 1 month before them on December 14 of 1911. The Norwegians had set up a black tent at the Pole which contained a letter addressed to the King of Norway, a

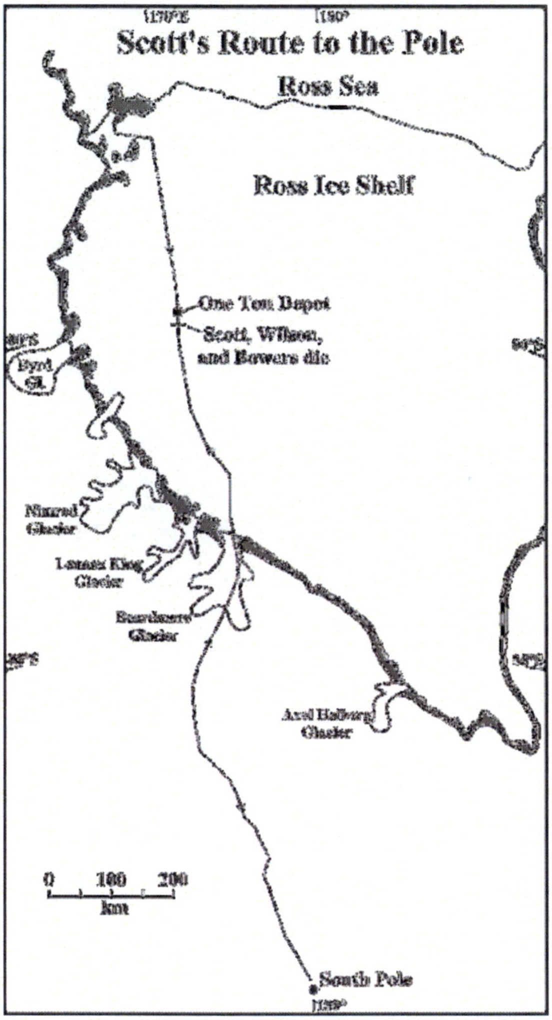

Fig. 1.15 Robert Scott and his team started their trek to the South Pole on November 3, 1911, and followed the route pioneered previously by Scott, Wilson, and Shackleton in 1902 and later by Shackleton, Marshall, Adams, and Wild in 1909. Scott and his men (Edward Wilson, Lawrence Oates, Edgar Evans, and Henry Bowers) reached the South Pole on January 18, 1912 about 1 month after Roald Amundsen. Scott, Wilson, and Bowers died in their tent only 1 day's travel from the One Ton Depot. Evans and Oates died earlier (Adapted from Stonehouse 2002)

sextant, a level, some clothing, and a plate with the names of Amundsen's crew. In addition, there was a letter addressed "For Robert F. Scott."

Scott wrote in his diary: "Great God! This is an awful place…." The group headed back to the Beardmore Glacier and descended to the Ross Ice Shelf. The men were cold, tired, starving, and suffering from vitamin deficiency (Huntford 1986a). Evans died on February 18, 1912, after a fall on the Beardmore Glacier. Oats sacrificed his life in mid-March by walking away from the tent during a snow storm. The three remaining men (Scott, Wilson, and Bowers) died in their tent a few days later during a snow storm only a short distance from their "One Ton Depot" marked in Fig. 1.15.

The death of Robert Scott and his four companions was caused by exhaustion, malnutrition, and exposure to the cold. Most observers have found fault with Robert Scott because he disliked using dogs and preferred instead to burden his team members with pulling heavy sleds up the Beardmore Glacier and across the polar plateau. Although Scott did not have food or the fuel to support five persons, he nevertheless added Bowers to his team thereby depriving the others of much-needed nourishment and space in the tent. If Scott had used dog teams, he could have carried more food and fuel and might have accomplished the trip in less time. Even if he did not win the race to the South Pole, he and his men might have survived the trip.

Scott disliked using dogs to pull sleds during his Antarctic expeditions because he thought that the work was too hard for them. Roald Amundsen did not have such sentimental feelings about dogs, although he treated them well, because he liked dogs and knew from prior experience in the Arctic that well-fed dogs work better than dogs that are malnourished.

When the polar party of Scott did not return by March 2 of 1912, the Terra Nova had to leave Cape Evans in order to avoid being trapped in the ice of McMurdo Sound. The naval surgeon Edward L. Atkinson sent out teams who searched in vain for Scott and his companions. The youngest member of the expedition, Apsley Cherry-Garrard, and a Russian dog handler, Dmitri Girev, drove dog teams all the way to the One-Ton Depot (Fig. 1.15) and waited there for Scott and his men to return as they had been instructed to do. Ironically, Scott and his men were lying in their tent only 11 miles farther south having run out of food and fuel. Cherry-Garrard waited for them for 6 days before he decided to turn back on March 12. He made it back to Cape Evans, but worried for the rest of his life that he should have driven beyond the One-Ton Depot, in which case he might have rescued Scott and his remaining companions (Huntford 1986a; Solomon 2001). Cherry-Garrard later published a detailed account of Scott's second Antarctic expedition, entitled "The worst journey in the world" (Cherry-Garrard 1922).

The remaining group of 12 men spent another winter at Cape Evans. In the spring of the following year, Atkinson led a large search party onto the Ross Ice Shelf in a final effort to find the bodies of Scott and his companions. When Charles Wright spotted the tent on November 12, 1912, Scott was lying in the center with Wilson on his left and Bowers on his right. The diaries and last letters they had written lay beside them. Atkinson took all the papers, rolls of exposed film, and rock samples that Wilson had collected along the Beardmore Glacier. The last words in Scott's diary were:

> For God's sake take care of our people.

Atkinson read a Burial Service after which the tent was collapsed and a large snow cairn was erected on top of it. The members of the search party returned to Cape Evans on November 26 and were greatly relieved that Campbell and his men had returned in their absence from their winter-over on Inexpressible Island. The Terra Nova arrived at Cape Evans on January 18, 1913, but before the remainder of Scott's second Antarctic expedition left Cape Evans, Atkinson and a group of men erected a cross on the summit of Observation Hill located directly behind McMurdo Station in order to commemorate the five men who died after reaching the geographic South Pole of Antarctica. The cross in Fig. 1.16 carries the names of the men who went to the pole (Captain Scott, Wilson, Bowers, Oats, and Evans) and an inscription selected by Cherry-Garrard (Bull and Wright 1993):

> To strive, to seek, to find, and not to yield.

The cross still stands on the summit of "Obs Hill" and new arrivals to McMurdo Station are encouraged to visit it at least once during their stay to pay their respects and to enjoy the view of McMurdo Station, Mt. Erebus, and the Royal Society Range on the mainland of Antarctica.

Sir Charles Wright (1887–1975), who was knighted in 1946, was in residence at McMurdo Station in February of 1964 when Gunter Faure returned from fieldwork in the Wisconsin Range of the Horlick Mountains. Sir Charles, who was 78 at the time, had a research project at Pole Station, but was prevented from working there himself for health reasons. During a social gathering one evening, Sir Charles reminisced freely about Scott's Terra Nova Expedition and its tragic end. However, when he was asked to describe what he had seen when Scott's tent was opened, he politely declined.

Fig. 1.16 The cross on the summit of Observation Hill overlooks McMurdo Station and is a popular destination for visitors to Antarctica. It was constructed of Australian Jarrah wood by the ship's carpenter of the Terra Nova. After some discussion, the survivors of Scott's expedition at Cape Evans decided to inscribe the cross with the concluding line of Tennyson's Ulysses: "To strive, to seek, to find and not to yield." When the cross was finished, Atkinson, Wright, Lashly, Crean, Debenham, Keohane, Cherry-Garrard, and the ship's carpenter transported the cross to Hut Point and then carried it to the top of Observation Hill where they erected it. At five P.M. on January 22, 1913, the job was done. The men gave three cheers and one more before returning to Cape Evans (Cherry-Garrard, 1922, p. 586/7) (Photo by G. Faure)

A memorial to Captain Robert F. Scott still stands in a pleasant park along the Avon River of Christchurch New Zealand. Many Antarctic travelers stop to view the greater-than-life statue in Fig. 1.17 which was carved by his wife, Kathleen. The inscription in the granite base of the statue was taken from Scott's diary:

> I do not regret this journey, which has shown that Englishmen can endure hardship, help one another and meet death with as great fortitude as ever in the past.

Fig. 1.17 The statue of Robert Scott in polar clothing strikes a heroic pose in the park along the Avon River in Christchurch, New Zealand. The statue was carved from a block of marble by Scott's wife, Kathleen (Photo by T.M. Mensing)

Fig. 1.18 Roald Amundsen of Norway was not a man who smiled a lot. He was instead driven by his passion for exploration of the polar regions. Being a thorough and methodical person, he trained himself and his companions in the art of polar travel and planned meticulously in order to assure success without injury to the members of his team. This portrait of Amundsen is displayed in Oslo in the museum that houses the Fram, the ship in which he and his men traveled to Antarctica for the purpose of reaching the South Pole

1.4.4 Amundsen's (Fram) Expedition of 1910–1912

Roald Engelbregt Gravning Amundsen (1872–1928) was a man of few words whose consuming passion was polar travel by land, by sea, and finally by air. He adopted the lifestyle of the Inuits whom he met during his successful voyage from Norway to Alaska by way of the Northwest Passage. The remarkable career of this expert polar traveler was described by Huntford (1986a). In addition, Huntford (1987) published Amundsen's photographs that illustrate the principal accomplishments of his career as a polar explorer: The Northwest Passage Expedition (1903–1906), the South-Pole Expedition (1910–1912), and the Maud (Northeast) Expedition (1918–1923). In spite of his reluctance to smile in Fig. 1.18, he was in fact a warm-hearted man who loved the dogs that helped him to reach the South Pole. Huntford (1987) described him this way: Behind a forbidding exterior there hid a sensitive man with an immense capacity for affection.

Amundsen first came to Antarctica as the second officer of the Belgica (1897–1899) and subsequently sailed through the Northwest Passage in a small fishing boat called the Gjoa. During this trip, which lasted from 1903–1906, Amundsen and his six companions spent two winters in the Arctic during which he established good working relationships with the local Inuits and traveled with them to the magnetic pole of the Arctic and to the settlement of Eagle in Alaska from where he telegraphed a message concerning his progress. The Gjoa arrived in San Francisco in October of 1906 (Stonehouse 2002).

After Amundsen returned to Norway, he began preparations for an expedition to the North Pole using the Fram which **Fridtjof Nansen** (1861–1930) had agreed to lend him for this purpose. However, when Dr. Frederick Cook and Robert Peary both claimed to have reached the North Pole in late 1908 (Cook) and early 1909 (Peary), Amundsen decided to attempt to reach the South Pole instead. He announced this change of plans only after he had left Oslo in August of 1910. When he reached the island of Madeira, he informed his crew and asked his brother to send a telegram to Robert Scott who had left London on June 1 of 1910.

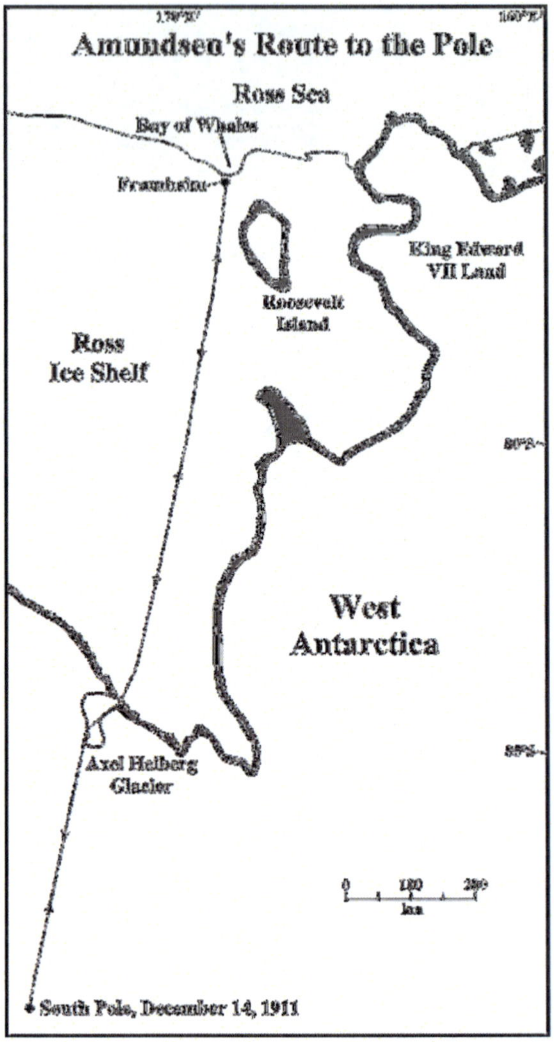

Fig. 1.19 The route taken by Amundsen and his men led directly from Framheim at the Bay of Whales to the mouth of the Axel Heiberg glacier which they used to reach the polar plateau and hence the South Pole. They started the trip on October 20, 1911, with four sleds, each of which was pulled by 13 dogs. Amundsen selected experienced men to drive the sleds and made rapid progress across the Ross Ice Shelf. The ascent of the Axel Heiberg glacier was difficult because it is steep and heavily crevassed. The group required 5 days to traverse the glacier and to reach the polar plateau. They arrived at the South Pole on December 14 of 1911 (Adapted from Stonehouse 2002)

Amundsen sailed directly to the Bay of Whales along the edge of the Ross Ice Shelf and set up a base camp in January of 1911 about three kilometers inland of the ice edge. His camp consisted of the hut "Framheim" and of several tents. He and his companions had all traveled extensively in the Arctic and were skilled dog handlers. They went to work immediately laying a series of caches of food and fuel across the Ross Ice Shelf. Even before the end of the Winter on September 8, 1911, Amundsen tried to get an early start, but he was forced to return to Framheim by the extremely cold and windy weather.

He tried again on October 20 taking four men (Olav Bjaaland, Helge Hassel, Helmer Hanssen, and Oscar Wisting) and headed south across the Ross Ice Shelf in Fig. 1.19. Each of his companions drove a sled pulled by 13 Greenland huskies who were in excellent condition. Consequently, they crossed the Ross Ice Shelf with ease traveling about 32 km per day and building cairns at fixed intervals to guide their return. They reached the Transantarctic Mountains on November 11 only 22 days after leaving Framheim. Amundsen decided to cross the mountains by following the Axel Heiberg Glacier to the polar plateau. The ascent to the plateau was very difficult because of the steep gradient and the abundance of deep crevasses in the ice of this glacier. After 5 days of intense effort in cold and windy weather, they reached the polar plateau at an elevation of about 10,000 ft or 3,050 m. From there, the run to the South Pole was relatively easy. They arrived at the Pole on December 14, 1911, where all five men participated in planting the Norwegian flag.

Helmer Hanssen described the arrival at the South Pole in his diary (Huntford 1987, p. 131):

It was… "a solemn moment for us all. As always, Amundsen thought of his companions, and when we planted the Norwegian flag at the South Pole, he let us all hold the bamboo stick with the flag, when it was fixed in the snow…."

Oscar Wisting recalled the occasion this way:

Roald Amundsen asked us to gather round to plant the flag. 'It is not the privilege of one man alone to carry out this ceremony. It is the privilege of all those,' he said, 'who have risked their lives for this cause.' Each man gripped the flagpole, and together we planted Norway's flag at the South Pole, where no human had yet set foot.

Amundsen himself recorded in his diary that:

[T]he Norwegian flag, with the Fram pennant underneath, are flying from the top of the tent pole. In the tent, I have placed various things: my sextant with a glass artificial horizon, a hypsometer, 3 reindeer fur boot warmers, some kimikks and mittens, and a few small things. In a folder, I left a letter for the King and some words to Scott who, I must assume will be the first to visit the place after us.

Amundsen spent 3 days making sure that they had correctly located the Pole. Before they departed, they set up the tent that was later found by Scott and his companions. The trip back to Framheim was uneventful and ended on January 25, 1912. Amundsen's prediction came true when Robert Scott and his companions discovered the black tent and the Norwegian flag at the South Pole on January 18, 1912. Their profound disappointment is evident in Fig. 1.20.

While Amundsen and his men were on their way to the South Pole, the Fram had carried out an extensive oceanographic cruise in the South Atlantic Ocean. After Amundsen had returned from the Pole, all members of the expedition boarded the ship and sailed away on January 30, 1912. They arrived in Hobart on February 7, 1912 (Amundsen 1912; Victor 1964; McPherson 1975; Reader's Digest 1985,1990; Huntford 1986a, 1987; Stonehouse 2002).

Fig. 1.20 Robert Scott and his companions reached the South Pole on January 18, 1912, where they found the black tent and Norwegian flag left by Amundsen and his men on December 14, 1911. They were profoundly disappointed to have lost the race to the South Pole as they stood around dejectedly. The individuals shown here from left to right are: Scott, Oates, Wilson, and Evans. The picture was taken by Henry (Birdie) Bowers (Photo by the Scott Polar Research Institute, University of Cambridge, used here with permission)

1.4.5 Shackleton's Second (Endurance) Expedition, 1914–1917

After Amundsen and Scott had reached the South Pole, Sir Ernest Shackleton proposed to cross Antarctica from the Weddell Sea to the Ross Sea, a feat that had previously been attempted by the German explorer **Wilhelm Filchner** (1877–1957) during the German Antarctic (Deutschland) Expedition of 1911–1913 (Stonehouse 2002). Accordingly, Shackleton organized the Imperial Trans-Antarctic Expedition of 1914–1917 which involved the use of two ships. The Endurance commanded by Captain Frank Worsley was scheduled to set up a base camp along the Luitpold Coast of the southern coast of the Weddell Sea (Worsley 1931). Shackleton and his men planned to use dogs to travel from this base across the continent to the Ross Ice Shelf and from there to Ross Island. At the same time, the Aurora commanded by Captain Aeneas Mackintosh, was supposed to return to Shackleton's hut at Cape Royds on Ross Island in order to establish a chain of depots across the Ross Ice Shelf to the base of the Beardmore Glacier for use by Shackleton and his party.

The Endurance left South Georgia Island on December 5, 1914, and headed south into the Weddell Sea where it was trapped by pack ice in early January of 1915. Shackleton was forced to winter-over in his ship as it slowly moved around the Weddell Sea until November 15, 1915, when the ship was crushed by the ice and ultimately sank. Shackleton set up tents on the pack ice for his 28 men. In April of 1916 the camp had been transported by the moving ice to the northern edge of the pack ice and was located within sight of open water. Therefore, Shackleton put his three lifeboats into the water and headed for Elephant Island (61°10'S, 055°14'W). All three boats reached the island on April 14, 1916, where they set up the camp in Fig. 1.21 by inverting two of the lifeboats and by

Fig. 1.21 Shackleton's camp at Cape Wild on Elephant Island which he and his men reached after their ship, the Endurance, was crushed by pack ice in the Weddell Sea on November 15, 1915. The shelter consisted of two lifeboats turned up-side-down on the beach. On April 24, 1916, Shackleton and five of his men sailed away in the lifeboat, James Caird, hoping to reach the whaling station on South Georgia Island. The remaining 22 men survived in their improvised shelter until August 30 when Shackleton returned on the Chilean Trawler, Yelcho, to rescue them. All of the men survived (Photo by the Scott Polar Research Institute, University of Cambridge, used here with permission)

setting them up on low walls made of beach cobbles. Shackleton instructed the ship's carpenter to make the third lifeboat (James Caird) sea-worthy and then used it with five men to sail 1,120 km across the open ocean to South Georgia Island. After 17 days at sea, they landed at King Hakon Bay on the uninhabited south side of South Georgia Island. Shackleton and two of his men decided to walk across the mountainous island and reached a whaling station at Stromness Bay on the following day.

The station manager immediately sent a boat to the other side of the island to pick up the three men Shackleton had left on the beach. Another ship with Shackleton and his two companions on board attempted to reach Elephant Island, but could not land because the island was surrounded by pack ice. Therefore, the ship took the three men to the Falkland Islands. Shackleton continued his efforts to reach the 22 men that were still camping on the beach at Elephant Island and eventually succeeded on August 30 of 1916 by means of a Chilean steam-powered tugboat. It is easy to imagine the anxiety of the men who had been waiting on Elephant Island since April 14 in hopes of being rescued without knowing whether Shackleton had reached South Georgia Island.

Meanwhile, another drama was unfolding on Ross Island during the summer of 1915 because a solid sheet of ice prevented the Aurora from landing at Cape Royds and forced Captain Mackintosh to land at Cape Evans instead. Before all of the supplies had been unloaded, the Aurora drifted away and was trapped in the pack ice for a period of 10 months. Eventually, Lieutenant J.R. Stenhouse managed to free the ship and returned it to New Zealand.

The group of ten men at Cape Evans lacked the supplies and equipment they needed to set up the depots for Shackleton and his men. This caused a great deal of concern for Captain Mackintosh and his men because they did not know that the Endurance had sunk and that Shackleton was not traveling across the continent as planned. In spite of lacking essential equipment including polar clothing, Captain Mackintosh and his men made heroic efforts to set up the six depots on the Ross Ice Shelf that Shackleton needed to complete his crossing of Antarctica. They actually reached Mount Hope at 83°31'S, 171°16'E near the mouth of the Beardmore Glacier. This futile effort cost three lives: Arnold Spencer-Smith died of scurvy while on the trail and Captain Mackintosh and Victor Hayward died

Fig. 1.22 This memorial at Cape Evans was erected by the Antarctic Division of D.S.I.R. of New Zealand in 1962–1963. The inscription reads: "Sacred to the memory of Captain A.L.A. Mackintosh, R.N.R. and V.G. Hayward who perished on the sea-ice in a blizzard about May 8, 1916, and of Reverend A.P. Spencer-Smith, B.A. who died on the Ross Barrier about March 6, 1916" (Photo by G. Faure)

when they broke through the ice between Hut Point and Cape Evans. The memorial in Fig. 1.22 was erected by New Zealanders and bears an inscription composed by A.K. Jack who survived the ordeal. The seven survivors at Cape Evans were finally rescued on January 10 of 1917 (Shackleton 1920; Joyce 1929; Worsley 1931; Huntford 1986b).

The scientists and crew of the Endurance who had survived on Elephant Island were reunited with Shackleton in Punta Arenas, Argentina. In spite of unimaginable hardships and great emotional strain, Shackleton had once again brought all of the men under his command back alive. His great determination as an explorer and his concern for the well-being of his men were praised by Roald Amundsen:

> Sir Ernest Shackleton's name will for evermore be engraved with letters of fire in the history of Antarctic exploration.

1.5 The Modern Era Begins

The exploration of Antarctica up to and including Shackleton's attempt to travel from the Weddell Sea to the Ross Sea were handicapped by the lack of reliable communication and by having to travel either by man-hauling or by dogsleds. The record of these expeditions assembled by Reader's Digest (1985, 1990) and by Stonehouse (2002) shows that fixed-wing aircraft were first used in 1928 by the Australian explorer George H. Wilkins (1888–1958) and by the American Richard E. Byrd (1888–1957) in Fig. 1.23. At about the same time, short-wave transmitters and receivers permitted long-distance communication by means of Morse code and by voice. If Nordenskjőld had been able to communicate with his ship and with its shipwrecked crew on Paulet Island, some of the suffering they endured could have been avoided. Therefore, the introduction of aircraft, wireless communications, and reliable snowmobiles has greatly facilitated the exploration of Antarctica and the on-going study of its geology and biology.

Fig. 1.23 Richard E. Byrd (1888–1957) was an American naval aviator who distinguished himself by flying to the North Pole in 1927 and subsequently to the South Pole in 1929. He organized and led two large-scale expeditions to Antarctica in 1928–1930 and 1933–1935. He set up base camps on the Ross Ice Shelf which he named Little America I and II. Admiral Byrd pioneered the use of fixed-wing aircraft for the exploration of the interior of Antarctica while at the same time sending the geologist Laurence Gould on the longest journey ever undertaken with dog sleds to explore the Queen Maud Range of the Transantarctic Mountains (Photo by T. M. Mensing taken at McMurdo Station)

1.5.1 Byrd's Little America Expedition, 1928–1930

The First World War (1914–1918) stimulated the development and use of fixed-wing aircraft in military operations over Europe which motivated the US Navy to establish the naval air service. **Richard Evelyn Byrd, Jr**. (1888–1957), who had graduated from the Naval Academy in Annapolis in 1912, joined the naval air service and distinguished himself as a planner in support of long-distance flights in the Arctic (Anderson 1974). In 1926 Richard Byrd and Floyd Bennet used a tri-motor Fokker aircraft to fly from Svalbard (Spitsbergen) to the North Pole and returned safely after flying 2,100 km over featureless sea ice. Byrd also crossed the Atlantic Ocean in 1927 by flying from Newfoundland to the village of Ver-sur-Mer in France.

Thus encouraged, Byrd organized an expedition to Antarctica in order to fly to the South Pole. This objective received support from the US government, the US Navy, the news media, and the general public all of whom contributed to the financing of the expedition. As a result, Byrd's first Antarctic Expedition (1928–1930) included two ships, the City of New York and the Eleanor Bolling, the latter of which was named after his mother. He carried abundant supplies, 95 dogs, several radio transmitters, a tractor, and three planes. These supplies were unloaded at the Bay of Whales on the Ross Ice Shelf and moved away from the ice edge a distance of 11 km where Byrd set up his base camp called Little America I, which resembled a small village where 42 men and the sled dogs wintered over. On January 27 of 1929 Byrd and the pilot Brent Balchen flew to Edward VII Land first seen by Robert

Scott in 1902 (Section 1.4.1) and discovered the Rockefeller Mountains (78°00'S, 155°00'W) of West Antarctica from the air.

While Byrd was making several additional short reconnaissance flights in February and early March of 1929, several dog teams carried supplies across the Ross Ice Shelf for a sledging expedition to be led by Laurence Gould. On November 4, 1929, Gould started the longest sledging journey in Antarctica to explore the Queen Maud Range (86°00'S, 160°00'W) located 800 km from Little America (Gould 1931).

Three weeks later, on November 28, 1929, Byrd and three of his companions including Brent Balchen took off in the tri-motor aircraft and reached the South Pole in less than 10 h. After circling the Pole, they returned to Little America I in about 6 h. Byrd had made the trip in less than 1 day that had taken Amundsen 90 days. For many years, Richard E. Byrd remained the only man who had flown across both geographic poles of the Earth (Byrd 1931).

1.5.2 Byrd's Second Expedition, 1933–1935

For the second expedition to Antarctica Byrd, once again, had two ships (Bear of Oakland and Jacob Ruppert) which arrived off the coast of Edward VII Land on December 20, 1933. After penetrating the pack ice, the ships entered the Bay of Whales on January 17, 1934 and unloaded their cargos including 150 sled dogs, four aircraft (including a helicopter), several tractors, three Guernsey cows, and a calf who were to provide fresh milk for 56 men. None of the land-based expeditions of Robert Scott, Ernest Shackleton, and Roald Amundsen was as well equipped or included so many men. Nevertheless, both of Byrd's expeditions relied on traditional as well as on modern means of transportation.

Byrd set up a weather station at 80°08'S and 163°57'W on the Ross Ice Shelf and moved there in March of 1934 in order to spend the winter making a continuous series of weather observations. At first, all appeared to be well until his team members at Little America II noticed that Byrd was beginning to behave erratically. They quickly organized a small tractor train in August of 1934 and brought Byrd back to Little America II where he recovered from what was later diagnosed as carbon monoxide poisoning caused by a defective heater.

During the 1934/35 field season Byrd's teammates explored large areas of West Antarctica and measured the thickness of the Ross Ice Shelf (90 m) as well as the depth of the water in which it floated (600 m). All of the various field parties returned to Little America II in January of 1935 and were picked up by the ships in February. Although Byrd and his men had explored large areas in the Ross-Sea sector of Antarctica, the US government did not formally claim these territories as other nations had done (e.g., Britain, New Zealand, Norway, and France).

After Byrd's second expedition, the exploration of Antarctica by means of ships, aircraft, and dog sleds continued by the American, British, and German expeditions listed in Appendix 1.10.1 (Byrd 1935; Victor 1964; Reader's Digest 1985, 1990; Stonehouse 2002).

1.5.3 International Geophysical Year (IGY), 1957–1958

At a dinner party on April 5 of 1950 in the home of Dr. James Van Allen in Silver Springs, Maryland, the suggestion was made that the time was ripe for another International Polar Year (IPY) even though the next one was not scheduled until 1982–1983. The first IPY had occurred in 1882–1883 followed 50 years later by the second IPY in 1932–1933. The proposal for a third IPY was submitted to the International Council of Scientific Unions which approved it in 1952 but expanded its scope to include geophysical studies of the whole Earth and of the space around the Earth. The resulting International Geophysical Year (IGY) extended from July 1 of 1957 to December 31 of 1958 and coincided with the maximum of the Sun-spot cycle.

The study of the Sun during the IGY coincided and may have influenced the timing of the first artificial satellites launched by the Soviet Union (Sputnik 1, October 4 and Sputnik 2, November 2, 1957) followed by Explorer 1 of the USA on January 31, 1958. At the request of James van Allen, the American satellite carried a Geiger Counter that detected the inner radiation belt that encircles the Earth (Victor 1964; Faure and Mensing 2007).

During the IGY 12 nations established 55 research stations in Antarctica where more than 5,000 scientists made measurements of various geophysical parameters (Crary 1982). This level of participation was made possible because all of the nations that had claimed territory in Antarctica agreed not to enforce these claims and to cooperate fully not only with the claimant but also with the non-claimant nations:

Claimant nations: Argentina, Australia, Chile, France, New Zealand, Norway, United Kingdom.

Non-claimant nations: Belgium, Japan, South Africa, Soviet Union, and the United States of America.

This arrangement not only permitted the non-claimant nations to conduct research in Antarctica but it also encouraged the claimant nations to cooperate with each other.

The history and evolution of the US Antarctic Research Program (USARP), later renamed the US Antarctic Program (USAP), and its implementation by Operation Deep Freeze of the US Navy were described by Crary (1982) in the Antarctic Journal of the United States which was established in 1966. Operation Deep Freeze, commanded by Rear Admiral George Dufek, constructed several research stations during the austral summers of 1955/56 and 1956/57 for use by American scientists during the IGY: Little America on the Ross Ice Shelf, Byrd Station in Marie Byrd Land in West Antarctica, Hallett Station in northern Victoria Land, and South Pole Station, later renamed Amundsen–Scott Station. In addition, McMurdo Station at the tip of the Hut Point Peninsula on Ross Island was established to support the movement of cargo from New Zealand to Antarctica by means of large transport planes such as the C-124 Globemasters which were flown by the pilots and crews of Air Squadron VXE-Six of the US Navy. These pilots and crews also flew the planes that carried cargo and civilian passengers to their destinations within the continent. This operation lasted 44 years from 1955 to 1999 (Gillespie 2005).

During the IGY the USA also maintained Ellsworth Station (77°43'S, 041°08'W) on the Filchner Ice Shelf from 1957 to 1962. A personal account of life at this station was later published by Behrendt (1998). Following the IGY, the USA opened Siple Station (75°33'S, 083°33'W) in 1969 and occupied the site, which is located 2,450 km from McMurdo Station, until 1989 when the last version of Siple Station was closed. Eights Stations at 75°14'S, 077°10'W in southwestern Palmer Land was operated by the USA from 1961 to 1965 in support of seismic surveys and studies of glaciology and upper atmospheric physics. In 1959 Hallet Station in northern Victoria Land was transferred to joint US/New Zealand management until 1964 when the principal laboratory buildings were destroyed by fire. The station was subsequently refurbished in 1984 for intermittent use. Palmer Station at the south end of Anverse Island at 64°46'S, 064°03'W was established in 1965, 7 years after the end of the IGY and has been occupied continuously ever since (Stonehouse 2002).

1.5.4 Commonwealth Trans-Antarctic Expedition, 1955–1958

During the IGY, the interior of Antarctica was explored by means of tractor trains that traversed the continent while making geophysical, glaciological, and meteorological observations (Schulthess 1960). For example, the Commonwealth Trans-Antarctic Expedition (1955–1958) led by Vivian E. Fuchs and Edmund Hillary crossed Antarctica from the edge of the Filchner Ice Shelf to Scott Base on Ross Island. The main body of the expedition departed on November 24 of 1957 from Shackleton Base at the edge of the Filchner Ice Shelf and arrived at South Pole Station (Amundsen–Scott) on January 19, 1958. The tractors used by this expedition were far more sophisticated than the primitive vehicles Scott tried out during the Terra Nova Expedition in Fig. 1.24. The modern tractors of the Trans-Antarctic Expedition overcame serious obstacles such as the large crevasse in Fig. 1.25. In the meantime, Hillary and his men had driven from Scott Base on Ross Island up the Skelton Glacier and arrived at Pole Station on January 4, 1958 where they waited for the arrival of the tractor train led by Fuchs. The two groups merged and left Pole Station on January 24 and arrived at Scott Base on March 2 of 1958. The group led by Fuchs and Hillary had traveled 3,453 km in 99 days and had accomplished the goal of Shackleton's Imperial Trans-Antarctic Expedition of 1914–1917 (Fuchs and Hillary 1958; Fuchs 1982, 1990).

This famous traverse across Antarctica is one of many examples of successful international cooperation that achieved its objectives. In many other ways, the IGY in Antarctica was a huge success not only in

Fig. 1.24 Robert Scott started his trek to the South Pole with two tractors both of which broke down within 3 days after leaving Cape Evans on November 3, 1911 (Photo by the Scott Polar Research Institute, University of Cambridge, used here with permission)

terms of its scientific results but also in terms of the international cooperation among participating nations (Crary et al. 1956; Crary 1962, 1982; Odishaw and Ruttenberg 1958).

After the IGY, many nations that did not previously have an Antarctic research program decided to join the community of the Antarctic nations in order to benefit politically. As a result, the number of nations that have set up permanent or temporary research bases on the mainland of Antarctica and on the off-shore islands listed in Appendix 1.10.2 has increased from 12 during the IGY to more than 28 in 2002 (Stonehouse 2002).

The benefits that have been realized transcend the results of geographic exploration of the continent and the acquisition of a large body of numerical data on subjects that range across the entire spectrum of science. The most far-reaching consequence of the IGY and its aftermath is the harmonious working relationship that has developed among the nations that continue to participate in Antarctic research in accordance with the Antarctic Treaty that was first negotiated in 1958 (McKenzie 1984; Beck 1986; Fifield 1987; Berkman 2002; Joyner and Theis 1997).

1.6 Antarctic Treaty

The peaceful cooperation of nations with overlapping territorial claims in Antarctica and with conflicting priorities elsewhere in the world was an unexpected benefit of the IGY (Sullivan 1957; Gould 1958; Siple 1959). Therefore, the 12 nations that had participated in the exploration of Antarctica during the IGY (i.e., Argentina, Australia, Belgium, Chile, France, Japan, New Zealand, Norway, South Africa, USSR, and USA) negotiated a treaty in order to continue to use Antarctica exclusively for peaceful purposes. The preamble of the resulting Antarctic Treaty identifies its objectives (Stonehouse 2002, Appendix E):

Fig. 1.25 One of the modern tractors used by the Commonwealth Trans-Antarctic Expedition (1955–1958) led by V.E. Fuchs nearly fell into a large crevasse. In spite of its precarious situation in this photograph, the tractor was recovered and the tractor train continued to South Pole Station and eventually reached Scott Base on Ross Island (Photo ©Royal Geographical Society, reproduced by permission)

"*Recognizing* that it is in the interest of all mankind that Antarctica shall continue forever to be used exclusively for peaceful purposes and shall not become the scene or object of international discord;

Acknowledging the substantial contributions to scientific knowledge resulting from international cooperation in scientific investigation in Antarctica;

Convinced that the establishment of a firm foundation for the continuation and development of such cooperation on the basis of freedom of scientific investigation in Antarctica as applied during the International Geophysical Year accords with the interests of science and the progress of all mankind;

Convinced also that a treaty ensuring the use of Antarctica for peaceful purposes only and the continuance of international harmony in Antarctica will further the purposes and principles embodied in the Charter of the United Nations."

The provisions of the Antarctic Treaty are specified in 14 Articles that treat various aspects of the agreement such as:

1. The use of Antarctica for peaceful purposes (Article 1)
2. Promotion of international cooperation in scientific research including the exchange of scientists (Articles 2 and 3)
3. Refusal to become involved in current or future claims of sovereignty (Article 4)
4. Prohibition against nuclear experimentation and disposal of radioactive waste (Article 5)
5. Agreement that the treaty shall apply to the area south of 60°S latitude including ice shelves but excluding the high seas (Article 6)
6. Inspection of stations by representatives of signatory nations and exchange of information concerning their operation (Article 7)
7. Provision that observers and scientists working in Antarctica are subject to the laws of their own countries (Article 8)
8. Scheduling of consultative meetings for the discussion of issues and disputes (Article 9)
9. Responsibility of signatory nations to enforce the provisions of the Antarctic Treaty (Articles 10 and 11)
10. Modification of the Treaty by means of amendments and major review after 30 years (Article 12)
11. Mechanism to permit other states to accede to the Treaty (Article 13)
12. Selection of the official languages of the Treaty: English, French, Russian, and Spanish (Article 14)

After the signing of the Antarctic Treaty by the original 12 signatory nations, many other nations "acceded" to the treaty and some were actually granted full membership with "consultative" status. The number of full-member nations listed in Appendix 1.10.3 has increased to 30 including India and China which are the most populous nations in the world. In addition, 15 countries have become associate members which supports the claim that more than 70% of the people of the Earth have accepted the principles embodied in the Antarctic Treaty.

1.7 Scientific Meetings and Publications

Whereas the treaty regulates the governance of Antarctica, the International Council of Scientific Unions (ICSU) set up the Scientific Committee on Antarctic Research (SCAR) in order to initiate, promote, and coordinate scientific research in Antarctica. The members of this committee are appointed by the scientific academies of the nations that have achieved consultative status. In this way, SCAR helps to implement the scientific research in Antarctica within the guidelines provided by the Antarctic Treaty. SCAR has formed several Working Groups that provide technical information and advice to the Consultative Meetings of the nations that have full voting privileges. SCAR also schedules scientific symposia of Antarctic research in different countries identified in Appendix 1.10.4 (Ford 2006).

Another recurring series of international conferences are the Gondwana Symposia which publish Proceedings Volumes (Appendix 1.10.5). In addition, the American Geophysical Union of Washington, D.C., has published monographs in the Antarctic Research Series that started during the IGY (Appendix 1.10.6).

In 1966 the National Science Foundation and the US Naval Support Force, Antarctica, launched the Antarctic Journal of the Unites States in order to provide "... a common outlet for information on the logistic and scientific aspects of the national effort to a broad audience of participants and interested observers" (Jones and Bakutis 1966, p. 1). In 1977, the Antarctic Journal started publishing Annual Review issues (vol. 12, no. 4) that contained brief preliminary (i.e., unreviewed) reports submitted by the research groups that were active in Antarctica with financial and logistical support by NSF and by Operation Deep Freeze. These special issues contain a wealth of information about the ongoing scientific research in Antarctica. The series ended in 1997 (vol. 32, no. 5) and publication of the Antarctic Journal was officially discontinued in 2002.

The rapid growth of the scientific literature pertaining to Antarctica caused the National Science Foundation to establish an office within the Library of Congress whose mission was to bring this literature "under bibliographic control." The resulting Cold Regions Bibliography Section of the Science and Technology Section of the Library of Congress republished English translations of the abstracts of scientific papers in 13 subject areas that originated in 30 countries and were originally written in 13 different languages. The first volume of the resulting Antarctic Bibliography was issued in 1965 and the series of volumes eventually ended with volume 25 (Doumani 1965). These books have excellent indexes of both authors and subjects and remain a valuable source of information pertaining to a wide range of subjects including geological and biological sciences, ice and snow studies, meteorology, and even political geography.

Scientific journals that are either devoted to the publication of Antarctic science or regularly contain reports on this subject include:

Antarctic Science
Polar Record
South African Journal of Antarctic Research
British Antarctic Survey Bulletin
New Zealand Journal of Geology and Geophysics
Polish Polar Research
Antarctic Record

Several other scientific journals occasionally publish peer- reviewed scientific reports on the topic of Antarctic science (e.g., Geology, Geological Society of America Bulletin, Journal of Geology, Tectonics, Geophysical Research Letters, Precambrian Research, Journal of Geophysical Research, Science, Nature, etc.).

1.8 Popular Books About Antarctica

The history of exploration of Antarctica was beautifully illustrated in two large books published by Reader's Digest (1985, 1990) while McPherson (1975) presented pictures of the landscapes of the Transantarctic Mountains where geologists now travel and work. The wildlife in the context of the surreal landscapes of Antarctica was captured in pictures by the incomparable Elliot Porter (Porter 1978). In addition, the calendars published annually by Colin and Betty Monteath of New Zealand emphasize the serene beauty of Antarctica when the Sun happens to be shining (Monteath and Monteath 1990–2004).

Other Antarctic photographers who have published their pictures of Antarctica include, but are not limited to,

Emil Schulthess, Ann Hawthorne, Margaret Bradshaw, Walt Seelig, Guy Guthridge, and many others.

The most recent trend in the popular Antarctic literature is the publication of specialized encyclopedias including those of McGonigal and Woodworth (2001), Stonehouse (2002), Trewby (2002), and a magnificent two-volume set edited by Riffenburg (2007). These encyclopedias contain explanatory text on a multitude of subjects pertaining to Antarctica, they are illustrated by a large number of photographic images and detailed maps, and they contain extensive lists of references.

There is no shortage of Antarctic adventure stories such as the reports by Siple (1931, 1936) about his participation in Byrd's two expeditions to Antarctica, the excellent account of life at Ellsworth Station during the IGY by Behrendt (1998), and the story by Dewart (1989) of how in 1960 he spent 1 year with the Russians at the Mirnyy research station. Another book by a geologist is the account by Doumani (1999) of the trials and tribulations of scientists and military personnel who were trying to live and work together in Antarctica from 1958 to 1965. In addition, Neider (1974) published an astute account of life in McMurdo Station during the austral summer of 1970/71 that included a helicopter crash near the summit of Mt. Erebus. Other eye-witness accounts of travel to and within Antarctica were published by Neider (1972), Parfit (1985), and Pyne (1986).

The successes and failures of the pioneers in the exploration of Antarctica are discussed in the literature: Fisher and Fisher (1957), Lansing (1959), Fuchs (1982, 1990), and Solomon (2001). Even Admiral Bellingshausen was memorialized in a book edited by Debenham (1945). Other books that describe the occupation of Antarctica by humans were published by Chapman (1965), Bertrand (1971), Baughman (1994), Rowell (1995), Walton and Atkinson (1996), and Green (2008). Lest we forget, Gillespie (2005) reminded us that the success of the US Antarctic Program was made possible by the men and women of the US Navy Detachment known as VXE-Six.

Antarctica also continues to attract more than its share of adventurers who must rely on their own resources to reach Antarctica and to achieve their personal objectives. The policy of the National Science Foundation is not to offer support unless their very lives are in imminent danger (e.g., Mears and Swan 1987; Messner 1990; Steger and Bowermaster 1992; Arnesen and Bancroft 2003; Andrews 2007).

1.9 Summary

The exploration of Antarctica started in the eighteenth and nineteenth centuries by means of daring voyages by brave men in wooden sailing ships who managed to penetrate the pack ice and who survived the cold and stormy weather of the Southern Ocean. Several of the early explorers acted on orders by their governments (e.g., Cook, Bellingshausen, Dumont d'Urville, Wilkes, Ross, etc.) to discover new land and to assess the opportunities for commercial enterprises such as fishing, and the "taking" of fur seals and whales. The discoveries they made attracted a large number of professional seal hunters and whalers into the Southern Ocean some of whom explored the coast of Antarctica on their own initiative in order to discover new hunting grounds (e.g., Weddell and Palmer). These commercial enterprises were so successful that processing plants were set up on some of the islands in the Southern Ocean near the Antarctic Peninsula.

When the ships of some of the scientific expeditions were trapped by the pack ice, the explorers were forced to winter-over in their ships (e.g., de Gerlache, Drygalski, Nordenskjöld, etc.) while others deliberately allowed their ships to be frozen into the ice (Charcot) or came prepared to winter-over on land (Borchgrevink). The lessons learned during these accidental or deliberate attempts to survive the winter in Antarctica encouraged their successors to explore the continent from base camps on land and ultimately to try to reach the geographic South Pole (e.g., Scott, Shackleton, and Amundsen).

The race to reach the South Pole during the first decade of the twentieth century was won by Roald Amundsen and his team from Norway who reached the South Pole on December 14, 1911, followed about 1 month later by Scott and his four companions on January 18, 1912. Amundsen was successful because he and his men used Greenland huskies to pull their sleds, whereas Scott and his companions had to man-haul the sled containing their camping gear, food, and fuel up the Beardmore Glacier and across the polar plateau. Although they did reach the pole, Scott and his men died during the return trip because of malnutrition, exhaustion, and hypothermia aggravated by unusually cold and stormy weather. Shackleton faced similar problems but elected to turn back before reaching the South Pole and thus saved his own life and those of his companions. The respective merits of the different styles of operation exemplified by the Antarctic journeys of Amundsen, Scott, and Shackleton continue to be debated by historians of Antarctic exploration.

The exploration of Antarctica was revolutionized by the introduction of fixed-wing aircraft, motorized sleds, and wireless communication used by Richard Byrd in the 1930s. The availability of these labor-saving devices and technological improvements greatly facilitated the exploration of the interior of the continents from the air and by means of motorized vehicles operating from temporary camps on land. These techniques enabled several nations to carry out comprehensive geophysical surveys and geographic exploration of Antarctica during the International Geophysical Year (IGY) (1957–1958). The success of these projects prompted 12 nations to enter into the Antarctic Treaty which continues to regulate the use of Antarctica for scientific research while it prohibits commercial exploitation of the natural resources of the continent and in the surrounding ocean south of 60° latitude.

The Antarctic Treaty has become a model for international cooperation among nations whose objectives elsewhere in the world may conflict with those of other Treaty Nations. The diplomatic precedents established by the Antarctic Treaty are at least as important as the results of the on-going scientific research in Antarctica. Among other benefits, the Antarctic Treaty has served as a model for the "Law of the Sea" (Joyner 1992) and for the treaty "... Governing the Activities of States in the Exploration and Use of Outer Space" adopted by the United Nations in 1963 (Faure and Mensing 2007).

*The dust jacket of P.-E. Victor's book contains a photograph by Emil Schulthess which shows "an American geologist returning from the ascent of Mount Glossopteris" in the Ohio Range. The person in the photograph is Dr. William Long who later became a Ph.D. candidate at The Ohio State University (Appendix 2.12.1).

1.10 Appendices

1.10.1 Exploration of Antarctica Following Byrd's Second Expedition (Stonehouse 2002)

Expedition	Dates
Ellsworth Expeditions:	1934–1936
Lincoln Ellsworth (1880–1951) and Sir Hubert Wilkins (1888–1951) attempted several long-distance overflights of West Antarctica and the Antarctic Peninsula.	
British Graham Land Expedition:	1934–1937
Explored and mapped the area south and east of Marguerite Bay on the Antarctic Peninsula.	
German Antarctic (Schwabenland) Expedition:	1938–1939
Explored the area east of the Weddell Sea in order to support a territorial claim of an area later claimed by Norway (i.e., Dronning Maud Land).	
US Antarctic Service Expedition:	1939–1941
Set up permanent bases at the Bay of Whales and on Stonington Island (68°11'S, 067°00'W) from which West Antarctica was mapped by aerial photography combined with ground truths established by dog sled teams.	
Operation Tabarin:	1943–1944
Precursor of the Falkland Island Dependencies Survey intended to strengthen British territorial claim to the Falkland Islands and the adjacent sector of Antarctica.	
US Naval Operation Highjump:	1946–1947
Naval task force that was organized to map the coastline of Antarctica and to train military personnel in cold-weather operations.	
US Naval Operation Windmill:	1947–1948
Continuation of Operation Highjump.	
Falkland Island Dependencies Aerial Survey Expedition:	1955–1957
Exploration and mapping by long-range aircraft of the Antarctic Peninsula and associated islands.	
Commonwealth Trans-Antarctic Expedition:	1957–1958
Traversed the Antarctic mainland by tractor train and dog sleds from the Filchner Ice Shelf to McMurdo Sound (Section 1.5.4).	
International Geophysical Year:	1957–1958
Twelve nations established more than 50 stations in Antarctica and undertook a coordinated program of research.	

1.10.2 Principal Research Stations in Antarctica (Stonehouse 2002)

Nation	Station	Location
Argentina	Primero de Mayo (Decepcion)	62°57'S, 060°38'W
	Teniente Jubany	62°14'S, 058°40'W
	Petrel	63°28'S, 056°17'W
	Melchior (1947)	64°20'S, 062°59'W
	General San Martin	68°07'S, 067°06'W
	Almirante Brown	64°53'S, 062°53'W
	Esperanza	63°24'S, 057°00'W
	Teniente Cámara	62°36'S, 059°54'W
	General Belgrano I	77°43'S, 038°04'W
	General Belgrano II	77°52'S, 034°37'W
	General Belgrano III	77°54'S, 045°59'W
	Primavera	64°09'S, 060°57'W
Australia	Casey (Wilkes)	66°17'S, 110°32'E
	Davis	68°35'S, 077°59'E
	Mawson	67°36'S, 062°52'E
Belgium	Roi Baudouin (The Netherlands)	70°26'S, 024°19'E
Brazil	Commandante Ferraz	62°05'S, 058°23'W
Bulgaria	St. Kliment Ohridiski	62°38'S, 060°22'W
Chile	Capitán Arturo Prat	62°30'S, 059°41'W
	General Bernardo O'SHiggins	63°12'S, 058°58'W
	Esperanza	63°24'S, 057°00'W
	General Ramon Cañas Montalva	63°32'S, 066°48'W
	Presidente Eduardo Frei	62°11'S, 058°57'W
	Presidente González Videla	64°49'S, 062°51'W
	Escudero (Teniente Rodolfo Marsh)	62°12'S, 057°00'W
	Teniente Carvajal (Adelaide)	67°46'S, 068°55'W
China	Great Wall	62°13'S, 058°58'W
	Zhongshan	69°22'S, 076°22'E
Ecuador	Pedro Vicente Maldonado	62°27'S, 059°43'W
Finland	Aboa (summer only)	73°03'S, 013°25'W
France	Dumont d'Urville	66°40'S, 140°01'E
	Concordia (Italy)	75°00'S, 125°00'E
Germany	Georg von Neumayer	70°38'S, 008°16'W
	Dallmann Laboratory	62°14'S, 058°40'W
	Lillie Marleen (hut)	71°12'S, 164°31'E
German Democratic Republic	George Forster	70°46'S, 011°50'E
India	Maitri	70°46'S, 011°44'E
Italy	Terra Nova Bay	74°41'S, 164°07'E
	Concordia	75°00'S, 125°00'E
Japan	Syowa	69°00'S, 139°35'E
	Asuka	71°31'S, 024°08'E
	Dome Fuji	77°30'S, 037°30'E
	Mizuho	70°41'S, 044°54'E
Korea	King Sejong	62°13'S, 058°47'W
The Netherlands	Roi Baudouin (Belgium)	70°26'S, 024°19'E
New Zealand	Scott Base (Ross Island)	77°30'S, 168°00'E

(continued)

1.10.2 (continued)

Nation	Station	Location
Norway (Sweden, Finland)	Norway	70°30'S, 002°32'W
	Troll	72°01'S, 020°32'E
	Tor	71°53'S, 005°09'E
Peru	Machu Picchu	62°05'S, 058°28'W
Poland	Henryk Arctowski	62°09'S, 058°27'W
Russia/USSR	Mirnyy	66°33'S, 093°01'E
	Pionerskaya (IGY only)	69°44'S, 095°30'E
	Oazis (Poland)	66°16'S, 100°44'E
	Kosomal'skaya (IGY only)	74°05'S, 097°29'E
	Vostok	78°28'S, 106°48'E
	Sovetskaya (IGY only)	78°23'S, 087°32'E
Soviet Union	Lazarev (1959)	70°00'S, 013°00'E
	Novolazarevskaya (1961)	70°46'S, 011°50'E
	Molodezhnaya (1962)	67°40'S, 045°51'E
	Bellingshausen (1968)	62°11'S, 058°58'W
	Leningradskaya (1970)	69°30'S, 159°23'E
	Druzhnaya (1976)	77°34'S, 040°13'W
	Russkaya (1979)	74°42'S, 136°51'W
	Soyoz (1982)	70°35'S, 068°47'E
	Progress (1989)	69°23'S, 076°23'E
South Africa	SANAE (1960)	70°41'S, 002°50'W
	Borga (1969, summer)	72°58'S, 003°48'W
	Sarie Marais	72°03'S, 002°49'W
Spain	Juan Carlos Island	62°39'S, 060°23'W
	Gabriele de Castilla	62°58'S, 060°41'W
Sweden	Wasa	73°03'S, 013°25'W
(Norway, Finland)	Svea	74°35'S, 011°13'W
Ukraine (UK)	Akademik Vernadsky (Faraday, UK)	65°15'S, 064°16'W
United Kingdom	Bird Island	54°00'S, 038°03'W
	Halley	74°35'S, 026°30'W
	Rothera	67°34'S, 068°07'W
	Signy (summer)	60°43'S, 045°36'W
	Fossil Bluff (1961–63, 1969–75)	71°20'S, 068°17'W
	Grytviken (King Edward Point)	54°17'S, 036°30'W
	Horseshoe Island	67°49'S, 067°18'W
USA	Byrd (1957)	80°01'S, 119°32'W
	Wilkes (Australia)	66°15'S, 110°32'E
	Ellsworth (1957–1962)	77°43'S, 041°08'W
	Hallett (New Zealand)	72°18'S, 179°16'E
	Plateau (1965–1968)	79°28'S, 040°35'E
	Amundsen-Scott	90°S
	McMurdo	77°51'S, 166°40'E
	Palmer (1965)	64°46'S, 064°03'W
	Siple (1969)	75°33'S, 083°33'W
	Eights (1961–1965)	75°14'S, 077°10'W
Uruguay	Artigas (1984)	62°11'S, 058°54'W
	Teniente Puperto Elichiribehety	63°24'S, 056°59'W

1.10.3 Member Nations of the Scientific Committee on Antarctic Research (SCAR) (Stonehouse 2002)

	Dates of Joining	
Nation	Associate Member	Full Member
Argentina	March 5, 1995	Feb. 3. 1958
Australia	Sept. 5, 1994	Feb. 3. 1958
Belgium	July 23, 1990	Feb. 3. 1958
Brazil	Sept. 12, 1988	Oct. 1, 1984
Bulgaria	June 15, 1992	July 27, 1998
Canada	July 1, 1988	Feb. 3. 1958
Chile		June 23, 1986
China, R.R.		June 15, 1992
Colombia[a]		July 23, 1990
Ecuador		March 3, 1958
Estonia		May 22, 1978
Finland		
France		
Germany[b]		
India		Oct. 1, 1984
Italy	May 19, 1987	Sept. 12, 1988
Japan		Feb. 3, 1958
Korea (South)	Dec. 8, 1987	July 23, 1990
The Netherlands	May 20, 1987	July 23, 1990
New Zealand		Feb., 3, 1958
Norway		Feb. 3. 1958
Pakistan	June 15, 1992	
Peru	April 14, 1987	
Poland		May 22, 1978
Russia/USSR[c]		Feb. 3. 1958
South Africa		Feb. 3. 1958
Spain	Jan. 15, 1987	July 23, 1990
Sweden	March 24, 1987	April 24, 1984
Switzerland	June 16, 1987	
Ukraine	Sept. 5, 1994	
UK		Feb. 3. 1958
USA		Feb. 3. 1958
Uruguay	July 29, 1987	

[a]Colombia withdrew from SCAR on July 3, 1995.
[b]The German Democratic Republic originally joined SCAR on September 9, 1981. Germany was later unified on October 3, 1990.
[c]The USSR was replaced by Russia in December of 1991.

1.10.4 International Symposia of Antarctic Research Organized by SCAR (Ford 2006)

Year	Country	City	Proceedings
1963	South Africa	Cape Town	Adie (1964)
1970	Norway	Oslo	Adie (1972)
1977	USA	Madison	Craddock (1982)
1982	Australia	Adelaide	Oliver et al. (1983)
1987	UK	Cambridge	Thomson et al. (1991)
1891	Japan	Tokyo	Yoshida et al. (1992)
1995	Italy	Siena	Ricci (1997)
1999	New Zealand	Wellington	Gamble et al. (2002)
2003	Germany	Potsdam	Fűtterer et al. (2006)
2008	Russia	St. Petersburg	

1.10.5 Gondwana Conferences and Their Proceedings Volumes

Year	Country	City	Editor(s)
1. 1967	Argentina	Mar Del Plata	Anonymous (1970)
2. 1970	South Africa	Cape Town and Johannesburg	Haughton (1970)
3. 1973	Australia	Canberra	Campbell (1975)
4. 1977	India	Calcutta	Laskar and Raja Rao (1979a, b)
5. 1980	New Zealand	Wellington	Creswell and Vella (1981)
6. 1985	USA	Columbus, OH	McKenzie (1987a, b)
7. 1988	Brazil	Sao Paulo	Ulbrich and Rocha Campos (1991)
8. 1991	Australia	Hobart	Findlay et al. (1993)
9. 1994	India	Hyderabad	Anonymous (1994)
10. 1998	South Africa	Cape Town	Almond and DeWitt (1999)
11. 2002	New Zealand	Christchurch	Storey et al. (1999) Gondwana Res., 7 (1)
12. 2005	Argentina	Mendoza	
13. 2008	China	Dali	

1.10.6 Selected Volumes of the Antarctic Research Series of the American Geophysical Union (AGU) of Washingtion, DC (All Publications Listed Here Are Also Included in Section 1.10)

Volume	Year	Editor(s)	Title
2	1964	Mellor, M.	Antarctic snow and ice studies
4	1965	Waynick, A.T.	Geomagnetism and aeronomy; studies in the ionosphere, geomagnetism and atmospheric radio noise
6	1965	Hadley, J.B.	Geology and paleontology of the Antarctic
8	1966	Tedrow, J.C.F.	Antarctic soils and soil-forming processes
9	1966	Rubin, M.J.	Studies in Antarctic meteorology
15	1971	Reid, J.L.	Antarctic oceanology I
16	1971	Crary, A.P.	Antarctic snow and ice studies II
19	1972	Hayes, D.E.	Antarctic oceanology II: The Australian-New Zealand sector
21	1974	Foster, M.W.	Recent Antarctic and subantarctic brachiopods
22	1974	Gunderson, E.K.E.	Human adaptability to Antarctic conditions
25	1977	Businger, J.	Meteorological studies at Plateau Station
29	1978	Lanzerotti, L.J. and C.G. Park	Upper atmosphere research in Antarctica
33	1981	McGinnis, L.D.	Dry Valley drilling project
36	1982	Turner, M.D. and J.F. Splettstoesser	Geology of the central Transantarctic Mountains
42	1990	Bentley, C.R. and D.E. Hayes	The Ross Ice Shelf: Glaciology and geophysics
43	1985	Jacobs, S.	Oceanology of the Antarctic continental shelf
46	1986	Stump, E.	Geological investigations in Northern Victoria Land
48	1990	LeMasurier, W.E. and J.W. Thomson	Volcanoes of the Antarctic plate and southern ocean
50	1990	Elliot, D.H.	Contributions to Antarctic research I
51	1990	Splettstoesser, J.F. and G.A.M. Dreschhoff	Mineral resources potential of Antarctica
53	1991	Elliot, D.H.	Contributions to Antarctic research II
54	1991	Hayes, D.E.	Marine geological and geophysical atlas of the circum-Antarctic to 30°S
55	1992	Stilwell, J.D. and W.J. Zinsmeister	Molluscan systematics and biostratigraphy of the Lower Tertiary La Meseta Formation
56	1992	Kennett, J.P. and D.A. Warnke	The Antarctic paleoenvironement; A perspective on global change, Part I
57	1992	Elliot, D.H.	Contributions to Antarctic research III
59	1993	Green, W.J. and E.I. Friedmann	Physical and biochemical processes in Antarctic lakes

(continued)

1.10.6 (continued)

Volume	Year	Editor(s)	Title
60	1993	Kennett, J.P. and D.A. Warnke	The Antarctic paleoenvironment: A perspective on global change, Part II
61	1994	Bromwich, D.H. and C.R. Stearns	Antarctic meteorology and climatology: Studies based on automatic weather stations
62	1994	Weiler, C.S. and P.A. Penhale	Ultraviolet radiation in Antarctica; Measurements and biological effects
66	1995	Kyle, P.R.	Volcanological and environemental studies of Mt. Erebus, Antarctica
67	1995	Elliot, D.H. and G.L. Blaisdell	Contributions to Antarctic research IV
68	1995	Cooper, A.K. et al.	Geology and seismic stratigraphy of the Antarctic margin, Part 1
71	1997	Barker, P.F. and A.K. Cooper	Geology and seismic stratigraphy of the Antarctic margin, Part 2
72	1998	Priscu, J.C.	Ecosystem dynamics in a polar desert: The McMurdo dry valleys, Antarctica
74	1998	Jeffries, M.O.	Antarctic sea ice: Physical processes, interactions, and variability
75	1998	Jacobs, S. and R. Weiss	Ocean, ice, and atmosphere; Interactions at the Antarctic continental margin
76	2000	Stilwell, J.D. and R.M. Feldmann	Paleobiology and paleoenvironments of Eocene rocks, McMurdo Sound, East Antarctica
77	2001	Alley, R.B. and R.A. Bindschadler	The West Antarctic ice sheet: Behavior and environment
78	2004	DiTullio, G.R. and R.B. Dunbar	Biogeochemistry of the Ross Sea
79	2003	Domack, E.W., A. Deventer, A. Burnett, R. Bindschadler, P. Convey, and M. Kirby	Antarctic Peninsula climate variability: Historical and paleoenvironmental perspectives

1.10.7 Memoirs, Special Papers, and Maps of the Geological Society of America (GSA) Relevant to Antarctica

Number	Year	Authors/editors	Titles
A. Special papers			
336	1999	Ramos and Keppie	Laurentia-Gondwana connections before Pangea
339	1999	Dressler and Sharpton	Large meteorite impacts and planetary evolution II
356	2002	Koeberl and MacLeod	Catastrophic events and mass extinctions: Impacts and beyond
384	2005	Kenkman et al.	Large meteorite impacts III
B. Memoirs			
170	1992	Webers et al.	Geology and paleontology of the Ellsworth Mountains, West Antarctica
187	1995	Veevers and Tewari	Gondwana master basin of peninsular India between Tethys and the interior of the Gondwanaland Province of Pangea
192	1999	Crowell	Pre-Mesozoic ice ages; Their bearing on understanding the climate system
Maps and charts			
MCH 093F (folded)	2005	Davis and Blankenship	Geology of the Scott-Reedy glaciers area, southern Transantarctic Mountains, Antarctica

References (Including those listed in Appendices 1.10.4, 1.10.5, 1.10.6, and 1.10.7)

Adie RJ (ed) (1964) Antarctic geology. North Holland, Amsterdam, The Netherlands

Adie RJ (ed) (1972) Antarctic geology and geophysics. Universitetsforlaget, Oslo, Norway

Allen AD (1962) Geological investigations in Southern Victoria Land, Antarctica. New Zealand J. Geol. Geophys. 5:278–294

Amundsen R (1912) The South Pole: an account of the Norwegian Antarctic expedition in the Fram, 1910–1912. Two volumes. John Murray, London

Anderson PJ (1974) Richard Evelyn Byrd: polar explorer. The Iron Worker 28(4):2–13

Andrews S (2007) In cold pursuit. St. Martin's Minotaur, New York

Anonymous (1970) Symposium on Gondwana stratigraphy. First Gondwana Symposium, Mar del Plata, Argentina. Rijka Geologische Dienst, Haarlem, The Netherlands

Anonymous (1994) Gondwana nine; Ninth International Gondwana Symposium, vols. 1 and 2. Oxford/IBH Publishing, New Delhi, India

Anonymous (1997) The United States in Antarctica; Report of the US Antarctic Program External Panel. National Science Foundation, Washingtion, DC

Anonymous (ed) (2008) Gondwana 13 Conference, Dali, Yunnan Province, China

Arnesen L, Bancroft A (2003) No horizon is so far. Da Capo, Cambridge, MA

Barker PF, Cooper AK (eds) (1997) Geology and seismic stratigraphy of the Antarctic margin, Part 2. Antarctic Research Series, vol. 71. American Geophysical Union, Washington, DC

Baughman T (1994) Before the heroes came: Antarctica in the 1890s. University of Nebraska Press, Lincoln, NE

Beaglehole JC (ed) (1955) The journals of Captain James Cook on his voyages of discovery. I. The voyage of the Endeavour 1768–71. Hakluyt Society, London

Beaglehole JC (ed) (1961) The journals of Captain James Cook on his voyages of discovery. II. The voyage of the Resolution and Adventure 1772–75. Hakluyt Society, London

Beaglehole JC (ed) (1962) The Endeavour journals of Joseph Banks. 2 vols. Angus and Robertson, Sydney, Australia

Beaglehole JC (ed) (1967) The journals of Captain James Cook on his voyages of discovery. III. The voyage of the Resolution and Disovery 1776–1780. Hakluyt Society, London

Beck P (1986) The international politics of Antarctica. Croom Helm, London

Behrendt JC (1998) Innocents on the Ice; A memoir of Antarctic exploration, 1957. University of Colorado Press, Niwot, CO

Bellec F (2000) Unknown lands; the log books of the great explorers. Overlook, Woodstock, GA/New York

Bentley CR, Hayes D (eds) (1990) The Ross Ice Shelf: glaciology and geophysics. Antarctic Research Series, vol. 42. American Geophysical Union, Washington, DC

Berkman PA (2002) Science into policy: global lessons from Antarctica. Academic, San Diego, CA

Bertrand, K.J., 1971. Americans in Antarctica 1775–1948. American Geographical Society, New York

Bromwich DH, Stearns CR (eds) (1994) Antarctic meteorology and climatology: studies based on automatic weather stations. Antarctic Research Series, vol. 61. American Geophysical Union, Washington, DC

Bull C, Wright PF (eds) (1993) Silas: the Antarctic diaries and memoir of Charles S. Wright. Ohio State University Press, Columbus, OH

Businger J (ed) (1977) Meteorological studies of Plateau Station, Antarctica. Antarctic Research Series, vol. 25. American Geophysical Union, Washington, DC

Byrd RE (1931) Little America; aerial exploration in the Antarctic. Putnam, New York

Byrd RE (1935) Discovery: the story of the second Byrd Antarctic expedition. Putnam, New York

Byrd RE (1938) Alone. Putnam, London

Campbell KSW (ed) (1975) Gondwana geology; papers presented at the Third Gondwana Symposium. Australian National University Press, Canberra, ACT

Chapman W (ed) (1965) Antarctic conquest: the great explorers in their own words. Bobbs-Merrill, New York

Cherry-Garrard A (1922) The worst journey in the world. Penguin Books (Republished in 1994 by Picador, London)

Collerson KD, McCulloch MT (1983) Nd and Sr isotope geochemistry of leucite-bearing lavas from Gaussberg, East Antarctica. In: Oliver RL, James PR, Jago JB (eds) Antarctic earth science, Australian Academy of Science, Canberra, ACT, pp 676–680

Cook FA (1909) Through the first Antarctic night 1898–1899. Doubleday, New York

Cooper AK, Barker PF, Brancolini G (eds) (1995) Geology and seismic stratigraphy of the Antarctic margin. Antarctic Research Series, vol. 68. American Geophysical Union, Washington, DC

Craddock C (ed) (1982) Antarctic geoscience. University of Wisconsin Press, Madison, WI

Crary AP, Gould LM, Hurlbut EO, Odishaw H, Smith WE (eds) (1956) Antarctica in the International Geophysical Year. Geophysical Monograph No. 1. American Geophysical Union, Washington, DC

Crary AP (1962) The Antarctic. Scientific American 207(3):2–14

Crary AP (ed) (1971) Antarctic snow and ice studies II. Antarctic Research Series, vol. 16. American Geophysical Union, Washington, DC

Crary AP (1982) International geophysical year: its evolution and US participation. Ant J US 17(4):1–6

Creswell MM, Vella P (eds) (1981) Gondwana Five: papers and abstracts of papers presented a the Fifth International Gondwana Symposium. Balkema, Rotterdam, The Netherlands

Crowell JC (1999) Pre-Mesozoic ice ages: their bearing on understanding the climate system. Memoir 192, The Geological Society of America, Boulder, CO

Davis MB, Blankenship DD (2005) Geology of the Scott-Reedy glaciers area, southern Transantarctic Mountains, Antarctica. MCH 093F (folded). The Geological Society of America, Boulder, CO

Debenham F (ed) (1945) The voyage of Captain Bellingshausen in the Antarctic Seas, 1819–1821. 2 vols. Hakluyt Society, London
Dewart G (1989) Antarctic comrades; An American with the Russians in Antarctica. Ohio State University Press, Columbus, OH
DiTullio GR, Dunbar RB (eds) (2003) Biochemistry of the Ross Sea. Antarctic Research Series, vol. 78. American Geophysical Union, Washington, DC
Domack E, Deventer A, Burnett A, Bindschadler R, Convey P, Kirby M (2003) Antarctic Peninsula climate variability: historical and paleoenvironmental perspectives. Antarctic Research Series, vol. 79. American Geophysical Union, Washinton, DC
Doumani GA (ed) (1965) Antarctic bibliography, vol. 1. Office of Antarctic Programs, National Science Foundation, Washington, DC
Doumani GA (1999) The frigid mistress; Life and exploration in Antarctica. Noble House, Baltimore, MD
Dressler BO, Sharpton VL (eds) (1999) Large meteorite impacts and planetary evolution. Special Paper 339. The Geological Society of America, Boulder, CO
von Drygalski E (1912) Der Gaussberg; Seine Kartierung und seine Formen. Sűdpolar-Expedition 1901–1903. Geographie und Geologie 11(1):1–46
Duncan RA (1981) Hot spots in the southern ocean – an absolute frame of reference for the motion of Gondwana continents. Tectonophysics 74:29–42
Elliot DH (ed) (1990) Contributions to Antarctic research I. Antarctic Research Ser., vol. 50. American Geophysical Union, Washington, DC
Elliot DH (ed) (1991) Contributions to Antarctic research II. Antarctic Research Ser., vol. 53. American Geophysical Union, Washington, DC
Elliot DH (ed) (1992) Contributions to Antarctic research III. Antarctic Research Ser., vol. 57. American Geophysical Union, Washington, DC
Elliot DH, Blaisdell GL (eds) (1995) Contributions to Antarctic research IV. Antarctic Research Ser., vol. 67. American Geophysical Union, Washington, DC
Emiliani C (1992) Planet Earth; Cosmology, geology, and the evolution of life and environment. Cambridge University Press, Cambridge
Faure G (1998) Principles and applications of geochemistry, 2nd edn. Prentice Hall, Upper Saddle River, NJ
Faure G (2001) Origin of igneous rocks; The isotopic evidence. Springer, Heidelberg, Germany
Faure G, Mensing TM (2007) Introduction to planetary science; The geological perspective. Springer, Dordrecht, The Netherlands
Fifield R(1987) International research in the Antarctic. Cambridge and Oxford University Press, SCAR/ICSU
Findlay RH, Unrug R, Banks MR, Veevers JJ (eds) (1993) Gondwana eight: assembly, evolution, and dispersal. A.A. Balkema, Rotterdam, The Netherlands
Fisher M, Fisher J (1957) Shackleton. Barrie, London
Ford AB (2006) The road to Gondwana via the early SCAR symposia. In: Fűtterer DK, Damaske D, Kleinschmidt G, Miller H, Tessensohn F (eds) Antarctica: contributions to global earth science, 3–6. Springer, Heidelberg, Germany
Foster MW (ed) (1974) Recent Antarctic and subantarctic brachiopods. Antarctic Research Series, vol. 21. American Geophysical Union, Washington, DC
Fuchs VE (1982) Of ice and men. Anthony Nelson, Oswestry
Fuchs VE (1990) A time to speak. Anthony Nelson, Oswestry
Fuchs VE, Hillary E (1958) The crossing of Antarctica. Cassell, London
Fűtterer DK, Damaske D, Kleinschmidt G, Miller H, Tessensohn F (eds) (2006) Antarctica: contributions to global earth science. Springer, Heidelberg, Germany
Gamble JA, Skinner DNB, Henrys S (eds) (2002) Antarctica at the close o f a millennium. Royal Soc New Zealand, Bull 35
Gillespie N (2005) Courage, sacrifice, devotion: the history of the United States Navy Antarctic Air Squadron VXE-Six. Infinity Publishing, 1094 New Dehaven St., Suite 100, West Conshohocken, PA, pp 19428–2713
Gould LM (1931) Cold: the record of an Antarctic sledge journey. Warren & Putnam, New York
Gould LM (1958) The polar regions in their relation to human affairs. American Geographical Society, New York
Green B (2008) Water, ice, and stone; Science and memory on the Antarctic lakes. Bellevue Literary Press, New York
Green WJ, Friedmann EI (eds) (1993) Physical and biogeochemical processes in Antarctic lakes. Antarctic Research Series, vol. 59. American Geophysical Union, Washington, DC
Gunderson EKE (ed) (1974) Human adaptability to Antarctic conditions. Antarctic Research Series, vol. 22. American Geophysical Union, Washington, DC
Hadley JB (ed) (1965) Geology and paleontology of the Antarctic. Antarctic Research Series, vol. 6. American Geophysical Union, Washington, DC
Haughton SH (ed) (1970) Second Symposium on Gondwana stratigraphy and paleontology. Council for Scientific and Industrial Research, Geological Society of South Africa, Marshalltown, South Africa
Hayes DE (ed) (1972) Antarctic oceanology II: the Australian-New Zealand sector. Antarctic Research Series, vol. 19. American Geophysical Union, Washington, DC
Hayes DE (ed) (1991) Marine geological and geophysical atlas of the circum-Antarctic to 30°S; Bathymetry map (rolled) and booklet. Antarctic Research Series, vol. 54. American Geophysical Union, Washington, DC
Huntford R (1986a) The last place on Earth, revised edn. Athenum, New York
Huntford R (1986b) Shackleton. Athenum, New York
Huntford R (ed) (1987) The Amundsen photographs. The Atlantic Monthly Press, New York
Huxley L (1914) Scott's last expedition, 2 vols. Smith Elder, London
Jacobs S (ed) (1985) Oceanology of the Antarctic continental shelf. Antarctic Research Series, vol. 43. American Geophysical Union, Washington, DC
Jacobs S, Weiss R (eds) (1998) Ocean, ice, and atmosphere: interactions at the Antarctic continental margin. Antarctic Research Series, vol. 75. American Geophysical Union, Washington, DC
Jones TO, Bakutis FE (1966) Introduction to the Antarctic Journal of the United States. Ant J US 1(1):1
Joyce EEM (1929) South polar trail: the log of the Imperial Trans-Antarctic Expedition. Duckworth, London

Joyner CC (1992) Antarctica and the Law of the Sea. Martinus Nijhoff, Dordrecht, The Netherlands/Boston, MA/London

Joyner CC, Theis ER (1997) Eagle over the ice; The U.S. in the Antarctic. University Press of New England, Hanover, NH

Kenkmann ThF Hőrz, Deutsch A (eds) (2005) Large meteorite impacts III. Special Paper 384. The Geological Society of Americal Bulletin, Boulder, CO

Kennett JP, Warnke DA (eds) (1992) The Antarctic paleoenvironment; A perspective of global change, Part 1. Antarctic Research Series, vol. 56. American Geophysical Union, Washington, DC

Koeberl C, MacLeod KG (eds) (2002) Catastrophic events and mass extinctions; Impacts and beyond. Special Paper 356. The Geological Society of American Bulletin, Boulder, CO

Kyle PR (ed) (1995) Volcanological and environmental studies of Mount Erebus, Antarctica. Antarctic Research Series, vol. 66. American Geophysical Union, Washington, DC

Lansing A (1959) Endurance, Shackleton's incredible voyage. McGraw-Hill, New York

Lanzerotti LJ, Park CG (eds) (1978) Upper atmosphere research in Antarctica. Antarctic Research Series, vol. 29. American Geophysical Union, Washington, DC

Laskar B, Raja Rao CS (eds) (1979) Fourth International Gondwana Symposium (vols. 1 and 2). Hindustan, Delhi

LeMasurier WE, Thomson JW (eds) (1990) Volcanoes of the Antarctic Plate and Southern Oceans. Antarctic Research Series, vol. 48. American Geophysical Union, Washington, DC

Lűdecke C (2006) Exploring the unknown: history of the first German south polar expedition 1901–1903. In: Fűtterer DK, Damaske D, Kleinschmidt G, Miller H, Tessensohn F (eds) (2006) Antarctica; Contributions to Global Earth Sciences. Springer, Heidelberg, Germany, pp 7–11.

Mawson D (1915) The home of the blizzard. 2 vols. Heinemann, London

McGinnis LD (ed) (1981) Dry Valley drilling project. Antarctic Research Series, vol. 33. American Geophysical Union, Washington, DC

McGonigal D, Woodworth L (2001) Antarctica and the Arctic; The complete encyclopedia. Firefly Books, Global Book Publishing, Auckland, New Zealand

McKenzie GD (1984) National security and the austral crescent. Quarterly Report 8(4):1–8. Mershon Center, The Ohio State University, Columbus, OH

McKenzie GD (ed) (1987a) Gondwana six; Structure, tectonics, and geophysics. American Geophysics Union, Geophysical Monograph 40, Washington, DC

McKenzie GD (ed) (1987b) Gondwana six; Stratigraphy, sedimentology, and paleontology. American Geophysical Union, Geophysical Monograph 41, Washington, DC

McPherson JG (1975) Footprints on a frozen continent. Hicks Smith, Wellington, New Zealand

Mears R, Swan R (1987) A walk to the Pole. Crown, New York

Mellor M (ed) (1964) Antarctic snow and ice studies. Antarctic Research Series, vol. 2. American Geophysics Union, Washington, DC

Messner R (1990) Antarktis; Himmel und Hőlle zugleich. Piper, Műnchen, Germany

Monteath C, Monteath B (1990–2004) Antarctica. Annual calendars, Hedgehog House, Christchurch, New Zealand

Neider C (ed) (1972) Antarctica: authentic accounts of life and exploration in the world's highest, driest, windiest, coldest, and most remote continent. Random House, New York

Neider C (1974) Edge of the world; Ross Island, Antarctica. Doubleday, Garden City, NJ

Nordenskjőld NOG, Andersson JG, Skottsberg C, Larsen CA (1905) Antarctica; Or two years amongst the ice of the South Pole. Hurst & Blackett, London

Odishaw H, Ruttenberg S (eds) (1958) Geophysics and the IGY. Geophysical Monograph No. 2, American Geophysical Union, Washington, DC

Oliver RL, James PR, Jago JB (eds) (1983) Antarctic earth science. Australian Academy of Science, Canberra

Parfit M (1985) South light: a journey to the last continent. Macmillan, New York

Philippi E (1912a) Geologische Beschreibung des Gaussbergs. Deutsche Sűdpolar Expedition 1901–1903. Geographie und Geologie 11(1):47–71

Phillippi E (1912b) Die Grundproben der deutschen Sűdpolar Expedition 1901–1903. Deutsche S dpolar Expedition 2(6):411–616

Porter E (1978) Antarctica. E.P. Dutton, New York

Priestley RE (1914) Antarctic adventure; Scott's northern party. T. Fisher Unwin, London

Priscu JC (ed) (1998) Ecosystem dynamics in a polar desert: the McMurdo dry valleys, Antarctica. Antarctic Research Series, vol. 72. American Geophysical Union, Washington, DC

Pyne SJ (1986) The ice; A journey to Antarctica. University of Iowa Press.

Ramos VA, Keppie JD (eds) (1999) Laurentia-Gondwana connections before Pangea. Special Paper 336. The Geological Society of America, Boulder, CO

Reader's Digest (1985) Antarctica; Great stories from the frozen continent. Reader's Digest Services, Surry Hill, New South Wales, Australia

Reader's Digest (1990) Antarctica: the extraordinary story of man's conquest of the frozen continent, 2nd edn. Reader's Digest, Sydney, Australia

Reid JL (ed) (1971) Antarctic oceanology I. Antarctic Research Series, vol. 15, American Geophysical Union, Washington, DC

Rheinisch R (1912) Petrographische Beschreibung der Gaussberg-Gesteine. Deutsche Sűdpolar-Expedition 1901–1903. Geographie und Geologie 11(1):73–87

Ricci CA (ed) (1997) The Antarctic region: geological evolution and processes. Terra Antarctica Pub., Siena, Italy

Riffenburgh B (ed) (2007) Encyclopedia of the Antarctic, vols. 1 and 2. Rutledge, Taylor & Francis, New York

Ross JC (1847) A voyage of discovery and research in the southern Antarctic regions during the years 1839–43. John Murray, London

Rowell G (1995) Poles apart: parallel visions of the Arctic and Antarctic. University California Press, Los Angeles, CA

Rubin MJ (ed) (1966) Studies in Antarctic meteorology. Antarctic Research Series, vol. 9. American Geophysical Union, Washington, DC

Schulthess E (1960) Antarctica; A photographic survey. Simon & Schuster, New York

Scott RF (1905) The voyage of the Discovery. Smith Elder, London

Shackleton EH (1909) The heart of the Antarctic, 2 vols. Heinemann, London
Shackleton EH (1920) South. Macmillan, New York
Sheraton JW, Cundari A (1980) Leucitites from Gaussberg, Antarctica Contrib Mineral Petrol 71:417–427
Siple PA (1931) A boy scout with Byrd. Putnam, New York
Siple PA (1936) Scout to explorer. Back with Byrd to the Antarctic. Putnam, New York
Siple P (1959) 90° South: the story of the American South Pole conquest. Putnam, New York
Solomon S (2001). The coldest march. Scott's fateful Antarctic expedition. Yale University Press, New Haven, CT
Splettstoesser JF, Dreschhoff GAM (eds) (1990) Mineral resources potential of Antarctica. Antarctic Research Series, vol. 51. American Geophysical Union, Washington, DC
Steger W, Bowermaster J (1992) Crossing Antarctica. Knopf, New York
Stilwell JD, Zinsmeister WJ (eds) (1992) Molluscan systematics and biostratigraphy of the Lower Tertiary La Meseta Formation. Antarctic Research Series, vol. 55. American Geophysical Union, Washington, DC
Stilwell JD, Feldmann RM (eds) (2000) Paleobiology and paleoenvironments of Eocene rocks, McMurdo Sound, East Antarctica. Antarctic Research Series, vol. 76. American Geophysical Union, Washington, DC
Stonehouse B (ed) (2002) Encyclopedia of Antarctica and the southern ocean. Wiley, Chichester, England
Storey BC, Rubridge BS, Cole DI, deWit MJ (eds) (1999) Gondwana 10; Event stratigraphy of Gondwana, vols. 1, 2, and 3. Elsevier, Oxford
Strange C, Bashford A (2008) Griffith Taylor; visionary environmentalist explorer. National Library of Australia, Canberra, ACT
Stump E (ed) (1986) Geological investigations in Northern Victoria Land. Antarctic Research Series, vol. 46. American Geophysical Union, Washington, DC
Sullivan W (1957) Quest for a continent. McGraw-Hill, New York
Tedrow JCF (ed) (1966) Antarctic soils and soil-forming processes. Antarctic Research Series, vol. 8. American Geophysical Union, Washington, DC
Thomson MRA, Crame JA, Thomson JW (eds) (1991) Geological evolution of Antarctica. Cambridge University Press, Cambridge
Tingey RJ (1990) Gaussberg. In: LeMasurier WE, Thomson JW (eds) Volcanoes of the Antarctic Plate and Southern Oceans. Antarctic Research Series, vol. 48. American Geophysical Union, Washington, DC, pp 446–448
Tingey RJ, McDougall I, Gleadow AJW (1983). The age and mode of formation of Gaussberg, Antarctica. J Geol Soc Aust 30:241–246
Trewby M (ed) (2002) Antarctica; An encyclopedia from Abbott Ice Shelf to Zooplankton. Firefly Books, Auckland, New Zealand
Turner MD, Splettstoesser JF (eds) (1982) Geology of the central Transantarctic Mountains. Antarctic Research Series, vol. 36. American Geophysical Union, Washington, DC
Veevers JJ, Tewari RC (1995) Gondwana master basin of peninsular India between Tethys and the interior of the Gondwana Province of Pangea. Memoir 187, The Geological Society of America, Boulder, CO
Victor* P-E (1964). Man and the conquest of the poles. Hamish Hamilton, London
Walton K, Atkinson R (1996) Of dogs and men: fifty years in the Antarctic. Images, Malvern, PA
Waynick AH (ed) (1965) Geomagnetism and aeronomy; studies in the ionosphere, geomagnetism and atmospheric radio noise. Antarctic Research Series, vol. 4. American Geophysical Union, Washington, DC
Webers GF Craddock C, Splettstoesser JF (eds) (1992) Geology and paleontology of the Ellsworth Mountains, West Antarctica. Memoir 170, The Geological Society of America, Boulder, CO
Weiler CS, Penhale PA (eds) (1994) Ultraviolet radiation in Antarctica; Measurement and biological effects. Antarctic Research Series, vol. 62. American Geophysical Union, Washington, DC
Wexler H, Rubin MJ, Caskey JE Jr (eds) (1962) Antarctic research: the Matthew Fontaine Maury Memorial Symposium. Monograph No. 7. American Geophysical Union, Washington, DC

Chapter 2
Antarctica: The Continent

The area of Antarctica is 13.97×10^6 km^2 making it the fifth largest of the seven continents (Stonehouse 2002). It is conventionally oriented on maps as shown in Fig. 2.1 and is subdivided into East Antarctica, West Antarctica, the Antarctic Peninsula, and certain islands that rise more than 500 m above sea level (i.e., Alexander, Bear, Berkner, Roosevelt, Ross, and Thurston). In addition, Antarctica is surrounded by the Ross, Ronne, Filchner, Riiser-Larsen, Fimbul, and Amery floating ice shelves as well as by the Larsen ice shelf located along the east coast of the Antarctic Peninsula. Except for the northernmost tip of the Antarctic Peninsula, the continent lies within the Antarctic Circle at latitudes greater than 62.5° south.

The continuing exploration of Antarctica has uncovered several natural phenomena that have not been recognized elsewhere on the Earth. All of these phenomena, which are mentioned here in passing, are described in detail in the appropriate chapters of this book:

1. The stratosphere over Antarctica is depleted in ozone because of the release of chlorine-bearing gases (CFCs) by humans in the mid-latitudes of the Earth.
2. Large meltwater lakes that exist at the base of the East Antarctic ice sheet may be linked by subglacial streams of liquid water.
3. Subglacial evaporite deposits form as a result of refreezing of meltwater at the base of the East Antarctic ice sheet in response to decreases in pressure.
4. Numerous lakes and ponds occur in the ice-free regions of southern Victoria Land and in the Bunger Hills of East Antarctica. Some of these ponds contain highly concentrated brines that do not freeze even at –50°C.
5. Cold and unusually salty water forms in the circum-Antarctic oceans and then sinks to the bottom of the ocean basins where it flows north into the South Atlantic, South Pacific, and the South Indian oceans.
6. Meteorite specimens that land on the East Antarctic ice sheet are transported to the margins of the ice sheet and may be exposed on the bare-ice fields in the zone of ablation adjacent to the Transantarctic Mountains, and at the Yamato and Grove mountains of East Antarctica. In addition, several dozen rock samples from the Moon and from Mars have been collected in Antarctica.

The conventional orientation of the continent of Antarctica in Fig. 2.1 causes East Antarctica to be located *west* of an observer facing north anywhere within the Transantarctic Mountains. This contradiction arises because East Antarctica is positioned on the right-hand side of a map when the continent is oriented in the conventional manner as in Fig. 2.1. All geological maps in this book are oriented with north at the top in order to avoid any ambiguity about the directions of east and west.

2.1 Topography

The rocks that form the crust of Antarctica are almost completely hidden under a thick ice sheet that covers 97.6% of the entire land mass. The average elevation of the surface of the ice sheet is more than 2,000 m above sea level (a.s.l.). The highest elevations occur in a series of ice domes that are roughly aligned along a semi-circle in the interior of East Antarctica: Dome Fuji (3,807 m at 77°00′S, 046°00′E), Dome Argus

G. Faure and T.M. Mensing, *The Transantarctic Mountains: Rocks, Ice, Meteorites and Water*,
DOI 10.1007/978-90-481-9390-5_2, © Springer Science+Business Media B.V. 2010

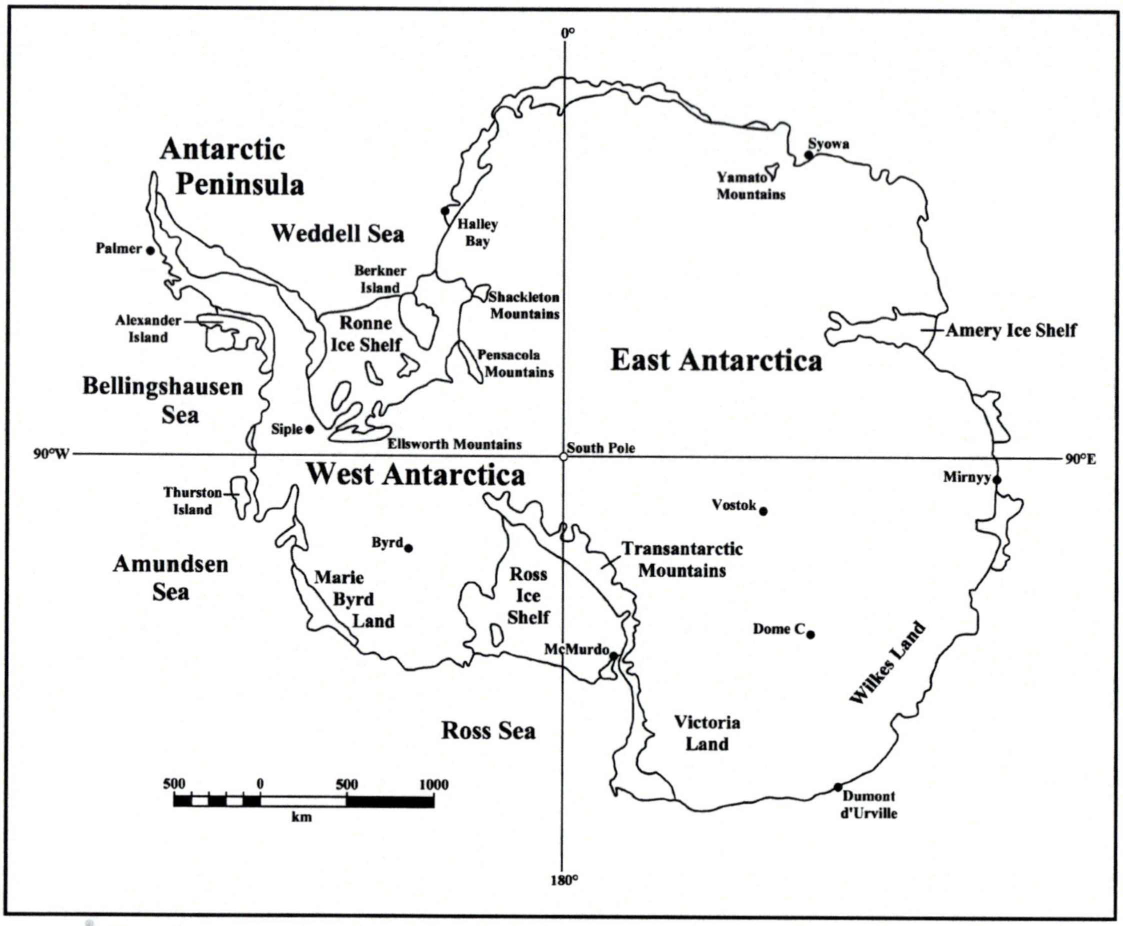

Fig. 2.1 Antarctica is conventionally oriented as shown and is subdivided into East Antarctica, West Antarctica, and the Antarctic Peninsula. The Transantarctic Mountains extend from northern Victoria Land along the Ross Sea and the Ross Ice Shelf toward the Pensacola Mountains adjacent to the Ronne Ice Shelf. Some of the research stations mentioned in the text are indicated on this map. A complete list of these stations is provided in Appendix 1.11.2 (Adapted from the Antarctic Journal of the US, volume 13, No. 4, October 1978)

(4,030 m at 81°00′S, 077°00′E), and Dome Charlie (3,206 m at 75°00′S, 125°00′E). The South Pole is located on the polar plateau at an elevation of 2,835 m. The average thickness of the Antarctic ice cap is 2,160 m and its volume has been estimated to be 30.11×10^6 km^3, which amounts to about 90% of the ice on the Earth (Stonehouse 2002). The greatest thickness of ice occurs in a small area in Wilkes Land at 69°30′S, 135°00′E where the bedrock is below sea level (National Geographic Society 1990). This site may be a large meteorite impact crater buried by the ice (Section 18.4.1).

The rocks of the Antarctic crust are exposed primarily on the Antarctic Peninsula, in the Transantarctic Mountains of East Antarctica, in the Ellsworth Mountains of West Antarctica, and in the extinct volcanoes of Marie Byrd Land. In addition, small mountain ranges project through the East Antarctic ice sheet in Queen (or Dronning) Maud Land, in Enderby Land, in Mac. Robertson Land, and in a few places in Wilkes Land (e.g., Gaussberg, Section 1.3.3).

The Sentinel Range of the Ellsworth Mountains in West Antarctica (Fig. 2.1) includes the Vinson Massif which contains the highest peak in Antarctica at 4,901 m (Stonehouse 2002). The highest peaks of the Transantarctic reach an elevation of 4,528 m in the Queen Alexandra Range along the north side of the Beardmore Glacier (National Geographic Society 1990).

Geophysical data indicate that the average thickness of the continental crust of East Antarctica is about 40 km. However, the crust thickens appreciably to about 55 km under the Transantarctic Mountains and under the subglacial Gamburtsev Mountains located in East Antarctica between 70° and 80°E longitude at about 80°S latitude (Bentley 1983; Kadmina et al. 1983).

The present elevation of the bedrock surface of some parts of East and West Antarctica is actually below sea level. For example, two large subglacial basins in Wilkes Land of East Antarctica lie below sea level. In addition, the surface of subglacial Lake Vostok at 78°28′S and 106°48′E is below sea level by about 200–300 m. The bedrock surface of most of West Antarctica is also below sea level partly because of the mass of the overlying ice sheet (Drewry et al. 1983; Bentley and Robertson 1982; Bentley et al. 1982). Consequently, the incursion of seawater into the subglacial basins could result in the break-up of the West Antarctic ice sheet and of the ice in the Wilkes basin of East Antarctica.

2.2 Volcanoes

Antarctica truly is the land of fire and ice because it contains a large number of volcanoes that were active during the Tertiary Period (LeMasurier 1990). At the present time only Mt. Erebus on Ross Island in Fig. 2.2a, b is still active. The summit of Mt. Erebus, which rises to a height of 3,794 m a.s.l., contains a large crater which intermittently fills with alkali-rich (basanite and kenyte) basalt lava (Kyle 1995).

Mt. Erebus was discovered in January of 1841 by Captain James Ross who named it after the ship he and his crew had sailed to Antarctica (Section 1.1). The volcano was emitting plumes of steam when it was discovered and it was still active in March of 1908 when it was climbed by William E. David and his five companions of Shackleton's Nimrod Expedition. When it was climbed a second time in December of 1912 by Raymond Priestley and other members of Scott's Terra Nova Expedition, it was still active. The activity has continued into the modern era and actually intensified in 1982. The volcano became violent in December of 1984 when it ejected incandescent lava bombs accompanied by thunderous explosions and by the emission of colored steam and other gases.

The volcanoes of Antarctica occur not only in the Transantarctic Mountains and on the islands off the coast of Victoria Land, but also in Marie Byrd Land of West Antarctica, on the Antarctic Peninsula and on its off-shore islands, on the South Sandwich Islands, in East Antarctica, and on the islands of the Southern Oceans. All of these volcanoes were described in a book edited by LeMasurier and Thomson (1990). The descriptions of the volcanoes include photographs, maps, chemical analyses of the rocks, and interpretations of these analyses. Even the Gaussberg on the coast of East Antarctica (Section 1.3.3) is included in this compilation of Antarctic volcanoes. The book also

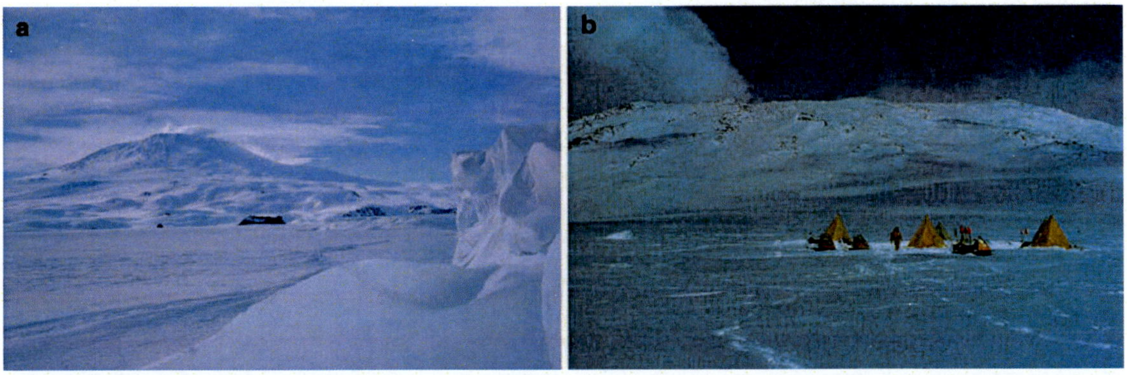

Fig. 2.2 (**a**) Mt. Erebus on Ross Island as seen from the Erebus Glacier tongue which projects into McMurdo Sound. A small plume of steam and other gases was rising from the summit of the volcano in January of 1985 (Photo by G. Faure) (**b**) Tent camp on the Fang Glacier below the summit of Mt. Erebus where scientists who need to work on the volcano spend at least one night in order to adjust to the high elevation (3,794 m above sea level). A large plume of steam was rising from the summit in the background during the 1982/83 field season (Photo courtesy of J.M. Palais)

contains references to the extensive literature on the volcanoes of Antarctica.

The volcanoes of the Transantarctic Mountains occur between Cape Hallett in northern Victoria Land and Mount Early (87°04′S, 153°46′W), which is located only 300 km from the South Pole. The volcanic mountains and cinder cones in the Transantarctic Mountains define the McMurdo Volcanic Province which is subdivided into four regions (Kyle 1990). These volcanoes were active in Late Tertiary time starting less than 25 million years ago and continuing to the present as in the case of Mount Erebus. All of them have extruded silica-undersaturated and alkali-rich lavas in marked contrast to the Ferrar Dolerite and Kirkpatrick Basalts of Middle Jurassic age both of which consist largely of silica-saturated tholeiites. These two suites of volcanic rocks are products of quite different petrogenetic processes that nevertheless occurred at different times in the lithospheric mantle underlying the Transantarctic Mountains. The petrogenesis of these different suites of volcanic rocks is the subject of Chapter 12 (Kirkpatrick Basalt) and Chapter 13 (Ferrar Dolerite), and Chapter 16 (Cenozoic Volcanoes).

The volcanoes in the Transantarctic Mountains and in Marie Byrd Land of West Antarctica erupted lava flows and pyroclastic ash that was deposited on the surface of the ice sheets. The ash was subsequently buried by snow and was thereby incorporated into the ice. The resulting ash layers now serve a useful purpose in the study of the ice sheets because they are unique "event horizons" whose age can be determined by isotopic methods (e.g., Folco et al. 2007). In addition, these horizons have preserved a record of the deformation of the ice sheets that is revealed by mapping their outcrop patterns on the bare-ice surfaces in the ablation zones. The chemical composition of the ash has been used to identify the volcanoes from which certain ash layers were erupted, while the sulfate concentration and the acidity (pH) of the ice above an ash layer provide clues to the amount of sulfuric acid that was injected into the stratosphere (Palais 1985). The volcanic dust and sulfuric acid in the stratosphere can cause temporary cooling of the global climate as demonstrated by the eruptions of Krakatau (Indonesia) in 1883, Mount St. Helens (Washington) in 1980, El Chichon (Mexico) in 1982, and Mount Pinatubo (Philippines) in 1992 (Holland and Petersen 1995; Thompson and Mosley-Thompson 1981; Kyle et al. 1981; Self et al. 1981).

2.3 Climate

The weather of Antarctica is proverbially hostile to life but actually varies widely depending on the season, the latitude, the elevation, and on local factors. The average monthly temperatures at McMurdo and South Pole stations in Fig. 2.3 display the range of seasonal temperature variations at these locations in relation to their latitude, elevation, and topographic setting:

McMurdo, 77°50′S, 24 m, coastal
South Pole, 90°S, 2,835 m, polar plateau

The highest average monthly temperature at the South Pole is only −32.3°C which occurs in December, whereas the highest average monthly temperature at McMurdo Station of −3.1°C occurs in December and January (Stonehouse 2002, p. 61). The highest average monthly temperatures at both locations in Antarctica are below the freezing temperature of water.

The "local conditions" that affect the weather alluded to above include the katabatic wind that blows from the polar plateau through the valleys of the outlet glaciers in the Transantarctic Mountains. The wind is generated when the layer of cold air that forms directly above the surface of the polar plateau drains downslope toward the coast. The speed of the wind on a gentle slope

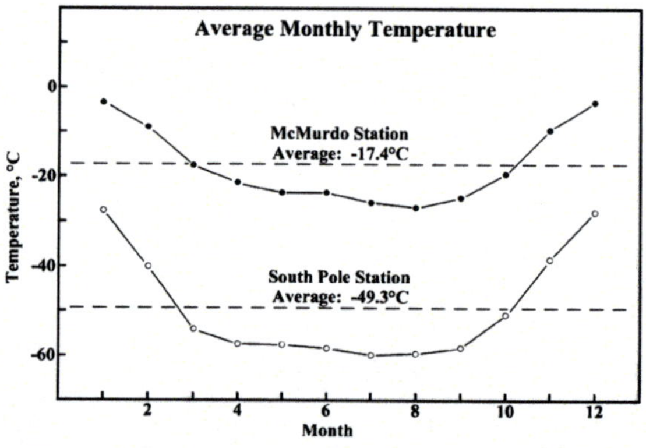

Fig. 2.3 The average monthly temperatures at McMurdo Station range from −26.9°C (August) to −3.1°C (December and January) compared to only −59.9°C (July) and −27.7°C (December) at the South Pole Station. The low seasonal temperature profile of the South Pole is partly attributable to its high southern latitude (90°S) and the high elevation of this site (2,835 m above sea level), whereas McMurdo Station is located at 77°50′S latitude at sea level on Ross Island. Nevertheless, the average annual temperatures at both sites are below the freezing temperature of water (Data from Stonehouse 2002)

is generally less than 36 km/h, but it can accelerate to hurricane strength when it is compressed in the narrow valleys through which ice from the plateau flows to the coast. For example, Mawson (1915) reported that on August 16, 1912, the katabatic wind at Cape Denison in Adelie Land of East Antarctica reached 129 km/h. Strong and long-lasting katabatic winds can cause frostbite in humans and can severely limit visibility because of the amount of snow that is transported in the atmosphere. Windy days during the summer months on the polar plateau and in the Transantarctic Mountains are, in many cases, accompanied by unusually low temperatures (e.g., –20°C to –30°C), clear skies, and low surface resolution because of blowing snow.

During cloudy weather a condition known as *whiteout* may occur when surface definition deteriorates because of the absence of shadows and because the horizon disappears. These conditions are dangerous because they can lead to accidents that result from poor visibility and associated disorientation. Whiteout conditions in Fig. 2.4 can also occur during dense fog and may be caused by ground blizzards when strong katabatic winds inject snow into the air. Travel on the polar plateau and in the mountains is "not indicated" under whiteout conditions.

Locations close to the coast on Ross Island and in Victoria Land occasionally do register temperatures above 0°C during the austral summer in December and January. In spite of such rare warm summer days along the coast, meteoric precipitation in the Transantarctic Mountains occurs in the form of snow rather than rain. However, the presence of morphologically preserved beech leaves and twigs in till of the Dominion Range less than 500 km from the South Pole indicates that during the Pliocene Epoch a temperate climate prevailed in that part of the Transantarctic Mountains (McKelvey et al. 1991; Hill et al. 1991; Mercer 1987; Webb and Harwood 1987). The meteorology of Antarctica is the subject of books by Rubin (1966), Businger (1977), Schwerdtfeger W. (1984), Bromwich and Stearns (1994), and King and Turner (1997).

Fig. 2.4 During whiteout conditions in the Transantarctic Mountains the horizon becomes invisible and surface definition is lost. In the case shown here, the whiteout conditions resulted from high wind associated with snowfall near the coast of northern Victoria Land. The Scott tent, still used by geological fieldparties in the Transantarctic Mountains and on the polar plateau, is well suited for use in cold and windy weather, especially when it is set up in well-packed snow or névé. Mills Valley, Pain Mesa, northern Victoria Land, December 1982 (Photo by G. Faure)

2.4 Cold-Weather Injuries

The effects of the low air temperatures in the Transantarctic Mountains are magnified by wind which accelerates the loss of heat from the human body. This phenomenon is expressed quantitatively by the *wind-chill scale* (Rees 1993) that converts the measured temperature into an equivalent wind-chill temperature. For example, the arrow in Fig. 2.5a indicates that the measured air temperature of –10°C at a wind speed of 8 m/s corresponds to a wind-chill temperature of –20°C. In addition, a wind speed of 8 m/s in Fig. 2.5b is equivalent to a speed of 28.8 km/h. The wind-chill temperature also permits the definition of the discomfort index in Table 2.1. Accordingly, a wind-chill temperature of –20°C is perceived as being "bitterly cold." Such conditions are not unusual during the austral summer on the polar plateau and in the Transantarctic Mountains, except along the coast.

The discomfort experienced by humans because of the cold and windy conditions that characterize the summer weather in the Transantarctic Mountains and on the polar plateau is aggravated by the extremely low humidity of the air. Humans working out-of-doors under these conditions must protect themselves by wearing appropriate cold-weather clothing. Their ability to cope with the stressful environmental conditions improves when they are physically fit, well rested, well nourished, healthy, and when they are highly motivated to accomplish their mission (Gunderson 1974).

The scientists who do research in Antarctica are, in most cases, highly motivated to accomplish their objectives and therefore can tolerate the adverse

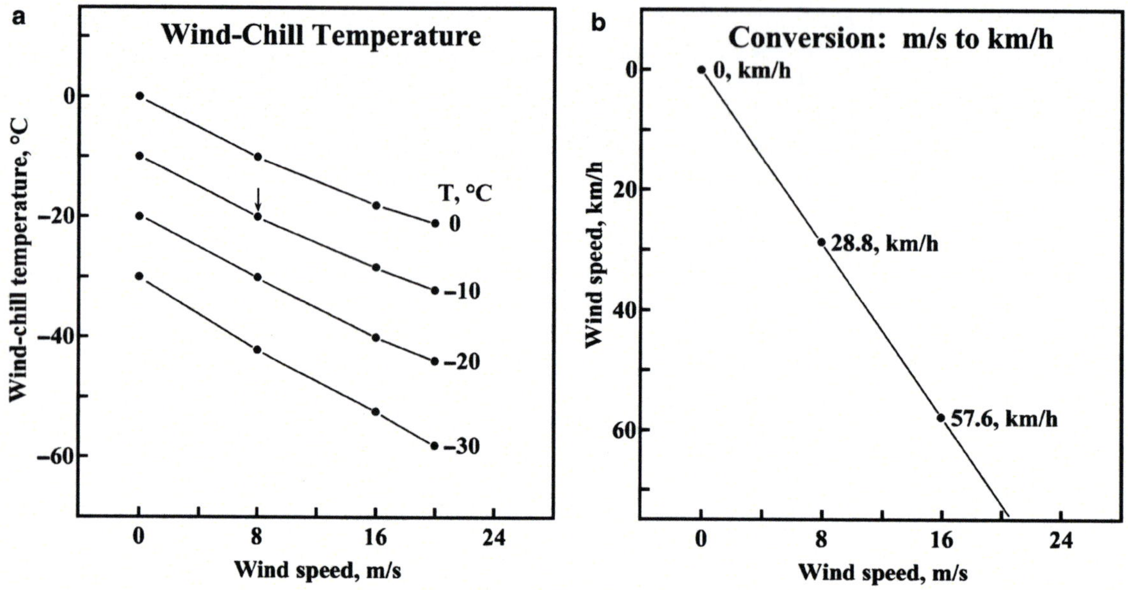

Fig. 2.5 (**a**) Conversion of the measured air temperature in degrees Celsius and the wind speed in meters per second into the wind-chill temperature which includes the effect of accelerated heat loss of the human body with increasing wind speed. The *arrow* indicates that a measured temperature of −10°C at a wind speed of 8 m/s corresponds to a wind-chill temperature of −20°C (**b**) Wind speeds measured in meters per second are converted into the equivalent unites of kilometers per hour by the relations: km/h = 3,600 m/1,000 s (Data from Stonehouse 2002, p. 293)

Table 2.1 Discomfort scale based on the wind-chill temperature derived in Fig. 2.5a from the measured temperature and the wind speed (Stonehouse 2002, p. 293)

Discomfort level	Range of wind-chill temperatures (°C)
Cool	+15 to +10
Very cool	+10 to 0
Cold	0 to −10
Very cold	−10 to −15
Bitterly cold	−15 to −25
Freezing cold	<−25

Table 2.2 Health hazards in Antarctica

(A) Weather-related
Hypothermia
Fire (burns and loss of shelter)
Frostbite
Snow blindness
Sunburn
Dehydration
Skin rashes
(B) Topographic hazards
Crevasses
Ice cliffs
Rock falls
Rock-climbing accidents
Drownings
Skiing/sledding accidents
(C) Aircraft and surface vehicles
Plane crashes[a]
Automobile accidents
Snowmobile accidents

[a]See Anderson (1974)

working conditions. However, some individuals may suffer emotional distress caused by loneliness, anxiety, sleeplessness, home sickness, constant daylight, anger, boredom, and the like. These kinds of emotional problems can also arise in individuals who are wintering over at one of the research stations which are cut off from the outside world for up to 6 months during the Antarctic winter.

The injuries caused by the harsh environmental conditions can be avoided in many cases by preventive measures taken in the field. If left untreated, minor bodily ailments can result in debilitating conditions that may require medical attention. The potential injuries and environmental hazards listed in Table 2.2 are not only weather-related but may also result from hazards associated with travel across mountainous terrain and on glaciers and sea ice.

Although hypothermia, sunburn, frostbite, and snow blindness do occur in Antarctica, these conditions rarely affect experienced field scientists. The most *common* weather-related health problems arise because of *dehydration* which not only affects kidney function and blood chemistry but also causes uncomfortable cracking and irritation of the skin. The most *dangerous* weather-related hazard is *fire* which can cause painful and potentially life-threatening burns and can also result in the loss of shelter, such as a dormitory building in McMurdo Station or a Scott tent in the field. The fire hazard is enhanced by the low humidity of the air and by the scarcity of liquid water with which to douse the flames.

The field manual that is issued by the Office of Polar Programs of the National Science Foundation to all groups that deploy through McMurdo Station contains the following admonition:

> Anyone deploying to remote locations in Antarctica should have a strong background in cold-weather survival or, at the very least, [should] employ a safety-survival guide with previous Antarctic experience. Antarctica is not a place to learn cold-weather skills. (Anonymous 1994)

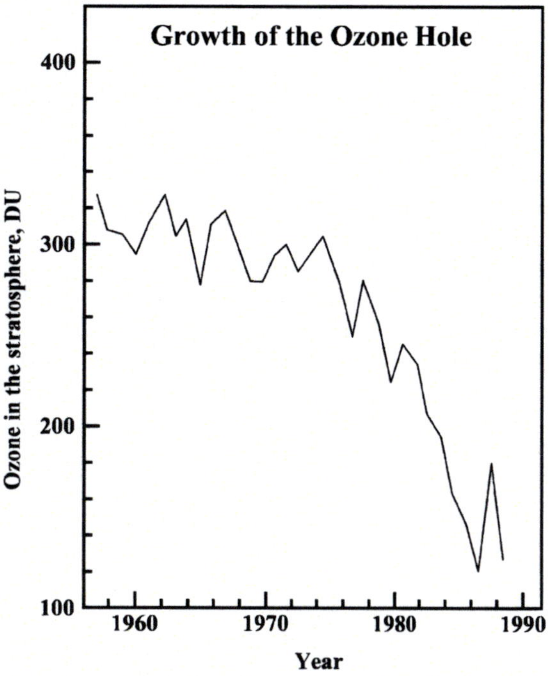

Fig. 2.6 Average ozone inventory in the stratosphere above the British research station at Halley Bay in October from 1957 to 1989. During that time interval the monthly averages for October decreased from about 320 Dobson units (DU) in the late 1950s to about 120 DU in the late 1980s. The decline of the average ozone inventory in October continued throughout the 1990s into the early twenty-first century (Adapted from Graedel and Crutzen 1993, Fig. 1.1)

2.5 The Ozone Hole

The hazardous working conditions in Antarctica arise not only from the harsh climate and the rugged surface environment, but also from exposure to excessive ultraviolet (UV) radiation caused by the destruction of ozone in the stratosphere above the continent (Appendices 2.11.2–2.11.5).

The seasonal loss of ozone from the stratosphere over Antarctica was first reported by Farman et al. (1985). Their data in Fig. 2.6 indicate that the amount of ozone in the atmosphere above the British Antarctic research station at Halley Bay on the east coast of the Weddell Sea (Appendix 1.10.2) declined from about 330 Dobson Units (DU) in October of 1957 to about 220 DU in October of 1984 (Appendix 2.11.4). In subsequent years, the average amount of atmospheric ozone in October over Antarctica continued to decline and approached 120 DU in 1989 followed by progressively lower values throughout the 1990s. The seasonal deficit in the amount of ozone over Antarctica is called the "Ozone Hole." The lowest amount of ozone of 90 DU was recorded in 1998 when the loss of ozone extended from about 12–24 km above the surface of the continent. In the same year, the ozone hole covered an area of 26.2×10^6 km^2 which is larger than the area of the North American continent.

Under normal circumstances, the oxygen molecules (O_2) of the air in the stratosphere, located 10–50 km above the surface of the Earth, are broken up by ultraviolet radiation emitted by the Sun and subsequently recombine to form ozone (O_3) (Appendix 2.11.4). The ozone of the stratosphere absorbs ultraviolet radiation thereby preventing it from reaching the surface of the Earth. This protection breaks down in Antarctica during September and October because, in the early spring, the stratosphere above the continent contains much less ozone than is present at other times of the year.

As the season progresses, the air in Antarctica warms up and the ozone hole gradually closes, but reappears in the spring of the following year. Measurements in

Fig. 2.6 demonstrate that this phenomenon has been occurring every spring since about 1970. It has been attributed to the release of anthropogenic chlorine-bearing gases such as chlorofluorohydrocarbons (CFCs) primarily in the northern hemisphere of the Earth.

The ozone hole in Fig. 2.7 is expected to fade away slowly and to disappear completely by the year 2068 because CFCs are no longer manufactured and therefore the amounts that are discharged into the atmosphere worldwide are decreasing annually (Cook-Andersen 2006).

We now know that the annual appearance of the ozone hole over Antarctica is caused by the circulation and low temperature of the stratosphere during the southern winter. The CFCs and other anthropogenic chlorine compounds, that are released into the atmosphere in the mid-latitudes of the Earth, are transported into the polar regions by the circulation of the atmosphere. During the winter in Antarctica (June, July, August, and September) the chlorine-bearing air of the stratosphere is isolated by strong westerly winds that form the circum-polar vortex. As a result, the temperature of the stratosphere within the vortex decreases to less than −80°C, which causes water vapor in the air to condense to form microscopic crystals of ice. Atoms of chlorine and other halogens are stored within the vortex by being sorbed to the surfaces of the ice crystals. When the vortex breaks down at the end of the Antarctic winter, warmer air enters the stratosphere and causes the ice crystals to sublime. The atoms of chlorine and other halogens that are thereby released accelerate the destruction of ozone in the stratosphere above Antarctica.

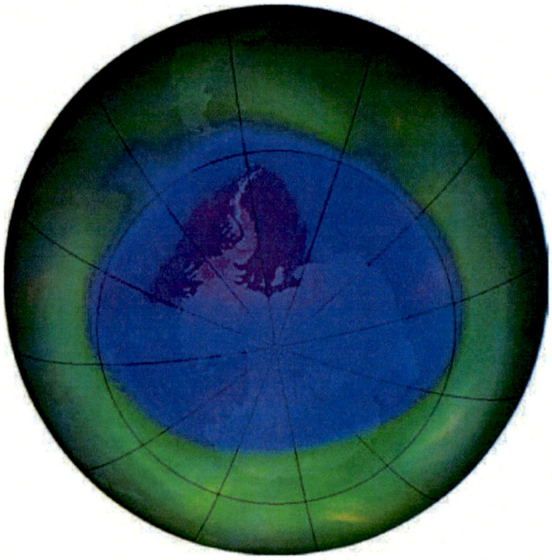

Fig. 2.7 The ozone hole over Antarctica in September of 2005 covered an area of more than 24 million square kilometers which is approximately equal to the area of North America. The ozone hole has been forming in the stratosphere over Antarctica since about 1970 as a result of the release of anthropogenic CFC gases into the atmosphere. Computer models now predict that the size of the ozone hole will begin to decline in 2018 and will stop forming altogether in 2068. This is good news because the ozone of the stratosphere absorbs ultraviolet radiation which can cause skin cancer and eye damage in humans and is harmful to marine organisms. The ozone content of the stratosphere over Antarctica has declined annually by about 70% below normal during September and October whereas the decline over the USA has only been between 3% and 6% (Cook-Andersen 2006; Photo courtesy of NASA)

The deterioration of the stratospheric ozone layer over Antarctica permits excessive amounts of solar ultraviolet radiation (UV) to reach the surface of the continent (Appendix 2.11.5). The increase in the intensity of UV radiation in Antarctica affects organisms on land and in the oceans as described in a book edited by Weiler and Penhale (1994). A summary by Graedel and Crutzen (1993) of the health effects of UV radiation indicates that it causes skin cancer in light-skinned individuals. Studies referred to by them indicate that a 1% reduction in the inventory of stratospheric ozone increases the effective UV dose by 2%, which causes a 4% increase in the incidence of the basal-cell carcinomas and about a 6% increase in the incidence of squamous-cell carcinomas. Therefore, unless special precautions are taken, geologists and other people who usually work out-of-doors in Antarctica during September and October may be exposed to higher doses of UV radiation than persons elsewhere on the Earth. The energy spectrum of solar UV radiation is discussed in Appendix 2.11.3.

Ozone also occurs as a contaminant in the air of large cities such as Los Angeles, and Mexico City. This low-altitude ozone persists for 25 days on average and can be transported for long distances by prevailing winds. The ozone in city smog is a byproduct of the combustion of gasoline and diesel fuel by automobile engines and can cause respiratory difficulties in humans. Ozone is also produced by lightning discharge in the air of the troposphere.

2.6 McMurdo Station

American scientists and support personnel who are scheduled to work in the Transantarctic Mountains or on the polar plateau of East Antarctica, in Marie Byrd Land in West Antarctica, and at South Pole Station will, in most cases, depart from Christchurch, New Zealand, and fly to McMurdo Station which is located at the tip of the Hut Point Peninsula on Ross Island in Fig. 2.8 (Section 1.2). This site was originally selected in 1955 by Admiral George J. Dufek as a logistics base for Operation Deep Freeze in preparation for research to be carried out by American scientists during the IGY (1957–1958). The site was chosen because it is located in a broad basin adjacent to a deep harbor where supply ships can unload cargo either onto a floating ice dock or directly to the shore. These favorable conditions also caused Robert Scott in 1901 to select this site for his winter-over base where he set up his Discovery Hut in Fig. 2.9 which still contains some of the equipment and supplies that he and his men left behind (Section 1.4.1).

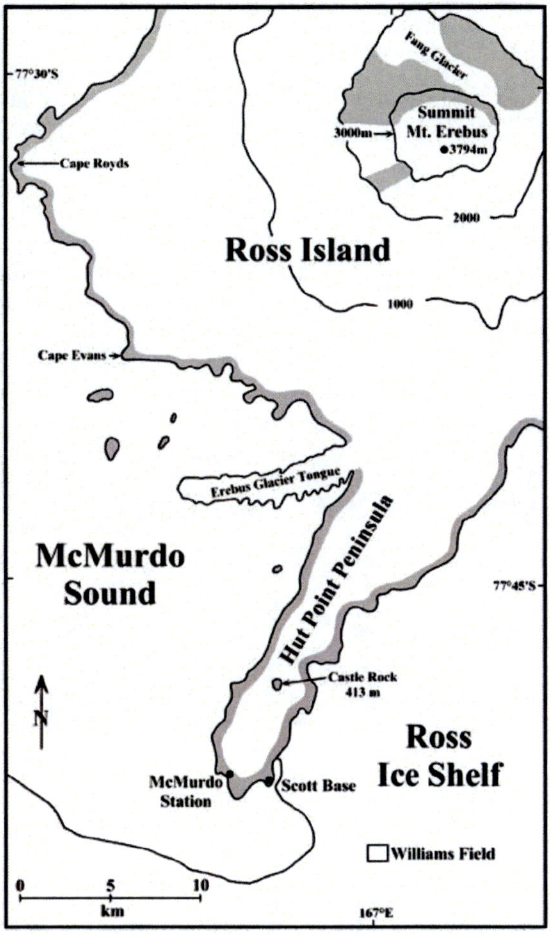

Fig. 2.8 McMurdo Station of the USA and Sott Base of New Zealand are both located at the southern tip of the Hut Point Peninsula on Ross Island. Castle Rock is a mass of hyaloclastite that rises to 413 m above sea level (a.s.l.) and is located about 6 km northeast of McMurdo Station. Cape Evans and Cape Royds contain the huts from which Robert Scott and Ernest Shackleton started their treks to the geographic South Pole, respectively. The Erebus Glacier Tongue is a prominent landmark for scientists traveling from McMurdo Station to Cape Evans and Cape Royds. The summit crater of Mt. Erebus is at 3,794 m a.s.l. (Adapted from the topographic map of Ross Island, Antarctica (ST 57–60/6, 1970, US Geol. Survey))

During the IGY, McMurdo Station was operated by the US Navy in support of American civilian scientists and technicians. At that time and for several years thereafter, McMurdo Station consisted mainly of olive-green canvass-covered Jamesway huts that served as bunk houses for the enlisted men and for some of the transient scientists prior to their deployment. Neider (1974) wrote that in 1970/71 McMurdo Station looked like a military supply base. Partly for that reason female scientists were not allowed to work in Antarctica because the facilities in McMurdo were considered to be inadequate (Chipman 1986; Rothblum et al. 1995; Arnesen and Bancroft 2003). McMurdo, which is now administered by a civilian contractor, has evolved into a substantial village that can accommodate a summer-time population of about 1,200 scientists, technicians, and field assistants who live in modern dormitory buildings. Women now make up more than one third of the population of McMudo Station. During the winter (March–August) the population of McMurdo drops to about 250. The support staff returns in mid-August on several special flights from New Zealand (Operation Winfly) in order to prepare for the arrival of scientists starting in October.

The US Antarctic Research Program (USAP) is administered by a representative of the National Science Foundation (NSF Rep) who occupies a prominent building called "The Chalet." The original Chalet in Fig. 2.10a was a Jamesway hut that had been painted red and had been dressed up by adding a wooden porch and gable. This hut was later replaced by a handsome wooden building in Fig. 2.10b that was built on the same site where it provides a spectacular view of the Royal Society Range across McMurdo Sound.

Fig. 2.9 View of McMurdo Station from Hut Point toward the Discovery Hut of Robert Scott (1901–1904) and Observation Hill in the background on the right. Arrival Heights is the basalt plateau on the left. The green building half-way up the slope of "Obs Hill" used to house the nuclear reactor that has since been dismantled and returned to the continental US (Conus). The sea ice in the "harbor" of McMurdo had not yet broken up when this picture was taken in January of 1983. Scott's ship, the RRS Discovery, was tied up next to the hut from 1902 to 1904 (Photo by G. Faure)

Fig. 2.10 (**a**) The Chalet in 1964 was a Jamesway Hut that had been painted red and that had the distinction of having a small porch (Photo by G. Faure). (**b**) Twelve years later in 1978, the offices of the NSF Rep were housed in a large permanent building that continues to be used for this purpose (Photo by G. Faure)

All activities related to on-going scientific research are supervised by the NSF Rep and his or her staff who have offices in the modern Chalet.

"Old Antarctic Explorers" (OAEs) will recall other landmarks of McMurdo such as Obs Hill (Fig. 2.9 and Section 1.4.3), the Berg Field Center (BFC), the Crary

Science and Engineering Center (Crary Lab), the Chow Hall, the Mammoth Mountain Inn (MMI), and the Hotel California (where you can check out anytime, but you can never leave). They will also remember Scott Base of New Zealand, Scott's second hut at Cape Evans, and Shackleton's hut at Cape Royds where the Adelie penguins come to incubate their eggs and to raise their chicks. Many OAEs have guided their field assistants to Castle Rock (Fig. 2.8) along the backbone of the Hut-Point Peninsula and have climbed to the top of this landmark in order to enjoy the view of Mt. Erebus and the Royal Society Range.

Former residents of McMurdo Station will also recall the windstorms that suddenly reduce visibility to less than 100 ft (30.48 m), with wind speeds in excess of 55 knots (101.9 km/h), and wind-chill temperatures that drop below −100°F (−73.3°C). Under these conditions, all personnel in McMurdo are required to remain in-doors (Condition I) and all travel is suspended for the duration of the storm that may last several days. The classification of weather conditions detailed in Table 2.3 is the responsibility of the McMurdo Weather Office (Mac Weather). Condition I storms can occur anywhere within the Transantarctic Mountains during the austral spring and fall especially on major outlet glaciers, such as the Reedy Glacier in Fig. 2.11, that channel the katabatic winds from the polar plateau to the Ross Ice Shelf or to the Ross Sea. Field parties that experience such storms learn first-hand that weather still rules in Antarctica.

A large base such as McMurdo Station requires a large amount of water for drinking, cooking, washing and, last but not least, for fighting fires. During the IGY and the years that followed, liquid water was obtained by melting snow that was scraped off the sides of the hills that surround McMurdo Station. The snow was trucked to the snow melter which was located adjacent to the original Chow Hall. This procedure was tedious, inefficient, and inadequate because, after a few years, the snow was used up and was not replaced by new snow.

In addition, the seasonal population of McMurdo Station continued to grow which increased the demand for water. For that reason, the US Congress in 1960 authorized the construction of a nuclear-fission reactor in order to provide power for the desalination of seawater. The components arrived on December of 1961 and were installed in a building that was erected at a site on the slope of Observation Hill above the station (Fig. 2.9). This reactor, which was put into operation in March of 1962, provided the power required to operate a desalination plant that converted seawater into fresh water (Neider 1974). However, in spite of the technological superiority of this process, water continued to be in short supply and had to be rationed. Matters came to a head when the representatives of the Antarctic Treaty Nations determined that the nuclear reactor violated the Treaty and therefore had to be shut down, dismantled, and all parts of it had to be removed from Antarctica. The Office of Polar Programs (OPP) did what was required and all radioactive waste was shipped to California. The nuclear installation was replaced by a desalination plant that is energized by fuel oil. The capacity of the present facility based on reverse osmosis is sufficient to provide an adequate

Table 2.3 Classification of weather emergencies at McMurdo Station and vicinity that are declared by the McMurdo Station weather office (Mac Weather) (Anonymous 1994)

Weather category	Description of weather conditions
Condition III	Wind speed up to 48 knots, wind chill down to −75°F, and visibility more than 1,320 ft (0.25 miles). Travel and all other activities are permitted although severe weather conditions may occur within 12 h.
Condition II	Wind speed between 48 and 55 knots, wind chill temperature between −75 and −100°F, and visibility between 100 and 1,320 ft (0.25 miles). Only pedestrian travel between buildings in McMurdo is permitted. Travel outside of McMurdo is restricted to marked trails and roads in radio-equipped vehicles.
Condition I	Wind speed in excess of 55 knots, wind chill temperatures less than −100°F, and visibility less than 100 ft. These parameters indicate extreme weather conditions. All personnel must remain inside buildings or in the nearest available shelter.

$$C^\circ = \frac{F^\circ - 32}{9} \times 5$$

1 mile = 5280 ft
1 ft = 12 in.
1 in. = 2.54 cm
1 knot = 1.852 m per h (Navy unit)
Wind-chill scale in Fig. 2.5

Fig. 2.11 Strong katabatic wind on the Reedy Glacier caused near whiteout conditions in this geological field camp in November of 1964 (Photo by G. Faure)

supply of water for the summer residents of McMurdo Station although shortages do occur and some of the water must be stored in case of fire.

The irony of the water shortage in McMurdo is that Antarctica contains 90% of the ice that exists on the Earth and 70% of the fresh water. However, mining of ice and snow on Ross Island for use in McMurdo Station is prohibitively expensive and may violate the Antarctic Treaty because of potential environmental issues. For these reasons, the demand for fresh water in the station continues to be met by desalination of seawater, although this process is also expensive and does impact the environment in McMurdo Sound because of the discharge of hypersaline brine by the desalination plant and of wastewater by the wastewater treatment plant.

Another way to make use of the fresh water that is stored in the glaciers of Antarctica is to tow icebergs to other regions of the world where water is in short supply. Each year large tabular icebergs break off from the Ross and Ronne ice shelves. These icebergs slowly disintegrate and melt as they are carried north by ocean currents. However, feasibility studies have demonstrated that icebergs cannot be towed because most of their volume is below the surface of the water and they tend to break apart by the stress of the tow cables (Husseiny 1978; Schwerdtfeger P. 1979; Holdgate 1980).

2.7 Amundsen-Scott South-Pole Station

The geographic South Pole was first reached in 1911 by Roald Amundsen and his companions and 1 month later in January of 1912 by Robert Scott and his team (Sections 1.4.3, 1.4.4). Nearly 17 years passed until 1929 when Admiral Richard Byrd flew over the South Pole in a Ford tri-motor airplane. Finally, in 1956 Admiral George J. Dufek landed at the South Pole in an LC-47 aircraft to prepare the way for the construction of an American research station to be used during

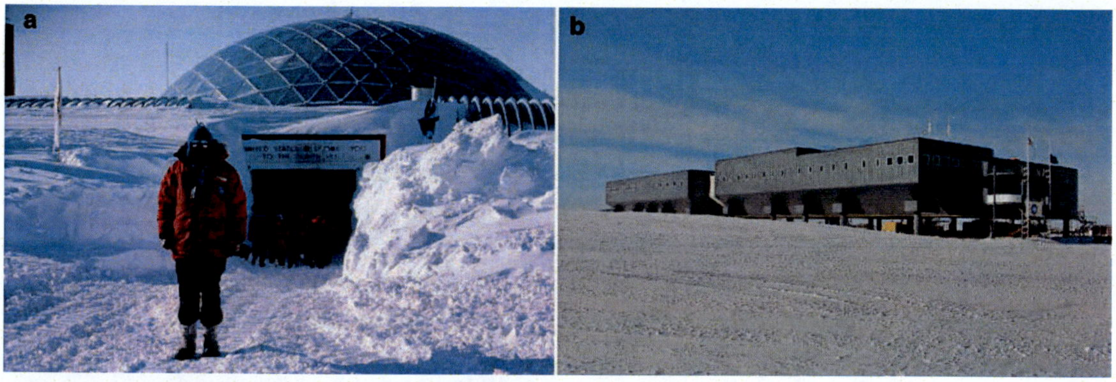

Fig. 2.12 (a) Entrance to the geodesic dome at Amundsen-Scott South Pole Station in December of 1982 (Photo by T.M. Mensing). (b) The new building at South Pole Station contains dorm rooms, laboratories, office space, a cafeteria, and recreational facilities. It is elevated above the surface to prevent snow drifts from accumulating (Photo by Keith Vanderlinde, National Science Foundation, taken on January 12, 2008)

the IGY (1957–1958). The first leader of Pole Station was Paul Siple who originally went to Antarctica as a Boy Scout with Byrd's first expedition in 1928 (Section 1.5.1).

The original building of South Pole Station was replaced in 1975 by a large geodesic dome located about 350 m from the actual pole. The dome in Fig. 2.12a, which was nearly 16 m high and 50 m in diameter, sheltered the science facilities, the galley, living quarters, communications equipment, and other essential services. The garage, the power plant, fuel bladders, and medical facilities were housed in a 252-m long tunnel covered by steel arches.

The famous geodesic dome that had sheltered the South-Pole Station for 30 years was replaced in 2007 because it was buried by snow that drifted around it. The new building in Fig. 2.12b is elevated on stilts in order to avoid the problems caused by the accumulation of snow. The geodesic dome will be dismantled and its parts will be returned to the continental US (Conus). The new station can accommodate a population of about 200 persons working on research projects in astronomy, upper atmosphere studies, meteorology, geophysics, and glaciology during the austral summer (October-February). Most of the scientists and technicians return home at the end of the summer. Only a small group of scientists and technicians stays to maintain the station during the winter.

Although Pole Station is more than 1,600 km south of Ross Island, it continues to be resupplied during the summer by ski-equipped C-130 aircraft flying out of McMurdo Station. The winter-over crew at Pole Station is resupplied only once in August by Winfly when mail and other supplies are dropped by parachute from a C-130 aircraft circling overhead in total darkness. Occasionally, aircraft have landed during the winter to evacuate persons requiring emergency medical treatment.

In addition to other science facilities, the South Pole Station includes a seismological observatory that has operated since 1957. This facility was greatly improved in January of 2003 when a new observatory was placed into service about 8 km from the main building. The sensor of the seismograph at the new facility is located in a borehole at a depth of 300 m below the surface where it is not affected by vibrations caused by the diesel-powered electric generators and other activities at the main station. As a result, the South Pole seismological observatory is now virtually free of locally-generated seismic noise and therefore is ideally suited to detect earthquakes that occur elsewhere on the Earth.

Visitors to Pole Station like to have their picture taken beside a barber pole topped by a silver globe surrounded by the flags of the Treaty Nations (Fig. 2.13). This "pole" is not at the precise location of the geographic South Pole because the East Antarctic ice sheet is sliding on its base which means that the true position of the South Pole must be redetermined annually.

Fig. 2.13 Gunter Faure (*left*) and Teresa M. Mensing (*right*) at Amundsen-Scott South Pole Station in December of 1982 (Photo by G. Faure)

2.8 Fieldwork in Antarctica

A voyage to Antarctica in a wooden sailing ship used to be a dangerous undertaking that lasted 2 years in most cases. Even steam-powered ships exposed travelers to severe wind storms, high waves, as well as to potentially dangerous icebergs and to the pack-ice that forms around the continent as the sea ice breaks apart in the spring.

The early explorers also used dogsleds to travel in Antarctica (Sections 1.3–1.5) which required skill in the handling of dogs and caused many dogs to die of malnutrition and exhaustion. In addition, dogs were routinely killed during extended trips in order to feed the remaining dogs. For example, in 1911 Roald Amundsen and his men left their base at the Bay of Whales (Framheim) with 55 dogs, but when they returned 89 days later after a journey of 2,976 km (1,860 miles), only 14 dogs remained alive (Section 1.4.4). Robert Scott and Ernest Shackleton also used dogs on their expeditions but preferred ponies and attempted to use primitive tractors to pull sleds (Sections 1.4.1, 1.4.2). Even Richard Byrd took a large number of dogs to Antarctica in addition to tractors and fixed-wing aircraft (Sections 1.5.1, 1.5.2). After the end of World War II, New Zealanders continued to use dog teams in Antarctica where they maintained a group of specially bred huskies (Fig. 2.14) at Scott Base on Ross Island (e.g., Herbert 1962, Figs. 4 and 5). However, the dog teams were gradually replaced by snowmobiles, helicopters, and fixed-wing aircraft. All sled dogs had to be removed from Antarctica prior to

Fig. 2.14 Antarctic huskies who were bred at Scott Base on Ross Island were used by New Zealand fieldparties to travel on the glaciers and snow fields of the Transantarctic Mountains. All dogs were removed from Antarctica in 1994 (Photo by G. Faure, November 1964)

April 1, 1994, in compliance with the Antarctic Treaty (Walton and Atkinson 1996; Stonehouse 2002).

In the modern era, most American scientists fly to Christchurch, New Zealand, on commercial airliners and are then transported to McMurdo Station by ski-equipped C-130 and wheeled C-141 cargo planes. A one-way trip from Christchurch to McMurdo Station takes 6–8 h depending on the type of aircraft, on the amount of cargo being transported, and on weather conditions. Most of the flight path crosses the open water of the South Pacific Ocean and of the Ross Sea where the water temperature is close to −2°C. Immersion in water at this low temperature is survivable by humans for only 2 or 3 min. Even though thousands of flights have crossed this dangerous stretch of water in the 50 years since the IGY, none have had to ditch in the ocean. However, several flights from Christchurch to McMurdo Station each year are forced to turn back before passing the point-of-no-return because of adverse weather conditions in McMurdo.

During the stop-over in Christchurch the Antarctic travelers are issued their cold-weather clothing at the Clothing Distribution Center (CDC) located near the airport. All Antarctic travelers need to try on the clothing they are issued to make sure that all items fit

Fig. 2.15 A helicopter of the New Zealand Air Force on Shapeless Mountain of southern Victoria Land is being loaded with the camping gear of an American fieldparty in December of 1994. This example of international cooperation is characteristic of operations in Antarctica (Photo by G. Faure, December 1994)

Fig. 2.16 This ski-equipped C-130 aircraft, flown by pilots of the US Navy, landed on Evans Névé in northern Victoria Land in 1982 to drop off a geological fieldparty (Photo by T.M. Mensing)

comfortably. Some OAEs prefer to wear their own clothing in Antarctica, but first-timers should use the clothes they are issued and should learn how to wear them correctly.

Geologists preparing to work in the Transantarctic Mountains or on the polar plateau spend about 10 days in McMurdo checking out camping equipment and food at the Berg Field Center (BFC). In addition, all persons who intend to live and work outside of the station are required to attend a 2-day snow-craft course taught by experienced American mountaineers (Yankielun 2007). Although helicopter support such as in Fig. 2.15 is available to field parties working within about 240 km (150 miles) of McMurdo Station, groups that work on the polar plateau, in most cases, prefer to use snowmobiles which can be transported into the field in the cargo holds of ski-equipped C-130 aircraft shown in Fig. 2.16 or suspended from helicopters. Field parties also receive a short-wave radio which they are required to use for daily check-ins with the field-party communications center in McMurdo or at South Pole Station. Most groups use two-burner Coleman stoves to heat their tents and to cook their food. Great care and skill in the use of these and other kinds of stoves is required to avoid burns and to prevent fires.

After field parties have checked out all of the necessary equipment from the BFC in McMurdo, including snowmobiles, they are ready for an overnight "shakedown trip" to a location of their choice on Ross Island, such as Cape Royds or Cape Evans. This trip, illustrated in Fig. 2.17, is the first opportunity for the members of a group to practice their skills in loading Nansen sleds, driving snowmobiles, setting up tents, preparing food, sleeping in their tents, and communicating with McMurdo Station by short-wave radio.

Fig. 2.17 Most geological field parties take an overnight trip to Cape Evans or Cape Royds on Ross Island in order to practice driving snowmobiles, setting up their Scott tents, cooking a meal, and communicating with McMurdo Station on their short-wave field radio (This photo was taken at Cape Royds with Mt. Erebus in the background in 1992 by T.M. Mensing)

The large displacement of the magnetic south pole from the geographic pole (i.e., 2,820 km in 1990) and the steep inclination of the magnetic-field lines detract from the utility of the magnetic compass in Antarctica. The angle subtended by the magnetic and geographic poles for an observer at McMurdo Station is 130°. In other words, the magnetic-compass needle points 130° east of the geographic south pole. Therefore, the *magnetic declination* at this location is 130° east. In addition, the needles of magnetic compasses used in Antarctica must be specially balanced in order to allow them to swing freely because of the steep inclination of the magnetic field in the vicinity of the magnetic pole. The use of the magnetic compass for travel within the continental area of Antarctica and in the adjacent oceans has been superseded by the Global Positioning System (GPS) which is based on a set of satellites the positions of which are known precisely.

2.9 Preservation of the Environment

The large number of persons who annually pass through the American research stations at McMurdo, South Pole, and Palmer as well as those who live and work outside of these stations, generate large amounts of waste that must be collected, sorted, packaged, and removed from the continent. The disposal of waste generated in Antarctica was originally mandated by the Antarctic Treaty which came into force in 1961 and is now required by US Public Law 95–541 also known as the Antarctic Conservation Act (ACA) of 1978. This Act specifically protects all mammals, birds, and plants in Antarctica, as well as designated historical sites and other places of interest. In addition, the Protocol on Environmental Protection of the Antarctic Treaty, which was signed by 26 nations in 1991, has guided the activities of the US Antarctic Program (USAP) ever since. The *rules* by means of which USAP assures compliance with the Protocol were published in 1993. Beginning on August 15 of 1993, all American citizens working in Antarctica are required to follow these rules. Violators can be fined or charged in courts of law in the state where they reside (Anonymous 1993).

The basic rule concerning waste disposal is that anything that is shipped or flown in must eventually be shipped out. Accordingly, the amount of material that is brought in must be *reduced* to the essentials, items that are no longer needed should be *re-used* if possible, and the rest must be collected for shipment to Conus for *recycling* and/or safe disposal. These guidelines are essential for the preservation of the environment in Antarctica because the cold and dry conditions prevent most materials from decomposing such as organic waste, paper, metals, wood, plastics, clothing, etc. Certain items are banned from Antarctica altogether:

1. Polystyrene chips (called peanuts) that are used as filler in packing boxes
2. Nonindigenous plants and animals such as house plants, flowers, pets, tropical fish, and nonsterile soil
3. Pesticides (except for hygienic or scientific use), and
4. Polychlorinated biphenols (PCBs)

Certain other substances are considered to be hazardous pollutants that must be handled safely in order to prevent injury to humans. These hazardous materials are generally labeled as being flammable, corrosive, reactive, toxic, or radioactive. Any waste that contains these kinds of materials is labeled "Antarctic hazardous waste." Pollutants that are required for research purposes either at one of the stations or in the field require a permit before they can be imported to Antarctica. Therefore, USAP each years applies for a comprehensive permit to import such potentially hazardous materials.

The rules concerning the disposal of waste in Antarctica are implemented by requiring that it be sorted at the source into the appropriate waste streams. USAP has defined 16 different kinds of waste and provides separate color-coded containers for each including not only glass, paper, plastics, and metal cans, but also batteries, cardboard, clothing, construction debris, domestic combustibles, food-contaminated waste, hazardous waste, heavy metals, light metals, wood, and miscellaneous waste. As a result, of these efforts, the three American research stations have become models of environmental preservation and even remote field camps are required to participate. Everything that is taken into the field must eventually be removed and they really do mean *everything*.

The transformation of McMurdo Station from a temporary military supply base to its present squeaky-clean appearance did not happen overnight and required a change in the local culture. However, after the problems were graphically described and condemned in the popular press (e.g., Lemonick 1990) and with prodding from Greenpeace (e.g., May 1988; Anonymous 1990), the activities of the US Antarctic Program continue to comply with ACA and the rules arising from it.

2.10 Summary

Antarctica is a large continent with a highly diversified topography and weather conditions. Sweeping generalizations about its weather have little value because the conditions on the surface of Antarctica depend critically on the latitude, the elevation above sea level, the distance to the nearest coast, the time of year, and the local topography. Similarly, the topography of Antarctica includes large and elevated continental ice sheets, rugged mountains, magnificent valley glaciers, ice-free valleys containing meltwater streams and lakes filled with brine, and coastal regions where primitive plants grow on land during the austral summer and where an abundant fauna and flora thrive in the ocean (Mastro and Wu 2004). In spite of the cold and dry conditions that characterize the Transantarctic Mountains, primitive plants such as algae, lichens, and mosses have been found there (Williams 1995).

The low temperatures and high winds that commonly occur in the Transantarctic Mountains and on the adjacent polar plateau of East Antarctica combine to cause wind-chill conditions that can result in injury to humans. Cold-related injuries are avoided by the excellent polar clothing and shelters that USAP provides to groups living and working in the field. Injuries may also result from dehydration in the extremely dry air and from fires that cause burns and the loss of shelter. Other sources of injuries include accidents related to mountaineering, travel on glaciers and sea ice, as well as plane crashes, and accidents that occur during construction and repair of buildings, and from accidental immersion in seawater. Work in Antarctica is not without risk. The motto is: "Plan ahead, stay alert, keep in touch, and never walk alone."

Another potential source of personal injury arises from the high flux of solar ultraviolet radiation that reaches the surface of Antarctica in September and October because of the destruction of ozone in the stratosphere. The loss of ozone results from CFCs and other chlorine-bearing anthropogenic gases that are primarily released by people in the northern hemisphere. These gases are temporarily trapped by the south-polar vortex that develops over Antarctica during the austral winter. When the vortex breaks down in the spring, the chlorine atoms are released and convert ozone (O_3) back into its diatomic form (O_2). The destruction of up to 70% of the ozone in the stratosphere over Antarctica causes the so-called ozone hole and permits excessive UV radiation to reach the surface. Prolonged and unprotected exposure to UV radiation can cause skin cancer and cataracts in humans. However, no acute cases have been reported that are attributable to the depletion of ozone in the stratosphere over Antarctica.

The USA maintains three research stations in Antarctica: at McMurdo on Ross Island, at the South Pole, and Palmer Station on Anvers Island in the Palmer Archipelago. These stations are a major component of the US Antarctic Program that is administered by the Office of Polar Programs of the National Science Foundation which is funded directly by the Congress of the USA. McMurdo and South Pole stations have been operated continuously since the IGY on a year-round basis, whereas Palmer Station opened in 1965. These stations provide facilities for research to scientists who either work at the stations or in the field outside the stations.

The large number of people that annually pass through McMurdo or live and work at South Pole and Palmer stations generate large amounts of waste of different kinds that must be removed from Antarctica in order to avoid contamination of the continent. The emphasis on the protection of the pristine environment of Antarctica has greatly reduced the impact American scientists and technicians have had on the areas where they have worked since the IGY.

2.11 Appendices

2.11.1 Exploration of Antarctica by Tractor Train

On November 1, 1958, a tractor train left Byrd Station in West Antarctica for a systematic study of snow stratigraphy, seismic profiling of the ice and underlying crust of Antarctica, and for a geological reconnaissance of the Horlick Mountains (Schulthess 1960). The group of six was led by Dr. Charles Bentley of the University of Wisconsin in Madison and included William E. Long, William Chapman, Fred Darling, Jack Long, and Leonard LeSchack. The departure of the tractor train was preceded in October of 1958 by a reconnaissance flight from Byrd Station to the Wisconsin Range of the Horlick Mountains where the

Fig. 2.18 The tractor train of the Marie-Byrd-Land oversnow traverse left Byrd station in West Antarctica on November 1, 1958, as part of the exploration of Antarctica during the IGY. The tractor train consisted of three Sno-cats each of which pulled a sled. The leading tractor, which was driven by Willam E. Long, carried a crevasse detector that projected from the front of the vehicle. The tractor train stopped at regular intervals to carry out glaciological and geophysical research. In addition, four members of the research team, including Bill Long, climbed Mt. Glossopteris in the Ohio Range and collected samples of rocks and fossils (Photo by Emil Schulthess reproduced by permission of Matthias Kamm, administrator of the photo archive of Emil Schulthess)

flight path turned east to the Ohio Range and continued to the Thiel Mountains where the aircraft turned northwest to the Whitmore Mountains in West Antarctica before returning to Byrd Station. Bill Long who was on this plane noted that a thick sequence of stratified rocks occurred in the Ohio Range and decided to make a collection of the stratified rocks he had seen in the Ohio Range.

The tractor train that left Byrd Station in Fig. 2.18 consisted of three Sno-cats each of which pulled a sled. The leading tractor, known as the "Sally Jeanne," was driven by Bill Long assisted by Fred Darling. This tractor carried a crevasse detector that is visible in Fig. 2.18. Occasional mechanical breakdowns of the vehicles, bad weather, and difficulties with crevasses limited the daily progress of the tractors to less than 36 km. The tractor train was resupplied from Byrd Station by a Dakota R4D aircraft that brought fuel for the tractors, food and mail for the crew, and, at least once, a dentist who treated one of the scientists. The elevation of the polar plateau in the path of the tractor train was in excess of 2,100 m above sea level and the weather was cold and windy most of the time with temperatures ranging from −20°C to −25°C.

Eventually the tractor train arrived at Station 414 at the foot of the Horlick Mountains and was delayed there by lack of fuel. The resupply plane from Byrd Station had to turn back several times because of bad weather, problems with radio transmissions, and mechanical problems. While the group waited for several days for the plane, Bill Long and three of his companions decided to climb Mt. Glossopteris which is one of the highest mountains in that part of Antarctica at 2,867 m above sea level.

In spite of bad weather, the group consisting of Bill Long, Fred Darling, Charles Bentley, and Jack Long set out in the morning and disappeared into the blowing snow and swirling clouds on the mountain. During the day a fog bank obscured the tractors parked about

Fig. 2.19 This picture taken by Emil Schulthess, a Swiss photographer who spent some time in the field with the Byrd-Station traverse group, captures the moment Bill Long returned to the Sno-cats at Station 414 after he and his companions had successfully climbed Mt. Glossopteris in the Ohio Range. The climbers were exhausted but unhurt after spending 12 h on the mountain shrouded in fog at this time (Photo by Emil Schulthess. Reproduced by permission of Matthias Kamm, administrator of the photo archive of Emil Schulthess)

2 miles from the foot of the mountain. After 12 h, Bill Long suddenly appeared out of the fog in Fig. 2.19 carrying a large backpack full of rock samples and a variety of fossils including marine invertebrates and fossilized Glossopteris leaves. The other climbers also returned utterly exhausted but unhurt.

This episode in the exploration of Antarctica had positive consequences for Bill Long whose study of the samples he collected on Mt. Glossopteris earned him a Master's Degree in geology at The Ohio State University (Long 1959, 1961). The results of his work gained the support of the Office of Polar Programs of the National Science Foundation in Washington, D.C. which provided the logistical support for geological fieldwork in the Ohio Range in 1960/61 and 1961/62. Several other scientists who joined Bill Long contributed their expertise to the study of all aspects of this remote mountain range, including: G.A. Doumani, J.H. Mercer, J.M. Schopf, S.B. Treves, R.L. Oliver, A.J. Boucot, L.L. Lackey, M.D. Higgins, J. Ricker, and C.J. Skinner. Bill Long earned a Ph.D. in Geology at The Ohio State University for his dissertation on the stratigraphy and environment of deposition of the sedimentary rocks in the Ohio Range (Long 1965).

The geology of the Ohio Range as it is presently understood is presented in this book in Section 7.5.1 (basement rocks) and in Section 10.5.2 (Beacon Supergroup). In addition, Fig. 10.22 depicts the adit of the Dirty Diamond Coal Co. from which Bill

Long and his companions extracted an unweathered sample of coal from a seam in the Mt. Glossopteris Formation.

2.11.2 Structure of the Atmosphere

The structure of the atmosphere of the Earth in Fig. 2.20 is defined in terms of the variations of pressure and temperature that occur with increasing altitude above the surface of the Earth. The major structural units are:

Troposphere, 0–10 km
Stratosphere, 10–50 km
Mesosphere, 50–100 km
Thermosphere, >100 km

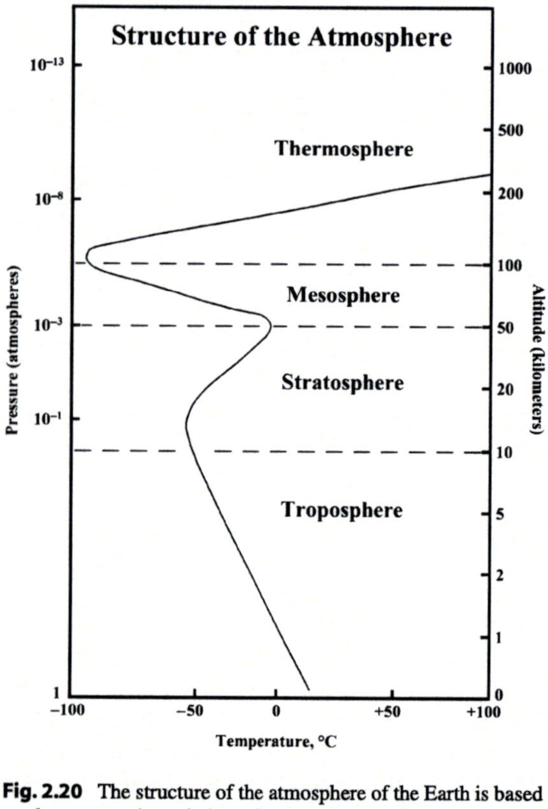

Fig. 2.20 The structure of the atmosphere of the Earth is based on the systematic variation of temperature (in degrees Celsius) and pressure (in atmospheres). The major subdivisions are the troposphere (0–10 km), stratosphere (10–50 km), mesosphere (50–100 km), and thermosphere (>100 km). Ozone (O_3) occurs in the lower part of the stratosphere between 10 and 25 km above the surface of the Earth (Adapted from Graedel and Crutzen 1993, Fig. 3.5)

2.11.3 Energy Spectrum of UV Radiation

Electromagnetic radiation emitted by the Sun is subdivided into three types defined by their wavelength (λ):

Infrared, $\lambda > 750$ nanometers (nm)
Visible light, $\lambda = 400$–750 nm
Ultraviolet, $\lambda < 400$ nm

Ultraviolet light (UV) in Fig. 2.21 is itself subdivided into:

UV-A, $\lambda = 400$–320 nm
UV-B, $\lambda = 320$–290 nm
UV-C, $\lambda < 290$ nm

The energy (E) of electromagnetic radiation is inversely proportional to its wavelength according to the equation:

$$E = hc/\lambda \tag{2.1}$$

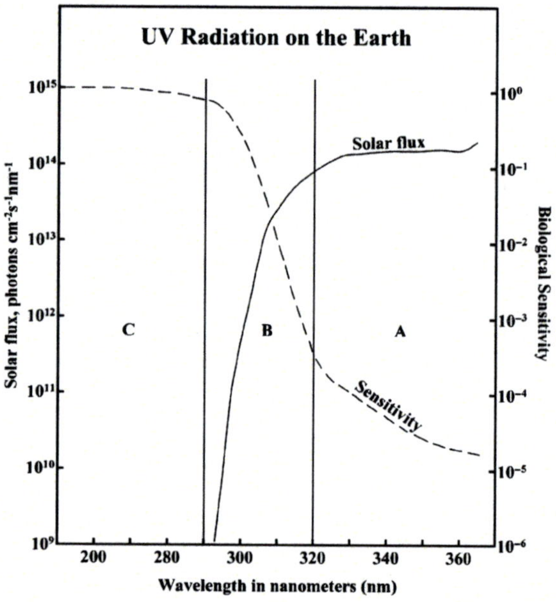

Fig. 2.21 Flux (*solid line*) and biological sensitivity of DNA (*dashed line*) of solar ultraviolet (UV) radiation on the surface of the Earth. UV radiation is subdivided into UV-A, UV-B, and UV-C with decreasing wavelength expressed in nanometers (nm). The solid line represents the UV-flux that reaches the surface of the Earth when the Sun is at a zenith angle of 39° and is expressed by the number of UV photons/cm²/s/nm. The biological sensitivity is scaled in terms of factors of 10^{-1} starting at 1.0 which is the maximum sensitivity and lowest tolerance of DNA molecules for UV radiation of different wavelengths (Adapted from Graedel and Crutzen 1993, Fig. 13.7)

where h = Planck's constant = 6.62517×10^{-27} erg/s and c = speed of light = 2.99792×10^{8} m/s

Accordingly, UV radiation is more energetic than visible light and UV-C is the most energetic form of UV radiation.

2.11.4 *Formation and Destruction of Ozone*

The most energetic UV radiation (i.e., UV-C) causes O_2 molecules in the stratosphere to dissociate:

$$O_2 + E \rightarrow O + O \tag{2.2}$$

The free oxygen atoms react with diatomic molecules of O_2 to form ozone (O_3):

$$O + O_2 + M \rightarrow O_3 + M \tag{2.3}$$

where M is a neutral molecule such as N_2 or O_2 which absorbs the energy released by reaction 2.3 and disperses it into the environment.

The ozone molecules preferentially absorb UV-C with a wavelength of $\lambda = 250$ nm (Hartley absorption band) and are dissociated in the process:

$$O_3 + E \rightarrow O_2 + O \tag{2.4}$$

$$O_3 + O \rightarrow 2O_2 \tag{2.5}$$

for a net reaction of:

$$2O_3 \rightarrow 3O_2 \tag{2.6}$$

Reactions 2.2–2.6 produce and destroy ozone and thereby maintain its inventory in the atmosphere. However, when certain trace constituents are introduced into the stratosphere, such as: NO, HO, Cl, Br, etc., the natural balance is disturbed because, in the presence of these catalysts, ozone is destroyed more rapidly than it is produced (Holland and Petersen 1995).

The amount of ozone in the atmosphere is expressed in terms of the Dobson Unit (DU) which is the number of ozone molecules that would form a layer of pure ozone that has an area of 1.0 cm² and is 0.01 mm thick at a pressure of 1.0 atmosphere and a temperature of 0°C. According to this definition, a column of air with an ozone concentration of 1 DU would contain 2.69×10^{16} molecules of O_3 per square centimeters. Correspondingly, an ozone concentration of 100 DU represents a layer of ozone that is 1.0 mm thick and has an area of 1.0 cm² at the temperature and pressure specified above. The average amount of ozone in the stratosphere of the equatorial region of the Earth is about 250 DU.

The principal anthropogenic contaminants that contribute to the destruction of ozone are the chlorofluorocarbons (CFCs) that were once widely used as propellants in aerosol cans containing hair spray and as coolants in refrigerators and air conditioners. These compounds are related to methane (CH_4) in which hydrogen has been replaced by the halogens fluorine, chlorine, and bromine. Therefore, the molecules of CFCs have formulas such as: $CFCl_3$, CF_2Cl_2, and CF_3Cl. Most of these compounds are gases at room temperature and their solubility in water is very low.

When CFC gases are released into the atmosphere, they gradually rise into the lower stratosphere because they are not washed out of the troposphere by rain. In the stratosphere the chlorine atoms are released when the CFC molecules are broken up by energetic UV-C radiation as indicated by the reaction:

$$2CFCl_3 + E + \frac{3}{2}O_2 + H_2O \rightarrow 2CO_2 + 2HF + 6Cl \tag{2.7}$$

The chlorine permits reactions to occur that destroy the ozone of the stratosphere:

$$Cl + O_3 \rightarrow ClO + O_2 \tag{2.8}$$

$$O_3 + E \rightarrow O + O_2 \tag{2.9}$$

$$ClO + O \rightarrow Cl + O_2 \tag{2.10}$$

When reactions 2.8–2.10 are summed, they add up to:

$$2O_3 \rightarrow 3O_2 \tag{2.11}$$

which is identical to Eq. 2.6. In this way, chlorine atoms in the stratosphere act as "ozone killers" and thereby decrease the ozone inventory of the global stratosphere.

Although the production of CFCs has been halted and they are no longer released into the atmosphere, these compounds continue to be the principal source of chlorine in the stratosphere. Natural sources, such as: sea spray, volcanic eruptions, and methyl chloride gas (CH_3Cl) released by plants are minor sources of atmospheric chlorine. Even though CFCs were outlawed, the chlorine concentration of the stratosphere is still five times higher than normal because of delayed releases of anthropogenic CFCs and other volatile halogen-bearing industrial compounds such as the cleaning solvent methyl chloroform (Graedel and Crutzen 1993, p. 143).

Although both the area and the depth of the ozone hole over Antarctica have varied annually, the size of the hole has not *decreased* appreciably even after the manufacture and release of the ozone-destroying gases were prohibited in 1987 by the Montreal Protokol. However, the computer models predict that the area and depth of the ozone hole will begin to decline in 2018 and that it will disappear by the end of the present century. If that happens, we will have successfully reversed the effects of this case of anthropogenic contamination of the atmosphere (Morell 2007; Stonehouse 2002; Faure 1998; Holland and Petersen 1995; Graedel and Crutzen 1993; Stolarski 1988; Molina and Rowland 1974).

2.11.5 Effect of UV Radiation on the Biosphere

The reduction of the ozone content of the stratosphere of the Earth as a whole, and of the Antarctic stratosphere in particular, has significantly *increased* the amount of UV radiation that reaches the surface of the Earth. The ozone layer in the stratosphere shields the Earth from the most energetic UV-C radiation ($\lambda < 290$ nm) by means of the ozone-destroying reaction of equation 2.4 (i.e., the Hartley absorption band). In addition, N_2O and O_2 also adsorb UV-C and prevent it from reaching the surface of the Earth. The amount of UV-B ($\lambda = 290$–320 nm) that passes through the atmosphere in Fig. 2.21 rises with increasing wavelength (and decreasing energy), whereas the low-energy UV-A radiation ($\lambda = 320$–400 nm) is not absorbed by the atmosphere and therefore all of it reaches the surface. Consequently, plants and animals have evolved a tolerance for UV-A and for the low-energy part of the UV-B spectrum. However, the decrease of the ozone content of the stratosphere of the Earth has caused an undesirable increase in the flux of energetic UV-B and UV-C radiation that reaches the surface. The problem is especially acute in Antarctica during the early spring in October.

Biological organisms on the surface of the Earth cannot survive prolonged exposure to UV-C ($\lambda < 290$ nm) because it destroys organic molecules. The relation between the solar flux of UV radiation and the wavelength of that radiation in Fig. 2.21 indicates that the flux of UV-C at the surface of the Earth is less than 10^{-5} times the flux of UV-B at $\lambda = 320$ nm. The flux of UV-B that reaches the Earth increases by a factor of more than 10^5 as the wavelength increases from 290 to 320 nm. In contrast to UV-B, the flux of UV-A is nearly independent of the wavelength from $\lambda = 320$ to 400 nm.

The biological sensitivity of DNA molecules also depends on the wavelength of the UV radiation. According to the data in Fig. 2.21, the sensitivity of DNA to UV-C radiation is 1.0 (i.e., UV-C is lethal). The sensitivity of DNA molecules to UV-B radiation decreases with increasing wavelength from 290 to 320 nm (i.e., long-wavelength UV-B radiation is tolerated better than short-wavelength UV-B radiation). DNA molecules tolerate UV-A best of all because organisms that did not adapt to this radiation did not survive on the surface of the Earth.

References

Anderson PJ (1974) United State aircraft losses in Antarctica. Antarctic J US 9(1):1–21

Anonymous (1990) Scientific report: Greenpeace Antarctic expedition. Greenpeace International, Amsterdam, The Netherlands

Anonymous (1993) Waste management in the United States Antarctic program. NSF93–128. Washington, DC, 20 p

Anonymous (1994) Field manual for the United States Antarctic program. Office of Polar Programs, National Science Foundation, Washington, DC

Arnesen L, Bancroft A (2003) No horizon so far. Da Capo Press, Cambridge, MA

Bentley CR (1983) Crustal structure of Antarctica from geophysical evidence: a review. In: Oliver RL, James PR, Jago JB (eds) Antarctic earth science. Australian Academy of Science, Canberra, ACT, Australia, pp 491–497

Bentley CR, Robertson JD (1982) Isostatic gravity anomalies in West Antarctica. In: Craddock C (ed) Antarctic geoscience. University of Wisconsin Press, Madison, WI, pp 949–954

Bentley CR, Robertson JD, Greischar LL (1982) Isostatic gravity anomalies on the Ross Ice Shelf. In: Craddock C (ed) Antarctic geoscience. University of Wisconsin Press, Madison, WI, pp 1077–1081

Bromwich DH, Stearns CR (1994) Antarctic meteorology and climatology: studies based on automated weather stations. Antarctic Research Series, vol. 61, American Geophysical Union, Washington, DC

Businger J (1977) Meteorological studies at Plateau Station. Antarctic Research Series, vol. 25, American Geophysical Union, Washington, DC

Chipman E (1986) Women on the ice. Melbourne University Press, Melbourne, Australia

Cook-Andersen G (2006) NASA study finds clock ticking slower on ozone-hole recovery. Earth Observ 18(4):33

Drewry DJ, Jankowski EJ, Steed, RHN (1983) Isostatic displacement of Antarctic lithosphere. In: Oliver RL, James PR, Jago JB (eds) Antarctic earth science. Australian Academy of Science, Canberra, ACT, Australia, 503 pp

Farman JC, Gardiner BG, Shanklin JD (1985) Large losses of total ozone in Antarctica reveal seasonal interaction. Nature 315:207–210

Faure G (1998) Principles and applications of geochemistry, 2nd edn. Prentice-Hall, Upper Saddle River, NJ

Folco L, Curzio P, Laurenzi MA, Mellini M, Zeoli A (2007) The age of ice at the Frontier Mountain blue ice field (Antarctica) constrained by $^{40}Ar/^{39}Ar$ chronology of englacial tephra. Meteor Planet Sci 42(Suppl):A49

Graedel TE, Crutzen PJ (1993) Atmospheric change: an earth-system perspective. W.H. Freeman, New York

Gunderson EKE (ed) (1974) Human adaptability to Antarctic conditions. Antarctic Research Series, vol. 22, American Geophysical Union, Washington, DC

Herbert WW (1962) The Axel Heiberg glacier. In: Collins BW (ed) Special Antarctic Issue, 681–706. New Zealand J Geol Geophys 5(5):671–913

Hill RS, Harwood DM, Webb PN (1991) Last remnant of Antarctica's Cenozoic flora: Nothofagus of the Sirius Group, Transantarctic Mountains. Abstracts, 8th international symposium on Gondwana, p. 43. Hobart, Tasmania

Holdgate MW (1980) Some environmental questions: an after-dinner address. Ann Glaciolo 1:135–136

Holland HD, Petersen U (1995) Living dangerously, Princeton Un. Press, Princeton, New Jersey

Husseiny AA (ed) (1978) Iceberg utilization. Pergamon, Oxford

Kadmina IN, Kurinin RG, Masolov VN, Grikurov GE (1983) Antarctic crustal structure from geophysical evidence: a review. In: Oliver RL, James PR, Jago JB (eds) Antarctic earth science. Australian Academy of Science, Canberra, ACT, Australia, pp 498–502

King JC, Turner J (1997) Antarctic meteorology and climatology. Cambridge University Press, Cambridge

Kyle PR (1990) McMurdo volcanic group: western ross embayment; Introduction. In: LeMasurier WE, Thomson JW (eds) Volcanoes of the Antarctic plate and southern oceans, Antarctic Research Series, vol. 48. American Geophysical Union, Washington, DC, pp 19–25

Kyle PR (ed) (1995) Volcanological and environmental studies of Mount Erebus, Antarctica. Antarctic Research Series, vol. 66, American Geophysical Union, Washington, DC

Kyle PR, Jezek PA, Mosley-Thompson E, Thompson LG (1981) Tephra layers in the Byrd Station ice core and the Dome C ice core, Antarctica, and their climatic importance. J Volcan Geotherm Res 11:29–39

LeMasurier WE (1990) Late Cenozoic volcanism on the Antarctic plate. In: LeMasurier WE, Thomson JW (eds) Volcanoes of the Antarctic plate and southern oceans, Antarctic Research Series, vol. 48. American Geophysical Union, Washington, DC, pp 1–17

LeMasurier WE, Thomson JW (eds) (1990) Volcanoes of the Antarctic plate and southern oceans, Antarctic Research Series, vol. 48. American Geophysical Union, Washington, DC

Lemonick MD (1990) Once inaccessible and pristine, the white continent is now threatened by spreading pollution, budding tourism and the world's thirst for oil. Antarctica: Time, 135(3):56–62, Jan. 15

Long WE (1959) Preliminary report of the geology of the central range of the Horlick Mountains, Antarctica. Ohio State Univ Res Found Rept 825–2, Part 7:1–23

Long WE (1961) Byrd Station and Marie Byrd Land traverse, 1958/59. Ohio State Univ Res Found Rept 825–2, Part 11:1–296

Long WE (1965) Stratigraphy of the Ohio range, Antarctica. In: Hadley JB (ed) Geology and paleontology of the Antarctic. Antarctic Research Series, vol. 6. American Geophysical Union, Washington, DC, pp 71–116

Mastro J, Wu N (2004) Under Antarctic ice. University of California Press, Los Angeles, CA

Mawson D (1915) The home of the Blizzard, vols. 1 and 2. Heinemann, London

May J (1988) The Greenpeace Book of Antarctica: a review of the seventh continent. Dorling Kindersley, London

McKelvey BC, Webb P-N, Harwood DM, Mabin, MCG (1991) The Dominion range Sirius group; a late Pliocene-early Pleistocene record of the ancestral Beardmore Glacier. In: Thomson MRA, Crame JA, Thomson JW (eds) Geological evolution of Antarctica. Cambridge University Press, Cambridge, pp 675–682

Mercer JH (1987) The Antarctic ice sheet during the late Neogene. In: Coetzee JA (ed) Paleoecology of Africa and the surrounding islands. Balkema, Rotterdam, The Netherlands, pp 21–33

Molina MJ, Rowland FS (1974) Stratospheric sink for chlorofluoromethanes: chlorine-atom-catalyzed destruction of ozone. Nature 249:810–812

Morell V (2007) Ahead in the clouds; Susan Solomon helped patch the ozone hole. Smithsonian, February:83–85

National Geographic Society (1990) Atlas of the world, 6th edn. Washington, DC

Neider C (1974) The edge of the world; Ross Island Antarctica. Doubleday, Garden City, NJ

Palais JM (1985) Tephra layers and ice chemistry in the Byrd-Station ice core, Antarctica. Ph.D. dissertation, Dept. Geol. Mineral., The Ohio State University, Columbus, OH

Rees WG (1993). A new wind-chill nomogram. Pol Rec 29(170):229–234

Rothblum ED, Morris JF, Weinstock JS (1995) Women in the Antarctic: risk-taking and social consequences. World Psychol 1(1):83–112

Rubin MJ (1966) Studies in Antarctic meteorology. Antarctic Research Series, vol. 9. American Geophysical Union, Washington, DC

Schulthess E (1960) Antarctica; A photographic survey. Artemis Verlags-AG, Zűrich, Switzerland

Schwerdtfeger P (1979) Review on icebergs and their uses. Cold Reg Sci Technol 1(1):59–79

Schwerdtfeger W (1984) Weather and climate of the Antarctic. Elsevier, Amsterdam, The Netherlands

Self, S, Rampino MR, Barbera JA (1981) The possible effects of large 19th and 20th century volcanic eruptions on zonal and hemispheric surface temperatures. J Volcan Geotherm Res 11:41–60

Stolarski RS (1988) Changes in ozone over the Antarctic. In: Rowland FS, Isaksen ISA (eds) The changing atmosphere. Wiley, New York, pp 109–119

Stonehouse B (ed) (2002) Encyclopedia of Antarctica and the southern ocean. Wiley, Chichester, England

Thompson LG, Mosley-Thompson E (1981). Temporal variability of microparticle properties in polar ice sheets. J Volcan Geotherm Res 11:11–27

Walton K, Atkinson R (1996) Of dogs and men: fifty years in the Antarctic. Images, Malvern

Webb PN, Harwood DM (1987) Terrestrial flora of the Sirius formation: its significance for late Cenozoic glacial history. Antarctic J US 22(4):7–11

Weiler CS, Penhale PA (eds) (1994) Ultraviolet radiation in Antarctica; Measurement and biological effects, Antarctic Research Series, vol. 62. American Geophysical Union, Washington, DC

Williams TD (1995) The Penguins: Spheniscidae. Oxford University Press, Oxford

Yankielun NE (2007) How to build an Igloo and other snow shelters. W.W. Norton, New York

Part II
The Basement Rocks

Chapter 3
Southern Victoria Land; Basement Rocks

The geological architecture of the Transantarctic Mountains in Table 3.1 is deceptively simple because it consists of only four suites of rocks which are separated by unconformities and intrusive contacts. The most pervasive and longest erosional interval is represented by the Kukri Unconformity, also called the Kukri Peneplain. This erosional surface separates the rocks of the underlying basement complex from the overlying sedimentary rocks of the Beacon Supergroup of Early Devonian to Late Triassic age and the dolerite sills of the Ferrar Group that intruded the Beacon rocks during the Middle Jurassic. The Beacon rocks are locally overlain by the sheet-like flows of the Kirkpatrick Basalt which formed where Ferrar magmas were erupted through fissures. The most recent episode of volcanic activity started during the Miocene, less than about 25 million years ago, when alkali basalt lava and pyroclastics of the McMurdo Volcanic Group were erupted through vents at numerous sites in the Transantarctic Mountains from northern Victoria Land (Cape Adare) to Mt. Early located close to the South Pole (LeMasurier and Thomson 1990). Mt. Erebus on Ross Island is the last of these Tertiary volcanoes to remain active at the present time (Kyle 1995).

The basement complex which underlies the Transantarctic Mountains along their entire length is exposed in a broad region along the coast of Victoria Land and along the edge of the Ross Ice Shelf all the way to the Horlick Mountains and beyond (Grindley 1981; Laird and Bradshaw 1982). It consists of sedimentary and volcanic rocks that were intensely folded, regionally metamorphosed, and intruded by granitic rocks of the Granite Harbor Intrusives (Table 3.1). The metamorphism of the sedimentary and volcanic rocks that occurred during the Ross Orogeny was sufficiently severe to convert these rocks locally into paragneisses and to generate the granitic magmas that intruded the volcano-sedimentary complex between 530 and 470 million years ago (i.e., Early Cambrian to Middle Ordovician) based on the Geologic Time Scale of the International Union of Geological Sciences issued in 2002 (IUGS 2002).

The rocks of the basement complex formed and were subsequently altered while Antarctica was an integral part of the supercontinent Gondwana, which contained all of the southern continents we know today. The lithologic and structural diversity of the basement rocks is compounded by their fragmentary exposure along the Transantarctic Mountains. Therefore, we have chosen to describe the basement rocks in each of the five segments defined in Fig. 3.1 and we will also use these subdivisions of the Transantarctic Mountains to describe the rocks that overlie the basement complex.

Rocks of Precambrian age are also exposed along the coast of East Antarctica in Fig. 3.2 and under the ice sheet of East Antarctica. The rocks of the subglacial crust are known only in terms of their geophysical properties that have been measured during surface traverses as well as by airborne and satellite surveys. The Precambrian rocks of the subglacial crust and of the coastal areas of East Antarctica have not been correlated with the basement rocks of the Transantarctic Mountains except in terms of their ages as measured by radiogenic-isotope geochronometry (Faure and Mensing 2005).

The geology of all parts of Antarctica was compiled by Craddock (1969a) and a group of 24 collaborators in the form of color-coded geological maps and associated text. This important work contains 18 geological maps that include all major outcrop areas of Antarctica. The geology of the Transantarctic Mountains is presented in six of these maps (i.e., 5, 13, 14, 15 16, and 17). Although these maps were published more

G. Faure and T.M. Mensing, *The Transantarctic Mountains: Rocks, Ice, Meteorites and Water*,
DOI 10.1007/978-90-481-9390-5_3, © Springer Science+Business Media B.V. 2010

Table 3.1 Simplified geological architecture of the continental crust of the Transantarctic Mountains in southern Victoria Land

Rock types	Geologic unit	Geologic time, Ma[a]
Alkali-rich volcanic rocks	McMurdo Volcanic Group	Miocene to Holocene 25 to 0.0
Dolerite sills and basalt flows	Ferrar and Kirkpatrick groups	Middle Jurassic 176.8 ± 1.8[b]
Sandstone, shale, tillite, coal	Beacon Supergroup	Early Devonian to Late Triassic 417 to 205
~~	Kukri Unconformity	~~~~~~~~
Granitic plutons and late-stage dikes	Granite Harbor Intrusives	Late Cambrian to Middle Ordovician
Metamorphosed and folded volcano-sedimentary complexes	Ross Supergroup Koettlitz and Skelton groups	Precambrian to Late Cambrian

[a] Ma = millions of years ago (mega anna)

[b] Heimann et al. (1994)

than 35 years ago and although they are now available only in private collections and certain libraries, they remain a useful resource for geologists who are interested in the geology of Antarctica.

In addition, the proceedings volumes of SCAR Meetings and of International Gondwana Symposia listed in Appendices 1.10.4 and 1.10.5 contain a wealth of information about the geology of the continent, as do the volumes of the Antarctic Research Series of the American Geophysical Union (Appendix 1.10.6) and the publications of the Geological Society of America (Appendix 1.10.7). The accumulated information about the geology of Antarctica has been summarized and interpreted by Gunn (1963), Ford (1964), Fairbridge (1975), Rowley (1983), Swithinbank (1988), Tingey (1991, 1996), Gamble et al. (2002), and others.

The Transantarctic Mountains have been mapped by geologists from New Zealand, the United Kingdom, USA, Australia, Germany, the USSR, and Italy. Several of these nations have established research stations on Ross Island and in Victoria Land. The fieldwork during the IGY (1957/58) and in the decades that followed was done by use of dogsleds (Herbert 1962) and snowmobiles, as well as by man-hauling sleds (Nichols 1963; Wade et al. 1965), and even by backpacking (McKelvey and Webb 1962) until helicopters and fixed-wing aircraft became available for support of geological field parties. The results of studies by scientists from New Zealand were published in ten special issues of the New Zealand Journal of Geology and Geophysics (Collins 1962, 1963, 1965, 1967; Bodley 1968). All of the publications arising from work done under the aegis of the New Zealand Antarctic Research Programmes between 1956 and 1964 were listed by Quartermain (1963, 1965). The post-IGY review papers of the geology of the Transantarctic Mountains include the work of Harrington (1958), Gunn and Warren (1962), Gunn (1963), Grindley and Warren (1964), and Schmidt (1966).

The amount of information that has accumulated about the geology of the Transantarctic Mountains is very large. Even when only the basement rocks are considered, the information in published reports is overwhelming and is difficult to fit into a coherent tectonic model, partly because the stratigraphy, structural deformation, and metamorphic grade of the rocks vary regionally and because the tectonic environment of the area of deposition of the basement rocks is obscured by the present configuration of Antarctica. Nevertheless, several authors have proposed explanations for the tectonic evolution of Antarctica in the context of the assembly of Gondwana (e.g., Hamilton 1967; Elliot 1975; Laird 1981; Grikurov 1982; Findlay et al. 1984, 1993; Allibone et al. 1993a, b; Stump 1995; Encarnación and Grunow 1996; Fűtterer et al. 1996).

The most successful synthesis of the data pertaining specifically to the basement rocks of the Transantarctic Mountains was published in a book by Stump (1995) whose tectonic model is based on the insight that the basement rocks were deposited along a passive margin of Gondwana after the break-away of Laurentia from Rodinia during the Neoproterozoic (Bell and Jefferson 1987; Moores 1991).

According to this concept, the paleo-Pacific margin of Gondwana in Fig. 3.3 was activated when a subduction zone developed along it (Ferracioli et al. 2002). The resulting subduction of the oceanic crust caused compression of the sediment and volcanics that had accumulated during the passive-margin phase and resulted in magmatic activity that started around 600 Ma. The intrusion of syntectonic plutons into the Ross

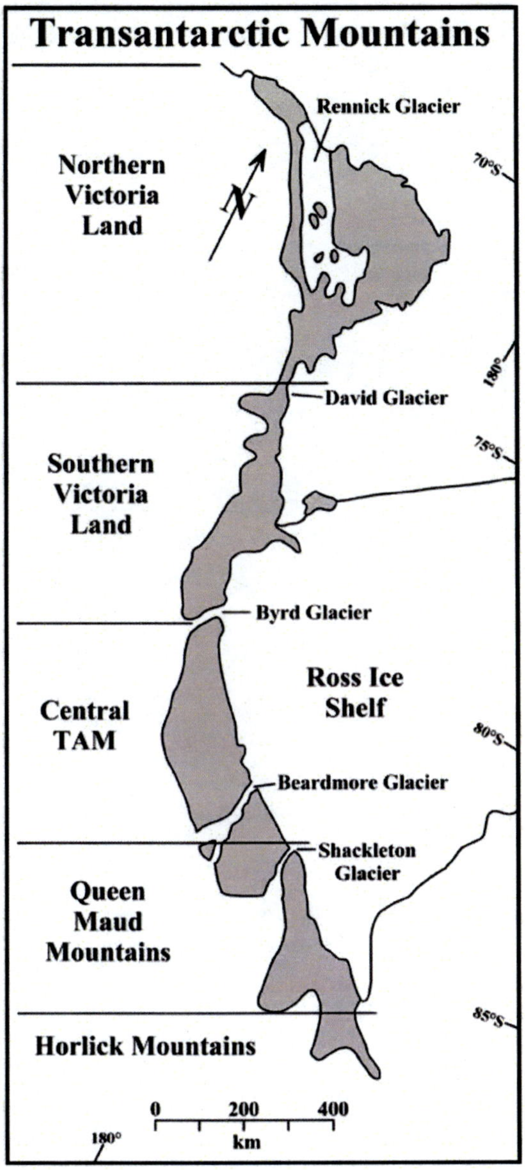

Fig. 3.1 For the purposes of this presentation the Transantarctic Mountains have been subdivided into five segments which are defined in this diagram. The shaded outcrop areas expose the basement rocks that are the subject of the following series of chapters. North in this and other maps in this book is at the top (Adapted from Stump et al. 2006)

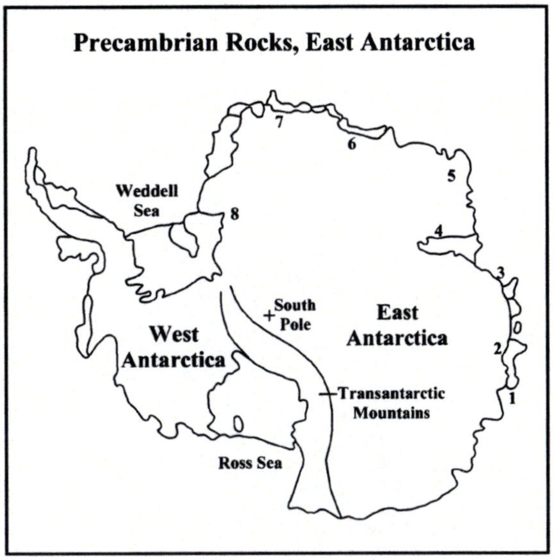

Fig. 3.2 Precambrian basement rocks are exposed along the coast of East Antarctica and in the mountain ranges along the edge of the ice sheet. The localities that are identified by number include: 1.Vincennes Bay, 2. Bunger Hills (Queen Mary Coast), 3. Vestfold and Larsemann Hills 4. Prince Charles Mountains (MacRobertson Land) 5. Enderby Land; 6. Sor Rondane Mountains 7. New Schwabenland (Queen Maud Land), and 8. Shackleton Range. Precambrian rocks also form the crust of East Antarctica that is covered by the ice sheet (Adapted from Stonehouse 2002; Craddock 1982)

Mountains ended at about 500 Ma and was followed by post-tectonic magmatic activity until about 460 Ma. Grunow et al. (1996) suggested that the magmatic activity along the Transantarctic Mountains coincided with and may be related to the Pan-African deformation event that occurred in Africa, southern India, Sri Lanka, and those parts of South America that were joined to Africa prior to the opening of the Atlantic Ocean.

The model of the origin and tectonic evolution of the Ross orogen described by Stump (1995), Grunow et al. (1996), Encarnación and Grunow (1996), Borg and DePaolo (1991, 1994), and Dalziel (1991) places the Transantarctic Mountains into the context of the assembly and subsequent dispersal of Gondwana which began when Laurentia split from Rodinia. Many details about the subsequent evolution of the resulting rift margin of Gondwana are still open for discussion, including the accretion of "suspect" terranes (Bradshaw et al. 1985) and the deposition of marine carbonate rocks along a coast that was presumably experiencing compression and magmatic activity resulting from subduction of oceanic crust under the continental margin of Gondwana (Stump 1995). These and other topics related to the formation of the Transantarctic Mountains will be discussed in subsequent chapters of this book.

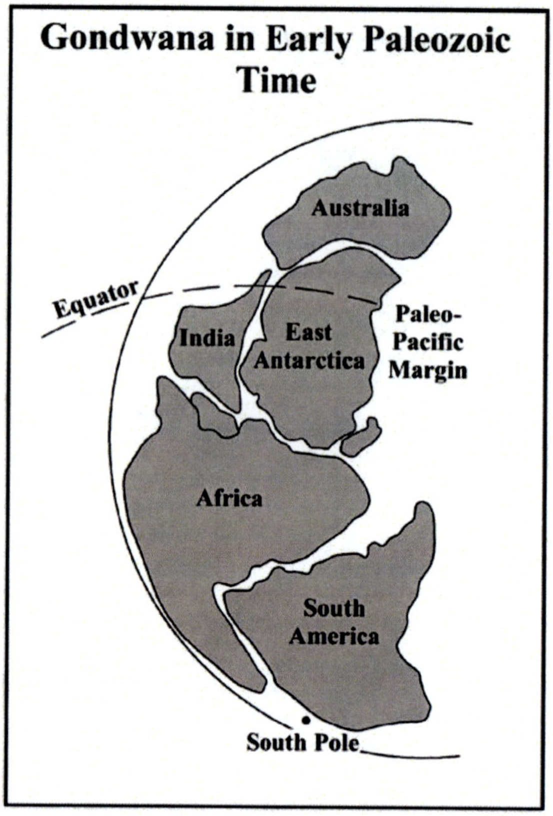

Fig. 3.3 The rifting of the supercontinent Rodinia during the Neoproterozoic Era split the future East Antarctic craton and caused Laurentia (the continental fragment that split from East Antarctica) to drift away. The newly formed paleo-Pacific coast of Gondwana, including East Antarctica, thereby became a passive margin along which detrital sediment was deposited by turbidity currents. Subsequently, the turbidites were folded and metamorphosed when an active subduction zone developed along the coast of Gondwana in early Paleozoic time. The deformation of the sedimentary rocks and the formation of granitic magma formed the Ross orogen which later became the basement upon which the sedimentary rocks of the Beacon Supergroup were deposited (Adapted from Encarnación and Grunow 1996)

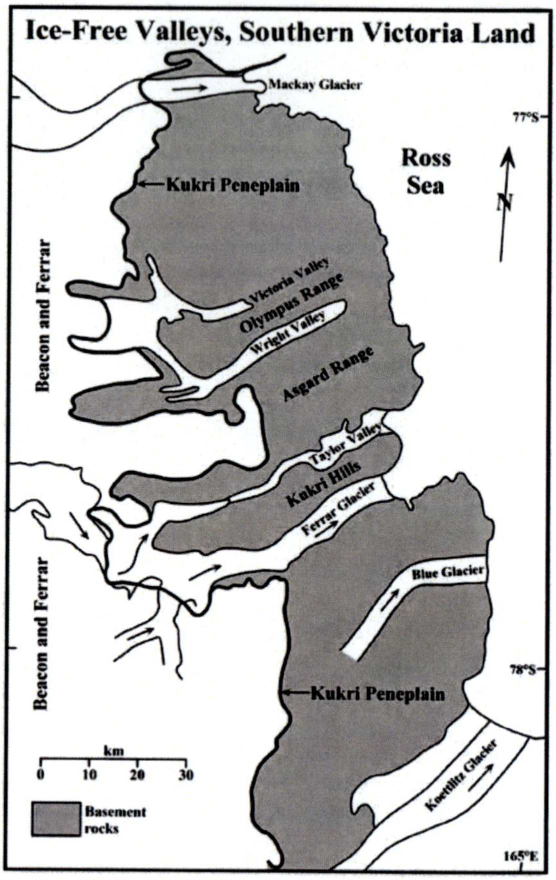

Fig. 3.4 Outcrop belt of the basement rocks along the coast of southern Victoria Land in the area of the ice-free valleys and adjacent to the Koettlitz Glacier. The trace of the Kukri Peneplain divides the area where basement rocks are exposed to the east from the area to the west where the basement rocks are overlain by the sandstones of the Beacon Supergroup (Devonian to Triassic) and by the sills of the Ferrar Dolerite (Middle Jurassic). The basement rocks in this part of the Transantarctic Mountains consist of the quartzo-feldspathic metasediments (Neoproterozoic) and marine carbonates (Cambrian) of the Ross Supergroup which was intruded by the Granite Harbor Intrusives consisting of syntectonic and post-tectonic granitic rocks ranging in composition from granite to granodiorite and diorite (Adapted from the geologic map of Craddock (1969a) and Warren (1969))

The basement rocks of *southern Victoria Land* defined in Fig. 3.1 are exposed in an irregular belt along the coast and consist primarily of the Granite Harbor Intrusives of Cambrian age which intruded the folded metasedimentary rocks of the Ross Supergroup of Cambrian and Neoproterozoic age (Warren 1969). This outcrop belt extends from the David Glacier south to the Byrd Glacier and includes the basement rocks that are exposed in the ice-free Victoria, Wright, and Taylor valleys in Fig. 3.4. Another large exposure of basement rocks included in Fig. 3.4 occurs between the Ferrar and the Koettlitz glaciers. A third area of exposure of Granite Harbor Intrusives in the Brown Hills between the Carlyon and Darwin glaciers and in the Britannia Range, located between the Koettlitz and Byrd glaciers (Grindley and Laird 1969) is not shown in Fig. 3.4.

3.1 Ice-Free Valleys

The ice-free valleys of southern Victoria Land in Fig. 3.4 were the first part of the Transantarctic Mountains to be studied by geologists during Scott's Discovery Expedition (1901–1904) which included the geologist Hartley T. Ferrar and the surveyor and navigator Albert B. Armitage. A few years later, Shackleton's Nimrod Expedition (1907–1909) included the geologists T. William E. David, Douglas Mawson, Raymond E. Priestley, and Sir Philip L. Brocklehurst. After returning from the Nimrod Expedition with Shackleton, Raymond Priestley joined Scott's Terra Nova Expedition (1910–1913) along with Frank Debenham, Charles S. Wright, and T. Griffith Taylor.

3.1.1 Topography

The ice-free valleys in Fig. 3.4 consist of several parallel ice-carved valleys that are separated from each other by mountain ranges. The central area of this region is occupied by the Wright and Taylor valleys which are separated from each other by the Asgard Range. North of Wright Valley is a large ice-free area which includes the Victoria, Barwick, Balham, and McKelvey valleys which are separated from the Wright Valley by the Olympus Range. The valley of the Ferrar Glacier, located south of Taylor Valley, is separated from it by the Kukri Hills.

Wright Valley in Fig. 3.5 is ice-free because only a small amount of ice from the polar plateau is presently entering it via the Airdevronsix Icefall which feeds the Wright Upper Glacier. Even smaller amounts of ice are flowing into Barwick Valley by way of the Webb Glacier and into Victoria Valley via the Victoria Upper Glacier in Fig. 3.6. The starvation of the outlet glaciers that once occupied Wright Valley and the Victoria Valleys in the past was caused by a decrease of the thickness of the East Antarctic ice sheet combined with the presence of subglacial bedrock obstructions along the ice-covered western slope of the Transantarctic Mountains in this area (Calkin 1974; Studinger et al. 2004). In addition, Bull (1966) demonstrated that the ice-free areas of southern Victoria Land absorb more solar heat than neighboring areas which are still ice covered. He concluded that the increased heat-absorption, combined with the small amount of annual precipitation, is currently preventing snow from accumulating in the ice-free valleys of southern Victoria Land. However, the "U-shaped" profiles of these valleys and the presence of glacial deposits within them leaves no doubt that they were occupied by glaciers in the past (Péwé 1960; Denton et al. 1970; Mayewski and Goldthwait 1985). The aridity of the climate in the ice-free (or "dry") valleys was previously demonstrated by the legendary Robert Nichols, Professor of Geology, at Tufts University of Massachusetts who introduced many of his students to the glacial geology and geomorphology of southern Victoria Land (Nichols 1963). Taylor Valley in Fig. 3.7 actually does contain a glacier that still flows from the polar plateau and terminates within that valley, most of which is ice-free, whereas the Ferrar Glacier continues to transport ice from the polar plateau to New Harbor on McMurdo Sound.

The rugged topography of the Transantarctic Mountains of southern Victoria Land originated by erosion of the landscape by glaciers that have sculpted the mountains into horns separated by cirque basins. For example, the summit plateau of the Olympus Range in Fig. 3.5 contains a large number of glacial horns including: Mt. Theseus (1,829 m), Mt. Peleus (1,790 m), Mt. Jason, Mt. Hercules, Mt. Aeolus, Mt. Boreas (2,180 m), Mt. Dido (2,070 m), Mt. Circe, and Mt. Electra. All of these peaks as well as the mountain range itself (i.e., Olympus) were named after characters in Greek mythology (Houtzager 2003). The glacial horns of the Asgard Range recall the names of deities in Norse mythology who had residences in Asgard: Odin, Thor, Baldr, the Valkyries, and Loki (spelled "Loke" on the USGS map ST 57–60/6*). The highest peaks of the ice-free valleys occur along the edge of the polar plateau, including Mt. Feather (2,985 m), Shapeless Mountain (2,739 m), Tabular Mountain (2,700 m), and Mt. Bastion (2,530 m).

Certain places in the ice-free valleys have been designated Sites of Special Scientific Interest (SSSIs), in order to protect ongoing scientific research or to preserve them in pristine condition for possible future study. This action was taken by the seventh Antarctic Treaty Consultative Meeting (ATCM VII) in 1972 and is contained in Recommendation VII-3 (Stonehouse 2002, p. 369). The SSSI sites in the ice-free valleys are: Barwick Valley (Fig. 3.6), the area between the Canada Glacier and Lake Fryxell (Taylor Valley), and the Linnaeus Terrace

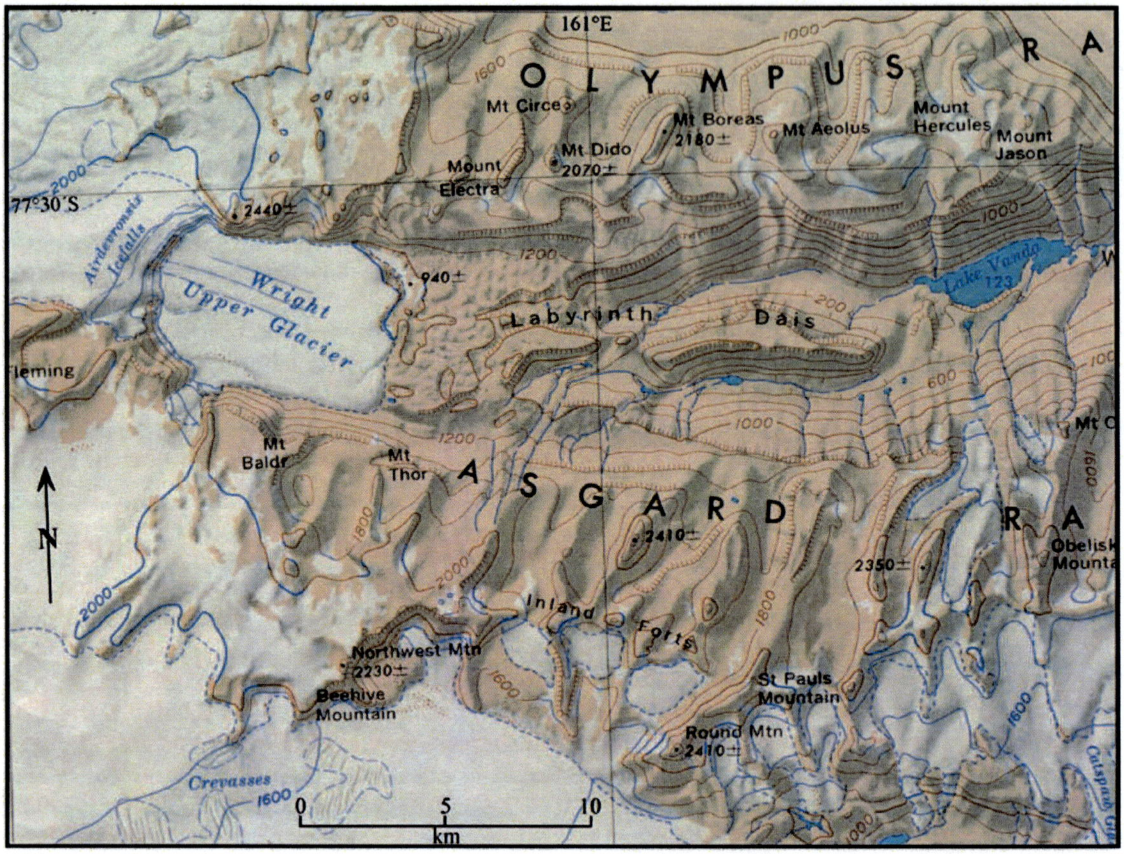

Fig. 3.5 The western end of Wright Valley in southern Victoria Land is framed by the Olympus Range in the north and the Asgard Range in the south. Ice from the East Antarctic ice sheet of the polar plateau enters the valley via the Airdevronsix icefall and forms the Wright Upper Glacier. The Labyrinth at its base was carved by meltwater streams flowing under pressure at the base of a glacier that once filled the Wright Valley. Lake Vanda is perennially covered by ice except for a moat of open water that forms during the summer months in December and January. In contrast to its icy surface, Lake Vanda contains a layer of hot brine at the bottom that is heated by sunlight which penetrates the ice and overlying clear water (Excerpt from the Taylor Glacier, Antarctica topographic map (ST 57–69/5; 77198-S1-TR-250; revised 1988) published by the U.S. Geological Survey, Denvery, Colorado)

(77°36'S, 161°05'E,). The Linnaeus Terrace was protected because of the presence of endolithic algae and lichens in the Beacon Sandstone at that site (Friedmann et al. 1988).

3.1.2 Geology

The basement rocks of the ice-free valleys of southern Victoria Land, including the area between the Ferrar and the Koettlitz glaciers farther south in Fig. 3.4, exemplify the structural complexity, lithologic diversity, regional metamorphism, and magmatic activity of the Ross orogen. Therefore, the geology of this area is presented in sufficient detail to serve as the type example for the basement rocks exposed elsewhere along the Transantarctic Mountains.

The geology of the Victoria and Wright valleys in Fig. 3.8 was mapped by P.N. Webb and B.C. McKelvey during the austral summers of 1957 and 1958, respectively (Webb and McKelvey 1959; McKelvey and Webb 1962). Their work, together with that of Harrington (1958), Gunn and Warren (1962), and Grindley and Warren (1964), set the stage for the ongoing study of the geology of the Transantarctic Mountains (e.g., Findlay et al. 1984; Allibone et al. 1993a, b; Encarnación and Grunow 1996).

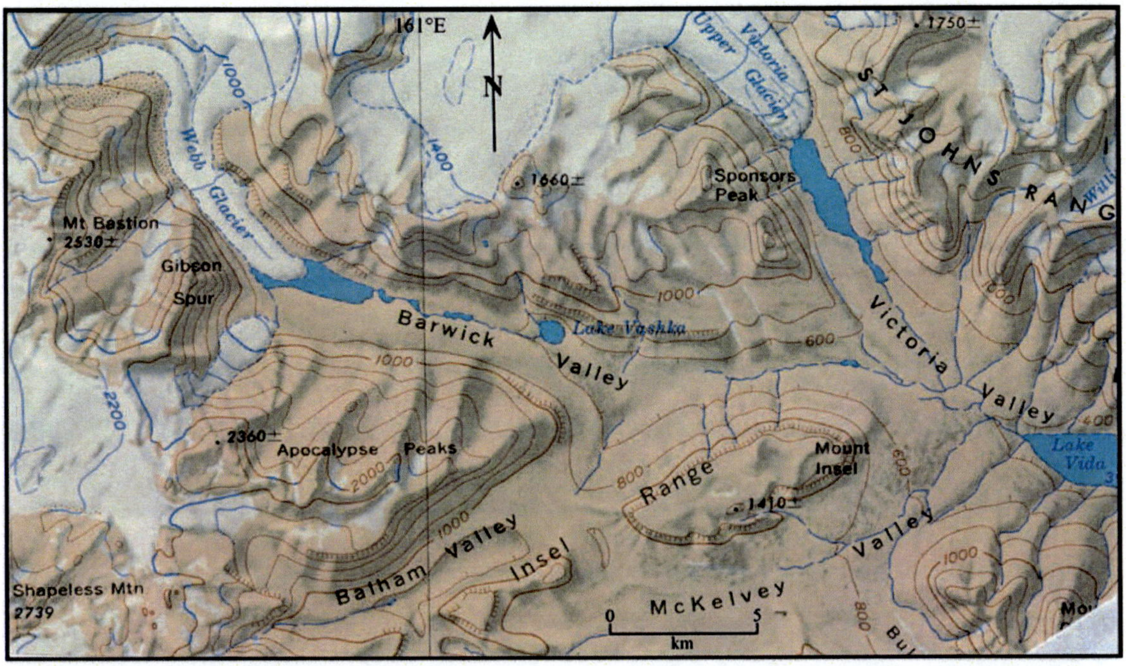

Fig. 3.6 The Balham, Barwick, Victoria, and McKelvey valleys of southern Victoria Land are ice-free because the flow of ice from the polar plateau via the Webb Glacier and the Upper Victoria Glacier is currently restricted. These valleys therefore expose metamorphic and granitic igneous rocks of the basement complex overlain unconformably by the Lower Devonian sandstones of the Beacon Supergroup. The Vida Granite in this area is named after Lake Vida in Victoria Valley. Meltwater lakes occur at the foot of the Webb and Upper Victoria Glaciers. The Barwick Valley also contains Lake Vashka (Excerpt from the Taylor Glacier, Antarctica topographic map (ST 57–69/5; 77198-S1-TR-250; revised 1988) published by the U.S. Geological Survey, Denver, Colorado)

McKelvey and Webb (1962) reported that large parts of *Wright Valley* are underlain by tightly folded metasedimentary rocks of the Asgard Formation which they assigned to the Neoproterozoic-Early Cambrian Skelton Group of the Ross orogen. Although Findlay et al. (1984) later abolished the Asgard Formation, it remains a valid local stratigraphic unit. The metasedimentary rocks of the Asgard Formation were intruded by the syntectonic Wright Intrusives consisting of: Theseus Granodiorite, Loke Microgranite, Dais Granite, and Olympus Granite-Gneiss. After the main phase of compressive deformation (i.e., the Ross Orogeny), the Asgard Formation and the Wright Intrusives were invaded by the Vida Granite and dikes of Vanda Lamprophyre and Porphyry which together constitute the suite of Victoria Intrusives of McKelvey and Webb (1962). In a larger context, the Wright and Victoria intrusives are the local representatives of the Granite Harbor Intrusives of southern Victoria Land.

More recently, Allibone et al. (1993a, b) and Cox (1993) identified 15 major granitic plutons in southern Victoria Land and proposed a new set of names for them because the lithologic character of the plutons is sufficiently diverse to resemble several of the intrusives previously named by McKelvey and Webb (1962). The classification of Allibone et al. (1993a, b) consists of three groups of intrusives:

1. Elongate, *concordant* plutons, which range from monzodiorite to granodiorite, are relatively undeformed and contain aligned K-feldspar megacrysts, hornblende, biotite, and mafic inclusions. These plutons were emplaced between 589 and 490 million years ago at deep levels during the metamorphism to the upper amphibolite facies of the metasedimentary rocks of the Asgard Formation.

   The names of these concordant plutons are: Bonney, Denton, Cavendish, and Wheeler.

Fig. 3.7 Taylor Valley in southern Victoria Land still contains a glacier that discharges meltwater into Lake Bonney. Ice from the East Antarctic ice sheet enters both the Taylor and Ferrar valleys but only the Ferrar Glacier actually reaches McMurdo Sound. The Rhone Glacier descends from the valley wall on the right. A thick sill of the Ferrar Dolerite is exposed farther up the valley (Photo by Rebecca Witherow reproduced by permission)

2. Elongate, *discordant* plutons are composed of equigranular, homogeneous biotite granodiorite and granite, and were intruded at shallow depth in the crust at 490 million years ago.

   The names of these discordant plutons are: Hedley, Valhalla, St. Johns, and Suess.

3. Discordant ovoid plutons that cut and are themselves cut by the Vanda dikes of mafic and felsic porphyry. These plutons were emplaced at shallow depth between 486 and 477 million years ago and range in composition from monzonite to granite. Some of these plutons contain K-feldspar megacrysts that resemble the texture of the oldest concordant plutons.

   The names of these youngest discordant plutons are: Pearse, Nibelungen, Orestes, Brownworth, Swinford, and Harker.

The new terminology for the magmatic rocks of the ice-free valleys was adopted by Encarnación and Grunow (1996) and was evaluated by Stump (1995) in the light of the nomenclature used in the literature that preceded the publications of Allibone et al. (1993a, b). These authors tried to improve the mapping of the ice-free valleys of southern Victoria Land by examining the contacts of individual plutons and by applying geochemical criteria that were not available to the geologists who worked in the valleys during the IGY and in the decade that followed. The new geologic map of Allibone et al. (1993a, b) is based on interpolations

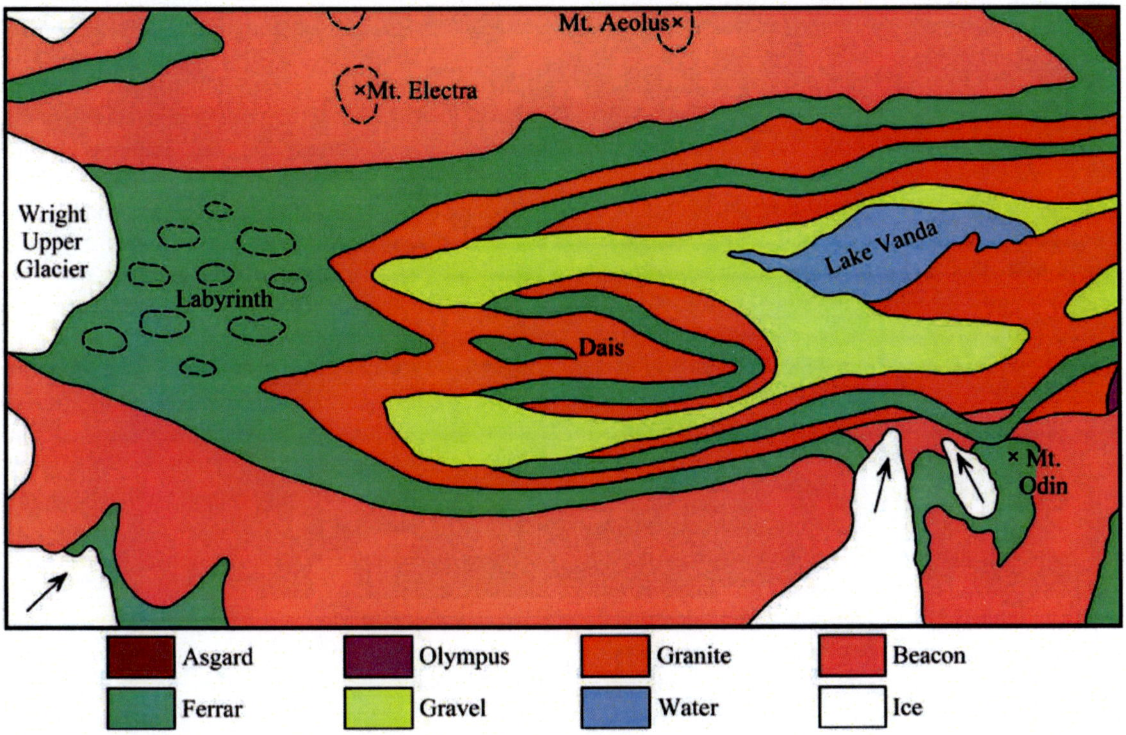

Fig. 3.8 Geologic map of the western Wright Valley in southern Victoria Land. This area is included in Fig. 3.5 (Adapted from McKelvey and Webb 1962)

between outcrops of plutonic basement rocks, which are extensively covered by the sandstones of the Beacon Supergroup and by the sills of the Ferrar dolerite. These interpolations between outcrops can be tested in the future by diamond drilling at selected sites in order to confirm the continuity of the underlying plutons beneath the cover rocks and glacial deposits.

The *Asgard Formation* of Wright Valley described by McKelvey and Webb (1962) consists of interbedded marbles, hornfels, and schists that grade locally into paragneiss and migmatite close to contacts with plutons of the Wright Intrusives. The type locality of this formation is the south wall of Wright Valley between Mt. Loke and Mt. Valkyrie.

The marbles are coarse-grained white rocks that disintegrate readily and therefore are rarely found in talus or moraines. They are composed of twinned grains of calcite with minor graphite and diopside. The metamorphic rocks classified as hornfels by McKelvey and Webb (1962) are composed of red garnet (13 mm) and fine-grained green diopside with minor amounts of quartz. Some specimens contain scapolite, diopside, and sphene in contact with the marble layers. The most common schist layers are composed of quartz and feldspar, but the schist locally contains plagioclase, biotite, and hornblende. The metasedimentary rocks of the Asgard Formation contain intercalated intrusive granitic orthogneisses (Allibone et al. 1993a) which form the Olympus Granite-Gneiss of McKelvey and Webb (1962). These orthogneisses are prominent along the north wall of Wright Valley east of Bull Pass. In this exposure, the foliation of the gneiss is aligned parallel with the strike of the Asgard Formation, inclusions of which occur in the transition zone within the gneiss. The relationship of the Olympus Granite-Gneiss to the other lithologies of the basement complex is uncertain although McKelvey and Webb (1962) considered the Olympus Granite-Gneiss to be one of the Wright Intrusives.

The development of augen gneiss by metasomatism of quartzo-feldspathic gneiss which formed by regional metamorphism during intense deformation of greywacke and sandstone was described by Smithson et al. (1971a, b) as an explanation for the gradational relationship between the Olympus Granite-Gneiss and the interbedded metasedimentary rocks of the Asgard Formation. Smithson et al. (1971c) also proposed that amphibolite reaction rims can form by metasomatic introduction of iron and magnesium coupled with the loss of calcium from diopside granofels, which is a medium- to coarse-grained granoblastic metamorphic rock with little or no foliation or lineation (Jackson 1997). The conversion of metasedimentary rocks into crystalline gneiss complexes is a characteristic of the Ross orogen that forms the basement complex of the Transantarctic Mountains (Grindley 1971, 1981).

The other members of the Wright Intrusives described by McKelvey and Webb (1962) are distinguished primarily by their texture, mineral composition, and bulk chemistry. The relation of these units to the terminology of Allibone et al. (1993a, b) is indicated in Table 3.2. Until the new terminology of the plutonic rocks in the ice-free valleys has been tested in the field and is confirmed by drilling and/or additional mapping, we will continue to use the traditional nomenclature. However, we consider that the Asgard Formation of McKelvey and Webb (1962) is largely correlative with the amphibolite-grade metasediments north of the Skelton Glacier and we will therefore refer to it as the Asgard Formation (Koettlitz Group) or simply as Koettlitz Group metasediments.

Table 3.2 Relation of the terminology of Allibone et al. (1993a, b) to the classification of the igneous and metamorphic rocks in Victoria, Wright, and Taylor valleys by Allen and Gibson (1962), McKelvey and Webb (1962), and Haskell et al. (1965a), respectively

Allibone et al. (1993a, b)	Previous classification
Bonney pluton	Olympus Granite-Gneiss and Dais Granite (Wright Valley) Larsen Granodiorite (Taylor Valley)
Denton pluton	Olympus Granite-Gneiss (Wright Valley)
Wheeler pluton	Larsen Granodiorite (Taylor Valley) Dais Granite (Victoria Valley)
Catspaw pluton	Irizar Granite (Taylor Valley)
St. Johns pluton	Vida Granite (Victoria Valley)
Hedley pluton	Vida Granite (Ferrar Valley)
Suess pluton	Larsen Granodiorite
Valhalla pluton	Olympus Granite and Dais granite (Wright Valley)
Biotite granitoid dikes	Theseus Granodiorite (Wright and Taylor Valleys)

The *Dais Granite* is exposed in the so-called Dais west of Lake Vanda close to the western (inland) end of Wright Valley. It is coarsely foliated parallel to the strike of the Koettlitz-Group metasediments and of the Olympus Granite-Gneiss. The Dais Granite is characterized by being porphyritic with large phenocrysts of orthoclase in a coarse matrix of oligoclase, hornblende, biotite, and quartz. Allanite, zircon, iron oxide, and apatite are present as accessory minerals. The Dais Granite has the composition of an adamellite that grades locally into alkali granite (Appendix 3.7.1).

The *Loke* (or Loki) *Microdiorite* occurs as dikes up to 2.5 m wide that intrude the Olympus Granite-Gneiss at Mt. Loke (Loki) and east of Mt. Theseus. The microdiorite contains up to 25% biotite and 15% green hornblende, as well as plagioclase, orthoclase, and quartz (Appendix 3.6.1).

Dikes of *Theseus Granodiorite* intrude the metasedimentary rocks of the Koettlitz Group and all of the other members of the Wright Intrusives. The type locality of the Theseus Granodiorite is on the north wall of Wright Valley between Mt. Theseus and Bull Pass. Similar intrusives of grey biotite granite have been reported at Granite Harbor and in the Kukri Hills.

The post-orogenic *Vida Granite* of the Victoria Intrusives intrudes Dais Granite and Olympus Granite-Gneiss in the western Asgard and Olympus mountains. The type locality is south of Lake Vida on the northern slope of the Olympus Range in Victoria Valley. The Vida Granite is a pink, hornblende-bearing, equigranular rock composed of orthoclase, oligoclase, quartz, biotite, and green hornblende with fine-grained interstitial quartz. Similar massive and unfoliated granitic intrusives have been reported from the valley of the Ferrar Glacier and from the coast of southern Victoria Land at Granite Harbor and Cape Irizar (Kurasawa et al. 1974).

All of the Wright and Victoria intrusives are cut by thin dikes of the *Vanda Lamprophyre and Porphyry*. At the type locality just east of Lake Vanda a swarm of these dikes intrudes the Dais and Vida granites and can be traced from there through Bull Pass into Victoria Valley. The lamprophyre is a dark aphanitic rock containing zoned plagioclase, green hornblende, chlorite, quartz, iron oxide, and accessory sphene and apatite. A typical porphyry dike at Bull Pass, about 6 m wide, contains phenocrysts of pink and white feldspar, greenish black hornblende, biotite, and quartz in a purplish aphanitic matrix.

The geology of Wright Valley mapped by McKelvey and Webb (1962) was extended by Allen and Gibson (1962) into the *Victoria Valley*. They described the Asgard Formation (Koettlitz Group) as a thick sequence of isoclinally folded marbles which are interbedded with layers of paragneiss, granulite, and quartzo-feldspathic schist. The metasedimentary rocks grade into Olympus Granite-Gneiss which passes gradationally into Dais Granite. The Asgard Formation (Koettlitz Group) and the Dais Granite are cut by the Vida Granite, and all of the basement rocks contain late-stage dikes of Vanda Lamprophyre and Porphyry.

The geology of *Taylor Valley* was initially examined by McKelvey and Webb (1959), Hamilton and Hayes (1960), and by Angino et al. (1960, 1962). Subsequently, Haskell et al. (1965a) published a map of the geology of Taylor Valley and described the metasediments of the Koettlitz Group which occur in four parallel north-trending belts, most prominent of which is the Middle-Taylor-Valley belt which is about 150 m wide. The metasediments of the Middle belt exposed along the south side of Taylor Valley consist of interbedded white marble, metaquartzite, magnesium-rich schist, calc-silicate hornfels, and quartzo-feldspathic schist. The mineral assemblages are consistent with the amphibolite facies and higher grades of regional metamorphism.

A second belt of Koettlitz-Group metasediments forms a south-plunging syncline in the *Nussbaum Riegel* which is a prominent topographic feature that projects from the Kukri Hills northward into Taylor Valley. According to Haskell et al. (1965a), the rocks of this feature consist of interbedded marble, metaquartzite, and quartz-labradorite-biotite schist. These rocks are correlative with the Asgard Formation (Koettlitz Group) of Wright and Victoria valleys (McKelvey and Webb 1962; Allen and Gibson 1962). In addition, Haskell et al. (1965a) correlated the rocks of Nussbaum Riegel with the basal part of the *Hobbs Formation* which occurs in the area between the Ferrar and the Koettlitz glaciers in Fig. 3.4. The structure and regional metamorphism of the metasedimentary rocks of Nussbaum Riegel were later described in great detail by Williams et al. (1971).

The metasedimentary rocks of the Koettlitz Group in Taylor Valley were intruded by several facies of the Granite Harbor Intrusives, including the highly foliated Olympus Granite-Gneiss, the Larsen Granodiorite which resembles the Dais Granite in Wright Valley, the Theseus Granodiorite, and the Irizar Granite. All of the different types of the Granite Harbor Intrusives and the metasedimentary rocks of the Koettlitz Group in Taylor Valley are cut by thin dikes of pegmatite, aplite, lamprophyre, porphyry, microdiorite, and ultramafic rocks.

The *Larsen Granodiorite* in Taylor Valley contains an orbicular phase located along the north side of the Taylor Glacier about 2.6 km from its terminus (Haskell et al. 1965a, Fig. 4). This occurrence was later investigated by Palmer et al. (1967) and by Dahl and Palmer (1981, 1983). Additional occurrences of orbicular textures were described by Allibone et al. (1993a).

A pluton of the post-tectonic *Irizar Granite*, exposed on Mt. Falconer (77°34'S, 163°08'E,) near the mouth of Taylor Valley, was studied by Ghent and Henderson (1968). These authors provided detailed petrographic descriptions of the various rock types exposed on Mt. Falconer and suggested that the Falconer pluton was originally emplaced at a depth of less than 8–10 km below the surface of the Earth. Parts of the pluton were exposed by erosion of the overlying rocks before sediment of the Beacon Supergroup was deposited on the Kukri Peneplain during the Early Devonian between 417 and 391 million years ago (IUGS 2002). McDougall and Ghent (1970) later reported K-Ar dates of biotite and hornblende from the Irizar Quartz Monzonite (adamellite) of the Mt. Falconer pluton, from the metasedimentary rocks of the Koettlitz Group, and from dikes that intruded the latter.

3.2 Koettlitz and Skelton Groups

The area between the Ferrar and the Skelton glaciers in Fig. 3.9 has been difficult to map because it is mountainous, snow-covered, and windy with occasional storms that can deposit up to 1.2 m of snow (Murphy et al. 1970; Flory et al. 1971; Rees et al. 1989). In spite of the active glaciation of most of this area, several ice-free valleys occur along the south side of the mountain range located between the Blue and Koettlitz glaciers (e.g., Garwood, Marshall, Miers, and Hidden valleys) which are identified on a map published by Bull (1962). The geology of this area was mapped by Blank et al. (1963) following Gunn and Warren (1962). In addition, Williams et al. (1971)

Ice-Free Valleys, Koettlitz Glacier, Southern Victoria Land

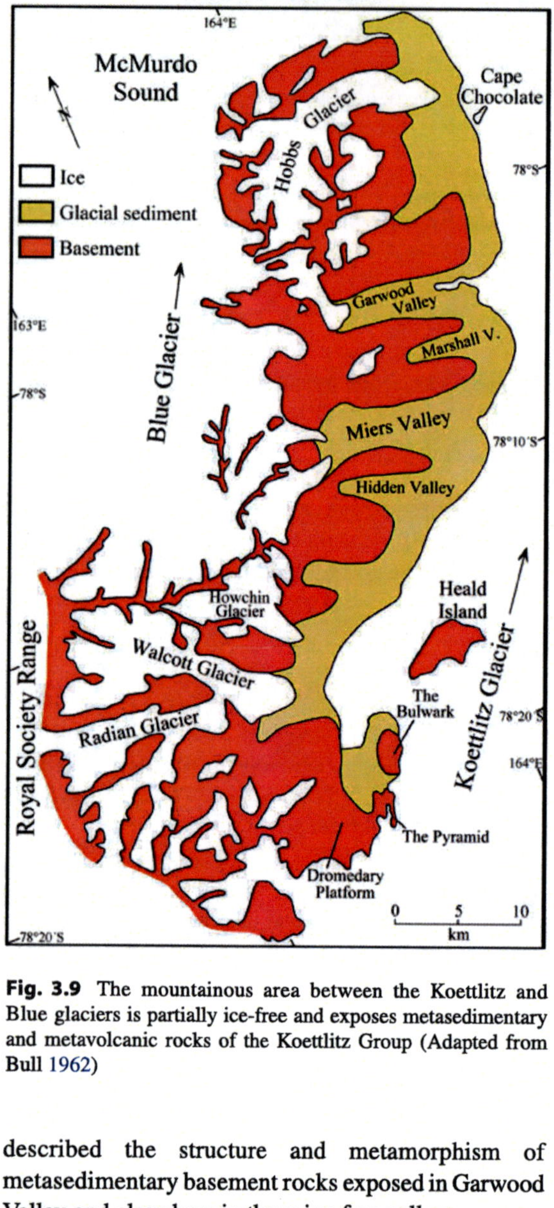

Fig. 3.9 The mountainous area between the Koettlitz and Blue glaciers is partially ice-free and exposes metasedimentary and metavolcanic rocks of the Koettlitz Group (Adapted from Bull 1962)

described the structure and metamorphism of metasedimentary basement rocks exposed in Garwood Valley and elsewhere in these ice-free valleys.

The multiply folded metasedimentary rocks that dominate the basement complex of this area were originally assigned to the Skelton Group by Gunn and Warren (1962). However, Grindley and Warren (1964) later placed only the low-grade metasedimentary rocks of the greenschist facies into the Skelton Group and assigned amphibolite-grade metasediments in the basement of southern Victoria Land to the Koettlitz Group.

Rocks of the Skelton Group are exposed on Teall Island, on both sides of the Skelton Glacier, and near Mt. Cocks between the Skelton and the Koettlitz Glaciers. Gunn and Warren (1962) divided the rocks of the Skelton Group into the upper *Teall Formation* and the underlying *Anthill Formation*, but Skinner (1982) later questioned the existence of the Teall Formation because the rocks at the type locality on Teall Island at the mouth of the Skelton Glacier resemble calc-silicate rocks that occur within the Anthill Formation. In place of the Teall Formation, Skinner (1982) defined the *Cocks Formation* based on metagraywacke and metavolcanic rocks that unconformably overlie the Anthill Formation (limestone) at Cocks Glacier (78°41'S, 162°00'E,).

The *Anthill Formation* is about 3,300 m thick (Stump 1995, p. 99) and consists of well-bedded white to grey limestone interbedded with lesser amounts of mudstone, siltstone, and quartzite. The limestone appears to be unfossiliferous. The metamorphic grade of the Anthill Formation increases from the greenschist facies on the south side of the Skelton Glacier to the amphibolite facies farther north between the Skelton and the Koettlitz glaciers.

The *Cocks Formation* contains a porphyritic pillow lava that was dated by Rowell et al. (1993) by means of the samarium-neodymium (Sm-Nd) method (Appendix 3.6.6). The rock yielded a Sm-Nd model date of 700–800 Ma, which indicates that the flow was extruded during the Neoproterozoic Era. The metamorphic grade of the Cocks Formation also increases in a northerly direction.

The variation of metamorphic grade of the metasedimentary basement rocks of southern Victoria Land has caused confusion because the distinction between of the Skelton and Koettlitz groups is based on metamorphic grade rather than on lithology or age of the rocks. Although the metasedimentary rocks of the ice-free valleys of southern Victoria Land were originally assigned to the Skelton Group, their metamorphic grade is more consistent with the Koettlitz Group which contains rocks in the amphibolite facies. Blank et al. (1963) defined five formations in the area between the Koettlitz and the Blue glaciers and assigned them to the Koettlitz Group with increasing stratigraphic age:

Hobbs Formation (youngest)
Salmon Marble
Garwood Lake Formation
Miers Marble
Marshall Formation (oldest)

However, Mortimer (1981) and Findlay et al. (1984) who also worked in the area did not agree with the stratigraphic sequence proposed by Blank et al. (1963). Mortimer (1981) concluded that the Salmon and Miers Marble were the same layer of rock and that the Garwood Lake and the Marshall Formation were also the same unit. Findlay et al. (1984) who remapped basement rocks throughout southern Victoria Land subdivided the Koettlitz Group into the Marshall Formation, the Salmon Marble, and the Hobbs Formation but did not place them in stratigraphic order.

In spite of the effort that has been made to identify mappable rock units in the Ferrar-Blue-Koettlitz-Skelton glaciers area, the relationship of the Skelton and Koettlitz groups remains in doubt. The only point of agreement among the investigators is that the rocks in both groups were deposited in a marine environment (Laird 1981). Moreover, both groups contain clastic metasediments, calc-silicates, marbles, conglomerates, volcanic rocks, and amphibolites. Therefore, the Skelton and Koettlitz groups could be regarded as a single complex of sedimentary and volcanic rocks of Neoproterozoic to Cambrian age whose metamorphic grade increases regionally from greenschist along the Skelton Glacier in the south to amphibolite in the Koettlitz-Blue Glacier area farther north. However, Stump (1995) recommended caution in the interpretation of the evidence which is still fragmentary and controversial (Rees et al. 1989).

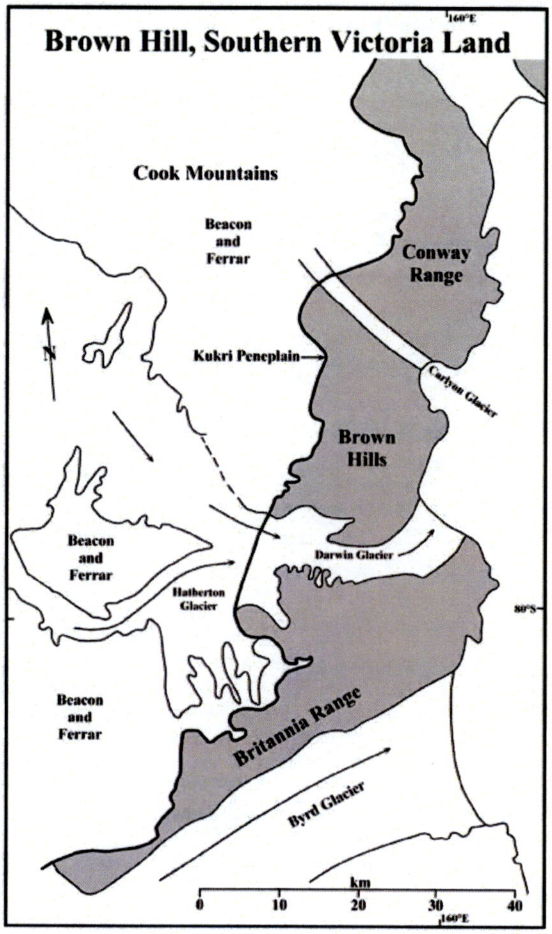

Fig. 3.10 The basement rocks that are exposed along the coast of the ice-free valleys of southern Victoria Land extend south into the Conway Range, the Brown Hills, and the Britannia Range. These rocks consist almost entirely of the Middle to Late Cambrian Granite Harbor Intrusives, which are subdivided into the Hope Granite, Skelton Granodiorite, Mt. Rich Granite, and the Carlyon Granodiorite. The older metasedimentary rocks of the Goldie Formation occur only in the form of inclusions within the granitic rocks of the Britannia Range (Adapted from Grindley and Laird 1969; Craddock 1969a)

3.3 Brown Hills

The basement rocks of southern Victoria Land in Fig. 3.10 extend south along the coast of the Ross Ice Shelf from the Skelton and Mullock glaciers to the Byrd Glacier. This segment of the Transantarctic Mountains contains the Conway Range, the Brown Hills, and the Britannia Range, all of which expose almost exclusively the granitic plutons of the Granite Harbor Intrusives (Grindley and Laird 1969). The granitic basement rocks of the Brown Hills in Fig. 3.11, which Encarnación and Grunow (1996) called the Brown Hills pluton, include the syntectonic Carlyon Granodiorite and the Mt. Rich Granite as well as the post-tectonic Hope Granite and dikes of leucocratic granite, pegmatite, lamprophyre, and meladiorite that cut the Carlyon and Mt. Rich plutons (Haskell et al. 1964, 1965b).

Metasedimentary rocks which are common in the ice-free valleys and in the Koettlitz-Skelton Glacier areas occur in the Brown Hills only as scattered inclusions within the Carlyon Granodiorite in the form of dark quartz-biotite-hornblende schist and light-colored metaquartzite. A large raft of metasediment, about 10,000 m^2 in area, occurs in the valley between the

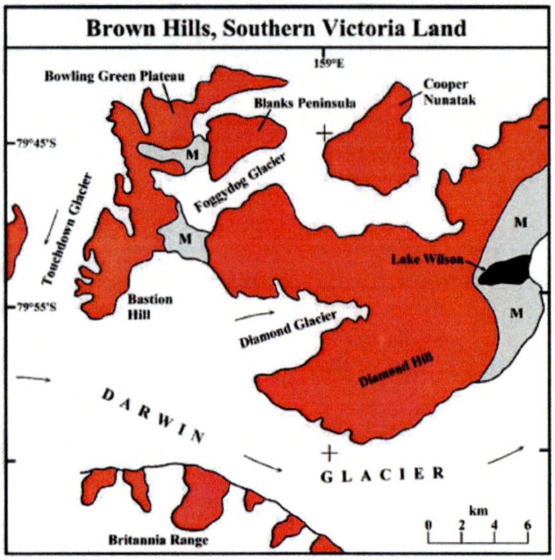

Fig. 3.11 The basement rocks exposed in the Brown Hills of southern Victoria Land consist almost exclusively of representatives of the Granite Harbor Intrusives (red) which were subdivided by Haskell et al. (1965b) into the synkinematic Carlyon Granodiorite and the Mt. Rich Granite followed by the post kinematic Hope Granite and a variety of dikes composed of leucocratic granite, pegmatite, lamprophyre, and meladiorite. Metasedimentary rocks consisting of quartz-biotite-hornblende schist occur only as scattered inclusions, the largest of which has an area of 10,000 m^2 and is located between the east end of Diamond Glacier and Lake Wilson (M = Moraine) (Adapted from Felder and Faure 1990, after Haskell et al. 1965b)

Fig. 3.12 Inclusion of metasedimentary rocks in the Granite Harbor Intrusives of the Brown Hills being investigated by Robert Felder during the 1978/79 Antarctic field season (Photo by G. Faure)

Diamond Glacier and Lake Wilson on the coast of the Ross Ice Shelf in Fig. 3.12. The strike of the metasedimentary rocks and the foliation of the Carlyon Granodiorite are parallel and the contact between them is gradational.

The Carlyon Granodiorite is medium to coarse grained, foliated to gneissic, and porphyritic in places. It grades into the Mt. Rich facies which is less foliated but more porphyritic than the Carlyon facies. Both intrusives are syntectonic and are composed of quartz, plagioclase, K-feldspar, biotite, hornblende, and sphene with accessory apatite, zircon, magnetite, and pyrite. The leucocratic dikes have higher concentrations of plagioclase and K-feldspar but lower concentrations of quartz and biotite than the Carlyon and Mt. Rich plutons and lack hornblende and sphene (Felder 1980). The average chemical analyses of the three major representatives of the Granite Harbor Intrusives published by Felder and Faure (1990) suggest that each was differentiated by fractional crystallization. However, differences in their initial $^{87}Sr/^{86}Sr$ ratios indicate that they are not the products of magmatic differentiation of a single parent magma.

A pluton composed of Hope Granite was mapped by Haskell et al. (1965b) on the Blanks Peninsula and the Bowling Green Plateau in Fig. 3.11. It is a fine-to-medium grained, equigranular, leucocratic granite composed of quartz, microcline, orthoclase, oligoclase, biotite, red garnet, and accessory chlorite. It resembles the Hope Granite described by Gunn and Walcott (1962) at the foot of Mt. Markham (82°51'S, 161°21'E,) located along the Lowery Glacier which is a southern tributary of the Nimrod Glacier (Section 5.2.3). The stratigraphic position of the Hope Granite is similar to that of the Vida Granite of the Victoria-Wright valleys and of the Irizar Granite in Taylor Valley. However, a genetic relationship among these post-kinematic granite plutons of southern Victoria Land has not yet been demonstrated.

3.4 Age Determinations

The geologic history of the basement rocks of southern Victoria Land started with the crystallization of crustal rocks of Gondwana which later became the source of sediment that was deposited along the passive paleo-Pacific margin of the continent. The sediment was subsequently folded and metamorphosed during the Ross Orogeny after the passive margin became a compressive subduction zone. During this active phase of the orogeny, the synkinematic plutons of the Granite Harbor Intrusives formed by crystallization of magma that originated by partial melting of accumulated sedimentary and volcanic rocks and by local recrystallization of the orogen (i.e., by granitization). After the main phase of the orogeny had ended, the magmatic activity continued with the intrusion of post-kinematic plutons and of various kinds of dikes of porphyry, lamprophyre, pegmatites, and mafic rocks. Although the sequence of events can be reconstructed by the interpretation of evidence in the field, the timing of the rock-forming events must be determined in the laboratory by radiogenic-isotope geochronometry (Faure and Mensing 2005). Summaries of the principal methods of dating igneous and metamorphic rocks are provided in Appendices 3.6.2–3.6.6.

Age determinations of rocks from the Transantarctic Mountains were included in several compilations of isotopic dates of Antarctic rocks. These compilations include, but are not necessarily limited to: Webb (1962), Picciotto and Coppez (1962, 1964a, b), Angino and Turner (1963), Ravich and Krylov (1964), Webb and Warren (1965), Craddock (1969b), Stuiver and Braziunas (1985), and others. Stuiver and Braziunas (1985) listed not only radiogenic-isotope dates, but also carbon-14 dates that record environmental processes during the most recent past extending to about 30,000 years before the present (BP).

3.4.1 K-Ar Dates

The first age determination of an Antarctic rock was reported by Goldich et al. (1958) who used the K-Ar method (Appendix 3.6.2) to date biotite from a sample of granite gneiss collected by P.N. Webb at Gneiss Point (77°24'S, 163°44'E,) on the coast of southern Victoria Land. Subsequently, additional K-Ar dates of biotite from the ice-free valleys were reported by Angino et al. (1962), Pearn et al. (1963), and McDougall and Ghent (1970). Although the numerical values of the decay constants of ^{40}K were changed after these reports were published, only the date reported by Goldich et al. (1958) is significantly altered when it is recalculated using the revised constants of Steiger and Jäger (1977).

The K-Ar dates of biotite in the basement rocks of Wright and Taylor valleys range from 425 to 520 million years (Ma) and have a slightly skewed but unimodal distribution in Fig. 3.13a. These dates record the time when the biotite cooled through its blocking temperature of 300 ± 25°C for radiogenic ^{40}Ar, whereas the blocking temperature of hornblende is 550 ± 25°C based on a compilation of data from the literature by Dallmeyer et al. (1981). Therefore, hornblende in a cooling body of rocks starts to retain radiogenic ^{40}Ar before biotite does, which causes K-Ar dates of hornblende to be *older* than K-Ar dates of biotite in the same body of rocks. The closure temperature of biotite (370 ± 20°C) reported by Berger and York (1981) agrees fairly well with that of Dallmeyer et al. (1981), but the closure temperature for hornblende (685 ± 55°C) of Berger and York (1981) is about 100° higher than that of Dallmeyer et al. (1981).

Although the K-Ar dates of biotites in the Granite Harbor Intrusives of the ice-free valleys in Fig. 3.13a post-date the time of crystallization of these rocks and form a "metamorphic veil" that obscures the crystallization age of the rocks, other methods of dating are available (to be discussed below) that do record the time of original crystallization of igneous and metamorphic rocks in spite of diffusion of radiogenic atoms during slow cooling of the rocks (i.e., the U-Pb method).

3.4.2 Rb-Sr Dates

Biotites in the granitic basement rocks of southern Victoria Land have also been dated by the Rb-Sr method which is based on the decay of naturally occurring radioactive ^{87}Rb to stable radiogenic ^{87}Sr (Appendix 3.6.3). The resulting Rb-Sr dates for biotite are just as sensitive to the loss of radiogenic ^{87}Sr during slow cooling as K-Ar dates of biotite. Therefore, Rb-Sr

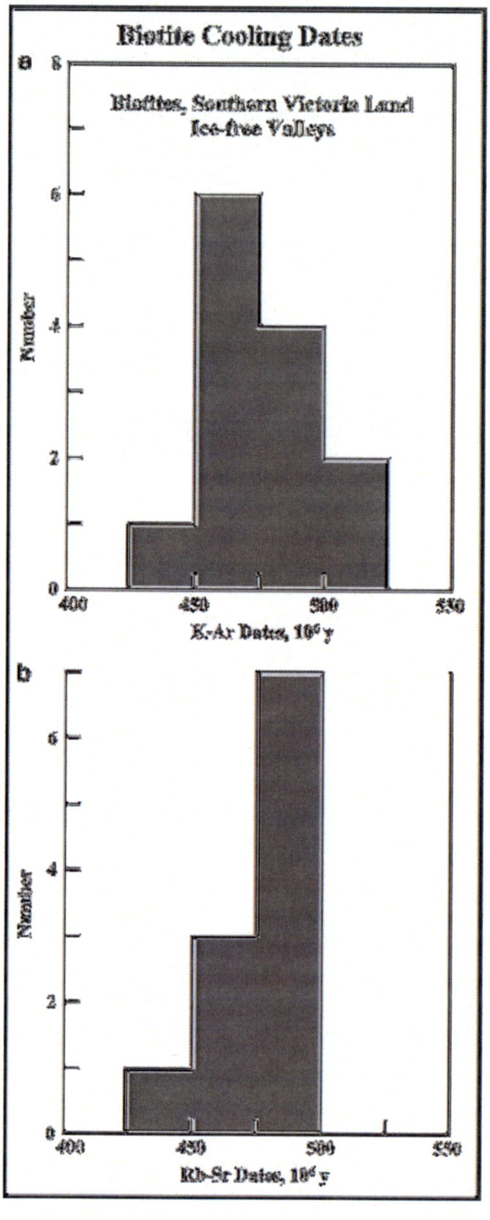

Fig. 3.13 (**a**) Histogram of K-Ar dates of biotite in the metasedimentary rocks (Asgard Formation, Koettlitz Group) as well as from the Granite Harbor Intrusives and late-stage mafic and lamprophyre dikes in Wright and Taylor Valley of southern Victoria Land. (Data from Goldich et al. 1958; Angino et al. 1962; Pearn et al. 1963; McDougall and Ghent 1970). (**b**) Histogram of Rb-Sr dates pf biotite in rocks of the basement complex described above. All dates were recalculated to $\lambda = 1.42 \times 10^{-11}$ year^{-1} for the decay constant of ^{87}Rb (Data from Deutsch and Webb 1964)

dates of biotite also record the time when plutonic igneous and high-grade metamorphic rocks cooled through the blocking temperature of Sr in biotite. In addition, Rb-Sr dates of single specimens of rocks and minerals (so-called "model dates") are based on assumed values of the initial $^{87}Sr/^{86}Sr$ ratio as explained in Appendix 3.6.3 and by Faure and Mensing (2005). This requirement does not apply to the calculation of K-Ar dates because the number of ^{40}Ar atoms that were incorporated into biotite or hornblende at the time these minerals crystallized from a magma is assumed to be zero.

The Rb-Sr dates of biotite extracted from rocks of the Asgard Formation (Koettlitz Group) and from the granitic intrusives in the ice-free valleys reported by Deutsch and Webb (1964) have a unimodal but skewed distribution in Fig. 3.13b. One biotite sample from the Koettlitz Group analyzed by those authors yielded an anomalously low Rb-Sr date of only 334 Ma (recalculated to $\lambda = 1.42 \times 10^{-11}$ year^{-1} for ^{87}Rb). Deutsch and Webb (1964) reported that this sample was collected "a few hundred feet" from a dolerite dike of the Middle Jurassic Ferrar Group and probably lost radiogenic ^{87}Sr as a result of heating during contact metamorphism. When this date is excluded, the Rb-Sr dates of biotite reported by Deutsch and Webb (1964) range from 425 to 500 million years and do not differ significantly from the K-Ar dates of biotites in Fig. 3.13a which were measured on a different set of samples collected in Wright and Taylor valleys.

A mixture of biotite and hornblende from a post-kinematic *porphyry dike* dated by Deutsch and Webb (1964) yielded a Rb-Sr date of 467 ± 15 Ma (recalculated to $\lambda = 1.42 \times 10^{-11}$ year^{-1}). This Middle Ordovician date may approach the crystallization age of this rock because the porphyry dikes in the basement of southern Victoria Land are thin (less than 1 m) and therefore cooled more rapidly than the much more voluminous Granite Harbor Intrusives (e.g., Olympus, Dais, Theseus, Vida, and Irizar).

Deutsch and Webb (1964) also reported an anomalously *old date* of 922 ± 80 Ma for a feldspar concentrate and 979 ± 80 Ma for a whole-rock sample from the same porphyry dike. These dates taken at face value suggested that this porphyry dike had crystallized about 1,000 million years ago, which implied that the basement rocks in the ice-free valleys had formed in Precambrian time more than 1,000 million years ago. In order to test this conclusion, Jones and Faure (1967) dated two porphyry dikes in Wright Valley by the Rb-Sr isochron method. The results in Fig. 3.14 yielded a well-constrained date of 470 ± 7 Ma (recalculated to $\lambda = 1.42 \times 10^{-11}$ year^{-1}). This date is a reliable determination of the age of the porphyry dikes in Wright Valley and confirms that their age is Middle Ordovician rather than Neoproterozoic.

The whole-rock Rb-Sr dates of the granitic intrusives in the ice-free valleys of southern Victoria Land in Table 3.3 range from 478 ± 4 to 490 ± 14 Ma (Allibone et al. 1993a; Faure et al. 1974). A well-defined Rb-Sr isochron for leucocratic granite dikes in the Brown Hills of 484 ± 6 Ma also fits in that range (Felder and Faure 1990). Several other whole-rock Rb-Sr isochron dates listed in Table 3.3 have large uncertainties caused by scatter of the initial $^{87}Sr/^{86}Sr$ ratios within individual plutons. The dates measured by Stuckless and Erickson (1975), Graham and Palmer (1987), and Allibone et al. (1993a) by a combination of whole-rock samples and constituent minerals range from 493 ± 17 Ma (Bonney pluton, Kukri Hills) to 472 ± 15 Ma (hornblende-biotite orthogneiss, Wright Valley). Two dates in this set are marginally younger (452 and 459 Ma, quartz monzonite, Granite Harbor) perhaps because biotite in these rocks lost radiogenic ^{87}Sr.

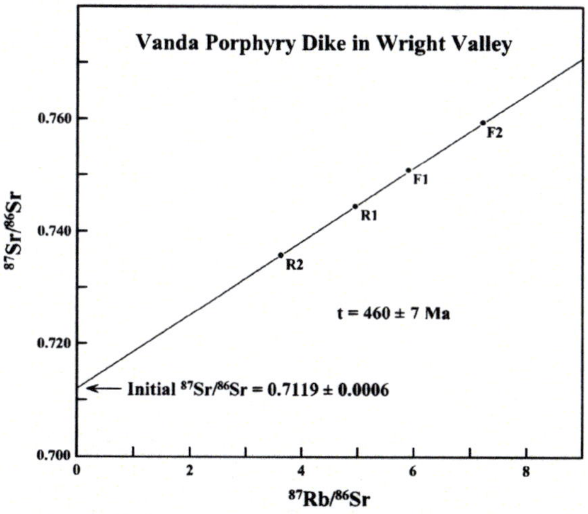

Fig. 3.14 Rb-Sr isochron defined by whole-rock samples of two Vanda Porphyry dikes (R1 and R2) and K-feldspar concentrates (F1 and F2). The date of 460 ± 7 Ma is the crystallization age of these dikes which were intruded into the metasedimentary rocks of the Asgard Formation, and into the plutonic igneous rocks of the Granite Harbor Intrusives in Wright Valley during the Middle Ordovician Epoch after the Ross Orogeny (Plotted from data by Jones and Faure 1967)

The slopes of Rb-Sr mineral isochrons are controlled primarily by biotite which has much higher $^{87}Rr/^{86}Sr$ ratios than K-feldspar, hornblende, or whole-rock samples (e.g., $^{87}Rb/^{86}Sr$ = 24.383 to 122.491). Consequently, the statistical precision of such dates is improved by inclusion of biotite on the isochron, but the dates calculated from the slope of the isochrons may be lowered if some of the accumulated radiogenic ^{87}Sr was lost from biotite during slow cooling and/or reheating of the rocks after crystallization from magma. Therefore, Rb-Sr isochron dates based on whole-rock samples and constituent minerals (e.g., biotite) need to be evaluated cautiously.

Most of the granitic rocks in southern Victoria Land that have been dated by the Rb-Sr whole-rock or whole-rock/mineral isochron methods in Table 3.3 have elevated initial $^{87}Sr/^{86}Sr$ ratios ranging from 0.70861 ± 0.00010 to 0.7121 ± 0.0011 compared to values between 0.7035 and 0.7045 for basalt that originated by decompression melting in the subcrustal mantle of the Earth (Faure 2001). Therefore, the granitic magmas that intruded the Ross orogen in southern Victoria Land and elsewhere in the Transantarctic Mountains contained *excess* radiogenic ^{87}Sr of crustal origin. Such magmas may form by partial melting during orogenic deformation and metamorphism of volcano-sedimentary complexes.

The age determinations of biotite and of whole-rock samples of the granite intrusives of southern Victoria Land reveal the timing of significant events in the evolution of the Ross orogen in this area. However, all of these dates *post-date* the deposition of the sediment along the paleo-Pacific coast of Gondwana. Adams and Whitla (1991) attempted to fill this gap in the history of the Ross orogen by using the whole-rock Rb-Sr method to date samples of the Asgard Formation (Koettlitz Group) collected in Wright and Victoria valleys. The time of deposition of clastic sedimentary rocks is difficult to determine by this method because the initial $^{87}Sr/^{86}Sr$ ratios of whole-rock samples depend on their lithologic composition and provenance. Therefore, whole-rock samples of sedimentary rocks in many cases scatter above and below Rb-Sr isochrons, which increases the uncertainty of dates derived from them and may introduce systematic errors. Adams and Whitla (1991) assumed that the isotopic composition of strontium in different lithologies

Table 3.3 Summary of radiogenic-isotope dates of plutonic igneous and metasedimentary rocks of the basement rocks of southern Victoria Land

Geologic unit	Date, Ma	$(^{87}Sr/^{86}Sr)_i$	References
Whole-rock Rb-Sr isochrons			
Marble, Asgard Fm., Koettlitz Gp.	840 ± 30	0.7081	1
Quartz-feldspar schist, Asgard Fm.	~670	0.707–0.713	1
Mica schist, Asgard Fm.	>615 ± 15	0.727	1
Olympus Granite Gneiss, Wright Valley	488 ± 43	0.7109 ± 0.0007	2
Vida Granite and Vanda Porphyry	471 ± 44	0.7104 ± 0.0008	2
Vanda Porphyry dikes	460 ± 7	0.7119 ± 0.0006	3
Harker pluton, St. Johns Range	478 ± 4	0.70952 ± 0.00012	4
St. Johns pluton, St. Johns Range	490 ± 14	0.70881 ± 0.00010	4
Granite, Lion Island, Cape Archer, Gregory Island	478 ± 35	0.70870 ± 0.00084	4
Carylon Granodiorite, Brown Hills	568 ± 9	0.7122 ± 0.00015	10
Mt. Rich Granite, Brown Hills	593 ± 238	0.7084 ± 0.0014	10
Leucocratic Granite dikes, Brown Hills	484 ± 6	0.7119 ± 0.0006	10
Whole-rock plus mineral Rb-Sr isochrons			
Vida Granite, Lake Vida	476 ± 14	0.7098 ± 0.0007	5
Avalanche Bay 77°01'S, 162°44'E,	452 ± 6	0.70892 ± 0.00005	6
Couloir Cliffs 77°01'S, 162°44'E,	459 ± 4	0.70903 ± 0.00056	6
Robertson Ridge 77°24'S, 162°12'E,	480 ± 14	0.70879 ± 0.00010	6
Hornblende-biotite orthogneiss, Wright Valley	472 ± 15	0.70861 ± 0.00010	4
Bonney pluton, Kukri Hills	493 ± 17	0.70893 ± 0.00008	4
Bonney pluton, Cathedral Rocks	479 ± 15	0.71016 ± 0.00013	4
$^{40}Ar/^{39}Ar$ plateau dates			
Biotite, CarylonGranodiorite, Brown Hills	513 ± 6, 504 ± 5, 508 ± 5		10
Hornblende, Carylon Granodiorite, Brown Hills	515 ± 6, 534 ± 6		10
Biotite, Mt. Rich Granite, Brown Hills	515 ± 5		10
$^{40}Ar/^{39}Ar$ total-gas release dates			
Biotite, Carylon Granodiorite, Brown Hills	476–510		10
Biotite, Mt. Rich Granite, Brown Hills	486–519		10
Hornblende, Carlyon Granodiorite, Brown Hills	531		10
U-Pb dates of zircon			
Couloir Cliffs, Granite Harbor	498 ± 4		7
Porphyry dike, Wright Valley	484 ± 7		7
Bonney pluton, Miers Ridge	505 ± 2		7
Granite, Cocks-Skelton Glacier	551 ± 4		7
Hornblende-biotite granite, Brown Hills	515 ± 8		7
Vida Granite, zircon	447 ± 34		8
Olympus Granite-Gneiss, zircon	462 ± 6		8
Olympus Granite-Gneiss, provenance	2554 ± 330		8
Olympus Granite-Gneiss	610 Upper intercept		9
Quartz syenite, Skelton-Cocks Glacier	551 ± 4		11
Sm-Nd model date			
Basalt, Cocks Fm., Skelton Glacier	809 ± 10		11

1 Adams and Whitla (1991); 2. Faure et al. (1974); 3. Jones and Faure (1967); 4. Allibone et al. (1993a); 5. Stuckless and Erickson (1975); 6. Graham and Palmer (1987) modified by Allibone et al. (1993a); 7. Encarnaciön and Grunow (1996); 8. Vocke and Hanson (1981); 9. Deutsch and Grőgler (1966); 10. Felder and Faure (1980, 1990); 11. Rowell et al. (1993) recalculated by Faure and Mensing.

of the Asgard Formation was homogenized during regional metamorphism and that the $^{87}Sr/^{86}Sr$ ratios of the pelitic layers had approached the values that existed in the limestone layers. However, the results do not support this assumption.

The measured $^{87}Rb/^{86}Sr$ ratios of the 37 rock samples analyzed by Adams and Whitla (1991) ranged widely from 0.086 to 11.88 depending on the presence of biotite, which is the principal Rb-bearing mineral in these rocks followed by K-feldspar. In addition to quartz-biotite-feldspar schists, the analyzed rock samples included marbles, amphibolites, and biotite-hornblende schists. The $^{87}Sr/^{86}Sr$ and $^{87}Rb/^{86}Sr$ ratios of the samples analyzed by Adams and Whitla (1991) form one coherent array (not shown) with the exception of two biotite schist samples from Wright Valley that have high $^{87}Rb/^{86}Sr$ ratios of >4.0. Adams and Whitla (1991) grouped these data on the basis of their lithologic composition and calculated separate Rb-Sr isochron dates for each group ($\lambda = 1.42 \times 10^{-11}$ year^{-1}): Mica schist, >615 ± 15 Ma; Quartz-feldspar schist, ~670 Ma; Marbles, 840 ± 30 Ma.

The authors concluded that the mica schists were regionally metamorphosed prior to 615 ± 15 Ma and that the isotopic composition of strontium in the quartz-feldspar schists was incompletely homogenized during a preceding event at about 670 Ma. A well-defined isochron of three samples of impure marble in Fig. 3.15 records the time when these rocks were deposited at 840 ± 30 Ma during the Neoproterozoic Era. The initial ratio of the marbles (0.7081) is higher than values of this ratio for unaltered marine carbonate rocks at 840 ± 30 Ma (Faure and Mensing 2005, p. 450). The chemical analysis published by Adams and Whitla (1991) for one of the marble samples suggest that the apparent enrichment of the marbles in ^{87}Sr is probably attributable to the presence of Sr-bearing impurities that were included in the carbonate minerals at the time of deposition (e.g., illite).

3.4.3 $^{40}Ar/^{39}Ar$ Dates

Felder and Faure (1990) dated biotite and hornblende in the granitic basement rocks of the Brown Hills by the $^{40}Ar/^{39}Ar$ method (Faure and Mensing 2005). This method is superior to the conventional K-Ar method because it is capable of reliably dating minerals even when they have lost some of the radiogenic ^{40}Ar that formed by the decay of ^{40}K (Appendix 3.6.4). The step-wise release $^{40}Ar/^{39}Ar$ dates of three biotites in the Carlyon Granodiorite reported by Felder and Faure (1990) are tightly clustered between 504 ± 5 and 513 ± 5 Ma with a mean of 508 ± 5 Ma, whereas two hornblendes in the Carlyon Granodiorite have partial release dates of 515 ± 6 and 534 ± 6 Ma. The older of the two hornblendes (534 ± 6 Ma) has a well developed partial-release plateau in Fig. 3.16 that represents the time when this mineral cooled through its blocking temperature for radiogenic ^{40}Ar. The other hornblende fraction contained biotite which explains its lower partial-release date (i.e., 515 ± 6 Ma). Therefore, the Carlyon Granodiorite cooled through the blocking temperature of hornblende (550 ± 25°C) at 534 ± 6 Ma and reached the blocking temperature of biotite (300 ± 25°C) at about 508 ± 5 Ma which yields a cooling rate of 9.6°C/Ma. These age determinations suggest that the crystallization age of the Carlyon Granodiorite and of the Mt. Rich Granite is more than 534 ± 6 Ma, which is in satisfactory agreement with the whole-rock Rb-Sr

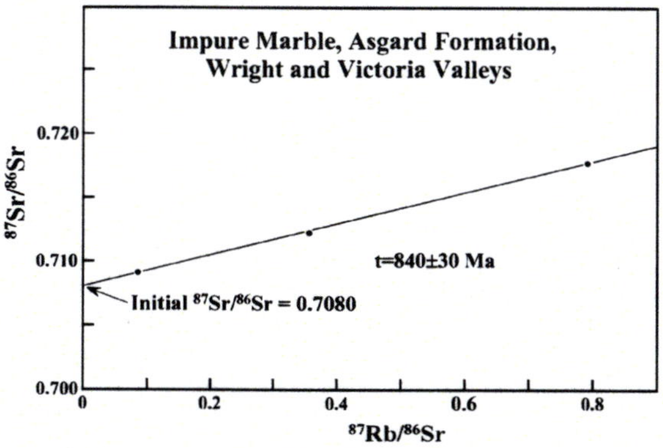

Fig. 3.15 Three whole-rock samples of marble in the Asgard Formation (Koettlitz Group) define a straight line on the Rb-Sr isochron diagram that yields a date of 840 ± 30 Ma and an initial $^{87}Sr/^{86}Sr$ ratio of 0.7080. Date and initial $^{87}Sr/^{86}Sr$ ratio calculated by Adams and Whitla (1991) from their own measurements.

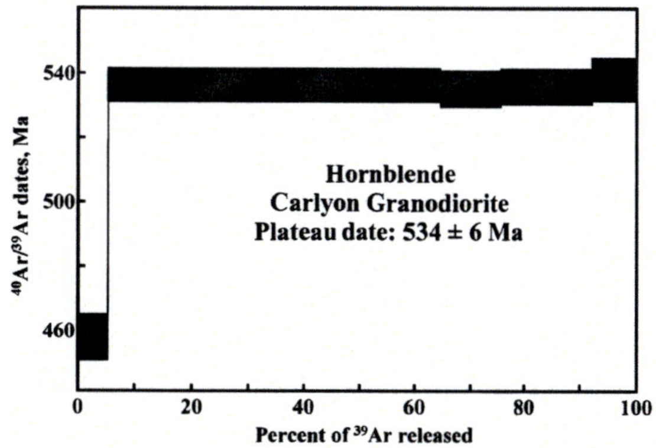

Fig. 3.16 The $^{40}Ar/^{39}Ar$ ratios of argon released by step-wise heating of a powdered sample of neutron-irradiated hornblende were used to calculate a series of dates which define a plateau at 534 ± 6 Ma. This date is a reliable estimate of the crystallization age of the Carlyon Granodiorite in the Brown Hills, southern Victoria Land (Replotted from data by Felder and Faure 1990)

date of the Carlyon Granodiorite (568 ± 34 Ma) listed in Table 3.3.

Felder and Faure (1990) also reported total-gas dates of biotites and hornblende from the Carlyon Granodiorite (biotite, 501–510 Ma; hornblende, 531 Ma) and from the Mt. Rich Granite (biotite, 486–519 Ma). These dates are equivalent to conventional K-Ar dates. Nevertheless, the total gas date of the hornblende (531 Ma) is indistinguishable from the partial-release plateau date (534 ± 6 Ma) of the same sample.

More recently, Encarnación and Grunow (1996) reported a U-Pb date of 515 ± 8 Ma for zircon in the equigranular and unfoliated granite of the Cooper Nunatak in the Brown Hills (see below). This date is compatible with the $^{40}Ar/^{39}Ar$ dates of biotites and hornblende reported by Felder and Faure (1990) and further constrains the time of intrusion and crystallization of the granite plutons in the Brown Hills.

3.4.4 U-Pb Dates

The U,Th-Pb methods of dating igneous and metamorphic rocks is based on the decay of the isotopes of uranium and thorium to stable isotopes of lead: $^{238}U \rightarrow {}^{206}Pb$; $^{235}U \rightarrow {}^{207}Pb$; $^{232}Th \rightarrow {}^{208}Pb$ in U, Th-bearing minerals such as zircon, sphene, apatite, or even rutile (Appendix 3.6.5). This method was used by Deutsch and Grőgler (1966) to date zircon grains extracted from a sample of Olympus Granite-Gneiss collected at 77°25' and 162°05'E, along the south shore of Lake Vida in Victoria Valley. The dates calculated by the authors for three zircon size-fractions are *discordant*, which implies that the zircons lost varying amounts of radiogenic lead or gained uranium and/or thorium (Faure and Mensing 2005). An interpretation of these data on a *concordia diagram* (Appendix 3.6.5) by the authors indicates that the zircons crystallized at about 610 Ma (or lost all pre-existing radiogenic lead) and subsequently lost varying amounts of radiogenic lead right up to the present. A later re-interpretation of the data by Vocke and Hanson (1981) yielded an upper-intercept date of 638 ± 92 Ma on a U-Pb concordia diagram and a lower-intercept date of 200 ± 193 Ma which is indistinguishable from zero.

Vocke and Hanson (1981) also dated zircons which they extracted from samples of the Vida Granite and the Olympus Granite-Gneiss in core 6 recovered by the Dry Valley Drilling Project (DVDP) at Lake Vida in Victoria Valley. An examination of the zircon grains in the Olympus Granite-Gneiss by cathodoluminescence revealed that these grains contain old cores with more recent overgrowths. Therefore, the data points representing different size-fractions define a chord on the concordia diagram in Fig. 3.17 that has an upper intercept at 2554 ± 330 Ma and a lower intercept at 462 ± 6 Ma. The most plausible interpretation of these dates in the context of the geology of southern Victoria Land is that the zircon cores are detrital grains that originally crystallized in Late Archean/Paleoproterozoic time. Accordingly, the upper-intercept date is the time of crystallization of the zircons in the *source area* from which they were liberated by erosion (i.e., it is the age of provenance). After deposition along the passive paleo-Pacific margin of Gondwana, the zircon grains

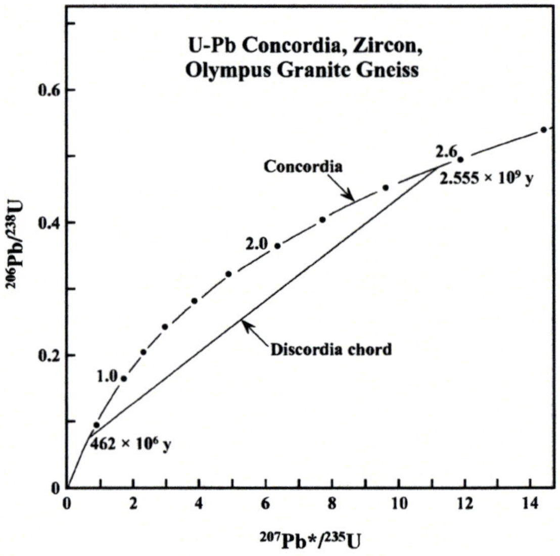

Fig. 3.17 The U-Pb concordia curve is plotted by solving Eqs. 3.16 and 3.17 for the same value of t in order to obtain the $^{206}Pb^*/^{238}U$ and $^{207}Pb^*/^{235}U$ ratios of points that represent concordant U-bearing minerals (see Faure and Mensing 2005, Table 10.3, p. 220). The discordant zircon crystals of the Olympus Granite-Gneiss analyzed by Vocke and Hanson (1981) define the discordia chord which intersects concordia at two points. The coordinates of these two points yield concordant dates by solving Eq. 3.18 or 3.19 (or both). Lower intercept, t = 462 ± 6 Ma; upper intercept, t = 2555 ± 330 Ma. The upper intercept yields the original crystallization age of the provenance of the zircon. The lower intercept is the time that has elapsed since the abraded zircon grains acquired fresh overgrowths during the Ross Orogeny. A more comprehensive interpretation was provided by Vocke and Hanson (1981) and by Vocke et al. (1978)

developed overgrowths at 462 ± 6 Ma by regional metamorphism of the sediment during the Ross Orogeny. Although the zircons analyzed by Vocke and Hanson (1981) have old cores, those of Deutsch and Grőgler (1966) do not, even though both originated from the Olympus Granite-Gneiss in Victoria Valley. A more comprehensives study of zircons in the granitic plutons of the ice-free valley could shed light on the provenance of the zircons and on the age of the plutons in which they now reside.

Different size fractions of the zircons extracted by Vocke and Hanson (1981) from the *Vida Granite* in DVDP6 define a chord that intersects the concordia curve at 447 ± 34 Ma and at 0 Ma. The upper intercept is the crystallization age of zircon in the Vida Granite and the lower intercept (0 Ma) indicates that the zircons may have been losing radiogenic lead by continuous diffusion as a result of chemical weathering.

Size-fractions of zircon extracted from a pluton of unfoliated quartz diorite which intrudes the intensely folded metasedimentary rocks of the *Anthill and Cocks formations* in the Skelton Glacier-Cocks Glacier area at 78°39' and 161°00'E, were dated by Rowell et al. (1993) by the U-Pb method. The data define a chord that intersects concordia at a point that corresponds to a well-constrained date of 551 ± 4 Ma. Zircon from a second pluton, composed of granite and located only 1 km from the quartz-diorite pluton, yielded an identical date of 551 ± 4 Ma (Encarnación and Grunow 1996). This date not only establishes the age of the two plutons, but also sets a lower limit to the time of deposition and deformation of the metasedimentary rocks that were intruded by these plutons.

3.4.5 Sm-Nd Dates

The most reliable way to determine the depositional age of unfossiliferous sedimentary (or metasedimentary) rocks is to date mafic volcanic rocks that may be interbedded with them. The Asgard Formation (Koettlitz Group) of the Wright and Victoria valleys as well as the Cocks Formation of the Skelton Glacier area contain *interbedded basalt*. The volcanic rocks in the *Asgard Formation* (Koettlitz Group) of Wright and Victoria valleys are represented by amphibolite and hornblende schist both of which have low initial $^{87}Sr/^{86}Sr$ ratios of 0.7032 ± 0.0002 and 0.70475 ± 0.00025, respectively (Adams and Whitla 1991). The *Cocks Formation* in the Skelton-Glacier area contains pillow basalts which were analyzed by Rowell et al. (1993) for dating by the Sm-Nd method (Appendix 3.6.6).

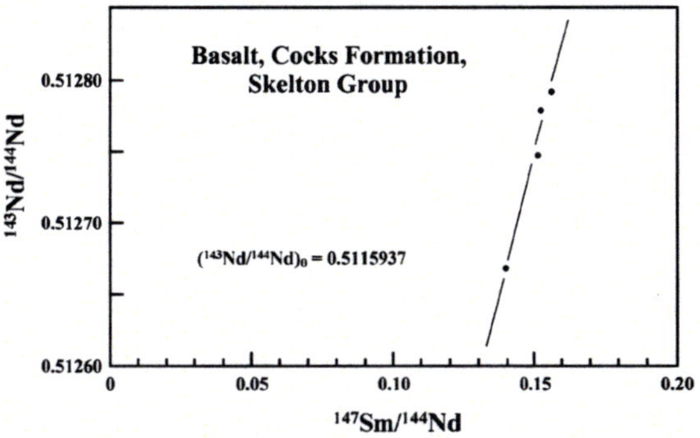

Fig. 3.18 The $^{147}Sm/^{144}Nd$ and $^{143}Nd/^{144}Nd$ ratios of a whole-rock sample and magnetic fractions of a pillow basalt in the Cocks Formation, Skelton Glacier area, define a straight line that was used to determine the initial $^{143}Nd/^{144}Nd$ ratio of this rock. The value was used in Appendix 3.6.6 to calculate a model date of 809 Ma based on the isotopic evolution of neodymium in CHUR (Plotted from data by Rowell et al. 1993)

We used the $^{143}Nd/^{144}Nd$ and $^{147}Sm/^{144}Nd$ ratios of one whole-rock basalt sample and three magnetic splits reported by Rowell et al. (1993) to determine their initial $^{143}Nd/^{144}Nd$ ratio by means of the Sm-Nd isochron diagram in Fig. 3.18. An unweighted least-squares regression of the data yields a value of 0.511594 for the initial $^{143}Nd/^{144}Nd$ ratio which we used to calculate the time in the past when the $^{143}Nd/^{144}Nd$ ratio of the chondritic uniform reservoir (CHUR) of the Earth in Fig. 3.19 had this value. The resulting Sm-Nd model date of 809 ± 10 Ma is similar to the date obtained by Rowell et al. (1993) who reported t = 700 to 800 Ma.

The Sm-Nd model date calculated above is the time when the pillow basalt in the *Cocks Formation* could have been erupted and was subsequently buried by sediment. If this conjecture is correct, then the age of the basalt is equal to the time of deposition of the sediment of the Cocks Formation. The Sm-Nd isotope systematics of mafic volcanic rocks are sufficiently robust to remain undisturbed by structural deformation and regional metamorphism as exemplified by volcanic rocks of Archean age (e.g., komatiite flows, Onverwacht Group, South Africa dated by Hamilton et al. 1979).

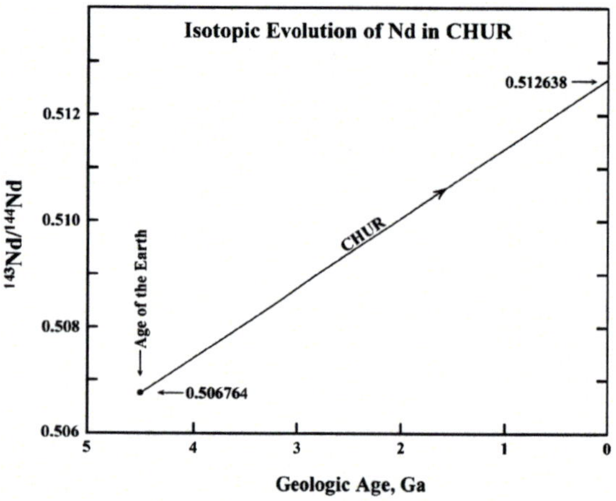

Fig. 3.19 The $^{143}Nd/^{144}Nd$ ratio of neodymium in the Chondritic Uniform Reservoir (CHUR) which represents the mantle of the Earth, increases as a function of time because of the alpha-decay of naturally-occurring radioactive ^{147}Sm (samarium). The initial $^{143}Nd/^{144}Nd$ ratio of CHUR at t = 4.5 × 10^9 years was 0.50674 and the present value of this ratio at t = 0 years is 0.512638

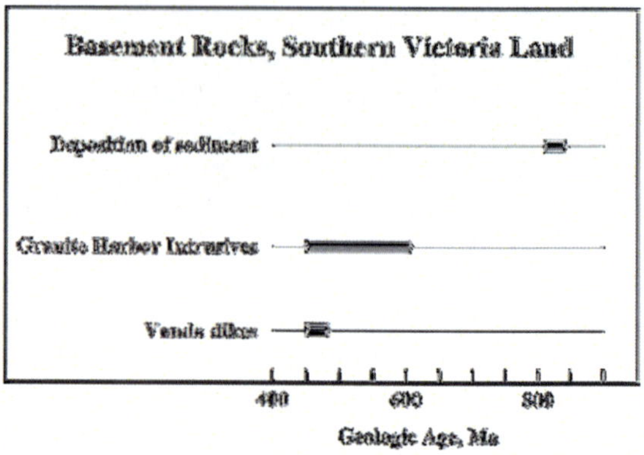

Fig. 3.20 The isotopic age determinations reviewed in this chapter were used to reconstruct the geologic history of the basement rocks of the Transantarctic Mountains in southern Victoria Land

3.5 Geologic History of Southern Victoria Land

The age determinations compiled in Table 3.3 can be used to reconstruct the geologic history of the Ross Orogen in southern Victoria Land with due regard to several self-evident considerations:

1. The geologic processes by means of which the rocks of the orogen originated acted slowly in the course of tens of millions of years.
2. The geologic processes did not occur everywhere at the same time but acted randomly in terms of time and space.
3. The radiogenic isotope systematics of K-Ar, Rb-Sr, and U-Pb were altered during tectonic deformation of sedimentary rocks, during slow cooling of granitic plutons and of high-grade metamorphic complexes.
4. The alteration of the isotope systematics occurred not only at high temperature and pressure at depth in the crust, but also at low temperature and pressure at shallow depth during the erosion of overlying strata of the Ross Mountains which were reduced to the Kukri Peneplain.
5. The sedimentary rocks were originally saturated with water which was squeezed out during tectonic deformation and regional metamorphism. The resulting flux of aqueous fluids toward the surface of the Earth contributed to the loss of radiogenic isotopes from minerals at depth in the crust.

The geologic history of the Ross Orogen in Fig. 3.20 started when the supercontinent Rodinia split into Gondwana and Laurentia which subsequently drifted apart. The resulting passive rift-margin of Gondwana permitted sediment and interbedded basalt lava to accumulate off-shore. The interval of time during which sediment and pillow lavas accumulated in the ocean off the passive margin of Gondwana is indicated by the whole-rock Rb-Sr isochron date of 840 ± 30 Ma for marbles in Wright and Victoria Valleys (Adams and Whitla 1991). Another date from that time period is the model Sm-Nd date of pillow basalt in the Cocks Formation of the Skelton-Glacier area (809 ± 10 Ma) based on data by Rowell et al. (1993). Consequently, the break-up of Rodinia occurred *prior* to these dates.

The clastic sediment that was deposited along the passive rift margin of Gondwana originated from the source regions that had crystallized in Neoarchean/Paleoproterozoic time at about 2554 ± 330 Ma (Vocke and Hanson 1981). Some time after 809 ± 10 Ma, the passive rift margin of Gondwana was transformed into a compressive subduction zone following unspecified changes in the pattern of convection of the mantle beneath the paleo-Pacific Ocean. As a result, the sedimentary-volcanic complex was compressed against the Gondwana craton, dewatered, extensively folded, and heated, which caused regional metamorphism that ranged from the greenschist to the amphibolite facies. Parts of the sedimentary-volcanic complex may have been subducted beneath the continental crust of Gondwana, which caused uplift of the continental margin. The intensity of regional metamorphism of the wet sediment was sufficient to cause large volumes of granitic magma to form by partial melting which intruded the overlying rocks while they were being deformed.

The resulting batholiths and plutons of the Granite Harbor Intrusives crystallized as they were being deformed by tectonic stress. The magmatic activity started during the latest Neoproterozoic to Early Cambrian time as exemplified by the Carlyon granodiorite/Mt. Rich granite of the Brown Hills which was dated by a whole-rock Rb-Sr isochron at 568 ± 54 Ma, by a $^{40}Ar/^{39}Ar$ plateau date of hornblende of 534 ± 6 Ma, and by a total Ar-release date of 531 Ma (Felder and Faure 1990). In addition, Encarnación and Grunow (1996) reported a U-Pb concordia date of zircon of 551 ± 4 Ma for a granite in the Cocks-Skelton Glacier area, and Rowell et al. (1993) obtained an identical U-Pb zircon date for a nearby quartz-syenite pluton. Deutsch and Grőgler (1966) reported an even older U-Pb concordia date of 638 ± 92 Ma for zircon in the Olympus Granite-Gneiss in Victoria Valley (recalculated by Vocke and Hanson 1981).

The magmatic activity continued for several tens of millions of years until it ended at about 460 ± 7 Ma with the intrusion of the Vanda Porphyry dikes in Wright Valley (Jones and Faure 1967). At this time in the geologic history of southern Victoria Land, a chain of high mountains existed along the former rift margin of Gondwana. The so-called Ross Mountains were subsequently eroded to form the Kukri Peneplain on which the sandstones and other clastic sedimentary rocks of the Beacon Supergroup were deposited starting at about 415 Ma in Early Devonian time.

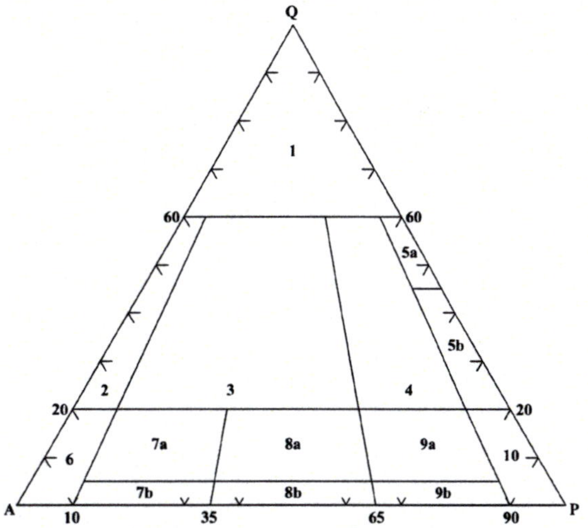

Fig. 3.21 The classification of plutonic igneous rocks by Streckeisen (1967, 1976) is based on the modal abundances of quartz (Q), alkali feldspar (including albite, An < 5%) (A), and plagioclase (An > 5%) (P). The fields have been numbered in order to identify the rock types that correspond to them: 1. quartz, 2. alkali granite, 3. granite, 4. granodiorite, 5a. quartz diorite (trondhjemite), 5b. quartz diorite (tonalite), 6. alkali syenite, 7a. quartz syenite, 7b. syenite, 8a. quartz monzonite (adamellite), 8b. monzonite, 9a. monzodiorite, 9b. monzogabbro, 10. diorite/gabbro (Adapted from Streckeisen 1967)

3.6 Appendices

3.6.1 *Classification of Plutonic Rocks of Granitic Composition*

Plutonic rocks of "granitic" composition are classified on the basis of the *modal* abundances of quartz, alkali feldspar including albite (An 0–5%), and plagioclase (An > 5%). The resulting triangular Q-A-P diagram was subdivided by Streckeisen (1967, 1976) into fields in Fig. 3.21 and numbered consecutively from 1 to 10. The plutonic rocks that occupy each of these fields are identified in the caption to Fig. 3.21. Four of the principal fields are subdivided as indicated by the letters *a* and *b*. For example, 7a contains quartz syenites while 7b is occupied by regular syenites. Streckeisen's classification has been widely used to classify granitic rocks of the Granite Harbor Intrusives in the Transantarctic Mountains.

3.6.2 *K-Ar Method*

Potassium has three naturally occurring isotopes:

$$^{39}_{19}K = 93.26\%, {}^{19}_{40}K = 0.0117\%, \text{and } {}^{19}_{41}K = 6.73\%.$$

Potassium-40 is radioactive and decays by branched decay to ^{40}Ar and to ^{40}Ca. The number of radiogenic decay products that accumulate as a function of time is expressed by an equation derived from the Law of Radioactivity (Faure and Mensing 2005).

$$^{40}Ar^* + ^{40}Ca^* = ^{40}K\left(e^{\lambda t} - 1\right) \quad (3.1)$$

where λ is the decay constant of ^{40}K and the asterisk identifies the radiogenic isotopes of K and Ca. The decay constant (λ) of ^{40}K is the sum of the decay constants that govern each of the two branches of the decay:

λ_e: electron-capture decay of ^{40}K to form $^{40}Ar^*$

λ_β: beta decay of ^{40}K to form $^{40}Ca^*$and the total decay constant of ^{40}K is:

$$\lambda = \lambda_e + \lambda_\beta \quad (3.2)$$

The decay of ^{40}K to ^{40}Ar is expressed by the equation:

$$^{40}Ar^* = \frac{\lambda_e}{\lambda}\,^{40}K\left(e^{\lambda t} - 1\right) \quad (3.3)$$

where λ_e/λ is the fraction of the ^{40}K atoms that decay to $^{40}Ar^*$. This equation can be used to date a K-bearing mineral, such as biotite or muscovite, by measuring the number of ^{40}Ar atoms that have accumulated in a unit weight of the mineral in the time that has elapsed since the mineral cooled sufficiently to allow the atoms of radiogenic ^{40}Ar to accumulate in its crystal lattice. In addition, the number of ^{40}K atoms that remain in the same unit weight of the mineral can be calculated from its measured potassium concentration.

The K-Ar method of dating arises from Eq. 3.3 which can be used to calculate the length of time in the past during which $^{40}Ar^*$ has accumulated:

$$e^{\lambda t} = \left(\frac{^{40}Ar^*}{^{40}K}\right)\left(\frac{\lambda}{\lambda_e}\right) + 1$$

$$t = \frac{1}{\lambda}\ln\left[\left(\frac{^{40}Ar^*}{^{40}K}\right)\left(\frac{\lambda}{\lambda_e}\right) + 1\right] \quad (3.4)$$

Equation 3.4 can now be used to illustrate the calculation of the K-Ar date of a biotite extracted from a hand specimen of paragneiss collected at Gneiss Point along the coast of southern Victoria Land at 77°24'S, 163°44'E,. Analyses reported by Goldich et al. (1958) of this biotite sample yield:

$$^{40}Ar^* / ^{40}K = 0.0333$$

The ratio of the decay constants has the numerical value:

$$\frac{\lambda}{\lambda_e} = \frac{5.543 \times 10^{-10}}{0.581 \times 10^{-10}} = 9.9540$$

Substituting into Eq. 3.4 yields:

$$t = \frac{1}{5.543 \times 10^{-10}}\ln\left[0.0333 \times 9.540 + 1\right]$$

$$t = \frac{\ln 1.31768}{5.543 \times 10^{-10}} = \frac{0.2758}{5.543 \times 10^{-10}} = 500 \times 10^6\,\text{years}$$

This date is the amount of time that has elapsed since the rock sample cooled sufficiently to allow radiogenic ^{40}Ar to accumulate in the biotite. This date is less than the time since crystallization of the mineral. Therefore, the rock is older than 500 × 10^6 years by an unknown amount of time depending on the cooling rate.

The date calculated above differs from the value of 520 million years reported by Goldich et al. (1958) because the values of the decay constant of ^{40}K used by these authors have been replaced by more accurate values in use at the present time. Nevertheless, this was the first age determination of a rock in Antarctica based on radioactivity. The date reported by Goldich et al. (1958) more than 50 years ago indicated that high-grade metamorphism of the basement rocks of the Transantarctic Mountains occurred during Middle Cambrian time (IUGS 2002).

3.6.3 Rb-Sr Method

Rubidium is an alkali metal that has two naturally occurring isotopes (^{85}Rb and ^{87}Rb), one of which is radioactive and decays to a stable isotope of strontium (Faure and Mensing 2005):

$$^{87}Rb \rightarrow ^{87}Sr^* + \beta^- + \bar{\nu} + Q \quad (3.5)$$

where β^- is a beta particle, $\bar{\nu}$ is an antineutrino, and Q is the energy liberated by the decay of one atom of ^{87}Rb. The number of $^{87}Sr^*$ atoms that form by decay of ^{87}Rb in a closed system is expressed by means of the Law of Radioactivity:

$$^{87}Sr^* = ^{87}Rb\left(e^{\lambda t} - 1\right) \quad (3.6)$$

where the decay constant $\lambda = 1.42 \times 10^{-11}$ year^{-1}. The total number of ^{87}Sr atoms in a unit weight of Rb-bearing mineral (e.g., biotite or muscovite) is:

$$^{87}Sr = {}^{87}Sr_i + {}^{87}Sr^* \qquad (3.7)$$

where $^{87}Sr_i$ is the number of atoms of this isotope that were incorporated into the Rb-bearing mineral at the time of its formation. Therefore,

$$^{87}Sr = {}^{87}Sr_i + {}^{87}Rb(e^{\lambda t} - 1) \qquad (3.8)$$

Each term of Eq. 3.8 is divided by the number of stable and non-radiogenic ^{86}Sr atoms of "ordinary" strontium:

$$\frac{^{87}Sr}{^{86}Sr} = \left(\frac{^{87}Sr}{^{86}Sr}\right)_i + \left(\frac{^{87}Rb}{^{86}Sr}\right)(e^{\lambda t} - 1) \qquad (3.9)$$

This equation can be used to date Rb-bearing minerals by measuring their $^{87}Sr/^{86}Sr$ and $^{87}Rb/^{86}Sr$ ratios and by solving Eq. 3.9 for t based on the known value of λ and an assumed value of $(^{87}Sr/^{86}Sr)_i$:

$$t = \frac{1}{\lambda} \ln \left[\frac{^{87}Sr/^{86}Sr - (^{87}Sr/^{86}Sr)_i}{(^{87}Rb/^{86}Sr)} + 1 \right] \qquad (3.10)$$

The resulting dates may be systematically in error in cases where the measured $^{87}Sr/^{86}Sr$ ratio of the mineral has about the same magnitude as the initial $^{87}Sr/^{86}Sr$ ratio. In addition, radiogenic ^{87}Sr may escape from Rb-bearing minerals as a result of slow cooling or by subsequent contact metamorphism or by isotope exchange with Sr in brines that may permeate the rocks. The loss of radiogenic ^{87}Sr by these or other processes lowers Rb-Sr mineral dates and causes them to underestimate the crystallization age of the host rock.

These problems are avoided by the whole-rock Rb-Sr isochron method of dating because the radiogenic ^{87}Sr that may diffuse out of Rb-rich minerals is quantitatively retained in most cases within the grain boundaries and by other minerals in a sample of the whole rock. When Eq. 3.9 is applied to a suite of rock samples all of which have the same age t, had the same initial $^{87}Sr/^{86}Sr$ ratio at the time of formation, and neither gained nor lost radiogenic ^{87}Sr, their measured $^{87}Sr/^{86}Sr$ and $^{87}Rb/^{86}Sr$ ratios form points on a straight line. This line is an isochron because all the rock or mineral samples that plot on this line have the same age.

Another way to recognize this phenomenon is to re-examine Eq. 3.9. If t is a constant, then $e^{\lambda t}$-1 is also a constant and the equation defines a straight line of the form:

$$y = b + mx \qquad (3.11)$$

where $y = {}^{87}Sr/^{86}Sr$, $x = {}^{87}Rb/^{86}Sr$, $b = (^{87}Sr/^{86}Sr)_i$, and $m = e^{\lambda t}-1$ is the slope of the line.

The Rb-Sr isochron method was used by Jones and Faure (1967) in Section 3.4.2 to date whole-rock samples and separated alkali feldspars of two Vanda porphyry dikes in Wright Valley. A least-squares regression of the data points in Fig. 3.14 yields a slope (m):

$$m = e^{\lambda t} - 1 = 0.006553$$

$$t = \frac{1}{\lambda} \ln(0.006553 + 1)$$

$$t = \frac{0.006531}{1.42 \times 10^{-11}} = 460 \times 10^6 y$$

Any sample of rock or mineral that does not satisfy the preconditions stated above does not plot on the isochron defined by a suite of cogenetic rocks that had the same initial $^{87}Sr/^{86}Sr$ ratio when they formed. Therefore, the goodness of fit of data points that define a Rb-Sr isochron is indicative of how well the assumptions of dating by this method are satisfied.

3.6.4 $^{40}Ar/^{39}Ar$ Partial-Release Dates

In order to date a potassium-bearing mineral by the $^{40}Ar/^{39}Ar$ method the sample is first irradiated by energetic neutrons in the core of a nuclear reactor (Faure and Mensing 2005). The neutrons interact with the nuclei of ^{39}K atoms and transform them into nuclei of ^{39}Ar by a nuclear reaction of the form:

$$^{39}_{19}K + \text{neutron} \rightarrow {}^{39}_{18}Ar + \text{proton} \qquad (3.12)$$

The resulting concentration of ^{39}Ar in the irradiated sample can be used to calculate the concentration of ^{39}K it contains. After the irradiation, the sample is heated stepwise in a vacuum and the isotopic composition of the Ar that is released at each step is measured with a gas-source mass spectrometer. The measured ratio of the argon isotopes is related to the age of the sample by the equation:

$$\frac{^{40}Ar^*}{^{39}Ar} = \frac{e^{\lambda t} - 1}{J} \quad (3.13)$$

where λ is the total decay constant of potassium (Appendix 3.6.2) and J is a constant that depends on the flux and energy distribution of the neutrons used in the irradiation. The "J-factor" is determined by irradiating a flux monitor which is a mineral whose age was determined by another radiogenic-isotope geochronometer. Therefore, the measured $^{40}Ar^*/^{39}Ar$ ratio of the mineral to be dated can be used to solve Eq. 3.13 for t which yields a date in the past.

When the sample to be dated is heated stepwise at increasing temperatures, the $^{40}Ar/^{39}Ar$ ratios of the argon that is released can be used to calculate a series of dates which form a spectrum when they are plotted versus the temperature of each heating step. Experience has shown that the $^{40}Ar/^{39}Ar$ ratios in many cases increase with increasing temperature because the gas that is released at low temperature has lost some of its radiogenic ^{40}Ar while the mineral cooled after it crystallized. The argon that is released at higher temperatures originates from the interiors of mineral grains and therefore has not lost radiogenic ^{40}Ar. Therefore, the $^{40}Ar/^{39}Ar$ dates form a flat-topped plateau that records the time when the mineral cooled through its blocking temperature.

The $^{40}Ar/^{39}Ar$ method of dating was used by Felder and Faure (1990) to determine the age of hornblende crystals from a sample of the Carlyon Granodiorite in the Brown Hills of southern Victoria Land. The spectrum of dates in Fig. 3.16 has a well-developed plateau that yields an average date of 534 ± 6 Ma.

3.6.5 U-Pb Methods

The long-lived and naturally occurring isotopes of uranium (^{238}U and ^{235}U) and thorium (^{232}Th) each decay to a different stable radiogenic isotope of lead through lengthy chains of intermediate radioactive daughters:

$$^{238}U \rightarrow {}^{206}Pb, \lambda_1 = 1.55125 \times 10^{-10} y^{-1}$$

$$^{235}U \rightarrow {}^{207}Pb, \lambda_2 = 9.8485 \times 10^{-10} y^{-1}$$

$$^{232}Th \rightarrow {}^{208}Pb, \lambda_3 = 4.9475 \times 10^{-11} y^{-1}$$

The growth of radiogenic isotopes of lead in minerals that contain uranium and thorium is expressed by equations derivable from the Law of Radioactivity (Faure and Mensing 2005):

$$\frac{^{206}Pb}{^{204}Pb} = \left(\frac{^{206}Pb}{^{204}Pb}\right)_i + \frac{^{238}U}{^{204}Pb}(e^{\lambda_1 t} - 1) \quad (3.14)$$

$$\frac{^{207}Pb}{^{204}Pb} = \left(\frac{^{207}Pb}{^{204}Pb}\right)_i + \frac{^{235}U}{^{204}Pb}(e^{\lambda_2 t} - 1) \quad (3.15)$$

$$\frac{^{208}Pb}{^{204}Pb} = \left(\frac{^{208}Pb}{^{204}Pb}\right)_i + \frac{^{232}Th}{^{204}Pb}(e^{\lambda_3 t} - 1) \quad (3.16)$$

where ^{204}Pb is the only non-radiogenic stable isotope of lead. In addition, the ratio of radiogenic ^{207}Pb to ^{206}Pb is expressed by combining Eqs. 3.14 and 3.15 where the asterisk identifies the radiogenic origin:

$$\left(\frac{^{207}Pb}{^{206}Pb}\right)^* = \frac{\frac{^{207}Pb}{^{204}Pb} - \left(\frac{^{207}Pb}{^{204}Pb}\right)_i}{\frac{^{206}Pb}{^{204}Pb} - \left(\frac{^{206}Pb}{^{204}Pb}\right)_i} = \frac{^{235}U}{^{238}U}\left(\frac{e^{\lambda_2 t} - 1}{e^{\lambda_1 t} - 1}\right) \quad (3.17)$$

The dates calculated from these equations based on analytical data of U-bearing minerals (e.g., zircon) are *discordant* in most cases either because of continuous or episodic loss of atoms of radiogenic lead or gain of uranium after crystallization. The oldest and most reliable dates are obtained from the ratio of radiogenic $^{207}Pb/^{206}Pb$ (Eq. 3.17) because the loss of radiogenic lead does not alter this ratio as much as the $^{206}Pb/^{204}Pb$ and $^{207}Pb/^{204}Pb$ ratios. Minerals containing thorium can be dated by the Th-Pb method (Eq. 3.16) although analytical difficulties discourage its use.

The loss of radiogenic lead atoms formed by decay of uranium is circumvented by an ingenious mathematical procedure that was devised by Wetherill (1963):

Equations 3.14 and 3.15 are each written in the form:

$$\frac{^{206}Pb^*}{^{238}U} = e^{\lambda_1 t} - 1 \quad (3.18)$$

$$\frac{^{207}Pb^*}{^{235}U} = e^{\lambda_2 t} - 1 \qquad (3.19)$$

These equations are used to calculate compatible sets of $^{206}Pb^*/^{238}U$ and $^{207}Pb^*/^{235}U$ ratios for selected values of t. The resulting values of these ratios are the coordinates of points that define the *concordia* curve which is the locus of all points whose coordinates yield concordant U-Pb dates. Any point that represents a mineral whose measured $^{206}Pb^*/^{238}U$ and $^{207}Pb^*/^{235}U$ ratios do not plot on concordia yield discordant U-Pb dates which are not valid age determinations.

Experience has shown that the $^{206}Pb^*/^{238}U$ and $^{207}Pb^*/^{235}U$ ratios of zircons extracted from the same body of rock define a straight-line *discordia* chord that intersects the concordia curve at two points. The coordinates of these points yield concordant dates because they lie on concordia. Several alternative explanations have been proposed to explain why zircons that yield discordant U-Pb dates nevertheless define a straight-line chord on the concordia diagram (Faure and Mensing 2005). The simplest but not necessarily the only explanation is that the discordance of U-bearing minerals is caused either by loss of radiogenic lead or by gain or loss of uranium. In any case, the dates calculated from the coordinates of the upper intercept are considered to be either the time of original crystallization of the U-bearing mineral or the time when the mineral was so severely altered that all radiogenic lead was lost. The lower intercept, in some cases, dates an episode of partial loss of radiogenic lead which should be confirmed by K-Ar and Rb-Sr dates of coexisting mica minerals (Appendices 3.6.2 and 3.6.3). If the chord intersects concordia at the origin, then the loss of radiogenic lead is occurring as a result of chemical weathering at the present time. The concordia diagram in Fig. 3.17 illustrates the case presented by Vocke and Hanson (1981) for zircons they extracted from the Olympus Granite-Gneiss in DVDP core 6 drilled at Lake Vida in Victoria Valley of southern Victoria Land.

3.6.6 Sm-Nd Method

Samarium (Sm) and neodymium (Nd) are rare earth elements (REE) that are widely distributed in rocks composed of silicate minerals (Faure and Mensing 2005). One of the naturally occurring isotopes of Sm is radioactive and decays to a stable isotope of Nd by emitting alpha particles (nuclei of ^4_2He):

$$^{147}_{62}Sm \rightarrow ^{143}_{60}Nd + ^4_2He + \text{energy} \qquad (3.20)$$

The halflife of ^{147}Sm is 1.06×10^{11} years which corresponds to a decay constant of $\lambda = 6.54 \times 10^{-12}$ year^{-1}.

The growth of radiogenic ^{143}Nd is expressed by an equation derived from the Law of Radioactivity:

$$\frac{^{143}Nd}{^{144}Nd} = \left(\frac{^{143}Nd}{^{144}Nd}\right)_i + \left(\frac{^{147}Sm}{^{144}Nd}\right)(e^{\lambda t} - 1) \qquad (3.21)$$

This equation is used for dating suites of Sm-bearing rocks by the isochron method used also for dating comagmatic igneous rocks by the Rb-Sr method.

The Sm-Nd method can be used to date mafic igneous rocks (basalt and gabbro) which are not suitable for dating by the Rb-Sr method. Magma of basaltic chemical composition originates by decompression melting in the mantle of the Earth which contains radiogenic ^{143}Nd that has formed by decay of ^{147}Sm. The isotopic evolution of Nd in the mantle of the Earth is represented by a model in Fig. 3.19 that is based on the isotope composition of neodymium in chondrite meteorites and which is therefore known as the "Chondritic Uniform Reservoir" (CHUR). The present value of the $^{143}Nd/^{144}Nd$ ratio in CHUR is:

$$\left(\frac{^{143}Nd}{^{144}Nd}\right)^0_{CHUR} = 0.512638$$

where the superscript "0" indicates that it applies to the present time (i.e., t = 0). The $^{147}Smd/^{144}Nd$ ratio of CHUR is:

$$\left(\frac{^{147}Sm}{^{144}Nd}\right)^0_{CHUR} = 0.1967$$

and the initial $^{143}Nd/^{144}Nd$ ratio of CHUR at t = 4.5 × 10^9 years ago is:

$$\left(\frac{^{143}Nd}{^{144}Nd}\right)^t_{CHUR} = 0.506764$$

The increase of the $^{143}Nd/^{144}Nd$ ratio of CHUR as a function of time is indicated in Fig. 3.19.

The model for the isotopic evolution of Nd in the mantle of the Earth (i.e., CHUR) can be used to calculate the time when the Nd in a crustal rock was separated from the "chondritic reservoir." This calculation is made by adapting Eq. 3.21:

$$0.512638 = \left(\frac{^{143}Nd}{^{144}Nd}\right)_i + 0.1967(e^{\lambda t} - 1) \quad (3.22)$$

which can be solved for t based on the initial $^{143}Nd/^{144}Nd$ ratio obtained from a Sm-Nd isochron of a suite of comagmatic igneous rocks:

$$t = \frac{1}{6.54 \times 10^{-12}} \ln\left[\frac{0.512638 - (^{143}Nd/^{144}Nd)_i}{0.1967} + 1\right] \quad (3.23)$$

The analytical data of Rowell et al. (1993) for a whole-rock sample and magnetic fractions of a pillow basalt in the Cocks Formation in the Skelton Glacier area define a Sm-Nd isochron in Fig. 3.18 which yields an initial $^{143}Nd/^{144}Nd$ ratio of 0.511594. Substituting this value into Eq. 3.23 yields:

$$t = \frac{1}{6.54 \times 10^{-12}} \ln\left[\frac{0.512638 - 0.511594}{0.1967} + 1\right]$$

$$t = \frac{\ln 1.0053075}{6.54 \times 10^{-12}} = 809 \times 10^6 \text{ y}$$

The procedure by which this date was obtained is more transparent than the method used by Rowell et al. (1993) who obtained dates between 700 and 800 Ma.

References

Adams CJ, Whitla PF (1991) Precambrian ancestry of the Asgard Formation (Skelton Group): Rb-Sr age of basement metamorphic rocks in the Dry Valley region, Antarctica. In: Thomson MRA, Crame JA, Thomson JW (eds) Geological evolution of Antarctica. Cambridge University Press, Cambridge, pp 129–135

Allen AD, Gibson GW (1962) Geological investigations in southern Victoria Land, Antarctica. Part 6: Outline of the geology of the Victoria Valley region. New Zealand J Geol Geophys 5:234–242

Allibone AH, Cox SC, Graham IJ, Smillie RW, Johnstone RD, Ellery SG, Palmer K (1993a) Granitoids of the Dry Valleys area, southern Victoria Land, Antarctica: Plutons, field relationships, and isotopic dating. New Zealand J Geol Geophys 36:281–297

Allibone AH, Cox SC, Smillie RW (1993b) Granitoids of the Dry Valleys area, southern Victoria Land: Geochemistry and evolution along the early Paleozoic Antarctic craton margin. New Zealand J Geol Geophys 36:299–316

Angino EE, Turner MD (1964) Antarctic orogenic history from absolute age dates. In: Adie RJ (ed) Antarctic geology. North-Holland, Amsterdam, The Netherlands, pp 552–556

Angino EE, Turner MD, Zeller EJ (1960) Reconnaissance geology of the Mt. Nussbaum area, Taylor Dry Valley, Victoria Land, Antarctica. Geol Soc Amer Bull 71(12):1816

Angino EE, Turner MD, Zeller EJ (1962) Reconnaissance geology of lower Taylor Valley, Victoria Land, Antarctica. Geol Soc Amer Bull 73:1553–1562

Bell RT, Jefferson CW (1987) An hypothesis for an Australian-Canadian connection in the Late Proterozoic and the birth of the Pacific Ocean. Proceedings of the Pacific-Rim Congress, 1987, Parkville, Victoria, 39–50, Australian Inst. Mining and Metallurgy

Berger GW, York D (1981) Geothermometry from $^{40}Ar/^{39}Ar$ dating experiments. Geochim Cosmochim Acta 45:795–812

Blank HR, Cooper RA, Wheeler RH, Willis IAG (1963) Geology of the Koettlitz-Blue Glacier region, southern Victoria Land, Antarctica. Trans Roy Soc New Zealand Geol 2(5):79–100

Bodley FA (ed) (1968) Fifth special Antarctic issue. New Zealand J Geol Geophys 11(4):791–1076

Borg SG, DePaolo DJ (1991) A tectonic model of the Antarctic Gondwana margin with implications for southeast Australia: Isotopic and geochemical evidence. Tectonophys 196: 339–358

Borg SG, DePaolo DJ (1994) Laurentia, Australia, and Antarctica as a Late Proterozoic supercontinent: Constraints from isotopic mapping. Geol 22:307–310

Bradshaw JD, Weaver SD, Laird MD (1985) Suspect terranes and Cambrian tectonics in northern Victoria Land, Antarctica. In: Howell DG (ed) Tectonostratigraphic Terranes of the Circum Pacific Region. Circum-Pacific Council for Energy Mineral Resources, Houston, TX, pp 467–479

Bull C (1962) Gravity observations in the Koettlitz Glacier area, southern Victoria Land, Antarctica. New Zealand Geol Geophys 5:810–819

Bull C (1966) Climatological observations in the ice-free areas of southern Victoria Land, Antarctica. In: Rubin MJ (ed) Studies in Antarctic Meteorology. Antarctic Research Series, vol. 9. American Geophysical Union, Washington, DC, pp 177–194

Calkin PE (1974) Subglacial geomorphology surrounding the ice-free valleys of southern Victoria Land, Antarctica. J Glaciol 13(69):415–429

Collins BW (ed) (1962) Special Antarctic issue. New Zealand J Geol Geophys 5(5):671–672

Collins BW (ed) (1963) Special Antarctic issue. New Zealand J Geol Geophys 6(3):305–306

Collins BW (ed) (1965) Third special Antarctic issue. New Zealand J Geol Geophys 8(2):163–390

Collins BW (ed) (1967) Fourth special Antarctic issue. New Zealand J Geol Geophys 10(2):307–623

Cox SC (1993) Inter-related plutonism and deformation in southern Victoria Land, Antarctica. Geol Mag 130:1–14

Craddock C (ed) (1969a) Geologic maps of Antarctica. Antarctic Folio Series, Folio 12. American Geographic Society, New York

Craddock C (1969b) Radiometric age map of Antarctica. Antarctic Map Folio Series, Folio 12, Plate 19, American Geographical Society, New York

Craddock C (ed) (1982) Antarctic geoscience. University of Wisconsin Press, Madison, WI

Dahl PS, Palmer DF (1981) Field study of orbicular granitic rocks in Taylor Valley, southern Victoria Land. Ant J US 16(6):47–49

Dahl PS, Palmer DF (1983) The petrology and origin of orbicular tonalite from western Taylor Valley, southern Victoria Land, Antarctica. In: Oliver RJ, James PR, Jago JB (eds) Antarctic Earth Science. Australian Academy of Science, Canberra, ACT, pp 156–159

Dallmeyer RD, Hess JR, Whitney JA (1981) Post-magmatic cooling of the Elberton Granite; Bearing on the Paleozoic tectonothermal history of the Georgia inner piedmont. J Geol 89:585–600

Dalziel IWD (1991) Pacific margin of Laurentia and East Antarctica-Australia as a conjugate rift pair: Evidence and implications for an Eocambrian supercontinent. Geology 19:598–601

Denton GH, Armstrong RL, Stuiver M (1970) Late Cenozoic glaciation in Antarctica; The record in the McMurdo Sound region. Antarctic J US 5:15–21

Deutsch S, Webb PN (1964) Sr/Rb dating on basement rocks from Victoria Land; Evidence for a 1000 million year old event. In: Adie RJ (ed) Antarctic geology. Wiley, New York, pp 557–562

Deutsch S, Grőgler N (1966) Isotopic age of Olympus Granite–Gneiss (Victoria Land, Antarctica). Earth Planet. Sci. Letters, 1:82–84

Elliot DH (1975) Tectonics of Antarctica: A review. Amer J Sci 275-A:45–106

Encarnaci6n J, Grunow A (1996) Changing magmatic and tectonic styles along the paleo-Pacific margin of Gondwana and the onset of early Paleozoic magmatism in Antarctica. Tectonics 15(6):1325–1341

Fairbridge RW (1975) Antarctica. In: Fairbridge RW (ed) The Encyclopedia of World Regional Geology, Part 1; Western Hemisphere (Including Antarctica and Australia). Dowden, Hutchinson & Ross, Stroudsburg, PA, pp 2–13

Faure G (2001) Origin of igneous rocks; The isotopic evidence. Springer, Heidelberg, Germany

Faure G, Mensing TM (2005) Isotopes; Principles and applications. Wiley, Hoboken, NJ

Faure G, Jones LM, Owen LB (1974) Isotopic composition of strontium and geologic history of the basement rocks of Wright Valley, southern Victoria Land, Antarctica. New Zealand J Geol Geophys 17(3):611–627

Felder RP (1980) Geochronology of the Brown Hills, Transantarctic Mountains. M.Sc. thesis, Department of Geology and Mineral Industries, The Ohio State University, Columbus, Ohio

Felder RP, Faure G (1980) Rubidium-strontium age determinations of part of the basement complex of the Brown Hills, central Transantarctic Mountains. Antarctic J US 15(5):16–17

Felder RP, Faure G (1990) Age and petrogenesis of the granitic basement rocks, Brown Hills, Transantarctic Mountains. In: Miller H (ed) Workshop on Antarctic Geochronology. Zentralblatt für Geologie und Paläontology, Teil I, vol. 1990 (1/2), pp 45–62

Ferracioli F, Bozzo E, Capponi G (2002) Aeromagnetic and gravity constraints for an early Paleozoic subduction system of Victoria Land, Antarctica. Geophys Res Lett 29(10):1406

Findlay RH, Skinner DNB, Craw D (1984) Lithostratigraphy and structure of the Koettlitz Group, McMurdo Sound, Antarctica. New Zealand J Geol Geophys 27:513–536

Findlay RH, Unrug R, Banks MR, Veevers JJ (eds) (1993) Gondwana Eight; Assembly, evolution, and dispersal. Balkema, Rotterdam, The Netherlands

Flory RF, Murphy DJ, Smithson SB, Houston RS (1971) Geologic studies of basement rocks in southern Victoria Land. Antarctic J US 6(4):119–120

Ford AB (1964) Review of Antarctic geology. Trans Amer Geophys Union 45:363–381

Friedmann EI, Hua M, Ocampo-Friedmann R (1988) Cryptoendolithic and cyanobacterial communities of the Ross Desert, Antarctica. Polarforschung 58:251

Fűtterer DK, Damaske D, Kleinschmidt G, Miller H, Tessensohn F (1996) Antarctica; Contributions to global Earth sciences. Springer, Berlin/Heidelberg, Germany

Gamble JA, Skinner DNB, Henrys S (eds) (2002) Antarctica at the close of the millennium. Roy Soc New Zealand Bull 35:251–259

Ghent ED, Henderson RA (1968) Geology of the Mt. Falconer pluton, lower Taylor Valley, south Victoria Land, Antarctica. New Zealand J Geol Geophys 11(4):851–880

Goldich SS, Nier AO, Washburn AL (1958) $^{40}Ar/^{40}K$ age of gneiss from McMurdo Sound, Antarctica. Trans Amer Geophys Union 39(8):956–958

Graham IJ, Palmer K (1987) New precise Rb-Sr mineral and whole-rock dates for I-type granitoids from Granite Harbor, south Victoria Land, Antarctica. New Zealand Antarctic Record 8(1):72–80

Grikurov GE (1982) Structure of Antarctica and outline of its evolution. In: Craddock C (ed) Antarctic geoscience. University of Wisconsin Press, Madison, WI, pp 791–804

Grindley GW (1971) Polyphase deformation of the Precambrian Nimrod Group, central Transantarctic Mountains. In: Adie RJ (ed) Antarctic geology and geophysics, Universitetsforlaget, Oslo, Norway

Grindley GW (1981) Precambrian rocks of the Ross Sea region. J Roy Soc New Zealand 11(4):411–423

Grindley GW, Laird MG (1969) Geology of the Shackleton Coast. In: Craddock C (ed) Geologic maps of Antarctica, Folio 12, Sheet 15. American Geographical Society, New York

Grindley GW, Warren G (1964) Stratigraphic nomenclature and correlation in the western Ross Sea region. In: Adie RJ (ed) Antarctic geology. North- Holland, Amsterdam, The Netherlands, pp 314–333

Grunow A, Hanson R, Wilson TJ (1996) Were aspects of Pan-African deformation linked to Iapetus opening? Geol 24(12):1063–1066

Gunn BM (1963) Geological structure and stratigraphic correlation in Antarctica. New Zealand J Geol Geophys 6(3):423–443

Gunn BM, Walcott RI (1962) The geology of the Mount Markham region, Ross Dependency, Antarctica. New Zealand J Geol Geophys 5(3):407–426

Gunn BM, Warren G (1962) Geology of Victoria Land between the Mawson and the Mullock Glaciers, Antarctica. New Zealand Geol Surv Bull 71:1–157

Hamilton PJ, Evensen NM, O'Nions RK, Smith HS, Erlank AJ (1979) Sm-Nd dating of Onverwacht Group volcanics, Southern Africa. Nature 279:25–28

Hamilton W (1967) Tectonics of Antarctica. Tectonophysics 4(4/6):555–568

Hamilton W, Hayes PT (1960) Geology of the Taylor Glacier-Taylor dry valley region, south Victoria Land, Antarctica. US Geol Surv Prof. Paper 400-B:376–377

Harrington HJ (1958) Nomenclature of rock units in the Ross Sea region, Antarctica. Nature 182:290

Haskell TR, Kennett JP, Prebble WM (1964) Basement and sedimentary geology of the Darwin Glacier area, Antarctica. In: Adie RJ (ed) Antarctic geology. North-Holland, Amsterdam, The Netherlands, pp 348–351

Haskell TR, Kennett JP, Prebble WM, Smith G, Willis IAG (1965a) The geology of the middle and lower Taylor Valley of south Victoria Land, Antarctica. Trans Roy Soc New Zealand 2(12):169–186

Haskell TR, Kennett JP, Prebble WM (1965b) Geology of the Brown Hills and Darwin Mountains, southern Victoria Land. Trans Roy Soc New Zealand 2(15):231–248

Heimann A, Fleming TH, Elliot DH, Foland KA (1994) A short interval of Jurassic continental flood basalt volcanism in Antarctica as demonstrated by $^{40}Ar/^{39}Ar$ geochronology. Earth Planet Sci Lett 121:19–41

Herbert WW (1962) The Axel-Heiberg Glacier. New Zealand J Geol Geophys 5(5):681–706

Houtzager G (2003) The complete encyclopedia of Greek mythology. Chartwell Books, Edison, NJ

IUGS (2002) International stratigraphic chart. Commission on the Geologic Map of the World, UNESCO, United Nations, New York

Jackson JA (ed) (1997) Glossary of geology, 4th edn. American Geological Institute, Alexandria, VA

Jones LM, Faure G (1967) Age of the Vanda Porphyry dikes in Wright Valley, southern Victoria Land, Antarctica. Earth Planet. Sci. Letters, 3:321–324

Kurasawa H, Yoshida Y, Mudrey MG Jr (1974) Geological log of the Lake Vida core-DVDP 6. In: Dry Valley Drilling Project, Bulletin 3. DeKalb, Northern Illinois University, DeKalb, IL, pp 92–108

Kyle PR (ed) (1995) Volcanological and environmental studies of Mt. Erebus, Antarctica. Antarctic Research Series, vol. 66. American Geophysical Union, Washington, DC

Laird MB (1981) Lower Paleozoic rocks of Antarctica. In: Holland CH (ed) Lower Paleozoic of the Middle East, Eastern and Southern Africa, and Antarctica. Wiley, London, pp 257–314

Laird MG, Bradshaw JD (1982) Uppermost Proterozoic and lower Paleozoic geology of the Transantarctic Mountains. In: Craddock C (ed) Antarctic Geoscience. University of Wisconsin Press, Madison, WI, pp 525–533

LeMasurier WE, Thomson JW (eds) (1990) Volcanoes of the Antarctic plate and southern Oceans. Antarctic Research Series, vol. 48. American Geophysical Union, Washington, DC

Mayewski PA, Goldthwait RP (1985) Glacial events in the Transantarctic Mountains: A record of the East Antarctic ice sheet. In: Turner MD, Splettstoesser JF (eds) Geology of the Transantarctic Mountains. Antarctic Research Series, vol. 36, Paper 12. American Geophysical Union, Washington, DC, pp 275–324

McDougall I, Ghent ED (1970) Potassium-argon dates on minerals from the Mt. Falconer area, lower Taylor Valley, south Victoria Land, Antarctica. New Zealand J Geol Geophys 13(4):1026–1029

McKelvey BC, Webb PN (1959) Geological investigations in south Victoria Land, Antarctica. Part 2: Geology of the upper Taylor Glacier region. New Zealand J Geol Geophys 2(4):718–728

McKelvey BC, Webb PN (1962) Geological investigations in southern Victoria Land, Antarctica. New Zealand J Geol Geophys 5:143–162

Moores EM (1991) Southwest U.S.-East Antarctic (SWEAT) connection: A hypothesis. Geol 19:425–428

Mortimer G (1981) Provisional report on the geology of the basement complex between Miers and Salmon Valley, McMurdo Sound, Antarctica. New Zealand Antarctic Record 3(2):1–8

Murphy DJ, Flory RF, Houston RS, Smithson SB (1970) Geological studies of basement rocks in south Victoria Land. Antarctic J US 5(4):102

Nichols RL (1963) Geologic features demonstrating aridity of the McMurdo Sound area, Antarctica. Amer J Sci 261:20–31

Palmer DF, Bradley J, Prebble WM (1967) Orbicular granodiorite from Taylor Valley, south Victoria Land, Antarctica. Geol Soc Amer Bull 78:1423–1428

Pearn WC, Angino EE, Stewart D (1963) New isotopic age measurements from the McMurdo Sound area, Antarctica. Nature 199:685

Péwé TL (1960) Multiple glaciation in the McMurdo Sound region, Antarctica: A progress report. J Geol 68:498–514

Picciotto E, Coppez A (1962) Bibliographie des mesures d'ages absolus en Antarctique. Annales Soc Geol Belg 85(8):263–308

Picciotto E, Coppez A (1964a) Bibliographie des mesures d'ages absolus en Antarctique (addendum, aout 1963). Ann Soc Geol Belg 85(4):115–128

Picciotto E, Coppez A (1964b) Bibliography of absolute age determinations in Antarctica (addendum). In: Adie RJ (ed) Antarctic geology. North-Holland, Amsterdam, The Netherlands, pp 563–569

Quartermain LB (1963) Publications resulting from work done under the aegis of the New Zealand Antarctic Research Programmes 1956–1962. New Zealand J Geol Geophys 6(3):348–360

Quartermain LB (1965) Publications resulting from work done under the aegis of the New Zealand Antarctic Research Programmes 1963–1964. New Zealand J Geol Geophys 8(2):371–379

Ravich MG, Ya Krylov A (1964) Absolute ages of rocks from East Antarctica. In: Adie RJ (ed) Antarctic geology. North-Holland, Amsterdam, The Netherlands, pp 491–494

Rees MN, Duebendorfer EM, Rowell AJ (1989) The Skelton Group, southern Victoria Land. Antarctic J US 24(5):21–24

Rowell AJ, Rees MN, Duebendorfer EM, Wallin ET, VanSchmus WR, Smith EI (1993) An active Neoproterozoic margin: Evidence from the Skelton Glacier area, Transantarctic Mountains. J Geol Soc London 150:677–682

Rowley PD (1983) Developments in Antarctic geology during the past half century. In: Boardman SJ (ed) Revolution of the earth sciences; advances in the past half-century. Kendall/ Hunt, Dubuque, IA

Schmidt DL (1966) The Transantarctic Mountains. Geotimes (November):15–16

Skinner DNB (1982) Stratigraphy and structure of low-grade metasedimentary rocks of the Skelton Group, southern Victoria Land; Does the Teall graywacke really exist? In: Craddock C (ed) Antarctic geoscience. University of Wisconsin Press, Madison, WI, pp 555–563

Smithson SB, Murphy DJ, Houston RS (1971a) Development of an augen-gneiss terrain. Contrib Mineral Petrol 33: 184–190

Smithson SB, Fikkan PR, Murphy DR, Houston RS (1971b) Development of augen-gneiss in the ice-free valley area, south Victoria Land. In: Adie RJ (ed) Antarctic geology and geophysics. Universitetsforlaget, Oslo, Norway, pp 293–298

Smithson SB, Fikkan PR, Houston RS (1971c) Amphibolitization of calc-silicate metasedimentary rocks. Contrib Mineral Petrol 31:228–237

Steiger RH, Jäger E (1977) Subcommission on geochronology: Convention on the use of decay constants in geo- and cosmochronology. Earth Planet Sci Lett 36:359–362

Stonehouse B (ed) (2002) Encyclopedia of Antarctica and the Southern Oceans. Wiley, Chichester

Streckeisen AL (1967) Classification and nomenclature of igneous rocks. Neues Jahrbuch Miner. Abhandlungen 107(2/3):144–240

Streckeisen AL (1976) To each plutonic rock its proper name. Earth Sci Rev 12:1–33

Stuckless JS, Ericksen RL (1975) Rb-Sr ages of basement rocks recovered from borehole DVDP 6, southern Victoria Land, Antarctica. Antarctic J US 10(6):301–307

Studinger M, Bell RE, Buck R, Karner GD, Blankenship DD (2004) Sub-ice geology inland of the Transantarctic Mountains in light of new aerogeophysical data. Earth Planet Sci Lett 220:391–408

Stuiver M, Braziunas TF (1985) Compilation of isotopic dates from Antarctica. Radiocarbon 27(2A):117–304

Stump E (1995) The Ross Orogen of the Transantarctic Mountains. Cambridge University Press, New York

Swithinbank C (1988) Antarctica. U.S. Geol. Surv. Prof. Paper 1386-B. U.S. Government Printing Office, Washington, DC

Tingey RJ (ed) (1991) The geology of Antarctica. Clarendon, Oxford

Tingey RJ (1996) How the south was won: A review of the first 150 years of Antarctic geologic exploration. Terra Antartica 3:1–10

Vocke RD Jr, Hanson GN (1981) U-Pb zircon ages and petrogenetic implications for two basement units from Victoria Valley, Antarctica. In: McGinnis LD (ed) Dry Valley drilling project. Antarctic Research Series, vol. 33. American Geophysical Union, Washington, DC, pp 247–255

Vocke RD Jr, Hanson GN, Stuckless JS (1978) Ages of the Vida granite and the Olympus granite gneiss, Victoria Valley, southern Victoria Land. Antarctic J US 13:15–17

Wade FA, Yeats VL, Everett JR, Greenlee DW, LaPrade KE, Shenk JC (1965) The geology of the central Queen Maud Range, Transantarctic Mountains, Antarctica. Antarct. Rept. Ser., 65–1. Texas Tech. Coll., Lubbock, TX

Warren G (1969) Geology of the Terra Nova Bay-McMurdo Sound area, Victoria Land. In: Craddock C (ed) Geologic maps of Antarctica, Folio 12, Sheet 14, American Geographical Society, New York

Webb PN (1962) Isotope dating of Antarctic rocks. A summary, I. New Zealand J Geol Geophys 5(5):790–796

Webb PN, McKelvey BC (1959) Geological investigations in south Victoria Land, Antarctica. Part 1: Geology of Victoria Valley. New Zealand J Geol Geophys 2(1):120–136

Webb PN, Warren G (1965) Isotope dating of Antarctic rocks, II. New Zealand J Geol Geophys 8(2):221–230

Wetherill GW (1963) Discordant uranium-lead ages, part 2: Discordant ages resulting from diffusion of lead and uranium. J Geophys Res 68:2957–2965

Williams PF, Hobbs BE, Vernon RH, Anderson DE (1971) The structural and metamorphic geology of basement rocks in the McMurdo Sound area, Antarctica J Geol Soc Australia 18(2):127–142

Chapter 4
Northern Victoria Land

Northern Victoria Land in Fig. 4.1 extends from the David Glacier northward to the Pennell and Oates coasts. The width of the Transantarctic Mountains of this area increases to more than 400 km measured at the latitude of Cape Adare. The area is traversed by the Lillie, Rennick, and Matusevich glaciers all of which flow northwest and discharge ice into the South Pacific Ocean. Also, the Tucker, Borchgrevink, Mariner, Aviator, Campbell, and Priestley glaciers (not all of which are identified in Fig. 4.1) flow southeast into the Ross Sea.

The mountains east of the Rennick Glacier are deeply dissected by narrow valleys which are occupied by a network of glaciers that flow toward the coast. The six highest peaks along the east coast reach elevations in excess of 3,000 m including Mt. Minto (71°47′S, 168°45′E) which at 4,163 m is even higher than Mt. Erebus (3,794 m) on Ross Island and Mt. Lister (4,025 m) in the Royal Society Range of southern Victoria Land.

The highest peak in the mountains west of the Rennick Glacier (e.g., the Daniels Range of the USARP Mountains) is Roberts Butte (72°39′S, 160°08′E) at 2,828 m, followed by Welcome Mountain (2,494 m) and Mt. Southard (2,402 m). The elevation of the polar plateau west of the Daniels Range rises gradually to about 2,500 m.

The basement rocks of northern Victoria Land in Table 4.1 consist of paragneisses of the Wilson Group, the metavolcanic and metasedimentary rocks of the Bowers Group, and the graywackes and argillites of the Robertson Bay Group. The rocks of the Wilson Group were intruded by plutons of Granite Harbor Intrusives of Late Cambrian to Middle Ordovician age, whereas the Bowers Group and the Robertson Bay Group were intruded by the younger Admiralty Intrusives of Devonian to Carboniferous age. These younger intrusives are genetically related to the rhyolites of the Gallipoli Porphyries in the Freyberg Mountains (72°15′S, 163°45′E) located in the part of the Wilson Terrane that lies east of the Rennick Glacier.

The rocks of northern Victoria Land are cut by several northwest-trending fault zones including the Leap-Year, Lanterman, and Wilson faults in Fig. 4.2. These fault zones define the Wilson, Bowers, and Robertson Bay terranes. The juxtaposition of unrelated geologic features on opposite sides of these boundary faults suggests that they are major dislocations in the continental crust of northern Victoria Land and that they may represent sutures between accreted terranes.

Northern Victoria Land also contains comparatively small remnants of the Beacon Supergroup and of the Ferrar Dolerite sills that intruded them. In addition, the Mesa Range at the head of the Rennick Glacier is a large dissected plateau composed of sheet-like flows of the Kirkpatrick Basalt. The youngest igneous rocks in northern Victoria Land are the alkali-rich basalt flows of the McMurdo Volcanic Group of Neogene age which crop out primarily along the coast of the Ross Sea (Hamilton 1972).

4.1 Exploration

The history of the geological exploration of northern Victoria Land was described by Stump (1995) who has recorded the contributions made by geologists from many countries including Great Britain, New Zealand, USA, Germany, Italy, the Soviet Union, and Australia.

The geologic mapping of northern Victoria Land began during the International Geophysical Year (IGY)

G. Faure and T.M. Mensing, *The Transantarctic Mountains: Rocks, Ice, Meteorites and Water*,
DOI 10.1007/978-90-481-9390-5_4, © Springer Science+Business Media B.V. 2010

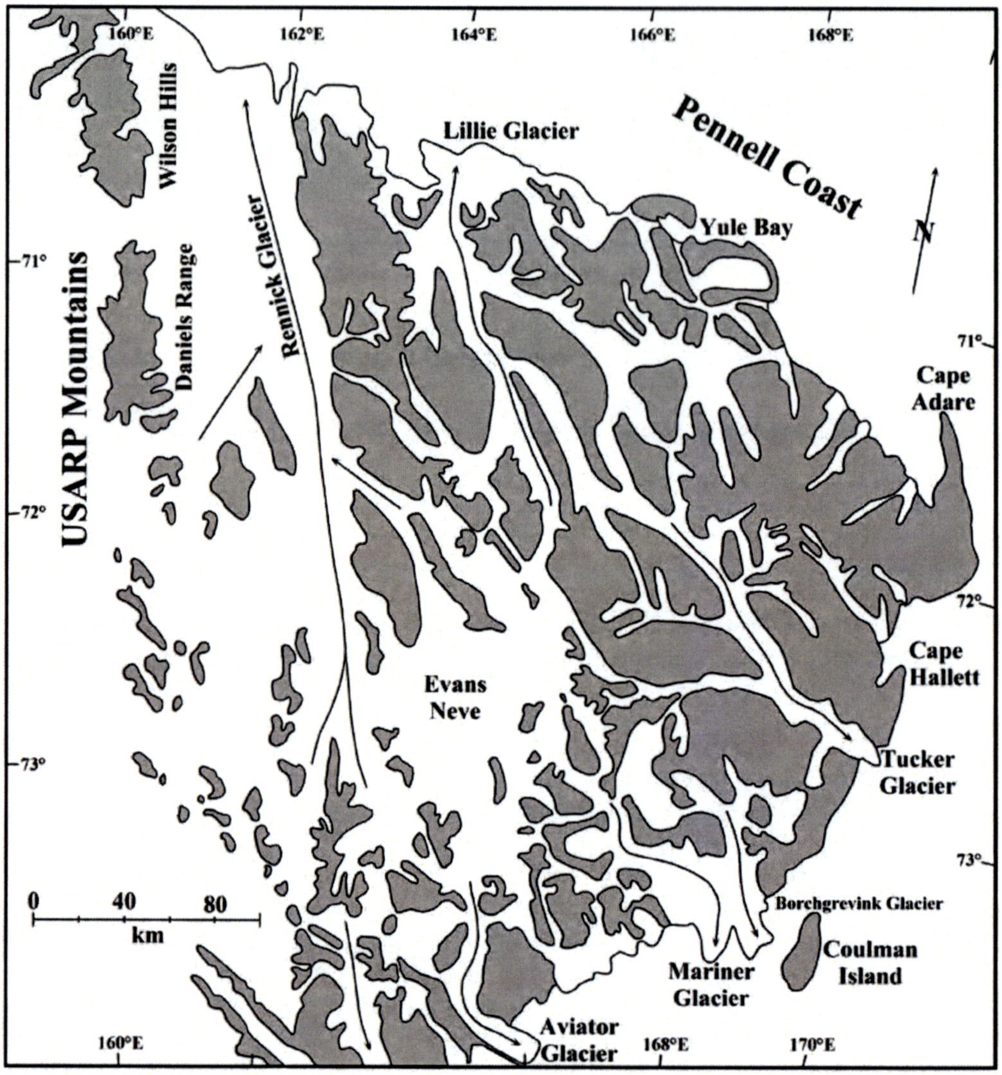

Fig. 4.1 Northern Victoria Land consists of irregularly shaped nunataks separated by narrow valleys through which glaciers wind their way to the coast (e.g., Lillie, Tucker, Borchgrevink, Mariner, and Aviator). The only exception is the Rennick Glacier which occupies a wide fault basin. Several mountains along the Ross-Sea coast have elevations in excess of 3,000 m (e.g., Mt. Minto, 4,163 m; Mt. Sabine, 3,719 m; Mt. Supernal, 3,660 m; Mt. Murchison, 3,500 m; Mt. Herschel, 3,335 m; and Mt. Phillips, 3,036 m). The elevations of the mountains west of the Rennick Glacier reach only 2,828 m (Roberts Butte), 2,494 m (Welcome Mountain), 2,402 m (Mt. Southard), and 2,257 m (Mt. Weihaupt). The polar plateau west of the Daniels Range rises to an elevation of about 2,500 m above sealevel (Adapted from Borg et al. 1986; Gair et al. 1969; National Geographic Society 1990)

in 1957/58 when a team of geologists from New Zealand worked on the Hallett Peninsula and along the Tucker Glacier (Harrington et al. 1964). In the decades after the IGY, New Zealand geologists continued to map large parts of northern Victoria Land using dog sleds for transportation. The publications that arose from this important activity were listed by Stump (1995, p. 12).

During the summer of 1958/59 geologists of the Soviet Union used the survey ship *Ob* to map the Oates and King George V coasts of northern Victoria Land. They also discovered and named Sputnik, Znamenskiy, and Babushkin islands as well as several topographic features along the Oates Coast including Ob' Bay and Lunik, Yermak, and Belousov points. The Russians

Table 4.1 Stratigraphy of the rocks exposed in northern Victoria Land (Gair et al. 1969)

Geologic unit	Geologic age (Ma)	Rock types
McMurdo Volcanic Group	Miocene to Holocene 25–0.0 Ma	Alkali-rich volcanic rocks
Ferrar and Kirkpatrick Basalt groups	Middle Jurassic 176.8 ± 1.8 Ma	Dolerite sills and basalt flows
Beacon Supergroup	Early Devonian to Late Triassic 417–205 Ma	Sandstone, shale, tillite, coal
Admiralty Intrusives in the Robertson Bay and Bowers terranes	Devonian to Carboniferous	Post-tectonic granites and granodiorites
Gallipoli Porphyries	Devonian to Carboniferous	Porphyritic rhyolite
	Kukri Unconformity	
Granite Harbor Intrusives in the Wilson Terrane	Late Cambrian to Early Ordovician	Syn- and post-tectonic granites, granodiorites, quartz diorites
	Intrusive contact	
Robertson Bay Group	Late Cambrian	Graywacke and argillite
	Leap Year Fault	
Bowers Group	Cambrian to Ordovician	Mafic volcanic tuff, argillite, sandstone, graywacke, quartzite, limestone,conglomerate
	Lanterman Fault	
Wilson Group in the Wilson Terrane	Neoproterozoic	Biotite-garnet paragneiss, marble, biotite schist, and migmatite
	Base not exposed	

also named features on the mainland such as the Zykov Glacier, the Mt. Belolikov and Kavrayskiy hills, and the Lazarev Mountains at the mouth of the Matusevich Glacier (Klimov and Soloviev 1958; Soloviev 1960).

Australian geologists (e.g., Douglas Mawson) have had a long-standing interest in Antarctica because the Oates Coast is located only about 2,600 km south of Hobart, Tasmania, measured along the meridian at 148°E. Consequently, two Australian research vessels, the Thala Dan and the Magga Dan, visited the Oates Coast during the summers from 1959 to 1962 and made observations that were published by McLeod (1964) and McLeod and Gregory (1967).

During the summer of 1964/65 two American geologists, Warren Hamilton and Dwight Crowder, used helicopters out of Hallett Station to examine rocks of a large part of northern Victoria Land between the Rennick Glacier and the coast of the Ross Sea (Hamilton 1967; Crowder and Hamilton 1967; Crowder 1968). The modern era of geological research started with the International Northern Victoria-Land Project during the 1981/82 field season supported by US helicopters operating from a field camp on the Evans Névé at 72°45′S, 164°30′E in Figs. 4.3a and b (Stump 1986).

Starting in 1979/80, geologists from Germany have made important contributions by means of the German Northern Victoria Land Expeditions which they abbreviated "GANOVEX." This name is a humorous reference to the word "Ganove" which is German slang for a habitual criminal. The first group of German geologists in northern Victoria Land set up a small prefabricated shelter in the Everett Range at the confluence of the Lillie and Ebbe Glaciers at 71°S, 163°34′E. The name of this shelter was Lillie Marleen after the song made famous by Marlene Dietrich (Fig. 4.4). The next German field season in 1981/82 had to be cancelled because the ship, Polarstern, that was supposed to support GANOVEX II was crushed by the pack ice in Yule Bay along the Pennell Coast. On that occasion, Americans from McMurdo Station helped to unload the ship before it sank and then transported the German crew and scientists to McMurdo by helicopter where they boarded C-130 aircraft for the flight to Christchurch in New Zealand. In 1982/82 the Germans built Gondwana Station in Fig. 4.5 at Gerlache Inlet in Terra Nova Bay and continued to work in northern Victoria Land (Fűtterer et al. 2006).

In 1987 the GANOVEX team published a comprehensive geological map of northern Victoria Land based on the accumulated results of their field work (Tessensohn et al. 1981) and a preliminary map of the structure of the Wilson, Bowers, and Robertson Bay terranes (GANOVEX-ITALIANTARTIDE 1991) in cooperation with Italian investigators.

In 1985/86 a team of Italian geologists started field work in Terra Nova Bay on the coast of the Ross Sea and set up Terra Nova Bay Station at Gerlache Inlet (74°41′S, 164°07′E) located only about 10 km southwest

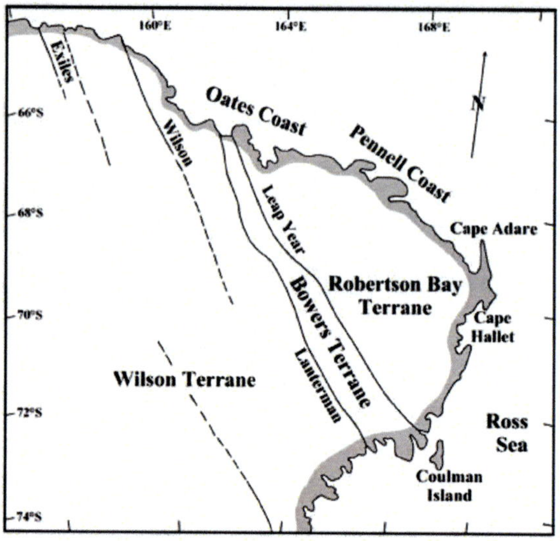

Fig. 4.2 Northern Victoria Land consists of three terranes that are separated from each other by the Leap Year and Lanterman faults. Each terrane is composed of characteristic assemblages of metasedimentary and metavolcanic rocks of Neoproterozoic and Cambrian ages. These basement rocks were locally intruded by plutons of the Granite Harbor Intrusives and by the younger Admiralty Intrusives. Rocks of the Beacon Supergroup occur only in a few places in northern Victoria Land including in the Morozumi Range and Helliwell Hills, in the adjacent Freyberg Mountains, and in the Lichen Hills. The Kukri Peneplain, which is so prominent in southern Victoria Land, is only preserved in these locations (Adapted from Stump 1995)

Fig. 4.4 The geologists of GANOVEX (1979/80) set up a small shelter called Lillie Marleen adjacent to the Lillie Glacier in northern Victoria Land (Fig. 4.1). The shelter was completed on January 14, 1980, and was used to refuel helicopters and as a refuge in case the weather deteriorated (Adapted from Fig. 4 of Kothe et al. 1981)

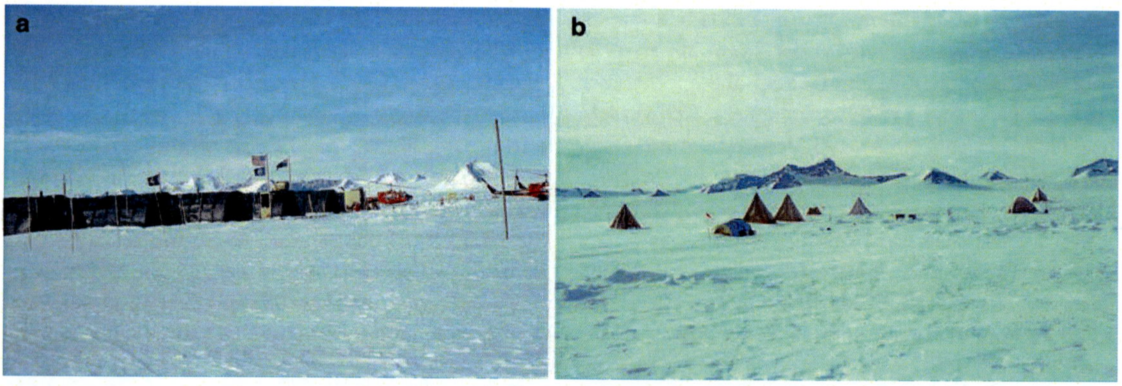

Fig. 4.3 (a) The camp of the International Northern Victoria-Land project 1981/82 consisted of a series of Jamesway huts on the on the Evans Névé at 72°45′S and 164°30′E. The geological field work was supported by helicopters of the U.S. Navy (Photo by G. Faure). **(b)** In addition to the Jamesway huts, the northern Victoria Land camp included several Scott tents used by geologists who preferred to camp in the style to which they were accustomed (Photo by G. Faure)

Fig. 4.5 Gondwana Station, the principal support base for scientists working with GANOVEX, is located at Gerlache Inlet of Terra Nova Bay on the Ross-Sea coast of northern Victoria Land (Photograph © 2010 BGR)

of the German Gondwana Station. The locations of both stations are indicated in Fig. 4.6.

American geologists returned to northern Victoria Land in 1988/89 and 1989/90 working in the area of the upper Tucker Glacier and along the coast of the Robertson Bay Terrane (Wright and Dallmeyer 1991; Dallmeyer and Wright 1992). Scientists from several other countries have continued to work in northern Victoria Land supported by helicopters and fixed-wing aircraft.

4.2 Terra Nova Bay

The Ross Orogen of southern Victoria Land extends across the David Glacier into the southern part of the Wilson Terrane of northern Victoria Land. The geology of this part of the Transantarctic Mountains in Fig. 4.6 includes a basement composed mainly of the Granite Harbor Intrusives and of the metasedimentary rocks of the Priestley Formation. The overlying Beacon Sandstones, the Ferrar Dolerite and Kirkpatrick Basalt, and McMurdo Volcanics are not shown in Fig. 4.6. The Terra-Nova-Bay area is mountainous and includes numerous peaks with elevations between 1,000 and 3,000 m (e.g., Mt. Levick, 2,733 m; Mt. Nansen, 2,738 m; Mt. Melbourne, 2,733 m; Mt. Baxter, 2,621 m; Mt. Mackintosh, 2,468 m; Mt. Dickason, 2,033 m; Mt. Abbott, 1,023 m; and Mt. Browning, 760 m). The area is also traversed by several large outlet glaciers that transport ice from the polar plateau to the coast of the Ross Sea (e.g., the Aviator, Campbell, Priestley, Reeves, Larsen, and David glaciers). Some of the mountains and glaciers of this area were given the names of the men of Scott's Terra Nova Expedition who wintered on Inexpressible Island in 1912: Victor

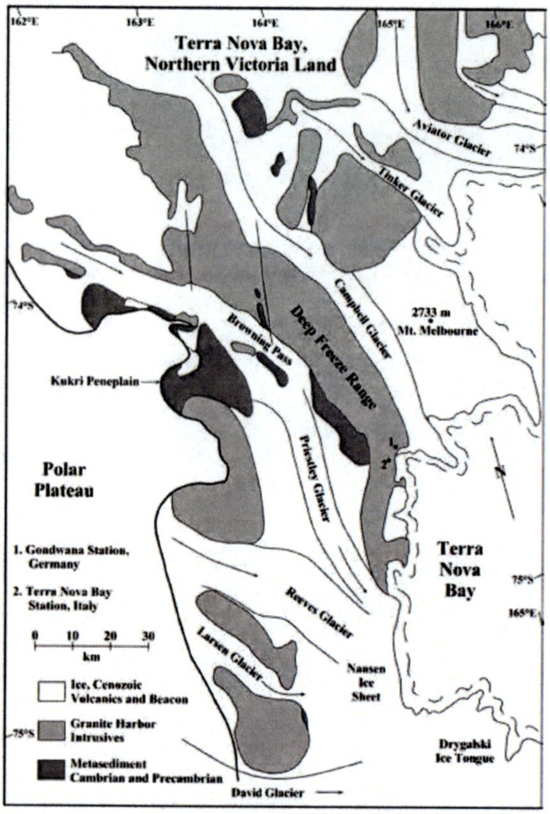

Fig. 4.6 The basement rocks of southern Victoria Land extend northward across the David Glacier into the Terra-Nova-Bay region of the Wilson Terrane of northern Victoria Land (NVL). The basement rocks of this region consist mainly of Granite Harbor Intrusives and of metasedimentary rocks of Neoproterozoic to Cambrian age. The rocks above the Kukri Peneplain are not shown (i.e., Beacon Supergroup, Ferrar Dolerite, and McMurdo Volcanics). This area contains several prominent mountains including Mt. Melbourne which is an extinct volcano of the Cenozoic McMurdo Volcanics (Adapted from Warren 1969; Gair et al. 1969; Skinner 1983)

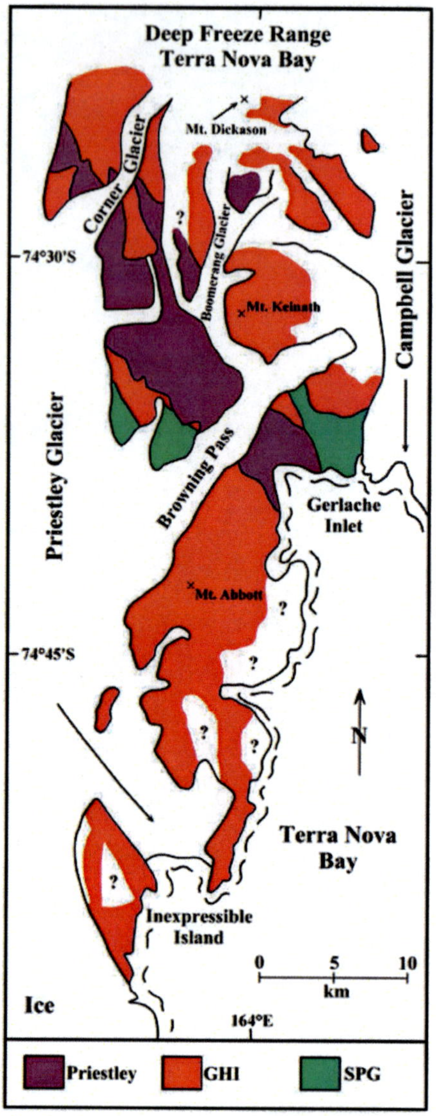

Fig. 4.7 The rocks exposed in the Deep Freeze Range between the Campbell and Priestley Glaciers include the folded metasedimentary rocks of the Priestley Schist which was intruded by the composite Terra Nova batholith of the Granite Harbor Intrusives (GHI). The metamorphic grade of the Priestley Schist increases until it grades into the Snowy Point Gneiss (SPG) which is a paragneiss composed of coarse and lenticular layers of quartz, microcline, oligoclase, biotite, and muscovite (Adapted from Skinner 1983)

Campbell, Raymond Priestley, Murray Levick, Frank Browning, Harry Dickason, and George Abbott (Section 1.4.3).

The geology of the area between the Campbell and the Aviator glaciers in Fig. 4.6 was mapped by Gair (1964, 1967), Ricker (1964), and by Skinner and Ricker (1968). Gair et al. (1969) and Warren (1969) compiled the available information into geologic maps of Victoria Land which includes the Terra-Nova-Bay area. The exploration of this area was continued by Nathan (1971a, b), Adamson (1971), Nathan and Skinner (1972), and Skinner (1972).

Skinner (1983) published a summary of the geology of the Deep Freeze Range between the Campbell and Priestley glaciers in Fig. 4.7. This mountain range is primarily composed of the Granite Harbor Intrusives which form the Terra Nova batholith. Skinner (1983) recognized 12 different mappable facies of this batholith which enclose remnants of the Priestley Schist and the Snowy Point Paragneiss. The granitic rocks of the Deep Freeze Range are foliated and cataclastically deformed indicating that granitic magmas were intruded sequentially and crystallized during the Ross Orogeny. Skinner (1983) stated that none of the granitic rocks of the Terra Nova batholith are post-kinematic in contrast to the granitic basement rocks of southern Victoria Land.

The Terra Nova batholith contains the Browning Mafites which are inclusions of mafic plutonic rocks consisting of foliated biotite-rich olivine- and pyroxene-diorites, and gabbros described by Skinner (1972, 1983). These mafic rocks occur both as rafted blocks and in the form of dikes and small plutons that are not distinguished in Fig. 4.7. A pluton of hypersthene-diopside-hornblende monzogabbro in the Browning Pass area grades upward through mafic granodiorite and granite into porphyritic quartz-rich microcline leucogranite containing the amphibole riebeckite. Another two-pyroxene mafic diorite along the Boomerang Glacier in the Deep Freeze Range differentiated into several pyroxenite facies including biotite-bearing, olivine-bearing, and plagioclase-hornblende-olivine-bearing varieties. These rocks are cut by pegmatite veins and were enriched in potassium and silica by metasomatism causing phlogopite, hornblende, biotite, quartz, and microcline to form.

The metasedimentary rocks of the Priestley Formation defined by Skinner and Ricker (1968) and shown in Fig. 4.7 are composed of pelitic, thinly bedded argillites interbedded with subordinate quartzo-feldspathic greywacke, fine-grained limestone, and quartzite (Skinner 1983). The metamorphic grade of the Priestley Formation in the Deep Freeze Range increases from the north to the south such that the metasedimentary rocks grade into the Snowy Point Paragneiss which crops out north and south of Browning Pass. According to Skinner (1983), this unit consists of coarsely foliated, lenticular quartz-microcline-oligoclase-biotite-muscovite paragneiss that contains layers rich in cordierite, garnet, and sillimanite. The lenticular bodies are boudinaged and consist of quartz-garnet-tourmaline pegmatite. In addition, the Snowy Point Paragneiss contains thin layers of marble and quartzite as well as amphibolite lenses that form prominent ridges along Gerlache Inlet where the gneiss is strongly migmatized and contains granite neosomes that are oriented parallel to the foliation.

The Priestley Formation of the Deep Freeze Range does not correlate with the metasediments of the Robertson Bay Terrane, contrary to a proposal by Nathan and Skinner (1972) who, at that time, had not yet visited the type locality of the Robertson Bay Group in Edisto Inlet (Skinner 1983). It seems more likely that the Priestley Formation is the local representative of the Koettlitz Group of southern Victoria Land because it occupies the same stratigraphic position as the Asgard Formation in the ice-free valleys of southern Victoria Land (McKelvey and Webb 1962; Section 3.1.2) and because both formations contain sedimentary carbonate beds. However, this conjecture has not been confirmed or even tested.

Isotopic age determinations of the basement rocks of the Terra-Nova-Bay area are listed in Appendix 4.8.1 based primarily on the work of Kreuzer et al. (1981, 1987), Vetter et al. (1984), Borsi et al. (1988), Armienti et al. (1990), and Adams (2006). The whole-rock Rb-Sr dates reported by Armienti et al. (1990) prove that the age of the Granite Harbor Intrusives of the Deep Freeze Range is Middle Cambrian based on the time scale of the IUGS (2002). All of the whole-rock samples they analyzed define a straight line in Fig. 4.8 that yields a date of 507 ± 7 Ma and an initial $^{87}Sr/^{86}Sr$ ratio of 0.71034 ± 0.00022.

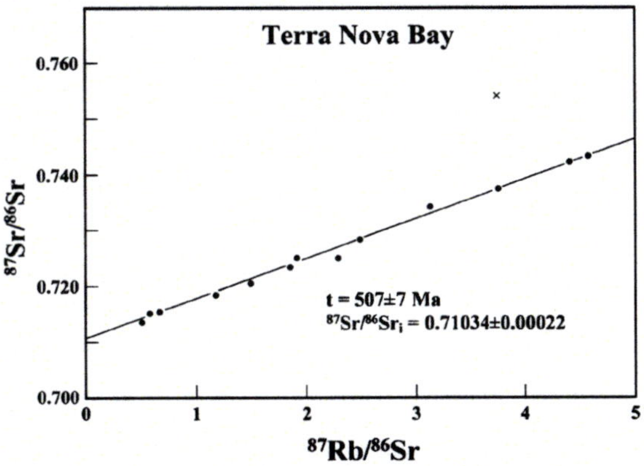

Fig. 4.8 Whole-rock Rb-Sr isochron of Granite Harbor Intrusives (GHI) of the Terra Nova batholith in the Deep Freeze Range and from Timber Peak, Ogden Heights, and Mt. Nansen in the Terra-Nova-Bay area of the Wilson Terrane. All of the samples of the GHI lie close to the straight line that was fitted by linear least-squares regression. The slope (m = 0.00722569) yields a date of 507.0 ± 7 Ma and an initial $^{87}Sr/^{86}Sr$ ratio of 0.71034 ± 0.00022 ($\bar{\sigma}$). A migmatite (×) that was intruded by the Mt. Abbott pluton in the Deep Freeze Range (Fig. 4.7) is either much older (about 820 Ma) or has a higher initial ration (0.72709) than the rocks that define the straight line in this diagram (Data from Armienti et al. 1990)

The age of the Granite Harbor Intrusives reported by Armienti et al. (1990) implies that the sedimentary rocks of the Priestley Formation were deposited prior to about 510 Ma and that their age is probably Neoproterozoic. This hypothesis is confirmed by whole-rock K-Ar dates measured by Adams (2006) of slate from outcrops of the Priestley Formation along the Foolsmate Glacier (74°01′S, 161°55′E) of the upper Priestley Glacier. These dates, which range from 513 to 556 Ma (Appendix 4.8.1) are illustrated in the form of histograms in Fig. 4.9. Although K-Ar dates of slates are difficult to interpret quantitatively, they are in most cases younger than the age of deposition of the original clay-rich sediment. In the simplest case, the K-Ar clock is reset to zero when all of the inherited and the in-situ produced radiogenic ^{40}Ar is lost during burial, deformation, metamorphism, and subsequent cooling of the metasedimentary rocks (Faure and Mensing 2005). However, even in the simplest case, varying fractions of ^{40}Ar produced by decay of ^{40}K following burial, metamorphism, and cooling may escape because slates do not retain ^{40}Ar quantitatively. In the case at hand, the whole-rock K-Ar dates of Adams (2006) in Fig. 4.9 support the conclusion that the sediment of the Priestley Formation was deposited before 560 Ma during the Neoproterozoic Era.

The display of dates in Fig. 4.9 also demonstrates that the whole-rock K-Ar dates of the Priestley Schist are lower than those of the Priestley Slate and that they overlap the K-Ar and Rb-Sr dates of biotite of the Granite Harbor Intrusives of the Terra-Nova-Bay area. This observation confirms the hypothesis that the rocks of the Priestley Schist were outgassed during the magmatic activity that accompanied the Ross Orogeny. The Granite Harbor Intrusives and the Priestley Schist subsequently cooled at about the same rate until the Middle Ordovician at about 450 Ma.

The initial $^{87}Sr/^{86}Sr$ ratios of the Granite Harbor Intrusives of the Terra-Nova-Bay area listed in Appendix 4.8.1 range from 0.7094 to 0.71049 based primarily on the data of Armienti et al. (1990) and Borsi et al. (1988). These values demonstrate that the granitic magmas contained excess radiogenic ^{87}Sr presumably because they formed by partial melting (i.e., anatexis) of crustal rocks during the Ross Orogeny. These results and conclusions are consistent with the Rb-Sr dates and initial $^{87}Sr/^{86}Sr$ ratios of the Granite Harbor Intrusives of southern Victoria Land (Table 3.3, Section 3.4.2 and Appendix 3.6.3).

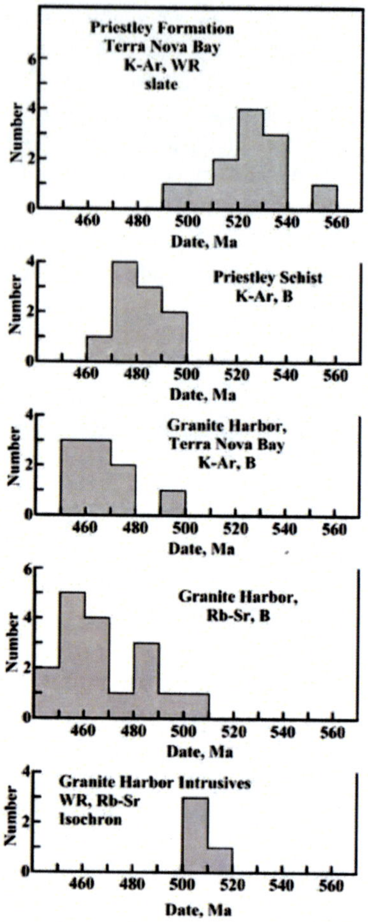

Fig. 4.9 The slates and schists of the Priestley Formation in the Terra-Nova-Bay area were intruded by the Granite Harbor Intrusives of the Terra Nova batholith in the Deep Freeze Range. The whole-rock K-Ar dates of slightly metamorphosed slates (490–560 Ma) indicate that these rocks were probably deposited in Neoproterozoic time and that they lost varying proportions of radiogenic ^{40}Ar during the intrusion of the granitic magmas of the Granite Harbor Intrusives in the course of the Ross Orogeny. The K-Ar dates of biotite in the Priestley Schist record the time in the past when they cooled to their blocking temperature following the Ross Orogeny. The K-Ar and Rb-Sr dates of micas in the Granite Harbor Intrusives of the TNB area likewise indicate that these rocks cooled slowly into the Ordovician Period after crystallizing during the Middle Cambrian Epoch (IUGS, 2002) (Data from Adams 2006; Kreuzer et al. 1981, 1987; Vetter et al. 1984; Nathan 1971b; Armienti et al. 1990; Borsi et al. 1988; Faure and Gair 1970)

4.3 Wilson Terrane

The boundary between the Terra-Nova-Bay area and the main part of the Wilson terrane is not defined because the Ross Orogen continues from southern Victoria Land through the Terra-Nova-Bay area into the Wilson Terrane of northern Victoria Land without discernible interruption. The Wilson Terrane extends from the Terra-Nova-Bay area north to the Oates Coast of Antarctica and is separated from the Bowers Terrane by the Lanterman fault zone in Fig. 4.2. The Wilson Terrane extends east across the Rennick Glacier and includes the Lanterman and Salamander ranges which form its "eastern province."

The basement rocks of the Wilson Terrane are well exposed in the Wilson Hills and in the Daniels Range of the USARP Mountains as well as in the Emlen Peaks, on Welcome Mountain, and at Mt. Southard (Gair et al. 1969; Sturm and Carryer 1970; Tessensohn et al. 1981). Even though these mountain ranges are located adjacent to the polar plateau, Archean and Paleoproterozoic granulites of the East Antarctic craton have not been recognized in the Wilson Terrane or anywhere else in northern Victoria Land.

The stratigraphy of the basement rocks of the Wilson Terrane located west of the Rennick Glacier consists of six units identified in Table 4.2: Berg Group, Rennick Schist, Priestley Formation, Morozumi Phyllite, Wilson Gneisses, and the Granite Harbor Intrusives. We have already discussed the Priestley Formation of the Terra-Nova-Bay area and therefore turn now to the Berg Group and the Rennick Schist. The stratigraphic units listed in Table 4.2 consist of variably metamorphosed and highly deformed shales, graywackes, sandstones, and limestones that were deposited in marine environments. Although they have not been correlated into a coherent picture, they are the key to understanding the complicated structure of the Wilson Terrane.

4.3.1 Berg Group, Oates Land

The metasedimentary rocks of the Berg Group occur in the Berg Mountains of Oates Land located west of the Matusevich Glacier along the George V Coast at 69°11′S, 155°58′E. These rocks were first described by Ravich et al. (1965) who visited the area from a Russian ship during the summer of 1958/59. However, Gair et al. (1969) did not include these outcrops on the geologic map of northern Victoria Land because the report of Ravich and his colleagues was not available to them. Thirty years later, in the summer of 1988/89, the area west of the Matusevich Glacier was visited by geologists of GANOVEX V including David Skinner

Table 4.2 Stratigraphy of the Wilson Terrane located west of the Rennick Glacier (Gair et al. 1969; Stump 1995)

Geologic unit	Age	Description
Granite Harbor Intrusives	Cambro-Ordovician	Biotite granite, granodiorite, and tonalite
	Intrusive contact	
Wilson Gneisses	Neoproterozoic to Cambrian	Biotite-garnet paragneiss, minor marble, quartz-biotite schist, and migmatite
	Gradational contact	
Rennick Schist[a]	Neoproterozoic	Fine-grained pelitic and psammitic schist and gneiss with calc-silicate lenses, marble, and carbonatite
Berg Group[a]	Neoproterozoic	Fine-grained graywackes and phyllite, marine turbidites containing sedimentary carbonate lenses and beds
Priestley Formation[a]	Neoproterozoic	Pelitic and thinly bedded argillites interbedded with quartzo-feldspathic graywacke, fine-grained limestone, and quartzite
Morozumi Phyllite and Spotted Schist	?	Graywacke and pelite including carbonate at the tops of some pelite beds

[a]The Priestley Formation, Rennick Schist, Morozumi Phyllite, and the Berg Group occur in different areas of the Wilson Terrane. Although all four units appear to be of Neoproterozoic age and although their lithologic compositions are similar, they have not been correlated with each other

who subsequently described the sedimentary rocks of the Berg Group illustrated by means of two measured stratigraphic sections (Skinner 1992; Skinner et al. 1996; Schűssler 1996).

We now know that the low-grade metasediments of this area are tightly folded and are cut by granite plutons (Olesch et al. 1996). The quartzofeldspathic graywackes, siltstones, and phyllites of the Berg Group contain carbonate nodules, lenses, and discrete layers of limestone. These sedimentary rocks are considered to be turbidites with graded bedding, stacked climbing ripples, flute casts, load casts, and laminations (Stump 1995). The phyllites of the Berg Group are composed primarily of quartz, biotite, sericite, and chlorite which facilitates age determinations by the K-Ar and Rb-Sr methods (Adams 1996).

The whole-rock K-Ar dates of slates of the Berg Group range from 478 to 492 Ma. The Berg Granite yielded a Rb-Sr whole-rock date of 489 ± 11 Ma in Fig. 4.10 with an initial $^{87}Sr/^{86}Sr$ ratio of 0.7139 ± 0.00076 (Adams 1996). The Archangel Granite, located 30 km southeast of the Berg Mountains at about 69°28′S, 156°26′E, appears to be older as indicated by its Rb-Sr whole rock isochron date of 573 ± 10 Ma with a low initial $^{87}Sr/^{86}Sr$ ratio of 0.7032 ± 11 (Adams 1996). The Berg Granite is a typical representative of the Granite Harbor Intrusives in the Transantarctic Mountains. However, the Archangel Granite has the lowest initial $^{87}Sr/^{86}Sr$ ratio among all of the granitic rocks in northern Victoria Land that have been analyzed to date. The samples of the Archangel Granite that Adams (1996) analyzed range in composition from leucogranodiorites to leucogranites and were collected from an outcrop that measured only 30 m in diameter without xenoliths or cross-cutting dikes or sills. Therefore, Adams (1996) concluded that the Archangel Granite is an early representative of the Granite Harbor Intrusives with an unusually low initial $^{87}Sr/^{86}Sr$ ratio. The low initial ratio implies that the granitic magma formed by partial melting of mantle-derived plutonic or volcanic rocks soon after these source rocks arrived in the crust of Antarctica.

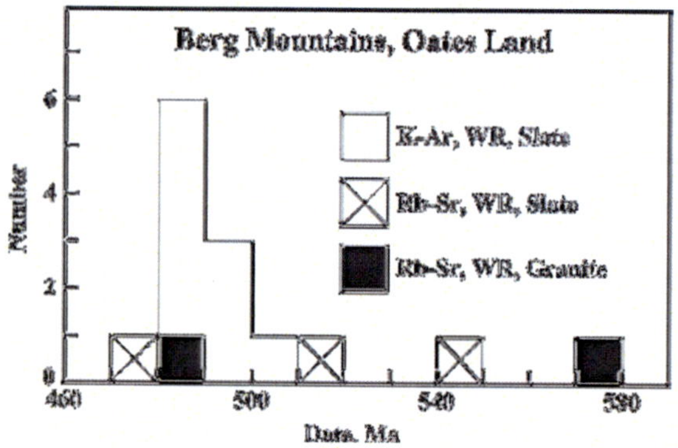

Fig. 4.10 The granitic intrusives of the Berg and Archangel Mountains in Oates Land have whole-rock Rb-Sr dates of 489 ± 11 and 573 ± 10 Ma, respectively. Both plutons intruded the folded but low-grade metasediments of the Berg Group. K-Ar dates of biotite and muscovite in the granites and whole-rock samples of slate range from 480 to 510 Ma which dates the time of uplift and cooling of the igneous and metasedimentary rocks. Two suites of slates of the Berg Group actually define whole-rock Rb-Sr isochrons with elevated initial $^{87}Sr/^{86}Sr$ ratios of 0.7173 that are significantly higher than the initial ratios of the granite plutons (These data from Adam (1996, 2006) are also compiled in Appendix 4.8.2)

The sedimentary rocks of the Berg Group contain acritarchs (a type of microfossil) of middle Neoproterozoic (Riphean) age (Iltchenko 1972). In addition, Skinner (1991) proposed that the Berg Group may be correlated with the metasedimentary rocks of the Priestley Formation (Section 4.2) mainly on the basis of lithologic similarities including the presence of sedimentary carbonate beds described by Skinner (1983). Marble beds also occur in the type section of the Rennick Schist in the Sequence Hills (Gair 1967). Therefore, sedimentary carbonate beds appear to be present in several sequences of metasedimentary rocks: the Skelton and Koettlitz groups of southern Victoria Land, and the Priestley Formation, Rennick Schist, and Berg Group in northern Victoria Land. A regional correlation among these sedimentary sequences listed in Table 4.2 is consistent with the active-subduction model for the origin of the Ross Orogen and therefore should be investigated (Stump 1995).

4.3.2 Rennick Schist

The Rennick Schist is composed of pelitic and psammitic metasedimentary rocks that have been intensely deformed and regionally metamorphosed to the amphibolite facies. This unit was first described by Gair (1967) from outcrops in the Sequence Hills (73°03′S, 161°15′E) where it is interbedded with marble. The quartz-biotite schist that makes up the bulk of the Rennick Schist also contains layers of granitic rocks thus forming lit-par-lit injection gneiss. In addition, the Granite Harbor Intrusives of the Wilson Terrane contain xenoliths composed of remnants of the Rennick Schist ranging from a few centimeters to hundreds of meters in length.

Later workers have recognized the Rennick Schist in nunataks and mountain ranges north of the Sequence Hills all the way to the Allegro Valley in the Daniels Range in Fig. 4.11 (Plummer et al. 1983; Babcock et al. 1986;

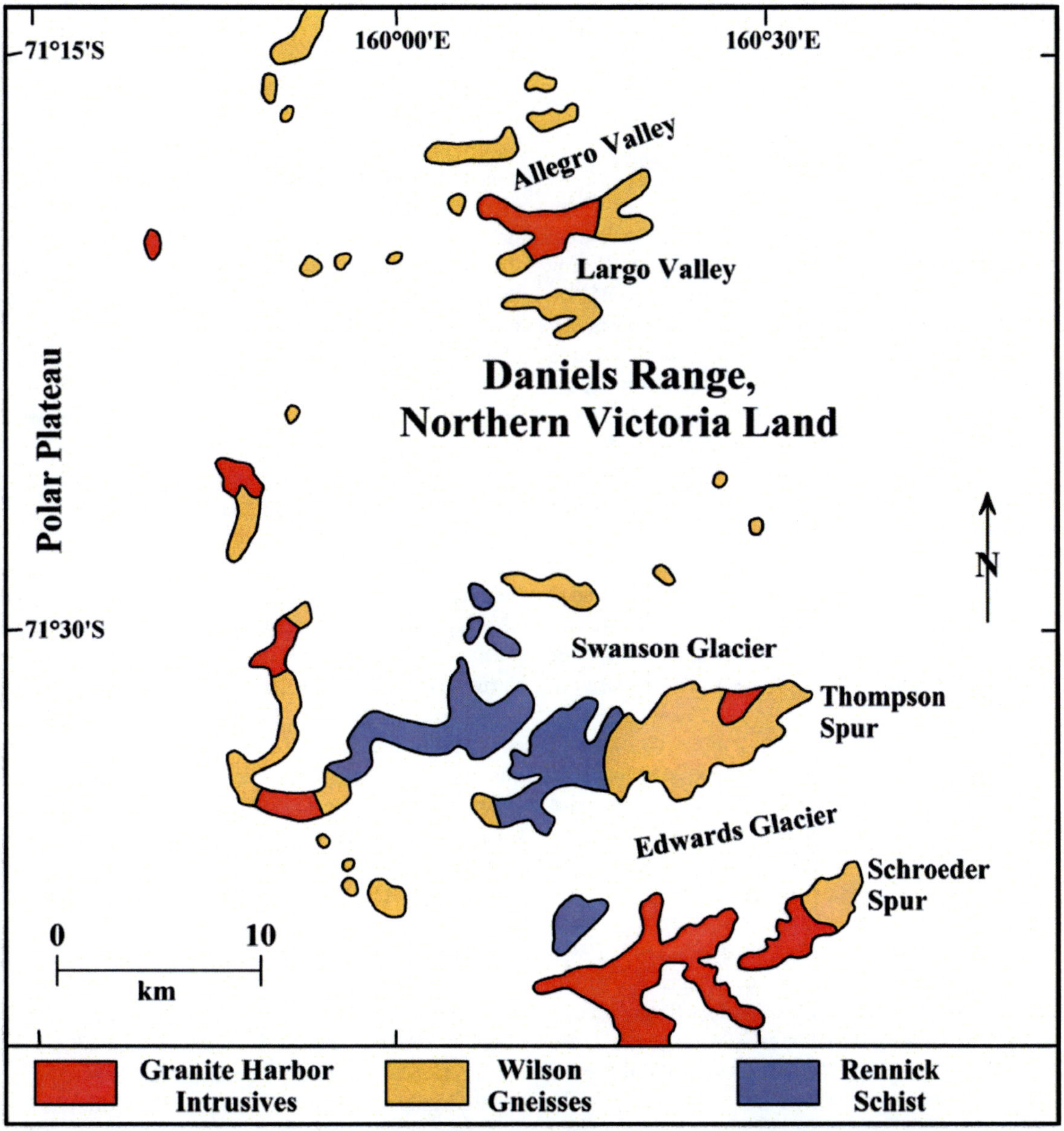

Fig. 4.11 The Wilson Gneisses and Rennick Schist of the Wilson Group in the southern part of he Daniels Range (Fig. 4.1), USARP Mountains, northern Victoria Land, were intruded by granitic plutons of the Granite Harbor Intrusives. This area exemplifies the complex relationships among the metamorphic gneisses and biotite schists that characterize the basement rocks of the western province of the Wilson Terrane that includes the Wilson Hills and the Daniels Range of the USARP Mountains as well as the Emlen Peaks, Welcome Mountain, Mt. Southard and lesser nunataks farther south (Adapted from Adams 1986 with information from Gair et al. 1969)

Roland et al. 1989; Stump 1989). Most recently, the composition and occurrence of the Rennick Schist was summarized by Stump (1995) who included a picture of a large recumbent fold of Rennick Schist on Welcome Mountain (72°14′S, 160°12′E). He also pointed out that the southernmost exposures of the Rennick Schist are only about 50 km north of the northernmost outcrops of the Priestley Formation.

4.3.3 Daniels Range and Wilson Hills

The western province of the Wilson Terrane is characterized by amphibolite-facies gneisses which Klimov and Soloviev (1958) originally assigned to the Wilson Group based on exposures in the Wilson Hills. Grindley and Warren (1964) later extended this term to describe all of the gneisses in the USARP Mountains including the Daniels Range. Sturm and Carryer (1970) subsequently subdivided the Wilson Group into the Wilson Gneisses (orthogneiss and migmatite) and the Rennick Schists as defined by Gair (1967). They also excluded plutons of granitic rocks that intrude the Wilson Gneisses in the USARP Mountains and classified them as Granite Harbor Intrusives. Accordingly, the stratigraphy of the basement rocks of the Daniels Range and the Wilson Hills of the USARP Mountains consists of Rennick Schist, the Wilson Gneisses, and the Granite Harbor Intrusives.

These three stratigraphic units are well exposed in the Daniels Range of the USARP Mountains in Fig. 4.11, especially along the Thompson and Schroeder spurs and in the Allegro and Largo valleys. At these locations, the Rennick Schist grades into Wilson Gneisses which supports the conjecture that the gneisses formed by regional metamorphism and partial melting of the Rennick Schist (Babcock et al. 1986; Tessensohn et al. 1981). If this conjecture is correct, then the Rennick Schist of the Daniels Range and in the upper Rennick-Glacier area is a remnant of the metasedimentary protoliths from which the different kinds of Wilson Gneisses formed.

The first age determinations by the whole-rock K-Ar method of gneisses and granites in the Wilson Hills and on several small off-shore islands by Ravich and Krylov (1964) ranged from 564 to 434 Ma. A sample of biotite from Aviation Island (also called Ostrova Polyarnoy) at 69°16′S, 158°47′E, yielded an even lower date of 420 Ma (Gair et al. 1969). These dates have been set aside because their relation to recognizable geologic events is uncertain. Nevertheless, the dates reported by Ravich and Krylov (1964) were thought to demonstrate that the Wilson Gneisses cooled in Neoproterozoic to early Paleozoic time and that these rocks are not an extension of the Archean craton of East Antarctica (see also Fig. 4.31 in Appendix 4.8.6).

Babcock et al. (1986) and Tessensohn et al. (1981) agreed that the Wilson Gneisses of the Daniels Range consist of syntectonic granites, migmatites, and closely related metamorphic rocks all of which formed by partial melting of associated metasedimentary rocks of the Rennick Schist. Babcock et al. (1986) subdivided the so-called Wilson plutonic complex into xenolithic intrusives and layered intrusives which occur in virtually every nunatak of the Daniels Range in Fig. 4.11.

The Granite Harbor Intrusives of the Daniels Range occur as dikes, sills, and plutons having diameters that range up to 10 km. These intrusives are composed primarily of granite or leucogranite in addition to granodiorite and quartz monzonite all of which have well-defined contacts with the Wilson Gneisses and the Rennick Schist. They are generally unfoliated and therefore post-tectonic, although evidence for the flow of magma is preserved by the alignment of feldspar phenocrysts and of biotite crystals. The Granite Harbor Intrusives of the Daniels Range are aluminum-rich and have been classified as S-type granites by Babcock et al. (1986) and by Borg et al. (1986, 1987) following the twofold classification of granitic rocks proposed by Chappell and White (1974) and by McCulloch and Chappell (1982).

The basement rocks of the Daniels Range were dated by K-Ar, Rb-Sr, and U-Pb methods and the resulting dates are compiled in Appendix 4.8.3 and are visualized by means of the nested histograms in Fig. 4.12.

Whole-rock Rb-Sr isochron dates indicate that the Rennick Schist of the Daniels Range crystallized at about 532 ± 20 Ma (Early Cambrian; IUGS 2002) and that the sandy (psammitic) and clay-rich (pelitic) facies had elevated initial $^{87}Sr/^{86}Sr$ ratios of 0.7187 ± 0.0009 and 0.7243 ± 0.0009, respectively (Adams 1986). A Rb-Sr date of a single whole-rock sample reported by Faure and Gair (1970) was recalculated to 548 ± 33 Ma based on initial ratios obtained by Adams (1986) and using the decay constant of ^{87}Rb recommended by

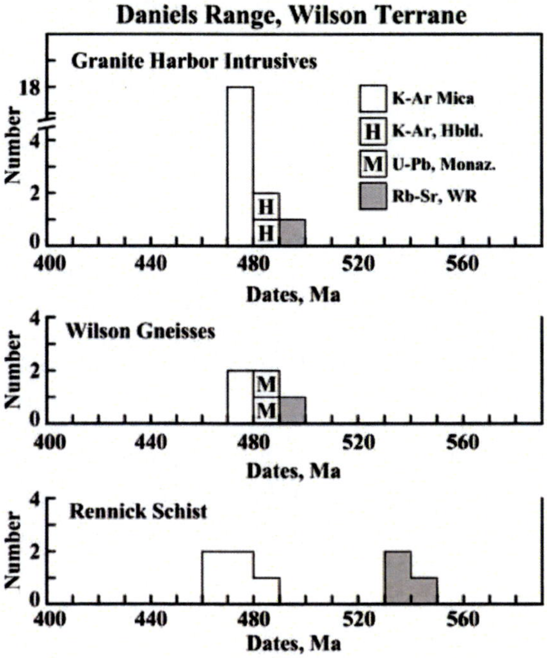

Fig. 4.12 Whole-rock Rb-Sr dates of the Rennick Schist in the Daniels Range of the Wilson Terrane between 530 and 550 Ma refer to the time of isotopic homogenization of strontium during regional metamorphism of these rocks. This event predates the Ross orogeny during which protoliths of Rennick Schist were converted into the Wilson Gneiss of the Daniels Range between 490 and 500 Ma. This gneiss was simultaneously intruded by granitic magma which formed the syntectonic Granite Harbor Intrusives. The initial $^{87}Sr/^{86}Sr$ ratios of the Granite Harbor Intrusives (0.7125) are lower than those of the Rennick Schist (0.7187 and 0.7243) and of the Wilson Gneiss (0.7205). The K-Ar dates of biotite and muscovite of all three rock units are similar (470–480 Ma) because the entire complex cooled at about the same rate to the blocking temperature of mica minerals for radiogenic ^{40}Ar (Data from Adams 1986; Faure and Gair 1970; Kreuzer et al. 1981, 1987; Adams 2006; Adams et al. 1982; Adams and Hőhndorf 1991). See also Appendix 4.8.3

Steiger and Jäger (1977). This date is consistent with the isochron dates of Adams (1986) and supports the conclusion that the isotopic composition of strontium in the Rennick Schist was homogenized as a result of regional metamorphism during the Early Cambrian Epoch. Therefore, the Rennick sediment was presumably deposited during the Neoproterozoic Era. A K-Ar date of biotite reported by Kreuzer et al. (1981) suggests that the Rennick Schist cooled slowly for about 60 million years to 472 ± 3 Ma when it reached the blocking temperature of biotite for radiogenic ^{40}Ar (300 ± 25°C).

The whole-rock Rb-Sr date of the Wilson Gneisses reported by Adams (1986) is 490 ± 30 Ma and its initial $^{87}Sr/^{86}Sr$ ratio of 0.7205 ± 0.0016 is almost as high as that of the Rennick Schist. Therefore, the Wilson Gneisses could have been derived by partial melting of the Rennick Schist. The K-Ar dates of biotite and muscovite of the Wilson Gneisses at the Schroeder and Thompson Spurs of the Daniels Range reported by Kreuzer et al. (1981) indicate that these rocks cooled to the blocking temperature of biotite at 474 ± 2.0 Ma which is in good agreement with the cooling history of the Rennick Schist at this location. Two U-Pb dates of monazite reported by Adams (1986), 484 and 488 Ma, confirm and refine the whole-rock Rb-Sr date of the Wilson Gneiss (490 ± 30Ma).

The Granite Harbor Intrusives of the Daniels Range have whole-rock Rb-Sr dates between 495 ± 10 and 510 ± 36 Ma (Adams 1984; Vetter et al. 1983; Kreuzer et al. 1981). The initial $^{87}Sr/^{86}Sr$ ratios of these rocks are lower than those of the Rennick Schist and of the Wilson Gneisses. Consequently, the magma of the Granite Harbor Intrusives of the Wilson Terrane did not originate from the Rennick Schist. In addition, the whole-rock samples of the Granite Harbor Intrusives scatter above and below their Rb-Sr errorchrons because the isotope composition of strontium they contain was not homogeneous at the time of crystallization. In this respect, the Granite Harbor Intrusives of the Wilson Terrane resemble those of the ice-free valleys of southern Victoria Land (Section 3.4.2).

Biotite and muscovite of the Granite Harbor Intrusives and biotite of a tonalite dike cooled to about 300°C at about 475 ± 3 Ma based on K-Ar dates measured by Kreuzer et al. (1981). In contrast to the wide range of whole-rock K-Ar dates of gneisses in the Wilson Hills reported by Ravich and Krylov (1964), the K-Ar dates of the basement rocks in the Daniels Range measured by Kreuzer et al. (1981) vary only from 471.5 to 478.5 Ma with a mean of 474.3 ± 2.4 Ma for ten samples of biotite and muscovite.

Borg et al. (1986, 1987) made a thorough study of the Granite Harbor Intrusives of northern Victoria Land in order to determine the origin of these rocks and to detect regional trends in their chemical compositions. The initial $^{87}Sr/^{86}Sr$ ratios (at 550 Ma) of 21 samples range from 0.7066 (Lanterman Range) to 0.7190 (Lichen Hills) and collectively exceed the $^{87}Sr/^{86}Sr$ ratios of mantle-derived plutonic and volcanic igneous rocks. These results confirm that the Granite

Harbor Intrusives of northern Victoria Land contain varying amounts of excess radiogenic ^{87}Sr of crustal origin (see also Section 3.4.2). The initial $^{143}Nd/^{144}Nd$ ratios of five of the 21 specimens range from 0.510685 (Daniels Range) to 0.510830 (Aviator Glacier) and are all lower than the $^{143}Nd/^{144}Nd$ ratio of the chondritic uniform reservoir (CHUR) at the present time (Fig. 3.19, Section 3.4.5, and Appendix 3.6.6). The deficiency of radiogenic ^{143}Nd in these rocks (at 550 Ma) also implies that they contain Nd of crustal origin because the rocks of the continental crust have lower Sm/Nd ratios and therefore have lower $^{143}Nd/^{144}Nd$ ratios than the rocks of the lithospheric mantle (Faure and Mensing 2005, Chapter 17).

The initial $^{87}Sr/^{86}Sr$ and $^{143}Nd/^{144}Nd$ ratios of these samples of Granite Harbor Intrusives constrain a mixing hyperbola in Fig. 4.13 between mantle-derived basalt and granitic rocks of the continental crust. The distribution of the data points in quadrant four of the Sr-Nd mixing diagrams confirms and refines the evidence that the magmas of the Granite Harbor Intrusives of northern Victoria Land contained Sr and Nd of crustal origin. However, these magmas did not form from a protoliths like the Rennick Schist because of the difference in their initial $^{87}Sr/^{86}Sr$ ratios.

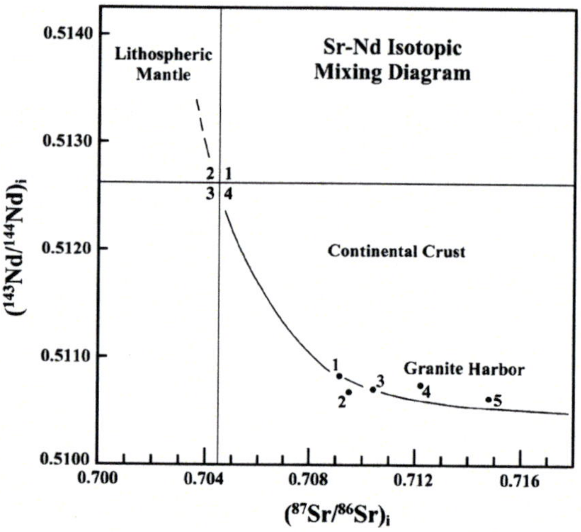

Fig. 4.13 The initial $^{87}Sr/^{86}Sr$ and $^{143}Nd/^{144}Nd$ ratios (at 550 Ma) of samples of Granite Harbor Intrusives in the Wilson Terrane of northern Victoria Land loosely define a Sr-Nd mixing hyperbola that extends from quadrant 4 (crustal rocks) to quadrant 2 (mantle-derived rocks). Therefore, the magmas of these Granite Harbor Intrusives originated primarily by partial melting of preexisting crustal rocks. The locations of the Granite Harbor Intrusives are identified by number: 1. Mt. Murchison, 2. Daniels Range, 3. Alamein Range, 4. Aviator Glacier, 5. Daniels Range (Data from Borg et al. 1987)

These isotopic data combined with the chemical compositions published by Borg et al. (1986) strongly support the conclusion that the Granite Harbor Intrusives of northern Victoria Land are S-type granitoids. This conclusion applies also to the Wright and Vanda Intrusives of the ice-free valleys of southern Victoria Land and perhaps to some of the Cambro-Ordovician granitic plutons elsewhere in the Transantarctic Mountains.

In the next section, the geology of the Morozumi Range and of the Lanterman and Salamander Ranges will be discussed separately because their tectonic settings differ from that of the Daniels Range in the western province of the Wilson Terrane.

4.3.4 Morozumi Range

The Morozumi Range in Fig. 4.14 is located within the Rennick Graben and therefore is separated from the Wilson-Group rocks of the USARP Mountains (Daniels Range and Wilson Hills) by the Rennick fault identified in Fig. 4.15. The basement rocks of the Morozumi Range consist of low-grade metasedimentary rocks including greywacke-shale that were originally assigned to the Robertson Bay Group (Gair et al. 1969; Sturm and Carryer 1970; Dow and Neall 1974). The northern end of the Morozumi Range is occupied by a pluton of Granite Harbor Intrusives called the Adamellite Massif. The metasedimentary rocks were recrystallized and were mapped as the Wilson Gneiss (Engel 1984). More recently, these metasedimentary rocks have been described as the phyllites and spotted schists of the "Morozumi Phyllite" in Table 4.2 (Stump 1995) (see also Fig. 4.33 in Appendix 4.8.6).

Closer examination indicates that the Morozumi Phyllites consist of alternating beds of graywacke and pelite that range in thickness from 10 cm to 2 m. The tops of the clay-rich shale units are calcareous. In addition, graded bedding, cross bedding, and laminations indicate that the metasedimentary rocks were originally deposited as turbidites. The rocks are folded and display well developed slaty cleavage. Stump (1995) stated that similar metasedimentary rocks occur in the

Fig. 4.14 View of the northern end of the Morozumi Range where light-colored adamellite is in contact with black metasedimentary rocks that were assigned to the Robertson Bay Group by Gair et al. (1969) and by Tessensohn et al. (1981). The metasedimentary rocks were intruded by vein-like bodies of granitic rock. The outcrop pattern of one such body forms the light-colored arch in this picture (Adapted from Tessensohn et al. 1981, Fig. 5, p. 50)

eastern Helliwell Hills, in the eastern Kavrayskiy Hills at the mouth of the Rennick Glacier, and at Lonely One Nunatak (71°12′S, 161°18′E). Apparently, the Morozumi Phyllite is not restricted to the Morozumi Range and its lithologic composition relates it to other metasedimentary rocks listed in Table 4.2.

Age determinations by the whole-rock Rb-Sr method listed in Appendix 4.8.3.10 of granites in the Morozumi Range have yielded dates of 515 ± 28 Ma $(^{87}Sr/^{86}Sr)_i = 0.7136 \pm 0.0022$ (Middle Cambrian) for the Adamellite Massif at the north end of the range and a lower date of 478 ± 14 Ma $(^{87}Sr/^{86}Sr)_i = 0.7092 \pm 0.0008$ for granitic rocks in the Jupiter Amphitheater and in Unconformity Valley (Vetter et al. 1983; Kreuzer et al. 1987). The K-Ar dates of biotite and muscovite in the Granite Harbor Intrusives of the Morozumi Range in Fig. 4.16 reached the blocking temperature between 460 and 490 Ma at essentially the same time as the basement rocks of the Daniels Range in Fig. 4.11 (Kreuzer et al. 1981, 1987). The initial $^{87}Sr/^{86}Sr$ ratios of the two suites are similar to the initial $^{87}Sr/^{86}Sr$ ratios of Granite Harbor Intrusives in the southern Daniels Range and elsewhere in northern Victoria Land. Therefore, the Granite Harbor pluton in the Morozumi Range formed from a sedimentary protoliths without achieving complete homogenization of the isotopic composition of strontium.

The Morozumi Range has preserved a record of past fluctuations of the volume of the ice in the Rennick Glacier in late Tertiary time (Denton et al. 1986). The evidence presented by Mayewski (1975, 1976), Mayewski and Attig (1978), and by Mayewski et al. (1979) indicates that the Rennick Glacier covered all

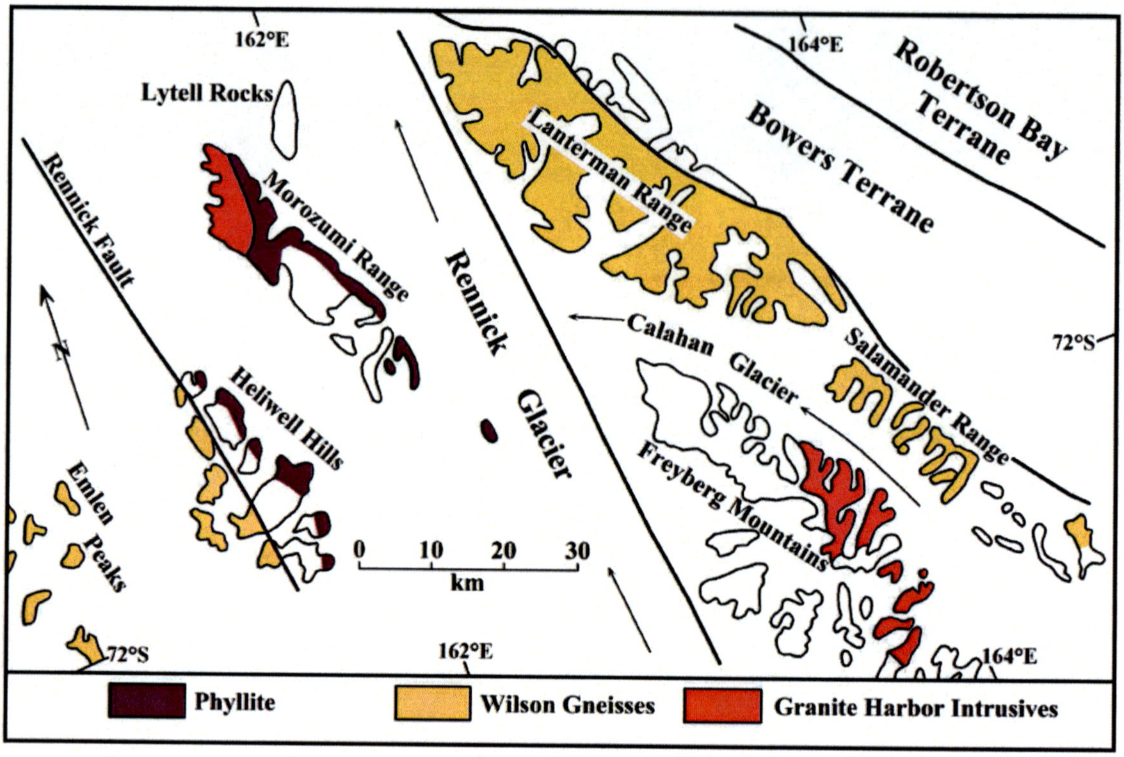

Fig. 4.15 The Morozumi, Lanterman, and Salamander ranges extend from the western province of the Wilson Terrane across the Rennick Glacier to the Bowers and Robertson Bay Terranes. The principal lithologic units of the basement complex are identified by colors. Outcrop areas in white expose sandstone of the Beacon Supergroup and of the Kirkpatrick Basalt. The major boundary faults are indicated by heavy black lines (Adapted from the geologic map of Gair et al. 1969)

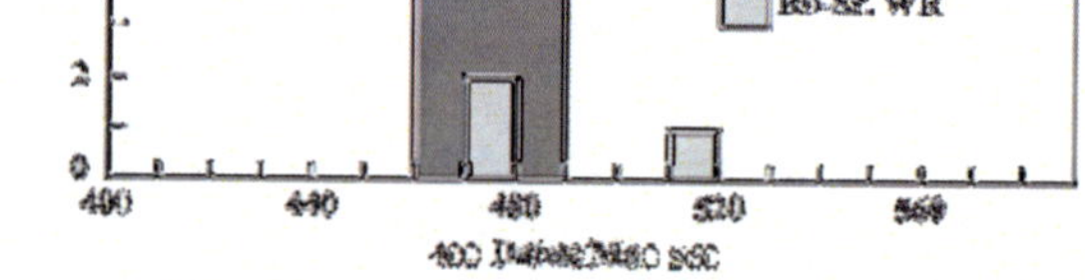

Fig. 4.16 The Granite Harbor Intrusives of the Morozumi Range in the Wilson Terrane crystallized between 510 and 520 Ma and then reached the blocking temperature of biotite and muscovite between 460 and 490 Ma after cooling for about 35 million years from 515 to 480 Ma (Data from Kreuzer et al. 1981, 1987)

but the highest peaks along its course during the Evans Glaciation to which they assigned an age of 4.2 Ma (Pliocene). During this high-stand, the glacier deposited basal till at 71°37′S, 162°01′E in the Morozumi Range several hundred meters above its present level. The till contains grains of detrital feldspar which yielded a Rb-Sr date of 469 ± 24.9 Ma (Middle Ordovician) and an initial $^{87}Sr/^{86}Sr$ ratio of 0.71319 ± 0.00074 (Faure 1986).

The comparison of the initial $^{87}Sr/^{86}Sr$ ratios in Fig. 4.17 supports the conclusion that the feldspar in the till of the Morozumi Range originated primarily from the Granite Harbor Intrusives (GHI) and not from the Rennick Schist (RS) and the Wilson Gneiss (WG) which are exposed in the Daniels Range and in the adjacent Lanterman and Salamander ranges in Fig. 4.14.

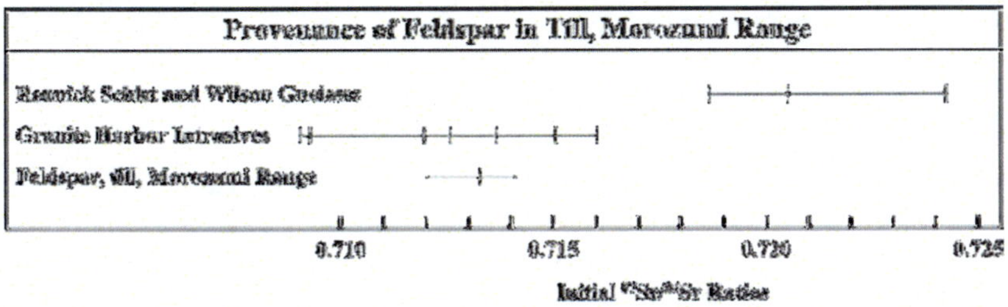

Fig. 4.17 A comparison of initial $^{87}Sr/^{86}Sr$ ratios supports the conclusion that feldspar grains in Neogene (late Tertiary) till in the Morozumi Range originated primarily from outcrops of Granite Harbor Intrusives (GHI) and not from the Rennick Schist (RS) and the Wilson Gneiss (WG) of the Daniels and Lanterman/Salamander ranges (Data from Kreuzer et al. 1981; Vetter et al. 1983; Adams 1986; Faure 1986)

4.3.5 Lanterman and Salamander Ranges

The Lanterman and Salamander ranges in Fig. 4.15 are located on the east side of the Rennick Glacier about 75 km east of the Thompson Spur in the Daniels Range. The Lanterman Range is composed of Wilson Gneisses overlain unconformably by remnants of the Beacon Supergroup. The Wilson Gneisses of the Lanterman Range consist primarily of biotite schists that are permeated by granitic dikes that coalesce locally into bodies of granite choked with rafts of biotite schist and smaller xenoliths. The Wilson Gneisses of the Lanterman Range also include hornblende gneisses, amphibolites, and metaconglomerates that were metamorphosed to the amphibolite facies like the gneisses of the Daniels Range (Tessensohn et al. 1981). However, Grew et al. (1984) determined that the rocks of the Daniels Range are characterized by the transition of andalusite to sillimanite in cordierite-muscovite-quartz assemblages and have steep metamorphic gradients, whereas the basement gneisses of the Lanterman and Salamander ranges contain transitions from kyanite to sillimanite, staurolite occurs in pelitic as well as mafic rocks, and ultramafic rocks composed of spinel-olivine pyroxenite and garnet-olivine bearing pyroxenite occur in this part of the Wilson Terrane (DiVincenzo et al. 1997). The plutonic rocks of the eastern province of the Wilson Terrane consist primarily of I-type granitoids rather than the S-type rocks that dominate in the western province (Borg et al. 1987).

The K-Ar dates of the Wilson Gneisses of the Lanterman Range in Appendix 4.8.3.12 and in Fig. 4.18 indicate that biotite, muscovite, and hornblende reached their blocking temperature for radiogenic ^{40}Ar between 451 and 502 Ma after crystallizing at 550 ± 20 Ma when the $^{87}Sr/^{86}Sr$ ratio of the Wilson Gneisses was 0.7155 ± 0.0003 (Adams and Hőhndorf 1991; Appendix 4.8.3.13).

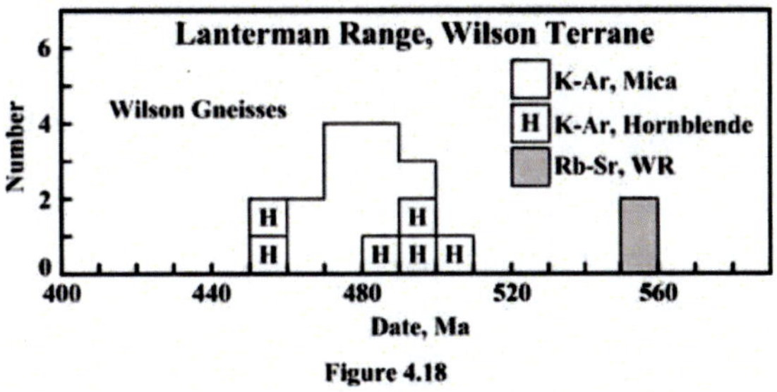

Fig. 4.18 Whole-rock Rb-Sr dates of Wilson Gneisses in the Lanterman and Salamander ranges are 550 ± 20 and 555 ± 36 Ma, respectively. These dates indicate the time of regional metamorphism during which the strontium in these rocks was isotopically homogenized. The K-Ar dates of biotite, muscovite, and hornblende are cooling ages (Data from Kreuzer et al. 1981, 1987; Adams 1982; Adams and Hőhndorf 1991). See also Appendix 4.8.3.12.

The Salamander Range is located southeast of the Lanterman Range and is separated from the Freyberg Mountains by the upper Canham Glacier. The northern part of the Salamander Range consists of Wilson Gneisses which, in the southernmost part, is overlain by residual outcrops of the Beacon Supergroup and of the Kirkpatrick Basalt.

Adams and Hőhndorf (1991) reported a Rb-Sr whole-rock isochron date of 555 ± 36 Ma $(^{87}Sr/^{86}Sr)_i$ = 0.7158 ± 0.0004 (Appendix 4.8.3.14 and Fig. 4.18). The dates and initial $^{87}Sr/^{86}Sr$ ratios of the Wilson Gneisses of the Salamander Range are within the errors of the Rb-Sr dates and initial $^{87}Sr/^{86}Sr$ ratios of the Wilson Gneisses in the Lanterman Range. Moreover, the initial $^{87}Sr/^{86}Sr$ ratios of both suites of Wilson Gneisses are higher than the initial $^{87}Sr/^{86}Sr$ ratios of Granite Harbor Intrusives in northern and southern Victoria Land.

The Salamander Granite that occurs at Mount Tukotok (72°17′S, 164°43′E) in the southern end of the Salamander Range was originally assigned to the Admiralty Intrusives by Gair et al. (1969). This assignment seemed to be confirmed when Laird et al. (1974) reported a K-Ar date of 337 ± 7 Ma for biotite that was measured by Richard L. Armstrong of Yale University. Stump (1995) later obtained a Rb-Sr whole-rock date of 319 ± 5 Ma. However, the genetic affiliation of the Salamander Granite remains in doubt because the Rb-Sr isochron date is actually younger than the cooling ages of Admiralty Intrusives in the Bowers and Robertson Bay terranes. In addition, Borg et al. (1986) showed that the chemical composition and petrography of the Salamander Granite differ from those of the Admiralty Intrusives.

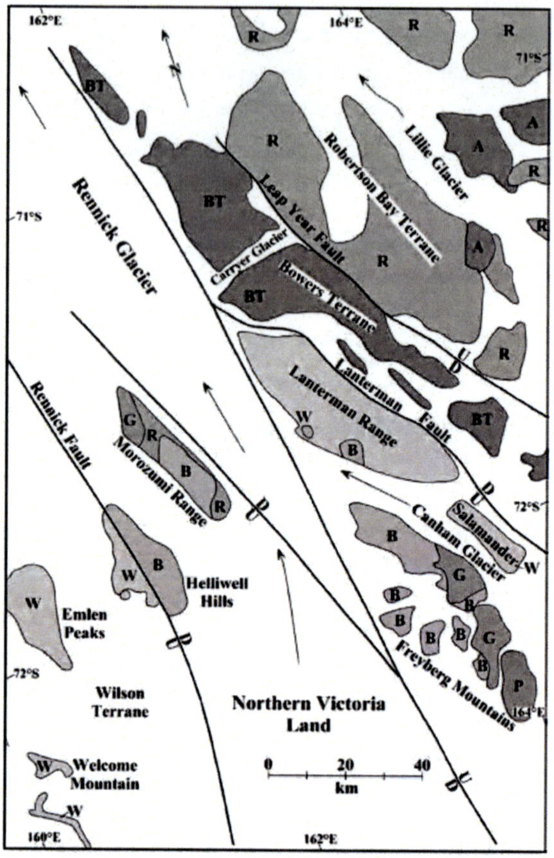

Fig. 4.19 The Bowers Terrane occupies the central area of northern Victoria Land flanked by the Wilson Terrane in the west and the Robertson Bay Terrane in the east. The major rock types of this segment of the Transantarctic Mountains are identified by capital letters in alphabetical order: A = Admiralty Intrusives, B = Beacon Supergroup, BT = Bowers Terrane, G = Granite Harbor Intrusives, P = Gallipoli Porphyries, R = Robertson Bay Group, W = Wilson Group. The Kirkpatrick Basalt has been omitted from this map for the sake of clarity and the Ferrar Dolerite sills are included with the Beacon Supergroup. The Lanterman and Salamander ranges constitute the eastern province of the Wilson Terrane (Adapted from Gair et al. 1969)

4.4 Bowers Terrane

The Bowers Terrane is a narrow fault-bounded block of the continental crust of northern Victoria Land that is located between the Wilson Terrane on the west and the Robertson Bay Terrane on the east as shown in Figs. 4.2 and 4.19. It extends for about 350 km from the Rennick Bay on the Oates Coast southeast to the Lady Newness Bay south of the mouth of the Mariner Glacier on the Borchgrevink Coast. Although it widens to 50 km at its southern end, its greatest width adjacent to the Salamander Range is only about 30 km. The Bowers Terrane is traversed by several small glaciers that are tributaries of the Rennick Glacier (i.e., the Carryer and Sledgers glaciers) and by the Black Glacier which joins the Lillie Glacier in the Robertson Bay Terrane.

The topography of the Bowers Terrane is mountainous with steep slopes and the glaciers are heavily crevassed. The principal mountain ranges listed sequentially from the Oates Coast to the head of the Mariner Glacier include the Frolov Ridge and the Explorers Range of the Bowers Mountains, as well as the eastern part of the Lanterman Range, the Leitch Massif, and the King Range (Gair et al. 1969; Tessensohn et al. 1981; Wodzicki and Robert 1986; Stump 1995).

The Bowers Terrane consists of volcanic and sedimentary rocks of Cambrian to Ordovician age which together constitute the Bowers Supergroup (Wodzicki and Robert 1986). After 2 decades of fieldwork by geologists from New Zealand, Germany, and the USA,

Table 4.3 Stratigraphy of the Bowers Supergroup in the Bowers Terrane of northern Victoria Land (Laird et al. 1982; Wodzicki and Robert 1986; Stump 1995)

Group	Formation	Age
Leap Year 3000+, m	Camp Ridge/ Reilly Carryer	Ordovician to late Late Cambrian
Mariner 2000+, m	Eureka Edlin/ Spurs	Late to Middle Cambrian
Sledgers 3500+, m	Glasgow/Molar	Early Cambrian to Neoproterozoic
370+, m	Husky Conglomerate	Neoproterozoic?

the Bowers Supergroup has been subdivided into three groups identified in Table 4.3.

The rocks of the Bowers Supergroup are structurally deformed but only slightly metamorphosed (prehnite-pumpellyite or lowermost greenschist facies) and their ages are indicated by fossils such as acritarchs, trace fossils, as well as trilobites, brachiopods, and mollusks (Laird et al. 1972, 1974; Dow and Neall 1974; Shergold et al. 1976; Andrews and Laird 1976; Cooper et al. 1976, 1982).

The rocks of the Bowers Supergroup are locally overlain by erosional remnants of the Beacon Supergroup (Permian to Triassic) which includes sills of the Ferrar Dolerite (Middle Jurassic). In addition, erosional remnants of the Kirkpatrick Basalt plateau (Middle Jurassic) are present adjacent to the Lanterman Range and in the Leitch Massif (Gair et al. 1969).

4.4.1 Sledgers Group

The most widely distributed rocks of the Bowers Terrane belong to the Sledgers Group which occurs intermittently from the Explorers Range in the north to the Mariner Glacier in the south. The Sledgers Group consists of interbedded volcanic and sedimentary rocks composed of sandstones and mudstones. Laird et al. (1982) placed these rocks into two separate formations and by doing so raised the former Sledgers Formation to the status of a group consisting of the Glasgow Formation (volcanics) and the Molar Formation (sediments). An alternative procedure, preferred by Tessensohn et al. (1981), would have been to consider the volcanic and sedimentary rocks to be lithologic facies of the Sledgers Formation. The approach taken by Laird et al. (1982) has caused uncertainty in the stratigraphic position of the Molar Formation relative to the Glasgow Volcanics. For example, in the western Molar Massif (71°38′S, 163°45′E) sedimentary rocks of the Molar Formation underlie the volcanics of the Glasgow Formation, whereas in the upper Carryer Glacier (Fig. 4.19), 800 m of Molar sediments interfinger with Glasgow Volcanics (Wodzicki and Robert 1986). In addition, in the area around Mt. Soza (71°10′S, 162°34′E) the Glasgow Formation is sandwiched between 1,000 m of Molar sediment below and 450 m above (Stump 1995). Evidently, parts of the Molar Formation, as presently defined, are both older, equivalent in age, and younger than the Glasgow Volcanics. Although such stratigraphic complications are not specifically excluded by the definition of "formations" (Jackson 1997), the two lithologies could have been accommodated as facies of a conventionally defined Sledgers Formation.

The Glasgow Formation consists primarily of volcanic breccia which reaches a thickness of 2,500 m at the type locality on Mt. Glasgow (71°08′S, 162°55′E). The composition of the volcanic rocks ranges from basalt to andesite but does include rhyolites. The volcanic breccias grade into debris flows and tuffaceous sediment of the Molar Formation. The presence of pillow basalt and marine fossils indicates that the eruptions occurred primarily in a subaqueous and marine environment. The volcanic rocks were altered by seawater to form spilites and keratophyres composed of quartz, albite, chlorite, epidote, actinolite, and varying amounts of calcite and leucoxene (Stump 1995).

The Sledgers Group was intruded by a layered gabbro located at the southern end of Spatulate Ridge (73°28′S, 167°13′E) in Lady Newness Bay. This body was first mentioned by Gibson et al. (1984) and was later described by GANOVEX (1987). It consists of cumulate layers composed of clinopyroxene, calcic plagioclase, orthopyroxene, and accessory olivine. The layers dip 60°SW and are cut by pegmatite dikes composed of hornblende and plagioclase. The intrusive is surrounded by a contact-metamorphic aureole that is 4.5 km wide where temperatures and pressures reached 690–730°C and 2–4 kilobars, respectively. Kreuzer et al. (1987) reported a K-Ar date of 521 ± 10 Ma (Middle Cambrian) for hornblende in this intrusive. Kleinschmidt and Tessensohn (1987) considered this gabbro to be part of an ophiolite complex while Bradshaw (1989) suggested that it is a subvolcanic pluton related to the Glasgow Volcanics. In any case, the K-Ar date of 521 ± 10 Ma reported by Kreuzer et al. (1987) confirms an Early Cambrian to Neoproterozoic age for the Sledgers Group.

The Molar Formation is composed of interbedded sandstone, mudstone, and some conglomerate. Wodzicki and Robert (1986) reported that at its type locality in the western Molar Massif (71°38′S, 163°45′E) more than 2,000 m of sedimentary rocks of the Molar Formation underlie volcanic rocks of the Glasgow Formation. The Molar Formation consists primarily of dark gray mudstone that is interbedded with fine grained and laminated graywacke sandstone which is cross-bedded in places and contains ripples. The graywacke sandstones contain feldspar and lithic grains composed of mudstone, limestone, and volcanic fragments as well as pieces of plutonic and metamorphic rocks. In addition to the fine-grained clastic sediment, the Molar Formation contains beds of coarse grained sandstone that have conglomerate at their base and fine upward. The clasts in the conglomerates consist of volcanic rocks, limestone, quartzite, granitoids, and muscovite schist (Wodzicki et al.). Limestone beds do occur in the Sledgers Group, but the presence of clasts composed of granitoids and muscovite schist indicates that plutonic and metamorphic rocks were exposed in the source area of the detrital sediment during the time of deposition of the Molar Formation.

The age of the Molar Formation is thought to be late Middle Cambrian based on the occurrence of polymerid trilobites in a limestone boulder of a conglomerate in the upper Molar Formation at a locality east of the Neal Massif (Cooper et al. 1983). This age assignment implies that no significant interruption occurred between the deposition of the Sledgers Group and the overlying Middle-to-Late Cambrian Mariner Group. If the Sledgers and Mariner groups were deposited without interruption, then the Glasgow Volcanics may have been erupted in Early-to-Middle Cambrian time rather than during the Neoproterozoic Era. The uncertainty of the age of the Glasgow Volcanics could be eliminated by dating them by the Sm-Nd method.

K-Ar dates of slate and of volcanic rocks of the Sledgers Group in Fig. 4.20 and Appendix 4.8.4.1 range from 332 ± 7 to 467 ± 3 Ma (Adams et al. 1982; Adams 2006). Two whole-rock samples of volcanic rocks yielded K-Ar dates of 328 ± 7 Ma (Lower Carboniferous) and 360 ± 6 Ma (Late Devonian). All of the K-Ar dates of the Sledgers Group are lower than the late Middle Cambrian age (i.e., ~500 Ma, IUGS 2002) indicated by polymerid trilobites. The evident loss of radiogenice ^{40}Ar from the slates and from the volcanic rocks is attributable not only to the limited retentivity of the rocks, but also to episodic increases of the temperature (Kreuzer et al. 1987, p. 43).

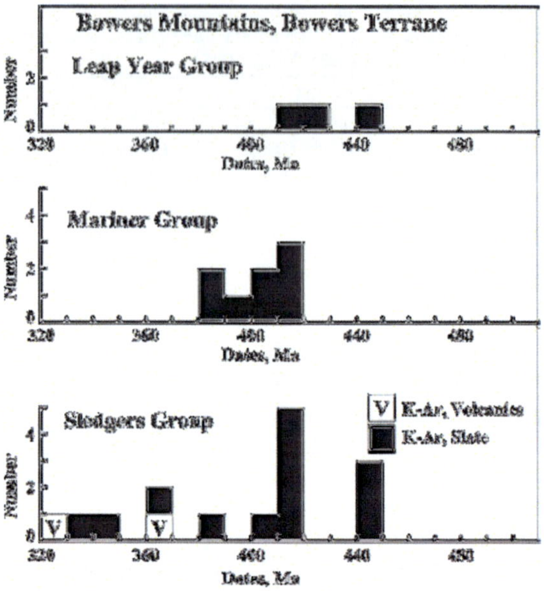

Fig. 4.20 The whole-rock K-Ar dates of slates in the Bowers Terrane vary widely from less than 470 Ma to greater than 330 Ma. The numerical values of these dates depend on several factors, including the extent of outgassing of inherited ^{40}Ar, possible contact metamorphic effects, and the retentivity of slate for radiogenic ^{40}Ar at low temperature at or below the surface of the Earth. Therefore, the sedimentary and volcanic rocks of the Bowers Group were probably deposited more than about 470 million years ago prior to the start of the Middle Ordovician Epoch (Data from Adams et al. 1982; Adams 2006)

One final problem in the stratigraphy of the Sledgers Group is the presence of the Husky Conglomerate in the eastern Lanterman Range adjacent to the high-grade metamorphic rocks of the Wilson Gneisses. According to Laird et al. (1982), the Husky Conglomerate consists of a sequence of dark-green amphibolitic conglomerates interbedded with sandstone and mudstone. These authors originally reported that the clasts are dominantly composed of amphibolite although the conglomerate itself was metamorphosed only to the greenschist facies. Even though the contact with the Wilson Gneisses is covered by snow, Laird et al. (1982) considered it to be unconformable. The top of the Husky Conglomerate is also obscured by snow but appears to be faulted against eastward-dipping and overturned layers of mudstone and volcaniclastic sandstone of the Sledgers Group.

Most workers now regard the Husky Conglomerate to be a basal facies of the Sledgers Group that does not warrant the status of a formation (Cooper et al. 1983; Bradshaw and Laird 1983; Wodzicki and Robert 1986). The discussion in the literature concerning the Husky Comglomerate was reviewed by Stump (1995).

4.4.2 Mariner Group

The Mariner Group in Table 4.3 consists of the Edlin, Spurs, and Eureka formations which have a combined thickness of greater than 2,500 m (Stump 1995). The Edlin Formation is interbedded with the Spurs Formation and both are conformably overlain by the Eureka Formation at the type locality on Eureka Spur (72°42′S, 116°00′E) near the head of the Mariner Glacier (Andrews and Laird 1976). The Mariner Group rests conformably on volcanics of the Sledgers Group, but it is separated from the overlying Leap Year Group by an unconformity (Laird and Bradshaw 1983).

At the type locality the Spurs Formation consists of more than 2,500 m of fissile mudstone containing lenses and thin beds of fine-grained sandstone and conglomerate. The rocks are fossiliferous (trilobites and brachiopods) and contain ripple marks and cross bedding. At Reilly Ridge the Spurs Formation contains conglomerate beds interbedded with coarse-grained sandstone. Some of these conglomerates contain blocks of limestone up to 500 m in length which appear to have been deposited by mass flow or sliding (Stump 1995). Other conglomerates at Reilly Ridge are polymict and contain clasts composed not only of sandstone, limestone, and volcanic rocks, but also of two-mica S-type granites.

The type section of the Edlin Formation, which is laterally equivalent to parts of the Spurs Formation, is located 1 km north of Mt. Glasgow. The Edlin Formation is composed of fine-grained sandstone and mudstone containing lenticular bodies of oosparite limestone up to 5 m thick and less than 50 m long. It also contains tuff beds, conglomerate, and reddish-brown mudstones which contain probable mudcracks, indicating that the redbed sediment may have been exposed at the surface (Laird and Bradshaw 1983).

The Eureka Formation at Eureka Spur consists of wavy-bedded sandstone, mudstone with lenses of limestone, and brownish-gray quartz sandstone. Minor mudstone layers interbedded with the sandstone contain trace fossils and mudcracks suggesting an intertidal environment of deposition (Andrews and Laird 1976). The Eureka Formation is fossiliferous and was deposited during the late Middle to late Late Cambrian (Cooper et al. 1982) in a sequence of environments that evolved from shallow water on a continental shelf (sands and muds of the Spurs Formation) to near-shore intertidal lagoons (reddish-brown muds and near-shore sands of the Eureka Formation). The top of the Eureka Formation was eroded prior to the deposition of the Leap Year Group (Laird and Bradshaw 1983).

Whole-rock K-Ar dates of slate and siltstones of the Mariner Group in Appendix 4.8.4.2, reported by Adams et al. (1982), range from 417 ± 3 (Lower Devonian) to 380 ± 3 Ma (Middle Devonian) and are lower than the age indicated by fossils (late Middle to late Late Cambrian), although they do overlap with the K-Ar dates of the Sledgers Group in Fig. 4.20.

4.4.3 Leap Year Group

The primarily fluvial rocks of the Leap Year Group have been subdivided into the Carryer Conglomerate, the Reilly Conglomerate, and the Camp Ridge Quartzite in order of decreasing age (Table 4.3), although the Reilly Conglomerate may be a coarser facies of the Camp-Ridge Quartzite (Laird and Bradshaw 1983). These formations were deposited on an erosion surface that cuts progressively older rocks of the Mariner and Sledger groups in a northwesterly direction within the Bowers Terrane.

The Carryer Conglomerate, which is at least 800 m thick, occurs about 15 km north of the mouth of the Carryer Glacier (Fig. 4.19) in the northern part of the Bowers Terrane. It is a polymict conglomerate composed of well-rounded clasts up to 3 m in diameter. In the area between the mouth of the Carryer Glacier and Mt. Soza (71°10′S, 162°34′E) and located just north of this glacier, the Carryer Conglomerate includes fine to medium-grained quartz arenite (Laird and Bradshaw 1983). The sandstones that are interbedded with conglomerate beds are ripple marked and cross-bedded (Dow and Neall 1974; Laird et al. 1982). The cobbles of the conglomerate consist of sandstone and mudstone as well as of volcanic and plutonic rocks (Wodzicki and Robert 1986). The volcanic rocks include pyroxene-bearing andesites, andesite breccias, and dacite porphyries. The plutonic rocks consist of two-mica

S-type granites, granodiorite containing hornblende and biotite, quartz monzonites, and alaskite (i.e., alkali-rich quartz-feldspar granite). The lithologic composition of the clasts indicates that volcanic and plutonic rocks were exposed to weathering and erosion a short distance from the site of deposition in the Bowers basin.

The Camp Ridge Quartzite and Reilly Conglomerate consist of quartz sandstone and conglomerate. The Camp Ridge Quartzite was the first formation to be named in the Bowers Terrane (LeCouteur and Leitch 1964) and was subsequently examined by several other field parties identified by Stump (1995). According to Laird and Bradshaw (1983), the thickness of the Camp Ridge Quartzite at the Leitch Massif is 4,000 m. This enormous lithologic unit has been traced for 300 km from the head of the Sheehan Glacier (70°56′S, 162°24′E) in the north to the mouth of the Mariner Glacier in the eastern outcrop belt of the Leap Year Group, whereas the Reilly Conglomerate occurs only on the western side of that belt. Laird and Bradshaw (1983) admitted that the sandy conglomerates that occur on Reilly Ridge (71°32′S, 163°18′E) may be a coarse-grained facies of the Camp Ridge Quartzite. The Reilly Conglomerate rests directly on the Carryer Conglomerate and has a thickness of at least 300 m.

The whole-rock K-Ar dates of two samples of slate and one of siltstone of the Leap Year Group in Appendix 4.8.4.3 range from 448 ± 9 (Late Ordovician) to 421 ± 3 Ma (Late Silurian) (Adams et al. 1982; Adams 2006). These dates shown Fig. 4.20 are also younger than the probable Cambrian depositional age of the rocks.

4.5 Robertson Bay Terrane

The rocks of the Robertson Bay Terrane of northern Victoria Land were originally described by Rastall and Priestley (1921) based on fieldwork by the Northern Party which camped at Cape Adare from 1911 to 1912 during Scott's second Antarctic Expedition (Section 1.4.3). Harrington et al. (1964) later referred to the rocks between Cape Hallett and the lower Tucker Glacier in Fig. 4.1 as the "Robertson Bay Group" and designated Edisto Inlet (72°20′S, 170°05′E) as the type locality because it was easier to reach from Hallett Station, where Harrington and his party were staying, than Robertson Bay itself. Three years later, Harrington (1967) published a comprehensive description of the geology and topography of the Cape Hallett area.

The Robertson Bay Group consists of rhythmically interbedded graywackes and slates or phyllites that were deposited as turbidites and were subsequently folded. Estimates of the thickness of these rocks range widely from 2,000 to 6,000 m (Stump 1995) because neither the top nor the bottom of these beds has been identified in the field. The rocks were folded along northwest trending axes and slightly metamorphosed by the crystallization of chlorite and sericite. The metamorphic grade and tightness of the folding increases gradually from the east to the west across the Robertson Bay Terrane and both trends are thought to be the result of a single orogenic event (Kleinschmidt 1981, 1983) (see also Fig. 4.32 in Appendix 4.8.6.).

The rocks of the Robertson Bay Group are uniformly dark gray or greenish in color with few distinctive marker beds such as lava flows, limestone beds, or conglomerates. The graywackes are fine-to-medium grained and are composed of quartz, minor feldspar, and lithic grains including siltstone, pelite, chert, mafic igneous rocks, and calcite. This clast assemblage suggests that the sediment was derived by erosion of metamorphic rocks in a continental area. Because of its apparent homogeneity, the Robertson Bay Group has not been subdivided into formations except for the Handler Formation and the Millen Schist.

4.5.1 The Handler Formation

The age of the Robertson Bay Group is Late Cambrian to Early Ordovician based on conodonts in a limestone boulder that was found in the upper part of the Robertson Bay Group along Handler Ridge (72°30′S, 167°00′E) (Burrett and Findlay 1984; Buggisch and Repetski 1987). Therefore, the age of the Robertson Bay Group is not Proterozoic as had been previously suggested on the basis of acritarchs (Iltchenko 1972). The upper part of the Robertson Bay Group was later named the Handler Formation by Wright and Brodie (1987) based on its exposure on Handler Ridge which is located along the Trafalgar Glacier near the Millen Range located less than 20 km east of the Leap Year Fault in Fig. 4.19.

The lower part of the Handler Formation (500–1,000 m) is composed primarily of the same alternating dark gray slate and graywacke that characterize the bulk of the Robertson Bay Group. However, in the Handler Formation red and green slates and siltstones are more abundant and individual beds are thicker than in the main body of the Robertson Bay Group. In addition, conglomerates composed of pea-sized quartz pebbles and quartz-rich rock fragments are present in the lower Handler Formation.

The upper part of the Handler Formation which is at least 500 m thick contains a jumble of blocks composed of limestone, quartz-pebble conglomerate, and quartz sandstone (Wright and Findlay 1984). The diameters of these blocks range from a few centimeters to several meters. The limestone block containing Late Cambrian-Early Ordovician conodonts, mentioned above, originated from this part of the Handler Formation which is considered to be a debris flow that moved from a source area where limestone was deposited in shallow water. One of the limestone blocks that has been examined contained a solution cavity that was filled with quartz-pebble conglomerate. This feature and the red color of some of the slates suggests that the source area on the continental shelf had developed karst topography and had accumulated terra rosa (i.e., a reddish brown residual soil on limestone bedrock as in the karst areas around the Adriatic Sea) (Jackson 1997).

4.5.2 Millen Schist

The rocks of the Robertson Bay Group located close to the Leap Year Fault were deformed during several episodes of tectonic activity and were metamorphosed to the lower greenschist facies. Volcanic and sedimentary rocks that resemble the Sledgers Formation of the Bowers Supergroup have also been recognized in this area. Apparently, rocks of the Robertson Bay Group and the Bowers Supergroup were mixed tectonically along the boundary fault that separates the two terranes.

These rocks were originally referred to as the Millen Range Schists by Findlay and Field (1983) but are now known as the Millen Schist (GANOVEX 1987) based on their occurrence in the Millen Range (72°20′S, 166°15′E) east of the Leap Year Fault in the Robertson Bay Terrane. The Millen Schist has been traced all the way to the most northerly part of the Bowers Mountains and southeast to the Malta Plateau (72°58′S, 167°18′E). The rocks consist of quartz-feldspar schist and phyllite derived from sedimentary protoliths of the Robertson Bay Group, from the Molar Formation (Sledgers Group), and from metavolcanic rocks of mafic composition. In contrast to the rocks of the Robertson Bay Group, the Millen Schist was multiply deformed (Bradshaw et al. 1982; Jordan et al. 1984; Findlay 1986) presumably during tectonic activity along the boundary fault.

4.5.3 Admiralty Intrusives

The sedimentary rocks of the Robertson Bay and the Bowers terranes were intruded by numerous plutons predominantly composed of granodiorite but ranging in composition from monzogranites to tonalites and diorites. Harrington (1958) originally assigned these plutons to the Admiralty Intrusives and extended this term to include all of the granite plutons in the Transantarctic Mountains. However, when isotopic age determinations indicated that some of the plutons in northern Victoria Land are younger than those in southern Victoria Land, Grindley and Warren (1964) divided the granitic plutons of the Transantarctic Mountains into the Devonian Admiralty Intrusives and the Cambro-Ordovician Granite Harbor Intrusives. We now know that the Admiralty Intrusives occur only in northern Victoria Land, whereas the Granite Harbor Intrusives occur throughout the Transantarctic Mountains from northern Victoria Land to the Pensacola Mountains adjacent to the Ronne Ice Shelf.

The Admiralty Intrusives of northern Victoria Land occur predominantly in the Robertson Bay Terrane and only in a few places in the Bowers Terrane, including on Znamenskiy Island at the mouth of the Rennick Glacier (Gair et al. 1969). The status of certain plutons that occur in the eastern province of the Wilson Terrane remains uncertain because it is not clear whether they are Admiralty or Granite Harbor Intrusives. For example, Tessensohn (1984) suggested that the Salamander Granite on Mt. Tukotok (72°17′S, 164°43′E), which is a Carboniferous pluton and is located close to the fault between the Wilson and Bowers terranes, is a "terrane stitching" pluton that penetrated the boundary between these terranes (see also Section 4.3.5 and 4.14). Tessensohn (1984) also considered that a pluton on Mt. Supernal (73°04′S, 165°42′E) straddles the boundary between the Bowers and the Wilson terranes. However, Borg et al. (1987) disputed these suggestions based on petrographic, geochemical, and geological information (some of which was unpublished) and concluded that the Admiralty Intrusions do not occur in the Wilson Terrane. This is an important point to be considered in the reconstruction of the tectonic history of northern Victoria Land.

The Admiralty plutons in the Robertson Bay Terrane are sharply discordant with chilled contacts against the country rocks but their contact-metamorphic aureoles are only a few tens of meters wide. They are also unfoliated except close to contacts and are thought to have been intruded at shallow depth after the tectonic deformation of the country rocks.

Chemical and isotopic data by Borg et al. (1986, 1987) indicate that the Admiralty Intrusives are primarily I-type granitoids in contrast to the Granite Harbor Intrusives which include both I-type and S-type rocks. Nevertheless, regional trends in the chemical compositions of the Admiralty Intrusives suggest that crustal rocks of sialic composition contributed to their petrogenesis and that the involvement of crustal rocks increased in a northeasterly direction across the Robertson Bay Terrane.

Kreuzer et al. (1981, 1987) measured a large number of precisely determined K-Ar dates of biotite, muscovite, and hornblende from several Admiralty Intrusives listed in Appendix 4.8.5.3. Tessensohn et al. (1981) also compiled a list of K-Ar dates of Admiralty Intrusives reported prior to 1975. We used the K-Ar and Rb-Sr dates of Kreuzer et al. (1987) to construct the set of histograms in Fig. 4.21 that constrain the cooling ages of several Admiralty Intrusives. Each of the 25 K-Ar dates of biotites or muscovites in part "a" of Fig. 4.21 is the average obtained for coarse and fine sieve fractions, which required a large investment of analytical labor but enhanced the reliability of this data set. The K-Ar dates of biotite and muscovite range from older than 345 Ma to younger than 375 Ma and have a well-developed unimodal distribution that peaks between 355 and 365 Ma in the Famennian Stage of the Late Devonian Epoch (IUGS 2002).

The K-Ar dates of hornblende in Fig. 4.21b range similarly from older than 345 Ma to less than 370 Ma and are also concentrated in the Famennian Stage. The good agreement between K-Ar dates of mica and

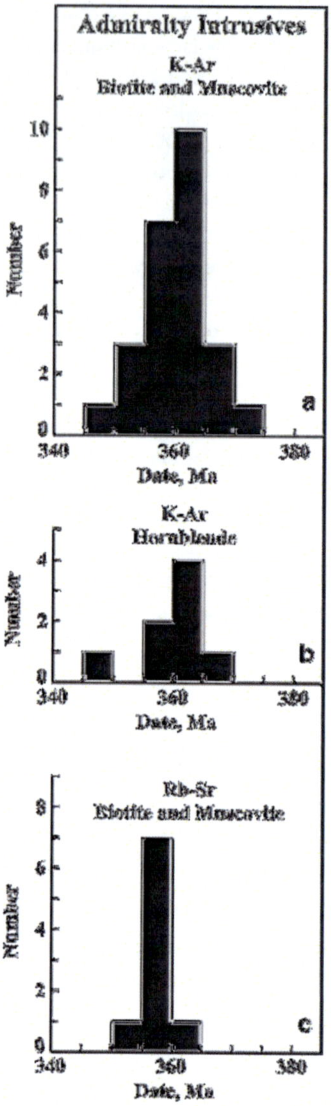

Fig. 4.21 K-Ar and Rb-Sr dates of bitotite, muscovite, and hornblende of the Admiralty Intrusives in the Robertson Bay Terrane. These dates sharply define the time of crystallization between 355 and 365 Ma during the Famennian Stage of the Late Devonian Epoch (IUGS 2002). The good agreement of the K-Ar biotite and hornblende indicates that the rocks cooled rapidly presumably because they were intruded at shallow depth (Data from Kreuzer et al. 1987)

hornblende indicates that the Admiralty plutons cooled rapidly presumably because they were intruded at shallow depth.

A small number of Rb-Sr dates of micas of Admiralty plutons at Cooper Spur (70°38′S, 165°03′E) and

Gregory Bluffs (70°44′S, 165°03′E) in the ANARE Mountains along the Pennell Coast range from 350 to 365 Ma with a strong peak in Fig. 4.21c in the Famennian Stage in excellent agreement with the K-Ar dates of micas. Kreuzer et al. (1987) also cited discrepant whole-rock Rb-Sr dates of 525 ± 26 Ma for Cooper Spur and 388 ± 13 Ma for Gregory Bluffs reported by Vetter et al. (1984).

The whole-rock Rb-Sr method has had limited success for dating the Admiralty Intrusives because data points scatter on the isochron diagrams. For example, only four of the nine whole-rock samples of the Yule Bay Batholith (Pennell Coast) in Fig. 4.22a analyzed by Kreuzer et al. (1987) define a straight line that yields a date of 364 ± 5 Ma and an initial $^{87}Sr/^{86}Sr$ ratio of 0.71495 ± 0.00015. The other five rock samples scatter above and below the line because their initial $^{87}Sr/^{86}Sr$ ratios at 360 Ma range from 0.7125 to 0.7184.

Nine whole-rock samples of the Admiralty Intrusive in the Everett Range at the confluence between the Lillie and Ebbe glaciers in the northern region of the Robertson Bay Terrane form a poorly defined straight line (called a "scatterchron") in Fig. 4.22b. The slope of the line yields a date of 358 ± 15 Ma and an initial $^{87}Sr/^{86}Sr$ ratio of 0.7116 ± 0.0004. This date is similar to the average of the three hornblende K-Ar dates of 362 ± 2 Ma of the same intrusive. This is noteworthy because the Rb-Sr whole-rock date of this Admiralty pluton should be older than its K-Ar hornblende date.

The reason for the scatter of data points on Rb-Sr isochron diagrams is that the strontium in whole-rock samples of seemingly homogeneous Admiralty plutons did not have the same isotopic composition at the time of crystallization. The initial $^{87}Sr/^{86}Sr$ ratios at 360 Ma calculated by Kreuzer et al (1987) for samples of several Admiralty Intrusives range widely from 0.7051 to 0.7172. Figure 4.23 shows that the initial $^{87}Sr/^{86}Sr$ ratios of the Yule and Everett plutons overlap only partly and that Admiralty plutons in the southern part of the Robertson Bay Terrane have lower initial $^{87}Sr/^{86}Sr$ ratios than the northern plutons. This trend is more specifically expressed by the contours of average initial $^{87}Sr/^{86}Sr$ ratios at 360 Ma in Fig. 4.24. The regional variation of the initial $^{87}Sr/^{86}Sr$ ratios as well as the range of chemical compositions reported by Borg et al. (1987) both suggest that the involvement crustal rocks in the petrogenesis of the Admiralty Intrusives increases from south to north-northeast across the Robertson Bay Terrane. This conclusion is confirmed by the initial $^{143}Nd/^{144}Nd$ ratios (at 400 Ma) reported by Borg et al. (1987) which decrease regionally from south to north in the opposite sense to the initial $^{87}Sr/^{86}Sr$ ratios because of the way samarium and neodymium are partitioned when magmas form by partial melting of silicate minerals in the crust and in the mantle of the Earth (Faure and Mensing 2005).

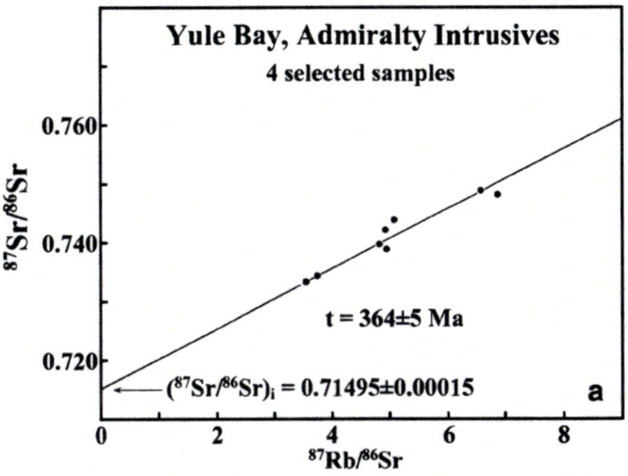

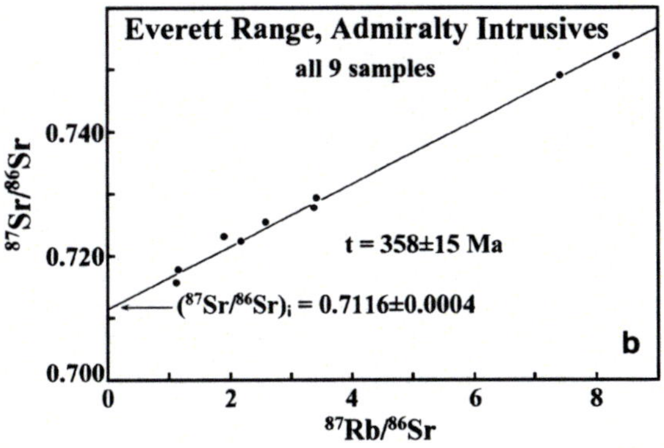

Fig. 4.22 (a) Four selected data points representing whole-rock samples of the Yule Bay granite batholith of the Admiralty Intrusives closely define a straight line in Rb-Sr coordinates: slope = 0.0051836 and initial $^{87}Sr/^{86}Sr$ = 0.71495 ± 0.00015, derived by unweighted linear regression. The slope yields a date of 364 ± 5 Ma (Appendix 4.8.4). The remaining five samples, one of which plots off-scale, scatter above and below the straight line because their initial $^{87}Sr/^{86}Sr$ ratios at 360 Ma range widely from 0.7125 to 0.7184 (Data from Kreuzer et al. 1987). (b) Nine whole-rock samples of the Admiralty pluton in the Everett Range at the confluence of the Lillie and Ebbe Glaciers in the northern part of the Robertson Bay Terrane define a straight line having a slope m = 0.0050967 and initial $^{87}Sr/^{86}Sr$ = 0.71162 ± 0.00037. The slope of the line yields a date of 358 ± 15 Ma (Appendix 4.8.4) (Data from Kreuzer et al. 1987)

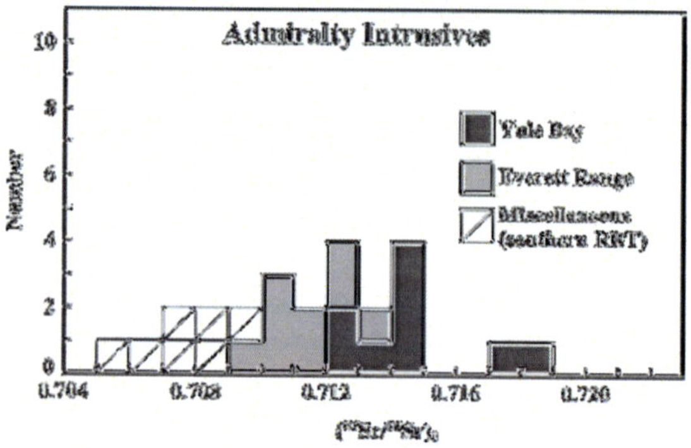

Fig. 4.23 The initial $^{87}Sr/^{86}Sr$ ratios at 360 Ma of Admiralty Intrusives in the Robertson Bay Terrane range widely depending on the geographic distribution of the plutons. Whole-rock samples of the Yule Bay batholith on the Pennell Coast have the highest initial $^{87}Sr/^{86}Sr$ ratios (0.7125–0.7184). The magmatic rocks from the Everett Range at the confluence of the Lillie and Ebbe Glaciers have somewhat lower initial $^{87}Sr/^{86}Sr$ ratios (0.7096–0.7133). The Admiralty Intrusives in the southern region of the Robertson Bay Terrane have the lowest initial $^{87}Sr/^{86}Sr$ ratios (0.7051–0.7099). The regional variation of these ratios indicates that pre-existing crustal rocks contributed increasingly from south to north to the formation of the granitic magmas of the Admiralty Intrusives (Data from Kreuzer et al. (1987) and Borg et al. (1987) after the $^{87}Sr/^{86}Sr$ ratios of the latter were recalculated to 360 Ma)

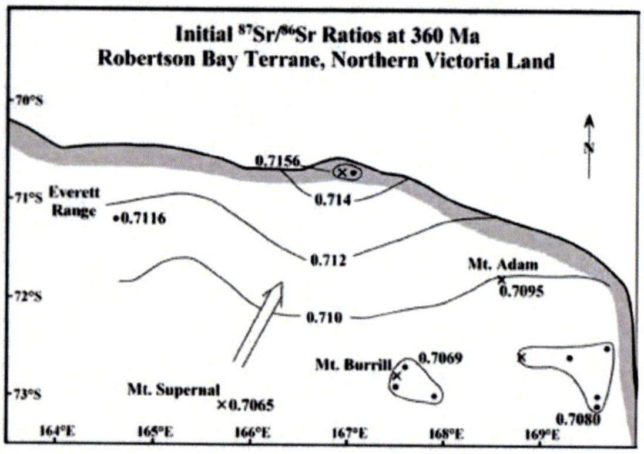

Fig. 4.24 The initial $^{87}Sr/^{86}Sr$ ratios at 360 Ma of the Admiralty Intrusives in the Robertson Bay Terrane have been contoured in order to express the regional increase of this ratio from south to north-northeast. The data by Borg et al. (1987) indicated by crosses were recalculated to 360 Ma to be compatible with data by Kreuzer et al. (1987) indicated by solid circles. Clusters of data from closely spaced collecting sites were averaged as shown. The contour interval is 0.002. The large arrow was aligned between Mt. Supernal to the south and the Yule Bay batholith on the Pennell Coast.

The isotope ratios of strontium and neodymium of five Admiralty Intrusives constrain a mixing hyperbola in Fig. 4.25 when the measured isotope ratios are recalculated to 400 Ma. All of the data points lie in quadrant 4 which is populated by crustal rocks that have elevated $^{87}Sr/^{86}Sr$ and low $^{143}Nd/^{144}Nd$ ratios. The alignment of data points along a mixing hyperbola indicates that the granitic magmas originated primarily by partial melting of mixtures of crustal rocks which contained varying proportions of mantle-derived rocks of basaltic composition that populate quadrant 2. Figure 4.13 contains a similar Sr-Nd mixing hyperbola for Granite Harbor Intrusives at 550 Ma in the Wilson Terrane. Such mixing arrays are observed in the study of volcanic rocks that were erupted in oceanic and crustal environments (Faure 2001; Section 1.7.1).

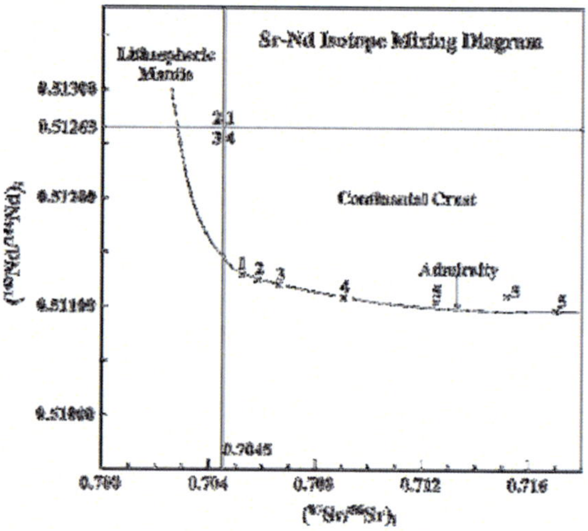

Fig. 4.25 The initial $^{87}Sr/^{86}Sr$ and $^{143}Nd/^{144}Nd$ ratios at 400 Ma of whole-rock samples of the Admiralty Intrusives in the Robertson Bay and Bowers Terranes define an isotopic mixing hyperbola. Although the chemical criteria identify the Admiralty Intrusives as I-type granitoids, the isotopic ratios of Sr and Nd reveal that they formed primarily by partial melting of crustal rocks that were enriched in radiogenic ^{87}Sr and depleted in radiogenic ^{143}Nd because crustal rocks are typically enriched in rubidium and depleted in samarium relative to the ultramafic rocks of the lithospheric mantle (Faure 2001). The Admiralty Intrusives are identified by number: 1. Mt. Supernal, 2. Mt. Burrill, 3. Lower Tucker Glacier, 4. Mt. Adams, and 5. Yule Bay. The alignment of data points along the mixing hyperbola indicates that the granite of Mt. Supernal (1) contains a higher proportion of mantle-derived Sr and Nd than the Yule Bay granite (5) (Data from Borg et al. (1986, 1987). The classification of granitic rocks is by Chappell and White (1974) and by McCulloch and Chappell (1982))

4.5.4 Gallipoli Porphyries and Carboniferous Volcanics

The intrusion of the Admiralty plutons was accompanied by the eruption of porphyritic rhyolites at Gallipoli Heights (72°26′S, 163°48′E) in the southern Freyberg Mountains located in the eastern province of the Wilson Terrane in Fig. 4.19 (Gair et al. 1969).

The age of the felsic volcanics at Gallipoli Heights was determined by Faure and Gair (1970) by means of a two-point whole-rock Rb-Sr age determination based on samples collected by S.J. Carryer. The results yielded a Middle-to-Late Devonian date of 370 ± 40 Ma (recalculated to $\lambda = 1.42 \times 10^{-11}$ $year^{-1}$) and an initial $^{87}Sr/^{86}Sr$ ratio of 0.70548. This value of the initial $^{87}Sr/^{86}Sr$ ratio is typical for continental volcanic rocks (e.g., rhyolites of Central America and Mexico; Faure 2001), but it is lower than the initial $^{87}Sr/^{86}Sr$ ratio (at 360 Ma) of most of the 27 specimens of Admiralty Intrusives analyzed by Kreuzer et al. (1987).

Volcanic rocks composed of andesite, rhyolite, and rhyolite pyroclastics occur also at Lawrence Peaks (72°50′S, 166°20′E) at the head of the Mariner Glacier in the Bowers Terrane. Lava flows, agglomerates, and breccias composed of dark green amygdaloidal basaltic andesite and pyroxene-hornblende feldsparphyric andesite or dacites also rest unconformably on Robertson Bay metasediments at Mt. Black Prince (71°47′S, 168°15′E) near Mt. Minto between the Tucker Glacier and Pennell Coast (Findlay and Jordan 1984; Grindley and Oliver 1983).

Andesites of the Black Prince Volcanics have whole-rock K-Ar dates of 323 ± 6 and 375 ± 6 Ma (Late Devonian to Early Carboniferous) reported by Adams et al. (1986). These dates are in good agreement with age determinations of the Admiralty Intrusives in Fig. 4.21. The lava flows at Mount Black Prince are interbedded with sedimentary rocks that contain plant fossils of Middle Devonian to Early Carboniferous age (Findlay and Jordan 1984).

Adams et al. (1986) also dated six dikes of mafic to intermediate composition from a variety of locations in the Robertson Bay Terrane. The dates range from 300 ± 5 (andesite dike, Millen Range, 4 km NE of Turret Peak; 72°16′S, 166°06′E) to 382 ± 9 Ma (altered mafic dike at the head of Shipley Glacier; 71°26′S, 169°12′E). Additional occurrences of mafic and felsic dikes in northern Victoria Land were described by Crawford et al. (1984). Some of these dikes are genetically related to the Late Tertiary Hallett Volcanic Province (Hamilton 1972), while others are composed of the Middle Jurassic Ferrar Dolerite.

We conclude the description of the geology of the Robertson Bay Terrane by comparing the K-Ar dates of low-grade metasediments and mica minerals in all three terranes of northern Victoria Land (Adams et al. 1982; Adams 2006). Figure 4.26 shows that the basement

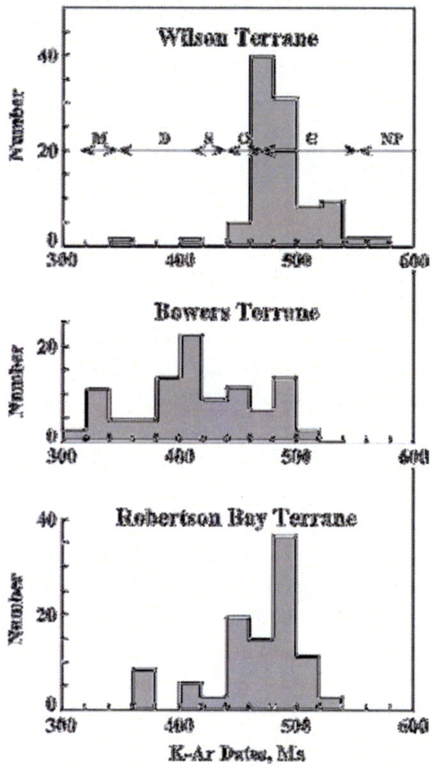

Fig. 4.26 The K-Ar dates of low-grade metasediments and of mica minerals of plutonic rocks record the time when these rocks cooled to the blocking temperature for radiogenic ^{40}Ar. Therefore, these dates can be interpreted as uplift ages of the three tectonic terranes of northern Victoria Land. The K-Ar dates of the rocks in the Wilson and Robertson Bay Terranes are strongly clustered between 440 and 540 Ma (Late Ordovician to Early Cambrian). In marked contrast, the K-Ar dates of the Bowers Terrane range widely from 300 to 500 Ma (Late Carboniferous to Late Cambrian). These comparatively young cooling dates provide evidence that the rocks of the Bowers Terrane may have been partially outgassed by a thermal event during the Cretaceous Period (Kreuzer et al. 1987). Alternatively, the Bowers Terrane may have been uplifted more recently than the adjacent Wilson and Robertson Bay Terranes in accordance with a tectonic model by Kleinschmidt and Tessensohn (1987) (Data by Adams et al. 1982; Adams 2006). NP = Neoproterozoic

rocks of the Wilson and Robertson Bay terranes cooled primarily between 440 (Late Ordovician) and 540 Ma (Early Cambrian). However, the slates of the Bowers Terrane cooled between 300 (Late Carboniferous) and 500 Ma (Late Cambrian). Kreuzer et al. (1987) suggested that the rocks of the Bowers Terrane were partially outgassed by an episodic thermal event during the Cretaceous Period. Alternatively, the low cooling ages may indicate that the Bowers Terrane was uplifted later than the adjacent terranes in accordance with a tectonic model proposed by Kleinschmidt and Tessensohn (1987).

4.6 Tectonics

The key to the tectonic history of northern Victoria Land are the large northwest-striking faults that divide this region into the three terranes described in this chapter. The fold axes of metasedimentary rocks also generally strike northwest and the folds are overturned to the southwest which implies that the rocks in all three terranes were compressed by forces directed from northeast to southwest. The compression of the terranes, combined with the strike-slip motions along the boundary faults, can account for the differences in lithology, tectonic deformation, and metamorphic grade of adjacent terranes. Our present understanding of the tectonic evolution is based on the work of many geologists who have contributed to the mapping of this region, including LeCouteur and Leitch (1964), Crowder (1968), Gair et al. (1969), Dow and Neall (1974), Tessensohn et al. (1981), Weaver et al. (1984), Gibson and Wright (1985), Wodzicki and Robert (1986), Vetter and Tessensohn (1987), Kleinschmidt and Tessensohn (1987), Flőttmann and Kleinschmidt (1991), and many others whose contributions were identified by Stump (1995).

The tectonic history of northern Victoria Land is constrained by the need to provide plausible answers to certain questions concerning the geology of this region:

- Which of the three terranes contains the Ross Orogen of southern Victoria Land?
- Where was the source of the detrital sediment that formed the rocks of the Bowers and Robertson Bay terranes?
- Where in northern Victoria Land is the paleo-Pacific coastline that existed after the rifting of Rodinia and the departure of Laurentia?
- What relative motions occurred along the boundary faults that define the three terranes?

One fact stands out among all the speculations and conjectures concerning the tectonic evolution of northern Victoria Land: In Neoproterozoic to early Paleozoic time, the continent of Antarctica did not exist in the form it has today. After Laurentia split from Rodinia during the Neoproterozoic Era, the East Antarctic craton remained an intergral part of Gondwana to which West Antarctica and the Antarctic Peninsula were added later. Therefore, the evolution of northern Victoria Land took place as part of the evolution of Gondwana before it broke up during the Mesozoic Era. In particular, the land masses of Australia, Tasmania, and New Zealand had not yet separated from Gondwana and therefore could have played a role in the tectonic evolution of northern Victoria Land (Grindley and Davey 1982; Findlay 1987). The separation of the southern continents from Gondwana, the subsequent assembly of the continent of Antarctica, and its drift to the south-polar region of the Earth will be discussed in Chapter 9.

The geological evidence that has been recorded along the Lanterman and Leap Year faults indicates that both thrusting and strike-slip motions have occurred. Therefore, two kinds of tectonic models have been proposed:

- Westward subduction of oceanic crust of the paleo-Pacific plate along the active margin of Gondwana
- Accretion of "suspect" or allochthonous terranes involving significant strike-slip motion as well as thrusting along the terrane boundaries

4.6.1 Subduction Model

The model of westward subduction (Kleinschmidt and Tessensohn 1987) divides the process into three stages starting at a time prior to the Ross Orogeny and ending at 550 Ma during Stage I, which was followed by Stages IIa and IIb that lasted from the main phase of the Ross Orogeny to 500 and 480 Ma, ending with Stage III at about 360 Ma.

During Stage I (Fig. 4.27) two parallel westward-dipping subduction zone existed along the paleo-Pacific

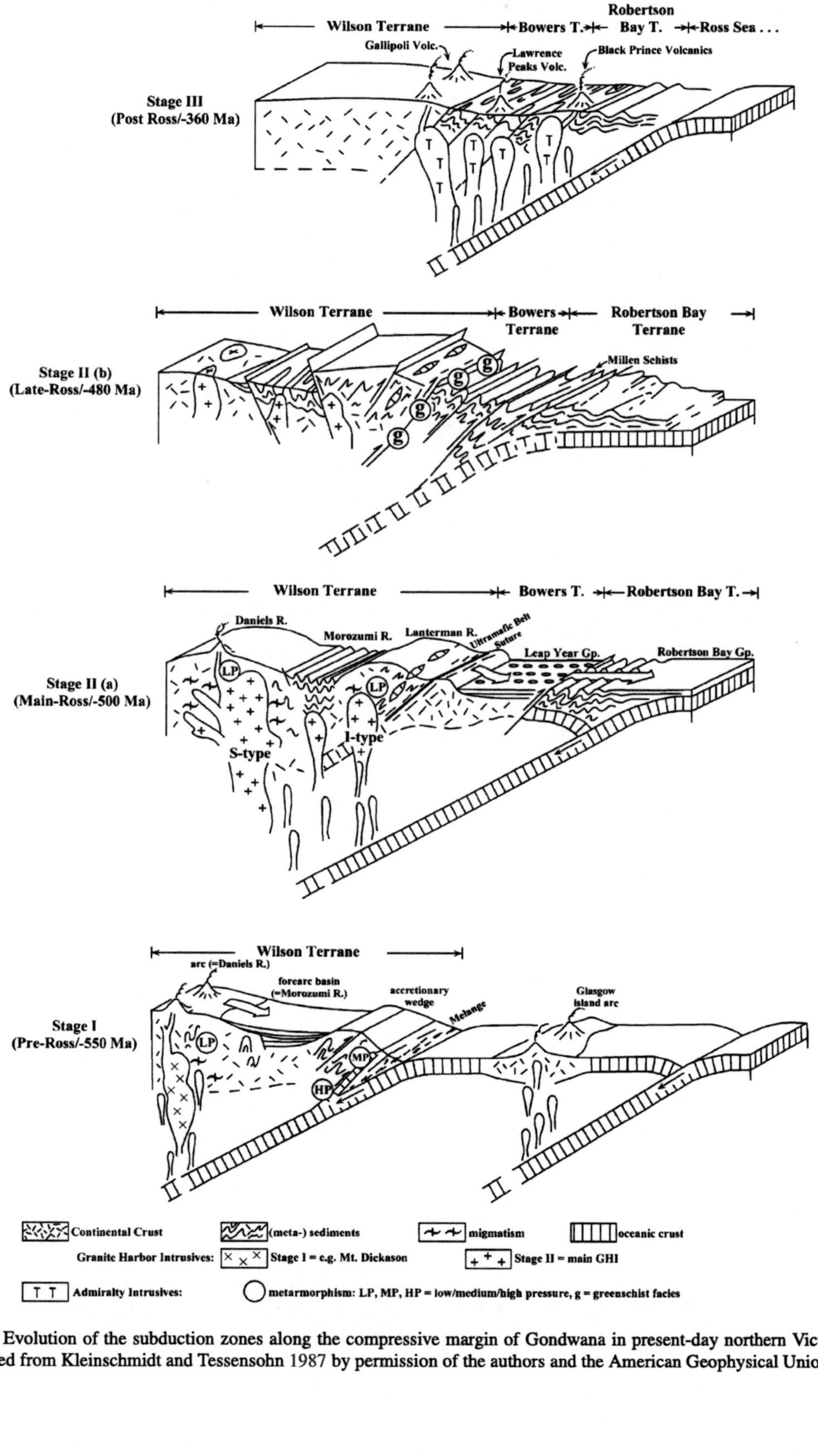

Fig. 4.27 Evolution of the subduction zones along the compressive margin of Gondwana in present-day northern Victoria Land (Reproduced from Kleinschmidt and Tessensohn 1987 by permission of the authors and the American Geophysical Union)

margin of Gondwana which caused partial melting of the overlying mantle wedge above each of the two down-going slabs. The resulting volcanic arc along the eastern edge of the craton evolved into the Daniels Range and the Wilson Hills of the Wilson Terrane. The second volcanic arc, located some distance off-shore, formed the Glasgow arc which evolved into the volcano-sedimentary complex of the Sledgers Group of the Bowers Terrane.

In Stage IIa the near-shore subduction zone died out because the accreted wedge of tightly folded sediment was "cratonized" in the form of the Morozumi and Lanterman ranges which shed sediment eastward into a forearc basin that developed between the eastern subduction zone and the proto-Lanterman Range. During Stage IIb these sediment layers were folded to form the Leap Year Group of the Bowers Terrane. In addition, a westward-dipping reverse fault (Leap Year Fault) separated the folded sedimentary rocks of the Bowers Terrane from the folded turbidites of the Robertson Bay Terrane. The Millen Schist developed at this stage from sediment of the Robertson Bay Group as well as from the Bowers Supergroup along the Leap Year Fault. According to the model of Kleinschmidt and Tessensohn (1987), all three terranes of northern Victoria Land had formed at the end of Stage IIb about 480 million years ago.

During Stage III between 480 and 360 Ma volcanic activity resulted in the eruption of lava flows of the Gallipoli Heights, Lawrence Peak, and Mt. Black Prince. At about the same time, westward subduction of oceanic crust continued and caused granitic plutons of the Admiralty Intrusives to form within the overlying folded and metamorphosed sedimentary rocks of the Bowers and Robertson Bay terranes. The diagram for Stage III in Fig. 4.27 shows that Kleinschmidt and Tessensohn (1987) postulated that the Admiralty Intrusives cut across both the Lanterman and the Leap Year Faults and thereby inhibited further motions along them.

4.6.2 Tectonics of the Wilson Terrane

The Wilson Hills and the Pomerantz Tableland in the northern Wilson Terrane contain the Wilson Thrust (WT) and the Exile Thrust (ET) as well as the Lazarev Thrust (LT), all of which strike north-northwest but dip in opposite directions. The Wilson Thrust in Fig. 4.28 dips west at about 30°, whereas the Exile Thrust (ET) and Lazarev Thrust (LT) dip east. Therefore, these faults define a crustal block in the Wilson Terrane that has a roughly triangular cross-section as illustrated in Fig. 4.29. Flőttmann and Kleinschmidt (1991) pointed out that compressive motions along these faults may have caused this block to be uplifted and to be thrust both east and west over folded low-grade metasedimentary rocks of the Paleozoic subduction zone that was previously defined by Kleinschmidt and Tessensohn (1987) in Fig. 4.27.

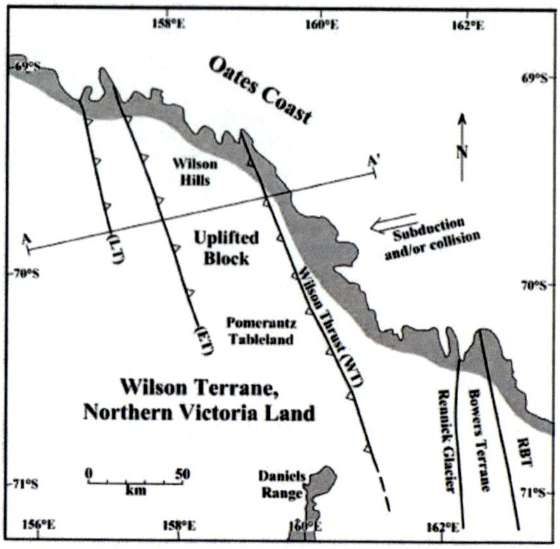

Fig. 4.28 The Wilson Terrane of northern Victoria Land (NVL) contains three thrust faults which define a crustal block that was uplifted as a result of compression from the east. The plane of the Wilson Thrust (WT) dips west at about 30° while the Exile Thrust (ET) and the Lazarev Thrust (LT) dip east. The uplifted block includes the Wilson Hills and the Pomerantz Tableland. The compressive force that activated the thrust faults originated either from the westward subduction, or by the collision of "suspect" terranes (i.e., BT and RBT) with the Wilson Terrane or both (Adapted from Adams 2006)

The hypothesis of Flőttmann and Kleinschmidt (1991) was subsequently supported by Adams (2006) who demonstrated in Fig. 4.30 that the K-Ar dates of biotites in the Uplifted Block in Fig. 4.29 are significantly lower on average than K-Ar dates of slates and biotite in the adjacent part of the Wilson Terrane:

Uplifted Block (UB): 470 ± 4 Ma (17 samples)
West of UB: 490 ± 4 Ma (6 samples)
East of UB: 500 ± 5 Ma (7 samples)

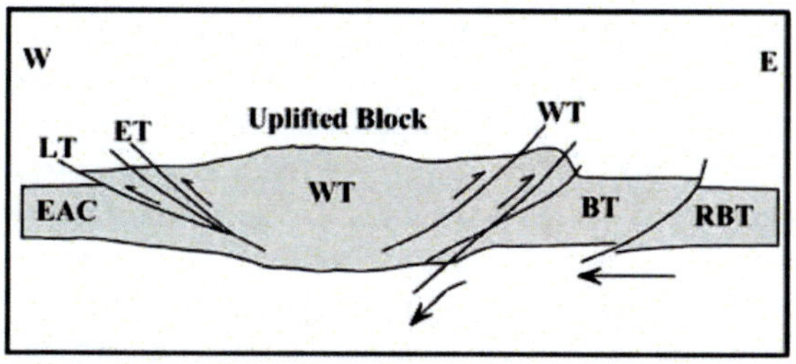

Fig. 4.29 The Wilson Thrust (WT) and the Exile and Lazarev Thrusts (ET and LT, respectively) in the northern Wilson Terrane (Fig. 4.28) dip in opposite directions and thus carve out a wedge-shaped block of the crust of the Wilson Terrane. West-directed forces arising from subduction of oceanic crust illustrated in Fig. 4.27, or the collision of the Bowers Terrane (BT) and the Robertson Bay Terrane (RBT) with the Wilson Terrane caused the crustal block to be uplifted relative to the East Antarctic Craton (EAC) and the Bowers Terrane. As a result, high-grade metamorphic rocks of the Uplifted Block were thrust over folded metasedimentary rocks located east and west of the Uplifted Block (Adapted from Fl ttmann and Kleinschmidt 1991)

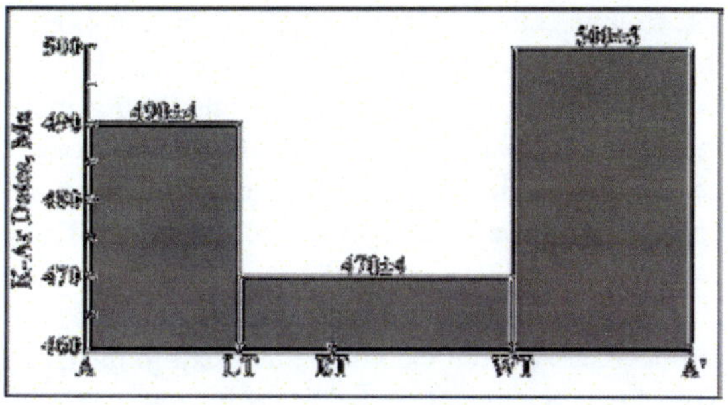

Fig. 4.30 The average K-Ar dates of biotite and slate of the Wilson Terrane, when projected to the line A-A' in Fig. 4.28, vary systematically as shown. The average K-Ar date of 17 biotites in the Uplifted Block (Fig. 4.29) is 470 ± 4 Ma which is significantly lower than the averages of the K-Ar dates of the adjacent crust. The apparent rejuvenation of the rocks in the Uplifted Block is attributable both to cooling of the rocks after uplift and to the loss of radiogenic ^{40}Ar in response to heat generated by the friction along the faults. The movement along the thrust faults and the resulting uplift of the fault block was caused by compression of the Wilson Terrane either as a result of subduction or because of collision of the Bowers and Robertson Bay terranes, or both. LT = Lazarev Thrust, ET = Exile Thrust, and WT = Wilson Thrust. The errors are one standard deviation of the mean based on the number of samples in each average indicated in the text. The tectonic model was proposed by Flőttmann and Kleinschmidt (1991); the K-Ar dates are by Adams (2006)

Evidently, the tectonic activity that was proposed by Flőttmann and Kleinschmidt (1991) can explain the significant losses of radiogenic ^{40}Ar from biotite in the Uplifted Block.

The tectonic history of this region combined with the K-Ar dates supports the hypothesis that all of the basement rocks of the Wilson Terrane were originally metamorphosed during the Cambro-Ordovician Ross Orogeny which occurred prior to 490 Ma. The uplift of the fault block caused rocks from depth to be lifted above the isotherm of the blocking temperature of radiogenic ^{40}Ar in biotite. After erosion of the rocks that had previously cooled following the Ross Orogeny, the K-Ar dates of biotite in the rocks that are presently exposed at the surface of the Uplifted Block record the time that elapsed since the biotite cooled through the blocking temperature of ^{40}Ar. Given that uplift preceded cooling, the uplift occurred prior to 470 ± 4 Ma (Middle Ordovician) but after the Ross Orogeny ended at 490 ± 4 Ma.

The Ross Orogeny was presumably the culmination of the westward subduction of oceanic crust and of the overlying sediment that had previously accumulated along the paleo-Pacific coast of Gondwana. If the Ross Orogeny ended at about 490 ± 4 Ma as implied by the data in Fig. 4.30, an explanation is needed for what triggered the westward compression of the Wilson Terrane that activated the thrust faults and caused the uplift of the crustal block defined by the Wilson Thrust and the Exile-Lazarev thrusts. Adams (2006) suggested that the uplift of the fault block was triggered by the collision of the Bowers and Robertson Bay terranes with the Wilson Terrane. This explanation supports the view that the Wilson Terrane formed by subduction and by the subsequent cratonization of a volcanic arc, whereas the Bowers Terrane and the Robertson Bay Terrane were added later by collisions of "suspect" terranes with Gondwana.

Alternatively, the heat generated by friction along the thrust faults could also have caused radiogenic ^{40}Ar to be lost from biotite in the rocks of the Uplifted Block. This conjecture is supported by the fact that biotite in a sample of granodiorite collected close to the Exile Thrust at 69°58′S, 158°04′E has the lowest K-Ar date (412 ± 8) reported by Adams (2006) for rocks of the Uplifted Block. In this case, the Exiles Thrust would have been active in the Early Devonian Epoch assuming that most of the radiogenic ^{40}Ar was lost from the biotite at that time. If the rejuvenation of the Uplifted Block was caused, at least in part, by frictional heat generated by the thrust faults, and if these faults were activated by the collision of the Bowers and Robertson Bay terranes with Gondwana, then the collision of these terranes occurred during the Early Devonian Epoch (IUGS 2002).

Adams (2006) considered that the evident loss of radiogenic ^{40}Ar from the sample adjacent to the Exile Thrust is probably not related to the extensive rejuvenation of the rocks of the Bowers Terrane but was apparently caused by a localized thermal event that occurred less than 300 million years ago in the Late Carboniferous (Pennsylvanian) Period.

The tectonic evolution of the Wilson Terrane was revisited by Läufer et al. (2006) who presented evidence that high-grade metamorphic rocks of the Wilson Terrane were thrust onto the East Antarctic Craton as far west as the Mertz and Ninnis glaciers in George V Land at about 145°E longitude.

4.6.3 Accreted-Terrane Model

The possibility that northern Victoria Land is a composite of accreted terranes was first suggested by three experienced New Zealand geologists, S.D. Weaver, J.D. Bradshaw, and M.G. Laird (Weaver et al. 1984) based on their study of the chemical composition of the Glasgow Volcanics of the Sledgers Group in the Bowers Terrane. These authors concluded that the Glasgow Volcanics and the interbedded volcanogenic sediment of the Molar Formation, which have a combined thickness of more than 3.5 km, could not have been deposited in their present continental setting located over 200 km west of the margin of Antarctica. The chemical composition of the Glasgow Volcanics instead favors the interpretation that they are part of an intra-oceanic island arc that could not fit into the present Bowers Terrane which is only 60 km wide on average. Therefore, the Bowers Terrane is only a small part of a larger volcano-sedimentary complex.

Another objection to the subduction model cited by Weaver et al. (1984) is that the boundary zone of the Lanterman fault does not contain crustal fragments of the subduction-zone complex. In addition, they noted that plutons of the S-type Granite Harbor Intrusives, that characterize the Wilson Terrane, occur only 6 km west of the tholeiite basalts of the Glasgow Volcanics in the Bowers Terrane. The juxtaposition in the Ross orogen of the Wilson Terrane and the volcano-sedimentary complex of the intra-oceanic island arc of the Bowers Terrane suggests that the intervening crustal segment was lost either by telescoping or by strike-slip faulting.

Weaver et al. (1984) considered four different tectonic models to explain the juxtaposition of the geologically incompatible terranes of northern Victoria Land. Three of the models they considered involved westward-dipping subduction zones that developed at the margin of the Wilson Terrane which had been cratonized previously during the Cambro-Ordovician Ross Orogeny. The fourth model they proposed assumed that the present arrangement of terranes is the result of strike-slip movement along the Lanterman and Leap Year fault zones which caused allochthonous terranes to be assembled in their present positions and which can explain the absence of the intervening crust. Weaver et al. (1984) concluded that the Bowers and Robertson Bay terranes had cooled independent of each other prior to collision with the Wilson Terrane and therefore had developed different regional patterns of K-Ar dates. After these two terranes had collided with the Wilson Terrane, they were intruded by plutons of the Admiralty Intrusives which occur in both terranes but are absent from the Wilson Terrane. Vetter and Tessensohn (1987) suggested that the Admiralty Intrusives, which are I-type granitoids, formed as a result of decompression melting triggered by uplift following the collision of the Bowers and Robertson Bay terranes by basin closure. However, the process by which the magma of the Admiralty Intrusives formed remains uncertain partly because of the delay of about 50 million years between the collision of the Robertson Bay and Bowers terranes (412 Ma) and the crystallization of the Admiralty Intrusives (360 Ma). The apparent time gap would, however, be shorter if the Admiralty Intrusives crystallized at about 400 Ma as postulated by Borg et al. (1987).

The accretion model has been supported by many authors, including Bradshaw et al. (1985), Bradshaw (1987,1989), Gibson (1987), Dallmeyer and Wright (1992), as well as by Capponi et al. (1999) who refer to the Lanterman Fault as a "suture zone," and by Wodzicki and Robert (1986) in their detailed study of the geology of the Bowers Mountains.

4.7 Summary

The geology of northern Victoria Land is complicated by the juxtaposition of three terranes that were assembled during the Early Devonian after they had evolved elsewhere. The only indigenous component is the Wilson Terrane because it formed from the volcano-sedimentary complex of a subduction zone along the compressive margin of the East Antarctic craton. This complex was deformed and recrystallized during the Ross Orogeny in Cambro-Ordovician time. Remnants of metasedimentary and metavolcanic rocks are preserved in the Priestley Formation and the derivative Snowy Point Paragneiss of the Terra-Nova-Bay area, in the Rennick Schist of the Sequence Hills and the USARP Mountains, in the metasediments of the Berg Group of Oates Land, and in the phyllites of the Morozumi Range. The metasedimentary rocks of the USARP Mountains (i.e., Wilson Hills, Daniels Range, Emlen Peaks, and Welcome Mountain) were transformed into the Wilson Gneisses that characterize the Wilson Terrane. The Granite Harbor Intrusives which intruded the Wilson Gneisses and permeated them in the form of dikes and sills contain excess radiogenic ^{87}Sr indicating that the granitic magmas formed by partial melting of crustal rocks. However, the granitic magmas that intruded the Wilson Terrane did not form by melting of the Rennick Schist because their initial $^{87}Sr/^{86}Sr$ ratios do not match.

After the Ross Orogeny, the Wilson Terrane was tectonically deformed by thrust faulting that may have been caused by the collision of the Bowers and Robertson Bay terranes that also caused uplift of the Wilson Hills and the Pomerantz Tableland. Still later, the Rennick Graben formed by normal faulting after relaxation of the compressive stress.

The Bowers Terrane is located east of the Wilson Terrane and is separated from it by the Lanterman fault zone which has been traced from the Bowers Mountains on the Oates Coast to the Lady Newness Bay on the Borchgrevink Coast of the Ross Sea. The Bowers Terrane appears to be a sliver of a much larger landmass that originally included an oceanic island-arc system. It is composed of oceanic tholeiites interbedded with and overlain by fossiliferous sedimentary rocks that were deposited in a marine environment. The rocks were compressed by westward-directed forces that caused the rocks to be folded and faulted. However, the metamorphic grade is lowermost greenschist facies. The eastern boundary of the Bowers Terrane is the Leap Year fault zone which separates it from the Robertson Bay Terrane.

The Robertson Bay Terrane consists of a thick succession of interbedded slate and graywacke that formed as turbidites in a marine environment seaward of a second subduction zone that developed in early Paleozoic time after the Wilson Terrane had been cratonized. The thickness of the sedimentary rocks of the Robertson Bay Group is unknown because neither the top nor bottom have been recognized. The age of the Robertson Bay Group is Late Cambrian to Early Ordovician based on conodont fossils in limestone boulders in the Handler Formation that is exposed along Handler Ridge. The sedimentary rocks are folded but only slightly metamorphosed indicated by the recrystallization of clay minerals to sericite. The metamorphic grade of the Robertson Bay Group increases adjacent to the Leap Year fault which contains a mixture of fault blocks of Bowers Supergroup and the Robertson Bay Group. The rocks in the Millen Range close to the Leap Year Fault were intensively deformed and recrystallized to form the Millen Schist. This zone of deformation has been traced along the entire boundary region between the Bowers and Robertson Bay terranes.

The Robertson Bay Terrane includes numerous batholiths and plutons of post-tectonic granitic rocks of the Admiralty Intrusives of Late Devonian to Carboniferous age. The initial $^{87}Sr/^{86}Sr$ ratios of the Admiralty Intrusives increase in a northeasterly direction from low values in the south to high values in the Yule Bay batholith along the Pennell Coast. Evidently, crustal rocks having elevated $^{87}Sr/^{86}Sr$ ratios contributed increasingly to the magmas of the Admiralty Intrusives in the northern part of the Robertson Bay Terrane. Admiralty Intrusives also occur in the Bowers Terrane, but appear to be absent from the Wilson Terrane, except perhaps in the Salamander Range along the Lanterman Fault. In addition, rhyolites and porphyry eruptives of Late Devonian age occur at Gallipoli Heights in the Freyberg Mountains which are located within the Wilson Terrane. Nevertheless, these Gallipoli Porphyries are considered to be the volcanic equivalent of the Admiralty Intrusives of the Robertson Bay Terrane. If the Gallipoli Porphyries are genetically related to the Admiralty Intrusives of the Robertson Bay Group, then the Late Devonian-Early Carboniferous magmatic activity of the Robertson Bay Terrane apparently did affect the Wilson Terrane.

4.8 Appendices

4.8.1 Age Determinations of Basement Rocks of the Terra-Nova-Bay Area

Field number	Rock type/location	Material	Date (Ma)	Reference
Priestley Formation and Priestley Schist, K-Ar				
ABB5	Mica schist (PS), Tethys Bay	B	468 ± 9	4
F08	Slate (PF), Foolsmate Glacier, S	WR	530 ± 11	4
F015	Slate (PF), Foolsmate Glacier, S	WR	517 ± 10	4
F025	Slate (PF), Foolsmate Glacier, W	WR	522 ± 10	4
F036	Slate (PF), Foolsmate Glacier, W	WR	524 ± 10	4
F045	Slate (PF), Foolsmate Glacier, névé	WR	556 ± 11	4
F047	Slate (PF), Foolsmate Glacier, névé	WR	513 ± 10	4
F048	Slate (PF), Foolsmate Glacier, névé	WR	538 ± 11	4
F049	Slate (PF), Foolsmate Glacier, névé	WR	530 ± 11	4
F055	Slate (PF), Foolsmate Glacier, névé	WR	522 ± 10	4
F061	Mica schist (PS), Wasson Rock	WR	524 ± 10	4
F071	Slate (PF), O'SKane Canyon	WR	494 ± 10	4
F079	Mica schist (PS), Simpson Crags	WR	500 ± 10	4
B09	Mica schist (PS), Wishbone Ridge	B	495 ± 10	4
B018	Mica schist (PS), Wishbone Ridge	B	472 ± 9	4
B031	Mica schist (PS), Wishbone Ridge	B	495 ± 10	4
B037	Mica schist (PS), Tourmaline Plateau	B	472 ± 9	4
B044	Mica schist (PS), Mt. Dickason	B	482 ± 10	4
B065	Biotite paragneiss (PS), Cape Sastrugi	B	483 ± 10	4
B068	Biotite paragneiss (PS), Snowy Point	B	475 ± 9	4
B045	Biotite paragneiss (TNF), Mt. Dickason	B	481 ± 10	4
B052	Mylonite (TNF), Mt. Emison	WR	474 ± 9	4
G02	Granodiorite gneiss (TNF), Cape Moebius	B	473 ± 9	4
Granite Harbor Intrusives, K-Ar				
4250	Granite, 6 km NE of Mt. Dickason	B	458.5 ± 2	2
4251	Granite, 6 km NE of Mt. Dickason	B	456.5 ± 2	2
4252	Granite, 4 km W of Mt. Dickason	B	462 ± 2.5	2
4255	Granite, ridge S of Gerlache Inlet	B	455.5 ± 2.5	2
4256	Granodiorite, S spur B of Black Ridge	B	475 ± 2.5	2
B042A	Pegmatite, Black Ridge	M	472 ± 9	4
FM1	Granite, Frontier Mountain	B	467 ± 9	4
		M	491 ± 10	4
Granite Harbor Intrusives, Rb-Sr, Minerals				
L1–10	Syenogranite. Mt. Abbott	B	443 ± 13	6
L16	Syenogranite, Mt. Abbott	B	450 ± 13	6
B5	Syenogranite, Mt. Abbott	B	472 ± 14	6
B2 (21–1–86)	Syenogranite, Mt. Abbott	B	464 ± 14	6
L21	Syenogranite, Mt. Abbott	B	457 ± 14	6
B2 (18–1–86)	Syenogranite, Mt. Abbott	B	491 ± 14	6
B5	Syenogranite, Mt. Abbott	B	506 ± 15	6
B6	Syenogranite, Mt. Abbott	B	583 ± 16	6
FP51	Monzogranite, Deep Freeze Range	B	450 ± 13	5
AB5c	Monzogranite Deep Freeze Range	B	457 ± 13	5
		M	461 ± 14	
G18	Monzogranite, Timber Peak	B	484 ± 14	5
B3	Monzogranite, Mr. Nansen	B	487 ± 14	5

(continued)

4.8.1 (continued)

Field number	Rock type/location	Material	Date (Ma)	Reference
L13	Migmatite, Terra Nova Bay	B	456 ± 13	5
		M	458 ± 20	
G16b	Quartz monzodiorite, Deep Freeze Range	B	463 ± 13	5
AB2	Granodiorite, Deep Freeze Range	B	462 ± 14	5
FP21	Monzogranite, Ogden Heights	B	488 ± 14	5
Granite Harbor Intrusives, Rb-Sr, Whole Rocks				
L3, L5, L11, L1–10, L16	Granitoids, Mt. Abbott	5 WR +2 Fs	508 ± 5 0.71020 ± 0.00010	6,5
FP51, AB5c, G16b, AB2	Granitoids, Deep Freeze Range	4 WR	504 ± 28 0.71042 ± 0.00065	5[a]
FP21, G18, B3	Monzogranites, TNB area	3 WR	508 ± 26 0.71049 ± 0.00049	5[a]
	Granitoids, Mt. Dickason	WR	548 ± 18 0.7094	3,2
	Granitoid, Illusion Hills	WR	519 ± 13	7

1. Kreuzer et al. (1981); 2. Kreuzer et al. (1987); 3. Adams (1986); 4. Adams (2006); 5. Armienti et al. (1990); 6. Borsi et al. (1988); 7. Faure and Gair (1970)

B = biotite, M = muscovite, WR = whole rock, PF = Priestley Fm., PS = Priestley Schist, TNF = Terra-Nova-Bay Fm

aCalculated by Faure and Mensing

4.8.2 Age Determinations of Granites and Metasediments of the Berg Group, Oates Land (Adams 1996, 2006)

Rock Type	Material	K-Ar, Ma	Rb-Sr, Ma	Reference
Berg Granite, Berg Mountains				
Granites	WR		489 ± 11 0.7139 ± 0.0076	1
Granite (GXV135)	B	507 ± 5		1
	M	489 ± 5		1
Berg Group, Berg Mountains, K-Ar				
Slate (GXV186A)	WR	491 ± 10		2
Slate (GXV113)	WR	487 ± 10		2
Slate (GXV116)	WR	483 ± 10		2
Slate (GXV120)	WR	489 ± 10		2
Archangel Granite, Outrider Nunatak				
Granite	WR		573 ± 10 0.7032 ± 0.00011	1
Granite (GXV120)	WR	506 ± 10		2
Granite (GXV110)	B	484 ± 5 483 ± 10		1 2
Berg Metasediment (East Ridge)				
Slate (330 m)	WR		476 ± ? 0.7174 ± 0.0008	1
Slate (GXV086A)	WR	492 ± 7 491 ± 10		1 2
Berg Metasediment (West Ridge)				
Slate (640 m)	WR		515 ± 7 0.7172 ± 0.00010	1
Slate (GXV120)	WR	490 ± 7 489 ± 10		1 2
Slate (GXV116)	WR	484 ± 7 483 ± 10		1 2
Slate (GXV)	WR	488 ± 7		1
Berg Metasediment (Ob Cairn)				
Marble	WR		548 ± 22 0.7151 ± 0.0002	1
Archangel Nunataks				
Slate (GXV113)	WR	487 ± 10		2

1. Adams (1996); 2. Adams (2006).

4.8.3 Age Determinations of Basement Rocks of the Lazarev Mountains, Daniels, Morozumi, and Lanterman Ranges, Wilson Terrane

Field number	Rock type/Location	Material	Date, Ma	Reference
4.8.3.1 Rennick Schist, Daniels Range, Rb-Sr				
	Metapelites	WR	531 ± 17 0.7243 ± 0.0009	3
	Metapsammites	WR	534 ± 24 0.7187 ± 0.0009	3
47 (Gair)	Quartz-biotite-schist	WR	548 ± 24 0.7187 and 0.7243 recalculated	4
DA13	Metasediment	B	472 ± 3 (K-Ar)	1
4.8.3.2 Rennick Schist, Lazarev Mountains, K-Ar				
GXV60	Granodiorite gneiss	B	465 ± 9	6
GXV68	Mica schist	B	472 ± 9	6
GVX25	Granodiorite gneiss, Williamson Head	B	465 ± 9	6
GXV107	Mica schist, Celestial Peak	B	484 ± 10	6
4.8.3.3 Wilson Gneisses, Daniels Range, Various Methods as Indicated				
9 Samples	Granodiorite gneiss	WR	490 ± 33 (Rb-Sr) 0.7205 ± 0.0016	3
RG73	Granodiorite	Mon	483.6 (U-Pb)	3
RG77	Granodiorite	Mon	487.5 (U-Pb)	3
GXV24	Amphibolitic gneiss, Williamson Head, Oates Coast	B	476 ± 10 (K-Ar)	6
GXV45	Granodiorite gneiss, Drake Head, Davis Bay	B	473 ± 9 (K-Ar)	6
4.8.3.4 Granite Harbor Intrusives, Schroeder Spur, Daniels Range, K-Ar (Minerals) Except as Indicated				
DA5, 11, 12 DA10 DA3 DA4	Granites Pegmatite Leucogranite Granodiorite	WR	495 ± 12 (Rb-Sr) 0.7125 ± 0.0011	2
DA3	Migmatite, Leucogranite	M B B	473.5 ± 2.5 473 ± 2.5 471.5 ± 2	1,2 1 2
DA5	Granite	M B	478.5 ± 1.5 471.5 ± 3	1
DA9	Tonalite dike	B H B	472 ± 2.5 >485 ± 5 479 ± 3	1 2 2
DA11	Granite	M	477 ± 2.5	2
DA12	Granite	M B	476.5 ± 2 473.5 ± 2	2 2
DA4	Granodiorite	B	471 ± 3	2
DA10	Pegmatite	M B	474.5 ± 2 472 ± 4	2 2
4.8.3.5 Granite Harbor Intrusives, Thompson Spur, Daniels Range, K-Ar				
DA1	Diorite	H B	>483 ± 6 470.5 ± 3	2 2
DA7	Granodiorite	B	473.5 ± 2	2
DA8	Migmatite	M B B	476 ± 2.5 472 ± 2 474.5 ± 2.5	1,2 2 1
4.8.3.6 Granite Harbor Intrusives, Renirie Rocks, K-Ar				
RE1	Granite	M B	479.5 ± 2.5 478 ± 2	2 2
TES115/1	Granite	M	479.5 ± 2.5	2
TES115/2	Granite	B	471 ± 2.5	2
RE2	Aplite dike	B	475 ± 3.5	2
4.8.3.7 Granite Harbor Intrusives, Matusevich Glacier, Harald Bay, and Davis Bay, K-Ar				
GXV26	Pegmatite, Harald Bay	M	471 ± 9	6
GXV31	Granodiorite, Exiles Nunatak	B	412 ± 8	6
GXV32	Granodiorite, Mt. Blowaway	B	463 ± 9	6
GXV36	Pegmatite, Thompson Peak	B	466 ± 9	6
GXV102	Granodiorite, Harald Bay	B	468+9	6
GXV50	Leucogranite, Davis Bay	B	464 ± 9	6
GXV826A	Granodiorite, Aviation Island	B	482 ± 10	6
4.8.3.8 Granite Harbor Intrusives, Pomerantz Tableland, K-Ar				
GXV12	Granodiorite, Armstrong Platform	B	477 ± 10	6
GXV14	Granodiorite, MacPherson Peak	B	476 ± 10	6
GXV15	Granodiorite, MacPherson Peak	B	462 ± 9	6
GXV17	Granodiorite, Mt. Harrison	B	463 ± 9	6
GXV19	Granodiorite, Robilliard Glacier	B	497 ± 10	6

(continued)

4.8.3 (continued)

Field number	Rock type/Location	Material	Date, Ma	Reference
4.8.3.9 Granite Harbor Intrusives, Kavraysky Hills, K-Ar				
GXV20	Granite, Yermak Point	B	507 ± 10	6
GXV140	Adamellite, Serrat Glacier	B	509 ± 10	6
GXV141	Granite, Kavraysky Hills	B	513 ± 10	6
GXV1445	Granodiorite gneiss, Kavraysky Hills	B	585 ± 10	6
GXV156	Granodiorite, Kavraysky Hills	B	502 ± 10	6
4.8.3.10 Granite Harbor Intrusives, Morozumi Range, Rb-Sr				
M01,2,3,5,6 M04	Adamellite leucogranite, Adamellite Massif	WR	515 ± 28 0.7136 ± 0.0022	8,2
M08, 4185, 4181,4182, 4184	Leucogranite adamellite, Jupiter Amphitheater	WR	470 ± 18 0.7114 ± 0.0013	2
M07, 4190 to 4194	Leucogranite, adamellite, Granite, Unconformity Valley	WR	475 ± 12 0.7097 ± 0.0007	8,2
4.8.3.11 Granite Harbor Intrusives, Morozumi Range, K-Ar				
M01	Adamellite Adamellite Massif	B	476 ± 2	2
M01	Granite Adamellite Massif	B	478.5 ± 2	1
M02	Adamellite Adamellite Massif	B	468 ± 2.5	2
M03	Granite Adamellite Massif	B	467 ± 2	1
M05	Adamellite Adamellite, Massif	B	478.5 ± 3	2
M06	Adamellite Adamellite Massif	B B	478 ± 2 478.5 ± 2.5	2 2
M06	Granite Adamellite Massif	M B	479 ± 2 479 ± 2	1 1
M08	Leucogranite Jupiter Amphitheater	M B	482.5 ± 3 476 ± 2.5	2 2
4181	Adamellite Jupiter Amphitheater	M	486 ± 3	2
4182	Adamellite Jupiter Amphitheater	M B	483 ± 3 469 ± 3	2 2
4183	Adamellite Jupiter Amphitheater	B	380.5 ± 2.5	2
4184	Adamellite Jupiter Amphitheater	B	463.5 ± 2.5	2
4185	Aeucogranite Jupiter Amphitheater	M	483.5 ± 3	2
4190	Adamellite Unconformity Valley	M	488 ± 3.5	2
4191	Granite Unconformity Valley	M B	482 ± 3.5 482 ± 3	2 2
4192	Leucogranite Unconformity Valley	M B	486 ± 4 478.5 ± 3.5	2 2
4193	Adamellite, greisenized Unconformity Valley	M	483 ± 3	2
4194	Granite, greisenized Unconformity Valley	M	489 ± 3	2
4.8.3.12 Wilson Gneisses, Lanterman Range, K-Ar				
LA1	Granodiorite 3 km SSW Carnes Crag	M	484 ± 3	2
LA2	Amphibole-rich schlieren	H	488 ± 7	2
LA2	Diorite	B	479.5 ± 2.5	1
LA3	Diorite	B	490.5 ± 2.5	1
LA3	Diorite South Hoshka Glacier	H B	502 ± 4 489 ± 2.5	2 2
2133	Orr Granite Amphibolite facies 5 km SSE Mt. Moody	B	487.5 ± 2.5	2
W1	Granodiorite gneiss	B,M	483 ± 3	5
W2	Granodiorite gneiss	B	466 ± 3	5
W3	Amphibolitic gneiss	H	453 ± 3	5
W4a	Granodiorite gneiss	M	476 ± 3	5
W7	Granodiorite gneiss	B	474 ± 3	5
W11	Amphibolitic gneiss	H	355 ± 3(1)	5
W13	Amphibolitic gneiss	H	451 ± 3	5
W15	Hornblende gneiss	B	479 ± 3	5
RG250	Diorite pegmatite	H	498 ± 6	7
W12	Granodiorite gneiss	M	467 ± 3	5

(1) Retrogressive metamorphism close to Lanterman Fault (Adams et al. 1982)

(continued)

4.8.3 (continued)

Field number	Rock type/Location	Material	Date, Ma	Reference
4.8.3.13 Wilson Gneisses, Lanterman Range, Rb-Sr				
W1, W2, W4a, W7, W15	Granodiorite gneisses	WR	550 ± 20 0.7155 ± 0.0003	7
4.9.3.14 Wilson Gneisses, Salamander Range, Rb-Sr				
RG230, RG231, RG232, RG233, RG239, RG241, RG242, RG245, RG248, RG252, RG269, RG270	Quartz-biotite schist ±Feldspar, ±Ca-silicate	WR	555 ± 36 0.7158 ± 0.0004	7

1. Kreuzer et al. (1981); 2. Kreuzer et al. (1987); 3. Adams (1986); 4. Faure and Gair (1970); 5. Adams et al. (1982); 6. Adams (2006); 7. Adams and Hőhndorf (1991); 8. Vetter et al. (1983)

B = biotite, M = muscovite, H = hornblende, Mon = monazite, WR = whole rock

4.8.4 Age Determinations of the Basement Rocks in the Bowers Terrane

Field number	Rock type/location	Material	Date (Ma)	Reference
4.8.4.1 Sledgers Group, K-Ar				
S13	Slate Molar Massif	WR	467 ± 3	5
S19A	Slate Sledgers Glacier	WR	401 ± 3	5
S29	Slate Carryer Glacier	WR	414 ± 3	5
S30	Slate Carryer Glacier	WR	413 ± 3	5
S32	Slate Carryer Glacier	WR	416 ± 3	5
2M83	Slate Molar Massif	WR	450 ± 3	5
2M85	Slate, Molar Massif	WR	453 ± 3	5
S2	Siltstone Molar Massif	WR	441 ± 3	5
GXV50	Andesite Glasgow Volcanics	WR	360 ± 6	6
GXV7	Metavolcanic rock Glasgow Volcanics	WR	328 ± 7	6
MS3509	Slate, Mt. Bruce	WR	417 ± 8	6
MS3514	Slate, Mt. Bruce	WR	382 ± 8	6
MS3519	Slate, Mt. Sheila	WR	445 ± 9	6
MS3523	Slate Winkler Glacier	WR	424 ± 8	6
MS3536	Slate Arruiz Glacier	WR	365 ± 7	6
MS3538	Slate Arruiz Glacier	WR	332 ± 7	6
MA15	Slate, Index Point	WR	340 ± 7	6
4.8.4.2 Mariner Group				
S44	Slate Sledgers Glacier	WR	413 ± 3	5
S49	Slate Sledgers Glacier	WR	402 ± 3	5
S52	Slate Sledgers Glacier	WR	417 ± 3	5
S55	Slate Sledgers Glacier	WR	384 ± 3	5
S57	Slate Sledgers Glacier	WR	380 ± 3	5
S59	Slate Sledgers Glacier	WR	400 ± 3	5
H8	Siltstone Leap Year Glacier	WR	413 ± 3	5
S43	Siltstone Helix Pass	WR	394 ± 3	5

(continued)

4.8.4 (continued)

Field number	Rock type/location	Material	Date (Ma)	Reference
4.8.4.3 Leap Year Group, K-Ar				
P5	Red siltstone Evans Névé	WR	421 ± 3	5
MA37	Slate, Mt. Hayton	WR	434 ± 9	6
MA38	Slate, Pyramid Peak	WR	448 ± 9	6

5. Adams et al. (1982); 6. Adams (2006)

4.8.5 Age Determinations of the Robertson Bay Group and Admiralty Intrusives, Robertson Bay Terrane

Field number	Rock type/location	Material	Date (Ma)	Reference
4.8.5.1 Robertson Bay Group, Millen Range-Leap Year Glacier, K-Ar				
MI2	Slate, Mt. McDonald	WR	486 ± 10	5
MI21	Slate, Mt. Hancox	WR	505 ± 10	5
MI22	Slate, Mt. Hancox	WR	458 ± 9	5
MI40	Slate, Jato Nunatak	WR	490 ± 10	5
MI48	Slate, Gless Peak	WR	495 ± 10	5
MI53	Slate, Mueller Glacier	WR	469 ± 9	5
MI60	Mica schist, Mueller, Glacier	WR	444 ± 9	5
4.8.5.2 Robertson Bay Group, K-Ar				
MI12	Slate, Mt. Hancox	WR	493 ± 10	5
MI16	Slate, Mt. Hancox	WR	499 ± 10	5
MI33	Slate, Mt. Hussey	WR	419 ± 8	5
MI54	Slate, Mueller Glacier	WR	439 ± 9	5
MI63	Slate, Lensen Glacier	WR	515 ± 10	5
MI69	Slate, Lensen Glacier	WR	525 ± 11	5
MGL/RB2	Argillite, Mt. McCarthy	WR	476 ± 3	4
P13	Siltstone, Pyramid Peak	WR	499	4
RB1	Slate, Leap Year Glacier	WR	502 ± 4	4
RB4	Slate, Leap Year Glacier	WR	447 ± 3	4
RB5	Slate, Leap Year Glacier	WR	495 ± 4	4
RB6	Slate, Leap Year Glacier	WR	468 ± 3	4
RB7	Phyllite, Champness Glacier	WR	448 ± 3	4
RB8	Slate, Champness Glacier	WR	490 ± 3	4
RB9	Slate, Champness Glacier	WR	491 ± 4	4
RB10	Slate, Leap Year Glacier	WR	476 ± 3	4
RB13	Slate, Graveson Glacier	WR	474 ± 3	4
RB14	Slate, Leap Year Glacier	WR	482 ± 3	4
RB15	Slate, Leap Year Glacier	WR	491 ± 4	4
J9–17	Slate, Mirabito Peak	WR	492 ± 4	4
J10–7	Slate, Mirabito Peak	WR	491 ± 4	4
J13–8	Slate, Mirabito Peak	WR	361 ± 3	4
4.8.5.3 Admiralty Intrusives, K-Ar				
GXV8	Granite, Znamenskiy Island	B	368 ± 7	5
GXV1	Granite, Ob Bay	B	367 ± 7	5
GXV2	Granite, Sputnik Island	B	366 ± 7	5
MA23	Granodiorite, Emerging Island	B	328 ± 7	5
MA25	Granodiorite, Mt. Montreuil	B	385 ± 8	5
MA28	Granodiorite, Lawrence Peaks	B	375 ± 8	5

(continued)

4.8.5 (continued)

Field number	Rock type/location	Material	Date (Ma)	Reference
4.8.5.4 Admiralty Intrusives, Yule Batholith, Yule Bay, Pennell Coast, K-Ar				
TES102	Novosad Island	B	361 ± 1.5	1,2
4010	South of O'SHara Glacier	B	365.5 ± 2	2
4012	Tapsell Foreland	B	364.5 ± 2.5	2
4014	Tapsell Foreland	B	361.5 ± 2	2
4015	Birthday Ridge	B	360 ± 2	2
4016	Fine-grained, biotite-hornblende-plagioclase xenolith	B	362 ± 2	2
4018	Fine-grained, biotite-hornblende-plagioclase xenolith, Birthday Ridge	B	363.5 ± 2	2
TS2	Cape Moore	B	363 ± 2	1
TS103	Hughes Island	B	367 ± 1.5	1
TS104	Sentry Rocks	B	366.5 ± 1.5	1
TS106	Missen Head	B	365.6 ± 1.5	1
TH1	Thala Island	B	366 ± 2	1
	Yule Bay	Rb-Sr WR	393 ± 20	3
4.8.5.5 Admiralty Intrusives, Everett Range, Lillie and Ebbe Glaciers, K-Ar				
EV1	Granite	B	362 ± 2	1
EV6	Granite, "810-Ridge"	B H H	358 ± 3 363.5 ± 4 366 ± 2.5	1 1 2
EV8	Granite, 6 km N of Pt. 1670	H B	360 ± 3 360.5 ± 1.5	1,2 1
EV11	Granite, 2 km NW of Pt. 1775	H B H	359 ± 4 359 ± 4 363 ± 3.5	2 1 1
TS1	Granite, Football Saddle	B B	357.5 ± 2 355.5 ± 2	1 2
TU1	Granodiorite, lower Tucker Glacier	B	363 ± 2	1
	Lillie Glacier	Rb-Sr WR	360 0.710–0.713	3
4.8.5.6 Admiralty Intrusives, Gregory Bluffs (70°45′S, 166°04′E), K-Ar				
4120	Athurson Bluff	B	363 ± 2	2
4123	4 km SSW of Athurson Bluff	B	365.5 ± 1.5	2
4127	4 km E of Mt. Harwood	M B	364 ± 1.5 359 ± 2	2 2
4128	4 km E of Mt. Harwood	M	370.5 ± 2	2
4.8.5.7 Admiralty Intrusives, Gregory Bluffs (70°45′S, 166°04′E), Rb-Sr				
4120	Athurson Bluff	WR-B	358 ± 7	2
4123	4 km SSW of Athurson Bluff	WR-B	358 ± 7	2
4127	4 km E of Mt. Harwood	WR-B WR-M	356 ± 7 357 ± 7	2 2
4128	4 km E of Mt. Harwood	WR-B WR-M	351 ± 7 359 ± 7	2 2
		WR	388 ± 13	3
4.8.5.8 Admiralty Intrusives, Granitoids, Lillie and Champness Glaciers Area (71°25′S, 164°22′E), K-Ar				
TES110	Platypus Ridge, Lower Lillie Glacier	H	362 ± 4 363 ± 2	2 2
4050	Griffith Ridge Champness Glacier area	B	354 ± 2	2
4054	Griffith Ridge Champness Glacier area	B	354.5 ± 2	2
4055	Biotite schlieren, Griffith Ridge Champness Glacier area	B	359.5 ± 2	2
4056	Radspinner Ridge Champness Glacier area	H B	356 ± 4 359.5 ± 2	2 2
4059	Radspinner Ridge Champness Glacier area	H B	348.5 ± 3 359 ± 2.5	2 2
4060	Copperstain Champness Glacier area	B	358 ± 2	2
4.8.5.9 Admiralty Intrusives, area of Tucker Inlet (72°37′S, 169°18′E)				
TU14	Tucker Inlet area	B	356 ± 2	2
4.8.5.10 Admiralty Intrusives, Upper Hand Glacier and Malta Plateau, K-Ar				
3228/4411	3 km NNW of Mt. Alberts	B	350.5 ± 2	2
3273/4416	South side of Upper Hand Glacier	B	348 ± 2.5	2
3279/4417	3 km NNE of Mt. Hussey	H B	363 ± 3.5 368 ± 2.5	2 2
4.8.5.11 Admiralty Intrusives, Cooper Spur (70°38′S, 165°03′E)				
4160		WR-B	359 ± 7	2
4161		WR-B WR-M	362 ± 7 359 ± 7	2 2
		WR	525 ± 26	3

1. Kreuzer et al. (1981); 2. Kreuzer et al. (1987); 3. Adams (1986); 4. Adams (1982); 5. Adams (2006)

B = biotite, M = muscovite, H = hornblende, WR = whole-rock.

Fig. 4.31 The Wilson Gneiss in Thompson Spur of the Daniels Range in Fig. 4.11 consists of black biotite schist that was intruded by a network of granitic veins. The granitic magma presumably originated from a center of migmatization at depth (Adapted from Tessensohn et al. 1981, Fig. 11, p. 62)

Fig. 4.32 This folded sequence of turbidite beds of the Robertson Bay Group is located on the north side of Colbeck Bay (71°38′S, 170°05′E) in the southwest corner of Robertson Bay (Adapted from Wright 1981, Fig. 2, p. 131)

Fig. 4.33 Sandstones of the Beacon Supergroup lie unconformably on basement rocks composed of black phyllites in the Unconformity Valley of the Morozumi Range. The Beacon rocks (Permian to Triassic) were intruded by a thick sill of the Ferrar Dolerite (Middle Jurassic) which now caps the mountain (Adapted from Tessensohn et al. 1981, Fig. 7, p. 52)

4.8.6 Additional Photographs of the Geology of Northern Victoria Land

Figure 4.31 Wilson Gneiss, Daniels Range
Figure 4.32 Turbidites, Robertson Bay Group
Figure 4.33 Beacon Sandstone, Morozumi Range

References

Adams CJ (1986) Age and ancestry of metamorphic rocks of the Daniels Range, USARP Mountains, Antarctica. In: Stump E (ed) Geological investigations in North Victoria Land. Antarctic Research Series, vol. 46. American Geophysical Union, Washington, DC, pp 25–38

Adams CJ (1996) Geochronological evolution of the western margin of northern Victoria Land: Rb-Sr and K-Ar dating of the Berg Group/Archangel Granites. In: Roland N (ed) German Antarctic North Victoria Land Expedition 1990/91 (GANOVEX V), 2. Geologisches Jahrbuch, B 89:179–194

Adams CJ (1997) Initial strontium isotopic signature of the late Precambrian-early Paleozoic metasediments from northern Victoria Land, East Antarctica. In: Ricci CA (ed) The Antarctic Region: Geological Eevolution and processes. Terra Antartica Publication, Siena, Italy, pp 227–231

Adams CJ (2006) Style of uplift of Paleozoic terranes in northern Victoria Land, Antarctica: Evidence from K-Ar age patterns. In: Fütterer DK, Damaske D, Kleinschmidt G, Miller H, Tessensohn F (eds) Antarctica: Contributions to Global Earth Sciences. Springer, Berlin, pp 205–214

Adams CJ, Hőhndorf A (1991) Age of the metamorphic basement of the Salamander and Lanterman ranges, northern Victoria Land, Antarctica. In: Thomson MR, Crame JA, Thomson JW (eds) Geological evolution of Antarctica. Cambridge University Press, Cambridge, pp 149–153

Adams CJ, Kreuzer H (1984) Potassium-argon age studies of slates and phyllites from the Bowers and Robertson Bay terranes, north Victoria Land, Antarctica. Geologisches Jahrbuch, Reihe B 60:265–288

Adams CJD, Gabites JE, Wodzicki A, Laird MG, Bradshaw JD (1982) Potassium-argon geochronometry of the Precambrian-Cambrian Wilson and Robertson Bay groups and Bowers Supergroup, northern Victoria Land, Antarctica. In: Craddock

C (ed) Antarctic geoscience. University of Wisconsin Press, Madison, WI, pp 543–548

Adams CJ, Whitla PF, Findlay RH, Field BF (1986) Age of the Black Prince Volcanics in the central Admiralty Mountains and possibly related hypabyssal rocks in the Millen Range, northern Victoria Land, Antarctica. In: Stump E (ed) Geological investigations in Northern Victoria Land. Antarctic Research Series, vol. 46. American Geophysical Union, Washington, DC, pp 203–210

Adamson RG (1971) Granitic rocks of the Campbell-Priestley divide, northern Victoria Land, Antarctica. New Zealand J Geol Geophys 14:486–503

Andrews PB, Laird MG (1976) Sedimentology of a Late Cambrian regressive sequence (Bowers Group), northern Victoria Land, Antarctica. Sed Geol 16:21–44

Armienti P, Ghezzo C, Innocienti F, Manetti P, Rocchi S, Tonarini S (1990) Granite Harbor Intrusives from north Victoria Land between David and Campbell glaciers: New geochronological data. In: Miller H (ed) Workshop on Antarctic Geochronology, Munich 1989. Zentralblatt Geol Paläont, Teil I, H. 1/2, 63–74

Babcock RS, Plummer CC, Sheraton JW, Adams CJ (1986) Geology of the Daniels Range north Victoria Land, Antarctica. In: Stump E (ed) Geological investigations in Northern Victoria Land. Antarctic Research Series, vol. 46. American Geophysical Union, Washington, DC, pp 1–24

Berger GW, York D (1981) Geothermometry from $^{40}Ar/^{39}Ar$ dating experiments. Geochim Cosmochim Acta 45:795–811

Borg SG, Stump E, Holloway JR (1986) Granitoids of northern Victoria Land, Antarctica: A reconnaissance study of field relations, petrography, and geochemistry. In: Stump E (ed) Geological investigations in Northern Victoria Land. Antarctic Research Series, vol. 46. American Geophysical Union, Washington, DC, pp 115–188

Borg SG, Stump E, Chappell BW, McCulloch MT, Wyborn D, Armstrong RL, Holloway JW (1987) Granitoids of northern Victoria Land, Antarctica: Implications of chemical and isotopic variations to regional crustal structure and tectonics. Amer J Sci 287:127–169

Borsi L, Ferrara G, Tonarini S (1988) Geochronological data on Granite Harbor Intrusives from Terra Nova Bay and Priestley Glacier, Victoria Land, Antarctica. Proceedings of Geosciences in Victoria Land, Antarctica, Sept. 1987, Siena, Italy

Bradshaw JD (1987) Terrane boundaries and terrane displacement in northern Victoria Land, Antarctica; Some problems and constraints. In: Leitch EC, Schreibner E (eds) Terrane accretion and orogenic belts. Geophysical Monograph, vol. 42. American Geophysical Union, Washington, DC, pp 199–205

Bradshaw JD (1989) Terrane boundaries in north Victoria Land. Memorie Della Societa Geologica Italiana 33:9–15

Bradshaw JD, Laird MG, Wodzicki A (1982) Structural style and tectonic history in northern Victoria Land. In: Craddock C (ed) Antarctic geoscience. University Wisconsin Press, Madison, WI, pp 809–816

Bradshaw JD, Weaver SD, Laird MG (1985) Suspect terranes in north Victoria Land, Antarctica. In: Howell DC, Jones DL, Cox A, Nur A (eds) Circum-Pacific Terrane Conference, Proceedings, 36–39. Stanford University Press, Stanford, CA

Buggisch W, Repetski JE (1987) Uppermost Cambrian (?) and Tremadocian conodonts from Handler Ridge, Robertson Bay Terrane, north Victoria Land, Antarctica. Geologisches Jahrbuch 66:145–185

Burrett CF, Findlay RH (1984) Cambrian and Ordovician conodonts from the Robertson Bay Group, Antarctica, and their tectonic significance. Nature 307:723–726

Capponi G, Crispini L, Meccheri M (1999) Structural history and tectonic evolution of the boundary between the Wilson and Bowers terranes, Lanterman Range, northern Victoria Land, Antarctica. Tectonophysics 312:249–266

Capponi G, Castorina F, DiPisa A, Meccheri M, Petrini R, Villa IM (2002) The meta-igneous rocks of the Barber Glacier area (northern Victoria Land): A clue to the enigmatic Borchgrevink orogeny? In: Gamble J Skinner D, Henrys S (eds) Antarctica at the close of the millennium. Royal Soc New Zealand Bull 35: 99–104

Chappell BW, White AJR (1974) Two contrasting granite types. Pacific Geology 8:173–174. The Royal Society of New Zealand, Wellington

Cooper RA, Jago JB, MacKinnon DI, Simes JE, Braddock PE (1976) Cambrian fossils from the Bowers Group, northern Victoria Land, Antarctica. New Zealand J Geol Geophys 19(2):283–288

Cooper RA, Jago JB, MacKinnon DI, Shergold JH, Vidal G (1982) Late Precambrian and Cambrian fossils from northern Victoria Land and their stratigraphic implications. In: Craddock C (ed) Antarctic Geoscience. University Wisconsin Press, Madison, WI, pp 629–633

Cooper RA, Jago JB, Rowell AJ, Braddock P (1983) Age and correlation of the Cambrian-Ordovician Bowers Supergroup, northern Victoria Land. In: Oliver RL, James PR, Jago (eds) Antarctic earth science. Australian Academy of Science, Canberra, ACT, pp 128–131

Crawford AJ, Green DH, Findlay RH (1984) A preliminary petrographical- geochemical survey of the dyke rocks from northern Victoria Land, Antarctica. Geologisches Jahrbuch, Reihe B 60:153–165

Crowder DF (1968) Geology of a part of northern Victoria Land, Antarctica. Prof. Paper 600-D, 95–107. US Geological Survey, Reston, VI

Dallmeyer RD, Wright TO (1992) Diachronous cleavage development in the Robertson Bay Terrane, northern Victoria Land, Antarctica: Tectonic implications. Tectonics 11:437–448

Denton GH, Bockheim JG, Wilson SC, Schlüchter C (1986) Late Cenozoic history of Rennick Glacier and Talos Dome, northern Victoria Land, Antarctica. In: Stump E (ed) Geological investigations in Northern Victoria Land. Antarctic Research Series, vol. 46.. American Geophysical Union, Washington, DC, 339–375

DiVincenzo G, Palmeri R, Talarico F, Andriesen PAM, Ricci C (1997) Petrology and geochronology of eclogites from the Lanterman Range, Antarctica J Petrol 38:1391–1417

Dow JAS, Neall VE (1974) Geology of the lower Rennick Glacier, northern Victoria Land, Antarctica. New Zealand J Geol Geophys 17:659–714

Engel S (1984) Petrogenesis of contact schists in the Morozumi Range, north Victoria Land. Geologisches Jahrbuch, B 60:167–186

Engel S (1987) Contact metamorphism by the layered gabbro at Spatulate Ridge and Apostrophe Island, north Victoria Land, Antarctica. Geologisches Jahrbuch 66:275–301

Faure G, Gair HS (1970) Age determinations of rocks from northern Victoria Land, Antartica. New Zealand J. Geol. Geophys. 13(4):1024–1026

Faure G (1986) Provenance of feldspar in till from the Morozumi Range, northern Victoria Land. In: Stump E (ed) Geological investigations in Northern Victoria Land. Antarctic Research Series, vol. 46. American Geophysical Union, Washington, DC, pp 377–381

Faure G (2001) Origin of igneous rocks; The isotopic evidence. Springer, Heidelberg, Germany

Findlay RH (1986) Structural geology of the Robertson Bay and Millen terranes, northern Victoria Land, Antarctica. In: Stump E (ed) Geological investigations in Northern Victoria Land. Antarctic Research Series, vol. 46. American Geophysical Union, Washington, DC, pp 91–114

Findlay RH (1987) A review of the problems important for the interpretation of the Cambro-Ordovician paleogeography of northern Victoria Land (Antarctica), Tasmania, and New Zealand. In: McKenzie GD (ed) Gondwana Six: structure, tectonics, and geophysics. Geophysical Monograph, vol. 40. American Geophysical Union, Washington, DC, pp 49–66

Findlay RH, Field BD (1982) Black Prince Volcanics, Admiralty Mountains: Reconnaissance observations. New Zealand Antarctic Rec 4(2):11–14

Findlay RH, Field BD (1983) Tectonic significance of deformations affecting the Robertson Bay Group and associated rocks, northern Victoria Land, Antarctica. In: Oliver RL, James PR, Jago JB (eds) Antarctic earth science. Australian Academy of Science, Canberra, ACT, pp 107–112

Findlay RH, Jordan H (1984) Volcanics at Mount Prince and Lawrence Peaks, northern Victoria Land, Antarctica. Geologisches Jahrbuch, Reihe B 60:143–151

Flőttmann T, Kleinschmidt G (1991a) Opposite thrust systems in northern Victoria Land, Antarctica: Imprints of Gondwana's Paleozoic accretion. Geology 19:45–47

Flőttmann T, Kleinschmidt G (1991b) Kinematics of major structures of north Victoria and Oates lands, Antarctica. Mem Soc Geol Ital 46:283–289

Fűtterer DK, Damaske D, Kleinschmidt G, Miller H, Tessensohn F (2006) Antarctica; contributions to global earth science. Springer, Berlin

Gair HS (1964) Geology of the upper Rennick Glacier to the coast, northern Victoria Land, Antarctica. In: Adie RJ (ed) Antarctic geology. North-Holland, Amsterdam, The Netherlands, pp 188–198

Gair HS (1967) The geology from the upper Rennick Glacier to the coast, northern Victoria Land, Antarctica. New Zealand J Geol Geophys 10:309–344

Gair HS, Sturm A, Carryer SJ, Grindley GW (1969) The geology of northern Victoria Land. In: Craddock C (ed) Geologic maps of Antarctica. Antarctic Map Folio Series, Folio 12, Sheet 13. American Geographical Society, New York

GANOVEX-ITALIANTARTIDE (1991) Preliminary geological-structural map of Wilson, Bowers, and Robertson Bay terranes in the area between Aviator and Tucker glaciers (northern Victoria Land, Antarctica). Memorie della Societá Geologica Italiana 46:267–272

GANOVEX Team (1987) Geological map of northern Victoria Land, Antarctica, 1:500,000; Explanatory notes. Geol Jahrbuch B 66:7–79

Gibson GM (1987) Metamorphism and deformation in the Bowers Supergroup; Implications for terrane accretion in northern Victoria Land, Antarctica. In: Leitch EC, Schreibner E (eds) Terrane accretion and orogenic belts. Geophysical Monograph, vol. 41. American Geophysical Union, Washington, DC, pp 207–219

Gibson GM, Wright TO (1985) Importance of thrust faulting in the tectonic development of northern Victoria Land, Antarctica. Nature 315:480–483

Gibson GM, Tessensohn F, Crawford AJ (1984) Bowers Supergroup rocks west of the Mariner Glacier and possible greenschist facies equivalents. Geologisches Jahrbuch 60:289–318

Grew ES, Kleinschmidt G, Schubert W (1984) Contrasting metamorphic belts in north Victoria Land, Antarctica. Geologisches Jahrbuch B 60:253–263

Grindley GW, Davey FJ (1982) The reconstruction of New Zealand, Australia, and Antarctica. In: Craddock C (ed) Antarctic geoscience. University of Wisconsin Press, Madison, WI, pp 15–29

Grindley GW, Oliver PJ (1983) Post-Ross orogenic cratonisation of northern Victoria Land. In: Oliver RL, James PR, Jago JB (eds) Antarctic earth science. Australian Academy of Science, Canberra, ACT, pp 133–139

Grindley GW, Warren G (1964) Stratigraphic nomenclature and correlation in the western Ross-Sea region. In: Adie RJ (ed) Antarctic geology. North- Holland, Amsterdam, The Netherlands, pp 314–333

Hamilton WB (1967) Tectonics of Antarctica. Tectonophysics 4:555–568

Hamilton WB (1972) The Hallett volcanic province. US Geol. Surv. Prof. Paper, 456-C:1–62

Harrington HJ (1958) Nomenclature of rock units in the Ross Sea region, Antarctica. Nature 182:290

Harrington HJ (1967) Topography and geology of the Cape Hallett district, Victoria Land, Antarctica Bull 80, New Zealand Geological Survey, Lower Hutt, New Zealand

Harrington HJ, Wood BL, McKellar IC, Lensen GJ (1964) The geology of Cape Hallet-Tucker Glacier district. In: Adie RJ (ed) Antarctic geology. North Holland, Amsterdam, The Netherlands, pp 220–228

Iltchenko LN (1972) Late Precambrian Achritarca of Antarctica. In: Adie RJ (ed) Antarctic geology and geophysics. Universitetsforlaget, Oslo, Norway, pp 599–602

IUGS (2002) International stratigraphic chart. International Commission on Stratigraphy. UNESCO, United Nations, New York

Jackson JA (ed) (1997) Glossary of geology, 4th edn. American Geological Institute, Alexandria, VA

Jordan H, Findlay RH, Mortimer G, Schmidt-Thomé M, Crawford AJ, Műller P (1984) Geology of the northern Bowers Mountains, north Victoria Land, Antarctica. Geologisches Jahrbuch B 60:57–81

Kleinschmidt G (1981) Regional metamorphism in the Robertson Bay Group area and in the southern Daniels Range, north Victoria Land, Antarctica: A preliminary comparison. Geologisches Jahrbuch B 41:201–208

Kleinschmidt G (1983) Trends in regional metamorphism and deformation in northern Victoria Land, Antarctica. In: Oliver RL, James PR, Jago JB (eds) Antarctic earth science. Australian Academy Science, Canberra, ACT, pp 119–122

Kleinschmidt G, Skinner DNB (1981) Deformation styles in the basement rocks of north Victoria Land, Antarctica. Geologisches Jahrbuch B 41:155–199

Kleinschmidt G, Tessensohn F (1987) Early Paleozoic westward subduction at the Pacific margin of Antarctica. In: McKenzie GD (ed) Gondwana Six: structure, tectonics, and geophysics. Geophysical Monograph, vol. 40. American Geophysical Union, Washington, DC, pp 89–105

Kleinschmidt G, Roland NW, Schubert W (1984) The metamorphic basement complex of the Mountaineer Range, north Victoria Land, Antarctica. Geologisches Jahrbuch B 60:213–251

Klimov LV, Soloviev DS (1958) Some features of the geological structure of Wilkes Land, King George V Land, and Oates Coast (East Antarctica). Doklady of the Academy of Sciences of the USSR, Geological Sciences Section 123(1–6):889–892

Kothe J, Tessensohn F, Thonhauser W, Wendebourg R (1981) The expedition and its logistics. Geol. Jahrbuch, Reihe B, Heft 41:3–30

Kreuzer H, Hőhndorf A, Lenz H, Vetter U, Tessensohn F, Műller P, Jordan H, Harre W, Besang C (1981) K/Ar and Rb/Sr dating of igneous rocks from northern Victoria Land, Antarctica. Geologisches Jahrbuch B 41:267–273

Kreuzer H, Delisle G, Fromm K, Hőhndorf A, Lenz H, Műller P, Vetter U (1987) Radiometric and paleomagnetic results from northern Victoria Land, Antarctica. In: McKenzie GD (ed) Gondwana Six: structure, tectonics, and geophysics. Geophysical Monograph, vol. 40. American Geophysical Union, Washington, DC, pp 31–47

Laird MG, Bradshaw JD (1983) New data on the lower Paleozoic Bowers Supergroup, northern Victoria Land. In: Oliver RL, James PR, Jago JB (eds) Antarctic earth science. Australian Academy of Science, Canberra, A.C.T, pp 123–126

Laird MG, Andrews PB, Kyle PR, Jennings P (1972) Late Cambrian fossils and the age of the Ross Orogeny, Antarctica. Nature 238:5358

Laird MG, Andrews PB, Kyle PR (1974) Geology of the northern Evans Névé, Victoria Land, Antarctica. New Zealand J Geol Geophys 17(3):587–601

Laird MG, Bradshaw JD, Wodzicki A (1982) Stratigraphy of the Upper Precambrian and Lower Paleozoic Bowers Subgroup, northern Victoria Land, Antarctica. In: Craddock C (ed) Antarctic geoscience. University of Wisconsin Press, Madison, WI, pp 535–542

Läufer AL, Kleinschmidt G, Rossetti F (2006) Late-Ross structures in the Wilson Terrane in the Rennick Glacier area (northern Victoria Land, Antarctica). In: Fűtterer DK, Damaske D, Kleinschmidt G, Miller H, Tessensohn F (eds) Antarctica: Contributions to global earth sciences. Springer, Berlin/Heidelberg, Germany, pp 195–204

LeCouteur PC, Leitch EC (1964) Preliminary report on the geology of an area south-west of the Upper Tucker Glacier, northern Victoria Land, Antarctica. In: Adie RJ (ed) Antarctic geology. North-Holland, Amsterdam, The Netherlands, pp 229–236

Mayewski PA (1975) Glacial geologic investigation of the upper Rennick Glacier region, northern Victoria Land. Antarctic J US 10(4):164–166

Mayewski PA (1976) Past levels and present state of northern Victoria Land glaciers. Antarctic J US 11(4):277–279

Mayewski PA, Attig JW Jr (1978) A recent decline in available moisture in northern Victoria Land, Antarctica. J Glaciol 20(84):593–594

Mayewski PA, Attig JW Jr, Drewry DJ (1979) Pattern of ice-surface lowering for the Rennick Glacier, northern Victoria Land, Antarctica. J Glaciol 22(86):53–65

McCulloch MT, Chappell BW (1982) Nd isotopic characteristics of S- and I-type granites. Earth Planet Sci Lett 58:51–64

McLeod IR (1964) Geological observations in Oates Land. In: Adie RJ (ed) Antarctic geology. North-Holland, Amsterdam, The Netherlands, pp 482–486

McLeod IR, Gregory CM (1967) Geological investigations along the Antarctic coast between longitudes 108°E and 166°E. Geology and Geophysics Report 78. Australian Bureau of Mineral Resources, Canberra, ACT

Nathan S (1971a) Geology and petrology of the Campbell-Aviator divide, northern Victoria Land, Antarctica. Part 2: Paleozoic and Precambrian rocks. New Zealand J Geol Geophys 14(3):564–596

Nathan S (1971b) K-Ar dates from the area between the Priestley and Mariner glaciers, northern Victoria Land, Antarctica. New Zealand J Geol Geophys 14:504–511

Nathan S, Skinner DNB (1972) Recent advances in the pre-Permian geology of northern Victoria Land. In: Adie RJ (ed) Antarctic geology and geophysics. Universitetsforlaget, Oslo, Norway, pp 333–337

National Geographic Society (1990) Atlas of the world, 6th edn. National Geographic Society, Washington, DC

Olesch M, Roland NW, Fenn G, Krauss U (1996) Petrogenesis of granitoid rocks of Oates Land, Antarctica. Geologisches Jahrbuch B 89:195–245

Plummer CC, Babcock RS, Sheraton JW, Adams CJ, Oliver RL (1983) Geology of the Daniels Range, northern Victoria Land, Antarctica: A preliminary report. In: Oliver RL, James PR, Jago JB (eds) Antarctic earth science. Australian Academy of Science, Canberra, A.C.T., pp 113–117

Prior GT (1899) Petrographical notes on the rock specimens collected in Antarctic regions during the voyage of HMS Erebus and Terror under Sir James Clark Ross in 1839–43. Mineral Mag 12:69–91

Prior GT (1902) Report on rock specimens collected by the Southern Cross Antarctic Expedition. Report on Southern Cross Expedition. British Museum, London, pp 321–332

Rastall RH, Priestley RE (1921) The shale-graywacke formation of Robertson Bay. British Antarct Terra Nova Exped, 1919. Nat Hist Report, Geol 1(4b):121–129. British Museum of Nat Hist, London

Ravich MG, Krylov AJ (1964) Absolute ages of rocks from East Antarctica. In: Adie RJ (ed) Antarctic geology. North-Holland, Amsterdam, The Netherlands, pp 579–589

Ravich MG, Klimov LV, Soloviev DS (1965) The Precambrian of East Antarctica. Trans. Scient. Research Institute of the Geology of the Arctic of the State Geological Committee, Moscow

Ravich MG, Klimov LV, Soloviev DS (eds) (1968) The Precambrian of East Antarctica. Israel Program for Scientific Translation, Jerusalem, Israel

Reader'Ss Digest (1985) Antarctica; Great stories from the frozen continent. Reader'Ss Digest Services, Surry Hills, New South Wales, Australia

Ricker J (1964) Outline of the geology between Mawson and Priestley glaciers, Victoria Land. In: Adie RJ (ed) Antarctic geology. North-Holland, Amsterdam, The Netherlands, pp 265–275

Roland NW, Gibson GM, Kleinschmidt G, Schubert W (1984) Metamorphism and structural relations of the Lanterman Metamorphics, north Victoria Land, Antarctica. Geologisches Jahrbuch B 60:319–361

Roland NW, Olesch M, Schubert W (1989) Geology and petrology of the western border of the Transantarctic Mountains between Outback Nunataks and Reeves Glacier, north Victoria Land, Antarctica. Geologisches Jahrbuch E 38:119–141

Schüssler U (1996) Metamorphic rocks in the northern Wilson Terrane, Oates Coast, Antarctica. Geologisches Jahrbuch B 89:247–269

Shergold JH, Cooper RA, MacKinnon DI, Yochelson EL (1976) Late Cambrian brachiopoda, mollusca, and trilobita from northern Victoria Land, Antarctica. Paleontology 19(2):38–42

Skinner DNB (1972) Differentiation source for the mafic and ultramafic rocks ("enstatite peridotites") and some porphyritic granites of Terra Nova Bay, Victoria Land. In: Adie RJ (ed) Antarctic geology and geophysics. Universitetsforlaget, Oslo, Norway, pp 299–303

Skinner DNB (1983) The geology of Terra Nova Bay. In: Oliver RL, James PR, Jago JB (eds) Antarctic earth science. Australian Academy of Science, Canberra, ACT, pp 150–155

Skinner DNB (1991) The Priestley Formation, Terra Nova Bay, and its regional significance. In: Thomson MRA, Crame JA, Thomson JW (eds) Geological evolution of Antarctica. Cambridge University Press, Cambridge, UK, pp 137–141

Skinner DNB (1992) Metasedimentary rocks of the western Wilson Terrane (Victoria Land-Oates Land) and Gondwana connections to Australia. In: Yoshida Y, Kaminuma K, Shiraishi K (eds) Recent progress in Antarctic earth science. Terra Scientific, Tokyo, pp 219–226

Skinner DNB, Ricker J (1968) The geology of the region between the Mawson and Priestley glaciers, northern Victoria Land, Antarctica. Part 1: Basement metasedimentary and igneous rocks. New Zealand J Geol Geophys 11:1009–1040

Skinner DNB, Jordan H, Schmidt-Thomé M (1996) The Berg Group of Oates Land, East Antarctica. Geologisches Jahrbuch B 89:271–293

Soloviev DS (1960) The lower Paleozoic metamorphic slates of the Oates Coast. Nauchoissledovatelsky Institut Geologii Antarktiki 113:147–158

Steiger RH, Jäger E (1977) Subcommission on geochronology: Convention on the use of decay constants in geo- and cosmochronology. Earth Planet Sci Lett 36:359–362

Stump E (ed) (1986) Geological investigations in northern Victoria Land. Antarctic Research Series, vol. 46. American Geophysical Union, Washington, DC

Stump E (compiler) (1989) Reconnaissance geologic map of the Welcome Mountain quadrangle, Transantarctic Mountains, Antarctica. US Geological Survey, US Antarctic Research Program, Reston, VA

Stump E (1995) The Ross orogen of the Transantarctic Mountains. Cambridge University Press, Cambridge

Stump E, Gootee B, Talarico F (2006) Tectonic model for development of the Byrd-Glacier discontinuity and the surrounding regions of the Transantarctic Mountains during the Neoproterozoic-early Paleozoic. In: Fütterer DK, Damaske D, Kleinschmidt G, Miller H, Tessensohn F (eds) Antarctica: Contributions to global earth sciences. Springer, Berlin/Heidelberg, Germany, pp 181–190

Sturm A, Carryer SJ (1970) Geology of the region between the Matusevich and Tucker glaciers, North Victoria Land, Antarctica. New Zealand J Geol Geophys 13:408–435

Tessensohn F (1984) Geological and tectonic history of the Bowers structural zone, north Victoria Land, Antarctica. Geologisches Jahrbuch B 60:377–396

Tessensohn F, Duphorn K, Jordan H, Kleinschmidt G, Skinner DBN, Vetter U, Wright TO, Wyborn D (1981) Geological comparison of basement units in north Victoria Land, Antarctica. Geologisches Jahrbuch B 41:31–88

Vetter U, Tessensohn F (1987) S- and I-type granitoids of north Victoria Land, Antarctica, and their inferred geotectonic setting. Geologische Rundschau 76(1):233–243

Vetter U, Roland NW, Kreuzer H, Höhndorf A, Lenz H, Besang C (1983) Geochemistry, petrography, and geochronology of the Cambro-Ordovician and Devonian-Carboniferous granitiods of northern Victoria Land, Antarctica. In: Oliver RL, James PR, Jago JB (eds) Antarctic earth science. Australian Academy of Science, Canberra, ACT, pp 140–143

Vetter U, Lenz H, Kreuzer H, Besang C (1984) Pre-Ross granites at the Pacific margin of the Robertson Bay Terrane, north Victoria Land, Antarctica. Geologisches Jahrbuch, Reihe B 60:363–369

Warren G (1969) Geology of the Terra Nova Bay-McMurdo Sound area, Victoria Land. In: Craddock C (ed) Geologic Map Folio Series, Folio 12, Sheet 14. American Geographical Society, New York

Weaver SD, Bradshaw JD, Laird MG (1984) Geochemistry of Cambrian volcanics of the Bowers Supergroup and implications for the early Paleozoic tectonic evolution of northern Victoria Land, Antarctica. Earth Planet Sci Lett 68: 128–140

Wodzicki A, Robert R Jr (1986) Geology of the Bowers Supergroup, central Bowers Mountains, northern Victoria Land. In: Stump E (ed) Geological investigations in Northern Victoria Land. Antarctic Research Series vol. 46. American Geophysical Union, Washington, DC, pp 39–68

Wodzicki A, Bradshaw JD, Laird MG (1982) Petrology of the Wilson and Robertson Bay groups and Bowers Supergroup, northern Victoria Land, Antarctica. In: Craddock C (ed) Antarctic geoscience. University of Wisconsin Press, Madison, WI, pp 549–554

Wright TO, Brodie C (1987) The Handler Formation, a new unit of the Robertson Bay Group, northern Victoria Land, Antarctica. In: McKenzie GD (ed) Gondwana Six: Structure, tectonics, and geophysics. Geophysical Monograph, vol. 40. American Geophysical Union, Washington, DC, pp 20–29

Wright TO, Dallmeyer RD (1991) The age of cleavage development in the Ross orogen, northern Victoria Land, Anarctica; Evidence from whole-rock slate ages. J Struct Geol 13:677–690

Wright TO, Findlay RH (1984) Relationships between the Robertson Bay Group and the Bowers Supergroup: New progress and complications from the Victoria Mountains, north Victoria Land. Geologisches Jahrbuch 60:105–116

Chapter 5
Central Transantarctic Mountains

We now return to the Brown Hills of southern Victoria Land (Section 3.3) and get ready to move south across the Byrd Glacier into the central sector of the Transantarctic Mountains which extends from the Byrd Glacier to the Shackleton Glacier for a distance of about 600 km. This part of the Transantarctic Mountains is traversed by several large outlet glaciers that flow from the East Antarctic ice sheet to the Ross Ice Shelf. The names of the principal outlet glaciers in Fig. 5.1 from north to south are: Byrd, Nimrod, Lennox-King, Beardmore, and Shackleton. The mountain ranges through which these glaciers flow include the Churchill Mountains and the Holyoake Range between the Byrd and Nimrod glaciers, the Geologist and Miller ranges at the head of the Nimrod Glacier, the Queen Elizabeth and the Holland ranges between the Nimrod and Lennox-King glaciers, the Queen Alexandra Range between the Lennox-King and Beardmore glaciers, and several smaller mountain ranges between the Beardmore and the Shackleton glaciers (e.g., the Commonwealth and Separation ranges, the Hughes and Sullivan ranges, and the Bush Mountains). In addition, we include the Duncan Mountains in this chapter even though they are located east of the Shackleton Glacier in the Queen Maud Mountains which are the topic of Chapter 6. Note also that the Transantarctic Mountains gradually swing east such that the Beardmore Glacier and all other outlet glaciers that follow down-range flow from south to north.

The basement rocks of the central Transantarctic Mountains are exposed primarily along the coast, except for the Geologist and the Miller Ranges which are located at the edge of the polar plateau more than 80 km inland of the coast. The basement rocks along the coast are unconformably overlain by the sedimentary rocks of the Beacon Supergroup which were intruded by sills of the Ferrar Dolerite and are locally overlain by remnants of the Kirkpatrick Basalt.

The highest peaks in the area consist of Beacon Sandstone and Ferrar Dolerite, capped in a few cases by Kirkpatrick Basalt as, for example, in the Queen Alexandra Range which contains Mt. Kirkpatrick (4,530 m), Mt. Elizabeth (4,480 m), Mt. Mackellar (4,297 m), Mt. Falla (3,825 m), and Tempest Peak (3,412 m). Other noteworthy peaks are Mt. Markham (4,351 m), Mt. Miller (4,160 m), Mt. Wexler (4,024 m), Mt. Odishaw (3,965 m), and Mt. Macdonald (3,930 m). Mt. Kirkpatrick is the fourth highest mountain in Antarctica exceeded only by the Vinson Massif (5,140 m), Mt. Tyree (4,965 m), and Mt. Shinn (4,801 m) all of which are located in the Sentinel Range of the Ellsworth Mountains in West Antarctica.

The geology of the central Transantarctic Mountains is depicted on maps edited by Craddock (1969) with descriptions by Grindley and Laird (1969) and by McGregor and Wade (1969). The geology of this sector was also investigated by Grindley (1963, 1967, 1972), Grindley et al. (1964), Laird (1963, 1964), Laird et al. (1971), McGregor (1965a, b), Skinner (1964), and Wade et al. (1965), Laird and Bradshaw (1982).

The stratigraphy of the basement rocks in Table 5.1 includes three groups of metasedimentary rocks in order of decreasing age: The Nimrod Group (Proterozoic), the Beardmore Group (Neoproterozoic), and the Byrd Group (Cambrian). All of these metasedimentary rocks were intruded by the Hope Granite of the Granite Harbor Intrusives (Cambro-Ordovician). The names of the formations in Table 5.1 were taken from the map of Grindley and Laird (1969) of the area between the Byrd and Beardmore glaciers which encompasses more than 80% of the central sector of the Transantarctic Mountains.

G. Faure and T.M. Mensing, *The Transantarctic Mountains: Rocks, Ice, Meteorites and Water*,
DOI 10.1007/978-90-481-9390-5_5, © Springer Science+Business Media B.V. 2010

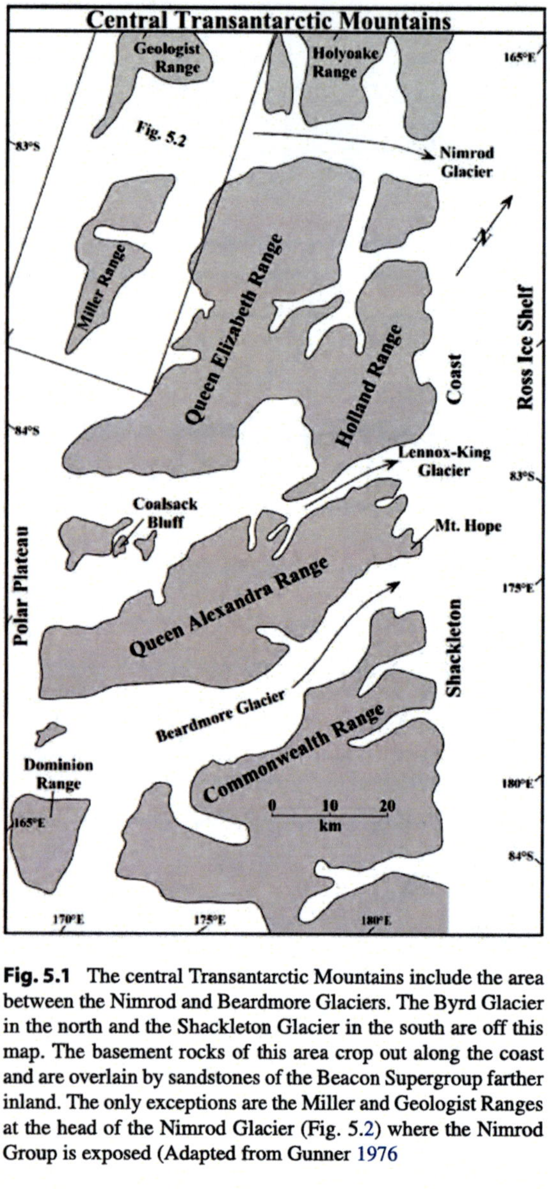

Fig. 5.1 The central Transantarctic Mountains include the area between the Nimrod and Beardmore Glaciers. The Byrd Glacier in the north and the Shackleton Glacier in the south are off this map. The basement rocks of this area crop out along the coast and are overlain by sandstones of the Beacon Supergroup farther inland. The only exceptions are the Miller and Geologist Ranges at the head of the Nimrod Glacier (Fig. 5.2) where the Nimrod Group is exposed (Adapted from Gunner 1976

We begin the description of the geology of the central Transantarctic Mountains with the Miller and Geologist Ranges at the head of the Nimrod Glacier in Fig. 5.1 because the Nimrod Group, which is composed of the oldest metasedimentary rocks in the basement of the Transantarctic Mountains, is only exposed in this area, whereas the rocks of the Beardmore and the Byrd Groups form the coastal mountains from the Byrd Glacier to the Shackleton Glacier and beyond.

5.1 Nimrod Group, Miller Range

The Miller Range at 83°15′S and 157°00′E in Fig. 5.2 is an uplifted fault block that reaches an elevation of 2,680 m in the Martin Dome. It is flanked on the east by the Marsh Glacier, which is a southern tributary of the Nimrod Glacier, and on the west by the polar plateau of the East Antarctic ice sheet. The related Geologist Range at 82°30′S and 155°30′E in Fig. 5.2 is located north of the Nimrod Glacier at the edge of the polar plateau. Both mountain ranges are composed of the amphibolite-grade gneisses of the Nimrod Group and of the plutons of the Granite Harbor Intrusives.

5.1.1 Geology of the Miller Range

The Miller Range was first mapped during the 1961/62 field season by geologists of the Geological Survey of New Zealand who named the Nimrod Group and defined the five formations listed in Table 5.1 of which it is composed (Grindley et al. 1964). In subsequent reports, Grindley and Laird (1969) and Grindley (1972) postulated that the Miller Formation had been emplacd on top of the Aurora Orthogneiss by the Endurance Thrust which disrupted the basal and upper parts of the Miller and Aurora formations, respectively. Grindley (1972) also published the most detailed map of the Miller Range which identifies some of the principal folds and faults as well as the outcrop areas of the five formations of the Nimrod Group.

The rocks of the Nimrod Group are lithologically so diverse and have been so intensely deformed that the formations are difficult to recognize in the field. Therefore, Gunner (1969) identified the major rock types that occur south of the Argosy Glacier without assigning them to specific formations. When Goodge et al. (1990) later mapped the Geologist Range, they found the same kind of rocks that occur in the Miller Range but mapped them only in terms of lithologic units (e.g., amphibolitic gneiss, orthogneiss, metasediments, and granite).

According to summaries by Stump (1995), the Worsley Formation occurs at the base of the Nimrod Group between Aurora Heights north of the Argosy Glacier and the Skua Glacier in Fig. 5.2. It contains tremolite-bearing marble and garnet-biotite schist for

Table 5.1 Stratigraphy of the basement rocks of the central Transantarctic Mountains (Grindley and Laird 1969; McGregor and Wade 1969; Stump1981, 1982

Geologic unit	Geologic age	Description
~~	Kukri Peneplain	~~~~~~
Granite Harbor Intrusives	Cambrian to Ordovician	Hope Granite: granite, adamellite, granodiorite, tonalite, gabbro
――――	Intrusive Contact	――――
Byrd Group	Cambrian	Douglas, Dick, Starshot: sandstone, conglomerate, shale, and volcanics Taylor Formation[a]: quartzite, hornfels, marble Shackleton Limestone & Selborne Marble: Archaeocyatha limestone, marble, conglomerate, quartzite
――――	Conformable Contact (Goodge 1997	――――
Beardmore Group	Neoproterozoic	Goldie Formation[b]: graywacke, argillite, phyllite, hornfels Cobham Formation: biotite schist, marble, quartzite, conglomerate
~~	Unconformity	~~~~~~
Nimrod Group	Paleoproterozoic (Goodge and Fanning 1999)	Miller Formation: marble, amphibolite, semipelitic gneiss Argo Gneiss: diorite, orthogneiss with eclogite lenses Aurora Formation: feldspathic gneiss, psammitic gneiss Argosy Formation: semipelitic gneiss, garnet schist, quartzite Worsley Formation: marble, dolomite, schist, skarn rock, hornfels

[a]The Taylor Formation is not recognized in the mountains north of the Beardmore Glacier but is a prominent formation of the Byrd Group south of the Beardmore Glacier between the Kosco and Gough Glaciers located north and south of the Shackleton Glacier which marks the southern limit of the central Transantarctic Mountains (McGregor and Wade 1969)
[b]Most of the rocks of the Goldie Formation are actually Cambrian in age and have been reassigned to the Starshot Formation of the Byrd Group as explained in Section 5.3.2 (Goodge et al. 2002; Myrow et al. 2002)

a total thickness of about 700 m (Grindley et al. 1964). The Worsley Formation was intruded by the Hope Granite which crops out both east and west of the Skua Glacier. The contact metamorphic aureole adjacent to the plutons includes diopside and cummingtonite in Worsley marble and andalusite-sillimanite in the schist layers.

The Argosy Formation occurs both north and south of the Argosy Glacier. The rocks are predominantly composed of garnet-biotite schist interbedded with psammitic (arenaceous) gneiss, quartzite, calc-silicate schist, marble, and amphibolite. Gunner (1969) observed relict cross-bedding, laminations, and relict bedding on Kreiling Mesa south of the Argosy Glacier. According to Grindley et al. (1964), the Argosy Formation is approximately 1,000 m thick.

The Aurora Formation was originally described by Grindley et al. (1964) as feldspathic, lenticular-banded, biotite gneiss derived from sedimentary rocks, although they did observe that this formation contains diorite-orthogneiss grading into granodiorite. Borg et al. (1986) interpreted the rocks to be mylonitized granodiorite that had intruded schists of the Nimrod Group. Although Borg et al. (1990) referred to it as the Camp Ridge Granodiorite, Goodge et al. (1991) preferred to use the original name and called it the Aurora Orthogneiss (Stump 1995).

The Nimrod Group contains tectonic blocks of mafic and ultramafic rocks which Grindley and Laird (1969) named the Argo Gneiss. The cores of these blocks contain eclogitic mineral assemblages described by Peacock and Goodge (1995). Goodge et al. (1993a) determined that the eclogitic mineral assemblages had formed at a temperature of 600°C and a pressure of 12–25 kilobars (kb). The blocks are roughly spherical in shape, have diameters between 0.5 and 50 m, and are rimmed by hornblende + plagioclase ± garnet ± biotite consistent with upper amphibolite to granulite facies metamorphism.

The Miller Formation was described by Stump (1995) as a layered sequence of polydeformed gneisses

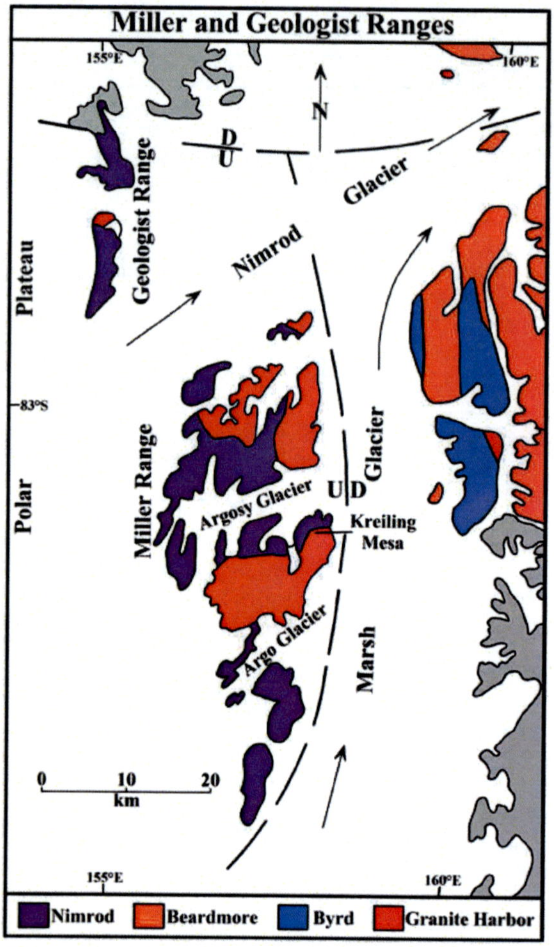

Fig. 5.2 The Geologist and Miller Ranges at the head of the Nimrod Glacier (Fig. 5.1) contain the only known outcrops of the Nimrod Group which is the oldest geologic unit in the Central Transantarctic Mountains. The outcrop areas colored gray expose the sandstones of the Beacon Supergroup which was deposited on the Kukri Peneplain. The trace of the peneplain is not shown on this map because the Beacon Supergroup has been removed by erosion from the Miller Range and occurs only in the northernmost Geologist Range. The dashed line represents inferred faults; however, not all of the faults are shown. D = downthrown, U = upthrown (Adapted from Grindley and Laird 1969

and schists interbedded with quartzite, marble, calc-silicate, and amphibolite. This formation is exposed along the western edge of the Miller Range at the head of the Argosy Glacier and south of the Argo Glacier (Fig. 5.2). The Miller Formation may be older than the other formations in the Miller Range based on several lines of evidence: It was thrust from the west on top of the Aurora Orthogneiss by the Endurance Thrust, it was metamorphosed to a higher grade, and it appears to be more highly deformed than the other formations. Goodge et al. (1991) reported that the gneisses of the Miller Formation contain hornblende, plagioclase, and biotite as well as accessory garnet, epidote, and clinopyroxene, whereas the pelitic schists contain kyanite and minor sillimanite indicating peak metamorphic temperatures between 680°C and 750°C and pressures of 8–14 kb.

A structural cross-section drawn by Grindley et al. (1964) indicates that the rocks generally dip west and that they are cut by high-angle thrust faults that also dip west such that the upthrown sides are on the western side of the faults. The Endurance Fault is a low-angle thrust fault along the western edge of the Miller Range. The fault plane was folded but dips west at low angles. The geologic sketch map in Fig. 5.2 does not contain all of the faults that appear on the map by Grindley et al. (1964).

5.1.2 Age Determinations

The first K-Ar age determinations by McDougall and Grindley (1965) demonstrated that the Hope Granite north of the Argosy Glacier in the Miller Range cooled during the Ordovician Period at 463 and 478 Ma (IUGS 2002). K-Ar dates of biotite and muscovite in the metamorphic rocks of the Nimrod Group range similarly from 456 to 633 Ma. The oldest date of 633 Ma is a clue that the rocks of the Nimrod Group in the southern Miller Range were metamorphosed prior to about 633 Ma and that the gneisses of the Nimrod Group were later intruded by the Hope Granite during the Ross Orogeny in Cambro-Ordovician time. Alternatively, apparent pre-Ross Orogeny K-Ar dates of minerals in the Nimrod Group may also be attributable to the presence of excess ^{40}Ar that could have been released from K-rich minerals during the Ross Orogeny.

In a subsequent paper, Grindley and McDougall (1969) reported K-Ar dates of hornblende from the Miller and Argosy Formations that range from 504 ± 5 to 1043 ± 16 Ma. These dates form a small cluster between 1000 and 1050 Ma in the histogram in Fig. 5.3. Although the authors considered the possibility that the hornblende contains excess ^{40}Ar, they tentatively suggested that the rocks of the Nimrod Group were

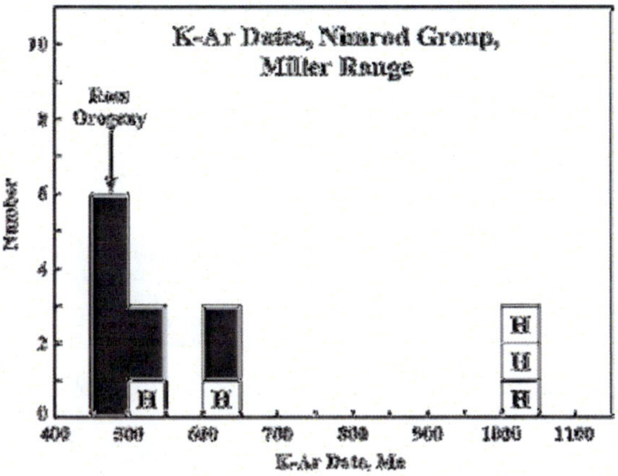

Fig. 5.3 The K-Ar dates of biotite and muscovite (gray) of the Nimrod-Group metasediments refer to the time elapsed after these minerals cooled following the Ross Orogeny and after the intrusion of plutons of Hope Granite into the metasedimentary rocks. The K-Ar dates of hornblende (H) record the end of a preceding orogeny that Grindley and McDougall (1969) named the Nimrod Orogeny, which occurred prior to about 1020 Ma (Data from McDougall and Grindley 1965; Grindley and McDougall 1969

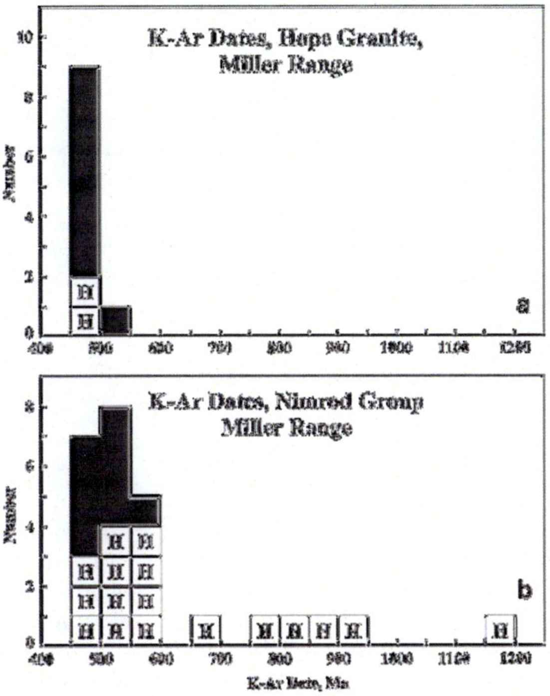

Fig. 5.4 a) The K-Ar dates of biotite (gray) and hornblende (H) of the Hope Granite in the Miller Range display the narrow range (450–550 Ma) observed previously by McDougall and Grindley (1965) and by Grindley and McDougall (1969) in Fig. 5.3. The good agreement between K-Ar dates of biotite and hornblende implies that the Hope plutons were intruded at shallow depth and cooled in less than 5 million years to the respective blocking temperatures of biotite and hornblende (Data from Adams et al. 1982). (**b**) The K-Ar dates of biotite (gray) and hornblende (H) from different formations of the Nimrod Group in the Miller Range are strongly clustered between 450 and 600 Ma. Most of the K-Ar dates were reset by loss of radiogenic ^{40}Ar during the intrusion of the Hope Granite close to the end of the Ross Orogeny. However, Adams et al. (1982) recognized the Beardmore Orogeny on the basis of the K-Ar hornblende dates between 550 and 600 Ma and concluded that there is no direct evidence for the Nimrod Orogeny proposed by Grindley and McDougall (1969). Adams et al. (1982) considered that the older K-Ar hornblende dates (>650 to <1200 Ma) are the result of contamination with ^{40}Ar during the Beardmore Orogeny (Data from Adams et al. 1982

deformed and metamorphosed prior to about 1020 Ma during the Nimrod Orogeny which, at that time, was thought to be the oldest orogeny that had affected the basement rocks of the Transantarctic Mountains.

The existence of the Nimrod Orogeny prior to about 1020 Ma was challenged by Adams et al. (1982). Although their K-Ar dates of biotite and hornblende from the Miller Range in Fig. 5.4a and b confirm that the Hope Granite cooled between 450 and 550 Ma exactly as demonstrated by Grindley and McDougall (1969), Adams et al. (1982) found no evidence for the occurrence of the Nimrod Orogeny because their K-Ar hornblende dates scatter widely from greater than 600 Ma to less than 1200 Ma without clustering between 1000 and 1050 Ma as expected. Instead, Adams et al. (1982) obtained several hornblende dates between 550 and 600 Ma which precede the intrusion of the Hope Granite in the Miller Range and thereby indicate that the rocks of the Nimrod Group were deformed and metamorphosed during the Beardmore Orogeny which they placed at about 600–630 Ma. Adams et al. (1982) suggested that some hornblendes as well as biotites in the rocks of the Nimrod Group were contaminated by excess radiogenic ^{40}Ar causing these minerals to have invalid old K-Ar dates ranging up to 1150 Ma as seen in Fig. 5.4b. The petrogenesis of the Hope Granite in the central Transantarctic Mountains is discussed in Section 5.4.

After the publication of the paper by Grindley and McDougall (1969), Gunner (1971a) and Gunner and Faure (1972) dated whole-rock samples and minerals of the Nimrod Group in the Miller Range by the Rb-Sr method. Nineteen whole-rock samples of the Nimrod Group scattered widely on the Rb-Sr geochronometry diagram and failed to define a straight line from which the age and initial $^{87}Sr/^{86}Sr$ ratio of the rocks could have been determined. The scatter of data points implies that the rock samples either had a range of different initial $^{87}Sr/^{86}Sr$ ratios or that their isotope systematics and their chemical compositions were altered during the Ross Orogeny and as a consequence of the intrusion of the Hope Granite. In order to eliminate samples that may have been altered in this way, Gunner and Faure (1972) selected samples that satisfied three criteria:

1. They originated from a small area.
2. They were collected at least 12 km from the nearest outcrop of Hope Granite.
3. Their field relations and petrography indicated that they had not been subjected to melting.

Only 8 of the 19 samples qualified for consideration, but even these samples did not define a straight line on the Rb-Sr chronometry diagram. Nevertheless, five of the eight preselected samples defined two parallel lines the slopes of which yielded similar dates (recalculated to $\lambda = 1.42 \times 10^{-11}$ year^{-1}) but different initial $^{87}Sr/^{86}Sr$ ratios (not shown):

I. $t = 1984 \pm 38$ Ma, $(^{87}Sr/^{86}Sr)_i = 0.7276 \pm 0.0021$
II. $t = 1789 \pm 141$ Ma, $(^{87}Sr/^{86}Sr)_i = 0.7059 \pm 0.0042$

The slope of line I is defined by only two samples one of which contained the rubidium-rich chromium mica fuchsite, the reliability of which as a geochronometer remains untested. Line II is defined by three typical metamorphic rocks of the Nimrod Group:

517: Epidote-hornblende schist
522: Biotite-microcline-oligoclase augengneiss
552: Andesine-biotite-hornblende schist

These samples define the straight line in Fig. 5.5 which also includes an internal mineral isochron for sample 522 indicating a date of 446 ± 14 Ma (recalculated to $\lambda = 1.42 \times 10^{-11}$ year^{-1}) and an initial $^{87}Sr/^{86}Sr$ ratio of 0.7511 ± 0.0007.

The whole-rock Rb-Sr date of 1789 ± 141 Ma (Paleoproterozoic) is the time when these rocks were recrystallized during regional metamorphism to amphibolite grade and the strontium they contained was isotopically homogenized so that all three specimens had the same initial $^{87}Sr/^{86}Sr$ ratio of 0.7059 ± 0.0041. This value indicates that their protoliths contained strontium derived from the lithospheric mantle in the form of volcanic rocks or their weathering products.

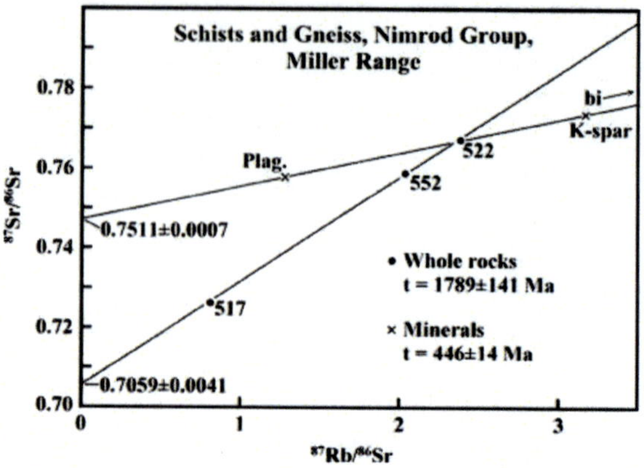

Fig. 5.5 Three selected whole-rock samples of the Nimrod Group in the Miller Range define a straight line that corresponds to a Rb-Sr date of 1789 ± 141 Ma ($\lambda = 1.42 \times 10^{-11}$ year^{-1}) and an initial $^{87}Sr/^{86}Sr$ ratio of 0.7059 ± 0.0042. In addition, the minerals of one of the rock samples (#522) form a line that yields a date of only 446 ± 14 Ma ($\lambda = 1.42 \times 10^{-11}$ year^{-1}) and an initial $^{87}Sr/^{86}Sr$ ratio of 0.7511 ± 0.0007. The whole-rock date records the time of isotopic homogenization of strontium in the rocks during a major orogenic event during which the rocks were structurally deformed and regionally metamorphosed. The minerals date the time of isotopic re-equilibrium of strontium in the minerals during the Ross Orogeny in Cambro-Ordovician time (Data from Gunner and Faure 1972)

The minerals of rock sample 522 date the time when the strontium they contained was re-equilibrated to an $^{87}Sr/^{86}Sr$ ratio of 0.7511 ± 0.0007. The Rb-Sr date of this event (i.e., 446 ± 14 Ma) agrees with the K-Ar dates of biotite from the Hope Granite in Fig. 5.4.

Gunner and Faure (1972) concluded from the evidence that the rocks of the Nimrod Group were a part of the East Antarctic craton and that they represent a volcano-sedimentary complex that was accreted to the craton during the Nimrod Orogeny that occurred about 1.79×10^9 years ago. According to this interpretation, the Nimrod Orogeny occurred about 740 million years earlier than originally proposed by Grindley and McDougall (1969).

Gunner and Mattinson (1975) later tested these conclusions concerning the geologic history of the Nimrod Group by measuring U-Pb dates of zircon and sphene in a sample of Hope Granite collected in the Miller Range. The U-Pb dates of sphene were found to be concordant, which means that the ^{238}U-^{206}Pb date and the ^{235}U-^{207}Pb date are identical within analytical error at 476 and 477 Ma, respectively. The most reliable date of 484 ± 40 Ma was obtained from the ratio of radiogenic ^{207}Pb to radiogenic ^{206}Pb in the sphene (Faure and Mensing 2005). This date confirmed that the sphene in the Hope Granite started to retain lead quantitatively during the Ordovician Period as indicated previously by K-Ar and Rb-Sr dates of minerals. Although two size-fractions of zircon analyzed by Gunner and Mattinson (1975) yielded discordant U-Pb dates, their radiogenic $^{207}Pb/^{206}Pb$ dates were in agreement at 537 ± 5 Ma (Early Cambrian). This date is distinctly older than the Ross Orogeny which suggests that the zircons are either detrital or that they contain old cores with young overgrowths.

The secret of the Nimrod zircons was eventually revealed by Goodge and Fanning (1999) and Goodge et al. (2001) by dating zircons using a SHRIMP mass spectrometer at the Australian National University that was designed and built by Professor William Compston (Faure and Mensing 2005). The acronym of this mass spectrometer stands for "Sensitive High-Resolution Ion Micro-Probe" (Compston and Williams 1984; Compston 1999).

Goodge and Fanning (1999) extracted zircon crystals from a sample of banded hornblende-biotite-gneiss collected at Camp Ridge at the head of the Argosy Glacier in the Miller Range (Fig. 5.2). Most of these zircon grains shown in Fig. 5.6 contain prismatic cores

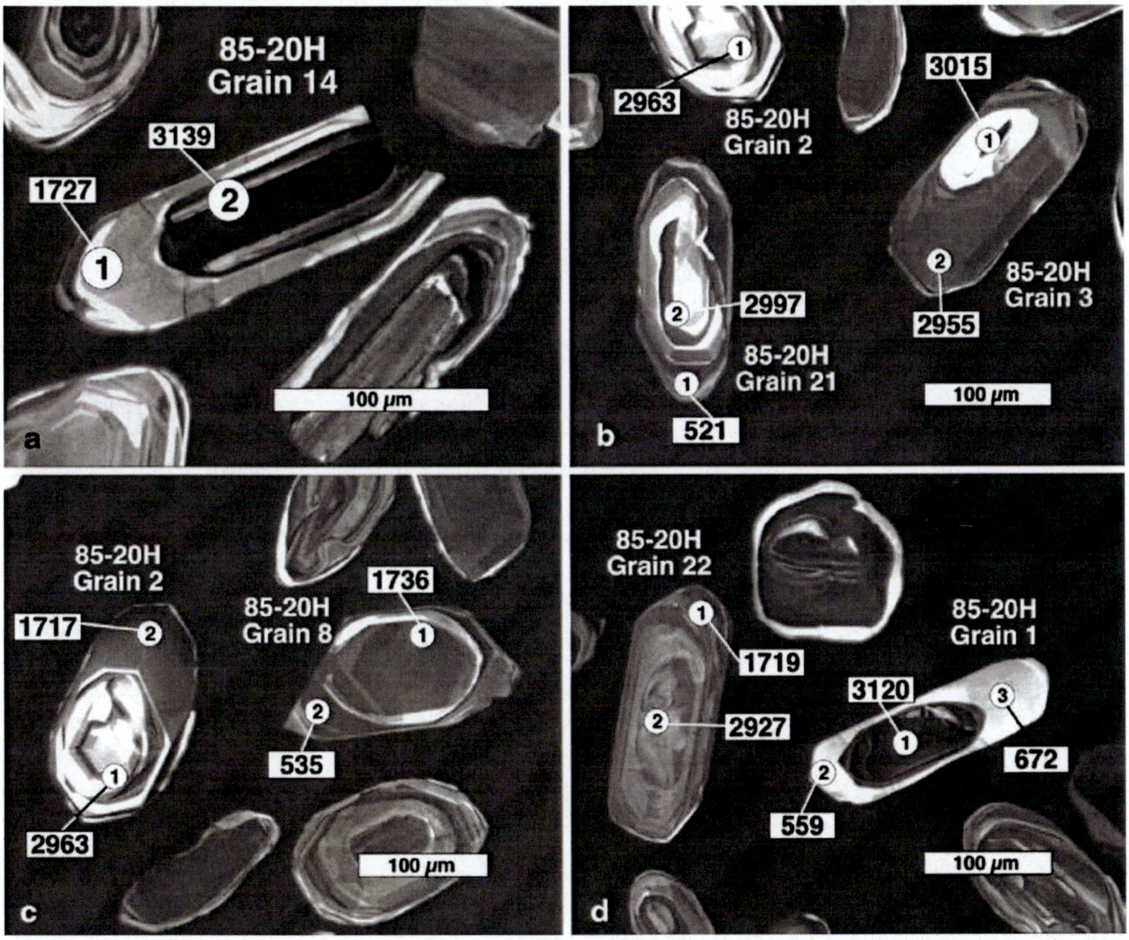

Fig. 5.6 Cathodoluminescence images of zircon grains in sample 85–20H (banded hornblende-biotite gneiss) which was collected along Camp Ridge at the head of the Argosy Glacier in the Miller Range. The zircon grains contain prismatic cores with younger overgrowths. The areas in each grain that were analyzed by the microprobe mass spectrometer are identified by numbered circles and the resulting U-Pb dates are expressed in the rectangular areas in units of Ma. The dates of about 3000 and 1700 Ma were calculated from the measured $^{207}Pb/^{206}Pb$ ratios of the radiogenic lead, whereas dates of about 530 Ma are based on $^{206}Pb/^{238}U$ ratios. The dates are listed in Table 5.2 which also contains representative averages (Reproduced by permission of the Geological Society of America and of John W. Goodge; Goodge and Fanning 1999)

Table 5.2 Summary of U-Pb SHRIMP dates of cores and overgrowths of zircon grains in a sample of gneiss of the Nimrod Group in the Miller Range of the central Transantarctic Mountains (Goodge and Fanning 1999)

	U-Pb dates (Ma[a])		
Grain number	Cores	Overgrowths	
1	3120	–	(672)
			559
2	2963	1717	–
3	3015	–	–
	2955		
8	–	1736	535
14	3139	1727	–
21	2997	–	521
22	2927	1719	–
Average all grains	3017 ± 31[b]	1725 ± 5	529[c] ± 7

[a]The dates were calculated from the radiogenic $^{207}Pb/^{206}Pb$ ratios and were taken from Fig. 2 of Goodge and Fanning (1999)
[b]The errors of the average dates are one standard deviation of the mean
[c]Weighted mean calculated by Goodge and Fanning (1999)

covered by overgrowths of younger zircon, in contrast to the zircon crystals which Gunner and Mattison (1975) had previously recovered from a sample of the Hope Granite in the Miller Range which did not have these cores. Goodge and Fanning (1999) were able to measure $^{206}Pb/^{238}U$ and $^{207}Pb/^{235}U$ ratios of the cores and the overgrowths by means of the ion-microprobe mass spectrometer (SHRIMP) and interpreted these measurements on a U-Pb concordia diagram (Appendix 3.6.5). The resulting $^{207}Pb/^{206}Pb$ dates of the cores and overgrowths in Table 5.2 clearly demonstrate that the zircon cores originally crystallized at 3017 ± 31 Ma during the Mesoarchean Era (2800–3200 Ma), whereas the overgrowths formed during the Paleoproterozoic Era at about 1725 ± 5 Ma and during the Cambrian Period at 529 ± 7 Ma. The 672 Ma date for the overgrowth of grain #1 was omitted from the average in Table 5.2 because Goodge and Fanning (1999) suspected that it may represent a mixture of overgrowth with core-zircon. The evidence for magmatic activity during the Mesoarchean Eon obtained by Goodge and Fanning (1999) by U-Pb dating of zircon cores is consistent with Sm-Nd isotopic evidence for the formation of a continental crust in the central Transantarctic Mountains at this time reported by Borg et al. (1990). In addition, Goodge and Fanning (1999) emphasized that the subsequent orogeny in Paleoproterozoic time included magmatic activity at 1730 Ma and was followed by intense regional metamorphism from 1720 to 1700 Ma. The episode of zircon growth at about 530 Ma coincides with the U-Pb dates of sphene and zircon from the Hope Granite in the Miller Range reported by Gunner and Mattinson (1975).

Goodge et al. (2001) later expanded the scope of their work by dating zircons from other rock samples of the Nimrod Group, including another sample of layered gneiss, a relict block of eclogite (Peacock and Goodge 1995), and a granodiorite orthogneiss. The results of these age determinations confirmed the conclusion of Goodge and Fanning (1999) that the cores of the zircons in the rocks of the Nimrod Group crystallized during the Mesoarchean Era and that the overgrowths on the original zircons grains formed when the rocks were tectonically deformed and metamorphosed during the Paleoproterozoic Era and then again during Cambrian Period. The authors recommended that the episode of deep-seated metamorphism at 1725 ± 5 Ma is, in fact, the Nimrod Orogeny that Grindley and McDougall (1969) originally defined on the basis of K-Ar hornblende dates which turned out to be unreliable because the hornblende contains varying amounts of excess ^{40}Ar (Adams et al. 1982).

The studies of Goodge and Fanning (1999) and of Goodge et al. (2001) brought into focus the seemingly contradictory evidence concerning the geologic history of the Nimrod Group (e.g., Gunner and Faure 1972; Borg and DePaolo 1986; Walker and Goodge 1991; Goodge and Dallmeyer 1992, 1996; Goodge et al. 1992, 1993a, b; Bennett and Fanning 1993). We now know that zircon grains in the rocks of the Nimrod Group in the Miller and Geologist ranges originally crystallized at 3017 ± 31 Ma during the Mesoarchean Era. They were eventually released by weathering of their original host and were redeposited in a marine basin with other detrital sediment. The accumulated sedimentary rocks were buried, tectonically deformed, and metamorphosed during the Nimrod Orogeny which ended at 1725 ± 5 Ma during late Paleoproterozoic time. The same rocks were later deformed a second time and reheated during the Ross Orogeny during the Cambrian followed by slow cooling into the Ordovician Period (495–440 Ma). Still later, perhaps during the Mesozoic Era when Gondwana broke up, a block of continental crust was uplifted and dissected by erosion to form the mountains of the Miller and Geologist ranges. The topography of these mountains was modified by the East Antarctic ice sheet which caused the development of the present geomorphology that was described by Grindley (1967).

5.2 Beardmore Group, Queen Elizabeth Range

According to the master hypothesis for the origin of the Ross orogen, the sedimentary rocks of the Beardmore Group were deposited in the paleo-Pacific ocean along the Antarctic coast of Gondwana (Stump 1995). The sediment consisted of weathering products and detrital grains of rocks and minerals derived from the Paleoproterozoic Nimrod Mountains that were slowly eroding for about 1 billion years following the Nimrod Orogeny at 1725 ± 5 Ma. The master hypothesis leads us to believe that the sediment of the Beardmore Group was subsequently compressed and regionally metamorphosed as a result of subduction of oceanic crust against and under the Antarctic craton. We may assume that the deformation and metamorphism of the Beardmore sediment took place during the Cambrian Period and ended in Cambro-Ordovician time with the intrusion and cooling of granitic plutons. However, this conjecture leaves us with the question: Are we dealing with two separate but closely spaced orogenies (i.e., the Beardmore and the Ross orogenies) or was there only one prolonged Beardmore/Ross Orogeny? We will try to answer this question by studying the rocks in the field and in the laboratory.

5.2.1 *Conventional Stratigraphy*

The basement rocks of the central Transantarctic Mountains consist largely of low-grade metamorphic rocks of the Beardmore Group which is exposed from the mouth of the Starshot Glacier south to the Ramsey Glacier. The Beardmore Group is younger than the Nimrod Group and is separated from it by an inferred unconformity which represents the time that elapsed between the Nimrod Orogeny and the onset of Beardmore sedimentation.

The Beardmore Group in Table 5.1 has been divided into the lower Cobham Formation and the upper Goldie Formation both of which were presumably deposited during the Neoproterozoic Era (Grindley and Laird 1969). The Cobham Formation at the base of the Beardmore Group is exposed in the Cobham Range north of the Nimrod Glacier. It is composed of interbedded schists, calc-schists, marbles, and quartzite that have an aggregate thickness of about 510 m. The top of the Cobham Formation is defined as the top of the highest marble bed, but the base is not exposed. The rocks have been regionally metamorphosed to almandine-biotite schist and tremolite-bearing calc-schists. The rocks were later reheated, presumably by the intrusion of granitic magma and now have a granoblastic texture. The original sedimentary rocks were deposited in a shallow-water marine environment on a continental shelf (Laird et al. 1971).

According to a summary by Stump (1995), the Goldie Formation is composed of a monotonous sequence of interbedded slates, phyllites, and graywackes (or litharenites). The rocks are typically dark gray in color but have developed a brownish weathered surface (Grindley 1963). The clasts in the graywacke are subangular to subrounded and are composed primarily of quartz with minor feldspar and muscovite. Trace amounts of detrital apatite, sphene, zircon, epidote, tourmaline, and hornblende occur as accessory constituents (Gunn and Walcott 1962). In addition, Gunner (1976) reported the presence of lithic fragments composed of mica schist, granite, and chert in several localities. The clasts are generally matrix-supported although the volume of matrix decreases locally to qualify the rock as an arenite. The matrix of the graywacke contains calcite in which case the rocks grade into a calcarenite. Even rare limestone and fine-grained conglomerate beds have been seen in the Goldie Formation by several field investigators (e.g., Gunn and Walcott 1962; Laird et al. 1971; Oliver 1972; Gunner 1976). The Goldie Formation does not contain detrital grains of volcanic or pelitic rocks and appears to have been derived from an area where crystalline rocks were exposed.

The most widely observed sedimentary structures of the graywacke beds are graded bedding and, more rarely, ripple marks and cross bedding. The layers of graywacke grade upward into pelites which are overlain by another graded graywacke bed in a nearly-endless succession. These sedimentary features indicate that the Goldie Formation was deposited by turbidity currents flowing off the continental shelf and slope of the paleo-Antarctic coast.

Stump et al. (1988) reported that four layers of diamictite occur in the Goldie Formation at Panorama Point at the north end of The Palisades (82°50′S, 159°10′E) located at the confluence of the Marsh and Nimrod Glaciers shown in Fig. 5.7. The thickest of the diamictite beds (about 6 m) is overlain by a large

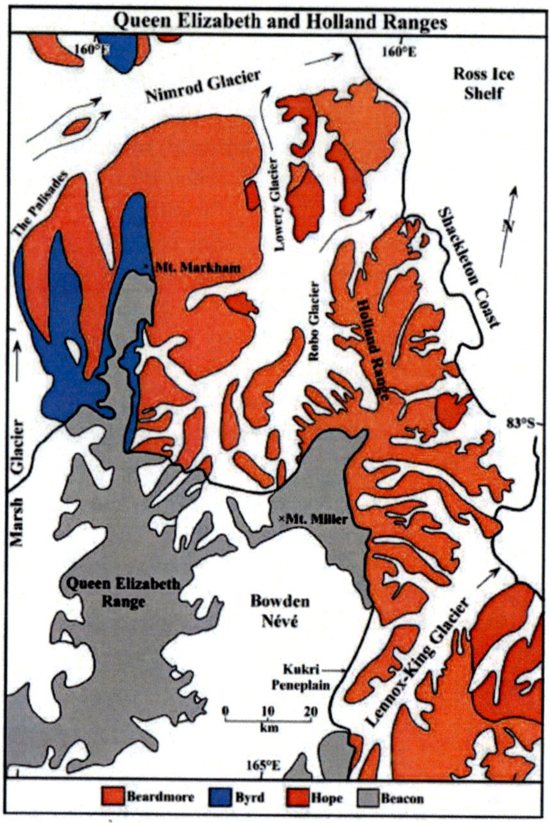

Fig. 5.7 The basement rocks of the Queen Elizabeth and Holland Ranges in the central Transantarctic Mountains consist of the low-grade metasedimentary rocks of the Beardmore Group which is locally overlain by the Byrd Group. Both groups of rocks were intruded by plutons of the Hope Granite of Cambro-Ordovician age. The flat-lying sedimentary rocks of the Beacon Supergroup (Devonian to Triassic) were subsequently deposited on the Kukri Peneplain. The white areas on the map are the ice of the mountain glaciers and of the Ross Ice Shelf (Adapted from Grindley and Laird 1969)

deposit of pillow basalt. The other diamictite units are less than 1 m thick and are located down-section from the pillow basalt. The diameters of clasts of the thickest diamictite layers range from sand-size particles up to about 40 cm and are composed of quartzite, calcarenite, dolomite, and quatz-muscovite schist. The matrix consists of clay. Although the clasts are not faceted or striated and no dropstones were found, Stump et al. (1988) suggested that these diamictites could be glacigenic. However, the origin of the diamictites is still not understood.

Volcanic rocks of various kinds have been noted in the Goldie Formation by several authors (e.g., Wade and Cathey 1986; Oliver 1972). The pillow basalt at Panorama Point (The Palisades) is associated with gabbro which yielded a Sm-Nd date of 762 ± 24 Ma based on separated plagioclase and pyroxene combined with a sample of pillow basalt (Borg et al. 1990). This date is the crystallization age of the basalt and gabbro and it may also indicate that the Goldie Formation exposed at this locality was deposited during the Neoproterozoic Era. In addition, Adams et al. (1982) reported K-Ar dates of 477 and 465 Ma for whole-rock samples of slate of the Goldie Formation. These dates record the time elapsed since cooling of the slates at the end of the Ross Orogeny and therefore imply that the Goldie sediment was deposited prior to 477 Ma.

5.2.2 Revision of the Goldie Formation

The geologic history of the paleo-Pacific margin of Antarctica that arises from the conventional model of the stratigraphy of the Beardmore Group, though plausible, has been revised in light of new age determinations and a refinement of the tectonic evolution of the rift margin of Gondwana after the departure of Laurentia. These revisions affect not only the stratigraphy of the Goldie Formation but also involve the overlying formations of the Byrd Group. For this reason, we present first the new insights concerning the origin of the Goldie Formation and postpone the synthesis of the depositional history of both groups until we have described the conventional stratigraphy of the Byrd Group in Section 5.3.1.

Contrary to several previous investigators, Goodge (1997) concluded from field evidence that the rocks of the Goldie Formation in the vicinity of the Nimrod Glacier "… contain only a single set of geometrically related regional structures associated with the development of upright, large-scale and small-scale flexure-slip folds." In addition, Goodge (1997) concluded, in agreement with Rowell et al. (1986) and Rees et al. (1989), that the contact between the top of the Goldie Formation and the base of the overlying Shackleton Limestone of the Byrd Group is conformable with some local displacement caused by folding of beds having differing competence. If the contact between the Goldie Formation and the Shackleton Limestone is conformable, then the change in lithology signals a change in the depositional environment without a major hiatus in the deposition of sediment. The evidence that the rocks of the Goldie Formation were not folded several times and that they were deposited continuously in a changing environments removes the requirement that the Ross Orogeny was preceded by the Beardmore Orogeny. Goodge et al. (1997) therefore concluded that the sedimentary rocks of the Beardmore and Byrd groups were deformed only by the Ross Orogeny which lasted longer and was a more complex tectonic event than previous investigators had recognized. Actually, the existence of the Beardmore Orogeny had been questioned previously by Stump et al. (1991), by Pankhurst et al. (1988), and by others. Even the K-Ar dates of hornblende in the Nimrod Group in Fig. 5.4b, on the basis of which Adams et al. (1982) defined the Beardmore Orogeny (600–550 Ma), vary continuously from 600 to 450 Ma and overlap the dates that characterize the Ross Orogeny. Therefore, these dates do not contradict the proposal that the orogenic activity was continuous and is attributable exclusively to the Ross Orogeny.

Goodge (1997) further pointed out that the rocks of the Beardmore Group were folded less than 600 million years ago, as indicated by U-Pb dates of detrital zircons (Walker and Goodge 1994), but more than 500 million years ago, based on age determinations of the post-kinematic Hope Granite. The evidence for a continuous spectrum of dates within the Ross orogen from Late Neoproterozoic to Ordovician is best accounted for by one continuous and extended tectonic process encompassed within the Ross Orogeny.

The last step in the reconstruction of the geologic history of the paleo-Pacific coast of Antarctica is to explain the cause for the westward compression of the Cobham and Goldie Formations. The compression cannot be attributed to the collision of crustal fragments with the Antarctic craton because, in contrast to northern Victoria Land, no such fragments have been recognized in the central Transantarctic Mountains, although Borg et al. (1990) did propose the existence of a Beardmore microcontinent. For this and other reasons, Goodge (1997) outlined a four-stage process that implements the assumptions of the master hypothesis for the origin of the Ross orogen, namely that it resulted from the transformation of the passive rift-margin of East Antarctica into a compressive subduction zone:

Stage 1: Rifting of Rodinia (About 700–600 Ma)

The supercontinent Rodinia, which included what later became East Antarctica, developed a rift that widened as Laurentia began to move away. As a result, the edge of the East Antarctic craton was thinned by the development of parallel "down-to-the-east" normal faults. Carbonate-shale deposits began to form on this subsiding shelf which extended eastward into a deepening continental slope where fine-grained clastic sediment was deposited by turbidity currents. The near-shore carbonate-shale sequence evolved into the Cobham Formation, whereas the deep-water turbidites became the Goldie Formation of the Beardmore Group. The sediment contained in these formations originated from sources in East Antarctic without additions from Laurentia which continued to move away. This stage in the tectonic evolution probably started at around 650 Ma and lasted for about 50–100 million years.

Stage 2: Compression of Sediment (About 600–550 Ma)

The Beardmore sediment that had accumulated in the off-shore basin began to be compressed as a result of seafloor spreading directed westward away from the spreading ridge. The normal faults that had formed along the Antarctic rift margin propagated upward into the accumulated sediment which was compressed into open folds verging toward the craton.

Stage 3: Start of Subduction (About 550 Ma)

Continued widening of the ocean that now separated Gondwana and Laurentia ultimately triggered subduction of oceanic crust because of the requirement that seafloor spreading must be balanced by subduction. Sediment continued to be compressed both on the margin of the continent and in the wedge between the trench and the craton, but the direction of the compressive force may have changed from being at right angles to the original rift to becoming oblique. Carbonate rocks continued to be deposited in shallow water along the coast.

Stage 4: Steady-State Subduction (About 550–450 Ma)

Ongoing subduction continued and thereby transformed the former passive rift-margin of Gondwana into an active subduction zone. This stage represents the culmination of the Ross Orogeny with magmatic activity and regional metamorphism not only of the supracrustal sediment of the Beardmore Group but affecting also the deeply buried rocks of the Nimrod Group farther inland.

The crystalline basement rocks of the Antarctic craton experienced higher temperatures and pressures compared to the supracrustal rocks which caused the two geologic units to respond differently to the applied stresses that originated from the subduction zone.

Goodge (1997, p. 698) summed up the presentation of his model for the tectonic evolution of the passive rift margin of East Antarctic into an active subduction zone which resulted in the prolonged Beardmore + Ross Orogeny with these words:

> This model successfully integrates structural, stratigraphic, and geochronological data from several areas along the Transantarctic Mountains… and it is consistent with the independent conclusion that Beardmore Group deformation is but one manifestation of a protracted plate-tectonic continuum referred to as the Ross Orogeny.

To which he later added:

> The Ross orogenic cycle constitutes a broad and diachronous sequence of geologic events during the latest Neoproterozoic to early Paleozoic manifested by different strain histories in rocks occupying different crustal levels.

Future generations of geologists may want to fine-tune this model; but for the time being it is a realistic synthesis of the geological processes that produced the Ross orogen.

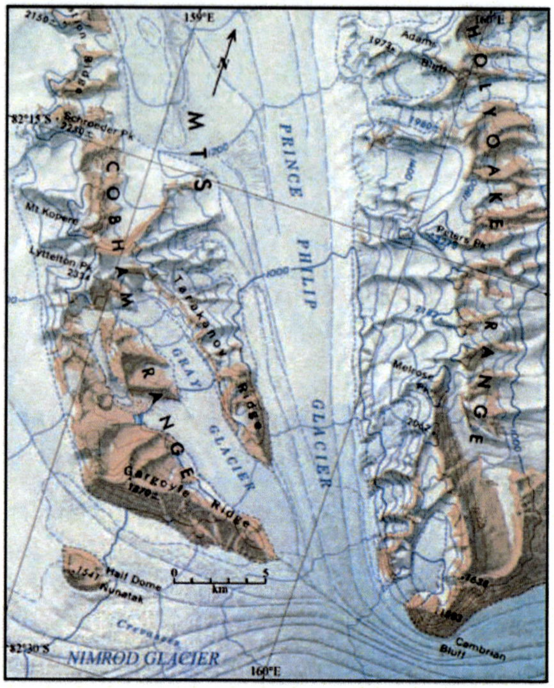

Fig. 5.8 The Holyoake Range and the Cobham Range of the Churchill Mountains are composed of the Shackleton Limestone of the Byrd Group and of the Goldie Formation of the underlying Beardmore Group. A recent restudy of the geology of this area has caused a significant revision of the stratigraphy and environment of deposition of these rocks (Section 5.3.2) (Excerpt from the topographic map: Nimrod Glacier, Antarctica (SU 56–60/10) published in 1966 by the US Geological Survey)

5.3 Byrd Group, Churchill Mountains

Outcrops of the Byrd Group occur in the Churchill Mountains in Fig. 5.8 and extend south from the Byrd Glacier toward the Nimrod Glacier and into the Queen Alexandra Range. The Byrd Group is in stratigraphic contact with the Beardmore Group in the Holyoake Range north of the Nimrod Glacier and in the Cotton Plateau of the Queen Elizabeth Range south of it (Grindley and Laird 1969). Rocks of the Byrd Group are also present adjacent to the Shackleton Glacier and in small areas farther south where it has been extensively intruded by plutons of the Granite Harbor Intrusives (McGregor and Wade 1969). The rocks of the Byrd Group, like those of the underlying Beardmore Group, are folded and are now considered to be conformable with the Goldie Formation of the Beardmore Group (Goodge 1997) although many previous investigators concluded that the contact is an unconformity.

In the northern part of the central Transantarctic Mountains between the Byrd and Nimrod Glaciers, Laird (1963), Skinner (1964, 1965), and Grindley and Laird (1969) described the stratigraphy of the Byrd Group in terms of the formations listed in Table 5.1:

> Dick, Douglas, and Starshot Formations:
>
> Sandstone, conglomerate, shale, volcanics
>
> Shackleton Limestone and Selborne Marble:
>
> Archaeocyatha limestone, marble, conglomerate, and quartzite

All investigators agree that the rocks of the Byrd Group were deposited during the Cambrian Period because the Shackleton Limestone as well as the Taylor Formation (in the Shackleton Glacier area) contain fossils of Cambrian age (Laird and Waterhouse 1962; Hill 1964a, b; Palmer and Gatehouse 1972; Yochelson and Stump 1977; Debrenne and Kruse 1986).

Rowell et al. (1988a) published excellent photographs of the shelly fauna and of eodiscoid and polymeroid trilobites of the Shackleton Limestone in the Churchill Mountains and at the head of the Beardmore Glacier where outcrops of the Shackleton Limestone were described by Rowell and Rees (1989) from an area south of Mt. Bowers (85°00′S, 164°05′E) on Buckley Island.

5.3.1 Conventional Stratigraphy

The Selborne Marble in Table 5.1 consists of highly deformed and metamorphosed scapolite-bearing marbles interbedded with biotite schist and hornfels that occur only at Cape Selborne and at the north side of Couzens Bay, both of which are located south of the mouth of the Byrd Glacier. The relation of the Selborne Marble to the Shackleton Limestone is uncertain because the two units are not in contact. Skinner (1964, 1965) suggested that the Selborne Marble is either a highly altered part of the Shackleton Limestone or is a marble-amphibolite complex of the Precambrian Nimrod Group. Stump (1980) who worked on the Shackleton Limestone at nearby Mount Madison favored the explanation that the Selborne Marble is a highly metamorphosed member of the Shackleton Limestone.

The Shackleton Limestone at the base of the Byrd Group consists of fine-grained, multi-colored, mostly unmetamorphosed but tightly folded limestone (Grindley and Laird 1969). According to Laird (1963, 1964), the thickness of the Shackleton limestone exceeds 9,000 m. However, later workers have questioned this estimate because the apparent thickness of the Shackleton Limestone is augmented by folding and faulting (e.g., Burgess and Lammerink 1979; Stump et al. 1979; Rees et al. 1985,1987; Rowell et al., 1988b, 1992; Rowell and Rees, 1991) and it probably does not exceed 2,000 m (Stump 1995).

The Shackleton Limestone in the Holyoake Range (Fig. 5.8) contains intertidal cryptalgal laminites, intraformational conglomerates, and archaeocyathid-algal reefs that formed in shallow water on a carbonate reef in depositional environments that ranged from intertidal to near storm-wave base (Rees et al. 1985, 1988, 1989; Rowell et al. 1988a, b; Rowell and Rees 1991). These shallow-water deposits grade laterally into oolitic limestone and thin-bedded peloidal wackestone (mud-supported sedimentary carbonate rocks containing more than 10% grains having diameters greater than 20 μm; Jackson 1997) that formed in deeper water farther off shore.

Archaeocyathid-bearing limestone boulders were first collected on the Beardmore Glacier by members of Shackleton's Nimrod Expedition (1907–1909) as well as by Scott' Ss Terra Nova Expedition (1910–1912). The first archaeocyathid-bearing rocks in situ were found by Laird and Waterhouse (1962) near Mt. Hunter in the Holyoake Range. These authors also suggested that the archaeocyathid-bearing boulders on the Beardmore Glacier originated from a small exposure of the Shackleton Limestone on the south end of Buckley Island. A few years later, Hill (1964a) found additional limestone boulders containing archaeocyathids at Plunket Point at the northern tip of the Dominion Range located only about 20 km from the limestone outcrop on Buckley Island. The presence of archaeocyathids indicated for the first time that the depositional age of the Shackleton Limestone is Early to Middle Cambrian (Laird 1963). This conclusion was a milestone in the exploration of the basement rocks of the Transantarctic Mountains and caused Grindley (1963) to name the Shackleton Limestone in honor of Ernest Shackleton who led the Nimrod Expedition (Section 1.4.2). Rees et al. (1985) also reported the presence of fossil algae (Epiphyton and Renalcis) as well as shelly fossils including mollusks, brachiopods, and trilobites of Early Cambrian age.

The Shackleton Limestone is overlain by clastic sedimentary rocks that have been assigned to the Dick, Douglas, and Starshot formations. The Dick Formation is composed of siltstone, argillite, sandstone, and silty limestone. Rees et al. (1988) reported that sandstones of the Dick Formation are ripple-marked and cross-bedded and that interbedded shales contain rare mudcracks. In addition, an outcrop of the Dick Formation located east of Mt. Dick (89°49′S, 159°32′E) and about 20 km south of the Byrd Glacier contains a lava flow that is 9 m thick and is composed of spilite (Skinner 1964, 1965). Rees et al. (1988) later interpreted this lava flow as a pillow basalt and noted that the contact between it and sedimentary rocks of the Byrd Group was covered by snow. They also reported that they had obtained a whole-rock K-Ar date of 586 ± 20 Ma (Neoproterozoic), which is incompatible with the presence of Early Cambrian fossils in the underlying Shackleton Limestone. Therefore, Rees et al. (1988)

concluded that the volcanic rocks at Mt. Dick are not part of the Byrd Group. The age and affiliation of this lava flow remain in doubt because it is extensively altered to chlorite, sericite, and granular epidote (Skinner 1964, 1965), which caused Stump (1995) to question the reliability of the age determination reported by Rees et al. (1988). In addition, rocks of Neoproterozoic age are not known to be present near this locality. However, the eruption of lava flows during the deposition of the sedimentary rocks of the Byrd Group is not necessarily excluded.

The Douglas Formation consists of polymict conglomerates and breccias that were deposited on an erosion surface of folded Shackleton Limestone. However, Stump (1995) reported that, at its type locality, the Douglas Formation conformably overlies the finer-grained clastic rocks of the Dick Formation. The conglomerates have a sandy matrix that has been altered to epidote, albite, calcite, chlorite, and limonite, which together define the quartz-albite-muscovite-chlorite subfacies of the greenschist facies (Skinner 1964, 1965). The boulders of the conglomerate are composed of quartz and limestone with minor argillite, siltstone, metaquartzite, arkosic sandstone, diorite, two-mica granite, gneissic granite, and amygdaloidal spilite. The Douglas Conglomerate reaches a thickness of more than 500 m at a locality called Castle Crags (82°01′S, 159°12′E) where it lies unconformably on karsted Shackleton Limestone. Rees and Rowell (1991) interpreted the clastic rocks of the Dick and Douglas Formations to be alluvial fan deposits that may have formed in a marine environment. The field evidence cited by Rees and Rowell (1991) indicates that the Shackleton Limestone was folded at least once prior to being exposed to erosion and prior to contributing most of the boulders preserved in the Douglas Conglomerate which itself was deformed after deposition.

The Starshot Formation of calcareous sandstone and shale with minor pebble conglomerate was mapped by Laird (1963) and Laird et al. (1971) on the eastern side of the Starshot Glacier. At the type locality at Mt. Ubique (81°30′S, 160°32′E) the formation has a thickness of about 470 m. The sedimentary rocks are interbedded with several orange-colored layers of altered volcanic rocks tentatively identified as rhyolite by Laird (1963). These volcanic rocks predate the Granite Harbor Intrusives which are post-tectonic in origin. The rocks of the Starshot Formation were deposited, at least in part, in shallow water and were extensively folded.

5.3.2 Revision of the Stratigraphy

The depositional age of the Goldie Formation of the Beardmore Group was assumed to be Neoproterozoic based on a Sm-Nd isochron date of 762 ± 24 Ma of gabbro and basalt that occur within the Goldie Formation at the Cotton Plateau south of the Nimrod Glacier (Borg et al. 1990). Most investigators also agreed that the Goldie Formation was deposited by turbidity currents in deep water off the passive margin of Antarctica soon after the breakup of the supercontinent Rodinia. This scenario was questioned by Myrow and Goodge (1999) whose study of the Goldie Formation indicated that most, but perhaps not all, of the rocks were deposited in shallow water in a marine environment. In addition, Goodge et al. (2002) reported a U-Pb concordia date of 668.1 ± 1.2 Ma for zircons they extracted from the gabbro in the Goldie Formation at the Cotton Plateau. This date is younger and more precise than the Sm-Nd date obtained by Borg et al. (1990) for gabbro and basalt collected at the same site.

Goodge et al. (2002) also dated large numbers of detrital zircons in samples of fine-grained sandstone of the Goldie and Cobham Formations by means of SHRIMP mass spectrometers. They preferentially dated the clear overgrowths and avoided the cores of the detrital zircon grains because the overgrowths formed after deposition as a result of subsequent metamorphism during deep burial, deformation, and intrusion of granite plutons. The results of this study were a complete surprise because they showed that the Goldie Formation does not consist of a monotonous sequence of deep-water turbidites of Neoproterozoic age. Zircon grains in samples of the Cobham Formation in the Cobham Range and in the Goldie Formation from the Cotton Plateau were found to have polymodal distributions of U-Pb dates ranging from 3290 (Paleoarchean) to 1065 ± 65 Ma (Mesoproterozoic) in accordance with the time scale of the IUGS (2002). Evidently, these zircon grains originated from igneous, metamorphic, and pre-existing sedimentary source rocks of the East Antarctic craton.

The U-Pb dates of detrital zircons in samples of the Goldie Formation collected from outcrops closer to the present coast near Mt. Markham in the Queen Elizabeth Range and in the Commonwealth Range along the south side of the Beardmore Glacier have a completely different distribution. Although the U-Pb dates of these zircons also have a polymodal distribution, they are strongly clustered between 645 and 520 Ma

(Neoproterozoic to Early Cambrian) which characterizes the igneous rocks that formed during the Ross Orogeny. In addition, U-Pb dates of zircon populations in these samples have prominent peaks at 1100 and 1085 Ma which coincide with the Grenvillian Orogeny of North America. This may seem strange, but it actually corroborates a radical hypothesis concerning the present location of Laurentia (Moores 1991; Dalziel 1992, 1997).

The youngest zircon grains in the Goldie Formation of the Queen Elizabeth and Commonwealth ranges have U-Pb dates of 552 ± 18 Ma and 525 ± 6 Ma, respectively. These age determinations demonstrate that the sediment of the Goldie Formation in the present coastal ranges of the Transantarctic Mountains was largely derived from the Ross orogen in contrast to rocks of the Beardmore Group that are exposed farther inland in the Cobham Range and the Cotton Plateau which are composed of sediment derived from Archean and Proterozoic sources located in the hinterland on the East Antarctic craton. Goodge et al. (2002) referred to the two parts of the Goldie Formation as the "inboard assemblage" and the "outboard assemblage." The age of deposition of the inboard sediment is less than 1143 ± 15 Ma (Cobham Formation) and less than 1065 ± 65 Ma (Goldie Formation, Cotton Plateau). In fact, the U-Pb date of zircon in the gabbro of the Cotton Plateau indicates that the "inboard" parts of the Goldie Formation were deposited as late as 668 ± 1 Ma. The outboard part of the Goldie Formation was deposited less than about 520 million years ago (Middle Cambrian or younger).

The early Paleozoic age of the outboard section of the Goldie Formation is supported by the discovery of trace and body fossils of apparent Cambrian age at the collecting sites of the outboard samples (Goodge et al. 2002). In other words, the outboard section of the Goldie Formation was deposited during and after the Ross Orogeny at the same time as sediment of the Byrd Group. This important insight has required a major revision of the stratigraphy of the Beardmore and the Byrd groups.

Based on the new information, Goodge et al. (2002) and Myrow et al. (2002) revised the stratigraphy of the Beardmore and Byrd groups in the central Transantarctic Mountains by including the outboard part of the Goldie Formation of the Beardmore Group with the Starshot Formation of the Byrd Group. This assignment is valid because the outboard part of the Goldie Formation and the Starshot Formation are both composed of clastic sediment that was deposited during the same interval of time along the paleo-Pacific coast of East Antarctica and because both rock sequences contain sediment derived from magmatic rocks that intruded the rocks of the Beardmore Group during the Ross Orogeny. Accordingly, Goodge et al. (2002) revised the geologic map in Fig. 5.9 by combining the outcrop areas of outboard Goldie, Starshot, and Dick Formations into one continuous belt that extends along the coast from the Byrd Glacier all the way to the Commonwealth Range south of the Beardmore Glacier. The entire assemblage of clastic sedimentary rocks of the newly-defined upper Byrd Group formed by erosion of the Ross Mountains which began to shed sediment at about 540 Ma when the tectonic activity started that culminated in the Ross Orogeny. Contrary to previously held views, only the Cobham Formation and the inboard Goldie Formation formed close to the Neoproterozoic rift margin by deposition of sediment that originated from Archean and Paleoproterozoice rocks of the East Antarctic craton.

An alternative hypothesis favored by Borg et al. (1990), Encarnación and Grunow (1996), and Grunow and Encarnación (2000) proposes that the basement rocks of the coastal mountains formed as result of the accretion of a microcontinent to the paleo-Pacific coast of East Antarctica. However, Goodge et al. (2002) and Myrow et al. (2002) emphasized that all of the sediment of the Beardmore and Byrd groups was transported by currents that flowed from west to east and that the sources of sediment were located on the East Antarctic craton and in the Ross Mountains that were being uplifted and deformed along the former Neoproterozoic rift margin.

In addition, the U-Pb date of zircons in the gabbro of the Cotton Plateau (668 ± 1 Ma) implies that Rodinia broke up somewhat later and that the rift valley between Laurentia and the East Antarctic craton began to trap sediment later than the previous interpretation of the evidence had indicated. The reassignment of the bulk of the Goldie Formation to the Starshot Formation of the Byrd Group also means that the contact between the inboard part of the Goldie Formation and the overlying Shackleton Limestone is now presumed to be an unconformity.

According to Myrow et al. (2002), the transition from the carbonate rocks of the Shackleton Limestone to the clastic rocks of the upper Byrd Group in the Holyoake-Cotton Plateau area (i.e., the Douglas, Dick, and Starshot formations) records the change from sedimentation along a passive continental margin to

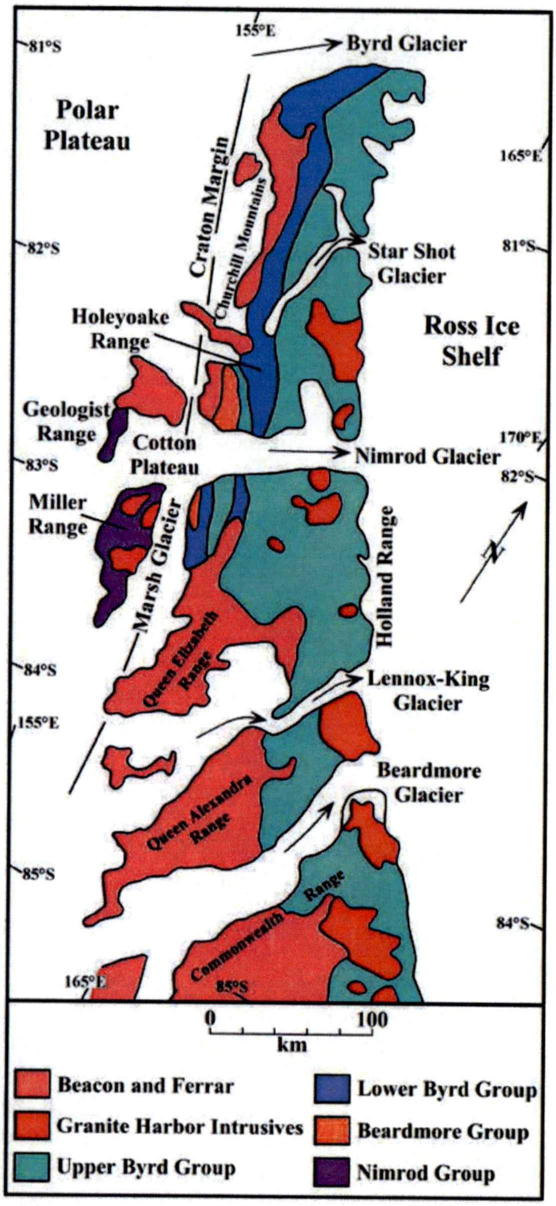

Fig. 5.9 In this revised geologic map of the central Transantarctic Mountains the "outboard" part of the Goldie Formation of the Beardmore Group has been reassigned to the Starshot Formation of the overlying Byrd Group. Consequently, the coastal mountain ranges are composed of the upper Byrd Group, whereas the "inboard" part of the Beardmore Group, consisting of the Cobham Formation and the basal Goldie Formation, occurs only in the Cobham Range and the Cotton Plateau of the Queen Elizabeth Range. The rocks of the Nimrod Group are exposed only in the Miller and Geologist Ranges which are located on the East Antarctic craton (Adapted from Goodge et al. 2002)

more active sedimentation of weathering products generated by the uplift and erosion of the Ross Mountains. Myrow et al. (2002) observed that the Shackleton Limestone is capped by archaeocyathan bioherms (i.e., mounds of carbonate rock built up by sedentary organisms enclosed by rocks of a different lithology) up to 40 m high relative to the seafloor. These bioherms are covered by a phosphate layer which was deposited as the depth of water increased prior to the start of the Ross Orogeny. The carbonate reefs were subsequently buried by black shale and by a mixture of nodular carbonate and shale which constitute the newly defined Holyoake Formation (Myrow et al. 2002).

The Holyoake Formation grades upward into the trilobite- and hyolithid-bearing calcareous siltstones of the Starshot Formation and into the alluvial fan deposits of the Douglas Conglomerate (the hyolitha are an extinct taxon of Paleozoic invertebrates that had a conical shell with a single aperature; Jackson 1997). The trilobites in the basal siltstones of the Starshot Formation indicate that the transition from passive- to active-margin sedimentation occurred in latest Early Cambrian (late Botomian). The transition from carbonate reefs to clastic sedimentation was followed by the deposition of upward-coarsening siliciclastic rocks which prograded eastward across the continental shelf. The resulting clastic formations of the upper Byrd Group were deposited simultaneously in different environments such that the coarse-grained alluvial-fan deposits formed adjacent to and contemporaneously with the more distal-shelf siltstones of the Starshot Formation. The transport direction of sediment was from west to east (or southwest to northeast). This revised stratigraphy of the pre-Devonian basement rocks is summarized in Table 5.3.

5.3.3 Pegmatite, Mt. Madison

The Shackleton Limestone of the Churchill Mountains south of the Byrd Glacier is tightly folded and was intruded by "… small plutons of hornblende granite, pyroxene-hornblende diorite or gabbro, and dikes of quartz porphyry, all probably related to the Hope Granite" (Grindley and Laird 1969). Faure and Felder (1984) described and sampled a complex pegmatite that intruded the Shackleton Limestone on one of two

Table 5.3 Revised stratigraphy of the basement rocks of the central Transantarctic Mountains (Myrow et al. 2002)

Geologic age	Group	Formations
late Middle to late Early Cambrian	Upper Byrd	Dick Conglomerate[a] Starshot/Goldie (outboard Douglas Conglomerate
Early Cambrian		Holyoake
~~	Disconformity	~~~~~~
Early Cambrian	Lower Byrd	Shackleton Limestone
~~	Unconformity	~~~~~~
Neoproterozoic	Beardmore	Goldie (inboard Cobham
~~	Unconformity	~~~~~~
Paleoproterozoic	Nimrod	Schists, gneisses, and marble

[a]The Douglas, Starshot, Goldie (outboard), and Dick Formations are different facies of sedimentary rocks deposited simultaneously in different environments

spurs on the north face of Mt. Madison marked by an "x" in Fig. 5.10a. The larger context of this area is apparent by reference to Fig. 5.9.

The physical dimensions of the pegmatite are approximately 60 × 90 × 5 m, giving it a volume of approximately 27,000 m^3. The pegmatite in Fig. 5.10b contains large crystals (up to 10 cm long) of spodumene ($LiAlSi_2O_6$) as well as lepidolite ($(K_2Li_3Al_4Si_7O_{21})OH,F)_3$), quartz, albite, K-feldspar, muscovite, and green tourmaline. Thin quartz veins extending from the pegmatite contain bismuthinite (Bi_2S_3) and small amounts of stibnite (Sb_2S_3), galena (PbS), and an unidentified copper-sulfide mineral. The pegmatite has a chilled contact against the Shackleton Limestone which has recrystallized to form coarse-grained and bleached marble in the contact aureole. Two hand specimens of pegmatite containing spodumene and lepidolite and one concentrate of bismuthinite and other sulfide minerals from a quartz vein were analyzed by A.S. McCreath and Son, Inc. of Harrisburg, PA. The results in Table 5.4 indicate that the two pegmatite samples Li-1 and Li-2 contain 1.28% and 2.35% Li_2O, respectively. The concentrate of bismuthinite and other sulfide minerals, analyzed by semi-quantitative optical spectroscopy, contains "major" concentrations of Bi, as well as 0.5–5% Sb, 0.3–3% each of Cu and Pb, as well as smaller concentrations (0.03–0.3%) of Cr and Fe and 0.02–0.2% of V.

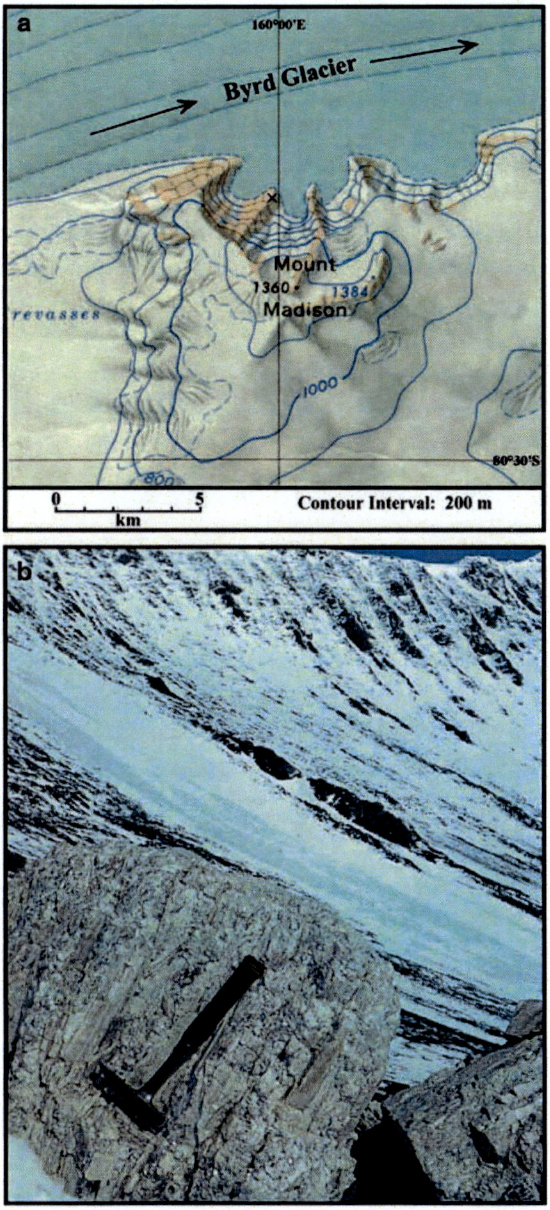

Fig. 5.10a Mt. Madison (80°27′S, 160°02′E) is located on the south side of the mouth of the Byrd Glacier. Its north-facing slope contains three small cirques that are separated from each other by two bedrock spurs which are composed of the Shackleton Limestone of the Byrd Group. **Fig. 5.10b** The crest of the western spur contains an outcrop of pegmatite composed of spodumene, lepidolite, bismuthinite, and tourmaline (Faure and Felder, 1984). The spur on which the pegmatite is located is marked by an "x." This map is an excerpt of the topographic map: Cape Selborne, Antarctica (SU 56–60/2) published in 1963 by the US Geological Survey in Washington, DC Photo: G. Faure

Table 5.4 Concentrations of Li-oxide and other metallic elements in samples of the pegmatite on Mt. Madison in Fig. 5.10 (Faure and Felder 1984)

Sample	Analyte	Concentration (wt%)
Li-1	Li-oxide	1.38
Li-2	Li-oxide	2.35
Bi-1[a]	Bismuth	"Major"
	Antimony	0.5–5
	Lead	0.3–3
	Copper	0.3–3
	Chromium	0.3–3
	Iron	0.03–0.3
	Vanadium	0.02–0.2

McCreath and Sons, Inc., Harrisburg, PA, chemical analysts
[a]Sample Bi-1 was analyzed by semi-quantitative optical spectrography. In addition to the metals listed in the table, sample Bi-1 contained detectable concentrations of Mn, B, Sn, Ag, and Mg

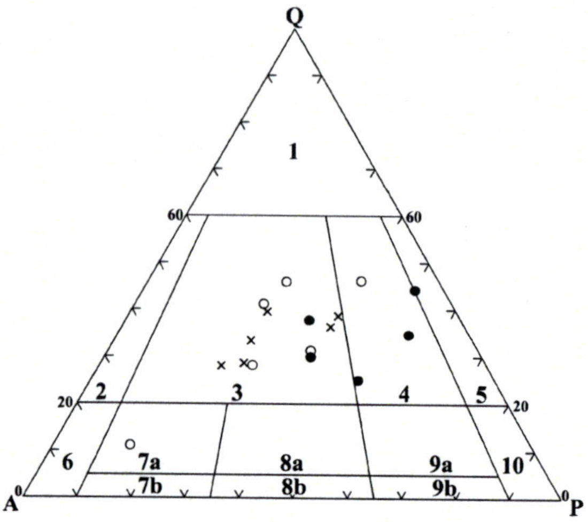

Fig. 5.11 The plutonic granitic rocks that intruded the metasedimentary rocks of the Nimrod Group in the Miller Range (solid circles) and of the Beardmore and Byrd Groups (open circles) along the Shackleton Coast between the Nimrod and Beardmore Glaciers consist primarily of granite (field #3) and granodiorite (field #4). The Ida Granite (x), which is a variant of the Hope Granite on Mt. Ida (83°35′S, 170°29′E) located on the west side of the mouth of the Beardmore Glacier, is also composed of granite in the classification of Streckeisen (1967) in Appendix 3.6.1. Modal analyses by Gunner (1976)

Faure and Felder (1984) also reported a whole-rock Rb-Sr date of 490 ± 9 Ma for sample Li-2 based on Rb = 1,444.7 ± 12.1 parts per million (ppm), Sr = 15.55 ± 0.15 ppm, $^{87}Sr/^{86}Sr = 3.009 \pm 0.005$, and assuming that the initial $^{87}Sr/^{86}Sr$ ratio was 0.7100 and $\lambda = 1.42 \times 10^{-11}$ year^{-1}. The date represents the time elapsed since cooling of the pegmatite and is consistent with K-Ar and Rb-Sr dates of biotite of the Hope Granite in the central Transantarctic Mountains (McDougall and Grindley 1965; Grindley and McDougall 1969; Adams et al. 1982).

The Sr concentration of the Shackleton Limestone adjacent to the pegmatite is 179.3 ppm and its $^{87}Sr/^{86}Sr$ ratio is 0.70996 ± 0.00031. The comparatively high value of the $^{87}Sr/^{86}Sr$ ratio of the Shackleton Limestone is consistent with the increase in this ratio in the oceans during the Cambrian Period which peaked during the late Middle Cambrain at about 0.7093 (Faure and Mensing 2005).

5.4 Petrogenesis of the Hope Granite

The plutons of Hope granite that intruded the metamorphic rocks of the Nimrod Group in the Miller Range and the Beardmore Group in the Queen Elizabeth Range are composed primarily of granite and granodiorite and do not appear to be differentiates of a more primitive parent magma (Fig. 5.11). These plutons did not form simultaneously although their intrusion and subsequent cooling did occur in the course of the Ross Orogeny during the Cambrian and Ordovician periods. The Hope Granite plutons that intruded the rocks of the Beardmore Group are post-tectonic and have well-defined contacts, whereas the Hope Granite plutons in the Miller Range display lit-par-lit injection and are in some cases syntectonic.

The investigation of the petrogenesis of the Hope Granite in the Miller and Queen Elizabeth Ranges was motivated by the following hypothesis posed by Gunner (1976):

> Granitic rocks that intrude metamorphic rocks in orogenic belts may have formed by crystallization of magma that originated by anatexis of volcano-sedimentary complexes at depth in the root zone of an orogen. In that case, the initial $^{87}Sr/^{86}Sr$ ratios of such anatectic granites are genetically related to the metamorphic rocks into which they were intruded.

Evidence bearing on the origin of the Hope Granite was originally presented by Gunner (1971a, b, 1974, 1976) and Gunner and Mattinson (1975) in the form of a whole-rock Rb-Sr diagram that indicated that the Hope plutons in the Miller and Queen Elizabeth ranges had crystallized at nearly the same time from magmas that had quite different $^{87}Sr/^{86}Sr$ ratios. A later study by Borg et al. (1990) confirmed this result and both data sets are presented in Fig. 5.12a and b. The whole-rock Rb-Sr dates and initial $^{87}Sr/^{86}Sr$ ratios of the Hope Granite in Table 5.5 also include the results for granitic rocks in the Gabbro Hills (84°42′S, 173°00′W) located on the coast east of the mouth of the Shackleton Glacier.

The results of these two studies are in excellent agreement and support three important conclusions about the petrogenesis of the Hope Granite in the central Transantarctic Mountains:

1. The average crystallization age of the plutons of the Hope Granite is 510 ± 9 Ma (1σ) indicating that they formed within a period of about 20 million years during the Ross Orogeny.
2. The Hope Granites that intruded the metasedimentary rocks of the Nimrod Group have higher initial $^{87}Sr/^{86}Sr$ ratios than plutons that intruded the Beardmore/Byrd Group metasediments.
3. The low initial $^{87}Sr/^{86}Sr$ ratio of the Hope Granite in the Gabbro Hills indicates that the granitic magma contained strontium derived from mafic volcanic rocks in the underlying lithospheric mantle.

The difference in the initial $^{87}Sr/^{86}Sr$ ratios of the Hope Granites suggests, but does not demonstrate convincingly, that the rocks of the Nimrod and Beardmore Groups had different average $^{87}Sr/^{86}Sr$ ratios at the time of magma formation by anatexis. Therefore, Gunner (1971a, 1976) prepared and analyzed seven composite rock samples of the Nimrod Group and six composites of the Beardmore Group taken from the same mountains from which the samples of Hope Granite originated. He reported that the $^{87}Sr/^{86}Sr$ ratios of the Beardmore composites (at the time of crystallization of the Hope Granite) ranged narrowly between 0.71257 and 0.71897, whereas the initial $^{87}Sr/^{86}Sr$ ratios of the Nimrod composites range more widely from 0.7065 to 0.74238. Nevertheless, these isotope ratios did not match the initial $^{87}Sr/^{86}Sr$ ratios of the Hope Granites in Table 5.5 well enough to prove a genetic connection between the metasedimentary rocks and the granite plutons that intruded them (Gunner, 1982).

Therefore, we have enlarged the data set by including initial $^{87}Sr/^{86}Sr$ ratios at 500 Ma of all of the individual rock samples of the two groups that were analyzed by Gunner and Mattinson (1975) and by Borg et al. (1990) in Fig. 5.12a and b. The results in Fig. 5.13a demonstrate that the samples of Hope Granite from plutons that intruded Beardmore-Group metasediments have lower initial $^{87}Sr/^{86}Sr$ ratios at 500 Ma than samples of Hope Granite that intruded Nimrod Group metasediments. This result strengthens the conclusion derived from Fig. 5.12a and b because the data of Gunner and Borg were merged to construct the histograms in Fig. 5.13a.

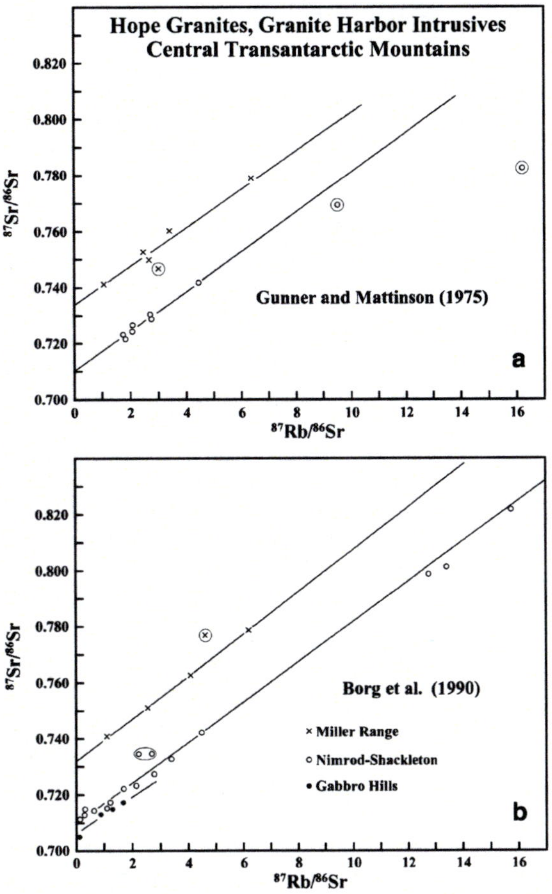

Fig. 5.12 **a** and **b** The whole-rock Rb-Sr data points of Hope Granites in different regions of the central Transantarctic Mountains (T.A.M.) constrain parallel straight lines that yield statistically identical dates but different initial $^{87}Sr/^{86}Sr$ ratios listed in Table 5.5. Data points that were excluded are identified by being circled. The data published by Gunner and Mattinson (1975) were recalculated to $\lambda = 1.42 \times 10^{-11}$ year^{-1} and by excluding samples that had high Rb/Sr ratios which appear to have lost some of the in-situ produced radiogenic ^{87}Sr. The data sources are reference within the diagram

The initial $^{87}Sr/^{86}Sr$ ratios of metasedimentary rocks at 500 Ma of the Beardmore Group in Fig. 5.13b lie within the range of the initial $^{87}Sr/^{86}Sr$ ratios of Hope

Table 5.5 Whole-rock Rb-Sr age determinations of Hope Granite from plutons that intruded the Nimrod Group metasediments in the Miller Range and the Beardmore and Byrd Groups in the coastal mountains between the Nimrod and the Shackleton Glaciers (Gunner 1971a, b, 1974, 1976; Gunner and Matttinson 1975; Borg et al. 1990) (Numbers in parentheses represent the number of samples analyzed)

Parameter	Gunner's data	Borg's data
Miller Range (Nimrod Group)		
Rb-Sr date, Ma	511 ± 18[a]	516 ± 6 (4)
$(^{87}Sr/^{86}Sr)_i$	0.73167 ± 0.00082[a]	0.73226 ± 0.00027
Coastal Mountains (Nimrod to Beardmore Glaciers)		
Rb-Sr date, Ma	501 ± 13[a]	500 ± 10 (14)
$(^{87}Sr/^{86}Sr)_i$	0.70864 ± 0.00047[a]	0.70995 ± 0.00058
Gabbro Hills[b] (84°42′S, 173°00′W)		
Rb-Sr date, Ma	–	521 ± 35 (4)
$(^{87}Sr/^{86}Sr)_i$	–	0.70484 ± 0.00029

[a]Recalculated to $\lambda = 1.42 \times 10^{-11}$ year^{-1} after excluding samples that are circled in Fig. 5.12a

[b]Located along the Dufek Coast and east of the mouth of the Shackleton Glacier

plutons that intruded Beardmore-Group metasediments. Similarly, the initial $^{87}Sr/^{86}Sr$ ratios of Nimrod metasediments in Fig. 5.13c range from greater than 0.705 to less than 0.770 and encompass the initial ratios of the Hope Granites in the Miller Range.

Therefore, we conclude that the granitic magmas that formed the plutons of Hope Granite could have formed by anatexis of the metasedimentary rocks of the Ross orogen which these magmas intruded. The isotopic composition of strontium in the metasedimentary rocks of the Nimrod Group was heterogeneous at the time the granite magmas formed because it is composed of a wide variety of rock types having a wide range of Rb/Sr ratios including marble, amphibolite, biotite schist, and quartz-feldspar gneiss. In addition, these rocks had been variously enriched in radiogenic ^{87}Sr by decay of ^{87}Rb between 1725 ± 5 Ma (Table 5.2) and the time of anatexis at 500 Ma. The Goldie Formation of the Beardmore Group has a more uniform lithologic composition than the Nimrod Group and was deposited at 668 ± 1.2 Ma only about 170 million years before granitic magmas formed by partial melting. The demonstration of a probable genetic link between granite plutons and the country rocks they intrude has relevance elsewhere in the world. Therefore, this insight is an important contribution to the petrogenesis of intrusive granites.

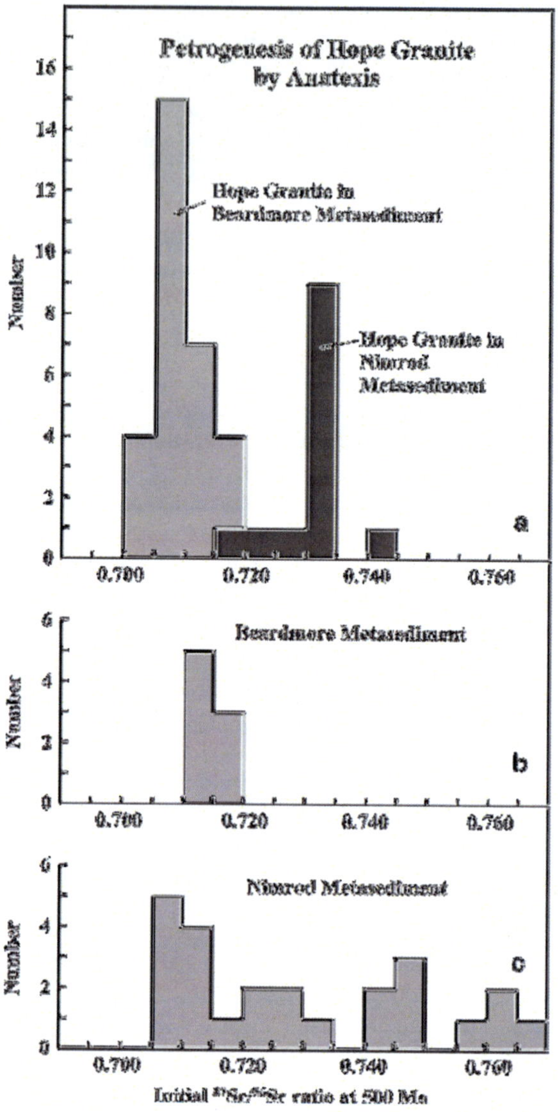

Fig. 5.13 **a)** Plutons of Hope Granite in the central Transantarctic Mountains that intruded the metasedimentary rocks of the Beardmore Group (late Neoproterozoic) have lower initial $^{87}Sr/^{86}Sr$ ratios at 500 Ma than Hope plutons that intruded the Nimrod metasediments of the Miller Range. **b)** The initial $^{87}Sr/^{86}Sr$ ratios of Beardmore metasediments match those of the Hope plutons that intruded them. **c)** The intial $^{87}Sr/^{86}Sr$ ratios of the schists and gneisses of the Nimrod Group range widely above and below the intial $^{87}Sr/^{86}Sr$ ratios of the Hope plutons that intruded them. These data demonstrate that the granitic magmas of the Hope Granites in the central Transantarctic Mountains could have formed during the Ross Orogeny by anatexis (partial melting) of the metasedimentary rocks they intruded (Data from Gunner 1971c, 1976; Borg et al. 1990)

We also refer here to the isotope compositions of neodymium and the Sm/Nd ratios of Hope plutons in the Gabbro Hills, Nimrod-Shackleton Glacier area, and in the Miller Range reported by Borg et al. (1990). These data scatter on the Sm-Nd geochronometry diagram (not shown) and therefore cannot be used to date these rocks. Instead, Borg et al. (1990) used the isotope ratios of Sr and Nd at 500 Ma to plot a Sr-Nd isotope mixing diagram similar to Fig. 5.14. The systematic difference of the initial isotope ratios of Sr and of Nd of Hope Granites that intruded metasedimentary rocks of the Beardmore and Nimrod Groups supports the hypothesis that the granitic magmas formed locally by partial melting of the adjacent country rocks. In other words, the clustering of data points in Fig. 5.14 arises because the sediment that formed the protoliths of the granite magmas originated from sources of differing age and chemical composition.

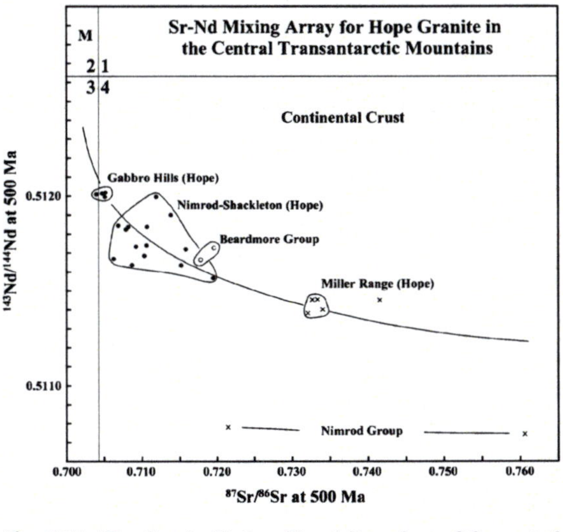

Fig. 5.14 The Granite Harbor (Hope) Intrusives of the central Transantarctic Mountains (T.A.M.) loosely constrain a hypothetical Sr-Nd isotopic mixing hyperbola at 500 Ma. All of the data points lie in quadrant 4 which is the domain of rocks in the continental crust. The rock samples from the Gabbro Hills between the Shackleton and Liv Glaciers contain more Sr and Nd of mantle origin than the Hope Granites that intruded the Beardmore and Nimrod Groups. The systematic differences between the initial $^{87}Sr/^{86}Sr$ and $^{143}Nd/^{144}Nd$ ratios of the Hope Granites confirm the existence of a genetic link between the isotope composition of Sr and Nd of the granite magmas and the ages and compositions of the metasedimentary protoliths from which they formed. The $^{143}Nd/^{144}Nd$ ratios were renormalized to $^{146}Nd/^{144}Nd = 0.7219$ as explained by Faure and Mensing (2005: 198–199) (The data were published by Borg et al. 1990)

5.5 Beardmore to Shackleton Glaciers

The coastal mountains of the southern region of the central Transantarctic Mountains in Fig. 5.15 are composed of Goldie Formation (Beardmore Group) and of turbidites and metavolcanic rocks (Byrd Group) which occur only adjacent to the Shackleton Glacier and which have not been correlated with the formations of the Byrd Group in the Churchill Mountains. The metasedimentary basement rocks of the southern region were intruded during the Ross Orogeny by large plutons of the Hope Granite (McGregor and Wade 1969; Wade and Cathey 1986; Stump 1985, 1995).

The Shackleton Glacier (84°35′S, 176°20′W) marks the southern limit of the central Transantarctic Mountains as defined in Fig. 3.1. It is also the northern limit of the Queen Maud Mountains which extend along the strike of the Transantarctic Mountains to the Reedy Glacier and to the Wisconsin Range of the Horlick Mountains located at about 86°S and 125°W.

The principal outlet glaciers of this segment of the Transantarctic Mountains include the Canyon, Ramsey, and Kosco glaciers in addition to the Beardmore and Shackleton glaciers, all of which are identified in Fig. 5.15. These glaciers flow from south to north as they transport ice of the East Antarctic ice sheet to the Ross Ice Shelf. The coastal mountains are deeply dissected by glacial valleys and include two peaks with elevations in excess of 4,000 m: Mt. Wexler (4,024 m; 84°30′S, 175°01′E) located at the head of the Canyon Glacier and Mt. Wade (4,084 m, 84°51′S, 174°19′W) east of the Shackleton Glacier. In addition, the elevations of six peaks between the Beardmore and the Shackleton Glaciers exceed 3,000 m. The resulting rugged topography enhances the scenic beauty of this area but presents a challenge to field geologist even though they can now reach the summits of the alpine peaks by using helicopters.

5.5.1 Beardmore to Ramsey Glaciers

The basement of the coastal mountains between the Beardmore and the Ramsey Glaciers in Fig. 5.15 consists of a thick sequence of clastic sedimentary rocks that resemble the Goldie Formation of the Churchill

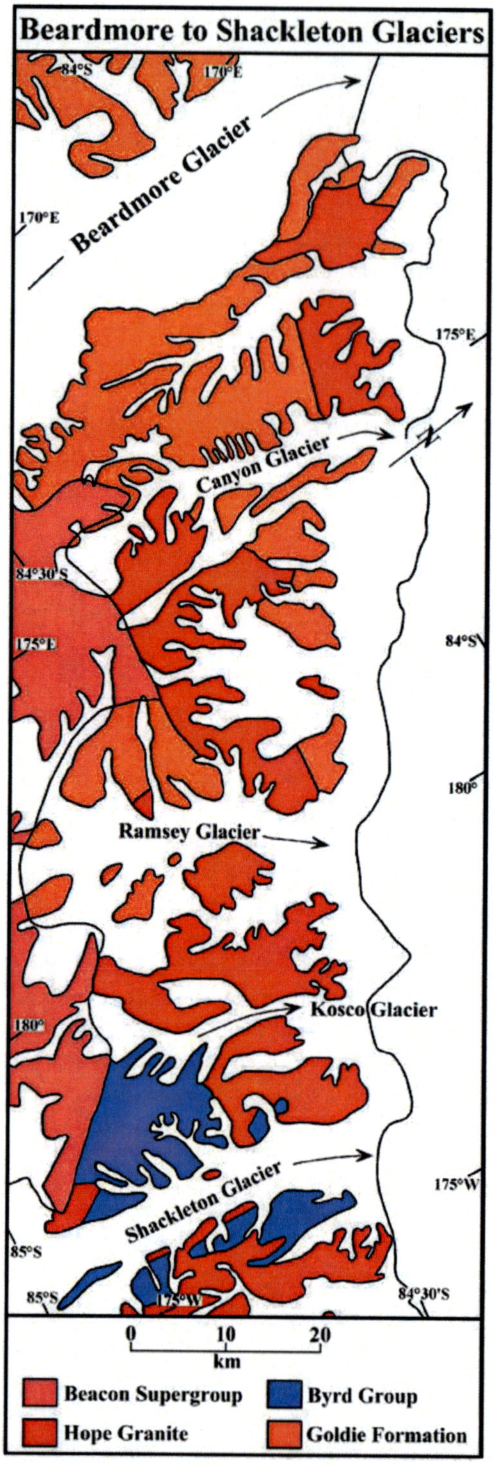

Mountains and the Queen Elizabeth Range. The Goldie Formation in the Beardmore-Ramsey Glacier area is composed of interbedded metagraywackes and phyllites that strike north-south parallel to the valley of the Ramsey Glacier and dip nearly vertically (Wade and Cathey 1986). They were intruded by mafic sills and may be isoclinally folded. The graywackes are coarse grained and are composed predominantly of quartz, orthoclase, microcline, and albite. Muscovite and biotite are also present, as well as accessory chlorite, apatite, zircon, tourmaline, and opaques. Similar rocks are exposed on the eastern side of the Ramsey Glacier. Although the sedimentary graywacke-phyllite complex adjacent to the Ramsey Glacier was assigned to the Goldie Formation, we do not yet know whether these rocks represent the "outboard" Goldie (Byrd Group) or the "inboard" Goldie (Beardmore Group) as defined by Goodge et al. (2002) and by Myrow et al. (2002).

5.5.2 Shackleton Glacier Area

The basement rocks adjacent to the Shackleton Glacier in Table 5.6 consist of interbedded quartzites, volcanic rocks, and marbles that have been subdivided into several formations described by Wade (1974), Stump (1985), and by Wade and Cathey (1986):

Henson marble
Taylor formation
Greenlee formation

The Taylor Formation was shown to be of Early to Middle Cambrian age by Yochelson and Stump (1977) and by Encarnación et al. (1999).

Fig. 5.15 The basement rocks of the southern region of the central Transantarctic Mountains between the Beardmore and Shackleton glaciers consist primarily of the Goldie Formation which was intruded by plutons of the Hope Granite. These units are unconformably overlain by the Beacon Supergroup. The trace of the Kukri Peneplain is indicated as a heavy line at the base of the Beacon rocks. The Byrd Group adjacent to the Shackleton Glacier consists of turbidites and metavolcanic rocks that were intruded by plutons of the Hope Granite. White areas represent snow and ice of the outlet glaciers and of the Ross Ice Shelf (Adapted from McGregor and Wade 1969)

Table 5.6 Stratigraphy of the basement rocks in the coastal mountains between the Ramsey and Shackleton Glaciers in the southern-most part of the central Transantarctic Mountains (Wade and Cathey 1986; Stump 1985; McGregor and Wade 1969; Mirsky 1969)

Age		Geologic units
Devonian to Triassic		Beacon Supergroup
∼∼∼∼∼	Kukri Peneplain	∼∼∼∼∼
Cambrian to Ordovician		Hope Granite of the Granite Harbor Intrusives
———	Intrusive Contact	———
Early to Middle Cambrian		Henson Marble Taylor Formation
———	Conformable or Faulted	———
Neoproterozoic to Early Cambrian[a]		Greenlee Formation

[a]Stump (1985) assigned the Taylor/Henson and Greenlee Formations of the Shackleton Glacier area to the Liv Group. The correlation of the Goldie Formation (Beardmore Group) to the Greenlee Formation is uncertain

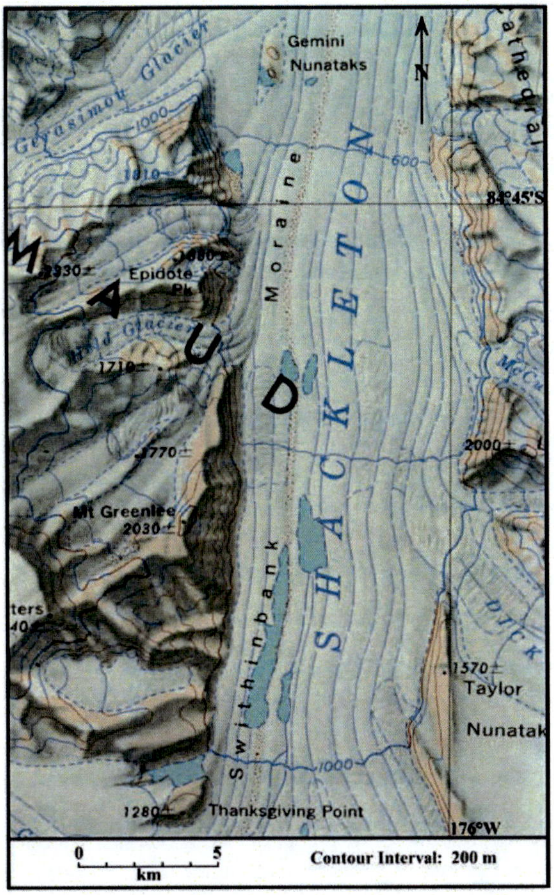

Fig. 5.16 The Gerasimou Glacier, Epidote Peak, Mt. Greenlee, and Taylor Nunatak in the central part of the Shackleton Glacier are the principal landmarks that identify the type section of the Greenlee and Taylor formations and of the Henson Marble which constitute the Byrd Group of this area. Excerpt of the topographic map entitled Shackleton Glacier, Antarctica, SV1–10/1; S 8400-W-16500/1 × 5; published by the US Geological Survey, Washington, DC

The Greenlee Formation in Table 5.6 is exposed at the northern end of Mt. Greenlee in Fig. 5.16 and continues north for about 15 km along the western side of the Shackleton Glacier. It was originally composed of silty muds, argillaceous silts, and impure sands. Limestone is present near the top of the formation. The grain size increases up section from silt near the base to very fine sand in the mid section followed by still coarser sand near the top of the formation. Wade and Cathey (1986) also mentioned the presence of layers of graywacke and of rare basaltic lava flows and ashfall tuffs.

The sediment was deposited near-shore in shallow water of a basin that was slowly subsiding. The sediment was derived from an area of low relief as weathering products of granitic rocks. The rocks of the Greenlee Formation dip south at 45° to nearly 90° and may form the limb of a major fold, the axis of which presumably trends east-west. Wade and Cathey (1986) estimated that the thickness of the Greenlee Formation exceeds 13 km provided that the exposure along the Shackleton Glacier is a continuous sequence without repetitions. However, Stump (1986) stated that the thickness of the Greenlee Formation along the west side of the Shackleton Glacier does not exceed 1,000 m.

The original shales of the Greenlee Formation were first converted to slates, phyllites, and schists during an episode of low-temperature tectonic deformation during which the limestones partly crystallized to marble. Subsequently, the rocks were heated during the intrusion of plutons of the Hope Granite (Granite Harbor Intrusives) which caused them to be converted to hornfels and amphibolites containing thin conformable layers of tonalite and pegmatite dikes.

The overlying Taylor Formation consists of ashfall and ashflow tuffs, felsic and mafic lava flows, and marbles. Wade and Cathey (1986) placed the contact of the Taylor Formation with the underlying Greenlee Formation at about 6 km south of the northern tip of Mt. Greenlee (Fig. 5.16). The contact is obscured by the brecciation and hydrothermal alteration of the Greenlee Formation manifested by the crystallization of epidote.

The beds of the Taylor Formation exposed along the southern spur of Mt. Greenlee dip south at 40–70° and appear to reach a thickness of about 5.7 km. Additional outcrops of the Taylor Formation occur on Taylor Nunatak (84°44′S, 176°10′W) and on Waldron Spur which is located on the east side at the mouth of the Shackleton Glacier (Wade and Cathey 1986).

The ashfall tuff that dominates the Taylor Formation contains angular to subrounded porphyroclasts of quartz, plagioclase, K-feldspar, and lithic fragments. Many porphyroclasts are corroded, embayed, and fragmented. Mafic minerals are not present among the porphyroclasts. The matrix is very fine grained and generally pink, gray, green, or tan in color. It is composed primarily of quartz with minor flakes of sericite as well as small grains of an opaque mineral (magnetite?), secondary calcite, and epidote. Wade and Cathey (1986) suggested that the ashfall tuffs of the Taylor Formation may correlate with the Wyatt Formation which occurs along the Scott Glacier (Stump 1985; Faure et al. 1968; Murtaugh 1969).

Taylor Nunatak in Fig. 5.16 extends parallel to the ice-flow direction of the Shackleton Glacier for about 6 km and the elevation of its summit is 1,570 m. The rocks exposed in its nearly vertical slope that faces the glacier consist of ashfall tuffs, ashfall flows, and marble. The rocks have been folded into a large anticline indicated by a reversal of dip from 70° north at the northern end of the nunataks to 70° south at its southern end. However, the lithologic compositions of the north and south limbs do not match, perhaps because the structure is a fault rather than an anticline. The crest of the nunatak is covered by a sill of Ferrar Dolerite (Middle Jurassic) which is about 12 m thick. The Taylor Formation on Taylor Nunatak is composed of approximately equal proportions of marble and pyroclastics that do not correlate with the rocks of the Taylor Formation on Mt. Greenlee only 7.5 km distant, presumably because the nunatak is located on the opposite side of a fault that extends along the valley of the Shackleton Glacier.

5.5.3 Age of the Taylor Formation

Yochelson and Stump (1977) found a trilobite fragment and what appear to be worm tubes in a bed of limestone breccia that is interbedded with chert and felsic volcanics near the summit of Taylor Nunatak (Fig. 5.16). The tubes have oval-shaped cross-sections that resemble fossils from the Lower Cambrian rocks of San Juan Province in Argentina which Yochelson and Herrera (1974) identified as *Cloudina?* borrelloi. The presence of *Cloudina?* and of trilobite fragments indicates that the age of the Taylor Formation is Early Cambrian in agreement with the age of the Shackleton Limestone of the Churchill Mountains (Hill 1964a, b) and with the Middle Cambrian age of the Leverett Formation in the Scott-Glacier area which also contains trilobite fragments that were identified by Palmer and Gatehouse (1972).

The age of the Taylor Formation was further refined by Encarnación et al. (1999) who collected trilobites from a carbonate bed near the summit of Taylor Nunatak including specimens of *Amphoton* oatesi and *Nelsonia* schesis both of which are endemic to Antarctica and were identified by Palmer and Gatehouse (1972) in the Nelson Limestone of the Neptune Range in the Pensacola Mountains (83°45′S, 055°00′W). These authors tentatively concluded that the *N.* schesis faunule is of late Middle Cambrian age, whereas the *A.* oatesi faunule indicates an early late-Middle Cambrian age. By comparing the occurrence of *Amphoton* trilobites in the Taylor Formation of Antarctica to the biostratigraphic zonation of this species in northern China and Australia, Encarnación et al. (1999) concluded that the fossiliferous limestone bed just below the summit of Taylor Nunatak is Undillan (Middle Cambrian) in age.

Encarnación et al. (1999) also dated zircons in samples of airfall tuff and welded tuff breccia which they collected near the summit of Taylor Nunatak at sites above the trilobite-bearing carbonate. They also collected a third sample of zircon north of the summit and about 15 m beneath the fossiliferous carbonate bed. The authors calculated U-Pb dates from the $^{207}Pb/^{206}Pb$ ratio of radiogenic lead in these zircons because this ratio is insensitive to recent loss of lead. The resulting error-weighted mean age for all three zircon samples is 505.1 ± 1.3 Ma. This date lies within the Middle Cambrian in the timescale of the IUGS (2002) in excellent agreement with the Australian time scale used by Encarnación et al. (1999). These authors also recalculated a U-Pb date reported previously by Van Schmus et al. (1997) for a zircon from a metarhyolite of the Taylor Formation at Lubbock Ridge on the east side of the Shackleton Glacier. This zircon yielded a $^{207}Pb/^{206}Pb$ date of 506 ± 5 Ma and is therefore in good agreement with the date of 505.1 ± 1.3 Ma for the three zircons from Taylor Nunatak analyzed by Encarnación et al. (1999). Consequently, radiogenic isotope age determinations and fossil evidence both confirm a Middle Cambrian age for the Taylor Formation in the Shackelton Glacier area.

5.5.4 Queen Maud Batholith

The rock outcrops in the valley of the Shackleton Glacier and along the coast between the Kosco and Ramsey glaciers in Fig. 5.15 expose a cross-section of the Queen Maud batholith that extends from the Shackleton glacier to the Wisconsin Range of the Horlick Mountains. Wade and Cathey (1986) published modal analyses of 37 specimens of these granitic rocks which have been used to classify them in the Streckeisen diagram in Fig. 5.17 (Streckeisen 1967, 1976). The distribution of data points indicates that the composition of the Queen Maud batholith ranges from granite (field #3) to granodiorite (field #4) and to quartz diorite (field #5).

Wade and Cathey (1986) also reported isotopic age determinations of feldspar and biotite of the Hope Granite by the K-Ar and Rb-Sr methods. However, feldspars do not retain radiogenic ^{40}Ar and the initial $^{87}Sr/^{86}Sr$ ratios used to calculate the Rb-Sr dates were not provided. Therefore, the reliability of these dates is uncertain. However, two Rb-Sr dates of biotite from Longhorn Spurs (84°40′S, 174°41′W) and Den Hartog Peak (84°20′S, 178°52′E) indicate that the granitic rocks at these locations cooled sufficiently to retain radiogenic ^{87}Sr at 470 ± 25 Ma during the Ordovician Period.

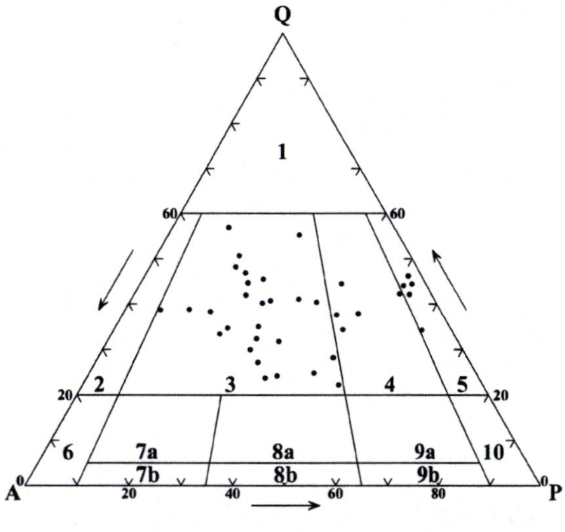

Fig. 5.17 The plutons of Hope Granite that intruded the metasedimentary basement rocks of the Shackleton-Glacier area are collectively referred to as the Queen Maud batholith. The modal concentrations of Quartz (Q), alkali feldspar (A), and plagioclase (P) are used to classify a representative collection of the Hope Granite from this area (see also Fig. 3.21, Appendix 3.6.1). The distribution of data points indicates that the Hope Granite consists of granite (field #3), granodiorite (field #4), and quartz diorite (field #5) (Data by Wade and Cathey (1986) and classification of plutonic igneous rocks by Streckeisen (1967))

5.6 Summary

The exploration of the geology of the central Transantarctic Mountains has progressed through several stages and has, in some cases, achieved a remarkably sophisticated level of understanding. The geologists who first mapped the rocks of this area traveled by dogsleds on highly crevassed glaciers and scrambled on foot up the steep slopes of the mountainous terrain. In spite of such formidable obstacles, these pioneers recorded what they saw and tried to organize their observations into a coherent stratigraphy (e.g., Gould 1935; Grindley 1963, 1967; Grindley et al. 1964; Gunn and Walcott 1962; Laird 1963, 1964; Laird et al. 1971; McGregor 1965a, b; Skinner 1964, 1965; Wade et al. 1965; and many others).

These pioneers were followed by a second group of geologists who benefitted from the use of snowmobiles and helicopters and who were able to build on the foundation that had been established by their predecessors. The improved logistical support enabled specialized studies to be carried out by experts working in selected area where important geological problems could be addressed.

The insights gained by these specialized studies of the central Transantarctic Mountains must be examined critically to determine whether the master hypothesis concerning the formation of the Ross orogen presented by Stump (1995) is still valid and still explains satisfactorily the deposition of detrital sediment along a passive rift margin which later evolved into an active subduction zone that caused the deformation and regional metamorphism of the rocks that now form the basement complex of the central Transantarctic Mountains.

References

Adams CJ, Gabites JE, Grindley GW (1982) Orogenic history of the central Transantarctic Mountains: New K-Ar age data on the Precambrian-lower Paleozoic basement. In: Craddock C (ed) Antarctic geoscience. University of Wisconsin Press, Madison, WI, pp 817–826

Bennett VC, Fanning CM (1993) A glimpse of the cryptic Gondwana shield: Archean and Proterozoic ages from the central Transantarctic Mountains. Abstract, Geol Soc Amer. Abstracts with Programs 25(5):49

Borg SG, DePaolo DJ (1986) Geochemical investigations of lower Paleozoic granites of the Transantarctic Mountains. Antarctic J US 21(5):41–43

Borg SG, Goodge JW, DePaolo DJ, Mattinson JM (1986) Field studies of granites and metamorphic rocks: Central Transantarctic Mountains. Antarctic J US 21(5):43–45

Borg SG, DePaolo DJ, Smith BM (1990) Isotopic structure and tectonics of the central Transantarctic Mountains. J Geophys Res 95:6647–6667

Burgess CJ, Lammerink W (1979) Geology of the Shackleton Limestone (Cambrian) in the Byrd Glacier area. New Zealand Antarctic Record 2(1):12–16

Compston W (1999) Geological age determination by instrumental analysis: The 29th Halmond Lecture. Mineral Mag 63:297–311

Compston W, Williams IS (1984) U-Pb geochronology of zircons from lunar breccia 73217 using a sensitive mass-resolution ion microprobe. J Geophys Res 89(Suppl):B525–B534

Craddock C (ed) (1969) Geologic maps of Antarctica. Antarctic Maps Folio Series, Folio 12, American Geographical Society, New York

Dalziel IWD (1992) Antarctica: A tale of two supercontinents? Ann Rev Earth Planet Sci 20:501–526

Dalziel IWD (1997) Neoproterozoic-Paleozoic geography and tectonics: Review, hypothesis, environmental speculation. Geol Soc Amer Bull 109:16–42

Debrenne F, Kruse PD (1986) Shackleton Limestone archaeocyathids. Alcheringa 10:235–278

Encarnación J, Grunow A (1996) Changing magmatic and tectonic styles along the paleo-Pacific margin of Gondwana and the onset of early Paleozoic magmatism in Antarctica. Tectonics 15:1325–1341

Encarnación J, Rowell AJ, Grunow AM (1999) A U-Pb age for the Cambrian Taylor Formation, Antactica: Implications for the Cambrian time scale. J Geol 107:487–504

Faure G, Felder RP (1984) Lithium-bearing pegmatite and bismuth-antimony-lead-copper-bearing veinlets on Mount Madison, Byrd Glacier area. Antarctic J US 19(5):13–14

Faure G, Mensing TM (2005) Isotopes: Principles and applications, 3rd edn. Wiley, Hoboken, NJ

Faure G, Murtaugh JG, Montigny RJE (1968) The geology and geochronology of the basement complex of the central Transantarctic Mountains. Canadian J Earth Sci 5:555–560

Goodge JW (1997) Latest Neoproterozoic basin inversion of the Beardmore Group, central Transantarctic Mountains, Antarctica. Tectonics 16:682–701

Goodge JW, Dallmeyer RD (1992) $^{40}Ar/^{39}Ar$ mineral age constraints and the Paleozoic tectonothermal evolution of high-grade basement rocks within the Ross orogen, central Transantarctic Mountains. J Geol 100:91–106

Goodge JW, Dallmeyer RD (1996) Contrasting thermal evolution within the Ross orogen, Antarctica: Evidence from mineral $^{40}Ar/^{39}Ar$ dates. J Geol 104:435–458

Goodge JW, Fanning CM (1999) 2.5 billion years of punctuated Earth history as recorded in a single rock. Geology 27:1007–1010

Goodge JW, Hansen VL, Peacock SM, Smith BK (1990) Metamorphic rocks in the Geologist and Miller ranges, Nimrod Glacier area, central Transantarctic Mountains. Antarctic J US 25(5):35–36

Goodge JW, Borg SG, Smith BK, Bennett VC (1991) Tectonic significance of Proterozoic ductile shortening and translation along the Antarctic margin of Gondwana. Earth Planet Sci Lett 102:58–70

Goodge JW, Hansen VL, Peacock SM (1992) Multiple petrotectonic events in high-grade metamorphic rocks of the Nimrod Group, central Transantarctic Mountains. In: Yoshida Y, Kaminuma Y, Shiraishi K (eds) Recent progress in Antarctic earth science. Terra Scientific, Tokyo, Japan, pp 203–209

Goodge JW, Hansen VL, Peacock SM, Smith BK, Walker NW (1993a) Kinematic evolution of the Miller Range shear zone, central Transantarctic Mountains, Antarctica, and implications for Neoproterozoic to early Paleozoic tectonics of the East Antarctic margin of Gondwana. Tectonics 12:1460–1478

Goodge JW, Walker NW, Hansen VL (1993b) Neoproterozoic-Cambrian basement-involved orogenesis within the Antarctic margin of Gondwana. Geology 21:37–40

Goodge JW, Fanning CM, Bennett VC (2001) U-Pb evidence of ~1.7 Ga crustal tectonism during the Nimrod Orogeny in the Transantarctic Mountains, Antarctica: Implications for Proterozoic plate reconstructions. Precambrian Res 112:261–288

Goodge JW, Myrow PM, Williams IS, Bowring S (2002) Age and provenance of the Beardmore Group, Antarctica: Constraints on the Rodinia supercontinent breakup. J Geol 110(4):393–406

Gould LM (1935) Structure of the Queen Maud Mountains. Geol Soc Amer Bull 46:937–984

Grindley GW (1963) The geology of the Queen Alexandra Range, Beardmore Glacier, Ross Dependency, Antarctica; with notes on the correlation of Gondwana sequences. New Zealand J Geol Geophys 6(3):307–347

Grindley GW (1967) The geomorphology of the Miller Range, Transantarctic Mountains, with notes on the glacial history and neotectonics of East Antarctica. New Zealand J Geol Geophys 10(2):557–598

Grindley GW (1972) Polyphase deformation of the Precambrian Nimrod Group, central Transantarctic Mountains. In: Adie RJ (ed) Antarctic geology and geophysics. Universitetsforlaget, Oslo, Norway, pp 313–318

Grindley GW, Laird MG (1969) Geology of the Shackleton Coast. In: Craddock C (ed) Geologic Map of Antarctica, Sheet 15. American Geographical Society, New York

Grindley GW, McDougall I (1969) Age and correlation of the Nimrod Group and other Precambrian rock units in the central Transantarctic Mountains, Antarctica. New Zealand J Geol Geophys 12(2/3):391–411

Grindley GW, McGregor VR, Walcott RI (1964) Outline of the geology of the Nimrod-Beardmore-Axel Heiberg Glaciers region, Ross Dependency. In: Adie RJ (ed) Antarctic geology. North-Holland, Amsterdam, The Netherlands, pp 206–219

Grunow AM, Encarnación J (2000) Terranes or Cambrian polar wander: New data from the Scott Glacier area, Transantarctic Mountains, Antarctica. Tectonics 19:168–191

Gunn BM, Walcott RI (1962) The geology of the Mt. Markham region, Ross Dependency, Antarctica. New Zealand J Geol Geophys 5:407–426

Gunner JD (1969) Petrology of metamorphic rocks from the Miller Range, Antarctica. Institute of Polar Studies, Report No. 32, The Ohio State University, Columbus, Ohio

Gunner JD (1971a) Age and origin of the Nimrod Group and of the Granite Harbour Intrusives, Beardmore Glacier region, Antarctica. Ph.D. dissertation, The Ohio State University, Columbus, OH

Gunner JD (1971b) Anatectic origin of late orogenic granitic intrusions, central Transantarctic Mountains, Antarctica (Abstract). Eos, Trans Amer Geophys Union 52:373

Gunner JD (1971c) Ida Granite: a new formation of the Granite Harbor Intrusives, Beardmore Glacier region. Antarctic J US 6(5):194–196

Gunner JD (1974) Investigations of lower Paleozoic granites in the Beardmore Glacier region. Antarctic J US 9(3): 76–81

Gunner JD (1976) Isotopic and geochemical studies of the pre-Devonian basement complex, Beardmore Glacier region, Antarctica. Institute of Polar Studies, Report No. 41, The Ohio State University, Columbus, OH

Gunner JD (1982) Basement geology of the Beardmore Glacier region. In: Turner MD, Splettstoesser JF (eds) Geology of the Central Transantarctic Mountains, Antarctic Research Series, vol. 36. American Geophysical. Union, Washingtion, DC, pp 1–9

Gunner JD, Faure G (1972) Rb-Sr geochronology of the Nimrod Group, central Transantarctic Mountains. In: Adie RJ (ed) Antarctic geology and geophysics. Universitetsforlaget, Oslo, Norway, pp 305–311

Gunner JD, Mattinson JM (1975) Rb-Sr and U-Pb ages of granites in the central Transantarctic Mountains. Geol Mag 112:25–31

Hill D (1964a) Archaeocyatha from loose material at Plunket Point at the head of the Beardmore Glacier. In: Adie RJ (ed) Antarctic geology. North-Holland, Amsterdam, The Netherlands, pp 619–622

Hill D (1964b) Archaeocyatha from the Shackleton Limestone of the Ross System, Nimrod Glacier area, Antarctica. Roy Soc New Zealand Trans, Geol 2:137–146

IUGS (2002) International stratigraphic chart. Commission on the Geologic Map of the World, UNESCO, United Nations, New York

Jackson JA 3d (1997) Glossary of geology, 4th edn. American Geological Institute, Alexandria, VA

Laird MG (1963) Geomorphology and stratigraphy of the Nimrod Glacier-Beaumont Bay region, southern Victoria Land, Antarctica. New Zealand J Geol Geophys 6(3):465–484

Laird MG (1964) Petrography of rocks from the Nimrod Glacier-Starshot Glacier region, Ross Dependency. In: Adie RJ (ed) Antarctic geology. North-Holland, Amsterdam, The Netherlands, pp 463–472

Laird MG, Bradshaw JD (1982) Uppermost Proterozoic and lower Paleozoic geology of the Transantarctic Mountains. In: Craddock C (ed) Antarctic geoscience. University of Wisconsin Press, Madison, WI, pp 525–534

Laird MG, Waterhouse JB (1962) Archaeocyathine limestone in Antarctica. Nature 194:861

Laird MG, Mansergh GD, Chappell JMA (1971) Geology of the central Nimrod Glacier area, Antarctica. New Zealand J Geol Geophys 14(3):427–468

McDougall I, Grindley GW (1965) Potassium-argon dates on micas from the Nimrod-Beardmore-Axel Heiberg region, Ross Dependency, Antarctica. New Zealand J Geol Geophys 8(2):304–313

McGregor VR (1965a) Notes on the geology of the area between the heads of the Beardmore and Shackleton Glaciers, Antarctica. New Zealand J Geol Geophys 8(2):278–291

McGregor VR (1965b) Geology of the area between the Axel Heiberg and Shackleton Glaciers, Queen Maud Range, Antarctica. New Zealand J Geol Geophys 8(2):314–343

McGregor VR, Wade FA (1969) Geology of the western Queen Maud Mountains. In: Craddock C (ed) Geologic maps of Antarctica. Antarctic Map Folio Series, Folio 12, Sheet 16. American Geographical Society, New York

Mirsky A (1969) Geology of the Ohio Range-Liv Glacier area. In: Craddock C (ed) Geologic map of Antarctica, Sheet 11, Plate XVI, Folio 12. American Geographical Society, New York

Moores EM (1991) Southwest U.S.-East Antarctica (SWEAT) connection: a hypothesis. Geology 19:425–428

Murtaugh JG (1969) Geology of the Wisconsin Range batholith, Transantarctic Mountains. New Zealand J Geol Geophys 12(2/3):526–550

Myrow PM, Goodge JW (1999) Reinterpretation of depositional and tectonic setting of Neoproterozoic strata, Transantarctic Mountains. In: Skinner DNB (ed) 8th International Symposium on Antarctic Earth Sciences. Programme and Abstracts. Royal Society of New Zealand, Wellington, New Zealand, p 222

Myrow PM, Pope MC, Goodge JW, Fischer W, Palmer AR (2002) Depositional history of pre-Devonian strata and timing of Ross orogenic tectonism in the central Transantarctic Mountains. Geol Soc Amer Bull 114(9):1070–1088

Oliver RL (1972) The geology of an area near the mouth of the Beardmore Glacier, Ross Dependency, Antarctica. In: Adie RJ (ed) Antarctic geology and geophysics. Universitetsforlaget, Oslo, Norway, pp 379–385

Palmer AR, Gatehouse CG (1972) Early and Middle Cambrian trilobites from Antarctica. US Geol Surv Prof. Paper 456D, pp 1–37

Pankhurst RJ, Storey BC, Millar IL, Macdonald DIM, Vennum WR (1988) Cambrian-Ordovician magmatism in the Thiel Mountains, Transantarctic Mountains, and implications for the Beardmore Orogeny. Geology 16(3):246–249

Peacock SM, Goodge JW (1995) Eclogite facies metamorphism preserved in tectonic blocks from a lower crustal shear zone, central Transantarctic Mountains, Antarctica. Lithos 36:12–13

Rees MN, Rowell AJ (1991) The pre-Devonian Paleozoic clastics of the central Transantarctic Mountains: Stratigraphy and depositional setting. In: Thomson MRA, Crame JA, Thomson JW (eds) Geological evolution of Antarctica. Cambridge University Press, Cambridge, pp 187–192

Rees MN, Rowell AJ, Pratt BR (1985) The Byrd Group of the Holyoake Range, central Transantarctic Mountains. Antarctic J US 19(5):3–5

Rees MN, Girty GH, Panttaja SK, Braddock P (1987) Multiple phases of early Paleozoic deformation in the central Transantarctic Mountains. Antarctic J US 22(5):33–35

Rees MN, Rowell AJ, Cole ED (1988) Aspects of the Late Proterozoic and Paleozoic geology of the Churchill Mountains, southern Victoria Land. Antarctic J US 23(5):23–25

Rees MN, Pratt BR, Rowell AJ (1989) Early Cambrian reefs, reef complexes, and associated lithofacies of the Shackleton Limestone, Transantarctic Mountains. Sedimentology 36:341–361

Rowell AJ, Rees MN (1989) Early Paleozoic history of the upper Beardmore Glacier area: Implications for a major Antarctic structural boundary within the Transantarctic Mountains. Antarctic Sci 1:249–260

Rowell AJ, Rees MN (1991) Setting and significance of the Shackleton Limestone, central Transantarctic Mountains. In: Thomson MRA, Crame JA, Thomson JW (eds) Geological evolution of Antarctica. Cambridge University Press, Cambridge, pp 171–175

Rowell AJ, Rees MN, Braddock P (1986) Pre-Devonian Paleozoic rocks of the central Transantarctic Mountains. Antarctic J US 21:48–50

Rowell AJ, Evans KR, Rees MN (1988a) Fauna of the Shackleton Limestone. Antarctic J US 23(5):13–14

Rowell AJ, Rees MN, Cooper RA, Pratt BR (1988b) Early Paleozoic history of the central Transantarctic Mountains: Evidence from the Holyoake Range, Antarctica. New Zealand J Geol Geophys 31:397–404

Rowell AJ, Rees MN, Evans KR (1992) Evidence of major Middle Cambrian deformation in the Ross orogen, Antarctica. Geology 20:31–34

Skinner DNB (1964) A summary of the geology of the region between the Byrd and Starshot glaciers, south Victoria Land. In: Adie RJ (ed) Antarctic geology. North-Holland, Amsterdam, The Netherlands, pp 284–292

Skinner DNB (1965) Petrographic criteria of the rock units between the Byrd and Starshot Glaciers, south Victoria Land, Antarctica. New Zealand J Geol Geophys 8(2):292–303

Streckeisen AL (1967) Classification and nomenclature of igneous rocks. Neues Jahrbuch Miner. Abhandlungen 107(2/3): 144–240

Streckeisen AL (1976) To each plutonic rock its proper name. Earth Sci Rev 12:1–33

Stump E (1980) Two episodes of deformation at Mt. Madison, Antarctica. Antarctic J US 15(5):13–14

Stump E (1981) Observations on the Ross orogen, Antarctica. In: Cresswell MM, Vella P (eds) Gondwana Five. Balkema, Rotterdam, The Netherlands, pp 205–208

Stump E (1982) The Ross Supergroup in the Queen Maud Mountains. In: Craddock C (ed) Antarctic geoscience. University of Wisconsin Press, Madison, WI, pp 565–570

Stump E (1985) Stratigraphy of the Ross Supergroup, central Transantarctic Mountains. In: Turner MD, Splettstoesser JF (eds) Geology of the Central Transantarctic Mountains. Antarctic Research Series, vol. 36. American Geophysical Union, Washington, DC, pp 225–274

Stump E (1995) The Ross orogen of the Transantarctic Mountains. Cambridge University Press, Cambridge

Stump E, Sheridan MF, Borg SG, Lowery PH, Colbert PV (1979) Geological investigations in the Scott Glacier and Byrd Glacier area. Antarctic J US 14(5):39–40

Stump E, Miller MG, Korsch RJ, Edgerton DG (1988) Diamictite from Nimrod Glacier area, Antarctica; Possible Proterozoic glaciation on the seventh continent. Geology 16(3):225–228

Stump E, Korsch RJ, Edgerton DG (1991) The myth of the Beardmore and Nimrod orogenies. In: Thomson MRA, Crame JA, Thomson JW (eds) Geological evolution of Antarctica. Cambridge University Press, Cambridge, pp 143–147

Van Schmus WR, McKenna LW, Gonzales DA, Fetter AH, Rowell AJ (1997) U-Pb geochronology of parts of the Pensacola, Thiel, and Queen Maud mountains, Antarctica. In: Ricci A (ed) The Antarctic Region: Evolution and Processes. Terra Antartica, Siena, Italy, pp 187–200

Wade FA (1974) Geological surveys of Marie Byrd Land and the central Queen Maud Range. Antarctic J US 9(5):241–242

Wade FA, Cathey CA (1986) Geology of the basement complex, western Queen Maud Mountains, Antarctica. In: Turner MD, Splettstoesser JF (eds) Geology of the Central Transantarctic Mountains. Antarctic Research Series, vol. 36. American Geophysical Union, Washington, DC, pp 429–453

Wade FA, Yeats VL, Everett JR, Greenlee DW, LaPrada KE, Shenk JC (1965) Geology of the central portion of the Queen Maud Range, Transantarctic Mountains. Science 150:1808–1809

Walker NW, Goodge JW (1991) Significance of Late Archean-Early Proterozoic U-Pb ages of individual Nimrod Group detrital zircons and Cambrian plutonism in the Miller Range, Transantarctic Mountains (Abstract). Geol Soc Amer, Abstracts with Programs 231(5):306

Walker NW, Goodge JW (1994). Tectonic significance of 650–1100 Ma detrital zircons from the Neoproterozoic Goldie Formation, Beardmore Group, central Transantarctic Mountains, Antarctica (Abstract). Geol Soc Amer, Abstracts with Programs 26:49

Yochelson EL, Herrera HE (1974) Un fossil enigmatico del Cambrico Inferior de Argentina. Ameghiniana 11:283–294

Yochelson EL, Stump E (1977) Discovery of Early Cambrian fossils at Taylor Nunatak, Antarctica. J Paleontol 51: 872–875

Chapter 6
The Queen Maud Mountains

The Transantarctic Mountains were a formidable barrier which the "pole seekers" had to overcome in order to reach the polar plateau (Section 1.4). Ernest Shackleton and Robert Scott forced their way up the Beardmore Glacier, whereas Roald Amundsen climbed the Axel Heiberg Glacier which he named after one of his supporters in Norway. After 4 days of intense effort he and his men, with the help of 42 dogs, reached the polar plateau on November 21, 1911 (Huntford 1987). Amundsen also named the Queen Maud Mountains after the reigning Queen of Norway (Section 1.4.4).

Although Amundsen did not specify the geographic limits, the Queen Maud Mountains as they are now defined, extend from the Shackleton Glacier to the Reedy Glacier, a distance of about 400 km. The Transantarctic Mountains widen from about 60 km at the Axel Heiberg Glacier to more than 200 km measured along the Scott Glacier. In fact, Mt. Howe at 87°22′S, 149°30′W at the head of the Scott Glacier is the most southerly mountain in all of Antarctica because it is located only 300 km from the South Pole. The Queen Maud Mountains are traversed by the Barrett, Liv, Strom, Axel Heiberg, Amundsen, Scott, Leverett, and Reedy glaciers. The Axel Heiberg Glacier is flanked by Mt. Fridtjof Nansen (3,989 m) on the west and by Mt. Don Pedro Christophersen (3,766 m) on the east. Many other peaks in the Queen Maud Mountains also have elevations greater than 3,000 m.

The geology of the Queen Maud Mountains was first explored by Laurence Gould during Richard Byrd'Ss first Antarctic expedition (1928–1929). Gould used dogsleds and had to travel 800 km across the Ross Ice Shelf just to reach the Queen Maud Mountains (Section 1.5.2; Gould 1935). The geological maps of McGregor and Wade (1969) and of Mirsky (1969) indicate that the Queen Maud Mountains consist primarily of metasedimentary and metavolcanic rocks of the basement complex which were intruded by the Granite Harbor Intrusives of the Queen Maud batholith.

The exposures of metasedimentary and metavolcanic rocks, which occur extensively and almost continuously in the coastal mountains of the central Transantarctic Mountains, are more scattered in the Queen Maud Mountains except adjacent to the Scott Glacier and in the Harold Byrd Mountains along the Leverett Glacier. However, when viewed as a whole, the geology of the Queen Maud Mountains is dominated by the Granite Harbor Intrusives of the Queen Maud batholith that extends for about 220 km from the eastern side of the Scott Glacier to the Reedy Glacier (Mirsky 1969).

The basement rocks of the Queen Maud Mountains are unconformably overlain by the sedimentary rocks of the Beacon Supergroup, remnants of which occur primarily at the head of the Shackleton, Liv, and Axel Heiberg glaciers (Barrett 1965; McGregor 1965a). Farther east, scattered outcrops of Beacon rocks occur on the Nilsen Plateau (86°20′S, 158°00′W), on Mt. Weaver (86°58′S, 153°50′W), and Mt. Saltonstall at the head of the Scott Glacier, at Mt. Blackburn (86°17′S, 147°16′W), and at the head of the Leverett Glacier (Stump et al. 1985).

The Transantarctic Mountains between the Shackleton and Axel Heiberg glaciers in Fig. 6.1 are composed primarily of the Granite Harbor Intrusives. Metasedimentary and metavolcanic basement rocks occur only between the Barrett and the Gough glaciers and in the Duncan Mountains located on the coast between the Liv and Strom glaciers. For this reason, we start the presentation of the geology of the

G. Faure and T.M. Mensing, *The Transantarctic Mountains: Rocks, Ice, Meteorites and Water*,
DOI 10.1007/978-90-481-9390-5_6, © Springer Science+Business Media B.V. 2010

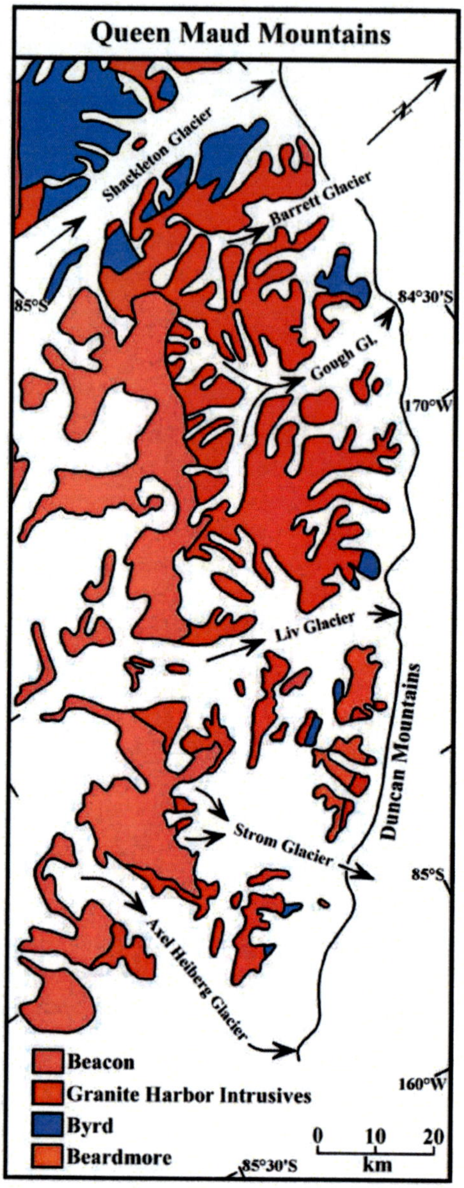

Fig. 6.1 The Queen Maud Mountains extend southeast from the Shackleton Glacier to the Reedy Glacier of the Wisconsin Range in the Horlick Mountains. The segment of the Queen Maud Mountains on this map includes the Duncan Mountains on the coast between the Liv and Strom glaciers. This small mountain range contains exposures of metasedimentary rocks that have been assigned to the Beardmore and Byrd groups. Most of this segment of the Queen Maud Mountains is composed of the Granite Harbor Intrusives which form the Queen Maud batholith. The sedimentary rocks of the Beacon Group (including sills of Ferrar Dolerite) rest unconformably on the basement rocks (Adapted from McGregor and Wade 1969)

Queen Maud Mountains in the Duncan Mountains which will serve as a transition from the geology of the Shackleton Glacier area to the metasedimentary and metavolcanic basement rocks that occur west of the Scott Glacier.

6.1 Duncan Mountains

The Duncan Mountains in Fig. 6.2 are only about 40 km long and are dominated topographically by Mt. Fairweather (1,190 m) and by Morris Peak (980 m). The west side of the mouth of the Liv Glacier is occupied by the Mayer Crags which include Mt. Henson (903 m) made famous by a thick deposit of the Henson Marble that was described by McGregor (1965b). The stratigraphy of the basement complex of the Duncan Mountains in Table 6.1 consists of the Duncan Formation, the Fairweather Formation, and the Henson Marble in ascending stratigraphic sequence. These formations were correlated by Wade and Cathey (1986) with the Greenlee and Taylor formations of the Shackleton Glacier area as shown in Table 6.1. However, Stump (1985) correlated the Duncan Formation with the Goldie Formation of the Queen Elizabeth Range. The principal references to the geology of the Duncan Mountains are the reports of McGregor (1965b), McGregor and Wade (1969), Stump (1981,1985), and Wade and Cathey (1986). In addition, Grindley et al. (1964), Linder et al. (1965), Wade et al. (1965), Wade (1974), Stump (1975, 1981), and Stump et al. (1976) worked in this area and Stump (1995) included a description of the geology of the Duncan Mountains in his book.

The Duncan Formation consists of dark-colored schists and hornfelses that crop out along the northern spurs of the Duncan Mountains and at the base of Mt. Henson (Stump 1981, 1995). The rocks were originally composed of well-bedded graywacke that was recrystallized to hornfels by heat that emanated from the Queen Maud batholith (i.e., Granite Harbor Intrusives). Near intrusive contacts the Duncan Formation consists of contorted gneisses and migmatites (McGregor 1965b). According to Stump (1985, 1995), the Duncan Formation is at least 4.6 km thick, provided that the thickness of the rocks is not increased by folding. The presence of calc-silicate layers indicates

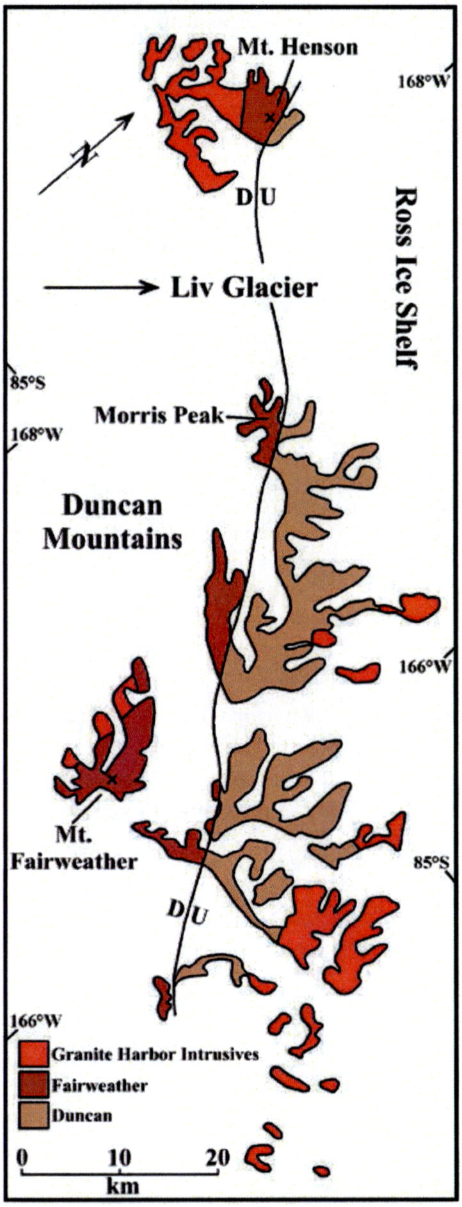

Fig. 6.2 The basement rocks that are exposed in the Duncan Mountains consist of the Duncan and Fairweather formations both of which were intruded by the Granite Harbor Intrusives. The contact between the Duncan and Fairweather formations is a fault along which the upthrown side is composed of the Duncan Formation. The Henson Marble on Mt. Henson is considered to be a member of the upper Fairweather Formation (Adapted from Stump 1995)

that the sediment originally contained carbonate beds. The age of the Duncan Formation is assumed to be Neoproterozoic because no diagnostic fossils have been found in these rocks. Stump (1985) suggested that the Duncan Formation was originally a turbidite.

Table 6.1 Stratigraphy of the metasedimentary basement rocks in the Shackleton Glacier area and in the Duncan Mountains (Adapted from Wade and Cathey 1986)

Age	Shackleton Glacier	Duncan Mountains
Devonian to Triassic	Beacon Supergroup	Beacon Supergroup
~~	Kukri Peneplain	~~~~~~
Cambrian to Ordovician	Granite Harbor Intrusives	Granite Harbor Intrusives
———	Intrusive Contact	———
Cambrian	Henson Marble Taylor Formation	Henson Marble Fairweather Formation
Neoproterozoic to Early Cambrian	Greenlee Formation	Duncan Formation
———	Base not exposed	———

The Fairweather Formation is younger than the Duncan Formation and was originally described by McGregor (1965b) as a crushed quartz-sandstone. Stump et al. (1976) classified it as a massive, silicic, porphyritic volcanic rock including nonporphyritic varieties and volcanic breccias. In addition, the Fairweather Formation contains thin beds of marble and chert. Wade and Cathey (1986) considered that the lithologic composition of the Fairweather Formation resembles the Taylor Formation in the Shackleton Glacier area and therefore correlated the two formations as indicated in Table 6.1.

The Henson Marble overlies the Fairweather Formation at Mt. Henson (Fig. 6.2). Wade and Cathey (1986) recalled that a similar bed of white marble crops out on the summit of Epidote Peak and in the cirque at the head of the Gerasimou Glacier in the Shackleton Glacier area (Section 5.5.2). This marble, like the Henson Marble on Mt. Fairweather, appears to be the youngest stratigraphic unit of the local basement complex. Consequently, Wade (1974) tentatively correlated the marble on Epidote Peak with the Henson Marble on Mt. Henson and Wade and Cathey (1986) later correlated both marble deposits with the Early Cambrian Shackleton Limestone of the Holyoake and Churchill mountains. However, the relation of the Greenlee and Duncan formations to the Goldie Formation remains in doubt following the reorganization of the stratigraphy of the Goldie Formation into "inboard" and "outboard" components by Goodge et al. (2002) and Myrow et al. (2002).

6.2 O'Brien Peak

The next significant exposure of metasedimentary basement rocks along the coast of the Ross Ice Shelf occurs at O'Brien Peak which is located in Fig. 6.3 between the mouths of the Amundsen and Scott glaciers at 85°28′S, 156°42′W. The geologic map of Mirsky (1969) indicates that this mountain is composed of the Granite Harbor Intrusive although undifferentiated basement rocks extend along the west side of the Scott Glacier and reach the coast at O'Brien Peak.

The area was first visited in 1929 by members of Gould's sledging group and the rock samples they collected were later described by Stewart (1934). Twenty four years later, Linder et al. (1965) mentioned O'Brien Peak in their report. The rock samples they collected were dated by Craddock et al. (1964) who reported Rb-Sr whole-rock and mineral dates between 520 ± 30 (microcline) and 450 ± 20 Ma (biotite) which reflect the crystallization and subsequent cooling of Granite Harbor Intrusives during and after the Ross Orogeny (Mirsky 1969).

The most detailed information on the geology of O'Brien Peak was provided by Katz and Waterhouse (1970a) who visited the area during the 1969/70 field season. They reported that O'Brien Peak is composed of a sequence of high-grade metamorphic rocks, including marbles, which were intruded locally by a small pluton of the Granite Harbor Intrusives. The metasedimentary rocks that form O'Brien Peak include interbedded hornfels, marble, schist, paragneiss, amphibolite, migmatite, and granite. A summary of the geology of this area is provided by Fig. 6.4 based on a diagram in the book by Stump (1995).

Although the presence of marble and calc silicate units is reminiscent of the basement complex of southern Victoria Land, the isolated occurrence of the outcrops on O'Brien Peak prevents correlation with the rocks of the Nimrod, Beardmore, or Byrd groups. However, the Rb-Sr dates of Craddock et al. (1964) do indicate that the metasedimentary rocks exposed on O'Brien Peak were deposited prior to the Ross Orogeny and were subsequently folded, metamorphosed, and intruded by granite magma followed by cooling to the closure temperature for radiogenic ^{87}Sr in biotite at 450 ± 20 Ma.

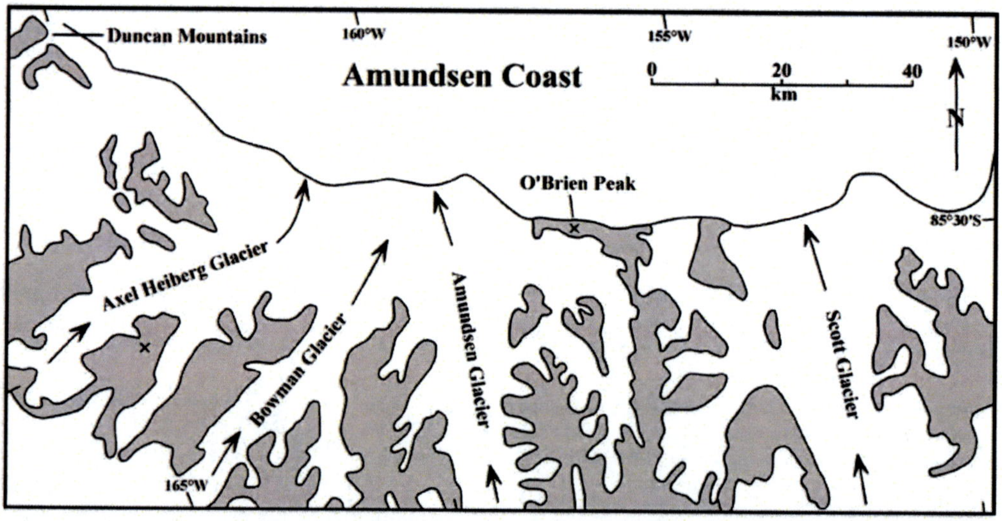

Fig. 6.3 O'Brien Peak is located on the Amundsen Coast between the mouths of the Amundsen and Scott glaciers about 90 km east of the Duncan Mountains. The area was first visited in 1929 by members of Gould's sledging party who collected rock samples that were described by Stewart (1934). O'Brien Peak was later mapped by Katz and Waterhouse (1970a) who discovered a complex of metamorphic rocks including marble, schist, paragneiss, and granite (Adapted from Mirsky 1969)

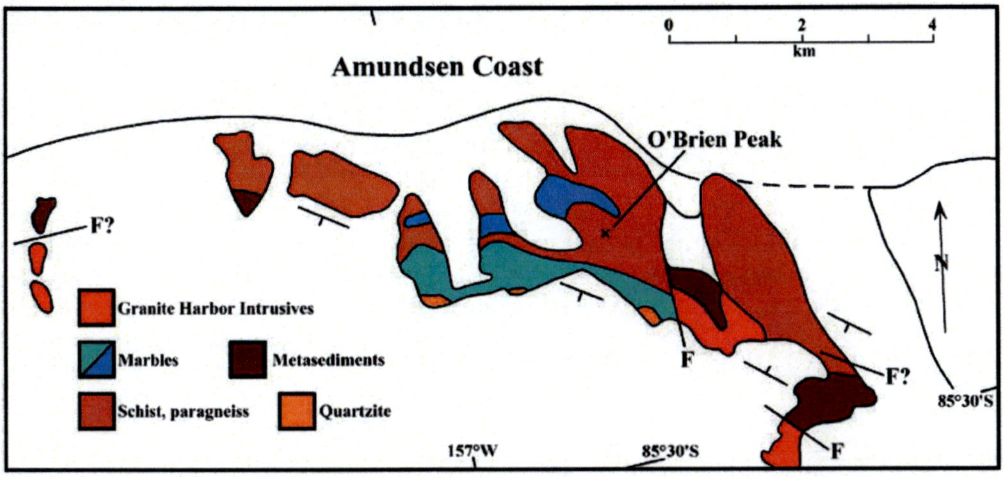

Fig. 6.4 O'Brien Peak on the Amundsen Coast at 85°28′S and 156°42′W consists of metasedimentary rocks including beds of pure marble (dark blue) and carbonate-rich rocks (light blue) which are interbedded with schists, paragneiss, amphibolite, migmatite, and hornfels. These, rocks as well as other kinds of undifferentiated metasediments and quartzites, were mapped by Katz and Waterhouse (1970a) (Adapted from Stump 1995)

6.3 Nilsen Plateau

We next move up the Amundsen Glacier to the Nilsen Plateau at 86°20′S, 158°00′W. This mountain range, which is also known as the Thorvald Nilsen Mountains, is located east of the head of the Amundsen Glacier about 100 km south of the Amundsen Coast of the Ross Ice Shelf (Davis and Blankenship 2005). The plateau includes Mt. Kristensen (3,460 m), Lindstrom Peak (2,640 m), and Kutchin Peak (2,360 m). The basement rocks of the Nilsen Plateau are exposed in the walls of the cirques that face the Amundsen Glacier and in the glacial valleys of the northern part of the plateau. The area was investigated during the 1963/64 field season by W.E. Long, D. McLelland, C.J. Skinner, and R.P. Goldthwait of the Institute of Polar Studies, The Ohio State University. The geology of the Nilsen Plateau was described by Mirsky (1969) who also listed age determinations published by McLelland (1967).

6.3.1 Granite Harbor Intrusives (Southern Nilsen Plateau)

The basement of the southern part of the Nilsen plateau in Fig. 6.5, mapped by McLelland (1967), consists of the Granite Harbor Intrusives which are unconformably overlain by the sedimentary rocks of the Beacon Supergroup. Samples of these granitic rocks were subsequently dated by Eastin (1970) by the whole-rock Rb-Sr method. The metasedimentary and metavolcanic basement rocks of the northern part of the Nilsen Plateau were later described by Stump (1985, 1995) and are not shown on McLelland's map in Fig. 6.5.

McLelland (1967) subdivided the Granite Harbor Intrusives of the southern part of the Nilsen Plateau into four locally-defined phases: Lonely Ridge Granodiorite, Cougar Canyon Quartz Monzonite, North Quartz Monzonite, and South Quartz Monzonite. These phases are not separately identified in Fig. 6.5. The Lonely Ridge Granodiorite, the Cougar Canyon Quartz Monzonite, and the South Quartz Monzonite are cataclastically deformed (i.e., they appear to be syntectonic), whereas the North Quartz Monzonite is not appreciably deformed and may therefore be younger than the other facies.

The Lonely Ridge Granodiorite is a dark, medium-grained rock on Lonely Ridge in Fig. 6.5 and is composed of plagioclase (An_{33} to An_{45}), quartz, biotite, microcline-perthite, and minor hornblende. It contains bands of gneissic texture containing augen of quartz and feldspar up to 10 mm in length. These blastomylonitic layers strike N40°W ± 10° and have a near-vertical dip (McLelland 1967). The Lonely Ridge

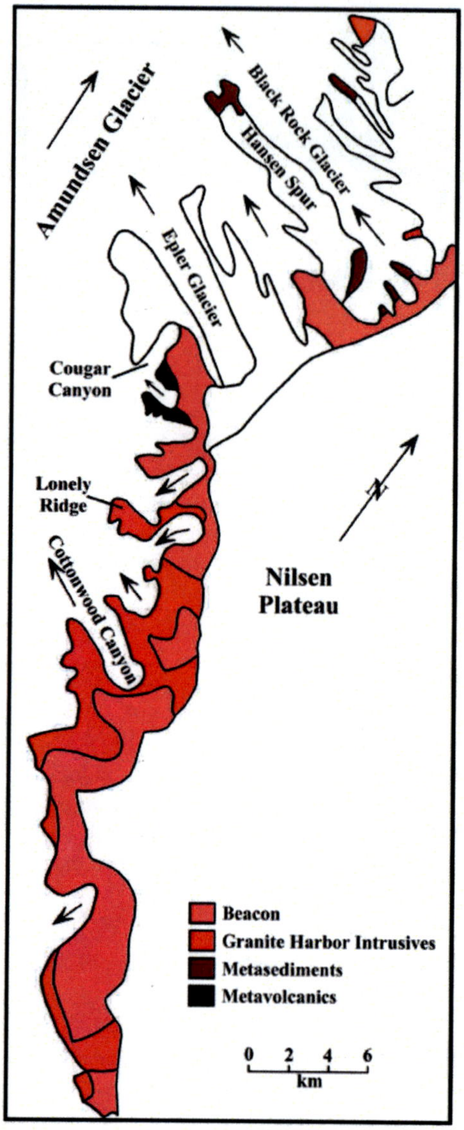

Fig. 6.5 The western edge of the southern part of the Nilsen Plateau along the Amundsen Glacier contains small outcrops of metavolcanic and metasedimentary rocks of the basement complex that were intruded by the Granite Harbor suite. McLelland (1967) divided the granitic basement rocks into the Cougar Canyon Quartz Monzonite, the Lonely Ridge Granodiorite, the North Quartz Monzonite, and the South Quartz Monzonite which are not differentiated on this map. The basement rocks are unconformably overlain by sandstones of the Beacon Supergroup and sills of Ferrar Dolerite (Adapted from McLelland 1967)

Granodiorite was intruded by dikes of the South Quartz Monzonite which also contains xenoliths of Lonely Ridge Granodiorite. These observations indicate that the Lonely Ridge Granodiorite is older than the South Quartz Monzonite.

McLelland (1967) reported two isotopic age determinations for biotites of the Lonely Ridge Granodiorite (Mirsky 1969):

K-Ar: 472 ± 10 Ma; Rb-Sr: 846 ± 35 Ma

The Rb-Sr date is anomalous and was later corrected by Felder and Faure (1979) who obtained a Rb-Sr date of 470 ± 17 Ma for biotite from a specimen of the Lonely Ridge Granodiorite collected by McLelland (McLelland #208). Consequently, McLelland's date (846 ± 35 Ma) should not be cited in the literature.

6.3.2 Volcano-Sedimentary Complex (Northern Nielsen Plateau)

The northern part of the Nilsen Plateau in Fig. 6.6, where the Amundsen Glacier turns east around the Olsen Crags, is dissected by the Epler Glacier and by the Blackwall Glacier which McLelland (1967) called the "Black Rock Glacier" in Fig. 6.5 and which is not identified at all on the topographic map of the Nilsen Plateau published by the US Geological Survey.

Figure 6.6 contains the highest point of the Nilsen Plateau at 3,940 m, as well as Crown Mountain (3,830 m), Mt. Kristens (3,460 m), Mt. Sundbeck (3,030 m), Mt. Stubbered (2,970 m), Beck Peak (2,650 m), and Lindstrom Peak (2,640 m) in addition to several as yet unnamed prominent topographic high-points. The ridges and spurs of the northern Nilsen Plateau in Fig. 6.7 consist almost exclusively of the volcano-sedimentary rocks of the basement complex of the Queen Maud Mountains. The stratigraphy of these rocks is summarized in Table 6.2 based on a correlation of the basement rocks by Stump (1985, 1995). Hansen Spur in Fig. 6.6 and 6.7 is composed of the metasedimentary rocks of the La Gorce Formation, whereas the ridges on the east side of Blackwall Glacier expose the low-grade meta-igneous rocks of the Wyatt Formation. Both the La Gorce Formation and the Wyatt Formation were named by Minshew (1967) based on exposures in the La Gorce Mountains and on Mt. Wyatt of the Scott Glacier area, respectively. The petrology and geochemistry of the rocks of the Wyatt Formation were later described by Borg (1980).

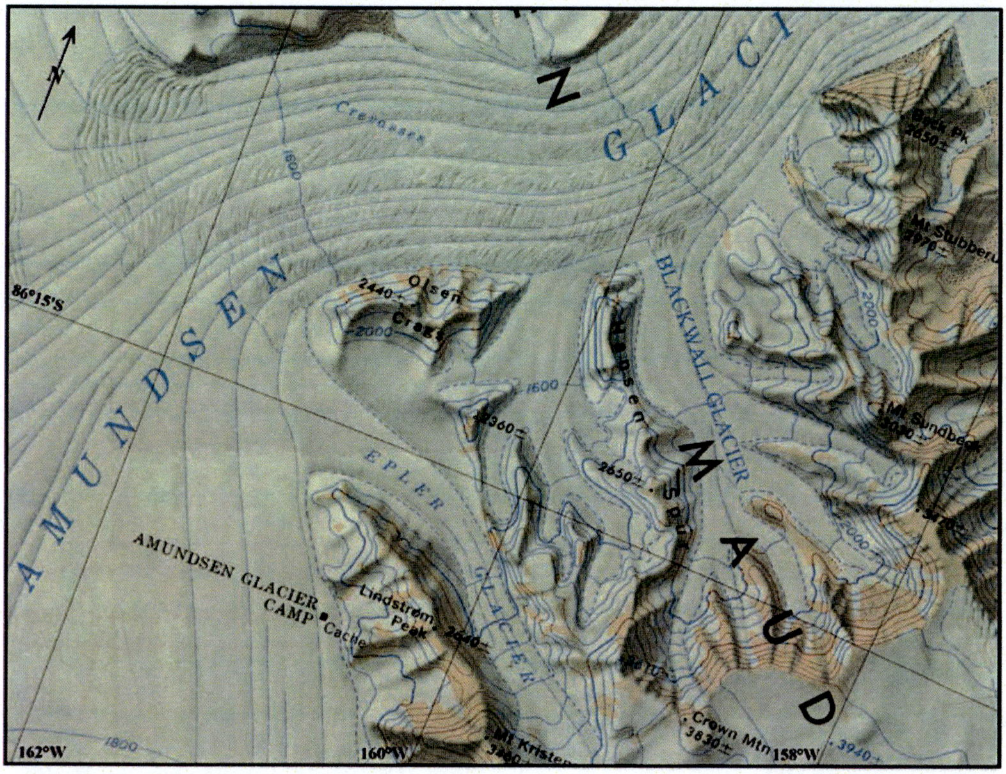

Fig. 6.6 The northwestern margin of the Nilsen Plateau at 86°20′S and 158°00′W of the Queen Maud Mountains is dissected by cirques and alpine valley glaciers that drain ice which has accumulated on the top of the plateau. Rock outcrops in the cirques and valleys of the northern Nilsen Plateau expose metasedimentary rocks of the La Gorce Formation and metavolcanic rocks of the Wyatt Formation which was intruded by small plutons of Granite Harbor Intrusives (Fig. 6.7). A cache of food and miscellaneous equipment was left at the Amundsen Glacier camp that was occupied during the 1963/64 field season. Excerpt of the topographic map SV1–10/10; S-8600–15,000/1 × 15 (Nilsen Plateau, Antarctica) of the US Geological Survey, Washington, DC

The valley of the Blackwall Glacier in Fig. 6.7 also contains several small outcrops of cross-bedded quartzite and shale which were described by Stump (1985). The relation of these rocks to the La Gorce and Wyatt formations is uncertain. All of the basement rocks of the Nilsen Plateau are unconformably overlain by the sedimentary rocks of the Beacon Supergroup (Devonian to Triassic) described by Long (1965). The Beacon rocks were intruded by sills of Ferrar Dolerite which are not shown in Fig. 6.5 and 6.7.

According to McLelland (1967), the La Gorce Formation consists of interbedded metagraywacke and slate which were asymmetrically folded along horizontal axes that strike north. The wavelength of the folds is about 1.5 km. Stump (1985) stated that the thickness of the La Gorce Formation on Hansen Spur exceeds 2,000 m. The beds of metagraywacke are between 0.5 and 3.0 m thick and consist of subangular grains of quartz and albite with accessory K-feldspar and epidote. The matrix is composed of biotite, sericite, and minor chlorite. The slate beds are less than 15 cm thick and contain very fine-grained biotite, sericite, quartz, and euhedral crystals of pyrite.

The rocks of the overlying Wyatt Formation are composed of quartz, albite, and biotite with phenocrysts of K-feldspar and albite (1–5 mm) in a fine-grained matrix of quartz, albite, and K-feldspar. The albite phenocrysts and the groundmass are partially altered to chlorite and sericite, whereas the matrix also contains epidote, calcite, and magnetite (McLelland 1967; Eastin 1970).

The Wyatt Formation occurs in conjunction with the La Gorce Formation not only in the northern part of the Nilsen Plateau but also on the west side of the

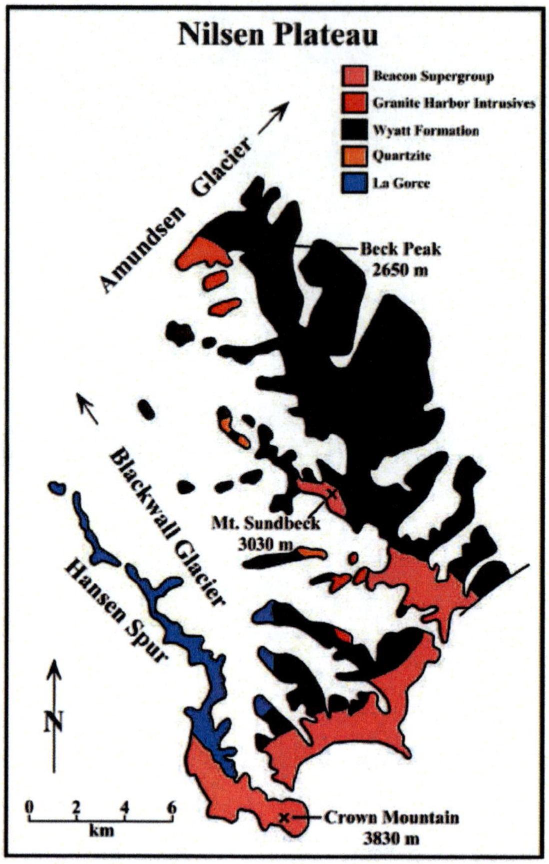

Fig. 6.7 The basement complex of the northern part of the Nilsen Plateau consists of the La Gorce and Wyatt formations and of the Granite Harbor Intrusives. The volcano-sedimentary rocks predominate in the northern part of the Nilsen Plateau, whereas Granite Harbor Intrusives dominate the basement rocks of the southern part in Fig. 6.5. The Blackwall Glacier was referred to as the Black Rock Glacier by McLelland (1967). The Kukri Peneplain that divides the basement from the overlying Beacon Supergroup is not flat in this area but has elevation differences of about 34 m (Adapted from Stump 1985, 1995)

Scott Glacier between Mt. Gardiner and Mt. Wyatt and in the La Gorce Mountains east of the Scott Glacier. The Wyatt Formation is also present on Metavolcanic Mountain near the head of the Reedy Glacier (Murtaugh 1969). Stump (1985, 1995) classified the rocks of the Wyatt Formation as a silicic porphyry that formed both as a hypabyssal intrusive and as a volcanic eruptive. The Wyatt Formation of the Nilsen Plateau intrudes the La Gorce Formation at Blackwall Glacier and is itself intruded by plutons of the Granite Harbor Intrusives.

Table 6.2 Stratigraphy of the basement rocks in the Duncan Mountains and the Nilsen Plateau of the Queen Maud Mountains (Stump 1985)

Group	Duncan Mountains	Nilsen Plateau
Granite Harbor Intrusives	Granite Harbor Intrusives	Granite Harbor Intrusives
	Intrusive Contact	
Liv	Fairweather and Henson Marble	Not present
Beardmore	Not present	Wyatt Formation[a]
	Dunan Formation	LaGorce Formation

[a]A measured section of cross-bedded quartzite, shale, and fine- to medium-grained sandstones that crop out in the valley of Blackwall Glacier (Fig. 6.7) was published by Stump (1985). The relation of these rocks to the La Gorce and Wyatt formations is not clear

6.3.3 Age Determinations

The basement rocks of the Nilsen Plateau were dated by Eastin (1970) who used the whole-rock Rb-Sr isochron method. The resulting dates in Table 6.3 have been recalculated to $\lambda = 1.42 \times 10^{-11}$ year^{-1}. All of the rock specimens dated by Eastin (1970) were collected by Douglas McLelland or other members of the group that mapped the Nilsen Plateau during the 1963/64 field season and were provided to Eastin by M.J. Hibbard, University of Nevada in Reno. The dates were calculated by the method of York (1966, 1969). The analytical data are tabulated in Appendix 6.7.1 and are interpreted there in Figs. 6.8–6.11.

The whole-rock Rb-Sr date of 713 ± 27 Ma of the La Gorce Formation of the Nilsen Plateau in Fig. 6.8, line A, records the time when the isotopic composition of strontium in these rocks was homogenized to an average $^{87}Sr/^{86}Sr$ ratio of 0.7117 ± 0.0005 (Eastin 1970). The sediment of the La Gorce Formation was probably deposited prior to 713 ± 27 Ma during the Neoproterozoic Era. The most Rb-rich samples of the La Gorce Formation (line B, Fig. 6.8, Appendix 6.7.1.1) and in Table 6.3 did not begin to retain in situ produced radiogenic ^{87}Sr until 475 ± 27 Ma (Early to Middle Ordovician) and their $^{87}Sr/^{86}Sr$ ratio at the time was 0.7173 ± 0.0005.

The metavolcanic rocks of the Wyatt Formation in the Nilsen Plateau were extruded or intruded into the rocks of the La Gorce Formation after the sedimentary

Table 6.3 Whole-rock Rb-Sr dates of the basement rocks exposed in the Nilsen Plateau at the head of the Amundsen Glacier (Data from Eastin 1970 are listed in Appendix 6.7.1 with Rb-Sr isochron diagrams in Figs. 6.8, 6.9, 6.10, and 6.11)

Geological unit	Number of samples	Initial $^{87}Sr/^{86}Sr$	Date[a] (Ma)
South Quartz Monzonite[b]	5 of 6	0.7248 ± 0.0037	443 ± 14
Lonely Ridge Granodiorite[c]	4 of 7	0.7116 ± 0.0003	598 ± 17
Wyatt Formation[d]	7 of 9	0.7150 ± 0.0007	475 ± 10 (reset)
La Gorce Formation	4 of 6	0.7117 ± 0.0005	713 ± 27
La Gorce Formation, Rb-rich samples[e]	2	0.7173 ± 0.0005	475 ± 27 (Reset)

[a]All dates have been recalculated to $\lambda = 1.42 \times 10^{-11}$ year^{-1}

[b]One sample that has a high $^{87}Rb/^{86}Sr$ ratio of 134.7 ± 2.9 plots below the Rb-Sr array defined by the other samples

[c]The three samples that do not fit the data array all have high $^{87}Rb/^{86}Sr$ ratios of 2.398, 71.52, and 270.7 and appear to have lost radiogenic ^{87}Sr during slow cooling

[d]The whole-rock Rb-Sr date of the Wyatt Formation appear to have been reset by contact metamorphism caused by the North Quartz Monzonite (Eastin 1970)

[e]The date is strongly controlled by a sample of phyllite that has a high $^{87}Rb/^{86}Sr$ ratio of 19.07 ± 0.29 and was collected from a xenolith within the South Quartz Monzonite at Lonely Ridge. Therefore, the date refers to the time at which the granite host-rock cooled sufficiently to allow radiogenic ^{87}Sr to accumulate in the xenolith

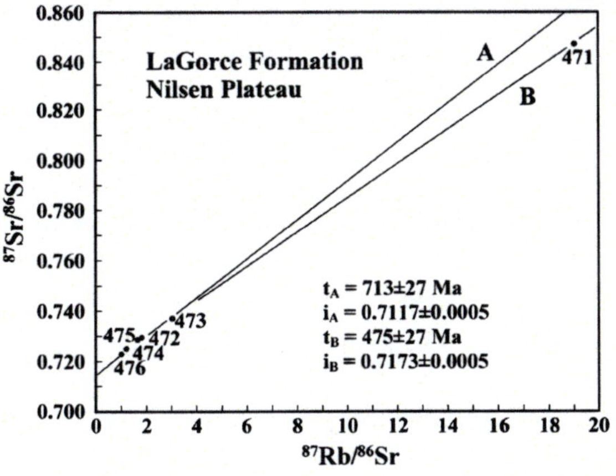

Fig. 6.8 Rb-Sr age determinations of metasedimentary rocks of the LaGorce Formation from the valley of the Blackwall Glacier, Nilsen Plateau (Fig. 6.7). The dates and initial $^{87}Sr/^{86}Sr$ ratios were calculated by Eastin (1970) by the methods of York (1966, 1969) and have been adjusted to $\lambda = 1.42 \times 10^{-11}$ year^{-1}. Isochron A is based on four colinear data points (472, 474, 475. and 476). Line B is defined by samples 471 and 473 which have the highest $^{87}Rb/^{86}Sr$ ratios and have lost some in situ produced radiogenic ^{87}Sr. Plotted from data by Eastin (1970)

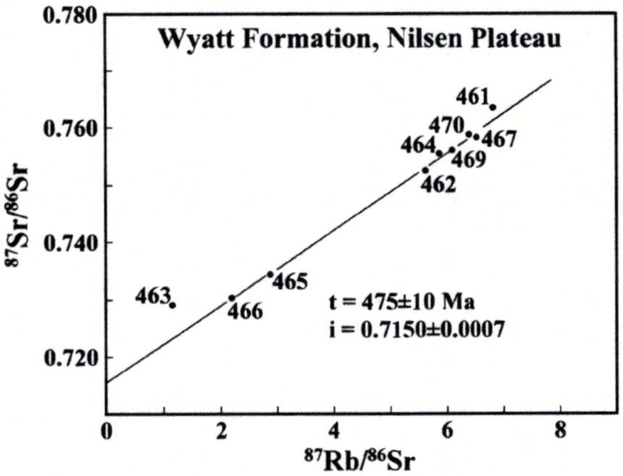

Fig. 6.9 Rb-Sr age determination of metavolcanic rocks of the Wyatt Formation in the valley of the Blackwall Glacier, Nilsen Plateau (Fig. 6.7). The date and initial $^{87}Sr/^{86}Sr$ ratio were calculated by Eastin (1970) by the methods of York (1966, 1969) and have been adjusted to $\lambda = 1.42 \times 10^{-11}$ year^{-1}. Samples 461 and 463 were excluded because they deviate significantly from the line fitted to the remaining data points. Plotted from data by Eastin (1970)

rocks were deposited prior to 713 ± 27 Ma. The rocks of the Wyatt Formation subsequently cooled slowly until 475 ± 10 Ma (Early to Middle Ordovician) when the average $^{87}Sr/^{86}Sr$ ratio of the metavolcanic rocks was 0.7150 ± 0.0007 (Table 6.3, Appendix 6.7.1.2, Fig. 6.9). The evidence for slow cooling implies that the metavolcanic rocks were intruded into the metasedimentary rocks of the La Gorce Formation and cooled at the same rate as the rocks of the basement complex of the Nilsen Plateau.

The rocks of the La Gorce and of the Wyatt formations of the Nilsen Plateau were intruded by syntectonic plutons of the Granite Harbor Intrusives represented here by the Lonely Ridge Granodiorite which crystallized at about 598 ± 17 Ma with an $^{87}Sr/^{86}Sr$ ratio of 0.7116 ± 0.0003 (Table 6.3 and Appendix 6.7.1.3, Fig. 6.10). This date may be more uncertain than indicated by the error limits (±17 Ma) because only four of seven samples analyzed by Eastin (1970) satisfy the assumptions required for dating by the Rb-Sr method. Therefore, the magma of the Lonely Ridge Granodiorite could have crystallized either during the Neoproterozoic or Early Cambrian time and subsequently cooled until 472 ± 10 Ma (K-Ar date, biotite; McLelland 1967; and Rb-Sr date, biotite; Felder and Faure 1979).

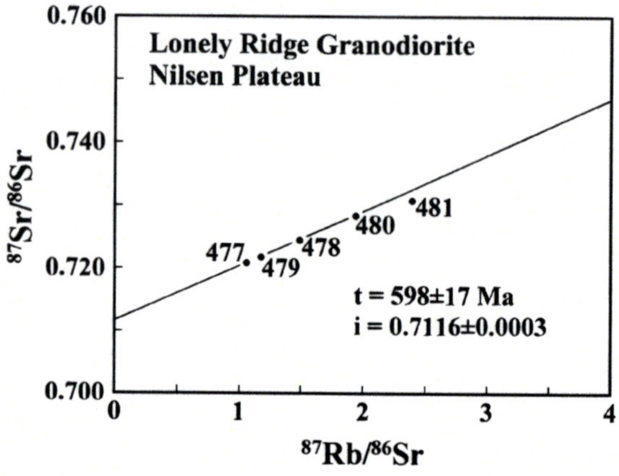

Fig. 6.10 Rb-Sr age determination of whole-rock samples of the Lonely Ridge Granodiorite of Lonely Ridge, Nilsen Plateau (Fig. 6.5). The date and initial $^{87}Sr/^{86}Sr$ ratio were calculated by Eastin (1970) by the methods of York (1966, 1969) and have been adjusted to $\lambda = 1.42 \times 10^{-11}$ year^{-1}. Samples 481, 482, and 483 were excluded from the calculation of the Rb-Sr date because they did not remain closed to radiogenic ^{87}Sr. Plotted from data by Eastin (1970)

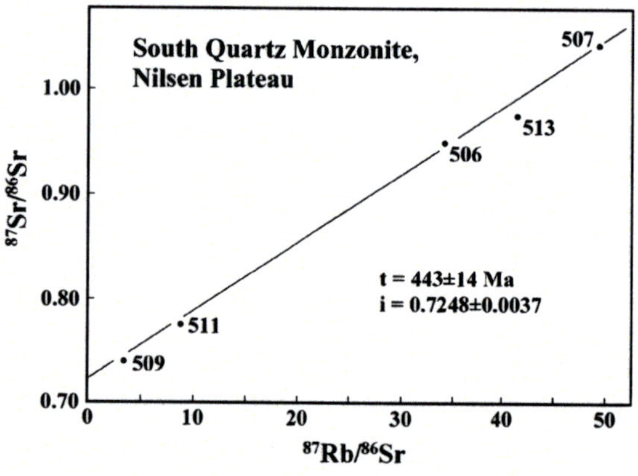

Fig. 6.11 Rb-Sr age determination of whole-rock samples of the South Quartz Monzonite from the southwest wall of Cottonwood Canyon (506, 511), from Windy Ridge (507, 508, and 509), and from South Ridge (513) of the Nilsen Plateau (Fig. 6.5). The date and initial $^{87}Sr/^{86}Sr$ ratio were calculated by Eastin (1970) by the methods of York (1966, 1969) and have been adjusted to $\lambda = 1.42 \times 10^{-11}$ year^{-1}. Sample 508 was excluded from the calculation of the Rb-Sr date because it lost radiogenic ^{87}Sr. Plotted from data by Eastin (1970).

The South Quartz Monzonite was intruded into plutons of the Lonely Ridge Granodiorite, but the whole-rock Rb-Sr date based on data in Appendix 6.7.1.4 and Fig. 6.11 indicates only that the isotope composition of strontium was homogenized at 443 ± 14 Ma (Late Ordovician) to a high $^{87}Sr/^{86}Sr$ ratio of 0.7248 ± 0.0037. Therefore, this is a cooling date that post-dates the time of original crystallization of the granitic magma.

According to the Rb-Sr dates of Eastin (1970), the Ross orogen began to form prior to about 713 ± 27 Ma with the deposition of clastic sedimentary rocks along the passive rift margin of Gondwana. The passive rift margin was subsequently transformed into a compressive subduction zone (Goodge et al. 2001), which caused the accumulated sediment of the La Gorce Formation to be folded and was followed by the deposition of the igneous rocks of the Wyatt Formation. As the Ross Orogeny continued, granitic magma of the Lonely Ridge Granodiorite intruded the early-formed volcano-sedimentary complex and crystallized at about 598 ± 17 Ma. The Ross Orogeny continued with the intrusion of the South Quartz Monzonite and other facies of the Granite Harbor Intrusives. At the end of the Ross Orogeny all parts of the Ross orogen in the Nilsen Plateau cooled to the retention temperature of radiogenic ^{87}Sr and ^{40}Ar:

South Quartz Monzonite, Rb-Sr whole rock:	443 ± 14 Ma
Lonely Ridge Granodiorite, Rb-Sr biotite:	470 ± 17 Ma
Lonely Ridge Granodiorite, K-Ar, biotite:	472 ± 10 Ma
La Gorce Formation, Rb-rich, Rb-Sr whole-rock:	475 Ma
Wyatt Formation, Rb-Sr whole rock:	476 ± 10 Ma

After a lengthy period of uplift and erosion, sedimentary rocks of Permian age described by Long (1965) were deposited on the Kukri Peneplain which formed by erosion of the rocks of the Ross orogen.

6.4 Scott Glacier Area

We now move east from the Nilsen Plateau into the drainage of the Scott Glacier in Fig. 6.12 formerly known as the Thorne Glacier. Virtually the entire area drained by the Scott Glacier and its tributaries exposes rocks of the basement complex consisting of the La Gorce Formation, the Wyatt Formation, the Ackerman Formation, and the

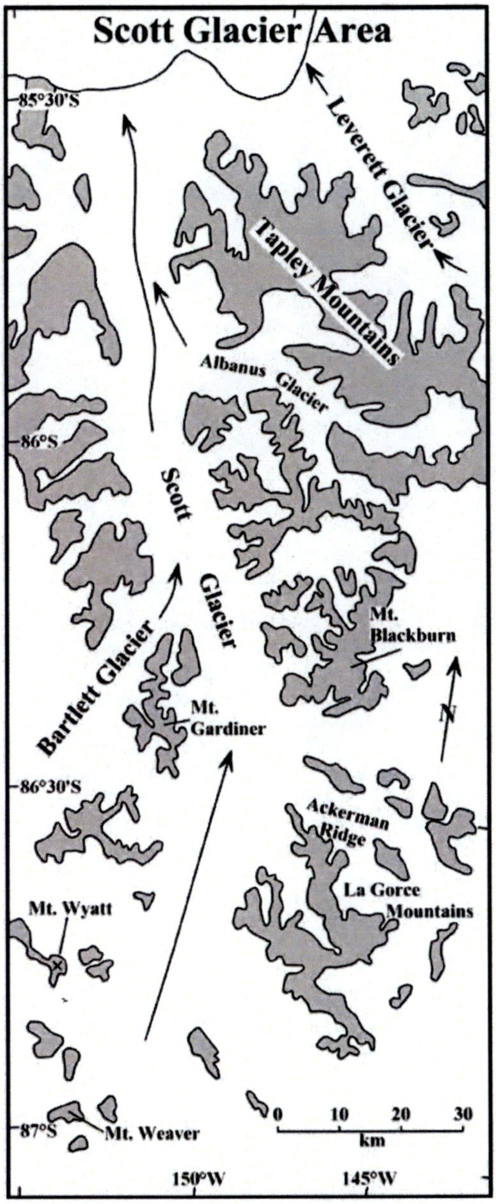

Fig. 6.12 The Scott Glacier flows north from the polar plateau and discharges ice into the Ross Ice Shelf. The Bartlett and Albanus glaciers are its principal tributaries. The major mountain ranges adjacent to the Scott Glacier are the La Gorce Mountains at its head and the Tapley Range east of its mouth. In addition, the map identifies Mt. Weaver, Mt. Wyatt, Mt. Gardiner, and Mt. Blackburn. The valley of the Leverett Glacier is located east of the Scott Glacier and drains the Harold Byrd Mountains to be discussed in Section 6.5 (Adapted from Mirsky 1969)

Granite Harbor Intrusives (Davis and Blankenship 2005). The principal tributaries of the Scott Glacier are the Bartlett and Albanus glaciers as well as the Van Reeth and Poulter glaciers (not shown in Fig. 6.12). The highest mountains located west of the Scott Glacier are Mt. Astor (3,711 m), Mt. Bowser (3,654 m), Mt. Crockett (3,479), Mt. Vaughan (3,140 m), and Mt. Griffith (3,097 m), all of which occur within a small area measuring 10 × 20 km at about latitude 86°S. Farther south but still west of the Scott Glacier we find Mt. Saltonstall (2,937 m), Mt. Wyatt (2,930 m), and Mt. Gardiner (2,479 m). The tallest peak in the La Gorce Mountains east of the Scott Glacier is Mt. Paine (3,330 m), followed by Johansen Peak (3,310 m), Mt Grier (3,037 m), and Mt. Mooney (2,850 m). Turning north from the La Gorce Mountains we find Mt. Blackburn (3,220 m). Still farther north in the Tapley Mountains are Mt.Gould (named after the Antarctic explorer) and Mt. Bushnell (editor of the American Geographical Society). Near the mouth of the Scott Glacier in Fig. 6.12, we note the Leverett Glacier which flows past the Harold Byrd Mountains (not shown in Fig. 6.12).

6.4.1 La Gorce Mountains

The La Gorce Mountains at the head of the Scott Glacier in Fig. 6.12 are about 45 km wide in an east-west direction and include several high peaks some of which were named above. The La Gorce Mountains also contain several cirques the largest of which in the center of the range contains ice-cored rock debris as do several smaller cirques that face the Scott Glacier and the Robinson Glacier on the north and east, respectively. The accumulation of rock debris on the ice in the largest cirque of the La Gorce Mountains implies that the ice is stagnant and is not actively supporting the flow of the Scott Glacier at the present time. However it probably did so in the past.

The Scott Glacier area was first visited by geologists of Byrd's second Antarctic Expedition of 1933–1935 (Blackburn 1937). Members of this group traveled by dogsleds and reached the northern part of the La Gorce Mountains where they recorded the presence of black porphyritic rocks (Stump et al. 1986). Subsequently, the geology of the southern mountain ranges of the Scott Glacier area, including the La Gorce Mountains, was investigated by Velon H. Minshew who traveled by helicopter from a camp at the base of Mt. Weaver (86°58′S, 153°50′W) located on the west side of the

Table 6.4 Stratigraphy of the basement complex of the La Gorce Mountains and the Leverett-Glacier area of the Queen Maud Mountains (Stump et al. 1986; Stump 1995; Davis and Blankenship 2005)

Group	La Gorce Mountains	Leverett Glacier
Granite Harbor Intrusives	Granite Harbor Intrusives	Granite Harbor Intrusives
————	Intrusive Contact	————
Liv		Leverett Party
Beardmore	Ackerman	
	Wyatt	
	La Gorce	
————	Base not exposed	————

Scott Glacier (Fig. 6.12). Minshew (1967) named the La Gorce and Wyatt formations and determined from the difference in the elevations of their outcrops that the Wyatt Formation is younger than the La Gorce Formation (Table 6.4). This conjecture was later confirmed by Stump et al. (1981) who prepared the geologic map of the La Gorce Mountains in Fig. 6.13. The center of attention in these later studies was Ackerman Ridge which was originally mapped as La Gorce Formation along its entire length (Mirsky 1969). However, Katz and Waterhouse (1970b) reported that part of the Ackerman Ridge is actually composed of the Wyatt Formation as indicated in Fig. 6.13.

The La Gorce Formation on Ackerman Ridge and Kessen Peak consists of alternating feldspar- and mica-bearing graywacke and metapelites. The rocks were tightly folded and metamorphosed to the greenschist facies characterized by the mineral assemblage quartz + feldspar + biotite ± muscovite ± chlorite. Cordierite and andalusite are present locally. The sedimentology of the La Gorce Formation was described by Smit and Stump (1986) who concluded that the graywacke-mudstone sequence was deposited by high-density turbidity currents. The La Gorce Formation also contains Bouma sequences composed of layers A, C, and E of the turbidite model of Bouma (1962) described in Appendix 6.7.2 and illustrated in Fig. 6.20. Smit and Stump (1986) concluded from the sedimentology of the La Gorce Formation that it was deposited as a submarine fan on a broad shelf-rise. The persistent presence of trace minerals such as garnet, muscovite, epidote, clinopyroxene, sphene, and zircon indicates that the sediment originated, at least in part, by erosion of high-grade metamorphic and felsic plutonic rocks. This conclusion by Smit and

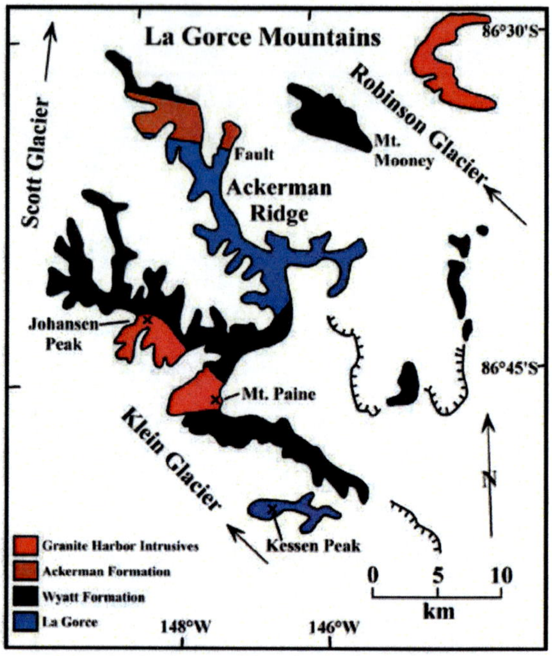

Fig. 6.13 The rocks exposed in the La Gorce Mountains include all of the stratigraphic units that form the basement complex of the Scott Glacier area. This mountain range is located near the head of the Scott Glacier at the edge of the polar plateau. Mt. Paine (3,330 m) and Johansen Peak (3,310 m) are the highest points of the La Gorce Mountains followed closely by Mt. Grier (3,037 m) which is not identified on Fig. 6.13. The geology of this mountain range was mapped during the 1980/81 field season (Stump et al. 1981) (Adapted from Stump 1983, 1995)

Stump (1986) is supported by the occasional presence of apatite, rutile, corundum, kyanite, and tourmaline. Stump (1995) also noted that the La Gorce Formation differs from the Goldie Formation of the central Transantarctic Mountains (Section 5.2) by the absence of carbonate beds except for calcite veins near faults.

The Wyatt Formation in Fig. 6.13 is widely exposed in the La Gorce Mountains and on Mt. Mooney and its petrography was described by Stump (1985). Stump et al. (1986) reported that the silicic porphyries of the Wyatt Formation intruded folded rocks of the La Gorce Formation at a newly-discovered site in the La Gorce Mountains. The rocks of the Wyatt Formation are dark gray to brown in outcrop with an aphanitic matrix containing phenocrysts of plagioclase and embayed quartz that range in abundance from 20% to 70% by volume. Most specimens also contain mafic phenocrysts that have been altered to chlorite. Borg (1980) reported the occurrence of unaltered hypersthene in rocks of the

Wyatt Formation near Mt. Wyatt on the west side of the Scott Glacier (Fig. 6.12).

The mineral composition (mode) of the Wyatt Formation in the Scott-Glacier area was determined by Minshew (1967) and is listed in Appendix 6.7.3.1 followed by chemical analyses in Appendix 6.7.3.2. The rocks of the Wyatt Formation in the La Gorce Mountains were altered either by the original magmatic fluid or during later metamorphism. In extreme cases, plagioclase was completely altered to sericite or saussurite (Stump et al. 1986). Saussurite is a mineral aggregate composed of albite and zoisite or epidote with variable amounts of calcite, sericite, and prehnite (Jackson 1997).

The mode of emplacement of the Wyatt Formation has been obscured by the lack of internal structure of the rocks and by the alteration of the groundmass. Minshew (1967) considered it to be a volcanic rock of pyroclastic origin based on the presence of zoned plagioclase, abundance of a fine-grained matrix, and the absence of flow structures. Murtaugh (1969) agreed with Minshew because the presence of broken phenocrysts suggested to him that the rocks were erupted explosively; but he also considered the possibility that parts of the Wyatt Formation formed as a shallow intrusive. Stump (1976) later found evidence in the Nilsen Plateau that the rocks of the Wyatt Formation had intruded the La Gorce Formation without extensive contact metamorphism (i.e., as a shallow intrusive that cooled rapidly), whereas Borg (1980) observed the presence of lenses of felsic rocks in the Wyatt Formation near Mt. Wyatt which he interpreted as flattened fragments of pumice thus implying a volcanic-eruptive origin. Evidently, the available evidence indicates that the Wyatt Formation was emplaced as a shallow intrusive in some areas and as a volcanic eruptive in others.

The Ackerman Formation of the La Gorce Mountains (Table 6.4) consists of a sequence of interbedded sedimentary and volcaniclastic rocks that crop out on the northern part of Ackerman Ridge in Fig. 6.13. Green shale at the base of the Ackerman Formation was deposited conformably on massive rocks of the Wyatt Formation. The top of the Ackerman Formation is faulted against the sedimentary rocks of the La Gorce Formation. The volcanic rocks of the Ackerman Formation resemble the Wyatt Formation in color, texture, phenocrysts content, and bulk chemical composition. The rocks were sericitized and mafic minerals were converted to chlorite, but the extent of alteration of the Ackerman volcanics is less that that of the Wyatt porphyries.

The chemical composition of the volcanic rocks of the Ackerman and Wyatt formations in Fig. 6.14 are similar but not identical. The high concentrations of silica and the reported presence of quartz indicate that the rocks of both formations are oversaturated with respect to silica according to the classification of Shand

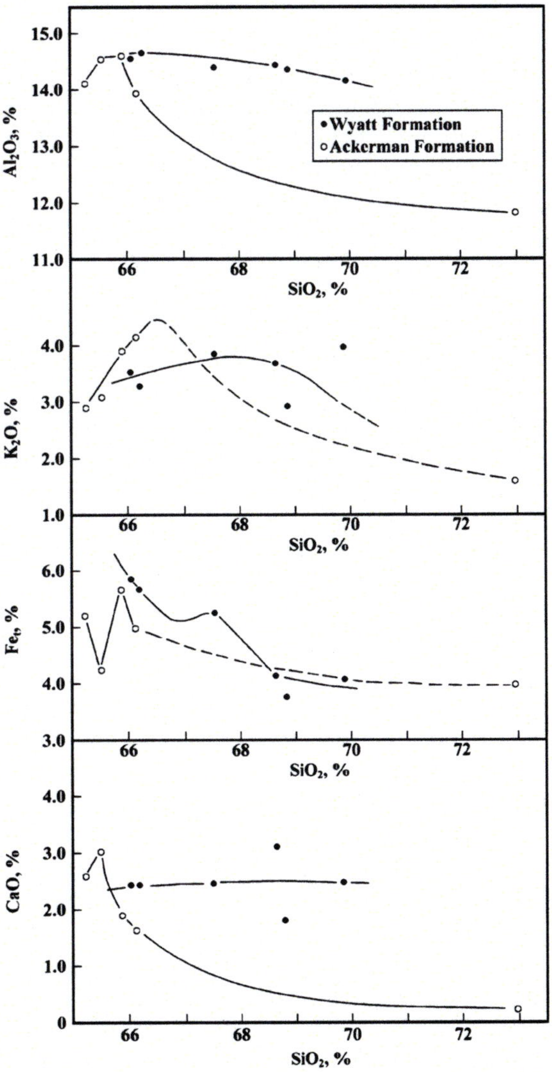

Fig. 6.14 Silica-variation diagrams of the silicic porphyries of the Wyatt Formation and of the Ackerman Formation in the La Gorce Mountains and in the upper Scott Glacier area (Data by Stump et al. 1986; Minshew 1967)

(1947). In addition, the rocks of both formations are peraluminous because their molar ratios of Al_2O_3/($K_2O + Na_2O + CaO$) are greater than 1.0 in all samples analyzed by Stump et al. (1986). The silica-variation diagrams in Fig. 6.14 suggest that the concentrations of the major rock-forming elements were affected by fractional crystallization of the respective parent magmas.

6.4.2 Age Determinations

The igneous rocks of the Wyatt Formation occur at many locations within the Queen Maud Mountains, as well as in the Horlick and Thiel Mountains. In spite of these extensive exposures, the origin of the Wyatt Formation is still uncertain because in some places it clearly intruded folded metasedimentary rocks of the LaGorce Formation, whereas in other places it overlies them conformably. In addition, hydrothermal alteration of the rocks and the effects of regional metamorphism described by Minshew (1967) have disturbed the isotopic composition of strontium of the rocks, which has caused data points to scatter on Rb-Sr isochron diagrams and caused large uncertainties in age determinations of the Wyatt Formation listed in Table 6.5.

Faure et al. (1979) analyzed five samples of the Wyatt Formation collected by Minshew between Mt. Wyatt and Mt. Gardiner in the upper reaches of the Scott Glacier. The results in Appendix 6.7.3.3 were originally used only to state an upper limit of 788 Ma (Neoproterozoic) for the age of the Wyatt Formation. A least-squares linear regression of these data (plus one duplicate) defines line A in Fig. 6.15 which results in a calculated date of 524 ± 13 Ma (Early Cambrian) with an initial $^{87}Sr/^{86}Sr$ ratio of 0.711919 ± 0.000547 for $\lambda = 1.42 \times 10^{-11}$ year^{-1}. Three of the samples in Fig. 6.15 precisely define line B which corresponds to a date of 802 ± 2 Ma and an initial $^{87}Sr/^{86}Sr$ ratio of 0.69775 ± 0.0000997. This date is invalid because the initial $^{87}Sr/^{86}Sr$ ratio of line B is less than that of primordial strontium (i.e., BABI or basaltic achondrites best initial, which has a value of 0.69897 ± 0.00003; Faure and Mensing 2005).Therefore, the Wyatt Formation was deposited during the Cambrian Period.

A more recent attempt to date rocks of the Wyatt Formation in the Scott Glacier area by Stump et al. (1986) was also frustrated by the scatter of data points.

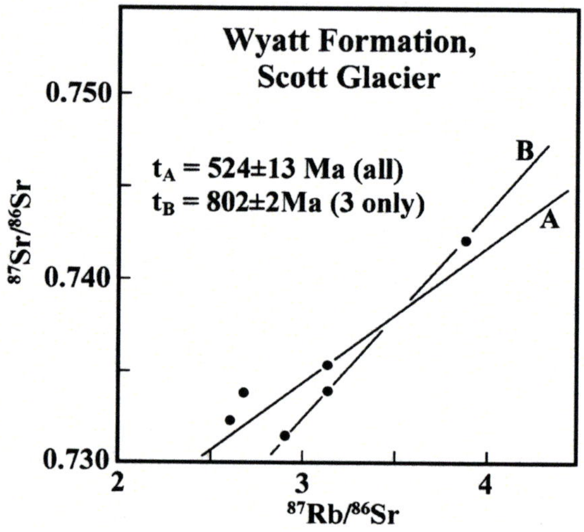

Fig. 6.15 A suite of five whole-rock samples of the Wyatt Formation collected by V.H. Minshew from outcrops between Mt. Wyatt and Mt. Gardiner in the Scott Glacier area define two straight lines labeled A and B derived by least-squares linear regression. Line A which includes all five samples plus one duplicate analysis yields a date of 524 ± 13 Ma and a high initial $^{87}Sr/^{86}Sr$ ratio of 0.711919 ± 0.000547 (1σ). Line B is defined by three selected samples and corresponds to a precise date of 802 ± 2 Ma; but it has an impossibly low initial $^{87}Sr/^{86}Sr$ ratio of 0.69775. We conclude that the igneous rocks of the Wyatt Formation are older than 524 ± 13 Ma but younger that 802 ± 2 Ma. These previously unpublished data by G. Faure are presented in Appendix 6.7.3.3. In addition, Appendices 6.7.3.1 and 6.7.3.2 contain modal analyses and chemical compositions of rocks of the Wyatt Formation from Minshew (1967)

Table 6.5 Age determinations of rocks in the La Gorce Mountains and other parts of the Scott Glacier area (Data from Mirsky 1969; Stump et al. 1986; Van Schmus et al. 1997; Faure unpub. [Appendix 6.7.3.3])

Rock unit	Method	Date (Ma)	Reference
Granite Harbor Intrusive, Mt. Wilbur	K-Ar, biotite	479 ± 14	Minshew (1969)
Wyatt Formation, Scott Glacier	Rb-Sr, whole rock	524 ± 13	Faure et al. (unpublished) (Appendix 6.7.3.3)
Wyatt Formation, west of upper Scott Glacier	Rb-Sr, whole rock	>545 ± 59	Stump et al. (1986)
Wyatt Formation, near Mt. Wyatt	U-Pb, zircon	520 – 540	Van Schmus et al. (1997)
Wyatt Formation	U-Pb, zircon	526 ± 2 Ma	Encarnación and Grunow (1996)

Nevertheless, the slope of a straight line defined by their data yielded a date of 545 ± 59 Ma with and initial $^{87}Sr/^{86}Sr$ ratio of 0.7114 ± 0.0030. Stump et al. (1986) concluded that the date had been reset and therefore is an underestimate of the crystallization age of the Wyatt Formation. Even the U-Pb method of dating zircon failed to provide a reliable date of the Wyatt Formation because the data points derived from the zircon fractions analyzed by Van Schmus et al. (1997) did not define a discordia curve. Even so, Van Schmus et al. (1997) inferred Early Cambrian dates between 520 and 540 Ma for this formation. The most precise age determination for the Wyatt Formation is a U-Pb date of 526 ± 2 Ma reported by Encarnaci6n and Grunow (1996). Therefore, the age of the Wyatt Formation appears to be Early Cambrian.

The story we take away from the age determinations of basement rocks in the Scott Glacier area listed in Table 6.5 starts with the deposition of graywacke-shale of the La Gorce Formation along the passive rift margin of Gondwana during the Neoproterozoic Era. These sediments were subsequently folded by compression followed by the eruption/intrusion of the Wyatt Formation during an early phase of the Ross Orogeny (Stump et al. 1991).

After the igneous rocks of the Wyatt Formation had been extruded and/or injected, the granitic plutons of the Granite Harbor Intrusives crystallized from granitic magmas that had formed by anatexis of the volcano-sedimentary rocks. These plutons cooled slowly into the Ordovician Period as indicated by the K-Ar dates of biotite in the granite of Mt. Wilbur (479 ± 14 Ma, Table 6.5). This hypothetical geologic history of the basement complex of the Scott Glacier area in the Queen Maud Mountains is consistent with the "standard model" for the geologic history of the basement rocks of the Transantarctic Mountains of Stump (1995) and outlined in the introduction to Chapter 3.

6.4.3 Queen Maud Batholith

The granitic rocks of the Granite Harbor Intrusives occur in the core of the Ross orogen from the Wilson Terrane of northern Victoria Land to the Ohio Range of the Horlick Mountains and beyond. This group of rocks is characterized by a wide range of chemical compositions and textures depending on the protoliths and on the pressure and temperature to which they were subjected during the Ross Orogeny. The resulting ignio-metamorphic rocks typically occur in the cores of orogenic mountains not only in Antarctica, but also on all of the continents. For this reason, the rocks of the Queen Maud batholith merit close attention because they contain a record of the evolution of one of the major mountain ranges of the Earth.

The origin of the granitic rocks of the Granite Harbor Intrusives of the Queen Maud batholith was investigated by Borg (1983), Borg et al. (1986a, 1987a, 1990), Gunner (1971, 1976), Gunner and Faure (1972), and by Goodge and Borg (1987). In addition, Borg (1984) and Borg et al. (1986b, 1987b) worked extensively on the Granite Harbor Intrusives of northern Victoria Land.

An important issue in the study of orogenic granitoids is to determine whether they formed primarily from clay-rich sedimentary rocks (S-type) or primarily from volcanic rocks of basaltic-to-andesitic composition (I-type), based on their chemical compositions (Chappell and White 1974). Alternatively, the provenance of orogenic granitoids is indicated by the numerical values of the initial $^{87}Sr/^{86}Sr$ and $^{143}Nd/^{144}Nd$ ratios of the rocks by means of the Sr-Nd isotopic mixing diagram (Faure and Mensing 2005; McCulloch and Chappell 1982; Faure and Hurley 1963).

The *chemical* criteria were used by Borg et al. (1986b) to classify the granitoids of the Queen Maud batholith into the two types defined by Chappell and White (1974) and White et al. (1977). The *isotopic* criteria were used by McCulloch and Chappell (1982) to show that the granitoids of the Berridale and Kosciusko batholiths of southeastern Australia form overlapping mixing arrays in quadrant 4 of the Sr-Nd isotopic mixing diagram. Therefore, orogenic granitoids are best regarded as products of granitization of volcano-sedimentary complexes composed of mixtures of volcanic and sedimentary protoliths in a wide range of proportions.

The granitoids of the Queen Maud batholith include plutons of diorite, quartz diorite, quartz monzonite, granodiorite, and granite. The rocks include both foliated and unfoliated varieties which indicate that they range from being synorogenic (foliated) to post-orogenic (unfoliated). Nevertheless, all parts of the Queen Maud batholith are considered to be products of the Ross Orogeny. The foliated varieties contain phenocrysts of pink microcline in a matrix of fine-grained feldspar, brown biotite (or hornblende) and quartz. Borg (1983) stated that the foliation generally parallels the margins of individual plutons and does not extend beyond their margin. The undeformed varieties are medium to coarse grained, contain pink microcline phenocrysts, and are also composed of plagioclase, biotite, and quartz. Sphene and allanite are common accessory minerals. The chemical compositions of 31 specimens analyzed by Borg (1983, Table 1) vary widely as indicated in Table 6.6.

Borg (1983) classified the granitoids of the Queen Maud batholith between the Scott and Reedy Glaciers into the I- and S-types based on their chemical compositions. The results identified an S-I boundary that separates the granitic rocks in the Tapley Mountains, the Harold Byrd Mountains, and Berry Peaks in the north from the rocks at the head of the Scott Glacier in the south. The granitoids north of the boundary are exclusively I-type while south of the boundary they are I-type and S-type. Borg (1983) suggested that the S-I boundary separates the oceanic environment in the north from the continental environment south of the boundary.

The granitic rocks in the segment of the Transantarctic Mountains between the Nimrod-Beardmore-Shackleton glaciers include three crustal provinces that were identified by Borg et al. (1990) on the basis of isotopic systematics of Rb-Sr, Sm-Nd, and oxygen. These provinces occur in the Miller Range south of the Nimrod Glacier, in the Beardmore Glacier area between the Marsh and Shackleton glaciers, and in the Gabbro Hills at the mouth of the Shackleton Glacier. The average initial $^{87}Sr/^{86}Sr$ ratios at 500 Ma of the granitoid rocks and gabbro of these areas are:

Table 6.6 Range of chemical composition of 31 granitoids in the Queen Maud batholith between the Scott and Reedy glaciers (Data from Borg 1983)

Oxide	Range (%)
SiO_2	58.86–75.72
Al_2O_3	12.05–18.26
Fe_2O_3[a]	1.64–7.91
MgO	0.50–3.92
CaO	0.76–7.14
Na_2O	2.29–4.20
K_2O	1.14–5.38
P_2O_5[b]	0.00–0.34

[a]Total iron
[b]MnO and loss on ignition are not shown in the table

Gabbro Hills: 0.70515 (0.70446–0.70589)
Beardmore area: 0.71091 (0.70684–0.71906)
Miller Range: 0.73481 (0.73243–0.74174)

The granitoids of the Miller Range represent the Paleoproterozoic crust (2000 Ma) of the East Antarctic craton (Gunner and Faure 1972). The granitoids of the Beardmore area belong to the Granite Harbor suite of orogenic origin, although Borg et al. (1990) considered that they represent a hypothetical Beardmore microcontinent. Similarly, we regard the mafic rocks of the Gabbro Hills as the descendants of the mafic lavas that were originally erupted along the compressive margin of East Antarctica prior to the culmination of the Ross Orogeny.

6.5 Leverett Glacier Area

The area north of the Leverett Glacier in Fig. 6.12 contains a scattering of nunataks which collectively form the Harold Byrd Mountains, the Bender Mountains, the Berry Peaks, and the Fallone Nunataks in addition to Mt. Webster (85°40′S, 144°24′W), Price Peak (85°43′S, 142°24′W), Racine Nunatak (85°28′S, 136°18′W), Langford Peak (85°33′S, 135°23′W), and other nunataks identified on the USGS topographic map of the Leverett Glacier, Antarctica (SV1–10/7, 1968). The history of the geological exploration of this large area of about 3,600 km² was summarized by Stump (1995).

6.5.1 History of Exploration

The Harold Byrd Mountains in Fig. 6.16 include Supporting Party Mountain where Laurence Gould and his team of geologists first set foot upon the rocks of the Transantarctic Mountains during Richard Byrd's first expedition to Antarctica from 1928 to 1930 (Section 1.5.1). The chain of nunataks of the Harold Byrd Mountains extends eastward from the Supporting Party Mountain for a distance of about 30 km and includes several noteworthy peaks including Mt. Graham, Mt. Nichols, Mt. Manke, as well as Cressey Peak and Fadden Peak located off-scale in Fig. 6.16 east of Supporting Party Mountain. The rock samples collected by Gould during his first landfall were later

Fig. 6.16 The Harold Byrd Mountains are located north of the Leverett Glacier which flows in a northwesterly direction and discharges ice into the Ross Ice Shelf just east of the mouth of the Scott Glacier (Fig. 6.12). This mountain range extends east from Supporting Party Mountain for about 30 km and includes Cressey Peak (870 m) and Fadden Peak (920 m) which lie off-scale east of Mt. Manke. The Harold Byrd Mountains consist of highly deformed and regionally metamorphosed mainly pelitic sedimentary rocks believed to be of Neoproterozoic age (Stump 1995; Davis and Blankenship 2005). Excerpt from the topographic map of the Leverett Glacier, Antarctica, SV1–10/7 published in 1968 by the US Geological Survey in Washington, DC Contour interval 200 m

identified by Stewart (1934) as biotite schist and biotite-hornblende schist.

During the 1964/65 field season Velon H. Minshew landed on Mt. Webster in a US Army helicopter that was stationed at the time at a geological field camp (Camp Ohio II) on the Reedy Glacier (Caloplaca Hills, Antarctica, SV1–10/12, US Geological Survey, 1968, Washington, DC). Mt. Webster in Fig. 6.17 is the principal nunatak in the area north of the Leverett Glacier and is located about 23 km south of Mt. Manke in the Harold Byrd Mountains. Minshew measured a stratigraphic section of the sedimentary and volcanic rocks at Mt. Webster and used these rocks to define the Leverett Formation (Minshew 1967; Mirsky 1969).

The stratigraphic section on Mt. Webster, measured by Minshew in 1965, was continued by Edmund Stump during the 1978/79 field season in the course of a geological reconnaissance. Stump (1995) later used the results of this work to define the Party Formation which is composed primarily of clastic sedimentary rocks exemplified by the rocks exposed along a ridge south of Mt. Nichols in Fig. 6.16 (Stump et al. 1978; Stump 1995; Lowry 1980; Heintz 1980).

The geologic mapping of the nunataks north of the Leverett Glacier was expanded during the 1992/93 field

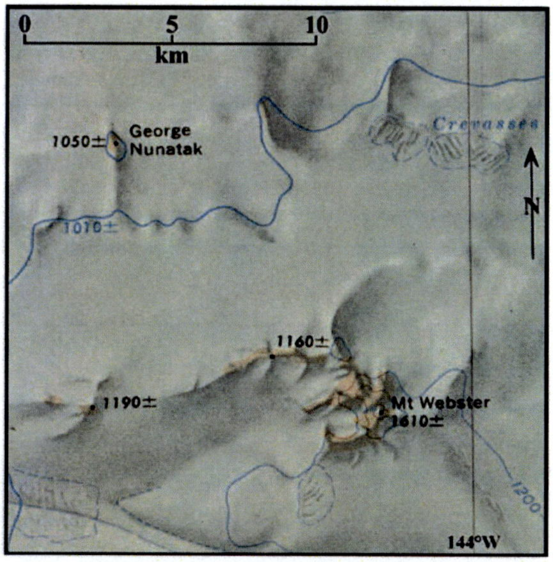

Fig. 6.17 Mount Webster is located south of the Harold Byrd Mountains (Fig. 6.16) and north of the Leverett Glacier at 85°40′S and 144°24′W. It is the type locality of the Cambrian Leverett Formation defined by Minshew (1967) and is composed of cross-bedded sandstones overlain by shale, limestones, and marble which in turn are overlain by silicic volcanic rocks. The limestones contain fragmental trilobite fossils (Minshew 1967) and archaeocyathids (Rowell et al. 1997). Excerpt from the topographic map of the Leverett Glacier, Antarctica, SV1–10 published in 1968 by the US Geological Survey, Washington, DC Contour interval 200 m

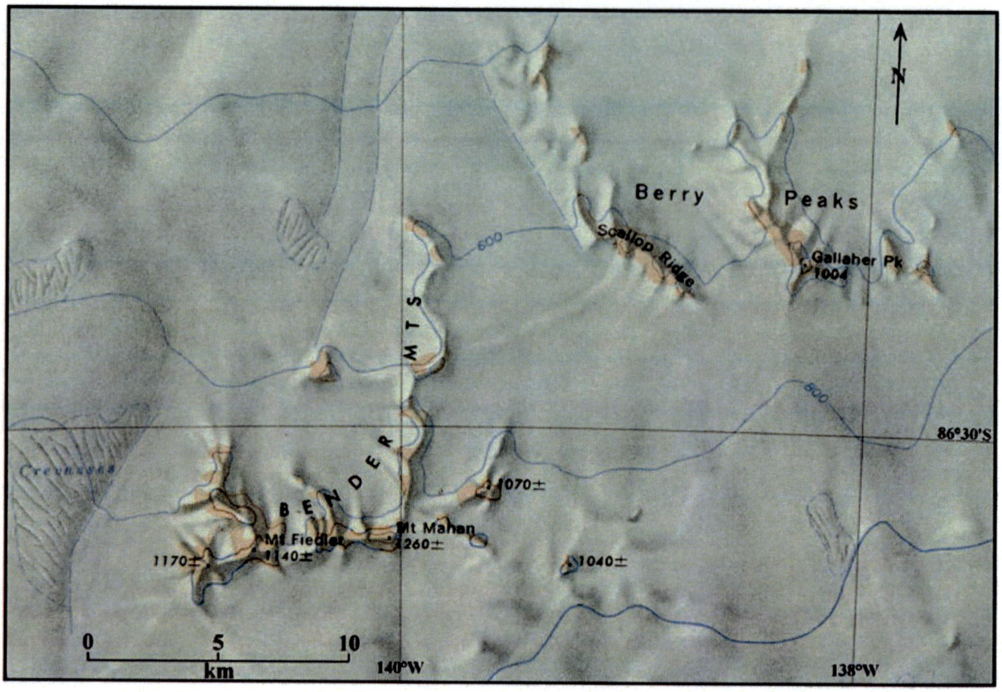

Fig. 6.18 The Bender Mountains and the Berry Peaks are located north of the Leverett Glacier and east of the Harold Byrd Mountains in Fig. 6.16. According to the geologic map of Davis and Blankenship (2005), the Bender Mountains are composed of the metasedimentary rocks of the Leverett Formation intruded by several small plutons of Granite Harbor Intrusives, whereas all of the Berry Peaks and the chain of small Fallone Nunataks north of the Harold Byrd Mountain consist predominantly of granitic rocks of the Granite Harbor suite. Excerpt from the topographic map of the Leverett Glacier, Antarctica SV1–10/7 published in 1968 by the US Geological Survey, Washington, DC Contour interval 200 m

season by A.J. Rowell who confirmed the presence of felsic volcanic rocks on Mt. Fiedler in the Bender Mountains in Fig. 6.17 and reported that the limestone exposed on Mt. Mahan contains archaeocyathids of Early Cambrian age (Rowell et al. 1993, 1997). Rowell also observed that the Berry Peaks and the Fallone Nunataks, which are located about 15 km north of the Harold Byrd Mountains (but which were not included in Fig. 6.16) are composed of the Granite Harbor Intrusives.

The most recent geologic map of the Queen Maud Mountains by Davis and Blankenship (2005) indicates that the western half of the Harold Byrd Mountains consists of rocks of the Party Formation and of Granite Harbor Intrusives, whereas the eastern nunataks (Mt. Manke, Cressey Peak, and Fadden Peak) consist of the rocks of the Leverett Formation). Similarly, the Bender Mountains in Fig. 6.18 are composed primarily of the Leverett Formation (e.g., Mt. Fiedler), whereas Mt. Mahan of the Bender Mountains contains a pluton of Granite Harbor Intrusives.

6.5.2 Stratigraphy

The basement rocks exposed in the nunataks north of the Leverett Glacier are composed of sedimentary and volcanic rocks of Cambrian age that were extensively deformed and variously metamorphosed during the Ross Orogeny. These rocks were subsequently intruded by plutons of granitic rocks of the Granite Harbor Intrusives which then cooled slowly until the Middle Ordovician.

Rocks of the La Gorce and Wyatt formations which are prominent in the basement complex of the Scott Glacier area (e.g., Fig. 6.13) have not been recognized in the nunataks north of the Leverett Glacier. In addition, rocks of the Beacon Supergroup occur only on Mt. Beazley (86°51′S, 142°51′W) where the ice of the Leverett Glacier descends from the Watson Escarpment on its way to the coast. The map of Davis and Blankenship (2005) also indicates the presence of small outcrops of unidentified metamorphic basement rocks (amphibolite, gneiss, schist, quartzite, and phyllite) in the Watson Escarpment east of Mt. Beazley as well as on the Phlegar Dome and Mt. Ratcliff.

The Party Formation in Table 6.7 was named by Stump (1995) after its occurrence on Supporting Party Mountain. It is composed primarily of pelitic schists with minor quartzite and rare beds of marble exemplified by the rocks exposed on the ridge south of Mt. Nichols in Fig. 6.16. However, no stratigraphic section has yet been measured and Stump (1995) did not designate a type locality except as mentioned above (i.e., Mt. Nichols).

The rocks of the Party Formation are tightly folded and consist of schists that contain abundant biotite and muscovite, as well as hornblende and garnet which occur locally. The Party Formation also contains diopside-bearing calc-schist that ranges in color from tan to brown and even green. Beds of fairly pure quartzite are also present at a place informally known as "Tongue Hills" (Stump 1995). The Party Formation in the vicinity of Mt. Graham (Fig. 6.16) is composed mostly of quartzofeldspathic gneisses interbedded with thin layers of mafic minerals and a thick bed of white marble (30–40 m) that crops out about 2 km south southeast of Mt. Graham. Stump (1995) attributed the deformation of these rocks to two separate events based on evidence in the field.

Because diagnostic sedimentary features were largely obliterated by structural deformation and metamorphic recrystallization of the rocks, Stump (1995) could not identify the original environment of deposition of the Party Formation with any degree of confidence. The presence of carbonate beds may indicate a shallow marine environment while the original interbedded sand and shale sequences suggest that the rocks are former turbidites. Stump (1995) favored the latter alternative because even deep-water turbidites like the rocks of the Goldie Formation may contain carbonate beds (Section 5.2.1).

The Leverett Formation in Table 6.7 was defined by Minshew (1967) based on a heterogeneous assemblage of sedimentary rocks including cross-bedded sandstones, shale, and limestone overlain by silicic volcanic rocks that occur on Mt. Webster in Fig. 6.17. He subdivided the lower part of the section exposed on this nunatak into seven members labeled A to G and estimated that the total thickness of the Leverett Formation is greater than about 2,000 m, although the top and bottom of the Leverett Formation are not exposed. The seven members of the lower part of the Leverett Formation on Mt. Webster described by Minshew (1967) are identified and briefly described in Appendix 6.7.4.1.

Minshew (1967) collected trilobite fragments from limestone Member E which were identified by Palmer (1970) and by Palmer and Gatehouse (1972). In a written

Table 6.7 Stratigraphic column of the sedimentary and volcanic rocks that form the basement complex of the Harold Byrd Mountains, the Bender Mountains, the Berry Hills, and the Falone Nunataks in the area north of the Leverett Glacier (Minshew 1967; Rowell et al. 1993; Stump 1995; Davis and Blankenship 2005)

Geologic unit	Age	Occurrence
Granite Harbor Intrusives Group	Cambrian to Ordovician	Western Harold Byrd Mountains Mt. Webster, Bender Mountains Berry Peaks, Fallone Nunataks (Figs. 6.16, 6.17, 6.18)
	Intrusive Contact	
Leverett Formation	Early to Middle Cambrian	Mt. Webster (Fig. 6.13) eastern Harold Byrd Mountains including Mt. Manke, Bender Mountains (Figs. 6.16, 6.13)
	Undefined	
Party Formation	Neoproterozoic (?) Early Cambrian	Western Harold Byrd Mountains including Mt. Fridovich Supporting Party Mountain, Mt. Graham, Mt. Nichols (Fig. 6.16)
	Undefined	

communication to Minshew (November 19, 1965) Palmer related the trilobites to the genera *Mapania*, *Lisamiella*, and *Sunaspis*. He noted the Asiatic aspect of this faunule and suggested a Middle Cambrian age for the Leverett Formation. Stump (1995) later reported that he was unable to find additional trilobite fragments in the E Member of the Leverett Formation on Mt. Webster.

Rowell et al. (1997) agreed that the presence of trilobites and archaeocyathids indicates that the age of the limestones of the Leverett Formation on Mt. Webster and Mt. Mahan is Early to Middle Cambrian (i.e., 545–500 Ma, IUGS 2002). However, the age of the Party Formation is not well constrained because it is not in contact with the Leverett Formation, because it does not contain fossils, and because the rocks have not been isotopically dated.

6.5.3 Granite Harbor Intrusives

The metasedimentary and metavolcanic rocks of the area north of the Leverett Glacier were intruded by plutons of the Granite Harbor Intrusives. Granitic rocks also form the bedrock of the Watson Escarpment which extends from the head of the Albanus Glacier (eastern tributary of the Scott Glacier) to Mt. Beazley (along the Leverett Glacier) and from there eastward to the Phleger Dome, Dzema Peak, and Mt. Ratcliff along the northern valley wall of the Kansas Glacier (western tributary of the lower Reedy Glacier).

The plutons of the Granite Harbor Intrusives that intruded the Party Formation in the Harold Byrd Mountains have well-defined contacts indicating that they are post-tectonic in origin. The heat released by the cooling magma caused porphyroblasts of biotite and andalusite to form in the contact zones of the country rocks. The southeast face of the summit of the Supporting Party Mountain also contains an unusual sequence of concordant aplite sills that apparently formed during the magmatic activity that followed the deformation and regional metamorphism that occurred during the Ross Orogeny.

In a regional summary of the geology of the basement rocks of the Queen Maud Mountains, Stump (1995) stated that the composition of the Granite Harbor Intrusives varies widely from granite to gabbro but noted that intermediate rocks (e.g., granodiorite and quartz monzonite) are most common. The granitic basement rocks of the Scott Glacier area were described by Borg (1983), whereas Murtaugh (1969) worked on the granitic rocks of the neighboring Wisconsin Range.

6.5.4 Isotopic Age Determinations

A whole-rock Rb-Sr age determination of rhyolite samples collected by Minshew on Mt. Webster was reported by Faure et al. (1968a, b, 1977) based on analyses of four samples that define a satisfactory straight line on the isochron diagram in Fig. 6.19. The resulting date is 483 ± 9 Ma (Early Ordovician) after recalculating the date to the modern value of the decay constant of ^{87}Rb ($\lambda = 1.42 \times 10^{-11}$ year^{-1}). The initial $^{87}Sr/^{86}Sr$ ratio (0.7153 ± 0.0005) is anomalously high presumably because the isotopic composition of strontium in these samples was homogenized some time after the rocks had crystallized. In that case, the Rb-Sr date underestimates the age of these rocks which probably crystallized during the Cambrian Period as indicated by the

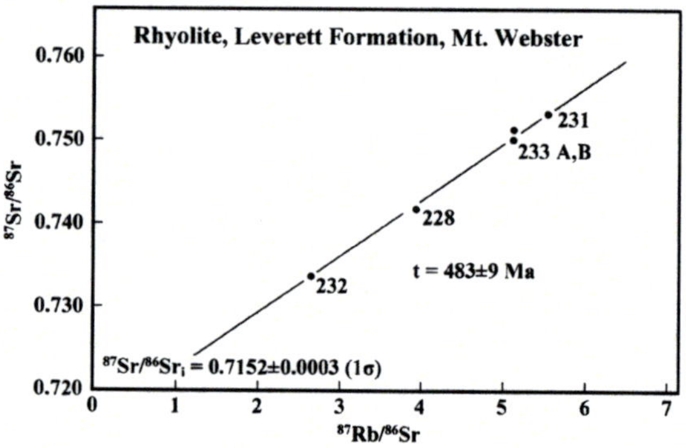

Fig. 6.19 Rb-Sr isotope ratio diagram of rhyolite in the Leverett Formation on Mt. Webster in the area north of the Leverett Glacier, Queen Maud Mountains. The analytical data pertaining to these samples are listed in Appendix 6.7.4.3 because they have not been published previously. The slope of the straight line yields an Early Ordovician date of 483 ± 9 Ma (recalculated to $\lambda = 1.42 \times 10^{-11}$ year^{-1}), but the high initial $^{87}Sr/^{86}Sr$ ratio suggests that the isotopic composition of strontium of these samples was homogenized at a later time following their initial crystallization. Therefore, the age of the Leverett Formation is probably greater than the date based on the Rb-Sr method and may well be Cambrian as indicated by trilobites and archaeocyathids in the limestone beds. The diagram is based on unpublished data by G. Faure. The date and initial $^{87}Sr/^{86}Sr$ ratio were included in publications by Faure et al. (1968a, b) and by Faure et al. (1979)

trilobite fragments in Member E. The chemical composition and isotope ratios of rubidium and strontium of the felsic pyroclastics of the Leverett Formation are tabulated in Appendices 6.7.4.2 and 6.7.6.3.

The age of the Leverett Formation was determined by Van Schmus et al. (1997) by the U-Pb method using samples collected during the 1992/93 field season. Suites of zircons from the volcanic rocks on Mt. Fiedler in the Bender Mountains (Fig. 6.18) and from schist on Cressey Peak in the Harold Byrd Mountains (Rowell et al. 1993) yielded upper-intercept dates on U-Pb concordia diagrams of 515 ± 4 and 519 ± 4 Ma, respectively, with lower intercepts at the origin. These patterns are consistent with continuous loss of radiogenic lead and indicate that the zircons crystallized during the Middle Cambrian (IUGS 2002) in good agreement with the paleontological evidence. The U-Pb dates of zircon in Table 6.8 are more precise and more accurate than the Rb-Sr date reported by Faure et al. (1968a, 1979), but they confirm the conjecture of these authors that the volcanic rocks are older than 483 ± 9 Ma.

Van Schmus et al. (1997) also dated zircons and monazites from the Granite Harbor Intrusives of the Leverett Glacier area. The results in Table 6.8 generally confirm that the intrusion of these plutons occurred after the eruption of the volcanic rocks of the Leverett Formation. The analytical data also indicate that some of the samples contain old xenocrystic zircons that formed prior to about 1.0 Ga. In such cases, VanSchmus et al. (1997) calculated dates from concordant zircon fractions that plotted close to or near the concordia curve.

The U-Pb dates of zircons in granites on Mt. Webster and in the Berry Peaks (494 ± 3 and 480 ± 5 Ma, respectively) indicate that the age of these granites is Early Ordovician and that they crystallized at the end of the Ross Orogeny. The gneisses in the Fallone Nunatak contain old xenocrystic zircons that may have been derived from Neoproterozoic "Grenville" basement that could have existed in the crustal magma sources off-shore of the paleo-Pacific coast of the present Transantarctic Mountains (McKenna et al. 1994; Dalziel 1992).

6.6 Summary

The Queen Maud Mountains form a major part of the Transantarctic Mountains that extends from the Shackleton Glacier to the Reedy Glacier. The basement rocks exposed in these mountains consist primarily of the granitic gneisses of the Granite Harbor Intrusives but also include metasedimentary and metavolcanic rocks of Neoproterozoic and Cambrian ages of the Ross orogen.

The geologic history of the basement rocks of the Queen Maud Mountains is recorded by the volcano-sedimentary complex which has been preserved at several

Table 6.8 U-Pb dates of zircon and monazite in rhyolites of the Leverett Formation and the Granite Harbor Intrusives in the nunataks north of the Leverett Glacier, Queen Maud Mountains (Van Schmus et al. 1997; Rowell et al. 1993; Davis and Blankenship 2005)

		Concordia Intercept dates (Ma)	
Geologic unit	Location	Upper	Lower
Quartz-porphyry metarhyolite, Leverett Formation	Cressey Peak, Harold Byrd Mountains	519 ± 4	31 ± 53
Felsic metavolcanic rock, Leverett Formation	Mt. Fiedler, Bender Mountains	516 ± 4	40 ± 44
Granite, Granite Harbor Intrusives	Mt. Webster, Harold Byrd Mountains	494 ± 3 (concordant U-Pb data)	–
Granite, Granite Harbor Intrusives	Berry Peaks	480 ± 5 (concordant U-Pb data)	
Granite, Granite Harbor Intrusives (zircon)	Fallone Nunataks	909 ± 85[a]	481 ± 35
Migmatitic biotite gneiss[b] (zircon)	Fallone Nunataks	1061 ± 138[a]	527 ± 64
Granite and migmatite combined	Fallone Nunataks	1016 ± 71[a]	508 ± 30
Migmatitic gneiss,[b] (monazite)	Fallone Nunataks	490	–

[a]These dates reflect the presence of old xenocrystic zircons that may have been derived from "Grenville-age" crust that may underlie the Fallone Nuntaks (McKenna et al. 1994)

[b]The geological affiliation of the biotite gneiss is unclear. It may be an early synorogenic phase of the Granite Harbor Intrusives

key localities that are surrounded by a "sea" of granitic rocks of the Queen Maud Batholith. The localities we have examined in this chapter include from west to east: The Duncan Mountains, O'Brien Peak, Nilsen Plateau, La Gorce Mountains of the Scott Glacier area, and the nunataks of the Harold Byrd Mountains. At each of these localities the basement consists of intensely folded and regionally metamorphosed sedimentary and volcanic rocks separated from the volcano-sedimentary complexes of neighboring areas by large volumes of intrusive granitic gneiss. The lack of continuity of the outcrops of the volcano-sedimentary complexes caused by the extensive "overburden" of snow and ice makes correlations of these complexes difficult, especially because they are not easy to date reliably by isotopic methods and because they contain very few fossils.

In spite of these difficulties, the basement rocks of the Queen Maud Mountains support the master hypothesis concerning the geological evolution of the Transantarctic Mountains by the rifting of the supercontinent Rodinia during the Proterozoic Eon and by the subsequent conversion of the passive rift margin of Gondwana into an active subduction zone (Stump 1995; Goodge et al. 2002). This hypothesis concerning the origin and evolution of the Ross orogen remains the organizing principle that allows us to recognize the process that has resulted in the formation of the Transantarctic Mountains that we know today.

6.7 Appendices

6.7.1 Rb-Sr Systematics, Nilsen Plateau (Eastin 1970; Fig. 6.5)

6.7.1.1 La Gorce Formation, Blackwall Glacier

Sample number	Description	$^{87}Rb/^{86}Sr$	$^{87}Sr/^{86}Sr$
471	Phyllite[a]	19.07±0.29	0.8464±0.0012
472	Metagraywacke	1.798±0.028	0.7297±0.0012
473	Metagraywacke	3.012±0.046	0.7377±0.0026
474	Quartzite	1.208±0.019	0.7244±0.0024
475	Metagraywacke	1.703±0.027	0.7293±0.0012
476	Quartzite	1.074±0.017	0.7226±0.0004

[a]Xenolith in the Lonely Ridge Granodiorite which was recrystallized by the heat emanating from the magma as indicated in Fig. 6.8

6.7.1.2 Wyatt Formation, Cougar Canyon

Sample number	Description	$^{87}Rb/^{86}Sr$	$^{87}Sr/^{86}Sr$
461	Metavolcanic	6.806±0.107	0.7636±0.0006
462	Metavolcanic	5.628±0.087	0.7527±0.0016
463	Metavolcanic	1.147±0.018	0.7292±0.0018
464	Metavolcanic	5.873±0.093	0.7554±0.0028
465	Metavolcanic	2.887±0.044	0.7344±0.0014
466	Metavolcanic	2.199±0.035	0.7307±0.0032
467	Metavolcanic	6.539±0.102	0.7583±0.0028
469	Metavolcanic	6.047±0.094	0.7560±0.0022
470[a]	Metavolcanic	6.399±0.111	0.7589±0.0014

[a]South Ridge, Nilsen Plateau

6.7.1.3 Lonely-Ridge Granodiorite, Lonely Ridge

Sample number	Description	$^{87}Rb/^{86}Sr$	$^{87}Sr/^{86}Sr$
477	Granodiorite	1.052±0.017	0.7206±0.0018
478	Granodiorite	1.487±0.025	0.7242±0.0018
479	Granodiorite	1.161±0.019	0.7216±0.0018
480	Granodiorite	1.946±0.031	0.7284±0.0036
481	Granodiorite	2.398±0.038	0.7305±0.0009
482	Granodiorite	71.52±1.42	1.164±0.000
483	Granodiorite	270.7±6.6	2.192±0.066
Biotite[a] Rb/Sr date 470 ± 17 Ma			

[a]Analyzed by Felder and Faure (1979) who calculated this date relative to an assumed initial $^{87}Sr/^{86}Sr$ ratio of 0.7040 and $\lambda = 1.42 \times 10^{-11}$ year^{-1}

6.7.1.4 South Leuco-Quartz Monzonite, Cottonwood Canyon and Windy Ridge

Sample number	Description	$^{87}Rb/^{86}Sr$	$^{87}Sr/^{86}Sr$
506[a]	Quartz Monzonite	34.37±0.54	0.9483±0.0042
507[b]	Quartz Monzonite	49.59±0.76	1.042±0.004
508[b]	Quartz Monzonite	134.7±2.9	1.464±0.090
509[b]	Quartz Monzonite	3.346±0.052	0.7397±0.0041
511[a]	Quartz Monzonite	7.931±0.123	0.7760±0.0015
513[c]	Quartz Monzonite	41.64±0.68	0.9737±0.0041

[a]Southwest wall of Cottonwood Canyon
[b]Windy Ridge
[c]Half way to South Ridge

6.7.2 *Bouma Cycles*

The Dutch sedimentologist Arnold H. Bouma defined a characteristic of sedimentary layers that are deposited by fast-moving bottom currents containing suspended sediment. The sediment is deposited in deep water and consist of a basal layer of coarse sand which fines upward and is overlain by layers of fine sand, silt, and mud in a characteristic pattern labeled from A to E as shown in Fig. 6.20.

The shaly layer that ends the sequence is overlain by the A layer of the next sequence which displays sole markings (i.e., scoured depressions in the underlying shaly layer filled with sand). Not all of the layers in a Bouma cycle are necessarily present in all cases (e.g., the sequence may be B-C-E, A-E, or others). Sediment deposited by episodically active gravity-powered bottom flows form what are commonly described as "interbedded alternating and laterally persistent layers of sandstone and shale" (Friedman and Sanders 1978).

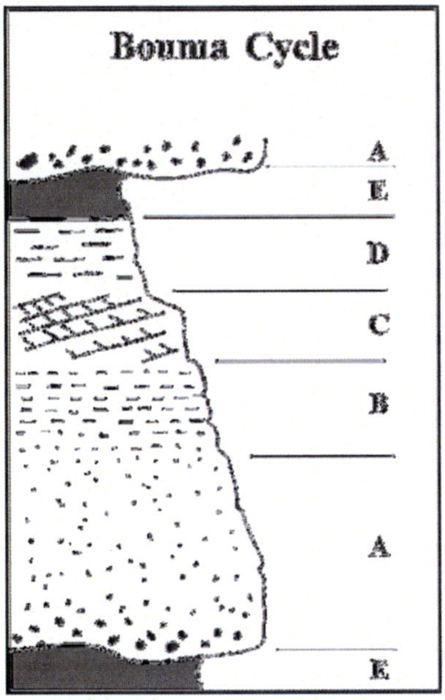

Fig. 6.20 Sequence of sediment deposited by gravity-powered bottom flow of sediment-laden water. The layers are identified by letters A to E (Bouma 1962; Friedman and Sanders 1978).

A. Graded sandstone that fines upward, but may also be massive, that was deposited on the underlying shaley layer (E). The sandstone was deposited by rapidly flowing water containing sediment in suspension
B. Parallel laminated sandstone that was deposited by fast-flowing current after the coarser particles had settled out to form layer A
C. Ripple cross-laminated fine-grained sandstone that formed at lower current velocities than layers A and B
D. Faintly laminated mudstone formed at low current velocity
E. Shaley layer deposited as the final fall-out from the tail of the turbidity current

6.7.3 *Modal and Chemical Compositions and Rb-Sr Systematics of the Wyatt Formation, Scott-Glacier Area and Wisconsin Range, Transantarctic Mountains*

6.7.3.1 Modal Analyses of the Wyatt Formation Collected by V.H. Minshew in the Scott Glacier Area (All Data in This Table Are from Minshew 1967; Fig. 6.12)

	Modal analysis, percent by volume				
Number	Quartz	Plagioclase	Biotite	K-feldspar	Muscovite
236 (64–226)[a]	52.5	12	18.5	12	5
	Rock fragments: 0; chlorite: 0; opaques: 0; matrix: 0				
237 (64–128)	15	21	5	0.5	0
	Rock fragments: 14.5; opaques: 1.0; matrix: 43				
328 (64–244)	14.5	18	0	4	0
	Rock fragments: trace; chlorite: 3.5; opaques: trace; matrix: 58				
329 (64–236)	2	20.5	0	6	0
	Rock fragments: 0; chlorite: 12; opaques: 0.5; epidote: 0.5; matrix: 58.6				
330 (64–218)	15	21	5	0.5	0
	Rock fragments: 14.5; opaques: 1.0; matrix: 43				

[a]Minshew's field numbers are indicated in parentheses

6.7.3.2 Chemical Compositions of Whole-Rock Samples of the Wyatt Formation from the Scott Glacier Area (Minshew, 1967) and from the Wisconsin Range of the Horlick Mountains in Percent by Weight (Analyst: Sam Botts, US Geological Survey)

Oxide	A	B	C	D	E
SiO_2	68.8	70.3	67.8	71.9	69.7
Al_2O_3	14.4	14.4	14.6	14.3	14.4
Fe_2O_3	1.6	1.1	1.8	1.1	1.4
FeO	3.4	2.2	3.2	1.8	2.6
MgO	2.0	1.1	2.0	0.99	1.5

(continued)

(continued)

Oxide	A	B	C	D	E
CaO	1.8	1.8	1.9	2.0	1.9
Na_2O	3.2	2.7	2.6	2.6	2.8
K_2O	2.9	3.9	3.9	4.1	3.7
H_2O+	1.3	1.4	1.4	1.3	1.3
H_2O-	0.05	0.03	0.13	0.03	0.06
TiO_2	0.78	0.54	0.82	0.43	0.64
P_2O_5	0.15	0.18	0.18	0.17	0.17
MnO	0.04	0.02	0.04	0.00	0.02
CO_2	0.04	0.06	0.02	0.01	0.03
Total	100.4	99.73	100.4	100.7	100.2

A: 235, Wyatt Formation, Mt. Gardiner, Scott Glacier
B: F-64–61, Wyatt Formation, Metavolcanic Mountain, Wisconsin Range
C: F-64–108, Wyatt Formation, Wisconsin Plateau, east of Olentangy Glacier
D: F-64–129, Wyatt Formation, Metavolcanic Mountain, Wisconsin Range
E: Average of A to D

6.7.3.3 Rb-Sr Systematics, Wyatt Formation, Scott-Glacier Area, Queen Maud Mountains (Faure, Unpublished)

The data listed below are previously unpublished because they scatter on the isochron diagram and were judged to be unsuitable for dating the Wyatt Formation by the whole-rock Rb-Sr method. Faure et al. (1979) used this data set only to set an upper limit of 788 Ma for the age of the Wyatt Formation.

Number	Location	Rb[a] (ppm)	Sr[a] (ppm)	$^{87}Rb/^{86}Sr$	$^{87}Sr/^{86}Sr^b$
236	Mt Gardiner	139.0	104.2	3.875 ± 0.058	0.7421
237	Mt. Wyatt	144.3	155.9	2.686 ± 0.032	0.7338
328A	Nunataks	137.1	127.1	3.130 ± 0.038	0.7351
328B	Nunataks	137.1	127.1	3.130 ± 0.038	0.7338
329	Nunataks	130.7	137.4	2.950 ± 0.064	0.7314
330	Mt. Wyatt	132.7	148.3	2.597 ± 0.053	0.7324

[a]Rb and Sr concentrations were determined by x-ray fluorescence using the rock standards of the US Geological Survey (BCR-1, AGV-1, GSP-1, G-2, W-1). Each value is the average of three replicate determinations

[b]The isotope composition of Sr was measured on a solid-source, single-filament (Ta) mass spectrometer relative to 0.70806 ± 0.00025 for the $^{87}Sr/^{86}Sr$ ratio of the Eimer and Amend $SrCO_3$ isotope standard. The measurement error of $^{87}Sr/^{86}Sr$ is 0.1% of the reported value.

All dates in Fig. 6.15 were calculated using $\lambda = 1.42 \times 10^{-11}$ year^{-1} for the decay constant of ^{87}Rb

The scatter of data points in the Rb-Sr isotope-ratio diagram is evident in Fig. 6.15. Nevertheless, a least-squares, linear, and unweighted regression of all six data points yields straight line A with a slope m_A = 0.00746354, initial $^{87}Sr/^{86}Sr$ = 0.711919 ± 0.000547 (one standard deviation of the mean, s.d.m.). The date corresponding to the slope is t_A = 524 ± 13 Ma (λ = 1.42×10^{-11} year^{-1}). The elevated value of the initial $^{87}Sr/^{86}Sr$ ratio implies that the isotope composition of Sr in these rocks was re-equilibrated some time after crystallization of these samples.

Inspection of Fig. 6.15 indicates that three samples (236, 328B, and 329) are highly collinear and define straight line B which is characterized by slope m_B = 0.011456458, initial $^{87}Sr/^{86}Sr$ = 0.69775 ± 0.0000997 (1 s.d.m.), and a linear correlation coefficient of 0.99952. The slope of line B yields a precise date of 802 ± 2 Ma, but the initial $^{87}Sr/^{86}Sr$ ratio is below that of achondrite meteorites referred to as BABI (basaltic achondrites best initial) where BABI = 0.69897 ± 0.00003 (Faure and Mensing 2005). The current theory of the isotopic evolution of Sr in the Earth specifies that BABI is the $^{87}Sr/^{86}Sr$ ratio of Sr that was incorporated into the Earth at the time of its formation 4.56×10^9 years ago. Therefore, the strontium in all terrestrial rocks throughout geologic time has had $^{87}Sr/^{86}Sr$ ratios greater than the value of BABI. Accordingly, the initial $^{87}Sr/^{86}Sr$ ratio of line B must be increased which lowers the slope and hence the date when the rocks of the Wyatt Formation crystallized. Therefore, we conclude that the age of the Wyatt Formation must be less than 802 ± 2 Ma.

6.7.4 *Leverett Formation, Mt. Webster, Harold Byrd Mountains*

6.7.4.1 Stratigraphy of the Lower Part of the Leverett Formation on Mt. Webster (Minshew 1967; Fig. 6.17)

Member	Description
G	Pyroclastic rocks of rhyolitic composition interbedded with beds of conglomeratic sandstones
F	White, massive, cliff-forming limestone, 100–150 m thick, grading upward into unit G through tuffaceous limestone and calcareous tuff
E	Thin-bedded, dark gray argillaceous limestone, more than 175 m thick. This unit contains fragmented trilobites examined by A.R. Palmer (Section 6.5.2)
D	Interbedded limestone and shale (~150–175 m)

(continued)

(continued)

Member	Description
C	Thin-bedded, silty green and brown shale intruded by pegmatite dikes. Stringers of gypsum permeate the unit (~125 m). No fossils were found
B	Pebbly mudstone (~100 m) containing angular grains of metamorphic and volcanic rocks in a red clay-sericite matrix cut by malachite- and azurite-bearing pegmatite dikes
A	Cross-bedded, coarse, red arkosic sandstone (~300 m) containing about 15% microcline on average, quartz, and rock fragments

6.7.4.2 Chemical Analysis of Sample 228 Felsic Pyroclastic, Leverett Formation, Mt. Webster (Minshew 1967) (Analysis by A.S. McCreath and Son, Inc)

Oxide	Concentration (%)	Oxide	Concentration (%)
SiO_2	66.48	H_2O+	2.82
Al_2O_3	13.52	H_2O-	–
Fe_2O_3	0.54	TiO_2	0.80
FeO	4.34	P_2O_5	0.124
MgO	1.57	MnO	–
CaO	1.90	CO_2	1.44
Na_2O	2.76	Total	99.85
K_2O	4.10		

6.7.4.3 Rb-Sr Systematics of Felsic Pyroclastics of Unit G of the Leverett Formation, Mt. Webster, Harold Byrd Mountains (Faure, Unpublished)

Sample number	Rb (ppm)[a]	Sr (ppm)[a]	$^{87}Rb/^{86}Sr$	$^{87}Sr/^{86}Sr^{b}$
228	171.1	126.4	3.932 ± 0.065	0.7420
231	196.6	102.9	5.556 ± 0.115	0.7534
232	127.3	138.5	2.667 ± 0.050	0.7337
233	210.8	119.3	5.136 ± 0.096	0.7514
				0.7501

[a]Rb and Sr concentrations were determined by x-ray fluorescence (Eastin 1970; Ray 1973)

[b]$^{87}Sr/^{86}Sr^{b}$ by solid-source mass spectrometry with a reproducibility of ± 0.1% of the measured value and relative to $^{87}Sr/^{86}Sr$ = 0.70806 ± 0.00025(1σ) for the Eimer and Amend $SrCO_3$ isotope standard

References

Barrett PJ (1965) Geology of the area between the Axel Heiberg and Shackleton glaciers, Queen Maud Range, Antarctica. Part 2: Beacon Group. New Zealand J Geol Geophys 8:344–363

Blackburn QA (1937) The Thorne Glacier section of the Queen Maud Mountains. Geogr Rev 27: 598–614

Borg SG (1980) Petrology and geochemistry of the Wyatt Formation and the Queen Maud batholith, upper Scott Glacier area, Antarctica. Unpublished MS thesis, Arizona State University, Tempe, AZ

Borg SG (1983) Petrology and geochemistry of the Queen Maud batholith, central Transantarctic Mountains, with implications for the Ross Orogeny. In: Oliver RL, James PR, Jago JB (eds) Antarctic earth science. Australian Academy of Science, Canberra, pp 165–169

Borg SG (1984) Granitoids of northern Victoria Land, Antarctica. PhD dissertation, Arizona State University, Tempe, AZ

Borg SG, Goodge JW, DePaolo DJ, Mattinson J (1986a) Field studies of granites and metamorphic rocks: Central Transantarctic Mountains, Antarctica. Antarctic J US 21(5):43–45

Borg SG, Stump E, Holloway JR (1986b) Granitoids of northern Victoria Land, Antarctica: a reconnaisance study of field relations, petrography, and geochemistry. In: Stump E (ed) Geological investigations in Northern Victoria Land. Antarctic Research Series, vol. 46. American Geophysical Union, Washington, DC, pp 115–188

Borg SG, Goodge JW, Bennett VC, DePaolo DJ, Smith BK (1987a) Geochemistry of granites and metamorphic rocks, central Transantarctic Mountains. Antarctic J US 22(5):21–23

Borg SG, Stump E, Chappell BW, McCulloch MT, Wyborn D, Armstrong RL, Holloway JR (1987b) Granitoids of northern Victoria Land, Antarctica: implications of chemical and isotopic variations to regional crustal structure and tectonics. Amer J Sci 287:127–169

Borg SG, DePaolo DJ, Smith BM (1990) Isotopic structure and tectonics of the central Transantarctic Mountains. J Geophys Res 95(B5):6647–6667

Bouma AH (1962) Sedimentology of some flysch deposits. Elsevier, Amsterdam, The Netherlands

Chappell BW, White AJR (1974) Two contrasting granite types. Pacific Geology 8:173–174

Craddock C, Gast PW, Hanson GN, Linder H (1964) Rubidium-strontium ages from Antarctica. Geol Soc Amer Bull 75:237–240

Dalziel IWD (1992) Antarctica: A tale of two supercontinents? Ann Rev Earth Planet Sci 20:501–526

Davis MB, Blankenship DD (2005) Geology of the Scott and Reedy glaciers area, southern Transantarctic Mountains, Antarctica. Map and Chart Series. Geological Society of America, Boulder, CO. doi: 10. 1130/2005. MCH 093

Eastin R (1970) Geochronology of the basement rocks of the central Transantarctic Mountains. Ph.D. dissertation, The Ohio State University, Columbus, OH

Encarnación JP, Grunow AM (1996) Changing magmatic and tectonic styles along the paleo-Pacific margin of Gondwana and the onset of early Paleozoic magmatism in Antarctica. Tectonics 15:1325–1341

Faure G, Hurley PM (1963) The isotopic composition of strontium in oceanic and continental basalts: Application to the origin of igneous rocks. J Petrol 4(1):31–50

Faure G, Mensing TM (2005) Isotopes: principles and applications. Wiley, Hoboken, NJ

Faure G, Murtaugh JG, Montigny RJE (1968a) The geology and geochronology of the basement complex of the central Transantarctic Mountains. Canadian J Earth Sci 5:555–560

Faure G, Hill RL, Eastin R, Montigny RJE (1968b) Age determinations of rocks and minerals from the Transantarctic Mountains. Antarctic J US 3(5):173–175

Faure G, Eastin R, Ray PT, McLelland D, Schultz CH (1979) Geochronology of igneous and metamorphic rocks, central Transantarctic Mountains. In: Laskar B, Raja Rao CS (eds) Fourth International Gondwana Symposium Papers. Hindustan, New Delhi, pp 805–813

Felder RP, Faure G (1979) Investigation of an anomalous date for Lonely Ridge granodiorite, Nilsen Plateau, Transantarctic Mountains. Antarctic J US 14(5):24

Friedman GM, Sanders JE (1978) Principles of sedimentology. Wiley, New York

Goodge JW, Borg SG (1987) Metamorphism and crustal structure of the Miller Range, central Transantarctic Mountains. Abstracts, 5th international symposium on Antarctic earth science, SCAR, Cambridge, UK

Goodge JW, Fanning CM, Bennett VC (2001) U-Pb evidence of ~ 1.7 Ga crustal tectonism during the Nimrod Orogeny in the Trans-Antarctic Mountains., Antarctica. Precamb, Res., 112:261–288

Goodge JW, Myrow P, Williams IS, Bowring SA (2002) Age and provenance of the Beardmore Group, Antarctica: Constraints on Rodinia supercontinent break- up. J Geol 110:393–406

Gould LM (1935) Structure of the Queen Maud Mountains, Antarctica. Geol Soc Amer Bull 46:973–984

Grindley GW, McGregor VR, Walcott RI (1964) Outline of the geology of the Nimrod-Beardmore-Axel Heiberg glaciers region, Ross Dependency. In: Adie RJ (ed) Proceedings of the first international symposium on Antarctic geology. North-Holland, Amsterdam, The Netherlands, pp 206–219

Gunner J (1971) Age and origin of the Nimrod Group and of the granite Harbour Intrusives, Beardmore Glacier region, Antarctica. Ph.D. dissertation, The Ohio State University, Columbus, OH

Gunner J (1976) Isotopic and geochemical studies of the pre-Devonian basement complex, Beardmore Glacier region, Antarctica. Institute of Polar Studies, Report 41, The Ohio State University, Columbus, OH

Gunner J, Faure G (1972) Rb-Sr geochronology of the Nimrod Group, central Transantarctic Mountains. In: Adie RJ (ed) Antarctic geology and geophysics. Universitetsforlaget, Oslo, Norway, pp 305–312

Heintz GM (1980) Structural geology of the Leverett Glacier area, Antarctica. MSc thesis, Arizona State University, Tempe, AZ

Huntford R (1987) The Amundsen photographs. The Atlantic Monthly Press, New York

IUGS (2002) International stratigraphic chart. International Commission on Stratigraphy. UNESCO, New York

Jackson JA (ed) (1997) Glossary of geology, 4th edn. American Geological Institute, Alexandria, VA

Katz HR, Waterhouse BC (1970a) The geologic situation at O'Brien Peak, Queen Maud Range, Antarctica. New Zealand J Geol Geophys 13(4):1038–1047

Katz HR, Waterhouse BC (1970b) Geological reconnaissance of the Scott Glacier area, southeastern Queen Maud Range, Antarctica. New Zealand J Geol Geophys 13(4):1030–1037

Linder H, de Hills SM, Thiel EC, 1965. Basement complex in the Queen Maud Mountains. In: Hadley JB (ed) Geology and paleontology of the Antarctic. Antarctic Research Series, vol. 6. American Geophysical Union, Washington, DC, pp 141–144

Long WE (1965). Stratigraphy of the Thorvald Nilsen Mountains, Queen Maud Range, central Antarctica. Geol Soc Amer, Special Paper 82, Abstracts for 1964, p. 124

Lowry PH (1980) The stratigraphy and petrography of the Cambrian Leverett Formation, Antarctica. MSc thesis, Arizona State University, Tempe, AZ

McCulloch MT, Chappell BW (1982) Nd isotopic characteristics of S- and I-type granites. Earth Planet Sci Lett 59(8):51–64

McGregor VR (1965a) Notes on the geology of the area between the heads of the Beardmore and Shackleton glaciers. New Zealand J Geol Geophys 8(2):278–291

McGregor VR (1965b) Geology of the area between the Axel Heiberg and Shackleton glaciers, Queen Maud Range, Antarctica. New Zealand J Geol Geophys 8(2):314–343

McGregor VR, Wade FA (1969) Geology of the western Queen Maud Mountains. In: Craddock C (ed) Geologic map of Antarctica. Folio 12, Sheet 16. American Geographical Society, New York

McKenna LW, Van Schmus WR, Rowell AJ, Seik GS (1994) Neoproterozoic continental crust and an attenuated continental margin outboard of the Gabbro Hills block in the Transantarctic Mountains. Abstract Geol Soc Amer Abstracts with Programs, A504

McLelland D (1967) Geology of the basement complex, Thorvald Nilsen Mountains, Antarctica. MSc thesis, University of Nevada, Reno, NV

Minshew VH (1965) Potassium-argon age from a granite at Mt. Wilbur, Queen Maud Range, Antarctica. Science 150:741–743

Minshew VH (1967) Geology of the Scott Glacier and Wisconsin Range areas, central Transantarctic Mountains, Antarctica. Ph.D. dissertation, The Ohio State University, Columbus, OH

Mirsky A (1969) Geology of the Ohio Range-Liv Glacier area. In: Craddock C (ed) Geologic map of Antarctica, Folio 12, Sheet 17. American Geographical Society, New York

Murtaugh JC (1969) Geology of the Wisconsin Range batholith, Transantarctic Mountains. New Zealand J Geol Geophys 12:526–550

Myrow PM, Pope MC, Goodge JW, Fischer W, Palmer AR (2002) Depositional history of pre-Devonian strata and timing of Ross orogenic tectonism in the central Transantarctic Mountains, Antarctica. Geol Soc Amer Bull 114(9):1070–1088

Palmer AR (1970) Early and Middle Cambrian trilobites from Antarctica. Antarctic J US 5(5):162

Palmer AR, Gatehouse CG (1972) Early and Middle Cambrian trilobites from Antarctica. US Geol Surv Prof Paper, 456 D, Reston, VA

Ray PT (1973) Rubidium-strontium age determinations of intrusive aplite and pegmatite dikes in the Wisconsin Range of the Horlick Mountains, Antarctica. MSc thesis, The Ohio State University, Columbus, OH

Rowell AJ, Evans KR, McKenna LW III (1993) Cambrian and possibly Proterozoic strata in the Transantarctic Mountains north of the Leverett Glacier. Antarctic J US 28(5):35–37

Rowell AJ, Gonzales DA, McKenna LW, Evans KR, Stump E, Van Schmus WR (1997) Lower Paleozoic rocks in the Queen Maud Mountains, revised ages and significance. In: Ricci CA (ed) The Antarctic region: geological evolution and processes. Terra Antartica, Siena, Italy, pp 201–207

Shand SJ (1947) Eruptive rocks. Wiley, New York

Smit JH, Stump E (1986) Sedimentology of the La Gorce Formation, La Gorce Mountains, Antarctica. J Sed Pet 56(5):663–668

Stewart D (1934) The petrography of some rocks from south Victoria Land. Proceedings of the American Philosophical Society, vol. 74, pp 307–310

Stump E (1975) Geology of the Duncan Mountains. Antarctic J US 10(4):179–180

Stump E (1976) On the late Precambrian–early Paleozoic metavolcanic and metasedimentary rocks of the Queen Maud Mountains, Antarctica. Just. Polar Studies, Rept. 62, The Ohio State Un., Columbus., Ohio, 1–212pp

Stump E (1981) Structural relationships in the Duncan Mountains, central Transantarctic Mountains, Antarctica. New Zealand J Geol Geophys 24:87–93

Stump E (1983) Type locality of the Ackerman Formation, La Gorce Mountains, Antarctica. In: Oliver RL, James PR, Jago JB (eds) Antarctic earth science. Australian Academy of Science, Canberra, ACT, pp 170–174

Stump E (1985) Stratigraphy of the Ross Supergroup, central Transantarctic Mountains. In: Turner MD, Splettstoesser JF (eds) Geology of the Central Transantarctic Mountains. Antarctic Research Series, vol. 36. American Geophysical Union, Washington, DC, pp 225–274

Stump E (1995) The Ross orogen of the Transantarctic Mountains. Cambridge University Press, Cambridge

Stump E, Corbató CE, Browning A (1976) Stratigraphy of the Fairweather Formation. Abstract Geol Soc Amer Abstracts with Programs 8:511–512

Stump E, Lowry PH, Heintz-Stocker GM, Colbert PV (1978) Geological investigations in the Leverett Glacier area. Antarctic J US 13(4):3–4

Stump E, Sheridan MF, Borg SG, Sutter JF (1980) Early Miocene subglacial basalts, the East Antarctic ice sheet and uplift of the Transantarctic Mountains. Science 207:757–759

Stump E, Self S, Smit JH, Colbert PV, Stump TM (1981) Geological investigations in the La Gorce Mountains and central Scott Glacier area. Antarctic J US 16(5):55–57

Stump E, Smit JH, Self S (1985) Reconnaissance geologic map of the Mt. Blackburn Quadrangle, Antarctica. Map A-11. US Geol Survery, US Antarctic Research Program, Reston, VA

Stump E, Smit JH, Self S (1986) Timing of events during the Late Proterozoic Beardmore Orogeny, Antarctica: Geological evidence from the La Gorce Mountains. Geol Soc Amer Bull 97:953–965

Stump E, Korsch RJ, Edgerton DG (1991) The myth of the Nimrod and Beardmore orogenies. In: Thomson RA, Crame JA, Thomson JW (eds) Geologic evolution of Antarctica. Cambridge University Press, Cambridge, pp 143–147

Van Schmus WR, McKenna LW, Gonzales DA, Fetter AH, Rowell AJ (1997) U-Pb geochronology of parts of the Pensacola, Thiel, and Queen Maud mountains, Antarctica. In: Ricci CA (ed) The Antarctic region: Geological evolution and process. Terra Antartica, Siena, Italy, pp 187–200

Wade FA (1974) Geological surveys of Marie Byrd Land and the central Queen Maud Range. Antarctic J US 9(5):241–242

Wade FA, Cathey CA (1986) Geology of the basement complex, western Queen Maud Mountains, Antarctica. In: Turner MD, Splettstoesser JF (eds) Geology of the Central Transantarctic Mountains. Antarctic Research Series, vol. 36. American Geophysical Union, Washington, DC, pp 429–453

Wade FA, Yeats VL, Everett JR, Greenlee DW, La Prade KE, Shenk JC (1965) Geology of the central portion of the Queen Maud Range, Transantarctic Mountains. Science 150:1808–1809

White AJR, Williams IS, Chappell BW (1977) Geology of the Berridale 1:100,000 sheet. Geological Survey, 1–138. New South Wales, Australia

York D (1966) Least-squares fitting of a straight line. Can J Phys 44:1079–1086

York D (1969) Least-squares fitting of a straight line with correlated errors. Earth Planet Sci Lett 5:320–324

Chapter 7
Horlick Mountains

The Horlick Mountains east of the Reedy Glacier include the Wisconsin Range, the Long Hills, and the Ohio Range, which together extend for about 235 km from the mouth of the Reedy Glacier to Iversen Peak which is the most easterly nunatak of the Ohio Range. The geologic map of the Horlick Mountains published by Mirsky (1969) includes the Wisconsin Range, the Long Hills, and the Ohio Range, whereas the map of Davis and Blankenship (2005) covers the eastern Queen Maud and the main body of the Wisconsin ranges but does not extend to the Long Hills and the Ohio Range.

7.1 Exploration of the Wisconsin Range

The mountains of the Wisconsin Range were first photographed from the air during Operation Highjump (1946–1947) carried out by the US Navy Taskforce 68. Subsequently in 1958, a temporary station was established on the Ross Ice Shelf by members of an oversnow tractor train that originated from Byrd Station in West Antarctica. William E. Long and F. Darling walked from this camp to the north-facing escarpment of the Wisconsin Range (Fig. 7.1) and collected several samples of granitic rocks at Mt. LeSchack. Long (1965) later reported that sedimentary rocks occur unconformably at the top of the escarpment at a height of 450–600 m above its base (Long 1965; Minshew 1967). Additional preliminary reports about the geology of the Wisconsin Range and the Scott Glacier area were published by Long (1962), Minshew and Summerson (1963), and Hadley (1965).

For the purpose of this presentation we include the mountains west of Reedy Glacier (i.e., the Watson Escarpment south of the Kansas Glacier, the Quartz Hills, Caloplaca Hills, and the Cleveland Mesa) with the main body of the Wisconsin Range which extends east from the Reedy Glacier to the Long Hills. The Wisconsin Range as defined in this way occupies an area of about 8,600 km^2. Because of its large area, only a small part of the Wisconsin Range centered on the Olentangy Glacier is included in Fig. 7.2.

The geology of the Wisconsin Range and of its eastern extension to the Long Hills was investigated during the 1694/65 field season by a group of geologists from the former Institute of Polar Studies (now the Byrd Polar Research Center) of the Ohio State University. The group included Velon H. Minshew, John G. Murtaugh, James T. Teller, Courtney J. Skinner, and Harry S. Gair (New Zealand) and was jointly led by Gunter Faure and John H. Mercer. The group set up a Jamesway Hut (Camp Ohio II in Fig. 7.3) and traveled primarily by snowmobile to rock outcrops along the Olentangy Glacier and on Metavolcanic Mountain. The arrival of three US Army helicopters on 17 December, 1964, put all parts of the Wisconsin Range and the Long Hills within reach of the geologists. It also enabled V.H. Minshew to land on Mt. Webster north of the Leverett Glacier (Section 6.5.2) and to return to Mt. Weaver at the head of the Scott Glacier for some follow-up field work in that area. The availability of helicopters also enabled Arthur Boucot and George Doumani, who had arrived at Camp Ohio II on 29 December, 1964, to fly to the Ohio Range in order to collect fossils. Teller (1966) later published a summary of his experiences during the field season in the Wisconsin Range.

A second group of geologists from the Ohio State University led by David Harwood later returned to the

G. Faure and T.M. Mensing, *The Transantarctic Mountains: Rocks, Ice, Meteorites and Water*,
DOI 10.1007/978-90-481-9390-5_7, © Springer Science+Business Media B.V. 2010

Fig. 7.1 The north-facing escarpment of the Wisconsin Range of the Horlick Mountains is composed primarily of granitic basement rocks. This photograph was taken on January 9, 1965, at about 10 p.m. local time on a rare day when the weather was sunny and calm (Photo by G. Faure)

Wisconsin Range for a study of the Neogene deposits of till on the Wisconsin Plateau and along the walls of the Reedy Glacier. This project was motivated by the discovery of marine diatoms in Neogene till of the Sirius Group in the Transantarctic Mountains, including in the Wisconsin Range (Harwood 1983, 1986a, b; Webb et al. 1983, 1984). The study of diatoms in glacial deposits of the Transantarctic Mountains led to a far-reaching hypothesis concerning fluctuations of the volume of the East Antarctic ice sheet in late Tertiary time. This hypothesis, proposed by D.M. Harwood and P.N. Webb, resulted in a spirited discussion of the evidence concerning the dispersal of diatoms in Antarctica by Harwood and Webb (1998), Stroeven et al. (1998), and Miller and Mabin (1998). In the same year, the stratigraphy of the Neogene glacial deposits in the Wisconsin Range was defined by Wilson et al. (1998). The discovery by Webb and Harwood (1987) of well preserved leaves and twigs of beech (Nothofagus) in late Pliocene till in the Dominion Range at the head of the Beardmore Glacier proves that during the Pliocene Epoch the climate in that part of the Transantarctic Mountains was not as cold and dry as it is at the present time (Hill et al. 1991). This episode in the study of the Transantarctic Mountains involving the dispersal of diatoms and the preservation of plant remains is one of the most significant discoveries that has yet been made in Antarctica. It will come up for formal discussion in the context of Chapter 19.

The next science team to work in the Wisconsin Range collected samples of basement rocks during the 1990/91 field season by using a Twin Otter aircraft from the camp of CASERTZ (Corridor Aerogeophysics of the Southeastern Ross Transect Zone) located on the West Antarctic Ice Sheet (Borg et al. 1991). The objective of this group was to characterize the basement rocks of the Transantarctic Mountains by means of isotopic compositions of neodymium, strontium, and oxyen in order to extend previous work by these authors in the central Transantarctic Mountains (Borg et al. 1990).

7.2 Basement Rocks, Wisconsin Range

The basement rocks of the Wisconsin Range in Fig. 7.4 are similar to those of the Queen Maud Mountains (Chapter 6) and are generally correlated with the LaGorce and Wyatt formations and with the Granite Harbor Intrusives. Sills and dikes of the Ferrar Dolerite are rare in the Wisconsin Range, in contrast to the Transantarctic Mountains between southern Victoria Land and the Amundsen Glacier where numerous dolerite sills occur within the Beacon Supergroup.

7.2.1 LaGorce Formation

Metasedimentary rocks composed of gray to black phyllites, metagraywackes, and impure quartzites occur in the Quartz Hills south of the Kansas Glacier and in the low ridges of the Ford and Gierloff Nunataks located north of the north-facing escarpment of the Wisconsin Plateau in Fig. 7.4. Murtaugh (1969) assigned these rocks to the LaGorce Formation of the Scott Glacier area (Section 6.4.1) based on the similarity of their lithologic composition and stratigraphic position (i.e., they were intruded by the granitic basement rocks). Another small outcrop of the metasedimentary rocks occurs east of Mt. Huckaby at the head of the Olentangy Glacier (Fig. 7.2).

The thickness of the LaGorce Formation in the Wisconsin Range is unknown because of folding and faulting. Murtaugh (1969) considered that the thickness of the LaGorce Formation could be more than about 1.5 km. In spite of the evident structural deformation and low-grade regional metamorphism, the rocks of the LaGorce Formation contain a variety of primary sedimentary structures, including bedding, graded bedding, cross-bedding, fluting, and sole markings.

Fig. 7.2 Excerpt of the central region of the Wisconsin Range including the Olentangy Glacier and Tillite Spur which contains a large remnant of the Beacon Supergroup that was studied by Minshew (1967). The map also includes Mims Spur and the adjacent Jamesway Hut of Camp Ohio II that was set up in November of 1964 by a team of geologists from The Ohio State University and which continues to serve as a potential refuge. Faure Peak (3,941 m) and Mt. Minshew (3,895 m) at the northern edge of the Wisconsin Plateau are the highest peaks in the area. Other mountains worth mentioning are Murtaugh Peak (2,921 m; 85°41'S, 130°15'W) and Teller Peak (3,530 m; 85°57'S, 135°28'W) all of which were named after members of the first field party that mapped the geology of the Wisconsin Range in 1964/65. Taken from the topographic maps of the Wisconsin Range published in 1968 by the US Geological Survey, Wisconsin Range, Antarctica SV1-10.8 and Caloplaca Hills, Antarctica SV1-10/12

Fig. 7.3 Camp Ohio II on the McCarthy Glacier which is a tributary to the Reedy Glacier in the Wisconsin Range of the Horlick Mountains. The person standing next to the Polaris snowmobile is Courtney Skinner. The Jamesway Hut still stands on the glacier as a potential shelter (Photo by G. Faure)

However, no fossils were found. The beds strike east-west and dip steeply in most places (Murtaugh 1969).

The rocks of the LaGorce Formation in the Ford and Gierloff Nunataks were intruded by plutons of unfoliated porphyritic granitic rocks of the Wisconsin Range Batholith. The granitic intrusives are chilled against the country rocks and caused contact metamorphism in the phyllites of the LaGorce Formation. Both rock types contain pegmatite dikes and quartz veins which apparently formed during the waning stage of the Ross Orogeny.

7.2.2 *Wyatt Formation*

The metavolcanic rocks of the Wyatt Formation occur primarily on Metavolcanic Mountain and on Shapley Ridge of the Cleveland Plateau of the Watson Escarpment west of Reedy Glacier and opposite Metavolcanic Mountain in Figs. 7.4. Murtaugh (1969) described the dark gray, metamorphosed igneous rocks and assigned them to the Wyatt Formation which was defined by Minshew (1967) based on the type locality on Mt. Wyatt in the Scott Glacier area (Section 6.3.2). Murtaugh (1969) also reported that an outcrop of metavolcanic rocks east of Mt. Huckeby in Fig. 7.2 is topographically higher than metasedimentary rocks of the LaGorce Formation and concluded that the Wyatt Formation is younger than the LaGorce Formation in agreement with Minshew (1967).

The rocks of the Wyatt Formation consist of fragments of quartz, feldspar, and rounded grains of blue quartz (?) in a fine-grained matrix of quartz and feldspar. The matrix is recrystallized and contains varying amounts of biotite and muscovite which locally express a strong secondary foliation of these rocks. The grains of blue quartz are highly strained and partially resorbed. These grains resemble the mineral cordierite which was identified by Ford (1964) in the so-called "cordierite-bearing hypersthene-quartz-monzonite porphyry" in the Thiel Mountains.

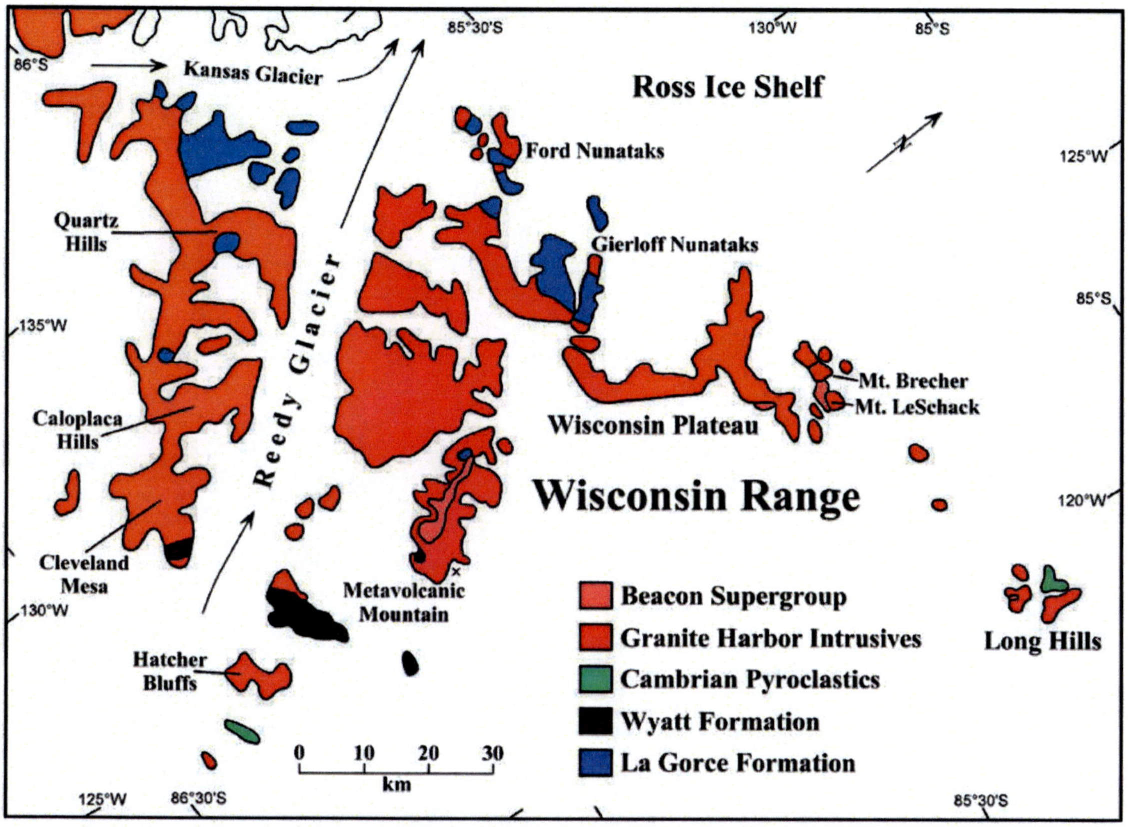

Fig. 7.4 The geology of the Wisconsin Range and of the Long Hills is dominated by the Granite Harbor Intrusives of the Wisconsin Range Batholith. These granitic rocks intruded the metasedimentary rocks of the LaGorce Formation and the metavolcanic rocks of the Wyatt Formation, both of which also occur in the Nilsen Plateau (Section 6.3.2) and in the LaGorce Mountains (Section 6.4.1). The granitic basement is locally overlain by Permian sedimentary rocks of the Beacon Supergroup but sills of Ferrar Dolerite are rare (Adapted from Mirsky 1969)

The rocks of the Wyatt Formation on Metavolcanic Mountain contain no primary structures and neither their thickness nor their attitude are known. Along Shapley Ridge of the Cleveland Mesa in Fig. 7.4 the rocks of the Wyatt Formation grade into gneissic granite and are cut by dikes of the same granite.

The chemical analyses of the Wyatt Formation in Appendix 7.8.1 demonstrate that the rocks have high silica concentrations ranging from 67.46% to 70.3%. Although the Wyatt Formation appears to be a deposit of volcanic ash that was extruded on the surface of the Earth, the pattern of decreasing concentrations of CaO + MgO, total Fe, and increasing Na_2O + K_2O with increasing silica concentration in Fig. 7.5 resembles the effects of fractional crystallization of magma. In addition, the pattern of variation of the chemical compositions of the Wyatt Formation in Fig. 7.5 appears to originate from a point that represents the composition of the porphyritic quartz monzonite facies of the Wisconsin Range Batholith.

7.2.3 Southern Nunataks

A cluster of small nunataks southwest of Metavolcanic Mountain in Fig. 7.6 includes Hatcher Bluffs, as well as Savage, Strickland, and Spear nunataks. These nunataks are composed of several different kinds of rocks that were recorded by Murtaugh (1969): Hatcher Bluffs and the Savage Nunatak consist of granitic gneisses. Strickland Nunatak exposes light-colored argillaceous quartzites and siltstones that differ markedly from the dark gray phyllites of the LaGorce

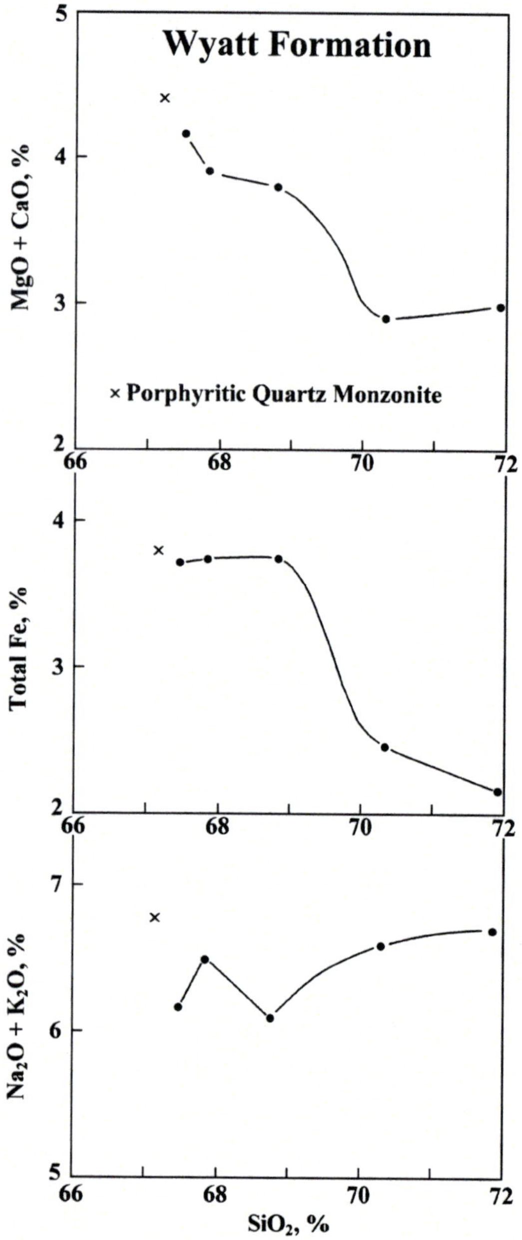

Fig. 7.5 Silica-variation diagram of whole-rock samples of the Wyatt Formation in the Wisconsin Range of the Horlick Mountains. The "x" represents a sample of the porphyritic quartz monzonite (Data from Minshew 1967; Faure unpublished in Appendix 7.8.1)

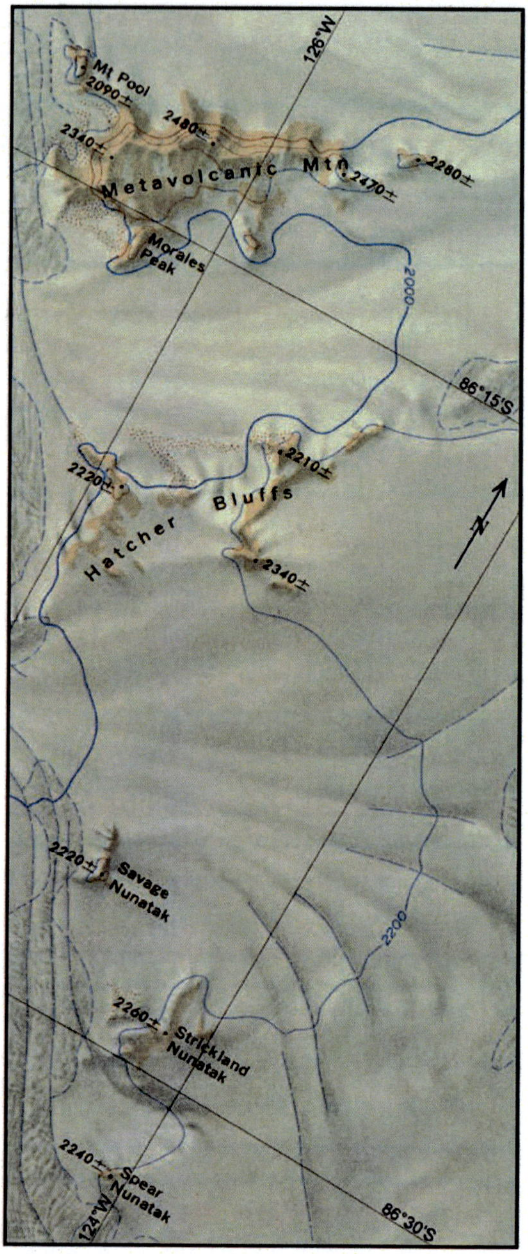

Fig. 7.6 The nunataks south of Metavolcanic Mountain in the Wisconsin Range include the Hatcher Bluffs and Savage, Strickland, and Spear nunataks. The Hatcher Bluffs and Savage Nunatak expose granitic gneisses of the basement complex. Strickland Nunatak consists of light-colored argillaceous quartzite and siltstone that appear to be younger than the phyllites of the LaGorce Formation. Spear Nunatak is composed of hornblende migmatite (Fig. 7.7) that may be the oldest rock unit of the Wisconsin Range. Excerpt of the US Geological Survey topographic map Caloplaca Hills, Antarctica, SV1-10/12 (1969) with geological information from Murtaugh (1969)

Formation. The beds dip to the northwest at about 50° and contain ripple marks. The quartzite beds are cut by quartz-feldspar veins that contain specular hematite. The age of these rocks is not known because they do

Fig. 7.7 The hornblende migmatite that occurs on Spear Nunatak south of Metavolcanic Mountain in Fig. 7.6 may be the oldest rock unit in the Wisconsin Range, but its stratigraphic position cannot be determined in the field because it is not in contact with other rocks and it has not been dated by isotopic methods (Murtaugh 1969). An attempt by Faure (unpublished) to date this specimen failed because its Rb/Sr ratio was too low (Photo by G. Faure)

not contain fossils and the isolated occurrence and small size of Strickland Nunatak prevent correlation of these rocks to the Ackerman Formation in the LaGorce Mountains (Section 6.4.1, Table 6.4 or to the Leverett Formation on Mt. Webster in the Harold Byrd Mountains (Section 6.5.2, Table 6.6).

Even more problematic is the hornblende migmatite on Spear Nunatak which is composed of layers of hornblende and of quartz-plagioclase (andesine) that have been folded and metamorphosed to the amphibolite facies (Fig. 7.7). Murtaugh (1969) considered the migmatites to be the oldest rocks in the Wisconsin Range and Minshew (1967) tentatively correlated them to the Nimrod Group of the Miller Range (Section 5.1.1).

7.2.4 Wisconsin Range Batholith

The granitic basement rocks of the Wisconsin Range Batholith are the dominant lithology in the Quartz Hills and Caloplaca Hills west of Reedy Glacier (Fig. 7.4). They also form the main part of the exposed basement rocks east of the Reedy Glacier and extend east as far as the Long Hills. These granitic basement rocks intruded the metasedimentary rocks of the LaGorce Formation as well as the metavolcanic rocks of the Wyatt Formation (Murtaugh 1969; Mirsky 1969; Davis and Blankenship 2005). Therefore, the granitic rocks of the Wisconsin Range Batholith are not the continental crust upon which the sedimentary and volcanic rocks of the Horlick Mountains were deposited prior to the Ross Orogeny.

The Wisconsin Range Batholith contains several textural varieties of granitic rocks which Murtaugh (1969) described as:

Foliated porphyroblastic
Massive porphyroblastic
Equigranular
Tourmaline granite
Porphyritic

The contacts between these facies are generally gradational but are intrusive in some cases. The foliated granitoids are characterized by microcline porphyroblasts which are mantled in the manner of rapakivi granites described by Eskola (1963) from the Precambrian of Finland. Evidently, the foliated facies of the Wisconsin Range Batholith are syntectonic and may have experienced metasomatism during the Ross Orogeny, whereas the unfoliated (i.e., massive) varieties are post-tectonic.

The bulk of the Wisconsin Range Batholith is composed of foliated and massive porphyroblastic facies. The tourmaline-bearing granite occurs on Savage Nunatak (Fig. 7.6) and is illustrated in Fig. 7.8. The tourmaline occurs in the form of ovoid rosettes up to

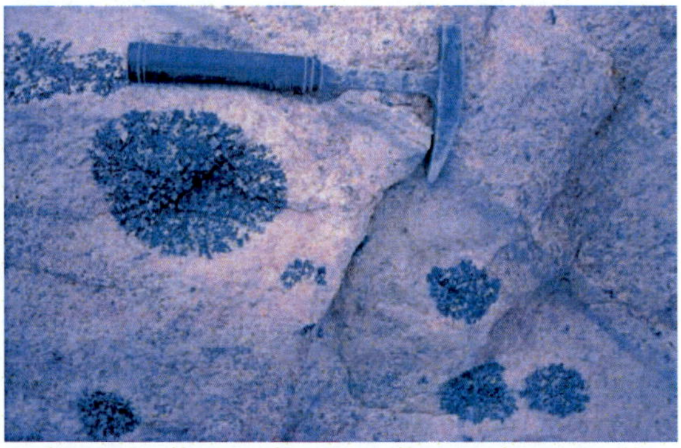

Fig. 7.8 Equigranular and unfoliated granitoid of the Wisconsin Range Batholith on Savage Nunatak containing rosettes of black tourmaline surrounded by haloes of light-colored granitic rock (Photo by G. Faure)

15 cm in diameter which are surrounded by light-colored haloes (see also Fig. 17 of Murtaugh 1969).

The porphyritic facies is composed of quartz monzonite which contains euhedral phenocrysts of pink microcline. The phenocrysts are 2–3 cm in length. This facies occurs on Mt. Brecher and Mt. LeSchack as well as along the eastern part of the north-facing escarpment of the Wisconsin Plateau and is similar to the granitic basement rocks in the Ohio Range described by Treves (1965).

The granitoides of the Wisconsin Range Batholith contain xenolithic inclusions that range in diameter from a few centimeters up to about 1 m. The xenoliths in Fig. 7.9 were originally composed of sedimentary and volcanic rocks that were recrystallized and partly resorbed by the granitic magma in which they were suspended. In many cases, the xenoliths are aligned parallel to the foliation of the granitic rocks (e.g., Fig. 16 of Murtaugh 1969).

In addition to various structural and textural varieties of granitic rocks, the Wisconsin Range Batholith contains pegmatite and aplite dikes as well as quartz veins both within the granitic rocks and in the metasedimentary and metavolcanic rocks of LaGorce and Wyatt formations, respectively. The pegmatite dikes are generally less than 1 m wide and are composed of quartz (including amethyst and rose varieties), microcline, plagioclase, biotite, muscovite, black tourmaline (schorlite), red garnet, and magnetite. In many cases, the pegmatite dikes are zoned and are the youngest igneous rocks in the Wisconsin Range except for andesite dikes, small plutons of hornblendite, and one dolerite dike north of Mims Spur in Fig. 7.2 where it intersects pegmatites described by Murtaugh (1969).

The granitic basement rocks which form the spurs that project into the valley of the Olentangy Glacier in Fig. 7.2 are unconformably overlain by the *Buckeye Tillite* of Permian age which is the basal unit of the Beacon Supergroup in the Wisconsin Range (Minshew 1966, 1967). The unconformity in Fig. 7.10 is the important Kukri Peneplain which formed by erosion of the Ross Mountains prior to renewed sedimentation that started during the Devonian Period. Remnants of the formerly more extensive sedimentary rocks of the

Fig. 7.9 Large angular xenoliths embedded in the equigranular facies of the Wisconsin Range Batholith in the valley of the Reedy Glacier. See also Fig. 16 of Murtaugh (1969) (Photo by G. Faure)

Fig. 7.10 The Kukri Peneplain is exposed on Tillite Spur in the valley of the Olentangy Glacier (Fig. 7.2) where Permian tillites rest unconformably on unfoliated and porphyritic facies of the Wisconsin Range Batholith. John G. Murtaugh is pointing to the unconformity a this site (Photo by G. Faure)

Beacon Supergroup have been preserved along the Olentangy Glacier in Fig. 7.4 and at the eastern end of the Wisconsin Plateau near Mt. Brecher (2,098 m) and Mt. LeSchack (2,265 m).

7.3 Age Determinations, Wisconsin Range

With the probable exception of the hornblende migmatite on Spear Nunatak (Section 7.2.3), the phyllites and metagraywackes of the LaGorce Formation are the oldest rocks in the basement of the Wisconsin Range.

7.3.1 Age of the LaGorce Formation

The first Rb-Sr age determination of rocks of the LaGorce Formation by Faure et al. (1968a) yielded a date of only 460 ± 16 Ma ($\lambda = 1.39 \times 10^{-11}$ year^{-1}) based on eight samples. In a subsequent report, Faure et al. (1968b) stated the age as 462 ± 17 Ma which they considered to be a lower limit. The samples that Faure et al. (1968a, b) analyzed originated from a site in the Ford Nunataks where the LaGorce Formation was intruded by a pluton of porphyritic quartz monzonite which had undoubtedly raised the temperature in the contact zone. Therefore, the samples were reanalyzed by Montigny and Faure (1969) in order to determine the extent to which the isotopic composition of strontium in the phyllites of the LaGorce Formation was homogenized by the intrusion and cooling of the granitic rocks.

The analytical data derived from eight whole-rock samples of the LaGorce Formation in Appendix 7.8.2.1 define a straight line in Fig. 7.11 which was fitted to the data points by unweighted linear regression. The corresponding Rb-Sr date ($\lambda = 1.42 \times 10^{-11}$ year^{-1}) and initial $^{87}Sr/^{86}Sr$ ratio are:

$$t = 450 \pm 15\,\text{Ma}, (^{87}\text{Sr}/^{86}\text{Sr})_i = 0.7170 \pm 0.0032\,(1\sigma)$$

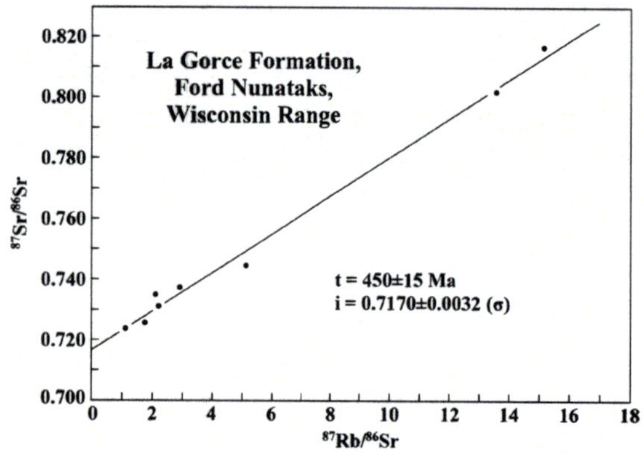

Fig. 7.11 Whole-rock Rb-Sr age determinations of the LaGorce Formation in the basement complex of the Wisconsin Range of the Horlick Mountains (Data from Montigny and Faure 1969; Faure unpublished in Appendix 7.8.2.1)

The high average initial $^{87}Sr/^{86}Sr$ ratio of these rock samples (0.7170 ± 0.0032) and the stratigraphic position of the rocks within the basement complex of the Wisconsin Range support the view that these rocks were deposited prior to about 450 ± 15 Ma and that the memory of this event was subsequently erased by the homogenization of the isotope composition of strontium in these rocks during the intrusion and subsequent cooling of granitic magma. Therefore, the Rb-Sr date of this suite of phyllites of the LaGorce Formation was reset and is an underestimate of the age of these rocks. However, the Rb-Sr systematics of the phyllites of the LaGorce Formation also demonstrate that contact metamorphism can cause partial isotopic homogenization of strontium in fine-grained clastic sedimentary rocks even though the mineral composition of the affected rocks in hand specimens is not detectably altered. We can also say that the whole-rock Rb-Sr date of the LaGorce Formation adjacent to the granitic pluton is equal to the cooling age of that intrusive. Unfortunately, we do not have a K-Ar or Rb-Sr date of biotite from the intrusive and can only cite the K-Ar biotite date of 470 ± 14 Ma reported by Minshew (1965) for granite on Mt. Wilbur at the head of the Scott Glacier. Additional K-Ar and Rb-Sr dates of biotite and muscovite in granitic basement rocks of the Queen Maud and Horlick Mountains are listed in Table 7.1 based on a compilation of Mirsky (1969). All of these dates are Ordovician cooling age of igneous rocks that crystallized during the Cambrian Period.

7.3.2 Age of the Wyatt Formation

The Wyatt Formation was initially assigned a Neoproterozoic date of 633 ± 13 Ma ($\lambda = 1.39 \times 10^{-11}$ year^{-1}) based on the Rb-Sr systematics of four samples from Metavolcanic Mountain (Faure et al. 1968a; Montigny and Faure 1969). This date was later restated as 630 ± 14 Ma by Faure et al. (1968b). After analyzing additional samples, Faure et al. (1977) reported a date of 543 ± 48 Ma (Early Cambrian) when the decay constant of ^{87}Rb is taken to be $\lambda = 1.42 \times 10^{-11}$ year^{-1}.

The analytical data of seven samples of the Wyatt Formation from Metavolcanic Mountain, the Cleveland Mesa, and the Wisconsin Plateau in Appendix 7.9.2.2 constrain a straight line in Fig. 7.12 which defines the time of crystallization of these rocks:

$$t = 514 \pm 8\,\text{Ma}, (^{87}\text{Sr}/^{86}\text{Sr})_i = 0.7115 \pm 0.0014(1\sigma)$$

The date indicates that these rocks were erupted during the Middle Cambrian Epoch. This revision of the geologic history of the Transantarctic Mountains indicates that the volcanic rocks of the Wyatt Formation were erupted at the surface of the Earth or intruded at shallow depth during the active phase of the Ross Orogeny and that the rocks probably cooled rapidly compared to the deep-seated granitic plutons of the Wisconsin

Table 7.1 R-Sr and K-Ar dates of biotite and muscovite in granites of the Queen Maud and Horlick mountains (Mirsky 1969)

Location	Mineral	Rb-Sr (Ma)	K-Ar (Ma)	Reference
Mt. Wilbur, Q. Maud Mtns.	Biotite		470 ± 14	Minshew (1965)
Discovery Ridge, Ohio Range	Biotite		472 ± 24	Treves (1965)
	Feldspar	471 ± 49		Treves (1965)
Lonely Ridge, Nilsen Plateau	Biotite		472 ± 10	McLelland (1967)
O'Brien Peak, Q. Maud Mtns.	Muscovite	460 ± 20		Craddock et al. (1964)
	Biotite		450 ± 20	
Zigzag Bluff, Axel Heiberg/Strom Glaciers	Muscovite		445 ± 13	McDougall and Grindley (1965)

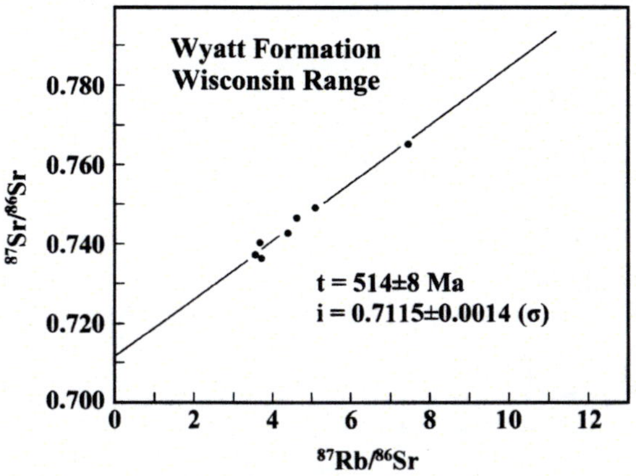

Fig. 7.12 Whole-rock Rb-Sr age determinations of the Wyatt Formation discussed in the text (Data from Montigny and Faure (1969); Faure unpublished in Appendix 7.8.2.2)

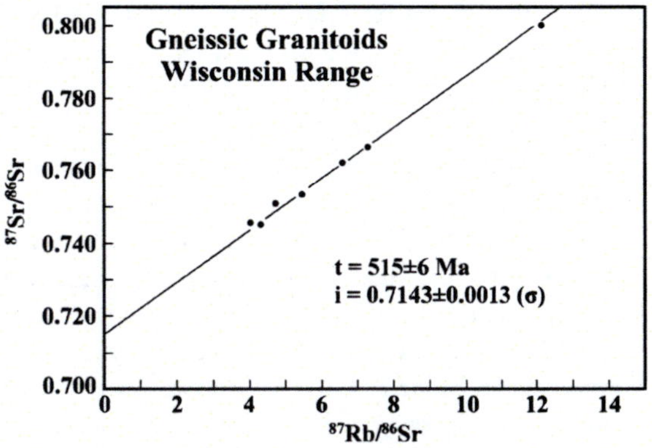

Fig. 7.13 Whole-rock Rb-Sr age determinations of granitiod basement rocks in the Wisconsin Range (Data from Faure unpublished in Appendix 7.8.2.3)

Range Batholith which continued to lose radio genic ^{87}Sr (and ^{40}K) from mica minerals (Table 7.1) as they cooled slowly until the Middle Ordovician.

7.3.3 Age of the Wisconsin Range Batholith

The Wisconsin Range Batholith is primarily composed of gneissic granitoids of syntectonic origin. A suite of seven specimens of these rocks from Mims and Polygon spurs in the valley of the Olentangy Glacier (Appendix 7.8.2.3) form a straight line in Fig. 7.13 from which the age and initial $^{87}Sr/^{86}Sr$ of these rocks was determined:

$$t = 515 \pm 6\,\text{Ma}, (^{87}\text{Sr}/^{86}\text{Sr})_i = 0.7143 \pm 0.0013\,(1\sigma)$$

The Middle Cambrian age and elevated initial $^{87}Sr/^{86}Sr$ ratio are typical of the Granite Harbor Intrusives of the Transantarctic Mountains.

A second group of granitic basement rocks in the Wisconsin Range and in the Long Hills in Appendix 7.8.2.4 define a straight line in Fig. 7.14 that yields a distinctly younger date and a lower initial $^{87}Sr/^{86}Sr$ ratio than the granite gneisses in Fig. 7.13:

$$t = 494 \pm 5\,\text{Ma}, (^{87}\text{Sr}/^{86}\text{Sr})_i = 0.7055 \pm 0.0014\,(1\sigma)$$

All specimens in this group are unfoliated (i.e., post-tectonic) and equigranular or porphyritic with euhedral phenocrysts of pink microcline. These

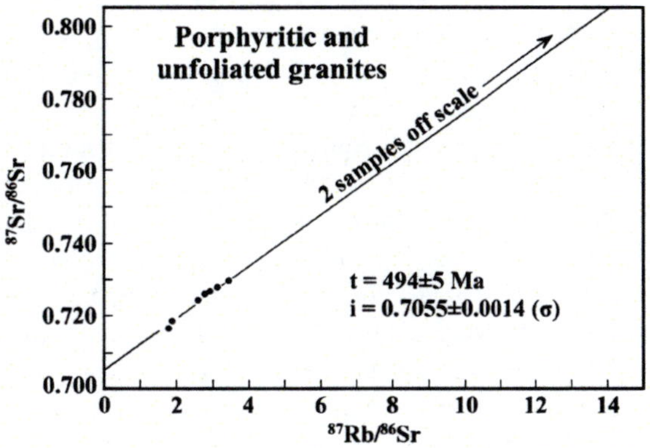

Fig. 7.14 Whole-rock Rb-Sr age determinations of porphyritic and unfoliated granitic basement rocks in the Wisconsin Range (Data from Montigny and Faure 1969; Faure unpublished in Appendix 7.8.2.4)

samples originated from Gratton Nunatak at the confluence of the Olentangy and McCarthy Glaciers, from Mt. Huckaby and Red Spur in the valley of the Olentangy Glacier (Fig. 7.2), from the Ford Nunataks and from Mt. LeSchack along the north-facing escarpment of the Wisconsin Plateau (Fig. 7.4), and from Peters Butte and Todd Ridge in the Long Hills.

The Rb-Sr date (494 ± 5 Ma) and low initial $^{87}Sr/^{86}Sr$ ratio (0.7055 ± 0.0014) indicate that the rocks of the post-tectonic facies of the Wisconsin Range Batholith crystallized during the Late Cambrian to Early Ordovician from magma that was not a derivative of the syntectonic magmas that formed the gneissic granitoids. Moreover, the wide distribution of the post-tectonic facies in the Wisconsin Range and in the Long Hills implies that the intrusion of post-tectonic granitic magma was a major event at the end of the compressive phase of the Ross Orogeny.

These insights concerning the history of magmatic activity in the Wisconsin Hills remind us that McLelland (1967) divided the Granite Harbor Intrusives in the southern part of the Nilsen Plateau into four distinctive facies (Section 6.3.1). Three of these facies are cataclastically deformed (i.e., syntectonic), whereas the North Quartz Monzonite is not foliated (i.e., post-tectonic) and may be younger than the other facies.

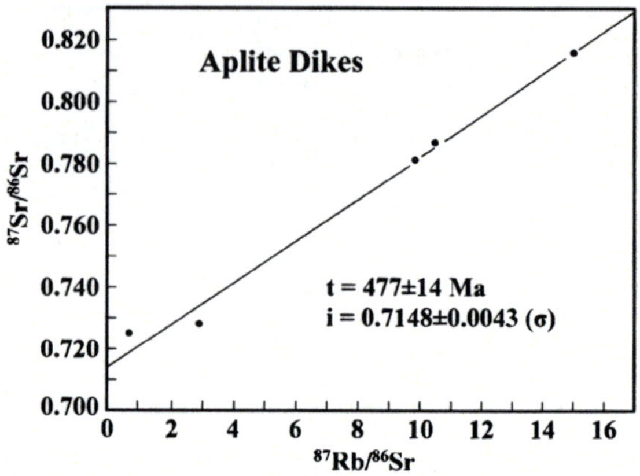

Fig. 7.15 Whole-rock Rb-Sr age determination of aplite dikes in the Wisconsin Range (Data from Ray 1973 in Appendix 7.8.2.5)

7.3.4 Age of Aplite and Pegmatite Dikes

The final stage in the development of the Wisconsin Range Batholith was the intrusion of dikes of aplite and pegmatite. The aplite dikes used for dating in this study originated from Mims, Polygon, and Tillite spurs where they intruded foliated granitic rocks. All aplite dikes are fine-grained, equigranular, and their mineral compositions are similar to those of quartz monzonite: quartz, microcline, plagioclase, biotite, as well as muscovite, augite, and apatite. Plagioclase is altered to sericite and calcite, and biotite grains contain iron oxide on cleavage planes. However, the extinctions of quartz on the petrographic microscope are sharp and the absence of foliation indicates that the aplite dikes were intruded after the orogenic stresses had relaxed.

The concentrations of rubidium and strontium in whole rock samples of aplite in Appendix 7.8.2.5 range widely causing their Rb/Sr ratios to vary between wide limits from 0.231 to 126. The corresponding measured $^{87}Sr/^{86}Sr$ ratios also range widely from 0.7253 to 3.957 (Ray 1973). However, a plot of the data reveals a great deal of scatter far in excess of analytical errors. The scatter of data points is attributable to the alteration of the minerals by chemical weathering which may have been accelerated by the large surface area of the small crystals of which the aplites are composed. Nevertheless, the whole-rock Rb-Sr date and initial $^{87}Sr/^{86}Sr$ ratio of the aplite dikes in Fig. 7.15 are:

$$t = 477 \pm 14\,\text{Ma}, (^{87}\text{Sr}/^{86}\text{Sr})_i = 0.7148 \pm 0.0043 (1\sigma)$$

This result confirms that the aplite dikes began to retain in-situ produced radiogenic ^{87}Sr during the Middle Ordovician (IUGS 2002) although they were probably intruded into the granite gneisses of Mims and Polygon spurs prior to that time. The magma from which the aplite dikes formed was a late-stage differentiate of the granitic magma of the Wisconsin Range Batholith parts of which were enriched in rubidium and depleted in strontium as a result of fractional crystallization feldspar.

The pegmatite dikes on Mims Spur are composed of coarse-grained quartz, microcline, plagioclase, muscovite, biotite, black tourmaline, and red garnet. Ray (1973) analyzed minerals from four of these pegmatite dikes and two samples of granite gneiss. The data in Appendices 7.8.2.6 and 7.8.2.7 yielded highly irregular dates between 118 and 828 Ma.

However, the feldspar concentrates derived from four pegmatite dikes define a straight line in Fig. 7.16 that yields:

$$t = 478 \pm 6\,\text{Ma}, (^{87}\text{Sr}/^{86}\text{Sr})_i = 0.7175 \pm 0.0024 (1\sigma)$$

The date is consistent with the Rb-Sr and K-Ar dates of biotite in the Queen Maud and Wisconsin Range batholiths listed in Table 7.1 and refers to the time of cooling of the pegmatite dikes. The initial $^{87}Sr/^{86}Sr$ ratio (0.7175 ± 0.0024) is high because it includes the radiogenic ^{87}Sr that formed in the pegmatites during the cooling interval.

The muscovites in the four pegmatite dikes and one sample of granite gneiss scatter widely and therefore are not shown in Fig. 7.16. The issues raised by the redistribution of radiogenic ^{87}Sr (and ^{40}Ar) during slow cooling or as a result of contact metamorphism verge toward thermochronology which was dealt with in books by Bernet and Spiegel (2004) and McDougall and Harrison (1988) as well as in papers by Berger and York (1981), Harrison and McDougall (1982), and Dodson (1973). Additional references to thermochronology were listed by Faure and Mensing (2005, Section 7.9).

7.4 Geologic History, Wisconsin Range Basement

The dates derived from the basement rocks of the Wisconsin Range in Table 7.2 outline a hypothetical geological history for this segment of the Transantarctic Mountains that is illustrated in Fig. 7.17. The results generally confirm the master hypothesis concerning the origin of the Ross orogen presented by Stump (1995).

The greywacke/shale complex of the LaGorce Formation was deposited prior to the start of the Ross Orogeny during which these rocks were folded and intruded by plutons of the Granite Harbor Intrusive. The metavolcanic rocks of the Wyatt Formation were erupted at the surface or intruded at shallow depth during the Middle Cambrian at about the time when the gneissic basement granites crystallized at depth. The Wisconsin Range Batholith also includes a post-tectonic facies of granitic rocks that crystallized in Late Cambrian to Early Ordovician time. This suite of granitic rocks is widely distributed in the basement of the Wisconsin Range and the Long Hills and may also occur in the basement of the Nilsen Plateau. The magmatic activity of the Ross Orogeny ended with the intrusion of aplite and pegmatite dikes into fractures in the gneissic granite of the Wisconsin Range Batholith. These rocks subsequently cooled and became closed systems at about 478 Ma during the Middle Ordovician.

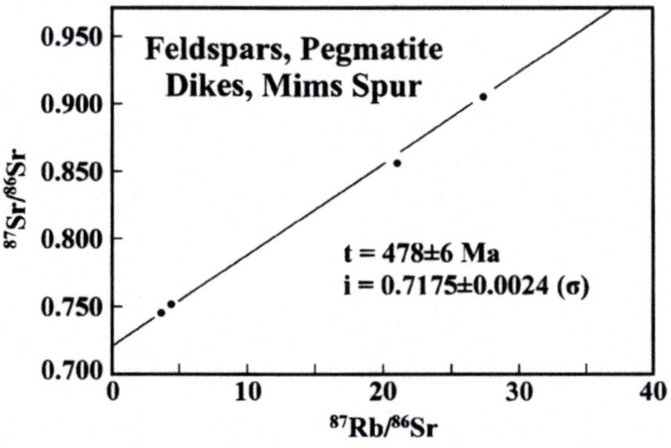

Fig. 7.16 Age determinations of feldspar from pegmatite dikes in the basement complex of the Wisconsin Range (Data from Ray 1973 in Appendices 7.8.2.6 and 7.8.2.7)

7.5 Long Hills

The Long Hills in Fig. 7.18 consist of a small cluster of nunataks located about 45 km east of Mt. LeSchack at the eastern limit of the main body of the Wisconsin

Table 7.2 Summary of Rb-Sr dates of whole-rocks and mineral concentrates from the basement rocks of the Wisconsin Range and the Long Hills in the Horlick Mountains. Recalculated to λ (^{87}Rb) = 1.42×10^{-11} year^{-1} from unpublished analytical data by Faure et al. (1968a, b, 1977) and Montigny and Faure (1969) (Appendices 7.8.2.1 to 7.8.2.7)

Stratigraphic unit	Locality	Date (Ma)	Initial ^{87}Sr/^{86}Sr
Feldspar, pegmatite dikes	Mims Spur, Wisconsin Range	478 ± 6	0.7175 ± 0.0024 (1σ)
Aplite dikes	Mims, Polygon, Tillite spurs, Wisconsin Range	477 ± 14	0.7148 ± 0.054 (1σ)
Porphyritic granite	Wisconsin Range and Long Hills	497 ± 4	0.7055 ± 0.0012 (1σ)
Granite gneiss, Granite Harbor	Mims and Polygon spurs, Wisconsin Range	515 ± 6	0.7143 ± 0.0013 (1σ)
Wyatt Fm.	Metavolcanic Mtn., Wisconsin Range	514 ± 8	0.7115 ± 0.0014 (1σ)
LaGorce Fm.	Ford Nunataks, Wisconsin Range	450 ± 15 (R)	0.7170 ± 0.0032 (1σ)

The dates and initial ^{87}Sr/^{86}Sr ratios were recalculated for this presentation by unweighted linear regression

Concentrations of Rb and Sr in whole-rock samples were measured by x-ray fluorescence using pressed pellets prepared from powdered samples (Eastin 1970; Ray 1973)

Isotope analyses were made on a Nuclide Corporation 60°-sector mass spectrometer from a single filament composed of Ta

The measurement precision of ^{87}Sr/^{86}Sr ratios was 0.1% for ^{87}Sr/^{86}Sr < 0.800 and 0.3% for ^{87}Sr/^{86}Sr > 0.800

The average ^{87}Sr/^{86}Sr ratio of the Eimer and Amend Sr carbonate isotope standard was 0.70806 ± 0.00025 (1σ)

R = reset

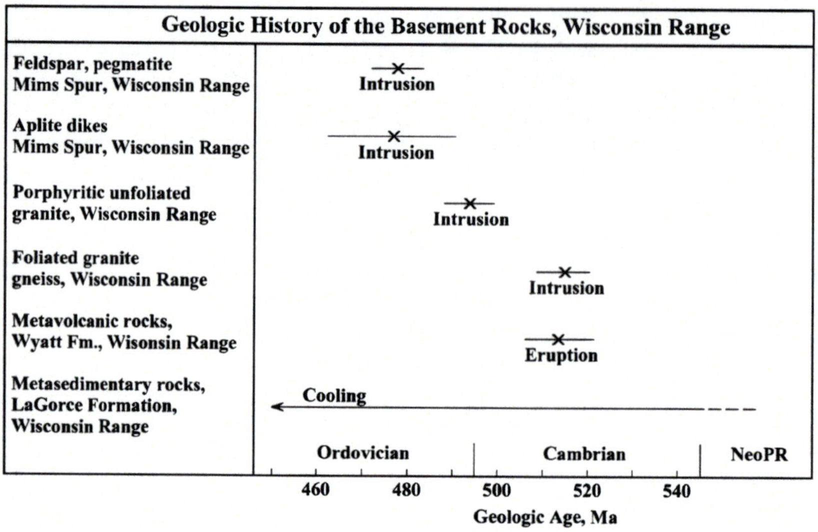

Fig. 7.17 Summary of age determinations by the Rb-Sr method of basement rocks of the Wisconsin Range in the Horlick Mountains. Dates from Figs. 7.11–7.16

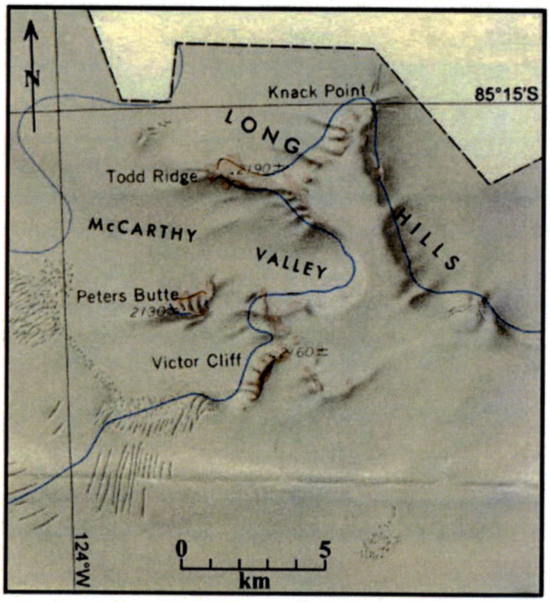

Fig. 7.18 The Long Hills of the Horlick Mountains are located about 45 km east of Mt. LeSchack (85°25'S, 124°00'W) connected to the Wisconsin Range by an ice-covered escarpment marked by the Spencer and Widich nunataks both of which expose granitic basement rocks of the Granite Harbor Intrusives (not shown). The Long Hills are noteworthy because Todd Ridge consists of rhyolite pyroclastics overlain by an erosional remnant of the Beacon Supergroup (Mirsky 1969). Excerpt of the topographic map Long Hills, Antarctica S8500-W11000/1 × 15, 1960 published by the US Geological Survey, Washington, DC in 1960

Range. The principal nunataks are Peters Butte (2,130 m), Todd Ridge (2,190 m), Victor Cliff (2,160 m), and Knack Point. In addition, Spencer Nunatak (2,007 m) and Widich Nunatak (1,986 m), which are not included in Fig. 7.18, occur along the ice-covered escarpment that extends east from Mt. LeSchack to the Long Hills.

The geologic map of Mirsky (1969) indicates that Peters Butte, Victor Cliff, and Knack Point as well as the Spencer and Widich nunataks consist of granitic rocks of the Wisconsin Range Batholith, whereas Todd Ridge is composed of felsic pyroclastic rocks which resemble the volcanic rocks of the Leverett Formation on Mt. Webster (Section 6.5.2). The volcanic rocks on Todd Ridge are porphyritic and appear massive in hand specimen. The phenocrysts of feldspar are embedded in an aphanitic groundmass, but the rocks are so extensively altered that the only remaining primary mineral is partly resorbed quartz and accessory minerals such as zircon, apatite, and magnetite. Feldspar has altered to fine-grained sericite, zeolite, carbonate, clay, and epidote. The matrix appears to be composed of devitrified glass. The volcanic rocks contain small xenoliths of siltstone, graywacke, and plutonic rocks, which suggests that they represent a former pyroclastic deposit that was cataclastically deformed and hydrothermally altered after deposition. The volcanic rocks on Todd Ridge as well as the granitic basement rocks on Victor Cliff are unconformably overlain by sedimentary rocks of the Beacon Supergroup.

7.5.1 Todd Ridge Pyroclastics, Chemical Composition

The chemical compositions of the felsic pyroclastics on Todd Ridge are interpreted by means of silica-variation diagrams in Fig. 7.19 based on the chemical analyses in Appendix 7.8.3. The concentrations of CaO + MgO decrease with increasing silica concentration as expected, but a single sample of rhyolite of the Leverett Formation on Mt. Webster does not fit the pattern, which does not support, although it does not invalidate, the conjecture that the felsic pyroclastics on Todd Ridge in the Long Hills represent the Leverett Formation. The variation of the concentrations of total Fe is inconsistent with the effects of fractional crystallization of magma and the concentrations of $Na_2O + K_2O$ scatter erratically in Fig. 7.19, thereby revealing the effects of hydrothermal alteration of these rocks.

7.5.2 Age of the Todd Ridge Pyroclastic Rocks

The Todd Ridge pyroclastic rocks may have been intruded by the porphyritic facies of the Wisconsin Range Batholith in which case the felsic pyroclastic rocks should be older than 497 ± 4 Ma (Table 7.2). Actually, Faure et al. (1968b) reported a whole-rock Rb-Sr date of 521 ± 38 Ma (Early Cambrian) for six specimens of the Todd Ridge pyroclastics (recalculated to $\lambda = 1.42 \times 10^{-11}$ year^{-1}).

The analytical data for eight samples of the Todd Ridge pyroclastics in Appendix 7.8.4 define a straight line in Fig. 7.20 from which we calculate a Rb-Sr date and initial $^{87}Sr/^{86}Sr$ ratio:

$$t = 534 \pm 5\,\text{Ma}; i = 0.7107 \pm 0.0023\,(1\sigma)$$

This Early Cambrian date is compatible with the stratigraphy and age determinations of the basement rocks of the Wisconsin Range. However, the severe alteration of the Todd Ridge pyroclastics casts doubt on the reliability of this date.

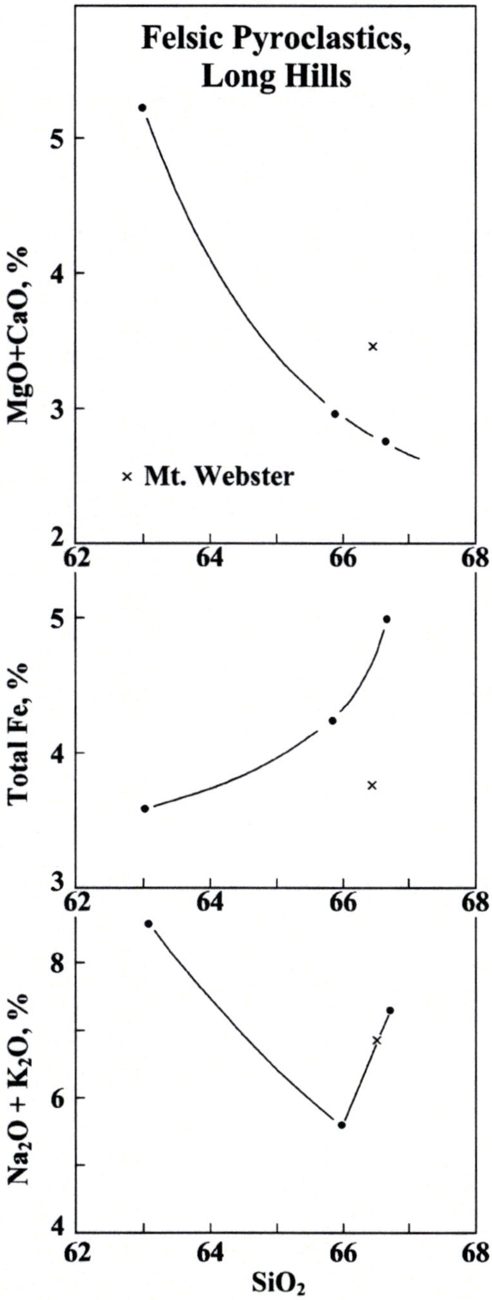

Fig. 7.19 Silica-variation diagram of rhyolite pyroclastic rocks on Todd Ridge of the Long Hills, Horlick Mountains. One sample of rhyolite from Mt. Webster in the Harold Byrd Mountains is represented by the "×" (Data from Minshew 1967; Faure unpublished in Appendix 7.8.3)

7.6 Ohio Range

The Ohio Range is located about 55 km northeast of the Long Hills and forms the most easterly part of the Horlick Mountains. The topographic map in Fig. 7.21 identifies the principal components of this mountain range starting with Lackey Ridge (2,380 m) at its western end followed by Mt. Glossopteris (2,867 m) and Mt. Schopf (2,990 m) at the east end of the range. The Ohio Range, like the Wisconsin Range, consists of a partly dissected plateau called the Buckeye Table which forms a north-facing escarpment along most of its length of about 40 km.

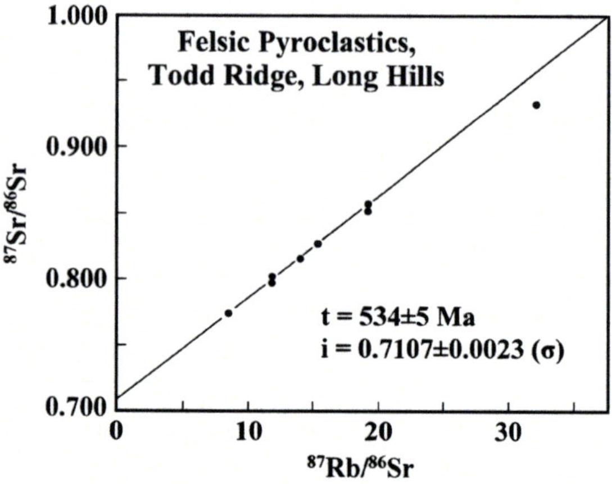

Fig. 7.20 Whole-rock Rb-Sr age determination of rhyolite pyroclastics in the basement complex of the Long Hills (Data from Faure unpublished in Appendix 7.8.4)

The Ohio Range was first explored during the 1958/59 field season in the course of an oversnow traverse that started on November 1, 1958, from Byrd Station in West Antarctica (Long 1965). The tractor train first advanced to the Wisconsin Range where Bill Long and Fred Darling collected samples of granitic basement rocks from the Wisconsin Escarpment. In December of 1958 the tractor train reached the Ohio Range and established a temporary camp at Mile 414 about 5 km north of Mt. Glossopteris. On December 6, 1958, a group of four men led by Charles Bentley and including Bill Long, Fred Darling, and Jack Long climbed to the summit of Mt. Glossopteris in order to collect rock specimens and fossils (Long 1961).

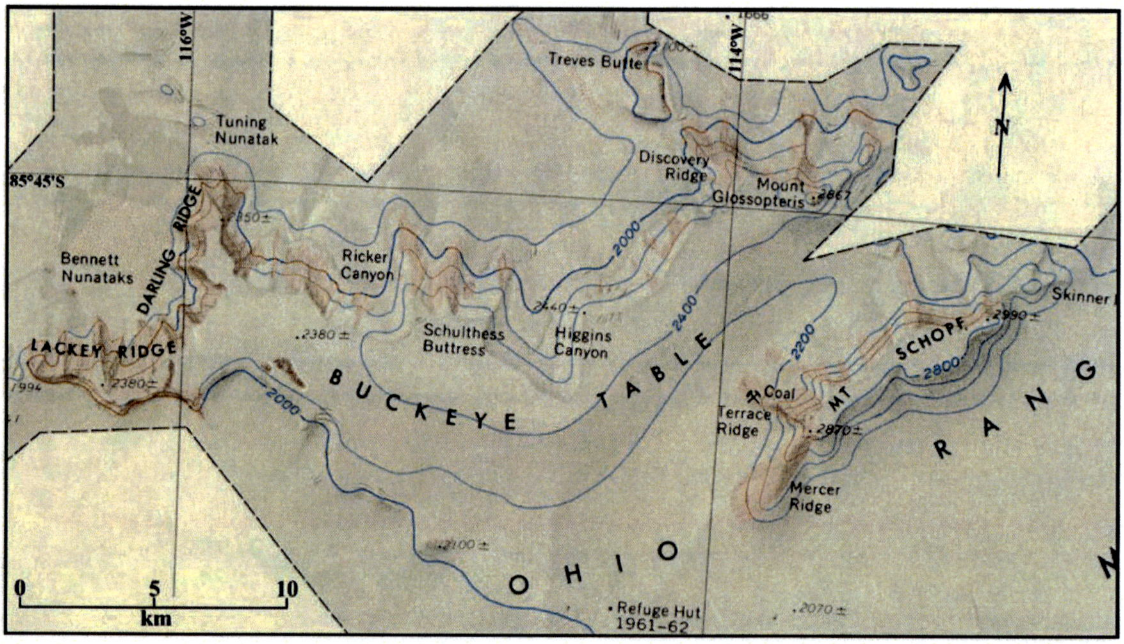

Fig. 7.21 Excerpt of the central part of the Ohio Range in the Horlick Mountains. Mirsky Ridge and Iversen Peak in the eastern part of the range are not shown. Several small nunataks (Knox, Vann, and Eldridge peaks) west of Lackey Ridge have also been excluded from this map. (Adapted from the topographic map: Ohio Range, Antarctica SV11-20/1 published by the US Geological Survey, Washington, DC)

7.6.1 Geology of the Ohio Range

The geology of the Ohio Range consists of a basement composed of porphyritic quartz monzonite and granodiorite unconformably overlain by about 990 m of sedimentary rocks of the Beacon Supergroup described by Long (1961, 1962, 1964, 1965). The erosion surface of the granitic basement is grooved, striated, and polished and is overlain by residual sandstone of the *Horlick Formation* (Early Devonian) followed by about 275 m of the *Buckeye Tillite* of Permian age. The upper part of the Beacon Supergroup in the Ohio Range contains several coal seams and is capped by a sill of the Ferrar Dolerite (about 190 m thick) which occurs on the summit of Mt. Schopf and its extension into Mercer Ridge. According to Long (1965), this is the only dolerite sill in the Ohio Range (Fig. 7.22).

The granitic basement and the overlying rocks of the Beacon Supergroup as well as the Ferrar Dolerite sill are cut by normal faults of post-Middle Jurassic age (Long 1965). However, the location of these faults is obscured by snow and ice except locally on the slopes of Mt. Glossopteris. The existence of normal faults is also indicated by the vertical displacement of the Kukri Peneplain between Treves Butte and Mt. Glossopteris (Long 1965, Fig. 7.22). The sedimentary rocks of the Beacon Supergroup are flat-lying with a slight regional dip to the south (see also Fig. 10.23).

7.6.2 Granitic Basement

The basement rocks of the Ohio Range are exposed along the north-facing escarpment that extends along the entire length of this mountain range. The basement consists primarily of unfoliated and porphyritic quartz monzonite and of slightly foliated granodiorite described by Treves (1965). Metasedimentary rocks like those of the LaGorce Formation and metavolcanic rocks of the Wyatt Formation have not been recognized in the Ohio Range but may be represented by xenoliths in the granitic rocks.

The first description of the granitic rocks of the basement complex by Long (1961) was illustrated by

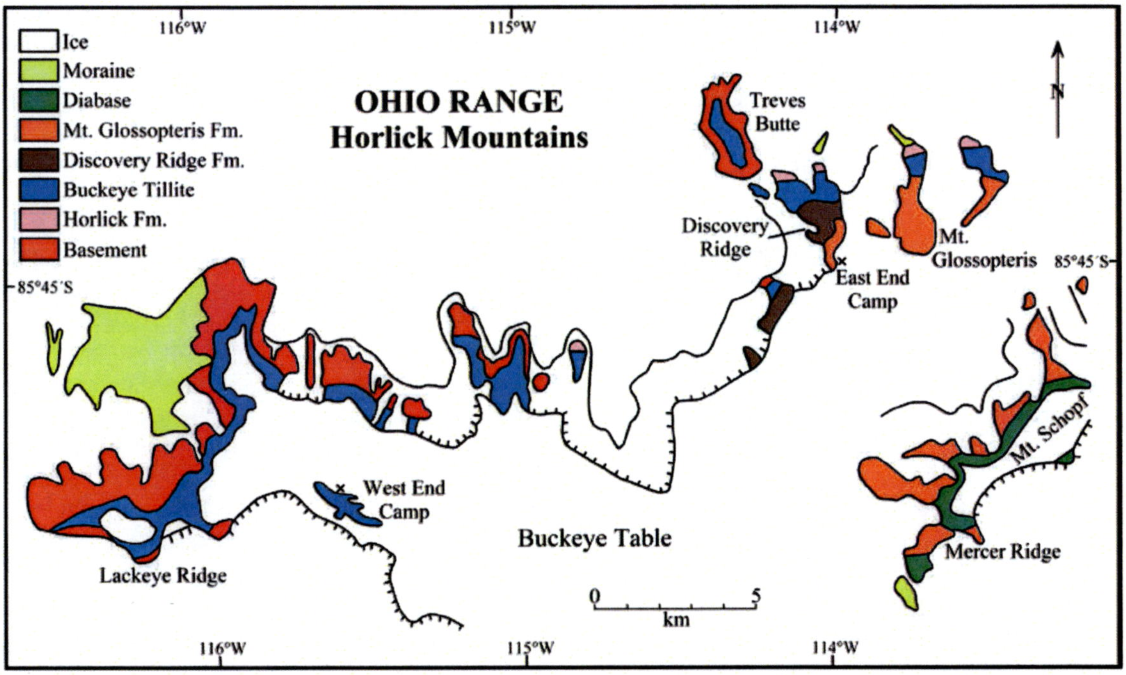

Fig. 7.22 The basement rocks of the Ohio range consist of granitic rocks which are overlain by flat-lying sedimentary rocks of the Beacon Supergroup. Mt. Schopf is capped by a sill of the Ferrar Dolerite (Adapted from Long 1965)

several excellent photographs and included modal analyses listed in Appendix 7.8.5. In addition, a chemical analysis of the quartz monzonite cited by Long (1961) is reproduced in Appendix 7.8.6. Treves (1965) reported that the granodiorite contains more plagioclase and biotite than the quartz monzonite but less quartz and microcline. He considered the granodiorite to be an early border facies of the local granitic pluton and suggested that its chemical composition may have been altered by the assimilation of older rocks. The contact between the two facies is gradational in most places; however, locally the quartz monzonite does intrude the granodiorite facies.

Isotopic age determinations by the Isotope Geology Branch of the US Geological Survey reported by Treves (1965) indicate cooling ages of 472 ± 24 Ma (K-Ar, biotite, Discovery Ridge) and 471 ± 49 Ma (Rb-Sr, feldspar, Treves Butte) and a crystallization age of 516 ± 72 Ma (Rb-Sr, whole rock, Treves Butte) for the porphyritic quartz monzonite. These dates are compatible with the geologic history of the basement rocks of the Wisconsin Range in Fig. 7.17.

7.7 Summary

The basement rocks of the Horlick Mountains consists of gneissic granites and of younger porphyritic quartz monzonites which are an eastward extension of the Queen Maud Batholith. In addition, the basement of the Wisconsin Range contains remnants of metasedimentary rocks of the LaGorce Formation and of metavolcanic rocks of the Wyatt Formation both of which were defined by Minshew (1967) based on outcrops near the head of the Scott Glacier (Sections 6.3.2 and 6.4.1). The rocks of the LaGorce Formation were deposited prior to the onset of the Ross Orogeny during which these rocks were folded, faulted, and regionally metamorphosed to the greenschist facies. The volcanic rocks of the Wyatt Formation were extruded on the surface of the Earth and in the form of shallow intrusives at about the same time during the Ross Orogeny when gneissic granitic rocks of the Wisconsin Range Batholith were crystallizing from granite magma at the depth in the crust.

The granitic rocks of the Wisconsin Range Batholith include a porphyritic quartz monzonite facies that is younger than the gneissic facies. Both facies are recognized throughout the Horlick Mountains from the valley of the Reedy Glacier to the east end of the Ohio Range. However, the LaGorce and Wyatt formations of the Wisconsin Range do not occur in outcrop in the basement of the Long Hills and of the Ohio Range.

Todd Ridge in the Long Hills exposes a suite of highly altered felsic pyroclastic rocks of apparent Early Cambrian age. The stratigraphic affiliation of these rocks remains in doubt, although they may be related to the Leverett Formation on Mt. Webster in the Harold Byrd Mountains (Section 6.5). The most southerly nunatak of the Wisconsin Range is composed of an amphibolite gneiss or migmatite that has not been seen anywhere else in the Horlick Mountains. The amphibolite grade of regional metamorphism and the severe deformation cause this rock to resemble the Nimrod Group of the Miller Range (Section 5.1).

The sedimentary rocks of the Beacon Supergroup are prominent in the Ohio Range as documented by the work of Long (1961, 1962, 1964, 1965), but only small remnants are preserved in the Long Hills and in the Wisconsin Range where they were studied by Minshew (1966, 1967). Likewise, sills of the Ferrar Dolerite which intruded the Beacon Supergroup throughout most of the Transantarctic Mountains are virtually absent from the Wisconsin Range and from the Long Hills but are represented by one voluminous sill that caps Mt. Schopf and Mercer Ridge of the Ohio Range.

7.8 Appendices

7.8.1 Chemical Composition of Metavolcanic Rocks, Wyatt Formation, Scott Glacier Area and Wisconsin Range in Weight Percent (Faure unpublished; Minshew 1967)

	A	B	C	D	E	F
SiO_2	67.46	68.8	70.3	67.8	71.9	67.15
Al_2O_3	15.62	14.4	14.4	14.6	14.3	14.13
Fe_2O_3	1.12	1.6	1.1	1.8	1.1	1.27
FeO	3.81	3.4	2.2	3.2	1.8	3.72
MgO	1.91	2.0	1.1	2.0	0.99	1.27
CaO	2.23	1.8	1.8	1.9	2.0	3.12
Na_2O	2.96	3.2	2.7	2.6	2.6	3.27
K_2O	3.20	2.9	3.9	3.9	4.1	3.51
TiO_2	0.68	0.78	0.54	0.82	0.43	0.72
P_2O_5	–	0.15	0.18	0.18	0.17	0.192
MnO	0.10	0.04	0.02	0.04	0.00	–
H_2O+	0.26	1.3	1.4	1.4	1.3	1.17
H_2O-	0.20	0.05	0.03	0.13	0.03	–
CO_2	0.34	0.04	0.06	0.02	0.01	–
Sum	99.89	100.46	99.73	100.4	100.7	99.52

A: Mt. Gardiner, Scott Glacier (#226)
B: Mt. Gardiner, Scott Glacier (# 236)
C: Metavolcanic Mountain, Wisconsin Range (F-64-61)
D: Wisconsin Plateau, Wisconsin Range (F-64-108)
E: Metavolcanic Mountain, Wisconsin Range (F-64-129)
Analyses by Sam Bolt, US Geological Survey (A to E)
F: Porphyritic quartz monzonite Peters Butte, Long Hills
Analysis by A.S. McCreath and Son, Inc. (F only)

7.8.2 Rb-Sr Data, Basement Rocks, Wisconsin Range (Montigny and Faure 1969; Ray 1973; Faure unpublished)

7.8.2.1 LaGorce Formation, Ford Nunataks, Wisconsin Range

Sample No	Rb, ppm	Sr, ppm	$^{87}Rb/^{86}Sr$	$^{87}Sr/^{86}Sr$
F-64-103A	131.6	173.9	2.195	0.7315
B	78.6	199.1	1.144	0.7237
C	225.9	48.79	13.53	0.8023, 0.8024
D	114.9	158.2	2.109	0.7357
E	222.0	125.8	5.128	0.7445
F	114.3	199.5	1.662	0.7253
G	253.7	48.88	15.20	0.8166
H	91.8	91.58	2.911	0.7375

7.8.2.2 Wyatt Formation, Metavolcanic Mountain, Wisconsin Range

Sample No	Rb, ppm	Sr, ppm	$^{87}Rb/^{86}Sr$	$^{87}Sr/^{86}Sr$
F-64-45	167.8	110.4	4.415 ± 0.076	0.7426
F-64-46	227.4	89.06	7.433 ± 0.152	0.7655
F-64-61	185.8	117.6	4.591 ± 0.059	0.7465
F-65-129	186.1	106.1	5.098 ± 0.027	0.7496
F-65-130	185.0	149.4	3.595 ± 0.042	0.7373
Wyatt Formation, Cleveland Mesa, Wisconsin Range				
F-65-119C	161.3	129.0	3.629 ± 0.065	0.7362
Wyatt Formation, Wisconsin Plateau, Wisconsin Range				
F-65-108	189.3	149.5	3.677 ± 0.063	0.7404

7.8.2.3 Granitic Gneiss, Olentangy Glacier, Wisconsin Range

Sample No	Rb, ppm	Sr, ppm	$^{87}Rb/^{86}Sr$	$^{87}Sr/^{86}Sr$
F-64-9	175.7	125.1	4.081 ± 0.095	0.7459, 0.7462
F-64-11	189.9	115.8	4.767 ± 0.091	0.7507
F-64-13	231.8	101.3	6.660 ± 0.210	0.7627
F-64-14	181.4	126.0	4.252 ± 0.054	0.7450
F-64-16	311.2	74.91	12.14 ± 0.26	0.8058, 0.8033
F-64-32	254.2	100.8	7.342 ± 0.087	0.7665
F-64-81	174.1	92.62	5.466 ± 0.120	0.7536

7.8.2.4 Unfoliated Porphyritic Granitic Rocks, Gratton Nunatak and Olentangy Glacier

Sample No	Rb, ppm	Sr, ppm	$^{87}Rb/^{86}Sr$	$^{87}Sr/^{86}Sr$
F-64-69	174.1	278.0	1.814 ± 0.032	0.7174
F-64-70	183.7	273.8	1.944 ± 0.039	0.7184
F-64-76	572.9	58.5	28.89 ± 0.48	0.9094, 0.9110
F-64-84	196.6	194.6	2.930 ± 0.083	0.7272
Porphyritic granite, Ford Nunataks and Mt. LeSchack Wisconsin Range				
F-64-102	243.9	41.0	17.40 ± 0.87	0.8263
240	193.4	192.5	2.914 ± 0.14	0.7275
Quartz Monzonite, Todd Ridge and Peters Butte, Long Hills				
F-64-307WR	193.3	178.8	3.217 ± 0.024	0.7282
F-64-227WR	197.6	218.3	2.624 ± 0.087	0.7248, 0.7259
227B	711.7	38.4	55.49 ± 2.52	1.055
227F	311.6	257.1	3.516 ± 0.076	0.7299

7.8.2.5 Aplite Dikes, Mims, Polygon, Tillite Spurs, Wisconsin Range

Sample No	Rb, ppm	Sr, ppm	$^{87}Rb/^{86}Sr$	$^{87}Sr/^{86}Sr$
F-64-6	176.3	52.2	9.847 ± 0.338	0.7810
F-64-19	153.1	153.5	2.892 ± 0.102	0.7266, 0.7282
F-64-22	324.7	2.78	446.3 ± 28.5	3.822, 3.957
F-64-23	349.0	2.76	476.6 ± 47.5	3.673, 3.693
F-64-25	96.93	418.0	0.6723 ± 0.0217	0.7253
F-64-27	209.9	23.6	25.90 ± 1.36	0.886, 0.8934
F-64-41	190.4	53.1	10.46 ± 0.51	0.7866

7.8.2.6 Mineral Concentrates, Granitic Rocks and Pegmatites, Mims Spur

Sample No	Rb, ppm	Sr, ppm	$^{87}Rb/^{86}Sr$	$^{87}Sr/^{86}Sr$
F-64-11B	756.6	11.76	206.9 ± 15.1	1.841
F-64-13M	445.6	45.9	29.51 ± 2.08	1.219
F-64-13B	882.0	9.75	308.9 ± 19.2	2.304, 2.507

7.8.2.7 Mineral Concentrates, Pegmatites, Mims Spur, Wisconsin Range

Sample No	Rb, ppm	Sr, ppm	$^{87}Rb/^{86}Sr$	$^{87}Sr/^{86}Sr$
F-64-8M	295.1	5.73	180.3	2.714
F-64-8F	189.5	20.4	27.45 ± 1.25	0.9067
F-64-10M 76.9	3.12	76.97	1.513	
F-64-10F	332.3	46.5	20.97 ± 0.23	0.8568
F-64-10T 1.32	5.69	0.673		–
F-64-15M 238.4	1.80	409.9	1.412	
F-64-15B	1080.9	12.8	257.6	1.258
F-64-15F	121.9	75.5	4.693 ± 0.100	0.7502
F-64-17M	265.5	4.99	173.5	2.008
F-64-17F	64.07	41.4	4.501 ± 0.114	0.7484

All Rb and Sr concentrations of mica minerals were determined by isotpe dilution including blank corrections for Sr = 0.1 μg and $^{87}Sr/^{86}Sr$ = 0.709 (Ray 1973)

WR = whole rock; B = biotite; M = muscovite, F = feldspar, T = tourmaline

7.8.3 Chemical Analyses of Felsic Pyroclastic Rocks on Todd Ridge, Long Hills, and on Mt. Webster, Harold Byrd Mountains, in Weight Percent (Faure unpublished; Minshew 1967)

	A	B	C	D
SiO_2	65.93	66.72	63.05	66.48
Al_2O_3	14.33	13.55	13.70	13.52
Fe_2O_3	0.01	0.69	0.54	0.54
FeO	5.43	5.81	4.13	4.34
MgO	1.32	1.71	1.24	1.90
CaO	1.62	1.03	4.00	1.57
Na_2O	0.25	2.02	3.26	2.76
K_2O	5.33	5.31	5.32	4.10
TiO_2	0.87	0.83	0.82	0.80
P_2O_5	0.176	0.170	0.199	
MnO	–	–	–	
H_2O+	2.82	1.66	1.65	2.28
H_2O-	–	–	–	–
CO_2	1.37	0.40	1.55	1.44
Sum	99.85	99.46	99.46	99.73

A: Todd Ridge, Long Hills (F-64-128,2)

B: Todd Ridge, Long Hills (F-64-128,4)

C: Todd Ridge, Long Hills (F-64-128,10)

D: Mt. Webster, Harold Byrd Mtns (#228)

Analyses by A.S. McCreath and Son, Inc.

7.8.4 Rb-Sr Data, Felsic Pyroclastic Rocks, Todd Ridge, Long Hills

Sample No	Rb, (ppm)	Sr, (ppm)	$^{87}Rb/^{86}Sr$	$^{87}Sr/^{86}Sr$
F-64-128(2)	236.7	21.64	32.36 ± 2.03	0.9333
128(3)	239.1	45.80	15.29 ± 0.14	0.8277
128(4)	228.7	57.62	11.59 ± 0.34	0.7987
128(5)	172.7	42.9	11.75 ± 0.27	0.7997
128(6)	286.7	43.88	19.19 ± 0.90	0.8526, 0.8592
128(9)	243.1	50.80	14.00 ± 0.46	0.8198
128(10)	200.8	68.25	8.573 ± 0.18	0.7748

7.8.5 Modal Compositions of the Granitic Basement Rocks, Ohio Range (Long 1961; Treves 1965)

	Concentration (%)		
Minerals	A (1)	A (2)	B (2)
Microcline	38.9	27.5	20.1
Quartz	26.6	35.8	15.2
Plagioclase	28.0	30.7	45.4
Biotite	5.1	3.8	10.1
Magnetite	0.7	0.3	4.1
Sphene	0.3	–	–
Apatite	0.2	–	5.1
Myrmekite	0.3	–	
Allanite	0.1	0.2	
Zircon	–	–	
Epidote	–		

1. Analysis by Warren Hamilton cited by Long (1961)
2. Analysis by Treves (1965)

A: quartz monzonite

B: granodiorite

7.8.6 Chemical Analysis of a Porphyritic Quartz Monzonite, Mt. Glossopteris, Ohio Range (W.W. Brannock, US Geological Survey, Reported by Long 1961)

Oxide	Concentration (%)
SiO_2	72.9
Al_2O_3	13.8
Fe_2O_3	0.9
FeO	1.2
MgO	0.57
CaO	1.8
Na_2O	2.5
K_2O	5.2
H_2O	0.78
TiO_2	0.26
P_2O_5	0.10
MnO	0.04
CO_2	0.05
Sum	100.00

References

Berger GW, York D. (1981) Geothermometry from $^{40}Ar/^{39}Ar$ dating experiments. Geochim Cosmochim Acta 45:795–811

Bernet M, Spiegel C (2004) Detrital thermochronology - provenance analysis, exhumation, and landscape evolution of mountain belts. Geological Society of America, Boulder, CO

Borg SG, DePaolo DJ, Smith BM (1990) Isotopic structure and tectonics of the central Transantarctic Mountains. J Geophys Res 95(B5):6647–6667

Borg SG, DePaolo DJ, Daly EE, Sims KWW (1991) Studies of granitic and metamorphic rocks, Horlick and Whitmore Mountains area. Antarctic J US 26(5):24–25

Craddock C (ed) (1970) Geologic map of Antarctica. Antarctic Map Folio Series, Folio 12. American Geographical Society, New York

Craddock C, Gast PW, Hanson GN, Linder H (1964) Rubidium-strontium ages from Antarctica. Geol Soc Amer Bull 75:237–240

Davis MB, Blankenship DD (2005) Geology of the Scott and Reedy Glaciers area, southern Transantarctic Mountains, Antarctica. Geol Soc Amer, Map and Chart Series, MCH093

Dodson MH (1973) Closure temperature in cooling geochronological and petrological systems. Contrib Mineral Petrol 40:259–274

Eastin R (1970) Geochronology of the basement rocks of the central Transantarctic Mountains, Antarctica. PhD dissertation, Department of Geology and Mineralogy, The Ohio State University, Columbus, OH

Eskola P (1963) The Precambrian of Finland. In: Rankama K (ed) The Precambrian, vol. 1. Wiley, Interscience, New York, pp 145–263

Faure G, Mensing TM (2005) Isotopes: Principles and applications, 3rd edn. Wiley, Hoboken, NJ

Faure G, Hill RL, Eastin R, Montigny RJE (1968a) Age determination of rocks and minerals from the Transantarctic Mountains. Antarctic J US 3(5):173–175

Faure G, Murtaugh JG, Montigny RJE (1968b) The geology and geochronology of the basement complex of the central Transantarctic Mountains. Canadian J Sci 5:555–560

Faure G, Eastin R, Ray PT, McLelland D, Schultz CH (1977) Geochronology of igneous and metamorphic rocks, central Transantarctic Mountains. In: Laskar B, Raja Rao RS (eds) Fourth international Gondwana symposium, vol. 2. Geological Survey of India, Calcutta, India, pp 805–813

Ford AB (1964) Cordierite-bearing, hypersthene-quartz monzonite porphyry and its regional importance. In: Adie RJ (ed) Antarctic geology. North-Holland, Amsterdam, The Netherlands, pp 429–441

Hadley JB (ed) (1963) Geology and paleontology of the Antarctic. Antarctic Research Series, vol. 6. American Geophysical Union, Washington, DC

Harrison TM, McDougall I (1982) The thermal significance of potassium feldspar K-Ar ages inferred from $^{40}Ar/^{39}Ar$ age spectrum results. Geochim Cosmochim Acta 46:1811–1820

Harwood DM (1983) Diatoms from the Sirius Formation, Transantarctic Mountains. Antarctic J US 18(5):98–100

Harwood DM (1986a) Diatom biostratigraphy and paleoecology with a Cenozoic history of Antarctic ice sheets. Ph.D. dissertation, Department of Geology and Mineralogy, The Ohio State University, Columbus, OH

Harwood DM (1986b) Recycled siliceous microfossils from the Sirius Formation. Antarctic J US 21:101–103

Harwood DM, Webb PN (1998) Glacial transport of diatoms in the Antarctic Sirius Group: Pliocene refrigerator. GSA Today 8(4):1, 4–8

Hill RS, Harwood DM, Webb PN (1991) Last remnant of Antarctica's Cenozoic flora: Nothofagus of the Sirius Group, Transantarctic Mountains. Eighth international symposium on Gondwana, Abstracts, Hobart, Tasmania, p. 43

Long WE (1961) Geology of Mount Glossopteris, Central Range of the Horlick Mountains, Antarctica. MSc thesis, Department of Geology and Mineralogy, The Ohio State University, Columbus, OH

Long WW (1962) Sedimentary rocks of the Buckeye Range, Horlick Mountains, Antarctica. Science 136(3513):319–321

Long WE (1964) The stratigraphy of the Ohio Range, Antarctica. PhD dissertation, Department of Geology and Mineralogy, The Ohio State University, Columbus, OH

Long WW (1965) Stratigraphy of the Ohio Range, Antarctica. In: Hadley JB (ed) Geology and paleontology of the Antarctic, vol. 6. Antarctic Research Series. American Geophysical Union, Washington, DC, pp 71–116

McDougall I, Grindley GW (1965) Potassium-argon dates on micas from the Nimrod-Beardmore-Axel Heiberg region. Ross Dependency, Antarctica. New Zealand J Geol Geophys 8:304–313.

McDougall I, Harrison TM (1988) Geochronology and thermochronology by the $^{40}Ar/^{39}Ar$ method. Oxford University Press, New York

McLelland D (1967) Geology of the basement complex, Nilsen Plateau, Antarctica. MSc thesis, University of Nevada, Reno, NV

Miller MF, Mabin M (1998) Antarctic Neogene landscapes-In the refrigerator or in the deep freeze. GSA Today 8(4):1–3

Minshew VH (1965) Potassium-argon age from a granite at Mt. Wilbur, Queen Maud Range, Antarctica. Science 150:741–743

Minshew VH (1966) Stratigraphy of the Wisconsin Range, Horlick Mountains, Antarctica. Science 152:637–638

Minshew VH (1967) Geology of the Scott Glacier and Wisconsin Range areas, central Transantarctic Mountains. Ph.D. dissertation, Department of Geology and Mineralogy, The Ohio State University, Columbus, OH

Minshew VH, Summerson CH (1963) Sedimentary section of the Robert Scott Glacier area, Queen Maud Range, Antarctica. Geol Soc Amer, Spec Paper 76:312

Mirsky A (1969) Geology of the Ohio Range-Liv Glacier area. In: Craddock C (ed) Geologic map of Antarctica, Sheet 17, Folio 12. American Geographical Society, New York

Montigny RJE, Faure G (1969) Contribution au problème de l'homogénéisation isotopique du strontium des roches totales au cours du métamorphisme: Cas du Wisconsin Range, Antarctique. Compte Rondu Acad Sci Paris 268:1012–1015

Murtaugh JG (1969) Geology of the Wisconsin Range batholith, Transantarctic Mountains. New Zealand J Geol Geophys 12(213):526–555

Ray PT (1973) Rubidium-strontium age determinations of intrusive aplite and pegmatite in the Wisconsin Range of the Horlick Mountains, Antarctica. MSc thesis, Department of Geology and Mineralogy, The Ohio State University, Columbus, OH

Stroeven AP, Burckle L, Kleman J, Prentice ML (1998) Atmospheric transport of diatoms in the Antarctic Sirius Group: Pliocene deep freeze. GSA Today 8(4):1, 4–5

Stump E (1995) The Ross orogen of the Transantarctic Mountains. Cambridge University Press, Cambridge

Teller JT (1966) In a land where nothing lives. Cincinnati Alumnus, March, 11–13 and 29–30

Treves SB (1965) Igneous and metamorphic rocks of the Ohio Range, Horlick Mountains, Antarctica. In: Hadley JB (ed) Geology and paleontology of the Antarctic. Antarctic Research Series, vol. 6. American Geophysical Union, Washington, DC, pp 117–125

Webb PN, Harwood DN (1987) Terrestrial flora of the Sirius Formation: Its significance for late glacial history. Antarctic J US 22(4):7–11

Webb PN, Harwood DM, McKelvey BC, Mercer JH, Stott DL (1983) Late Neogene and older Cenozoic microfossils in high-elevation deposits of the Transantarctic Mountains: Evidence for marine sedimentation and ice volume variation of the East Antarctic craton. Antarctic J US 18(5):96–97

Webb PN, Harwood DM, McKelvey BC, Mercer JH, Stott LD (1984) Cenozoic marine sedimentation and ice volume variation on the East Antarctic craton. Geology 12: 287–291

Wilson GS, Harwood DM, Askin RA, Levy RH (1998) Late Neogene Sirius Group strata in Reedy Valley, Antarctica: A multiple-resolution record of climate, ice sheet, and sea-level events. J Glacial 44(148):437–447

Chapter 8
The Far-Eastern Mountains

The Transantarctic Mountains extend to the east beyond the Ohio Range for at least 1,300 km. This far-eastern segment of the Transantarctic Mountains in Fig. 8.1 includes the Thiel Mountains, the Pensacola Mountains, the Argentina Range, the Wichaway Nunataks, the Shackleton Range, and the Theron Mountains. These mountain ranges are separated from each other by large distances covered by the East Antarctic ice sheet. In spite of the remote location of these far-eastern mountains, all of them have been studied in the field by teams of geologists.

The Thiel and Pensacola mountains were mapped by geologists of the US Geological Survey (Schmidt and Ford 1969). The geologic map of the Shackleton Range was published by Clarkson (1982) and by Clarkson et al. (1995) based on field work of geologists from many countries. The Theron Mountains were discovered and described by members of the Commonwealth Trans-Antarctic Expedition (1955–1958; Section 1.5.4; Stephenson 1966).

8.1 Thiel Mountains

The Thiel Mountains are located at 85°15′S and 091°00′W about 200 km east of the Ohio Range. They were first sighted in 1958/59 by members of the over-snow exploratory traverse that traveled from Byrd Station to the Horlick Mountains (Section 7.1). In the following year, a geophysical-survey party landed at the eastern end of the Thiel Mountains and collected rock samples which were later dated by Craddock et al. (1964b). The entire mountain range was subsequently mapped during the 1960/61 and 1961/62 field seasons by geologists of the US Geological Survey (Ford and Aaron 1962) and their findings were incorporated by Schmidt and Ford (1969) into the set of geologic maps edited by Craddock (1969 a, b). More recently in 1983/84, a joint US-British airborne geology field party landed at the Thiel Mountains and made observations which resulted in publications by Storey and Dalziel (1987), Storey and Macdonald (1987), and Pankhurst et al. (1988).

The Thiel Mountains consist of several flat-topped plateaus that are being dissected to form a long and convoluted escarpment that faces north to northeast. The Ford Massif at the western end of the Thiel Mountains in Fig. 8.2 is surrounded by a steep cliff which is more than 400 m high. This cliff contains several ice-carved valleys that define rocky spurs between them including Reed Ridge, as well as Walker and Janulis spurs. The cliff that outlines the Ford Massif merges into the Bermel Escarpment which connects the Massif to a string of nunataks that form the southeastern end of the Thiel Mountains. These nunataks include Mount Walcott (2,154 m), Mount Wrather (2,095 m), Mendenhall Peak (2,131 m), and Smith Knob (2,135 m). The system of cliffs, escarpments, and ice falls measured from the Johnson Nunataks in the west to Nolan Pillar in the east is more than 145 km long.

The geologists of the US Geological Survey who were mapping the Thiel Mountains during the 1961/62 field season discovered two pieces of a meteorite on the ice near Mt. Wrather. The meteorite weighed 28.6 kg and was identified as a pallasite composed of olivine, troilite (FeS), and schreibersite [$(FeNi)_3P$] in a matrix of metallic Fe-Ni alloy (Chapter 18). At the time this specimen was collected, it was the largest of only four meteorites discovered in Antarctica and the only pallasite known from that continent (Ford and Tabor 1971).

G. Faure and T.M. Mensing, *The Transantarctic Mountains: Rocks, Ice, Meteorites and Water*,
DOI 10.1007/978-90-481-9390-5_8, © Springer Science+Business Media B.V. 2010

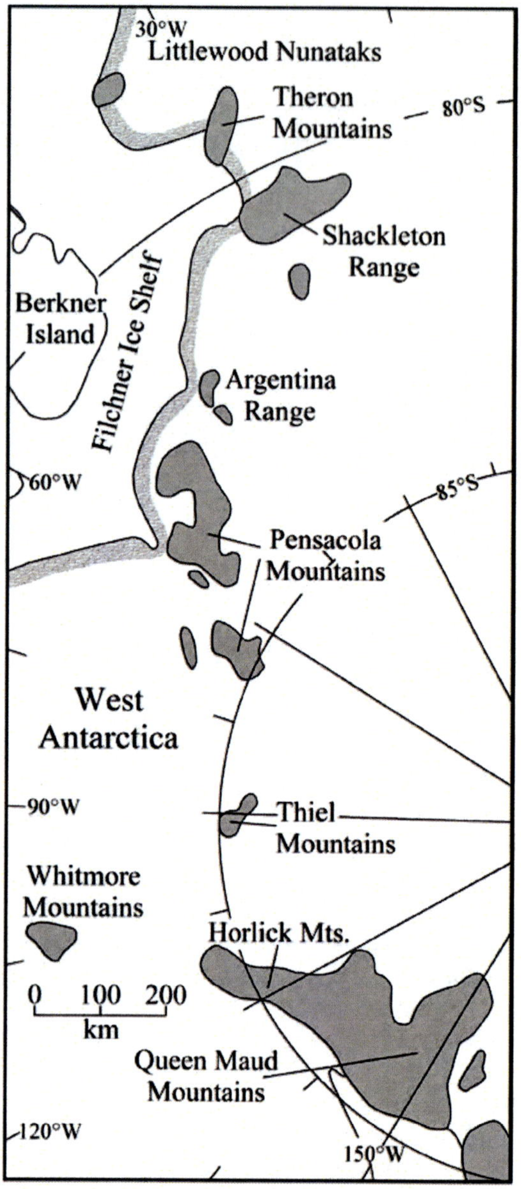

Fig. 8.1 The eastern end of the Transantarctic Mountains consists of the Horlick Mountains, the Thiel Mountains, the Pensacola Mountains, the Shackleton Range, and the Theron Mountains. These mountain ranges are separated from each other by large gaps covered by the East Antarctic ice sheet and in this way occupy about 1,300 km of the total length of the Transantarctic Mountains. All of these distant mountains have been mapped by geologists (e.g., Mirsky 1969; Davis and Blankenship 2005; Schmidt and Ford 1969; Clarkson et al. 1995) (Adapted from Craddock 1982)

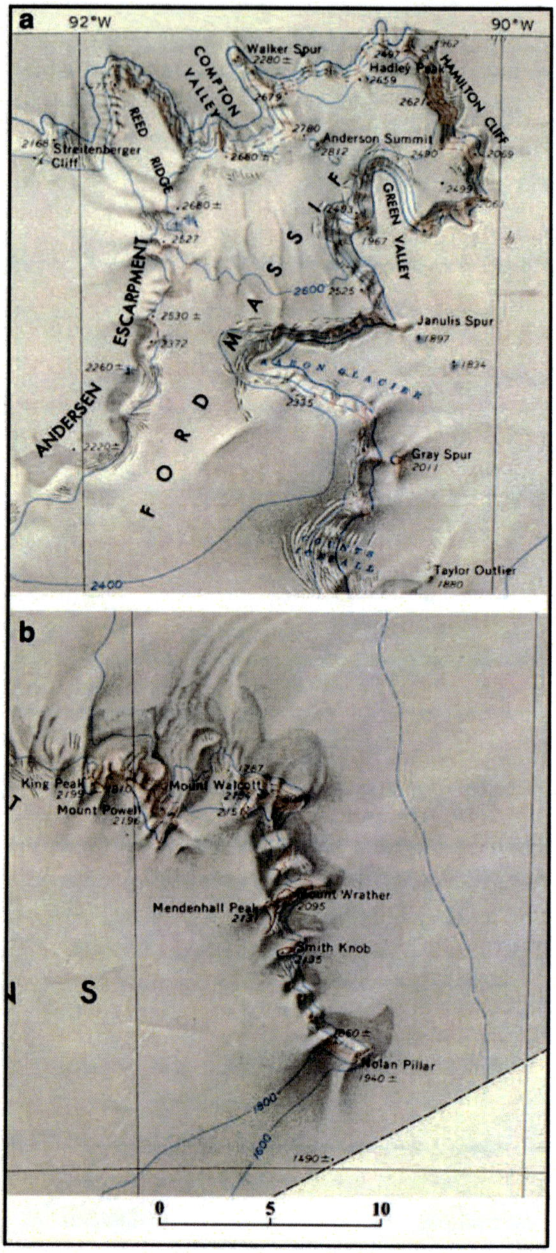

Fig. 8.2 (**a**) Topography of the Ford Massif at the western end of the Thiel Mountains which consists of a long and convoluted system of rocky cliffs and ice falls that extends for about 145 km from northwest to southeast. (**b**) The southeastern end of the Thiel Mountains consists of several peaks that are aligned along an escarpment including Mt. Walcott, Mt. Wrather, Smith Knob, and Nolan Pillar. Excerpted from the topographic map: Thiel Mountains Antarctica, S8500-W8000/1 × 15, 1959 published by the US Geological Survey, Washington, DC

8.1.1 Geology

The rocks exposed in the Thiel Mountains belong to the basement complex of the Transantarctic Mountains. The sedimentary rocks of the Beacon Supergroup that unconformably overlie the basement rocks elsewhere in the Transantarctic Mountains do not occur in the Thiel Mountains either because they have been removed by erosion or because they were not deposited there. The stratigraphy of the basement rocks of the Thiel Mountains in Table 8.1 is deceptively simple compared to the basement rocks elsewhere in the Transantarctic Mountains (e.g., the Queen Maud Mountains and the Wisconsin Range).

8.1.1.1 Mt. Walcott Formation

The oldest rocks in the Thiel Mountains belong to the *Mt. Walcott Formation* which occurs only on Mt. Walcott and Mt. Wrather at the southeastern end of the range. It consists of flat-lying, thin-bedded, volcaniclastic sandstone, pebbly sandstone, and subordinate amounts of mudstone, breccia, and limestone with a thickness of about 100 m (Schmidt and Ford 1969; Storey and Macdonald 1987; Pankhurst et al. 1988; Stump 1995). The section contains a basal volcaniclastic breccia which may have formed when lava that was erupted near the coast flowed into the ocean. The Mt. Walcott Formation was intruded by several dacite sills that are about 15 m thick. The sills exposed on the slopes of Mt. Wrather transgressed up-section by following steeply dipping faults. The sedimentary rocks are also interbedded conformably with a lava flow composed of porphyritic dacite. In spite of the evidence for the eruption of large volumes of volcanic rocks not only in the Mt. Walcott Formation but also in the overlying Thiel Mountain Porphyry, the metamorphic grade of the sedimentary rocks is low.

Schmidt and Ford (1969) mentioned that a thin layer of marble in the Mt. Walcott Formation contains stromatolites. However, Storey and Macdonald (1987) and Pankhurst et al. (1988) later stated that these stromatolites are actually composed of interlaminated calcareous siltstone and limestone that underwent soft-sediment deformation. The same authors did find abundant echinoderm plates and sponge spicules similar to *Protospongia fenestrata* which occurs in the Middle Cambrian Nelson Limestone of the Pensacola Mountains (Weber and Fedorov 1980, 1981). The Mt. Walcott Formation also contains trace fossils (*Planolites* and *Chondrites*). Although *Planolites* has been found in rocks that are 1,000 million years old, the earliest appearance of *Chondrites* is in rocks of Cambrian age. Therefore, Pankhurst et al. (1988) concluded that the paleontological evidence supports a Phanerozoic age for the Mt. Walcott Formation, whereas Schmidt and Ford (1969) had previously assigned it a probable Neoproterozoic age.

The sandstones in the lower part of the Mt. Walcott Formation are interbedded with mudstones; but higher in the section the sandstones are interbedded with pebbly sandstone layers some of which contain lenses of conglomerate. Some of the sandstone beds are cross-bedded and have symmetrical ripple marks on their top surfaces. The uppermost part of a stratigraphic section measured by Storey and Macdonald (1987) is composed of dark silty mudstone containing abundant calcareous concretions and thin silty sandstone beds. These authors concluded from the stratigraphic evidence that the rocks of the Mt. Walcott Formation were deposited in a volcanically active area along the coast that was gradually flooded because of subsidence or because of a rise in sea level. The sedimentary rocks of Mt. Walcott Formation were subsequently covered by a thick layer of the *Thiel Mountain Porphyry* which is up to about 1 km thick. Storey and Macdonald (1987)

Table 8.1 Stratigraphy of the basement rocks of the Thiel Mountains (Schmidt and Ford 1969; Pankhurst et al. 1988; Stump 1995)

Formation	Description
Reed Ridge Granites	Light gray and pink, coarse-grained, biotite granitic rocks in plutons and smaller intrusives
Intrusive contact	
Thiel Mountains Porphyry	Dark-gray, cordierite-and hypersthene-bearing porphyry probably forming a thick sill
Intrusive contact	
Mt. Walcott Formation	Flat-lying, thin-bedded clastic sedimentary rocks of low metamorphic grade interbedded with thin dacite flows or sills. Total thickness about 100 m

as well as Storey and Dalziel (1987) proposed that the Mt. Walcott Formation of the Thiel Mountains is equivalent to the Ackerman Formation in the LaGorce Mountains of the Queen Maud Range (Section 6.4.1), partly because the rocks of the overlying Thiel Mountain Porphyry closely resemble rocks of the Wyatt Formation which we have already encountered in the Nilsen Plateau (Section 6.3.2), in the Scott Glacier area (Section 6.4.1), and in the Wisconsin Range (Section 7.1.2).

Perhaps the most noteworthy aspect of the Mt. Walcott Formation is that it is not folded and therefore seems not to have been subjected to the compression and deep burial that characterize the metasedimentary and metavolcanic rocks of the Ross orogen (Stump 1995). We have seen that the LaGorce Formation in the Wisconsin Range was deformed during the Ross Orogeny and that even the felsic pyroclastic rocks on Todd Ridge of the Long Hills were folded. We assume that the basement rocks of the Ohio Range were also deformed during the Ross Orogeny even though they consist of granitic rocks. Therefore, the Ross orogen that underlies the Transantarctic Mountains from the Wilson Terrane of northern Victoria Land to the Ohio Range of the Horlick Mountains may not underlie the Thiel Mountains.

8.1.1.2 Thiel Mountains Porphyry

The Ford Massif in Fig. 8.2a of the Thiel Mountains consists of cordierite-bearing hypersthene-quartz-monzonite-porphyry described by Ford (1964) which is now known as the Thiel Mountains Porphyry. This rock unit is also exposed intermittently along the Bermel Escarpment and is the principal rock type in the mountains at the southeastern part of the Thiel Mountains in Fig. 8.2b. It also forms Chastain Peak (2,255 m) along the Moulton Escarpment (85°10′S, 94°45′W) which is located 25 km west of the Andersen Escarpment that forms the western edge of the Ford Massif. The Thiel Mountains Porphyry even occurs on Sonntag Nunatak (84°53′S,86°42′W) more than 60 km north of Smith Knob. Accordingly, the Thiel Mountains Porphyry may underlie an area of about 5,000 km^2 most of which is presently covered by the East Antarctic ice sheet. The rock is composed of coarse angular plagioclase crystals in a dark and very fine-grained matrix. In addition, the rock contains phenocrysts of quartz, as well as smaller phenocrysts (or xenocrysts) of potassium feldspar, hypersthene, and cordierite which together comprise about 60% of the rock on average. The matrix is holocrystalline and contains anhydral quartz and alkali feldspar.

Table 8.2 Average abundances of minerals (modal composition) of 11 specimens of the Thiel Mountains Porphyry determined by Ford (1964)

Minerals	Average abundance (%)	Range (%)
Major minerals[a]		
Potassium feldspar	31	24–35
Quartz	31	25–41
Plagioclase	27	24–34
Chlorite	5	1–10
Biotite	4	1–8
Hypersthene	2	0–7
Cordierite	<1	0–1
Accessory minerals		
Zircon, apatite, iron oxide, sphene, and rutile		

[a]Determined by point-counting of thin sections and based on 1,700–2,000 points per slide. The results include the phenocrysts as well as the groundmass

Modal analyses of 11 specimens prepared by Ford (1964) in Table 8.2 identify the Thiel Mountains Porphyry as a quartz monzonite even though it is not a plutonic granitic rock. The plagioclase phenocrysts display oscillatory zoning in angular crystals some of which are fragmented. Quartz crystals are partly resorbed and rounded. Biotite and chlorite are alteration products of hypersthene, whereas cordierite (1–3 mm) occurs at low concentrations throughout the porphyry as rounded phenocrysts, as irregularly-shaped crystals associated with feldspar, and as small anhedral grains in xenoliths of hypersthene-cordierite-labradorite granulite.

Ford (1964) reported that the Thiel Mountains Porphyry contains several types of granulite xenoliths most of which are less than 2 cm in diameter; but one large xenolith of gneissic granulite measured 150 by 60 cm. The presence of xenoliths composed of hypersthene-cordierite granulites is an important clue to the origin of the Porphyry.

The Thiel Mountains Porphyry is essentially a massive body of igneous rocks with locally-developed faint stratification which Ford (1964) attributed to the flow of magma that began to crystallize at depth before it was either intruded at shallow depth in the crust or was erupted at the surface where it cooled rapidly. Although Ford (1964) originally favored the explanation that the Porphyry formed from magma that was intruded at shallow depth, Ford and Sumsion (1971) later concluded that the Thiel Mountains Porphyry is a crystal-rich silicic tuff.

The presence of granulite xenoliths suggests that the magma formed by partial melting of high-grade metamorphic rocks of the East Antarctic craton. According to this hypothesis proposed by Ford (1964), hypersthene and cordierite are xenocrysts derived from granulite or charnockite protoliths. The provenance of cordierite and hypersthene crystals in the Thiel Mountains Porphyry was later confirmed by Ford and Himmelberg (1976) because cordierite is a refractory mineral during magma formation by anatexis and because both cordierite and hypersthene occur in granulite xenoliths embedded within the Porphyry. This hypothesis implies that the Thiel Mountains may be located on the East Antarctic craton in contrast to the Ross orogen which is believed to be located east of the craton for most of its length with the possible exception of the Nimrod Group in the Miller/Geologist ranges.

Ford and Himmelberg (1976) also published the average chemical composition of 27 specimens of the Thiel Mountains Porphyry, which they classified as a rhyodacite, as well as chemical analyses of the cordierite and the orthopyroxene all of which are listed in Appendix 8.5.1. The chemical composition of the Thiel Mountains Porphyry closely resembles the chemical compositions of the metavolcanic rocks of the Wyatt Formation on Mt. Gardiner in the Scott Glacier area and on Metavolcanic Mountain at the head of the Reedy Glacier (Appendix 7.8.1).

Many investigators who have worked on the rocks of the Wyatt Formation (Sections 6.3.2, 6.4.1, and 7.1.2) and on the Thiel Mountains Porphyry have commented on the similarities of their chemical composition, their texture, their massive structure, their modes of emplacement, and their relation to underlying clastic sedimentary rocks (e.g., Vennum and Storey 1987a; Pankhurst et al. 1988; Stump 1995). Even Ford and Sumsion (1971) pointed out that the Thiel Mountains Porphyry and the rocks of the Wyatt Formation are both of pyroclastic origin and formed during an episode of silicic volcanic activity on an "immense scale" because the volume of pyroclastic rocks in the Thiel Mountains alone seems to have exceeded 5,000 km^3.

However, even if the Thiel Mountains Porphyry and the pyroclastic rocks of the Wyatt Formation formed during the same episode of silicic volcanic/magmatic activity, cordierite has only been identified in the Thiel Mountains Porphyry. The presence of this mineral as xenocrysts and in small granulite xenoliths supports the hypotheses proposed by Ford (1964) that the magma of the Thiel Mountain Porphyry formed by anatexis of high-grade metamorphic rocks of the East Antarctic craton. In that case, the Thiel Mountains may be located on a "salient" of the East Antarctic craton which explain why the underlying sedimentary rocks of the Mt. Walcott Formation were not folded during the Ross Orogeny.

8.1.1.3 Reed Ridge Granite

The Thiel Mountains Porphyry of the Ford Massif and of the area near Smith Knob in Fig. 8.2a, and b was intruded by plutons of the Reed Ridge Granite which is a light-gray, coarse grained, porphyritic, and biotite-rich granitic rock that ranges in composition from granodiorite to quartz monzonite. These granite intrusives were named after Reed Spur which extends north from the Ford Massif west of Compton Valley in Fig. 8.2a. The Porphyry is also cut by dikes of aplite and pegmatite that appear to be related to the granitic plutons (Schmidt and Ford 1969).

The petrology and chemical composition of the Reed Ridge Granite in Appendix 8.5.2 were investigated by Vennum and Storey (1987a) in the context of a comprehensive study of the granitic rocks that form the nunataks of the Ellsworth-Whitmore Mountains crustal block. These authors concluded that the Reed Ridge Granites of the Thiel Mountains are representatives of the Granite Harbor Intrusives that were intruded in Cambro-Ordovician time in the course of the Ross Orogeny. The tectonic setting of the Thiel Mountains raises the question of how this mountain range relates to the Patuxent and Neptune ranges of the Pensacola Mountains (Fig. 8.1) which are considered by many investigators to be an extension of the Transantarctic Mountains.

8.1.2 Age Determinations

The first age determinations of the Thiel Mountains Porphyry and of the Reed Ridge Granites published by Ford et al. (1963), Aaron and Ford (1964), Ford (1964), and Schmidt and Ford (1969) were based in part on the lead-alpha method of dating that was described by Gottfried et al. (1959) and summarized by Faure (1986). The Pb-alpha dates of zircon in the Thiel Mountains

Porphyry ranged from 670 ± 50 to 530 ± 60 Ma, which caused Schmidt and Ford (1969) to assign the Thiel Mountains Porphyry a Neoproterozoic age.

The average of three lead-alpha dates of zircon in granodiorites (513 ± 45 Ma, 1σ) confirmed the field evidence that indicated that the granite plutons are younger than the Porphyry. The K-Ar dates of biotite in the granitic rocks reported by Schmidt and Ford (1969) range from 511 ± 26 to 484 ± 24 Ma (i.e., Cambro-Ordovician) and are cooling ages typical of granitic basement rocks in the Ross Orogen of the Transantarctic Mountains. The same authors also reported whole-rock Rb-Sr dates of 634 ± 85, 558 ± 70, and 499 ± 30 Ma for the Reed Ridge Granites (recalculated to $\lambda = 1.42 \times 10^{-11}$ year^{-1}). These dates range widely from Neoproterozoic to Late Cambrian and have large uncertainties arising from the scatter of data points on the Rb-Sr geochronometry diagram. Craddock et al. (1964b) reported a Rb-Sr date of 528 ± 20 Ma for muscovite in a boulder of pegmatite collected near Smith Knob (recalculated to $\lambda = 1.42 \times 10^{-11}$ year^{-1}). The specimen dated by Craddock et al. (1964b) was collected in December of 1959 by E.C. Thiel and E.S. Robinson who landed in the Thiel Mountains (Fig. 8.2b) while carrying out an airborne geophysical and geological reconnaissance survey (Thiel 1961).

The lead-alpha dates of zircon in the Thiel Mountains Porphyry appeared to be confirmed by Eastin (1970) who obtained a whole-rock Rb-Sr date of 632 ± 102 Ma and an initial $^{87}Sr/^{86}Sr$ ratio of 0.7086 ± 0.0059 for specimens of the Porphyry. This date and the initial $^{87}Sr/^{86}Sr$ was derived by an error-weighted regression procedure applied to the analytical data in Appendix 8.5.3. The same data were used by Faure et al. (1977) to calculate a date of 660 ± 79 Ma and an initial ratio of 0.7069. The large uncertainty (± 102 × 10^6 year) of the date calculated by Eastin (1970) arose because the data points do not constrain the slope and intercept of a straight line well enough to calculate a geologically useful crystallization date of the Thiel Mountains Porphyry.

Eastin (1970) also analyzed two samples of biotite granite and three of quartz monzonite from the Thiel Mountains for dating by the Rb-Sr method (Appendix 8.5.3). He used these data to determine the crystallization date of the granite but was unable to fit a straight line to all five data points. Therefore, we excluded a biotite-rich sample of quartz monzonite from Smith Knob (499; A.B. Ford # 7.12·61.64) which had lost radiogenic ^{87}Sr. The four remaining samples of granitic basement rocks closely define an isochron in Fig. 8.3 that yields a date of 530 ± 6 Ma and an $^{87}Sr/^{86}Sr$ ratio of 0.71153 ± 0.00057 (1σ). This date and the initial $^{87}Sr/^{86}Sr$ ratio associate the granitic rocks of the Thiel Mountains with the Granite Harbor Intrusives and thereby indicate that this remote mountains range is an extension of the Transantarctic Mountains.

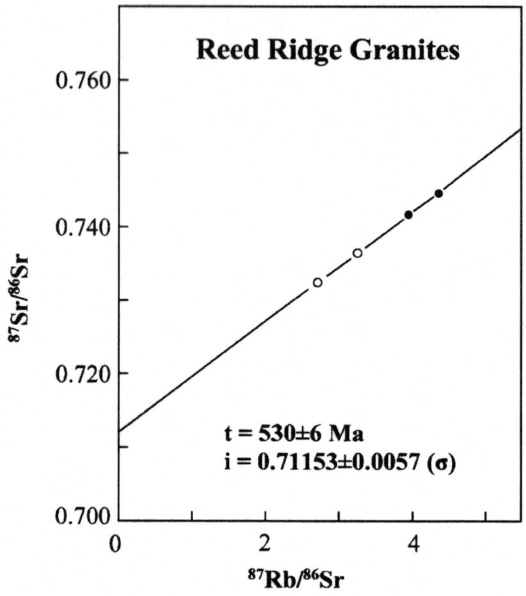

Fig. 8.3 The crystallization age of the granitic basement rocks of the Thiel Mountains was calculated from analytical data of Eastin (1970) listed in Appendix 8.5.3. The rock samples were provided by Arthur B. Ford. Solid circles = quartz monzonite; open circles = biotite granite. One of the five samples (#499) analyzed by Eastin (1970) was excluded because it lost radiogenic ^{87}Sr. The value of the decay constant of ^{87}Rb used to calculate the date from the slope of the isochron was $\lambda = 1.42 \times 10^{-11}$ year^{-1}. The slope and intercept of the isochron were determined by unweighted linear regression

The most recent and most successful whole-rock Rb-Sr age determinations of the basement rocks of the Thiel Mountains were published by Pankhurst et al. (1988) based on analyses of 45 specimens of igneous and sedimentary rocks. Appendix 8.5.4 includes these age determinations and initial $^{87}Sr/^{86}Sr$ ratios as well as the results published earlier by Schmidt and Ford (1969) and by Eastin (1970). However, the lead-alpha dates of zircon and whole-rock Rb-Sr dates of Schmidt and Ford (1969) were excluded from this tabulation.

The whole-rock Rb-Sr dates Appendix 8.5.4 have larger error limits up to about 13% because the rocks used for dating scatter on the isochron diagram more than is consistent with analytical errors. Therefore, the U-Pb date of zircon reported by Van Schmus et al. (1998) for a sample of the Thiel Mountains Porphyry is a valuable check on the whole-rock Rb-Sr date of this rock determined by Pankhurst et al. (1988). The zircon fractions analyzed by Van Schmus et al. (1998) define a discordia line that intersects concordia at two points (Section 3.7.5). The lower intercept at 500 ± 4 Ma presumably refers to the time of crystallization of zircon overgrowths, whereas the upper intercept of 1093 ± 23 Ma is the approximate average age of xenocrystic zircon cores. The lower-intercept date is identical to the whole-rock Rb-Sr date derived by Pankhurst et al. (1988) by regressing all 45 data points they analyzed of the igneous and sedimentary rocks from the Thiel Mountains. These authors considered the possibility that the Rb-Sr systematics of the Mt. Walcott Formation and of the Thiel Mountains Porphyry had been reset when these rocks were reheated during the intrusion of the Reed Ridge Granite. However, the wide spacing of the collecting sites makes such large-scale isotopic homogenization unlikely. Therefore, we are left with the suggestion that the basement rocks of the Thiel Mountains formed during an interval of about ten million years about 500 million years ago.

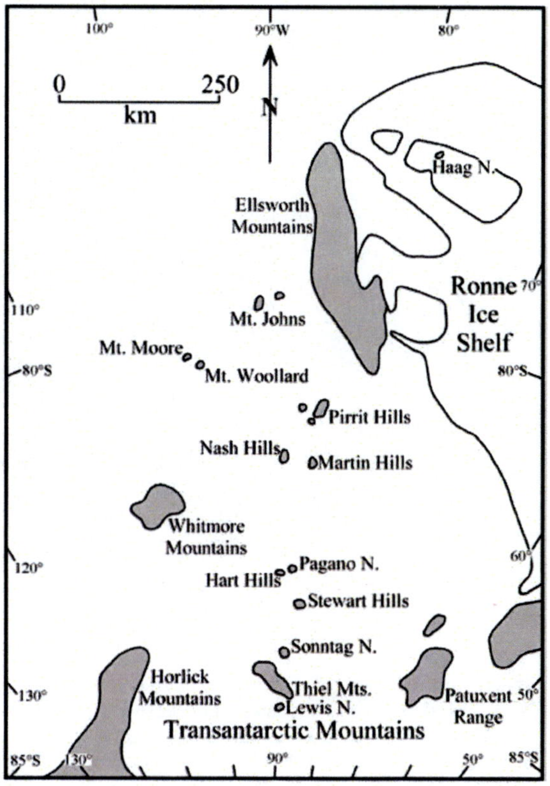

Fig. 8.4 The granitic plutons of the nunataks located in a belt in West Antarctica between the Thiel Mountains and the Ellsworth Mountains were intruded during the Jurassic Period. The dates of these plutons are listed in Appendix 8.6.5 (Adapted from Vennum and Storey 1987a)

8.1.3 Ellsworth-Whitmore Mountains

The hypothetical salient of the East Antarctic craton may also have played a role in the formation of small magmatic centers in the region between the Thiel Mountains and the Ellsworth Mountains identified in Fig. 8.4 that were studied by Vennum and Storey (1987a, b), Grunow et al. (1987), and by Dalziel et al. (1987). Vennum and Storey (1987a) concluded that the granitic intrusives of the nunataks in this area are cupolas composed of highly evolved granitic rocks of much larger bodies of less evolved granitic rocks of Mesozoic age that are present at depth in the crust. The whole-rock Rb-Sr dates reported by Millar and Pankhurst (1987) of the granitic rocks of these nunataks range from 195 to 173 Ma (Fig. 8.5).

The only exception are the aplogranites and microgranites of the Haag Nunataks (77°00′S, 078°18′W) in West Antarctica which crystallized at 1058 ± 53 and 1003 ± 18 Ma, respectively. A similar whole-rock Rb-Sr date of 980 ± 16 Ma (recalculated to $\lambda = 1.42 \times 10^{-11}$ year^{-1}) was reported by Eastin and Faure (1971) for rhyolites in the Littlewood and Bertrab nunataks at 77°53′S, 034°10′W in Coats Land on the east coast of the Weddell Sea (Section 9.4). However, the relationship between the Haag Nunataks and the small nunataks of Coats Land can only be understood in the context of the rifting of Rodinia and the departure of Laurentia which are the subjects of Chapter 9.

The Ellsworth Mountains of West Antarctica in Fig. 8.4 are composed of a thick sequence of highly folded and low-grade metasedimentary and metavolcanic rocks which range in age from Proterozoic to Permian (Craddock et al. 1964a, 1986; Craddock 1969 a; Schopf 1969; Webers et al. 1992). Although volcanic rocks do occur in the Heritage Group of Neoproterozoic

Fig. 8.5 Historical photograph of a US Navy R4D-8 (modified DC-3 on skis) that landed on January 23, 1960 in the Nash Hills at 81°53′S, 89°23′W in West Antarctica. The scientists on that flight were Edward Thiel, Campbell Craddock, Edwin Robinson, and Henry Brecher (This photo by Henry Brecher was published with his permission)

to Cambrian age (Craddock 1969), the Ellsworth Mountains were not intruded by granite plutons of Neoproterozoic age (e.g., the Haag Nunataks) or of Jurassic age (e.g., the Whitmore Mountains and related nunataks of the Ellsworth-Whitmore Mountains block).

8.1.4 Summary

The Thiel Mountains are part of the far-eastern mountains that extend from the Ohio Range of the Horlick Mountains to the east coast of the Weddell Sea. The rocks exposed along the north-facing escarpment of the Thiel Mountains consist of undeformed sedimentary rocks of low metamorphic grade (Mt. Walcott Formation) that are overlain by the cordierite-bearing hypersthene-quartz-monzonite porphyry of the Thiel Mountains Porphyry. This unit is a massive deposit of crystal-rich silicic tuff that contains cordierite and hypersthene as xenocrysts and in small granulite xenoliths. The presence of cordierite in this volcanic or shallow intrusive body is significant because this mineral occurs primarily in high-grade metamorphic rocks rather than in pyroclastic or intrusive igneous rocks. However, cordierite is a refractory mineral that can occur in magma that formed by partial melting of high-grade metamorphic rocks. Therefore, the presence of cordierite in the Thiel Mountains Porphyry is an indication that the magma that was erupted to form the Porphyry originated by partial melting of granulite in the underlying East Antarctic craton. According to this hypothesis proposed by Ford (1964), the rocks of the Thiel Mountains formed on the East Antarctic craton and therefore were not deformed during the Ross Orogeny.

The Thiel Mountains also contain plutons of granitic rocks that intruded the basal sedimentary rocks and the overlying pyroclastic porphyry during the Cambrian Period. The stratigraphic position, chemical composition, crystallization age, and initial $^{87}Sr/^{86}Sr$ ratio identify the Reed Ridge Granite as a member of the Granite Harbor Instrusives that characterize the basement rocks of the Transantarctic Mountains along their entire length.

8.2 Pensacola Mountains

The Pensacola Mountains in Fig. 8.6 are located along the southern coast of the Ronne Ice Shelf and about 175 km northeast of the Thiel Mountains. This large and complex range of mountains is about 375 km long and includes the Patuxent, Neptune, and Forrestal ranges as well as the Dufek Massif.

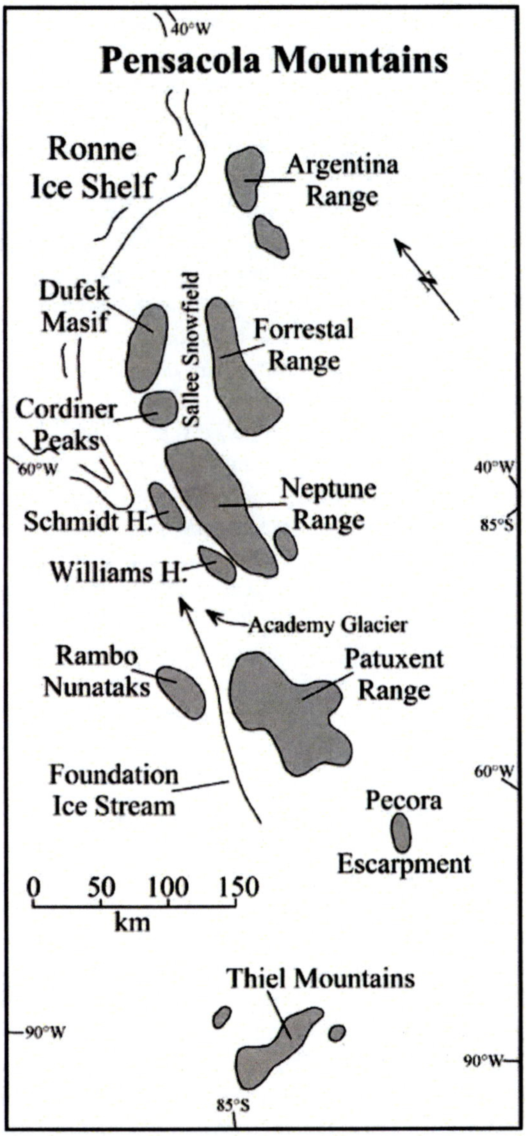

Fig. 8.6 The Patuxent Range of the Pensacola Mountains is separated from the Thiel Mountains by a distance of about 200 km that is covered by the East Antarctic Ice Sheet. The only other mountains close to the Patuxent Range are the Pecora Escarpment and the Rambo Nunataks. The northern part of the Pensacola Mountains includes the Neptune Range, the Forrestal Range, and the Dufek Massif (Adapted from the Geologic Map of Antarctica in Craddock 1982)

8.2.1 Discovery and Exploration

The Pensacola Mountains were discovered on January 13, 1956, during a transpolar reconnaissance flight by the US Navy. They were observed again on October 23, 1957, by Captain Finn Ronne of the US Navy and Dr. Edward C. Thiel during a flight from Ellsworth Station to the Dufek Massif.

The first geological observations of the Dufek Massif were made in December of 1957 by Aughenbaugh (1961) and Walker (1961) during a traverse of a tractor train under the auspices of the IGY. Subsequently, all parts of the Pensacola Mountains were mapped by geologists of the US Geological Survey during the field seasons of 1962/63, 1963/64, 1965/66 (Schmidt and Ford 1963, 1966; Schmidt 1964, 1969; Schmidt et al. 1964, 1965). The work of these American geologists culminated in the publication of geological maps by Schmidt and Ford (1969), Schmidt et al. (1978), and Ford et al. (1978a, b). More recently, several other geological field parties have worked in the Pensacola Mountains:

German Democratic Republic and USSR: 1978/79, 1979/80
US–UK (airborne): 1987/88
US (ground based): 1989/90
US (ground based): 1993/94

The publications arising from these studies were listed by Stump (1995) in a review of the geology of the Pensacola Mountains

8.2.2 Topography

The Patuxent Range in Fig. 8.6 consists of a large number of nunataks that project through the East Antarctic Ice Sheet in an area of about 6,000 km^2.

The area between the Thiel Mountains and the Patuxent Range in Fig. 8.6 is covered by the East Antarctic ice sheet which flows through this gap into the Ronne Ice Shelf. The only other mountain ranges associated with the Patuxent Range are the Pecora

Escarpment and the Rambo Hills identified in Fig. 8.6. The Rambo Hills are separated from the Patuxent Range by the Foundation Ice Stream which is about 40 km wide at this location.

The rocks exposed in the Pecora Escarpment consist of Permian siltstone and shale of the Pecora Formation which was intruded by Middle Jurassic sills of the Ferrar Dolerite. The Rambo Nunataks are composed of argillaceous sandstone and shale of the Patuxent Formation which is the principal lithology of the basement complex of the Patuxent Range (Schmidt and Ford 1969).

The Neptune Range is located 60 km northeast of the Patuxent Range and is separated from it by the Academy Glacier which flows from the polar plateau and joins the Foundation Ice Stream before it reaches the Ronne Ice Shelf. The main body of the Neptune Range covers an area of about 4,900 km^2 and is surrounded by the Schmidt Hills and the Williams Hills on its western side and by numerous clusters of nunataks on the eastern side. The topographic maps published by the US Geological Survey indicate that the elevations of the peaks in the Patuxent Range and of the Neptune Range are generally less than 2,000 m above sea level. The most noteworthy topographic feature of the Neptune Range is the Washington Escarpment which extends for about 90 km along its eastern edge and connects the main body of the Neptune Range to the Cordiner Peaks located south of the Dufek Massif.

The northern Pensacola Mountains are dominated by the Dufek Massif which consists of a very large body of layered gabbro and ultramafic rocks of Jurassic age (Ford 1970; Ford and Kistler 1980). This body extends to the southeast across the Sallee Snowfield in Fig. 8.6 to the Forrestal Range where it covers the folded sedimentary rocks of Paleozoic age that also occur in the Neptune Range. The structure and origin of the Dufek Massif is presented in Chapter 13.

8.2.3 Stratigraphy

The geologic map of the Pensacola Mountains prepared by Schmidt and Ford (1969) lists an extensive sequence of sedimentary and interbedded volcanic rocks identified in Table 8.3. Schmidt and Ford (1969) observed that the oldest rocks in all parts of the Pensacola Mountains are interbedded layers of graywacke and shale which they named the Patuxent Formation. The rocks of the Patuxent Formation are tightly folded throughout the Pensacola Mountains and, in the eastern Neptune Range, are unconformably overlain by the fossiliferous Nelson Limestone of Middle Cambrian age. The stratigraphic position of the Patuxent Formation at the base of the succession in the Neptune Range, its structural deformation, absence of fossils, and its similarity to other volcano-sedimentary complexes in the basement of the Transantarctic Mountains (e.g., the Goldie and LaGorce formations) suggested that it was deposited during the Neoproterozoic Era. However, age determinations by Millar and Storey (1995) of felsites interbedded with the Patuxent Formation eventually indicated that the rocks of the Patuxent Formation in the nunataks of the Schmidt and Williams hills west of the main body of the Neptune Range are younger than the rocks of this formation in the eastern Neptune Range.

Rowell et al. (2001) resolved this dilemma by placing the sedimentary basement rocks of the eastern Neptune Range into the Hanna Ridge Formation (pre-Middle Cambrian) and by restricting the former Patuxent Formation (post-Middle Cambrian) to the sedimentary and volcanic basement rocks of the Schmidt Hills, the Williams Hills, the Rambo Nunataks, and the Patuxent Range. Accordingly, the base of the Hannah Ridge Formation may be Neoproterozoic as proposed by Schmidt and Ford (1969), whereas the redefined Patuxent Formation is Late Cambrian as postulated by Rowell et al. (1994). Therefore, the Hannah Ridge Formation and the Patuxent Formation are both listed in Table 8.3.

The metasedimentary rocks of the Patuxent Formation in the Schmidt and Williams hills are locally interbedded with lava flows of pillow basalt and felsite and were intruded by diabase sills. The metasedimentary rocks of the Hannah Ridge Formation in the eastern Neptune Range are not interbedded with basalt and felsite but are also composed of alternating layers of graywacke and slate that are difficult to distinguish in the field from the rocks of the Patuxent Formation. In spite of the lithologic similarity, the Hannah Ridge Formation is **older** than the redefined Patuxent Formation (Table 8.3).

8.2.3.1 Patuxent Range

The basement rocks of the Patuxent Range in Fig. 8.7 consists almost exclusively of the isoclinally folded metasedimentary rocks of the Patuxent Formation (redefined) and are **not** interbedded with felsite and basalt flows; but they are cut by thin porphyritic

Table 8.3 Stratigraphy of the Paleozoic sedimentary and volcanic rocks of the Patuxent, Neptune, and southern Forrestal ranges (Schmidt and Ford 1969; Barrett et al. 1986; Millar and Storey 1995; Van Schmus et al. 1997; Rowell et al. 2001)

Rock unit	Age	Description	Occurrence
Ferrar Dolerite sills	Jurassic	Intruded Permian siltstone and shale	Pecora Escarpment
Ferrar Gabbro intrusion	Jurassic	Layered gabbro and ultramafic rocks	Dufek Massif and Forrestal Range
Intrusive Contact			
Pecora Fm.	Permian	Siltstone and shale with coaly layers	Pecora escarpment, southern Forrestal Range
Gale Mudstone	Permian	Tillite, black and homogeneous	Neptune Range, several sites
Disconformity?			
Sandstone	Devonian	Quartz sandstone interbedded with carbonaceous beds	Patuxent Range (south end)
Dover Sandstone	Devonian	Quartz sandstone, medium to coarse	Neptune Range Cordiner Pks. Forrestal Range (southern)
Disconformity?			
Neptune Group	Devonian	Heiser Sandstone Elbow Fm. Elliott Sandstone Brown Ridge Congl.	Neptune Range Cordiner Pks. Forrestal Range (southern)
Unconformity			
Granitic rocks (Granite Harbor Instrusives?)	Cambro/Ordovicain	Gray or pink, coarse, biotite granitic plutons and dikes	Neptune Range (Washington Escarp.) also Thiel Mtns.
Mt. Hawkes Rhyolite	Cambro/Ordovician	Rhyolite porphyry sills and hypabyssal plutons	Neptune Range (southern)
Wiens Fm.	Cambrian	Dark green and reddish brown shale, siltstone, oolitic ls.	Neptune Range (southern)
Gambacorta Fm.	Cambrian	Rhyolite flows, pyroclastic rocks, sandstone, and conglomerate	Neptune Range (southern)
Unconformity?			
Felsic and mafic volcanic rocks in the Patuxent Fm.	Cambrian	Felsic mega-breccias and pillow basalt	Schmidt Hills, Williams Hills, Patuxent Range (southern)
Patuxent Formation	Cambrian	Argillaceous sandstone/shale	Schmidt Hills, Williams Hills, Rambo Nunataks, Patuxent Range
Hannah Ridge Formation	Early Cambrian to Neoproterozoic	Argillaceous sandstone/shale	Eastern Neptune Range
Base not exposed			

[a] Schmidt and Ford (1969) estimated that the total thickness of the sedimentary rocks and lava flows in the Pensacola Mountains is between 14.3 and 16.3 km

olivine-clinopyroxene-biotite lamprophyre dikes. Three K-Ar dates of biotite in these dikes, determined by S.S. Goldich and reported by Schmidt and Ford (1969), average 232 Ma (Middle Triassic; IUGS 2002).

8.2.3.2 Schmidt and Williams Hills

The geology of the Schmidt and Williams hills differs from the geology of the main body of the Neptune Range in three significant ways (Schmidt and Ford 1969):

1. The Patuxent Formation (redefined) in the Schmidt and Williams hills contains interbedded lava flows of pillow basalt and felsite (e.g., the Gorecki Felsite Member), whereas the Hannah Ridge Formation in the eastern Neptune Range does not contain interbedded volcanic rocks.
2. The Hannah Ridge Formation in the eastern Neptune Range is unconformably overlain by the fossiliferous Middle Cambrian Nelson Limestone followed by a succession of Paleozoic formations identified in Table 8.3. The Nelson Limestone and the overlying Paleozoic formations are not present in the Schmidt and Williams hills.
3. The main body of the Neptune Range contains plutons of the Serpan Granite and Gneiss which do not occur in the nunataks west of the main part of the Range.

These abrupt differences in the geology suggests that the original distance between western nunataks

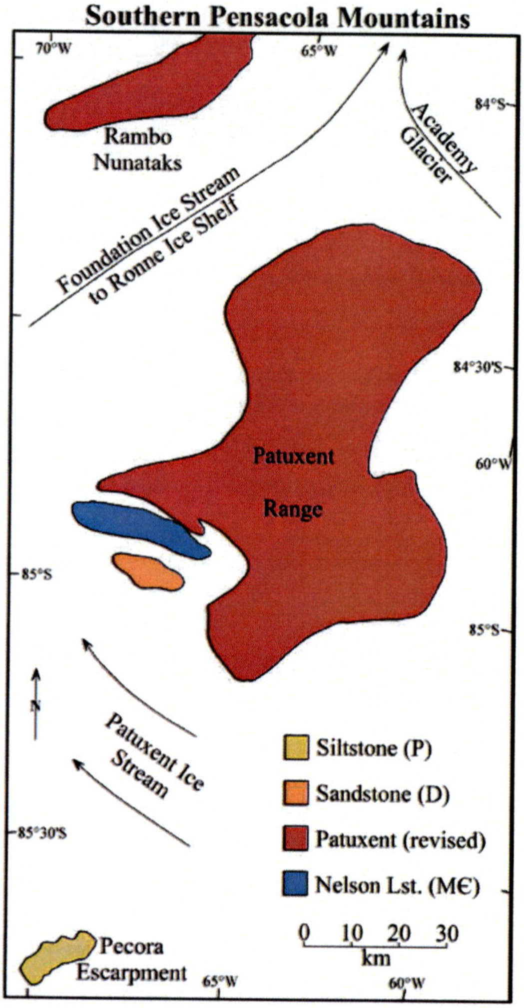

Fig. 8.7 The Patuxent Range forms the most southerly part of the Pensacola Mountains. It consists of nunataks composed of the Patuxent Formation (as revised by Rowell et al. 2001) which forms the local basement complex. The metasedimentary rocks of the Patuxent Formation are isoclinally folded and consist of a repetitious sequence of graywacke and slate. These rocks are younger than the fossiliferous Nelson Limestone which is late Middle Cambrian. The map area also contains outcrops of Devonian sandstone south of the Patuxent Range and of Permian siltstone and shale of the Pecora Formation in the Pecora Excarpment (Adapted from Schmidt and Ford 1969 with information from Rowell et al. 2001)

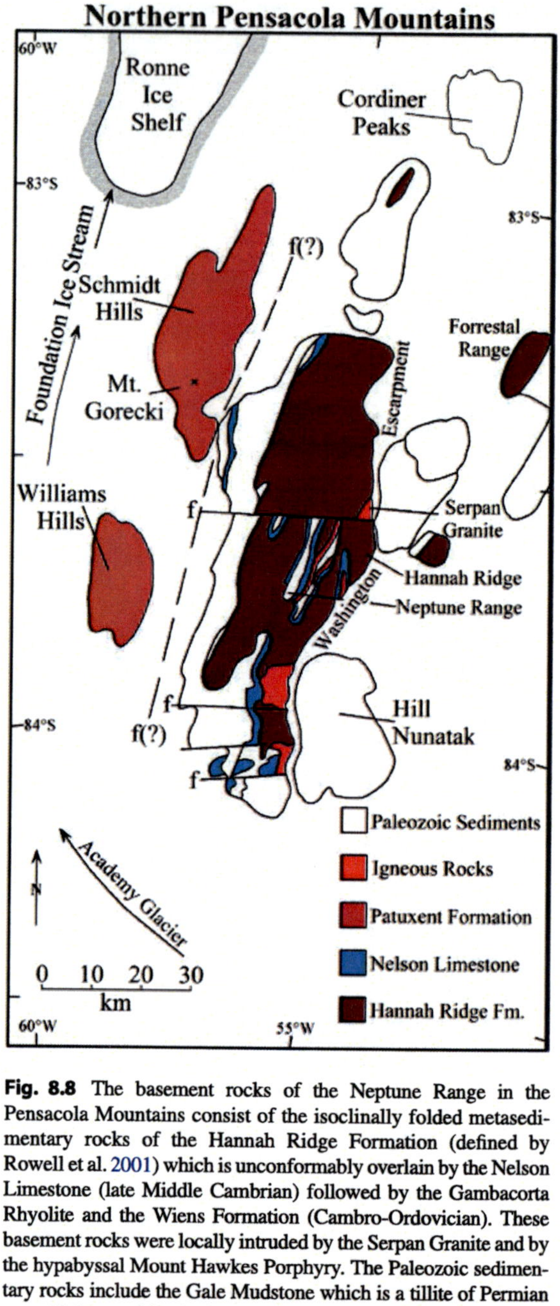

Fig. 8.8 The basement rocks of the Neptune Range in the Pensacola Mountains consist of the isoclinally folded metasedimentary rocks of the Hannah Ridge Formation (defined by Rowell et al. 2001) which is unconformably overlain by the Nelson Limestone (late Middle Cambrian) followed by the Gambacorta Rhyolite and the Wiens Formation (Cambro-Ordovician). These basement rocks were locally intruded by the Serpan Granite and by the hypabyssal Mount Hawkes Porphyry. The Paleozoic sedimentary rocks include the Gale Mudstone which is a tillite of Permian age which resembles the Buckeye Tillite of the Beacon Supergroup in the Ohio Range of the Horlick Mountains (Adapted from Schmidt and Ford 1969 with information from Stump 1995)

and the main body of the Neptune has decreased by movement along a fault in the Roderick Valley which presently separates these crustal blocks in Fig. 8.8 (Schmidt et al. 1978).

8.2.3.3 Neptune Range

The sedimentary rocks of the Hannah Ridge Formation in the Neptune Range were folded and eroded prior to the deposition of the marine Nelson Limestone during the Middle Cambrian. The limestone is interbedded with the overlying Gambacorta Formation consisting of rhyolitic lava flows, pyroclastic deposits, and beds of volcaniclastic sandstone and conglomerate. The thick pile of volcanic rocks of the Gambacorta Formation is followed by Cambrian red shale, siltstone, and fine-grained sandstone with minor oolitic limestone of the Wiens Formation. This sequence of Cambrian sedimentary and volcanic rocks was intruded by the hypabyssal Mount Hawk Rhyolite Porphyry and by a pluton of the Serpan Granite and Gneiss of Cambro-Ordovician age.

After an interval of erosion, the Cambrian limestone, rhyolite flows and related sedimentary rocks of the Gambacorta and Wiens formations were overlain by Ordovician or younger sedimentary formations of the Neptune Group (Table 8.3) which consists of the Brown Ridge Conglomerate, the Elliott Sandstone, the Elbow Formation, and the Heiser Sandstone in order of decreasing age and increasing stratigraphic position (Schmidt and Ford 1969). The composition of the post-Cambrian formations of the Neptune Group was described by Williams and Ford (1968) and by Williams (1969).

The formations of the Neptune Group are succeeded by the Devonian Dover Sandstone and other thick-bedded quartz sandstones and quartzite which are disconformably overlain by the Permian Gale Mudstone and by siltstones and shales of the Pecora Formation. Schmidt and Ford (1969) described the Gale Mudstone as a black, homogeneous tillite containing scattered cobbles of igneous and sedimentary rocks. This unit was deposited over the entire area of the Neptune Range, the Forrestal Range, and the Cordiner Peaks in the Pensacola Mountains. It resembles the tillites that were deposited by the Permian continental ice sheets in the Horlick Mountains and elsewhere in the Transantarctic Mountains (Gunn and Warren 1962; Long 1965).

The Gale Mudstone is stratigraphically overlain by well-bedded tan-colored siltstones and shales of Permian age of the Pecora Formation including coaly layers that contain fossilized *Glossopteris* leaves similar to those that have been described in the rocks of the Beacon Supergroup elsewhere in the Transantarctic Mountains (Schopf 1964). Evidently, the Devonian Sandstones, Gale Mudstone (tillite), and the Pecora Formation of the Neptune and Forrestal ranges resemble the rocks of the Beacon Supergroup. However, all of the sedimentary and volcanic rocks of the Neptune Range that rest unconformably on the Hannah Ridge Formation were folded during the Late Permian Weddell (or Gondwanian) Orogeny (Ford 1972). Even the isoclinally folded sedimentary rocks of the Hannah Ridge Formation of the Neptune Range were deformed during this event. However, the Permian siltstones and shales in the Pecora Escarpment located only about 80 km south of the Patuxent Range in Fig. 8.6 were not folded, which suggests that the Weddell Orogeny was concentrated in the area of the Weddell Sea. The rocks of the Beacon Supergroup to be discussed in Chapter 10 are likewise not structurally deformed except locally along faults.

The extensive literature on the geology of the Pensacola Mountains includes contributions by Schmidt et al. (1963, 1965, 1969), Schopf (1964, 1968, 1969), Huffman and Schmidt (1966), Schmidt and Williams (1967, 1969), Williams and Ford (1968), Williams (1969), Schmidt and Ford (1969), Beck and Schmidt (1971), Ford (1972), Behrendt et al. (1966a, b, 1973), Cameron and Ford (1974), Ford et al. (1978a, b), Schmidt et al. (1978), Weber and Fedorov (1980, 1981), Weber (1982), Frischbutter (1981, 1982), Frischbutter and Vogler (1985), Boyd (1991), Storey et al. (1992, 1996), Rowell et al. (1994, 2001), Stump (1995), Millar and Storey (1995), Van Schmus et al. (1997), Leat et al. (2001), and Curtis (2002).

8.2.4 Age Determinations (Rb-Sr Method)

The Pensacola Mountains originally posed a problem concerning the age and stratigraphic position of the Patuxent Formation which consists of turbidites that were presumably deposited along the passive rift margin of Gondwana much like the folded metasedimentary rocks that dominate the basement complex throughout the Transantarctic Mountains (Storey et al. 1992). Therefore, the age of the Patuxent Formation was assumed to be Neoproterozoic supported by the fact that it is unconformably overlain by the Nelson Limestone which contains trilobites of late Middle Cambrian age (Schmidt and Ford 1969; Palmer and Gatehouse 1972; Rowell et al. 1992, 2001).

8.2.4.1 Patuxent/Hannah Ridge Formations

The rocks of the *Patuxent/Hannah Ridge* formations have been difficult to date by the whole-rock Rb-Sr method because the interbedded graywacke and shale layers did not have the same initial $^{87}Sr/^{86}Sr$ ratios at the time of deposition and because radiogenic ^{87}Sr that formed by decay of ^{87}Rb may have been redistributed during subsequent metamorphism and tectonic deformation of the rocks.

Nevertheless, Eastin (1970) attempted to measure the ages of the siliciclastic sedimentary basement rocks and of the igneous rocks that are associated with it by the whole-rock Rb-Sr method using rock samples provided by Dwight L. Schmidt of the US Geological Survey. Eastin's data and brief descriptions of the samples are compiled in Appendices 8.5.5 and 8.5.6 and his interpretations are illustrated by diagrams included in the text. All of the dates calculated by Eastin (1970) have been recalculated to the modern value of the decay constant of ^{87}Rb ($\lambda = 1.42 \times 10^{-1}$ year^{-1}) and therefore supersede dates that were quoted previously by Faure et al. (1968, 1977).

Eastin (1970) analyzed 11 samples of laminated siltstone, slate, and subgraywacke of the Patuxant Formation that were collected by D.L. Schmidt from outcrops in the Patuxent Range that had not been altered by contact metamorphism. Nevertheless, the data points representing these samples scatter far beyond analytical error above and below line A in Fig. 8.9 that was fitted by Eastin (1970) to all of the samples by the method of York (1966, 1967). Line B is defined only by samples 437 and 444, whereas line C is constrained by samples 439, 441, 436, 440, and 446. The dates and initial $^{87}Sr/^{86}Sr$ ratios that Eastin (1970) derived from these straight lines are:

Line	Date, Ma	$(^{87}Sr/^{86}Sr)_i$
A	387 ± 10 (11)	0.7169 ± 0.0024
B	427 ± 12 (2)	0.7173 ± 0.0006
C	375 ± 7 (5)	0.7154 ± 0.0008

Eastin (1970) concluded that these Rb-Sr dates were reset by the last metamorphic event that affected the rocks of the Patuxent Range and that none of the dates represent the time of deposition these sedimentary rocks.

Another way to determine the age of the Patuxent Formation is to date the felsic lava flows that are interbedded with the sedimentary rocks of the Patuxent Formation in the Schmidt and Williams hills. However, the Rb-Sr data for eight samples of felsic igneous rocks in Appendices 8.5.8 and 8.5.9 scatter widely in Fig. 8.10. Eastin (1970) interpreted these data by fitting three straight lines based on geometrical considerations. The dates and initial $^{87}Sr/^{86}Sr$ ratios that he derived from these lines are:

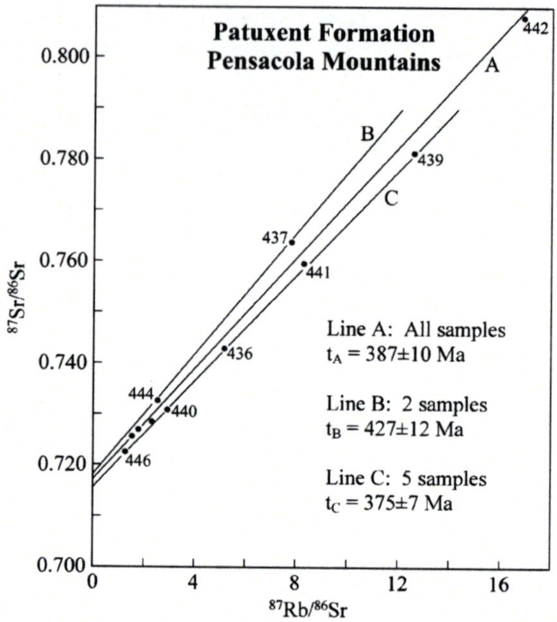

Fig. 8.9 Samples of graywacke and slate of the Patuxent Formation in the Patuxent Range scatter on the Rb-Sr geochronometry diagram. Line A which includes all 11 samples yields: t_A = 387 ± 10 Ma, $(^{87}Sr/^{86}Sr)_i$ = 0.7169. Line B is defined by samples 437 and 444 yields: t_B = 427 ± 12 Ma, $(^{87}Sr/^{86}Sr)_i$ = 0.7173 ± 0.0006. Line C is constrained by samples 439, 441, 436, 440, and 446 corresponds to: t_C = 375 ± 7 Ma, $(^{87}Sr/^{86}Sr)_i$ = 0.71555 ± 0.00079. All three dates are less than the time of deposition of the Patuxent Formation which must be older than Middle-to-Late Cambrian based on the presence of trilobite fossils in the overlying nNelson Formation (Palmer and Gatehouse 1972; Rowell et al. 1992; Evans et al. 1991). The samples of the Patuxent Formation were collected by D.L. Schmidt and were analyzed by Eastin (1970) whose results and sample descriptions are listed in Appendices 8.5.6 and 8.5.7, respectively. The decay constant of ^{87}Rb was $\lambda = 1.42 \times 10^{-11}$ year^{-1}

Line	Date, Ma	$(^{87}Sr/^{86}Sr)_i$
A	465 ± 2 (2)	0.7074 ± 0.0002
B	489 ± 8 (2)	0.7171 ± 0.0001
C	1,184 ± 76 (4)	0.7052 ± 0.0008
A&B	478 ± 6	

The weighted average date derived by pooling the slopes of lines A and B is 478 ± 6 Ma. This Early Ordovician date (IUGS 2002) differs markedly from the much older date of 1184 ± 76 Ma that is derivable from

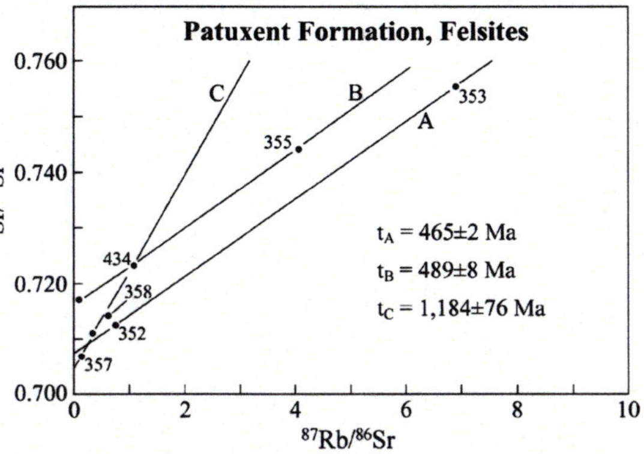

Fig. 8.10 Felsic flows and intrusive plugs in the Patuxent Formation of the Schmidt and Williams hills scatter widely on the Rb-Sr geochronometry diagram. Nevertheless, Eastin (1970) fitted three lines to the date points labeled A, B, and C. The dates and the initial ($^{87}Sr/^{86}Sr$) ratios indicated by these lines are: A: 489 ± 8 Ma, 0.7074 ± 0.0002, B: 465 ± 2 Ma, 0.7171 ± 0.0001, C: 1184 ± 76, 0.7052 ± 0.0008

line C defined by four samples two of which originated from the Schmidt Hills (354 and 358) and two from the Williams Hills (356 and 434). There is no apparent connection between these samples, which raises doubts about the validity of the date indicated by line C.

Three felsite samples from the Gorecki Rhyolite (352, 354, 358) which is interbedded with the Patuxent Formation on Mount Gorecki in the Schmidt Hills in Fig. 8.8 (Rowell et al. 1994) are not suitable for dating by the Rb-Sr method because the data points scatter irregularly in coordinates of $^{87}Rb/^{86}Sr$ and $^{87}Sr/^{86}Sr$.

The Patuxent Formation in the Schmidt and Williams hills (but not in the main body of the Neptune Range) contains voluminous basalt flows and diabase sills. These rocks provide yet another opportunity to date the time of deposition of the Patuxent Formation. Eastin (1970) analyzed six samples of diabase (Schmidt Hills) and four of basalt (Williams Hills) and attempted to use the results in Appendices 8.5.10 and 8.5.11 to date these rocks even though their low Rb/Sr ratios make them unfavorable for dating by the whole-rock Rb-Sr method. Four of the diabase samples originated from different levels within a 55-m thick sill on Mount Gorecki.

One of the samples from this sill (# 421) has the highest Rb/Sr ratio (0.241) among these samples, perhaps because it was collected only 1.8 m below the top of the sill. Eastin (1970) used this sample to define two straight lines (A and B, not shown) from which he derived two dates:

Line	Date, Ma	($^{87}Sr/^{86}Sr)_i$
A	762 ± 59 (4)	0.7065 ± 0.0003
B	1240 (2)	0.7020

Line A is constrained by the four samples of diabase derived from the sill on Gorecki Peak, whereas line B is based on only two samples. A third possible interpretation of the Rb-Sr data of these mafic rocks is based on what may be a fortuitous alignment of two diabase samples (419, 421) from the sill on Mount Gorecki in the Schmidt Hills and one basalt (415) from Nunatak #8 in the Williams Hills. These three samples closely constrain a straight line in Fig. 8.11 which yields a date of 542 ± 3 Ma (Early Cambrian), and an initial $^{87}Sr/^{86}Sr$ ratio of 0.70688 ± 0.000007 (1σ). We conclude that neither the whole-rock Rb-Sr dates of the metasediments nor the felsic and mafic igneous rocks indicate unequivocally whether the age of the Patuxent Formation in the Schmidt and Williams hills is Neoproterozoic or Cambrian.

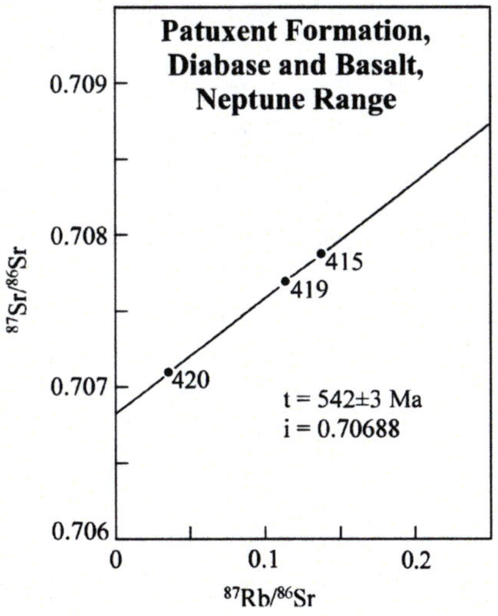

Fig. 8.11 The data points representing diabase sills and basalt flows of the Patuxent Formation analyzed by Eastin (1970) scatter in a tight cluster between 0.7060 and 0.7080 ($^{87}Sr/^{86}Sr$) and between 0.03 and 0.24 ($^{87}Rb/^{86}Sr$). However, three data points in that cluster closely define an isochron that yields a date of 542 ± 3 Ma and an initial $^{87}Sr/^{86}Sr$ ratio of 0.70688 ± 0.000007 (1σ). The goodness of fit of this isochron may be fortuitous and gives only tentative support to the conclusion that the diabase sills and basalt flows of the Patuxent Formation crystallized during the Early Cambrian Epoch. Eastin (1970) preferred a different line the slope of which was based largely on sample 421 (not shown) that yielded a date of 762 ± 59 Ma and supported the Neoproterozoic age of the Patuxent Formation that was also favored by Schmidt and Ford (1969) (Data from Eastin 1970)

Twenty two years after Eastin (1970) completed his work, Storey et al. (1992) were also unable to derive meaningful whole-rock Rb-Sr dates for felsites, mafic sills, and pillow basalts associated with the Patuxent Formation in the Schmidt and Williams hills. They expressed their conclusion succinctly by stating:

> New Rb-Sr data on the igneous rocks we have studied... further illustrate the extreme postcrystallization disturbance of their isotopic systems and the consequent difficulty of deriving meaningful geochronological results from the Patuxent Formation. (Storey et al. 1992, p. 1397)

Even the concentrations of samarium (Sm) and neodymium (Nd) and the $^{143}Nd/^{144}Nd$ isotope ratios of these rocks measured by Storey et al. (1992) scatter widely and fail to yield reliable Sm-Nd dates. Therefore, this attempt to determine the age of the Patuxent Formation, as redefined by Rowell et al. (2001), by means of the Sm-Nd method also failed.

8.2.4.2 Gambacorta Formation

The *Gambacorta Formation* of the Neptune Range consists of a large volume of volcanic rocks that may be amenable to dating by the Rb-Sr method because these rocks have high Rb/Sr ratios. Eastin (1970) analyzed 13 samples of the Gambacorta Formation from the southern Neptune Range and Hill Nunatak made available to him by D.L. Schmidt. The results in Appendices 8.5.12 and 8.5.13 loosely define a straight line in Fig. 8.12. Eastin (1970) excluded sample 290 which is a grayish-red rhyolite with prominent flow banding. Quartz phenocrysts in this sample are highly embayed, feldspar crystals have altered to sericite, and the rock contains inclusions of altered pumice. The remaining 12 samples yielded a date of 556 ± 39 Ma (late Neoproterozoic; IUGS 2002) and an initial $^{87}Sr/^{86}Sr$ ratio of 0.7052 ± 0.0015. This date is unacceptable because the Gambacorta Formation conformably overlies the Nelson Limestone which is late Middle Cambrian in age (520–500 Ma) based on fossil evidence. Accordingly, stratigraphic considerations indicate that the rhyolites of the Gambacorta Formation are younger than about 520 Ma based on the timescale of the IUGS (2002).

We have attempted to refine Eastin's interpretation of his data by excluding sample 389 on the grounds that it has the highest Rb/Sr ratio of this suite of samples and therefore exerts a strong influence on the slope of line A

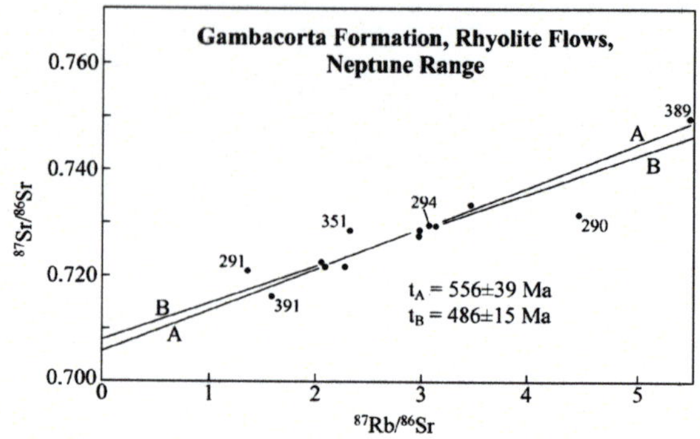

Fig. 8.12 The data points representing rhyolite flows, rhyodacite, and felsic tuffs of the Gambacorta Formation in the Neptune Range scatter widely primarily because the samples were altered after crystallization and because of differences in their initial $^{87}Sr/^{86}Sr$ ratios. Eastin (1970) excluded sample 290 which is a highly altered rhyolite (Appendix 8.5.13) and regressed the remaining 12 samples (Line A) to obtain a date of 556 ± 39 Ma and $(^{87}Sr/^{86}Sr)_i$ = 0.7052 ± 0.0015. An alternative date of 486 ± 15 Ma and $(^{87}Sr/^{86}Sr)_i$ = 0.7078 ± 0.0018 arises from Line B after sample 389 is also excluded. Although this date (Early Ordovician) is consistent with the stratigraphic position of the Gambacorta Formation, it arises from the arbitrary exclusion of a sample that appears to be unaltered (Based on data by Eastin 1970)

in Fig. 8.12. The alternative line B was fitted by unweighted linear regression which yields a date of 486 ± 15 Ma. In summary, the whole-rock Rb-Sr dates and initial $^{87}Sr/^{86}Sr$ ratios of the Gambacorta Formation are:

Line	Date, Ma	$(^{87}Sr/^{86}Sr)_i$
A	556 ± 39 (12)	0.7052 ± 0.0015
B	486 ± 15 (11)	0.7078 ± 0.0018

Although the date of 486 ± 15 Ma is compatible with the stratigraphic position of the Gambacorta Formation and with the fossil evidence for a Middle Cambrian age of the underlying Nelson Limestone, this age determination is contrived to fit independent evidence and therefore is not convincing.

Eastin (1970) also reported the chemical analyses in Appendix 8.5.14 of seven samples of the Gambacorta Formation provided by D.L. Schmidt. Silica-variation diagrams in Fig. 8.13 indicate that the concentrations of alkali metals, alkaline earths, and oxides of iron vary regularly with increasing silica concentrations from 72% to 78%. The only sample that deviates from the smooth trends (#294) is a pyroclastic rhyolite clast from the basal part of the Elliott Sandstone (Pemberton Section) 1.6 km west of Wiens Peak. The Elliott

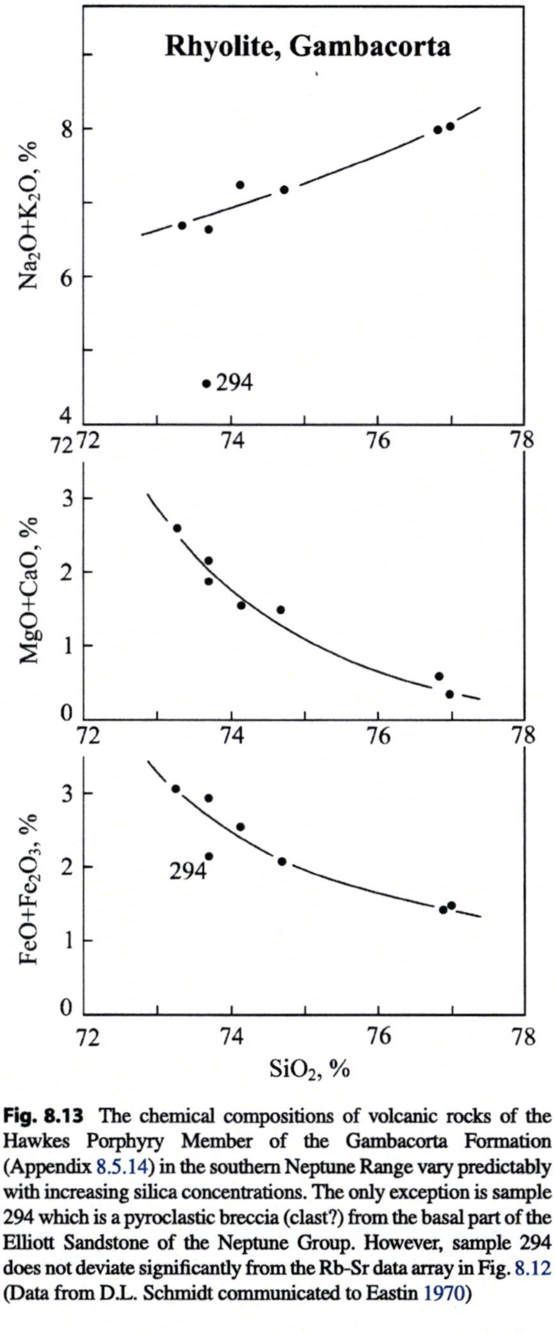

Fig. 8.13 The chemical compositions of volcanic rocks of the Hawkes Porphyry Member of the Gambacorta Formation (Appendix 8.5.14) in the southern Neptune Range vary predictably with increasing silica concentrations. The only exception is sample 294 which is a pyroclastic breccia (clast?) from the basal part of the Elliott Sandstone of the Neptune Group. However, sample 294 does not deviate significantly from the Rb-Sr data array in Fig. 8.12 (Data from D.L. Schmidt communicated to Eastin 1970)

Sandstone is part of the Neptune Group which was deposited unconformably on the Wiens and Gambacorta formations in Table 8.3. Therefore, this sample may have been derived as a clast from the Gambacorta Formation and was altered during transport and after redeposition in the Elliott Sandstone.

8.2.4.3 Serpan Granite and Gneiss

The granite pluton on *Serpan Peak* (Fig. 8.8) intruded the Hannah Ridge Formation of the eastern Neptune Range (Rowell et al. 2001). These granitic rocks occur as a coarse-grained equigranular facies (white or red) and as foliated granitic gneisses containing varying amounts of hornblende, biotite, and sphene. Schmidt (written communication to Eastin 1970) suggested that the Serpan *Gneiss* is a contaminated border facies of the Serpan *Granite*. In that case, the two suites of granitic rocks have the same age but may have different initial $^{87}Sr/^{86}Sr$ ratios in which case they should define two parallel straight lines on the Rb-Sr isochron diagram. The chemical compositions of these rocks in Appendices 8.5.16, 8.5.17 and the silica-variation diagrams in Fig. 8.14 confirm that the *Serpan Gneiss* may have formed by contamination of granite magma which caused the concentrations of the alkali metals to decrease and the concentration of the alkaline earths to increase. The concentrations of Rb and Sr responded similarly such that the average Rb/Sr ratio of the *Serpan Gneiss* is significantly lower than the Rb/Sr ratio of the *Serpan Granite*:

Serpan Gneiss: Rb/Sr = 0.291
Serpan Granite: Rb/Sr = 4.53

As a consequence of this difference, the samples analyzed by Eastin (1970) should form two clusters of data points on the Rb-Sr isochron diagram (not shown). The line connecting these data clusters is a mixing line whose slope cannot be used to date these rocks.

Eastin (1970) analyzed two samples of Serpan Granite and eight samples of Serpan Gneiss for dating by the Rb-Sr method (Appendix 8.5.15). In addition, D.L. Schmidt contributed Rb-Sr data for a third sample of granite identified as "USGS".

The samples of the *Serpan Gneiss*, taken separately, form two parallel straight lines (A and B) in Fig. 8.15 from which Eastin (1970) calculated a pooled date of 514 ± 15 Ma (Middle Cambrian; IUGS 2002). The initial $^{87}Sr/^{86}Sr$ ratios of line A is 0.7084 ± 0.0002 and that of line B is 7057 ± 0.0006.

The samples of the *Serpan Granite* define straight line C in Fig. 8.15 which yielded a date of 519 ± 34 Ma (Middle Cambrian) and an initial $^{87}Sr/^{86}Sr$ ratio of 0.7064 ± 0.00061. Evidently, the Rb-Sr date of the *Serpan Granite* is indistinguishable from the pooled date of the *Serpan Gneiss*:

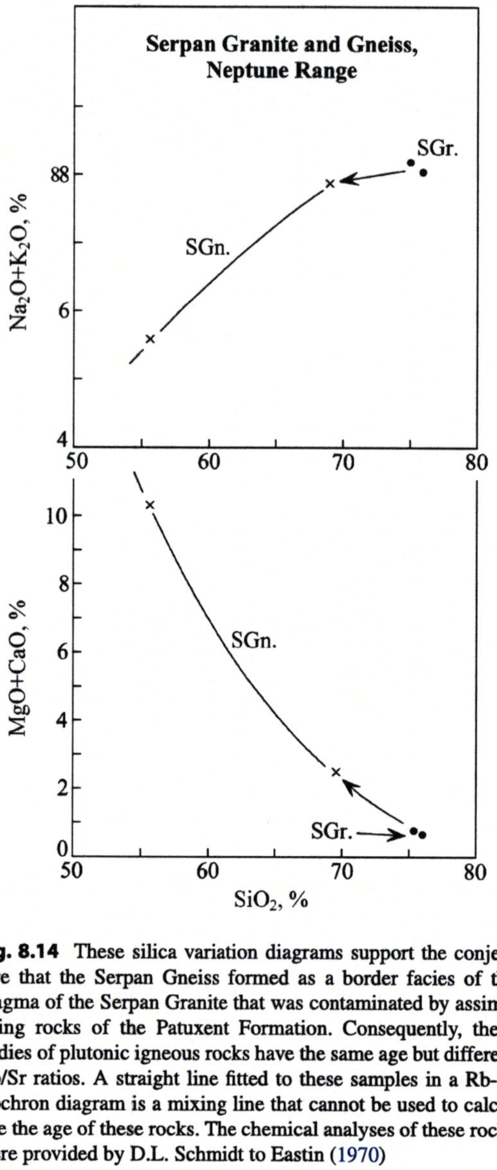

Fig. 8.14 These silica variation diagrams support the conjecture that the Serpan Gneiss formed as a border facies of the magma of the Serpan Granite that was contaminated by assimilating rocks of the Patuxent Formation. Consequently, these bodies of plutonic igneous rocks have the same age but different Rb/Sr ratios. A straight line fitted to these samples in a Rb-Sr isochron diagram is a mixing line that cannot be used to calculate the age of these rocks. The chemical analyses of these rocks were provided by D.L. Schmidt to Eastin (1970)

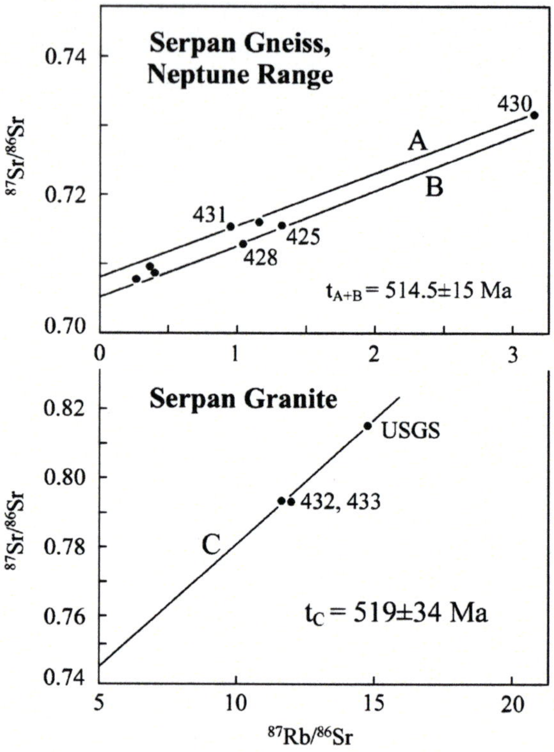

Fig. 8.15 The samples of the Serpan Gneiss analyzed by Eastin (1970) form two quasi-parallel lines A and B which yield a pooled Rb-Sr date of 514 ± 5 Ma (Middle Cambrian). The Serpan Granites have higher $^{87}Rb/^{86}Sr$ ratios than the Gneiss and define line C from which Eastin (1970) calculated a date of 519 ± 34 Ma. These results unequivocally indicate a Middle Cambrian age for the Serpan Granite and Serpan Gneiss. Therefore, the pluton on Serpan Peak was intruded during the Ross Orogeny at about the same time as the Granite Harbor Intrusives elsewhere in the Transantarctic Mountains (Data from Eastin 1970 and D.L. Schmidt [personal communication to Eastin 1970])

Line	Date, Ma	$(^{87}Sr/^{86}Sr)_i$
Serpan Gneiss		
A	505 ± 14	0.7084 ± 0.0002
B	520 ± 5	0.7057 ± 0.0006
A+B	514 ± 15	0.7057 to 0.7084
Serpan Granite		
C	519 ± 34	0.7064 ± 0.0006

This summary confirms that the Serpan Granite and Gneiss crystallized at the same time within the statistical errors of the calculated dates during the Middle Cambrian.

8.2.5 Age Determinations (U-Pb Method)

The age determinations by the Rb-Sr method of rocks in the Pensacola Mountains have been extended and refined by U-Pb dates of separated and hand-picked zircon crystals. The U-Pb measurements of zircon and of other U-bearing minerals are facilitated by use of concordia diagrams (Faure and Mensing 2005), which are also briefly described in Appendix 3.6.5. Another important advantage of the U-Pb method of dating is that the accessory U-bearing minerals in igneous and metamorphic rocks are more refractory than Rb-bearing minerals. Therefore, these minerals (e.g., zircon, monazite, and others) retain the accumulated isotopes of radiogenic lead better than Rb-bearing minerals (e.g., biotite, microcline, and others) retain radiogenic strontium. Consequently, U-Pb dates of zircon are, in many cases, more precise than Rb-Sr dates of whole-rock samples and therefore have provided more useful information about the geologic history of the rocks being dated.

The first U-Pb dates of hand-picked zircon crystals of the Gorecki Felsite and of rhyolites in the Gambacorta Formation, published by Millar and Storey (1995), had far-reaching consequences for our understanding of the geologic history of the Pensacola Mountains. These authors reported that three fractions of zircon from the Gorecki Felsite cluster close to concordia and indicate a date of 500 ± 8 Ma. Zircons from the Gambacorta Formation yielded a similar date of 501 ± 3 Ma. Accordingly, both felsites crystallized within a short interval of time during the Late Cambrian to Early Ordovician.

The U-Pb dates of the Gorecki Felsite and the Gambacorta Formation reported by Millar and Storey (1995) were confirmed by Van Schmus et al. (1997) who obtained a date of 502 ± 3 Ma for zircons in felsite collected at Mt. Gorecki (Schmidt Hills) and at Teeny Rock (Williams Hills). In addition, Van Schmus et al. (1997) reported U-Pb dates for zircons in the Gambacorta Rhyolite at Brown Ridge, Mt. Hawkes, and Jones Valley all of which are located in the Neptune Range:

Brown Ridge and Mt. Hawkes: 500 ± 5 Ma
Q-F Porphyry, Jones Valley: 495 ± 5 Ma

They also analyzed zircon fractions from granodiorite at Serpan Peak (Neptune Range) which define two discordia lines that converge to a lower intercept on concordia at 448 ± 6 Ma. The dates of the upper intercepts are 981 ± 31 Ma and ~1700 Ma. These upper-intercept dates represent the approximate average age of zircon crystals which the granite magma inherited from its source.

Van Schmus et al. (1997) also reported a similar upper-intercept date of 1093 ± 23 Ma for zircons in the Thiel Mountains Porphyry, also known as the "cordierite-bearing hypersthene-quartz monzonite" (Section 8.1.1.2). The same authors reported another example of an upper-intercept date of about 1100 Ma of zircons in granite and gneiss in the Fallone Nunataks north of the Leverett Glacier (Section 6.5.4, Table 6.7). These results suggest that igneous and/or high-grade metamorphic rocks of Mesoproterozoic age (1000–1600 Ma) occur in the continental crust underlying the Transantarctic Mountains. The age determinations summarized in Appendix 8.5.18 were used by Rowell et al. (2001) to propose significant revisions in the stratigraphy and geologic history of the Pensacola Mountains.

8.2.6 Revised Stratigraphy

The best available evidence for the age of the Patuxent Formation (as redefined by Rowell et al. 2001) arises from age determinations of the Gorecki Rhyolite in the Schmidt Hills. If the Gorecki Rhyolite consists of lava flows or other kinds of *volcanogenic surface deposits* that are interbedded with the Patuxent Formation at Mount Gorecki and elsewhere, then age determinations of these rhyolites also date the time of deposition of the local Patuxent Formation. On the other hand, if the felsic magma of the Gorecki Rhyolite *intruded* the Patuxent Formation in the form of sills or small plutons, then the Patuxent Formation is **older** than isotopic dates of the felsites in the Schmidt Hills.

Schmidt et al. (1978) considered the Gorecki Felsite in the Schmidt Hills to be interbedded with the rocks of the Patuxent Formation and Rowell et al. (1994) asserted emphatically that the age of the Patuxent Formation in the Schmidt Hills is early Paleozoic and not Neoproterozoic because the felsite at Mount Gorecki contains fragments of fossiliferous limestone derived from the Nelson Limestone:

> The Gorecki Felsite Member… consists predominantly of oligomictic megabreccias that include large blocks of felsitic tuffs,…, together with locally abundant limestone blocks…. We interpret them [i.e., the mega breccias] as a volcanic apron deposit formed by submarine-driven mass movement that probably was associated with a caldera event. A diversity of carbonate rock types is present representing marine depositional settings from peritidal to below storm-wave base. A few of the limestone blocks are fossiliferous, and their sparse fauna includes hyolithids and sponge spicules…. The Gorecki Felsite Member is clearly Phanerozoic, not Neoproterozoic! (Rowell et al. 1994, p. 43)

The U-Pb date of zircon from the Gorecki Felsite (500 ± 8 Ma) confirmed that the age of the Patuxent Formation in the Schmidt Hills is in fact, Middle to Late Cambrian according to the geologic timescale of the IUGS (2002).

However, Millar and Storey (1995) also recalled that the Patuxent Formation in the *eastern* Neptune Range (as originally defined) is unconformably *overlain* by the Nelson Limestone which means that, in this area, the sedimentary basement rocks are ***older*** than the Nelson Limestone. This apparent contradiction in the age of the sedimentary basement rocks was resolved by Rowell et al. (2001) who reassigned the folded graywacke-slate complex of the eastern Neptune Range to the newly-defined Hannah Ridge Formation. Accordingly, the original Patuxent Formation is now restricted to the basement rocks of the Schmidt and Williams hills, the Rambo Nunataks, and to the Patuxent Range.

The U-Pb date of the Gorecki Felsite measured by Millar and Storey (1995) implies that the age of the pillow-basalt flows in the Schmidt and William hills is also Late Cambrian to Early Ordovician. Storey et al. (1992) previously proposed that these pillow basalts were erupted during an episode of the rifting of the continental crust during which the Thiel Mountains Porphyry and the basaltic rocks in the Ellsworth Mountains were also erupted (Section 8.1.6). According to Millar and Story (1995), the episode of extensional tectonics in the far-eastern sector of the Transantarctic Mountains may have been caused by the development of a back-arc basin in response to the active subduction that was taking place at that time along the paleo-Pacific coast of Antarctica.

Another point to consider is that the rhyolite lavas of the Gambacorta Formation have the same age (Late Cambrian to Early Ordovician) as the Gorecki Felsite (Eastin 1970; Millar and Storey 1995; Van Schmus et al. 1997). Therefore, the rhyolites of the Gorecki Member of the Patuxent Formation and of the Gambacorta Formation in the eastern Neptune Range were erupted during the same interval of time between 488 ± 4 and 501 ± 3 Ma (Table 8.3).

The revised stratigraphy of the Pensacola Mountains is illustrated in Fig. 8.16. The diagram helps to visualize that in the eastern Neptune Range the folded graywacke-slate complex of the Hannah Ridge Formation is unconformably overlain by the Middle Cambrian Nelson Limestone, whereas in the Schmidt and Williams hills the graywacke-slate of the Patuxent Formation overlies the Nelson Limestone. Although the Patuxent and the Hannah-Ridge formations are composed of similar interbedded graywacke and slate, the Patuxent Formation in the western hills and in the Patuxent Range is post-Middle Cambrian in age, whereas the newly defined Hannah-Ridge Formation in the eastern Neptune Range is pre-Middle Cambrian in age.

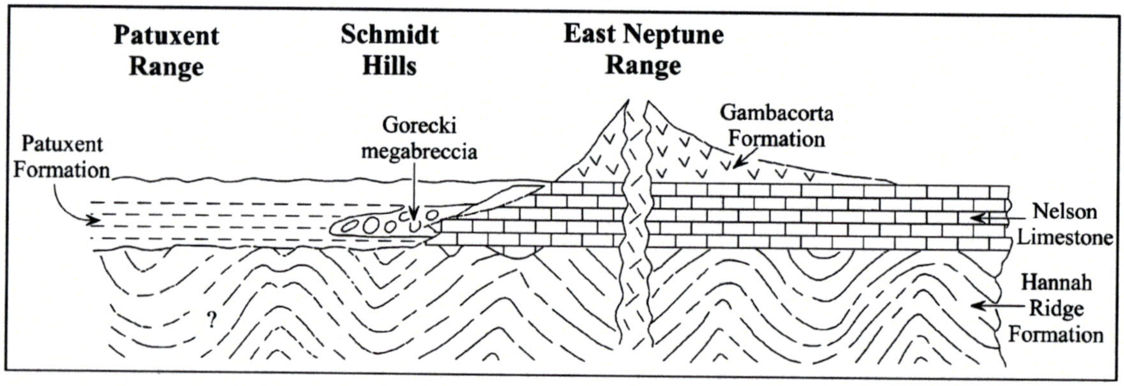

Fig. 8.16 The hypothetical cross-section of the Pensacola Mountains at about 500 Ma extends from the eastern Neptune Range to the Schmidt Hills and the Patuxent Range. The diagram offers a plausible explanation why turbidites that are older than the Nelson Limestone in the Neptune Range appear to be younger that the Nelson Limestone in the Schmidt Hills and in the Patuxent Range. The solution to this riddle is that the turbidites in the Schmidt Hills and in the Neptune Range look alike in the field but were actually deposited at different times. Therefore, Rowell et al. (2001) assigned the turbidites in the Neptune Range to the newly-defined Hannah Ridge Formation, whereas the turbidites of the Schmidt Hills, Williams Hills, and the Patuxent Range remain in the redefined Patuxent Formation. The distinction between the redefined Patuxent Formation and the newly defined Hannah Ridge Formation was made in Table 8.3 of this chapter and has been maintained throughout the foregoing discussion of the stratigraphy and isotopic age determinations of the Pensacola Mountains (Adapted from Rowell et al. 2001)

8.2.7 *Summary*

The Pensacola Mountains consist of the Patuxent, Neptune, and Forrestal ranges which are associated with several clusters of nunataks including the Pecora Escarpment and the Rambo Hills in the vicinity of the Patuxent Range, the Schmidt and Williams hills west of the Neptune Range, and the Cordiner Peaks north of the Neptune Range. The northern Pensacola Mountains consist of the Forrestal Range, a large part of which is buried under the layered gabbros and ultramafic rocks of Dufek intrusion, which is one of the largest stratified mafic bodies in the world comparable to the Bushveld Complex of South Africa, the Stillwater Complex of Montana, and the Muskox Complex of northern Canada (Faure 2001).

The Patuxent Range consists of folded and regionally metamorphosed turbidites composed of the alternating layers of graywacke and shale of the Patuxent Formation. At the southern tip of the Patuxant Range the turbidites contain sills of diabase and are in contact with Cambrian limestone of the Nelson Formation. Still farther south, outcrops of Devonian sandstone exist, but the relationship of the Nelson Limestone and the Devonian sandstone to the turbidites of the Patuxant Formation is not clear at this locality.

In the Schmidt Hills the turbidites of the Patuxant Formation contain diabase sills, pillow basalt, and felsic porphyry. In the Williams Hills the Patuxant Formation contains basalt flows and felsic porphyry, but diabase sills are absent. In spite of its close proximity to the Schmidt and Williams hills, the turbidite basement complex of the main body of the Neptune Range does not contain basaltic or felsic volcanic rocks. The difference has been attributed to the presence of a fault in the Roderick Valley that lies between the Neptune Range and the Schmidt and William hills and that may have moved these parts of the Pensacola Mountains closer together.

In addition, the turbidite basement complex of the Neptune Range is unconformably overlain by the Middle Cambrian Nelson Limestone which is absent from the Schmidt and Williams hills. Therefore, the age of the turbidites that underlie the Nelson Limestone in the Neptune Range must be "pre-Middle Cambrian", whereas the turbidites in the Schmidt and Williams Hills where the Nelson Limestone are absent are "post-Middle Cambrian" in age as confirmed by isotopic age determinations. This confusing situation was resolved by assigning the turbidites of the Neptune Range to the Hannah Ridge Formation which is Neoproterozoic, whereas the turbidites of the Schmidt and Williams hills and of the Patuxant Range continue to be referred to the post-Middle Cambrian Patuxant Formation.

The Nelson Limestone of the Neptune Range is overlain by a succession of volcanic and sedimentary rocks of Paleozoic age including the Gambacorta Rhyolites and the Mount Hawkes Rhyolite Porphyry of Cambro-Ordovician age which occur only in the Neptune Range of the Pensacola Mountains. These volcanic and hypabyssal rocks are overlain by a variety of sandstones and thin shale layers of Devonian age leading up to the Permian tillite of the Gale Mudstone. The youngest rocks in this sequence are tan-colored siltstones and black shale (Pecora Formation) containing coaly layers and Glossopteris plant fossils. In addition, the siltstones were intruded by sills of the Ferrar Dolerite of Jurassic age which is genetically related to the gabbros of the Dufek Massif.

A unique feature of the Pensacola Mountains is that Paleozoic formations of the Neptune and the Forrestal ranges were folded during the Weddell Orogeny of late Paleozoic age. This event distinguishes the Paleozoic formations of the Pensacola Mountains from the sedimentary rocks of the Beacon Supergroup elsewhere in the Transantarctic Mountains which have low dips of less than 10° and were folded only locally along faults.

8.3 Argentina Range

The Argentina Range is an extension of the Pensacola Mountains along the coast of the Ronne Ice Shelf. This small mountain range consists of three clusters of nunataks located at about 82°20′S and 042°00′W which form the Schneider Hills, Mt. Ferrara, and the Panzarini Hills in Fig. 8.17. The Schneider Hills are separated from the Dufek Massif and the Forrestal Range of the Pensacola Mountains by the Support Force Glacier which channels the East Antarctic Ice Sheet into the Ronne Ice Sheet.

The geology of the Argentina Range in Fig. 8.17 includes a complexly folded sequence of interbedded graywacke and slate that Schmidt and Ford (1969) tentatively assigned to the Patuxent Formation of Neoproterozoic age. However, Rowell et al. (2001)

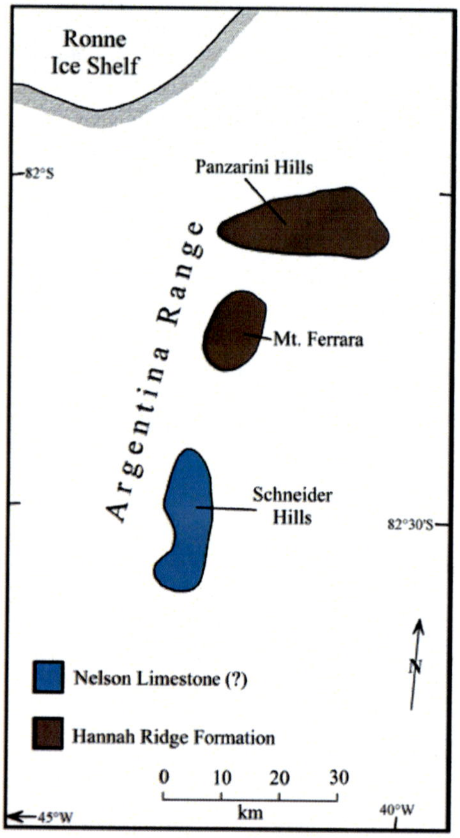

Fig. 8.17 The Argentina Range is located northeast of the Pensacola Mountains along the coast of the Ronne Ice Shelf. It consists of three clusters of nunataks which form the Schneider Hills, Mt. Ferrara, and the Panzarini Hills. These nunataks are composed of interbedded graywackes and slate of the Hannah Ridge Formation and of limestone of uncertain stratigraphic position (i.e., either Nelson Limestone or a facies of the Hannah Ridge Formation) (Adapted from the map of Schmidt and Ford 1969 with stratigraphic information from Rowell et al. 2001)

later correlated these basement rocks of the Argentina Range with the Hannah Ridge Formation of the Neptune Range where it is unconformably overlain by the Middle Cambrian Nelson Limestone.

Limestones that occur in the Schneider Hills of the Argentina Range strike northeast and dip steeply to the east. Schmidt and Ford (1969) placed them into the Nelson Limestone, whereas Rowell et al. (2001) considered them to be a facies of the Early Cambrian Hannah Ridge Formation. Schmidt and Ford (1969) reported that the limestone exposed at Ruthven Bluff in the Schneider Hills contains abundant archaeocyathids and that the limestone was intruded by sills of Jurassic Ferrar Dolerite. Sills of the Ferrar Dolerite are not common elsewhere in the Pensacola Mountains except, of course, in the Dufek Massif where the gabbros have been correlated with the sills of the Ferrar Dolerite.

8.4 Shackleton Range and Theron Mountains

The Shackleton Range and the Theron Mountains are the last of the far-eastern ranges of the Transantarctic Mountains. The *Theron Mountains* at 79°05′S and 028°15′W are located in Coats Land north of the Shackleton Range (Stephenson 1966). Aughenbaugh et al. (1965) reported that the Theron Mountains consists of flat-lying late Paleozoic and early Mesozoic sedimentary rocks, which are presumably assignable to the Beacon Supergroup. The sedimentary rocks of the Theron Mountains were deposited on a metamorphic basement complex of Neoproterozoic to early Paleozoic age.

The *Shackleton Range* at 80°40′S and 026°00′W is located south of the Slessor Glacier and east of the Filchner Ice Shelf. It extends for about 170 km from west to east and about 90 km from north to south and thus occupies an area of 15,300 km^2. The Shackleton Range consists of several clusters of nunataks including the Read Mountains, the Pioneers Escarpment, the Herbert Mountains, the La Grange Nunataks, the Haskard Highlands, the Otter Highlands, and the Stephenson Bastion all of which are grouped around the Fuchs Dome in Fig. 8.18 in counter-clockwise order. The northern and southern limits of the Shackleton Range are the Slessor and Recovery glaciers, respectively. Both ice streams flow from east to west and occupy fault zones that have caused the Shackleton Range to be uplifted as a horst that is regionally tilted to the east. Accordingly, the western end of the Shackleton Range is an eroding fault scarp whereas the eastern end is submerged by the East Antarctic Ice Sheet (Clarkson, 1972, 1982; Clarkson et al. 1995). The nunataks that form the Shackleton Range generally have low elevations compared to the mountains of the Pensacola Mountains. For example, the summit of Mt. Provender in the Haskard Highlands in Fig. 8.18 is only 901 m above sea level (Grew and Halpern 1979).

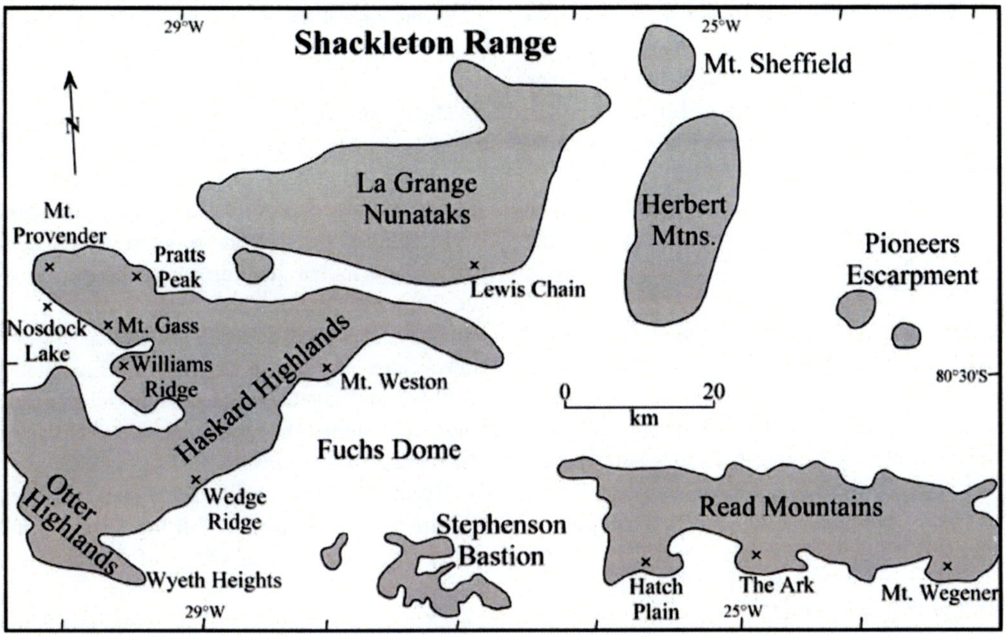

Fig. 8.18 The nunataks of the Shackleton Range along the coast of the Filchner Ice Shelf form several mountain ranges that surround the central Fuchs Dome. Only a few of the largest nunataks are identified by name because they are mentioned in the description of the geology of the Shackleton Range. The strike of the schists and gneisses that form the basement of this range is east-west because these rocks were compressed during the collision of the East Antarctic craton with South Africa. The Slessor and Recovery glaciers along the northern and southern limits of the Shackleton Range, respectively, flow west in fault zones (not shown) (Adapted from Pankhurst et al. 1983; Clarkson 1982; Grew and Halpern 1979)

The strike of the Shackleton Range is roughly east-west and thus differs from the southwest-to-northeast strike of the Pensacola Mountains. The difference is significant because it reveals that the rocks of the Shackleton Range formed by compression from the north rather than from the east as in the Neptune Range of the Pensacola Mountains. Tessensohn (1997), Tessensohn et al. (1999a), and Talarico et al. (1999) attributed the stress that deformed the crystalline basement rocks of the Shackleton Range to the collision of the East Antarctic craton (East Gondwana) with the Mozamibique belt of Africa (West Gondwana) which resulted in the development of the Filchner domain at right angles to the trend of the Ross Orogen.

In this way, the Shackleton Range completes a trend that began in the Thiel Mountains and continued in the Pensacola Mountains both of which are increasingly displaced from the subduction zone along which the Ross Orogen formed by compression from east to west at about 500 Ma. However, Clarkson (1982) reported that the Cambro-Ordovician rocks at the western end of the Shackleton Range strike north and south, presumably because these rocks form the eastern border of the Ross Orogen. Evidently, the rocks of the Shackleton Range have been deformed as a result of two overlapping tectonic regimes which caused various stratigraphic units to be combined into tectonostratigraphic blocks that were displaced relative to each other (see also Dalziel et al. 1994).

8.4.1 Discovery and Mapping

The Shackleton Range was first seen from the air by an Argentine reconnaissance flight in 1955 (Pujato 1977) and was visited in February of 1956 by members of the British Commonwealth Trans-Antarctic Expedition (Fuchs and Hillary 1959; Fuchs 1969; Stephenson 1966). In the following year, David Stratton and Kenneth Blaiklock reached this mountain range by means of dog sleds from Halley Bay located on the Caird Coast of the Weddell Sea more than 500 km from the Shackleton Range. The long commute from their home base made it impractical for British geologists and surveyors to work in the Shackleton Range using dog sleds. Therefore, several British surveyors and geologists with 27 dogs were airlifted in 1968 by an American LC-130 aircraft from Halley Bay to the Shackleton Range. The plane landed safely on the Recovery Glacier far from their intended destination because of cloudy weather. In spite of injuries sustained by two members of the team while working in the mountains, Ken Blaiklock and Tony True made good progress with their topographic survey, while Peter Clarkson and the other geologists mapped parts of the Herbert and Read mountains. The group was picked up in late January of 1969 by an American LC-130 which returned them to their base at Halley Bay. The mapping of the Shackleton Range by British surveyors and geologists with logistical support by American LC-130 aircraft continued for two more years during the 1969/70 and 1970/71 field season (Fuchs 1970; Clarkson 1971).

Several years later, an international group of scientists worked in the Shackleton Range during the 1976/77 field season as part of a 5-year program of study of the Weddell-Sea area by USSR. The group consisted of 133 people including Edward Grew and Arthur Ford (Grew 1977; Ford 1977) from the USA. The expedition (22 SAE) was supported by a base on the northern edge of the Filchner Ice Shelf called Druzhnaya (Friendly). Transportation in the field was provided by four fixed-wing aircraft, one helicopter, and a skimobile named "Buran". Edward Grew worked in the Mt. Provender area of the Haskard Highlands and later published isotopic age determinations of rocks from the Metamorphic Complex in that area (Grew and Halpern 1979; Grew and Manton 1980).

Ford (1977) who worked in the central Read Mountains and in the La Grange Nunataks pointed out that the Shackleton Range does not contain early Paleozoic Granite Harbor Intrusives and that the youngest sedimentary rocks (Blaiklock Glacier Group) were not folded during the Weddell Orogeny which did cause folding of the Paleozoic rocks in the Neptune Range of the Pensacola Mountains.

The insights into the geology of the Shackleton Range that were gained by the efforts of many geologist from different countries were summarized by Clarkson (1995) who wrote on page 7 of his report:

> [T]he Shackleton Range lies at a geological cross roads between the craton of East Antarctica, the Transantarctic Mountains and the enigmatic tectonic provinces that together constitute West Antarctica. (Cited by Tessensohn and Thomson 1999, p. 175)

The most recent study of the geology and structure of the Shackleton Range in the context of the assembly of the supercontinent Gondwana was carried out in 1994/95 by a group of 14 European geologists and a large number of support personnel from German, British, and Italian government agencies (Tessensohn and Thomson 1999). The science team of this large-scale project, called EUROSHACK, was composed of experienced investigators some of whom had previously worked together in northern Victoria Land, the Pensacola Mountains, and in the Shackleton Range. The reports of these investigators in issue 3 or volume 6 of Terra Antartica published in Siena, Italy, shed new light on the geologic history of the Shackleton Range and on its role in the assembly of Gondwana by the merger of East Gondwana (consisting of East Antarctica, India, and Australia) with West Gondwana (composed of Africa and South America).

8.4.2 Geology

The initial progress in mapping the rocks of the Shackleton Range was reported by Stephenson (1966), Thomson (1972), and in a series of publications by Clarkson (1971, 1972, 1981, 1982, 1983), Clarkson et al. (1979, 1995), Grew and Halpern (1979), Grew and Manton (1980), Hofmann et al. (1980, 1981), Marsh (1983a, b) and by Hofmann and Paech (1983). The stratigraphy of the Shackleton Range consists of three units that are identified in Table 8.4:

> The Shackleton Range Metamorphic Complex, the Turnpike Bluff Group, and the Blaiklock Glacier Group.

The Shackleton Range Metamorphic Complex consists of basal orthogneiss, migmatite, and intrusive granodiorite and overlying schist, marble, and amphibolite. The Turnpike Bluff Groups is unconformable to the Metamorphic Complex and consists of clastic sedimentary rocks that were folded and metamorphosed to the greenschist facies. The Blaiklock Glacier Group includes red and gray clastic sedimentary rocks of Middle Cambrian age which are undeformed and unmetamorphosed. All of the stratigraphic units were intruded of dolerite sills of Ordovician and Carboniferous age (Spaeth et al. 1995; Storey 1996).

Table 8.4 A stratigraphy of the Shackleton Range based on groups of formations that were defined by Clarkson (1972, 1981, 1982, 1983)

Group	Description and age	Outcrop area[a]
Dolerite sills	Ordovician and Carboniferous	Haskard Highlands
Intrusive contact		
Blaiklock Glacier	Sandstone and shale (red and gray), pebble conglomerate, slightly warped to form a regional anticline	Otter and Haskard Highlands, La Grange Nunataks
Unconformity		
Turnpike Bluff	Quartzite, sandstone, siltstone, argill. silt-stone, pebble congol-merate. Intensively folded, Neoproterozoic or Cambrian	Wyeth Heights of the Otter Highlands, Stephenson Bastion, Flett Crags, and Mt. Wegener of the Read Mountains
Unconformity		
Shackleton Range Metamorphic Complex (upper part)	Schist, marble, and amphibolite. Precambrian	Haskard Highlands, La Grange Nunataks, Herbert Mountains, Pioneers Escarpment
Shackleton Range Metamorphic Complex (lower part)	Granitic gneisses and granodiorite of the East Antarctic craton. Precambrian	Read Mountains Haskard Highlands

[a]The outcrop areas are identified in Fig. 8.19

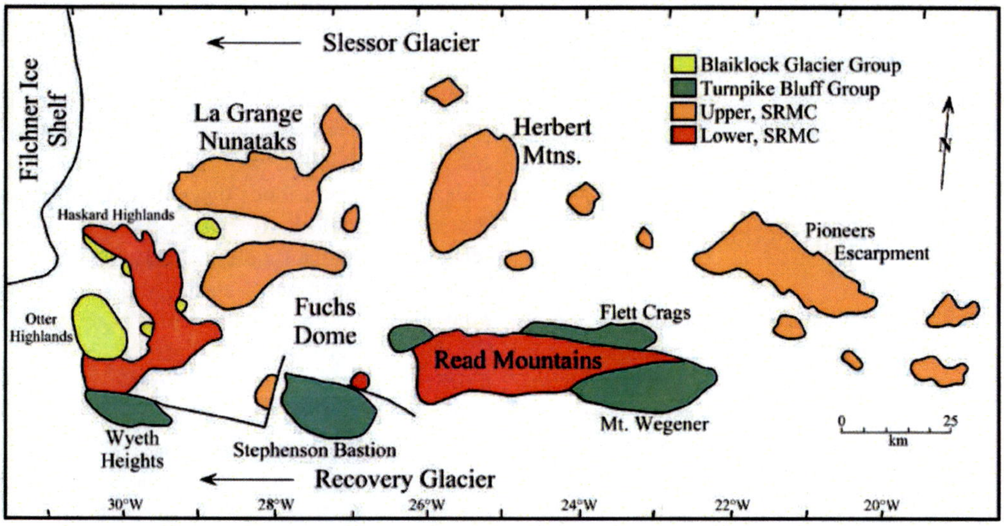

Fig. 8.19 Geological sketch map of the Shackleton Range by Clarkson (1982) based primarily on outcrops of three principal components in order of increasing age: Blaiklock Glacier Group of unmetamorphosed and undeformed sedimentary rocks of Cambrian age; Turnpike Bluff Group of metasedimentary rocks of Neoproterozoic to Cambrian age; Shackleton Range Metamorphic Complex composed of an upper sequence of middle-amphibolite grade schists and a lower sequence of upper amphibolite- and granulite-grade gneisses, migmatites, and granitoid intrusives of Precambrian age (Adapted from Clarkson 1982)

According to Clarkson (1982), the Metamorphic Complex of the Shackleton Range crops out primarily in the Read Mountains and in the Haskard Highlands in Fig. 8.19 (Olesch et al. 1995). The overlying metasedimentary schists in the Haskard Highlands were subdivided by Marsh (1983a) into the Nostoc Lake, Mount Gass, and Williams Ridge formations (Braun 1995; Zeh et al. 1999).

Although these formations were defined for rocks at specific localities in the Haskard Highlands, similar metasedimentary sequences occur elsewhere in the northern Shackleton Range where they form the Herbert Group defined by Hofmann and Paech (1980). In addition, Marsh (1983b) noted that the Watts Needle Formation, composed of quartzite and purple shales which grade upwards into thinly bedded sandstone and quartzite interbedded with gray limestone beds, rests unconformably on gneisses of the Metamorphic Complex along the southern edge of the Read Mountains. Low-grade metasedimentary rocks of the Turnpike Bluff Group occur in the western Haskard Highlands, in the Herbert Mountains, in the northern Read Mountains, and in the Stephenson Bastion. The unmetamorphosed sedimentary rocks of the Blaiklock Glacier Group occur in the western Haskard Highlands and in the topographic gap between the Haskard Highlands and the La Grange Nunataks.

The sedimentary rocks of the Shackleton Range contain fossils including stromatolites and microfossils (Golovanov et al. 1979, 1980; Weber 1991) as well as invertebrates, and trace fossils (Thomson 1972; Solov'ev and Grikurov 1978; Clarkson et al. 1979; Popov and Solov'ev 1981; Clarkson and Wyeth 1983; Thomson et al. 1995). Although the Cambro-Ordovician age of the Blaiklock Glacier Group was indicated by the fossil invertebrates (Thomson and Weber 1999), fossils have not provided enough information to unravel the geologic history and tectonics the Shackleton Range.

8.4.3 *Age Determinations*

The complicated geologic history of the Shackleton Range could not have been reconstructed based only on the evidence provided by field work and by the sparse fossil assemblages. Therefore, several geochronologists have dated representative collections of rocks from the Shackleton Range by isotopic methods in order to provide a time scale for the geologic history of this region. The literature concerning the *geochronology* of the rocks in the Shackleton Range can be divided based on the major mountains ranges:

Haskard Highlands: Rex (1972), Grew and Halpern (1979), Grew and Manton (1980), Pankhurst et al. (1983), Zeh et al. (1999).

La Grange Nunataks: Pankhurst et al. (1983), Brommer et al. (1999).

Herbert Range: Hofmann et al. (1980, 1981) Pankhurst et al. (1983), Talarico et al. (1999), and Brommer and Henjes-Kunst (1999).

Read Mountains: Rex (1972), Hofmann et al. (1980), and Pankhurst et al. (1983).

8.4.3.1 Haskard and Otter Highlands

The rocks of the Haskard and Otter highlands encompass all of the major stratigraphic units of the Shackleton Range listed in Table 8.4 and in Fig. 8.19: The Precambrian basement gneisses of the Metamorphic Complex, the sedimentary rocks of the Turnpike Bluff Group, and the unmetamorphosed sedimentary rocks of the overlying Blaiklock Glacier Group. These two mountain ranges are also located closer than any other part of the Shackleton Range to the Ross Orogen.

As expected by Clarkson (1982), the granitic gneisses of the Metamorphic Complex in the Wedge Ridge area of the Haskard Highland in Fig. 8.18 have yielded the oldest whole-rock Rb-Sr date of 2700 ± 100 Ma (Neoarchean) with an initial $^{87}Sr/^{86}Sr$ ratio of 0.700 ± 0.004 (Pankhurst et al. 1983). The low initial $^{87}Sr/^{86}Sr$ ratio indicates that the strontium in these rocks originated from the mantle without significant contamination by radiogenic ^{87}Sr from crustal rocks. However, the separated minerals (K-feldspar and biotite) and the whole-rock sample of one of these gneisses form a good Rb-Sr isochron that yields a date of only 504 ± 6 Ma with a high initial $^{87}Sr/^{86}Sr$ ratio of 0.8820 ± 0.0003. This result demonstrates that the Archean gneisses in the southern Haskard Highlands were thoroughly reheated during the Cambrian Period which caused the isotopic composition of strontium in the constituent minerals to be homogenized to a high value of the $^{87}Sr/^{86}Sr$ ratio. A Rb-Sr model date of 1700 ± 50 Ma of a Rb-rich muscovite pegmatite in the crystalline complex of the Haskard Highlands suggests that the pegmatite either crystallized or was reheated at that time.

Additional age determinations of the rocks from the Mt. Provendar area (Fig. 8.18) in the western Haskard Highlands by Grew and Halpern (1979) and Grew and Manton (1980) yielded dates of 583 ± 48 and 656 ± 66 Ma (Rb-Sr, whole rock) and about 500 Ma (U-Pb, zircon), respectively. In addition, Grew and Halpern (1979) calculated a model Rb-Sr date of 519 ± 15 Ma for biotite in the feldspathic augen gneiss that crystallized at 583 ± 48 Ma. Rex (1972) reported the first dates from the Haskard Highlands including whole-rock K-Ar dates of 1446 ± 60 Ma (Mesoproterozoic) for a granodiorite dike as well as 457 ± 18 Ma (Late Ordovician) and 297 ± 12 Ma (Carboniferous) for two dolerite dikes.

The northern Haskard Highlands contain two crustal blocks separated by the Northern Haskard Fault (Zeh et al. 1999). The northern block is composed of metadiorite and metasedimentary rocks of the Nostoc Lake Formation. The southern block includes the Mt. Weston Gneiss, the Mount Gass Formation, and the Williams Ridge Formation (Marsh 1983b).

Age determinations in Table 8.5 by Zeh et al. (1999) of the Mt. Weston Gneisses in the southern block reveal their long and complex geologic history:

Sedimentary rocks in the Haskard Highlands, dated by Pankhurst et al. (1983) by the whole rock Rb-Sr method yielded Cambro-Ordovician dates:

Mica schist, Turnpike Bluff Grp.: 520 ± 24 Ma
Williams Ridge Fm., Rb-Sr: 0.7134 ± 0.0006
Shales, Blaiklock Glacier Grp.: 475 ± 42 Ma
Mount Provender Fm., Rb-Sr: 0.7159 ± 0.0036

Both groups of sedimentary rocks are of early Paleozoic age. The clustering of dates in this time interval (Appendix 8.5.19) reflects the effects of the Ross Orogeny on the rocks of the western Haskard Highlands followed by uplift and cooling.

8.4.3.2 La Grange Nunataks

Isotopic dates of the rocks in the western La Grange Nunataks in Table 8.6, by Pankhurst et al. (1983) and by Brommer et al. (1999), reveal the geologic history of this part of the Shackleton Range (Appendix 8.5.19). The high initial $^{87}Sr/^{86}Sr$ ratio of granitic gneisses in the La Grange Nunataks dated by Pankhurst et al. (1983) implies either that the isotopic composition of strontium was homogenized at the indicated date (2310 ± 126 Ma) or that the magma formed by partial melting of crustal rocks that were previously enriched in radiogenic ^{87}Sr by decay of ^{87}Rb. In either case, these granitic gneisses of the western La Grange Nunataks are older than their Rb-Sr date. The U-Pb (zircon) dates of the gneiss at Mathys Bank in the La Grange Nunataks obtained by Brommer et al. (1999) (2328 ± 47, 1715 ± 6 Ma) are in good agreement with the Rb-Sr dates of Pankhurst et al. (1983) for granitic gneisses in the same area (2310 ± 126 Ma).

Table 8.5 Summary of isotopic age determinations by Zeh et al. (1999) of the Haskard Highlands, Shackleton Range

Mineral	Method	Date (Ma)	Interpretation
Mt. Weston Gneiss			
Zircon	U-Pb	1810 ± 2	Emplacement
Monazite	U-Pb	1737 ± 3	Emplacement
Garnet	Sm-Nd	1665 ± 60	Granulite metamorphism 700°C, 8 kbars
Zircon	Pb loss	574 ± 29	Ross Orogeny
Biotite	Rb-Sr	513 ± 6	Cooling to closure temp.
Nostoc Lake Formation			
Zircon	U-Pb	532 ± 18	Emplacement
Whole rock	Sm-Nd	506 ± 6	Granulite metamorphism 750°C, 10 kbars
Biotite (Pankhurst et al. 1983)	Rb-Sr	500 ± 10	Cooling to closure temp.
Pratts Peak			
Pyroxenite (bi., ap., clino)	Sm-Nd	505 ± 140	
Mt. Gass Fm.			
Zircon	U-Pb	~1000	Provenance (?)
Biotite	Rb-Sr	449 ± 5	Cooling to closure temp.

bi = biotite, ap = apatite, clino = clinopyroxene

Table 8.6 Summary of isotopic age determinations from the La Grange Nunataks, Shackleton Mountains

Rock type	Method	Date (Ma)	Reference
Granitic gneiss	Rb-Sr, w.r.	2310 ± 126	1
		0.7219 ± 0.0039	
Schist, Lewis Chain	Rb-Sr, w.r.	505 ± 18	1
		0.7141 ± 0.0016	
Gneiss, Mathys Bank	U-Pb, zircon	2328 ± 47	2
Gneiss, Mathys Bank	U-Pb, zircon	1715 ± 6	2
Gneiss, Mathys Bank	Sm-Nd, garnet	535 ± 22	2

1. Pankhurst et al. (1983) w.r. = whole rock
2. Brommer et al. (1999)

8.4.3.3 Herbert Mountains

The geologic history of the Herbert Mountains and of nearby Mt. Sheffield was investigated by Hofmann et al. (1980, 1981), Hofmann and Paech (1983), and Brommer and Henjes-Kunst (1999). The rocks include migmatite and pyroxene-biotite gneisses, garnet-kyanite schists as well as minor quartzite, marble, and calc-silicates. Mount Sheffield in the northern Herbert Mountains (Fig. 8.18) even contains post-tectonic basalt that may be related to the Jurassic Ferrar Dolerite (Brommer and Henjes-Kunst 1999).

K-Ar dates of whole rock samples from the Sumgin Buttress (80°18′S, 025°44′W) and the Charpentier Pyramid (80°16′S, 025°37′W), measured by Hofmann et al. (1980) and listed by Hofmann and Paech (1983), scatter widely from 268 ± 21 Ma (fuchsite, quartz schist, Sumgin Buttress) to 434 ± 35 Ma (whole rock, amphibolite, Sumgin Buttress). Judging from the context of the geologic history of the Shackleton Range, all of these dates have been lowered by loss of radiogenic ^{40}Ar. Two dolerite dikes at Sumgin Buttress and Charpentier Pyramids in the Herbert Mountains yielded whole-rock K-Ar dates of 391 ± 31 and 417 ± 33 Ma, respectively.

Hofmann et al. (1981) reported that two samples of garnet-bearing mica schist from a nunatak 4 km south southeast of Sumgin Buttress yielded a Rb-Sr date of 1414 ± 184 Ma, whereas three additional samples indicated 470 ± 36 Ma. These authors suggested that the former date refers to an episode of regional metamorphism as does a whole-rock K-Ar date of 1401 ± 70 Ma of a porphyroblastic granitoid in the Read Mountains reported by Hofmann et al. (1980).

Brommer and Henjes-Kunst (1999) measured K-Ar dates of amphibole, biotite, and muscovite in rocks of the Herbert Range that range from 598 ± 14 (amphibole, Mt. Sheffield) to 495 ± 3 Ma (biotite, Maclaren Monolith). The average K-Ar amphibole date (580 ± 25 Ma) is significantly older than the average K-Ar biotite and muscovite date (500 ± 4 Ma). The difference indicates that these rocks cooled to the respective blocking temperatures of amphibole and mica at the end of the last episode of regional metamorphism and tectonic deformation. The prior geologic history of these rocks remains unknown.

Fieldwork during the EUROSHACK expedition in 1994/95 led to the discovery by Talarico et al. (1999) of the remnants of an ophiolite complex in the Bernhardi Heights of the Herbert Mountains. The ophiolite complex is composed of amphibolites, metagabbro, and serpentinites and is tectonically intercalated between two high-grade gneiss complexes. The K-Ar dates of amphiboles in the ophiolite complex range from 510 ± 13 to 491 ± 13 Ma. In addition, the isotope ratios of strontium and neodymium of the ophiolite at 540 Ma form points in quadrant 2 of a Sr-Nd mixing diagram (not shown), whereas the basement gneisses in the Herbert Mountains lie in quadrant 4 which is the field for rocks of the continental crust (Faure 2001; Faure and Mensing 2005). The strontium in the ophiolites is enriched in radiogenic ^{87}Sr relative to the mantle array as a result of alteration of the rocks by seawater (Talarico et al. 1999).

8.4.3.4 Read Mountains

The Read Mountains in the southern Shackleton Range (Figs. 8.18 and 8.19) consist primarily of granitic gneisses of the Shackleton Range Metamorphic Complex overlain along their northern and southern borders by metasedimentary rocks of the Turnpike Bluff Group (Clarkson 1982).

Crystalline basement rocks at Hatch Plain in the southwestern Read Mountains crystallized at 1599 ± 38 Ma (0.714 ± 0.001) and at 1763 ± 21 Ma (0.7041 ± 0.0012). Additional samples of granitic gneiss suggest a date of 1820 ± 160 Ma (0.705 ± 0.003), whereas granodiorite dikes that cut the granitic gneiss of the Read Mountains yield dates between 1900 and 1300 Ma (Rb-Sr, whole rock; Pankhurst et al. 1983). Rex (1972) reported a whole-rock K-Ar date of 1454 ± 60 Ma (recalculated by Pankhurst et al. 1983) for one of these granodiorite dikes. Age determinations by Hofmann et al. (1980) that were reviewed by Hofmann and Paech (1983) include a whole-rock K-Ar date of 1401 ± 70 Ma for porphyroblastic granitoids from the Read Mountains.

The isotopic composition of strontium in metasedimentary rocks of the Mt. Wegener Formation along the southern rim of the Read Mountains (Figs. 8.18 and 8.19) was homogenized at 526 ± 6 Ma (0.7152 ± 0.0005). Pankhurst et al. (1983) considered this date to be the time of diagenesis of the sediment. In that case, the Read Mountains are the only part of the Shackleton Range that was *not* affected by the Ross Orogeny or by the Pan-African orogenic event.

One sample of purple shale of the Watts Needle Formation has a Rb-Sr model date of 720 Ma assuming an initial ^{87}Sr/^{86}Sr ratio of 0.715. Limestone beds in the Watts Needle Formation contain Riphean stromatolites (Golovanov et al. 1979, 1980) consistent with a Proterozoic age (1650–800 Ma; Harland et al. 1990) for this formation.

Summaries of the isotopic dates in Fig. 8.20 and Appendix 8.5.19 reveal that the geologic history of the Shackleton Range started during the Archean Eon more than 2,700 million years ago. The initial formation of continental crust was followed by an episode of magmatic activity, tectonic deformation, and regional metamorphism between 1600 and 1800 Ma. A second less well developed episode of regional metamorphism may have reset the isotopic chronometers between 1400 and 1500 Ma. A third episode of magmatic activity occurred during the Neoproterozoic between 1000 and 540 Ma followed by resetting of the isotopic dates during the Ross Orogeny and/or during the Pan-African orogenic event. The only exception are the granitic basement rocks of the Read Mountains which were not affected by this event as far as is known at the present time.

8.4.4 Tectonics

In spite of the efforts by many geologists to organize the stratigraphy of the Shackleton Range into a coherent picture, the geologic map in Fig. 8.18 does not convey information concerning the structure of the Shackleton Range which is required for a complete understanding of the origin of this mountain range. Similarly, the isotopic dates measured by Pankhurst et al. (1983) and by other geochronologists identified in Fig. 8.20 do not form coherent patterns on the map of the Shackleton Range because the structure of this mountain range is too complex for such patterns to be recognized (Kleinschmidt and Buggisch, 1994).

Therefore, Clarkson et al. (1995) and Tessensohn et al. (1999a) reorganized the stratigraphy of the Shackleton Range into tectono-stratigraphic units defined in Table 8.7 which were displaced relative to each other along low-angle thrust faults and by the movement of nappes. In addition, the presence of the ophiolite complex discovered by Talarico et al. (1999) and the evidence for significant displacement of tectono-stratigraphic blocks are clues that the Shackleton Range is a collisional orogen and therefore may contain a suture between two continental fragments that merged during the assembly of Gondwana.

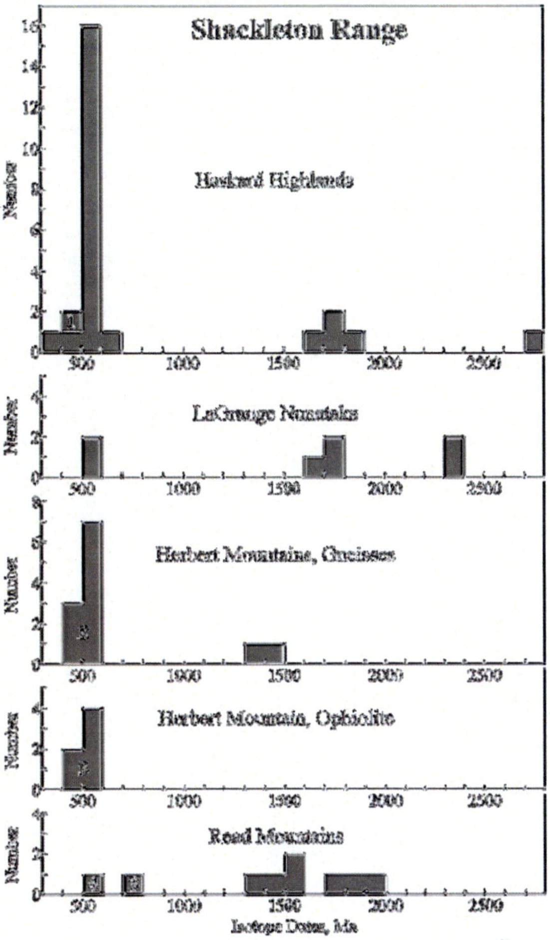

Fig. 8.20 Isotopic age determinations of rocks and minerals by U-Pb (zircon and monazite), and by Rb-Sr, Sm-Nd, and K-Ar (whole-rock and minerals) of the major mountain ranges in the Shackleton Range: (1) Rb-Sr, whole-rock, Blaiklock Group, Haskard Highlands; (2) K-Ar, minerals, granitic gneisses, Herbert Mountains; (3) K-Ar, amphibole, Ophiolite Complex, Herbert Mountains; (4) Rb-Sr, whole-rock, Mt. Wegener Formation, Turnpike Bluff Group; (5) Rb-Sr, whole-rock, Watts Needle Formation, Turnpike Bluff Group (Data from Rex 1972; Grew and Halpern 1979; Grew and Manton 1980; Hofmann et al. 1980, 1981; Pankhurst et al. 1983; Zeh et al. 1999; Brommer et al. 1999; Brommer and Henjes-Kunst 1999; Talarico et al. 1999)

Clarkson et al. (1995) and Tessensohn et al. (1999a) used the tectono-stratigraphic units defined in Table 8.7 and depicted diagrammatically in Fig. 8.21 to construct a new geologic map of the Shackleton Range reproduced in Fig. 8.22. According to this model, the gneisses and migmatites that form the core of the Read Mountains

Table 8.7 Tectono-stratigraphic units of the bedrock in the Shackleton Range (Tessensohn et al. 1999a; Clarkson et al. 1995; Buggisch et al. 1994)

Unit	Name	Description
I	Remobilized Northern Basement	High-grade metamorphic rocks of the Kalahari (Africa) craton of Precambrian age that were remobilized during the Pan-African orogenic event which occurred about the same time as the Ross Orogeny of the Transantarctic Mountains (Tessensohn et al. 1999a identified at least four additional subunits)
II	Ophiolite Complex	Remnant of oceanic crust older than 500 Ma and younger than 1000 Ma. It is composed of mid-ocean ridge basalt (MORB), oceanic-island basalt (OIB), meta-gabbro, ultramafic rocks, and interbedded meta-carbonates, metapelites, and probably meta-chert (Talarico et al. 1999)
III	Metasedimentary rocks and fossiliferous Cambrian sediments	In the northern Shackleton Range this unit consists of quartzites, marbles, and metapelites deposited along a passive continental margin partly in shallow water. In the Pioneers Escarpment the rocks of this unit resemble those of the Mt. Wegener Formation in the southern Shackleton Range where fossils, including Archeocyathids, and isotopic age determinations indicate a Cambrian age (Buggisch and Henjes Kunst 1999; Buggisch et al. 1990)
IV	Southern Precambrian Foreland	Metamorphic and magmatic rocks of the basement complex of the Read Group of Precambrian age are without evidence for reheating during the Ross Orogeny. The basement rocks of the Read Mountains are unconformably overlain by undeformed and unmetamorphosed Neoproterozoic quartzites and calcarcerous cover rocks containing stromatolites of the Watts Needle Formation (Buggisch et al. 1994). The crystalline basement rocks of the Read Group and the overlying sedimentary cover rocks belong to the East Antarctic craton. This unit was overridden by Unit III which was thrust-faulted over them from the north (Buggisch et al. 1994)

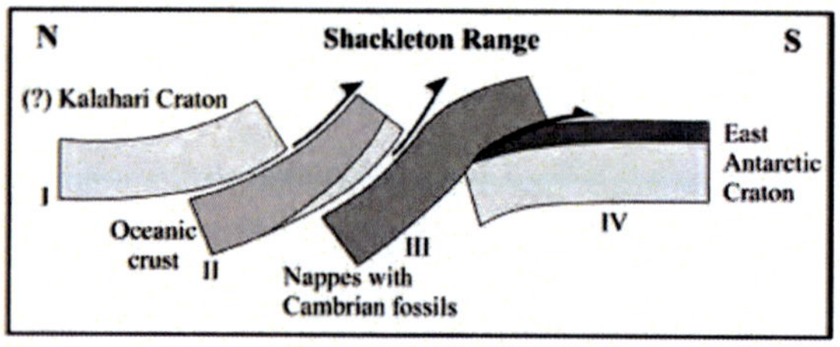

Fig. 8.21 The structure of the Shackleton Range is the result of compression of the basement rocks and overlying sediment in the Filchner domain between the Mozambique belt of southeastern Africa and the East Antarctic craton. This illustration of the continental-collision model demonstrates that nappes composed of sedimentary rocks containing Cambrian fossils (Unit III) were thrust south onto the East Antarctic craton (Unit IV). The nappes of sedimentary rocks were themselves covered by oceanic crust in the form of an ophiolite complex (Unit II) and by polymetamorphic granitic basement rocks of the Kalahari craton of southern Africa (Unit I) (Adapted from Tessensohn et al. 1999a)

(Unit IV in Table 8.7) were covered by nappes of fossiliferous sedimentary rocks (Unit III) of Cambrian age. The slab containing the ophiolite (Unit II) was thrust south onto Unit III, and the high-grade metamorphic rocks of the northern Shackleton Range (Unit I) originated from the Kalahari craton of southern Africa which moved south during the collision of the East Antarctic and African cratons during the Pan-African orogenic event. In other words, the Shackelton Range is a stack of tabular slices of crustal rocks that piled up on the margin of the East Antarctic craton as a result of the collision between Africa and East Antarctica. These events occurred in the so-called Filchner domain at about the same time during which the Ross Orogen formed by subduction and compression along the paleo-Pacific margin of the East Antarctic craton. The collision model explains the east-west strike of the Shackleton Range and suggests that the north-south fabric of the Blaiklock Glacier Group reflects the influence of the Ross Orogeny on the rocks in the western Shackleton Range. Although the Shackleton Range is not a product of the Ross Orogeny, it was, nevertheless, marginally affected by it.

The tectonic map of the Shackleton Range in Fig. 8.22 reveals that the Read Mountains are a window that exposes the granitic gneisses and granitoid intrusives of the East Antarctic craton (Talarico and Kroner 1999). These rocks are unconformably overlain by the undeformed and unmetamorphosed sedimentary rocks of the Watts Needle Formation which contains Neoproterozoic stromatolites. The crystalline basement rocks of the Read Mountains and their sedimentary cover-rocks (Unit IV) were over-ridden by the metasedimentary rocks of Unit III which were originally mapped by a Clarkson (1982) as the Turnpike

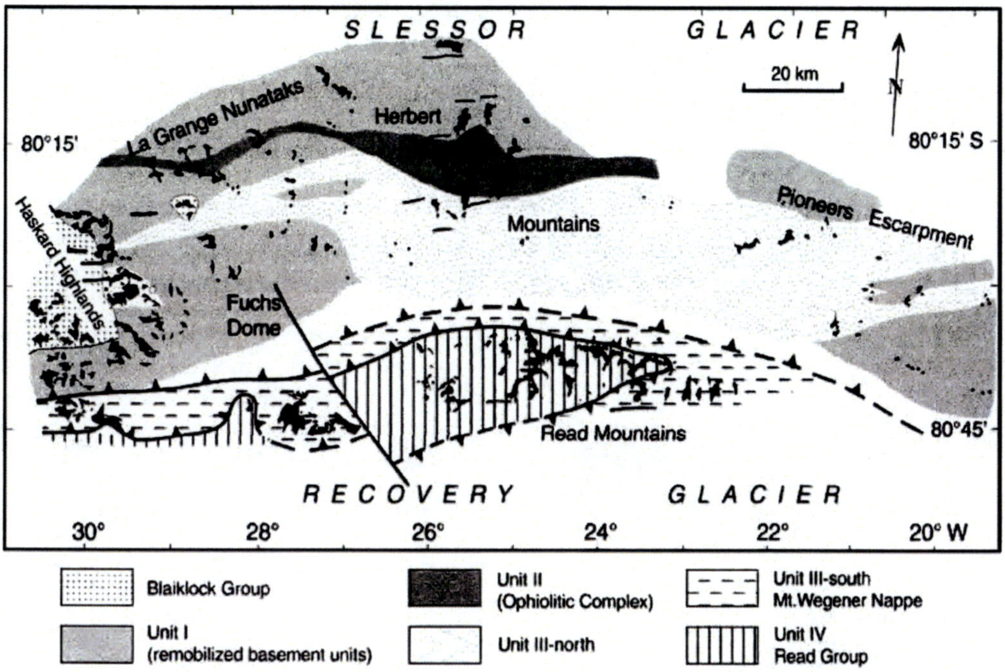

Fig. 8.22 The Shackleton Range consists of several thrust slices and nappes that were stacked on the margin of the East Antarctic craton as a result of its collision with the continent of Africa at about 500 Ma (Tessensohn et al. 1999a). According to this interpretation of the geology, the Read Mountains are a window that exposes the gneisses and migmatites of the East Antarctic craton (Unit IV). These basement rocks were over-ridden by the Mt. Wegener nappe (Unit III, south), which was itself partly covered by a slice of rocks (Unit III, north), which included an ophiolite (Unit II). The northern Shackleton Range is composed of remobilized basement rocks derived from the Kalahari craton of southern Africa (Unit I) (Adapted from Tessensohn et al. 1999a)

Bluff Group (Buggisch and Henjes-Kunst 1999). Outcrops of the sedimentary rocks of this group extend along the southern margin of the Shackleton Range in Fig. 8.19 from the Wyeth Heights of the Otter Highlands east to Stephenson Bastion, and to the Mt. Wegener area in the southeastern corner of the Read Mountains (Buggisch and Kleinschmidt 1999). Clarkson (1982) also mapped the sedimentary rocks in the Flett Crags along the northern border of the Read Mountains as the Turnpike Bluff Group (Fig. 8.19). The new map of the Shackleton Range in Fig. 8.22 indicates that the sedimentary rocks of Unit III (north) also occur in the La Grange Nunataks, in the Herbert Mountains, and in the Pioneers Escarpment.

The continental-collision model developed by Clarkson et al. (1995) and Tessensohn et al. (1999a) was supported by Schmädike and Will (2006) who determined that the ultramafic rocks of the ophiolite complex (Unit II) were subjected to high pressures (20–23 kilobars) and high temperatures (710–810°C) corresponding to the eclogite facies of regional metamorphism.

Pressures and temperatures of this magnitude are expected to occur during continental collision (Coward and Ries 1996) and thereby support the hypothesis that the northern Shackleton Range contains a suture between the East Antarctic craton (East Gondwana) and southern Africa (West Gondwana) which collided in a subduction zone that was active during the Pan-African Orogeny (Shackleton 1996). Eclogite-facies metamorphic rocks also occur in the Nimrod Group of the Miller Range (Section 5.1.1; Peacock and Goodge 1995) and in the Lanterman Range of northern Victoria Land (Section 4.3.5; Di Vincenzo et al. 1997). The merger of the two continents contributed to the assembly of the supercontinent of Gondwana which came into existence during the Paleozoic and which disintegrated during the Mesozoic into the southern continents of the Earth.

The paleogeographic reconstruction of the southern continents during the collision between East and West Gondwana in Fig. 8.23 demonstrates that the Shackleton Range is an extension of the Mozambique mobile belt of east Africa where the former Mozambique ocean

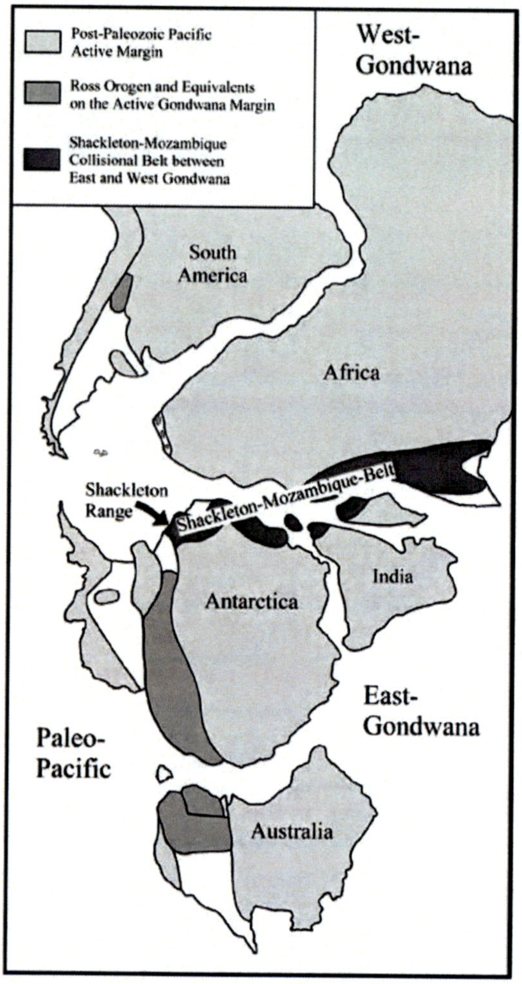

Fig. 8.23 The continental collision model considers that the structure of the Shackleton Range is the result of compression caused by the collision of the East Antarctic craton with the African continent. The resulting closure of the Mozambique ocean is recorded by the Mozambique fold belt. The Ross Orogen, which underlies the Transantarctic Mountains formed by compression of sedimentary and volcanic rocks in an active subduction zone followed by intrusion of the anatectic granitoids of the Granite Harbor Intrusives. Both tectonic processes affected the deposition and subsequent deformation and metamorphism of the basement rocks of the Shackleton Range (Adapted from Tessensohn et al. 1999a)

closed as a result of subduction of oceanic crust. This reconstruction is based in part on the east-west strike of the Shackleton Range which is not the result of rotation of a small crustal fragment of East Antarctica but arises from the alignment of the Shackleton Range with the Mozambique fold belt. Similarly, the Ross Orogen (or fold belt) extends from the Wilson terrane of northern Victoria Land into eastern Australia (Section 4.3) and from the Filchner domain into South America via the South African fold belt. The Shackleton Range is located close to the intersection of these two-fold belts. The undeformed fossiliferous sedimentary rocks of the Blaiklock Glacier Group were deposited along the eastern margin of the Ross Orogen in Cambro-Ordovician time (Buggisch et al. 1999; Thomson and Weber 1999). These investigations largely confirm and extend observations concerning the geology of the Blaiklock Glacier Group published by Stephenson (1966), Clarkson and Wyeth (1983), Marsh (1984), and Buggisch et al. (1994).

8.4.5 Glaciation

The location of the Shackleton Range at the edge of the East Antarctic craton and between the Filchner Ice shelf and the East Antarctic Ice Sheet suggests that it should have been glaciated during the Permian Period as well as during the Plio/Pleistocene ice age.

8.4.5.1 Permian Glaciation

Tillites of Permian age, which are part of the Beacon Supergroup elsewhere in the Transantarctic Mountains, were discovered during the EUROSHACK expedition of 1994/95 on the most easterly nunatak of the Shackleton Range (Tessenshon et al. 1999b). The tillite (5 m) at this location is overlain by about 14 m of light-colored siltstone capped by a thin layer of sandstone about 1 m thick (Pecora Formation?). The tillite contains angular clasts composed of sandstone, quartzite, phyllite, and mica schist which are faceted and striated. Some of the clasts were metamorphosed during the intrusion and cooling of sills of the Ferrar Dolerite (Jurassic), but such sills do not occur at this location. The basement on which the tillite was deposited is not exposed but may consist of biotite schist which crops out a few kilometers west of the nunatak where the glacial sediment was discovered. No fossils of any kind where found in the tillite and the overlying sedimentary rocks.

Insofar as is known, the outcrop described by Tessensohn et al. (1999b) is the only occurrence of sedimentary rocks of the Beacon Supergroup on the Shackleton Range. Beacon rocks do occur in the Theron Mountains north of Shackleton Range and in the Wichaway Nunataks to the south. In addition, tillites of Permian age occur in the Gale Mudstone of the Neptune Range in the Pensacola Mountains (Nelson 1981) and form the Buckeye Tillite of the Ohio Range (Long 1962).

The discovery of tillite proves that the mountains of the Shackleton Range were over-ridden by Permian continental ice sheets that once covered the Transantarctic Mountains before they were uplifted. The evidence for the Permian glaciation of the Shackleton Range draws attention to the possibility that the erosion surface on the present-day summit plateaus is the local representative of the Kukri Peneplain and that this surface was not necessarily formed by erosion during the Plio/Pleistocene glaciation.

The uplift history of the Shackleton Range that arises from the presence of the Kukri Peneplain on the summit plateaus of the mountain ranges (e.g., the Read Mountains) was discussed by Tessensohn et al. (1999a) and is the subject of a study by Lisker et al. (1999) who used fission-track dates (see Faure and Mensing 2005) of apatite and zircon to identify three stages in the exhumation of the Shackleton Range:

1. Early Jurassic, prior to the break-up of Gondwana and the intrusion of the Ferrar Dolerite sills
2. Late Cretaceous, when about 2 km of rock were removed by erosion during tectonic readjustments within Gondwana
3. Early Tertiary, during uplift of the Transantarctic Mountains

These considerations remind us that the Transantarctic Mountains did not exist during the Permian glaciation of the East Antarctic craton and that the eventual uplift of the mountains may have between initiated by large-scale rifting of Gondwana which was accompanied by the eruption of flood basalts during the Jurassic Period.

8.4.5.2 Plio/Pleistocene Glaciation

The Plio/Pleistocene glaciation of the Shackleton Range was caused by ice that originated from three potential sources: The East Antarctic Ice Sheet, the Filchner Ice Shelf, and the ice caps that formed locally on the high plateaus. Geomorphic evidence demonstrates that the Fuchs Dome is part of an ice divide that extends east-west across the Shackleton Range and reaches an elevation of more than 1,600 m above sea level (Kerr and Hermichen 1999). The present-day glaciers north of the divide flow north into the Slessor Glacier, whereas glaciers south of the divide flow south into the Recovery Glacier. The Blaiklock and Stratton glaciers of the Haskard Range flow west directly into the Filchner-Ronne Ice Shelf.

A study of the provenance of glacial erratics by Höfle and Buggisch (1993) suggested that till and erratics in the northwestern part of the Shackleton Range were deposited during the most recent glacial maximum when the Filchner Ice Shelf thickened because it was grounded following a decrease in sea level. Höfle and Buggisch (1993) also concluded that the Shackleton Range was over-ridden in the late Miocene by ice from the East Antarctic ice sheet.

The report by Kerr and Hermichen (1999) contains topographic contour maps of parts of the Haskard Range and of Stephenson Bastion that were published in 1980 by unnamed Russian investigators from the Institute of Okeangeologia in St. Petersburg, Russia. The Plio/Pleistocene glaciation of the Shackleton Range was also discussed by Stephenson (1966), Skidmore and Clarkson (1972) Elverhoi (1981), and Meer et al. (1992).

8.4.6 Summary

The Shackleton Range is remote even by Antarctic standards, which has retarded the geological exploration of this important mountain range. Present knowledge indicates that the Shackleton Mountains are located in the Filchner domain and that the basement rocks were affected both by the Ross Orogeny and by the Pan-African Orogeny. The latter caused thrust slices and nappes to be piled up along the edge of the East Antarctic craton.

Isotopic age determinations (Appendix 8.5.19) indicate that parts of the granitic gneisses of the Haskard Highlands crystallized during the Late Archean Era and that the crystalline basement rocks were intruded by granitic magma of anatectic origin and were subjected to high grades of regional metamorphism in Early

Proterozoic time between 1600 and 1800 Ma. The current understanding of the structure of the Shackleton Range indicates that the Read Mountains are a window that was over-ridden by a nappe composed of metasedimentary rocks from the north. The tectonic assemblage of the Shackleton Mountains also includes remnants of an ophiolite complex in the Herbert Mountains and in the La Grange Nunataks. This ophiolite complex represents oceanic crust that was mobilized during the collision between the East Antarctic craton and the continent of Africa. The merger of two large crustal blocks contributed to the assembly of Gondwana during the early Paleozoic Era.

The only erosional remnant of the Beacon Supergroup is a small outcrop of Permian tillite and overlying siltstone in the eastern Shackleton Range. Beacon rocks do occur in the Wichaway Nunataks to the south and in the Theron Mountains north of the Shackleton Range. The presence of flat-lying sedimentary cover rocks of the Beacon Supergroup relate the far-eastern mountain ranges to the Transantarctic Mountains. However, the pre-Devonian geologic histories of the Thiel Mountains, the Pensacola Mountains, and especially of the Shackleton Range deviate from the history of the Ross Orogen.

The Shackleton Range was glaciated during the Permian Period by continental ice sheets that covered large areas of Gondwana. More recently, the Shackleton Range was over-ridden in post-Miocene time by ice from the East Antarctic and from the Filchner ice sheets which deposited glacial sediment and modified the landscape.

8.5 Appendices

8.5.1 Chemical Analyses of Whole-Rock Samples of the Thiel Mountains Porphyry and of Xenocrysts of Cordierite and Hypersthene (Ford and Himmelberg 1976)

Component	Whole-rock samples (27)	Cordierite (4)	Orthopyroxene (9)
SiO_2	69.8	47.2	47.4
Al_2O_3	14.5	33.9	3.39
Fe_2O_3	1.1	–	–
FeO	3.0	8.2	33.6
MgO	1.2	7.7	13.7
CaO	2.2	0.18	0.24
Na_2O	2.5	0.18	0.01
K_2O	3.8	0.15	–
H_2O+	1.1	1.25	–
TiO_2	0.69	0.01	0.25
P_2O_5	0.18	–	–
MnO	0.02	0.35	0.60
CO_2	0.05	0.25	–
Sum	100.1	99.37	99.2

The number of specimens analyzed is indicated in parentheses

8.5.2 Chemical Analyses of the Reed Ridge Granite (A and B) and of the Thiel Mountains Poprhyry (C and D) in Percent by Weight and Parts Per Million (ppm)as Indicated (Vennum and Storey 1987 a, b)

Component	A	B	C
Major elements (%)			
SiO_2	71.12	75.27	69.78
TiO_2	0.45	0.11	0.63
Al_2O_3	14.07	12.81	14.13
Fe_2O_3	0.67	0.42	0.95
FeO	2.30	0.77	3.10
MnO	0.06	0.04	0.07
MgO	0.81	0.17	1.05
CaO	1.63	0.33	1.95
Na_2O	2.77	2.57	2.69
K_2O	4.73	6.66	4.01
P_2O_5	0.13	0.13	0.17
H_2O	0.95	0.67	1.58
Sum	99.69	99.95	100.11
Trace elements (ppm)			
Ba	430	148	495
Cr	13	~0	11
Nb	12	8	16
Ni	6	~0	5
Sr	101	40	144
Rb	247	281	167

Component	A	B	C
Th	17	9	19
V	35	3	35
Y	49	29	57
Zr	184	65	333
Rare earths (ppm)			
La	31.79	7.54	47.12
Ce	65.89	17.09	98.67
Pr	7.95	2.10	11.65
Nd	31.10	7.92	46.95
Sm	7.00	2.59	9.88
Eu	1.17	0.41	1.59
Gd	6.60	2.49	9.10
Dy	6.93	3.70	8.03
Ho	1.37	0.72	1.59
Er	4.07	2.32	4.51
Yb	3.58	2.61	3.80
Lu	0.57	0.41	0.61

A = granodiorite, B = alaskite, C = porphyry (average of two analyses): Note that Appendix 8.6.1 contains the average chemical analysis of 27 specimens of the Thiel Mountains Porphyry

8.5.3 Rubidium-Strontium Systematics of the Thiel Mountains Porphyry and Reed Ridge Granite (Data from Eastin 1970)

Sample Number	Rb (ppm)	Sr (ppm)	$^{87}Rb/^{86}Sr$	$^{87}Sr/^{86}Sr$
Thiel Mountains Porphyry				
493	197.2	127.4	4.191 ± 0.065	0.7447 ± 0.0027
494	191.6	130.1	3.941 ± 0.060	0.7427 ± 0.0018
495	211.6	112.6	5.134 ± 0.078	0.7544 ± 0.0036
496	184.4	126.4	3.835 ± 0.059	0.7433 ± 0.0018
Biotite Granite				
497	153.7	154.4	2.722 ± 0.043	0.7325 ± 0.0017
498	185.2	155.6	3.276 ± 0.0050	0.7355 ± 0.0026
Quartz Monzonite				
499	303.7	66.0	12.62 ± 0.20	0.7913 ± 0.0028
500	147.2	91.9	4.386 ± 0.069	0.7446 ± 0.0014
501	147.5	102.3	3.930 ± 0.060	0.7417 ± 0.0007

8.5.4 Summary of Isotopic Age Determinations of the Basement Rocks of the Thiel Mountains (Schmidt and Ford 1969; Eastin 1970; Pankhurst et al. 1988)

Rock unit	Material	Date (Ma)	Reference
Rb-Sr Dates			
Pegmatite	Muscovite	528 ± 20[a]	2
Pegmatite	Muscovite	499 ± 20[a]	4
Reed Ridge Granite	Whole rocks	491 ± 12 0.7153 ± 0.0014	1
Reed Ridge Granite	Whole rocks	558 ± 70[a] 634 ± 85[a]	4
Reed Ridge Granite	Feldspar	632 ± 85[a]	4
Reed Ridge Granite	Whole rocks	580 ± 6 0.71153 ± 0.00057	3
Thiel Mtns. Porphyry	Whole rocks	493 ± 24 0.7143 ± 0.0014	1
Mt. Walcott Formation	Dacite sill	480 ± 18 0.7150 ± 0.0010	1
Mt. Walcott Formation	Sedimentary Rocks	495 ± 6 0.7138 ± 0.0003	1
U-Pb Dates			
Thiel Mtns. Porphy. (lower concordia-intercept)	Zircon	500 ± 4	5
Thiel Mtns. Porph. (upper concordia-intercept)	Zircon	1093 ± 23	5
K-Ar Dates			
Reed Ridge Granite	Biotite	500	4
		491	4
		484	4
		511	4
		504	4

1. Pankhurst et al. (1988); 2 Craddock et al. (1964b); 3. Eastin (1970; recalculated); 4. Schmidt and Ford (1969); 5. Van Schmus et al. (1998)

[a]Recalculated to λ (^{87}Rb) = 1.42×10^{-11} year^{-1}

8.5.5 Isotopic Age Determinations of Granitic Rocks in Whitmore Mountains and Other Nunataks of the Ellsworth-Whitmore Mountains Block

Location	Rock type	Method (Ma) K-Ar Rb-Sr	Reference
Whitmore Mountains			
Mt. Seelig	Seelig Granite	195 ± 21	1
		0.70728 ± 0.00146	
Mt. Seelig	Biotite (recalculated)	195 ± 8	3
Mt. Seelig	Microgranite	182 ± 5	1
		0.72237 ± 0.00078	
Mt. Chapman	Granite: coarse and fine, minerals	173	2
		0.7148	
Mt. Radlinski			
Linck Nunatak	Biotite	179 ± 5 (recalculated)	3
	Other Nunataks		
Pirrit Hills	Granite	173 ± 3	1
		0.70703 ± 0.00158	
Nash Hills	Granite	175 ± 8	1
		0.71224 ± 0.00077	
Pagano Nun.	Biotite	175 ± 8	1
		0.71571 ± 0.00135	
Pagano Nun.	Biotite (recalculated)	179 ± 4	3
Haag Nun.	Granite gneiss	1176 ± 76	1
		0.70308 ± 0.00060	
Haag Nun.	Aplogranite	1003 ± 18	1
		0.70360 ± 0.00023	
Haag Nun.	Microgranite	1058 ± 53	1
		0.70293 ± 0.00050	
Haag Nun.	Hornblende	1018 ± 28	4
E. 4690.1	Biotite	991 ± 22	
	K-feldspar	745 ± 18	
	Plagioclase	628 ± 18	
Haag Nun.	Hornblende	1031 ± 14	4
E. 4690.2	Biotite	1002 ± 24	
	K-feldspar	731 ± 18	
	Plagioclase	596 ± 16	

1. Millar and Pankhurst (1987); 2. Kovach and Faure (1978); 3. Webers et al. (1982); 4. Clarkson and Brook (1977)

8.5.6 Patuxent Formation in the Patuxent Range (Data by Eastin (1970))

Sample Number	Rb (ppm)	Sr (ppm)	$^{87}Rb/^{86}Sr$	$^{87}Sr/^{87}Sr$
436	195.0	103.6	5.103 ± 0.079	0.7430 ± 0.0019
437	184.8	65.2	7.704 ± 0.118	0.7639 ± 0.0026
438	102.9	193.2	1.452 ± 0.023	0.7253 ± 0.0025
439	228.7	50.0	12.498 ± 0.193	0.7815 ± 0.0038
440	106.8	98.4	2.936 ± 0.046	0.7309 ± 0.0011
441	256.5	87.1	8.009 ± 0.123	0.7596 ± 0.0020
442	225.5	36.6	16.735 ± 0.261	0.8087 ± 0.0008
443	–	–	1.671 ± 0.029	0.7272 ± 0.0026
444	153.6	176.0	2.471 ± 0.038	0.7328 ± 0.0016
445	–	–	2.253 ± 0.035	0.7283 ± 0.0036
446	115.5	229.2	1.369 ± 0.021	0.7225 ± 0.0022

8.5.7 Patuxent Formation: Description of Samples from the Patuxent Range

436 (1 Sa.2.12.65) Interbedded siltstone and slate, Far West Peak.
437 (5S. 12.4.62) Laminated slate, Seminole Nunatak.
438 (65S.2.12.62) Subgraywacke, Winnebago Nunatak.
439 (9S.2.12.65) Slate, Pueblo Nunatak area.
440 (1S.11.30.62) Subgraywacke, Dash Nunatak.
441 (1S.1.24.62a) Interbedded siltstone and slate, Mohawk, "A" Nunatak.
442 (2XS.11.13.62) Slate, Snowbank Nunatak.
443 (95Sb.22.12.65) Subgraywacke, Black Pyramid.
444 (3Sa.11.16.62) Laminated siltstone and subgraywacke, Corner Glacier area.
445 (S.1.2.63) Laminated siltstone, Bannock Ridge.
446 (65F-1.31.12.62) Slate, Ogalalla Ridge

[a]The samples were provided by D.L. Schmidt whose field numbers are enclosed in parentheses

8.5.8 Patuxent Formation Samples of Felsic Flows and Plugs (Data by Eastin 1970)

Sample Number	Rb[a] (ppm)	Sr[a] (ppm)	$^{87}Rb/^{86}Sr$	$^{87}Sr/^{86}Sr$
352	77.71	300.8	0.7483 ± 0.0157	0.7126 ± 0.0008
353	264.4	111.4	6.903 ± 0.029	0.7555 ± 0.0024
354	36.00	318.2	0.3276 ± 0.0041	0.7111 ± 0.0010
355	115.2	81.75	4.094 ± 0.071	0.7442 ± 0.0006
357	1.82	40.49	0.1299 ± 0.0384	0.7070 ± 0.0036
358	102.6	484.4	0.6136 ± 0.0077	0.7145 ± 0.0014
434	44.8[b]	116.5[b]	1.048 ± 0.017	0.7231 ± 0.0008
435	1.5[b]	81.6[b]	0.0557 ± 0.00464	0.7174 ± 0.0023

[a]Rb and Sr concentrations determined by isotope dilution unless otherwise indicated
[b]Rb and Sr concentration determined by x-ray fluorescence

8.5.9 Felsites of the Patuxent Formation, Schmidt and Williams Hills

352 (12Sa.11.1.66) Rhyolite, east prong of Gorecki Peak, Schmidt Hills. Light gray vitric-crystal tuff (?). Quartz and K-feldspar grains about 60%. Black angular inclusions up to 2.5 cm long.

353 (32 Sa.7.2.64) Rhyolite plug (about 30 m in diameter) intruding the Patuxent Formation, Williams Hills. K-feldspar phenocrysts (15%) in an aphanitic groundmass.

354 (12.Sb.11.1.6.6) Rhyolite, Gorecki Peak, Schmidt Hills. Similar to #352.

355 (32 Sd.7.2.64) Rhyolite, Williams Hills. Similar to #353, but from a different outcrop.

357 (18.Sb.5.2.64) Felsic tuff, east-central Williams Hills. Numerous black angular inclusions (25%) in aphanitic groundmass.

358 (W44a.23.1.64) Rhyolite, Gorecki Peak, Schmidt Hills. Similar to 352 and 354.

434 (1Se.6.12.65) Rhyolite, Williams Hills, 2.4 km west of collecting site of samples 353 and 355. From a sheared lense in folded Patuxent Fm. Phenocrysts of quartz, plagioclase, and minor. K-feldspar. Abundant sericite (?) in the groundmass.

435 (1 Si.6.12.65) Rhyolite, Williams Hills. Collected 2.4 km west of samples 353 and 355. Similar to 434.

8.5.10 Patuxent Formation: Samples of Diabase and Basalt (Data by Eastin 1970)

Sample Number	Rb[a] (ppm)	Sr[a] (ppm)	$^{87}Rb/^{86}Sr$	$^{87}Sr/^{86}Sr$
Diabase				
418	12.1	198.0	0.1768 ± 0.0030	0.7079 ± 0.0013
429	8.0	203.6	0.1131 ± 0.0068	0.7077 ± 0.0014
420	2.7	223.8	0.0342 ± 0.0011	0.7071 ± 0.0011
421	66.1	273.9	0.6647 ± 0.0107	0.7138 ± 0.0013
422	9.4	212.9	0.1208 ± 0.0030	0.7061 ± 0.0019
423	23.4	267.4	0.2370 ± 0.0058	0.7062 ± 0.0020
Basalt				
414	8.1[b]	100.3[b]	0.1830 ± 0.0070	0.7078 ± 0.0011
415	10.1[b]	207.6[b]	0.1371 ± 0.0081	0.7079 ± 0.0018
416	13.0	244.1	0.1415 ± 0.0041	0.7061 ± 0.0024
417	11.1	194.5	0.1655 ± 0.0114	0.7077 ± 0.0016

[a]Rb and Sr concentrations determined by x-ray fluorescence unless otherwise indicated

[b]Rb and Sr concentrations determined by isotope dilution

8.5.11 Patuxent Formation: Description of Diabase and Basalt Samples from the Neptune Range

Diabase

418 (10 Sa. 11.166) Gorecki Peak, Schmidt Hills. Medium grained, collected near the base of a 55-m-thick sill that intruded the Patuxent Formation. The diabase sill is located about 30 m below a rhyolite flow.

419 (10Sb.11.1bb) Gorecki Peak, Schmidt Hills. Collected 27m above sample 418 from the same sill.

420 (10 Sc.11.1.66) Gorecki Peak, Schmidt Hills. Coarse grained, collected 42 m above sample 18 from the same sill.

421 (10Sd.11.1.66) Gorecki Peak, Schmidt Hills. Fine grained, collected about 1.8 m below the top of the same sill as samples 418–420.

422 (B30.164.BGP.11.D - 3) Medium grained, Nunatak #10, Schmidt Hills. Collected frm the interior of a sill estimated to be between 91 to 122 m thick.

423 (B 30.1.64.GBP.11.D-4) Moderately coarse-grained from the same locality as sample # 422.

Basalt

414 (S23-24B.2.7.64) Interior of a flow 30 m thick, Nunatak #6, Williams Hills.

415 (S5.2.4.64) Moderately amygdaloidal, composed of chlorite with some calcite. Nunatak #8, Williams Hills.

416 (1S.1.13.64) Nunatak #2, Williams Hills.

417 (B4.15.1.64) Pillow basalt, Nunatak #1, Williams Hills. Collected from the interior of a basalt pillow 30 cm in diameter.

8.5.12 Gambacorta Formation: Felsic Volcanic Rocks of the Hawkes Porphyry Member of the Pensacola Mountains (Data by Eastin 1970)

Sample Number	Rb[a] (ppm)	Sr[a] (ppm)	$^{87}Rb/^{86}Sr$	$^{87}Sr/^{86}Sr$
290	261.9	168.7	4.502 ± 0.023	0.7315 ± 0.0008
291	114.7	243.0	1.367 ± 0.014	0.7207 ± 0.0009
292	130.0	166.0	2.270 ± 0.028	0.7214 ± 0.0018
293	144.3	136.5	3.064 ± 0.038	0.7286 ± 0.0012
294	140.6	132.2	3.085 ± 0.038	0.7292 ± 0.0014
348	131.1	121.2	3.136 ± 0.039	0.7291 ± 0.0012
349	121.7	170.5	2.066 ± 0.026	0.7216 ± 0.0010
350	128.5	107.3	3.474 ± 0.043	0.7333 ± 0.0013
351	106.3	112.6	2.736 ± 0.034	0.7286 ± 0.0010
356	136.8	132.7	2.987 ± 0.037	0.7274 ± 0.0008
389	158.7	83.90	5.498 ± 0.069	0.7495 ± 0.0008
390	135.8	193.2	2.035 ± 0.025	0.7223 ± 0.0007
391	94.57	173.1	1.582 ± 0.020	0.7160 ± 0.0005

[a]Rb and Sr concentrations were determined by isotope dilution

8.5.13 Gambacorta Formation and Hawkes Porphyry in the Neptune Range (Rock Descriptions by D.L. Schmidt)

290 (W21d.12.27.63) Rhyolite, grayish red with prominent red-brown flow banding. Porphyritic texture, highly embayed quartz, and pseudomorphs of sericite after feldspar with inclusions of altered pumice. Red-Brown Member, Bragg Valley, southern Neptune Range.

291 (N1.12.24.63) Latite, aphanitic, greenish-gray, with dark greenish-gray streaks. Lower Member, Upper Jackson Valley, southern Neptune Range.

292 (N2.1.12.64) Porphyritic quartz latite, phenocrysts of K-feldspar, plagioclase, and quartz in a dark-red groundmass. Hawkes Porphyry Member, west side of "Hawkes camp" valley, southern Neptune Range.

293 (S25.12.29.64) Porphyritic quartz latite like 292 containing more phenocrysts (33%). Hawkes-Porphyry Member, middle Bragg Valley, southern Neptune Range.

294 (S2-11.12.63) Pyroclastic rhyolite, autolithic breccia, from basal part of the Elliott Sandstone (Neptune Grp., Ordovician?). Pemberton section, 1.6 km west of Wiens Peak, southern Neptune Range.

348 (45Sa.13.12.65) Rhyodacite containing 20–25% phenocrysts of quartz and altered K-feldspar in a greeniwth-gray aphanitic groundmass. Hawkes-Porphyry Member, Hill Nunatak.

349 (73S.17.12.65) Rhyodacite, similar to #348. Hawkes-Porphyry Member, from the east face of the Mount Hawkes.

350 (46Sa.13.12.65) Rhyodacite, altered olive-gray vitric-crystal tuff. Quartz = 30%, alkali feldspar = 10%, plagioclase = 15%. Hawkes-Porphyry Member, Hill Nunatak.

351 (46Sb.13.12.65) Vitric-crystal tuff similar to #350. Hawkes-Porphyry Member, Hill Nunatak.

356 (47Wa.28.1.64) Porphyritic felsite with phenocrysts of quartz, K-feldspar, and plagioclase and black angular inclusions. Hypabyssal porphyry intrusive, 37 m wide, into Patuxent Formation at Pope Nunatak.

389 (23S.28.12.63) Vitric-crystal tuff (?) containing quartz and plagioclase phenocrysts in an aphanitic groundmass. Red-Brown Member, Bragg Valley.

390 (21Wa.27.12.63) Vitric-crystal tuff (?) similar to #389 but containing light and dark bands. Red-Brown Member, Bragg Anticline.

391 (5N.12.1.64) Black ash flow tuff, containing phenocrysts (33%) of quartz and plagioclase. Johnston Porphyry Member, eastern Hood Ridge.

8.5.14 Chemical Compositions of Rock Samples from the Hawkes-Porphyry Member of the Gambacorta Formation Provided by D.L. Schmidt to R. Eastin, in Weight Percent

Element	292	293	294	348	349	350	351
SiO_2	76.98	76.88	73.71	74.71	73.27	73.71	74.10
Al_2O_3	12.06	11.95	14.12	12.68	12.63	12.65	12.46
Fe_2O_3	1.22	1.08	1.52	0.94	0.85	1.82	1.31
FeO	0.27	0.38	1.40	1.15	2.20	0.61	1.24
MgO	0.17	0.08	0.38	0.44	0.54	0.49	0.39
CaO	0.19	0.55	1.79	1.31	2.05	1.38	1.15
Na_2O	2.21	3.31	0.91	3.47	3.36	3.33	3.79
K_2O	5.88	4.75	3.64	3.72	3.34	3.32	3.50
H_2O+	0.52	0.32	1.67	0.89	1.08	1.29	0.85
H_2O-	0.08	0.05	0.27	0.05	0.05	0.12	0.04
TiO_2	0.11	0.28	0.11	0.24	0.34	0.23	0.26
P_2O_5	0.02	0.07	0.02	0.05	0.06	0.02	0.02
MnO	0.01	0.02	0.05	0.05	0.08	0.07	0.06
CO_2	0.01	0.02	0.15	0.27	0.07	0.75	0.59
Cl	0.01	0.01	0.02	0.00	0.01	0.01	0.00
F	0.2	0 03	0.02	0.03	0.08	0.04	0.03
Sum	99.76	99.76	99.77	100.00	100.01	99.84	99.79

292: Porphyritic quartz latite, Hawkes Porphyry
293: Porphyritic quartz latite, Hawkes Porphyry
294: Pyroclastic rhyoltie, basal Elliot sandstone
348: Rhyodacite, Hawkes Prophyry
349: Rhyodacite, Hawkes Prophyry
351: Vitric-crystal tuff, Hawkes Prophyry

8.5.15 Serpan Granite and Gneiss: Rb and Sr Concentrations and $^{87}Sr/^{86}Sr$ Ratios (Measured by Eastin 1970)

Sample Number	Rb[a] (ppm)	Sr[a] (ppm)	$^{87}Rb/^{86}Sr$	$^{87}Sr/^{86}Sr$
Serpan Granite, Serpan Peak				
432	280.2	66.6	11.64 ± 0.19	0.7933 ± 0.0030
433	266.2	60.9	11.87 ± 0.18	0.7934 ± 0.0026
USGS[b]	278.0	54.5	14.74	0.8154
Serpan Gneiss, Serpan Peak				
424	161.2	452	0.9623 ± 0.0149	0.7153 ± 0.0002
425	156.6	314.5	1.347 ± 0.021	0.7157 ± 0.0008
426	87.2	823.2	0.2842 ± 0.0044	0.7078 ± 0.0002
427	98.4	654.7	0.4096 ± 0.0066	0.7087 ± 0.0022
428	164.0	440.6	1.020 ± 0.016	0.7130 ± 0.0018
429	89.2	628.1	0.3826 ± 0.0061	0.7094 ± 0.0014
430	205.5	175.2	3.155 ± 0.048	0.7311 ± 0.0014
431	170.2	397.3	1.157 ± 0.018	0.7162 ± 0.0017

[a]Determined by x-ray fluorescence unless otherwise indicated
[b]Data provided by D.L. Schmidt by written communication to R. Eastin

8.5.16 Serpan Granite and Gneiss from Serpan Peak in the Neptune Range of the Pensacola Mountains (Data by Eastin 1970)

Serpan Granite
432 (18S. 2.12.64R-1) Granite, coarse grained, red variety.
433 (18S.2.12.64W-3) Granite, coarse grained, white variety.
Serpan Gneiss
424 (W66a.11.2.64) Granite gneiss, biotite (?) altered to chlorite.
425 (W66aa.11.2.64) Granite gneiss, chlorite common.
426 (W66c.11.2.64) Hornblende-biotite-sphene gneiss. Biotite is partly altered to chlorite.
427 (W66cc11.2.64) Hornblende-biotite-sphene gneiss. Biotite was altered to chlorite.
428 (W66d.11.2.64) Granite gneiss.
429 (W66e.11.2.64) Hornblende-biotite-sphene gneiss. Contains a dark alteration lenticule and biotite is chloritized.
430 (W66f.11.2.64A) Granite gneiss.
431 (W66f.11.2.64B) Granite gneiss.

8.5.17 Chemical Compositions of Rock Samples from the Serpan Granite and Serpan Gneiss on Serpan Peak, Northern Neptune Range, Pensacola Mountains, in Weight Percent

	Serpan Granite		Serpan Gneiss	
	432	433	427	431
SiO_2[a]	75.96	75.25	55.41	69.86
Al_2O_3	12.87	12.92	16.71	14.37
Fe_2O_3	0.50	0.55	3.32	0.85
FeO	1.03	1.27	4.97	2.21
MgO	0.09	0.09	3.54	1.06
CaO	0.59	0.65	6.76	1.41
Na_2O	3.29	3.26	2.99	3.50
K_2O	4.78	4.92	2.60	4.39
H_2O+	0.50	0.59	1.38	1.21
H_2O-	0.08	0.09	0.04	0.07
TiO_2	0.10	0.12	1.30	0.50
P_2O_5	0.01	0.01	0.49	0.15
MnO	0.04	0.05	0.15	0.08
CO_2	0.02	0.03	0.05	0.17
Cl	0.01	0.02	0.02	0.03
F	0.02	0.02	0.16	0.06
Sum	99.89	99.84	99.89	99.92

[a]D.L. Schmidt written communication to Eastin (1970)

8.5.18 Selected Isotopic Age Determinations of Rocks and Minerals in the Pensacola Mountains in Units of 10^6 Years (Ma)

K-Ar, biotite	Rb-Sr Whole rocks	U-Pb, zircon and monazite	Reference
Lamprophyre, Patuxent Range			
244, 219			1
233			
Serpan Granite and Gneiss, Neptune Range			
	514 ± 15		2
	519 ± 34		2
		488 ± 6 (l.i.)	4
		981 ± 31 (u.i.)	4
		~1700 (u.i.)	4
	510 ± 30		6
265 ± 13			6
Hawkes Porphyry, Gambacorta Rhyolites			
		488 ± 4 (l.i.)	4
		1013 ± 131 (u.i.)	4
		495 ± 10	4
		500 ± 5	4
	486 ± 15		2
		501 ± 3	3
Patuxent Formation, Diabase and Basalt			
	542 ± 3		2
	762 ± 59 (4 samples, one sill)		2
Patuxent Formation, Gorecki Rhyolite and Felsites			
	478 ± 6	491 ± 6 (l.i.)	4
		1180 ± 140 (u.i.)	4
		502 ± 3	4
		500 ± 8	3
Patuxent Formation, Patuxent Range, Graywacke-Slate			
	>387 ± 10 (R)		2
	>427 ± 12 (R)		2
	>375 ± 7 (B)		2
		<496 ± 12	5
Hannah Ridge Formation, Eastern Neptune Range (Defined by Rowell et al. 2001)			
		<561 ± 2	5

1. Schmidt and Ford (1969); 2. Eastin (1970); 3. Millar and Storey (1995); 4. Van Schmus et al. (1997); 5. Rowell et al. (2001); 6. Beck and Schmidt (1971). li = lower intercept on a U-Pb concordia; ui = upper intercept on a U-Pb concordia

8.5.19 Summary of Isotopic Age Determinations of Rocks and Minerals in the Principal Mountain Ranges of the Shackleton Range Expressed in Ma

Ordovician 440–495	Cambrian 495–545	Neoproterozoic 540–1000	Mesoproterozoic 1000–1600	Paleoproterozoic 1600–2500	Neoarchean 2500–2800
Haskard Highlands					
	504 ± 6	537 ± 36	>1550	1700 ± 50	2700 ± 100
	500 ± 10	582 ± 57	<1500	1810 ± 2	
	520 ± 24	>900	1446 ± 60	1737 ± 3	
475 ± 42	500	574 ± 29		1665 ± 60	
457 ± 18	513 ± 5	~1000			
(297 ± 12)[a]	532 ± 18	656 ± 66			
449 ± 5	506 ± 6				
	500 ± 10				
	505 ± 14				
	519 ± 15				
La Grange Nunataks					
	505 ± 18			2310 ± 12	
	535 ± 22			2328 ± 47	
				1715 ± 6	
Herbert Mountains (Gneiss)					
(434 ± 35)	495 ± 3	598 ± 14	1414 ± 194		
(268 ± 21)	496 ± 3	551 ± 12			
(351 ± 28	505 ± 3	588 ± 13			
(399 ± 32)	504 ± 3				
(391 ± 31)	500 ± 3				
(417 ± 33)	499 ± 5				
470 ± 36					
Herbert Mountains (Ophiolite)					
491 ± 13	508 ± 13				
492 ± 13	510 ± 13				
	504 ± 12				
	504 ± 15				
Read Mountains					
	526 ± 6	720	1599 ± 38	1763 ± 21	
			>1300	1820 ± 160	
			1454 ± 60	<1900	
			1401 ± 70		

[a]Dates that are less than 440 to 495 Ma are identified by parentheses

References: Rex (1972), Grew and Halpern (1979), Grew and Manton (1980), Hofmann et al. (1980, 1981), Pankhurst et al. (1983), Zeh et al. (1999), Brommer et al. (1999), Brommer and Henjes-Kunst (1999), and Talarico et al. (1999)

References

Aaron JM, Ford AB (1964) Isotope age determinations in the Thiel Mountains, Antarctica. Geological Society of America, Special Paper 76, Abstracts for 1963, p. 1

Aughenbaugh NB (1961) Preliminary report on the geology of the Dufek Massif. Reports of Antarctic Geologic Observations, 1956–1960. IGY Glaciol. Rept. Ser., No. 4: 155–193. IGY World Data Center A, Glaciology, American Geographical Society, New York

Aughenbaugh NB, Lounsbury RW, Behrendt JC (1965) The Littlewood Nunataks. Antarctica J Geol 73(6):889–894

Barrett PJ, Elliot DE, Lindsay JF (1986) The Beacon Supergroup (Devonian-Triassic) and Ferrar Group (Jurassic) in the Beardmore Glacier area, Antarctica. In, MD Turner and JF Splettstoesser, eds, Geology of the Central Transantarctic Mountains. Antarctic Research Series, 36:339–428. Amer, Geophys. Union, Washington, DC

Beck ME Jr, Schmidt DL (1971) Reset direction of remanent magnetization for Upper Cambrian rhyodacite welded tuff, Pensacola Mountains, Antarctica. US Geol Surv Prof. Paper 750-C:174–178

Behrendt JC, Henderson JR, Meister LJ (1966a) Geophysical reconnaissance in the Pensacola Mountains. Antarctic J US 1(4):125–126

Behrendt JC, Henderson JR, Meister LJ (1966b) Airborne geophysical study in the Pensacola Mountains of Antarctica. Science 153:1373–1376

Behrendt JC, Rambo W, Henderson JR, Wanous RE, Meister L (1973) Simple Bouguer gravity and aeromagnetic maps of the Davis Valley Quadrangle and part of the Cordiner Peaks Quadrangle and vicinity, Antarctica. Geophys. Investigatons, Map GP-887, US Geological Survey, Washington, DC

Boyd WW (1991) Crustal evolution in the Pensacola Mountains: Inferences from chemistry and petrology of the igneous rocks and nodule-bearing lamprophyre dykes. In: Thomson MRA, Crame JA, Thomson JW (eds) Geological evolution of Antarctica. Cambridge University Press, Cambridge, pp 557–562

Braun HM (1995) Structural investigations in Proterozoic to Lower Palaeozoic rocks in the Read Mountains and Haskard Highlands of the Shackleton Range, Antarctica. Polarforschung 63(2/3):63–99

Brommer A, Millar IL, Zeh A (1999) Geochronology structural geology and petrology of the northern La Grange Nunataks, Shackleton Range, Antarctica. Terra Antartica, 6(3):

Brommer A, Henjes-Kunst F (1999) Structural investigations and K-Ar geochronology of the northern Herbert Mountains and Mt. Sheffield, Shackleton Range, Antarctica. Terra Antartica 6(3):279–291

Buggisch W, Henjes-Kunst (1999) Stratigraphy, facies, and provenance analyses of the Lower Cambrian Mt. Wegner Formation of the Shackelton Range, Antarctica. Terra Antartica 6(3):211–228

Buggisch W, Kleinschmidt G (1999) New evidence for nappe tectonics in the southern Shackleton Range, Antarctica. Terra Antartica 6(3):203–210

Buggisch W, Kleinschmidt G, Kreuzer H, Krumm H (1990) Stratigraphy, metamorphism, and nappe tectonics in the Shackleton Range (Antarctica). Geodätisch-geophysikalische Veröffentlichungen, Reihe I 15(6):64–86

Buggisch W, Kleinschmidt G, Kreuzer H, Krumm H (1994) Metamorphic and structural evolution of the southern Shackleton Range during the Ross Orogeny. Polarforchung 63(1):33–56

Buggisch W, Bachtadse V, Henjes-Kunst F (1999) Lithostratigraphy, facies, geo-chronology, and paleomagnetic data from the Blaicklock Glacier Group, Shackleton Range, Antarctica. Terra Antartica 6(3):229–239

Cameron RE, Ford AB (1974) Baseline analysis of soils in the Pensacola Mountains. Antarctic J US 9(4):116–119

Clarkson PD (1971) Shackleton Range geological survey. Antarctic J US 6:121–122

Clarkson PD (1972) Geology of the Shackleton Range: A preliminary report. British Antarctic Surv Bull 31:1–15

Clarkson PD (1981) Geology of the Shackleton Range, I: The Shackleton Range metamorphic complex. British Antarctic Surv Bull 51:257–283

Clarkson PD (1982) Tectonic significance of the Shackleton Range. In: Craddock C (ed) Antarctic geoscience. University of Wisconsin Press, Madison, WI, pp 835–839

Clarkson PD (1983) Geology of the Shackleton Range, II: The Turnpike Bluff Group. British Antarctic Surv Bull 52:109–124

Clarkson PD (1995) Introduction. In: Clarkson PD, Tessensohn F, Thomson JW, et al. (eds) Geological map of the Shackleton Range, Antarctica. BAS GEOMAP Series, Sheet 4, 1:250,000, with supplementary text. British Antarctic Survey, Cambridge, 79 pp

Clarkson PD, Brook M (1977) Age and position of the Ellsworth Mountains crustal fragment. Antarctica Nature 265:615–616

Clarkson PD, Wyeth RB (1983) Geology of the Shackleton Range: III. The Blaiklock Glacier Group. British Antarctic Surv Bull 53:233–244

Clarkson PD, Hughes CP, Thomson MRA (1979) Geological significance of a Middle Cambrian fauna from Antarctica. Nature 279:791–792

Clarkson PD, Tessensohn F, Thomson JW, et al. (1995) Geologic map of the Shackleton Range, Antarctica: BAS GEOMAP Series, Sheet 4. British Antarctic Survey, Cambridge

Coward MP, Ries AC (eds) (1996) Collision tectonics. Special Publication 76. Geological Society of London, London

Craddock C (1969a) Geology of the Ellsworth Mountains. In: Craddock C (ed) Geologic map of Antarctica, Sheet 4, Folio 12. American Geographical Society, New York

Craddock C (1969b) Geologic maps of Antarctica. Folio 12. American Geographical Society, New York

Craddock C (ed) (1982) Antarctic geoscience. University of Wisconsin Press, Madison, WI

Craddock C et al. (1986) The geology of the Ellsworth Mountains. One colored map, 33" × 59". Geological Society of America, Map and Chart Series, MC–77

Craddock C, Anderson JJ, Webers GF (1964a) Geologic outline of the Ellsworth Mountains. In: Adie RJ (ed) Antarctic geology. North-Holland, Amsterdam, The Netherlands, pp 155–170

Craddock C, Gast PW, Hanson GN, Linder H (1964b) Rubidium-strontium ages from Antarctica. Geol Soc Amer Bull 75:237–240

Curtis M (2002) Palaeozoic to Mesozoic polyphase deformation of the Patuxent Range, Pensacola Mountains, Antarctica. Antarctic Sci 14(2):175–183

Dalziel IWD, Garrett SW, Grunow AM, Pankhurst RJ, Storey BC, Vennum WR (1987) The Ellsworth-Whitmore Mountains block: Its role in the tectonic evolution of West Antarctica. In: McKenzie GD (ed) Gondwana Six: Structure, tectonics, geophysics. Geophysical Monograph, vol. 40. American Geophysical Union, Washington, DC, pp 173–182

Dalziel IWD, Helper MA, Hutson FE, Grimes SW (1994) Geologic investigations in the Shackleton Range and Coats Land nunataks, Antarctica. Antarctic J US 29(5):4–6

Di Vincenzo G, Palmeri R, Talarico F, Andriesen PAM, Ricci C (1997) Petrology and geochronology of eclogites from the Lanterman Range, Antarctica. J Petrol 38:1391–1417

Eastin R (1970) Geochronology of the basement rocks of the central Transantarctic Mountains, Antarctica. PhD dissertation, Department of Geology and Mineralogy, The Ohio State University, Columbus, OH

Eastin R, Faure G (1971) The age of the Littlewood Volcanics of Caots Land, Antarctica. J Geol 79:241–245

Elverhoi A (1981) Evidence for a late Wisconsin glaciation of the Weddell Sea. Nature 293:641–642

Evans KR, Rowell AJ, Rees MN (1991) Sea-level fluctuations and the evolution of a Middle Cambrian carbonate ramp in the Neptune Range. Antarctic J US 26(5):47–49

Faure G (1986) Principles of isotope geology, 2nd edn. Wiley, New York

Faure G (2001) Origin of igneous rocks; the isotopic evidence. Springer, Heidelberg, Germany

Faure G, Mensing TM (2005) Isotopes: Principles and applications, 3rd edn. Wiley, Hoboken, NJ

Faure G, Hill RL, Eastin R, Montigny RJE (1968) Age determinations of rocks and minerals from the Transantarctic Mountains. Antarctic J US 3(5):173–175

Faure G, Eastin R, Ray PT, McLelland D, Schultz CH (1977) Geochronology of igneous and metamorphic rocks, central Transantarctic Mountains. In: Laskar B, Raja Rao CS (eds) Fourth international Gondwana symposium. Hindustan, Delhi-110007, India, pp 805–813

Ford AB (1964) Cordierite-bearing hypersthene-quartz-monzonite-porphyry in the Thiel Mountains and its regional importance. In: Adie RJ (ed) Antarctic geology. North-Holland, Amsterdam, The Netherlands, pp 429–441

Ford AB (1970) Development of the Layered Series and capping Granophyre of the Dufek Intrusion of Antarctica. In: Symposium on the Bushveld Igneous Complex and Other Layered Intrusions. Special Publication, vol. 1. The Geological Society of South Africa, Capetown, South Africa, pp 492–510

Ford AB (1972) Weddell Orogeny; Latest Permian to early Mesozoic deformation at the Weddell Sea margin of the Transantarctic Mountains. In: Adie RJ (ed) Antarctic geology and geophysics. Universitetsforlaget, Oslo, Norway, pp 419–425

Ford AB (1977) Geological comparison of the Shackleton Range and the Pensacola Mountains during the 22nd Soviet Antarctic Expedition. Antarctic J US 12(4):88–90

Ford AB, Aaron JM (1962) Bedrock geology of the Thiel Mountains, Antarctica. Science 137:751–752

Ford AB, Himmelberg GR (1976) Cordierite and orthopyroxene megacrysts in late Precambrian volcanic rocks of the Thiel Mountains. Antarctic J US 11(4):260–262

Ford AB, Kistler RW (1980) K-Ar age, composition, and origin of Mesozoic mafic rocks related to the Ferrar Group, Pensacola Mountains, Antarctica. New Zealand J Geol Geophys 23:371–390

Ford AB, Sumsion RS (1971) Late Precambrian silicic pyroclastic volcanism in the Thiel Mountains, Antarctica. Antarctic J US 6(5):185–186

Ford AB, Tabor RW (1971) The Thiel Mountains pallasite of Antarctica. Geol Surv Prof. Paper 750-D:56–60

Ford AB, Hubbard HA, Stern TW (1963) Lead-alpha ages of zircon in quartz-monzonite porphyry, Thiel Mountains, Antarctica: A preliminary report. US Geol Surv Prof. Paper 450E:105–107

Ford AB, Schmidt DL, Boyd WW Jr, Nelson WH (1978a) Geologic map of the Saratoga Table Quadrangle, Pensacola Mountains, Antarctica. U.S. Antarctic Research Program, Map A-9. US Geological Survey, Washington, DC

Ford AB, Schmidt DL, Boyd WW Jr (1978b) Geologic map of the Davis Valley Quadrangle and part of the Cordiner Peaks Quadrangle, Pensacola Mountains, Antarctica. U.S. Antarctic Research Program, Map A-10. US Geological Survey, Washington, DC

Frischbutter A (1981) Gliederung, Bau and Entwicklung des Transantarktischen Gebirges im Bereich der Neptune Range (Antarktis). Teil 1: Stratigraphie und regional Einbindung. Zeitschrift für Geologische Wissenschaften 9:817–833

Frischbutter A (1982) Gliederung, Bau und Entwicklung des Transantarktischen Gebirges im Bereich der Neptune Range (Antarktis). Teil 2: Deformation, Metamorphose und Gesamtentwicklung. Zeitschrift für Geologische Wissenschaften 10:165–180

Frischbutter A, Vogler P (1985) Contributions to the geochemistry of magmatic rocks in the upper Precambrian lower Paleozoic profile of the Neptune Range, Transantarctic Mountains, Antarctica. Zeitschrift für Geologische Wissenschaften 13:345–357

Fuchs R, Sir (1970) Shackleton Range traverse, 1969–1970. Antarctic J US 5(4):107–108

Fuchs V, Sir (1969) Shackleton Range traverse. Antarctic J US 4(4):139–140

Fuchs VE, Hillary E (1959) The crossing of Antarctica; The Commonwealth Trans-Antarctic Expedition, 1955–58. London

Golovanov NP, Mil'shteyn VE, Mikhaylov VM, Shulyatin OG (1979) Stromatoliths and microphytoliths of the Shackleton Range (Western Antarctica). Doklady Akademii Nauk, SSSR 249(4):977–979 (In Russian)

Golovanov NP, Mikhailov VM, Stulyatin OG (1980) First diagnostic stromatolites from Antarctica and their biostratigraphical significance. Antarktica 19:152–159

Gottfried D, Jaffe HW, Senftle RE (1959) Evaluation of the lead-alpha (Larsen) method for determining ages of igneous rocks. US Geol Surv Bull 1097-A:63

Grew ES (1977) With the Soviets in Antarctica, austral summer 1976–1977. Antarctic J US 12(4):86–88

Grew ES, Halpern M (1979) Rubidium-strontium dates from the Shackleton Range Metamorphic Complex in the Mount Provender area, Shackleton Range, Antarctica. J Geol 87:325–332

Grew ES, Manton WI (1980) Uranium-lead ages of zircons from Mount Provender, Shackleton Range, Transantarctic Mountains. Antarctic J US 15(5):45–46

Grunow AM, Dalziel IWD, Kent DV (1987) Ellsworth-Whitmore Mountains crustal block, western Antarctica: New paleomagnetic results and their tectonic significance. In: McKenzie GD (ed) Gondwana Six: Structure, tectonics, and geophysics. Geophysical Monograph, vol. 40. American Geophysical Union, Washington, DC, pp 161–171

Gunn BM, Warren G (1962) Geology of Victoria Land between the Mawson and Mulock Glaciers, Antarctica. New Zealand Geol Survey Bull 71:1–157

Harland WB et al. (1990) A geologic time scale 1989. Cambridge University Press, Cambridge, UK

Höfle HC, Buggisch W (1993) Glacial geology and petrography of erratics in the Shackleton Range, Antarctica. Polarforschung 63(2/3):183–201

Hofmann J, Paech HJ (1980) Zum strukturgeologischen Bau am Westrand der ostantarktischen Tafel. Zeitschrift für Geologische Wissenschaften 8:425–537

Hofmann J, Paech H-J (1983) Tectonics and relationships between structural stages in the Precambrian of the Shackleton Range, western margin of the East Antarctic craton. In: Oliver RL, James PR, Jago JG (eds) Antarctic earth science. Australian Academy of Science, Canberra, A.C.T., pp 183–189

Hofmann J, Kaiser G, Klemm W, Paech H-J (1980) K/Ar Alter von Doleriten und Metamorphiten der Shackleton Range und der Wichaway Nunataks, Ost- und Südostumrandung des Filchner Eisschelfs (Antarktis). Zeitschrift für Geologische Wissenschaften 8(9):1227–1232

Hofmann J, Pilot HJ, Schlichting M (1981) Das Rb-Sr Alter von Metamorphiten der Herbert Mountains, Shackelton Range, Antarctica. Zeitschrift für Geologische Wissenschaften 9(8):835–842

Huffman JW, Schmidt DL (1966) Pensacola Mountains project. Antarctic J US 1(4):123–124

IUGS (2002) International stratigraphic chart. International Commission on Stratigraphy, UNESCO, New York

Kerr A, Hermichen WD (1999) Glacial modification of the Shackleton Range, Antarctica. Terra Antartica 6(3):353–360

Kleinschmidt G, Buggisch W (1994) Plate tectonic implications of the structure of the Shackleton Range, Antarctica. Polarforschung 63:57–62

Kovach J, Faure G (1978) Rubidium-strontium geochronology of granitic rocks from Mt. Chapman, Whitmore Mountains, West Antarctica. Antarctic J US 13(4):17–18

Leat PT, Riley TR, Storey BC, Kelley SP, Millar IL (2000) Middle Jurassic ultramafic lamprophyre dyke within the Ferrar magmatic province, Pensacola Mountains. Antarctica Mineral Mag 64:95–111

Lisker F, Schäfer T, Olesch M (1999) The uplift/denudation history of the Shackleton Range (Antarctica) based on fission-track analyses. Terra Antartica 6(3):345–352

Long WE (1962) Sedimentary rocks of the Buckeye Range, Horlick Mountains, Antarctica. Science 136(3513):319–321.

Long WE (1965) Stratigraphy of the Ohio Range, Antarctica. In: Hadley JB (ed) Geology and paleontology of the Antarctic. Antarctic Research Series, vol. 6. American Geophysical Union, Washington, DC, pp 71–116

Marsh PD (1983a) The late Precambrian and early Paleozoic history of the Shackleton Range, Coats Land. In: Oliver RL, James PR, Jago JB (eds) Antarctic earth science. Australian Academy of Science, Canberra, ACT, pp 190–193

Marsh PD (1983b) The stratigraphy and structure of the Haskard Highlands and Otter Highlands of the Shackleton Range. British Antarctic Surv Bull 60:23–43

Marsh PD (1984) The stratigraphy and structure of Lagrange Nunataks, northern Fuchs Dome and Herbert Mountains of the Shackleton Range. British Antarctic Surv Bull 63:19–40

Meer JJ van der, Mücher HJ, Höfle HC (1992) Micromorphological observations on till samples from the Shackleton Range and North Victoria Land, Antarctica. Polarforschung 62(1): 57–65

Millar IL, Pankhurst RJ (1987) Rb-Sr geochronology of the region between the Antarctic Peninsula and the Transantarctic Mountains. In: McKenzie GD (ed) Gondwana Six: Structure, tectonics, and geophysics. Geophysical Monograph, vol. 40. American Geophysical Union, Washington, DC, pp 151–160

Millar IA, Storey BC (1995) Early Paleozoic rather than Neoproterozoic volcanism and rifting withing the Transantarctic Mountains. J Geol Soc London 152:417–460

Nelson WH (1981) The Gale Mudstone: A Permian (?) tillite in the Pensacola Mountains and neighboring parts of the Transantarctic Mountains. In: Hambrey MJ, Harland WB (eds) Earth's Pre-Pleistocene glacial record. Cambridge University Press, Cambridge, pp 227–229

Olesch M, Braun H-M, Kamenev EN, Kameneva GI, Schubert W (1995) The Read Group. In: Clarkson PD, Tessensohn F, Thomson JW, et al. (eds) Geological map of the Shackleton Range, Antarctica. BAS GEOMAP Series, Sheet 4, 1:250,000, with supplementary text. British Antarctic Survey, Cambridge, 79 pp

Palmer AR, Gatehouse CG (1972) Early and Middle Cambrian trilobites from Antarctica. US Geol Surv, Prof. Paper 456-D

Pankhurst RJ, Marsh PD, Clarkson PD (1983) A geochronologic investigation of the Shackleton Range. In: Oliver RL, James PR, Jago JB (eds) Antarctic earth science. Australian Academy of Science, Canberra, A.C.T., pp 176–182

Pankhurst RJ, Storey BC, Millar IL, Macdonald DIM, Vennum WR (1988) Cambro-Ordovician magmatism in the Thiel Mountains, Transantarctic Mountains, and implication for the Beardmore Orogeny. Geol 16:246–249

Peacock SM, Goodge JW (1995) Eclogite facies metamorphism preserved in tectonic blocks from a lower crustal shear zone, central Transantarctic Mountains, Antarctica. Lithos 36:12–13

Popov LE, Solov'ev LA (1981) Middle Cambrian inarticulate brachiopoda, chancelloria, coniconchs, and trilobites of western Antarctica (Shackleton and Argentina ranges). Antarktika 20:64–72

Pujato H (1977) Los vuelos precursors del General Pujato. Antartida 8:32–39.

Rex DC (1972) K-Ar age determinations on volcanic and associated rocks form the Antarctic Peninsula and Dronning Maud Land. In: Adie RJ (ed) Antarctic geology and geophysics. Universitetsforlaget, Oslo, Norway, pp 133–136

Rowell AJ, Rees MN, Evans KR (1992) Evidence of major Middle Cambrian deformation in the Ross orogen, Antarctica. Geol 20:31–34.

Rowell AJ, Van Schmus WR, McKenna LW III, Evans KR (1994) Early Paleozoic continental-rise deposition off East Antarctica: The Patuxent Formation of the Pensacola Mountains. Antarctic J US 29(5):42–44

Rowell AJ, Van Schmus WR, Storey BC, Fetter AH, Evans KR (2001) Latest Neoproterozoic to Mid-Cambrian age for the main deformation phases of the Transantarctic Mountains: New stratigraphic and isotopic constraints from the Pensacola Mountains, Antarctica. J Geol Soc London 158:295–308

Schmädike E, Will TM (2006) First evidence of eclogite facies metamorphism in the Shackleton Range, Antarctica: Trace of a suture between East and West Gondwana? Geol 34(3):133–136

Schmidt DL (1964) Geology of the Pensacola Mountains. Bull US Antarctic Projects Officer 5(10):98–101

Schmidt DL (1969) Precambrian and lower Paleozoic igneous rocks, Pensacola Mountains, Antarctica. Antarctic J US 4(5):203–204

Schmidt DL, Ford AB (1963) U.S. Geological Survey in the Patuxent Mountains. Bull US Antarctic Projects Officer 4(8):20–24

Schmidt DL, Ford AB (1966) Geology of the northern Pensacola Mountains and adjacent areas. Antarctic J US 1(4):125

Schmidt DL, Ford AB (1969) Geology of the Pensacola and Thiel Mountains. In: Craddock C et al. (eds) Geologic maps of Antarctica. Antarctic Map Folio Series, Sheet 5, Folio 12. American Geographical Society, New York

Schmidt DL, Williams PL (1967) Continental glaciation of late paleozoic age, Pensacola Mountains, Antarctica. In, AJ Amos, ed., Gondwana Stratigraphy, 2:91–106. UNESCO, Paris

Schmidt DL, Williams PL (1969) Continental glaciation of late Paleozoic age Pensacola Mountains, Antarctica. In, Gondwana Stratigraphy. IUGS Symposium, Buenos Aires, 1967. UNESCO (Paris), Earth Sci Ser 2:617–649

Schmidt DL, Dover JH, Ford AB, Brown RD (1964) Geology of the Patuxent Mountains. In: Adie RJ (ed) Antarctic geology. North-Holland, Amsterdam, The Netherlands, pp 276–283

Schmidt DL, Williams PL, Nelson WH, Ege JR (1965) Upper Precambrian and Paleozoic stratigraphy and structure of the Neptune Range, Antarctica. US Geol Surv Prof. Paper 525-D:112–119

Schmidt DL, Williams PL, Nelson WH (1978) Geologic map of the Schmidt-Hills quadrangle and part of the Gambacorta quadrangle, Pensacola Mountains, Antarctica. US Antarctic Research Program, Map A-8. US Geological Survey, Reston, VA

Schopf JM (1964) Paleobotanical studies in Antarctica. Geological Society of America, Special Paper 76, Abstracts for 1963, p 317

Schopf JM (1968) Studies in Antarctic paleobotany. Antarctic J US 3(5):176–177

Schopf JM (1969) Ellsworth Mountains: Position in West Antarctica due to sea-floor spreading. Science 164:63–66

Shackleton RM (1996) Precambrian collision tectonics in Africa. In: Coward MP, Ries AC (eds) Collision tectonics. Special Publication, vol. 76. Geological Society London, London, pp 345–362

Skidmore MJ, Clarkson PD (1972) Physiography and glacial geomorphology of the Shackleton Range. British Antarctic Surv Bull 30:69–80

Solov'ev IA, Grikurov GE (1978) First finds of Middle Cambrian trilobites in the Shackleton Range (Antarctica). Antarktika Doklady Komissii 17:187–198. Izdatelstvo "Nauka", Moscow, Russia

Spaeth G, Hotten R, Peters M, Techmer K (1995) Mafic dykes in the Shackleton Range, Antarctica. Polarforschung 63(2/3):101–121

Stephenson PJ (1966) Geology. 1. Theron Mountains, Shackleton Range, and Wichaway Nunataks. Commonwealth Trans-Antarctic Expedition 1955–1958. Science Rept No 8:1–79

Storey BC (1996) Microplates and mantle plumes in Antarctica. Terra Antartica 3(2):91–102

Storey BC, Dalziel IWD (1987) Outline of the structural and tectonic history of the Ellsworth Mountains – Thiel Mountains Ridge. In: McKenzie GD (ed) Gondwana Six: Structure, tectonics, and geophysics. Geophysical Monograph, vol. 40. American Geophysical Union, Washington, DC, pp 117–128

Storey BC, Macdonald DIM (1987) Sedimentary rocks of the Ellsworth - Thiel Mountains Ridge and their regional equivalents. British Antarctic Surv Bull 76:21–49

Storey BC, Alabaster T, Macdonald DIM, Millar IL, Pankhurst RJ, Dalziel IWD (1992) Upper Proterozoic rift-related rocks of the Pensacola Mountains, Antarctica: Precursors to supercontinent breakup? Tectonics 11:1392–1405

Storey BC, Macdonald DIM, Dalziel IWD, Isbell JL, Millar IL (1996) Early Paleozoic sedimentation, magmatism, and deformation in the Pensacola Mountains, Antarctica: The significance of the Ross Orogeny. Geol Soc Amer Bull 108:685–707

Stump E (1995) The Ross orogen of the Transantarctic Mountains. Cambridge University Press, Cambridge

Talarico F, Kroner U (1999) Geology and tectono-metamorphic evolution of the Read Group, Shackleton Range: A part of the East Antarctic craton. Terra Antartica 6(3):183–202

Talarico F, Kleinschmidt G, Henjes-Kunst F (1999) An ophiolite complex in the northern Shackleton Range, Antarctica. Terra Antartica 6(3):293–315

Tessensohn F (1997) Shackleton Range, Ross orogen and SWEAT hypothesis. In: Ricci CA (ed) The Antarctic Region: Geological evolution and processes. Proceedings of the seventh international symposium on Antarctic Science, Terra Antartica, Siena, Italy, pp 5–12

Tessensohn F, Thomson MRA (1999) The EUROSHACK Project: A brief outline. Terra Antartica 6(3):175–182

Tessensohn F, Kleinschmidt G, Talarico F, Buggisch W, Brommer A, Henjes-Kunst F, Kroner U, Millar IL, Zeh A (1999a) Ross-age amalgamation of East and West Gondwana: Evidence from the Shackleton Range, Antarctica. Terra Antartica 6(3):317–325

Tessensohn F, Kleinschmidt G, Buggisch W (1999b) Permo-Carboniferous glacial beds in the Shackleton Range. Terra Antartica 6(3):337–344

Thiel E (1961) Antarctica, one continent or two? Polar Rec 10(67):335–348

Thomson MRA (1972) Inarticulate brachiopoda from the Shackleton Range and their stratigraphic significance. Bull Brit Antarct Surv 31:17–20

Thomson MRA, Weber B (1999) Discovery of an Ordovician invertebrate fauna in the Blaiklock Glacier Group, Shackleton Range, Antarctica. Terra Antartica 6(3):241–248

Thomson MRA, Solov'ev IA, Buggisch W (1995) Trilobite shales. In: Clarkson PD, Tessensohn F, Thomson JW, et al. (eds) Geological map of the Shackleton Range, Antarctica. BAS GEOMAP Series, Sheet 4, 1:250,000, with supplementary text. British Antarctic Survey, Cambridge, 79 pp

Van Schmus WR, McKenna LW, Gonzales DA, Fetter AH, Rowell AJ (1998) U-Pb geochronology of the Pensacola, Thiel, and Queen Maud Mountains, Antarctica. In: Ricci CA (ed) The Antarctic Region: Geological evolution and processes. Terra Antartica, Siena, Italy, pp 187–200

Vennum WR, Storey BC (1987a) Petrology, geochemistry, and tectonics setting of granitic rocks from the Ellsworth-Whitmore Mountains crustal block and Thiel Mountains, West Antarctica. In: McKenzie GD (ed) Gondwana Six: Structure, tectonics, and geophysics. Geophysical Monograph, vol. 40. American Geophysical Union, Washington, DC, pp 139–150

Vennum WR, Storey BC (1987b) Correlation of gabbroic and diabasic rocks from the Ellsworth Mountains, Hart Hills, and Thiel Mountains, West Antarctica. In: McKenzie GD (ed) Gondwana Six: Structure, tectonics, and geophysics. Geophysical Monograph, vol. 40. American Geophysical Union, Washington, DC, pp 129–138

Walker PT (1961) Study of some rocks and minerals from the Dufek Massif, Antarctica. Reports of Antarctic Geological Observations, 1956–1960. IGY Glaciol. Rept. Ser., No. 4:195–213. IGY World Data Center A, Glaciology, American Geographical Society, New York

Weber K (1991) Microfossils in Proterozoic sediments from the southern Shackleton Range, Antarctica: A preliminary report. Zeitschrift für Geologische Wissenschaften 19(2):185–197

Weber W (1982) Beitrag zur Geologie des Pensacola-Gebirges (Antarktika). Freiberger Forschungshefte 371:41–96

Weber W, Fedorov LV (1980) Zur Geologie des nördlichen Teiles des Neptune Range/Pensacola Gebirges (Antarctika). Gedätische und Geophysikalische Veröffentlichungen R1(8):68–94

Weber W, Fedorov LV (1981) Geology of the northern Neptune Range, Pensacola Mountains. Geodätische und Geophysikalische Veröffentlichungen. Ser 1(8):68–94

Webers GF, Craddock C, Rogers MA, Anderson JJ (1982) Geology of the Whitmore Mountains. In: Craddock C (ed) Antarctic geoscience. University of Wisconsin Press, Madison, WI, pp 841–847

Webers GF, Craddock C, Splettstoesser JF (eds) (1992) Geology and paleontology of the Ellsworth Mountains. West Antarctica Geological Society of America, Memoir 170, Boulder, CO

Williams PL (1969) Petrology of upper Precambrian and Paleozoic sandstones in the Pensacola Mountains, Antarctica. J Sed Pet 39:1455–1465

Williams PL, Ford AB (1968) Pensacola Mountains project. Antarctic J US 3(5):181–182

York D (1966) Least squares fitting of a straight line. Canadian J Phys 44:1079–1086

York D (1967) The best isochron. Earth Planet Sci 2:479–482

Zeh A, Millar IL, Kroner U, Görz I (1999) The structural and metamorphic evolution of the northern Haskard Highlands, Shackleton Range, Antarctica. Terra Antartica 6:249–268

Chapter 9
From Rodinia to Gondwana

The standard model for the origin of the Transantarctic Mountains postulates that during the Neoproterozoic Era a supercontinent existed called Rodinia which included what later became the East Antarctic craton. This supercontinent split into two large continental fragments called Laurentia and Gondwana. The continental fragment we call Laurentia drifted away (was displaced) and eventually became the North American continent. Gondwana originally included the land areas of Australia, East Antarctica, and India. The rift valley that split Rodinia widened into an ocean which was bordered by passive rift margins that formed the coasts of East Antarctica and of Laurentia. Along both coasts turbidite sequences composed of alternating layers of graywacke and shale were deposited. At the end of the Neoproterozoic the passive rift margin of East Antarctica was transformed into an active subduction zone along which the siliciclastic turbidites and locally developed platform carbonates and volcanic rocks were folded by compression against the mainland and, in the process, were regionally metamorphosed and intruded by granitic magma of anatectic origin (Stump 1995). The details of this process are the subject of Chapters 3–8 of this book.

Before we continue to discuss the evolution of the Transantarctic Mountains, we elaborate what is known about the rifting of Rodinia, the fate of Laurentia that drifted away, and the assembly of Pangea which formed in the late Paleozoic and then broke up during the Mesozoic Era. The resulting continental fragments drifted apart and thereby formed the present continents of the Earth. That, in a few words, is how the continents we know today came into existence. The details of this process are complex and still under discussion at the periodic Gondwana conferences and SCAR meetings: Laskar and Raja Rao (1979), Craddock (1982), Oliver et al. (1983), McKenzie (1987), Ulbrich and Rocha Campos (1990), Thomson et al. (1991), Findlay et al. (1993), Ricci (1997), Gamble et al. (2002), and Fütterer et al. (2006).

9.1 Continental Drift

Even a casual perusal of the map of the Earth suggests that Africa and South America fit together. This observation was elaborated by Alfred Wegener in a book published in 1912: "Die Entstehung der Kontinente und Ozeane". In this book, Wegener presented an impressive amount of geologic evidence to support his hypothesis that all of the continents of the Earth are fragments of a supercontinent he named Pangea which was fragmented during the Mesozoic.

Pangea (meaning "all earth") came into existence during the Paleozoic by the assembly of fragments of continental crust which were scattered over the surface of the Earth (Siever 1983; Emiliani 1987, 1992; Press and Siever 1986). Figure 9.1 indicates that Pangea consisted of a northern part called Laurasia and a southern part called Gondwana. The name Laurasia is a contraction of Laurentia (Canadian shield) and Eurasia. The name of Gondwana was coined by the Austrian geologist Eduard Suess (1831–1914) after the land of the Gonds in India where tillite and coal deposits occur like those that occur on all of the southern continents. The name which Suess originally proposed was "Gondwanaland", but "Gondwana" is now preferred because it means the "land of Gonds" in the native language.

Wegener's hypothesis was rejected by most American geologists because he could not explain why continents appear to drift (be displaced) on the surface of the Earth. Nevertheless, Wegner's book went through at least four

G. Faure and T.M. Mensing, *The Transantarctic Mountains: Rocks, Ice, Meteorites and Water*,
DOI 10.1007/978-90-481-9390-5_9, © Springer Science+Business Media B.V. 2010

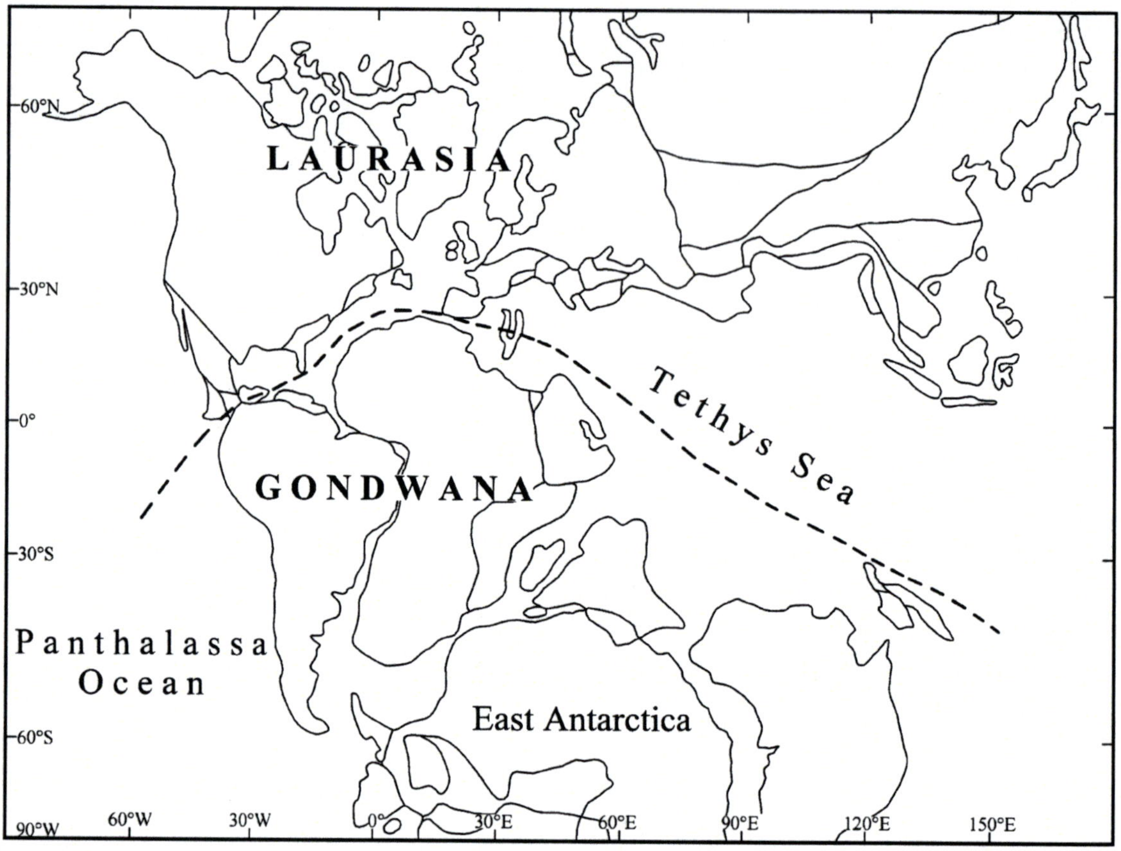

Fig. 9.1 The supercontinent Pangea was composed of Laurasia and Gondwana during the Jurassic Period when it began to disintegrate into the continents and islands we know today. The gap between Laurasia and Gondwana was occupied by the Tethys Sea which evolved into the Mediterranean Sea. Note that Antarctica was not located at the Geographic South Pole at that time and that the southern continents and islands (Australia, India, Ceylon, Madagascar, South Africa, and South America) were clustered around it. The basement rocks and the Beacon Supergroup of the Transantarctic Mountains had already been deposited when Gondwana broke up (Adapted from Fig. 10.11 of Smith 1992)

editions in Europe (Wegener 1929) and was eventually reprinted in the USA (Wegener 1966).

The hypothesis concerning continental drift was supported by Alex L. du Toit in South Africa who became convinced that the southern continents drifted to their present positions after the supercontinent Pangea broke up during the Mesozoic (du Toit 1937).

An important piece of evidence in du Toit's reconstruction was the distribution of tillites of Permian age on the southern continents including Antarctica, South America, South Africa, India, and Australia. The geographic distribution of tillite on the southern continents and the ice-flow direction derived from striations on the glaciated rock surfaces seemed to require the improbable conclusion that the source of the Permian ice sheet was located somewhere in the southern oceans. However, when the continents were reassembled into a coherent body in Fig. 9.2, the areal distribution of tillite deposits indicated that the center of the ice sheet was on Gondwana.

Continental drift was eventually taken seriously after it became possible in the 1950s to track the past movement of the continents by measuring the declination and the inclination of the magnetic field that was embedded in basalt flows at the time they crystallized (van Andel 1992; Smith 1992). The pattern of magnetic anomalies of the ocean floor that was revealed by Vine and Matthews (1963), combined with the recognition of island arcs by Wilson (1963a, 1965, 1973), supported the concept of sea-floor spreading

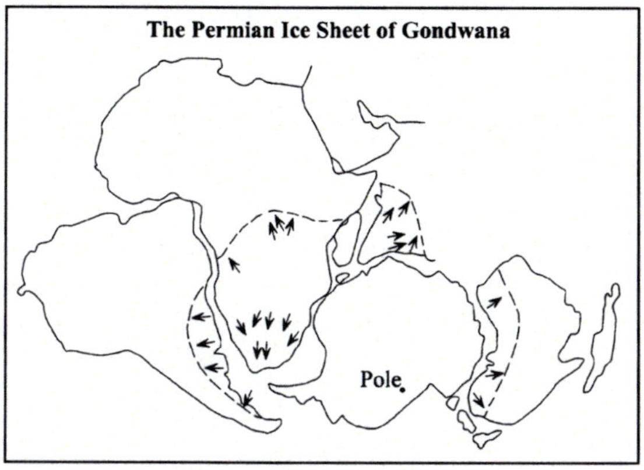

Fig. 9.2 The regional distribution of Permian tillite on the southern continents and the ice-flow directions indicate the size of the Permian ice sheets when the continents are reassembled into Gondwana. This interpretation also indicates that Gondwana broke up after the Permian glaciation and that the continental fragments subsequently drifted to their present positions on the southern hemisphere of the Earth (Adapted from Smith (1992) based on the reconstruction by du Toit (1937))

that was proposed by Dietz (1961) and by Hess (1962). These insights soon led to the theory of plate tectonics (Vine and Hess 1970; Wyllie 1976; Windley 1977) which has become part of the standard curriculum in the Earth Sciences (Press and Siever 1986).

As soon as continental drift (or displacement) was recognized as a consequence of sea-floor spreading in the context of plate tectonics, the re-assembly of past supercontinents became a "hot topic" in geology. The reconstruction of Gondwana by du Toit (1937) was endorsed by Hamilton (1963, 1967) who confirmed that the folded metasedimentary basement rocks of the Transantarctic Mountains extend across the continent in a continuous belt that can be traced from eastern Australia, across Antarctica, to the fold belt at the southern tip of South Africa, and into South America. The evident success of du Toit's reconstruction of Gondwana encouraged many other investigators to attempt to reassemble the continents of the Earth and thereby to reconstruct the geography of the world prior to the disintegration of Pangea (Cox and Hart 1986).

The tectonics of Antarctica in the context of Gondwana were discussed again by Hamilton (1967), as well as by Ford (1972), Elliot (1972, 1975) and others while Frakes and Crowell (1970) used their observations of Permian tillites to refine the positions of Antarctica in Gondwana. At about the same time, Dietz and Sproll (1970), Dietz and Holden (1970), and Dietz et al. (1972) employed computers to match the shapes of continents at the 1,000 fathom depth (i.e., 1 fathom = 1.83 m; defined as the distance from the tip of the middle finger on one hand to the tip of the middle finger on the other hand of the fully extended arms of a large man). All of these and other studies repeatedly confirmed the reconstruction of Gondwana as originally postulated by Alfred Wegener and Alex du Toit.

Hurley and Rand (1969) reassembled the continents by tracing orogenic belts which they identified by means of geological information and by isotopic age determinations of basement rocks. The distribution of isotopic dates revealed that all of the present continents in Fig. 9.3 contain nuclei composed of high-grade metamorphic and plutonic igneous rocks that crystallized prior to about 1,700 Ma. These continental nuclei are surrounded by younger mountain belts composed of folded and metamorphosed volcano-sedimentary complexes intruded by anatectic granitic plutons. Hurley and Rand (1969) noted that in the pre-drift reconstruction of Pangea the continental nuclei cluster tightly within Laurasia and Gondwana. Therefore, they concluded that the nuclei had remained in their present locations from 1,700 to 200 Ma and that they had been overprinted and surrounded by orogenic fold belts during that interval of geologic time. Hurley and Rand (1969) expressed their interpretation of Fig. 9.3 by the statement:

> It seems unlikely that during the time between 1700 and 200 million years ago the continents were scattered and drifting, only to be reassembled with this degree of ordering at 200 million years ago. Instead, there appear to have been nonmoving ancient nuclei and continental accretion up to the time of the great rift.

We proposed an alternative interpretation that the continental nuclei are fragments of a supercontinent that formed spontaneously in the course of Precambrian time. The hypothetical Precambrian supercontinent was broken up by rifting and the fragments were dispersed by movement of the tectonic plates in which they were embedded. The microcontinents became surrounded by belts of folded volcano-sedimentary complexes that piled up around them when the continental nuclei got stuck in oceanic subduction zones. The driving force that causes rifting of continents originates from plumes of hot rock that rise through the asthenosphere and spread out against the underside of the continental lithosphere. As a result, fractures form

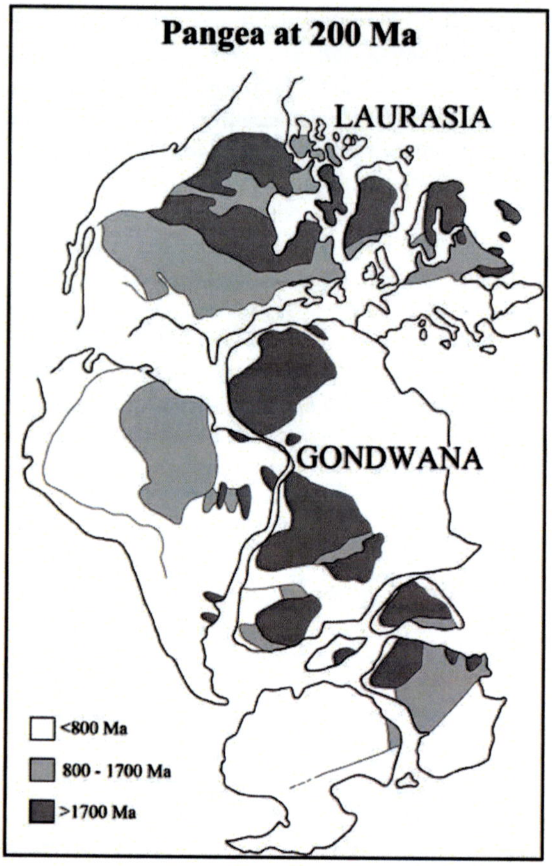

Fig. 9.3 When the continents are assembled into the supercontinent of Pangea, each continent is seen to contain a nucleus of high-grade metamorphic and plutonic rocks that crystallized more than 1,700 million years ago. Even East Antarctica contains such a nucleus although this was not known at the time this map was first assembled in the late 1960s. The interpretation of this information is problematical. Hurley and Rand (1969) concluded that "... there was no significant fragmentation or scattering of the continental nuclei prior to the great drift episode." An alternative interpretation is suggested in the text (Adapted from Hurley and Rand 1969)

in the overlying lithosphere and propagate upward toward the surface. The development of fractures causes decompression of the plume heads and of the surrounding lithospheric mantle, which triggers partial melting on a large scale leading to the eruption of volcanic rocks at the surface. Evidence of the interaction of mantle plumes with the lithospheric mantle was presented in a book by Faure (2001) which included, among many other examples, a detailed account of igneous petrogenesis in the volcanoes of Hawaii.

According to present knowledge, Pangea formed during the Paleozoic Era from fragments of continental crust that collided with the growing supercontinent. These fragments presumably originated by the fragmentation of previously existing continents. In other words, before large continents like Gondwana and Laurasia could form, pre-existing continents had to break up to produce the fragments. In this way, the growth of Pangea by the accretion of microcontinents implies the earlier existence of a land mass that disintegrated.

9.2 The SWEAT Hypothesis

The study of the structure and tectonic evolution of Antarctica was stimulated by a proposal that the East Antarctic craton was originally connected to North America (Moores 1991; Dalziel 1991; Hoffman 1991). This startling hypothesis was preceded by reports of Hartnady (1986, 1991) and was prompted by the work of the Bell and Jefferson (1987) who had previously correlated the late Precambrian rocks of western Canada with similar rocks in South Australia. These authors concluded that a rift had formed between western Canada and Australia at about 1,200 Ma and that both continental fragments had developed complementary passive rift margins by about 750 Ma.

The bold proposal of Bell and Jefferson (1987) was followed by the even more provocative hypothesis of Moores (1991) who postulated that the East Antarctic craton was once connected to the Yavapai-Mazatzal orogenic belt of southwestern USA where basement rocks crystallized between 2,100 and 1,500 Ma. The title of the Moores (1991) paper was: "Southwest U.S.-East Antarctic connection..." which he abbreviated to the catchy acronym SWEAT. The orogenic belts of North America, Australia, India, Africa, South America, and Eurasia that enabled Moores to propose his continental reconstruction had been previously delineated by Hurley and Rand (1969). Although the SWEAT hypothesis was unorthodox, it was actually based on a large amount of geological, structural, and geophysical evidence, including the insight that the Cordilleran Mountains of North America and the Transantarctic Mountains had both formed along passive rift margins.

The re-alignment of East Antarctica, Australia, India, and North America proposed by the SWEAT hypothesis

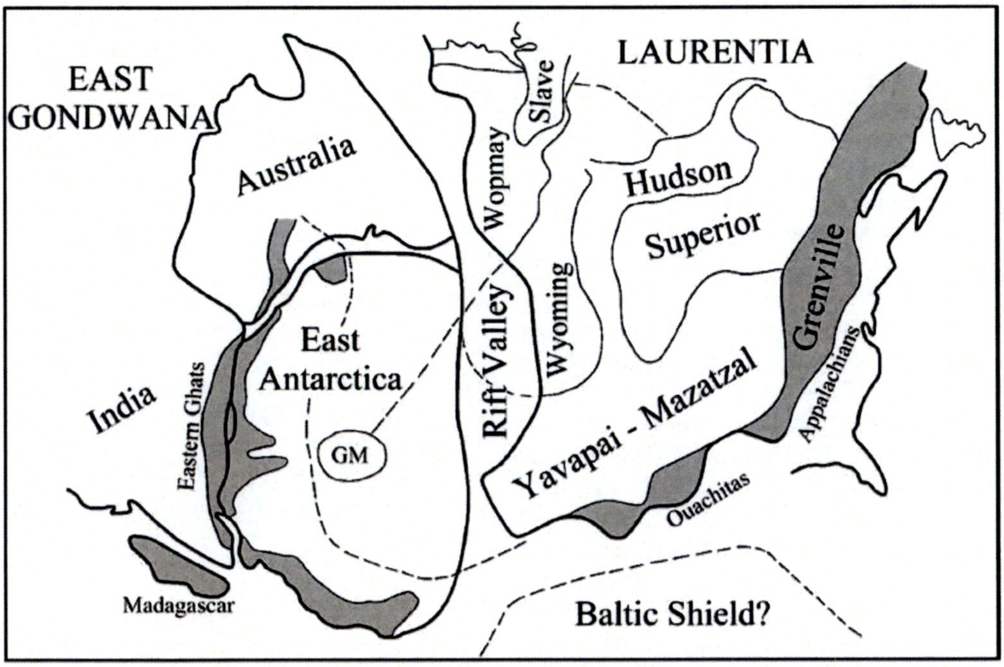

Fig. 9.4 During the Neoproterozoic Era the supercontinent Rodinia split into East Gondwana and Laurentia. As the rift valley widened into an ocean, graywacke-shale sequences were deposited along the passive rift margins of both continental fragments. The turbidites that were deposited along the coast of East Gondwana became part of the Transantarctic basement complex that was compressed and metamorphosed during the Ross Orogeny. The sediments that accumulated along the rift margin of Laurentia are now represented by the rocks of the Belt-Purcell Group (Middle to Late Proterozoic) of Canada and Montana and by the Windermere Group (Neoproterozoic to early Paleozoic) which can be traced from northern Canada into California (Price 1964; Dickinson 1977; Winston 1986). The Transantarctic Mountains and the Cordillera are not shown in this diagram because they formed after the deposition of turbidites that occurred along the respective rift margins (Adapted from Moores 1991)

in Fig. 9.4 reveals several interesting relationships, one of which concerns the Grenville structural province of eastern North America (1,500–800 Ma). This province extends from eastern Canada south into Texas and then continues around the Precambrian core of East Antarctica. In the area where India is joined to East Antarctica, Grenville-age rocks occur in the Eastern Ghats Mountains and in Madagascar. The Genville belt continues from East Antarctica into the southern region of Western Australia. In other words, the cratons of all three continents are connected to each other by a broad belt of Grenville-age schists, gneisses, mafic complexes, and anorthosites. This consequence of the SWEAT hypothesis certainly challenges the Principle of Least Astonishment, but it has stimulated research into the tectonic evolution of the continents which has already increased our understanding of global tectonics.

Another interesting consequence of the SWEAT reconstruction is that the Yavapai-Mazatzal orogenic belt of North America (1,500–2,100 Ma) extends into the ice-covered interior of the East Antarctic craton and therefore provides a clue to the geology of this part of Antarctica about which we still know very little, in spite of conjectures by Grikurov et al. (1972) and Grikurov (1982). Similarly, Moores (1991) projected the Wopmay orogenic belt of Canada into the East Antarctic craton and noted that it converges in the subglacial Gamburtsev Mountains with the extension of the Yavapai-Mazatal tectonic province. He also noted that the boundary between the Wyoming and Wopmay orogenic belts, known as the Snowbird Zone in northwest Canada, has a distinctive geophysical signature which may be detectable in the ice-covered interior of East Antarctica by appropriately designed geophysical surveys of magnetism and gravity. Alternatively, the Wyoming craton (>2,100 Ma) may continue into the East Antarctic craton, whereas the Wopmay belt extends into Australia. These and other consequences of the SWEAT hypothesis have stimulated a lively debate in the geological literature which has already produced new insights into the tectonic relationship of East Antarctica and the other southern continents: (e.g., Dalziel 1992; Storey et al. 1994; Tessensohn 1997; Harley et al. 1997; Gose et al. 1997; Unrug 1997; Jacobs et al. 1999; Goodge 2001; Goodge et al. 2002, 2008; and many others).

The rifting of Rodinia depicted in Fig. 9.4 occurred after about 750 Ma and caused sediment to be deposited along the rift margin of East Gondwana which subsequently was compressed and metamorphosed during the Cambro-Ordovician Ross Orogeny and now forms the metasedimentary component of the basement of the Transantarctic Mountains. The sediment that accumulated along the rift margin of Laurentia is represented by the Belt-Purcell Group of Montana and by the Windermere Group that extends south from northwestern Canada into California (Price 1964; Dickinson 1977; Winston 1986). Evidently the SWEAT hypothesis offers a plausible explanation for the simultaneous deposition of similar sedimentary sequences in similar tectonic and environmental settings that now occur on different continents in opposite hemispheres of the Earth.

During the Ross Orogeny, East Gondwana, consisting of Australia, East Antarctica, and India, collided with West Gondwana (Africa and South America) as evidenced by the geology of the Shackleton Range (Section 8.4). In addition, the size of Gondwana continued to grow by the accretion of several continental fragments such as the Bowers and Robertson Bay terranes of northern Victoria Land (Sections 4.4 and 4.5) and of several as yet poorly known microcontinents that assembled to form West Antarctica and the Antarctic Peninsula (e.g., Jankowski and Drewry 1981; Dalziel and Elliot 1982; Millar and Pankhurst 1987, and many other papers too numerous to list here).

The assembly of both Gondwana and Laurentia continued during the Paleozoic until the two supercontinents merged at about 200 Ma to form Pangea which contained most of the continental crust existing at that time. Pangea was surrounded by a global ocean, called Panthalassa, which contained a large embayment that was occupied by the Tethys Sea. The world we know today began to take shape during the Jurassic Period when Gondwana and Laurasia began to be broken up into the present-day continents which drifted apart and were separated from each other by oceans (Wilson 1963b.) The fragmentation of Gondwana to form the southern continents was accompanied by large-scale eruption of basalt and the intrusion of diabase sills not only along the Ross orogen of Antarctica (i.e., the Ferrar Dolerite) but also in South Africa, in Brazil, in northwest Africa, and in North America (Faure 2001).

9.3 Coats Land

The reconstruction of Rodinia proposed by Moores (1991) relies partly on the extension of the Grenville structural province from North America to Coats Land located in East Antarctica along the east coast of the Weddell Sea in Fig. 9.5. The only rock outcrops of Coats Land are the Littlewood, Bertrab, and Moltke

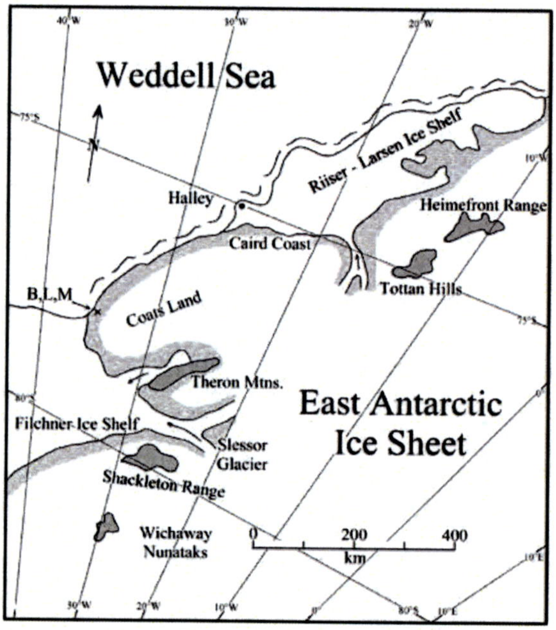

Fig. 9.5 Coats Land is located along the east coast of the Weddell Sea in East Antarctica. It consists of a low ice-covered ridge about 580 km in length north of the Filchner Ice Shelf. The only rock outcrops in Coats Land are the Littlewood, Bertrab, and Moltke nunataks (B, L, M) all of which are inaccessible and have been rarely visited by geologists. These nunataks are composed of rhyolite and granophyre which crystallized at 1,076 ± 7 Ma (Storey et al. 1994). This date as well as other age determinations in Table 9.1 confirm that these nunataks lie within the belt of Grenville-age rocks of the East Antarctic craton in agreement with the SWEAT hypothesis of Moore (1991) (Adapted from Anonymous 1990)

nunataks located at 77°53'S and 34°38'W (Marsh and Thompson 1984; Hutson et al. 1997). The rocks of these nunataks (rhyolite porphyry and granophyre) differ markedly from the rocks exposed in the Shackleton Range (Section 8.4).

The Littlewood Nunataks consist of four low, lichen-covered ridges about 50 m wide located between the Schweitzer and Lerchenfeld glaciers. These nunataks were visited for the first time on January 28, 1959, by Dr. J.C. Behrendt and Lt (J.G.) Erickson of the U.S. Navy who landed in a helicopter from the ice breaker U.S.S. Edisto. Aughenbaugh et al. (1965) later published a whole-rock K-Ar date of 840 ± 30 Ma for a specimen of rhyolite from one of the Littlewood Nunataks. The analyst at Geochron Laboratories, Cambridge, Massachusetts, where the specimen was dated, considered the K-Ar date to be an underestimate of the age of the rock that may have crystallized between 1,000 and 1,100 Ma. Additional rock specimens from the nearby Bertrab Nunataks were described by Capurro (1955) and Cordini (1959). The specimens consist of rhyolite porphyry, granophyre, and of dioritic lamprophyre that occurs in dikes on the Bertrab Nunataks but not on the Littlewood Nunataks. Capurro (1955) also described a specimen of carbonate rock composed of calcite crystals with altered rock fragments composed of chlorite and of plagioclase stained by iron oxide. Provided that this specimen is not an erratic, the presence of limestone containing fragments of a fine-grained volcanic rock may indicate that the rocks of Bertrab Nunataks are part of a marine volcanic sequence (Capurro 1955), whereas Aughenbaugh et al. (1965) referred to these rocks as the Littlewood Volcanics. In any case, the chemical compositions of the rhyolite porphyry on Littlewood Nunatak and of the granite porphyry on the Bertrab Nunataks are virtually indistinguishable (Augenbaugh et al. 1965).

A small suite of specimens from the Littlewood Nunataks and from the Bertrab Nunataks was dated by Eastin and Faure (1971) who obtained a whole-rock Rb-Sr date of 980 ± 16 Ma (recalculated to $\lambda = 1.42 \times 10^{-11} y^{-1}$). When that age determination was published in 1971, the relation of the Littlewood Volcanics to the geology of Antarctica was not understood because felsic volcanic rocks that crystallized at about 1,000 Ma were not known in the Transantarctic Mountains or in Dronning Maud Land. Although the date appeared to be "Grenvillian", felsic volcanic rocks are not typical of the rocks of the Grenville province of North America either. For the next 20 years the Littlewood Volcanics remained in limbo because their place in the geologic history of Antarctica was not understood.

Moores (1991) eventually suggested that the small nunataks of Coats Land are the stepping stones that permit the Grenville province of North America to be extrapolated into Antarctica. Consequently, efforts were made to establish the precise locations of these nunataks and to study the rocks of which they are composed (Toubes Spinelli 1983; Marsh and Thompson 1984; Dalziel et al. 1994; Hutson et al. 1997). The age determinations by Storey et al. (1994) and by Gose et al. (1997) in Table 9.1 confirm that the rocks of the Littlewood and Bertrab nunataks crystallized between 1,100 and 1,000 Ma.

The SWEAT hypothesis of Moores (1991) also positions the Haag Nunatak, which is presently located close to the crustal block of the Ellsworth-Whitmore Mountains (Section 8.1.3), within the Grenville belt in the gap between Coats Land and the Yavapai-Mazatzal province of North America in Fig. 9.6. In this way, the Haag Nunatak has found a home that is consistent with

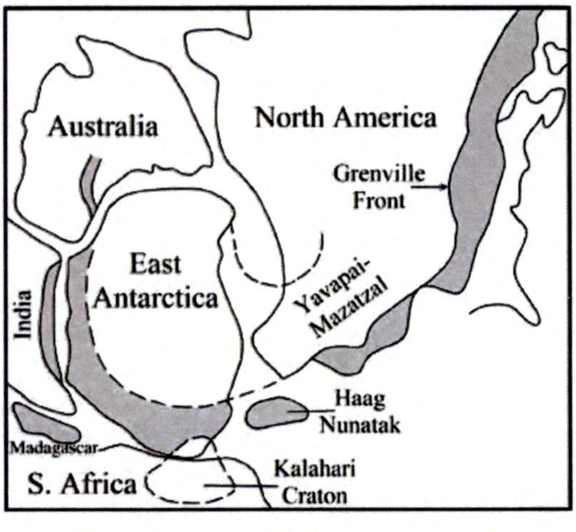

Fig. 9.6 The Grenville belt of eastern North America extends into Queen Maud Land of East Antarctica, into India, Madagascar, and into southern Australia where it appears to die out. The Haag Nunatak of the Ellsworth-Whitmore mountain block (Section 8.1.3) bridges the gap in the Grenville belt between North America and Coates Land (Maslanyj and Storey 1990). The Grunehogna structural province contains high-grade metamorphic rocks of the Kalahari craton of South Africa which was thrust over younger rocks in the Shackleton Range (Section 8.4). The placement of the Haag Nunatak within the Grenville belt confirms and refines the reconstruction of Rodinia proposed by Moores (1991)

Table 9.1 Isotopic age determinations of the Littlewood and Bertrab nunataks in Coats Land and of the Haag Nunatak of West Antarctica

Rock/mineral	K-Ar	Rb-Sr	Sm-Nd	U-Pb	References
			Ma		
Littlewood and Bertrab Nunataks					
Whole rock	840 ± 30				1
Whole rock		980 ± 16 (1σ) 0.7042 ± 0.0011			2
Whole rock		1,076 ± 7 0.7023 ± 0.0002			3
Whole rock			1,003 ± 38 0.51564 ± 0.00027 (1σ)		4
Zircon				1,112 ± 4	5
Titanite				1,106 ± 3	5
Haag Nunatak (77°00'S, 078°18'W)					
Granodiorite gneiss		1,176 ± 76 0.70308 ± 0.00060			6
Microgranite sheets		1,058 ± 53 0.70293 ± 0.00050			6
Aplogranites		1,003 ± 18 0.70360 ± 0.00023			6
Biotite and hornblende	991–1,031				7

1 Aughenbaugh et al. (1965)
2 Eastin and Faure (1971)
3 Storey et al. (1994)
4 Recalculated from data by Storey et al. (1994)
5 Gose et al. (1997)
6 Millar and Pankhurst (1987)
7 Clarkson and Brook (1977)

See also: Kleinschmidt, G. and S.D. Boger 2009. The Bertrab, Littlewood and Moltke nunataks in Coats Land: Enigma of East Antarctic geology. Polarforschung, 78 (3): 95–104

the age of the rocks listed in Table 9.1 as reported by Clarkson and Brook (1977) and Millar and Pankhurst (1987).

The age determinations of Aughenbaugh et al. (1965) and Eastin and Faure (1971) originally indicated that the Littlewood Volcanics were not metamorphosed during the Ross orogeny at 500 Ma. Therefore, Dalziel et al. (1994) suggested that these rocks should have retained their primary magnetization. Subsequently, Hutson et al. (1997) located the magnetic paleo-pole at 23.9°S and 258.5°E by measuring 84 oriented samples they collected at six sites on two of the Bertrab Nunataks and at three sites on one of the Littlewood Nunataks all of which yielded identical results. After rotating the East Antarctic craton into the position it occupied in the SWEAT hypothesis, the magnetic pole of Coats Land plotted on the apparent polar wander path of Laurentia. This result therefore supported the SWEAT hypothesis (Hutson et al. 1997).

Additional support for the SWEAT hypothesis has come from new age determinations published by Storey et al. (1994), Gose et al. (1994, 1997), and Goodge et al. (2002, 2008). Storey et al. (1994) reported a whole-rock Rb-Sr date of 1,076 ± 7 Ma with an initial $^{87}Sr/^{86}Sr$ ratio of 0.7023 ± 0.0002 for a suite of samples from one of the Bertrab Nunataks including three rhyolites, three granophyres, and two dolerites. The low initial $^{87}Sr/^{86}Sr$ ratio means that the magma formed by partial melting of mantle-derived basaltic rocks without significant input of crustal strontium. The same authors also published Sm-Nd model dates of 1,076 Ma (rhyolite), 1,129 Ma (dolerite), and 1,055 Ma (granophyre) from the Bertrab Nunatks. These results leave no room for doubt that the nunataks of Coats Land crystallized during the Grenvillian orogeny and have not been metamorphosed since then.

U-Pb dates of zircon by Gose et al. (1997) indicated a date of 1,112 ± 4 Ma for a Littlewood rhyolite and 1,113.5 + 18,- 6 Ma of a Bertrab granophyre. A sample of titanite at Bertrab has a U-Pb date of 1,106 ± 3 Ma which Gose et al. (1997) interpreted as a cooling age and which indicates that the rocks cooled rapidly after crystallization from magma.

The SWEAT hypothesis continues to be tested in various ways. Many studies have supported the SWEAT hypothesis even though the timing of the break-up of Rodinia and the distribution of continental fragments have been questioned or refined (e.g., Tessensohn 1997; Gose et al. 1997; Fitzsimons 2000). In addition, other research groups have refined our understanding of the structure and geochronology of Queen Maud Land located north of Coats Land (e.g., Moyes and Groenewald 1996; Jacobs et al. 1999).

9.4 Summary

Continental drift (or displacement) is one of the most important phenomena in geology because it has helped us to learn how the Earth works. We have progressed from reconstructing supercontinents by the shapes of existing continents and islands to explaining continental displacements in terms of a process in the mantle of the Earth that causes rifting of continents, sea-floor spreading, subduction of sediment, and the resulting formation of mountain belts and volcanic island arcs.

The tectonic history of Antarctica, insofar as we know it today, started in the Proterozoic Eon when the East Antarctic craton was embedded in a large continent we call Rodinia. During the Neoproterozoic Era this supercontinent split into two large fragments. The fragment called Laurentia drifted away from East Gondwana which included the land areas of Australia, East Antarctica, and India. The rift valley between Laurentia and East Gondwana widened into an ocean in which sediment was deposited along the passive rift margins of the two continental fragments. The alternating greywacke and shale layers of the turbidites that were deposited along the coast of East Gondwana were subsequently compressed against the craton when the passive rift margin became an active subduction zone in Late Cambrian to Early Ordovician time. As a result, the turbidite complex and locally extruded volcanic rocks were folded, metamorphosed, and intruded by granitic magma of anatectic origin. The resulting Ross Mountains were welded to the former rift margin of East Gondwana and became the roots of the present Transantarctic Mountains. The Ross fold belt can be traced from eastern Australia, through Antarctica, across the southern tip of Africa, and even into the South America. Therefore, it establishes a bond between these continents which together formed the supercontinent of Gondwana along with India and Madagascar.

The rifting of Rodinia and the formation of Laurentia and Gondwana is the subject of the SWEAT hypothesis which postulates that the western side of North America was joined to East Antarctica and Australia when all three land masses were united in Rodinia. Whereas the turbidites that were deposited along the rift margin of East Antarctica now make up the Ross orogen, the turbidites that accumulated along the passive rift margin of Laurentia have been preserved as the Belt-Purcell and Windermere sequences of western U.S.A and Canada, respectively. In addition, North America and East Antarctica are stitched together in the SWEAT hypothesis by the Grenville orogenic belt which extends from northeastern Canada to Texas and then continues to the coastal mountains of East Antarctica, to the eastern Ghat Mountains of India, to Madagascar, and finally into southern West Australia. This bold proposal to merge East Antarctica and North America defies the Principle of Least Astonishment as well as Ocam's Razor. Nevertheless, the SWEAT hypothesis has been supported by several investigators who set out to test its validity.

One such test has involved the ages of rhyolite and granophyre in the Littlewood and Bertrab nunataks in Coats Land on the coast of the Weddell Sea. This is the area where the Grenville belt of North America enters East Antarctica according to the SWEAT hypothesis. Isotopic age determinations have repeatedly and convincingly indicated that the rocks of the Littlewood and Bertrab nunataks crystallized during the Grenville Orogeny in North America (1,200–1,000 Ma) and that these rocks were not metamorphosed during the Ross orogeny around 500 Ma. Therefore, the rifting of Rodinia presumably occurred after the crystallization of the Littlewood Volcanics and before the deposition of the turbidites in the basement of the Transantarctic Mountains. The consequences of the SWEAT hypothesis are revolutionizing our understanding of the geologic history of the Earth. North America, which was once connected to East Antarctica when both were parts of Rodinia, ended up in the northern hemisphere of the Earth while East Antarctica was displaced to the geographic South Pole.

References

Andel TH van (1992) Seafloor spreading and plate tectonics. In: Brown GC, Hawkesworth CJ, Wilson RCL (eds) Understanding the Earth, 167–186. Cambridge University Press, Cambridge

Anonymous (1990) Atlas of the World,, 6th edn. National Geographic Society, Washington, DC

Aughenbaugh NB, Lounsbury RW, Behrendt JC (1965) The Littlewood Nunataks, Antarctica. J Geol 73(6):889–894

Bell RT, Jefferson CW (1987) An hypothesis for an Australian-Canadian connection in the Late Proterozoic and the birth of the Pacific Ocean. In: Proceedings, Pacific Rim congress.. Australian Institute of Mining and Metallurgy, Parkville, Victoria, Australia, pp 39–50

Capurro LRA (1955) Expedicion Argentina al Marde Weddell. Minist. Marina, Dir. Gen., Naveg. Hidrograf., pp. 129–131

Clarkson PD, Brook M (1977) Age and position of the Ellsworth Mountains crustal fragment Antarctica. Nature 265: 615–616

Cordini IR (1959) Elcomocimiento geologico de la Antartica. Just. Ant. Argent, Pub. No. 6, pp 141–145

Cox A, Hart RB (1986) Plate tectonics: how it works. Blackwell Scientific, Oxford

Craddock C (ed) (1982) Antarctic geoscience. University of Wisconsin Press, Madison, WI

Dalziel IWD (1991) Pacific margin of Laurentia and East Antarctica/Australia as a conjugate rift pair: evidence and implications for an Eocambrian supercontinent. Geol 19(6):598–601

Dalziel IWD (1992) Antarctica: a tale of two supercontinents? Ann Rev Earth Planet Sci 20:501–526

Dalziel IWD, Elliot DH (1982) West Antarctica: problem child of Gondwanaland. Tectonics 1:3–19

Dalziel IWD, Helper MA, Hutson FE, Grimes SW (1994) Geologic investigations in the Shackleton Range and Coats Land nunataks, Antarctica. Antarctic J US 29(5):4–6

Dickinson WR (1977) Paleozoic plate tectonics and the evolution of the Cordilleran continental margin. In: Stewart JH, Stevens CH, Fritsche AE (eds) Paleozoic paleogeography of the western United States. Society of Economic Paleontologists, Pacific Coast Paleogeography Symposium vol 1, pp 137–155

Dietz RS (1961) Continent and ocean basin evolution by spreading of the sea floor. Nature 190:854–857

Dietz RS, Holden JC (1970) Reconstruction of Pangaea, its breakup and dispersion of continents: permian to present. J Geophys Res 75:4939–4956

Dietz RS, Sproll WP (1970) Fit between Africa and Antarctica: a continental drift reconstruction. Science 167(3925):1612–1614

Dietz RS, Holden JC, Sproll WP (1972) Antarctica and continental drift. In: Adie RJ (ed) Antarctic geology and geophysics. Universitetsforlaget, Oslo, Norway

Du Toit AL (1937) Our wandering continents. Oliver & Boyd, London

Eastin R, Faure, G. (1971) The age of the Littlewood volcanics of Coats Land, Antarctica. J Geol 79:241–245

Elliot DH (1972) Aspects of Antarctic geology and drift reconstructions. In: Adie RJ (ed) Antarctic geology and geophysics. Universitetsforlaget, Oslo, Norway, pp 849–858

Elliot DH (1975) Tectonics of Antarctica: a review. Am J Sci 275A:45–106

Emiliani C (1987) Dictionary of the physical sciences. Oxford University Press, Oxford

Emiliani C (1992) Planet earth; cosmology, geology, and the evolution of life and environment. Cambridge University Press, New York

Faure G (2001) Origin of igneous rocks: the isotopic evidence. Springer, Heidelberg, Germany

Findlay RH, Unrug R, Banks MR, Veevers JJ (eds) (1993) Gondwana eight: assembly evolution, and dispersal. Balkema, Rotterdam, The Netherlands

Fitzsimons ICW (2000) Grenville-age basement provinces in East Antarctica: evidence for three separate collisional orogens. Geology 28:879–882

Ford AB (1972) Fit of Gondwana continents. 24th IGC, Sect. 3: 113–121

Frakes LA, Crowell JC (1970) Geologic evidence for the place of Antarctica in Gondwanaland. Antarctic J US 5(3):67–69

Fütterer DK, Damaske D, Kleinschmidt G, Miller H, Tessensohn F (eds) (2006) Antarctica; contributions to global earth science. Springer, Berlin/Heidelberg, Germany

Gamble JA, Skinner DNB, Henry S (eds) (2002) Antarctica at the close of a millennium. Proceedings of 8th International Symposium. Antarctic Earth Science, Wellington, 1999. Royal Society of New Zealand, Bulletin 35, Wellington, New Zealand

Goodge JW (2001) From Rodinia to Gondwana: supercontinent evolution in the Transantarctic Mountains. In: Gamble J, Skinner DA (eds) Proceedings of 8th International Symposium on Antarctic Earth Science, Royal Society of New Zealand, Bulletin 35, Wellington, New Zealand

Goodge JW, Myrow P, Williams IS, Bowring SA (2002) Age and provenance of the Beardmore Group, Antarctica: constraints on Rodinia supercontinent breakup. J Geol 110(4):393–406

Goodge JW, Verwoort JD, Fanning CM, Brecke DM, Farmer GL, Williams IS, Myrow PM DePaolo DJ (2008) A positive test of East Antarctica-Laurentia juxtaposition within the Rodinia supercontinent. Science 321(5886):235–240

Gose WA, Dalziel IWD, Helper MA, Hutson FE, Grimes S (1994) A positive test of the SWEAT hypothesis: new paleomagnetic data from the "Grenvillian" rocks of Coats Land, Antarctica. Geol Soc Am Abst Prog 26:503–504

Gose WA, Helper MA, Connelly JN, Hutson FE, Dalziel IWD (1997) Paleomagnetic data and U-Pb isotopic ages from Coats Land, Antarctica: implications for Late Proterozoic plate reconstructions. J Geophys Res 102(B4):7887–7902

Grikurov GE (1982) Structure of Antarctica and outline of its evolution. In: Craddock C (ed) Antarctic geoscience. University of Wisconsin Press, Madison, WI, pp 791–804

Grikurov GE, Ravich MG, Soloviev DS (1972) Tectonics of Antarctica. In: Adie RJ (ed) Antarctic geology and geophysics. Universitetsforlaget, Oslo, Norway, pp 457–468

Hamilton W (1963) Antarctic tectonics and continental drift. In: Polar wandering and continental drift. Society of Economic Paleontologists and Mineralogists, Special Publications 10, pp 74–93

Hamilton W (1967) Tectonics of Antarctica. Tectonophyics 4(4/6):555–568

Harley SL, Hensen BJ, Jacobs J (1997) Antarctica in the amalgamation of Gondwana. In: Ricci CA (ed) The Antarctic

region; geological evolution and processes. Terra Antartica, Siena, Italy
Hartnady CJH (1986) Was North America ("Laurentia") part of south-western Gondwanaland during the Late Proterozoic Era? Suid-Afr Tydskr Wet 82:251–254
Hartnady CJH (1991) About turn for supercontinents. Nature 352:476–478
Hess HH (1962) History of ocean basins. In: Engel AEJ, James HL, Leonard BF (eds) Petrologic studies. Geological Society of America, Buddington Volume, pp 599–620
Hoffman PF (1991) Did the breakout of Laurentia turn Gondwanaland inside-out? Science 252:1409–1412
Hurley PM, Rand JR (1969) Pre-drift continental nuclei. Science 164(3885):1229–1242
Hutson FE, Helper MA, Dalziel IWD, Grimes SW (1997) Initial results of geologic investigations in the Shackleton Range and southern Coats Land nunataks, Antarctica. Antarctic J US 32(5):234–237
Jacobs J, Hansen BT, Henjes-Kunst F, Thomas RJ, Weber K, Bauer W, Armstrong RA, Cornell DH (1999) New age constraints on the Proterozoic/Lower Palaeozoic evolution of Heimefontjella, East Antarctica, and its bearing on Rodinia/ Gondwana correlations. Terra Antartica 6(4):377–389
Jankowski EJ, Drewry DJ (1981) The structure of West Antarctica from geophysical studies. Nature 291:17–21
Laskar B, Raja Rao CS (eds) (1979) Fourth international Gondwana symposium, vols. 1 and 2, Hindustan Publishing Corporation, Delhi, India
Marsh PD, Thompson JW (1984) Location and geology of nunataks in north-western Coats Land. British Antarct. Surv Bull 65:33–39
McKenzie GD (ed) (1987) Gondwana six: structure, tectonics, and geophysics. Geophysics Monograph 40, American Geophysical Union, Washington, DC
Millar IL, Pankhurst RJ (1987) Rb-Sr geochronology of the region between the Antarctic Peninsula and the Transantarctic Mountains: Haag nunataks and Mesozoic granitoids. In: McKenzie GD (ed) Gondwana six: structure, tectonics, and geophysics. American Geophysical Union, Washington, DC, pp 151–160
Moores EM (1991) The southwest U.S.-East Antarctica (SWEAT) connection: a hypothesis. Geology 19(5):452–428
Moyes AB, Groenewald PB (1996) Isotopic constraints on Pan-African metamorphism in Dronning Maud Land, Antarctica. Chem Geol 129:247–256
Oliver RL, James PR, Jago JB (eds) (1983) Antarctic earth science. Australian Academy of Science, Canberra, ACT
Press F, Siever R (1986) Earth, 4th edn. Freeman, New York
Price RA (1964) The Precambrian Purcell system in the Rocky mountains of southern Alberta and British Columbia. Bull Canadian Petrol Geol 12:399–426
Ricci CA (ed) (1997) The Antactic region: geological evolution and processes. Terra Antartica Siena, Italy
Siever R (1983) The dynamic earth. Sci Am 249(3):46–55
Smith GA (1992) Plate tectonics and continental drift. In: Brown GC, Hawkesworth CJ, Wilson RCL (eds) Understanding the earth. Cambridge University Press, Cambridge, pp 187–203
Storey BC, Pankhurst RJ, Johnson AC (1994) The Grenville Province within Antarctica: a test of the SWEAT hypothesis. J Geol Soc London 151:1–4
Stump E (1995) The Ross orogen of the Transantarctic Mountains. Cambridge Un Press, Cambridge
Tessensohn F (1997) Shackleton Range, Ross Orogen, and SWEAT hypothesis. In: Ricci CA (ed) The Antarctic region; geological evolution and processes. Terra Antartica, Siena, Italy, pp 5–12
Thomson MRA, Crame JA, Thomson JW (ed) (1991) Geological evolution of Antarctica. Cambridge University Press, Cambridge
Toubes Spinelli RO (1983) Geology of the Bertrab Nunatak, Argentinian sector of Antarctica. Contribucion Instituto Antartico Argentino, 296:1–9 (in Spanish)
Ulbrich H, Rocha Campos AC (eds) (1990) Gondwana seven proceedings. Instituto de Geociencias, São Paulo, Brazil
Unrug R (1997) Rodinia to Gondwana: the geodynamic map of Gondwana supercontinent assembly. GSA Today 7(1):1–6
Vine FJ, Hess HH (1970) Sea floor spreading. In: Maxwell AE (ed) The sea, vol. 4. Wiley, New York, pp 587–622
Vine FJ, Matthews PM (1963) Magnetic anomalies over ocean ridges. Nature 199:947–949
Wegener A (1912) Die Entstehung der Kontinente. Petermanns Mitteilungen 185–195, 253–256, 305–309
Wegener A (1929) Die Entstehung der Kontinente und Ozeane, 4th edn. Vieweg und Sohn, Braunschweig, Germany
Wegener A (1966). The origin of the continents and oceans. Reprinted by Dover, New York
Wilson JT (1963a) Evidence from islands on the spreading of ocean floor. Nature 197:536–538
Wilson JT (1963b) Continental drift. Sci Am 208(4): 86–100
Wilson JT (1965) A new class of faults and their bearing on continental drift. Nature 207:343–347
Wilson JT (1973) Continental drift, transcurrent and transform faulting. In: Maxwell AE (ed) The sea, vol. 4. Wiley, New York, pp 623–644
Windley BF (1977) The evolving continents. Wiley, New York
Winston D (1986) Sedimentation and tectonics of the Middle Proterozoic Belt basin and their influence on Phanerozoic compression and extension in western Montana and northern Idaho. In: Peterson JA (ed) Paleozoic tectonics and sedimentation in the Rocky mountains region, United States of American Association of Petroleum Geologists, Mem. vol. 41, pp 3–20
Wyllie PJ (1976) The way the earth works. Wiley, New York

Part III
Gondwana: Growth and Disintegation

Chapter 10
The Beacon Supergroup

The geologists who accompanied Robert Scott and Ernest Shackleton to Antarctica were amazed when they entered the mountains of southern Victoria Land and discovered the ice-free valleys such as Wright Valley in Fig. 10.1 (Sections 1.4, 2.3, and 3.1). The modern traveler entering the ice-free valleys by helicopter from McMurdo is similarly affected because the wide U-shaped valleys present a familiar view of sandy plains dotted with lakes and ponds filled with liquid water and of meltwater streams that flow from alpine glaciers descending from the high mountain ranges that frame these valleys.

The early explorers on Scott's first Antarctic Expedition (1901–1904) took note of the massive cliffs of sandstone and shale that are interlayered with thick sills of diabase. Ferrar (1907) proposed to place the sandstones of southern Victoria Land in the Beacon Formation because of their occurrence at a spectacular mountain that Lieutenant Albert B. Armitage had named the Beacon Heights in Fig. 10.2 The name was later adopted by Harrington (1958) in his seminal paper on the nomenclature of rocks in the Ross-Sea region. A proposal by Mirsky (1964) to reconsider the naming of the Beacon Formation was not supported by his colleagues.

The first scientific descriptions of the Beacon rocks in the ice-free valleys were published by Ferrar (1907), David and Priestley (1914), and Debenham (1921) followed after the IGY by Webb and McKelvey (1959), McKelvey and Webb (1959, 1961, 1962), Hamilton and Hayes (1960, 1963), and Zeller et al. (1961). These reports were followed by the papers of Gunn and Warren (1962), Angino et al. (1962), Allen (1962), Allen and Gibson (1962) Harrington and Speden (1962), Shaw (1962), and Vialov (1962). In the following year, Webb (1963) published the first comprehensive description of the Beacon rocks in southern Victoria Land and Haskell et al. (1965a) followed with a treatment of the geology of Taylor Valley in Fig. 10.3 including the basement rocks, the Beacon rocks, and the sills of Ferrar Dolerite.

The Beacon rocks occur not only in southern Victoria Land in Fig. 10.4 but characterizes the Transantarctic Mountains from northern Victoria Land to the Ohio Range of the Horlick Mountains. They continue from there to the far-eastern mountains described in Chapter 8 including the Pensacola Mountains, the Shackleton Range, the Theron Mountains, and even the Kottas Mountains on the coast of the Weddell Sea in Queen Maud Land (Poscher 1989; Barrett 1991).

Flat-lying sandstones, siltstones, and coal seams of Permian and Triassic ages also occur in the Lambert Graben at the head of the Amery Ice Shelf in East Antarctica. These rocks have been assigned to three formations which together constitute the Amery Group:

Jetty Member, 60 m, Late Triassic
Flagstone Bench Formation, 400 m, Late Permian
Bainmedart Coal Measure, 1,100 m, Late Permian
Radock Conglomerate, 400 m, Middle to Late Permian

The sedimentary rocks of the Lambert Group were mapped by Webb and Fielding (1993) who referred to additional studies of these rocks in the Lambert Graben.

The Beacon rocks in the Transantarctic Mountains attract attention because they are the products of geological processes that occurred in Gondwana during the late Paleozoic and early Mesozoic eras. In addition, they contain fossilized remains of plants and animals that inhabited the landscape at this time in history of the Earth. The Beacon rocks also contain coal

G. Faure and T.M. Mensing, *The Transantarctic Mountains: Rocks, Ice, Meteorites and Water*,
DOI 10.1007/978-90-481-9390-5_10,

Fig. 10.1 Wright Valley in southern Victoria Land is ice-free and contains the Onyx River which flows inland from Lake Brownworth at the base of the Wright Lower Glacier near the Ross Sea to Lake Vanda located 50 km from the coast. The peaks of the Asgard Range appear in the background (Photo by R. Montigny)

deposits, tillites, calcite concretions, and radioactive minerals. In the sections that follow, we first review the occurrence of the Beacon Supergroup in certain key areas of the Transantarctic Mountains and then consider the special topics mentioned above.

10.1 Southern Victoria Land

The topography of southern Victoria Land (S.V.L.) was described in Section 3.1.1 and is depicted in the contoured maps published by the US Geological Survey available on scales of 1:250,000 and 1:50,000. In addition, the geology of southern Victoria Land is depicted in Fig. 10.4 and was described by Warren (1969) on Sheet 14 of Folio 12 edited by Craddock (1969). We start the description of the Beacon Supergroup in southern Victoria Land because it is the most accessible part of the Transantarctic Mountains located only 60 km from McMurdo Station on Ross Island.

10.1.1 Ice-Free Valleys

The type section of the Beacon rocks is located on West Beacon Heights (2,420 m) adjacent to the upper part of the Taylor Glacier between the Arena and Beacon valleys in Fig. 10.2 (Harrington and Speden 1962). Hamilton and Hayes (1963) later assembled a composite section of more than 1,700 m from exposures on the Terra Cotta Hills, New Mountain, West Beacon Heights, and Beacon Valley. The Beacon sandstones at these locations were intruded by several dolerite sills having a total thickness of at least another 1,200 m according to estimates by Hamilton and Hayes (1963).

The geologic map of Warren (1969) indicates that Beacon sandstones occur widely along the edge of the polar plateau from the Koettlitz Glacier in the south to the David Glacier in the north. Some of the principal outcrop areas include the Royal Society Range with Mt. Lister at 4,025 m, the Beacon Heights area (Fig. 10.2), the Willett Range which

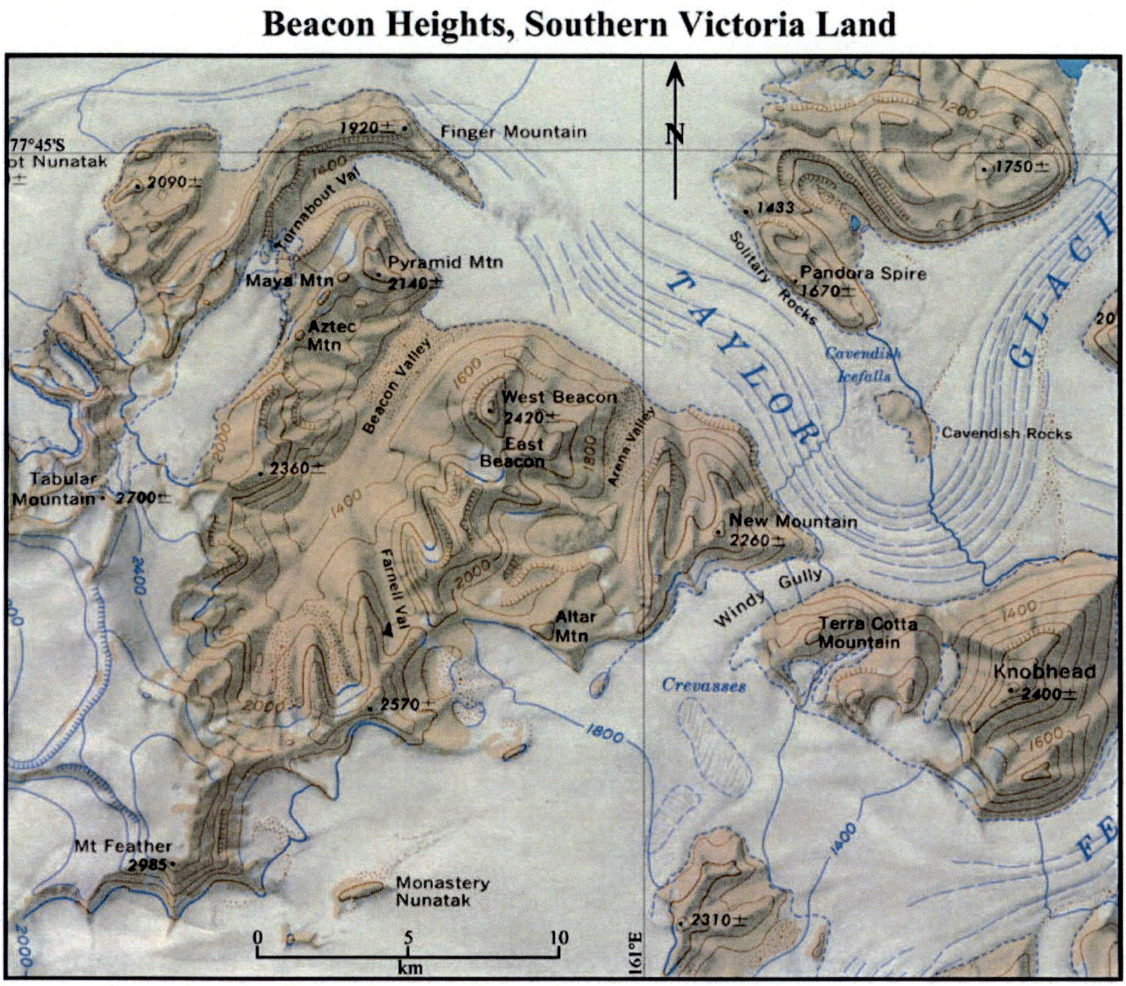

Fig. 10.2 The mountain known as West Beacon adjacent to the upper Taylor Glacier in southern Victoria Land is the type section of the Beacon Supergroup (Harrington and Speden 1962). Hamilton and Hayes (1963) subsequently assembled a composite section of more than 1,799 m from outcrops at Terra Cotta Hills and New Mountain, at West Beacon, and at Beacon Valley all of which are identified on this topographic map. The Terra Cotta and Windy-Gully parts of the section are from Zeller et al. (1961), whereas the Beacon Valley section is from McKelvey and Webb (1959). Excerpt of the Taylor Glacier topographic map (ST 57-60/5) published in 1967 by the US Geological Survey, Washington, DC

contains Shapeless Mountain (2,739 m), and the Convoy Range north of the Mackay Glacier featuring Mt. Brooke (2,675 m) and including the well-known Allan Hills (Fig. 10.4).

The sandstones and related sedimentary rocks of the Beacon Supergroup rest everywhere on the Kukri erosional surface. In addition, the rocks are flat-lying and undeformed except locally near faults and where the sandstones were intruded by sills of Ferrar Dolerite (McElroy et al. 1967). Webb (1963) reviewed the stratigraphic terms that were proposed by some of the early workers and he defined four formations listed in Table 10.1 which have an aggregate thickness of about 1,220 m:

Weller Sandstone (youngest)
Aztec Siltstone
Beacon Heights Orthoquartzite
Odin Arkose (oldest)

The type locality of the *Odin Arkose* is Mt. Boreas (77°28'S, 161°07'E) in the Olympus Range north of Wright Valley in Fig. 10.5. The unit is about 90 m thick

Taylor Valley, Southern Victoria Land

Fig. 10.3 Taylor Valley in southern Victoria Land is flanked by the Asgard Range to the north and by the Kukri Hills to the south. Both mountain ranges consist of the sandstones of the Beacon Supergroup which rest unconformably on igneous and metamorphic basement rocks (Section 3.1.2) and were intruded by sills of the Ferrar Dolerite of Jurassic age. Excerpt of the topographic map: Ross Island, Antarctica (ST 57-60/6*) published in 1960 by the US Geological Survey, Washington, DC

and includes the *Boreas Subgraywacke Member* which differs from the main body of the Odin Arkose in color, lithology, induration, and sedimentary structure. The Odin Arkose resembles the *Windy Gully* and *Terra Cotta Mountain* members which occur, like the Odin Arkose, at the base of the Beacon Group at these respective locations (Zeller et al. 1961) in Table 10.1.

The *Beacon Heights Orthoquartzite* is a thick-bedded cliff-forming body of rocks that resembles the Fortress Sandstone mapped in the Victoria Valley by Allen (1962) and the so-called Lower Arenite described by Gunn and Warren (1962) from Alligator Peak (78°28'S, 158°45'E) in the Boomerang Range located south of the Ferrar Glacier in Fig. 10.6. The Beacon Heights Orthoquartzite is characterized by festoon cross-bedding and contains trace fossils such as worm casts and trails. In addition, the rocks contain ripple marks and ferruginous concretions up to about 90 cm in diameter. The sandstone beds are locally interbedded with greenish siltstone and shale up to 60 cm thick.

The *Aztec Siltstone* is named for a deposit of red and green siltstones that overlie the Beacon Heights Orthoquartzite on Aztec Mountain on the west wall of Beacon Valley, upperTaylor Glacier, in Fig. 10.2. The Aztec Formation also occurs south of the Taylor Glacier on Maya Mountain, Mt. Weller, and Knobhead Mountain in Fig. 10.2. Some maroon-colored siltstone beds contain desiccation polygons and small grayish white carbonate concretions.

The *Weller Sandstone* is exposed in the western wall of Beacon Valley (Fig. 10.2). It consists of at least 152 m of quartz sandstone and conglomerate, carbonaceous sandstone, siltstone, and shale. The sandstones are coarse grained, cross-bedded, and contain abundant carbonaceous plant fragments. The Weller Formation may be correlated with the Mt. Bastion Coal Measures (Permian) in Victoria Valley described by Allen (1962).

The *Beacon rocks* in Wright and Taylor valley contain a variety of trace fossils described by Vialov (1962), Webb (1963), Webby (1968), Gevers et al. (1971),

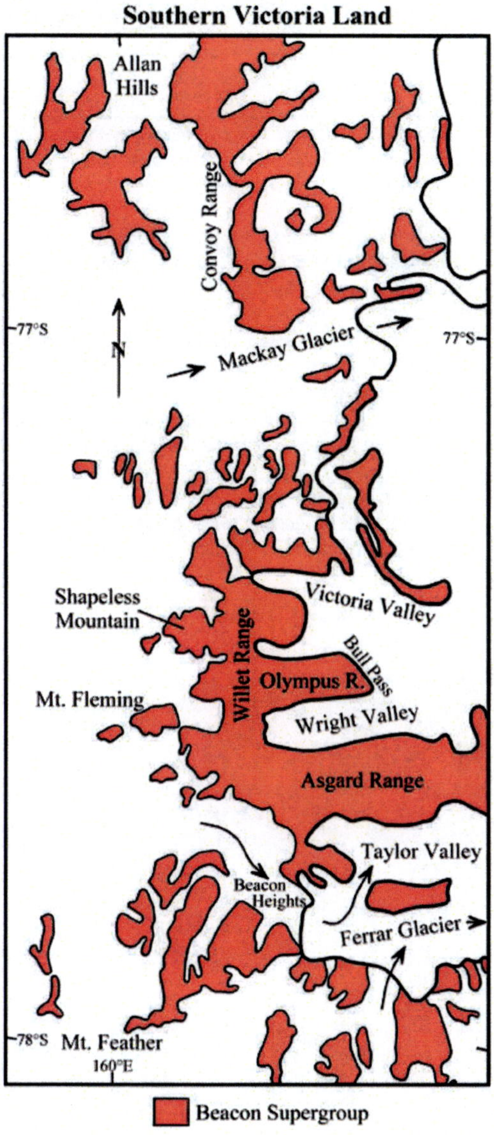

Fig. 10.4 Outcrops of the Beacon Supergroup in the mountains of southern Victoria Land (Adapted from Craddock 1969, Sheet 14, Folio 12)

Gevers and Twomey (1982). The presence of plant and trace fossils in the Beacon rocks of southern Victoria Land suggests that the climate at the time of deposition of the Beacon sandstones was temperate and supported an abundant flora and fauna even though conditions at certain times and in certain places in Gondwana were arid as suggested by the presence of mud-cracked surfaces of shale layers and calcite concretions. The geologic history from the Early Devonian to the Early Jurassic in the Ross-Sea region of the Transantarctic Mountains was reviewed by Barrett (1981).

The sandstones also contain resistant accessory minerals, including apatite, zircon, tourmaline, garnet, and rutile, some of which can be dated by the U-Pb method (see also Stewart 1934; Shaw 1962; Mirsky et al. 1965). Webb (1963) suggested that the sediment could have been derived primarily from the underlying metamorphic and igneous rocks of the basement complex. Alternatively, the resistant mineral grains could also have been derived from the high-grade metamorphic and plutonic igneous rocks of the East Antarctic craton or even from the cratons of Australia, India, and South Africa which surrounded East Antarctica in Gondwana.

10.1.2 Expanded Stratigraphy

The work of Webb (1963) is a point of departure for the studies of the Beacon Supergroup in southern Victoria Land that followed. Seven years later, McKelvey et al. (1970) published an expanded stratigraphic column derived from the area between Olympus Range (Fig. 10.5 in Wright Valley) and the Boomerang Range in Fig. 10.6 located about 140 km southwest of the Olympus Range in the Skelton Névé. By this time, the stratigraphy of the Beacon rocks in Table 10.2 had expanded into the lower Taylor Group and the upper Victoria Group and therefore merited the designation *Beacon Supergroup*.

The *Taylor Group* consists of texturally mature rocks that contain little carbonaceous material, although the Devonian fish and other kinds of fossils occur in the upper stratigraphic units of this group. The *Victoria Group* is composed of a heterogeneous assemblage of rocks containing abundant carbonaceous material including leaves of *Glossopteris* in the lower units (Permian) and leaves of *Dicroidium* in the upper parts (Triassic).

The Taylor Group of southern Victoria Land includes the basal *New Mountain Sandstone* which is a light-colored quartz sandstone containing pebbly sandstone, siltstone, mudstone, and breccia/conglomerate layers near its base (Plume 1978; Wizevich 1997). The New Mountain Sandstone thins northward from 220 m at Mt. Handsley located a short distance south of Knobhead (Fig. 10.2) to 46 m near Mt. Boreas in the Olympus Range (Fig. 10.5) the top of this formation is truncated by the *Heimdall erosion surface*.

Table 10.1 Stratigraphy of the Beacon Supergroup in Wright and Taylor valleys of southern Victoria Land (Adapted from Webb 1963)

Formation Name/location	Thickness (m)	Description
Weller Sandstone Mt. Weller, western wall of Beacon Valley (Fig. 10.2)	152 +	Quartz sandstone and conglomerate, carbonaceous sandstone, siltstone, and shale. The sandstone is massive, coarse grained, and cross-bedded. Fossilized plant debris, coal seams, and worm casts in siltstone
Aztec Siltstone, Aztec Mountain upper Taylor Gl. (Fig. 10.2)	21	Red and green siltstone, found also at Maya Mtn., Mt. Webber, Beacon Heights, and Knobhead. Desiccation polygons, small carbonate concretions, and abundant white worm tubes aligned vertically in red siltstone ("Scolithus")
Beacon Heights Orthoquartzite, West Beacon Heights, upper Taylor Gl. (Fig. 10.2)	762–914	Homogeneous white to tan colored orthoquartzite. Massive festoon cross-bedded with worm casts, trails, ripple marks, and ferruginous concretions
Odin Arkose Mt. Boreas, Olympus Range (Fig. 10.5)	91	Alternating beds of breccia, conglomerate, sandstone, siltstone, and shale. Graded bedding, cross-bedding, ripple marks, and worm casts are present
Boreas Subgraywacke Member (Fig. 10.5)		Distinctive basal layer present only at Mt. Boreas. Composed of angular fragments of igneous rocks, quartzite, and chert in a matrix of sandstone and siltstone
	Kukri Peneplain	
Basement rocks including the Asgard Formation and the Granite Harbor Intrusives (McKelvey and Webb 1962)		

Fig. 10.5 The Wright Valley in southern Victoria Land slopes west, away from the coast, and contains the Onyx River which flows from Lake Brownworth at the foot of the Wright Lower Glacier to Lake Vanda at the western end of the valley (Fig. 3.5). The Asgard Range south of Wright Valley supports several valley glaciers which were named by Professor R.L. Nichols after his field assistants. Wright Valley has a U-shaped cross-section indicating that it was once occupied by an outlet glacier that flowed from the polar plateau to the Ross Sea. Excerpt from the topographic map: Ross Island, Antarctica, ST 57-6-/6* (162–170°E) published by the US Geological Survey in 1960 and revised 1970

The overlying *Altar Mountain Formation*, which is 162 m thick at Altar Mountain (Fig. 10.2), contains the Odin Arkose Member (30 m, Table 10.1) which grades upward into quartz sandstone and maroon-to-green siltstones. The *Arena Sandstone* that conformably overlies the Altar Mountain Formation is a buff-colored slope-forming sandstone about 385 m thick in contrast to the overlying *Beacon Heights Orthoquartzite* (Table

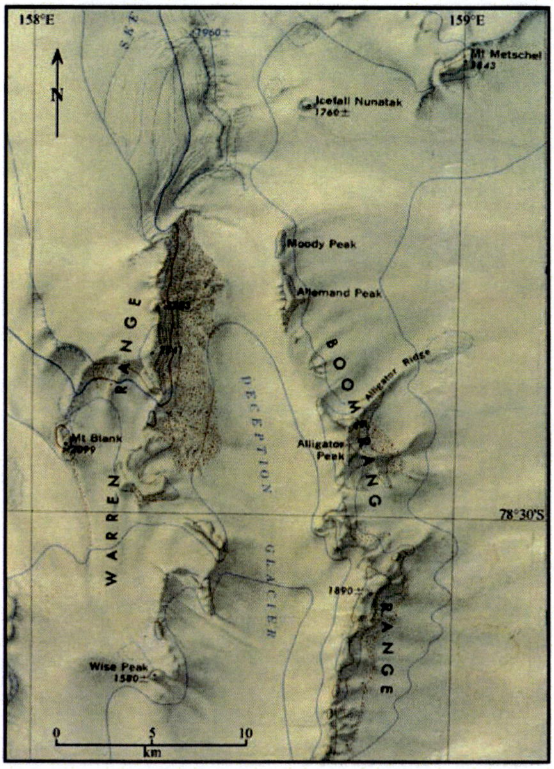

Fig. 10.6 The Boomerang and Warren ranges in the area of the Skelton Névé in southern Victoria Land consist primarily of sandstones of the Beacon Supergroup that were intruded by sills of the Ferrar Dolerite (Jurassic). Note the location of Mt. Metschel (upper right corner) which contains an outcrop of the Metschel Tillite (Permian). The rocks of the Boomerang Range were first examined by Gunn and Warren (1962) and by McKelvey et al. (1970) who used them to refine the stratigraphy of the Beacon Supergroup outlined in Table 10.2. Excerpt from the Mount Harmsworth sheet (ST 57-60/9; S7800-E15600/1 × 6) published in 1965 by the US Geological Survey, Washington, DC

Table 10.2 Expanded stratigraphy of the Beacon Supergroup between the Olympus Range (Wright Valley) and the Boomerang Range (Skelton Névé) in southern Victoria Land (McKelvey et al. 1970)

Age	Group	Formation
Triassic to Permian	Victoria	Lashly
		Fleming
		Feather Conglomerate
		Weller Coal Measures
	Pyramid disconformity	
Permian	Victoria	Metschel Tillite
	Maya disconformity	
Middle-Late Devonian	Taylor	Aztec Siltstone
Early-Middle Devonian		Beacon Heights Orthoquartzite
		Arena Sandstone
		Altar Mountain
		Odin Arkose Member
	Heimdall disconformity	
Devonian or older	Taylor	New Mountain Sandstone
		1. Terra Cotta Siltstone M.
		2. Windy Gully Sandstone M.
		3. Boreas Subgraywacke M.
		(3 = 2 + 1)
	Kukri Peneplain	
Cambro/Ordovician		Granite Harbor Intrusives
Neoproterozoic		Skelton Grp.

10.1) which forms spectacular white cliffs more than 300 m high. The Beacon Heights Formation grades upward into the red and green siltstones of the *Aztec Formation* (Table 10.1) which thin northward from 138 m in the Boomerang Range (Fig. 10.6) to only 40 m at Aztec Mountain (Fig. 10.2).

McKelvey et al. (1970) collected abundant fragmentary fish fossils in the rocks of the Aztec Formation of the northern Boomerang Range (78°30'S, 158°45'E) following their original discovery in 1957/58 by Guyon Warren. The fish fragments that McKelvey et al. (1970) collected during the 1968/69 field season included jaws, teeth, skulls, dermal plates, and fin-spines of freshwater fish of late Middle Devonian age. The Aztec Formation also contains a small number of plant fossils and conchostracans.

The environment of deposition of the Taylor Group of the Beacon Supergroup in southern Victoria Land has become a controversial issue because the scarcity of diagnostic fossils has caused differences of opinion to develop among investigators. The only point of agreement is that the Aztec Formation, youngest in the Taylor Group, is non-marine as indicated by abundant fresh-water fish and conchostracans (McPherson 1978; Grindley et al. 1980).

The absence of diagnostic fossils in the sandstones of the Taylor Group may be caused by the poor preservation of chitinous exoskeletons, although the presence of an abundant fauna is indicated by traces of arthropods and other animals capable of locomotion. For this reason, several authors have favored a shallow-water marine origin for these sandstones (e.g., Gevers and Twomey 1982; Allen 1962; Webby 1968), whereas Barrett (1979) and Gunn and Warren (1962) favored a non-marine origin for the rocks of the Taylor Group on sedimentological grounds.

Bradshaw (1981) identified 26 ichnotaxa (trace fossils) in the Taylor Group of southern Victoria Land most of which occur in the New Mountain Sandstone and its two members (i.e., the Terra Cotta Siltstone and the Windy Gully Sandstone). Trace fossils are less abundant in the Altar Mountain and Arena formations and are rare in the Beacon Heights Orthoquartzite and in the Aztec Siltstone (Table 10.2). Bradshaw (1981) acknowledged the opposing points of view but concluded on the basis of the abundance of trace fossils that the Windy Gully, Terra Cotta, New Mountain, and Altar formations are marine and that the Arena, Beacon Heights, and Aztec formations are probably non-marine.

Although glacial beds were at first thought to be absent in the mountains of southern Victoria Land, McKelvey et al. (1970) reported that tillites are present at Aztec and Maya mountains in Fig. 10.2 and in the Boomerange Range (Fig. 10.6). They assigned the tillites to the *Metschel Formation* after Mt. Metschel (Fig. 13.3), a small nunatak about 18 km north of the northern end of the Boomerang Range in the Skelton Névé (Fig. 10.6). The Metschel Tillite was deposited on the *Maya erosion surface* which cuts the top of Aztec Formation as shown in Fig. 10.7a. Subsequently, the glacial deposits were removed in many places by the *Pyramid erosion event* which not only removed the tillite but also cut through the Maya erosion surface into the Aztec beds in Fig. 10.7b. In such places, the *Weller Formation* was deposited on the Pyramid erosion surface that was cut into the Aztec Formation. In places where the remnants of the Metschel Tillite were preserved, they rest on the Maya erosion surface and are disconformably overlain by the Weller Formation which was deposited on the Pyramid erosion surface. Therefore, the Metschel Tillite has been preserved in southern Victoria Land only as scattered erosional remnants sandwiched between the Maya and Pyramid erosion surfaces (e.g., on Mt. Feather, Pinet et al. 1967).

The Metschel Formation consists of tillite, varved siltstones, and other kinds of stratified sediment and thickens to the south from 27 m at Mt. Metschel to 38 m in the Boomerang Range (Barrett and Askin 1975). The tillite is massive in most places with a sandy gray-green matrix and contains clasts of granitic, gneissic, and metasedimentary rocks. Some clasts are up to 1.5 m in diameter. The Maya erosion surface at the base of Metschell Tillite on Aztec Mountain contains smoothly rounded grooves and ridges aligned northwest-to-southeast which are believed to be of glacial origin (McKelvey et al. 1970).

The *Weller Coal Measures* lie on the Pyramid erosion surface and consist of cross-bedded carbonaceous sandstone (170 m) and siltstones with occasional conglomerate lenses. The Weller Formation contains seams of coal of bituminous to anthracite rank that are up to 1 m thick. The lower part of the Weller Formation is composed of arkosic rocks and contains a thin basal conglomerate with clasts that resemble those of the Metschel Tillite even though the tillite may not be present.

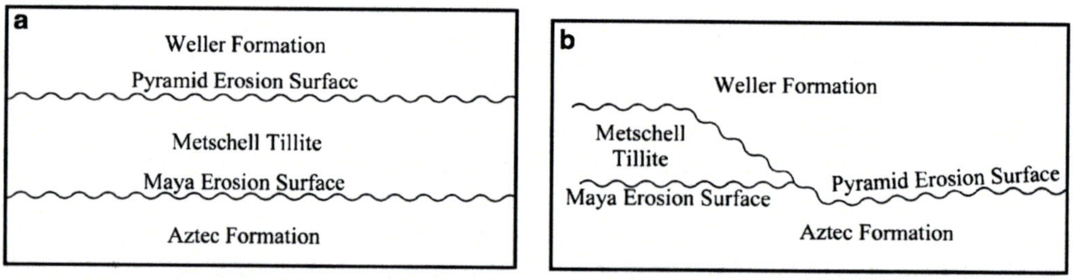

Fig. 10.7a The Metschel Tillite is bounded above and below by the Maya and Pyramid erosion surfaces which are parallel to each other **b** In places where the Pyramid erosion surface cuts through the Metschel Tillite and through the Maya erosion surface into the Aztec Formation, the Weller Formation lies disconformably on the Aztec Formation and the Maya erosion surface is thereby eliminated.

Fossilized leaves of *Glossopteris* and *Gangamopteris* of Permian age are present in this formation.

The coal-bearing Weller Formation is overlain by the *Feather Conglomerate* which is prominent at Mt. Feather (Fig. 10.2), but which also occurs on Mt. Fleming (Fig. 10.5) and on Portal Mountain located at the edge of the polar plateau. The Feather Conglomerate is 160 m thick, according to McKelvey et al. (1970), and was deposited by braided streams during the Permian Period (Barrett and Fitzpatrick 1985).

The Feather Conglomerate is conformably overlain by the *Fleming Formation* which consists of 45 m of coarse-grained and cross-bedded orthoquartzite and interbedded greenish-yellow siltstone on Mt. Fleming located the head of Wright Valley and at the edge of polar plateau. The geology of Mt. Fleming was described by Pyne (1984). The Fleming Formation is followed up-section by the *Lashly Formation* which contains 300 m of sandstone, shale, and carbonaceous beds at its type section on Mt. Feather (Fig. 10.2) and elsewhere (e.g., Mt. Fleming and Portal Mountain; McKelvey et al. 1970). The Lashly Formation contains plant microfossils of late Middle Triassic age.

The stratigraphy of the *Beacon Supergroup* proposed by McKelvey et al. (1970) is applicable throughout the Transantarctic Mountains with local modifications as necessary. In a later report, McKelvey et al. (1977) expanded the study area from the Olympus Range in Wright Valley northward to the Mackay Glacier (Fig. 10.4). In addition, Whitby et al. (1983) remapped the formations of the Beacon Supergroup in southern Victoria Land with special attention to the Weller Coal Measures. The geology of the Beacon Supergroup in southern Victoria was reviewed by Barrett et al. (1971, 1972), Korsch (1974), and by Barrett (1981).

10.1.3 Darwin Mountains

We now move south from the ice-free valleys and the Boomerang Range to the Darwin Mountains of southern Victoria Land adjacent to the Darwin and Hatherton glaciers and north of the Byrd Glacier in Fig. 10.8. This segment of the Transantarctic Mountains also includes the Brown Hills (Section 3.3) where granitic rocks of the local basement complex are exposed. The Beacon rocks of the Darwin Mountains were originally described by Gunn and Warren (1962) and were later mapped by Haskell et al. (1964, 1965b). The geology of the Darwin Mountains was also included on the geologic map of Grindley and Laird (1969) and was reviewed by Anderson (1979).

The Beacon rocks of the Darwin Mountains in Fig. 10.9 include both the Taylor and Victoria groups which consist of formations that closely resemble those of the Olympus Boomerang ranges described by McKelvey et al. (1970). The stratigraphic column in Table 10.3 is based on information provided by Haskell et al. (1965b).

The *Brown Hills Conglomerate* which is the oldest Beacon formation in the Darwin Mountains was deposited on the Kukri erosion surface by streams that transported sediment derived by erosion of nearby basement rocks. The pebbles are up to 4 cm in diameter, in part well rounded, and are composed of vein quartz and granitic rocks in a coarse sandy matrix. The accessory minerals include sphene, augite, black tourmaline, and magnetite. The conglomerate is locally cemented by quartz which was deposited in optical continuity with detrital quartz. The conglomerate contains carbonaceous matter formed from plants which inhabited the flood plain of the mature stream that transported the sediments. The Brown Hills Conglomerate is at least 34 m thick at its type locality on Bastion Hill (70°40'S, 158°34'E) at the confluence of the Touchtown Glacier and the Darwin Glacier, but it is not present elsewhere in the Darwin Mountains.

The lower *Hatherton Sandstone* is a well-sorted, flat-lying, texturally mature quartz arenite cemented by secondary quartz. The upper part of this formation contains rare grains of feldspar and enough biotite (less than 1.0%) to cause the weathered rock to be stained by secondary iron oxide. The accessory minerals consist of sphene, apatite, and zircon. At its type section on the north wall of Misthound Cirque (79°45'30"S, 156°15'E) in Fig. 10.10 the Hatherton sandstone is 450 m thick. An even thicker section is exposed along the Hatherton Glacier in the eastern Darwin Mountains. The upper part of the Hatherton Sandstone contains trace fossils including *Beaconites antarcticus* Vialov and others described and pictured by Gevers et al. (1971).

Splettstoesser and Jirsa (1985) reported that an outcrop of the Hatherton Sandstone in the Britannia Range

Fig. 10.8 The Hatherton Glacier flows from the polar plateau between the Darwin Mountains on the right and the Britannia Range on the left. Both mountain ranges in this area consist of sedimentary rocks of the Beacon Supergroup that were intruded by sills of Jurassic Ferrar Dolerite. The Westhaven Nunatak at the edge of the polar plateau is composed of the Jurassic Kirkpatrick Basalt. Aerial photograph from US Geological Survey, Earth Explorer, US Antarctic Indexes, TMA 1011, roll 00031L, frame 43, November 13, 1962

(Fig. 10.9) exhibits columnar jointing up to 1 m from the contact with a sill of the Ferrar Dolerite (50 m) which itself displays columnar jointing. This kind of jointing also occurs in a sandstone bed at the top of the Mackellar Formation (Permian) in the Queen Alexandra Range (Barett et al. 1986).

The *Darwin Tillite* is a massive unsorted sandstone containing erratic boulders. It has a thickness of more than 60 m and contains layers of sand and silt. A stratigraphic column of this tillite was published by Frakes et al. (1968). The type locality is at Colliseum Ridge in Fig. 10.10, but it occurs also along Haskell Ridge, at the base of Midnight Plateau, and in the eastern Darwin Mountains. It was divided into three members by Haskell et al. (1965b) based on lithologic differences. The middle member (6 m of sand and silt with graded

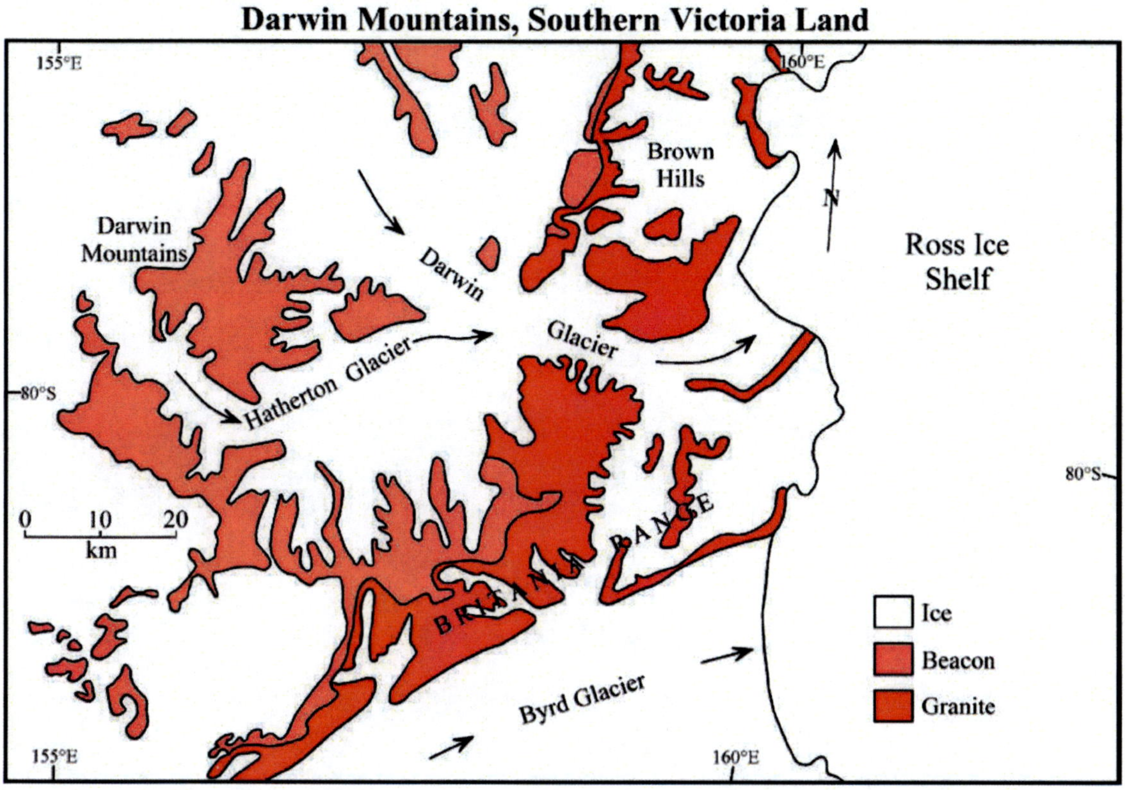

Fig. 10.9 Outcrops of the Beacon Supergroup in the Darwin Mountains adjacent to the Darwin and Hatherton glaciers. The Beacon rocks were deposited on the Kukri Peneplain and lie unconformably on the Granite Harbor Intrusives of the basement complex described in Section 3.3. The trace of the erosion surface is not shown on this map (Adapted from Grindley and Laird 1969)

Table 10.3 Stratigraphy of the Beacon Supergroup in the Darwin Mountains in the southernmost part of the southern Victoria Land located north of the Byrd Glacier (Fig. 10.9) (Adapted from Haskell et al. 1965b)

Age	Group[a]	Formation
Permian	Victoria	Ellis
Permian	Victoria	Misthound Coal Measures
		Disconformity
Permian	Victoria	Darwin Tillite
		Disconformity
Devonian	Taylor	Upper Hatherton Sandstone
Devonian	Taylor	Lower Hatherton Sandstone
Devonian	Taylor	Brown Hills Congl
		Kukri Unconformity
Basement Complex (Section 3.3)		

[a]The formations in this table were assigned to the Victoria and Taylor groups in accordance with information by McKelvey et al. (1970)

bedding) contains ripple marks and small drag folds caused by the movement of the overrding ice. The Darwin Tillite contains rare unfossilized wood fragments which are the only plant remains in this formation (Barrett and Askin 1975). Striated boulders are not common presumably because the striations have been obscured by weathering. Frakes et al. (1968) reported that the sediment-transport direction was southeast.

The *Misthound Coal Measures* disconformably overlie the Darwin Tillite. Haskell et al. (1965b) estimated that about 20 m of the tillite was eroded before the overlying sandstone was deposited. The Misthound Coal Measures are 93 m thick at their type locality in Misthound Cirque (Fig. 10.10). The formation consists of cross-bedded sandstone, carbonaceous sandstone,

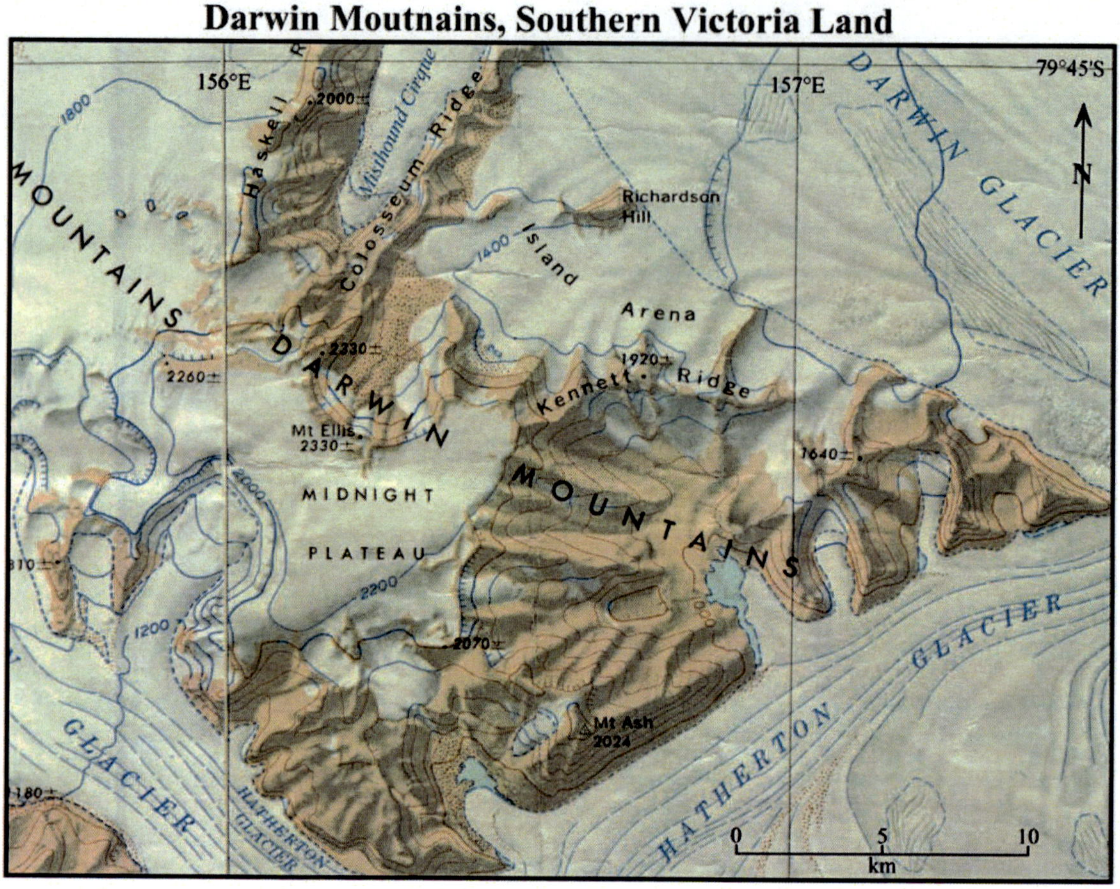

Fig. 10.10 The Darwin Mountains of southern Victoria Land are located at the confluence of the Darwin and Hatherton glaciers. They consist of the sandstones of the Beacon Supergroup and were mapped by Haskell et al. (1965a). Many topographic features they used to describe the geology of the mountain range can be found on this map. Excerpt from the topographic map: Carlyon Glacier, Antarctica (ST 57-60/13*; 156–163°E; S7900-E15600/1 × 7) published in 1966 by the US Geological Survey, Washington, DC

conglomerate, shale, and coal seams up to 20 cm thick. The lower part of the formation is a massive sandstone that contains leaves and stems of *Gangamopteris* as well as pyrite concretions.

The *Ellis Formation* consists of a 30 m sequence of sandstone, siltstone, and intraformational conglomerates that overlie the Misthound Coal Measures. The contact between these formations is obscured by a sill of the Ferrar Dolerite that is also 30 m thick. At its type section beneath the western end of the Midnight Plateau the Ellis Formation consists of well-bedded yellow to white sandstone including thin layers of green sandstone, white siltstone, and quartz-pebble conglomerates. The sandstones are predominantly composed of angular and poorly sorted quartz grains. Rock fragments and grains of orthoclase and plagioclase are less abundant. The accessory minerals include zircon, tourmaline, sphene, and magnetite. The minerals in the Ellis Formation have been altered by contact metamorphism caused by the dolerite sill. Haskell et al. (1965b) concluded that the Ellis Formation is late Permian or Early Triassic in age. Barrett et al. (1986) placed this formation into the Triassic System.

The stratigraphies of the *Beacon Supergroup* in the Olympus-Boomerang area and in the Darwin Mountains of southern Victoria Land are compared in Table 10.4.

Table 10.4 Comparison of the stratigraphic columns of the Beacon Supergroup in the Olympus-Boomerang area (McKelvey et al. 1970) and in the Darwin Mountains (Haskell et al. 1965b)

Age	Group	Formations: Olympus/ Boomerang	Formations: Darwin Mtns.
Triassic	Victoria	Lashly	Ellis
Late Permian		Fleming	
Permian		Feather	
Permian		Weller	Misthound
	Pyramid Disconformity		
Permian	Victoria	Metschel	Darwin
	Maya Disconformity		
Devonian	Taylor	Aztec	
		Beacon Heights	Hatherton
		Arena	
		Altar Mtn.	
		Odin Arkose	
	Heimdall Disconformity		
Devonian or older	Taylor	New Mountain Basal members	Brown Hills
	Kukri Disconformity		
Cambro/ Ordvician		Granite Harbor	
Neoproterozoic		Skelton Grp.	

The rocks of the Beacon Supergroup in both areas were deposited in the same interval of time between the Early Devonian until the Middle Triassic. The difference is that the Beacon rocks in the Olympus-Boomerang area have been more highly subdivided into 11 formations compared to the Beacon rocks in the Darwin Mountains which have been divided into only five formations. However, the rocks in both areas resemble each other in lithology, mineral composition, structure, environment of deposition, and fossil content. Evidently the environmental conditions in the depositional basin in this part of Gondwana remained stable for about 170 million years (Early Devonian to Middle Triassic) except for an interval of continental glaciation during the Permian.

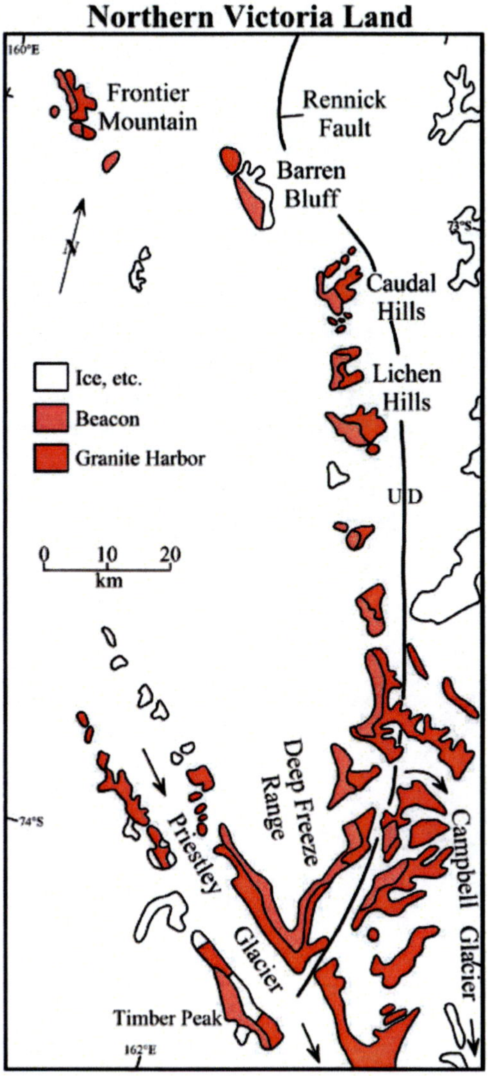

Fig. 10.11 Erosional remnants of the Beacon Supergroup occur on the nunataks in the Rennick basin and on its western edge along the polar plateau in northern Victoria Land. The formations are of Permian and Triassic age and belong to the Victoria Group. Rocks of the Devonian Taylor Group are not present in northern Victoria Land (Adapted from Gair et al. 1969)

10.2 Northern Victoria Land

The Beacon rocks of northern Victoria Land in Figs. 10.11 and 4.33 occur only as erosional remnants scattered within the Rennick Graben and in nunataks at the edge of the polar plateau. These rocks have been assigned to three formations that have a total thickness of only a few hundred meters and belong to the *Victoria Group* of Permian and Triassic ages. The older *Taylor Group* of Devonian age has not been recognized in northern Victoria Land and presumably was not deposited there. The Beacon rocks in the Lanterman Range of northern

Victoria Land were first noticed by Crowder (1968) and then described by Laird et al. (1977) and Walker (1983).

The Beacon rocks on the rim of the Rennick basin in Fig. 10.11 occur on Timber Mountain along the Priestley Glacier, in the Deep Freeze Range (Fig. 4.6 in Section 4.2), and its northward extension into the Lichen and Caudal hills, Barren Bluff, Frontier Mountains, and Roberts Butte.

A second cluster of outcrops of Beacon rocks in Fig. 10.12 within the Rennick basin includes the Freyberg Mountains, the Morozumi Range, the Helliwell Hills, as well as the Lanterman and Salamander ranges. The Beacon rocks of northern Victoria Land lie uncomformably on the metasedimentary rocks of the basement complex (Sections 4.3.4 and 4.3.5) including the Granite Harbor Intrusives (e.g., in the Freyberg Mountains). All but one of the outcrops of Beacon rocks in northern Victoria Land are located in the Wilson Terrane. The only exception are the Beacon rocks of the Neall Massif which is located east of the Salamander Range in the Bowers Terrane (Section 4.4). The Beacon rocks do not occur in the Robertson Bay Terrane (Gair et al. 1969).

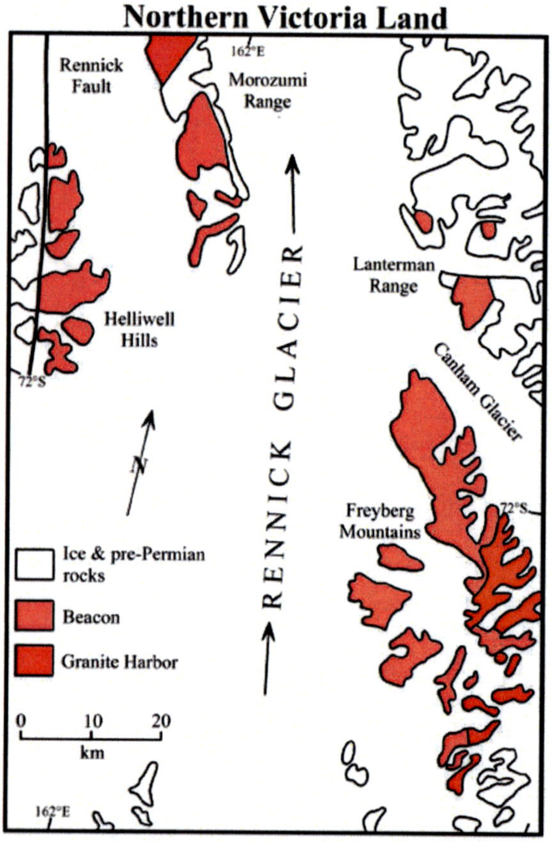

Fig. 10.12 Beacon rocks of the Victoria Group within the Rennick basin in northern Victoria Land. All of the outcrops of Beacon rocks rest unconformably on the Kukri erosion surface which is not shown. The Rennick Fault defines the western edge of the Rennick basin. The outcrops of Beacon rocks along this margin are identified in Fig. 10.11 (Adapted from Gair et al. 1969)

The stratigraphy of the Beacon rocks in Table 10.5 includes three formations defined by Collinson et al. (1986):

Section Peak (youngest)
Takrouna/basal breccia
Diamictite (glacial, oldest)

In addition, Elliot et al. (1986) described and named the Exposure Hill Formation based on exposures in the Mesa Range, at Agate Peak, and near Mt. Carson in the upper Aeronaut Glacier. The rocks at those localities are composed of volcanic breccias containing a mixture of angular clasts composed of the overlying Jurassic Kirkpatrick Basalt and sandstone of the Triassic Section Formation. The Exposure Hill Formation also contains lapilli tuff, tuff, and tuffaceous sandstone. The presence of abundant basalt clasts in the Exposure Hill Formation indicates that it formed by explosive volcanic activity at a time when the eruption of Kirkpatrick Basalt in the Mesa Range had already commenced. Therefore, the Exposure Hill Formation is an early manifestation of the magmatic activity that produced the rocks of the Jurassic Ferrar Group.

The Exposure Hill Formation of northern Victoria Land occupies the same stratigraphic position as the Mawson Formation in the Allan and Coombs hills of southern Victoria Land. Both formations are composed of volcanic breccias and related pyroclastic rocks that formed at the start of the eruption of the Kirkpatrick Basalt and the intrusion of sills of the Ferrar Dolerite of the Middle Jurassic Ferrar Group which is the subject of Chapter 12.

10.2.1 Glacial Diamictite

The unnamed *glacial diamictite* occurs in a down-faulted block of the western Lanterman Range in Fig. 10.12. It is about 250 m thick (McKelvey 1983;

McKelvey and Walker 1983) and includes glacial outwash as well as glaciolacustrine and fluvioglacial deposits (Laird and Bradshaw 1981; Skinner 1981). Collinson et al. (1986) and McKelvey and Walker (1983) correlated the glacial beds of northern Victoria Land with the *Metschel Tillite* of southern Victoria.

Each outcrop of Beacon rocks is unconformable to underlying pre-Permian basement rocks and the three formations identified in this table are not in contact with each other. The only exception is the basal breccia which grades into the overlying Takrouna Formation in the Neall Massif and at the DeGoes Cliff at the southern end of the Morozumi Range in Fig. 10.12.

Collinson et al. (1986) also reported that *metasedimentary breccias* occur sporadically at the base of the Takrouna Formation at DeGoes Cliff (northern Morozumi Range) and in the Neall Massif. The breccia in the Neall Massif is 56 m thick and contains clasts of mudstone, fine-grained sandstone, argillite, granite, rhyolite (?), and chlorite schist. The matrix is composed of sandy mudstone. Collinson et al. (1986) reported that the clasts are not striated and that there is no evidence supporting a glacial origin for the breccias. Instead, the breccias grade up-section into a carbonaceous mudstone of the Takrouna Formation containing fossilized leaves of *Gangamopteris* (Permian). Therefore, the breccias are a basal facies of the Takrouna Formation.

10.2.2 Takrouna Formation

The *Takrouna Formation* was named by Dow and Neall (1974) for a sequence of sandstone and carbonaceous shale at Takrouna Bluff (71°59'S, 163°23'E) in the Alamein Range of the northern Freyberg Mountains in Fig. 10.12 (Collinson and Kemp 1983; Collinson et al. 1986). The Takrouna Formation is composed of quartz sandstone interbedded with silty mudstones and coal seams. The exact thickness of this formation is not known because the top has been eroded. At the DeGoes Cliff in the southern Morozumi Range the Takrouna Formation unconformably overlies the *Rennick Schist* (Section 4.3.2) with a basal breccia layer similar to the exposure in the Neall Massif (Table 10.5). The exposed section of the Takrouna Formation at this locality is 280 m thick. The sandstones are ripple-marked, cross-bedded, laminated, and contain fining-upward cycles. The grain size of the sandstones of the Freyberg Mountains increases eastward toward the Neall Massif and decreases westward toward the Helliwell Hills and the Morozumi Range. Collinson et al. (1986) reported that the sandstones are composed of quartz, feldspar, biotite, muscovite, tourmaline, epidote, zircon, and garnet. All detrital mineral grains were derived from the local basement rocks, which included plutonic rocks of granitic composition and

Table 10.5 Stratigraphy of the Victoria Group of the Beacon Supergroup in northern Victoria Land (Collinson et al. 1986)

Age	Formation	Locations	Description
Triassic	Section Peak	Section Peak	Sandstone and conglomerate
		Vantage Hills	
		Roberts Butte	Fossil wood at Section Peak
		Unconformity	
Permian	Takrouna	DeGoes Cliff	Sandstone, coarse-to medium- to fine-grained, conglomerate, siltstones, carbonaceous mudstone, coal seams, fossil wood
		Boggs Balley	
		Jupiter Amph.	
		Alamein Range	
		Neall Massif	
		Monte Cassino	
		Moawhango	
		Smiths Bench	
		Unconformity	
Permian	Diamictite	Lanterman R.	Tillite and other glacial deposits
Late Carbon-iferous	Breccia	DeGoes Cliff (Morozumi R.)	Non glacial, clasts of local origin
		Neall Massif	
		Unconformity	

low-grade metasedimentary rocks. The Takrouna Formation is equivalent to the *Weller Coal Measures* of southern Victoria Land (Table 10.2) on the basis of lithologic similarity, the presence of coal seams, and a Permian age indicated by *Glossopteris*, *Gangamopteris*, *Paracalamites*, and *Vertebraria* described of Dow and Neall (1974). Additional information on plant fossils in the Takrouna Formation was provided by Hammer (1986).

The Takrouna Formation was deposited in a northwest trending intra-cratonic basin that is now occupied by the Rennick Glacier. The presence of glacial deposits in a down-faulted block of the Lanterman Range indicates that, prior to the deposition of sand and gravel, the area was glaciated. Collinson et al. (1986) interpreted the non-glacial breccias at the base of the Takrouna Formation as debris flows that formed along the walls of glacial valleys. The sediment was derived from the pre-Permian rocks of the Wilson Terrane that were exposed in the Rennick basin. The area of deposition of the Takrouna Formation was limited by the Rennick Basin which was more than 100 km wide and as much as 650 m deep. The floor of this valley was drained by braided streams that flowed northwest. The coarse texture and immature character indicates that the sediment was derived from local sources and was not transported over long distances.

10.2.3 Section Peak Formation

The *Section Peak Formation* was named for the outcrop of Beacon sandstones on Section Peak (73°14'S, 161°57'E) in the Lichen Hills on the edge of the polar plateau in Fig. 10.11. The Beacon rocks at this locality are capped by a sill of Ferrar Dolerite which was initially misidentified as a flow of the Kirkpatrick Basalt. As a result, the age these Beacon rocks was thought to be Jurassic (Gair 1967). Plant fossils described by Gair et al. (1965) and Norris (1965) actually indicate a Triassic age for the Section Peak Formation. Additional outcrops of the Section Peak Formation occur at Timber Peak and at several groups of nunataks along the edge of the polar plateau in Fig. 10.11 up to and including Roberts Butte (72°38'S, 160°08'E). The thickness of the Section Peak Formation increases from about 50 m in the Lichen Hills to 160 m at Timber Peak in the south (Fig. 10.11).

The Section Peak Formation consists of sandstone and conglomerate with abundant carbonaceous plant remains including coalified wood and calcite concretions. The rocks were deposited by braided streams much like the Takrouna Formation. The sandstones contain angular grains of quartz, feldspar, biotite, muscovite, tourmaline, granite, and epidote. Lithic grains are present and are composed primarily of volcanic rocks. Fragments of plutonic and metasedimentary rocks are rare. The grains of volcanic rocks are felsic in composition and consist of microcrystalline mosaics of anhedral quartz and feldspar.

The Section Peak Formation was deposited on the margins of Rennick basin after that basin was filled by sediment of the Takrouna Formation. The rocks of the Section Peak Formation resemble sedimentary rocks that occur at the top of the *Lashly Formation* in the Allan Hills (Table 10.2). Collinson et al. (1986) suggested that the Section Peak and Lashly formations are similar to the Late Triassic Falla Formation in the Beardmore and Shackleton glacier areas.

The volcanic detritus in the Section Peak Formation presumably originated from an off-shore subduction zone along the Pacific margin of East Antarctica. Similar volcanic sediment occurs at the base of the Permian System in the Ellsworth Mountains (Section 8.1.3), in the Late Permian rocks of the Beardmore Glacier area, and in the Early Triassic rocks of southern Victoria Land (Collinson et al. 1986). The correlation of the stratigraphic sections of Beacon rocks in northern and southern Victoria Land is indicated in Table 10.6.

Table 10.6 Correlation of the Victoria Group of the Beacon Supergroup in northern Victoria Land with the Beacon rocks of southern Victoria Land (Collinson et al. 1986; compiled from sources in the literature)

Northern Victoria Land	Southern Victoria Land	Age
Section Peak	Lashly	Triassic
60–220 ± m[a]	520 ± m	
Erosional gap	Feather	–
	250 m	
Takrouna	Weller	Permian
300 + m[b]	250 m	
Glacial beds	Metschel	–
		L. Carboniferous
0–650 m	0–70 m	
	Kukri Peneplain	

[a]± m: thickness uncertain

[b]+ m: thickness may be greater than indicated

10.3 Central Transantarctic Mountains

The central sector of the Transantarctic Mountains extends from the Byrd Glacier south to the mouth of the Shackleton Glacier (Chapter 5; Fig. 3.1). This segment contains thick deposits of Beacon rocks which form several large mountain ranges at the edge of the polar plateau between the major outlet glaciers:

Byrd to Nimrod glaciers:	Churchill Mountains
Nimrod to Lennox-King:	Queen Elizabeth Range
Lennox-King to Beardmore:	Queen Alexandra Range
Beardmore to Shackleton:	Queen Maud Mountains

The elevations of the summits of six mountains in this area exceed 4,000 m and 16 peaks have elevations between 3,000 and 4,000 m above sea level. Clearly, the mountain ranges in Fig. 10.13 rival the Alps of Europe, the Rocky Mountains of North America, and the Andes of South America.

The geology of the central Transantarctic Mountains is depicted on the maps prepared by Grindley and Laird (1969) and by McGregor and Wade (1969). In addition, Barrett et al. (1970), Lindsay et al. (1973), Barrett and Elliot (1973), and Elliot et al. (1974) published outcrop maps of the mountains in the Beardmore Glacier area. The rocks of the Beacon Supergroup in the Beardmore Glacier area were described by Grindley (1963), Grindley and Warren (1964), Grindley et al. (1964), McGregor (1965a, b), Barrett et al. (1967, 1968a), Barrett (1969, 1972), Lindsay (1969), Collinson et al. (1981), Collinson and Isbell (1986), and Barrett et al. (1986). The evolution of the Transantarctic basin during the Permian and Triassic was summarized in a paper by Collinson et al. (1994) which was published in a monograph on the geologic history of the Panthalassan margin of Gondwana edited by Veevers and McA. Powell (1994).

An important event in the mapping of the Beacon rocks in the Beardmore Glacier area was the discovery of an amphibian bone in December of 1967 at Graphite Peak (Barrett et al. 1968b). Subsequently, many more reptilian and amphibian bones were collected at Coalsack Bluff (Elliot et al. 1970) and in the Cumulus Hills along the McGregor Glacier (Kitching et al. 1972; Hammer and Cosgriff 1981; Colbert 1982). The McGregor Glacier, located at 85°08'S and 174°50'W, is an eastern tributary of

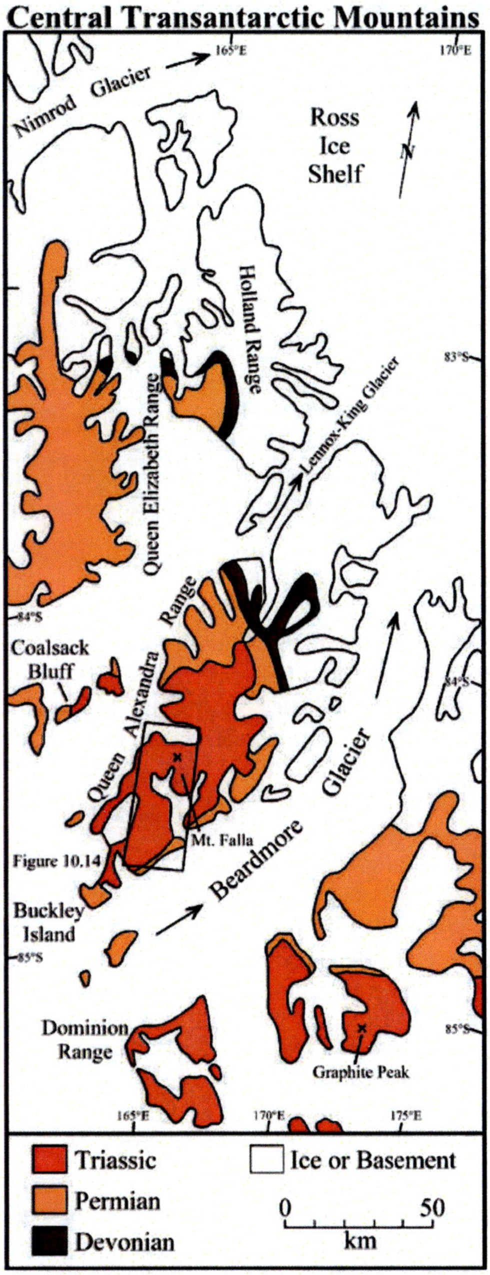

Fig. 10.13 The Beacon rocks in the Queen Elizabeth and Queen Alexandra ranges form high peaks in the central Transantarctic Mountains between the Nimrod and Beardmore glaciers. These rocks were deposited during the Devonian, Permian, and Triassic periods; but the extent of the Devonian Alexandra Formation of the Taylor Group is limited as shown (Adapted from Barrett et al. 1986)

Table 10.7 Stratigraphy of the Beacon Supergroup in the Beardmore Glacier area (Barrett et al. 1986)

Age	Group	Formations	
		Beardmore Gl.[a]	Darwin Mtns.[b]
Triassic	Victoria	Falla (160–530 m)	Not present
		Fremouw (650 m)	Ellis (180)
	Disconformity		
Permian	Victoria	Buckley (750 m)	Misthound (150 m)
		Fairchild (130–220 m)	Not present
		Mackellar (60–140 m)	Disconformity
		Pagoda (126–395 m)	Darwin (27–82 m)
	Disconformity		
Devonian	Taylor	Alexandra (0–320 m)	Hatherton (450 + m)
		Not present	Brown Hills (34 + m)
	Kukri Peneplain		
Cambro-Ordovician		Granite Harbor	Intrusives
Neoproterozoic-Cambrian		Metasedimentary rocks of the Goldie Formation	

[a]Beardmore section by: Grindley (1963), Barrett et al. (1968a), Barrett (1969), Lindsay (1969)
[b]Darwin section by: Haskell et al. (1965b), Barrett et al. (1971)

the Shackleton Glacier. The locations of Graphite Peak and Coalsack Bluff are indicated in Fig. 10.13. The discovery of fossil vertebrates in Antarctica aroused a great deal of interest in the geology of the Transantarctic Mountains and spurred further exploration of the Beacon rocks. This topic is presented in Section 11.1.

The stratigraphy of the Beacon rocks in the Beardmore Glacier area and in the Darwin Mountains in Table 10.7 was compiled from the literature by Barrett et al. (1986). Both sections include the Taylor and Victoria groups, although the Beacon rocks in the Beardmore area are considerably thicker than in the Darwin Mountains (i.e., Beardmore: 1976–2255 m; Darwin: 841–896 m). In spite of differences in total thickness, the stratigraphy of the Beacon rocks extends from southern Victoria Land into the Beardmore Glacier area with only a few changes, thereby confirming that the Beacon basins of Gondwana persisted in time and space for nearly 170 million years (Elliot 1975; Collinson et al. 1994). This period of geological tranquillity of Gondwana ended during the Jurassic Period when volcanic activity broke out followed by rifting and the intrusion of dolerite sills. These sills greatly increased the total thickness of the Beacon sections and caused contact metamorphism in the adjacent sedimentary rocks (Vavra et al. 1981).

10.3.1 Alexandra Formation

The *Alexandra Formation* of the Taylor Group in Table 10.7 is the oldest Beacon formation in the Beardmore Glacier area in Fig. 10.13. It is a mature, light-colored, thick- bedded, and cliff-forming quartz sandstone that was deposited during the Devonian Period on the *Kukri* erosion surface cut into Granite Harbor Intrusive and into the tightly folded Goldie Formation. The sandstones are coarse-grained and conglomeratic in some places. The quartz grains and pebbles are well rounded and cemented by secondary quartz. Some exposures are greenish-gray because of the presence of abundant garnet (e.g., 20 km west of Clarkson Peak). The sandstones are coarse bedded and yield paleocurrent directions that scatter widely. Based on this and other evidence, Barrett et al. (1986) concluded that the sandstones of the Alexandra Formation were deposited by streams flowing across a coastal plain.

The Alexandra Formation is present only in the northern Alexandra and Holland ranges as shown in Fig. 10.13; but it does occur in a few areas of the Queen Elizabeth Range and in Holyoake and Cobham ranges of the Churchill Mountains north of the Nimrod Glacier (Laird et al. 1971; Fig. 5.8). The Alexandra Formation correlates with the Hatherton Sandstone of the Darwin Mountains (Table 10.7) and with the New Mountain sandstone in the Olympus-Boomerang area of southern Victoria Land (Table 10.4); but it does not occur in northern Victoria Land (Section 10.2).

The sandstones of the Alexandra Formation, like those of Beacon Heights Orthoquartzite and the Hatherton Sandstone, contain only a few trace fossils. Nevertheless, the Devonian age of the Alexandra Formation is indicated by its correlation to the Beacon Heights Orthoquartzite which contains impressions of *lycopod* stems (Plumstead 1962), while the overlying Aztec Siltstone contains abundant fossil fish and plant microfossils which constrain a Devonian age (McKelvey et al. 1970, 1972; Helby and McElroy 1969).

10.3.2 Pagoda Tillite

The *Pagoda Tillite*, which is the oldest formation of the Victoria Group in Table 10.7, was deposited during the Late Carboniferous/Permian on the eroded surface of the Alexandra Formation. The erosion surface is equivalent to the Maya disconformity in the Beacon rocks of southern Victoria Land. In places where the Alexandra Formation was eroded or was not deposited, the Pagoda Formation was deposited directly on the Kukri erosion surface which was polished and striated by the continental ice sheet.

The Pagoda Tillite is composed of massive, poorly sorted sandstone containing scattered pebbles and cobbles. It is interbedded with layers of sandstone and shale studied by Lindsay (1968, 1969, 1970a, b) as well as by Frakes et al. (1971). In addition, Coates (1972, 1985) considered the Pagoda Tillite of the Queen Alexandra Range in the context of glacial deposits elsewhere in the central Transantarctic Mountains.

The Pagoda Formation is exposed in the vicinity of Pagoda Peak (83°56'S, 166°45'E) and at Mt. Miller in the northern Alexandra Range and it also occurs in the Queen Elizabeth Range. The thickness of the Pagoda Formation in the Queen Alexandra Range varies from 126 to 174 m most of which consists of tillite. The glacial beds in the Queen Elizabeth Range are considerably thicker (i.e., 390 m), but they thin to only 110–10 m north of the Nimrod Glacier (Fig. 10.13). Barrett et al. (1986) noted that the increase in thickness of the Pagoda Formation in the Queen Elizabeth Range is accompanied by a decrease in the thickness of Alexandra Formation and suggested that the Pagoda Formation filled glacial valleys cut into the Alexandra Sandstone.

The tillite beds range in thickness from 2 to 12 m, they are greenish-gray to dark gray in color, and contain clasts most of which are less than 10 cm in diameter, although some of the largest clasts range up to 2.8 m. The clasts originated from the local basement rocks. Fewer than 1% of the small clasts, but about 5% of the larger clasts, are striated. Sandstones were deposited in channels, streams, deltas, eskers, frost wedges, and during occasional episodes of subglacial melting (Lindsay 1970a) and comprise from 6% to 70% of the Pagoda Formation. In the area of the Tillite Glacier on the west side of the Queen Alexandra Range, the Pagoda Formation contains dark gray to black shale which includes thin beds of dark gray to black limestone which are about 20 cm thick and can be traced for considerable distances.

The ice-flow direction is indicated by grooved and striated bedrock surfaces at the base of the Pagoda Tillite and by the orientation of the long axes of clasts. Lindsay (1970a) reported an ice-flow direction of 168 ± 24° (southeast) which is in good agreement with the ice-flow direction of 155° reported by Frakes et al. (1968) for the Darwin Tillite in the Darwin Mountains of southern Victoria Land.

Although no diagnostic fossils have been found in the tillite, the Pagoda Formation is probably Late Carboniferous to Early Permian in age because the Buckley Formation, which occurs stratigraphically above it, contains glossopterid plant fossils of Permian age. Accordingly, the Pagoda Formation was deposited at about the same time as the Darwin and Metschel tillites of southern Victoria Land, although small differences in the depositional ages of these formations cannot be ruled out (Frakes et al. 1971). Kemp (1975) and Askin and Schopf (1982) placed the glacial beds of the Transantarctic Mountains into Stage 2 (Early Permian) of the Australian palynostratigraphic zonation (Schopf and Askin 1980).

The glacial sediments of the Pagoda Formation were deposited during the retreat of the large continental ice sheet in Gondwana. The tillites formed directly by ablation of basal ice, whereas the sandstones, siltstones, shales, and thin carbonate beds formed in ice-marginal lakes and meltwater streams. Some well-sorted sediment could also have been deposited by water flowing under the ice sheet. The tillite beds may record advances of the ice margin over previously deposited ice-marginal deposits. Lindsay (1968) suggested that the ice front advanced at least three times across the area that is now included in the central Transantarctic Mountains.

10.3.3 Mackellar Formation

The *Mackellar Formation* is composed of dark carbonaceous shales interbedded with sandstones that were deposited conformably on the Pagoda Formation (Grindley 1963). The Mackellar Formation is overlain by the Fairchild Formation; but the contact was intruded by a sill of the Ferrar Dolerite. The thickness of the Mackellar Formation in the Beardmore Glacier area varies irregularly from 55 m at Clarkson Peak (83°19'S, 164°34'E) in the Holland Range to 140 m on Buckley Island (Fig. 10.13). The formation occurs not only in the Queen Alexandra, Queen Elizabeth, and Holland ranges but it also extends into the Holyoake Range north of the Nimrod Glacier (Laird et al. 1971) and south into the area of the Shackleton Glacier (LaPrade 1970).

The sandstone beds of the shale-sandstone couplets are cross-bedded and up to 40 cm thick. The beds rarely extend for more than 10 m. In the Queen Elizabeth Range the Mackellar Formation contains lenticular bodies of fine- to medium-grained sandstones that are 5–25 m thick, fine upward, and contain shale fragments at their base. Barrett et al. (1986) suggested that these sandstone bodies formed in stream channels. Near Mt. Mackellar the top of the formation includes an unstratified layer of fine-grained and olive-gray sandstone that is 20 m thick. This bed contains white spheroidal concretions about 1 m in diameter. Some of these concretions contain calcite crystals and at least one was found to contain massive pale-green prehnite (zeolite). This sandstone bed has columnar jointing which extends downward from an overlying dolerite sill and cuts the concretions, which proves that the concretions formed before the sill was intruded. A similar occurrence was reported by Splettstoesser and Jirsa (1985) in the Hatherton Sandstone (Devonian) of the Britannia Range north of the Byrd Glacier (Section 10.1.2).

In the type area near Mt. Mackellar the Mackellar Formation contains several white and nonfossiliferous limestone beds up to 30 cm thick which extend for several meters. Brown-weathering gray limestone beds also occur in the lower part of the formation in the Queen Elizabeth Range. Elsewhere, limestone beds are not common in the Mackellar Formation. The strontium in the lenticular limestone beds is enriched in radiogenic ^{87}Sr (Faure and Barrett 1973) which indicates that they were deposited by aqueous solutions of non-marine origin.

Lindsay (1969) described a bed of pebbly sandstone (6.6 m thick) at Mount Angier (83°21'S, 161°00'E) near Mt. Markham in the northern Queen Elizabeth Range. About 3% of this rock consists of clasts whose diameter is greater than 1 cm. The lithologic composition of the clasts is similar to that of the Pagoda Tillite and some of the clasts are actually striated. This bed can be traced for more than 1 km before it pinches out. Lindsay (1968, 1969) concluded that the pebbly sandstone was a mudflow that originated from glacial deposits at an elevated site north of the area of deposition.

The Mackellar Formation contains only a few trace fossils including gastropod trails and worm burrows. A Permian age is, nevertheless, indicated by its conformable relation to the overlying and *Glossopteris*-bearing Fairchild Formation. The Mackellar Formation was deposited in a large body of anoxic water which Elliot (1975) has called the Mackellar Sea. However, the enrichment of the carbonate beds in radiogenic ^{87}Sr indicates that the beds were not deposited from seawater, although the Mackellar Sea may have had a connection to the open ocean like the Baltic Sea does at the present time. The increase in the sandiness of the formation in the western Queen Elizabeth Range and north of the Nimrod Glacier suggest that the margin of the basin was located in that direction. Sandstones of the Mackellar Formation in the western Queen Elizabeth Range (Fig. 10.13) actually contain ripple marks, mudcracks, and trails suggesting that they were deposited in shallow water close to the beach and that the water sometimes receded.

Lindsay (1970a) suggested that the Mackellar Sea formed as a result of isostatic depression caused by the Early Permian continental ice sheet and that its extent decreased as the crust rebounded until the Mackellar Sea eventually evolved into an alluvial plain on which the overlying Fairchild Formation was deposited.

The Mackellar Formation does not occur in the Darwin Mountains and in the Olympus-Boomerang area because the Mackellar Sea existed only south of these areas (see also Horner and Krissek 1991).

10.3.4 Fairchild Formation

The *Fairchild Formation* is exposed on a ridge that extends from Fairchild Peak at the mouth of Tillite Glacier toward the summit of Mt. Mackellar on the west side of the Queen Alexandra Range (Barrett 1969). It consists of a fine- to medium-grained cross-bedded sandstone that is conformably overlain by the Buckley Coal Measures. The Fairchild Formation is 145 m thick at its type locality on the northwest face of Mt. Mackellar. It grades upward from cliff-forming beds of light gray sandstone to fissile siltstones and shale which are recessive. The formation contains scattered quartzite pebbles 1–10 cm in diameter that are rounded as a result of being transported by streams. Current directions derived from large crossbeds by Barrett et al. (1986) indicate transport to the southeast at 168 ± 34°.

The sandstone is primarily composed of quartz grains some of which are well rounded and may have been derived from the Alexandra Formation of the Devonian Taylor Group. Grains of microcline and sodic plagioclase occur throughout but decrease in abundance toward the top of the formation. In addition, the sandstones contain lithic grains (schist and phyllite), biotite, muscovite, and pink garnet. The matrix consists of sericite, chlorite, quartz, and calcite which are post-depositional.

The Fairchild Formation contains coalified plant fragments and impressions of *Gangamopteris* leaves of Permian age. Sandstones that resemble the Fairchild Formation of the Queen Alexandra Range were reported by Laird et al. (1971) from the Holyoake Range north of the Nimrod Glacier and by other investigators from locations in the Queen Maud Mountains and the Horlick Mountains (LaPrade 1970; Doumani and Minshew 1965; Long 1965).

The sediment was deposited by braided streams of low sinuosity. The presence of feldspar grains and the generally poor rounding of quartz grains suggest that the sediment was derived from granitic and metasedimentary basement rocks, was transported less than about 200 km, and was deposited without subsequent reworking. The quartzite pebbles presumably originated from the Neoproterozoic Cobham Formation in the Cobham Range north of the Nimrod Glacier (Section 5.2.1, Fig. 5.7). This formation consists primarily of quartzite with lesser amounts of metagraywacke, argillite, and marble. The wide distribution of the Fairchild Formation in the Transantarctic Mountains suggest that it was deposited in a basin that was at least 1,000 km long which received sediment from the highlands on the margins of the basin and that the sand was distributed by streams flowing to the southeast (Barrett et al. 1986).

10.3.5 Buckley Coal Measures

The *Buckley Formation* contains coal beds and was therefore referred to by Grindley (1963) as the Buckley Coal Measures (Young and Ryburn 1968). The name was taken from Buckley Island near the head of the Beardmore Glacier where Frank Wild discovered coal seams during Shackleton's trek to the South Pole (1907–1909) (Section 1.4.2). The lower part of the Buckley Formation consists of sandstone which Barrett (1969) redefined as the Fairchild Formation described above. The Buckley Formation is overlain disconformably by the sandstones and noncarbonaceous mudstones of the Fremouw Formation of Triassic age (Table 10.7).

The Buckley Formation forms a gently plunging syncline which is the dominant structure of Mesozoic age in the Beardmore region. Although a complete section has not been found, the thickness of the Buckley Formation is close to 745 m (Barrett et al. 1986). It consists primarily of thick-bedded sandstones interbedded with layers of carbonaceous shale. In addition, the formation contains coal beds, conglomerate lenses, and occasional thin beds of limestone. The abundance of shale increases up-section such that the upper 100 m of the Buckley Formation are composed almost entirely of shale. The sandstone beds range from 1 to 10 m in thickness and, in many cases, contain shale fragments at their base. These basal segments as well as the main bodies of the sandstone beds contain coalified trunks and branches of trees from 5 to 30 cm in diameter.

The shale layers, composed of gray mudstone, range in thickness from 1 to 5 m in the lower part of the

Table 10.8 Chemical composition of coal samples from the Buckley and Fremouw formations in the Beardmore Glacier area in percent by weight (Barrett 1969)

Sample	H	C	N	O	S	Ash	BTU[a]
Buckley Formation[b]							
A321	2.6	81.0	2.0	8.5	0.5	8.2	12,880
C203	2.8	82.0	2.0	7.7	0.8	7.1	13,240
R005	2.4	79.8	2.0	9.8	0.7	8.0	12,670
R008	2.6	79.4	2.0	7.7	1.1	10.8	12,700
R014	0.85	39.7	0.7	3.3	0.2	55.2	5,990
G012	1.6	61.5	1.4	20.2	0.4	22.3	8,660
Fremouw Formation							
F328	2.9	72.8	1.9	14.4	0.5	11.5	11,580

[a]1 BTU = 1,055 J or 252 g calories

[b]The samples were submitted by J.M. Schopf to the Coal Analysis Section of the US Bureau of Mines

formation but thicken to 10 m in the upper part and reach 100 m at one location. Some carbonaceous shale beds are highly fissile and contain leaves of *Glossopteris*. The upper part of the Buckley Formation contains coal seams which are 1–2 m thick and occur both within and at the tops of the shale beds. The thickest coal seam (10.7 m) observed by Barrett et al. (1986) occurs on Mount Picciotto (83°46'S, 163°00'E) on the east side of the Queen Elizabeth Range adjacent to the Bowden Névé. The coal has a low ash and sulfur content and was classified by Barrett (1969) as low-volatile bituminous. The chemical analyses in Table 10.8 confirm that four samples of coal from the Buckley Formation have low ash content (7.1–10.8%), high carbon concentration (79.4–82.0%), and correspondingly high heat content (12,670–13,240 BTU). Two additional coal samples from Graphite Peak have higher ash contents (22.3–55.2%), lower carbon concentrations (61.5–39.7%), and lower heat contents (5,990–8,660 BTU). A single sample of coal of the Triassic Fremouw Formation does not quite measure up to the best coal of Buckley Formation.

The sandstone beds have large-scale trough cross-beds and are ripple-laminated as are the siltstone layers. The flow direction on the western side of the synclinal axis is to the east, whereas on the eastern side it is to the south. In general, the transport direction is to the southeast. Barrett et al. (1986) concluded that the sediment of the Buckley Formation originated partly from the East Antarctic craton located west the basin and partly from sources to the north where the Ross Ice Shelf is presently located.

The sandstones of the lower part of the Buckley Formation are mainly arkosic and grade up-section into lithic varieties in which more than 10% of the grains are volcanic. The sandstones of both kinds are fine to medium grained and fairly well sorted. Quartz, feldspar, and lithic grains of the arkosic and lithic sandstones are primarily angular to slightly rounded. The feldspar grains consist of microcline (7%) and plagioclase (13%). The abundances of different feldspar compositions vary regionally which suggests that they were derived from separate sources (i.e., plagioclase from the north and microcline from the west). The lithic sandstones contain more plagioclase (25%) and lithic grains (35%) but less quartz (10%) and K-feldspar (1%) than the arkosic sandstones. The lithic grains are composed of volcanic rocks of pyroclastic origin or are weathering products of lava flows. However, lava flows of Permian or Triassic age do not occur in the Beacon Supergroup of the central Transantarctic Mountains. The lithic grains of the volcanic sandstone were altered to zeolites by the heat that emanated from sills of Ferrar Dolerite.

The average of the chemical analyses of four samples of volcaniclastic sandstone of Buckley Formation in Table 10.9 indicates a rhyolitic composition, although these specimens were mixtures of lithic grains (30–44%), plagioclase (20–36%), and quartz (4–10%). The chemical composition of these rocks is characterized by high SiO_2 (68.55%), Al_2O_3 (15.49%), Na_2O (3.98%), and K_2O (2.83%) with notably low concentrations of FeO (0.89%), Fe_2O_3, (1.28%), MgO (0.65%), CaO (1.92%), and TiO_2 (0.87%). The presence of volcaniclastic sandstones in the upper Buckley Formation is a signal that volcanic rocks of felsic composition were erupted in the East Antarctic region of Gondwana during Late Permian. This process continued

Table 10.9 Average chemical composition to four volcaniclastic sandstones from the upper Buckley Formation (Permian) in the central Transantarctic Mountains (Barrett et al. 1986; Table 11)

Oxide	Concentration (%)
SiO_2	68.55[a]
TiO_2	0.87
Al_2O_3	15.49
Fe_2O_3	1.28
FeO	0.89
MnO	0.10
MgO	0.65
CaO	1.92
Na_2O	3.98
K_2O	2.83
H_2O+	2.48
H_2O-	0.44
P_2O_5	0.27
CO_2	0.29
Sum	100.4

[a]Analyst A.S. McCreath and Son, Inc.

during the Triassic Period and is recorded by the volcaniclastic sandstones of the overlying Fremouw and Falla formations. The suite of heavy minerals in the sandstones of the Buckley Formation differs from that of the Fairchild Formation and includes not only zircon, tourmaline, and garnet but also muscovite and biotite both of which are detrital.

Plant fossils are abundant in the Buckley Formation and include leaves of *Glossopteris* and *Vertebraria* as well as permineralized peat (Schopf 1970, 1971, 1982). In addition, Barrett et al. (1986) reported trace fossils and possible coprolites (ovoid structures). The Buckley Formation and its correlatives occur intermittently all the way to the Ohio Range about 1,000 km southeast of the Beardmore Glacier area. The Buckley Formation may also correlate in age with the coal-bearing Misthound Formation of the Darwin Mountains although that formation was deposited disconformably on the Darwin Tillite (Table 10.7) and the Mackellar and Fairchild formations of the Beardmore area were not deposited in the Darwin Mountains. The age of the Buckley Formation ranges from Early Permian (i.e., *Gangamopteris* and *Glossopteris*) to Middle and Late Permian (i.e., *Glossopteris* only).

The sandstones of the Buckley Formation were deposited on a slowly aggrading alluvial plain, whereas the shales and mudstones accumulated in lakes and ponds (Barrett 1969; Barrett et al. 1986). The volcaniclastic component of the sandstones was probably erupted penecontemporaneously as pyroclastic ash but no beds of volcanic tuff have been found in the Beardmore area.

10.3.6 Fremouw Formation

The *Fremouw Formation* in Table 10.7 is the oldest unit of Triassic age in the Beardmore area of the central Transantarctic Mountains. It was named by Barrett (1969) for Fremouw Peak which is located south of the Prebble Glacier in the Queen Alexandra Range in Fig. 10.14. The Fremouw Formation is composed of sandstones and mudstones that rest disconformably on the Buckley Formation (Barrett et al. 1986). The basal beds of the Fremouw Formation are composed of quartz sandstone and greenish-gray mudstone. This part of the formation contains vertebrate fossils that were discovered and described by Barrett et al. (1968b), Elliot et al. (1970), Kitching et al. (1972), and Hammer et al. (1986). The vertebrate fossils include bones of both reptiles and amphibians that occur as channel lag deposits in the basal Fremouw at Coalsack Bluff identified in Fig. 10.13.

The middle part of the Fremouw Formation is composed predominantly of mudstone that is overlain by the volcanic sandstone of the upper part of the formation. The middle and upper parts of the Fremouw Formation contain plant fossils but no fossil vertebrates have been found. The mudstone and volcanic sandstone are recessive and are overlain by the more resistant cliff-forming quartz sandstone of the Falla Formation.

The Fremouw Formation is widely distributed in the central Transantarctic Mountains and occurs in the Queen Alexandra and Queen Elizabeth ranges as well in the Dominion and Supporters ranges at the head of the Beardmore Glacier (McGregor 1965a, b). In addition, Collinson and Elliot (1984) mapped the Fremouw Formation in the Shackleton Glacier area of the Queen Maud Mountains. Barrett et al. (1986) estimated that the total thickness of the Fremouw Formation lies between 670 and 800 m.

The sandstones are light gray in color, fine grained, well sorted, faintly laminated, and trough cross-bedded. The sandstone beds fine upward and grade into overlying mudstone layers which are several meters thick and grade up-section into fissile carbonaceous shale and beds of coal (Table 10.8). The mudstone layers contain branching sets of vertical tubes that may have formed from the roots of plants that grew in swamps. The drainage in the Fremouw basin was to the northwest contrary to the southeasterly sediment-transport direction of the underlying Buckley Formation.

The lower part of the Fremouw Formation contains fossilized bones of tetrapods of the *Lystrosaurus* assemblage which indicate an Early Triassic age (Kitching et al. 1972). Therefore, the disconformity between the upper Buckley and the lower Fremouw formations includes the transition from the Permian to the Triassic periods. The age of the upper part of the Fremouw Formation is Middle to Late Triassic because of the occurrence of leaves of *Dicroidium odontopteroides* in rocks about 30 m below the top of the formation (Barrett et al. 1986). The abundance of plant fossils in the rocks indicates that the basin contained vegetation which sheltered a variety of reptiles and amphibians. Tree trunks up to 1 m thick contain widely-spaced tree rings that suggest a moderately moist but strongly seasonal climate.

Barrett et al. (1986) speculated that the well-rounded sand grains of the lower part of the Fremouw Formation originated from Devonian sandstones of the Pensacola Mountains described by Schmidt and Ford (1969). The sandstone of the middle part of the Fremouw Formation contains angular to slightly rounded grains of quartz, microcline, and plagioclase that originated by mechanical weathering of granitic rocks. The third kind of sediment in the upper part of the Fremouw Formation consist of grains of volcanic rocks and sodic plagioclase which originated by erosion of pyroclastic deposits similar to those in the underlying Buckley Formation.

The Fremouw Formation of the Beardmore Glacier area extends north as far as the Mackay Glacier of southern Victoria Land where it is correlated with the Lashly Formation. It also extends southeast to the Nilsen Plateau (Barrett et al. 1972; Elliot 1975; Collinson et al. 1994). However, sedimentary rocks of Triassic age do not occur in the Scott Glacier area (Doumani and Minshew 1965; Minshew 1966, 1967), in the Ohio Range (Long 1962, 1965), and in the Pensocola Mountains (Schmidt et al. 1965; Schmidt and Ford 1969).

10.3.7 Falla Formation

The *Falla Formation* was originally defined by Grindley (1963), but Barrett (1969) later subdivided Grindley's Falla Formation into the Fremouw Formation and redefined the Falla Formation as the overlying sequence of cliff-forming sandstones and shales capped by tuffaceous sandstone. The Falla Formation crops out on Mt. Falla (84°22'S, 164°55'E) in the Queen Alexandra Range (Fig. 10.14) where its thickness is 530 m (Barrett et al. 1986). The Falla Formation is overlain by the Prebble Formation which is composed of laharic deposits and pyroclastic breccias of Early to Middle Jurassic age equivalent to the Mawson Formation of the Allan Hills in southern Victoria Land (Section 10.1.3). The Falla Formation thins laterally and only a few outcrops are know outside of the southern to central Queen Alexandra Range (Elliot et al. 1974; Kitching et al. 1972; Collinson and Elliot 1984; Barrett et al. 1986; Elliot 1996).

The lower part of the Falla Formation consists of 12 fining-upward cycles. Each of these cycles was deposited on an erosion surface and each is composed of reddish-brown weathering medium- to fine-grained sandstone containing shale fragments at its base. The cycles range from 5 to 54 m in thickness and contain coalified plant remains. The tops of the sandstone cycles are composed of carbonaceous shale from 1 to 3 m thick. The shale beds contain coaly layers up to 1 m thick, but no actual coal seams were found by Barrett et al. (1986). The shale contains plant fragments including the leaves of *Dicroidium*. The upper part of the Falla Formation consists almost entirely of greenish-gray mudstone composed of volcanic ash (i.e., it is a tuff). The tuff in the upper part of the formation contains accretionary lapilli (Barrett 1969) that could have formed in clouds of ash ejected during volcanic eruptions.

The chemical composition of the tuff in the Falla Formation in Table 10.10 resembles the composition of volcaniclastic sandstone of the Buckley Formation (Table 10.9). Both rock types are silica- and alumina-rich with low concentrations of FeO, MgO, and CaO and comparatively high concentrations of alkali oxides. Barrett et al. (1986) concluded that the tuff of the Falla Formation has a rhyolitic composition. Table 10.10 also contains the chemical analysis of a trachyte pebble from the Falla Formation on Mt. Falla.

Fig. 10.14 The southern part of the Queen Alexandra Range in Fig. 10.13 exposes the Fremouw and Falla formations of the Beacon Supergroup as well as the overlying Prebble Formation and Kirkpatrick Basalt of the Ferrar Group. This part of the Queen Alexandra Range is located west of the Beardmore Glacier and east of the Walcott Névé. Excerpt of the Buckley Island (SV 61-60/3) and Cloudmaker (SV 51-60/4) topographic maps of Antarctica. US Geological Survey, 1967, Washington, DC

Silica-variation diagrams (not shown) reveal irregularities in the concentration of Na_2O, K_2O, and CaO of the samples of tuff which were presumable caused by alteration of the samples during diagenesis or by the subsequent intrusion of dolerite sills. Modal analyses by Barrett et al. (1986) indicate that several samples of sandstone contain zeolites and calcite cement. The samples of tuff range from rhyolite to dacite in the classification diagram of Wilson (1989) in Figure 10.15 although one of the samples is depleated in alkali elements. The single pebble of volcanic rock analyzed by Barrett et al. (1986) is intermediate between trachyte and trachyandesite. This pebble and others like it may have been eroded from the same felsic flows and pyroclastic deposits that also provided pebbles to the Fremouw Formation.

The Falla Formation of the Beardmore and Shackleton glacier areas is time-equivalent to the Lashly Formation of southern Victoria Land because both formations contain leaves of *Dicroidium*. However, the tuff beds at the top of the Falla Formation occur only in the Beardmore Glacier area (Barrett et al. 1986). The presence of *Dicroidium odontopteroides* indicates a Middle to Late Triassic age for the Falla Formation (Townrow 1967). The fossil plants in the Beacon rocks of the Beardmore area were described by Taylor and Smoot (1985a, b), Stubblefield and Taylor (1985), and Pigg and Taylor (1985).

10.3.8 Age Determinations, Falla Formation

The layers of the tuff near the top of the Falla Formation and a trachyte pebble collected on Mt. Falla were dated by isotopic methods. The pebble (F218A-19) was analyzed by the whole-rock K-Ar method which yielded a date of 197.7 ± 2.7 Ma (Barrett and Elliot 1972). A chemical analysis of this pebble is contained in Table 10.10 and Fig. 10.15. Barrett et al. (1986) recalculated the K-Ar date of this pebble to 203 ± 3 Ma by using the constants of Steiger and Jäger (1977). Barrett and Elliot (1972) also reported a whole-rock K-Ar date of 183.2 ± 10 Ma (recalculated to the constants of Steiger and Jäger 1977) for a dolerite boulder in a lenticular lahar deposit 120 m below the lowest flow of

Table 10.10 Chemical analysis of four samples of tuff of rhyolitic composition and of one pebble of volcanic rock from the Falla Formation in the Beardmore Glacier area (Barrett et al. 1986)[a]

	Concentration (%)	
Oxide	Rhyolitic tuff[b]	Trachyte pebble[c]
SiO_2	69.92	61.57
TiO_2	0.42	0.86
Al_2O_3	12.63	19.41
Fe_2O_3	2.51	0.53
FeO	0.57	0.13
MnO	0.04	0.08
MgO	1.08	0.32
CaO	2.00	2.36
Na_2O	1.75	7.41
K_2O	3.02	1.70
H_2O+	4.50	1.18
H_2O-	0.42	0.95
P_2O_5	0.24	0.23
CO_2	0.27	0.00
SO_3	0.00	1.79
Sum	99.37	98.52

[a]Analyses by A.S. McCreath and Son, Inc.

[b]The samples of tuff are identified by the numbers K020, K023, F220, and F247

[c]The sample number of the pebble is F218A-19

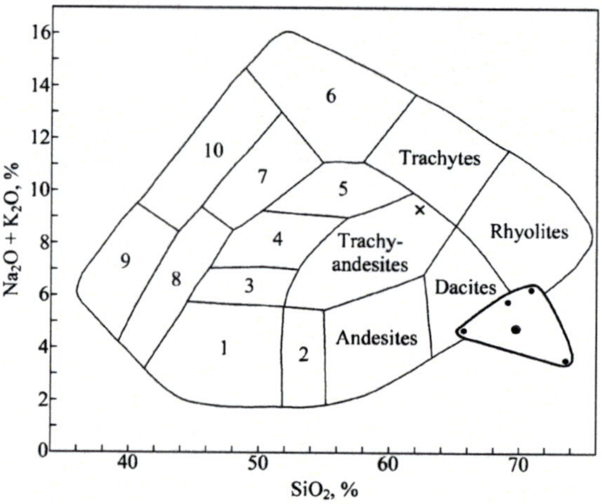

Fig. 10.15 The classification of volcanic rocks, based on their concentrations of Na_2O + K_2O and SiO_2 in weight percent, indicates that the tuff of the Falla Formation ranges in composition from dacite to rhyolite, whereas a pebble of volcanic rock is intermediate in composition between trachyte and trachyandesite. The numbered fields are identified as follows: (1) Basalt, (2) Basaltic andesite, (3) Hawaiite, (4) Mugearite and Trachybasalt, (5) Benmoreite, (6) Phonolite, (7) Phonolite tephrite, (8) Basanite and Tephrite, (9) Nephelinite, and (10) Phonolitic nephelinite. Solid circles: Falla Tuff; solid circle inside circle: Average of four samples of tuff; Cross: "trachyte" pebble (Adapted from Wilson 1989; Faure 2001; data from Barrett et al. 1986)

Kirkpatrick Basalt (Ferrar Group) at the Otway Massif (85°27'S, 172°00'E). The K-Ar dates of the trachyte pebble and the basalt boulder constrain the age of the Falla Formation to be younger than 203 ± 3 Ma but older than 183.2 ± 10 Ma.

Five whole-rock samples of tuff collected by Peter Barrett from the type section of the Falla Formation on the northwest face of Mt. Falla (293–414 m above the base of the Falla Formation) initially yielded a whole-rock Rb-Sr date of 190 ± 9 Ma based on $\lambda = 1.39 \times 10^{-11}$ year^{-1} (Faure and Hill 1973). A plot of the data in Fig. 10.16 indicates that sample F240 (Barrett's number) has lost radiogenic ^{87}Sr or has gained Rb. We have therefore recalculated the date based on an unweighted linear regression of the remaining four samples which yielded 196.4 ± 12.2 Ma and an initial $^{87}Sr/^{86}Sr$ ratio of 0.71220 ± 0.00129 for λ (^{87}Rb) = 1.42×10^{-11} year^{-1}. The Rb-Sr date of the Falla tuff and the K-Ar dates of the two pebbles in Fig. 10.17 constrain the age of the tuff to be Early Jurassic, although the uncertainty of the Rb-Sr dates extends into the Late Triassic. The high initial $^{87}Sr/^{86}Sr$ ratio of the tuff in Fig. 10.16 indicates that the strontium in the rhyolitic pyroclastics probably originated from crustal rocks of granitic composition.

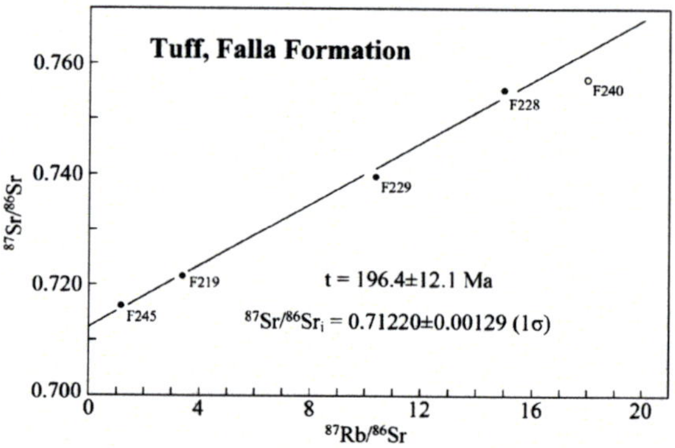

Fig. 10.16 Whole-rock samples of tuff from the upper part of the Falla Formation define a straight line on the Rb-Sr isochron diagram. The slope and intercept of the line yield a date of 196.4 ± 12.2 Ma ($\lambda = 1.42 \times 10^{-11}$ year^{-1}) and initial $^{87}Sr/^{86}Sr$ ratio of 0.71220 ± 0.00129, respectively. Sample F240 was excluded from the isochron because it appears to have lost radiogenic ^{87}Sr or gained rubidium. The data points are identified by the sample numbers of P.J. Barrett (Data from Faure and Hill 1973)

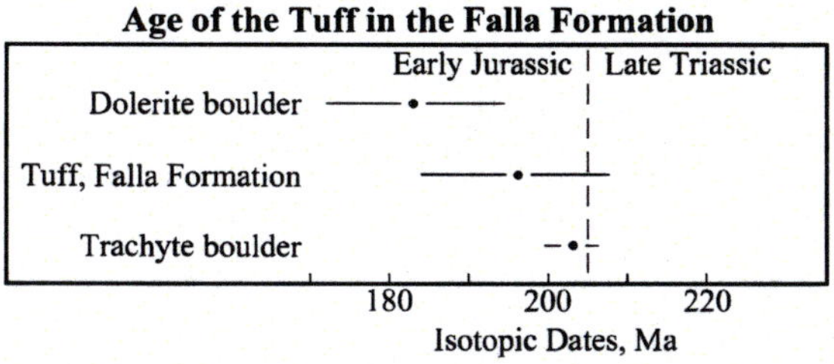

Fig. 10.17 The whole-rock K-Ar dates of a trachyte boulder from Mt. Falla and of a Ferrar Dolerite boulder from the Otway Massif, reported by Barrett and Elliot (1972), constrain the age of the tuff in the upper part of the Falla Formation. Both K-Ar dates were recalculated to the constants recommended by Steiger and Jäger (1977). The whole-rock Rb-Sr date (Fig. 10.16) of the Falla tuff (196.4 ± 12.1 Ma, $\lambda = 1.42 \times 10^{-11}$ year^{-1}), calculated from data by Faure and Hill (1973), indicates an Early Jurassic to Late Triassic age for the upper part of the Falla Formation. The boundary between the Late Triassic and Early Jurassic at 205.1 Ma is based on the time scale of the IUGS (2002)

10.4 Queen Maud Mountains

The Beacon Supergroup continues southeast of the Beardmore Glacier as far as the Axel Heiberg Glacier of the Queen Maud Mountains. This large area contains extensive deposits of the Beacon Supergroup which were described by McGregor (1965a, b) and Barrett (1965). At about the same time, a team of US geologists led by F.A. Wade divided the Beacon rocks of the central Queen Maud Range into four formations ranging in age from Devonian (or Carboniferous) to Triassic (Wade et al. 1965). During the next 20 years, the Beacon rocks of the Shackleton Glacier area were studied by Kirby LaPrade (1970, 1982, 1984). In addition, the Beacon rocks of the Queen Maud Mountains were described by McGregor and Wade (1969) on the geologic map of the area edited by Craddock (1969).

10.4.1 Cumulus Hills, Shackleton Glacier

The Beacon rocks exposed in the Cumulus Hills along the Shackleton Glacier attracted attention because they contain fossilized bones of large quadruped amphibians and reptiles to be presented in Section 11.1 (Kitching et al. 1972; Hammer and Cosgriff 1981; Colbert 1982; Hammer et al. 1996). The rocks of the Beacon Supergroup in the Cumulus Hills on the east side of the Shackelton Glacier in Fig. 10.18 were described by Collinson et al. (1978, 1981), Vavra et al. (1981), and by Vavra (1982). The formations that occur in this area were defined by Barrett et al. (1986) based on exposures in the Queen Alexandra Range of the Beardmore Glacier area. However, Devonian sandstones of the Alexandra Formation (Taylor Group) are not present in the Cumulus Hills. In addition, LaPrade (1982) combined the Falla Formation with the Fremouw Formation. Subsequently, Collinson and Elliot (1984) reviewed the stratigraphy of the Beacon Supergroup in the Cumulus Hills and recognized both the Triassic Fremouw and Falla formations which contain the tetrapod bones:

Triassic: Falla, 170+ m
Fremouw, 735 m
Permian: Buckley ~450 m
Fairchild ~200 m
Mackellar 160 m
Pagoda 4–190 m

The *Pagoda Tillite* in the Cumulus Hills was deposited on the eroded surface of the local basement complex consisting primarily of the Granite Harbor Intrusives (Chapter 6, Table 6.1). The Pagoda Formation is composed of tillite interbedded with sandstone containing microfloras of Early Permian (or Late Carboniferous) age (Askin and Schopf 1982). The *Mackellar Formation* which overlies the Pagoda Tillite consists of black shale and fine grained sandstone. It was succeeded by the sandstones of the *Fairchild Formation*, followed by the *Buckley Coal Measures*.

The Triassic rocks of the *Fremouw* and *Falla formations* consist of interbedded sandstone and mudstone some of which are carbonaceous. The volcaniclastic rocks of the *Prebble Formation* top the sequence and are overlain by lava flows of the *Kirkpatrick Basalt* of Jurassic age. The Beacon rocks in the Shackleton Glacier area were intruded by voluminous sills of the *Ferrar Dolerite*.

The sandstone beds of the middle and upper members of the *Fremouw Formation* in the Cumulus Hills contain lithic grains of igneous, metamorphic, and sedimentary origin. Most of the igneous grains consists of felsic volcanic rocks and are composed of microcrystalline aggregates of quartz and feldspar. Other lithic grains consists of granitic rocks, metasiltstone, metaquartzite, phyllite, and quartz-mica schist. The matrix is composed of phyllosilicates and zeolite (laumontite) described by Vavra (1982). In general, the

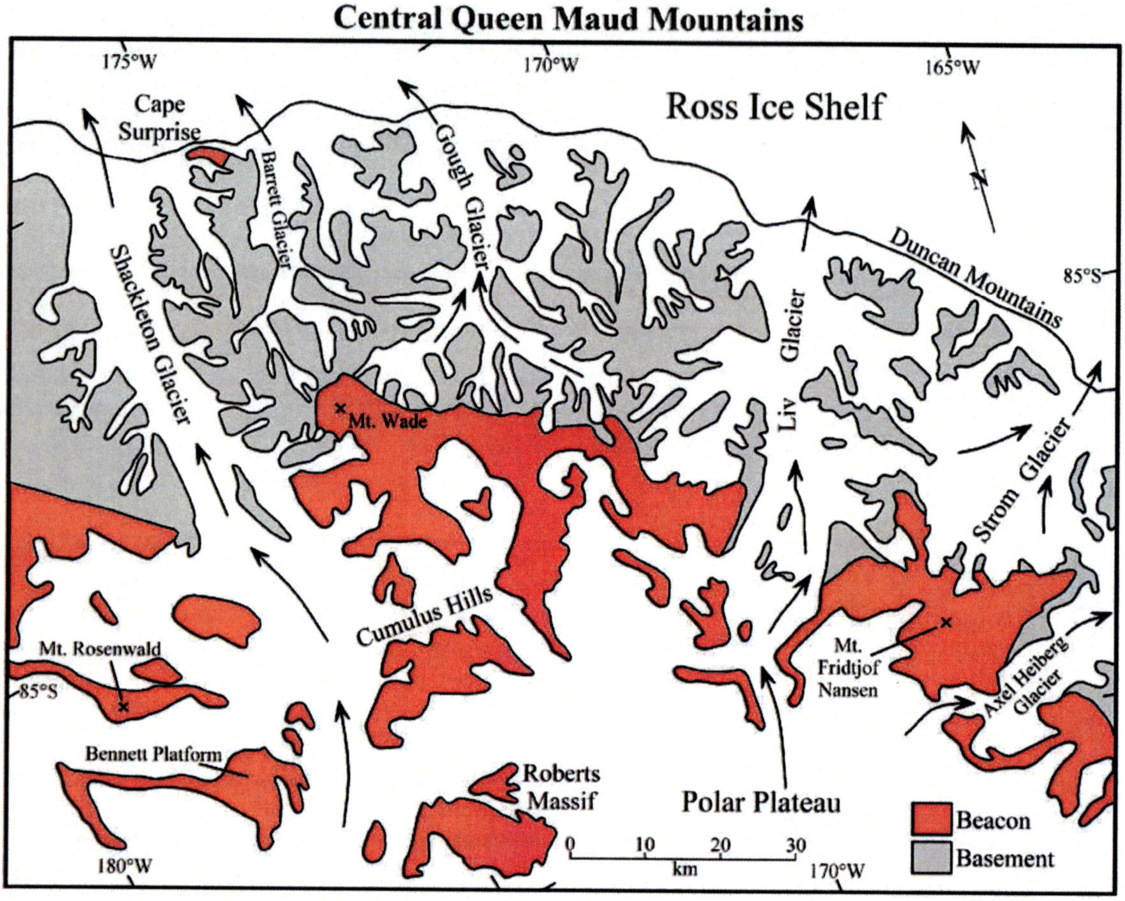

Fig. 10.18 Rocks of the Beacon Supergroup unconformably overlie the basement rocks of the coastal mountains in the central Queen Maud Mountains between the Shackleton and Axel Heiberg glaciers. The geology of the basement complex of this area is depicted in Fig. 6.1 of Chapter 6 which includes a discussion of the geology of the Duncan Mountains in Section 6.1 and Fig. 6.2 Barrett (1965) first published stratigraphic sections of the Beacon rocks of Mt. Fridtjof Nansen, Mt. Wade, and Cape Surprise which are identified on this map (Adapted from the map of McGregor and Wade 1969)

rocks of the Fremouw Formation in the Cumulus Hills are similar to the same formation in the Queen Alexandra Range. However, the paleocurrent directions measured by Collinson and Elliot (1984) in the Fremouw Formation of the Cumulus Hills average 210° (southwest), whereas Barrett et al. (1984) reported a mean current direction of 319° (northwest) for the Fremouw Formation in the Beardmore Glacier area.

The *Falla Formation* is composed of sandstone, shale, and tuffaceous beds as defined by Barrett et al. (1986) in the Queen Alexandra Range. In the Shackleton Glacier area only the lower part of the formation is present composed primarily of quartz sandstones which are bluff-forming and weather reddish brown in contrast to the greenish-gray and slope-forming sandstones of the upper Fremouw Formation. The sandstone units in the lower part of the Falla Formation fine upward into silty mudstones that contain casts of plant roots, but no other plant remains were found. Quartz (52%), feldspar (18%), and lithic grains (26%) are most abundant with minor amounts of mica (1%). Grains of volcanic origin are rare in the sandstones of the lower Falla. Collinson and Elliot (1984) concluded that the Early Triassic strata in the Shackleton Glacier area were deposited on an alluvial plain by streams of low sinuosity which evolved into sand-bearing braided streams during the Middle to Late Triassic.

10.4.2 Mt. Weaver, Scott Glacier

Southeast of the Axel Heiberg Glacier in the eastern Queen Maud Mountains only a few scattered remnants of the Beacon Supergroup have been preserved including on the Nilsen Plateau (Section 6.3), at the head of the Scott Glacier, on Mt. Blackburn, and on the Watson Escarpment where the Leverett Glacier descends to the coast of the Ross Ice Shelf. The geology of this large region was summarized by Mirsky (1964) on the geologic map of Antarctica edited by Craddock (1969).

The geology of Mt. Weaver and of the neighboring mountains was investigated during 1962/63 field season by a group of geologists from the Institute of Polar Studies of The Ohio State University led by George Doumani and including V.H. Minshew, L.L. Lackey, and C.J. Skinner. This group subdivided the Beacon rocks on Mt. Weaver into three formations listed in Table 10.11. These formations do not conform to the stratigraphy of the Beardmore-Shackleton glacier areas defined by Barrett et al. (1986), but they do correlate with the Beacon stratigraphy of the Wisconsin Range (Minshew 1966) and of the Ohio Range (Long 1965). The formations on Mt. Weaver were initially designated in the field by capital letters from A to E (Doumani and Minshew 1965) but were later given formational names (Minshew 1966, 1967). The total thickness of the Beacon rocks on Mt. Weaver is about 660 m.

At *Mt. Saltonstall* and Mt. Innes-Taylor in Fig. 10.19 Doumani and Minshew (1965) found boulders of greenish tillite in the moraines on the northern slopes of these mountains. The tillite contains faceted and striated pebbles and cobbles of metamorphic, granitic, and sedimentary rocks and resembles the Buckeye Tillite of the Horlick Mountains. These boulders were presumably transported and deposited by the Permian ice sheet although tillite is not present in outcrop at these locations.

Sunny Ridge located about 5 km southwest of Mt. Weaver in Fig. 10.19 contains a thick unit of black shale with interbedded sandstone and coal. These sedimentary rocks are about 160 m thick and dip southwest at 40°. The shale contains leaves of *Glossopteris* and petrified tree trunks up to 20 cm thick. The relation of the Sunny-Ridge section to the stratigraphy of Mt. Weaver is not understood.

The Beacon rocks on *Mt. Blackburn*, located close to the confluence of the Van Reeth Glacier with the Scott Glacier north of the LaGorce Mountains, occur at three sites labeled A, B, and C by Doumani and Minshew (1965). At locality C a sequence of sandstone, conglomerate, and black shale is *capped* by a layer of massive till about 15 m thick. This till

Table 10.11 Stratigraphy of the Beacon Supergroup on Mt. Weaver and on neighboring mountains in the area of the upper Scott Glacier (Doumani and Minshew 1965)

Unit	Formation	Description
E	Queen Maud (youngest)	Cyclical deposits of sandstone, shale and coal. Each cycle starts with a massive sandstone bed featuring large-scale trough cross-bedding and shale clasts at the base. Fossil leaves of Glossopteris and petrified tree trunks up to 60 cm in diameter are common, many are still upright. A quartz-pebble conglomerate occurs at the base of this formation
		Disconformity
D	Weaver (upper)	Massive sandstone with large-scale cross-bedding and ripple marks. The sandstone contains intraformational conglomerates composed shale clasts and burrows of crustacians or worms (Vialov 1962). This unit grades upward into shale and contains two coal beds in the upper 15 m
C	Weaver (middle)	Thin bedded fine-grained sandstone interbedded with thin beds of carbonaceous shale. Some interbeds are conspicuously brown and green
B	Weaver (lower)	Massive and coarse-grained sandstone containing limonitic concretions. The sandstone grades into thin-bedded black shale which contains muscovite on bedding planes. Carbonate stringers and large concretions up to 38 cm in diameter were noted as were animal trails and cone-in-cone structures in the shale
A	Scott Glacier (0–20, oldest)	Basal conglomerate composed of weathered granitic boulders in a matrix of sandstone, siltstone, and shale. The conglomerate rests on an erosion surface of the granitic basement. The relief on the peneplain is about 15 m. A bed of massive, gray mudstone occurs locally between the conglomerate and the erosion surface. This basal conglomerate overlies the basement rocks on Mt. Wilbur and elsewhere in this area where the contact of the Beacon rocks with the underlying basement is exposed

contains boulders of quartzite, black phyllite, gneiss, and diabase which are faceted and striated. Doumani and Minshew (1965) pointed out that the presence of diabase boulders suggests that this tillite is Jurassic or younger in age. Jurassic till is not known from other parts of the Transantarctic Mountains and this till is not present at nearby locality A. The total thickness of Beacon rocks on Mt. Blackburn is more than 600 m.

Beacon rocks are also exposed at *Mt. Howe* and *D'Angelo Bluff* located about 50 km south of Mt. Weaver beyond the southern limit of Fig. 10.19. Shales at Mt. Howe contain abundant *Glossopteris* leaves. The sedimentary rocks were intruded by diabase sills which converted the sedimentary rocks to hornfels and altered the coal to graphite.

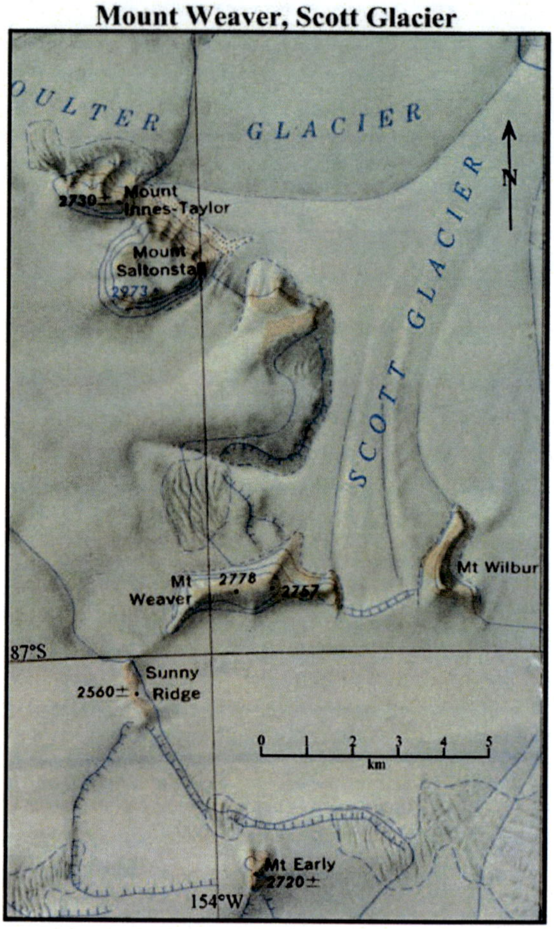

Fig. 10.19 Mt. Weaver (86°58'S and 153°50'W) is located at the head of the Scott Glacier which drains ice from the polar plateau to the coast of the Ross Ice Shelf. Sedimentary rocks of the Beacon Supergroup occur on Mt. Weaver, Sunny Ridge, Mt. Saltonstall, Mt. Innes-Taylor, and Mt. Wilbur all of which are identified on this map. In addition, D'Angelo Bluff and Mt. Howe, which are located still closer to the South Pole but are not included on this map, are also composed of Beacon rocks. Mt. Early is an extinct olivine basalt volcano of Cenozoic age. The geology of the Mt. Weaver area was described by Doumani and Minshew (1965) and by Minshew (1966, 1967). Excerpt of the topographic maps: Nilsen Plateau, Antarctica (SV 1-10/10), 1967, and D'Angelo Bluff, Antarctica (SV 1-10/15*), 1968, published by the US Geological Survey, Washington, DC

10.5 Horlick Mountains

The Horlick Mountains include the Wisconsin Range, the Long Hills, and the Ohio Range. The Wisconsin Range is largely composed of granitic basement rocks (Section 7.1) overlain in a few places by Glossopteris-bearing Permian rocks of the Beacon Supergroup. The Long Hills (Section 7.4) also contain small outliers of Beacon rocks on Todd Ridge and Victor Cliff. However, the most extensive exposure of the Beacon rocks occurs in the Ohio Range (Section 7.5) which were studied by W.E. Long in the early 1960s.

10.5.1 Wisconsin Range

The Beacon rocks on Tillite Spur in the valley of the Olentangy Glacier (Fig. 7.2) were divided into the same three formations that Doumani and Minshew (1965) described on Mt. Weaver in the Scott Glacier area:

Queen Maud
Weaver
Buckeye

Minshew (1966, 1967) reported that the *Buckeye Formation* in the Wisconsin Range is predominantly a tillite, 80–110 m thick, that rests unconformably on the granitic basement which is striated, grooved, and polished (Fig. 7.10). The lower part of the Buckeye Tillite is a locally stratified and well bedded but poorly sorted glacio-lacustrine deposits. The overlying tillite is greenish-gray, devoid of bedding, and contains abundant striated and faceted clasts. The matrix of the tillite is composed of mixture of clay and silt.

The Buckeye Tillite grades up-section into the Weaver Formation, the lower part of which is composed of shale with animal trails and ice-rafted pebbles. The middle part consists of interbedded siltstone and shale containing trace fossils. The upper part is a massive sandstone topped by a thin shale bed. The sandstone is characterized by the presence of animal burrows while the shale contains abundant *Glossopteris* leaves.

Minshew (1966) considered that the rocks of the Weaver Formation are a regressive sequence starting with shale that was deposited in a periglacial lake adjacent to the terminus of the Permian ice sheet. Subsequently, silt was transported into the basin and was interbedded with shale in the middle member. Ultimately, the sandstone of the upper member was deposited on a beach or in a near-shore environment. The sequence ended when the lake basin was transformed into a swamp in which black shale was deposited together with leaves of *Glossopteris*. Minshew (1966) questioned whether the large regional extent of the Weaver Formation is compatible with deposits in a periglacial lake which must have been several hundred kilometers in length. Therefore, he favored the idea that the Weaver Formation in the Wisconsin Range and its correlatives in the Queen Alexandra Range formed in a marine embayment or in brackish water.

The *Queen Maud Formation* is a coarse grained conglomeratic sandstone that lies unconformably on the rocks of the Weaver Formation. This unit is about 25 m thick and is equivalent to the basal quartz-pebble conglomerate of the Queen Maud Formation on Mt. Weaver. On Tillite Ridge in the Wisconsin Range the Queen Maud Formation is overlain by indurated till more than 30 m thick. This till is probably of Pleistocene or even of Pliocene age and thus is one of the many late Cenozoic glacial deposits in the Transantarctic Mountains that are collectively included in the Sirius Group.

10.5.2 Ohio Range

The Ohio Range was first seen from the air in October 1958 during a geological reconnaissance flight from Byrd Station in West Antarctica to the Wisconsin Range, the Ohio Range, the Thiel Mountains, and finally to the Whitmore Mountains before returning to Byrd Station. During this flight, W.E. Long observed that a thick section of stratified rocks was present in the Ohio Range. An opportunity presented itself to examine these rocks when a tractor train left Byrd Station for the Marie-Byrd Land oversnow traverse and, after passing the Wisconsin Range, made a temporary stop a short distance from the base of Mt. Glossopteris (Fig. 7.21). On this occasion, Bill Long and three companions from the traverse crew climbed Mt. Glossopteris of the Ohio Range in Fig. 10.20 and collected rock samples and fossils including leaves of *Glossopteris* which were later described by Schopf (1962; Long 1959).

The geologic map in Fig. 10.21 indicates that the Ohio Range consists of a high plateau called the Buckeye Table topped by Mt. Schopf (2,990 m) and by Mt. Glossopteris (2,867 m). These mountains expose a thick section of the Beacon Supergroup that was deposited unconformably on the early Paleozoic granitic basement complex (Section 7.5). The Beacon rocks of the Ohio Range were studied during 1960/61 and 1961/62 by two geological field parties from the Institute of Polar Studies of The Ohio State University led by W.E. Long who was supported by G.A. Doumani, J.H. Mercer, and others. Bill Long worked on the stratigraphy and sedimentology of the Beacon rocks in the Ohio Range (Long 1962, 1965), Mercer (1963) interpreted the glacial geomorphology, and Cridland (1963) discussed the *Glossopteris* flora. In addition, Boucot et al. (1963) described Early Devonian terebratuloid brachiopods from the Ohio Range and Doumani et al. (1965) wrote an extensive report about the Early Devonian marine invertebrate fauna in the basal Horlick Formation of the Ohio Range. This fauna correlates with a similar fauna in the Crashsite Group of the Ellsworth Mountains that was described by Webers et al. (1992).

Long (1962, 1965) defined four formations of the Beacon Supergroup in the Ohio Range which are identified and briefly described in Table 10.12. The ages of these rocks range from Early Devonian (Horlick Formation) to Late Permian (Mt. Glossopteris Formation). Rocks of Carboniferous (Mississippian and Pennsylvanian) and Triassic ages are not present. In other words, the Fremouw and Falla formations do not occur in the Ohio Range, nor do they occur in the Wisconsin Range and on Mt. Weaver in Scott Glacier area (Minshew 1966, 1967; Doumani and Minshew 1965).

Fig. 10.20 Mt. Glossopteris (2,867 m) of the Ohio Range is composed of 1,220 m of flat-lying sedimentary rocks of the Beacon Supergroup. These rocks were first examined in December of 1958 by four American geologists who reached the Ohio Range during the Marie-Byrd Land oversnow-traverse and climbed the mountain. Rock samples and fossils collected by the group were later described by Long (1959) (The photograph is reproduced here by permission of Dr. Margaret Bradshaw, Canterbury Museum, Christchurch, New Zealand)

The *Horlick Formation* is the oldest Beacon formation in the Ohio Range (Long 1965). It is composed of feldspathic and quartzitic sandstones which are locally cross-bedded and interbedded with beds of shale and siltstone. The poorly sorted sandstones contain marine invertebrate fossils, whereas the dark shale and siltstones contain fossilized primitive plants. The Horlick Formation is comparatively thin (0–46 m), but it occurs widely in the Ohio Range although it was eroded in some places before the Permian glaciation. In such cases, the Buckley Tillite was deposited directly on the striated and polished erosion surface of the granitic basement. The invertebrate fossils include brachiopods, pelecypods, bryozoans, trilobites, cephalopods, and *Tentaculites*. Boucot et al. (1963) assigned most of the brachiopods to the genus *Pleurothyrella* in the family Terebratulidae. These fossils establish an Early Devonian age for the Horlick Formation.

During the 1979/80 field season a group of New Zealand geologists led by M.A. Bradshaw returned to the Ohio Range in order to examine more closely the environment of deposition of the Early Devonian Horlick Formation. The group confirmed that brachiopods, mollusks, trilobites, crinoids, fish bones, and trace fossils occur at all levels in this formation and that it was indeed deposited in a marine environment in contrast to most sedimentary rocks of the Beacon Supergroup which are predominantly of non-marine origin (Bradshaw and McCartan 1983).

The *Buckeye Tillite* in Table 10.12 is an unsorted boulder clay composed of boulders, cobbles, and pebbles in a fine-grained bluish-green matrix. Long (1965) published an excellent close-up picture of this tillite in his Fig. 8. The tillite is interbedded with water-deposited layers of sandstone and shale. The Buckeye Tillite occurs along the whole length of the northern escarpment and is also present on Treves

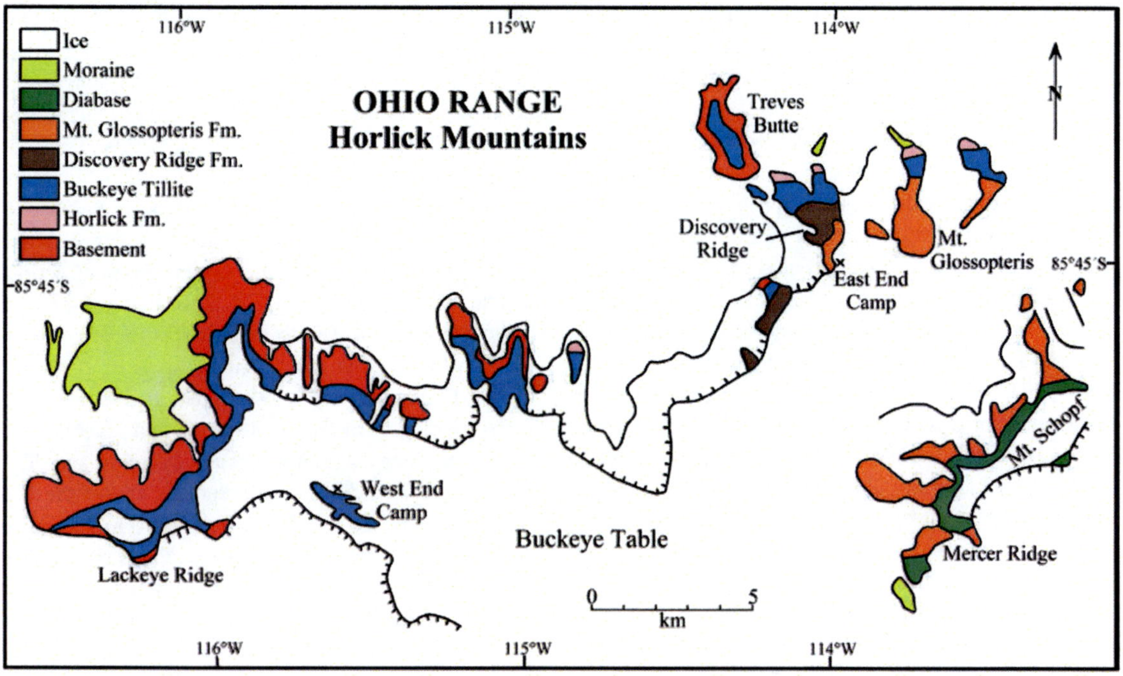

Fig. 10.21 The geology of the Ohio Range includes a basement of Cambrian granitic rocks overlain by 700 m of flat-lying sedimentary rocks of the Beacon Supergroup which contains four formations identified in the diagram. The Horlick Formation consists of Early Devonian marine sandstones and shales containing invertebrate fossils. The overlying Buckeye Tillite was deposited during the Permian glaciation of Gondwana. The Discovery Ridge Formation is composed of shales and siltstones of Permian age. It is overlain by the Glossopteris Formation of sandstones, shales, and coal seams containing *Glossopteris* flora, petrified tree trunks, and conchostrachans. Sedimentary rocks of Triassic age do not occur in the Ohio Range (Adapted from Long 1965)

Table 10.12 Stratigraphy of the Beacon Supergroup in the Ohio Range of the Horlick Mountains (Long 1962, 1965)

Age	Formation and thickness (m)	Description
Jurassic	Ferrar Dolerite 183	Dolerite (diabase)
	Intrusive contact	
Late Permian	Mt. Glossopteris 700	Arkose, sandstone shale, coal, *Glossopteris* leaves, *Dadoxylon*, *Leaia*
Permian	Discovery Ridge 122 (upper)	Carbonaceous shale, cone-in-cone structures, limestone
	Discovery Ridge 46 (lower)	Platy shale
	Disconformity	
Permian	Buckeye Tillite 274	Tillite, striated pebbles and striated pavements, with intercalated beds of sandstone and shale
	Disconformity	
Early Devonian	Horlick 0.54	Sandstone, shale with marine fossil assemblages
	Kukri Unconformity	
Cambrian	Granite Harbor	Granitic rocks
		(a) Porphyritic biotite granite with pink phenocrysts of K-feldspar
		(b) Medium- to coarse-grained granite (Treves 1965)

Butte (Fig. 7.21). The thickness of the tillite ranges from 256 to 313 m and it is separated from overlying shales of the Discovery Ridge Formation by a disconformity (Long 1965, Fig. 10).

The clasts are composed of sedimentary (72%), igneous (22%), and metamorphic (6%) rocks. They are in part angular, subangular, subrounded, and only rarely rounded. Twelve percent of the clasts are striated and 58% are faceted. Striated and grooved surfaces occur at many levels in the Buckeye Tillite (Long 1965, Figs. 11, 12, and 14) and indicate that the Permian ice sheet moved from west to east (i.e., N75°E to N90°E). Although sedimentary rocks that resemble tillite may form in several non-glacial environments (e.g., landslides, mudflows, volcanic eruptions, etc.), there is no longer any doubt that the formations containing tillite in the Beacon Supergroup were deposited by a continental sheet that covered a large area of Gondwana during the Permian Period.

The *Discovery Ridge Formation* consists of a succession of shales which Long (1965) subdivided into lower and upper members (Table 10.12). Both members occur only in the eastern part of the Ohio Range because, in the western part, all rocks above the Buckeye Tillite have been eroded. The lower member is a dark-colored, platy shale containing up to 40% silt. The upper member consists of a soft carbonaceous shale interbedded with lenticular layers of gray carbonate about 15 cm thick. These carbonate lenses do not contain fossils and are not of biogenic origin. Cone-in-cone structures composed of calcite are common in these beds. The Discovery Ridge Formation does not contain fossils although different kinds of animal trails are present (Long 1965, Fig. 15). The age of this formation is assumed to be Permian because the overlying Mt. Glossopteris Formation and the underlying Buckeye Tillite are both Permian in age.

The *Mt. Glossopteris Formation* is a cyclical deposit of arkose, feldspathic sandstone, siltstone, shale, and coal. The shale beds contain fossil leaves of *Glossopteris*. This formation occurs only on Mt. Glossopteris and Mt. Schopf in the eastern Ohio Range. Some of the sandstone beds form cliffs up to 30 m high which are not easy to climb. The formation is 700 m thick making it by far the thickest of the four Beacon formations in the Ohio Range.

The Mt. Glossopteris Formation contains fossilized leaves and related parts of *Glossopteris*. In addition, the formation contains fossilized wood in the form of small chips and complete logs up to 7.3 m long. One log pictured by Long (1965, Fig. 19) is more than 30 cm thick and contains 34 annual growth rings which are about 1 cm thick. The presence of fossilized tree trunks in the sandstone beds indicates that the Mt. Glossopteris Formation was deposited by meandering streams lined by forests. Many logs were transported as driftwood because they are aligned in a northeasterly direction parallel to the current direction. Several kinds of trees appear to have been present including *Dadoxylon* and a generic species that Schopf (1962) referred to as *Antarctixylon* sp. (i.e., Antarctic wood). Cridland (1963) identified five species of *Glossopteris*, one of *Gangamopteris*, and several other species of plants, which indicate a Permian age. The absence of *Dicroidium* means that the age of the Mt. Glossopteris Formation is pre-Triassic. The only animal fossils in this formation are conchostracans that Doumani and Tasch (1963, 1965) assigned to the new species *Leaia* and *Cyzicus*.

The Mt. Glossopteris Formation contains several coal beds which range in thickness from 1.2 to 3.6 m and amount to a total of about 23 m. The coal ranges in rank from bituminous to semi-anthracite but has a low grade because of its high ash content (Schopf and Long 1960; Schopf 1962). Long and his companions dug an adit into the lowest coal seam on Terrace Ridge (Fig. 10.22) in order to recover a sample of unweathered coal from a depth of about 5 m. The resulting adit was named the *Dirty Diamond Mine* and aroused much public interest after pictures of the "mine" appeared in the National Geographic Magazine. The fresh sample of the dry coal was found to have a carbon concentration of 80.7% with less than 20% of ash. The coal seam contained a lens of pyrite ("sulfur ball") measuring 15 × 41 cm and fractures in the coal near the surface were typically lined with yellowish-brown iron oxyhydroxide. Some of the coal seams in the Ohio Range extend laterally for about 1.2 km while others appear to be lenticular and their thickness varies along strike.

The environment of deposition of the Glossopteris Formation was non-marine and the sediment was deposited by streams that flowed from west to east. The striations in the underlying Buckeye Tillite indicate that the ice also flowed from west to east following the regional slope of the land. At the time of deposition of the Mt. Glossopteris Formation the climate was sufficiently humid to cause feldspar to weather while the topography of the highlands in the west was rugged

Fig. 10.22 An adit was dug into the outcrop of the lowest large coal seam in the Mt. Glossopteris Formation (Late Permian) on Terrace Ridge of Mt. Schopf in the Ohio Range. A composite sample of coal from this excavation contained 80.7% carbon (dry and free of mineral matter) which places it in the class of low-volatile bituminous coal (Long 1965). This picture of the "coal miners" of the Dirty Diamond Coal Co. was provided by Dr. William E. Long and is reproduced with his permission

enough to permit rapid erosion by streams. The scarcity of conglomerates indicates that the site of deposition in the Ohio Range consisted of floodplains beyond the range of alluvial fans that would have formed at the foot of the western mountains.

The Glossopteris Formation was intruded by a thick diabase sill (177 m) of the Ferrar Dolerite which still caps Mt. Schopf in Fig. 10.21. All sedimentary rocks above the sill have been eroded. Therefore, the top of the Mt. Glossopteris Formation is no longer present in the Ohio Range. Long (1965) considered that the Mt. Glossopteris Formation of the Ohio Range is equivalent in age and facies to coal-bearing sandstone-shale sequences in southern Victoria Land, in the central Transantarctic Mountains, and in the Queen Maud Mountains.

The structure of the Ohio Range is dominated by normal faults which define these mountains. However, the boundary faults cannot be located because the escarpments have been eroded and no longer coincide with the original fault zones. Long (1965) stated that the principal faults in the Ohio Range strike northeast-southwest. For example, a fault in Fig. 10.23 raised the Kukri Peneplain on Treves Butte by nearly 200 m relative to its level in the main body of the Ohio Range. Long (1965) also suggested that Mt. Glossopteris was uplifted by faulting relative to Mt. Schopf because the elevation of the summit of Mt. Glossopteris is greater than the elevation of the base of the diabase sill that caps Mt. Schopf. Geophysical evidence derived from seismic and gravity measurements by Bentley and Ostenso (1961) during the Byrd-Station traverse in 1958/59 indicates that the difference in elevation between the summit of Mt. Glossopteris and the bedrock surface at traverse station 414 is of the order of 3,000 m even though that station was located only about 6 km from the foot of the mountains.

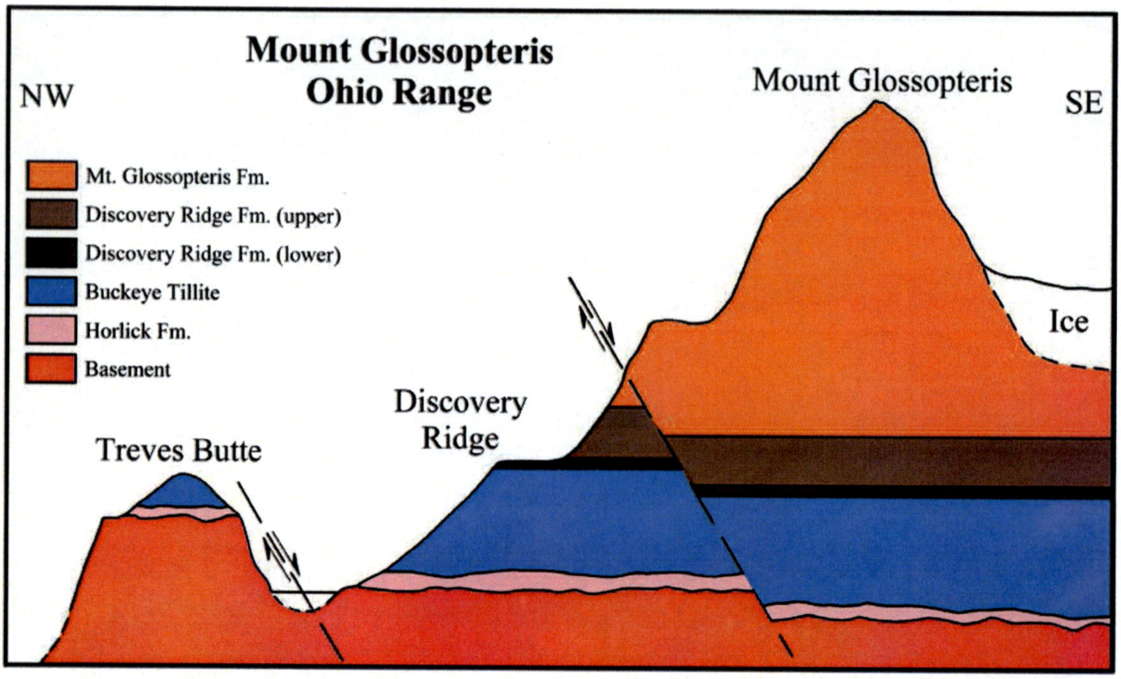

Fig. 10.23 The structural cross-section through Mount Glossopteris and Treves Butte of the Ohio Range emphasizes the importance of normal faults that displace the flat-lying sedimentary rocks of the Beacon Supergroup as well as the granitic rocks of the basement complex. The Discovery Ridge Formation in this cross-section consists of the upper and lower members (Adapted from Long 1965)

Table 10.13 Comparison of stratigraphies of the Beacon Supergroup in the Transantarctic Mountains (Barrett et al. 1986; Elliot 2000)

Geologic age	S. Victoria Land	Darwin Gl.	Beardmore Gl.	Ohio R.	Pensacola Mtns.
Jurassic	Mawson		Prebble		
			Hanson		
Triassic			Falla		
	Lashly	Ellis	Fremouw		
Permian	Feather	Misthound	Buckley	Mt. Glossopteris	Pecora
	Weller		Fairchild		
			Mackellar	Discovery R.	
	Metschel	Darwin	Pagoda	Buckey Till.	Gale
Devonian	Aztec			Horlick	Dover
	Beacon Hts.				
	Arena	Hatherton	Alexandra		Heiser
	Altar				Elbow
	New Mtn.	Brown			Elliott
					Brown Ridge
			Unconformity		

Ross orogen: Measedimentary and metavolcanic rocks intruded by the Granite Harbor Intrusives.

10.6 The Far Eastern Mountains

The compilation of stratigraphic units in Table 10.13 suggests that the sedimentary rocks of the Beacon Supergroup in the Ohio Range are an extension of the Beacon rocks that characterize the main body of the Transantarctic Mountains. Farther east, the geologic history of the mountains changes. For example, Beacon rocks do not occur in the Thiel Mountains. In the Pensacola Mountains, the basement rocks are overlain by sedimentary rocks of the Neptune Group and by sandstones to which Barrett et al. (1986) assigned a Devonian age. In that case, these rocks may be coeval with the Devonian sandstones of southern Victoria Land, although the basin in which they were deposited may not have been coextensive with the basin in which the sandstones of the Taylor Group were deposited. These Devonian formations in the Pensacola Mountains are disconformably overlain by the Gale Mudstone which is a tillite of Permian age presumed to be correlative with similar tillites throughout the Transantarctic Mountains. The Gale Mudstone is overlain by siltstones and carbonaceous shale of the Pecora Formation which also contains coaly layers (Section 8.2.3). These rocks contain *Glossopteris* and resemble the Discovery Ridge/Mt. Glossopteris formations of the Ohio Range. Sedimentary rocks of Triassic age are not present in the Pensacola Mountains. Although sedimentary rocks of the Beacon Supergroup are not present in the Argentina Range, such rocks do occur in the Wichaway Nunataks, in the Shackleton Range, and in the Theron Mountains (Craddock 1982). Still farther afield in Queen Maud Land of East Antarctica, Beacon-like sedimentary rocks occur in the Kraul Mountains, in the Tottan Hills, and in the Kuven Hills. However, these locations are far removed from the Transantarctic Mountains and the relation of the sedimentary rocks to the Beacon Supergroup of the Transantarctic Mountains is uncertain.

10.7 Summary

The sedimentary rocks of the Beacon Supergroup range in age from Devonian to Triassic and occur, with interruptions, throughout the entire length of the Transantarctic Mountains from northern Victoria Land to the Theron Mountains on the coast of the Weddell Sea. In fact, the Beacon rocks characterize the Transantarctic Mountains because they form the peaks and ridges of this enormous mountain range. The Transantarctic Mountains are also shaped by the numerous and voluminous sills of the Ferrar Dolerite and by the related Kirkpatrick Basalt.

Although the Devonian sandstones in southern Victoria Land, in the Beardmore Glacier area, and in the Ohio Range were deposited in shallow marine basins, most of the Permian and Triassic rocks formed in continental basins of Gondwana.The only exception may be the Mackellar Formation which was deposited in a stagnant body of water left behind after the Permian ice sheets had retreated from East Antarctica.

Another noteworthy feature of the Beacon Supergroup are the Permian tillites that formed from glacial deposits that accumulated during the retreat of the continental ice sheets that covered a large area of Gondwana. These rocks closely resemble the Dwyka Tillite and other Permian glacial deposits of southern Africa and India.

The tillites are overlain by sandstones, siltstones, and shale including substantial seams of bituminous coal. The presence of abundant plant fossils (e.g., *Glossopteris*, *Gangamopteris*, *Diocroidium*, etc.) as well as petrified tree trunks and coalified plant debris required temperate and humid climatic conditions that were radically different from the cold and dry climate that prevails in the Transantarctic Mountains at the present time. During this time, Gondwana was also inhabited by aquatic and terrestrial vertebrates such as fish, amphibians, and reptiles all of which are represented in the fossil record of the Transantarctic Mountains.

References

Allen AD (1962) Geological investigations in south Victoria Land, Antarctica. Part VII: Formations of the Beacon Group in the Victoria Valley region. New Zealand J Geol Geophys 5(2):278–294

Allen AD, Gibson GW (1962) Geological investigations in southern Victoria Land, Antarctica. Part 6. Outline of the geology of the Victoria Valley region. New Zealand J Geol Geophys 5(2):234–242

Anderson JM (1979) The geology of the Taylor Group, Beacon Supergroup, Byrd Glacier area, Antarctica. New Zealand Antarctic Record 2(1):6–11

Angino EE, Turner MD, Zeller EJ (1962) Reconnaissance geology of lower Taylor Valley, Victoria Land, Antarctica. Geol Soc Amer Bull 73:1553–1562

Askin RA, Schopf JM (1982) Permian and Triassic palynostratigraphy of the Victoria Group, Transantarctic Mountains. In: Craddock C (ed) Antarctic geoscience. University of Wisconsin Press, Madison, WI, pp 649–659

Barrett PJ (1965) Geology of the area between the Axel Heiberg and Shackleton glaciers, Queen Maud Range, Antarctica. Part 2: Beacon Group. New Zealand J Geol Geophys 8(2):344–363

Barrett PJ (1969) Stratigraphy and petrology of the mainly fluviatile Permian and Triassic Beacon rocks, Beardmore Glacier area, Antarctica. Institute of Polar Studies, Report No. 34, 132 p

Barrett PJ (1972) Stratigraphy and petrology of the mainly fluviatile Permian and Triassic part of the Beacon Supergroup, Beardmore Glacier area. In: Adie RJ (ed) Antarctic geology and geophysics. Universitetsforlaget, Oslo, Norway, pp 365–372

Barrett PJ (1979) Nonmarine character of the Taylor Group (Devonian) in southern Victoria Land. In: Laskar B, Raja Rao CS (eds) Fourth international Gondwana symposium, vol. 2. Hindustan, India, pp 478–480

Barrett PJ (1981) History of the Ross Sea region during the deposition of the Beacon Supergroup 400–180 million years ago. J Roy Soc New Zealand 11:447–458

Barrett PJ (1991) The Devonian-to-Jurassic Beacon Supergroup of the Transantarctic Mountains and correlatives in other parts of Antarctica. In: Tingey RJ (ed) The geology of Antarctica. Clarendon, Oxford, pp 120–152

Barrett PJ, Askin RA (1975) The Early Permian glacial beds of south Victoria Land and the Darwin Mountains. In: Campbell KSW (ed) Gondwana geology. Australian National University Press, Canberra, A.C.T., pp 333–346

Barrett PJ, Elliot DH (1972) The early Mesozoic volcaniclastic Prebble Formation, Beardmore Glacier area. In: Adie RJ (ed) Antarctic geology and geophysics. Unversitetsforlaget, Oslo, Norway, pp 403–409

Barrett PJ, Elliot DH (1973) Reconnaissance geologic map of the Buckley Island Quadrangle, Transantarctic Mountains, Antarctica. US Antarctic Research Program, Antarctic Geologic Map A-3. US Geological Survey, Washington, DC

Barrett PJ, Fitzpatrick PG (1985) Deposition of the lower Feather Conglomerate, a Permian braided-river deposit in southern Victoria Land, Antarctica, with notes on the regional paleogeography. Sediment Geol 45:189–208

Barrett PJ, Elliot DH, Lindsay JF (1967) Geology of the Beardmore Glacier area. Antarctic J US 2(4):110–112

Barrett PJ, Elliot DH, Gunner J, Lindsay JF (1968a) Geology of the Beardmore Glacier area, Transantarctic Mountains. Antarctic J US 3(4):102–106

Barrett PJ, Baillie RJ, Colbert EH (1968b) Triassic amphibian from Antarctica. Science 161(3840):460–462

Barrett PJ, Lindsay JF, Gunner J (1970) Reconnaissance geologic map of the Mount Rabot Quadrangle, Transantarctic Mountains, Antarctica. US Antarctic Research Program, Antarctic Map No. 1. US Geological Survey, Washington, DC

Barrett PJ, Kohn BP, Askin RA, McPherson JG (1971) Preliminary report of Beacon Supergroup studies between the Hartherton and Mackay glaciers, Antarctica. New Zealand J Geol Geophys 14:605–614

Barrett PJ, Grindley GW, Webb PN (1972) The Beacon Supergroup of East Antarctica. In: Adie RJ (ed) Antarctic geology and geophysics. Universitetsforlaget, Oslo, Norway, pp 319–332

Barrett PJ, Elliot DH, Lindsay JF (1986) The Beacon Supergroup (Devonian-Triassic) and Ferrar Group (Jurassic) in the Beardmore Glacier area, Antarctica. In: Turner MD, Splettstoesser JF (eds) Geology of the Central Transantarctic Mountains. Antarctic Research Series, vol. 36. American Geophysical Union, Washington, DC, pp 339–428

Bentley CR, Ostenso NA (1961) Glacial and subglacial topography of West Antarctica. J Glaciol 3(29):882–911

Boucot AJ, Caster KE, Ives D, Talent JA (1963) Relationships of a new Lower Devonian terebratuloid (Brachiopoda) from Antarctica. Bull Amer Paleo 46(207):77–151

Bradshaw MA (1981) Palaeoenvironmental interpretations and systematics of Devonian trace fossils from the Taylor Group (lower Beacon Supergroup), Antarctica. New Zealand J Geol Geophys 24:615–652

Bradshaw MA, McCartan LM (1983) The depositional environment of the Lower Devonian Horlick Formation, Ohio Range. In: Oliver RL, James PR, Jago JB (eds) Antarctic earth science. Australian Academy of Science, Canberra, ACT, pp 238–241

Coates DA (1972) Pagoda Formation: Evidence of Permian glaciation in the Central Transantarctic Mountains. In: Adie RJ (ed) Antarctic geology and geophysics. Unversitetsforlaget, Oslo, Norway, pp 359–364

Coates DA (1985) Late Paleozoic glacial patterns in the Central Transantarctic Mountains, Antarctica. In: Turner JD, Splettstoesser JF (eds) Geology of the Transantarctic Mountains. Antarctic Research Series, vol. 36. American Geophysical Union, Washington, DC, pp 25–338

Colbert EH (1982) Triassic vertebrates in the Transantarctic Mountains. Ant. Res. Ser., 36:11–35. Amer. Geophys. Union, Washington, DC

Collinson JW, Elliot DH (1984) Triassic stratigraphy of the Shackleton Glacier region, Transantarctic Mountains. In: Turner MD, Splettstoesser JF (eds) Geology of the Central Transantarctic Mountains. Antarctic Research Series, vol. 36. American Geophysical Union, Washington, DC, pp 103–117

Collinson JW, Isbell JL (1986) Permian-Triassic sedimentology of the Beardmore Glacier region. Antarctic J US 21(5):29–30

Collinson JW, Kemp NR (1983) Permian-Triassic sedimentary sequence in northern Victoria Land, Antarctica. In: Oliver RL, James PR, Jago JB (eds) Antarctic earth science. Australian Academy of Science, Canberra, ACT, pp 221–225

Collinson JW, Stanley KO, Vavra CL (1978) Stratigraphy and sedimentary petrology of the Fremouw Formation (Lower Triassic), Cumulus Hills, central Transantarctic Mountains. Antarctic J US 13(4):21–22

Collinson JW, Stanley KO, Vavra CL (1981) Triassic fluvial depositional systems in the Fremouw Formation, Cumulus Hills, Antarctica. In: Cresswell MM, Vella PP (eds) Gondwana Five. Balkema, Rotterdam, The Netherlands, pp 141–148

Collinson JW, Chapman Pennington D, Kemp NR (1986) Stratigraphy and petrology of Permian and Triassic fluvial deposits in northern Victoria Land, Antarctica. In: Stump E (eds) Geological investigation in Northern Victoria Land. Antarctic Research Series, vol. 46. American Geophysical Union, Washington, DC, pp 211–242

Collinson JW, Isbell JL, Elliot DH, Miller MF, Miller JMG, Veevers JJ (1994) Permian-Triassic Transantarctic basin. In: Veevers JJ, Powell CMcA (eds) Permian-Triassic Pangean Basin and Foldbelts along the Panthalassan Margin of Gondwana. Geological Society of America, Memoir, vol. 184. Boulder, CO, pp 173–222

Craddock C (ed) (1969) Geological map of Antarctica. Folio 12. American Geographical Society, New York

Craddock C (1982) Geologic map of Antarctica. In: Craddock C (ed) Antarctic geoscience. University of Wisconsin Press, Madison, WI

Cridland AA (1963) A Glossopteris flora from the Ohio Range, Antarctica. Amer J Bot 50(2):186–195

Crowder DF (1968) Geology of a part of northern Victoria Land, Antarctica. US Geol Surv Prof. Paper 600D, 95–107

David TWE, Priestley R (1914) Glaciology, physiography, stratigraphy, and tectonic geology of south Victoria Land. Brit. Ant. Exp. 1907–1909. Repts Sci Invest, Geology, Part 1:1–319

Debenham F (1921) The sandstone, etc. of the McMurdo Sound, Terra Nova Bay, and Beardmore Glacier regions. Brit Mus Nat Hist, Nat Hist Rept, Geology 1(4a):103–109

Doumani GA, Minshew VH (1965) General geology of the Mount Weaver area, Queen Maud Mountains, Antarctica. In: Hadley JB (ed) Geology and paleontology of the Antarctic. Antarctic Research Series, vol. 6. American Geophysical Union, Washington, DC, pp 127–139

Doumani GA, Tasch P (1963) Leaiid Conchostracan zone in Antarctica and its Gondwana equivalents. Science 142:591–592

Doumani GA, Tasch P (1965) A Leaiid *Conchotracan* zone (Permian) in the Ohio Range, Horlick Mountains, Antarctica. In: Hadley JB (ed) Geology and paleontology of the Antarctic. Antarctic Research Series, vol. 6. American Geophysical Union, Washington, DC, pp 229–239

Doumani GA, Boardman RS, Rowell AJ, Boucot AJ, Johnson JG, McAlester AL, Saul J, Fisher DW (1965) Lower Devonian fauna of the Horlick Formation, Ohio Range, Antarctica. In: Hadley JB (ed) Geology and paleontology of the Antarctic. Antarctic Research Series, vol. 6. American Geophysical Union, Washington, DC, pp 241–281

Dow JAS, Neall VE (1974) Geology of the lower Rennick Glacier, northern Victoria Land, Antarctica. New Zealand J Geol Geophys 17:659–714

Elliot DH (1975) Gondwana basins in Antarctica. In: Campbell KSW (ed) Gondwana geology. Australian National University Press, Canberrra, ACT, pp 493–536

Elliot DH (1996) The Hanson Formation: A new stratigraphical unit in the Transantarctic Mountains. J African Earth Sci 31:77–89

Elliot DH, Barrett PJ, Mayewski PA (1974) Reconnaissance geologic map of the Plunket Point Quadrangle, Transantarctic Mountains, Antarctica. Institute of Polar Studies, Ohio State University. Antarctic Map A-4. US Geological Survey, Reston, VA

Elliot DH, Colbert EH, Breed WJ, Jensen JA, Powell JS (1970) Triassic tetrapods from Antarctica. Science, 169:1197–1201

Faure G (2001) Origin of igneous rocks; The isotopic evidence. Springer, Heidelberg, Germany

Faure G, Barrett PJ (1973) Strontium isotope compositions of non-marine carbonate rocks from the Beacon Supergroup of the Transantarctic Mountains. J Sed Petrology 43(2):447–457

Faure G, Hill RL (1973) The age of the Falla Formation (Triassic), Queen Alexandra Range, Antarctica. Antarctic J US 8(5):264–266

Ferrar HT (1907) Report on the field geology of the region explored during the "Discovery" Antarctic Expedition: 1901–1904. Nat Antarctic Exp, Nat Hist Repts 1(1):1–100

Frakes LA, Marzolf JE, Edwards LN, Gevers TW, Crowell JC (1968). Sedimentological aspects of the Darwin Tillite in the Darwin Mountains. Antarctic J US 3:109–110

Frakes LA, Matthews JL, Crowell JC (1971) Late Paleozoic glaciation, III, Antarctica. Geol Soc Amer Bull 82(6):1581–1604

Gair HS (1967) The geology from the upper Rennick Glacier to the coast, northern Victoria Land, Antarctica. New Zealand J Geol Geophys 10(2):309–344

Gair HS, Norris G, Ricker J (1965) Early Mesozoic microfloras from Antarctica. New Zealand J Geol Geophys 8:231–235

Gair HS, Sturm A, Carryer SJ, Grindley GW (1969) The geology of northern Victoria Land. In: Craddock C (ed) Geologic map of Antarctica, Folio 12, Sheet 13. American Geographical Society, New York

Gevers TW, Twomey A (1982) Trace fossils and their environments in Devonian (Silutian) lower Beacon strata in the Asgard Range, Victoria Land, Antarctica. In: Craddock C (ed) Antarctic geoscience. University of Wisconsin Press, Madison, WI, pp 639–647

Gevers TW, Frakes LA, Edwards LN, Marzolf JE (1971) Trace fossils in the lower Beacon sediments (Devonian), Darwin Mountains, southern Victoria Land, Antarctica. J Paleotol 45(1):81–94

Grindley GW (1963) The geology of the Queen Alexandra Range, Beardmore Glacier, Ross Dependency, Antarctica, with notes on the correlation of Gondwana sequences. New Zealand J Geol Geophys 6:307–347

Grindley GW, Laird MG (1969) Geology of the Shackleton Coast. In: Craddock C (ed) Geologic map of Antarctica, Folio 12, Sheet 15. American Geographical Society, New York

Grindley GW, Warren G (1964) Stratigraphic nomenclature and correlation in the western Ross Sea region. In: Adie RJ (ed) Antarctic geology. North-Holland, Amsterdam, The Netherlands, pp 314–333

Grindley GW, McGregor VR, Walcott RI (1964) Outline of the geology of the Nimrod-Beardmore-Axel Heiberg glaciers region, Ross Dependency. In: Adie RJ (ed) Antarctic geology. North-Holland, Amsterdam, The Netherlands, pp 206–219

Grindley GW, Mildenhall DC, Schopf JM (1980) A mid-Late Devonian flora from the Ruppert Coast, Marie Byrd Land, West Antarctica. J Roy Soc New Zealand 10(3):271–285

Gunn BM, Warren G (1962) Geology of the basement rocks between the Mawson and Mullock glaciers, south Victoria Land, Antarctica. New Zealand Geol Surv Bull No. 71

Hamilton W, Hayes PT (1960) Geology of Taylor Glacier, Taylor dry valley region, south Victoria Land, Antarctica. US Geol Surv Prof. Paper 400-B

Hamilton W, Hayes PT (1963) Type section of the Beacon Sandstone of Antarctica. US Geol Surv Prof. Paper 456-A:1–18
Hammer WR (1986) Takrouna Formation fossils of northern Victoria Land. Ant. Res. Set. 46:243–247. Amer. Geophys Union, Washington, DC
Hammer WR, Cosgriff JW (1981) Myosaurus gracilis an anomodont reptile from the Lower Triassic of Antarctica and South Africa. J Paleo 55(2):410–424
Hammer WR, Ryan WJ, Tamplan JW, DeFauw SL (1986) New vertebrates from the Fremouw Formation (Triassic) Beardmore Glacier, Antarctica. Ant J US 21(5):24–26
Hammer WR, Hickerson WJ, Collinson JW (1996) Preliminary analysis of Triassic vertebrates from the Shackleton region. Antartic J. US, 31(2):8–9
Harrington HJ (1958) Nomenclature of rock units in Ross Sea region of Antarctica. Nature 182:290
Harrington HJ, Speden IG (1962) Section through the Beacon Sandstone at Beacon Height West, Antarctica. New Zealand J Geol Geophys 5(5):707–717
Haskell TR, Kennett JP, Prebble WM (1964) Basement and sedimentary geology of the Darwin Glacier area, Antarctica. In: Adie RJ (ed) Antarctic geology. North-Holland, Amsterdam, The Netherlands, pp 348–351
Haskell TR, Kennett JP, Prebble WM, Smith G, Willis IAG (1965a) The geology of the middle and lower Taylor Valley of south Victoria Land, Antarctica. Trans Roy Soc New Zealand 2(12):169–186
Haskell TR, Kennett JP, Prebble WM (1965b) Geology of the Brown Hills and the Darwin Mountains, southern Victoria Land, Antarctica. Trans Roy Soc New Zealand 2(15):231–248
Helby RJ, McElroy CT (1969) Microfloras from the Devonian and Triassic of the Beacon Group, Antarctica. New Zealand J Geol Geophys 12:376–382
Horner TC, Krissek LA (1991) Contributions of sedimentologic, thermal alteration, and organic carbon data to paleoenvironmental interpretation of fine grained Permian clastics from the Beardmore Glacier region, Antarctica. In: Elliot DH (ed) Contributions to Antarctic Research II. Antarctic Research Series, vol. 53. American Geophysical Union, Washington, DC, pp 33–65
Kemp EM (1975) The palynology of late Paleozoic glacial deposits of Gondwana. In, Campbell KSW, ed, Gondwana Geology, 397–414. Aust. Nat. Un. Press, Canberra
Kitching JW, Collinson JW, Elliot DH, Colbert EH (1972) Lystrosaurus zone (Triassic) fauna from Antarctica. Science 175(4021):524–527
Korsch RJ (1974) Petrographic comparison of the Taylor and Victoria groups (Devonian to Triassic) in south Victoria Land, Antarctica. New Zealand J Geol Geophys 17:523–541
Laird MG, Bradshaw JD (1981) Permian tillites of north Victoria Land, Antarctica. In: Hambrey MJ, Harland WB (eds) Earth's Pre-Pleistocene glacial record. Cambridge University Press, Cambridge, pp 237–240
Laird MG, Mansergh GD, Chappell JMA (1971) Geology of the central Nimrod Glacier area, Antarctica. New Zealand J Geol Geophys 14(3):427–468
Laird MG, Cooper RA, Jago JB (1977) New data on the lower Palaeozoic sequence of northern Victoria Land, Antarctica, and its significance for Australian-Antarctic relations in the Palaeozoic. Nature 265:107–110
LaPrade KE (1970) Permian-Triassic Beacon Group of the Shackleton Glacier area, Queen Maud Range, Transantarctic Mountains. Ged Soc Amer Bull 891(5):1403–1410
LaPrade KE (1982) Petrology and petrography of the Beacon Supergroup, Shackleton Glacier area, Queen Maud Range, Transantarctic Mountains. In: Craddock C (eds) Antarctic geoscience. University of Wisconsin Press, Madison, WI, pp 581–590
LaPrade KE (1984) Climate, geomorphology, and glaciology of the Shackleton Glacier area, Queen Maud Mountains, Transantarctic Mountains, Antarctica. In: Turner MD, Splettstoesser JF (eds) Geology of the Central Transantarctic Mountains. Antarctic Research Series, vol. 36. American Geophysical Union, Washington, DC, pp 163–196
Lindsay JF (1968) Stratigraphy and sedimentation of the lower Beacon rocks of the Queen Alexandra, Queen Elizabeth, and Holland ranges, Antarctica, with emphasis on the Paleozoic glaciation. Ph.D. dissertation, Department of Geology and Mineralogy, The Ohio State University, Columbus, OH
Lindsay JF (1969) Stratigraphy and sedimentation of lower Beacon rocks in the central Transantarctic Mountains, Antarctica. Institute of Polar Studies, Rept. 33. The Ohio State University, Columbus, OH, 58 p
Lindsay JF (1970a) Depositional environment of Paleozoic glacial rocks in the central Transantarctic Mountains. Geol Soc Amer Bull 81(4):1149–1172
Lindsay JF (1970b) Clast fabric strength of tillite. J Geol 78(5):597–603
Lindsay JF, Gunner J, Barrett PJ (1973) Reconnaissance geologic map of the Mount Elizabeth and Mt. Kathleen quadrangles, Transantarctic Mountains, Antarctica. US Antarctic Research Program, Antarctic Geologic Map A-2. US Geological Survey, Washington, DC
Long WE (1959) Preliminary report of the geology of the central range of the Horlick Mountains, Antarctica. Institute of Polar Studies, Research Foundation Report 825–2, Part 7. The Ohio State University, Columbus, OH, 23 pp
Long WE (1962) Sedimentary rocks of the Buckeye Range, Horlick Mountains, Antarctica. Science 136(3513):319–321
Long WE (1965) Stratigraphy of the Ohio Range, Antarctica. In: Hadley JB (ed) Geology and paleontology of the Antarctic. Antarctic Research Series, vol. 6. American Geophysical Union, Washington, DC, pp 71–116
McElroy CT, Rose G, Bryan JH (1967) A unique occurrence of deformed sedimentary rocks of the Beacon Group, Antarctica. Antarctic J US 2:241–244
McGregor VR (1965a) Notes on the geology of the area between the heads of the Beardmore and Shackleton glaciers, Antarctica. New Zealand J Geol Geophys 8(2):278–291
McGregor VR (1965b) Geology of the area between the Axel Heiberg and Shackleton glaciers, Queen Maud Range, Antarctica. Part 1: Basement complex, structure, and glacial geology. New Zealand J Geol Geophys 8(2):314–343
McGregor VR, Wade FA (1969) Geology of the Western Queen Maud Mountains. In: Craddock C (ed) Geologic map of Antarctica. Folio 12, Sheet 16. American Geographical Society, New York
McKelvey BC (1983) A palaeogeographic comparison of the late Palaeozoic Lanterman Range tillite of northern Victoria Land, Antarctica, with equivalent tillites in eastern Australia. Geol Soc Amer, Abstracts with Programs 15(4):215

McKelvey BC, Walker BC (1983) Late Palaeozoic glacigene strata in northern Victoria Land. In: Oliver RL, James PR, Jago JB (eds)Antarctic earth science, 226. Australian Academy of Science, Canberra, ACT
McKelvey BC, Webb PN (1959) Geological investigations in south Victoria Land, Antarctica. Part II, Geology of the upper Taylor Glacier region. New Zealand J Geol Geophys 2(4):718–728
McKelvey BC, Webb PN (1961) Geological reconnaissance in Victoria Land, Antarctica. Nature 189(4764):545–547
McKelvey BC, Webb PN (1962) Geological investigations in southern Victoria Land, Antarctica, Part 3: Geology of Wright Valley. New Zealand J Geol Geophys 5(1):143–162
McKelvey BC, Webb PN, Gorton MP, Kohn BP (1970) Stratigraphy of the Beacon Supergroup between the Olympus and Boomerang ranges, Victoria Land. Nature 227(5363):1126–1128
McKelvey BC, Webb PN, Gorton MP, Kohn BP (1972) Stratigraphy of the Beacon Supergroup between the Olympus and Boomerang ranges, Victoria Land. In: Adie RJ (ed) Antarctic geology and geophysics. Unversitetsforlaget, Oslo, Norway, pp 345–352
McKelvey BC, Webb PN, Kohn BR (1977) Stratigraphy of the Taylor and lower Victoria groups (Beacon Supergroup) between the Mackay Glacier and the Boomerang Range, Antarctica. New Zealand J Geol Geophys 20:813–863
McPherson JG (1978) Stratigraphy and sedimentology of the Upper Devonian Aztec Siltstone, southern Victoria Land, Antarctica. New Zealand J Geol Geophys 21:667–683
Mercer JH (1963) Glacial geology of the Ohio Range, central Horlick Mountains, Antarctica. Institute of Polar Studies, Rept. 8. The Ohio State University, Columbus, OH, 13 pp
Minshew VH (1966) Stratigraphy of the Wisconsin Range, Horlick Mountains, Antarctica. Science 152(3722): 637–638
Minshew VH (1967). Geology of the Scott Glacier and Wisconsin Range areas, central Transantarctic Mountains, Antarctica. Ph.D. dissertation, The Ohio State University, Columbus, OH
Mirsky A (1964) Reconsideration of the Beacon as a stratigraphic name in Antarctica. In: Adie RJ (ed) Antarctic geology, North-Holland, Amsterdam, The Netherlands
Mirsky A, Treves SB, Calkin PE (1965) Stratigraphy and petrography, Mount Gran area, southern Victoria Land, Antarctica. In: Hadley JB (ed) Geology and paleontology of the Antarctic. Antarctic Research Series, vol. 6. American Geophysical Union, Washingtn, DC, pp 145–175
Norris G (1965) Triassic and Jurassic miospores and acritarchs from the Beacon and Ferrar groups, Victoria Land, Antarctica. New Zealand J Geol Geophys 8(2):236–277
Pigg KB, Taylor TN (1985) Anatomically preserved Glossopteris from the Beardmore Glacier area of Antarctica. Antarctic J US 19(5):8–10
Pinet PR, Matz DB, Hayes MO (1967) Petrology of the upper division of the Beacon Sandstone. Antarctic J US 2(4):108–109
Plume RW (1978) A revision of the existing stratigraphy of the New Mountain Sandstone (Beacon Supergroup), south Victoria Land, Antarctica. New Zealand J Geol Geophys 21:167–175
Plumstead EP (1962) Fossil floras of Antarctica (with an appendix on Antarctic fossil wood by R. Krausel). Trans-Antarctic Expedition 1955–1958. Sci Rept Geol 9:154
Poscher G (1989) Sedimentpetrographische Charakterisierung von Diamiktiten der Beacon Super-Group, Kottas Berge, Heimefrontfjella (Antarktis) Zeitschrift für Gletscherkunde und Glazialgeologie 25(1):1–16
Pyne AR (1984) Geology of the Mt. Fleming area, south Victoria Land, Antarctica. New Zealand J Geol Geophys 27:505–512
Schmidt DL, Ford AB (1969) Geology of the Pensacola and Thiel Mountains. In: Craddock C (ed) Geologic map of Antarctica, Folio 12, Sheet 5. American Geographical Society, New York
Schmidt DL, Williams PL, Nelson WH, Ege JR (1965) Upper Precambrian and Paleozoic stratigraphy and structure of the Neptune Range, Antarctica. US Geol Survey Prof. Paper 525-D:112–119
Schopf JM (1962) A preliminary report on plant remains and coal of the sedimentary section in the central Horlick Mountains, Antarctica. Inst Polar Studies, Rept. 2. The Ohio State University, Columbus, OH, 61 pp
Schopf JM (1970) Petrified peat from a Permian coal bed in Antarctica. Science 169(3942):274–277
Schopf JM (1971) Notes on plant tissue preservation and mineralization in a Permian deposit of peat from Antarctica. Amer J Sci 271(5):522–543
Schopf JM (1982) Forms and facies of Vertebraria in relation to Gondwana coal. In: Turner MD, Splettstoesser JF (eds) Geology of the Central Transantarctic Mountains. Antarctic Research Series, vol. 36. American Geophysical Union, Washington, DC, pp 37–62
Schopf JM, Askin RA (1980) Permian and Triassic floral biostratigraphic zones of southern land masses. In: Dilcher DL, Taylor TN (eds) Biostratigraphy of fossil plants. Dowden, Hutchinson, and Ross, Stroudsburg, PA, pp 119–152
Schopf JM, Long WE (1960) Antarctic coal geology. Abstr Geol Soc Amer Bull 71(12) Part 2:1967
Shaw SE (1962) Petrography of Beacon Sandstone samples from Beacon Height West, Upper Taylor Glacier, Antarctica. New Zealand J Geol Geophys 5(5):733–739
Skinner DNB (1981) Possible Permian glaciation in north Victoria Land, Antarctica. Geologisches Jahrbuch, Reihe B 41:261–266
Splettstoesser JF, Jirsa MA (1985) Columnar jointed sandstone in Beacon Supergroup, Britannia Range, Antarctica. New Zealand J Geol Geophys 28:761–764
Steiger RH, Jäger E (1977) Subcommission on geochronology: Convention on the use of decay constants in geo- and cosmochronology. Earth Planet Sci Lett 36:359–362
Stewart D (1934) The petrography of the Beacon Sandstone of South Victoria Land. Amer Mineral 19:351–359
Stubblefield SP, Taylor TN (1985) Fossil fungi in Antarctic wood. Antarctic J US 19(5):7–8
Taylor TN, Smoot EL (1985a) A new Triassic cycad from the Beardmore Glacier area of Antarctica. Antarctic J US 19(5):5–7
Taylor TN, Smoot EL (1985b) Plant fossils from the Ellsworth Mountains. Antarctic J US 19(5):48–49
Townrow JA (1967) Fossil plants from Allan and Carapace nunataks, and from the upper Mill and Shackleton glaciers, Antarctica. New Zealand J Geol Geophys 10(2):456–473

Treves SB (1965) Igneous and metamorphic rocks of the Ohio Range, Horlick Mountains. In: Hadley JB (ed) Geology and paleontology of the Antarctic. Antarctic Research Series, vol. 6. American Geophysical Union, Washington, DC, pp 117–125

Vavra CL (1982) Provenance and alteration of the Triassic Fremouw and Falla formations central Transantarctic Mountains, Antarctica. Ph.D. dissertation, Department of Geology and Mineralogy, The Ohio State University, Columbus, OH

Vavra CL, Stanley KO, Collinson JW (1981) Provenance and alteration of Triassic Fremouw Formation, central Transantarctic Mountains. In: Cresswell MM, Vella P (eds) Gondwana Five. A.A. Balkema, Rotterdam, The Netherlands, pp 149–153

Veevers JJ, Powell CMcA (eds) (1994) Permian-Triassic Pangean basins and foldbelts along the Panthalassan margin of Gondwana. Geological Society of America, Memoir, vol. 184, Boulder, CO

Vialov OS (1962) Problematica of Beacon Sandstone at Beacon Height West, Antarctica. New Zealand J Geol Geophys 5(5):718–732

Wade FA, Yeats VL, Everett JR, Greenlee DW, LaPrade KE, Shenk JC (1965) The geology of the central Queen Maud Range, Transantarctic Mountains, Antarctica. Science 150:1808–1809

Walker BC (1983) The Beacon Supergroup of northern Victoria Land, Antarctica. In: Oliver RL, James PR, Jago JB (eds) Antarctic earth science. Australian Academy of Science, Canberra, A.C.T., pp 211–214

Warren G (1969) Geology of the Terra Nova Bay - McMurdo Sound area, Victoria Land. In: Craddock C (ed) Geologic map of Antarctica, Folio 12, Sheet 14. American Geographical Society, New York

Webb JA, Fielding CR (1993) Permo-Triassic sedimentation within the Lambert Graben, northern Prince Charles Mountains, East Antarctica. In: Findlay RH, Unrug R, Banks MR, Veevers JJ (eds) Gondwana Eight: Assembly, evolution, and dispersal. Balkema, Rotterdam, The Netherlands, pp 357–370

Webb PN (1963) Geological investigations in southern Victoria Land, Antarctica. Part 4: Beacon Group of the Wright Valley and Taylor Glacier region. New Zealand J Geol Geophys 6(3)361–387

Webb PN, McKelvey BC (1959) Geological investigations in southern Victoria Land, Antarctica. Part I. Geology of Victoria dry valley. New Zealand J Geol Geophys 2(1):120–136

Webby BD (1968) Devonian trace fossils from the Beacon Group of Antarctica. New Zealand J Geol Geophys 11(4):1001–1008

Webers GF, Glenister B, Pojeta J Jr, Young G (1992) Devonian fossils from the Ellsworth Mountains, West Antarctica. In: Webers GF, Craddock C, Splettstoesser JF (eds) Geology and paleontology of the Ellsworth Mountains, West Antarctica. Geological Society of America, Memoir, vol. 170. Boulder, CO, pp 269–278

Whitby KJ, Rose G, McElroy CT (1983) Formational mapping of the Beacon Supergroup type area with special reference to the Weller Coal Measures, south Victoria Land, Antarctica. In: Oliver RL, James PR, Jago JB (eds) Antarctic earth science. Australian Academy of Science, Canberra, A.C.T., pp 228–232

Wilson M (1989) Igneous petrogenesis. Unwin Hyman, London

Wizevich MC (1997) Fluvial-eolian deposits in the Devonian New Mountain Sandstone, Table Mountain, southern Victoria Land, Antarctica: Sedimentary architecture, genesis, and stratigraphic evolution. In: Ricci CA (ed) The Antarctic region: Geological evolution and processes. Terra Antartica, Siena, Italy, pp 933–944

Young DJ, Ryburn RJ (1968) The geology of Buckley and Darwin islands, Beardmore Glacier, Ross Dependency, Antarctica. New Zealand J Geol Geophys 11(4):922–929

Zeller EJ, Angino EE, Turner MD (1961) Basal sedimentary section at Windy Gully, Taylor Glacier, Victoria Land, Antarctica. Geol Soc Amer Bull 72:781–786

Chapter 11
Beacon Supergroup; Special topics

The sedimentary rocks of the Beacon Supergroup contain a record of the geologic history of Gondwana from the Early Devonian to the Late Triassic. The preceding chapter contains a review of the stratigraphy, some comments about the petrography, and statements about the environment of the deposition of the Beacon rocks. We next consider certain special topics that deserve to be included in the presentation of the Beacon rocks of the Transantarctic Mountains.

11.1 Isotopic Studies of Carbonate Rocks

The Beacon Supergroup contains thin lenticular layers of carbonate minerals that are interbedded with virtually all of its formations. In addition, carbonate minerals occur as concretionary nodules, especially in siltstone and shale layers, and as calcite "cleats" in joints and on bedding planes in coal seams and in the overlying sandstones of Permian and Triassic age.

The origin of such carbonate rocks is not clear because some may have formed at the time of deposition of the sediment while others originated during initial burial and subsequent diagenesis of the sediment. The difference is important because carbonates that are biogenic or formed as chemical precipitates at the water-sediment interface contain a record of the depositional environment, whereas diagenetic carbonates reflect the properties of the porewater in the sediment after burial (Schmidt and Friedman 1974).

The carbonate minerals that occur in the sedimentary rocks of the Beacon Supergroup consist primarily of calcite ($CaCO_3$) which contains varying concentrations of strontium, magnesium, iron, and manganese as well as oxygen and carbon. The isotopic composition of *strontium* in calcite in sedimentary rocks can be used to determine whether this mineral was deposited from seawater in a marine environment (Faure and Mensing 2005). The isotope composition of *oxygen* in calcite indicates whether the mineral precipitated in equilibrium with seawater or with meteoric water. The isotopes of *carbon* are fractionated during photosynthesis which identifies carbon that has been processed by green plants during photosynthesis (Hoefs 1997; Sharp 2007).

11.1.1 Strontium

The isotopic composition of strontium (Section 3.4.2 and Appendix 3.7.3) in the crust of the Earth changes with time because of the decay of naturally-occurring radioactive ^{87}Rb to stable, radiogenic ^{87}Sr with a halflife of 48.8×10^9 years corresponding to a decay constant of 1.42×10^{-11} year^{-1} (Sections 3.4.2 and 3.7.3). Consequently, the $^{87}Sr/^{86}Sr$ ratios of rocks and minerals increase with time at rates that depend on their $^{87}Rb/^{86}Sr$ ratios which are calculated from their Rb/Sr concentration ratios (Eq. 3.9, Appendix 3.6.3).

The old *granitic basement rocks* of the continental crust have elevated Rb/Sr ratios and therefore have been enriched in radiogenic ^{87}Sr. Their $^{87}Sr/^{86}Sr$ ratios typically have values of about 0.725 ± 0.005 but may range up to 1.0 or higher. *Volcanic rocks of basaltic composition* originate from the subcrustal mantle which has lower Rb/Sr ratios than granitic rocks in the continental crust. Therefore, the Sr in basaltic rocks has low $^{87}Sr/^{86}Sr$ ratios of only 0.704 ± 0.002 (Faure 2001). The value of the $^{87}Sr/^{86}Sr$ ratio of strontium in *seawater* during the Phanerozoic has been intermediate (0.7080 ± 0.0015) because the oceans contain a

G. Faure and T.M. Mensing, *The Transantarctic Mountains: Rocks, Ice, Meteorites and Water*,
DOI 10.1007/978-90-481-9390-5_11, © Springer Science+Business Media B.V. 2010

mixture of strontium derived from old granitic rocks and from young basaltic rocks, both of which are exposed to weathering on the continents. The basaltic rocks that form along spreading ridges in the oceans are an additional source of unradiogenic strontium in solution in seawater. Furthermore, marine limestones exposed to weathering on the continents release strontium having an intermediate isotope composition that acts to stabilize the value of the $^{87}Sr/^{86}Sr$ ratio of seawater. Nevertheless, the $^{87}Sr/^{86}Sr$ ratio of the oceans has varied with time depending on the changing amounts of strontium that have originated from granitic rocks in the continental crust and from young basaltic rocks that are erupted on the continents and in the ocean basins. These time-dependent fluctuations of the $^{87}Sr/^{86}Sr$ ratio in the oceans, shown graphically in Fig. 11.1, have been measured with great precision by analyzing the isotopic composition of strontium in unaltered marine limestones of Phanerozoic age. The relevant references were listed by Faure and Mensing (2005).

The use of the $^{87}Sr/^{86}Sr$ ratio to determine the environment of deposition of an unknown carbonate sample hinges on the assumption that the isotopic composition of strontium of the carbonate sample and of the aqueous solution from which it formed are identical. In other words, we assume that strontium isotopes are not *fractionated* during the transfer of the element from the solution into the solid phase. Actually, all measurements of the isotopic composition of strontium are corrected for isotope fractionation to $^{86}Sr/^{88}Sr$ = 0.11940, which means that isotope fractionation can be excluded from consideration in the interpretation of data. Therefore, the well-known $^{87}Sr/^{86}Sr$ ratios of marine limestones of Phanerozoic age in Fig. 11.1 provide a basis for assessing the environment of deposition of the carbonate layers in the Beacon Supergroup.

In cases where the $^{87}Sr/^{86}Sr$ ratio of a carbonate rock has the same value as marine limestones of corresponding age, the unknown carbonate sample *could* have formed from seawater. If the $^{87}Sr/^{86}Sr$ ratio of the carbonate of unknown origin is higher (or lower) than that of seawater at the time deposition, then the unknown carbonate sample *did not* form in a marine environment. In that case, the sample may have formed in a non-marine environment or it may have been precipitated from pore water during burial and diagenesis of the sediment.

The validity of this criterion is corroborated by measurements of the $^{87}Sr/^{86}Sr$ ratios non-marine limestones. For example, Clauer and Tardy (1972) reported

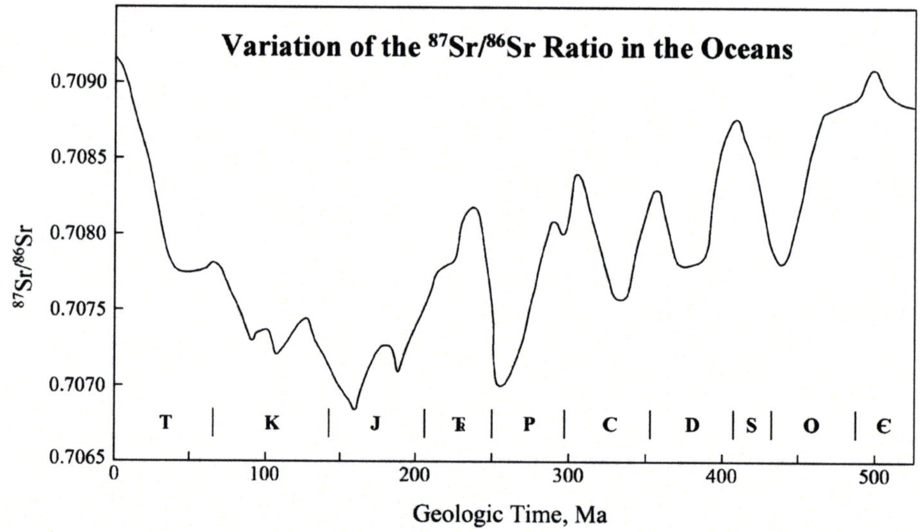

Fig. 11.1 The isotopic composition of strontium in the oceans has varied systematically during the Phanerozoic. The present $^{87}Sr/^{86}Sr$ ratio of seawater in the open ocean is 0.70918 ± 0.00001 (2$\bar{\sigma}$). The value of this ratio at any given time in the past was controlled by the proportions of strontium that entered the oceans from different sources including granitic rocks of Precambrian age of the continental crust and young volcanic rocks derived from the subcrustal mantle of the Earth by volcanic eruptions on the continents and in the ocean basins. The $^{87}Sr/^{86}Sr$ ratios of all carbonate rocks that formed in the oceans during the Phanerozoic fit this curve (Adapted from Faure and Mensing 2005)

$^{87}Sr/^{86}Sr$ ratios ranging from 0.7161 to 0.7225 for lacustrine carbonates of Oligocene age in the Massif Central of France. These values are far higher than the $^{87}Sr/^{86}Sr$ ratio of seawater at the time of deposition of these non-marine limestones. Similarly, Faure et al. (1963) measured $^{87}Sr/^{86}Sr$ ratios ranging from 0.712 to 0.726 in contemporary mollusk shells in lakes and rivers of the Precambrian Shield of Canada. These results demonstrate that the strontium in solution in these lakes and rivers is enriched in radiogenic ^{87}Sr which is released into the water by weathering of Precambrian rocks that have elevated Rb/Sr concentration ratios.

Faure and Barrett (1973) and Lord et al. (1988) used measurements of the $^{87}Sr/^{86}Sr$ ratios to determine the environment of deposition of carbonate rocks from the Devonian, Permian, and Triassic formations of the Beacon Supergroup in the Transantarctic Mountains. Most of the samples were selected by Peter Barrett who also provided information about the distance of the collecting site of each sample to the nearest sill of the Ferrar Dolerite and about the thickness of the nearest sill. This information permits a qualitative estimate of the thermal history (maximum temperature and duration) to which the carbonate samples were exposed. The concern about the thermal history arises because all the carbonate samples contained insoluble residues consisting of Rb-bearing clay minerals, iron oxide, and grains of rock-forming silicates (e.g., quartz, feldspar, mica, etc.) When these impure carbonate rocks were heated following the intrusion of dolerite sills, some of the radiogenic ^{87}Sr that had formed by decay of ^{87}Rb in the impurities may have diffused into the carbonate matrix in which the impurities were embedded.

The carbonate samples in Table 11.1 include representatives of the Taylor and theVictoria groups from southern Victoria Land and from the central Transantarctic Mountains. The results are presented in Fig. 11.2 which permits a comparison of the $^{87}Sr/^{86}Sr$ ratios of the calcite samples with the $^{87}Sr/^{86}Sr$ ratio of seawater at the times of deposition of the Beacon rocks. The carbonate rocks include samples of Devonian age from the Alexandra and Aztec formations of southern Victoria Land and samples from the Permian formations of the central Transantarctic Mountains. In addition, two carbonate samples from the Triassic Fremouw Formation were analyzed. The results in Fig. 11.2 clearly demonstrate that the $^{87}Sr/^{86}Sr$ ratios of all carbonate samples from the Beacon Supergroup are *higher* than the seawater-values of this ratio. Therefore, these carbonate rocks were not precipitated in contact with seawater. Instead, the data indicate that the carbonates were either deposited in non-marine environments or formed during diagensis of the host rocks by precipitating from pore water.

The $^{87}Sr/^{86}Sr$ ratios of two Cambrian limestone formations of known marine origin (i.e., the Nelson Limestone, Pensacola Mountains, and limestone of the Leverett Formation, Mt. Webster, Harold Byrd Mountains) are both consistent with the $^{87}Sr/^{86}Sr$ ratio of Cambrian seawater. In addition, the $^{87}Sr/^{86}Sr$ ratio of one carbonate sample from the Triassic Fremouw Formation plots close to the seawater curve

Table 11.1 Average $^{87}Sr/^{86}Sr$ ratios of *calcite* in the sedimentary rocks of the Beacon Supergroup in the central Transantarctic Mountains (Faure and Barrett 1973; Lord et al. 1988, unpublished data)

Formation	Age	Aver. $^{87}Sr/^{86}Sr$ and number of sples	Range
Fremouw	Triassic	0.7100(2)	0.7084–0.7116
Buckley	Permian	0.7171(2)	0.7150–0.7194
Fairchild	Permian	0.7185(1)	
Mackellar	Permian	0.7177(9)	0.7145–0.7218
Pagoda	Permian	0.7194(6)	0.7178–0.7228
Weaver, Scott Gl.	Permian	0.7150(1)	
Buckeye, Reedy Gl.	Permian	0.7168(2)	0.7163–0.7172
Aztec	Devonian	0.7272(3)	0.7260–0.7291
Alexandra	Devonian	0.7177(1)	
Leverett[a]	Cambrian	0.7089(1) } marine	
Nelson[b]	Cambrian	0.7093(1) } marine	

[a] Mt. Webster, Harold Byrd Mountains

[b] Neptune Range, Pensacola Mountains

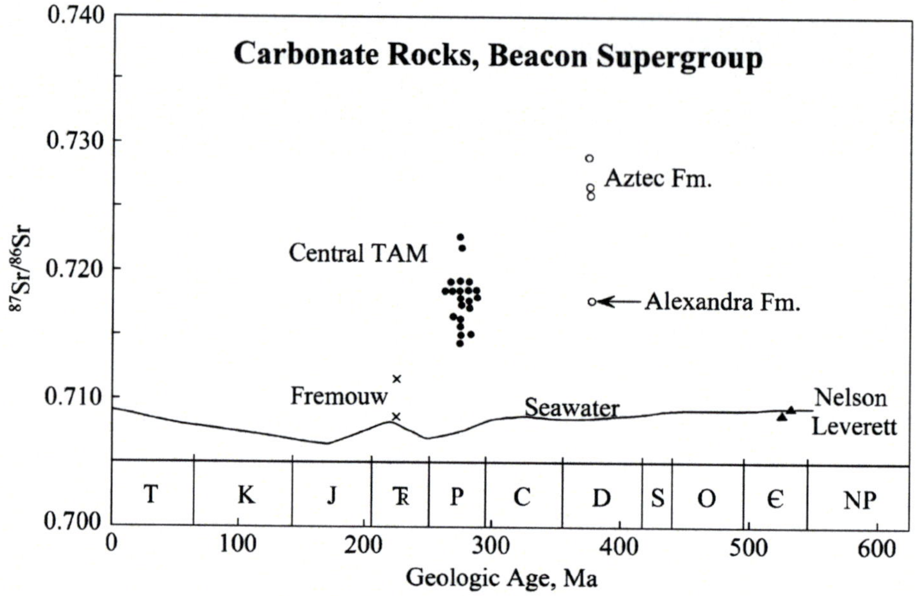

Fig. 11.2 The $^{87}Sr/^{86}Sr$ ratios of calcite in carbonate layers and concretions of the Beacon Supergroup in southern Victoria Land and in the central Transantarctic Mountains are higher than those of seawater at their respective depositional ages. The only exceptions are the Nelson Limestone of the Pensacola Mountains and the limestone of the Leverett Formation on Mt. Webster in the Harold Byrd Mountains both of which are known to be of marine origin. These results clearly demonstrate that the carbonate rocks analyzed in this study did not form in contact with seawater but are either non-marine or diagenetic in origin (Data from Faure and Barrett 1978)

in Fig. 11.2 and therefore could contain marine strontium. However, we consider it more likely that this sample (F055B) precipitated from pore water that contained strontium derived from pyroclastics that occur in the upper part of the Fremouw Formation.

In summary, the $^{87}Sr/^{86}Sr$ ratios of calcite in carbonate rocks of Triassic, Permian, and Devonian age indicate that this mineral did not precipitate from seawater. Instead, the calcite probably precipitated from pore water percolating through the sediment after deposition. The strontium in solution in the pore water originated by incongruent dissolution of detrital grains of K-feldspar, biotite, and other Rb-bearing minerals in the sediment. We conclude that even marine sandstones (e.g., the Alexandra Formation) may contain diagenetic carbonate lenses and concretions that formed after deposition and therefore did not form in a marine environment.

These conclusions were later confirmed by Lord et al. (1988) following additional measurements of the $^{87}Sr/^{86}Sr$ ratios of calcite in carbonate rocks of Permian age from the Beacon rocks in the central Transantarctic Mountains. These authors also measured the $^{87}Sr/^{86}Sr$ ratios and Rb/Sr ratios of the insoluble residue fractions of the carbonate rocks they had analyzed. The $^{87}Sr/^{86}Sr$ ratios of the residues are higher than those of the coexisting calcites and the $^{87}Sr/^{86}Sr$ ratios of the non-carbonate (silicate and oxide) fractions correlate positively with their Rb/Sr ratios. This observation is illustrated in Fig. 11.3 which shows that the non-carbonate fractions of Permian carbonates loosely constrain a straight line in coordinates of $^{87}Sr/^{86}Sr$ and $^{87}Rb/^{86}Sr$. The slope of this line corresponds to a date of 297 ± 137 Ma. Although this is by no means a reliable age determination, it is a clue that the non-carbonate fractions of non-marine carbonate rocks may convey useful information about the provenance and post-depositional history of their host rocks.

11.1.2 Oxygen

When calcium carbonate precipitates from an aqueous solution, the isotopes of oxygen are fractionated depending on the temperature. As a result, the isotope

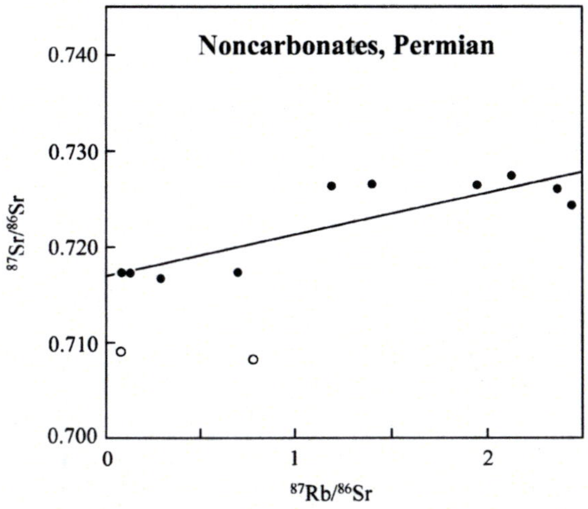

Fig. 11.3 The $^{87}Sr/^{86}Sr$ and $^{87}Rb/^{86}Sr$ ratios of non-carbonate fractions of carbonate rocks from the Permian Pagoda, Mackellar, and Fairchild formations in the Queen Elizabeth and Queen Alexandra ranges are loosely correlated. The straight line fitted to these data points has a slope of m = 0.004229 and an intercept of $^{87}Sr/^{86}Sr = 0.71729 \pm 0.00246$ (1σ) which yield a nominal date of 297 ± 137 Ma. The non-carbonate fractions are composed of mixtures of quartz, plagioclase, K-feldspar, diopside, muscovite, and garnet as well as alteration products, including prehnite, montmorillonite, and chlorite or wollastonite. The open circles are non-carbonate fractions of carbonate rocks in the Buckley Formation from Coalsack Bluff (Data from Lord et al. 1988)

composition of oxygen in calcite differs from the isotopic composition of oxygen in the water from which it formed. The isotope composition of oxygen in calcite is expressed as the permil difference (δ) between the measured $^{18}O/^{16}O$ ratio and the value of this ratio in a standard known as PDB which is the calcite of the internal skeleton of a Cretaceous belemnite (cephalopod) in the Pee Dee Formation of South Carolina. Based on this convention, the isotope composition of oxygen in calcite is expressed in terms of:

$$\delta^{18}_{Spl.}O(PDB) = \left[\frac{\left(^{18}O/^{16}O\right)_{Spl.} - \left(^{18}O/^{16}O\right)_{PDB}}{\left(^{18}O/^{16}O\right)_{PDB}} \right] \times 10^3 \text{‰} \quad (11.1)$$

The $\delta^{18}O$ parameter can be positive, negative, or equal to zero. If $\delta^{18}O$ is positive, the oxygen in the calcite is enriched in ^{18}O relative to PDB. If $\delta^{18}O$ is negative, the calcite is depleted in ^{18}O relative to PDB; and if $\delta^{18}O = 0$, the oxygen in the calcite has the same isotope composition as the oxygen in PDB.

Table 11.2 Average isotopic composition of oxygen in calcite of carbonate rocks in Permian formations of the Queen Elizabeth and Queen Alexandra ranges, central Transantarctic Mountains (Lord et al. 1988; Faure and Lord 1990)

Formation	Age	$\delta^{18}_C O^a_{PDB}$ ‰	T[b] (°C)	$\delta^{18}_W O^c_{SMOW}$ ‰
Fremouw (2)	Triassic	−25.0	15	−25.4
Buckley (2)	Permian	−28.4	15	−28.9
Fairchild (1)	Permian	−28.1	15	−28.6
Mackellar (10)	Permian	−26.2 ± 2.8	5	−29.1
Pagoda (6)	Permian	−25.6 ± 3.0	0	−29.8
Alexandra (1)	Devonian	−24.5	20	−23.8

[a] Average measured $\delta^{18}O$ values of two sets of calcite samples published by Lord et al. (1988) and Faure and Lord (1990). Both sets of measurements were made by Richard Reesman at Geochron Laboratories (aka Krueger Enterprises, Inc.) Cambridge, Massachusetts

[b] The temperatures estimates are based on an average global temperature of 15°C (Fremouw, Buckley, Fairchild), 5°C for the Mackellar carbonates, 0°C for the Pagoda Formation, and 20°C for the Devonian Alexandra Formation

[c] Calculated $\delta^{18}_W O$ values of water in equilibrium with the calcite at the assumed temperature and expressed on the SMOW scale

A similar set of measurements of $\delta^{18}O$ and $\delta^{13}O$ values for calcites from the Gale Mudstone (tillite) in the Pensacola Mountains was published by Schmidt and Friedman (1974)

The isotope composition in water is expressed relative to the $^{18}O/^{16}O$ ratio in Standard Mean Ocean Water which is abbreviated: SMOW. Evidently, the isotope composition of oxygen in calcite and water are expressed on different scales which are related by the equations (Faure and Mensing 2005):

$$\delta^{18}O_{PDB} = 0.97002\,\delta^{18}O_{SMOW} - 29.98 \quad (11.2)$$

$$\delta^{18}O_{SMOW} = 1.03091\,\delta^{18}O_{PDB} + 30.91 \quad (11.3)$$

Given that the PDB standard is composed of calcite that formed in a *marine environment*, all calcites of marine origin are characterized by having $\delta^{18}_C O_{PDB}$ values close to zero.

The average $\delta^{18}O$ values on the PDB scale of calcite in samples of carbonate rocks from the Beacon Supergroup in the central Transantarctic Mountains in Table 11.2 range from −24.5‰ (Alexandra Formation) to −28.4‰ (Buckley Formation). The evident depletion in ^{18}O relative to oxygen in marine calcite is strong evidence that these calcites were *not* precipitated from seawater. This conclusion applies even to the calcite in the Alexandra Formation which contains trace fossils that indicate that the rocks of this formation were deposited in a marine environment (Bradshaw 1981; Gevers and Twomey 1982; Bradshaw and McCartan 1983; Barrett et al. 1986). The explanation for the discrepancy between paleontological evidence and the isotopic composition of strontium and oxygen is that the calcite in the carbonate rocks of the Beacon Supergroup formed from pore water which was depleted in ^{18}O. Such water occurs in the polar regions of the Earth where meteoric precipitation is depleted in ^{18}O relative to seawater as a result of isotope fractionation by progressive condensation of water vapor in air masses which causes rain or snow in the polar regions to have strongly negative $\delta^{18}O$ values on the SMOW scale (Hoefs 1997; Faure and Mensing 2005; Sharp 2007).

The isotope composition of oxygen in calcite is related to the isotope composition of liquid water by an equation derived by O'Neil et al. (1969) and restated by Friedman and O'Neil (1977):

$$\delta^{18}O_C - \delta^{18}O_W = \frac{2.78\times10^6}{T^2} - 2.89 \qquad (11.4)$$

where T = temperature on the Kelvin scale, the δ–values must be expressed on the same scale (PDB or SMOW), and the oxygen in the calcite must have been in isotopic equilibrium with the oxygen of the water in which it formed. Therefore, Eq. 11.4 can be used to calculate the $\delta^{18}O$ values of the water from which the calcites in the Beacon rocks were precipitated.

The six calcite samples from the Pagoda Tillite in Table 11.2 analyzed by Lord et al. (1988) and by Faure and Lord (1990) have an average $\delta_C^{18}O_{PDB} = -26.4 \pm 0.5‰$. Converting this value to the SMOW scale by means of Eq. 11.3 yields:

$$\delta_C^{18}O_{SMOW} = 1.0391(-26.4) + 30.91 = 3.69‰$$

We substitute this value into Eq. 11.4 and assume a temperature of 0°C (273.15 K) because the calcites presumably formed in glacial meltwater in a periglacial or sub-glacial environment:

$$+3.69 - \delta_W^{18}O_{SMOW} = \frac{2.78\times10^6}{(273.15)^2} - 2.89$$

from which it follows that $\delta_W^{18}O_{SMOW} = -30.7‰$.

The most plausible interpretation of the result of this calculation is that carbonate concretions in the Pagoda Tillite formed in contact with glacial meltwater of the continental ice sheets that covered a large area of Gondwana during the Permian Period. This conclusion is not significantly affected by the assumed temperature. For example if T = 5°C (278.15 K), Eq. 11.4 yields $\delta_W^{18}O = -29.4‰$ and if T = 10°C (283.5 K), $\delta_W^{18}O = -28.1‰$. All reasonable temperature estimates lead to the conclusion that the water from which the calcites in the Pagoda Formation precipitated was strongly depleted in ^{18}O, which is characteristic of snow and ice in the polar regions including the present ice sheets in Antarctica and Greenland (e.g., Faure et al. 1992, 1993; Lorius et al. 1979).

Repeating the same calculations for calcites in the other Beacon formations in Table 11.2 leads to the unexpected result that all of the calcites were precipitated in water that was depleted in ^{18}O even though the assumed temperatures in Gondwana before and after the Permian glaciation were modulated based on geological evidence. Faure and Lord (1990) suggested that glacial meltwater continued to dominate the alluvial plains where the Beacon sediment was deposited after the Permian glaciation had ended. However, it is unlikely that this condition could have persisted for more than 40 million years from Early Permian to Early Triassic.

Alternatively, the rain water that fell in the Beacon basin may have been highly fractionated by repeated condensation of water vapor in the atmosphere as the air masses traveled from a distant source to the Beacon basins in Gondwana. Such a situation actually exists at the present time on the tropical Quelccaya Ice Cap in the Andes of Peru where the $\delta^{18}O_{SMOW}$ of snow ranges from −31‰ to −9‰ with an average of −21‰ (Thompson and Dansgaard 1975). The ^{18}O – depletion of the snow was unexpected because the average annual air temperature at the site is close to 0°C and the latitude of the site is only 14°S, although the elevation is 5,500 m a.s.l. The explanation for this apparent anomaly is that the air masses that deposit snow at Quelccaya originate from the Atlantic Ocean, rather from the nearby Pacific Ocean and travel a distance of about 3,000 km up the valley of the Amazon River. During this transit, the air masses preferentially lose water molecules that contain ^{18}O and gain water molecules by re-evaporation of surface water that are depleted in ^{18}O. As a result, the water vapor that arrives over the site of the Quelccaya Ice Cap is much more depleted in ^{18}O than expected from its low latitude and comparatively warm average annual temperature (Thomson and Dansgaard 1975; Thompson et al. 1979, 1985, 1986; Grootes et al. 1988).

11.1.3 Carbon

The isotopic composition of carbon is expressed relative to the $^{13}C/^{12}C$ ratio of the PDB standard:

$$\delta^{13}C_{spl} = \left[\frac{\left(^{13}C/^{12}C\right)_{spl.} - \left(^{13}C/^{12}C\right)_{PDB}}{\left(^{13}C/^{12}C\right)_{PDB}} \right] \times 10^3 \text{‰} \qquad (11.5)$$

Carbon isotopes are fractionated primarily (but not exclusively) during photosynthesis in green plants which depletes the carbon in ^{13}C. Therefore, carbon that was processed by green plants has negative $\delta^{13}C$ values on the PDB scale.

The $\delta^{13}C$ values of calcites in the Beacon formations of the central Transantarctic Mountains in Table 11.3 reported by Lord et al. (1988) and Faure and Lord (1990) are, in most cases, negative and imply the presence of varying amounts of biologically-processed carbon.

The only exception to this generalization are calcites in the glacial deposits of the Pagoda Formation. Five of the nine calcites of this formation actually have positive $\delta^{13}C$ values although the average of all nine samples is −0.2‰. On average, the $\delta^{13}C$ values of these calcites indicate that only a small fraction of the bicarbonate ions in the periglacial lake originated from carbon dioxide that formed by oxidation of plant debris. Lord et al. (1988) therefore concluded that the general absence of biologically processed carbon is consistent with the lack of vegetation in and around the periglacial lake in which the sandstone and shales of the Pagoda Formation were deposited.

In marked contrast to the calcites in the Pagoda Formation, 10 of the 14 calcite samples of the overlying Mackellar Formation have $\delta^{13}C$ values that range from +3.4‰ to −14.1‰ with a mean of −4.6‰. The ^{13}C-depletion of these calcites indicates that organic matter was oxidizing in the water of the Mackellar Sea to form carbon dioxide which reacted with water molecules to form bicarbonate ions in solution. The prevalence of biologically-processed carbon in the calcites of the Mackellar Formation is consistent with the high concentration of organic carbon in the shale beds of this formation and with the presence of trace fossils which indicate that the Mackellar Sea contained both plant and animal life and that the surrounding mudflats were vegetated (Horner and Krissek 1991).

The $\delta^{13}C$ values of calcite in the Fairchild and Buckley formations in Table 11.3 are negative although only three samples from these formations were analyzed. Nevertheless, this evidence indicates that plant debris was buried with the sediment and was partly converted into carbon dioxide that dissolved in the pore water.

Two calcites from the Triassic Fremouw Formation have identical $\delta^{13}C$ values of −17.9‰ and contain a major proportion of biogenic carbon. Apparently, vegetation was abundant and supported amphibians and reptiles whose bones were found in the sandstones of the Fremouw Formation in the area of the Beardmore and Shackleton glaciers.

In summary, the isotopic compositions of strontium and oxygen of calcite samples from the Beacon Supergroup indicate that this mineral formed by diagenesis from pore water after deposition of the sediment. Even the carbonate rocks of the Devonian Aztec and Alexandra formations formed this way although the sandstones and shales of these formations are probably marine. The isotope composition of carbon in the calcites in the glacial sediment of Pagoda Formation is largely unfractionated which is consistent with the expected lack of vegetation in the periglacial lake and its surroundings adjacent to the ice sheet. However, the isotope composition of carbon implies that abundant plant and animal life existed in the Mackellar Sea and continued into the Triassic Fremouw Formation.

Table 11.3 Average isotope composition of carbon in calcites from the Beacon Supergroup of the central Transantarctic Mountains (Faure and Lord 1990; Lord et al. 1988, unpublished data)

Formation and number of sples.	Age	$\delta^{13}C$ (PDB) ‰	Range‰
Fremouw (2)	Triassic	−17.9	−17.9
Buckley (2)	Permian	−5.5	−5.5 to −5.6
Fairchild (1)	Permian	−4.0	
Mackellar (13)	Permian	−4.6	+3.4 to −14.1
Pagoda (9)	Permian	−0.2	+2.8 to - 6.1
Alexandra (1)	Devonian	−4.4	

11.1.4 Calcite Cleats in Coal

Coal deposits in the Transantarctic Mountains contain carbonate minerals (calcite and siderite) in the form of cleats that formed on bedding planes and in joints within the coal and in the sandstones that overlie the coal seams. Carbonate minerals may also occur as concretions and coal balls that formed during deposition and diagenesis (Weber and Keith 1962; Weber et al. 1964; Anderson et al. 1980; Orr et al. 1982; Faure and Szabo 1983). Calcite cleats are thought to have formed by precipitation from the ambient groundwater *after* deposition of the rocks. Coal may also contain sulfide minerals (e.g., pyrite or marcasite) and silicates (clay, zeolites, etc.) (Shieh and Suter 1979). The coal in the Transantarctic Mountains has a low average sulfur concentration of about 0.7% because it formed in non-marine environments (Whitby et al. 1983).

Samples of calcite and the associated coal from several locations in southern Victoria Land were analyzed by Faure and Botoman (1984) for the isotope compositions of carbon, oxygen, and strontium. The objectives of this study were to derive information about the environmental conditions in which the calcite cleats formed and to determine the provenance of the carbon and strontium they contain. The results are listed in Table 11.4.

The $\delta^{13}C$ values of the coal samples range from $-23.5‰$ to $-26.7‰$ on the PDB scale which means that the coal was strongly *depleted* in ^{13}C by photosynthesis of the plants from which the coal formed. The carbon in the calcite is also depleted in ^{13}C as indicated by $\delta^{13}C$ values ranging from $-15.6‰$ to $-16.9‰$. Evidently, most but not all, of the carbon in the calcite originated from the coal. Faure and Botoman (1984) suggested that methane gas in the coal seams may have been oxidized to carbon dioxide gas (Eq. 11.6), which reacted with water molecules to form bicarbonate ions (Eq. 11.7), which then reacted with calcium ions to form calcium carbonate, which crystallized as calcite (Eq. 11.8):

$$\underset{methane}{CH_4} + 2O_2 \rightarrow CO_2 + 2H_2O \qquad (11.6)$$

$$CO_2 + H_2O \rightarrow HCO_3^- + H^+ \qquad (11.7)$$

$$HCO_3^- + Ca^{2+} \rightarrow \underset{calcite}{CaCO_3} + H^+ \qquad (11.8)$$

The bicarbonate ions that were derived from the coal *mixed* with bicarbonate ions that had formed from atmospheric carbon dioxide which has $\delta^{13}C = -7.7‰(PDB)$. The fraction of carbon in the calcite can be estimated from mixing theory (Faure and Mensing 2005):

$$\delta^{13}C_{cal.} = f\,\delta^{13}C_{coal} + (1-f)\,\delta^{13}C_{atm.} \qquad (11.9)$$

Taking the $\delta^{13}C$ values of the calcite and coal in the Allan Hills (Table 11.4) yields:

$$-16.9 = f(-26.0) + (1 - f)(-7.7)$$
$$f = 0.50$$

which means that half of the carbon in the calcite cleats in the Allan Hills originated from the coal.

The $\delta^{18}O$ values of oxygen in the calcite cleats range from $+9.14‰$ *to* $+13.06‰$ on the SMOW scale. The $\delta^{18}O$ value of calcite is related to the $\delta^{18}O$ value of

Table 11.4 Isotopic composition of carbon, oxygen, and strontium in calcite cleats of coal seams in southern Victoria Land (Faure and Botoman 1984)

Location	Material	$\delta^{13}C$ (PDB) ‰	$\delta^{18}O$ (SMOW) ‰	$^{87}Sr/^{86}Sr$
Allan Hills	Coal	−26.0	–	–
	Calcite	−16.9	+11.82	0.71318
Shapeless Mt.	Coal	−23.5	–	–
	Calcite	−16.8	+13.06	0.72392
Mt. Fleming	Coal	−23.5	–	–
	Calcite[b]	−2.8	+9.14	0.71602
Mt. Feather	Coal	−26.7	–	–
	Calcite	−15.6	–	0.71233

[a]The isotope compositions of carbon and oxygen were determined by Krueger Enterprises of Cambridge, MA. The $^{87}Sr/^{86}Sr$ ratios were measured in the Laboratory for Isotope Geology and Geochemistry (Isotopia) at The Ohio State University relative to $^{87}Sr/^{86}Sr = 0.70806 \pm 0.00007$ (1σ) for the Eimer and Amend $SrCO_3$

[b]Sparry calcite lens interbedded with the sandstone of the Mt. Fleming Formation (Permian) and was *not* associated with coal. The lenticular calcite bed was about 5 cm thick and several tens of meters wide

water from which the calcite formed by Eq. 11.4 of Friedman and O'Neil (1977). This equation includes the temperature of the environment which is unknown, but presumable must be greater than 0°C but less than 25°C. The limiting temperature estimates yield corresponding $\delta^{18}O$ values of the water in which calcite cleats in the Allan Hills could have been precipitated:

$$T = 0^{\circ}C\,(273.15K), \delta^{18}O_w = -22.5‰$$

$$T = 25^{\circ}C\,(298.15K), \delta^{18}O_w = -19.4‰$$

Regardless of the assumed temperature, the water from which the calcite cleats in the Allan Hills formed was *depleted* in ^{18}O compared to seawater $(\delta^{18}O_{SMOW} = 0‰)$ or meteoric water in the mid-latitudes $(\delta^{18}O_{SMOW} \sim -8‰)$. The other calcite cleats in Table 11.4 yield very similar results.

The water which formed the calcite cleats in the Permian coal deposits of southern Victoria Land is depleted in ^{18}O almost as much as the water that formed the carbonate rocks in the Permian sandstones and shales in the central Transantarctic Mountains (Table 11.2). Therefore, the cleats may have formed during the Permian Period in glacial meltwater that was depleted in ^{18}O. Alternatively, the cleats could have formed much more recently from meltwater of the present-day snow adjacent to the Transantarctic Mountains which also has $\delta^{18}O$ values between $-20‰$ and $-30‰$ on the SMOW scale. The age of the calcite cleats in the Permian coal of the Transantarctic Mountains is presently unknown.

The $^{87}Sr/^{86}Sr$ ratios of the calcite cleats in Table 11.4 (0.71233–0.72392) are far higher than those of seawater at any time in the geologic history of the Earth (Fig. 11.1). They are, however, similar to the average $^{87}Sr/^{86}Sr$ ratios of carbonate rocks in the Permian formations of the central Transantarctic Mountains in Table 11.1. Therefore, the calcite cleats were very probably formed by the ambient groundwater which contained radiogenic ^{87}Sr derived by weathering of K-feldspar, biotite, and other Rb-bearing mineral grains in the Permian sandstones of the Beacon Supergroup. The Rb-bearing minerals in the sandstones were derived from the granitic and metamorphic basement rocks of the Transantarctic Mountains and of the East Antarctic craton. The distribution of coal deposits in Antarctica and their economic potential are discussed in Section 11.5.4.

11.2 The Glaciation of Gondwana

Glacial deposits of Permo-Carboniferous age and the associated Glossopteris flora enabled du Toit (1937) to reassemble the southern continents into the supercontinent Gondwana. His successful reconstruction of the paleogeography of the southern hemisphere of the Earth supported the hypothesis of Wegener (1929) concerning the origin of continents and ocean basins as a result of continental displacement (Section 9.1). The reconstruction of Gondwana in Fig. 11.4 by Smith and Hallam (1970) summarizes the evidence that existed in the 1960s concerning the extent of the Permo-Carboniferous glaciation in Gondwana. The existence of this supercontinent permitted land-based amphibians and reptiles as well as plants to spread across all of the southern continents during the Paleozoic Era until continental rifts began to form during the Jurassic which then slowly widened into the oceans that now separate the continents of the southern hemisphere.

Caputo and Crowell (1985), Veevers and McA. Powell (1987), and Crowell (1999), among others, have assembled the geological evidence for the Paleozoic glaciations of Gondwana in South Africa, South America, Australia, Antarctica, and India. This evidence includes glacial striations and grooves on polished bedrock surfaces, faceted clasts embedded in a fine-grained and highly-indurated matrix, crescentic gouges, boulder pavements, dropstones in both lacustrine and marine fine-grained and laminated shale, and roches moutonées.

The glaciation of Gondwana during the Permo-Carboniferous left a record not only on the southern continents, but it also affected the entire Earth by lowering sealevel and changing the global climate. Large-scale glaciations have occurred many times in the geological history of the Earth. Crowell (1999) recognized that continental glaciations occurred during the Neoproterozoic in South Australia, Western Australia, China, Namibia, South America, Norway, and Greenland. In addition, a Paleoprotorozoic glaciation of North America is indicated by the Gowganda Tillite exposed along the north shore of Lake Huron in Canada.

The most recent continental glaciation which occurred during the Pleistocene may not yet have ended. For example, Carlson et al. (2008) have recently modeled the disintegration of the Laurentide ice sheet of North America and Siddall and Kaplan (2008)

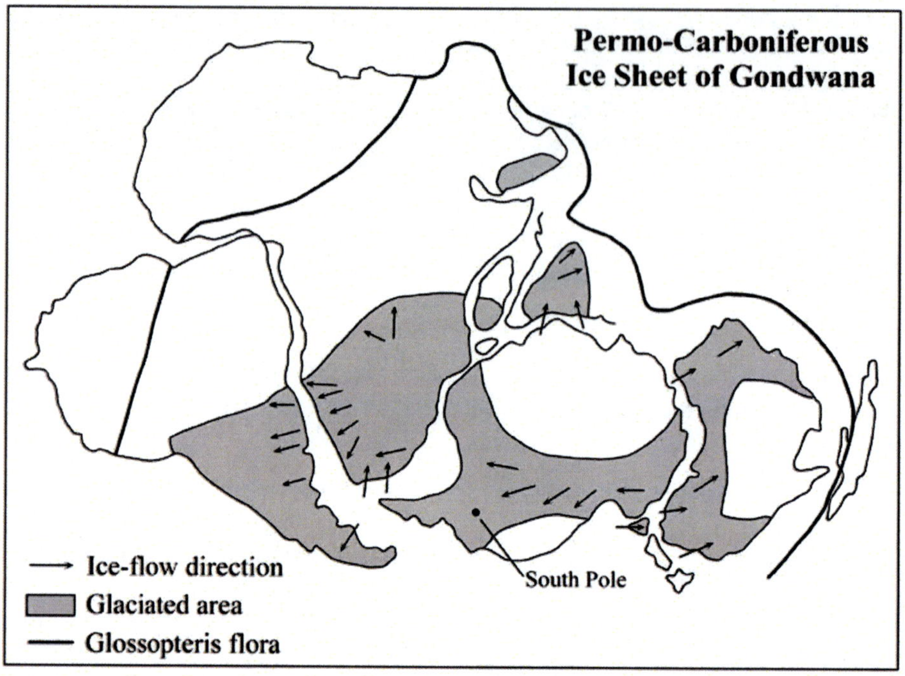

Fig. 11.4 The Permo-Carboniferous ice sheet deposited tillite on all of the southern continents at a time when these continents were united in the supercontinent Gondwana. The arrows indicate ice flow directions which suggest that in the Early Permian the center of the accumulation zone of the ice sheet was located in Victoria Land of Antarctica in the vicinity of the geographic South Pole. A that time, the ice flowed from Antarctica to Australia, India, and southern Africa from where it flowed off the African continent to South America across what later became the South Atlantic Ocean. The *Glossopteris* flora occurred around the edge of the ice sheet in what is now northern South America, central Africa, and northern India as well as on southern Timor (Adapted from Windley (1977) based on Smith and Hallam (1970), with information from Crowell and Frakes (1970), Seyfert and Sirkin (1973), and Plumstead (1973))

considered the fate of the Greenland ice sheet and the effect of its future disintegration on the human population.

The evidence left behind by the Permo-Carboniferous glaciation in the southern hemisphere was described and interpreted in a book edited by Fielding et al. (2008a) which includes a contribution by Isbell et al. (2008) about the Early Permian glacial deposits of the Transantarctic Mountains. The book also contains a review of the current understanding of global climate patterns by the same authors (Fielding et al. 2008b).

Erosional remnants of Early Permian tillites occur at many sites throughout the Transantarctic Mountains from southern Victoria Land to the Ohio Range of the Horlick Mountains and beyond to the Pensacola and the Shackleton ranges. The lithologic similarity, stratigraphic position, and Late Carboniferous to Early Permian age strongly suggest that these glacial deposits record the former presence of a continental ice sheet in the area of the present Transantarctic Mountains when Antarctica was an integral part of Gondwana. For example, Mirsky (1969) correlated the tillites that occur in the Beacon Supergroup at Mt. Fridtjof Nansen (Liv Glacier) with similar rocks on the Nilsen Plateau (Amundsen Glacier), at Mt. Weaver (Scott Glacier), at Tillite Ridge (Wisconsin Range), and at Mt. Glossopteris (Ohio Range). The tillite at these sites rests unconformably on granitic and metamorphic basement rocks except in the Ohio Range where the Buckeye Tillite was deposited on the Devonian Horlick Formation. Schmidt and Williams (1969) and Schmidt and Friedman (1974) described the tillite of the Gale Mudstone in the Pensacola Mountains and stated that it is probably correlative with tillite occurrences as far west as the Darwin Range and with tillite in the Ellsworth Mountains. Subsequently, Barrett et al. (1986) also correlated the Permian tillites throughout the Transantarctic Mountains from the Gale Mudstone of the Pensacola Mountains all the way to the Metschel Tillite in the Boomerang Range of southern Victoria Land. The most northerly deposits of tillite occurs in the Lanterman Range of northern Victoria Land where it is overlain by sedimentary rocks of the Takrouna Formation which contains a *Glossopteris-Gangamopteris* flora of Permo-Carboniferous age. Collinson et al. (1986) therefore correlated this tillite with other tillite deposits of Early Permian age in Antarctica and Tasmania.

The Pagoda Tillite in the central Transantarctic Mountains has been described by several investigators starting with Grindley (1963), followed by Lindsay (1970), Coates (1972, 1985), Barrett et al. (1986) and, more recently, by Miller (1989), Isbell (1999), and Isbell et al. (2003). Coates (1985) noted that the thickness of the glacial deposits in the central Transantarctic Mountains decreases systematically along the strike of the mountains from the Queen Elizabeth Range (300 m) to Mt. Fridtjof Nansen (0–1.5 m) adjacent to the Liv Glacier. The glacial deposits in the Queen Maud Mountains are thin (0–20 m) from the Liv Glacier to the Reedy Glacier where the thickness increases to about 50 m in the Wisconsin Range and to 250–300 m in the Ohio Range. Coates (1985) also noted that the elevation of the erosion surface of the underlying basement rocks is high in the area where tillite is thin. The ice-flow directions in the region of thin-till vary widely with a strong component to the south but swing to the west into the Beardmore basin and to the east into the Horlick/Queen Maud basin on the flanks of this ice-divide. The western end of the Beardmore basin was located in the Holland and Queen Elizabeth ranges where the prevailing ice-flow direction was to the southeast.

The observations by Coates (1985) were confirmed and elaborated by Isbell et al. (2008) in the most recent summary of the Permian glaciation of the Transantarctic Mountains. The cross-section of Isbell et al. (2008) drawn parallel to the strike of the Transantarctic Mountains in Fig. 11.5 clearly shows the thinning of the tillite in the Queen Maud Mountains and the thickening of the section in the areas of the Beardmore-Nimrod glaciers and in the Ohio Range.

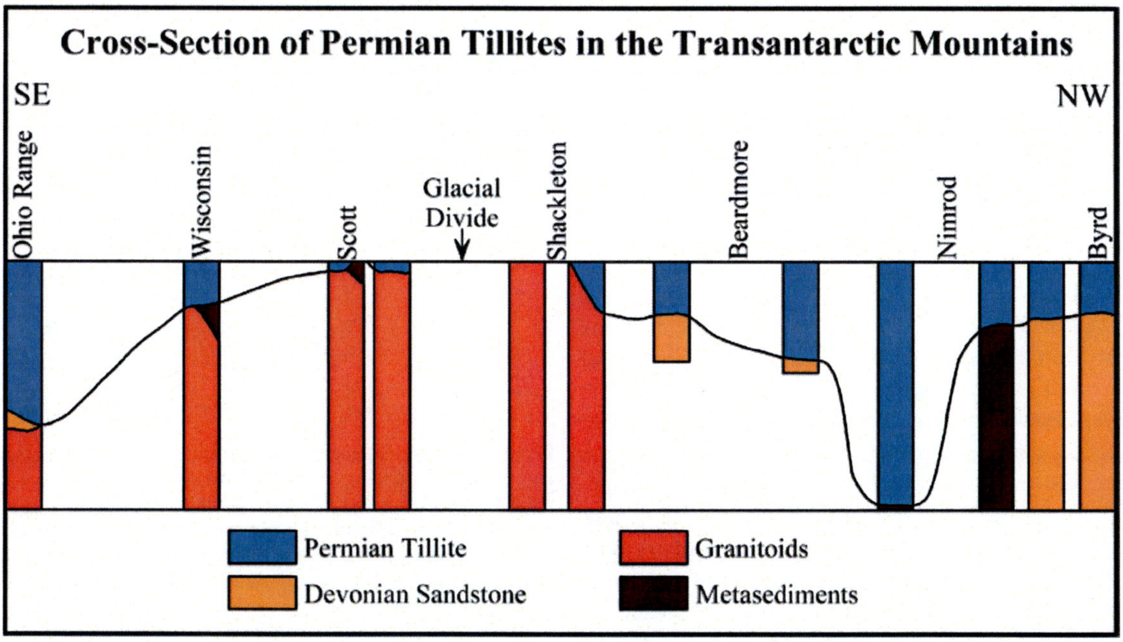

Fig. 11.5 The thickness of the Permian tillites varies along the strike of the Transantarctic Mountains from the Byrd-Glacier area to the Ohio Range. In addition, the tillites were deposited on Devonian sandstones (Byrd and Beardmore glaciers), on folded metasedimentary basement rocks (Nimrod Glacier), and on granitic basement rocks in the Queen Maud Mountains between the Shackleton Glacier and the Wisconsin Range. In the Ohio Range, the Buckeye Tillite rests disconformably on the Devonian Horlick Formation (Adapted from Isbell et al. 2008)

These trends were expressed by Isbell et al. (2008) by means of an isopach map of the tillites in the Transantarctic Mountains. The zero-thickness contour in Fig. 11.6 outlines the Beardmore basin and the Horlick/Queen Maud basin where the thickness of the tillite exceeds 300 m. The basins are separated by a glacial divide where only a thin layer of tillite was deposited by basal ice. However, questions remain whether the ice sheet was wet-based, in which case glacial meltwater may have played an important role in the deposition of sediment. The presence of water (fresh or marine) is indicated by the *Conchostracans* in the glacigenic sediment of the Shackleton Glacier that were described by Babcock et al. (2002), whereas palynomorphic and other carbonaceous plant fossils were damaged by the heat that emanated from the sills of Ferrar Dolerite.

Isbell et al. (2001, 2008) concluded that the tillites in the Darwin and the central Transantarctic Mountains formed in a glaciomarine rather than a continental setting and that Antarctica was *not* covered by a single, massive continental ice sheet. These conclusions differ from the views expressed earlier by Lindsay (1970), Barrett and Askin (1975), Barrett et al. (1986), and Miller (1989) who considered that lodgment and melt-out till were deposited in Antarctica during the waxing and waning of a continental ice sheet. These authors, as well as Frakes et al. (1971), concluded that the multiple tillite layers interbedded with stratified deposits of sandstone and shale of Permian age formed during four to six advances and retreats of a land-based ice sheet.

The history of Paleozoic glaciations of Gondwana has been assembled not only from evidence in the Transantarctic Mountains but also from rock outcrops on the other southern continents (e.g., Caputo and Crowell 1985; Veevers and McA. Powell 1987; McA. Powell and Veevers 1987). According to Crowell (1999), the glaciation of Gondwana started during the Late Devonian with the development of alpine glaciers in the area that later became South America. These local valley glaciers coalesced into a coherent ice sheet at about 340 Ma during the Early Carboniferous (Visean). The resulting large-scale glaciation of Gondwana lasted until about 265 Ma in the Late Permian (Kazanian). According to these estimates, the glaciation of Gondwana lasted for about 75 million years and, during that time, caused sealevel to be significantly depressed.

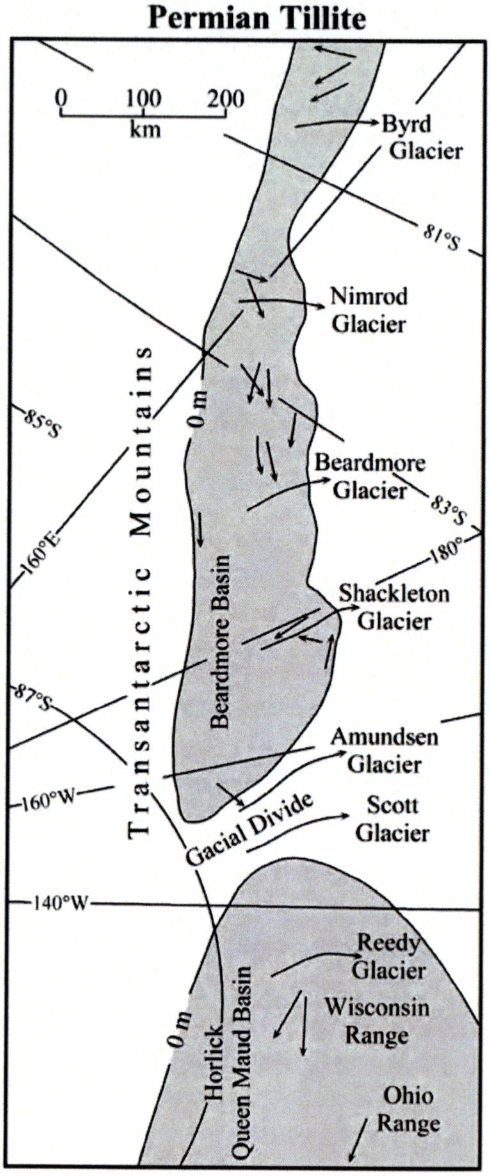

Fig. 11.6 The zero isopach contour outlines the Beardmore and the Horlick-Queen Maud glacial basins which are separated by a glacial divide where the tillite is absent or very thin (<1.5 m). The measured ice flow directions are indicated by short and straight arrows, whereas the long and curved arrows mark the locations of the major outlet glaciers (Adapted from Isbell et al. 2008)

Although Powell and Veevers (1987) attributed the start of the glaciation of Gondwana to uplift during the Early Carboniferous in Australia and South American, the major cause was climate change resulting from the displacement of Gondwana toward the south geographic pole. Crowell (1999) cited sources in the literature which indicate that during the Late Ordovician and Early Silurian the magnetic pole passed through South America, looped north into Africa from Early Devonian to Late Carboniferous and reached the vicinity of Antarctica during the Early Permian.

Veevers and Powell (1987) placed Antarctica between latitudes 0° to 30°S during the Early Devonian at 400 Ma. Forty million years later during the Late Devonian, Antarctica was between latitudes 30°S and 60°S. Still later, at 325 Ma during the Early Carboniferous, most of the future territory of Antarctica was located inside the Antarctic circle. During the latest Carboniferous at 290 Ma, the South Pole was located close to Victoria Land and then passed into eastern Australia were it stayed until Late Permian/Early Triassic before it moved into the basin of the Pacific Ocean.

According to this sequence of events, the glaciation of Gondwana passed through three episodes which are identified in Fig. 11.7 based on the work of Veevers and Powell (1987). Episodes I and II consisted of alpine glaciations in South America (Brazil) and adjacent areas of Africa. Episode III was the major glaciation of Gondwana by a continental ice sheet. Part A of episode III occurred primarily in South America, eastern Australia, and possibly in Antarctica during late Early Carboniferous (i.e., Late Mississippian, Serpukhovian). This phase was initially followed by a reduction of ice volume, but was succeeded later by the main phase of the glaciation (B) during the Late Carboniferous (i.e., Late Pennsylvanian, Kazimovian and Gzhelian). Veevers and Powell (1987) and Fielding et al. (2008b) considered that the glaciation ended in part C of episode III in the Early Permian (Sakmarian/Artinskian), whereas Crowell (1999) shows that it continued into the Late Permian. According to Veevers and Powell (1987), the final deposits were mainly glaciomarine perhaps because isostatic depression of parts of Gondwana caused local transgressions of seawater onto the continent.

When viewed in the larger context of Gondwana, the tillites of the Transantarctic Mountains and the interbedded sandstones and mudstones were deposited close to the end of the glaciation of Gondwana when the ice sheet was ablating and meltwater streamed into local basins and into the ocean. As a result, tillites and water-deposited sedimentary rocks were locally interbedded. Another insight provided by the larger view is that the glaciation of Gondwana did not start in Antarctica and that Victoria Land was not the center of the Permian ice sheet in

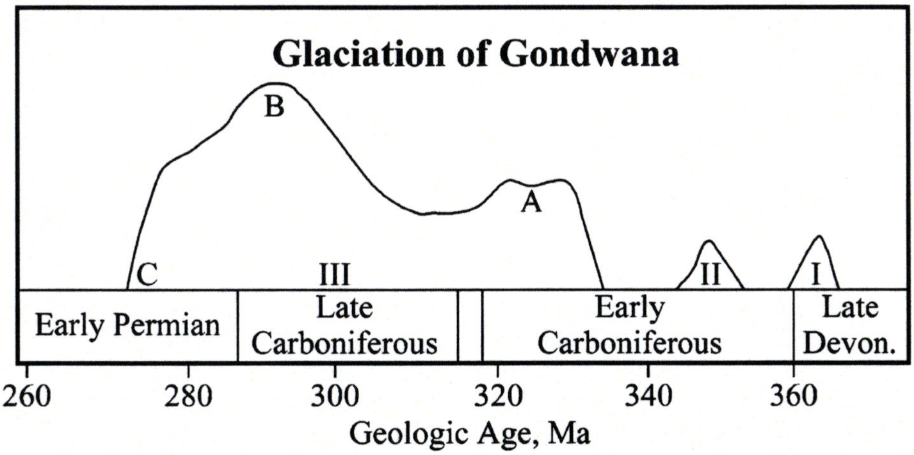

Fig. 11.7 The glaciation of Gondwana occurred primarily because parts of this large continent moved across the south geographic pole of the Earth from the Late Devonian to Early Permian. After two initial episodes of alpine glaciation (I and II) in the area of what is now South America, a large continental ice sheet (IIIA) developed during the Early Carboniferous (i.e., Mississippian) and spread across a large area of Gondwana including territories that are today South Africa, Australia, Antarctica, and India. The volume of this ice sheet reached its peak (B) in the Late Carboniferous (i.e., Pennsylvanian) and then declined during the Early Permain (C). The glacial deposits in the Transantarctic Mountains date from the Late Carboniferous/Early Permian and include glaciomarine sediment deposited at the margin of the ice sheet presumably because seawater transgressed locally onto Gondwana due to isostatic depression (Adapted from Veevers and Powell (1987))

Antarctica except perhaps temporarily when remnants of the ice may have flowed into the area presently occupied by the Transantarctic Mountains. Moreover, the Permo-Carboniferous ice sheet may have occupied parts of East Antarctica that are presently ice-covered, such as the Gamburtsev Mountains. However, Isbell et al. (2003) asserted that the Transantarctic Mountains were ice-free during the Carboniferous and that this area was covered only by the *margin* of the ice sheet during the Early Permian. In that case, the Gondwanian ice sheet may not have penetrated into the interior of the East Antarctic craton although the Gamburtsev Mountains probably were covered by local ice caps.

11.3 Tetrapod Fauna

The rocks of the Beacon Supergroup of Gondwana are noteworthy because they were deposited on *land* rather than in the coastal ocean of Gondwana. In addition, the sediment was deposited after plants and animals had made the transition from the ocean to life on land. Even more fortunate is the fact that the rocks of the Beacon Supergroup have preserved a record of the evolution of air-breathing tetrapods because the sediment that was deposited on alluvial plains was not removed by erosion even before it was lithified. For these reasons, the sedimentary rocks of the Beacon Supergroup can provide information that is not available in rocks that were deposited in the oceans at the same time.

When the siltstones of the Aztec Formation were deposited during the Late Devonian, fish already existed in the rivers and lakes of Gondwana. The recovery of fossilized fish in the Boomerang Range of southern Victoria Land was described in Section 10.1.2 (White 1968; McKelvey et al. 1972; McPherson 1978; Grande and Eastman 1986; Young 1987).

We now turn to the fossilized bones of amphibians and reptiles that have been found in the Early Triassic Fremouw Formation of the central Transantarctic Mountains. The Fremouw basin was inhabited by a variety of tetrapods including labyrinthodont amphibians and therapsid reptiles which are believed to be the Triassic ancestors of Cretaceous mammals (Colbert 1949). These animals were widely distributed in East Antarctica and on the other continents of the southern hemisphere that once were part of Gondwana (Hammer 1990). Therefore, the presence of these vertebrates in East Antarctic, South Africa, and India strengthens the evidence for the break-up of Gondwana and the displacement of the continental fragments to their present positions in the southern hemisphere of the Earth (Taylor and Taylor 1990).

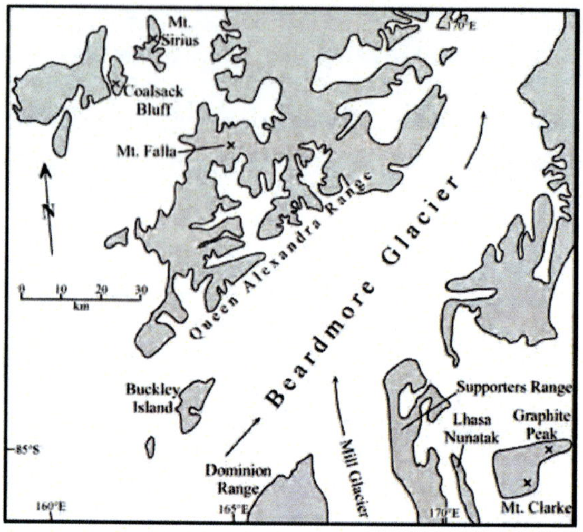

Fig. 11.8 Graphite Peak and Coalsack Bluff east and west, respectively, of the upper Beardmore Glacier are the sites where fossilized bones of tetrapod vertebrates were first discovered in 1967 and 1969. The bone fragment found by Peter Barrett at Graphite Peak was part of the lower jaw of a labyrinthodant amphibian. The bones found subsequently at Coalsack Bluff belong to several species of tetrapods including the therapsid *Lystrosaurus* which is regarded as an Early Triassic ancestor of Cretaceous mammals (Colbert 1949) (Adapted from Collinson and Elliot (1984) with modification of the Graphite Peak area from the topographic map of Plunket Point Antarctica (SV 51-60/8*; 162°–180°E) published in 1967 by the US Geological Survey, Washington, DC)

Fossilized bones of tetrapods have been recovered from the Fremouw Formation at Graphite Peak and Coalsack Bluff east and west, respectively, of the upper Breadmore Glacier in Fig. 11.8, in the Cumulus Hills along the east side of the Shackleton Glacier, and in Gordon Valley in the Queen Alexandra Range identified in Fig. 10.14 (Colbert 1971, 1975, 1977, 1978, 1979, 1982a, b).

11.3.1 Graphite Peak

Graphite Peak in Fig. 11.9a is located near Mt. Clarke (3,211 m) east of Lhasa Nunatak and east of the Supporters Range adjacent to the Mill Glacier. The sedimentary

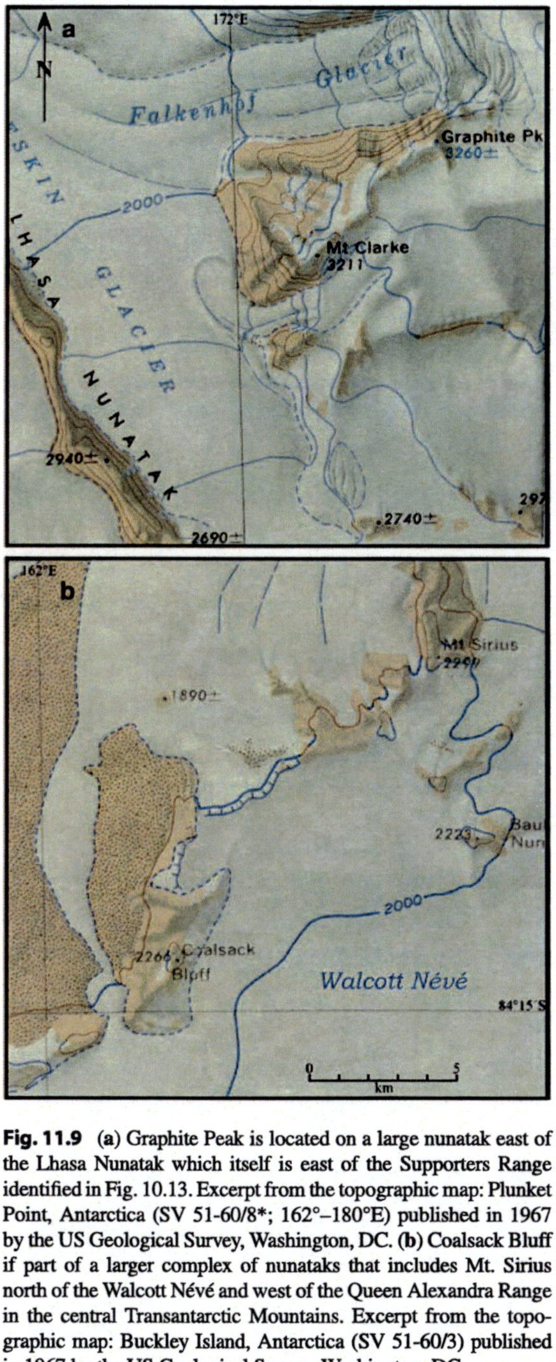

Fig. 11.9 (a) Graphite Peak is located on a large nunatak east of the Lhasa Nunatak which itself is east of the Supporters Range identified in Fig. 10.13. Excerpt from the topographic map: Plunket Point, Antarctica (SV 51-60/8*; 162°–180°E) published in 1967 by the US Geological Survey, Washington, DC. (b) Coalsack Bluff if part of a larger complex of nunataks that includes Mt. Sirius north of the Walcott Névé and west of the Queen Alexandra Range in the central Transantarctic Mountains. Excerpt from the topographic map: Buckley Island, Antarctica (SV 51-60/3) published in 1967 by the US Geological Survey, Washington, DC

Fig. 11.10 *Glossopteris* leaves are preserved in most cases as films of amorphous carbon on bedding planes of fine-grained sedimentary rocks. Although the shape of the leaves and the venetation are clearly visible, the internal anatomy and the surface features of the leaves are lost (Photo by T.N. Taylor and E.L. Taylor, reproduced with permission)

rocks at this location include sandstones of the upper *Buckley Formation* (Permian) which is disconformably overlain by the *Fremouw Formation* (Early Triassic) composed of greenish-gray mudstone, quartz sandstone, and volcaniclastic sandstone. The sandstones of the Buckley Formation contain carbonized plant stems and silicified wood as well as graphitic coal and *Glossopteris* leaves like those in Fig. 11.10. The mudstones of the overlying Fremouw Formation contain *Dicroidium* odontopteroides and carbonized plant stems which Barrett et al. (1968) tentatively identified as *Neocalamites*.

The first bone fragment was found on December 28, 1967, by Peter J. Barrett in a pebbly quartz-sandstone of the lower Fremouw Formation on a ridge about 4 km west of Graphite Peak in Fig. 11.9a. This bone was identified as the posterior part of the left mandibular ramus of a labyrinthodont amphibian even though the bone fragment was only about 7 cm long. Barrett et al. (1968) published a detailed technical description of the osteological characteristics of this bone and Colbert (1982a) identified it as *Austrobrachyops jenseni*.

Hammer et al. (1986) revisited Graphite Peak during the 1985–1986 field season and collected additional specimens of tetrapod bones. Although most of these bones belonged to *Lystrosaurus murrayi*, the new collection also included a mandible of the carnivorous cynodont reptile *Phrinaxodon* (Hammer 1990b).

11.3.2 Coalsack Bluff

The second discovery of tetrapod bones occurred 2 years later on November 23, 1969, at Coalsack Bluff

(84°14'S, 162°25'E) in Figs. 11.8 and 11.9b where David Elliot recovered bone fragments from a coarse sandstone of the Fremouw Formation. Subsequently, more than 400 specimens of bone belonging to labyrinthodont amphibians and to thecodont as well as to therapsid reptiles, including *Lystrosaurus*, were collected at this site. The bones were found at the base of sandstone beds that filled stream channels in the Early Triassic landscape (Elliot et al. 1970).

The geology of Coalsack Bluff was described by Collinson and Elliot (1984) who also published a large-scale map of the nunatak which is only 5 km long and 2 km wide. The rocks exposed at Coalsack Bluff include the upper part of the Buckley Formation and the lower part of the overlying Fremouw Formation for a total thickness of approximately 500 m. The sedimentary rocks of Coalsack Bluff were intruded by three sills of Ferrar Dolerite. Four whole-rock K-Ar dates, cited by Collinson and Elliot (1984), range from 170.1 ± 2.0 to 182.3 ± 3.2 Ma and confirm a Middle Jurassic age for these rocks (IUGS 2002).

Glacial deposits of Quaternary age on the northwestern side of Coalsack Bluff contain cobbles and boulders of locally derived Ferrar Dolerite, clasts of igneous and metamorphic rocks, as well as fragments of limestone containing well-preserved archaeocyathids of Cambrian age. These limestone clasts could have originated from an outcrop of Shackleton Limestone that occurs at the south end of Buckley Island located about 95 km southeast of Coalsack Bluff (Section 5.3.1).

11.3.3 Cumulus Hills, Shackleton Glacier

During the field season of 1969–1970 abundant fossil bones were recovered from the Fremouw Formation of the Cumulus Hills in the McGregor Glacier area in Fig. 11.11a (Colbert 1971; Kitching et al. 1972; Cosgriff et al. 1978; Cosgriff and Hammer 1979, 1981; Cosgriff 1983; Hammer and Cosgriff 1981). The area was revisited during the 1985–1986 field season by Hammer et al. (1996) who found additional fossil bones on Collinson Ridge, Shenk Peak, and Halfmoon Bluff of the Cumulus Hills and at several other sites along the Shackleton Glacier (Fig. 11.11b).

The McGregor Glacier is an eastern tributary of the Shackleton Glacier in the Queen Maud Mountains.

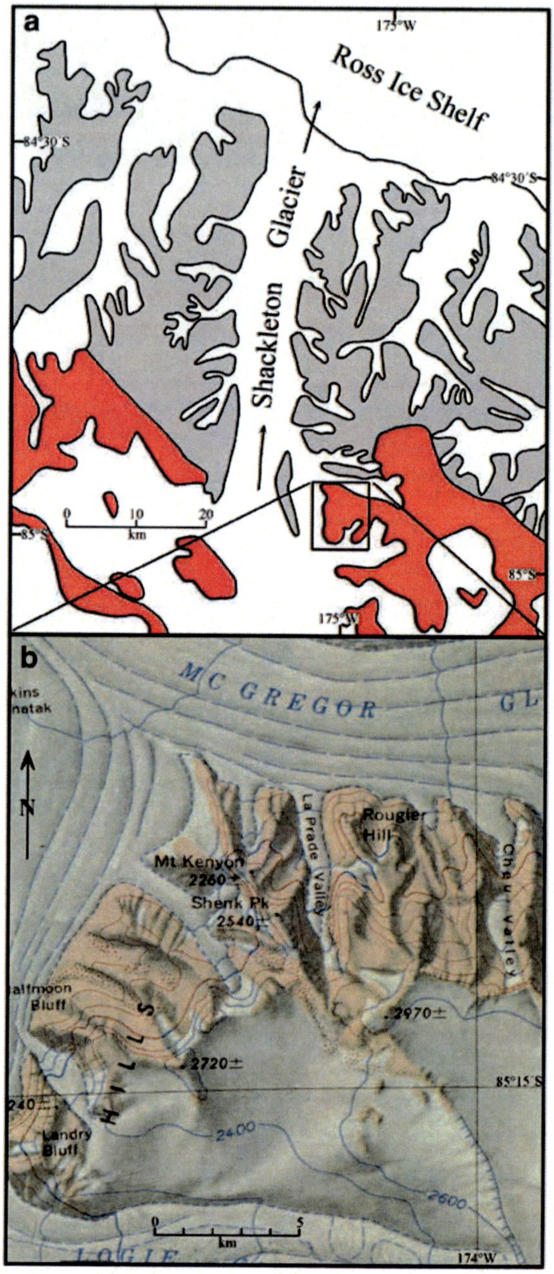

Fig. 11.11 (a) Shackleton Glacier area of the Queen Maud Mountains including Beacon rocks (pink) south of the McGregor Glacier. The coastal mountains expose the Granite Harbor intrusives of the basement complex (Adapted from McGregor and Wade 1969). (b) Rock outcrops of the Beacon Supergroup south of the McGregor Glacier where Kitching et al. (1972) and Colbert (1971) recovered bones of Early Triassic vertebrate tetrapods including a terapsid reptile of the genus *Lystrosaurus*. Excerpt from the topographic map: Liv Glacier, Antarctica (SV 1-10/5) published in 1968 by the US Geological Survey, Washington, DC

The collecting sites are located south of the McGregor Glacier and east of the Shackleton Glacier. The rocks of the Beacon Supergroup in this area were described by LaPrade (1970) who referred the Triassic strata to the Mt. Kenyon Formation named after Mt. Kenyon in Fig. 11.11b. However, these rocks were later placed in the Fremouw and Falla formations described by Barrett et al. (1986) in the Beardmore Glacier area (LaPrade 1970). Accordingly, the bones in the McGregor Glacier area, like those at Graphite Peak and Coalsack Bluff, occur in the lower part of the Fremouw Formation which consists of cyclical deposits of sandstone grading into overlying mudstone. The coarse sandstones at the base of each cycle contain disarticulated bones which were probably derived by erosion of the mudflats in the depositional basin. The layers of mudstone contain more complete skeletal remains than the coarse grained sandstones.

The bones that were recovered in the McGregor Glacier area included *Lystrosaurus murrayi* and other members of the Lystrosaurus fauna including the theriodont reptile *Phrinaxodon liorhinus*, the cotylosaurian *Procolophon trigoniceps* (Colbert 1977), and the skull of a small labyrinthodont amphibian pictured by Colbert (1971) and by Kitching et al. (1972).

The Triassic vertebrates recovered from these three sites in the Transantarctic Mountains were described and evaluated in comprehensive reports by Colbert (1982a, b) in which he reviewed all of the fossils that were collected at Coalsack Bluff, at Graphite Peak, and at the McGregor Glacier. The long list of species that have been identified include a dipnoan fish, several labyrinthodent amphibians and a large number of reptiles (Colbert 1982a, p. 32).

11.3.4 Gordon Valley and Mt. Kirkpatrick

Bones of vertebrate animals were also recovered from the upper Fremouw Formation in Gordon Valley in the Queen Alexandra Range (Fig. 10.14) and from the upper Falla Formation on Mt. Kirkpatrick (Hammer et al. 1987). The Fremouw Formation in Gordon Valley actually contains foot prints of vertebrates which were described and photographed by Macdonald et al. (1991). The bones recovered by W.R. Hammer and his associates were described and identified in a series of reports (Hammer 1990a, b; Hammer et al. 1990, 1991, 1995, 1996; Hammer and Hickerson 1992, 1993, 1994a, b). The stratigraphy and sedimentology of the bone-bearing beds were worked out by Elliot et al. (1972), Isbell and Macdonald (1991), Krissek et al. (1992), Retallack and Hammer (1996), and Krissek and Horner (1991).

11.3.5 Lystrosaurus

Lystrosaurus belongs to the Subclass Synapsida, Order Therapsida, and is a member of the Family Lystrosaurida. The reconstruction of Lystrosaurus in Fig. 11.12 was made from bones recovered at several of the sites described by Kitching (1968), Colbert (1973, 1974) Colbert and Kitching (1977), and Cosgriff et al. (1982).

Lystrosaurus was a heavily built, mammal-like dicynodont reptile about 90 cm long. The skull was short and toothless except for two large tusks that grew on each side of the skull of many individuals. The jaws resembled the beaks of modern turtles and presumably had sharp edges used for biting off vegetation. The structure of the shoulders and hip joints indicates that *Lystrosaurus* had a sprawling gait and was capable of digging burrows for nesting. The occurrence of this reptile in rocks of Early Triassic age indicates that it, or its ancestors, survived the Permo-Triassic extinction event. The reasons for the survival of this genus are unknown partly because the causes of this extinction event are still not clearly understood.

11.3.6 Permo-Triassic Extinction Event

The end of the Permian Period is defined by the extinction of about 90% of all life forms that existed on the Earth at that time. This catastrophe was even more severe than the extinction event that defines the end of the Cretaceous Period which resulted from the profound environmental disturbance caused by the impact of an asteroid at Chicxulub on the Yucatán Peninsula of Mexico. The cause of the Permo-Triassic extinction could have been a dramatic fluctuation of the global climate, or of sealevel, or the impact of an asteroid or comet, or severe volcanic activity, or all of the above.

Fig. 11.12 This plaster cast of a *Lystrosaurus* reptile resides in the Byrd Polar Research Center of the Ohio State University in Columbus, Ohio (Photo by G. Faure. Image preparation by Ikon)

For example, the impact of an asteroid upon the Earth could have triggered large-scale volcanic eruptions at the antipode which further aggravated the disturbance of the global climate by filling the atmosphere with dust that blocked sunlight for several years or even decades. The explanation of the Permo-Triassic extinction event is currently a "hot" topic in the Earth sciences (e.g., Wayt Gibbs 2001; Kaiho et al. 2001, 2002; Koeberl et al. 2002; McAllister Rees 2002; Ryskin 2003; Twitchett et al. 2004). Although several large meteorite impact craters that formed close to the time of the Permian-Triassic boundary have been considered, none has yet been identified as the cause for the Permo-Triassic extinction event (see also Section 18.4.1).

Retallack et al. (1998) briefly reviewed the literature concerning the causes of the Permo-Triassic extinction event as an introduction to their own study to determine whether boundary clays at Graphite Peak between the Buckley and Fremouw formations contain shocked quartz and anomalously high concentrations of iridium of meteoritic origin. Although they found shocked quartz and enhanced iridium concentrations both at Graphite Peak and at Mt. Crean in southern Victoria Land, the evidence is inconclusive because the claystones at the Permo-Triassic boundary at these locations consist of redeposited soil rather than of impact ejecta.

In a related study, Basu et al. (2003) reported finding meteorite fragments and metallic grains in rock samples from Graphite Peak. However, in the absence of an impact crater and other geomorphic evidence that is characteristic of impact events, the search for the causes of the Permo-Triassic extinction event continues.

11.4 Plant Fossils

The survival of amphibians and reptiles in Gondwana was facilitated by the presence of plants that grew on the coastal plains which provided them with food and shelter. The leaves, stems, and roots of these plants have been preserved in the sedimentary rocks of the Beacon Supergroup as carbonaceous films and by per-mineralization. Plants were scarce during the Devonian and during the Permo-Carboniferous glaciation only cold-adapted plants survived. When the glaciation finally ended in the Early Permian, the climate improved and allowed abundant vegetation to flourish.

When the geologists who accompanied Robert Scott and Ernest Shackleton first entered the Transantarctic Mountains (Sections 1.4.1, 1.4.2., and 1.4.3), they were amazed to find evidence that plants and animals had once thrived in Antarctica (e.g., Seward 1914; David and Priestley 1914; Debenham 1921). Doumani and Long (1962) reported that these discoveries caused Charles Hedley to write in 1912:

> Of rippling brooks, of singing birds, of blossoming flowers, and forest glades in the heart of Antarctica.

Although we know that birds and flowers evolved much later, Antarctica was a wetter and warmer place in the Permo-Triassic than it is at the present time (Collinson 1997). In addition, the existence of very similar floras in the sedimentary rocks of Permo-Triassic age on all of the southern continents can only be explained by the break-up of Gondwana followed by the displacement of the continental fragments. This aspect of the paleobotany of the southern continents was the theme of a book edited by Taylor and Taylor (1990). In 1962, Doumani and Long could only speculate that the climate change implied by the fossil record required either that the poles had wandered or the continents had drifted.

The technical descriptions of the plant fossils of the Transantarctic Mountains form a large body of literature with major contributions by Plumstead (1962, 1964, 1973), Townrow (1967), and a remarkable series of papers by James M. Schopf (1965, 1966a, b, 1967, 1968, 1969, 1970a, b, c, 1971, 1973, 1975, 1977, 1978; Schopf and Long 1966; Rigby and Schopf 1969). The study of plant fossils in the Beacon Supergroup was continued by Rigby (1969, 1985), by Tom and Edith Taylor and their graduate students (Taylor and Smoot 1983, 1984, 1985a, b; Taylor et al. 1986a, b, 1990, 1992; Taylor and Taylor 1988, 1990, 1992) leading up to a comprehensive treatise on paleobotany by Taylor, Taylor, and Krings (2009).

The plant fossils of the Beacon Supergroup occur primarily in the Permo-Triassic rocks of the Victoria Group which is present throughout the Transantarctic Mountains from northern Victoria Land (e.g., Gair et al. 1965; Tessensohn and Mädler 1987) to southern Victoria Land (e.g., Edwards 1928; Taylor et al. 1989a, b, 1990) and beyond, to the Ohio Range (Doumani and Long 1962; Cridland 1963) and the Ellsworth Mountains (e.g., Craddock et al. 1965). Plant fossils from the far-eastern Transantarctic Mountains were described by Plumstead (1962, 1964, 1973).

We also take note that Doumani and Tasch (1963, 1965) described two species of conchostracans (*Leaia* gondwanella, n.sp. and Cyzicus (Lioestheria) doumanii n. sp.) which occur in a thin layer of black carbonaceous-shale (10–15 cm thick) in the Permian Mt. Glossopteris Formation of the Ohio Range. These conchostracans appear to have lived in small ponds in a swampy environment with variable weather conditions that permitted them to survive for 21–48 days.

The flora that has been preserved in the Beacon Supergroup of the Transantarctic Mountains and in Australia, South Africa, India, and South America contain leaves of *Glossopteris*, *Gangamopteris*, and *Dicroidium*, as well as gymnospermous trees (Meyer-Berthaud and Taylor 1989), cycads, ferns, and palynomorphs (pollen, spores, and acritarchs). The fossilized remains of these plants occur primarily in the Buckley, Fremouw, and Falla formations.

A great deal has been learned about these plants from the study of blocks of silicified peat on Fremouw Peak (Fig. 10.14) and on Mt. Augusta (84°48'S, 163°06'E) in the Marshall Mountains of the Queen Alexandra Range. The Mt. Augusta locality of Schopf (1970b, 1971, 1977, 1978) was re-investigated by Taylor et al. (1986b) who reported that the silicified peat actually occurs at the southeastern end of Skaar Ridge about 4.8 km from Mt. Augusta. The fossils at the Skaar-Ridge site occur in the upper Buckley Formation and are Late Permian in age. In addition, well-preserved plant fossils have been found in a shale layer on Mt. Falla. The descriptions that follow are based primarily on fossil plants recovered at these three sites although the same species occur at many other locations in the Transantarctic Mountains and elsewhere on the Gondwana continents.

11.4.1 *Glossopteris*

The Permian flora of Gondwana was characterized by the tree-like *Glossopteridales* pictured by Schopf (1970a). The stem of this hypothetical tree was com-

Fig. 11.13 *Glossopteris* leaves in Permian sandstone of the Beaon Supergroup in the Transantarctic Mountains (Photo by T.N. Taylor and E.L. Taylor, reproduced with permission. This image was published in *Paleobotany* by Thomas N. Taylor, Edith L. Taylor, and Michael Krings, second edition. Figure 14.174, p. 600. Copyright Elsevier, 2009)

posed of *Araucarioxylon*-type wood, the roots have been assigned to *Vertebraria*, and it had long and slender leaves known as *Glossopteris* with reticulate venetion that is clearly visible in Fig. 11.13. In spite of the wide-spread occurrence of *Glossopteris* in the Permian rocks of Gondwana, these leaves have provided little information about the biology of the plant itself because, in most cases, the leaves are preserved only as impressions in fine-grained sedimentary rocks (Taylor and Smoot 1985b).

The genus *Glossopteris* was first defined by A. Brongniart in 1828 ..."for simple, sublanceolate leaves with a prominent midrib, entire margin, and net venetion" (Taylor and Taylor 1992). This definition has not been particularly helpful because most fossilized specimens of *Glossopteris* have not preserved the cuticle which is critically important in plant taxonomy. In addition, the relation of *Glossopteris* to *Gangamopteris* is unclear because these leaves had similar anatomies. *Glossopteris* occurs primarily in rocks of Permo-Triassic age, but a few specimens of similar leaves have been found in sedimentary rocks of Jurassic age. Pigg and Taylor (1985) described the cellular anatomy of well-preserved *Glossopteris* in the silicified peat associated with the Permian rocks on Mt. Augusta and with the Triassic rocks on Fremouw Peak. The only other locality where *Glossopteris* leaves were preserved with structural integrity is in the Bowen Basin of Queensland in Australia (Gould and Delevoryas 1977).

Taylor and Taylor (1992) later identified 16 different species of *Glossopteris* in the Polarstar Formation of the Ellsworth Mountains. In addition, these authors identified two species of *Gangamopteris* and one of *Vertebraria* australis, as well as the equisetum *Paracalamites* australis and *Raniganjia* bengalensis. The flora in the Polarstar Formation of the Ellsworth Mountains closely resembles the floras in the Beacon rocks of the Ohio Range (Cridland 1963), in the Theron Mountains, and the Wichaway Nunataks (Plumstead 1964). The similarity of the floras strengthens the evidence that the Ellsworth Mountains are closely related to the Transantarctic Mountains and may have been moved out of place by tectonic activity as suggested by Schopf (1969).

Additional information on the structure of *Glossopteris* is available in the books by Taylor (1981), and by Taylor and Taylor (1990), in the treatise on paleobotany by Taylor, Taylor and Krings (2009), and in the paleontological literature including papers by Doumani and Long (1962), Craddock et al. (1963), Plumstead (1964), Schopf (1966a, 1967, 1973, 1976), Rigby (1969), Rigby and Schopf (1969), Gould and Delevoryas (1977), Taylor and Smoot (1985), Taylor and Taylor (1992), Smoot and Taylor (1986), and others.

11.4.2 Dicroidium

Although Dicroidium in Fig. 11.14 is an important Triassic index fossil, the biology of the plant is not

Fig. 11.14 Fronds of *Dicroidium* in sandstone of Triassic age in the Beacon Supergroup of the Transantarctic Mountains (Photo by T.N. Taylor and E.L. Taylor, reproduced with permission. This image was published in *Paleobotany* by Thomas N. Taylor, Edith L. Taylor, and Michael Krings, second edition. Figure 15.18, p. 628. Copyright Elsevier, 2009)

well known. Pigg and Taylor (1987) and Pigg (1990) described the internal structure of *Dicroidium* recovered from the silicified peat on Fremouw Peak (Fig. 10.14). In addition, Taylor and Taylor (1988) collected well-preserved specimens of *Dicroidium* from a layer of black shale in the Falla Formation on Mt. Falla. These specimens included reproductive structures which had not been seen before.

The literature on *Dicroidium* includes papers by Townrow (1957), Pigg (1990), Taylor et al. (1992), and Taylor and Taylor (1990, 2009).

11.4.3 Cycads and Other Gymnosperms

The so-called cycads are gymnospermous trees having compound leaves and naked seeds borne separately on sporophylls or in simple cones. They evolved during the Permian Period and thrived during the Mesozoic Era. Several genera of the order Cycadales have survived to the present. The cycads are palm-like trees with tall trunks composed of a large pithy core surrounded by soft woody tissue. They have crowns of large leaves that grow from the top of the stem in a way that is still characteristic of palm trees today. The cycad stems on Fremouw Peak were described by Taylor and Smoot (1984, 1985a), Smoot et al. (1985), and Taylor et al. (1986a).

In addition to the cycads, the collecting sites at Skaar Ridge and on Fremouw Peak contain silicified wood of gymnospermous trees that reached a height of up to 22 m (Barrett 1969). A silicified trunk of a gymnospermous tree in Fig. 11.15 is partly buried in sand and Fig. 11.16 shows the upright stump of a tree in the Gordon Valley of the Queen Alexandra Range that has weathered out of the sedimentary rocks in which it was embedded. This tree stump is part of a "forest" of about

Fig. 11.15 The silicified trunk of a tree lies partly buried in sandstone of a former stream channel in the Queen Alexandra Range of the Transantarctic Mountains (Photo by T.N. Taylor and E.L. Taylor, reproduced with permission. This image was published in *Paleobotany* by Thomas N. Taylor, Edith L. Taylor, and Michael Krings, second edition. Figure 1.4, p. 3. Copyright Elsevier, 2009)

Fig. 11.16 A silicified tree stump in the growth position was found in the upper Fremouw Formation (Middle Triassic) at the head of the Gordon Valley at 84°11'S, 164°54'E in the Queen Alexandra Range. The height of the stump is about 1 m. This photograph was first published in 1991 on the cover of the Antarctic Journal of the United States (Vol. 16, No. 5) and in the report by Taylor et al. (1991) (Photo by R. Cuneo and T.N. Taylor reproduced with permission. This image was published in *Paleobotany* by Thomas N. Taylor, Edith L. Taylor, and Michael Krings, second edition. Figure 1.54, p. 29. Copyright Elsevier, 2009)

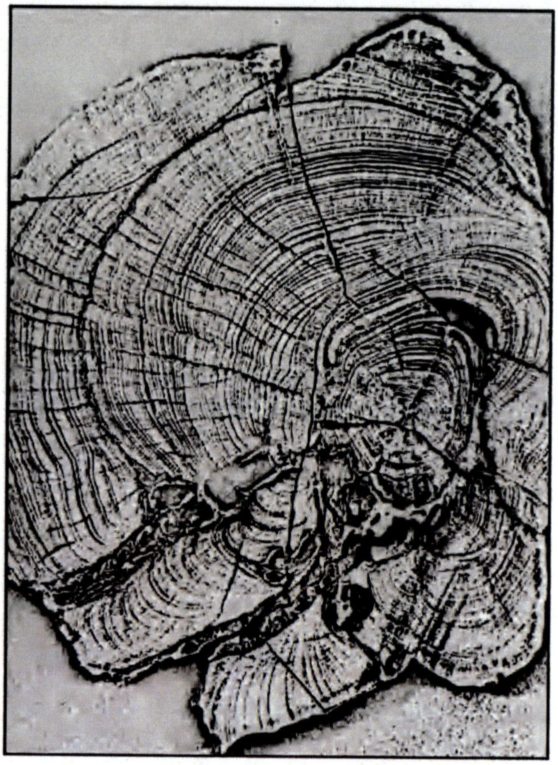

Fig. 11.17 The cross-section of a stem of *Antarctixylon* sp. collected by P.J. Barrett from the Early Triassic Fremouw Formation on Fremouw Peak in the Queen Alexandra Range. The width and shape of the rings constitute a climate record that is difficult to interpret in this case. High-contrast reproduction of a photo of the etched surface of specimen (CB 365-3 (B-2) by Schopf (1973)

100 tree stumps in an area of about 3,900 m^2. The diameters of these trunks range from 13 to 61 cm and the largest tree had 86 growth rings. The bedding plane around the trees in this forest is littered with *Dicroidium* which probably grew on these trees. The only other forest of petrified trees in Antarctica occurs in the Early Cretaceous rocks of Alexander Island (71°00'S, 70°00'W) located off the coast of western Palmer Land on the Antarctic Peninsula (Jefferson 1982).

Jefferson and Taylor (1983), Taylor and Smoot (1983), and Taylor et al. (1983) reported that the cells and growth rings of the wood of petrified trees in the Queen Alexandra Range were anatomically preserved. A microscopic examination revealed that the trees on Fremouw Peak were damaged by fire in the Early Triassic before they were petrified. In addition, Stubblefield and Taylor (1985, 1986) later reported that wood of *Araucarioxylon*-type and *Vertebraria* at Fremouw Peak and at Skaar Ridge (Mt. Augusta) was attacked by fungi that left holes in the wood. An additional example of damage to the stem of a gymnospermous tree was uncovered by Schopf (1973) in Fig. 11.17.

Even more important to the advancement of botany is evidence which suggests that seed ferns may be the "missing link" between the gymnosperms and the angiosperms. The critical evidence takes the form of a small seed-bearing capsule in the silicified peat at

Fremouw Peak described by Taylor and Taylor (1987). The capsule containing the seeds is open on one side, indicating that it was part of a gymnospermous plant. However, the seeds were at least partly covered suggesting that the plant was at least partly angiospermous. Therefore, the seed-bearing ferns at Fremouw Peak described by Taylor and Taylor (1987) and by Millay et al. (1987) may be the evolutionary ancestors of the flowering plants as postulated by Zavada and Taylor (1986). However, Retallack and Dilcher (1981) proposed that the angiosperms evolved from the glossopterids.

11.4.4 Palynomorphs

Taylor et al. (1988) also recovered a large number of palynomorphs (pollen, spore, and acritarchs) from shale and coal of the Falla Formation. These palynomorphs are unusually well preserved compared to the material previously studied by Askin and Fasola (1978) and by Askin and Schopf (1982). The microflora recovered by Taylor et al. (1988) indicates a Late Triassic (Norian) age for the Falla Formation. These palynomorphs may be used to define a biostratigraphy which could provide better stratigraphic control than is possible with plant megafossils. This possibility enhances the relevance of the studies by paleobotanists who have recovered and classified small organic particles in the rocks of the Beacon Supergroup (e.g., Norris 1965; Helby and McElroy 1969; Kemp 1972, 1975; Askin 1977; Askin and Schopf 1977, 1978, 1982; Askin and Fasola 1978; Tasch 1977, 1978; Farabee et al. 1989).

11.4.5 Permo-Triassic Climate

The climate of Antarctica during the Permo-Triassic was at least partly determined by its proximity to the geographic south pole which affected the average annual temperature and the duration of darkness during the winter. Nevertheless, the geological evidence indicates that sand and mud were deposited by liquid water that flowed freely across alluvial plains. In addition, the *flora* implies that the sedimentary rocks of the Victoria Group were deposited under temperate climatic conditions.

For example, Jefferson (1983) relied on the size and form of *Glossopteris* leaves in the Permo-Triassic rocks of the Transantarctic Mountains to conclude that these plants had grown during a warm and frost-free growing season. In addition, the well-defined growth rings of *Araucarioxylon*, up to 8 mm wide, suggest a warm climate and long growing season. Collinson (1997) agreed that the sedimentological and paleontological evidence indicates that the climate in the Beacon basins of Antarctica warmed significantly from cold-glacial in the Early Permian, to cool-humid during the Middle and Late Permian, to warm-with-seasonal-moisture during the Triassic.

These considerations constrain the paleolatitudes of the Beacon basins during the Permo-Triassic to be *lower* than they are at present. For example, the present climate of the Queen Alexandra Range, which is located between latitudes 84°S and 85°S, does *not* sustain vegetation, which *did* grow there during the Permo-Triassic. Therefore, the flora that has been preserved in the Beacon rocks of the Queen Alexandra Range is *inconsistent* with the present latitude of the area and, instead, requires a warmer climate and hence a lower latitude. Although the climate of an area does not depend on its geographic latitude alone, we note that trees do not grow in the tundra of the Arctic regions of North America and Asia north of about 70°N latitude. Therefore, the Permo-Triassic latitude was probably less than about 70°S.

Crowell (1999) used paleomagnetic data of Bachtadse and Briden (1990) and of Chen et al. (1993) to construct a polar wander path for the Paleozoic which shows that during the Early Permian the magnetic pole was off the coast of Victoria Land. Subsequently, the magnetic pole moved farther away from the Antarctic mainland which might have caused the climate in the Beacon basins to improve, provided that the magnetic pole remained in the vicinity of the geographic pole. This is not always the case because, at the present time, the magnetic pole has been moving away from the geographic south pole at a rapid rate (Cullington 1963; Burrows 1963). When the magnetic pole of Antarctica was discovered during the British Antarctic (Nimrod) Expedition of 1907–1909, it was located at 72°25'S, 155°16'E near the Talos Dome of the East Antarctic Ice Sheet adjacent to northern Victoria Land (Section 1.4.2). At that time, it was located about 2,000 km from the geographic south pole. In 1914 the magnetic pole had moved to 71°S, 150°E and in 1990 it was located at 65°S, 139°E in the ocean off the Adélie Coast of East

Antarctica and was more than 2,800 km from the geographic pole (Stonehouse 2002). However, it is not clear how long the separation of the magnetic pole from the geographic pole can last because the magnetic pole is constrained by the rotation of the Earth to remain in the vicinity of the geographic pole.

Jefferson (1983) attempted to explain the apparent polar warming during the Permo-Triassic by postulating that the inclination of the axis of rotation relative to the ecliptic plane (i.e., the axial obliquity) had decreased. If true, the area circumscribed by the Antarctic circle would be reduced. As a result, the amount of Sun light received by the polar regions of the Earth would increase, thus warming the polar regions (Faure and Mensing 2007).

Collinson (1997) attributed the appearance of amphibians and reptiles in Antarctica during the Triassic to global warming caused by the Milankovitch cycles resulting from periodic changes in the eccentricity of the orbit of the Earth, the tilt of the axis of rotation (axial obliquity), and the longitude of perihelion all of which have periodicites of about 100,000, 41,000, and 23,000 years, respectively. Collinson (1997) also noted that global climate models fail to explain the warm and moist climate of the Beacon basins of Antarctica during the Triassic because they do not take into account the warming of the atmosphere by the water of the Panthalassa Ocean.

Perhaps we can invoke Uniformitarianism by pointing out that North America and northern Eurasia were in the grip of continental ice sheets only about 10,000–15,000 years ago and that the climate of these areas has warmed up and now supports abundant vegetation without requiring a major change in latitude because the circulation of the oceans and of the atmosphere distribute solar heat and thereby determine the climate in different parts of the Earth.

11.5 Mineral Deposits

Antarctica undoubtedly contains mineral deposits that formed in the course of its long geologic history. The vast amount of ice on the continent (Section 2.6) and the coal seams that are interbedded with the rocks of the Beacon Supergroup have already been mentioned in Section 10.3.5. In addition, Ravich et al. (1982) described Precambrian iron deposits in the Prince Charles Mountains located west of the Lambert Glacier and the Amery Ice Shelf of East Antarctica, and Ford (1983) evaluated the potential mineral resources of the Dufek Massif in the Pensacola Mountains. The existence of these and presumably other potentially valuable mineral deposits in Antarctica forced the Treaty Nations to make plans about how they should respond in case mining companies decided to search for mineral deposits in Antarctica, especially if they wanted to start full-scale mining operations of large ore deposits. Another cause for alarm was the proposal to use Antarctica as a repository of highly radioactive waste produced by nuclear power reactors and by the nuclear weapons industry (e.g., Angino and Zeller 1977).

This issue aroused concern because mining operations or the construction of a nuclear-waste repository site in Antarctica would undoubtedly disturb the environment (Anonymous 1977; Zumberge 1977, 1979a, b; Elliot 1977), because questions of sovereignty could arise again (Triggs 1987), and because nations which had not signed the Antarctic Treaty demanded that the benefits of the natural resources of Antarctica be shared among all nations of the world.

11.5.1 CRAMRA

Negotiations concerning exploration and exploitation of mineral deposits in Antarctica were started in 1982 during the Fourth Special Antarctic Treaty Consultative Meeting in Wellington, New Zealand, and continued for 7 years. The resulting *Convention on the Regulation of Antarctic Mineral Resource Activities* (CRAMRA) was finally presented in 1988 at a Consultative Meeting of the Treaty Nations. CRAMRA contained 67 articles which attempted to regulate mining operations with a strong emphasis on environmental protection. Nevertheless, it was attacked both by environmentalist, who wanted to ban all mining operations in Antarctica, and by representatives of the mining industry who considered CRAMRA to be overly complicated and unwieldy. In addition, CRAMRA provided no assurance or incentives to mining companies that might consider investing large sums of money for exploration and production of mineral resources in Antarctica. As result, American mining companies were not interested in exploring for minerals in Antarctica, although some interest has persisted in the exploration for oil and gas in the Ross Sea and on the continental shelf of Antarctica (St. John 1990).

CRAMRA was also unpopular because only the nations that had signed the Antarctic Treaty would be allowed to mine in Antarctica. Consequently, the United Nations voted without dissent in favor of making Antarctica a protected wilderness area where no nation is allowed to search for and to exploit mineral deposits. At about the same time, an international forum consisting of diplomatic representatives of 35 nations voted unanimously against mineral exploration and in favor of declaring Antarctica to be a wilderness area. In addition, the Kerry/Conte bill enacted by the US Congress makes it unlawful for an American citizen to carry out commercial mineral activities in Antarctica. Although the proponents of CRAMRA claimed that its principal purpose was to protect the environment, the environmentalists stated that the only way to protect the environment of Antarctica is to ban mining and other dangerous activities altogether.

The vigorous discussion of the issues raised by CRAMRA made many people aware of the value Antarctica has for scientists and tourists and of the environmental damage that would result from mining operations no matter how well they were regulated. Two of the original treaty nations (France and Australia) refused to ratify CRAMRA and thereby prevented it from being adopted.

Some of the provisions of CRAMRA that dealt with environmental protection were used to frame the *Protocol of Environmental Protection to the Antarctic Treaty*. Article 7 of the Protocol prohibits all activities relating to mineral resources except for the purpose of scientific research. In this way, the Treaty Nations have been prohibited from mining and drilling for oil since 1991 when the Protocol to the Antarctic Treaty went into effect. In addition, the Protocol now guides all activities undertaken by the US Antarctic Program (Section 2.9). The text of the Protocol on Environmental Protection can be read in the book published by Stonehouse (2002, Appendix D).

11.5.2 Inventory of Mineral Deposits

The first descriptions of Antarctic mineral resources by Chalmers (1957), Mueller (1964), Stewart (1964), McLeod (1965), Potter (1969), and Smith (1972) was followed by the publication of a comprehensive list of Antarctic mineral deposits by Wright and Williams (1974). At about the same time, Goodell et al. (1971) published an influential paper on the ferromanganese deposits in the South Pacific, the Drake Passage, and the Scotia Sea. Manganese nodules on the ocean floor are a potential source of nickel, cobalt, and copper although they are composed of oxides of manganese and iron (Glasby 1976). In spite of great expectations, manganese nodules are not currently recovered in the Southern Ocean or anywhere else because of environmental issues, excessive cost, questions of sovereignty, and lack of sufficient demand.

The world-wide gasoline shortage in the 1970s also prompted questions in the press (e.g., the Wall Street Journal) about finding oil or gas in Antarctica (e.g., Spivak). Such questions were stimulated by reports of hydrocarbon gas in cores taken in the Ross Sea (e.g., McIver 1975). Soon thereafter, Holdgate and Tinker (1979) published a book on the occurrence of oil and gas in Antarctica and Ivanhoe (1980) speculated in the Oil and Gas Journal about the operating conditions for the recovery of oil in Antarctica. Although an impressive amount of geophysical information has been gathered off the coast of the Antarctic Peninsula and in the Ross Sea, no giant oil or gas fields have been discovered that could motivate an effort to re-open the debate about exploitation of natural resources of Antarctica (Behrendt 1983; St. John 1990).

The radar-echo surveys of the Antarctic ice sheet by Drewry (1975) provided information about the sub-ice topography but did not detect mineral deposits. More recently, Behrendt et al. (1980) combined the radar-echo sounding technique of Drewry (1975) with an airborne magnetometer survey which revealed that a large part of the Dufek intrusion in the Pensacola Mountains is covered by the East Antarctic ice sheet. The enlargement of the area of this intrusion increased the amount of potential ore of iron, titanium, and vanadium in this differentiated body of gabbro, pyroxenite, and anorthosite.

Prior to the adoption in 1991 of the Environmental Protocol, which prohibits the search for mineral deposits and their exploitation, a large number of papers were published about metallic mineral deposits in all parts of Antarctica and about the occurrence of oil and gas off-shore. These publications are listed in Appendix 11.7.1 under the following headings:

(a) Antarctic Peninsula and off-shore islands
(b) Ellsworth Mountains, West Antarctica
(c) Transantarctic Mountains (excluding coal)
(d) Petroleum: Ross Sea, ice shelves, and TAM
(e) Antarctica, whole continents

11.5.3 Radioactivity Surveys

Sedimentary rocks that contain potassium, as well as uranium and thorium and their respective radioactive daughters emit gamma rays as the unstable atoms of these elements decay (Faure and Mensing 2005). Uranium and thorium occur in the "heavy minerals" that are included in sandstones and conglomerates at the time of deposition. In addition, uranium (but not thorium) may be deposited from groundwater in case the sediment contains organic matter that can reduce hexavalent uranyl ions $\left(UO_2^{+2}\right)$ to the tetravalent state (U^{+4}). Compounds of U^{+4} are deposited with the sediment because they are far less soluble than compounds of UO_2^{+2}. The uranium in ground water is released into solution by chemical weathering of plutonic and volcanic igneous rocks. The fixation of tetravalent uranium in sedimentary rocks of the Beacon Supergroup became possible because land-based plants that evolved during the Devonian Period provided organic matter that was buried with the sediment.

LeRoux and Toens (1987) classified the uranium deposits of the late Paleozoic sedimentary rocks of Gondwana into four types: Sandstone-hosted, shale-hosted, coal-hosted, and vein-hosted. Sandstone-hosted uranium deposits occur in Argentina and Brazil, in the Karoo basin of South Africa (Blackstroem 1974), in Madagascar, India, and Australia. The uranium concentrations of shale and coal in the sedimentary rocks of Gondwana, in most cases, are low and do not reach ore-grade (Gilbert et al. 1984). Vein-hosted uranium deposits are associated with granitic intrusives and with complex pegmatites on all of the southern continents including Antarctica.

The sandstones of the Beacon Supergroup include several kinds of "heavy minerals" that contain uranium and thorium either as primary constituents or by replacement of other elements. For example, zircon ($ZrSiO_4$) always contains ions of uranium and thorium because they can substitute for Zr^{+4} (Faure 1998). Other heavy minerals that commonly contain uranium and thorium include monazite, apatite, rutile, sphene, etc. These uranium- and thorium-bearing minerals in the Beacon sandstones originated as accessory minerals in granitic rocks that were exposed to weathering and erosion in the areas adjacent to the basins of deposition. Uranium and thorium are concentrated during the fractional crystallization of granitic magma and can reach high concentrations in small volumes of residual magma that crystallize as late-stage pegmatite dikes or quartz veins.

Because of the wide-spread occurrence of uranium and thorium in all kinds of rocks and because uranium is used to fuel nuclear fission reactors, a thorough evaluation of the natural resources of Antarctica should include a search for deposits of uranium that can be detected by the gamma-ray flux that emanates from uranium- and thorium- bearing rocks. This task was undertaken by Ed Zeller and Gisella Dreschhoff of the University of Kansas in Lawrence starting in 1977 and continuing well into the 1980s. Preliminary reports of their results were published in the Antarctic Journal of the US and in other technical journals: Zeller et al. (1977, 1979, 1986), Dreschhoff et al. (1980, 1981, 1982, 1983a, b), Dreschhoff and Zeller (1986), and Zeller and Dreschhoff (1980, 1987).

Zeller et al. (1982) described the gamma-ray detector they used in the field which employed a large crystal of thallium-bearing sodium iodide (1,835 cm^3) (Lövborg et al. 1976). The equipment was mounted in one of the helicopters (UH-1/N) stationed in McMurdo (Darnley 1971). The results of an airborne traverse over Nussbaum Riegel in Taylor Valley in southern Victoria Land closely matched measurements taken on the ground and reflected changes in the composition of the rocks along the traverse.

The average counting rates of different rock types in Table 11.5 indicate that pegmatites and lit-par-lit granitoids containing closely-spaced pegmatite dikes have the highest average gamma-ray activity of 4.1 relative to ice = 1.0. The pegmatites (and lit-par-lit granitic gneisses) are followed by late-stage Vanda Porphyry dikes (3.7), augen gneisses (3.0), and Irizar Granite (2.6). Metasedimentary rocks (2.4) and marble (2.2) have the lowest average gamma-ray counting rate among the basement rocks. Basalt of the McMurdo Volcanics and the Ferrar Dolerite also have low average counting rates of 2.0 and 1.9, respectively. These mafic igneous rocks are known to have low concentrations of potassium, uranium, and thorium. Trachytes are enriched in potassium and uranium compared to cogenetic basalt which accounts for their high gamma counting rate of 3.7 compared to ice = 1.0.

The counting rates of the Beacon rocks appear to depend on the concentration of organic matter they contain. For example, the average gamma-ray counting rate of Devonian rocks of the Beacon

Table 11.5 Average gamma-ray counting rates relative to ice based on airborne measurements in the Transantarctic Mountains (Zeller et al. 1982)

Rock type	Average counting rate relative to ice = 1.0
McMurdo Volcanics	
Basalt	2.0
Trachyle	3.7
Ferrar Dolerite	
Dolerite	1.9
Beacon Supergroup	
Triassic rocks	2.3
Perm. coal meas.[a]	2.1
Aztec/fishbeds	2.6
Devon. Rocks	1.8
Basal Beacon	2.5
Pegmatites[b]	4.1
Vanda Porphyry	3.7
Irizar Granite	2.6
Metasediments	2.4
Marble	2.2
Augen gneiss	3.0

[a]Radiological analyses of samples of Permian Triassic coal from the central Transantarctic Mountains and from the Queen Maud Range were published by Gilbert et al. (1984)

[b]Included lit-par-lit injection granitic gneisses

Supergroup is only 1.8, whereas the counting rate of the fishbeds of Devonian Aztec Formation is 2.6. The high counting rate of the basal Beacon rocks (2.5) in Table 11.5 may be caused by an abnormally high abundance of heavy minerals in the basal sandstones and conglomerates directly above the Kukri Peneplain. The Permian coal measures and overlying Triassic rocks have comparatively low counting rates of 2.1 and 2.3, respectively.

The surveys of radioactivity of Zeller and Dreschhoff included areas in southern and northern Victoria Land (Dreschhoff et al. 1983a; Zeller et al. 1986), the Ellsworth Mountains (Dreschhoff et al. 1980, 1992), and the Scott Glacier area of the Queen Maud Mountains (Dreschhoff et al. 1983b). In spite of the large areas that were surveyed in each region, no noteworthy uranium deposits were found.

Although the gamma-ray flux emanating from glacial *ice* is a convenient reference for radiological measurements of rocks, it is not actually constant. Zeller et al. (1977) reported that the counting rate of snow exposed in a pit at Williams Field on the Ross Ice Shelf near McMurdo declined with depth below the surface but increased at 1.83 and 2.44 m. The authors suggested that the increase was caused by radioactive fallout from the testing of nuclear bombs by France in 1970–1971 at Mururoa Island in the South Pacific.

In the following year, Dreschhoff and Zeller (1978) reported a similar result from a snow pit in Marie Byrd Land in West Antarctica where they found a peak in the activity profile at a depth of approximately 3.3 m. Subsequently, they tested the snow in a pit near South Pole Station which was more than 10 m deep. The resulting gamma-ray activity profile in Fig. 11.18

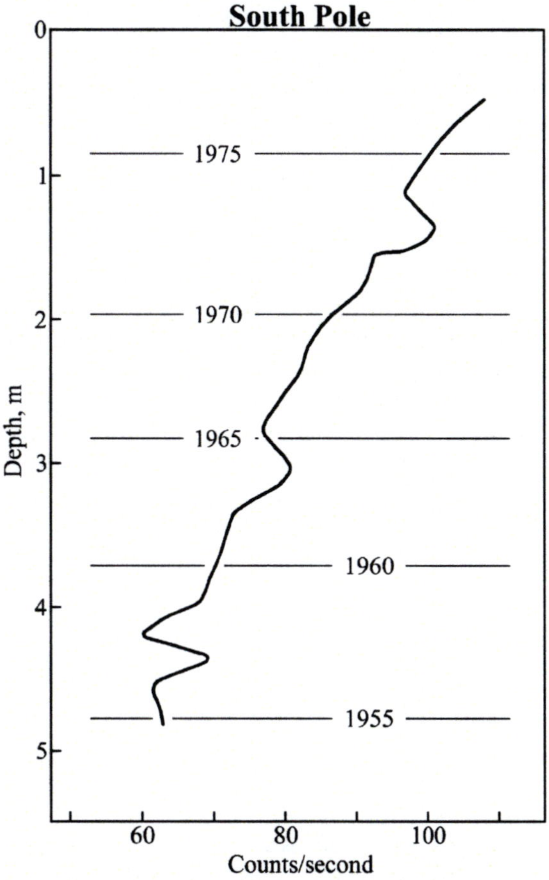

Fig. 11.18 The gamma-ray counting rate of snow and firn in a pit close to South Pole Station decreases with depth but rises at 1.4, 3.1, and 4.4 m. These peaks represent fallout from the testing of nuclear weapons between 1970/71 at Mururoa Island (1.4 m), thermonuclear detonations in September of 1961 (3.1 m), and tests that ended in 1958 (4.4 m) (Picciotto and Wilgain 1963). The background activity decreased with increasing depth because the detector was increasingly shielded from cosmic rays. The year of deposition of the snow was determined by counting layers in the walls of the pit (Adapted from Dreschhoff and Zeller 1978)

has three peaks at depths of 1.4, 3.1, and 4.4 m below the surface. Dreschhoff and Zeller (1978) attributed the peak at 1.4 m to fallout resulting from nuclear weapons tests during 1970/71 at Mururoa Island, the peak at 3.1 m to the explosion of thermonuclear devices, and the peak at 4.4 m to nuclear-bomb tests that ended in 1958. The decrease of the background radiation with depth, evident in Fig. 11.18, was caused by increased shielding of the detector from cosmic rays at increasing depth in the snow pit.

Fission products were also used by Picciotto and Wilgain (1963) as a reference for measuring snow accumulations in Antarctica. Additional fission products were deposited in Antarctica after the accidental explosion of a nuclear reactor in Chernobyl, Ukraine, on April 26 of 1986 (Faure et al. 1997).

11.5.4 Bituminous Coal

Coal occurs throughout the Transantarctic Mountains in the Victoria Group of the Beacon Supergroup as well as in all of the southern continents that were once part of Gondwana (Section 10.3.5). The processes by which coal forms from accumulated plant debris were described by Schopf (1948), who also provided a precise definition of coal (Schopf 1956). He recommended procedures for sampling coal beds (Schopf 1960), and specified its different evolutionary stages consisting of peat, lignite, bituminous coal, and graphite (Schopf 1966b). After W.E. Long collected plant fossils and coal on Mt. Glossopteris in the Ohio Range in 1958 (Long 1959), Schopf (1962) described the fossils and subsequently teamed up with W.E. Long to describe the alteration of coal by contact metamorphism (Schopf and Long 1966).

The coal deposits in the Gondwana basins of India were extensively discussed during the Fourth International Gondwana Symposium in 1977 (Laskar and Raja Rao 1979) which included a paper by Sanyal and Subramanian (1979) on the petrology of coal deposits of India. The environment of deposition of the Indian coal measures was discussed by Ghosh (1975) and by Chandra and Taylor (1975). Additional reports about the coal deposits of Brazil, South Africa, and Australia were presented at the Seventh International Gondwana Symposium in 1988 (Ulbrich and Rocha Campos 1991) and by Hobday (1987). Coal deposits of Permian age were also included in the proceedings volumes of subsequent Gondwana symposia edited by Oliver et al. (1983), Findlay et al. (1993), Ricci (1997), and Gamble et al. (2002). The petrology of coal was explained by Stach et al. (1982) and its importance as a source of plant fossils was detailed by Schopf (1982) and by Taylor, Taylor, and Krings (2009).

The coal deposits of Antarctica were identified by Splettstoesser (1970, 1980) based partly on the compilation of Antarctic natural resources (including coal) by Wright and Williams (1974). The list of locations presented by Splettstoesser (1980) includes 20 sites, including northern Victoria Land (Walker 1983; Collinson and Kemp 1983), southern Victoria Land (Mulligan et al. 1963a, b; McKelvey et al. 1977; Whitby et al. 1983), the Buckley (Permian) and Fremouw (Triassic) formations of the Beardmore Glacier area (Barrett et al. 1986), and the Ohio Range of the Horlick Mountains (Long 1964, 1965, Fig. 10.22). In addition, minor amounts of coal occur in the Permian Polarstar Formation in the Ellsworth Mountains (Webers et al. 1992; Collinson et al. 1992), in the Pensacola Mountains (Schmidt and Ford 1969), in the Theron Mountains (Brown and Taylor 1960), in the Permian Amelang Formation in western Queen Maud Land (Bauer et al. 1997), and along the west coast of the Amery Ice Shelf adjacent to the Prince Charles Mountains of East Antarctica (Bennett and Taylor 1972). All of these coal deposits formed during the Permo-Triassic after the end of the Permo-Carboniferous glaciations of the Gondwana. An unusually thick deposit of Permian coal can be seen in Fig. 11.19. The Devonian formations of the Taylor Group in southern Victoria Land, in the Beardmore Glacier area, and in the Pensacola Mountains do not contain coal (Wizevich 1997; Barrett et al. 1986; Schmidt and Ford 1969).

The potential for mining coal in the Transantarctic Mountains and perhaps in the Prince Charles Mountains of East Antarctica has been discussed by several authors including Splettstoesser (1980), Kameneva and Grikurov (1983), and Rose and McElroy (1987). All have mentioned the logistical difficulties imposed by the harsh weather, long periods of total darkness in the winter, the remote location, and the need for long-distance transport of the coal. These factors would increase the cost of mining and cause it to exceed the commercial value of the product, which makes mining coal in Antarctica economically unattractive. Nevertheless, the rising demand for energy in the world, especially by nations that do not have indigenous deposits

Fig. 11.19 An unusually thick deposit of bituminous coal of Permian age in the Beardmore Glacier area of the Transantarctic Mountains. The coal at this location is 2.7 m thick and is overlain by light-colored sandstone of the Buckley Formation. The chemical composition of coal from several sites in the Transantarctic Mountains is given in Section 10.3.5 (Photo by T.N. Taylor and E.L. Taylor, reproduced with permission. This image was published in *Paleobotany* by Thomas N. Taylor, Edith L. Taylor, and Michael Krings, second edition. Figure 1.30, p. 19. Copyright Elsevier, 2009)

of fossil fuel, continues to generate interest in the coal deposits of Antarctica. Most important of all, Article 7 of the Protocol on Environmental Protection to the Antarctic Treaty states (Stonehouse 2002, Appendix D, p. 315):

> Any activity relating to mineral resources, other than scientific research, shall be prohibited.

The intention of the Protocol to the Antarctic Treaty is very clear. Any activity designed to find, evaluate, and exploit a mineral resource in Antarctica is a violation of the Antarctic Treaty.

11.6 Summary

The Permian and Triassic rocks of the Victoria Group of the Transantarctic Mountains were deposited on land rather than in the oceans and have therefore preserved a record of land-based plants and animals. Although the Beacon Supergroup consists predominantly of clastic sedimentary rocks, carbonates in the form of lenticular beds and concretions occur at most stratigraphic levels. Isotopic compositions of strontium, oxygen, and carbon indicate that these carbonate rocks were not deposited by seawater. Instead, the carbonate rocks formed during burial and diagenesis of the sediment from pore water which contained strontium in solution that was enriched in radiogenic ^{87}Sr. The water molecules were depleted in ^{18}O and the bicarbonate ions in the pore water were variously depleted in ^{13}C. Calcite also occurs in the form of cleats that formed in joints and on bedding planes of coal seams in the overlying layers of sandstone.

The study of glacial deposits on the southern continents indicates that glaciers first formed during the Early Carboniferous in the area of Gondwana that later became South America. The resulting ice sheets expanded and retreated as Gondwana drifted south until the coast of Victoria Land was located close to the geographic South Pole. The area that is now occupied by the Transantarctic Mountains consisted of several basins that were located along the margins of the ice sheet. When the ice sheet began to disintegrate during the Early Permian, till deposited by the ice sheet was interbedded with sandstone and shale that formed in periglacial meltwater lakes and marine embayments. During the waning stages of the glaciation of Gondwana the ice flowed southeast from a center in Victoria Land into the Beardmore basin, but the glaciation of Gondwana did not start in Antarctica and the ice sheet may not have spread onto the East Antarctic craton.

After the glaciation had ended, the climate of Antarctica improved and sustained an abundant flora of gymnospermous trees, cycads, ferns, and horsetails (calamites) which grew on the swampy alluvial plains along the coast. The flora is characterized by leaves of *Glossopteris*, *Gangamopteris*, and *Dicroidium* which were preserved as impressions in fine-grained sediment and as carbonized films on bedding planes. The stems of tree-like plants, some of which are still in the upright growth position, were preserved by silicification. Two deposits of silicified peat in the Queen Alexandra Range have provided valuable anatomical information about the Permo-Triassic plants.

The Early Triassic Fremouw Formation has yielded fossil bones of a wide variety of land-based amphibians and reptiles that inhabited the forests on the coastal plain of East Antarctica. The collecting sites in the Beardmore and Shackleton glacier areas of the central Transantarctic Mountains include Graphite Peak, Coalsack Bluff, the Cumulus Hills, and the Gordon Valley. The most famous Triassic reptile is *Lystrosaurus* which was a plump little animal about 90 cm long that seems to have been well adapted to digging burrows in the soil.

Lystrosaurus and most of the plants and animals that inhabited the Triassic forests of East Antarctica were the descendants of ancestors who survived the mass extinction at the end of the Permian Period. The cause of that catastrophe during which 90% of all life forms perished is still unknown. The best explanation is that the extinction resulted from the consequences of the impact of an asteroid which triggered large-scale volcanic eruptions at the antipode opposite the impact point. The volcanic eruptions that followed prolonged the disturbance of the global climate because the dust shielded the surface of the Earth from sunlight and thereby prevented photosynthesis by green plants and caused cooling of the global climate.

The harsh climate and remote location of present-day Antarctica have shielded it from the exploitation of its natural resources except for whales, seals, and fish in the Southern Ocean. The natural resources of Antarctica include the ice that covers most of the continent, large deposits of bituminous coal, and as yet unexplored reservoirs of petroleum and natural gas. In addition, Antarctica is known to contain large deposits of iron ore in the Prince Charles Mountains and in the Dufek Massif which may also contain titanium, vanadium, and chromium. None of these deposits has yet been exploited because the cost of production is likely to exceed the value of the materials that are recovered, because of the danger of serious accidents in a hostile environment, and because of the unavoidable but undesirable contamination of the environment. Last but not least, the governments of the Antarctic Treaty Nations have prohibited all activities relating to mineral resources in Antarctica.

11.7 Appendix

11.7.1 List of Publications Concerning Metallic Mineral Deposits in Different Regions of Antarctica by Year of Publication

A. Antarctic Peninsula and Off-Shore Islands

1958: Caballero, M.A. and N.H. Fourcade. Obsevaciones geologicas en Caleta Copper Mine, Islas Shetland del Sur., Antartida Argentina. Inst. Antart. Argentino, Contr. 23:1–39

1974: del Valle, R., J. Morelli, and C. Rinaldi. Manifestacion cupro-plumbifera "Don Bernabe", Isla Livingston, Islas Shetland del Sur, Antartida Argentino. Inst. Antart. Argentino, Contr. 175:1–35

1975: Rowley, P.D., P.L. Williams, D.L. Schmidt, R.L. Reynolds, A.B. Ford, A.H. Clark, E. Farrar, and S.L. McBride. Copper mineralization along the Lassiter Coast of the Antarctic Peninsula. Econ. Geol., 70(5):982–987

1977: Rowley, P.D., P.L. Williams, and D.L. Schmidt. Geology of an Upper Cretaceous copper deposit in the Andean Province, Lassiter Coast, Antarctic Peninsula. US Geol. Surv., Prof. Paper 984:1–36

1978: Littlefair, M.J. The "quartz-pyrite" rock of the South Shetland Islands, western Antarctic Peninsula. Econ. Geol., 73:1184–1189

1979: Vennum, W.R. Evaporite encrustations and yellow and green surficial salts from the Orville Coast and eastern Ellsworth Land. Antarctic J. US, 14(5):22–23

1980: Rowley, P.D. and W.R. Vennum, eds. Studies of the geology and mineral resources of the southern Antarctic Peninsula and eastern Ellsworth Land, Antarctica. US Geol. Surv. Prof. Paper 1351

1980: Vennum, W.R. Evaporite encrustations and sulphide oxidation products from the southern Antarctic Peninsula. New Zealand J. Geol. Geophys., 23(4):499–505

1980: Cox, C., R. Ciocanelea, and D.E. Pride. Genesis of mineralization associated with Andean intrusions, northern Antarctic Peninsula region. Antarctic J. US, 15(5):22–23

1981: Pride, D.E., S. Moody, and M. Rosen. Metallic mineralization, South Shetland Islands, Gerlache Strait, and Palmer Station. Antarctic J. US, 16(5):13–14

1981: Hawkes, D.D. and M.F. Littlefair. An occurrence of molybdenum, copper, and iron in the Argentine Islands, West Antarctica. Econ. Geol., 76:898–904

1982: Rowley, P.D. and D.E. Pride. Metallic mineral resources of the Antarctic Peninsula. In, C. Craddock, ed. Antarctic Geoscience, 859–870. University of Wisconsin Press, Madison, WI

1982: Vieira, C., B. Alarcón, J. Ambrus, and L. Olcay. Metallic mineralization in the Gerlache Strait region, Antarctica. In, C. Craddock, ed. Antarctic Geoscience, 871–876. University of Wisconsin Press, Madison, WI

1983: Rowley, P.D., E. Farrar, P.E. Carrara, W.R. Vennum, and K.S. Kellogg. Porphyry-type copper deposits and K-Ar ages of plutonic rocks of the Orville Coast and eastern Ellsworth Land, Antarctica. In, R.L. Oliver, P.R. James, and J.B. Jago, eds. Antarctic Earth Sci., 437. Australian Acad. Sci., Canberrra, A.C.T.

B. Ellsworth Mountains, West Antarctica

1981: Vennum, W.R. and J.M. Nishi. New Antarctic mineral occurrences. Antarctic J. US, 16(5):14–15

1992: Vennum, W.R. and J.M. Nishi. Chemical weathering of Cu, Fe, and Pb sulfides, southern Ellsworth Mountains, West Antarctica. In, G.F. Webers, C. Craddock, and J.F. Splettstoesser, eds. Geology and Paleontology of the Ellsworth Mountains. West Antarctica. Geol. Soc. Amer., Memoir 170:433–441

1992: Splettstoesser, J.F. Economic resources of the Ellsworth Mountains, West Antarctica. In, G.W. Webers, C. Craddock, and J.F. Splettstoesser, eds. Geology and Paleontology of the Ellsworth Mountains, West Antarctica. Geol. Soc. Amer., Memoir 170: p. 443, Boulder, CO

C. Transantarctic Mountains (Excluding Coal)

1983: Ford, A.B. The Dufek intrusion of Antarctica and a survey of its minor metals and possible resources. In, J.C. Behrendt, ed. Petroleum and Minerals Resources of Antarctica. US Geol. Surv., Circular 909:51–75. Alexandria, VA

1983: Ford, A.B., R.E. Mays, J. Haffty, and B.P. Fabbi. Reconnaissance of minor metal abundances and possible resources of the Dufek intrusion, Pensacola Mountains. In, R.L. Oliver, P.R. James, and J.B. Jago, eds. Antarctic Earth Science, 433–436. Australian Acad. Sci., Canberra, A.C.T

1984: Faure, G. and R.P. Felder. Lithium-bearing pegmatite and bismuth-antimony-lead-copper-bearing veinlets on Mount Madison, Byrd Glacier area. Antarctic J. US, 19(5):13–14

D. Petroleum: Ross Sea, shelves, and TAM

1974: Anonymous Antarctica: World hunger for oil spurs Security Council review. Science, 184(4138):776–779

1975: McIver, R.D. Hydrocarbon gases in canned core samples from Leg 28 sites 271, 272, and 273, Ross Sea. Initial Reports of the Deep Sea Drilling Project, 28:815–819. US Government Printing Office, Washington, DC

1979: Behrendt, J.C. Speculations on petroleum resource potential of Antarctica. Abstract. Amer. Assoc. Petroleum Geol., 63(3):418

1983: Behrendt, J.C. Petroleum and mineral resources of Antarctica. US Geol. Surv., Circular 909, Alexandria, VA

1983: Behrendt, J.C. Geophysical and geological studies relevant to the assessment of the petroleum resources of Antarctica. In, R.L. Oliver, P.R. James, and J.B. Jago, eds. Antarctic Earth Science 423–428. Australian Acad. Sci., Canberra, A.C.T.

1983: Behrendt, J.C. and C.D. Masters. Speculations on the petroleum resources of Antarctica. Abstract. US Geol. Surv., Polar Research Symposium, Circular 911:3–4

1983: Behrendt, J.C. Are there petroleum resources in Antarctica? In, J.C. Behrendt, ed. Petroleum and Mineral Resources of Antarctica, 1–24. US Geol. Surv., Circular 909, Alexandria, VA

1989: Anonymous. Oil in Antarctica. Nature, 337:490

1990: St. John, B., ed. Antarctica as an exploration frontier: Hydrocarbon potential, geology, and hazards. Amer. Assoc. Petroleum Geol., Studies in Geology, 31:1–154

1990: Anderson, J.B., P.G. Pope, and M.A. Thomas. Evolution and hydrocarbon potential of the northern Antarctic Peninsula continental shelf. In, B. St. John, ed. Hydrocarbon Potential, Geology, and Hazards. Amer. Assoc. Petroleum Geol., Studies in Geology, 31:1–12

1990: Jeffers, J.D. and J.B. Anderson. Sequence stratigraphy of the Bransfield Basin, Antarctica: Implications for tectonic history and hydrocarbon potential. In, B. St. John, ed. Hydrocarbon Potential, Geology, and Hazards. Amer. Assoc. Petroleum Geol., Studies in Geology, 31:13–29

1990: Reid, D.E. and J.B. Anderson. Hazards to Antarctic exploration and production. In, B. St. John, ed. Hydrocarbon Potential, Geology, and Hazards. Amer. Assoc. Petroleum Geol., Studies in Geology, 31:31–43

1990: Cooper, A.K., F.J. Davey, and K. Hinz. Geology and hydrocarbon potential of the Ross Sea, Antarctica. In, B. St. John, ed. Hydrocarbon Potential, Geology, and Hazards. Amer. Assoc. Petroleum Geol., Studies in Geology, 31:47–67

1990: Behrendt, J.C. Multichannel seismic reflection surveys over the Antarctic continental margin relevant to petroleum resource studies. In, B. St. John, ed. Hydrocarbon Potential, Geology, and Hazards. Amer. Assoc. Petroleum Geol., Studies in Geology, 31:69–75

1990: Wanneson, J. Geology and petroleum potential of the Adelie Coast margin, East Antarctica. In, B. St. John, ed. Hydrocarbon Potential Geology, and Hazards. Amer. Assoc. Petroleum Geol., Studies in Geology, 31:77–87

1990: Royer, J.-Y., L.M. Gahagan, L.A. Lawver, C.L. Mayes, D.T. Nürnberg, C.R. Sandwell. Scotese, A tectonic chart for the Southern Ocean derived from Geosat altimeter data. In, B. St. John, ed. Hydrocarbon Potential, Geology, and Hazards. Amer. Assoc. Petroleum Geol., Studies in Geology, 31:89–99

1990: Macdonald, D.I.M. and P.J. Butterworth. The stratigraphy, setting, and hydrocarbon potential of the Mesozoic sedimentary basins of the Antarctic Peninsula. In, B. St. John, ed. Hydrocarbon Potential, Geology, and Hazards. Amer. Assoc. Petroleum Geol., Studies in Geology, 31:101–125

1990: Gambôa, L.A.P. and P.R. Maldonado. Geophysical investigations in the Bransfield Strait and in the Bellingshausen Sea, Antarctica. In, B. St. John, ed. Hydrocarbon Potential, Geology, and Hazards. Amer. Assoc. Petroleum Geol., Studies in Geology, 31:127–141

1990: Collen, J.D. and P.J. Barrett. Petroleum geology from CIROS-1 drill hole, McMurdo Sound: Implications for the potential of the Victoria Land basin, Antarctica. In, B. St. John, ed. Hydrocarbon Potential, Geology, and Hazards. Amer. Assoc. Petroleum Geol., Studies in Geology, 31:143–151

E. Antarctica, Whole Continent

1965: McLeod, I.R. Antarctica: Geology and mineral occurrences. In, J.M. Dew, ed. Handbook, Australia and New Zealand. Eighth Commonwealth Mining and Metallurgical Congress, 165–166. Australasian Institute of Mining and Metallurgy, Melbourne, Victoria, Australia

1970: Runnels, D.D. Continental drift and economic minerals in Antarctica. Earth Planet. Sci. Letter, 8:400–402

1976: Erickson, G.E. Metallogenic provinces of the southeastern Pacific region. In, M.T. Halbouty, J.C. Maher, and H.M. Lilian, eds. Circum-Pacific Energy and Mineral Resources. Amer. Assoc. Petroleum Geol., Memoir 25:527–558

1976: Splettstoesser, J.F. Mining in Antarctica: Survey of mineral resources and possible exploitation methods. In, Proceedings, Third Internat. Conf., Port and Ocean Engineering under Arctic Conditions, 1137–1155. Un. Alaska Press, Fairbanks, Alaska

1976: Wade, F.AS. Antarctica: An unprotected, unexploited continent: Summary. In, M.T. Halbouty, J.C. Maher, and H.M. Lian, eds. Circum-Pacific Energy and Mineral Resources. Amer. Assoc. Petroleum Geol., Memoir 25:74–79

1976: Schofield, E.A. Antarctica up for grabs. Bull., Sierra Club, 61(6):17–19

1978: Dugger, J.A. Exploiting Antarctic mineral resources: Technology, economics, and the environment. Un. Miami Law Review, 33:315–339.

1979: Lovering, J.F. and J.R.V. Prescott, Last of lands: Antarctica. Melbourne University Press, Melbourne, Victoria, Australia

1979: Zumberge, J.H. Mineral resources and geopolitics in Antarctica. Amer. Scientist, 67:68–77

1979: Rowley, P.D. and P.L. Williams. Metallic mineral resources of Antarctica: Development far in the future. Abstract. Amer. Assoc. Petroleum Geol. Bull., 63(3):518

1983: Rowley, P.C. and D.E. Pride. Mineral occurrences of Antarctica. In, J.C. Behrendt, eds. Petroleum and Mineral Resources of Antarctica. US Geol. Surv., Circular 909, 25–49. Alexandria, VA

1983: Rowley, P.D., P.C. Williams, and D.E. Pride. Metallic and nonmetallic mineral resources of Antarctica. In, J.F. Splettstoesser, ed. Mineral Resource Potential of Antarctica. University Texas Press, Austin, TX

1983: Rowley, P.D., A.B. Ford, P.L. Williams, and D.E. Pride. Metallogenic provinces of Antarctica. In, R.L. Oliver, P.R. James, and J.B. Jago, eds. Antarctic Earth Science, 414–419. Australian Acad. Sci., Canberra, A.C.T.

1983: Splettstoesser, J.F. Mineral resource potential in Antarctica: Review and predications. Abstract. In, R.L. Oliver, P.R. James, and J.B. Jago, eds. Antarctic Earth Science, 413. Australian Acad. Sci., Canberra, A.C.T.

1983: Rowley, P.D., P.L. Williams, and D.E. Pride. Mineral resources of Antarctica. Abstract. US Geol. Surv., Polar Research Symposium, Circular 911:3

1983: Kameneva, G.I. and G.E. Grikurov. A metallogenic reconnaissance of Antarctic major structural provinces. In, R.L. Oliver, P.R. James and J.B. Jago, eds. Antarctic Earth Science, 420–422. Australian Acad. Sci., Canberra, A.C.T.

1985: Splettstoesser, J.F. Antarctic geology and minerals resources. Geology Today 1(2):41–45

1986: Shapley, D. The seventh continent: Antarctica in a resource age. Resources for the future, 1616 P Street NW, Washington, DC

1985: Weber, W. Nutzbare Mineralisationen und Rohstoffvorkommen in der Antarktis. Freiberger Forschungshefte, C 409. Geowissen-schaften: Geologie

1988: Shabecoff, P. Development seen for the minerals of all Antarctica. New York Times, June 8, 1988

1988: Molnia, B.F. Regulating Antarctic mineral development. Geotimes, 33(11):3–5

1988: Anonymous. Antarctic pact could open the way for mining. Science, 240:1612

1989: Ivanov, V.L. Evolution of Antarctic prospective sedimentary basins. Antarctic Science 1(1):51–56

1990: Splettstoesser, J.F. Antarctic geology, minerals, and politics: Implications for Treaty Review in 1991. Unpublished abstract. Geol. Soc. Amer., Forum, People, Rocks, and Penguins: Antarctic Mineral Resource Policy. Oct. 28, 1990 Hyatt Regency Hotel, Dallas, TX

1990: Splettstoesser, J.F. and G.A.M. Dreschhoff, eds. Mineral resources potential of Antarctica. Antarctic Research Series, vol. 51. Amer. Geophys. Union, Washington, DC

References

Anderson TF, Brownlee ME, Phillips TL (1980) A stable-isotope study on the origin of permineralized peat zones in the Herrin coal. J Geol 88:713–722

Angino EE, Zeller EJ (1977) Burial of high level radioactive wastes. Antarctic J US 12(4):125–127

Anonymous (1977) Antarctic resources: Effects of mineral exploration. Polar Rec 18(17):631–636

Askin (Kyle) RA (1977) Palynostratigraphy of the Victoria Group of south Victoria Land, Antarctica. New Zealand J Geol Geosphys 20(6):1081–1102

Askin (Kyle) RA, Fasola A (1978) Triassic palynology of the Beardmore Glacier area of Antarctica. In: La Palinologia, num. extraord. 1:313–139

Askin RA, Schopf JM (1977) Palynomorph preservation in the Beacon Supergroup of the Transantarctic Mountains. Antarctic J US 12(4):121–122

Askin RA, Schopf JM (1978) Palynologic studies in the Transantarctic Mountains. Antarctic J US 13(4):18–19

Askin RA, Schopf JM (1982) Permian and Triassic palynostatigraphy of the Victoria Group, Transantarctic Mountains. In: Craddock C (ed) Antarctic geoscience. University of Wisconsin Press, Madison, WI, pp 649–660

Babcock LE, Isbell JL, Miller MF, Hasiotis ST (2002) New late Pateozoic Conchostracan (Crustacea: Branchipoda) from the Shackleton Glacier area, Antarctica; Age and paleoenvironmental implications. J Paleont 76:70–75

Bachtadse V, Briden JC (1990) Paleomagnetic constraints on the position of Gondwana during Ordovician and Devonian times. In: McKerrow WS, Scotese CR (eds) Paleozoic palaeogeography and biogeography. Geological Society London, Memoir, 12. London, England, pp 43–48

Barrett PJ (1969) Stratigraphy and petrology of the mainly fluviatile Permian and Triassic Beacon rocks, Beardmore Glacier area, Antarctica. Institute of Polar Studies, Report 34. The Ohio State University, Columbus, OH

Barrett PJ, Askin RA (Kyle) (1975) The Early Permian glacial beds of southern Victoria Land and the Darwin Mountains. In: Campbell KSW (ed) Gondwana geology. Australian National University Press, Canberra, ACT, pp 333–346

Barrett PJ, Baillie RJ, Colbert EH (1968) Triassic amphibian from Antarctica. Science 161:460–462

Barrett PJ, Elliot DH, Lindsay JF (1986) The Beacon Supergroup (Devonian-Triassic) and Ferrar Group (Jurassic) in the Beardmore Glacier area, Antarctica. In: Turner MD, Splettstoesser JF (eds) Geology of the Central Transantarctic Mountains. Antarctic Research Series, vol. 36. American Geophysical Union, Washington, DC, pp 339–428

Basu AR, Petaev MI, Poreda RJ, Jacobsen SB, Becker L (2003) Chondritic meteorite fragments associated with the Permian-Triassic boundary in Antarctica. Science 302:1388–1392

Bauer W, Hagemann HW, Poscher G, Sachsenhofer RF, Spaeth G (1997) Permian coals from Western Dronning Maud Land: Composition, environment, and the influence of Jurassic magmatism on their maturity. In: Ricci CA (ed) The Antarctic region: Geological evolution and processes. Terra Antartica, Siena, Italy, pp 945–951

Behrendt JC, Drewry DJ, Jankowski E, Grim MS (1980) Aeromagnetic and radio-echo ice sounding measurements show much greater area of the Dufek intrusion, Antarctica. Science 209:1014–1017

Bennett AJR, Taylor GH (1972) Coals from the vicinity of the Prince Charles Mountains. In: Adie RJ (ed) Antarctic geology and geophysics. Universitetsforlaget, Oslo, Norway, pp 591–598

Blackstroem JW von (1974) Uranium deposits in the Karroo Supergroup near Beufort West, Cape Province, South Africa. In: Anonymous (ed) Formation of uranium ore deposits. International Atomic Energy Agency, SM-183/48. Vienna, Austria, pp 419–424

Brongniart A (1828) Prodrome d'une histoire des vegetaux fossiles. Dictionaire des Sciences Naturelles 57:1–223

Bradshaw MA (1981) Palaeoenvironmental interpretations and systematics of Devonian trace fossils from the Taylor Group, Antarctica. New Zealand J Geol Geophys 24:615–652

Bradshaw MA, McCartan LM (1983) The depositional environment of the Lower Devonian Horlick Formation, Ohio Range. In: Oliver RL et al. Antarctic Earth Science. Australian Acad. Sci. 238–241

Brown HR, Taylor GH (1960) Metamorphosed coal from the Theron Mountains. Trans-Antarctic Commonwealth Expedition, 1955–1958. Sci Repts No 12:1–11

Burrows AL (1963) Location of the south magnetic pole. New Zealand J Geol Geophys 6(3):454–464

Caputo MV, Crowell JC (1985) Migration of glacial centers across Gondwana during the Paleozoic Era. Geol Soc Amer Bull 96:1020–1036

Carlson AE, Legrande AN, Oppo DW, Came RE, Schmidt GA, Anslow FS, Licciardi JM, Obbink EA (2008) Rapid early Holocene deglaciation of the Laurentide ice sheet. Nat Geosci 1:620–624

Chalmers RO (1957) Mineral possibilities of the Antarctic. Australian Museum Mag 12(8):252–255

Chandra D, Taylor GH (1975) Gondwana coals. In: Stach E et al. (eds) Coal petrology. Borntraeger, Berlin, pp 139–159

Chen Z, Li ZX, Powell CM, Balme BE (1993) Palaeomagnetism of the Brewer Conglomerate in central Australia, and fast movement of Gondwanaland during the Late Devonian. Geophys J Internat 115:564–574

Clauer N, Tardy Y (1972) Discrimination between continental and marine sedimentary environments from the isotopic composition of strontium. Acad Sci Comptes Rendu t:273

Coates DA (1972) Pagoda Formation: Evidence of Permian glaciation in the central Transantarctic Mountains. In: Adie RJ (ed) Antarctic geology and geophysics. Universitetsforlaget, Oslo, Norway, pp 359–364

Coates DA (1985) Late Paleozoic glacial patterns in the central Transantarctic Mountains, Antarctica. In: Turner MD, Splettstoesser JF (eds) Geology of the Central Transantarctic Mountains. Antarctic Research Series, vol. 36. American Geophysical Union, Washington, DC, pp 325–338

Colbert EH (1949) The ancestors of mammals. Sci Amer 806:2–5

Colbert EH (1971) Triassic tetrapods from the McGregor Glacier. Antarctic J US 6(5):188–189

Colbert EH (1973) Antarctic Lystrosaurus defined. Antarctic J US 8(5):273–274

Colbert EH (1974) Lystrosaurus from Antarctica. Amer Mus Novitates 2535:1–44

Colbert EH (1975) Further determinations of Antarctic Triassic tetrapods. Antarctic J US 10(5):250–252

Colbert EH (1977) Cynodont reptiles from the Triassic of Antarctica. Antarctic J US 12(4):119–120

Colbert EH (1978) Proto-lizards from the Triassic of Antarctica. Antarctic J US 13(4):20–21

Colbert EH (1979) Triassic scaloposaurian reptiles. Antarctic J US 14(5):32–33

Colbert EH (1982a) Triassic vertebrates in the Transantarctic Mountains. In: Turner MD, Splettstoesser JF (eds) Geology of the Central Transantarctic Mountains. Antarctic Research Series, vol. 36. American Geophysical Union, Washington, DC, pp 11–35

Colbert EH (1982b) Mesozoic vertebrates of Antarctica. In: Craddock C (ed) Antarctic geosciences. University of Wisconsin Press, Madison, WI, pp 619–628

Colbert EH, Kitching JW (1977) Triassic cynodont reptiles from Antarctica. Amer Mus Novitates 2611:1–30

Collinson JW (1997) Paleoclimate of Permo-Triassic Antarctica. In: Ricci CA (ed) The Antarctic region; Geological evolution and processes. Terra Antartica, Siena, Italy, pp 1029–1034

Collinson JW, Elliot DH (1984) Geology of Coalsack Bluff, Antarctica. In: Turner MD, Splettstoesser JF (eds) Geology of the Central Transantarctic Mountains. Antarctic Research Series, vol. 36. American Geophysical Union, Washington, DC, pp 97–102

Collinson JW, Kemp NR (1983) Permian-Triassic sedimentary sequence in northern Victoria Land, Antarctica. In: Oliver RL, James PR, Jago JB (eds) Antarctic earth sciences. Australian Academy of Science, Canberra, Australia, pp 221–225

Collinson JH, Pennington DC, Kemp NR (1986) Stratigraphy and petrology of Permian and Triassic fluvial deposits in northern Victoria Land. In: Stump E (ed) Geological investigations in northern Victoria Land. Antarctic Res Ser 46:211–242

Collinson JW, Vavra CL, Zawiskie JM (1992) Sedimentology of the Polarstar Formation (Permian), Ellsworth Mountains, West Antarctica. In: Webers GF, Craddock C, Splettstoesser JF (eds) Geology and paleontology of the Ellsworth Mountains, West Antarctica. Geological Society of America, Memoir, 170. Boulder, CO, pp 63–79

Cosgriff JW (1983) Large thecodont reptiles from the Fremouw Formation. Antarctic J US 18(5):52–55

Cosgriff JW, Hammer WR (1979) New species of Dicynodontia from the Fremouw Formation. Antarctic J US 14(5):30–32

Cosgriff JW, Hammer WR (1981) New skull of Lystrosaurus curvatus from the Fremouw Formation. Antarctic J US 16(5):52–53

Cosgriff JW, Hammer WR, Zawiskie JM, Kemp NR (1978) New Triassic vertebrates from the Fremouw Formation of the Queen Maud Mountains. Antarctic J US 13(4):23–24

Cosgriff JW, Hammer WR, Ryan WJ (1982) The Pangean reptile, *Lystrosaurus* maccaigi in the Lower Triassic of Antarctica. J Paleontology 56(2):371–385

Craddock C, Bastien TW, Rutford RH, Anderson JJ (1965) *Glossopteris* discovered in West Antarctica. Science 148:634–637

Cridland AA (1963) A *Glossopteris* flora from the Ohio Range, Antarctica. Amer J Botany 50:186–195

Crowell JC (1999) Pre-Mesozoic ice ages: Their bearing on understanding the climate system. Geol Soc Amer, Memoir 192. Boulder, CO, pp 1–106

Crowell JC, Frakes LA (1970) Phanerozoic glaciation and the causes of ice ages. Amer J Sci 268:193–224

Cullington AL (1963) The geomagnetic secular variation in the Ross Dependency. New Zealand J Geol Geophys 6(3):444–453

Darnley AG (1971) Airborne gamma-ray survey techniques. Department of Energy, Mines, and Resources. Geol Surv Canada, Ottawa, Canada

David TWE, Priestley RE (1914) Glaciology, physiography, stratigraphy, and tectonic geology of south Victoria Land. Rept. Brit. Antarct. Exped. 1907–1909. Geology 1:1–319

Debenham F (1921) The sandstone, etc., of the McMurdo Sound, Terra Nova Bay and Beardmore Glacier regions. Natural History Report British Antarctic (Terra Nova) Expedition, 1910. Geology 1(4a):103–119

Doumani GA, Long WE (1962) The ancient life of the Antarctic. Sci Amer 863, September:2–14

Doumani GA, Tasch P (1963) Leaiid conchostracan zone in Antarctica and its Gondwana equivalents. Science 142:591–592

Doumani GA, Tasch P (1965) A leaid conchostracan zone (Permian) in the Ohio Range, Horlick Mountains, Antarctica. In: Hadley JB (ed) Geology and paleontology of the Antarctic. Antarctic Research Series, vol. 6. American Geophysical Union, Washington, DC, pp 229–239

Dreschhoff GAM, Zeller EJ (1978) Gamma-ray activity from radioactive fallout at the South Pole. Antarctic J US 13(4):193–194

Dreschhoff GAM, Zeller EJ (1986) Uranium mineralization in southern Victoria Land, Antarctica. Antarctic J US 21(5):11–13

Dreschhoff GAM, Zeller EJ, Thoste V, Bulla K (1980) Resource and radioactivity survey in the Ellsworth Mountains. Antarctic J US 15(5):32

Dreschhoff GAM, Zeller EJ, Busch K, Bulla K (1981) Resource and radioactivity survey in southern Victoria Land. Antarctic J US 16(5):38–39

Dreschhoff GAM, Zeller EJ, Kropp WR (1982) Study of radioelement concentrations in northern Victoria Land by gamma-ray spectrometry. Ant J US 17(5):4–5

Dreschhoff GAM, Zeller EJ, Kropp WR (1983a) Radiometric survey in northern Victoria Land. In: Oliver RL, James PR, Jago JB (eds) Antarctic earth science. Australian Academy of Science, Canberra. A.C.T., pp 429–432

Dreschhoff GAM, Zeller EJ, Schmid H, Bulla K, Morency M, Tremblay A (1983b) Radioactive mineral occurrence at Szabo Bluff, Transantarctic Mountains. Antarctic J US 18(5):48–49

Dreschhoff GAM, Zeller EJ, Thoste V (1992) Radioelement distribution in the sedimentary sequence of the Ellsworth Mountains, West Antarctica. In: Webers GF, Craddock C, Splettstoesser JF (eds) Geology and paleontology of the Ellsworth Mountains of West Antarctica. Geological Society of America, Memoir 170—. Boulder, CO, pp 325–332

Drewry DJ (1975) Radio-echo sounding map of Antarctica (90°E–180°). Polar Rec 17:359–374

Du Toit AL (1937) Our wandering continents. Oliver and Boyd, London, England

Edwards WN (1928) The occurrence of Glossopteris in the Beacon Sandstone of the Ferrar Glacier, South Victoria Land. Geol Mag 65(7):323–327

Elliot DH (1977) A framework for assessing environmental impacts of possible Antarctic mineral development. Part 1. U.S. Dept. Commerce, Nat. Tech. Info. Service, PB-262 750

Elliot DH, Colbert EH, Breed WJ, Jensen JA, Powell JS (1970) Triassic tetrapods from Antarctica. Evidence for continental drift. Science 169(3951):1197–1201

Elliot DH, Collinson JW, Powell JS (1972) Stratigraphy of Triassic tetrapod-bearing beds of Antarctica. In: Adie RJ (ed) Antarctic geology and geophysics. Universitetsforlaget, Oslo, Norway, pp 265–271

Farabee MJ, Taylor TN, Taylor EL (1989) Pollen and spore assemblages from the Falla Formation (Upper Triassic), central Transantarctic Mountains, Antarctica. Rev Palaeobot Palynology 61:101–138

Faure G (1998) Principles and applications of geochemistry, 2nd edn. Prentice Hall, Upper Saddle River, NJ
Faure G (2001) Origin of igneous rocks; the isotopic evidence. Springer, Heidelberg, Germany
Faure G, Barrett PJ (1973) Strontium isotope compositions of non-marine carbonate rocks from the Beacon Supergroup of the Transantarctic Mountains. J Sed Pet 43(2):447–457
Faure G, Botoman G (1984) Origin of epigenetic calcite in coal from Antarctica and Ohio. Isotope Geoscience 2:313–324 (Chem Geol 46(4):313–324)
Faure G, Lord BK (1990) Oxygen and carbon isotope compositions of calcite in carbonate concretions in the Beacon rocks of Antarctica: Products of the Gondwana ice sheet. In: Ulbrich H, Rocha Campos AC (eds) Gondwana Seven Proceedings. Instituto de Geociencias, Universidade de São Paulo, Brazil, pp 303–315
Faure G, Mensing TM (2005) Isotopes: Principles and applications, 3rd edn. Wiley, Hoboken, NJ
Faure G, Mensing TM (2007) Introduction to planetary science; The geological perspective. Springer, Dordrecht, The Netherlands
Faure G, Szabo Z (1983) The origin of secondary calcite in coal deposits of south Victoria Land, Antarctica, and in the Berea sandstone of Ohio. Geol Soc Amer Abstracts with Program 15(6):570
Faure G, Hurley PM, Fairbairn HW (1963) An estimate of the isotopic composition of strontium in rocks of the Precambrian shield of North America. J Geophys Res 68:2323–2329
Faure G, Grootes P, Buchanan D, Hagen EH (1992) Oxygen isotope study of the ice fields surrounding the Reckling Moraine on the East Antarctic ice sheet. In: Elliot DH (ed) Contributions to Antarctic research III. Antarctic Research Series, vol. 57. American Geophysical Union, Washington, DC, pp 15–26
Faure G, Wehn KS, Montello JM, Hagen EH, Strobel ML, Johnson KS (1993) Isotope composition of the ice and sub-glacial geology near the Allan Hills, Victoria Land, Antarctica. In: Findlay RH (ed) Gondwana Eight; Assembly, evolution, and dispersal. Balkema, Rotterdam, The Netherlands, pp 485–495
Faure G, Sipp JA, Lee G (1997) Search for anthropogenic cesium-137 in a soil profile in Beacon Valley, southern Victoria Land. Antarctic J US 32(5):17–18
Fielding CR, Frank TD, Isbell JL (eds) (2008a) Resolving the late Paleozoic ice ages in time and space. Geological Society of America, Special Paper 441, Boulder, CO
Fielding CR, Frank TD, Isbell JL (2008b) The late Paleozoic ice age: A review of current understanding and synthesis of global climate patterns. In: Fielding CR, Frank ID, Isbell JL (eds) Resolving the late Paleozoic Ice Age in time and space. Geological Society of America, Special Paper 441. Boulder, CO, pp 343–354
Findlay R, Unrug R, Banks MR, Veevers JJ (1993) Gondwana Eight; Assembly, evolution, and dispersal. Balkema, Rotterdam, The Netherlands
Ford AB (1983) The Dufek Intrusion of Antarctica and a survey of its minor metals and possible resources. In: Behrendt JC (ed) Petroleum and mineral resources of Antarctica. US Geological Survey, Circular 909. Alexandria, VA, pp 51–75
Frakes LA, Matthews JL, Crowell JC (1971) Late Paleozoic glaciation, III, Antarctica. Geol Soc Amer Bull 82:1581–1603
Friedman I, O'Neil JR (1977) Compilation of stable isotope fractionation factors of geochemical interest. In: Fleischer M (ed) Data of geochemistry, Prof. Paper 440-KK, US Geological Survey, Washington, DC
Gair HS, Norris G, Ricker J (1965) Early Mesozoic micro-floras from Antarctica. New Zealand J Geol Geophys 8(2):231–235
Gamble JA, Skinner DNB, Henry S (2002) Antarctica at the close of a millennium. Royal Society of New Zealand, Bulletin 35, Royal Society of New Zealand, Wellington, New Zealand
Gevers TW, Twomey A (1982) Trace fossils and their environment in Devonian (Silurian) lower Beacon strata in the Asgard Range, Victoria Land, Antarctica. In: Craddock C (ed) Antarctic geoscience. University of Wisconsin Press, Madison, WI, pp 639–648
Ghosh PK (1995) The environment of coal formation in the peninsular Gondwana basins of India. In: Campbell KSW (ed) Gondwana geology. Australian National University Press, Canberra, A.C.T., pp 221–231
Gilbert GE, Gasella VR, Bishop CT (1984) Activities of isotopes of uranium, thorium, radium, and lead of Antarctic coals, 1–9. Monsanto Research Corp., MLM-3151, Miamisburg, OH
Glasby GP (1976) Manganese nodules in the South Pacific: A review. New Zealand J Geol Geophys 19(5):707–736
Goodell HG, Meyland MA, Grant B (1971) Ferromanganese deposits of the South Pacific Ocean, Drake Passage, and Scotia Sea. In: Reid JL (ed) Antarctic oceanology I. Antarctic Research Series, vol. 15. American Geophysical Union, Washington, DC, pp 27–92
Gould RE, Delevoryas T (1977) The biology of *Glossopteris*: Evidence from petrified seed-bearing and pollen-bearing organs. Alcheringa 1:387–399
Grande L, Eastman JT (1986) A review of Antarctic ichtyofaunas in the light of new fossil discoveries. Palaeontology 29:113–137
Grindley GW (1963) The geology of the Queen Alexandra Range, Beardmore Glacier, Ross Dependency, Antarctica; With notes on the correlation of Beacon sequences. New Zealand J Geol Geosphys 6(3):307–347
Grootes PM, Stuiver M, Thompson LG, Mosley-Thompson E (1988) Oxygen isotope ratio changes in tropical ice, Quelccaya, Peru. J Geophys Res 94(D1):1187–1194
Hammer WR (1990a) Triassic terrestrial vertebrate faunas of Antarctica. In: Taylor TN, Taylor EL (eds) Antarctic paleobiology. Springer, New York
Hammer WR (1990b) Thrinaxodon from Graphite Peak, central Transantarctic Mountains. Antarctic J US 25(5):26–27
Hammer WR, Cosgriff JW (1981) Myosaurus gracilis, and anomodont reptile from the Lower Triassic of Antarctica and South Africa. J Paleontology 55(2):410–424
Hammer WR, Hickerson WJ (1992) Comments on the fossil vertebrates from the Falla Formation Jurassic, Beardmore Glacier region, Antarctica. Antarctic J US 27(5):1–2
Hammer WR, Hickerson WJ (1993) A brief review of Mesozoic terrestrial vertebrates from the Transantarctic Mountains including evidence of a scavenging theropod. Antarctic J US 28(5):33–35
Hammer WR, Hickerson WJ (1994a) Dinosaur assemblage from the Transantarctic Mountains. Antarctic J US 29(5):31–33

Hammer WR, Hickerson WJ (1994b) A crested theropod dinosaur from Antarctica. Science 264(5160):828–830

Hammer WR, Ryan WJ, Tamplin JW, DeFauw SL (1986) New vertebrates from the Fremouw Formation (Triassic), Beardmore Glacier region, Antarctica. Antarctic J US 21(5):24–26

Hammer WR, Ryan WJ, DeFauw SL (1987) Comments on the vertebrate fauna from the Fremouw Formation (Triassic), Beardmore Glacier region, Antarctica. Antarctic J US 22(5):32–33

Hammer WR, Collinson JW, Ryan WJ III (1990) A new Triassic vertebrate fauna from Antarctica and its depositional setting. Antarctic Sci 2(2):163–167

Hammer WR, Hickerson WJ, Krippner S, Tamplin J (1991) Therapsid, temnospondyls, and dinosaurs from the Fremouw and Falla formations, Beardmore Glacier region, Antarctica. Antarctic J US 26(5):19–20

Hammer WR, Hickerson WJ, Slaughter RW (1995) Evidence of theropod behavior from the Jurassic of Antarctica. Antarctic J US 30(5):55–56

Hammer WR, Hickerson WJ, Collinson JW (1996) Preliminary analysis of Triassic vertebrates from the Shackleton Glacier region. Antarctic J US 31(2):8–9

Helby RJ, McElroy CT (1969) Microfloras from the Devonian and Triassic of the Beacon Group, Antarctica. New Zealand J Geol Geophys 12(2/3):376–382

Hobday DK (1987) Gondwana coal basins of Australia and South Africa; Tectonic setting, depositional systems and resources. In: Scott AC (ed) Coal and coal-bearing Strata: Recent advances. Geological Society of London, Spec Pub 32:219–233

Hoefs J (1997) Stable isotope geochemistry, 4th edn. Springer, Heidelberg, Germany

Holdgate MW, Tinker J (1979) Oil and other minerals in the Antarctic. House of Print, London

Horner TC, Krissek LA (1991) Contributions to sedimentologic, thermal alteration and organic carbon data to paleoenvironmental interpretation of fine-grained Permian clastics from the Beardmore Glacier region, Antarctica. In: Elliot DH (ed) Contributions to Antarctic research II. Antarctic Research Series, vol. 53. American Geophysical Union, Washington, DC, pp 33–65

Isbell JL (1999) The Kukri erosion surface; A reassessment of its relationship to rocks of the Beacon Supergroup in the central Transantarctic Mountains, Antarctica. Antarctic Sci 11:228–238

Isbell JL, MacDonald DIM (1991) Lithofacies analysis of the Triassic Fremouw Formation of the Gordon Valley vertebrate site, Antarctica. Antarctic J US 26(5):15–16

Isbell JL, Miller MF, Babcock LE, Hasiotis ST (2001) Ice-marginal environment and escosystem prior to initial advance of the late Paleozoic ice sheet in the Mount Butters area of the central Transantarctic Mountains, Antarctica. Sedimentology 48:953–970

Isbell JL, Lenaker PA, Askin RA, Miller MF, Babcock LE (2003) Reevaluation of the timing and extent of late Paleozoic glaciation in Gondwana: Role of the Transantarctic Mountains. Geology 31:977–980

Isbell JL, Koch ZJ, Szablewski GM, Lenaker PA (2008) Permian glacigenic deposits in the Transantarctic Mountains, Antarctica. In: Fielding CR, Frank TD, Isbell JL (eds) Resolving the late Paleozoic Ice Age in time and space. Geological Society of America, Special Paper 441. Boulder, CO, pp 71–82

IUGS (2002) International stratigraphic chart. International Commission on Stratigraphy, International Union of Geological Sciences

Ivanhoe LF (1980) Antarctic-operating conditions and petroleum prospects. Oil Gas J 78(52):212–220

Jefferson TH (1982) Fossil forests from the Lower Cretaceous of Alexander Island, Antarctica. Palaeontology 25(4):681–708

Jefferson TH (1983) Palaeoclimatic significance of some Mesozoic Antarctic fossil floras. In: Oliver RL, James PR, Jago JB (eds) Antarctic earth science, Australian Academy of Science, Canberra, A.C.T., pp 593–598

Jefferson TH, Taylor TN (1983) Permian and Triassic wood from the Transantarctic Mountains: Paleoenvironmental indicators. Antarctic J US 18(5):55–57

Kaiho K, Kajiwara Y, Nakano T, Miura Y, Kawahata H, Tazaki K, Ueshima M, Chen Z, Shi GR (2001) End-Permian catastrophe by a bolide impact: Evidence of a gigantic release of sulfur from the mantle. Geology 29(9):815–818

Kaiho K, Kajiwara Y, Miura Y (2002) Reply to Koeberl et al. Geology 30(9):856

Kameneva GI, Grikurov GE (1983) A metallogenic reconnaissance of Antarctic major structural provinces. In: Oliver RL, James PR, Jago JB (eds) Antarctic earth science. Australian Academy of Science, Canberra, Australia, pp 420–422

Kemp EM (1972) Lower Devonian palynomorphs from the Horlick Formation, Ohio Range, Antarctica. Palaeontographica 139B:105–124

Kemp EM (1975) The palynology of late Paleozoic glacial deposits of Gondwana. In: Campbell KSW (ed) Gondwana geology. Australian National University Press, Canberra, A.C.T., pp 397–414

Kitching JW (1968) On the Lystrosaurus zone and its fauna with special reference to some immature Lystrosauridae. Paleontologica Africana 11:61–76

Kitching JW, Collinson JW, Elliot DH, Colbert EH (1972) Lystrosaurus zone (Triassic) fauna from the Antarctica. Science 175:524–527

Koeberl C, Gilmour I, Reimold WU, Claeys P, Ivanov B (2002) End-Permian catastrophe by bolide impact: Evidence of a gigantic release of sulfur from the mantle: Comment. Geology 30(9):855–856

Krissek LA, Horner TC (1991) Sedimentology of a vertebrate bone-bearing bed in the Triassic Fremouw Formation at Gordon Valley, Beardmore Glacier region, Antarctica. Antarctic J US 26(5):17–19

Krissek LA, Horner TC, Elliot DH, Collinson JW (1992) Stratigraphy and sedimentology of vertebrate bone-bearing beds in the Triassic (and Jurassic?) Fremouw and Falla formations, Beardmore Glacier regions, Antarctica. In: Yoshida Y (ed) Recent progress in Antarctic earth science. Terra Scientific, Tokyo, Japan, pp 249–256

LaPrade KE (1970) Permian-Triassic Beacon Group of the Shackleton Glacier area, Queen Maud Range, Transantarctic Mountains, Antarctica. Geol Soc Amer Bull 81(5): 1403–1410

Laskar B, Raja Rao CS (1979) Fourth International Gondwana Symposium: Papers 1:223–344

LeRoux JP, Toens PD (1987) The Permo-Triassic uranium deposits of Gondwana. In: McKenzie GD (ed) Gondwana

Six: Stratigraphy, sedimentology, and paleontology. Geophysical Monograph, vol. 41. American Geophysical Union, Washington, DC, pp 139–146
Lindsay JF (1970). Depositional environment of Paleozoic glacial rocks in the central Transantarctic Mountains. Geol Soc Amer Bull 81:1149–1172
Long WE (1959) Preliminary report of the geology of the central range of the Horlick Mountains, Antarctica. Ohio Research Foundation Rept. 825–2, part 7, pp 1–23
Long WE (1964) The stratigraphy of the Horlick Mountains. In: Adie RJ (ed) Antarctic geology. North-Holland, Amsterdam, The Netherlands, pp 352–363
Long WE (1965) Stratigraphy of the Ohio Range, Antarctica. In: Hadley JB (ed) Geology and paleontology of the Antarctic. Antarctic Research Series, vol. 6. American Geophysical Union, Washington, DC, pp 71–116
Lord BK, Jones LM, Faure G (1988) Evidence for the existence of the Gondwana ice sheet in the ^{18}O depletion of carbonate rocks in the Permian formations of the Transantarctic Mountains. Chem Geol 72:163–171
Lorius C, Merlivat L, Jouzel J, Pourchet M (1979) A 30,000-yr isotope climate record from Antarctic ice. Nature 280: 644–648
Lövborg L, Kirkegaard P, Mose Christiansen E (1976) Design of NaI (Tl) scintillation detectors for use in gamma-ray surveys of geological sources. In: Anonymous (ed) Exploration for uranium ore deposits. International Atomic Energy Agency, Proceedings Series, Vienna, Austria, pp 127–148
Macdonald DIM, Isbell JL, Hammer WR (1991) Vertebrate trackways from the Triassic Fremouw Formation, Queen Alexandra Range. Antarctic J US 26(5):20–21
McAllister Rees P (2002) Land-plant diversity and the end-Permian mass extinction. Geology 30(9):827–830
McIver RD (1975) Hydrocarbon gases in canned core samples from Leg 28 sites 271, 272, and 273, Ross Sea. Initial Reports of the Deep Sea Drilling Project, 28:815–819. US Government Printing Office, Washington, DC
McKelvey BC, Webb PN, Gorton MP, Kohn BP (1972) Stratigraphy of the Beacon Supergroup between the Olympus and Boomerang ranges, Victoria Land. In: Adie RJ (ed) Antarctic geology and geophysics. Universitetsforlaget, Oslo, Norway, pp 345–352
McKelvey BC, Webb PN, Kohn BR (1977) Stratigraphy of the Taylor and lower Victoria group (Beacon Supergroup) between the Mackay Glacier and the Boomerang Range. New Zealand J Geol Geophys 20:813–863
McLeod IR (1965) Antarctica: Geology and mineral occurrences. In: Dew JM (ed) Handbook, Australia and New Zealand. Eighth Commonwealth mining and metallurgical congress. Australasian Institute of Mining and Metallurgy, Melbourne, Victoria, Australia, pp 165–166
McPherson JG (1978) Stratigraphy and sedimentology of the Upper Devonian Aztec Siltstone, southern Victoria Land, Antarctica. New Zealand J Geol Geophys 21:667–683
Meyer-Berthaud B, Taylor TN (1989) The structure and affinities of woody stems from the Triassic of Antarctica. Amer J Botany 76(Supplement):170–171
Millay MA, Taylor TN, Taylor EL (1987) Studies of Antarctic fossil plants: An association of ferns from the Triassic of Fremouw Peak. Antarctic J US 22(5):31–32
Miller JMG (1989) Glacial advance and retreat sequence in a Permo-Carboniferous section, central Transantarctic Mountains. Sedimentology 36:419–430
Mirsky A (1969) Geology of the Ohio Range-Liv Glacier area. In: Craddock C (ed) Geologic map of Antarctica, Folio 12, Sheet 17. American Geographical Society, New York
Mueller G (1964) Some notes on mineralization in Antarctica. In: Adie RJ (ed) Antarctic geology. North-Holland, Amsterdam, The Netherlands, pp 393–394
Mulligan JJ et al. (1963a) Mount Gran coal deposits, Victoria Land, Antarctica. US Bur. Mines, Rept. Investig. 6218:1–66
Mulligan JJ et al. (1963b) Willett Range coal deposits, Machay Glacier area, Victoria Land, Antarctica. US Bur. Mines, Rept. Investig. 6331:1–67
Norris G (1965) Triassic and Jurassic miospores and acritarchs from the Beacon and Ferrar Group, Victoria Land, Antarctica. New Zealand J Geol Geophys 8:236–277
Oliver RL, James PR, Jago JB (1983) Antarctic Earth Science. Australian Acad. Sci., Canberra, ACT, pp 1–697
O'Neil JR, Clayton RN, Mayeda TK (1969) Oxygen isotope fractionation in divalent metal carbonates. J Chem Phys 51:5547–5558
Orr SR, Faure G, Botoman G (1982) Isotopic study of a siderite concretions, Tuscarawas County, Ohio. Ohio J Sci 82:52–54
Picciotto E, Wilgain S (1963) Fission products in Antarctic snow, a reference level for measuring accumulation. J Geophys Res 68(21):5965
Pigg KB (1990) Anatomically preserved Dicroidium foliage from the central Transantarctic Mountains. Rev Palaeobot Palyn 66:127–145
Pigg KB, Taylor TN (1985) Anatomically preserved *Glossopteris* from the Beardmore area of Antarctica. Antarctic J US 20(5):8–10
Pigg KB, Taylor TN (1987) Anatomically preserved *Dicroidium* from the Transantarctic Mountains. Antarctic J US 22(5):28–29
Plumstead EP (1962) Fossil floras in Antarctica. Trans. Antarctic Exped. (1955–1958) Sci Rept 9:1–154
Plumstead EP (1964) Palaeobotany of Antarctica. In: Adie RJ (ed) Antarctic geology. North-Holland, Amsterdam, The Netherlands, pp 637–654
Plumstead EP (1973) The late Palaeozoic Glossopteris flora. In: Hallam A (ed) Atlas of palaeobiogeography. Elsevier, Amsterdam, The Netherlands, pp 187–206
Potter N (1969) Natural resource potentials of the Antarctic. Amer Geog Soc, Occasional Pub 4:1–97
Powell C McA, Veevers JJ (1987) Namurian uplift in Australia and South America triggered the main Gondwana glaciation. Nature 326:177–179
Ravich MG, Fedorov LV, Tarutin DA (1982) Precambrian iron deposits of the Prince Charles Mountains. In: Craddock C (ed) Antarctic geoscience. University of Wisconsin Press, Madison, WI, pp 853–858
Retallack G, Dilcher DL (1981) Arguments for a glossopterid ancestry of angiosperms. Paleobiol 7:54–67
Retallack GJ, Hammer WR (1996) Paleoenvironment of the Triassic therapsid Lystrosaurus in the central Transantarctic Mountains, Antarctica. Antarctic J US 31(2):33–35
Retallack GJ, Seyedolali A, Krull ES, Holser WT, Ambers CP, Kyte FT (1998) Search for evidence of impact at the Permian-Triassic boundary in Antarctica and Australia. Geology 26(11):979–982

Ricci CA (ed) (1997) The Antarctic region: Geological evolution and processes. Terra Antartica, Siena, Italy
Rigby JF (1969) Permian sphenopsids from Antarctica. US Geol Surv Prof. Paper 613-F. US Government Printing Office, Washington, DC
Rigby JF (1985) Some Triassic (middle Gondwana) floras from south Victoria Land, Antarctica. Hornibrook Symposium Abstracts. DSIR, Christchurch, New Zealand, pp 78–79
Rigby JF, Schopf JM (1969) Stratigraphic implications of Antarctic paleobotanical studies. In: Amos AJ (ed) Gondwana stratigraphy. UNESCO, Paris, France, pp 91–106
Rose G, McElroy CT (1987) Coal potential of Antarctica. Department of Resources and Energy, Resource Rept 2:1–19
Ryskin G (2003) Methane-driven oceanic eruptions and mass extinction. Geology 31(9):741–744
Sanyal SP, Subramanian CS (1979) Petrology of Gondwana coals of India: A comparative study. In: Laskar B, Raja Rao CS (eds) Fourth International Gondwana Symposium Papers 1:305–315
Schmidt DL, Ford AB (1969) Geology of the Pensacola Mountains. In: Craddock C (ed) Geologic map of Antarctica, Folio 12, Sheet 5. American Geological Society, New York
Schmidt DL, Friedman I (1974) Continental deposition of Antarctic tillite indicated by carbon and oxygen isotopes. J Research US Geol Surv 2(6):711–715
Schmidt DL, Williams PL (1969) Continental glaciation of late Paleozoic age, Pensacola Mountains, Antarctica. In: Amos AJ (ed) Gondwana stratigraphy (IUGS symposium, Buenos Aires, 1967). Earth Science Series, vol. 2. UNESCO, Paris, pp 617–649
Schopf JM (1948) Variable coalification: The processes involved in coal formation. Econ Geol 43:207–225
Schopf JM (1956) A definition of coal. Econ Geol 51(6): 521–527
Schopf JM (1960) Field description and sampling of coal beds. Bull US Geol Surv 1111-B:1–70
Schopf JM (1962) A preliminary report on plant remains and coal of the sedimentary section in the central Horlick Mountains, Antarctica. Institute of Polar Studies, Rept. 2—. The Ohio State University, Columbus, OH, pp 1–62
Schopf JM (1965) Anatomy of the axis in *Vertebraria*. Antarctic Research Series, vol. 6. American Geophysical Union, Washington, DC, pp 217–228
Schopf JM (1966a) Antarctic paleobotany and palynology. Antarctic J US 1(4):135
Schopf JM (1966b) Definitions of peat and coal and of graphite that terminates the coal series (graphocite). J Geol 74(5) Part I:584–592
Schopf JM (1967) Antarctic fossil plant collecting during the 1966–1967 season. Antarctic J US 2:114–116
Schopf JM (1968) Studies in Antarctic paleobotany. Antarctic J US 3(5):176–177
Schopf JM (1969) Ellsworth Mountains: Position in West Antarctica due to sea-floor spreading. Science 164:63–66
Schopf JM (1970a) Gondwana paleobotany. Antarctic J US 5(3):62–66
Schopf JM (1970b) Petrified peat from a Permian coal bed in Antarctica. Science 169:274–277
Schopf JM (1970c) Relation of floras of the southern hemisphere to continental drift. Taxon 15(5):657–674
Schopf JM (1971) Notes on plant tissue preservation and mineralization in a Permian deposit of peat from Antarctica. Amer J Sci 271:522–543
Schopf JM (1973) The contrasting plant assemblages from Permian and Triassic deposits in southern continents. In: Logan A, Hills LV (eds) The Permian and Triassic system and the mutual boundary. Canadian Society of Petroleum Geologists, Memoir 2., Calgary, Alberta, Canada, pp 379–397
Schopf JM (1975) Modes of fossil preservation. Rev Palaeobot Palynol 20:27–53
Schopf JM (1976) Morphologic interpretation of fertile structures in glossopterid gymnosperms. Rev Palaeobot Palynol 21:25–64
Schopf JM (1977) Coal forming elements in permineralized peat from Mt. Augusta (Queen Alexandra Range). Antarctic J US 12(4):110–112
Schopf JM (1978) An unusual osmundaceous specimen from Antarctica. Canadian J Botany 56:3083–3095
Schopf JM (1982) Forms and facies of *Vertebraria* in relation to Gondwana coal. In: Turner MD, Splettstoesser JF (eds) Geology of the Central Transantarctic Mountains. Antarctic Research Series, vol. 36. American Geophysical Union, Washington, DC, pp 37–62
Schopf JM, Long WE (1966) Coal metamorphism and igneous associations in Antarctica. In: Gould RF (ed) Coal science, Adv. in Chemistry Ser., No. 55:156–195. Amer. Chem. Soc.
Seward AC (1914) Antarctic fossil plants. British Antarctic (Terra Nova) Expedition, 1910. Natural History Report, Geology 1(1):1–149. British Museum of Natural History, London, England
Seyfert CK, Sirkin LA (1973) Earth history and plate tectonics. Harper & Row, New York
Sharp Z (2007) Principles of stable isotope geochemistry. Prentice Hall, Upper Saddle River, NJ
Shieh YN, Suter TG (1979) Formation conditions of authigenic kaolinite and calcite in coals by stable isotope determinations. Clays Clay Miner 27:154–156
Siddall M, Kaplan MR (2008) A tale of two ice sheets. Nat Geosci 1:570–571
Smith AG, Hallam A (1970) The fit of the southern continents. Nature 225:139–144
Smith PJ (1972) The resource potential of Antarctica. Comments Earth Sci (geophysics) 3(1):23–28
Smoot EL, Taylor TN (1986) Evidence of simple polyembryony in Permian seeds from Antarctica. Amer J Botany 73:1079–1081
Smoot EL, Taylor TN, Delevoryas T (1985) Structurally preserved plants from Antarctica. I. Antarticycas gen. nov., a Triassic cycad stem from the Beardmore Glacier area. Amer J Botany 72:1410–1423
Spivak J (1974) Now the energy crisis spurs idea of seeking oil at the South Pole. Wall Street J, 21 February:1
Splettstoesser JF (1970) Coal in Antarctica. Abstract Amer Assoc Petroleum Geol Bull 63(3):532
Splettstoesser JF (1980) Coal in Antarctica. Econ Geol 75:936–942
Stach E, Mackowsky MT, Teichmüller M, Taylor GH, Chandra D, Teichmüller R (1982) Textbook of coal petrology, 3rd edn. Borntraeger, Berlin, Germany
Stewart D (1964) Antarctic mineralogy. In: Adie RJ (ed) Antarctic geology. North-Holland, Amsterdam, The Netherlands

St. John B (1990) Antarctica as an exploration frontier: Hydrocarbon potential, geology, and hazards. Amer. Assoc. Petroleum Geol., Studies in Geology, 31:1–154

Stonehouse B (ed) (2002) Encyclopedia of Antarctica and the Southern Ocean. Wiley, Chichester, UK

Stubblefield SP, Taylor TN (1985) Fossil fungi in Antarctic wood. Antarctic J US 20(5):7–8

Stubblefield SP, Taylor TN (1986) Wood decay in silicified gymnosperms from Antarctica. Botany Gaz 147:116–125

Tasch P (1977) Intercontinental correlations by conchostracans and palynomorphs from Antarctica, western Australia, India, and Africa. Antarctic J US 12(4):121

Tasch P (1978) Permian palynomorphs (Coalsack Bluff, Mt. Sirius, Mt. Picciotto) and other studies. Antarctic J US 13(4):19–20

Taylor TN (1981) Paleobotany. McGraw-Hill, New York

Taylor TN, Smoot EL (1983) Structurally preserved plants from the Beardmore Glacier area. Antarctic J US 18(5):57–58

Taylor TN, Smoot EL (1984) Fossil plants from the Beardmore Glacier area. Antarctic J US 19(5):12

Taylor TN, Smoot EL (1985a) A new Triassic cycad from the Beardmore Glacier area of Antarctica. Antarctic J US 20:5–7

Taylor TN, Smoot EL (1985b) Plant fossils from the Ellsworth Mountains. Antarctic J US 20(5):48–49

Taylor EL, Taylor TN (1988) Late Triassic flora from Mt. Falla, Queen Alexandra Range. Antarctic J US 23(5):2–3

Taylor TN, Taylor EL (eds) (1990) Antarctic paleobiology: Its role in the reconstruction of Gondwana. Springer, New York

Taylor TN, Taylor EL (1992) Permian plants from the Ellsworth Mountains, West Antarctica. In: Webers GF, Craddock C, Splettstoesser JR (eds) Geology and paleontology of the Ellsworth Mountains, West Antarctica. Geological Society of America, Memoir 170. Boulder, CO, pp 285–294

Taylor TN, Smoot EL, Delevoryas T (1983) Structurally preserved plants from Antarctica, A. Triassic cycad stem. Amer J Botany 70:80

Taylor TN, Taylor EL, Collinson JW (1986a) Paleoenvironment of Lower Triassic plants from the Fremouw Formation, Antarctica. Antarctic J US 21(5):26–27 (*Formerly E.L. Smoot)

Taylor EL, Taylor TN, Collinson JW, Elliot DH (1986b) Structurally preserved Permian plants from Skaar Ridge, Beardmore Glacier region. Antarctic J US 21(5):27–28

Taylor TN, Taylor EL, Farabee MJ (1988) Palynostratigraphy of the Falla Formation (Upper Triassic) Beardmore Glacier region. Antarctic J US 23(5):8–9

Taylor EL, Taylor TN, Isbell JL, Cúneo NR (1989a) Fossil floras of southern Victoria Land: 1. Aztec Mountain. Antarctic J US 24(5):24–26

Taylor EL, Taylor TN, Isbell JL, Cúneo NR (1989b) Fossil floras of southern Victoria Land: 2. Kennar Valley. Antarctic J US 24(5):26–30

Taylor EL, Taylor TN, Meyer-Berthaud B, Isbell JL, Cúneo NR (1990) A Late Triassic flora from the Allan Hills, southern Victoria Land. Antarctic J US 25(5):20–21

Taylor EL, Boucher LD, Taylor TN (1992) Dicroidium foliage from Mt. Falla, central Transantarctic Mountains. Antarctic J US 27(5):2–3

Taylor EL, Cúneo R, Taylor TN (1991) Permian and Triassic fossil forests from the central Transantarctic Mountains. Antarctic J US 26(5):23–24

Taylor TN, Taylor EL, Krings M (2009) Paleobotany; The biology and evolution of fossil plants. Academic, Elsevier, Amsterdam, The Netherlands

Tessensohn F, Mädler K (1987) Triassic plant fossils from north Victoria Land, Antarctica. Geologisches Jahrbuch B66:187–201

Thompson LG, Dansgaard W (1975) Oxygen isotope and microparticle studies of snow samples from the Quelccaya Ice Cap, Peru. Antarctic J US 10(1):24–26

Thompson LG, Hastenrath S, Morales Arnao B (1979) Climatic ice-core records from the tropical Quelccaya Ice Cap. Science 203(4386):1240–1243

Thompson LG, Mosley-Thompson E, Bolzan JF, Koci BR (1985) A 1500-year record of tropical precipitation recorded in ice cores from the Quelccaya Ice Cap, Peru. Science 229(4717):971–973

Thompson LG, Mosley-Thompson E, Dansgaard W, Grootes PM (1986) The "Little Ice Age" as recorded in the stratigraphy of the tropical Quelccaya Ice Cap. Science 234:361–364

Townrow JA (1957) On *Dicroidium*, probably a pteridospermous leaf, and other leaves now removed from this genus. Trans Geol Soc Africa 60:21–60

Townrow JA (1967) Fossil plants from Allan and Carapace Nunataks, and from the upper Mill and Shackleton Glaciers, Antarctica. New Zealand J Geol Geophys 10:456–473

Triggs GD (1987) The Antarctic Treaty regime. Cambridge University Press, Cambridge, UK

Twitchett RJ, Krystyn K, Baud A, Wheeley JR, Richoz S (2004) Rapid marine recovery after the end-Permian mass-extinction event in the absence of marine anoxia. Geology 32(9): 805–808

Ulbrich H, Rocha Campos AC (eds) (1991) Gondwana Seven proceedings. Instituto de Geociencias, University of Sao Paulo, Sao Paulo, Brazil

Veevers JJ, Powell C McA (1987) Late Paleozoic glacial episodes in Gondwanaland reflected in transgressive-regressive depositional sequences in Euramerica. Geol Soc Amer Bull 98:475–487

Walker BC (1983) The Beacon Supergroup of northern Victoria Land, Antarctica. In: Oliver RL, James PR, Jago JB (eds) Antarctic earth science. Australian Academy of Science, Canberra, Australia, pp 211–214

Wayt Gibbs W (2001) On the termination of species. Scientific American, November, 40–49

Weber JN, Keith ML (1962) Carbon-isotope composition and the origin of calcareous coal balls. Science 138:900–902

Weber JN, Williams EG, Keith ML (1964) Paleoenvironmental significance of carbon isotopic composition of siderite nodules in some shales of Pennsylvanian age. J Sediment Petrol 34:814–818

Webers GF, Craddock C, Splettstoesser JF (1992) Geologic history of the Ellsworth Mountains, West Antarctica. In: Webers GF, Craddock C, Splettstoesser JF (eds) Geology and paleontology of the Ellsworth Mountains, West Antarctica. Geological Society of America, Memoir 170. Boulder, CO, pp 1–8

Wegener A (1929) Die Entstehung der Kontinente und Ozeane, 4th edn. Viehweg und Sohn, Braunschweig, Germany

Whitby KJ, Rose G, McElroy CT (1983) Formational mapping of the Beacon Supergroup type area with special references to the Webber Coal Measures, South Victoria Land, Antarctica

White EI (1968) Devonian fishes of the Mawson-Mullock area, Victoria Land, Antarctica. Trans-Antarctic Exped., 1955–1958, Sci. Repts. 1960–1968, 16:1–26

Windley BF (1977) The evolving continents. Wiley, London

Wizevich MC (1997) Fluvial-eolian deposits in the Devonian New Mountains Sandstone, Table Mountain, southern Victoria Land, Antarctica: Sedimentary architecture, genesis, and stratigraphic evolution. In: Ricci CA (ed) The Antarctic region: Geological evolution and processes. Terra Antartica, Siena, Italy, pp 933–944

Wright NA, Williams PL (1974) Mineral resources of Antarctica. US Geol Surv Circular 705:1–29

Young GC (1987) Devonian vertebrates of Gondwana. In: McKenzie GD (ed) Gondwana Six: Stratigraphy, sedimentology, and paleontology. Geophysical Mongraph vol. 4. American Geophysical Union, Washington, DC, pp 1–50

Zavada MS, Taylor TN (1986) The role of self incompatibility as a mate choice mechanism in the gymnosperm-angiosperm transition: A hypothesis. Amer Naturalist 128:538–550

Zeller EJ, Dreschhoff GAM (1980) Evaluation of uranium resources in Antarctica. In: Uranium evaluation and mining techniques. International Atomic Energy Agency, Vienna, Austria, pp 381–390

Zeller EJ, Dreschhoff GAM (1987) Radioactive minerals and the pre-Beacon erosion surface, Antarctica. In: McKenzie GD (ed) Gondwana Six: Stratigraphy, sedimentology, and paleontology. Geophysical Monograph, vol. 41. American Geophysical Union, Washington, DC, pp 63–72

Zeller EJ, Dreschhoff G, Crisler K (1977) Resource and radioactivity survey in southern Victoria Land. Antarctic J US 12(4):112–113

Zeller E, Dreschhoff G, Thoste V, Kropp WR (1979) Radioactivity survey in Antarctica 1978/79. Antarctic J US 14:38–39

Zeller EJ, Dreschhoff G, Crisler K, Tessensohn F (1982) Resource and radioactivity survey in Antarctica by airborne gamma-ray spectrometry. In: Craddock C (ed) Antarctic earth science. University of Wisconsin Press, Madison, WI, pp 877–883

Zeller EJ, Dreschhoff GAM, Kropp W-R (1986) Evaluation of the uranium resource potential of northern Victoria Land. In: Stump E (ed) Geological investigations in Northern Victoria Land. Antarctic Research Series, vol. 46. American Geophysical Union, Washington, DC, pp 383–391

Zumberge JH (1977) A preliminary assessment of the environmental impact of mineral exploration/exploitation in Antarctica. SCAR Secretariat, Scott Polar Research Institute, Lensfield Road, Cambridge, England

Zumberge JH (ed) (1979a) Environmental effects of mineral exploration and exploitation in Antarctica. Sci. Comm. Antarctic Research, University Library, University of Cambridge, Cambridge, UK

Zumberge JH (1979b) Mineral resources and geopolitics in Antarctica. Amer Scientist 67(1):68–77

Chapter 12
The Ferrar Group: Kirkpatrick Basalt

The rocks of the Ferrar Group consist of the sills of the Ferrar Dolerite, of the flows of the Kirkpatrick Basalt, and of phreatomagmatic volcanic breccias that underlie the basalt plateaus. These rocks formed in a short interval of time during the Middle Jurassic in a setting of extensional tectonics. The basalt magmas originated at depth in the subcrustal mantle and intruded the basement of the Transantarctic Mountains and the overlying sedimentary rocks of the Beacon Supergroup by means of deep crustal rifts that started the ultimate break-up of Gondwana.

Flood basalts and diabase sills and dikes were also erupted on all of the southern continents during the break-up of Gondwana, including southern Africa, India, South America, and Australia (Faure 2001). The sills and dikes of Jurassic dolerite in Tasmania and basalts on Kangaroo Island off the coast of South Australia are of special interest because they are extensions of the Ferrar magmatic province of Antarctica and because they were studied by geochemists and petrologists before the importance of the Ferrar rocks to the tectonic evolution of Gondwana was recognized. The investigations of these rocks helped to define the issues that warranted the subsequent studies of the basaltic rocks of the Ferrar Group in the Transantarctic Mountains.

12.1 Wisanger Basalt, South Australia

The flows of Jurassic basalt on Kangaroo Island at the mouth of the Gulf of the Vincent in South Australia are composed of tholeiite and have been assigned to the Wisanger Formation which appears to be a remnant of a more extensive deposit of flood basalt. Milnes et al. (1982) published chemical analyses of 12 specimens and reported concentrations of rubidium and strontium as well as the $^{87}Sr/^{86}Sr$ ratios of three samples of the Wisanger Basalt. Their initial $^{87}Sr/^{86}Sr$ ratios calculated at 170 Ma range from 0.71062 to 0.71165 which indicates the presence of excess radiogenic ^{87}Sr. The Wisanger Basalt was dated by McDougall and Wellman (1976) by the K-Ar method. The age, chemical composition, and elevated initial $^{87}Sr/^{86}Sr$ ratio relate the Wisanger Basalt on Kangaroo Island to the Tasmanian Dolerite and to the flood basalt and dolerite of the Ferrar Group in the Transantarctic Mountains.

12.2 Tasmanian Dolerite

The Tasmanian Dolerite intruded flat-lying clastic sedimentary rocks of the Parmeener Supergroup which consist of a varied assemblage of Late Carboniferous to Triassic age with an aggregate thickness of about 2,000 m (Forsyth 1987). These rocks rest unconformably on a basement of Late Devonian granites and older folded sedimentary rocks. The Lower Division of the Parmeener Supergroup is composed of glaciomarine strata (Banks and Clarke 1987; Hand 1983; Domack et al. 1993), whereas the Upper Division (Late Permian to Triassic) contains *non-marine* sandstones, siltstones, mudstone, and coal measures as well as lenticular beds of conglomerate and *tuff*. The fossil flora of the Upper Division of the Parmeener Supergroup includes *Glossopteris*, *Gangamopteris*, *Vertebraria* and several kinds of palynomorphs. Even fossilized bones of labyrinthodont amphibians have been found in Triassic sandstones of Tasmania (Cosgriff 1974; Cosgriff and Hammer 1983).

The similarity of the Upper Division of the Parmeener Supergroup to the Permo-Triassic rocks of the Beacon Supergroup is unmistakable, but correlations

G. Faure and T.M. Mensing, *The Transantarctic Mountains: Rocks, Ice, Meteorites and Water*,
DOI 10.1007/978-90-481-9390-5_12,

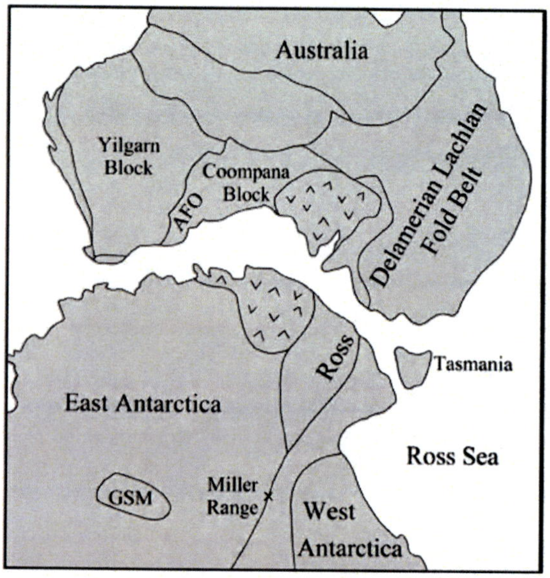

Fig. 12.1 In this continental reconstruction by Lawver et al. (2003), Tasmania is located close to the coast of northern Victoria Land and within the Ross-Delamerian-Lachlan fold belt. Therefore, the geologic history of Tasmania was similar to that of the Ross orogen and the Transantarctic Mountains from early Paleozoic to the break-up of Gondwana in the late Mesozoic. AFO = Albany-Fraser orogen; the "V" patterned areas identify an igneous province of Paleoproterozoic and Mesoproterozoic age; GSM = Gamburtsev Subglacial Mountains; the Miller Range was discussed in Section 5.1.1; the lines within Australia and East Antarctica are tectonic boundaries (Adapted from Fig. 4.1-2 by Finn et al. (2006))

of specific formations are not feasible. Rather, the similarities of lithologic composition and of the fossil record arise because the rocks were deposited in similar topographic and climatic environments during the same interval of geologic time when Tasmania in Fig. 12.1 was still a part of Gondwana in close proximity of East Antarctica (Collinson and Kemp 1983; Finn et al. 2006).

The sedimentary rocks of the Upper Permeenar Group were intruded by sills and dikes of the Tasmanian Dolerite and the Red Hill Granophyre. K-Ar age determinations of the sills by Schmidt and McDougall (1977) yielded an average date of 170.5 ± 8.0 Ma which corresponds to an early Middle Jurassic age.

The Tasmanian Dolerite forms several thick transgressive sills and dikes. The dike at Red Hill is up to 1,600 m wide and grades from tholeiite at depth to granphyre at the top. McDougall (1962) used chemical, petrographic, and mineralogical evidence to show that the granophyre formed by fractional crystallization of basalt magma. He also concluded that the basalt magma formed by partial melting of rocks in the lithospheric mantle at a depth of about 100 km and a temperature of about 1,400°C. The magma was intruded rapidly and without assimilating crustal rocks. The subsequent differentiation of the magma caused the rocks to become enriched in quartz and alkali feldspar with increasing felsic index, whereas the abundances of plagioclase and pyroxene declined although the abundance of "iron ore" (i.e., magnetite) increased slightly.

In a subsequent study of the Tasmanian Dolerites Heier et al. (1965) reported that the concentrations of uranium, thorium, and potassium increase steadily from the chilled margins of the dolerite to the granophyre and that the U/K, Th/K, and the Th/U ratios are virtually constant throughout this highly differentiated sequence of rocks. Even more remarkably, the numerical values of these ratios differ from the values observed in basaltic rocks elsewhere and, instead, resemble the values of these ratios in granitic rocks. Moreover, the homogeneity of these ratios and the large mass of the dolerite and granophyre argue against contamination of the magma by assimilation of crustal rocks. Heier et al. (1965) also reported that the K/Rb ratios of the Tasmanian Dolerite is anomalously low (214 ± 12) and is identical to that of the Red Hill Granphyre (205 ± 12). The K/Rb ratios of both rock units are lower than the average K/Rb ratios of basalt in Brazil (403 ± 59) and in India (385 ± 159) measured by Erlank and Hofmeyr (1968). However, the average K/Rb ratios of the Tasmanian Dolerite and Granophyre are identical to the average K/Rb ratio of dolerite and basalt of the Ferrar Group in the Transantarctic Mountains (208 ± 34). Erlank and Hofmeyr (1968) pointed out that the K/Rb ratios of the Jurassic dolerite in Tasmania and Antarctica are similar to the crustal average which is 230. The concentration ratios of uranium, thorium, and potassium as well as the K/Rb ratios raise interesting questions about the source of the basalt magmas. Taken at face value, the geochemical evidence favors the conclusion that the magmas formed by partial melting of crustal rocks. Alternatively, mantle-derived basalt magmas may have assimilated large amounts of crustal rocks. However, the internal homogeneity of geochemical parameters of the Tasmanian Dolerites makes this an unlikely alternative.

Even more startling was the revelation by Heier et al. (1965) that the Tasmanian Dolerite and Granophyre contain excess radiogenic ^{87}Sr indicated by an average initial $^{87}Sr/^{86}Sr$ ratio of 0.7115 ± 0.0007. Mantle-derived basalts generally have significantly lower $^{87}Sr/^{86}Sr$ ratios than rocks in the continental crust because the ultramafic rocks of the mantle have lower Rb/Sr ratios than granitic rocks of the continental crust. For example, the $^{87}Sr/^{86}Sr$ ratio of basalt on the island of Hawaii is 0.7043 (Faure 2001).

The "crustal" characteristics of the Jurassic dolerites of Antarctica and Tasmania relate these rocks to each other and set them apart from Jurassic dolerite in southern Africa and Brazil. This point was emphasized by Compston et al. (1968) in a direct comparison of Rb-Sr systematics of Mesozoic basaltic rocks in southern Victoria Land, Tasmania, South Africa, and South America. In addition, these authors reported K-Ar dates for plagioclase and pyroxene from a pegmatoid in the Lake Vanda sill in Wright Valley of southern Victoria Land. The results indicated a Middle Jurassic age for the Ferrar Dolerite, in agreement with the age of the Tasmanian Dolerite and the K-Ar dates of the Wisanger Basalt on Kangaroo Island of South Australia reported later by McDougall and Wellman (1976).

The principal contribution by Compston et al. (1968) was to demonstrate that dolerites from several sills of Ferrar Dolerite in the Transantarctic Mountains have elevated initial $^{87}Sr/^{86}Sr$ ratios similar to the initial $^{87}Sr/^{86}Sr$ ratios of the Tasmanian Dolerite and the Wisanger Basalt on Kangaroo Island. These studies alerted other geologists to the challenge posed by the dolerite sills and basalt flows of the Transantarctic Mountains, Tasmania, and Kangaroo Island. The principal objectives of the research that followed were to:

1. Identify the magma sources and to explain the "crustal" geochemical signature of the basalt and dolerite of the Transantarctic Mountains.
2. Explain the homogeneity of the chemical and isotopic compositions of these rocks.
3. Explain how magma was transported in the subsurface while showing little evidence of contamination by crustal rocks (e.g., they contain few if any crustal xenoliths).
4. Determine the precise date of intrusion/eruption and the duration of this magmatic episode.
5. Relate the formation of basalt magma to the rifting that caused Gondwana to break up.

12.3 Diamictites, Transantarctic Mountains

The Beacon Supergroup is overlain by deposits of diamictites composed of angular clasts in a fine-grained matrix interbedded with stratified pyroclastic rocks. The clasts are composed of basalt, dolerite, and sedimentary rocks of the underlying Beacon rocks. These deposits were formed during the early stages of eruption of the Kirkpatrick Basalt and the intrusion of sills of Ferrar Dolerite into the Beacon Supergroup. The diamictites occur in different parts of the Transantarctic Mountains and have been assigned to three formations identified in Table 12.1.

12.3.1 Mawson Formation

The diamictites and pyroclastic rocks of the Mawson Formation in Fig. 12.2 occur in the Allan and Coombs hills and at Carapace Nunatak in Fig. 12.3 as well as in some of the nunataks of the Prince Albert Mountains located north of the Allan Hills (Elliot 2002; Elliot et al. 2006). Gunn and Warren (1962) first described these breccias in the Allan Hills and considered them to be a tillite. However, Ballance et al. (1965) questioned the glacial origin of the deposit and tentatively suggested a volcanic origin. Still later, Borns and Hall (1969) and Hall et al. (1982) reinterpreted the Mawson Formation as a debris flow following its extrusion as a volcanic explosion breccia. More recently, the origin of the Mawson Formation in the Coombs Hills was explained by White and McClintock (2001) as the

Table 12.1 Deposits of diamictites and pyroclastics of the Ferrar Group in the Transantarctic Mountains

Formation	Location
Mawson	Allan Hills
	Coombs Hills
	Prince Albert Mtns., in southern Victoria Land
Prebble	Queen Alexandra Range
	Otway Massif
	Grosvenor Mtns., in the central Transantarctic Mountains
Exposure Hill	Mesa Range and nearby nunataks
	Deep Freeze Range
	northern Victoria Land

Fig. 12.2 Outcrop of Mawson Diamictite in the Allan Hills. The rocks of this formation are unstratified and are composed of dolerite and sandstone clasts in a fine-grained matrix that contains zeolites and other products of hydrothermal alteration (Photo by G. Faure)

product of multiple phreatomagmatic volcanic explosions. Elliot et al. (2006) subsequently emphasized that the diamictites in the Allan Hills intruded the sedimentary Beacon rocks consisting of the Weller, Feather, and Lashly formations and published a geologic map of the area reproduced in Fig. 12.4. The age of the Mawson Formation in the Allan Hills was determined by J.F. Sutter by means of a whole-rock $^{40}Ar/^{39}Ar$ date of 185.7 ± 3.4 Ma of a basalt dike that intruded the diamictite (Hall et al. 1982). This date indicates that the age of the Mawson Formation is Early Jurassic.

The phreatomagmatic origin of the Mawson Formation, demonstrated by White and McClintock (2001), was caused by explosions in the subsurface when large amounts of water interacted with basalt magma that had entered rifts in the basement of the Transantarctic Mountains. The water that caused the phreatomagmatic explosion originated primarily from pore water in the Beacon sandstones. In addition, rift valleys that may have formed at the surface may have contained lakes (Hanson and Elliot 1996) which could have augmented the water available in subsurface. The violent explosions that followed formed funnel-shaped bodies of breccia composed of clasts of sandstone and basalt in a matrix of powdered rocks. The schematic diagram of such a breccia body in Fig. 12.5 indicates that it extends from depth toward the surface and thus appears to intrude the local country rocks, as emphasized by Elliot et al. (2006), even though the breccia diatreme is not composed of plutonic igneous rocks.

White and McClintock (2001) divided the breccia pipe (or diatreme) in Fig. 12.5 into three sections which they identified as:

UD = Upper Diatreme
LD = Lower Diatreme, and
RZ = Root Zone

The *root zone* includes a feeder dike, incompletely homogenized shattered country rock, basalt fragments, pulverized rock debris, and contorted basalt dikes.

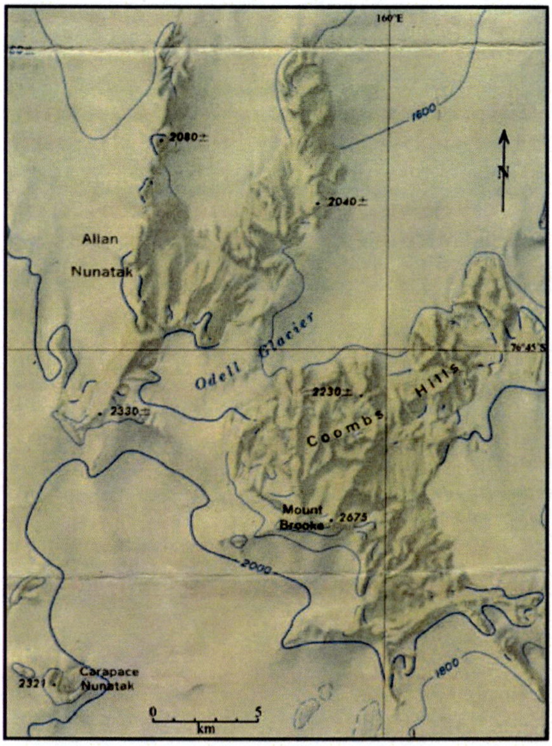

Fig. 12.3 The Allan Hills, the Coombs Hills, and Carapace Nunatak are located at the edge of the polar plateau in southern Victoria Land. The sedimentary rocks of the Beacon Supergroup in this area were intruded by sills of the Ferrar Dolerite and are locally overlain by the Kirkpatrick Basalt. In addition, a large complex of volcanic breccia and related pyroclastics of the Mawson Formation are exposed here which formed by violent phreatomagmatic explosions that preceded and partially overlapped the eruption of the Kirkpatrick Basalt. Excerpt of the topographic map: Convoy Range, Antarctica, ST-57–60/1, published by the US Geological Survey in 1965

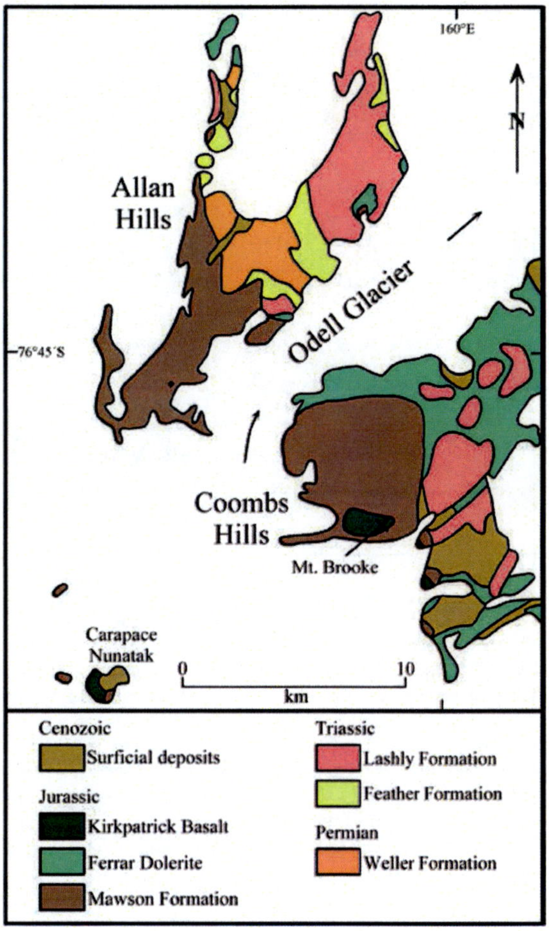

Fig. 12.4 The Allan Hills and Coombs Hills at the edge of the polar plateau in southern Victoria Land expose the Mawson Formation as well as the underlying formations of the Beacon Supergroup and the overlying Kirkpatrick Basalt and Ferrar Dolerite sills. The Mawson Formation consists of laharic deposits and volcanic breccias interbedded with layers of tuffaceous sandstone. This formation represents a transition from felsic pyroclastics, that characterize the Triassic Lashly Formation, and the large-scale eruption of tholeiitic basalt in Gondwana during the Jurassic (Adapted from Elliot et al. 2006)

The *lower diatreme* contains an unstratified mixture of crushed country rock, basalt clasts, and blocks of stratified rim-tuff as well as basalt dikes that originated in the root zone. The *upper part* of the diatreme features a shallow maar crater that is partially filled with well-bedded post-eruptive sediment including pyroclastic rocks.

The phreatomagmatic volcanic eruptions expelled both compressed gases and rock fragments and formed shallow maar craters at the surface surrounded by a tephra ring. The superheated steam that was discharged caused pervasive hydrothermal alteration of the breccia pipes and associated pyroclastics, including the deposition of zeolites in the open spaces of the phreatomagmatic breccia (Ballance 1967; Ballance and Watters 2001).

White and McClintock (2001) considered that the root zone of a phreatomagmatic diatreme can migrate sideways as a result of multiple explosions that shatter the country and thereby expand the root zone. This process can enlarge the volume of brecciated country and cause additional diatremes to form. Accordingly, the Mawson Formation in the Coombs Hills consists of

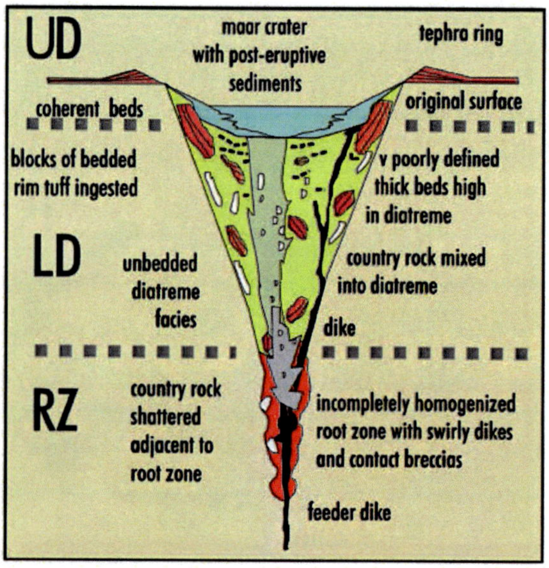

Fig. 12.5 The phreatomagmatic volcanic diatremes were subdivided by White and McClintock (2001) into a root zone (RZ), a lower diatreme section (LD), and an upper diatreme section (UD). The explosions that break up the country rock occur in the root zone where large volumes of water interact with basalt magma rising from depth. The cone shaped breccia pipe is composed of angular fragments of country rock and basalt in a matrix of powdered rocks and is capped by a maar crater that vents steam and other gases. The crater is surrounded by a ring of tephra and rock fragments ejected by the explosions (Reproduced by permission of J.D.L White and the Geological Society of America)

several coalesced phreatomagmatic diatremes. White and McClintock (2001) considered that the Mawson Formation in the Coombs Hills is composed of the root zones and lower-diatreme facies of several diatremes having a volume in excess of 10 km^3. The surface deposits may have been removed by lahars and by the East Antarctic ice sheet which has repeatedly overridden the western flanks of the Transantarctic Mountains during the Cenozoic.

The Mawson Formation in the Coombs Hills is locally overlain by the Kirkpatrick Basalt and was intruded by dikes and sills of the Ferrar Dolerite. This evidence could mean that the Mawson Formation predates the Ferrar Group. However, the presence of basalt and dolerite clasts in the Mawson Diamictite indicates that basalt was already erupting and that sills of dolerite had already intruded the underlying Beacon rocks when the phreatomagmatic explosions occurred. Therefore, the Mawson Formation is part of the Ferrar Group and it is not the youngest formation of the Beacon Supergroup.

Phreatomagmatic volcanic eruptions have also occurred on some of the Cenozoic volcanoes of Marie Byrd Land in West Antarctica (Palais 1985; Palais and Sigurdsson 1989) and in the Eifel volcanic province of Germany (Wörner and Schmincke 1984). In addition, volcanic breccias of phreatomagmatic origin occur at the base of flood basalts in the Karoo volcanic province of southern Africa (Marsh and Skilling 1998), in the Deccan Traps of India (Srinivasan et al. 1998), and in the flood-basalt province of Siberia (Zolotukhin and Al'Mukhamedov 1988). At about the same time, rifting of Laurasia in the northern part of Pangea was accompanied by eruptions of basalt along the eastern coast of North America and in Liberia on the coast of West Africa (Mauche et al. 1989). The petrogenesis of the flood basalts in these areas and in the Transantarctic Mountains was discussed by Faure (2001) based on isotopic and chemical data.

12.3.2 Prebble Formation

In the Queen Alexandra Range of the Central Transantarctic Mountains diamictites and pyroclastic rocks of the Prebble Formation occur below the Kirkpatrick Basalt. The Prebble Formation overlies the felsic tuffs of the upper Falla Formation which Elliot (1996) renamed the Hanson Formation. Whole-rock samples of the tuff yielded a Rb-Sr date of 186 ± 9 Ma corresponding to an Early Jurassic age (Faure and Hill 1978) which was later confirmed by Hammer and Hickerson (1994) based on the bones of tetrapod vertebrates.

The Prebble Formation occurs not only in the Queen Alexandra Range, but also in the Otway Massif, at Mt. Pratt located 30 km east of the Massif, and in the scattered nunataks of the Grosvenor Mountains all of which are located close to the head of Mill Glacier on the edge of the polar plateau. The Prebble Formation at these locations is composed of lahar deposits, pyroclastic breccias, and tuff. The type section of this formation is on the northwest face of Mt. Kirkpartrick at the head of the Prebble Glacier (Fig. 10.14). The laharic deposits contain clasts of basalt, dolerite, and sedimentary rocks of the underlying Beacon rocks. They also contain clasts of felsic volcanic rocks and they are interbedded with layers of felsic tuff (Barrett and Elliot 1972; Barrett et al. 1986; Elliot and Larsen 1993). Evidently, the felsic and mafic volcanic rocks are not sharply divided between the Kirkpatrick Basalt and the Hanson Formation. The thickness of the Prebble Formation in the Queen Alexandra Range changes from about 150 m on Mt. Kirkpatrick to only 3 m on Lindsay Peak but reaches 460 m in the Otway Massif (Barrett et al. 1986). The Falla and the overlying Hanson formations in the Queen Alexandra Range were cut by intrusive bodies of breccia (Elliot 2000) which suggests that these breccias formed by a similar explosive phreatomagmatic process that formed the breccias of the Mawson Formation in the Coombs Hills (White and McClintock 2001).

12.3.3 Exposure Hill Formation

The Exposure Hill Formation occurs at the base of the sheet-like lava flows of Kirkpatrick Basalt that form the Mesa Range of northern Victoria Land and on nearby nunataks most of which are located in the Rennick graben in Fig. 12.6. The deposits of breccia and related pyroclastic rocks occur at Exposure Hill at the southern end of Gair Mesa and at Mt. Fazio and Scarab Peak on Tobin Mesa as well as on Mt. Carson adjacent to the Aeronaut Glacier in Fig. 12.7. In addition, Elliot et al. (1986a) described a deposit of breccia on Agate Peak located about 26 km east of the northern tip of Pain Mesa (Fig. 12.6).

The thickness of the Exposure Formation varies locally between 10 and 150 m. It is composed of breccia containing clasts of altered basalt, vesicular basalt, dolerite, sandstone, and beds of lapilli tuff, tuff, and tuffaceous ripple-marked sandstone (Elliot et al. 1986a; Wörner 1992; Roland and Wörner 1996).

The matrix of the breccias and lapilli tuffs of the Exposure Hill Formation consists of grains of quartz, feldspar, amphibole, pyroxene, garnet, biotite, and opaques (Elliot et al. 1986a). Angular quartz grains are most abundant, followed by plagioclase (oligoclase), whereas alkali feldspar (sanidine) is rare. The matrix of the breccias and lapilli tuffs also contains blocky grains of altered basalt and shards of silicic volcanic glass that have been replaced by zeolites. Except for the grains of basalt, all of the mineral grains as well as the shards of silicic glass occur in the underlying

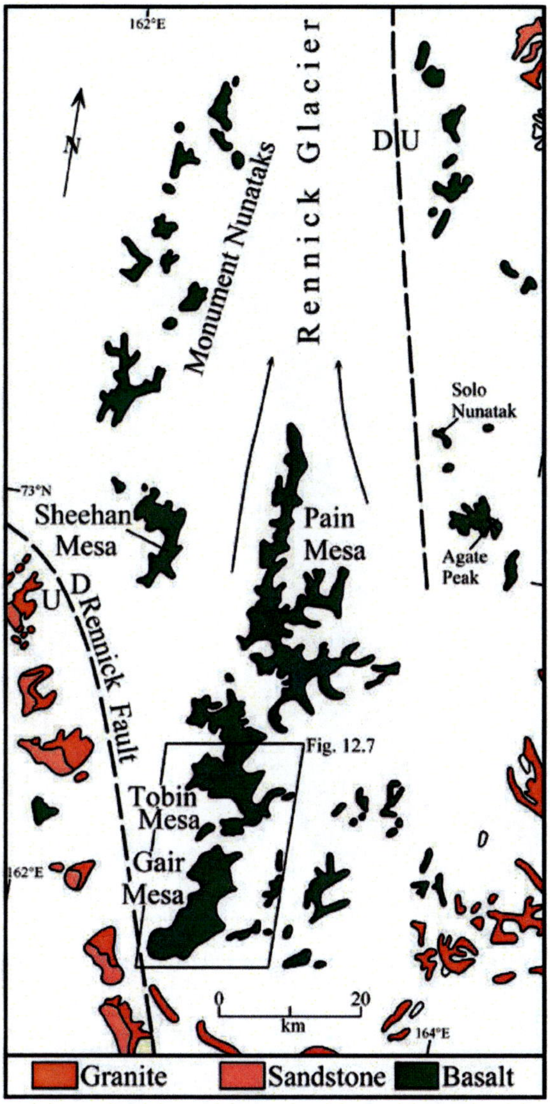

Fig. 12.6 The geology of the Mesa Range includes the Kirkpatrick Basalt which is underlain by the sandstones of the Beacon Supergroup. The Beacon rocks were deposited unconformably on the granitic rocks of the local basement complex (Adapted from Gair et al. 1969 with information from Elliot et al. 1986a)

Section Peak Formation. Therefore, the breccias of the Exposure Hill Formation could have formed as a result of phreatomagmatic explosions in the subsurface like those that produced the Mawson Diamictite in the Coombs Hills (White and McClintock 2001).

The age of Exposure Hill Formation is constrained by its stratigraphic position to be older than

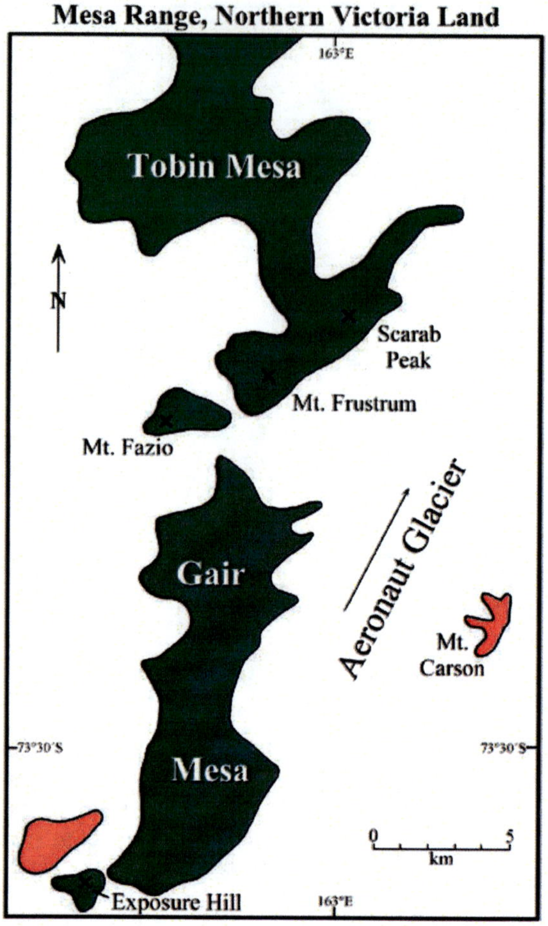

Fig. 12.7 Outcrops of the Exposure Hill Formation in the Mesa Range of northern Victoria Land occur at Exposure Hill south of Gair Mesa and on Tobin Mesa at Mt. Fazio and Scarab Peak, as well as on Mt. Carson along the Aeronaut Glacier. Gair and Tobin mesas are composed of Kirkpatrick Basalt (green) which was deposited on the sandstones of the Section Peak Formation (pink) of the Beacon Supergroup (Adapted from Gair et al. 1969, Elliot et al. 1986a), and based on the topographic map : Mt. Murchison, Antarctica; SS 58–50/5, published in 1970 by the US Geological Survey, Washington, DC)

the Section Peak Formation and younger than the Kirkpatrick Basalt. The age of the Section Peak Formation is Late Triassic according to palynomorphs identified by Norris (1969). The whole-rock K-Ar dates of the Kirkpatrick Basalt in northern Victoria Land range widely from 113 to 178 Ma (Elliot and Foland 1986). These authors considered that only the oldest date accurately records the age of these rocks, whereas the other eight samples they dated had lost radiogenic ^{40}Ar. A date of 178 ± 3 Ma corresponds to an early Middle Jurassic age for the Kirkpatrick Basalt. A more precise age determination by the $^{40}Ar/^{39}Ar$ method of 176.6 ± 0.7 Ma was later published by Heimann et al. (1994). On the basis of this evidence, the age of the Exposure Hill Formation is early Middle Jurassic.

12.4 Kirkpatrick Basalt, Northern Victoria Land

The outcrops of the Kirkpatrick Basalt identified in Table 12.2 are remnants of much larger deposits that occur from the Litell Rocks and Mesa Range of northern Victoria Land to the Otway Massif and Grosvenor Mountains at the head of the Mill Glacier in the central Transantarctic Mountains, a distance of 1,600 km.

The Kirkpatrick Basalt does not occur in the Queen Maud, Horlick, Thiel, and Pensacola mountains. Exposures of basalt in the Shackleton Range have not been correlated with the Kirkpatrick Basalt. However, sills of Ferrar Dolerite do occur in the Whichaway Nunataks (Brewer 1989), in the Theron Mountains (Blundell and Stephenson 1959; Brewer et al. 1987, 1992). In addition, basalt magma pooled in a large magma chamber in the crust beneath the Pensacola Mountains where it differentiated into the layered gabbro complex of the Dufek intrusion (Ford and Kistler 1980).

12.4.1 Mesa Range

The Mesa Range in Fig. 12.8 is a large and deeply dissected plateau consisting of flat-lying layers of

Table 12.2 Principal outcrops of the Kirkpatrick Basalt in the Transantarctic Mountains

Mesa Range, Monument Nunataks, Litell Rocks, and related nunataks in northern Victoria Land
Prince Albert Mountains, Brimstone Peak, Allan Hills, and Carapace Nunatak in southern Victoria Land
Westhaven Nunatak at the head of Hatherton Glacier in the Britannia Range
Mt. Kirkpatrick, Mt. Falla, Tempest Peak, Blizzard Peak in the Queen Alexandra Range
Otway Massif, Grosvenor Mountains at the head of Mill Glacier

Fig. 12.8 Aerial view of the Mesa Range in northern Victoria Land consisting of Gair Mesa in the foreground followed by Tobin and Pain mesas farther north. The Rennick Glacier west of the Mesa Range flows north to the Oates Coast. The flat-lying layers of the Kirkpatrick Basalt are well exposed in this view of the western slope of Gair Mesa (Photo courtesy of the US Geological Survey, US Antarctic Resource Center, TMA 872, right-looking oblique camera (33R), frame 166, taken November 5, 1961)

Kirkpatrick Basalt interbedded with occasional sedimentary interbeds of lacustrine and volcaniclastic origin. The Mesa Range consists of the Gair, Tobin, and Pain mesas and extends for nearly 80 km from the southern end of Gair Mesa to the northern tip of Pain Mesa. The highest elevation of 3,280 m occurs at the southern tip of Gair Mesa. The elevations of the summit plateaus decrease in a northerly direction toward Mt. Masley on Pain Mesa which has an elevation of 2,605 m.

Flows of Kirkpatrick Basalt are also exposed on the Sheehan Tableland, Sculpture Mountain, the Monument Nunataks, and on many small nunataks located east of the Mesa Range, including Solo Nunatak (at 72°50′S, 163°35′E). The most northerly outcrop of Kirkpatrick Basalt occurs at the Litell Rocks (71°24′S, 162°00′E) adjacent to the northeast side of the Morozumi Range (Skinner et al. 1981; Hornig 1981). The Mesa Range and the mountains between the upper Rennick Glacier and the Ross-Sea coast were originally mapped by Gair (1967) and by Nathan and Schulte (1968). Gair et al. (1969) later published the first comprehensive geologic map of the area. The Mesa Range was revisited during the 1981/82 field season as part of the International Northern Victoria Land Project with logistical support by the US Antarctic Program (Elliot et al. 1982; Stump 1986).

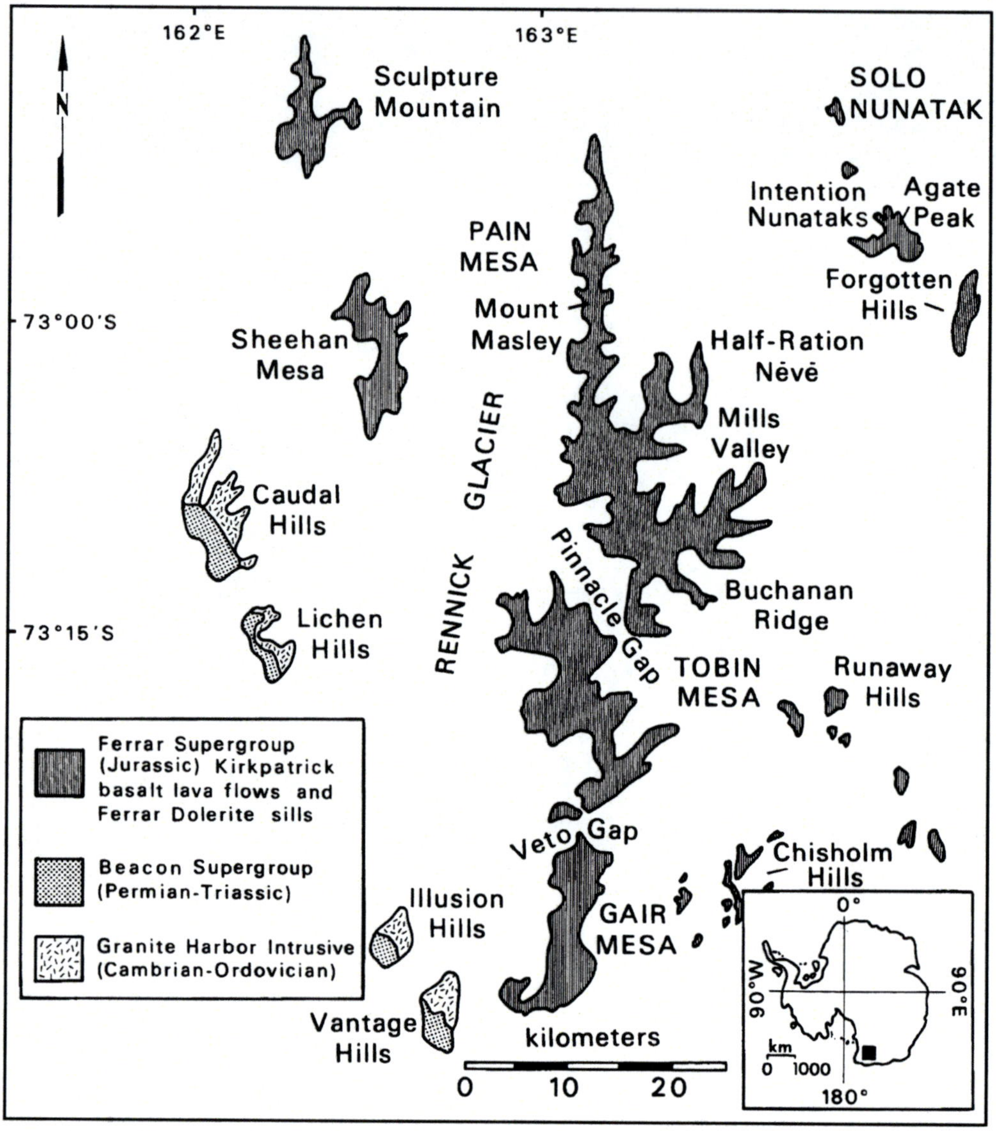

Fig. 12.9 The Mesa Range of northern Victoria Land is located along the upper reaches of the Rennick Glacier which flows north to the Oates Coast. The mesas and associated nunataks are composed primarily of basalt flows of the Kirkpatrick Basalt. The map identifies the location of Mt. Masley on Pain Mesa and of Solo Nunatak which are described in the text (Reproduced from Mensing 1987)

12.4.2 Tobin Mesa

The flow stratigraphy, petrography, and chemical composition of the basalt flows on *Tobin Mesa* in Fig. 12.9 were described by Siders (1983), Haban (1984), Siders and Elliot (1985), Elliot et al. (1986b), and Fleming et al. (1995). This mesa is composed of about 40 flows that have a total stratigraphic thickness of about 900 m. The thickness of the flows ranges from less than 1 m to more than 120 m. All flows have amygdaloidal zones at their bases and tops and the vesicles are filled with zeolites, quartz, and calcite. The thickest flows are coarse grained and have diabasic texture. Tobin Mesa is capped by a black flow composed of aphanatic to glassy basalt.

The basalt flows on Tobin Mesa are interbedded with thin layers of sedimentary rocks most of which are volcanogenic, but some of which are lacustrine in origin. Elliot et al. (1986b) described a locality on the southwest ridge of Gair Mesa where a bed of finely laminated sediment, 1.85 m thick, occurs beneath a thick basalt flow. The lower part of this bed consists of sandstone and mudstone containing mudcracks and mud-flake breccias. The upper 10–15 cm of this bed consist of very fine grained sediment containing conchostracans and other unspecified freshwater fossils. Elliot et al. (1986b) also mentioned that large conchostrachan-bearing clasts occur between pahoehoe toes at Agate Peak and they photographed an upright tree trunk in the basalt at Mt. Fazio on Tobin Mesa. Some of the sedimentary interbeds are actually fossil soils that formed by weathering of the underlying lava flows and they contain volcanic ash of both basaltic and silicic composition (e.g., Elliot et al. 1988, 1991). Layers of volcaniclastic sediment up to 4 m thick that occur at Mt. Fazio and Haban Spur on Tobin Mesa provide evidence of local environmental conditions during the eruption of the Kirkpatrick Basalt.

Siders and Elliot (1985) published chemical analyses of 18 basalt specimens most of which originated from Siders Bluff on the northwest corner of Tobin Mesa. The rocks are silica-rich but alkali-poor such that most specimens from Tobin Mesa analyzed by Siders and Elliot (1985) are classifiable in Fig. 12.10 as basaltic andesite and even andesites. However,

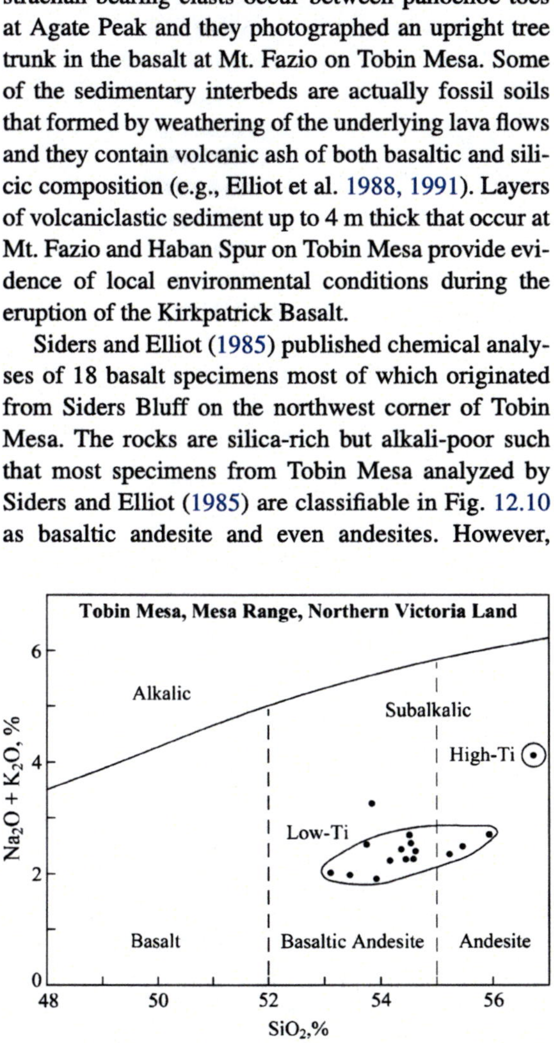

Fig. 12.10 The lava flows on Tobin Mesa are subalkaline basaltic andesites and even stray into the andesite field in this alkali-silica diagram (Wilson 1989). The phenocryst assemblage of plagioclase, pyroxene, and magnetite identifies these rocks as tholeiitic basalt in spite of their high silica concentrations (Data from Siders and Elliot 1985)

these rocks are described as basalts because the phenocrysts are composed of Ca-rich plagioclase (labradorite and bytownite), pyroxene (augite, pigeonite, and hypersthene), and opaques (titanomagnetite) (Siders and Elliot 1985). The flow that caps Tobin Mesa has high concentrations of TiO_2 (1.93%), SiO_2 (56.66%) $Na_2O + K_2O$ (4.31%), and P_2O_5 (0.27%). In addition, the capping flow on Tobin Mesa has exceptionally high concentrations of certain trace elements (e.g., Ba = 398 ppm, V = 391 ppm, Zr = 252 ppm, Rb = 71 ppm, Y = 60 ppm, Nb = 13 ppm, and thorium = 10 ppm.) However, the concentration of strontium (128 ppm) does not differ appreciably from the average strontium concentration of 17 low-Ti flows on Tobin Mesa (124 ± 23 ppm, 1σ). Additional information on the lava flows on Tobin Mesa was included in a report by Siders (1983).

Elliot and Foland (1986) reported whole-rock K-Ar dates of different size fractions of seven samples of basalt from Tobin Mesa. Although the dates ranged widely from 178 ± 3 to 113 ± 2 Ma, the authors considered the age of the lava flows to be greater than 178 Ma and speculated that all of the samples lost radiogenic ^{40}Ar either continuously since crystallization or as a result of a geologic event during the Cretaceous Period. The first reliable age determination of the Kirkpatrick Basalt on Tobin Mesa of 174.2 ± 1.0 Ma was reported by McIntosh et al. (1986) based on the $^{40}Ar/^{39}Ar$ incremental-heating method (Appendix 3.6.4). This date established the Middle Jurassic age of the Kirkpatrick Basalt in the Mesa Range.

Subsequently, Foland et al. (1993) reported $^{40}Ar/^{39}Ar$ dates of 176.8 ± 0.8 and 176.1 ± 0.6 Ma for two plagioclase fractions of the youngest basalt flow on Mt. Frustrum of Tobin Mesa which confirmed the result published earlier by McIntosh et al. (1986). Foland et al. (1993) also successfully dated a sample of basalt glass from southern Victoria Land by the $^{40}Ar/^{39}Ar$ method and obtained a date of 176.6 ± 0.5 Ma. These dates were the first reliable indication that the eruption of basalt in northern and southern Victoria Land occurred virtually at the same time during the Middle Jurassic.

The first measurements of the isotopic composition of strontium in the Kirkpatrick Basalt flows on Tobin Mesa were made by L.M. Jones and interpreted by Hill (1969). The samples were collected by H.S. Gair and included three specimens of the high-Ti flow that caps Scarab Peak, one sample of low-Ti basalt taken about

Table 12.3 Analytical data for Kirkpatrick Basalt on Scarab Peak, Tobin Mesa, NVL (Collected by H.S. Gair, analyzed by L.M. Jones, and interpreted by Hill 1969)

Sample & flow type	Rb (ppm)	Sr	K (%)	Rb/Sr	$^{87}Rb/^{86}Sr$	$^{87}Sr/^{86}Sr$
296[a] hi-Ti	70.4	131.5	1.49	0.535	1.546	0.7158
297[b] hi-Ti	64.7	130.4	1.52	0.496	1.434	0.7140
298[c] hi-Ti	80.8	130.0	1.74	0.621	1.796	0.7139
299[d] lo-Ti	28.0	173.6	0.627	0.161	0.466	0.7140
300[e] lo-Ti	39.5	141.2	0.905	0.279	0.808	0.7155

[a]Scarab Peak, top of dark flow (61 m) that forms the top of Tobin Mesa (Gair's sample 1)

[b]Scarab Peak, 2.50 m below sample 296 (Gair's sample 5)

[c]Scarab Peak, 61 m below sample 296 from the base of the top flow (Gair's sample 9)

[d]Scarab Peak, 640 m below the top of Tobin Mesa (Gair's sample #17)

[e]First-Climb Peak (Gair's name), Ferrar Dolerite, 305 m below the top of the mountain (Gair's sample #25)

640 m below the summit of that peak, and a sample of a dolerite sill taken 305 m below the top of "First-Climb Mountain" (probably Gair Mesa). The results are compiled in Table 12.3.

The samples of high-Ti basalt have comparatively high concentrations of rubidium (72 ± 8 ppm) and potassium (1.58 ± 0.14%) which yield a low average K/Rb ratio of 219 characteristic of sialic rocks of the continental crust. The low-Ti basalt and dolerite have lower concentrations of both rubidium (34 ± 8 ppm) and potassium (0.766 ± 0.196%) which yields a K/Rb ratio of 225 that is characteristic of the Kirkpatrick Basalt and the Ferrar Dolerite of the Transantarctic Mountains.

The $^{87}Sr/^{86}Sr$ ratios of the rocks in Table 12.3 range from 0.7140 to 0.7160 but are widely scattered and therefore do not permit an age determination. The average initial $^{87}Sr/^{86}Sr$ ratio at 175 Ma ($\lambda = 1.42 \times 10^{-11}$ year^{-1}) is 0.71162 ± 0.00075 $(1\bar{\sigma})$. Evidently, the Kirkpatrick Basalt on Tobin Mesa is enriched in radiogenic ^{87}Sr and may well be genetically related to the Wisanger Basalt (Section 12.1) and the Tasmanian Dolerite (Section 12.2).

12.4.3 Pain Mesa

Pain Mesa in Fig. 12.11 is composed of flat-lying layers of Kirkpatrick Basalt much like Tobin Mesa described in the previous section. These basalt flows were investigated by Mensing (1987) and by Mensing et al. (1991). Mount Masley on Pain Mesa is composed of more than 20 basalt flows that were correlated to the flows on Solo Nunatak (Fig. 12.9) by Mensing et al. (1991) by means of a prominent marker flow ("Lower Black Marker"). The combined sequence consists of more than 40 flows with a total thickness greater than 800 m.

Mensing (1987) divided the flows on Pain Mesa into two groups based on differences in their chemical compositions. The flows of the upper suite on Mt. Masley in Fig. 12.12 have higher concentrations of SiO_2, TiO_2, P_2O_5, and $FeO + Fe_2O_3$, but lower concentrations of MgO, CaO, and Al_2O_3 than the flows of the lower suite. Accordingly, the upper suite is composed of high-Ti flows, whereas the lower suite consists of low-Ti flows. In spite of the pronounced difference in chemical composition, all of the flows on Pain Mesa are supersaturated with respect to silica and are therefore classifiable as "tholeiites". The high-Ti flows also have high concentrations of zirconium and yttrium which distinguish them from the low-Ti flows in Fig. 12.13.

The basalt flows on Pain Mesa are hypocrystalline with phenocrysts of plagioclase, pyroxene, and iron oxide in a quenched glassy matrix (Mensing 1987). Olivine was not found in any of the samples. The low-Ti flows are more coarse grained than the high-Ti flows and contain Ca-rich plagioclase (labradorite) as well as clinopyroxene (augite and pigeonite) and orthopyroxene (hypersthene) which is rimmed by clinopyroxene. The high-Ti flows are aphanitic and contain quenched plagioclase crystals with "belt buckle" and "swallow-tail" morphology composed of Na-rich andesine. The high-Ti flows contain only clinopyroxene (augite and

Fig. 12.11 Mt. Masley at the northern end of Pain Mesa is composed of a sequence of lava flows of low-Ti tholeiite basalt and the overlying darker high-Ti flows of the Kirkpatrick Basalt. Profiles of the chemical compositions of the flows on Mt. Masley are shown in Fig. 12.12 (Photo by T.M. Mensing)

pigeonite). The matrix, which contains disseminated opaque crystals (magnetite), comprises about 5% by volume of the low-Ti rocks and as much as 75% of the high-Ti flows. Haban (1984) reported that the iron oxide crystals in the flows of both suites on Tobin Mesa are composed of titanomagnetite.

The classification of the Kirkpatrick Basalt in the Mesa Range is ambiguous because the silica concentrations of the high-Ti flows exceed the range characteristic of basalt. The total-alkali-silica diagram of Cox et al. (1979) and Wilson (1989) in Fig. 12.14 for flows of Pain Mesa illustrates this discrepancy. Although all of the samples analyzed by Mensing (1987) are subalkaline, the silica concentrations extend from 49.5% in the basalt field across the basaltic andesite into the andesite field at 56.0%. Therefore, the low-Ti flows range from basalt to basaltic andesite in Fig. 12.14, whereas the high-Ti flows, range from basaltic andesites to andesites. Nevertheless, the phenocryst mineralogy allows both suites to be classified as tholeiite basalt.

The summary of the chemical analyses reported by Mensing (1987) in Table 12.3 demonstrates that the chemical composition of the high-Ti basalt flows that cap Pain Mesa at Mills Valley, Buchanan Ridge, and at Mt. Masley is distinctly different from the chemical composition of the low-Ti flows that make up the main body of Pain and Tobin mesas. The high-Ti flows overlie, but are *not* interbedded with, the low-Ti flows. The "Lower Black Marker" as well as the "Upper Black Marker" both of which are interbedded with low-Ti flows in Mills Valley, on Buchanan Ridge, and on Mt. Masley of Pain Mesa are actually composed of low-Ti basalt in spite of their appearance (Mensing 1987; Mensing et al. 1991, Fig. 4). The transition from the lower to the upper suites appears to be abrupt because it is not marked by layers of paleosoil or sedimentary interbeds. This point is significant because it justifies the assumption that the age of the high-Ti flows is only slightly less than the $^{40}Ar/^{39}Ar$ date of 174.2 ± 1.0 Ma reported by McIntosh et al. (1986) for a whole-rock samples of a low-Ti basalt from the base of the Mesa Range.

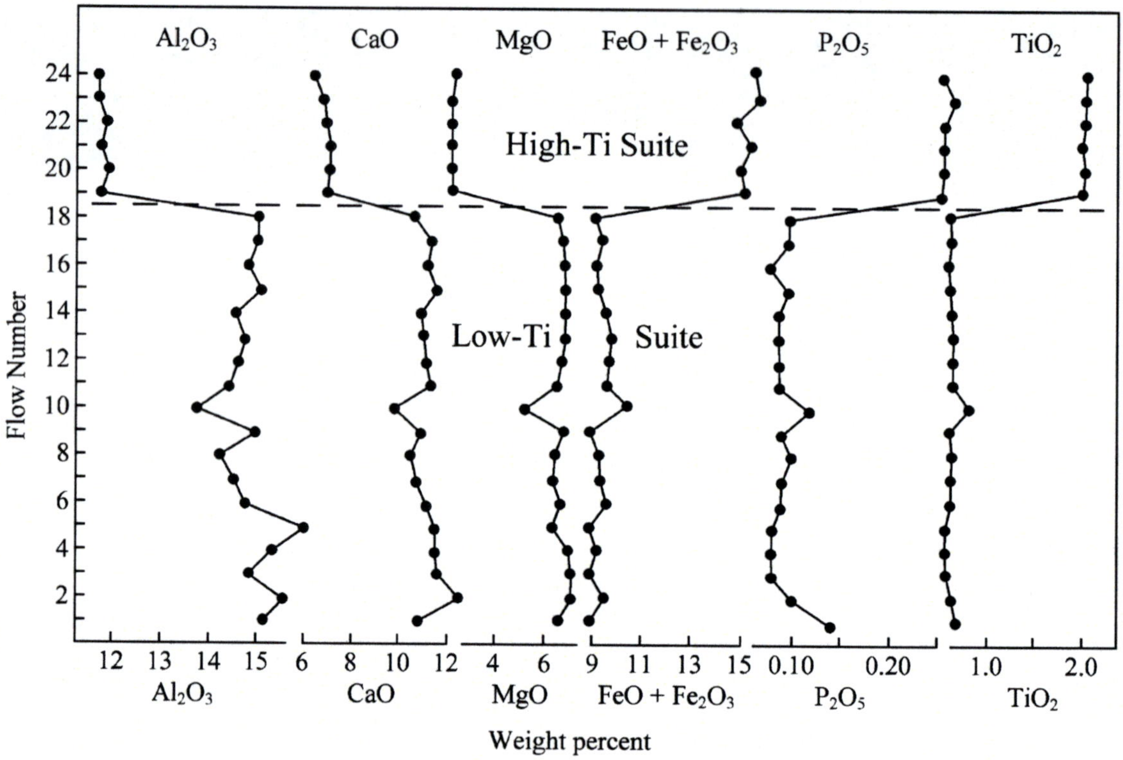

Fig. 12.12 The lava flows that form Mt. Masley of Pain Mesa consist of two suites that have distinctly different major-element concentrations, including SiO_2 and certain trace elements (e.g., zirconium and yttrium). The rocks of both suites have been classified as tholeiitic basalts based on the mineralogy of phenocrysts (Ca-rich plagioclase, two pyroxenes, and magnetite in a fine-grained matrix). The matrix of the high-Ti flows is aphanitic and black, whereas the matrix of the low-Ti flows is fine grained and dark grey (Mensing 1987)

The expectation that the age of the high-Ti flows on Pain Mesa is only slightly less than 175 Ma is drawn into question by the whole-rock Rb-Sr dates of the two types of flows. For example, the high-Ti flows in Mills Valley analyzed by Mensing (1987) yield a Rb-Sr date of only 82 ± 9 Ma in Fig. 12.15 with an initial $^{87}Sr/^{86}Sr$ ratio of 0.71162 ± 0.00016 compared to 169 ± 17 Ma for the low-Ti flows and an initial $^{87}Sr/^{86}Sr$ ratio of 0.71041 ± 0.00017. Additional Rb-Sr dates of the two types of flows at Buchanan Ridge and Mt. Masley in Table 12.5 confirm that the high-Ti flows crystallized during the Late Cretaceous while the low-Ti flows were extruded during the Middle Jurassic. Both types of flows have high initial $^{87}Sr/^{86}Sr$ ratios that are characteristic of Precambrian basement rocks of the continental crust and are *incompatible* with magmas derived from the litherospheric mantle (Faure 2001). The initial $^{87}Sr/^{86}Sr$ ratios of the Kirkpatrick Basalt are similar to the high initial $^{87}Sr/^{86}Sr$ ratios of the Wisanger Basalt of southern Australia (Section 12.1) and the Tasmanian Dolerite (Section 12.2) both of which were not only extruded at about the same time but also formed from magma of similar composition and origin.

The evident enrichment of the Kirkpatrick Basalt in radiogenic ^{87}Sr has been explained by two alternative hypotheses:

1. The magma originated from heterogeneous sources in the subcontinental mantle (Kyle 1980).
2. Mantle-derived basalt magma was contaminated by assimilating crustal rocks (Faure 1981).

These opposing alternatives were reconciled by an ingenious proposal by Hergt (1987). The petrogenesis of the basalt and dolerite of the Ferrar Group in Antarctica is an important topic in petrology because it is closely related to the rifting of Gondwana and the subsequent displacement of the continental fragments. Therefore, we postpone the discussion of the

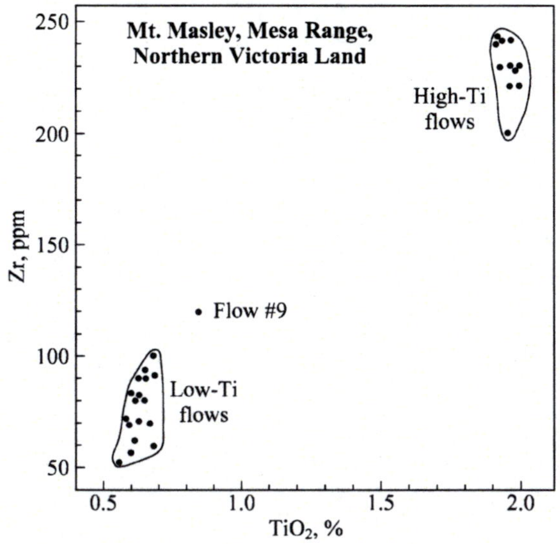

Fig. 12.13 The lava flows that form Mt. Masley (Fig. 12.9) consist of low-Ti basalt capped by six flows of high-Ti basalt (Fig. 12.12) which also have elevated concentrations of zirconium and yttrium. The difference of the chemical compositions between the high-Ti and low-Ti flows implies that they may have formed from different magmas (Mensing 1987)

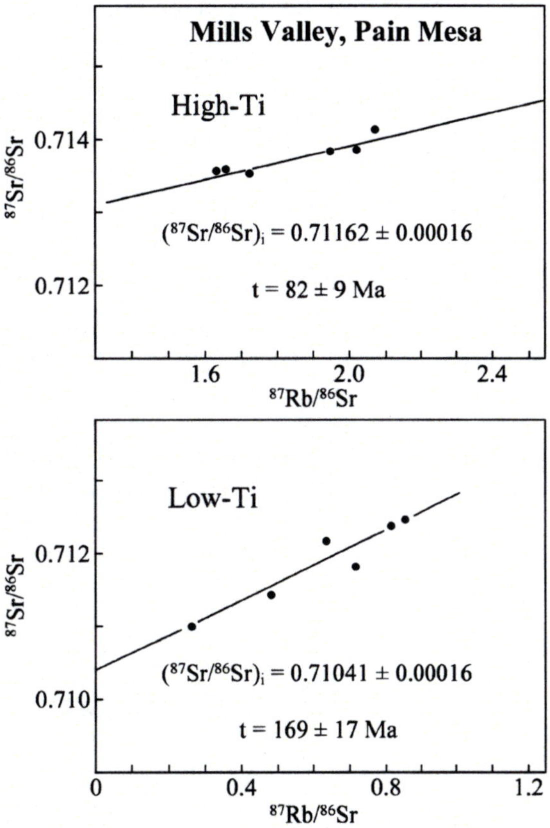

Fig. 12.15 The low-Ti flows of the Kirkpatrick Basalt in Mills Valley of Pain Mesa yield a whole-rock Rb-Sr date of 169 ± 7 Ma (Middle Jurassic) and an initial $^{87}Sr/^{86}Sr$ ratio of 0.71041 ± 0.00017, whereas the high-Ti flows that cap the mesa yield a much lower date of 82 ± 9 Ma (Late Cretaceous) and a higher initial $^{87}Sr/^{86}Sr$ ratio of 0.71162 ± 0.00016. This unexpected result means either the high-Ti flows were erupted much later than the low-Ti flows (contrary to evidence in the field) or that the isotopic composition of strontium of the high-Ti flows was homogenized by a thermal pulse during the Late Cretaceous as first suggested by Mensing (1987, 1991)

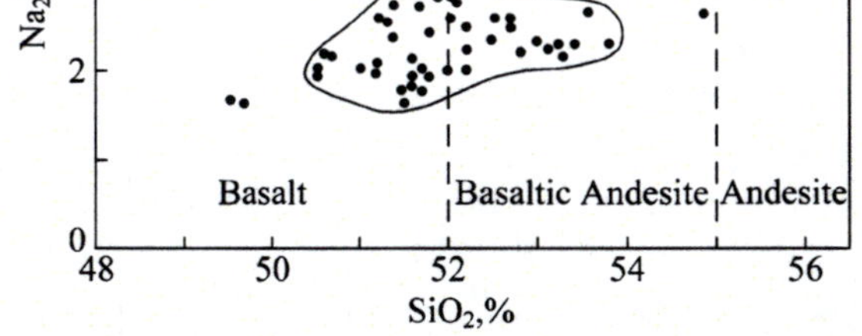

Fig. 12.14 The flows of Kirkpatrick Basalt on Pain Mesa have a wide range of silica concentrations from 49.5% to 56.0% which causes them to extend from basalt to basaltic andesite in the chemical classification of volcanic rocks by Cox et al. (1979) and Wilson (1989). The chemical classification based on this diagram is in conflict with the classification based on the mineralogy of the phenocrysts on the basis of which these rocks are tholeiite basalts (Mensing 1987)

petrogenesis of the Kirkpatrick Basalt until after we have described the basalt flows in the Alexandra Range of the central Transantarctic Mountains.

The apparent Late Cretaceous isotopic homogenization of strontium of the high-Ti flows that cap Pain Mesa is a separate issue that is related to the tectonic history of northern Victoria. The most likely explanation for this phenomenon is illustrated in Fig. 12.16. The idea is that the high-Ti flows were extruded soon after the eruption of the low-Ti basalt at about 170 Ma and that the anomalously low date of these flows is the result of isotopic homogenization of strontium during

Table 12.4 Average chemical analyses of low-Ti and high-Ti lava flows at Mills Valley, Buchanan Ridge, and Mt. Masley on Pain Mesa, Mesa Range, northern Victoria Land (Mensing 1987)

	High-Ti	Low-Ti[a]
	(%) ± 1σ (N = 28)	(%) ± 1σ (N = 44)
SiO_2	55.0 ± 0.5	52.6 ± 4.9
Al_2O_3	11.5 ± 0.1	14.5 ± 0.50
CaO	6.76 ± 0.18	10.7 ± 0.60
MgO	2.14 ± 0.05	6.56 ± 0.57
Na_2O	2.13 ± 0.23	1.67 ± 0.16
K_2O	1.66 ± 0.26	0.68 ± 0.27
Fe_2O_3	16.0 ± 0.28	9.77 ± 0.42
MnO	0.20 ± 0.01	0.17 ± 0.01
TiO_2	1.95 ± 0.03	0.62 ± 0.05
P_2O_5	0.25 ± 0.01	0.09 ± 0.01
LOI	1.61 ± 0.59	2.73 ± 1.49
Sum	99.2	100.09
Trace elements, ppm (μg/g)		
Chronium	12.4 ± 4.3	108 ± 21
Rubidium	79.5 ± 13.9	28.4 ± 11.6
Strontium	128.0 ± 3.4	116.9 ± 28.3
Yttrium	43.4 ± 10.1	22.5 ± 15.3
Zirconium	2.26 ± 13	76.6 ± 14.9
Niobium	28.2 ± 8.6	20.4 ± 7.2
Barium	444 ± 23 (7)	–

[a]Excluding the Lower Black Marker on Mt. Masley of Pain Mesa

Table 12.5 Summary of Rb-Sr dates and initial $^{87}Sr/^{86}Sr$ ratios of Kirkpatrick Basalt in Pain Mesa, Mesa Range, NVL (Data and interpretation by Mensing 1987, 1991)

	Date (Ma)	Initial $^{87}Sr/^{86}Sr$
	Mills Valley, low-Ti basalt	
	169 ± 17	0.71041 ± 0.00017 (1σ)
	Mills Valley, high-Ti basalt	
	82 ± 9	0.71162 ± 0.00016 (1σ)
	Buchanan Ridge, low-Ti basalt	
	164 ± 11	0.71033 ± 0.00017 (1σ)
	Mt. Masley, low-Ti basalt	
A	183 ± 6	0.70932 ± 0.000055 (1σ)
B	172 ± 10	0.71035 ± 0.00013 (1σ)
C	177 ± 10	0.71073 ± 0.00011 (1σ)
	Mt. Masley, high Ti-basalt	
	88 ± 5	0.71163 ± 0.00020 (1σ)
	Average Rb-Sr dates	
	Low-Ti basalt:	173 ± 8 Ma ($2\bar{\sigma}$)
	High-Ti basalt:	85 ± 6 Ma ($2\bar{\sigma}$)

a geologic event in the Early Cretaceous. The nature of the event will become evident in Chapter 15 when we discuss the rifting of Gondwana that started during

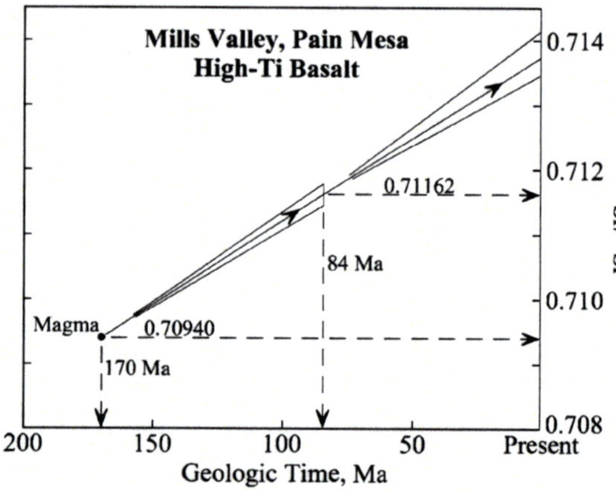

Fig. 12.16 The strontium in the high-Ti flows at Mills Valley of Pain Mesa, northern Victoria Land, could have evolved from magma that was erupted at 170 Ma with an initial $^{87}Sr/^{86}$ Sr ratio of 0.70940. Subsequently, the $^{87}Sr/^{86}Sr$ ratio in each flow increased at rates depending on their Rb/Sr (i.e., $^{87}Rb/^{86}Sr$) ratios. At 84 Ma, when the average $^{87}Sr/^{86}Sr$ ratio of the high-Ti flows had reached 0.71162, the $^{87}Sr/^{86}Sr$ ratios of all high-Ti flows were equilibrated (reset) to a value of 0.71162, thereby yielding the errorchron shown in Fig. 12.15 (Mensing 1987, 1991)

Early to Middle Jurassic and continued well into the Cretaceous Period.

12.4.4 Solo Nunatak

Solo Nunatak in Figs. 12.9 and 12.17 is an erosional remnant of the lava plateau that forms the Mesa Range. It is located 22 km east northeast of the northern tip of Pain Mesa and is composed of 24 flows of low Ti-basalt that were correlated with the low-Ti flows on Pain Mesa (Mensing 1987; Mensing et al. 1991). The flows contain phenocrysts of plagioclase (labradorite) and pyroxene (augite, pigeonite, and hypersthene) in a fine-grained to aphanitic matrix containing disseminated crystals of magnetite. This assemblage of phenocrysts identifies the rocks as tholeiite basalt. However, the concentrations of $Na_2O + K_2O$ and SiO_2 reported by Mensing (1987) and by Mensing et al. (1984, 1991) indicate that these rocks extend into the field of basaltic andesites in Fig. 12.18.

The majority of the flows in Fig. 12.19 have nearly uniform chemical compositions (e.g., SiO_2 ~ 53%, MgO + CaO ~ 18%). These profiles lack the discontinuities caused by the differences in chemical composition of the low-Ti and high-Ti flows on Pain Mesa in Fig. 12.12. Therefore, all of the flows on Solo Nunatak belong to the low-Ti suite of the Kirkpatrick Basalt with concentrations of TiO_2 between 0.54% and 0.79% (Mensing 1987).

The initial $^{87}Sr/^{86}Sr$ ratios of the low-Ti flows on Solo Nunatak reported by Mensing et al. (1984) range from 0.70985 to 0.71207 with an average of 0.71069 ± 0.00001 ($1\bar{\sigma}$) for 37 samples from 24 flows recalculated to 175 Ma using $\lambda = 1.42 \times 10^{-11}$ year^{-1}. The average initial $^{87}Sr/^{86}Sr$ ratio of the Kirkpatrick Basalt on Solo Nunatak is virtually identical to the average initial $^{87}Sr/^{86}Sr$ ratios of the low-Ti flows on Pain Mesa,

Fig. 12.17 Solo Nunatak at 72°50′S and 163°35′E is an erosional remnant located 22 km east northeast of the northern tip of Pain Mesa (Fig. 12.9). Is it composed of 23 flows of the Kirkpatrick Basalt of the low-Ti suite. Mensing (1987, 1991) correlated the flows on Solo Nunatak with the flows on Mt. Masley of Pain Mesa by means of a prominent black marker flow that occurs in both sections (Photo by G. Faure)

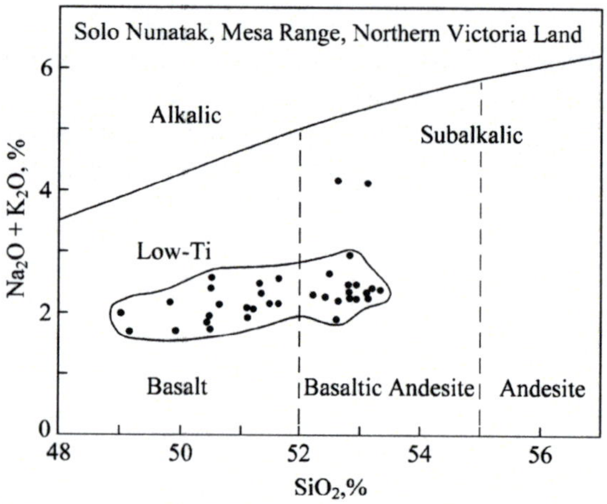

Fig. 12.18 On the alkali-silica diagram of Cox et al. (1979) and Wilson (1989) the flows of Kirkpatrick Basalt of Solo Nunatak extend from basalt to basaltic andesite much like the lava flows on Pain Mesa (Fig. 12.14) and on Tobin Mesa (Fig. 12.10). However, the mineral composition of the phenocrysts cause all of the flows to be classified as subalkalic tholeiite basalt. In addition, all of the flows are composed of low-Ti basalt because the high-Ti basalt flows that cap Tobin and Pain mesas of the Mesa Range are not present and may have been eroded (Mensing 1987)

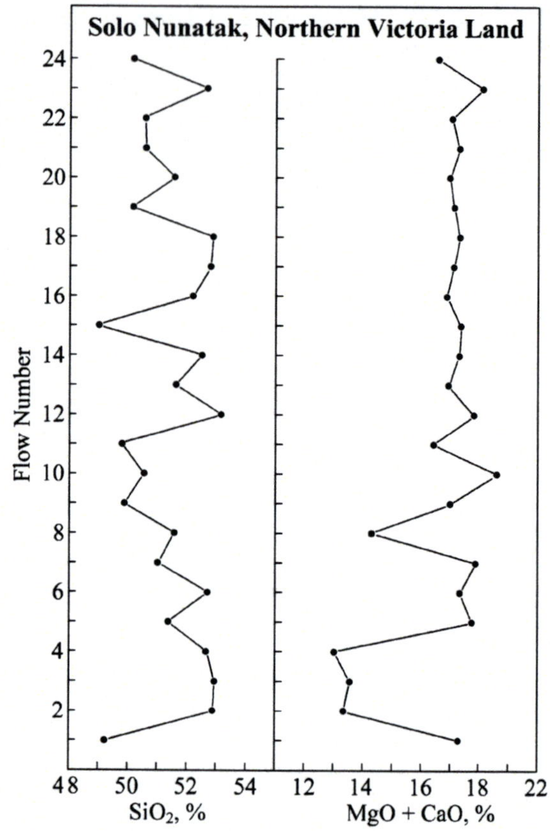

Fig. 12.19 The concentration profiles of SiO_2 and of MgO + CaO of the flows on Solo Nunatak of the Mesa Range reveal compositional irregularities of flows 2, 3, 4, and 8. However, all of the flows at this location belong to the low-Ti suite of the Kirkpatrick Basalt (TiO_2 = 0.54 to 0.79%) (Mensing 1987; Mensing et al. 1984)

the Wisanger Basalt (Section 12.1), and the Tasmanian Dolerite (Section 12.2).

The profile of initial $^{87}Sr/^{86}Sr$ ratios of the flows in Solo Nunatak in Fig. 12.20 vary systematically within narrow limits. The $^{87}Sr/^{86}Sr$ and $^{87}Rb/^{86}Sr$ ratios of the flows on Solo Nunatak measured by Mensing et al. (1984) scatter widely in Fig. 12.21 and do not define a straight line. Therefore, we have drawn a reference isochron corresponding to an age of 175×10^6 years and an initial $^{87}Sr/^{86}Sr$ ratio of 0.71069 ± 0.00007 ($1\bar{\sigma}$) calculated from these measurements. The differences in the $^{87}Sr/^{86}Sr$ ratios of the flows on Solo Nunatak indicate that the magma was contaminated with radiogenic ^{87}Sr either at the source or during its transit through the continental crust. This hypothesis is supported by the loosely defined fan-shaped array of data points in Fig. 12.22. In this interpretation of the data, the contaminant had a wide range of high $^{87}Sr/^{86}Sr$ ratios between about 0.711 and 0.713, whereas the $^{87}Sr/^{86}Sr$ ratio of the magma was between 0.709 and 0.710 which implies that the magma formed by partial melting of crustal rocks. These hypothetical deductions echo the conclusions expressed by McDougall (1962), Heier et al. (1965), and Compston et al. (1968) about the crustal characteristics of the Tasmanian Dolerite.

Additional evidence favoring the contamination hypothesis is provided by the isotope composition of oxygen in the flows on Solo Nunatak published by Mensing et al. (1984). Eleven $\delta^{18}O$ values of ten flows range from +6.0‰ to +9.3‰ which means that these rocks are variously enriched in the heavy isotope of oxygen compared to continental flood basalt which have δ^{18} O values between +4.3‰ and 6.5‰ according to a compilation by Harmon and Hoefs (1995). The systematics of oxygen isotopes are explained in Section 11.1.2 and by Faure and Mensing (2005, Chapter 26). The $\delta^{18}O$ values and initial $^{87}Sr/^{86}Sr$ ratios at 175 Ma of four flows on Solo Nunatak in Fig. 12.23 constrain a straight-line segment of a Sr-O isotopic mixing hyperbola. Seven additional samples are clustered separately. The apparent correlation of the isotopic compositions of strontium and oxygen strengthens the

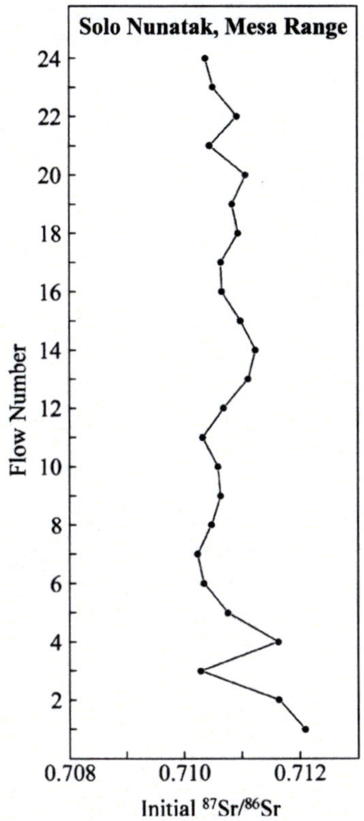

Fig. 12.20 The initial $^{87}Sr/^{86}$ Sr ratios of the 24 basalt flows on Solo Nunatak in the Mesa Range vary systematically within narrow limits between 0.70985 and 0.71207. These ratios are typical of the isotopic composition of strontium in the low-Ti flows of the Kirkpatrick Basalt of the Mesa Range. The high initial $^{87}Sr/^{86}$ Sr ratios indicate that the magma was either derived from a heterogeneous source in the mantle or that it was contaminated by rocks of the continental crust. The measured $^{87}Sr/^{86}$ Sr ratios were corrected for decay of ^{87}Rb to t = 175 Ma and $\lambda = 1.42 \times 10^{-11}$ year^{-1} (Mensing et al. 1984)

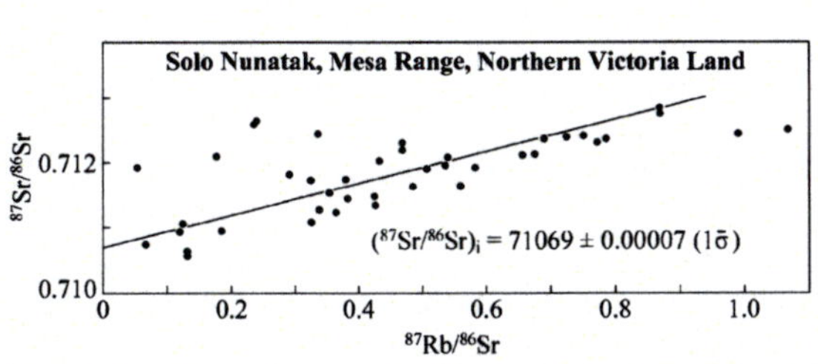

Fig. 12.21 The data points representing the flows on Solo Nunatak of the Mesa Range scatter above and below a 175-Ma reference isochron drawn through the average initial $^{87}Sr/^{86}$ Sr ratio at 0.71069 ± 0.00007 (1$\bar{\sigma}$). The failure of the data points to define a straight-line isochron indicates either that these flows had different initial $^{87}Sr/^{86}$ Sr ratios or that their Rb/Sr ratios were altered after crystallization, or both (Mensing et al. 1984)

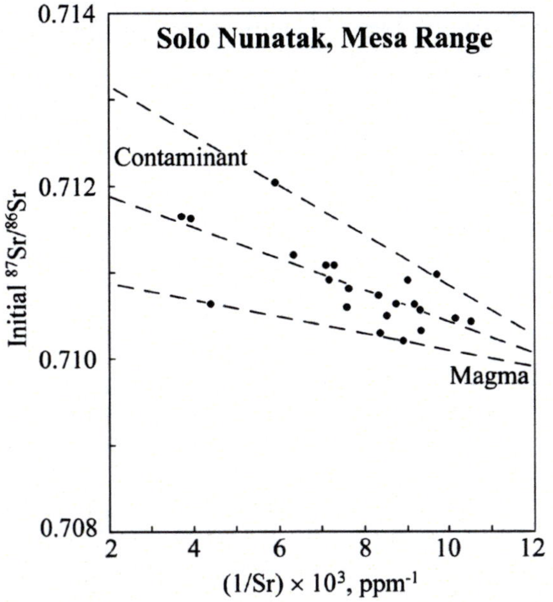

Fig. 12.22 The initial $^{87}Sr/^{86}Sr$ ratios at 175 Ma ($\lambda = 1.42 \times 10^{-11}$ year^{-1}) of the basalt flows on Solo Nunatak vary inversely with the reciprocals of their strontium concentrations. The fan-shaped array of data points supports the hypothesis that the magma was contaminated by strontium that was variably enriched in radiogenic ^{87}Sr formed by decay of ^{87}Rb in the Precambrian rocks of the crust of Antarctica (Mensing et al. 1984)

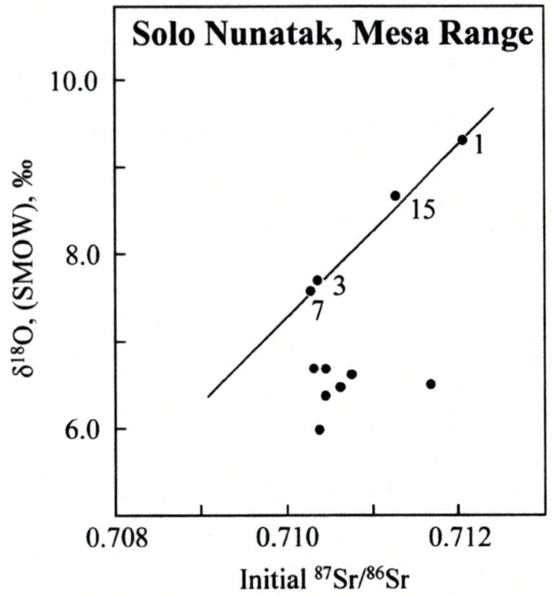

Fig. 12.23 Four flows (1, 3, 7, and 15) on Solo Nunatak define a straight line in coordinates of the initial $^{87}Sr/^{86}Sr$ ratio (at 175 Ma) and the isotope composition of oxygen expressed as the $\delta^{18}O$ value relative to SMOW. The other samples cluster between $\delta^{18}O = +6.0$ and +6.7‰. The linear relationship between isotope compositions of strontium and oxygen is attributable to mixing of basalt magma with a crustal component having elevated $^{87}Sr/^{86}Sr$ ratios and high $\delta^{18}O$ values (Faure and Mensing 2005; Mensing et al. 1984)

magma-contamination hypothesis because quartz and feldspar in igneous and metamorphic rocks of the continental crust are enriched in ^{18}O.

12.5 Brimstone Peak, Prince Albert Mountains

The Prince Albert Mountains in the northern part of southern Victoria Land consist of a large number of nunataks located on the polar plateau at the heads of Mawson, Harbord, and David glaciers all of which flow into the Ross Sea. Several of these nunataks are composed of sills of Ferrar Dolerite although flows of Kirkpatrick Basalt occur on Brimstone Peak (75°48′S, 158°33′E). Molzahn et al. (1996) reported isotope ratios of strontium, neodymium, and osmium as well as trace-element concentrations of five dolerite samples from the nunataks of the Prince Albert Mountains and two basalt samples from Brimstone Peak, including rubidium, strontium, samarium, neodymium, rhenium (Re) and osmium (Os). They also measured isotope ratios of strontium, neodymium, and lead in clinopyroxene and plagioclase of these rocks.

The Re-Os systematics of the whole-rock samples yielded a date of 172 ± 5 Ma and an initial $^{187}Os/^{188}Os$ ratio of 0.194 ± 0.023 (Faure and Mensing 2005). The Re-Os date reported by Molzahn et al. (1996) confirms the validity of the $^{40}Ar/^{39}Ar$ dates reported by McIntosh et al. (1986) and of Foland et al. (1993) for Kirkpatrick Basalt in the Mesa Range of northern Victoria Land. Molzahn et al. (1996) also concluded that the magma that formed the Ferrar rocks in the Prince Albert Mountains originated in the lithospheric mantle that contained subducted sediment in agreement with Hergt et al. (1989).

12.6 Kirkpatrick Basalt, Queen Alexandra Range

We now move to the Queen Alexandra Range of the central Transantarctic Mountains. By doing so we pass by the outcrops of Kirkpatrick Basalt on Westhaven Nunatak at the head of the Hatherton Glacier which have not been studied intensively (Grindley and Laird 1969). We also omit a discussion of the Kirkpatrick Basalt on the Otway Massif and the Grosvenor Mountains on the polar plateau at the head of the Mill Glacier (McGregor 1965; McGregor and Wade 1969; Elliot 1970a; Barrett et al. 1986).

The Kirkpatrick Basalt of the Queen Alexandra Range forms the summits of Mt. Kirkpatrick and Mt. Falla in the Queen Alexandra Range in Fig. 12.24 as well as of Tempest and Blizzard peaks in the Marshall Mountains (Grindley and Laird 1969). These rocks were named by Grindley (1963) and were subsequently described by Elliot (1970a, b, 1972), Elliot et al.

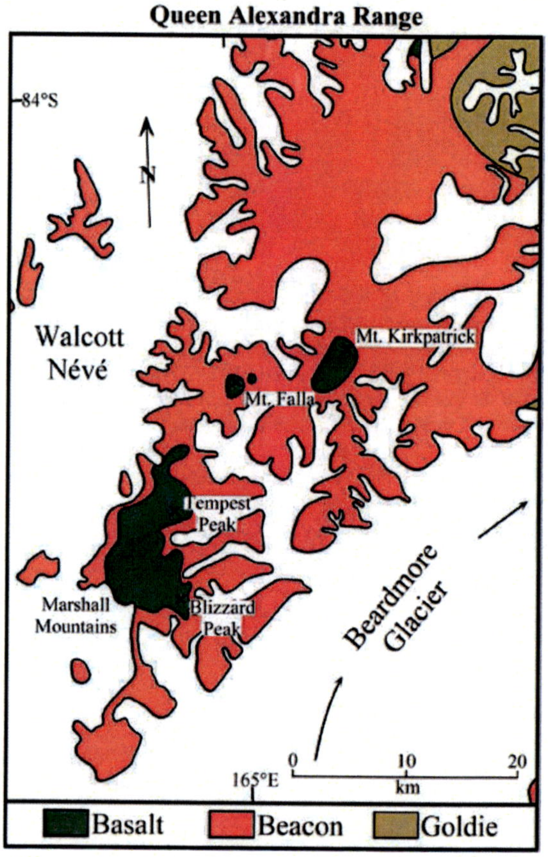

Fig. 12.24 The Kirkpatrick Basalt in the Queen Alexandra Range occurs primarily on Mt. Kirkpatrick (4,528 m) and on Mt. Falla (3,825 m). In addition, basalt flows form the summits of the Marshall Mountains including Tempest Peak (3,347 m), Storm Peak (3,280 m), Blizzard Peak (3,379 m), and Mt. Marshall (3,160 m) identified in Fig. 10.14. The basalt flows of the Kirkpatrick Basalt were deposited on the sedimentary rocks of the Beacon Supergroup (Chapter 10) which rest unconformably on the metasedimentary rocks of the Goldie Formation (Beardmore Group) (Adapted from Grindley and Laird (1969))

(1985), Fleck et al. (1977), Hoefs et al. (1980), and Faure et al. (1972, 1974, 1982, 1984). The descriptions of the Kirkpatrick Basalt in the Beardmore Glacier area by Elliot (1970a) and Barrett et al. (1986) include photographs of Mt. Falla (Queen Alexandra Range), of Peterson Ridge on Storm Peak (Marshall Mountains), and of Mt. Bumstead (Grosvenor Mountains).

12.6.1 Stratigraphy and Chemical Composition

Elliot (1970a) measured 12 stratigraphic sections of the Kirkpatrick Basalt in the Queen Alexandra Range and Marshall Mountains including Mt. Falla in Fig. 12.25 as well as Mt. Kirkpatrick, Blizzard Peak, and Tempest Peak. The section on Storm Peak in the Marshall Mountains consists of numerous basalt flows of varying thickness that add up to about 530 m.

Fig. 12.25 Lava flows of the Kirkpatrick Basalt form the summit of Mt. Falla in the Queen Alexandra Range (Photo by G. Faure)

Most of the flows are amygdaloidal, some have diabasic texture, and one flow contains pillow-like bodies (i.e., pahoehoe toes). In addition, two flows contain petrified three trunks and a sedimentary interbed near the top of the section. Elliot (1970a) also published a stratigraphic section of the basalt flows on Mt. Bumstead in the Grosvenor Mountains near the Otway Massif.

The chemical compositions of the flows on Storm Peak vary up-section with discontinuities between flows 1 and 2 and flows 10–11 which divide the section in Fig. 12.26 into three intervals. The first flow to be erupted has low concentrations of SiO_2 (53.14%) and total iron as FeO (9.07%), but high concentrations of Al_2O_3 (13.98%) and MgO (7.14%) compared to all other flows on Storm Peak. The concentrations of SiO_2 and total iron decrease up section from flow 2 to 10 whereas the concentration of Al_2O_3 and MgO increase. The last flows (11 and 12) in Fig. 12.26 differ markedly from the chemical compositions of the underlying flows.

The evident stratigraphic variation of the chemical composition of the flows on Storm Peak indicates that the magma was chemically heterogeneous presumably as a result of differentiation by fractional crystallization. Alternatively, the basalt magma may have been contaminated by mixing with a partial melt of rocks in the continental crust or because the magma originated from a mechanical mixture of crustal rocks embedded in the ultramafic rocks of the subcrustal mantle (i.e., a heterogeneous mantle). The alternative hypotheses cannot be evaluated based only on the major-element concentrations and will be discussed in the light of the isotopic compositions of strontium, neodymium, and oxygen.

The silica-alkali diagram in Fig. 12.27 demonstrates that the flows on Mt. Falla and Storm Peak have a wide range of compositions from basaltic andesite to andesite. Nevertheless, the mineralogy of the phenocrysts (Ca-rich plagioclase (labradorite), clinopyroxene (augite and pigeonite), and magnetite) as well as the high silica concentrations also allow these rocks to be classified as tholeiite basalts (Barrett et al. 1986). The minerals mentioned above display quench textures like those observed in the Mesa Range and Solo Nunatak (Mensing 1987). The matrix (mesostasis) consists of an intergrowth of quartz and alkali-feldspar which increases the concentrations of silica and alkali metals of these rocks.

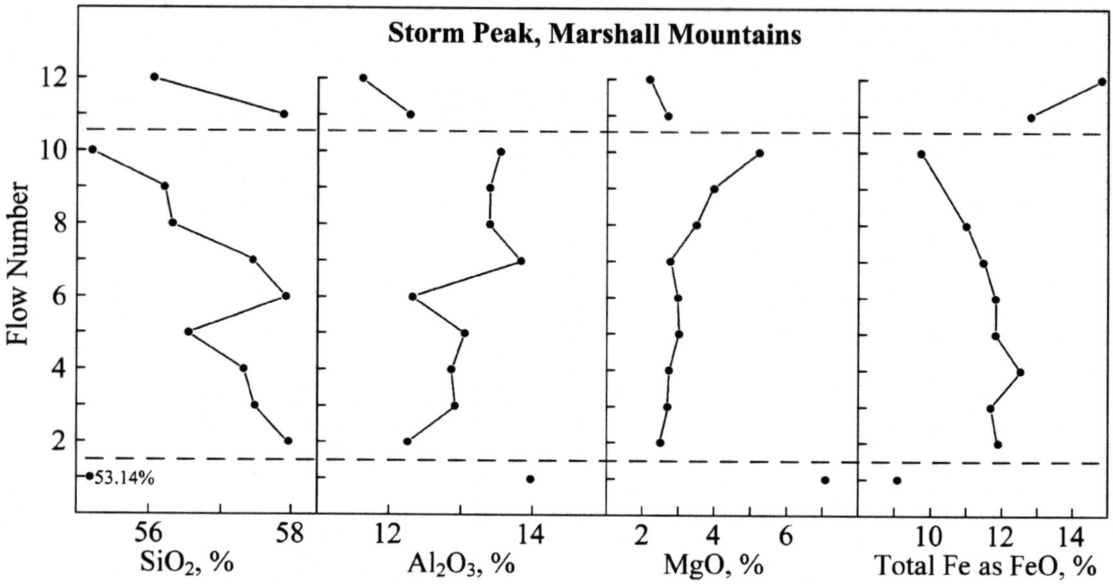

Fig. 12.26 The concentrations of all major elements in the flows of Kirkpatrick Basalt on Storm Peak in the Marshall Mountains vary sequentially up-section. The sequence includes discontinuities between flows 1 and 2 and between 10 and 11 which divide the section into three parts. The stratigraphic variation of the chemical compositions of the flows indicates that the sequentially erupted flows are samples of a heterogeneous magma. The magma my have been differentiating by fractional crystallization, or by mixing with a silica-rich partial melt of crustal rocks, or by having formed by partial melting of a heterogeneous mixture of sialic and ultramafic rocks. The chemical compositions of basalt flows on Mt. Falla also vary stratigraphically although individual flows are not easy to correlate with those on Storm Peak at a distance of about 30 km (Data from Elliot 1970,1972 with additions from Bowman 1971)

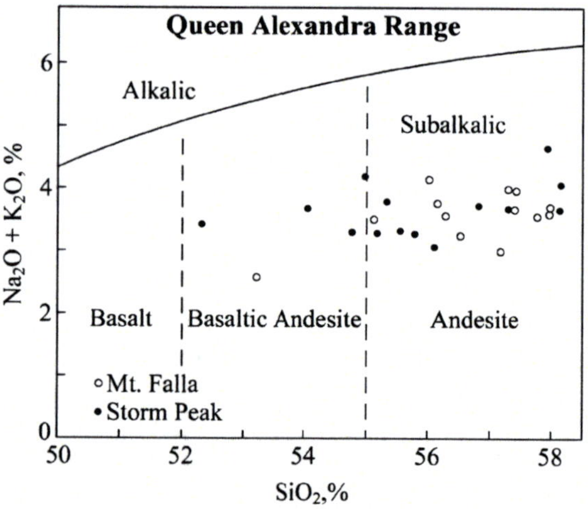

Fig. 12.27 Samples of the Kirkpatrick Basalt on Mt. Falla (Queen Alexandra Range) and Storm Peak (Marshall Mountains) range in composition from subalkalic tholeiite basaltic andesites to andesites. Alternatively, the phenocryst compositions identify these rocks as tholeiitic basalts. The high silica concentrations of these rocks presumably arise from the aphanitic matrix. The chemical compositions of the Kirkpatrick Basalt in the Beardmore Glacier area are similar to the basalt flows of the Mesa Range in northern Victoria Land (Data from Elliot 1970a, 1972;1974 Faure et al., 1982)

The concentrations of TiO_2 of the basalt flows on Storm Peak in Fig. 12.28 range from greater than 0.4% to less than 2.4% and include both low-Ti and high-Ti varieties. Most of the flows on Storm Peak have intermediate TiO_2 concentrations between 1.0% and 1.6%. In this regard, the flows on Storm Peak in the Queen Alexandra Range differ from the Kirkpatrick Basalt in the Mesa Range of northern Victoria Land where the TiO_2 concentrations have a distinctly bimodal distribution seen in Fig. 12.28.

The lava flows in the Queen Alexandra Range and in the Marshall Mountains are interbedded in a few places with layers of tuff and lacustrine sediment. The lake sediment consists largely of finely laminated volcanic ash, partly replaced by calcite, and containing abundant organic debris including conchostracan, fish remains, and fragments of wood and plants. The composition of the volcanic ash and of the lacustrine sediment appears to be felsic, indicating that silicic ash was apparently erupted contemporaneously with the basalt lava flows. The presence of lacustrine sediment means that, at certain times, the volcanic eruptions ceased

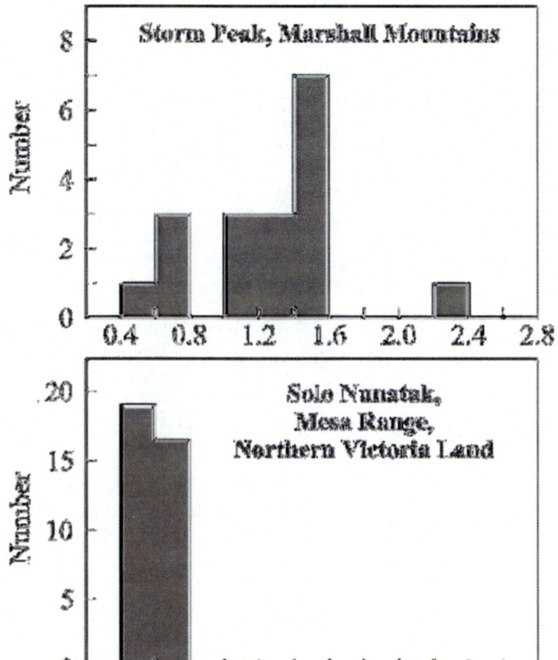

Fig. 12.28 The concentrations of TiO_2 of the Kirkpatrick Basalt on Storm Peak in the Marshall Mountains range from greater than 0.4% to less that 2.4% but are not as sharply divided into low-Ti and high-Ti suites as the basalt flows on Solo Nunatak and Pain Mesa of northern Victoria Land (Data from Elliot 1970a, 1972; Faure et al. 1982; Mensing et al. 1984; Mensing 1987)

and that vegetation grew in soils that formed by chemical weathering of the basalt flows and volcanic ash.

12.6.2 Fossils Among the Lava Flows

The sedimentary interbeds were deposited in local ponds and lakes during breaks in the eruption of basalt flows. These interbeds contain a variety of non-marine fossils including palynomorphs and acritarcs described by Norris (1965). In addition, the anatomy of petrified tree trunks in the growth position in the flows of Kirkpatrick Basalt of the Mesa Range was studied by Jefferson et al. (1983). This work was preceded by a report published by Jefferson (1980) about fossil angiosperms in volcanogenic shales in Antarctica.

An especially well-preserved fauna and flora occurs at Carapace Nunatak near the Allan Hills in southern Victoria Land (Bradshaw 1987). This locality was described by Ball et al. (1979) who identified syncarids, ostracodes, conchostracans, notostracans, isopods, and insects. They also recovered about 50 specimens of insects belonging to the orders Ephemeroptera, Odonata, Blattaria, and Hemiptera. The plant fossils consist primarily of ferns, including *Cladophlebis* and *Coniopteris*, and the conifer *Nothodacrium* as well as the cycadophyte (*Dictyozamites*). All the plant fossils and the insects are of Jurassic age. These authors also published an aerial photograph of Carapace Nunatak and excellent pictures of the rocks including the tegmen of a fossil cockroach.

The sedimentary interbeds of Kirkpatrick Basalt in the Queen Alexandra Range contain fossil fish that were collected by D.H. Elliot and were classified by Schaeffer (1971, 1972). The specimen in Fig. 12.29 is a bony ray-finned fish belonging to the order of Pholidophoriformes that belongs to the family Archaeomenidae of Jurassic age found in New South Wales, Australia. The similarity of this specimen to fossil fish in Australia is significant because Antarctica and Australia were still connected during the Jurassic Period. Schaeffer (1972) subsequently identified the fish in Fig. 12.29 as a new genus and new species named *Oreochima elliotí*. The name is a reference to Storm Peak in the Queen Alexandra Range and is derived from the Greek words "oreos" (mountain) "cheima" (Storm). More than 60 fragmentary specimens of *Oreochima ellioti* were recovered and several of them were described and pictured by Schaeffer (1972).

The pholidophoriform fishes evolved into the teleosts which are the most common and most highly diversified kind of fish that exist at the present time.

Fig. 12.29 This fossil of a 176 million-year-old fish was found in a layer of sedimentary rocks interbedded with flows of the Kirkpatrick Basalt on Storm Peak in the Marshall Mountains. The specimen is about 6 cm long and was given the name *Oreochima ellioti*. It lived in a pond or small lake during a temporary interruption in the extrusion of basalt lava. The pond in which the fish lived was also inhabited by conchostracans, ostracods, insects, and algae (Reproduced from Schaeffer 1971)

Richter and Thomson (1989) reported finding the first specimen of a teleost fish in a block of Late Jurassic tuffaceous mudstone embedded in the Middle Cretaceous Whisky Bay Formation on James Ross Island on the east side of the Antarctic Peninsula.

The conchostracans in the sedimentary interbeds belonging to the genera *Cyzicus* and *Paleolimnadia* were pictured by Tasch and Gifford (1984). The same authors also described trace fossils, fish fragments, insect wings, corniids, ostracodes, and algae. Additional papers by Paul Tasch on the subject of fossils in non-marine sedimentary environments in Antarctica include Tasch (1968, 1969, 1973, 1976, 1977, 1987).

The fossil fauna in the sedimentary interbeds between flows of Kirkpatrick Basalt at Carapace Nunatak in southern Victoria Land and in the Marshall Mountains of the Queen Alexandra Range was described by Stigall et al. (2008). The authors of this study were especially interested in documenting the preservation of fossils in these lacustrine environments. They examined not only conchostracans but also ostracodes, insect nymphs, actinopterygian fish, and plant remains. The evidence indicates that in places where microbial mats are absent the carapaces of conchostracans and fish-remains are preserved by calcium phosphate, whereas ostracode carapaces are preserved as calcium carbonate in agreement with the skeletal mineralogy of these organisms. In the presence of microbial mats, conchostracans are replaced by silica more frequently than in other kinds of substrates suggesting that silicification was aided by bacteria.

The study of fossils in the sedimentary interbeds of Jurassic lava flows in Antarctica is motivated by the "preservation paradox" that arises because the most abundant life forms (e.g., insects, spiders, scorpions, nematodes, and others) that lack solid shells or skeletons are vastly under-represented in the fossil record which favors the preservation of species that have "hard" parts but make up less than 20% of animal species (Babcock et al. 2006).

12.6.3 K-Ar Dates

Although the Middle Jurassic age of the Kirkpatrick Basalt in the Queen Alexandra Range and elsewhere in the Transantarctic Mountains is no longer in doubt (McDougall 1963), the whole-rock K-Ar dates of these rocks are, in many cases, inaccurate because of the loss of radiogenic ^{40}Ar and, in a few cases, because of the presence of extraneous ^{40}Ar especially in pyroxene crystals. For example, the whole-rock K-Ar dates of basalt flows (Storm Peak, Mt. Cecily, Mt. Spohn) and a dolerite sill (Tillite Glacier) reported by Elliot (1970a) range from 163 ± 10 to 179 ± 7 Ma (Geochron Laboratories, Cambridge, Massachusetts). The dates reported later by Elliot et al. (1985) for basalt flows and dolerite sills of the Ferrar Group in the Queen Alexandra Range, the Grosvenor Mountains, on Mt. Piccioto, Coalsack Bluff, and other sites in the central Transantarctic Mountains range even more widely from 114.0 ± 1.5 to 204.8 ± 5.6 Ma. Even the whole-rock K-Ar dates of well crystallized dolerite samples on Mt. Picciotto range from 142.1 ± 1.6 to 204.8 ± 5.6 Ma. Perhaps it is necessary to recall that McDougall (1963) dated separated plagioclase and pyroxene rather than whole-rock samples in his successful attempt to date the basalt and dolerite of the Ferrar Group.

When the K-Ar dates reported by Elliot (1970a), Elliot et al. (1985), and Fleck et al. (1977) are combined in Fig. 12.30, a bimodal distribution emerges with peaks between 130 to 140 Ma and 160 to 180 Ma.

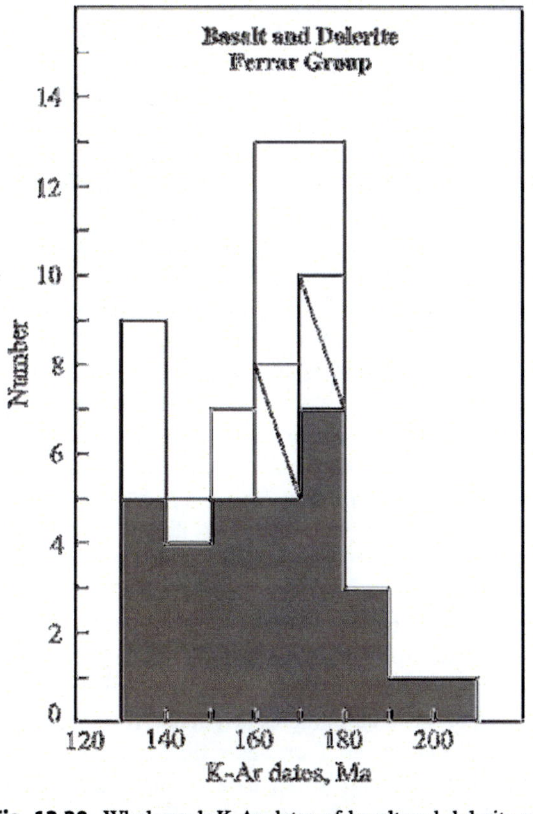

Fig. 12.30 Whole-rock K-Ar dates of basalt and dolerite of the Ferrar Group in the Queen Alexandra Range and elsewhere in the central Transantarctic Mountains. Shaded: Elliot et al. (1985), diagonal line: Elliot (1970a), white: Fleck et al. (1977). Twenty six dates between 160 and 180 Ma have an average value of 171 ± 2 Ma (2$\bar{\sigma}$). The dates in the interval 130–140 Ma (Early Cretaceous) indicate that ^{40}Ar was lost due to slow cooling and diffusion at low temperature. One date at 114 ± 1.4 Ma is not shown

The average of the 26 dates that lie between 160 and 180 Ma is 171 ± 2 Ma (2$\bar{\sigma}$). This date is at least representative of the available K-Ar dates of the Middle Jurassic basalt and dolerite of the Ferrar Group in the central Transantarctic Mountains. The dates between 130 and 140 Ma have been lowered by varying losses of radiogenic ^{40}Ar during slow cooling of dolerite sills and by diffusion at low temperature during devitrification and chemical weathering of basalt flows.

12.6.4 ^{40}Ar/^{39}Ar Dates

The effect of ^{40}Ar loss during devitrification of the glassy matrix of Kirkpatrick Basalt flows was investigated by Fleck et al. (1977) by means of the step-wise heating technique of the ^{40}Ar/^{39}Ar method. In preparation of this study, Fleck et al. (1977, p. 19) adopted the following definition of the "plateau date":

> A 'plateau' is that part of an age-spectrum diagram composed of contiguous gas fractions that together represent more than 50% of the total ^{39}Ar released from the sample and for which no difference in age can be detected between *any* two fractions at the 95% level of confidence

This definition of the plateau date effectively removed the subjective judgment of investigators from the interpretation of the spectra of ^{40}Ar/^{39}Ar partial-release dates and is still in use worldwide.

The results in Table 12.6 indicate that sample 64.01 (Mt. Bumstead) yields a plateau date of 170.4 ± 1.4 Ma which is statistically identical to 169.4 ±1.1 Ma derived from an Ar-isotope correlation diagram (Faure and Mensing 2005). These dates are more accurate and more precise than the conventional whole-rock K-Ar dates of 188.6 ± 2.2 and 149.7 ± 2.1 Ma of this sample. Two samples (64.10, Mt. Bumstead and 27.17, Storm Peak) did not form ^{40}Ar/^{39}Ar plateaus, but the Ar-isotope ratios did define isochrons with dates of 167.5 ± 3.8 and 161.8 ± 8.6 Ma, respectively. Sample 27.46 (Storm Peak) produced a plateau-date of 129.0 ± 5.0 and a 126 ± 3.8 Ma isochron date both of which indicate major loss of ^{40}Ar consistent with the high abundance (85%) of devitrified matrix.

The effect of the devitrified matrix on the retention of radiogenic ^{40}Ar is graphically illustrated in Fig. 12.31 based on the data of Fleck et al. (1977). This diagram emphasizes the point that the loss of ^{40}Ar rises with increasing abundance of the devitrified matrix.

Table 12.6 Age determinations of basalt flows by the conventional whole-rock K-Ar and the $^{40}Ar/^{39}Ar$ partial release method (Fleck et al. 1977)

Locality & number	Matrix (%)	K-Ar (Ma)	$^{40}Ar/^{39}Ar$ (Ma) Plateau	Isochron
Mt. Bumstead[a] 64.01	17.8	188.6 ± 2.2 149.7 ± 2.1	170.4 ± 1.4	169.4 ± 1.1
Mt. Bumstead 64.10	45.2	173.1 ± 1.9	No plateau	167.5 ± 3.8
Storm Peak[b] 27.17	62.2	160.5 ± 4.8 134.6 ± 6.9	No plateau	161.8 ± 8.6
Storm Peak 27.46	85.0	139.5 ± 4.0 126.3 ± 4.4	129.0 ± 5.0	126.1 ± 3.8

[a]Grosvenor Mountains
[b]Marshall Mountains

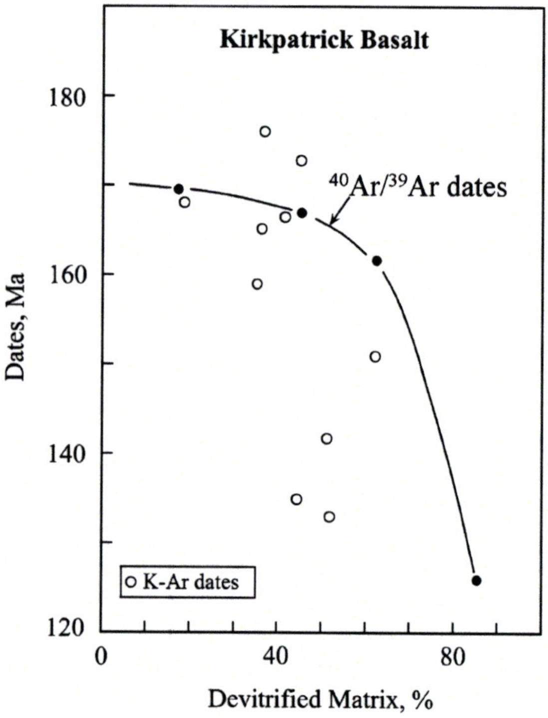

Fig. 12.31 The $^{40}Ar/^{39}Ar$ dates of four specimens of Kirkpatrick Basalt from Storm Peak (Marshall Mountains) and Mt. Bumstead (Grosvenor Mountains) decrease smoothly with increasing abundance of devitrified matrix. This result indicates that the basalt crystallized close to 170 Ma and that the $^{40}Ar/^{39}Ar$ dates of samples containing more than about 20% devitrified matrix increasingly underestimate the age of these rocks. The conventional whole rock K-Ar dates range from being older to being younger than the age of the basalt and are not reliable (Data from Fleck et al. 1977 and Table 12.3)

The resulting pattern of variation suggests that the crystallization age of the Kirkpatrick Basalt is greater than 170 Ma and that samples containing devitrified matrix consistently underestimate the age of the rocks even by the $^{40}Ar/^{39}Ar$ method. Fleck et al. (1977, p. 18), described the devitrified matrix as:

> Microcrystalline to cryptocrystalline intergrowth of quartz, alkali feldspar, plagioclase, and pyroxene with minor amounts of glass, chlorophaeite, zeolite (?), and chlorite, probably formed by devitrification of glass.

The conventional whole-rock K-Ar dates of the Kirkpatrick Basalt in Fig. 12.31 literally vary all over the place and range from being too old to being too young. (Chlorophaeite is chlorite that fills cavities or fractures in mafic igneous rocks).

The loss of ^{40}Ar by the devitrified matrix of the Kirkpatrick Basalt that was documented by Fleck et al. (1977) caused Heimann et al. (1994) to date only feldspar separated from basalt samples collected at Storm Peak in the Marshall Mountains, from Carapace Nunatak in southern Victoria Land, and from the Mesa Range in northern Victoria Land. In addition, Heimann et al. (1994) dated basalt clasts from the Prebble, Mawson, and Exposure Hill formations by the $^{40}Ar/^{39}Ar$ method. Eleven samples yielded plateau dates, whereas six samples had complex and discordant spectra. The results in Table 12.7 are remarkably consistent, which is all the more surprising because the samples originated from a 1,600-km segment of the Transantarctic Mountains. The average $^{40}Ar/^{39}Ar$ plateau date of the Kirkpatrick Basalt according to Heimann et al. (1994) is 176.6 ± 1.8 Ma (including uncertainty of "J" parameters). This date remains the most precise and the most representative age determination of the Kirkpatrick Basalt in the Transantarctic Mountains. It means that more than 500,000 km³ of basalt were erupted in a comparatively short time of less than one million years, according to Heimann et al. (1994). The rapid rate of eruption of basalt implied by this result has been observed also in other continental basalt provinces, including the Siberian Traps (244–242 Ma), the Paraná Basalt of Brazil (135–130 Ma), the Deccan Traps of India (66–64 Ma) and the Columbia River Basalt of Oregon and Washington (17–15 Ma) (see also Faure 2001).

Table 12.7 $^{40}Ar/^{39}Ar$ plateau dates on samples of Kirkpatrick Basalt from the Transantarctic Mountains (Heimann et al. 1994)

Site	$^{40}Ar/^{39}Ar$ Plateau date (Ma)
Mesa Range, NVL	
Mt. Frustrum, Tobin Mesa	176.8 ± 0.8
Pain Mesa	(174.1)[a]
Pain Mesa	176.4 ± 0.9
Exposure Hill, Exposure Hill Fm., basalt clast	(177.3)
Southern Victoria Land	
Carapace Nunatak	176.4 ± 0.5
Carapace Nunatak	176.4 ± 0.4
Allan Hills, Mawson Fm., basalt clast	176.4 ± 0.4
Central Transantarctic Mtns.	
Storm Peak	176.8 ± 0.5
Storm Peak	177.2 ± 0.5
Storm Peak	176.5 ± 0.4
Mt. Pratt, east of Otway Massif. Prebble Fm., basalt clast	176.4 ± 0.4
Average age of the Kirkpatrick Basalt in the Transantarctic Mtns.	176.6 ± 1.8[b]

[a]Parentheses mean "not adopted" by Heimann et al. (1994)
[b]Includes the uncertainty of the "J" parameters

12.6.5 Rb-Sr Dates

Samples of basalt from flows 1 to 6 (from the bottom) on Mt. Falla analyzed by Pace (1977) define a straight line in Fig. 12.32 which has an intercept on the y-axis of 0.7128 ± 0.00008 ($2\bar{\sigma}$) and yields a date of 170 ± 5 Ma ($\lambda = 1.42 \times 10^{-11}$ year^{-1}). The upper flows on Mt. Falla (7–14) scatter in Fig. 12.32 and are not shown.

The flows of Kirkpatrick Basalt on Storm Peak analyzed by Bowman (1971) also scatter widely on the Rb-Sr isochron diagram and do not permit a geologically meaningful age determination. Nevertheless, the age of the Kirkpatrick Basalt in the Transantarctic Mountains is closely defined by the available $^{40}Ar/^{39}Ar$ and Rb-Sr dates between 170 and 176 Ma which defines the Bajocian Stage of the Middle Jurassic Epoch (IUGS 2002).

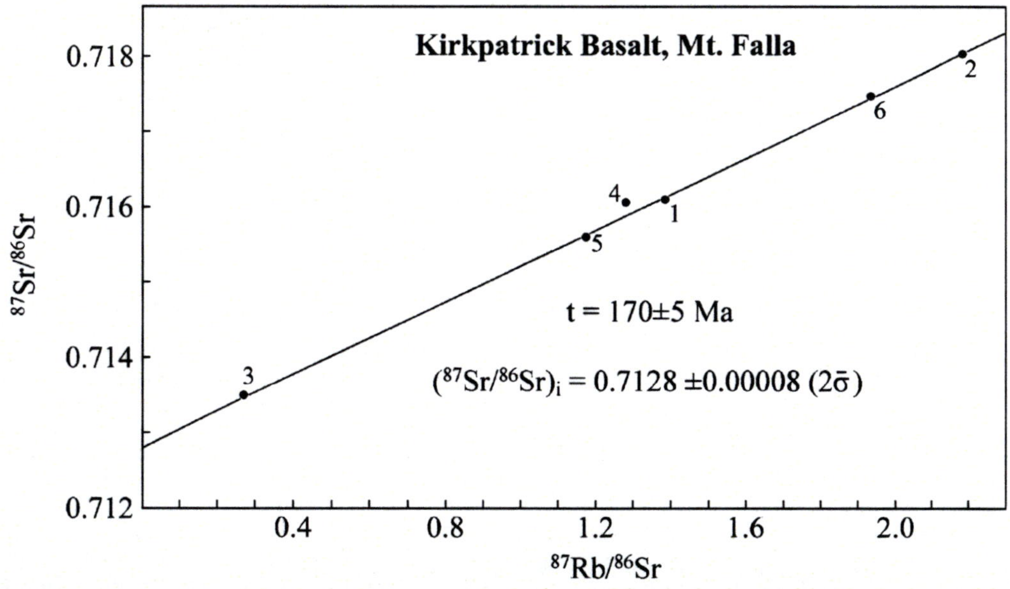

Fig. 12.32 Samples of the Kirkpatrick Basalt in flows 1–6 (from the bottom) on Mt. Falla define a straight line that yields an initial $^{87}Sr/^{86}Sr$ ratio of 0.7128 ± 0.00008 ($2\bar{\sigma}$) and a date of t = 170 ± 5 Ma. Samples from flows 7 to 14 are not shown because they scatter widely on this diagram (Data by Pace (1977) reported by Faure et al. (1982))

12.6.6 Intial $^{87}Sr/^{86}Sr$ Ratios

The isotope composition of strontium is an important parameter because it conveys information about the petrogenesis of igneous rocks such as the basalt and dolerite of the Ferrar Group. The first measurements of the initial $^{87}Sr/^{86}Sr$ ratios of the Ferrar Dolerite by Compston et al. (1968) revealed that the strontium in these rocks is anomalously enriched in radiogenic ^{87}Sr. In the following year, Hill (1969) confirmed that the Kirkpatrick Basalt in the Mesa Range of northern Victoria Land and in the Queen Alexandra Range also has elevated initial $^{87}Sr/^{86}Sr$ ratios. In addition, Hill (1969) reported that the K/Rb ratios of the basalt flows and dolerite sills have low average values of 223 ± 9 (basalt) and 262 ± 11 (dolerite) that are typical of granitic rocks in the continental crust.

Subsequent reports by Faure et al. (1972, 1974, 1982) added data for basalt flows on Storm Peak analyzed by Bowman (1971) and on Mt. Falla analyzed by Pace (1977). The initial $^{87}Sr/^{86}Sr$ ratios of basalt specimens from these locations form a prominent peak in Fig. 12.33 between 0.7100 and 0.7130 with an average 0.7117 ± 0.0004 ($2\bar{\sigma}$). This value is similar to the initial $^{87}Sr/^{86}Sr$ ratios of the Granite Harbor Intrusives in the basement of the Transantarctic Mountains and it differs profoundly from the $^{87}Sr/^{86}Sr$ ratios of mantle-derived basalt in the oceans and continents of the Earth (Faure 2001).

The initial $^{87}Sr/^{86}Sr$ of the sequentially erupted flows on Mt. Falla and on Storm Peak in Fig. 12.34 vary stratigraphically. The first six flows on Mt. Falla actually have a nearly constant initial $^{87}Sr/^{86}Sr$ ratio at 170 Ma which has permitted these flows to be dated by the Rb-Sr method in Fig. 12.32. The initial $^{87}Sr/^{86}Sr$ ratios of the upper flows on Mt. Falla (7–14) vary systematically between 0.7100 and 0.7121. These variations imply that the magma had a range of $^{87}Sr/^{86}Sr$ ratios and that the sequentially erupted lava flows are samples drawn from different parts of the magma in its subsurface reservoir.

The lava flows on Storm Peak also have a range of initial $^{87}Sr/^{86}Sr$ ratios that vary stratigraphically. The lower stratigraphic section on Storm Peak in Fig. 12.34 (flows 1–7) consists of interbedded flows that have alternating high and low initial $^{87}Sr/^{86}Sr$ ratios with a consistent trend toward higher values up-section. Following a discontinuity between flow 7 and 8, the initial $^{87}Sr/^{86}Sr$ ratios of the upper flows (8–12) also alternate between high and low values but with a decreasing trend up-section.

The evident heterogeneity of the isotopic composition of strontium in the magma cannot be caused by crystallization of an isotopically homogeneous magma in a closed magma chamber because the isotopes of strontium are not fractionated by this process. Therefore, the isotopic heterogeneity of strontium of the magma

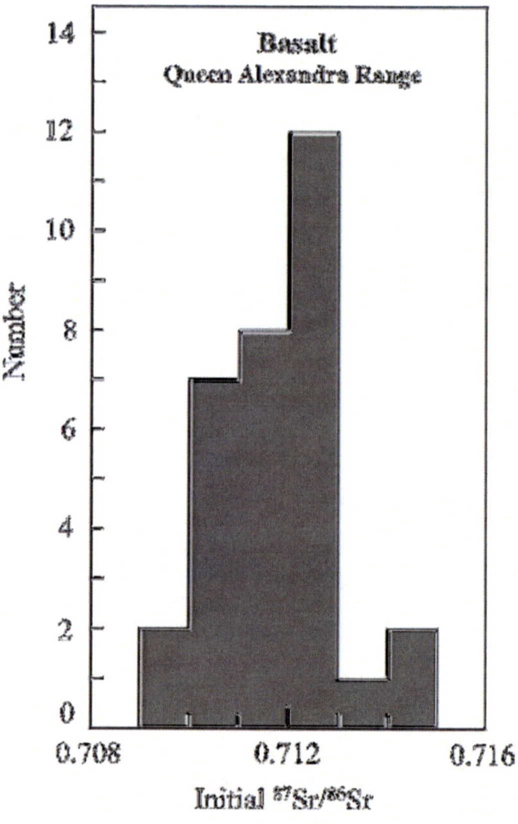

Fig. 12.33 The initial $^{87}Sr/^{86}Sr$ ratios of the Kirkpatrick Basalt on Storm Peak and Mt. Falla in the Queen Alexandra Range at 170 Ma ($\lambda = 1.42 \times 10^{-11}$ year^{-1}) range widely but cluster between 0.7100 and 0.7130 with an average of 0.7117 ± 0.0004 ($2\bar{\sigma}$) for 27 specimens. This value resembles the initial ratio of the Granite Harbor Intrusives in the basement of the Transantarctic Mountains and is not compatible with basalt magma derived from the mantle. The isotope composition of strontium in sills of Ferrar Dolerite was not included in this data set by Bowman (1971) and Pace (1977) reported by Faure et al. (1974, 1982). The data of Hill (1969) were included in the report by Faure et al. (1972)

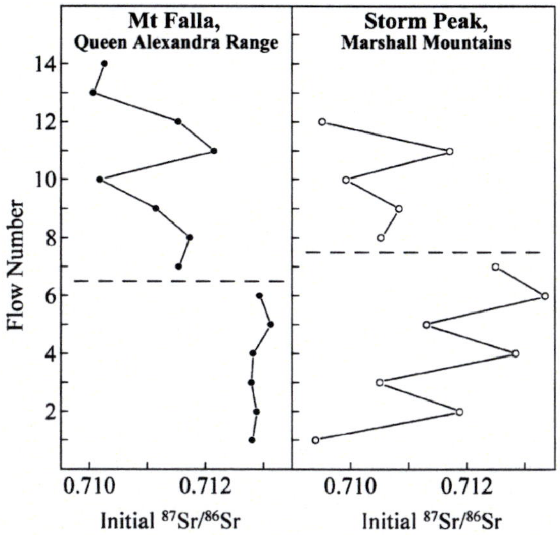

Fig. 12.34 The initial $^{87}Sr/^{86}Sr$ ratios (170 Ma) of sequentially erupted lava flows on Mt. Falla and Storm Peak vary between wide limits from 0.7094 to 0.7133. The pattern of variation on Mt. Falla consists of a lower section (flows 1–6) having a nearly constant initial $^{87}Sr/^{86}Sr$ ratio of 0.7128 ± 0.00008 followed by an upper section (flows 7–14) where the initial $^{87}Sr/^{86}Sr$ ratios vary systematically up-section. The lower flows on Storm Peak (1–7) alternate between having high and low initial $^{87}Sr/^{86}Sr$ ratios with an increasing trend up-section. The upper flows continue the pattern of alternating high and low initial $^{87}Sr/^{86}Sr$ ratios with a decreasing trend up-section similar to the upper flows on Mt. Falla. The isotopic heterogeneity of strontium in the Kirkpatrick Basalt indicates that these flows formed from an isotopically heterogeneous magma. Data for Storm Peak from Bowman (1971) and for Mt. Falla from Pace (1977) (These data were reported by Faure et al. 1974, 1982)

that gave rise to the Kirkpatrick Basalt is attributable to one or the other alternative explanations:

1. The magma formed by partial melting of isotopically and chemically heterogeneous source rocks of crustal origin, or
2. An isotopically homogenerous mantle-derived basalt magma was contaminated by assimilating rocks of the continental crust that lined the fractures through which the magma rose to the surface.

12.6.7 Magma Mixing

When magmas having different $^{87}Sr/^{86}Sr$ ratios and different strontium concentrations mix in differing proportions, the resulting mixtures define a hyperbola in coordinates of the $^{87}Sr/^{86}Sr$ ratio and the strontium concentration (Faure and Mensing 2005). The equation of the hyperbola can be transformed into the equation of a straight line by redefining the x-variable to be the reciprocal of the strontium concentration. Therefore, a plot of initial $^{87}Sr/^{86}Sr$ ratios and reciprocals of the strontium concentration of two-component mixtures define a straight line.

The data for the Kirkpatrick Basalt on Storm Peak and Mt. Falla do, in fact, define straight lines in Fig. 12.35. Least-squares regressions of the data points yield linear correlation coefficients of –0.9435 (Storm Peak) and –0.9633 (Mt. Falla), where the minus sign indicates that the slopes of the lines are negative. According to this interpretation of the data, the magma that formed the flows at Storm Peak had an $^{87}Sr/^{86}Sr$ ratio less than 0.7090 and a strontium concentration less than 125 ppm, whereas the $^{87}Sr/^{86}Sr$ ratio of the crustal contaminant was greater than 0.7135 and its strontium concentration was greater than 140 ppm. A similar statement applies to the flows on Mt. Falla. Such contaminants could have formed by partial melting of the volcano-sedimentary rocks of the Ross orogen or of rocks in the underlying continental crust of Antarctica. Partial melting of the crustal rocks would have been facilitated by the presence of hydrous minerals (e.g., clay minerals, mica, and amphiboles, etc.) and by the depth-dependent temperature of the crustal rocks before the intrusion of basalt magma from greater depth. Partial melting of crustal rocks caused by the heat emanating from basalt magma derived either from the lithospheric mantle or from the greater depth in the crust has caused the co-eruption of basalt and rhyolite in large igneous provinces elsewhere on the Earth (Faure 2001).

Magma mixing also causes the isotope composition of strontium of the resulting mixtures to be correlated with the concentrations of other components such as SiO_2, MgO, and others depending on the chemical composition of the contaminant. Such correlations of isotopic and chemical parameters were first reported by Faure et al. (1972, 1974, 1982) for flows of the Kirkpatrick Basalt in the Transantarctic Mountains and of the Kirwan Volcanics in Queen Maud Land (Chapter 14).

The data of Bowman (1971) and Pace (1977) in Fig. 12.36 demonstrate that the initial $^{87}Sr/^{86}Sr$ ratios at 170 Ma of the Kirkpartick Basalt on Storm Peak and on Mt. Falla both rise with *increasing* concentrations of SiO_2.

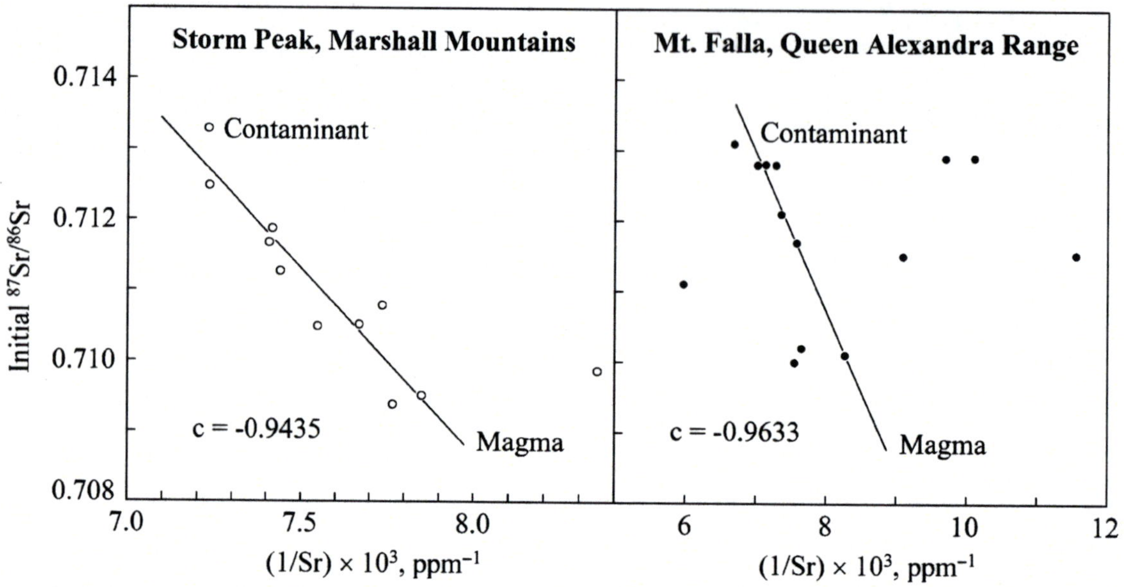

Fig. 12.35 The initial $^{87}Sr/^{86}Sr$ ratios at 170 Ma and reciprocal strontium concentrations of the Kirkpatrick Basalt on Storm Peak and Mt. Falla define straight lines that result from mixing of two components in varying proportions (i.e., the basalt magma and a crustal contaminant). The contaminant my have formed by partial melting of the rocks of the Ross orogen or of the continental crust of Gondwana in response to heat emanating from the basalt magma which originated from greater depth in the crust or from the underlying lithospheric mantle. Note that some data points scatter far off the mixing lines especially in the case of Mt. Falla (Data from Bowman (1971) for Storm Peak and Pace (1977) for Mt. Falla and reported by Faure (1974, 1982))

The correlation of these parameters is consistent with mixing of basalt magma with a silica-rich partial melt of crustal rocks with elevated $^{87}Sr/^{86}Sr$ ratios depending on their Rb/Sr ratio and age. A second example of this phenomenon in Fig. 12.37 demonstrates that the increase of the initial $^{87}Sr/^{86}Sr$ ratios at 170 Ma on Storm Peak with *decreasing* concentrations of MgO. The pattern of variation supports the hypothesis that the basalt magma mixed with a crustal melt that had a low concentration of MgO and a high $^{87}Sr/^{86}Sr$ ratio. Additional correlations of initial $^{87}Sr/^{86}Sr$ ratios and other major elements were reported by Faure et al. (1974) for the flows on Storm Peak analyzed by Bowman (1971). The positive correlation of initial $^{87}Sr/^{86}Sr$ ratios of the basalt flows on Solo Nunatak (Section 12.4.1.3) was described by Mensing et al. (1983).

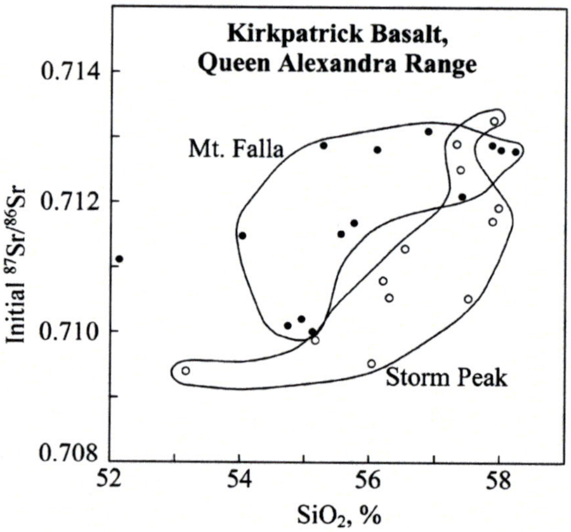

Fig. 12.36 The initial $^{87}Sr/^{86}Sr$ ratios at 170 Ma of the flows of Kirkpatrick Basalt on Mt. Falla and Storm Peak both rise with *increasing* concentrations of SiO_2. Such correlations of isotopic and chemical parameters cannot occur by differentiation of magma in a closed chamber, but are expected as a result of magma mixing, or by magma formation by partial melting of mixtures of sialic and ultramafic rocks (Data from Bowman 1971; Pace 1977 reported by Faure et al. 1974, 1982)

12.7 Petrogenesis: Isotopic Evidence

All of the evidence concerning the chemical and isotopic composition of the Tasmanian Dolerite and of the Kirkpatrick Basalt in the Transantarctic Mountains

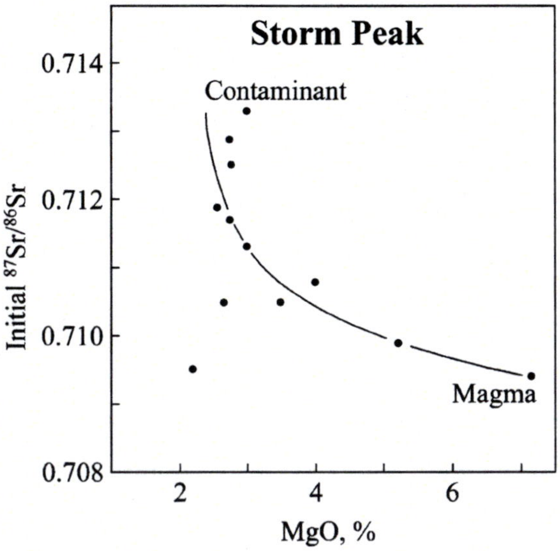

Fig. 12.37 The initial $^{87}Sr/^{86}Sr$ ratios at 170 Ma of basalt flows on Storm Peak (Marshall Mountains) rise with *decreasing* concentrations of MgO. This pattern of variation supports the hypothesis that the basalt magma mixed with a crustal melt that had a low concentration of MgO but a high $^{87}Sr/^{86}Sr$ ratio (Data from Bowman 1971; Faure et al. 1974)

points to the crustal characteristics of these major igneous rocks. Hergt et al. (1989) expressed this aspect of the petrogenesis of these rocks by stating:

> Many of the trace element and isotopic characteristics compare more closely with those of continental crustal rocks than with typical mantle-derived materials, indicating the dominance of a crustal component in the petrogenesis of these tholeiites.

The evidence we have already presented includes the anomalously high initial $^{87}Sr/^{86}Sr$ ratios (Figs. 12.21, 12.33, 12.34), the Sr-isotopic mixing lines (Figs. 12.23, 12.35), and the correlation of initial $^{87}Sr/^{86}Sr$ ratios with major-element concentrations including SiO_2 (Fig. 12.36) and MgO (Fig. 12.37).

The most direct way to recognize the crustal character of these rocks is by means of the initial isotope ratios of strontium and neodymium in Fig. 12.38. The Sr-Nd plane in this diagram is divided into four quadrants by the $^{143}Nd/^{144}Na$ ratio of the Chondritic Uniform Reservoir (CHUR) and by the corresponding $^{87}Sr/^{86}Sr$ ratio: $^{143}Nd/^{144}Nd = 0.512638$ and $^{87}Sr/^{86}Sr = 0.70450$. The isotope ratios of strontium and neodymium in igneous, metamorphic, and sedimentary rocks of the continental crust form points in quadrant 4 whereas mantle-derived mafic igneous rocks occupy quadrant 2 where they define the "Mantle Array". When mantle-derived magmas mix with crustal magmas, the isotope ratios of strontium and neodymium lie along isotopic mixing hyperbolas that connect the two end-member components (Faure and Mensing 2005).

12.7.1 Tasmanian Dolerite

The initial $^{87}Sr/^{86}Sr$ and $^{143}Nd/^{144}Nd$ ratios of 21 samples of the chilled margins of Tasmanian Dolerite analyzed by Hergt et al. (1989) form a tight cluster of points in quadrant 4 of the Sr-Nd isotope-mixing diagram in Fig. 12.38. In addition, five samples of "contact sediment" define a mixing hyperbola that passes through the cluster of dolerite samples in quadrant 4. The placement of the Tasmanian Dolerite in the "crustal" quadrant of the Sr-Nd isotope-mixing diagram is consistent with conclusions expressed previously by McDougall (1962), Heier et al. (1965), and Compston et al. (1969) in Section 12.2. The challenge these data pose is to explain the mechanism whereby the Tasmanian Dolerite acquired the crustal characteristics revealed by its chemical and isotopic compositions.

The process that caused the Jurassic magmas of Tasmania and the Transantarctic Mountains to acquire the chemical and isotopic characteristics of rocks in the continental crust has been explained by two alternate hypotheses mentioned repeatedly in this chapter (e.g., Section 12.4):

1. Mantle-derived magmas were contaminated by mixing with crustal melts or by assimilating crustal rocks from the walls of the fractures through which they moved toward the surface.
2. Magmas formed in a heterogeneous mantle that had been altered prior to rifting and had acquired crustal characteristics.

Although both hypotheses are plausible, they do not account for the enormous volume and for the chemical and isotopic homogeneity of magma that was produced in a short time in Tasmania and along the entire length of the Transantarctic Mountains.

Hergt et al. (1989) used trace-element modeling to show that the chemical and isotopic compositions of the

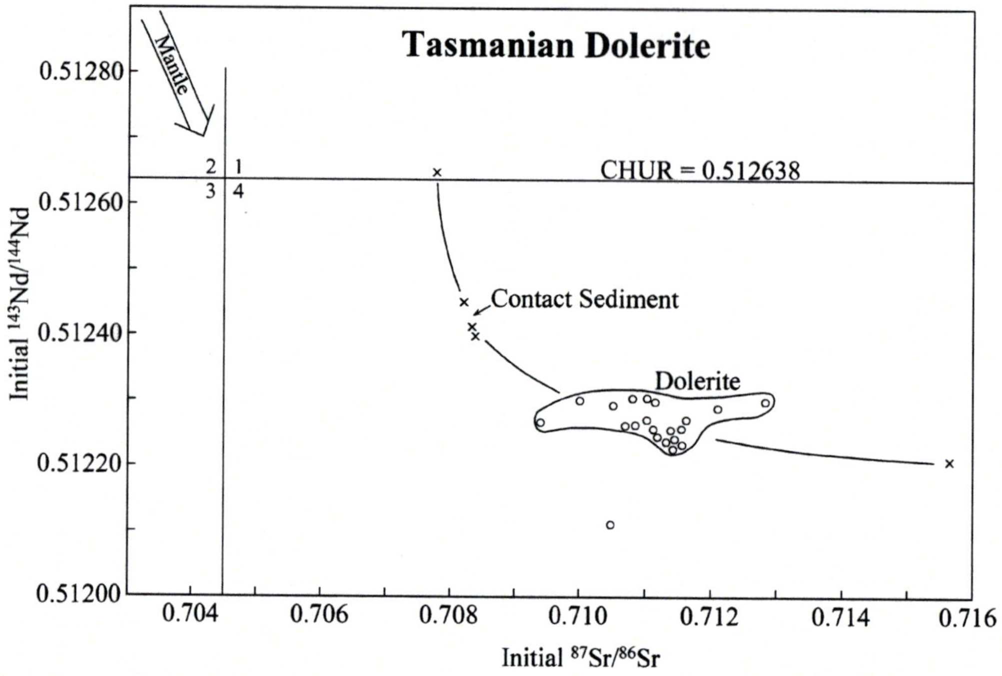

Fig. 12.38 The initial $^{87}Sr/^{86}Sr$ and $^{143}Nd/^{144}Nd$ ratios of the Tasmanian Dolerites (open circles) cluster in quadrant 4 of the Sr-Nd isotopic mixing diagram which is partitioned based on the isotopic composition of the Chondritic Uniform Reservoir (CHUR) described by Faure and Mensing (2005). The diagram also contains data points (crosses) of "contact sediment" which define an isotopic mixing hyperbola that passes through the cluster of data points represented by the Tasmanian Dolerite. Quadrant 4 is the locus of points representing rocks in the continental crust that have elevated $^{87}Sr/^{86}Sr$ ratios and low $^{143}Nd/^{144}Nd$ ratios. The measured $^{143}Nd/^{144}Nd$ ratios were multiplied by 1.00154 in order to make them compatible with CHUR = 0.512638 (Data by Hergt et al. 1989)

Tasmanian Dolerite could be explained by the presence of less than 3% weight percent of sediment in the subcrustal mantle. In other words, the contamination of the basalt magma occurred *at the source* rather than after the magma intruded the crust. In this way, the hypothesis of Hergt et al. (1989) involved both of the earlier proposals advocated by Faure (1981) and by Kyle (1980). The new proposal gained strength from the likelihood that the sedimentary component was introduced into the subcrustal mantle as a result of *subduction* which occurred along the paleo-Pacific coast of East Antarctica in early Paleozoic time. The compressive tectonic regime that originated from the subduction caused the Ross Orogeny which formed the igneous and metamorphic rocks of the basement complex that underlies the present Transantarctic Mountains. We recall here again that the initial $^{87}Sr/^{86}Sr$ ratios of the Tasmanian Dolerite and of the Kirkpatrick Basalt are similar to the initial $^{87}Sr/^{86}Sr$ ratios of the Granite Harbor Intrusives of the Transantarctic Mountains.

The convergence of several lines of evidence gives strong support to the hypothesis of Hergt et al. (1989) and elevates it to the status of a *theory*. According to the theory of Hergt et al. (1989), the formation of basalt magma in the subcrustal mantle during the Middle Jurassic is the culmination and conclusion of the sequence of geologic events we have described on the pages of this book:

1. Rifting of Rodinia and the departure of Laurentia during the Neoproterozoic
2. Deposition of turbidites along the passive margin of East Antarctica until about Early Cambrian
3. Folding and metamorphism of the turbidites by compression resulting from the development of a subduction zone along the paleo-Pacific margin of East Antarctica
4. Formation of granitic magma by anatexis of the volcano-sedimentary rocks of the Ross orogen and intrusion into the orogen as syntectonic and posttectonic plutons and batholiths

5. Relaxation of the compressive stress and cooling of the orogen until about the Late Ordovician
6. Erosion of the Ross Mountains and peneplanation until Early Devonian
7. Deposition of the clastic sedimentary rocks of the Beacon Supergroup between the Early Devonian and the Late Triassic
8. Initiation of rifting of the crust underlying the area of the present Transantarctic Mountains and large-scale eruption and intrusion of basalt magma with crustal characteristics during the Middle Jurassic
9. Break-up Gondwana and displacement of the continental fragments accompanied by the final assembly of Antarctica during the Cretaceous

12.7.2 Kirkpatrick Basalt, Mesa Range

The petrogenesis of the Kirkpatrick Basalt and the Ferrar Dolerite in the Transantarctic Mountains was once the subject of an intense debate. The contributions by various research groups were concerned with:

1. The explanation for the crustal character of the rocks of the Ferrar Group in Antarctica
2. The origin of the distinctly different high-Ti and low-Ti suites, and
3. The widespread occurrence of continental flood basalt and comagmatic dolerite sills/dikes of Middle Jurassic to Early Cretaceous age in Gondwana

The debate concerning these important problems subsided after Hergt (1987) and Hergt et al. (1989, 1991) presented convincing evidence that the magmas that formed these rocks were generated by extensive partial melting in the subcrustal mantle of Gondwana which contained a small amount of sediment that was introduced by subduction along the paleo-Pacific margin during the early Paleozoic. The principal protagonists in this debate included Rhodes and Bornhorst (1986), Brooks and Hart (1978), Allègre et al. (1982), Kyle (1977, 1980), Kyle et al. (1981, 1983, 1987), Hoefs et al. (1980), Faure (1981), Faure et al. (1972, 1974, 1982, 1984), Mensing et al. (1984), Fleming and Elliot (1988), Brotzu et al. (1988), and Hornig (1993).

The initial $^{87}Sr/^{86}Sr$ and $^{143}Nd/^{144}Nd$ ratios of Kirkpatrick Basalt in the Mesa Range of northern Victoria Land in Fig. 12.39 occupy an area in quadrant 4 close to the data cluster of the Tasmanian Dolerite. Therefore, this Sr-Nd isotope mixing diagram confirms the close genetic relationship of the Kirkpatrick Basalt and the Tasmanian Dolerite (Fig. 12.38) which suggests that the magma of the basalt in the Mesa Range originated from the same source in the mantle as the magma that formed the Tasmanian Dolerite.

12.8 Continental Tectonics

Some of the flows on Pain and Tobin mesas in the Mesa Range of northern Victoria Land appear to have been altered because they yield Cretaceous whole-rock Rb-Sr dates. Mensing (1987) first reported that samples of the high-Ti flows on Pain Mesa scatter above and below a Rb-Sr reference isochron corresponding to a date of 92 Ma. Mensing and Faure (1996) subsequently calculated a Rb-Sr date of 100 ± 15 Ma and an initial $^{87}Sr/^{86}Sr$ ratio of 0.711289 for 28 whole-rock samples of high-Ti basalt on Pain Mesa. Fleming et al. (1992, 1993) likewise obtained a low Rb-Sr date of 103 Ma for samples of high-Ti basalt from Tobin Mesa. However, two plagioclase concentrates of a high-Ti basalt yielded $^{40}Ar/^{39}Ar$ plateau dates of 176.8 ± 0.8 and 176.1 ± 0.6 Ma. These dates confirm that the *age* of the high-Ti basalt is Jurassic in agreement with Rb-Sr, $^{40}Ar/^{39}Ar$, and Re-Os dates of the Kirkpatrick Basalt in the Transantarctic Mountains.

The conventional K-Ar dates of the basalt on Tobin Mesa reported by Elliot and Foland (1986) for basalt flows on Tobin Mesa range widely between 178 ± 3 and 113 ± 2 Ma which the authors attributed to loss of radiogenic ^{40}Ar. Faure and Mensing (1993) published another set of whole-rock K-Ar dates of the high-Ti flows on Pain Mesa which also ranged widely from 174 ± 7 to 103 ± 4 Ma. The two sets of K-Ar dates both indicated that the flows originally crystallized at about 176 ± 2 Ma and that they stopped losing radiogenic ^{40}Ar during the Cretaceous Period at about 108 ± 5 Ma. In other words, the Rb-Sr and K-Ar dates of the basalt flows in the Mesa Range suggest that their Rb-Sr and K-Ar systematics were disturbed during the Cretaceous Period.

The high-Ti flows appear to be conformable to the underlying low-Ti flows on Pain Mesa and there is no evidence for an extended hiatus between the eruption of the two sets of flows. In addition, the $^{40}Ar/^{39}Ar$ plateau dates of plagioclase in the high-Ti flow on Tobin Mesa demonstrate that the age of this flow is Middle Jurassic and not Cretaceous. The loss of ^{40}Ar and the homogeni-

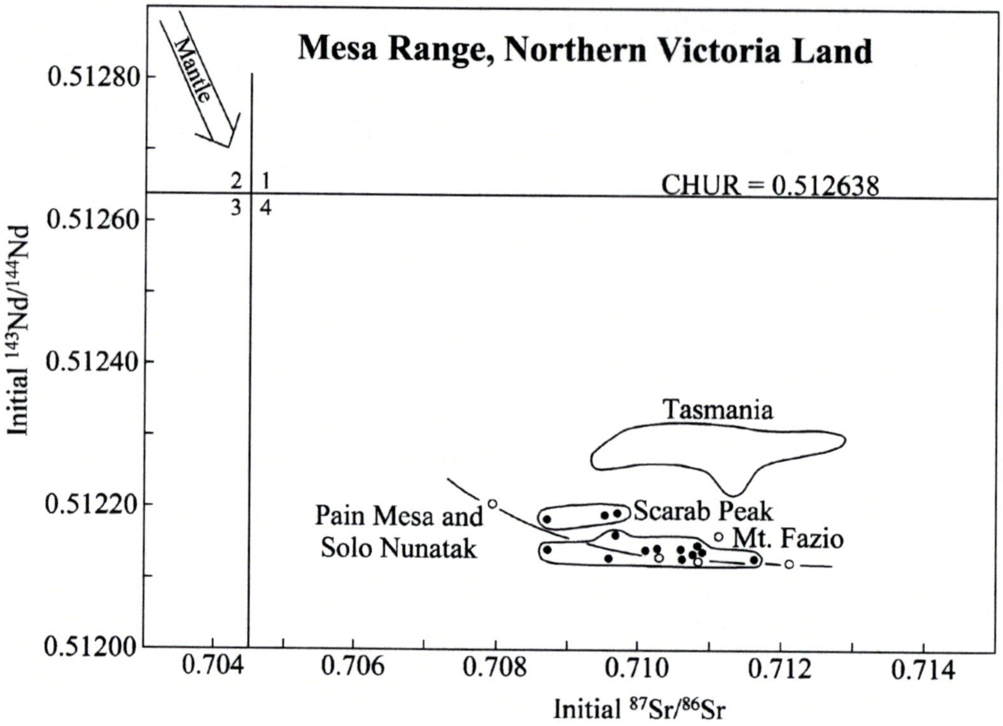

Fig. 12.39 The initial $^{87}Sr/^{86}Sr$ and $^{143}Nd/^{144}Nd$ ratios of samples of the Kirkpatrick Basalt in the Mesa Range of northern Victoria Land define points in quadrant 4 of the Sr-Nd isotope mixing diagram. Rocks from Scarab Peak and Mt. Fazio (Tobin Mesa) cluster adjacent to each other and occupy the same area in the diagram as rocks from Pain Mesa and Solo Nunatak (open circles). A hypothetical isotopic-mixing hyperbola is indicated for the samples from Pain Mesa and Solo Nunatak. The close relationship of the Tasmanian Dolerites and the Kirkpatrick Basalt of the Mesa Range is unmistakable. The presence of crustal strontium and neodymium in all of these rocks supports the theory of Hergt et al. (1989) (Data for Mt. Fazio and Scarab Peak from Fleming et al. (1995), for Pain Mesa and Solo Nunatak from Mensing and Mukasa (unpublished) and for the Tasmanian Dolerite from Hergt et al. (1989))

zation of the isotopic composition of strontium of the high-Ti flows on Tobin and Pain mesas could have been caused by an increase in temperature and by hydrothermal activity caused by the thermal event. The hypothetical increase of the temperature may have been caused by *tectonic* activity because neither the Mesa Range nor any part of northern Victoria Land was affected by magmatic activity during the Cretaceous Period. The high concentration of devitrified glass in the matrix of the high-Ti flows may have made them more susceptible to alteration by the thermal event than the underlying low-Ti flows which contain more crystals and less matrix. However, some samples of high-Ti basalt actually did not lose ^{40}Ar and have yielded Jurassic K-Ar dates. These conjectures concerning the cause for the alteration of the isotopic systematics of the flows in the Mesa Range point to a tectonic event that primarily affected northern Victoria Land.

This train of thought has led us to a "hot topic" in the geologic history of the Transantarctic Mountains concerning the way rifting along the former plaeo-Pacific coast of Antarctica eventually caused the disintegration of Gondwana. This topic has been debated extensively in the geological literature with contributions by Elliot (1972, 1974, 1983, 1991, 1992), Wilson (1992, 1993, 1995), Storey and Alabaster (1991), Behrendt et al. (1991), Schmidt and Rowley (1986), and many others.

12.9 Virtual Geomagnetic Poles

Measurements of the position of the virtual geomagnetic pole (VGP) have supported the hypothesis that the Rb-Sr and K-Ar dates of certain flows of the Kirkpatrick Basalt in the Mesa Range were lowered by a Cretaceous thermal event that was caused by tectonic activity in northern Victoria Land. For examples, Burmester and Anderson (1984) reported that the VGP of the Admiralty Intrusives in the Salamander Range (360–367 Ma, Kreuzer et al. 1981) deviates from the polar wander path of Thompson and Clark (1982) for the Paleozoic in a way that suggests "post-Jurassic motion between northern Victoria Land and East Antarctica." In a subsequent study of the paleomagnetism of the Bowers Supergroup, Schmierer and Burmester (1986, pp. 87–88) noted that the direction of magnetization of the rocks of the Bowers Supergroup does not match the direction expected for Devonian or Jurassic thermal events and concluded that:

> The apparent duration of the remagnetization event and the high precision of the mean direction from the Bowers Supergroup are most consistent with locking in of the remanence during the Cretaceous normal polarity superchron.

A study of the paleomagnetism by Delisle (1983) of the basalt and dolerite in the Litell Rocks provided additional evidence for a Cretaceous thermal event. Skinner et al. (1981) who described these rocks in the field, considered them to be representatives of the Ferrar Group of Middle Jurassic age, although Kreuzer et al. (1981) reported Cretaceous K-Ar dates between 90 and 120 Ma for specimens that appeared to be hydrothermally altered. Three samples of basalt from the Litell Rocks measured by Delisle (1983) indicated a well-defined Cretaceous pole position. Therefore, Delisle (1983, p. 149) concluded that:

> [I]f this material was emplaced during the Jurassic, then it must have been effectively magnetically overprinted during the Cretaceous.

Additional measurements of the paleomagnetism by Delisle and Fromm (1984, 1989) confirmed that the position of the VGP of Ferrar Dolerite sills in the Rennick Graben deviates significantly from its position during the Jurassic at 55°S and 220′E. The VGP reported by Delisle and Fromm (1989) for a dolerite sill on Roberts Butte located west of the Rennick Graben does not coincide with the Jurassic VGP but matches the Cretaceous VGP of the Litell Rocks obtained by Delisle (1983). Based on this new evidence Delisle and Fromm (1989, p. 149) concluded that:

> The discovery of a unit on Roberts Butte with a paleomagnetic record very similar to that of the basaltic flow at Litell Rocks,..., is additional evidence in favor of the assumption of a thermal event in northern Victoria Land during the Cretaceous, resetting in places the paleomagnetic record. This event could well be associated with the initial splitting of Antarctica from Australia.

This conclusion was endorsed by Kreuzer et al. (1987, p. 34) who stated:

> [P]aleomagnetic rejuvenations in Mesozoic volcanics on both sides of the Rennick Graben (Delisle and Fromm 1984) point to a Cretaceous thermal event.

The average VPG of 15 flows of Kirkpatrick Basalt in the Mesa Range analyzed by McIntosh et al. (1986) at 64°S and 210°E lies south of the Jurassic VGP. These authors attributed the discrepancy to incomplete averaging of the paleosecular variation of the magnetic field. However, Elliot and Foland (1986) noted that the K-Ar dates of the flows on Tobin Mesa range from 178 ± 3 (Middle Jurassic) to 113 ± 2 Ma (Early Cretaceous) and speculated that radiogenic ^{40}Ar could have been lost during a low-level Cretaceous heating event associated with the formation of the Rennick Graben. Alternatively, the heating event may have been caused by the separation of Australia from Antarctica.

Additional VGPs of the Kirkpatrick Basalt and Ferrar Dolerite in the Prince Albert Mountains, the ice-free valleys southern Victoria Land, and in the Queen Alexandra Range were reported by Lanza and Zanella (1993), Hoffman et al. (1984), Funaki (1984), McIntosh et al. (1982), Cherry (1981), Ostrander (1971), Briden and Oliver (1963), Bull et al. (1962), Bull and Irving (1960), and Turnbull (1959). Paleomagnetic measurements in the Dufek Massif in the Pensacola Mountains were reported by Beck (1972), Beck et al. (1979), Burmester and Sheriff (1980), and Burmester and Beck (1981). The VGP of Jurassic basalt and dolerite in the Theron Mountains was determined by Blundell and Stephenson (1959) while Blundell (1964) and Lövlie (1979, 1980) measured VGPs for the Jurassic basalt flows in the nunataks of Vestjella in Queen Maud Land (Chapter 14).

The paleomagnetism of the Tasmanian Dolerite was determined by Schmidt and McDougall (1977) while Valencio et al. (1979), Valencio (1979), and Vilas and Valencio (1979) compared the paleomagnetic properties of Mesozoic and Cenozoic igneous rocks in Antarctica, South Africa, and South America. A memorial to Dr. Daniel A. Valencio was published by Anonymous (1991).

The measurements of VGPs of igneous rocks of Mesozoic age in Antarctica and on the other conti-

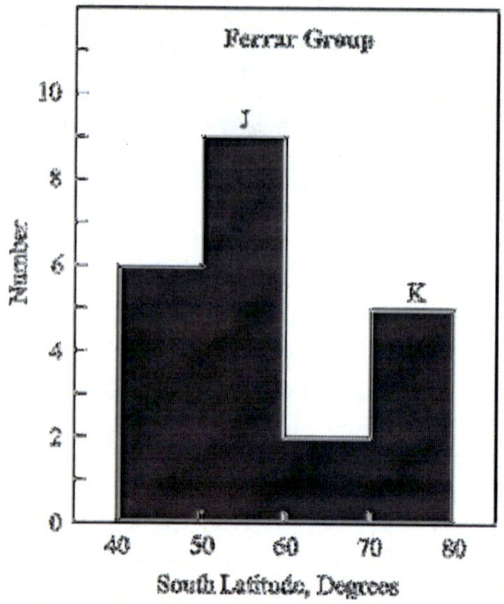

Fig. 12.40 The south latitudes of the virtual geomagnetic pole (VGP) derived from basalt and dolerite of the Ferrar Group in Antarctica have a bimodal distribution. The *average* latitude of Jurassic VGPs range from 51° to 57° whereas the *average* Cretaceous VGPs are located between 65°S and 78°S. This compilation of average VGPs documents that certain basalt flows and dolerite sills of Jurassic age were remagnetized during a thermal event during the Cretaceous (Data from Lanza and Zanella 1993)

nents of the southern hemisphere has provided vital information concerning the break-up of Gondwana and the subsequent displacement of the continental fragments to their present positions. The eruption of large volumes of basalt flows and the intrusion of dolerite sills along the paleo-Pacific margin of Antarctica combined with evidence for a thermal event in northern Victoria Land during the Cretaceous Period both imply that deep crustal rifts formed along strike of the present Transantarctic Mountains. Schmidt and Rowley (1986) proposed that the rifts originated as transform faults and that the landmasses of West Antarctica and the Antarctic Peninsula were displaced by right-lateral strike-slip motion in Fig. 12.40 from their original position adjacent to Victoria Land to their present positions.

12.10 Summary

The eruption of lava flows of the Kirkpatrick Basalt and the intrusion of dolerite sills in Tasmania and in the area of the Transantarctic Mountains was caused by rifting of the continental crust along the paleo-Pacific margin of Antarctica during the Middle Jurassic. The resulting release of pressure caused basalt magma to form by decompression melting in the subcrustal mantle which contained less than about 3% of subducted sediment. The presence of subducted sedimentary rocks in the subcrustal mantle caused the basalt magma to be enriched in radiogenic ^{87}Sr and to have a pronounced crustal character in its major-element and trace-element composition.

The eruptions occurred rapidly along the entire length of Transantarctic Mountains from northern Victoria Land to the Theron Mountains and into Queen Maud Land. The large volume of basalt that was erupted and the aggregate thickness of the cogenetic dolerite sills that were intruded into the sedimentary rocks of the Beacon Supergroup contributed substantially to the uplift of the Transantarctic Mountains. The formation of crustal rifts and the resulting magmatic activity in Antarctica and Tasmania were a prelude to the ultimate break-up of Gondwana. In fact, the magmatic activity that occurred in Gondwana during the Middle Jurassic was part of the global tectonic disintegration of Pangea that created the geography of the Earth we know today.

References

Allègre CJ, Dupré B, Richard P, Rousseau D (1982) Subcontinental versus suboceanic mantle, II. Nd-Sr-Pb isotopic comparison of continental tholeiites with mid-ocean ridge tholeiites, and the structure of the lithosphere. Earth Planet Sci Lett 57:25–34

Anonymous (1991) In Memoriam Prof. Daniel A. Valenico. In: Ulbrich H, Rocha Campos AC (eds) Gondwana Seven. Instituto de Geociencias, Sao Paulo, Brazil

Ball HW, Borns HW Jr, Hall BA, Brooks HK, Carpenter FM, Delevoryas T (1979) Biota, age, and significance of lake deposits, Carapace Nunatak, Victoria Land, Antarctica. In: Laskar B, Raja Rao CS (eds) Fourth international Gondwana symposium papers, vol. 1. Hindustan, Delhi, India, pp 166–175

Ballance PF (1965) A reinterpretation of the Mawson Tillite, Antarctica. Aust J Sci 28(8):312

Ballance PR (1967) Stratigraphy and hydrothermal mineralization of the Beacon Group in the Allan Hills-Carapace Nunatak area, Victoria Land, Antarctica. Aust J Sci 30(11):446

Ballance P, Watters WA (1971) The Mawson Diamictite and the Carapace Sandstone formations of the Ferrar Group at Allan Hills and Carapace Nunatak, Victoria Land, Antarctica. New Zealand J Geol Geophys 14:512–527

Ballance P, Watters WA (2001) Hydrothermal alteration, contact metamorphism, and authigenesis in Ferrar Supergroup and Beacon Supergroup rocks, Carapace Nunatak, Allan Hills, and Coombs Hills, Victoria Land, Antarctica. New Zealand Geol Geophys 45(1):71–84

Banks MR, Clarke MJ (1987) Changes in the geography of the Tasmania Basin in the late Paleozoic. In: McKenzie GD (ed) Gondwana Six: stratigraphy, sedimentology, and paleontology. Geophysical Monograph, vol. 41. American Geophysical Union, Washington, DC, pp 1–4

Barrett PJ, Elliot DH (1972) The early Mesozoic volcaniclastic Prebble Formation, Beardmore Glacier area. In: Adie RJ (ed) Antarctic geology and geophysics. Universitetsforlaget, Oslo, Norway, pp 403–409

Barrett PJ, Elliot DH, Lindsay JF (1986) The Beacon Supergroup (Devonian-Triassic) and Ferrar Group (Jurassic) in the Beardmore Glacier area, Antarctica. In: Turner MD, Splettstoesser JF (eds) Geology of the Central Transantarctic Mountains. Antarctic research series, vol. 36. American Geophysical Union, Washington, DC, pp 339–428

Beck ME (1972). Paleomagnetism and magnetic polarity zones in the Jurassic Dufek intrusion, Pensacola Mountains, Antarctica. Geophys J Roy Astron Soc 28(1):49–64

Behrendt JC, LeMasurier WE, Cooper AK, Tessensohn F, Tréhu A, Damaske D (1991) Geophysical studies of the West Antarctic rift system. Tectonics 10(6):1257–1273

Blundell DJ (1964) Orientated dolerite sample from Dronning Maud Land. British Antarctic Surv Bull 4:57

Blundell DJ, Stephenson PJ (1959) Paleomagnetism of some dolerite intrusions from the Theron Mountains and Whichaway nunataks, Antarctica. Nature 184:1860

Borns HW, Hall BA (1969) Mawson "Tillite" in Antarctica: Preliminary report of a volcanic deposit of Jurassic age. Science 166:870–872

Bowman JR (1971) Use of the isotope composition of strontium and SiO_2 content in determining the origin of Mesozoic basalt from Antarctica. MSc thesis, The Ohio State University, Columbus, OH

Bradshaw MA (1987) Additional field interpretation of the Jurassic sequence at Carapace Nunatak and Coombs Hills, south Victoria Land, Antarctica. New Zealand J Geol Geophys 30:37–49

Brewer TS (1989) Mesozoic dolerites from Whichaway Nunataks. Antarctic Sci 1(2):97–108

Brewer TS, Palacz Z, Hawkesworth CJ (1987) Mesozoic tholeiites from Coats Land and Dronning Maud Land, Antarctica: Further pieces in the Gondwanaland jigsaw. Terra Cognita 7:604

Brewer TS, Hergt JM, Hawkesworth CJ, Rex D, Storey BC (1992) Coats Land dolerites and the generation of Antarctic continental flood basalts. In: Storey BC, Alabaster T, Pankhurst RJ (eds) Magmatism and the causes of continental break-up. Geol Soc London, Spec Pub 68:185–208

Briden JC, Oliver RL (1963) Paleomagnetic results from the Beardmore Glacier region, Antarctica. New Zealand J Geol Geophys 6:388–394

Brooks C, Hart SR (1978) Rb-Sr mantle isochrons and variations in the chemistry of Gondwanaland's lithosphere. Nature 271(5642):220–223

Brotzu P, Capaldi G, Civetta L, Melluso L, Orsi G (1988) Jurassic Ferrar dolerites and Kirkpatrick basalts in northern Victoria Land (Antarctica); Stratigraphy, geochronology, and petrology. Mem Soc Geol Italiana 43:97–116

Bull C, Irving E (1960) The paleomagnetism of some hypabyssal intrusive rocks from south Victoria Land, Antarctica. Geophys J Roy Astron Soc 3:211–214

Bull C, Irving E, Willis I (1962) Further paleomagnetic results from south Victoria Land, Antarctica. Geophys J Roy Astron Soc 6:320–336

Burmester RF, Anderson JK (1984) Did northern Victoria Land collide with East Antarctica in the Cretaceous? Antarctic J US 19(5):31–33

Burmester RF, Beck ME (1981) Paleomagnetism of the Dufek intrusion. Antarctic J US 16(5):58–60

Burmester RF, Sheriff SD (1980) Paleomagnetism of the Dufek intrusion, Pensacola Mountains, Antarctica. Antarctic J US 15(5):43–45

Cherry EM (1981) A paleomagnetic investigation of Jurassic Kirkpatrick Basalt flows from south Victoria Land, Antarctica. BA thesis, The Ohio State University, Columbus, OH

Cherry EM, Noltimier HC (1982) Paleomagnetic results: Kirkpatrick basalts at Brimstone Peak and Gorgon Peak, Antarctica. Eos Trans AGU 62:555

Collinson JW, Kemp NR (1983) Permian-Triassic sedimentary sequence in northern Victoria Land, Antarctica. In: Oliver RL, James PR, Jago JB (eds) Antarctic geoscience. Australian Academy of Science, Canberra, ACT, pp 221–225

Compston W, McDougall J, Heier KS (1968) Geochemical comparison of the Mesozoic basaltic rocks of Antarctica, South Africa, South America, and Tasmania. Geochim Cosmochim Acta 32:129–149

Cosgriff JW (1974) Lower Triassic temnospondyli of Tasmania. Geol Soc Amer Spec Paper 149:1–134

Cosgriff JW, Hammer WR (1983) The labyrinthodont amphibians of the earliest Triassic from Antarctica, Tasmania, and South Africa. In: Oliver RL, James PR, Jago JB (eds) Antarctic earth science. Australian Academy of Science, Canberra, ACT, pp 590–592

Cox KG, Bell JD, Pankhurst RJ (1979) The interpretation of igneous rocks. Allen & Unwin, London

Dalziel IWD, Elliot DH (1981) West Antarctica: Problem child of Gondwanaland. Tectonics 1:3–19

Delisle G (1983) Results of paleomagnetic investigations in northern Victoria Land, Antarctica. In: Oliver RL, James PR, Jago JB (eds) Antarctic earth science. Australian Academy of Science, Canberra, ACT, pp 146–149

Delisle G, Fromm K (1984) Paleomagnetic investigations of Ferrar Supergroup rocks, north Victoria Land. Geologisches Jahrbuch 41:41–55

Delisle G, Fromm K (1989) Further evidence for a Cretaceous thermal event in northern Victoria Land. Geologisches Jahrbuch E38, Hannover:143–151

Domack EW, Burkley LA, Domack CR, Banks MR (1993) Facies analysis of glacial marine pebbly mudstone in the Tasmania Basin: Implications for regional paleoclimates during the late Paleozoic. In: Findlay RH, Unrug R, Banks MR, Veevers JJ (eds) Gondwana Eight; Assembly, evolution, and dispersal. A.A. Balkema, Rotterdam, The Netherlands, pp 471–484

Elliot DH (1970a) Jurassic tholeiites of the central Transantarctic Mountains, Antarctica. In: Gilmour EH, Stradling D (eds) Proceedings of the second Columbia River Basalt sympo-

sium. Eastern Washington State College Press, Cheney, Washington, DC, pp 301–325
Elliot DH (1970b) Beardmore Glacier investigations: Narrative and geological report. Antarctic J US 5(4):83–85
Elliot DH (1972a) Major oxide chemistry of the Kirkpatrick Basalt, central Transantarctic Mountains. In: Adie RJ (ed) Antarctic geology and geophysics. Universitetsforlaget, Oslo, Norway, pp 413–418
Elliot DH (1972b) Aspects of Antarctic geology and drift reconstructions. In: Adie RJ (ed) Antarctic geology and geophysics. Universitetsforlaget, Oslo, Norway, pp 849–858
Elliot DH (1974) The tectonic setting of the Jurassic Ferrar Group, Antarctica. In: Gonzalez-Ferran O (ed) Proceedings of the symposium on Andean and Antarctic volcanology problems. International Association of Volcanology and Chemistry of the Earth's Interior, Special Series, pp 357–372
Elliot DH (1991) Triassic-Early Cretaceous evolution of Antarctica. In: Thomson MRA, Crame JA, Thomson JW (eds) Geological evolution of Antarctica. Cambridge University Press, Cambridge, pp 541–548
Elliot DH (1992) Jurassic magmatism and tectonism associated with Goundwanaland break-up: An Antarctic perspective. In: Storey B, Alabaster T, Pankhurst RJ (eds) Magmatism and the causes of continental break-up. Geological Society of London, Special Publication, vol. 68. London, pp 165–184
Elliot DH (2000) Stratigraphy of Jurassic pyroclastic rocks in the Transantarctic Mountains. J African Earth Sci 31:77–89
Elliot DH (2002) Paleovolcanological setting of the Mawson Formation: Evidence from the Prince Albert Mountains, Victoria Land. In: Gamble JA, Skinner DNB, Henrys S (eds) Antarctica at the close of the Millennium. The Royal Society of New Zealand Bulletin, vol. 35. Wellington, New Zealand, pp 185–192
Elliot DH, Foland KA (1986) Potassium-argon age determinations of the Kirkpatrick Basalt, Mesa Range. In: Stump E (ed) Geological investigations in Northern Victoria Land. Antarctic research series, vol. 46. American Geophysical Union, Washington, DC, pp 279–288
Elliot DH, Hanson RE (2001) Origin of widespread, exceptionally thick basaltic phreatomagmatic tuff breccia in the Middle Jurassic Prebble and Mawson Formations, Antarctica. J Volcanol Geotherm Res 111:183–201
Elliot DH, Larsen D (1993) Mesozoic volcanism in the central Transantarctic Mountains, Antarctica: Depositional environment and tectonic setting. In: Findlay RH, Unrug R, Banks MR, Veevers JJ (eds) Gondwana Eight: Assembly, evolution, and dispersal. A.A. Balkema, Rotterdam, The Netherlands, pp 397–410
Elliot DH, Siders M, Faure G, Taylor KS (1982) The Kirkpatrick Basalt, Mesa Range, north Victoria Land. Antarctic J US 17(5):19–20
Elliot DH, Fleck RJ, Sutter JF (1985) Potassium-argon age determinations of Ferrar Group rocks, central Transantarctic Mountains. In: Turner MD, Splettstoesser JF (eds) Geology of the Central Transantarctic Mountains. American Geophysical Union, Washington, DC, pp 197–224
Elliot DH, Haban MA, Siders MA (1986a) The Exposure Hill Formation Mesa Range. In: Stump E (ed) Geological investigations in Northern Victoria Land. Antarctic research series, vol. 46. American Geophysical Union, Washington, DC, pp 267–278
Elliot DH, Siders MA, Haban MA (1986b) Jurassic tholeiites in the region of the Upper Rennick Glacier, north Victoria Land. In: Stump E (ed) Geological investigations in Northern Victoria Land, Antarctic research series, vol. 46. American Geophysical Union, Washington, DC, pp 249–265
Elliot DH, Bigham J, Jones FS (1988) Weathering profiles in the Jurassic basalt sequence, Beardmore Glacier region. Antarctic J US 23(5):17–19
Elliot DH, Bigham J, Jones FS (1991) Interbeds and weathering profiles in the Jurassic basalt sequence, Beardmore Glacier region, Antarctica. In: Ulbrich H, Rocha Campos (eds) Proceedings of the Seventh Gondwana symposium. University of Sao Paulo, Sao Paulo, Brazil, pp 289–302
Elliot DH, Fleming TH, Kyle PR, Foland KA (1999) Long distance transport of magmas in the Jurassic Ferrar Large Igneous Province. Earth Planet Sci Lett 167:89–104
Elliot DH, Fortner EH, Grimes CB (2006) Mawson breccias intrude Beacon strata at Allan Hills, South Victoria Land: Regional implications. In: Fütterer DK, Damaske D, Kleinschmidt G, Miller H, Tessensohn F (eds) Antarctica: Contributions to global science. Springer. Berlin/Heidelberg, Germany, pp 291–298
Faure G (1981) Strontium isotope composition of volcanic rocks: Evidence for contamination of the Kirkpatrick Basalt, Antarctica. In: O'Connell RJ, Fyfe WS (eds) Evolution of the earth, Geodynamics series, vol. 5. American Geophysical Union, Washington, DC, pp 75–81
Faure G (1998) Principles and applications of geochemistry, 2nd edn. Prentice Hall, Upper Saddle River, NJ
Faure G (2001) Origin of igneous rocks; The isotopic evidence. Springer, Heidelberg, Germany
Faure G, Hill RL (1963) Age of the Falla Formation (Triassic), Queen Alexandra Range. Antarctic J US 8(5):264–266
Faure G, Mensing TM (1993) K-Ar dates and paleomagnetic evidence for Cretaceous alteration of Mesozoic basaltic lava flows, Mesa Range, northern Victoria Land, Antarctica. Chem Geol 109:305–315
Faure G, Mensing TM (2005) Isotopes: principles and applications, 3rd edn, Principles of isotope geology. Wiley, Hoboken, NJ
Faure G, Hill RL, Jones LM, Elliot DH (1972) Isotope composition of strontium and silica content of Mesozoic basalt and dolerite from Antarctica. In: Adie RJ (eds) Antarctic geology and geophysics. Universitetsforlaget, Oslo, Norway, pp 617–624
Faure G, Bowman JR, Elliot DH, Jones LM (1974) Strontium isotope composition and petrogenesis of the Kirkpatrick Basalt, Queen Alexandra Range, Antarctica. Contrib Mineral Petrol 58(3):153–169
Faure G, Pace KK, Elliot DH (1982) Systematic variations of $^{87}Sr/^{86}Sr$ ratios and major element concentrations in the Kirkpatrick Basalt of Mt. Falla, Queen Alexandra Range, Transantarctic Mountains. In: Craddock C (ed) Antarctic geoscience. University of Wisconsin Press, Madison, WI, pp 715–723
Faure G, Hoefs J, Mensing TM (1984) Effect of oxygen fugacity on sulfur isotope compositions and magnetite concentrations in the Kirkpatrick Basalt, Mount Falla, Queen Alexandra Range, Antarctica. Isotope Geosci (Chem Geol) 2:301–311

Finn CA, Goodge JW, Damaske D, Fanning CM (2006) Scouting craton's edge in paleo-Pacific Gondwana. In: Fütterer DK, Damaske D, Kleinschmidt G, Miller H, Tessensohn F (eds) Antarctica: Contributions to global earth science. Springer, Heidelberg, Germany, pp 165–174

Fleck RJ, Sutter JF, Elliot DH (1977) Interpretation of discordant $^{40}Ar/^{39}Ar$ age-spectra of Mesozoic tholeiites from Antarctica. Geochim Cosmochim Acta 41:15–32

Fleming TH, Elliot DH (1988) Iron-rich tholeiitic rocks of the Kirkpatrick Basalt, Beardmore Glacier region. Antarctic J US 23(5):15–17

Fleming TH, Elliot DH, Jones LM, Bowman JR, Siders MA (1992) Chemical and isotopic variations in an iron-rich lava flow from the Kirkpatrick Basalt, north Victoria Land, Antarctica: Implications for low-temperature alteration. Contrib Mineral Petrol 111:440–457

Fleming TH, Elliot DH, Foland KA, Jones LM, Bowman JR (1993) Disturbance of Rb-Sr and K-Ar isotopic systems in the Kirkpatrick Basalt, north Victoria Land, Antarctica: Implications for middle Cretaceous tectonism. In: Findlay RH, Unrug R, Banks MR, Veevers JJ (eds) Gondwana Eight: Assembly, evolution, and dispersal. A.A. Balkema, Rotterdam, The Netherlands, pp 411–424

Fleming TH, Foland KA, Elliot DH (1995) Isotopic and chemical constraints on the crustal evolution and source signature of Ferrar magmas, north Victoria Land, Antarctica. Contrib Mineral Petrol 121:217–236

Foland KA, Fleming TH, Heimann A, Elliot DH (1993) Potassium-argon dating of fine-grained basalt with massive Ar loss: Application of the $^{40}Ar/^{39}Ar$ technique to plagioclase and glass from the Kirkpatrick Basalt, Antarctica. Chem Geol 107:173–190

Ford AB, Kistler RW (1980) K-Ar age, composition, and origin of Mesozoic mafic rocks related to Ferrar Group, Pensacola Mountains, Antarctica. New Zealand J Geol Geophys 23:371–390

Forsyth SM (1987) Review of the Upper Parmeener Supergroup. Report 1987/01. Tasmania Department of Mines, Hobarth, Tasmania, pp 1–38

Funaki M (1984) Paleomagnetic investigation of McMurdo Sound region, southern Victoria Land, Antarctica. Mem Nat Inst Polar Res, Series C, Earth Sci, No 16:1–81. Tokyo, Japan

Gair HS (1967) The geology from the upper Rennick Glacier to the coast, northern Victoria Land, Antarctica. New Zealand J Geol Geophys 10(2):309–344

Gair HS, Sturm A, Carryer SJ, Grindley GW (1969) The Geology of northern Victoria Land. In: Craddock C (ed) Geologic map of Antarctica, Folio 12, Sheet 13. American Geographical Society, New York

Grapes RH, Reid DL, McPherson JG (1974) Shallow dolerite intrusion and phreatic eruption in the Allan Hills region, Antarctica. New Zealand J Geol Geophys 17:563–577

Grindley WG (1963) The geology of the Queen Alexandra Range, Beardmore Glacier, Ross Dependency, Antarctica; With notes on the correlation of Gondwana sequences. New Zealand J Geol Geophys 6(3):307–347

Grindley GW, Laird MG (1969) Geology of the Shackleton Coast. In: Craddock C (ed) Geologic map of Antarctica, Folio 12, Sheet 15. American Geographical Society, New York

Grindley GW, Warren G (1964) Stratigraphic nomenclature and correlation in the western Ross Sea region. In: Adie RJ (ed) Antarctic geology. North-Holland, Amsterdam, The Netherlands, pp 314–333

Haban MA (1984) The mineral chemistry and petrogenesis of the Ferrar Supergroup, north Victoria Land, Antarctica. MSc thesis, The Ohio State University, Columbus, OH

Hall BA, Sutter JF, Borns HW Jr (1982) The inception and duration of Mesozoic volcanism in the Allan Hills-Carapace Nunatak area, Victoria Land, Antarctica. In: Craddock C (ed) Antarctic geoscience. University Wisconsin Press, Madison, WI, pp 709–714

Hammer WR, Hickerson WJ (1994) A crested therapod dinosaur from Antarctica. Science 264:828–830

Hand SJ (1993) Paleogeography of Tasmania's Permo-Carboniferous glacigenic sediments. In: Findlay RH, Unrug R, Banks MR, Veevers JJ (eds) Gondwana Eight; assembly, evolution, and dispersal. A.A. Balkema, Rotterdam, The Netherlands, pp 459–469

Hanson RE, Elliot DH (1996) Rift-related Jurassic basaltic phreatomagmatic volcanism in the central Transantarctic Mountains: Precursory stage to flood-basalt effusions. Bull Volcanol 58:327–347

Harmon RS, Hoefs J (1995) Oxygen isotope heterogeneity of the mantle deduced from global ^{18}O systematics of basalts from different geotectonic settings. Contrib Mineral Petrol 120:95–114

Heier KS, Compston W, McDougall I (1965) Thorium and uranium concentrations, and the isotopic composition of strontium in the differentiated Tasmanian dolerites. Geochim Cosmochim Acta 29:643–659

Hergt JM (1987) The origin and evolution of the Tasmanian Dolerites. Ph.D. dissertation, Australian National University, Canberra, Australia

Hergt JM, Chappell BW, McCulloch MT, McDougall I, Chivas AR (1989) Geochemical and isotopic constraints on the origin of the Jurassic Dolerites of Tasmanian. J Petrol 30(4):841–883

Hergt JM, Peate DW, Hawkesworth CJ (1991) The petrogenesis of Mesozoic Gondwana low-Ti flood basalts. Earth Planet Sci Lett 105:134–148

Hill RL (1969) Strontium isotope composition of basaltic rocks in the Transantarctic Mountains, Antarctica. MSc thesis, Department of Geology, The Ohio State University, Columbus, OH

Hoefs J, Faure G, Elliot DH (1980) Correlation of $\delta^{18}O$ and initial $^{87}Sr/^{86}Sr$ ratios in Kirkpatrick Basalt on Mt. Falla, Transantarctic Mountains. Contrib Mineral Petrol 75:199–203

Hoffman J, Nairn AEM, Peterson DN (1984) The paleomagnetic investigation of flows and sills from the Queen Alexandra Range, Antarctica. In: Turner MD, Splettstoesser JF (eds) Geology of the Central Transantarctic Mountains. Antarctic research series, vol. 36. American Geophysical Union, Washington, DC, pp 63–74

Hornig I (1993) High-Ti and low-Ti tholeiites in the Jurassic Ferrar Group, Antarctica. In: Danske D, Fritsch J (eds) German Antarctic North Victoria Land Expedition 1988/89 (GANOVEX V) Geologisches Jahrbuch, Reihe E 47:335–369

Houghton BF, Smith RT (1993) Recycling of magmatic clasts during explosive eruptions: Estimating the true juvenile content of phreatomagmatic volcanic deposits. Bull Volcanol 55:414–420

Ishizaka K, Yanagi T, Hayatsu K (1977) A strontium isotopic study of the volcanic rocks of the Myoko Volcano Group, central Japan. Contrib Mineral Petrol 63:295–307

IUGS (2002) International stratigraphic chart. International Commission of Stratigraphy

Jefferson TH (1980) Angiosperm fossils in supposed Jurassic volcanogenic shales, Antarctica. Nature 285:157–158

Jefferson TH, Siders MA, Haban MA (1983) Jurassic trees engulfed by lavas of the Kirkpatrick Basalt Group, northern Victoria Land. Antarctic J US 18(5):14–16

Kreuzer H, Höhndorf A, Lenz H, Vetter U, Tessensohn F, Muller P, Jordan H, Harre W, Besang C (1981) K/Ar and Rb/Sr dating of igneous rocks from northern Victoria Land. Antarctica. Geologisches Jahrbuch 41:267–273

Kreuzer H, Delisle G, Fromm K, Höhndorf A, Lenz H, Müller P, Vetter U (1987) Radiometric ages of pre-Mesozoic rocks from northern Victoria Land, Antarctica. In: McKenzie GD (eds) Gondwana Six: Structure, tectonics, and geophysics. Geophysical Monograph, vol. 40. American Geophysical Union, Washington, DC, pp 31–47

Kyle PR (1977) Petrogenesis of Ferrar Group rocks. Antarctic J US 12(4):108–110

Kyle PR (1980) Development of heterogeneities in the subcontinental mantle: Evidence from the Ferrar Group. Contrib Mineral Petrol 73:89–104

Kyle PR, Elliot DH, Sutter JF (1981) Jurassic Ferrar Supergroup tholeiites from the Transantarctic Mountains, Antarctica, and their relationship to the initial fragmentation of Gondwana. In: Greswell MM, Vella P (eds) Gondwana V, pp 283–287

Kyle PR, Pankhurst R, Bowman JR (1983) Isotopic and chemical variations in Kirkpatrick Basalt Group rocks from southern Victoria Land. In: Oliver RL, James PR, Jago JB (eds) Antarctic earth sciences. Australian Academy of Science, Canberra, ACT, pp 234–237

Kyle PR, Pankhurst P, Moorbath S, Bowman JR (1987) Nature and development of an enriched mantle source for Jurassic Ferrar tholeiites, Transantarctic Mountains. Abstract Terra Cognita 7:618

Lanza R, Zanella E (1993) Paleomagnetism of the Ferrar dolerite in the northern Prince Albert Mountains, Victoria Land, Antarctica. Geophys J Int 114:501–511

Larsen D (1988) The petrology and geochemistry of the volcaniclastic upper part of Falla Formation and Prebble Formation, Beardmore Glacier area, Antarctica. MSc thesis, The Ohio State University, Columbus, OH

Lawver LA, Dalziel IWD, Gahagan LM, Martin KM, Campbell DA (2003) The 2003 atlas of plate reconstructions (750 Ma to present day). Plates Progress Report 280-0703, UTIG Tech. Report 190

Lorenz V (1986) On the growth of maars and diatremes and its relevance to the formation of tuff rings. Bull Volcanol 48:265–274

Lorenz V (2000) Formation of maar-diatreme volcanoes. Terra Nostra 2000:284–291

Lövlie R (1979) Mesozoic palaeomagnetism in Vestfjella, Dronning Maud Land, East Antarctica. Geophys J Roy, Astron Soc 59:529–537

Lövlie R (1988) Evidence for deuteric magnetization in hydrothermally altered Mesozoic basaltic rocks from East Antarctica. Phys Earth Planet Interiors 52(3/4):352–364

Marsh JS, Skilling I (1998) Field excursion guide book A3:Karoo volcanic and intrusive rocks, Eastern Cape, South Africa. IAVCEI International Volcanology Conference

McDougall I (1962) Differentiation of the Tasmanian Dolerites: Red Hill Dolerite – Granophyre association. Geol Soc Amer Bull 73:279–316

McDougall I (1963) Potassium-argon age measurements on dolerite from Antarctica and South Africa. J Geophys Res 68(5):1535–1545

McDougall I, Wellman P (1976) Potassium-argon ages for some Australian Mesozoic igneous rocks. J Geol Soc Australia 23:1–9

McGregor VR (1965) Notes on the geology of the area between the heads of the Beardmore and Shackleton glaciers, Antarctica. New Zealand J Geol Geophys 8:278–291

McGregor VR, Wade FA (1969) Geology of the Western Queen Maud Mountains. In: Craddock C (ed) Geologic map of Antarctica, Folio 12, Sheet 16. American Geographical Society, New York

McIntosh WC, Kyle PR, Cherry EM, Noltimier HC (1982) Paleomagnetic results from the Kirkpatrick Basalt Group, Victoria Land. Antarctic J US 17(5):20–22

McIntosh WC, Kyle PR, Sutter JF (1986) Paleomagnetic results from the Kirkpatrick Basalt Group, Mesa Range, north Victoria Land, Antarctica. In: Stump E (ed) Geological investigations in Northern Victoria Land. Antarctic Research Series, vol. 46—. American Geophysical Union, Washington, DC, pp 289–303

McKelvey BC, Webb PN (1959) Geology of upper Taylor Glacier region, Pt. 2 of Geological investigations in southern Victoria Land, Antarctica. New Zealand J Geol Geophys 2(4):718–728

Mensing TM (1987) Geology and petrogenesis of the Kirkpatrick Basalt, Pain Mesa and Solo Nunatak, northern Victoria Land, Antarctica Ph.D. dissertation, The Ohio State University, Columbus, OH

Mensing TM, Faure G (1996) Cretaceous alteration of Jurassic volcanic rocks, Pain Mesa, northern Victoria Land, Antarctica. Chem Geol 129:153–161

Mensing TM, Faure G, Taylor KS, Jones LM (1983) Petrogenesis of the Kirkpatrick Basalt, Solo Nunatak, northern Victoria Land. Antarctic J US 18(5):12–14

Mensing TM, Faure G, Jones LM, Bowman JR, Hoefs J (1984) Petrogenesis of the Kirkpatrick Basalt, Solo Nunatak, northern Victoria Land, Antarctica, based on isotopic compositions of strontium, oxygen, and sulfur. Contrib Mineral Petrol 87:101–108

Mensing TM, Faure G, Jones LM, Hoefs J (1991) Stratigraphic correlation and magma evolution of the Kirkpatrick Basalt in the Mesa Range, northern Victoria Land, Antarctica. In: Ulbrich H, Rocha Campos AC (eds) Gondwana Seven proceedings. Instituto de Geociencias, Universidade de São Paulo, Brazil, pp 654–667

Milnes AR, Cooper BJ, Cooper AJ (1982) The Jurassic Wisanger Basalt of Kangaroo Island, South Australia. Trans Roy Soc South Australia 106(Pt. 1):1–13

Molzahn M, Reisberg L, Wörner G (1996) Os, Sr, Nd, Pb, O isotope and trace element data from the Ferrar flood basalts, Antarctica: Evidence for an enriched subcontinental lithospheric source. Earth Planet Sci Lett 144:529–546

Nathan S, Schulte FJ (1968) Geology and petrology of the Campbell-Aviator divide, northern Victoria Land, Antarctica.

Part I. Post-Paleozoic rocks. New Zealand J Geol Geophys 11:940–975

Norris G (1965) Triassic and Jurassic miospores and acritarchs from the Beacon and Ferrar groups, Victoria Land, Antarctica. New Zealand J Geol Geophys 8(2):236–277

Ostrander JH (1971) Paleomagnetic investigations of the Queen Alexandra Range, Antarctica. Antarctic J US 6(5):183–185

Pace KK (1977) Interpretation of $^{87}Sr/^{86}Sr$ ratios and chemical compositions of the Kirkpatrick Basalts, Mt. Falla, Queen Alexandra Range, Antarctica. MSc thesis, The Ohio State University Columbus, OH

Palais JM (1985) Tephra layers and ice chemistry in the Byrd-Station ice core, Antarctica. PhD dissertation, The Ohio State University, Columbus, OH

Palais JM, Sigurdsson H (1989) Petrologic evidence of volatile emissions from major historic and pre-historic eruptions. In: Berger A et al. (eds) Volcanic volatile emissions. Geophysical Monograph, vol. 7. American Geophysical Union, Washington, DC, pp 31–53

Pankhurst RJ, Leat PT, Sruoga P, Rapela CW, Marquez M, Storey BC, Riley TR (1998) The Chon Aike province of Patagonia and related rocks in West Antarctica: A silicic large igneous province. J Volcanol Geotherm Res 81:113–136

Rhodes RC, Bornhorst ThJ (1976) Petrologic provinces in Jurassic tholeiites of Gondwanaland. Geol Rundschau 65:930–938

Richter M, Thomson MRA (1989) First Aspidorhynchidae (Pisces: Teleostei) from Antarctica. Antarctic Sci 1(1):57–64

Roland NR, Wörner G (1996) Kirkpatrick flows and associated pyroclastics: New occurrences, definition, and aspects of a Jurassic Transantarctic rift. Geologisches Jahrbuch, Reihe B89:97–121

Schaeffer B (1971) Jurassic fishes from Antarctica. Antarctic J US 6(5):190–191

Schaeffer B (1972) A Jurassic fish from Antarctica. Amer Museum Novitates, No. 2492, June 30:1–16

Schmidt DL, Rowley PD (1986) Continental rifting and transform faulting along the Jurassic Transantarctic Rift, Antarctica. Tectonics 5(2):279–291

Schmidt PW, McDougall I (1977) Palaemagnetic and potassium-argon dating studies of the Tasmanian Dolerites. J Geol Soc Australia 25(Pt. 6):321–328

Schmierer K, Burmester R (1986) Paleomagnetic results from the Cambro-Ordovician Bowers Supergroup, northern Victoria Land, Antarctica. In: Stump E (ed) Geological investigations on Northern Victoria Land. Antarctic Research Series, vol. 46. American Geophysical Union, Washington, DC, pp 69–90

Siders MA (1983) Intraflow variability, chemical stratigraphy, and petrogenesis of the Kirkpatrick Basalt from the Mesa Range, north Victoria Land, East Antarctica. MSc thesis, The Ohio State University, Columbus, OH

Siders MA, Elliot DH (1985) Major and trace element geochemistry of the Kirkpatrick Basalt, Mesa Range, Antarctica. Earth Planet Sci Lett 72:54–64

Skinner DNB, Ricker J (1968) The geology of the region between the Mawson and Priestley glaciers, north Victoria Land, Antarctica; 2, Upper Paleozoic to Quaternary Geology. New Zealand J Geol Geophys 11(4):1041–1075

Skinner DNB, Tessensohn F, Vetter U (1981) Lavas in the Ferrar Group of Litell Rocks, north Victoria Land, Antarctica. Geol Jahrbuch B 41:251–259

Srinivasan R, Jaffri SH, Rao GV, Reddy GK (1998) Phreatomagmatic eruptive center from the Deccan Trap Province, Jabalpur, central India. Current Sci 74:878–890

Stigall AL, Babcock LE, Briggs DEG, Leslie SA (2008) Taphonomy of lacustrine interbeds in the Kirkpatrick Basalt (Jurassic) Antarctica. Palaios 23:344–355

Storey BC, Alabaster T (1991) Tectonomagnetic controls on Gondwana break-up models: Evidence form the proto-Pacific margin of Antarctica. Tectonics 10(6):1274–1288

Storey BC, Hole MJ, Pankhurst RJ, Millar IL, Vennum W (1988) Middle Jurassic within-plate granites in West Antarctica and their bearing on the break-up of Gondwanaland. J Geol Soc London 145:999–1007

Tappert R, Foden J, Stachel T, Muehlenbachs K, Tappert M, Wills K (2009) Deep mantle diamonds from South Australia: A record of Pacific subduction at the Gondwana margin. Geol 37(1):43–46

Tasch P (1968) Quantitative paleolimnology and fossil conchostrancans. Antarctic J US 3(5):179–180

Tasch P (1969) Branchipoda. In: Moore RC (ed) Treatise on invertebrate paleontology. The University of Kansas, Lawrence, Kansas, pp R129–R185

Tasch P (1973) Antarctic and other Gondwana non-marine deposits. Antarctic J US 8(5):272–273

Tasch P (1976) Jurassic non-marine trace fossils (Transantarctic Mountains) and the food web. J Paleontol 50(4):754–758

Tasch P (1977) Ancient Antarctic freshwater ecosystems. In: Llano GA (ed) Adaptation within Antarctic Ecosystems. Gulf Publishing, Houston, TX

Tasch P (1987) Fossil conchostracans of the southern hemisphere and continental drift. Geological Society of America, Memoir 165. Boulder, CO, pp 1–290

Tasch P, Gifford EL Jr (1984) Central Transantarctic Mountains nonmarine deposits. In: Turner MD, Splettstoesser JF (eds) Geology of the Central Transantarctic Mountains. Antarctic Research Series, vol. 36. American Geophysical Union, Washington, DC, pp 75–96

Thompson R, Clark RM (1982) A robust, least-squares Gondwana apparent polar wander path and the question of paleomagnetic assessment of Gondwana reconstructions. Earth Planet Sci Lett 57:152–158

Turnbull G (1959) Some paleomagnetic measurements in Antarctica. Arctic 12:151–157

Valencio DA (1979) Correlation of late Palaeozoic-early Mesozoic rocks from South America and Africa on the basis of their magnetic remanences. In: Laskar B, Raja Rao CS (eds) Fourth international Gondwana symposium, vol. 2. Pp 917–922

Valencio DA, Mendia JE, Vilas JF (1979) Paleomagnetism and K-Ar age of Mesozoic and Cenozoic igneous rocks from Antarctica. Earth Planet Sci Lett 45:61–68

Vilas JF, Valencio DA (1979) Palaeomagnetism of South American rocks and the Gondwana continent. In: Laskar B, Raja Rao CS (eds) Fourth international Gondwana symposium, vol. 2. pp 923–930

White JDL (1991) Maar-diatreme phreatomagmatism at Hopi Buttes, Navajo Nation (Arizona), USA. Bull Volcanol 53:239–258

White JDL (1996) Impure coolants and interaction dynamics of phreatomagmatic eruptions. J Volcanol Geothermal Res 74:155–170
White JDL, McClintock MK (2001) Immense vent complex marks flood-basalt eruption in a wet, failed rift: Coombs Hills, Antarctica. Geology 29(10):935–938
Wilson M (1989) Igneous petrogenesis; a global tectonic approach. Unwin Hyman, London
Wilson TJ (1992) Mesozoic and Cenozoic kinetic evolution of the Transantarctic Mountains. In: Yoshida Y et al. (eds) Recent progress in Antarctic earth science. Terra Scientific, Tokyo, Japan, pp 304–314
Wilson TJ (1993) Jurassic faulting and magmatism in the Transantarctic Mountains. Implications for Gondwana break-up mechanisms. In: Findlay RH, Unrug R, Banks MR, Veevers JJ (eds) Gondwana Eight: Assembly, evolution, and dispersal. A.A. Balkema, Rotterdam, The Netherlands, pp 563–572
Wilson TJ (1995) Cenozoic transtension along the Transantarctic Mountains - West Antarctic rift boundary, southern Victoria Land, Antarctica. Tectonics 14(2):531–545
Wörner G (1992) Kirkpatrick lavas, Exposure Hill Formation and Ferrar sills in the Prince Albert Mountains, Victoria Land, Antarctica. Polarforschung 60(2):87–90
Wörner G, Schmincke H-U (1984) Petrogenesis of the zoned Laacher See tephra sequence. J Petrol 25:836–851
Yanagi T, Ishizaka K (1978) Batch fractionation model for the evolution of volcanic rocks in an island arc: An example from central Japan. Earth Planet Sci Lett 40:252–262
Zolotukhin VV, Al'Mukhamedov AI (1988) Traps of the Siberian platform. In: Macdougall JD (ed) Continental flood Basalts. Kluwer, Dordrecht, The Netherlands, pp 273–310

Chapter 13
Ferrar Group: Dolerite Sills and the Dufek Intrusion

During Robert Scott's first expedition to Antarctica from 1901–1904, a group of men led by Albert Armitage and including the geologist Hartley Ferrar crossed McMurdo Sound and explored the mountains of Victoria Land. They discovered the Ferrar Glacier and ascended it to the edge of the polar plateau (Section 1.4.1). During this trip, Hartley Ferrar photographed diabase sills that intruded the Beacon sandstones and later described the geology of the region (Ferrar 1907). The diabase samples he collected plus others that were collected during Shackleton's first expedition from 1907–1909 were later described by Prior (1907) and Benson (1916), respectively. Several other geologists worked in the ice-free valleys during Scott's "Terra Nova" expedition, but only the report by Smith (1924) provided information about the diabase sills. That was pretty much the extent of the information on the diabase sills of southern Victoria Land until the IGY (1957/58) when several geologists from New Zealand and the USA began to map the geology of the Transantarctic Mountains. British geologists prefer the term "dolerite" for rocks which American geologists call "diabase". We will use the British term out of respect for the New Zealand geologists who continue to work in Antarctica.

Important contributions were made by Harrington (1958) who named the Ferrar Dolerite and by Grindley (1963) who discovered the Kirkpatrick Basalt in the Beardmore Glacier area. Subsequently, Grindley and Warren (1964) considered the Ferrar Dolerite and the Kirkpatrick Basalt to be Formations which together constitute the Ferrar Group. This usage has prevailed until the present time (Barrett et al. 1986).

13.1 Southern Victoria Land

The Ferrar Dolerite sills in Wright Valley of southern Victoria Land were included on the geological map of the area prepared by McKelvey and Webb (1962). The cross-section in Fig. 13.1 of the Olympus Range north of Wright Valley identifies the three dolerite sills that intrude the granite of the basement complex, the Kukri Peneplain, and the sedimentary rocks of the Beacon Supergroup. The Basement sill was injected into exfoliation joints in the basement complex, and the Peneplain sill was intruded along the unconformity between the basement and the overlying sedimentary rocks. The Beacon sills were forcefully intruded along bedding planes of the sandstones of the Beacon Supergroup and rose up-section by cutting across thick layers of sedimentary rocks as illustrated in Fig. 13.2.

The dolerite sills in Taylor Valley were mapped by Warren Hamilton in 1958 who published excellent photographs of Taylor and Wright valleys including the sills on Finger Mountain in the upper Taylor Valley. Hamilton (1965) also published a geologic map of the Beacon Heights area that identifies the outcrops of the three dolerite sills. The combined thickness of the dolerite sills locally reaches 2,000 m which significantly increased the elevation of the Beacon rocks which had been deposited on alluvial plains close to sea level (Chapter 10). In this sense, the eruption of the Kirkpatrick Basalt and the intrusion of the sills of the Ferrar Dolerite initiated the uplift of the Transantarctic Mountains. The sills did not emanate from identifiable magma centers but permeated the entire complex of

G. Faure and T.M. Mensing, *The Transantarctic Mountains: Rocks, Ice, Meteorites and Water*,
DOI 10.1007/978-90-481-9390-5_13,

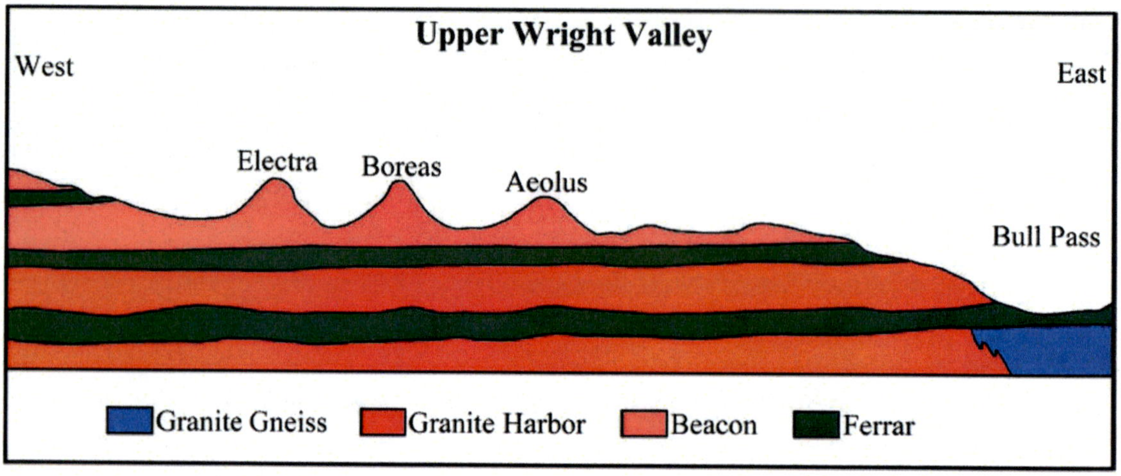

Fig. 13.1 The Wright Valley in southern Victoria Land contains three sills of Ferrar Dolerite: The Basement sill intruded the Granite Harbor Intrusives of the local basement, the Peneplain sill lies between the granitic basement and the overlying sandstones of the Beacon Supergroup, and the uppermost sill was injected into the Beacon rocks. The cross-section extends from Bull Pass to the polar plateau and shows the glacial horns of the Olympus Range. The view is to the north (Adapted from McKelvey and Webb 1962)

Fig. 13.2 The sills of Ferrar Dolerite were forcefully intruded into the sandstones of the Beacon Supergroup throughout the Transantarctic Mountains from northern Victoria Land to the Theron Mountains on the coast of the Weddell Sea. In this view, a dolerite sill has intruded sandstone of the Buckley Formation at the foot of Mt. Achernar in the Queen Alexandra Range (Photo by T.M. Mensing)

the Beacon Supergroup throughout the Transantarctic Mountains by following bedding planes and forcing their way up-section through fissures and pre-existing fractures or faults.

The sills have also affected the topography of the Transantarctic Mountains by protecting the underlying sedimentary rocks of the Beacon Supergroup from erosion, as for example on Mt. Metschel (Section 10.1.2) in Fig. 13.3 and on Mt. Schopf in the Ohio Range (Section 10.5.2). In addition, the sills have caused contact metamorphism especially where they have intruded volcaniclastic sandstones, shale, and coal of the Beacon Supergroup. The alteration of the Triassic Fremouw and Falla formations in the Queen Alexandra Range was documented by Vavra et al. (1981) and Vavra (1982).

Starting during the IGY, the pioneering studies of Gunn (1962, 1963, 1965, 1966) provided information about the internal structure of the dolerite sills, their petrography, and chemical compositions while Gunn and Warren (1962) recorded the wide-spread occurrence of the sills in southern Victoria Land. Gunn (1966) recognized three types of dolerite depending on the presence of distinctive microphenocrysts in the chilled margins of the sills: pigeonite, hypersthene, and olivine.

Fig. 13.3 Mount Metschel near the Boomerang Range of southern Victoria Land is capped by a sill of Ferrar Dolerite that intruded sedimentary rocks of the Beacon Supergroup (Section 10.1.2). The picture shows a geological field party on the move with snowmobiles pulling Nansen sleds which carry Scott tents, food, and fuel (Photo by Peter Barrett reproduced by permission)

13.1.1 Isotopic Dating

The age of the Ferrar Dolerite in the Transantarctic Mountains was first determined by McDougall (1963) who published K-Ar dates of plagioclase and pyroxene of dolerite specimens collected by B.M. Gunn and E. Irving. The results ranged from 147 to 163 Ma from which McDougall correctly concluded that the age of the Ferrar Dolerite is Middle Jurassic based on the geologic timescale of Kulp (1961) and in agreement with the timescale of the IUGS (2002). McDougall (1963) also derived several conclusions about the interpretation of the K-Ar dates he had measured that have withstood the test of time:

1. The K-Ar dates of plagioclase and pyroxene agree closely with each other which suggests that pyroxene does not contain excess ^{40}Ar and that plagioclase has not lost argon by diffusion [exceptions do occur].
2. The K-Ar dates of the Basement sill in Victoria Valley agree with the dates of the Peneplain sill indicating that the two sills were intruded almost simultaneously.
3. The petrology and ages of the dolerite sills in Antarctica and Tasmania are strikingly similar.

Subsequently, Compston et al. (1968) published additional K-Ar dates of plagioclase and pyroxene from a dolerite sill near Lake Vanda in Wright Valley. These dates ranged from 153.9 to 169.2 Ma and were combined in Fig. 13.4 with the K-Ar dates published by McDougall (1963).

Compston et al. (1968) also measured ^{87}Sr/^{86}Sr and Rb/Sr ratios and used them to date the Ferrar Dolerite sills by the Rb-Sr method. The Rb/Sr ratios of the samples ranged widely from 0.0173 for plagioclase to 2.50 for sanidine from a pegmatoid, but the ^{87}Sr/^{86}Sr ratios did not vary consistently. Using 11 samples of hypersthene-bearing dolerites and pegmatitic rocks, Compston et al. (1968) calculated a date of 151 ± 18 Ma with an initial ^{87}Sr/^{86}Sr ratio of about 0.711. We recalculated the date by least-squares regression of nine data points using $\lambda = 1.42 \times 10^{-11}$ year^{-1} and obtained a date of 151 ± 15 Ma with an initial ^{87}Sr/^{86}Sr ratio of 0.71174 ± 0.00042 ($2\bar{\sigma}$). The date is compatible with the Middle Jurassic age of the Ferrar Dolerite based on the K-Ar method.

A second set of Rb-Sr measurements of dolerites in southern Victoria Land, published by Jones et al. (1973), yielded a date of 176 ± 15 Ma and an initial ^{87}Sr/^{86}Sr ratio of 0.71195 ± 0.00015 in Fig. 13.5. The date and initial ^{87}Sr/^{86}Sr ratio agree with results for Kirkpatrick Basalt on Mt. Falla in the Queen Alexandra Range in Fig. 12.32 of Section 12.6.5.

The average initial ^{87}Sr/^{86}Sr ratio at 170 Ma of the Ferrar Dolerites analyzed by Compston et al. (1968) and Jones et al. (1973a) is very nearly identical to the average initial ^{87}Sr/^{86}Sr ratio of the Tasmanian Dolerite in Fig. 13.6 reported by Heier et al. (1965) and Hergt et al. (1989a). Both suites have unimodal distributions although the Ferrar Dolerite samples analyzed by Compston et al. (1968) included pegmatoids from the upper parts of several sills which, in some cases, appear

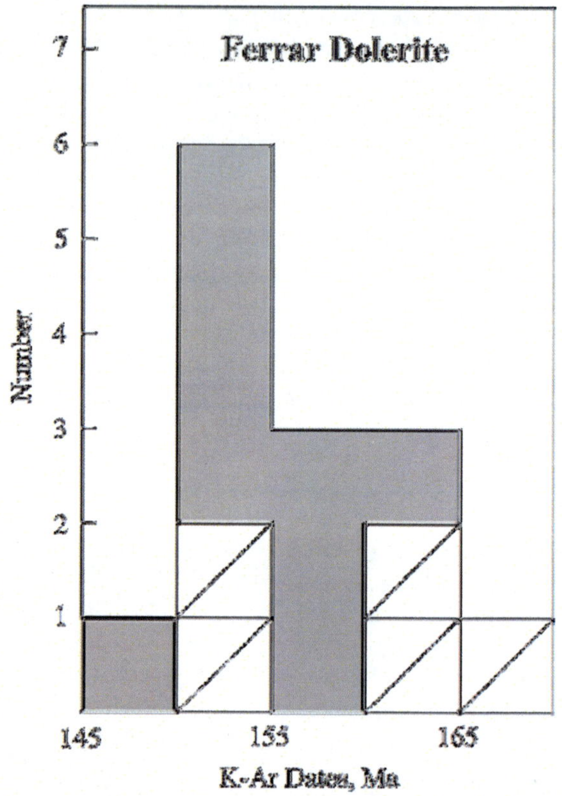

Fig. 13.4 The K-Ar dates of plagioclase and pyroxene (diagonal lines) of dolerite sills in southern Victoria Land measured by McDougall (1963) ranged from 147 to 163 Ma. Subsequently, Compston et al. (1968) published additional K-Ar dates of these minerals (shaded) between 153.9 and 169.2 Ma. The oldest dates indicate that the age of the Ferrar Dolerite is Middle Jurassic (160–180 Ma) based on the timescales of Kulp (1961) and of the IUGS (2002). These results also agree with the K-Ar dates of the Tasmanian Dolerites reported earlier by McDougall (1961, 1962)

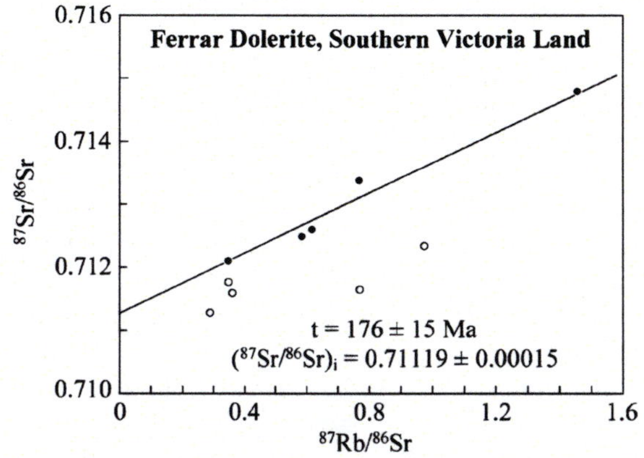

Fig. 13.5 Dolerite samples from sills in southern Victoria Land scatter on the Rb-Sr geochronometry diagram. However, five of the ten samples (solid circles) constrain a straight line that indicates a date of 176 ± 15 Ma and an initial $^{87}Sr/^{86}Sr$ ratio of 0.71119 ± 0.00015 at 95% confidence based on $\lambda = 1.42 \times 10^{-11}$ year^{-1}. The other samples (open circles) were excluded because they do not fit the line (Data from Jones et al. 1973)

to be contaminated with radiogenic ^{87}Sr because the dolerite magma may have assimilated rocks from the roof above the sills. The high initial $^{87}Sr/^{86}Sr$ ratios of the dolerite magma in Tasmania and in the Transantarctic Mountains are characteristic of sialic rocks of the continental crust and have not been seen in mantle-derived mafic igneous rocks elsewhere in the world (Faure 2001).

The initial $^{87}Sr/^{86}Sr$ ratios of the dolerites in the Transantarctic Mountains and in Tasmania in Fig. 13.6 raise questions about the origin of the magma from which they formed. These questions are relevant to the geologic history of Gondwana because the eruption of the Kirkpatrick Basalt (Chapter 12) and the intrusion of the Ferrar Dolerite throughout the Transantarctic Mountains was a prelude to the rifting of Gondwana and the subsequent displacement of the continental fragments.

Compston et al. (1968) also measured certain trace-element ratios in Table 13.1 which further strengthen the genetic link between the Jurassic dolerites of Tasmania and the Transantarctic Mountains, but which also distinguish them from the Karoo Dolerites of South Africa. The trace-element ratios of the Tasmanian and Antarctic dolerites are characteristic of crustal rocks, whereas the South African dolerites look more like mantle-derived igneous rocks (Heier and Rogers 1963; Heier et al. 1964; Gunn 1965; Erlank and Hofmeyr 1966, 1968).

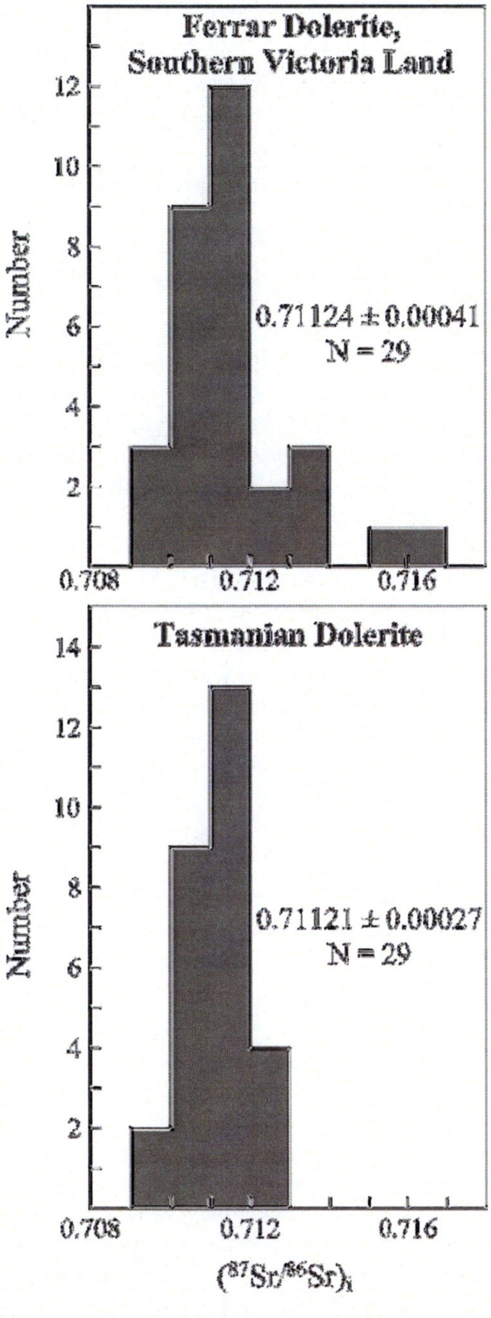

Fig. 13.6 The initial $^{87}Sr/^{86}Sr$ ratios of the Ferrar Dolerite in the Transantarctic Mountains have a mean of 0.71124 ± 0.00041 (1σ) that is identical to the average initial $^{87}Sr/^{86}Sr$ ratio of the Tasmanian Dolerite when both data sets are corrected to 170 Ma and $\lambda = 1.42 \times 10^{-11}$ year^{-1}. These isotope ratios are characteristic of granitic rocks in the continental crust in spite of the fact that the chemical composition of the dolerites indicates that the basalt magma from which they formed originated in the mantle beneath the continental crust (Data for the Ferrar Dolerite from Compston et al. 1968; Jones et al. Tasmanian Dolerite data from Heier et al. 1965; Hergt et al. 1989a)

Table 13.1 Average trace-element ratios of Jurassic dolerites and granophyre (Data from Compston et al. 1968; Heier et al. 1965)

Location	Th/U	Th/K × 10^4	U/K × 10^4	K/Rb
Antarctica	3.3(4) ± 0.7	5.6(4) ± 1.3	1.8(4) ± 0.9	226(3) ± 37
Tasmania	4.1(5) ± 0.2	5.1(5) ± 0.2	1.25(5) ± 0.05	209(6) ± 13
South Africa[a]	5.4(10) ± 2.2	2.8(10) ± 0.8	0.57± 0.20	407(3) ± 16

[a]Cited by Compston et al. (1968) from the literature.

13.1.2 Chemical Compositions

The reports of Gunn (1962, 1966) and of Hamilton (1965) include chemical analyses of dolerite samples collected from the stratigraphic sections of several sills including the Basement sill that intruded quartz monzonite in the Kukri Hills which form the south wall of the Taylor Valley. The graphical representation in Fig. 13.7 of the chemical analyses published by Hamilton (1965) strongly suggests that the Basement sill is internally differentiated and contains zones of diorite pegmatite (D) and granophyre (G) near the top of the sill which is 207 m thick at this location. Petrologists traditionally assume that the chemical compositions of the lower (L) and upper (U) chilled zones of internally differentiated mafic instrusives approach the chemical composition of the magma. The concentrations of SiO_2 and Na_2O + K_2O increase sharply in the granophyre zone (G) near the top of the sill in Fig. 13.7, while the concentration of iron expressed as FeO peaks in the zone of diorite pegmatite (D). The concentrations of TiO_2 (0.43–1.3%) and P_2O_5 (0.03–0.24%) are low except in the diorite pegmatite and the granophyre where both reach peak values. The rocks in the chilled zones and in the main body of the Basement sill are low-Ti dolerites.

The concentrations of Na_2O + K_2O and SiO_2 of the sills of Ferrar Dolerite for which Hamilton (1965) provided chemical analyses (except pegmatites and granophyre) plot in the field of quartz dolerite and diorite in Fig. 13.8. The concentrations of silica are high and range from 53% to 58%, whereas the concentrations of alkali metals (Na_2O + K_2O) cluster around 3.0% (<2.0–4.0%) with one high value at 5.3% in Fig. 13.8. Accordingly, the sills of Ferrar Dolerite in the Transantarctic Mountains are subalkalic quartz dolerites and diorites based on the chemical classification of Cox et al. (1979) and Wilson (1989).

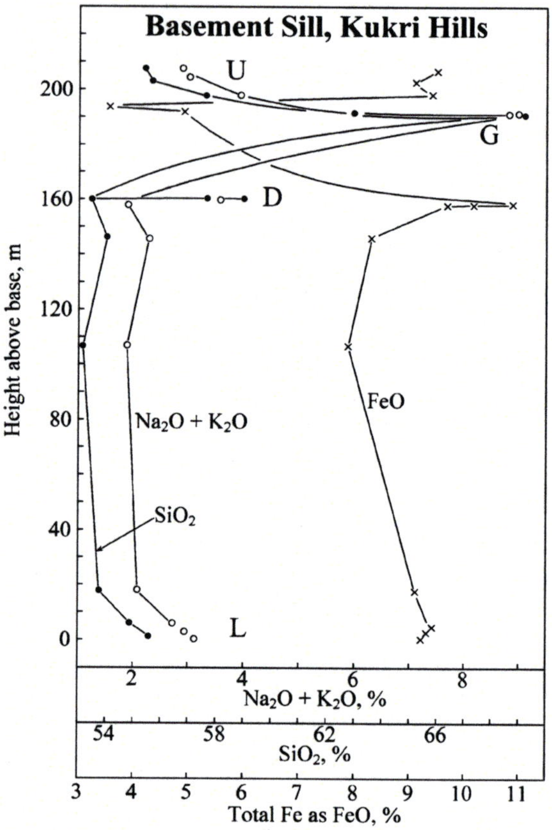

Fig. 13.7 The Basement sill of the Ferrar Dolerite in the Kukri Hills south of the Taylor Glacier is chemically stratified as illustrated by the concentrations of Na_2O + K_2O, SiO_2, and total iron as FeO in percent. L = lower chilled zone, D = diorite pegmatite schlieren, G = granophyre zone, and U = upper chilled zone. The rocks of the granophyre zones are enriched in alkali metals (Na_2O + K_2O) and silica (SiO_2), but depleted in iron (FeO) (Data from Hamilton 1965)

13.1.3 Mineral Stratigraphy

The most abundant mineral in the rocks of the main body of the Basement sill studied by Hamilton (1965) is calcic plagioclase feldspar (labradorite-bytownite).

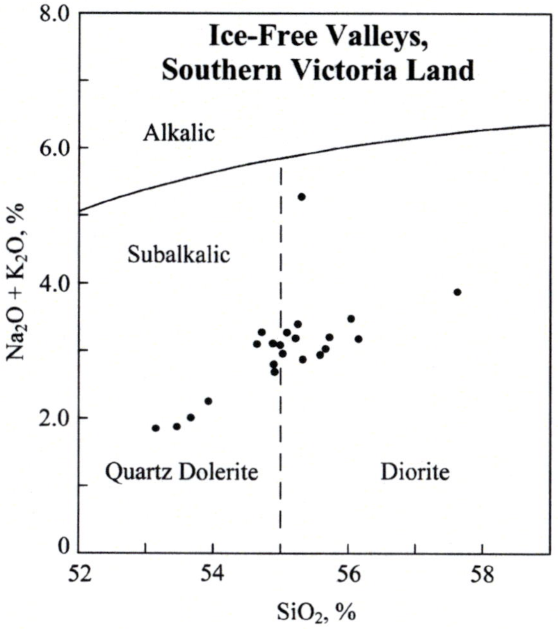

Fig. 13.8 The concentrations of silica and of alkali metals of the sills of Ferrar Dolerite in the ice-free valleys of southern Victoria Land place these rocks into the quartz dolerite and diorite fields. The high concentrations of silica and low content of alkali metals also identify these rocks as subalkalic quartz-normative tholeiites. The rocks are classified as "dolerites" because they are primarily composed of calcic plagioclase and pyroxenes with ophitic texture. The Ferrar Dolerite closely resembles the Kirkpatrick Basalt in terms of chemical composition and mineralogy, but differs by its ophitic texture and intrusive mode of occurrence. Accordingly, dolerite sills and basalt flows are assumed to have formed from the same mafic magma (Data from Hamilton 1965)

The crystals are generally euhedral and unaltered. The dolerite also contains two and, in some cases, even three pyroxenes: augite, pigeonite, and hypersthene. In addition, all specimens contain opaque minerals (magnetite and ilmenite). Quartz and alkali feldspar are prominent in the rocks of the diorite pegmatite and granophyre zones.

Hamilton (1965) reported that the Peneplain and other sills and dikes that intruded the Beacon rocks on New Mountain and Pyramid Mountain along the upper Taylor Glacier are likewise composed of plagioclase, pyroxenes, and opaques with minor quartz and alkali feldspar as well as chlorite and other alteration products. In addition, the Peneplain sill, located east of the mouth of Arena Valley, is also differentiated and includes a pegmatitic zone near its top. The chemical compositions and mineralogy of all of the sills and dikes of Ferrar Dolerite in Taylor Valley and elsewhere in southern Victoria Land identify these rocks as *subalkalic tholeiitic dolerites*.

The petrogenesis of the sills of Ferrar Dolerite in the Transantarctic Mountains has two aspects:

1. The source and origin of the magma
2. The crystallization of the magma accompanied by crystal-liquid fractionation, segregation of crystals, contamination of the magma after intrusion, and hydrothermal alteration

Gunn (1966) investigated the second aspect of the petrogenesis by documenting the stratigraphic variation of the mineral composition of several sills in southern Victoria Land and by relating the trace-element concentrations of the sills to the mineral stratigraphy. He concluded that sills having characteristic mineral compositions were intruded sequentially. Olivine tholeiite sills were intruded first followed by hypersthene tholeiite and pigeonite tholeiite sills. The chemical compositions of each of these sills vary in a characteristics manner in accordance with their mineral stratigraphy. Examples of each mineralogical type identified by Gunn (1966) are listed in Appendix 13.6.1. Some of the Basement sills that were examined by Gunn (1966) were contaminated by granite xenoliths. For example, he stated that the upper part of the Basement sill at Salina Pond "is contaminated by digestion of granite." He also described the contamination of the Terra Cotta dome:

> Lenses and dikes of granophyre up to 20 cm in thickness criss-cross the dolerite and some transgress both dolerite and overlying granite. They appear to originate in partially fused granite xenoliths. (Gunn 1966, p. 884)

A third example of wall-rock contamination occurs in the Peneplain sill at Solitary Rocks adjacent to the upper Taylor Glacier where Gunn (1966, p. 884) reported:

> A granophyre dike about 50 cm wide and several kilometers long cuts the sill.

These reports of wall-rock contamination of dolerite sills indicate that the pegmatoids and granophyres that occur near the top of certain dolerite sills did not necessarily form by differentiation of the basalt magma.

The three mineralogical varieties of dolerite also differ from each other in terms of both major-element and trace-element compositions. Gunn (1966) reported that the rare *olivine*-bearing dolerites are slightly undersaturated with respect to silica, whereas the hypersthene-bearing dolerites contain 8–9% normative silica, which could crystallize as quartz under favorable conditions. The pigeonite-bearing dolerites are even more silica-rich with 10–12% normative quartz. The hypersthene tholeiites contain orthopyroxene, augite, and plagioclase. The pigeonite tholeiites are composed of clinopyroxene (pigeonite, subcalcic augite, and augite) as well as plagioclase.

The abundances of the principal rock-forming minerals of the Vanda hypersthene-sill in Fig. 13.9 vary stratigraphically and define a central mafic zone, an overlying felsic zone, and a pegmatoid zone near the top of the sill which is 332 m thick. Hypersthene occurs exclusively in the mafic zone, whereas augite and pigeonite as well as plagioclase occur at all stratigraphic levels. Plagioclase feldspar becomes dominant in the felsic zone which also contains a layer of micropegmatite about 270 m above the base. The abundance of micropegmatite rises near the top of the sill and defines the pegmatoid zone which overlaps with the upper of the felsic zone. The pegmatoid rock in Fig. 13.10 was described by Gunn (1962, p. 830) in the Basement sill of the Kukri Hills (Appendix 13.6.1):

> A coarse graphic intergrowth of quartz and potash feldspar of small optic angle that appears to be sanidine is typical. Quartz is also found in graphic intergrowth with more sodic extensions of plagioclase.... Pegmatite differentiates are common in all thick basic sheets and those of the Ferrar Dolerites display the usual features of such bodies, as described by Walker (1953).

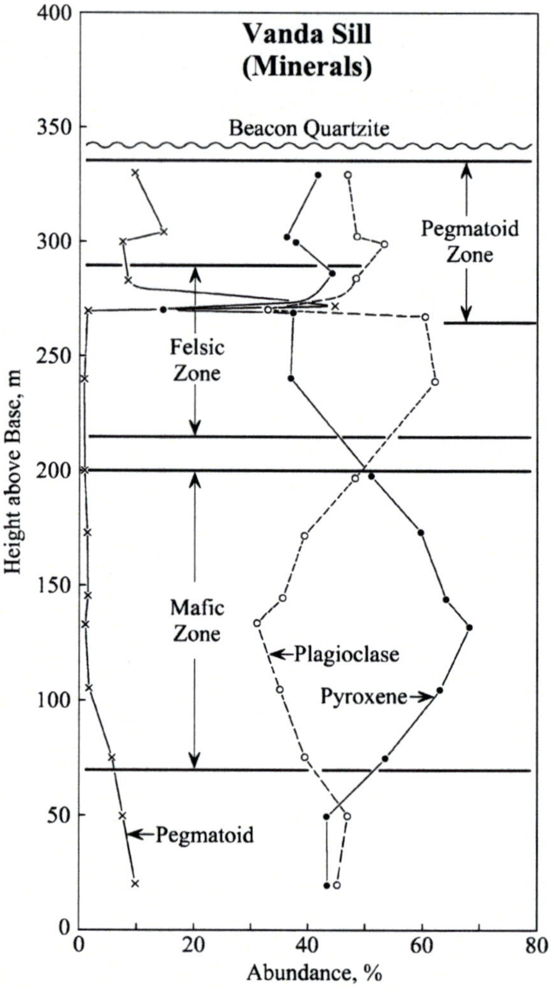

Fig. 13.9 The stratigraphic variations of the major rock-forming minerals of the Vanda hypersthene sill define a central mafic zone, an overlying felsic zone, and a pegmatoid zone near the top of the sill. The pegmatoid zone partly overlaps the felsic zone and includes a thin layer at about 270 m above the base where the pegmatoid assemblage reaches 44.5%. The stratigraphic variation of the abundances of the rock-forming minerals is evidence for the internal differentiation of the Vanda sill by fractional crystallization (Data from Gunn 1966)

13.1.4 Trace Elements

Trace element concentrations of the Vanda hypersthene sill measured by Gunn (1966) include chromium, nickel, copper, zinc, rubidium, strontium, zirconium, and barium. Gunn (1966) also reported concentrations of the same trace elements in the New Mountains pigeonite sill and in the Thumb Point olivine sill. The profiles of nickel and copper concentrations in Fig. 13.11 reveal that the rocks of the mafic zone of the Vanda hypersthene-sill are enriched in nickel and depleted in copper. The concentration of nickel decreases up-section in the felsic zone and reaches a minimum value of 32 ppm in the pegmatoid layer at 270 m above the base of the sill. The concentration of copper in the felsic zone actually increases and has a maximum value of 159 ppm in the pegmatoid layer which is also enriched in zinc (118 ppm), rubidium (34 ppm), strontium (137 ppm), zirconium (142 ppm), and barium (395 ppm).

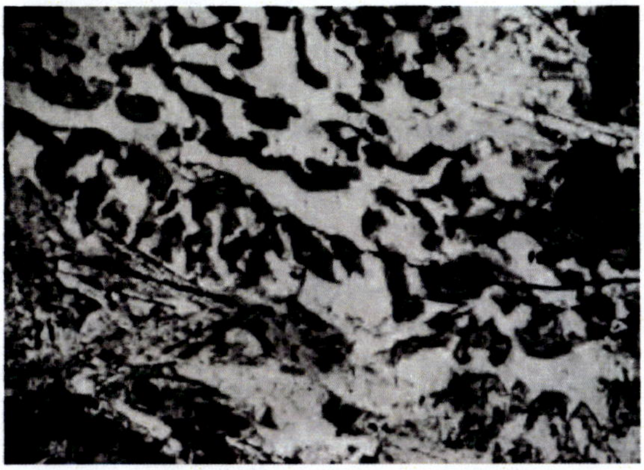

Fig. 13.10 The pegmatoid facies of the Basement sill in the Kukri Hills is composed of graphic intergrowth of quartz and potassium feldspar. This rock type occurs at all levels in the sills of Ferrar Dolerite and becomes abundant in layers near the tops of most thick sills regardless of their stratigraphic position or the characteristic composition of pyroxenes (Adapted from Gunn 1962)

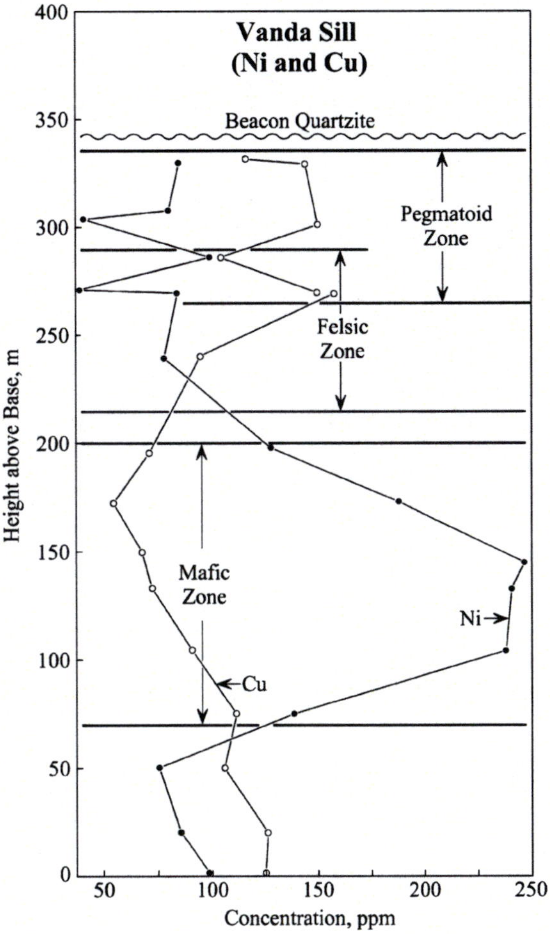

Fig. 13.11 The concentrations of nickel and copper in the Vanda hypersthene sill vary stratigraphically in the opposite sense. The mafic zone of this sill is strongly enriched in nickel and depleted in copper, whereas the felsic zone is enriched in copper and depleted in nickel. The lowest concentrations of nickel and the highest concentrations of copper occur in the layer of pegmatoid at 270 m above the base of the sill. Nickel is closely correlated to magnesium indicating that it resides in pyroxenes, whereas copper and zinc probably occur as sulfides in these rocks (Data from Gunn 1966)

The anticorrelation of nickel and copper indicates that these elements reside in different minerals. Nickel is closely correlated with magnesium (not shown) which indicates that it resides in pyroxene, whereas copper is correlated with zinc (not shown) and both elements may occur in sulfide minerals.

Gunn (1965) reported K/Rb and K/Ba ratios for each of the three mineralogical types of sills in southern Victoria Land. The average K/Rb and K/Ba ratios in Table 13.2 of the rocks in each type of dolerite are remarkably uniform and low. The average K/Rb ratio of 41 specimens of dolerite and cogenetic rocks is 237 ± 22 compared to 203 ± 57 for granitic basement rocks from the Transantarctic Mountains. Cenozoic, mantle-derived alkali-rich volcanic rocks from New Zealand, Antarctica, and Heard Island have a distinctly higher average K/Rb ratio of 335 ± 52. These results by Gunn (1965) helped to draw attention to the crustal character of the Ferrar Dolerite sills of Antarctica. The K/Ba ratios of the Antarctic dolerites are also uniformly low in contrast to the same ratios in alkali-rich volcanic rocks of Cenozoic age in Antarctica, New Zealand, and Heard, the K/Ba ratios of which range widely from 8 to 1,197.

Gunn (1966) concluded that the stratigraphic variation of the chemical composition, and hence of the mineral assemblages, support the view that the sills of Ferrar Dolerite formed by multiple injection of magma of differing compositions and that they did not form by internal differentiation of an initially homogeneous magma. This hypothesis may account for the apparent heterogeneity of the isotope composition of strontium (e.g., Compston et al. 1968; Jones et al. 1973). Multiple injection of mafic magma may also have occurred during the formation of the Dufek intrusion.

Table 13.2 Average K/Rb and K/Ba ratios of rocks in the three types of Ferrar Dolerite sills of southern Victoria Land, including all of lithologic varieties that occur there (Data from Gunn 1965)

Type of sills	K/Rb	K/Ba
Olivine dolerite (7)	235 ± 18[a]	25 ± 3
Hypersthene dolerite (21)	240 ± 25	28 ± 3
Pigeonite dolerite (13)	234 ± 19	29 ± 3
Granitic rocks[b] (5)	203 ± 57	53 ± 37
Alkali volcanics[c] (11)	335 ± 52	28 ± 18[d] (9)

[a]One standard deviation.
[b]Primarily from the basement of the Transantarctic Mountains.
[c]Cenozoic volcanic rocks from New Zealand, Antarctica, and Heard Island.
[d]Omitted two samples having K/Ba = 720 (alk. trachyte, Antarctica) and 1197 (phonolite, New Zealand).
The number of samples included in each average is indicated in parentheses.

13.1.5 Flowage Differentiation

When magma containing entrained crystals is erupted rapidly to the surface, the crystals congregate in the center of the conduit. If the rate of flow is sufficiently fast, the crystals are carried upward even though they are sinking through the magma. The largest crystals are concentrated in the center of the entrained mush and are surrounded by smaller crystal. In this way, crystals are segregated from the residual magma which moves ahead of the crystals. When magma flows horizontally, as during the intrusion of a sill, the coarse crystals are concentrated in the center of a vertical cross-section and the flow regime is revealed by the horizontal distribution of phenocrysts (Bhattacharji and Smith 1964).

This phenomenon is well known in fluid mechanics (Leal 1980) and was applied by Simkin (1967) to the intrusion of the mafic magma of dikes on the Isle of Skye in Scotland and by Upton and Wadsworth (1967) to the intrusion of a small sill on the volcano Piton des Neiges on Réunion. The textures of igneous rocks that result from flow differentiaton of mafic magma have been described by Marsh (1996a, b).

Marsh and Philipp (1996) reported that the concentrations of MgO in vertical profiles of the Basement sill in the ice-free valleys of southern Victoria Land increase up-section toward peak values before decreasing higher up in the sill. Gunn (1966) had previously observed this pattern and noted that it was caused by the vertical distribution of hypersthene which occurs only in the center of the Basement sill, whereas augite and pigeonite are present throughout. Marsh and Philipp (1996) proposed that the crystals of hypersthene had been concentrated in the central section of the Basement sill by flowage differentiation.

The Basement sill which is exposed in the north wall of Wright Valley thins progressively from 350 m at Bull Pass to about 0.05 m east of Bull Pass. In the thinnest part of the Basement sill crystals of orthopyroxene (opx) are absent and the MgO concentration is only about 7%. The MgO concentration increases to 15% in the direction of Bull Pass as the thickness of the opx tongue in the center of the sill increases to 75–80% of the total thickness of the sill. The MgO profiles of the Basement sill in Fig. 13.12 demonstrate that the opx-tongue extends north from Bull Pass into Victoria Valley and south into the Kukri Hills. The regional variation of the thickness of the Basement sill suggest that the magma spread laterally from an area near Lake Vanda and advanced in all directions by lifting the overlying granitic and sedimentary rocks which had a combined thickness of about 4 km. In spite of the weight of the overlying rocks, the magma flowed sufficiently rapidly to cause the hypersthene crystals to be segregated from the residual liquid in which they were entrained. The authors suggested that the intrusion consisted of several pulses and that it spread preferentially in the north–south direction parallel to the trend of the present Transantarctic Mountains. These insights into the way the dolerite sills were intruded prepare the way for three-dimensional mapping of the opx bodies and the reconstruction the intrusion of these sills.

In the course of their fieldwork in Wright Valley, Marsh and Zieg (1997) discovered that the Basement sill exposed in the Dais between Lake Vanda and the Wright Upper Glacier is stratified. The thickness of the layers ranges from a few centimeters to tens of meters. On close inspection, the thick layers (10–30 m) consist of alternating thin layers of hypersthene (brownish green) and plagioclase (white). Two plagioclase-rich anorthosite layers (15–35 cm thick) near the base of the exposed section at the northeast side of the Dais extend for about 50–70 m.

The layering of differentiated mafic intrusions is an important phenomenon because the layers are a record of processes that occurred within the magma chamber, such as:

1. Settling of crystals from a cooling magma like snow flakes from a cloud

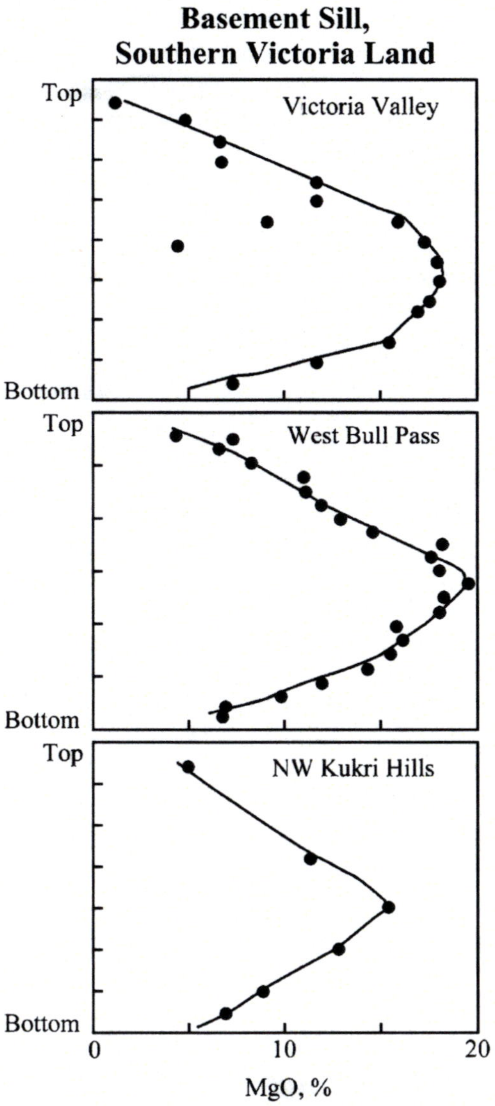

Fig. 13.12 The central part of the Basement sill in the ice-free valleys of southern Victoria Land is enriched in MgO. The Mg-enrichment coincides with the distribution of hypersthene ($(Mg,Fe)_2Si_2O_6$) in addition to Mg-bearing augite ($Ca(Mg,Fe,Al)(Al,Si)_2O_6$) and pigeonite (Ca,Mg,Fe silicate) (Gunn 1966). The accumulation of coarsely crystalline hypersthene in the middle of the Basement sill was attributed by Marsh and Philipp (1996) to flowage differentiation (Adapted from Fig. 2 of Marsh and Philipp 1996)

2. Preferential nucleation of certain minerals
3. Convection of magma with entrained crystals that are deposited against the walls of the chamber

The stratification of the Basement sill anticipates the much more extensive layering of the Dufek intrusion in the Pensacola Mountains to be presented in Section 13.4. The geology of the principal differentiated gabbro intrusives of the Earth was described by Faure (2001) and by Cawthorn (1996).

13.1.6 Magma Transport

The large volume of mafic magma and the forceful way in which it was intruded raise a question about the way the magma was transported that enabled it to penetrate the rocks of the Beacon Supergroup and to reach the surface where it formed large basalt plateaus. This question was considered by Cartwright and Möller-Hansen (2006) for the case of two sill complexes off the coast of Norway. The sills are discordant to the stratification of the country rocks and intersect each other at depths ranging from 8 to 12 km below the sediment-water interface. The authors used seismic data to show that the sills are connected with each other in such a way that they:

> ...acted as a single throughgoing magmatic plumbing system, transporting mafic melts from mid-lower crustal depths to near surface.

Accordingly, Cartwright and Möller-Hansen (2006) concluded that sills can act as conduits for the efficient transport of large volumes of magma through the continental crust and can thereby contribute to the build-up of large lava plateaus on the surface.

The sills of Ferrar Dolerite in the Transantarctic Mountains are known to be transgressive and to be interconnected. Therefore, the sills could have been the channels through which dolerite magma moved through the rocks of the Beacon Supergroup to be erupted at the surface. In other words, the sills and dikes that permeate the rocks of the Beacon Supergroup and the underlying basement are the *feeders* through which basalt magma reached the surface.

Elliot et al. (1999) addressed the question of magma transport in the Transantarctic Mountains and arrived at the same conclusion as Cartwright and Möller-Hansen (2006). Elliot et al. (1999) discussed various options for dispersal of basalt magma from several sources or from a single source. The most likely scenario seems to be that the magma originated from a source or sources and was subsequently distributed by the subsurface plumbing system to all of the sites where these rocks now occur. The authors conjectured that the low-Ti flows, which form the bulk of the basalt plateaus, and the high-Ti flows, which cap these plateaus, originated from the same source. The source of the magma, and the basic cause for the ultimate break-up of Gondwana, may have been the arrival of a large plume from depth in the mantle which caused the development of rifts between Africa and Queen Maud Land of East Antarctica (Storey and Kyle 1997).

13.1.7 Petrogenesis

In spite of the fact that the basalt flows and dolerite sills of the Ferrar Group are all products of the same magmatic event, the stratigraphic variations of their chemical and isotopic compositions are difficult to reconcile with fractional crystallization of homogeneous magma (e.g., Gunn 1966). Therefore, Morrison and Reay (1995) considered a wide range of petrogenetic processes in their attempt to explain the origin of a thick dolerite sill on Terra Cotta Mountain in Taylor Valley based on its chemical composition, including major and trace elements. The sill on Terra Cotta Mountain was intruded along the unconformity between the basement complex and the overlying sedimentary rocks of the Beacon Supergroup. Because of the stratigraphic position of this sill, Hamilton (1965) referred to it as the Peneplain sill, whereas Gunn (1966) identified it as the New Mountains sill (Appendix 13.6.1).

The petrogenetic processes which Morrison and Reay (1995) considered in their study of the origin of the Terra Cotta sill included:

1. Crustal contamination of the magma
2. Varying degrees of partial melting
3. Fractional crystallization in a closed system
4. Magma mixing
5. Formation of magma from heterogeneous source rocks in the subcrustal mantle, and
6. Subsolidus hydrothermal alteration

The authors acknowledged that some aspects of the chemical composition suggest that fractional crystallization played an important part in the petrogenesis of the Terra Cotta sill. However, subsolidus alteration and the heterogeneity of the source rocks also contributed to the petrogenesis. Morrison and Reay (1995) questioned whether crustal contamination of mantle-derived magma can explain the regional similarity of isotopic and trace-element compositions of the basalt flows and dolerite sills in northern and southern Victoria Land and concluded that:

> "…the trace element and isotopic characteristics are enherited from a uniformly enriched mantle source region.

However, Morrison and Reay (1995) did not suggest how the magma sources in the mantle were "uniformly enriched." The most plausible explanation for the crustal characteristics of the basalt and dolerite of the Ferrar Group is the widely accepted proposal by Hergt et al. (1989a) that the magma sources in the subcrustal mantle contained a small amount of subducted sediment.

Many other investigators have contributed to the debate about the petrogenesis of the Jurassic tholeiites of Gondwana, including: Faure et al. (1974), Rhodes and Bornhorst (1976), Kyle (1977, 1980), Faure (1981), Kyle et al. (1983), Brotzu et al. (1988), Hergt et al. (1991), Brewer et al. (1992), Fleming et al. (1995), and Elliot and Fleming (2004).

Kyle (1980) pointed to the high average initial $^{87}Sr/^{86}Sr$ of 0.7115 ± 0.0012 of the basalt and dolerite of the Ferrar Group in the Transantarctic Mountains and noted that the high initial $^{87}Sr/^{86}Sr$ ratios of these rocks remained constant over a linear distance of more than 3,500 km and that they characterize a volume of 400,000 km^3 of magma. He also acknowledged that the stratigraphic variation of the concentrations of the major elements are compatible with fractional crystallization of the magma and that the rocks are enriched in the light rare earths compared to mantle-derived mid-ocean ridge basalt. Although Kyle (1980) admitted that the concentrations of SiO_2 and CaO are correlated with the initial $^{87}Sr/^{86}Sr$ ratios, which is a sign of crustal contamination of magma by assimilation or magma mixing, he preferred the hypothesis that the magma originated from the subcrustal mantle that was enriched in rubidium and in the light rare earths. He postulated that the enrichment of the mantle could have occurred by one of several processes prior to the formation of magma during the Jurassic Period:

1. Addition of rocks from the continental crust by subduction into the mantle

2. Metasomatic enrichment of the mantle by volatile elements originating from the lower mantle, and
3. Accumulation of radiogenic ^{87}Sr by radioactive decay of ^{87}Rb in the mantle having a high Rb/Sr ratio as a result of alternatives 1 and 2 above

Kyle (1980) concluded:

> The interpretation of $^{87}Sr/^{86}Sr$ ratios in continental tholeiites is ambiguous. Presently available data on the Ferrar Group do not give strong support to models involving either selective contamination by crustal-derived components or generations of the tholeiites in a heterogeneous mantle.

In a subsequent report, Kyle et al. (1983) wrote that the basalts on Gorgon Peak (Prince Albert Mountains):

> ...show strong covariations between initial $^{87}Sr/^{86}Sr$ ratios and major elements.

The correlations of major-elements with initial $^{87}Sr/^{86}Sr$ ratios are positive for SiO_2 and TiO_2, and negative for Al_2O_3, MgO, and CaO. In addition, the concentrations of the light rare earth elements and strontium show excellent positive correlations with initial $^{87}Sr/^{86}Sr$ ratios. Nevertheless, Kyle et al. (1983) argued that the initial $^{87}Sr/^{86}Sr$ ratio of the Ferrar magmas was about 0.709 and that the addition of crustal strontium raised the value only incrementally to 0.7115. However, the evidence that sills of Ferrar Dolerite have a distinctive crustal character compelled Kyle et al. (1983) to conclude:

> ...mantle contamination may have occurred as a result of a large scale recycling of crustal material back into the mantle by subduction.

This conclusion by Kyle et al. (1983) is the compromise that led to the solution of the petrogenesis problem. The Ferrar magmas did not acquire their crustal characteristics by assimilating crustal rocks or by mixing with crustal melts. Instead, the magmas formed by partial melting of mixed source rocks in the subcrustal mantle which contained sediment and/or crustal rocks that were introduced into the mantle by subduction along the paleo-Pacific coast of Gondwana during early Paleozoic time. This proposal by Kyle et al. (1983) was later elaborated and tested by numerical modeling by Hergt et al. (1989a) in the study of the petrogenesis of the Tasmanian Dolerite (Section 12.6.1). In this way, the petrogenesis of the basalt flows and dolerite sills of the Ferrar Group is integrated into the tectonic history of Gondwana.

13.2 Roadend Nunatak, Touchdown Glacier

The sedimentary rocks of the Beacon Supergroup north and south of the ice-free valleys of southern Victoria Land contain an abundance of intrusive dolerite sills of the Ferrar Group. The sills are not popular with stratigraphers and paleontologists because they interrupted the continuity of the exposures of the Beacon rocks and caused contact metamorphism of reactive sedimentary rocks that contain diagnostic fossils.

Brief statements about the occurrence of basalt flows and dolerite sills of the Ferrar Group were included in the text that accompanied the geologic maps of the Transantarctic Mountains edited by Craddock (1969). For example, Grindley and Laird (1969), who mapped the geology of the Shackleton Coast of the central Transantarctic Mountains, stated:

> The Beacon Group is intruded by a vast quantity of tholeiitic dolerite (Ferrar Dolerite) as sills, dikes, and plugs... Individual sills may be 500 m thick, and the total thickness of dolerite is about 2000 m... The thicker sills and flows are strongly differentiated.

13.2.1 Stratigraphy

Roadend is a small nunatak at the confluence of the Touchdown and Darwin glaciers at 79°48′S and 158°02′E between the Brown Hills (Section 3.3) and the Darwin Mountains (Section 10.1.3) in Fig. 13.13. This nunatak contains the Basement sill which intrudes the Hope Granite (Section 3.3) and the Peneplain sill which forms the summit of the nunatak in Fig. 13.14. The Peneplain sill is underlain by a thin layer of the Brown Hills Conglomerate (Section 10.1.3) which was deposited directly on the Kukri Peneplain that truncates the Hope Granite at this site.

The exposed thickness of the Basement sill is 150 m but its true thickness is unknown because the lower contact is covered by the ice of the Touchdown Glacier. The thickness of the Peneplain sill is 148 m but its original thickness is also unknown because the top of the sill has been eroded. The dolerite exposed on the summit of Roadend Nunatak is deeply weathered and is not covered by glacial erratics or till.

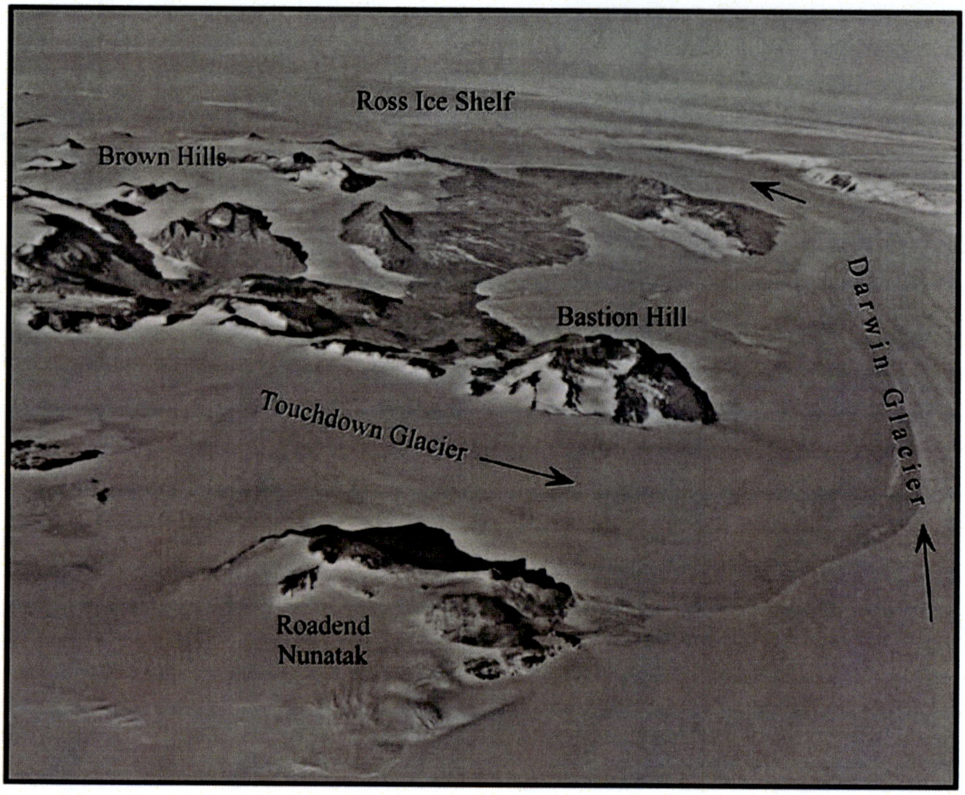

Fig. 13.13 The Roadend Nunatak is located at the confluence of the Touchdown and Darwin glaciers close to Bastion Hill in the Brown Hills. This nunatak contains partial sections of the Basement sill and of the Peneplain sill. The Basement sill intruded the Hope Granite of the Granite Harbor Intrusives. The Peneplain sill is underlain by a sliver of the Brown Hills Conglomerate of the Beacon Supergroup. The top of the Peneplain sill on the summit of Roadend Nunatak is deeply weathered but no glacial erratics or other evidence of glaciation of the summit of the nunatak was found. Aerial photograph from US Geological Survey, Earth Explorer, US Antarctic Indexes, TMA 1011, roll 00033R, frame 4b, November 13, 1962

The rocks of the lower part of the Basement sill on Roadend Nunatak contain hypersthene, pigeonite, augite, Ca-rich plagioclase, and accessory opaques in addition to micrographic intergrowth of K-feldspar and quartz (pegmatoid). The anorthite-content of the plagioclase in the Basement sill decreases up-section from about 80% (bytownite) to 60% (labradorite). Hypersthene is not present in the *upper* part of the Basement sill where augite is the dominant pyroxene.

The pyroxenes of the Peneplain sill consist entirely of pigeonite and augite (i.e., hypersthene is absent). Plagioclase is more abundant and more sodic than the plagioclase of the Basement sill. The micrographic intergrowth of quartz and K-feldspar is also more abundant (10–14%) in the Peneplain sill.

13.2.2 Chemical Compositions

The chemical analyses in Appendix 13.6.2 were used in Fig. 13.15 to show that the rocks of both sills on Roadend Nunatak are subalkalic dolerites similar to the dolerites in the ice-free valleys in Fig. 13.8. The normative mineral compositions in Table 13.3, calculated from the analyses in Appendix 13.6.2, include quartz, K-feldspar, plagioclase, and pyroxenes with accessory magnetite, ilmenite, and apatite. The Peneplain sill differs from the Basement sill by containing more quartz, alkali feldspar, magnetite, ilmenite, and apatite in the norm; but less anorthite and pyroxene. Accordingly, the plagioclase of the Peneplain sill is more sodium-rich than

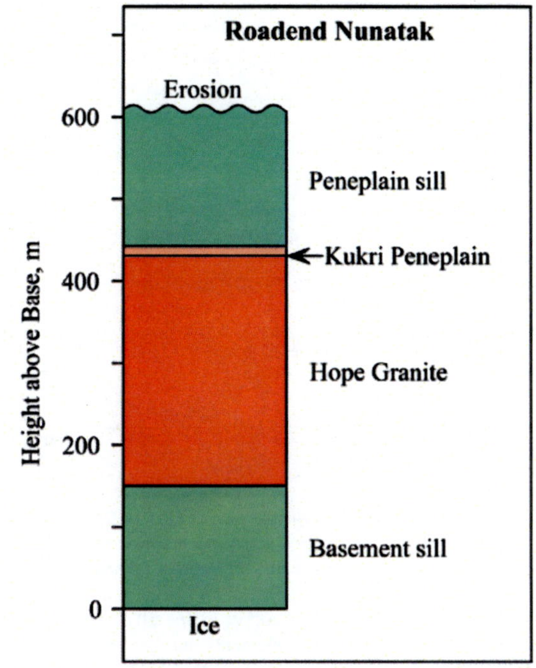

Fig. 13.14 The rocks exposed on Roadend Nunatak consist of the Basement sill of the Ferrar Dolerite which intrudes the Hope Granite of the local basement complex. The Kukri Peneplain is overlain by a sliver of the Brown Hills Conglomerate of the Beacon Supergroup which was intruded by the Peneplain sill. The Roadend Nunatak is now capped by the Peneplain sill because the overlying Beacon rocks were eroded. This section was measured during the 1978/79 field season by G. Faure and R.P. Felder

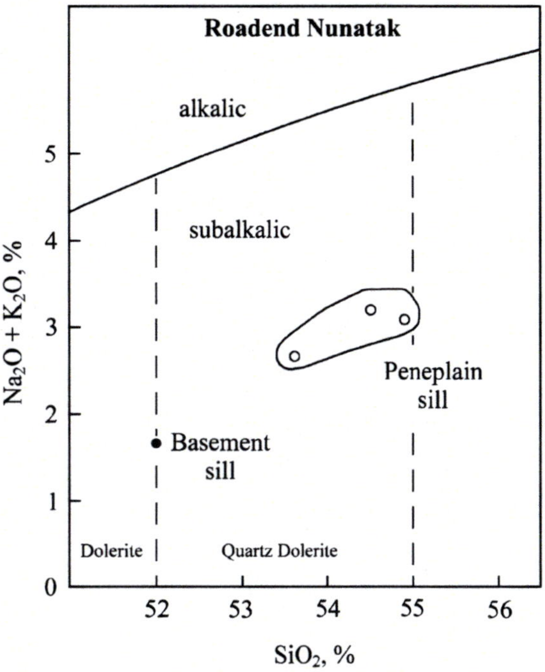

Fig. 13.15 The Basement and Peneplain sills on Roadend Nunatak are composed of subalkalic dolerite and quartz dolerite, respectively. The chemical analyses of Kibler (1981) indicate that the Peneplain sill is more evolved than the Basement sill Data in Appendix 13.6.2

Table 13.3 Normative mineral compositions of the Basement and Peneplain sills on Roadend Nunatak calculated from the chemical analyses in Appendix 13.6.2 and expressed in percent (Kibler 1981)

	Basement sill	Peneplain sill		
Height above base (m)	91.4	475.5	512.1	585.2
Quartz	2.6	9.1	8.4	6.6
K-feldsfar	2.1	7.2	6.9	5.1
Albite	10.9	16.7	15.9	15.4
Anorthite	34.6	26.3	27.7	31.6
Plagioclase	An_{76}	An_{61}	An_{64}	An_{67}
Pyroxenes	46.9	34.5	36.5	36.9
Magnetite	1.5	2.0	1.9	1.7
Ilmenite	0.7	1.5	1.4	1.1
Apatite	0.14	0.30	0.28	0.23
Sum	99.4	97.6	99.0	98.6

the plagioclase of the Basement sill. Based on this evidence, the Peneplain sill appears to be more evolved than the Basement sill.

The basal zone of the Basement sill on Roadend Nunatak Fig. 13.16 is enriched in MgO (Kibler 1981), presumably because of the presence of three pyroxenes including hypersthene. This zone is not present in the Peneplain sill which has a uniformly low MgO concentration of less than 7.0%. The Mg-enrichment of the Basement sill on Roadend Nunatak is similar to the MgO enrichment of the hypersthene-bearing sills in the ice-free valleys of southern Victoria Land reported by Gunn (1962, 1966) which was later attributed by Marsh and Philipp (1996) to flowage differentiation.

The Mg-enriched zone of the Basement sill in Fig. 13.17 is also enriched in nickel (299 ppm) and depleted in copper (103 ppm) (Kibler 1981). The resulting concentration profile of the Basement sill on Roadend Nunatak is similar to the profile in Fig. 13.11 of the concentrations of nickel and copper in the Vanda hypersthene- bearing sill (Gunn 1966).

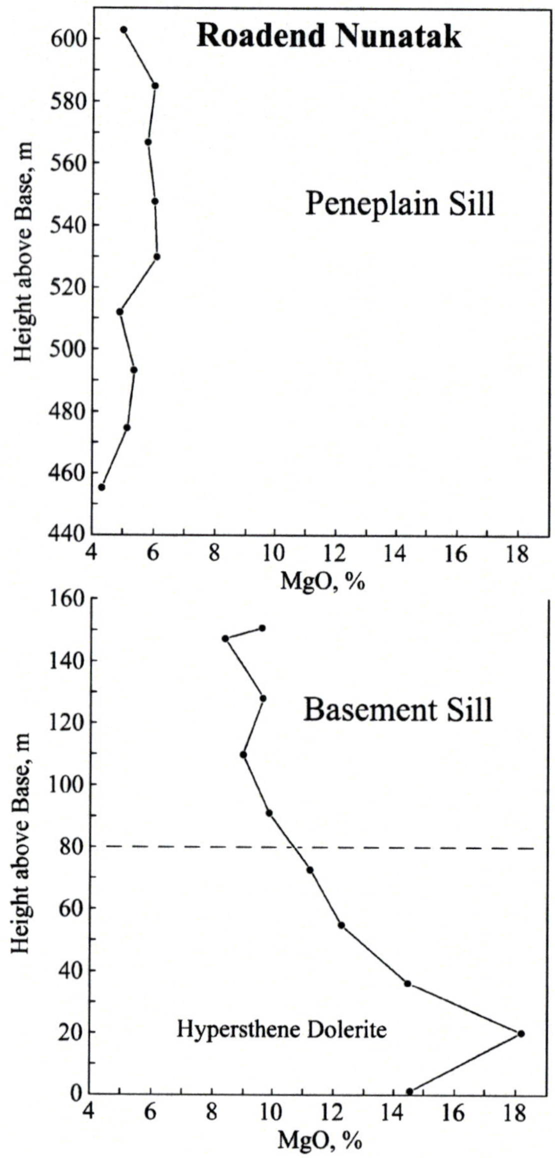

Fig. 13.16 The lower part of the Basement sill on Roadend Nunatak has elevated concentrations of MgO up to 18.1% caused by the presence of hypersthene, pigeonite, and augite. Marsh and Philipp (1996) attributed the enrichment of the central sections of the Basement sill in the ice-free valleys of southern Victoria Land to flowage differentiation which concentrated the hypersthene crystals because of their greater size. The concentrations of MgO of the Peneplain sill are uniformly low at less than 7.0% arising from the presence of the clinopyroxenes (cpx), pigeonite and augite, whereas hypersthene (opx) is absent (Data from Kibler 1981)

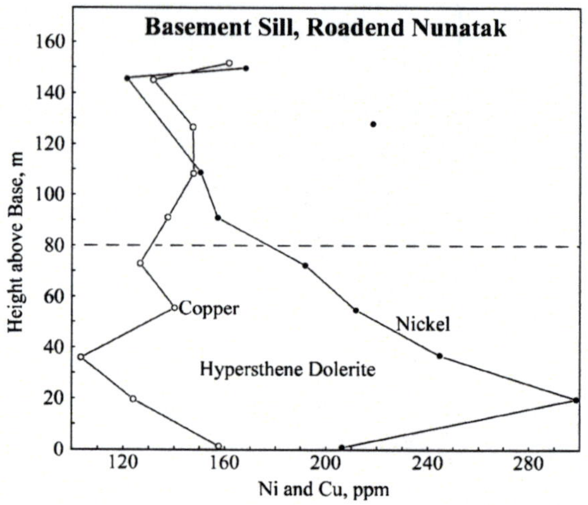

Fig. 13.17 The concentrations of nickel closely follow MgO and reach a maximum of 299 ppm in the lower hypersthene dolerite zone of the Basement sill on Roadend Nunatak where copper reaches a minimum concentration of 103 ppm. The concentration profiles of nickel and copper in this sill closely resemble the distribution of these elements in the Vanda sill in Fig. 13.11 based on the data of Gunn (1966). The lower contact of the Basement sill of Roadend Nuntak is covered by the ice of the Touchdown Glacier (Data from Kibler 1981)

13.2.3 Rb-Sr Dating

The Rb-Sr systematics of the sills on Roadend Nunatak (Appendix 13.6.3) are interpreted in Fig. 13.18 to determine the age of the two sills. Eight of nine whole-rock samples of the Peneplain sill loosely constrain a straight line that yields a date of 182 ± 9 Ma and an initial $^{87}Sr/^{86}Sr$ ratio of 0.71167 ± 0.00014 (1σ). Ten whole-rock samples of the Basement sill have more scattered isotope ratios such that only six approach a straight line in Fig. 13.18. The date derivable from this line is 187 ± 20 Ma and the initial $^{87}Sr/^{86}Sr$ ratio is 0.71048 ± 0.00017 (1σ).

The whole-rock Rb-Sr dates of the two sills on Roadend Nunatak are not distinguishable from each other and are both compatible with the average of the $^{40}Ar/^{39}Ar$ plateau dates (176.8 ± 0.5 Ma) of the Kirkpatrick Basalt reported by Heimann et al. (1994). However, the sills have distinctly *different* initial $^{87}Sr/^{86}Sr$ ratios. In addition, the initial $^{87}Sr/^{86}Sr$ ratios of four whole-rock samples of the Basement sill range from 0.7092 to 0.7100 and are significantly lower than

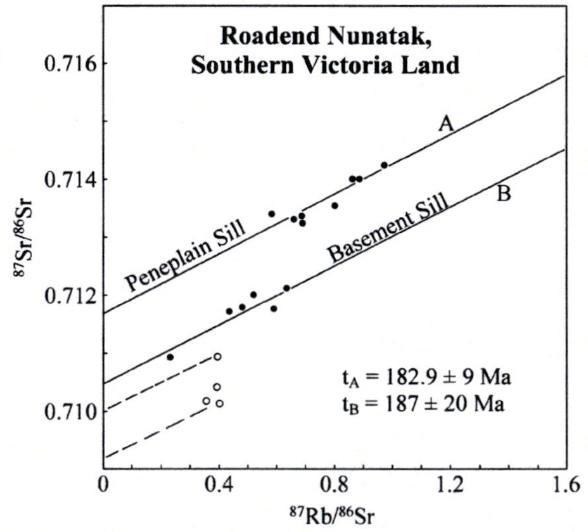

Fig. 13.18 The whole-rock Rb-Sr dates and initial $^{87}Sr/^{86}Sr$ ratios of the sills of Ferrar Dolerite on Roadend Nunatak: Basement sill: t = 187 ± 20 Ma; 0.710475 ± 0.00017 (1σ) Peneplain sill: t = 182 ± 9 Ma; 0.711673 ± 0.00014 (1σ) Note that the two sills have different initial $^{87}Sr/^{86}Sr$ ratios and that four samples were excluded from the age determination of the Basement sill because their initial $^{87}Sr/^{86}Sr$ ratios are low and range from 0.7092 to 0.7100. The Rb-Sr dates of the two flows have overlapping errors and do not differ from the average $^{40}Ar/^{39}Ar$ plateau date of 176.8 ± 5 Ma reported by Heimann et al. (1994) for samples of the Kirkpatrick Basalt (Data from Kibler 1981; Faure et al. 1991)

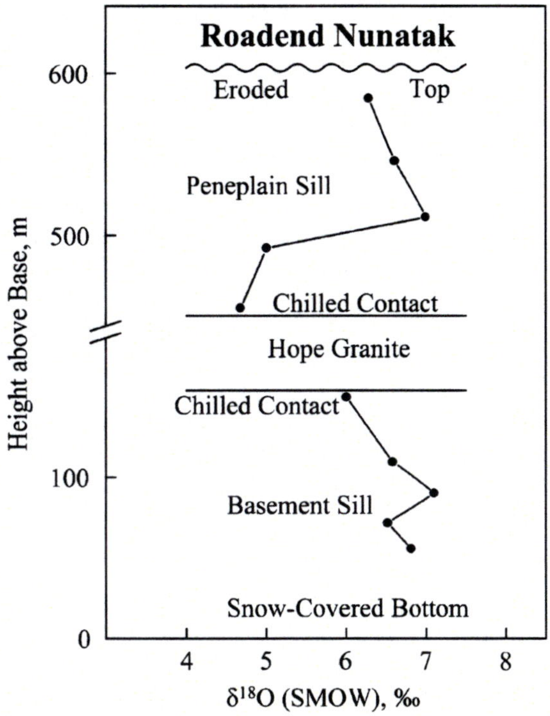

Fig. 13.19 The $\delta^{18}O$ (SMOW) values of the Peneplain and Basement sills vary stratigraphically such that the chilled contacts have low $\delta^{18}O$ values and the samples from the interiors of the sills have higher values. This pattern of variation is attributed in the text to the effects of hydrothermal alteration (Data from Kibler 1981)

the intercept of the errorchron line of that sill. The anomalous samples originated from the top of the Basement sill and therefore could have been contaminated during the intrusion of the dolerite magma into the Hope Granite. Nevertheless, the difference between the intercepts of the Rb-Sr errorchron lines demonstrates that the two sills formed from magma that had different $^{87}Sr/^{86}Sr$ ratios at the time of intrusion. Evidently, the initial $^{87}Sr/^{86}Sr$ ratios of the sills of Ferrar Dolerite are *not* constant everywhere in the Transantarctic Mountains.

13.2.4 Oxygen

The isotope compositions of oxygen of the Peneplain and Basement sills of Roadend Nunatak vary stratigraphically and have $\delta^{18}O$ (SMOW) values between +4.7‰and +7.1%. This range of oxygen-isotope compositions is well within the range of $\delta^{18}O$ values for basalt in different geologic settings (Hoefs 1997, p. 82). However, the stratigraphic variation of this parameter in Fig. 13.19 indicates that the isotope compositions of oxygen in the rocks of the sills may have been altered. The most probable cause for the observed pattern of variation is hydrothermal alteration of the rocks at mesothermal to epithermal temperatures.

The modal analyses by Kibler (1981) indicate that the rocks contain secondary biotite (2–6%) which formed by alteration of pyroxene, that ilmenite was altered to leucoxene plus magnetite, and that all rock specimens contain micrographic intergrowths of quartz and alkali feldspar (2–12%). When silicate minerals or volcanic glass interact with water, the solid phases are enriched in ^{18}O relative to the oxygen in the water depending on the temperature and on the water-to-rock ratio (Faure and Mensing 2005)

A test of this hypothesis in Fig. 13.20 suggests that the $\delta^{18}O$ values and strontium concentrations of the rocks of the Peneplain and Basement sills on Roadend Nunatak are negatively correlated. The data in Appendix 13.6.3 indicate that the chilled contact of the Peneplain sill has the highest concentration of strontium (142.48 ppm) and the lowest $\delta^{18}O$ value (+4.7‰). A sample from the center of this sill (at 512.1 m) has the lowest strontium concentration (122.50 ppm) and the highest $\delta^{18}O$ value of +7.0‰. The same relationship applies to the Basement sill because the chilled top (at 150.0 m) has the highest concentration of strontium (100.40 ppm) and the lowest $\delta^{18}O$ value (+6.0‰) for that sill. All of the other samples from the interior parts of the Basement sill have lower strontium concentrations (<100.40 ppm) and higher $\delta^{18}O$ values (>6.0‰). We note that this interpretation of the data implies that interior parts of the sills were altered by heated groundwater that flowed through joints in the sills while the fine-grained contact rocks record the original composition of the magma.

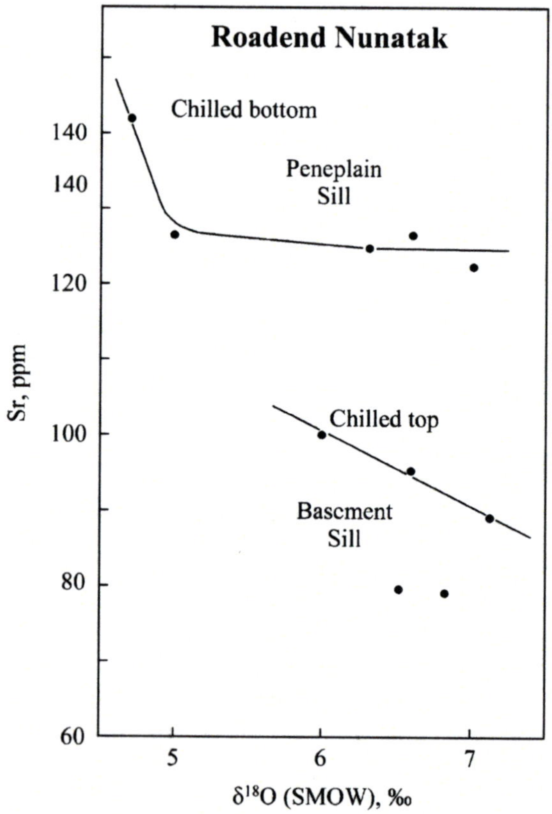

Fig. 13.20 The $\delta^{18}O$ values of the Peneplain and Basement sills on Roadend Nunatak appear to increase with decreasing concentrations of strontium. This relationship may be caused by hydrothermal alterations which enriched the rocks in ^{18}O and depleted them in strontium. Note that the chilled bottom of the Peneplain sill and that the chilled top of the Basement sill have the lowest $\delta^{18}O$ values and the highest strontium concentrations in each sill. The most highly altered samples (low Sr, high $\delta^{18}O$) originate from the central regions of both sills. Strontium concentrations by Kibler (1981) and $\delta^{18}O$ measurements by J. Hoefs (personal communication 1986)

13.3 Central Transantarctic Mountains

The rocks of the Beacon Supergroup in the central Transantarctic Mountains are permeated with intrusive sills and dikes of the Ferrar Dolerite. For example, the area around Mt. Falla in the Queen Alexandra Range contains five dolerite sills within the Beacon Supergroup; but there may be more because the Basement and Peneplain sills are not exposed. These sills were combined with the sedimentary rocks of the Beacon Supergroup on the geologic map of Grindley and Laird (1969) but were separately identified on the map of Barrett and Elliot (1973). The dolerite sills of the adjoining areas of the central Transantarctic Mountains were included on the geologic maps of Barrett et al. (1973) and Lindsay et al. (1973). In addition, the paleomagnetism of Jurassic basalt flows and dolerites sill of the area were investigated by Hoffman et al. (1984).

13.3.1 Mt. Achernar, MacAlpine Hills

The MacAlpine Hills in Fig. 13.21 are located along the south side of the upper Law Glacier which flows into the Bowden Névé between the Queen Elizabeth and Queen Alexandra ranges. Mt. Achernar is located at the junction of the MacAlpine Hills and the Lewis Cliff in Fig. 13.22. It consists of flat-lying sandstones and coal seams of the Permian Buckley Formation which was intruded by three sills of the Ferrar Dolerite. The elevation of the summit plateau of Mt. Achernar is 2,691 m a.s.l. while its base is at 2,000 m. The sills are exposed on the steep slopes of two large cirques that face the Law Glacier in Fig. 13.23.

The thickness of Sill #1 (lowest) is 58 m, Sill #2 is 126 m thick, and the remaining thickness of Sill #3 (highest) is 67 m. The base of Sill #1 is covered by

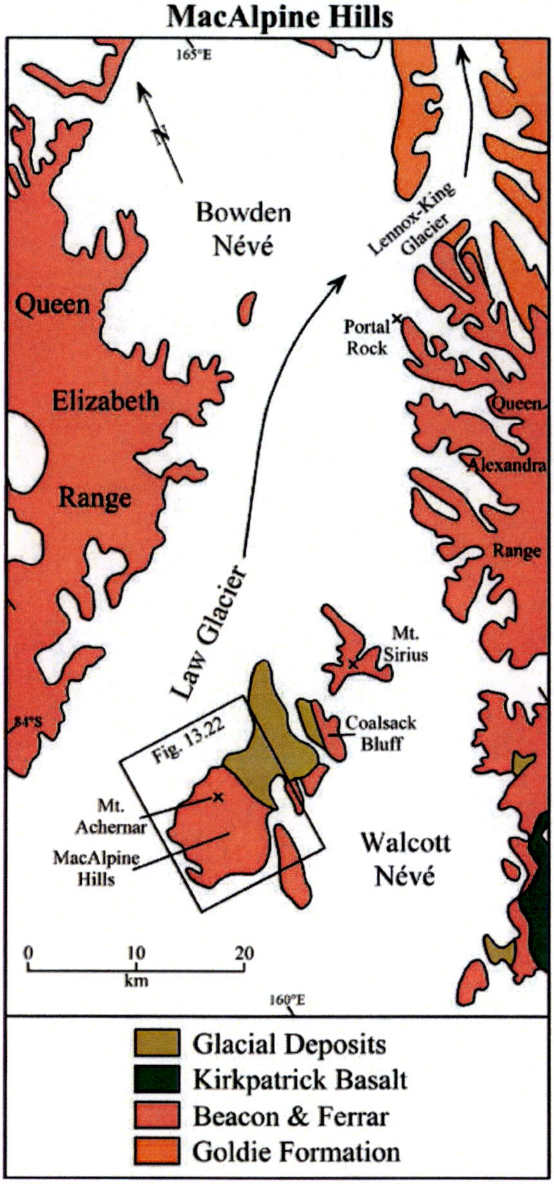

Fig. 13.21 The MacAlpine Hills of the central Transantarctic Mountains are located at the head of the Law Glacier between the Queen Alexandra Range and the Queen Elizabeth Range. The ice leaves the Bowden Névé via the Lennox-King Glacier which flows into the Ross Ice Shelf. Mt. Achernar is located at the junction between the MacAlpine Hills and Lewis Cliff. The sills of Ferrar Dolerite are not distinguished from the sedimentary rocks of the Beacon Supergroup in this diagram (Adapted from the map of Grindley and Laird 1969)

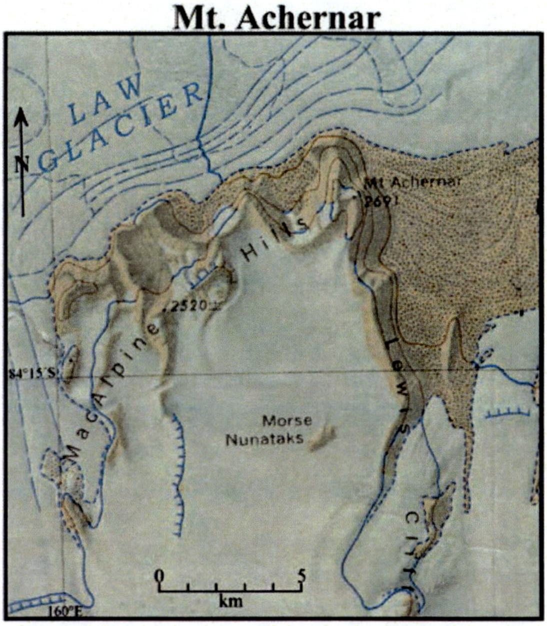

Fig. 13.22 Mount Achernar at 84°12′S and 160°56′E contains three sills of Ferrar Dolerite which intruded sandstones of the Permian Buckley Formation. The underlying basement rocks are not exposed in this area. Mount Achernar is located a short distance west of Coalsack Bluff (Section 11.2.2; Fig. 11.9B) and of Mt. Sirius which contains a deposit of Neogene till on its summit. Excerpt of the topographic map: Buckley Island, Antarctica; SV 51–60/3 published in 1967 by the US Geological Survey

snow and the top of Sill #3 has been eroded. The sills have caused contact metamorphism (hornfels) in the sandstones and siltstones of the Buckley Formation, which contains three coal seams in the interval between sills 2 and 3. The dolerite in Fig. 13.24 is strongly jointed with evidence of hydrothermal activity at the base and between 63 and 87 m above the base of Sill #2. The dolerite of Sill #2 is fine grained at the top and bottom and coarse grained in the central part.

The chemical analyses of 35 dolerite samples from the three sills on Mt. Achernar in Appendix 13.6.4 display a wide range of silica concentrations (51.5–55.1%), but the low concentrations of alkali metals in Fig. 13.25 (1.9–3.8%) classify most of the rocks as quartz dolerite (Wilson 1989).

The profile in Fig. 13.26 reveals that the concentrations of MgO in Sill #3 rise to a maximum at 21 m above the bottom of the sill. This pattern of variation is presumably caused by the presence of a hypersthene zone like those described by Marsh and Philipp (1996) in the Basement sill of Wright Valley. The concentrations of MgO in sills 1 and 2 on Mt. Achernar are all less than 7.0%.

Fig. 13.23 Mount Achernar in the Queen Alexandra Range contains three sills of the Ferrar Dolerite which intruded the Buckley Formation of the Beacon Supergroup. The base of the lowest sill is covered by talus and the upper part of the top sill has been eroded. A coal seam is visible in the Buckley Formation between sills #1 and #2 (Photo by T.M. Mensing)

The Rb-Sr systematics in Appendix 13.6.5 were used in Fig. 13.27 to date the three sills on Mt. Achernar by the whole-rock Rb-Sr method. The data points of all three sills scatter above and below errorchron lines which yield dates with overlapping errors that confirm the Middle Jurassic age of the Ferrar Dolerite:

Sill # 3: $t = 166 \pm 9$ Ma, 0.710298 ± 0.000112 (1σ)
Sill # 2: $t = 180 \pm 11$ Ma, 0.708088 ± 0.000092 (1σ)
Sill # 1: $t = 180 \pm 16$ Ma, 0.708732 ± 0.000160 (1σ)

The initial $^{87}Sr/^{86}Sr$ ratios in Fig. 13.28 were derived entirely from the intercepts of the errorchrons in Fig. 13.18 (Roadend) and in Fig. 13.27 (Mt. Achernar). The samples that were excluded from the age determinations were also excluded from the determination of the initial $^{87}Sr/^{86}Sr$ ratios which are significantly *lower* than the average initial $^{87}Sr/^{86}Sr$ ratios of the Jurassic dolerites of southern Victoria Land and of Tasmania in Fig. 13.6. The initial $^{87}Sr/^{86}Sr$ ratios of sills 1 and 2 on Mt. Achernar are also lower than the initial $^{87}Sr/^{86}Sr$ ratios of the sills on Roadend Nunatak in Fig. 13.18. The differences of the initial $^{87}Sr/^{86}Sr$ ratios of the dolerites sills on Roadend Nunatak and Mt. Achernar are a potentially important clue concerning the petrogenesis of the sills of Ferrar Dolerite because only samples that were used to constrain the errochrons were used to define them.

The $\delta^{18}O$ values of whole rock samples of Sill #1 in Fig. 13.29 vary up-section from +4.4‰ near the lower contact to higher values in the midsection ranging from +5.20‰ to +6.47‰ and then decline toward +4.35%at the upper contact. These data are within the range of $\delta^{18}O$ values of continental basalt elsewhere in the world (Hoefs 1997; Faure 2001). The stratigraphic variation of $\delta^{18}O$ within Sill #1 is attributable to differences in the mineral composition of the rocks or to the enrichment of the rocks in ^{18}O as a result of isotope exchange with heated groundwater.

Figure 13.30 is intended to test the hypothesis that the rocks of Sill #1 may have been enriched in ^{18}O and lost strontium as a result of interacting with heated

Fig. 13.24 Sill #2 on Mt. Achernar, like most Ferrar Dolerite sills in the Transantarctic Mountains, is heavily jointed which permitted widespread hydrothermal alteration even of the interior parts of thick sills (Photo by G. Faure)

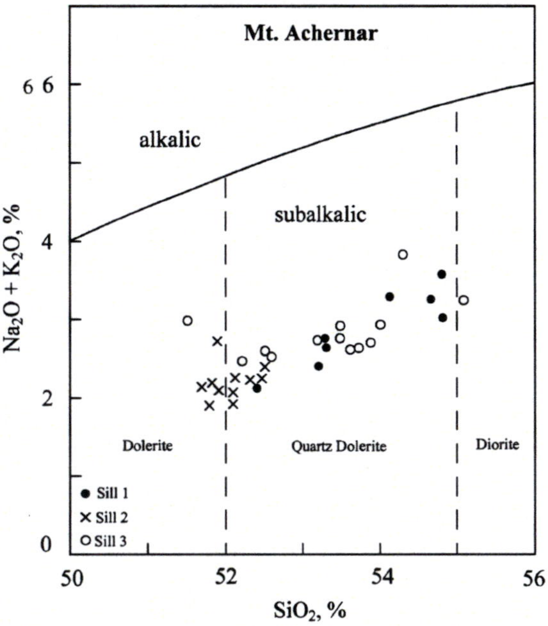

Fig. 13.25 The concentrations of SiO_2 and $Na_2O + K_2O$ of the sills on Mt. Achernar indicate that the rocks are subalkalic quartz-normative dolerites most of which occupy the field of quartz dolerites. The chemical analyses in Appendix 13.6.4 were used by Faure et al. (1991) to calculate average compositions of each sill on Mt. Achernar

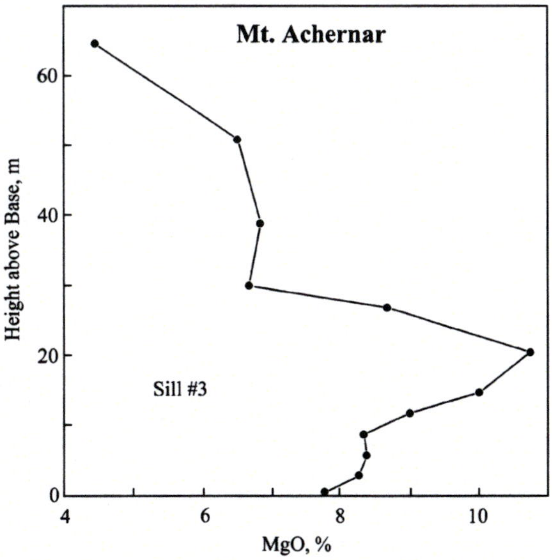

Fig. 13.26 The concentrations of MgO in Sill #3 at the top of Mt. Achernar rise up-section and peak at 10.7%. The Mg-rich layer of this sill is more than 30 m thick and is presumably caused by the presence of hypersthene in addition to augite and pigeonite (Section 13.1.5, Fig. 13.12). Plotted from data in Appendix 13.6.4

groundwater as suggested in Fig. 13.20. Although the evidence is equivocal, the diagram does suggest an inverse correlation between the abundance of ^{18}O and the concentration of strontium. Two samples from the chilled margin of Sill #1 deviate from the trend of data points that represent the central part of the sill.

The $\delta^{18}O$ parameter of cogenetic plagioclase and pyroxene in Sill #2 also vary stratigraphically in Fig. 13.31a based on measurements listed in Appendix 13.6.6. The *difference* between the $\delta^{18}O$ values of the two minerals is related to the temperature at which the isotope composition of oxygen they contain was equilibrated. The relationship is expressed by an equation explained by Faure and Mensing (2005):

$$\Delta_2^1 = \delta_1 - \delta_2 = \frac{A \times 10^6}{T^2} \qquad (13.1)$$

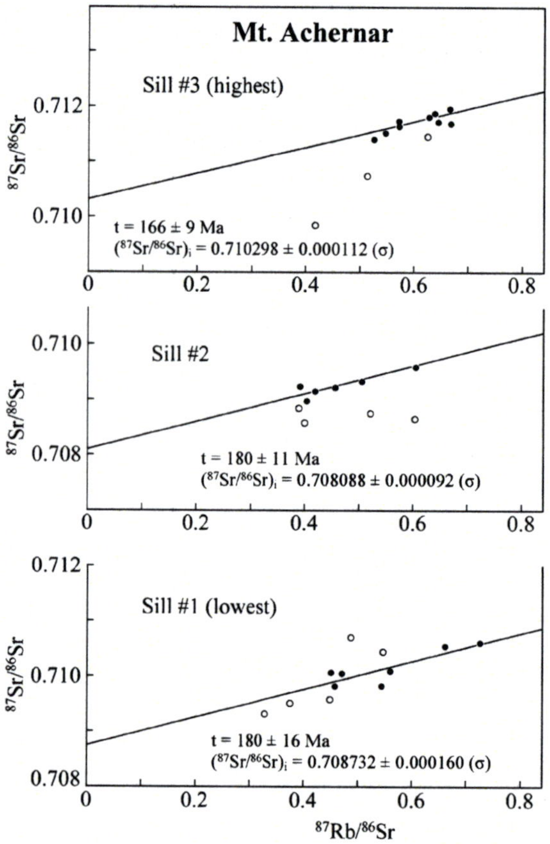

Fig. 13.27 The ^{87}Sr/^{86}Sr and ^{87}Rb/^{86}Sr ratios of whole-rock samples of the three dolerite sills on Mt. Achernar form errochrons that yield Middle Jurassic Rb-Sr dates based on $\lambda = 1.42 \times 10^{-11}$ year^{-1}. However, several samples from each sill scatter above and below their respective errorchrons. The open circles are samples that were excluded from the age determination. Each sill has a different initial ^{87}Sr/^{86}Sr ratio all of which are lower than the initial ^{87}Sr/^{86}Sr ratios of the sills on Roadend Nunatak (Fig. 13.18) and in the ice-free valleys of southern Victoria Land (Fig. 13.6) (Data from Faure et al. 1991)

where δ_1 and δ_2 pertain to two cogenetic minerals in isotopic equilibrium with the same oxygen reservoir at the temperature T in kelvins. The value of the coefficient (*A*) depends on the identity of the cogenetic minerals and is derivable both by a priori calculations (Bottinga and Javoy 1975) and by experimentation in the laboratory (Chiba et al. 1989; Chacko et al. 2001). The temperature estimates based on the measurements of δ^{18}O values of plagioclase and pyroxene in Sill #2 are explained in Appendix 13.6.6 and have been plotted stratigraphically in Fig. 13.31b.

The temperatures estimated by means of the thermometry equation of Chiba et al. (1989) are 54° higher on average than those obtained by the equation of

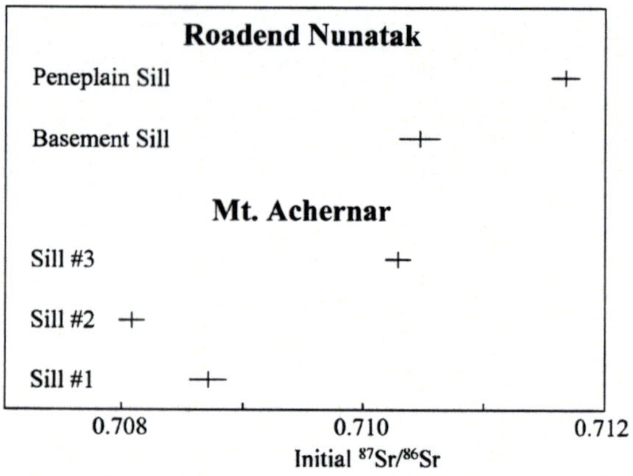

Fig. 13.28 The initial ^{87}Sr/^{86}Sr ratios derived from the errorchrons of the Ferrar Dolerite sills on Roadend Nunatak and on Mt. Achernar are not constant as expected, but differ significantly. This potentially important observation concerning the petrogenesis of the dolerite sills needs to be confirmed by further studies. These values were taken from Figs. 13.18 and 13.27 based on data listed in Appendices 13.7.3 (Roadend Nunatak) and 13.7.4 (Mt. Achernar)

Bottinga and Javoy (1975); but both sets of values in Appendix 13.6.6 are reasonable estimates of the oxygen-isotopic equilibration temperature of mafic igneous rocks that crystallized from magma. Actually, the temperatures indicated by the oxygen-isotope thermometers are probably *less* than the crystallization temperature of the silicate minerals.

When the lowest and the highest temperatures are excluded, the average isotopic-equilibration temperatures are:

Bottinga and Javoy (1975): 872 ± 112°C (1σ)
Chiba et al. (1989): 926 ± 117°C (1σ)

By combining both averages, we obtain a grand average of 900 ± 30°C. This value is well above the retention temperature of plagioclase and pyroxene for radiogenic ^{40}Ar and therefore explains why K-Ar dates underestimate the age of the Ferrar basalts and dolerites depending on the rate of cooling.

13.3.2 Portal Rock, Queen Alexandra Range

Portal Rock (83°50′S, 165°36′E) is a small mountain at the confluence of the Tillite Glacier with the Lennox-King Glacier in Fig. 13.32. The location of the

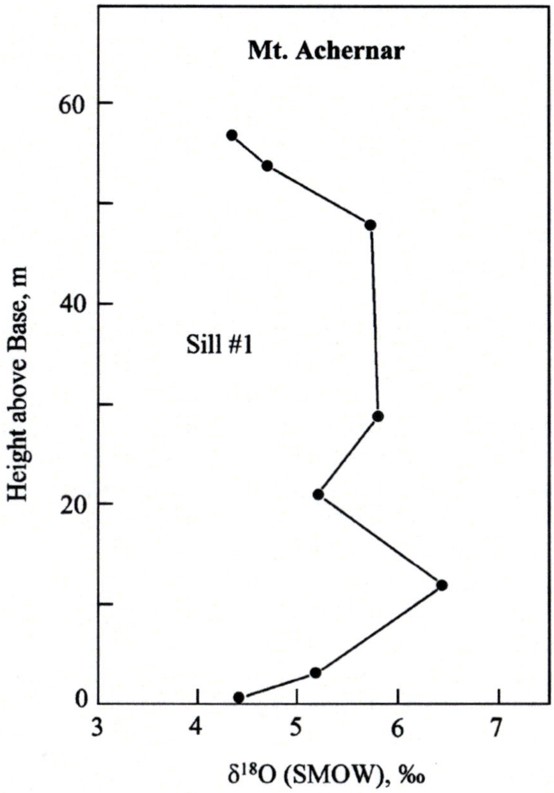

Fig. 13.29 The $\delta^{18}O$ values of whole-rock samples of the lowest dolerite sill (Sill #1) on Mt. Achernar increase up section from +4.4‰ close to the basal contact to higher values (+5.20‰ to +6.47‰) in the mid-section and then decline again to +4.35‰ at the upper contact. The increase in the abundance of ^{18}O is unexplained, but may be caused by an increase in the abundance of plagioclase or by low-temperature isotope fractionation during interaction with heated groundwater. Data in Appendix 13.6.5 by J. Hoefs at the University of Göttingen, Germany

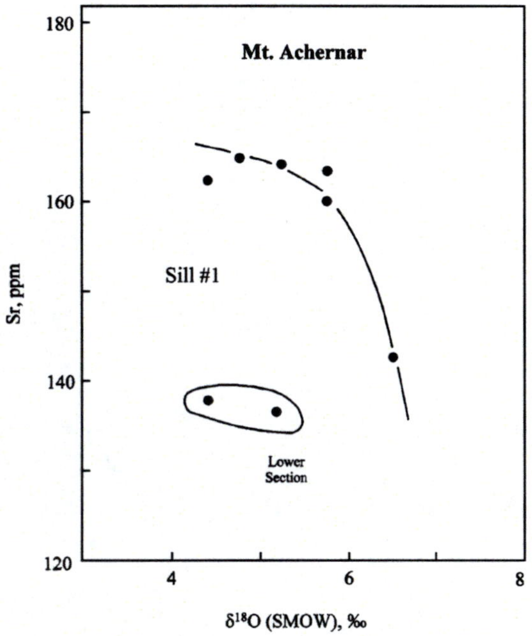

Fig. 13.30 The $\delta^{18}O$ values of whole-rock samples taken from a measured section of Sill #1 on Mt. Achernar give only limited support to the hypothesis that the rocks were enriched in ^{18}O and depleted in strontium by interacting with heated groundwater. Two samples from the base of the section deviate markedly from the data points that represent samples from the center of the sill. Data in Appendix 13.6.5

mountain is also marked on the map of the western Queen Alexandra Range in Fig. 13.21. Portal Rock (aka Portal Peak), which rises 390 m above the level of the surrounding ice in Fig. 13.33, consists of the Permian Pagoda Tillite and the overlying Mackellar Formation which was intruded by a sill of the Jurassic Ferrar Dolerite (Barrett et al. 1986). A stratigraphic section in Fig. 13.34 indicates that the thickness of this sill is 129 m and that it has chilled margins at the top and bottom against sedimentary rocks of the Beacon Supergroup. The sedimentary rocks beneath this sill and the bottom part of the sill itself (10 m above the base) were altered hydrothermally. The main body of the sill consists of medium to coarse grained dolerite that is well jointed.

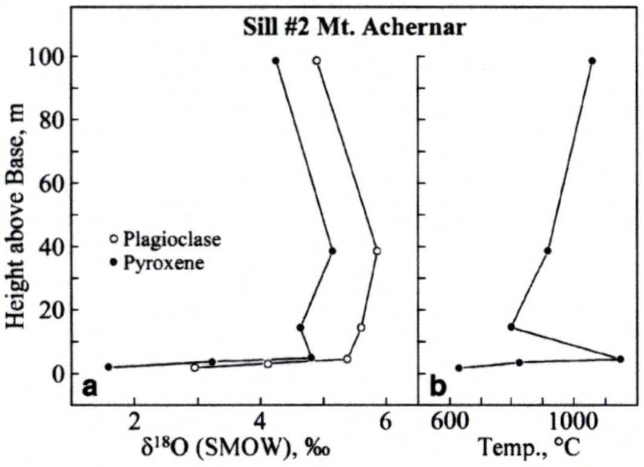

Fig. 13.31 (**a**) The $\delta^{18}O$ values of cogenetic plagioclase and pyroxene in Sill #2 on Mt. Achernar vary stratigraphically from low values at the lower contact of the sill to values between +5‰ and +7‰ about 5 m above the base. The $\delta^{18}O$ values of plagioclase are consistently higher than those of coexisting pyroxene. The difference is an indicator of the temperature at which the isotope composition of oxygen was locked in. (**b**) This temperature is estimated to be about 900 ± 30°C (872–926°C). The temperatures plotted in the diagram are averages for each sample based on the isotopic thermometry equation of Bottinga and Javoy (1975) and Chiba et al. (1989)

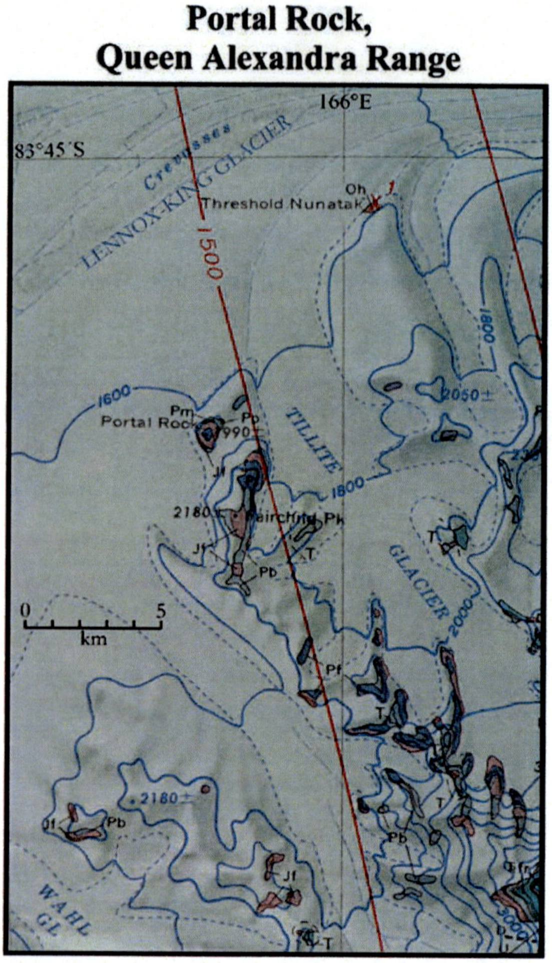

Fig. 13.32 Portal Rock (aka Portal Peak) at the mouth of the Tillite Glacier in the Queen Alexandra Range exposes the Permian Mckellar Formation (Pm) which overlies the Pagoda Tillite (Pp). The Mackellar Formation at this location was intruded by a sill of the Jurassic Ferrar Dolerite (Jf). The geology of the Queen Alexandra Range was described by Barrett et al. (1986). The elevation of the summit of Portal Rock is 1,990 m a.s.l. which causes it to stand 390 m above the surface of the Lennox-King Glacier. Excerpt of the reconnaissance geologic map by Lindsay et al. (1973)

The chemical composition and petrogenesis of the dolerite sill of Portal Rock were discussed by Hergt et al. (1989b) in the context of a regional overview of the basalt and the dolerite of the Ferrar Group in Gondwana. The concentrations of SiO_2 and alkali metals reported by Hergt et al. (1989b) in Fig. 13.35 demonstrate that most of the sill is composed of subalkalic, low- TiO_2 (0.36–0.94%) quartz dolerite. The data points on the silica-alkali metal diagram are tightly clustered suggesting that the sill on Portal Rock is chemically homogeneous and therefore undifferentiated. The lower chilled margin appears to be enriched in K_2O (5.44%), Al_2O_3 (21.45%), TiO_2 (1.12%), P_2O_5 (0.23%), and H_2O+ (4.93%); but it is depleted in SiO_2 (50.80%), MgO (4.09%), and CaO (0.70%) and is *not* representative of the chemical composition of the dolerite magma.

The profiles of SiO_2, MgO, and Na_2O in Fig. 13.36 reveal that the concentrations of these elements vary only within narrow limits, except for the lower chilled contact described above. The concentration of SiO_2 decreases up-section from about 54% to 52%, but the trend reverses at about 100 m above the base of the sill and the concentrations of SiO_2 increase in the upper 30 m toward 54%. The concentration of MgO ranges from about 6% to 7% and includes only local variations. This profile indicates that the sill at Portal Rock does not contain a hypersthene zone as do the dolerite sills in the ice-free valleys of southern Victoria Land (Gunn 1962 1966). The concentrations of Na_2O remain virtually constant throughout the body of this sill contrary to the expectation that the crystallization of early-formed anorthite enriched the residual magma in Na_2O and caused the plagioclase to become enriched in the albite component.

The Rb-Sr and Sm-Nd systematics of six samples analyzed by Hergt et al. (1989b) scatter widely in Fig. 13.37 and do not converge to their respective Rb-Sr and Sm-Nd isochrons. Nevertheless, three samples do form a Rb-Sr errorchron in part A that yields a date of 169 ± 16 Ma and an initial $^{87}Sr/^{86}Sr$ ratio of 0.70988 ± 0.00010 (σ). The other samples in this set scatter widely, probably because their Rb/Sr ratios were altered and *not* because the magma from which they formed had a range of different isotope compositions of strontium. Although this conjecture needs to be tested, we prefer to base conclusions about petrogenesis only on the intercepts of errorchron or isochron lines.

The Sm-Nd date in part B of Fig. 13.37 approximate an errorchron based on four of the same six samples analyzed by Hergt et al. (1989b) for dating by the Rb-Sr method. The slope of the Sm-Nd errochron corresponds to a date of 185 ± 7 Ma and the intercept is 0.510602 ± 0.0000058 (σ). The $^{143}Nd/^{144}Nd$ ratios measured by Hergt et al. (1989b) were multiplied by 1.00154 in order to make them compatible with $^{146}Nd/^{144}Nd = 0.7219$ (the value of the decay constant of ^{147}Sm is 6.54×10^{-12} $year^{-1}$). The Rb-Sr and Sm-Nd dates have overlapping

Fig. 13.33 Portal Rock at 83°33′S and 165°36′E at the mouth of the Tillite Glacier contains a sill of the Ferrar Dolerite which intruded the Mackellar Formation of Permian age (Photo by T.M. Mensing)

errors and therefore are not distinguishable. However, whole-rock Rb-Sr dates are lower than Sm-Nd dates of the same specimens, in many cases, because rubidium and strontium are more mobile in the subsurface environment than samarium and neodymium.

The initial $^{87}Sr/^{86}Sr$ and $^{143}Nd/^{144}Nd$ ratios at 175 Ma of the dolerite sill on Portal Rock plot in quadrant 4 in the Sr-Nd isotope-mixing diagram in Fig. 13.38. Evidently, the isotopic compositions of strontium and neodymium are mixtures of these elements that originated from the subcrustal mantle and the continental crust. We recall that the Tasmanian Dolerite in Fig. 12.38 and the flows of Kirkpatrick Basalt in the Mesa Range of northern Victoria Land in Fig. 12.39 also plot in quadrant 4 of this diagram. The data points representing the sill on Portal Rock in Fig. 13.38 form a segment of the isotopic mixing hyperbola (Faure and Mensing 2005). We also note that the initial $^{87}Sr/^{86}Sr$ ratio (0.70988 ± 0.00010) of the sill on Portal Rock is similar to the initial $^{87}Sr/^{86}Sr$ ratio of Sill #3 on Mt. Achernar in Fig. 13.28. Therefore, the intercept values of Rb-Sr and Sm-Nd errorchrons can perhaps be used to correlate sills of the Ferrar Dolerite on a regional basis. Rock samples that scatter away from such errorchrons represent cases of local contamination or alteration and therefore should be excluded from the determinations of initial isotopic ratios of strontium and neodymium.

The basalt flows and dolerite sills of the Ferrar Group in Antarctica are part of a larger petrogenetic province in Fig. 13.39 that extended across Gondwana from Australia, Tasmania, and the Transantarctic Mountains as far as the Theron Mountains in Coats Land (Hergt et al. 1991). These rocks are related to the Jurassic basalt and dolerite of the Kirwan Volcanics in Queen Maud Land and to the rocks of the Karoo Group in southern Africa. The belt continues with the Etendeka Volcanics in Namibia and the Paraná Basalt of Brazil. The ages of these rocks range from 180 Ma (Early/Middle Jurassic) for the Karoo-Kirwan subprovince, to 175 Ma (Middle Jurassic) for the rocks of the Ferrar Group in the Transantarctic Mountains, and to

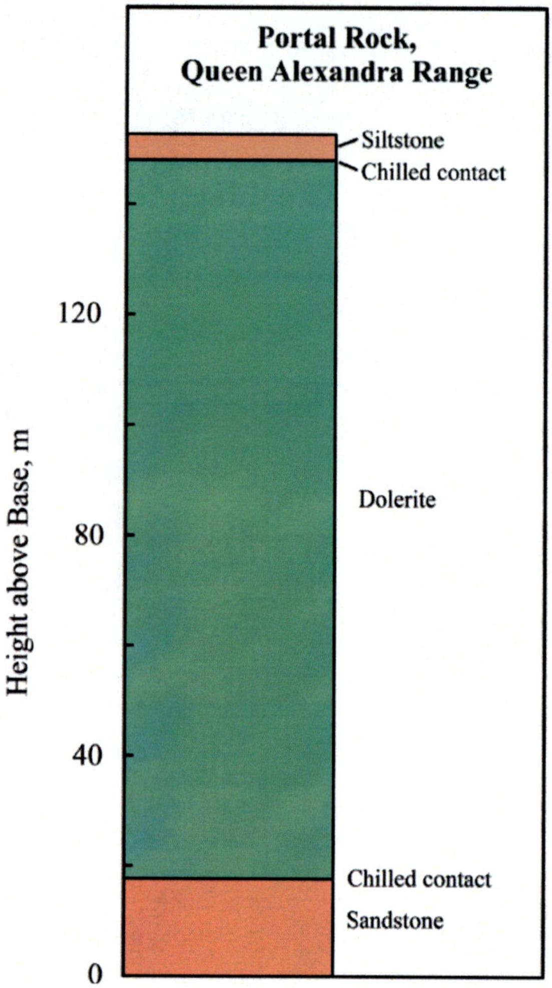

Fig. 13.34 The dolerite sill on Portal Rock is 129 m thick. It has chilled contacts at its top and bottom and has caused contact metamorphism of the sedimentary rocks above and below. Hydrothermal alteration is evident in the sediment at the base of the sill and within the sill at 10 m above the base. Section measured by G. Faure and T.M. Mensing on November 22, 1985

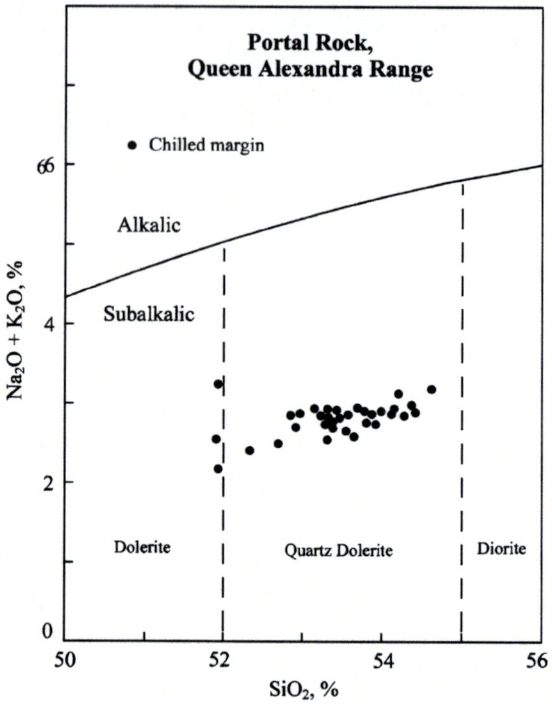

Fig. 13.35 The dolerite sill on Portal Rock in the Queen Alexandra Range is a subalkalkic low-Ti quartz dolerite. The rocks of this sill make a small footprint on this diagram because their chemical compositions are remarkably homogeneous. The only deviation is the chilled margin at the base of the sill (sample P-1-3C) which has a low concentration of SiO_2 (50.8%) and high concentration of K_2O (5.44%). This specimen is not representative of the chemical composition of the dolerite magma. Data in Appendix 13.6.7 by J.M. Hergt (personal communication 1987)

130 Ma (Cretaceous) for the Paraná-Etendeka province. The isotopic characteristics of these rocks were described by Faure (2001).

13.4 Dufek Intrusion

The Dufek intrusion is a layered deposit of mafic and ultramafic plutonic rocks that form the Dufek Massif and parts of Forrestal Range of the northern Pensacola Mountains (Schmidt and Ford 1969; Ford et al. 1977; Ford and Himmelberg 1991). The Forrestal Range was first viewed from the air on January 13, 1956, during a transpolar reconnaissance flight by the US Navy from McMurdo to the Filchner Ice Shelf. On October 23, 1957, Captain Finn Ronne (US Navy) and Dr. Ed Thiel flew from Ellsworth Station to the Dufek Massif but did not land. In December of the same year, the Ellsworth-Station IGY Traverse arrived at the Dufek Massif where two American geologist (N.B. Aughenbaugh and P.T. Walker) measured a 400-m stratigraphic section of the lower part of the Dufek intrusion (Aughenbaugh 1961; Walker 1961). In the field seasons of 1962/63 and 1963/64 the Dufek Massif and Forrestal Range were mapped by A.B. Ford and teams of geologists of the US Geological Survey who reached the area by snowmobile from the Patuxent and Neptune ranges of the Pensacola Mountains, respectively. Finally, in 1965/66 a large group of

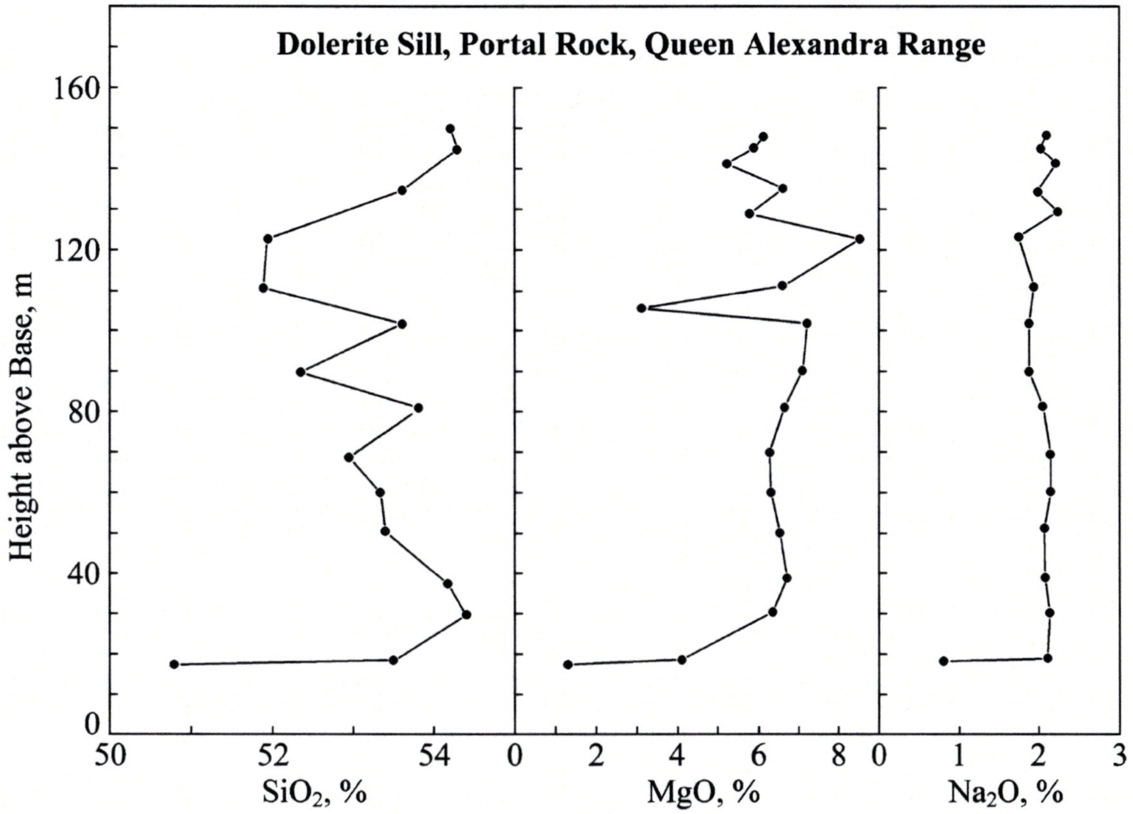

Fig. 13.36 The concentration profiles of the dolerite sill on Portal Rock in the Queen Alexandra Range reflect the homogeneity of its chemical composition. The principal exception is the lower chilled margin which has low concentrations of SiO_2, MgO, Na_2O, and CaO (0.70%) while Al_2O_3 (21.45%) and H_2O+ (4.93%) are unusually high. The overall impression is that the magma did not differentiate by fractional crystallization in situ in contrast to the sills in the ice-free valleys in southern Victoria Land (e.g., the Basement sill, Kukri Hills, Fig. 13.7). Data from Appendix 13.6.7

geologists and geophysicists of the US Geological Survey worked in the Pensacola Mountains with helicopter support. Subsequently, during 1976/77 field season, England and Nelson (1977) carried out geophysical traverses on the ground from the Dufek Massif to the Cordiner Peaks, to the Brown Nunataks, and north toward the Ronne Ice Shelf (England et al. 1979). At the same time, A.B. Ford and a group of geologists continued the geological fieldwork (Ford et al. 1977).

13.4.1 Topography

The Dufek intrusion in Fig. 13.40 extends from the Dufek Massif (82°36′S, 052°30′W) across the Sallee Snowfield to the Forrestal Range (33°00′S, 049°30′W). Geophysical Surveys by Behrendt et al. (1981) indicate that the Dufek intrusion is a continuous tabular body that extends north from the Dufek Massif and from the Forrestal Range under the Filchner Ice Shelf for about 250 km measured from the northern tip of the Dufek Massif. The resulting area occupied by the Dufek intrusion is larger than 50,000 km^2, which places it in second place behind the Bushveld Complex of South Africa in the hierarchy of stratified mafic intrusions (Ford 1990). The large size of the Dufek intrusion has been questioned by Ferris et al. (1998) who reduced its area to only 6,600 km^2. Even if that estimate is confirmed in the future, the Dufek intrusion remains in second place behind the Bushveld Complex.

The *Dufek Massif* in Fig. 13.41 is a rugged mountain range composed almost entirely of layered gabbro and related mafic igneous rocks. The main body of this mountain range is aligned from southwest to northeast and extends for 20 km followed by scattered nunataks to

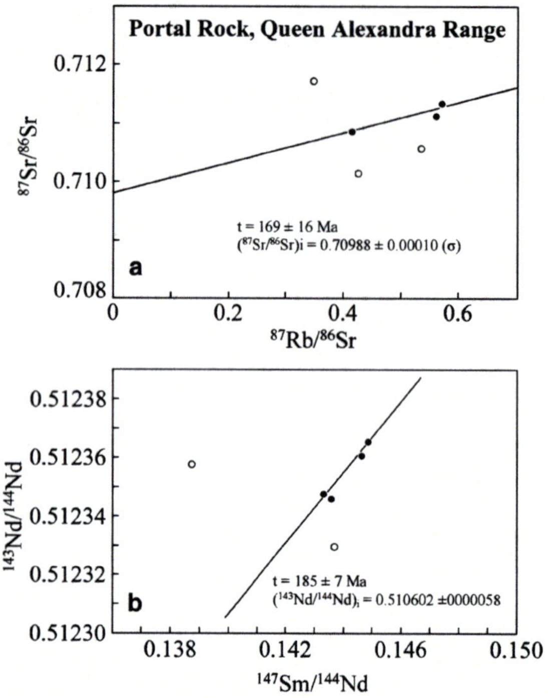

Fig. 13.37 The Rb-Sr data constrains an errorchron that corresponds to a date of t = 169 ± 16 Ma and has an intercept at $^{87}Sr/^{86}Sr$ = 0.70988 ± 0.00010 (λ = 1.42×10^{-11} year^{-1}). The Sm-Nd data define an errorchron that yields a date of 185 ± 7 Ma and has an intercept at $^{143}Nd/^{144}Nd$ = 0.510602. The $^{143}Nd/^{144}Nd$ ratios were multiplied by 1.00154 to correct for isotope fractionation to $^{146}Nd/^{144}Nd$ = 0.7129 (λ = 6.54×10^{-12} year^{-1}). Rb-Sr and Sm-Nd data by Hergt et al. (1989b)

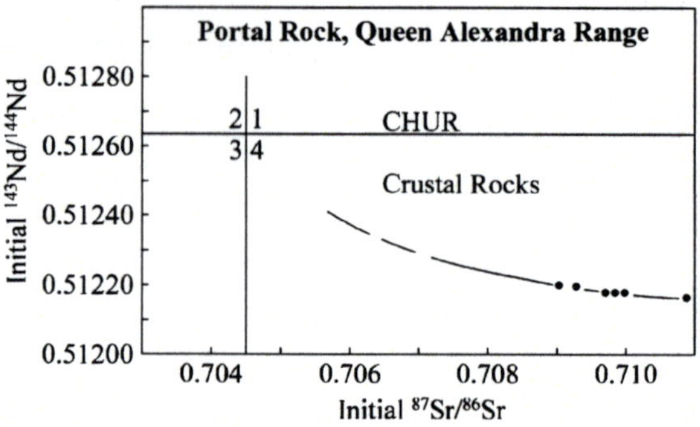

Fig. 13.38 The initial $^{87}Sr/^{86}Sr$ and $^{143}Nd/^{144}Nd$ ratios at 175 Ma of six specimens of quartz dolerite from the sill on Portal Peak, Queen Alexandra Range, form a linear array in quadrant 4 of the Sr-Nd isotope mixing diagram. The placement of the data points in quadrant 4 signifies that the isotopic compositions of strontium and neodymium in the dolerite samples are mixtures of these elements that originated from the subcrustal mantle and from the continental crust. In this respect, the petrogenesis of the sill at Portal Rock conforms to the model that has been developed by Hergt (1987) for rocks of the Tasmanian Dolerite (Section 13.1.7). The $^{143}Nd/^{144}Nd$ ratios measured by Hergt et al. (1989b) were corrected for isotope fractionation to $^{146}Nd/^{144}Nd$ = 0.7219 (see also Fig. 13.37)

Rankine Rock located about 30 km northeast along a northwest-facing escarpment. The Dufek intrusion is cut by normal faults, called the Enchanted Valley faults, which extend parallel to the Dufek Massif with the downthrown side on the north. In other words, the Dufek Massif is the mountain range that is composed of the Dufek intrusion. The Dufek Massif contains several jagged peaks including Walker, Neuberg, Aughenbaugh, and Hannah. Other noteworthy locations are Spear and Frost spurs, Davis Valley, and Rankin Rock. The total area occupied by the Dufek Massif is approximately 700 km^2 (Ford and Boyd 1968).

The *Forrestal Range* in Fig. 13.42 is separated from the Dufek Massif by the Sallee Snowfield which is about 75 km wide and is entirely covered by ice and snow. The Forrestal Range is composed of the Lexington Table and the Saratoga Table both of which are flat-topped plateaus or mesas separated by May Valley. The rocks of the Dufek intrusion are primarily exposed along the steep cliffs of the plateaus and on isolated peaks and spurs along their edges, such as: Mt. Stephens, Mt. Lechner, and Mt. Zirzow, as well as on Mathis Spur, and Sorna Bluff. The Forrestal Range extends to the northeast toward the Kester Peaks and Mt. Malville about 90 km from the southern edge of the Saratoga Table (Schmidt and Ford 1969). The total area occupied by the Forrestal Range is approximately 2,600 km^2 (Ford and Boyd 1968).

The *Cordiner Peaks* are located about 15 km south of the southern end of the Dufek Massif (Fig. 13.41). This cluster of nunataks is composed of gently folded sedimentary rocks that characterize the Neptune Range of the Pensacola Mountains (Section 8.2.3) including the Gale Mudstone, the Dover Sandstone, and the sedimentary rocks of the Neptune Range of Paleozoic age (Schmidt and Ford 1969). These rocks were intruded by sills of the Ferrar Dolerite.

Sills of Ferrar Dolerite also intruded the Nelson Limestone (Cambrian) in the *Schneider Hills* of the *Argentina Range* (Section 8.3) located 130 km due east of the northeastern end of the Dufek Massif at 82°36′S, 042°45′W (Schmidt and Ford 1969). The dolerite sills

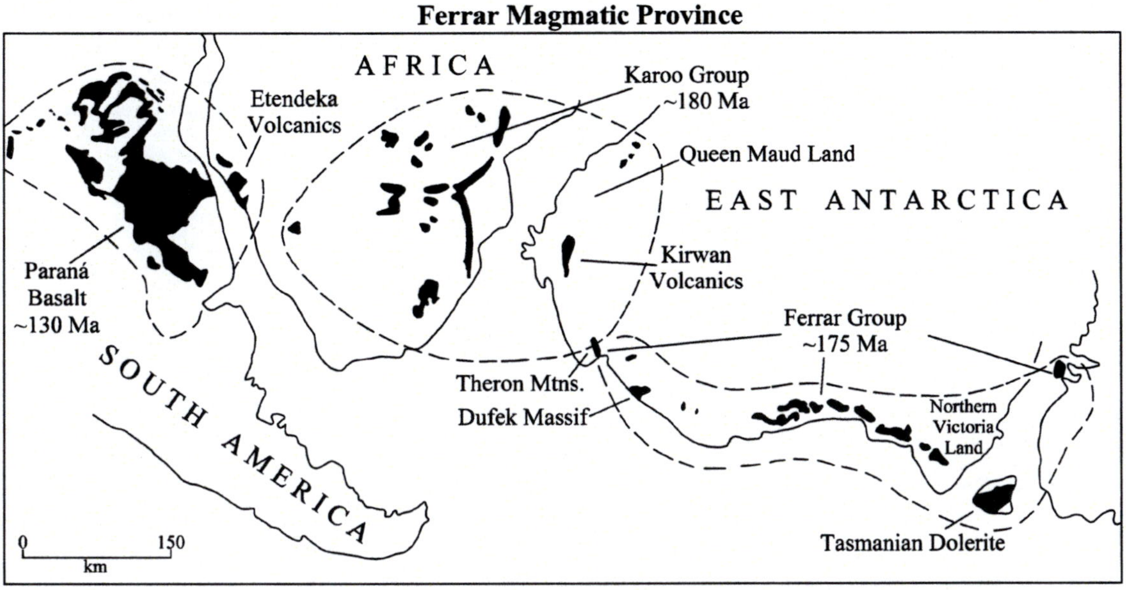

Fig. 13.39 The magmatic activity that was initiated by the rifting of Gondwana during the Mesozoic Era occurred in a belt that now extends across four continents of the southern hemisphere. The chemical and isotopic compositions of the resulting continental flood basalts and intrusives depend on the make-up of the source rocks from which the magmas formed by decompression melting. This view relates the igneous rocks of the Ferrar Group of Antarctica to a global tectonic event that resulted in the fragmentation of Gondwana and in the opening of the Atlantic Ocean by the break-up of Laurasia (Adapted from Hergt et al. 1991)

that occur in the areas north and south of the Dufek Massif and the Forrestal Range may be genetically related to the Dufek intrusion, in which case they can provide estimates of the chemical and isotopic composition of the Dufek magma at the time of intrusion.

13.4.2 Stratigraphy

The preliminary investigations of the well-bedded and flat-lying rocks of the Forrestal Range initially failed to recognize their magmatic origin (Ford and Boyd 1968). This was not the first time in the history of geology that stratified gabbro intrusions were mistakenly identified from a distance as sedimentary rocks (e.g., Wager and Deer (1939) who mapped the Skaergaard intrusion of East Greenland).

Ford and Boyd (1968) divided the Dufek intrusion into several descriptive series listed in Table 13.2. The intrusion is underlain by lightly folded sedimentary rocks of Paleozoic age which rest unconformably on the tightly folded low-grade metamorphic rocks of the Patuxent Formation (Section 8.2.3). The older sequence of metasediments contains interbedded basalt flows, whereas the overlying sedimentary rocks contain rhyolite flows and pyroclastic rocks of the Gambacorta Formation (Section 8.2). However, Ford and Boyd (1968) stated empathically that volcanic rocks related to the Dufek intrusion are not present in the Pensacola Mountains. Ford (1976) later assigned the Upper Exposed Series of the Forrestal Range and the Lower Exposed Series of the Dufek Massif to the *Forrestal Gabbro Group* and subdivided it into the *Saratoga Gabbro* (Forrestal Range) and the *Aughenbaugh Gabbro* (Dufek Massif). These formations each contain several members identified in Table 13.3.

The *Forrestal Gabbro* encompasses all of the layered gabbros in the Forrestal Range and in the Dufek Massif (Ford 1976). It consists not only of gabbro but also includes mafic and felsic differentiates ranging in composition from pyroxenite and magnetite rocks to leucogabbro, anorthosite, and granophyre. Although neither the top nor the bottom of the Forrestal Gabbro Group is exposed, Ford (1976) estimated that the total thickness of the Dufek intrusion is 8 to 9 km including the Middle Hidden Series (covered by the Sallee Snowfield) and the Lower Hidden Series at the base of the intrusion.

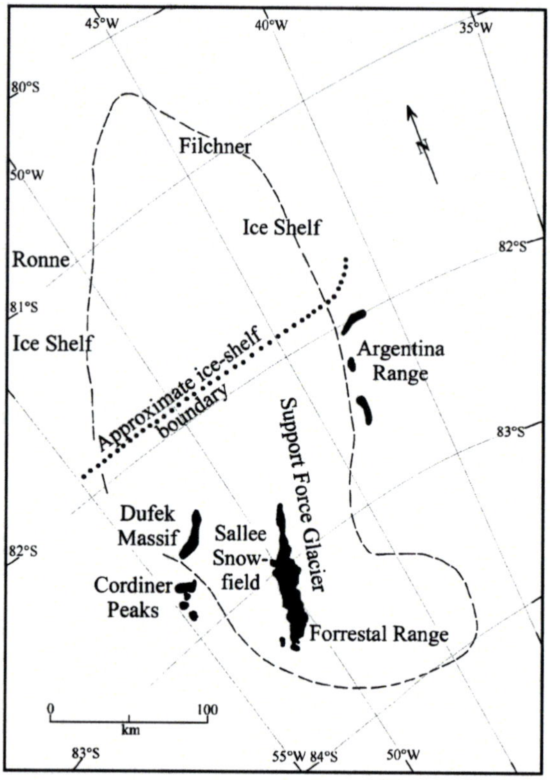

Fig. 13.40 The Dufek intrusion in the northern Pensacola Mountains extends from the Dufek Massif across the Sallee Snowfield to the Forrestal Range. Geophysical surveys by Behrendt et al. (1974, 1979, 1980, 1981) indicate that the intrusion is a continuous tabular body that extends north about 250 km from the northern tip of the Dufek Massif. Ferris et al. (1998) questioned this extrapolation of the area of the Dufek intrusion and reduced it from about 50,000 to only 6,600 km². Even in that case, the area of the Dufek intrusion is exceeded only by the Bushveld Complex of South Africa (Adapted from Ford 1990)

Table 13.2 Stratigraphy of the Dufek intrusion in the Dufek Massif and the Forrestal Range (Ford and Boyd 1968)

Stratigraphic units
A. Upper Missing series, eroded, Forrestal Range
B. Upper Exposed Series, 1,750 m, Forrestal Range
Granophyre (top of the section)
Upper gabbro zone
Upper inclusion zone
Upper layered zone
Lower inclusion zone
Lower layered zone
C. Middle Hidden Series
Covered by Sallee Snowfield
D. Lower Exposed Series, 1,825 m, Dufek Massif
Upper layered zone
Brown zone
Middle gabbro
Lower brown zone
Lower gabbro zone
Anorthosite zone
E. Lower Hidden Series
Not exposed (bottom of the section)

Table 13.3 Refined stratigraphy of the Dufek intrusion in the Forrestal Range and the Dufek massif (Ford 1976)[a]

Group	Formations	Members
Forrestal Range	(Upper Exposed Series)	
Forrestal Gabbro	Lexington Granophyre	
	Saratoga Gabbro	
		Upper inclusion
		Lower inclusion
		Stephens Anorthosite
Dufek Massif	(Lower Exposed Series)	
Forrestal Gabbro	Aughenbaugh Gabbro	
		Spear Anorthosite
		Frost Pyroxenite
		Neuberg Pyroxenite
		Lower Anorthosite
	Walker Anorthosite	

[a]The Middle Hidden Series of Ford and Boyd (1968) is covered by the Sallee Snowfield and the Lower Hidden Series is overlain by the rocks of the Upper and Lower Exposed Series.

The *Walker Anorthosite* in Table 13.3 underlies the Aughenbaugh Gabbro in the Dufek Massif and was named after Walker Peak in Fig. 13.41 where a 230-m section of this formation is exposed. It is composed of a medium-grained plagioclase cumulate (An_{75} to An_{85}). The anorthosite is white to cream colored or gray and is spotted because of the presence of coarse poikilitic postcumulus crystals of pyroxene.

The *Aughenbaugh Gabbro* (Table 13.3) is a mafic cumulate that concordantly overlies the Walker Anorthosite and extends throughout the Dufek Massif from Walker Peak in the southwest for about 50 km to Rankine Rock which is the most northeasterly nunatak of the Dufek Massif in Fig. 13.41. The thickness of this formation is at least 1,600 m. The Aughenbaugh Gabbro is composed of medium-grained cumulates composed of plagioclase (An_{60} to An_{70}), augite, and inverted pigeonite with accessory magnetite in the upper 200 m of the formation. The rocks are light to medium gray in color and are layered on a scale of millimeters to centimeters. The Aughenbaugh Gabbro

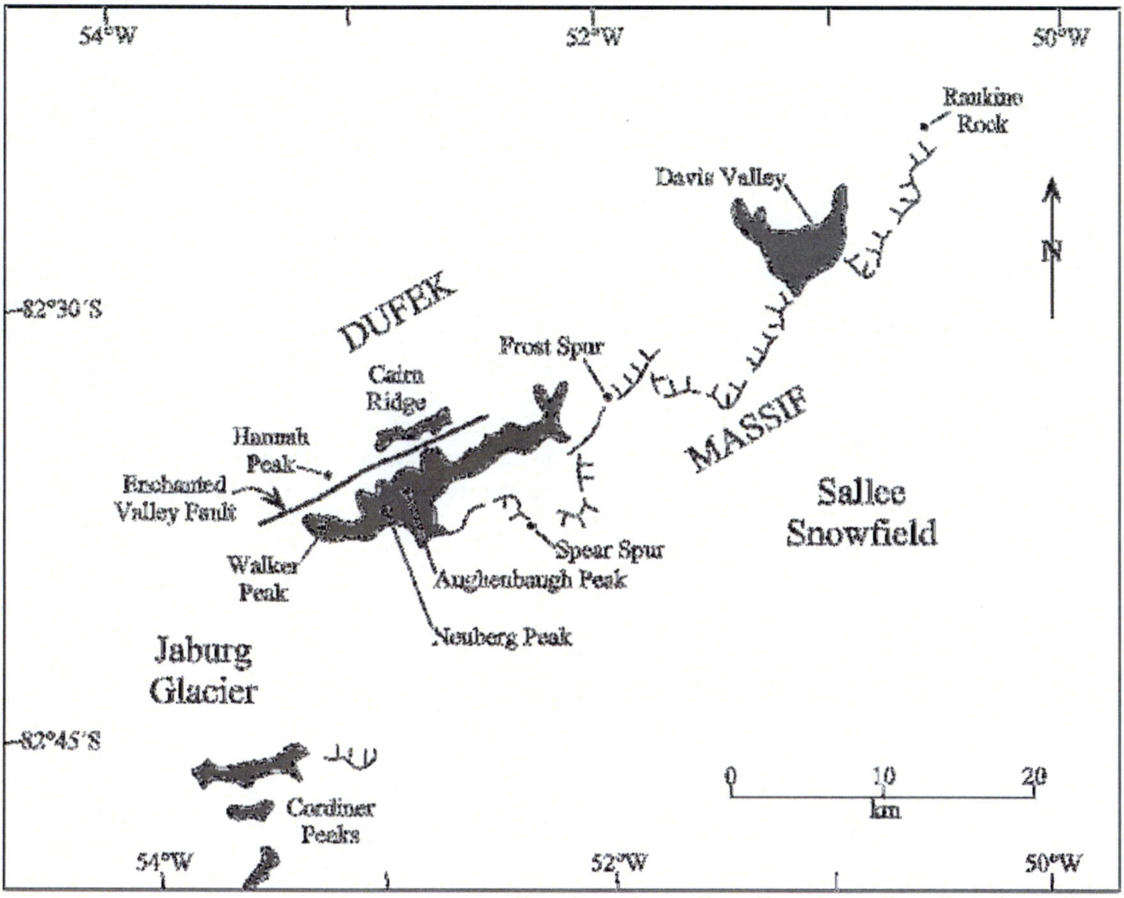

Fig. 13.41 The Dufek Massif is composed almost entirely of the stratified gabbroic cumulates of the Dufek intrusion which is genetically related to the Middle Jurassic Ferrar Dolerite. This mountain range consists of steep-sided peaks and ridges that extend about 20 km between Walker Peak and Frost Spur. The geology of this mountain range was mapped by geologists of the US Geological Survey under the leadership of Arthur B. Ford (Adapted from Fig. 1 of Ford 1976)

contains four members identified in Table 13.3 and described below.

The Lower Anorthosite Member is a white to cream colored plagioclase cumulate about 40 m above the base of the Aughenbaugh Gabbro in the Dufek Massif. The layer is about 12 m thick and has sharply defined contacts at the top and bottom. Like the Walker Anorthosite, the Lower Anorthosite Member contains clots of postcumulus pyroxenes that give the rock a spotted appearance. This layer is absent in some areas where the lower part of the Aughenbaugh Gabbro is exposed.

The Neuberg Pyroxenite Member is a conspicuous black layer of pyroxene cumulate located stratigraphically about 200 m above the base of the Aughenbaugh Gabbro. The layer is about 5 m thick. It has a sharply defined base but grades upward into the plagioclase-rich gabbro of the Aughenbaugh Formation. The pyroxene, which is coarse grained and occurs in clusters 5 to 7 cm in diameter, originally crystallized as pigeonite which settled out of the cooling magma and later inverted to orthopyroxene (Himmelberg and Ford, 1976). The Neuberg Member can be traced for about 11 km before it passes out of sight under snow and talus.

The *Frost Pyroxenite Member* occurs near the top of Neuberg Peak about 800 m above the base of the Aughenbaugh Gabbro (Figure 13.41). The north-easterly dip of the Dufek intrusion causes the Frost Pyroxenite to crop out in the Davis Valley at the level of the lower ice plateau. This unit consists of six or more layers which have sharply defined bases and grade upward into gabbro as the abundance of plagioclase increases. Three of these layers are composed of dark brown pyroxenite, two of which occur near the base of this unit and one at the top. The total thickness of the Frost Pyroxenite is about 43 m and it extends for about 35 km along nearly the entire length of the Dufek massif.

The *Spear Anorthosite Member* consists of three layers of cream colored to light gray anorthosite cumulate interbedded with dark gray cumulate gabbro for a total thickness of about 105 m in the upper part of the Aughenbaugh Gabbro. The thickness of the plagioclase cumulate layers is about 46 m, 12 m, and more than 5 m from the lowest to the highest. These anorthosite layers are interbedded with layers of pyroxene-plagioclase-magnetite cumulate which are about 18 m (lower) and 24 m thick (upper). The anorthosite layers contain clusters of pyroxene crystals which give them a spotted appearance similar to the Walker Anorthosite.

The *Saratoga Gabbro* is a well-layered gabbro cumulate that is exposed throughout the *Forrestal Range* (Ford 1976) and underlies most of the snow covered Saratoga and Lexington tables in Fig. 13.42. The unit has been traced for at least 130 km from its most southwesterly outcrops to the northeastern nunataks of the Forrestal Range. Geophysical evidence indicates that the Saratoga Gabbro may continue under the ice for another 100 km in a northeasterly direction (Behrendt

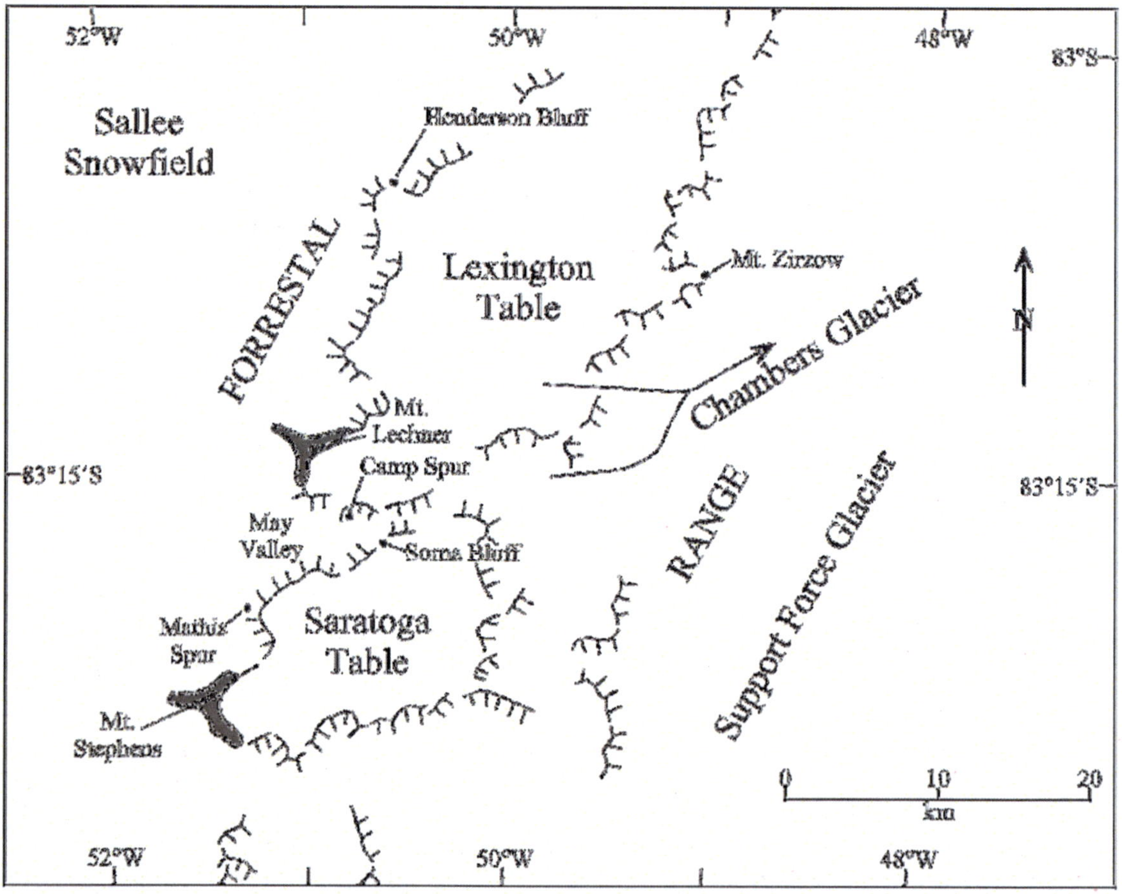

Fig. 13.42 The Dufek intrusion extends from the Dufek Massif under the Sallee Snowfield to the Forrestal Range located about 75 km to the southeast. The Forrestal Range consists primarily of the Saratoga Table and the Lexington Table both of which are plateaus or mesas composed of the layered gabbros of the Dufek intrusion. These rocks are primarily exposed along the edges of the snow-covered plateaus at Mt. Stephens, Mt. Lechner, and Sorna Bluff (Adapted from Fig. 1 of Ford 1976)

et al. 1974). The total exposed thickness of the Saratoga Gabbro is approximately 1,400 m. The type section on the western spur of Mt. Stephens contains about 600 m of layered rocks composed of pyroxene-plagioclase-magnetite cumulates and of locally developed layers of cumulate plagioclase and magnetite. The upper part of the Saratoga Gabbro of about 730 m is not present on Mt. Stephens but sections measured down-dip indicate that its lithologic composition is similar to the lower part of this formation. The base of the Saratoga Gabbro is not exposed but the top is conformably overlain by the Lexington Granophyre at Sorna Bluff, Mt. Lechner, and Mt. Zirzow in Fig. 13.42.

The Saratoga Gabbro differs from the Augenbaugh Gabbro of the Dufek Massif by being more conspicuously layered although individual layers are not as continuous as those of the Aughenbaugh Gabbro. The Saratoga Gabbro also contains more magnetite which causes it to be darker than the Aughenbaugh Gabbro, whereas plagioclase is more sodic (e.g., An_{50} to An_{60}). In addition, the Saratoga Gabbro contains apatite and, more rarely, Fe-rich olivine (fayalite) both of which are not present in the Aughenbaugh Gabbro. The Saratoga Gabbro contains three members defined by Ford (1976) and described briefly below (Table 13.3).

The *Stephens Anorthosite Member* forms the base of the Saratoga Gabbro. It is 300 m thick and is composed of alternating layers of white to light gray plagioclase cumulates (7 to 15 m thick) interbedded with dark gray gabbro composed of pyoxene, plagioclase, and magnetite. The anorthosite layers have well-defined bases but grade upwards into the gabbro layers as pyroxene becomes more abundant. The Stephens Anorthosite is exposed only at Mt. Stephens in Figure 13.42 and is ice-covered elsewhere.

The *Lower Inclusion Member* is composed of dark gabbroic cumulate that contains light colored rounded masses of quartz-bearing leucogabbro and lesser amounts of anorthosite and granophyre one to three meters in diameter. Ford (1976) and Ford and Boyd (1968) concluded that these inclusions originated from the roof of the magma chamber where late-stage differentiated magma was crystallizing and that they were transported laterally by magma currents. The base of the Lower Inclusion Member is exposed on the western spur of Mt. Stephens at an elevation of about 2020 m and is located about 518 m above the base of the Saratoga Gabbro. This unit is 46 m thick and extends for about 10 km north of Mt Stephens.

The *Upper Inclusion Member* is exposed at Sorna Bluff at the north end of the Saratoga Table in Figure 13.42 and along the southern rim of the Lexington Table at Mt. Lechner and in May Valley. Inclusions of anorthosite and leucogabbro are embedded in dark gabbro. The unit is about 100 m thick and is located 1,195 m above the base of the Saratoga Gabbro.

The *Lexington Granophyre* is the highest formation in the Forrestal Gabbro Group and does not occur in the Dufek Massif. The granophyre is exposed along the escarpment of the Lexington Table and on the north rim of the Saratoga Table. The type section of the formation is located on Mt. Zirzow where its thickness is about 238 m. Good exposures of the granophyre also occur on Mt. Lechner and Sorna Bluff. The lower contact of the Lexington Granophyre is sharp and appears to be concordant; however, the upper part of the granophyre and the overlying roof rocks of the intrusion have been eroded. The granophyre is light gray to cream colored and is composed of fine-grained myrmekitic intergrowths of a quartz and K-feldspar. The rocks also contain microphenocrysts of zoned andesine plagioclase and pale green augite which has altered to green amphibole and mica. The modal abundance of pyroxene and its alteration products ranges from 1 to 5%. Acicular crystals of apatite are ubiquitous but make up less than one percent of the mode.

Ford (1970) reported that the granophyre has a uniform appearance and contains no evidence of crystal settling. It does not contain xenoliths of silicic country rocks that might have contaminated the magma. The volume of granophyre is relatively small and makes up only 3–4% of the entire intrusion. Therefore, the Lexington Granophyre probably crystallized from highly differentiated magma of the Dufek intrusion.

13.4.3 Chemical Composition and Mineralogy

The average chemical compositions of the various differentiated facies of the Dufek intrusion in Appendix 13.6.8 and Fig. 13.43 show the effects of fractional crys-

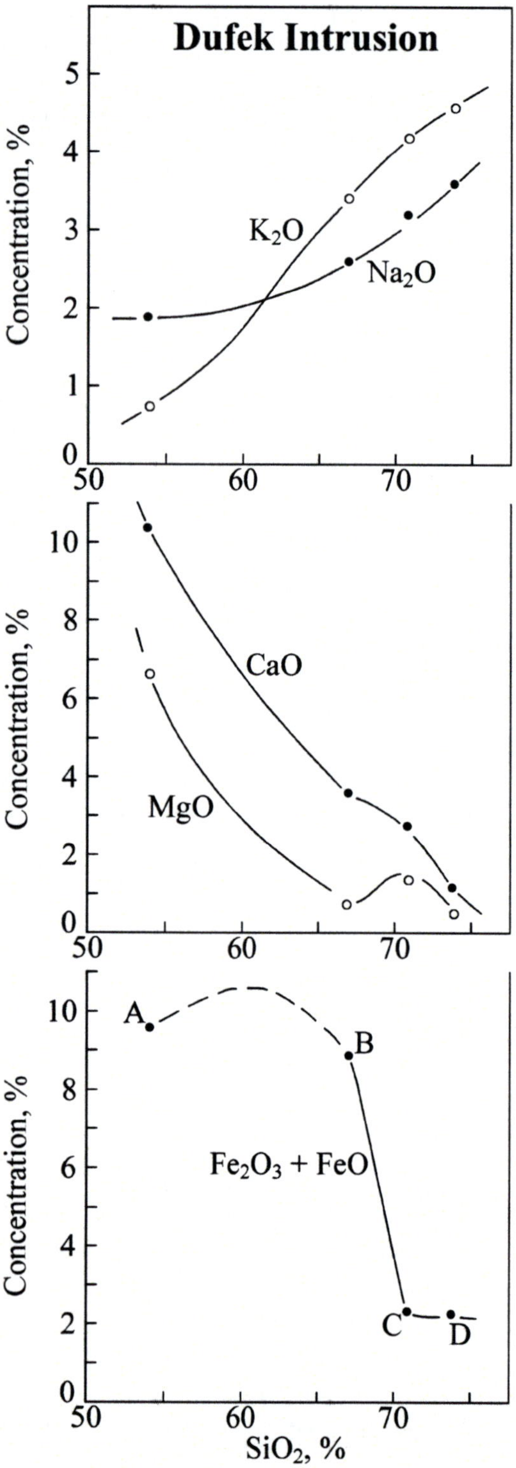

Fig. 13.43 The average concentrations of major elements in the rocks of the Dufek intrusion vary with increasing concentrations of SiO_2. The chilled margin of the Dufek intrusion exposed at Mt. Lechner in the Forrestal Range is enriched in iron and is not representative of the parental magma. Therefore, the chemical composition of "A" is based on tholeiite basalt in dikes at Cordiner Peaks (Ford 1974). The concentrations of sodium and potassium increased in the residual magma, whereas the concentrations of calcium, magnesium, and iron decreased by crystallization of Ca-rich plagioclase, pyroxene, and magnetite, respectively. The concentration of potassium actually rises above that of sodium because sodium was removed from the magma by crystallization of plagioclase, whereas potassium remained in the magma. Data from Ford (1970) and Ford and Kistler (1980) tabulated in Appendix 13.6.8

tallization of the Dufek magma. The concentrations of sodium and potassium rise with increasing silica concentrations, whereas the concentrations of calcium, magnesium, and iron decline. The data reveal that the concentration of potassium actually rises above that of sodium which is noteworthy because the crustal abundance of potassium is less than that of sodium (Faure 1998, Chapter 4). The reason for the rise of potassium over sodium is that sodium was removed from the magma by the crystallization of plagioclase (Abel et al. 1979; Haensel et al. 1986), whereas potassium remained in the magma until orthoclase began to crystallize from the residual magma that formed the granophyre. Calcium and magnesium were also removed from the mafic magma by crystallization of Ca-rich plagioclase and by pyroxenes, respectively. Iron was initially retained in the residual magma; but eventually left the magma by crystallization of pyroxene followed by magnetite in the upper part of the Dufek intrusion. The aplite dikes are the most highly differentiated part of the Dufek intrusion and contain only about 2% FeO and Fe_2O_3 combined. These progressive changes of the chemical compositions of the gabbro and the overlying granophyre were discussed and illustrated by Ford and Boyd (1969) and Ford (1969, 1970).

These authors distinguished a "mafic" trend defined by gabbros of the layered series (i.e., the Aughenbaugh and Saratoga gabbros in Table 13.3) from a "felsic" trend for interbedded anorthosites and leucogabbros. The two trends converge in the granophyre. Leucocratic dikes (i.e., aplites) in Fig. 13.43 are extreme differentiates which have the highest average concentrations of SiO_2 (72.3%), K_2O (4.4%), and Na_2O (3.4%) but low concentrations of FeO + Fe_2O_3 (1.2%), MgO (0.95%), CaO (2.0%), TiO_2 (0.21%), P_2O_5 (0.09%), and MnO (0.065%). The concentrations of these components differ significantly from the chemical compositions of sills of Ferrar Dolerite in the Cordiner Peaks located about 12 km south of Walker Peak in the Dufek Massif in Fig. 13.41. The layered gabbros as well as the pyroxenite, magnetitites, and anorthosites of the Dufek intrusion formed by the settling of crystals from the overlying magma and by deposition from convection currents. These currents apparently eroded channels in the previously deposited crystal mush which were subsequently filled with later-formed crystals of another mineral.

The rocks of the Dufek intrusion have a wide range of densities which correlate positively with the abundance of pyroxene. This relationship in Fig. 13.44 was worked out by Ford and Nelson (1972, Fig. 1) who measured the densities of approximately 600 rock samples of the Dufek intrusion. The results indicate that the different rock types can be characterized by their densities in grams per cubic centimeters:

Granophyre: 2.65
Anorthosite: 2.70
Gabbro: 2.80–3.20
Pyroxenite: 3.30
Magnetitite: 3.50

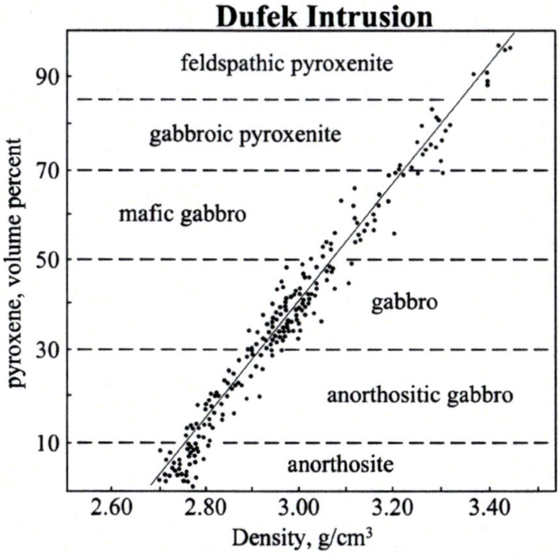

Fig. 13.44 The densities of rocks of the Dufek intrusion increase linearly with the rising abundance of pyroxene. Accordingly, the density of the rocks can be used to classify rocks of the Dufek intrusion. In addition, the densities of the layered gabbros in the Dufek Massif and the Forrestal Range increase up-section because of the rising abundance of magnetite. As a result, the average density of the rocks in the section of the Forrestal Range (3.03 g/cm³) is greater than the average density of the underlying section of the Dufek Massif (2.95 g/cm³). The average density of the entire body including the unexposed series in Table 13.2 is 2.98 g/cm³ (Data from Ford and Nelson 1972)

The linear relationship between the density of different kinds of rocks and their pyroxene concentrations can be used to classify rocks of the Dufek intrusion. In addition, the densities of the gabbros of the Dufek Massif and of the Forrestal Range increase up-section. As a result, the rocks of the Forrestal Range have higher densities than the rocks in the Dufek Massif because they contain more magnetite. In this way, the densities of the rocks confirm that the initial crystallization of Mg-rich pyroxenes from the Dufek magma enriched the residual magma in iron until magnetite began to form which reduced the iron concentration of the remaining magma and caused the aplite dikes in Fig. 13.43 to have low concentrations of iron.

The progressive iron enrichment of the cooling Dufek magma is reflected also by the composition of the cumulate pyroxenes (Himmelberg and Ford 1973, 1976) and of the iron-titanium oxides (Himmelberg and Ford 1975, 1977). The concentrations of iron as FeO in Ca-rich pyroxenes (augite) in Fig. 13.45 demonstrate the iron enrichment in the pyroxenes of the Dufek intrusion caused by progressive fractional crystallization. The concentrations of MgO display the opposite trend (not shown). The effects of fractional crystallization on the composition of plagioclase were investigated by Abel et al. (1979) and Haensel et al. (1986), while Drinkwater et al. (1986) studied apatite in the upper part of the Dufek intrusion.

13.4.4 Age and Petrogenesis

The youngest rocks intruded by the Dufek magma are Glossopteris-bearing sedimentary rocks of Permian age. Therefore, Ford and Boyd (1968) suggested that the age of the Dufek intrusion is Jurassic by analogy with the dolerite sills of the Ferrar Group which were dated by McDougall (1963). Several years later, Ford and Kistler (1980) published K-Ar and total-release $^{40}Ar/^{39}Ar$ dates of the Dufek intrusion, of dikes in the Cordiner Peaks, and of the dolerite sills in the Pecora Escarpment. These age determinations confirmed the Jurassic age of all of these rocks:

Dufek intrusion: 172 ± 2.3 Ma (3)
Cordiner Peaks: > 169 ± 4 Ma (1)
Pecora Escarpment: 179 ± 5 Ma (2)

Subsequently, Brewer et al. (1996) reported $^{40}Ar/^{39}Ar$ plateau dates of 182.1 ± 2.4 and 182.9 ± 2.5 Ma for plagioclase separates from the Dufek intrusion and Minor and Mukasa (1997) dated zircons in a large sample (70 kg) of the Lexington Granophyre from the top of Sorna Bluff in the Forrestal Range and a hornblende in a granitic dike that intruded layered gabbro just below the granophyre at Mt. Lechner. The zircon fractions from the Lexington Granophyre were strongly abraded to remove surface alteration and then yielded concordant U-Pb dates in Table 13.4 with an average of 183.9 ± 0.3 Ma which Minor and Mukasa (1997) interpreted as the crystallization age of the Dufek intrusion. The zircon fractions extracted from the granitic dike also yielded concordant dates averaging 182.7 ± 0.4 Ma. The age of the Dufek intrusion determined by Minor and Mukasa (1997) is indistinguishable from the U-Pb date of 183.6 ± 1.0 Ma reported by Encarnación et al. (1996) for zircon and baddeleyite from two sills of the Ferrar Dolerite. Encarnación et al. (1996) also obtained a U-Pb date of 183.7 Ma for a Karoo sill in South Africa. These results strongly support the hypothesis that the Ferrar Dolerite and the Dufek that the Ferrar Dolerite and the Dufek intrusion of East Antarctica crystallized at the same time as Karoo sills in South Africa.

Although the Kirkpatrick Basalt may be a product of the same magma that formed the sills of Ferrar Dolerite,

Table 13.4 Average U-Pb dates of zircon and $^{40}Ar/^{39}Ar$ plateau dates of hornblende extracted from the Lexington Granophyre and from a granite dike that intruded the layered gabbro of the Dufek intrusion (Data from Minor and Mukasa 1997)

Unit	U-Pb (Ma)	$^{40}Ar/^{39}Ar$ plateau (Ma)
Granophyre	183.9 ± 0.3	174.1 ± 0.8 (A)
		175.6 ± 0.8 (B)
Granite dike	182.7 ± 0.4	180.0 ± 0.8 (C)
		178.1 ± 1.1 (D)

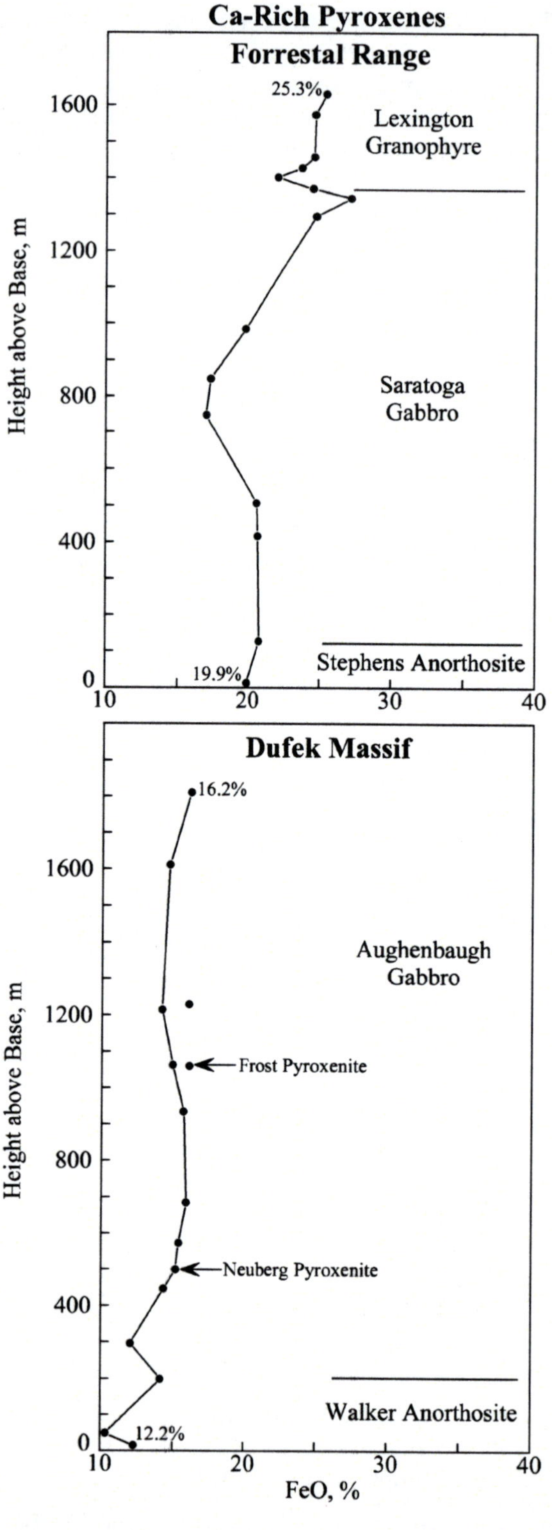

Fig. 13.45 The concentrations of FeO (total iron) of Ca-rich pyroxenes (augite) *increase* up-section in the Dufek Massif from 12.2% to 16.2% and in the Forrestal Range from 19.9% to 25.3%. The concentration of MgO (not shown) in the same stratigraphic interval *decreases* up-section from 15.4% to 12.8% (Dufek Massif) and from 11.5% to 8.4% (Forrestal Range). The Ca-poor pyroxenes (pigeonite inverted to Fe-rich hypersthene) have higher concentrations of FeO and MgO than the Ca-rich pyroxenes, but they vary similarly up-section in the Dufek Massif from 21.7% to 25.4% (FeO) and from 20.0% to 16.4% (MgO). The data for the Forrestal Range are incomplete. The stratigraphic variations of FeO and MgO in the pyroxenes of the Dufek intrusion are unambiguous evidence of the evolution of the Dufek magma by fractional crystallization (Data from Himmelberg and Ford 1976)

the age determinations of the Kirkpatrick Basalt do not yet confirm a genetic relationship with the Ferrar Dolerite. The best average $^{40}Ar/^{39}Ar$ plateau date of plagioclase (176.6 ± 1.8 Ma) published by Heimann et al. (1994) may need to be revised because the age of the flux monitor Mmhb-1 they used was recalibrated by Renne et al. (1994, 1998) from 513.5 to 523.5 Ma. Use of the revised age of the flux monitor increases the $^{40}Ar/^{39}Ar$ plateau date of the Kirkpatrick Basalt from 176.6 ± 1.8 (Heimann et al. 1994) to 180.4 ± 1.8 Ma (Minor and Mukasa 1997). Similarly, the $^{40}Ar/^{39}Ar$ plateau dates of hornblende from the Dufek intrusion in Table 13.4 would also rise (Minor and Mukasa 1997):

Granophyre: (A) 175.2 ± 0.8 Ma; (B) 176.8 ± 0.8 Ma
Granite dike: (C) 181.2 ± 0.8 Ma; (D) 179.3 ± 1.1 Ma

In spite of the uncertainty of the isotopic age determinations, the mafic igneous rocks that were formed during the initial split between East Antarctica and Africa constitute the largest known volume of rift-related magma in the world.

Although the Dufek intrusion is known to have formed by differentiation of a large volume of mafic magma, the origin of this magma and its relation to the Ferrar Dolerite are less well established. Ford and Kistler (1980) reported that a dolerite sill in the Pecora Escarpment located south of the Pensacola Mountains (Section 8.2.2, Fig. 8.6), and a basalt dike in the Cordiner Peaks (Fig. 13.40) have high initial $^{87}Sr/^{86}Sr$ ratios of 0.7104 and 0.7126, respectively. These values clearly associate these rocks with the sills of Ferrar Dolerite on Roadend Nunatak and Mt. Achernar in Fig. 13.28. Two samples of pyroxenite collected by N.B. Aughenbaugh from the Dufek intrusion in 1957 and analyzed by Faure et al. (1972) also have high initial $^{87}Sr/^{86}Sr$ ratios of 0.7097 and 0.7088 at 175 Ma and using $\lambda = 1.42 \times 10^{-11}$ year^{-1}. More recently, Ford et al. (1986) published four additional initial $^{87}Sr/^{86}Sr$ ratios of pyroxene and plagioclase from the Dufek intrusions that range from 0.70763 to 0.70918. Even with these additions, the number of measurements of the isotopic composition of strontium in the Dufek intrusion is insufficient to fully explain the petrogenesis of this exceptionally large and complex body of plutonic igneous rocks. However, the available initial $^{87}Sr/^{86}Sr$ ratios in Fig. 13.46 overlap those of the Ferrar Dolerite and thereby indicate that the Dufek magma, like the Ferrar magma, was enriched in radiogenic ^{87}Sr, presumably because it formed by decompression melting in the subcrustal mantle that contained subducted crustal rocks that had higher Rb/Sr ratios than the ultramafic rocks of the lithospheric mantle.

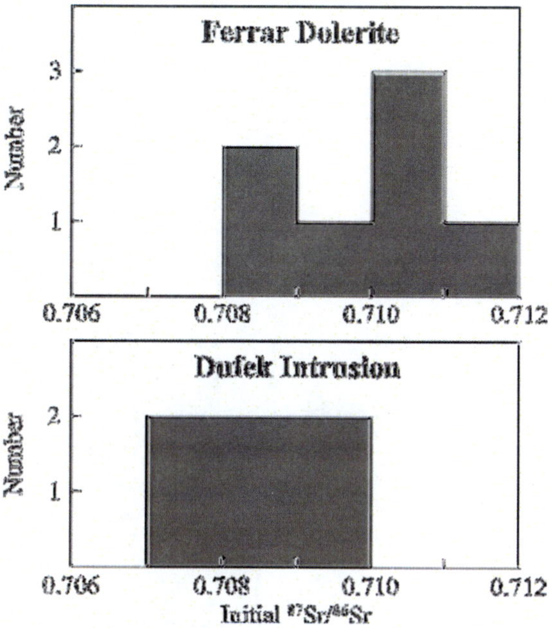

Fig. 13.46 The initial $^{87}Sr/^{86}Sr$ ratios of the Dufek intrusion overlap the range of initial $^{87}Sr/^{86}Sr$ ratios of the Ferrar Dolerite indicating that both bodies of rock contain excess radiogenic ^{87}Sr compared to mantle-derived basalts in the oceans and on the continents (Faure 2001). Evidently, the Dufek magma could have originated from sources in the subcrustal mantle that contained subducted crustal rocks having higher Rb/Sr ratios than the rocks of the mantle. The initial $^{87}Sr/^{86}Sr$ ratios of the Ferrar Dolerite are the intercept values of Rb-Sr errorchrons for sills in Wright Valley (Fig. 13.5), Roadend Nunatak (Fig. 13.8), Mt. Achernar (Fig. 13.27), and Portal Rock (Fig. 13.37). The initial $^{87}Sr/^{86}Sr$ ratios of the Dufek intrusion are based on single samples analyzed by Faure et al. (1972) and Ford et al. (1986) whose Rb/Sr ratios could have been altered after crystallization

The $\delta^{18}O$ values of rocks from the *Dufek Massif* reported by Ford et al. (1986) in Fig. 13.47 range narrowly from +5‰ to +7‰ with one low value of less than +4‰. These values are "normal" for mafic igneous rocks that crystallized from mantle-derived magma without extensive assimilation of crustal rocks or hydrothermal alteration. The $\delta^{18}O$ values of rocks from the Forrestal Range vary from "normal" values of +6‰ to low values between 0‰ and +1‰. Several prominent rock types in the Forrestal Range have distinctive $\delta^{18}O$ values (Ford et al. 1986):

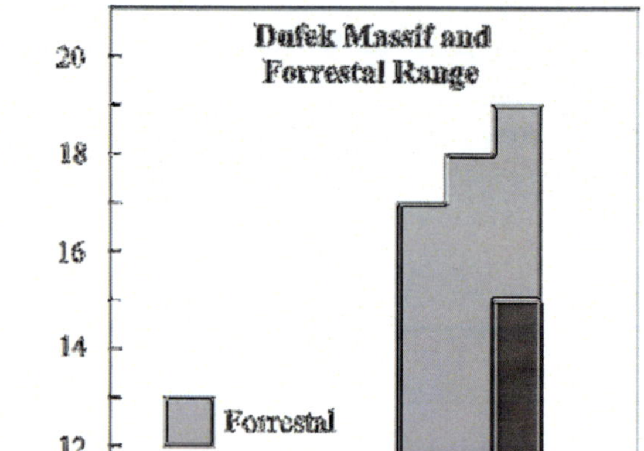

Fig. 13.47 The $\delta^{18}O$ (SMOW) values of whole-rock samples of the Dufek intrusion vary widely. The $\delta^{18}O$ values of the lower part of the intrusion in the Dufek Massif range in most cases from greater than +5‰ to less than +7‰ which is "normal" for mantle-derived mafic igneous rocks. However, the $\delta^{18}O$ of the upper part of the intrusion in the Forrestal Range decrease from "normal" values to low values between 0‰ and +1‰. The low $\delta^{18}O$ values indicate that the magma assimilated crustal rocks or that the rocks were altered after crystallization, or both (Data from Ford et al. 1986)

Grabbro: +6.2‰ to 0.0‰
Contact gabbro: +2.6‰ to +2.2‰
Granophyre: +6.0‰ to +3.8‰
Inclusions: +6.2‰ to +0.6‰
Felsic dike: +4.8‰ to +3.6‰

Assuming that $\delta^{18}O$ values between +5.8 and +6.2 are "normal", rocks having low $\delta^{18}O$ values were altered either by contamination of the magma (e.g., contact gabbro) or after crystallization by interaction with ^{18}O-depleted meteoric water in the subsurface, or both.

The Dufek intrusion formed from a very large volume of magma and therefore may have been the source of the magma that formed the sills of the Ferrar Dolerite and the flows of the Kirkpartick Basalt. Elliot et al. (1999) assumed a thickness of 9 km and an area of 6,600 km^2 (Ferris et al. 1998) which yields an estimated volume of 59,400km^3. However, in spite of its impressive size, the volume of the Dufek intrusion is probably *less* than the total volume of the sills of the Ferrar Dolerite in the Transantarctic Mountains which is about 100,000 km^3 (Kyle et al. 1981) or between 110,000 and 170,000 km^3 (Fleming et al. 1997). Therefore, Elliot et al. (1999) considered that the volume of the Dufek intrusion is too small to have been the central source of all Ferrar magmas. A more attractive alternative is that the Ferrar magmas formed at many places in the subcrustal mantle along the trend of the Transantarctic Mountains and did not originate from the chamber of the Dufek intrusion which is located at the far end of the Ferrar magmatic province. The areal extent and volume of the Dufek intrusion remain in dispute to be settled in the future.

13.4.5 Natural Resources

The potential mineral resources of the Transantarctic Mountains, including metallic and radioactive minerals as well as coal in the Beacon Supergroup, were discussed in Section 11.5. We now turn to the Dufek intrusion which contains several important metals: Iron, titanium, chromium, vanadium, copper, cobalt, nickel, and members of the platinum group of elements. Although the Dufek intrusion is an attractive target for mineral exploration, Ford (1990) emphasized that no minable deposits of any kind are currently known to exist there. The scientific merits of drilling the Dufek intrusion were discussed in 1979 by the Committee on Energy and Natural Resources of the US. Senate, but these discussions were terminated because of concern about contamination of the environment.

The most abundant ore mineral in the Dufek intrusion is magnetite which also contains titanium and vanadium (Rowley et al. 1983). The magnetite occurs in cumulate layers that also contain disseminated grains of sulfide minerals including pyrite, chalcopyrite, pyrrhotite, and perhaps bornite (Walker 1961; Drinkwater et al. 1985; Ford 1990).

The layers of magnetite in the Dufek Massif are only a few centimeters thick and occur primarily in the upper 200 m of the Aughenbaugh Gabbro. Magnetite is much more abundant in the Saratoga Gabbro of the Forrestal Range where the modal concentration of this mineral ranges from 2% to 12% and where the magnetite layers are up to 1 m thick. The magnetite layers contain ferrian ilmenite and vanadium tri-oxide (Himmelberg and Ford 1977; Ford 1990).

The rocks of the Forrestal Range have a more complex mineralogy than the rocks of the Dufek Massif because all specimens analyzed by Ford et al. (1983) contain varying concentrations of oxide (2–34%), pyroxene (11–46%), and plagioclase (14–75%) in addition to trace amounts of apatite and visible sulfide especially in the upper part of the section. The concentrations of sulfur and of eight metals in 24 whole-rock samples from measured sections of the Dufek Massif and of the Forrestal Range reported by Ford et al. (1983) are listed in Appendix 13.6.9B and are summarized in Table 13.5. In most cases, the average concentrations of metals in the rocks of the Forrestal Range are higher than in the Dufek Massif. The only exceptions are chromium and nickel which have higher average concentrations in the Dufek Massif. The three most abundant metals (excluding titanium and iron) in the rocks of the Dufek Massif are vanadium (198 ppm), chromium (174 ppm), and nickel (76 ppm), whereas the most abundant metals in the Forrestal Range are vanadium (691 ppm), copper (330 ppm), and cobalt (61 ppm). Sulfur is strongly correlated with copper

Table 13.5 Average concentrations of metals in whole-rock samples of the Dufek intrusion in the Dufek Massif and in the Forrestal Range (Data from Ford et al. 1983)

	Dufek M.	Forrestal R.
Metal	13 samples	11 samples
Pt (ppb)	13 (3)[a]	22 (4)[a]
Pd (ppb)	8 (2)	16 (5)
S (ppm)	101 (13)	677 (10)
Cu (ppm)	29 (13)	330 (11)
V (ppm)	198 (13)	691 (11)
Co (ppm)	50 (13)	61 (11)
Cr (ppm)	174 (13)	25 (8)
Ni (ppm)	76 (13)	51 (9)
TiO_2 (%)	0.33 (13)	3.72 11)

[a]The number of samples with measurable concentrations is indicated in parentheses.

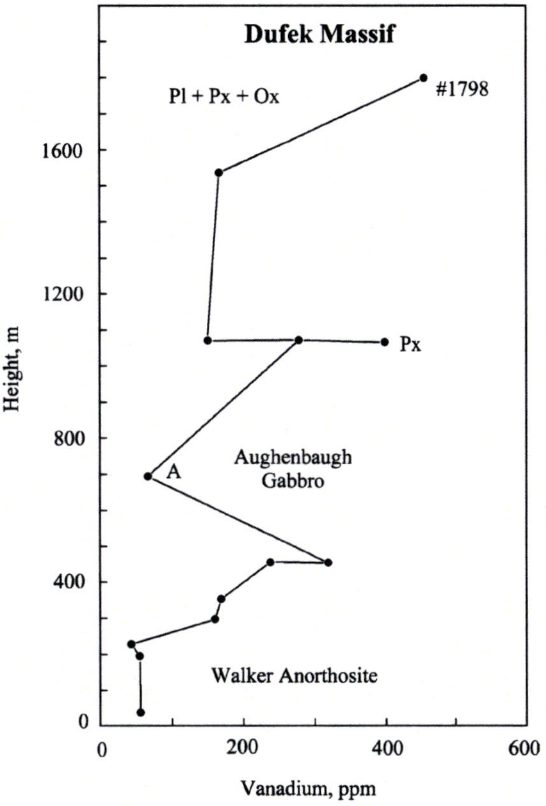

Fig. 13.48 The concentrations of vanadium in the rocks of the Dufek Massif are closely tied to the stratigraphic variation of the mineral composition. Pyroxene and oxide minerals have high concentrations of vanadium, whereas plagioclase contains less than 100 ppm of vanadium. Quantitative estimates of the vanadium concentrations in the text are based on Fig. 13.49 and calculations in Appendix 13.6.10 (Data by Ford et al. 1983)

(r = 0.9946) but more weakly with nickel (r = 0.5808) and with cobalt (r = 0.3550). Vanadium is related to TiO_2 (r = 0.7747) but also seems to follow sulfur (r = 0.6318), perhaps because the sulfide minerals are associated with titaniferous magnetite.

The platinum-group elements that were determined by Ford et al. (1983; Ford 1983a, b) have low concentrations, expressed in nanograms per gram or parts per billion (ppb), and are below the limit of detection in most samples. For example, rhodium (12 ppb) was measureable in only one out of 24 samples, whereas palladium and platinum concentrations were each measured in 7 of 24 samples. All three platinum group metals are more abundant in the Saratoga Gabbro of the Forrestal Range than in the Aughenbaugh Gabbro of the Dufek Massif.

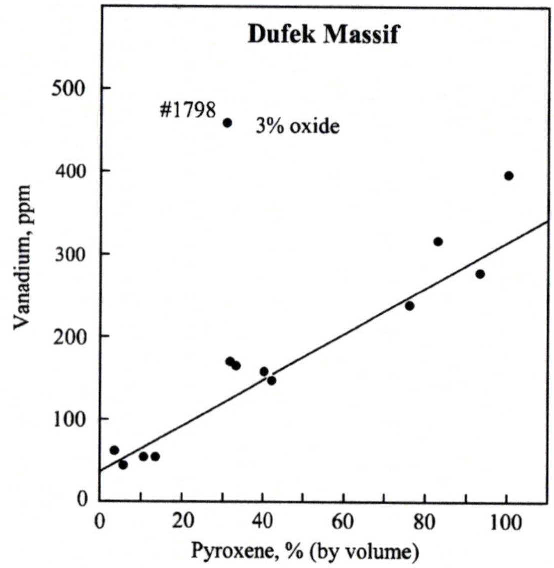

Fig. 13.49 The concentration of vanadium in the rocks of the Dufek Massif increases linearly with increasing modal concentrations of pyroxene. The relationship is expressed by Eq. 13.1 in the text. The highest sample in the stratigraphy of the Dufek intrusion at 1,798 m has an excessive vanadium concentration of 460 ppm because it contains titaniferous magnetite at a modal concentration of 3% by volume. The concentration of vanadium in the oxide is 0.63% by weight (Appendix 13.6.10) (Data by Ford et al. 1983)

The vanadium concentrations of the *Dufek intrusion* in Fig. 13.48 vary stratigraphically depending primarily on the modal concentration of pyroxene in the host rocks. This relation in Fig. 13.49 is expressed by the linear equation:

$$V = 356 + 317(\text{pyr}) \quad (13.2)$$

where V = vanadium concentration in ppm, pyr = modal pyroxene concentration in volume percent, and the linear correlation coefficient r = 0.9639. The intercept at V = 35.6 ppm is the average concentration of vanadium in plagioclase. One sample at 1,798 m, which contains 3% of modal oxide minerals (i.e., titaniferous magnetite) has an exceptionally high vanadium concentration of 460 ppm which indicates that the oxide minerals contain even more vanadium than pyroxenes. Data in Appendix 13.6.9 and the calculation in Appendix 13.6.10 leads to the estimates of the vanadium concentrations of the minerals in Dufek Massif: V_{ox} = 6305 ppm, V_{pyr} = 320 ppm, V_{plag} = 36 ppm.

The concentrations of chromium and nickel in these rocks (Appendix 13.6.10, Fig. 13.50a and b)

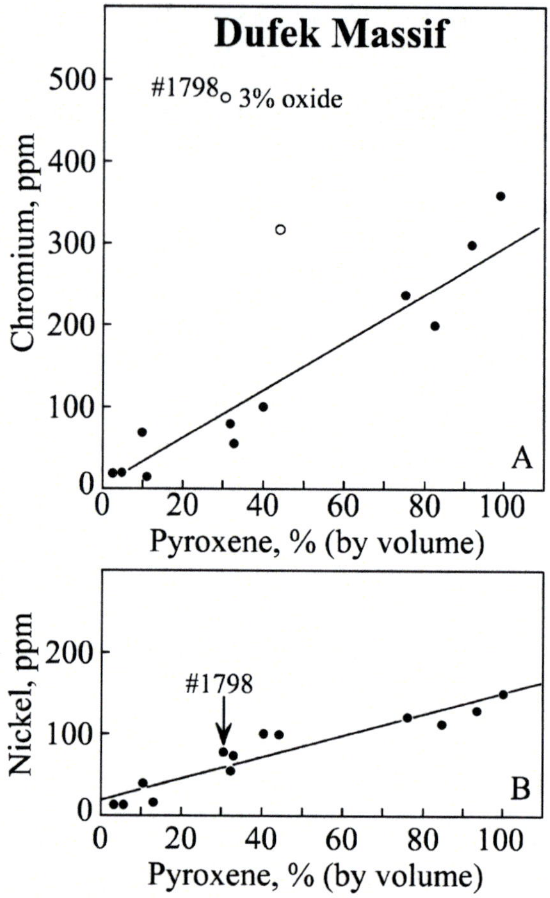

Fig. 13.50 (**a**) The concentrations of chromium in the rocks of the Dufek Massif are linearly related to the modal concentration of pyroxene (solid circles). Two samples deviate from the trend one of which (1,798 m) contains oxide minerals which contain a higher concentration of chromium than pyroxene. The chromium concentration of the oxide minerals in this sample is 7,100 ppm or 0.71% by weight. (**b**) The concentration of nickel in the rocks of the Dufek Massif also rise with increasing modal concentrations of pyroxene. Sample #1798 does not deviate from the array because the oxide minerals in this rock sample do not contain nickel (Data from Ford et al. 1983)

likewise rise with the modal pyroxene concentrations. Chromium, like vanadium, is concentrated in the oxide minerals but nickel is not.

The presence of sulfide minerals in the Saratoga Gabbro of the *Forrestal Range* is reflected by higher average concentrations of sulfur, copper, and cobalt, whereas the greater abundance of oxide minerals has enriched these rocks in vanadium and TiO_2 (Table 13.5). Figure 13.51 demonstrates that copper is concentrated in layers composed of pyroxene + oxide +

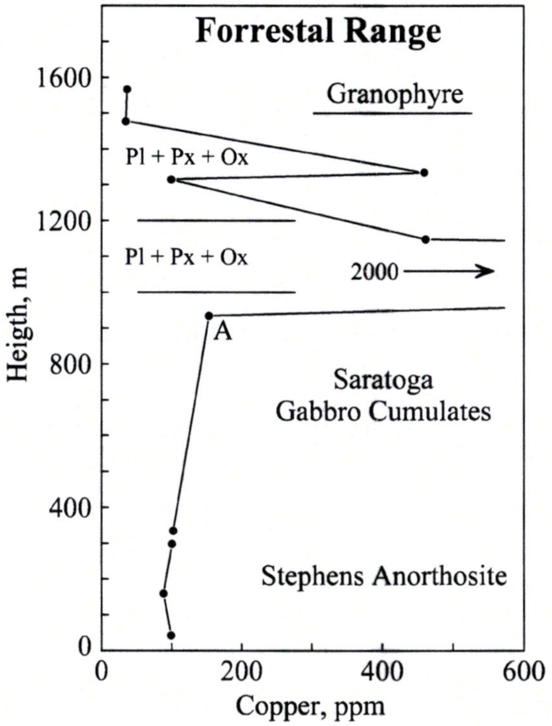

Fig. 13.51 The concentrations of copper in the Forrestal Range vary stratigraphically and reach high values in two layers at 1,080 m (Cu = 2,000 ppm) and at 1,343 m (Cu= 460 ppm) both of which have high modal pyroxene concentrations (71% at 1,080 m; 46% at 1,343 m, respectively). The copper concentrations of the Stephens Anorthosite and the Lexington Granophyre are less than 100 ppm. The association between copper and pyroxene is caused by the association of sulfide minerals with layers of pyroxenite in the Forrestal Range (Data from Ford et al. 1983)

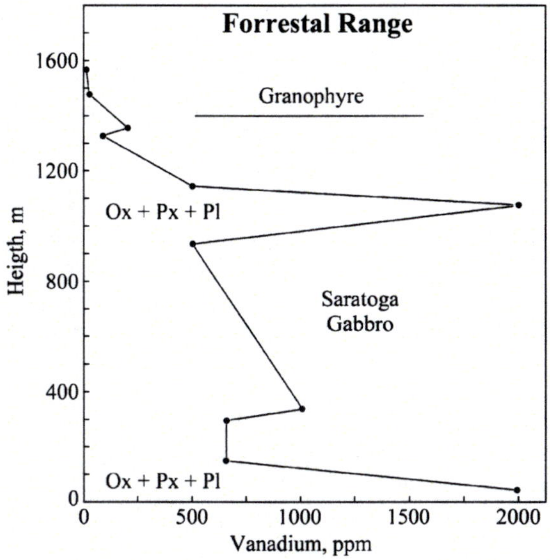

Fig. 13.52 Two layers in the Forrestal Range that are composed of cumulates of oxide minerals, pyroxene, and plagioclase have high concentrations of vanadium of 2,000 ppm. The basal pyroxenite + oxide layer also contains elevated concentrations of cobalt, nickel and TiO_2, but sulfur, copper, and chromium in the basal layer are not enriched (Appendix 13.6.9). In general, the concentrations of metals in the rocks of the Forrestal Range do not correlate with the modal concentrations of pyroxene and oxide minerals in contrast to the Dufek Massif (e.g., Fig. 13.49) (Data from Ford et al. 1983)

plagioclase. The highest concentrations of copper and sulfur (Cu = 2,000 ppm; S = 4,700 ppm) occur in rocks at 1,080 m which contain 71% pyroxene, 18% oxides, and 9% plagioclase by volume (Ford et al. 1983). The geochemical coherence of copper and sulfur in the rocks of the Forrestal Range is expressed by an equation that was derived by regression of the data in Appendix 13.6.9:

$$\mathrm{Cu} = 32.0 + 0.4206\,\mathrm{S} \qquad (13.3)$$

where Cu and S are concentrations in ppm and the linear correlation coefficient r = 0.9958. The layers of pyroxene + oxide + plagioclase of the Forrestal Range in Fig. 13.52 are enriched in vanadium as well as in TiO_2, nickel, and cobalt (not shown). However, the average chromium and nickel concentrations of the Forrestal Range in Table 13.5 are lower than those of the Dufek Massif. Rock layers that are enriched in metals are known in other stratified gabbro intrusions including the Bushveld Complex (Merensky Reef; Kruger and Marsh 1982) and the gold-bearing sulfides in the Skaergaard Intrusion of East Greenland (Bird et al. 1986, 1988).

In general, the distribution of metals (Fe, Ti, Cr, V, Cu, Co, Ni, PGEs) as well as of sulfur in the Dufek intrusion appears to depend on the mineral composition of the host rocks. The highest concentrations occur in layers composed of pyroxene-oxide-plagioclase cumulates, whereas the lowest concentrations of metals and sulfur occur in plagioclase cumulates. The stratigraphic distribution of metals in the rocks of the Forrestal Range is not related to the mineral composition of the rocks as clearly as in the Dufek Massif. The distribution of metals in other layered intrusion (e.g., the Skaergaard intrusion of East Greenland) was affected by hydrothermal redistribution after crystallization of the minerals (Faure 2001).

13.5 Summary

The sills of *Ferrar Dolerite* permeate the rocks of the Beacon Supergroup in large areas of the Transantarctic Mountains. The intrusion of these sills initiated the uplift of the Transantarctic Mountains by as much or more than 1,000 m. The magma that was intruded crystallized to form plagioclase, pyroxene, and oxide minerals which caused the remaining liquid to be enriched in silica and alkali metals. Crystals of the orthopyroxene *hypersthene* were segregated during the rapid injection of the magma and were concentrated in the centers of thick sills in southern Victoria Land. Chemical profiles reveal that the cores of these sills have high concentrations of MgO leading upward to pegmatoid zones. The age of the sills is 176.4 ± 0.5 Ma based on the $^{40}Ar/^{39}Ar$ plateau dates of the Kirkpatrick Basalt which is the extrusive equivalent of the intrusive sills. The initial $^{87}Sr/^{86}Sr$ ratios of the Ferrar Dolerite sills range from 0.708 to 0.712 and are higher than those of mantle-derived basalt in the ocean basin and on the continents. In addition, certain trace-element ratios of the dolerite resemble the values of these ratios in granitic rocks of the continental crust. The crustal character of the Ferrar Dolerite indicates that the magma formed by decompression melting along deep crustal rifts that penetrated into the subcrustal mantle which contained subducted oceanic sediment.

The *Dufek intrusions* in the Dufek Massif and in the Forrestal Range of the northern Pensacola Mountains is a large tabular body of layered gabbro second in volume only to the Paleoproterozoic Bushveld Complex of southern Africa. The rocks of the Dufek intrusion are cumulates that form a thick sequence of layers of gabbro interbedded with anorthosite, pyroxenite, magnetite, and capped by a thick layer of light-colored granophyre. The lower part of the intrusion in the Dufek Massif extends for about 75 km across the Sallee Snowfield to the Forrestal Range where the upper part of the intrusion is exposed.

The Dufek intrusion is Middle Jurassic in age and resembles the Ferrar Dolerite in its chemical composition and its initial $^{87}Sr/^{86}Sr$ ratios. The layers of pyroxenite and magnetite contain not only iron and titanium, but also vanadium, chromium, as well as platinum and palladium in low concentrations. The layers composed of pyroxene and titaniferous magnetite also contain disseminated sulfides which contain copper, cobalt, and nickel.

13.6 Appendices

13.6.1 Mineralogical Types of Ferrar Dolerite Sills in Southern Victoria Land (Gunn 1966)

Name/Location	Thickness (m)	Country Rock
Hypersthene sills		
Vanda sill, 1,500 m above the floor of Wright Valley	332	Granite and gneiss, basement complex
Basement sill, Salina Pond, 410 m beneath the Vanda sill, Wright Valley	300	Granite
Basement sill, Solitary Rocks, upper Taylor Glacier	262	Irizar Granite
Basement sill, west end of Kukri Hills, lower Taylor Glacier	253	Irizar Granite
Terra Cotta dome, at the base of Mt. Terra Cotta, upper Ferrar Glacier	Not provided	Granite
Emmanuel sill, on east side of Emmanuel Glacier tributary of Ferrar Glacier	430	Irizar Granite
Finger Mountain, upper Taylor Glacier	280	Not provided
Pigeonite sills		
Peneplain sill, Kukri Hills, western end of the range	433	Beacon Sandstone above the sill
Peneplain sill, Solitary Rocks, upper Taylor Glacier, cut by granophyre dike, 50 cm wide	320	Beacon Sandstone above the sill
Finger Mountain sill, upper Taylor Glacier	150	Beacon Sandstone above the sill
New Mountain sill, west side of upper Taylor Glacier	207	Beacon Sandstone above, granite beneath
Northwest Mountain sill, caps the mountain, west end of the Asgard Range	430	Beacon Sandstone beneath
Olivine sills		
Thumb Point sill, Prince Albert Mtns., north of Mawson Glacier	420	Base covered, top eroded
Painted Cliffs sill, near Beardmore Glacier. Grindley (1963)	Not provided	Not provided

13.6.2 Chemical Analyses of Dolerite Sills on Roadend Nunatak, Southern Victoria Land

	Basement	Peneplain	Peneplain	Peneplain
Height above base (m)	91.4	475.5	512.1	585.2
Major elements (%)				
SiO_2	52.0	54.5	54.9	53.6
TiO_2	0.38	0.81	0.72	0.58
Al_2O_3	15.2	14.2	14.5	15.5
Fe_2O_3[a]	7.93	10.5	10.0	9.14
MnO	0.14	0.16	0.16	0.14
MgO	9.89	5.05	5.78	5.96
CaO	12.3	9.03	9.65	10.9
Na_2O	1.29	1.97	1.88	1.82
K_2O	0.36	1.21	1.16	0.86
P_2O_5	0.06	0.13	0.12	0.10
LOI[b]	0.70	1.31	1.08	0.93
Sum	100.4	99.0	100.0	99.6
Trace elements (ppm)				
Chromium	666	110	151	120
Yttrium	18	30	23	10
Zirconium	31	120	115	80
Niobium	10	10	16	20
Barium	160	280	263	200

[a]Total iron as Fe_2O_3.
[b]Loss on ignition (LOI).
Analyses by x-ray fluorescence by X-Ray Assay Lab, Don Mills, Ontario, Canada.

13.6.3 Rb-Sr Systematics of the Dolerite Sills on Roadend Nunatak at the Confluence of the Touchdown and Darwin Glaciers, Southern Victoria Land

Height above base (m)	Rb (ppm)	Sr (ppm)	$^{87}Rb/^{86}Sr$	$^{87}Sr/^{86}Sr$[a]	$\delta^{18}O$[b] (‰)
Peneplain sill					
603.5	36.78[c]	120.62[c]	0.8828	0.71404	–
585.2	28.56	125.10	0.6609	0.71335	+6.3
566.9	28.90	122.78	0.6814	0.71327	–
548.6	28.62	122.30	0.6774	0.71340	–
530.4	35.08	126.54	0.8025	0.71358	+6.6
512.1	37.58	122.50	0.8881	0.71400	+7.0
493.8	25.52	126.76	0.5828	0.71342	+5.0
475.5	40.80	121.80	0.9698	0.71427	–
455.4	54.42	142.48	1.1062	0.71809(?)	+4.7
Basement sill[d]					
150.0	12.24	100.40	0.3528	0.71021	+6.0
146.3	13.84	100.32	0.3992	0.71015	–
128.0	14.30	106.54	0.3884	0.71044	–
109.7	15.64	95.24	0.4753	0.71183	+6.6
91.4	12.16	89.04	0.3952	0.71095	+7.1
73.2	16.50	79.72	0.5990	0.71183	+6.5
54.9	11.68	79.08	0.4275	0.71176	+6.8
36.6	13.54	76.46	0.5125	0.71202	–
19.8	12.86	58.60	0.6352	0.71214	–
0	5.86	74.78	0.2268	0.71094	–

[a]Measured by L.M. Jones. The average of four replicate determinations of the Eimer and Amend isotope standard is $^{87}Sr/^{86}Sr = 0.708053 \pm 0.000037$ (1σ). The $^{87}Sr/^{86}Sr$ ratio of the NBS 987 isotope standard is 0.71024 ± 0.00001.
[b]Measured by J. Hoefs relative to SMOW.
[c]Rubidium and strontium were measured by Kibler (1981) by x-ray fluorescence.
[d]The section was measured by G. Faure and R.P. Felder during the 1978/79 field season.

13.6.4 Major-Element Analyses of Whole-Rock Samples, Ferrar Dolerite Sills, Mt. Achernar, Queen Alexandra Range, in Percent by Weight

Height above base (m)	SiO_2	Al_2O_3	CaO	MgO	Na_2O	K_2O	Fe_2O_3	MnO	TiO_2	P_2O_5	LOI	Sum
Sill # 1 (lowest)												
57 (top)	53.3	14.3	10.0	5.53	1.89	0.85	10.8	0.17	0.80	0.14	2.00	99.9
48	54.1	14.1	9.43	4.55	2.13	1.16	11.6	0.17	0.95	0.17	1.70	100.1
39	54.8	14.1	8.81	3.54	2.21	1.35	11.8	0.17	1.03	0.18	1.62	99.7
27	53.3	14.9	10.8	5.53	1.85	0.77	10.5	0.16	0.72	0.12	1.62	100.3
21	54.8	14.6	9.93	4.90	1.98	0.99	10.4	0.17	0.72	0.14	1.54	100.3
15	54.7	13.9	9.12	4.57	2.08	1.18	11.6	0.17	0.98	0.18	1.62	100.6
6	53.2	14.9	11.5	6.37	1.69	0.71	9.54	0.16	0.58	0.10	1.54	100.4
0 (bottom)	52.4	16.1	12.7	7.09	1.59	0.52	8.10	0.14	0.44	0.08	1.23	100.5
Sill # 2												
126 (top)	51.5	15.8	7.95	5.14	1.88	1.11	10.7	0.25	0.74	0.13	4.54	99.8
124.5	53.9	14.8	10.5	5.65	1.90	0.80	10.1	0.16	0.68	0.12	1.54	100.2
123	53.7	14.7	10.7	5.75	1.83	0.78	9.96	0.16	0.66	0.12	1.31	99.7
111	53.6	15.0	11.0	5.96	1.82	0.79	9.62	0.16	0.64	0.12	1.31	100.1
105	52.6	15.8	11.8	6.52	1.82	0.66	8.84	0.15	0.53	0.09	1.00	99.9
93	55.1	13.6	8.94	4.06	2.12	1.08	12.4	0.19	0.93	0.17	1.70	100.4
81	54.0	14.3	10.1	5.48	2.01	0.89	10.6	0.17	0.72	0.13	1.62	100.1
69	52.2	15.5	11.2	6.03	1.76	0.70	9.18	0.15	0.60	0.10	1.77	99.3
57	52.5	15.5	11.2	6.08	1.83	0.74	9.16	0.15	0.59	0.10	1.39	99.3
45	53.2	15.5	10.9	6.02	1.94	0.79	9.54	0.16	0.61	0.11	1.54	100.4
27	53.5	15.0	10.3	6.39	2.05	0.87	9.62	0.16	0.64	0.11	1.16	99.9
12	53.5	15.1	10.7	6.16	1.92	0.81	9.77	0.16	0.65	0.11	1.16	100.1
4.5	53.9	14.9	10.6	5.68	1.92	0.73	9.99	0.16	0.67	0.12	1.62	100.4
3	53.7	14.9	10.7	5.81	1.94	0.75	9.82	0.17	0.66	0.12	1.70	100.3
2.25	53.3	16.0	7.34	5.08	2.75	1.09	9.51	0.17	0.73	0.13	3.70	99.9
0 (bottom)												
Sill # 3 (highest)												
64.5 (top)	51.9	18.7	11.7	4.37	1.94	0.74	8.69	0.13	0.58	0.08	1.23	100.0
51	52.5	16.4	11.8	6.48	1.70	0.70	8.90	0.15	0.54	0.08	1.08	99.8
39	52.1	15.8	11.6	6.80	1.56	0.68	9.22	0.15	0.56	0.08	1.08	100.5
30	52.3	16.8	11.9	6.61	1.55	0.67	8.62	0.14	0.52	0.08	1.16	100.2
27	52.4	14.4	10.6	8.64	1.48	0.67	9.73	0.16	0.55	0.08	0.85	100.1
21	52.1	14.2	10.6	10.7	1.35	0.57	9.43	0.17	0.47	0.07	0.77	100.3
15	51.8	15.2	10.7	10.0	1.36	0.54	9.18	0.15	0.47	0.08	0.85	100.0
12	51.7	15.5	10.7	8.94	1.48	0.67	9.52	0.15	0.55	0.08	0.93	99.7
9	51.8	15.5	10.8	8.26	1.52	0.68	9.61	0.16	0.56	0.08	1.08	100.3
6	52.1	15.4	10.9	8.36	1.41	0.67	9.67	0.16	0.56	0.08	1.16	99.8
3	51.9	15.3	10.8	8.23	1.45	0.67	9.49	0.16	0.55	0.08	1.00	100.3
0 (bottom)	52.2	15.3	10.8	7.85	1.45	0.68	9.61	0.16	0.57	0.08	1.08	100.0

Analyses of X-Ray Assay Laboratories, Don Mills, Ontario.

13.6.5 Rb-Sr Systematics of the Sills of Ferrar Dolerite on Mt. Achernar, Queen Alexandra Range (84°12′S, 160°56′E)

Height above base (m)	Rb (ppm)	Sr (ppm)	$^{87}Rb/^{86}Sr$	$^{87}Sr/^{86}Sr$	$\delta^{18}O$ (SMOW) (‰)
Sill # 1 (lowest)					
57 (top)	25.33	162.2	0.4522	0.71006	+4.35
54	26.77	165.5	0.4682	0.71005	+4.72
48	34.22	180.7	0.5482	0.71048	+5.73
45	36.67	218.1	0.4867	0.71071	–
39	41.01	163.6	0.7253	0.71061	+5.75
27	24.74	157.3	0.4552	0.70982	–
21	31.91	164.7	0.5607	0.71012	+5.20
15	36.52	161.2	0.6558	0.71053	–
12	26.82	142.4	0.5452	0.70983	+6.47
6	22.22	141.5	0.4545	0.70961	–
3	17.74	136.5	0.3762	0.70949	+5.16
0 (bottom)	15.56	137.8	0.3268	0.70932	+4.44
Sill # 2 (middle)					
126 (top)	32.16	179.3	0.5190	0.70877	
123	22.13	160.0	0.4002	0.70897	
105	77.36	130.8	1.712	0.71159	
99	31.52	151.4	0.6023	0.70866	
93	32.19	155.1	0.6005	0.70963	
51	23.17	159.9	0.4193	0.70918	
27	29.06	167.0	0.5034	0.70936	
21	24.13	187.2	0.3730	0.70900	
12	23.75	152.2	0.4517	0.70924	
6	20.77	156.5	0.3841	0.70887	
3	21.79	160.9	0.3919	0.70926	
2.25	33.91	246.8	0.3975	0.70858	
0 (bottom)					
Sill # 3 (highest)					
64.5 (top)	26.75	142.6	0.5430	0.71154	
51	26.56	122.6	0.6269	0.71186	
39	25.49	117.0	0.6307	0.71188	
30	24.28	122.8	0.5722	0.71176	
27	23.69	103.1	0.6648	0.71198	
21	20.13	101.9	0.5717	0.71168	
15	20.58	112.6	0.5289	0.71145	
12	25.54	145.4	0.5085	0.71078	
9	26.34	185.3	0.4115	0.70983	
6	24.55	110.6	0.6423	0.71171	
3	24.63	113.8	0.6265	0.71151	
0 (bottom)	26.23	113.7	0.6676	0.71168	

The average $^{87}Sr/^{86}Sr$ ratio of four replicate determinations of the Eimer and Amend Sr-isotope standard is 0.708053 ± 0.000037 (1σ).

The $^{87}Sr^{86}Sr$ ratio of NBS 987 is 0.71024 ± 0.00001.

The base of the lowest sill (#1) is not exposed and the upper contact of this sill is covered by scree. At 59 m above the base of sill # 1 the sediment of the Buckley Formation is "baked."

The base of sill # 2 starts at 2.25 m above the zero point. The dolerite between 2.25 and 3.0 m is hydrothermally altered. The rock at 126 m above base is the fine-grained top of the sill.

The base of sill # 3 at 0 m is fine grained. The upper part of sill #3 was removed by erosion.

The section was measured during 1985/86 field season by G. Faure and T. M. Mensing.

The $^{87}Sr^{86}Sr$ ratios were measured by L.M. Jones and the $\delta^{18}O$ values of sill #1 were measured by J. Hoefs.

13.6.6 $\delta^{18}O$ Values of Plagioclase and Pyroxene in Dolerite Samples of Sill # 2 on Mt. Achernar and Estimates of the Isotope Equilibration Temperature

Height above base (m)	Plag. (‰)	Pyr. (‰)	Δ Plag. - Pyr. (‰)	Temperature (°C) (1)	Temperature (°C) (2)
2.25	+3.94	+2.57	1.37	613	655
3.00	+5.15	+4.22	0.93	802	853
4.50	+6.36	+5.81	0.55	1,126	1,192
15	+6.57	+5.59	0.98	775	824
39	+6.87	+6.07	0.80	887	941
99	+5.85	+5.21	0.64	1,023	1,085

1. Bottinga and Javoy (1975).
2. Chiba et al. (1989).
The $\delta^{18}O$ values were measured by Jochen Hoefs, University of Göttingen, Germany.

The geothermometry equation of Bottinga and Javoy (1975) is:

$$\Delta_2^1 = \frac{(1.70-1.04\beta)10^6}{T^2} \quad (13.4)$$

where β is the mole fraction of anorthite in the plagioclase and A = (1.70 – 1.04 β). Assuming that the plagioclase in Sill # 2 on Mt. Achernar is labradorite defined as An_{50} to An_{70}, we set β = 0.6 which yields A = 1.076. Substituting this value into Eq. 13.4 and setting Δ_2^1 = 0.98 (15 m above base) we obtain:

$$0.98 = \frac{1.076\times10^6}{T^2}$$

$$T = \left(\frac{1.076\times10^6}{0.98}\right)^{1/2} = 1047.8K$$

$$T = 1047.8 - 273.15 = 774.6 \sim 775°C$$

The resulting temperature estimates in column (1) range from 613°C to 1126°C. Omitting the lowest and the highest values yields a mean of 872 ± 112°C (1σ)

The oxygen-isotope fractionation factors reported by Chiba et al. (1989) for feldspar-pyroxene are: Albite-Diopside = 1.81, Anorthite-Diopside = 0.76. For labradorite (An = 60%, Ab = 40%) the numerical value of A is:

$$A = 0.6 \times 0.76 + 0.4 \times 1.81 = 1.18$$

and Eq. 13.4 takes the form:

$$\Delta_2^1 = \frac{1.18\times10^6}{T^2}$$

Setting Δ_2^1 = 0.98, yields T = 824°C. Omitting the lowest and highest temperature estimates yields an average of 926 ± 117°C (1σ)

13.6.7 Chemical Analyses of Rock Samples from a Measured Section of the Sill of Ferrar Dolerite on Portal Rock, Queen Alexandra Range (J.M. Hergt personal communication to G. Faure, April 27, 1987)

Sample #	P-1-2S	P-1-3C	P-1-4D	P-1-5D	P-1-6D	P-1-7D	P-1-8D	P -1-10D	P-1-11D	P-12D
Height above base (m)	18	18.5	19	19.5	21	24	27	30	33	36
Major elements (%)										
SiO_2	82.48	50.80	53.49	53.21	53.66	53.43	53.78	54.40	54.41	53.57
TiO_2	0.46	1.12	0.77	0.74	0.74	0.73	0.73	0.71	0.71	0.70
Al_2O_3	7.99	21.45	14.41	14.23	14.35	14.30	14.38	14.58	14.75	14.69
Fe_2O_3	0.50	1.13	2.12	2.04	2.06	2.18	2.32	2.30	2.49	2.08
FeO	2.11	8.37	7.28	7.45	7.56	7.36	7.20	7.13	6.89	7.31
MnO	0.05	0.07	0.13	0.15	0.16	0.15	0.16	0.16	0.16	0.16
MgO	1.20	4.09	6.02	6.04	6.19	6.29	6.35	6.32	6.47	6.50
CaO	0.82	0.70	10.52	10.57	10.63	10.51	10.56	10.49	10.63	10.77
Na_2O	2.98	0.82	2.10	2.07	2.12	2.01	2.06	2.10	2.15	2.05

(continued)

13.6.7 (continued)

Sample #	P-1-2S	P-1-3C	P-1-4D	P-1-5D	P-1-6D	P-1-7D	P-1-8D	P -1-10D	P-1-11D	P-12D
Height above base (m)	18	18.5	19	19.5	21	24	27	30	33	36
K_2O	0.27	5.44	0.74	0.78	0.83	0.82	0.86	0.91	0.85	0.83
P_2O_5	0.18	0.23	0.13	0.12	0.12	0.12	0.12	0.12	0.11	0.12
S	0.01	0.01	0.02	0.01	0.01	0.01	0.01	0.00	0.03	0.02
H_2O+	1.30	4.93	1.51	1.35	1.17	1.10	1.07	1.06	0.92	1.13
H_2O–	0.11	0.51	0.56	0.50	0.39	0.29	0.35	0.35	0.27	0.36
CO_2	0.09	0.11	0.09	0.08	0.11	0.10	0.10	0.12	0.10	0.14
Total	100.55	99.78	99.89	99.34	100.10	99.40	100.05	100.75	100.95	100.43
Trace elements (ppm)										
Rb	15	307	20	23.5	25.5	27	28.5	30	27.5	29
Sr	232	355	142	145	142	145	153	144	152	146
Pb	20	38	5	5	6	4	5	4	4	5
Y	28	37	22	22	21	19	20	20	19	20
Zr	214	112	105	101	99	98	98	97	92	86
Nb	8	20	5	5	5	4.5	5	5	5	5
Ga	7	32	16.5	17	16	16.5	16.5	16	16.5	16

13.6.7 (continued) Part 2

Sample #	P-1-13D	P-1-14D	P-1-15D	P-1-16D	P-1-17D	P-1-18D	P-1-19D	P-1-20D	P-1-21D
Height above base (m)	39	42	45	48	51	54	57	60	63
SiO_2 (%)	54.13	53.49	54.13	53.95	53.38	54.00	54.41	53.35	53.81
TiO_2	0.69	0.69	0.68	0.68	0.69	0.70	0.66	0.69	0.75
Al_2O_3	14.79	14.54	14.82	15.01	14.93	14.67	14.83	14.89	14.21
Fe_2O_3	1.80	2.11	2.00	1.91	1.96	2.07	1.72	1.77	1.88
FeO	7.39	7.16	7.12	7.13	7.24	7.34	7.49	7.47	7.84
MnO	0.16	0.16	0.15	0.16	0.16	0.16	0.15	0.15	0.16
MgO	6.59	6.62	6.60	6.66	6.53	6.45	6.56	6.28	6.65
CaO	10.95	10.81	10.88	11.10	11.05	10.77	10.95	10.90	10.66
Na_2O	2.03	2.09	2.12	2.01	2.03	2.03	2.10	2.12	2.00
K_2O	0.82	0.76	0.78	0.75	0.76	0.86	0.72	0.75	0.81
P_2O_5	0.11	0.12	0.11	0.12	0.11	0.11	0.11	0.13	0.15
S	0.02	0.04	0.01	0.01	0.03	0.02	0.01	0.02	0.01
H_2O+	0.96	1.05	1.03	0.96	0.96	1.11	0.85	0.85	0.97
H_2O-	0.34	0.35	0.31	0.34	0.40	0.39	0.33	0.31	0.35
CO_2	0.19	0.18	0.08	0.13	0.11	0.16	0.09	0.13	0.10
Total	100.97	100.17	100.82	100.92	100.34	100.84	99.98	9.81	100.35
Rb (ppm)	26.5	24	24	24	25	29.5	21.5	21.5	26
Sr	148	149	144	143	145	148	145	145	144
Pb	5	5	4	5	4	4	4	4	5
Y	19	19	19	19	19	19	18	19	22
Zr	92	87	88	86	86	92	82	89	96
Nb	4	5	4	4.5	4	4	3.5	5	4.5
Ga	16	16	16	16.5	16	16.5	17	17	16.5

13.6.7 (continued) Part 3

Sample #	P-1-22D	P-1-23D	P-1-24D	P-1-25D	P-1-26D	P-1-27D	P-1-28D	P-1-29D	P-1-30D
Height above base (m)	66	69	72	75	78	81	84	87	90
SiO_2 (%)	53.18	52.97	53.28	52.83	53.29	53.80	53.33	52.88	52.34
TiO_2	0.74	0.72	0.69	0.71	0.68	0.65	0.66	0.64	0.58
Al_2O_3	14.79	14.88	14.69	14.59	14.41	15.12	14.76	14.78	15.47
Fe_2O_3	2.01	1.99	2.13	1.96	2.03	1.82	1.75	1.61	1.52
FeO	7.48	7.50	7.37	7.43	7.45	7.30	7.43	7.32	6.91
MnO	0.16	0.16	0.16	0.15	0.16	0.15	0.16	0.15	0.14
MgO	6.23	6.26	6.39	6.37	6.62	6.64	6.72	6.96	7.05
CaO	10.65	10.80	10.70	10.77	10.86	11.12	11.11	11.18	11.74
Na_2O	2.17	2.14	2.13	2.06	2.02	2.05	2.02	1.97	1.88
K_2O	0.78	0.75	0.78	0.79	0.74	0.71	0.68	0.72	0.54
P_2O_5	0.14	0.12	0.11	0.11	0.12	0.10	0.10	0.10	0.09
S	0.00	0.00	0.04	0.06	0.03	0.00	0.00	0.00	0.01
H_2O+	0.95	1.02	0.98	0.91	0.87	0.99	1.06	1.02	1.49
H_2O-	0.31	0.39	0.44	0.39	0.40	0.40	0.32	0.48	0.59
CO_2	0.07	0.10	0.15	0.19	0.13	0.11	0.07	0.09	0.32
Total	99.66	99.80	100.04	99.32	99.81	100.96	100.17	99.80	100.67
Rb (ppm)	24	22.5	25	24.5	22	21	19.5	21.5	15.5
Sr	150	150	149	148	146	145	144	145	146
Pb	4	5	4	5	4	4	4	4	3
Y	20	19	19	18	19	17	17	18	16
Zr	94	89	87	88	88	82	80	83	65
Nb	4	4.5	4	4.5	5	4	4	3.5	4
Ga	16.5	17	17	16	17	16.5	16.5	16	16

13.6.7 (continued) Part 4

Sample #	P-1-31D	P-1-32D	P-1-33D	P-1-34D	P-1-35D	P-1-36D	P-1-37D	P-1-38D	P-1-39-D
Height above base (m)	93	96	99	102	105	111	117	123	129
SiO_2 (%)	52.69	53.65	53.29	53.60	51.94	51.89	53.53	51.96	54.20
TiO_2	0.55	0.61	0.62	0.68	0.55	0.71	0.62	0.36	0.79
Al_2O_3	15.59	15.44	15.13	14.55	21.64	14.39	14.58	15.35	14.12
Fe_2O_3	1.40	1.56	1.49	1.90	1.71	2.47	1.82	1.23	2.55
FeO	7.08	7.05	6.98	6.97	4.53	6.67	7.40	6.86	7.74
MnO	0.15	0.15	0.15	0.15	0.10	0.16	0.16	0.15	0.17
MgO	7.37	7.28	7.20	7.14	3.12	6.58	6.98	8.56	5.77
CaO	11.37	11.47	11.66	11.22	12.16	12.27	11.17	12.83	10.20
Na_2O	1.89	1.93	1.87	1.86	2.42	1.94	1.95	1.75	2.23
K_2O	0.60	0.65	0.68	0.69	0.83	0.60	0.67	0.43	0.89
P_2O_5	0.09	0.09	0.10	0.11	0.09	0.12	0.10	0.05	0.12
S	0.05	0.00	0.01	0.00	0.01	0.00	0.02	0.00	0.00
H_2O+	0.88	0.82	0.98	0.91	0.93	2.03	1.01	0.81	1.07
H_2O-	0.35	0.34	0.31	0.32	0.20	0.57	0.30	0.19	0.35
CO_2	0.22	0.05	0.13	0.08	0.15	0.11	0.13	0.12	0.10
Total	100.28	101.09	100.60	100.18	100.38	100.51	100.44	100.65	100.30
Rb (ppm)	17.5	19	21.5	21.5	31.5	17	21	15	28.5
Sr	144	139	141	136	206	135	144	143	160
Pb	3	4	4	4	4	3	4	2	5
Y	14	16	17	19	14	19	18	10	21
Zr	69	72	77	85	71	78	79	35	96
Nb	2.5	4	4.5	4.5	4	5.5	3.5	2	5.5
Ga	15	16	16	15	19	16	15.5	14.5	17.5

13.6.7 (continued) Part 5

Sample #	P-1-40D	P-1-41D	P-1-42D	P-1-43C	P-1-44C	P-1-45S	P-1-46S
Height above base (m)	135	141	145.5	147	148	149.5	150
SiO_2 (%)	53.59	54.61	54.27	54.41	54.21	63.17	62.77
TiO_2	0.62	0.94	0.78	0.82	0.78	0.73	0.58
Al_2O_3	15.09	13.65	14.33	14.22	14.31	16.15	12.07
Fe_2O_3	2.19	2.91	2.13	2.11	1.74	0.68	0.52
FeO	6.91	8.25	7.66	8.02	8.11	5.11	2.95
MnO	0.15	0.17	0.16	0.16	0.16	0.06	0.54
MgO	6.50	5.18	5.86	5.80	6.05	2.79	2.19
CaO	10.99	9.76	10.43	10.31	10.46	0.67	11.06
Na_2O	2.00	2.19	2.04	2.13	2.11	1.36	2.50
K_2O	0.84	1.05	0.80	0.78	0.81	4.48	3.42
P_2O_5	0.11	0.15	0.13	0.14	0.13	0.18	0.17
S	0.00	0.00	0.01	0.03	0.02	0.00	0.01
H_2O+	0.99	1.17	1.09	1.21	1.31	3.37	1.12
H_2O-	0.40	0.36	0.32	0.41	0.39	0.49	0.22
CO_2	0.11	0.13	0.14	0.14	0.16	0.15	0.39
Total	100.49	100.52	100.15	100.69	100.75	99.39	100.51
Rb (ppm)	29.5	31	22.5	20.5	19	209	116
Sr	153	160	152	153	159	239	182
Pb	3	6	4	5	5	26	21
Y	18	25	22	23	22	34	30
Zr	81	119	106	106	101	238	253
Nb	3.5	6.5	5.5	6.5	5	14	10
Ga	16	17	17	17	17	22	14.5

13.6.8 Average Chemical Analyses of the Lexington Granophyre and Other Felsic Differentiates of the Dufek Intrusion in the Forrestal Range and Dufek Massif, in Weight Percent (Ford 1970; Ford and Kistler 1980)

Unit and oxide	A	B	C	D
SiO_2 (%)	53.9	67.0	70.8	73.7
Al_2O_3	15.7	12.3	14.9	13.8
Fe_2O_3	1.52	3.8	0.55	1.7
FeO	8.1	5.1	1.8	0.60
MgO	6.7	0.84	1.4	0.49
CaO	10.4	3.6	2.8	1.2
Na_2O	1.92	2.6	3.2	3.6
K_2O	0.73	3.4	4.2	4.6
TiO_2	0.66	0.86	0.19	0.23
P_2O_5	0.13	0.37	0.09	0.09
MnO	0.13	0.14	0.08	0.05
Sum	99.89	100.01	100.01	100.06

A. Average of 12 samples of little differentiated basalt and dolerite dikes of the Ferrar Dolerite in the Cordiner Peaks less than 15 km southeast of the southernmost nunataks of the Dufek Massif (corrected to $H_2O+ = 0\%$).
B. Average of nine samples of the Lexington Granophyre in the Forrestal Range.
C. Average of three samples of aplite dikes in the Dufek Massif.
D. Average of five samples of aplite dikes in the Forrestal Range.

13.6.9 Modal Concentrations of Minerals in the Rocks of the Forrestal Range and the Dufek Massif (Data from Ford et al. 1983)

	Oxides	Pyroxene	Plagioclase	Ap.[a]	Sulf.[b]
Height (m)	% by volume				
Forrestal Range					
1,564 (top)	2	12	21	tr.[c]	0
1,483	2	11	14	tr.	0
1,343	12	46	42	tr.	tr.
1,320	13	38	46	3	tr.
1,149	11	39	50	tr.	tr.
1,080	18	71	9	tr.	2
939	4	21	70	0	0
340	9	39	52	0	0
298	8	17	75	0	0
165	7	12	61	0	tr.
37	34	31	35	0	tr.
Dufek Massif					
1,798 (top)	3	30	67	0	0
1,540	tr.	32	68	0	0
1,075	0	42	58	0	0
1,070	0	93	7	0	0
1,070	0	99	1	0	0
700	tr.	4	96	0	0
460	0	82	18	0	0
460	tr.	76	24	0	0
380	0	33	67	0	0
300	0	40	60	0	0
230	0	5	95	0	tr.
200	0	13	87	0	tr.
29	tr.	10	90	0	tr.

[a]Ap = apatite.
[b]Sulf. = visible sulfide.
[c]tr. = trace.

13.6.9B Concentrations of Metals in Whole-Rock Samples of the Dufek Intrusion (Ford et al. 1983)

	Pt	Pd	S	Cu	V	Co	Cr	Ni	TiO_2
Height (m)	(ppb)	(ppb)	(ppm)	(ppm)	(ppm)	(ppm)	(ppm)	(ppm)	(%)
Forrestal Range (11 samples)									
1,564			130	32	6	7	2		1.6
1,483			140	36	15	8	2		0.68
1,343				460	200	90		13	4.4
1,320			320	100	80	70	8	2	5.4
1,149	25		800	460	500	85		24	4.8
1,080?	13	6	4,700	2,000	2,000	100	44	65	7.5
939			90	150	500	38	10	32	1.3
340	14	44	170	100	1,000	55		50	2.9
298		11	100	100	650	60	50	50	2.5
166		12	100	90	650	38	65	32	1.2
37	35	6	220	100	2,000	120	20	190	8.6
Average	22(4)	16.(5)	677(10)	330(11)	691(11)	61(11)	25(8)	51(9)	3.72(11)
Dufek Massif (13 samples)									
1,798		10	130	36	460	70	480	80	0.87

(continued)

13.6.9B (continued)

Height (m)	Pt (ppb)	Pd (ppb)	S (ppm)	Cu (ppm)	V (ppm)	Co (ppm)	Cr (ppm)	Ni (ppm)	TiO_2 (%)
1,540			190	36	170	44	80	55	0.39
1,075			40	12	150	70	320	100	0.22
1,070			110	35	280	85	300	130	0.36
1,070	15		90	24	400	90	360	150	0.46
700			70	20	65	10	20	12	0.30
460			160	40	320	65	200	110	0.41
460			130	40	240	80	240	120	0.41
380	15		50	24	170	55	55	75	0.22
300			100	30	160	55	100	100	0.28
230	10		60	26	44	8	20	12	0.06
200			40	20	55	8	14	12	0.21
20		5	140	34	55	14	70	38	0.17
Average	13(3)	8(2)	101(13)	29(13)	198(13)	50(13)	174(13)	76(13)	0.33(13)

13.6.10 Concentrations of Vanadium in the Oxide Minerals of the Dufek Intrusion in the Dufek Massif (Ford et al. 1983)

The vanadium concentrations of whole-rock samples of the Dufek intrusion in the Dufek Massif in Fig. 13.49 form a straight line represented by Eq. 13.4:

$$V = 3.56 + 3.17(\text{pyr}) \tag{13.4}$$

where V is the vanadium concentration of a rock in the Dufek Massif in ppm and (pyr) is the modal concentration of pyroxene in volume percent in that rock. One of the samples (at 1,798 m) in Fig. 13.49 has an anomalously high vanadium concentration presumably because this rock contains oxides as well as pyroxene and plagioclase (Ford et al. 1983).

The vanadium concentration of the oxide minerals can be estimated by means of Eq. 13.5:

$$aV_{ox} + bV_{pyr} + CV_{plag.} = V_{rock} \tag{13.5}$$

where a, b, and c are weighting factors expressed as decimal fractions of the abundances of the minerals in weight percent (i.e., a + b + c = 1.0). The modal concentrations of the minerals in volume percent must be converted to percent by weight because the concentrations of vanadium are expressed in weight units. The conversion is based on Eq. 13.6:

$$\text{Mass} = \text{Volume} \times \text{Density} \tag{13.6}$$

The modal concentrations of the minerals in volume percent in the anomalous rock sample are: Plagioclase = 67%, pyroxene = 30%, and oxides = 3% (Ford et al. 1983). The densities of these minerals are: Plagioclase = 2.70 g/cm^3, pyroxene = 3.30 g/cm^3, and oxides = 5.18 g/cm^3. Therefore, the masses of the minerals and their abundances in this rock are:

Plagioclase: 0.67 × 2.70 = 1.809 g = 61.23 wt. %
Pyroxene: 0.30 × 3.30 = 0.99 g = 33.50 wt. %
Oxides: 0.03 × 5.18 = 0.1554 g = 5.25 wt. %
Sums = 2.9544 g = 99.98 wt. %

The concentration of plagioclase is 1.809 × 100/2.9544 = 61.23 wt. % and similarly for the other components. By substituting into Eq. 13.5 and solving for V_{ox} we obtain: $0.0525\,V_{ox} + 0.335 \times 320 + 0.6123 \times 35.6 = 460$

$$V_{ox} = \frac{460 - 0.335 \times 320 + 0.6123 \times 35.6}{0.0525} = 6305\,ppm$$

or V_{ox} = 0.63% by weight.

The concentrations of chromium (Fig. 13.50a) and nickel (Fig. 13.50b) in the rocks of the Dufek intrusion are also positively correlated with modal pyroxene concentrations and define straight lines:

Cr = –2.06 + 3.11 (pyr),[1] r = 0.9781
Ni = 20.1 + 1.310 (pyr), r = 0.9328

where r = linear correlation coefficient.

[1]Forced through the origin by adding 15 zeros.

References

Abel KD, Himmelberg GR, Ford AB (1979) Petrologic studies of the Dufek intrusion: Plagioclase variation. Antarctic J US 14(5):6–8

Aughenbaugh NB (1961) Preliminary report on the geology of the Dufek Massif. IGY World Data Center, A. Glaciol Glac Rept 4:155–193

Barrett PJ, Elliot DH (1973) Reconnaissance geologic map of the Buckley Island Quadrangle, Transantarctic Mountains, Antarctica. Antarctic geologic map A-3. US Geological Survey, Washington, DC

Barrett PJ, Elliot DH, Lindsay JF (1986) The Beacon Supergroup (Devonian-Triassic) and Ferrar Group (Jurassic) in the Beardmore Glacier area, Antarctica. In: Turner MD, Splettstoesser JF (eds) Geology of the Central Transantarctic Mountains. Antarctic Research Series, vol. 36. American Geophysical Union, Washington, DC, pp 339–428

Behrendt JC, Henderson JR, Meister L, Rambo WL (1974) Geophysical investigations of the Pensacola Mountains and adjacent glacierized areas of Antarctica. US Geol Surv Prof. Paper 844:1–28

Behrendt JC, Drewry DJ, Jankowski E, England AW (1979) Revision of the known area of the Dufek intrusion. Antarctic J US 14:6

Behrendt JC, Drewry DJ, Jankowski E, Grim MS (1980) Aeromagnetic and radio echo ice-sounding measurements show much greater area of the Dufek intrusion, Antarctica. Science 209:1014–1017

Behrendt JC, Drewry DJ, Jankowski E, Grim MS (1981) Aeromagnetic and radio echo ice-sounding measurements over the Dufek intrusion, Antarctica. J Geophys Res 86:3014–3020

Benson WN (1916) Report on the petrology of the dolerites collected by the British Antarctic expedition, 1907–1909. British Antarctic Exped., 1907–09. Scientific Investigations, Rept. 2 (Geology), Pt. 9:153–160

Bhattacharji S, Smith C (1964) Flowage differentiation. Science 145:150–153

Bird DK, Rogers RD, Manning CE (1986) Mineralized fracture systems of the Skaergaard intrusion. Medd Grönland Geosci 16:1–68

Bird DK, Manning CE, Rose NM (1988) Hydrothermal alteration of Tertiary layered gabbros, East Greenland. American J Sci 288:405–407

Bottinga Y, Javoy M (1975) Oxygen isotope partitioning among the minerals of igneousand metamorphic rocks. Rev Geophys Space Phys 13:401–418

Brewer TS, Hergt JM, Hawkesworth CJ, Rex D, Storey BC (1992) Coats Land dolerites and the generation of Antarctic continental flood basalts. In: Storey BC, Alabaster T, Pankhurst RJ (eds) Magmatism and the causes of continental break-up. Geol Soc London, Spec Pub 68:185–208

Brewer TS, Rex D, Guise PG, Hawkesworth CJ (1996) Geochronology of Mesozoic tholeiitic magmatism in Antarctica: Implications for the development of the failed Weddell-Sea rift systems. In: StoreyBC et al. (eds) Weddell Sea tectonics and Gondwana break-up. Geol Soc London, Spec Paper 108:45–61

Brotzu P, Capaldi G, Melluso L Orsi G (1988) Jurassic Ferrar Dolerites and Kirkpatrick Basalts in northern Victoria Land (Antarctica): Stratigraphy, geochronology, and petrology. Mem Soc Geol Italiana 43:97–116

Cartwright J, Möller-Hansen D (2006) Magma transport through the crust via interconnected sill complexes. Geology, 34(11):929–932

Cawthorn RG (1996) Layered intrusions. Elsevier, Amsterdam, The Netherlands

Chacko T, Cole DR, Horita J (2001) Equilibrium oxygen, hydrogen, and carbon isotope fractionation factors applicable to geologic systems. In: Valley JW, Cole DR (eds) Rev Mineral Geochem 43:1–81

Chiba H, Chacko T, Clayton RN, Goldsmith JR (1989) Oxygen isotope fractionations involving diopside, forsterite, magnetite, and calcite: Applications to geothermometry. Geochim Cosmochim Acta 53:2985–2995

Compston W, McDougall I, Heier KS (1968) Geochemical comparison of the Mesozoic basaltic rocks of Antarctica, South Africa, South America, and Tasmania. Geochim Cosmochim Acta 32:129–149

Cox KG, Bell JD, Pankhurst RJ (1979) The interpretation of igneous rocks. Allen & Unwin, London

Craddock C (ed) (1969) Geologic map of Antarctica. Antarctic map Folio Series, Folio 12. American Geographical Society, New York

Drinkwater JL, Ford AB, Czamanske GK (1985) Study of sulfide mineral distribution in the Dufek intrusion. Antarctic J US 20(5):50–51

Drinkwater JL, Ford AB, Czamanske GK (1986) Apatites of the Dufek Intrusion: A preliminary study. Antarctic J US 21(5):66–69

Elliot DH, Fleming TH (2004) Occurrence and dispersal of magmas in the Jurassic Ferrar Large Igneous Province, Antarctica. Gondwana Res 7:225–239

Elliot DH, Fleming TH, Kyle PR, Foland KA (1999) Long-distance transport of magmas in the Jurassic Large Igneous Province, Antarctica. Earth Planet Sci Lett 167:89–104

Encarnación J, Fleming TH, Elliot DH (1996) Synchronous emplacement of Ferrar and Karoo dolerites and the early break up of Gondwana. Geology 24:535–538

England AW, Nelson WH (1977) Geophysical studies of the Dufek intrusion, Pensacola Mountains, Antarctica, 1976–1977. Antarctic J US 12(4):93–94

England AW, Cooke JE, Hodge SM, Watts RD (1979) Geophysical investigations of Dufek intrusion, Pensacola Mountains. Antarctic J. US 14:4–5

Erlank AJ, Hofmeyr PK (1966) K/Rb and K/Cs ratios in Karroo dolerites from South Africa. J Geophys Res 71:5439

Erlank AJ, Hofmeyr PK (1968) K/Rb ratios of Mesozoic and tholeiites from Antarctica, Brazil, and India. Earth Planet Sci Lett 4:33–38

Faure G (1981) Strontium isotope composition of volcanic rocks: Evidence for contamination of the Kirkpatrick Basalt, Antarctica. In: O'Connell RJ, Fyfe WS (eds) Evolution of the earth. Geodynamics Series, vol. 5. American Geophysical Union, Washington, DC, pp 75–81

Faure G (1998) Principles and applications of geochemistry, 2nd edn. Prentice Hall, Upper Saddle River, NJ

Faure G (2001) The origin of igneous rocks: The isotopic evidence. Springer, Heidelberg, Germany

Faure G, Mensing TM (2005) Isotopes: Principles and applications, 3rd edn. Wiley, Hoboken, NJ

Faure G, Hill RL, Jones LM, Elliot DH (1972) Isotope composition of strontium and silica content of Mesozoic basalt and

dolerite from Antarctica. In: Adie RJ (ed) Antarctic geology and geophysics. Universitetsforlaget, Oslo, Norway, pp 617–624
Faure G, Bowman JR, Elliot DH, Jones LM (1974) Strontium isotope composition and petrogenesis of the Kirkpatrick Basalt, Queen Alexandra Range, Antarctica. Contrib Mineral Petrol 58(3):153–169
Faure G, Mensing TM, Jones LM, Hoefs J, Kibler EM (1991) Isotopic and geochemical studies of Ferrar Dolerite sills in the Transantarctic Mountains. In: Ulbrich H, Rocha Campos AC (eds) Gondwana Seven Proceedings. Instituto Geociencias, Universidade de São Paulo, São Paulo, Brazil, pp 669–683
Ferrar HT (1907) Report on the field geology of the region explored during the "Discovery" Antarctic expedition, 1901–05. National Antarctic Exped., vol. 1 (Geology):1–100. London
Ferris J Johnson A, Storey B (1998) Form and extent of the Dufek intrusion, Antarctica, from newly compiled aeromagnetic data. Earth Planet Sci Lett 154:185–202
Fleming TH, Foland KA, Elliot DH (1995) Isotopic and chemical constraints on the crustal evolution and source signature of Ferrar magmas, north Victoria Land, Antarctica. Contrib Mineral Petrol 121:217–236
Fleming TH, Elliot DE, Heimann A, Foland KA (1997) $^{40}Ar/^{39}Ar$ geochronology of Ferrar Dolerite sills from the Transantarctic Mountains, Antarctica: Implications for the age and origin of the Ferrar magmatic province. Geol Soc Amer Bull 109:533–546
Ford AB (1969) Chemical trends in the Dufek intrusion, Pensacola Mountains. Antarctic J US 4(5):202–203
Ford AB (1970) Development of the layered series and capping granophyre of the Dufek intrusion of Antarctica. In: Visser DJL, von Gruenewaldt G (eds) Symposium on the Bushveld igneous complex and other layered intrusion. Geol Soc South Africa, Spec Pub 1:492–510
Ford AB (1974) Basalt dikes of the Cordiner Peaks: Satellite bodies of the Dufek intrusion? Antarctic J US 9(4):149–152
Ford AB (1976) The stratigraphy of the Dufek intrusion, Antarctica. US Geol Surv Bull 1405D:1–36
Ford AB (1983a) The Dufek Intrusion of Antarctica and a survey of its minor metals related to possible resources. Abstract US Geological Survey Polar Research Symposium, Circular 911. Denver, CO, pp 7–10
Ford AB (1983b) The Dufek intrusion of Antarctica and a survey of its minor metals and possible resources. In: Behrendt JC (eds) Petroleum and mineral resources of Antarctica. US Geological Survey, Circular 909. Denver, CO, pp 51–75
Ford AB (1990) The Dufek intrusion of Antarctica. In: Splettstoesser JF, Dreschoff GAM (eds) Mineral resources of Antarctica. Antarctic Research Series, vol. 51. American Geophysical Union, Washington, DC, pp 15–32
Ford AB, Boyd WW Jr (1968) The Dufek intrusion, a major stratiform gabbroic body in the Pensacola Mountains, Antarctica. 23rd International Geological Congress, Proceedings, vol. 2. Prague, pp 213–228
Ford AB, Himmelberg GR (1991) Geology and crystallization of the Dufek intrusion. In: Tingey RJ (eds) Geology of Antarctica. Oxford University Press, Oxford, UK
Ford AB, Kistler RW (1980) K-Ar age, composition and origin of Mesozoic mafic rocks related to the Ferrar Group, Pensacola Mountains, Antarctica. New Zealand J Geol Geophys 23:371–390
Ford AB, Nelson SW (1972) Density of the stratiform Dufek intrusion, Pensacola Mountains, Antarctica. Antarctic J US 7(5):147–149
Ford AB, Carlson C, Czamanske GK, Nelson WH, Nutt CJ (1977) Geological studies of the Dufek intrusion, Pensacola Mountains, 1976–1977 Antarctic J US 12(4):90–92
Ford AB, Mays RE, Haffty J, Fabbi BP (1983) Reconnaissance of minor metal abundances and possible resources of the Dufek Intrusion, Pensacola Mountains. In: Oliver RL, James PR, Jago JB (eds) Antarctic earth science. Australian Academy of Science, Canberra, A.C.T., pp 433–436
Ford AB, Kistler RW, White LD (1986) Strontium and oxygen isotope study of the Dufek Intrusion. Antarctic J US 21:63–66
Grindley WG (1973) The geology of the Queen Alexandra Range, Beardmore Glacier, Ross Dependency, Antarctica: With notes on the correlation of Gondwana sequences. New Zealand J Geol Geophys 6(3):307–347
Grindley GW, Laird MG (1969) Geology of the Shackleton Coast. In: Craddock C (ed) Geologic map of Antarctica, Folio 12, Sheet 15. American Geographical Society, New York
Grindley GW, Warren G (1964) Stratigraphic nomenclature and correlation in the western Ross Sea region. In: Adie RJ (ed) Antarctic geology. North-Holland, Amsterdam, The Netherlands, pp 314–333
Gunn BM (1962) Differentiation in Ferrar Dolerites, Antarctica. New Zealand J Geol Geophys 5(5):820–863
Gunn BM (1963) Layered intrusions in the Ferrar Dolerites, Antarctica. Mineral Soc Amer, Spec Paper 1:124–133
Gunn BM (1965) K/Rb and K/Ba ratios in Antarctic and New Zealand tholeiites and alkali basalts. J Geophys Res 70(24):6241–6247
Gunn BM (1966) Modal and element variation in Antarctic tholeiites. Geochim Cosmochim Acta 30(9):881–920
Gunn BM, Warren G (1962) Geology of Victoria Land between the Mawson and Mulock glaciers, Antarctica. New Zealand Geol Surv Bull 71:1–157
Haensel JM Jr, Himmelberg GR, Ford AB (1986) Plagioclase compositional variations in anorthosites of the lower part of the Dufek intrusion. Antarctic J US 21(5):61–63
Hamilton WB (1965) Diabase sheets of the Taylor Glacier region, Victoria Land, Antarctica. US Geol Surv Prof. Paper 456-B:1–71
Harrington HJ (1958) Nomenclature of rock units in the Ross Sea region, Antarctica. Nature 182:209
Heier KS, Rogers JJW (1963) Radiometric determination of thorium, uranium, and potassium in basalts and in two magmatic differentiation series. Geochim Cosmochim Acta 27:137–154
Heier KS, McDougall I, Adams JAS (1964) Thorium, uranium, and potassium concentrations in Hawaiian lavas. Nature 201:254–256
Heier KS, Compston W, McDougall I (1965) Thorium and uranium concentrations, and the isotopic compositions of strontium in the differentiated Tasmanian dolerites. Geochim Cosmochim Acta 29:643–659
Heimann A, Fleming TH, Elliot DH, Foland KA (1994) A short interval of Jurassic continental flood basalt volcanism in Antarctica as demonstrated by $^{40}Ar/^{39}Ar$ geochronology. Earth Planet Sci Lett 121:19–41
Hergt JM (1987) The origin and evolution of the Tasmanian Dolerites. PhD dissertation. The Australian National University, Canberra, Australia

Hergt JM, Chappell BW, McCulloch MT, McDougall I, Chivas AR (1989a) Geochemical and isotopic constraints on the origin of the Jurassic dolerites of Tasmania. J Petrol 30(4):841–883

Hergt JJ, Chappell BW, Faure G, Mensing TM (1989b) The geochemistry of Jurassic dolerites from Portal Peak, Antarctica. Contrib Mineral Petrol 102:298–305

Hergt JM, Peate DW, Hawkesworth CJ (1991) The petrogenesis of Mesozoic Gondwana low-Ti flood basalts. Earth Planet Sci Lett 105:134–148

Himmelberg GR, Ford AB (1973) Pyroxene compositional trends in the Dufek intrusion, Pensacola Mountains. Antarctic J US 8(5):260–263

Himmelberg GR, Ford AB (1975) Petrologic studies of the Dufek intrusion, Pensacola Mountains: Iron-titanium oxides. Antarctic J US 10(5):241–244

Himmelberg GR, Ford AB (1976) Pyroxenes of the Dufek intrusion, Antarctica. J Petrol 17(2):219–243

Himmelberg GR, Ford AB (1977) Iron-titanium oxides of the Dufek intrusion. Amer Mineral 62:623–633

Hoefs J (1997) Stable isotope geochemistry, 4th edn. Springer, Heidelberg, Germany

Hoffman J, Nairn AEM, Peterson DN (1984) The paleomagnetic investigation of flows and sills from the Queen Alexandra Range, Antarctica. In: Turner MD, Splettstoesser JF (eds) Geology of the Central Transantarctic Mountains. Antarctic Research Series, vol. 36, paper 4

IUGS (2002) International stratigraphic chart. International Commission of Stratigraphy

Jones LM, Walker RL, Hall BA, Borns HW Jr (1973) Origin of the Jurassic dolerite and basalts of southern Victoria Land. Antarctic J US 8(5):268–270

Kibler EM (1981) Petrogenesis of two Ferrar Dolerite sills, Roadend Nunatak, Transantarctic Mountains, Antarctica. MSc thesis, The Ohio State University, Columbus, OH

Kruger FJ, Marsh JS (1982) Significance of $^{87}Sr/^{86}Sr$ ratios in the Merensky cyclic unit of the Bushveld complex. Nature 298:53–55

Kulp JL (1961) Geologic time scale. Science 133:1105–1114

Kyle PR (1977) Petrogenesis of the Ferrar rocks. Antarctic J US 12(4):108–110

Kyle PR (1980) Development of heterogeneities in the subcontinental mantle: Evidence from the Ferrar Group, Antarctica. Contrib Mineral Petrol 73:89–104

Kyle PR, Elliot DH, Sutter JF (1981) Jurassic Ferrar Supergroup tholeiites from the Transantarctic Mountains, Antarctica, and their relationship to the initial fragmentation of Gondwana. In: Cresswell VM, Vella P (eds) Gondwana. Balkema, Rotterdam, The Netherlands, pp 283–287

Kyle PR, Pankhurst RJ, Bowman JR (1983) Isotopic and chemical variations in Kirkpatrick Basalt Group rocks from southern Victoria Land. In: Oliver RL, James PR, Jago JB (eds) Antarctic earth science. Australian Academy of Science, Canberra, A.C.T., pp 234–237

Leal LG (1980) Particle motions in a viscous fluid. Ann Rev Fluid Mechanics 12:435–476

Lindsay JF, Gunner J, Barrett PJ (1973) Reconnaissance geologic map of the Mount Elizabeth and Mount Kathleen Quadrangles, Transantarctic Mountains, Antarctica. Antarctic geologic map A-2. US Geological Survey, Washington, DC

Marsh BD (1996a) Solidification fronts and magmatic evolution. Mineral Mag 60:5–40

Marsh BD (1996b) Sill sequence as piecemeal layered intrusions and links to volcanism: Evidence from the Ferrar dolerites of Antarctica. EOS, Trans Amer Geophys Union 77:S292

Marsh BD, Philipp JR (1996) Three-dimensional magmatic filling of Basement sill revealed by unusual crystal concentrations. Antarctic J US 31(2):39–40

Marsh BD, Zieg MJ (1997) The Dais layered intrusion: A new discovery in the Basement sill of the McMurdo dry valleys. Antarctic J US 32(5):18–20

McDougall I (1961) Determination of the age of a basic igneous intrusion by the potassium argon method. Nature 190:1184–1186

McDougall I (1962) Differentiation of the Tasmanian Dolerites: Red Hill Dolerite-Granophyre association. Geol Soc Amer Bull 73:279–316

McDougall I (1963) Potassium-argon age measurements on dolerites from Antarctica and South Africa. J Geophys Res 68(5):1535–1545

McKelvey BC, Webb PN (1962) Geology of Wright Valley. Pt. 3 of Geological investigations in southern Victoria Land, Antarctica. New Zealand J Geol Geophys 5(3):225–251

Minor D, Mukasa S (1997) Zircon U-Pb and hornblende ^{40}Ar-^{39}Ar ages for the Dufek layered mafic intrusion, Antarctica: Implications for the age of the Ferrar large igneous province. Geochim Cosmochim Acta 61:2497–2504

Morrison AD, Reay A (1995) Geochemistry of Ferrar Dolerite sills and dykes at Terra Cotta Mountain, south Victoria Land, Antarctica. Antarctic Sci 7(1):73–85

Prior GT (1907) Report on the rock specimens collected during the "Discovery" Antarctic Expedition, 1901–1904. Nat Antarct Exped, Nat History, vol. 1, Geology:101–140, London

Renne PR, Deino AL, Walter RC, Turrin BD, Swisher CC III, Becker TA, Curtis GH, Sharp WD, Jasouni AR (1994) Intercalibration of astronomical and radioisotope time. Geology 22:783–786

Renne PR, Swisher CC, Deino AL, Karner DB, Owen TL,. DePaolo DJ (1998) Intercalibration of standards, absolute ages, and uncertainties in $^{40}Ar/^{39}Ar$ dating. Chem Geol 154:117–152

Rhodes RC, Bornhorst TJ (1976) Petrologic provinces in Jurassic tholeiites of Gondwanaland. Geologische Rundschau 65:930–937

Rowley PD, Ford AB, Williams PL, Pride DE (1983) Metallogenic provinces of Antarctica. In: Oliver RL, James PR, Jago JB (eds) Antarctic earth science. Australian Acad. Sci., Canberra, A.C.T., pp 414–419

Schmidt DT, Ford AB (1969) Geology of the Pensacola and Thiel mountains. In: Craddock C (ed) Geologic map of Antarctica, Antarctic Folio Series, Folio 12, Sheet 5. American Geographical Society, New York

Simkin T (1967) Flow differentiation in the picritic sills of North Skye. In, P.J. Wyllie, ed., Ultramafic and Related Rocks. Wiley, New York

Smith WC (1924) The plutonic and hypabyssal rocks of south Victoria Land. British Antarctic "Terra Nova" Exped. 1910–1913. Nat Hist Rep, Geology 1(6):167–227

Storey BC, Kyle PR (1997) An active mantle mechanism for Gondwana break-up. South African J Geol 100:283–290

Upton BGJ, Wadsworth WJ (1967) A complex basalt-mugearite sill in Piton des Neiges volcano, Reunion. Amer. Mineral. 52:1475–1492

Vavra CL (1982) Provenance and alteration of the Triassic Fremouw and Falla formations, central Transantarctic Mountains, Antarctica. PhD dissertation, Department of Geology and Mineralogy, The Ohio State University, Columbus, OH

Vavra CL, Stanley KO, Collinson JW (1981) Provenance and alteration of the Triassic Fremouw Formation, Central Transantarctic Mountains. In: Cresswell MM, Vella P (eds) Gondwana Five. A.A. Balkema, Rotterdam, The Netherlands, pp 149–153

Wager LR, Deer WA (1939) The petrology of the Skaergaard Intrusion, Kangerdlugssuaq East Greenland. Medd om Grönland 105:1–352

Walker F (1953) The pegmatitic differentiates of basic sheets. Amer J Sci 251:41–60

Walker PT (1961) Study of some rocks and minerals from the Dufek Massif, Antarctica. IGY World Data Center, A. Glaciol, Glac Rep 4:195–213

Wilson M (1989) Igneous petrogenesis. Unwin Hyman, London

Chapter 14
Kirwan Volcanics, Queen Maud Land

Western Queen (Dronning) Maud Land in Fig. 14.1 is not considered to be part of the Transantarctic Mountains because the basement rocks were not deformed and regionally metamorphosed by the Ross Orogeny. Nevertheless, some of the nunataks that project through the East Antarctic Ice Sheet in the area outlined in Fig. 14.1 and enlarged in Fig. 14.2 consist of Jurassic tholeiite basalt of the *Kirwan Volcanics* that superficially resemble the Kirkpatrick Basalt in the Transantarctic Mountains. In addition, flat-lying sedimentary rocks of the Permian *Amelang Formation* resemble the sandstones and shales of the Beacon Supergroup. The groups of nunataks in Fig. 14.2 where the Kirwan Volcanics are exposed include Vestfjella (including Plogen and Basen), Fossilryggen, Mannefallknausane, Heimefrontfjella, Bjőrnnutane, Sembberget, and the Kirwan Escarpment. The first information about the geology of the Kirwan Volcanics was published by Brunn (1964) and Blundell (1964). The latter collected oriented samples of dolerite in the southern part of Vestfjella for a study of paleomagnetism (see also Lövlie 1979, 1988). These samples were later described by Brown (1967) and dated by Rex (1967).

Vestfjella has continued to be the most frequently visited mountain range in western Queen Maud Land because it is located adjacent to the Riiser-Larsen Ice Shelf and is accessible from the British research station at Halley Bay. Hjelle and Winsnes (1972) reported that the nunataks of Vestfjella are composed of a large number of lava flows that are between 1 and 25 m thick and dip to the northwest at low angles of 10–25°. The highest peak in Vestfjella has an elevation of 1,132 m. The whole-rock K-Ar dates of dolerites from Vestfjella (168 ± 6 and 172 ± 6 Ma) reported by Rex (1967) indicate a Middle Jurassic age for the Kirwan Volcanics. In a later report, Rex (1972) reported additional K-Ar dates for basalt and dolerite from Queen Maud Land, the Theron Mountains, and the Shackleton Range. Still more K-Ar dates of basalt flows and dikes in Vestfjella were later published by Furnes and Mitchell (1978), Furnes et al. (1987), and Peters et al. (1991). All of these dates are listed in Appendix 14.5.1.

Juckes (1968, 1969) published the first chemical analyses of Jurassic dolerites on the nunataks of Mannefallknausane (Fig. 14.2) and Vestfjella. He observed that the silica concentrations of these rocks range widely from 43.89% to 51.19% and that the rocks are deficient in potassium (K_2O = 0.11% to 0.95%) compared to the basalts and dolerites of the Ferrar Group of Victoria Land in the Transantarctic Mountains. He also reported that the K/Rb ratios of eight dolerite specimens range from 83 to 380 with an average of 224. Juckes (1968 therefore concluded that the chemical compositions of the Jurassic dolerites in Queen Maud Land range more widely than those of the Ferrar Group in the Transantarctic Mountains.

Subsequently, Faure and Elliot (1971) reported that the initial $^{87}Sr/^{86}Sr$ ratios of ten samples of basalt and dolerite from nunataks at Bjőrnnutane, Sembberget, Heimefrontfjella, and Vestfjella range from 0.7037 to 0.7075 (at t = 170 Ma) and are consistently *lower* than those of the Kirkpatrick Basalt and the Ferrar Dolerite. They also reported that the rubidium concentrations of the Kirwan Volcanics are lower but the strontium concentrations are higher than those of the Jurassic basalts in the Transantarctic Mountains that were analyzed by Hill (1969). These initial studies suggested that the Kirwan Volcanics of Queen Maud Land *differ* from the basalt and dolerite of the Ferrar Group in the Transantarctic Mountains and in Tasmania in terms of their chemical and Sr-isotope compositions, although both suites of rocks appear to have formed at about the same time during the Jurassic Period and both suites are quartz-normative tholeiite basalts.

G. Faure and T.M. Mensing, *The Transantarctic Mountains: Rocks, Ice, Meteorites and Water*,
DOI 10.1007/978-90-481-9390-5_14,

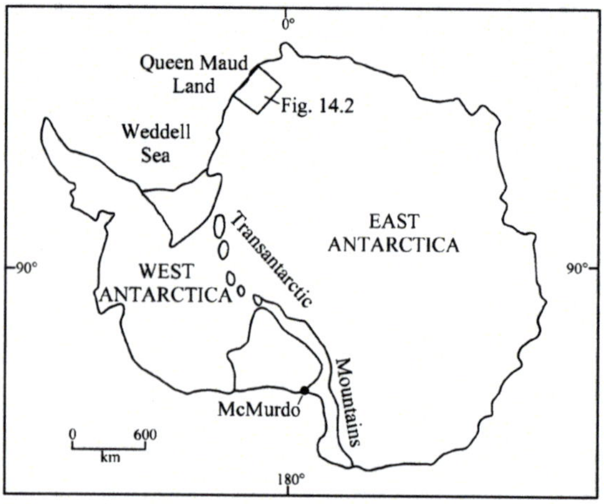

Fig. 14.1 Queen (or Dronning) Maud Land is located in East Antarctica along the east coast of the Weddell Sea. It is not considered to be an extension of the Transantarctic Mountains, although it does include flat-lying sandstones of the Permian Amelang Formation which are overlain by continental flood basalts of the Jurassic Kirwan Volcanics. The chemical and isotopic compositions of the basalt flows and sills resemble the Karoo Basalt of southern Africa which was adjacent to Queen Maud Land before the break-up of Gondwana (Adapted from Furnes et al. (1987))

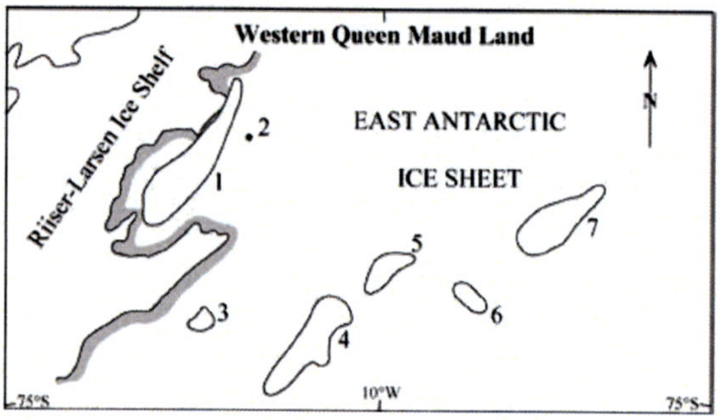

Fig. 14.2 Queen (or Dronning) Maud Land contains groups of nunataks composed of the Jurassic Kirwan Volcanics. The groups of nunataks are identified by number: 1, Vestfjella; 2, Fossilryggen; 3, Mannefallknausane; 4, Heimefrontfjella; 5, Bjőrnnutane; 6, Sembberget; and 7, Kirwan Escarpment. These groups of nunataks rise above the surface of the East Antarctic ice sheet which covers the entire area on this map (Adapted from Fig. 1 of Furnes et al. (1987))

14.1 Kirwan Escarpment

The Kirwan Escarpment of Queen Maud Land at 73°25′S and 03°30′W in Fig. 14.2 consists of a large number of nunataks that are aligned along a northwest-facing escarpment for a length of about 70 km (Faure et al. 1979). The nunataks in Fig. 14.3 were first mapped by Aucamp et al. (1972) of the Antarctic Division of the Geological Survey of South Africa who referred to them as "Kirwanveggen," whereas Norwegian geologists who have worked in Queen Maud Land (e.g., Hjelle and Winsnes 1972) call these nunataks "Kirwanryggen." We will follow the British geologists who use the name "Kirwan Escarpment" (e.g., Juckes 1968) even though Neethling (1972a) reported that these nunataks are actually the peaks of a high mountain range that is buried in the East Antarctic ice sheet which is about 2.4 km thick on the inland side of the Kirwan nunataks.

The nunataks that are composed of basalt flows are identified by the letters A to J (Harris et al. 1990). Some of the basalt nunataks were named by Wolmarans and Kent (1982): A = Turnaround Ridge, C = Muskeg Cliff, D = Mount Alex du Toit, E = Concretion Point, F = Frostbite Bluff, G = Petrel Peak, I = Ladneset, J = Tunga. Harris et al. (1990) published stratigraphic sections of the basalt flows on these nunataks and identified the underlying rock formations. The basement consists of the granitic gneisses, quartzite, amphibolite, and pegmatite of the Sverdrup Group (Roots 1969), except at Tunga where the basalt flows are underlain by early Paleozoic quartzite of the Urfjell Group (Aucamp et al. 1972).

During the summers of 1984/85 and 1985/86 the Kirwan Escarpment was visited by Chris Harris of the University of Cape Town who measured stratigraphic sections of the basalt flows and collected samples for a study of the petrogenesis of the Kirwan Volcanics. Harris et al. (1990) subsequently published major-element and trace-element analyses of 27 basalt specimens as well as isotopic compositions of strontium, neodymium, and oxygen. The concentrations of the major-element oxides (Fig. 14.4) vary only slightly in contrast to the dolerites at Mannefallknausane (Juckes 1968), Heimefrontfjella (Juckes 1972), and Vestfjella (Hjelle and Winsnes 1972; Furnes et al. 1987; Luttinen and Siivola 1997; Luttinen et al. 1998). In addition, Harris et al. (1990) emphasized the similarity of the chemical compositions of the Kirwan Volcanics to the Karoo Basalt of southern Africa. The age of the Sverdrup Group, according to Roots (1969) is Precambrian (?). We can only add that volcanic rocks of the Trollkjellrygg Group at the mouth of the Jutulstraumen Glacier in Queen Maud Land yielded a

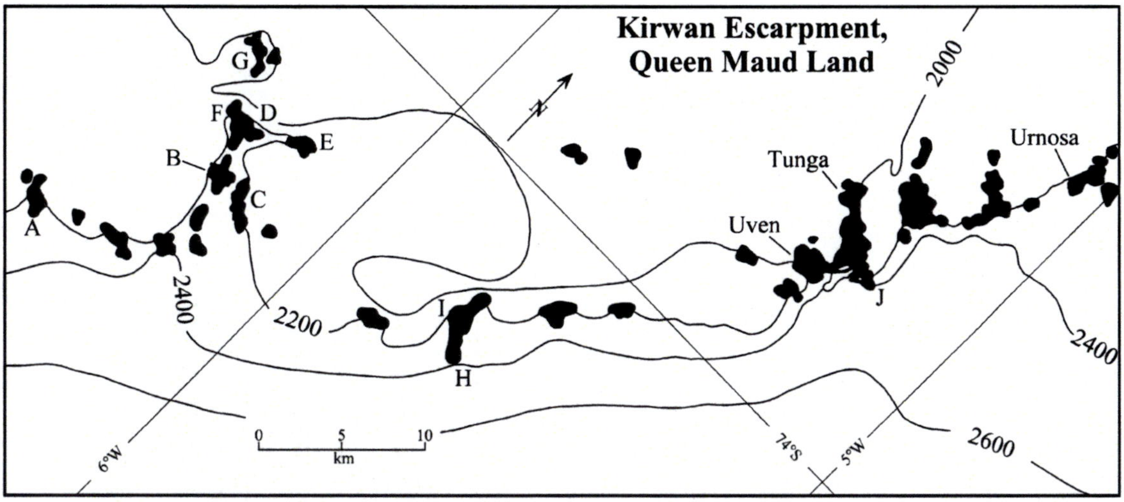

Fig. 14.3 The Kirwan Escarpment of Queen Maud Land in East Antarctica (#7 in Fig. 14.2) consists of a large number of nunataks. Some of the nunataks which are composed of the Jurassic lava flows of he Kirwan Volcanics are identified by letters A to J. Faure et al. (1979) analyzed 12 samples of basalt from ten flows on nunatak B and one sample from Tunga. The samples were collected by geologists of the South African Expedition SANAE 25 during the 1969/70 field season (Aucamp et al. 1972; Wolmarans and Kent 1982) (Adapted from Faure et al. (1979) and Harris et al. (1990))

whole-rock Rb-Sr date of 838 ± 28 Ma (Eastin et al. 1970; recalculated to $\lambda = 1.42 \times 10^{-11}$ year^{-1}) and an initial $^{87}Sr/^{86}Sr$ ratio of 0.7097 ± 0.0009. This date confirms the Neoproterozoic age of the Trollkjellrygg Group assigned to it by Roots (1969). Additional age determinations of rocks in Queen Maud Land were published by Allsopp and Neethling (1970).

In some places (e.g., Mt. Alex du Toit, Concretion Point, Frostbite Bluff, and Petrel Peak), the basalt flows are underlain by the Permian sedimentary rocks of the *Amelang Formation* (Neethling 1971). Lindsay et al. (2002) reported that the formation is up to 80 m thick and is composed of a basal tillite overlain by fine-grained sandstones which are interbedded with carbonaceous shale and thin discontinuous coal seams. The sandstones coarsen upward and are interbedded with thin conglomerate beds and plant-bearing shales. Lindsay et al. (2002) suggested that the Amelang Formation could be divided into the lower Frostbite Bluff and the upper Concretion Point members. The Amelang Formation in Heimefrontfjella was studied by Larsson et al. (1990), Guy-Ohlson and Lindstrom (1994), and Bauer et al. (1997). The basalt flows of the Kirwan Escarpment are also interbedded with thin layers of volcanic ash (e.g., at Tunga, Ladneset, Concretion Point, and Turnaround Ridge). The greatest thickness of basalt flows observed by Harris et al. (1990) is 460 m.

The samples of basalt from nunatak B and from Tunga analyzed by Faure et al. (1979) were collected by A.P.H. Aucamp. The trace-element concentrations of these samples were later determined by Furnes et al. (1987) who also published trace-element concentrations of five basalt samples from Bjőrnnutane and three from Sembberget that were previously analyzed by Faure and Elliot (1971) and by Faure et al. (1972).

The initial $^{87}Sr/^{86}Sr$ ratios (corrected for decay of ^{87}Rb to 170 Ma) of the basalt flows on nunataks B and Tunga measured by Faure et al. (1979) range from 0.7027 to 0.7063 and have a poly-modal distribution in Fig. 14.5 with an average of 0.7044 ± 0.0010 (1σ). This value is significantly lower than the average $^{87}Sr/^{86}Sr$ ratio of the Kirkpatrick Basalt of the Ferrar Group in the Queen Alexandra Range of the Transantarctic Mountains.

Three samples from nunatak B (B07, B03, and B014) define a Rb-Sr isochron in Fig. 14.6 which corresponds to a date of 181.9 ± 0.4 Ma and an initial $^{87}Sr/^{86}Sr$ ratio of 0.70418 ± 0.0000005. The other samples scatter above and below this errorchron indicating

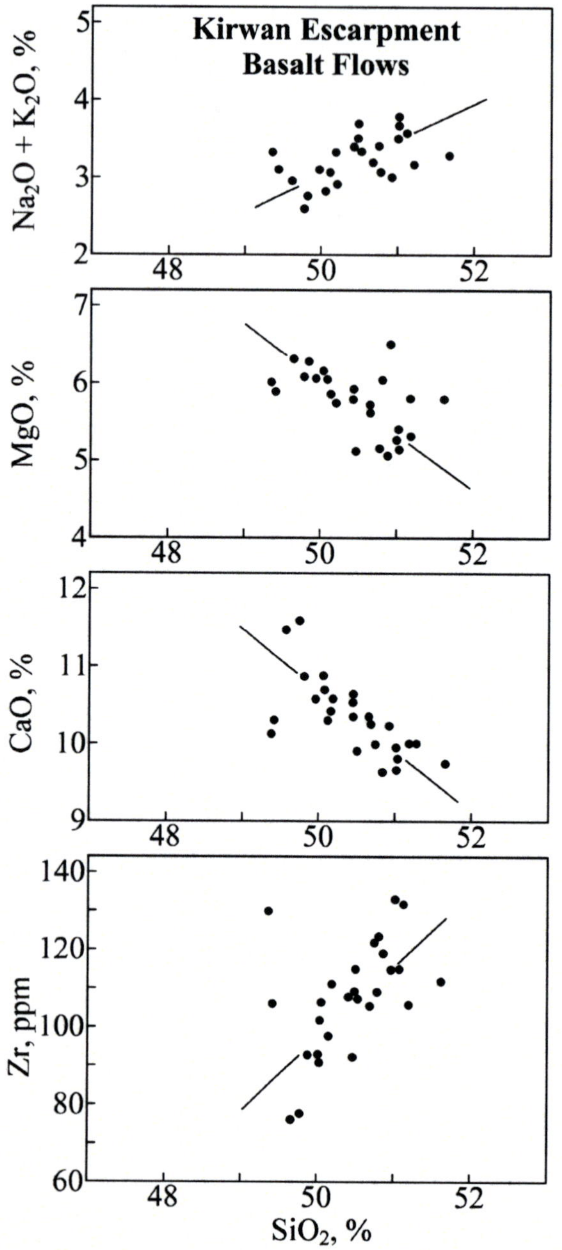

Fig. 14.4 The variation of the concentrations of alkali metals and alkaline earths with increasing silica concentrations in the basalt flows of the Kirwan Escarpment is consistent with the effects of crystallization of Ca-rich plagioclase, olivine, and pyroxene from basalt magma. The concentrations of zirconium rise with increasing silica concentration as a result of progressive fractional crystallization in the magma chamber (Data from Harris et al. (1990))

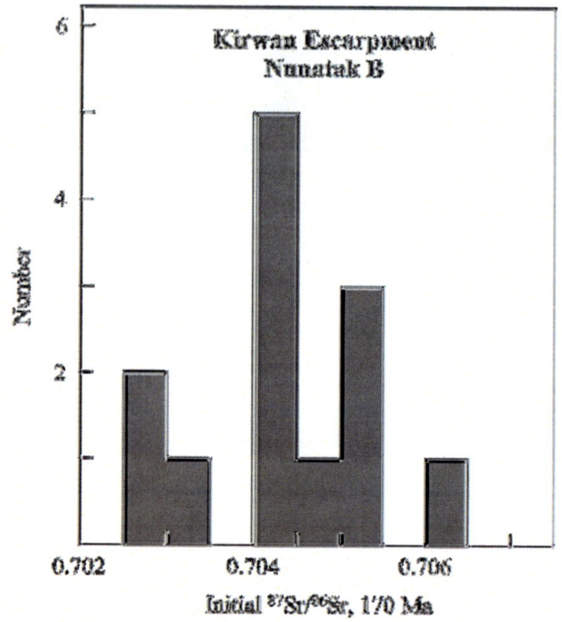

Fig. 14.5 The initial $^{87}Sr/^{86}Sr$ ratios of the lava flows on nunataks B and Tunga of the Kirwan Escarpment have a polymodal distribution at 170 Ma. The principal mode is between 0.7040 and 0.7045. The wide range of variation of the initial $^{87}Sr/^{86}Sr$ ratios and their numerical values are relevant to the origin of these rocks (Data from Faure et al. (1979))

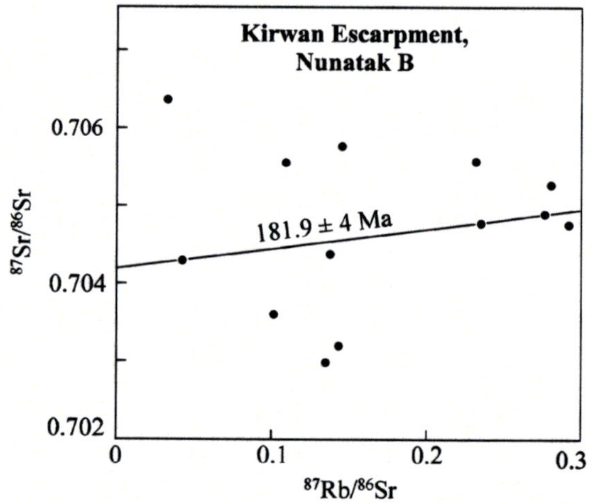

Fig. 14.6 Most of the basalt samples from nunataks B and Tunga in the Kirwan Escarpment scatter widely on the Rb-Sr isochron diagram. The only exceptions are three samples that define a straight line that yields a date of 181.9 ± 0.4 Ma and an initial $^{87}Sr/^{86}Sr$ ratio of 0.70418 ± 0.00000026 ($2\bar{\sigma}$). The samples that scatter above and below the line either formed from a heterogeneous magma that assimilated strontium from the Precambrian basement complex or they were altered after crystallization (Data from Faure et al. (1979))

that they either had different initial $^{87}Sr/^{86}Sr$ ratios (i.e., the magma was variously contaminated) or they were altered after crystallization. Aucamp et al. (1972) reported a whole-rock K-Ar date of 172 ± 10 Ma for the basalt sample (T-1-1) from Tunga that was analyzed by Faure et al. (1979).

The range of initial $^{87}Sr/^{86}Sr$ ratios of the basalt flows on nunatak B and the associated variation of their strontium concentrations may have been caused by additions of a Sr-bearing contaminant to the mantle-derived magma. Mixing theory presented by Faure (1998, Chapter 18) indicates that two-component mixtures form a straight line in coordinates of the $^{87}Sr/^{86}Sr$ ratios and the reciprocals of the strontium concentrations. The data published by Faure et al. (1979) loosely define such a mixing line in Fig. 14.7 (correlation coefficient: 0.81). Assuming that the basalt magma had an initial $^{87}Sr/^{86}Sr$ ratio of about 0.7030 at t = 170 Ma, a contaminant that originated from the Precambrian granitic gneisses of the continental crust should have had an elevated $^{87}Sr/^{86}Sr$ ratio consistent with the Rb/Sr ratios and the age of the rocks. The mixing line in Fig. 14.7 appears to support this hypothesis. However, the negative slope of the mixing line requires that the contaminant had a higher strontium concentration than the basalt magma. For example, if the 1/Sr ratio of the contaminant is 40×10^{-4} ppm^{-1}, its Sr concentration is 250 ppm. A 1/Sr ratio of 70×10^{-4} ppm^{-1} of the uncontaminated magma is equivalent to a Sr concentration of 145 ppm, which is lower than expected for continental flood basalt. In this way, the test of the mixing hypothesis has raised more questions about the origin of the Kirwan Volcanics on Nunatak B of the Kirwan Escarpment.

The initial $^{87}Sr/^{86}Sr$ ratios reported by Harris et al. (1990) have a somewhat higher average than the basalt on nunatak B reported by Faure et al. (1979). Nevertheless, the initial $^{87}Sr/^{86}Sr$ and 1/Sr ratios of the eight freshest samples selected by Harris et al. (1990) out of a total of 26 also form a mixing line with a negative slope and a correlation coefficient of 0.79. These authors noted that a similar phenomenon occurs in the basalts of the Sabie River Formation (Karoo Group) in southern Lebombo (Cox and Bristow 1984).

The most persuasive evidence for the contamination of the basalt flows on the Kirwan Escarpment arises from a Sr-Nd isotopic mixing diagram in Fig. 14.8 based on data by Harris et al. (1990). All data points were corrected for decay of the respective parent

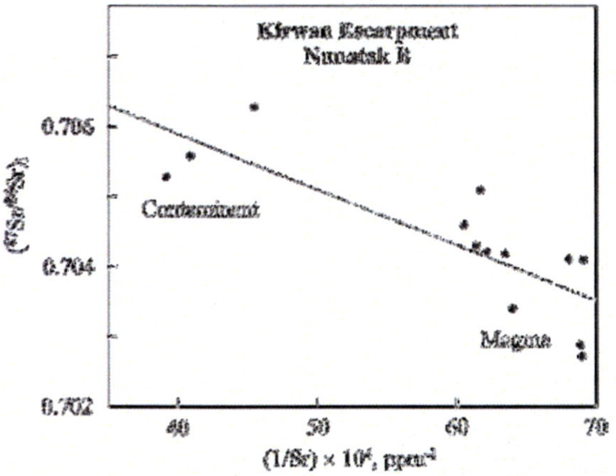

Fig. 14.7 The basalt flows on the nunataks B and Tunga of the Kirwan Escarpment loosely constrain a straight line in coordinates of the initial $^{87}Sr/^{86}Sr$ ratios (at t = 170 Ma) and the reciprocals of their strontium concentrations (correlation coefficient: 0.81). Such linear arrays of data points characterize mixtures of two components in differing proportions including volcanic rocks that formed from magma that assimilated strontium from the rocks of the continental crust. The negative slope of the mixing line is unusual because it implies that the contaminant had a higher strontium concentration than the basalt magma (Data from Faure et al. (1979))

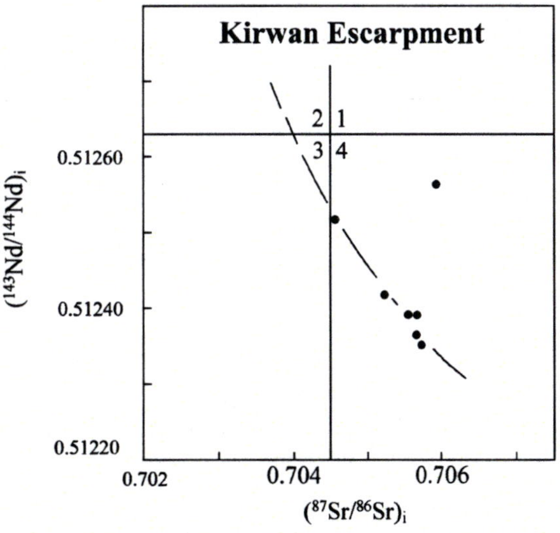

Fig. 14.8 The initial $^{87}Sr/^{86}Sr$ and $^{143}Nd/^{144}Nd$ ratios at 172 Ma of seven basalt samples from the Kirwan Escarpment define a segment of a Sr-Nd isotope mixing hyperbola located in quadrant 4, which indicates that the samples contain varying amounts of Sr and Nd from the continental crust. One sample (URF-7) is anomalous but also lies in quadrant 4 (Data from Harris et al. (1990))

isotope to an age of 172 Ma. Six of the seven samples analyzed by Harris et al. (1990) define a segment of an isotopic mixing hyperbola in quadrant 4 which is populated by rocks of the continental crust. Accordingly, these isotopic data indicate that the basalt flows on the Kirwan Escarpment contain crustal strontium and neodymium. The contamination of these rocks by strontium and neodymium from the Precambrian gneisses of the continental crust underlying Queen Maud Land could have occurred in three different ways:

1. During the formation of the magma by partial melting of a mechanical mixture of mantle-derived rocks and crustal rocks (e.g., gabbro dikes that intruded the deep continental crust)
2. By mixing of the basalt magma with a partial melt formed in granitic basement rocks that lined a magma chamber or the walls of conduits through which basalt magma was moving
3. Deposition of crustal strontium and neodymium by hydrothermal fluids after eruption and crystallization of basalt lava

In fact, all three processes are likely to have contributed to affect the isotopic compositions of strontium and neodymium and the concentrations of the respective parent and daughter elements. Although the chemical composition of the basalt magma was undoubtedly changed both by simultaneous fractional crystallization and assimilation of wall rock, the isotopic composition of strontium and neodymium is altered *only* by the assimilation of a contaminant and not by fractional crystallization.

The Karoo Basalt, like the Kirwan Volcanics, formed from mantle-derived magma that differentiated by fractional crystallization and concurrent contamination in an open system in contact with granitic gneisses of Precambrian age. The proximity of Queen Maud Land to southern Africa in pre-Jurassic time supports the hypothesis that the Kirwan Volcanics are more closely related to the Karoo Basalt than to the basalt and dolerite of the Ferrar Group (Neethling 1972b; Harris et al. 1991; Faure 2001, Fig. 5.67).

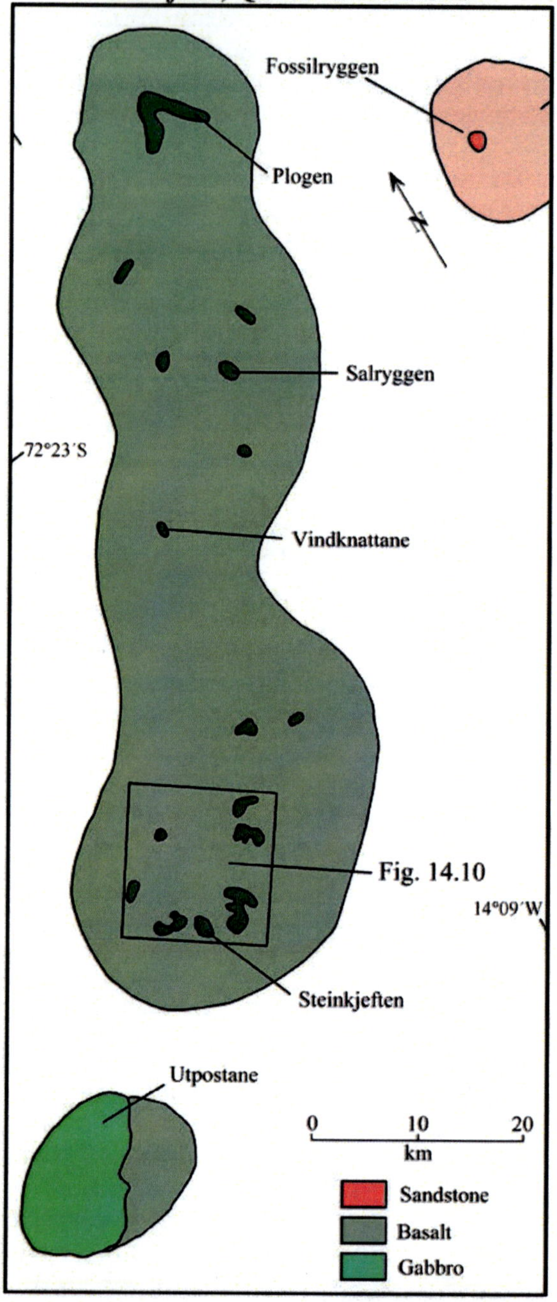

Fig. 14.9 The nunataks of the main body of Vestfjella consist of basalt flows and dolerite dikes of the Kirwan Volcanics (Jurassic). The nunataks called Plogen and Basen at the northeastern end of Vestfjella are off scale. Fossilryggen located east of Plogen is composed of Permian *Glossopteris*-bearing sandstones of the Amelang Formation. Utpostane of the southwestern end of Vestfjella is an olivine-bearing gabbro which is genetically related to the lava flows of the Kirwan Volcanics (Adapted from Hjelle and Winsnes (1972))

14.2 Vestfjella

Vestfjella in Fig. 14.9 consists of a series of isolated nunataks that occur along an ice-covered rise that extends in a northeasterly direction from Utpostane for about 130 km to the nunatak called Basen (not included in Fig. 14.9). The basaltic lava flows of Vestfjella are underlain by Permian sandstones of the Amelang Formation which is exposed only in Fossilryggen and which contains *Glossopteris* (Hjelle and Winsnes 1972). The geology of Vestfjella was first described by Hjelle and Winsnes (1972) who published photographs of the rocks taken in the field and major-element analyses of ten samples of basalt and dolerite. They also estimated that the thickness of the basalt flows at Steinkjeften is about 2,000 m, which was later revised to 2,500 m (Furnes and Mitchell 1978). The flows vary in texture, color, and degree of alteration. The basal parts of the flows contain plagioclase phenocrysts, whereas the upper parts contain amygdules filled with calcite, epidote, quartz, and prehnite. The surfaces of the flows are oxidized and cracked with pahoehoe flow structures. Hjelle and Winsnes (1972) also reported that the rocks on north-facing cliffs are coated with white salts that form by evaporation of meltwater. Basalt pillows at Vindknattane (Fig. 14.9) indicate that the lava flows at that locality were extruded into ponds or lakes.

14.2.1 Chemical Compositions

Hjelle and Winsnes (1972) reported chemical analyses of basalt from several nunataks in Vestfjella including one sample of olivine gabbro from Utpostane (Fig. 14.9). In addition, Juckes (1968, 1969) published a chemical analysis of basalt from the "VA Nunatak" which may be "Nunatak A" in Fig. 14.10 according to Fig. 1 of Furnes and Mitchell (1978). Later workers (e.g., Furnes and Mitchell 1978; Furnes et al. 1987; Luttinen et al. 1998) published a large number of additional major-element and trace-element analyses from the nunataks of Vestfjella, including Plogen and Basen (Peters et al. 1989; Luttinen and Siivola 1997; Luttinen et al. 1994, 1998).

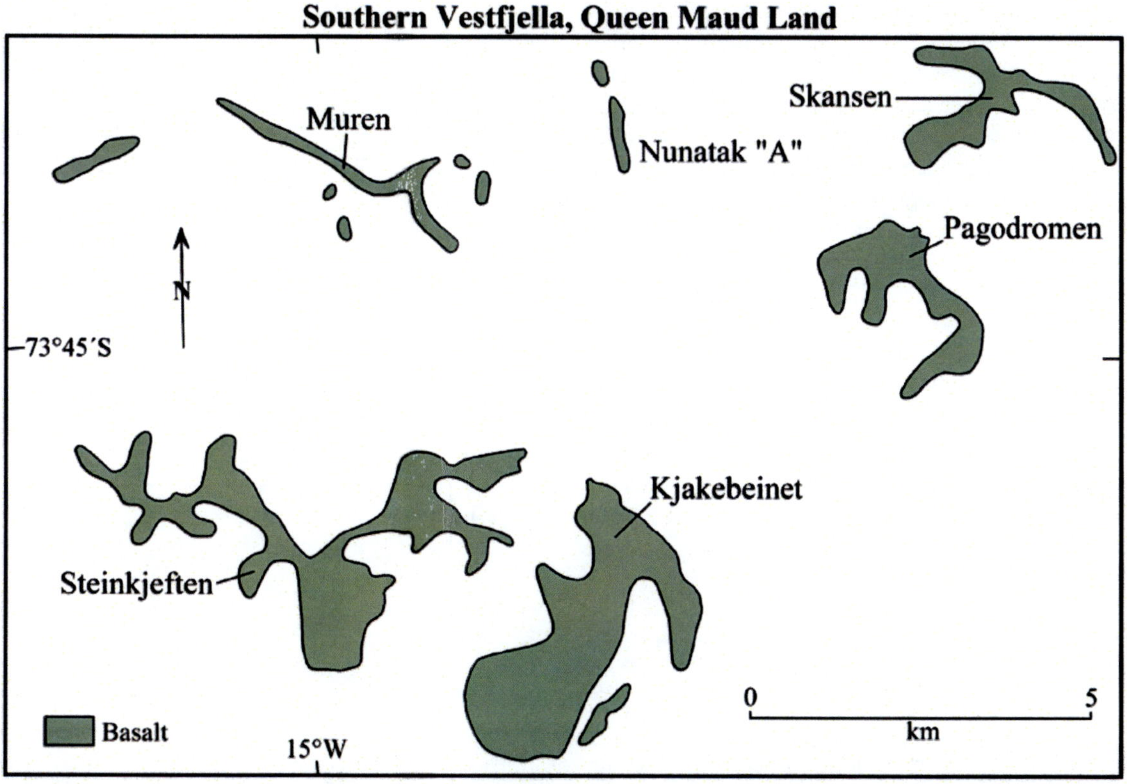

Fig. 14.10 Several large nunataks at the southwestern end of Vestfjella are composed of basalt flows and dolerite dikes of the Kirwan Volcanics (Jurassic). The flows dip northwest at 9–18°. Furnes et al. (1987) published whole-rock and plagioclase K-Ar dates that scatter widely from 171 ± 2 to 695 ± 11 Ma. Nevertheless, all of the dated samples are considered to be of Jurassic age (Adapted from Furnes et al. (1987))

The chemical compositions of the flows in south Vestfjella reported by Furnes and Mitchell (1978) in Fig. 14.10 vary only within narrow limits and closely resemble the chemical compositions of the flows on the Kirwan Escarpment in Fig. 14.4. For example, the silica concentrations of most of the basalt flows at both locations vary between 49% and 52%. The only exceptions are two olivine-bearing flows at Muren and Kjakebeinet in Vestfjella which have comparatively low silica concentrations (46.2%) and elevated MgO concentrations (14.0%). Similar high-magnesium flows occur on Pagodromen and East Steinkjeften (Furnes et al. 1987) and in a dike (14-KHG) from northern Vestfjella (Luttinen et al. 1998). The chemical compositions of these flows are affected by the presence of olivine phenocrysts which also occur in the olivine gabbro at Utpostane which contains 45.12% SiO_2 and 23.07% MgO and is correspondingly depleted in $Na_2O + K_2O$ = 1.55% and CaO = 7.22%. The relation of the olivine gabbro to the basalt flows is not yet certain although a cogenetic origin seems probable.

The silica concentrations of the dikes that intruded the flows in southern Vestfjella range more widely in Fig. 14.11 than those of the flows (Furnes and Mitchell 1978). This is true also of zirconium which is a refractory trace element that is not susceptible to hydrothermal alteration or chemical weathering.

The concentrations of SiO_2 and $Na_2O + K_2O$ in weight percent serve to classify the Kirwan Volcanics in southern Vestfjella into subalkalic tholeiite basalts and into alkali basalts (Wilson 1989). Figure 14.12 contains all of the basalt and dolerite specimens of southern Vestfjella that were analyzed by Hjelle and Winsnes (1972), Furnes and Mitchell (1978), and Furnes et al. (1987). Most of the data points are located in the field of subalkalic basalt and only a few stray into the alkali basalt territory and even fewer are classifiable as basaltic andesites. In this regard, the Kirwan Volcanics differ from the Kirkpatrick Basalt in the

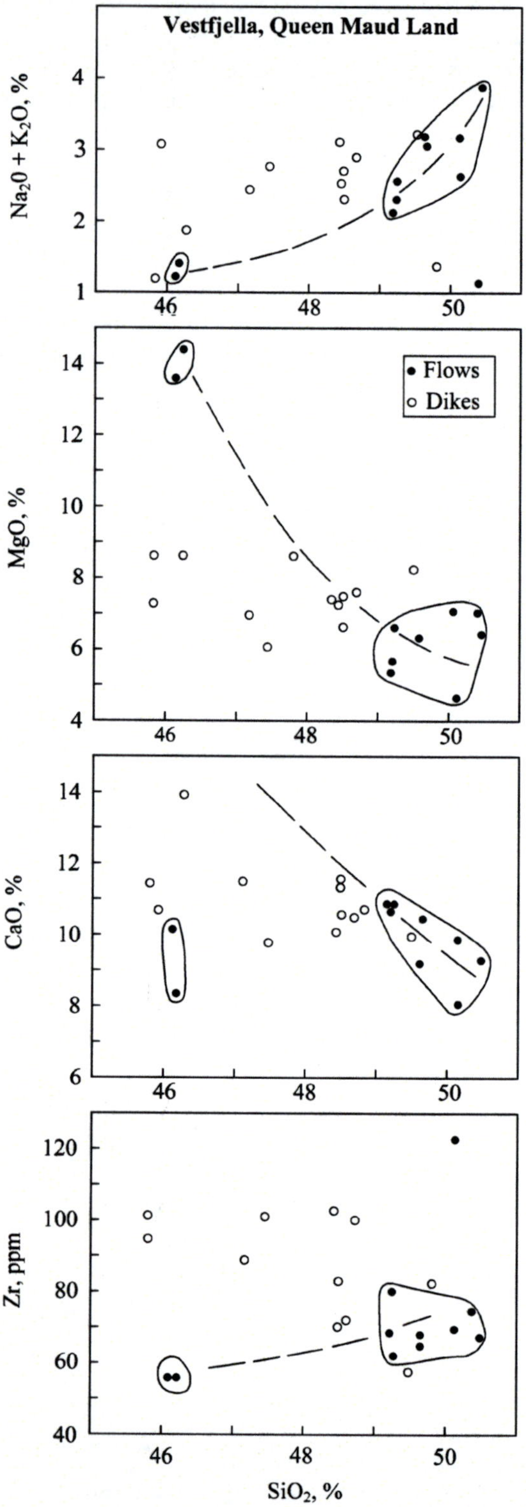

Fig. 14.11 The concentrations of alkali metals, alkaline earths, silica, and zirconium of most basalt flows in southern Vestfjella in Fig. 14.10 vary only within narrow limits. Two olivine-rich flows from Muren and Kjakebeinet have low silica concentrations (46.2%) and are correspondingly enriched in altered olivine (MgO = 14.0%). The chemical composition of the other flows from Vestfjella resemble the composition of the flows of the Kirwan Excarpment in Fig. 14.4 (Data from Furnes and Mitchell (1978))

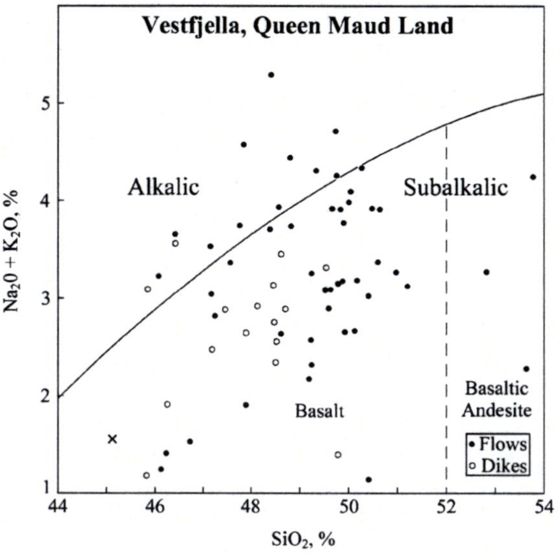

Fig. 14.12 The basalt flows and dolerite dikes/sills of the Kirwan Volcanics in southern Vestfjella scatter widely on the alkali-silica diagram. However, most remain in the subalkalic field and only a few transgress into alkalic territory (Data from Hjelle and Winsnes (1972), Furnes and Mitchell (1978), and Furnes et al. (1987))

Transantarctic Mountains most of which are classified as basaltic andesites or even as andesites on the silica-alkali metals diagram.

14.2.2 Isotopic Age Determinations

The whole-rock K-Ar dates of basalt dikes and lava flows in the nunataks of Vestfjella in Fig. 14.13 vary widely from 90 to 695 Ma (Appendix 14.5.1). Dates based on plagioclase and groundmass reported by Peters et al. (1991) also scatter. Nevertheless, more than half of the dates are between 160 and 180 Ma, which defines the Middle Jurassic Epoch. These dates support the general conclusion that the Kirwan Volcanics of Queen Maud Land were erupted at about the same time as the Kirkpatrick Basalt and the Ferrar Dolerite in the Transantarctic Mountains and Tasmania (Chapters 12 and 13). The basalt *dikes* dated by Rex (1967), Furnes and Mitchell (1978), and Peters et al. (1991) yielded slightly younger K-Ar dates than the lava *flows*:

Dikes: 166.4 ± 2.8 ($2\bar{\sigma}$), N = 19
Flows: 179.2 ± 4.4 ($2\bar{\sigma}$), N = 10

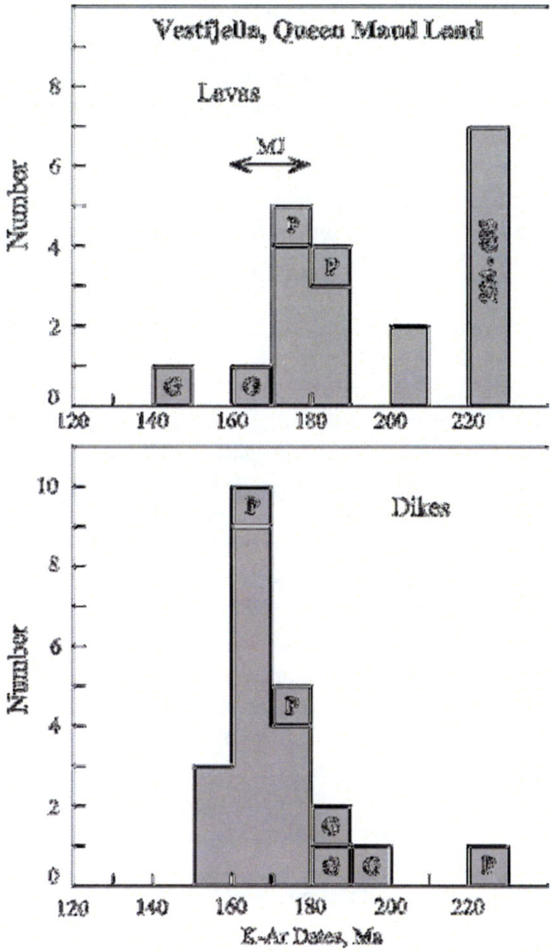

Fig. 14.13 Whole-rock K-Ar dates of basalt flows and dikes of the Kirwan Volcanics in the nunataks of Vestfjella in western Queen Maud Land. MJ = Middle Jurassic, P = plagioclase, G = ground mass (Data from Rex (1967), Furnes and Mitchell (1978), Furnes et al. (1987), and Peters et al. ((1991) listed in Appendix 14.5.1)

The K-Ar dates of basalt and dolerite in Milorgfjella (Heimefrontfjella) reported by Rex (1972) range from 162 to 179 Ma with a mean of 172 ± 6 Ma (Middle Jurassic). However, a pre-Permian dolerite dike in the Proterozoic basement of Milorgfjella yielded a K-Ar date of 455 ± 15 Ma (Ordovician). Two dolerite dikes at Mannefallknausane also formed during the Ordovician magmatic event although one was altered and yielded a low K-Ar date of 258 ± 10 Ma, while the other has a low potassium concentration (0.14%) and yielded a poorly defined date of 564 ± 100 Ma. Furnes et al. (1987) reported anomalously old K-Ar dates in excess of 300 Ma for basalt from West Steinkjeften and West Muren in Vestfjella. These older dates are probably caused by the presence of excess radiogenic ^{40}Ar that entered the basalt magma while it was in transit through the Precambrian rocks of the continental crust. An example of this phenomenon occurs in the Jurassic diabase dikes in Liberia, West Africa, that intruded the Precambrian basement rocks of this region. Dalrymple et al. (1975) reported that the K-Ar dates of pyroxene and plagioclase of Liberian diabase dikes range from 177 ± 4 to 1236 ± 36 Ma. The same authors confirmed the presence of excess radiogenic ^{40}Ar in these minerals by obtaining stepwise-release $^{40}Ar/^{39}Ar$ spectra. The initial $^{87}Sr/^{86}Sr$ ratios of the Liberian diabase dikes also scatter widely on a Rb-Sr isochron diagram (Mauche et al. 1989).

Unfortunately, the basalt flows on the southern nunataks of Vestfjella (Fig. 14.10) cannot be dated by the whole-rock Rb-Sr method. The samples that were analyzed by Furnes et al. (1987) scatter on the Rb-Sr isochron diagram in Fig. 14.14a and do not define a unique straight line. Instead, we have drawn a reference line based on an assumed age of 170 Ma and an initial $^{87}Sr/^{86}Sr$ ratio of 0.70416 ± 0.00021 ($2\bar{\sigma}$) which is the average of the initial ratios of the specimens in this set. This value is identical to the initial ratio of (0.70418) of the basalt flows on nunatak B of the Kirwan Escarpment in Fig. 14.6. The failure of the lava flows at both locations to define Rb-Sr isochrons indicates that either the isotope composition of strontium in the respective magmas was not homogeneous, or that the rocks gained or lost rubidium and strontium after crystallization, or both.

Although the basalt dikes that intrude the flows of southern Vestfjella are reported to be unaltered, the samples analyzed by Furnes et al. (1982) also scatter on the isochron diagram in Fig. 14.14b. Nevertheless, the Jurassic age of these rocks is assured because they overlie *Glossopteris*-bearing sandstones of the Amelang Formation which crops out in Fossilryggen. However, basalt flows are not present at that locality.

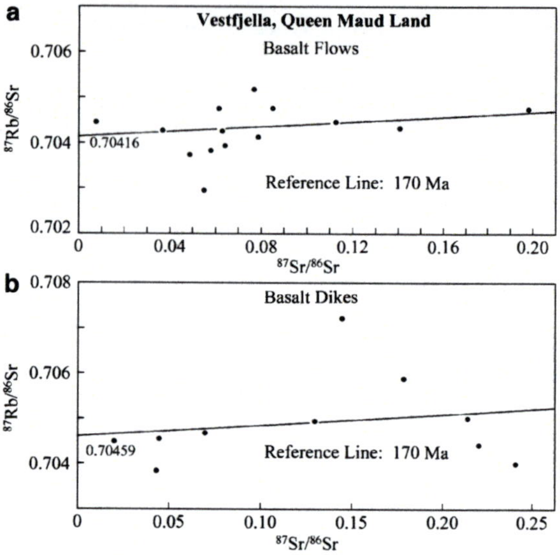

Fig. 14.14 (**a**) The data points representing basalt flows on the southern nunataks of Vestfjella scatter above and below a 170 Ma reference line with an intercept of 0.70416 ± 0.00021 ($2\bar{\sigma}$). The data points do not define an isochron because the rocks were either altered after they crystallized, or because the $^{87}Sr/^{86}Sr$ ratio of the magma was not uniform, or both. Therefore, this group of rocks cannot be dated by the whole-rock Rb-Sr method (Data from Furnes et al. (1987)). (**b**) The basalt dikes that intruded the flows in the southern nunataks of Vestfjella also deviate from the 170 Ma reference line and cannot be dated by the Rb-Sr method. The initial $^{87}Sr/^{86}Sr$ ratio of the dikes at 170 Ma is 0.70459 ± 0.00061 ($2\bar{\sigma}$) (Data by Furnes et al. (1982))

14.2.3 Petrogenesis

Several research groups have contributed to the ongoing effort to explain the origin of the basalt magma that was erupted in Queen Maud Land and to relate these volcanic rocks to the Karoo Basalt of southern Africa (Furnes and Mitchell 1978; Faure et al. 1979; Furnes et al. 1982, 1987; Lövlie 1988; Harris et al. 1990; Luttinen and Siivola 1997; Luttinen et al. 1994, 1998).

The concentrations of major elements of the basalts in southern Vestfjella vary with silica concentrations in Fig. 14.11 in ways that are at least partly attributable to fractional crystallization in a closed magma chamber. In addition, the magma seems to have been contaminated by assimilating partial melts that formed in the Precambrian basement rocks through which the basalt magma was rising toward the surface. These contaminants changed the chemical composition of the magma and altered its $^{87}Sr/^{86}Sr$ ratio. The resulting contaminated magma was erupted through deep crustal rifts and formed lava flows which reacted with hydrothermal fluids consisting largely of heated groundwater. Several authors have commented about the evident

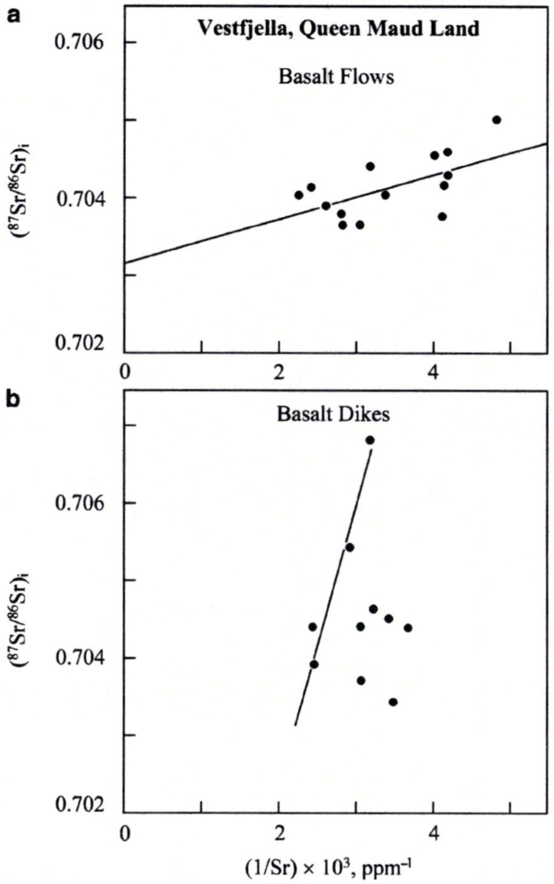

Fig. 14.15 (**a**) The data points representing the basalt flows on the nunataks of Vestfjella scatter in coordinates of the reciprocal Sr concentrations and the corresponding initial $^{87}Sr/^{86}Sr$ ratios corrected for decay of ^{87}Rb for 170×10^6 years. If these samples are mixtures of a mantle-derived magma with varying amounts of contaminant having a well-defined Sr concentration and $^{87}Sr/^{86}Sr$ ratio at 170 Ma, then all of the resulting mixtures must form a straight line on this diagram (Faure 1998, Section 18.5). The scatter of the data points means either that the contaminant had a range of $^{87}Sr/^{86}Sr$ ratios and Sr concentrations, or that the rocks gained or lost Sr during hydrothermal alteration after they had crystallized. The straight line fitted to the data points by least-squares regression has a correlation coefficient of 0.60 and a slope of 0.2893 (Data from Furnes et al. (1987)). (**b**) The initial $^{87}Sr/^{86}Sr$ and 1/Sr ratios of basalt dikes in Vestfjella also scatter and most are not simple two-component mixtures. Four samples appear to define a mixing line between a hypothetical magma ($^{87}Sr/^{86}Sr \sim 0.7035$, Sr ~ 435 ppm) and a contaminant ($^{87}Sr/^{86}Sr > 0.707$, Sr < 315 ppm) (Data from Furnes et al. (1982))

state of alteration of the Kirwan Volcanics in Queen Maud Land (Hjelle and Winsnes 1972; Furnes and Michell 1978; Harris et al. 1990). Lövlie (1988) investigated the effect of the epidote-prehnite facies of low-grade regional metamorphism on the magnetic properties of the flows.

The numerous basalt dikes which intruded the flows of southern Vestfjella are *unaltered* in contrast to the flows (Furnes and Mitchell 1978). Nevertheless, Furnes et al. (1982) reported that the initial $^{87}Sr/^{86}Sr$ ratios of ten dikes range from 0.70347 to 0.70687 and average 0.70460 ± 0.00061 ($2\bar{\sigma}$). This result reinforces the evidence that the initial $^{87}Sr/^{86}Sr$ ratios of basalt flows and dikes of the Kirwan Volcanics in Queen Maud Land are not constant and that average values of this ratio are *lower* than the initial $^{87}Sr/^{86}Sr$ ratios of the basalt flows and dolerite sills of the Ferrar Group in the Transantarctic Mountains.

The variation of the chemical compositions and initial $^{87}Sr/^{86}Sr$ ratios of the Jurassic basalt flows and dikes of Queen Maud Land may have been caused by contamination of the basalt magma. We tested this hypothesis by plotting the initial $^{87}Sr/^{86}Sr$ ratios of flows and dikes from Vestfjella at 170 Ma versus the reciprocals of their strontium concentrations. Although mixing theory indicates that two-component mixtures form straight lines in coordinates of 1/Sr and $^{87}Sr/^{86}Sr$, when a homogeneous magma assimilates varying amounts of contaminants that have a range of $^{87}Sr/^{86}Sr$ ratios and strontium concentrations, the resulting mixtures scatter and form a fan-shaped array of mixing lines that converge to the coordinates of the magma.

The data for basalt *flows* from the southern nunataks of Vestfjella published by Furnes et al. (1987) scatter above and below the best-fit line in Fig. 14.15a. This line has a correlation coefficient of 0.60 and a positive slope of 0.2893. Nevertheless, the scattering of the data points means that these flows are not two-component mixtures. Instead, either they formed from magma that assimilated contaminants that had a range of $^{87}Sr/^{86}Sr$ ratios and strontium concentrations, or the rocks were altered after they crystallized, or both.

The basalt *dikes* in Vestfjella analyzed by Furnes et al. (1982) also scatter in Fig. 14.15b, although four samples approach a mixing line between a hypothetical magma and a contaminant having a higher $^{87}Sr/^{86}Sr$ ratio than the magma (>0.707) and a lower strontium concentration (<315 ppm). If the basalt dikes were not altered by hydrothermal solutions, then the scatter of data points is presumably the result of magma contamination alone.

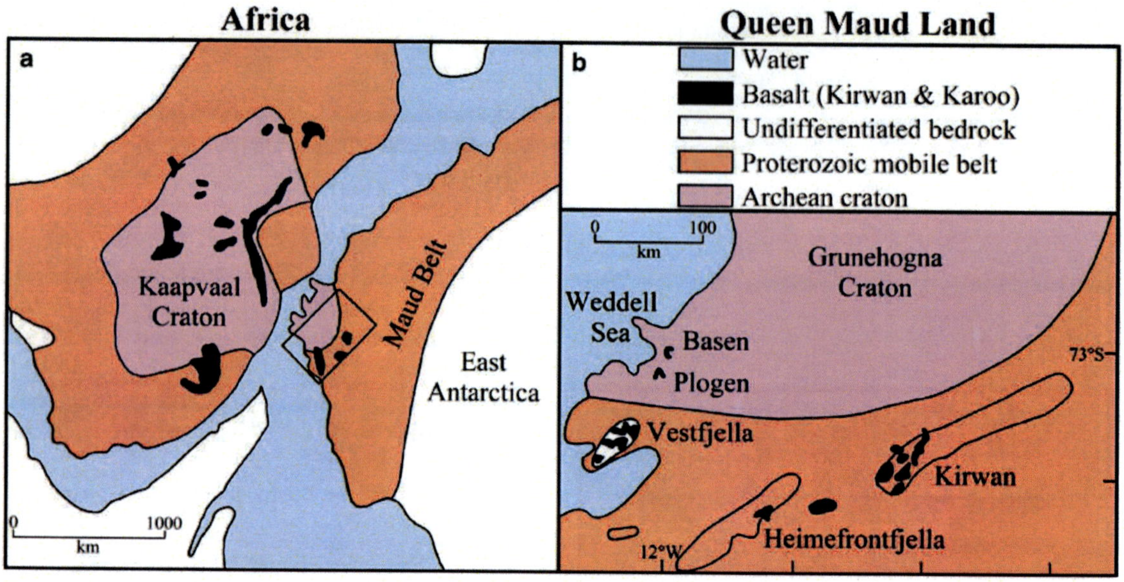

Fig. 14.16 (**a**) The Archean gneisses of the Grunehogna craton in Queen Maud Land of East Antarctica are an extension of the Kaapvaal craton of southern Africa when both are restored to the positions they occupied when the Kirwan Basalt and the Karoo Basalt were extruded during the Jurassic Period. (**b**) The continental flood basalts of the Kirwan Volcanics are exposed on the nunataks of Plogen and Basen which continue the northeasterly trend of the nunataks of Vestfjella (Adapted from Luttinen et al. (1998))

The Rb-Sr systematics indicate that the flows as well as the dikes of Vestfjella originated in an open magmatic system in which the chemical composition of the magma was altered by interaction with the Precambrian basement rocks that were characterized by having higher $^{87}Sr/^{86}Sr$ ratios but lower strontium concentration than the mantle-derived basalt magma.

14.2.4 Permian Sedimentary Rocks

Hjelle and Winsnes (1972) described sedimentary rocks of the Amelang Formation (Section 14.1) at Fossilryggen which consists of a narrow ridge, 2 km long, and a small nunatak located about 30 km east of the main trend of Vestfjella (Fig. 14.9). The sedimentary rocks are 43 m thick and consist of cross-bedded quartz sandstone (25 m) overlain by interbedded siltstones and quartz sandstone some of which contain coarse worm tracks and large sandy concretions. The exposed section is capped by dark-colored clay-shale (up to 10 m). The sedimentary rocks contain *Glossopteris*, *Gangamopteris*, and *Vertebraria* of Early Permian age. The uppermost shale layer exposed at Fossilryggen is a coal shale containing abundant plant fragments. These sedimentary rocks dip northwest at low angles and were intruded by sills and dikes of dolerite. Hjelle and Winsnes (1972) indicated that flat-lying Late Carboniferous or Permain sedimentary rocks occur also in Heimefrontfjella and in some of the basalt nunataks of the Kirwan Escarpment as confirmed later by Harris et al. (1990).

The presence of Permian sedimentary rocks at Fossilryggen demonstrates that non-marine Beacon-like sedimentary rocks were deposited in Queen Maud Land and that these kinds of rocks are not restricted to the Beacon Supergroup of the Transantarctic Mountains. Therefore, the sedimentary rocks at Fossilryggen fill a gap between the Beacon Supergroup of Antarctica and the Karoo Group in southern Africa and thereby reinforce the juxtaposition of these continents prior to the break-up of Gondwana.

14.3 Plogen and Basen

The northeasterly trend of the nunataks of Vestfjella is continued by two large nunataks called Plogen and Basen. The nunataks in Figs. 14.16a and b are located on the Grunehogna craton which is considered to be an extension of Kaapvaal craton of southern Africa. The geology and tectonics of the Grunehogna craton and the surrounding Proterozoic rocks of the Maud fold-belt were described by Groenewald et al. (1995), Jacobs et al. (1993), and Wolmarans and Kent (1982). Additional information about the geology of this region was provided by Neethling (1969) and by Roots (1969).

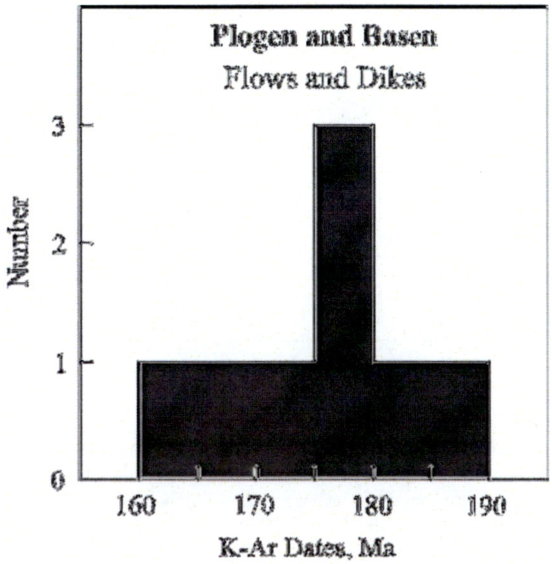

Fig. 14.17 Thirteen K-Ar dates of basalt flows and dikes on the nunataks Plogen and Basen in northernmost Vestfjella listed in Appendix 14.5.1 range from 90 to 352 Ma. Eight of these dates lie between 160 and 190 Ma and three form the peak between 175 and 180 Ma in this diagram. The average of these three dates is 178 ± 2 Ma (1σ). On the basis of this evidence, the age of the lava flows and dikes of northern Vestfjella is early Middle Jurassic (IUGS 2002) (Data from Peters et al. (1991))

14.3.1 Geology and Geochemistry

The nunataks Plogen and Basen consist of basalt flows of the Kirwan Volcanics cut by basalt dikes that strike northeast. The flows are interbedded with minor layers of pluvial sediment and the total stratigraphic thickness of the flow and sediment layers is greater than 900 m on Plogen and greater than 300 m on Basen (Luttinen and Siivola 1997).

The K-Ar dates of the flows and dikes on these nunataks reported by Peters et al. (1991) range widely from 90 to 352 Ma based on analyses of plagioclase, groundmass, and whole-rock samples. Eight dates that range only from 160 to 190 Ma are included in Fig. 14.17. The three dates that occupy the interval from 175 to 180 Ma have a mean of 178 ± 2 Ma (1σ) which corresponds to the early Middle Jurassic on the IUGS (2002) geologic time scale.

The Plogen nunatak consists of about 30 basalt flows most of which are porphyritic with phenocrysts of plagioclase, olivine, and pyroxene in a groundmass of plagioclase laths, clinopyroxene, Fe-Ti oxides, and mesostasis. The plagioclase phenocrysts of many flows are altered to sericite and, in some cases, have been converted to pseudomorphs of saussurite (mixture of alteration products containing chlorite, calcite, albite, and epidote). Olivine was also altered, but clinopyroxene is pristine. In contrast to the pervasive alteration of the flows, the basalt dikes are unaltered. The flow tops are typically amygdaloidal and the former vesicles are filled with chlorite, quartz, prehnite, and calcite.

The Ti/Y ratios in Fig. 14.18 identify the flows and dikes of Plogen and Basen as "low-Ti" basalt.

Luttinen and Siivola (1997) divided the flows and dikes on Plogen and Basen into three chemical types (CT) based on the concentration ratios of certain incompatible elements: Ti/P, Nb/Ta, Ti/Zr, Ti/Y, and the magnesium number defined as the molar ratio of Mg/(Mg + Fe^{2+}). The three chemical types are differentiated from each other in Figs. 14.19a and b in coordinates of Ti/P and Nb/Ta versus Ti/Zr. Luttinen and Siivola (1997) noted that the CT-1 and CT-2 flows and dikes resemble the Karoo Basalt of southern Lebombo and the Rooi Rand Dolerites of southeastern Africa which demonstrates that the Karoo volcanic province of Africa extends into western Queen Maud Land. The CT-3 flows are similar to the basalt flows in southern Vestfjell but are not represented in southern Africa. The element ratios that characterize the different groups of flows and dikes are not the result of fractional crystallization of basalt magma but probably reflect the chemical heterogeneity of the magma sources.

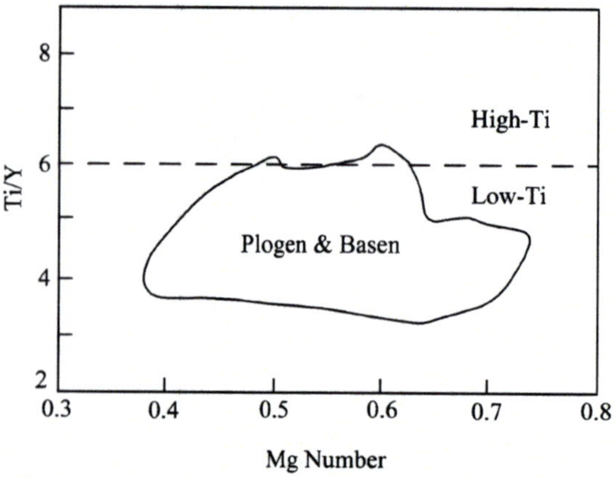

Fig. 14.18 The flows and dikes on Plogen and Basen have titanium/yttrium (Ti/Y) ratios less than six which is characteristic of low-titanium basalts. The magnesium number of these rocks, defined as the atomic (Mg/Mg+Fe^{2+}) ratio, varies widely from about 0.4 to 0.7 (Adapted from Luttinen and Siivola (1997))

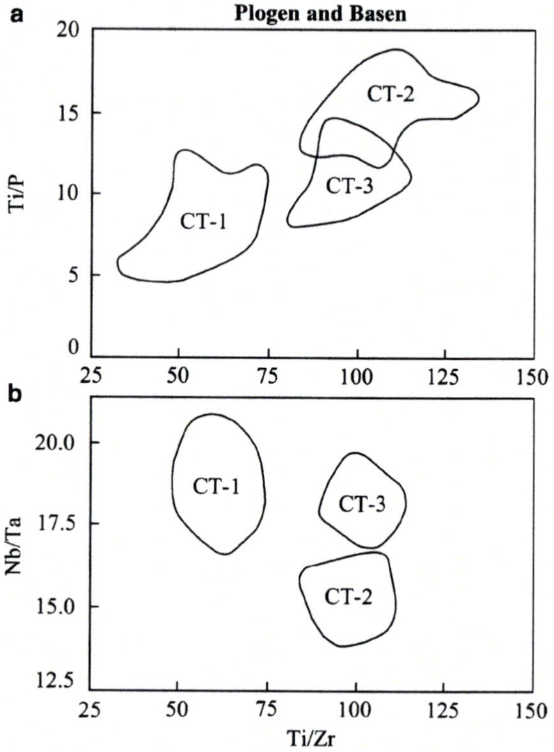

Fig. 14.19 The chemical types (CT) of flows and dikes on Plogen and Basen are defined in terms of concentration ratios of Ti/P vs. Ti/Zr (**a**) and Nb/Ta vs. Ti/Zr (**b**). The chemical compositions of CT-1 and CT-2 resemble certain basalt flows of the Karoo Group in southern Africa, whereas the CT-3 samples resemble the Kirwan Volcanics on the nunataks of southern Vestfjella. A fourth chemical type (CT-4) was later added by Luttinen et al. (1998) (Data and interpretation from Luttinen and Siivola (1997))

14.3.2 Isotopic Compositions

In a subsequent work, Luttinen et al. (1998) measured isotopic compositions of strontium and neodymium and added a fourth magma type (CT-4) based on the chemical compositions of two dikes on Basen which are subalkaline tholeiites but have higher Ti/Zr and Ti/Mg# than the CT-3 samples. They also have higher concentrations of MgO and iron than most of the flows and dikes on Plogen and Basen probably because they contain excess olivine.

The initial $^{87}Sr/^{86}Sr$ and $^{143}Nd/^{144}Nd$ ratios at 180 Ma of the flows and dikes of Plogen and Basen range widely from 0.70401 to 0.70794 and from 0.511596 to 0.51259, respectively. These isotope ratios are not correlated with the reciprocals of the concentrations for strontium and neodymium, respectively, which effectively rules out magma contamination. In addition, secondary alteration of the flows is unlikely because neodymium is not a mobile element in weathering environments. Although the data points do not form mixing lines, Fig. 14.20 demonstrates that the initial $^{143}Nd/^{144}Nd$ ratios (at 180 Ma) and the 1/Nd ratios cluster into the same chemical types that were identified by Luttinen and Siivola (1997). The strontium-mixing diagram (not shown) also breaks up into the CT clusters.

The isotopic ratios of strontium and neodymium scatter on their respective isochron diagrams and do not record the crystallization age of these rocks. Although the Rb-Sr systematics of four CT-1 samples approach a straight line (not shown), the date derivable from it (309 Ma) is geologically meaningless. The Sm-Nd data in Fig. 14.21 reproduce the CT groups on the isochron diagram and thereby confirm that the $^{143}Nd/^{144}Nd$ ratios were inherited from the magma sources and do not depend on the time elapsed since crystallization. The complete failure of the Sm-Nd (and Rb-Sr) systematics to indicate the age of the Kirwan Volcanics on Plogen and Basen cannot be explained by isotope fractionation during fractional crystallization of the magma because isotope fractionation effects are eliminated by the fractionation

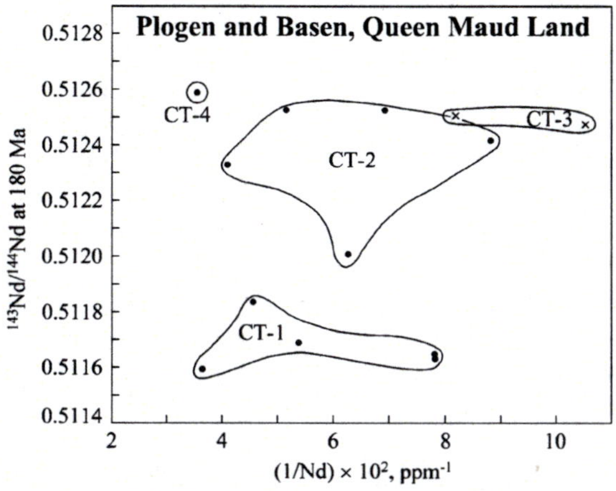

Fig. 14.20 The initial $^{143}Nd/^{144}Nd$ and 1/Nd ratios do not form mixing lines on this diagram as expected but form clusters that correspond to the chemical types (CT-1, CT-2, CT-3, and CT-4) defined by Luttinen and Siivola (1997) on the basis of chemical criteria. These data therefore demonstrate that the heterogeneity of the initial $^{143}Nd/^{144}Nd$ ratios of the "CT-magmas" are not the result of mixing of magma with crustal contaminants. Instead, the heterogeneous isotopic composition of neodymium was apparently inherited from the magma sources (Isotopic data from Luttinen et al. (1998))

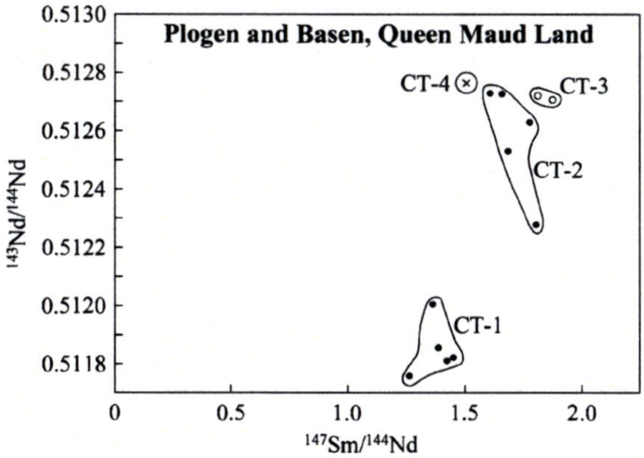

Fig. 14.21 The data points representing the flows and dikes on Plogen and Basen scatter widely in coordinates of $^{147}Sm/^{144}Nd$ and $^{143}Nd/^{144}Nd$ and fail to form the expected linear array from which the age of crystallization can be determined. Instead, the data form four groups that correspond to the chemical types (CT-1, CT-2, CT-3, and CT-4). The failure of the data points to define a whole-rock Sm-Nd isochron confirms yet again the heterogeneity of the $^{143}Nd/^{144}Nd$ ratios of magma source of these rocks (Data from Luttinen et al. (1998))

corrections that are routinely made to measured values of the respective isotope ratios (Faure and Mensing 2005).

14.3.3 Petrogenesis

The most direct evidence concerning the petrogenesis of the flows and dikes on Plogen and Basen is derivable from the Sr-Nd isotope-mixing diagram in Fig. 14.22. The initial isotope ratios (at 180 Ma) of these rocks form points in quadrant 4 which contains rocks and minerals of the continental crust that have high $^{87}Sr/^{86}Sr$ and low $^{143}Nd/^{144}Nd$ ratios (Faure and Mensing 2005). Part 4a of quadrant 4 is occupied by granitic rocks of Paleozoic age, whereas part 4b contains granitic gneisses of Precambrian age (Faure 1986, Fig. 13.8). The initial isotope ratios of the volcanic rocks on Plogen and Basen form clusters that correspond to the CT groups defined by Luttinen and Siivola (1997) on the basis of *chemical* criteria. The CT-1 flows and dikes form an elongated array in area 4b of Fig. 14.22 which suggests that the magma of this group of rocks formed primarily by partial melting of Precambrian rocks at depth in the continental crust of East Antarctica. The magma sources of the CT-1 group may have been Archean granitic basement rocks of the Grunehogna craton which crystallized at 2.0 Ga according to a Rb-Sr date reported by Halpern (1970).

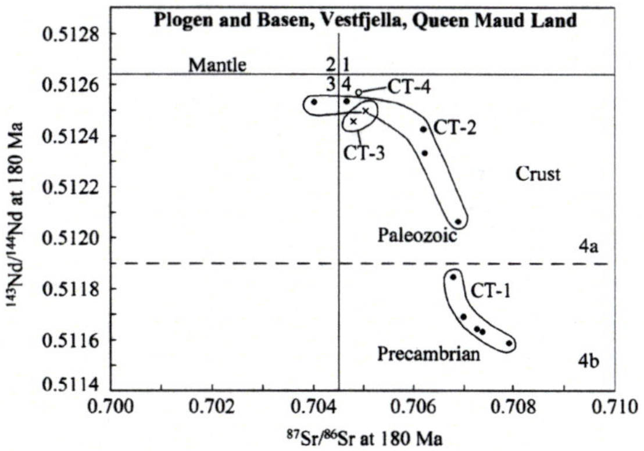

Fig. 14.22 The basalt flows and dikes of Middle Jurassic age on the nunataks Plogen and Basen of Vestfjella have initial $^{87}Sr/^{86}Sr$ and $^{143}Nd/^{144}Nd$ ratios at 180 Ma that place them in quadrant 4 of the Sr-Nd mixing diagram. The data points form the same clusters that were originally defined by Luttinen and Siivola (1997) based on *chemical* criteria. The data clusters result from the contamination of mantle-derived basalt magma with varying amounts of crustal rocks of Paleozoic and Precambrian ages. The magmatic activity and petrogenesis may have been a consequence of a plume from the asthenospheric mantle that exerted extensional stresses on the underside of the lithospheric plate which caused large volumes of basalt magma to form by decompression melting of the rocks in the plume head and adjacent lithosphere (Data from Luttinen et al. (1998))

The CT-2 flows and dikes plot in area 4a and extend toward the part of Fig. 14.22 that contains mantle-derived basalts on oceanic islands and mid-ocean ridges. Accordingly, the CT-2 flows and dikes formed from mantle-derived basalt magma that assimilated granitic rocks of Paleozoic age at intermediate depth in the continental crust.

The assimilation of crustal rocks by hot mantle-derived basalt magma is facilitated in cases where the temperature of the crust approaches the melting point of granitic rocks. Under these circumstances a partial melt may form by the additional heat emanating from the basalt magma that is rising from depth. Such silica-rich partial melts move with the basalt magma toward the surface and are mixed with it while both are in transit.

The CT-3 and CT-4 magmas originated primarily in the subcrustal mantle with only limited input from the continental crust. These magmas as well as those of the CT-2 group of flow and dikes, could have originated by decompression melting in the head of a plume that rose from depth in the mantle (i.e., the asthenosphere) until it impinged on the underside of the lithospheric mantle in which East Antarctica and Africa were embedded. The heads of such plumes spread out laterally and thereby exert extensional stresses on the overylying lithospheric plate. The resulting fractures (triple junction) propagate upward toward the surface and may widen into rift valleys that split continents as they open into ocean basins (e.g., the Red Sea). The decrease in pressure causes large-scale decompression melting of the rocks in the head of the plume and in the adjacent lithospheric mantle. The resulting basalt magma rises toward the surface by following the rifts and may be erupted through crustal fissures without forming volcanic mountains. In this way, the plume hypothesis can account for the petrogenesis of the Kirwan Volcanics of Queen Maud Land and of continental flood basalts elsewhere in the world (Faure 2001).

14.4 Summary

Several clusters of nunataks in western Queen Maud Land are composed of basalt flows and dikes which are collectively known as the Kirwan Volcanics. These flows are underlain by flat-lying sandstones and shales of the Permian Amelang Formation which overlies Neoproterozoic basement rocks. The basalt flows of the Kirwan Volcanics are composed of subalkaline tholeiites similar to the Kirkpatrick Basalt. However, the $^{87}Sr/^{86}Sr$ ratios of the Kirwan Volcanics are consistently lower than those of the Kirkpatrick Basalt even though both suites of basalt flows crystallized at about the same time during the Jurassic Period. Hence, the magma of the Kirwan Volcanics originated by a different petrogenetic process or from a different source than the magma of the Ferrar Group in the Transantarctic Mountains.

The age of the Kirwan Volcanics is known to be post-Permian because the sedimentary rocks of the underlying Amelang Formation contain *Glossopteris*. The K-Ar dates of the basalt flows range widely because these rocks lost varying proportions of radiogenic ^{40}Ar although several samples contain excess ^{40}Ar. In addition, the data points representing flows and dikes of the Kirwan Volcanics scatter on a Rb-Sr geochronometry diagram. Nevertheless, reasonable interpretations of the K-Ar dates support a value of about 179 ± 4.4 Ma for the crystallization age of the basalt flows of the Kirwan Volcancs.

The comparatively low $^{87}Sr/^{86}Sr$ ratios of the basalt flows of the Kirwan Volcanics, as well as subtle aspects of their chemical compositions, associate these rocks with certain basalt flows and dolerite sills of the Jurassic Karoo Group of southern Africa. The simultaneous eruption of basalt during the Jurassic Period in Queen Maud Land and parts of southern Africa resulted from the separation of Africa from East Antarctica by the formation of a rift between them. This rift was part of a triple junction caused by the stress exerted by a mantle plume on the continental crust and lithospheric mantle of Gondwana. The rifting caused decompression melting within the crust/upper mantle of Gondwana and the resulting basalt magma was erupted through the same rifts that caused the magmas to form. In this way, the Jurassic magmatic activity in Queen Maud Land and southern Africa was the start of the break-up of Gondwana which ended during the Late Cretaceous with the separation of Australia from East Antarctica.

14.5 Appendices

14.5.1 K-Ar Age Determinations of Jurassic Basalt Flows and Dikes in Queen Maud Land

Description		K-Ar date, Ma	Reference
Milorgfjella (Heimefrontfjella)			
Basalt	74°36′S, 010°00′W	172 ± 7	2
Basalt	74°28′S, 008°15′W	162 ± 6	2
Basalt	74°36′S, 010°05′W	173 ± 7	2
Lava	74°37′S, 010°02′W	173 ± 5	2
Dolerite	74°19′S, 009°50′W	179 ± 7	2
Dolerite	74°17′S, 009°45′W	455 ± 15	2
Mannefallknausane[a]			
Dolerite	74°36′S, 014°25′W	258 ± 10 (altered)	2
Dolerite	74°36′S, 014°20′W	564 ± 100	2
$\lambda_e = 0.584 \times 10^{-10}$ year^{-1}, $\lambda_\beta = 4.72 \times 10^{-10}$ year^{-1}, $^{40}K/K = 1.19 \times 10^{-2}$ atom percent			
Vestfjella (lavas)			
Basalt	Skansen	234 ± 10	3
Basalt	Pagodromen	180 ± 4	3
Basalt	Pagodromen	186 ± 4	3
Basalt	Pagodromen	185 ± 4	3
Basalt	Kjakebeinet	171 ± 2	3
Basalt	Kjakebeinet	172 ± 3	3
Basalt	E. Steinkjeften	185 ± 3	3
Basalt	E. Steinkjeften	205 ± 4	3
Basalt	W. Steinkjeften	224 ± 7	3
Basalt	W. Steinkjeften	225 ± 2	3
Basalt	W. Steinkjeften	317 ± 10	3
Basalt	Muren	205 ± 6	3
Basalt	W. Muren	328 ± 5	3
Plagioclase	W. Muren	330 ± 11	3
Basalt	W. Muren	695 ± 11	3
$\lambda_e = 0.581 \times 10^{-10}$ year^{-1}, $\lambda_\beta = 4.962 \times 10^{-10}$ year^{-1}, $^{40}K/K = 1.167 \times 10^{-2}$ atom percent			
Vestfjella (dikes)			
Dolerite A	73°50′S, 015°00′W	168 ± 6	1
Dolerite B	73°50′S, 015°00′W	172 ± 6	1
$\lambda_e = 0.584 \times 10^{-10}$ year^{-1}, $\lambda_\beta = 4.72 \times 10^{-10}$ year^{-1}, $^{40}K/K = 1.19 \times 10^{-2}$ atom percent			
Basalt	VF44	156 ± 4	4
Basalt	VF41	158 ± 3	4
Basalt	VF45	158 ± 2	4
Basalt	VF110	162 ± 2	4
Basalt	VF34	164 ± 2	4
Basalt	VF82	164 ± 3	4
Basalt	VF120	164 ± 2	4
Basalt	VF83	166 ± 2	4
Basalt	VF30	169 ± 5	4
Basalt	VF32	165 ± 2	4
Basalt	VF116	168 ± 2	4
Basalt	VF40	171 ± 3	4
Basalt	VF128	170 ± 2	4
Basalt	VF129	172 ± 2	4

(continued)

14.5.1 (continued)

Description		K-Ar date, Ma	Reference
Milorgfjella (Heimefrontfjella)			
$\lambda_e = 0.584 \times 10^{-10}$ year^{-1}, $\lambda_\beta = 4.72 \times 10^{-10}$ year^{-1}, ^{40}K/K = 1.19×10^{-2} atom percent			
Kirwan Escarpment			
		172 ± 10	6
Decay constants and ^{40}K/K were not specified			
Plogen and Basen (northern Vestfjella)			
B1	Flow, whole rock	176 ± 10	5
B2	Flow, whole rock	352 ± 2	5
B3	Dike, plagioclase	295 ± 19	5
B3	Dike, groundmass	197 ± 10	5
B6	Flow, groundmass	148 ± 10	5
B6	Flow, whole rock	189 ± 10	5
B9	Flow, groundmass	90 ± 9	5
B10	Flow, plagioclase	179 ± 13	5
B10	Flow, groundmass	169 ± 13	5
B13	Dike, groundmass	180 ± 11	5
B16	Dike, plagioclase	174 ± 13	5
B16	Dike, groundmass	183 ± 20	5
B17	Dike, plagioclase	160 ± 16	5
Decay constants and ^{40}K/K were not specified			

1. Rex (1967), 2. Rex (1972), 3. Furnes et al. (1987), 4. Furnes and Mitchell (1978), 5. Peters et al. (1991), 6. Aucamp et al. (1972).

[a]Called Milorgknausane prior to 1967 (Juckes 1968).

References

Allsopp HL, Neethling DC (1970) Rb-Sr isotopic ages of Precambrian intrusives from Queen Maud Land, Antarctica. Earth Planet Sci Lett 8:66–70

Aucamp APH, Wolmarans LG, Neethling DC (1972) The Urfjell Group, a deformed (?) early Paleozoic sedimentary sequence, Kirwanveggen, western Dronning Maud Land. In: Adie RJ (ed) Antarctica geology and geophysics. Universitetsforlaget, Oslo, Norway, pp 557–561

Bauer W, Hagemann HW, Porscher G, Sachsenhofer RF, Spaeth G (1997) Permian coals from western Dronning Maud Land: Composition, environment, and the influence of Jurassic magmatism on their maturity. In: Ricci CA (ed) The Antarctic Region: Geological evolution and processes. Terra Antartica, Siena, Italy, pp 945–951

Blundell DJ (1964) Orientated dolerite sample from Dronning Maud Land. British Antarctic Survey Bull 4:57

Brown JW (1967) Jurassic dolerites from the Falkland Islands and Dronning Maud Land. British Antarctic Surv Bull 13:89–92

Brunn V von (1964) Note on some basic rocks in western Dronning Maud Land. In: Adie RJ (ed) Antarctic geology. North-Holland, Amsterdam, The Netherlands, pp 415–418

Cox KG, Bristow JW (1984) The Sabie River Basalt Formation of the Lebombo Monocline and south east Zimbabwe. Geol Soc South Africa, Spec Pub 13:125–148

Dalrymple GB, Grommé CS, White RW (1975) Potassium-argon age and paleomagnetism of diabase dikes in Liberia: Initiation of central Atlantic rifting. Geol Soc Amer Bull 86:399–411

Eastin R, Faure G, Neethling DC (1970). The age of the Trollkjellrygg Volcanics of western Queen Maud Land. Antarctic J US 5(5):157–158

Faure G (1986) Principles of isotope geology, 2nd edn. Wiley, New York

Faure G (1998) Principles and applications of geochemistry, 2nd edn. Prentice Hall, Upper Saddle River, NJ

Faure G (2001) Origin of igneous rocks; The isotopic evidence. Springer, Heidelberg, Germany

Faure G, Elliot DH (1971) Isotope composition of strontium in Mesozoic basalt and dolerite from Dronning Maud Land. British Antarctic Surv Bull 25:23–27

Faure G, Mensing TM (2005) Isotopes: Principles and applications, 3rd edn. Wiley, Hoboken, NJ

Faure G, Hill RL, Jones LM, Elliot DH (1972) Isotope composition of strontium and silica content of Mesozoic basalt and dolerite from Antarctica. In: Adie RJ (ed) Antarctic geology and geophysics. Universitetsforlaget, Oslo, Norway, pp 617–624

Faure G, Bowman JR, Elliot DH (1979) The initial $^{87}Sr/^{86}Sr$ ratios of the Kirwan Volcanics of Dronning Maud Land: Comparison with the Kirkpatrick Basalt, Transantarctic Mountains. Chem Geol 26:77–90

Furnes H, Mitchell JG (1978) Age relationships of Mesozoic basalt lava and dykes in Vestfjella, Dronning Maud Land, Antarctica. Norsk Polarinstitutt, Skrifter Nr. 169:45–68. Oslo, Norway

Furnes H, Neumann ER, Sundvoll B (1982) Petrology and geochemistry of Jurassic dykes from Vestfjella, Dronning Maud Land, Antarctica. Lithos 15:295–304

Furnes H, Vad E, Austrheim H, Mitchell JG, Garmann LB (1987) Geochemistry of basalt lavas from Vestfjella and adjacent areas, Dronning Maud Land, Antarctica. Lithos 20:337–356

Groenewald PM, Moyes AP, Grantham GH, Krynauw JR (1995) East Antarctic crustal evolution: Geological constraints and modeling in western Dronning Maud Land. Precambrian Res 75:231–250

Guy-Ohlson D, Lindstrom S (1994) Palaeoecology of the Early Permian strata at Heimefrontjella, Dronning Maud Land, Antarctica. Antarctic Sci 6(4):507–515

Halpern M (1970) Rubidium-strontium date of possibly 3 billion years for a granitic rock from Antarctica. Science 169:977–978

Harris C, Marsh JS, Duncan RA, Erland AJ (1990) The petrogenesis of the Kirwan Basalt of Dronning Maud Land, Antarctica. J Petrol 31(2):341–369

Hill RL (1969) Strontium isotope composition of basaltic rocks in the Transantarctic Mountains, Antarctica. MSc thesis, Department of Geology, The Ohio State University, Columbus, OH

Hjelle A, Winsnes T (1972) The sedimentary and volcanic sequence of Vestfjella, Dronning Maud Land. In: Adie RJ (ed) Antarctic geology and geophysics. Universitetsforlaget, Oslo, Norway, pp 539–546

IUGS (2002) International stratigraphic chart. International Commission of Stratigraphy, UNESCO

Jacobs J, Thomas RJ, Weber K (1993) Accretion and indentation tectonics at the southern edge of the Kaapvall craton during the Kibaran (Grenville) orogeny. Geology 21:203–206

Juckes LM (1968) The geology of Mannefallknausane and part of Vestfjella, Dronning Maud Land. British Antarctic Surv Bull 18:65–78

Juckes LM (1969) Trace-element values for dolerites from western Dronning Maud Land. British Antarctic Surv Bull 22:95–98

Juckes LM (1972) The geology of northeastern Heimefrontfjella, Dronning Maud Land. British Antarctic Surv Bull 65:1–45

Larsson K, Lindstrom S, Guy-Ohlson D (1990) An early palynoflora from Milorgfjella, Dronning Maud Land, Antarctica. Antarctic Sci 2(4):331–334

Lindsay P, Croaker MC, Millsteed B (2002) Geology of the Amelang Formation, Western Dronning Maud Land. In: Antarctica at the close of the Millenium. Roy Soc New Zealand Bull 35:193–197

Lövlie R (1979) Mesozoic palaeomagnetism in Vestfjella, Dronning Maud Land, East Antarctica. Geophys J Roy Astron Soc 59:529–537

Lövlie R (1988) Evidence for deuteric magnetization in hydrothermally altered Mesozoic basaltic rocks from East Antarctica. Phys Earth Planet Inter 52(3/4):352–364

Luttinen AV, Siivola JU (1997) Geochemical characteristics of Mesozoic lavas and dikes from Vestfjella, Dronning Maud Land: Recognition of three distinct chemical types. In: Ricci CA (ed) The Antarctic Region: Geological evolution and processes. Terra Antartica, Siena, Italy, pp 495–503

Luttinen AV, Grind KH, Siivola JU, Räisänen MH (1994) The mafic igneous rocks of Vestfjella, wetern Queen Maud Land, Antarctica. Antarctic Rep Finland 4:12–19

Luttinen AV, Rämo OT, Huhma H (1998) Neodymium and strontium isotopic and trace element compositions of a Mesozoic CFB suite from Dronning Maud Land, Antarctica; Implications for lithosphere and asthenosphere contributions to Karoo magmatism. Geochim Cosmochim Acta 62(15):2701–2714

Mauche R, Faure G, Jones LM, Hoefs J (1989) Anomalous isotopic compositions of Sr, Ar, and O in the Mesozoic diabase dikes of Liberia, West Africa. Contrib Mineral Petrol 101:12–18

Neethling DC (1969) Geology of the Ahlmann Ridge, western Queen Maud Land. In: Craddock C (ed) Geologic map of Antarctica, Folio 12, Sheet 7. American Geographical Society, New York

Neethling DC (1971) South African Earth Science exploration of western Queen Maud Land, Antarctica. PhD dissertation, University of Natal, Natal, South Africa

Neethling DC (1972a) South African Antarctic Earth Science Programme 1959–1969: Solid Earth geophysics, oceanography, and glaciology. South African J Antarctic Res 2:2–15

Neethling DC (1972b) Comparative geochemistry of Proterozoic and Paleo-Mesozoic tholeiites of western Dronning Maud Land. In: Adie RJ (ed) Antarctic geology and geophysics. Universitetsforlaget, Oslo, Norway, pp 603–616

Peters M (1989) Die Vulkanite im westlichen und mittleren Neuschwabenland, Vestfjella, und Ahlmannryggen, Antarktika. Berichte zur Polarforschung, Alfred Wegener-Institut für Polar-und Meeresforschung, Bremerhaven, pp 1–186

Peters M, Haverkamp B, Emmermann R, Kohnen H, Weber K (1991) Palaeomagnetism, K-Ar dating, and geodynamic setting of igneous rocks in western and central Neuschwabenland, Antarctica. In: Thomson MRA, Crame JA, Thomson JW (eds) Geological evolution of Antarctica. Cambridge University Press, Cambridge, UK, pp 549–555

Rex DC (1967) Age of dolerite from Dronning Maud Land. British Antarctic Surv Bull 11:101

Rex DC (1972) Potassium-argon age determinations on volcanic and associated rocks from the Antarctic Peninsula and Dronning Maud Land. In: Adie RJ (ed) Antarctic geology and geophysics. Universitetsforlaget, Oslo, Norway

Roots EF (1969) Geology of western Queen Maud Land. In: Craddock C (ed) Geologic map of Antarctica, Folio 12, Sheet 6. American Geographical Society, New York

Wilson M (1989) Igneous petrogenesis, A global tectonic approach. Unwin Hyman, London, UK

Wolmarans LG, Kent LE (1982) Geological investigations in western Dronning Maud Land, Antarctica: A synthesis. South African J Antarctic Res Suppl 2:1–93

Chapter 15
Break-up of Gondwana and Assembly of Antarctica

Antarctica, as we know it, is a large landmass that is completely surrounded by the Atlantic, Indian, and Pacific oceans. Seen from space, the ice sheets that cover East and West Antarctica gleam by the sunlight they reflect. Actually, the interior of West Antarctica is a large tectonic rift that lies below sea level. Even the subglacial bedrock surface of the Wilkes Basin of East Antarctica is below sea level. In the bathymetric image in Fig. 15.1, Antarctica is depicted as an elevated continent in hues of brownish red and yellow because the elevation of the East Antarctic ice sheet increases from sea level along the coast to about 4,000 m at Dome A in the interior. Even the elevation of the West Antarctic ice sheet exceeds 2,000 m in the Whitmore Mountains and in parts of Marie Byrd Land (e.g., Executive Committee and Flood ranges). Although the geologic history of East Antarctica extends more than three billion years into the Archean Eon, the present appearance of Antarctica dates from the Cenozoic Era that started at the end of the Cretaceous Period 65 million years ago.

In the preceding chapters of this book we have dropped hints that the magmatic activity during the Jurassic Period was a symptom of the tectonic processes that ultimately caused Gondwana to break up, followed by the displacement of the resulting continental fragments to their present positions in the southern hemisphere. The break-up of Gondwana, the concurrent assembly of Antarctica, and the uplift of the Transantarctic Mountains continue to be debated in the geological literature and at international conferences. The data relevant to this spectrum of topics have come from measurements of paleomagnetism combined with isotopic age determinations, from the geographic distribution of fossils, from observations of the displacements along fault zones, and from modeling based on the physical properties of rocks and on the release of heat by elements that have long-lived radioactive isotopes (e.g., uranium, thorium, and potassium). The former connections between the continents of the southern hemisphere are also indicated by the continuity of geological structures and by the similarities of rock units that formed before the continents of the southern hemisphere separated (e.g., the Kirwan Volcanics of Queen Maud Land and the Karoo Basalt of southern Africa). The break-up of Gondwana was a global phenomenon that affected all of Pangea and resulted in a complete rearrangement of the continents in both hemispheres of the Earth.

15.1 The Plume Hypothesis

A plausible explanation for the occurrence of this global upheaval arises from the *plume hypothesis* which we outline briefly here without documentation because that would exceed the scope of this monograph. The plume hypothesis relates the formation of magma to a tectonic process; namely, plumes of hot rocks rise from depth in the asthenospheric mantle until they impinge on the underside of the rigid lithospheric shell of the Earth. After a plume has reached the underside of the lithospheric plate of a continent, the head of the plume expands laterally and may reach a diameter of 1,000 km (i.e., the Afar plume, Gulf of Aden; Schilling 1973). The radial expansion of the plume-head exerts stress on the underside of the lithosphere which causes it to fracture. These fractures propagate upward into the overlying continental crust in the form of rifts which, in time, may widen into oceans (e.g., the Red Sea).

The fractures in the lithospheric mantle reduce the pressure which causes the rocks of the plume head and the adjacent lithosphere to melt by decompression.

G. Faure and T.M. Mensing, *The Transantarctic Mountains: Rocks, Ice, Meteorites and Water*,
DOI 10.1007/978-90-481-9390-5_15,

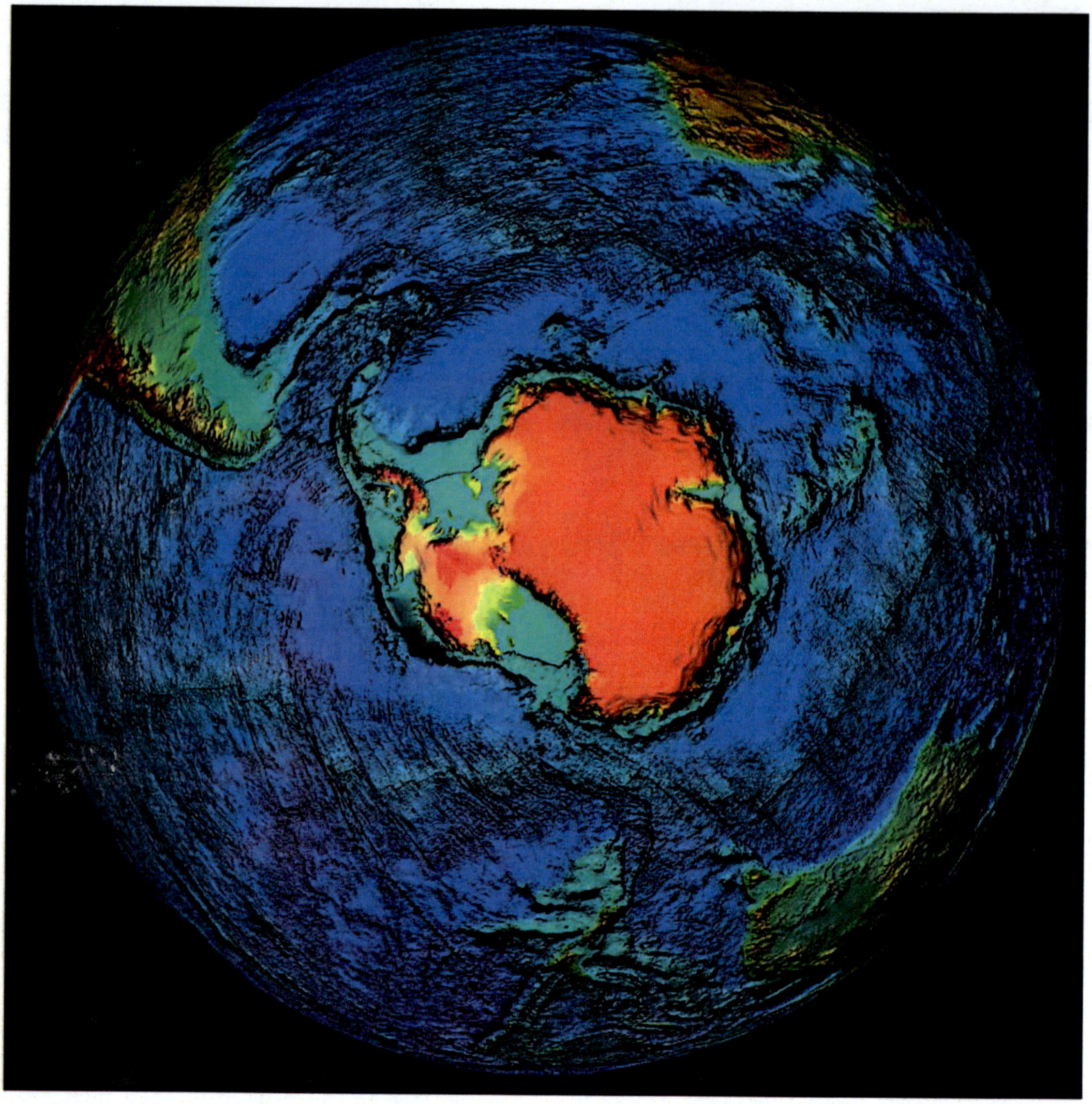

Fig. 15.1 This bathymetric image of Antarctica was prepared by the National Geophysical Data Center (NGDC) and was last revised July 23, 2008. The coloration of Antarctica reflects the high elevation of the East Antarctic ice sheet which reaches 4000 m above sea level at Dome Argus (81°00′S, 077°00′E). Even the elevation of the West Antarctic ice sheet exceeds 2,000 m above sea level in the area of the Whitmore Mountains and in parts of Marie Byrd Land (Image from: <http://www.ngdc.noaa.gov/mgg/global/global.html> The site is maintained by: <carla.j.moore@noaa.gov>)

The basalt magma that is thereby produced intrudes the fractures and rises toward the surface. If the magma reaches the surface, it is erupted to form large plateaus of tholeiite basalt (e.g., the Mesa Range in northern Victoria Land). Part of the magma may also intrude near-surface rocks in the form of sills and dikes (e.g., sills of Ferrar Dolerite), and it may form large tabular intrusives composed of layered gabbro and other products of fractional crystallization (e.g., the gabbros of the Dufek intrusion and related pyroxenites, anorthosites, and granophyres). Note, however, that in some cases the granophyres that cap large layered gabbro intrusives are either recrystallized roof rocks or are heavily contaminated by assimilating roof rocks (e.g., the Lebowa Granite, Bushveld Complex, South Africa). The petrogenesis of mafic igneous rocks from magma that formed by decompression melting in the head of a plume was discussed by Cox (1978, 1988), Macdougall (1988), White and McKenzie (1989), Wilson (1989), Campbell and Griffiths (1990), Duncan and Richards (1991), Turner and Hawkesworth (1995), Ebinger and Sleep (1998), and in additional reports referenced by Faure (2001, Section 2.6).

The basalt plateaus that form by the extrusion of lava through deep fractures in the crust may be torn apart when the rifts widen into oceans (e.g., the Kirwan Volcanics of Queen Maud Land and the Karoo Basalt of Southern Africa). Other examples of paired basalt plateaus that are now separated by an ocean are the Edendeka Basalt of Namibia and the Paraná Basalt of Brazil as well as the diabase dikes of Liberia in West Africa and the diabase and basalt along the east coast of North America. Additional, examples were described by Faure (2001).

The plumes which play such an important role in the plate tectonics of the Earth derive their energy from the heat that is released by the radioactive decay of uranium, thorium, and potassium. The ultramafic rocks of the mantle of the Earth have only low concentrations of these heat-producing elements which, instead, have been concentrated in the granitic gneisses of the continental crust. However, oceanic crust and the overlying marine sediment that are subducted into the mantle have higher concentrations of the heat-producing elements than the pristine rocks of mantle (Hofmann and White 1982). Therefore, the temperature of subducted oceanic crust and marine sediment in the asthenospheric mantle increases as they age. The increase in temperature causes them to expand which lowers their density and generates a buoyant force that causes them to rise slowly toward the surface. The resulting diapir or plume in Fig. 15.2 develops a "head" followed by a "tail" that may consist of entrained asthenospheric rocks (Richards et al. 1989). The time that elapses between the initial subduction and the ultimate return to the top of the asthenosphere (i.e., the incubation time of the subducted oceanic crust) is of the order of several hundred million years, but the temperature differential between the plume and the adjacent mantle is no more than 100–200°C. Nevertheless, the plumes are transporting heat from the interior of the Earth into the overlying continental crust which expands causing uplift of the surface. The heat emanating from the head

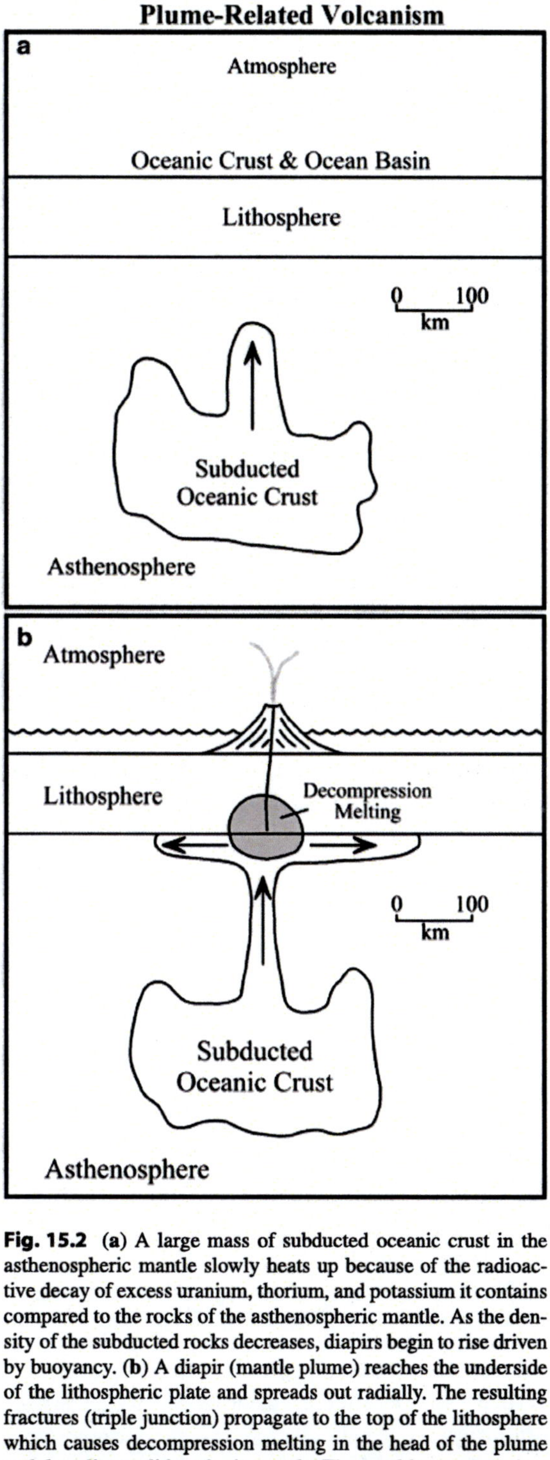

Fig. 15.2 (**a**) A large mass of subducted oceanic crust in the asthenospheric mantle slowly heats up because of the radioactive decay of excess uranium, thorium, and potassium it contains compared to the rocks of the asthenospheric mantle. As the density of the subducted rocks decreases, diapirs begin to rise driven by buoyancy. (**b**) A diapir (mantle plume) reaches the underside of the lithospheric plate and spreads out radially. The resulting fractures (triple junction) propagate to the top of the lithosphere which causes decompression melting in the head of the plume and the adjacent lithospheric mantle. The resulting magma rises through the fractures and is erupted at the surface producing either a shield volcano (as shown) or a lava plateau (such as the Kirkpatrick Basalt)

of the plume and from basalt magma may also cause partial melting of the granitic basement rocks. The resulting silicic melts may mix with the mantle-derived basalt magma, or they may be erupted together with the basalt flows (e.g., the Snake River Plain of Idaho and Yellowstone Caldera of Wyoming, USA). This tectonic process has operated throughout the history of the Earth from the Eoarchean Era more than 3.6 billion years ago. It is still working at the present time and it will maintain the geological activity of the Earth for hundreds of millions of years into the future.

According to this hypothesis, the geological activity that is observable on the surface of the Earth (displacement of continents, subduction, volcanic activity, earthquakes, uplift, and orogenies) are all consequences of the movement of plumes from depth in the mantle until they reach the underside of the rigid lithospheric shell of the Earth. This global tectonic process has caused lithospheric plates to collide at certain times in the history of the Earth leading to the assembly of large continents like Rodinia, Gondwana, and Pangea. At other times, this process has broken up the supercontinents and caused the fragments to be dispersed. In this way, the break-up of Gondwana and the displacement of the continental fragments can be regarded as a manifestation of deep-Earth tectonics involving plumes.

15.2 The Weddell-Sea Triple Junction

We begin with the assertion that the break-up of Gondwana is an established fact that no longer needs to be demonstrated by citing the evidence for its occurrence. However, the aspects of this geological phenomenon that still attract attention are:

1. The timing of the separations of the southern continents from East Antarctica
2. The concurrent tectonic activity within Antarctica
3. The uplift of the Transantarctic Mountains

These aspects of the break-up of Gondwana are all part of the same tectono-magmatic process which relates them to each other. In other words, the tectonic processes that caused Gondwana to break up also caused certain crustal fragments to assemble in West Antarctica; even the uplift of the Transantarctic Mountains is probably a consequence of the tectonic activity that dominated the geologic processes of Antarctica during the late Mesozoic and Cenozoic. Each of the topics we have identified above has been extensively researched by workers with specialized expertise who have used their results to formulate hypotheses that have, in some cases, conflicted with hypotheses expressed by their colleagues. These matters will continue to be debated at scientific meetings and in the relevant technical journals until eventually a consensus is reached. At the present time, West Antarctica continues to be "the problem child of Gondwanaland" (Dalziel and Elliot 1982) and the reasons for the uplift of the Transantarctic Mountains remain unsettled.

The break-up of Gondwana is easier to research and to comprehend than its original assembly because the outcome of the break-up is known. It was difficult originally to assemble Gondwana by reuniting the continents of the southern hemisphere because the very existence of such a supercontinent was in doubt and because the processes that caused it to break up and caused the continental fragments to drift apart were not understood. We now know that continents may be broken up by plumes that rise from depth in the asthenosphere of the mantle and by transform faults that displace midocean ridges along which new seafloor is produced by volcanic activity (Wilson 1965; Wyllie 1971).

The fractures in the lithospheric plate above the head of a large plume, in many cases, radiate from a central point in a pattern called a *triple junction* (Burke and Dewey 1973). A good example of this phenomenon is the Red Sea, Gulf of Aden, and the Afar Depression (Windley 1977). The rifts that emanate from a triple junction do not necessarily develop equally; some may die out while others become dominant and extend for thousands of kilometers. For example, the Red Sea rift extends for more than 2,300 km north from the Gulf of Aden.

Several authors have related the eruption of continental basalt in Gondwana to the arrival of a large plume which caused the formation of a triple junction in the area of the Weddel Sea between Queen Maud Land of East Antarctica and southern Africa (Storey 1995; Storey and Kyle 1997; Storey et al. 1996; Encarnación et al. 1996; Elliot and Fleming 2000). It is by no means certain that the crustal rifts, through which the Ferrar magmas in the Transantarctic Mountains were erupted, originated from a triple junction in the area between Queen Maud Land and southern Africa, nor is it certain that the magma originated from that area (Elliot et al. 1999). It seems that the principles of tectono-magmatic processes are understood, but their application to the petrogenesis of the Jurassic basalt in the Transantarctic Mountains is still uncertain.

15.3 Subduction Along the Paleo-Pacific Coast

During the early part of the Mesozoic Era a subduction zone in Fig. 15.3 stretched from the west coast of South America to the paleo-Pacific coast of East Antarctica. This subduction zone dipped under the Antarctic Peninsula, Thurston Island, Marie Byrd Land, Southern New Zealand, and Tasmania all of which were later moved into the positions they occupy at the present time (Elliot 1991). The subduction of oceanic crust caused the eruption of large amounts of rhyolitic ignimbrites in the Chon Aike province of Patagonia and in the Antarctic Peninsula in Fig. 15.3, making it the largest igneous province of silicic volcanic rocks on Earth (Pankhurst et al. 1998). In retrospect, the volcanic eruptions along this subduction zone may have been the source of rhyolitic volcanic ash in the sedimentary rocks of the Beacon Supergroup in the central Transantarctic Mountains and elsewhere (e.g., the Buckley, Fremouw, and Falla formations, Section 10.3). If so, then the subduction along the paleo-Pacific coast of Gondwana was active from the Late Permian, or before, until the Late Jurassic, or later. The compression of the coastal region of Gondwana may also have caused the folding associated with the Gondwanide orogeny in the Pensacola Mountains described by Ford (1972)

The rocks that form the Antarctic Peninsula consist almost entirely of batholithic granitoids and volcano-sedimentary complexes that formed as a result of the igneous activity associated with this subduction zone (Storey and Alabaster 1991). The rocks of the so-called Antarctic Peninsula Volcanic Group (APVG) were

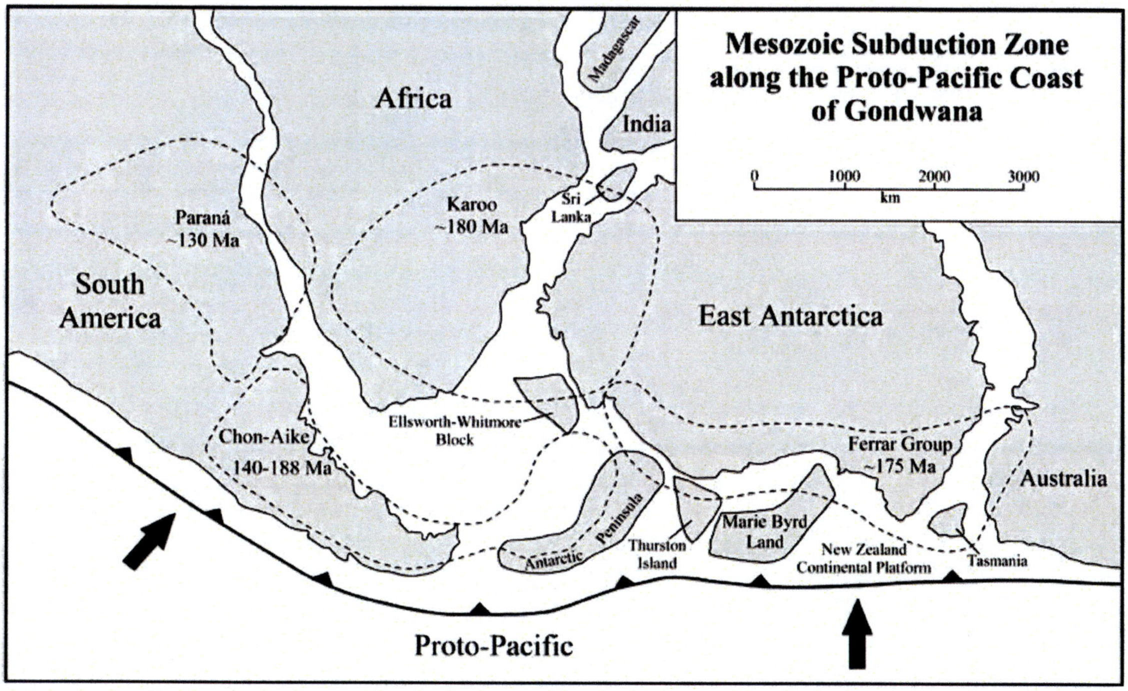

Fig. 15.3 A major subduction zone existed along the paleo-Pacific coast of Gondwana during the Mesozoic and Cenozoic eras. The compression as well as extension caused by this active tectonic feature affected, and may have caused, the eruption of ignimbrites and mafic volcanics in the Chon Aike province of Patagonia and along the Antarctic Peninsula (Adapted from Tappert et al. (2009) and Pankhurst et al. (1998))

described by Pankhurst et al. (1998) who also cited references to the literature dealing with the geology of the Antarctic Peninsula. The felsic magmas that caused the volcanic eruptions along the paleo-Pacific subduction zone formed by partial melting of rocks in the lower crust as a result of heat transferred from depth in the mantle by basalt magma, only small amounts of which were erupted.

An important insight that arises from this work is that the granitic plutons of the Ellsworth-Whitmore block of West Antarctica in Fig. 15.3 were formed by the intrusion of granitic magma that originated in the off-shore subduction zone (Section 8.1.3, Fig. 8.4, Appendix 8.5.5). The ages of these plutons in Fig. 15.4 overlap the ages of the Ferrar Dolerite and the Kirkpatrick Basalt in the Transantarctic Mountains (Storey et al. 1988; Halpern 1966). The synchronous formation of felsic and mafic igneous rocks in different parts of Antarctica was noted by Professor M.G. Ravich, Vice Director of the Research Institute of Arctic Geology in St. Petersburg, Russia (Ravich and Grikurov 1976). These magmatic events may be related based on the following conjecture.

The surface of the subglacial bedrock underlying West Antarctica in Fig. 15.5 is below sea level as a result of thinning of the continental crust. The resulting Byrd Subglacial Basin is the back-arc basin associated with the subduction zone along the paleo-Pacific coast of Gondwana in Fig. 15.6 (Elliot 1974; LeMasurier 2008). The extension of the crust of West Antarctica caused the formation of horsts and grabens separated by normal faults. These faults run parallel to the trace of the subduction zone and to the Ross-Sea coast of East Antarctica. Therefore, these faults could have cut deeply through the Ross orogen and triggered partial melting in the subcrustal mantle and perhaps within the lower crust itself. Alternatively, the back-arc faults may have facilitated the movement of magmas that originated in Queen Maud Land. In either case, the faulting in the back-arc basin of the paleo-Pacific subduction zone permits a genetic connection between the granitic plutons of West Antarctica and the dolerite sills and basalt flows of the Ferrar Group in the Transantarctic Mountains.

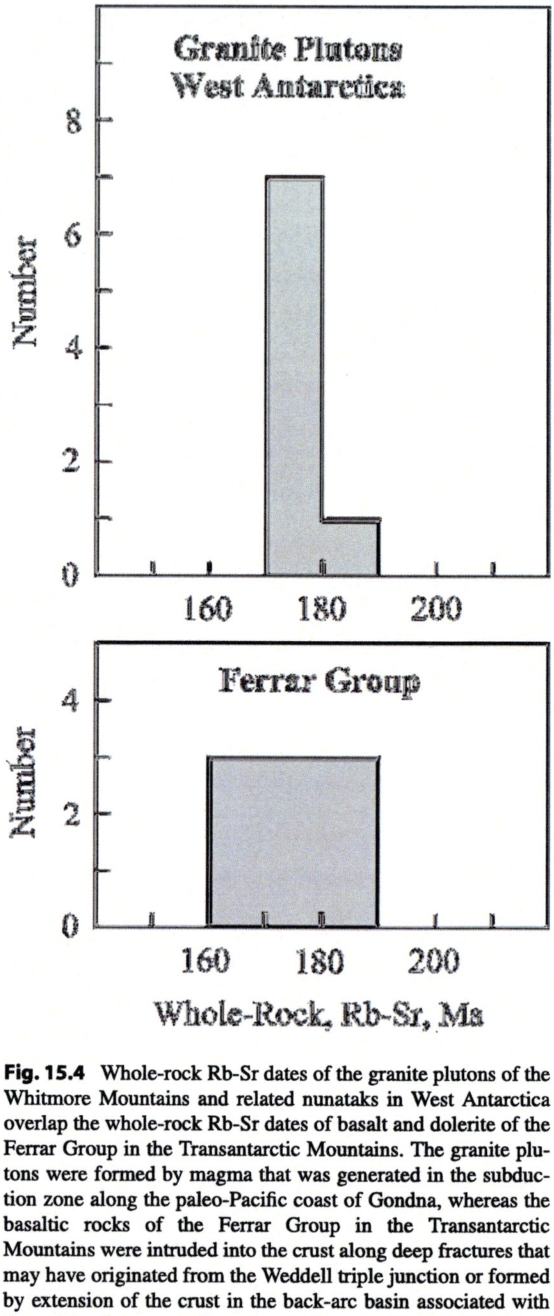

Fig. 15.4 Whole-rock Rb-Sr dates of the granite plutons of the Whitmore Mountains and related nunataks in West Antarctica overlap the whole-rock Rb-Sr dates of basalt and dolerite of the Ferrar Group in the Transantarctic Mountains. The granite plutons were formed by magma that was generated in the subduction zone along the paleo-Pacific coast of Gondna, whereas the basaltic rocks of the Ferrar Group in the Transantarctic Mountains were intruded into the crust along deep fractures that may have originated from the Weddell triple junction or formed by extension of the crust in the back-arc basin associated with the subduction zone (Data from Chapters 12 and 13 (Ferrar Group) and Chapter 8 (Whitmore Mountains))

Additional information about the Mesozoic rocks of West Antarctica is available in the reports of:

Halpern (1966, 1967, 1968), Stump (1973), Dalziel and Elliot (1982), Webers et al. (1982, 1983), Grindley and Oliver (1983), Jankowski et al. (1983), Adams (1986), Millar and Pankhurst (1987), Storey and Dalziel (1987), Grunow et al. (1987), Dalziel et al. (1987a, b), Vennum and Storey (1987a, b), Garrett et al. (1987),

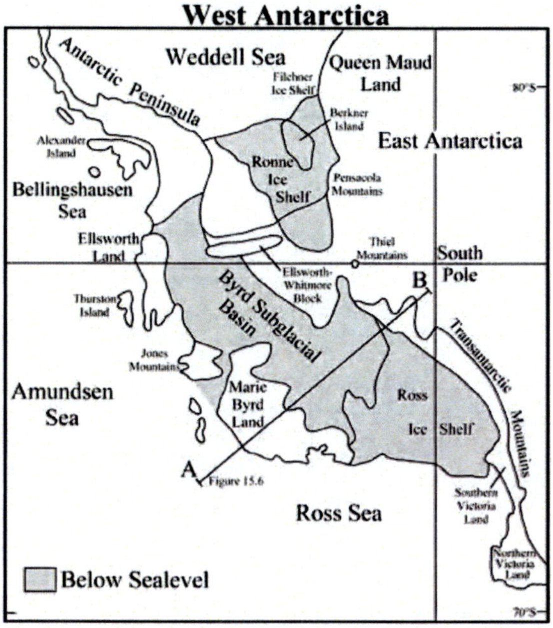

Fig. 15.5 West Antarctica is a mosaic of several crustal fragments that are arranged about the Byrd Subglacial Basin, the bedrock surface of which is below sea level. The crustal fragments include: the Ellsworth-Whitmore Block, Ellsworth Land, Thurston Island, the Jones Mountains, and Marie Byrd Land. The Byrd Subglacial Basin is roughly parallel to the Transantarctic Mountains and their extension to the Thiel and Pensacola mountains. The line A-B is the approximate position of the cross section of the subduction zone on the paleo-Pacific coast of Gondwana in Fig. 15.6 (Adapted from Elliot (1974) and LeMasurier (2008))

LeMasurier and Rex (1989), Storey (1991) Pankhurst et al. (1991), Grunow et al. (1991), Elliot (1991), Weaver et al. (1991), Pankhurst et al. (1993), Behrendt et al. (1991a, b), Adams et al. (1995), Siddoway et al (2004), Winberry and Anandakrishna (2004).

15.4 Assembly and Break-Up of Gondwana

Many investigators have proposed chronologies for the break-up of Gondwana and have plotted the positions of the continental fragments until they arrived at their present locations. Interest in this topic was first aroused by Wegener (1912) followed by du Toit (1937) who actually assembled the southern continents into a supercontinent (Section 9.1, Fig. 9.3 and Section 11.2, Fig. 11.4). The reconstruction of Gondwana was taken up by Sproll and Dietz (1969), Dietz and Sproll (1970), Dietz et al. (1972), Vilas and Valencio (1970), and by Smith and Hallam (1970) who fitted the continents together based primarily on the shapes of their submerged continental shelves. The most satisfactory reconstruction of Gondwana in Fig. 15.7 was published by Lawver and Scotese (1987) and by Lawver et al. (1991).

In all of these reconstructions, Antarctica turned out to be the keystone around which the other continents could be fitted even though the Antarctic Peninsula and West Antarctica seemed to get in the way because of unexpected overlaps with South America and New Zealand. Elliot (1972) suggested a solution to these problems by detaching the Antarctic Peninsula and Marie Byrd Land from the main body of East Antarctica. He also prepared a set of time-lapse diagrams that showed how Gondwana was broken up and how the fragments "drifted" apart to their present positions in the southern hemisphere. The entire process lasted about *125 million years* from the Late Jurassic (140 Ma) to the middle Miocene (15 Ma), according to the time scale of the IUGS (2002).

Many other attempts to reconstruct Gondwana have been published based on information from geology, paleontology, and paleomagnetism: Frakes and Crowell (1970), Craddock (1970), Colbert (1970), Schopf (1969), and McElhinny and Luck (1970). These early works were followed by refined reconstructions of Gondwana many of which considered specific areas between Antarctica and adjacent continents:

Australia and Antarctica: Kennett et al. (1972), Weissel and Hayes (1974), Griffiths (1974), Norton and Molnar (1977), Falvey and Mutter (1981), Cande and Mutter (1982), Audley-Charles (1983), Stump et al. (1986), Squire and Miller (2003).

New Zealand and Antarctica: Weissel et al. (1977), Laird (1981), Grindley and Davey (1982), Katz (1982), Grindley and Oliver (1983), Walcott (1984), Stump et al. (1986), Lawver and Scotese (1987), Bradshaw (1991), Stock and Cande (2002), Mortimer et al. (2006).

South Africa and Antarctica: Cox (1988), Harris et al. (1991), Kristoffersen and Hinz (1991), Elliot (1992), Grunow (1993), Storey (1995), Storey et al. (1996), Encarnación et al. (1996), Storey and Kyle (1997), Elliot and Fleming (2000).

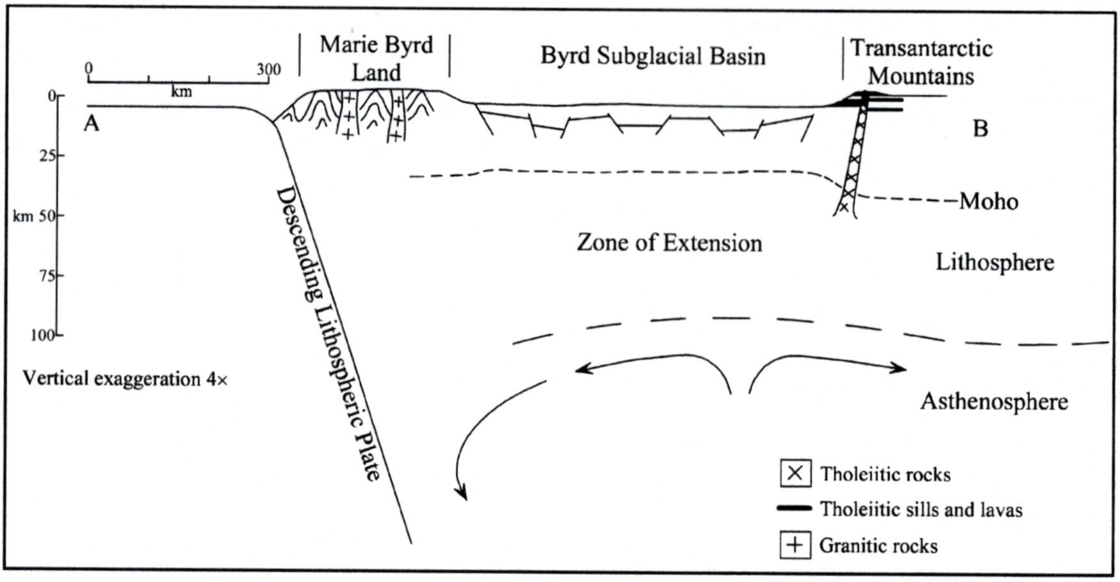

Fig. 15.6 The subduction of a lithospheric plate during the Mesozoic Era caused the eruption of felsic and mafic lavas in the Antarctic Peninsula and in Marie Byrd Land. The crust of Gondwana in the back-arc basin was stretched which caused block-faulting and subsidence of the Byrd Subglacial Basin. Crustal extension of the back-arc basin may also have caused rifts to form in the Ross orogen and in the overlying rocks of the Beacon Supergroup. In that case, the resulting decompression could have caused basalt magmas to form by partial melting of the mantle. These magmas subsequently intruded the overlying continental crust and formed the sills of the Ferrar Dolerite and the plateaus of the Kirkpatrick Basalt. The line of section A-B is marked in Fig. 15.5 (Adapted from Elliot (1974, 1992))

India, Sri Lanka, Madgascar, and Antarctica: Katz (1978), Katz and Premoli (1979), Norton and Sclater (1979), Rabinowitz et al. (1983), Lawver and Scotese (1987).

The involvement of southeast Asia, Tibet, and the parts of northern China in Gondwana was suggested by Stauffer and Gobbett (1972), Crawford (1974), and Audley-Charles (1983). The Falkland Plateau and the Falkland Islands were originally located between southern Africa and South America prior to the opening of the South Atlantic Ocean in the Early Cretaceous (Mitchell et al. 1986) and were later moved to their present position off the coast of Patagonia by the development of the Scotia Arc (Dalziel and Elliot 1971; Martin et al. 1981).

Summaries of the tectonic evolution of Antarctica that treat all parts of the continent were published by: Voronov (1964), Hamilton (1967), Grikurov et al. (1972), Elliot (1975), McGlynn et al. (1975), and Grikurov (1982). *Laurasia*, the northern part of Pangea was also broken up during the Mesozoic which caused a drastic rearrangement of the geography of the northern hemisphere (Nance and Linnemann 2008).

The break-up of Gondwana started during the Jurassic Period and proceeded in a series of steps that were outlined by Lawver et al (1991) in Fig. 15.8a–d:

Figure 15.8a: At 140 Ma (Late Jurassic to Early Cretaceous) Africa and South America had already separated from Queen Maud Land of East Antarctica, but all other parts of Gondwana were still attached to East Antarctica.

Figure 15.8b: Ten million years later, at 130 Ma (Early Cretaceous), the gap between Africa and East Antarctica had widened, Madagascar had separated from the African mainland, but still lay close to India. In addition, a gap had formed between South America and Africa meaning that the South Atlantic Ocean had begun to open up.

Figure 15.8c: At 118 Ma (Early Cretaceous) India, Sri Lanka, and Madagascar started to move away from East Antarctica, thereby opening the Indian Ocean. The South Atlantic Ocean was also getting wider.

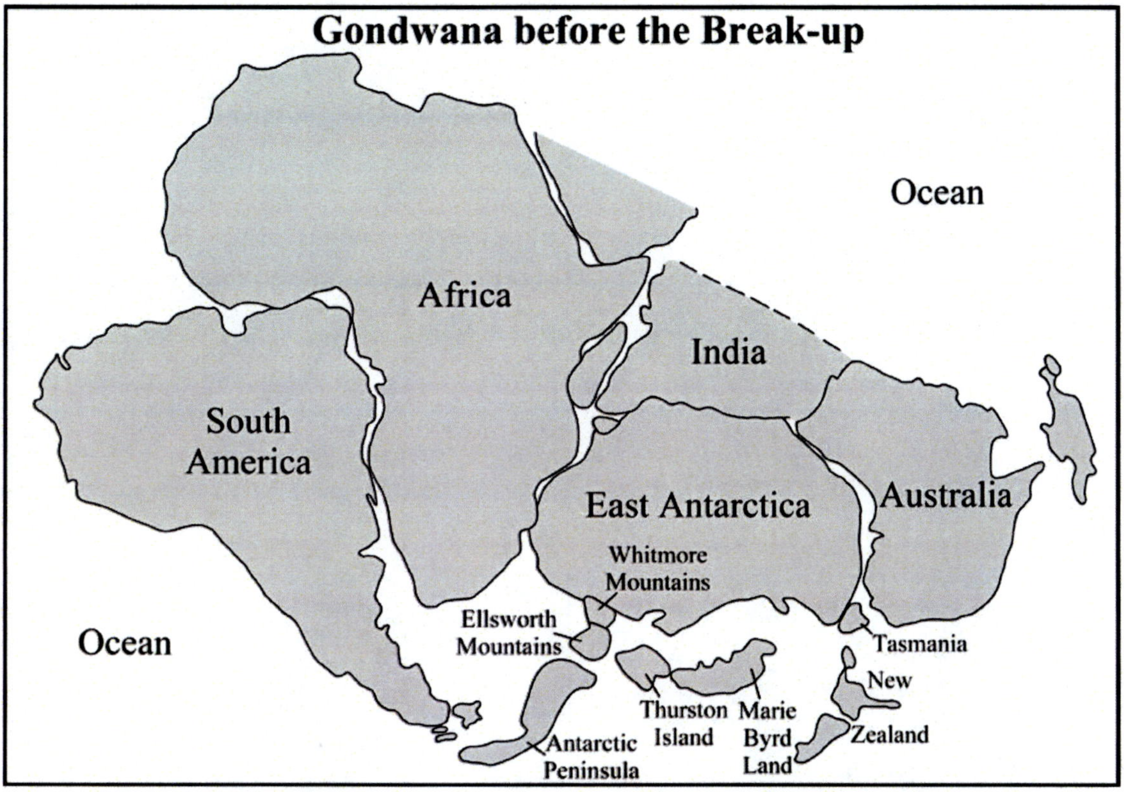

Fig. 15.7 Gondwana, fully assembled at the start of the Mesozoic Era at about 260 Ma, contained all of the present continents of the southern hemisphere with East Antarctica in the center. Several small land masses were not attached to East Antarctica at this time: the Antarctic Peninsula, the Ellsworth Mountains, the Whitmore Mountains, Thurston Island, Marie Byrd Land, the North and South islands of New Zealand, and Tasmania (Adapted from Lawver and Scotese (1987))

Australia was still attached to East Antarctica and the microcontinents (Antarctic Peninsula, the Ellsworth-Whitmore block, Marie Byrd Land, and New Zealand) continued in their original positions.

Figure 15.8d: At 70 Ma (latest Late Cretaceous) Africa and South America moved north and the gap between them widened. As a result, the Atlantic and Indian oceans were approaching their present areal dimensions. The islands of New Zealand also moved north into the Pacific Ocean, and Australia started to pull away from East Antarctica. The Antarctic Peninsula, the Ellsworth-Whitmore block, and Marie Byrd Land were amalgamated into West Antarctica.

Insofar as is known, Antarctica was not yet glaciated during the Late Cretaceous although ice may have begun to accumulate on the highest mountains. The subduction zone west of the Antarctic Peninsula died out during the early Pliocene (4 Ma); but it continues unabated along the west coast of South America where it has caused the uplift of the Andes Mountains (Faure 2001).

The break-up of Gondwana and the displacement of the continental fragments was caused by rifts that evolved into midocean ridges along which seafloor is generated by volcanic activity. These spreading ridges now girdle the globe and, for the most part, are still active. Therefore, the lithospheric plates that form the surface of the Earth are still being displaced relative to each other and the continents that are embedded in the plates are still "drifting". However, the plate that contains the continent of Antarctica in Fig. 15.9 is surrounded by spreading ridges that hold it in place, which means that Antarctica is stuck in its present position (Rülke and Dietrich 2007).

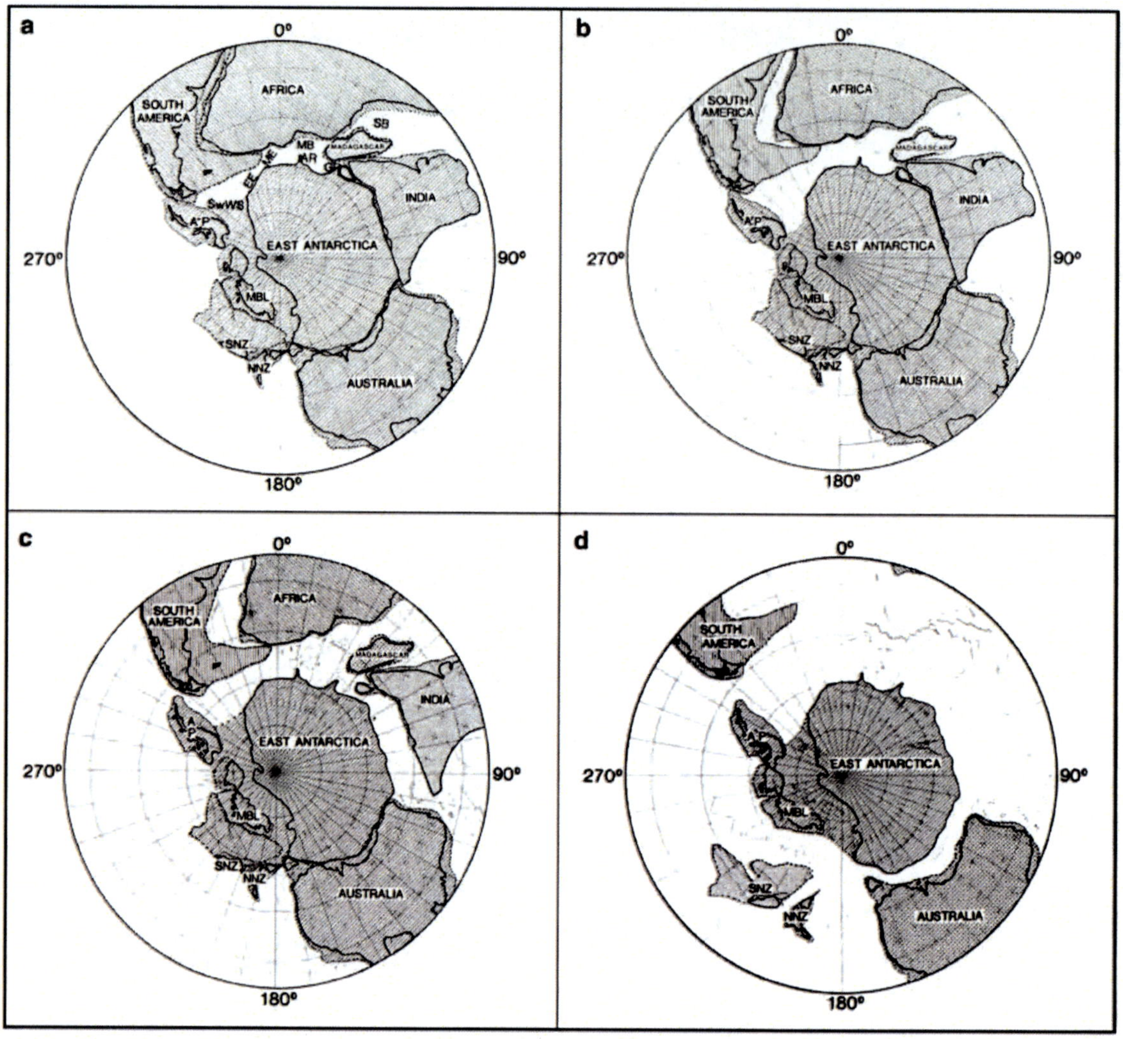

Fig. 15.8 The disintegration of Gondwana started at about 140 Ma (**a**) when Africa broke away from Queen Maud Land. At 130 Ma (**b**) the separation of Africa and EastAntarctica increased and the South Atlantic Ocean began to open up. Twelve million years later, at 118 Ma (**c**), India moved north and thereby enlarged the Indian Ocean. During the latest Cretaceous, at 70 Ma (**d**), Australia separated from East Antarctica, Marie Byrd Land merged into West Antarctica, and New Zealand moved north into the Pacific Ocean. By the middle Miocene at 15 Ma, Antarctica as we know it had emerged from Gondwana (Adapted from Lawver et al. (1991))

15.5 Uplift of the Transantarctic Mountains

The Transantarctic Mountains are one of the most spectacular topographic features in Antarctica. These mountains extend for 3,700 km from the Oats Coast in northern Victoria Land to the Theron Mountains on the coast of the Weddell Sea and their summits rise to elevations in excess of 4,000 m. This large mountain range is composed primarily of flat-lying sedimentary rocks of the Beacon Supergroup intruded by equally undeformed sills of the Ferrar Dolerite and capped locally by thick layers of the Kirkpatrick Basalt. These rocks, which form the visible part of the Transantarctic Mountains, were not folded and faulted by compression during an orogeny, in contrast to the Ross orogen on

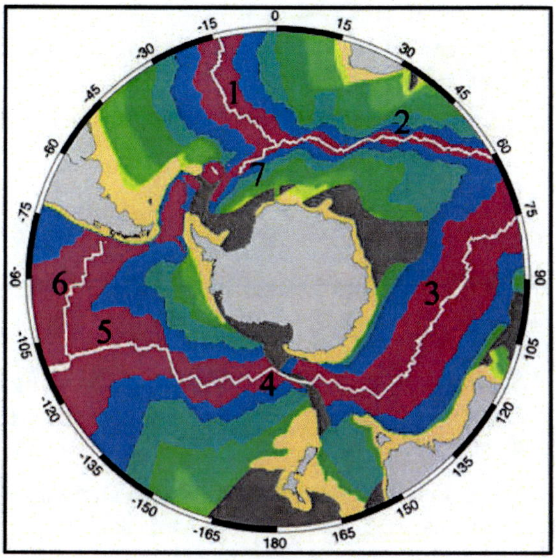

Fig. 15.9 Antarctica is almost completely surrounded by spreading ridges along which new seafloor is generated by volcanic activity. The *white line* that surrounds the continent is the trace of this system of spreading ridges. The age of the oceanic crust is color-coded such that red is the youngest, followed by two shades of blue, two shades of green, and yellow for the oldest oceanic crust. The ridges are numbered: 1 = Mid-Atlantic Ridge, 2 = Southwest-Indian Ridge, 3 = Southeast-Indian Ridge, 4 = Pacific-Antarctic Ridge, 5 = East Pacific Rise, 6 = Chile Rise, and 7 = American-Antarctic Ridge (Adapted from Hayes et al. (2009))

which the Beacon rocks were deposited. Mountain ranges of comparable magnificence elsewhere in the world formed by folding and thrust-faulting of sedimentary and volcanic rocks during orogenies that caused extensive metamorphic recrystallization of the affected rocks accompanied by intrusion of granitic magma formed by anatexis (e.g., the Appalachian Mountains in North America). The Transantarctic Mountains differ from other mountain ranges because they consist of uplifted and deeply dissected fault blocks composed of undeformed rocks. Although the reasons for the uplift of the fault blocks are not yet completely understood, the history of tectonic activity of Gondwana does suggest that the uplift of the Transantarctic Mountains occurred as a consequence of the break-up of Gondwana and the subsequent tectonic adjustments.

15.5.1 Crustal Structure of Antarctica

The thickness and composition of the crust of East and West Antarctica have been determined by measurements of gravity and magnetism by means of aircraft and satellites, complemented by seismic surveys carried out on the ground (Bentley and Clough 1972; Bentley 1973, 1983; Bentley and Robertson 1982; Behrendt et al. 1974, 1980; Robinson and Splettstoesser 1984). Much useful information was also obtained by airborne radio-echo sounding of large parts of Antarctica which indicated not only the thickness of the East Antarctic ice sheet, but also revealed the topography of the subglacial bedrock and the presence of meltwater lakes at the base of the ice sheet (Drewry 1972, 1975, 1976).

The results summarized by Bentley (1983) in Fig. 15.10 are based on gravity anomalies published by Groushinsky and Sazhina (1982). The diagram indicates that the thickness of the subglacial crust of

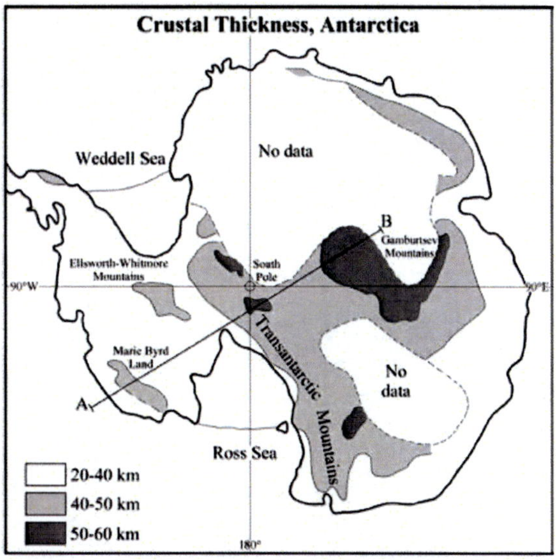

Fig. 15.10 The thickness of the continental crust of Antarctica ranges from 20 to 40 km in West Antarctica and from 40 to 60 km in East Antarctica. In general, the thickness of the crust of West Antarctica is at least 5 km less than the crust that underlies the Transantarctic Mountains. The boundary between East and West Antarctica is sharp and parallels the trend of the Transantarctic Mountains. The step-like increase of the crustal thickness suggests that the boundary between East and West Antarctica is a fault zone. The line A-B is the cross-section of crustal thickness shown in Fig. 15.11 (Adapted from Bentley (1983))

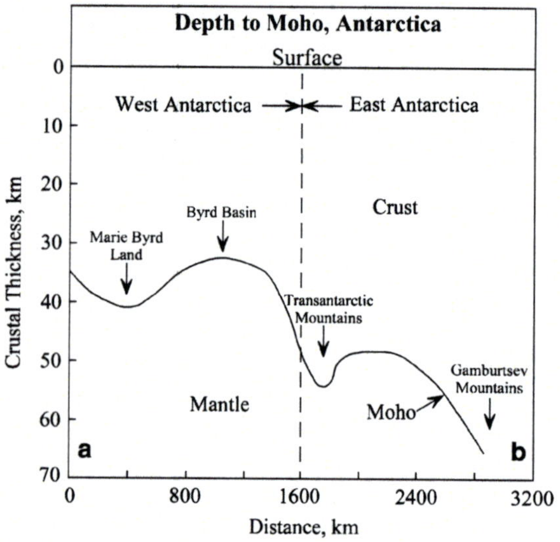

Fig. 15.11 This cross-section of the crust of West and East Antarctica was drawn along line A-B in Fig. 15.10 using the contours of crustal thickness of a diagram by Bentley (1983). The profile of crustal thickness indicates the depth to the Mohorovičić Discontinuity (Moho) which separates the continental crust of Antarctica from the lithospheric mantle below. This cross-section demonstrates that the crust of West Antarctica is thinner than the crust of East Antarctica and that the transition is abrupt which suggests the presence of a fault zone or suture. The vertical exaggeration is 40×. The crustal thicknesses were derived from gravity anomalies by Groushinsky and Sazhina (1982)

Antarctica increases stepwise from about 20 to 40 km in West Antarctica, to 40–50 km under the Transantarctic Mountains (TAM), and to more than 60 km under the Gamburtsev Mountains (GM). The principal exceptions in West Antarctica are the Ellsworth-Whitmore Mountains (EWM) and Marie Byrd Land (MBL) where the thickness of the crust rises to more than 40 km. The crustal thickness under the Transantarctic Mountains rises abruptly from 35 to about 45 km which suggests that this boundary is a fault zone or crustal suture.

The profile of the crustal thickness of parts of West and East Antarctica in Fig. 15.11 indicates the complementary depth to the Mohorovicic Discontinuity (Moho) and illustrates the increase of the crustal thickness at the boundary between West and East Antarctica. The structure of the crust of Antarctica was also investigated by Dewart and Toksöz (1965), Jankowski and Drewry (1981), Kurinin and Grikurov (1982), and by Kadmina et al. (1983). In spite of the insights gained by these and related studies, several important questions remain:

1. How was West Antarctica assembled from blocks of continental crust following the break-up of Gondwana?
2. How and when were the Transantarctic Mountains uplifted?
3. What is the origin of the accreted terranes of northern Victoria Land (Sections 4.3–4.6)?
4. How far do the sedimentary rocks of the Beacon Supergroup and the mafic igneous rocks of the Ferrar Group extend under the East Antarctic ice sheet?

These questions continue to motivate on-going research into the tectonic processes that shaped Antarctica following the break-up of Gondwana.

15.5.2 The Transantarctic Fault Zone

Robinson and Splettstoesser (1986) speculated that the Transantarctic fault zone, which separates East and West Antarctica, is the border between two lithospheric plates which were placed in juxtaposition in the course of the break-up of Gondwana. This hypothesis was supported by Scharnberger and Scharon (1972) who reported that the virtual geomagnetic poles derived from igneous rocks indicate that East and West Antarctica were originally located on different lithospheric plates.

Robinson and Splettstoesser (1986) also noted that the East and West Antarctic plates did not collide forcefully because the rocks of the Beacon Supergroup and of the Ferrar Group are not folded and thrust-faulted. In addition, the Transantarctic fault zone cannot be a "normal" fault because the geology of West Antarctica differs profoundly from the Transantarctic Mountains. Robinson and Splettstoesser (1986) also considered that the Transantarctic Mountains cannot be located on the passive margin of a continental rift that split the Antarctic plate because passive rift margins elsewhere do not reach the high elevations of the Transantarctic Mountains. Perhaps other processes have enhanced the elevation of the Transantarctic Mountains.

Assuming that the Transantarctic fault zone is neither a compressive nor a passive border between the East and West Antractic plates, Robinson and Splettstoesser (1986) concluded that it is a *transcurrent fault* zone along which the West Antarctic plate moved by strike-slip motion. In that case, the Transantarctic Mountains may have been uplifted by warping and block faulting of the edge of the East Antarctic plate caused by the motion of the West Antarctic plate. The complementary tectonic deformation of the West Antarctic plate is hidden by the West Antarctic ice sheet. The absence of volcanic activity and of earthquakes along the Transantarctic fault zone is an indication that East and West Antarctica are now strongly fused together within the pan-Antarctic plate.

Schmidt and Rowley (1986) assumed that a continental rift formed between East and West Antarctica and subsequently widened into the Transantarctic Rift System by extension and thinning of the continental crust. At about the same time, continental flood basalts were erupted and dolerite sills were intruded into the flat-lying sedimentary Beacon rocks of the Transantarctic Mountains on the East Antarctic side of the rift. This tectono-magmatic activity was an immediate precursor to the break-up of Gondwana.

These authors also assumed that in pre-Jurassic time *Marie Byrd Land* in Fig. 15.12 was located adjacent to *northern Victoria Land* and was later moved to its present position in West Antarctica by strike-slip motion along the Transantarctic rift. Schmidt and Rowley (1986) attributed the movement along the Transantarctic rift system to seafloor spreading between Queen Maud Land and Africa in Fig. 15.8a. In other words, the Transantarctic rift system may be a Late Jurassic transform fault of the Weddell-African rift.

Another interesting suggestion by Schmidt and Rowley (1986) was that during the Jurassic Period the Ellsworth Mountains in Fig. 15.12 were displaced by the strike-slip fault from their original position within the Transantarctic Mountains to their present position in West Antarctica. Seventeen years earlier, Schopf (1969) made a similar proposal based on the correlation of sedimentary rocks of the Ellsworth Mountains (Section 8.1.3) with the rocks of the Neptune Range in the Pensacola Mountains (Section 8.2.3) and of the Ohio Range in the Horlick Mountains (Section 10.5.2). The rocks of the Ellsworth Mountains and of the Neptune Range were folded during the Permian

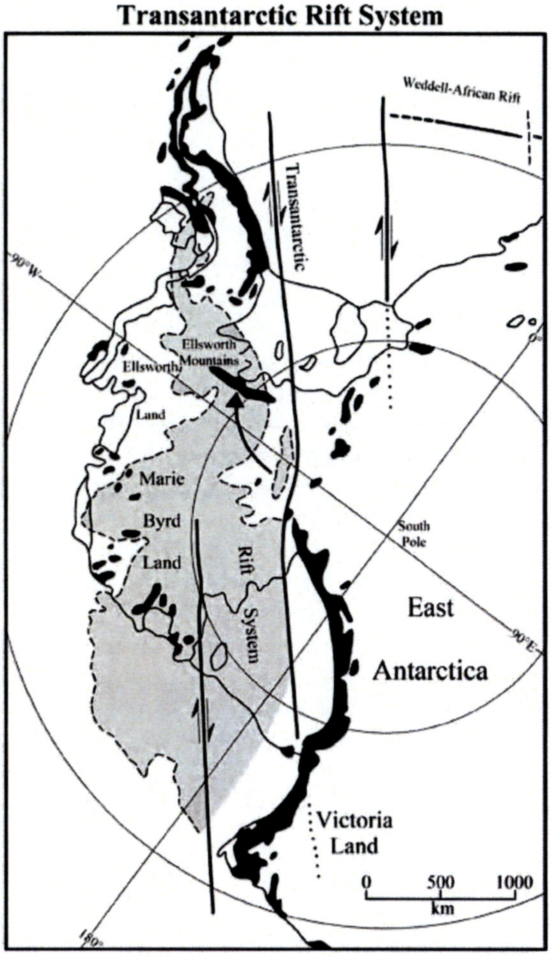

Fig. 15.12 According to a proposal by Schmidt and Rowley (1986), the Antarctic plate was split by a rift that developed during the Jurassic Period. This rift caused the eruption of basalt flows and the intrusion of dolerite sills of the Ferrar Group in the Transantarctic Mountains along the raised margin of the rift valley. Marie Byrd Land, which was originally located adjacent to northern Victoria Land, moved to its present position in West Antarctica by right-lateral motion along the Transantarctic rift system and the Ellsworth Mountain block was rotated clockwise out of alignment with the Transantarctic Mountains at about the same time. These displacements of blocks of continental crust were caused by transform faults that developed in the spreading ridge that separated Africa from Queen Maud Land of East Antarctica (Adapted from Fig. 1 of Schmidt and Rowley (1986))

Gondwanide orogeny which was presumably caused by compression of the subduction zone that was located along the paleo-Pacific coast of Gondwana.

15.5.3 Marie Byrd Land

The former association of Marie Byrd Land (MBL) with northern Victoria Land (NVL) was tested by isotopic age determinations of the Swanson Formation in the Ford Ranges located east of Edward VII Peninsula in MBL (Adams 1986; Adams et al. 1995). The Swanson Formation consists of folded metasedimentary rocks that were intruded by granitoids of Late Devonian age and by leucogranites of Jurassic to Cretaceous age. These plutons are surrounded by contact metamorphic aureoles of biotite hornfels that are 100–500 m wide. The pre-Devonian deformation of the Swanson Formation was accompanied by low-grade regional metamorphism of the rocks.

The thickness of the Swanson Formation is more than 1 km, but neither top nor bottom are exposed. It was deposited as a submarine fan on an east-facing paleoslope. Most of the rocks of this formation are slates composed of quartz, sericite, and chlorite. The rocks are unfossiliferous, although Bradshaw et al. (1983) reported finding a single trace fossil which may indicate that the depositional age of the Swanson Formation is post-Cambrian. K-Ar and Rb-Sr dates of biotite and feldspar in granitoids, granite gneiss, and schist in MBL compiled by Wade (1969) from the literature range from 143 ± 4 (Latest Jurassic) to 92 ± 5 Ma (Late Cretaceous).

Bradshaw et al. (1983) noted that the Swanson Formation resembles the rocks of the Robertson Bay Group in northern Victoria Land and that the two units may have overlapping ages. However, the paleocurrent directions of the Swanson Formation and the Robertson Bay Group diverge. Therefore, they concluded that the lithologic similarity is not sufficient to correlate the two units. Wade and Couch (1982) likewise noted both the similarity and differences of the two rock units but concluded that they are "totally unrelated."

Adams (1986) dated a large suite of samples of fine-grained pelitic rocks (well cleaved slates) from the Swanson and Mackay mountains and from Mt. Passel in the Ford Ranges of MBL by the K-Ar method. These dates in Fig. 15.13a have a distinctly bimodal distribution. The older suite reflects the low-grade metamorphism and deformation of the Swanson Formation during the Late Ordovician. The younger suite records the time when the rocks were heated by the intrusion of Late Devonian granites. Part B of Fig. 15.13 contains the whole-rock K-Ar dates of slates

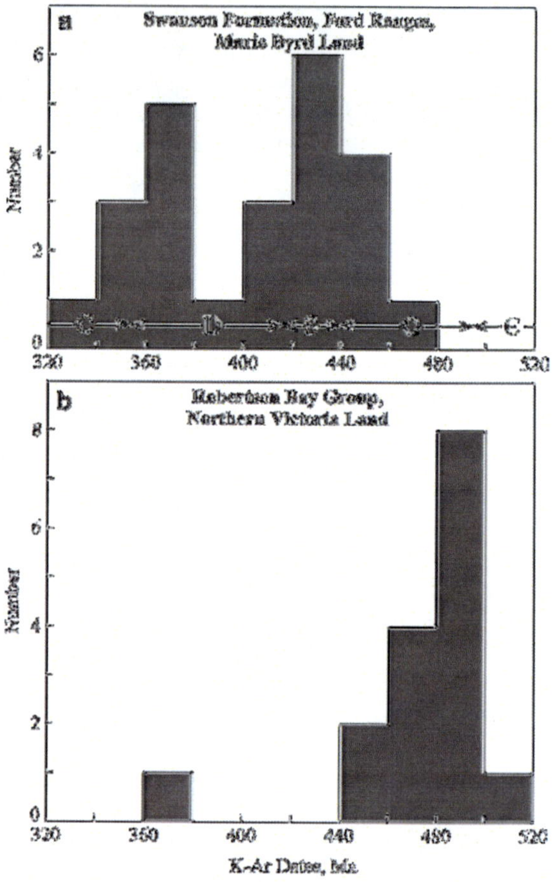

Fig. 15.13 (**a**) The whole-rock K-Ar dates of slates of the Swanson Formation in the Ford Ranges of Marie Byrd Land in West Antarctica have a bimodal distribution. The older suite (413–480 Ma) records the structural deformation and low-grade regional metamorphism of the sedimentary rocks during the Late Ordovician. The younger suite (336–384 Ma) reflects an episode of thermal metamorphism caused by the intrusion of Late Devonian granites (Data from Adams (1986)). (**b**) The whole-rock K-Ar dates of slates of the Robertson Bay Group in northern Victoria Land (Section 4.5) have a unimodal distribution that predates the deformation and low-grade metamorphism of the Swanson Formation. Therefore, the K-Ar dates do not support the hypothetical correlation of the Swanson Formation in West Antarctica with the Roberston Bay Group in northern Victoria Land. The lithologic similarities of these rocks presumably indicates that the rocks of both units were deposited in similar marine environments as distal turbidites (Data from Adams et al. (1982))

of the Robertson Bay Group in northern Victoria Land (Adams et al. 1982). In spite of the similar lithology, the K-Ar dates of the Robertson Bay Group are, for the most part, older than the K-Ar dates of the Swanson

Formation. Therefore, the low-grade metamorphism of the Robertson Bay slates occurred *before* the metamorphism of the slates of the Swanson Formation. In this way, the K-Ar dates also *do not* support the hypothesis that the Swanson Formation and the Robertson Bay Group are the same rock unit.

Adams (1986) also measured whole-rock Rb-Sr dates of fine-grained slates of the Swanson Formation based on samples derived from each of the three principal collecting sites (i.e., the Swanson and Mackay mountains and Mt. Passel in the Ford Ranges). The resulting dates in Table 15.1 are concordant and lie within the range of the older suite of K-Ar dates of the Swanson Formation in Fig. 15.13a. The whole-rock Rb-Sr dates of the slates represent the time elapsed since the isotopic composition of strontium was homogenized by low-grade regional metamorphism during the Late Ordovician. The rock samples from the three collecting sites have different initial $^{87}Sr/^{86}Sr$ ratios because the Rb/Sr and $^{87}Sr/^{86}Sr$ ratios of the detrital sediment at the time of deposition varied regionally and approached homogeneity only on a local scale.

The apparent failure of the correlation of the Swanson Formation in Marie Byrd Land and the Robertson Bay Group does not necessarily invalidate the hypothesis that Marie Byrd Land moved from a location near northern Victoria Land to its present position in West Antarctica. The paleomagnetic data of Scharnberger and Scharon (1972) as well as the evidence for strike-slip motion along the Transantarctic rift postulated by Robinson and Splettstoesser (1986) and by Schmidt and Rowley (1986) virtually require that Marie Byrd Land moved to its present position in West Antarctica in the course of the Late Jurassic to Early Cretaceous break-up of Gondwana.

Table 15.1 Whole-rock Rb-Sr dates of slates from the Swanson Formation in the Ford Ranges of Marie Byrd Land (Adams 1986)

Collecting site	Rb-Sr date Ma[a]	Initial $^{87}Sr/^{86}Sr$
Swanson Mountains	439 ± 30	0.712 ± 0.004
Mt. Passel	439 ± 6	0.718 ± 0.001
Mackay Mountains	440 ± 4	0.716 ± 0.001

[a] $\lambda = 1.42 \times 10^{-11}$ year^{-1}

15.5.4 The Accreted Terranes of NVL

The importance of strike-slip motion in the final assembly of Antarctica is also suggested by the accretion of the Bowers and Robertson Bay terranes to the Wilson terrane of northern Victoria Land (Sections 4.4–4.6). Although the provenance of these terranes and the tectonic process that caused them to move into juxtaposition with East Antarctica are as yet unexplained, the tectonic model of Schmidt and Rowley (1986) may offer a suggestion.

The break-away of Australia from East Antarctica between 100 and 65 Ma (Lawver et al. 1991) was initiated by a rift that subsequently evolved into the Southeast-Indian Ridge (Fig. 15.9). Transform faults associated with this spreading ridge may have caused crustal blocks from the Campbell Plateau adjacent to the South Island of New Zealand or from the Tasman Plateau located near Tasmania to be transported south into juxtaposition with the Wilson terrane of northern Victoria Land. The age of accretion of the Bowers and Robertson Bay terranes may have been recorded by the alteration of the Rb-Sr systematics of the high-Ti basalt flows that cap the Pain and Tobin mesas in the Mesa Range of northern Victoria Land (Sections 12.4–12.6).

15.5.5 Fission-Track Method of Dating

The rate of uplift of the Transantarctic Mountains has been measured by the fission-track method which is briefly reviewed in Appendix 15.7.1 and was described in detail by Faure and Mensing (2005). The fission-track method is well suited for this purpose because uranium-bearing minerals (e.g., apatite) retain fission tracks only after the minerals have cooled to the track-retention temperature which for apatite is 60°C (Gleadow and Fitzgerald 1987).

The interpretation of fission-track dates of apatite at increasing depths in a section of *stable* continental crust in Fig. 15.14 is based on the effect of temperature on the retention of fission tracks. This criterion is used to define three zones with increasing depth below the surface (Gleadow and Fitzgerald 1987):

1. The *complete retention zone* extends from the surface to a depth of 1.8 km where a normal geothermal

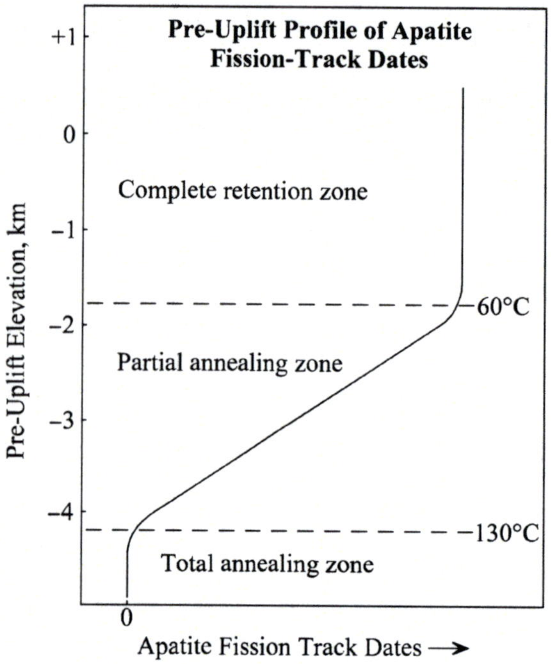

Fig. 15.14 The retention of fission tracks in apatite varies with depth below the surface depending on the geothermal gradient. In a stable region of the continental crust (i.e., before uplift occurs) all tracks are annealed in the *total annealing zone* at a depth greater than 4.2 km and at temperatures in excess of 130°C. At shallower depth, between 1.8 and 4.2 km and temperatures between 60°C and 130°C, in the *partial annealing zone*, track retention increases with decreasing depth. All tracks are retained in the *complete retention zone* at shallow depth from the surface to 1.8 km and temperatures less that 60°C. The resulting fission-track dates of apatite remain *constant* to a depth 1.8 km below the surface and then *decrease* at greater depth and approach 0 Ma at about 4.2 km because all tracks that form are rapidly annealed (Adapted from Gleadow and Fitzgerald (1987))

gradient of 30 ± 5°C/km yields an ambient temperature of 60°C. At this temperature, apatite retains all fission tracks and the fission-track dates of apatite in this zone are constant.

2. The *partial annealing zone* lies between depths of 1.8 and 4.2 km where the temperature rises from 60°C at 1.8 km to 130°C at 4.2 km. In this temperature interval the retention of fission tracks by apatite *decreases* with increasing temperature which means that the fission-track dates decrease with depth and approach 0 Ma at 4.2 km (130°C).
3. In the *total annealing zone* at temperatures greater than 130°C and at depths greater than 4.2 km, apatite does not retain fission tracks because they anneal rapidly after they form. Therefore, apatites in this zone yield track-dates of 0 Ma.

In an area where the crust is being *uplifted*, the fission-track dates of apatite increase with increasing elevations of the collecting sites and record the time that has elapsed since the samples cooled through the 60°C isotherm. All of the fission-track dates in the Transantarctic Mountains post-date the intrusion of the sills of Ferrar Dolerite which had already cooled before the apatites and other uranium-bearing minerals began to accumulate fission track.

Gleadow and Fitzgerald (1987) used fission-track dating of apatite in the granitic basement rocks of Wright Valley in southern Victoria Land to measure the local uplift rate on Mt. Jason in the Olympus Range (Fig. 3.5) and at Mt. Doorly adjacent to the Lower Wright Glacier. The results in Fig. 15.15a for Mt. Jason indicate that the fission-track dates of apatite in the Olympus Granite Gneiss increase linearly with increasing elevation of the collecting sites. The slope of this straight line is 15 m/10^6 year which is the local uplift rate between 120 and 40 Ma. In this time interval the elevation of the Olympus Range increased by 1,200 m.

The fission-track dates of apatite from Mt. Doorly form two line segments in Fig. 15.15b. The slope of the line from 100 to 50 Ma is 10 m/10^6 year. The uplift rate increased sharply at about 50 Ma (early Eocene). Gleadow and Fitzgerald (1987) estimated that the average post-Eocene uplift rate of Mt. Doorly was 96–106 m/10^6 year. The fission-track dates of apatite from Mt. Newall (not shown) confirmed the change in the uplift rate at about 50 Ma indicated by the results for Mt. Doorly.

The fission-track method has continued to be used to measure uplift rates in the Transantarctic Mountains and thereby to trace the evolution of the landscape during the Cenozoic (Gleadow et al. 1984; Fitzgerald et al. 1986; Gleadow and Fitzgerald 1987; Stump and Fitzgerald 1992; Fitzgerald and Stump 1997; Fitzgerald 1986, 1992, 2002).

The uplift of the Transantarctic Mountains has been explained by modeling of tectonic processes based on systematic measurements of the physical properties of the crust and underlying mantle (e.g., Bentley 1983; Behrendt and Cooper 1991; Coren et al. 1997; Della Vedova et al. 1997; Davey and DeSantis 2006;

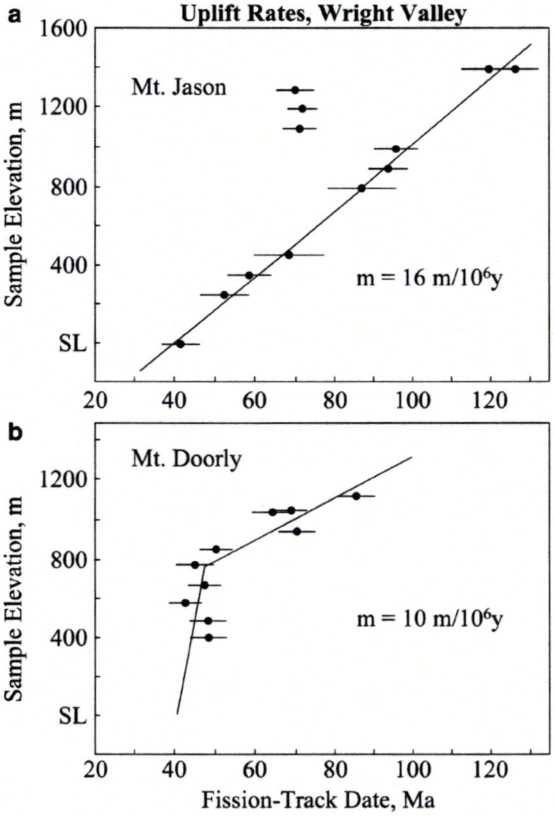

Fig. 15.15 (a) The fission-track dates of apatite in the Olympus Granite Gneiss on Mt. Jason in the Olympus Range of Wright Valley increase linearly with the elevation of the collecting sites above sea level (SL). The slope of the straight line (15 m/10^6 year) is the uplift rate recorded by the fission track dates of apatite. (b) The fission-track dates of apatite in granite basement rocks on Mt. Doorly located adjacent to the Lower Wright glacier form a "dogleg" which shows that the uplift rate increased at about 50 Ma (early Eocene) from 10 m/10^6 year to a much higher rate. The data indicate that the uplift of the Transantarctic Mountains accelerated during the Eocene when Mt. Doorly was uplifted by about 200 m into the complete track-retention zone in only a few million years (Data and interpretation by Gleadow and Fitzgerald (1987))

Fielding et al. 2006). The interpretation of geophysical data has been complemented by field-based studies of rocks in outcrop (e.g., Hamilton 1965; Wilson 1992a, b, 1993, 1995; Mortimer et al. 2002; Paulsen and Wilson 2009).

Behrendt et al. (1991a, b), in a summary of the evidence relevant to the uplift of the Transantarctic Mountains, concluded that the mountains have been rising since about 60 Ma at an average rate of 100 m per million years, but that the uplift after the middle Pliocene (about 3.6 Ma) has been episodic at rates of about 1 km per million years. However, different blocks of the Transantarctic Mountains have been uplifted at different rates and at different times. For example, Webb et al. (1994) reported uplift rates between about 350 m/10^6 year and 548 m/10^6 year based on the occurrence of Pliocene marine microfossils in the mountains of the Queen Alexandra Range adjacent to the Beardmore Glacier and in the Dominion Range.

15.5.6 Uplift of the Transantarctic Mountains

While the evidence in favor of uplift of the Transantarctic Mountains is irrefutable, the *causes* of the observed uplift are still a matter of debate. The problem seems to be that several different geophysical phenomena may have contributed to the uplift. An important proposal by Smith and Drewry (1984, p. 536) attributed the uplift:

> [T]o the delayed effect of the overriding by east Antarctica of anomalously hot asthenosphere forming under west Antarctica in the late Cretaceous.

These authors considered that an increase of 100°C in the temperature at the base of the lithosphere caused an uplift rate of 90 m per million years about 50 million years later.

A very different hypothesis was advanced by Stern and ten Brink (1989) who related the uplift of the Transantarctic Mountains to the depression of the subglacial Wilkes Basin west of the mountains and of the Victoria Land Basin located east of the mountains in the Ross Embayment. The Wilkes Basin is located 400–500 km west of the Transantarctic Mountains and reaches a depth of 1 km below sea level. The Victoria-Land Basin off the coast of Victoria Land contains at least 4–5 km of sediment of Cenozoic age. Stern and ten Brink (1989, p. 10, 315) stated:

> The wavelength and amplitude of these three structures can be accounted for by the elastic flexure of two cantilevered lithospheric plates if the boundary between East and West Antarctica is taken as a stress-free edge.

In subsequent papers Stern and Baxter (2002) and Stern et al. (2005) emphasized the effectiveness of isostatic rebound in response to the erosion of deep valleys into blocks of the Transantarctic Mountains (Stern et al. 2005, p. 221):

> In temperate climates, ~ 25% of peak elevations in mountain ranges can be created by isostatic rebound as a response to erosional incision. Significantly more relief generation and peak uplift are, however, possible for glacial erosion in a polar climate.

The models for the tectonic evolution of the Transantarctic Mountains and the assembly of West Antarctica are complemented by the field-based studies of T.J. Wilson and her students who have interpreted patterns of faults, fractures, dikes, and other structural elements in the rocks to reconstruct the tectonic history of Antarctica (Wilson 1992, 1993a, b; Demyanick 2006). Wilson's work has provided welcome ground truths that constrain the conflicting claims arising from models of the tectonic processes that have shaped Antarctica. For example, Wilson (1993a) investigated the relation between rifting and magmatic activity in the Transantarctic Mountains during the Middle Jurassic and concluded (Wilson 1993a, p. 563):

> Time relations between faults and Ferrar dolerites indicate that crustal extension occurred along the Transantarctic trend before and during the Ferrar magmatism. …Crustal extension followed by magmatism is consistent with recent models in which rifts form where extending lithosphere encounters a region of anomalously hot mantle, possibly associated with mantle plumes.

In another work, Wilson (1995) wrote:

> The Transantarctic Mountains are considered to be a rift-flank uplift, yet no substantial isostatic uplift is expected in a transtensional setting, and the mechanism of large-magnitude Cenozoic uplift of the mountains remains problematical.

These statements by Wilson remind us that the Transantarctic Mountains contain not only the silica-saturated basalt flows and dolerite sills of the Middle Jurassic Ferrar Group, but also late Tertiary alkali-rich volcanic rocks of the McMurdo Volcanics. Behrendt (1991a, b) combined the presence of these volcanic rocks with geophysical data to define the Cenozoic West Antarctic Rift system.

This rift system in Fig. 15.16 enters northern Victoria Land from the Indian Ocean and runs parallel to the Jurassic rift system in Fig. 15.12 (Schmidt and Rowley 1986) as far as the Horlick Mountains, where it turns and passes west of the Ellsworth-Whitmore mountains towards the Bellingshausen Sea. Accordingly, the Byrd Subglacial Basins of West Antarctica, defined by geophysical parameters in Fig. 15.5 (LeMasurier 2008) and by extension in a back-arc in Fig. 15.6 (Elliot 1992), is thereby identified as the "floor" of the West Antarctic rift system of Cenozoic age.

The Transantarctic Mountains are now widely regarded as the unusually high margin of the West Antarctic rift system (Behrendt and Cooper 1991). The Cenozoic alkali-rich volcanic rocks that were erupted along this rift occur not only along the Transantarctic Mountains but also characterize the surface geology of Marie Byrd Land and of Ellsworth Land (LeMasurier and Thomson 1990; LeMasurier 2006).

The tectonics of the West Antarctic rift zone will continue to be investigated in the years to come by indirect geophysical methods and by direct observation of rocks in outcrop. The evidence for crustal extension related to the West Antarctic rift is strong, but the study of the geology and tectonics of West Antarctica is made difficult by the cover of the West Antarctic ice sheet. Nevertheless, progress in understanding this part of Antarctica is exemplified by recent studies of Siddoway et al. (2004), Winberry and Anandakrishnan (2004), and LeMasurier (2006, 2008).

15.6 Summary

The break-up of Gondwana was initiated by the arrival of a mantle *plume* in the area between Queen Maud Land of East Antarctica and southern Africa. The stresses imposed on the lithosphere by this plume resulted in the development of a *triple junction* consisting of crustal rifts. The decompression melting of the rocks in the plume head and in the adjacent lithospheric mantle formed basalt magma which was erupted at the surface to form the *Kirwan Volcanics* of Queen Maud Land and parts of the *Karoo Basalt* of southern Africa. The rift between Africa and East Antarctica subsequently widened by seafloor spreading into the Weddell Sea and the Indian Ocean.

The break-up of Gondwana continued by means of crustal rifts which widened into the southern the oceans as the continental fragments were displaced to lower latitudes. The break-up of Gondwana started during the early Middle Jurassic and continued well into the

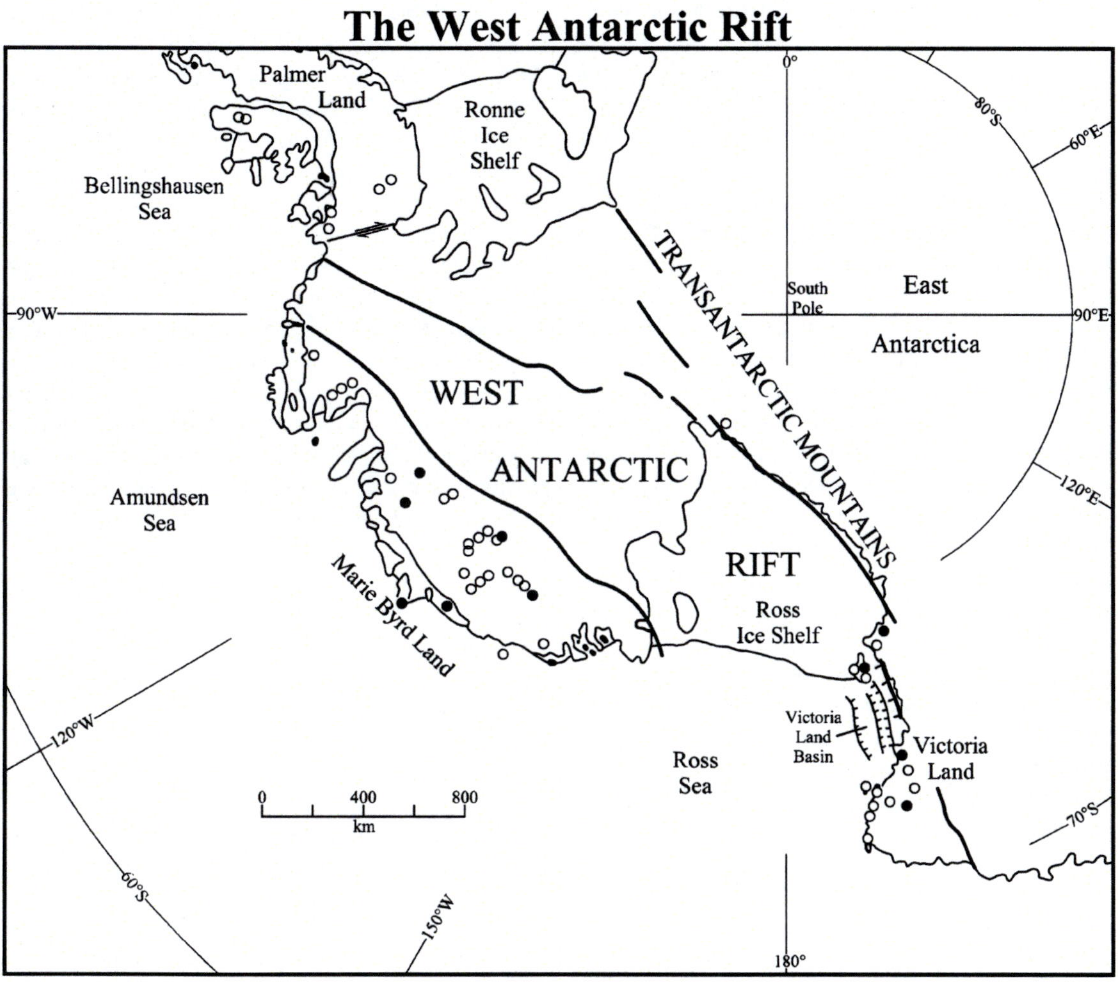

Fig. 15.16 The Transantarctic Mountains form part of the margin of the West Antarctic Rift which swings west near the southern tip of the Ross Ice Shelf and extends from there across West Antarctica toward the Bellingshausen Sea. The opposite margin of the West Antarctic Rift is located along the southern edge of Marie Byrd Land. Both margins of the West Antarctic Rift contain volcanoes of Cenozoic age which have erupted alkali-rich lavas. Mt. Erebus is presently the only one of these volcanoes that is still active. The broad floor of the West Antarctic Rift is the Byrd Subglacial Basin where the continental crust has been thinned to about 20 km by extension. *Solid circles*: Quaternary volcanic centers, *Open circles*: Older volcanic centers (Adapted from LeMasurier and Thomson (1990, p. 2))

Late Cretaceous when Australia was the last part of Gondwana to separate from East Antarctica. The tectonic plates of the southern hemisphere continue to move at the present time, except the Antarctic plate which is locked in place at the geographic south pole by spreading ridges that surround it almost completely.

Before the start of the break-up of Gondwana, a powerful subduction zone existed along its paleo-Pacific coast. This subduction zone compressed the volcano-sedimentary complex that had been deposited off-shore and caused the explosive eruption of felsic volcanic ash that was deposited in the Beacon basins of East Antarctic during the Late Permian and Triassic (i.e., in the Buckley, Fremouw, and Falla formations). The eruptions continued even into the Middle Jurassic during the early stages of the eruption of the Kirkpatrick Basalt (e.g., the Exposure Hill, Mawson, and Prebble formations). In addition, granitic magma was intruded into the crust of West Antarctica at about the same time while basalt magma formed the basalt flows and dolerite sills of the Ferrar Group in the Transantarctic Mountains. The subduction zone along the Antarctic Peninsula and West Antarctica has died out but continues to be active along the west coast of South America.

The present boundary between East and West Antarctica is a major tectonic discontinuity that has been variously interpreted to be a normal fault associated with the back-arc basin of the paleo-Pacific subduction zone, or a rift extending from the Weddell triple junction, or a transform fault with major strike-slip motion, or the passive margin of the West Antarctic rift zone, or a combination of two or more of these interpretations. In any case, rifts during the Middle Jurassic facilitated the transport of basalt magma of the Ferrar Group and may have caused partial melting by decompression of deep crustal and lithospheric rocks along their length. Later, during the Cenozoic, alkali-rich volcanic rocks were erupted along the Transantarctic Mountains and in West Antarctica in response to the formation of the West Antarctic rift system. The Transantarctic Mountains, from northern Victoria Land to the Horlick Mountains, are the uplifted margin of this rift system. The West Antarctic rift system turns west at the Horlick Mountains and continues across West Antarctica toward the Bellingshausen Sea.

The Transantarctic Mountains were initially uplifted by the intrusion of sills of the Ferrar Dolerite during the Middle Jurassic. The main phase of the uplift occurred during the late Cenozoic (Neogene) in response to the formation of the West Antarctic rift system. However, this explanation does not account for the unusually high elevation of the Transantarctic Mountains nor does it indicate why this rift formed in the first place. LeMasurier (2006) proposed that a dome on the southern slope of Marie Byrd Land may be the surface expression of a mantle plume. This hypothesis, if confirmed, would relate the origin of the West Antarctic rift system to the presence of one or several mantle plumes that have not yet been detected because the topography and geology of the floor of the West Antarctic rift valley (i.e., the Byrd Subglacial Basin) are obscured by the West Antarctic ice sheet and by the Ross Ice Shelf.

The origin of the Byrd Subglacial Basin is still uncertain because it has also been attributed to extensional thinning of continental crust in the back-arc associated with the subduction zone along the Paleo-Pacific coast of Gondwana, or to crustal thinning of the floor of the West Antarctic rift systems. West Antarctica has also been described as a collection of small fragments of continental crust which were merged into an apparent land mass by the cover of the West Antarctic ice sheet.

Although the tectonic evolution of the Transantarctic Mountains and of West Antarctica may be linked, the details of that relationship are still uncertain. The strike-slip motion along the boundary faults between the accreted terranes and the craton of East Antarctica is also uncertain. Perhaps we now have sufficient knowledge of the complex tectonic history of Antarctica to identify the areas where specialized studies may provide additional evidence we can use to replace uncertainty with understanding.

15.7 Appendix

15.7.1 The Fission-Track Method of Dating

Uranium is composed of three naturally occurring radioactive isotopes having the mass numbers: 238, 235, and 234 all of which decay by emitting alpha particles from their nuclei. All of the naturally occurring isotopes of uranium can also decay by *spontaneous fission*, but virtually all spontaneous fission events in a uranium-bearing mineral are due to ^{238}U which is by far the most abundant isotope of uranium at 99.2745% by number.

When a fission event occurs in a crystal of a uranium-bearing mineral (e.g., apatite, sphene, zircon, muscovite, epidote) or in a specimen of glass (e.g., volcanic, meteorite impact, or man-made), the fission fragments of the nucleus of ^{238}U collide forcefully with atoms of other elements in the crystal or glass. The damaged region along the tracks of the fission fragments can be enlarged by etching a polished interior surface of the specimen being dated. The solvents used for etching were listed by Faure and Mensing (2005, Table 22.1). The resulting fission tracks are then counted with an optical microscope to determine the number of tracks per unit area. The measured track density depends on the age of the crystal, on its uranium concentration, on the halflife of ^{238}U for spontaneous fission, and on the retention of the fission tracks which depends on the ambient temperature and therefore on the depth below the surface of the Earth where the samples were collected.

The temperature is an important variable because uranium-bearing minerals anneal fission tracks at elevated temperatures and retain them quantitatively only after the temperature has decreased below the level at which all tracks are preserved. Therefore, the fission track date of a mineral or glass is the time that has elapsed since the specimen cooled through the track-retention temperature for the last time. In other words, fission track dates are equal to the "age" of a specimen of mineral or glass *only* in cases when the specimen cooled to the track-retention temperature immediately after it was formed. The fission-track method appears to be straightforward in principle but is, in fact, quite labor intensive and demanding (Faure and Mensing 2005).

References

Adams CJ (1986) Geochronological studies of the Swanson Formation of Marie Byrd Land, West Antarctica, and correlation with northern Victoria Land, East Antarctica, and South Island, New Zealand. New Zealand J Geol Geophys 29:354–358

Adams CJ, Gabites JE, Wodzicki A, Laird MG, Bradshaw JD (1982) Potassium-argon geochronology of the Precambrian-Cambrian Wilson and Robertson Bay Groups and Bowers Supergroup, northern Victoria Land, Antarctica. In: Craddock C (ed) Antarctic geoscience. University of Wisconsin Press, Madison, WI, pp 543–548

Adams CJ, Seward D, Weaver SD (1995) Geochronology of Cretaceous granites and metasedimentary basement on Edward VII Peninsula (Ross margin), Marie Byrd Land, West Antarctica. Antarctic Sci 7:265–277

Audley-Charles MG (1983) Reconstruction of eastern Gondwanaland. Nature 306:48–50

Behrendt JC, Henderson JR, Meister L, Rambo W (1974) Geophysical investigations of the Pensacola Mountains and adjacent glacierised area of Antarctica. US Geol. Surv. Prof. Paper 844

Behrendt JC, Drewry DJ, Jankowski E, Grim MS (1980) Aeromagnetic and radio echo ice sounding measurements show much greater area of the Dufek intrusion, Antarctica. Science 209:1014–1017

Behrendt JC, Cooper AK (1991) Evidence of rapid Cenozoic uplift of the shoulder escarpment of the Cenozoic West Antarctic rift system and speculation on possible climate forcing. Geology 19:315–319

Behrendt JC, LeMasurier WE, Cooper AK, Tessensohn F, Tréhu A, Damaske D (1991a) Geophysical studies of the West Antarctic Rift. Tectonics 10(6):1257–1273

Behrendt JC, LeMasurier WE, Cooper AK, Tessensohn F, Tréhu A, Damaske D (1991b) The West Antarctic Rift system: A review of geophysical investigations. In: Elliot DH (ed) Contributions to Antarctic research II. American Geophysical Research, Washington, DC, pp 67–112

Bentley CR (1973) Crustal structure of Antarctica. Tectonophysics 20:229–240

Bentley CR (1983) Crustal structure of Antarctica from geophysical evidence: A review. In: Oliver RL, James PR, Jago JB (eds) Antarctic earth science. Australian Academy of Science, Canberra, A.C.T., pp 491–497

Bentley CR, Clough JW (1972) Antarctic subglacial structure from seismic refraction measurements. In: Adie RJ (ed) Antarctic geology and geophysics. Universitetsforlaget, Oslo, Norway, pp 683–691

Bentley CR, Robertson JD (1982) Isostatic gravity anomalies in West Antarctic. In: Craddock C (ed) Antarctic geoscience. University of Wisconsin Press, Madison, WI, pp 949–954

Bradshaw JD (1991) Cretaceous dispersion of Gondwana: Continental and oceanic spreading in the south-west Pacific-Antarctic sector. In: Thomson MRA, Crame JA, Thomson JW (eds) Geological evolution of Antarctica. Cambridge University Press, Cambridge, UK, pp 581–586

Bradshaw JD, Andrews PB, Field BD (1983) Swanson Formation and related rocks of Marie Byrd Land and a comparison with the Robertson Bay Group of northern Victoria Land. In: Oliver RL, James PR, Jago JB (eds) Antarctic earth science. Australian Academy of Science, Canberra, A.C.T., pp 274–279

Burke K, Dewey JF (1973) Plume-generated triple junctions: Key indicators in applying plate tectonics to old rocks. J Geol 81:406–433

Campbell IH, Griffiths RW (1990) Implications of mantle-plume structure for the evolution of flood basalts. Earth Planet Sci Lett 99:79–93

Cande SC, Mutter JC (1982) A revised identification of the oldest sea-floor spreading anomalies between Australia and Antarctica. Earth Planet Sci Lett 58(2):151–160

Coren F, Delisle G, Rolf C (1997) Evidence for Mesozoic-early Cenozoic tectonic lineations in the Ross Sea and Transantarctic Mountains, Antarctica. In: Ricci CA (ed) The Antarctic region: Geological evolution and processes. Terra Antartica, Siena, Italy, pp 577–584

Cox KG (1978) Flood basalts, subduction, and the break-up of Gondwanaland. Nature 274:47–49

Cox KG (1988) The Karoo province. In: Macdougall JD (ed) Flood Basalts. Kluwer Academic, Hingham, MA

Craddock C (1970) Antarctic geology and Gondwanaland. Antarctic J US 5(3):53–57

Crawford AR (1974) A greater Gondwanaland. Science 184:1179–1181

Dalziel IWD, Elliot DH (1971) Evolution of the Scotia Arc. Nature 233(5317):246–252

Dalziel IWD, Elliot DH (1982) West Antarctica-Problem child of Gondwanaland. Tectonics 1:3–19

Dalziel IWD, Garrett SW, Grunow AM, Pankhurst RJ, Storey BC, Vennum WR (1987a) The Ellsworth-Whitmore, Mountains crustal block: Its role in the tectonic evolution of West Antarctica. In: McKenzie GD (ed) Gondwana Six: Structure, tectonics, and geophysics. Geophysical Monograph, vol. 40. American Geophysical Union, Washington, DC, pp 173–182

Dalziel IWD, Storey BC, Garrett SW, Grunow AM, Herrod LDB, Pankhurst RJ (1987b) Extensional tectonics and fragmentation of Gondwanaland. In: Coward MP, Dewey JF, Hancock PL (eds) Continental extensional tectonics. Geol Soc London, Spec Pub 28:433–441

Davey FJ, De Santis L (2006) A multi-phase rifting model for the Victoria Land Basin, western Ross Sea. In: Fütterer DK, Damaske D, Kleinschmidt G, Miller H, Tessensohn F (eds) Antarctica: Contributions to global earth science. Springer, Heidelberg, Germany, pp 303–308

Della Vedova B, Pellis G, Trey H, Zhank J, Cooper AK, Markis J, et al. (1997) Crustal structure of the Transantarctic Mountains. In: Ricci CA (ed) The Antarctic region: Geological evolution and processes. Terra Antartica, Siena, Italy, pp 609–618

Demyanick E (2006) Structure of the Transantarctic Mountains from the Royal Society Range to the Churchill Mountains as determined from satellite image analysis. MSc thesis, School of Earth Sciences, The Ohio State University, Columbus, OH

Dewart G, Toksöz MN (1965) Crustal structure in East Antarctica from surface-wave dispersion. Geophys J 10:127–139

Dietz RS, Sproll WP (1970) Fit between Africa and Antarctica: A continental drift reconstruction. Science 167(3925): 1612–1614

Dietz RS, Holden JC, Sproll WP (1972) Antarctica and continental drift. In: Adie RJ (ed) Antarctic geology and geophysics. Universitetsforlaget, Oslo, Norway, pp 837–842

Drewry DJ (1972) Subglacial morphology between the Transantarctic Mountains and the south pole. In: Adie RJ (ed) Antarctic geology and geophysics. Universitetsforlaget, Oslo, Norway, pp 693–703

Drewry DJ (1975) Radio-echo sounding map of Antarctica (~90°E–180°). Polar Record 17(109):359–374

Drewry DJ (1976) Comparison of electromagnetic and seismic-gravity ice thickness measurements in East Antarctica. J Glaciol. 15(73):137–150

Duncan RA, Richards MA (1991) Hotspots, mantle plumes, flood basalts, and true polar wander. Rev Geophys 29:31–50

du Toit AL (1937) Our wandering continents. Oliver and Boyd, Edinburgh, Scotland

Ebinger CJ, Sleep NH (1998) Cenozoic magmatism throughout East Africa resulting from impact of a single plume. Nature 395:788–791

Elliot DH (1972) Aspects of Antarctic geology and drift reconstructions. In: Adie RJ (ed) Antarctic geology and geophysics. Universitetsforlaget, Oslo, Norway, pp 849–858

Elliot DH (1974) The tectonic setting of the Jurassic Ferrar Group, Antarctica. In: Gonzalez-Ferran O (ed) Andean and Antarctic volcanology problemsFrancesco Giannini and Sons, Naples, Italy, pp 357–372

Elliot DH (1975) Tectonics of Antarctica: A review. Amer J Sci 275-A:45–106

Elliot DH (1991) Triassic-Early Cretaceous evolution of Antarctica. In: Thomson MRA, Crame JA, Thomson JW (eds) Geological evolution of Antarctica. Cambridge University Press, Cambridge, UK, pp 541–548

Elliot DH (1992) Jurassic magmatism and tectonism associated with Gondwanaland break-up: An Antarctic perspective. In: Storey BC, Alabaster T, Pankhurst RJ (eds) Magmatism and the causes of continental breakup. Geol Soc London, Spec Pub 68:165–184

Elliot DH, Fleming TH (2000) Weddell triple junction: The principal focus of Ferrar and Karoo magmatism during initial break-up of Gondwana. Geology 28(6):539–542

Elliot DH, Fleming TH, Kyle PR, Foland KA (1999) Long-distance transport of magmas in the Jurassic large igneous province. Earth Planet Sci Lett 167:89–104

Encarnación J, Fleming TH, Elliot DH, Eales HV (1996) Synchronous emplacement of Ferrar and Karoo dolerites and the early breakup of Gondwana. Geology 24:535–538

Falvey DA, Mutter JC (1981) Regional plate tectonics and the evolution of Australia's passive continental margins. J Australian Geol Geophys 6(1):1–29

Faure G (2001) The origin of igneous rocks; The isotopic evidence. Springer, Heidelberg, Germany

Faure G, Mensing TM (2005) Isotopes: Principles and applications. Wiley Hoboken, New Jersey

Fielding CR, Henrys SA, Wilson TJ (2006) Rift history of the western Victoria Land Basin: A new perspective based on integration of cores with seismic reflection data. In: Fütterer DK, Damaske D, Kleinschmidt G, Miller H, Tessensohn F (eds) Antarctica: Contributions to global earth sciences. Springer, Heidelberg, Germany, pp 309–318

Fitzgerald PG (1986) Fission-track tectonic studies of the Transantarctic Mountains, Beardmore Glacier area. Antarctic J US 21(5):38–41

Fitzgerald PG (1992) The Transantarctic Mountains of southern Victoria Land: The application of apatite fission track analysis to a rift shoulder uplift. Tectonics 11:634–662

Fitzgerald PG (2002) Tectonics and landscape evolution of the Antarctic plate since the breakup of Gondwana, with an emphasis on the West Antarctic rift system and the Transantarctic Mountains. In: Gamble et al. (eds) Antarctica at the close of the Millennium. Roy Soc New Zealand Bull 35:453–469

Fitzgerald PG, Stump E (1997) Cretaceous and Cenozoic episodic denudation of the Transantarctic Mountains, Antarctica: New constraints from apatite fission track thermochronology in the Scott Glacier region. J Geophys Res 102:7747–7765

Fitzgerald PG, Sandiford M, Barrett PJ, Gleadow AJ (1986) Asymmetric extension associated with uplift and subsidence in the Transantarctic Mountains and Ross Embayment. Earth Planet Sci Lett 81(1):67–78

Ford AB (1972) Weddell orogeny: Latest Permian to early Mesozoic deformation at the Weddell-Sea margin of the Transantarctic Mountains. In: Adie RJ (ed) Antarctic geology and geophys. Universitetsforlaget, Oslo, Norway, pp 419–425

Frakes LA, Crowell JC (1970) Geologic evidence of the place of Antarctica in Gondwanaland. Antarctic J US 5(3):67–69

Garrett SW, Herrod LDB, Mantripp DR (1987) Crustal structure in the area of the Haag Nunataks, West Antarctica: New aeromagnetic and bedrock elevation data. In: McKenzie GD (ed) Gondwana Six: Structure, tectonics, and geophysics. Geophysical Monograph, vol. 40. American Geophysical Union, Washington, DC, pp 109–115

Gleadow AJW, Fitzgerald PG (1987) Uplift history and structure of the Transantarctic Mountains: New evidence from fission track dating of basement apatites in the Dry Valley area, southern Victoria Land. Earth Planet Sci Lett 82:1–14

Gleadow AJW, McKelvey BC, Ferguson KU (1984) Uplift history of the Transantarctic Mountains in the Dry Valleys area, southern Victoria Land, Antarctica, from apatite fission track ages. New Zealand J Geol Geophys 27(4):457–464

Griffiths JR (1974) Revised continental fit of Australia and Antarctica. Nature 249:336–338

Grikurov GE (1982) Structure of Antarctica and outline of its evolution. In: Craddock C (ed) Antarctic geoscience. University of Wisconsin Press, Madison, WI, pp 791–804

Grikurov GE, Ravich MG, Soloviev DS (1972) Tectonics of Antarctica. In: Adie RJ (ed) Antarctic geology and geophysics. Universitetsforlaget, Oslo, Norway, pp 457–468

Grindley GW, Davey FJ (1982) The reconstruction of New Zealand, Australia, and Antarctica. In: Craddock C (ed) Antarctic geoscience. University of Wisconsin Press, Madison, WI, pp 15–29

Grindley GW, Oliver PJ (1983) Palaeomagnetism of Cretaceous volcanic rocks from Marie Byrd Land, Antarctica. In: Oliver RL, James PR, Jago JB (eds) Antarctic earth science. Australian Academy of Science, Canberra, A.C. T., pp 573–578

Groushinsky NP, Sazhina NB (1982) Some features of Antarctic crustal structure. In: Craddock C (ed) Antarctic geoscience. University of Wisconsin Press, Madison, WI, pp 907–911

Grunow AM (1993) Creation and destruction of Weddell Sea floor in the Jurassic. Geology 21:647–650

Grunow AM, Dalziel IW, Kent DV (1987) Ellsworth-Whitmore Mountains crustal block, western Antarctica: New paleomagnetic results and their tectonic significance. In: McKenzie GD (ed) Gondwana Six: Structure, tectonics, and geophysics. Geophysical Monograph, vol. 40. American Geophysical Union, Washington, DC, pp 161–171

Grunow AM, Kent DV, Dalziel IWD (1991) New paleomagnetic data from Thurston Island: Implications for the tectonics of West Antarctica and Weddell Sea opening. J Geophys Res 96:17935–17954

Halpern M (1966). Rubidium-strontium date from Mt. Byerly, West Antarctica. Earth Planet Sci Lett 1:455–457

Halpern M (1967) Rubidium-strontium isotopic age measurements of plutonic igneous rocks in eastern Ellsworth Land and northern Antarctic Peninsula, Antarctica. J Geophys Res 72:5133

Halpern M (1968) Ages of Antarctic and Argentine rocks bearing on continental drift. Earth Planet Sci Lett 5:159–167

Hamilton W (1965) Diabase sheets of the Taylor Glacier region, Victoria Land, Antarctica. US Geol Survey, Prof. Paper, 456-B:1–71

Hamilton W (1967) Tectonics of Antarctica. Tectonophysics 4(4/6):555–568

Harris C, Erlank AJ, Duncan AR, Marsh JS (1991) The geochemistry of the Kirwan and other Jurassic basalts of Dronning Maud Land, and their significance for Gondwana reconstruction. In: Thomson MRA, Crane JA, Thomson JW (eds) Geological evolution of Antarctica. Cambridge University Press, Cambridge, UK, pp 563–568

Hayes DE, Zhang C, Weissel RA (2009) Modeling paleobathymetry to 30° South: Present to 75 my. EOS 90(19):165–166

Hofmann AW, White WM (1982) Mantle plumes from ancient oceanic crust. Earth Planet Sci Lett 57:421–436

IUGS (2002) International stratigraphic chart. International Commission on Stratigraphy. UNESCO, New York

Jankowski EJ, Drewry DJ (1981) The structure of West Antarctica from geophysical studies. Nature 291:17–21

Jankowski EJ, Drewry DJ, Behrendt JC (1983) Magnetic studies of upper crustal structure in West Antarctica and the boundary with East Antarctica. In: Oliver RL, James PR, Jago JB (eds) Antarctic earth science. Australian Academy of Science, Canberra, A.C.T., pp 197–203

Kadmina IN, Kurinin RG, Masolov VN, Grikurov GE (1983) Antarctic crustal structure from geophysical evidence: A review. In: Oliver RL, James PR, Jago JB (ed) Antarctic earth science. Australian Academy of Science, Canberra, A.C.T., pp 498–502

Katz MB (1978) Sri Lanka in Gondwanaland and the evolution of the Indian Ocean. Geol Mag 115:237–244

Katz HR (1982) West Antarctica and New Zealand: A geological test of the model of continental split. In: Craddock C (ed) Antarctic geology and geophysics. University Wisconsin Press, Madison, WI, pp 31–41

Katz MB, Premoli C (1979) India and Madagascar in Gondwanaland based on matching Precambrian lineaments. Nature 297:312–315

Kennett JP, Burns RE, Andrews JE, Churkin M Jr, Davies TA, Dumitrica P, Edwards AR, Galehouse JS, Packham GH, van der Lingen GJ (1972) Australian-Antarctic continental drift, palaeocirculation changes and Oligocene deep-sea erosion. Nature 239:51–55

Kristoffersen Y, Hinz K (1991) Evolution of the Gondwana plate boundary in the Weddell Sea area. In: Thomson MRA, Crame JA, Thomson JW (eds) Geological evolution of Antarctica. Cambridge University Press, Cambridge, UK, pp 225–230

Kurinin RG, Grikurov GE (1982) Crustal structure of part of East Antarctica from geophysical data. In, Craddock, C., ed. Antarctic Geoscience. Univ. Wisconsin Press, Madison, Wisconsin

Laird MG (1981) The late Mesozoic fragmentation of the New Zealand segment of Gondwana. In: Cresswell MM, Vella P (eds) Gondwana Five. A.A. Balkema, Rotterdam, The Netherlands, pp 311–318

Lawver LA, Scotese CR (1987) A revised reconstruction of Gondwanaland. In: McKenzie GD (ed) Gondwana Six: Structure, tectonics, and geophysics. Geophysical Monograph, vol. 40. American Geophysical Union, Washington, D.C, pp 17–23

Lawver LA, Royer JY, Sandwell DT, Scotese CR (1991) Evolution of the Antarctic continental margins. In: Thomson MRA, Crame JA, Thomson JW (eds) Geological evolution of Antarctica. Cambridge University, Press, Cambridge, UK, pp 533–539

LeMasurier WE (2006) What supports the Marie Byrd Land dome? An evaluation of potential uplift mechanisms in a continental rift system. In: Fütterer DK, Damaske D, Kleinschmidt G, Miller H, Tessensohn F (eds) Antarctica: Contributions to global earth science. Springer, Heidelberg, Germany, pp 299–302

LeMasurier WE (2008) Neogene extension and basin deepening in the West Antarctic rift inferred from comparisons with the East African rift and other analogs. Geology 36(3):247–250

LeMasurier WE, Rex DC (1989) Evolution of linear volcanic ranges in Marie Byrd Land, West Antarctica. J Geophys Res 94(B6):7223–7236

LeMasurier WE, Thomson JW (eds) (1990) Volcanoes of the Antarctic plate and southern oceans. Antarctic Research

Series, vol. 48. American Geophysical Union, Washington, DC

Macdougall JD (ed) (1988) Flood basalts. Kluwer Academic, Hingham, MA

Martin KA, Hartnady CJ, Goodland SW (1981) A revised fit of South America and south central Africa. Earth Planet Sci Lett 54:293–305

McElhinny MW, Luck GR (1970) Paleomagnetism and Gondwanaland. Science 168:830–832

McGlynn JC, Irving E, Bell K, Pullaiah G (1975) Palaeomagnetic poles and a Proterozoic supercontinent. Nature 255(5506): 318–319

Millar IL, Pankhurst RJ (1987) Rb-Sr geochronology of the region between the Antarctic Peninsula and the Transantarctic Mountains: Haag Nunataks and Mesozoic granitoids. In: McKenzie GD (ed) Gondwana Six: Structure, tectonics, and geophysics. Geophysical Monograph, vol. 40. American Geophysical Union, Washington, DC, pp 151–160

Mitchell C, Taylor GK, Cox KG, Shaw J (1986) Are the Falkland Islands a rotated microplate? Nature 319:131–134

Mortimer N, Forsyth PJ, Turnbull IM (2002) Reassessment of faults in the Wilson Piedmont Glacier area: Implications for age and style of Transantarctic Mountains uplift. In: Gamble JA et al. (eds) Antarctica at the close of the Millennium. Roy Soc New Zealand Bull 35:207–213

Mortimer N, Hoernle K, Hauff F, Palin JM, Dunlap WJ, Werner R, Faure K (2006) New constraints on the age and evolution of the Wishbone Ridge, southwest Pacific Cretaceous microplate, and the Zealandia-West Antarctic breakup. Geology 34(3):185–188

Nance RD, Linnemann U (2008) The Rheic Ocean: Origin, evolution, and significance. GSA Today 18(12):4–12

Norton IO, Molnar P (1977) Implications of a revised fit between Australia and Antarctica for the evolution of the eastern Indian Ocean. Nature 267:338–339

Norton IO, Sclater JG (1979) A model for the evolution of the Indian Ocean and the break-up of Gondwanaland. J Geophys Res 84(B12):6303–6330

Pankhurst RJ, Storey BC, Millar IL (1991) Magmatism related to the break-up of Gondwana. In: Thomson MRA, Crame JA, Thomson JW (eds) Geological evolution of Antarctica. Cambridge University Press, Cambridge, UK, pp 573–579

Pankhurst RJ, Millar IL, Grunow AM, Storey BC (1993) The pre-Cenozoic magmatic history of the Thurston Island crustal block, West Antarctica. J Geophys Res 98:11835–11849

Pankhurst RJ, Leat PT, Sruoga P, Rapela CW, Marquez M, Storey BC, Riley TR (1998) The Chon Aike province of Patagonia and related rocks in West Antarctica: A silicic large igneous province. J Volcanol Geotherm Res 81:113–136

Paulsen TS, Wilson TJ (2009) Structure and age of volcanic fissures on Mount Morning: A new constraint on Neogene to contemporary stress in the West Antarctic Rift, southern Victoria Land, Antarctica. Geol Soc Amer Bull 121:1071–1088

Rabinowitz PD, Coffin MF, Falvey D (1983) The separation of Madagascar and Africa. Science 220:67–69

Ravich MG, Grikurov GE (1974) Explanatary notes to the geologic map of Antarctica. Research Institute of Arctic Geology. St. Petersburg, Russia, pp 1–83

Richards MA, Duncan RA, Courtillot VE (1989) Flood basalts and hotspot tracks: Plume heads and tails. Science 246:103–107

Robinson ES, Splettstoesser JF (1984) Structure of the Transantarctic Mountains determined from geophysical surveys. In: Turner MD, Splettstoesser JF (eds) Geology of the Central Transantarctic Mountains. Antarctic Research Series 36:119–162

Rülke A, Dietrich R (2007) The SCAR GPS campaigns in the context of global reference system realization and geodynamic research in Antarctica: A keystone in a changing world. In: Cooper AK, Raymond CR, et al. (eds) Online Proceedings of the 10th ISAES. USGS Open-File Report 2007-1047. Extended Abstract 146:1–4

Scharnberger CK, Scharon L (1972) Palaeomagnetism and plate tectonics of Antarctica. In: Adie RJ (ed) Antarctic geology and geophysics. Universitetsforlaget. Oslo, Norway, pp 843–848

Schilling J-G (1973) Afar mantle plume: Rare-earth evidence. Nature 242:2–5

Schmidt DL, Rowley PD (1986) Continental rifting and transform faulting along the Jurassic Transantarctic rift, Antarctica. Tectonics 5(2):279–291

Schopf JM (1969) Ellsworth Mountains: Position in West Antarctica due to sea-floor spreading. Science 164:63–66

Siddoway CS, Baldwin SL, Fitzgerald PG, Fanning CM, Luyendyk BP (2004) Ross Sea mylonites and the timing of intracontinental extension within the West Antarctic Rift System. Geology 32(1):57–60

Smith AG, Drewry DJ (1984) Delayed phase change due to hot asthenosphere causes Transantarctic uplift? Nature 309:536–538

Smith AG, Hallam A (1970) The fit of the southern continents. Nature 225(5228):139–144

Sproll WP, Dietz RS (1969) Morphological continental drift fit of Australia and Antarctica. Nature 222(5191):345–348

Squire RJ, Miller JMcL (2003) Synchronous compression and extension in East Gondwana: Tectonic controls on world-class gold deposits at 440 Ma. Geology 31(12): 1073–1076

Stauffer PH, Gobbett DJ (1972) Gondwanaland possible involvement of southeast Asia. Nature 240:139–140

Stern TA, Baxter AB (2002) Glacial erosion, rock, and peak uplift within the Transantarctic Mountains. In: Gamble JA et al. (eds) Antarctica at the close of a Millennium. Roy Soc New Zealand Bull 35:471–478

Stern TA, ten Brink US (1989) Flexural uplift of the Transantarctic Mountains. J Geophys Res 94(B8):10315–10330

Stern TA, Baxter AK, Barrett PJ (2005). Isostatic rebound due to glacial erosion within the Transantarctic Mountains. Geology 33(3):221–224

Stock JM, Cande SC (2002) Tectonic history of Antarctic seafloor in the Australia-New Zealand-South Pacific sector: Implications for Antarctic continental tectonics. Roy Soc New Zealand Bull 35:351–259

Storey BC (1991) The crustal blocks of West Antarctica within Gondwana: Reconstruction and break-up model. In: Thomson MRA, Crame JA, Thomson JW (eds) Geological evolution of Antarctica. Cambridge University Press, UK, pp 587–592

Storey BC (1995) The role of mantle plumes in continental breakup: Case histories from Gondwanaland. Nature 337:301–308

Storey BC et al. (eds) (1996) Weddell Sea tectonics and Gondwana breakup. Geol Soc London, Spec Pub 108

Storey BC, Alabaster T (1991) Tectonomagmatic controls on Gondwana break-up models: Evidence from the proto-Pacific margin of Antarctica. Tectonics 10(6):1274–1288

Storey BC, Dalziel IWD (1987) Outline of the structural and tectonic history of the Ellsworth Mountains-Thiel Mountain ridge. In: McKenzie GD (ed) Gondwana Six: Structure, tectonics, and geophysics. Geophysical Monograph, vol. 40. American Geophysical Union, Washington, DC, pp 117–128

Storey BC, Kyle PR (1997) An active mantle mechanism for Gondwana break-up. South African J Geol 100:238–290

Storey BC, Hole MJ, Pankhurst RJ, Millar IL, Vennum WR (1988) Middle Jurassic within-plate granites in West Antarctica and their bearing on the break-up of Gondwanaland. J Geol Soc London 145:999–1007

Stump E (1973) Earth evolution in the Transantarctic Mountains and West Antarctica. In: Tarling DH, Runcorn SK (eds) Implications of continental drift to the earth sciences 2:909–924. Academic, London

Stump E, Fitzgerald PG (1992) Episodic uplift of the Transantarctic Mountains. Geology 20:161–164

Stump E, White AJR, Borg SG (1986) Reconstruction of Australia and Antarctica: Evidence from granites and recent mapping. Earth Planet Sci Lett 79(3/4):348–360

Tappert R, Foden J, Stachel T, Muehlenbachs K, Tappert M, Willis K (2009) Deep mantle diamonds from South Australia. A record of Pacific subduction at the Gondwana margin. Geology 37(1):43–36

Turner S, Hawkesworth C (1995) The nature of the sub-continental mantle: Constraints from the major-element composition of continental flood basalts. Chem Geol 120:295–314

Vennum WR, Storey BC (1987a) Petrology, geochemistry, and tectonic setting of granitic rocks from the Ellsworth-Whitmore mountains crustal block and Thiel Mountains, West Antarctica. In: McKenzie GD (ed) Gondwana Six: Structure, tectonics, and geophysics. Geophysical Monograph, vol. 40. American Geophysical Union, Washington, DC, pp 139–150

Vennum WR, Storey BC (1987b) Correlation of gabbroic and diabasic rocks from the Ellsworth Mountains, Hart Hills, and Thiel Mountains, West Antarctica. In: McKenzie GD (eds) Gondwana Six: Structure, tectonics, and geophysics. Geophysical Monograph, vol. 40. American Geophysical Union, Washington, DC, pp 129–138

Vilas JF, Valencio DA (1970) Paleogeographic reconstruction of the Gondwanic continents based on paleomagnetic and sea-floor spreading data. Earth Planet Sci Lett 7(5):397–405

Voronov PS (1964) Tectonics and neotectonics of Antarctica. In: Adie RJ (ed) Antarctic geology. North-Holland, Amsterdam, The Netherlands, pp 629–700

Wade FA (1969) Geology of Marie Byrd Land. In: Craddock C (ed) Geologic map of Antarctica. Folio 12, Sheet 18. American Geographical Society, New York

Wade FA, Couch DR (1982) The Swanson Formation, Ford Ranges, Marie Byrd Land: Evidence for and against a direct relationship with the Robertson Bay Group, northern Victoria Land. In: Craddock C (ed) Antarctic geoscience. University of Wisconsin Press, Madison, WI, pp 609–616

Walcott RJ (1984) Reconstruction of the New Zealand region for the Neogene. Paleo Paleo Paleo 46(2):217–231

Weaver SD, Bradshaw JD, Adams CJ (1991) Granitoids of the Ford Ranges, Marie Byrd Land, Antarctica. In: Thomson MRA et al. (eds) Geological evolution of Antarctica. Cambridge University Press, Cambridge, UK, pp 345–351

Webb P-N, Harwood DM, Mabin MGC, McKelvey BC (1994) Late Neogene uplift of the Transantarctic Mountains in the Beardmore Glacier region. Terra Antartica 1(2):463–467

Webers GF, Craddock C, Rogers MA, Anderson JJ (1982) Geology of the Whitmore Mountains. In: Craddock C (ed) Antarctic geoscience. University of Wisconsin Press, Madison, WI, pp 841–847

Webers GF, Craddock C, Rogers MA, Anderson JJ (1983) Geology of Pagano Nunatak and the Hart Hills. In: Oliver RL, James PR, Jago JB (ed) Antarctic earth science. Australian Academy of Science, Canberra, A.C.T., pp 251–255

Wegener A (1912) Die Entstehung der Kontinente. Geol Rundschau 3:276–292

Weissel JK, Hayes DE (1974) The Australian-Antarctic discordance: New results and implications. J Geophys Res 79(17):2579–2582

Weissel JK, Hayes DE, Herron EM (1977) Plate tectonics synthesis: The displacements between Australia, New Zealand, and Antarctica since the Late Cretaceous. Marine Geol 25:231–277

White R, McKenzie D (1989) Magmatism at rift zones: The generation of volcanic continental margins and flood basalts. J Geophys Res 94(B6):7685–7729

Wilson JT (1965) Transform faults, oceanic ridges, and magnetic anomalies south of Vancouver Island. Science 150:482–485

Wilson M (1989) Igneous petrogenesis; A global tectonic approach. Unwin Hyman, London

Wilson TJ (1992) Mesozoic and Cenozoic kinematic evolution of the Transantarctic Mountains. In: Yoshida Y et al. (eds) Recent progress in Antarctic earth science. Terra Scientific, Tokyo, Japan, pp 303–314

Wilson TJ (1993a) Jurassic faulting and magmatism in the Transantarctic Mountains: Implications for Gondwana break up. In: Findlay RH, Unrug R, Banks MR, Veevers JJ (eds) Gondwana Eight: Assembly, evolution and dispersal. A.A. Balkema, Rotterdam, The Netherlands, pp 563–572

Wilson TJ (1993b) Mesozoic and Cenozoic kinematic evolution of the Transantarctic Mountains. In: Yoshida Y et al. (ed) Recent progress in Antarctic earth science. Terra Scientific, Tokyo, Japan

Wilson TJ (1995) Cenozoic transtension along the Transantarctic – West Antarctic Rift boundary, southern Victoria Land, Antarctica. Tectonics 14:531–545

Winberry JP, Anandakrishnan S (2004) Crustal structure of the West Antarctic rift system and Marie Byrd Land hotspot. Geology 32(11):977–980

Windley BF (1977) The evolving continents. Wiley, London

Wyllie PJ (1971) The dynamic Earth. Wiley, New York

Part IV
Fire and Ice

Chapter 16
Cenozoic Volcanoes

The Transantarctic Mountains of northern and southern Victoria Land contain large volcanoes of Cenozoic (late Tertiary) age which have erupted lavas that are alkali-rich and silica undersaturated. In addition, the lavas contain inclusions of granulite from the deep crust and ultramafic rocks from the lithospheric mantle. The Cenozoic volcanoes in the Transantarctic Mountains are located on the East Antarctic rim of the West Antarctic rift system (Section 15.5.5). The opposite side of this rift in Marie Byrd Land also contains Cenozoic volcanoes in the Flood Range, the Erven Nunatak, the Executive Committee Range, and in the USAS Escarpment, all of which appear on the geologic map of the area by Wade (1969). The Executive Committee Range is so named because it consists of a linear array of four towering, equally-spaced volcanoes starting with Mt. Sidley (4,181 m) in the south and continuing with Mt. Hartigan (2,815 m), Mt. Cumming (2,615 m), and Mt. Hampton (3,223 m) in the north for a distance of 65 km.

Cenozoic volcanic rocks also occur on the Antarctic Peninsula and on the islands off its west coast (e.g., Deception Island; Faure et al. 1971) and on Gaussberg on the coast of East Antarctica (Section 1.3.3). All of the Cenozoic volcanoes on the Antarctic plate were individually described in a book edited by W.E. LeMasurier and J.W. Thomson assisted by P.E. Baker, P.R. Kyle, P.D. Rowley, J.L. Smellie, and W.J. Verwoerd. This book is an excellent source of authoritative information about all aspects of the geology of the Cenozoic volcanoes on the mainland of Antarctica and on the associated islands in the Southern Oceans (LeMasurier and Thomson 1990).

Mt. Erebus on Ross Island in Fig. 16.1 is the only volcano that is still active (Treves and Kyle 1973a). Most of the others are either dormant or extinct. However, fumarolic activity, ash deposits on ice, and Pleistocene lava flows having K-Ar dates less than 0.1 Ma have been reported from Mt. Melbourne and The Pleiades in northern Victoria Land (LeMasurier 1990). In addition, small basaltic cinder cones, lava flows, and pyroclastic deposits occur at scattered locations in northern and southern Victoria Land.

The Cenozoic volcanic rocks of the Transantarctic Mountains form the *McMurdo Volcanic Group* which is subdivided on the basis of geographic location into four provinces (LeMasurier and Thomson 1990):

1. Hallett volcanic province (NVL)
2. Melbourne volcanic province (NVL)
3. Erebus volcanic province (SVL)
4. Southern most volcanic province (QMM)

where NVL = northern Victoria Land, SVL = Southern Victoria Land, and QMM = Queen Maud Mountains.

16.1 Erebus Volcanic Province

We begin this chapter with descriptions of the lava flows that were erupted by the volcanoes of the Erebus volcanic province because they are accessible from McMurdo Station and Scott Base on Ross Island, because Mt. Erebus in Fig. 16.1 is the only active volcano in all of Antarctica, and because the volcanoes on Ross Island were discovered in January 1841 by James C. Ross and R.M. Crozier who named Mt. Erebus and Mt. Terror after their small wooden sailing ships. Even at that time, Mt. Erebus was active and has remained so to the present (Section 1.1). The first description of the volcanic activity on the summit of Mt. Erebus was provided by Ross (1847, p. 216, 220–221) on January 28, 1841:

G. Faure and T.M. Mensing, *The Transantarctic Mountains: Rocks, Ice, Meteorites and Water*,
DOI 10.1007/978-90-481-9390-5_16,

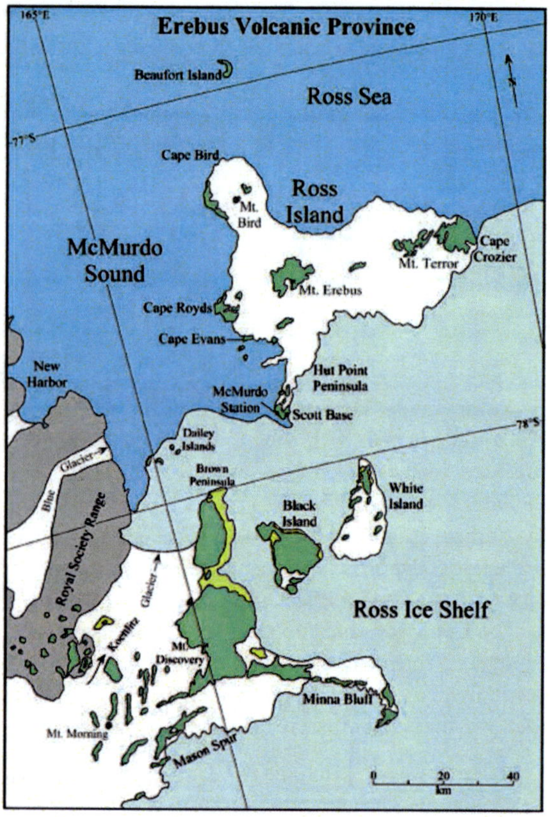

Fig. 16.1 The Cenozoic volcanoes of southern Victoria Land include Mt. Erebus (3,794 m), Mt. Terror (3,230 m), and Mt. Bird (1,766 m) on Ross Island as well as Mt. Discovery (2,681 m) and Mt. Morning (2,728 m) south of the Koettlitz Glacier. Cenozoic volcanic rocks also form the Hut Point Peninsula on Ross Island and Minna Bluff which is embedded in the Ross Ice Shelf. In addition, all of the islands in this area contain Cenozoic volcanic rocks including Black Island, White Island, and Beaufort Island. The volcanic rocks exposed on the volcanoes and associated islands in southern Victoria Land range in composition from basanites to phonolites (Adapted from Warren (1969) with information from LeMasurier (1990))

At 4 p.m. Mount Erebus was observed to emit smoke and flame in unusual quantities…The bright red flame that filled the mouth of the crater was clearly perceptible; and some of the officers believed they could see streams of lava pouring down its side….

Reported by Kyle et al. (1982a, p. 737).

The largest exposures of alkali-rich volcanic rocks in the Erebus volcanic province occur on Ross Island and in the Discovery subprovince (i.e., Mt. Discovery, Mt. Morning, Minna Bluff, and the Brown Peninsula in Fig. 16.1). The principal volcanoes on Ross Island are Mt. Erebus (3,794 m), Mt. Terror (3,230 m), and Mt. Bird (1,766 m). In addition, the Erebus volcanic province includes the Royal Society Range and the ice-free valleys (Taylor and Wright) of southern Victoria Land (Treves 1967).

Ross Island and the other nearby volcanic islands (e.g., White Island and Black Island) were described by geologists who were attached to the expeditions of Robert Scott and Ernest Shackleton (Section 1.4) (e.g., Ferrar 1907; David and Priestley 1914; Debenham 1923). These geologists also collected rock specimens that were later described by Prior (1907), Jensen (1916), Thomson (1916), and Smith (1954). Even Carsten E. Borchgrevink, who led the British Southern Cross Expedition (1898–1900), landed at Cape Crozier (Ross Island) in 1900 (Section 1.3.2) and collected rock specimens that were described by David et al. (1896) and Prior (1902). The latter also described the collections made by James C. Ross in 1941 (Prior 1899). The last explorers to visit Cape Crozier were E.A. Wilson, H.R. Bowers, and A. Cherry-Garrard who made the trip from Cape Evans in the winter of 1911 in order to collect eggs of Emperor Penguins who have a rookery at Cape Crozier (Section 1.4.3; Cherry-Garrard 1922).

These early reports were followed by investigations that were published after the IGY (1957–1958): Forbes (1963), Beck (1965), Cole and Ewart (1968), and Cole et al. (1971). Recent summaries of the geology, chemical composition, and isotope geochemistry of the Cenozoic volcanic rocks in the Erebus volcanic province are contained in books edited by McGinnis (1981), LeMasurier and Thomson (1990), and Kyle (1995).

Ross Island is located close to the southern end of the Victoria Land Basin in the western Ross Sea. This basin contains the Terror Rift which has been traced from Ross Island to Cape Washington and Mt. Melbourne in northern Victoria Land. Kyle and Cole (1974) pointed out that Mt. Bird, Mt. Terror, and the Hut Point Peninsula of Ross Island form a radial pattern at 120° angles around Mt. Erebus. A similar pattern occurs around Mt. Discovery in Fig. 16.1 where the Brown Peninsula, Minna Bluff, and Mt. Morning are also radially distributed around Mt. Discovery at 120° (Eggers 1979; Kyle et al. 1979). These apparent fracture patterns suggest doming of the crust, presumably by a mantle plume that provided the heat necessary to form the large volumes of lava that have been erupted by the volcanoes on Ross Island as well by Mt. Discovery and Mt. Morning (Kyle 1990a).

Fig. 16.2 Mount Erebus on Ross Island is the only active volcano in Antarctica. The crater at the summit contains a lava lake that vents gases which condense to form a plume. On this calm and sunny day the volcano was quiet. Castle Rock in the foreground on the Hut Point Peninsula is composed of basanite hyaloclastite that crystallized at 1.18 Ma (Kyle 1990b) (Photo by T.M. Mensing)

16.1.1 Petrology

The volcanic rocks of the Erebus volcanic province were described by Treves (1967, 1971) and by Goldich et al. (1975). The principal rock types recognized by them were: Basanitoid, trachybasalt, anorthoclase phonolite, mafic phonolite, and trachyte. Although the nomenclature has changed, the rock descriptions by Treves (1967) and Goldich et al. (1975) still convey useful information.

Basanitoids (aka basanite) are alkalic olivine basalts which contain more than 5% normative nepheline (Macdonald and Katsura 1964). Basanitoids are ubiquitous on Ross Island and have constant chemical compositions regardless of their stratigraphic position and location on the island. All have intersertal to granular textures and contain augite, plagioclase, nepheline, and opaque minerals in the groundmass. The phenocrysts generally consist of olivine (80% forsterite) and titaniferous augite. Plagioclase and kaersutite (amphibole) are less common. Basanitoids occur both as thin flows and as pyroclastic deposits with inclusions of sedimentary, and mafic, as well as ultramafic igneous rocks.

Trachybasalts have chemical compositions that are intermediate between basanitoids and phonolites in the classification of Goldich et al. (1975). These intermediate rocks are subdivided nowadays into tephrites, phonotephrites, and tephriphonolites in the nomenclature of Le Bas et al. (1986) because their chemical and mineralogical compositions are quite diverse. The trachybasalts are characterized by plagioclase phenocrysts as well as kaersutite and titaniferous augite. Olivine, opaques, and trace amounts of apatite are present as accessory phenocrysts. Some samples contain brown glass in the groundmass as well as plagioclase, augite, and finely disseminated opaques. The trachybasalt at Half Moon Crater, located 2 km from McMurdo on the Hut Point Peninsula, contains abundant inclusions of mafic igneous and sedimentary rocks.

Anorthoclase phonolite contains megacrysts of euhedral anorthoclase which distinguishes these rocks from mafic phonolites. Anorthoclase is a triclinic feldspar like microcline in which the Na-feldspar component is more abundant than the K-feldspar component (Boudette and Ford 1966). Anorthoclase also occurs in the alkali-rich rocks in the Oslo graben of Norway and on Mt. Kilimanjaro and Mt. Kenya in East Africa. For this reason, the anorthoclase phonolites on Mt. Erebus are also called Antarctic kenytes (Treves 1967). The chemical and mineralogical compositions of the anorthoclase phonolites on Ross Island are remarkably uniform. The lavas that form the summit of Mt. Erebus contain up to 40% subhedral to euhedral anorthoclase, whereas the matrix is composed of plagioclase, anorthoclase, apatite, opaques, and nepheline.

Treves (1967) reported that the *anorthoclase* crystals contain 16 mole percent orthoclase, 63 mole percent albite, and 21 mole percent anorthite. In addition, the crystals contain inclusions of apatite, glass, augite, opaques, and even olivine (Mason et al. 1982). When Mt. Erebus was especially active between 1972 and 1974, the volcano ejected bombs composed of brown glassy anorthoclase phonolite. The rim of the crater is littered with loose anorthoclase crystals more than 5 cm in length which appear to be weathering out of the "pumiceous" (i.e., glassy) phonolite porphyry. The crystallization of anorthoclase crystals from phonolite magma was discussed by Kyle et al. (1992) and by Dunbar et al. (1994).

Mafic phonolite occurs in the form of small plugs and flows throughout Ross Island and elsewhere within the Erebus volcanic province. For examples, Post Office Hill at Cape Crozier and Observation Hill at McMurdo Station are composed of this rock type. Their chemical compositions are quite variable. Normative nepheline can be as high as 30%. Most samples contain more augite and kaersutite than the anorthoclase phonolites.

Trachyte occurs at Mt. Cis between Cape Royds and Cape Evans. It is oversaturated with respect to silica (i.e., it is quartz normative) and contains abundant inclusions (Thomson 1916; Jensen 1916; Treves 1967). The inclusions consist of sanidinite, pyroxene granulite, diabase, and metamorphosed sandstone. These inclusion suggest that Ross Island may be underlain by granitic basement rocks, Beacon sandstones, and perhaps even by Ferrar Dolerite.

16.1.2 Chemical Compositions

The lavas of the Erebus volcanic province are alkali-rich and silica undersatured and are classified most effectively on the total alkali-silica diagram of Le Bas et al. (1986) in Fig. 16.3. However, Kyle et al. (1979) preferred the classification of Coombs and Wilkinson (1969) and identified most Cenozoic lavas in the Transantarctic Mountains as basanite, phonolite, or trachyte. Goldich et al. (1975, 1981), Sun and Hanson (1975, 1976), Stuckless and Ericksen (1976), Stuckless et al. (1981), and Weiblen et al. (1981) classified the lavas as trachybasalt and identified low-silica varieties as *basanitoids*. We will use the classification of Le Bas et al. (1986) which was also preferred by LeMasurier (1990).

The chemical compositions of volcanic rocks of the Erebus volcanic province reported by Goldich et al. (1975) and Stuckless et al. (1981) range widely in Fig. 16.4 from low-silica basanites (Bn) to tephrites (Tp), phonotephrites (Ptp), tephriphonolites (Tpp), benmoreites (Bm), and phonolites (P). The silica concentrations of these rocks range from 41.3% (basanite, Brown Peninsula) to 58.1% (benmoreite, Cape Royds, Ross Island). The basanites in Fig. 16.4 have low

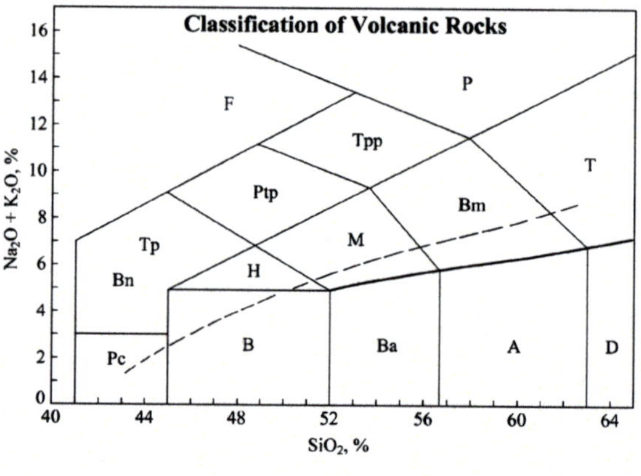

Fig. 16.3 The alkali-rich volcanic rocks of the Erebus volcanic province are classified on the total alkali-silica diagram of Le Bas et al. (1986). The rocks in this diagram can be arranged into three series with decreasing alkali content: (1) Basanite (Bn), tephrite (Tp), phonotephrite (Ptp), tephriphonolite (Tpp), and phonolite (P); (2) Hawaiite (H), mugearite (M), benmoreite (Bm), and trachyte (T); (3) Basalt (B), basaltic andesite (Ba), andesite (A), and dacite (D). The boundary between the alkali-rich and subalkalic rocks defined by Macdonald and Katsura (1964) is shown as a dashed line (Adapted from LeMasurier (1990) and Wilson (1989))

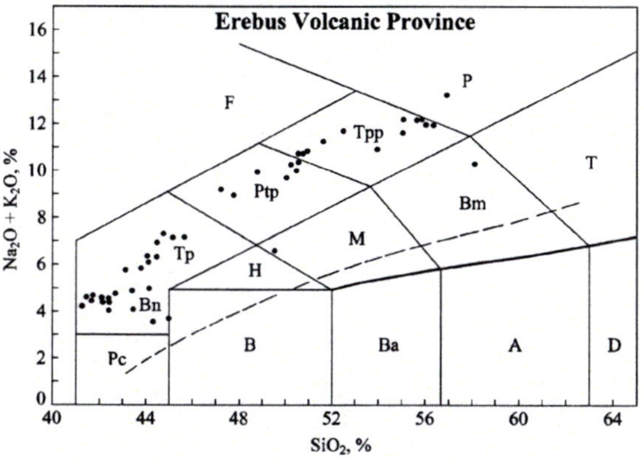

Fig. 16.4 The silica and total-alkali concentrations of Cenozoic lavas of the McMurdo Volcanics in the Erebus volcanic province range widely and are classifiable as basanite (Bn), tephrite (Tp), phonotephrite (Ptp), tephriphonolite (Tpp), phonolite (P), and benmoreite (Bm). All of the Cenozoic lavas in the diagram are alkali-rich and undersaturated with respect to silica. The alkali-rich Cenozoic lavas are associated with the West Antarctic rift system and occur not only in the Transantarctic Mountains but also in Marie Byrd Land and Ellsworth Land in West Antarctica (Data from Goldich et al. (1975) and Stuckless et al. (1981))

total-alkali concentrations (4–5%), whereas the phonolites are extremely alkali rich ($Na_2O + K_2O = 16.51\%$).

The chemical compositions of the Cenozoic lavas of the Erebus volcanic province in Fig. 16.4 differ profoundly from the subalkalic and quartz-normative basalts and dolerites of the Jurassic Ferrar Group. Evidently, the Cenozoic lavas formed by a different petrogenetic process even though they were erupted in the *same* area as the mafic rocks of the Ferrar Group. The contrast between these two suites of igneous rocks is presumably related to the difference in the tectonic regime in which they formed. The Ferrar magmas originated by large-scale decompression melting triggered by the arrival of a mantle plume which caused the formation of a triple junction between East Antarctica and Africa. The Cenozoic volcanics of the Transantarctic Mountains formed by low degrees of melting in a setting of extensional tectonics associated with the West Antarctic rift system and therefore resemble rift-related alkali-rich rocks elsewhere on the Earth, such as the East African rift valleys, as well as the Rhine graben of Germany and the Oslo graben of Norway (Faure 2001).

Stuckless et al. (1981) carried out a Q-mode factor analysis of the chemical compositions of the 49 rock samples in Fig. 16.4 and concluded that all of the samples were derived from one parent magma. They also concluded that the basanitoid magma began to crystallize 60–90 km below the surface. The phonolites and related alkali-rich lavas were derived from the basanitoid magma but began to crystallized at the base of the crust and therefore contain phenocrysts of *anorthoclase* and upper-crustal xenoliths, whereas the basanitoids contain ultramafic xenoliths from the lithsopheric mantle.

In the Erebus volcanic province a thick sequence of basanite flows is overlain by more alkali-rich lavas (i.e., tephrites, phonotephrite, and tephriphonolites) which are capped by flows of phonolite. The petrogenesis of the Cenozoic lavas of the Erebus volcanic province was also discussed by Stuckless et al. (1974), Goldich et al. (1975), Kurasawa (1975), Sun and Hanson (1975, 1976), Stuckless and Ericksen (1976), and Kyle and Rankin (1976).

16.1.3 K-Ar Dates

The ages of the Cenozoic volcanic rocks of the McMurdo Volcanic Group were determined by Armstrong (1978), Fleck et al. (1972), Forbes et al. (1974), Wright-Grassham (1987), and others by the whole-rock K-Ar method. Under ideal circumstances, these dates indicate the crystallization ages of late Cenozoic volcanic rocks that cooled rapidly after they were erupted on land, provided they were not hydrothermally altered and provided that all ^{40}Ar was outgassed from the lava at the time eruption.

Armstrong (1978) used the K-Ar method to date 91 specimens of the McMurdo Volcanics from *all parts* of Victoria Land including Ross Island and several other off-shore islands. This important work presented here in Fig. 16.5 indicates that volcanoes in this area have erupted continuously since the Miocene Epoch. The K-Ar dates also provided information about the glaciation of the Transantarctic Mountains (e.g., Armstrong et al. 1968; Denton and Armstrong 1968; Denton et al. 1969; Denton 1970, 1971). In addition, K-Ar dates of volcanic rocks in stratigraphic sections in Antarctica and elsewhere have provided a time scale for the reversals of the polarity of the magnetic field of the Earth (Cox 1969). The magnetic properties of the Cenozoic lavas of Erebus volcanic province were investigated by

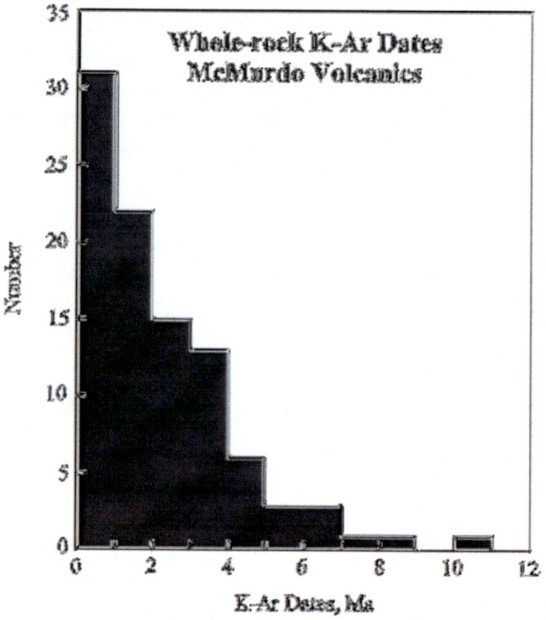

Fig. 16.5 Whole-rock K-Ar dates of the McMurdo Volcanics of the Erebus, Hallett, and Melbourne volcanic provinces form an unbroken sequence from the Holocene (0 Ma) to the middle Miocene (10 Ma). The oldest date is 18.0 ± 0.7 Ma (early Miocene) followed by 15.4 ± 0.5, 13.8 ± 0.2, and 13.2 ± 0.4 Ma, all of which are in the Miocene. The pattern of variation of K-Ar dates indicates that the volcanic activity started during the early Miocene and continued without interruption to the present. Early-formed volcanic rocks are removed by erosion and may also be covered by younger flows which explains why the frequency of K/Ar dates decreases with the increasing age of the rocks (Data from Armstrong (1978), Forbes et al. (1974), and Fleck et al. (1972))

Table 16.1 Whole-rock K-Ar dates of lavas exposed at the surface in the Erebus volcanic province (Kyle 1990a)

Locality	
Ross Island	Range of K-Ar dates (Ma)
Mt. Bird	3–4
Mt. Terror	0.6–2.0
Hut Point Peninsula	0.4–1.8
Mt. Erebus	<1.0
Discovery Subprovince	
Mt. Morning	<2.0
"Gandalf Ridge"	15.5–18.7
"Riviera Ridge	13.0–15.2
Mason Spur	12.4–13.5
Minna Bluff	
eastern tip	9.0–11.0
western part	7.3–8.3
Black Island	
northern end	10.9
Mt. Aurora	3.4–3.8
Mt. Discovery	4.4–5.3
Brown Peninsula	2.0–2.8
Mt. Morning	<2
White Island	0.2
Wright and Taylor Valleys	
Wright valley	2–4
Taylor valley	2–5

Cox (1966), McMahon and Spall (1974), Elston and Bressler (1980), and Mankinen and Cox (1988), among others.

The Cenozoic volcanic activity in the Transantarctic Mountains may have started even sooner than is indicated by the K-Ar dates of Armstrong (1978) because the first flows to be erupted were either eroded or buried by younger flows and therefore are no longer available in outcrop. Therefore, the onset of volcanic activity is more accurately recorded by volcanic rocks in the drill cores recovered by the Dry Valley Drilling Project (McGinnis 1981).

The available evidence indicates that volcanic activity in southern Victoria Land started in the early Oligocene at about 40 Ma based primarily on micropaleontological evidence (Barrett 1987). Volcanic rocks in the MSSTS-1 drill core in McMurdo Sound yielded K-Ar dates of 13.7 and 24.3 Ma (Gamble et al. 1986). The oldest volcanic rocks in the Discovery subprovince occur at "Gandalf Ridge" (15.5–18.7 Ma), at "Riviera Ridge" (13.0–15.2 Ma), and at Mason Spur (12.4–13.5 Ma) around the base of Mt. Morning in Fig. 16.1 which itself is composed of lava flows dated at less than 2.0 Ma (Kyle 1990a). The K-Ar dates of other geographic features of the McMurdo Sound area are listed in Table 16.1.

The results summarized in this table reveal that the ages of rocks on Mt. Bird on Ross Island (Fig. 16.1) range from 3 to 4 Ma, while the lavas on Mt. Terror and on the Hut Point Peninsula crystallized between 0.5 and 2.0 Ma. The lavas that form the cone of Mt. Erebus are even younger at less than 0.1 Ma. The lavas in the Discovery subprovince range from less than 2 to 14 Ma and are in part older than the lavas exposed on Ross Island.

16.1.4 Rb-Sr Systematics

Information concerning the origin of the alkali-rich lavas of the Erebus volcanic province is derivable from the isotopic composition of strontium and the concentrations of rubidium and strontium. The first measurements of the $^{87}Sr/^{86}Sr$ ratios of the Cenozoic volcanic rocks in the Erebus province by Jones and Walker (1972) ranged from 0.7022 to 0.7048 with a mean of 0.70365 ± 0.0024 ($2\bar{\sigma}$). These measurements demonstrated that the Cenozoic lavas are only moderately enriched in radiogenic ^{87}Sr in marked contrast to the basalt flows and dolerite sills of the Ferrar Group. These results were later confirmed by Stuckless and Ericksen (1976) whose measured $^{87}Sr/^{86}Sr$ ratios in Figs. 16.6a, b, and c have a unimodal distribution with a mean of 0.70332 ± 0.00005 ($2\bar{\sigma}$).

The alkali-rich lavas analyzed by Stuckless and Ericksen (1976) have high strontium concentrations that range from 640 to 2,000 ppm, whereas their rubidium concentrations range from only 30 to 173 ppm and the average Rb/Sr ratio of the alkali-rich lavas is 0.096. The only exception is a sample of trachyte which has a high $^{87}Sr/^{86}Sr$ ratio of 0.70501, Sr = 48 ppm, Rb = 104 ppm and Rb/Sr = 2.17.

The $^{87}Sr/^{86}Sr$ ratios of the *megacrysts* (kaersutite, titaniferous augite, diopside, and anorthoclase) in Fig. 16.6c range from 0.70326 to 0.70337 and average 0.703314 ± 0.00005 ($2\bar{\sigma}$), which is indistinguishable from the average $^{87}Sr/^{86}Sr$ ratio of the lavas. Evidently, the megacrysts crystallized in isotopic equilibrium with strontium in the magma and are therefore *phenocrysts* rather than *xenocrysts* in the rocks in which they occur.

The concentrations of strontium in these *minerals* range widely from 45 ppm in diopside to 2,400 ppm in the anorthoclase. The rubidium concentrations are

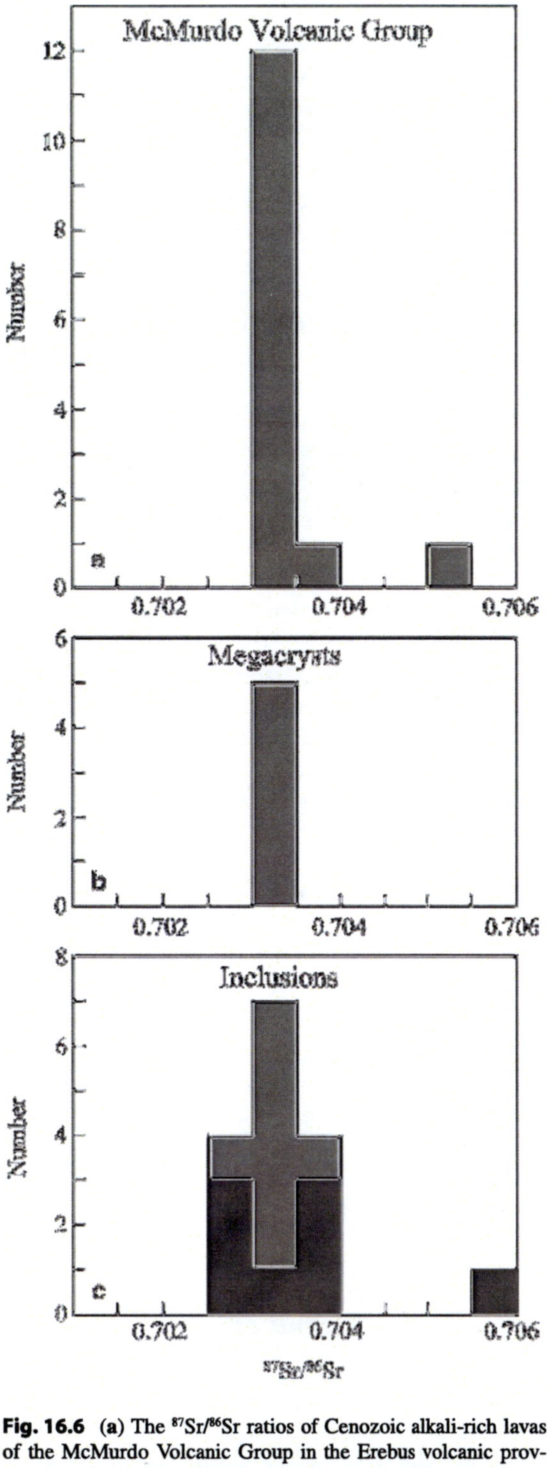

Fig. 16.6 (**a**) The $^{87}Sr/^{86}Sr$ ratios of Cenozoic alkali-rich lavas of the McMurdo Volcanic Group in the Erebus volcanic province have a *unimodal* distribution between 0.7030 and 0.7035. One sample of trachyte has an $^{87}Sr/^{86}Sr$ ratio of 0.70501 and a low Sr concentration of 48 ppm. (**b**) Megacrysts of kaersutite, titaniferous augite, diopside, and anorthoclase have the same $^{87}Sr/^{86}Sr$ ratios as the lavas in which they occur. (**c**) The $^{87}Sr/^{86}Sr$ ratios of granulite (*grey*) and ultramafic rocks (*black*) range more widely than those of the host lavas from 0.7025 to 0.7040. One coarse-grained inclusion of lherzolitic harzburgite from Cape Crozier has an elevated $^{87}Sr/^{86}Sr$ ratio of 0.70605 and a very low Sr concentration of 3.01 ppm. These $^{87}Sr/^{86}Sr$ ratios, measured by Stuckless and Ericksen (1976), were increased by 0.00012 to be compatible with the $^{87}Sr/^{86}Sr$ ratio of 0.71025 for the Sr-isotope standard SRM-987

extremely low with 0.045 ppm in the titaniferous augite and 13 ppm in the anorthoclase. The high strontium concentrations of the kaersutite and anorthoclase protect the isotopic composition of strontium in these minerals from alteration caused by addition of strontium from external sources.

The *inclusions* analyzed by Stuckless and Ericksen (1976) consist of granulites and of plutonic igneous rocks of ultramafic composition. The *granulite* inclusions are composed of pyroxene, kaersutite, plagioclase, and opaque oxides. In addition, several inclusions contain up to 65% olivine and small amounts of glass. The *ultramafic* inclusions consist of *wehrlite* (olivine and diopside) and *harzburgite* (olivine and enstatite or hypersthene). Several specimens analyzed by Stuckless and Ericksen (1976) were friable.

The $^{87}Sr/^{86}Sr$ ratios of the *granulite* inclusions in Fig. 16.6c range from 0.70289 to 0.70390 with a mean of 0.70323 ± 0.00019 ($2\bar{\sigma}$). This value is surprisingly low for rocks that presumable originated from the base of the continental crust. However, the granulites have a low average Rb/Sr ratio of 0.0145 ($^{87}Rb/^{86}Sr$ = 0.043) which causes their average $^{87}Sr/^{86}Sr$ ratio to increase by only 0.0061 per billion years. Although the granulite inclusions cannot be dated in this case, the Rb-Sr systematics permit their age to be early Precambrian.

The *ultramafic* inclusions in Fig. 16.6c have a bimodal distribution of the $^{87}Sr/^{86}Sr$ ratios and a low average Rb/Sr ratio of 0.030 ± 0.038 ($1\bar{\sigma}$). Three friable inclusions define a straight line on the Rb-Sr geochronometry diagram in Fig. 16.7 that corresponds to a date of 1,140 Ma ($\lambda = 1.42 \times 10^{-11}$ year^{-1}) and an initial $^{87}Sr/^{86}Sr$ ratio of 0.70345. The geological significance of this date is questionable because it relies entirely on the elevated $^{87}Sr/^{86}Sr$ ratio of sample CC-82 which has a low strontium concentration of only 3.01 ppm and is therefore susceptible to contamination by ground water or meltwater.

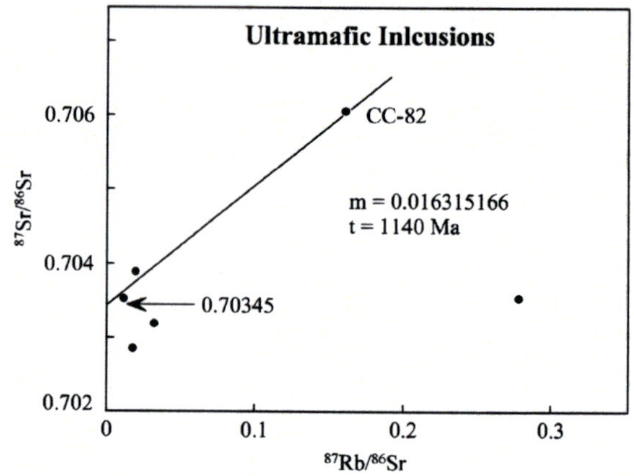

Fig. 16.7 The ultramafic inclusions that were erupted with the Cenozoic alkali-rich lavas at Cape Crozier (CC-82) and Cape Bird (CB-64) on Ross Island and at Brandau vent (BV-5g) in the Royal Society Range define a straight line that yields a date of 1140 Ma. This date is largely based on the sample of trachyte (CC-82) which contains only 3.01 ppm of strontium and is susceptible to contamination by radiogenic ^{87}Sr. Therefore, the geological significance of the Rb-Sr date is questionable (Data from Stuckless and Ericksen (1976))

16.1.5 U-Pb Systematics

The isotopic composition of lead in the alkali-rich lavas on Ross Island, on Mt. Morning and Mt. Discovery, and from Taylor Valley loosely define a straight line in Fig. 16.8 (Sun and Hanson 1975). The slope of this line is the ratio of radiogenic ^{207}Pb to radiogenic ^{206}Pb. The numerical value of this ratio is used to determine a date by interpolating in Table 10.3 of Faure and Mensing (2005). A linear regression of the $^{207}Pb/^{204}Pb$ and $^{206}Pb/^{204}Pb$ ratios of alkali-rich lavas from the Erebus volcanic province yields:

$$\left(\frac{^{207}Pb}{^{206}Pb}\right)^* = 0.08406, t = 1290\,\text{Ma}$$

Sun and Hanson (1975, p. 83) concluded that the date indicated by the isotope compositions of lead is about 1500 Ma and offered two alternative interpretations of its geological significance:

1. The basalts have been contaminated with basement material and the linear array is a result of mixing of two components.
2. The $^{207}Pb/^{206}Pb$ ratios represent primary variations in the mantle source which have existed for 1500 m.y.

We prefer the second alternative because the $^{87}Sr/^{86}Sr$ ratios of most Cenozoic lavas analyzed by Stuckless and Ericksen (1976) and by Jones and Walker (1972) provide no evidence that the magma was contaminated by strontium derived from the granitic basement rocks. The only exception is the sample of trachyte (CC-82, Fig. 16.7) which has a high $^{87}Sr/^{86}Sr$ ratio of 0.70501 (Stuckless and Ericksen 1976).

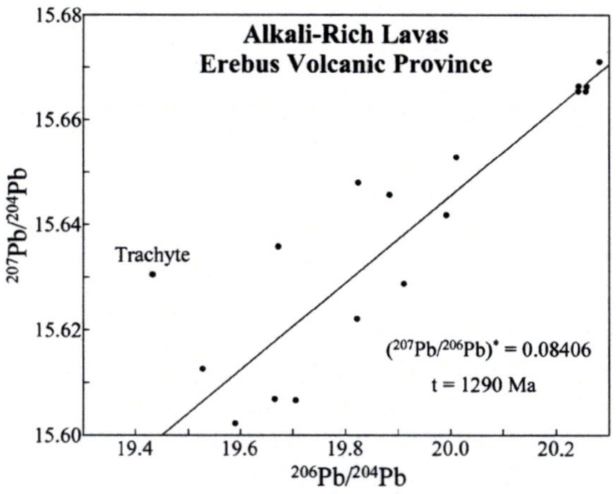

Fig. 16.8 The $^{206}Pb/^{204}Pb$ and $^{207}Pb/^{204}Pb$ ratios of Cenozoic lavas in the Erebus volcanic province loosely constrain a straight line the slope of which is the radiogenic ^{207}Pb-to-^{206}Pb ratio. This ratio varies with the passage of time by decay of the isotopes of uranium and therefore can be converted into a date by interpolating in Table 10.3 published by Faure and Mensing (2005). The date indicates how much time has elapsed since the isotope composition of lead was homogenized by a thermal event that affected the magma sources of the basanite magma. However, the geological significance of this event is presently unknown. The trachyte sample (CC-82) was excluded from the regression (Data from Sun and Hanson (1975))

16.1.6 Oxygen Isotopes

The isotope composition of oxygen in the Cenozoic lavas of Antarctica, measured by Wenner (1974), are mostly in the "normal" range for mantle-derived volcanic rocks (i.e., $\delta^{18}O$ = +5.0 to +6.5‰ relative to SMOW). Coexisting plagioclase phenocrysts also have $\delta^{18}O$ values in this range which suggests that the rocks have not been altered by interacting with water. However, two of ten whole-rock samples from the Hallett volcanic province have low $\delta^{18}O$ values of +1.6‰ and +3.9‰, although phenocrysts of plagioclase in the former and pyroxene in the latter have $\delta^{18}O$ values of +4.0‰ and +5.0‰, respectively. Wenner (1974) concluded that the magma from which these two samples crystallized had assimilated country rocks that were depleted in ^{18}O. In general, phenocrysts of feldspar, pyroxene, olivine, and quartz have slightly higher $\delta^{18}O$ values than the groundmass of the volcanic rock in which they occurred.

Quartz and feldspar of a *granite inclusion* in basalt ($\delta^{18}O$ ~ 5.2‰) are strongly *depleted* in ^{18}O with $\delta^{18}O$ = −4.4‰ (feldspar) and −8.6‰ (quartz). All of these results are consistent with the current state of knowledge of the oxygen isotope composition of igneous rocks (Taylor 1968; Anderson et al. 1971; Faure 1986; Hoefs 1997; Valley and Cole 2001; Sharp 2007).

16.1.7 Ultramafic Inclusions

The inclusions in the Cenozoic lavas of the Erebus volcanic province originated from the continental crust and from the underlying mantle and therefore can provide information about the chemical composition and structure of the subsurface lithosphere. For this reason, the inclusions have been collected and studied by many investigators including Kirsch (1981), Kalamarides et al. (1987), Kalamarides and Berg (1991), Gamble and Kyle (1987), Gamble et al. (1988), Kyle et al. (1987), Berg (1991), McGibbon (1991), and others.

Inclusions are especially abundant in Recent basanitic cinder cones in the Royal Society Range and in the ice-free valleys of southern Victoria Land (Kyle et al. 1987). The so-called Foster Crater in the Royal Society Range is unique because ultramafic phlogopite-rich clinopyroxenite and glimmerite inclusions have been collected at this site (McGibbon 1991). Some of the ultramafic inclusions at Foster Crater contain up to 80% of low-Ti phlogopite which occurs both in the form of veinlets and in disseminated form. The ultramafic inclusions at other sites in the Erebus volcanic province are composed of clinopyroxenite, spinel lherzolite, wehrlite, and dunite. The diameters of these inclusions range from a few centimeters to about 50 cm and many are rounded although some are blocky.

McGibbon (1991) determined that the spinel lherzolite inclusions record temperatures between 880°C and 1,100°C which implies that they originated from the mantle at depths of 35–70 km. The depth of origin of the clinopyroxenite and glimmerite inclusions is restricted by the stability of plagioclase to less than 30 km.

McGibbon (1991) also reported that the $^{87}Sr/^{86}Sr$ and $^{143}Nd/^{144}Nd$ ratios of spinel lherzolite and clinopyroxenite inclusions in the Erebus volcanic province place them in quadrant 2 (mantle array) of the Sr-Nd isotopic mixing diagram together with the host basanite. In addition, McGibbon (1991) reported that the clinopyroxenites and glimmerites at Foster Crater form a straight line in coordinates of $^{87}Sr/^{86}Sr$ and $^{87}Rb/^{86}Sr$ ratios which yields of Rb-Sr date of 439 ± 14.5 Ma. The geological significance of this date is uncertain because the colinearity of data points may be caused by mixing of clinopyroxene (low R/Sr, low $^{87}Sr/^{86}Sr$) and phlogopite (high Rb/Sr, high $^{87}Sr/^{86}Sr$). In fact, the $^{87}Sr/^{86}Sr$ ratios of separated phlogopite specimens range from 0.72622 to 0.77130, whereas the $^{87}Sr/^{86}Sr$ ratios of the clinopyroxenes are about 0.7085. Separated clinopyroxene and phlogopite also form a straight line in coordinates of $^{206}Pb/^{204}Pb$ and $^{207}Pb/^{206}Pb$ that corresponds to a date of 1095 ± 5 Ma. These results suggests that the mantle underlying the Erebus volcanic province was subjected to processes whose relation to geological events in the overlying crust is not yet clear.

16.1.8 Granulite Inclusions

The inclusions of crustal rocks in the lava flows of the Erebus volcanic province consist of lower-crustal granulites, upper-crustal basement rocks, and supracrustal rocks. Berg (1991) described inclusions from 17 sites in the Erebus volcanic province, 7 of which are located in the Royal Society Range (Brandau Vent, Roaring Valley, Foster Crater, Ward Valley, The Bulwark, Lower and Upper Radian Glacier) and five on Ross Island (Cape Crozier, Cape Barne, and at three places on the Hut Point Peninsula).

The *supracrustal inclusions* include dolerites of the Ferrar Group and partially fused and recrystallized sedimentary rocks (sandstones, calcareous sandstones, and sandy carbonate rocks). Dolerite inclusions are common at several sites in the Transantarctic Mountains (e.g., Taylor Valley), but are rare elsewhere in spite of systematic searches at every site visited by Berg (1991). The fused sedimentary rocks originally contained calcite and Ca-Mg carbonate that was recrystallized to diopside, coarse-grained carbonate, and wollastonite. Although these kinds of inclusions were originally attributed to the Beacon Supergroup by Thomson (1916), Berg (1991) concluded that they were probably derived from sedimentary rocks of Cenozoic age that occur as erratics on Black Island and on the Brown Peninsula (Vella 1969).

The boulders of Cenozoic rocks that occur in the Discovery subprovine and along the coast of McMurdo Sound were collected and dated by the fossils they contain (Harrington 1969; Stott et al. 1983; Stilwell et al. 1993).

The *upper-crustal inclusions* described by Berg (1991) consist of granitic and metamorphic rocks of the basement complex that underlies the Transantarctic Mountains. The metamorphic rocks are primarily quartzites, calc-silicate gneisses, and pelitic schists some of which contain sillimanite. The upper-crustal inclusions are particularly abundant at sites in the Transantarctic Mountains but also occur at other sites where inclusions are present.

The *lower-crustal inclusions* consist of pyroxene granulites and ultramafic rocks. Berg (1991) subdivided the granulite inclusions into the dominant two-pyroxene granulites and the less abundant clinopyroxene granulites. Both kinds of granulites contain plagioclase in contrast to the ultramafic inclusions discussed in Section 16.1.7. Berg and Herz (1986) reported that two-pyroxene granulites contain both orthopyroxene (hypersthene) and clinopyroxene (augite) along with lesser amounts olivine and spinel or garnet. Hornblende, quartz, ilmenite, apatite, biotite, sanidine, and scapolite occur as accessory minerals. Some of these minerals are present only in certain parts of the Erebus volcanic province. For example, garnet occurs only in two-pyroxene granulites in the Transantarctic Mountains, whereas olivine occurs only in granulite inclusions of the Ross Embayment (i.e., Ross Island and associated islands). In addition, spinel is abundant in two-pyroxene granulites on Black Island, but is rare elsewhere.

Berg and Herz (1986) and Berg (1991) estimated that the two-pyroxene granulites originated from depths between 17 and 45 km at pressures ranging from 5 to 14 kilobars. The various mineral assemblages that characterize these granulite inclusions indicate temperatures between 850°C and 1,000°C based on the work of Irving (1974) with similar inclusions in Australia. These conclusions require a steep geothermal gradient in the upper crust of 60°C/km to 100°C/km which should cause a high rate of heat flow at the surface. In fact, Decker and Bucher (1982) reported high heat-flow rates in drill holes of the Dry Valley Drilling Project (e.g., at depths of 30–135 m in hole #8 at the mouth of Taylor Valley adjacent to New Harbor). Decker and Bucher (1982, p. 887) concluded:

> …the Ross Island-dry valley region is a zone of high terrestrial heat flow. A zone of high heat flow in this part of Antarctica would imply near-melting conditions in the underlying crust and/or upper mantle.

High rates of heat flow in the Ross Sea and at Arrival Heights on the Hut Point Peninsula were also reported by Blackman et al. (1984) and by Risk and Hochstein (1974), respectively.

In conclusion, Berg (1991) pointed out that ultramafic inclusions that contain orthopyroxene resemble the two-pyroxene granulite inclusions and, in some cases, are interbedded with them in the largest specimens. However, as noted before, the ultramafic inclusions are *not* genetically related to the Ferrar Dolerite or to the Cenzoic alkalic lavas of the McMurdo Volcanic Group (McGibbon 1991).

16.1.9 Structural Discontinuity of the Deep Crust

Granulite inclusions occur throughout the Erebus volcanic province at sites where Cenozoic lavas and pyroclastics have been erupted. The abundances of certain accessory minerals (e.g., garnet, olivine, spinel, etc.) and textural varieties of the granulite inclusions appear to vary regionally within this province, which suggests that the underlying crust may be laterally inhomogeneous. This conjecture was confirmed by Kalamarides et al. (1987) who discovered significant differences in the isotopic compositions of oxygen and strontium in granulite inclusions from the Transantarctic Mountains and from the Ross Embayment (Ross Island and the Discovery subprovince).

The granulite inclusions from volcanic sites in the Transantarctic Mountains have calc-alkaline compositions, whereas the granulite inclusions in the Ross Embayment are basaltic. Average chemical analyses of these inclusions in Appendix 16.8.1 demonstrate the extreme heterogeneity of the two suites. One of the four granulites from the Transantarctic Mountains (# 82-220a) analyzed by Kalamarides and Berg (1991) approaches a felsic composition with SiO_2 = 62.91%, FeO = 3.58%, MgO = 1.83%, and K_2O = 2.36%. One of the three granulites from the Ross Embayment (#81.07) contains high Al_2O_3 (31.27%), low FeO (0.61%), low MgO (1.54%) and high CaO (15.97%).

The concentrations of trace elements are similarly irregular. Sample 82-220a in the Transantarctic suite contains 3,050 ppm Ba, 1,112 ppm Sr, 127 ppm Rb, but low concentrations of base metals (Co, Ni, Cu, and Zn). Sample #82-119 in the Ross suite is enriched in Cr (1,571.0 ppm), Ni (765.0 ppm), Co (160.0 ppm), Cu (57.0 ppm, and Zn (71.0 ppm).

Kalamarides and Berg (1991) demonstrated graphically that the granulites from the Transantarctic Mountains and the Ross Embayment form separate clusters with only minor overlap in coordinates of Al_2O_3 vs. P_2O_5, SiO_2 vs. FeO, and Ba vs. Ni (not shown). However, the most effective separation of the two suites of samples occurs in coordinates of the initial $^{87}Sr/^{86}Sr$ ratios vs. $\delta^{18}O$ values.

The granulite inclusion from sites in the Ross Embayment have lower $\delta^{18}O$ values and lower initial $^{87}Sr/^{86}Sr$ ratios at 900 Ma than granulite inclusions from the Transantarctic Mountains. These isotopic differences are evident in Fig. 16.9 which also demonstrates that the isotopic compositions of the Kirkpatrick

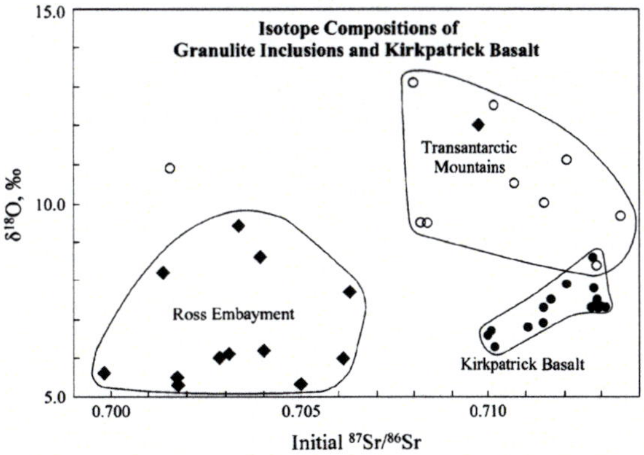

Fig. 16.9 The $\delta^{18}O$ values and initial $^{87}Sr/^{86}Sr$ ratios at 900 Ma of granulite inclusions in the Cenozoic volcanic rocks in the Ross Embayment and in the Transantarctic Mountains form separate clusters. These differences indicate that the inclusions originated from distinctly different crustal sources that are separated from each other by a discontinuity (Kalamarides et al. 1987). The isotope compositions of oxygen and strontium of the Kirkpatrick Basalt in the Transantarctic Mountains (Hoefs et al. 1980) define a third cluster having low $\delta^{18}O$ values and high initial $^{87}Sr/^{86}Sr$ ratios. The evidence in this diagram implies that a tectonic boundary exists in the deep crust underlying the Transantarctic Mountains and the Ross Embayment. The isotopic data also rule out a connection between the granulite inclusions in the Cenozoic lavas and the Kirkpatrick Basalt of the Transantarctic Mountains (Adapted from Fig. 1 of Kalamarides and Berg (1991))

Basalt of Middle Jurassic age differ systematically from those of the granulite inclusions. We note in passing that the flows of Kirkpatrick Basalt *do not* contain the kinds of inclusions that occur abundantly in the McMurdo Volcanics even though these different petrologic suites occupy overlapping geographic areas. In addition, the Cenozoic lavas in which the granulite inclusions occur have uniform $^{87}Sr/^{86}Sr$ ratios and $\delta^{18}O$ values both in the Transantarctic Mountains and in the Ross Embayment and do not contain evidence of the lateral inhomogeneity of the deep crust. Apparently, the basanite magma of the McMurdo Volcanic Group originated from the mantle *beneath* the crust from which the granulite inclusions originated.

The boundary between the deep crust of the Transantarctic Mountains and of the Ross Embayment in Fig. 16.10 extends along the present coast of southern Victoria Land. This boundary predates the development of the West Antarctic rift system during the Cenozoic although it may have predisposed the crust to fail along this line (Wilson 1995). Kalamarides and Berg (1991) and Kalamarides et al. (1987) considered that this crustal suture dates from the time of the Cambro-Ordovican Ross Orogeny and may have originated even earlier during the Precambrian. In addition, we note that the deep crustal boundary that was recognized by Kalamarides et al. (1987) supports the conjecture that East and West Antarctica were joined as a result of movement along the boundary as proposed by Schmidt and Rowley (1986).

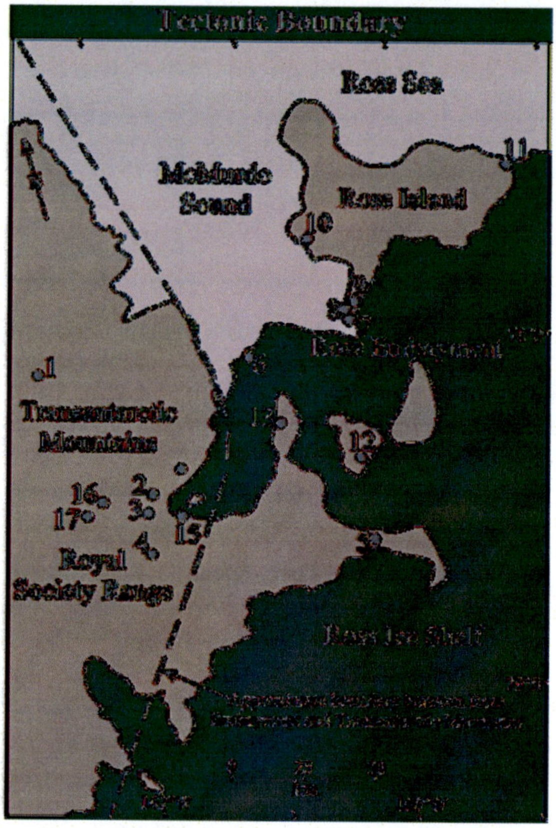

Fig. 16.10 Differences in the chemical and isotopic compositions of granulite inclusions that were brought to the surface by the eruption of Cenozoic alkali-rich lavas indicate that a deep-seated boundary exists between the Transantarctic Mountains and the Ross Embayment. This boundary extends parallel to the coast of Victoria Land and supports the view that the deep crust of East and West Antarctica were placed next to each other as a result of tectonic displacement of West Antarctica relative to East Antarctica. The numbers identify the sites where inclusions were collected by Kalamarides et al. (1987), Kalamarides and Berg (1991), and Berg (1991). (1) Taylor Valley, (2) Brandau Vent, (3) Roaring Valley, (4) Foster Crater, (5) Minna Bluff, (6) Dailey Islands, (7) Twin Crater, (8) Sulfur Cones, (9) Half Moon Crater, (10) Cape Barne, (11) Cape Crozier, (12) Black Island, (13) Brown Peninsula, (14) Ward Valley, (15) The Bulwark, (16) Lower Radian Glacier, (17) Upper Radian Glacier (Adapted from Kalamarides and Berg (1991) and Berg (1991))

16.2 Ross Island

Ross Island in Fig. 16.11 consists of the overlapping lava flows and pyroclastic deposits of Mt. Erebus and of several extinct volcanoes, including Mt. Bird, Mt. Terror, and Mt. Terra Nova which sits in the saddle between Erebus and Terror. Most of Ross Island is covered by ice that forms several glaciers that flow to the coast (e.g., the Erebus Glacier, the Barne Glacier, the Terror Glacier, the Aurora Glacier, and the Fang Glacier on the summit of Mt. Erebus).

The Erebus Glacier forms the Erebus Glacier Tongue which extends from the base of the Hut Point Peninsula into the McMurdo Sound for a distance of 11 km. The ice tongue is a convenient rest stop for geological field parties in Fig. 16.12 on their way to an overnight camp-out at Cape Royds. The ice tongue floats in the water of McMurdo Sound and is flexed by the daily tides. The fractures that form in the ice tongue as a result of the tidal deformation evolve locally into ice caves (Fig. 16.13) that could serve as emergency shelters for travelers who may be trapped on the sea ice by a sudden wind storm.

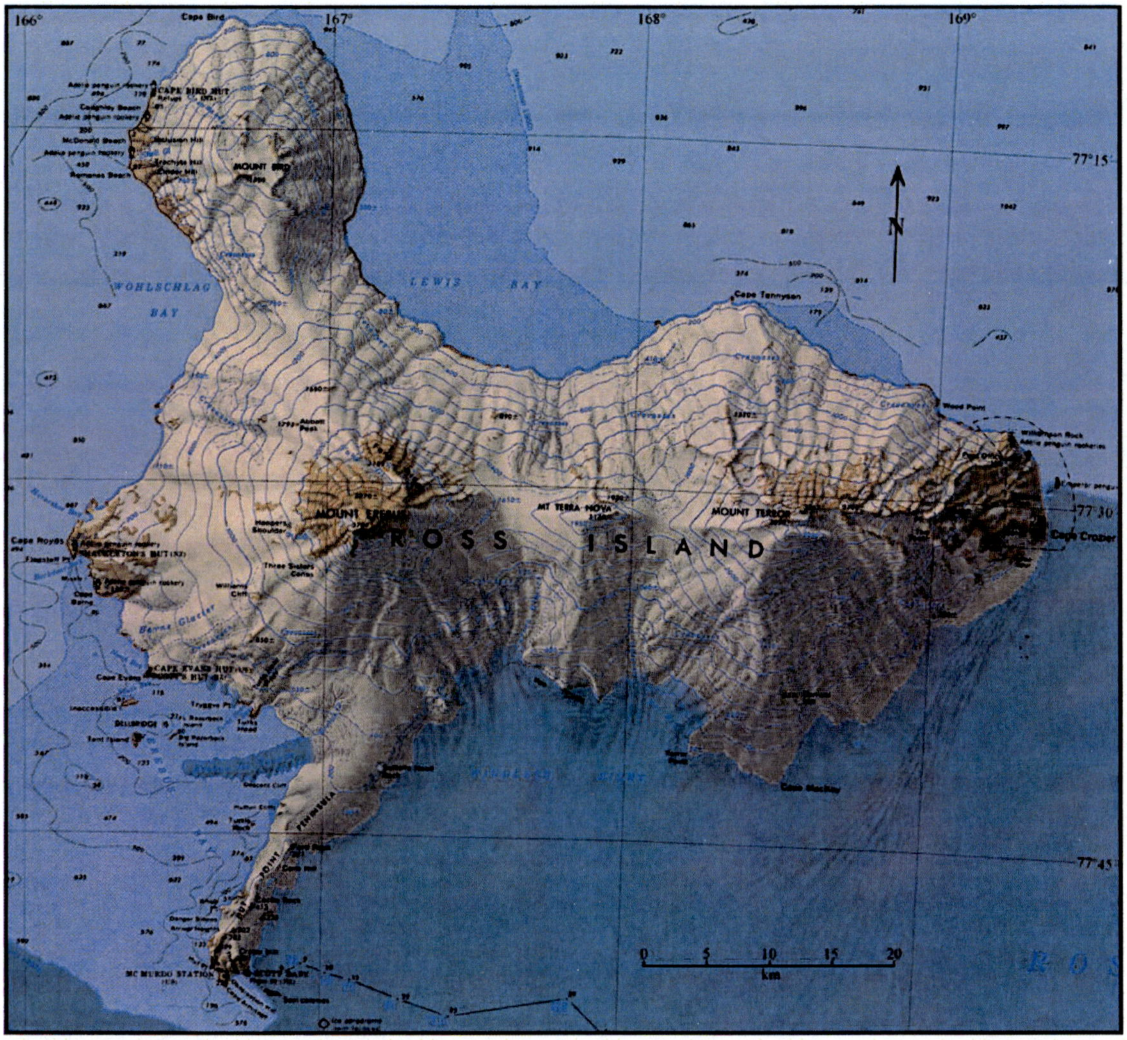

Fig. 16.11 Ross Island consists of volcanic rocks that were erupted by the three principal volcanoes: Mt. Erebus, Mt. Bird, and Mt. Terror. The island is separated from the mainland of Antarctica by McMurdo Sound where the sea ice breaks up in late December allowing cargo ships to dock at McMurdo Station on the southern tip of the Hut Point Peninsula. The south side of the island from Scott Base at the tip of the peninsula to Cape Crozier is in direct contact with the Ross Ice Shelf. Ross Island contains rookeries of Adelie penguins at Cape Royds, at Caughley Beach of Cape Bird, and at Williamson Rock at Cape Crozier. The rookery of the famous Emperor penguins is also located at Cape Crozier. Mount Erebus is the source of small seismic tremors caused by the movement of the phonolite magma within the cone. Excerpt of the topographic map: Ross Island, Antarctica, ST 57–60.6* published in 1960 by the US Geological Survey (Revised in 1970) (From the USGS Antarctic Resource Center, http://usarc.usgs.gov/)

The chemical compositions of the lava flows and pyroclastic deposits of Ross Island range widely from basanite to phonolite (Fig. 16.4). These rocks formed from basanite magmas which differentiated by fractional crystallization in subsurface magma chambers. Phonolites are the most highly differentiated products of this process and are probably less voluminous than the intermediate "trachybasalt" of Goldich et al. (1975) which are identified as tephrite, phonotephrites, and tephriphonolites in Fig. 16.4. In spite of their comparatively small volume, the phonolites were erupted last and therefore cover the intermediate lavas at the summit and on the slopes of Mt. Erebus and the other volcanoes on Ross Island.

Fig. 16.12 The Erebus Glacier Tongue extends from the base of the Hut Point Peninsula (located out of view on the right) into McMurdo Sound. The ice tongue is a place where geological field parties stop on their way from McMurdo Station to Cape Royds. The ice tongue contains small caves that form in fractures caused by the daily tides. The entrance to one of these caves is exposed in the ice cliff behind the snowmobiles. Mount Erebus looms large in the background. The notch to the left of the summit indicates the position of the Fang Glacier which is located behind the summit in this view (Photo by T.M. Mensing)

A noteworthy exposure of anorthoclase-phonolite porphyry in Fig. 16.14 occurs near Shackleton's Hut at Cape Royds. This lava flow may have been erupted by a subsidiary vent on the slopes of Mt. Erebus and flowed into the water of McMurdo Sound which caused the well-formed pillow structure displayed by this outcrop.

16.2.1 Hut Point Peninsula and Petrogenesis

The Hut Point Peninsula in Fig. 16.15 extends from the south side of Ross Island for a distance of more than 20 km. It was so named because Robert Scott's first hut is located at its southern tip. This site was later selected for the construction of McMurdo Station because cargo ships and icebreakers can tie up directly to shore (Section 2.6) in the deep-water bay at the tip of the peninsula.

The geology of the Hut Point Peninsula was mapped by Kyle (1981a, 1990b) based both on surface exposures of the volcanic rocks and on three cores which were recovered in 1973 by diamond drilling in the immediate vicinity of McMurdo Station. The drilling was carried out by the Dry Valley Drilling Project (Mudrey and McGinnis 1974; McGinnis 1981). The core of DVDP-1 reached a depth of 201.5 m and penetrated 40 lithologic units described by Treves and Kyle (1973b) and by Kyle (1981a) who published lithologic logs of DVDP 1, 2, and 3. In addition, Kyle (1981a) provided chemical analyses for 31 samples of volcanic rocks from DVDP-1 and described the petrography of each of the 40 stratigraphic units in this core. Kyle (1981a) and Goldich et al. (1981) also published chemical analyses of samples from cores DVDP-2 and 3 which were drilled 3 m apart and reached depths of 179.41 and 381.00 m, respectively.

Fig. 16.13 The ice caves in the Erebus Glacier Tongue are a target for exploration by geologists on their way to Cape Royds. The caves also provide shelter for travelers who may be exposed to a sudden wind storm. The ceiling of the ice cave supports long "ice stalactites" (i.e., icicles) and coarse crystals of ice that form by condensation of water vapor. Both persons in this image carry ice axes and are wearing polar clothing provided by the US Antarctic Program (Photo by T.M. Mensing)

The chemical composition of core DVDP-1 in Fig. 16.16 encompasses the full range of concentrations of silica and alkali metals that characterize the McMurdo Volcanics in Fig. 16.4. Accordingly, they are classified in order of increasing concentrations of alkali metals and silica into basanites, tephrites, phonotephrites and tephriphonolites. The alkali-metal concentrations in this sequence rise to 12.5% (tephriphonolites) but the samples from DVDP-1 analyzed by Kyle (1981a) do not include true phonolites.

The concentrations of silica and alkali metals in the core of DVDP-1 vary systemically up-section in Fig. 16.17 and generally increase to high levels in the upper 70 m of the core. The lava flows rest on a thick deposit of basanite hyaloclastite which occupies the bottom of the core from 203.1 to about 150 m. The hyaloclastite is overlain by alkali-rich flows of phonotephrite and tephriphonolite from 150 to 125 m, followed by another layer of basanite from 125 to 100 m. Above a depth of 100 m the alkali and silica concentrations increase up-section with a succession of tephrite and phonotephrite flows leading up to tephriphonolite at 60 m. The uppermost stratigraphic section consists of phonotephrites having fairly constant composition and extending from 60 m almost to the top of the core. The uppermost flow in DSDP-1 is an alkali-rich tephriphonolite that is about 5 m thick.

Based on information provided by Goldich et al. (1975, 1981), Sun and Hanson (1975, 1976), Stuckless et al. (1974, 1981), and Stuckless and Ericksen (1976), the entire sequence of alkali-rich volcanic rocks could have evolved from basanite parent magmas. In that case, successive sequences of lava flows and pyroclastic deposits may start with the eruption of basanite followed by more highly differentiated alkali-rich flows. Such a scenario seems to be supported by the

Fig. 16.14 The volcanic rocks exposed near Shackleton(s Hut on Cape Royds are phonolites which, in this case, formed pillows indicating that the lava flowed into the water of McMurdo Sound and was quenched instantly. Each lobe of the resulting rock is covered by a rind of volcanic glass which weathers preferentially and gives the outcrop an etched appearance. The cusps of the pillows point down thus indicating the top and bottom of the unit. The phonolite at Cape Royds and on the summit of Mt. Erebus contains large crystals of anorthoclase about 5 cm in length; therefore, this rock is referred to as an anorthoclase-phonolite porphyry. It can also be called an Antarctic kenyte because of its similarity to lava flows erupted by Mt. Kilimanjaro and Mt. Kenya in East Africa (Photo by G. Faure)

stratigraphy of the core of DVDP-1 in Fig. 16.17 where the two basanitelayers are overlain by more alkali-rich flow sequences.

A similar flow stratigraphy has been observed on certain oceanic islands including the Hawaiian islands (Faure 2001). MacDonald and Katsura (1964) and Clague and Frey (1982) recognized three suites of flows on the Hawaiian Islands on the basis of their chemical composition and stratigraphic position:

1. Nephelinitic suite (youngest)
2. Alkali suite
3. Tholeiitic suite (oldest)

This succession was explained by Chen and Frey (1983) by a gradual reduction of the extent of melting of the lithospheric mantle by heat emanating from the head of the large plume which is currently located beneath Mt. Kilauea on the island of Hawaii (i.e., the Big Island). The entire chain of Hawaiian volcanoes was formed from basalt magma that originated by partial melting caused by the plume which has remained stationary while the Pacific plate has moved over it in a northwesterly direction.

The comparison of the flow stratigraphy of the Erebus volcanic province with the flow sequence on the Hawaiian islands raises questions about the petrogenesis of the volcanic rocks on Ross Island:

1. Does the radial symmetry of the locations of volcanic vents about Mt. Erebus and Mt. Discovery indicate the existence of a mantle plume in the subsurface?
2. Did the Antarctic plate shift in such a way that the volcanic activity migrated from Mt. Discovery to Mt. Erebus?
3. Did the West Antarctic rift system form in response to the tectonic stress imposed on the continental crust by one or several mantle plumes?
4. Is Marie Byrd Land breaking away from the mainland of Antarctica by the widening of the West Antarctic rift system?

We cannot yet answer any of these questions except to assert in agreement with Kyle and Cole (1974) that the volcanic activity is a consequence of deep-seated tectonic processes which have caused patterns of crustal fractures that control the positions of the volcanoes in the Erebus volcanic province and elsewhere in Victoria Land, as well as in Marie Byrd Land (e.g., the Executive Committee Range).

The K-Ar dates of the lava flows on the Hut Point Peninsula in Table 16.2 indicate that they were erupted during the Pleistocene at a time when *Home erectus* was populating Europe and Asia.

Additional research reports about the geology and petrology of the volcanic rocks on the Hut Point Peninsula were published by Cole and Ewart (1968). Treves and Kyle (1973a, b), Kyle and Treves (1974a, b), Kyle and Rankin (1976), Treves et al. (1975), Kurasawa (1975, 1978), and Weiblen et al. (1981).

Fig. 16.15 The Hut Point Peninsula extends for more than 20 km from the south side of Ross Island. McMurdo Station and Scott Base are located at the southern tip of the peninsula. The volcanic rocks of the Hut Point Peninsula were penetrated by three diamond drill holes (DVDP-1, 2, and 3) which were sited in the immediate vicinity of McMurdo Station. These cores were logged by Treves and Kyle (1973) and Kyle and Treves (1974a, b). The stratigraphy and petrogenesis of the volcanic rocks were discussed by Kyle (1981a, b, 1990b, 1995) and by Goldich et al. (1981) (Aerial photography courtesy of the USGS TMA 550, image 216, F33)

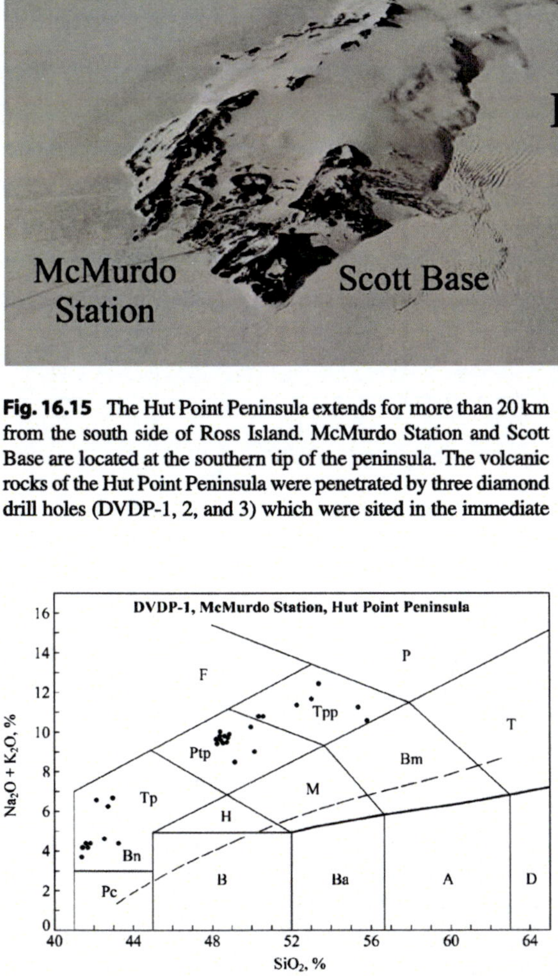

Fig. 16.16 The chemical compositions of volcanic rocks in DVDP-1 drilled at McMurdo Station on the Hut Point Peninsula range widely on the total alkali-silica diagram from basanite (Bn), to tephrite (Tp), to phonotephrite (Ptp), and to tephriphonolite (Tpp). Therefore, the rocks in this core (201.5 m) are almost as diversified as the McMurdo Volcanics in the entire Erebus volcanic province in Fig. 16.4. Samples of DVDP-2 and 3 are not shown but range just as widely as the rocks of DVDP-1 (Data from Kyle (1981a))

16.2.2 Mount Erebus, Summit

Mount Erebus is without a doubt one of the most photogenic mountains in Antarctica (e.g., Fig. 16.2) and has been seen by tens of thousands of men and women who have passed through McMurdo Station and Scott Base on their way to their assigned duty stations or research areas. A few researchers have even been able to live and work on the summit of this olympian volcano.

The summit of Mt. Erebus in Fig. 16.18 consists of a caldera at about 3,200 m which has filled with volcanic rocks composed of anorthoclase phonolite (Kyle 1977, 1990b). The active cone, which is located within this former caldera, contains the Main Crater which is elliptical in shape (500 × 600 m) and about 110 m deep. The Main Crater contains the Inner Crater which has a diameter of about 250 and is 100 m deep. The Inner Crater in Fig. 16.19 contains a lake composed of continuously

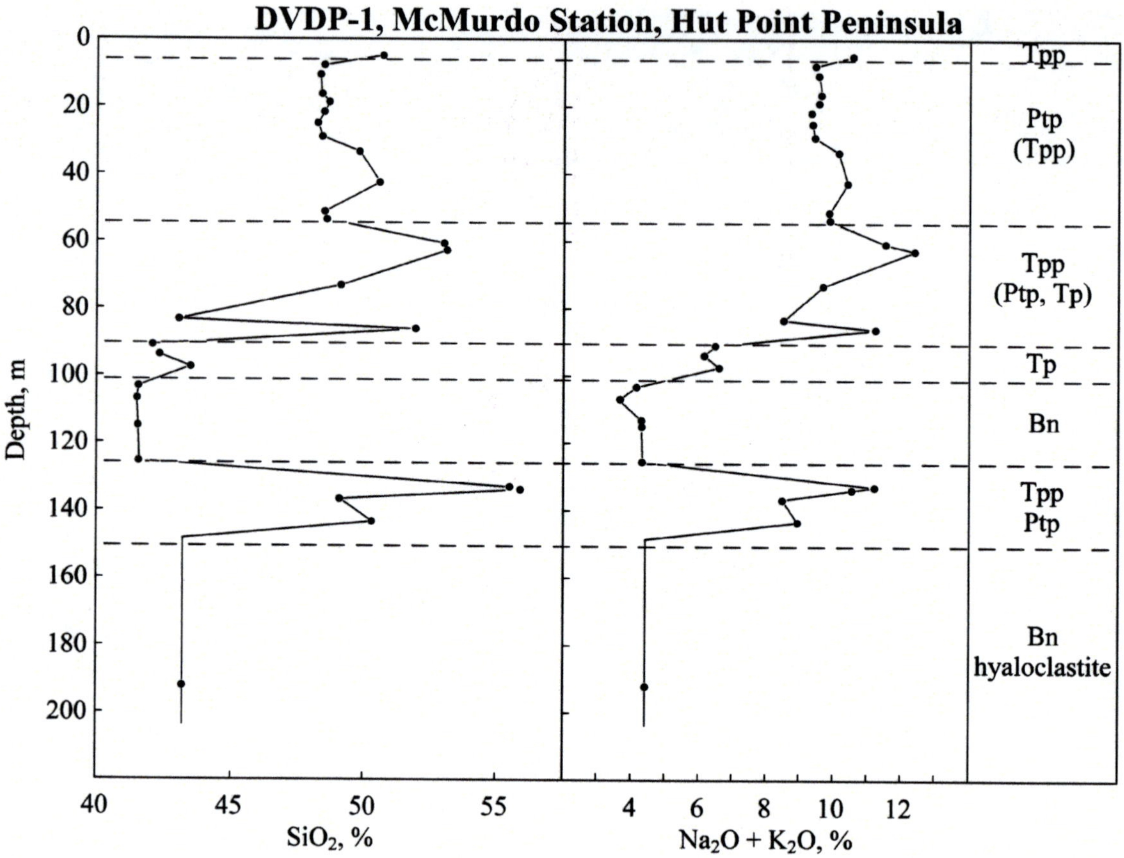

Fig. 16.17 The total alkali and silica concentrations of the DVDP-1 core define two stratigraphic sequences consisting of basanite pyroclastics and lava flows overlain, in each case, by alkali-rich flows ranging in composition from tephrite, to phonotephrite, to tephriphonolite. This stratigraphic pattern also occurs on the Hawaiian islands, where early-formed tholeiite basalts are overlain by alkali-rich lavas (Data from Kyle (1981a))

Table 16.2 Whole-rock K-Ar dates of lava flows on the Hut Point Peninsula of Ross Island compiled by Kyle (1981a)

Location	Date (Ma[a])	Reference
Basanite flows		
Black Knob	0.44 ± 0.1	Armstrong (1978)
SW of Black Knob	0.60 ± 0.06	Armstrong (1978)
Half Moon Crater	1.00 ± 0.2	Armstrong (1978)
Basanite dike		
Castle Rock	1.21 ± 0.05	Kyle (1981a)[b]
	Phonolite	
Observation Hill	1.21 ± 0.04	Forbes et al. (1974)

[a]Calculated from the new constants of Steiger and Jäger (1977).

[b]Measured by J.E. Gabites, Institute of Nuclear Science, D.S.I.R., Lower Hutt, New Zealand.

convecting liquid anorthoclase phonolite lava. The lava lake ejects lava bombs and discharges gases that form the plume that rises from the summit of Erebus.

Kyle et al. (1982a) recorded the volcanic activity on the summit of Erebus from January 1973 until December 1976. During this time, the lava lake increased substantially and strombolian gas eruptions occurred at the rate of one to three per day. Kyle et al. (1982a) also monitored the seismic activity of the summit area by means of seismographs that were installed in a warm ice cave originally described by Giggenbach (1976). The frequency of volcanic earthquakes in December of 1974 was about 1,0001984 per week (Dibble et al., 1989).

A noteworthy topographic feature on the summit of Mt. Erebus is Fang Ridge located northeast of the

Fig. 16.18 The summit of Mt. Erebus contains an old caldera that has filled with volcanic rocks to form a plateau. The plateau contains the active cone which includes the Main Crater of the volcano. The rocky slope in the foreground is Fang Ridge behind which the Fang Glacier is located where people who need to work on the summit of Mt. Erebus adjust to the elevation before they walk up to the summit plateau. Hut Point Peninsula is visible in the distance (Photo courtesy of the USGS, TMA 730, 0041, F31, November 1, 1960)

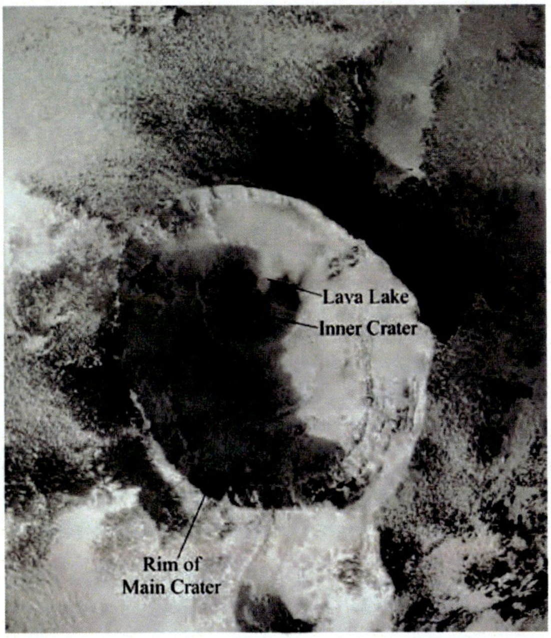

Fig. 16.19 The Main Crater at the summit of Mt. Erebus has an elliptical shape with diameters of 500 and 600 m. It is about 110 m deep and contains the Inner Crater which is 250 m in diameter, 100 m deep, and contains a continually convecting lake composed of liquid phonolite lava. The lava in this lake has not flowed over the rim of the Main Crater in historical time, but the lava lake does emit gases that form the plume that rises above the summit of the volcano. The lake has periodic gas eruptions that throw lava bombs out of the Main Crater (e.g. 1972/73, Giggenback et al. 1973) (Photo courtesy of the USGS, TMA 1214, 0074, F32, November 1, 1963)

summit and extending about 4 km in a northwest-southeast direction. The ridge forms a nearly vertical cliff, from 100 to 200 m high, facing Erebus. The space between Fang Ridge and the summit of Mt. Erebus is occupied by the Fang Glacier where tents have been set up to shelter people who need to adjust to the high elevation before proceeding to the summit of Mt. Erebus (Fig. 2.2b).

A K-Ar date of 0.81 ± 0.02 Ma was measured by Armstrong (1978) for the lowest flow on Fang Ridge and 0.73 ± 0.07 Ma for the uppermost flow. Armstrong (1978) also reported a date of 0.15 ± 0.05 Ma for an anorthclase phonolite collected from the rim of the summit plateau. However, anorthoclase phenocrysts from the summit cone yielded K-Ar dates ranging from 0.55 to 0.44 Ma. These dates are too old because the K-Ar dates of the anorthoclase crystals should be close to 0 Ma. Therefore, the anorthoclase crystals presumable contain excess radiogenic ^{40}Ar which may reside in the glass inclusions in these crystals (Moore and Kyle 1990). Kyle (1977) actually analyzed *glass* inclusions in three anorthoclase crystals collected from the summit region of Mt. Erebus and reported progressive changes in the composition of the glass from the center of one of the crystals to the rim. However, the composition of olivine, clinopyroxene, and magnetite inclusions in the anorthoclase are similar to the compositions of these minerals in the glassy matrix of the phonolites.

The presence of excess radiogenic ^{40}Ar in the glass inclusions of anorthoclase crystals was eventually confirmed by Esser et al. (1997) by means of the $^{40}Ar/^{39}Ar$ method of dating. They reported that the $^{40}Ar/^{39}Ar$ dates of an anorthoclase crystal that was erupted in 1984 decreased when part of the glass was removed from crushed samples by etching with hydrofluoric acid (HF). The $^{40}Ar/^{39}Ar$ dates in Fig. 16.20 decrease from 640 ± 30 (30% glass) to 179 ± 16 ka (10% glass)

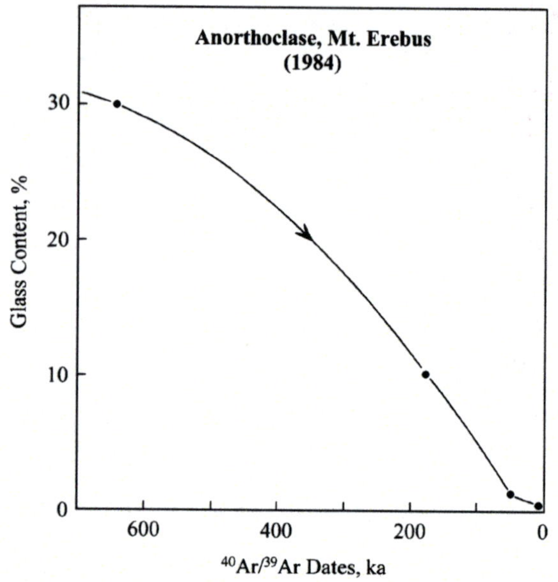

Fig. 16.20 The $^{40}Ar/^{39}Ar$ dates of anorthoclase crystals erupted on Mt. Erebus in 1984 decrease systematically after the inclusions of glass are removed by etching crushed samples with hydrofluoric acid. This result confirms that the melt inclusions contain excess ^{40}Ar. The lowest date of 8 ± 2 ka (1 ka = 10^3 years) was obtained by heating crushed anorthoclase crystals containing less than 1% glass at temperatures of less than 1,200°C (Data from Esser et al. (1997))

and finally to 48 ± 8 ka (1% glass). Esser et al. (1997) also demonstrated that melt inclusions (i.e., glass) in anorthoclase grains release argon at temperatures in excess of 1,200°C when the anorthoclase melts incongruently. Accordingly, samples of crushed anorthoclase containing less than 1% glass inclusions that were heated to less than 1,200°C yielded dates as low as 8 ± 2 ka.

16.2.3 Mount Erebus, Gas and Dust

The continuing activity of Mt. Erebus releases substantial amounts of volcanic particulates and gases into the atmosphere. In addition, a variety of salts are forming on rock surfaces and in the regolith in the area of the summit. The discharge of water vapor by fumaroles on the summit plateau has caused the growth of spectacular ice towers and warm ground that occurs south of the Main Crater and elsewhere within a radius of less than 1,000 m of its rim.

The gases vented by the summit of Mt. Erebus consists primarily of water vapor, sulfur dioxide, carbon dioxide, as well as chlorine and fluorine (Rose et al. 1985). These gases and the associated particles are the principal volcanic contaminants of the atmosphere over Antarctica and therefore have been studied by sampling from aircraft and by remote sensing using satellites(e.g., Rothery and Oppenheimer 1994).

The *fumaroles* on the summit plateau of Mt. Erebus discharge steam into the atmosphere which condenses rapidly in the cold air and the water then freezes to form hollow ice towers pictured in Fig. 16.21 (Holdsworth and Ugolini 1965). In addition *ice caves* have formed on the summit plateau where steam is discharged by subglacial vents. Some of these ice caves and the warm ground along the rim of the Main Crater have been used for recreational purposes.

The particulates in the Erebus plume were collected and analyzed on December 9 and 19, 1983, by Chuan et al. (1986) using a ten-stage cascade impactor with quartz-crystal microbalance impactor-plates (Chuan 1975). These investigators determined that the Erebus plume injects 21 ± 3 t of aerosol particles per day into the upper troposphere of Antarctica. The particles settle out and are deposited in snow and ice (Shaw 1982; Delmas et al. 1982, 1985). Chuan et al. (1986) reported that large aerosol particles having diameters of 5–25 micrometers (μm) in the Erebus plume are composed of elemental sulfur and silicates, a group of middle-sized particles consists of iron oxides, and small particles (less than 1 μm) are formed by complex liquids. The average mass concentration of particles of all sizes detected by Chuan et al. (1986) in the Erebus plume ranged from 70 to 370 μg/m³ and decreased exponentially for a distance of about 16 km from the summit.

Salts consisting of halite (Na_2Cl), thenardite (Na_2SO_4), and sylvite (KCl) identified by Keys (1980) in the Main Crater were also observed in the plume by Chuan et al. (1986). Sulfuric acid droplets and coatings of aerosol particles were found only on December 9, 1983, because they are discharged only during small gas eruptions of the lava lake. During these events, the plume also contains particles of native sulfur in crystalline and amorphous form in addition to sulfur dioxide (SO_2) and, possibly, hydrogen sulfide (H_2S). Mid-sized particles of iron oxide occurred in the plume of December 19 but not on December 9. In addition,

Fig. 16.21 Hollow ice towers on the summit plateau of Mt. Erebus form where fumaroles discharge steam into the cold air. In addition, ice caves also occur where warm ground or subglacial steam vents cause melting of glacial ice and snow. Similar ice towers occur on the summit of Mt. Melbourne on the coast of northern Victoria Land (Photo by J.M. Palais published with permission)

the mid-sized particles included Fe-Al silicates, aluminum sulfate, and silica (SiO_2) in the form of christobalite (?) that can form from silicon tetrafluoride (SiF_4) when it reacts with water in the atmosphere. The smallest particles are chlorine-rich and seem to have formed by evaporation of droplets of volatile liquids (e.g., water and dilute hydrochloric acid).

Given the variation in the speed and direction of the wind on Ross Island, the particles in the Erebus plume are dispersed widely such that their surface concentration is below detection. Deposits of volcanic ash that do occur in the East Antarctic ice sheet as well as locally on the surface of valley glaciers and ice caps formed during volcanic eruptions not only of Mt. Erebus but also of the other volcanoes and vents in the Transantarctic Mountains of Victoria Land and in Marie Byrd Land of West Antarctica (Keys et al. 1977; Palais et al. 1982, 1983a, b; Palais and Legrand 1985; Palais et al. 1989; Radke 1982; Kyle et al. 1982b).

16.2.4 Mt. Erebus, Soil Salts

The soils on Mt. Erebus were first described by Ugolini (1967). The rocks around the Main Crater of Mt. Erebus are coated with salts which also occur locally within the regolith. Keys (1980) and Keys and Williams (1981) identified the compounds listed in Table 16.3 which contain not only chlorine and fluorine, but also sulfate that forms by a reaction of sulfur dioxide gas with water vapor and oxygen of the atmosphere. In addition, the salts on the summit of Mt. Erebus contain aluminum, silicon, and iron (Keys and Williams 1981). Although some of the salt deposits in Fig. 16.22a and b are distinctly yellow, native sulfur is *not* present. Instead, the yellow deposits are probably composed of jarosite $[(K,Na)(Fe,Al)(SO_4)_2(OH)_6]$.

The volcanic origin of the salts on the summit of Mt. Erebus was confirmed by Jones et al. (1983) who demonstrated that the isotopic composition of strontium in yellow and tan-colored salts is indistinguishable from the isotopic composition of strontium in anorthoclase crystals. The average $^{87}Sr/^{86}Sr$ ratio of three samples of salt from the summit of Erebus is 0.70345 ± 0.00007 (1$\bar{\sigma}$) compared to 0.70330 ± 0.00003 for an anorthoclase crystal. The $^{87}Sr/^{86}Sr$ ratio of a carbonate sample from an ice cave on the summit is 0.70460 ± 0.00003. The authors considered that this sample contained a small amount of marine strontium ($^{87}Sr/^{86}Sr$ = 0.7090) that was dissolved in the meltwater from which this salt was precipitated.

Table 16.3 Compounds on the surface of rock clasts and in the regolith on the summit plateau of Mt. Erebus (Keys and Williams 1981)

Salt	
Positively identified	Formula
Halite	$NaCl$
Gypsum	$CaSO_4 2H_2O$
Alunite	$K_2Al_6(SO_4)_4(OH)_{12}$
Thenardite	Na_2SO_4
Sylvite	KCl
Mirabilite	$Na_2SO_4 1OH_2$
Calcite	$CaCO_3$
Tentatively identified	
Chloraluminate	$AlCl_6 6H_2O$
Alunogen	$Al_2(SO_4)_3 18H_2O$
Jarosite	$(K,Na)(Fe,Al)(SO_4)_2(OH)_6$
Malladrite	Na_2SiF_6
Sulphohalite	$Na_6ClF(SO_4)_2$
Al trifluoride[a]	AlF_3
Na-Al oxychloride[a]	$NaAl_4O_4Cl_5$
Hydromolysite[a]	$FeCl_3 6H_2O$
Ralstonite[a]	$NaMgAl(F,OH)_2H_2O$

[a]Tentatively identified by W. Rose (Michigan Technical University).

16.2.5 Soil Salts, Coast of Ross Island

Strontium derived from seawater is even more abundant in soil salts along the coast of Ross Island analyzed by Jones et al. (1983) and Faure and Jones (1989). The $^{87}Sr/^{86}Sr$ ratios of these salts in Fig. 16.23 and Appendix 16.8.2 range widely from 0.70344 ± 2 (calcite, Castle Rock, Hut Point Peninsula, elevation 413 m) to 0.70912 ± 3 (thenardite, Trachyte Hill, Cape Bird, elevation 2 m). The calcite on Castle Rock contains strontium of volcanic origin, whereas the strontium in thenardite along the beach at Cape Bird is of marine origin.

A compilation of all $^{87}Sr/^{86}Sr$ ratios of soil salts from Ross Island measured by Jones et al. (1983) and by Faure and Jones (1989) in Fig. 16.23 reveals a strongly bimodal distribution. The salts from the summit of Mt. Erebus have low $^{87}Sr/^{86}Sr$ ratios between 0.70339 ± 3 and 0.70353 ± 3, whereas the soil salts along the coast of Ross Island have consistently higher $^{87}Sr/^{86}Sr$ ratios that range from 0.70707 to 0.70912. These results support the hypothesis that the soil salts on Ross Island contain strontium of volcanic and marine origin in varying proportions. The salts that are forming in the vicinity of the active volcanic vent on

Fig. 16.22 (a) The inner crater of Mt. Erebus is the source of gases that are discharged by the lava lake. These gases are composed primarily of water vapor, sulfur dioxide, and lesser amounts of carbon dioxide, chlorine, and fluorine. (b) The gases react with the exposed rocks and cause the deposition of different kinds of chloride, sulfate, and fluoride salts listed in Table 16.3. The yellow coating on the rock clasts in this image is not composed of native sulfur, but is probably jarosite (Photos by J.M. Palais published with permission)

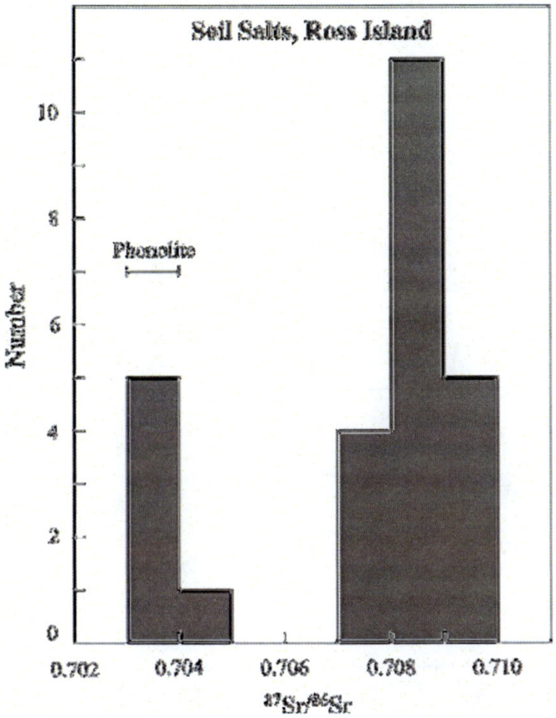

Fig. 16.23 The $^{87}Sr/^{86}Sr$ ratios of soil salts on Ross Island have a distinctly bimodal distribution. Salts on the summit of Mt. Erebus contain strontium that is derived from the volcanic rocks, whereas the strontium in soil salts along the coast originates both from seawater ($^{87}Sr/^{86}Sr = 0.7090$) and from volcanic rocks ($^{87}Sr/^{86}Sr = 0.7035$) (Data from Jones et al. (1983) and Faure and Jones (1989) listed in Appendix 16.8.2)

the summit of Mt. Erebus contain strontium of volcanic origin, whereas salts near the coast contain mixtures of marine and volcanic strontium.

This hypothesis was subsequently confirmed by measurements of the isotopic compositions of *sulfur* published by Faure and Jones (1989). The isotopic composition of sulfur is expressed by the $\delta^{34}S$ parameter which is defined as:

$$\delta^{34}S = \left[\frac{(^{34}S/^{32}S)_{spl} - (^{34}S/^{32}S)_{std}}{(^{34}S/^{32}S)_{std}} \right] \times 10^3 ‰ \quad (16.1)$$

The standard is the $^{34}S/^{32}S$ ratio of sulfur in troilite (FeS) in meteorites. The $\delta^{34}S$ value of sulfur in mantle-derived igneous rocks is 0‰ because the sulfur in meteorites and the Earth both originated from the solar nebula that contracted to form the Sun and the planets (Faure and Mensing 2007). The sulfate ion in present-day seawater has a $\delta^{34}S$ value of +20‰ which indicates that the sulfur in the oceans is enriched in ^{34}S (Hoefs 1997; Faure and Mensing 2005; Sharp 2007).

The strontium and sulfur in the soil salts along the coast of Ross Island are isotopic mixtures of these elements derived from volcanic and marine sources:

Strontium: $^{87}Sr/^{86}Sr$ from 0.7034 to 0.70906

Sulfur: $\delta^{34}S$ from 0.0 to 20.0‰

The shapes of the isotopic mixing hyperbolas, derived by Faure and Jones (1989) in Fig. 16.24, depend on the Sr/S concentration ratios of the marine and volcanic components: If $(Sr/S)_m > (Sr/S)_v$ (e.g., $K = (Sr/S)_m/(Sr/S)_v = 10$), then the mixing hyperbola in Fig. 16.24 is convex upward. If $(Sr/S)_m < (Sr/S)_v$, (e.g., $K = (Sr/S)_m/(Sr/S)_v = 0.1$), the mixing hyperbola is convex downward. These alternative mixing regimes were combined in Fig. 16.24 in order to define an area on the Sr-S isotopic mixing diagram that can accommodate the isotopic compositions of strontium and sulfur of all soil salts analyzed by Faure and Jones (1989). The construction of the mixing hyperbolas is explained and illustrated in Appendix 16.8.3.

The abundance of the marine component indicated by this model ranges from about 1% (Fang Ridge, summit of Erebus) to 98% for salt along the beach at Trachyte Hill at Cape Bird. The abundance of the marine component in the salts on the slope of Trachyte Hill decreases with increasing elevation to 81% at 130 m a.s.l. The trend reverses at higher elevations and reaches 97% marine at 390 m a.s.l. At Cape Royds the abundance of the marine component likewise decreases from 92% at 12 m to 75% at 129 m a.s.l with deviations at 60 and 284 m.

The patterns apparent at both locations indicate that the transport of marine strontium and sulfur decreases with increasing elevation and distance from the coast. The trend is reversed in areas where prevailing winds favor the transport of marine salt upslope or where excessive melting of snow releases marine salts into the environment. Faure and Jones (1989) noted that the water of McMurdo Sound at Cape Bird is ice-free somewhat longer than at Cape Royds which may explain why the marine component is more abundant at Trachyte Hill (Cape Bird) than at Cape Royds.

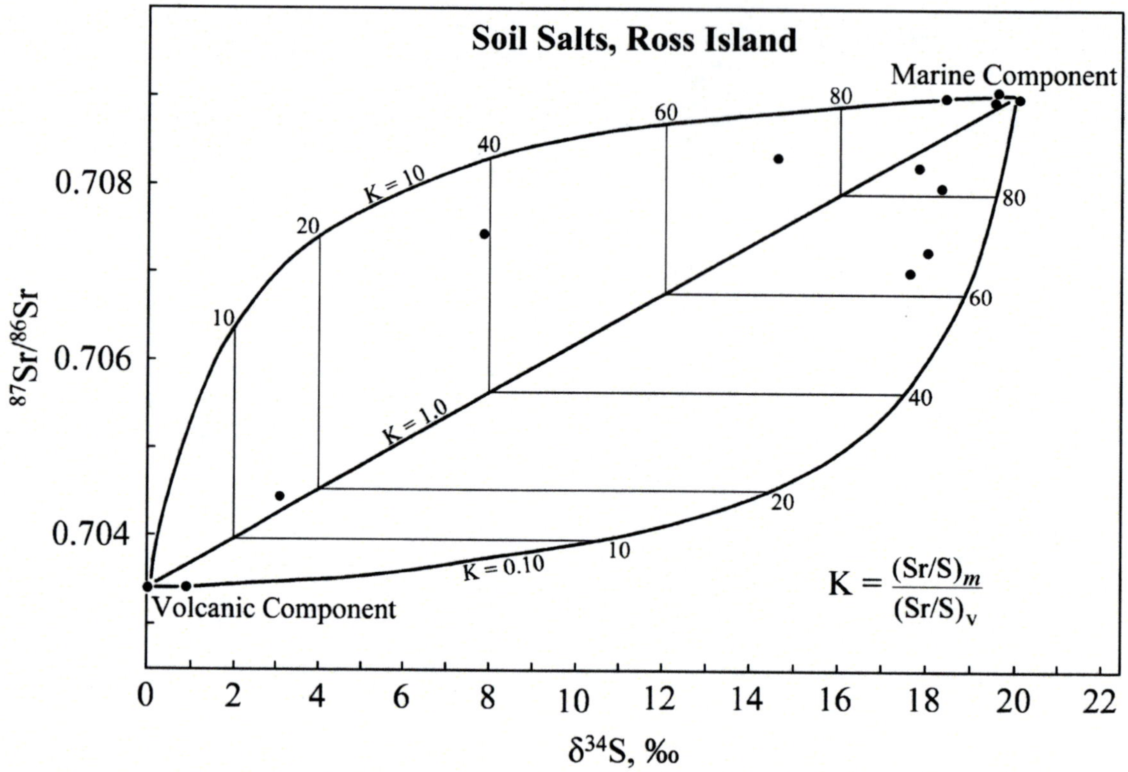

Fig. 16.24 The soil salts along the coast of Ross Island contain strontium and sulfur that originated from the volcanic rocks and from seawater. The isotopic compositions of strontium and sulfur in the soil salts are constrained by two isotopic mixing hyperbolas whose shapes depend on the assumed concentrations of strontium and sulfur expressed by the parameter K. The isotopic compositions of strontium and sulfur of the volcanic component are: $^{87}Sr/^{86}Sr = 0.7035$, $\delta^{34}S = 0.0‰$, whereas the marine strontium and sulfur are defined by $^{87}Sr/^{86}Sr = 0.7090$, $\delta^{34}S = +20‰$. The measured isotope compositions of these elements in soil salts analyzed by Faure and Jones (1989) fit this model. The data are listed in Appendix 16.8.2

16.3 Melbourne Volcanic Province

The Melbourne volcanic province of northern Victoria Land (Kyle 1990c) in Fig. 16.25 includes Mt. Melbourne, the Vulcan Hills, Mt. Overlord, The Pleiades, and the Malta Plateau all of which occupy an arcuate area on the mainland of northern Victoria Land together with numerous small volcanic vents. The geology of this area was originally described by Nathan and Schulte (1968) and was included in the geologic map of northern Victoria Land compiled by Gair et al. (1969). In addition, all of the volcanoes and vents of the Melbourne volcanic province were described in the book edited by LeMasurier and Thomson (1990). Geologists of GANOVEX IV mapped and sampled the rocks of the Melbourne volcanic province during the 1984/85 field season and carried out an extensive aeromagnetic survey of part of northern Victoria Land and the adjacent Ross Sea (Bosum et al. 1991; Behrendt et al. 1991a, b).

The K-Ar dates of the volcanic rocks from the Melbourne volcanic province measured by Armstrong (1978) range from 18.5 ± 0.7 Ma (Miocene) for the Nathan Hills to 0.003 ± 0.014 Ma (i.e., 3,000 years) for The Pleiades and 0.01 ± 0.02 Ma (i.e., 10,000 years) for Mt. Melbourne (Kyle 1990c), which suggests that The Pleiades and Mt. Melbourne may be dormant but are not yet extinct.

The volcanic rocks of the Melbourne volcanic province are alkali-rich like those of the Erebus volcanic province although silica concentrations in the Melbourne province range more widely. The representative chemical analyses compiled by the authors of the book edited by LeMasurier and Thomson (1990)

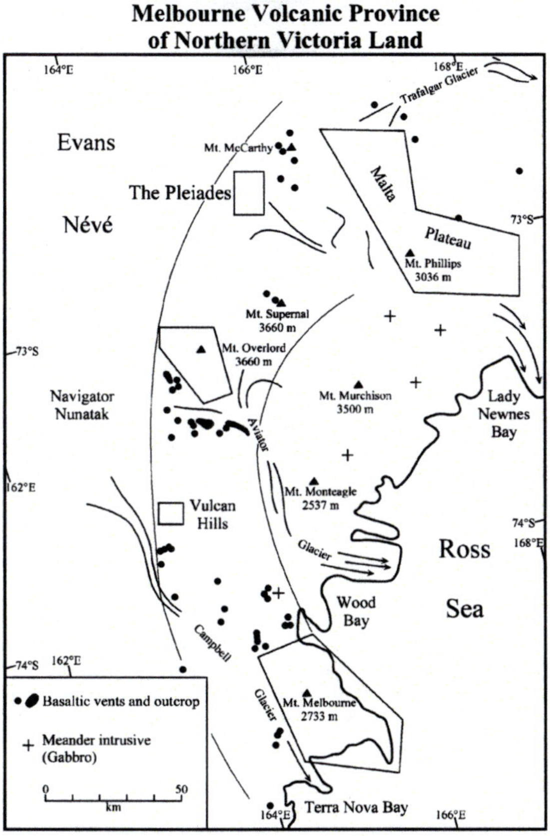

Fig. 16.25 The Melbourne volcanic province of northern Victoria Land forms an arc that extends from Mt. Melbourne in the south to the Vulcan Hills, Mt. Overlord, The Pleiades, and the Malta Plateau in the north. In addition, the Melbourne volcanic province contains numerous small volcanic vents where lava and volcanic ash of basaltic composition have been extruded. The Meander intrusive suite consists of gabbroic rocks whose presence was inferred by an aeromagnetic survey by Bosum et al. (1991) and Behrendt et al. (1991a, b) (Adapted from Kyle (1990f) and from Fig. 2 of Nathan and Schulte (1968))

are plotted in Fig. 16.26. The concentrations of silica range from 41% to 75% (Mt. Overlord) and the total alkali concentrations extend from 3% to 14%. The lavas of The Pleiades are the most alkali-rich and are assigned to the tephrite-phonolite-trachyte association. The volcanic rocks of Mt. Melbourne, the Vulcan Hills, Mt. Overlord, and of local vents are less alkali-rich and belong to the hawaiite-mugearite-trachyte suite. The Cenozoic volcanics of the Malta Plateau have a wide range of silica concentrations from 42% to 72% and range from basanite to trachyte and peralkaline rhyolites.

16.3.1 Mt. Melbourne

Mt. Melbourne (2,732 m) in Fig. 16.27 is located at the northern end of the Terror Rift within the broad Victoria Land Basin which is a local manifestation of the Cenozoic West Antarctic rift system. The southern end of the Terror Rift is occupied by Mt. Erebus indicating that these two volcanoes are related by this rift. The Victoria Land Basin and the Terror Rift contain several submarine volcanic vents marked in Fig. 16.27. Mt. Melbourne in Fig. 16.28 is located on the coast of northern Victoria Land north of Terra Nova Bay. It is accessible by a short helicopter flight from the German and Italian research stations which are located about 20–30 km south of Shield Nunatak at the foot of Mt. Melbourne. Several other volcanic centers are located in the vicinity of Mt. Melbourne, including Cape Washington, Edmonson Point, Baker Rocks, and the Random Hills north of the central volcano.

The geology of the Mt. Melbourne and some of the other volcanoes of the Melbourne volcanic province was first described by Gair (1967), Nathan and Schulte (1967, 1968), and by Riddolls and Hancox (1968). The most recent study of the geology of Mt. Melbourne and of the associated volcanic centers which together constitute the Melbourne volcanic field was published by Wörner and Viereck (1987) and Wörner et al. (1989). Geophysical surveys of this part of northern Victoria Land were carried out by Delisle and Thierbach (1989), Dürbaum et al. (1989), Bosum et al. (1991), and Behrendt et al. (1991a, b).

The first indication that Mt. Melbourne may have erupted quite recently was reported by Adamson and Cavaney (1967) who observed layers of volcanic ash in an ice cliff along the southwestern slope of the mountain and on the ice tongue of the Campbell Glacier. The continuing geothermal and volcanic activity of Mt. Melbourne was actually witnessed by Nathan and Schulte (1967) who climbed Mt. Melbourne from the north with the aid of snowmobiles. They observed many small areas of hot ground near the summit and small fresh-looking cones and mounds of volcanic rocks on the slopes and within the crater of the volcano. The areas of hot ground were venting steam which formed hollow ice pinnacles up to 1.80 m tall, similar to those on the summit plateau of Mt. Erebus. The ice pinnacles are connected to an extensive system of snow caves with chambers that are up to 7 m wide and 3 m high that continue for at least 300 m

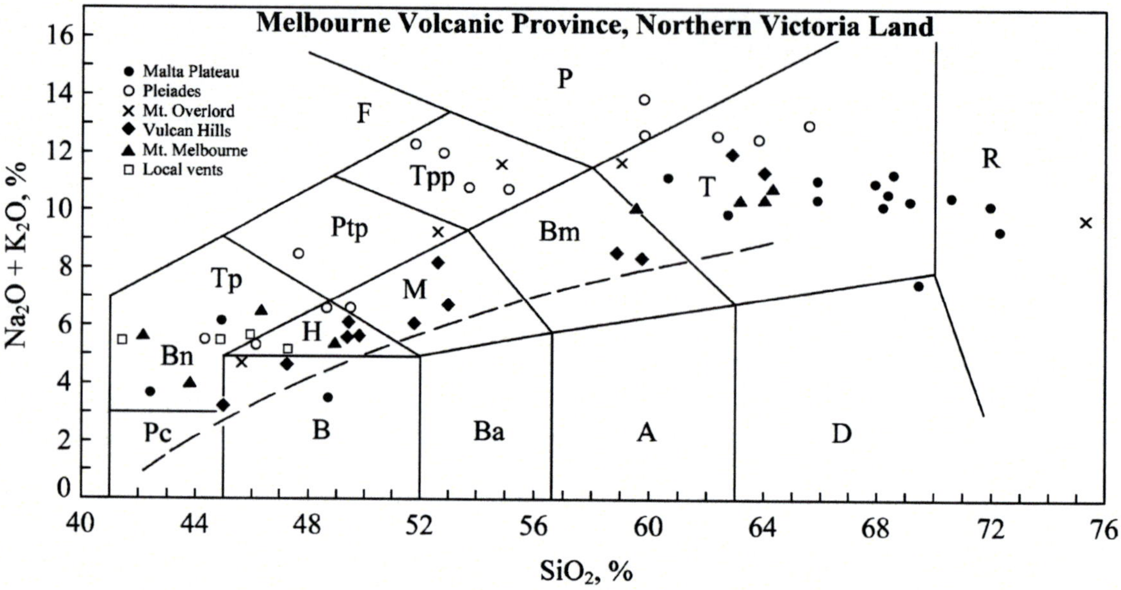

Fig. 16.26 The chemical compositions of the McMurdo Volcanics in the Melbourne volcanic province of northern Victoria Land range widely from basanites to phonolites, trachytes, and even rhyolites. Most of the volcanoes in this province include the full range of chemical compositions of lava flows. Bn = basanite, Tp = tephrite, Ptp = phonotephrite, Tpp = tephriphonolite, P = phonolite, F = foidite, H = hawaiite, M = mugearite, Bm = benmoreite, T = trachyte, R = rhyolite, Pc = picrobasalt, B = basalt, Ba = basaltic andesite, A = andesite, and D = dacite. The dashed line is the alkaline-subalkaline boundary of Macdonald and Katsura (1964) (Data from LeMasurier and Thomson (1990))

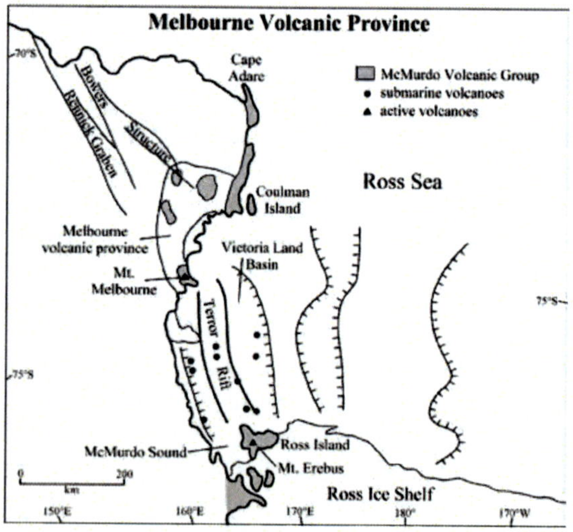

Fig. 16.27 Mounts Melbourne and Erebus are located on opposite ends of the Terror rift which is a local manifestation of the West Antarctic rift system. The summit of Mt. Melbourne displays fumaroles and high heat flow indicating that this volcano is not yet extinct. Mt. Erebus still maintains a lava lake in its Inner Crater. Small submarine vents have been detected by geophysical surveys. The alignment of Mt. Melbourne and Mt. Erebus and the presence of submarine volcanic vents within the Victoria Land Basin emphasize that the Cenozoic volcanic activity in the Transantarctic Mountains of Victoria Land is closely related to the tectonics of this region (Adapted from Kyle (1990) in LeMasurier and Thomson (1990))

(Delisle and Thierbach 1989). Nathan and Schulte (1967) also reported the smell of hydrogen sulfide within the ice pinnacles and noted the presence of small deposits of yellow salt which they considered to be native sulfur.

The temperature of the soil in the warm ground on the summit of Mt. Melbourne was measured in December of 1972 by Lyon and Giggenbach (1974) and in January of 1983 by Keys et al. (1983) who reported temperatures up to 59°C. Subsequently, Delisle and Thierbach (1989) carried out a thorough geothermal survey of the summit of Mt. Melbourne. Keys et al. (1983) observed that patches of moss are growing in the warm soil, indicating that the soil temperatures have remained constant for decades. Broady et al. (1987) later found several species of algae and one species of liverworts in addition to moss in the

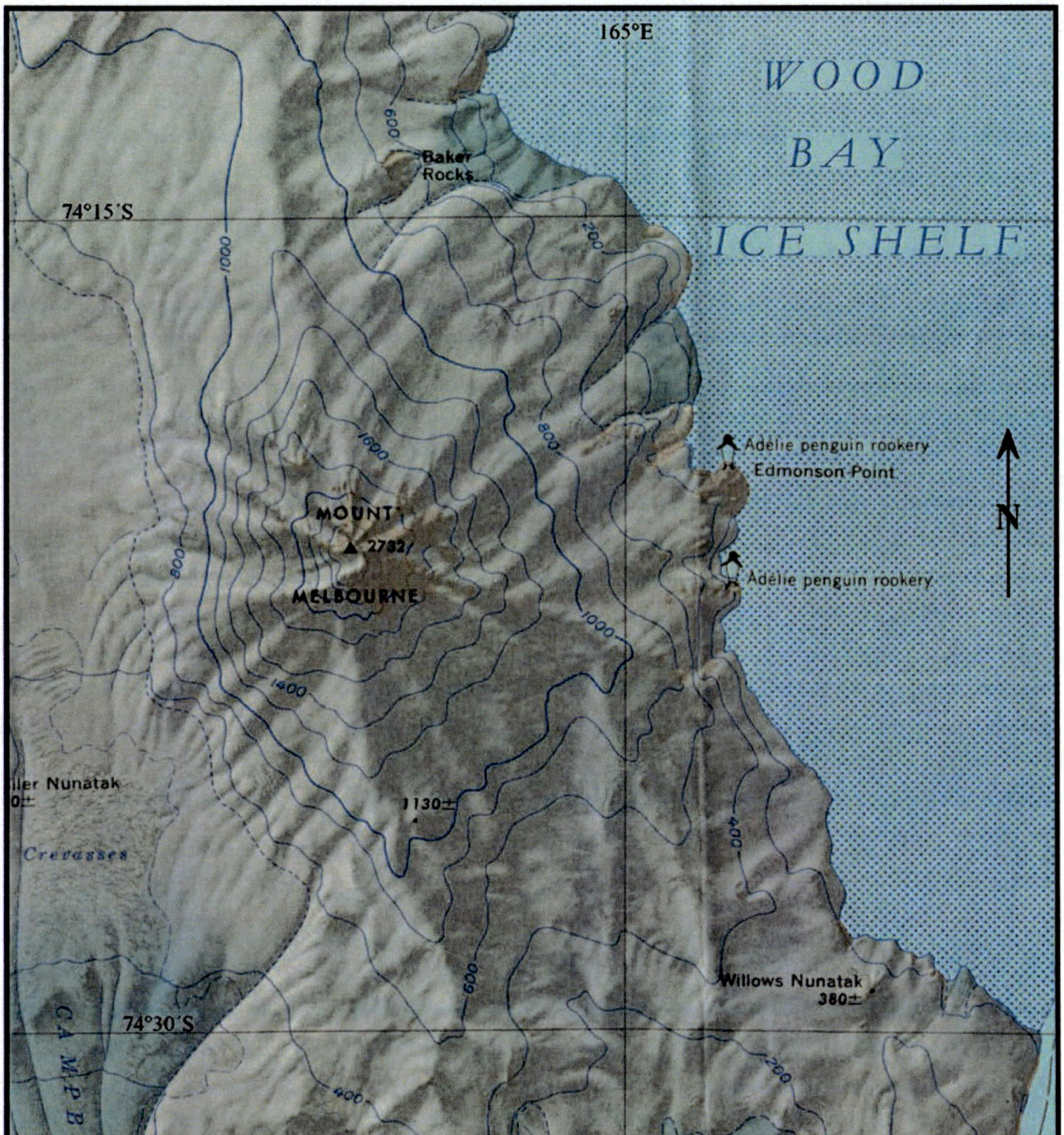

Fig. 16.28 Mount Melbourne on the Ross-Sea coast of northern Victoria Land is a stratovolcano composed of lava flows and pyroclastic deposits. The entire peninsula included in this map is underlain by volcanic rocks of the Melbourne volcanic field. This area contains several satellite centers of volcanic activity including Shield Nunatak at the mouth of the Campbell Glacier, Oscar Point, Cape Washington Ridge, Edmonson Point, and Baker Rocks. All of the volcanic rocks on Mt. Melbourne and at the associated centers are alkali-rich and range in composition from basanite to trachyte. Excerpt of the topographic map: Mount Melbourne, Antarctica, SS58–60/9. Published in 1968 by the US Geological Survey

warm ground on Mt. Melbourne. Keys et al. (1983) also confirmed the presence of volcanic ash interbedded with snow at a site about 2 km east of the summit caldera and published a photograph of this deposit. The deposit consists of lapilli tuff and contains black obsidian chips.

The volcanic rocks on Mt. Melbourne collected by Nathan and Schulte (1967) have the appearance of black vesicular basalt composed of plagioclase, pyroxene, olivine, and opaque minerals in a glassy matrix. However, the presence of alkali feldspar and of greenish clinopyroxene suggested that it is a trachyandesite. Numerous chemical analyses of volcanic rocks reported by Wörner and Viereck (1990) in Fig. 16.26 demonstrate that the rocks range from basanite to tephrite, hawaiite, and trachyte. In addition, Hornig et al. (1991) described inclusions of granulites from the lower crust, of mantle-derived lherzolite and harzburgites, and of tholeiitic cumulates genetically related to the lava flows in which they occur.

The concentrations of alkali metals, CaO, and MgO of the volcanic rocks that contain less than 48% SiO_2 vary *irregularly* in Fig. 16.29; but the concentrations of these elements vary *smoothly* in rocks that contain more than 48% of SiO_2. The smooth patterns of variation in silica-rich rocks are consistent with fractional crystallization of the magma, whereas the irregular variations of silica-undersaturated rocks may reflect variations in the extent of partial melting of the magma sources.

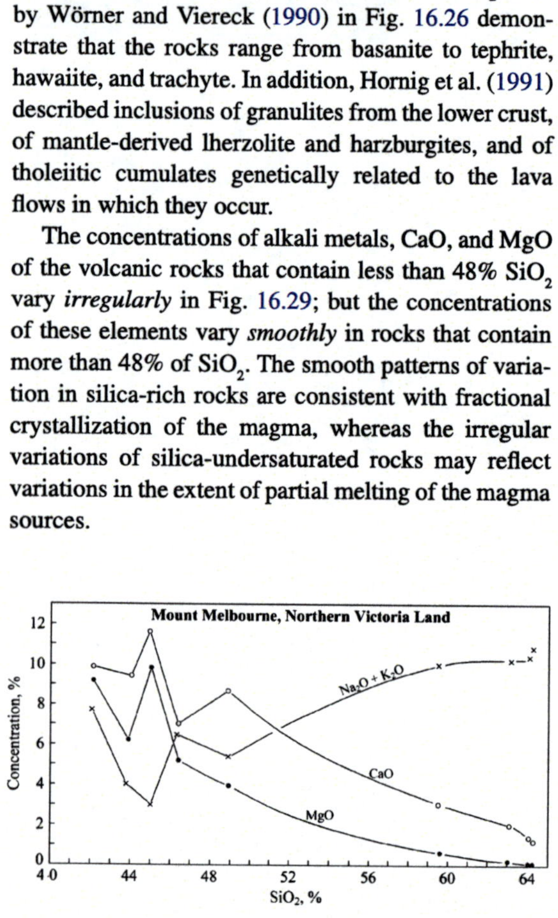

Fig. 16.29 The concentrations of $Na_2O + K_2O$, MgO, and CaO of rocks containing less than 48% SiO_2 vary irregularly perhaps because their chemical compositions reflect varying degrees for partial melting of the magma sources. The concentrations of the same elements in rocks containing more than 48% SiO_2 vary regularly with increasing silica concentrations consistent with fractional crystallization (Data from Wörner and Viereck (1990))

Wörner and Viereck (1990) noted that the $^{87}Sr/^{86}Sr$ ratios of the rocks range from 0.7029 to 0.7044 and suggested that magma was derived from different sources in the subcrustal mantle and/or that some magmas were contaminated. The comparatively low alkali-enrichment of the rocks of Mt. Melbourne implies a higher degree of partial melting than occurred in the magma sources of Mt. Erebus.

The internal activity of Mt. Melbourne is now being studied by Italian geophysicists who established a volcanological observatory on the mountain in 1988 (Armadillo et al. 2001). The results indicate that the east side of the volcano is affected by strike-slip motion and related transtension. Local seismic activity is also concentrated along the eastern flank and tilt meters indicate that the volcanic cone is tectonically active.

16.3.2 Petrogenesis (Mount Melbourne)

The origin of the volcanic rocks of the Melbourne volcanic field was investigated by Wörner and Viereck (1989) and by Wörner et al. (1989) by mapping the rocks in the field and by study in the laboratory. Wörner and Viereck (1989, p. 369) reported that the rocks consists of:

> …hyaloclastic mass flows, phreatomagmatic tuff ring deposits, auto-brecciated hydroclastic lava flows, flows invading water and wet sediment and one unwelded basaltic ignimbrite.

Lava lows that were erupted subaqueously formed pillows and pillow breccias, while scoria cones and pahoehoe flows are typical of subaerial eruptions. An older sequence of subaerially erupted flows is disconformably overlain by thick subglacial hyaloclastic deposits topped by subaerial tuff rings, scoria cones, and lava flows. These rock formations in turn are overlain by Holocene deposits formed by plinian, strombolian, and vulcanian eruptions.

Wörner et al. (1989) reported chemical analyses for 92 samples of volcanic rocks from ten subdivisions of the Melbourne volcanic field. The analyses included major and trace elements and the resulting normative mineral compositions. Without classifying each one of these specimens, it is sufficient to record that their chemical compositions range widely just as they do in the Erebus volcanic province of southern Victoria Land (Fig. 16.4). Wörner et al. (1989) also reported isotope compositions of strontium (17) and neodymium (13) of rock samples for which chemical analyses are available. These isotope analyses are listed in Appendix 16.8.5.

The measured $^{87}Sr/^{86}Sr$ and $^{143}Nd/^{144}Nd$ ratios primarily occupy quadrant 2 in the Sr-Nd isotopic mixing diagram in Fig. 16.30. This quadrant is the locus of rocks that originated from the mantle of the Earth. The data points loosely constrain a linear array which is a short segment of a mixing hyperbola between Sr and Nd in Mid-Ocean Ridge Basalt (MORB) and other components in the mantle that can contribute to the formation of magma. These components were defined by Zindler and Hart (1986) and Hart (1988) and are designated by the abbreviations (Faure 2001, p. 58):

MORB = mid-ocean ridge basalt
DMM = depleted MORB mantle
EM1 = enriched mantle 1
EM2 = enriched mantle 2, and
HIMU = mantle having a high $^{238}U/^{204}Pb$ ratio

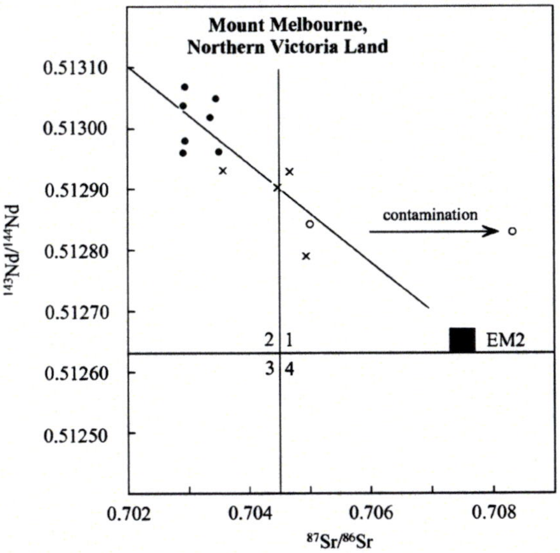

Fig. 16.30 The volcanic rocks of the Melbourne volcanic field loosely define a linear array in quadrant 2 of the Sr-Nd isotopic mixing diagram. The *solid circles* are alkali basalt and basanite, the crosses are mugearites and other rocks of intermediate alkali concentration, the *open circles* are trachytes. The data array is collinear with the EM2 mantle component which consists of crustal rocks that were subducted into the mantle (Hart 1988). One trachyte sample was contaminated with ^{87}Sr. The placement of the data points in quadrant 2 indicates that the magma originated by varying (but small) degrees of partial melting of mixed source rocks in the mantle (Data from Wörner et al. (1989))

The data array in Fig. 16.30 is collinear with EM 2 ($^{87}Sr/^{86}Sr = 0.7075$, $^{143}Nd/^{144}Nd = 0.51265$, $^{206}Pb/^{204}Pb = 19.2$). This component consists of continental rocks that were subducted into the mantle. The collinearity of the isotope ratios of strontium and neodymium of the volcanic rocks from the Melbourne field with EM 2 indicates that the magmas formed from sources in the mantle that contained subducted crustal rocks.

Wörner et al. (1989) observed that the $^{87}Sr/^{86}Sr$ ratios of the volcanic rocks cluster into three groups in Fig. 16.31 depending on their chemical compositions. Alkali basalt and basanite have the lowest $^{87}Sr/^{86}Sr$ ratios (0.7025–0.7040), mugearites, benmoreites, and trachyandesites have higher $^{87}Sr/^{86}Sr$ ratios (0.7035–0.7050), and trachytes have the highest $^{87}Sr/^{86}Sr$ ratios

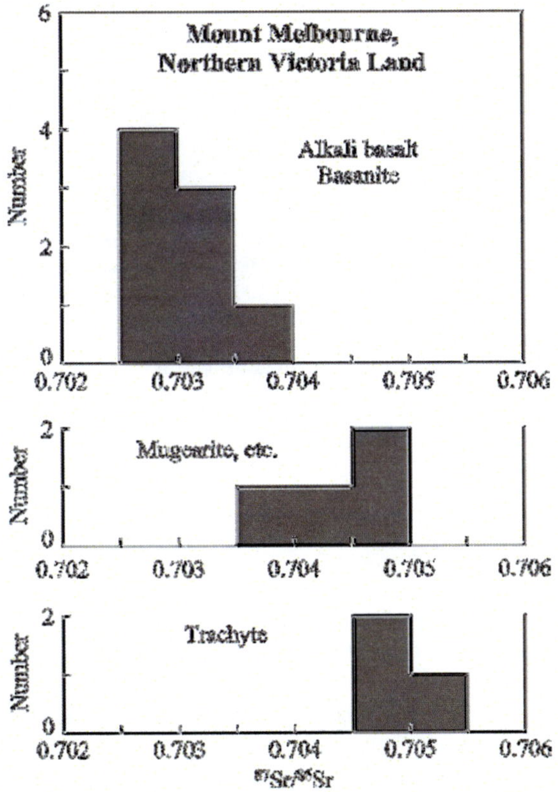

Fig. 16.31 The volcanic rocks of the Melbourne volcanic field form three groups based on their chemical compositions and $^{87}Sr/^{86}Sr$ ratios. One sample of trachyte ($^{87}Sr/^{86}Sr = 0.70829$) is not shown because it plots off scale. The apparent correlation of chemical and isotopic compositions is an important clue that these magmas originated from mixed sources or were contaminated by radiogenic ^{87}Sr of the continental crust (Data from Wörner et al. (1989))

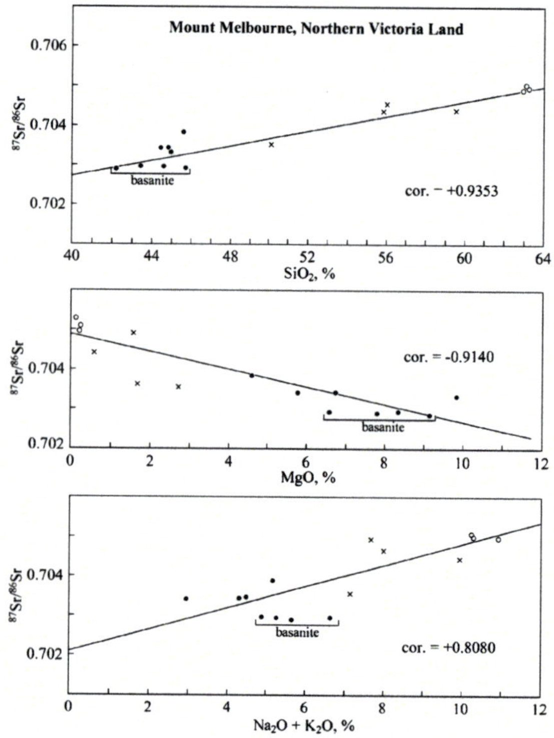

Fig. 16.32 The $^{87}Sr/^{86}Sr$ ratios of the volcanic rocks of the Melbourne volcanic field are strongly correlated with the concentrations of SiO_2, MgO, and $Na_2O + K_2O$. Such correlations cannot form by fractional crystallization of magma in a closed system and, instead, indicate that the magma was *contaminated* either by assimilating rocks from the continental crust or by forming from mixed source rocks in the mantle. The colinearity of data points in Fig. 16.30 with mantle component EM2 favors contamination at the magma source in the mantle (Data from Wörner et al. (1989))

(0.7045–0.7055) with one value of 0.70829 (off scale in Fig. 16.31). This anomalous trachyte from Edmonson Point deviates strongly from the linear data array in Fig. 16.30 because it was contaminated with ^{87}Sr from the continental crust.

The grouping of the rocks in the Melbourne volcanic field in Fig. 16.31 based on their chemical compositions and $^{87}Sr/^{86}Sr$ ratios is expressed in Fig. 16.32 by the quasi-linear correlation of the isotopic and chemical parameters. For example, the $^{87}Sr/^{86}Sr$ ratios are positively correlated with the silica concentrations (cor = +0.9353). The $^{87}Sr/^{86}Sr$ ratios also *decrease* with rising concentrations of MgO (cor = –0.9140), and they *increase* with increasing concentrations of alkali metals (cor = +0.8080). The correlation of the $^{87}Sr/^{86}Sr$ ratios with the chemical compositions of the rocks is evidence in favor of contamination of basanite magma either at the source or by assimilation of rocks of the continental crust.

Wörner et al. (1989) regarded the alkali basalt and basanites as the parental magmas and concluded that the crystallization and segregation of the phenocryst minerals (olivine, plagioclase, clinopyroxene, magnetite, apatite, and sanidine) could have caused the observed differentiation of the volcanic rocks in the Melbourne volcanic field. However, the wide range of isotopic compositions of these volcanic rocks cannot be caused by fractional crystallization *alone* because this process does not enrich the rocks in radiogenic ^{87}Sr nor does it deplete them in radiogenic ^{143}Nd. For example, the trachytes in Fig. 16.32 cannot have formed from either the alkali basalt or the basanite magma by fractional crystallization because the isotope ratios of strontium and neodymium of the trachytes differ significantly from those of the hypothetical parent magmas.

The collinearity of the Melbourne volcanic rocks with EM 2 in Fig. 16.30 suggests that the lavas formed by different degrees of partial melting of heterogeneous mantle rocks containing varying proportions of subducted crustal rocks. According to this hypothesis, the basanites formed by a larger degree of partial melting of nearly pure depleted mantle than the trachytes which originated by a small degree of partial melting of source rocks that included a subducted crustal component. The abundance of volcanic glass and the evidence of explosive eruptions indicate that the magmas moved rapidly toward the surface and did not differentiate during slow cooling in magma chambers. Therefore, we conclude reluctantly that the petrogenesis of the volcanic rocks of the Melbourne volcanic field remains unsettled. The geochemical evolution of magmas derived from a stratified source in the mantle beneath the Ross Sea was investigated by Rocholl et al. (1995).

16.3.3 The Pleiades

The Pleiades in Fig. 16.25 consist of a cluster of volcanic cones of late Cenozoic age located on the eastern edge of the Evans Névé and at the head of the Mariner Glacier. This location makes The Pleiades the most northerly volcanoes of the Melbourne volcanic province and places them 140 km west of the coast of northern Victoria Land. The location of The Pleiades

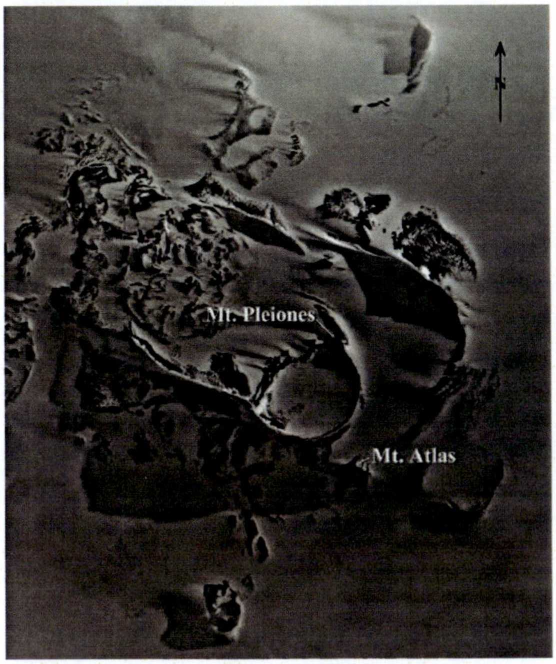

Fig. 16.33 Vertical view of Mt. Pleiones and Mt. Atlas in The Pleiades volcanic center adjacent to the Evans Névé in northern Victoria Land. The Taygete cone (not shown) has erupted as recently as 3,000 years ago (Data from Kyle (1990d); US Navy photograph TMA 1379 F32 279)

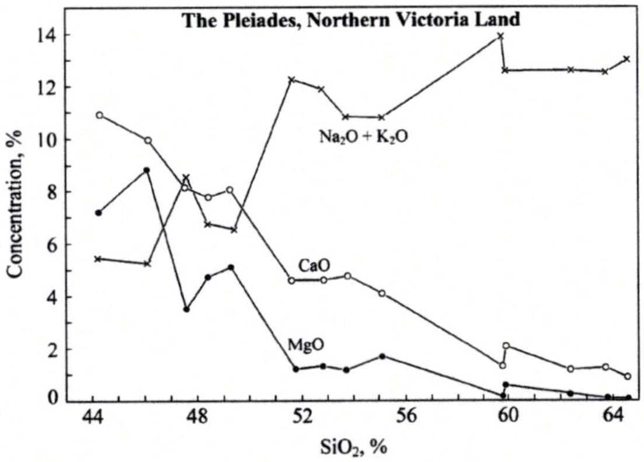

Fig. 16.34 The concentrations of MgO, CaO, and total alkalies vary irregularly with increasing concentrations of SiO_2 from about 44% to 65%. The concentrations of MgO and CaO decrease to very low values (MgO = 0.07%, CaO = 0.94%) at SiO_2 = 64.56% while the alkali elements rise to remarkably high values ($Na_2O + K_2O$ = 13.94%) at SiO_2 = 59.79%. The variation of the concentrations of these elements is more irregular than expected of magma that differentiated by fractional crystallization (Data by Kyle (1990d) in LeMasurier and Thomson (1990))

therefore deviates from the pattern of the coastal volcanoes that lie north of Mt. Melbourne and south of Cape Hallett.

The highest volcanic cones in The Pleiades are Mt. Atlas (~3,040 m) and Mt. Pleiones (~3,020 m) which project about 500 m above the surface of the Evans Névé in Fig. 16.33. Other noteworthy topographic features are Alcyone and Taygete cones located about 3.5 and 6.0 km north of Mt. Atlas, respectively.

The Pleiades were mentioned by Gair (1967) in his description of the geology of the upper Rennick Glacier area and they appear on the geologic map prepared by Gair et al. (1969). The Pleiades were also mentioned by Nathan and Schulte (1968) and by Riddolls and Hancox (1968) in their description of the regional geology. In addition, Kyle and Rankin (1976) included rock samples from The Pleiades in their study of the rare-earth geochemistry of the McMurdo Volcanic Group (Kyle and Rankin 1976). More recently, Kyle (1990d) presented a summary of the salient features of the geology of these volcanoes based on fieldwork in 1971/72 and 1981/82.

The chemical compositions of volcanic rocks from The Pleiades (Kyle 1990d) place them into the strongly alkalic basanite-tephrite-phonotephrite-phonolite suite in Fig. 16.26. In addition, Kyle (1990d) recognized a less alkali-rich suite of hawaiites, mugearites, benmoreites, and trachytes. The lavas on Mt. Pleiones contain xenoliths composed upper crustal granitoids as well cognate inclusions of syenite and essexite (i.e., feldspathoidal gabbro). The concentrations of MgO, CaO, and total alkali metals of volcanic rocks in The Pleiades in Fig. 16.34 vary irregularly with increasing silica concentrations (Kyle 1990d). The concentrations of a larger set of chemical analyses published by Kyle (1982, 1986) vary even more irregularly.

The K-Ar dates measured by Armstrong (1978) indicate that the volcanoes of The Pleiades erupted as recently as 3,000 years ago (Kyle 1990d). The low K-Ar dates of Armstrong (1978) were later confirmed by the $^{40}Ar/^{39}Ar$ step-wise release dates of Esser and Kyle (2002) which range from 847,000 to 6,000 years. These dates suggest that the volcanic eruptions of The Pleiades intensified about 100,000 years ago and that the cone of Mt. Atlas was formed about 65,000 years ago. At 45,000 years in the past, lava flows were erupted on Mt. Pleiones and near the summit Alcyone. The most recent activity occurred 6,000 years ago (i.e., 6 ± 6 ka) at Taygete Cone which erupted trachyte flows. The evidence for Holocene volcanic activity is confirmed by fresh hydrothermal alteration and by the deposition of pumice lapilli on the surface of parts of The Pleiades.

Lavas from the other volcanoes in the Melbourne volcanic province are also classified in Fig. 16.26 based on chemical analyses published by Schmidt-Thomé et al. (1990) for the Malta Plateau, by Kyle and Knoll (1990) for Mt. Overlord, by Kyle (1990e) for the Vulcan Hills, and by Kyle (1990f) for isolated volcanic vents throughout northern Victoria Land.

16.4 Hallett Volcanic Province

The volcanoes along the Ross-Sea coast of northern Victoria Land collectively form the Hallett volcanic province in Fig. 16.35. This province includes the Adare, Hallett, and Daniell peninsulas, and Coulman Island. Volcanic rocks of Cenozoic age also occur on the Possession Islands and at a few vents on the mainland west of the major volcanic complexes named above (McIntosh and Kyle 1990a).

16.4.1 Adare Peninsula

The Adare Peninsula was first visited by Carsten Borchgrevink and his men who wintered over in two huts on Ridley Beach (Section 1.3). In 1911/12 the huts were used again by Scott(s Northern Party led by Victor Campbell (Section 1.4.3). The remains of these huts still stand on Ridley Beach (McIntosh and Kyle 1990b). Descriptions of the geology of the Adare Peninsula and of the petrology of the volcanic rocks were published by Harrington et al. (1967), Hamilton (1967, 1972), Jordan (1981), and by LeMasurier et al. (1983).

Armstrong (1978) reported a K-Ar date of 2.27 ± 0.5 Ma corrected by McIntosh and Kyle (1990a) to the new decay constants of Steiger and Jäger (1977) using

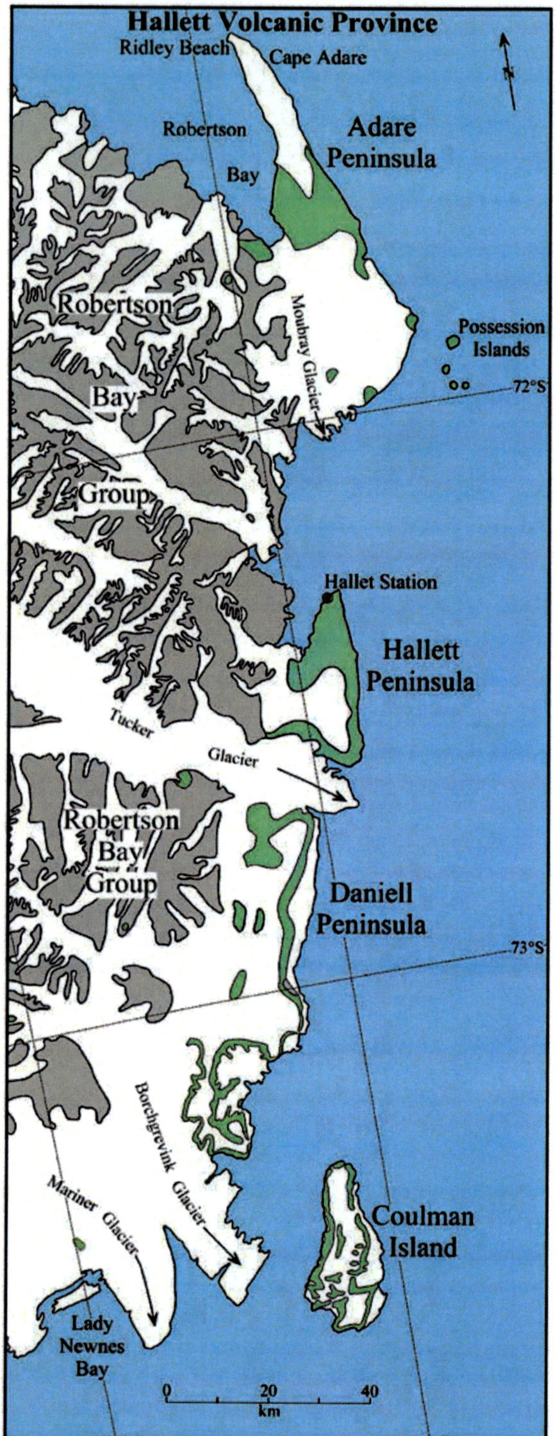

Fig. 16.35 The Hallett volcanic province consists of five regions along the Ross-Sea coast of northern Victoria Land: Adare Peninsula, Hallett Peninsula, Daniell Peninsula, Coulman Island, and the Possession Islands. The Cenozoic volcanic rocks in these areas are located adjacent to and may have been deposited on top of the low-grade metasedimentary rocks of the Robertson Bay Group of early Paleozoic age (Section 4.5) (Adapted from Gair et al. (1969))

the tables of Dalrymple (1979). Kreuzer et al. (1981) also published K-Ar dates of volcanic rocks collected from the western coast of the Adare Peninsula near the southern end of Robertson Bay. The dates ranged from 6.77 ± 0.05 to 13.24 ± 0.12 Ma (middle to late Miocene) with one anomalously low date of 1.14 ± 0.05 Ma.

The total-alkali and silica concentrations of three rock samples from the Adare Peninsula are classified in Fig. 16.36 as a basanite, a phonolite, and a trachyte (McIntosh and Kyle 1990b). Additional chemical analyses by Hamilton (1972), Smith (1964), David et al. (1896), and unpublished data by P.R. Kyle and W.C. McIntosh demonstrate a continuous spectrum of rock types on the Adare Peninsula from $SiO_2 = 42\%$ to 67%. The presence of large volumes of palagonitic tuff breccias and pillow breccias on the Adare Peninsula and elsewhere in the Hallett volcanic province led Hamilton (1972) to conclude that these rocks were erupted through ice at a time when the East Antarctic ice sheet had expanded. This interpretation was supported by the presence of abundant glacial erratics on the northern end of the Adare Peninsula (Priestley 1923).

An alterative explanation proposed by LeMasurier et al. (1983) and by McIntosh and Gamble (1991) is that the volcaniclastic rocks are subaerially erupted autoclastic breccias within unbrecciated lava flows. The subaerial eruption of lava *flows* was supported by several other geologists who also worked on the Adare Peninsula. For example, Jordan (1981) found a thick section (1,500 m) of subaerially erupted lava flows including reddish-brown hematite-stained scoria on the west coast of the Adare Peninsula. These flows are underlain by a basal unit (150–200 m) of palagonite breccia. LeMasurier et al. (1983) reported that crystalline lava flows interbedded with tuff occur along the *east coast* of the Adare Peninsula. They also noted the common occurrence of red and black cinders and volcanic bombs and the presence of extensive yellow and green products of hydrothermal alteration and of autobrecciated lava flows which were fragmented by the deformation of viscous lava.

Apparently, the volcanic rocks on the Adare Peninsula were erupted *subaerially* rather than *subglacially* although the basal flows may have come in contract with snow and meltwater in a periglacial environment. After the main phase of the eruption, the East Antarctic ice sheet probably did advance from the west and deposited the glacial erractics that cover the northern part of the Adare Peninsula.

16.4.2 Hallett Peninsula

The Hallett Peninsula in Fig. 16.35 contains several overlapping shield volcanoes which are covered by snow and ice. Rock outcrops occur primarily along the periphery of the Peninsula in the form of cliffs which are up to 1,400 m high (McIntosh and Kyle 1990c).

Hallett Station in Fig. 16.37 was used by the USA for research support during the IGY (Section 1.5.3). The station was transferred to joint US and New Zealand management in 1959; but the principal laboratory buildings burned down in 1964. The station was subsequently refurbished in1984 and is used intermittently by American and New Zealand scientists.

Chemical analyses in Fig. 16.36 classify the volcanic rocks of the Hallett Peninsula into the mildly alkaline basanite-hawaiite-mugearite-trachyte suite (McIntosh and Kyle 1990c). These authors also stated that the most alkali-rich rocks occur in the vicinity of volcanic vents where hydrothermal alteration is intense and that the evolved rock types (e.g., the trachytes) probably comprise less than 10% of the total volume of rocks on the Hallett Peninsula.

Armstrong (1978) reported two K-Ar dates of 5.49 ± 0.12 Ma and 6.6 ± 0.4 Ma (late Miocene) for rocks of the Hallett Peninsula. Three samples of basalt from Cape Hallett analyzed by Jones and Walker (1972) have $^{87}Sr/^{86}Sr$ ratios that range from 0.7037 to 0.7054 and are consistent with the $^{87}Sr/^{86}Sr$ ratios of the volcanic rocks of the Erebus and Melbourne volcanic provinces.

16.4.3 Daniell Peninsula

The Daniell Peninsula is located south of the Hallett Peninsula between the outflow of the Tucker Glacier in the north and the Borchgrevink Glacier at its southern end (Fig. 16.35). Outcrops of volcanic rocks occur primarily along the coast and along the sides of the two glaciers that define the Daniell Peninsula. The geology of this peninsula was described by Hamilton (1972) who visited six sites, two of which were later examined by LeMasurier et al. (1983). A summary of the geology of the Daniell Peninsula was prepared by McIntosh and Kyle (1990d).

The Daniell Peninsula has two summits formed by volcanic vents that are located along a rift that extend

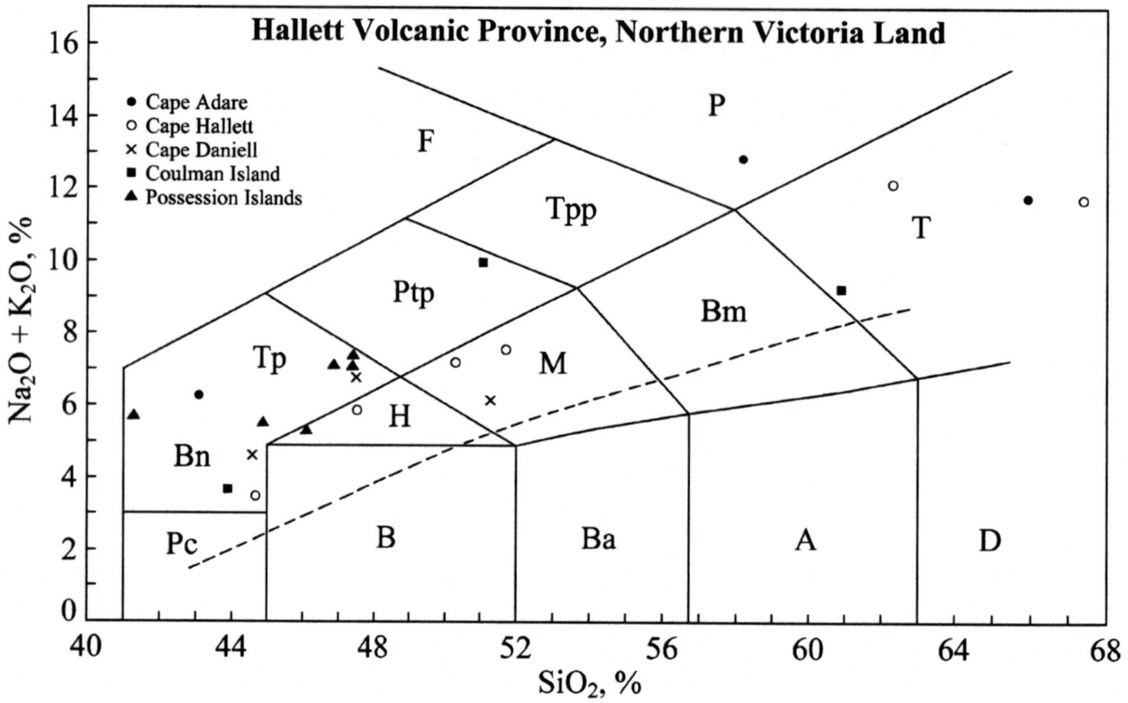

Fig. 16.36 The volcanic rocks in the Hallett volcanic province range from basanites to phonolites and trachytes. The principal volcanic centers represented on this diagram are Cape Adare, Cape Hallett, Cape Daniell, Coulman Island, and the Possession Islands. The rock types are identified by the letters in each field: Bn = basanite, Tp = tephrite, Ptp = phonotephrite, Tpp = tephriphonolite, P = phonolite, F = foidite, H = hawaiite, M = mugearite, Bm = benmoreite, T = trachyte, R = rhyolite, Pc = picrobasalt, B = basalt, Ba = basaltic andesite, A = andesite, and D = dacite (Data from McIntosh and Kyle (1990b, c, d, e, f) in LeMasurier and Thomson (1990))

north-to-south along the length of the Peninsula. One of these summits, Mt. Brewster (2,026 m), has primarily erupted lava flows of basanite and hawaiite. Outcrops in the northern part of the Peninsula along the Tucker Glacier consist of highly differentiated lavas (basanite to phonolite) and layers of tuff that were hydrothermally altered and intruded by dikes (McIntosh and Kyle 1990d). Three chemical analyses of rocks from the Daniell Peninsula (crosses) are identified in Fig. 16.36 as basanite, tephrite, and mugearite. The age of these rocks is Cenozoic (Miocene), but no age determinations have been published.

16.4.4 Coulman Island

Coulman Island in Fig. 16.35 is the most southerly volcanic center of the Hallet volcanic province and is located in the Ross Sea opposite the ice tongue of the Borchgrevink Glacier. The island is 35 km long and 15 km wide at its southern end which contains the Coulman caldera at 1,998 m. The caldera is about 5 km wide and at least 700 m deep. The lava flows that crop out north of the caldera are primarily basanite, basalt, and hawaiite. The rocks east of the caldera consist of a typical vent assemblage of hydrothermally altered alkali-rich lava flows (basanite to phonolite) and layers of tuff, all of which were intruded by dikes (McIntosh and Kyle 1990e).

The coast of Coulman Island is a steep cliff, 600–1,000 m high, consisting predominantly of lava flows, autoclastic breccias, and tuff. The breccias are holocrystalline and stained red by oxidation of iron which indicates that they were subaerially erupted. The topography and geology of Coulman Island were described by Harrington et al. (1967), Hamilton (1972), and McIntosh and Gamble (1989) following fieldwork by W.C. McIntosh in February of 1986.

The chemical compositions of three rock samples from Coulman Island are identified in Fig. 16.36 as

Fig. 16.37 Hallet Station on the northern tip of the Hallett Peninsula was refurbished in 1984 after a fire destroyed the principal laboratory buildings. The station is now administered by the USA and New Zealand and is used intermittently (Photo by Eugene Smith)

basanite, phonotephrite, and trachyte. Additional analyses compiled by McIntosh and Kyle (1990d) indicate that the silica concentrations of volcanic rocks on Coulman Island range from less than 45% to about 63% without significant gaps. The K-Ar date of a lava flow on the west side of Coulman Island analyzed by Armstrong (1978) is 7.2 ± 0.5 Ma based on the new constants of Steiger and Jäger (1977).

16.4.5 Possession Islands

The Possession Islands, between 71°53′S, 171°14′E and 71°58′S, 171°07′E, in Fig. 16.35 are located 8 km off the coast of Cape McCormick and Cape Roget on the southern end of the Adare Peninsula. They consist of a group of nine small islands and sea stacks composed of Cenozoic alkali-rich volcanic rocks described by Harrington et al. (1967), Waterhouse (1965), Hamilton (1972), and McIntosh and Kyle (1990f). In spite of their small size, all nine islands in Fig. 16.38 have been named from north to south: Possession Island, Dickson Pillar, Kristensen Rocks, Favreau Pillar, Foyn Island, Kemp Rock, Bull Island, and Heftye Island.

Possession (2.2 × 1.7 km) and Foyn (2.0 × 1.5 km, elevation 101 m) are the two largest islands followed by Bull and Heftey islands. The others are isolated pillar-like rocky islands detached from the larger islands by wave erosion augmented by weathering (i.e., they are sea stacks).

The Possession Islands were first visited by Captains James Ross and R.M. Crozier on January 11, 1841 (Section 1.2). Captain Ross dedicated the Possession Islands to Queen Victoria before continuing south along the coast of what is now Victoria Land until he reached a large island on January 24, 1841, later named

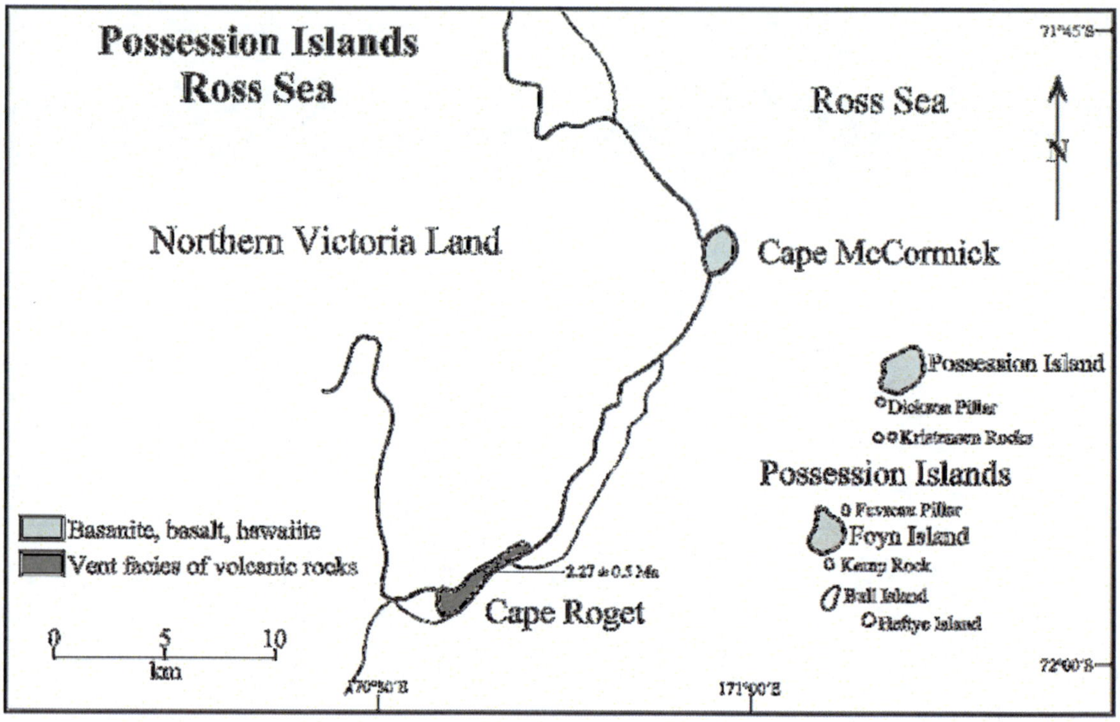

Fig. 16.38 The Possession Islands in the Ross Sea off the coast of northern Victoria Land were discovered by James Ross and R.M. Crozier on January 11, 1841, when Captain Ross took possession of the islands in the name of Queen Victoria. The Possession Islands are composed of basanite, basalt, and hawaiite which are less altered than the Cenozoic volcanic rocks on the mainland (Adapted from McIntosh and Kyle (1990b) in LeMasurier and Thomson (1990))

Ross Island. Captain Ross named the two volcanoes on Ross Island after the two small ships in his fleet (Victor 1963).

The volcanic rocks on Possession and Foyn islands in Fig. 16.36 are basanites and tephrites in the form of lava flows, volcanic bombs, and megapillows. McIntosh and Kyle (1990f) commented that the rocks of the Possession Islands are least altered of all the rocks in the Hallett volcanic province that have been analyzed.

16.5 The Balleny, Scott, and Peter I Islands

The Balleny Islands in Fig. 16.39 are located in the southern Pacific Ocean about 250 km north of the Pennell Coast of northern Victoria Land between 66°16′ to 67°38′S and 162°15′ to 164°44′E. They form a linear chain of volcanic islands between East Antarctica and Australia and resemble in this regard the volcanic island chains that characterize the South Pacific Ocean (Faure 2001).

Scott Island is a large submerged free-standing volcanic mountain that rises from the seafloor to a height of about 100 m above the surface of the ocean (Wörner and Orsi 1992). It is located at 67°30′S and 180°00′ about 500 km east of the Balleny Islands.

Peter I Island is a large shield volcano in the Bellingshausen Sea at 68°51′S and 90°35′W and about 280 km north of the coast of Ellsworth Land in West Antarctica. Its principal claim to fame is that it was discovered by Captain Thaddeus von Bellingshausen on January 10, 1821, during his circumnavigation of Antarctica.

All of these islands are Cenozoic volcanoes that erupted silica undersaturated and alkali-rich volcanic rocks that superficially resemble the volcanic rocks of the West Antarctica rift valley along the coast of Victoria Land and in Marie Byrd Land of West Antarctica. The existence of these volcanoes in the oceans close to Antarctica raises a question about

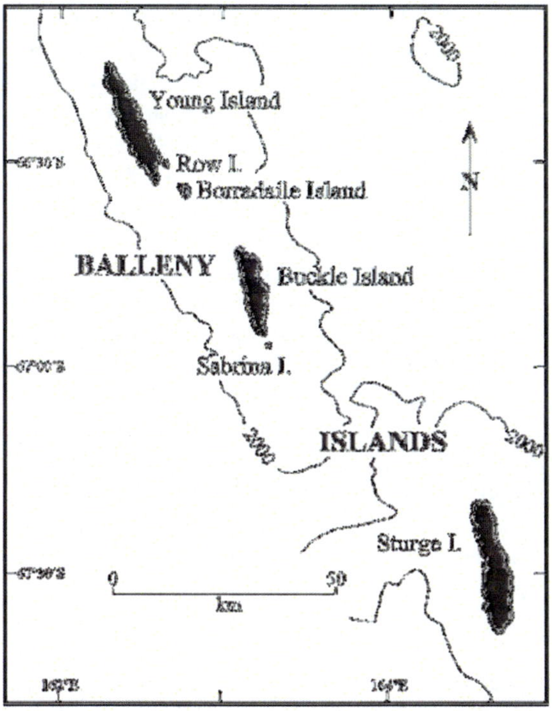

Fig. 16.39 The Balleny Islands are located about 250 km north of the Pennell Coast of northern Victoria Land. They are the summits of volcanoes on a submarine ridge that is aligned with the Tasman Fracture Zone located north of the Pacific-Antarctic sea-floor spreading ridge. Volcanic eruptions on Buckle Island were reported in 1839 and 1899. However, an infrared survey published in 1969 did not detect any thermal anomalies (Burge and Parker 1969) (Adapted from Wright and Kyle (1990a) in LeMasurier and Thomson (1990))

possible tectonic links to the Cenozoic volcanoes of the mainland of Antarctica.

16.5.1 Balleny Islands

The Balleny Islands in Fig. 16.39 consist of five named islands which are the summits of volcanoes on a submarine ridge that trends northwest and is aligned with the Tasman Fracture zone north of the Pacific-Antarctic sea-floor spreading ridge (Falconer 1972; Kyle and Cole 1974). The Balleny Islands extend for about 160 km and consists of: Young, Row, Borradaile, Buckle, Sabrina, and Sturge islands. All of them are composed of lava flows interbedded with layers of tuff and agglomerate scoria of probable late Cenozoic age (Wright and Kyle 1990a). The geologic history and topography of the Balleny Islands was also summarized by Quartermain (1964).

The Balleny Islands were discovered by J. Balleny on February 9, 1839. The coasts of the islands consist of steep cliffs and the interiors are covered by ice caps. Quartermain (1964) reported that volcanoes were erupting on Buckle Islands in 1839 and 1899; but a survey of infrared radiation by Burge and Parker (1969) did not detect thermal anomalies on any of the islands.

Petrographic studies by Mawson (1950) and Kyle (1976) identified primarily olivine trachybasalt and lesser amounts of olivine-augite basalt, whereas Waterhouse (1965) reported finding sugar-textured quartz xenoliths on Sabrina Island. The chemical analyses of five rock samples compiled by Wright and Kyle (1990a) identify four basanite and one hawaiite (Mawson 1950; Kempe 1973; Johnson et al. 1982). Deep-sea cores in the southern Pacific contain rhyolitic dust deposited at 2.5 Ma which probably originated from explosive volcanic eruptions on the Balleny Island (Huang et al. 1975). In general, the chemical compositions of the volcanic rocks on the Balleny Islands resemble the volcanic rocks of Hallett volcanic province (Wright and Kyle 1990a).

The Balleny Islands are frequently surrounded by pack ice until the end of the summer and, because of their remote location, they have been visited only rarely and no systematic geological field work has been done there.

16.5.2 Scott Island

Scott Island is only about 3 km wide and is bounded by a steep cliff (Wright and Kyle 1990b). The only landing was made in December 1992 by the Morning Expedition of Captain W. Colbeck. The rock sample that was collected on this occasion was later identified as a phonolite by Prior (1907) consistent with a chemical analyses that was included in the report of Wright and Kyle (1990b).

The chemical analyses of five dredge samples (Johnson et al. 1982) indicate that four of the analyzed samples are phonolites and one is mugearite (Campsie et al. 1983). The high concentrations of alkali metals (Na_2O + K_2O: 8.83–13.11%) and of silica (SiO_2: 50.93–58.76%) suggest that these rocks evolved by fractional crystallization of basanite magma (Wright and Kyle 1990b).

16.5.3 Peter I Island

Peter I Island in Fig. 16.40 is 19 km long and 11 km wide and rises to an elevation of 1,750 m above sea level. It has a steep coastline and is extensively covered by ice. The base of the island is circular and has a diameter of 64 km at a depth of 3,660 m. Therefore, Peter I Island is a large volcanic mountain that rises to a height of 5,240 m above the bottom of the surrounding ocean (Rowley 1990).

After its discovery in 1821, Peter I Island has been visited only a few times for brief periods of time. In January of 1927 the Norwegian ship *Odd I* circumnavigated the island and dredged a few rock samples described by Broch (1927). The first landing was made on February 2, 1929, by a group from the *Norvegia* commanded by Captain Nils Larsen (Holtedahl 1929). The geology of Peter I Island was described in some detail by Bastien and Craddock (1976a) who reached it in February 1960 from the U.S.S. Burton Island and spent 2.5 h examining rocks. In January and February 1987 a group of Norwegian surveyors spent 10 days on the island and collected rock samples at five locations (Rowley 1990).

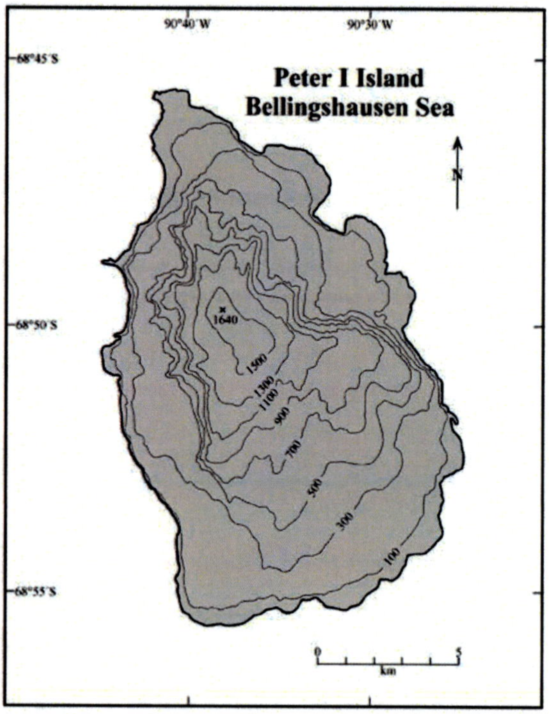

Fig. 16.40 Peter I Island was discovered by Captain Thaddeus von Bellingshausen on January 10, 1821, who circumnavigated Antarctica in two small ships named Vostok and Mirnyy (Section 1.1). The island is located in the Bellingshausen Sea about 280 km north of the coast of Ellsworth Land in West Antarctica. Peter I Island is a shield volcano that rises from the sea floor at 3,660 m below sea level to a summit at 1,750 m above sea level. Therefore, the island is a large volcanic mountain that rises 5,240 m from the bottom of the ocean. A K-Ar date of 12.8 ± 1.5 Ma reported by Bastien and Craddock (1976a) indicates that the lava flows on Peter I Island were erupted during the Miocene (Adapted from Rowley (1990) in LeMasurier and Thomson (1990))

The summit contains a small crater about 100 m wide. The volcanic rocks of Peter I Island consist of basalt and trachyte flows with minor amounts of volcanic breccia and lavas of intermediate composition. Some of the flows are highly vesicular. Pahoehoe structures have been reported, but no pillow lavas have been seen (i.e., the flows are subaerial). The lava sequences are locally cut by basalt dikes and plugs.

The basalt flows are gray to reddish brown, hypocrystalline to holocrystalline, and partly porphyritic with phenocrysts of olivine (9.8–17.2%), plagioclase (25.3–38.2%), and pyroxene in the matrix (30.0–41.6%). Chemical analyses of five rock samples were published by Broch (1927), Craddock et al. (1964), and Bastien and Craddock (1976a, b).

A K-Ar date of 12.8 ± 1.5 Ma (Miocene) of an olivine basalt collected near sea level on Peter I Island was reported by Bastien and Craddock (1976a). In addition, the lack of erosion of the volcanic cone supports the conclusion that Peter I Island formed during the late Cenozoic.

16.5.4 Petrogenesis (Balleny Islands)

The volcanic activity of the Balleny Islands and of Scott Island is related to their location between East Antarctica and Australia (Duncan and McDougall 1989; Lanyon et al. 1993, 1995). Partly for that reason, the $^{87}Sr/^{86}Sr$ ratios of the volcanic rocks on these islands are lower than the $^{87}Sr/^{86}Sr$ ratios of lavas on the mainland of northern Victoria Land. Hart and Kyle (1993) reported that the volcanic rocks of the Melbourne and Hallett volcanic provinces have higher $^{87}Sr/^{86}Sr$ ratios than the lavas on the off-shore islands including Franklin, Coulman, Possession, Scott, and the Balleny islands.

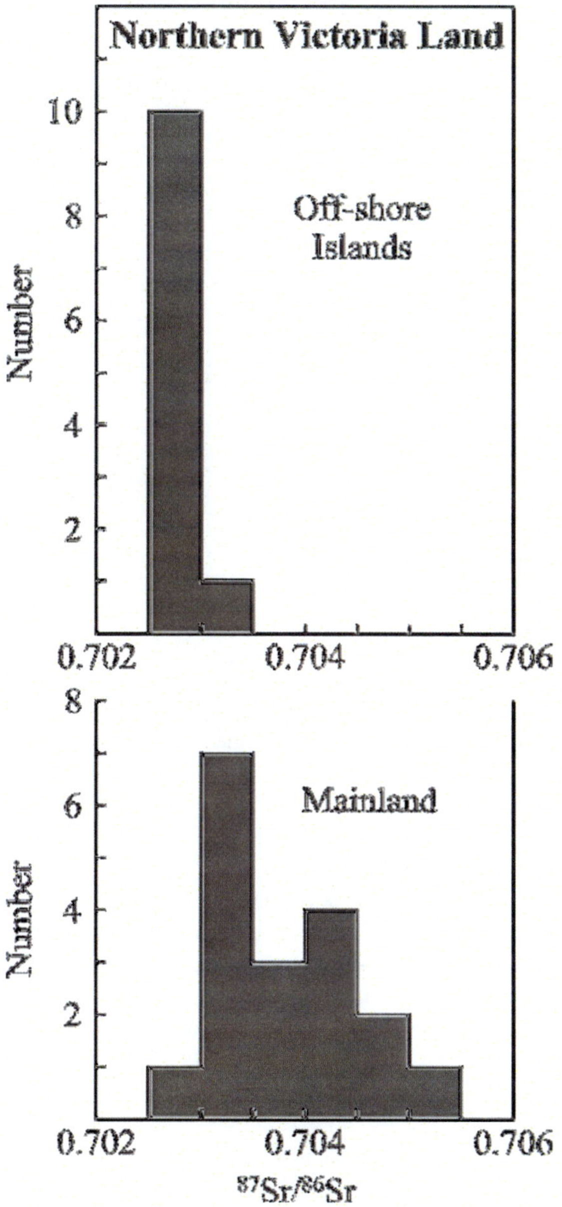

Fig. 16.41 The $^{87}Sr/^{86}Sr$ ratios of the Cenozoic volcanic rocks on the islands off the coast of Victoria Land are lower than those on the mainland. The islands included here are: Franklin, Coulman, Possession, Scott, and Balleny. The volcanoes on the mainland are Melbourne, Vulcan Hills, Overlord, and The Pleiades, as well as the Daniell, Hallett, and Adare peninsulas (Data from Jones and Walker (1972) and Hart and Kyle (1993))

The data in Appendix 16.8.5 and Fig. 16.41 demonstrate that the $^{87}Sr/^{86}Sr$ ratios of the off-shore islands range narrowly from 0.7025 to 0.7030, whereas the $^{87}Sr/^{86}Sr$ ratios of the lavas on the mainland have a bimodal distribution between 0.7025 and 0.7055 with abundance peaks between 0.7030–0.7035 and 0.7040–0.7045.

The low $^{87}Sr/^{86}Sr$ ratios of the Cenozoic volcanics on the off-shore islands extend the evidence presented earlier by Futa and LeMasurier (1983) that the $^{87}Sr/^{86}Sr$ ratios of certain Cenozoic volcanic rocks in the Marie Byrd Land are lower than those of oceanic island basalts (OIBs) and, instead, are similar to values characteristic of mid-ocean ridge basalt (MORBs). Therefore, the low $^{87}Sr/^{86}Sr$ ratios of the off-shore islands in Fig. 16.41 support the conjecture that a Rb-depleted mantle domain exists under the Ross Sea and Marie Byrd Land.

Hart and Kyle (1993, p. 15) summarized the presentation of their data with the hypothesis:

> …volcanics that erupt through significant continental crust become variably contaminated with radiogenic ^{87}Sr, whereas those erupted through extensionally thinned continental crust, or oceanic crust, are not significantly contaminated.

Hart and Kyle (1993) also measured isotopic compositions of lead (i.e., the $^{206}Pb/^{204}Pb$ ratios) in the Cenozoic lavas of the Hallett volcanic province, the Balleny Islands, as well as of Scott and Peter I islands. The $^{206}Pb/^{204}Pb$ and $^{87}Sr/^{86}Sr$ ratios of these rocks outline several overlapping areas in Fig. 16.42 that can be extrapolated to the HIMU component of the magma sources in the mantle (Zindler and Hart 1986; Hart 1988).

The HIMU component in the magma sources of the mantle has an elevated $^{206}Pb/^{204}Pb$ ratio presumably because it has a high $^{238}U/^{204}Pb$ ratio which isotope geochemists call "mu" (μ), hence the name HIMU. The apparent enrichment of the HIMU mantle component in uranium (or its depletion in lead) have been alternatively attributed to:

1. Transfer of lead from the mantle into the core thereby causing the U/Pb ratio of the affected mantle to rise
2. Recycling of old oceanic crust having an elevated U/Pb ratio by subduction into the mantle, and
3. Uranium enrichment of certain mantle domains by metasomatism (Hart 1988; Faure 2001)

Recycling of old oceanic crust (option 2 above) is the most attractive alternative because subduction of oceanic crust into the mantle is an important tectonic process that has been occurring on the Earth since the Archean Eon.

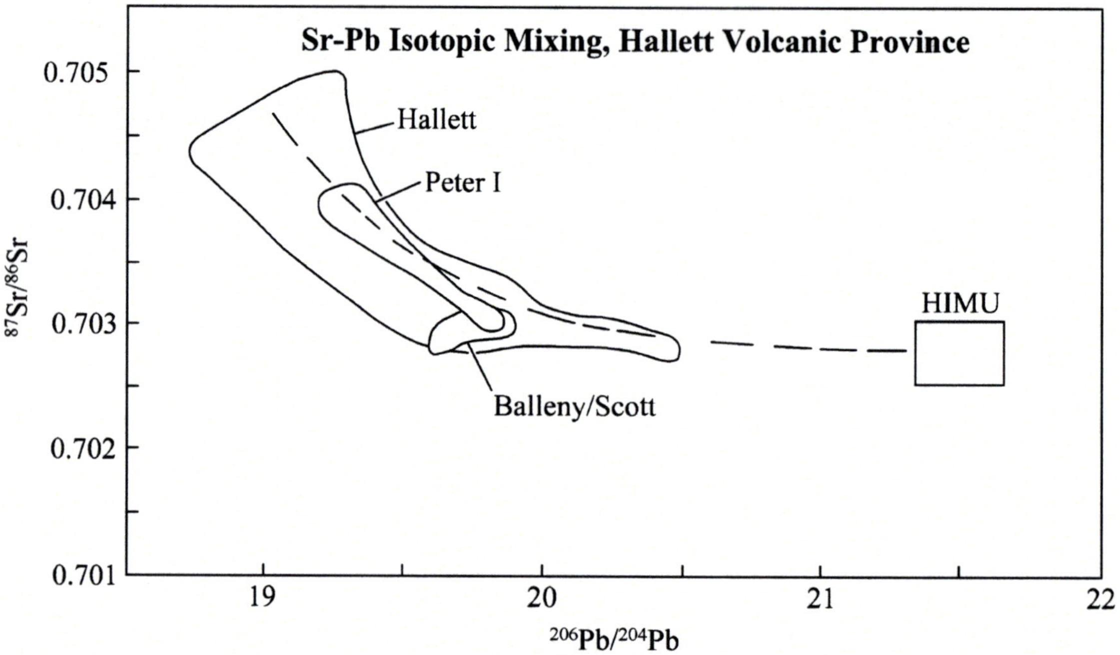

Fig. 16.42 The $^{206}Pb/^{204}Pb$ and $^{87}Sr/^{86}Sr$ ratios of the Cenozoic volcanic rocks on the Hallett Peninsula, the Balleny/Scott islands, and Peter I Island define data arrays that extrapolate to the HIMU component of magma sources in the mantle. The HIMU component is characterized by high $^{206}Pb/^{204}Pb$ ratios (20.0–21.5) and low $^{87}Sr/^{86}Sr$ ratios (0.7025–0.7030). The significance of these functional relationships between the isotope compositions of lead on strontium in these rocks is still uncertain (Data from Hart and Kyle (1993))

The HIMU component typically has a high $^{206}Pb/^{204}Pb$ ratio between 20.5 and 21.5 and a low $^{87}Sr/^{86}Sr$ ratio between 0.7025 and 0.7030 which places it in Fig. 16.42 on an extension of the hyperbolic Sr-Pb isotopic mixing line loosely defined by the Cenozoic volcanic rocks of the Hallett volcanic province. The volcanic rocks on the Balleny Islands and on Scott and Peter I islands have isotopic compositions of strontium and lead that overlap those of the Hallett volcanic province. However, the $^{87}Sr/^{86}Sr$ ratios of the Balleny/Scott islands are low (0.7025–0.7030), whereas those of Peter I Island (0.7038–0.7040) are normal for oceanic island basalts.

16.6 Mount Early and Sheridan Bluff, QMM

We now shift our attention from the Balleny Islands to the head of the Scott Glacier in the Queen Maud Mountains (QMM) (Sections 2.2, 6.4, and 10.4.2) because, in this area, Cenozoic volcanic activity appears to have interacted with the East Antarctic ice sheet. As a result, the rocks of this area provide information not only about the volcanic eruptions that occurred a scant 325 km from the geographic South Pole, but they also indicate the "age" of the East Antarctic ice sheet which is the subject of the next chapter.

Mt. Early (2,720 m) in Fig. 16.43 is located south of Mt. Weaver at the head of the Scott Glacier in the Queen Maud Mountains. It is the most southerly volcano on Earth located at 87°04'S and 153°46'W. Although Mt. Early was discovered in 1934, it was first described by G.A. Doumani and V.H. Minshew who worked in the area during the 1962/63field season (Section 10.4.2; Doumani and Minshew 1965). Mt. Early was visited again during the 1978/79 field season by a geological field party led by E. Stump (Stump et al. 1980).

Sheridan Bluff in Fig. 16.43 is located north of Mt. Weaver and about 20 km north of Mt. Early at 86°53'S, 153°30'W. This locality was briefly described by

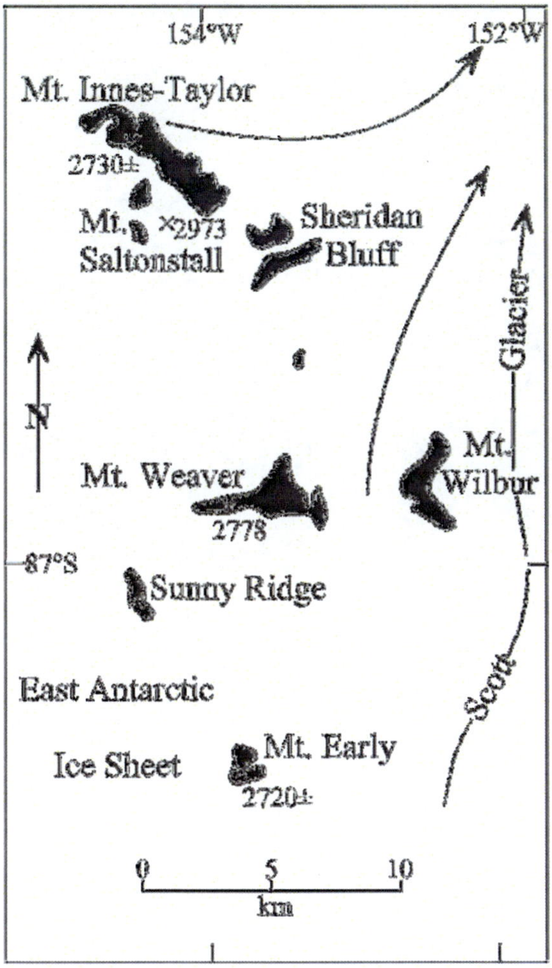

Fig. 16.43 Mount Early and Sheridan Bluff are located south and north, respectively, of Mt. Weaver at the head of the Scott Glacier in the Queen Maud Mountains. This area is only about 325 km from the geographic South Pole. The geology of this area was described in Sections 6.4 and 10.4.2 based on fieldwork by Doumani and Minshew (1965), Minshew (1967), and Stump et al. (1980) (Adapted from Stump et al. (1980))

Doumani and Minshew (1965), and Stump et al. (1980) measured a stratigraphic section of the volcanic rocks exposed on Sheridan Bluff. Both localities are identified on the geologic map of Mirsky (1969) and were described by Stump et al. (1990a, b).

These sites are significant in the geology of the Transantarctic Mountains because they fix the position of the southern rim of the West Antarctic Rift in Fig. 15.16 (LeMasurier and Thomson 1990). The alkali-rich magmas that formed in this rift caused the eruption of the lavas and pyroclastic deposits of the McMurdo Volcanic Group in all of Victoria Land, at the head of the Scott Glacier, in Marie Byrd Land, and in Ellsworth Land. Mt. Early also marks the spot where the south rim of the West Antarctic Rift first deviates from the trend of the Transantarctic Mountains and enters West Antarctica. Therefore, the location of Mt. Early and Sheridan Bluff is undoubtedly tectonically controlled, but the specific reasons for their existence are still obscure.

The rocks on Mt. Early in Fig. 16.44 are mildly alkaline olivine-bearing basalt containing less than 5% normative nepheline. The black pillow lavas which form the core of the volcanic cone are overlain by pillow breccia which grades up-section into fine grained, bright yellow, palagonite breccia (Stump et al. 1990a). The rocks exposed in Sheridan Bluff consist of a basal pillow breccia that was deposited on the eroded surface of the local basement complex (Granite Harbor Intrusive). The pillow breccia grades up-section into hyaloclastite breccia which consists of a matrix of yellow palagonite containing angular fragments of black pillow lava some of which have preserved parts of the glassy rind. The hyaloclastite at Sheridan Bluff is 85 m thick; but it thins in a northwesterly direction toward Mt. Saltonstall (Fig. 16.43). This hyaloclastic layer is overlain by a lava flow and by a second hyaloclastic layer 5–10 m thick. The top of the section consists of nine subaerially erupted lava flows with at total thickness of 110 m. The aggregate thickness of the volcanic rocks exposed at Sheridan Bluff is about 200 m (Stump et al. 1980).

Chemical analyses published by Minshew (1967) and Stump et al. (1990a, b) identify the olivine basalt on Mt. Early as a mugearite in the classification of Le Bas et al. (1986) in Fig. 16.45. The two rocks from Sheridan Bluff straddle the alkalic-subalkalic boundary of Macdonald and Katsura (1964) such that one is a hawaiite while the other is a subalkalic basalt.

A lava flow from the core of Mt. Early yielded a K-Ar date of 15.86 ± 0.30 Ma (Stump et al. (1980, 1990a) which is middle Miocene. K-Ar dates of three lava flows at Sheridan Bluff reported by Stump et al. (1980, 1990b) have a mean of 18.32 ± 0.35 Ma. The $^{40}Ar/^{39}Ar$ dates of two lava flows on Sheridan Bluff (19.43 ± 0.65 and 19.75 ± 1.57 Ma) are concordant.

Fig. 16.44 Mt. Early (2,720 m) is a large volcanic mountain of Miocene age located at the head of the Scott Glacier in Fig. 16.43. It is composed of pillow lavas, volcanic breccia, and palagonite (hydrated volcanic glass of subglacial or subaqueous origin). The age of the subglacially erupted lavas on Mt. Early and Sheridan Bluff (19.6 ± 1.7 Ma) indicates that glacial ice was already present at this site when the volcanoes erupted (Photo by C.H. Summerson published by Minshew (1967))

The average of the two $^{40}Ar/^{39}Ar$ dates is 19.59 ± 1.69 Ma (early Miocene). Another whole-rock K-Ar date of 22.2 ± 1.1 Ma of a volcanic rock from Sheridan Bluff was obtained by V.H. Minshew (unpublished). This date is not distinguishable from the average of the two $^{40}Ar/^{39}Ar$ plateau dates reported by Stump et al. (1986) because the errors overlap. A possible interpretation of these results is that the lavas at both sites were erupted at about the same time during early Miocene. The flow from the core of Mt. Early dated by Stump et al. (1980) (15.86 ± 0.30 Ma) may have lost ^{40}Ar during slow cooling after crystallization.

The presence of basal palagonite breccia, hyaloclastite, and pillow basalts on Mt. Early and Sheridan Bluff indicates that lavas at both vents were erupted through an ice sheet (Stump et al. 1980). Therefore, the age of these volcanic rocks tells us that glacial ice was present at the head of the Scott Glacier during the early Miocene.

Stump et al. (1980) also concluded that the topographic relief of the area at the head of the Scott Glacier at the time of the volcanic eruptions was about 900 m, which implies that the Transantarctic Mountains had been uplifted and that deep valleys had been carved prior to the eruption of the volcanoes.

Age determinations by LeMasurier and Rex (1982) of thick deposits of hyaloclastite indicate that the West Antarctic ice sheet formed in the middle Oligocene, or earlier, on a broad surface above sea level. After the ice sheet was in place, the underlying basement was subjected to extensional tectonics and concurrent subglacial volcanic eruptions during the Miocene. Therefore, the West Antarctic rift system and other grabens in West Antarctica formed *after* the ice sheet had already formed (Hughes 1973). The age of the subglacial volcanic activity at Mt. Early and Sheridan Bluff supports the interpretation of LeMasurier and Rex (1982).

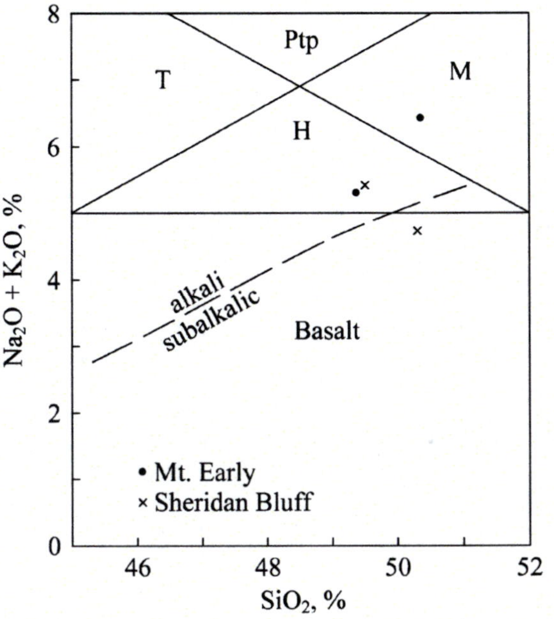

Fig. 16.45 The Miocene volcanic rocks that were erupted by Mt. Early and at Sheridan Bluff at the head of the Scott Glacier in the Queen Maud Mountains are classified on the total-alkali-silica diagram of Le Bas et al. (1986) as mildly alkali basalt, hawaiite (H), and mugearite (M) (Data from Minshew (1967) and Stump et al. (1990a, b))

16.7 Summary

Volcanic rocks of Miocene to Holocene age occur widely in the Transantarctic Mountains of Victoria Land and on the off-shore island including Ross Island. Most of the volcanoes are located in a belt along the coast which forms the western margin of the West Antarctic rift system. The lavas and pyroclastic deposits are predominantly alkali-rich and belong to the basanite-to-phonolite association. Subglacially erupted palagonite flows and hyaloclastite deposits indicate that ice had begun to accumulate on the Transantarctic Mountains at the time the volcanic eruptions began.

The Cenozoic volcanic rocks of the Transantarctic Mountains are collectively referred to the McMurdo Volcanic Group. These rocks were erupted by volcanoes located in four volcanic provinces described in this chapter. The Erebus volcanic province includes Mt. Erebus, Mt. Terror, and Mt. Bird on Ross Island as well as Mt. Discovery and Mt. Morning on the mainland of southern Victoria Land. Volcanic rocks of Cenozoic age occur widely on the other off-shore islands as well as on Minna Bluff, the Brown Peninsula, the Royal Society Range, and the ice-free valleys.

Mt. Erebus, which still maintains a lava lake in a crater on its summit, is the only active volcano in Antarctica. Mt. Melbourne of the Melbourne volcanic province also continues geothermal activity on its summit, and volcanoes in The Pleiades have erupted in the recent past. All other volcanoes in the Transantarctic Mountains are extinct, although geothermal activity still occurs on a few volcanoes in Mary Byrd Land.

The lava flows in the Erebus volcanic province contain inclusions of different rock types including granulites from the lower crust and ultramafic rocks from the subcrustal mantle. The granulite inclusions indicate the presence of a boundary that separates the crust underlying the Transantarctic Mountains from the crust beneath the Ross Embayment. Another boundary is defined by the isotopic composition of strontium in the volcanic rocks of northern Victoria Land and on the off-shore islands, including Franklin, Coulman, and Scott islands a well as the Possession and Balleny islands. These two boundaries extend along the coast of Victoria Land and appear to coincide with the western margin of the West Antarctic rift system.

Mt. Early is an extinct and partly dissected volcano at the head of the Scott Glacier located about 325 km from the geographic South Pole. The alkali-rich lavas and palagonitic hyaloclastite rocks were apparently erupted subglacially. Therefore, the average of the $^{40}Ar/^{39}Ar$ plateau dates (19.6 ± 1.7 Ma) of two lavas on Sheridan Bluff are a lower limit to the onset of the continental glaciation of East Antarctica.

16.8 Appendices

16.8.1 Average Chemical Compositions of the Granulite Inclusion from the Deep Crust Beneath the Transantarctic Mountains and the Ross Embayment (Kalamarides and Berg 1991)

	TAM		*RE*
Number of samples	3	1	3
Major Elements (%)			
Sample Numbers	#83–82, 82–160, 82–215	82–220a	#82–119, 81–7, 83–15
SiO_2	51.09 (48.18–53.53)	62.91	46.71 (43.52–48.81)
TiO_2	1.07 (0.8–1.55)	1.74	0.19 (0.09–0.26)
Al_2O_3	17.44 (15.72–19.92)	15.23	18.11 (10.04–31.27)
FeO	9.99 (8.44–11.0)	3.58	5.86 (0.61–11.71)
MnO	0.16 (0.14–0.19)	0.05	0.10 (0.01–0.19)
MgO	5.67 (4.98–6.05)	1.83	13.68 (1.54–24.44)
CaO	10.50 (9.89–11.14)	10.98	13.78 (9.04–16.32)
Na_2O	3.08 (2.59–3.42)	1.08	1.46 (0.77–2.54)
K_2O	0.79 (0.29–1.56)	2.36	0.11 (0.06–0.15)
P_2O_5	0.22 (0.13–0.28)	0.25	0.00 (0.00–0.01)
Trace Element, ppm			
Sc	22.4(18.3–27.9)	8.8	26.6 (12.8–41.0)
Ba	402.7 (106.0–745)	3050.0	26.6 (14.8–30.0)
Co	28.1 (25.3–30.0)	7.1	75.4 (6.3–160.0)
V	148.8 (123.0–117.0)	66.0	74.3 (23.0–109.0)
Cr	91.0 (61.0–125.0)	32.0	863.7 (81.0–1571.0)
Ni	27.4 (20.5–36.7)	11.6	339.7 (31.0–756.0)
Cu	21.7 (17.5–29.3)	5.6	32.2 (2.7–57.0)
Zn	104.5 (94.9–111.6)	37.0	37.3 (6.8–71.0)
Rb	6.63 (0.60–15.3)	127.0	2.70 (1.4–5.3)
Sr	670.7 (524.0–761.0)	1112.0	229.0 (126.0–397.0)
Y	29.8 (29.5–32.0)	17.8	12.5 (2.7–5.2)
Zr	70.0 (0.0–147.0)	39.0	11.9 (9.8–13.5)
Nb	9.57 (9.2–10.11)	13.9	5.25 (4.9–5.6)
Pb	3.57 (1.8–6.7)	4.2	1.53 (0.8–2.2)
Ga	21.0 (21.0)	–	7.4 (7.4)
Isotope composition			
$\delta^{18}O$, ‰	+10.2	+13.2	+6.1
$^{87}Sr/^{86}Sr$	0.7112	0.7113	0.7047
$^{87}Rb/^{86}Sr$	0.0420	0.2600	0.028

TAM = Transantarctic Mountains, RE = Ross Embayment.

16.8.2 Isotopic Compositions of Strontium and Sulfur in Soil Salts on Ross Island Including the Summit of Mt. Erebus (Jones et al.1983; Faure and Jones 1989)

Sample	Elevation (m, a.s.l)	Dominant mineral	$^{87}Sr/^{86}Sr$	$\delta^{34}S$ (‰)	Reference
Summit of Mt. Erebus					
81006	~3,600	Yellow salt	0.70353 ± 3		1
81007	~3,600	Yellow salt	0.70341 ± 2		1
810010	~3,600	Tan salt	0.70342 ± 2		1
81008[a]	~3,600	Na carbonates	0.70449 ± 3	+3.1	1,2
81400[b]	2,438	Thenardite	0.70339 ± 3	+0.8	2
G4	~3,600	Anorthoclase	0.70330 ± 3	1	
Hut Point Peninsula					
26	302	Thenardite	0.70729 ± 3	1,2	
Castle Rock					
29	413	Calcite	0.70344 ± 2	1	
Mt. Bird					
Bird 2[c]	1,800	Thenardite	0.70707 ± 2	+17.6	2
Cape Bird, Trachyte Hill					
13	2	Thenardite	0.70912 ± 3	+19.6	2
14	56	Thenardite	0.70825 ± 3	+17.8	2
15	130	Thenardite	0.70802 ± 3	+18.3	2
16	156	Thenardite	0.70845 ± 1		2
17	215	Thenardite	0.70854 ± 2		2
18	302	Thenardite	0.70879 ± 12		2
19	364	Thenardite	0.70906 ± 3		2
20	390	Thenardite	0.70904 ± 3	+19.4	2
Cape Royds					
3	12	Thenardite	0.70906 ± 3	+18.4	2
4	55	Thenardite	0.70902 ± 2		2
6	60	Thenardite?	0.70745 ± 4	+7.9	2
			0.70673 ± 4[d]		2
7	94	Thenardite?	0.70887 ± 2		2
			0.70788 ± 1[d]		2
8	129	Thenardite?	0.70838 ± 1	+14.6	2
			0.70789 ± 2		
10	191	Thenardite?	0.70819 ± 4		2
		0.70801 ± 3[d]	2		
11	253	Calcite?	0.70771± 2		2
12	284	Calcite	0.70834		2
25	15	An. phonolite	0.70363 ± 4		
Cape Barne					
	~10	Mirabilite		+20.3	2

[a]Collected by J.M. Palais.

[b]Collected by P.R. Kyle.

[c]Jones et al. (1983).

[d]Leached with cold dilute HCl

Eimer and Amend, $SrCO_3$, $^{87}Sr/^{86}Sr = 0.70811 \pm 2$ by Jones et al. (1983) and 0.70800 by Faure and Jones (1989).

16.8.3 Isotopic Compositions of Two-Component Mixtures (Faure and Jones1989)

The salts in the soil of Ross Islands contain strontium and sulfur both of which are mixtures of two components that have different isotope compositions. Such mixtures are described by an equation that was derived by Langmuir et al. (1978) and presented by Faure (1986). The procedure is to calculate the $^{87}Sr/^{86}Sr$ ratio and $\delta^{34}S$ value separately for the same mixing parameters. The results are the coordinates of points on the Sr-S mixing hyperbola which can be constructed by means of a smooth curve through the points calculated from the mixing equations:

$$R_M^X = \frac{R_A^X X_A f_A + R_B^X X_B(1-f_A)}{X_A f_A + X_B(1-f_A)} \quad (16.2)$$

R_M = the isotope ratio of an element X in the mixture M of components A and B,

X_A, X_B = the concentrations of element X in components A and B,

f_A = weight fraction of component A in any mixtures of A and B (i.e., $f_A = A/A + B$).

The curvature and attitude of the mixing hyperbola depend on the concentration ratios of the two elements X and Y which are related by the equation:

$$K = (X/Y)_A / (X/Y)_B \quad (16.3)$$

where X_A and Y_A are the concentrations of elements X and Y in component A and similarly for component B.

The hyperbolas in Fig. 16.24 are based on values of the parameter chosen from the data (Appendix 16.8.2). For the marine component:

$(^{87}Sr/^{86}Sr)_m = 0.70906$; $\delta^{34}S_m = +20.0‰$
For the volcanic component:
$(^{87}Sr/^{86}Sr)_v = 0.70340$; $\delta^{34}S = +0.0‰$

In the construction of Fig. 16.24 the concentrations of Sr and S were chosen such that K = 10 for one of the hyperbolas and K = 0.10 for the other. The concentration of S is fixed by the stoichiometry of the sulfate ion, which means that only the concentration of Sr is assumed to vary. Note also, that the mixtures form a straight line when K = 1.0

The $^{87}Sr/^{86}Sr$ ratio of a mixture of components of A and B is calculated from Eq. 16.2 for selected values of the mixing parameters f_A and K. Assume that $f_A = 0.40$ and the K = 10. In that case:

$$(Sr/S)_A = 10\ (Sr/S)_B$$

Let the marine component be A and the volcanic component B. If $S_A = S_B$, then

$$Sr_A = 10\ Sr_B$$

Therefore, we set $Sr_A = 100$ ppm and $Sr_B = 10$ ppm and substitute appropriate values into Eq. 16.2 to calculate the $^{87}Sr/^{86}Sr$ ratio:

$$\left(\frac{^{87}Sr}{^{86}Sr}\right)_M = \frac{0.70906 \times 100 \times 0.4 + 0.7034 \times 10(1-0.4)}{100 \times 0.4 + (1-0.4)}$$

$$\left(\frac{^{87}Sr}{^{86}Sr}\right)_M = \frac{28.3624 + 4.2204}{40+6} = 0.70832$$

Similarly for sulfur:

$$\delta^{34}S_M = \frac{20.0 \times S_A \times 0.4 + 0 \times S_B(1-0.4)}{S_A \times 0.4 + S_B(1-0.4)}$$

Since $S_A = S_B$,

$$\delta^{34}S_M = \frac{20 \times 0.4}{0.4+0.6} = 8.0$$

Therefore, the coordinates of the point on the mixing hyperbola at $f_A = 0.4$ are:

$$(^{87}Sr/^{86}Sr)_M = 0.70832$$

$$\delta^{34}S_M = 8.0‰$$

Additional points on the mixing hyperbola in Fig. 16.24 for K = 10 were calculated by varying f_A stepwise from 0.1, 0.2, 0.4, 0.6, 0.8, and 1.0.

16.8.4 Isotope Compositions of Strontium and Neodymium of Volcanic Rocks from the Mt. Melbourne Volcanic Field (Wörner et al. 1989)

Number	Rock type	$^{87}Sr/^{86}Sr$	$^{143}Nd/^{144}Nd$
SN 16-016	Alkali basalt	0.70338	0.51302
OR 09-035	Alkali basalt	0.70345	0.51296
NE 09-118	Alkali basalt	0.70387	0.51305
SE 04-160	Alkali basalt	0.70343	–
WR 01-041	Basanite	0.70297	0.51296
WR 13-140	Basanite	0.70288	0.51304
WR 21-152	Basanite	0.70293	0.51298
RH 07-215	Basanite	0.70295	0.51307
SN 10-010	Mugearite	0.70355	0.51293
EP 02-070	Benmoreite	0.70445	0.51290
MM 21 185	Trachyandesite	0.70466	0.51293
MM 21-238	Trachyandesite	0.70496	0.51279
MM 04-173-	Trachyte	0.70499	0.51284
MM 01-167 FSP	–	0.70506	–
EP 14-100-1	Trachyte	0.70829	0.51283
EP 14-100 FSP	–	0.70492	–
WR 18-148N	cpx	0.70345	–
E$A $SrCO_3$	–	0.70800	–
BCR-1		–	0.51266

Localities (Fig. 16.28): RH = Random Hills, BR = Baker Rocks, EP = Edmonson Point, NE = north Edmonson Point, SE = south Edmonson Point, MM = Mt. Melbourne WR = Washington Ridge, OR = Oscar Point, SN = Shield Nunatak, WCG = west Campbell Glacier.

16.8.5 Isotope Compositions of Strontium of the Cenozoic Lavas of Northern Victoria Land and Adjacent Islands

Location	$^{87}Sr/^{86}Sr$	Reference
Franklin Island	0.70289	1
	0.70292	1
	0.70294	1
Mt. Melbourne[a]	0.70328	1
	0.70287	1
Campbell Glacier	0.7034	2
	0.7041	2
Vulcan Hills	0.70392	1
	0.70439	1
Mt. Overlord	0.70334	1
	0.70330	1
	0.7044	2
	0.7047	2
Mt. Phillips	0.7049	1
The Pleiades	0.70364	1
	0.70350	1
	0.70326	1
Coulmann Island	0.70276	1
	0.70332	1
Daniell Peninsula	0.70326	1
Cape Hallett	0.7037	2
	0.7042	2
	0.7054	2
Possession Islands	0.70288	1
	0.70290	1
Cape Adare	0.70290	1
Scott Island	0.70285	1
Balleny Islands	0.70297	1
	0.70300	1

[a]See also Appendix 16.8.4 containing data by Wörner et al. (1989).

[b]Hart and Kyle (1993).

[c]Jones and Walker (1972).

References

Adamson RG, Cavaney RJ (1967) Volcanic debris layers near Mt. Melbourne, northern Victoria Land, Antarctica. New Zealand J Geol Geophys 10(2):418–421

Anderson AT, Clayton RN, Mayeda TK (1971) Oxygen isotopic thermometry of mafic igneous rocks. J Geol 19:715–729

Armadillo E, and 10 coauthors (2001) Geophysical features of the Mt. Melbourne area, Antarctica, and preliminary results from the integrated network for monitoring the volcano. In: Gamble JA et al. (eds) Antarctica at the close of a Millennium. Roy Soc New Zealand Bull 35:571–578

Armstrong RL (1978) K-Ar dating: Late Cenozoic McMurdo Volcanic Group and Dry Valley glacial history, Victoria Land, Antarctica. New Zealand J Geol Geophys 21(6):685–698

Armstrong RL, Hamilton W, Denton GH (1968) Glaciation in Taylor Valley, Antarctica, older than 2.7 million years. Science 159:187–189

Balleny J (1939) New land in the southern ocean. Literary World 2:185–187

Barrett PJ (1987) Oligocene sequence cored at CIROS-1, western McMurdo Sound. New Zealand Antarctic Rec 7:1–7

Bastien TW, Craddock C (1976a) The geology of Peter I Island. In: Initial Report of the Deep Sea Drilling Project, vol. 35. Government Printing Office, Washington, DC, pp 341–357

Bastien TW, Craddock C (1976b) Igneous rocks of Peter I Island. Antarctic J US 11(4):267–269

Beck AC (1965) A note on Mt. Erebus, Ross Island, Antarctica. New Zealand J Geol Geophys 8(2):180–185

Behrendt JC, Duerbaum HJ, Damaske D, Saltus R, Bosum W, Cooper AK (1991a) Extensive volcanism and related tectonism beneath the western Ross Sea continental shelf, Antarctica: Interpretation of an aeromagnetic survey. In: Thomson MRA, Crame JA, Thomson JW (eds) Geological evolution of Antarctica. Cambridge University Press, Cambridge, UK, pp 299–304

Behrendt JC, LeMasurier WE, Cooper AK, Tessensohn T, Trehu A, Damaske D (1991b) The West Antarctic rift system: A review of geophysical investigations. In: Elliot DH (eds) Contributions of Antarctic Research II, Antarctic Research Series, vol. 53. American Geophysical Union, Washington, DC, pp 67–112

Berg JH (1991) Geology, petrology, and tectonic implications of crustal xenoliths in Cenozoic volcanic rocks of southern Victoria Land. In: Thomson MRA, Crame CA, Thomson JW (eds) Geological evolution of Antarctica. Cambridge University Press, Cambridge, UK, pp 311–315

Berg JH, Herz DL (1986) Thermobarometry of two-pyroxene granulite inclusions in Cenozoic volcanic rocks of the McMurdo Sound region. Antarctic J US 21(5):19–20

Blackman DK, von Herzen RP, Lawver LA (1984) Heat flow and tectonics in the Ross Sea. Eos, Trans. Amer Geophys Union 65(45):1120

Bosum W, Damaske D, Behrendt JC, Saltus R (1991) The aeromagnetic survey of northern Victoria Land and the western Ross Sea during GANOVEX IV and a geophysical-geological interpretation. In: Thomson MRA, Crame JA, Thomson JW (eds) Geologic evolution of Antarctica. Cambridge University Press, Cambridge, UK, pp 267–272

Boudette EL, Ford AB (1966) Physical properties of anorthoclase from Antarctica. Amer Mineral 51:1374–1387

Broady P, Given D, Greenfield L, Thompson K (1987) The biota and environment of fumaroles on Mt. Melbourne, northern Victoria Land. Polar Biol 7:97–113

Broch OA (1927) Gesteine von der Peter I Insel, West Antrarktis. Avh. Nor. Vidensk. Akad. Oslo, Math. Naturvidensk, Klasse 9:1–41, Oslo, Norway

Burge W E, Parker DC (1969) Infrared survey in Antarctica: Imagery analysis. Rep. 1160–9-f, 1–53. Willow Run Lab., University of Michigan, Ann Arbor, MI

Campsie J, Neumann ER, Johnson GL (1983) Dredged volcanic rocks from the southern oceans: The Eltanin collection. New Zealand J Geol Geophys 26:31–45

Chen C-Y, Frey FA (1983) Origin of Hawaiian tholeiite and alkalic basalt. Nature 302: 785–789

Cherry-Garrard A (1922) The worst journey in the world; Antarctic 1910–1913. Penguin, London. Reissued in 1994 by Picador a division of Pan Macmillan, London

Chuan RL (1975) Fine particles; Aerosol generation, measurement, sampling and analysis. In: Liu BYH (ed). Academic, New York, pp 763–775

Chuan RL, Palais J, Rose WI, Kyle PR (1986) Fluxes, sizes, morphology, and compositions of particles in the Mt. Erebus volcanic plume, December 1983. J Atmos Chem 4:467–477

Clague DA, Frey FA (1982) Petrology and trace element geochemistry of the Honolulu Volcanics, Oahu: Implications for the oceanic mantle below Hawaii, J. Petrol. 23(3): 447–504

Cole JW, Ewart A (1968) Contributions to the volcanic geology of the Black Island, Brown Peninsula, and Cape Bird areas, McMurdo Sound, Antarctica. New Zealand J Geol Geophys 11(4):793–828

Cole JW, Kyle PR, Neall VE (1971) Contributions to the Quaternary geology of Cape Crozier, White Island, and Hut Point Peninsula, McMurdo Sound region, Antarctica. New Zealand J Geol Geophys 14:528–546

Coombs DS, Wilkinson JFG (1969) Lineages and fractionation trends in undersaturated volcanic rocks for the East Otago Volcanic Province (New Zealand) and related rocks. J Petrol 10:440–501

Cox A (1966) Paleomagnetic research on volcanic rocks of McMurdo Sound. Antarctic J US 1:136

Cox A (1969) Geomagnetic reversals. Science 163:237–245

Craddock C, Bastien TW, Rutford RH (1964) Geology of the Jones Mountains area. In, Adie RJ, ed. Antarctic Geology, 171–187. North-Holland, Amsterdam

Dalrymple GB (1979) Critical tables for the conversion of K-Ar ages from old to new constants. Geology 7:558–560

David TWE, Priestley RE (1914) Glaciology, physiography, stratigraphy and tectonic geology of south Victoria Land. Rept. British Antarct. Exped., 1907–1909. Geology 1:1–319, London

David TWE, Smeeth WF, Schofield JW (1896) Notes on Antarctic rocks collected byMr. C.E. Borchgrevink. Proc. Roy. Soc. New South Wales (for 1895) 29:461–492

Debenham F (1923) The physiography of the Ross archipelago. British (Terra Nova) Antarctic Exped., 1910–1913. Harrison, London

Decker ER, Bucher GJ (1982) Geothermal studies in the Ross Island-Dry Valley region. In: Craddock C (ed) Antarctic geoscience. University of Wisconsin Press, Madison, WI, pp 887–894

Delisle G, Thierbach R (1989) Radio echo sounding and geothermal investigations in the crater area of Mt. Melbourne, north Victoria Land, Antarctica. Geologisches Jahrbuch, Reihe E 38:435–453

Delmas RJ, Briat M, Legrand M (1982) Chemistry of south polar snow. J Geophys Res 87:4314–4318

Delmas RJ, Legrand M, Aristarain AL, Zanolini F (1985) Volcanic deposits in Antarctic snow and ice. J Geophys Res 90:12901–12920

Denton GH (1970) Late Cenozoic glaciations in Antarctica. Antarctic J US 5:15–21

Denton GH (1971) The late Cenozoic glacial history of Antarctica. In: Turekian KK (ed) The late Cenozoic Ice Ages. Yale University Press. New Haven, CT, pp 267–306

Denton GH, Armstrong RL (1968) Glacial geology and geochronology of the McMurdo Sound region. Antarctic J US 3:99–101

Denton GH, Armstrong RL, Stuiver M (1969) Historie glaciaire et chronologie de la region du detroit de McMurdo, sud de la Terre Victoria, Antarctide: Note preliminaire. Revue de Géographie Physique et de Géologie Dynamique 2 11(3):265–278

Dibble RR, Kienle J, Kyle PR, Shibuya K (1984) Geophysical studies of Erebus volcano, Antarctica, from 1974 December to 1982 January. New Zealand J Geol Geophys 27:425–456

Dibble RR, Barrett SID, Kaminuma K, Miura S, Kienle J, Rowe CA, Kyle PR, McIntosh WC (1989) Time comparisons between video and seismic signals from explosions in the lava lake of Erebus volcano. Bulletin of Disaster Prevention Research Institute, Kyoto Univ, Kyoto, Japan

Doumani GA, Minshew VH (1965) General geology of the Mount Weaver area, Queen Maud Mountains, Antarctica. In: Hadley JB (ed) Geology and paleontology of the Antarctic. Antarctic Research Series, vol. 6. American Geophysical Union, Washington, DC, pp 127–139

Dunbar NW, Cashman KV, Dupre R (1994) Crystallization processes of anorthoclase phenocrysts in the Mount Erebus magmatic system: Evidence from crystal composition, crystal size distribution, and volatile contents of melt inclusions. In: Kyle PR (ed) Volcanological and environmental studies of Mount Erebus, Antarctica. Antarctic Research Series, vol. 66. American Geophys Union, Washington, DC, pp 129–146

Duncan RA, McDougall I (1989) Volcanic time-space relationships. In: Johnson RW (ed) Intraplate Volcanism in Eastern Australia and New Zealand. Cambridge University Press, Cambridge, UK, pp 43–54

Dürbaum HJ, Druivenga G, Geipel H, Merkel G (1989) Gravity measurements along a traverse from Mt. Melbourne to the Polar Plateau in north Victoria Land, Antarctica. Geologisches Jahrbuch, Reihe E 38:231–243

Eggers AJ (1979) Scallop Hill Formation, Brown Peninsula, McMurdo Sound, Antarctica. New Zealand J Geol Geophys 22(3):353–361

Elston DP, Bressler SL (1980) Paleomagnetic investigation of Cenozoic glacigenic sediments, Taylor Valley and McMurdo Sound. Antarctic J US 15(5):9–10

Esser RP, Kyle PR (2002) $^{40}Ar/^{39}Ar$ chronology of the McMurdo Volcanic Group at The Pleiades, northern Victoria Land, Antarctica. In: Gamble JA et al. (eds) Antarctica at the close of the Millennium. Roy Soc New Zealand Bull 35:415–418

Esser RP, McIntosh WC, Heizler MT, Kyle PR (1997) Excess argon in melt inclusion in young (< 250 ka) anorthoclase feldspar from Mt. Erebus, Antarctica, as revealed by the $^{40}Ar/^{39}Ar$ method. Geochim. Cosmochim. Acta 61:3789–3801

Falconer RKH (1972) The Indian-Antarctic-Pacific triple junction. Earth Planet. Sci Lett 17:151–158

Faure G (1986) Principles of isotope geology, 2nd edn. Wiley, New York

Faure G (2001) The origin of igneous rocks; The isotopic evidence. Springer, Heidelberg, Germany

Faure G, Jones LM (1989) Distribution of marine salts along the west coast of Ross Island, Antarctica, based on isotopic compositions of strontium and sulfur. In: Augustithis SS (ed) Weathering; Its products and deposits. Theophrastus, S.A., Athens, Greece, pp 369–381

Faure G, Mensing TM (2005) Isotopes: Principles and applications, 3rd edn. Wiley, Hoboken, NJ

Faure G, Mensing TM (2007) Principles of planetary science: The geological perspective. Springer, Dordrecht, The Netherlands

Faure G, Schultz CH, Carwile RW (1971) Isotope composition of strontium in volcanic rocks from Deception Island. Antarctic J US 6(5):197–198

Ferrar HT (1907) Report on the field geology of the region explored during the Discovery Antarctic Expedition 1901–1904. Natural History Rept. Nat Antarctic Exped 1 (Geology):1–100

Fleck RJ, Jones LM, Behling RE (1972) K-Ar dates of the McMurdo volcanics and their relation to the glacial history of Wright Valley. Antarctic J US 7:244–246

Forbes RB (1963) Ultrabasic inclusions from the basalts of the Hut Point area, Ross Island. Bull Volcanol 26:13–21

Forbes RB, Turner DL, Carden JR (1974) Age of trachyte from Ross Island, Antarctica. Geol 2:297–298

Futa K, LeMasurier WE (1983) Nd and Sr isotopic studies on Cenozoic mafic lavas from West Antarctica: Another source for continental alkali basalts. Contrib. Mineral. Petrol. 83: 38–44

Gair HS (1967) The geology from the upper Rennick glacier to the coast, northern Victoria Land, Antarctica. New Zealand J Geol Geophys 10(2):309–344

Gair HS, Sturm A, Carryer SJ, Grindley GW (1969) The geology of northern Victoria Land. In: Craddock C (ed) Geologic map of Antarctica, Folio 12, Sheet 13. American Geographical Society, New York

Gamble JA, Kyle PR (1987) The origins of glass and amphibole in spinel-wehrlite xenoliths from Foster Crater, McMurdo Volcanic Group, Antarctica. J Petrol 28(5):755–780

Gamble JA, Barrett PJ, Adams CJ (1986) Basaltic clasts from Unit 8. New Zealand Dept. Sci Ind Res Bull 8, 237:145–152

Gamble JA, McGibbon F, Kyle PR, Menzies MA, Kirsch I (1988) Metasomatized xenotliths from Foster Crater, Antarctica: Implications for lithospheric structure and processes beneath the Transantarctic Mountain front. J Petrol, Special Lithosphere Issue:109–138

Giggenbach WF (1976) Geothermal ice caves on Mt. Erebus, Ross Island, Antarctica. New Zealand J Geol Geophys 19:365–372

Giggenbach WF, Kyle PR, Lyon GL (1973) Present volcanic activity on Mount Erebus, Ross Island. Geology 1:135–136

Goldich SS, Treves SB, Suhr NH, Stuckless JS (1975) Geochemistry of the Cenozoic volcanic rocks of Ross Island and vicinity, Antarctica. J Geol 83:415–435

Goldich SS, Stuckless JS, Suhr NH, Bodkin JB, Wamser RC (1981) Some trace element relationships in the Cenozoic volcanic rocks from Ross Island and vicinity, Antarctica. In: McGinnis LD (ed) Dry valley Drilling Project. Antarctic Research Series, vol. 33. American Geophysical Union, Washington, DC, pp 215–228

Hamilton W (1967) The Hallett Volcanic Province. Antarctic J US 2:177–178

Hamilton W (1972) The Hallett volcanic province, Antarctica. US Geol Surv Prof. Paper, 456C:1–62

Harrington HJ (1969) Fossiliferous rocks in moraines at Minna Bluff, McMurdo Sound. Antarctic J US 4(4):134–135

Harrington HJ, Wood BL, McKellar LC, Lensen GJ (1967) Topography and geology of the Cape Hallett district, Victoria Land, Antarctica. New Zealand Geol Surv Bull (new series) 80:1–100

Hart SR (1988) Heterogeneous mantle domains: Signatures, genesis, and mixing chronologies. Earth Planet Sci Lett 90:273–296

Hart SR, Kyle PR (1993) The geochemistry of McMurdo Group volcanic rocks. Antarctic J US 28:14–16

Hoefs J (1997) Stable isotope geochemistry, 4th edn. Springer, Heidelberg, Germany

Hoefs J, Faure G, Elliot DH (1980) Correlation of $\delta^{18}O$ and initial $^{87}Sr/^{86}Sr$ ratios in Kirkpatrick Basalt on Mt. Falla, Transantarctic Mountains. Contrib Mineral Petrol 75:199–203

Holdsworth G, Ugolini FC (1965) Fumarolic ice towers on Mount Erebus, Ross Island, Antarctic. J Glaciol 5:878–879

Holtedahl O (1929) On the geology and physiography of some Antarctic and subantarctic islands, with notes on the character and origin of fjords and strandflats of some northern lands, etc., Sci. Results Norw. Antarctic Expedition 1927–1928, vol. 3. Oslo, Norway, pp 1–172

Hornig IG, Wörner, Zipfel J (1991) Lower crustal and mantle xenoliths from the Mt. Melbourne volcanic field, northern Victoria Land, Antarctica. Mem Soc Geol Italiana 46:337–352

Huang TC, Watkins ND, Shaw DM (1975) Atmospherically transported volcanic glass in deep-sea sediments: Volcanism in subantarctic latitudes of the Pacific during late Pliocene and Pleistocene time. Geol Soc Amer Bull 86:1305–1315

Hughes T (1973) Is the West Antarctic ice sheet disintegrating? J Geophys Res 78:7884–7910

Irving AJ (1974) Geochemical and high-pressure experimental studies of garnet pyroxenite and pyroxene granulite xenoliths from the Delegate basaltic pipes, Australia. J Petrol 15:1–40

Jensen HI (1916) Report on the petrology of the alkaline rocks of Mt. Erebus, Antarctica. British Antarctic Expedition 1907–1909. Rept Scient Investig, Geology 2(7):93–128

Johnson GL, Kyle PR, Vanney JR, Campsie J (1982) Geology of Scott and Balleny islands, Ross Sea, Antarctica and morphology of adjacent seafloor. New Zealand J Geol Geophys 25:427–436

Jones LM, Walker RL (1972) Geochemistry of the McMurdo Volcanics, Victoria Land, Part 1:Strontium isotope composition. Antarctic J US 7(5):142–144

Jones LM, Faure G, Taylor KS, Corbató CE (1983) The origin of salts on Mount Erebus and along the coast of Ross Island, Antarctica. Isot Geosci (Chem Geol) 1:56–64

Jordan H (1981) Tectonic observations in the Hallett volcanic province, Antarctica. Geologisches Jahrbuch, Reihe G,41:111–125

Kalamarides RI, Berg JH, Hank RH (1987) Lateral isotopic discontinuity in the lower crust: An example from Antarctica. Science 237:1192–1195

Kalamarides RI, Berg JH (1991) Geochemistry and tectonic implications of lower-crustal granulites included in Cenozoic volcanic rocks of southern Victoria Land. In: Thomson MRA, Crane JA, Thomson JW (eds) Geological evolution of Antarctica. Cambridge University Press, Cambridge, UK, pp 305–310

Kempe DRC (1973) Rocks from Antarctica: The Discovery collection in the British Museum (Natural History). Bull British Mus Nat Hist, Mineral 2:337–376

Keys JR (1980) Salts and their distribution in the McMurdo region, Antarctica. PhD dissertation, Victoria University, Wellington, New Zealand

Keys JR, Williams K (1981) Origin of crystalline, cold desert salts in the McMurdo region, Antarctica. Geochim Cosmochim Acta 45:2299–2309

Keys JR, Anderton PW, Kyle PR (1977) Tephra and debris layers in the Skelton Névé and Kempe Glacier, south Victoria Land, Antarctica. New Zealand J Geol Geophys 20(5):971–1002

Keys JR, McIntosh WC, Kyle PR (1983) Volcanic activity of Mt. Melbourne, northern Victoria Land. Antarctic J US 18(5):10–11

Kirsch ID (1981) Evidence for mantle metasomatism in ultramafic inclusions from Foster Crater, Antarctica. MSc thesis, The Ohio State University, Columbus, OH

Kreuzer H, Höhndorf A, Lenz H, Vetter U, Tessensohn F, Muller B, Jordan H, Harre W, Besang C (1981) K/Ar and Rb/Sr dating of igneous rocks from North Victoria Land, Antarctica. Geologisches Jahrbuch, Reihe B 41:267–273

Kurasawa H (1975) Strontium isotopic studies of the Ross Island volcanics, Antarctica. In: Torii T (ed) Geochemical and geophysical studies of Dry Valleys, Victoria Land in Antarctica. Mem. Spec. Issue, vol. 4. Institute of Polar Research, Tokyo, Japan, pp 67–74

Kurasawa H (1978) Voclanism and volcanic rocks in Antarctica. Antarctic Record 58:204–234

Kyle PR (1976) Geology, mineralogy, and geochemistry of the late Cenozoic McMurdo Volcanic Group, Victoria Land. PhD dissertation, Victoria University, Wellington, New Zealand

Kyle PR (1977) Mineralogy and glass chemistry of Recent volcanic ejecta from Mt. Erebus, Ross Island, Antarctica. New Zealand J Geol Geophys 20(6):1123–1146

Kyle PR (1981a) Geologic history of Hut Point Peninsula as inferred from DVDP 1, 2, and 3 drillcores and surface mapping. In: McGinnis LD (ed) Dry Valley Drilling Project. Antarctic Research Series, vol. 33. American Geophysical Union, Washington, DC, pp 427–445

Kyle PR (1981b) Mineralogy and geochemistry of a basanite to phonolite sequence at Hut Point Peninsula based on core from Dry Valley Drilling Project drillholes 1, 2, and 3. J Petrol 22:451–500

Kyle PR (1982) Volcanic geology of The Pleiades, northern Victoria Land, Antarctica. In: Craddock C (ed) Antarctic geoscience. University of Wisconsin Press, Madison, WI, pp 747–754

Kyle PR (1986) Mineral chemistry of late Cenozoic McMurdo Volcanic Group from the Pleiades, northern Victoria Land. Antarctic Research Series, vol. 46. American Geophysical Union, Washington, DC, pp 305–337

Kyle PR (1990a) Erebus volcanic province: Summary. In: LeMasurier WE, Thomson JW (eds) Volcanoes of the Antarctic Plate and Southern Ocean. Antarctic Research Series, vol. 48. American Geophysical Union, Washington, DC, pp 81–88

Kyle PR (1990b) Hut Point peninsula. In: LeMasurier WE, Thomson JW (eds) Volcanoes of the Antarctic Plate and Southern Oceans. Antarctic Research Series, vol. 48. American Geophysical Union, Washington, DC, pp 109–112

Kyle PR (1990c) Melbourne Volcanic Province: Summary. In: LeMasurier W, Thomson W (eds) Volcanoes of the Antarctic Plate and Southern Oceans. Antarctic Research Series, vol. 48. American Geophysical Union, Washington, DC, pp 48–52

Kyle PR (1990d) The Pleiades. In: LeMasurier WE, Thomson JW (eds) Volcanoes of the Antarctic Plate and Southern Oceans. Antarctic Research Series, vol. 48. American Geophysical Union, Washington, DC, pp 60–64

Kyle PR (1990e) Vulcan Hills. In: LeMasurier WE, Thomson JW (eds) Volcanoes of the Antarctic Plate and Southern Oceans. Antarctic Research Series, vol. 48. Amer. Geophys. Union, Washington, D.C., pp 69–71

Kyle PR (1990f) Local suite basaltic rocks. In: LeMasurier WE, Thomson JW (eds) Volcanoes of the Antarctic Research

Series, vol. 48. American Geophysical Union, Washington, DC, pp 79–80

Kyle PR (ed) (1995) Volcanological and environmental studies of Mt. Erebus, Antarctica. Antarctic Research Series, vol. 66. American Geophysical Union, Washington, DC

Kyle PR, Cole JW (1974) Structural control of volcanism in the McMurdo Volcanic Group, Antarctica. Bull Volcanologique 38(1):16–25

Kyle PR, Knoll MR (1990) Mount Overlord. In: LeMasurier WE, Thomson JW (eds) Volcanoes of the Antarctic Plate and Southern Oceans. Antarctic Research Series, vol. 48. American Geophysical Union, Washington, DC, pp 65–68

Kyle PR, Rankin PC (1976) Rare earth element geochemistry of late Cenozoic alkaline lavas of the McMurdo Volcanic Group, Antarctica. Geochim Cosmochim Acta 40:1497–1507

Kyle PR, Treves SB (1974a) Geology of DVDP3, Hut Point Peninsula, Ross Island, Antarctica. Dry Valley Drilling Project Bull 3:13–18

Kyle PR, Treves SB (1974b) Geology of Hut Point peninsula, Ross Island. Antarctic J US 9(5):232–234

Kyle PR, Adams J, Rankin PC (1979) Geology and petrology of the McMurdo Volcanic Group at Rainbow Ridge, Brown Peninsula, Antarctica. Geol Soc Amer Bull 90:676–688

Kyle PR, Dibble RR, Giggenbach WF, Keys J (1982a) Volcanic activity associated with the anorthoclase phonolite lava lake, Mount Erebus, Antarctica. In:. Craddock C (ed) Antarctic geoscience. University of Wisconsin Press, Madison, WI, pp 735–745

Kyle PR, Palais JM, Delmas R (1982b) The volcanic record of Antarctic ice cores: Preliminary results and potential for future investigations. Ann Glaciol 3:172–177

Kyle PR, Wright AC, Kirsch I (1987) Ultramafic xenoliths in the late Cenozoic McMurdo Volcanic Group, Western Ross Sea segment, Antarctica. In: Nixon P (ed) Mantle Xenoliths. Wiley, New York, pp 287–293

Kyle PR, Moore JA, Thirlwall MF (1992) Petrologic evolution of anorthoclase phonolite lavas at Mount Erebus, Ross Island, Antarctica. J Petrol 33:849–875

Langmuir CH, Vocke RD Jr, Hanson GN, Hart SR (1978) A general mixing equation with applications to Icelandic basalts. Earth Planet Sci Lett 37:380–392

Lanyon R, Varne R, Crawford AJ (1993) Tasmanian Tertiary basalts, the Balleny plume, and the opening of the Tasman Sea (southwest Pacific Ocean). Geol 21:555–558

Lanyon R, Crawford AJ, Eggins SM (1995) Westward migration of Pacific Ocean upper mantle into the Southern Ocean region between Australia and Antarctica. Geol 23:511–514

Le Bas MJ, Le Maitre RW, Streckeisen A, Zanettin B (1986) A chemical classification of volcanic rocks based on the total alkali-silica diagram. J Petrol 27:745–750

LeMasurier WE (1990) Late Cenozoic volcanism on the Antarctic plate: An overview. In: LeMasurier WE, Thomson JW (eds) Volcanoes on the Antarctic Plate and Southern Oceans. Antarctic Research Series, vol. 48. American Geophysical Union, Washington, DC, pp 1–17

LeMasurier WE, Rex DC (1982) Volcanic record of Cenozoic glacial history in Marie Byrd Land and western Ellsworth Land: Revised chronology and evaluation of tectonic factors. In: Craddock C (ed) Antarctic geoscience. University of Wisconsin Press, Madison, WI, pp 725–746

LeMasurier WE, Thomson JW (eds) (1990) Volcanoes of the Antarctic plate and southern Oceans. Antarctic Research Series, vol. 48. American Geophysical Union, Washington, DC, pp 1–487

LeMasurier WE, McIntosh WC, Ellerman PJ, Wright AC (1983) USCGS Glacier cruise 1, December 1982-January 1983; Reconnaissance of hyaloclastites in the western Ross Sea region. Antarctic J US 18(5):60–61

Lyon GL (1986) Stable isotope stratigraphy of ice cores and the age of the last eruption at Mount Melbourne, Antarctica. New Zealand Geol Geophys 29(1):135–138

Lyon GL, Giggenbach WF (1974) Geothermal activity in Victoria Land, Antarctica. New Zealand J Geol Geophys 17(3):511–521

Macdonald GA, Katsura T (1964) Chemical composition of Hawaiian lavas. J Petrol 1:172–177

Mankinen EA, Cox A (1988) Paleomagnetic investigation of some volcanic rocks from the McMurdo volcanic province, Antarctica. J Geophys Res 93:11599–11612

Mason RA, Smith JV, Dawson JB, Treves SB (1982) A reconnaissance of trace elements in anorthoclase megacrysts. Mineral Mag 46:7–11

Mawson D (1950) Basaltic lavas of the Balleny Islands. ANARE report. Trans Roy Soc Australia 73:223–231

McGibbon FM (1991) Geochemistry and petrology of ultramafic xenoliths of the Erebus Volcanic Province. In: Thomson MRA, Crame JA, Thomson JW (eds) Geological evolution of Antarctica. Cambridge University Press, Cambridge, UK, pp 317–322

McGinnis LD (1981) Dry Valley Drilling Project. Antarctic Research Series, vol. 33. American Geophysical Union, Washington, DC

McIntosh WC, Gamble JA (1989) A subaerial eruptive environment for the Hallett Coast volcanoes. In, Thomson MRA, Crame JA, Thomson JW, eds. Geological Evolution of Antarctica. Cambridge Un. Press, New York

McIntosh WC, Kyle PR (1990a) Hallett volcanic province; Summary. In: LeMasurier WE, Thomson JW (eds) Volcanoes of the Antarctic Plate and Southern Oceans. Antarctic Research Series, vol. 48. American Geophysical Union, Washington, DC, pp 26–31

McIntosh WC, Kyle PR (1990b) Adare Peninsula. In: LeMasurier WE, Thomson JW (eds) Volcanoes of the Antarctic Plate and Southern Oceans. Antarctica Research Series, vol. 48. American Geophysical Union, Washington, DC, pp 32–35

McIntosh WC, Kyle PR (1990c) Hallett Peninsula. In: LeMasurier WE, Thomson JW (eds) Volcanoes of the Antarctic Plate and Southern Oceans. Antarctic Research Series, vol. 48. American Geophysical Union, Washington, DC, pp 36–39

McIntosh WC, Kyle PR (1990d) Daniell Peninsula. In: LeMasurier WE, Thomson JW (eds) Volcanoes of the Antarctic Plate and Southern Oceans Antarctic Research Series, vol. 48. American Geophysical Union, Washington, DC, pp 40–42

McIntosh WC, Kyle PR (1990e) Coulman Island. In: LeMasurier WE, Thomson JW (eds) Volcanoes of the Antarctic Plate and Southern Oceans. Antarctic Research Series, vol. 48. American Geophysical. Union, Washington, DC, pp 43–45

McIntosh WC, Kyle PR (1990f) Possession Islands. In: LeMasurier WE, Thomson JW (eds) Volcanoes of the Antarctic Plate and

Southern Oceans. Antarctic Research Series, vol. 48. American Geophysical Union, Washington, DC, pp 46–47

McMahon BE, Spall H (1974) Paleomagnetic data from unit 13, DVDP hole 2, Ross Island. Antarctic J US 9(5):229–232

Minshew VH (1967) Geology of the Scott Glacier and Wisconsin Range areas, Central Transantarctic Mountains, Antarctica. PhD dissertation, The Ohio State University, Columbus, OH

Mirsky A (1969) Geology of the Ohio Range - Liv Glacier area. In: Craddock C (ed) Geologic map of Antarctica, Folio 12, Sheet 17. American Geographical Society, New York

Moore JA, Kyle PR (1990) Mount Erebus. In: LeMasurier WE, Thomson JW (eds) Volcanoes of the Antarctic Plate Southern Oceans. Antarctic Research Series, vol. 48. American Geophysical Union, Washington, DC, pp 103–108

Mudrey MG, McGinnis LD (1974) Antarctica geologic history investigated by diamond drilling. Geology 2:291–294

Nathan S, Schulte FJ (1967) Recent thermal and volcanic activity on Mt. Melbourne, northern Victoria Land, Antarctica. New Zealand J Geol Geophys 10:422–430

Nathan S, Schulte FJ (1968) Geology and petrology of the Campbell-Aviator divide, northern Victoria Land, Antarctica. Part 1: Post Paleozoic rocks. New Zealand J Geol Geophys 11:940–975

Palais JM, Legrand M (1985) Soluble impurities in the Byrd-Station ice core, Antarctica; their origin and sources. J Geophys Res 90(C1):1143–1154

Palais JM, Whillans IM, Bull C (1982) Snow stratigraphic studies at Dome C., East Antarctica. An investigation of depositional and diagenetic processes. Ann Glaciol 3:239–242

Palais JM, Kyle PR, Delmas R (1983a) Detailed studies of tephra layers in the Byrd-Station ice core: Preliminary results and interpretation. Antarctic J US 18(5):109–110

Palais JM, Delmas R, Briat M, Jouzel J (1983b) Liquid conductivity of a 44-meter firn core, McMurdo ice shelf. Antarctic J US 18(5):106–107

Palais JM, Chuan R, Spencer MJ (1989) Soluble and insoluble impurities in snow samples from Ross Island, Antarctica. Ant. J. US 24(5):89–91

Priestley RE (1923) Physiography (Robertson Bay and Terra Nova regions). In: British Antarctic "Terra Nova" Expedition, 1910–1913. Harrison, London, pp 1–87

Prior GT (1899) Petrographical notes on the rock specimens collected in Antarctic regions during the voyage of H.M.S. Erebus and Terror under Sir James Clark Ross in 1839–1843. Mineral Mag 12:69–91

Prior GT (1902) Report on the rock specimens collected by the Southern Cross Antarctic Expedition. In: Sharpe RB (eds) Report on the Collections of Natural History during the Voyage of the "Southern Cross". British Museum, London, pp 321–332

Prior GT (1907) Report on the rock specimens collected during the Discovery Antarctic Expedition 1901–1904. Nat History 1, Geology 101–140

Quartermain LB (1964) The Balleny Islands: A descriptive and historical outline, compiled for use of the New Zealand Antarctic Research Expedition Balleny Islands Reconnaissance Party 1963–1964. Antarctic Div., New Zealand, D.S.I.R. Wellington, New Zealand, pp 1–42

Radke L (1982) Sulphur and sulphate from Mr. Erebus. Nature 299:710–712

Riddolls BW, Hancox GT (1968) The geology of the upper Mariner Glacier regions, north Victoria Land, Antarctica. New Zealand J Geol Geophys 11:881–899

Risk GF, Hochstein MP (1974) Heatflow at Arrival Heights, Ross Island, Antarctica. New Zealand J Geol Geophys 17(3):629–644

Rocholl A, Stine M, Molzahn M, Hart SR, Wörner G (1995) Geochemical evolution of rift magmas by progressive tapping of a stratified mantle source beneath the Ross Sea rift, northern Victoria Land, Antarctica. Earth Planet Sci Lett 131:207–224

Rose WI, Chuan RL, Kyle PR (1985) Rate of sulphur dioxide emission from Erebus volcano, Antarctica, December 1983. Nature 316:710–711

Rothery DA, Oppenheimer C (1994) Monitoring Mount Erebus by satellite remote sensing. In: Kyle PR (ed) Volcanological and environmental studies of Mount Erebus, Antarctica. Antarctic Research Series, vol. 66. American Geophysical. Union, Washington, DC, pp 51–56

Rowley PD (1990) Peter I. Island. In: LeMasurier WE, Thomson JW (eds) Volcanoes of the Antarctic Plate and Southern Ocean. Antarctic Research Series, vol. 48. American Geophysical Union, Washington, DC, pp 454–456

Schmidt DL, Rowley PD (1986) Continental rifting and transform faulting along the Jurassic Transantarctic rift, Antarctica. Tectonics 5(2):179–291

Schmidt-Thomé M, Mueller P, Tessensohn F (1990) Malta Plateau. In: LeMasurier WE, Thomson JW (eds) Volcanoes of the Antarctic Plate and Southern Oceans. Antarctic Research Series, vol. 48. American Geophysical Union, Washington, DC, pp 53–59

Sharp Z (2007) Principles of stable isotope geochemistry. Prentice Hall, Upper Saddle River, NJ

Shaw G (1982) On the residence time of the Antarctic ice sheet sulfate aerosol. J Geophys Res 87:4309–4313

Smith WC (1954) The volcanic rocks of the Ross Archipelago. Natural History Rept. British Antarctic Terra Nova Expedition, 1910. Geology 2(1):1–107

Smith WC (1964) Volcanic rocks of Cape Adare and erratics from the Terra Nova Region. Natural History Report, British Antarctic Terra Nova Expedition, 1910. Geology 2:151–206

Steiger RH, Jäger E (1977) Subcommission on geochronology: Convention on the use of decay constants in geo-and cosmochronology. Earth Planet Sci 36:359–362

Stilwell JD, Levy RH, Harwood DM (1993) Preliminary paleontological investigation of Tertiary glacial erratics from the McMurdo Sound region, East Antarctica. Antarctic J US 28(5):16–19

Stott LD, McKelvey BC, Harwood DM, Webb PN (1983) A revision of the ages of Cenozoic erratics at Mount Discovery and Minna Bluff, McMurdo Sound. Antarctic J US 18(5):36–38

Stuckless JS, Ericksen RL (1976) Strontium isotope geochemistry of the volcanic rocks and associated megacrysts and inclusions from Ross Island and vicinity, Antarctica. Contrib Mineral Petrol 58:111–126

Stuckless JS, Weiblen PW, Schulz KJ (1974) Magma evolution for the alkalic rocks of the Ross Island area, Antarctica. EOS, 55:474. American Geophysical Union, Washington, DC

Stuckless JS, Miesch AT, Goldich SS, Weiblen PW (1981) A Q-mode factor for the petrogenesis of the volcanic rocks from Ross Island and vicinity, Antarctica. In: McGinnis LD (ed) Dry Valley Drilling Project. Antarctic Research Series, vol. 33. American Geophysical Union, Washington, DC, pp 257–280

Stump E, Sheridan MF, Borg SG, Sutter JF (1980) Early Miocene subglacial basalts, the East Antarctic ice sheet, and uplift of the Transantarctic Mountains. Science 207:757–758

Stump E, Smit JH, Self S (1986) Timing of events during the late Proterozoic Beardmore Orogeny, Antarctica. Geol. Soc. Amer. Bull. 97(8):953–965

Stump E, Borg SG, Sheridan MR (1990a) Mount Early. In: LeMasurier MR, Thomson JW (eds) Volcanoes of the Antarctic Plate and Southern Oceans. Antarctic Research Series, vol. 48. American Geophysical Union, Washington, DC, pp 138–139

Stump E, Borg SG, Sheridan MF (1990b) Sheridan Bluff. In: LeMasurier WE, Thomson JW (eds) Volcanoes of the Antarctic Plate and Southern Oceans. Antarctic Research Series, vol. 48. American Geophysical Union, Washington, DC, pp 136–137

Sun S-S, Hanson GN (1974) Genesis of McMurdo Volcanics on Ross Island. Antarctic J US 9(4):234–236

Sun SS, Hanson GN (1975) Origin of Ross Island basanitoids and limitations upon the heterogeneity of mantle sources for alkali basalt and nephelinites. Contrib Minerals Petrol 52:77–106

Sun SS, Hanson GN (1976) Rare earth element evidence of the differentiation of McMurdo Volcanics, Ross Island, Antarctica. Contrib Mineral Petrol 54:139–155

Taylor HP Jr (1968) The oxygen isotope geochemistry of igneous rocks. Contrib Mineral Petrol 19:1–71

Thomson JA (1916) Report of the inclusion of the volcanic rocks of the Ross archipelago. British Antarctic Expedition, 1907–1909. Rept Scient Investig, Geology 2(8):129–148

Treves SB (1967) Volcanic rocks from the Ross Island, Marguerite Bay, and Mt. Weaver areas, Antarctica. Japanese Antarctic Research Expedition Scientific Report, vol. 1. National Institute of Polar Research, Tokyo, Japan, pp 136–149

Treves SB (1971) Geology of the volcanic rocks of the Ross Island area, Antarctica. Antarctic J US 6(5):193–194

Treves SB, Kyle PR (1973a) Renewed volcanic activity of Mt. Erebus, Antarctica. Antarctic J US 8:156

Treves SB, Kyle PR (1973b) Geology of DVDP 1 and 2, Hut Point Peninsula, Ross Island, Antarctica. Dry Valley Drilling Project ull 2:11–82

Treves SB, Barnes CG, Stillwell DP (1975) Geology and petrography of rocks from the Ross Sea near Ross Island and the mouth of Taylor Valley. Antarctic J US 10:297–302

Ugolini FC (1967) Soils of Mount Erebus, Antarctica. New Zealand J Geol Geophys 10(2):431–442

Valley JW, Cole DR (eds) (2001) Stable isotope geochemistry. Review in mineralology and geochemistry, vol. 43. Mineralogical Society of America, Blacksburg, VA

Vella P (1969) Surficial geological sequence, Black Island and Brown Peninsula, McMurdo Sound, Antarctica. New Zealand J Geol Geophys 12(4):761–770

Victor P-E (1963) Man and the conquest of the Poles. Hamish Hamilton, London

Wade FA (1969) Geology of Marie Byrd Land. In: Craddock C (ed) Geologic map of Antarctica, Folio 12, Sheet 18. American Geographical Society, New York

Warren G (1969) Geology of the Terra Nova Bay-McMurdo Sound area, Victoria Land. In: Craddock C (ed) Geologic map of Antarctica, Folio 12, Sheet 14. American Geographical Society, New York

Waterhouse BC (1965) Western Ross Sea-Balleny Islands Expedition: January-March, 1965. Report, D.S.I.R. and New Zealand Geological Survey, Wellington, New Zealand

Weiblen PW, Hunter WC, Stuckless JS, Schulz KJ, Mudrey MG Jr (1981) Correlation of clinopyroxene compositions with environment of formation based on data from Ross Island volcanics, Antarctica. In: McGinnis DL (ed) Dry Valley Drilling Project. Antarctic Research Series, vol. 33. American Geophysical Union, Washington, DC, pp 229–246

Wenner DB (1974) Oxygen isotopes in Cenozoic volcanic rocks in Antarctica. Antarctic J US 9(5):243–244

Wilson M (1989) Igneous petrology. A global tectonic approach. Unwin Hyman, London

Wilson TJ (1995) Cenozoic transtension along the Transantarctic Mountains-West Antarctic rift boundary, southern Victoria Land, Antarctica. Tectonics 14(2):531–545

Wörner G, Orsi G (1992) Volcanic observations on Scott Island in the Antarctic Ocean. Polarforschung 60:82–83

Wörner G, Viereck L (1987) Subglacial to emergent volcanism at Shield Nunatak, Mt. Melbourne volcanic field, Antarctica. Polarforschung 57(1,2):27–41

Wörner G, Viereck L (1990) Mount Melbourne. In: LeMasurier WE, Thomson JW (eds) Volcanoes of the Antarctic plate and Southern Oceans. Antarctic Research Series, vol. 48. American Geophysical Union, Washington, DC, pp 72–80

Wörner G, Viereck L, Hertogen J, Niephaus H (1989) The Mt. Melbourne volcanic field (Victoria Land, Antarctica). II. Geochemistry and magma genesis. Geologisches Jahrbuch E38:395–433

Wright-Grassham AC (1987) Volcanic geology, mineralogy, and petrogenesis of the Discovery volcanic subprovince, southern Victoria Land, Antarctica. PhD dissertation, New Mexico Institute of Mining and Technology, Socorro, New Mexico

Wright AC, Kyle PR (1990a) Balleny Islands. In: LeMasurier WE, Thomson JJ (eds) Volcanoes of the Antarctic Plate and Southern Oceans. Antarctic Research Series, vol. 48. American Geophysical Union, Washington, DC, pp 449–450

Wright AC, Kyle PR (1990b) Scott Island. In: LeMasurier WE, Thomson JW (eds) Volcanoes of the Antarctic Plate and Southern Oceans. Antarctic Research Series, vol. 48. American Geophysical Union, Washington, DC, pp 452–453

Zindler A, Hart SR (1986) Chemical geodynamics. Ann Rev Earth Planet Sci 14:493–571

Chapter 17
The East Antarctic Ice Sheet

Antarctica is the fifth largest continent on the Earth with an area of 14,200,000 km². About 98% of the surface of Antarctica is covered by an ice sheet that is 2,000 m thick on average. The volume of the ice sheet has changed in the course of time and reached a maximum during the late Wisconsin (Weichselian) ice age about 20,000 years ago. At that time, the volume of the Antarctic ice sheet was 37.7×10^6 km³, of which 24.2×10^6 km³ covered the East Antarctic craton and 13.5×10^6 km³ overlay the bedrock of West Antarctica and the associated islands (Hughes et al. 1981). Taken together, the ice sheets of Antarctica constitute about 90% of the ice that exists on the Earth. The ice cap that covers Greenland makes up about 9% of the total and all other ice caps and glaciers on the Earth contribute only 1% of the ice on the Earth (Hughes et al. 1981; Bindschadler 2004).

The ice sheets in Antarctica store a very large amount of water which has depressed the level of water in the oceans and has thereby enlarged the surface areas of the continents and oceanic island where a large proportion of the human population lives. If the ice in Antarctica and Greenland were to melt in the future, sealevel would rise by 65 m or more. Although that is not likely to happen in the foreseeable future, any increase in the amount of glacial meltwater that is discharged into the oceans causes sealevel to rise. Evidently, there is a close connection between the volume of the ice sheets in Antarctica and Greenland, the level of water in the oceans, and the amount of living space available to humans.

The ice sheets of the Earth are also affecting the global climate because they increase the *albedo* of the Earth (i.e., the fraction of sunlight that is reflected into space) and thereby lower the average global temperature. The present albedo of the Earth is 39% (Hartmann 2005). An increase in the average global temperature caused by the greenhouse effect will cause additional snow and ice to melt which will *decrease* the albedo of the Earth and cause the average global temperature to rise still more. In this way global warming can trigger a positive feed-back loop because a reduction in the annual snow cover decreases the albedo of the Earth, which leads to more warming and causes sea level to rise when the additional melt-water flows into the oceans. These issues emphasize the importance of the ice sheets of Antarctica to the global climate and hence to the environment in which we live.

17.1 Dynamics of the Antarctic Ice Sheets

The elevation of the present ice sheet in East Antarctica in Fig. 17.1 rises from sea level along the coast to more than 4,000 m in the interior at Dome Argus (18°00′S and 77°00′E). The ice moves under the influence of gravity from the interior of Antarctica toward the coast where it either forms icebergs or coastal ice shelves at the outflows of outlet glaciers and ice streams (e.g., the Ross, Renne, Filchner, Amery and other ice shelves).

The ice shelves grow in size by the continuous addition of ice from the interior of the continent until large tabular icebergs break off and float north into the open ocean where they gradually break up and melt. Occasionally, icebergs run aground along the coast of Antarctica and release sediment that was entrained in the basal ice. An iceberg that separated from the Ross Ice Shelf in March of 2000 was about 300 m thick and had an area of nearly 13,000 km² comparable to the

G. Faure and T.M. Mensing, *The Transantarctic Mountains: Rocks, Ice, Meteorites and Water*,
DOI 10.1007/978-90-481-9390-5_17, © Springer Science+Business Media B.V. 2010

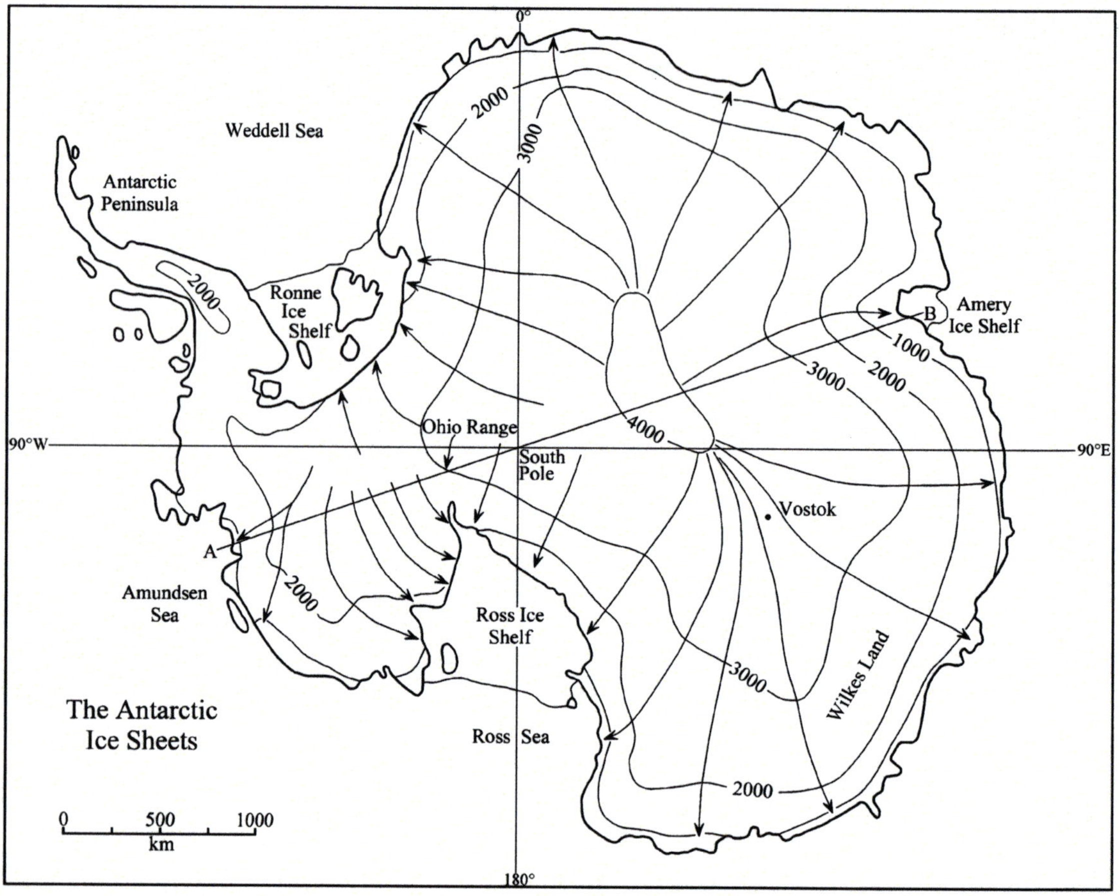

Fig. 17.1 The ice sheet that covers Antarctica rises to an elevation of more than 4,000 m above present sea level and slopes away from this central dome toward the coast. The ice flows under the influence of gravity at right angles to the topographic contour lines. The local thickness of the ice sheet varies irregularly depending on the subglacial topography of the bedrock and on its elevation above sea level. The line A–B identifies the position of the topographic profile of the ice sheet in Fig. 17.2. The approximate position of the Ohio Range locates the Transantarctic Mountains which separate East and West Antarctica in the region between the Ross and Ronne ice shelves (Adapted from Hughes 1981; National Geographic 1990)

area of the state of Connecticut. This iceberg became a floating ice island in the Ross Sea on which planes could land to refuel or in case of mechanical difficulties. This iceberg eventually ran aground in the Ross Sea about 300 km from its point of origin, but was refloated by the ground motions caused by a large earthquake (magnitude 8.4) in Peru on June 23, 2000 (Science News 2005, 167:45).

The potential use of large icebergs as a source of freshwater has been considered but was found to be impractical because icebergs are difficult to tow. The reason is that only about 15% of the thickness of an iceberg extends above the surface of the water depending on the density of the ice, the thickness of the berg, and the density of the seawater. A graph expressing the freeboard of icebergs as a function of their thickness was published by Schwerdtfeger (1979) who also provided information, originally compiled by Nazarov (1962), on the locations of icebergs in the southern oceans between 1773 and 1960. The engineering aspects of towing icebergs were investigated by Job (1975, 1978a, b) by means of a computer simulation. Other reports about the feasibility of using Antarctic icebergs as a source of freshwater were included in a book edited by Husseiny (1978). Additional reports were

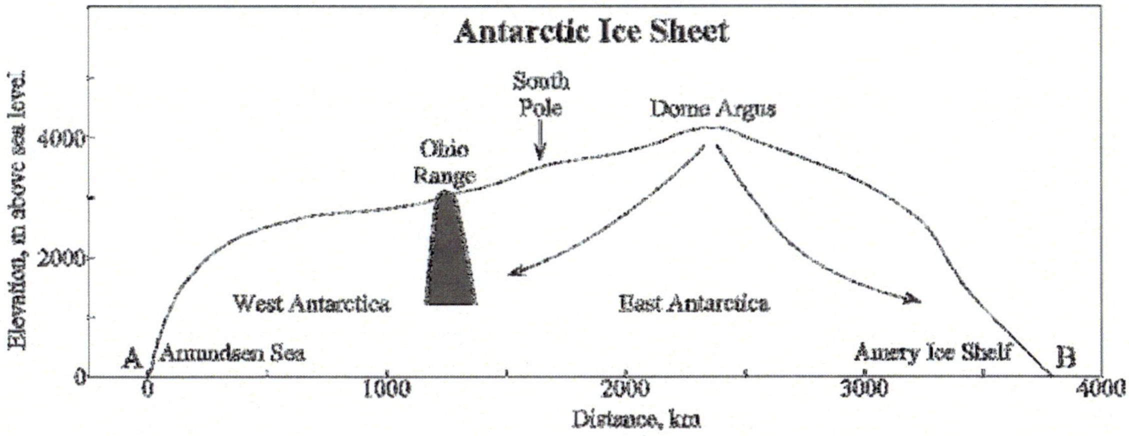

Fig. 17.2 The topographic profile of the Antarctic ice sheet along line A–B in Fig. 17.1 demonstrates that the ice flows from the region of Dome Argus toward the coast. The Ohio Range identifies the position of the Transantarctic Mountains along the line of section and thereby divides the ice sheet into the East and West Antarctic components. The East Antarctic ice sheet is underlain by rocks of the continental crust, whereas the West Antarctic ice sheet bridges the structural basins that formed by extension of the crust caused by rifting during the Cenozoic. The bedrock surfaces are not shown. The vertical exaggeration is 250 times (Drawn from Figure 6.10 of Hughes 1981)

published by Swithinbank (1969), Hult and Ostrander (1973), Weeks and Campbell (1973), Huppert and Turner (1978), Weeks and Mellor (1978), and Holdgate (1980).

Returning to the Antarctic ice sheets and the ice shelves that are the sources of icebergs, we note in Fig. 17.2 that the summit of the ice sheet (Dome Argus) is approximately centered over East Antarctica. We also note that the West Antarctic ice sheet is distinctly thinner than the ice sheet that covers East Antarctica even though they are connected to each other in Fig. 17.1 through the gap between the Ohio Range and the Pensacola Mountains. The apparent thickening of the East Antarctica icesheet is primarily caused by the Transantarctic Mountains which block the outward flow of the ice and force it to flow through comparatively narrow valleys that cross the Transantarctic Mountains. The names of outlet glaciers in the Transantarctic Mountains have been mentioned in the preceding chapters of this book (e.g., the Byrd, Nimrod, Beardmore, Shackleton, Scott, and Reedy glaciers in the central Transantarctic Mountains as well as the Ferrar, Mackay, David, Mariner, and Rennick glaciers in southern and northern Victoria Land).

The profile of the East Antarctic ice sheet in Fig. 17.2 is a sensitive function of several factors that maintain its shape and thickness. The ice in the interior of Antarctica forms by the accumulation of snow at the rate of only 5–10 mm of water-equivalent per year (Budd et al. 1971; Giovinetto and Bull 1987). The rate of precipitation in coastal areas ranges from 20 to 60 mm of water equivalent per year, and is up to about 12 times higher than the precipitation rate in the interior of the continent. Therefore, large parts of the interior of East Antarctica are a polar desert. In spite of the low rate of meteoric precipitation, the thickness of the ice in large areas is greater than 3,000 m (Hughes et al. 1981, Figure 6.16).

The thickness of the East Antarctic ice sheet is augmented by snow that is transported by wind from the coastal regions to the interior. Snow that is deposited in the central region of East Antarctica accumulates without melting and is gradually converted into *firn* composed of small ice pellets and then into solid *ice*. The only loss of ice and snow occurs by sublimation, by the flow of ice under the influence of gravity, and by melting at the base of the ice sheet. The ice sheet "flows" by sliding on its base and by plastic deformation. Cold-based ice sheets and valley glaciers that are frozen to their bed flow by plastic deformation of the ice, whereas wet-based glaciers slide on their base. The temperature at the base of the East Antarctic ice sheet near its margin exceeds the melting point of the ice and causes the basal ice to melt (Hughes et al. 1981). Therefore, the marginal parts of the East Antarctic ice sheet move by sliding, whereas the ice in the center of the Antarctic ice sheet flows by plastic deformation because it is frozen to its bed (see Fig. 17.34).

The average regional slope (s) between Dome Argus, (4,000 m) and the coast of East Antarctica located at a distance of 1,425 km is:

$$s = \frac{4000 \times 100}{1{,}425 \times 10^3} = 0.28\%$$

A gradient of this magnitude is not perceptible, which conveys the impression that the East Antarctic ice sheet is flat. However, the surface of the ice sheet along the Transantarctic Mountains and along the coast is far from being flat and has quite appreciable topography with elevation differences of the order of 100 m or more. The topographic relief of the ice sheet occurs where a thin layer of ice flows over hills and valleys in the underlying bedrock. Deep moats tend to form adjacent to nunataks which channel the catabatic wind around them and which radiate the heat they absorb from sunlight and thereby accelerate the rate of sublimation of ice and snow. Large funnel shaped depressions at the edge of the polar plateau occur where outlet glaciers draw down the ice. These kinds of depression are, in some cases, up to 100 km wide and several hundred meters below the regional level of the ice sheet. Wind also sculpts the surface snow into *sastrugi* which tend to be aligned along prevailing wind directions. Sustrugi may be up to 100 cm high and constitute a hazard for snowmobiles, tracked vehicles, and for aircraft landing or taking off on the ice sheet.

The greatest danger for persons traveling on the surface of the ice sheet arises from *crevasses* that form in areas where the ice sheet is subjected to extensional stress which occurs where the surface slope changes or where valley glaciers are forced to flow around a bend in their channels. Under these conditions, the brittle ice at the top of the ice sheet forms sets of fractures that may be up to 10 m wide and extend to depths of 100 m or more. These fractures (or crevasses) are hidden in most cases by snow bridges which may collapse without warning under the weight of a snowmobile or any kind of tracked oversnow vehicle. Persons walking on the ice sheet who fall into a crevasse rarely survive due to injury during the fall, hypothermia, or asphyxiation caused by compression of the chest.

The surface of the East Antarctic ice sheet consists of a central area of snow accumulation and a marginal region of ablation. In the *ablation zone* adjacent to the western slope of the Transantarctic Mountains the ice is exposed and ablates by sublimation directly into the air without melting. The color of the ice is pale blue because it absorbs the wavelength of red light. Hence these exposures of ice are called "blue ice" areas.

The blue ice areas are of great interest because they expose old ice that formed in the central accumulation area that may be up to 1,000 km distant in the "up-ice" direction. Moreover, meteorites, cosmic spherules, fly-ash, and other kinds of object or particles that fell on the surface of the ice sheet in the area of accumulation emerge from the ice in the ablation area and can be recovered for scientific study (Chapter 18). Blue ice is typically slippery and rippled by the katabatic winds which polish its surface. Many unwary travelers have slipped on the ice and fallen unexpectedly, sometimes with painful consequences.

17.2 Cenozoic Glaciation of Antarctica

The break-up of Gondwana during the Mesozoic Era altered the distribution of the continents in the southern hemisphere and exposed the East Antarctic craton to the southern ocean which still surrounds it. The tectonic processes that caused Gondwana to break up also affected East Antarctica by contributing to the uplift of the Transantarctic Mountains, by the formation of the West Antarctic rift system, and by rearranging the location of continental fragments (e.g., Marie Byrd Land, the Ellsworth Mountains, and the Antarctic Peninsula).

The accumulation of snow and the growth of local ice caps in East Antarctica started during the Cenozoic Era after the Transantarctic Mountains had begun to rise at about 50 Ma (early Eocene). The uplift of the Transantarctic Mountains was related to the development of the West Antarctic rift valley which caused the eruption of alkali-rich volcanic rocks (Chapter 16). In several cases, these rocks consist of subglacially erupted palagonites and hyaloclastites indicating that ice was present at the time of the eruption. The oldest subglacially erupted volcanic rocks at Sheridan Bluff and on Mt. Early at the head of the Scott Glacier crystallized at 19.59 ± 1.69 Ma (Section 16.6; Stump et al. 1990).

Even more direct evidence for the presence of glaciers in Antarctica was provided by the core recovered at site 270 in the Ross Sea by the Deep Sea Drilling Project (DSDP). This core contains a thick layer of till overlying a basal layer of glauconite-bearing sandstone. The K-Ar date of 26.0 ± 0.4 Ma (Oligocene) of the glauconite *predates* the deposition of the water-laid till. These results indicate that ice caps and valley glaciers were present in the Transantarctic after 26.0 ± 0.4 Ma and before 19.59 ± 1.69 Ma. In other words, the glaciation of the Transantarctic Mountains probably started in the late Oligocene and was widespread in early Miocene time.

These results support a hypothesis proposed by Drewry (1975) and contradict those who had previously postulated that the start of glaciation of Antarctica coincided with the start of the continental glaciation in the northern hemisphere. Additional support for the middle-to-late Cenozoic glaciation of Antarctica is provided by geomorphic evidence in the Transantarctic Mountains and by the ages of the microfossils and biological materials (leaves and wood) in till of the Sirius Group to be presented in Chapter 19. The apparent coincidence of the uplift of the Transantarctic Mountains and the growth of the East Antarctic ice sheet supports the proposals of Bull et al. (1962), Mercer (1968), and Calkin and Nichols (1971) that ice caps formed first in the Transantarctic Mountains and then expanded into the continental ice sheet that eventually covered the entire continent. An alternative proposal by Denton et al. (1971) holds that storms from the southern ocean penetrated into the interior of the continent and deposited snow on the Gamburtsev Mountains that were discovered in 1968 by the Russian geophysicist A.P. Kapitsa. The resulting glaciers then expanded from the Gamburtsev Mountains to form an ice sheet that eventually covered the entire continent.

These proposals are not mutually exclusive because the ice sheet may have started both along the Transantarctic Mountains and in the Gamburtsev Mountains. In fact, the radio-echo surveys of Drewry (1971, 1972, 1973) indicate that an extensive network of subglacial valleys exists along the western flank of the Transantarctic Mountains as well as in the Gamburtsev Mountains in Fig. 17.3. Subglacial mountains in Queen Maud Land and New Schwabenland were also favorably positioned along the coast and may have developed centers of glaciation that contributed to the growth of the East Antarctic ice sheet.

According to Drewry (1975), a decrease of the average annual temperature during the Oligocene at about 26 Ma caused snowfields and ice caps to form along the eastern side of the Transantarctic Mountains. The ice caps fed valley glaciers which flowed east toward the Ross Embayment and west into the interior lowlands of East Antarctica. The ice that flowed to the coast of the Ross Sea formed icebergs which deposited the marine till containing striated dropstones that was recovered in the DSDP core at site 270. The valley glaciers that flowed west from the Transantarctic Mountains coalesced to form piedmont glaciers in the East Antarctic lowlands. The piedmont glaciers gradually thickened until the ice-flow direction was reversed and outlet glaciers crossed the Transantarctic Mountains from west to east and built up the Ross Ice Shelf within the Ross Embayment.

As time passed, the volume of the East Antarctic ice sheet increased sufficiently to bury the Transantarctic Mountains during the early Miocene until only the highest peaks remained ice-free as nunataks. At this time, the ice deeply eroded the valleys of the outlet glaciers and imposed on them the typical U-shaped cross-sections that are evident in the ice-free valleys of southern Victoria Land and elsewhere in the Transantarctic Mountains. More recently, the East Antarctic ice sheet thinned causing previously ice-covered areas of the Transantarctic Mountains to become ice-free and exposing till of the Sirius Group that was deposited by the ice sheet during its maximum extent.

The sequence of events that led to the glaciation of East Antarctica have been widely discussed by many authors including Hollin (1962), Crowell and Frakes (1970), Kvasov and Verbitsky (1981), Wellman and Tingey (1981), Denton and Hughes (1981), Labeyrie et al. (1986), Robin (1988), Denton et al. (1989), Webb (1990), Moriwaki et al. (1992), Mayewski et al. (1995), and others. In addition, the advances and retreats of the ice sheet across the Transantarctic Mountains have been investigated in selected regions of the Transantarctic Mountains. The results of these regional studies will be presented in Chapter 19.

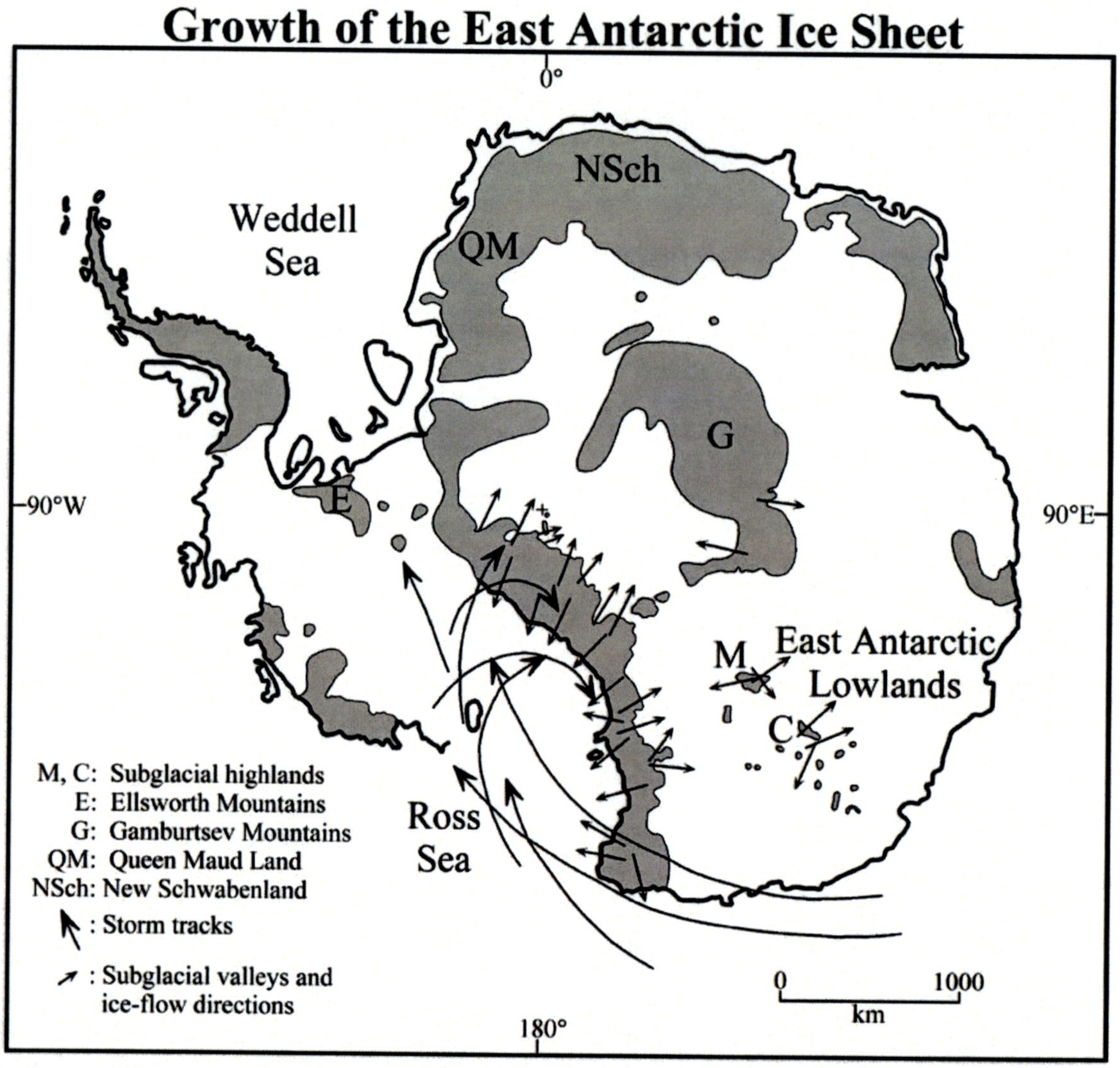

Fig. 17.3 The subglacial mountains of East Antarctica and the Transantarctic Mountains are outlined at the 500 m contour based on the results of Kapitsa (1968), Bentley (1972), and Drewry (1973). The *small arrows* indicate the locations of major valleys through which ice may have flowed into the adjacent lowlands. The diagram also indicates the tracks of low-pressure air masses (*large arrows*), which rotate in a clockwise direction in the southern hemisphere and therefore turn right in the Ross Sea toward the Transantarctic and Gamburtsev mountains (Adapted from Figure 5 of Drewry 1975)

The complex interaction of many factors that can cause continental glaciations was succinctly expressed by Crowell and Frakes (1970, p. 193):

> A basic cause of glaciation is…the arrangement of continental masses and their effect on ocean circulation. Additional significant factors…are the elevation of continents, the elevation and extent of highlands, the availability of moisture sources, the freezing and thawing of polar sea, the effect of snow and ice cover and frozen seas on the Earth's albedo and the mean Earth temperature, the positioning and swinging of storm tracks along the polar front with the development of strong air-mass contrast, and others. When many of these facts reinforce, continental glaciation ensues.

Another factor that favored the formation of the continental ice sheet of Antarctica during the Cenozoic was the displacement of the Antarctic plate into the south-polar region which caused climatic cooling.

17.3 The Elephant Moraine

The geology of the East Antarctica is still largely unknown because almost all of it is covered by the ice sheet. Some information about the geology of the region is derivable from regional airborne geophysical surveys, but actual samples of the rocks that underlie the ice sheet are only obtainable by drilling through the ice sheet or by study of rock clasts in the moraines that have accumulated along the edge of the ice sheet or on its surface.

The ice sheet was penetrated by drilling at Vostok Station (Russia) on the East Antarctic ice sheet and at Byrd Station (USA) in West Antarctica. The resulting cores contain valuable records of past climatic conditions, but only the core at Byrd Station recovered samples of the subglacial bedrock. The climate record contained in these cores and in the core that was recently recovered at Dome C will be featured in a later section of this chapter.

Samples of the bedrock of East Antarctica have accumulated in the terminal and recessional moraines that have been deposited by the ice sheet along the western slope of the Transantarctic Mountains. In addition, supraglacial moraines that are forming on the surface of the ice sheet contain rock clasts that originated from the bedrock in the path of the ice. The Elephant Moraine is such a supraglacial moraine located in a blue-ice area of the East Antarctic ice sheet west of the Transantarctic Mountains of southern Victoria Land. The Elephant Moraine, as well as the nearby Reckling Moraine, are composed of bands of clasts and fine-grained sediment that are released by sediment-laden "dirty" ice that is exposed at the surface. The bands of sediment extend parallel to the ice-flow direction and are composed of rock debris that originated from the bedrock at the base of the ice sheet.

The rock clasts of both moraines were classified in order to identify the rocks that underlie the ice sheet and thereby to test the hypothesis that the ice sheet west of the Transantarctic Mountains flows over granitic and high-grade metamorphic rocks of the East Antarctic craton.

17.3.1 Physical Dimensions

The Elephant Moraine (informal name) in Fig. 17.4 is located at 76°17.58′S and 157°20.082′E about 70 km west of the Allan Hills in southern Victoria Land. This area of the ice sheet is within the drainage basin of the David Glacier which crosses the Transantarctic Mountains north of the latitude of the Elephant Moraine. Accordingly, the ice sheet that underlies the Elephant Moraine is flowing in a northeasterly direction toward the Prince Albert Mountains.

The moraine is about 5.6 km long and about 1.7 km wide at its widest part. It originates in a large horseshoe-shaped basin open to the northeast. The basin is defined by a system of ice ramps that carry crevasses and ice pinnacles at their crests. The difference in the elevation from the polar plateau to the bottom of the basin is 120 m.

The Elephant Moraine was discovered by William MacDonald, Chief of the Branch of International Activities of the U.S. Geological Survey, during an airborne photographic survey of the mountains of southern Victoria Land sponsored by the U.S. Geological Survey and flown by crews of the U.S. Navy. It was named Elephant Moraine because, when viewed from the air, its shape resembles an Elephant with a long trunk, a massive body with a short tail, and stumpy legs. It even has an eye approximately in the right place.

17.3.2 Lithologic Composition of Rock Clasts

The Elephant Moraine is composed primarily of rock clasts and large boulders. Fine grained sediment is preserved only in areas where it is protected by lag gravel. The clasts and residual fines occur in bands on the surface of bare (blue) ice. Snow and firn occur only locally depending on the time that has elapsed since the most recent snowstorm.

The clasts and fine-grained sediment lie on sediment-laden ice in Fig. 17.5 from which they are derived by

Fig. 17.4 The supraglacial moraine, known informally as the Elephant Moraine, is located about 70 km northwest of the Allan Hills in a large blue-ice basin of the East Antarctic ice sheet. The ice sheet is flowing northeast at this locality into the ice-drainage basin of the David Glacier. The moraine consists of clasts and fine-grained sediment that are released by the ablation of sediment-laden ice from the base of the ice sheet. The clasts and fines consist primarily of dolerite, basalt, and diamictite of the Ferrar Group and of sandstone, siltstone, shale, and coal of the Beacon Supergroup. Granitic basement rocks were not found among the 47,687 clasts that were classified at 230 surveyed sites in the moraine (Faure and Taylor 1985; Taylor 1986) (Photo by G. Faure)

sublimation of the ice. Taylor (1986) mapped the entire moraine by counting and classifying 47,687 clasts at 230 surveyed sites of know area identified on Fig. 17.6. The abundances of clasts of different lithologies were expressed as the number of clasts per square meters. The clasts in Fig. 17.7a consist primarily of dolerite and basalt of the Ferrar Group with lesser abundances in Fig. 17.7b of sandstone, siltstone, shale, hornfels, and coal of the Beacon Supergroup (Faure et al. 1984; Mensing 1991). About 20% of the clasts are faceted, striated, and polished (Taylor 1986).

Clasts of diamictite of the Mawson Formation which is exposed in the Allan and Coombs hills (Section 12.3.1) occur at 60% of the surveyed sites at abundances greater than zero but less than about 8% by number in Fig. 17.7c. The abundances of diamictite clasts at four sites ranged from 14% to 26%, presumabley because diamictite boulders had disintegrated at these locations. The presence of diamictite boulders in the Elephant Moraine implies that they originated from subglacial deposits located in the path of the ice southwest of the moraine.

The Elephant Moraine also contains pellets of molded clay-rich sediment as well as clasts of black micritic calcite and opaline silica cementing pebbly sediment (Faure et al. 1986). No meteorite specimens were found within the moraine, although numerous meteorite specimens were recovered on the blue ice that surrounds the Elephant Moraine (Cassidy et al. 1983). In addition, no clasts of granitic or high-grade metamorphic basement rocks were identified among the large number of clasts that were examined (Taylor 1986).

Fig. 17.5 A band of sediment-laden ice in the Elephant Moraine is covered by clasts and fine-grained sediment that were released by the sublimation of the ice. Such bands of dirty ice can be traced for several kilometers along the length of the moraine and presumably are basal ice that was forced to the surface of the ice sheet by the mountainous topography of the bedrock over which the ice is flowing. The moraine is still forming at the present time depending on the rate of sublimation of the ice. The Elephant Moraine is located at the base of a topographic ramp in the surface of the blue ice area (Photo by G. Faure)

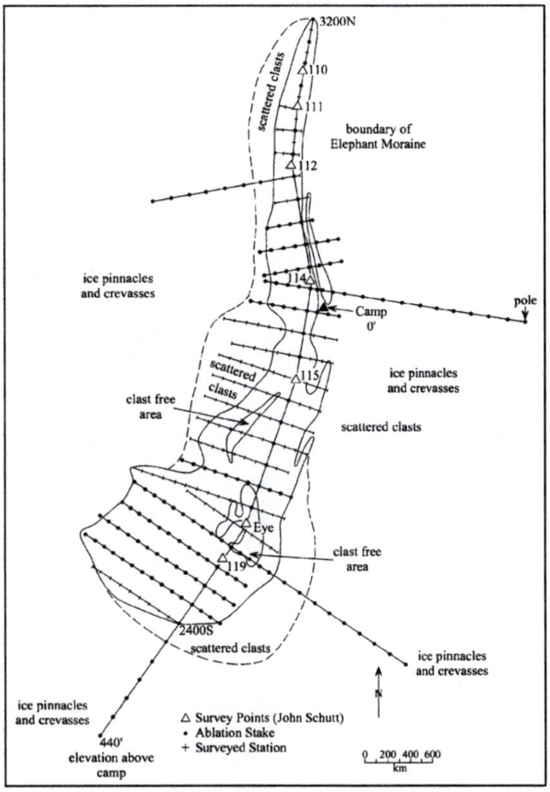

Fig. 17.6 The system of surveyed lines within the Elephant Moraine define the locations of stations at which clasts within the moraine were counted and classified by Taylor (1986). Locations identified by black circles were also used to measure ablation rates of the ice (Section 17.3.7). The baseline extends between survey stations established by John Schutt which are identified by white triangles numbered from 110 to 119. The "Eye" is an ice hill containing a survey marker by the U.S. Geological Survey. Clast-free areas within the moraine are outlined within the moraine. The Elephant Moraine is located within a basin, open to the northeast, formed by ice ramps with crevasses and pinnacles at their crests (Adapted from Faure and Sutton 1985 and Taylor 1986)

17.3.3 Origin of the Elephant Moraine

The high abundance of dolerite and basalt clasts of the Ferrar Group and the ubiquitous presence, but low abundance, of sedimentary rocks of the Beacon Supergroup demonstrates that the subglacial bedrock in the path of the ice southwest of the Elephant Moraine consists of sandstone and shale of the Beacon Supergroup capped by dolerite sills and related basalt flows. These rocks presumably cover the granitic and metamorphic rocks of the East Antarctic craton. The topographic relief of the ice and the very existence of the Elephant Moraine both suggest that the ice is flowing over mountainous topography of the bedrock at its base, perhaps consisting of mesas composed of sandstone and shale that are capped by dolerite sills and separated from each other by steep-sided valleys.

This conjecture was confirmed by a ground-based radio-echo survey (RES) carried out by Delisle and Sievers (1991) on the ice fields adjacent to the Allen Hills and, more recently, by Damm (1996) who used RES equipment mounted in a helicopter to measure the ice thickness near the David Glacier. Still more

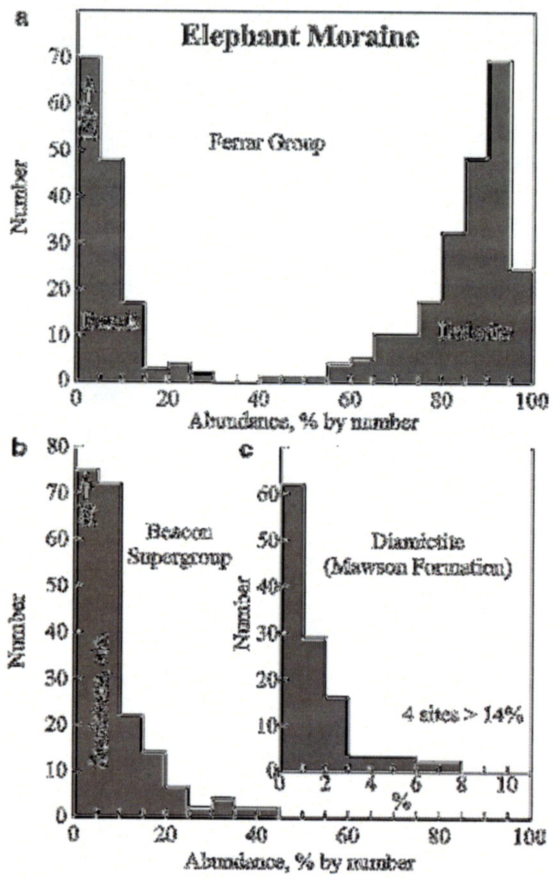

Fig. 17.7 (**a**) The Elephant Moraine is forming by ablation of bands of sediment-laden ice that occur in a large area of blue ice. The clasts consist primarily of dolerite of the Ferrar Group (90–95% by number), whereas basalt clasts of the Kirkpatrick Basalt are much less abundant (0–10% by number). (**b**) Sandstone, siltstone, shale, coal, and hornfels of the Beacon Supergroup are present at all 230 study sites on the moraine, but make up only 0–10% by number. (**c**) Clasts of diamictite of the Mawson Formation (Ferrar Group; Figure 22 of Taylor, 1986) are present at about 60% of the study sites on the moraine at abundances from >0% to <8% by number. Higher abundances of diamictite clasts, between 14% and 26%, occur at four locations. Clasts of granitic and high-grade metamorphic rocks of the basement complex and meteorite specimens were not found in the moraine (Data from Taylor 1986 and field notes by G. Faure)

recently, Coren et al. (2003) used synthetic aperture radar (SAR) to survey an area of 10,000 km^2 including the ice fields near the Allan Hills.

Damm (1996) demonstrated that the thickness of the ice in the drainage basins of the David Glacier varies widely as it flows over mountainous bedrock toward the coast of Victoria Land. The profile in Fig. 17.8 along the Antarctic Crustal Profile-1 (ACRUP-1) demonstrates that the relief of the subglacial bedrock approaches 1,000 m and that the topography of the ice that flows over this mountainous bedrock reflects the irregular profile of the underlying bedrock. The topographic profiles of the bedrock obtained by Damm (1996) indicate that the valley of the David Glacier extends to a depth of about 800 m *below* sea level, but they do not indicate whether supraglacial moraines are forming on some of the steps in the topographic profile of the ice sheet.

The profiles of ice thickness and bedrock topography obtained by Damm (1996) support a model for the formation of the Elephant Moraine as a consequence of thinning of the ice sheet that is flowing over a bed rock ridge. Figure 17.9a is a two-dimensional representation of the way a layer of sediment-laden basal ice in a *thick* ice sheet flows up the backside of a bedrock ridge without becoming exposed at the surface. However, the presence of the ridge does cause a ramp to form at a surface of the ice sheet along which crevasses and ice pinnacles develop.

If the ice sheet is thin as in Fig. 17.9b, basal ice does reach the surface and forms a supraglacial moraine as exemplified by the Elephant Moraine. The clasts and fine-grained sediment that were originally embedded in the basal ice form a supraglacial moraine that continues to increase in length at a rate depending on the flow velocity of the ice sheet. Consequently, the length of the moraine is related to its age and thus indicates the time that has elapsed since the ice sheet thinned sufficiently to cause basal ice to come to the surface. This model of ice-sheet dynamics therefore provides a method of dating the climate change that caused the thinning of the ice sheet (Taylor 1986).

17.3.4 Dating Supraglacial Moraines

Although these concepts have not been quantified, they support the conjecture that the Elephant Moraine began to form after the ice sheet had thinned sufficiently to cause basal ice to be exposed at the surface. Moreover, the limited length of the moraine (i.e., 5.6 km) indicates that the moraine is not old enough to have reached the David Glacier toward which the ice is moving. Accordingly, the age of Elephant Moraine can be approximated by the relation:

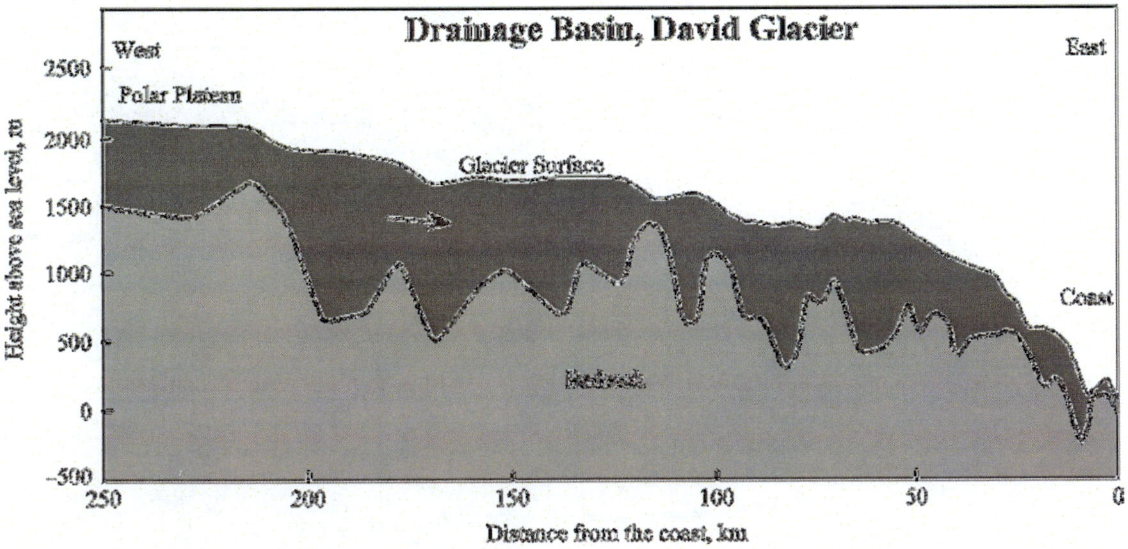

Fig. 17.8 The subglacial bedrock in the drainage basin of the David Glacier has mountainous topography with elevation differences of up to 1,000 m between peaks and adjacent valleys. The thickness of the ice that flows over this irregular bedrock surface varies widely from about 1,300 m on the polar plateau to a few hundred meters near the coast. In addition, the surface of the ice sheet has a step-like profile in response to the complex bedrock topography. The report by Damm (1996) does not mention whether supraglacial moraines are present along this traverse where basal ice may have been forced to the surface by the turbulent flow of the ice sheet (Adapted from Figure 6 of Damm 1996)

$$t \sim \left(\frac{l}{v}\right) \quad (17.1)$$

where t = age in years, l = length in cm, and v = velocity in cm/year.

An alternative method of dating the Elephant Moraine and similar supraglacial moraines arises from the storage of energy which nonconducting crystalline solids absorb from solar and/or nuclear radiation. The stored energy can be quantitatively released by heating samples of the material in the laboratory to a temperature of about 500°C. The radiation emitted by the specimens is recorded in the form of a glow curve. The amount of the thermoluminesence that is released by this procedure can be related to the duration of the irradiation which dates the exposure of the specimen at the surface of the ice sheet (Faure and Mensing 2005, Section 22.4).

Exploratory measurements reported by Faure and Sutton (1985) indicate that a sample of quartz-sandstone with prolonged sun exposure in the Elephant Moraine has a more intense glow curve than a sample that has had no sun exposure. Therefore, the exposure age of quartz sandstone clasts in the Elephant Moraine may be measurable after the method has been properly calibrated.

17.3.5 Micropaleontology of Molded Till Pellets

The pellets of clay-rich sediment that occur in the Elephant Moraine in Fig. 17.10a were apparently molded into spherical shapes as the ice sheet flowed over wet deposits of basal till of late Cenozoic age. Montello (1986) reported that the till pellets in the Elephant Moraine contain quartz, K-feldspar, plagioclase, and kaolinite which were presumably derived from the subglacial bedrock in the path of the ice sheet. The presence of K-feldspar (microcline) in the till pellets is noteworthy because this mineral commonly occurs in granitic and high-grade metamorphic rocks (Montello 1986). Four grain-size fractions of K-feldspar extracted from a till pellet yielded an approximate Rb-Sr date of 660 ± 33 Ma ($\lambda = 1.42 \times 10^{-11}\,year^{-1}$). Microcline in a second pellet yielded a spectrum of Rb-Sr dates between 677 and 453 Ma. These samples

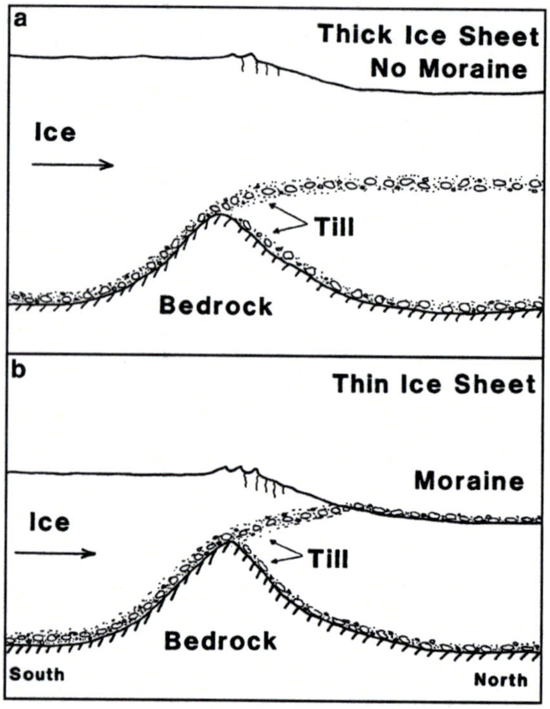

Fig. 17.9 (**a**) When a thick ice sheet flows across a topographic obstruction of the bedrock at its base, the surface of the ice sheet develops a "ramp" with crevasses and ice pinnacles at its crest. However, the sediment-laden ice at the base of the ice sheet does not become exposed at the surface and a supraglacial moraine does not form in this case. (**b**) After the ice sheet has thinned sufficiently caused by a change in climate or ice-flow regime, basal ice does reach the surface. In this case, a supraglacial moraine begins to form and is extended by the continuing flow of the ice over the subglacial bedrock ridge. The length of the moraine becomes a measure of the time elapsed since the ice sheet thinned sufficiently to allow basal ice to reach the surface (Adapted from Taylor 1985, Figure 36)

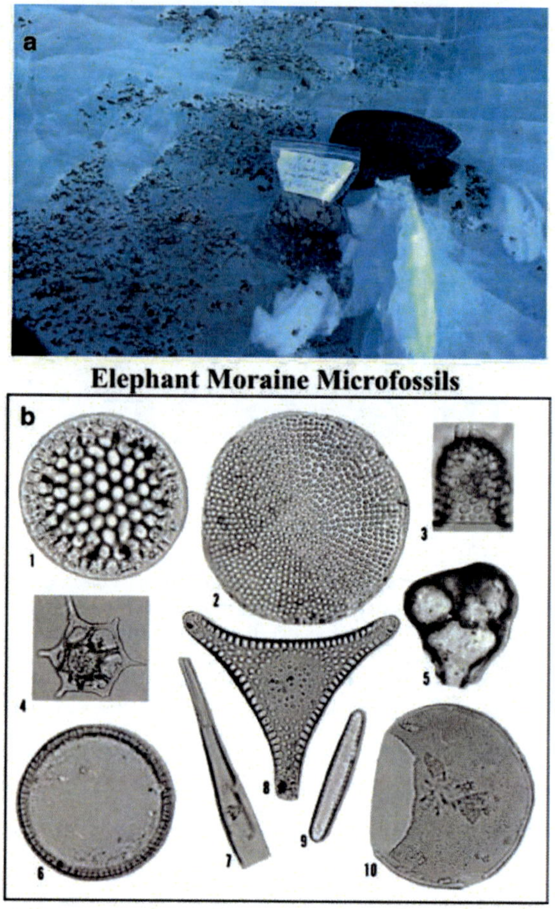

Fig. 17.10 (**a**) Molded pellets of clay-rich sediment are released by sublimation of bands of "dirty" ice in the Elephant Moraine. These pellets are samples of till from the base of the ice sheet which appears to be wet because the pellets are molded rather than abraded. The age of the till is assumed to be late Cenozoic in which case it could be assigned to the Sirius Group (Photo by G. Faure). (**b**) The till pellets that weather out of the ice in the Elephant Moraine contain a suite of reworked marine diatoms and other microfossils of Miocene and Pliocene ages identified by D.M. Harwood. Microfossils containing these same species were identified by Harwood (1983, 1986) in Neogene till in the Wisconsin Range, the Dominion Range, and on Mt. Feather (southern Victoria Land). The microfossils depicted in this image are identified by number: 1. *Stephanopyxis grunowii*, 2. *Thalassiosira lentiginosa*, 3. *Stephanopyxis turris*, 4. *Distephanus speculum* (silicoflagellate), 5. *Pseudoammodochium* sp. cf. *P. dictyoides* (ebridian), 6. *Paralia sulcata*, 7. Rhizosolenia sp., 8. *Trinacria excavate*, 9. *Thallassionema nitzschioides*, 10. *Stellarima microtrias* (Adapted from Faure and Harwood 1990)

are composed of mixtures of grains of different ages which means that the Rb-Sr dates may not be the age of a geological event. Nevertheless, these dates can be taken as evidence that a component of Precambrian feldspar is present in the till pellets of the Elephant Moraine.

The till pellets also contain a varied assemblage of reworked marine microfossils some of which are depicted in Fig. 17.10b. The flora includes 23 species of diatoms, as well as silicoflagellates, ebridians, cysts, radiolarians, sponge spicules, and palynomorphs identified by D.M. Harwood (Faure and Harwood 1990). Similar assemblages of marine microfossils occur in Cenozoic till of the Sirius Group at many places in the Transantarctic Mountains. This evidence suggests that the till of the Sirius Group was deposited in a marine embayment that existed during the Pliocene on the East Antarctic craton and it implies that the volume of the East Antarctic ice sheet decreased significantly at this time. This interpretation of the micropaleontological evidence by Webb et al. (1983, 1984), Harwood (1983, 1985, 1986), and Harwood and Webb (1990) was critically reviewed by Mercer (1985) and caused a controversy that has still not been resolved to everyone's satisfaction (Denton et al. 1993; Sugden et al. 1993; Wilch et al. 1993). This matter is more fully presented in Chapter 19.

17.3.6 Ablation Rates

The blue-ice areas surrounding the Elephant Moraine are part of the broad ablation belt of the East Antarctic ice sheet located west of the Transantarctic Mountains. In this area the eastward flow of the ice sheet is blocked except for the ice that squeezes through the narrow valleys of the outlet glaciers. Therefore, ablation dominates the glaciological processes that are occurring in this areas and the *rate* of ablation is an important factor in the mass balance of the ice sheet.

We expect that the ablation rate of the Elephant Moraine is reduced by the layer of rock clasts of which it is composed and by the discontinuous patches of snow that migrate across it driven by the prevailing wind. The presence of rock clasts may *reduce* the ablation rate because the clasts cover the surface of the ice. They may also *increase* the ablation rate because the clasts, which are primarily composed of dark-brown weathered dolerite, absorb solar radiation. The resulting increase in the temperature of the air above the moraine accelerates ablation by sublimation and even causes melting of the ice, especially under boulders that are warmed by solar radiation.

The ablation rate of the blue-ice fields surrounding the Elephant Moraine was measured at regular intervals along four surveyed lines that started from the surveyed baseline in Fig. 17.6 and extended up to 1.75 km beyond the edge of the moraine (Faure and Buchanan 1987, 1991). At each station a wooden dowel rod was placed into a hole drilled into the ice with an ice screw. The holes were 10–15 cm deep and had diameters of 1.0 cm. The level of the ice was recorded by a notch cut into the dowel and the location of each rod was inscribed in order to facilitate future identification. When these stations were revisited 2 years later, the level of the ice was recorded be cutting a second notch into the dowels. The distance between the two notches on each dowel that was measured with dial calipers calibrated to within 2.54×10^{-5} m. In some cases, the dowels were revisited at the end of the field season in order to obtain a measurement of the summer ablation rate for comparison to the annual rate. In addition, each site was described in order to assess the effect of snow and entrained sediment on the annual ablation rate of the ice.

Table 17.1 Average annual ablation rates and ranges at different environmental sites in and around the Elephant Moraine (Faure and Buchanan, 1991)

Environmental Site	Average annual Moraine	Ablation rate (cm/year) Ice field
Ice, snow-free	2.61 ± 0.15 (40)[a]	4.12 ± 0.15 (19)
	0.6–4.5	2.9–5.3
Ice, snow-covered	1.96 ± 0.24 (10)	3.24 ± 0.21 (2)
	0.5–3.9	
Firn (ice-field)	–	2.84 ± 0.45 (3)
		2.4– 3.7
"Dirty" ice, no Snow	2.73 ± 0.55 (6)	–
	0.4–4.5	
"Dirty" ice snow-covered	1.23 ± 0.22 (7)	–
	0.24–1.93	

[a]One standard deviation of the mean and number of sites in parentheses.

The results in Table 17.1 were derived from 112 of 167 sites for an overall recovery rate of 67% after a 2-year exposure time. The missing dowels were either buried in migratory snow drifts or were blown away by high winds. The results in Table 17.1 and Fig. 17.11 demonstrate that the ablation rates of the snow-free ice fields outside of the Elephant Moraine ranged from 2.9 to 5.3 cm/year with an average of 4.12 ± 0.15 cm/year for 19 determination. The ablation rates of snow-free ice within the moraine ranged from only 0.6 to 4.5 cm/year and average 2.61 ± 0.15 cm/year for 40 sites. Ice within the moraine that was covered by snow at the time the dowels were recovered ablated even more slowly at an average rate of 1.96 ± 0.24 cm/year at 10 sites.

In summary, the ice fields adjacent to the Elephant Moraine ablated more rapidly than the ice within the moraine, contrary to the expectation that the warming of clasts by solar radiation increases the ablation rate of ice in the moraine.

On the whole, the ablation rate of ice within the Elephant Moraine is appreciably reduced by migrating snow drifts and is enhanced much less than expected

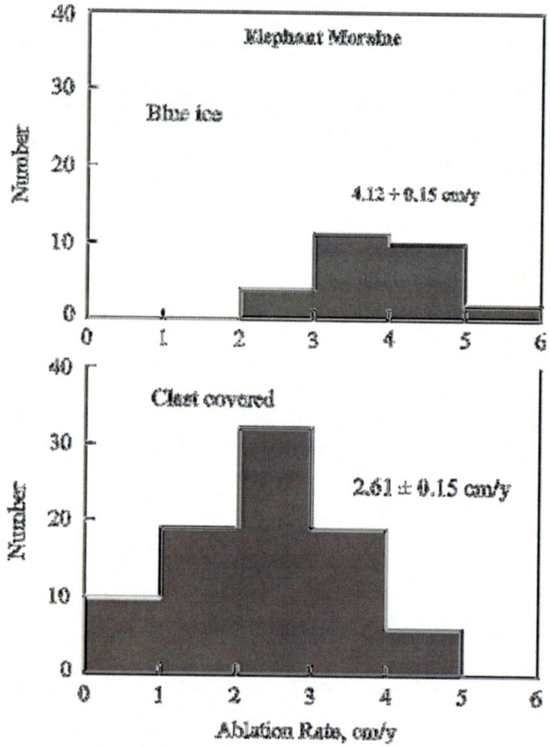

Fig. 17.11 The annual ablation rate of snow-free blue ice surrounding the Elephant Moraine ranges from 2.9 to 5.3 cm/year and has an average value of 4.1 ± 0.1 cm/year based measurements at 20 sites. The ablation rate of ice within the moraine is lower on average than the ablation rate of blue ice without a cover of snow. The average ablation rates at different environmental sites in and around the Elephant Moraine are listed in Table 17.1 (Data from Faure and Buchanan 1991)

by the solar heating of dark-colored rock clasts. Snow-free areas of blue ice adjacent to the moraine ablate more rapidly than bare ice within the moraine. As a result, parts of the Elephant Moraine are slightly higher topographically than the surrounding ice fields.

17.3.7 Subglacial Calcite and Opaline Silica

Clasts of black calcite and opaline silica occur at a few places in the Elephant Moraine where large clasts have disintegrated into smaller fragments. The black calcite in Fig. 17.12 is crudely stratified and was initially identified by Sipiera and Landis (1986) as a fragment of a stromatolite (i.e., an organosedimentary structure produced by colonies of micro-organisms such as cyanobacteria that trap or precipitate sediment as a result of their metabolic activity).

The fragments of black calcite are locally associated with clasts of fine-grained pebbly sediment (till?) that is permineralized with opaline silica. These kinds of clasts in Fig. 17.13 also contain thin-bedded black calcite. The co-occurrence in the moraine and in hand specimens of the two kinds of materials suggests that they are cogenetic.

The data in Table 17.2 indicate that the oxygen in *calcite* of sample 582 is strongly depleted in ^{18}O ($\delta^{18}O_c = -20.5 \pm 0.4‰$, SMOW). The fractionation of oxygen isotopes between water (w) and calcite (c) is expressed by the equation:

$$\delta^{18}O_c - \delta^{18}O_w = \frac{2.78 \times 10^6}{T^2} - 2.89 \qquad (17.2)$$

(Friedman and O'Neil 1977). Setting $\delta^{18}O_c = -20.5$ ‰ and assuming a temperature of 0°C (T = 273.15 K) yields $\delta^{18}O_w = -54.9‰$ on the SMOW scale. If the calcite had precipitated from seawater at 0°C, its $\delta^{18}O$ value, according to equation 17.2, would have been + 34.4 ‰ on the SMOW scale. The measured $\delta^{18}O$ value (−20.5‰) is obviously not consistent with a marine origin of the calcite clasts in the Elephant Moraine.

On the other hand, the extreme depletion in ^{18}O of the *water* in isotopic equilibrium with the black calcite at 0°C is consistent with the $\delta^{18}O$ values of the ice exposed in the Elephant Moraine which range from −42.6‰ to −49.7‰ (Faure et al. 1993). Similar calculations for all of the calcites in Table 17.2 yield $\delta^{18}O$ values of the water between −47.3‰ and −54.9‰. The evident depletion of the water in ^{18}O is characteristic of meteoric precipitation in the polar regions and is the result of isotope fractionation at low temperature during multiple cycles of evaporation of water from the surface of the Earth and condensation of water vapor in the atmosphere. This process can be modeled by the Rayleigh equations (Hoefs 1997, pp. 9–10).

In contrast to the isotope composition of the calcite, the oxygen of the *opaline silica* in Table 17.2 is *enriched* in ^{18}O ($\delta^{18}O_{sil} = +13.03‰$) relative to SMOW. The fractionation of oxygen isotopes between water and various forms of silica at low temperatures was expressed by equations compiled by Chacko et al. (2001). Equations 1–3 in Table 17.3 yield consistent $\delta^{18}O_w$ values for the water assuming that the temperature

Fig. 17.12 This specimen superficially resembles a stromatolite, but it is actually composed of calcite that formed from cryogenic brines at the base of the East Antarctic ice sheet. It was collected during the 1986/87 field season in the Elephant Moraine by David Buchanan. The calcite in a similar sample analyzed by Faure et al. (1988) is strongly depleted in ^{18}O ($\delta^{18}O$ SMOW = -20.5 ± 0.4‰) and ^{13}C ($\delta^{13}C$ PDB = -22.6 ± 0.1‰). In addition, the strontium in this sample is enriched in radiogenic ^{87}Sr ($^{87}Sr/^{86}Sr$ = 0.71417 ± 0.00002) relative to 0.70809 ± 0.00001 for the Eimer and Amend Sr-isotope standard. These data indicate that the calcite precipitated from glacial meltwater, that the carbon is of biogenic origin, and the strontium was derived from crustal rocks rather than from seawater (Photo by G. Faure)

was T = 273.15 K (0.0°C) and that oxygen in the silica was in isotopic equilibrium with the oxygen in the water. The average $\delta^{18}O_w$ of the aqueous phase calculated from equations 1–3 in Table 17.3 is:

$$\delta^{18}O_w = -29.9 \pm 0.8\ (1\sigma)‰$$

Equation 4 by Brandriss et al. (1998) yields a $\delta^{18}O$ value for the water of −23.0‰ (SMOW). Nevertheless, all of the oxygen-isotope fractionation equations in Table 17.3 indicate that the water that was in isotopic equilibrium with the opaline silica was somewhat *less* depleted in ^{18}O than the water from which the calcite precipitated.

The difference in the isotopic composition of oxygen indicates that the silica precipitated from water that may have gained ^{18}O by isotope exchange with silicate minerals of the bedrock underlying the ice sheet, whereas the calcite precipitated from glacial meltwater at the base of the ice sheet. In other words, the opaline silica appears to have precipitated from groundwater that was locally discharged at the base of the ice sheet. The discharge of water by springs under the ice sheet may have been caused by convection of groundwater that was heated by a local-volcanic center like those that existed in the ice-free valleys and in the Royal Society Range of southern Victoria Land (Faure et al. 1988; Section 16.1 of this book).

The carbon of the calcites in Table 17.2 is strongly depleted in ^{13}C. For example, the calcite of sample 582 has $\delta^{13}C$ = −22.6 ± 0.1‰ on the PDB scale. On this scale, carbon of marine calcite has $\delta^{13}C$ close to 0‰ because the PDB standard is a marine belemnite of the Cretaceous Peedee Formation in South Carolina. The evident ^{13}C depletion of all of the calcite samples from the Elephant Moraine in Table 17.2 is therefore *inconsistent* with a marine origin of the calcites.

Carbon isotopes are strongly fractionated during photosynthesis of green plants which are depleted in ^{13}C and have negative $\delta^{13}C$ values ranging from −22‰

Fig. 17.13 Clasts of pebbly sediment (till?) in the Elephant Moraine are extensively impregnated by opaline silica. These clasts are locally associated with fragments of black calcite (Fig. 17.12) and both presumably originated from a deposit at the base of the ice sheet. This particular specimen contains some layered black calcite suggesting that the calcite and opaline silica are cogenetic (Photo by T.M. Mensing)

Table 17.2 Isotopic composition of oxygen, carbon and strontium of black calcite and opaline silica from the Elephant Moraine on the East Antarctic ice sheet (Faure et al. 1988)

Sample	Composition	$\delta^{18}O_{SMOW}$	$\delta^{13}C_{PDB}$	$^{87}Sr/^{86}Sr$
Number		‰	‰	
582	Calcite	−20.5 ± 0.4 (1a,b)	−22.6 ± 0.1 (1a,b)	0.71417 ± 2² (1b)
583	Calcite	−14.2 (1a)	−23.05 (1a)	0.71402 ± 4 (1b)
584	Calcite	–	–	0.71383 ± 3 (1b)
EMA-2	Calcite	−12.91 (1c)	−22.06(1c)	0.71488 ± 3 (1b)
	Opal. silica	+13.03 (1b)	–	–
F-84-6A	Calcite	−13.7 (1a)	−22.0 (1a)	0.71484 ± 1 (1b)
E&A Sr isotope std.	SrCO3	–	–	0.70801 ± (1b)

1a. Analyzed by Krueger Enterprises, Cambridge, MA; 1b. L.M. Jones at Conoco Inc., Ponca City, OK; 1c. J. Hoefs, Geochemical; Institute, Göttingen, Germany.

2. Error in the last digit of the ratio (e.g. ± 2 = 0.00002).

to −33‰ (PDB) for plants that fix atmospheric CO_2 by pathways described by Calvin (1968) (i.e., C_3 plants). Bituminous coal, which is derived from plant remains, is similarly depleted in ^{13}C. For example, the $\delta^{13}C$ values (PDB) of bituminous coal in the Beacon rocks of the Transantarctic Mountains range from −23.5‰ to −26.0‰ (Faure and Botoman 1984). The similarity of the carbon-isotope compositions of the calcite samples in the Elephant Moraine and coal in the Beacon Supergroup is permissive evidence that

Table 17.3 Isotope fractionation of oxygen between various forms of solid silica and water at low temperatures (Chacko et al. 2001)

Silica	Equation T (kelvins)	Temperature (°C)	Reference
1. Quartz	$3.306\ (106/T^2) -2.71$	180–550	Zhang et al. (1989)
2. Amorph. sil.	$3.52\ (106/T^2) -4.35$	34–93	Kita et al. (1985)
3. Biog. sil.	$41.2-0.25\ t^a$	4–27	Labeyrie (1974)
4. Biog. sil.	$15.56\ (103/T)-20.92$	3.6–20.0	Brandriss et al. (1998)

^{a}t = temperature in °C.

the carbon in the calcites originated from coal at the base of the ice sheet.

The $^{87}Sr/^{86}Sr$ ratios of the calcite (0.71383–0.71488) are higher than the values of this ratio in seawater at any time in the history of the Earth (Figures 19.1 and 19.8 of Faure and Mensing 2005). Consequently, the strontium in the calcite samples did not originate from seawater or by weathering of marine limestone. Instead, the strontium was probably released into solution by weathering of detrital grains of microcline and of the Ferrar dolerite and basalt in the sediment at the base of the ice sheet. Feldspar size-fractions separated from two till pellets in the Elephant Moraine by Montello (1986) have $^{87}Sr/^{86}Sr$ ratios between 0.71825 and 0.73327, whereas the $^{87}Sr/^{86}Sr$ ratio of Ferrar Dolerite in southern Victoria Land range from 0.7116 to 0.7148 (Jones et al. 1973a). Therefore the strontium in the calcite clasts in the Elephant Moraine probably originated from *both* sources.

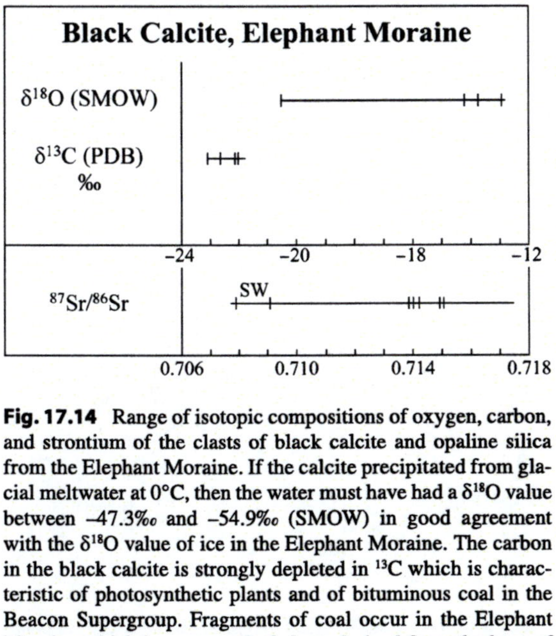

Fig. 17.14 Range of isotopic compositions of oxygen, carbon, and strontium of the clasts of black calcite and opaline silica from the Elephant Moraine. If the calcite precipitated from glacial meltwater at 0°C, then the water must have had a $\delta^{18}O$ value between −47.3‰ and −54.9‰ (SMOW) in good agreement with the $\delta^{18}O$ value of ice in the Elephant Moraine. The carbon in the black calcite is strongly depleted in ^{13}C which is characteristic of photosynthetic plants and of bituminous coal in the Beacon Supergroup. Fragments of coal occur in the Elephant Moraine which is composed of clasts derived from the base of the ice sheet. The strontium in the black calcite is strongly enriched in radiogenic ^{87}Sr compared to seawater (SW) of Phanerozoic age

The isotopic data summarized in Fig. 17.14 support the conclusion that the calcite in the Elephant Moraine was deposited from glacial meltwater at the base of the ice sheet, whereas the opaline silica precipitated from warm groundwater discharged by springs at the base of the ice sheet. These conclusions raise questions about the existence of liquid water at the base of the East Antarctic ice sheet and about the enrichment of glacial meltwater and groundwater in the ions that allowed calcite and opaline silica to precipitate.

17.4 Reckling Moraine and Allan Hills

The Reckling Moraine (informal name) at 76°16′S and 159°15′E is located in a large complex of ice fields in Fig. 17.15 that extends west from Reckling Peak for about 90 km and includes the Elephant Moraine. The ice fields are flanked along their south side by a system of ice ramps that reflect the mountainous topography of the rock surface over which the ice sheet isflowing. The Reckling Moraine is supraglacial in origin like the Elephant Moraine and is forming by ablation of basal ice that is exposed in the basin defined by the ice ramps.

The moraine in Fig. 17.16 is about 6 km long and 1.5 km wide and has the shape of a horse shoe opening to the northeast. The interior of the moraine in Fig. 17.16 contains an irregularly-shaped clast-free area which resembles the "Gingerbread Man" of Mother Goose fame. The southern branch of the moraine contains numerous parallel bands of sediment-laden ice. Such bands are less abundant in the northern branch.

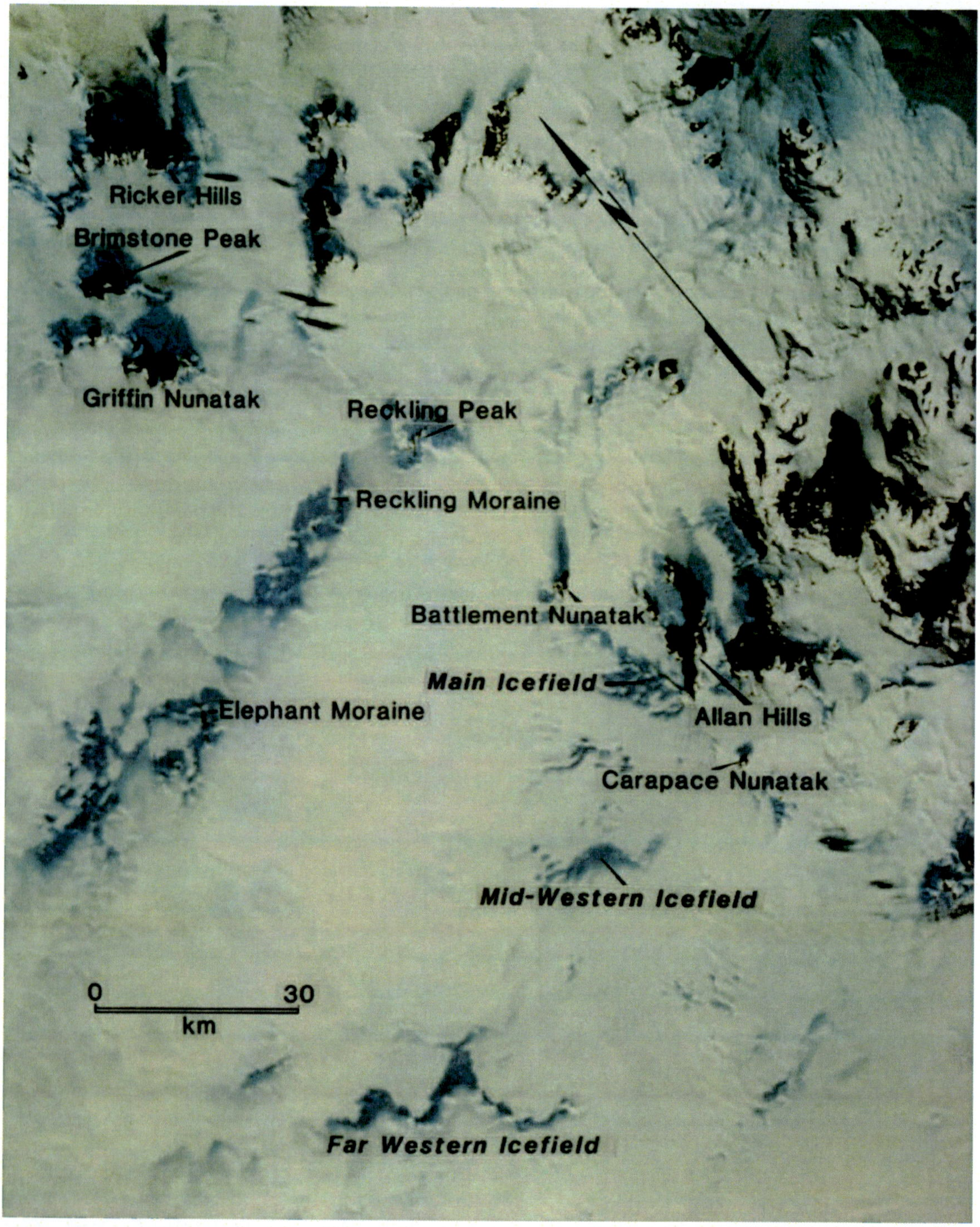

Fig. 17.15 The Reckling and Elephant moraines occur within a series of ice fields that extend for 90 km west from Reckling Peak which is the only outcrop of bedrock in this belt of blue ice. The ice fields are generally located north of a complicated system of ice ramps that has not been mapped topographically, but which is characterized by relief of 100 m or more. The ice fields adjacent to the Reckling and Elephant moraines exposed hundreds of meteorite specimens that are released by sublimation of the ice in which they were embedded (Whillans and Cassidy, 1983 and Chapter 18) (Photo by Landsat)

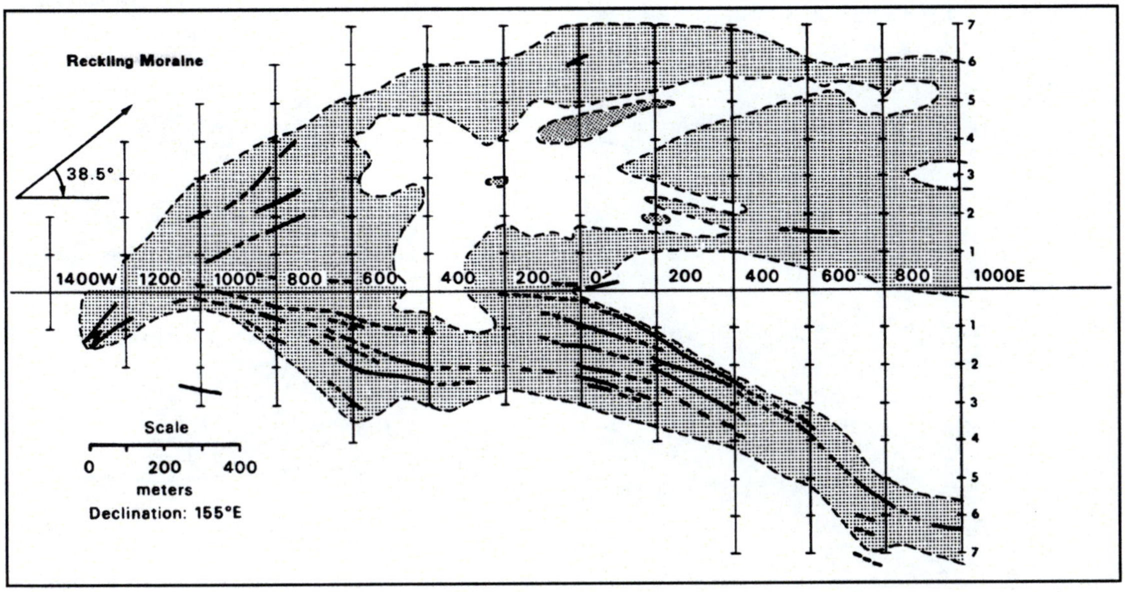

Fig. 17.16 The Reckling Moraine has the shape of a horse shoe open to the northeast. It contains an irregularly-shaped clast-free area that resembles a human figure (e.g., the Gingerbread Man). The southern branch of the moraine contains numerous parallel bands of sediment-laden basal ice. Similar bands of sediment-laden ice occur in the northern branch but appear to be less numerous. The bands of basal ice of both arms converge to a "W" at the western end of the moraine (Adapted from Figure 2 of Faure et al. 1987)

17.4.1 Lithology of Rock Clasts

The Reckling Moraine in Fig. 17.17 was mapped by counting and classifying 21,346 clasts greater than 1.6 cm in diameter at 78 surveyed stations within a system of picket lines in Fig. 17.16 (Faure et al. 1987, 1992, 1993; Faure 1990a, b, c, d).

The results in Fig. 17.18 indicate that clasts composed of the Ferrar Dolerite and sedimentary rocks of the Beacon Supergroup are equally abundant at about 49% each. The remaining 2% of the clasts consist of basalt, diamictite, coal, and molded till pellets. Clasts composed of granitic and high-grade metamorphic rocks and of black calcite and opaline silica are absent from the Reckling Moraine.

A clast of coarse grained quartz monzonite that was dated by Faure et al. (1984) probably originated from a moraine near the Allan Hills rather than from the Reckling Moraine. The Rb-Sr date of ($\lambda = 1.42 \times 10^{-11}$year^{-1}) of this specimen is 496 ± 4.6 Ma (Cambro-Ordovician) in good agreement with isotopic dates of the Granite Harbor suite elsewhere in the Transantarctic Mountains (Section 3.4.2).

17.4.2 Ablation Rates

The ablation rates of ice within the Reckling Moraine and in the surrounding ice fields were determined by Faure and Buchanan (1991). The results in Table 17.4 also include ablation rates of the main ice field adjacent to the Allan Hills where several hundred meteorite specimens have been recovered by personnel of Project ANSMET (Antarctic Search for Meteorites) (Score and Lindstrom 1990).

The ablation rates of ice within the Reckling Moraine and on the adjacent ice fields were measured by means of 77 wooden dowel rods that were "planted" in December 1986, 63 of which were recovered in December of 1988. The results in Table 17.4 demonstrate that the average ablation rates of the ice fields adjacent to the Reckling Moraine and Allan Hills are 4.11 ± 0.17 cm/year and 4.29 ± 0.24 cm/year, respectively. These values are indistinguishable from the average ablation rate of the ice fields around the Elephant Moraine in Table 17.1 (i.e., 4.12 ± 0.15 cm/year). The ablation rate of ice within the Reckling Moraine (1.51 ± 0.60 cm/year) is markedly lower than

Fig. 17.17 The Reckling Moraine is composed of unsorted clasts and fine-grained sediment that accumulate as the ice ablates by sublimation. The accumulated sediment in the foreground lies on a band of "dirty" ice that can be traced for several kilometers. The sediment-free ice in the distance is part of the system of ice ramps that define the basin within which the Reckling Moraine is forming (Photo by G. Faure)

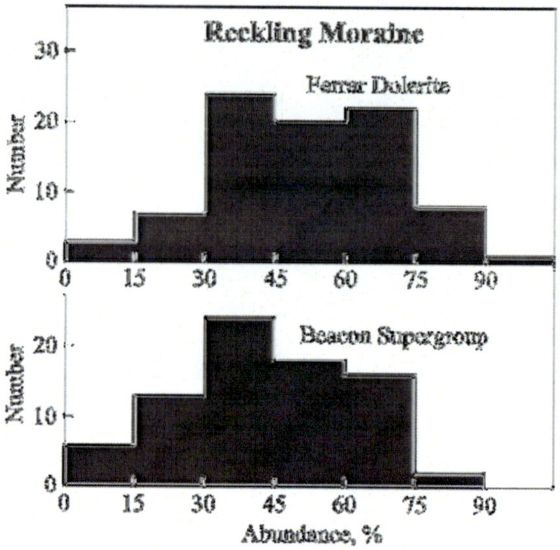

Fig. 17.18 Clasts of dolerite of the Ferrar Group and sedimentary rocks (sandstone, siltstone, and shale) of the Beacon Supergroup are about equally abundant in the Reckling Moraine. All other rock types (i.e., basalt, diamictite, coal, and molded till pellets) together comprise about 2% of the clasts. Neither clasts of granite, high-grade metamorphic rocks, nor of calcite and opaline silica were found in the Reckling Moraine. The clasts were counted and classified by Erik Hagen and Michael Strobel during the 1986/87 field season. The abundances of dolerite clasts within the Reckling Moraine were contoured on a map published by Faure et al. (1987)

Table 17.4 Average annual ablation rates of the ice fields adjacent to the Reckling Moraine and near the Allan Hills (Faure and Buchanan 1991).

Site description	Reckling Moraine (cm/year)	Allan Hills (cm/year)
Ice field, snow-free	4.11 ± 0.17 (50)[a]	4.29 ± 0.24 (15)
Moraine, snow-free	1.36 ± 0.47 (2)	–
Moraine, snow-covered	1.67 ± 0.33 (10)	–
Firn	1.98 (1)	–

[a]The number of sites is indicated in parentheses.

the ablation rate of bare ice regardless of whether the ice is covered by clasts or by snow. Consequently, the Reckling Moraine locally rises several meters above the level of the surrounding ice basin.

17.4.3 Oxygen Isotopes

The isotope composition of oxygen in the East Antarctic ice sheet varies regionally depending primarily on the distance from the coast and on the temperature in the accumulation area. Therefore, the $\delta^{18}O$ values of vertical cores in the accumulation areas have preserved a record of climatic conditions in the past. This record is disturbed when the ice flows from the accumulation zone to the margin of the ice sheet where it is exposed in a horizontal surface. In the ideal case, the $\delta^{18}O$ values of ice samples collected along a surveyed line across the exposed surface of the ice sheet might reveal the climate history of the area in the interior of the continent where the ice formed. Alternatively, such a "horizontal core" may reveal the structure of the ice sheet by the disruption of the inherited pattern of climate-controlled $\delta^{18}O$ values caused by thrust faults and folds within the ice sheet.

In order to test this hypothesis Faure et al. (1992) determined $\delta^{18}O$ values of ice collected along surveyed lines that extended up to 3 km from the baseline in the Reckling Moraine into the adjacent ice fields both north and south of the moraine. In addition, eight vertical ice cores were augered to a depth of about 1 m each, within a 2-km distance along one of the surveyed lines that crossed the moraine. The oxygen isotope composition of the ice in these cores was analyzed at 1-cm intervals in the Quaternary Isotope Laboratory at the University of Washington in Seattle.

The $\delta^{18}O$ values of the ice cores (not shown) range widely from −37.5 to about −50.0‰ (SMOW), but the patterns do not correlate between adjacent cores. Even the $\delta^{18}O$ profiles of two cores drilled 1 m apart do not resemble each other. Evidently, the ice sheet at the Reckling Moraine was severely deformed as it flowed from the accumulation area and the record of climate change it may have contained was lost.

The $\delta^{18}O$ values of ice collected 5–10 cm below the surface along surveyed lines at 100 m intervals also vary irregularly. Even samples collected at 10-m intervals at selected sites do not provide intelligible information about the climate history and structure of the ice sheet. These results therefore demonstrate that the $\delta^{18}O$ values of ice collected systematically along the surface of the ice sheet do not constitute a "horizontal core" from which the structure of the ice sheet and a climate record can be determined.

Some of the results suggest that surface ice may have been contaminated by recently-deposited snow that infiltrated into small fractures in the ice. Snow that fell in the Reckling Moraine on December 19 of 1986 had a $\delta^{18}O$ value of −32.1‰ (SMOW). The presence of this contaminant was detected in the ice cores, described above, to a depth of about 10 cm, but it did not affect the $\delta^{18}O$ values of ice collected at greater depth below the surface.

A graphical presentation in Fig. 17.19a shows that the $\delta^{18}O$ values of ice in and around the Reckling Moraine are strongly clustered in the interval from −42 to −52‰. A small number of samples have lower $\delta^{18}O$ extending to −36‰, presumably because of contamination by recently deposited snow. In general, the $\delta^{18}O$ values of the ice underlying the Reckling Moraine are far more negative than the $\delta^{18}O$ value of recently deposited snow ($\delta^{18}O$ = −32.1‰), which means that the ice could not have formed at the present location of the Reckling Moraine. Therefore, the isotopic composition of oxygen supports the hypothesis that the ice formed farther from the coast and at

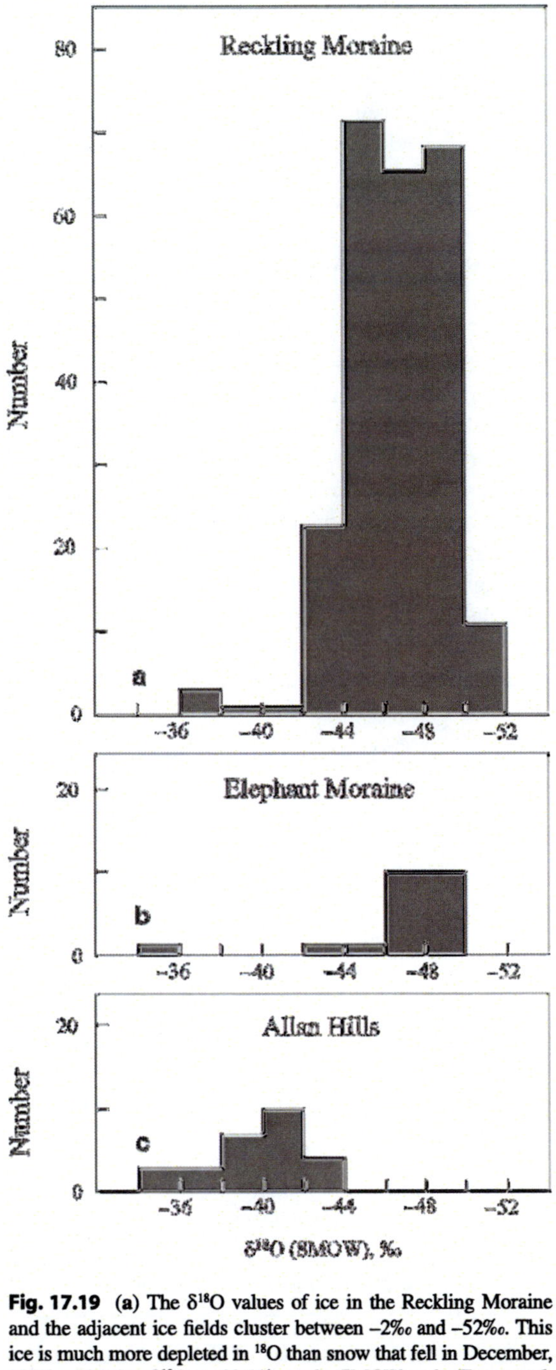

Fig. 17.19 (a) The $\delta^{18}O$ values of ice in the Reckling Moraine and the adjacent ice fields cluster between −2‰ and −52‰. This ice is much more depleted in ^{18}O than snow that fell in December, 1986, which has $\delta^{18}O$ = −32.1‰ on the SMOW scale (Faure et al. 1992). (b) The ice near the Elephant Moraine has similar $\delta^{18}O$ values as the ice at the Reckling Moraine. Therefore, the ice at both locations may have originated from the same source area in the interior of the continent (Faure et al. 1993). (c) The $\delta^{18}O$ values of the ice adjacent to the Allan Hills differ from the $\delta^{18}O$ values of ice in the Reckling and Elephant moraines by being less negative. Therefore, the source area of the Allan Hills ice field may have been closer to the coast and was not as cold as the source areas of the Elephant and Reckling moraines (Faure et al. 1993)

lower climatic temperatures than exist presently at the Reckling Moraine and that it flowed from the zone of accumulation in the interior of East Antarctica to its present location.

The $\delta^{18}O$ values of ice in and around the Elephant Moraine in Fig. 17.19b cluster in the same range as the $\delta^{18}O$ values in the Reckling Moraine. The evident similarity of oxygen isotope compositions supports the suggestion that the ice at both sites originated from the same area in the zone of accumulation.

The $\delta^{18}O$ values of ice from the Allan Hills in Fig. 17.19c differ markedly from the $\delta^{18}O$ values in the Reckling and Elephant moraines. Therefore, the ice exposed in the ice field adjacent to the Allan Hills originated from a different source area than the ice of the two moraines. This conclusion is consistent with the results of a triangulation survey of the Main Allan-Hills ice field in Fig. 17.20 by Annexstad and Nishio (1979) and Annexstad and Schultz (1982) who reported that the velocity of ice flowing east from the plateau into the Main Allan-Hills ice field decreases from about 1 m per year to near zero close to the Allan Hills. Therefore, the ice west of the Allan Hills is stagnant and is not flowing in the direction of the Reckling Moraine.

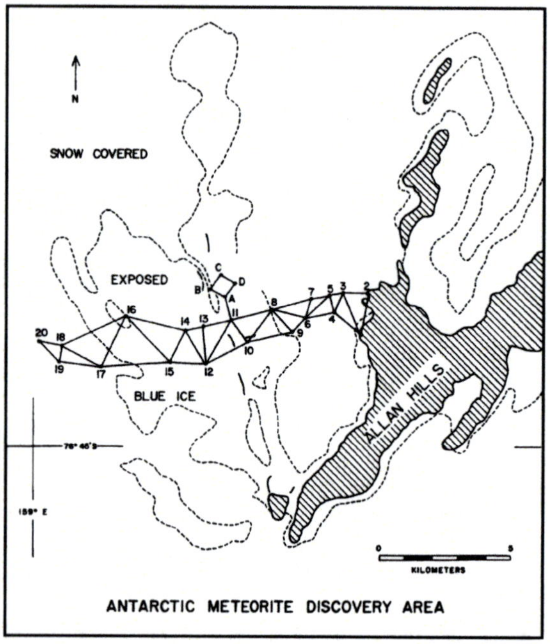

Fig. 17.20 The direction and rate of movement of a large area of blue ice located west of the Allan Hills was investigated by means of a triangulation network first surveyed in 1978/79 and later repeated in 1981/82. The results indicated that the ice moves from west to east and that the flow velocity decreases from about 1 m per year at the western end of the triangulation network to near 0 m per year close to the rock outcrops of the Allan Hills (Adapted from Annextad and Nishio 1979)

17.4.4 Bedrock Topography

The topography of the bedrock in the vicinity of Reckling Moraine and the Allan Hills was determined by measuring the thickness of the ice using a portable radar-echo device (Faure and Buchanan 1987; Faure 1990c, Faure et al. 1993). An example of the results for the Reckling Moraine in Fig. 17.21 demonstrates that the ice is flowing across a flat-topped ridge and that the moraine is located a short distance downstream of the crest of this subglacial ridge. The thickness of the ice varies from about 100 m above the ridge to more than 400 m in the adjacent valley. Two additional profiles across the Reckling Moraine (not shown) demonstrate that the subglacial bedrock ridge extends for at least 2 km parallel to the ice ramp south of the moraine.

A radar-echo profile of the edge of the ice sheet in the Allan Hills in Fig. 17.22 also reveals the presence of a subglacial ridge. The ice flowing across this ridge

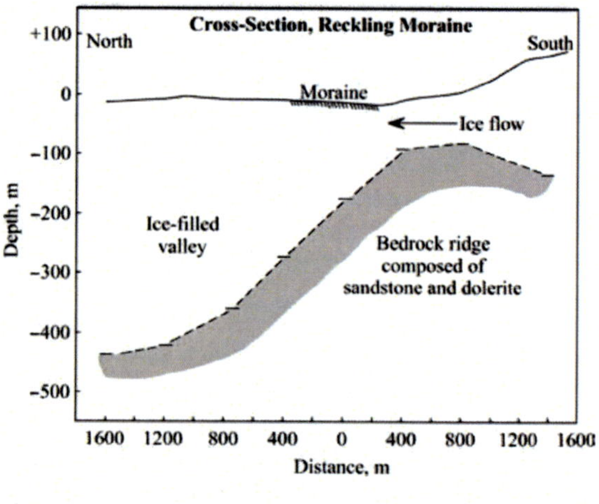

Fig. 17.21 Radar-echo measurements across the Reckling Moraine at 1,000 W (Fig. 17.16) reveal that the ice is flowing across a bedrock ridge which apparently deflects basal ice to the surface where it forms a supraglacial moraine. The presence of this ridge was confirmed by two additional profiles across the Reckling Moraine. The thickness of the ice over the ridge is only about 100 m, whereas the adjacent valley is filled with ice that is more than 400 m thick (Faure and Buchanan 1987; Faure 1990c; Faure et al. 1993)

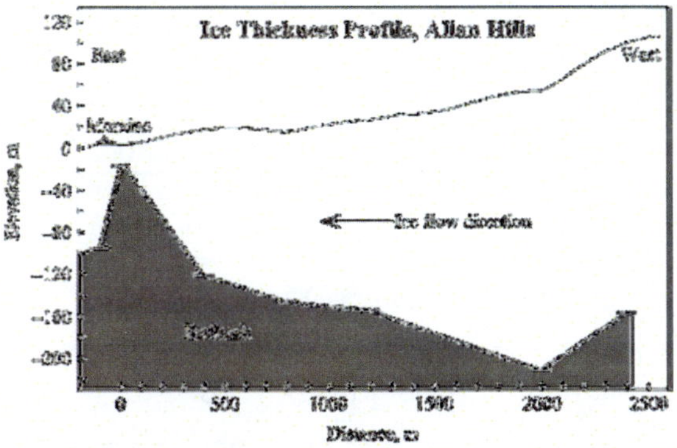

Fig. 17.22 The topography of the ice sheet adjacent to the Allan Hills is strongly affected by the bedrock at the base of the ice sheet. This profile of the ice sheet indicates that a bedrock ridge over which the ice if flowing forces basal ice to the surface which ablates to form a small supraglacial moraine on the surface of the ice. The Elephant and Reckling moraines probably form in the same way as a result of the interaction between the flow of the ice sheet and the mountainous topography of the subglacial bedrock (Adapted from Faure and Buchanan 1987)

thins from 260 m to only 20 m over the crest of this ridge which actually crops out about 500 m south of the profile (Faure and Buchanan 1987). As in the Elephant and Reckling moraines, the subglacial ridge in the Allan Hills forces sediment-rich basal ice to the surface where it forms a supraglacial moraine.

17.5 Accumulation Rates of Snow and Condensation Temperatures

The isotope composition of hydrogen and oxygen in snow deposited on the East Antarctic ice sheet contains information about the temperature at the time and place of deposition. The temperature dependence of the isotope fractionation factors causes the $\delta^{18}O$ values of snow to vary seasonally such that summer snow is less depleted in ^{18}O than winter snow. Similarly, the isotopic composition of hydrogen of snow, expressed as the δD value, also varies seasonally where δD value is defined by equation 17.3:

$$\delta D = \left(\frac{(D/H_{spl.} - (D/H)_{std.}}{(D/H)_{std.}} \right) \times 10^3 \text{‰} \quad (17.3)$$

where D = deuterium and the standard is SMOW (Faure and Mensing 2005).

The isotope compositions of oxygen and hydrogen in meteoric precipitation are related by the equation of a straight line (meteoric-water line) that was fitted to a large number of $\delta^{18}O$ and δD values of rain and snow from all over the Earth (Rozanski et al. 1993):

$$\delta D = (7.96 \pm 0.02)\, \delta^{18}O + 8.86 \pm 0.17 \quad (17.4)$$

This equation permits us to calculate δD values from measured values of $\delta^{18}O$ in snow and ice.

17.5.1 Accumulation Rates

The $\delta^{18}O$ and δD values of snow collected at increasing depth below the surface identify layers of snow deposited during successive summers and winters (Epstein et al. 1965). The layers of summer snow in Fig. 17.23 are *less depleted* in ^{18}O and D than the layers of winter snow. The resulting stratigraphic profile can be used to date snow layers by counting backward from the present. Epstein et al. (1965) concluded that the average rate of snow accumulation at the South Pole in the 6-year period between 1958 and 1963 was 7 cm of water per year.

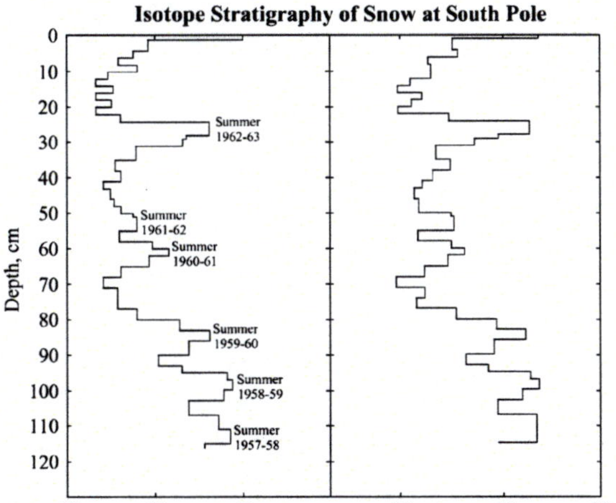

Fig. 17.23 The $\delta^{18}O$ and δD values of firn at the South Pole vary stratigraphically and identify snow deposited during the summers (December and January) and winter (July and August). The thickness of these layers was converted into an annual accumulation rate of water of 7 g/cm²/year which agreed with the rate of deposition indicated by accumulation stakes (Adapted from Epstein et al. 1965)

This technique of dating ice cores by counting annual layers, identified either visually or isotopically (or both), has been widely used to develop climate histories for the ice sheets in East and West Antarctica. However, the method is limited to about 1,000 years by the progressive thinning of annual layers with depth in the ice sheet.

The rates of snow accumulation in a large area of East Antarctica were subsequently measured during the South Pole-Queen Maud Land Traverse which made measurements at 75 stations during three successive summers from 1964 to 1968 (Picciotto et al. 1971). The participants of this traverse identified annual layers visually in snow pits and by detecting the fission-product radionuclides released by nuclear weapons tests in the summer of 1954/55. In addition, the decay of ^{210}Pb in snow layers was used at some of the stations (Crozaz et al. 1964). The accumulation rates based on the beta decay of fission products and by counting annual layers visually appear to be well correlated in most cases. The deposition rates derived from both methods ranged from 0 to 7 g/cm²/year and averaged 3.7 g/cm²/year for all sites where measurements were made. The average annual temperatures measured in 25-m bore holes in the ice included −58.4°C at Plateau Station (79°15′S, 40°30′E), −57.2°C at the former research station of the U.S.S.R. (82°07′S, 55°06′E) near the Pole of Inaccessibility (85°50′S, 65°47′E), −50.5°C close to the South Pole (89°36′S, 60°00′E), and −38.0° in Queen Maud Land (78°42′S, 6°52′W).

In a related study, Picciotto et al. (1968) compared different measurement strategies to determine the rate of accumulation at the former research station of the U.S.S.R. near the Pole of Inaccessibility. The resulting snow-accumulation rates in Table 17.5 based on stakes, on the beta activity of firn deposited in 1955, and on measurements of ^{210}Pb are in excellent agreement at about 3 g/cm²/year. The $\delta^{18}O$ values of summer snow near the Pole of Inaccessibility rose to about −50.5‰ in summer snow of 1965 and declined to about −61‰ in the winter of 1964. However, the distinction between winter and summer snow is not this obvious in all cases.

Table 17.5 Snow accumulation rates near the Pole of Inaccessiblity (82°07'S, 55°06'E) determined by different methods (Picciotto et al. 1968)

Method	Accumulation rate (g/cm²/year)
1. Accumulation stake	3.1 ± 0.5, 1965
	3.6, 1959–1965
2. Visual firn stratigraphy	<12
3. Oxygen isotope stratigraphy of upper 235 cm of snow	8.6 ± 0.9
4. Beta activity of 1955 snow	3.0 ± 0.1
5. Variation of ^{210}Pb with depth, 80-year average	3.1 ± 0.3

[1]The Pole of Inaccessibility is farthest from any point on the coast and has an elevation of 3,700 m above sea level. The U.S.S.R. established a temporary research station nearby from December 14–26 in 1956 at the coordinates indicated above. The Pole of Inaccessibility is actually located at 85°50'S, 65°47'E (Stonehouse 2002).

17.5.2 Temperature Estimates (Oxygen)

The seasonal variations of $\delta^{18}O$ and δD values of snow at the South Pole in Fig. 17.23 imply that the isotope compositions of oxygen and hydrogen in snow deposited on the East Antarctic ice sheet depend on the condensation temperature of water vapor in the atmosphere (Picciotto et al. 1960; Gonfiantini et al. 1963; Gow 1965). This hypothesis was subsequently investigated by Aldaz and Deutsch (1967) who measured $\delta^{18}O$ values of freshly deposited snow at the South Pole from November 1964 to October 1965 and recorded the surface temperature for 89 days in the 12-month period. The surface temperatures in Table 17.6 ranged from −23°C to −67°C with an average of −43.6°C, whereas $\delta^{18}O$ values of snow that fell in the same period decreased from −32.5‰ in the summer of 1964/65 to −62.8‰ and averaged −52.0‰. The extreme ^{18}O depletion of the snow at South Pole Station also gives credence to the low $\delta^{18}O$ of the ice in the Reckling and Elephant moraines in Fig. 17.19.

The surface temperatures and $\delta^{18}O$ values of the snow at the South Pole during 1964/65 are positively correlated and approximate to line A in Fig. 17.24:

$$\delta^{18}O = 04177\ t_s - 32.41\ (\text{line A}) \qquad (17.5)$$

where $\delta^{18}O$ is in permille on the SMOW scale, t_s is the surface temperature in °C, and the linear correlation coefficient $r^2 = 0.7564$ for 85 sets of data. This equation applies only to snow that was deposited in the area around the South Pole and to the firn and ice that formed from that snow.

A more serious restriction is that isotope fractionation is not governed by the temperature at the surface where the snow accumulated, but by the temperature

Table 17.6 Surface temperatures and $\delta^{18}O$ values of freshly deposited snow at the South Pole from November 1964 to October 1965 (Aldaz and Deutsch 1967)

Parameter	Max	Min	Average
Surf. Temp. (°C)	−23	−67	−43.5
$\delta^{18}O$ (SMOW) (‰)	−32.5	−62.8[a]	−52.0

[a]This is one of the lowest $\delta^{18}O$ values ever measured in Antarctic snow or ice.

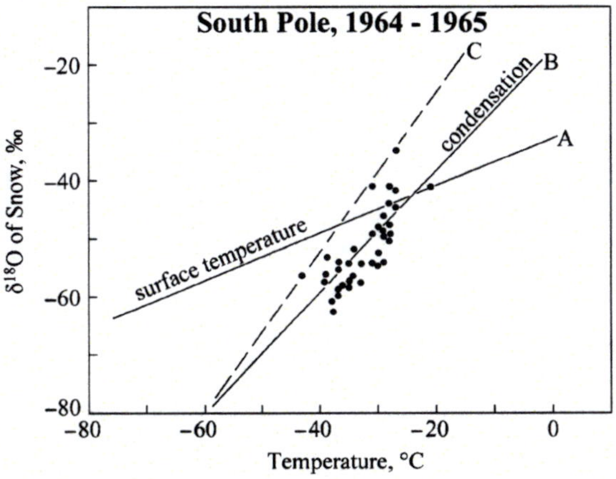

Fig. 17.24 The $\delta^{18}O$ values of freshly fallen snow at South Pole Station during the 12-month period from November 1964 to October 1965 are approximately correlated with surface temperatures (line A, equation 17.5) and with atmospheric temperatures (line B, equation 17.6). Only the condensation temperature at which snow flakes formed in the atmosphere above South Pole Station were plotted. Note that the data points based on surface temperatures are not shown. Line C (equation 17.7) is from Aldaz and Deutsch (1967) and is based on atmospheric temperatures. These quasi-linear relationships provide a basis for estimating surface temperatures as well as condensation temperatures from the $\delta^{18}O$ values of snow, firn, and ice (Data from Aldaz and Deutsch 1967)

in the clouds where the snow flakes grew by condensation of water vapor. Therefore, Aldaz and Deutsch (1967) also measured the atmospheric temperatures at different heights above the surface between 650 and 500 millibars by means of a "radiosonde". The highest temperatures at which water vapor in the clouds condensed to form snow flakes and the corresponding $\delta^{18}O$ values of snow in Fig. 17.24 define line B:

$$\delta^{18}O = 10.899\ t_c - 16.1165\ (\text{line B}) \quad (17.6)$$

where t_c is the maximum condensation temperature and the correlation coefficient $r^2 = 0.7829$ for 37 data pairs. Line C in Fig. 17.24 was defined by Aldez and Deutsch (1967):

$$\delta^{18}O = 1.4\ t_c + 4.0\ (\text{line C}) \quad (17.7)$$

Equations 17.5 (line A) and 17.6 (line B) can be used to estimate the temperature of condensation in the atmosphere and the temperature of deposition of the surface from the $\delta^{18}O$ values of snow deposited at the South Pole and, perhaps, elsewhere in the interior of the East Antarctic ice sheet. For example, if the measured $\delta^{18}O$ value of snow is −50.0‰, equation 17.5 (line A) yields t_s = −42°C and equation 17.6 (line B) indicates t_c = −31°C. The estimated surface temperature (t_s) is probably a more appropriate descriptor of the climate at South Pole, whereas the condensation temperature (t_c) is more relevant to the meteorological process of snow formation at the South Pole.

17.5.3 Temperature Estimates (Hydrogen)

The isotope composition of hydrogen in snow and ice of the East Antarctic ice sheet also contains a record of the temperature at which the snow flakes formed from water vapor in the atmosphere. The relation between δD of snow and the average annual temperature at the collecting sites was investigated by Lorius and Merlivat (1977) along a traverse from Dumont d'Urville on the coast of Adélie Land toward Dome C. The elevation along the traverse in Fig. 17.25 increased from sea level to about 3,100 m a.s.l. at a distance of nearly 900 km from the coast. As expected, the average annual temperature, measured at a depth of 10 m below the surface, decreased from about −20°C near the coast to − 54°C at an elevation of about 3,100 m at the end of the traverse.

The δD values of snow along this traverse decreased from about −150‰ at the coast to −360‰ at the highest elevation, at the lowest temperature, and at the greatest distance from the coast. The results reported by Lorius and Merlivat (1977) indicate that the δD values (in per mille) are linearly related to the mean annual temperature according to the equation:

$$\delta D\ (‰) = 6.04\ t\ (°C) - 51 \quad (17.8)$$

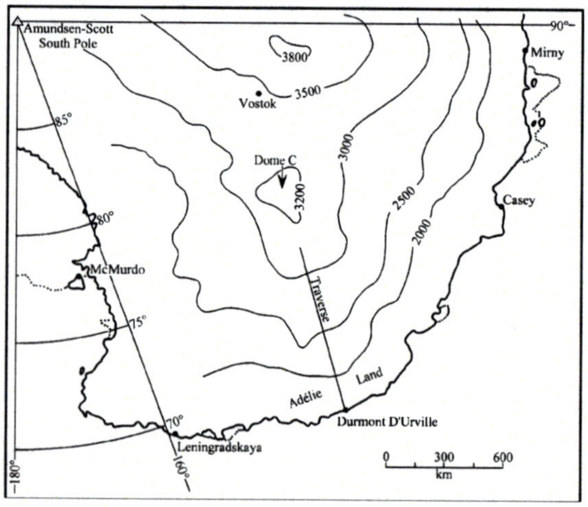

Fig. 17.25 The traverse by C. Lorius and L. Merlivat started from the French research station Dumont D'Urville on the Adélie Coast and proceeded toward Dome C. The purpose was to relate the δD values of firn to the elevation above sea level and to the average annual temperature of the stops along the way. The length of the traverse was close to 900 km. Dome C is located along the central ice divide of the East Antarctic ice sheet (Adapted from Figure 1 of Lorius and Merlivat 1977)

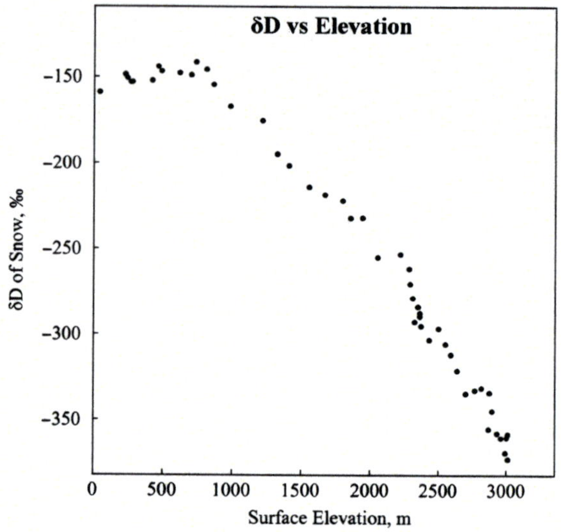

Fig. 17.26 The δD values of snow and firn collected by Lorius and Merlivat (1977) along the route of their traverse shown in Fig. 17.25 decrease with increasing elevation of the collecting sites as well as with decreasing average annual temperature measured 10 m below the surface. This relationship arises because the isotope-fractionation factor of hydrogen in water vapor in the atmosphere increases with decreasing temperature and because of the resulting progressive depletion of the water vapor in deuterium as the air masses move from the coast into the interior of the East Antarctic ice sheet (Adapted from Figure 3 of Lorius and Merlivat 1977)

Faure et al. (1993) used this equation to estimate the mean annual temperatures at the source of the ice that is now exposed in the Elephant and Reckling moraines. In order to do this, they first converted the $\delta^{18}O$ values of the ice to the corresponding δD values using the equation of the meteoric-water line in generalized form:

$$\delta D = 8\ \delta^{18}O + 10 \tag{17.9}$$

The resulting temperatures of the source of the ice ranged from −45°C to −57°C. At the present time, such low average annual temperatures occur only in the central region of the East Antarctic ice sheet and at the South Pole. The same procedure indicated that the ice adjacent of the Allan Hills formed at somewhat warmer average annual temperatures between −37°C and −47°C. Evidently, this ice either originated from an area that is closer to the Transantarctic Mountains than the source of the ice at the Elephant and Reckling moraines, or it formed during a previous interglacial stage (e.g., the Sangoman).

The data of Lorius and Merlivat (1977) in Fig. 17.26 also demonstrate a quasi-linear relationship between δD values of the snow and the elevation of the collecting site along their traverse. This relation yields elevations in excess of 3,000 m a.s.l. for the source of the ice in the Elephant and Reckling moraines but only about 2,700 m for the ice at the Allan Hills. The reliability of these estimates is confirmed by noting that the $\delta^{18}O$ value of snow that fell in the Reckling Moraine in December of 1986 (i.e., $\delta^{18}O = -32.3‰$) corresponds to an elevation of 2,100 m a.s.l. which agrees with the present elevation of this moraine.

In addition, the $\delta^{18}O$ values of snow that can be calculated from the δD values measured by Lorius and Merlivat (1977) rise from −48‰ to −20‰ with increasing average annual temperature from −60°C near Dome C and −20°C at the Adélie Coast. The resulting graph in Fig. 17.27 includes the $\delta^{18}O$ values of ice at the Elephant and Reckling moraines (E and R) and the Allan Hills (A). These empirical relationships relate the $\delta^{18}O$ and δD values of the ice to the climatic conditions at the time the snow was deposited. Although the climate history that is contained in the ice is destroyed by the deformation of the ice when it flows from its source to the area adjacent to the Transantarctic

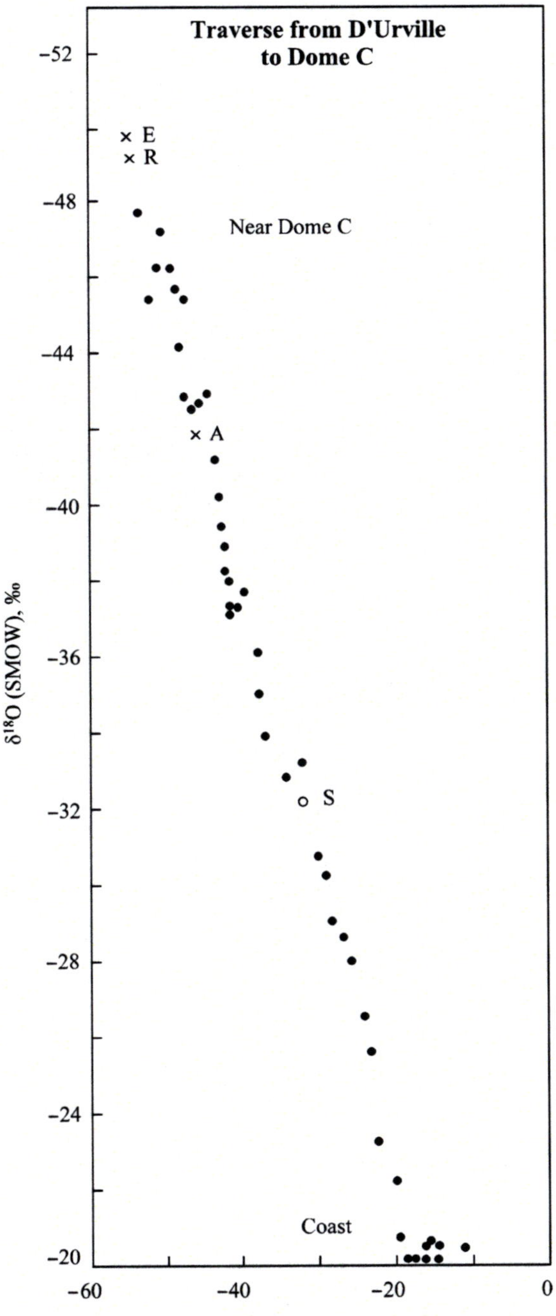

Fig. 17.27 The $\delta^{18}O$ values of *snow* and firn, calculated from the measured δD values in Fig. 17.26, decrease progressively with decreasing average annual temperature of the collecting sites. These $\delta^{18}O$ values refer to the snow and firn of the collecting sites and not to the underlying ice which is not exposed along the Adélie Coast of East Antarctica. However, the crosses are the measured $\delta^{18}O$ values of *ice* at the Elephant (E) and Reckling (R) moraines and at the ice fields adjacent to the Allan Hills (A). These measurements are used here to estimate the average annual temperatures of the source regions of the ice. The data point labeled S is the measured $\delta^{18}O$ value of snow deposited in the Reckling Moraine in December of 1986, which therefore indicates that the average annual temperature at the Reckling Moraine at the present time is –32.1°C (Based on data by Lorius and Merlivat 1977)

Mountains, the climate history is *preserved* in long cores that penetrate the ice sheet along the central divide such as at Dome C and in the vicinity of the Russian research station known as Vostok.

17.6 Climate Histories from Ice Cores

The ice that has accumulated along the central divide of the East Antarctic ice sheet contains a record of past climatic temperatures that is preserved in the isotope compositions of oxygen and hydrogen. In addition, the ice contains atmospheric gases that were trapped by the annual layers of snow. The ice also contains particles of terrestrial and extraterrestrial dust, as well as ions and molecular compounds that nucleated snow flakes. This record of past environmental conditions has been recovered by analysis of the ice in long cores drilled through the ice sheets of East and West Antarctica and of Greenland. As a result, we now have information about the history of the global climate and of the composition of the atmosphere extending up to 800,000 years into the past (Petit et al. 1999; EPICA 2004; Siegenthaler et al. 2005; Spahni et al. 2005; Loulergue et al. 2008; Lüthti et al. 2008). These studies of the past are motivated by the hope that knowledge of the past may be the key to the future (Palais 1986).

The ice cores that have already been drilled through the ice sheets in Greenland and Antarctica have not only provided information about the history of climate change, but have also indicated that climate changes during the Pleistocene glaciations have occurred more rapidly than expected (Alley and Bender 1998; Alley 2000, 2002, 2004). In addition, the ice cores have yielded information about the concentrations of carbon dioxide and methane in the ancient atmosphere. Variations in the concentrations of these greenhouse gases either caused global climate change or were affected by changes in the global climate that were caused by other factors.

17.6.1 The Vostok Core

Vostok is located on the high plateau of the East Antarctic ice sheet at 78°28′S and 106°48′E at an elevation of 3,488 m. The average annual temperature is −55.5°C and the rate of deposition of water in the form of snow ranges from 2.2 to 2.5 g/cm²/year (Young et al. 1982). The coldest day in Vostok occurred on July 21, 1983, when the temperature dropped to −89.6°C. The highest temperatures of about −15°C occur in January (Stonehouse 2002). Drilling at Vostok started in 1970 and proceeded in stages to a depth of 950 m which was reached in 1974. The $\delta^{18}O$ profile of this core encompassed the transition from the last ice age to the present interglacial (Barkov et al. 1974; Lorius et al. 1979, 1985).

Drilling resumed in 1980 and reached a depth of 2,083 m where the depositional age of the ice was estimated to be 150,000 years (Lorius et al. 1985). The $\delta^{18}O$ and δD values of the ice were measured at the Laboratoire de Geochemie Isotopique in Saclay, Gif-sur-Yvette, in France. The resulting $\delta^{18}O$ and δD profiles of the Vostok core were interpreted by Lorius et al. (1985) and by Jouzel et al. (1982), respectively. In addition, the concentrations of carbon dioxide and of chemical impurities in this core were measured in Russian and French laboratories.

The $\delta^{18}O$ profile of the Vostok core in Fig. 17.28 includes four intervals of low temperature when the ice was strongly depleted in ^{18}O and reached $\delta^{18}O$ values of −62‰ (SMOW). Lorius et al. (1985) used these cold intervals to subdivide the core into eight sections which they identified by the letters A-H. The depths in the core and the ages of these intervals are indicated in Table 17.7. Accordingly, the core extends from the Holocene (A) (Present) to the Wisconsinan glacial

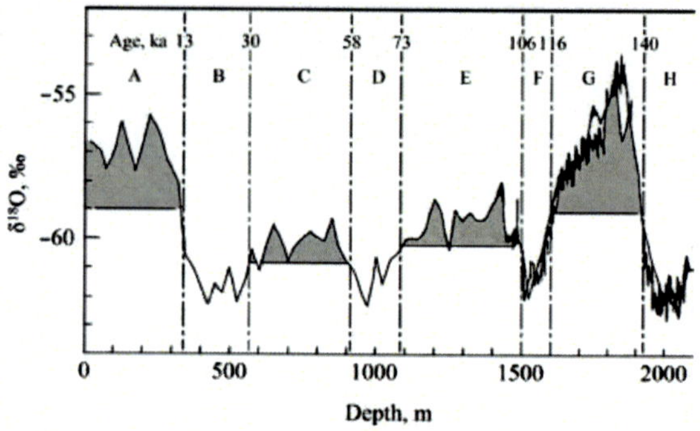

Fig. 17.28 The profile of $\delta^{18}O$ values of the 2083-m ice core drilled at Vostok (Fig. 17.25) records changes in the average annual temperature in late Pleistocene time that included two episodes of continental glaciation (Wisconsinan and Illinoian) and two interglacials (Holocene and Sangamonian). These geological events are identified by the letters: A = Holocene interglacial, B to F = Winconsinan glacial, G = Sangomonian interglacial, H = Illinoian glacial (Adpated from Lorius et al. 1985)

(B-F), the Sagoman interglacial (G), and the Illinoian (H) glacial stages.

The $\delta^{18}O$ values of *Holocene* ice range from −56‰ to −57.5‰, which corresponds to average surface temperatures between −56.5°C and −60.1°C based on equation 17.5 that was derived in Section 17.5.2 from the data of Aldaz and Deutsch (1967) at the South Pole. These temperature estimates are in general agreement with the average annual temperature of −55.5°C at Vostok (Lorius et al. 1985). The $\delta^{18}O$ values representing the *Wisconsinan* Stage (B-F) indicate that the average surface temperature declined to −70.8°C during three episodes (B, D, and F). These cold spells were separated by two intermediate intervals when the average annual temperature rose to −64.9°C (C) and −61.3°C (E). During the *Sangoman* interglacial (G) the average surface temperature rose to −51.7°C which is

Table 17.7 Warm and cold intervals indicated by the $\delta^{18}O$ values of ice in the long core (2,083m)[a] drilled at Vostok on the East Antarctic ice sheet (Lorius et al. 1985)

Interval	Warm/cold	Depth (m)	Age (year)	Glacial stage
A	Warm	0–330	0–13,000	Holocene
B	Cold	330–570	13,000–30,000	Wisconsinan
C	Intermediate	570–910	30,000–58,000	Wisconsinan
D	Cold	910–1,090	58,000–73,000	Wisconsinan
E	Intermediate	1,090–1,500	73,000–106,000	Wisconsinan
F	Cold	1,500–1,600	106,000–116,000	Wisconsinan
G	Warm	1,600–1,920	116,000–140,000	Sangoman
H	Cold	1,920–2,083	>140,000	Illinoian

[a]The total thickness of the ice at Vostok is approximately 3,700 m.

Table 17.8 Summary of environmental conditions during the past 150,000 years at Vostok on the East Antarctic ice sheet (Based on the work of Lorius et al. 1985)

Glacial stage	Duration (year)	Avg. Annual Temp. (°C)
Holocene (A)	13,000	−56.6 to −60.1
Wisconsinan (B-F)	103,000	−61.3 to −70.8
Sangoman (G)	24,000	−51.7 to −57.7
Illinoian (H)	>10,000	−68.4 to −70.8

warmer than the present mean annual temperature at Vostok. However, the *Illinoian* glaciation which preded the Sangoman interglacial was just as cold as the Wisconsin glaciation (i.e., −70.8°C).

The information derivable from the Vostok core is summarized in Table 17.8

1. The Holocene interglacial has lasted about 13,000 years which is about one half the duration of the Sangoman interglacial.
2. The average annual temperatures of the Wisconsin and Illinoian glaciations at Vostok were about 15°C colder than the present average annual temperature.
3. The Wisconsin glaciation in East Antarctica was interrupted by two intervals of partial warming, but it still lasted for more than 100,000 years.
4. If geologic history were to repeat itself, the next continental glaciation of the Earth may occur in about 10,000 years.

The climatic history derivable from the Vostok core was later extended by Petit et al. (1999) to 488,000 years.

17.6.2 Dating Ice

Several different methods have been used to date ice in long cores; but all of them have limitations. Methods based on the stratigraphy of the ice are unreliable because the thickness of annual layers decreases with depth in the core and because the isotope compositions of oxygen and hydrogen are homogenized by diffusion. The thinning of annual layers of ice in the Vostok core is illustrated in Fig. 17.29 which was derived from the ice-flow model that includes the thinning of annual layers and is based on the assumption that the rate of accumulation of snow on the polar plateau is strongly correlated with the surface temperature (Lorius et al. 1979, 1985).

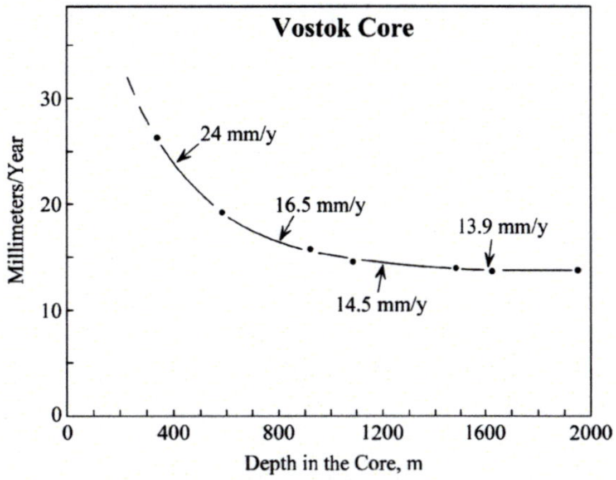

Fig. 17.29 The thickness of the annual layers of ice in the Vostok core (Fig. 17.28) decreases with increasing depth in the upper part of the core and approaches a constant value below a depth of about 1,600 m. This property causes the age of ice in a long core to vary non-linearly with depth (Derived from Figure 1 of Lorius et al. 1985)

This ice-flow model can be supplemented by certain other methods listed in Table 17.9. The application of methods based on natural or anthropogenic radioactive isotopes is limited because of low concentrations, irregular distribution, and because the basic assumptions of dating are not always satisfied.

The climate record revealed by the $\delta^{18}O$ and δD values of long ice cores from Antarctica and Greenland can also be compared in Fig. 17.30 to the cyclical variations of $\delta^{18}O$ (PDB) values of the calcium-carbonate shells of the foraminifer *Globigerina* sacculifer in piston cores from the Caribbean Sea that were dated by the uranium series disequilibrium method (Emiliani 1978, 1992). The Marine Isotope Stages (MIS) defined by the $\delta^{18}O$ (PDB) values of forminifera form a sequence of warm (up) and cold (down) intervals that are numbered from 1 (Holocene) to 20 (~750 × 10^3 year). The isotopic record in ice cores matches the marine record remarkably well which confirms the chronology of the ice cores and supports the hypothesis that both record variations of the global climate during the Pleistocene.

17.6.3 EPICA Core at Dome C

The success of the collaboration of Russian and French scientists who drilled and analyzed the long core at

Table 17.9 Methods of dating surface ice and long ice cores (Lorius et al. 1985; Faure and Mensing 2005)

Method	Application	Procedure
Stratigraphic methods		
Snow stratigraphy	Snow pits and short ice cores	Counting of annual layers
Ice-flow model	Long cores	Rate of deposition and thinning function
Volcanic dust	Long cores	Date anorthoclase and zircon by isotopic method
Fallout from nuclear weapons test in 1955	Snow pits and short ice cores	Rate of beta decay of ^{134}Cs, ^{137}Cs, ^{60}Co, ^{90}Sr, etc.
Isotopic methods		
Isotopic dating of snow and ice	Snow pits and short cores	^{14}C, ^{210}Pb, ^{230}Th, etc.
Cosmogenic radio-nuclides	Snow pits and short cores	^{10}Be, ^{26}Al, ^{36}Cl, ^{53}Mn, etc.
Comparison to marine isotopic stages		
Oxygen isotope record of foraminifera	Long ice cores	Matching of warm and cold periods of global climate

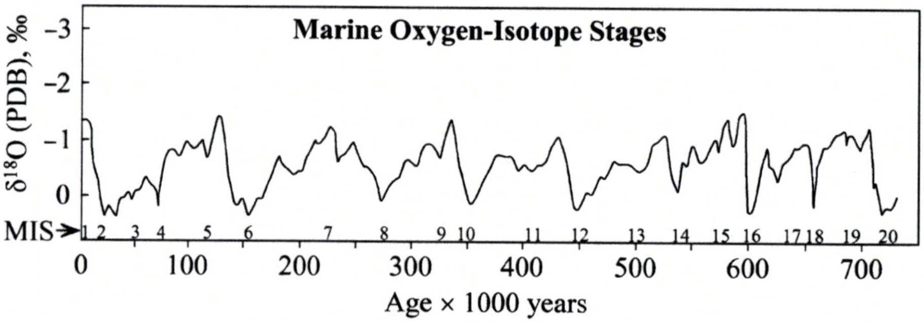

Fig. 17.30 The profile of $\delta^{18}O$ (PDB) of the carbonate shells of pelagic and benthic forams in piston cores of marine sediment from the Caribbean Sea define warm (up) and cold (down) intervals from the Holocene to about 750×10^3 years in the past. The so-called Marine Isotope Stages (MIS), which are numbered from 1 to 20, include eight episodes when the climate turned cold. The oxygen-isotope profile of ice cores recovered at Vostok and Dome C is similar to the marine record, but the correlation of the marine climate record to the land-based record is uncertain primarily because the land-based record is fragmentary (Emiliani 1978, 1992) (Adapted from Hoefs 1997)

Vostok prompted a group of European scientists to form a team for the purpose of drilling an even longer core at Dome C. The participants in the European Project of Ice Coring in Antarctica (EPICA) chose to drill at Dome C because it is located in the center of the snow-accumulation area where old ice should be present at depth and where the annual layers are not deformed by folding and faulting as they appear to be in the Elephant and Reckling moraines on the margin of the ice sheet (Sections 17.3 and 17.4).

The drill site chosen by EPICA is called *Concordia Station* and is located at 75°06′S and 123°21′E at an elevation of 3,233 m above sea level where the thickness of the East Antarctic ice sheet is 3,200 m. The consortium of scientists and logistics personnel from ten nations first drilled a hole at Kohnen Station in Queen Maud Land for comparison with the climate record contained in the ice cores recovered in Greenland. The ice core drilled at Dome C by EPICA (designated EDC) has been analyzed for $\delta^{18}O$, δD, and for the concentrations of carbon dioxide, methane, and nitrous oxide. The ice was dated by means of an ice-flow model that indicated an age of 800,000 years for the bottom of the core (EPICA 2004).

The temperature fluctuations in Fig. 17.31 recorded by the measured $\delta^{18}O$ values of the ice indicate that nine cold intervals occurred in the 800,000 years encompassed by the Dome-C core (Lüthti et al. 2008). The cold and warm intervals in this core are identified by the numbers of the corresponding marine oxygen-isotope stages in Fig. 17.30. As we noted before, the good agreement between the climate records of the marine foram shells and the EPICA ice core strengthens the interpretation of both and confirms that the

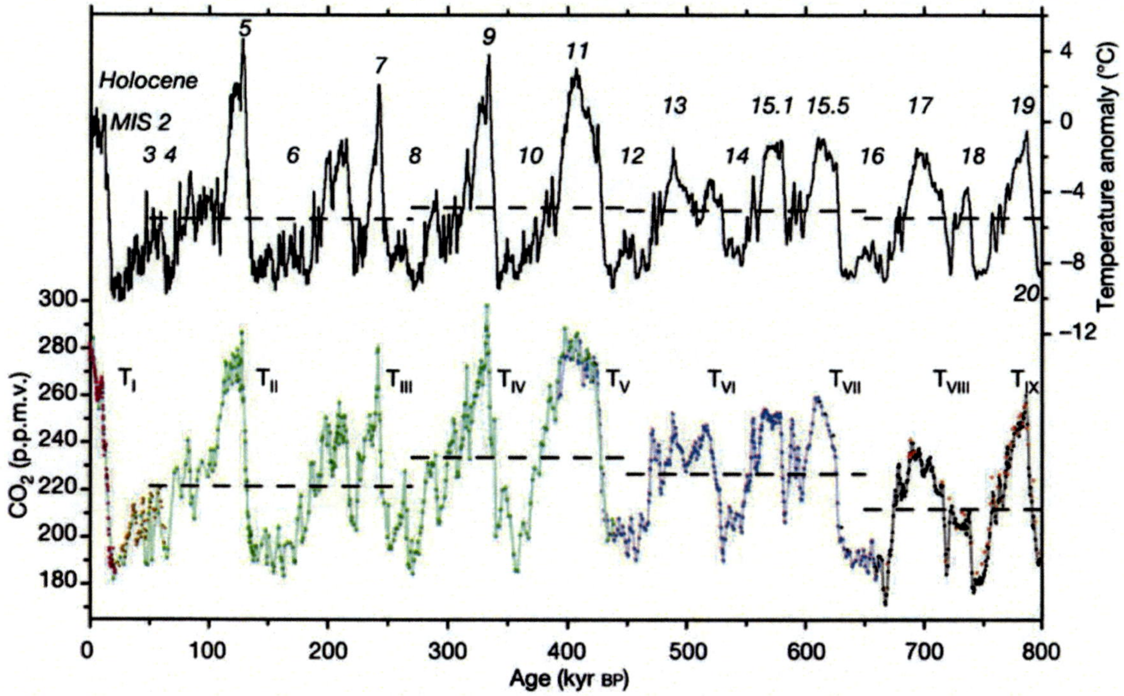

Fig. 17.31 The history of temperature fluctuations based on the isotope composition of oxygen and hydrogen in the ice core drilled by the European Project for Ice Coring in Antarctica (EPICA) at Concordia Station on Dome C (75°06′S, 123°21′SE) is closely correlated with the concentrations of carbon dioxide in the ice, both of which can be indexed to the marine isotope stages in Fig. 17.30. The correlation of these parameters confirms the scope of the climate fluctuations which include nine cold spells that terminated abruptly during the Pleistocene Epoch (i.e., T_I-T_{II}, etc.). The temperature profile (black curve) is based on measurements of δD and was plotted on the EDC3 timescale. The carbon dioxide concentrations of the ice at Dome C (*purple*, *blue*, and *black circles*) were measured at the University of Bern. The data for the Vostok core (*green*), Taylor Dome (*brown*), and Dome C (*red*) were measured at Grenoble. The horizontal lines are the mean values of temperature and carbon dioxide during the time intervals: 799–650, 650–450, 450–270, and 270–50 in thousands of years (ka) before the present (BP) (Reproduced from Lüthi et al. 2008 by kind permission of the senior author and by License Agreement 2310880793668, Nov. 16, 2009, with Nature Publishing Group)

climate of the Earth during the Pleistocene was even more unstable than suggested by the land-based record which contains only four glacial stages.

The concentrations of carbon dioxide in the ice of the Dome-C core range from about 300 parts per million by volume (ppmv) to 172 ppmv (Lüthi et al. 2008). These values are *lower* than the concentration of carbon dioxide in the present atmosphere, which has increased from about 340 ppmv in 1980 (Thompson and Schneider 1981) to about 385 ppmv in 2010 (Perkins 2008). Even more noteworthy is the close correlation of warm episodes (i.e., interglacials) and high concentrations of carbon dioxide in air trapped in the ice. Similarly, cold intervals (i.e., glaciations) correlate with low concentrations of carbon dioxide. The correlation of temperature fluctuations with variations of the concentrations of carbon dioxide raises the important question:

Was climatic warming caused by an increase in the carbon dioxide content of the atmosphere, or does climatic warming cause the concentration of atmospheric carbon dioxide to rise?

It could have gone either way before humans started to increase the carbon dioxide concentration of the atmosphere by combustion of fossil fuel, by large-scale burning of plants, and by other practices that benefit us in the short run, but which release excess carbon dioxide into the atmosphere.

The EPICA Dome-C core has also yielded concentrations of methane and nitrous oxide trapped in the ice

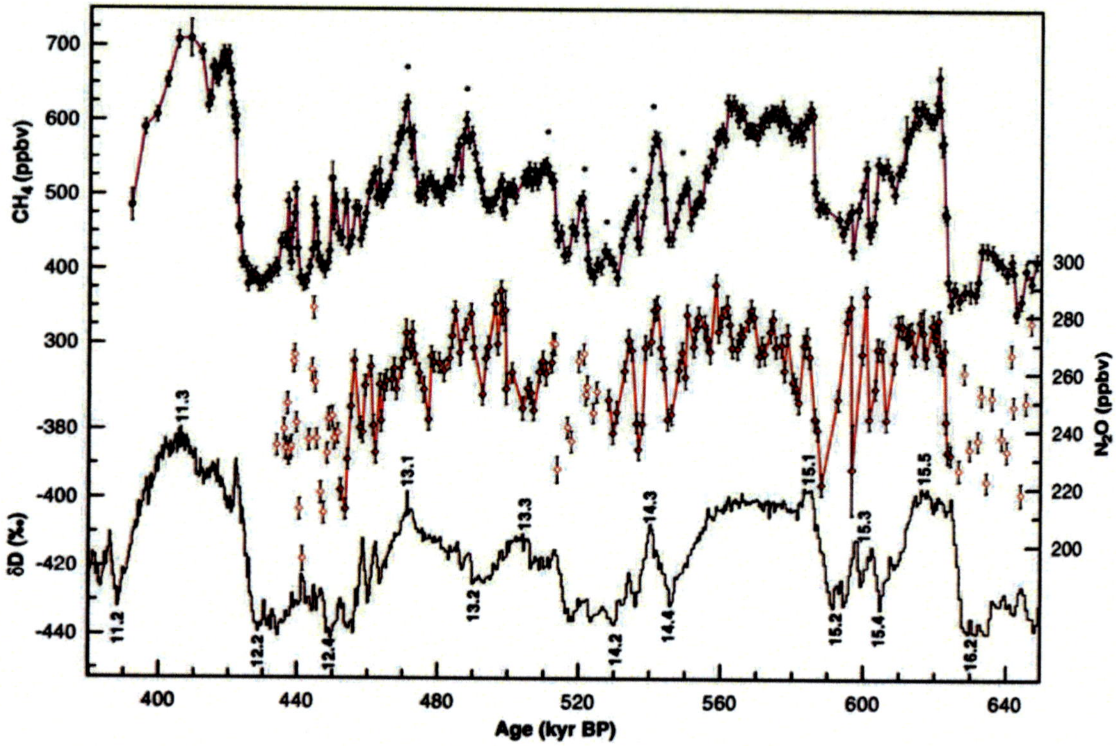

Fig. 17.32 The concentrations of methane in ice of the Dome-C core (EPICA) between 380 × 10³ years and 650×10³ years in the past vary in-step with the δD values and hence with the average annual surface temperatures at Dome C. The warm and cold climatic intervals are identified by the MIS (marine isotope stages) numbers from Fig. 17.30 where "up" means warm and "down" means cold. The profile of nitrous oxide concentrations also mirrors the fluctuations of δD. Data for the upper part of the core (0–220 ka) are available in Figure 1 of Spahni et al. (2005). The measurements were made at the Physics Institute of the University of Bern and at the Laboratorie de Gaciologie et Geophysique de l'Environnement (LGGE) in Grenoble. The data were plotted on the EDC2 time scale (2005) (Reproduced from Figure 2 of Spahni et al. 2005 by permission of Professor Dr. Thomas Stocker at the University of Bern and by the Copyright Clearance Center in Danvers, Massachusetts, USA)

(Spahni et al. 2005). The concentrations of both gases follow the δD values of ice in Fig. 17.32 between 380,000 and 650,000 years. The peaks and valleys of the δD profile are identified by MIS numbers placed on top for warm intervals (peaks) and on the bottom for cold intervals (valleys). The concentrations of methane in this time interval range from 375 parts per billion by volume (ppbv) to 725 ppbv which amounts to an increase by a factor of 1.9. Since methane, like carbon dioxide, is a "greenhouse gas" that absorbs infrared radiation, the variations of the methane concentrations correlate closely with variations in δD which tracks the average annual temperature of the East Antarctic ice sheet at Dome C.

Evidently, the analysis of ice cores from Antarctica and Greenland has already yielded important results that enhance our understanding of the environment on the surface of the Earth. The opportunity to participate in this important research has attracted scientists from other nations. For example, Japanese investigators recovered ice that is 340,000 years old in a core drilled at Dome Fuji located at 77°30′S and 37°30′E about 950 km south of the Japanese research station Syowa (Watanabe et al. 2003). In addition, China is preparing to drill at Dome Argus (80°22′S, 77°21′E) which is the highest area on the East Antarctic ice sheet at an elevation of 4,093 m. The thickness of the ice sheet at this location is uncertain because of the presence of the subglacial Gamburtsev Mountains. Nevertheless, the Chinese investigators and the International Partnerships in Ice Core Sciences (IPICS) are hoping to recover a long ice core that may extend even farther back in time than the EPICA core at Dome C.

17.7 Water Under the Antarctic Ice Sheet

The East Antarctic ice sheet appears to be a frozen wasteland of vast proportions. The seasonal temperatures at South Pole Station range from −21°C in the summer to −78°C in the winter with an average annual temperature of −50°C. Conditions at Vostok are even more severe. The thickness of the ice sheet ranges from less than 1,500 m above the peaks of the Gamburtsev Mountains to more than 4,500 m in the region between East and West Antarctica (Hughes et al. 1981). In spite of the low surface temperatures, large amounts of liquid water are present at the base of the East Antarctic ice sheet (Kiryukhin and Tolstikhin 1988; Bindschadler and Bentley 2002, Fricker et al. 2007; Bell et al. 2006, 2007; Bell 2008).

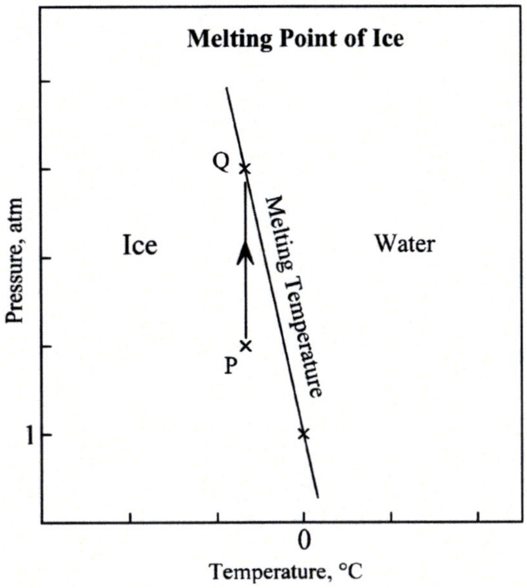

Fig. 17.33 The melting temperature of ice decreases with increasing pressure; therefore, ice melts at a lower temperature when pressure is exerted on it. The melting temperature (aka melting point) of ice at one atmosphere is 0°C (zero degrees Celsius is defined as the melting temperature of water at one atmosphere). When pressure is applied to ice at point P, without changing the temperature, the ice will melt at point Q because the increase in pressure favors the state of aggregation that has the higher density and, therefore, a smaller molar volume (i.e., liquid water is denser than ice). This phenomenon contributes to the melting of ice at the base of the East Antarctic ice sheet

17.7.1 Pressure-Melting Point

Liquid water exists at the base of the East Antarctic ice sheet because the melting point of ice is depressed by the pressure of the overlying ice and because the ambient temperature at the base of the ice sheet is at or slightly above the local pressure-melting point depending on the heat released by several sources:

1. Geothermal heat from the underlying crust of the East Antarctic craton and from the upper mantle beneath the crust.
2. Hot spots in the crust underlying the ice sheet resulting in the discharge of heated groundwater and/or the eruption of steam, volcanic pyroclastics, and lava flows.
3. Heat generated by the deformation of the ice sheet, by friction at the base of the ice sheet, and by exothermic chemical reactions.

The depression of the melting point of ice by pressure is a unique property of water and is a consequence of the fact that ice is less dense than water (i.e., ice floats in water). The reason why ice melts when pressure is applied to it is explained in Fig. 17.33.

When pressure is applied to ice, its melting temperature decreases along the line labeled "Melting Temperature" in Fig. 17.33. For example, when pressure is applied to ice existing at point P until the environmental conditions have changed to the coordinates of point Q (higher pressure, same temperature), the ice melts because point Q lies on the hypothetical melting curve of water. The basic reason for this unique behavior of ice arises because the water that forms at point Q occupies a smaller volume than the ice that melted, which has the effect of reducing the pressure. This phenomenon contributes to the melting of ice at the base of the East Antarctic ice sheet and applies equally to the ice in all continental ice sheets, ice caps, and valley glaciers.

The environmental conditions at the base of the East Antarctic ice sheet were classified by Hughes et al. (1981) into three categories:

1. Frozen, because the temperature at the base of the ice sheet is lower than the melting point of the ice at the prevailing pressure.
2. Melting, because the temperature at the base of the ice sheet is equal to or greater than the melting point of the ice at the prevailing pressure.

3. Freezing, because the temperature at the base of a wet ice sheet is less than the freezing temperature of the glacial meltwater.

The basal ice in the center of the East Antarctic ice sheet in Fig. 17.34 is completely frozen (dark blue) because the thickness of the ice sheet (1,500 m) is insufficient to exert enough pressure on the ice to lower its melting point to cause melting at the ambient temperature. This area is surrounded by a broad region where basal ice is melting (red) which grades into regions were water that has formed by melting in the red zone is refreezing in the pale green zone. The transition from melting to freezing causes the meltwater to refreeze as *regelation* ice which does not retain stratigraphic continuity with the overlying ice sheet. Therefore, the isotopic composition of oxygen and hydrogen as well as the chemical composition of the regelation ice are homogenized and the record of the conditions at the time the ice was originally deposited is lost, especially in case the regelation ice forms from meltwater that is flowing under the ice sheet in response to pressure gradients.

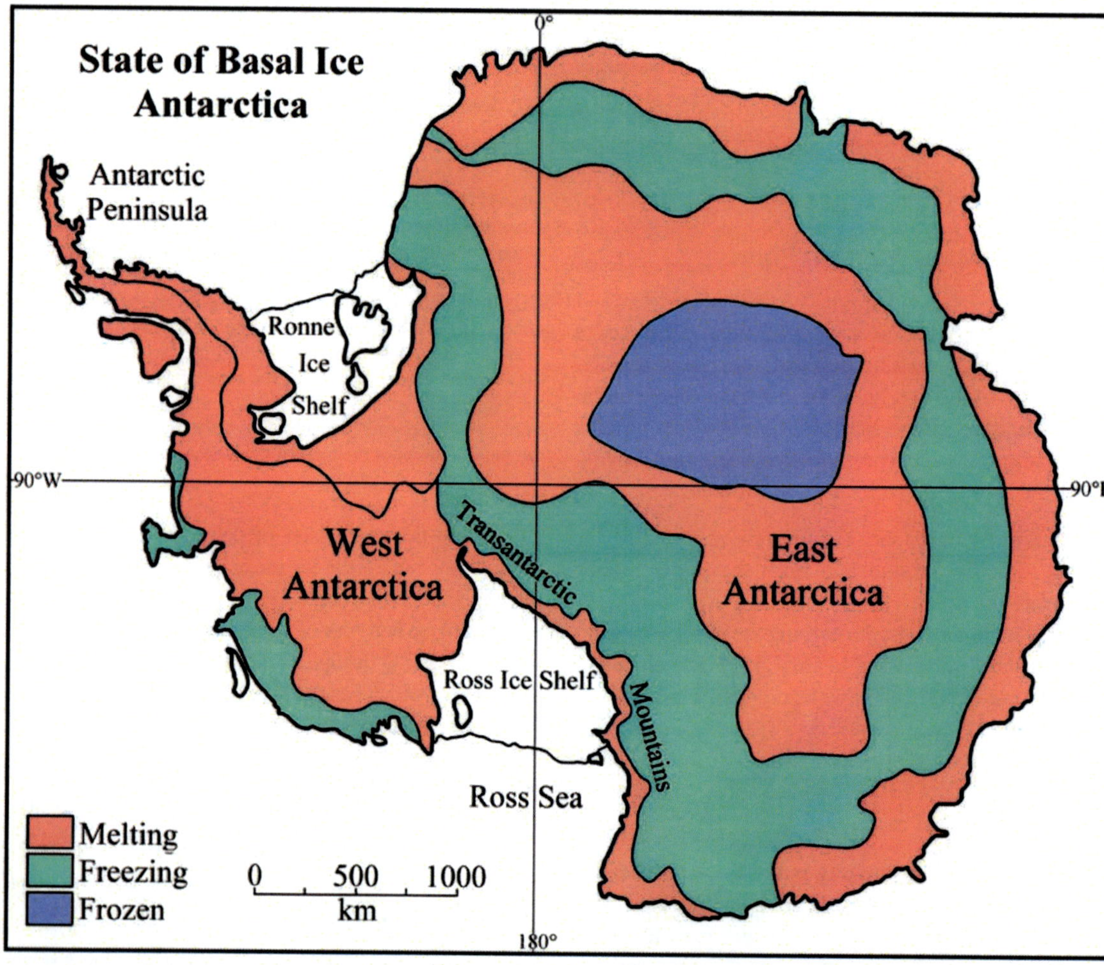

Fig. 17.34 The state of the ice at the bottom of the Antarctic ice sheets depends on the prevailing temperature and on the melting point of the ice which *decreases* with increasing pressure. As a result of the interplay of these factors, the bottom of the ice sheet is either frozen, melting, or freezing. Where the basal ice is melting, water in the liquid state is produced which flows along pressure gradients until it enters a freezing zone. When basal meltwater refreezes, dissolved salts are excluded and are concentrated in a residual brine from which salts may precipitate (e.g., carbonates, sulfates, chloride, etc.) (Adapted from Figure 6.7 of Hughes et al. 1981)

17.7.2 Lake Vostok

The water under the East Antarctic ice sheet has collected in more than 160 subglacial lakes which are connected by a network of channels (Cook-Anderson 2007; Bell et al. 2006, 2007; Fricker et al. 2007). Lake Vostok in Fig. 17.35 is the largest of these lakes with an area of about 13,000 square kilometers and a volume of 5,400 cubic kilometers comparable to the volume of Lake Michigan (Perkins 2006). The lake, which is overlain by 3,700 m of ice, is about 300 km long and about 50 km wide and appears to be located within a tectonic rift bounded by deep faults (Bell et al. 2006). The tectonic origin of the basin is supported by the occurrence of several earthquakes that have originated from epicenters near Lake Vostok (Bell 2008). Lake Sovetskaya and 90°E Lake, located about 400 km west of Lake Vostok in Fig. 17.35, may also occupy rift basins. The faults that presumable define the basin of Lake Vostok may heat groundwater which could feed hotsprings that discharge warm water into the lake.

Fig. 17.35 Lake Vostok in East Antarctica is the largest of about 160 subglacial lakes that have been discovered by radar echo surveys. The lake is about 300 km long and about 50 km wide. It is covered by 3,700 m of ice and occupies a tectonic rift that is bounded by deep faults. Lake Vostok was named after the Russian research station that is located close to the southern end of the subglacial lake. Lake Sovetskaya and 90°E Lake, which are located close to the 90°E Meridian may also occupy rift basins (Adapted from MODIS Mosaic of Antarctica acquired by NASA's satellites Terra and Aqua between November 20, 2003 and February 29, 2004)

Lake Vostok was named after the Russian research station that is located at the southern end of the lake. The hole that was drilled into the ice sheet at that station was stopped about 120 m above the lake because the water of the lake may contain organisms that have been isolated for millions of years. The journal Science News (October 9, 1999; p. 230) reported that bacteria had actually been found in the ice. The warm water that may be entering the lake via hotsprings, if confirmed, would strengthen the expectation that organisms have survived in the water of Lake Vostok.

Scientists at the Jet Propulsion Laboratory in Pasadena, California, have proposed to explore Lake Vostok by means of a "cryobot" which would could melt its way through the ice and could then release a "hydrobot" which would search for signs of life in the water (Wayt Gibbs 2001). The exploration of Lake Vostok was proposed as a test of the technology that could also be used to explore the ice-covered ocean of Europa, the second Galilean satellite of Jupiter (Faure and Mensing 2007). These plans have not been implemented because of the danger of contaminating Lake Vostok with modern life forms from the surface of the Earth, because the water in the lake contains carbon dioxide gas under pressure, and because of funding limits. Perhaps these technologies can be tested by exploring smaller and shallower lakes under the East Antarctic ice sheet that may also be more accessible from McMurdo Station than Lake Vostok. Several of these lakes and subglacial meltwater streams have been identified by naming them, including Lake Concordia, the Recovery lakes, and the Wingham River in East Antarctica as well as Lake Engelhardt, the Fricker cascading lakes, and Lake Ellsworth in West Antarctica (Bell 2008).

17.7.3 Ice Streams

The water under the East Antarctic ice sheet lubricates its motion and speeds up the delivery of ice to the southern ocean (Kohler 2007). That is especially true for the ice streams that drain the ice sheets of East and West Antarctica in Fig. 17.36 (Bindschadler and Bentley 2002). For example, meltwater from the four Recovery lakes, located about 500 km inland of the coast of the Filchner ice shelf, set in motion the fast-moving Recovery ice streams which discharge ice into

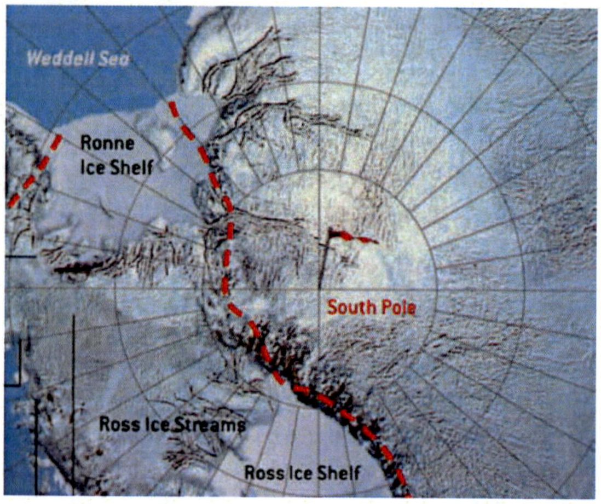

Fig. 17.36 The East Antarctic ice sheet loses mass not only by melting of basal ice, but also by means of ice streams and outlet glaciers. For example, the Recovery ice streams which are lubricated by water that originates from the subglacial Recovery lakes transports ice to the Filchner Ice Shelf, whereas the outlet glaciers which cross the Transantarctic Mountains discharge ice into the Ross Ice Shelf and directly into the Ross Sea. Similarly, ice streams that drain the West Antarctic ice sheet contribute most of the ice into the Ross Ice Shelf (Adapted from Bindschadler and Bentley 2002. By permission of the authors and the Editor of Scientific American)

the Filchner ice shelf (Bell et al. 2007; Bell 2008). In addition, five ice steams (A, B, C, D, and E) are draining the West Antarctic ice sheet into the Ross Ice Shelf (Bindschadler and Bentley 2002; Hughes 1987).

These ice streams play a vital role in the mass balance of the Antarctic ice sheets because 90% of the ice that enters the southern oceans is transported by ice streams which are up to about 2,000 m "deep", about 100 km wide, up to 800 km long, and move at the rate of 200–1,000 m/year, or more (Bell 2008). The ice streams flow much faster than most mountain glaciers especially when they are lubricated by the presence of water at their base. The ice is initially incorporated into the ice shelves that float in the water of the Ross Sea and the Weddell Sea. As these ice shelves grow in size, large tabular icebergs break off and melt as they float north.

17.7.4 Effect on Sea Level

The volume of the ice sheets in Antarctica and Greenland and the volume of water in the oceans are in delicate balance. If the global climate changes in such a way that the volume of glacial ice begins to decrease, the volume of water in the oceans must increase correspondingly. In that case, sea level rises and the coastal areas of the continents and low-lying islands are flooded. Therefore, the ice streams in Antarctica are being monitored to determine whether they are speeding up. Whereas the East Antarctic ice sheet appears to be stable at the present time, the West Antarctic and Greenland ice sheets are in greater jeopardy.

The presence of water at the base of the ice streams accelerates their rate of flow to the coast thereby increasing the volume of icebergs that enter the oceans annually. When icebergs derived from land-based ice sheets melt in the ocean, the volume of water in the ocean basins increases and sea level rises.

An increase of sea level is especially troubling for the stability of the West Antarctic ice sheet which rests on several subglacial islands that are separated by basins containing seawater. When sea level rises, the West Antarctic ice sheet is lifted off its foundations which may cause it to slide into the southern ocean. In that case, sea level would rise rapidly as the West Antarctic ice sheet melts. The Ross Ice Shelf, which holds the West Antarctic in its place, may also begin to deteriorate. In the worst case scenario, the entire West Antarctic ice sheet and the ice of the Ross Ice Shelf may slide into the southern ocean.

This series of events was originally considered by Hollin (1962) and has received much attention from the scientific community ever since. Terry Hughes at the University of Maine considered the effects of a rapid collapse of the ice sheets of Greenland and West Antarctica in a paper entitled (Hughes 1987):

Deluge II and the continent of doom: Rising sea level and collapsing Antarctic ice.

He was particularly concerned about the response of the ice streams that transport ice from the West Antarctic ice sheet into the Ross Ice Shelf because, as he put it: "Ice streams pull ice out of ice sheets" (Hughes 1987, p. 89). The connection between glaciation and sea level was also the topic of studies by Rutford et al. (1968) and Barrett et al. (1987), among others. The concern for the future of the West Antarctic ice sheet motivated studies by Bentley (1997), Bindschadler (1990, 1998), and Alley and Bindschadler (2001). Even more recently, Bindschadler and Bentley (2002) and Bentley et al. (2007) considered the changes that have occurred in the West Antarctic ice sheet and in the Ross Ice Shelf since the end of the last glaciation and speculated about the effects of future global warming on the stability of the West Antarctic ice sheet and the associated coastal ice shelves. They noted that the average annual temperature along the Antarctic Peninsula has already increased by two degrees Celsius since the 1950s and predicted that a complete collapse of the West Antarctic ice sheet would raise sea level by 5 m and flood low-lying coastal areas and permanently force about two billion people from their homes.

Whereas the West Antarctic ice sheet may disintegrate by sliding into the southern oceans, the ice sheet that covers Greenland is actually melting in response to global warming. Long (2009) reported that if the Greenland ice sheet continues to melt at the present rate, sea level would increase by more than 50 cm by the end of the present century which will have serious economic and social consequences. The estimates in Table 17.10 by Bell (2008) indicate that global sea level would rise by 7.3 m if all of the ice in Greenland melted. If all ice sheets were to melt, sea level would rise 64.9 m. If that ever happens, living space on the Earth would be drastically reduced and the human population would decline substantially.

At present time, sea level has been rising since the onset of the Industrial Revolution from about 1870 AD. The graph in Fig. 17.37 compiled by Long (2009) from data by Cazenave et al. (2008), Church and White (2006), and Jevrejeva et al. (2006) indicates that sea level increased at an average rate of about 0.6 mm/year from 1870 to 1930. The rate then increased to 2.1 mm/year from 1930 to 2000. At this rate, sea level will increase by only 20.7 cm in the course of the 21st century. However, if the Greenland ice sheet continues to melt and if the West Antarctic ice sheet surges, sea level will rise much more than that.

The stratigraphy of sedimentary rocks, evaluated by Vail and Hardenbol (1979), indicates that global sea level in Fig. 17.38 abruptly declined at about 29 Ma (late Oligocene) from +250 to −100 m relative to present sea level. Subsequently, sea level has fluctuated repeatedly with a high stand of +100 m during the

Table 17.10 Estimates of the global rise in sea level caused by the melting of the three continental ice sheets (Bell 2008, p. 63)

Ice sheet	Sea level (m)
East Antarctica	+51.8
West Antarctica	+5.8
Greenland	+7.3
All ice sheets	+64.9

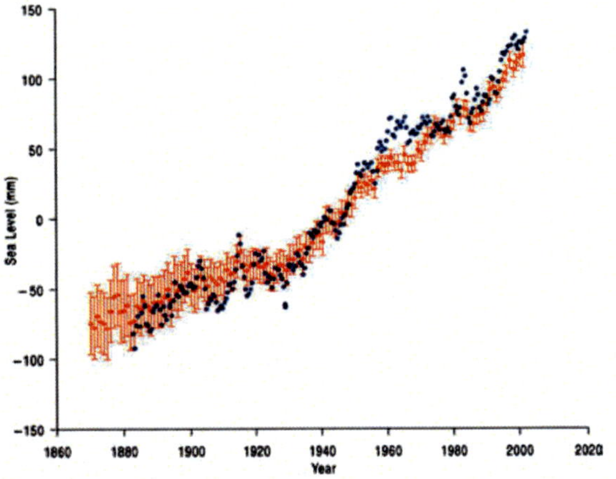

Fig. 17.37 Sea level has been rising between A.D. 1870 and 2000 at an accelerating rate from 0.6 mm/year (1870–1930) and 2.1 mm/year (1930–2000). At the current rate, sea level will rise about 21 cm during the 22nd century. The increase in sea level is partly attributable to an increase in the temperature of surface water in the oceans and to accelerated melting of valley glaciers and of the continental ice sheets of Greenland and Antarctica (Adapted from Long 2009)

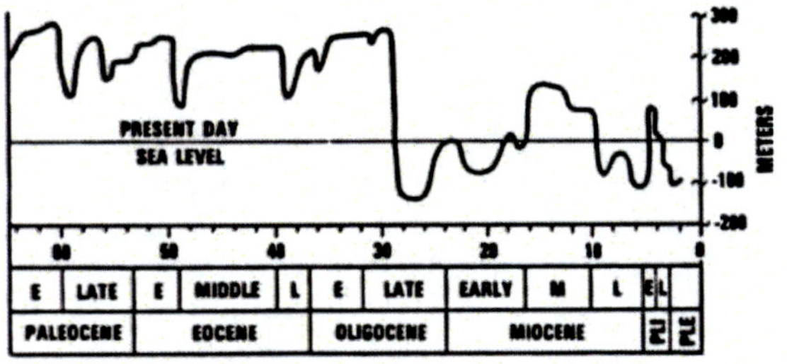

Fig. 17.38 Global sea level has varied by several hundred meters during the Cenozoic Era according to stratigraphic criteria. In general, sea level declined gradually during the Paleocene and Eocene but dropped abruptly at about 29 Ma during the late Oligocene. This decline of sea level occurred within one to two million years and amounted to a drop of about 350 m. After some minor fluctuations, sea level increased again to about 100 m above the present level during the middle Miocene and again during the Pliocene. The abrupt decrease of sea level during the Oligocene is attributable to the onset of the glaciation of Antarctica. The subsequent fluctuations of sea level are presumably caused by variations of the volume of ice in Antarctica. These variations of sea level *preceded* the history of glaciations that is recorded in the ice cores drilled at Vostok and Dome C. (Adapted from Vail and Hardenbol 1979)

Miocene followed by another decline at 10.8 Ma to −100 m. The authors cautioned that the rapid lowering of sea level during the Tertiary Period is controversial but concluded (Vail and Hardenbol 1979, p. 79):

"Glaciation is the only known mechanism that occurs rapidly enough to cause such rapid changes in sea level."

The decline of sea level at 29 Ma coincides approximately with evidence that the glaciation of Antarctica started prior to about 25 Ma. The curve also indicates that sea level increased significantly during the middle Miocene for about six million years and again during the Pliocene for one or two million years. These events occurred *before* the oldest ice at Dome C was deposited (EPICA 2004). Another point of approximate agreement is the estimate by Bell (2008) that sea level would rise by 64.9 m (Table 17.10) if the continental ice sheets of Antarctica and Greenland were to melt. In that case, the surface of the oceans would rise again to the level it attained during the Pliocene and Miocene in Fig. 17.38.

17.8 Cryogenic Brines and Evaporites

The freezing of glacial meltwater at the base of an ice sheet generates brines because the ions that are dissolved in the meltwater are excluded from the ice when the water refreezes. The removal of water from dilute aqueous solutions by freezing at the base of an ice sheet is analogous to the removal of water by evaporation in contact with the atmosphere. Therefore, we can speak about cryogenic brines that can form cryogenic evaporite deposits at the base of continental ice sheets. The fragments of black stratified calcite and opaline silica in the ElephantMoraine are examples of cryogenic evaporites.

The formation of cryogenic brines was discussed by Starinsky and Katz (2003) who pointed out that the freezing of dilute solutions not only increases the concentration of the solute, but also causes the chemical composition of the resulting brine to change depending on the precipitation of the salts that form from it. Brines enriched in calcium and chloride ions that are known to occur in the basement rocks of the Precambrian shields of Canada, Fennoscandia, and Bohemia were concentrated by freezing of seawater during the Pleistocene which caused their salinity to increase by factors of 25–30 relative to seawater as a result of cooling to −30°C.

The increase of the salinity of cryogenic brines causes minerals to precipitate depending on the chemical composition of the brine and the solubility of the minerals in question. The resulting cryogenic evaporites consist primarily of calcite and aragonite ($CaCO_3$), but may also contain a large number of sodium salts such as thenardite (Na_2SO_4), mirabilite ($Na_2SO_4 \cdot 10H_2O$), hydrohalite ($NaCl \cdot 2H_2O$), nahcolite ($NaHCO_3$), and others. The evolution of brines by evaporation of water at the surface has been described by Garrels and McKenzie (1967), Hardie and Eugster (1970), Wood (1975), Harvie and Weare (1980), Hardie (1984), Faure (1998a, Section 20.5), Klein-BenDavid et al. (2004), and Katz and Starinsky (2009). An example of subglacial formation of calcite and opaline silica in the Elephant Moraine was described in Section 17.3.6.

Subglacial deposits of calcite and aragonite from previously ice-covered areas in the world have been described by Bauer (1961), Ford et al. (1970), Page (1971), Magaritz (1973), Hallet (1976), Bjaerke and Dypvik (1977), Hanshaw and Hallet (1978), Peterson and Moresby (1979), Hillaire-Marcel et al. (1979), Lemmens et al. (1982), Souchez and Lemmens (1985), and Aharon (1988).

An excellent example of cryogenic brines and evaporite minerals was described by Fitzpatrick et al. (1990) from the Lewis Cliff ice tongue in the vicinity of Mt. Achernar along the Law Glacier (Section 13.3.1; Figure 13.22). The minerals occur within an ice-cored moraine north of the terminus of the Lewis Cliff ice tongue in Fig. 17.39 and on the eastern side of the ice tongue in the form of thin layers that strike north and dip steeply (75° ± 10°) to the west. A third deposit occurs around a supraglacial pond on the Lewis Cliff ice tongue about 1 m above the frozen surface of this pond. At a fourth locality in the moraine, a short distance from the northeast corner of the ice tongue, Fitzpatrick et al. (1990) found large blocks of friable borax and nahcolite up to 1.4 m long. All of the minerals that occur at these localities are either white or colorless and are easily overlooked when they occur on snow. Even the deposits in the moraine resemble patches of snow or ice.

The evaporite minerals described by Fitzpatrick et al. (1990) are noteworthy because they include salts of boron and potassium which were identified by x-ray diffraction supplemented by chemical analyses. The results, compiled in Table 17.11, indicate that the minerals are primarily carbonates, bicarbonates, sulfate, and borates of sodium. These salts have also formed in Searles Lake and Deep Springs Lake in California and in Lake Magadi in Kenya. The brines in these lakes

Fig. 17.39 A highly structured ice-marginal moraine has formed along the Law Glacier at the foot of Mt. Achernar (Section 13.3.1; Figure 13.22). The moraine ridges are ice-cored and the sediment that covers these ice ridges originated by sublimation of the underlying ice. Therefore, the moraine ridges appear to have been thrust up from the base of the Law Glacier. The brine which has been extruded along fault planes has precipitated the evaporite minerals that occur in a few places among the moraine ridges associated with the Lewis Cliff ice tongue located to the left out of view of this image (Photo by G. Faure)

originated by evaporation of meteoric water and are characterized by high pH values and low concentrations of calcium and magnesium (Hardie 1983; Eugster 1966). Fitzpatrick et al. (1990) determined that the brine that precipitated the evaporite minerals in the Lewis Cliff ice tongue could not have formed from water derived by melting of ice in the ice tongue.

Table 17.11 Evaporite minerals on the Lewis Cliff ice tongue and on the adjacent ice-cored moraine identified by Fitzpatrick et al. (1990)

Minerals	Composition
Nahcolite	$NaHCO_3$
Trona	$Na_2CO_3 \cdot NaHCO_3 \cdot 2H_2O$
Burkeite	$Na_6(CO_3)(SO_4)_2$
Thenardite	Na_2SO_4
Arcanite (?)	K_2SO_4
Tincalconite	$Na_2B_4O_5(OH)_4 \cdot 3H_2O$
Borax	$Na_2B_4O_5(OH)_4 \cdot 8H_2O$

Therefore, the brine probably formed from a mixture of meltwaters rather than from locally derived meltwater. The carbon dioxide that is required for the precipitation of nahcolite (Eugster 1966) may have originated either from atmospheric carbon dioxide that was released by melting of basal ice or by dissolution of carbonate rocks.

The carbon-14 dates of two samples of nahcolite on the Lewis Cliff ice tongue are concordant and have a mean value of 22,985 ± 518 years. However, the carbon-14 dates of three additional nahcolites are discordant and have values of 9,960 ± 100, 34,470 ± 710, and 37,730 ± 1,060 years. Fitzpatrick et al. (1990) also reported uranium-series disequilibrium dates of nahcolites based on the decay of ^{234}U to ^{230}Th (Faure and Mensing 2005; Section 20.1c). These dates range from 6,100 ± 100 to 150,000 ± 4,000 years and differ significantly from the carbon-14 dates. For example, the two nahcolite samples that yielded a combined carbon-14 date of 22,985 ± 518 years have apparent ^{230}Th - ^{234}U dates of 7,400 ± 100 and 6,100 ± 100 years for an average of 6,750 ± 143 years. The discrepancy between the carbon-14 and ^{230}Th - ^{234}U dates casts doubt on the reliability of these age determinations. If a significant fraction of carbon dioxide originated from limestone or other sources containing "dead" carbon, the carbon-14 date of the nahcolite is too old. On the other hand, if the nahcolite contains excess ^{234}U or is deficient in radiogenic ^{230}Th, the U-Th disequilibrium dates would be too low. In any case, the age determinations by both methods suggest that the cryogenic nahcolite in or near the Lewis Cliff ice tongue dates from the last ice advance of the Pleistocene Epoch (i.e., Wisconsinan or Weichselian).

The distribution of the collecting sites along the eastern margin of the Lewis Cliff ice tongue and within the adjacent ice-cored moraine ridges indicates that the minerals formed both by evaporation of brines that were extruded through near-vertical fractures in the ice as well as by subglacial precipitation. The salt deposits that occur in the soil of the ice-free valleys and associated with saline lakes and ponds in this area are presented in Chapter 19.

17.9 Chemical Composition of Antarctic Ice

The ice sheets of Antarctica and Greenland contain aerosol particles some of which are soluble in water. When ice containing water-soluble aerosol particles melts, the ions of which they are composed are released into the water. In addition, certain atmospheric gases (carbon dioxide, sulfur dioxide, nitrous oxide) that were trapped at the time of deposition of snow react with water molecules when the ice melts and form anions: HCO_3^-, CO_3^{2-}, Cl$^-$, SO_4^{2-}, and NO_3^-. The presence of these and other negatively charged ions (anions) in glacial meltwater is essential because all aqueous solutions must maintain a state of electrical neutrality which requires that the number of positive charges of the cations must equal the number of negative charges of the anions. The concentrations of ions in glacial meltwater are quite low which requires careful handling of ice samples in order to prevent their contamination (Legrand et al. 1982, 1984).

The principal sources of aerosol particles in the atmosphere are (Shaw 1989):

1. Soil-derived dust (100–1,000 million tons per year)
2. Oceanic sea salt (1,000 million tons per year)
3. Extraterrestrial sources
4. Oxidation products of algae, phytoplankton, and bacteria that produce sulfuric acid
5. Volcanic eruptions
6. Anthropogenic combustion products

Evidently, the chemical composition of Antarctic snow depends on the kind of water-soluble aerosol particles it contains, as well as on other factors such as the distance to the source of different kinds of aerosol particles, and on the rate of snow deposition.

Kumai (1976) identified the aerosol particles in snow flakes deposited at South Pole Station and Delmas et al. (1985) reviewed the deposition of volcanic dust in Antarctica. The formation of anions from occluded gases was discussed by Delmas (1982), Delmas et al. (1982a), Herron (1982), Legrand and Delmas (1984, 1987, 1988), whereas Hanappe et al. (1968), Ragone and Finelli (1972), Delmas et al. (1982b) reported chemical analyses of snow and ice in Antarctica.

The chemical composition of ice in the long cores that were recovered in East Antarctica at Dome C (74°39′S, 124°10′E), Vostok (78°30′S, 106°54′E) and South Pole (90°S) were determined by Petit et al. (1981), deAngelis et al. (1984), and Delmas et al. (1982a), respectively. The ice in the core drilled at Byrd Station in West Antarctica was analyzed by Palais (1985) and Palais and Legrand (1985).

The concentrations of sulfate and nitrate ions in the snow and ice of Antarctica and Greenland are of special interest because these ions form in part by chemical reactions in the atmosphere. For example, sulfate ions in the form of sulfuric acid (H_2SO_4) are deposited after volcanic eruptions, whereas the formation of nitrate (NO_3^-) has been attributed to lightning, solar flares, explosive impacts of large extraterrestrial objects, as well as to volcanic eruptions (Hitchcock et al. 1980; Delmas 1982; Delmas et al. 1982b; Herron 1982; Dreschhoff and Zeller 1994). The presence of electrically charged ions in ice gives rise to the electrical conductivity of the ice which can be measured routinely and conveniently after the ice has been melted (Dreschhoff and Zeller 1994).

17.9.1 Firn at Base Roi Baudouin

The chemical analyses of snow at Base Roi Baudouin (70°S,24°E), Amundsen-Scott Station (90°S), and Plateau Station (79°S,40°E) are listed in Appendix 17.12, based on analyses by Hanappe et al. (1968). The data demonstrate that firn at Base Roi Baudouin on the coast of East Antarctica has higher concentrations of sodium, magnesium, potassium, calcium, manganese, iron, and nickel than firn at South Pole and Plateau stations both of which are about 1,300 km from the nearest coast. The variation of the concentration ratios in Fig. 17.40 suggests that the concentrations of sea salts decrease in the sequence: Na (highest), Mg, K, Ca, Fe, and Mn. Nickel does not fit the pattern perhaps because it originates to some extent from extraterrestrial sources.

Hanappe et al. (1968) concluded from their data that the rate of deposition of inorganic particulate matter at Base Roi Baudouin is 10^{-4}g/cm^2/year, whereas at Plateau Station the rate of deposition is only 10^{-6}g/cm^2/year. In other words, the rate of deposition particulates at the coastal site is about 100 times higher than at the interior site. These authors also concluded that 98% of the solute at Plateau Station consists of terrestrial material and 2% is extraterrestrial dust containing about 1.3% nickel. If all the nickel in the firn at Plateau Station is assumed to be extraterrestrial, the rate of deposition of meteoritic dust on the surface of the Earth is about 100,000 metric tonnes per year.

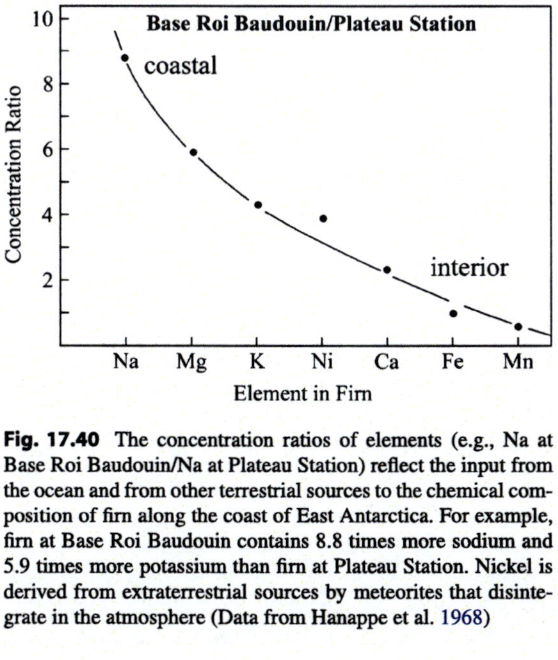

Fig. 17.40 The concentration ratios of elements (e.g., Na at Base Roi Baudouin/Na at Plateau Station) reflect the input from the ocean and from other terrestrial sources to the chemical composition of firn along the coast of East Antarctica. For example, firn at Base Roi Baudouin contains 8.8 times more sodium and 5.9 times more potassium than firn at Plateau Station. Nickel is derived from extraterrestrial sources by meteorites that disintegrate in the atmosphere (Data from Hanappe et al. 1968)

17.9.2 The Byrd-Station Ice Core

The Byrd-Station ice core was drilled through the West Antarctic ice sheet in 1968 and reached bedrock at a depth of 2,164 m (Gow et al. 1968). This core contains about 2,000 layers of volcanic ash or "tephra" (Gow and Williamson 1971; Gow 1971). The term includes all volcanic ejecta regardless of the diameters of the particles. These tephra were subsequently described by Kyle and Jezek (1978), Gow et al. (1979), and Kyle et al. (1981). In addition, Epstein et al. (1970) and Johnsen et al. (1972) demonstrated by means of the $\delta^{18}O$ values of the ice that the Byrd-Station core contains a record of the transition from the cold climate of the Wisconsinan ice age to the present Holocene interglacial.

The Byrd-Station ice core confirmed and extended conclusions reached previously from the study of the Camp-Century core drilled in northwestern Greenland in 1966 (Dansgaard et al. 1969, 1971; Johnsen et al. 1972; Mörner 1972, 1974; Langway et al. 1985). The ice of the Camp-Century core also contains tephra that originated from volcanic eruptions in Iceland (Hammer and Clausen 1981; Hammer 1984).

The concentrations of cations Na^+, Mg^{2+}, K^+, NH_4^+, and $H^+$and anions Cl^-, NO_3^-, and SO_4^{2-} in ice of the Byrd-Station core were determined by Palais (1985) and by Palais and Legrand (1985) who divided the core into four segments based on climate criteria derived from the $\delta^{18}O$ profile. The average concentrations of the ions in each of the segments are summarized in Table 17.12.

The data in Fig. 17.41 indicate that the highest concentrations occur in the ice of the late glacial stage (LG) between 1,350–1,700 m. The average concentrations of the ions subsequently decreased abruptly during the transition (T) between 1,050–1,350 m to the concentrations of the Holocene ice (0–1,050 m), which are consistently lower than the concentrations of ice during the pre-late glacial stage (PlG) of the Wisconsinan glaciation represented by the interval between 1,700 and 2,150 m.

Table 17.12 Average concentrations of cations and anions in four climatic intervals of the Byrd Station core (Palais and Legrand 1985)

Depth interval (m)	Na^+	NH_4^+	Mg^{2+}	K^+	Cl^-	NO_3^-	SO_4^{2-}
	microequilants per liter[a]						
Holocene interval							
0–1,050	1.074	0.122	0.268	0.041	1.194	0.610	0.817
n = 19[b]	±0.42	±0.03	±0.11	±0.02	±0.45	±0.16	±0.31
Isotopic transition (T)							
1,050–1,350	1.613	0.139	0.40	0.054	2.104	0.866	0.981
n = 4	±0.98	±0.07	±0.25	±0.04	±1.35	±0.15	±0.17
Late glacial interval (LG)							
1,350–1,700	2.287	0.133	0.57	0.067	2.848	0.79	1.294
n = 7	±0.43	±0.06	±0.11	±0.03	±0.52	±0.05	±0.33
Pre-late glacial interval (PlG)							
1,700–2,150	1.44	0.083	0.36	0.041	1.7969	0.697	1.044
n = 9	±0.37	±0.05	±0.09	±0.02	±0.43	±0.07	±0.22

[a]The equivalent weight of an ion is the mass of one mole of the ion (gram atomic weight) divided by the charge of the ion. If the ion has a charge ±2, one equivalent weight is equal to one half of a mole in grams.

[b]The number of samples included in each average is indicated by "n".

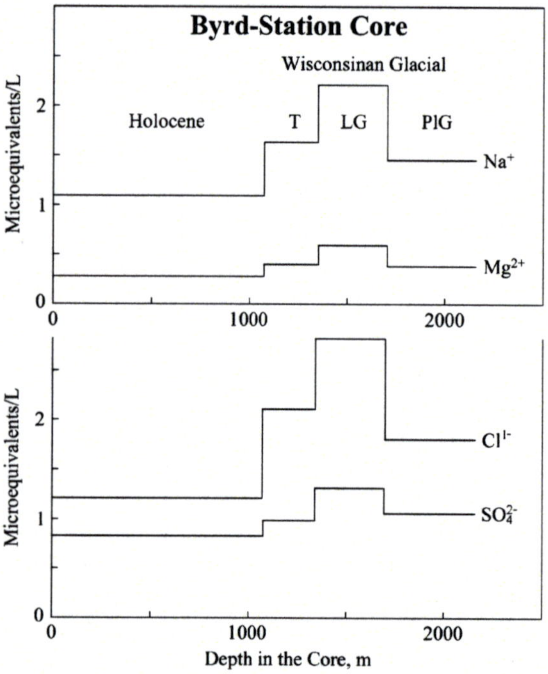

Fig. 17.41 The average concentrations of sodium, magnesium, chlorine, and sulfate ions in the Byrd-Station ice core vary incrementally with depth. The highest concentrations occur in the ice between 1,350 and 1,700 m that was deposited during the Late Glacial (LG) part of the Wisconsinan glacial stage. The ice in this interval also contains layers of volcanic ash (tephra). The concentrations of ions in the Holocene ice (0–1,050 m) are distinctly lower than the concentrations in the Pre-late Glacial (PlGO (1,700–2,150 m) interval. The transition (T) from (1,050–1,350 m) was a time of rapid change of the average global temperature from the Late Glacial Maximum to the Holocene (Data from Palais and Legrand 1985)

Palais and Legrand (1985) concluded that the ice of the Byrd Station core contains sodium chloride (NaCl) and sodium sulfate (Na_2SO_4) of marine origin. The interactions of ions during transport from the surface of the ocean to the interior of Antarctica permits elemental fractionation which causes an excess of Cl^- in some places and an excess of Na^+in others. The data in Table 17.12 indicate that the Byrd-Station ice core contains an excess of Cl^- over Na^+(not shown); but an excess of Na^+as Na_2SO_4 occurs in the ice of the Vostok core (deAngelis et al. 1984).

17.9.3 Nitrate and Sulfate Concentrations

Palais and Legrand (1985) also reported concentrations of nitrate NO_3^-, ammonium NH_4^+ and sulfate SO_4^{2-} in the ice of the Byrd-Station core. The nitrogen compounds originate from the atmosphere by chemical reactions of N_2 molecules which can be broken up by lightning (Huebert and Lazrus 1980a, b; Herron 1982; Delmas et al. 1982a, b). Parker et al. (1978) originally suggested that compounds of nitrogen can form in the atmosphere during auroral displays in response to solar activity. However, Herron (1982) later concluded that the nitrate concentrations of snow and ice in Greenland and Antarctica are not related to the activity of the Sun.

Subsequently, Dreschhoff and Zeller (1994) reported that the nitrate concentrations and liquid conductivities of a 122-m firn and ice core taken in June of 1992 in Greenland near the drill site of GISP 2 (Greenland Ice Sheet Project) also known as "Summit". Their results based on a total of 7,776 samples taken at 1.5 cm increments indicate that the profiles of annual nitrate concentrations and electrical conductivities are overlain by sporadic anomalies that record known volcanic eruptions in Iceland and elsewhere in the world (e.g., St. Augustine in 1986, Hekla in 1970 and 1947, Askja in 1961, Raikoke in 1924, Katla in 1918, Katmai in 1912, Krakatau in 1883, Tambora in 1815, etc.). The record even contains a major spike in 1908 when a large extraterrestrial object exploded in the atmosphere over Tunguska in Siberia. The anomalies caused by the volcanic eruptions and by the Tunguska Event were used together with the annual fluctuations of the electrical conductivity to construct a time scale for the core which extended 415 years into the past from 1992 to 1577 AD.

The nitrate concentrations and electrical conductivities also recorded the occurrence of a solar flare in 1859/60 which was observed in England. The scope of the study by Dreschhoff and Zeller (1994) is unprecedented and demonstrates that large-scale atmospheric events like volcanic eruptions, solar flares, and explosion of large objects from space are recorded by spikes of the nitrate concentrations and electrical conductivities of ice in Greenland and Antarctica.

The sulfate anion in Antarctic ice probably originates from several sources, including biogenic sources in the oceans and from seawater with major contributions from sporadic volcanic eruptions (Delmas et al. 1982a, b; Herron 1982). The segment of the Byrd-Station core between 1,300 and 1,650 m which includes the late-glacial maximum (LG in Fig. 17.41) contains tephra layers indicating that the atmosphere contained volcanic dust and sulfuric acid particles erupted by volcanoes in Marie Byrd Land and in northern Victoria Land (Kyle et al. 1981, 1982). The apparent loading of the atmosphere with sulfuric acid particles may have caused additional climatic cooling at about 25,000 years close to the end of the Wisconsin glacial stage.

Tephra layers in ice cores are typically overlain by ice that contains elevated sulfate concentrations and excess hydrogen ions (H^+). These chemostratigraphic relationships occur because the sulfur dioxide gas (SO_2) discharged during volcanic eruptions reacts with water vapor and oxygen of the atmosphere to form microscopic droplets of sulfuric acid that remain in the atmosphere longer than the silicate particles of volcanic ash which settle out soon after the volcanic eruptions. The resulting clouds of sulfuric acid droplets reflect sunlight and thereby cause temporary cooling of the global climate. The sulfuric acid droplets are eventually deposited on top of the complementary ash layer on the surfaces of the polar ice sheets.

The excess acidity and enhanced sulfate concentration of the ice associated with tephra-bearing layers in ice cores enabled Hammer et al. (1981a, b) to obtain a continuous profile of the acidity (pH) of ice in a core drilled in 1974 at Station Crete (71°N,37°W) in central Greenland. The profile confirmed that historical volcanic eruptions were represented by acidic layers in the Greenland ice sheet.

The effect of the volcanic eruptions during the 19th and 20th centuries on the temperature of the atmosphere was evaluated in a series of papers by Self et al. (1981), Newell (1981), Robok (1981), and Walker (1981), all of whom published their contributions in volume 11 of the Journal of Volcanology and Geothermal Research. The occurrence of tephra layers in the East Antarctic ice sheet is the subject of reports by Keys et al. (1977), Koeberl et al. (1987), Palais and Sigurdsson (1989), and Faure et al. (1994, with assistance from Mary Davis and Keith Henderson).

17.9.4 Lead in Continental Ice Sheets

The continental ice sheets of Antarctica and Greenland contain all kinds of solid particles and chemical compounds that occur in the atmosphere, including lead which is at least partly *anthropogenic* because it is released into the atmosphere by the combustion of coal, during the smelting of lead-bearing ores, by the use of lead oxides as a pigment, and by other anthropological activities. The most important source of atmospheric lead is its former use as an additive in gasoline which caused wide-spread lead contamination of the surface of the Earth.

The *natural* sources of environmental lead include (Nriagu 1978):

Windblown dust (85%)
Plant exudates (10%)

Forest fires, volcanic eruptions, and meteorite impacts (3%)

Radioactive decay and sea salt spray (2%)

Nriagu (1978) estimated that the natural sources contribute only about 4% of the lead discharged annually into the atmosphere by anthropogenic sources. The environmental geochemistry of lead is the subject of books by Cannon and Hopps (1971), Nriagu (1978), Thornton (1982), Salomons and Förstner (1984), Tessier and Turner (1995), and of numerous articles including those of Ghazi (1994), Fuge et al. (1996) and Faure (1998a, b).

The greatest contribution to our knowledge of the environmental geochemistry of lead was made by Professor C.C. Patterson and his associates at the California Institute of Technology. An entire issue of Geochimica et Cosmochimica Acta (vol. 58, No. 15, 1994) was devoted to Professor Patterson upon his retirement. The effect of environmental lead on public health was also considered in a major work on medical geochemistry edited by Selínus et al. (2005).

Analyses of ice in the core drilled at Camp Century in northwest Greenland revealed that the concentration of lead increased gradually from 0.011 μg/kg in 1753 A.D. to 0.068 μg/kg in 1945 A.D. Subsequently, the concentration of lead rose steeply and reached 0.16 μg/kg in 1960 (Murozumi et al. 1969). The dramatic increase in the concentration of lead in this ice core was attributed to the widespread use of gasoline that contained tetra-ethyl lead as an additive.

The accurate measurement of lead in ice cores requires elaborate precautions to avoid contamination (Wolff and Peel 1985; Peel 1986; Boutron 1990). Nevertheless, Boutron and Patterson (1983, 1986) detected lead in snow and firn in Antarctica even though this continent is far removed from industrial sources of lead. According to a review by Boutron et al. (1994), the average concentration of lead in Antarctic snow that was deposited between 10,000 and 5,000 years B.P. is 0.71 μg/kg. During the 20th century (1900–1960) the concentration of lead increased gradually to 3.6 μg/kg followed by a steep rise to 11.1 μg/kg in 1980. After the use of lead in gasoline was discontinued in North America and Europe in the early 1970s, the concentration of lead in surface water of the oceans and in snow in Greenland declined to low levels.

Ice in cores drilled through the East Antarctic ice sheet at Dome C and Vostok contains lead at widely varying concentration depending on climatic conditions (Boutron and Patterson 1986, 1987; Bolshov et al. 1998; Boutron et al. 1987, 1990, 1994). The pattern of variation of lead in Fig. 17.42 is similar to that of Na^+, Mg^{2+}, Cl^-, and SO_4^{2-} in Fig. 17.41. The concentration of lead in the ice deposited during the Glacial Maximum toward the end of the Wisconsinan glaciation rose to 30×10^{-12} g/g (0.03 μg/kg). A high level of lead also occurred at the end of the Illinoian glaciation more than 140,000 years ago. Ice of Wisconsinan age 115,000–20,000 years ago contains only about 2×10^{-12} g/g of lead and Holocene ice contains even less lead than that. The anthropogenic lead released in the 20th century does not register in Fig. 17.42 because the time scales are not compatible.

Nevertheless, the ice sheets of Antarctica and Greenland are archives that contain a recording of global climate change, volcanic eruptions and explosive impacts of exterrestrial objects. The ice also contains evidence of the contamination of the atmosphere by the human population of the Earth.

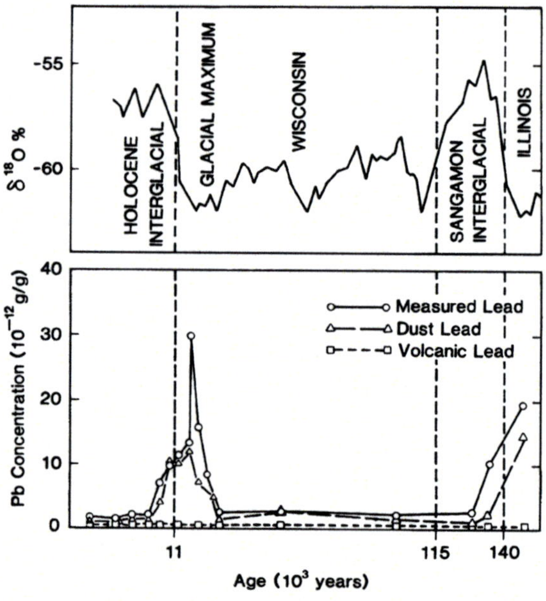

Fig. 17.42 The concentration of lead in ice cores drilled at Dome C and Vostok in East Antarctica varies systematically depending on climate. During the Glacial Maximum close to the end of the Wisconsinan glacial stage, the concentration of lead increased to 30×10^{-12} g/g (or 0.030 μg/kg). A similar increase occurred at the end of the Illinoian glaciation. The lead concentration during the main phase of the Wisconsinan glaciation was only about 2×10^{-12} g/g and decreased still more during the Holocene interglacial (Adapted from Patterson et al. 1987b with $\delta^{18}O$ values from Lorius et al. 1985)

17.10 Dust in the Ice Sheets of Antarctica

The ice cores that have been recovered in Antarctica and from Greenland contain several kinds of dust particles. Small particles having diameters ranging from less than 0.1 up to 2 μm are present at varying concentrations at all levels in these cores. Large particles with diameters greater than 2 μm and ranging up to 80 μm (e.g., tephra) are concentrated in discrete layers in the ice, but are much less abundant in the ice above and below these layers. The small particles are most numerous, but the large particles contribute most of the total mass. The small particles in the atmosphere determine its optical depth which affects the radiation balance of the Earth by the way they absorb and reflect solar radiation. The atmosphere also contains anthropogenic particles which include combustion products of fossil fuel called *fly ash* and other kinds of particles composed of synthetic materials, metals, etc., as well as extraterrestrial particles, such as micrometeorites and ablation spherules that form when meteorites are heated by friction as they pass through the atmosphere. These extraterrestrial particles are described in Chapter 18.

The microscopic particles are small enough to rise into the upper troposphere (up to 10 km above the surface of the Earth) and into the stratosphere (10–50 km above the surface) where they may reside for several years before they are eventually removed by meteoric precipitation. During their residence in the stratosphere the microscopic dust particles are transported widely over the Earth. The stratospheric dust particles in the Antarctic and Greenland *ice cores* are insoluble in water and are composed of clay minerals (Thompson 1977a, b; Kumai 1997) and x-ray amorphous oxides and hydroxides all of which were presumably derived from soil and regolith exposed in deserts and around receding ice sheets at the end of the Pleistocene.

The larger dust particles (2–80 μm) are composed primarily of volcanic glass and enclosed minerals that were erupted explosively by volcanoes in Antarctica, on the surrounding volcanic islands, and at more distant locations in Indonesia and in South and Central America. These kinds of particles occur in diffuse layers of volcanic ash that was deposited on the ice sheets of East and West Antarctica. The morphology and chemical composition of volcanic particles have been used to identify their sources and to obtain information about the petrology of the volcanic rocks and about the type of eruption that produced them. Under favorable circumstances, the volcanic particles can be dated by the $^{40}Ar/^{39}Ar$ laser-ablation method and by the U-Pb method using ion-probe mass spectrometers such as SHRIMP (sensitive high-resolution ion microprobes) (Compston 1999).

17.10.1 Stratospheric Dust

The first ice cores from East Antarctica that were available for the study of stratospheric dust were drilled at South Pole in 1974 (Thompson, E.M. 1979) and at Dome C in 1977/78. The South-Pole core was 101 m long and extended 900 years into the past. The seasonal accumulation and resulting stratigraphy of snow at South-Pole Station was described by Gow (1965), whereas Thompson and Mosley-Thompson (1981) published a profile of dust concentrations in the core.

The Dome C core was drilled in 1977/78 to a depth of 906 m as a contribution to the International Antarctic Glaciology Project (Lorius and Donnou 1978). The $\delta^{18}O$ profile of the ice in this core was measured and interpreted by Lorius et al. (1979). The temperatures of the ice at increasing depth in this hole were reported by Gillet and Rado (1979). The snow stratigraphy of Dome C, including both depositional and diagenetic processes, was investigated during the 1978/79 and 1979/80 field seasons by Palais (1980) and Palais et al. (1982). The study included the visible stratigraphy in a 3-m snow pit as well as the beta-activity and the stratigraphy of dust-particle concentrations.

The study of microparticles in Antarctic ice cores by L.G. Thompson and E. Mosley-Thompson at The Ohio State University (Thompson et al. 1975, 1981; Thompson 1977a, b; Mosley-Thompson 1979; Thompson and Mosley-Thompson 1981) indicated that the concentration of dust particles in ice cores from the South Pole and Dome C in East Antarctica, from Byrd Station in West Antarctica, as well as from Camp Century in northwestern Greenland have certain common characteristics identified in Table 17.13. The conclusions that are derivable from the data include:

1. The Holocene ice in Greenland is about three times dustier than ice of equivalent age in Antarctica.
2. Ice deposited during the Wisconinan (Weichselian) glaciation in Antarctica and Greenland contains

more dust particles than the corresponding Holocene ice.

3. The Wisconsinan ice at Camp Century in Greenland is about nine times dustier than Wisconsinan ice in Antarctica.

Table 17.13 Average concentration of microparticles having diameters greater than 0.6 μm in ice cores from Antarctica and Greenland expressed by the number of particles per 500 μL (Thompson and Mosley-Thompson 1981, Table 6)

	Antarctica		Greenland
Average Concentration	Dome C[a]	Byrd station	Camp century
Holocene	6,670	8,989	25,713
Wisconsinan	40,562	27,445	313,507
Wisconsinan Holocene	6	3	12
All core sections	20,796	15,369	119,411

[a]This is the 906-m core drilled in 1977/78.

Thompson and Mosley-Thompson (1981) also demonstrated in Fig. 17.43 that the greatest concentration of dust in all three cores occurred just prior to the transition from the Wisconsinan glacial to the Holocene interglacial age, as indicated by the $\delta^{18}O$ values of the ice at Dome C (Lorius et al. 1979), at Byrd Station (Epstein et al. 1970), and at Camp Century (Dansgaard et al. 1971). The concentration of dust in the EPICA core drilled at Dome C also increases prior to the transition from Wisconsinan to the Holocene, at the termination of the Illinoian glaciations, and it spikes repeatedly in the older ice presumably in conjunction with climate changes during transitions from glacial to interglacial stages (EPICA 2004).

The observed increase in the dust concentrations of ice just prior to the transitions implies that large areas

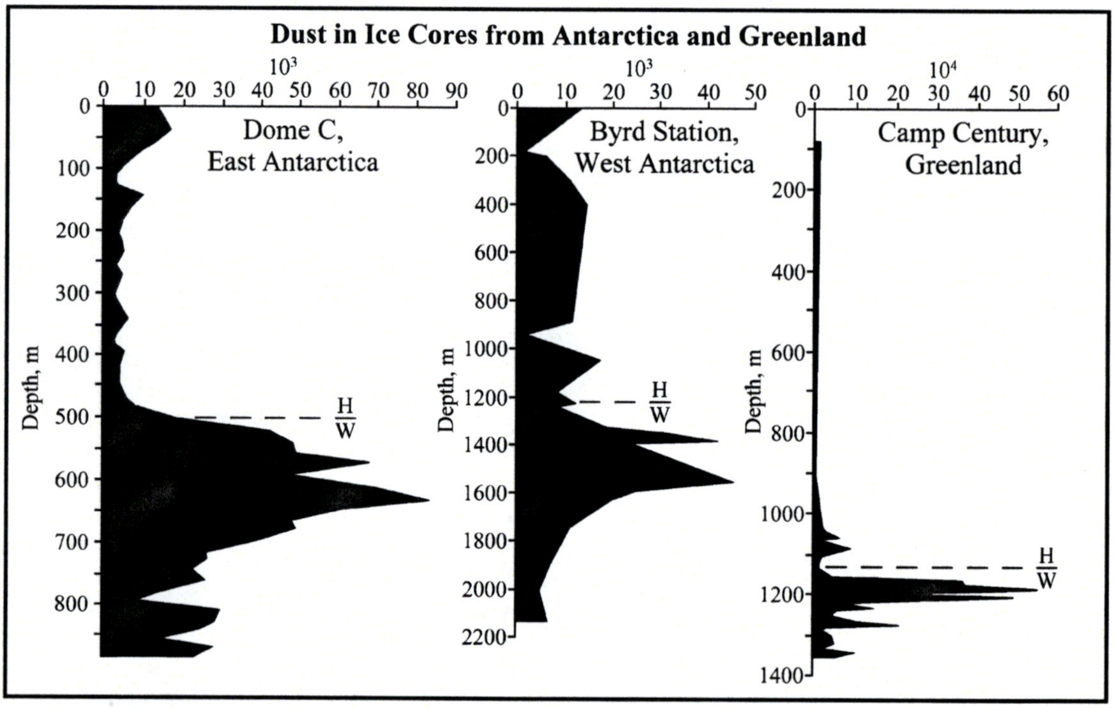

Fig. 17.43 The concentrations of microscopic dust particles in ice cores from Antarctica and Greenland increase sharply at the Holocen-Wisconsinan boundary indicated by a steep decrease of $\delta^{18}O$ values of the ice. The concentrations of dust particles having diameters greater than 0.6 μm are expressed by the number of particles per 500 μL of water. The concentrations were measured in a class 100 clean room by means of a Coulter counter (Thompson 1977a). The ice-core depth to the H/W (Holocene/Wisconsinan) transition occurs at different depths depending on the rate of accumulation of snow at each site. The H/W transitions in the Antarctic and Greenland cores occur at about the same time in the past which confirms the bipolar occurrence of continental glaciation (Adapted from Thompson and Mosley-Thompson 1981, Figure 6)

covered by glacial sediment were exposed along the margins of the receding ice sheets. The oxygen-isotope record combined with age determinations by means of ice-flow models indicates that the transition from glacial to interglacial conditions occurred at about the same time in the northern and southern hemispheres. In addition, this climate change occurred in a surprisingly short time. For example, Thompson and Mosley-Thompson (1981) reported that the transition from glacial to interglacial conditions in the Dome-C core occurred during at time interval of about 1,000 years. Alley (2000) documented even more rapid climate changes that have occurred during decades rather than centuries or millennia.

17.10.2 Tephra Layers in the Ice Sheets of Antarctica

Layers of ice containing disseminated particles of volcanic origin (i.e., tephra) are a common feature of blue ice areas and in the vicinity of volcanic vents in the Transantarctic Mountains. Volcanic tephra have also been found in the ice cores drilled at South Pole Station, Dome C, Vostock Station, and at Byrd Station, as well as in the glacial deposits that cover the floor of Wright Valley (Jones et al. 1973b; Boger and Faure 1988).

The chemical compositions of coarse tephra recovered from the core drilled at *Byrd Station* in West Antarctica are tightly clustered in the trachyte field of Fig. 17.44

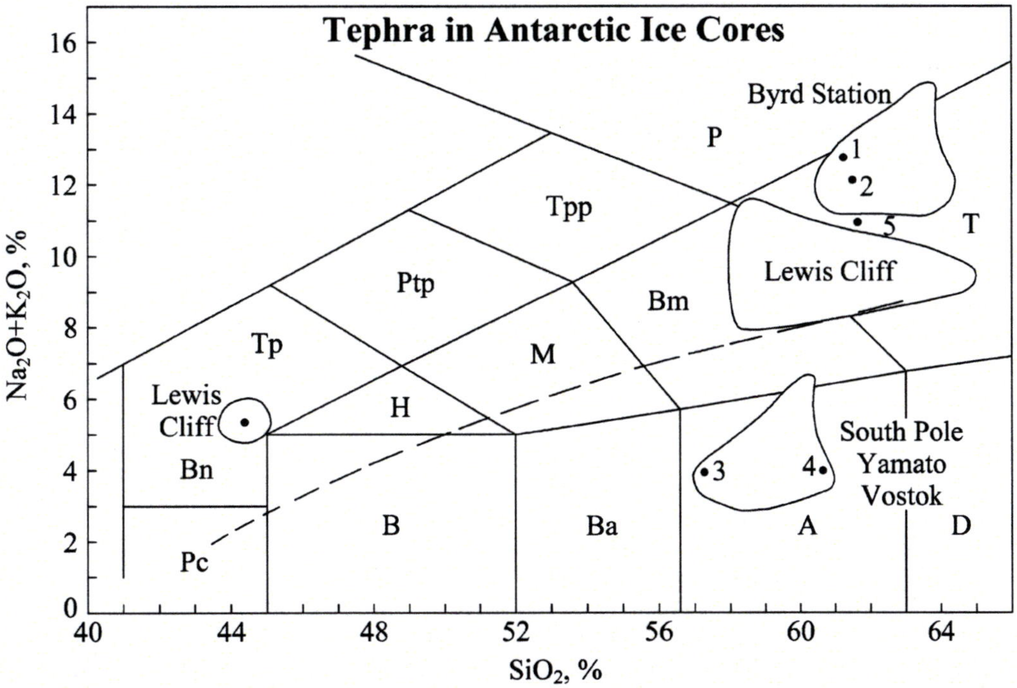

Fig. 17.44 The chemical compositions of *coarse* tephra at different depths in the Byrd-Station core classify them as alkali-rich trachytes/phonolites. The points labeled 1 and 2 are volcanic rocks from Mt. Takahe and Mt. Sidley, respectively, in Marie Byrd Land. The chemical compositions of the fine tephra in the Byrd-Station core scatter widely and are not shown on this diagram. The tephra from South Pole, the Yamato Mountains, and Vostok Station are andesites that could have originated from volcanoes in the South Sandwich Islands (3 = Candlemas Island, 4 = Bellingshausen Island). Tephra from the Lewis Cliff ice tongue are composed of benmoreites, trachytes, and basanites. These tephra may have originated from mixed sources including volcanoes in northern Victoria Land (e.g., The Pleiades, Section 16). Point 5 represents tephra in the 906-m core drilled at Dome C in 1977/78 (Kyle et al. 1981). The labels of the fields are explained in Figure 16.26. Data from Kyle et al. (1981; Byrd Station, Dome C, Takahe, and Sidley), Palais (1985a, b; coarse tephra from Byrd Station), Palais et al. (1987; South Pole, Vostok), Koeberl (1990; Lewis Cliff ice tongue), Nishio et al. (1985; Yamato Mountains), Baker (1978; Bellingshausen Island), Tomblin (1979; Candlemas Island)

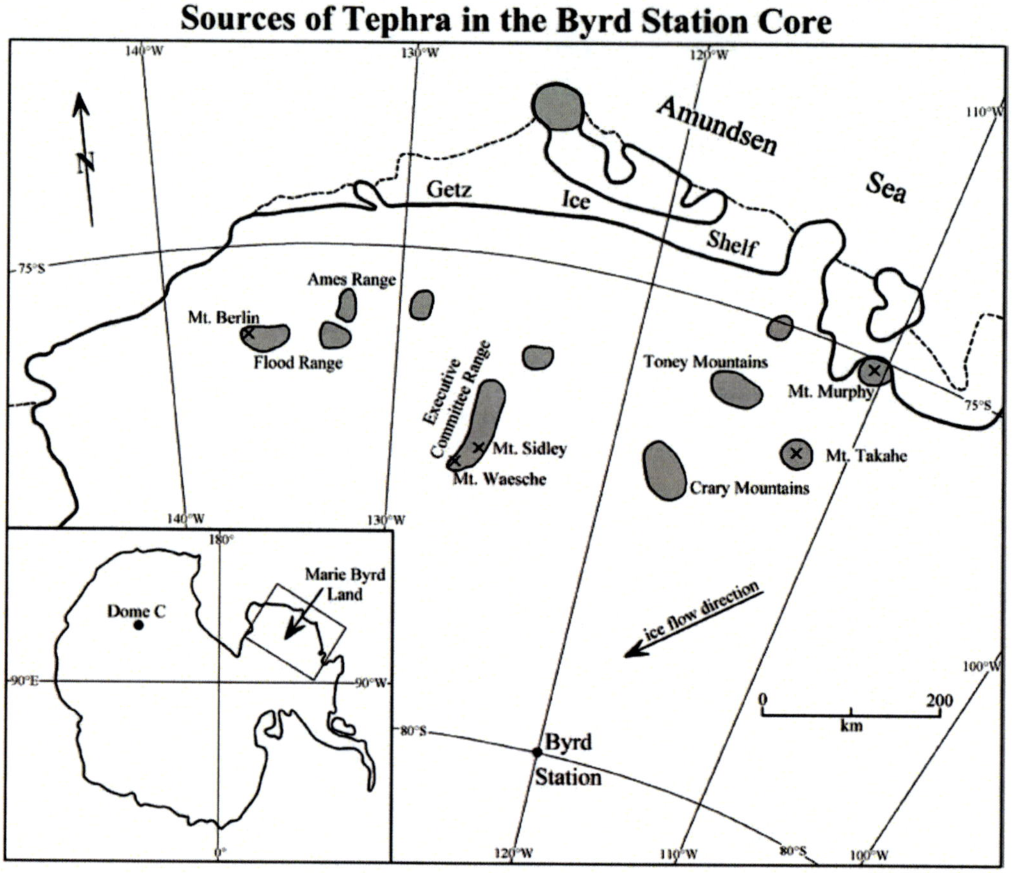

Fig. 17.45 Mt. Takahe in Marie Byrd Land is the probable source of tephra in the Byrd-Station ice core. The morphology of these tephra points to a phreatomagmatic eruption of Mt. Takahe, which implies that it was covered by the West Antarctic ice sheet and that the West Antarctic ice sheet may have been thicker at the time of the eruption (Adapted from Kyle et al. 1981)

and extend into the phonolite field (Kyle et al. 1981; Palais 1985a, b). The tephra could have originated from volcanoes in Marie Byrd Land, especially Mt. Takahe and Mt. Sidley in Fig. 17.45, both of which have erupted alkali-rich pyroclastics and lavas presented by Points 1 and 2 in Fig. 17.44, respectively (Kyle et al. 1981). The similarity of the chemical compositions of the volcanic rocks and the prevailing ice-flow direction of the West Antarctic ice sheet during the past 25,000 years, indicated by the black arrow in Fig. 17.45, make Mt. Takahe the probable source of the tephra in the Byrd Station core (Kyle and Jezek 1978; Kyle et al. 1991; Palais 1985a, b).

Mt. Takahe is a large shield volcano located about 500 km northeast of Byrd Station. It is composed of alkali-rich volcanic rocks ranging from basanite to trachyte. K-Ar dates reported by LeMasurier and Rex (1990) range from 0.31 ± 0.09 Ma to less than 0.1 Ma. Although no fumarolic activity has been observed on Mt. Takahe, this volcano may have erupted less than 30,000 years ago. Four other volcanoes in Marie Byrd Land (Mt. Berlin, Mt. Sidley, Toney Mountain, and Mt. Waesche) are also potential sources of tephra in the Byrd-Station core, but were disqualified by Kyle et al. (1981) and Palais (1985a, b). Mt. Sidley is too old (4–5 Ma), Mt. Waesche and Toney Mountains are not known to have erupted trachytes, and the volcanic rocks on Mt. Berlin are not a good match with the Byrd-Station tephra. The morphology of these tephra examined by Palais (1985a, b) suggests that they formed in a phreatomagmatic eruption from a subglacial volcanic vent.

The chemical composition of shards of volcanic glass from the 906-m core drilled during 1977/78 at *Dome C* (Lorius et al. 1979; Thompson et al. 1981) places them at point 5 in the trachyte field of Fig. 17.44. These particles originated from a depth of 726 m in the Dome-C core and were deposited about 25,000 years ago. Therefore, these tephra could also have originated from Mt. Takahe in spite of the great distance between Dome C and Mt. Takahe. The dispersion of tephra from Mt. Takahe may have been facilitated because the summit of this volcano is within 3,600 and 6,400 m of the boundary between the troposphere and the stratosphere. Therefore, the volcanic ash erupted by Mt. Takahe could have been injected directly into the stratosphere where it spread widely across Antarctica (Kyle et al. 1981).

The ice core drilled in 1983/84 at *South Pole* Station contained a 3-mm tephra layer at a depth of 303.44 m (Palais et al. 1987). The tephra are composed of vesicular glass with crystals of quartz, pyroxene, and plagioclase. The bulk chemical composition of these tephra plots close to the boundary between the andesite and benmoreite fields in Fig. 17.44 in strong contrast to the tephra in the cores at Byrd Station and Dome C (See also Koeberl 1990, Table 1).

The *Vostok tephra* are very similar in composition to the tephra in the South-Pole core and closely resemble the andesite in Candlemas and Bellingshausen islands of the South Sandwich Islands in Fig. 17.46 (57°45′S, 026°30′W) (Tomblin 1979; Baker 1978). The distance between these islands and Vostok station is about 4,500 km (Palais et al. 1987; Kyle et al.1984). In spite of the dispersion during transport through the atmosphere, the thickness of the tephra layer in the Vostok core is 5 cm at a depth of 101 m giving it an assigned age of 3,300 years. Therefore, the volcanic ash erupted by volcanoes in the South Sandwich Islands may have formed a large plume over East Antarctica. If so, this ash layer may be a useful stratigraphic marker in the East Antarctic ice sheet (Kyle et al. 1984).

Volcanic ash of andesitic composition also occurs in the ice adjacent to the *Yamato Mountains* in Fig. 17.46. Nishio et al. (1985) and Katsushima et al. (1984) analyzed 14 samples of tephra collected at three sites in the blue-ice area south of Massif A of the Yamato Mountains. All of these tephra are composed of subalkalic andesites similar in composition to the andesites erupted on the South Sandwich Islands.

The Lewis Cliff ice tongue 84°17′S and 161°05′E in Fig. 17.47 extends from the Walcott Névé north into an ice-marginal moraine that has formed along the Law Glacier in the Queen Maud Mountains (Figs. 13.22, 17.39). The chemical composition of three tephra from the Lewis Cliff ice tongue define a region in Fig. 17.44 that straddles the boundary between bemmoreites and trachytes. A fourth sample is a basanite. Koeberl (1990) compared the chemical compositions of the Lewis-Cliff tephra to the composition of volcanic rocks erupted in The Pleiades. Alternatively, these tephra could also be mixtures of alkali-trachyte tephra from Marie Byrd Land and andesite tephra from the South Sandwich Islands.

17.10.3 Tephra on the Allan Hills Ice Fields

The tephra layers exposed on the ice fields west of the Allan Hills were mapped by Dunbar (1995a, b) who also attempted to date anorthoclase crystals contained within them by the $^{40}Ar/^{39}Ar$ method. The outcrop patterns of these tephra layers in Fig. 17.48 demonstrate their linear continuity and the structural deformation of the ice in this area. The same investigators also sampled dust layers on other ice fields associated with the Allan Hills, such as the ice around the Elephant and Reckling moraines, the Far Western, Far Northern, and "Meteorite City" ice fields, ice adjacent to Carapace Nunatak, and other localities on the margin of the East Antarctic ice sheet in southern Victoria Land.

The tephra from these localities are classifiable as basanite, trachyte, phonolite, and even rhyolite. Dunbar et al. (1995a, b) attributed these tephra to volcanic eruptions of Mt. Erebus (Kyle et al. 1992; Dunbar et al. 1994), of Mt. Melbourne (Wörner et al. 1989), and The Pleiades (Kyle 1990).

Five samples of tephra in dusty ice collected by Marvin (1990) on the ice sheet adjacent to the Allan Hills in Fig. 17.49 (sites 1 and 2) are identified in

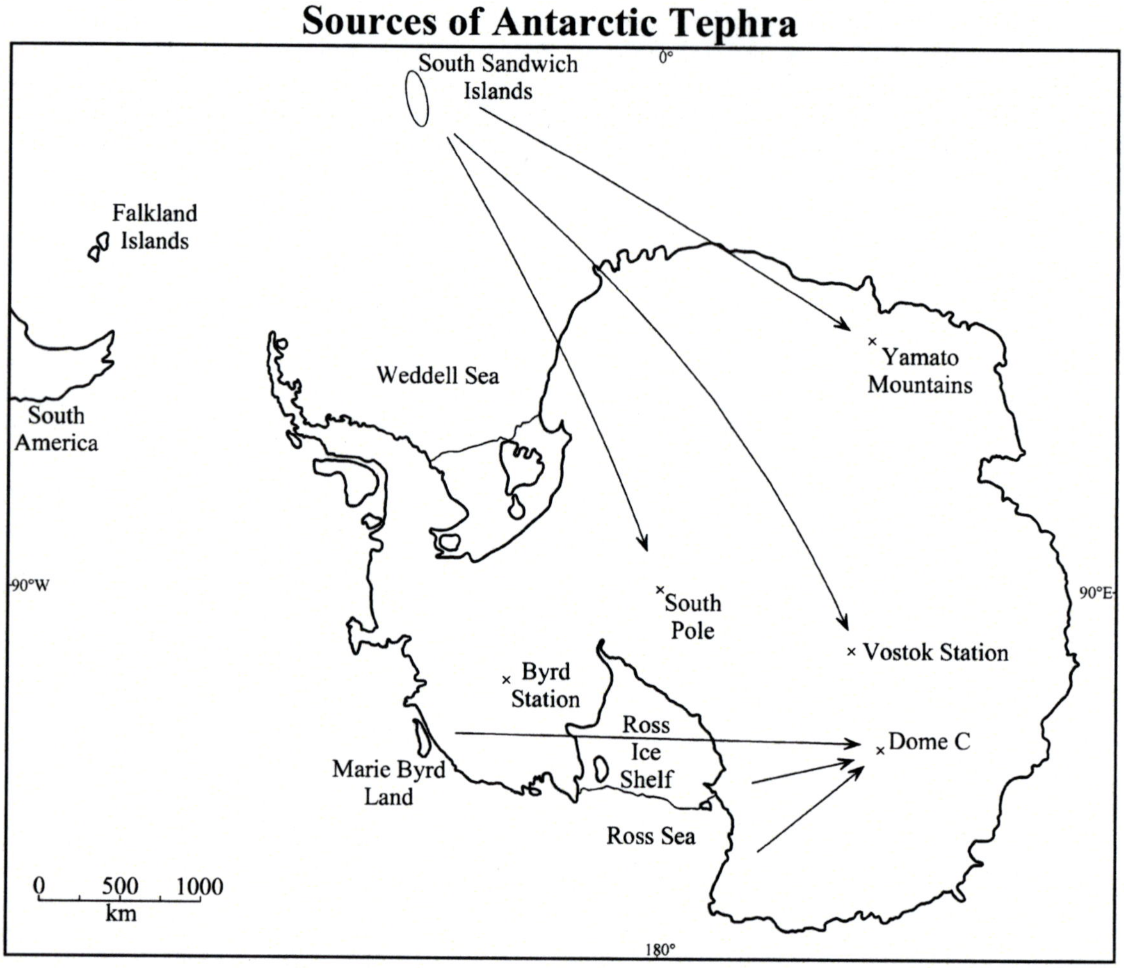

Fig. 17.46 The andesite volcanoes on Candlemas and Bellingshausen islands in the South Sandwich Islands are suspected of being the source of the tephra in the Yamato Mountains, at Vostok, and South Pole. The Dome-C tephra may have originated from the volcanoes in Marie Byrd Land or in northern and southern Victoria Land (Adapted from Palais et al. 1987)

Fig. 17.50 as trachytes (1a,b,c, and 2b) and dacite (2a). Three additional samples of tephra from ice in the "Cul de Sac" in Fig. 17.49 (sites 3 and 4) are classified as low-silica basanites (3) and basalt (4). Point 5 is the average of 12 tephra samples from the Allan Hills ice field analyzed by Nishio et al. (1985). The two groups of collecting sites sampled by Marvin (1990) are most likely at different stratigraphic levels in the ice sheet and therefore could have originated from different volcanic centers. The compositions of the tephra from the sites in the Cul de Sac resemble average "trachybasalt" from Ross Island published by Goldich et al. (1975). The tephra at sites 1 and 2 on the blue ice field in Fig. 17.49 could have originated from volcanoes of the Malta Plateau, The Pleiades, and Mt. Melbourne in Fig. 16.26.

17.10.4 Black Spherules, Allan Hills

A dust band discovered by Harvey et al. (1998) 5 km west of the western edge of the Allan Hills is composed of an unusual assemblage of dark and rounded particles. These spherical particles contain strongly zoned euhedral olivine crystals in a matrix of iron-rich glass

Fig. 17.47 The Lewis Cliff ice tongue contains several tephra layers and englacial clasts that emerge as the surrounding ice sublimates. A "dirt" layer occurs at the well-defined contact between pale blue ice and dark blue ice. The contact appears to be a fault in the ice (Photo by G. Faure)

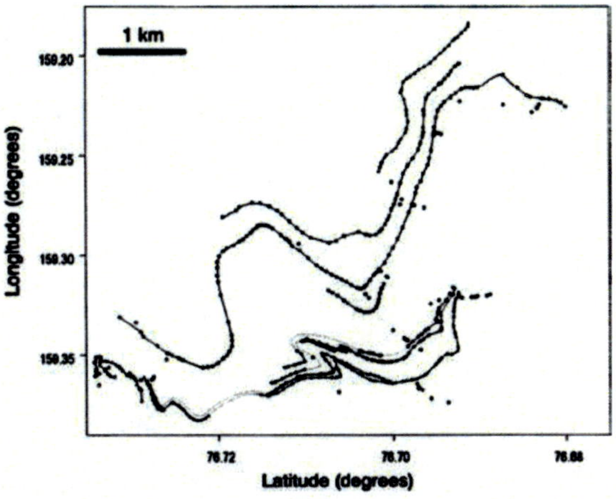

Fig. 17.48 Layers of dusty ice on the ice field west of the Allan Hills formed by deposition of volcanic ash and wind-blown dust particles on the surface of the East Antarctic ice sheet. The lower contacts of the dust layers are sharp, whereas the upper contacts are diffuse. The outcrop pattern of the dust layers and their variable dips reveal the structural deformation of the ice caused by flowage from the area of accumulation near the central ice divide to the margin of the ice sheet along the Transantarctic Mountains (Adapted from Dunbar et al. 1995)

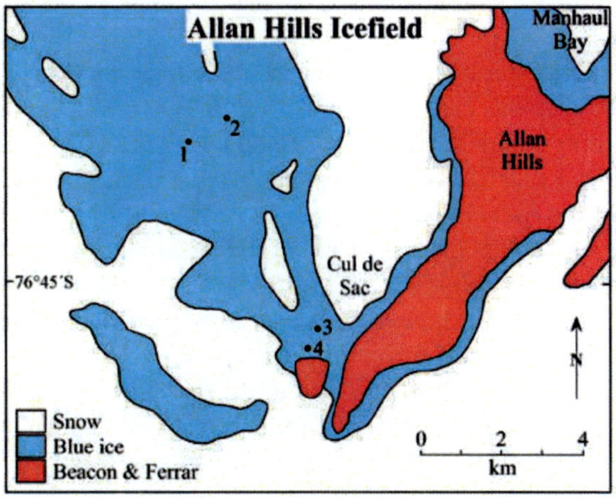

Fig. 17.49 Tephra from layers of dusty ice adjacent to the Allan Hills have a wide range of chemical compositions. Sites 1 and 2 contain tephra that range from trachyte to dacite in Fig. 17.50, whereas tephra at sites 3 and 4 in the "Cul de Sac" are composed of basanite and basalt. The ice at sites 3 and 4 in the Cul de Sac yielded U-series disequilibrium dates of 195,000 years (site 3) and 295,000 years (site 4). The ice at sites 1 and 2 is distinctly younger at 95,000 years (1) and 85,000 years (2) (Data from Marvin 1990)

with abundant crystals of Fe-Ni oxides as well as some metallic and sulfide inclusions. The concentrations of major-element oxides of the black spherules match those of average H-chondrite (i.e., stony) meteorites. The spherules also contain cosmogenic radionuclides (^{10}Be, ^{26}Al, and ^{36}Cl) which confirms that the spherules are of extraterrestrial origin and suggests that they fell from the sky between 2.3 and 2.7 million years ago. The apparent age of the spherules is similar to the terrestrial age (2.2 ± 0.4 Ma) of an H-chondrite specimen (ALH 88019) that was collected on the ice fields near the Allan Hills (Scherer et al. 1997). The similarity of chemical compositions supports the hypothesis that the black spherules originated by ablation of an H-chondrite that passed through the atmosphere. Harvey et al.

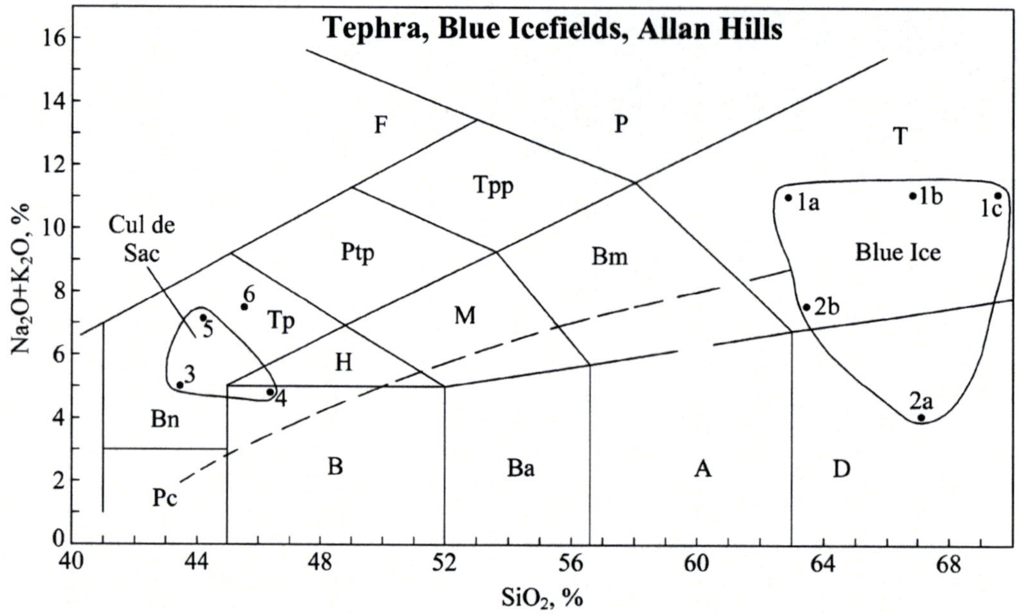

Fig. 17.50 The tephra at sites 1 and 2 on the ice adjacent to the Allan Hills (Fig. 17.49) are composed of trachyte (1a, b, c and 2b) and dacite (2a). The samples from sites 3 and 4 in the "Cul de Sac" consist of basanite (3) and basalt (4). Sample 5 is the average of 12 tephra samples from the Allan Hills ice field analyzed by Nishio et al. (1985). The chemical compositions of tephra 3, 4, and 5 are similar to average trachybasalt (6) on Ross Island (Goldich et al., 1975). The tephra at sites 1 and 2 are similar to volcanic rocks on the Malta Plateau, Mt. Melbourne, and The Pleiades in northern Victoria Land (Data from Marvin 1990; Nishio et al. 1985)

(1998) considered three impact events that are known to have occurred outside of Antarctica between two and three million years ago, but could not associate any of them with the ablation spherules in the icefield adjacent to the Allan Hills.

A final point to be made is that the spherules must have been deposited on the surface of the East Antarctic ice sheet, which means that East Antarctica was glaciated at the time (i.e., two to three million years ago). Therefore, the terrestrial age of the meteorite-ablation spherules is consistent with the evidence that Cenozoic volcanoes in the Transantarctic Mountains (e.g., Mt. Early and Sheridan Bluff) erupted through glacial ice from which it follows that the glaciation of East Antarctica started during the late Oligocene/early Miocene perhaps as early as 25 million years ago. However, none of the ice cores that have been drilled in East or West Antarctica have encountered ice that is as old as the meteorite ablation spherules in the icefield near the Allan Hills.

17.11 Summary

The East Antarctic ice sheet can only be described in terms of superlatives. It is the largest and thickest body of ice in the world, it reaches the highest elevation above sea level of any continental ice sheet, and it is the coldest area on the Earth. The East Antarctic ice sheet has stored most of the fresh water on the Earth because the ice formed from water that has passed through the distillation process of the hydrologic cycle. The ice sheet began to form during the late Oligocene and now covers almost the entire area of the East Antarctic craton.

The ice forms from snow that is deposited in the central region of the continent where it is converted to firn and then to ice by recrystallization in response to the weight of the overlying layers of snow. The ice moves from the zone of accumulation to the coast of East Antarctica and toward the Transantarctic Mountains, which force it to flow through narrow valleys in order to reach the Ross Ice Shelf and the Ross Sea from where it eventually returns to the southern oceans in the form of icebergs.

As the ice approaches the Transantarctic Mountains, it is forced to flow over rough terrain consisting of the flat-topped hills and ridges of the western foothills of the Transantarctic Mountains. The resulting structural deformation of the ice causes sediment-laden ice from the base of the ice sheet to be deflected to the surface where the clasts and fine-grained sediment accumulate as supraglacial moraines.

The Elephant and Reckling moraines are such supraglacial moraines west and northwest of the Allan Hills in southern Victoria Land. The clasts in these moraines are the only samples we have of the rocks over which the ice is flowing from the zone of accumulation toward the zone of ablation west of the Transantarctic Mountains. Much to our surprise, the clasts in these two moraines are composed entirely of sedimentary rocks of the Beacon Supergroup and of igneous rocks of the Ferrar Group. Clasts of high grade metamorphic rocks of the Precambrian shield of East Antarctica do not occur in these moraines.

The isotope compositions of oxygen and hydrogen in the ice record the temperature at the time and place of deposition. The relationship between isotope composition and the average annual temperature has been used to reveal the history of climate change in the accumulation areas from the present to 800,000 years in the past. In addition, the ice contains samples of air that was trapped between snow flakes at the time of deposition. Therefore, the climate history has been augmented by a history of the chemical composition of the atmosphere.

The temperature records in the long cores recovered at Vostok and Dome C indicate that the Pleistocene Epoch contained at least nine cold intervals separated by interglacials, compared to only four glaciations in North America. The concentrations of carbon dioxide and methane in the ice have fluctuated in step with the temperature such that the glacial episodes were accompanied by low concentrations of carbon dioxide and methane both of which are "greenhouse" gases. That raises the question whether, in the absence of human interference, cyclical changes in the composition of the atmosphere *caused* synchronous changes in the average global temperature of the Earth, or whether the changes in temperature *caused* changes in the composition of the atmosphere. Whatever drives the global climate machine, we marvel at the fact that we are able to reconstruct the composition of the air that *Homo antecessor* breathed 800,000 years ago.

Ice is virtually the only substance that is less dense than the liquid from which it forms. Hence ice floats in water, as we all know. An important consequence of this physical property is that the melting temperature of ice *decreases* with increasing pressure. The temperature at the base of the East Antarctica ice sufficiently high to cause the ice to melt. Radar-echo surveys have revealed that about 160 lakes of various sizes (e.g., Lake Vostok) exist in bedrock basins under several kilometers of overlying ice.

The meltwater at the base of the ice sheet reduces friction and thereby accelerates the delivery of icebergs to the oceans, where they melt and thus increase the volume of water in the oceans. The resulting increase of sea level threatens the stability of the West Antarctic ice sheet, parts of which are underlain by seawater. When sea level rise, the West Antarctic ice sheet is uplifted from its foundation, which may cause it to slide off the continent into the southern oceans. The resulting increase of sea level (5.8 m) will flood the coastal plains of the continents and will permanently displace the human populations from their homes.

Meanwhile, the meltwater under the ice sheet flows in response to pressure gradients and may refreeze in places where the pressure-melting point of the ice is higher than the ambient temperature. When the meltwater refreezes, the ions that are dissolved in it are excluded from the ice that forms. Therefore, the residual water becomes "salty" and may precipitate cryogenic evaporite minerals such as calcite, thenardite, and others. The cryogenic brine may also rise to the surface of the ice sheet through fractures along the edge of the ice sheet where it evaporates and where a variety of salts may be precipitated.

The snow that is deposited along the coast of East Antarctica contains elements that originated primarily from seawater in the form of microscopic aerosol particles that nucleate snow flakes. Snow in the interior of Antarctica also contains these elements but at much lower concentrations. The positive electrical charges of the metals (cations) must exactly balance the negative electrical charges of the non-nometals (anions) because all aqueous solutions (including snow and ice) must maintain a state of electrical neutrality. The negatively-charged anions in snow consist primarily of chloride, sulfate, nitrate, and bicarbonate which originate from seawater, from compounds emitted by plants and animals, by gases discharged by volcanic eruptions (e.g., sulfur dioxide), and by the naturally occurring gases in the atmosphere (e.g., molecular nitrogen and carbon dioxide). The nitrate ions in snow in Antarctica form when the di-atomic molecules of nitrogen (N_2) in the atmosphere are broken up by lightning, by solar flares, or by explosive impacts of asteroids and comets.

Long cores extracted from the ice sheets in East and West Antarctica contain layers of "dirty" ice composed of particles of volcanic ash called tephra. The tephra in the core drilled at Byrd Station in West Antarctica consist of alkali-rich volcanic ash almost certainly erupted by Mt. Takahe in Marie Byrd Land. The tephra in the ice at South Pole, Vostok, and the Yamato Mountains are composed of andesite that matches the composition of volcanic rocks erupted in the South Sandwich Islands 4,500 km from Vostok. Tephra in a core at Dome C are composed of trachyte like those in the Byrd-Station core and could have originated from the volcanoes in northern and southern Victoria Land or in Marie Byrd Land. The wide distribution of tephra in the ice sheets of East and West Antarctica suggests that they were injected into the stratosphere during the volcanic eruptions and were transported over great distances during the lengthy residence in that region of the atmosphere.

The ice fields adjacent to the Allan Hills, where a large number of meteorite specimens have been collected, also contain tephra layers. The outcrop patterns of these layers indicate that the ice was extensively deformed as it flowed over and around the subglacial foothills of the Transantarctic Mountains. A small deposit of black spherules embedded in the ice near the Allan Hills was attributed to the passage through the atmosphere of a stony meteorite. This deposit reminds us that the East Antarctic ice sheet contains extraterrestrial particles and specimens of meteorites in addition to terrigenous dust and volcanic ash.

17.12 Appendix

17.12.1 Chemical composition of snow at Base Roi Baudouin, Amundsen-Scott, and Plateau stations (Hanappe et al. 1968)

	Na	Mg	K	Ca	Mn	Fe	Ni
	(parts per billion per gram (ppb))						
Base Roi Baudouin (40 g/cm²/y)[a]							
1.	–	–	–	–	0.15 ± 0.03	11 ± 1	0.42 ± 0.08
2.	226 ± 11	27 ± 3	44 ± 7	27 ± 3	–	15 ± 3	0.80 ± 0.15
							0.74 ± 0.08
3.	–	–	–	–	0.19 ± 0.02	–	0.42 ± 0.04
4.	251 ± 12	31 ± 3	30 ± 4	18 ± 3	0.05 ± 0.01	2.6 ± 0.4	0.26 ± 0.04
						2.3 ± 0.3	0.29 ± 0.03
5.	265 ± 13	28 ± 3	16 ± 2	7.1 ± 1.5	0.12 ± 0.02	2.8 ± 0.3	0.13 ± 0.04
Avg.	247 ± 20	29 ± 2	30 ± 14	17 ± 10	0.13 ± 0.06	6.7 ± 5.9	0.44 ± 0.25
Amundsen-Scott Station (6 g/cm²/y)							
6.	9 ± 1	–	<15	–	0.75 ± 0.03	–	0.05 ± 0.02
7.	–	–	–	–	0.29 ± 0.03	–	0.13 ± 0.04
Avg.	9 ± 1	–	<15	–	0.52 ± 0.33	–	0.09 ± 0.06
Plateau Station (2.8 g/cm²/y)							
8.	18 ± 2	3.4 ± 0.4	4.5 ± 0.9	0.6 ± 0.3	<0.02	7.0 ± 0.7	<0.07
9.	35 ± 3	6.4 ± 0.6	13 ± 2	13 ± 2	0.24 ± 0.04	6.5 × 0.7	0.2 ± 0.1
					0.21 ± 0.03	–	0.17 ± 0.03
10.	–	–	–	–	0.27 ± 0.03	–	0.16 ± 0.03

(continued)

17.12.1 (continued)

	Na	Mg	K	Ca	Mn	Fe	Ni
	(parts per billion per gram (ppb))						
11.	32 ± 2	4.9 ± 1.5	3.5 ± 0.7	8.6 ± 0.9	0.41 ± 0.04	7.4 ± 0.7	0.062 ± 0.01
					0.37 ± 0.04	7.1 ± 0.4	0.063 ± 0.006
							0.064 ± 0.007
Avg.	28 ± 9	4.9 ± 1.5	7.0 ± 5.2	7.4 ± 6.3	0.25 ± 0.14	7.0 ± 0.4	0.113 ± 0.06

[a]Rate of meteoric precipitation of water.

References

Aharon P (1988) Oxygen, carbon, and U-series isotopes of aragonite from Vestfold Hills, Antarctica: Clues to geochemical processes in subglacial environments. Geochim Cosmochim Acta 52:2321–2331

Aldaz L, Deutsch S (1967) On a relationship between air temperature and oxygen isotope ratio of snow and firn in the South Pole region. Earth Planet Lett 3:267–274

Alley RB (2000) The two-mile time machine: ice cores, abrupt climate change, and our future. Amazon.com

Alley RB (ed) (2002) Abrupt climate change: inevitable surprises. Committee on Abrupt Climate Change, National Research Council, National Academy Press, Washington, D.C.

Alley RB (2004) Abrupt climate change. Scient Amer 291(5):62–69

Alley RB, Bender ML (1998) Greenland ice cores: frozen in time. Scient Amer 285(2):4–9

Alley RB, Bindschadler RA (eds) (2001) The West Antarctic ice sheet: behavior and environment. Antarctic Research Series, vol. 77. American Geophysical Union, Washington, DC

Annexstad JO, Nishio F (1979). Glaciological studies in Allan Hills, 1978–79. Antarctic J US 14(5):87–88

Annexstad JO, Schultz L (1982) Triangulation survey of the Allan Hills icefield, 1981–1982. Antarctic J US 17(5):57–58

Baker PE (1978) The South Sandwich Islands: III. Petrology of the volcanic rocks. British Antarctic Survey, Scient Rept 93:1–34

Barrett PJ, Elston DP, Harwood DM, McKelvey BC, Webb P-N (1987) Mid-Cenozoic record of glaciation and sea-level change on the margin of the Victoria Land Basin, Antarctica. Geology 15:634–637

Bauer VF (1961) Kalkabsätze unter Kalkalpengletschern und ihre Bedeutung für die Altersbestimmung heute gletscherfrei werdender Karrenformen. Zeitschrift für Gletscherkunde und Glazialgeologie 4:215–225

Bell RE (2008) The unquiet ice. Scient Amer 293 (February):60–67

Bell RE, Studinger M, Fahnestock MA, Shuman CA (2006) Tectonically controlled subglacial lakes on the flanks of the Gamburtsev subglacial mountains, East Antarctica. Geophys Res Lett 33:LO2504, 4p

Bell RE, Studinger M, Shuman CA, Fahnestock MA, Joughin I (2007) Large subglacial lakes in East Antarctica at the onset of fast-flowing ice streams. Nature 445:904–907

Bentley CR (1972) Subglacial topography. In: Heezen BC, Tharp M, Bentley CR (eds) Morphology of the Earth in the Antarctic and Subantarctic. Antarctic Map, Folio Series 16, Plate 7. American Geographysical Society, New York

Bentley CR (1997) Rapid sea-level rise soon from West Antarctic ice-sheet collapse? Science 275:1077–1078

Bentley CR, Thomas RH, Velicogna I (2007) Ice sheets. In: Global outlook for ice and snow. Section 6A:99–114. United Nations Environmental Programme, United Nations, New York

Bindschadler R (2004) The history and significance of ice Earth. In: Dasch P (ed) Ice worlds of the solar system. Cambridge University Press, Cambridge, pp 6–32

Bindschadler RA (ed) (1990) Sea rise: a multidisciplinary research initiative to predict rapid changes in global sea level caused by collapse of marine ice sheets. NASA Conference Pub 3075:1–47

Bindschadler, RA (1998) Future of the West Antarctic ice sheet. Science 282:428–429

Bindschadler RA, Bentley CR (2002) On thin ice. Scient Amer 287(6):98–105

Bjaerke T, Dypvik H (1977) Quaternary "stromatolitic" limestone of subglacial origin from Scandinavia. J Sed Petrol 47:1321–1327

Boger PD, Faure G (1988) The possible occurrence of volcanic ash in till from Victoria Land, Antarctica. Antarctic J US 23(5):29–30

Bolshov MA, Boutron CF, Zybin ZV (1989) Determination of lead in Antarctic ice at picogram-per-gram level by laser atomic fluorescence spectrometry. Anal Chem 61:1758–1762

Boutron CF (1990) A clean laboratory for ultralow concentration heavy metal analyses. Fresenius J Anal Chem 337:482–491

Boutron C, Patterson C (1983) The occurrence of lead in Antarctic recent snow, firn deposited over the last two centuries and prehistoric ice. Geochim Cosmcohim Acta 47(8):1355–1368

Boutron CF, Patterson CC (1986) Lead concentration changes in Antarctic ice during the Wisconsin Holocene transition. Nature 323:222–225

Boutron CF, Patterson CC (1987) Relative levels of natural and anthropogenic lead in recent Antarctic snow. J Geophys Res 92:8454–8464

Boutron CF, Patterson CC, Petrov VN, Barkov NI (1987) Preliminary data on changes of lead concentrations in Antarctic ice from 155,000 to 26,000 years BP. Atmos Environ 21:1197–1202

Boutron CF, Bolshov MA, Koloshnikov VG, Patterson CC, Barkov NI (1990) Direct determination of lead in Vostok Antarctic ancient ice by laser excited atomic fluorescence spectrometry. Atmos Environ 24A:1797–1800

Boutron CF, Candelone J-P., Hong SM (1994) Past and recent changes in the large scale tropospheric cycles of lead and other heavy metals as documented in Antarctic and Greenland snow and ice: a review. Geochim Cosmochim Acta 58(15):3217–3225

Bowser CJ, Rafter TA, Black RF (1970) Geochemical evidence for the origin of mirabilite deposits near Hobbs Glacier, Victoria Land, Antarctica Mineral. Soc Amer Spec Paper 3:261–272

Brandriss ME, O'Neil JR, Edlund MB, Stoermer EF (1998) Oxygen isotope fractionation between diatomaceous silica and water. Geochim Cosmochim Acta 62:1119–1125

Budd WF, Jenssen MJD, Radok U (1971) Derived physical characteristics of the Antarctic ice sheet. A.N.A.R.E. Interim Reports, Series A (N), Glaciology, 1–178 Antarctic Division, Melbourne, Australia

Bull C, McKelvey BC, Webb PN (1962) Quaternary glaciations in southern Victoria Land Antarctica. J Glaciol 4:63–78

Calkin PE, Nichols RL (1971) Quaternary studies in Antarctica. In: Adie RJ (ed) Antarctic geology and geophysics. Universitetsforlaget, Oslo, Norway, pp 625–644

Calvin M (1968) Chemical evolution. Oxford University Press, Oxford

Cannon HL, Hopps HC (eds) (1971) Environmental geochemistry in health and disease. Geol Soc Amer Mem 123:1–230

Cassidy WA, Meunier T, Buchwald V, Thompson C (1983) Search for meteorites in the Allan Hills/Elephant Moraine area, 1982–83. Ant J US 18:81–82

Cazenave A, Lombard A, Llovel W (2008) Present-day sea-level rise: a synthesis. Comptes Rendus Geosci 340:761–770

Chacko T, Cole DR, Horita J (2001) Equilibrium oxygen, hydrogen, and carbon isotope fractionation factors applicable to geological systems. In: Valley JW, Cole DR (eds) Stable isotope geochemistry. Reviews in mineralogy and geochemistry, vol. 43. Mineralogical Society of America, Blacksburg, Virginia, pp 1–81

Church JA, White NJ (2006) A 20th century acceleration in global sea-level rise. Geophys Res Lett 33. doi:10.1029/2006 GL026510

Claridge GGC, Campbell IB (1977) The salts in Antarctic soils, their distribution and relationship to soil processes. Soil Sci 123:377–384

Compston W (1999) Geological age determination by instrumental analysis. The 29th Halmond Lecture. Mineral Mag 63:297–311

Cook-Anderson G (2007) NASA satellites unearth clues to leaks in Antarctic "plumbing system". Earth Obser 19(2):32

Coren F, Delisle G, Sterzai P (2003) Ice dynamics of the Allan Hills meteorite concentration sites revealed by satellite aperture radar interferometry. Meteorit Planet. Sci 38(9):1319–1330

Crowell JC, Frakes LA (1970) Phanerozoic glaciation and the causes of ice ages. Amer J Sci 268:193–224

Crozaz G., Picciotto E, De Breuck W (1964) Antarctic snow chronology with Pb210. J Geophys Res 69(12):2597–2604

Damm V (1996) Subice morphology deduced by radio echo sounding (RES) in the area between David and Mawson glaciers, Victoria Land. Geol Jahrbuch B89:321–331

Dansgaard W, Johnsen SJ, Moller J, Langway CC (1969) One thousand centuries of climate record from Camp Century on the Greenland ice sheet. Science 166:377–381

Dansgaard W, Johnsen SJ, Clausen HP, Langway CC Jr (1971) Climatic record revealed by the Camp Century ice core. In: Turekin KK (ed) Late Cenozoic Glacial Ages. Yale University Press, New Haven, CT, pp 37–56

de Angelis M, Legrand M, Petit JR, Barkov NI, Korotkevitch YS, Kotlyakov VM (1984) Soluble and insoluble impurities along the 950-m deep Vostok core (Antarctica): climate implications. J Atmos Chem 1:215–239

Delisle G, Sievers J (1991) Sub-ice topography and meteorite finds near the Allan Hills and the Near Western ice field, Victoria Land, Antarctica. J Geophys Res 96(E1):15577–15587

Delmas R (1982) Antarctic sulphate budget. Nature 299:667–678

Delmas R, Barnola M, Legrand M (1982a) Gas derived aerosol in central Antarctic snow and ice: the case of sulfuric and nitric acids. Ann Glaciol 3:71–76

Delmas R, Briat M, Legrand M (1982b) Chemistry of South Polar snow. J Geophys Res 87(C13):4314–4318

Delmas R, Legrand M, Aristarain A, Zanolini F (1985) Volcanic deposits in Antarctic snow and ice. J Geophys Res 90:12,901–12,920

Denton GH, Hughes TJ (eds) (1981) The last great ice sheets. Wiley-Interscience, New York

Denton GF, Armstrong RF, Stuiver M (1971) The late Cenozoic glacial history of Antarctica. In: Turekian K (ed) Late Cenozoic Ice Ages. Yale University Press, New Haven, CT

Denton GH, Bockheim JG, Wilson SC, Stuiver M (1989) Late Wisconsin and Early Holocene glacial history, inner Ross Embayment, Antarctica. Quat Res 31(2):151–182

Denton GH, Sugden DE, Marchant DR, Hall BL, Wilch TI (1993) East Antarctic ice sheet sensitivity to Pliocene climatic change from a Dry Valleys, perspective. Geografiska Annaler 75A(4):155–204

Dreschhoff GAM, Zeller EJ (1994) 415-year Greenland ice-core record of solar proton events dated by volcanic eruptive episodes. Institute for Tertiary-Quaternary Studies-TER-QUA Symposium Series, 2:1–24

Drewry DJ (1971) Subglaical morphology between the Transantarctic Mountains and the South Pole. In: Adie RJ (ed) Antarctic geology and geophysics. Universitetsforlaget, Oslo, Norway, 693–704

Drewry DJ (1972) The contribution of radio-echo sounding to the investigation of Cenozoic tectonics and glaciation in Antarctica. In: Price RJ, Sugden DE (eds) Polar geomorphology. Inst Brit Geol Spec Publ 4:43–58

Drewry DJ (1973) Sub-ice relief and geology of East Antarctica. Ph.D. Dissertation University of Cambridge

Drewry DJ (1975) Initiation and growth of the East Antarctic ice sheet. J Geol Soc, Lond 131:255–273

Dunbar NW, Cashman KV, Dupre R (1994) Crystallization processes of anorthoclase phenocrysts in the Mount Erebus magmatic system. In: Kyle PR (ed) Volcanological Studies of Mount Erebus. Antarctic Research Series, vol. 60. American Geophysical Union, Washington, DC

Dunbar NW, Kyle PR, McIntosh WC, Esser RP (1995a). Tephra layers in blue ice, Allan Hills, Antarctica: a new source of glacial tephrochronlogical data. Antarctic J US 30(5):76–77

Dunbar NW, Kyle PR, McIntosh WC, Esser RP (1995b) Geochemical composition and stratigraphy of tephra layers in Antarctic blue ice: insights into glacial tephrochronology. 7th SCAR Symposium on Antarctic Sciences, Abstracts, p 115. Siena, Italy

Emiliani C (1978) The cause of ice ages. Earth Planet Sci Lett 37:349–352

Emiliani C (1992) Planet Earth: cosmology, geology, and the evolution of life and environment. Cambridge University Press, New York

EPICA (2004) Eight glacial cycles from an Antarctic ice core. Nature 429:623–628 (56 authors from 11 countries)

Epstein S, Sharp RP, Gow AJ (1965) Six-year record of oxygen and hydrogen isotope variations in South Pole firn. J Geophys Res 70(8):1809–1814

Epstein S, Sharp RP, Gow AJ (1970) Antarctic ice sheet; stable isotope analysis of Byrd Station cores and interhemispheric climatic implications. Science 168:1570–1572

Eugster HP (1966) Sodium carbonate-bicarbonate minerals as indicators of Pco_2. J Geophys Res 71:3369–3377

Faure G (1990a) Origin of stranding surfaces. Lunar Planet. Inst. Tech. Rep., 90–03:11–15. Houston, TX

Faure G (1990b) Physical description of the Elephant and Reckling moraines. Lunar Planet. Inst. Tech. Rep., 90–03:24–25. Houston, TXs

Faure G (1990c) Subice topography and the formation of supraglacial moraines. Lunar Planet. Inst. Tech. Rep., 90–03:32–33. Houston, TX

Faure, G. (1990d) Supraglacial moraines, meteorites, and climate change. Lunar Planet. Inst. Tech. Rep., 90–03:33–35. Houston, TX

Faure G (1998a) Principles and applications of geochemistry, 2nd edn. Prentice-Hall, Upper Saddle River, NJ

Faure G (1998b) Effect of environmental lead on human health. In: Faure G (ed) Principles and applications of geochemistry, 2nd edn. Prentice-Hall, Upper Saddle River, NJ, pp 485–505

Faure G, Botoman G (1984) Origin of epigenetic calcite in coal from Antarctica and Ohio based on isotope compositions of oxygen, carbon, and strontium. Isotope Geosci (Chem Geol) 2:313–324

Faure G, Buchanan D (1987) Glaciology of the East Antarctic ice sheet in the Allan Hills: a preliminary interpretation. Antarctic J US 22(5):74–75

Faure G, Buchanan D (1991) Ablation rates of the ice fields in the vicinity of the Allan Hills, Victoria Land, Antarctica. In: Elliot DH (ed) Contributions to Antarctic Research II. Antarctic Research Series vol. 53. American Geophysical Research, Washington, DC, pp 19–31

Faure G, Harwood DM (1990) Marine microfossils in till clasts of the Elephant Moraine on the East Antarctic ice sheet. Antarctic J US 25(5):23–25

Faure G, Mensing TM (2005) Isotopes: principles and applications, 3rd edn. Wiley, Hoboken, NJ

Faure G, Mensing TM (2007) Introduction to planetary science; the geological perspective. Springer, Dordrecht, The Netherlands

Faure G, Sutton S (1985) Thermoluminescence of sandstone clasts of the Elephant Moraine. Antarctic J US 19(5):12–14

Faure G, Taylor KS (1985) The geology and origin of the Elephant Moraine on the East Antarctic ice sheet. Antarctic J US 19(5):11–12

Faure G, Kallstrom ML, Mensing TM (1984) Classification and age of terrestrial boulders in the Elephant and Reckling moraines. Antarctic J US 19(5):28–29

Faure G, Taylor KS, Jones LM (1986) Hydrothermal calcite in the Elephant Moraine. Antarctic J US 21(5):21

Faure G, Strobel ML, Hagen EH, Buchanan D (1987) Glacial geology of the Reckling Moraine on the East Antarctic ice sheet. Antarctic J US 22(5):61–63

Faure G, Hoefs J, Jones LM, Curtis JB, Pride DE (1988) Extreme ^{18}O depletion in calcite and chert clasts from the Elephant Moraine on the East Antarctic ice sheet. Nature 332:352–354

Faure G, Grootes P, Buchanan D, Hagen EH (1992) Oxygen isotope study of the ice fields surrounding the Reckling Moraine on the East Antarctic ice sheet. In: Elliot DH (ed) Contributions to Antarctic Research. Antarctic Research Series, III, vol. 57. American Geophysical Research, Washington, DC, pp 15–26

Faure G, Wehn KS (Taylor), Montello JM, Hagen EH, Strobel ML, Johnson KS (1993) Isotopic composition of the ice and sub-glacial geology near the Allan Hills, Victoria Land, Antarctica. In: Findlay RH et al (eds) Gondwana eight: assembly, evolution, and dispersal. A.A. Balkema, Rotterdam, The Netherlands, pp 485–495

Faure G, Mensing TM, Manson VL, Place MC (1994) Geochemistry of a layer of volcanic ash in the ice near Brimstone Peak, southern Victoria Land. Antarctic J US 29(5):22–23

Fitzpatrick JJ, Muhs DR (1989) Borax in the supraglacial moraine of the Lewis Cliff, Buckley Island quadrangle - First Antarctic occurrence. Antarctic J US 24(5):63–65

Fitzpatrick JJ, Muhs DR, Jull AJT (1990) Saline minerals in the Lewis Cliff ice tongue, Buckley Island quadrangle, Antarctica. In: Elliot DH (ed) Contributions to Antarctic Research I. Antarctic Research Series, vol. 50. American Geophysical Union, Washington, DC, pp 57–69

Ford DC, Fuller PG, Drake JJ (1970) Calcite precipitates at the soles of temperate glaciers. Nature 226:441–442

Fricker HA, Scambos T, Bindschadler RA, Padman L (2007) An active subglacial water system in West Antarctica mapped from space. Science 315:1544–1548

Friedman I, O'Neil (1977) Compilation of stable isotope fractionation factors of geochemical interest. U.S. Geol. Surv. Prof. Paper 440-KK, Washington, DC

Fuge R, Billet M, Selinus O (1996) Environmental geochemistry. Appl Geochem 11(1/2):1–385

Garrels RM, McKenzie FT (1967) Origin of the chemical composition of some springs and lakes. In: Equilibrium concepts in natural water systems. American Chemical Society, Advances in Chemistry Series, vol. 67. Washington, DC, pp 222–242

Ghazi AM (1994) Lead in archaeological samples: an isotopic study by ICP-MS. Appl Geochim 9:627–636

Gibson GW (1962) Geological investigations in southern Victoria Land, Antarctica. Part 8: evaporite salts in the Victoria Valley region. New Zealand J Geol Geophys 5(3):361–374

Gillet F, Rado C (1979) A 180-meter core drilling at Dome C and measurements in the 905-meter drill hole. Antarctic J US 14(5):101

Giovinetto MB, Bull C (1987) Summary and analyses of surface mass balance compilations for Antarctica, 1960–1985. Report No. 1, Byrd Polar Research Center, Ohio State University, Columbus, OH (ISSN: 0896–2472):1–90

Goldich SS, Treves SB, Suhr NH, Stuckless JS (1975) Geochemistry of the Cenozoic volcanic rocks of Ross Island and vicinity, Antarctica J Geol 83(4):415–434

Gonfiantini R, Togliatti V, Tongiorgi E, DeBreuck W, Picciotto E (1963) Snow stratigraphy and oxygen isotopic variations in the glaciological pit of King Baudouin Station, Queen Maud Land, Antarctica. J Geophys Res 68(13): 3791–3798

Gow AJ (1965) On the accumulation and seasonal stratification of snow at the South Pole. J Glaciol 5(40):467–477
Gow AJ (1971) Analysis of Antarctic ice cores. Antarctic J US 6(5):205–206
Gow AJ, Williamson T (1971) Volcanic ash in the Antarctic ice sheet and its possible climate implications. Earth Planet Sci Lett 13:210–213
Gow AJ, Ueda HT, Garfield DE (1968) Antarctic ice sheet: preliminary results of the first core hole to bedrock. Science 161:1011
Gow AJ, Epstein S, Sheehy W (1979) On the origin of stratified debris in ice cores from the bottom of the Antarctic ice sheet. J Glaciol 23:185–192
Hallet B (1976) Deposits formed by subglacial precipitation of $CaCO_3$. Geol Soc Amer Bull 87:1003–1015
Hammer CU (1984) Traces of Icelandic eruptions in the Greenland ice sheet. Jokull 34:51–65
Hammer CU, Clausen HP, Dansgaard W (1981a) Greenland ice sheet evidence of post-glacial volcanism and its climatic impact. Nature 288:230–235
Hammer CU, Clausen HB, Dansgaard W (1981b) Past volcanism and climate revealed by Greenland ice cores. J Volcanol Geotherm Res 11:3–10
Hanappe F, Vosters M, Picciotto E, Deutsch S (1968) Chimie des neiges Antarctique et taux de deposition de matiere extraterrestre-deuxieme article. Earth Planet Sci 4:487–496
Hanshaw BB, Hallet B (1978) Oxygen isotope composition of subglacially precipitated calcite: possible paleoclimatic implications. Science 200:1267–1270
Hardie LA (1984) Evaporites: marine or nonmarine? Science 284:193–240
Hardie LA, Eugster HP (1970) The evolution of closed-system brines. Mineral Soc Amer Spec Paper 3:273–290
Hartmann WK (2005) Moons and planets. 5th edn., Brooks/Cole, Belmont California
Harvey RP, Dunbar NW, McIntosh WC, Esser RP, Nishiizumi K, Taylor S, Caffee MW (1998) Meteoritic event recorded in Antarctic ice. Geology 26(7):607–610
Harvie CE, Weare JH (1980) The prediction of mineral solubilities in natural water: The Na-K-Mg-Cl-SO_4-H_2O system from zero to high concentrations at 25°C. Geochim Cosmochim Acta 44:981–997
Harwood DM (1983) Diatoms from the Sirius Formation, Transantarctic J US 18(5):98–100
Harwood DM (1985) Late Neogene climatic fluctuation in the southern high latitudes: Implication of a warm Pliocene and deglaciated Antarctic continent. South African J Sci 81:239–241
Harwood DM (1986) Recycled siliceous microfossils from the Sirius Formation. Antarctic J US 21(5):101–103
Herron MM (1982) Impurity sources of F^-, Cl^-, NO_3^- and $SO_4^=$ in Greenland and Antarctic precipitation. J Geophys Res 87(C4):3052–3060
Hillaire-Marcel C, Soucy JM, Cailleux A (1979). Analyse isotopique de concretions sous-glaciaires de l' inlandsis laurentidien et teneus en oxygen-18 de la glace. Canadian J Earth Sci 16:1494–1498
Hitchcock DR, Spiller LL, Wilson WE (1980) Sulfuric acid aerosols and HCl release in coastal atmospheres: evidence of rapid formation of sulfuric acid particulates. Atmos Environ 14:165–182
Hoefs J (1997) Stable isotope geochemistry, 4th edn. Springer, Heidelberg, Germany
Holdgate MW (1980) Some environmental questions: an after-dinner address. Annal Glaciol 1:135–136
Hollin JT (1962) On the glacial history of Antarctica. J Glaciol 32:173–195
Huebert BJ, Lazrus AL (1980a) Tropospheric gas-phase and particulate nitrate measurements. J Geophys Res 85(C12):7322–7328
Huebert BJ, Lazrus AL (1980b) Bulk composition of aerosols in the remote troposphere. J Geophys Res 85(C12):7337–7344
Hughes TJ (1981) Numerical reconstruction of paleo-ice sheets. In: Denton GH, Hughes TJ (eds) The last great ice sheets. Wiley-Interscience, New York, pp 222–274
Hughes TJ (1987) Deluge II and the continent of doom; rising sea level and collapsing Antarctic ice. Boreas 16:89–100
Hughes TJ, Denton GH, Andersen BG, Schilling DH, Fastook JL, Lingle CS (1981) The last great ice sheets; a global view. In: Denton GH, Hughes TJ (eds) The last great ice sheets. Wiley-Interscience, New York, pp 275–317
Hult JL and Ostrander NC, 1973 Hult JL, Ostrander NC (1973) Antarctic icebergs as a global fresh water resource. The Rand Corp., R-1255-NFS
Huppert HE, Turner JS (1978) On melting icebergs. Nature 271:46–48
Husseiny AA (ed) (1978) Iceberg utilization: proceedings of the first international conference, Ames, Iowa, 2–6 October, 1977. Pergamon Press, Oxford
Jevrejeva S, Grinsted A, Moore JC, Holgate S (2006) Nonlinear trends and multi-year cycles in sea-level records. J Geophys Res. doi:10.1029/2005JC003229
Job JG (1975) Simulation of iceberg towing. Report to the Committee on Iceberg Utilization of the Australian Academy of Science, Canberra, A.C.T.
Job JG (1978a) Yields and energetics in moving unprotected icebergs to southern continents. In: Husseiny AA (ed) Iceberg utilization. Pergamon Press, Oxford, pp 339–349
Job (1978b) High efficiency iceberg propulsion systems. In: Husseiny AA (ed) Iceberg utilization. Pergamon Press, Oxford, pp 503–527
Johnsen SJ, Dansgaard W, Clausen HB, Langway CC (1972) Oxygen isotope profiles through the Antarctic and Greenland ice sheets. Nature 235:429–434
Jones LM, Walker RL, Hall BA, Borns HW Jr(1973a) Origin of the Jurassic dolerites and basalts of southern Victoria Land. Antarctic J US 12(5):268–270
Jones LM, Whitney JA, Stromer JC (1973b) A volcanic ash deposit, Wright Valley. Antarctic J US 8(5):270–272
Jouzel J, Merlivat I, Lorius C (1982) Deuterium excess in an East Antarctic ice core suggests higher relative humidity at the oceanic surface during the last glacial maximum. Nature 299:688–691
Kapitsa AR (1968) Podlednyy rel'yef Antarktidy. Moscow, USSR, pp 1–98
Katsushima T, Nishio F, Omae H, Ishikawa M, Takahashi S (1984) Composition of dirt layers in the bare ice areas near the Yamato Mountains in Queen Maud Land and the Allan Hills in Victoria Land, Antarctica. Mem. Nat. Inst. Pol Res Special Issue 34:174–187
Katz A, Starinsky A (2009) Geochemical history of the Dead Sea. Aquat Geochim 15:159–194

Keys JR, Williams K (1981) Origin of crystalline, cold desert salts in the McMurdo region, Antarctica. Geochim Cosmochim Acta 45:2299–2309

Keys JR, Anderton PS, Kyle PR (1977) Tephra and debris layers in the Skelton Névé and Kempe Glacier, South Victoria Land, Antarctica. New Zealand J Geol Geophys 20(5):971–1002

Kiryukhin VA, Tolstikhin NI (1988) The hydrology of Antarctica. Int Geol Rev 30(1):36–45

Kita I, Taguchi S, Matsubaya O (1985) Oxygen isotope fractionation between amorphous silica and water at 34–93°C. Nature 314:83–84

Klein-BenDavid O, Sass E, Katz A (2004) The evolution of marine evaporitic brines in inland basins: The Jordan-Dead Sea Rift valley. Geochim Cosmochim Acta 68(8):1763–1775

Koeberl C (1990) Dust bands in blue ice fields in Antarctica and their relationship to meteorites and ice. In: Cassidy WA, Whillans I (eds) Workshop of Antarctic meteorite stranding surfaces. Lunar Planet. Inst., Tech. Report. 90–03:70–74. Houston, TX

Koeberl C, Yanai K, Cassidy WA, Schutt JW (1987). Investigation of dust components from dust bands from blue ice fields in the Lewis Cliff (Beardmore) area, Antarctica. In: Nagata T (ed) 12th Symposium of Antarctic Meteorites. Nat. Inst. Polar Res., Tokyo, Japan

Kohler J (2007) Glaciology: lubricating lakes. Nature 445:830–831

Kumai M (1976) Identification of nuclei and concentrations of chemical species in snow crystal samples at South Pole. J Atmos. Sci 33:833–844

Kumai M (1977) Electron microscope analysis of aerosols in snow and deep cores from Greenland. In, Isotopes and Impurities in Snow and Ice. Proceedings of Grenoble Symposium, 1975. IAHS-AISH Pub. No 118:341–350

Kvasov DD, Verbitsky MYa (1981) Causes of Antarctic glaciation in the Cenozoic. Quatern Res 15:1–17

Kyle PR (1990) The Pleiades. In: LeMasurier WE, Thomson JW (eds) Volcanoes of the Antarctic plate and southern oceans. Antarctic Research Series, vol. 48. American Geophysical Union, Washington, DC, pp 60–64

Kyle PR, Jezek PA (1978) Compositions of three tephra layers from the Byrd-Station ice core, Antarctica. J Volcanol Geotherm Res 4:225–232

Kyle PR, Jezek PA, Mosley-Thompson E, Thompson LG (1981) Tephra layers in the Byrd Station ice core and the Dome C ice core, Antarctica, and their climatic importance. J Volcanol Geotherm Res 11:29–39

Kyle PR, Palais JM, Delmas R (1982) The volcanic record of Antarctic ice cores: Preliminary results and potential for future investigations. Annal Glaciol 3:172–177

Kyle PR, Palais JM, Thomas E (1984) The Vostok tephra: an important englacial stratigraphic marker? Antarctic J US 19:64–65

Kyle PR, Moore JA, Thirlwall MF (1992) Petrological evolution of anorthoclase phonolite lavas at Mount Erebus, Ross Island, Antarctica. J Petrol 33(4):849–875

Labeyrie L (1974) New approach to surface seawater paleotemperatures using $^{18}O/^{16}O$ ratios in silica of diatom frustules. Nature 248:40–41

Labeyrie LD, Pichon JJ, Labracherie M, Ippolito P, Duprat J, Duplessy JC (1986) Melting history of Antarctica during the past 60,000 years. Nature 322:701–705

Langway CC, JR, Oeschger H, Dansgaard W (eds) (1985) Greenland ice core: geophyics, geochemistry, and the environment. Geophysics Monograph, vol. 33. American Geophysical Union, Washington, DC

Legrand MR, Delmas RJ (1984) The ionic balance of Antarctic snow: a 10-year detailed record. Atmos Environ 18:1867–1874

Legrand MR, Delmas RJ (1987) A 220-years continuous record of volcanic H_2SO_4 in the Antarctic ice sheet. Nature 327:671–676

Legrand MR, Delmas RJ (1988) Formation of HCl in the Antarctic atmosphere. J Geophys Res 93(D6):7153–7168

Legrand MR, Aristarian AJ, Delmas RJ (1982) Acid titration of polar snow. Anal Chem 65:2446–2449

Legrand MR, deAngelis M, Delmas RJ (1984) Ion chromographic determination of common ions at ultratrace levels in Antarctic snow and ice. Anal Chem Acta 54:1336–1339

LeMasurier WE, Rex DC (1990) Mount Takahe. In: LeMasurier WE, Thomson JW (eds) Volcanoes of the Antarctic plate and southern oceans. Antarctic Research Series, vol. 48. American Geophysical Union, Washington, DC, pp 169–174

Lemmens M, Lorrain R, Haren J (1982) Isotopic composition of ice and subglacially precipitated calcite in an alpine area. Zeitschrift für Gletscherkunde und Glazialgeologie 18:151–159

Long AJ (2009) Back to the future: Greenland's contribution to sea-level change. GSA Today 19(6):4–10

Lorius C, Merlivat L (1997) Distribution of mean surface stable isotope values in East Antarctica: observed changes with depth in the coastal area. In: Rodda JC (ed) International symposium on isotopes and impurities in snow and ice. Internat. Assoc. Hydrol. Sci., Pub., vol. 118. Bartholomew Press, Surrey, pp 127–137

Lorius C, Donnou D (1978) A 905-meter deep core drilling at Dome C (East Antarctica) and related surface programs. Antarctic J US 13(4):50–51

Lorius C, Merlivat L, Jouzel J, Pourchet M (1979) A 30,000-yr isotope climatic record from Antarctic ice. Nature 280:644–648

Lorius C, Jouzel J, Ritz C, Merlivat L, Barkov NI, Korotkevich YS, Kotlyakov VM (1985) A 150,000 climate record from Antarctic ice. Nature 316:591–596

Loulergue L, Schilt A, Spahni R, Masson-Delmotte V, Blunier T, Lemieux B, Barnola J-M, Raynaud D, Stocker TF, Chappellaz J (2008) Orbital and millennial-scale features of atmospheric CH_4 over the past 800,000 years. Nature 453:383–386

Lüthi D, LeLoch M, Bereiter B, Blunier T, Barnola J-M, Siegenthaler U, Raynaud D, Jouzel J, Fischer H, Kawamura K, Stocker TF (2008) High-resolution carbon dioxide concentration record 650,000–800,000 years before the present. Nature 453:379–382

Magaritz M (1973) Precipitation of secondary calcite in glacier areas: carbon and oxygen isotopic compositions of calcites from Mt. Hermon, Israel, and the European Alps. Earth Planet. Sci Lett 17:385–390

Marvin UB (1990) Diverse components of dust bands in Allan Hills ice samples. In: Cassidy WA, Whillans IM (eds) Workshop on Antarctic meteorite stranding Surfaces. LPI Tech. Rept., vol. 90–03. Lunar and Planetary Institute, Houston, TX, pp 75–79

Mayewski PA, Lyons WB, Zielinski G, Twickler M, Whitlow S, Dibb J, Grootes P, Taylor K, Whung P-Y, Fosberry L, Wake C, Welch K (1995) An ice-core based, late Holocene history for the Transantarctic Mountains, Antarctica. In: Elliot DH (ed) Contributions to Antarctic geology IV. Antarctic Research Series, vol. 67. American Geophysical Union, Washington, DC, pp 33–46

Mensing TM (1991) High-titanium basalt and dolerite clasts from the Elephant and Reckling moraines. Antarctic J US 26(5):26–27

Mercer JH (1968) Glacial geology of the Reedy Glacier area, Antarctica. Geol Soc Amer Bull 79:471–486

Mercer JH (1985) When did open-marine conditions last prevail in the Wilkes and Pensacola basins, East Antarctica, and when was the Sirius Formation emplaced? South African J Sci 81:243–245

Mörner NA (1972) Time scale and ice accumulation during the last 125,000 years as indicated by the Greenland ^{18}O curve. Geol Mag 109:17–24

Mörner NA (1974) The Greenland ^{18}O curve: time scale and ice accumulation. Geol Mag 111:431–433

Montello JM (1986) The provenance of till at the Elephant Moraine on the East Antarctic ice sheet. M.Sc. Thesis, The Ohio State University, Columbus, OH

Moriwaki K, Yoshida Y, Harwood DM (1992) Cenozoic glacial history of Antarctica: a correlative synthesis. In: Yoshida Y (ed) Recent progress in Antarctic Earth Science. Terrapub., Tokyo, Japan, pp 773–780

Murozumi M, Chow TJ, Patterson CC (1969) Chemical concentrations of pollutant lead aerosols, terrestrial dusts, and sea salts in Greenland and Antarctic snow strata. Geochim Cosmochim Acts 33:1247–1294

National Geographic (1990) Atlas of the world, 6th edn. National Geographic Society, Washington, DC

Nazarov VS (1962) Ice of the Antarctic waters. Results of the I.G.Y. Okeanologiya, No. 6, Soviet Geophysics Committee, Academy of Sciences, Moscow, USSR

Newell RE (1981) Further studies of the atmospheric temperature change produced by the Mt. Agung volcanic eruption in 1963. J Volcanol Geotherm Res 11:61–66

Nishio F, Katsushima T, Ohmae H (1985 Volcanic ash layers in bare ice areas near the Yamato Mountains, Dronning Maud Land and the Allan Hills, Victoria Land, Antarctica. Anal Glaciol 7:34–41

Nriagu JO (ed) (1978) The biogeochemistry of lead in the environment. Parts A and B. Elsevier-North Holland, Amsterdam, The Netherlands

Page NR (1971) Subglacial limestone deposits in the Canadian Rocky Mountains. Nature 229:42–43

Palais JJ, Kyle PR, Mosley-Thompson E, Thomas E (1987) Correlation of a 3,200 year old tephra in ice cores form Vostok and South Pole stations, Antarctica. Geophys Res Lett 14(8):804–807

Palais JM (1980) Snow stratigraphic investigations at Dome C, East Antarctica: a study of depositional and diagenetic processes. M.Sc. Thesis, The Ohio State University, Columbus, OH

Palais JM (1985a). Tephra layers and ice chemistry in the Byrd-station ice core, Antarctica. Ph.D. Dissertation, The Ohio State University, Columbus, OH

Palais JM (1985b) Particle morphology, composition and associated ice chemistry of tephra layers in the Byrd ice core: evidence for hydrovolcanic eruptions. Anal Glaciol 7:42–48

Palais JM (1986/87) Polar ice cores. Oceanus 29(4):55–60

Palais JM, Legrand MR (1985) Soluble impurities in the Byrd Station ice core, Antarctica: their origin and sources. J Geophys Res 90(C1):1143–1154

Palais JM, Sigurdsson H (1989) Petrologic evidence of volatile emissions from major historic and pre-historic volcanic eruptions IUGG and AGU, Monograph, vol. 52, Washington, DC

Palais JM, Whillans IM, Bull C (1982) Snow stratigraphic studies at Dome C, East Antarctica; an investigation of depositional and diagenetic process. Anal Glaciol 3:239–242

Palais JM, Kyle PR, Delmas R (1983) Detailed studies of tephra layers in the Byrd Station ice core: preliminary results and interpretation. Antarctic J US 18(5):109–110

Parker BC, Heiskill LE, Thompson WJ, Zeller WJ, Zeller EJ (1978) Non-biogenic fixes of nitrogen in Antarctica and some ecological implications. Nature 271:651–652

Patterson C, Boutron C, Petrov VN (1987) Lead records in Antarctic ice: Changes in global atmospheric concentrations during the past 150,000 years. Antarctic J US 22(5):85–87

Peel D (1986) Is lead pollution of the atmosphere a global problem? Nature 323:200

Perkins S (2006) Cold and deep: Antarctica's Lake Vostok has two big neighbors. Science News 169 (February 4):69–70

Perkins, S (2008) Down with carbon. Science News, May 10:18–23

Peterson JA, Moresby JF (1979) Subglacial travertine and associated deposits in the Carstensz area, Irian Jaya, Republic of Indonesia. Zeitschrift für Gletscherkunde und Glazialgeologie 15:23–29

Petit JR, Briat M, Royer A (1981) Ice age aerosol content from East Antarctic ice core samples and past wind strength. Nature 293:391–394

Petit JR et al (1999) Climate and atmospheric history of the past 420,000 years from the Vostok ice core, Antarctica. Nature 399:429–436

Picciotto E, De Maere X, Friedman I (1960) Isotopic composition and temperature of formation of Antarctic snow. Nature 187:857

Picciotto E, Cameron R, Crozaz G, Deutsch S, Wilgain S (1968) Determination of the rate of snow accumulation at the Pole of Relative Inaccessibility, Eastern Antarctica: a comparison of glaciological and isotopic methods. J Glaciol 7(50):273–287

Picciotto E, Crozaz G, De Breuck W (1971) Accumulation on the South Pole-Queen Maud Land Traverse, 1964–1968. In: Crary AP (ed) Antarctic snow and ice studies II. Antarctic Research Series, vol. 16, pp 257–315

Robin G, de Q (1988) The Antarctic ice sheet, its history and response to sea level and climatic changes over the past 100 million years. Paleo Paleo Paleo 67:31–50

Robok A (1981) A latitudinally dependent volcanic dust veil index, and its effect on climate simulations. J Volcanol Geotherm Res 11:67–80

Rozanski K, Araguas-Araguas L, Conginatini R (1993) Isotopic patterns of modern global precipitation. In: Climate change in continental isotopic records. Geophysics Monograph, vol. 78. American Geophysical Union, Washington, DC, pp 1–36

Rutford RH, Craddock C, Bastien TW (1968) Late Tertiary glaciation and sea-level changes in Antarctica. Paleo Paleo Paleo 5:15–39

Scherer P et al (1997) Allan Hills 88019: an Antarctic H-chondrite with a very long terrestrial age. Meteoritics Planet Sci 32:769–773

Schwerdtfeger P (1979) On icebergs and their uses. Cold Regions Science and Technology, vol. 1. Hanover, NH, pp 60–79

Score R, Lindstrom MM (1990) Guide to the U.S. collection of Antarctic meteorites 1976–1988; Everything you wanted to know about the meteorite collection (so you don't have to ask).

Antarctic Met Newslett 13(1):1–135. NASA; Solar System Exploration Division, Planetary Science Branch, Publication 82, Lyndon B. Johnson Space Center, Houston, TX

Self S, Rampino MR, Barbera JJ (1981) The possible effects of large 19th and 20th century volcanic eruptions on zonal and hemispheric surface temperatures. J Volcanol Geothem Res 11:41–60

Shaw GE (1989) Aerosol transport from sources to ice sheets. In: Oeschger H, Langway CC Jr (eds) The environmental record in glaciers and ice sheets. Wiley-Interscience, Chichester

Siegenthaler U, Stocker TF, Monnin E, Lüthi D, Schwander J, Stauffer B, Raynaud D, Barnola J-M, Fischer H, Masson-Delmotte V, Jouzel J (2005) Stable carbon cycle-climate relationship during the late Pleistocene. Science 310:1313–1317

Sipiera PP, Landis CA (1986) A preliminary report of a possible stromatolite find from the Elephant Moraine, Antarctica: a potential directional indicator for ice movement. Lunar Planet Inst Tech Rept 86–01:101–103

Smith GJ (1965) Evaporite salts from the dry valleys of Victoria Land, Antarctica. New Zealand J Geol Geophys 8(2):381–382

Souchez RA, Lemmens M (1985) Subglacial carbonate deposition: an isotopic study of a present day case. Palaeo Palaeo Palaeo 51:357–364

Spahni R, Chappellaz J, Stocker TF, Loulergue L, Hausammann G, Kawamura K, Flückiger J, Schwander J, Raynaud D, Masson-Delmotte V, Jouzel J (2005) Atmospheric methane and nitrous oxide of the late Pleistocene form Antarctic ice cores. Science 310:1317–1321

Starinsky A, Katz A (2003) The formation of natural cryogenic brines. Geochim Cosmochim Acta 67(8):1475–1484

Stonehouse B (ed) (2002) Encyclopedia of Antarctica and the southern ocean. Wiley, Chichester

Stump E, Borg SG, Sheridan MR (1990) Sheridan bluff. In: LeMasurier WE, Thomson JW (eds) Volcanoes of the Antarctic plate and southern oceans. Antarctic Research Series, vol. 48. American Geophysical Union, Washington, DC, pp 136–137

Sugden DE, Marchant DR, Denton GH (1993) The case for a stable East Antarctic ice sheet: the background. Geografiska Annaler 75A(4):151–154

Swithinbank CWM (1969) Giant icebergs in Weddell Sea, 1967–1968. Pol Rec 14(91):477–478

Taylor KS (1986) Lithologies and distribution of clasts in the Elephant Moraine, Allan Hills, southern Victoria Land, Antarctica. M.Sc. Thesis, Kent State University, Kent, OH

Tessier A, Turner DR (1995) Metal speciation and bioavailability in aquatic systems. Wiley, Chichester

Thompson EM (Mosley-Thompson, E) (1979) 911-years of microparticle deposition at the South Pole: a climatic interpretation. Ph.D. Dissertation, The Ohio State University, Columbus, OH

Thompson LG (1977a) Microparticles, ice sheets and climate. Institute of Polar Studies, Report No. 64. The Ohio State University, Columbus, OH

Thompson LG (1977b) Variations in microparticle concentration, size distribution, and elemental composition found in Camp Century, Greenland, and Byrd Station, Antarctica, deep ice cores, Proceedings of Grenoble Symposium, 1975. IAHS-AISH Pub. No. 118:351–364

Thompson LG, Mosley-Thompson E (1981) Temporal variability of microparticle properties in polar ice sheets. J Volcanol Geotherm Res 11:11–27

Thompson LG, Hamilton WL, Bull C (1975) Climatological implications of microparticle concentrations in the ice core from Byrd Station, Western Antarctica. J Glaciol 14:433–444

Thompson LG, Mosley-Thompson E, Petit JR (1981) Glaciological and climatological interpretation of microparticle concentrations from the French 905-meter Dome C, Antarctica, core. In: Sea level, ice, and climate. IAHS Pub. No. 131:227–234

Thompson, S.L. and S.H. Schneider, 1981. Carbon dioxide and climate: ice and ocean. Nature, 290:9–10

Thornton I (ed) (1983) Applied environmental geochemistry. Academic Press, London

Tomblin JF (1979) The South Sandwich Islands II. The Geology of Candlemas Island. British Antarct. Survey, Science Report 92:1–33

Vail PR, Hardenbol J (1979) Sea-level changes during the Tertiary. Oceanus 22:71–79

Walker GPL (1981) Generations and dispersal of fine ash and dust by volcanic eruptions. J Volcanol Geotherm Res 11:81–92

Watanabe O et al (2003) Homogeneous climate variability across East Antarctica over the past three glacial cycles. Nature 422:509–512

Wayt Gibbs W (2001) Out in the cold. Scient Amer 284(3):16–17

Webb P-N (1990) The Cenozoic history of Antarctica and its global impact. Antarctic Sci 2:3–21

Webb P-N, Harwood DM, McKelvey BC, Mercer JH, Stott LD (1983) Late Neogene and older Cenozoic microfossils in high elevation deposits of the Transantarctic Mountains: Evidence for marine sedimentation and ice volume variation on the East Antarctic cratons. Antarctic J US 18(5):96–97

Webb P-N, Harwood DM, McKelvey BC, Mercer JH, Stott LD (1984) Cenozoic marine sedimentation and ice-volume variation on the East Antarctic craton. Geology 12(5):287–291

Weeks WF, Campbell WJ (1973) Icebergs as a freshwater source: an appraisal. J Glaciol 12:207–233

Weeks WF, Mellor M (1978) Some elements of iceberg technology. In: Husseiny AA (ed) Iceberg utilization. Pergamon Press, Oxford, pp 45–98

Wellman P, Tingey RJ (1981) Glaciation, erosion and uplift over part of East Antarctica. Nature 192:142–143

Wilch TI, Denton GH, Lux DR, McIntosh WC (1993) Limited Pliocene glacier extent and surface uplift in middle Taylor Valley, Antarctica. Geografiska Annaler 75A(4):331–351

Wörner G, Viereck L, Hertogen J, Niephaus H (1989) The Mt. Melbourne volcanic field. (Victoria Land, Antarctica) II: geochemistry and magma genesis. In: Damaske D, Durbaum HJ (eds) GANOVEX IV, 1984/85. Geologisches Jahrbuch, Reihe B, 38. Bundesanstalt für Geowissenschaften und Rohstoffe, Hannover, Germany

Wolff EW, Peel DA (1985) Closer to a true value for heavy metal concentrations in recent snow by improved contamination control. Anal Glaciol 7:61–69

Wood JR (1975) Thermodynamics of brine-salt equilibria I. The system $NaCl-KCl-CaCO_2-MgCl_2-H_2O$ and $NaCl-MgSO_4-H_2O$ at 25°C. Geochim Cosmochim Acta 39:1147–1163

Young NW, Pourchet M, Kotlyakov VM, Korolev PA, Dyugerov MB (1982) Accumulation distribution in the IAGP area, Antarctica: 90°E. Anal Glaciol 3:333–338

Zhang L, Liu J, Zhou H, Chen Z (1979) Oxygen isotope fractionation in the quartz-water system. Econ Geol 84:1643–1650

Chapter 18
Meteorites on Ice

The ice sheets of East and West Antarctica collect meteorites and other kinds of particles that fall to the Earth from space. Most of the meteorites that impact on the Earth fall into the oceans which cover just over 70% of its surface. Meteorites that land on the continents are destroyed by weathering or are buried and thus may be lost unless they happen to fall in the hot deserts of Africa, South America, and Asia where they may be preserved because of the low rates of weathering in these areas. Meteorites that fall in the zones of accumulation of the ice sheets of Antarctica are buried in the snow and are incorporated into the ice sheet where they are largely protected from weathering. These meteorites are transported toward the margins of the ice sheet where they emerge on its surface in the zone of ablation along the Transantarctic Mountains. Actually, a significant fraction of the meteorites in the East Antarctic ice sheet is lost at sea when the ice that has transported them to the coast breaks off and forms icebergs that melt as they float north in the southern oceans.

Meteorites have great scientific value because most of them are fragments of asteroids which are themselves remnants of larger bodies that formed in the solar system at the time the planets accreted from planetesimals in the protoplanetary disk that formed around the Sun at the time of its formation about 4.6×10^9 years ago. Therefore, meteorites are our principal source of information about the origin of Earth and of the other planets of the solar system.

Meteorites that fall in temperate and tropical climatic zones of the continents are rapidly altered by their exposure to water and to the terrestrial atmosphere. Therefore, only those meteorites that are observed to fall and which are immediately collected and preserved in specialized laboratories are suitable for study. The meteorites that have been entombed in the East Antarctic ice sheet are not "falls" strictly speaking, but they have been protected sufficiently well to be a reliable source of information about their origins, travel through space, and their arrival on the Earth. Meteorites that were collected at the time of their fall comprise only a small fraction of the total number of specimens in the museums of the world, which is why the meteorites collected on the East Antarctic ice sheet are a valuable addition to the available inventory of unaltered rock specimens from "outer space."

When the orbit of an asteroid is disturbed by gravitational interactions with Jupiter or by other causes, it may be broken up by colliding with another asteroid. The resulting fragments that cross the orbit of the Earth eventually collide with it. When a meteoroid enters the atmosphere of the Earth, it is heated by friction until it becomes a white-hot *fireball*. The frictional heating and the mechanical forces that are exerted by the atmosphere may cause the meteoroid to explode before it reaches the surface of the Earth. Therefore, the fall of a meteoroid on the Earth is a spectacular event. The fireball streaks across the sky at a high speed trailing a plume of dark smoke until it breaks up in a series of thunderous explosions. In most cases, the resulting fragments fall harmlessly to the surface of the Earth as a shower of meteorites. Small meteoroids may "burn up" in the atmosphere as streaks of light called *meteors* or "shooting stars" until they completely disintegrate into their constituent atoms and molecules.

18.1 Meteorites in Antarctica

The first meteorite discovered in Antarctica was found by a member of Douglas Mawson's expedition from 1911 to 1914 to the Adélie Coast (67°00′S, 139°00′E) of East Antarctica. The meteorite was collected in Commonwealth Bay about 43 km west of Cape Dennison

G. Faure and T.M. Mensing, *The Transantarctic Mountains: Rocks, Ice, Meteorites and Water*,
DOI 10.1007/978-90-481-9390-5_18, © Springer Science+Business Media B.V. 2010

where Mawson had set up a camp. The meteorite weighed about 1,000 g and was later identified as an "L5 chondrite." It is on display in the South Australian Museum in Adelaide in South Australia. Fifty years later in 1961, two pieces of an iron meteorite (named Lazarev) were found by a team of Russian geologists in the Humboldt Mountains (71°45′S, 011°30′E) along the Princess Astrid Coast of Queen Maud Land where the Soviet Union had established a base called Novalazarevskaya (Ravich and Revnov 1963). In the same year, Arthur Ford and his colleagues of the US Geological Survey collected two pieces of a stony-iron meteorite (pallasite) from the ice at the foot of Mt. Wrather in the Thiel Mountains (Section 8.1; Ford and Tabor 1971). A pallasite meteorite is depicted in Fig. 18.1. Three years later, in 1964, geologists of the US Geological Survey picked up another iron meteorite in the Neptune Range of the Pensacola Mountains (Section 8.2; Duke 1965). In other words, only five meteorite specimens were collected in Antarctica during a period of 53 years between 1911 and 1964. This meager harvest did not give any indication that the ice fields of Antarctica would later yield more than 25,000 meteorite specimens that have provided a great deal of information about the origin and subsequent evolution

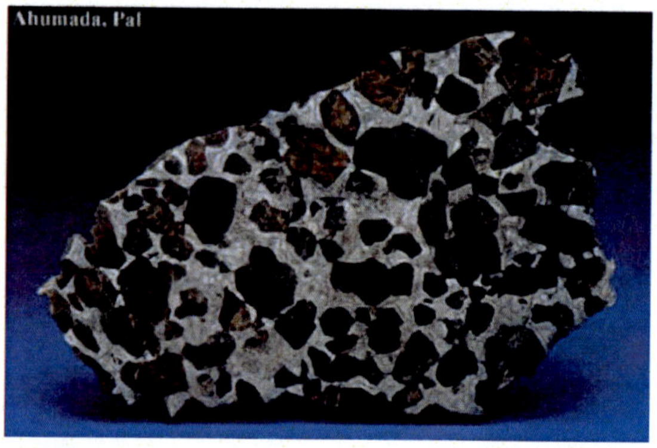

Fig. 18.1 The third meteorite to be collected in Antarctica was a pallasite that was recovered by A.B. Ford on December 7, 1961, in the Thiel Mountains (Ford and Tabor 1971). Pallasites are composed primarily of crystals of olivine with minor low-calcium pyroxene, chromite, and phosphates in a matrix of metallic iron-nickel alloy. The pallasite in this picture, known as Ahumada, illustrates the composition and texture of the Thiel Mountains pallasite although it was not recovered in Antarctica (Reproduced by permission of A.N. Krot from Krot et al. (2005, Fig. 25b, p. 114) and Elsevier, Inc. through Copyright Clearance Center, Inc. 222 Rosewood Drive, Danvers, MA 01932)

of the solar system (Cassidy 1991, 2003; Harvey 2003; Koeberl 1988a; Graham and Annexstad 1989).

The modern era of meteorite collection in Antarctica began during the 1969/70 field season when a team of Japanese glaciologists traveled from Syowa, the Japanese base at Lűtzow-Holm Bay (69°10′S, 037°30′W) on the coast of East Antarctica, to the Yamato Mountains located about 400 km from that base. While working on the large bare-ice area adjacent to the Yamato Mountains, Renji Naruse, one of the glaciologists, picked up a boulder that was lying on the ice far from any rock outcrops and correctly identified it as a meteorite. Therefore, Naruse and his companions spent the rest of that day searching an area of about 50 km^2 and found eight additional meteorite specimens. However, they did not attach much significance to their discovery and did not continue the search.

The nine meteorite specimens that were collected in the relatively small area of bare ice in the Yamato Mountains differed from each other and therefore represented nine different falls. When William A. Cassidy of the University of Pittsburgh became aware of this fact in 1973 during the annual meeting of the Meteoritical Society in Davos, Switzerland, it occurred to him that the meteorite specimens must have fallen in a much larger area and were subsequently concentrated by the flow of the ice sheet. In that case, Antarctica might be a good place to collect meteorites, which presumably occur not only on the bare-ice fields of the Yamato Mountains but also in other parts of the East Antarctic ice sheet where similar glaciological conditions occur. Although it was by no means easy to convince the skeptics, Bill Cassidy eventually received a grant from the National Science Foundation and, in 1976/77, started searching for meteorites on the ice sheet adjacent to the Transantarctic Mountains of southern Victoria Land. He was fortunate because Keizo Yanai, a Japanese meteoriticist who was a member of Cassidy's team, found two meteorite specimens only 20 min after the group had landed in a helicopter on a small bare-ice field at the edge of the polar plateau. However, Cassidy and his team did not find another meteorite for 6 weeks (Cassidy 2003).

Starting with that inauspicious beginning, the Antarctic Search for Meteorites (ANSMET) has been spectacularly successful and has recovered well over 10,000 specimens or about 350 per year on average. Bill Cassidy led ANSMET until 1994 when he turned the operation over to Ralph P. Harvey of Case Western

Reserve University in Cleveland who, in collaboration with John W. Schutt, has continued to collect large numbers of meteorite specimens on the East Antarctic ice sheet adjacent to the Transantarctic Mountains.

The meteorite specimens that have been collected annually by the ANSMET scientists are curated at the Johnson Space Center in Houston, Texas, where they are assigned official designations such as ALHA 81005. The letters "ALH" refer to the Allan Hills, the letter "A" indicates that the specimen was collected by ANSMET, "81" refers to the 1981/82 field season, and "005" tells us that it was the fifth Antarctic meteorite from that field season examined in Houston. The letter "A" was dropped after the 1981/82 fieldseason. The bare-ice (or blue-ice) areas adjacent to the Transantarctic Mountains are identified by the letter codes listed in Appendix 18.12.1. The origin of the blue-ice areas was discussed at a workshop in 1988 chaired by Cassidy and Whillans (1990) who referred to them as "stranding surfaces" (Faure 1990).

All of the meteorite specimens collected in Antarctica are listed in the Antarctic Meteorite Newsletters issued twice each year by the Meteorite Working Group at the Johnson Space Center in Houston, Texas. In addition, selected specimens are classified based on petrographic descriptions of thinsections. This information is also published in The Meteoritical Bulletins that are combined with the annual supplements of the journal Meteoritics and Planetary Science. The Meteoritical Bulletins list and describe meteorites recovered from all regions of the Earth including Antarctica.

The meteorites recovered by Japanese scientists from 1966 to 1994 were listed by Yanai and Kojima (1995) who also provided chemical analyses of a large number of specimens that originated from the Yamato Mountains, Asuka Station, Sör Rondane Mountains, Allan Hills, Bates Nunatak, Derrick Peak, and Mt. Baldr. Three of the specimens listed by Yanai and Kojima (1995) are lunar meteorites: Asuka-881757, Yamato-793274, and Yamato-793169. The compilation includes color photographs of 12 specimens of different kinds of stony meteorites including A-881757 from the Moon.

Antarctic meteorites were also described in the publications of Marvin and Mason (1980, 1982, 1984), Marvin and MacPherson (1989), Score et al. (1981, 1982), Score and Lindstrom (1990), and in the meteorite catalog of Grady (2000). In addition, Antarctic meteorites have been the subjects of numerous reports arising from special symposia and workshops including those of Bull and Lipschutz (1982), Annexstad et al. (1986), Cassidy and Whillans (1990), and Koeberl and Cassidy (1990, 1991). Antarctic meteorites were also included in the reviews by Lipschutz and Schultz (1999, 2006) and Harvey (2003). Cassidy (2003) provided an insightful and humorous behind-the-scenes account of his adventures on the polar plateau of East Antarctica including a great deal of information about the meteorites that he and his associates of ANSMET collected there. See also Delisle (1983, 1997) and Denton and Hughes (1981).

18.2 Classification of Meteorites

Meteorites have been collected for several centuries and about 2,000 specimens had accumulated in museums of the world prior to 1970 when meteorites were discovered in Antarctica. The meteorites in these museums include three different types: stones, irons, and stony irons. Each of these types has been subdivided into different groups depending on the minerals of which they are composed, their chemical compositions, the inclusions they contain, their textures, and the abundance of volatile matter. An abbreviated version of the classification of meteorites is contained in Table 18.1 (see also Davis 2005; Krot et al. 2005; Hartmann 2005; Lipschutz and Schultz 1999; McSween 1999; Taylor 1992).

Stony meteorites are most abundant and make up 94.2% by number of the so-called "falls" (i.e., meteorites that were collected soon after they fell). Irons are rare at only 4.5% and stony irons are even less common at 1.2%. In the modern classification in Table 18.1 all meteorites (falls only) are subdivided into:

Chondrites
Nonchondrites (differentiated)
Nonchondrites (undifferentiated)

The *chondrites* include the "ordinary" and the "carbonaceous" varieties of stony meteorites each of which is further split into different groups. Most chondrites contain *chondrules* which are spherical particles about 1 mm or less in diameter composed of silicate minerals such as olivine, pyroxene, plagioclase, as well as troilite, spinel, and metallic iron (Hewins 1997). The chondrules formed in the solar nebula by accretion of dust particles which were subsequently melted and

Table 18.1 Classification of meteorites based on falls only (Hartmann 2005; Krot et al. 2005)

Classes	Varieties	Groups
Chondrites	Carbonaceous	CI, CII, CIII
85.3%	4.7%	Enstatite (E)
		Bronzite (H)
	Ordinary	Hypersthene (L)
	80.6%	Amphoterite (LL)
		Others
Nonchondrites	Achondrites	Angrites
(differentiated)	8.9%	Aubrites
		Brachinites
		Ureilites
		HED meteorites:
		(howardites, eucrites, and diogenites)
		Others
	Martian	Shergottites
	0.0%	Nakhlites
		Chassignites
		Orthopyroxenites
	Lunar	Mare basalts
	0.0%	Highland anorthosites
		Regolith breccias
	Stony irons	Mesosiderites
	1.2%	Pallasites
	Irons	Octahedrites
	4.5%	Hexahedrites
		Ataxites
		Others
Nonchondrites		Acapulcoites
(undifferentiated, primitive)		Lodranites
		Winonaites
		Silicate inclusions
Totals	99.9%	Irons IAB & IIICD

The so-called stony meteorites include chondrites (carbonaceous and ordinary), as well as achondrites and the rare undifferentiated nonchondrites. Stony meteorites make up 94.2% of all falls, whereas irons (4.5%) and stony irons (1.2%) make up the rest. The undifferentiated nonchondrites and the martian and lunar meteorites are rare and have never been observed to fall.

then cooled quickly to form silicate glass "beads". The chondrules were later incorporated into the parent bodies of meteorites and therefore are among the oldest objects in the solar system (Whipple 1966; Hewins et al. 1996; Connolly and Desch 2004; Dodd 1971).

The differentiated *nonchondrites* in Table 18.1 consist primarily of *achondrites* which are stony meteorites that lack chondrules and contain evidence of magmatic differentiation, including the segregation of liquid iron and nickel which are insoluble in silicate melts. The metallic liquid presumably sank under the influence of gravity to form the cores of the meteorite parent-bodies. When the parent bodies were later broken up by collisions with other planetesimals, the fragments of the cores continued to orbit the Sun until they accidentally crossed the orbit of the Earth and were captured by its gravitational field. The resulting iron and stony iron meteorites make up, respectively, 4.5% and 1.2% by number of all falls, whereas the complementary achondrites constitute 8.9% by number.

The differentiated Nonchondrites in Table 18.1 also contain meteorites that originated from Mars and the Moon. These meteorites were ejected as a result of explosive impacts of solid objects (i.e., small asteroids or comets) and went into orbit around the Sun until they were captured by the gravitational field of the Earth. A significant number of the meteorites that have been recovered in Antarctica and in the hot deserts of the Earth originated from Mars, the Moon, and perhaps even from

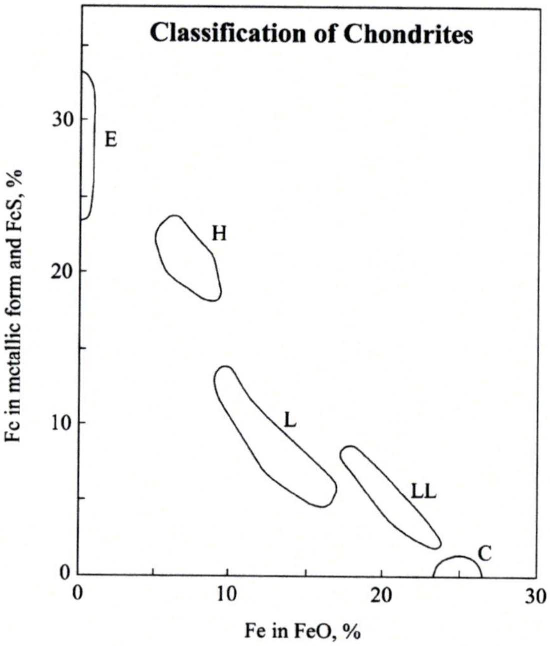

Fig. 18.2 The classification of chondrite meteorites is based on the distribution of iron between FeO and Fe + FeS combined with the fact that different classes of chondrites contain about the same amount of iron. This diagram demonstrates that as the concentration of oxidized iron (FeO) increases, the concentration of reduced forms of iron (Fe + FeS) decreases. E = enstatite chondrites, H = bronzite chondrites, L = hypersthene chondrites, LL = amphoterite (olivine-pigeonite) chondrites, C = carbonaceous chondrites (Adapted from Fig. 29 of Mason 1962, for falls only)

the planets Mercury or Venus, although no mercurian or venusian meteorites have yet been identified.

The undifferentiated nonchondrites are a rare class of stony meteorites in Table 18.1 that were never heated sufficiently to melt and therefore are composed of grains of silicate and oxide minerals that are not in chemical equilibrium with each other.

Each of the meteorites varieties in Table 18.1 consists of different groups of objects based on their bulk chemical and mineral compositions, textures, and other criteria. For example, the ordinary chondrites, which are the most abundant stony meteorites, are subdivided in Fig. 18.2 into groups identified by letter: E, H, L, and LL. The average chemical compositions of meteorites in these groups are listed in Table 18.2 together with the average chemical compositions of three groups of carbonaceous chondrites which comprise only 4.7% by number among all meteorite falls.

The *carbonaceous chondrites* are especially interesting because they contain organic matter of abiogenic origin. In other words, the organic matter they contain consists of complex molecules of carbon that is bonded to hydrogen, nitrogen, sulfur, phosphorus, and other elements. These organic compounds formed by chemical reactions in the solar nebula rather than as a result of biological metabolism of living organisms. The carbonaceous chondrites are designated by the letter C as in Table 18.2 and Fig. 18.2.

Table 18.2 Chemical compositions of different classes of ordinary and carbonaceous chondrites in weight percent (Lipschutz and Schultz 1999, Table III)

	Ordinary chondrites				Carbonaceous chondrites		
Component	E	H	L	LL	C1	C2M	C3V
SiO_2	38.47	36.60	39.72	40.60	22.69	28.97	34.00
TiO_2	0.12	0.12	0.12	0.13	0.07	0.13	0.16
Al_2O_3	1.78	2.14	2.25	2.24	1.70	2.17	3.22
Cr_2O_3	0.23	0.52	0.53	0.54	0.32	0.43	0.50
Fe_2O_3	–	–	–	–	13.55	–	–
FeO	0.23	10.30	14.46	17.39	4.63	22.14	26.83
MnO	0.02	0.31	0.34	0.35	0.21	0.25	0.19
MgO	21.63	23.26	24.73	25.22	15.87	19.88	24.58
CaO	1.03	1.74	1.85	1.92	1.36	1.89	2.62
Na_2O	0.64	0.86	0.95	0.95	0.76	0.43	0.49
K_2O	0.16	0.09	0.11	0.10	0.06	0.06	0.05
P_2O_5	Trace	0.27	0.22	0.22	0.22	0.24	0.25
H_2O+	0.34	0.32	0.37	0.51	10.80	8.73	0.15
H_2O-	?	0.12	0.09	0.20	6.10	1.67	0.10
Fe(m)	23.70	15.98	7.03	2.44	–	0.14	0.16
Ni(m)	1.78	1.74	1.24	1.07	–	–	0.29
Co(m)	0.12	0.08	0.06	0.05	–	–	0.01
FeS	8.09	5.43	5.76	5.79	9.08	5.76	4.05
C	0.32	0.11	0.12	0.22	2.80	1.82	0.43
S(el.)	–	–	–	–	0.10	–	–
NiO	0.11	–	–	–	1.33	1.71	–
CoO	–	–	–	–	0.08	0.08	–
NiS	–	–	–	–	–	–	1.72
CoS	–	–	–	–	–	–	0.08
SO_3	–	–	–	–	5.63	1.59	–
CO_2	–	–	–	–	1.50	0.78	–
Sum	98.77	99.99	99.99	99.92	98.86	99.82	99.84
Total Fe	29.01	27.45	21.93	19.63	18.85	21.64	23.60
$(Ni/Fe)_m$	0.075	0.108	0.176	0.438	–	–	1.812

(–) = not reported and assumed to be zero; H_2O- is loosely bound (adsorbed) water released at less than 110°C; H_2O+ is chemically bound water released only at T > 110°C. Total Fe includes all forms of reduced and oxidized Fe recalculated to the metallic state; m = metallic, el.= elementary, E = enstatite chondrites, Mason (1962, Table 6), H = bronzite chondrites, L = hypersthene chondrites, LL = pigeonite (amphoterite) chondrites, C1, C2, C3 = carbonaceous chondrites. C1= CI (Ivuna-like), C2M (Mighei-like), C3V (Vigarano-like) (see also Krot et al. 2005).

Another way to subdivide the chondrites is on the basis of the increasing extent of metamorphic equilibration of minerals, which is manifested primarily by the blurring of the boundaries of chondrules, by the presence or absence of glass, and by evidence for recrystallization of feldspar. The increasing metamorphic grade is expressed by numbers from 1 (unmetamorphosed) to 6 (most highly metamorphosed) (Van Schmus and Wood 1967). The mineralogical and metamorphic criteria of the classification of chondrites are combined in a rectangular grid in Table 18.3 (Wood 1968; Glass 1982). The entries in this table indicate that all enstatite chondrites (E) are variably metamorphosed but that carbonaceous chondrites are either unmetamorphosed (C1, C2) or only slightly metamorphosed (C3 and C4). The ordinary chondrites of the H, L, and LL groups are all variably recrystallized (grades 4–6), whereas those in metamorphic grade 3 are only slightly metamorphosed and are collectively referred to as the "unequilibrated ordinary chondrites." Table 18.3 also explains that the "L5 chondrite" collected by Douglas Mawson's team in Commonwealth Bay of East Antarctica was a hypersthene chondrite of metamorphic grade 5.

The concentrations of volatile compounds (e.g., H_2O^+, C, CO_2) of the carbonaceous chondrites in Table 18.3 decrease progressively from group C1 to group C4, which suggests that they lost increasing amounts of these compounds by heating. Consequently, the extent of alteration of the carbonaceous chondrites increases in the sequence indicated by their placement in Table 18.3. In other words, the C3 carbonaceous chondrites are more altered than those of C1. However, the carbonaceous chondrites are generally less altered than the ordinary chondrites in Groups E, H, L, and LL.

Table 18.3 Classification grid for chondrite meteorites based on mineralogical (chemical) criteria and metamorphic grade (Adapted from Glass 1982)

			Metamorphic grade			
	Not	Not	Slightly ⟶	Highest		
	1	2	3	4	5	6
E			E3	E4	E5	E6
C	C1	C2	C3	C4		
H			Unequilibrated	H4	H5	H6
L			Ordinary	L4	L5	L6
LL	Chondrites	LL4	LL5	LL6		

E = enstatite, C = carbonaceous, H = bronzite, L = hypersthene, LL = amphoterite (olivine-pigeonite) chondrites.

Most ordinary and carbonaceous chondrites are breccias because they contain angular fragments of other kinds of meteorites. The brecciation of chondrites indicates that they are products of collisions among asteroids of different mineral compositions. The mixed fragments of such collisions recombined under the influence of gravity to form a new body which could later collide with another asteroid to form objects that are composed of polymict breccias. The shock waves that were generated within two colliding asteroids caused a wide range of damage to the crystals of the common rock-forming minerals of which chondrites are composed. The resulting shock metamorphism of olivine, plagioclase, and orthopyroxene provides yet another set of properties that are used to describe and classify the chondrites.

The iron meteorites (differentiated nonchondrites in Table 18.1) are subdivided into 13 chemical groups that are identified by Roman numerals (I to IV) combined with capital letters (A, B, C, D, and E). In some cases, two letters are necessary to identify certain groups (e.g., IAB and IIICD). A significant number (i.e., 15%) of the known iron meteorites do not fit into any of the 13 defined groups. These are referred to as the *ungrouped* iron meteorites. If each group and each of the ungrouped iron meteorites originated from a separate parent body, then 20–30 such parent bodies may have existed in the asteroid belt (Scott 1979; Wasson 1990). Many iron meteorites have a characteristic texture shown in Fig. 18.3, called the Widmanstätten pattern, that consists of an intergrowth of *kamacite* lamellae (α-FeNi, <6% Ni) where the interstitial spaces are occupied by several high-Ni minerals including *taenite* (FeNi) and *awaruite* ($FeNi_3$). The width of the kamacite lamellae increases with decreasing cooling rate and varies widely among different groups of iron meteorites. The cooling rates can be used to infer the radii of the parent bodies from which iron meteorites of specific groups were derived. An alternative classification of iron meteorites based on structural criteria is presented in Table 18.4.

18.3 Antarctic Meteorites

The meteorite specimens collected by ANSMET and by other investigators include most of the groups of stony, stony-iron, and iron meteorites listed in Table 18.1.

Fig. 18.3 Iron meteorites have a characteristic texture which appears on polished and etched surfaces. This texture, called the Widmanstätten pattern, formed during slow cooling and crystallization of liquid iron-nickel in the core of a meteorite parent body. The tear-shaped inclusion is composed of iron sulfide which is immiscible in iron-nickel liquid. The meteorite in this image is Carbo, a IID iron meteorite, which was not collected in Antarctica (Reproduced by permission of H. Haack from Haack and McCoy (2005, Fig. 8, p. 337) and Elsevier, Inc. through the Copyright Clearance Center, Inc., 222 Rosewood Drive, Danvers, MA 01932)

Table 18.4 Structural classification of iron meteorites (Mason 1962)

Group	Minerals	Ni (%)	Widmanstätten
Hexahedrites[a]	Kamacite,[a] schreibersite, troilite, daubreelite, graphite	4–6	No
Ataxites, Ni-poor	Same as hexahedrites		No
Octahedrites	Kamacite, taenite,[b] awaruite, plessite,[c] schreibersite, troilite, cohenite (Fe_3C), graphite (diamond)	>6	Yes
Ataxites, Ni-rich	Plessite, taenite	>12	No

[a]Kamacite, αFe-Ni alloy, 5.5% Ni, body-centered cubic.
[b]Taenite, γFe-Ni alloy, 27–65% Ni, face-centered cubic.
[c]Plessite, eutectic intergrowth of kamacite and taenite.

The specimens also range widely in mass from fractions of 1 g up to 407 kg (ALH76009). All of the specimens are fragments of more massive meteoroids that exploded during their passage through the atmosphere. Therefore, small fragments collected in any given area are more numerous than large ones and low-mass specimens collected at a particular site in Antarctica outnumber those having masses greater than 1.0 kg.

Meteorite specimens are found primarily on the bare-ice fields where the East Antarctic ice sheet is actively ablating by sublimation rather than by melting. Some of the most productive collecting sites (e.g., the ice fields that surround the Elephant Moraine) are far removed from rock outcrops that could shed boulders of terrestrial rocks onto the adjacent ice. Nevertheless, some boulders of terrestrial rocks may be brought to the surface of the ice sheet by the internal deformation of the ice that is forced to flow over mountainous bedrock topography at its base (Section 17.3). Consequently, meteorites and terrestrial rocks may occur together on the same ice field and may not always be easy to distinguish in the field.

18.3.1 Physical Properties of Meteorites

By far the largest number of meteorite specimens that have been collected in Antarctica are ordinary chondrites of various compositions and metamorphic grades (Tables 18.1, 18.2, and 18.3). They are generally characterized by a fusion crust like the specimen in Fig. 18.4 that may range in color from brown (ordinary chondrites) to black (carbonaceous chondrites). Some specimens lack a fusion crust, either because it may have been removed by weathering or because such specimens are interior fragments of the original meteoroid. In some cases, the fusion crust is inconspicuous because it is light-colored.

The surfaces of freshly fallen meteorites contain evidence of aerodynamic sculpturing during their passage through the atmosphere and of selective ablation of inclusions such as troilite nodules that may leave cavities on the surface of the specimen. Most stony meteorite specimens have a smooth surface that may be modified in some cases by being nodular, striated, ribbed, porous, or even scoriaceous (Mason 1962, p. 44–45). The surfaces of many iron meteorites contain shallow depressions resembling thumb prints

Fig. 18.4 The Asuka-87251 stony meteorite is an LL5 chondrite that was collected by Japanese investigators in 1987 on the blue-ice fields near Asuka Station in the Sőr Rondane Mountains (72°S, 26°E) of Queen Maud Land south of the Princess Ragnhild Coast. This large stone weighs 46 kg which makes it the most massive meteorite specimen in the Asuka collection. The specimen has a well developed dark grey fusion crust and a nearly spherical shape resulting from ablation during its passage through the atmosphere of the Earth (Reproduced from Yanai and Kojima (1995, photo 6, p. 10) by permission of the Library Director of the National Institute of Polar Research in Tokyo, Japan)

called *regmaglypts*. However, iron meteorites may also have hackly surfaces like shrapnel in cases where the meteoroid had already begun to cool when it exploded in the atmosphere. Iron meteorites are relatively easy to identify in the field because of their high density (about 8 g/cm^3).

The densities of ordinary chondrites (3.3–3.9 g/cm^3) are higher than those of most terrestrial rocks because they contain grains of metallic iron (Mason 1962, p. 89–95). The densities of carbonaceous chondrites range widely depending on the concentration of volatile compounds they contain. Type C1 carbonaceous chondrites (Table 18.3) have low densities (2.2 g/cm^3 or less) and attract attention because they are black in color, may be friable, and lack metallic grains. However, they are strongly magnetic, presumably because of the presence of small grains of Fe-Ni spinel (Mason 1962, p. 96). The densities of C2 carbonaceous chondrites range from 2.5 to 2.9 g/cm^3, they are only weakly magnetic, and they contain less volatile material than C1 carbonaceous chondrites. The concentrations of volatile compounds of C3 carbonaceous chondrites are still lower than those of the C1 and C2 types and they do contain metallic Fe-Ni grains. Accordingly, the C3 carbonaceous chondrites have high densities between 3.4 and 3.5 g/cm^3 similar to the densities of some ordinary chondrites.

The presence of a fusion crust, the color and shape of the surface, the presence of metallic grains, and the unusual densities, in most cases, permit meteorite specimens to be distinguished from terrestrial boulders. These criteria apply to meteorite specimens that occur not only in the cold desert of Antarctica, but also to those that are collected in the hot deserts of the world, and in all other places where meteorites can be found. Even though tens of thousands of meteorite specimens have been collected on the East Antarctic ice sheet, Antarctica is not receiving a higher flux of meteorites compared to other areas of the Earth. The apparent abundance of meteorites is caused primarily by their better preservation in the ice and by the dynamics of the East Antarctic ice sheet.

The cardinal rule of meteorite collecting in Antarctica and elsewhere is: *When in doubt, collect it and let the experts decide whether it is a meteorite.* However, the scientific value of meteorites can be compromised if they are not collected properly. Therefore, ANSMET has recommended a simple procedure for collection meteorites in Antarctica and for documenting the collecting site:

1. Photograph the meteorite in place before collecting it.
2. Place the specimen into a clean plastic bag by using a clean knife or scissors without touching it with bare or gloved hands.
3. Identify the outside of the sample bag by a number and mark the collecting site and sample number on a map or aerial photograph.
4. Search the area for additional specimens because meteorites occur in clusters.
5. Place additional specimens in separate bags and give each a separate number. Make a sketch of the distribution of specimens in terms of distance and direction relative to the first find using the pace-and-compass method or GPS if available.
6. Keep the meteorites frozen and notify the Representative of the National Science Foundation in McMurdo who will arrange their transfer to ANSMET.
7. All meteorites collected by American investigators in Antarctica are shipped to the Meteorite Curation

Facility at the Johnson Space Center in Houston, Texas (French 1983).

8. Field parties that collect meteorite specimens are officially recognized in NASA publications. However, meteorites are not to be treated as personal souvenirs (Schmitt 2002).

One of the scientific benefits we have already derived from the study of Antarctic meteorites is that a significant number of the specimens that have been collected there are "ungrouped," which means that they do not fit into the arbitrary classification scheme that has been devised (Table 18.1). The ungrouped meteorite specimens expand the classification and provide new and additional insight into the origin and evolution of the solar system.

18.3.2 Transport and Exposure

The transport and exposure of meteorite specimens by the ice sheet adjacent to the Transantarctic Mountains was elegantly explained by Whillans and Cassidy (1983) in a paper entitled: "Catch a falling star: Meteorites and old ice." Their model assumes that meteorite specimens are deposited on the surface of the ice at a constant rate and that the rate of flow of the ice sheet is also constant. These assumptions lead to the conclusion that the meteorite specimens are uniformly distributed within the ice sheet. When the ice encounters the Transantarctic Mountains in Fig. 18.5, it is deflected upward causing the meteorite specimens to emerge at the surface as the ice sublimates. As time passes, meteorites accumulate on the stranding surface because they are released by the ice, because of direct infall to the ablation zones, and because of compression of the ice in the ablation zone which increases the concentration of meteorite specimens entrained in the ice sheet.

The Whillans-Cassidy model is based on an idealized conception of the interaction of the East Antarctic ice sheet with a mountain barrier. It assumes that the ice sheet is 3 km thick and that the bedrock surface over which it flows is relatively smooth. Actually, we know that the bedrock topography under the ice adjacent to the East Antarctic ice sheet consists of flat-topped mesas that are separated in some cases by deep valleys (Sections 17.3 and 17.4). The rugged subglacial bedrock topography causes meteorite stranding surfaces to form in unexpected places where the flow of the ice sheet is disrupted and where ablation is accelerated. We also know that the ice sheet is only about 100 m thick in places where it flows over the tops of the bedrock mesas (Faure and Buchanan 1987; Delisle et al. 1989; Delisle and Sievers 1991, Delisle 1997).

The model of Whillans and Cassidy (1983) in Fig. 18.5 assumes that meteorites that fall in the accumulation area of the East Antarctic ice sheet west of the Transantarctic Mountains, are buried and are then transported by the flow of the ice sheet until they emerge in the ablation zone adjacent to the Transantarctic Mountains. The flow lines of the Whillans-Cassidy model in Fig. 18.5 predict that the age of the ice in the ablation area decreases with increasing distance west of the Transantarctic Mountains. Therefore, the terrestrial ages of meteorite specimens (i.e., the time that has elapsed since the fall of a meteorite) should also decrease with increasing distance from the western slope of the mountains.

The West Antarctic ice sheet undoubtedly also contains meteorites that landed on its surface. However, the glaciological, climatic, and topographic conditions do not permit them to accumulate on blue-ice stranding surfaces.

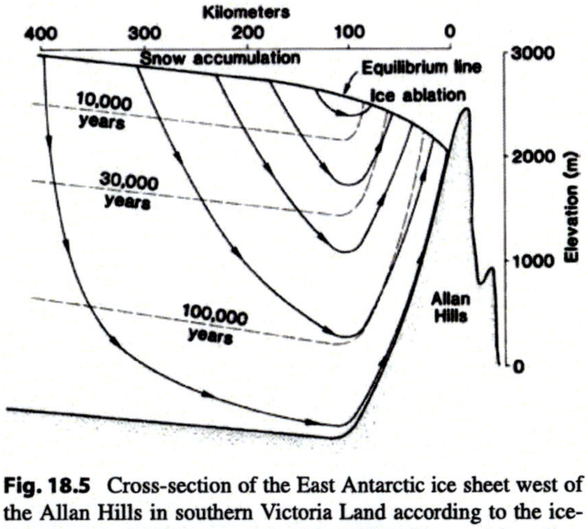

Fig. 18.5 Cross-section of the East Antarctic ice sheet west of the Allan Hills in southern Victoria Land according to the ice-flow model of Whillans and Cassidy (1983). The model predicts that the terrestrial ages of meteorite specimens that are transported by the ice sheet and are released when the ice sublimates in the ablation area should decrease with increasing distance west of the Allan Hills. Adapted from Whillans and Cassidy (1983). The view is to the north with west on the left and east on the right. The vertical exaggeration is 100× (Adapted from Whillans and Cassidy 1983)

18.3.3 Meteorite Collections from Antarctica

Experience has shown that ablation areas that expose bare ice occur in many places on the East Antarctic ice sheet adjacent to the entire length of the Transantarctic Mountains. Many of these areas have yielded a wide variety of meteorite specimens including rocks from the Moon and from Mars listed in Appendices 18.12.3 and 18.12.4. The ongoing search for meteorites by the personnel of ANSMET is one of the most successful programs of the United States Antarctic Program operated by the National Science Foundation. Several summary reports have been published by past and current participants in this enterprise, including especially the book by Cassidy (2003) and reports by Harvey (2003), Cassidy et al. (1992), Cassidy (1991), Cassidy and Harvey (1991), Koeberl (1988a), Marvin and Mason (1980, 1982, 1984), Marvin (1981, 1984, 1989), Marvin and McPherson (1989), Cassidy and Rancitelli (1982), and Cassidy et al. (1977).

By the end of 2006 the number of meteorite collecting sites along the Transantarctic Mountains where ANSMET has recovered meteorites had increased to 50 and continues to increase (Appendix 18.12.1). Meteorites have also been collected by Antarctic research organizations of other countries including Japan, Germany, Italy, China, South Korea, and others. The total number of Antarctic meteorite specimens that were collected by ANSMET up to and including 2006 stood at 15,029 (Weisberg et al. 2009). The Japanese collection of Antarctic meteorites which were derived mainly from the Yamato Mountains (71°30′S, 035°40′E), Asuka Station (72°S, 026°E), Belgica Mountains (72°35′S, 031°15′E), and from some of the ice fields adjacent to the Transantarctic Mountains were listed and described by Yanai and Kojima (1995). The Chinese Antarctic Research Expedition to the Grove Mountains (85°40′S, 175°00′E) in 2003 and 2006 recovered 950 meteorite specimens that were classified and briefly described in Meteoritical Bulletin No. 96 edited by Weisberg et al. (2009).

Most of the meteorite specimens that were recovered from the Grove Mountains in East Antarctica weigh less than 10 grams (g) and only 21 specimens weigh more than 100 g. The resulting mass distribution in Fig. 18.6 is fairly typical of meteorite specimens collected on other blue-ice areas on the East Antarctic ice sheet. Evidently, most of the specimens are fragments of meteoroids that exploded in the atmosphere.

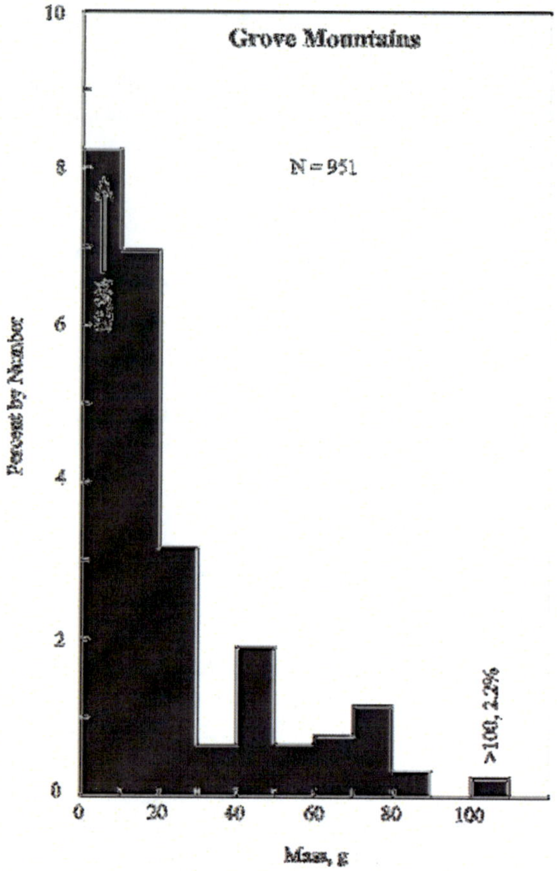

Fig. 18.6 The mass distribution of meteorite fragments recovered by the Chinese Antarctic Research Expedition in 2003 and 2006 in the Grove Mountains, East Antarctica, is weighted in favor of small specimens weighing less than 10 g which make up 82.3% by number of 951 specimens that were collected. Specimens weighing more than 100 g make up only 2.2% of the collection. The most massive specimen weighed 2,812 g. The high abundance of small meteorite specimens collected in the Grove Mountains is typical of meteorite collections elsewhere in East Antarctica (Data from Weisberg et al. 2009)

18.4 Meteorite-Impact Craters

The impact of large meteorites is an important geologic process that continues to affect the surface of the Earth although the frequency of such events has declined significantly since about 3.9 billion years in the past (Dressler et al. 1994; Koeberl and MacLeod 2002; Reimold and Gibson 2006; Faure and Mensing 2007). The shock-metamorphism of the target rocks caused by the explosive impact of meteorites has been illustrated and described by French (1998, 2004). As a rule of thumb, the diameter of a meteorite-impact crater is more than ten times greater than the diameter of the impactor (Melosh 1989). Some of the largest craters on the Earth (e.g., Sudbury, Ontario) have diameters in excess of 200 km, which implies that such large craters formed by the impact of asteroids close to 20 km in diameter. Some of the impact craters that have scarred the surface of the Moon have even larger diameters that exceed 1,000 km (e.g., the Mare Imbrium). Although most of the impact craters on the Earth have been destroyed by erosion or have been buried under sedimentary rocks and continental ice sheets, more than 160 such craters have been recognized on the continents and in the shallow water of the continental shelves (e.g., Meteor Crater, Winslow, Arizona; Shoemaker and Kieffer 1974). Since most of the known impact craters occur on land and since the surface area of the continents (excluding Antarctica) is about 139.1 million square kilometers, the average numerical crater density (d) on the continents of the Earth is:

$$d = \frac{160}{139.1 \times 10^6} \approx 1.2 \text{ craters per million square kilometers.}$$

Given that Antarctica has an area of 14.2 million square kilometers, it should contain about 16 meteorite impact craters like those that have been preserved on the other continents.

Although large meteorites have certainly impacted in Antarctica, only one such crater has been tentatively identified under the ice at about 71°30′S, 140°00′E in Wilkes Land south of the Adélie coast. Another more speculative meteorite impact crater may exist at Butcher Ridge (79°12′S, 155°48′E) in the Cook Mountains of southern Victoria Land.

18.4.1 Wilkes Land Impact Basin

The presence of a meteorite impact crater at 71.5°S and 140°E in Wilkes Land was first suggested by Schmidt (1962) as a potential source of tektites in Australia called *australites* based on the presence of a negative gravity anomaly that was recorded by the Victoria Land Traverse-2 in 1958/59. The presence of a meteorite impact crater in Wilkes Land was supported by Weihaupt (1961, 1976) on the basis of a seismic survey which indicated that the elevation of the bedrock surface at the site of the hypothetical crater is 415 below sea level. However, a reinterpretation of the seismic data by Crary (1963) lowered the elevation to 1,220 m below sea level which at least partially accounts for the low gravity anomaly of the site. Subsequently, Steed and Drewry (1982) carried out airborne radar-echo surveys in 1971/72 and 1974/75 which indicated the presence of a deep subglacial trench at about 70°S and 135°E in Wilkes Land. This trench reaches a depth of 1,120 m below sea level and is a tectonic feature rather than a meteorite impact crater.

The presence of a large subglacial meteorite impact basin at 70°S and 120°E in Wilkes Land was again proposed by von Frese et al. (2009). This site is located almost 800 km west of the hypothetical crater considered by Schmidt (1962) and Weihaupt (1976). The new site was identified from gravity measurements recorded by the GRACE satellite (Gravity Recovery and Climate Experiment) which indicate that a *positive* gravity anomaly occurs over a subglacial topographic basin located at the indicated site. The occurrence of a positive gravity anomaly over a large topographic basin in the crust of East Antarctica is compatible with the effects of a giant meteorite impact crater that is underlain by a plug of ultramafic rocks of the underlying mantle. The diameter of the impact basin in Fig. 18.7 is about 500 km which means that it is much larger than the famous Chicxulub crater (diameter = 170 km) on the Yucatán Peninsula of Mexico.

The Wilkes Land impact basin is antipodal to the Siberian flood basalts which were erupted between 248 and 250 Ma at the end of the Permian Period when nearly 90% of all species on Earth became extinct (see also Graphite Peak, Section 11.21.1). Consequently, the impact of a large meteorite in Wilkes Land may have triggered large-scale melting of the mantle underlying the crust of Siberia. In addition, the combined consequences of the impact of a large meteorite impact followed by the large-scale eruption of basalt lava may have disrupted the environmental conditions and thus caused the profound extinction event at the end of the Permian Period.

18.4.2 Butcher Ridge, Cook Mountains

Butcher Ridge in Fig. 18.8 is located at the very edge of the polar plateau in the Cook Mountains (79°25′S, 158°00′E) north of the Darwin Glacier in the

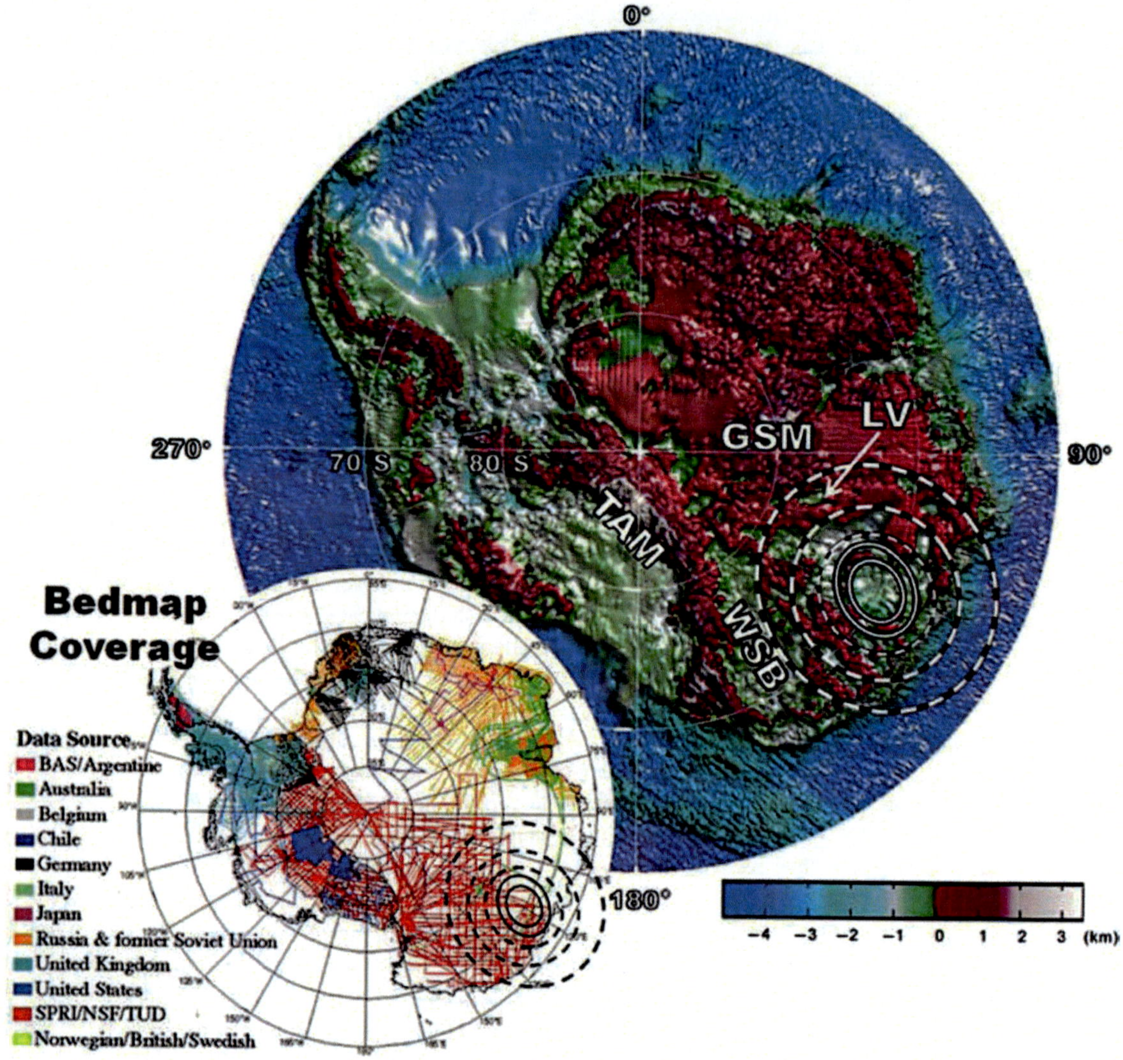

Fig. 18.7 The Wilkes Land subglacial meteorite-impact crater was identified on the basis of gravity data obtained by the GRACE satellite (Gravity Recovery and Climate Experiment). The location of the crater is indicated by a positive free-air gravity anomaly which occurs above a subglacial basin at 70°S and 120°E outlined by the solid white lines which delineate the rim of the crater. The dashed lines outline the hypothetical outer rings of this multi-ring basin. The coincidence of a positive gravity anomaly over a deep crustal basin is consistent with a large meteorite impact crater underlain by a plug of ultramafic rocks of the upper mantle. The Wilkes Land meteorite impact crater is even larger than the Chicxulub impact crater which caused a large-scale extinction event at 65 Ma. TAM = Transantarctic Mountains, GSM = Gamburtsev Subglacial Mountains, WSB = Wilkes Subglacial Basin, LV = Lake Vostok (Reproduced by permission of R.R.B. Von Frese and the American Geophysical Union from Von Frese et al. 2009, Fig. 3)

Transantarctic Mountains of southern Victoria Land. It consists of a semi-circular ridge with a steep east-facing cliff that extends north from Mt. Ayres (2,498 m, 79°20′S and 156°28′E) for about 23 km. The arc of Butcher Ridge in Fig. 18.9 partly encloses the Finger Ridges including Harvey Peak (2,120 m, 79°13′S and 157°01′E).

The geology of this part of the Transantarctic Mountains, described by Grindley and Laird (1969), consists of flat-lying quartz sandstones of the Beacon Supergroup that were intruded by sills of the Ferrar Dolerite. These rock units are underlain by the granitic basement rocks of the Granite Harbor Intrusives that are exposed in the Brown Hills along the coast of the

Fig. 18.8 Aerial view of Butcher Ridge of the Cook Mountains at the edge of the polar plateau. The view is to the south. Butcher Ridge is in the foreground (Photo by G. Faure)

Ross Ice Shelf (Section 3.3). Haskell et al. (1963, 1965) mapped the granitic basement rocks and overlying Beacon rocks in the Brown Hills and Darwin Mountains, but they did not include Butcher Ridge in their study area.

The geology of Butcher Ridge turned out to be unexpectedly complex with a wide range of rock types which Marshak et al. (1981, p. 54) described as:

- Diffusely layered to massive andesite
- Well-layered andesite with glass layers
- Well-layered rhyolite with glass layers
- Massive glass (1-km long)
- Rafts of sedimentary rocks

The same authors considered these rocks to be part of a hypabyssal intrusion which they assigned to the Ferrar Group even though these rocks:

> ...are quite unlike those of all known exposures of Ferrar Dolerite and may represent a previously undescribed type of igneous body.

Marshak et al. (1981) published photographs of well-layered rhyolite glass and of a large glass body. All of the rocks on Butcher Ridge are structurally contorted into noncylindrical disharmonic folds with amplitudes of about 20 m. The rocks are also pervasively layered with up to 100 distinct layers in several exposures. In some cases, crystalline layers are interbedded with black glass. Although Marshak et al. (1981) used the terminology of volcanic rocks to describe the geology of Butcher Ridge, there is no evidence that the layers are lava flows.

The authors concluded:

> We believe that the Butcher Ridge Igneous Complex is unique. It may represent mixing in a hypabyssal setting of two magmas, one silicic, derived by melting of crustal rocks, and one basaltic (precursor of Ferrar Dolerite) derived from the mantle.

Schellhorn (1982) added detailed petrographic descriptions of the principal rock types and a large number of chemical analyses (both major and trace elements),

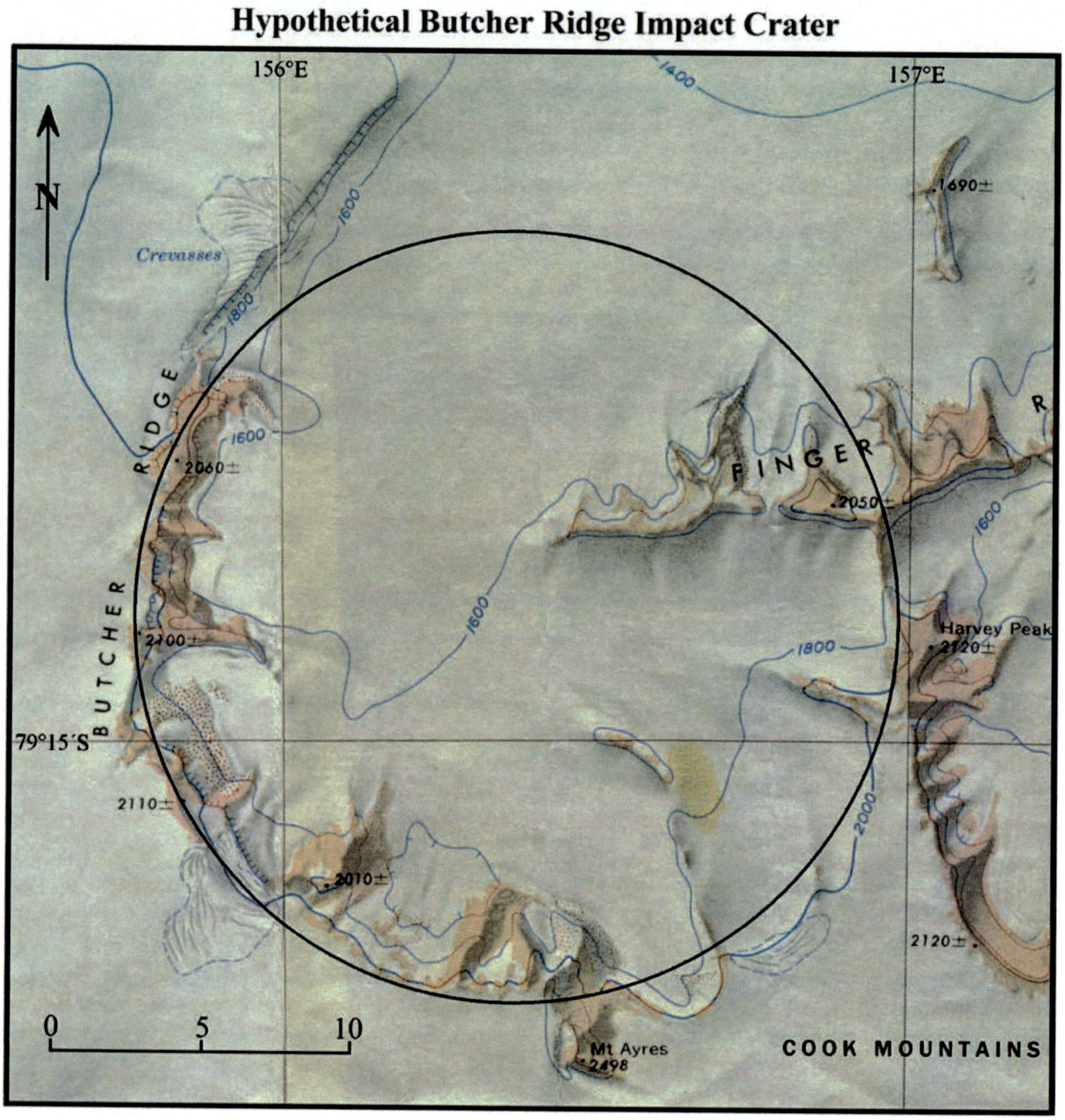

Fig. 18.9 Butcher Ridge at the edge of the polar plateau in the Cook Mountains of southern Victoria Land is composed of highly contorted and layered rocks consisting of glass and aphanitic rocks ranging from basalt to alkali-rich rhyolite. Although they resemble volcanic rocks, they formed by mixing of melts of basaltic and felsic composition followed by rapid cooling. The petrogenesis of these rocks is not understood. However, they may be meltrocks formed by the explosive impact of a large meteorite (1–2 km) into target rocks consisting of sedimentary Beacon rocks and sills of Ferrar dolerite of the Transantarctic Mountains (Adapted from the topographic map entitled Turnstile Ridge, Antarctica (ST53-56/16) published in 1966 by the US Geological Survey, Washington, DC)

as well as isotopic compositions of strontium and oxygen. He interpreted the data in terms of mixing of a basaltic magma of the Ferrar Group with a silica-rich melt of crustal rocks.

The Butcher-Ridge structure gives rise to a strong magnetic anomaly which is not compatible with a typical sill of Ferrar Dolerite. Instead, Behrendt et al. (1995) concluded that the magnetic profile, which peaks at 1,000 nT (nanoTesla), is comparable to the magnetic anomaly of the Dufek intrusion (Section 13.4). Their calculations showed that the thickness of the most magnetic rocks is greater than about 2 km, which implies that the entire body is probably even thicker.

However, the rocks at Butcher Ridge consist largely of *glass* and pseudo-volcanic rocks and cannot be described as *cumulates* that formed in a deep crustal magma chamber like the Dufek intrusion. We therefore conclude that the available geological and geophysical evidence is not compatible with the differentiation of basaltic magma in a crustal magma chamber.

Cassidy (2003) visited the area in 1978 to search for meteorites because Butcher Ridge is located close to the edge of the polar plateau. Although he did not find any meteorites, what he saw caused him to suspect that the igneous rocks exposed at Butcher Ridge are *melt-rocks* that formed when a large meteorite impacted on quartz-rich sandstones of the Beacon Supergroup and an interbedded sill of Ferrar Dolerite.

This alternative explanation is based on extensive information derived from the study of meteorite impact craters (e.g., Koeberl 1994; Dressler et al. 1994; Koeberl and MacLeod 2002). Unfortunately experts familiar with hypervelocity impacts of large meteorites have not had an opportunity to examine the rocks at Butcher Ridge.

18.5 Allan Hills Icefields

The Allan Hills are a relatively small group of sandstone and dolerite ridges on the edge of the polar plateau at 74°43′S and 159°40′E. The East Antarctic ice sheet adjacent to the Allan Hills in Fig. 18.10 contains several areas of blue ice where meteorite specimens have accumulated which are identified by the indicated letter codes (Appendix 18.12.1):

- Main, Near Western, Middle Western, and Far Western ice fields (ALH)
- Beckett Nunatak (BEC)
- Reckling Peak and Moraine (RKP)
- Elephant Moraine, Meteorite City, Texas Bowl, Northern Ice Patch (EET)
- Outpost Nunatak (OTT)
- David Glacier ice fields (DAV)
- MacKay Glacier ice fields (MCY)

The Main ice field of the Allan Hills is the site where in 1976/77 a joint American– Japanese team first collected nine meteorite specimens (ALHA76001 to ALHA76009). This collection consisted of seven ordinary chondrites, one iron, and one polymict eucrite (Score and Lindstrom 1990). In the following year, a systematic search by another American-Japanese team led by W.A. Cassidy collected 299 specimens. The next season (1978/79) yielded 260 specimens. After that collecting season, ANSMET had turned the corner and found "meteorite heaven" (Cassidy 2003, p. 34).

The progress in the recovery of "ALH" meteorites is recorded in Fig. 18.11. The spikes in the number of specimens collected identify the field seasons during which ANSMET carried out systematic searches in the blue-ice areas of the Allan Hills. The last major "harvest" during 1988/89 consisted of the 163 specimens that were collected by members of GANOVEX. The specimens collected in the 1976/77, 1977/78, and 1978/79 fieldseasons on the Allan Hills icefields were split and the USA and Japan shared half of each specimen. The catalog of Score and Lindstrom (1990) lists the original masses of these specimens, whereas Yanai and Kojima (1995) recorded the masses of the splits that were transferred to the National Institute of Polar Research (NIPR) in Tokyo, Japan.

The masses of the ALH meteorites listed in the catalog of Score and Lindstrom (1990) range from 0.1 g to 407 kg (ALH76009). The histogram in Fig. 18.12 indicates that 71% of all meteorite specimens recovered from the icefields of the Allan Hills weigh less than 100 g and that specimens weighing between 100 and 200 g have abundances of only 9.1%. An example of a small stony meteorite collected on the ice field surrounding the Reckling Moraine appears in Fig. 18.13. Specimens more massive than 1,000 g make up only 4.6% by number of all specimens that have been recovered on the Allan Hills ice fields. The masses of the Allan-Hills meteorites are larger on average than the masses of the Grove-Mountains meteorites in Fig. 18.6, but the abundances of both suites decrease exponentially with increasing mass. The distribution of the masses of meteorite specimens is consistent with their origin by the explosion of meteoroids during their passage through the atmosphere. In addition, stony meteorites may crack after landing on the ice sheet because of the expansion of ice that forms by freezing of liquid water in fractures (Section 18.7).

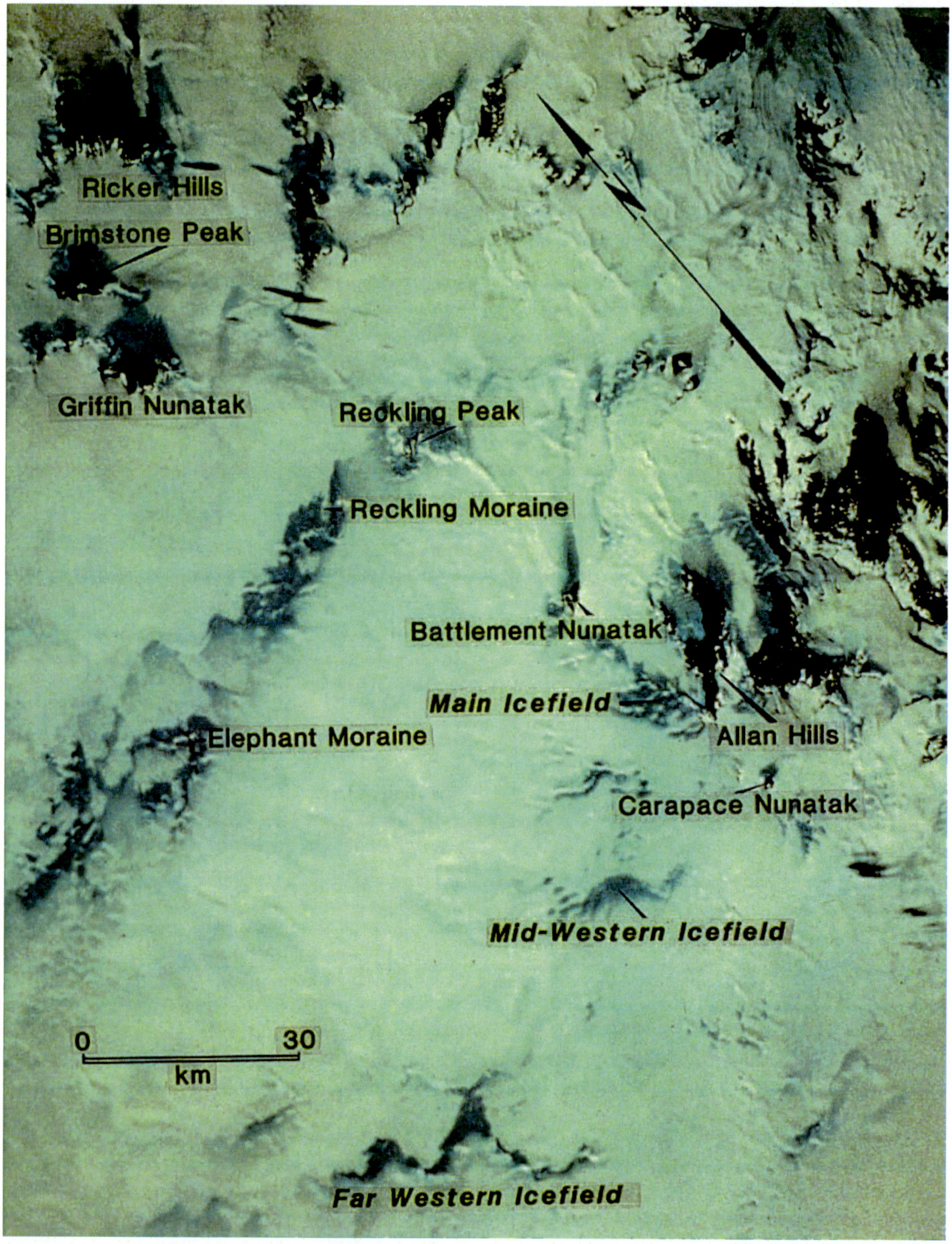

Fig. 18.10 Aerial view of the meteorite-collecting areas on the East Antarctic ice sheet located west of the Allan Hills in southern Victoria Land. The view also contains several mountains that project through the ice called "nunataks." The principal stranding surfaces of this area are listed in the text together with letter codes. The supraglacial Elephant and Reckling moraines were described in Chapter 17 (Courtesy of John W. Schutt who adapted it from Landsat)

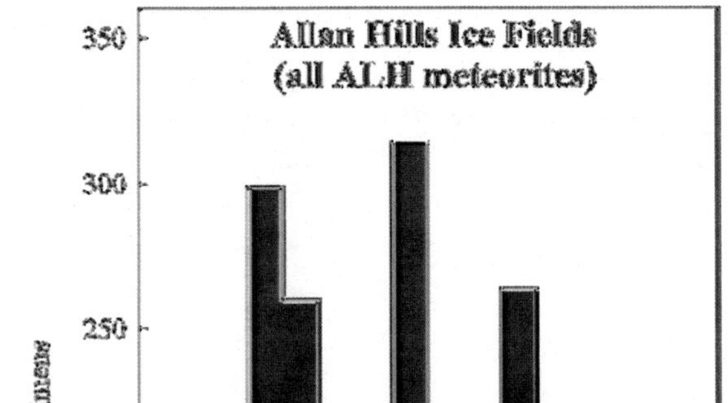

Fig. 18.11 During the 1976/77 field season a joint American-Japanese team collected nine meteorite specimens on the Main ice field adjacent to the Allan Hills. Subsequently, systematic searches by ANSMET yielded large numbers of specimens until the 1988/89 season when a team of American and German investigators collected additional specimens. More recently, geological field parties working in the area have continued to recover small numbers of additional specimens. The process that caused the "ALH" meteorite specimens to accumulate on the stranding surfaces of this area is continuing to deliver new specimens as the ice continues to ablate at the rate of about 5 cm/year (Data from Score and Lindstrom 1990, and from recent issues of the Antarctic Meteorite Newsletter)

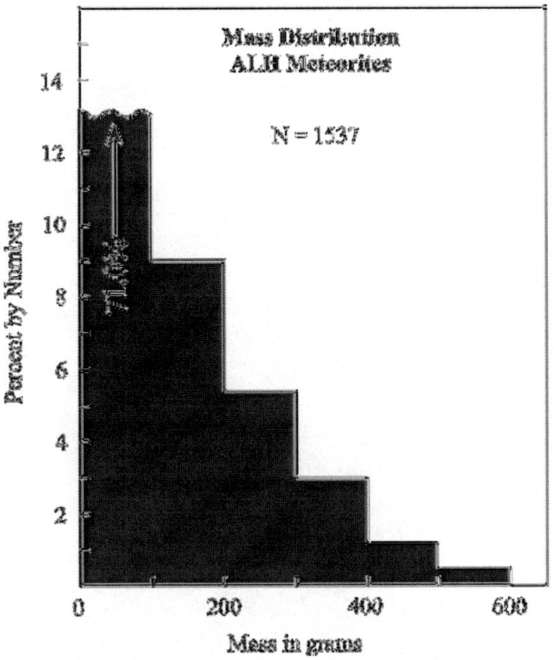

Fig. 18.12 The masses of most of the meteorite specimens that have been collected on the ice fields of the Allan Hills are less than 100 g each. In general, the abundance of meteorite specimens decreases as their masses increase. Specimens having masses greater than 1.0 kg are rare and make up only 4.6% by number compared to 71% for the low-mass fraction (<100 g). The observed distribution of the masses of meteorite specimens in the Allan Hills is consistent with the view that meteorites are fragments of larger meteoroids that exploded in the atmosphere and that these fragments may also have shattered during impact on the surface of the ice sheet and by the freezing of water in fractures (Data from Score and Lindstrom 1990)

The abundances of the different groups of Antarctic meteorites should resemble the abundances of *falls* because a representative fraction of all of the specimens that land on the accumulation area of the East Antarctic ice sheet ultimately emerge on the icefields of the ablation zone even though specimens in outflow glaciers are lost at sea. The abundances of meteorites that are *found* in areas where the climate is temperate or tropical are skewed because stony meteorites are destroyed by chemical weathering and those that remain are not easy to distinguish from terrestrial boulders, cobbles, and pebbles. The data in Table 18.5 confirm that the abundances of the Allan Hills meteorite specimens are similar to the abundances of *falls* and differ from the abundances of *finds*, although the ALH meteorite specimens are actually finds and not falls. However, stony irons (0.21%) and irons (1.0%) are *less* abundant in the ALH suite than *falls* elsewhere on the Earth and much less abundant than stony irons and irons among the *finds*, where iron and stony-iron meteorites are *dominant*" because they attract attention because of their high density and are therefore easy to identify.

The evident deficiency of iron and stony irons among the ALH meteorites may be a statistical aberration that will vanish when a larger population of specimens is considered. However, the abundance of

Fig. 18.13 The meteorite specimen discovered by David Buchanan on the ice near the Reckling Moraine was assigned the label RKP 86701 and was identified as an H-6 chondrite weighing 176.8 g. A thinsection of this specimen was analyzed by scanning electron microscopy by Boroughs (1992) and Boroughs et al. (1992) (Photo by G. Faure)

iron meteorites on *other* stranding surfaces in Antarctica is also low:

0.4% for the Lewis Cliff Ice Tongue and 1.5% for the Elephant Moraine (Score and Lindstrom 1990). A possible explanation for the low abundance of irons among Antarctic meteorites is that their high density (about 8 g/cm^3) causes them to sink more deeply into the ice sheet than chondrites. As a result, some of the irons and stony irons that fell onto the surface of the ice sheet may have been trapped in stagnant ice at the base of the ice sheet and are thereby prevented from returning to the surface in the ablation zone.

The meteorite specimens collected on the icefields of the Allan Hills include at least three rocks from Mars and one from the Moon. In addition, the collection includes rare specimens of certain achondrites and other nonchondrites including undifferentiated Acapulcoites (Table 18.1). Another noteworthy feature of Antarctic meteorites is the abundance of carbonaceous chondrites (Table 18.5) which are seriously under-represented among the *finds* elsewhere in the world because they are destroyed by chemical weathering in the presence of liquid water, carbon dioxide, and oxygen.

Table 18.5 Classification of meteorite specimens collected on the ice fields adjacent to the Allan Hills that carry the code ALH. Based on data from Score and Lindstrom (1990). Note that the abundances of the ALH meteorite specimens are similar to those of "falls" and differ from the abundances of "finds"

	Percent by number		
Group	ALH	Falls[a]	Finds[b]
Carbonaceous Chondrites			
C1	0		
C2	3.1		
C3	0.85		
C4	0.35		
All C Chondrites	4.3	4.7	0.30
Ordinary Chondrites			
H	58.0		
L	27.3		
LL	3.8		
E	1.2		
All Ordinary Chondrites	90.3	85.3	44.6
Nonchondrites (differentiated)			
Achondrites	3.6	8.9	1.0
Martian	0.21	–	
Lunar	0.07	–	
Stony irons	0.21	1.2	4.9
Irons	1.0	4.5	49.2
Nonchondrites (undifferentiated)			
Acapulcoites	0.21	–	
	99.9		

[a]Hartmann (2005).
[b]Glass (1982).

18.6 Meteorite Chronologies

The evolution of meteorites prior to their arrival on the Earth can be subdivided into a sequence of events that started with *nucleosynthesis* of atoms in stars followed by condensation of solid particles in the solar nebula and the accretion of the particles into the *meteorite parent-bodies* (Faure and Mensing 2007). These parent bodies were heated by the decay of unstable atoms which caused them to differentiate under the influence of their own gravitational fields into a metallic core, a silicate mantle, and a crust that was covered by a layer of regolith. The crystallization of minerals in the crust and mantle of the meteorite parent-bodies is recorded by the so-called *crystallization age* of stony meteorites, which can be determined by measuring the accumulated radiogenic daughter isotopes of long-lived radioactive parents (e.g., $^{87}Rb \rightarrow ^{87}Sr$, $^{147}Sm \rightarrow ^{143}Nd$, $^{238}U \rightarrow ^{206}Pb$, etc.; Faure and Mensing 2005). The results of age determinations of stony meteorites collected in Antarctica have repeatedly demonstrated that they crystallized between 4.55 and 4.65 billion years ago (Hartmann 2005; Misawa et al. 2005; Yamaguchi et al. 2001; Nyquist et al. 1986, 1994, 1997; Lugmair and Galer 1992; Goodrich et al. 1991).

After the parent bodies of the meteorites had crystallized, they were broken up by impacts and collisions. Fragments of various sizes that were ejected by the impacts went into orbit around the Sun in the form of meteoroids. The heat generated by the collisions also caused radiogenic ^{40}Ar to escape from the affected rocks, which is recorded by the low K-Ar dates of stony meteorites which are referred to as their *gas-retention ages*.

The meteoroids in their heliocentric orbits were irradiated by cosmic rays which consist primarily of high-energy cosmic rays that originate from sources in the galaxy. The exposure of meteoroids to cosmic rays in Fig. 18.14 caused the formation of stable as well as unstable "cosmogenic" product nuclides. The concentrations of certain *stable* cosmogenic nuclides in meteorites and their respective production rates are used to calculate the *cosmic-ray exposure ages* (CRE) which indicate how long a meteoroid orbited the Sun before it fell to the Earth (Herzog 2005; Eugster 2003).

After a meteorite has passed through the atmosphere of the Earth and has landed on its surface, it is shielded from cosmic rays by the magnetic field of the Earth and by the terrestrial atmosphere. Therefore, the production of all cosmogenic nuclides stops, but the existing cosmogenic *radionuclides* continue to decay. The length of time that has elapsed since a meteorite specimen fell to the surface of the Earth is calculated from the measured rate of decay of the remaining cosmogenic radionuclides, including ^{3}H, ^{10}Be, ^{26}Al, ^{36}Cl, ^{53}Mn, and several others. The result of this calculation is the so-called *terrestrial age* of a meteorite. In the case of meteorites that landed on the East Antarctic ice sheet, the terrestrial ages include the time during which the meteorite specimens were buried in the ice as well as the time they spent on the bare ice in the ablation zone of the ice sheet.

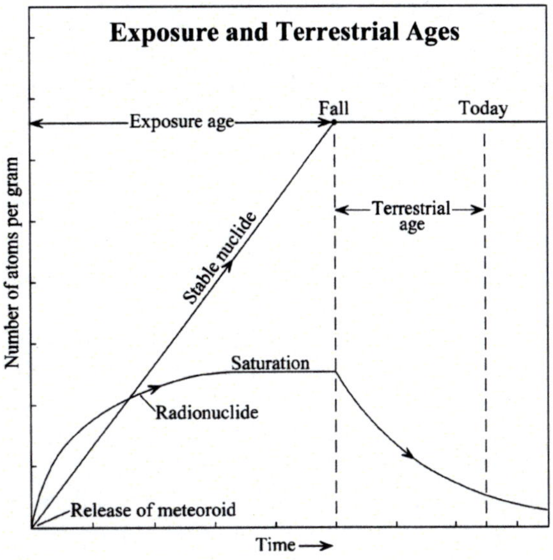

Fig. 18.14 Schematic representation of the production of stable and unstable nuclides in meteoroids by cosmic rays in space. The stable products accumulate at a constant rate until the meteoroid hits the Earth. The cosmogenic radionuclides decay as they form but may reach a state of saturation when the rate of decay is equal to the rate of production. When production of the cosmogenic nuclides stops at the time the meteoroid falls to the Earth, the concentration of the stable cosmogenic nuclides remains constant while the concentration of the unstable nuclides begins to decrease as they continue to decay (Adapted from Lipschutz and Schultz 1999)

18.6.1 Cosmic-Ray Exposure Ages

Cosmic rays generally *cannot* penetrate more than about 1 m of rocks composed of silicate minerals. Therefore, meteoroids that were excavated from a depth of more than 1 m below the surface of a meteorite parent-body are assumed to have been "pristine" in the sense that they did not contain cosmogenic nuclides at the time they were ejected into space. In addition, meteoroids are assumed to have had a simple history of irradiation without changes in shape that affected the shielding of the interior (Eugster 2003). In the simplest case, the concentration (s) of a *stable* cosmogenic nuclide in a meteorite is expressed in the form:

$$s = P_s t \tag{18.1}$$

where P_s = production rate of the nuclide and
t = cosmic-ray exposure (CRE) age.

The stable cosmogenic nuclides used for measuring CRE ages are ^{3}He, ^{21}Ne, ^{38}Ar, ^{83}Kr, or ^{126}Xe. The production rates of these nuclides can be calculated from the concentrations of the target elements. For example, stable ^{21}Ne is produced from the stable nuclei of magnesium by irradiation with cosmic-ray produced neutrons followed by emission of an alpha particle from the nucleus of the product:

$$^{24}_{12}Mg + ^{1}_{0}n \rightarrow ^{21}_{10}Ne + ^{4}_{2}He \qquad (18.2)$$

where $^{1}_{0}n$ is the neutron and $^{4}_{2}He$ is the alpha particle. The production rates of the cosmogenic nuclides were determined by Eugster (1988).

The CRE ages of *ordinary chondrites* that fell in Antarctica and elsewhere range from a few million years up to several tens of millions of years. Simulations of the celestial mechanics of meteoroids indicate that most of the exposure to cosmic rays occurred in the main asteroid belt from which they were ultimately ejected into Earth-crossing orbits when their periods of revolution came into resonance with that of Jupiter (Faure and Mensing 2007).

Carbonaceous chondrites tend to have short CRE ages of less than 2 million years. For example, the C4 meteorite EET83311 from the Elephant Moraine has a CRE age of only 40,000 years (Nishiizumi and Caffee 2002). Short exposure ages like this are difficult to measure using stable cosmogenic nuclides (e.g., ^{21}Ne) because their concentrations are low and because carbonaceous chondrites also contain trapped neon. Therefore, Ma et al. (2002) measured the CRE age of the carbonaceous chondrite EET96026 by determining not only stable ^{3}He and ^{21}Ne but also the unstable cosmogenic nuclides (e.g., ^{10}Be, ^{26}Al, ^{36}Cl, and ^{53}Mn). All of the data indicate that the CRE age of EET96026 is less than 0.52 million years.

In contrast to the short CRE ages of the two carbonaceous chondrites from the Elephant Moraine, the H chondrites FRO90174 and FRO90001 from the Frontier Mountain (72°59′S, 160°20′E) in northern Victoria Land have CRE ages of $(7.2 \pm 0.5) \times 10^6$ years and $(8 \pm 1) \times 10^6$ years, respectively (Welten et al. 2001).

The CRE ages of ordinary chondrites (H, L, LL, and E) range from less than 1 million years up to about 80 million years. The peaks in the spectra of CRE ages identify collisions in the asteroid belt that released large numbers of fragments (i.e., meteoroids) that were exposed to cosmic rays until they were forced to enter Earth-crossing orbits by gravitational interactions with Jupiter. For example, the spectrum of CRE ages of H chondrites in Fig. 18.15 has a prominent peak between 6 and 8 million years ago, which indicates that a collision occurred in the asteroid belt at that time. The other chondrite groups also originated as fragments due to collisions among different kinds of asteroids (Eugster 2003).

The iron meteorites stand out as a group because their exposure ages are far *older* than those of the chondrites and achondrites most of which were irradiated

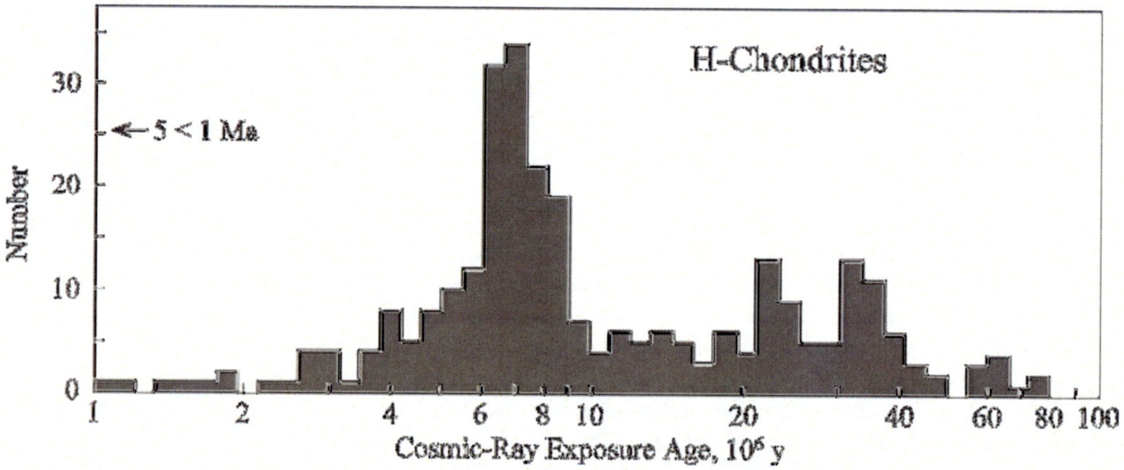

Fig. 18.15 The frequency distribution of the cosmic-ray exposure (CRE) ages of H chondrites has several peaks the largest of which occurs between 6 and 8 million years ago. Apparently a large number of H-chondrite meteoroids was produced by one or more collisions in the asteroid belt that caused the break-up of their parent body. The CRE ages of stony meteorites are generally less than 80 million years, whereas iron meteorites spend much more time in interplanetary space exceeding 1 billion years in some cases (Adapted from Fig. 1 of Eugster 2003)

for less than 80 million years before they impacted on the Earth, on the Moon, or on one of the other terrestrial planets (e.g., Mars, Venus, and Mercury). The CRE ages of iron meteorites range from less than 100 million years up to more than 2,200 million years and cluster at about 650 ± 60 million years for IIIAB irons (Lipschutz and Schultz 1999). The iron meteorites may have longer CRE ages than stony meteorites because they are not as easily broken up by collisions in space.

The interpretation of cosmic-ray exposure ages of all kinds of meteorites links them to the asteroids from which they originated but which we have not yet been able to sample directly. The study of meteorites indicates that some asteroids contain resources that humans may be able to recover in the future (Kargel 1994; Lewis 1996).

18.6.2 Terrestrial Ages

The production of radionuclides in meteoroids that are orbiting the Sun takes place by nuclear spallation and neutron-capture reactions with the atoms of the major elements (Bogard et al. 1995; Leya et al. 2000). The concentration of a particular radionuclide in a stony meteoroid exposed to cosmic rays in Fig. 18.14 initially increases with time until it reaches a state of equilibrium or saturation when its rate of decay is equal to its rate of production. When such a meteoroid enters the atmosphere of the Earth and explodes, the resulting meteorite specimens are assumed to be saturated with respect to the cosmogenic radionuclides they contain. After a meteorite has landed on the surface of the Earth, the production of radionuclides stops because the Earth is protected from cosmic rays by its magnetic field and by the atmosphere. Therefore, the rate of decay of cosmogenic radionuclides decreases with time as each nuclide continues to decay with its characteristic halflife. The terrestrial age of a meteorite specimen collected in Antarctica or anywhere else on the Earth is calculated from the rates of decay of the radionuclides (e.g., ^{36}Cl or ^{26}Al) that *remain* at the time of analysis (Jull 2001).

The rate of decay of cosmogenic radionuclides in meteorites is so low that this method of dating was not practical until after tandem-accelerator mass-spectrometers became available in the 1980s (Elmore and Phillips 1987). As a result of this development, the terrestrial ages of meteorites can now be measured by counting the atoms of selected cosmogenic radionuclides they contain. The most effective radionuclide for measuring the terrestrial ages of meteorites is ^{36}Cl which is produced by a nuclear spallation reaction in grains of metallic iron of ordinary chondrites. In addition, ^{10}Be, ^{14}C, ^{26}Al, ^{53}Mn, and ^{81}Kr have also been used successfully (Nishiizumi et al. 1981, 1983, 1989; Michlovich et al. 1995; Welten et al. 2001).

Chlorine-36 is favored by investigators because the grains of metallic iron are easily concentrated from crushed samples of stony meteorites and because the concentration of ^{36}Cl atoms per gram of metal can be determined routinely by accelerator mass-spectrometry. The corresponding rate of decay of ^{36}Cl, also called the *activity*, is calculated from the measured concentration by means of the law of radioactivity. The calculation of the terrestrial ^{36}Cl age of the LL6 chondrite ALH 78153 is demonstrated in Appendix 18.12.2.

The histogram of the terrestrial ages of meteorite specimens collected on the Allan-Hills ice fields in Fig. 18.16 demonstrates that more than 35% of the specimens that have been dated fell less than 100,000 years ago and that about 10% have been on the Earth for more than 500,000 years, including ALH88019

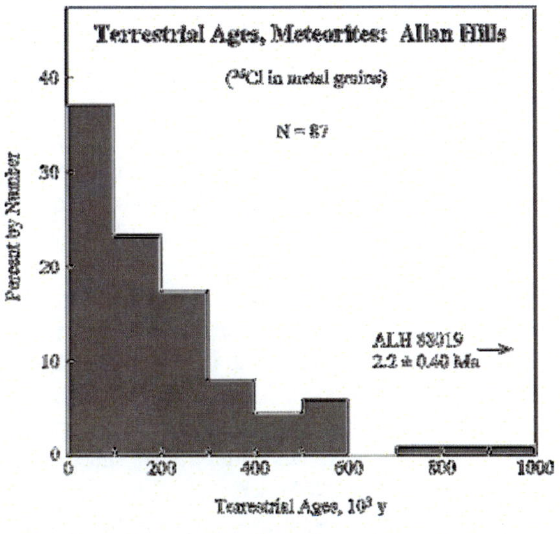

Fig. 18.16 The terrestrial ages of stony and iron meteorites from the ice fields adjacent to the Allan Hills range from <100 × 10^3 to 2,200 × 10^3 years. All dates were calculated from the rate of decay of cosmogenic ^{36}Cl in metallic Fe-Ni-Co grains measured by means of accelerator mass spectrometry (Data from Nishiizumi et al. 1989; Michlovich et al. 1995; Scherer et al. 1997)

which has a terrestrial age of 2.2 ± 0.4 Ma (Scherer et al. 1997). The terrestrial ages of stony and iron meteorites from other collecting sites on the East Antarctic ice sheet are *lower* on average than the terrestrial ages of ALH meteorites, although the terrestrial ages of four specimens from these sites are older than 1 million years (Nishiizumi et al. 1989; Jull 2001; Welten et al. 1997):

DRP 78001: 1.0 ± 0.1 Ma (iron)
LEW 86360: 2.35 ± 0.15 Ma
ILD 83500: 3.20 ± 0.20 Ma (iron)
Lazarev: 5.0 Ma (iron)

The sites are identified in Appendix 18.12.1.

The preservation of meteorite specimens that fell more than 1 million years ago is made possible by the slow rate of weathering in the cold and dry environment of the East Antarctic ice sheet. The survival of iron meteorites (i.e., DRP, ILD, and Lazarev listed above) is attributable to their low porosity. However, ALH88019 and LEW86360 both of which have long terrestrial ages, are ordinary chondrites. In addition, the long terrestrial ages of these meteorites support the geological evidence that the East Antarctic ice sheet formed as much as 25 million years ago even though its volume may have decreased temporarily during the Pliocene Epoch (Webb et al. 1984; Webb and Harwood 1991). However, if the ice retreated sufficiently from the Transantarctic Mountains to permit seawater to invade the continent, then some of the meteorite specimens that were recovered in the Allan Hills (e.g., ALH88019) and on the Lewis Cliff ice tongue (e.g., LEW86360) may have fallen into seawater and were later entrained in the ice sheet during its subsequent advance. However, no evidence has been reported that indicates that any Antarctic meteorites were altered by seawater. On the other hand, if the ice sheet was as extensive during the Pliocene Epoch as it is at present, in accordance with the "giant ice-sheet" hypothesis of Denton et al. (1993) and Marchant et al. (1993), then even the meteorites with the oldest terrestrial ages could have landed on the ice sheet or on rocks without falling into water.

The exponential decrease of the abundances of meteorite specimens with increasing terrestrial ages in Fig. 18.16 is noteworthy because it implies either that meteorites were progressively destroyed the longer they resided in the ice sheet or that they collect at the base of the ice sheet and are prevented from being exposed at the surface in the ablation areas. A third alternative that fewer meteorites fell prior to about 1 million years is not supported by planetary science (Faure and Mensing 2007).

18.6.3 Old Meteorites at the Allan Hills

The distribution of collecting sites of meteorites on the Main Icefield of the Allan Hills in Fig. 18.17 indicates that *all* 18 meteorite specimens having terrestrial ages greater than 300,000 years occur in a narrow band located along the margin of the ice sheet in this area. In addition, 80% of the specimens with terrestrial ages between 150,000 and 300,000 years also occur in this belt. The remarkable association of meteorite specimens that fell more than 150,000 years ago along the edge of the East Antarctic ice sheet confirms the meteorite-transport model of Whillans and Cassidy (1983) as noted by Harvey (1990). This model in Fig. 18.5 demonstrates that the transport times and the terrestrial residence ages of meteorite specimens rise with increasing depth of burial. Therefore, the prevalence of long residence ages among the meteorite specimens collected along the margin of active ice on the Main Allan Hills ice field implies that the meteorite specimens originated from depth in the ice sheet and that basal ice is forced to the surface by a bedrock barrier related to the Allan Hills.

Faure and Buchanan (1987) showed in Fig. 17.22 that this barrier is a subglacial ridge that extends north from a small outcrop at the southern end of the Allan Hills. The ridge underlies the prominent ice ramp that separates the active ice of the polar plateau from the mound of stagnant ice that is located between the ice ramp and the rock outcrops of the Allan Hills to the east. Most of the meteorite specimens with terrestrial ages between 150,000 and 300,000 years were collected along this ice ramp.

18.7 Chemical Weathering of Stony Meteorites

The long residence ages of meteorites on the ice fields of the Allan Hills and elsewhere in Antarctica imply that the rates of chemical weathering reactions are slow enough to allow meteorite specimens to survive for up to 1 million years (e.g., ALH78153) and more. Meteorites in the hot deserts of the world (e.g., North Africa, Oman, Chile, Australia) weather more rapidly than those that fall in the cold desert of the East Antarctic ice sheet and therefore do not survive as long (Bland 2001; Crozaz et al. 2003; Lee and Bland 2004). Stony meteorites that are exposed to the temperate climate conditions of North America and northern Europe disintegrate in a few thousand years. Iron meteorites survive longer than stony meteorites because their low porosity prevents water and other volatile compounds from penetrating into the interior of such specimens.

Meteoroids of all kinds are also altered in various ways by exposure to cosmic rays and to the solar wind while they revolve around the Sun (e.g., Section 18.6.1). Some specimens of stony meteorites even contain evidence that their chemical and mineral compositions were altered *before* they were released into

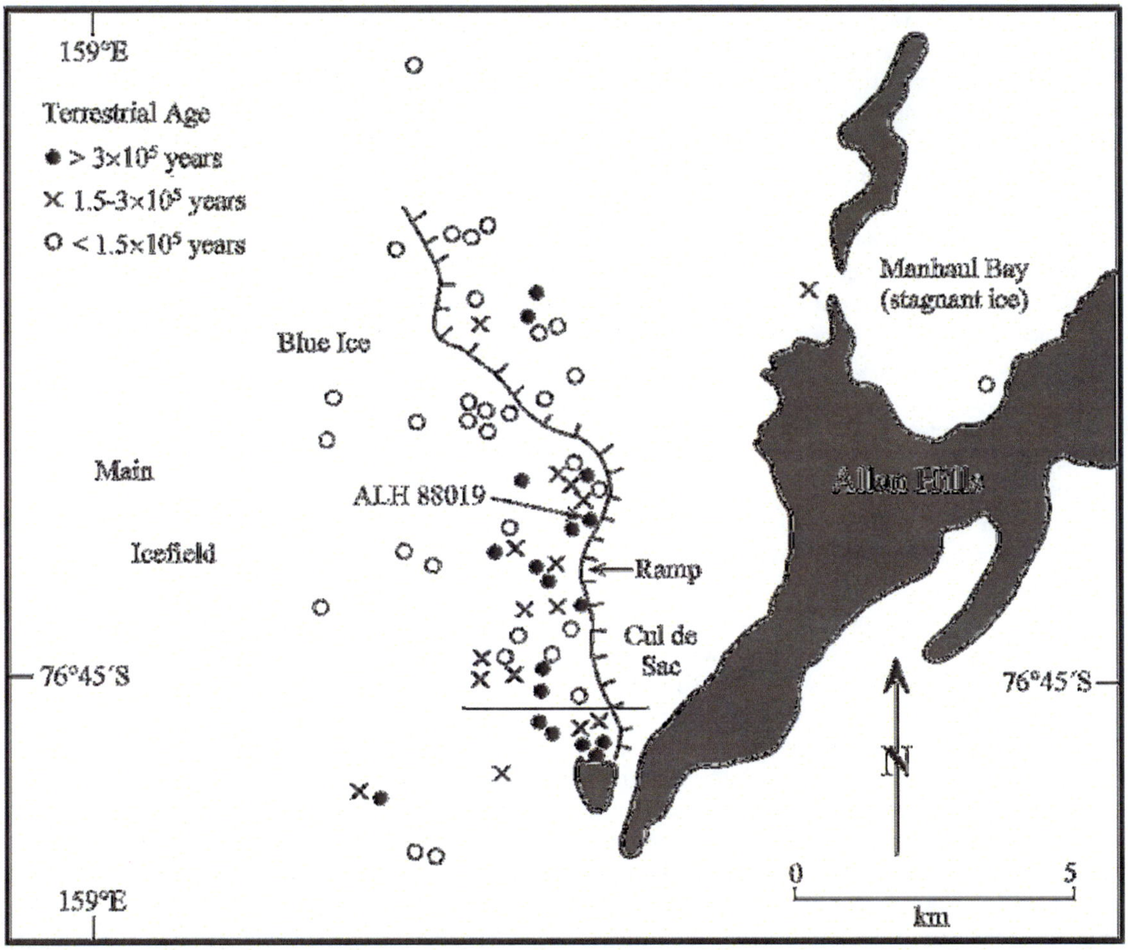

Fig. 18.17 The terrestrial ages of meteorites on the Main ice field of the Allan Hills range widely from greater than 300,000 years to less than 150,000 years. Specimens having terrestrial ages greater than 300,000 years (*solid circles*) are clustered along the ice ramp close to the rock outcrop of the Allan Hills. Specimens with terrestrial ages between 150,000 and 300,000 years (×) are also concentrated in this part of the Main ice field where the oldest ice from the base of the ice sheet is exposed (Whillans and Cassidy 1983; Fig. 18.5). Specimens that fell less than 150,000 years ago occur primarily west of the ice ramp on ice from higher stratigraphic levels in the ice sheet. The LL6 chondrite ALH 88019, which has the longest terrestrial age of 2.20 ± 0.40 Ma, also lies at the ice ramp with other specimens that fell more than 300,000 years ago. The line across the edge of the ice at the Cul de Sac is the radar-echo profile measured by Faure and Buchanan (1987) (Adapted from Nishiizumi et al. 1989; Scherer et al. 1997)

interplanetary space by the break-up of their parent asteroid (Gooding 1984).

At issue here is the chemical alteration of *stony meteorites* after they have been exposed on the ice-fields of Antarctica. The exposure of *stony* meteorites to water and oxygen of the terrestrial environment causes certain of their minerals to decompose into oxides and hydroxides of major elements. In addition, efflorescences of carbonates, sulfates, and other kinds of salts can form on the surfaces of meteorite specimens and can fill fractures in their interiors and certain chemical elements may also be deposited from the atmosphere (e.g., Cl, Br, I, Hg). The extent of weathering depends on the environmental conditions (e.g., temperature and humidity) and the presence of reactive minerals containing divalent iron (e.g., troilite, olivine, pyroxene). The porosity of stony meteorites and the presence of fractures accelerate the alteration and ultimate decomposition of stony meteorites. For this reason, the description of Antarctic meteorites by the curators of the Meteorite Working Group in Houston includes an assessment of the extent of fracturing and of the state of weathering (Score and Lindstrom 1990). The state of weathering is indicated by letters A, B, and C defined primarily by the presence and pervasiveness of rust halos.

The Antarctic meteorites are more extensively weathered and fractured than expected because of a misconception about the microclimatic conditions on the ice fields of the polar plateau. Although the air temperature remains below the freezing point of water, the temperature in the interior of a meteorite specimen exposed to solar radiation may rise sufficiently to melt snow on its surface. For example, Schultz (1990) showed in Fig. 18.18 that the internal temperature of a sample of the carbonaceous chondrite Allende placed on the ice of the Far Western ice field adjacent to the Allan Hills in December of 1985, was consistently higher than the air temperature by up to about 15°C and that the *temperature difference* increased as the wind speed and cloud cover *decreased*. The wind apparently cools meteorite specimens exposed on the icefields while the cloud cover modulates the amount of solar energy they receive. On several occasions, when the wind speed decreased to zero, the internal temperature of the test specimen monitored by Schultz (1990) actually rose to +5°C. Similar results were reported by Harvey (2003) who measured the temperature at the base of a 1-kg dolerite boulder in December and January of 1997/98 on the Mare Meteoriticus ice field (informal name). In this experiment, the temperature at the base of the boulder increased to +10°C, whereas the temperature of the ice beneath the boulder remained nearly constant at about −10°C (see also Miotke 1988).

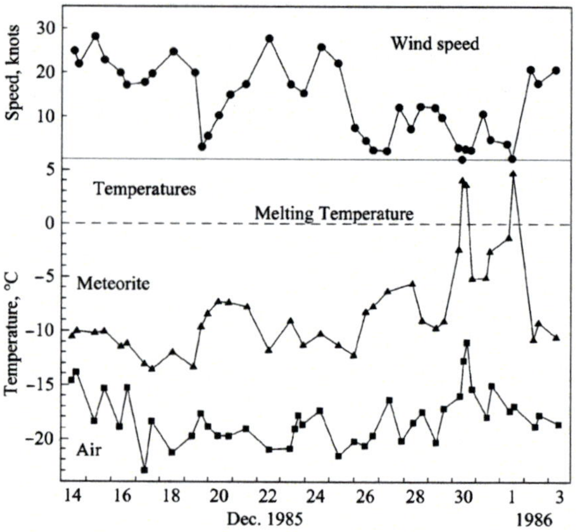

Fig. 18.18 The internal temperature of a meteorite specimen placed on the bare ice of the Far Western ice field of the Allan Hills during December and January of 1985/86 was consistently higher than the air temperature and, on 2 days, actually exceeded the melting temperature of ice. The highest internal meteorite temperatures occurred on days when the wind speed decreased to zero. At these times, meteorite specimens can melt the ice on which they lie and are thereby exposed to liquid water that fills the cup-shaped depressions that forms around them. The wind speed is here expressed in terms of "knots" used by the US Navy and defined as one nautical mile per hour which is equivalent to 1.852 km/h (Adapted from Schultz 1986, 1990)

The meltwater that stony meteorites absorb on calm and sunny days (and nights) during the austral summer subsequently freezes, causing fractures to widen until the specimens eventually disintegrate. The small meteorite fragments that result from these freeze-thaw cycles are transported by the wind and may collect in local depressions, in the lee of ice or bedrock ridges, and at the margins of icefields where the bare ice is covered by firn and snow which tend to trap small meteorite specimens. This phenomenon is exemplified in Fig. 18.19 by the distribution of meteorite collecting sites on the bare ice surrounding the Reckling Moraine (Section 17.4).

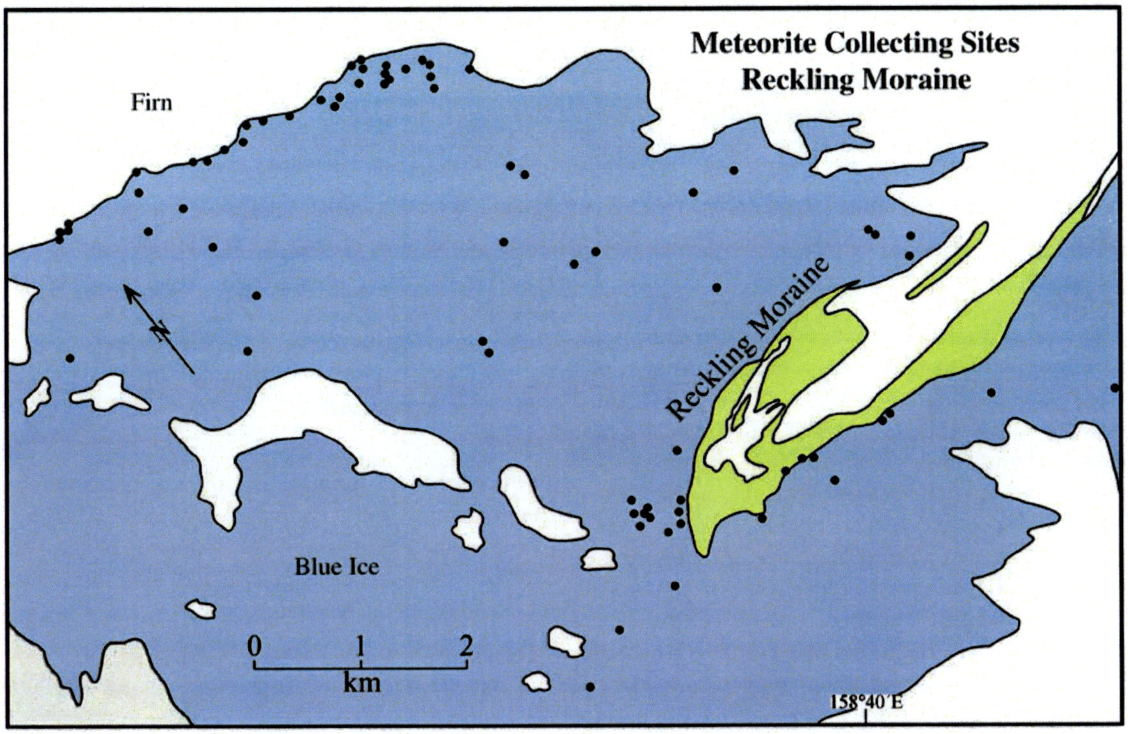

Fig. 18.19 Deposits of firn and glacial sediment of the Reckling Moraine trap small meteorite specimens which are moved by wind that blows primarily from the southwest to northeast (Adapted from J.W. Schutt, personal communication 2007)

18.7.1 Evaporite Minerals

The meltwater absorbed by stony meteorites also causes weathering which manifests itself by the formation of secondary minerals, by alteration of the isotopic composition of certain elements, and by the addition of halogens and similar compounds that are deposited by snow and from the air (Gooding 1986a, b, 1989; Lipschutz 1982, 1986; Velbel 1988; Jull et al. 1988; Velbel and Gooding 1990).

The stony meteorites collected during the 1977/78 field season on the icefields adjacent to the Allan Hills included seven specimens that contain deposits of white efflorescences on their surfaces and in cracks. Marvin (1980) reported that these deposits were composed of the hydrated carbonates and sulfates of magnesium and calcium listed in Table 18.6. In spite of the presence of these deposits, only two of the seven specimens have rust stains. The absence of heavy rust stains is difficult to reconcile with the prevalence of Mg-salts among the weathering products because olivine and pyroxene do not decompose until a significant fraction of the divalent iron they contain has been oxidized to

Table 18.6 Weathering products on stony meteorites collected during the 1977/78 field season on the ice fields adjacent to the Allan Hills (Marvin 1980)

Specimen #	Group	Mineral	Formula
ALHA 77257	Ureilite	Hydromagnesite	$Mg_4(OH)_2(CO_3)_3 \cdot 3H_2O$
ALHA 77307	C3	Epsomite	$MgSO_4 \cdot 7H_2O$
ALHA 77003	H3	Leonhardite[a]	$MgSO_4 \cdot 4H_2O$
ALHA 77294	H5	Amorphous	Mg-carbonate
ALHA 77231	L6	Nesquehonite	$Mg(HCO_3)(OH) 2H_2O$[b]
ALHA 77262	H4	Gypsum	$CaSO_4 \cdot 2H_2O$
ALHA 77140	L3	Not determined	

[a]This mineral should be called "starkeyite" because "leonhardite" is a variety of laumontite which is a zeolite mineral (Velbel 1988).
[b]The composition of nesquehonite can also be expressed by the formula $MgCO_3 \cdot 3H_2O$.

the trivalent state in the form of rust. In addition, Marvin (1980) noted that the kinds of Mg-salts that form on stony meteorites do not occur in the soils of Antarctica (Claridge and Campbell 1977).

Nevertheless, nesquehonite and other magnesium salts have also been found on a meteorite specimen from the Yamato Mountains (Yabuki et al. 1976) and from other localities adjacent to the Transantarctic Mountains. Velbel (1988) published a list of 94 specimens of weathered meteorites collected by ANSMET. The evaporite minerals formed either as a result of mobilization of elements within the meteorite specimens or by the addition of Mg, Ca, C, S, and O from outside sources. Velbel et al. (1991) concluded that the magnesium originated from olivine in the meteorites and that calcium was released by weathering of plagioclase which does occur in the matrix of stony meteorites. Regardless of whether the elements that form the evaporitic weathering products originate from internal or external sources (or both), the presence of evaporitic weathering products indicates that the affected meteorite specimens have been *altered*.

A study by Grady et al. (1989) confirmed an observation of Marvin (1980) that salt efflorescences are forming on Antarctic meteorite specimens while they are thawing in the laboratory at the Johnson Space Center in Houston, Texas. Marvin (1980) suggested that some Antarctic meteorites continue to be altered after collection because of the presence of small amounts of snow that were inadvertently included in the Teflon collecting bags. Evidently, the chemical weathering of Antarctic meteorites continues while they are stored in museums unless steps are taken to exclude water vapor, oxygen, and other volatile compounds that are present in the atmosphere.

The possibility that the alteration of meteorites occurred on the surface of their parent body in the asteroid belt (Gooding 1984) was directly addressed by Zolensky and Gooding (1986) by comparing the weathering products of the Antarctic carbonaceous chondrites ALHA 77003 to the weathering products of the carbonaceous chondrites Murchison and Allende, both of which are falls that did not come in contact with water or ice. The authors reported that ALHA 77003 contains jarosite, goethite, and small grains of magnetite that had formed by the progressive decomposition of the indigenous minerals in the presence of small amounts of water, whereas Murchison and Allende did not contain these or any other weathering products. Therefore, Zolensky and Gooding (1986) concluded that ALHA 77003 and other carbonaceous chondrites collected in Antarctica were altered *after* they landed on the East Antarctic ice sheet rather than on their parent bodies. In addition, Gounelle and Zolensky (2001) also concluded that sulfate veinlets in carbonaceous chondrites did *not* form on the meteorite parent body, but instead are terrestrial weathering products that could have formed during storage in museums.

18.7.2 Clay Minerals

The primary rock-forming silicate minerals in terrestrial rocks can react with liquid water to form various two-layer and three-layer clays (Faure 1998). For example, Na-plagioclase (i.e., albite) is transformed into kaolinte by the reaction:

$$\begin{aligned}2NaAlSi_3O_8 + 9H_2O + 2H^+ \rightarrow\ & Al_2Si_2O_5(OH)_4 \\ & + 2Na^+ + 4H_4SiO_4\end{aligned} \quad (18.3)$$

where water and hydrogen ions are provided by the environment and silicic acid takes the form of amorphous silica ($Si(OH)_4$). This transformation and others like it can occur in the soil of the ice-free valleys of Antarctica (Ugolini and Jackson 1982) as well as in stony meteorites that are exposed to solar radiation on the ice fields of the East Antarctic ice sheet.

The first evidence for the conversion of silicate minerals of stony meteorites in Antarctica was reported by Gooding (1986a) who detected the presence of clay mineraloids, gypsum, K-Fe sulfates (jarosite?), and "rust" on the surfaces and in cracks of achondrites and chondrites from Elephant Moraine (EET) and the Allan Hills. These weathering products had formed primarily from glass and plagioclase in the fusion crust and in cracks in the interiors of the meteorite specimens.

The formation of clay minerals in stony meteorites exposed to solar radiation on the ice fields of Antarctica is attributable to the episodic melting of ice and snow in contact with meteorite specimens that are warmed by sunlight. However, meteorites that were still embedded in the ice when they were collected also contained aluminosilicate weathering products (Gooding 1986b; Gow and Cassidy 1989; Gow 1990). Harvey and Score (1991) suggested that meteorites that were weathered while they were still embedded in the ice could be warmed by solar energy when they were less than 10 m below the surface of the ice sheet. The alteration of meteorites embedded in the ice may be facilitated by the presence of small amounts of sulfuric acid on grain boundaries between ice crystals of the East Antarctic ice sheet (Mulvaney et al. 1988).

18.7.3 Trace Elements

The evidence that stony meteorites collected in Antarctica are weathered implies that certain chemical elements are mobilized within the affected meteorite specimens. In addition, glacial meltwater and atmospheric carbon dioxide invade the affected meteorite specimens together with halogens, sulfur-bearing compounds, and organic molecules. Therefore, meteorites that fell on the East Antarctic ice sheet are altered mineralogically as well as chemically and, for that reason, their trace-element concentrations may differ from those of non-Antarctic meteorite falls.

The concentrations of potentially mobile trace elements in H5 and L6 chondrites of Antarctic and non-Antarctic origin (falls only) reported by Lipschutz (1989) permit the conclusion with 90% or higher confidence that the trace-element concentrations of Antarctic meteorites actually do *differ* from those of the non-Antarctic falls. However, the cause for this difference is still in doubt. On the one hand, the Antarctic meteorites may have originated from a different set of parent bodies in the asteroid belt, as postulated by Lipschutz (1989). On the other hand, even if the Antarctic chondrites did originate from different parent bodies, the concentrations of trace elements in Antarctic chondrites may also have been lowered by chemical weathering after they fell on the polar plateau of Antarctica. The distinction between the effects of terrestrial weathering of meteorites and the properties they inherited from their parent bodies remains a difficult issue in meteoritics.

18.7.4 Iodine Contamination

The concentrations of chlorine, bromine, and iodine of meteorites measured by Goles et al. (1967) raised the suspicion that *non-Antarctic* meteorites in museums had been contaminated by being handled with bare hands by collectors and because they had been stored for years in display cases through which air could circulate freely. Therefore, Dreibus and Wänke (1983) and Dreibus et al. (1986) measured the concentrations of halogens in *Antarctic* meteorites in order to document the contamination of *non-Antarctic meteorites*. The results in Table 18.7 demonstrated that Antarctic meteorites actually have unexpectedly *higher* concentrations of fluorine, chlorine, and iodine than similar meteorites of non-Antarctic origin. Even Antarctic martian meteorites have higher iodine concentrations than non-Antarctic meteorites from Mars.

Table 18.7 Concentrations of halogen elements in Antarctic and non-Antarctic stony meteorites in parts per million (Dreibus et al. 1986)

Sources	F	Cl	Br	I	I/Br
Eucrites					
Antarctic	79.9 (8)[a]	116 (9)	0.069 (9)	6.98 (9)	101
Non-Antarctic	32.6 (5)	75.2 (6)	0.352 (6)	0.151 (6)	0.42
Antarctic/non-Antarctic	2.4	1.5	0.19	46	
Howardites					
Antarctic	8.50 (2)	73.9 (4)	0.146 (4)	1.81 (4)	12.4
Non-Antarctic	1.75 (2)	4.1 (2)	0.034 (2)	0.015 (2)	0.44
Antarctic/non-Antarctic	4.9	18	4.3	120	
Shergottites (Martian meteorites)					
Antarctic	30.7 (3)	29.3 (3)	0.154 (3)	0.927 (3)	6.0
Non-Antarctic	41 (1)	108 (1)	0.891 (1)	0.036 (1)	0.04
Antarctic/non-Antarctic	0.75	0.27	0.17	25	
Seawater[b]	1.3	1.95×10^4	67	0.056	0.0008

[a]Number of specimens included in the arithmetic average.

[b]Taylor and McLennan (1985).

The apparent enrichment of Antarctic meteorites in iodine and, to a lesser extent, in chlorine is expressed in Table 18.7 by the Antarctic/non-Antarctic ratios of all four elements. The iodine ratios range from 25 in shergottites, to 46 in eucrites, and to 120 in howardites. Even Antarctic carbonaceous chondrites and ordinary chondrites (H) are enriched in iodine by factors of 24 and 2.5, respectively (not shown). These results demonstrate that the Antarctic meteorites are not pure and pristine as expected, but have been extensively contaminated by iodine and, to a lesser extent, by fluorine and chlorine.

Dreibus et al. (1986) suggested that the Antarctic meteorites may have been contaminated by marine evaporite deposits that were incorporated into the East Antarctic ice sheet. However, an analysis of the ice exposed near the Elephant Moraine on the polar plateau found only low concentrations of halogens (i.e., Cl = 0.31 ppm, Br = 0.015 ppm, I < 0.2 ppb) with a low I/Br ratio of less than 0.0130. Dreibus et al. (1986) also suggested that the iodine contamination of Antarctic meteorites may be caused by the deposition of aerosol particles from the atmosphere. The transport and size distribution of aerosol particles in the atmosphere above Antarctica was reviewed by Shaw (1988), while Duce et al. (1971, 1973) measured the halogen concentrations of particulates and air at McMurdo and South Pole Station. In addition, Heumann et al. (1987, 1990) and Shinonaga et al. (1994) demonstrated that halogens on the surfaces of Antarctic meteorites diffuse into their interiors especially at times when liquid water is present. They also showed that fluorine, chlorine, and bromine are derived from seaspray, whereas iodine occurs as methyl iodide (CH_3I) which is produced by marine organisms. This gaseous molecule is transported by the movement of the air without regard to distance from the coast.

These insights indicate that the presence of excess iodine, chlorine, fluorine and, to a lesser extent, bromine, on the surfaces of Antarctic stony meteorites is attributable to the deposition of aerosol particles (Cl, F, Br) and of methyl iodide from the atmosphere. In other words, the excess halogens on the surfaces of Antarctic meteorites are atmospheric *contaminants* rather than weathering products. Accordingly, the amount of excess iodine on a meteorite lying on an Antarctic ice field increases primarily as a function of time and is less dependent on the climatic conditions on the ice field and on its distance from the coast.

Langenauer and Krähenbűhl (1993) subsequently analyzed sequentially-removed surface layers of ordinary chondrites from four different locations along the Transantarctic Mountains in order to relate the extent of their contamination to their location on the ice fields of the Allan Hills, the Elephant Moraine, the Lewis Cliff ice tongue, and the Thiel Mountains. Their analytical data confirmed that the concentrations of iodine and of all other halogens *decrease* with depth below the surfaces of ordinary chondrites. The depth profile of iodine in ALHA 77258 (H6; terrestrial age (270 ± 70) × 10^3 years; Score and Lindstrom 1990; Nishiizumi et al. 1989) in Fig. 18.20 shows that the iodine concentration

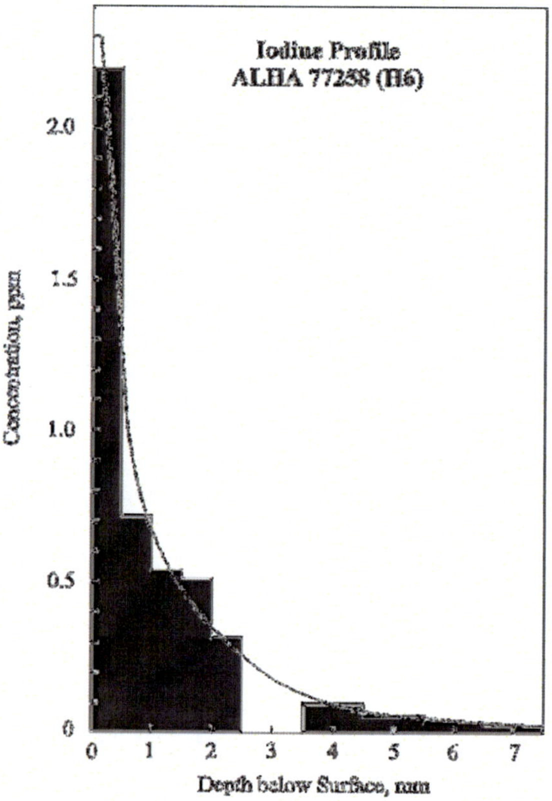

Fig. 18.20 The concentrations of iodine and of other halogens (F, Cl, Br) decrease with depth below the surface of the ordinary H6 chondrite ALHA 77258. The pattern is attributable to the deposition of CH_3I molecules from the atmosphere to the surface of this meteorite followed by diffusion of the iodine to a depth of about 6.5 mm below the surface. The amount of iodine contamination of the surface of a meteorite increases with increasing exposure time, whereas the contamination with F, Cl, and Br decreases with distance from the coast (Data from Langenauer and Krähenbűhl 1993)

Table 18.8 Apparent diffusion coefficients (D_a) for F^- and Cl^- ions in the surface layers of two ordinary chondrites from the ice fields of the Allan Hills (Langenauer and Krähenbühl 1993; Score and Lindstrom 1990)

Meteorite	D_a (F^-) (m^2/s)	D_a (Cl^-) (m^2/s)
ALHA 79025 (H5)	$>6.4 \times 10^{-19}$	$>4.4 \times 10^{-17}$
ALH 84082 (H6)	$>4.8 \times 10^{-20}$	$>2.7 \times 10^{-17}$

of this specimen decreases dramatically with increasing depth from 2.20 ± 0.05 ppm at the surface to 0.3 ± 0.01 ppm below a depth of 6.5 mm. All ten Antarctic meteorite specimens analyzed by Langenauer and Krähenbühl (1993) display similar depth-related decreases of the concentrations of all other halogens. The authors attributed these profiles to diffusion of ions through the network of available pore spaces and derived apparent diffusion coefficients (D_a) for fluorine and chlorine in two meteorite specimens in Table 18.8.

Fig. 18.21 Derrick Peak at 80°04′S and 156°23′E is the site where a small group of iron meteorites was found by a team of New Zealand geologists. The principal meteorite-discovery site is outlined in red. The Hatherton Glacier is a southern tributary to the Darwin Glacier which flows between the Brown Hills and the Britannia Range before merging into the Ross Ice Shelf (Adapted from U.S.G.S. map SU 56-60/1, Mount Olympus, Antarctica, with information from Shiraishi 1979)

18.8 Iron Meteorites: Derrick Peak

The iron meteorites are of special interest because this group of meteorites is under-represented among Antarctic meteorites (Table 18.5). Several large fragments of the Derrick Peak iron meteorite (DRP 78001–78009) fell about 1 million years ago and have lain on a substrate of soil and rock fragments ever since. Therefore, these specimens were not transported in the ice of the polar plateau to their present location like most other meteorites in Antarctica, but they actually landed on rocks where they were found.

Six specimens of this iron meteorite were discovered in December of 1978 by a group of New Zealand geologists led by Michael Selby of the Waikato University on the rocky slope of Derrick Peak in the Darwin Mountains at 80°04′S, 156°23′E adjacent to the Hatherton Glacier in Fig. 18.21. A photograph of the most massive of these specimens, weighing 160 kg, was published by Cassidy (2003). The ANSMET Group, working nearby, recovered ten additional specimens on the day following the original discovery. In subsequent years, searches by another New Zealand group and by ANSMET yielded 11 additional specimens bringing the total to 27.

When the geologists who made the original discovery reported by radio that they had found several iron meteorites on the slope of Derrick Peak, a helicopter was dispatched to transport the specimens to the camp on the nearby Darwin Glacier. The largest of the specimens was too heavy to be lifted into the helicopter and was therefore rolled onto a stretcher provided by the crew of the helicopter. However, when the stretcher was raised off the ground, the handles broke. (Medical stretchers are not designed to lift bodies weighing 160 kg). The specimen was eventually rolled into the helicopter on a heavy plank that happened to be available in the Darwin Camp.

Derrick Peak in Fig. 18.21 is a glacial horn whose summit at 2,070 m is about 900 m above the surface of the Hatherton Glacier. Most of the meteorite specimens were found on a flat bedrock bench along a northeast trending ridge, although two specimens appear to have slid or rolled off the ridge toward the Hatherton Glacier and one specimen was found more than 5 km southeast of the main collecting site.

The rocks exposed on Derrick Peak consist of a basal conglomerate overlain by quartz sandstones and orthoquartzites of the Devonian Taylor Group (Kamp and Lowe 1982). These rocks are unconformably overlain by the Permian tillite of the Darwin Formation and by sparsely carbonaceous sandstones of the Misthound Coal Measures that cap the summit of the mountain. The sedimentary rocks of the Taylor Group dip gently to the north at about 3° and were intruded by sills and dikes of the Jurassic Ferrar Dolerite. Consequently, the talus and lag gravels that cover the slope of Derrick Peak consist of a mixture of clasts of light-colored sandstone and of dark brown weathered dolerite.

All of the meteorite specimens on Derrick Peak are part of a shower of fragments that resulted from the explosion of a large iron meteoroid as it passed through the atmosphere. The upper surfaces of the specimens are dark brown which contrasted with the tan-colored sandstone and thus facilitated their discovery. The picture of one of the specimens in Fig. 18.22 demonstrates that the undersides of this and other specimens are deeply weathered because they were exposed to moisture in the ground for a long period of time. Nishiizumi et al. (1987) used cosmogenic radionuclides to determine that the Derrick Peak meteorite fell 1.0 ± 0.1 million years ago. Lazarev, the iron meteorite recovered by Russian geologists in the Humboldt Range of East Antarctica (Section 18.6.2) has an even older terrestrial age of 5 million years (Nishiizumi et al. 1989).

Fig. 18.22 This specimen of the Derrick Peak iron meteorite has a mass of 10 kg and is highly corroded as a result of chemical weathering after it fell about 1 million years ago. The upper surface is dark brown in color and is characterized by protruding crystals of schreibersite [$(Fe,Ni)_3P$] which are more resistant to chemical weathering than kamacite (Fe-Ni alloy). The original fusion crust and heat-altered surface layer are no longer present. The Derrick Peak meteorite is a member of group IIAB (coarse octahedrite) (Courtesy of NASA, JSC, Houston, Texas)

Table 18.9 Chemical composition of two specimens of the iron meteorite Derrick Peak (DRP78008 and DRP78009) (Clarke 1982)

Sample Number	78008.3[a]	78009.3[a]	78009.4[b]
Ni (%)	6.64	6.59	6.3
Co (%)	0.51	0.46	0.47
P (%)	0.35	0.34	–[c]
Cu (ppm)	–	–	110
Ga (ppm)	–	–	56
As (ppm)	–	–	9.6
W (ppm)	–	–	0.7
Ir (ppm)	–	–	<0.03
Au (ppm)	–	–	1.2

[a]E. Jarosewich, analyst.
[b]Daode et al. (1982).
[c]Not determined. See also Wasson et al. (2007).

The Derrick Peak meteorite specimens are composed of coarse blades of kamacite (5–8 mm wide) and contain inclusions of schreibersite and troilite which taken together identify this meteorite as a coarse octahedrite (Table 18.4). The chemical composition of two specimens in Table 18.9 indicates that the Derrick Peak meteorite is characterized by containing only about 6–7% Ni (Clarke 1982; Wasson et al. 2007). The schreibersite crystals protrude from the weathered surface of the specimens, which is unusual because this mineral ablates preferentially when iron meteorites are heated during their passage through the atmosphere. The explanation for the unusual appearance of the Derrick Peak specimens is that, after the fall, the iron has weathered faster than the schreibersite which has caused more than 1 cm of iron to be lost from the surface and explains why the schreibersite crystals are now sticking out of the surface (Clarke 1982).

The Derrick Peak iron meteorite was more extensively altered by chemical weathering than most other Antarctic meteorites which remained frozen until they were exposed by ablation of the ice in which they were embedded. The weathering products of DRP78009 (138.1 kg) include Cl-bearing *akaganéite* (β-FeOOH) and goethite (α-FeOOH) with smaller amounts of lepidocrocite (γ-FeOOH) and maghemite (γ-Fe_2O_3) (Buchwald and Clarke 1989).

Akaganéite was first described by MacKay (1962) from the Akagané limonite mine in Japan and was reported by Marvin (1983b) as a weathering product of iron meteorites and of a fragment of the Sputnik satellite launched into Earth orbit by the USSR in 1957. The formula of this mineral was subsequently worked out by Keller (1970). The akaganéite that formed on the Derrick Peak meteorite contains chlorine at surprisingly high concentrations ranging from 0.8% to 4.5%, as well as sulfur (0.2–0.5%), Si, Ca, and Mg. Buchwald and Clarke (1989) concluded that the chlorine did not originate by dissolution of meteoritic lawrencite ($FeCl_2$), but was sorbed by the meteorites from windblown snow and dust particles.

18.9 Lunar and Martian Meteorites

The importance of searching for meteorites on the East Antarctic ice sheet was widely recognized in 1983 after it became known that the small and inconspicuous rock sample in Fig. 18.23 that was collected by Ian Whillans and John Schutt on one of the ice fields

Fig. 18.23 This small rock sample was collected on January 18, 1982 by Ian Whillans and John Schutt on the Middle-Western ice field on the East Antarctic ice sheet west of the Allan Hills. It was subsequently identified as a lunar rock by scientists at the Johnson Space Center in Houston, Texas, and was assigned the identifying number ALHA 81005. The rock contains fragments of coarse-grained plagioclase feldspar in a fine-grained black matrix and has been described as an anorthosite breccia from the highlands of the Moon. (Courtesy of NASA/LPI))

west of the Allan Hills originated from the highlands of the Moon. Cassidy (2003) quoted the

description of this specimen from John Schutt's field notebook:

> #1422 - Strange meteorite. Thin, tan-green fusion crust, ~50%, with possible ablation features. Interior is dark grey with numerous white to grey breccia (?) fragments. Somewhat equidimensional at ~3 cm.

This was the first lunar rock to be collected in Antarctica or anywhere else on the Earth. However, the origin of the specimen was not recognized until it was examined by the planetary scientists of the Johnson Space Center in Houston who had been working with lunar rocks which the Apollo astronauts had collected on the surface of the Moon. These experts tentatively identified the specimen as a lunar rock, classified it as a polymict anorthosite breccia, and assigned its official name: ALHA 81005. At that time, the presence of a rock from the surface of the Moon on the polar plateau of Antarctica appeared to be highly unlikely because the experts doubted that rocks could be blasted into space from the surface of the Moon. However, when the origin of this meteorite was discussed in March of 1983 at the Lunar Planetary Science Conference in Houston, the verdict was unanimous: ALHA 81005 is a lunar meteorite that originated from the highlands of the Moon (Marvin 1983a; Keil and Papike 1983; Bogard 1983). We now know that the specimen was captured by the Earth more than 140,000 years ago after it had been exposed to cosmic rays on the surface of the Moon for 580 ± 180 million years (Eugster et al. 1986).

In subsequent years, the meteorite collectors of ANSMET have found many more lunar meteorites on the icefields adjacent to the Transantarctic Mountains while Japanese investigators recovered ten additional specimens of lunar rocks from the large icefields associated with the Yamato Mountains near the coast of East Antarctica. The lunar meteorites that have been collected in Antarctica are listed in Appendix 18.12.3 (see also Cassidy 2003; Table 7.1). The total mass of the lunar meteorites collected in Antarctica is greater than 2,580 g and continues to increase as additional specimens are recovered.

Prior to the recovery of ALHA 81005, the ANSMET group had collected two other unusual meteorites (ALHA 77005 and EETA 79001) one of which is shown in Fig. 18.24. The mineralogical composition

Fig. 18.24 The meteorite EETA 79001 was collected by ANSMET personnel during the 1979/80 field season on the ice fields surrounding the supraglacial Elephant Moraine (Chapter 7). It is an igneous rock of mafic to ultramafic composition that has been identified as originating from the surface of Mars because its mineral composition and texture resemble those of the so-called SNC (snic) meteorites defined by Shergotty, Nakhla, and Chassigny. The view of EETA 79001 shown here indicates that it has a black fusion crust and a smooth shape as a result of aerodynamic sculpturing during its passage through the atmosphere of the Earth (Courtesy of NASA/LPI)

of these specimens is identical to that of a strange achondrite known as Shergotty which together with Nakhla and Chassigny formed the so-called SNC (snic) group. These meteorites are mafic and ultramafic igneous rocks that crystallized much more recently than chondrites and achondrites from the asteroid belt. Therefore, they must have originated from a parent body that was large enough to have remained geologically active until long after the parent bodies of stony meteorites had cooled. The origin of SNC meteorites was hotly debated by meteoriticists until eventually a consensus was reached around 1979/80 that they originated from Mars. Therefore, by analogy with the SNC meteorites, ALHA 77005 and EETA 79001 were identified as rock samples from the surface of Mars. This conclusion has validated the on-going search for meteorites in Antarctica because the martian meteorites are contributing to the exploration of Mars which, to this day, we can only study by means of satellites in orbit and by robotic rovers (Faure and Mensing 2007; Carr 2006; McSween 2005; Papike 1998). The martian meteorites that have been recovered in Antarctica are listed in Appendix 18.12.4.

18.9.1 Lunar Rocks in Antarctica

The discovery of the first lunar meteorite (ALHA 81005) in 1981 on the Middle Western ice field adjacent to the Allan Hills was a milestone in planetary science. Subsequent study of this object (Fig. 18.23) established that it originated from the highlands of the Moon (Bogard 1983). The most plausible explanation is that it was excavated by the explosive impact on the Moon of a large meteoroid (or small asteroid) and then entered into an eccentric orbit around the Sun from which it was captured by the Earth. This explanation has far-reaching consequences because it suggests that large amounts of lunar rocks could have been transferred from the Moon to the Earth throughout the history of the solar system, especially during the period of intense lunar bombardment that ended about 3.8 billion years ago. The rocks that were ejected from the Moon and arrived on the Earth during that period were subsequently destroyed by weathering and were incorporated into the crust of the Earth. The process that caused rocks to be ejected from the surface of the Moon is presumably also capable of ejecting rocks from the surfaces of Mars and from the other terrestrial planets of the solar system. In this way, the discovery of the first lunar rock on the Earth supports the conjecture that the transfer of rock samples among the solid planets is a natural process that had escaped our notice until 1983 when ALHA 81005 was positively identified as a rock from the Moon (Gladman et al. 1996).

After the original discovery of ALHA 81005 in 1981/82, several additional lunar meteorites were recovered on the ice fields of Antarctica, but none were found elsewhere on the Earth. This created the erroneous impression that lunar meteorites have fallen only in Antarctica. Finally, a small stone about 3 cm in diameter and weighing only 19 g was collected in the desert of Australia at a place called Calkalong Creek. This specimen was examined by Hill et al. (1991) who identified it as a lunar meteorite based on certain diagnostic element ratios (e.g., Fe/Mn, K/La, Fe/Sc, and Ca/Al). These authors also reported that the name of the collecting site in Australia (i.e., Calkalong) is derived from the aboriginal phrase "Kalkalupilinguta" which means "Seven sisters went up in the sky chased by the Moon."

The number of lunar meteorites (also called "lunaites" by Warren et al. 2005) has increased from one in 1981 to 32 in 2005, including the 22 specimens in

Appendix 18.12.3 recovered in Antarctica (Zeigler et al. 2005). Several specimens are "paired" because they are either fragments of a lunar rock sample that broke up during its passage through the atmosphere or because they originated from the same impact crater on the Moon (i.e., source pairing, Warren 2005).

The lunar meteorites have aerodynamically rounded shapes and are covered by a fusion crust which they acquired while they passed through the atmosphere of the Earth. Lunar meteorites also contain cosmogenic radionuclides (e.g., ^{10}Be, ^{26}Al, ^{36}Cl, and ^{53}Mn) because they were exposed to cosmic rays on the surface of the Moon and while they were in transit from the Moon to the Earth. Many lunar meteorites are breccias and contain evidence of shock metamorphism. Like the other "nonchondrites" in Table 18.1, the lunar meteorites lack chondrules and do not contain grains of metallic Fe-Ni alloys. Because of their similarities to achondrite meteorites and to pebbles or cobbles of terrestrial rocks, lunar meteorites are not easy to identify in the field and, in some cases, were only collected in response to the ANSMET imperative: "When in doubt, collect…".

The chemical composition of ALHA 81005 in Table 18.10 was determined by Korotev et al. (1983) who used the method of instrumental neutron-activation analysis (INAA) because it is nondestructive, because it is well suited for the analysis of small samples, and because it can determine many elements simultaneously. For example, Korotev et al. (1983) analyzed seven small grains of the meteorite with a total weight of only 78 mg and obtained the concentrations of 33 elements from each of these granules. The final result in Table 18.10 is a weighted average of all subsamples. The subtitle of their report (*So Apollo 18 flew, but where did it sample?*) alluded to the fact that this Antarctic meteorite extended the exploration of the Moon that ended with the flight of Apollo 17, but the precise location on the Moon from which ALH 81005 originated is not known. The same is true for most, if not all, lunar rocks that have been recovered in Antarctica and elsewhere.

The chemical analysis of ALHA 81005 in Table 18.10 and the data contained in the report by Hill et al. (1991) define a field for lunar meteorites and lunar basalt in Fig. 18.25 in terms of the Fe/Mn, K/La, and Fe/Sc ratios. Data point 1 in that diagram identifies ALHA 81005 and point 2 is the Calkalong lunar meteorite. The locations of these data points within the cluster of lunar rocks (open circles) that were collected on the Moon by the Apollo astronauts leaves no doubt that ALHA 81005 and Calkalong are rocks from the Moon. The diagram also contains a field defined by HED achondrites (howardites, eucrites,

Table 18.10 Average chemical composition of eight fragments of the first lunar meteorite found on Earth (ALHA 81005) (Data from Korotev et al. 1983)

Major elements		Trace elements			
Oxide	Conc. (%)	Element	Conc. (ppm)	Element	Conc. (ppm)
TiO_2	0.23	Sc	8.81	Sr	141
Al_2O_3	25.1	V	23	Zr	19 ± 12
FeO_t	5.53	Cr	900	Sb	0.030
MgO	8.8	Mn	620	Cs	0.04 ± 0.02
CaO	0.3–0.4	Co	22.5	Ba	24
Na_2O	0.005	Ni	243	La	1.80
K_2O	0.01 to 0.02	Br	0.3	Nd	2.75
Total	40.03[a]	Rb	<6	Sm	0.855

[a]All elements were determines by INAA (instrumental neutron activation analysis) which is not capable of measuring SiO_2.

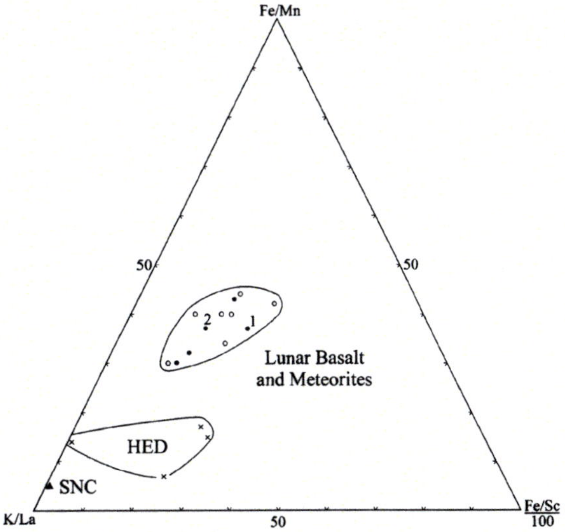

Fig. 18.25 Lunar basalt (open circles) collected by the Apollo astronauts and lunar meteorites (solid circles) collected in Antarctica occupy a well-defined area in coordinates of their Fe/Mn, K/La, and (Fe/Sc) × 0.01 ratios. The achondrites of the HED (Howardite, Eucrite, Diogenite) groups (crosses) lie in a separate field. The SNC (solid triangles) meteorites (Shergottites, Nakhlites, Chassignites) originated from Mars. The sample numbered 1 is ALHA 81005 and 2 is Calkalong which was recovered in the Australian desert. The Fe/Sc ratio was reduced by a factor or 0.01 in order to prevent the data points from crowding into the Fe/Sc corner of the triangle (Data for ALHA 81005 from Korotev et al. 1983; and for the other samples from Hill et al. 1991)

diogenites) in a different part of the diagram. In addition, the average chemical composition of the martian meteorites (Shergotty, Nakhla, and Chassigny) is indicated by the location of the solid triangle in Fig. 18.25.

The crystallization ages of lunar rocks collected by the Apollo astronauts range from about 4.55 billion to less than 3.0 billion years and define episodes of volcanic/magmatic activity on the Moon (Faure and Mensing 2005, p. 147). The lunar meteorites from Antarctica in Table 18.11 also crystallized in this interval of time (Misawa et al. 1993; Chen and Wasserburg 1985; Tera et al. 1974). In contrast to rocks from the Moon, the crystallization ages of meteorites from the asteroid belt in Fig. 18.26 are restricted to values between 4.55 and

Table 18.11 Crystallization ages of lunar meteorites collected in Antarctica in 10^9 years

Code No.	Sm-Nd	Rb-Sr	U-Th-Pb	$^{40}Ar/^{39}Ar$	K-Ar	Reference
Yamato/Asuka						
Y86032	–	–	–	–	3.87 ± 0.30	1
Y82192	4.30	3.90 ± 0.50	4.549 ± 0.006	4.24 ± 0.34	3.90 ± 0.20	1,5
Y82193	–	–	–	–	>3.0	1,5
Y791197	4.30	3.95 ± 0.24	4.25	4.07 ± 0.18	–	1,5
A8811757	3.87 ± 0.05	3.84 ± 0.03	3.85 ± 0.15	3.80 ± 0.012	–	2
			3.82 ± 0.29			
			3.93 ± 0.28			
Transantarctic Mountains						
ALHA 81005	–	–	4.50	–	4.30 ± 0.90	1,5
			3.90			
			4.60			
EET 96008	–	–	3.53 ± 0.27	–	–	3
			3.52 ± 0.10			
MAC 88104	–	–	–	–	3.55 ± 0.40	4
MAC 88105						
LAP 02205	2.992 ± 0.085	2.991 ± 0.014	–	–	–	6
LAP 02205	–	–	–	2.915 ± 0.010	–	7
	–	–	–	2.95 ± 0.04	–	8
LAP 02205	–	–	2.93 ± 0.15	–	–	9

1. Eugster (1989); 2. Misawa et al. (1993); 3. Anand et al. (2003); 4. Eugster et al. (1991); 5. Eugster and Niedermann (1988); 6. Rankenburg et al. (2007); 7. Fernandes and Burgess (2006); 8. Nyquist et al. (2005); 9. Anand et al. (2006).

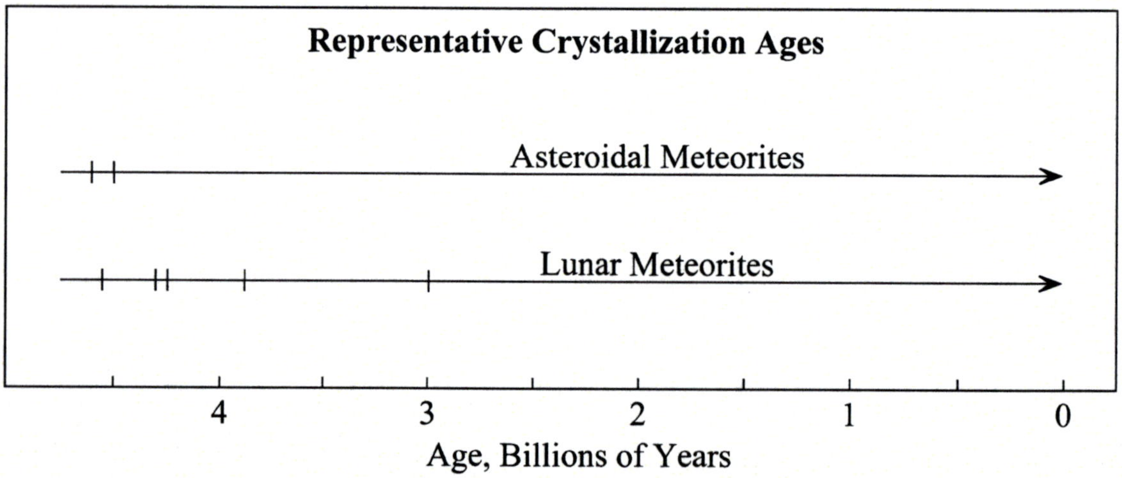

Fig. 18.26 The parent bodies of the asteroidal meteorites cooled and crystallized in about 100 million years between 4.50 and 4.60 billion years ago. The meteorites that originated from the Moon (also known as lunaites) crystallized during episodic igneous activity on the surface and in the crust of the Moon between about 4.55 and less than 3.0 billion years ago. The data are referenced in Table 18.11

4.65 billion years (Faure and Mensing 2007). The wide range of variation of the crystallization ages of lunar meteorites indicates that igneous activity continued on the Moon long after the parent bodies of asteroidal meteorites had cooled and crystallized.

Lunar meteorites collected in Antarctica have also yielded information about the frequency of impacts on the surface of the Moon. For example, the lunar meteorites from the MacAlpine Hills (MAC 88104 and MAC 88105) and from the Queen Alexandra Range (QUE 93069 and QUE 94269) are regolith breccias that contain small fragments of glass that formed by cooling of impact melts. The ages of these glass particles reported by Cohen et al. (2005) record the impacts of meteorites that occurred on the Moon prior to the ejection of the respective meteorites from the surface of the Moon. The samples, weighing between 4 and 121 μg, were dated by the $^{40}Ar/^{39}Ar$ method described by Faure and Mensing (2005). The best estimates of the dates obtained by Cohen et al. (2005) for glass clasts in Fig. 18.27 have a polymodal frequency distribution between 1.8 and 4.2 billion years. Impact glasses that formed more than 3.9 billion years ago have probably been destroyed by more recent impacts. Although the glass particles in regolith breccias are difficult to date and have average analytical errors of about ± 0.30 billion years, such measurements are capable of yielding valuable insights into the impact history of the Moon (see also Tera et al. 1974).

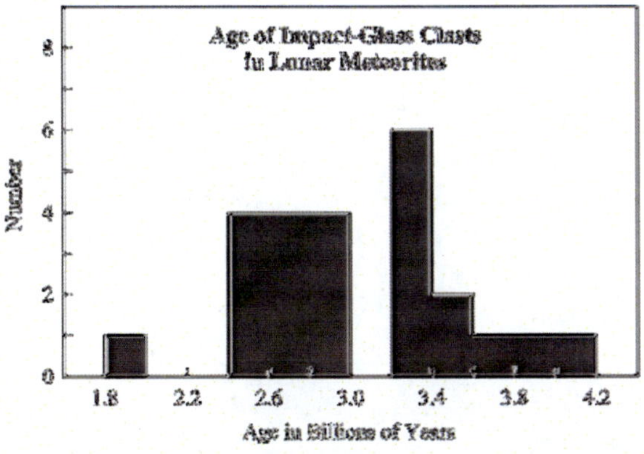

Fig. 18.27 The $^{40}Ar/^{39}Ar$ dates of small particles of impact-melt glass in lunar meteorites collected in Antarctica and in the Sahara Desert of Libya indicate the frequency of explosive impacts of meteoroids or asteroids on the Moon. The results derived from these samples yield a spectrum of dates ranging from >1.8 to <4.2 billion years. The low abundances of dates older than 3.4 billion years implies that some glass particles older than about 3.4 billion years were destroyed by more recent impacts on the Moon. The $^{40}Ar/^{39}Ar$ dates shown here were measured by Cohen et al. (2005)

Five lunar meteorites listed in Appendix 18.12.3 were discovered in 2002/3 on the LaPaz ice field in Antarctica (Appendix 18.12.1). The specimens are composed of low-Ti basalt and are paired with each other. This set of lunar meteorites includes LAP 02205 which has a mass of 1,230 g making it the most massive rock from the Moon ever found on the Earth. All of these lunar rocks crystallized 2.93 ± 0.25 billion years ago (Table 18.11) and fell to the Earth 55,000 ± 5,000 years ago (Nishiizumi et al. 2006).

18.9.2 Martian Rocks in Antarctica

The martian meteorites listed in Appendix 18.12.4 resemble Shergotty and Nakhla both of which are type examples of igneous rocks that were ejected from the surface of *Mars* (McSween 1999, 2005; Nyquist et al. 2001). The original shergottite is a stone of basaltic composition and texture that fell in 1865 at the village of Shergotty in India. Nakhla fell in 1911 near Alexandria in Egypt and was originally classified as a diopside-olivine achondrite. Shergotty, Nakhla, and Chassigny (a dunite) differ from other kinds of achondrites in Table 18.1. The shergottites are now subdivided into basaltic, lherzolitic, and olivine-phyric subgroups based on their grain size and mineral compositions.

The original Nakhla stone attracted attention when Papanastassiou and Wasserburg (1974) reported that its Rb-Sr crystallization age is only 1.37 ± 0.02 billion years, which was later confirmed by Gale et al. (1975) and by Nakamura et al. (1982). The anomalously low crystallization age distinguished Nakhla from most other meteorites and raised questions about its origin.

The so-called Shergotty achondrite is even more anomalous because its crystallization age of 165 ± 11 million years is much lower than that of Nakhla (Chen and Wasserburg 1986; Jagoutz and Wänke 1986; Nyquist et al. 1979; Bogard et al. 1979). Evidently, Nakhla and Shergotty originated from a body in the solar system that contained sufficient internal heat to form magma long after the parent bodies of the "normal" meteorites had cooled below the melting point of silicate minerals. The source of these meteorites was subsequently identified to be the planet Mars (Stolper and McSween 1979;

Cassidy 2003, p. 109). This conclusion was strengthened in 1983 when ALHA 81005 was identified as a lunar meteorite that was blasted off the surface of the Moon by the impact of a meteorite from the asteroid belt (Melosh 1989; Section 18.9.1). If it happened on the Moon, it could also happen on Mars.

The next challenge in this chain of forensic meteoritics was to identify the martian meteorites in the growing collection of stony meteorites that were being recovered in Antarctica. The "missing link" turned out to be the isotopic composition of the noble gases and of nitrogen in the SNC meteorites. Bogard and Johnson (1983) and Bogard et al. (1984) reported that certain isotope ratios of the noble gases in EET 79001 and ALHA 77005 are identical to those of the meteorite Zagami which had previously been identified as a member of the SNC group on the basis of its low crystallization age. The isotopic and elemental ratios of the noble gases in these meteorites differ consistently from noble gases in normal stony meteorites, in lunar rocks, and that exist in the terrestrial atmosphere; but they are *identical* to the isotopic compositions of the noble gases in the atmosphere of Mars measured by the Viking spacecraft that landed on Mars in 1976 (Klein 1978).

The Shergotty meteorite and two of its Antarctic relatives (e.g., EETA 79001 in Fig. 18.24) were intensively studied by the members of the so-called Shergotty Consortium (Laul 1986). The investigations included a wide range of topics, including their chemical compositions (Laul et al. 1986), the crystallization age of the Shergotty meteorite (Jagoutz and Wänke 1986; Chen and Wasserburg 1986; Jones 1986), the carbon-isotope composition of the shergottites (Wright et al. 1986), nitrogen and the noble gases they contain (Becker and Pepin 1986), as well as xenon and other noble gases (Swindle et al. 1986), their cosmic ray exposure ages (Nishiizumi et al. 1986; Bhandari et al. 1986), and their thermoluminescence (Hasan et al. 1986). In addition, Gooding and Muenow (1986) concluded that sulfate and other weathering products in EETA 79001 probably formed on Mars. The evidence for chemical weathering on Mars was later elaborated by the participants of a workshop organized by Burns (1993). More recently, Swindle and Olson (2004) used the $^{40}Ar/^{39}Ar$ method of dating to recognize the aqueous alteration of nakhlites and to date its occurrence. Another noteworthy contribution to the study of Mars was a list of the concentrations of the major and trace elements in eight martian meteorites published by Treiman et al. (1986).

The search for martian meteorites in Antarctica and in the hot deserts of the world continues because these meteorites are the only rocks from Mars that are presently available. The most recent information on the martian meteorites is contained in the books edited by Davis (2005) and by McFadden et al. (2006) and in the papers presented at a workshop on the topic of unmixing of SNCs (Treiman 2003).

Additional noteworthy contributions to the study of Martian meteorites include the papers by McCanta et al. (2004), Greshake et al. (2004), Jones (2003), Borg et al. (2003, 1997), Kiefer (2003), Goodrich (2002), Goodrich et al. (2003), Herd et al. (2002), Taylor et al. (2002), Eugster et al. (2002), Barrat et al. (2001), Sano et al. (2000), Hale et al. (1999), Kring et al. (2003), and Ash et al. (1996).

18.9.3 Life on Mars? (ALH 84001)

The most important meteorite discovered in Antarctica in more than 25 years of searching was picked up by Roberta Score of ANSMET during the 1984/85 field-season on the Far Western ice field near the Allan Hills (Cassidy 2003). This specimen in Fig. 18.28 was named ALH 84001 and was identified as a diogenite achondrite (Table 18.1) weighing in at 1,930.9 g (Score and Lindstrom 1990). Nearly 10 years passed before

Fig. 18.28 ALH 84001 was blasted from the surface of Mars by the impact of a meteorite 16 million years ago. It is a coarse-grained orthopyroxenite that crystallized at 4.5 ± 0.14 Ga (Mittlefehldt 1994; Nyquist et al. 1995). This specimen was collected by Roberta Score in 1984 on the Far Western ice field west of the Allan Hills (Courtesy of NASA, Johnson Space Flight Center, Houston, Texas)

David W. Mittlefehldt discovered that ALHA 84001 is actually "a coarse grained cataclastic orthopyroxenite related to the martian SNC meteorites" (Mittlefehldt 1994). In addition, Mittlefehldt reported the presence of two generations of carbonate minerals which could have formed at a moderately high temperature of about 700°C.

After ALHA 84001 had been recognized as a martian meteorite of unique composition, its crystallization age was measured with the expectation that its age would relate it either to the nakhlites (i.e., 1.3×10^9 years) or to the shergottites (i.e., 0.16×10^9 years). Instead, the crystallization age of ALH 84001 turned out to be 4.56 billion years (Nyquist et al. 2001). Evidently, this specimen is a sample of the primordial crust of Mars that contains a record of its long history including magma formation, cooling and crystallization at depth below the surface, followed by episodes of shock metamorphism which caused crystals of orthopyroxene and chromite to be crushed and which converted plagioclase crystals to maskelynite glass. Fractures within the rock were subsequently filled by hydrothermal Ca-Mg-Fe carbonate, magnetite, and euhedral pyrite. The rock was eventually excavated by the explosive impact of a meteorite and was exposed to cosmic rays in space for 14.7 ± 0.9 million years (Eugster 2003; Eugster et al. 2002; Goswami et al. 1997), whereas its terrestrial age is only 11,000 ± 1,000 years (Nishiizumi et al. 1989).

After ALH 84001 was identified as a unique martian meteorite, McKay et al. (1996) reported that fracture surfaces in the interior of this specimen contain polycyclic aromatic hydrocarbons (PAHs), as well as globules of carbonate minerals that contain fine-grained crystals of magnetite and iron sulfides. The authors conceded that the carbonate globules in Fig. 18.29 could have formed by inorganic processes, but noted that the presence of organic molecules favored a biogenic origin not only for the PAHs but also for the carbonate globules. Some of the carbonate globules are cut by fractures related to shock metamorphism that occurred either while this rock was still on Mars or while it was in space. In any case, the shock-faulting of the carbonate globules proves that they are not products of terrestrial weathering. McKay et al. (1996) described the molecular composition of the hydrocarbons and demonstrated that the PAHs in ALH 84001 are not of terrestrial origin. Instead, the molecular composition of the PAHs in this meteorite resembles those of carbonaceous chondrites but differs in some specific details.

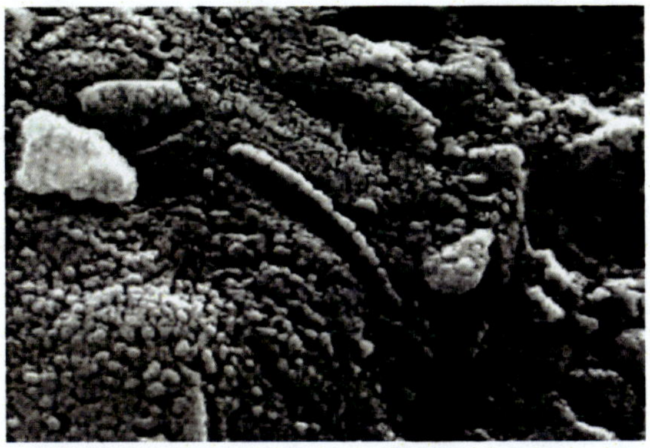

Fig. 18.29 Fracture surfaces in the martian meteorite ALH 84001 are lined with carbonate globules and tube-like structures which were tentatively identified by McKay et al. (1996) as evidence for the existence of micro-organisms on Mars. However, the evidence is not conclusive because an abiogenic origin of these features cannot be excluded (Courtesy of NASA and of the Johnson Space Flight Center, Houston, Texas)

The carbonate globules in Fig. 18.29 have an average diameter of about 50 μm and contain cores rich in calcium and manganese covered by alternating layers containing iron, magnesium, and sulfur. The iron-rich rims of the globules are composed of very fine grained magnetite crystals and lesser amounts of pyrrhotite. The co-existence of carbonates with magnetite and iron sulfide can be attributed to inorganic processes but requires a complex sequence of environmental conditions. The globules may also have formed by biological processes which can operate in a state of thermodynamic disequilibrium. In addition, magnetite particles of biogenic origin that resemble the magnetite particles in the globules of ALH 84001 do occur in different kinds of terrestrial sediment. These terrestrial magnetite particles are formed by certain anaerobic bacteria. McKay et al. (1996) concluded their presentation with the statement:

> Although there are alternative explanations for each of these phenomena taken individually, when they are considered collectively, particularly in view of their spatial association, we conclude that they are evidence for primitive life on early Mars.

This carefully worded conclusion aroused a great deal of public interest, which re-invigorated the exploration of the solar system and validated the collection of meteorites in Antarctica and elsewhere in the world. The paper by McKay et al. (1996) also started a debate among scientists whose comments were published under the heading: "Evaluating the evidence for past life on Mars" (Science, vol. 274, pp. 2119–2125). Although the evidence presented by McKay et al. (1996) did not convince the skeptics, the search for life on Mars and on other bodies of the solar system continues to receive a high priority by the mission planners at NASA. In spite of these efforts, no life forms of any kind have yet been found on Mars or anywhere else in the solar system. However, the absence of evidence is not evidence of absence. In fact, the conviction is spreading that self-replicating molecules will form anywhere liquid water is present:

> Ubi aqua, ibi vita.

After the publication of the reports by Mittlefehldt (1994) and McKay et al. (1996), the carbonate globules, magnetite and phosphate crystals, the atmospheric gases of Mars, and the hydrocarbons of ALH 84001 have continued to be studied by many science teams:

Carbonates: Jull et al. (1995, 1998), Harvey and McSween (1996), Shearer et al. (1996), Wadhwa and Lugmair (1996), Scott et al. (1997), Saxton et al. (1997), Leshin et al.(1997), Valley et al. (1997), Kirschvink et al. (1997), Kring et al. (1998), McSween and Harvey (1998), Kent et al. (2001), Greenwood and McSween (2001), Eiler et al. (2002a, b), Corrigan and Harvey (2004), Holland et al. (2005), and Niles et al. (2005).

Phosphate: Terada et al. (2003) and Greenwood et al. (2003).

Magnetite: Bradley et al. (1996, 1998), and Thomas-Keprta et al. (2000).

Atmospheric Gases: Miura et al. (1995), Swindle et al. (1995), Murty and Mohapatra (1997), Grady et al. (1998), Gilmour et al. (1998), Bogard and Garrison (1999), and Miura and Sugiura (2000).

Organic matter: Greenwood et al. (1997), Becker et al. (1997, 1999), McSween (1997), Thomas-Keprta et al. (1998), Zolotov and Shock (2000), and Stephan et al. (2003).

Miscellaneous: Warren and Kallemeyn (1996), Turner et al. (1997), and Treiman (1998).

Although the evidence derivable from ALH 84001 is not sufficient to prove that life existed on Mars, the search continues not only in the solar system but also in the Milky Way galaxy (Faure and Mensing 2007). The study of this martian rock from Antarctica has profoundly stirred the imagination of the people of planet Earth. What if there is life on Mars, or on Europa, or on Enceladus, … or on a planet orbiting one of the one-hundred-billion stars of the Milky Way galaxy?

Clark R. Chapman summed it up this way:

> The Mars rock mainly has proved to be a mirror in which scientists and laypeople alike see reflected their hopes or doubts about the prevalence of life in the universe. Chapman (1999)

18.10 Micrometeorites and Cosmic Spherules

The East Antarctic ice sheet transports not only meteorite specimens that have fallen onto its surface, but it also contains a mixture of small particles of terrestrial and extraterrestrial origin (Cassidy 1964). The *extraterrestrial* component of this sediment consists of unmelted micrometeorites (MMs), glassy cosmic spherules, and interplanetary dust particles (IDPs) which include the particles released by comets as they pass through the solar system (Faure and Mensing 2005, 2007).

Most of the micrometeorites, that fall to the Earth, land in the ocean and are detectable in deep-sea sediment. The particles that land on the continents are diluted by terrestrial rock and mineral particles and are subsequently decomposed by chemical weathering in soils. Fortunately, MMs that fall in Greenland and Antarctica are preserved by being incorporated into the ice sheets from which they are ultimately released when the ice melts or sublimates. Alternatively, the MMs can be recovered by melting the ice in-place and by collecting the sediment that accumulates in the melt cavity.

The *terrestrial* component of the dust particles embedded in the ice consists of volcanic ash, fine-grained dust derived from soil on the continents, carbon particles released by forest fires, biogenic particles (e.g., the skeletons of diatoms, seeds, and pollen grains), aerosol particles of atmospheric origin, including sea-spray particles that nucleate snow flakes (Section 17.10). In addition, the uppermost layer of snow and firn that was deposited after the start of the Industrial Revolution (i.e., post AD 1850) contains anthropogenic detritus such as flakes of metal, paint, and plastics, fly-ash particles and other combustion products, fibers (composed of wood, cotton, and synthetics), industrial contaminants (e.g., lead), and radioactive nuclides released by the testing of nuclear weapons and by the operation of nuclear reactors (Faure et al. 1997).

18.10.1 Discovery of Micrometeorites

The presence of terrestrial and extraterrestrial particles in the ice of Antarctica was originally suspected by Paul Siple (1908–1968) who first came to Antarctica in 1928–1930 as the representative of the Boy Scouts on Richard Byrd's first Antarctic expedition (Section 1.5.1). While he was working at one of Byrd's Little America stations on the Ross Ice Shelf, he asked to be lowered into the large cavity that had been produced by melting the ice below the surface. The sediment that was released by the ice collected at the bottom of the chamber from which water was pumped for use by the station personnel. This sediment was presumably composed not only of particles of volcanic rocks derived from the volcanoes of West Antarctica, but also contained particles of extraterrestrial origin. Unfortunately, the sediment was never examined because the jar in which it was stored can no longer be located (Cassidy 2003; Stonehouse 2002). The occurrence, morphology, and stratigraphy of tephra layers in the West Antarctic ice sheet have been described by LeMasurier and Thomson (1990), Palais (1985), Palais and Legrand (1985), Kyle and Jezek (1978), Kyle et al. (1981) and by others listed by them (Section 17.10).

The systematic study of MMs and other kinds of extraterrestrial particles became possible after 1986 when Michel Maurette and his associates discovered that large amounts of sediment (called cryoconite) have accumulated in the meltwater pools on the surface of the Greenland ice sheet (Maurette et al. 1986, 1987; Robin et al. 1990). This sediment serves as the substrate for algal mats that cover the bottoms of these ponds. In the following years, Maurette and his colleagues melted a large cavity into the East Antarctic ice sheet at Cap Prudhomme about 6 km from the French research station Dumont d'Urville on the Adélie Coast of East Antarctica. The ice that Maurette and his team melted had a mass of about 100 t and contained about 5,000 unmelted micrometeorites and a similar number of glassy spherules as well as an abundance of penguin feathers (Maurette et al. 1989). Therefore, the concentration of extraterrestrial particles (N) in the ice is (Cassidy 2003):

$$N = \frac{10{,}000}{100 \times 10^3} = 0.1 \text{ particles per kg of ice.}$$

The most productive sites for the recovery of large numbers of MMs and cosmic spherules on the East Antarctic ice sheet are located at:

- Cap Prudhomme, Adélie Coast of East Antarctica; Maurette et al. (1989, 1991)
- Astrolabe Glacier, 66°45′S, 139°55′E; Gounelle et al. (1999)
- Yamato Mountains, 71°30′S, 035°40′E; Terada et al. (2001)
- Dome C, 75°00(S, 125°00(E; Duprat et al. (2001, 2003)
- South Pole, water well, Taylor et al. (1998, 2000. 2007)
- Allan Hills, dust band, 76°45′S, 160°00′E; Harvey et al. (1998)
- Walcott Névé, 84°23′S, 162°40′E; Harvey and Maurette (1990)
- Dome Fuji, water well, Nakamura et al. (1999)

In addition, glassy spherules have been recovered from the margins of supraglacial moraines on the East Antarctic ice sheet (Cassidy 2003) and from deposits of till in the Transantarctic Mountains (Hagen 1988, 1995; Hagen et al. 1990; Koeberl et al. 1988; Koeberl and Hagen 1989; Koeberl et al. 1989; Harvey and Maurette 1990). Hagen et al. (1992) actually trapped glassy spherules from the wind blowing across the bare-ice surface of the Lewis Cliff ice tongue (84°17′S, 161°05′E). Additional occurrences of extraterrestrial particles in the ice of Antarctica have been described by Zolensky et al. (1988), Yiou et al. (1989), Yiou and Raisbeck (1987), Kumai et al. (1983), Wagstaff and King (1981), King and Wagstaff (1981), Mosley-Thompson (1980), Hodge et al. (1967), Shima (1966), Schmidt (1964), and Thiel and Schmidt (1961). Spherical particles of extraterrestrial origin have also been discovered in deep-sea sediment, in beach sand, and in a wide range of terrestrial environments listed by Taylor and Brownlee (1991) and by Taylor et al. (2000). The origin and significance of micrometeorites was discussed in a book published in 2006 by Michel Maurette (2006).

18.10.2 Origin and Composition

The study of extraterrestrial spherules in Fig. 18.30 (a, b) was initially retarded by their small size (less than 1 mm);

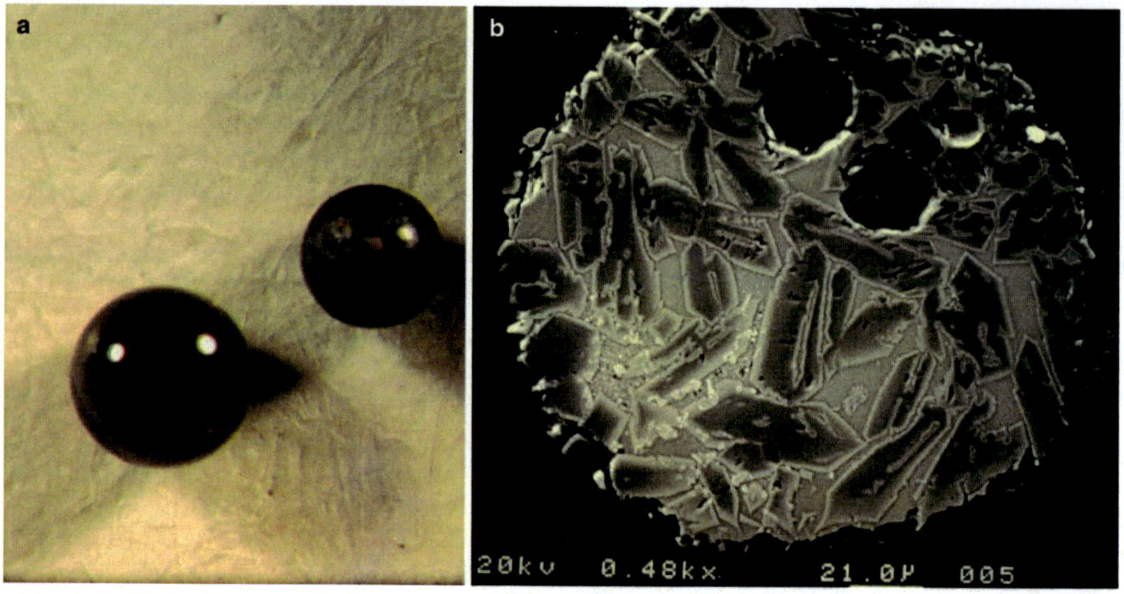

Fig. 18.30 (a) These shiny black spherules were extracted from till of the "Meteorite Moraine" near the Lewis Cliff ice tongue (Photo by E.M. Hagan). (b) The cross-section of a cosmic spherule from the "Meteorite Moraine" is composed of blocky olivine crystals in a glassy matrix. Back-scattered electron image by C. Koeberl

but these difficulties have been overcome by the use of modern analytical equipment. Prior to the discovery of large deposits of extraterrestrial particles on the ice sheets of Greenland and Antarctica, the number of particles available for study was small not only because the concentrations of MMs and cosmic spherules in deep-sea sediment and soil are low, but also because these particles are highly susceptible to chemical weathering in terrestrial environments. Micrometeorites and cosmic spherules that fall onto the surface of the ice sheets in Greenland and Antarctica have a much better chance to survive and therefore are more representative of all kinds of extraterrestrial particles that are deposited globally over the entire Earth.

Micrometeorites form as a consequence of collisions of meteoroids and asteroids in space (Genge et al. 2005). They have angular shapes and their chemical compositions range widely across the whole spectrum of stony meteorites including primarily carbonaceous and ordinary chondrites listed in Table 18.12 (Engrand et al. 1999). The unmelted micrometeorites contain the same rock-forming minerals as stony meteorites, including grains of metallic iron-nickel. Consequently, MMs are enriched in siderophile elements (e.g., iridium and gold) compared to crustal rocks of the Earth. For example, Tuncel and Zoller (1987) measured the iridium concentration of atmospheric particulates at the South Pole (9.1×10^{-17} g/m^3 of air) and used it to estimate the global meteorite-accretion rate (11,000 t/year). This value is somewhat lower than the global rate of deposition of "cosmic dust" (50–500 μm) of 30,000 ± 20,000 t/year obtained by Love and Brownlee (1993). However, both results imply that the global flux of cosmic dust to the Earth is *greater* than the flux of "hand-sized" meteorites. The cosmic dust is also an important source of organic molecules (Matrajt et al. 2004) and of certain trace elements in the surface environment of the Earth (Hodge et al. 1967; Hagen et al. 1990; Kurat et al. 1994).

Table 18.12 Estimates of the precursors of micrometeorites expressed as the percent by number. Based on a compilation from the literature by Taylor et al. (2007)

Precursors	Micrometeorites (%)
Carbonaceous chondrites	70
Ordinary chondrites	30
Iron meteorites	~1–2
HED (howardites, eucrites, diogenites)	~0.5

18.10.3 Classification

Micrometeorites are heated as they pass through the atmosphere depending on their velocity, mass, and entry angle (Toppani et al. 2001) and therefore display a wide range of textures from unmelted angular grains to glassy spherical particles. Kurat et al. (1994) described 23 micrometeorites from Cap Prudhomme and established a classification based on their textures and mineral compositions in order of *increasing* thermal alteration:

1. Unmelted micrometeorites (MMs): Characterized by a thin outer layer of platy magnetite containing not only iron but also magnesium, aluminum, silicon, and manganese.
   (a) Phyllosilicate MMs
   (b) Coarse-grained anhydrous MMs
   (c) Transitional MMs (a-b)
2. Scoriaceous MMs: Composed of highly vesicular glass with varying amounts of residual anhydrous minerals.
3. Cosmic spherules: Composed of quench-textured silicate glass with/without vesicles and residual anhydrous minerals.
4. Meteorite-ablation spherules: Form when the molten layer of stony meteoroids passing through the terrestrial atmosphere is stripped off and congeals into glassy spherules that fall to the surface of the Earth.

We have extended Kurat's classification for the purpose of this presentation (category 4 above) by adding "meteorite-ablations spherules" (Koeberl and Hagen 1989; Hagen et al. 1990; Hagen 1988, 1995; Harvey et al. 1998). Both kinds of spherules (i.e., 3 and 4 above) are present in the ice sheets of Antarctica and Greenland, but have not yet been distinguished from each other because both formed from stony meteoroids by melting in the terrestrial atmosphere and therefore have similar physical and chemical properties.

Cosmic spherules that formed by melting of micrometeorites can be distinguished from ablation spherules of stony meteorites by the observed concentrations of cosmogenic ^{10}Be, ^{26}Al, and ^{53}Mn they contain. Nishiizumi et al. (1991) measured the concentrations of ^{26}Al and ^{10}Be of a large number of cosmic spherules from deep-sea sediment and from the meltwater lakes on the Greenland ice sheet. The evidence indicated that the spherules had been exposed to galactic cosmic rays and to the solar wind for more than one hundred thousand, but less than 10 million, years while they revolved around the Sun in circular rather than highly elliptical orbits. Therefore, the concentrations of cosmogenic radionuclides of cosmic spherules considered by Nishiizumi (1983), Nishiizumi et al. (1991, 1995), and Raisbeck et al. (1986) support the view that the cosmic spherules were irradiated as small particles in space. Another way of expressing this conclusion is to say that the micrometeorites are "crumbs" (Taylor et al. 2007) that were produced as a result of collisions of asteroids such as Vesta and that the cosmogenic radionuclides they contain were produced as their orbits gradually decayed until they were captured by the Earth and were melted as they passed through the atmosphere.

The extraterrestrial sediment that is being deposited on the surface of the Earth also includes small interplanetary dust particles (IDPs) which are agglutinated clusters of crystals that have been collected by aircraft flying 20 km above the surface of the Earth (Bradley 2005; Warren et al. 1997; Zolensky et al. 1994; Brownlee 1985). These kinds of particles are probably deposited with snow accumulating on the surface of the ice sheets of Antarctica and Greenland; but they have not been recognized in the East Antarctic ice sheet because they are too fragile to be preserved and because their diameters (~10 μm) are smaller than those of most MMs and cosmic spherules. About 44% of these kinds of particles were classified as "extraterrestrial" by Warren et al. (1997), whereas 31% are anthropogenic-terrestrial and 25% are natural-terrestrial (Faure and Mensing 2007).

We also exclude from consideration the so-called *microtektites* which are small glassy particles that fall from the sky following the explosive impact of large meteoroids and asteroids. Although microtektites superficially resemble cosmic spherules and meteorite-ablation spherules, they are terrestrial in origin and therefore exceed the scope of this chapter (Faul 1966; Koeberl 1986, 1988c, 1990, 1994). The microtektites that are geographically associated with the tektites of Australia and Indochina have been described by Glass and Pizzuto (1994), Lee and Wei (2000), Prasad and Khedekar (2003), Glass et al. (2004), and by Kelly and Elkins-Taunton (2004).

18.10.4 Micrometeorites, Cap Prudhomme

The extraterrestrial particles from Cap Prudhomme analyzed by Gounelle et al. (2005) are composed

Table 18.13 Average major-element concentrations in micrometeorites and cosmic spherules (25–50 μm) collected at Cap Prudhomme, Antarctica, expressed in weight percent and recalculated to 100% (Gounelle et al. 2005)

	xtl (3)	fg (29)	sc (3)	cs (3)
SiO_2	50.6 ± 4.4	38.9 ± 8.2	42.0 ± 24.6	29.0 ± 13.5
TiO_2	0.08 ± 0.02	0.11 ± 0.04	0.07 ± 0.01	0.11 ± 0.03
Al_2O_3	0.92 ± 0.49	2.99 ± 1.67	3.92 ± 0.47	3.74 ± 2.22
Cr_2O_3	0.71 ± 0.09	0.61 ± 0.23	0.42 ± 0.12	1.28 ± 1.36
FeO	8.5 ± 6.5	31.9 ± 10.9	25.7 ± 20.5	42.9 ± 14.5
MnO	0.14 ± 0.02	0.32 ± 0.15	0.22 ± 0.10	0.26 ± 0.01
MgO	38.1 ± 2.7	21.7 ± 9.7	26.1 ± 11.3	21.4 ± 9.8
NiO	0.28 ± 0.35	0.30 ± 0.26	0.19 ± 0.16	0.34 ± 0.40
CaO	0.57 ± 0.24	0.73 ± 0.54	0.51 ± 0.06	0.70 ± 0.62
Na_2O	0.03 ± 0.03	0.58 ± 0.48	0.32 ± 0.07	0.10 ± 0.10
K_2O	<dl	0.20 ± 0.15	0.14 ± 0.6	0.15 ± 0.7
S	0.05 ± 0.05	0.90 ± 1.03	0.47 ± 0.46	0.52 ± 0.50
Total	99.98	99.24	100.06	100.50

xtl = crystalline MMs = coarse-grained anhydrous MMs, fg = fine grained MMs = phyllosilicate MMs, sc = scoriaceous MM, cs = cosmic spherules, < dl = less than the detection limit. The number of samples in each average is indicated in parentheses.
The analyses were made by microprobe at the University of Vienna and at the University of Paris.

Table 18.14 Average trace-element concentrations in micrograms per gram (ppm), unless otherwise indicated, of anyhydrous crystalline micrometeorites (xtl MMs) ranging in size from 100 to 400 μm collected at Cap Prudhomme, Antarctica (Kurat et al. 1994)[a]

Element	Concentration	Element	Concentration
Na (%)	1.10 ± 1.49	La	0.64 ± 0.34
K (%)	0.24 ± 0.47	C	1.33 ± 1.08
Sc	17.2 ± 13.1	Nd	<1.3 ± 0.5
Cr	6293 ± 8963	Sm	0.36 ± 0.18
Fe (%)	19.9 ± 8.9	Eu	0.10 ± 0.04
Co	254 ± 183	Gd	<0.35 ± 0.18
Ni	4469 ± 3076	Tb	0.08 ± 0.02
Zn	84 ± 17	Tm	<0.08 ± 0.02
Ga	3.94 ± 3.92	Yb	0.38 ± 0.18
As	4.79 ± 4.59	Lu	0.047 ± 0.018
Se	1.82 ± 0.76	Hf	0.42 ± 0.44
Br	1.32 ± 2.82	Ta	0.26 ± 0.43
Rb	5.96 ± 2.28	Os	0.34 ± 0.34
Sr	<95 ± 68	Ir	0.47 ± 0.69
Zr	<21 ± 15	Au	0.10 ± 0.07
Ru	0.98 ± 1.25	Th	0.122 ± 0.120
Sb	0.44 ± 0.91	U	0.130 ± 0.113
Ba	<22 ± 15		

[a]Determined by C. Koeberl using instrumental neutron activation analysis (INAA).

primarily of pyroxene and olivine with accessory amounts of magnetite and a few other iron-bearing minerals. Accordingly, the chemical compositions of these MMs in Table 18.13 are dominated by SiO_2, FeO, and MgO, whereas Al_2O_3, CaO, Na_2O, and K_2O have remarkably low concentrations. The concentrations of TiO_2, Cr_2O_3, MnO, NiO, and S are also less than 1.0% in most cases. Particles composed of metallic Fe-Ni, which occur in the Greenland suite (~5%), are absent in the Cap Prudhomme suite although small amounts of metallic Fe-Ni do occur in the crystalline MMs as an accessory phase (Gounelle et al. 2005).

Feldspar is not present in crystalline form though the composition of the glassy matrix of phyllosilicate (fg) MMs in Table 18.13 approaches Ca-rich feldspar.

The average trace-element concentrations of seven xtl MMs from the Cap Prudhomme collection are listed in Table 18.14 based on data published by Kurat et al. (1994). The enrichment or depletion of trace elements in the MMs relative to average rocks of the continental crust (Taylor and McLennan 1985) was expressed as the logarithm to the base 10 of the MM crustal-rock ratios. The resulting distribution of data points in Fig. 18.31 indicates that the MMs are strongly enriched in iridium and gold as well as in nickel, chromium, selenium, iron, cobalt, arsenic, and antimony relative to crustal rocks.

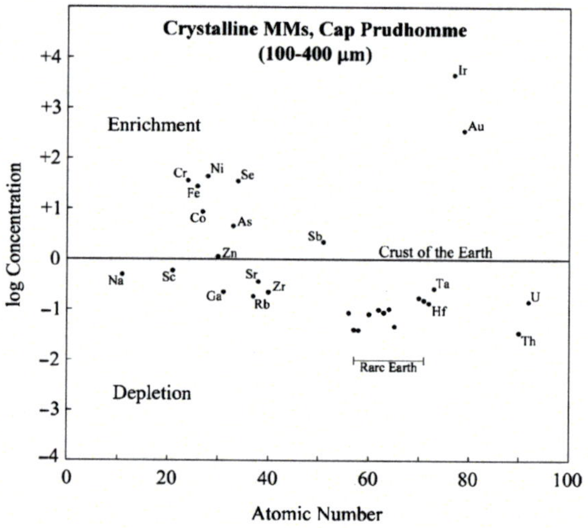

Fig. 18.31 The concentrations of chemical elements in the crystalline micrometeorites (MMs) at Cap Prudhomme, Antarctica, in the size range 100–400 μm deviate significantly from the chemical composition of the rocks in the crust of the Earth. The enrichment and depletion of the MMs is expressed as the logarithm to the base 10 of the ratios of the concentrations of the MMs divided by the concentrations in crustal rocks as reported by Taylor and McLennan (1985). The elements that are enriched in the MMs and their respective enrichment factors include primarily Ir (4700), Au (333), Ni (43), Se (36), Cr (34), Fe (28), Co (8.8), As (4.8), and Sb (2.2). The MMs were analyzed by C. Koeberl by instrumental neutron activation analysis (INAA)

The elements that plot below the dividing line are less abundant in the xtl MMs than in the crust of the Earth (i.e., sodium, scandium, gallium, rubidium, strontium, zirconium, barium, rare earths, hafnium, tantalum, thorium, and uranium). This comparison strengthens the hypothesis that MMs are fragments of stony meteoroids that originated from the asteroid belt (see also Hagen et al. 1990, Fig. 3). However, small particles derived from the surface of the Moon, Mars, or from some of the large asteroids (e.g., Vesta or Ceres) may also be present in the Cap Prudhomme suite of micrometeorites. Engrand and Maurette (1998) and Engrand et al. (1999) have linked MMs to carbonaceous chondrites.

18.10.5 Micrometeorites, South Pole

In 1995 several thousand extraterrestrial particles were recovered from the bottom of a cavity that had been formed by melting approximately 4,000 m^3 of ice at the South Pole (Taylor et al. 1997, 1998, 2000, 2007). The collection contained a large amount of anthropogenic detritus as well as cosmic spherules and unmelted or partially melted MMs. Taylor et al. (2000) reported that cosmic spherules constitute only 0.2 of the 175 g of sediment that was recovered from the bottom of the cavity and that the terrestrial component of that sediment included primarily both natural and anthropogenic sediment, such as iron oxide grains, wood fragments, paint chips, wire, aluminum flakes, copper spherules, and dark nonmagnetic "glue balls." Most of these particles originated from various kinds of equipment failures. For example, the iron-oxide grains were released by the failure of a pump in the water-supply system, the copper spherules originated during the repair of a soldered copper pipe, and the glue balls formed by melting of the insulation during a fire in an electrical cable.

Taylor et al. (2000) primarily described the cosmic spherules in the South-Pole suite of particles because the unmelted MMs were difficult to distinguish from the terrestrial particles. The ice from which the extraterrestrial particles of the South-Pole suite were recovered was deposited during a period of about 400 years between AD 1100 and 1500. The diameters of these cosmic spherules range from 50 to 1,000 μm and the grain-size distribution is assumed to be unbiased because all the particles that fell on the snow surface were preserved in the ice without appreciable losses by chemical weathering or attack by bacteria.

Table 18.15 Abundances of extraterrestrial spherules (50–800 μm) released by melting ice in a large cavity at the South Pole Station (Taylor et al. 2000)

Spherule	Abundance (%)	Description
Barred olivine	41	Lath-shaped olivine and small magnetite in interstitial glass
Glass	17	Mafic glass, highly spherical, some have scalloped edges
Cryptocrystalline	12	No visible crystals but crystalline in polarized light
Porphyritic olivine	11	Equidimensional olivine crystals and magnetite in interstitial glass
Relic-grain bearing	12	Relic olivine, metal/sulfide grains, incompletely melted
Scoriaceous	3	Vesicular glass, relic olivine, metal/sulfide grains, incompletely melted
I-type	2	Interlocking magnetite crystals with interstitial wűstite.[a]
Ca-Al-Ti (CAT)	1	High Mg/Si ratios of ~1.7. Oblate, white, and lack magnetite
G-Type	1	Iron-rich, magnetite dendrites in a glassy matrix

[a]Wűstite: FeO (metallic grey, cubic).

The South-Pole suite of extraterrestrial particles examined by Taylor et al. (2000) included several kinds of cosmic spherules based on differences in composition and on the extent of heating during their passage through the atmosphere. The classification of these particles in Table 18.15 contains brief descriptions of the principal types. In addition, the collection contained unmelted MMs rimmed by magnetite and composed of fine-grained metal, sulfide, and silicate grains giving them a chondritic bulk composition. Another type of unmelted MMs was composed of Fe-poor pyroxene and olivine embedded in an Fe-rich matrix.

18.11 Summary

When the search for meteorites in Antarctica started in 1976/77, few people expected that it would be successful because most of the continent is covered by an ice sheet of gigantic proportions. In retrospect, we now realize that meteorite specimens are concentrated on the bare-ice surfaces of the ablation zone adjacent to the Transantarctic Mountains where the flow of the ice is deflected upward by subglacial bedrock barriers.

The search for meteorites by the participants of ANSMET has not only provided a wealth of information about meteorites but has also enriched our understanding of the dynamics of the East Antarctic ice sheet.

The importance of meteorites in the context of planetary science arises because they are fragments of asteroids that revolve around the Sun primarily in the space between the orbits of Mars and Jupiter. Most asteroids are themselves fragments of larger bodies that originally formed by accretion of planetesimals between 4.5 and 4.6 billion years ago. Therefore, meteorites contain information about the origin of the solar system and about the composition of the tens of thousands of asteroids that continue to orbit the Sun. We know that asteroids have repeatedly collided with the Earth during its long history and have caused catastrophic climate change resulting in the extinction of many species of plants and animals.

The Antarctic meteorites have not only added specimens that fit within the previously existing classification, but also include samples that were ejected from the Moon and from Mars. The study of lunar meteorites has extended the information derived from lunar rocks collected by the astronauts of the Apollo Program of the USA and by the lunar landers launched by the USSR. The martian meteorites that have been collected in Antarctica and elsewhere on the Earth are the only samples of rocks from Mars that are available for study at the present time.

The properties of meteorites collected in Antarctica that have been measured include their chemical and mineralogical compositions as well as their textures, isotopic ages, and cosmic-ray exposure histories. The scientific value of the accumulated data is exemplified in this chapter by selected specimens of Antarctic meteorites such as the Derrick Peak iron, by the first lunar meteorite recovered on the icefields of the Allan Hills (ALHA 81005), and by the martian rock ALH 84001. The interpretation of data derived from these and other meteorites collected in Antarctica contribute to the on-going exploration of the solar system.

18.12 Appendices

18.12.1 Letter Codes and Locations of Collecting Sites of Meteorite Specimens (Antarctic Meteorite Newsletter, 29(2):3, 2006 and Gazetteer of the Antarctic (Fourth edition))

Code	Location name	Latitude & Longitude
ALH	Allan Hills	76°43′ S, 159°40′ E
BEC	Beckett Nunatak	76°02′ S, 160°11′ E
BOW	Bowden Névé	83°30′ S, 165°00′ E
BTN	Bates Nunataks	80°15′ S, 153°30′ E
CMS	Cumulus Hills	85°20′ S, 175°00′ W
CRA	Mt. Cranfield	79°56′ S, 158°40′ E
CRE	Mt. Crean	77°53′ S, 159°30′ E
CRS	Mt. Cranfield	
DAV	David Glacier	75°19′ S, 162°00′ E
DEW	Mt. DeWitt	77°12′ S, 159°50′ E
DNG	D'Angelo Bluff	87°18′S, 154°00′W
DOM	Dominion Range	85°20′ S, 166°30′ E
DRP	Derrick Peak	80°04′ S, 156°23′ E
EET	Elephant Moraine	76°11′ S, 157°10′ E
FIN	Finger Ridges	79°11′ S, 157°00′ E
FRO	Frontier Mountain	72°59′ S, 160°20′ E
GDR	Gardner Ridge	86°57′ S, 148°24′ W
GEO	Geologists Range	82°30′ S, 155°30′ E
GRA	Graves Nunatak	83°43′ S, 141°30′ W
GRO	Grosvenor Mountains	85°40′ S, 175°00′ E
GRV	Graves Nunatak	86°43′ S, 141°30′ W
HOW	Mt. Howe	87°22′ S, 149°30′ W

(continued)

18.12.1 (continued)

Code	Location name	Latitude & Longitude
ILD	Inland Forts	77°38′ S, 161°00′ E
KLE	Klein Ice Field	86°48′ S, 150°00′ W
LAP	LaPaz Ice Field[a]	85°55′ S, 63°00′ W
LAR	Larkman Nunatak	85°46′ S, 179°23′ E
LEW	Lewis Cliff	84°17′ S, 161°05′ E
LON	Lonewolf Nunataks	81°20′ S, 152°50′ E
MAC	MacAlpine Hills	84°13′ S, 160°30′ E
MBR	Mount Baldr	77°35′ S, 160°34′ E
MCY	MacKay Glacier	76°58′ S, 162°00′ E
MET	Meteorite Hills	79°40′ S, 155°36′ E
MIL	Miller Range	83°15′ S, 157°00′ E
ODE	Odell Glacier	76°44′ S, 159°55′ E
OTT	Outpost Nunataks	75°50′ S, 158°12′ E
PAT	Patuxent Range	84°43′ S, 064°30′ W
PCA	Pecora Escarpment	85°38′ S, 068°42′ W
PGP	Purgatory Peak	77°21′ S, 162°18′ E
PRA	Mt. Pratt	85°24′ S, 176°41′ E
PRE	Mt. Prestrud	86°34′ S, 165°07′ W
QUE	Queen Alexandra Range	84°00′ S, 168°00′ E
RBT	Roberts Massif	85°32′ S, 177°05′ W
RKP	Reckling Peak	76°16′ S, 159°15′ E
SAN	Sandford Cliffs	83°54′S, 159°17′ E
SCO	Scott Glacier	85°45′ S, 153°00′ W
STE	Stewart Hills	84°12′ S, 086°00′ W
TEN	Tentacle Ridge	79°37′ S, 157°15′ E
TIL	Thiel Mountains	85°15′ S, 091°00′ W
TYR	Taylor Glacier	77°44′ S, 162°10′ E
WIS	Wisconsin Range	85°45′ S, 125°00′ W
WSG	Mt. Wisting	86°27′ S, 165°26′ W

[a]The LaPaz icefield (informal name) is located in the vicinity of the Pecora Escarpment. It was named in memory of Dr. Lincoln LaPaz who, in 1945, founded the Institute of Meteoritics at the University of New Mexico and was its director until his retirement in 1966.
Dr. LaPaz was an inspiring teacher whose influence is still felt by his former students.

18.12.2 Calculation of the Terrestrial Age of the LL6 Chondrite ALH 78153 by the Decay of Cosmogenic ^{36}Cl and ^{26}Al

Chlorine-36 is a cosmogenic radioactive (i.e., unstable) nuclide that is produced by a nuclear spallation reaction in grains of metallic iron in stony meteorites. After a meteorite specimen has landed on the surface of the Earth, the production of all cosmogenic radionuclides stops and ^{36}Cl decays at a rate depending on its halflife. The terrestrial age of a meteorite is calculated from the remaining concentration of ^{36}Cl by an application of the law of radioactivity (Faure and Mensing 2005).

The relevant constants that govern the decay of ^{36}Cl in meteorites were determined by Nishiizumi et al. (1983). The rate of decay of ^{36}Cl in the metal grains of ordinary chondrites at the time of fall is 22.8 ± 3.1 dpm/kg (disintegrations per minute per kilogram of iron). The halflife of ^{36}Cl is 0.301 × 10^6 years which corresponds to a decay constant (λ) of 2.302 × 10^{-6} year^{-1}. The decay of ^{36}Cl takes place primarily by emission of beta particles (i.e., electrons) from its nucleus which transforms ^{36}Cl into stable atoms of $^{36}_{18}$Ar:

$$^{36}_{17}Cl \rightarrow ^{36}_{18}Ar + \beta^- + \overline{\nu} + Q \quad (18.4)$$

where β^- = electron emitted by the nucleus
$\overline{\nu}$ = antineutrino
Q = decay energy (0.7093 MeV).

The terrestrial age of meteorite specimens can be calculated from the measured decay rates (A) of cosmogenic radionuclides by an application of the law of radioactivity:

$$A_m = A_i e^{-\lambda t} \quad (18.5)$$

where A_m = measured decay rate of a cosmogenic radiocuclide

A_i = decay rate at the time of fall of the meteorite (assumed to be the decay rate at saturation)

λ = decay constant of the radionuclide (e.g., ^{26}Al, ^{36}Cl, etc.)

t = the time that has elapsed since the fall of the meteorite (i.e., its terrestrial age).

We use the activity of ^{36}Cl in the LL6 chondrite ALH78153 to demonstrate how its terrestrial age is calculated based on data by Nishiizumi et al. (1989):

$$A_m = {}^{36}Cl_m = 2.58 \pm 0.27 \text{ dpm/kg}$$

$$A_i = {}^{36}Cl_i = 22.8 \pm 3.1 \text{ dpm/kg}$$

$$\lambda({}^{36}Cl) = 2.302 \times 10^{-6} \text{ year}^{-1}$$

Solving Eq. 18.5 for t yields:

$$e^{\lambda t} = \frac{A_i}{A_m}$$

$$\lambda t = \ln\left(\frac{A_i}{A_m}\right) (\ln = \text{natural log})$$

$$t = \frac{1}{\lambda}\ln\left(\frac{A_i}{A_m}\right) \quad (18.6)$$

Substituting values for ^{36}Cl in ALH78513:

$$t = \frac{1}{2.302 \times 10^{-6}}\ln\left(\frac{22.8}{2.58}\right)$$

$$t = \frac{\ln 8.8372}{2.302 \times 10^{-6}} = \frac{2.1789}{2.302 \times 10^{-6}} = 0.9465 \times 10^6$$

$$t = (0.95 \pm 0.10) \times 10^6 \text{ years.}$$

The terrestrial age of ALH78153 can also be calculate from its activity of ^{26}Al (22 ± 1 dpm/kg) assuming that its activity at the time of fall was 56 ± 8 dpm/kg (Evans and Reeves 1987). Substituting into Eq. 18.6 yields:

$$t = \frac{1}{0.983 \times 10^{-6}}\ln\left(\frac{56}{22}\right) = (0.950 \pm 0.20) \times 10^6 \text{ years}$$

The terrestrial age of ALH78153 derived from ^{26}Al is in good agreement with its ^{36}Cl age because both methods indicate that this specimen fell onto the East Antarctic ice sheet about 1 million years ago. The concordance of the terrestrial ages of ALH78153 is an *exception* because of the uncertainty in the initial activity of ^{26}Al. Like all other meteorites that fell on the surface of the East Antarctic ice sheet, ALH78153 was subsequently transported by the ice sheet to the vicinity of the Allan Hills where it was left stranded on the ice of the ablation zone.

An even older terrestrial age of (2.2 ± 0.4) × 10^6 years was reported by Scherer et al. (1997) for the H5 chondrite ALH88019 which was collected on the Main Allan-Hills ice field close to an ice ramp where the active ice of the polar plateau merges into stagnant ice (see Fig. 18.17).

18.12.3 Lunar Meteorite Specimens Collected in Antarctica Including Paired Samples (Warren 2005; Papike et al. 1998; Eugster 1989; Bogard 1983)

Code	Mass (g)	Description	Reference
Highlands			
ALH 81005	31.4	Regolith breccia	Marvin (1983a)
MAC 88104	61.2	Polymict regol. breccia	Lindstrom et al. (1991a, b)
MAC 88105	662.5	Paired with MAC 88104	Lindstrom et al. (1991a, b), Cohen et al. (2005)
QUE 93069	21.4	Regolith breccia	Warren (2005), Cohen et al. (2005), Warren et al. (2005)
QUE 94269	3.1	Paired with QUE 93069	Warren (2005), Warren et al. (2005)
Y-791197	52.4	Regolith breccia	Yanai and Kojima (1995) Koeberl (1988b), Takahashi and Masuda 1987

(continued)

18.12.3 (continued)

Code	Mass (g)	Description	Reference
Y-983885	289	Paired with Y-791197 (?)	Warren (2005)
Y-82192	36.67	Polymict breccia	Yanai and Kojima (1995), Koeberl (1988b), Eugster and Niedermann (1988), Takahashi and Masuda (1987)
Y-82193	27.04	Paired with Y-82192	Yanai and Kojima (1995), Eugster and Niedermann (1988)
Y-86032	648.43	Paired with Y-82192	Yanai and Kojima (1995) Koeberl (1988)
Mare Basalts			
Asuka-81757	442.12	Gabbro, mare	Yanai and Kojima (1995)
EET 87521	31	Polymict breccia	Warren (2005)
EET 96008	53	Paired with EET 87521	Warren (2005), Anand et al. (2003)
QUE 94281	23.4	Regolith breccia, mare	Warren (2005)
Y-793169[a]	6.07	Gabbro, mare	Yanai and Kojima (1995)
Y-793274[a]	8.66	Regolith breccia, mare	Yanai and Kojima (1995)
Y-981031	186	Paired with Y-793274	Warren (2005), Warren et al. (2005)
LAP 02205	1,230	Low-Ti basalt	Zeigler et al. (2005)
LAP 02224	250	Paired with LAP 02205	Zeigler et al. (2005)
LAP 02226	240	Paired with LAP 02205	Zeigler et al. (2005)
LAP 02436	60	Paired with LAP 02205	Zeigler et al. (2005)
LAP 03632	90	Paired with LAP 02205	Zeigler et al. (2005)

Total mass: 4,453.39 g, total number of specimens: 22

[a]Although Y-793169 and Y-793274 were collected 2 years before ALHA 81005, they were not recognized as rocks from the Moon until 1990 (Y-793169) and 1987–1999 (Y-793169). Therefore, ALHA 81005 is regarded as the first lunar meteorite to be discovered on the Earth in 1981/82 and to be identified in 1983 (Warren 2005).

18.12.4 Partial List of Martian Meteorites Collected in Antarctica

Code	Mass (g)	Description	Reference
Transantarctic Mountains			
ALH 77005	482.5	Lherzolite shergottite	Score and Lindstrom (1990)
EETA 79001	7,942.0	Basalt. shergottite	Score and Lindstrom (1990)
ALHA 81313	0.5	Shergottite	Score and Lindstrom (1990)
ALHA 84001	1930.9	Orthopyroxenite	Score and Lindstrom (1990)
LEW 88516	13	Lherzolite shergottite	Score and Lindstrom (1990)
QUE 94201	12	Basalt. shergottite	Cassidy (2003)
GRV 99027	9.97	Lherzolite shergottite	Cassidy (2003)
MIL 03346		Nakhlite	Berczi et al. (2008)
RBT 04261	78.8	Olivine shergottite	Nagao and Park (2008)
RBT 04262	204.6	Olivine shergottite	Nagao and Park (2008)
LAR 06319	78.6	Olivine shergottite	Nagao and Park (2008)
Yamato Mountains			
Y 793605	13	Lherzolite shergottite	Cassidy (2003)
Y 000593	13,713	Nakhlite	Schwenzer et al. (2008)
Y 000749	1,283	Nakhlite	Cassidy (2003)
Y 980459		Shergottite	Koizumi et al. (2008)
Y 000027		Shergottite	Schwenzer et al. 2008
Y 000097		Shergottite	Schwenzer et al. (2008)
Y 000802		Nakhlite	Park et al. (2008)

References

Anand M, Taylor LA, Neal CR, Snyder GA, Patchen A, Sano Y, Terado K (2003) Petrogenesis of lunar meteorite EET 96008. Geochim Cosmochim Acta 67(18):3499–3518

Anand M, Taylor LA, Floss C, Neal CR, Terada K, Tanikawa S (2006) Petrology and geochemistry of LaPaz Icefield 02205: A new unique low-Ti mare-basalt meteorite. Geochim Cosmochim Acta 70(1):246–264

Annexstad JO, Schultz L, Wänke H (eds) (1986) Antarctic meteorites. LPI Tech. Rept. 86–01, Lunar and Planetary Science Institute, Houston, TX

Ash RD, Knott SF, Turner G (1996) A 4-Gyr shock age for a martian meteorite and implications for the cratering history of Mars. Nature 380:57–59

Barrat JA, Blichert-Toft J, Nesbitt RW, Keller F (2001) Bulk chemistry of Saharan shergottite Dar al Gani 476. Meteorit Planet Sci 36:23–29

Becker L, Glavin DP, Pada JL (1997) Polycyclic aromatic hydrocarbons (PAHs) in Antarctic martian meteorites, carbonaceous chondrites, and polar ice. Geochim Cosmochim Acta 61:475–481

Becker L, Bopp B, Rust T, Bada JL (1999) The origin of organic matter in martian meteorite ALH 84001. Earth Planet Sci Lett 167:71–79

Becker RH, Pepin RO (1986) Nitrogen and light noble gases. Geochim Cosmochim Acta 50(6):993–1000

Behrendt JC, McCafferty AE, Damaske D, Kyle PR (1995) High amplitude aeromagnetic anomaly over the Butcher Ridge igneous complex: Evidence of possible Jurassic cumulate rocks in the Transantarctic Mountains bordering the Ross Embayment. In: Elliot DE (ed) Contributions to Antarctic Research IV, Antarctic Research Series, vol. 67. American Geophysical Union, Washington, DC, pp 1–8

Berczi Sz et al. (2008) Recognition of planetary cumulate rock types for rover cameras. Meteorit Planet Sci 43(Supplement):A179

Bhandari N, Goswami JN, Jha R, Sengupta D, Shukla PN (1986) Cosmogenic effects in shergottites. Geochim Cosmochim Acta 50(b):1023–1030

Bland P (2001) Quantification of meteorite infall rates from accumulations in deserts, and meteorite accumulations on Mars. In: Peucker-Ehrenbrink B, Schmitz B (eds) Accretion of extraterrestrial matter throughout earth's history. Kluwer Academic/Plenum, New York, pp 267–298

Bogard DD (ed) (1983) A meteorite from the Moon. Geophys Res Lett 10:733–840

Bogard DD, Garrison DH (1999) Argon-39-argon-40 "age" and trapped argon in martianshergottites, Chassigny, and Allan Hills 84001. Meteorit Planet Sci 34:451–473

Bogard DD, Johnson P (1983) Martian gases in an Antarctic meteorite? Science 221:651

Bogard DD, Husain L, Nyquist LE (1979) ^{40}Ar-^{39}Ar age of the Shergotty achondrite and implications for its post-shock thermal history. Geochim Cosmochim Acta 43:1047–1055

Bogard DD, Nyquist LE, Johnson P (1984) Noble-gas contents of shergottites and implications for the Martian origin of SNC meteorites. Geochim Cosmochim Acta 48:1723–1739

Bogard DD, Nyquist LE, Bansal BM, Garrison DH, Herzog GF, Albrecht AA, Vogt S, Klein J (1995) Neutron-capture ^{36}Cl, ^{41}Ca, ^{36}Ar, and ^{150}Sm in large chondrites: Evidence for high fluences of thermalized neutrons. J Geophys Res 100:9401–9416

Borg LE, Nyquist LE, Taylor LA, Wiesmann H, Shih C-Y (1997) Constraints on martian differentiation processes from Rb-Sr and Sm-Nd isotopic analyses of the basaltic shergottite QUE 94201. Geochim Cosmochim Acta 61:4915–4931

Borg LE, Nyquist LE, Wiesmann H, Shih C-Y, Reese YD (2003) The age of Dar al Gani 476 and the differentiation history of the martian meteorites inferred from their radiogenic isotopic systematics. Geochim Cosmochim Acta 67(18): 3519–3536

Boroughs TJ (1992) Interpretation of the chemical composition of certain major minerals in a new chondritic meteorite from Antarctica. MSc thesis, The Ohio State University, Columbus, OH

Boroughs TJ, Faure G, Buchanan D (1992) Chemical compositions of minerals of the ordinary (H6) chondrite, RKP 86701, from the Reckling Moraine. Antarctic J US 27 (5):30–31

Bradley JP (2005) Interplanetary dust particles. In: Davis AM (ed) Meteorites, comets, and planets. Elsevier, Amsterdam, The Netherlands, pp 698–713

Bradley JP, Harvey RP, McSween HY Jr (1996) Magnetite whiskers and platelets in the ALH 84001 martian meteorite: Evidence of vapor-phase growth. Geochim Cosmochim Acta 60(24):5149–5155

Bradley JP, McSween HY Jr, Harvey RP (1998) Epitaxial growth of nanophase magnetite in martian meteorite Allan Hills 84001: Implications for biogenic mineralization. Meteorit Planet Sci 33:765–773

Brownlee DE (1985) Cosmic dust: Collection and research. Ann Rev Earth Planet Sci 13:147–173

Buchwald VF, Clarke RS Jr (1989) Corrosion of Fe-Ni alloys by Cl-containing akaganéite (β-FeOOH): The Antarctic meteorite case. Amer Mineralogist 74:656–667

Bull C, Lipschutz ME (eds) (1982) Workshop on Antarctic glaciology and meteoritics. LPI Tech. Rept. No. 82–03, Lunar and Planetary Science Institute, Houston, TX

Burns RG (ed) (1993) The MSATT workshop on chemical weathering on Mars. Geochim Cosmochim Acta 57:4551–4637

Carr MH (2006) Mars: Surface and interior. In: McFadden L-A, Weissman PR, Johnson TV (eds) (2006) Encyclopedia of the Solar System, 2nd edn. Academic, Elsevier, San Diego, CA

Cassidy WA (ed) (1964) Proceedings from the conference on cosmic dust. Ann New York Acad Sci 119

Cassidy WA (1991) Meteorites from Antarctica. In: Tingey RJ (ed) The geology of Antarctica. Clarendon, Oxford, UK, pp 652–666

Cassidy WA (2003) Meteorites, ice, and Antarctica. Cambridge University Press, Cambridge, UK

Cassidy WA, Harvey RP (1991) Are there real differences between Antarctic finds and falls meteorites? Geochim Cosmochim Acta 55:99–104

Cassidy WA, Rancitelli LA (1982) Antarctic meteorites. Amer Scientist 70:156–164

Cassidy WA, Whillans IM (eds) (1990) Antarctic meteorite stranding surfaces. LPI Tech. Rept. No. 90-03, Lunar and Planetary Science Institute, Houston, TX

Cassidy WA, Olsen E, Yanai K (1977) Antarctica: A deepfreeze storehouse for meteorites. Science 182:727–731

Cassidy WA, Harvey RP, Schutt JW, Delisle G, Yanai K (1992) The meteorite collecting sites of Antarctica. Meteoritics 27:490–525

Chapman CR (1999) News and views. The Planetary Report, July/August, p. 18.

Chen JH, Wasserburg GJ (1985) U-Th-Pb isotopic studies on meteorites ALHA 81005 and Ibitira. Lunar and Planetary Science Conference, vol. 26, pp 119–120

Chen JH, Wasserburg GJ (1986) Formation ages and evolution of Shergotty and its parent planet from U-Th-Pb systematics. Geochim Cosmochim Acta 50(6):955–968

Claridge GGC, Campbell IB (1977) The salts in Antarctic soils, their distribution and relationship to soil processes. Soil Science 123(6):377–384

Clarke RS Jr (1982) The Derrick Peak, Antarctica, iron meteorites. Meteoritics 17(3):129–134

Cohen BA, Swindle TD, Kring DA (2005) Geochemistry and $^{40}Ar/^{39}Ar$ geochronology of impact-melt clasts in feldspathic lunar meteorites: Implications for lunar bombardment history. Meteorit Planet Sci 40(5):755–777

Connolly HC Jr, Desch SJ (2004) On the origin of the "kleine Kügelchen" called spherules. Chemie der Erde 64:95–125

Corrigan CM, Harvey RP (2004) Multi-generational carbonate assemblages in martian meteorite Allan Hills 84001: Implications for nucleation, growth, and alteration. Meteorit Planet Sci 39(1):17–30

Crary AP (1963) Results of the United States traverse in East Antarctica, 1958–61. IGY Glaciology Reports No 7:1–144

Crozaz G, Floss C, Wadhwa M (2003) Chemical alteration and REE mobilization in meteorites from hot and cold deserts. Geochim Cosmochim Acta 67(2):4727–4741

Daode W, Malvin DJ, Wasson JT (1982) Classification of ten Chinese, eleven Antarctic and ten other iron meteorites. Lunar Planetary Science Conference, Abstract Volume. Houston, TX, p 139

Davis AM (ed) (2005) Meteorites, comets, and planets. Vol. 1, Treatise on Geochemistry. Elsevier, Amsterdam, The Netherlands

Delisle G (1993) Global change, Antarctic meteorite traps and the East Antarctic ice sheet. J Glaciol 39:397–408

Delisle G (1997) Sub-ice topography in selected areas of Victoria Land, Antarctica: Implications for its glacial erosion history. In: Barker P, Cooper AK (eds) Geology and seismic stratigraphy of the Antarctic Margin, Part 2. American Geophysical Union, Washington, DC, pp 127–135

Delisle G, Sievers J (1991) Sub-ice topography and meteorite finds near the Allan Hills and the Near Western Icefield, Victoria Land, Antarctica. J Geophys Res 96(E1):15577–15587

Delisle G, Sievers J, Schultz L (1989) Radio-echo sounding survey across the Allan Hills icefield. Ant J US 24(5):50–52

Denton GH, Hughes TJ (1981) The last great ice sheets. Wiley, New York

Denton GH, Sugden DE, Marchant DR, Hall BL, Wilch TI (1993) East Antarctic ice-sheet sensitivity to Pliocene climate change from a Dry Valleys perspective. Geogr. Annaler Stockholm, vol. 75A. Scandinavian University Press, Stockholm, Sweden, pp 155–204

Dodd RT (1971) The petrology of chondrules in the Sharps meteorite. Contrib Mineral Petrol 31:201–227

Dreibus G, Wänke (1983) Halogens in Antarctic meteorites. Meteoritics 18:291–292

Dreibus G, Wänke H, Schultz L (1986) Mysterious iodine-overabundance in Antarctic meteorites. In: Annexstad JO, Schultz L, Wänke H (eds) International Workshop on Antarctic Meteorites. LPI Tech. Rept. Number 86-01. Houston, TX, pp 34–36

Dressler BO, Grieve RAF, Sharpton VL (eds) (1994) Large meteorite impacts and planetary evolution. Special Paper 293. Geological Society of America, Boulder, CO

Duce RA, Zoller WH, Jones AG (1971) Atmospheric particle and gas sampling at McMurdo and South Pole stations. Ant J US 6(4):133–134

Duce RA, Zoller WH, Moyers JL (1973) Particulate and gaseous halogens in the Antarctic atmosphere. J Geophys Res 78:7802–7811

Duke MB (1965) Discovery of Neptune Mountains iron meteorite, Antarctica Meteorit Bull No. 34:2–3

Duprat J, Maurette M, Engrand C, Matrajt G, Immel G, Gounelle M, Kurat G (2001) An estimation of the contemporary micrometeorite flux obtained from surface snow samples collected in central Antarctica. Abstract Meteorit Planet Sci 36:A52

Duprat J, Engrand C, Maurette M, Gounelle M, Hammer C, Kurat G (2003) The CONCORDIA-collection: Pristine contemporary micrometeorites from central Anarctica surface snow. Abstract 1727. 34th Lunar and Planetary Science Conference. CD-ROM, Houston, TX

Eiler JM, Valley JW, Graham CM, Fournelle J (2002a) Two populations of carbonate in ALH 84001: Geochemical evidence for discrimination and genesis. Geochim Cosmochim Acta 66(7):1285–1303

Eiler JM, Kitchen N, Leshin L, Strausberg M (2002b) Hosts of hydrogen in Allan Hills 84001: Evidence for hydrous martian salts in the oldest martian meteorite? Meteorit Planet Sci 37:395–405

Elmore D, Phillips EM (1987) Accelerator mass spectrometry for measurement of long-lived radioisotopes. Science 236:543–550

Engrand C, Maurette M (1998) Carbonaceous micrometeorites from Antarctica. Meteorit Planet Sci 33:565–580

Engrand C, McKeegan KD, Leshin LA (1999) Oxygen isotopic compositions of individual minerals in Antarctica micrometeorites: Further links to carbonaceous chondrites. Goechim Cosmochim Acta 63(17):2623–2636

Eugster O (1988) Cosmic-ray production rates for ^{3}He, ^{21}Ne, ^{38}Ar, ^{83}Kr, and ^{126}Xe in chondrites based on ^{81}Kr-^{83}Kr exposure ages. Geochim Cosmochim Acta 52:1649–1662

Eugster O (1989) History of meteorites from the Moon collected in Antarctica. Science 245:1197–1202

Eugster O (2003) Cosmic-ray exposure ages of meteorites and lunar rocks and their significance. Chemie der Erde 63:3–30

Eugster O, Niedermann S (1988) Noble gases in lunar meteorites Y-82192 and Y-82193 and history of the meteorites from the Moon. Earth Planet Sci Lett 89:15–27

Eugster O, Geiss J, Krähenbühl U, Niedermann S (1986) Noble gas isotopic composition, cosmic ray exposure history, and terrestrial age of the meteorite Allan Hills A81005 from the Moon. Earth Planet SciLett 78:139–147

Eugster O, Beer J, Burger M, Finkel RC, Hofmann HJ, Krähenbühl U, Michel Th, Synal HA, Wölfli W (1991) History of paired lunar meteorites MAC 88104 and MAC 88105 derived from noble-gas isotopes, radionuclides, and

some chemical abundances. Geochim Cosmochim Acta 55:3139–3148
Eugster O, Busemann H, Lorenzetti S, Terribilini D (2002) Ejection ages from krypton-81-krypton-83 dating and pre-atmospheric sizes of martian meteorites. Meteorit Planet Sci 37:1345–1360
Evans JC, Reeves JH (1987) ^{26}Al survey of Antarctic meteorites. Earth Planet Sci Lett 82:223–230
Faul H (1966) Tektites are terrestrial. Science 152(3727): 1341–1345
Faure G (1990) Origin of stranding surfaces. In: Cassidy WA, Whillans IM (eds) Workshop on Antarctic meteorite stranding surfaces. Lunar Planet Sci (LPI) Tech. Rept. 90-03. Houston, TX, pp 11–15
Faure G (1998) Principles and applications of geochemistry, 2nd edn. Prentice Hall, Upper Saddle River, NJ
Faure G, Buchanan D (1987) Glaciology of the East Antarctic ice sheet at the Allan Hills: A preliminary interpretation. Ant J US 22(5):74–75
Faure G, Mensing TM (2005) Isotopes: Principles and applications, 3rd edn. Wiley, Hoboken, NJ
Faure G, Mensing TM (2007) Introduction to planetary science: The geological perspective. Springer, Dordrecht, The Netherlands
Faure G, Sipp JA, Lee G (1997) Search for anthropogenic cesium-137 in a soil profile in Beacon Valley, southern Victoria Land. Ant J US 32(5):17–18
Fernandes VA, Burgess R (2006) Ar-Ar studies of two lunar mare rocks: LAP 02205 and EET 96008. Lunar Planetary Science Conference, Abstract #1145 (CD-ROM)
Ford AB, Tabor RW (1971) The Thiel Mountains pallasite of Antarctica. US Geol Surv Prof. Paper 750-D:56–60
French BM (1983) Curation of meteorites in the NASA Johnson Space Center, Houston, Texas. Ant J US 18(5): 89–90
French BM (1998) Traces of catastrophe; a handbook of shock metamorphic effects in terrestrial meteorite impact structures. Lunar Planetary Science, Contribution 954. Houston, TX
French BM (2004) The importance of being cratered. Meteorit Planet Sci 39(2):169–197
Gale NH, Arden JW, Hutchison R (1975) The chronology of the Nakhla achondrite meteorite. Earth Planet Sci Lett 26:195–206
Genge MJ, Gileski A, Grady MM (2005) Chondrules in Antarctic micrometeorites. Meteorit Planet Sci 40(2):225–238
Gilmour JD, Whitby JA, Turner G (1998) Xenon isotopes in irradiated ALH 84001: Evidence for shock-induced trapping of ancient martian atmosphere. Geochim Cosmochim Acta 62:2555–2571
Gladman BJ, Burns JA, Duncan M, Lee P, Levison HF (1996) The exchange of impact ejecta between terrestrial planets. Science 271:1387–1392
Glass BP (1982) Introduction to planetary geology. Cambridge University Press, Cambridge, UK
Glass BP, Pizzuto JE (1994) Geographic variation in Australasian microtektite concentrations: Implications concerning the location and size of the source crater. J Geophys Res 99(E9):19075–19081
Glass BP, Huber H, Koeberl C (2004) Geochemistry of Cenozoic microtektites and clinopyroxene-bearing spherules. Geochim Cosmochim Acta 68(9):3971–4006
Goles GG, Greenland LP, Jérome DY (1967) Abundances of chlorine, bromine, and iodine in micrometeorites. Geochim Cosmochim Acta 31:1771–1787
Gooding JL (1984) Aqueous alteration on meteorite parent bodies: Possible role of "unfrozen" water and the Antarctic meteorite analogy. Meteoritics 19:228–229
Gooding JL (1886a) Clay-mineraloid weathering products in Antarctic meteorites. Geochim Cosmochim Acta 50: 2215–2223
Gooding JL (1986b) Weathering of stony meteorites in Antarctica. In: Annexstad JO, Schultz L, Wänke H (eds) International workshop on Antarctic meteorites. LPI Tech. Rept. 86-01. Houston, TX, pp 48–54
Gooding JL (1989) Significance of terrestrial weathering effects in Antarctic meteorites. In: Marvin UB, MacPherson GJ (eds) Field and laboratory investigations of meteorites from Victoria Land and the Thiel Mountains Region, Antarctica, 1982–1983 and 1083–1984. Smithsonian Contrib Earth Sci 28:93–98. Washington, DC
Gooding JL, Muenow DW (1986) Martian volatiles in shergottite EETA 79001: New evidence from oxidized sulfur and sulfur-rich aluminosilicates. Geochim Cosmochim Acta 50(6):1049–1059
Goodrich CA (2002) Olivine-phyric martian basalts: A new type of shergottite. Meteorit Planet Sci (Supplement) 37:B31–B34
Goodrich CA, Patchett PJ, Lugmair GW, Drake MJ (1991) Sm-Nd and Rb-Sr isotopic systematics of ureilites. Geochim Cosmochim Acta 55(3):829–848
Goodrich CA, Herd CDK, Taylor LA (2003) Spinels and oxygen fugacity in olivine-phyric and lherzolite shergottites. Meteorit Planet Sci 38(12):1773–1792
Goswami JN, Sinha N, Murty SVS, Mohapatra RK, Clement CJ (1997) Nuclear tracks and light noble gases in Allan Hills 84001: Pre-atmospheric size, fall characteristics, cosmic ray exposure duration and formation age. Meteorit Planet Sci 32:91–96
Gounelle M, Zolensky ME (2001) A terrestrial origin for the sulfate veins in CI1 chondrites. Meteorit Planet Sci 36:1321–1329
Gounelle M, Maurette M, Engrand C, Brandstätter F, Kurat G (1999) Mineralogy of the 1998 Astrolabe Antarctic micrometeorite collection. Meteorit Planet Sci 34:A46
Gounelle M, Engrand C, Maurette M, Kurat G, McKeegan KD, Brandstätter F (2005) Small Antarctic micrometeorites: A mineralogical and in-situ oxygen-isotopic study. Meteorit Planet Sci 40(6):917–932
Gow AJ (1990) In situ meteorites: Evidence for the imminent emergence of buried meteorites at the surface of the Antarctic ice sheet. In: Cassidy WA, Whillans IM (eds) Workshop on Antarctic meteorite stranding surfaces, LPI Tech. Rept. 90-03. Houston, TX, p. 85
Gow AJ, Cassidy WA (1989) The emerging meteorite: Crystalline structure of the enclosing ice. In: Marvin UB, MacPherson GJ (eds) Field and laboratory investigations of meteorites from Victoria Land and the Thiel Mountains Region, Antarctica, 1982–1983 and 1983–1984. Smithsonian Contrib Earth Sci 28:87–91. Washington, DC
Grady MM (2000) Catalogue of meteorites, 5th edn. Cambridge University Press, Cambridge, UK
Grady MM, Gibson EK Jr, Wright IP, Pillinger CT (1989) The formation of weathering products on the LEW 85320 ordinary

chondrite: Evidence from carbon and oxygen stable isotope compositions and implications for carbonates in SNC meteorites. Meteoritics 24(1):1–7

Grady MM, Wright IP, Pillinger CT (1998) A nitrogen and argon stable isotope study of Allan Hills 84001: Implications for the evolution of the martian atmosphere. Meteorit Planet Sci 33:795–802

Graham AL, Annexstad JO (1989) Antarctic meteorites. Antarctic Sci 1(1):3–14

Greenwood JP, McSween HP Jr (2001) Petrogenesis of Allan Hills 84001: Constraints from impact-melted feldspathic and silica glasses. Meteorit Planet Sci 36:43–61

Greenwood JP, Riciputi LR, McSween HY Jr (1997) Sulfide isotopic compositions in shergottites and ALH 84001, and possible implications for life on Mars. Geochim Cosmochim Acta 61:4449–4454

Greenwood JP, Blake RE, Coath CD (2003) Ion microprobe measurements of $^{18}O/^{16}O$ ratios of phosphate minerals in the martian meteorites ALH 84001 and Los Angeles. Geochim Cosmochim Acta 67(2):2289–2298

Greshake A, Fritz J, Stőffler D (2004) Petrology and shock metamorphism of the olivine-phyric shergottite Yamato 980459: Evidence for a two-stage cooling and a single-stage ejection history. Geochim Cosmochim Acta 68(10): 2359–2377

Grindley GW, Laird MG (1969) Geology of the Shackleton Coast. In: Bushnell VC, Craddock C (eds) Geologic maps of Antarctica. Antarctic Map Folio Series, Folio 12, Plate 14. American Geographical Society, New York

Haack H, McCoy TJ (2005) Iron and stony-iron meteorites. In: David AM (ed) Meteorites, comets, and planets. In: Holland HD, Turekian KK (eds) Treatise on geochemistry, vol. 1. Elsevier-Pergamon, Oxford, UK, pp 325–345

Hagen EH (1988) Geochemical studies of Neogene till in the Transantarctic Mountains; Evidence for an extraterrestrial component. MSc thesis, Department of Geology and Mineralogy, The Ohio State University, Columbus, OH

Hagen EH (1995) A geochemical and petrological investigation of meteorite ablation products in till and ice of Antarctica. PhD dissertation. The Ohio State University, Columbus, OH

Hagen EH, Koeberl C, Faure G (1990) Extraterrestrial spherules in glacial sediment, Beardmore Glacier area, Transantarctic Mountains. In: Elliot DH (ed) Contributions to Antarctic Research I, Antarctic Research Series, vol. 50—. American Geophysical Union, Washington, DC, pp 19–24

Hagen EH, Faure G, Buchanan D (1992) Recovery of extraterrestrial particles from the Lewis Cliff ice tongue using a passive collection system. Ant J US 27(5):31–33

Hale VPS, McSween HY Jr, McKay GA (1999) Re-evaluation of intercumulus liquid composition and oxidation state for the Shergotty meteorite. Geochim Cosmochim Acta 63(9):1459–1470

Hartmann WK (2005) Moons and planets, 5th edn. Brooks/Cole, Belmont, CA

Harvey RP (1990) Terrestrial age mapping of the Allan Hills Main Icefield and implications for the Whillans-Cassidy model of meteorite concentration. In: Cassidy WA, Whillans IM (eds) Workshop on Antarctic meteorite stranding surfaces. LPI Tech. Rept. 90-03. Houston, TX, pp 88–90

Harvey RP (2003) The origin and significance of Antarctic meteorites. Chemie der Erde 63:93–147

Harvey RP, Maurette M (1990) The best cosmic dust source in the world? The origin and significance of the Walcott Névé, Antarctica, micrometeorites. Lunar and Planetary Science Conference, vol. 21—. LPI, Houston, TX, pp 569–578

Harvey RP, McSween HY Jr (1996) A possible high-temperature origin for the carbonates in the martian meteorite ALH 84001. Nature 382:49–51

Harvey RP, Score R (1991) Direct evidence of in-ice or pre-ice weathering of Antarctic meteorites. Meteoritics 26:343–344

Harvey RP, Dunbar NW, McIntosh WC, Esser RP, Nishiizumi K, Taylor S, Caffee MW (1998) Meteoritic event recorded in Antarctic ice. Geology 26(7):607–610

Haskell TR, Kennett JP, Prebble WM (1963) Basement and sedimentary geology of the Darwin Glacier area. In: Adie RJ (ed) Antarctic geology. North-Holland, Amsterdam, The Netherlands, pp 348–351

Haskell TR, Kennett JP, Prebble WM (1965) Geology of the Brown Hills and Darwin Mountains, southern Victoria Land, Antarctica. Trans Roy Soc New Zealand Geology 2(15):231–247

Hassan FA, Haq M, Sears DWG (1986) Thermoluminescence and shock reheating history of meteorites III: The Shergottites. Geochim. Cosmochim. Acta, 50(6):1031–1038

Herd CDK, Borg LE, Jones JH, Papike JJ (2002) Oxygen fugacity and geochemical variations in the martian basalts: Implications for martian basalt petrogenesis and the oxidation state of the upper mantle of Mars. Geochim Cosmochim Acta 66(11):2025–2036

Herzog GF (2005) Cosmic-ray exposure ages of meteorites. In: Davis AM (ed) Meteorites, comets, and planets. Treatise on geochemistry, vol. 1. Elsevier, Amsterdam, The Netherlands, pp 347–380

Heumann KG, Gall M, Weiss H (1987) Geochemical investigations to explain iodine overabundances in Antarctic meteorites. Geochim Cosmochim Acta 51:2541–2547

Heumann KG, Neubauer J, Reifenhäuser W (1990) Iodine overabundances measured in the surface layers of an Antarctic stony and iron meteorite. Geochim Cosmochim Acta 54:2503–2506

Hewins RH (1997) Chondrules. Annu Rev Earth Planet Sci 25:61–83

Hewins RH, Jones RH, Scott ERD (eds) (1996) Chondrules and the protoplanetary disk. Cambridge University Press, Cambridge, UK

Hill DH, Boynton WV, Haag RA (1991) A lunar meteorite found outside the Antarctic. Nature 352:614–617

Hodge PW, Wright FW, Langway CC Jr (1967) Studies of particles of extraterrestrial origin, 5. Compositions of the interiors of spherules from Arctic and Antarctic ice deposits. J Geophys Res 72(4):1404–1406

Holland G, Saxton JM, Lyon IC, Turner G (2005) Negative $\delta^{18}O$ values in Allan Hills 84001 carbonate: Possible evidence for water precipitation on Mars. Geochim Cosmochim Acta 69(5):1359–1369

Jagoutz E, Wänke H (1986) Sr and Nd isotopic systematics of Shergotty meteorite. Geochim Cosmochim Acta 50(6): 939–953

Jones JH (1986) A discussion of isotopic systematics and mineral zoning in the shergottites: Evidence for a 180 m.y. igneous crystallization age. Geochim Cosmochim Acta 50(6):969–977

Jones JH (2003) Constraints on the structure of the martian interior determined from the chemical and isotopic systematics of SNC meteorites. Meteorit Planet Sci 38(12): 1807–1814

Jull AJT (2001) Terrestrial ages of meteorites. In: Peucker-Ehrenbrink B, Schmitz B (eds) Accretion of extraterrestrial matter throughout earth's history. Kluwer Academic/Plenum, New York, pp 241–266

Jull AJT, Cheng S, Gooding JL, Velbel MA (1988) Rapid growth of Mg-carbonate weathering products in a stony meteorite from Antarctica. Science 242:417–419

Jull AJT, Eastoe CJ, Xue S, Herzog GF (1995) Isotopic composition of carbonates in the SNC meteorites Allan Hills 84001 and Nakhla. Meteoritics 30:311–318

Jull AJT, Courtney C, Jeffery DA, Beck JW (1998) Isotopic evidence for a terrestrial source of organic compounds found in martian meteorite Allan Hills 84001 and Elephant Moraine 79001. Science 279:366–369

Kamp PJJ, Lowe DJ (1982) Geology and terrestrial age of the Derrick Peak meteorite occurrence, Antarctica. Meteoritics 17:119–127

Kargel JS (1994) Metalliferous asteroids as potential sources of precious metals. J Geophys Res 99(E10):21129–21144

Keil K, Papike JJ (eds) (1983) Meteorites from the Earth's Moon. Proceedings of the Lunar Planetary Science Conference, LPI, Contrib. 501, Houston, TX

Keller P (1970) Eigenschaften von $(Cl, F, OH)_{<2}Fe_8(O,OH)_{16}$ und Akaganéite. Neues Jahrbuch für Mineralogie, Abhandlungen 113:29–49

Kelly DC, Elkins-Tanton LT (2004) Bottle green microtektites from the South Tasman Rise: Deep-sea evidence for an impact event near the Miocene/Pliocene boundary. Meteorit Planet Sci 39(12):1921–1929

Kent AJR, Hutcheon ID, Ryerson FJ, Phinney DL (2001) The temperature of formation of carbonate in martian meteorite ALH 84001. Geochim Cosmochim Acta 65(2): 311–321

Kiefer WS (2003) Melting in the martian mantle: Shergottite formation and implications for present-day mantle convection on Mars. Meteorit Planet Sci 39(12):1815–1832

King EA, Wagstaff J (1981) Micrometeorites from Antarctic ice cores. Ant J US 16(5):92–93

Kirschvink JL, Maine AT, Vali H (1997) Paleomagnetic evidence of a low-temperature origin of carbonate in the martian meteorite ALH 84001. Science 275:1629–1633

Klein HP (1978) The Viking biological experiment on Mars. Icarus 34:666–674.

Koeberl C (1986) Geochemistry of tektites and impact glasses. Ann Rev Earth Planet Sci 14:323–350

Koeberl C (1988a) Antarktische Meteorite; Kosmische Bruchstücke im ewigen Eis. Die Geowissenschaften 6(4):106–114

Koeberl C (1988b) Trace element geochemistry of lunar meteorites Yamato-791197 and Y-82192. Proceedings of NIPR Symposium on Antarctic Meteorites, vol. 1. National Institute of Polar Research, Tokyo, pp 122–134

Koeberl C (1988c) The origin of tektites: A geochemical discussion. Proceedings of NIPR Symposium on Antarctic Meteorites, vol. 1. National Institute of Polar Research, Tokyo, pp 261–290

Koeberl C (1990) The geochemistry of tektites: An overview. Tectonophysics 171:405–422

Koeberl C (1994) Tektite origin by hypervelocity asteroidal or cometary impact: Target rocks, source craters, and mechanisms. In: Dressler BO, Grieve RAF, Sharpton VL (eds) Large meteorite impacts and planetary evolution. Geological Society of America, Special Paper 293. Boulder, CO, pp 133–151

Koeberl C, Cassidy WA (eds) (1990) Differences between Antarctic and non-Antarctic meteorites. LPI Tech. Rept. No. 90-01. Lunar and Planetary Science Institute, Houston, TX

Koeberl C, Cassidy WA (1991) Differences between Antarctic and non-Antarctic meteorites: An assessment. Geochim Cosmochim Acta 55:3–18

Koeberl C, Hagen EH (1989) Extraterrestrial spherules in glacial sediment from the Transantarctic Mountains, Antarctica: Structure, mineralogy, and chemical composition. Geochim Cosmochim Acta 53:937–944

Koeberl C, MacLeod KG (eds) (2002) Catastrophic events and mass extinctions: Impacts and beyond. SPE 356. Geological Society of America, Boulder, CO

Koeberl C, Hagen EH, Faure G (1988) Chemical composition and morphology of meteorite ablation spherules in Neogene till in the Dominion Range, Transantarctic Mountains. Lunar Planetary Science Conference, vol. 9(2). LPI, Houston, TX, pp 625–626

Koeberl C, Hagen EH, Faure G (1989) Extraterrestrial spherules from glacial sedimentin Antarctica: Internal structure, mineralogy, and chemical composition. Lunar Planetary Science Conference, vol. 20(2). LPI, Houston, TX, pp 528–529

Koizumi E, Mikouchi T, McKay G, Miyamo M (2008) Crystallization relationship of Yamato-980459 and QUE 94201 depleted shergottites. Meteorit Planet Sci 43(Supplement):A79

Korotev RL, Haskin LA, Lindstrom MM (1983) Lunar highlands breccia 81005 (ALHA): So Apollo 18 flew, but where did it sample? Fourteenth Lunar and Planetary Science Conference. LPI, Houston, TX, pp 12–13

Kring DA, Swindle TD, Gleason JD, Grier JA (1998) Formation and relative ages of maskelynite and carbonate in ALH 84001. Geochim Cosmochim Acta 62:2155–2166

Kring DA, Gleason JD, Swindle TD, Nishiizumi K, Caffee MW, Hill DH, Jull AJT, Boynton WV (2003) Composition of the first bulk melt sample from a volcanic region of Mars: Queen Alexandra Range 94201. MeteoritPlanet Sci 38(12):1833–1848

Krot AN, Keil K, Goodrich CA, Scott ERD, Weisberg MK (2005) Classification of meteorites. In: Davis AM (ed) Meteorites, comets, and lanets. In: Holland HD, Turekian KK (eds) Treatise of geochemistry, vol. 1. Elsevier-Pergamon, Oxford, UK, pp 83–128

Kumai M, Ackley SF, Clarke DB (1983) Elemental compositions and concentrations of microspherules in snow and pack ice from the Weddell Sea. Ant J US 18(5):128–131

Kurat G, Koeberl C, Presper T, Brandstätter F, Maurette M (1994) Petrology and geochemistry of Antarctic micrometeorites. Geochim Cosmochim Acta 58:3879–3904

Kyle PR, Jezek PA (1978) Composition of three tephra layers from the Byrd Station ice core, Antarctica. J Volcanol Geotherm Res 4:225–232

Kyle PR, Jezek PA, Mosley-Thompson E, Thompson LG (1981) Tephra layers in the Byrd Station ice core and the Dome C ice core, Antarctica, and their climatic importance. J Volcanol Geotherm Res 11:29–39

Langenauer M, Krähenbühl U (1993) Halogen contamination in Antarctic H5 and H6 chondrites and relation to sites of recovery. Earth Planet Sci Lett 120:431–442

Laul JC (1986) The Shergotty consortium and SNC meteorites: An overview. Geochim Cosmochim Acta 50(6):875–887

Laul JC, Smith MR, Wänke H, Jagoutz E, Dreibus G, Palme H, Spettel B, Burghele A, Lipschutz ME, Verkouteren RM (1986) Chemical systematics of the Shergotty meteorite and the composition of its parent body (Mars). Geochim Cosmochim Acta 50(6):909–926

Lee MR, Bland PA (2004) Mechanisms of weathering of meteorites recovered from hot and cold deserts and the formation of phyllosilicates. Geochim Cosmochim Acta 68(4):893–916

Lee M-Y, Wei K-Y (2000) Australasian microtektites in the South China Sea and the West Philippine Sea: Implications for age, size, and location of the impact crater. Meteorit Planet Sci 35:1151–1155

LeMasurier WE, Thomson JW (eds) (1990) Volcanoes of the Antarctic plate and southern oceans. Antarctic Research Series, vol. 48. American Geophysical Union, Washington, DC

Leshin L, McKeegan KD, Harvey RP (1997) Oxygen isotopic constraints on the genesis of carbonates from martian meteorite ALH 84001. Lunar Planetary Science Conference, vol. XXVIII—. Houston, TX, pp 805–806

Lewis JS (1996) Mining the sky, untold riches from asteroids, comets, and planets. Helix Books/Addison-Wesley, Reading, MA

Leya I, Lange H-H, Neumann S, Wieler R, Michel R (2000) The production of cosmogenic nuclides in stony meteoroids by galactic cosmic-ray particles. Meteorit Planet Sci 35:259–286

Lipschutz ME (1982) Weathering effects in Antarctic meteorites. In: Marvin UB, Mason B (eds) Catalog of meteorites from Victoria Land, Antarctica, 1978–1980. Smithsonian Contrib Earth Sci 24:67–69

Lipschutz ME (1986) Trace elements in Antarctic meteorites: Weathering and genetic information. In: Annexstad JO, Schultz L, Wänke H (eds) International workshop on Antarctic meteorites. LPI Tech. Rept. 86-01. Houston, TX, pp 68–70

Lipschutz ME (1989) Trace element variations between Antarctic (Victoria Land) and non-Antarctic meteorites. In: Marvin UB, MacPherson GJ (eds) Field and laboratory investigations of meteorites from Victoria Land and the Thiel Mountains Region, Antarctica, 1982–1983 and 1983–1984. Smithsonian Contributions to Earth Sciences, vol. 28. Washington, DC, pp 99–102

Lipschutz ME, Schultz L (1999) Meteorites. In: Weissman PR McFadden L-A, Johnson TV (eds) Encyclopedia of the solar system. Academic, San Diego, CA, pp 629–671

Lipschutz ME, Schultz L (2006) Meteorites. In: McFadden L-A, Weissman PR, Johnson TV (eds) Encyclopedia of the solar system, 2nd edn. Academic, San Diego, CA

Love SG, Brownlee DE (1993) A direct measurement of the terrestrial mass accretion rate of cosmic dust. Science 262:550–553

Lugmair GW, Galer SJG (1992) Age and isotopic relationships among angrites Lewis Cliff 86010 and Angra dos Reis. Geochim Cosmochim Acta 56:1673–1694

Ma P, Herzog GF, Schultz L, Weber HW, Knie K (2002) Cosmogenic radionuclides and light noble gases in EET 96026. Meteorit Planet Sci 37(7):A91

MacKay AL (1962) β-Ferric oxyhydroxite-akaganéite. Mineral Mag 33:270–280

Marchant DR, Denton GH, Swisher CC III (1993) Miocene-Pliocene-Pleistocene glacial history of Arena Valley, Quartermain Mountains, Antarctica. Geografiska Annaler Stockholm, vol. 75A. Scandinavian University Press, Stockholm, Sweden, pp 269–302

Marshak S, Kyle PR, McIntosh WC, Samsonov V, Schellhorn MA (1981) Butcher Ridge igneous complex, Cook Mountains, Antarctica. Ant J US 16(5):54–55

Marvin UB (1980) Magnesium carbonate and magnesium sulfate deposits on Antarctic meteorites. Ant J US 15(5):54–55

Marvin UB (1981) The search for Antarctic meteorites. Sky Telescope, 62 (November): 423–427

Marvin UB (1983a) The discovery and initial characterization of Allan Hills 81005: The first lunar meteorite. Geophys Res Lett 10:775–778

Marvin UB (1983b) Mineralogy of the oxidation products of the Sputnik 4 fragment and of iron meteorites. J Geophys Res 68:5059–5068

Marvin UB (1984) Meteorites on ice. Planet Rept 4(2) (March/April):12–14

Marvin UB (1989) Meteorite distributions of the Allan Hills Main Icefield and the pairing problem. In: Marvin UB, MacPherson GJ (eds) Field and laboratory investigations of meteorites from Victoria Land and the Thiel Mountains Region, Antarctica, 1982–1983 and 1983–1984. Smithsonian Contributions to the Earth Sciences, No. 28. Washington, DC, pp 113–119

Marvin UB, MacPherson GJ (eds) (1989) Field and laboratory investigations of meteorites from Victoria Land and the Thiel Mountains region, Antarctica, 1982–1983 and 1983–1984. Smithsonian Contributions to the Earth Sciences, No. 28. Smithsonian Institution Press, Washington, DC

Marvin UB, Mason B (eds) (1980) Catalog of Antarctic meteorites, 1977–1978. Smithsonian Contributions to the Earth Sciences, No. 23. Smithsonian Institution Press, Washington, DC

Marvin UB, Mason B (eds) (1982) Field and laboratory investigations of meteorites from Victoria Land, Antarctica. Smithsonian Contributions to the Earth Sciences, No. 24. Smithsonian Institution Press, Washington, DC

Marvin UB, Mason B (eds) (1984) Field and laboratory investigations of meteorites from Victoria Land, Antarctica. Smithsonian Contributions to the Earth Sciences, No. 26. Smithsonian Institution Press, Washington, DC

Mason B (1962) Meteorites. Wiley, New York

Matrajt G, Pizzarello S, Taylor S, Brownlee DE (2004) Concentrations and variability of the AIB amino acid in polar micrometeorites: Implications for the exogenous delivery of amino acids to the primitive Earth. Meteorit Planet Sci 39(11):1849–1858

Maurette M (2006) Micrometeorites and the mysteries of our origin. Springer, Heidelberg, Germany

Maurette M, Hammer C, Brownlee DE, Reeh N, Thomson HH (1986) Placers of cosmic dust in blue-ice lakes of Greenland. Science 233(August):869–872

Maurette M, Jehanno C, Robin E, Hammer C (1987) Characteristics and mass distribution of extraterrestrial dust from the Greenland ice cap. Nature 328:699–702

Maurette M, Pourchet M, Bonny Ph, de Angelis M, Siry P (1989) A new collection of micrometeorites extracted from

100 tonnes of artificially melted blue ice near Cap Prudhomme in Antarctica. Lunar Planet Sci 20(2):644–645

Maurette M, Olinger C, Michel-Levy MC, Kurat G, Pourchet M, Brandstätter F, Bourot-Denise M (1991) A collection of diverse micrometeorites recovered from 100 tonnes of Antarctic blue ice. Nature 351:44–47

McCanta MC, Rutherford MJ, Jones JH (2004) An experimental study of rare earth element partitioning between a shergottite melt and pigeonite: Implications for the oxygen fugacity of the martian interior. Geochim Cosmochim Acta 68(8): 1943–1952

McFadden L-A, Weissman PR, Johnson TV (eds) (2006) Encyclopedia of the solar system, 2nd edn. Academic, Elsevier, San Diego, CA

McKay DS, Gibson EK Jr, Thomas-Keprta KL, Vali H, Romanek CS, Clemett SJ, Chillier XDF, Maechling CR, Zare RN (1996) Search for past life on Mars: Possible relic biogenic activity in martian meteorite ALH 84001. Science 273:924–930

McSween HY Jr (1997) Evidence for life in a martian meteorite? GSA Today, vol. 7(7). Geological Society of America, Boulder, CO, pp 1–7

McSween HY Jr (1999) Meteorites. In: Beatty JK, Petersen CC, Chaikin A (eds) The new Solar System. Sky, Cambridge, MA, pp 351–363

McSween HY Jr (2005) Mars. In: Davis AM (ed) Meteorites, comets, and planets. Treatise on geochemsistry. Elsevier, Amsterdam, The Netherlands, pp 601–621

McSween HY Jr, Harvey RP (1998) An evaporation model for the formation of carbonates in the ALH 84001 martian meteorite. Int Geol Rev 40:774–783

Melosh HJ (1989) Impact cratering: A geological process. Oxford University Press, Oxford, UK

Michlovich ES, Wolf SF, Wang MS, Vogt S, Elmore D, Lipschutz ME (1995) Chemical studies of H chondrites 5. Temporal variations of sources. J Geophys Res 100(E2): 3317–3333

Miotke F-D (1988) Microclimate, weathering processes and salt within ice-free continental Antarctica. Polarforschung 58:201–209

Misawa K, Tatsumoto M, Dalrymple GB, Yanai K (1993) An extremely low U-Pb source in the Moon: U-Th-Pb, Sm-Nd, Rb-Sr, and $^{40}Ar/^{39}Ar$ isotopic systematics and age of lunar meteorite Asuka 881757. Geochim Cosmoshim Acta 57(19):4687–4702

Misawa K, Yamaguchi A, Kaiden H (2005) U-Pb and $^{207}Pb/^{206}Pb$ ages of zircon from basaltic eucrites: Implications for early basaltic volcanism on the eucrite parent body. Geochim Cosmochim Acta 69(24):5847–5861

Mittlefehldt DW (1994) ALH 84001, a cumulative orthopyroxenite member of the martian meteorite clan. Meteoritics 29:214–221

Miura YN, Sugiura N (2000) Martian atmosphere-like nitrogen in orthopyroxenite ALH 84001. Geochim Cosmochim Acta 64(3):559–572

Miura YN, Nagao K, Sugiura N, Sagawa H, Matsubara K (1995) Orthopyroxenite ALH 84001 and shergottite ALH 77005: Additional evidence for a martian origin of noble gases. Geochim Cosmochim Acta 59:2105–2113

Mosley-Thompson E (1980) 911 years of microparticle deposition at the South Pole: A climatic interpretation. Institute of Polar Studies, Report 73. Ohio State University, Columbus, OH

Mulvaney R, Wolff EW, Oates K (1988) Sulphuric acid at grain boundaries in Antarctic ice. Nature 331:247–249

Murty SVS, Mohapatra RK (1997) Nitrogen and heavy noble gases in ALH 84001: Signatures of ancient martian atmosphere. Geochim Cosmochim Acta 61:5417–5428

Nagao K, Park J (2008) Noble gases and cosmic-ray exposure ages of two martian shergottites, RBT 04262 and LAR 06319 recovered in Antarctica. Meteorit Planet Sci 43 (Supplement):A107

Nakamura N, Unruh DM, Tatsumoto M, Hutchison R (1982) Origin and evolution of the Nakhla meteorite inferred from the Sm-Nd and U-Pb systematics and REE, Ba, Sr, Rb, and K abundances. Geochim Cosmochim Acta 46:1555–1573

Nakamura T, Imae N, Noguchi T, and 12 additional persons (1999) Antarctic micrometeorites collected at the Dome Fuji Station. Antarctic Meteorite Research, vol. 12—. Tokyo, Japan, pp 183–198

Niles PB, Leshin LA, Guan Y (2005) Microscale carbon isotope variability in ALH 84001 carbonates and a discussion of possible formation environments. Geochim Cosmochim Acta 69(11):2931–2944

Nishiizumi K (1983) Measurement of ^{53}Mn in deep-sea iron and stone spherules. Earth Planet. Sci. Letters 63:223–228

Nishiizumi K, Caffee MW (2002) Exposure histories of C2 carbonaceous chondrites. Meteorit Planet Sci 37(7):A109

Nishiizumi K, Murell MT, Arnold JR, Elmore D, Ferraro RD, Gove HE, Finkel RC (1981) Cosmic ray produced ^{36}Cl and ^{53}Mn in Allan Hills-77 meteorites. Earth Planet Sci Lett 52:31–38

Nishiizumi K, Arnold JR, Elmore D, Ma X, Newman D, Gove HE (1983) ^{36}Cl and ^{53}Mn in Antarctic meteorites and ^{10}Be-^{36}Cl dating of Antarctic ice. Earth Planet Sci Lett 62:407–417

Nishiizumi K, Klein J, Middleton R, Elmore D, Kubik PW, Arnold JR (1986) Exposure history of shergottites. Geochim Cosmochim Acta 50:1017–1021

Nishiizumi K, Klein J, Middleton R, Arnold JR (1987) Long-lived cosmogenic nuclides in the Derrick Peak and Lazarev iron meteorites. Lunar Planetary Science Conference, vol. 18—. Houston, TX, pp 724–725

Nishiizumi K, Elmore D, Kubik PW (1989) Update on terrestrial ages of Antarctic meteorites. Earth Planet Sci Lett 93:299–313

Nishiizumi K, Arnold JR, Fink D, Klein J, Middleton R, Brownlee DE, Maurette M (1991) Exposure history of individual cosmic particles. Earth Planet Sci Lett 104:315–324

Nishiizumi K, Arnold JR, Brownlee DE, Caffee MW, Finkel RC, Harvey RP (1995) ^{10}Be and ^{26}Al in individual cosmic spherules from Antarctica. Meteoritics 30:728–732

Nishiizumi K, Hillegonds DJ, Welten KC (2006) Exposure and terrestrial histories of lunar meteorites LAP 02205/02224/02226/02436, MET 01210, and PCA02001. Lunar Planetary Science Conference, Abstract #2369 (CD-ROM). Houston, TX

Nyquist LE, Wooden JL, Bansal B, Wiesmann H, McKay GA, Bogard DD (1979) Rb-Sr age of the Shergotty achondrite and implications for metamorphic resetting of isochron ages. Geochim Cosmochim Acta 43:1057–1074

Nyquist LE, Takeda H, Bansal BM, Shih C-Y, Wiesmann H, Wooden JL (1986) Rb-Sr and Sm-Nd internal isochron ages of a subophitic basalt clast and a matrix sample from Y 75011 eucrite. J Geophys Res 91:8137–8150

Nyquist LE, Bansal B, Wiesmann H, Shih C-Y (1994) Neodymium, strontium, and chromium isotopic studies of the LEW 86010 and Angra dos Reis meteorites and the chronology of the angrite parent body. Meteoritics 29:872–885

Nyquist LE, Bansal B, Wiesmann H, Shih C-Y (1995) "Martians" young and old: Zagami and ALH 84001. Proceed Lunar Planet Sci Conf 31:1065–1066

Nyquist LE, Reese YD, Wiesmann H, Shih C-Y, Borg LE (1997) Samarium-neodymium age and manganese-chromium systematics of eucrite Elephant Moraine 90020. Meteorit Planet Sci 32(Supplement):A101-A102

Nyquist LE, Bogard DD, Shih C-Y, Greshake A, Stőffler D, Eugster O (2001) Ages and geologic histories of martian meteorites. Space Sci Rev 96:1050–1064

Nyquist LE, Shih C-Y, Reese YD, Bogard DD (2005) Age of lunar meteorite LAP 02205 and implications for impact sampling of planetary surfaces. Lunar Planetary Science Conference, Abstract #1374 (CD-ROM), Houston, TX

Palais JM (1985) Tephra layers and ice chemistry in the Byrd Station ice core, Antarctica. PhD dissertation, The Ohio State University, Columbus, OH

Palais JM, Legrand M (1985) Soluble impurities in the Byrd Station ice core, Antarctica: Their origin and sources. J Geophys Res 90:1143–1154

Papanastassiou DA, Wasserburg GJ (1974) Evidence for late formation and young metamorphism in the achondrite Nakhla. Geophys Res Lett 1(1):23–26

Papike JJ (ed) (1998) Planetary materials. Reviews in Mineralogy, vol. 36. Mineralogical Society of America, Washington, DC

Papike JJ, Ryder G, Shearer CK (1998) Lunar samples. In: Papike JJ (ed) Planetary materials, reviews in mineralogy, vol. 36, Chapter 5.. Mineralogical Society of America, Washington, DC, pp 1–234

Park J, Garrison DH, Bogard DD (2008) $^{40}Ar/^{39}Ar$ ages of nakhlites: A synthesis. Meteorit Planet Sci 43 (Supplement): A127

Prasad MS, Khedekar VD (2003) Impact microcrater morphology on Australasian microtektites. Meteorit Planet Sci 38(9):1351–1371

Raisbeck GM, Yiou F (1987) ^{10}Be and ^{26}Al in micrometeorites from Greenland ice. Meteoritics 22:485

Rankenburg K, Brandon AD, Norman MD (2007) A Rb-Sr and Sm-Nd isotope geochronology and trace element study of lunar meteorite LaPaz Icefield 02205. Geochim Cosmochim Acta 71:2120–2135

Ravich MG, Revnov BI (1963) Lazarev iron meteorite. Meteoritika 23:30–35 (In Russian) English translation in Meteoritica, vol. 23, 1965. Taurus, New York, pp 38–43

Reimold WU, Gibson RL (eds) (2006) Processes on the early Earth. Special Paper 405. Geological Society of America, Boulder, CO

Robin E, Christophe M, Bourot-Denise M, Jehanno C (1990) Crystalline micrometeorites from the Greenland blue lakes: Their chemical composition, mineralogy, and possible origin. Earth Planet Sci Lett 97:162–176

Sano Y, Terada K, Takeno S, Taylor LA, McSween HY Jr (2000) Ion microprobe uranium-thorium dating of Shergotty phosphates. Meteorit Planet Sci 35:341–346

Saxton JM, Lyon IC, Turner G (1997) Oxygen isotope ratio zoning in ALH 84001 carbonates (Abstract). In: Clifford S, Treiman A (eds) Workshop of early Mars—.. LPI Contribution No. 916. Lunar Planetary Science Institue, Houston, TX, pp 70–72

Scherer P, Schultz L, Neupert U, Knauer M, Neumann S, Leya I, Mokos MR, Lipschutz ME, Metzler K, Suter M, Kubik PW (1997) Allan Hills 88019: An Antarctic H-chondrite with a very long terrestrial age. Meteoritics 32:769–773

Schmitt DG (2002) The law of ownership and control of meteorites. Meteorit Planet Sci 37(Supplement): B5–B11

Schmidt RA (1962) Australites and Antarctica. Science 138(3538):443–444

Schmidt RA (1964) Microscopic extraterrestrial particles from the Antarctic Penninsula. In: Cassidy WA (ed) Cosmic dust. Ann NY Acad Sci 119:186–204

Schultz L (1986) Allende in Antarctica: Temperatures in Antarctic meteorites. Meteoritics 21(4):505

Schultz L (1990) Terrestrial ages and weathering of Antarctic meteorites. In: Cassidy WA, Whillans IM (eds) Workshop on Antarctic meteorite stranding surfaces. LPI Tech. Rept. 90-03. Houston, TX, pp 56–59

Schwenzer SP, Herrmann S, Ott U (2008) Noble gases in two shergottites and a nakhlite from Antarctica; Y-000027 and Y-000593. Meteoritics Planet. Sci., 43(Supplement): A140

Score R, Lindstrom MM (1990) Guide to the U.S. collection of Antarctic meteorites 1976–1988. Antarctic meteorite newsletter, vol. 13(1). Johnson Space Center, Houston, TX, 135 p

Score R, Schwartz CM, King TVV, Mason B, Bogard DD, Gabel EM (1981) Antarctic meteorite descriptions 1976-1977-1978-1979. Curatorial Branch Publications, 54, Johnson Space Center 17076. Houston, TX

Score R, Schwartz CM, Mason B, Bogard DD (1982) Antarctic meteorite descriptions 1980. Curatorial Branch Publications, 60, Johnson Space Center 18170, Houston, TX

Scott ERD (1979) Origin of anomalous iron meteorites. Mineral Mag 43:415–421

Scott ERD, Yamaguchi A, Krot AN (1997) Petrological evidence for shock melting of carbonates in the martian meteorite ALH 84001. Nature 387:377–379

Schellhorn MA (1982) The role of crustal contamination at the Butcher Ridge igneous complex, Antarctica. MSc thesis. New Mexico Institute of Mining and Technology, Socorro, New Mexico

Shaw GE (1988) Antarctic aerosols: A review. Rev Geophys 26:89–112

Shearer CK, Lane GD, Papike JJ, Spilde MN (1996) Sulfur isotope systematics in alteration assemblages in martian meteorite ALH 84001. Geochim Cosmochim Acta 60: 2921–2926

Shima M (1966) Glassy spherules (microtektite?) found in ice at Scott Base, Antarctica J Geophys Res 71(14):3595–3596

Shinonaga T, Endo K, Ebihara M, Heumann KG, Nakahara H (1994) Weathering of Antarctic meteorites investigated from contents of Fe^{3+}, chlorine, and iodine. Geochim Cosmochim Acta 59:3735–3740

Shoemaker EM, Kieffer SW (1974) Guidebook to the geology of Meteor Crater, Arizona. Pub. 17. Center for Meteorite Studies, Arizona State University, Tempe, AZ

Steed RHN, Drewry DJ (1982) Radio-echo sounging investigations of Wilkes Land, Antarctica. In: Craddock C (ed) Antarctic geoscience. University of Wisconsin Press, Madison, WI

Stephan T, Jessberger EK, Heiss CH, Rost D (2003) TOF-SIMS analysis of polycyclic aromatic hydrocarbons in Allan Hills 84001. Meteorit Planet Sci 38(1):109–116
Stolper EM, McSween HY (1979) Petrology and origin of the shergottite meteorite. Geochim Cosmochim Acta 43:1475–1498
Stonehouse B (ed) (2002) Encyclopedia of Antarctica and the Southern Oceans. Wiley, Chichester, UK
Swindle TD, Olson EK (2004) $^{40}Ar/^{39}Ar$ studies of whole rock nakhlites: Evidence for the timing and aqueous alteration on Mars. Meteorit Planet Sci 39(5):755–766
Swindle TD, Caffee MW, Hohenberg CM (1986) Xenon and other noble gases inshergottites. Geochim Cosmochim Acta 50(6):1001–1015
Swindle TD, Grier JA, Burkland MK (1995) Noble gases in orthopyroxenite ALH84001: A different kind of martian meteorite with an atmospheric signature. Geochim Cosmochim Acta 59:793–801
Takahashi K, Masuda A (1987) Two lunar meteorites Yamato 791197 and Yamato 82192: REE abundances and geochronological dating. Mem. Natl. Inst. Polar Res., Special Issue 46: 71–88
Taylor LA, Nazarov MA, Shearer CK, McSween HY Jr, Cahill J, Neal CR, Ivanova MA, Barsukova LD, Lentz RC, Clayton RN, Mayeda TK (2002) Martian meteorite Dhofar 019: A new shergottite. Meteorit Planet Sci 37:1107–1128
Taylor S, Brownlee DE (1991) Cosmic spherules in the geologic record. Meteoritics 26:203–211
Taylor S, Lever JH, Harvey RP, Govoni J (1997) Collecting micrometeorites from the South Pole water well. CREL Rept. 97-1. US Army Cold-Regions Research and Engineering Laboratory, Hanover, New Hampshire, England
Taylor S, Lever JH, Harvey RP (1998) Accretion rate of cosmic spherules measured at South Pole. Science 392:899–903
Taylor S, Lever JH, Harvery RP (2000) Numbers, types, and compositions of an unbiased collection of cosmic spherules. Meteorit Planet Sci 35:651–666
Taylor S, Herzog GF, Delaney JS (2007) Crumbs from the crust of Vesta: Achondritic cosmic spherules from the South Pole water well. Meteorit Planet Sci 42(2):223–233
Taylor SR (1992) Solar system evolution: a new perspective. Cambridge University Press, Cambridge, UK
Taylor SR, McLennan SM (1985) The continental crust: Its composition and evolution. Blackwell Scientific, Oxford, UK
Tera F, Papanastassiou D, Wasserburg GJ (1974) Isotopic evidence for a terminal lunar cataclysm. Earth Planet Sci Lett 22(1):1–21
Terada K, Yada T, Kojima H, and 21 additional persons (2001) General characterization of Antarctic micrometeorites collected by the 39th Japanese Antarctic Research Expedition: Consortium studies JARE AMMs (II). Antarctic Meteorite Res 14:89–107, Tokyo, Japan
Terada K, Monde T, Sano Y (2003) Ion microprobe U-Th-Pb dating of phosphates in martian meteorite ALH 84001. Meteorit Planet Sci 28(11):1697–1703
Thiel E, Schmidt RA (1961) Spherules from the Antarctic ice cap. J Geophys Res 66:307–310
Thomas-Keprta KL, McKay DS, Wentworth SJ, Stevens TO, Taunton AE, Allen CC, Coleman A, Gibson EK Jr, Romanek CS (1998) Bacterial mineralization patterns in basaltic aquifers: Implications for possible life in martian meteorite ALH 84001. Geology 26(11):1031–1034
Thomas-Keprta KL, Bazylinski DA, Kirschvink JL, Clemett SJ, McKay DS, Wentworth SJ, Vali H, Gibson EK Jr, Romanek CS (2000) Elongated prismatic magnetite crystals in ALH 84001 carbonate globules: Potential martian magnetofossils. Geochim Cosmochim Acta 64:4049–4083
Toppani A, Libourel G, Engrand C, Maurette M (2001) Experimental simulation of atmospheric entry of micrometeorites. Meteorit Planet Sci 36:1377–1396
Treiman AH (1998) The history of Allan Hills 84001 revisited: Multiple shock events. Meteorit Planet Sci 33:735–764
Treiman AH (ed) (2003) Unmixing of SNCs: Chemical, isotopic, and petrologic components of the martian meteorites. Meteorit Planet Sci 38(12):1711–1875
Treiman AH, Drake MJ, Janssens M-J, Wolf R, Ebihara M (1986) Core formation in the Earth and shergottite parent body (SPB): Chemical evidence from basalts. Geochim Cosmochim Acta 50(6):1071–1091
Tuncel G, Zoller WH (1987) Atmospheric iridium at the South Pole as a measure of the meteoritic components. Nature 329:703–705
Turner G, Knott FS, Ash RD, Gilmour JD (1997) Ar-Ar chronology of the martian meteorite ALH 84001: Evidence for the timing of the early bombardment of Mars. Geochim Cosmochim Acta 61:3835–3850
Ugolini FC, Jackson ML (1982) Weathering and mineral synthesis in Antarctic soils. In: Craddock C (ed) Antarctic geoscience. University of Wisoconsin Press, Madison, WI, pp 1101–1109
Valley JW, Eiler JM, Graham CM, Gibson EK Jr, Romanek CS, Stolper EM (1997) Low-temperature carbonate concretions in the martian meteorite ALH 84001: Evidence from stable isotopes and mineralogy. Science 275:1633–1638
Van Schmus WR, Wood JA (1967) A chemical-petrologic classification for the chondritic meteorites. Geochim Cosmochim Acta 31:747–765
Velbel MA (1988) The distribution and significance of evaportitic weathering products on Antarctic meteorites. Meteoritics 23:151–159
Velbel MA, Gooding JL (1990) Terrestrial weathering of Antarctic stony meteorites: Developments 1985–1989. In: Koeberl C, Cassidy WA (eds) Workshop on differences between Antarctic and non-Antarctic meteorites. LPI Tech. Rept. 90-01. Houston, Texas, pp 94–98
Velbel MA, Long DT, Gooding JL (1991) Terrestrial weathering of Antarctic stone meteorites: Formation of Mg-carbonates on ordinary chondrites. Geochim Cosmochim Acta 55:67–76
Von Frese RRB, Potts LV, Wells SB, Leftwich TE, Kim HR, Kim JW, Golynsky AV, Hernandez O, Gaya-Piqué LR (2009) GRACE gravity evidence for an impact in Wilkes Land, Antarctica. Geochim Geophsy, Geosystems 10(2):1–14. doi:10.1029/ 2008GC002149
Wadhwa M, Lugmair GW (1996) The formation age of carbonates in ALH 84001. Abstract Meteorit Planet Sci 31:A145
Wagstaff J, King EA (1981) Micrometeorites and possible cometary dust from Antarctic ice cores. Lunar Planetary Science Conference, vol. 13(3). LPI, Houston, Texas, pp 1124–1126

Warren JL, Zolensky ME, Thomas K, Dodson AL, Watts LA, Wentworth S (1997) Cosmic dust catalog, vol. 15, parts 1, 2, 3. Space and Life Sciences Directorate, Solar System Exploration Div., Office of the Curator, No. 93, JSC #27897. NASA, Johnson Space Center, Houston, TX

Warren PH (2005) The Moon. In: Davis AM (ed) Meteorites, comets, and planets. Treatise on geochemistry. Elsevier, Amsterdam, The Netherlands, pp 559–599

Warren PH, Kallemeyn GW (1996) Siderophile trace elements in ALH 84001, other SNC meteorites, and eucrites: Evidence of heterogeneity, possibly time-linked, in the mantle of Mars. Meteorit Planet Sci 31:97–105

Warren PH, Ulf-Moller F, Kallemeyen GW (2005) New lunar meteorites: Impact melt and regolith breccias and large-scale heterogeneities of the upper lunar crust. Meteorit Planet Sci 40(7):989–1014

Wasson JT (1990) Ungrouped iron meteorites in Antarctica: Origin of anomalously high abundance. Science 249:900–902

Wasson JT, Huber H, Malvin DJ (2007) Formation of IIAB iron meteorites. Geochim Cosmochim Acta 71(2):760–781

Webb PN, Harwood DM (1991) Late Cenozoic glacial history of the Ross Embayment, Antarctica. Quat Sci Rev 10:215–223

Webb PN, Harwood DM, McKelvey BC, Mercer JH, Stott LD (1984) Cenozoic marine sedimentation and ice-volume variation on the East Antarctic craton. Geology 12:287–291

Weihaupt JG (1961) Geophysical studies in Victoria Land, Antarctica. Research Report Series, Geophysics, Polar Research Center, vol. 1. University of Wisconsin, Madison, WI, pp 1–123

Weihaupt JG (1976) The Wilkes Land anomaly: Evidence for a possible hypervelocity impact crater. J Geophys Res 81(32):5651–5663

Weisberg MK, Smith C, Benedix G, Herd CDK, Righter K, Haack H, Yamaguchi A, Chennaoui Aoudjehane H, Grossman JN (2009) The Meteoritical Bulletin, No. 96. Meteorit Planet Sci 44(9):1355–1397

Welten KC, Alderliesten C, van der Borg K, Lindner L, Loeken T, Schultz L (1997) Lewis Cliff 86360: An Antarctic L-chondrite with a terrestrial age of 2.35 million years. Meteorit Planet Sci 32:775–780

Welten KC, Nishiizumi K, Caffee MW, Schäfer J, Wieler R (1999) Terrestrial ages and exposure ages of Antarctic H chondrites from Frontier Mountain, north Victoria Land. Antarct. Meteor. Res., 12:94–107

Welten KC, Nishiizumi K, Masarik J, Caffee MW, Jull AJT, Klandrud SE, Wieler R (2001) Cosmic-ray exposure history of two Frontier Mountain H-chondrite showers from spallation and neutron-capture products. Meteorit Planet Sci 36:301–317

Whillans IM, Cassidy WA (1983) Catch a falling star: Meteorites and old ice. Science 222:55–57

Whipple FL (1966) Chondrules: Suggestion concerning the origin. Science 153:54–56

Wood JA (1968) Meteorites and the origin of planets. McGraw-Hill, New York

Wright IP, Carr RH, Pillinger CT (1986) Carbon abundance and isotopic studies of Shergotty and other shergottite meteorites. Geochim Cosmochim Acta 50(6):983–991

Yabuki H, Okada A, Shima M (1976) Nesquehonite found on the Yamato 74371 meteorite. Scientific Papers, Institute of Physical and Chemical Research, vol. 70. Tokyo, Japan, pp 22–29

Yamaguchi A, Taylor GJ, Keil K, Floss C, Crozaz G, Nyquist LE, Bogard DD, Garrison DH, Reese YD, Weismann H, Shih C-Y (2001) Post-crystallization reheating and partial melting of eucrite EET 90020 by impact into the hot crust of asteroid 4 Vesta ~4.50 Ga ago. Geochim Cosmochim Acta 65(20):3577–3599

Yanai K, Kojima H (1995) Catalog of Antarctic meteorites, 1969–1994. National Institute of Polar Research, Tokyo, Japan

Yiou F, Raisbeck GM (1987) Cosmic spherules from an Antarctic ice core. Meteoritics 22:539

Yiou F, Raisbeck GM, Jéhanno C (1989) Influx of cosmic spherules to the Earth during the last ~10^5 years deduced from concentrations in Antarctic ice cores. Meteoritics 24:344

Zeigler RA, Korotev RL, Joliff BL, Haskin LA (2005) Petrography and geochemistry of the LaPaz Icefield basaltic lunar meteorite and source crater pairing with Northwest Africa 032. Meteorit Planet Sci 40(7):1073–1101

Zolensky ME, Gooding JL (1986) Aqueous alteration on carbonaceous-chondrite parent bodies as inferred from weathering of meteorites in Antarctica. Meteoritics 21: 548–549

Zolensky ME, Webb SJ, Thomas K (1988) The search for refractory interplanetary dust particles from pre-industrial aged Antarctic ice. Lunar Planetary Science Conference, vol. 18. LPI, Houston, TX, pp 599–605

Zolensky ME, Wilson TL, Rietmeijer FJM, Houston FGJ (eds) (1994) Analysis of inter-planetary dust, vol. 310. American Institute of Physics, Washington, DC

Zolotov M Yu, Shock EL (2000) An abiotic origin for hydrocarbons in the Allan Hills 84001 martian meteorite through cooling of magmatic and impact generated gases. Meteorit Planet Sci 35:629–638

Chapter 19
Glaciaton of Southern Victoria Land

The late Cenozoic (Neogene) history of the Transantarctic Mountains included several episodes when the East Antarctic ice sheet overflowed the mountains. The resulting glaciations of the Transantarctic Mountains caused major topographic changes which became apparent after the ice sheet subsided for the last time. Therefore, the landscape of the present Transantarctic Mountains has evolved as a result of the climate changes during the Neogene which caused the advance(s) and subsequent retreat(s) of the East Antarctic ice sheet and set the stage for the most recent geomorphic and hydrologic evolution of the ice-free valleys of southern Victoria Land.

The lowering of sea level which accompanied the continental glaciation of the northern and southern hemispheres during the Pleistocene caused the Ross Ice Shelf to be grounded. In other words, it was no longer floating on seawater but actually settled onto the sediment that had been deposited beneath it. As a result, the forward motion of the grounded ice slowed appreciably, whereas the input of ice from East and West Antarctica increased. Consequently, the grounded ice thickened substantially and buried the base of Ross Island (Forbes and Ester 1964). The grounding of the Ross Ice Shelf also blocked the valleys of the outlet glaciers in the Transantarctic Mountains and caused these glaciers to thicken far beyond their present levels. The ice also entered the valleys of southern Victoria Land and deposited terminal moraines (e.g., in the lower Wright Valley).

Evidently, the Neogene glaciation of the Transantarctic Mountains and of the McMurdo Sound was a complicated process that combined changes in climate and sea level with the resulting changes in the flow of the outlet glaciers and of the Ross Ice Shelf.

19.1 Neogene Sediment in McMurdo Sound

McMurdo Sound in Figs. 19.1 and 16.1 is positioned between the Transantarctic Mountains and Ross Island and therefore has accumulated sediment that was generated by the glaciation of the presently ice-free valleys of southern Victoria Land and by the volcanic activity in the Erebus volcanic province and in the Discovery subprovince (Sections 16.1 and 16.2). The stratigraphy of the sediment in McMurdo Sound has been investigated by geophysical surveys, by the study of sediment cores, and by geological mapping of Cenozoic sedimentary rocks on the islands off the coast of southern Victoria Land.

The initial geophysical surveys of McMurdo Sound by Robinson (1962, 1963, 1964) were continued by Pederson et al. (1981), Wong and Christoffel (1981), McGinnis et al. (1983, 1985), Stern (1984), and Davy and Alder (1989). The scope of the geophysical studies of McMurdo Sound was enlarged by Kim et al. (1986), Davey (1987), and by Cooper et al. (1987, 1991) to include the crustal structure and sediment fill of the Victoria Land Basin north of McMurdo Sound.

Our interest is focused on the age and provenance of the sediment in McMurdo Sound because the sediment records the erosion of the Transantarctic Mountains by glaciers and the concurrent volcanic activity during the Cenozoic. Both sediment sources are represented in the cores of the MSSTS-1 and CIROS-1 and CIROS-2 drill holes described in publications edited by Barrett (1986, 1989).

A reconstruction of the topography of McMurdo Sound by Kyle (1981) based on K-Ar dates of volcanic rocks indicates that at about 5 Ma (early Pliocene)

G. Faure and T.M. Mensing, *The Transantarctic Mountains: Rocks, Ice, Meteorites and Water*,
DOI 10.1007/978-90-481-9390-5_19,

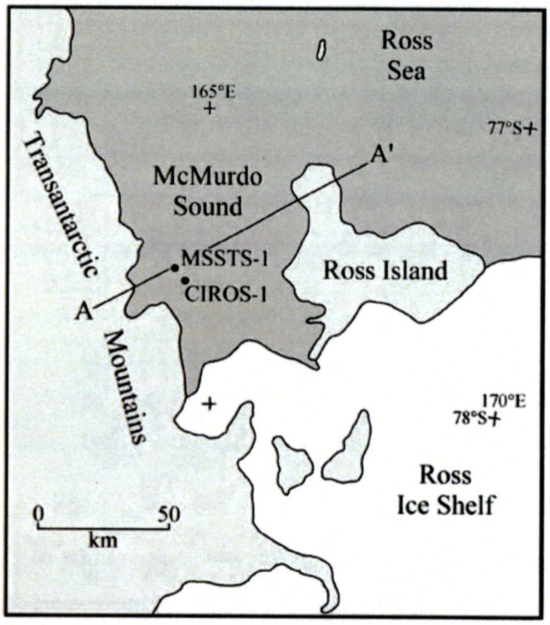

Fig. 19.1 CIROS-1 amd MSSTS-1 in McMurdo Sound were drilled in order to recover cores of the sediment that would reveal the history of glaciation of the Transantarctic Mountains in southern Victoria Land during the Cenozoic. A third core (CIROS-2) was drilled a short distance in front of the floating ice tongue of the Ferrar Glacier (not shown). CIROS-1 reached a depth of 702 m ending in beds of sand and conglomerate dated at 36.2 Ma (Eocene) (Adapted from Barrett et al. 1989)

only Mt. Morning (14–18 Ma), Black Island (10.9 Ma), Mt. Discovery (5.3 Ma), and Mt. Bird (4.5 ± 0.6 Ma) existed. In other words, Ross Island, White Island, the Brown Peninsula, and Minna Bluff were absent, allowing the Ross Ice Shelf to flow north unhindered and to enter the Ferrar, Taylor, and Wright valleys (Webb 1972).

Kyle (1981) also reported that the hyaloclastite of Castle Rock on the Hut Point Peninsula of Ross Island formed by a subglacial volcanic eruption at 1.21 ± 0.05 Ma because, at the time, the thickness of the Ross Ice Shelf had increased by more than 400 m compared to its present thickness. The increase in the thickness of the Ross Ice Shelf may have been caused partly by grounding and partly by the obstruction of its movement by Ross Island and Minna Bluff which had formed prior to 1.21 Ma.

The geologic history of McMurdo Sound also includes the deposition of fossiliferous sedimentary rocks which are now represented by small outcrops and scattered boulders on the islands and coast of McMurdo Sound (McIntyre and Wilson 1966; Wilson 1967; Vella 1969; Eggers 1976; Leckie and Webb 1979; Webb 1981; Stott et al. 1983). Even younger Quaternary deposits have been described by Speden (1962) and Kellogg et al. (1990).

The most detailed historical record of McMurdo Sound is contained in the drill cores that have been recovered from McMurdo Sound, which includes not only CIROS-1, CIROS-2, and MSSTS-1, but also DVDP 15 (Dry Valley Drilling Project; McGinnis 1981) and CRP-1, CRP-2/2A, and CRP-3 (Cape Roberts Project; Terra Antartica, 1998–2001).

19.1.1 CIROS-1 and MSSTS

The scientific reports arising from the study of the core of CIROS-1 in Fig. 19.1 were edited by Barrett (1989) and were summarized by Barrett et al. (1989, 1991). The CIROS-1 core contains sediment from two periods of deposition lasting from earliest Oligocene to early Miocene. The sediment was transported by rivers and glaciers from the Transantarctic Mountains which were rising at the time and had reached about half their present height (Barrett et al. 1989).

Temperate glaciers were discharging icebergs that deposited layers of pebbly mudstone (diamictite) off the coast of southern Victoria Land. The sediment contains palynomorphs and impressions of Nothofagus (beech) leaves indicating that forests existed in sheltered areas along the coast and that the climate was cool but distinctly warmer than at the present time. The ice may have originated from local ice caps in the Transantarctic Mountains or from a large ice sheet in the interior of East Antarctica. The sediment that was deposited during the late Oligocene/early Miocene records variations in sea level and corresponding changes in the extent of the ice. Barrett et al. (1989) concluded that:

> The glacial record from CIROS-1,..., substantiates the view that continent-wide glaciation of ice sheet proportions was a major feature of the Oligocene in Antarctica.

In addition, sand layers in the CIROS-1 core contain fragments of alkali-rich volcanic rocks of the McMurdo Volcanic Group, thereby extending the record of volcanic activity to the early Oligocene (George 1989). The geothermal gradient measured in CIROS-1 is 40°C/km (White 1989), which is consistent with the

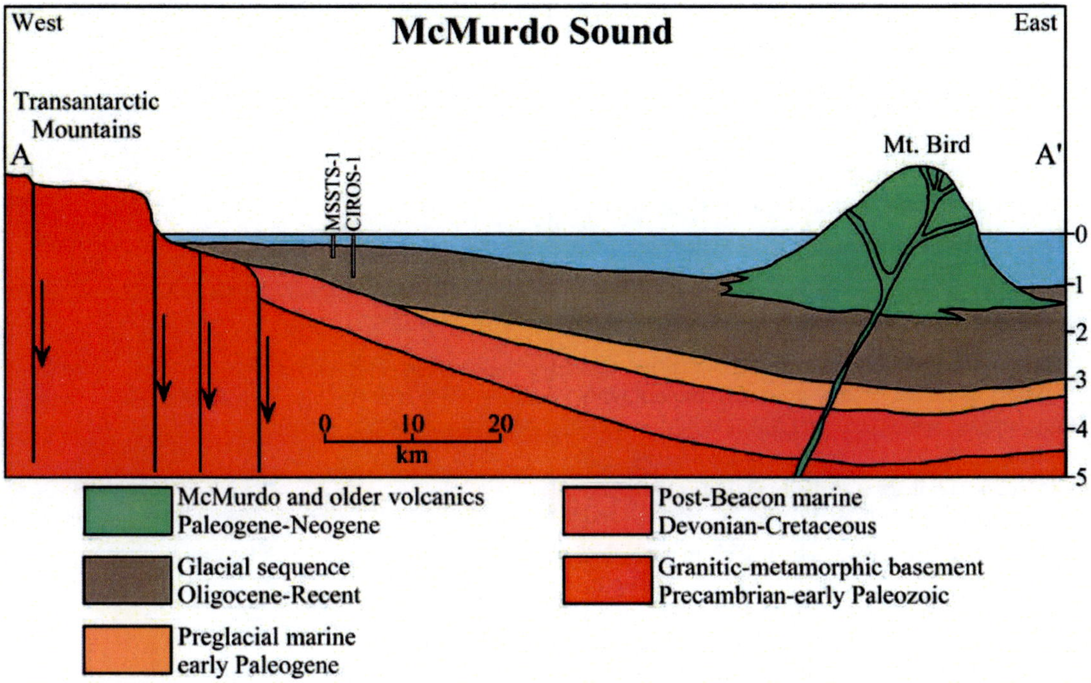

Fig. 19.2 The stratigraphy of the sediment and underlying granitic basement along line A-A′ in Fig. 19.1 is based on seismic refraction and reflection surveys by Wong and Christoffel (1981), McGinnis et al. (1983), and Davy and Alder (1989) in addition to the core logs of CIROS-1 and MSSTS-1. The uppermost unit consists of glacial diamictites and interbedded sandstone and mudstone extending in age from early Oligocene to Recent. The glacial deposits are underlain by marine sediment of Cretaceous to early Paleocene age (Adapted from Barrett et al. 1989, with input from Harwood et al. 1989; Webb 1989)

continuing activity of Mt. Erebus and suggests that the West Antarctic rift system is still alive. Hambrey et al. (1989) concluded that the lower sequence of CIROS-1 (366–702 m) consists of the pro-delta facies of mud interbedded with redeposited sand, whereas the upper sequence (27–366 m) represents a deltaic facies of mud, sand, and diamictite. Both facies constitute the "Glacial Sequence" in Fig. 19.2 which was adapted from Barrett et al. (1989). The underlying layers of "Preglacial Marine" and "Post-Beacon Marine" are interpretations of seismic surveys by Wong and Christoffel (1981), McGinnis et al. (1983), and Davy and Alder (1989). According to the geophysical data, McMurdo Sound is underlain by the granitic and high-grade metamorphic rocks that crop out in the mountains of southern Victoria Land.

The report edited by Barrett (1989) also contains a section in which the fossils of the CIROS-1 core are described. For example, Webb (1989) recovered 62 species of calcareous and agglutinated benthic foraminifera. Webb (1989) also integrated the sediment stratigraphies of CIRSO-1 with those of the MSSTS-1 and DSDP-270 (Deep Sea Drilling Project) cores. The oldest sediment recovered in the CIROS-1 core was dated by Barrera (1989) by means of the $^{87}Sr/^{86}Sr$ ratios of shells of mollusks deposited and preserved in situ in the sediment of CIROS-1. The age of this material is 34–36 Ma, which Webb (1989) placed within the lowest Oligocene, although the time scale of the IUGS (2002) suggests an uppermost Eocene (Priabonian) age.

The best time-resolution was achieved by Harwood et al. (1989) who used micropaleontologic, palynologic, magnetostratigraphic, and Sr-isotopic data to construct a biostratigraphic chronology of the sediment in CIROS-1. These authors identified a disconformity in the core between 30.5 and 34.5 Ma where 4 million years of sediment are missing. This disconformity divides the section into an upper part consisting of lower Miocene and upper Oligocene (22–30.5 Ma) and a lower part composed of lower Oligocene and uppermost Eocene (34.5–36.1 Ma). The sedimentation rate of the lower part of the section is constant at about 200 m per million years, whereas the sediment of the upper part was deposited at variable rates between 200 and about 15 m per million years.

The lower part of the core contains six layers of diamictite of glacial origin and thereby indicates that valley glaciers were reaching the coast and were forming icebergs that deposited clasts and glacial flour as they drifted into the Ross Sea. Note that Ross Island and the other volcanic islands that define McMurdo Sound (Fig. 19.1) did not exist during the lower Oligocene between 34.5 and 36.1 Ma. In the upper part of the CIROS-1 core glacial diamictites are the dominant lithology at about 56% and are interbedded with mudstone and sandstone.

The MSSTS-1 (McMurdo Sound Sediments and Tectonic Studies) hole was drilled inMcMurdo Sound only 4 km northwest of CIROS-1 (Davey and Christoffel 1984). It reached a depth of about 230 m and also intersected upper Oligocene and lower Miocene sediment (Barrett and McKelvey 1986; Barrett et al. 1987). The chronology of that core was subsequently revised by Harwood et al. (1989) which resulted in a better match of the sedimentation rates of the CIROS-1 and MSSTS-1 cores.

Evidently, ice was already flowing into the Ross Sea during the early Oligocene at about 35 Ma and increased in volume at about 30 Ma. These estimates refine the history of glaciation of Antarctica indicated by the K-Ar dates of subglacially erupted lava flows at Mt. Early and Sheridan Bluff (Section 16.6) and by the K-Ar date of greensand (glauconite) in the core of DSDP 270 in the Ross Sea reported by McDougall (1977) and revised by Leckie and Webb (1986). The DSDP-270 core was recovered in front of the Ross Ice Shelf about 500 km from the southern coast of Victoria Land.

19.1.2 Cape Roberts Project

In October of 1997 drilling started on fast sea ice at 77°0.48′S and 163°45.3′E approximately 16 km off Cape Roberts located south of Granite Harbor and about 30 km from the floating ice of the Mackay Glacier in Fig. 19.3. The drill site is underlain by 150 m of water near the margin of the Victoria Land Basin. Drilling had progressed to a depth of 148 m below the sea floor when operations were hurriedly suspended on October 24, 1997, because the sea ice began to break up after a severe windstorm (Cape Robert Science Team 1998).

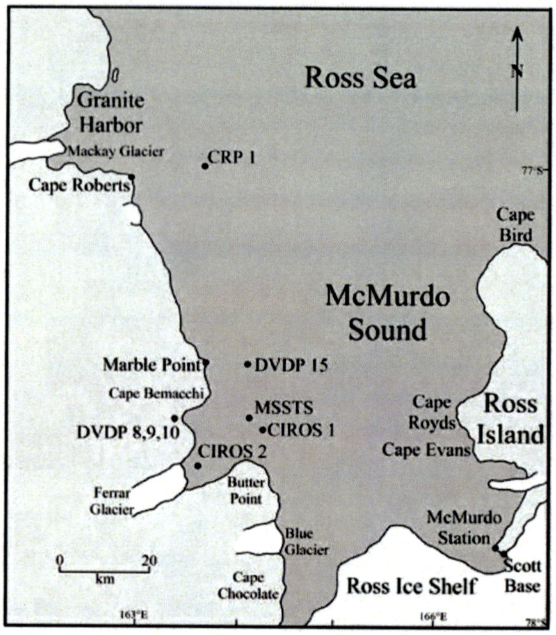

Fig. 19.3 The Cape Roberts Project drilled three holes through the sea ice about 16 km off the coast of southern Victoria Land at Cape Roberts. The cores are identified as CRP-1, CRP-2, 2A, and CRP-3 and were located close to CRP-1. The Cenozoic sediment at these sites dips away from the coast into the Victoria Land Basin. The cores consist of Oligocene to Miocene sediment composed of glacial diamictite interbedded with mudstone and sandstone. The sediment records the gradual cooling of the climate and increasing deposition of glacial sediments. CRP-3 bottomed in Devonian sandstone of the Beacon Supergroup (Adapted from Cape Roberts Science Team (1998) published in Terra Antartica vol. 5(1): 1–30, 1998)

The core recovered at CRP-1 consists of 43.55 mbsf (meters below sea floor) of glacial sediment of Quaternary age overlying early Miocene glacigenic sediment. The oldest sediment recovered at the bottom of the CRP-1 core is a mudstone dated at 22–24 Ma (early Miocene).

The drill site is located on the western edge of the Victoria Land Basin which extends to a depth of about 15 km below sea level and is filled with sedimentary and volcanic rocks on a basement of granitic gneisses. The sedimentary rocks are in large part of glacial origin. The Victoria Land Basin formed by crustal extension and subsidence related to the Cenozoic West Antarctic Rift (Section 15.5.5).

The core recovered at CRP-1 was subsequently described by a large number of scientists who published their results in Terra Antartica (Vol. 5, No. 3:255-713, 1998). These reports included papers by McIntosh (1998) on the $^{40}Ar/^{39}Ar$ dating of volcanic clasts, by Lavelle (1998) on the strontium-isotope stratigraphy, and by Hart and Webb (1998) on amino-acid geochronology. The results summarized by Barrett (1998a, b) yielded a date of 1.1 ± 0.3 Ma by the $^{87}Sr/^{86}Sr$ ratios of fragmental mollusk shells in the Quaternary part of the section from 28 to 33 m below the sea floor. This date is consistent with the $^{40}Ar/^{39}Ar$ dates of volcanic clasts and with the biostratigraphy of radiolarians (Bohaty et al. 1998).

The isotopic composition of strontium of mollusc shells in the early Miocene section of the CRP-1 core (46–62 mbsf) indicated dates of 16.6 ± 0.4 and 18.7 ± 0.3 Ma in agreement with the diatom biostratigraphy of Harwood et al. (1998). The $^{40}Ar/^{39}Ar$ date of a volcanic clast (18.4 ± 1.1 Ma) measured by McIntosh (1998) confirms the early Miocene (Burdigalian) age of the sediment.

Drilling resumed in 1998 at a nearby site and reached 624 m below the sea floor. The upper 27 m of the core of CRP-2,2A consists of poorly consolidated sediment of Pliocene and Quaternary age, followed by sediment of early Miocene (19 Ma) to early Oligocene (31 Ma) age. The geochronology of CRP-2/2A is sufficiently precise to suggest that the advances and retreats of the East Antarctic ice sheet were caused by Milankovitch cycles (i.e., based on the orbital mechanics of the Earth). In addition, the microfossils indicate that the climate deteriorated from cool-temperate during the Oligocene to subpolar in early Miocene. The persistence of a cold-adapted flora from the Oligocene into the Miocene and, perhaps, into the Pliocene is remarkable and implies a slow rate of cooling of the climate along the coast of the Transantarctic Mountains. Remnants of this flora are preserved in the land-based deposits of till of the Sirius Group (e.g., in the Dominion Range near the head of the Beardmore Glacier). The interpretations of data derived from CRP-2,2A were published, as before, in Terra Antartica (7(3):211–412, 2000 and 7(4/5): 413–644, 2000).

The third and final hole was completed in 1999 and reached sandstone of the Beacon Supergroup of Devonian age. The final depth of CRP-3 was 934 mbsf. The reports arising from the study of the third core were published in Terra Antartica (vol. 7 (1/2):3–210, 2000), (vol. 8(3):123–308, 2001), and vol. 8(4):313–620, 2001).

The study of the cores recovered at Cape Roberts has revealed the gradual cooling of the climate from the Oligocene to the Miocene, which presumably favored the growth of the East Antarctic ice sheet during that time. In spite of the periglacial conditions, a sparse flora persisted in sheltered areas along the coast possibly as late as the Pliocene. The sediment in the CRP cores also records the uplift of the Transantarctic Mountains and the progressive erosion of the rocks that form this mountain range.

19.1.3 Ross Ice Shelf Project

Personnel of the Ross Ice Shelf Project melted a hole through the ice shelf at a site known as J9 (82°22.5′S, 168°87.5′S) and used it to recover several piston cores (Clough and Hansen 1979). The ice at that location is 360 m thick and is followed by 237 m of seawater. The reports arising from this project were published in Science (vol. 203, No. 4379, 1979).

Webb et al. (1979) reported that the sediment is of glaciomarine origin and contains fossils of middle Miocene age. Sediment of Pliocene and Pleistocene age is not present and may have been removed when sea level declined during the Pleistocene continental glaciations of the northern and southern hemispheres causing the Ross Ice Shelf to ground. The sediment contains a microflora of spores and pollen grains including Nothofagidites identified by Brady and Martin (1979). The pollen assemblage suggests that the Ross Shelf was not present during the Miocene and that beech trees or brush and other plants grew in sheltered locations between the glaciers along the coast of West Antarctica which was the principal source of the sediment (Shaffer and Faure 1976; Kovach and Faure 1977; Webb et al. 1979; Mercer 1981).

19.2 The Sirius Group

The glaciation of the Transantarctic Mountains is indicated by several features that are the characteristic products of glacial ice moving over mountainous bedrock:

- Deepening of stream valleys and development of U-shaped cross sections
- Erosion of bedrock ridges by the formation of cirque basins, glacial horns, and arêtes
- Striations and glacial polish on bedrock several hundred meters above the present level of the ice
- Glacial erratic boulders lying on a substrate of different lithology
- Development of lateral moraines along the walls of glacial valleys at several different levels above the surface of present glaciers in these valleys
- Deposits of till in the form of terminal moraines and ground moraines
- Disruption of the original drainage and development of undrained depressions
- Lacustrine deposits of periglacial meltwater lakes

The present low-stand of the East Antarctic ice sheet is indicated by the exposure of all of the geological features listed above. For example, cirques that were once the sources of tributary glaciers are now empty hanging valleys several tens or even hundreds of meters above the bottom of the valley that was once occupied by a valley glacier. Similarly, recessional moraines record the presence of glaciers that have since vanished. These observations have two possible explanations:

1. The thickness of the East Antarctic ice sheet has declined sufficiently to prevent it from overriding the Transantarctic ice sheet; or
2. The Transantarctic Mountains have been uplifted sufficiently to restrict the outflow of ice from the East Antarctic ice sheet through narrow fault valleys that cross the mountains.

The correct choice between these alternatives is not obvious and will emerge only as all of the relevant facts are recorded and integrated into a coherent and realistic explanation of the late Cenozoic geologic history of the Transantarctic Mountains.

The deposits of till and glacial outwash deposits in the Transantarctic Mountains are especially noteworthy because their very existence at sites well above the level of the present outlet glaciers indicates either the thickness of the glaciers that once flowed across the Transantarctic Mountains or the amount of uplift of the mountains relative to a more or less constant thickness of the East Antarctic ice sheet. Therefore, we approach this topic with questions about the age (or ages) of the glacial advances, about the rate and timing of uplift of the Transantarctic Mountains, about the presence of plant debris in the glacial sediment, and about the lithologic composition of the subglacial basement over which the ice flowed (Stroeven 1997).

19.2.1 Mt. Sirius

In one of the many seminal papers published by John Mercer he reported the presence of "thick coherent drift capping…peaks and ridges" in the Beardmore Glacier area in Fig. 19.4 (Mercer 1972). The till deposits Mercer described occur on Mt. Sirius (2,300 m, 84°08′S and 163°15′E) and in the Dominion Range (2,300 m, 85°20′S and 166°30′E) at the head of the Beardmore Glacier.

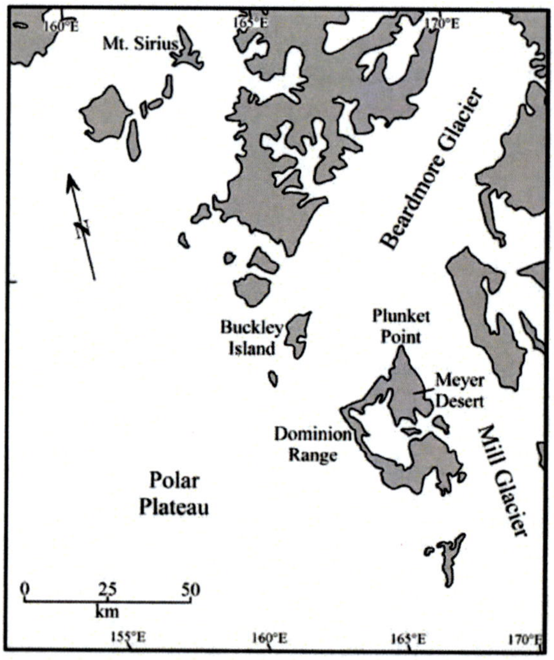

Fig. 19.4 Mount Sirius at 84°08′S and 163°15′E and the Dominion Range at 85°20′S and 166°30′E both contain large deposits of massive till composed of cobbles and boulders in a fine-grained matrix. The till on the summit of Mt. Sirius is about 100 m thick. The till in the Dominion Range reaches a thickness of up to 200 m in a deposit that extends for 6 km (Adapted from Mercer 1972, Fig. 1)

Fig. 19.5 (**a**) Mount Sirius in the Beardmore Glacier area (Figs. 11.8 and 11.9) is a mesa composed of sandstones of the Beacon Supergroup overlain by an intrusive sill of the Ferrar Dolerite. The summit consists of a massive deposit of Neogene till which is the type example of the Sirius Formation as originally defined by Mercer (1972). The elevation of Mt. Sirius is 2,300 m above sea level compared to 2,000 m on the Walcott Névé (Photo by G. Faure). (**b**) The Neogene lodgement till on the summit of Mt. Sirius was deposited on a striated surface of the dolerite sill (McKelvey et al. 1984). The till contains boulders of sandstone, siltstone, shale, and coal of the Beacon Supergroup and of dolerite and basalt of the Ferrar Group. The matrix is composed of sand and silt-size particles. The till also contains limestone clasts and grains of detrital pyrite and unreplaced wood (Hagen 1988; Hagen et al. 1989) as well as recycled Cretaceous and Cenozoic microfossils (Webb et al. 1984; McKelvey et al. 1984; Harwood 1984) (Photo by G. Faure)

Mt. Sirius in Fig. 19.5a is a steep-sided mesa composed of sandstones of the Fremouw Formation of the Beacon Supergroup overlain by a dolerite sill of the Ferrar Group which is covered by a thick deposit of the till first described by Mercer (1972) and subsequently by McKelvey et al. (1984) and by Hagen (1988). Mt. Sirius is the type locality of the Sirius Formation which was later elevated to the status of the Sirius Group that includes the numerous small deposits of late Cenozoic till that occur in the Transantarctic Mountains from Manhaul Bay in the Allan Hills of southern Victoria Land to the Wisconsin Range in the Horlick Mountains.

The till on Mt. Sirius was deposited on the glaciated surface of the dolerite sill that caps the mountain. The color of the fresh surface of the matrix of this till is light yellow to tan with large clasts composed of the local bedrock, including semi-lithified sediment fragments that contain microfossils. The till is faintly stratified and contains lenticular deposits of conglomerate and siltstone in the uppermost 30 m of the section where soft-sediment deformation is common (McKelvey et al. 1984).

The clasts are predominantly composed of dolerite and basalt of the Ferrar Group (76.5% by number), followed by sandstone and siltstone of the Beacon Supergroup (20.7%), volcanic breccia (0.8%), hornfels (0.8%) and trace amounts of coal. In addition, three clasts of limestone and an equal number of granitic gneiss were found among 776 clasts in six samples of till from Mt. Sirius (Hagen 1988).

One of the limestone clasts weighing 70 g is black in color and contains fractures filled with white calcite. This specimen contains fragmentary shells of brachiopods and/or pelecypods as well as possible archaeocyathids photographed in thinsections by Hagen (1988, p. 149–155). The isotopic compositions of strontium, oxygen, and carbon of selected unaltered material are:

$$^{87}Sr/^{86}Sr = 0.70811 \pm 0.00001$$

$$\delta^{18}O = +22.2‰ (SMOW)$$

$$\delta^{13}C = -1.2‰ (PDB)$$

The isotope compositions of oxygen and carbon identify this specimen as a marine limestone of Cambrian age most likely derived from the Shackleton Limestone (Veizer and Hoefs 1976). However, the $^{87}Sr/^{86}Sr$ ratio is lower than $^{87}Sr/^{86}Sr$ ratio of marine strontium of Cambrian age (Montañez et al. 1996) presumably because of contamination during the Ross orogeny.

Table 19.1 Chemical composition of the heavy-mineral fraction (1–0.0625 mm) in samples of till from Mt. Sirius and from the Dominion Range in micrograms per gram (Hagen 1988, p. 77)

	Mt. Sirius	Dominion R.
Element	(1)[a]	(3)[a]
Zn	261 ± 5[b]	193 ± 23[c]
Cu	178 ± 2	163 ± 19
Ni	138 ± 5	137 ± 7
Fe	169,317 ± 1618	139,830 ± 1917
Mn	4247 ± 47	3114 ± 76
Cr	876 ± 27	586 ± 26
Ti	29,437 ± 416	12,310 ± 571

[a]Number of samples.
[b]Analytical error.
[c]One standard deviation for three samples.

The matrix of the till on Mt. Sirius is composed of quartz, plagioclase, K-feldspar, kaolinite, montmorillonite, and illite (Hagen 1988, p. 22). The sand-size fractions (1.0–0.0625 mm) contain heavy minerals including pyroxene, hornblende, garnet, tourmaline, rutile, zircon, apatite, sphene, magnetite, ilmenite, hematite, pyrite, and chalcopyrite. A bulk sample of the heavy mineral fraction of till on Mt. Sirius in Table 19.1 is composed primarily of iron, titanium, and manganese with lesser concentrations of chromium, zinc, copper, and nickel (Hagen 1988). Additional analyses of the heavy-mineral fractions of till in the Transantarctic Mountains were published by Faure et al. (1995).

19.2.2 Marine Microfossils and Real Wood

Till of the Sirius Group at Mt. Sirius and at many sites in the Transantarctic Mountains contains a diverse assemblage of recycled marine diatoms, foraminifera, radiolaria, silicoflagellates, etc. that range in age from Eocene to late middle Pliocene (Webb et al. 1983; Harwood 1983, 1986a, b). These microfossils restrict the depositional age of the Sirius till to less than middle Pliocene (i.e., less than about 3 Ma). The marine microfossils are "reworked" which means that they were originally deposited in a marine environment and were then redeposited with the glacial sediment by an ice sheet that transported them to high elevations in the Transantarctic Mountains. Alternatively, Mt. Sirius may have been uplifted through the ice sheet after the till was deposited on the surface of the dolerite sill that now caps the mesa.

Webb et al. (1983) considered three possible explanations for how the marine microfossils in the till of the Sirius Group could have been deposited:

1. The Sirius till could have been deposited in situ at a time when the Transantarctic Mountains were below sea level.
2. Marine sediment from the Ross Embayment could have been redeposited in the Transantarctic Mountains by a greatly expanded West Antarctic ice sheet which invaded the Transantarctic Mountains from the east.
3. The East Antarctic ice sheet could have receded from the Transantarctic Mountains and allowed seawater to fill the Wilkes and Pensacola basins both of which are below sea level by as much as 1,000 m.

The third alternative requires that the climate of Antarctica during the Cenozoic warmed sufficiently to cause melting of the East Antarctic ice sheet, thus permitting marine sediment to be deposited in the basins on the East Antarctic craton. Subsequently, the climate cooled again, the ice readvanced, and deposited the marine sediment in the Transantarctic Mountains.

Webb et al. (1983) favored the third hypothesis partly because the first option seemed to require unrealistic rates of uplift of the Transantarctic Mountains and because the second alternative is contradicted by the prevailing eastward flow of the East Antarctic ice sheet. According to the favored hypothesis, part of the East Antarctic ice sheet melted during the Cenozoic because of temporary climatic warming which caused sea level to rise as noted by Vail and Hardenbol (1979) in Fig. 17.38.

The warming of the Antarctic climate during the Cenozoic is supported by the presence of leaf impressions of Nothofagus in the sediment of McMurdo Sound (CIROS-1, Section 19.1.1). In addition, small slivers of real wood were discovered by Hagen (1988) in a sample of till on Mt. Sirius and by D.M. Harwood in till of the Dominion Range. The isotope compositions of carbon of these wood fragments (Hagen 1988, p. 143) are:

$$\text{Mt. Sirius}: \delta^{13}C = -24.4‰ \text{ (PDB)}$$

$$\text{Dominion Range}: \delta^{13}C = -21.7‰ \text{ (PDB)}$$

The $\delta^{13}C$ values of these wood fragments are representative of modern wood which suggests that they originated from woody plants that grew in the vicinity of Mt. Sirius and the Dominion Range during the middle Pliocene. Subsequently, Webb and Harwood (1987) reported that they found leaves and twigs of Nothofagus in the till deposits of the Dominion Range.

We also recall that the till pellets weathering out of basal ice in the Elephant Moraine contain a suite of marine microfossils (Fig. 17.10b) including some of the same species that Harwood (1983) identified in till samples of the Sirius Group collected at Tillite Spur in the Wisconsin Range, from Metavolcanic Mountains in the Quartz Hills along Reedy Glacier, from Plunket Point in the Dominion Range at the head of the Beardmore Glacier, from Mt. Sirius, and from Mt. Feather in southern Victoria Land. Evidently, marine microfossils occur in deposits of Cenozoic till in a large part of the Transantarctic Mountains.

19.2.3 Pyrite Grains and Cosmic Spherules

The till of the Sirius Group in the Dominion Range and at Mt. Achernar also contains detrital grains of pyrite. The isotope composition of sulfur of these grains was determined by Hagen (1988, p. 162):

$$\delta^{34}S = +2.3, -1.4‰ \text{ (Dominion Range)}$$

$$\delta^{34}S = +3.1‰ \text{ (Mt. Achernar)}$$

These values suggest that the pyrite originated from igneous rocks which contain sulfur having $\delta^{34}S$ values near zero relative to the sulfur in troilite (FeS) of meteorites which is used as the standard. The most likely source of the pyrite grains are the basalt flows and dolerite sills of the Ferrar Group in the path of the ice that deposited the till.

Hagen (1988) also recovered several magnetic spherules in the heavy-minerals fractions of Sirius till from the Dominion Range. The density of these spherules is greater than 2.83 g/cm³ and their color ranges from shiny black to dull brown. The high-density spherules contrast with low-density spherules (<1.8 g/cm³) in the till of Mt. Sirius which are non-magnetic and green to white in color. The diameters of the magnetic and non-magnetic spherules recovered by Hagen (1988, p. 163) range from 70 to 130 μm. Smaller spherules may be present, but were not recovered. The magnetic spherules may be related to meteorites (Section 18.10), whereas the non-magnetic spherules probably include rounded mineral grains, volcanic lapillae, microtektites, or even fly-ash particles that were deposited during the twentieth century on the surface of till deposits in Antarctica. Hagen (1995) subsequently analyzed a large collection of cosmic spherules he extracted from Cenozoic till in the Transantarctic Mountains and from the blue-ice fields adjacent to the mountains (Koeberl et al. 1988, 1989; Koeberl and Hagen 1989; Hagen et al. 1990, 1992). In addition, Wight (1995) prepared photomicrographs and determined chemical compositions of magnetic and non-magnetic spherules from the "Meteorite Moraine" (informal name) adjacent to the Lewis Cliff ice tongue in the Beardmore Glacier area.

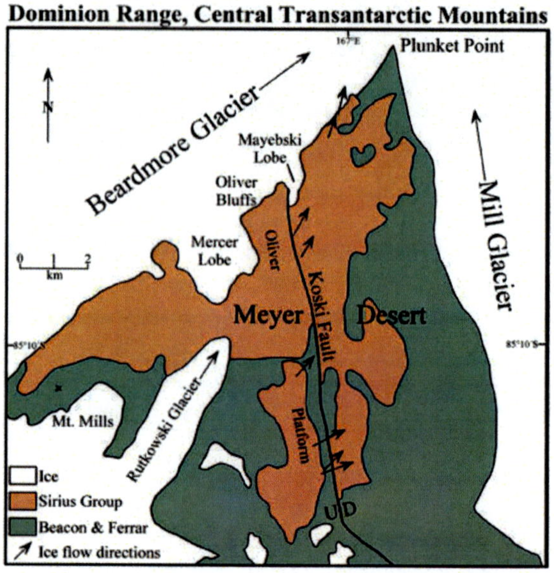

Fig. 19.6 The Dominion Range consists of a basement of Ferrar dolerite and Beacon sandstones overlain disconformably by erosional remnants of glacial deposits of the Sirius Group of middle-to-late Pliocene age. The glacial deposits and underlying rocks are cut by a normal fault (Koski Fault) with the upthrown side on the west. The arrows indicate that the Pliocene ice-flow directions are parallel to the flow of ice in the present Beardmore Glacier. The best exposure of the Meyer Desert Formation (Sirius Group) occurs at Oliver Bluffs where unreplaced twigs and leaves of *Nothofagus* (southern beech) have been found (Adapted from McKelvey et al. 1991)

19.3 Dominion Range

The Dominion Range in Fig. 19.6, located at the confluence of the Beardmore and Mill glaciers at 85°20′S and 166°30′E, was first visited by members of Robert Scott's and Ernest Shackleton's expeditions on their way to the South Geographyic Pole (Section 1.4). David and Priestley (1914) mentioned that members of Shackleton's party found several small limestone erratics on the summit of Mt. Hope on the western side of the mouth of the Beardmore Glacier and about 800 m above the present level of the ice. This observation was later confirmed by Oliver (1963) who reported that a lateral moraine on the northeast slope of Mt. Kyffin at the mouth of the Beardmore Glacier extends to an elevation of about 1,060 m above the present level of the Beardmore Glacier.

19.3.1 Stratigraphy

The glacial deposits in the Dominion Range occur on a series of benches that constitute the Oliver Platform in Fig. 19.7. The elevations of these benches increase stepwise from about 1,800 m at Oliver Bluffs at the western edge of the Dominion Range adjacent to the Beardmore Glacier to about 2,600 m at the southern end of the Upper Oliver Platform (Webb et al. 1987). The glacial deposits of the Sirius Group in Table 19.2 consist of the late Pliocene Meyer Desert Formation (150 m) and of the overlying early Pleistocene Mt. Mills Formation (50 m) (Webb and Harwood 1993).

The Meyer Desert Formation is composed of bouldery diamictite, exemplified by Fig. 19.8, interbedded with fluvio-glacial sandstone and siltstone and with thin-bedded glacio-lacustrine sandstone and mudstone (Passchier 2002). The lithological diversity of the Meyer Desert Formation indicates that it was deposited during several advances and retreats of the East Antarctic ice sheet in glacial and periglacial environments. The Meyer Desert Formation also contains unfossilized roots and stems of *Nothofagus* preserved in situ as well as twigs and leaves that were deposited in fluvial and lacustrine sediment. The fine-grained sediment contains indigenous and reworked palynomorphs as well as reworked siliceous and calcareous

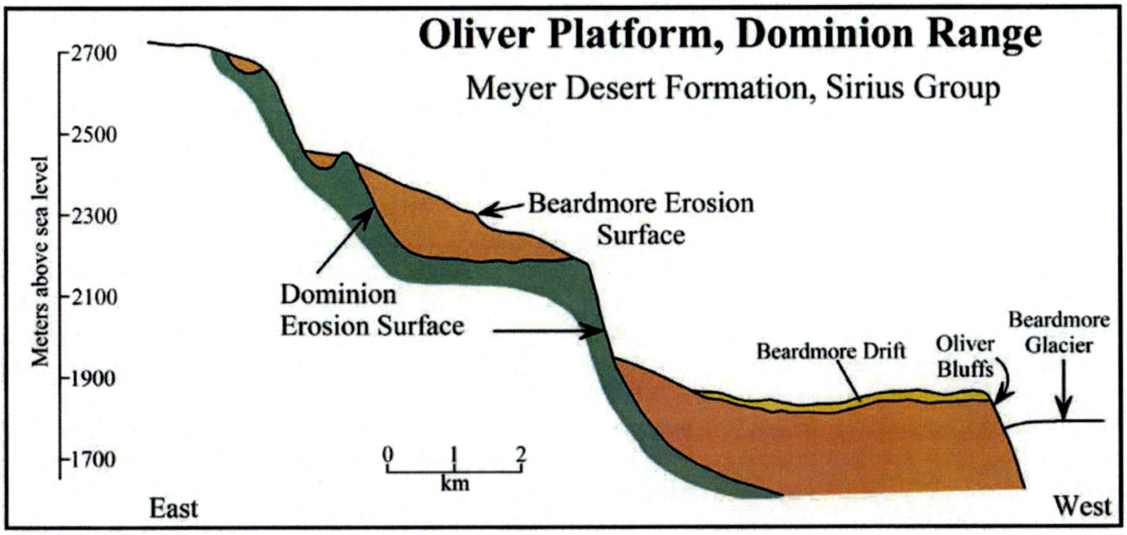

Fig. 19.7 The erosional remnants of the Meyer Desert Formation and the overlying Beardmore Drift of the Sirius Group are located on benches of the Dominion Erosion Surface. The glacial sediments (*orange*) were deposited on Ferrar dolerite and Beacon sandstones (*green*). The Beardmore Drift of unconsolidated gravel and till of Plio/Pleistocene age (*yellow*) was deposited on the Beardmore Erosion Surface on the lowest step of the topographic profile which was drawn parallel to the Koski Fault (Adapted from McKelvey et al. 1991)

Table 19.2 Stratigraphy of the Pliocene glacial deposits in the Dominion Range and adjacent areas in the Transantarctic Mountains (Webb and Harwood 1993)

Formation	Group	Age
Beardmore drift	III	Late Pleistocene
Beardmore drift	II	Late Pleistocene
Beardmore drift	I	Late Pleistocene
Beardmore Erosion Surface		
Mt. Mills, 50 m	Sirius	Early Pleistocene (?)
Meyer Desert, 150 m	Sirius	Late Pliocene <2.5–3.1 Ma
Dominion Erosion Surface		
	Ferrar	Jurassic
	Beacon	Triassic

microfossils of Cretaceous to Pliocene ages (e.g., diatoms, silicoflagellates, radiolarians, and calcareous foraminifera).

The Mt. Mills Formation which overlies the Meyer Desert Formation may be of early Pleistocene age although it is unfossiliferous and is composed of diamictite, breccia, conglomerate, and sandstone. The sediment of the Mills Formation was eroded by glaciers that subsequently occupied the Beardmore and Mill valleys. These glaciers deposited lateral moraines along the valley walls which constitute the Beardmore drift composed of unconsolidated gravel and till

Fig. 19.8 The Meyer Desert Formation of the Sirius Group in the Dominion Range has a fine grained yellowish-tan matrix of sand, silt, and clay that encloses angular boulders of dolerite of the Ferrar Group and of sedimentary rocks of the Beacon Supergroup. This facies of the glacial deposits closely resembles the diamictite on Mt. Sirius in Fig. 19.5b. However, the section in the Dominion Range also includes thin bedded and well-sorted fluvial and lacustrine sandstones and siltstones that formed in periglacial environments that existed between episodic advances of the East Antarctic ice sheet during the late Pliocene (Photo by G. Faure)

(Mercer 1972). These moraines record the former levels of the glaciers in the area, where Beardmore I is the oldest and highest. The different levels of Beardmore drift are distinguished by the extent of weathering of dolerite and sandstone clasts. The observations recorded by Oliver (1963) suggest that the highest drift on Mt. Kyffin belongs to the Beardmore I lateral moraine.

19.3.2 Pliocene Nothophagus

Oliver Bluffs in Fig. 19.9 is a large exposure of the Meyer Desert Formation. Webb and Harwood (1987) subdivided this outcrop into four units (not shown) and reported that Nothofagus wood occurs in thin horizons at the boundaries between units 1 and 2 and between units 3 and 4. The wood fragments in Fig. 19.10a are up to 20 cm long and 2 cm thick and occur in the upper 5 cm of the diamictite in units 1 and 3. The pieces of wood are overlain by gravel and well-sorted sand of the overlying units indicating that the beech shrubs grew in the diamictite and were subsequently buried by outwash deposits during the interglacial episodes.

Most of the pieces of wood in Fig. 19.10a are well preserved and all of them originated from a single species that closely resembles *Nothofagus betuloides* from Tierra Fuego in southern Argentina and southern Chile (Carlquist 1987). The wood also resembles *Nothofaugs gunnii* which is native to Tasmania. The best available information indicates that the wood in the diamictite at Oliver Bluffs is about 2 million years old. In spite of its almost unprecedented age, the wood floats in water and burns with a slight resinous odor (Webb and Harwood 1987). The presence of Nothofagus shrubs in the Dominion Range during the late Pliocene was confirmed by Askin and Markgraf (1986) who found Nothofagus pollen in sediment of unit 2 as well as other kinds of palynomorphs most of which appear to be reworked.

The till of the Sirius Group also contains well-preserved leaves of Nothogagus in the form of dense

Fig. 19.9 The till and interbedded lacustrine sediment of the Meyer Desert Formation is well exposed at Oliver Bluffs in the Dominion Range. The age of this formation is middle-to-late Pliocene based on the presence of marine diatoms of middle Pliocene age (Harwood 1986a, b). In addition, the glacial outwash in the Meyer Desert Formation contains unfossilized wood of the southern beech (*Nothofagus*) that was collected by D.M. Harwood and P.-N. Webb (Photo by David Harwood, published with his permission)

Fig. 19.10 (**a**) Shrubs of *Nothofagus* grew in the glacial deposits of the Dominion Range during the middle Pliocene about 2 million years ago. The wood still floats in water and burns with a bluish flame giving off a resinous odor. The presence of *Nothofagus* indicates that the climate in the Dominion Range during the Pliocene was warmer and more humid than the present climate which is cold and dry (Photo by David Harwood, published with his permission). (**b**) The leaves of *Nothofagus beardmorensis* were preserved in the lacustrine sediment of the Meyer Desert Formation in the Dominion Range (Hill et al. 1991). They were shed at the end of the austral summer and were buried by mud in a periglacial pond or lake. The venation and cell structure are resolvable with a scanning electron microscope. All of the leaves belong to the same species of *Nothofagus* and were not damaged by grazing animals or fungal infestations. The presence of unreplaced wood and leaves of *Nothofagus* in sediment of middle-to-late Pliocene age is a key fact in the glacial history of the Beardmore Glacier area of the Transantarctic Mountains (Photo by David Harwood, published with his permission)

mats within fluvioglacial sediment associated with one of the wood horizons (Hill et al. 1991). The leaves in Fig. 19.10b were apparently swept up by a stream and were deposited with fine-grained sediment. The leaves are well preserved and are up to 6 cm in length. All leaves are of the same species of *Nothofagus* although pollen of other angiosperms and of the conifer *Dacrydium* are present. In addition, the layers of sediment which contain the *Nothofagus* leaves also contain a moss cushion. Hill et al. (1991) noted that the leaves show no signs of animal grazing, fungal infestation, or physical damages. The large size of the leaves implies relatively cold climatic conditions or low intensity of light caused by shade or frequent cloud cover.

All of the evidence points to the fact that *Nothofagus* and a few other species of plants grew on the glacial deposits in the Dominion Range during the late Pliocene after the ice had temporarily receded and were subsequently buried by outwash deposits. Therefore, these plants survived in spite of the high apparent elevation of the Dominion Range and the long period of darkness that characterizes locations at latitude 85°S.

The climate of the Dominion Range during the late Pliocene may have been warmer than it is at the present time because of an episode of global warming and/or because its elevation during the Pliocene was lower than it is presently. Webb et al. (1994) demonstrated that the Dominion Range is located at a higher elevation (i.e., 1,800 m a.s.l.) and higher latitude (85°S) than all other sites in Australia, New Zealand, and South America where southern beech trees grow at the present time. Webb et al. (1994) considered several alternative explanations for the anomalous occurrence of *Nothofagus* during the Pliocene in the Dominion Range:

1. The physiology of the Pliocene Nothofagus in the Dominion Range was more tolerant of the harsh climatic conditions than modern species.
2. The climate in Antarctica during the Pliocene at high latitudes and at high elevations in Antarctica was exceptionally warm.
3. The Dominion Range and adjacent parts of the Transantarctic Mountains in the Beardmore Glacier area were uplifted at the end of the Pliocene to its present elevation.

The association of plant-bearing horizons with water-laid glacigenic outwash deposits, the presence of marine sediment interbedded with glacial deposits in the valley of the Beardmore Glacier, and evidence for faulting of the Sirius Group all favor the hypothesis that the Dominion Range was uplifted at the end of the Pliocene to its present elevation which caused the local climate to deteriorate and killed the vegetation.

19.3.3 The Beardmore Fjord

An important outcrop of the Sirius Group on the Cloudmaker located 75 km "down-stream" of the Dominion Range on the west side of the Beardmore Glacier contains *marine* glacial diamictite (141 m) overlain by terrestrial glacial deposits (103 m). The sediment immediately below the terrestrial beds contains marine foraminifera preserved in situ. The depth of water indicated by the forams is probably 100 m, whereas the present elevation of the foram beds is 1,231 m a.s.l. Therefore, the deposits of the Sirius Group at The Cloudmaker were uplifted by about 1,330 m relative to present sea level. If the forams were deposited at 3.1 Ma, the uplift rate was 1,330/3.1 = 429 m per million years, which amounts to 0.429 mm/year or 4.29 cm/century. In other words, the presence of forams in glacial deposits at The Cloudmaker indicates that the valley of the Beardmore Glacier was flooded by seawater during the Pliocene and therefore was a *fjord* (Webb et al. 1994).

The sediment of unit 1 of the Meyer Desert Formation at Oliver Bluffs also contains marine forams (McKelvey et al. 1991) indicating that the present Dominion Range was within the Beardmore Fjord during the Pliocene at least part of the time.

The evidence that the Dominion Range and The Cloudmaker were at sea level during the Pliocene helps to explain why beech shrubs could grow there in spite of the high latitude. In addition, an episode of global warming may have occurred during the Pliocene which caused melting of glaciers and raised sea level worldwide. Although the volume of the East Antarctic ice sheet may have declined, it is not yet clear whether the Pensacola and Wilkes basins were flooded with seawater and whether the marine microfossils in the Sirius Group originated from the marine sediment that was deposited in these basins. However, the evidence that the valley of Beardmore Glacier was a fjord during the Pliocene eliminates the need to postulate large-scale volume fluctuations of the East Antarctic ice sheet in order to explain why the ice sheet could override the Dominion Range (Webb et al. 1994).

We recall at this point that the Transantarctic Mountains consist of several large fault blocks which were uplifted at different rates and different times. Therefore, the differential uplift of the fault blocks may have affected their glacial histories, such that low-lying areas experienced overriding glaciations, whereas up-faulted blocks remained ice-free at the same time.

19.3.4 History of Glaciation of the Transantarctic Mountains

Lodgment till and associated periglacial outwash deposits occur widely in the Transantarctic Mountains. Although all of the Cenozoic glacial deposits are now included in the Sirius Group, these deposits are difficult to date because the microfossils they contain are reworked and because the various kinds of plant remains they may contain are not sufficiently time-sensitive to determine their depositional ages.

The task of describing and correlating the late Cenozoic glacial deposits of the Transantarctic Mountains and of integrating them into a coherent geologic history was undertaken by Mayewski (1972, 1975) and by Mayewski and Goldthwait (1985). The first compilation of reported occurrences of the Sirius Formation included nineteen deposits at elevations ranging from 1,750 (Allan Hills, southern Victoria Land) to 3,000 m (Mt. Blackburn, Scott Glacier area). Only one deposit at Mt. Roth in the Shackleton Glacier area is located at 800 m a.s.l. (Mayewski 1972).

In a subsequent report, Mayewski (1975) established a chronology for the glaciation of the Transantarctic Mountains based on deposits in the Queen Maud Mountains and their correlatives in southern Victoria Land. The time scale in Table 19.3 contains four glacial events. The estimated ages and durations of these glacial events reflect the state of knowledge in the 1970s.

The Queen Maud glaciation of Mayewski (1975) is based on the till on the summit of Mt. Sirius (>4.2 Ma) which he considered to be early Pliocene in age, whereas Webb et al. (1987) and Webb and Harwood (1987) preferred a late Pliocene age (<3.5 Ma) based primarily on the age of the youngest reworked microfossils (e.g., marine diatoms).

Table 19.3 Time scale of the successive glaciations of the Transantarctic Mountains (Mayewski 1975)

Event	Age (year or Ma)	Epoch
Amundsen	<9490 years	Holocene
Shackleton	49,000 years to 1.6 Ma	Pleistocene
Scott	2.1–2.4 Ma	Late Pliocene
Queen Maud	>4.2 Ma	Early Pliocene

The subsequent Scott, Shackleton, and Amundsen glaciations are based on the presence of lateral moraines in the valleys of the major outlet glaciers. Mayewski (1975) defined these moraines by their elevation and continuity relative to the present ice surface such that the highest moraine is the oldest and the lowest moraine is the youngest. He also noted the morphology, thickness, presence or absence of ice cores, and the state of weathering of rock clasts in the till. In addition, Mayewski (1975) considered certain properties of the soils on these moraines, including particle-size distribution, color, the abundance of secondary salts, acidity, and the presence of clay minerals (see also Mercer 1972).

Mayewski (1975) used the elevations of the lateral moraines to project the level of ice in the outlet glaciers to the East Antarctic ice sheet. The resulting ice-sheet reconstruction indicated that during the Queen Maud Glaciation the Transantarctic Mountains were almost completely covered by ice and that the ice sheet along the central divide in East Antarctica was about 350 m thicker than it is at the present time. In addition, the grounding line of the Ross Ice Shelf moved forward to a position 225 km north of the present edge of the ice shelf.

During the Pliocene "interglacial" that followed the Queen Maud Glaciation, the East Antarctic ice sheet at first thinned to less than its present volume. Subsequently, the East Antarctic ice sheet readvanced into the Transantarctic Mountains during the Scott Glaciation (2.1–2.4 Ma; late Pliocene) during which the highest and oldest of the lateral moraines was deposited. The highstand of the valley glaciers during the Scott Glaciation was followed by a step-wise recession of the ice in response to reductions of the level of the East Antarctic ice sheet which supported the Shackleton (49000–1.6 Ma; Plio/Pleistocene) and Amundsen glaciations (<9490 years; Holocene).

Mayewski's model of the glacial history of the Transantarctic Mountains successfully accounts for the information that was available in the 1970s by postulating changes in the volume of the East Antarctic ice sheet that were caused, presumably, by changes in the average annual temperature and by the rate of deposition of snow on the ice sheet. An alternative view advocated by P.-N. Webb and his associates is that the glaciation of the Transantarctic Mountains was modulated by the uplift of fault blocks rather than by changes in the volume of the East Antarctic ice sheet. Webb's model can explain the regional differences in the history of glaciation of the Transantarctic Mountains, although the flooding of the Pensacola and Wilkes basin that preceded the Pliocene glaciation of the Dominion Range remains controversial.

A third alternative to explain the glaciation of the Transantarctic Mountains arises from the response of the Ross Ice Shelf to changes in the global sea level. We mentioned in the introduction to this chapter that the lowering of sea level caused the Ross Ice Shelf to become grounded. As a result, the grounding line moved north to a position in front of the present edge of the Ross Ice Shelf and the outlet glaciers thickened because the ice was effectively dammed up. In this scenario, the Scott, Shackleton, and Amundsen glaciations reflect the stepwise unblocking of the outlet glaciers in response rising sea level. However, the lowering of sea level was presumably caused by an increase in the volume of the East Antarctic and other ice sheets in the world. Therefore, the Scott, Shackleton, and Amundsen glaciations could have been caused by the combined effects of the grounding of the Ross Ice Sheet, the contemporaneous increase in the thickness of the East Antarctic ice sheet, and the uplift of fault blocks of the Transantarctic Mountains. These complex interactions can perhaps be attributed to changes in the insolation of the Earth caused by cyclical changes in certain orbital parameters which alter the global climate in accordance with the theory of Milankovitch (Broecker et al. 1968).

19.3.5 Trouble with Diatoms

The presence of reworked marine diatoms of Cenozoic age in glacial deposits at high elevations in the Transantarctic Mountains (Harwood 1986a, b) seemed to require a source on the East Antarctic craton. In addition, the co-occurrence of wood fragments of Nothofagus and marine diatoms in the Dominion Range and on Mt. Sirius indicated a moderate climate in the area during the Pliocene. These facts are accommodated by the hypothesis that the East Antarctic ice sheet retreated from the Transantarctic Mountains during the Pliocene allowing the Pensacola and Wilkes topographic basins on the East Antarctic craton to be flooded by seawater. During the subsequent readvance of the ice sheet the marine sediment containing marine diatoms and other marine microfossils was redeposited as lodgment till at high elevations in the Transantarctic Mountains, not only in the Dominion Range and at Mt. Sirius, but also at many other sites throughout the Transantarctic Mountains.

As soon as this hypothesis was published, it was attacked by skeptical colleagues who could not reconcile it with their own observations. A major point of disagreement was the evidence that the climate in the mountains of southern Victoria Land remained cold and dry during the Pliocene. Another troublesome issue was that diatoms can be transported through the atmosphere and are not reliable indicators of the provenance of sediment deposited on land. This point was made repeatedly by Burckle et al. (1988), Burckle (1995), Burckle and Potter (1996), Burckle and Wasell (1995), and Burckle et al. (1997). Finally, Burckle and Delaney (1999) reported that they found marine diatoms in chondrite meteorites collected in the Allan Hills and the Queen Alexandra Range. This discovery emphasized that diatoms are *ubiquitous* and it cautioned that stony meteorites *found* on the surface the East Antarctic ice sheet, or elsewhere on the Earth, contain biogenic contaminants of terrestrial origin.

The discussion concerning the source of marine diatoms in Pliocene glacial deposits of the Transantarctic Mountains has continued and is still not completely resolved. The evidence that the valley of the Beardmore Glacier was a fjord and was uplifted only after the deposition of the Meyer Desert Formation explains the growth of Nothofagus on the till of units 1 and 3 at Oliver Bluffs. The differential uplift of fault blocks also offers a plausible explanation for the cold and dry climate of the mountains in southern Victoria Land to be presented in Section 19.4.

The occurrence of diatoms in glacial deposits of Antarctica was explained by Kellogg et al. (1997) following several studies of diatoms in the Sirius Formation and in ice cores at the South Pole by Kellogg and Kellogg (1984, 1996a, b). The process begins with the eolian deposition of diatoms and other small particles in the zone of accumulation of the East Antarctic ice sheet (Ram et al. 1988). The diatoms are then covered by snow and are buried in the ice sheet. The cross-section of the ice sheet in Fig. 19.11 outlines two alternative scenarios for the transport of diatoms within the ice sheet. If the basal ice is melting (part A), the diatoms and other particulates are transferred from the ice to the basal till. If the ice sheet is frozen to its base (part B), the diatoms stay within the ice until they reach the terminus where the ice melts and the diatoms end up in the till of a terminal moraine. Kellogg et al. (1997) also considered the case of an ice sheet (not shown) that is freezing at its base and that forms a floating ice shelf. The latter case represents the ice streams of the West Antarctic ice sheet which are depositing basal till containing diatoms (Scherer 1991). Continued basal melting of the floating ice tongue also releases additional diatoms from the ice into the glacigenic marine till.

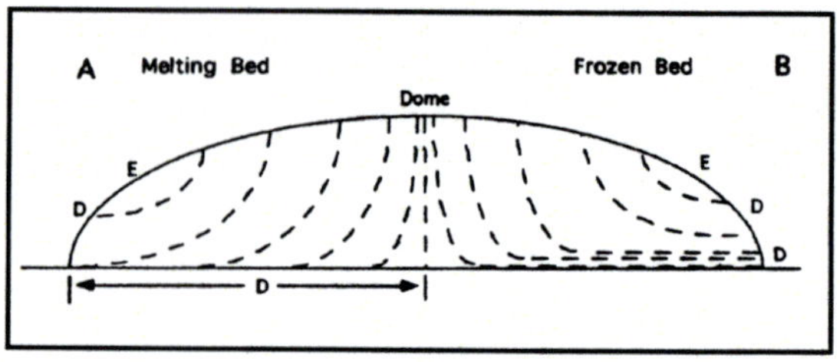

Fig. 19.11 Kellogg et al. (1997) proposed that marine diatoms of Cenozoic age are transported through the air and are deposited on the East Antarctic ice sheet. The subsequent transport of the diatoms depends on the way the ice sheet moves. (**a**) If the base of the ice sheet melts, the diatoms (D) are deposited in the basal till and may be brought to the surface in molded till pellets as in the Elephant Moraine (Section 17.3.5). (**b**) In cases where the base of the ice sheet is frozen to its bed, the diatoms (D) are transported to the edge of the ice sheet where they are deposited in end moraines. The authors also considered the cases (not shown) where the base of the ice sheet freezes to its bed and where the ice sheet forms an ice shelf floating in seawater (Adapted from Kellogg et al. 1997)

The proposed dispersal of diatoms by wind is based on the fact that diatoms have small shells (0.01–1.0 mm) composed of opaline silica. Marine diatoms date from the middle Cretaceous (about 100×10^6 years), whereas fresh-water species appeared during the Miocene (20×10^6 years). Both types became abundant and highly diversified. More than 12,000 species of diatoms have been described (Emiliani 1992). The small size and low density of the shells allow them to be transported by wind until they are deposited on rock outcrops, soil, or ice sheets anywhere on the Earth.

Therefore, the presence of marine diatoms in till of the Sirius Group does not necessarily require the conclusion that the diatoms originated from marine sediment that was deposited in topographic basins on the East Antarctic craton. Furthermore, the ice sheet did not necessarily melt to expose these basins to invasion by seawater. However, even when the extent of melting is scaled back, the presence of Nothofagus growing on till and outwash deposits of the Beardmore fjord indicates that temperate climatic conditions existed during all or part of the Pliocene along the coast of Antarctica.

19.3.6 The Outlet Glaciers

The glaciation of East Antarctica that started during the Oligocene has continued to the present time. Therefore, we are able to observe it in action and to determine how the different components are related to each other. The outlet glaciers that cross the Transantarctic Mountains are in integral part of this process. Several of the major outlet glaciers have already been studied to determine the past levels of the ice, the rate at which the ice flows, the thickness of the ice, and the position of the grounding line (e.g., Reedy Glacier, Mercer 1968; Wilson et al. 1998). In many cases, the glaciers were studied as part of an investigation of the glacial history of the adjacent mountains (e.g., Mercer (1963) in the Ohio Range; McGregor (1965) at the head of the Shackleton Glacier; Grindley (1967) in the Miller Range at the head of the Nimrod Glacier).

The rates of movement of seven of the major outlet glaciers that discharge ice into the Ross Ice Shelf are listed in Table 19.4 based on measurement by Swithinbank (1963). The observed rates range from 2.31 ± 0.02 m/d (meters per day) for the Byrd Glacier to only 0.03 ± 0.025 m/d for the Liv Glacier. The Byrd Glacier in Fig. 19.12 flows more than twice as fast as any of the other glaciers in Table 19.4 and contributes almost 50% of the total amount of ice discharged by the outlet glaciers on that list. The actual volume of ice that is discharged annually requires knowledge of the thickness of the glaciers at their outlets which was not available to Swithinbank (1963). Nevertheless, the comparison of the widths of these outlet glaciers conveys the impression that they are mighty rivers of ice. The five ice streams that transport ice from the West Antarctic ice sheet into the Ross Ice Shelf are even larger than the outlet glaciers in Table 19.4.

Table 19.4 Rate of movement of ice in the outlet glaciers that cross the Transantarctic Mountains and discharge into the Ross Ice Shelf (Swithinbank 1963)

Glacier	Rate of movement (m/d)[a]	Width (km)	Rate × Width (km^2/year)
Byrd	2.31 ± 0.02	25.5	21.5
Beardmore	0.998 ± 0.004	23.0	8.4
Mulock	1.06 ± 0.01	13.0	5.0
Nimrod	0.620 ± 0.005	18.5	4.2
Scott	0.70 ± 0.02	12.5	3.2
Amundsen	0.64 ± 0.03	11.0	2.6
Liv	0.30 ± 0.025	9.0	0.99

[a]m/d = meters per day.

The geologic map of Grindley and Laird (1969) of the Shackleton Coast indicates that the valleys of the Mullock, Byrd, Nimrod, and Beardmore glaciers are fault zones that define crustal blocks of the Transantarctic Mountains. For example, the valley of the Byrd Glacier in Fig. 19.13 separates the Britannia Range to the north from the Churchill Mountains on the southern side of the valley. The Britannia Range (Section 3.3) is composed of the Granite Harbor Intrusives overlain by rocks of the Beacon Supergroup, whereas the Churchill Mountains (Section 5.3) consist of the Shackleton Limestone and the overlying sandstones of the Beacon Supergroup. The rocks on both sides of the valley were intruded by dolerite sills of the Ferrar Group.

A large deposit of till at the base of Mt. Tuatara on the south side of the Byrd Glacier in Fig. 19.14 a few tens of meters above the present level of the glacier extends up the side of the mountains for about 100 m. The glacial polish of the Shackleton Limestone on Mt. Tuatara indicates that the level of the glacier in the past was several hundred meters above its present level. The elevation of the base of the deposit is about 300 m a.s.l.

The color of the weathered surface of the glacial deposit on the slope of Mt. Tuatara is yellowish but fresh surfaces are light grey. The till contains abundant striated and faceted clasts of limestone as well as of igneous and metamorphic rocks. In addition, the till is underlain by a lenticular deposit of coarse gravel and contains structures indicative of soft-sediment deformation. No plant or animal fossils were observed perhaps because the till was deposited at the base of the Byrd Glacier and was exposed only when the level of the glacier declined to its present level.

Fig. 19.12 The Byrd Glacier viewed from Mt. Madison on the south side of the valley is heavily crevassed and is virtually impassable on foot. The Britannia Range is visible on the far side of the Byrd Glacier. The surface of the Beardmore Glacier is similarly torn up although Robert Scott and Ernest Shackleton were able to manhaul their sleds along the north side of the valley (Photo by G. Faure)

19.3.7 Rb-Sr Dating of Feldspar in Till

Questions concerning the provenance of till deposited by continental ice sheets arise not only in the Transantarctic Mountains but also in North America and Europe and apply not only to deposits of late Cenozoic and Quaternary ages but also to tillite of Permian and older ages. Till that originated from granitic gneisses of Precambrian and/or Paleozoic ages contains detrital grains of K-feldspar (microcline) and plagioclase that can be dated by the Rb-Sr method (Taylor and Faure 1980). This method was used by Faure and Taylor (1981) to date size fractions of feldspar concentrates prepared from samples of the till exposed at Mt. Tuatara on the south side of the Byrd Glacier. The resulting data points in Fig. 19.15 define line A in coordinates of the $^{87}Sr/^{86}Sr$ and $^{87}Rb/^{86}Sr$ ratios. The slope of line A yields a date of 1100 ± 69 Ma and an intercept on the y-axis at 0.7050 ± 0.0021 ($\lambda = 1.42 \times 10^{-11}$ year^{-1}). This date exceeds the known age of the Granite Harbor Intrusives of the Transantarctic Mountains presumably because the till at Mt. Tuatara contains feldspar of Precambrian age derived from the East Antarctic craton (e.g., Halpern 1970; Grew and Manton 1979).

Feldspar in Cenozoic till on the summit of Table Mountain (77°57′S, 162°00′E) south of the upper Ferrar Glacier (Barrett and Powell 1981) yielded a Rb-Sr date of 524 ± 115 Ma from line B in Fig. 19.15 and an initial $^{87}Sr/^{86}Sr$ ratio of 0.7132 ± 0.0068. Both parameters are typical of the Granite Harbor Intrusive which locally underlies the rocks of the Beacon Supergroup. These and additional results from the Wisconsin Range and from Mt. Fleming in southern Victoria Land are compiled in Table 19.5. A component of Precambrian feldspar was detected only in the till on Mt. Tuatara on the south side of the Byrd Glacier.

Improvements in the technology of analytical instruments that have occurred since the measurements in Table 19.5 were made permit individual mineral grains to be dated by the Rb- Sr, $^{40}Ar/^{39}Ar$, and U-Pb methods. The crystallization ages of selected mineral grains in till

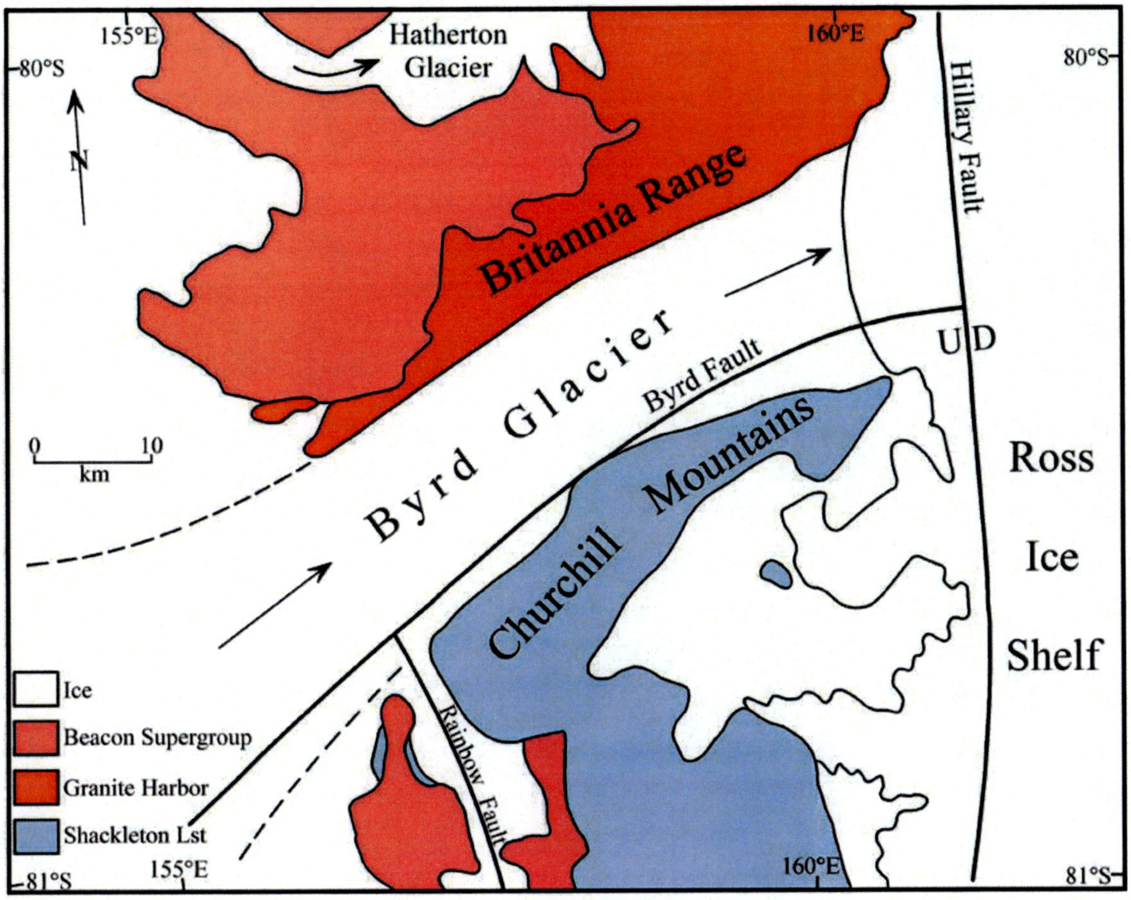

Fig. 19.13 The valley of the Byrd Glacier is a fault zone that separates two blocks of the Transantarctic Mountains. The northern block includes the Britannia Range which is composed of the Granite Harbor Intrusives overlain by sedimentary rocks of the Beacon Supergroup. The southern block is part of the Churchill Mountains which consist of the Shackleton Limestone and of the overlying Beacon Supergroup. The Byrd Glacier is about 30 km wide at Cape Selbourne and extends more than 200 km onto the polar plateau beyond the range of the existing topographic contour maps. Although the average gradient of the surface of the Byrd Glacier is only 0.8%, the ice flows at an average rate of 2.31 ± 0.02 m/day and discharges more ice into the Ross Ice Shelf than any other outlet glacier in the Transantarctic Mountains (Swithinbank 1963) (Adapted from Grindley and Laird (1969) with information derived from the US Geological Survey topographic maps: Cape Selbourne (SU 56-60/2) and Mount Olympus (SU 56-60/1) published in 1966).

are better indicators of provenance than the Rb-Sr dates of mixtures of grains derived from different sources.

19.4 Volcanic Activity, Southern Victoria Land

We now return to the region of McMurdo Sound where we started this chapter and to the topics we previously discussed in Sections 3.1, 10.1, 13.1, 16.1, 16.2, and 19.3. The ice-free valleys of southern Victoria Land in Fig. 19.16 consists not only of the Taylor, Wright, and the Victoria valleys, but include also the Garwood, Marshall, Miers, and Hidden valleys which are located along the north side of the Koettlitz Glacier. The geology of this area was discussed in Section 3.2 in connection with the Koettlitz and Skelton groups of metasedimentary rocks of the Ross orogen.

The history of glaciation of southern Victoria Land was presented and interpreted in a book edited by Denton and Hughes (1981) and in a special issue of

Fig. 19.14 A large deposit of yellowish till is exposed along the north-facing slope of Mt. Tuatara on the south side of the Byrd Glacier. The deposit extends from the present level of the glacier to about 100 m up the polished and striated slope of the Shackleton Limestone. The till was deposited by the Byrd Glacier at a time when it was several hundered meters thicker than it is at the present time. The glacier was wet-based at the time because the till displays features of soft-sediment deformation and includes a lenticular deposit of clasts at the base of the exposure indicating the action of glacial meltwater. The site was visited by G. Faure and R. Felder in December of 1978 (Photo by G. Faure)

Geografiska Annaler edited by Sugden, Marchant, and Denton (1993). The book edited by Denton and Hughes (1981) includes a report by Stuiver et al. (1981) on the history of the marine ice sheet of West Antarctica which deposited Ross Sea drift throughout the McMurdo-Sound area including all of the ice-free valleys of southern Victoria Land. The reports edited by Sugden et al. (1993) are concerned with the Cenozoic glaciation of the ice-free valleys and address the stability of the East Antarctic ice sheet during the Pliocene. This issue was discussed again in GSA Today (vol. 8(4): 1–8, 1998) which included an exposition of the problem by Miller and Mabin (1998a) and presentations of opposing interpretations of the diatom distributions by Harwood and Webb (1998) and Stroeven et al. (1998). The exchange was closed by Miller and Mabin (1998b)

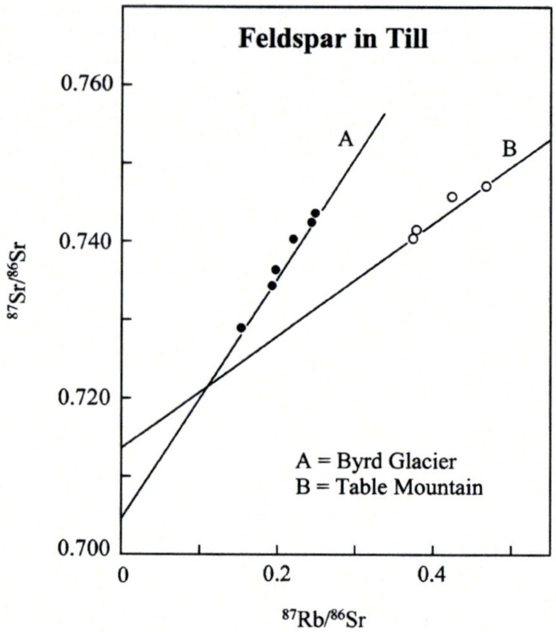

Fig. 19.15 Size fractions of K-feldspar and plagioclase in the till at the base of Mt. Tuatara south of the Byrd Glacier (line A) and in till of the Sirius Group on the summit of Table Mountains (line B) yield Rb-Sr dates and intercepts: Line A, t = 1100 ± 69 Ma, 0.7050 ± 0.0021; line B, t = 524 ± 115 Ma, 0.7132 ± 0.0068. These results indicate that the till on Mt. Tuatara contains feldspar of Precambrian age derived from the East Antarctic craton, whereas the till on Table Mountain originated from the Granite Harbor Intrusives (Adapted from Faure and Taylor 1981)

Table 19.5 Rb-Sr dates of size fractions of feldspar in till of late Cenozoic and Quaternary age in the Transantarctic Mountains

Location of till	Rb-Sr date (Ma)	Initial $^{87}Sr/^{86}Sr$	Interpretation and reference
Mt. Tuatara Byrd Glacier	1100 ± 69	0.7050 ± 0.0021	Paleozoic and Precambrian (1)
Table Mtn Ferrar Glacier	524 ± 115	0.7132 ± 0.0068	Granite Harbor Intrusives (1)
Wisconsin Range Reedy Glacier	480 ± 21	0.7144 ± 0.0030	Granite Harbor Intrusives (2)
Mt. Fleming Wright Valley	506	0.70779	Granite Harbor Intrusives (3)

1. Faure and Taylor (1981); 2. Faure et al. (1983); 3. Boger et al. (1987).

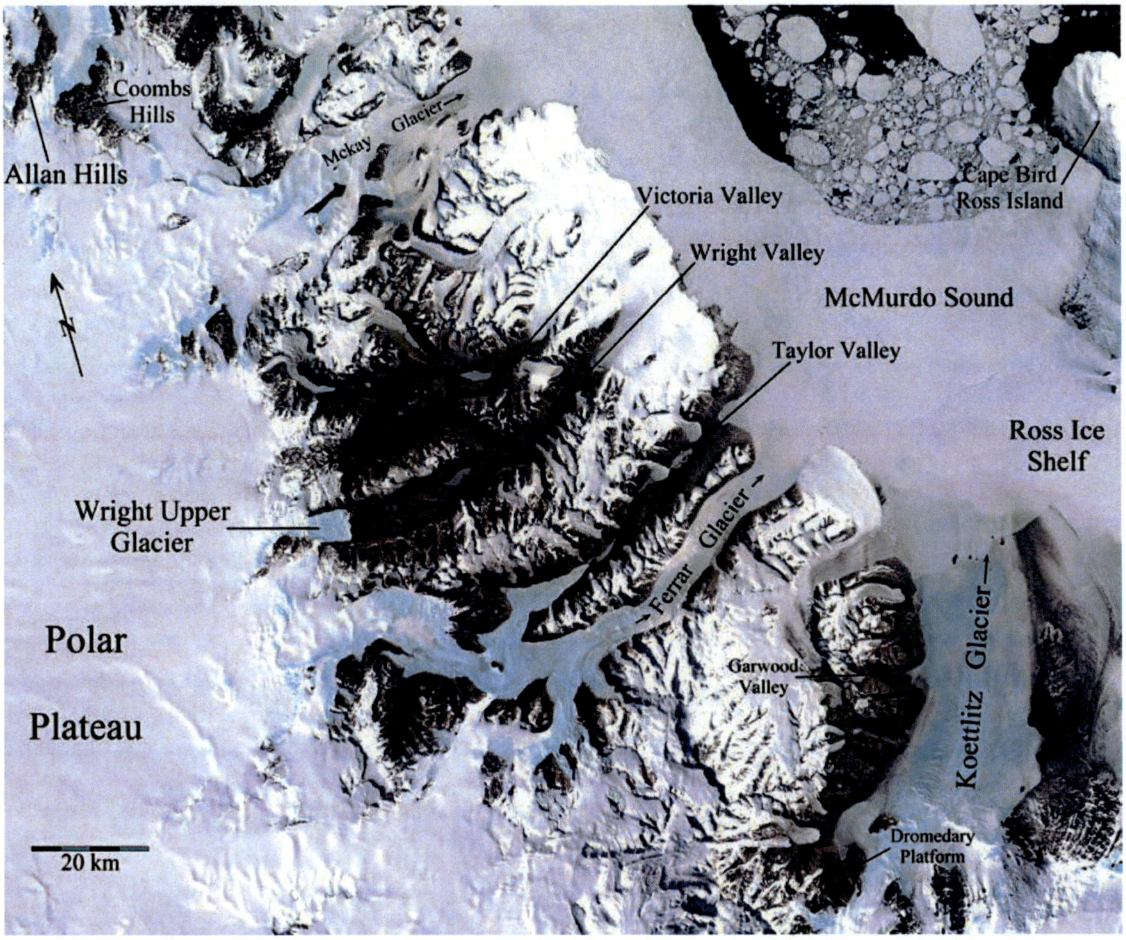

Fig. 19.16 The ice-free valleys of southern Victoria Land include the Victoria, Wright, and Taylor valleys. In addition, small ice-free valleys occur along the Koettlitz Glacier (i.e., the Garwood, Marshall, Miers, and Hidden valleys). This area also includes the Dromedary Platform, The Bulwark, and the Pyramid. Ice from the polar plateau still occurs in the Wright Upper and Taylor glaciers, but only the Ferrar and Koettlitz valleys contain continuous outlet glaciers that discharge ice from the polar plateau into McMurdo Sound (Adapted from the Landsat Image Mosaic of Antarctica (LIMA) http://lima.usgs.gov)

with specific questions that need to be answered before we can decide whether or not the East Antarctic ice sheet melted during the Pliocene.

A historical perspective of the exploration of Wright Valley was provided by the contributors to an issue of the Antarctic Journal of the United Stated (vol. 7(6):225–264, 1972) which was introduced by Bull (1972). This issue contains a seminal paper by Webb (1972) who announced the startling conclusion that Wright Valley was once a fjord.

One of the problems in understanding the geological history of the ice-free valleys of southern Victoria Land is to develop a consistent chronology of the many changes that have occurred here. These changes have been affected by numerous interacting factors, including but not limited to:

- Changes in the thickness of the East Antarctic ice sheet relative to the Transantarctic Mountains
- Uplift of the Transantarctic Mountains of southern Victoria Land relative to the thickness of the East Antarctic ice sheet
- Retreat of the outlet glaciers caused by thinning of the East Antarctic ice sheet, or by uplift of the Transantarctic Mountains, or by climate change, or all of the above

- Volcanic eruptions in the previously glaciated valleys
- Decrease in global sea level which caused the Ross Ice Shelf to be grounded and allowed ice to advance from the coast into the valleys that open onto McMurdo Sound
- Increase in sea level, retreat of the grounding line of the Ross Ice Shelf, and deglaciation of the valleys
- Flooding of the valleys by seawater to form fjords in which the shells of marine organisms were deposited
- Draining of the fjords by uplift of the continental margin
- Development of meltwater streams and lakes, formation of arctic soil impregnated with salts, and establishment of colonial algae, lichens, moss, and nematodes in the ice-free valleys

We do not know how many times the valleys were filled with ice from the polar plateau and the durations of these glaciations. The Ferrar Valley still contains an outlet glacier that discharges ice into the Ross Sea. The Taylor Valley also contains a former outlet glacier (i.e., the Taylor Glacier) which has retreated up the Taylor Valley (Fig. 3.7). Only Wright and the Victoria valleys are presently largely ice-free.

Persons entering the valleys of southern Victoria Land are emotionally affected by the grandeur of the scenery that confronts them. Several visitors have put their impressions into words, including Neider (1972, 1974), Nichols (1982), Lopez (1988a, b, c), and Green (2008). Many others have experienced the feeling of awe which continues to draw them to the ice-free valleys of southern Victoria Land.

19.4.1 Taylor Valley

The ice-free valleys and the Royal Society Range contain numerous volcanic vents where small lavas flows and volcanic cinders of the McMurdo Volcanic Group were erupted during the Cenozoic Era (Section 16.1). All cinder cones in Taylor Valley in Fig. 19.17 rest on glacially eroded rocks and therefore post-date the last through-going glaciation. The cinder cones and lava flows below on elevation of about 1,250 m were later deformed during advances of the Taylor Glacier which scattered clasts of the volcanic rocks and incorporated them into the glacial drift. Some cinder cones were also overrun by advances of local alpine glaciers. In spite of the apparent deformation of many cinder cones by the movement of local glaciers, the volcanic rocks undoubtedly formed *after* the last through-going glaciation of Taylor Valley. Therefore, the ages of the volcanic rocks constitute a *lower* limit of the age of the last glaciation that modified the shape of Taylor Valley and discharged ice into the Ross Sea (Armstrong et al. 1968).

Three specimens of vesicular and porphyritic basalt from Taylor Valley were dated by Armstrong et al. (1968) who used the whole-rock K-Ar method. The results for two of the samples indicate an average date of 2.7 ± 0.2 Ma. The date of the third sample (2.3 ± 0.8 Ma) has a larger analytical error than the dates of the other two samples. Therefore, Armstrong et al. (1968) concluded that the last valley-cutting glaciation of Taylor Valley occurred *prior* to 2.7 ± 0.2 Ma followed by lesser advances of the Taylor Glacier and of the alpine glaciers. Additional age determinations of volcanic rocks from Wright and Taylor valleys by Armstrong (1978) later confirmed that the eruptions in these valleys started at about 4.5 Ma and continued until about 1.5 Ma.

More recent age determinations of volcanic rocks in Taylor Valley by Wilch et al. (1993a, b) by the $^{40}Ar/^{39}Ar$ plateau method range from 1.50 ± 0.05 (east of the Rhone Glacier) to 3.87 ± 0.15 Ma (Sollas Glacier). These additional dates have the same range as the whole-rock K-Ar dates reported previously by Armstrong et al. (1968) and Armstrong (1978) for volcanic rocks in the middle Taylor Valley. Wilch et al. (1993a, b) concluded that the Taylor Valley was uplifted by only about 300 m since 2.57 Ma and that the Taylor Glacier advanced modestly four times during the Plio/Pleistocene reaching an elevation of 1,082 m a.s.l. between 2.97 and 2.71 Ma. Wilch et al. (1993b, p, 331) therefore concluded that:

> there has not been a significant change in the magnitude or character of glacial activity since mid-Pliocene time. All glacial deposits in middle Taylor Valley contain clasts with ventifaction and desert varnish, suggesting that desert conditions existed prior to 3.47 Ma.

In addition, they reported that all volcanic rocks in Taylor Valley were erupted subaerially rather than subaqueously or subglacially, which confirms that the Taylor Glacier did not expand sufficiently to cover the volcanic vents.

Fig. 19.17 Taylor Valley is not entirely ice-free because the Taylor Glacier descends from the polar plateau and terminates at Lake Bonney (Fig. 3.7). Taylor Valley also contains several additional meltwater lakes including Mummy Pond between the Matterhorn and Suess glaciers, lakes Chad and Hoare between the Suess and Canada glaciers, and Lake Fryxell between the Canada and Commonwealth glaciers (not shown), all of which are identified in Fig. 10.3. These lakes are fed by meltwater discharged by the alpine glaciers which line the walls of Taylor Valley. The cinder cones are located along the margins of Rhone, Matterhorn, Sollas, and Marr glaciers (Armstrong et al. 1968) (Adapted from the Landsat Image Mosaic of Antarctica (LIMA) http://lima.usgs.gov)

19.4.2 Wright and Ferrar Valleys

Volcanic cinder cones in Fig. 19.18 also occur along the southern wall of Wright Valley. The K-Ar dates of two samples of basalt from cinder cones and two basalt clasts from a later moraine at the terminus of the Meserve Glacier range from 2.4 ± 0.3 to 4.2 ± 0.2 Ma (Fleck et al. 1972). According to these dates, the last valley-cutting glaciations in Wright Valley occurred prior to 4.2 ± 0.2 Ma and the last volcanic eruptions occurred at 2.5 ± 0.3 Ma. In addition, the K-Ar dates of two basalt clasts from the lateral moraine near the terminus of the Meserve Glacier indicate that this glacier existed at 3.4 ± 0.1 Ma. These preliminary conclusions may be changed when additional samples of Cenozoic volcanic rocks in the ice-free valleys are dated. The presence of volcanic detritus in the glacial deposits of the ice-free valleys was reported by McCraw (1962), Everett (1971), Jones et al. (1973), Kurasawa (1978), Boger and Faure (1988), and others.

An additional point of reference in the glacial history of southern Victoria Land was provided by K-Ar and $^{40}Ar/^{39}Ar$ dates of volcanic glass and anorthoclase crystals from a bed of volcanic ash in the core of CIROS 2 drilled close to the edge of the terminus of the Ferrar Glacier (Barrett et al. 1992). The layer of volcanic ash is interbedded with glaciomarine sediment containing marine diatoms of middle Pliocene age (Webb et al. 1984; Webb and Harwood 1991). The volcanic ash is composed of detrital grains of volcanic glass, feldspar, quartz, and other minerals. The results of age determinations of different mineral fractions of

Fig. 19.18 The Wright Valley in southern Victoria Land is ice-free except for the glaciers at its upper and lower ends. In addition, Wright Valley contains numerous alpine glaciers along its south wall (e.g., Bartley, Meserve, Hart, etc.). The intermittent meltwater streams that are discharged by these glaciers during the summer (December and January) feed the Onyx River which flows west from the Wright Lower Glacier on the coast to Lake Vanda at the west end of the valley (Fig. 10.1). Small volcanic cones along the west side of the Meserve Glacier and below the terminus of the Goodspeed Glacier have erupted basalt which was dated by Fleck et al. (1972) by the K-Ar method. In addition, Hall et al. (1993) dated basalt clasts in glacial deposits of the east-central Wright Valley (Adapted from the Landsat Image Mosaic of Antarctica (LIMA) http://lima.usgs.gov)

the volcanic ash in the CIROS 2 core are summarized below (Barrett et al. 1992):

$$\text{Volcanic glass K} - \text{Ar } 3.36 \pm 0.17 \text{ Ma}$$

$$\text{Anorthoclase}^{40}\text{Ar}/^{39}\text{Ar } 2.77 \pm 0.03 \text{ Ma}$$

$$\text{Volcanic glass}^{40}\text{Ar}/^{39}\text{Ar } 2.91 \pm 0.11 \text{ Ma}$$

These dates confirm that the age of the diatoms in the CIROS-2 core is about 3.0 Ma which is the time when the East Antarctic ice sheet may have melted allowing seawater to flood the Pensacola Basin on the East Antarctic craton. However, the eolian transport of diatoms advocated by Kellogg et al. (1997) means that the presence of marine diatoms in glacial deposits no longer requires the conclusion that the East Antarctic ice sheet collapsed during the Pliocene and that the Pensacola Basin was flooded by seawater.

19.4.3 Koettlitz Glacier

Volcanic rocks of Cenozoic age occur also in the ice-free valleys located between the Koettlitz and Blue glaciers in Fig. 19.19 (Sections 3.2 and 16.1). Age determinations by Armstrong (1978) indicate that volcanic activity in this area occurred first at about 13.5 ± 0.3 Ma (middle Miocene) and resumed later at about 2.88 ± 0.15 Ma at The Bulwark during the Pliocene. The younger volcanic rocks occur primarily on the Dromedary Platform, at The Bulwark, and at The Pyramid in Fig. 19.19. The volcanic activity in this area occurred intermittently and

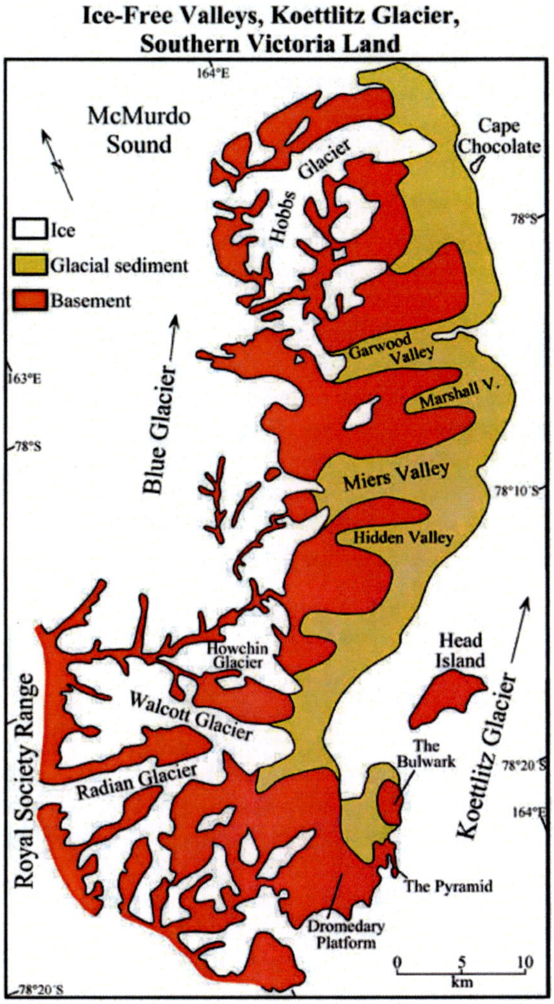

Fig. 19.19 The area between the Koettlitz and the Blue glaciers in southern Victoria Land consists of mountainous terrain containing several glaciers and snow fields as well as ice-free valleys (e.g., Garwood, Marshall, and Miers) that face the north side of the Koettlitz Glacier. This area contains volcanic vents and small lava domes on the Dromedary Platform, The Bulwark, The Pyramid, and along the Howchin, Walcott, and Radian glaciers (Adapted from Bull 1962a)

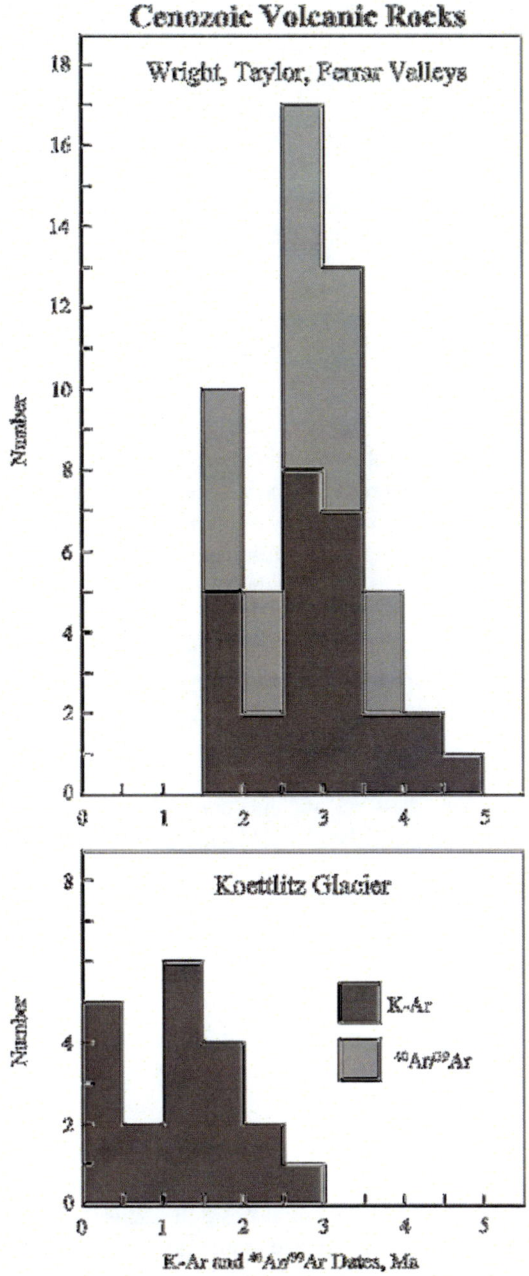

Fig. 19.20 The Cenozoic volcanic rocks of the McMurdo Volcanic Group in the Wright, Taylor, and Ferrar valleys of southern Victoria Land were erupted between 5.0 and 1.5 Ma (Pliocene), whereas the eruptions in the ice-free valleys north of the Koettlitz Glacier (Fig. 19.19) started at 3.0 Ma and lasted well into the Pleistocene. The eruptions in both areas were subaerial rather than subglacial and occurred *after* the Transantarctic Mountains of southern Victoria Land had been overridden by the East Antarctic ice sheet (Data from Armstrong 1968; Fleck et al. 1972; Barrett et al. 1992; Wilch et al. 1993b; Hall et al. 1993)

on a small scale until 0.08 ± 0.13 Ma (Pleistocene). The eruptions appear also to have been subaerial rather than subglacial suggesting that the area was not covered by an ice sheet at this time.

A summary of all age determinations of Plio/Pleistocene volcanic rocks in the ice-free valleys of southern Victoria Land in Fig. 19.20 indicates that the volcanic activity in the Koettlitz-Glacier area started later during the Pliocene than the eruptions in the Wright, Taylor, and Ferrar valleys. The frequency of eruptions in both areas was sporadic and gave rise to polymodal distribution patterns. The volcanic activity in both areas lasted about 3 million years; but the eruptions in the Koettlitz-Glacier area continued almost to the present, whereas the activity in the Wright, Taylor, and Ferrar valleys ended at about 1.5 Ma.

19.5 Arena Valley and Western Asgard Range

The isotopic dates of Pliocene/Pleistocene volcanic rocks that were erupted in the ice-free valleys of southern Victoria Land provide only a lower limit to the age of the last major overriding glaciation of the Transantarctic Mountains in this area. We therefore turn to deposits of volcanic ash in the Arena Valley of the Quartermain Mountains and in the western Asgard Range which are located on opposite sides of the upper Taylor Glacier near the edge of the polar plateau (Marchant et al. 1993a, b). The landscape evolution and the stratigraphy of glacial deposits combined with isotopic age determination of volcanic glass and sanidine crystals in the volcanic ash have extended the glacial history of southern Victoria well into the Miocene. A large and well preserved layer of white volcanic ash occurs also near the Hart Glacier in Wright Valley (Hall et al. 1993; Wilch et al. 1989).

19.5.1 Arena Valley

The Arena Valley in Fig. 19.21 is located on the south side of the upper Taylor Glacier between Windy Gully and Beacon Valley. The physical characteristics, geographic setting, stratigraphic position, and isotopic age determinations of till and colluvium in the Arena Valley were described by Marchant et al. (1993a). The glacial deposits consist of an older suite of Miocene age and a younger suite of Plio/Pleistocene age.

The Plio/Pleistocene deposits listed in Table 19.6 are lateral moraines associated either with the Taylor Glacier (i.e., Taylor II, III, IV) or with alpine glaciers (i.e., Alpine A, B, C, D). Both types of moraines are composed of boulders and cobbles in a matrix of sand and gravel. The boulders are composed of dolerite (~68%), sandstone (~31%), and granite (~1.0%) none of which display glacial polish and striations. The thickness of the Taylor and Alpine moraines ranges from 0.5 up to about 4.0 m. The Taylor Moraines occur in the lower Arena Valley which was once occupied by a lobe of ice of the Taylor Glacier. The ages of the Taylor and Alpine moraines are indicated in Table 19.6 based on isotopic age determinations of volcanic ash.

The Miocene deposits consist of till and associated colluvium most of which are older than 11.4 Ma. Both types of deposits are composed of gravelly sand, gravelly mud, or sandy gravel and their thickness ranges from less than 1.0 to about 3.0 m. The clasts are composed of dolerite, sandstone, and granite in about the same proportions as in the Plio/Pleistocene drift of Arena Valley. Only the clasts in the Arena till and the Altar till contain faceted and striated clasts. In addition, the Brawhm till and the Arena till were deposited on striated bedrock.

19.5.2 Western Asgard Range

The western Asgard Range in Fig. 19.22 is located between the upper Taylor Glacier and the upper Wright Valley including the Wright Upper Glacier, the Labyrinth, the Dais, and Lake Vanda. The names of the north-facing valleys of the Asgard Range are consistent with the theme of Nordic mythology and include Sessrumnir, Njord, Jotunheim, Nibelungen, Donner, Odin, Beowulf, and Fenrir valleys. The mountain peaks are also named after deities and heroes of nordic mythology such as Mt. Baldr, Mt. Thor, Mt. Odin, Mt. Valhalla, and Mt. Beowulf.

The deposits of Miocene till in the valleys of the western Asgard Range in Table 19.7 are composed of gravel, sand, and mud with clasts of dolerite, sandstone, and granite. The thickness of the glacial deposits ranges from less than 1 to 1.50 m, whereas the thickness of the rock glaciers exceeds 3 m. Faceted and striated clasts occur only in the Asgard, Inland Forts, and Sessrunmir till (Marchant et al. 1993b).

19.5.3 Volcanic Ash

The history of glaciation of southern Victoria Land as we know it relies heavily on isotopic dating of Cenozoic

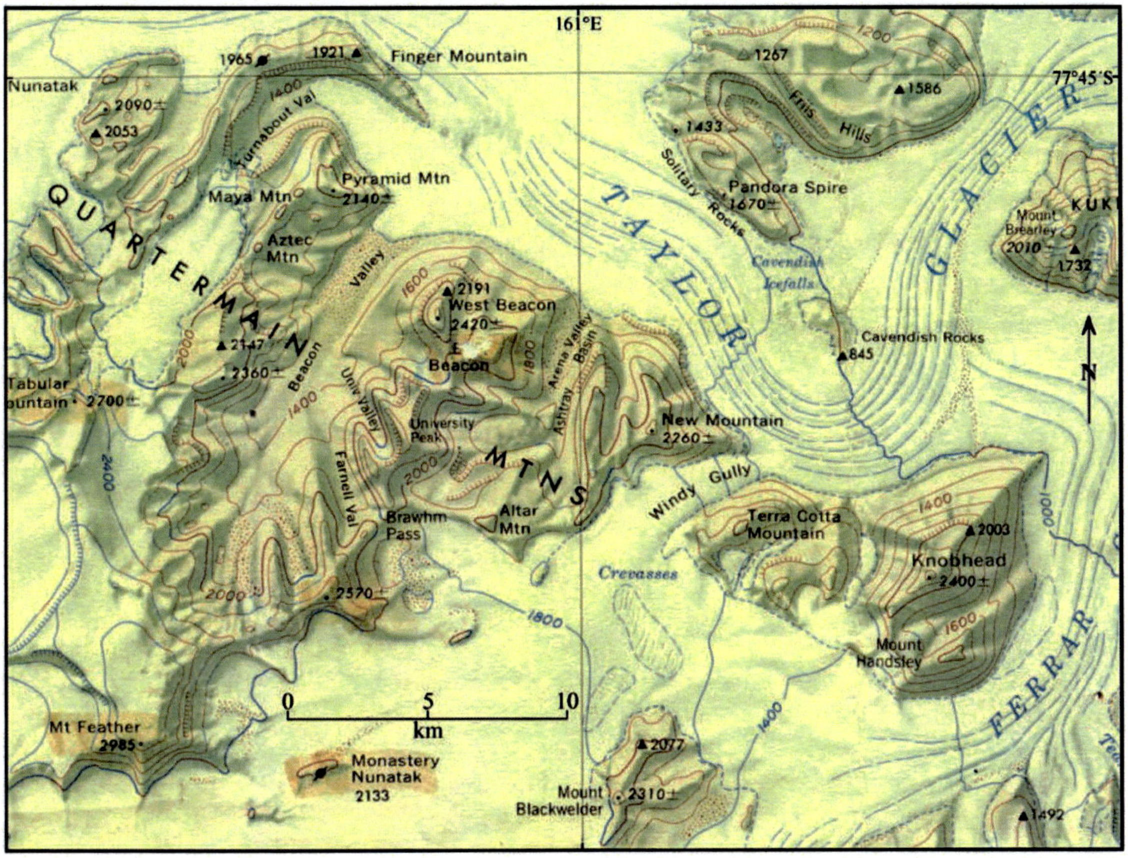

Fig. 19.21 The Arena Valley is located in the Quartermain Mountains on the south side of the Taylor Glacier between Windy Gully and Beacon Valley. Arena Valley contains glacial deposits of Miocene and Plio/Pleistocene ages that were dated by the $^{40}Ar/^{39}Ar$ method. The glacial deposits of Arena Valley were studied by Marchant et al. (1993a) who concluded: "The geologic record of Arena Valley glaciation and landscape evolution shows persistent cold-desert conditions and hence implies stability of the adjacent East Antarctic Ice Sheet for at least the last 11.3 Ma." Excerpt from the topographic map: Taylor Glacier, Antarctica, ST57-60/5 published by the US Geological Survey in 1989

volcanic rocks (Section 19.4) and of extensive deposits of volcanic ash in Arena Valley (Marchant et al. 1993a), in the western Asgard Range (Marchant et al. 1993b), and near the Hart Glacier in Wright Valley (Hall et al. 1993). The volcanic ash was deposited from the air during eruptions of some of the Cenozoic volcanoes along the coast (e.g., Mt. Discovery and Mt. Morning, Section 16.1) of the McMurdo volcanic province. The ash is composed of alkali-rich phonolitic glass shards with crystals of sanidine and anorthoclase.

In some cases, the volcanic ash in Arena valley was mixed into lobate achalanche deposits and into Quartermain I till. Ash also occurs in thin layers within colluvium and was deposited on the desert pavement of the Monastery colluvium in the central Arena valley. The undisturbed ash is more than 99% pure, whereas the reworked deposits contain more than 50% of non-volcanic contaminants. The ash appears to be unweathered and contains less than 5% of secondary clay indicating that the climate in this area has remained cold and dry since the middle Miocene (Marchant et al. 1993a).

Selected crystals of sanidine and shards of volcanic glass were dated by the $^{40}Ar/^{39}Ar$ method at the Berkeley Geochronology Center. In addition, exposure ages were determined by means of in-situ produced cosmogenic radionuclides (Brown et al. 1991; Brook et al. 1993). The $^{40}Ar/^{39}Ar$ dates derived from volcanic ash in Arena Valley and from sites in the western Asgard Range are listed in Appendix 19.9.1.

Table 19.6 Stratigraphy of the glacial deposits of Plio/Pleistocene and Miocene ages in the Arena Valley (Fig. 19.21) of southern Victoria Land (Marchant et al. 1993a)

Strat. Unit	Age (ka or Ma)
Plio/Pleistocene	
Taylor II[a]	>117 ka
Alpine A	
Taylor III	>200 ka
Alpine B	
Taylor IV[a]	>1.0 Ma
Alpine C	
Taylor IV[b]	>2.2, > 7.4 Ma
Alpine D	
Miocene	
Quartermain I till	<11.6, >4.4 Ma
Brawhm till	<11.6, >7.4 Ma
Undifferentiated[b] colluvium	>11.3 Ma
Arena till	>11.3 Ma
Slump Mountain diamicton	>11.3 Ma
Quartermain II till	>11.3 Ma
Monastery colluvium	>11.3 Ma
Altar till	>11.3 Ma

[a]Taylor drift was deposited by the Taylor Glacier whereas the Alpine drift is associated with alpine glaciers.

[b]Unsorted and unstratified sediment composed of boulders in a matrix of sand and gravel that formed by landslides and other forms of down-slope movement.

The $^{40}Ar/^{39}Ar$ dates of sanidine crystals and shards of volcanic glass in Arena Valley range from 4.33 ± 0.07 (Pliocene) to 12.90 ± 0.06 Ma (Miocene). These dates and the dates based on cosmogenic radionuclides support the ages that Marchant et al. (1993a) assigned to the stratigraphic units in Table 19.6. The dates and the apparent geomorphic stability of slopes and of the glacial deposits support the conclusion that the Arena Valley has been a cold desert for more than 11.3 Ma and was overrun by a northeast flowing ice sheet prior to that date.

The most recent activity in the Arena Valley involved the expansion of the Taylor Glacier which thickened by 475 m during the Pliocene and by 325 m during the Pleistocene. The persistence of cold and dry conditions in the Arena Valley even during the Pliocene is not consistent with the proposed melting of the East Antarctic ice sheet at about 3 Ma. However, this conclusion does not alter the fact that Nothofagus did grow in the Dominion Range in the Beardmore Valley during the Pliocene.

The volcanic ash in the western Asgard Range occurs in sand wedges that formed by contraction of frozen ground (Péwé 1959; Black 1976). Therefore, the deposits in which the sand wedges occur are older than the volcanic ash that occurs in these sand wedges. The isotopic dates of sanidine grains and glass shards in ash-filled sand wedges reported by Marchant et al. (1993b) and listed in Appendix 19.9.1 range from 10.08 ± 0.17 to 15.01 ± 0.02 Ma. Volcanic ash in rock glaciers, waterlain ash, and ash overlying desert pavement all formed during the middle Miocene extending back in time to 15.24 Ma. Marchant et al. (1993b) concluded that the Sessrumnir and Inland Forts till (Table 19.8) were deposited prior to 15.2 Ma from wet-based alpine glaciers. The Asgard, Jotunheimen, and Nibelungen tills were deposited between 13.6 and 14.8 (15.2) Ma from a thick northeast flowing ice sheet. There is no evidence that any of the glaciers in the western Asgard Range expanded after about 13.6 Ma. The absence of glacial deposits of the Sirius Group in the Asgard Range (and in Arena Valley) means that this part of the Transantarctic Mountains was not overrun by the East Antarctic ice sheet in late Pliocene time. However, Mt. Feather which is located only about 20 km from Arena Valley and about 40 km from the western Asgard Range apparently does contain deposits of Sirius till (Harwood 1986a).

The contradictory evidence concerning the history of glaciation of southern Victoria Land and the Beardmore area (Section 19.3) during the late Pliocene is perplexing. A possible solution of the problem may arise from differences in the elevations of these areas. The cold and dry climatic conditions of the Quartermain Mountains and Asgard Range imply a high elevation (1,800–2,000 m a.s.l.) during the Pliocene, whereas the Beardmore Valley was a fjord at sea level during the same time. Consequently, the climate in the Beardmore Valley permitted Nothofagus to grow on glacial deposits at the edge of the East Antarctic ice sheet while the climate in the mountains at the head of the Taylor and Ferrar glaciers was cold and dry.

19.5.4 Endolithic Plants

In spite of the present harsh climatic conditions of the western Asgard Range, or perhaps because of them, endolithic plants are presently growing in the sandstone exposed in the Linnaeus Terrace on the north-facing (i.e., sunny) slope of the Asgard Range in Fig. 19.22

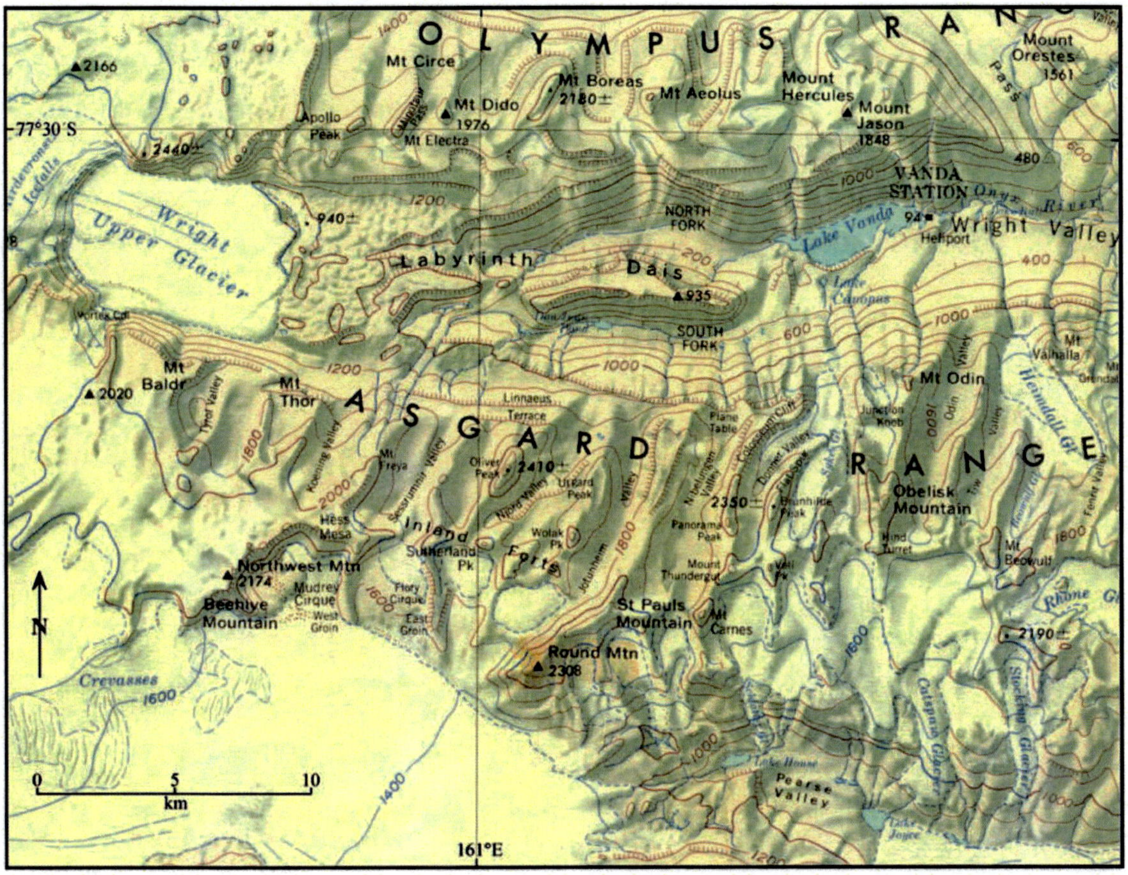

Fig. 19.22 The valleys of the western Asgard Range contain deposits of till and colluvium of Miocene age. Although these deposits are restricted to specific valleys, they nevertheless record environmental conditions that occurred in the Asgard Range as well as in Arena Valley during the Miocene and Plio/Pleistocene (Marchant et al. 1993a, b). Excerpt from the topographic map: Taylor Glacier, Antarctica (ST57-60/5) published by the US Geological Survey in 1989

Table 19.7 Stratigraphy of the glacial deposits of Miocene age in the western Asgard Range (Fig. 19.22) of southern Victoria Land (Marchant 1993b)

Strat, Unit	Age (Ma)
Rock glaciers	<12.5
Undiff, colluviums	>12.0
Jotunheimen till	>10.5
Nibelungen till	<15.2, >13.6
Asgard till	<15.2, >13.6
Koenig colluviums	>15.2
Inland Forts till	>13.5
Sessrumnir till	>15.2

(Friedmann and Ocampo 1976; Friedmann 1977, 1978, 1982, 1993; Gallikowski and Hirsch 1988).

Lichens and mites were also discovered at several locations in the Transantarctic Mountains by biologist from the Bernice P. Bishop Museum in Honolulu, Hawaii (Gressitt 1965). For example, Keith Wise of the Bishop Museum discovered lichens and mites in the Wisconsin Range in January 1965 (Sections 7.1–7.3). This is considered to be the southernmost location for these organisms. Lichen, moss, and other plants are well known from the coast of southern and northern Victoria Land (e.g., Marble Point, McMurdo Sound) and colonial algae grow profusely in the warm brines at the bottoms of several of the lakes in the ice-free valleys (e.g., Lake Vanda and Lake Bonney; Holm-Hansen 1964; Parker and Simmons 1978; Simmons et al. 1979).

The presence of endolithic cyanobacteria in rocks of the ice-free valleys was discovered in 1976 and led to a systematic study of the micro-organisms that live below the surface of porous rocks that are sufficiently translucent to permit photosynthesis. The rocks that meet these criteria are primarily sandstone, granite, and marble, whereas the dark and nonporous dolerite is not a suitable host (Friedmann 1982).

The most common endolithic plants in the ice-free valleys are lichen composed of filamentous fungi and unicellular green algae in a symbiotic relationship. Endolithic blue-green algae (cyanobacteria) are less common. The habitat of the endolithic lichen is layered with an upper black zone (about 1 mm), followed by a white zone (2–4 mm), and by a basal green zone which grades imperceptibly into the underlying rock matrix. Lichen may also grow in fractures in rocks that are not sufficiently porous, in which case they are referred to as chasmoendolithic. In addition, the lichen colonies contain as yet unidentified colorless bacteria which presumable decompose organic matter.

The lichen prefer rock surfaces that face north because they are warmed by solar radiation, as demonstrated by Miotke (1979a) in Taylor Valley. His results indicated that the temperature at a depth of 10 cm in marble and dolerite fluctuated daily and approached +10°C in the afternoon, provided that the sun was shining. Liquid water that forms by melting of snow can infiltrate the underlying rock where it supports the metabolic activity of the endolithic organisms (Friedmann 1978) and promotes chemical weathering (Miotke 1980). When the weathering rind flakes off, the endolithic lichen are exposed at the surface where they cannot survive.

The endolithic lichen and algae of Antarctica are a remarkable adaptation of these plants to the harsh climatic conditions, especially in the winter when the sun does not shine and the temperature can drop to about –60°C (Heinrich 1976).

The apparent similarity of the ice-free valleys of southern Victoria Land and the surface of Mars was recognized by Morris et al. (1972) and by Anderson et al. (1972) several years before the Viking spacecraft landed on Mars in 1976 (Faure and Mensing 2007, p. 211). Although the Viking landers did not find evidence that life exists in the soil at the landing sites, the possible existence of endolithic organisms on Mars has not yet been tested. Friedmann (1982, p. 1052) noted that endolithic communities of lichen and algae may have evolved in Antarctica

> when conditions for life deteriorated (For example, because of a glaciation) and organisms retreated into the protected niche provided by the inside of porous rocks.

The glaciation of Antarctica started millions of years ago judging by the evidence from the CIROS-1 core (Section 19.1.1) and from the age of subglacially erupted volcanic rocks at Sheridan Bluff and Mt. Early (Section 16.6). The work of Marchant et al. (1993a, b) leaves little room for doubt that the climate of the Quartermain Mountains and the western Asgard Range has been cold and dry for up to 15 million years. Presumably, lichen and algae have existed in endolithic shelters in this part of the Transantarctic Mountains ever since the climate deteriorated to the cold and dry conditions that exist at the present time. Perhaps colonies of endolithic micro-organisms have also survived on the surface of Mars in porous rocks that are exposed to sunlight in places where atmospheric pressure is sufficient to permit snow to melt (e.g., in the cliffs of Valles Marineris).

These considerations warrant further study to determine exposure ages of rock surfaces in the ice-free valleys of southern Victoria Land by means of cosmogenic nuclides (Schäfer et al. 1999; Summerfield et al. 1999) and to measure the abrasion rate of rocks in the ice-free valleys by natural processes (Miotke 1979a, b; Malin 1984). The endolithic lichen and algae have continued to attract the attention of microbiologists of North America and Europe (Friedmann and Ocampo 1976; Friedmann 1977; Hirsch et al. 1988; Meyer et al. 1988; Palmer and Friedmann 1988; Ocampo-Friedmann et al. 1988; Bonani et al. 1988).

19.6 Wright Valley

The glacial history of the Quartermain Mountains and of the western Asgard Range (Marchant et al. 1993a, b) suggests that the Wright Valley in Fig. 19.22 was also shaped prior to about 15 Ma (middle Miocene) when it was subjected to one of several glaciations by the East Antarctic ice sheet. Subsequently, the topography of Wright Valley was further modified by the alpine glaciers that still descend from the Agard Range and by the lobe of ice that entered the valley from the Ross Embayment.

The glaciation of the Wright Valley was initially investigated by Bull (1962b), Bull et al., 1962, 1964), Nichols (1964, 1965, 1966), and Calkin et al. (1970). These early studies were continued by Denton and Armstrong (1968), Denton et al. (1970) and led to a synthesis of the history of the Antarctic ice sheet by Denton et al. (1984) and, more recently, by Denton et al. (1993).

Other features of Wright Valley that have attracted attention are Lake Vanda and the associated ponds (Don Juan and Canopus), the origin of soil salts and brines, the Onyx River and related meltwater streams, the invasion of seawater into the valley, and the plants and animals that inhabit the soil and lakes of Wright Valley (Tedrow and Ugolini 1963).

19.6.1 Glaciation

The most recent contributions to the history of glaciation of Wright Valley are by Prentice (1982), Prentice et al., 1987), and Hall et al. (1993) which led to a new synthesis by Denton et al. (1993). Hall et al. (1993) recognized three kinds of glacial deposits in Wright Valley: Peleus till, Alpine moraines, and Ross Sea drift. The stratigraphy in Table 19.8 indicates that these deposits formed in a glaciated landscape which existed when the Peleus till was deposited at >3.8 Ma.

The evidence that Wright Valley was once occupied by an outlet glacier in pre-Pliocene time is corroborated in Fig. 19.23 by the Labyrinth at the foot of the Wright Upper Glacier (Shaw and Healy 1977). The Labyrinth consists of channels incised into a sill of the Ferrar Dolerite by glacial meltwater flowing under pressure at the base of a thick valley glacier that descended from the East Antarctic ice sheet into the Wright Valley. The Wright Upper Glacier is all that remains of the former Wright Glacier that flowed through the valley more than 3.8 million years ago (Hall et al. 1993).

Table 19.8 Stratigraphy of glacial deposits in Wright Valley, southern Victoria Land (Hall et al. 1993)

Strat. Unit	Age (Ma)
Eolian deposits	Modern
desert pavement	
fluvial deposits, etc.	
Alpine I drift	<3.3, Quaternary
Alpine II drift	
Onyx drift	<3.3, Quaternary
Wright drift	<3.4, Quaternary
Loop moraine	Quaternary
Alpine III drift	<3.5, Pliocene
Alpine IV drift	>3.7
Hart Ash	3.9, Pliocene
Peleus till	>3.8
Glacial landscape	> >3.8

The size of the Wright outlet glacier was reconstructed in Fig. 19.24 by Prentice et al., (1987) assuming that the ice at its base was at the pressure melting point and that the climate was warmer than at present. The authors determined by modeling that the ice of the Wright outlet glacier originated from the East Antarctic ice sheet which thickened sufficiently to overflow the rocky threshold at the head of the Wright Valley.

The glacial deposits of Wright Valley are interbedded with a layer of volcanic ash that was deposited from the air. The ash in Fig. 19.25 is especially well preserved at a location between the Hart and Goodspeed alpine glaciers in Fig. 19.18 at an elevation of 378 m a.s.l. (Hall et al. 1993). The ash is underlain by strongly oxidized colluvium and was preserved in-situ judging by the sharpness of the lower contact. It is overlain by a thin layer of lag gravel (desert pavement) of the Onyx or Wright drift. The ash is fine grained, white, and contains abundant glass shards. The deposit was overridden by westward flowing ice which distributed the upper part (20–30 cm) of the ash layer which, in some places, is up to 90 cm thick. The K-Ar date of the volcanic glass of the Hart Ash is 3.9 ± 0.3 Ma (Hall et al. 1993).

After the deposition of volcanic ash, the alpine glaciers formed sets of lateral moraines that were originally labeled by Calkin and Bull (1972) by means of Roman numerals such that IV is the oldest and I is the youngest.

The landscape evolution of Wright Valley was interrupted in late Pliocene/Quaternary time by the intrusion of ice from the grounded Ross Ice Shelf which deposited the Loop Moraine near the present terminus of the Goodspeed alpine glacier and other deposits identified as Onyx and Wright drift in Table 19.8. This episode ended when sea level rose sufficiently during the late Pleistocene to refloat the Ross Ice Shelf causing the ice to retreat from Wright Valley.

The Alpine glaciations continued (Alpine II) and are still going on at the present time (Alpine I). The present geological activity in the Wright Valley is manifested by the formation of small dunes, by deflation and the resulting development of desert pavement, by fluvial deposits associated with the Onyx River and with the seasonal meltwater streams issuing from the alpine glaciers, and by various forms of mass wasting (colluvium, rock glaciers, and debris flows.).

Fig. 19.23 The Labyrinth at the head of Wright Glacier in Fig. 19.22 was carved by meltwater flowing under pressure at the base of the former Wright Glacier that carried ice from the polar plateau to the Ross Sea (Photo by G.H. Denton reproduced with his permission from the cover of Geology (vol. 11, No. 2, 1983) through the courtesy of the Geological Society of America)

19.6.2 Peleus Till

The Peleus till covers the floor of east-central Wright Valley up to an elevation of about 1,150 m (Prentice 1982; Hall et al. 1993), but it does not form moraines or other kinds of glacial landforms. It consists of a pale yellow to yellowish brown diamicton with a silty or sandy matrix containing clasts of dolerite, granitoids, and metasedimentary rocks, as well as of orthoquartzite and porphyritic dike rocks, all of which occur within the Wright Valley. However, the Peleus till does not contain clasts of late Cenozoic basalt of the McMurdo Volcanic Group. Evidently the Peleus till was deposited before the start of volcanic eruptions in the area. The soil that has developed on the Peleus till is oxidized to a depth of about 30 cm and contains salt crystals and encrustations in the upper 20 cm (Bockheim 1978). The extent of weathering of the Peleus till generally corresponds to stage 4 of Campbell and Claridge (1987). Hall et al. (1993) concluded that:

> all available evidence indicates that Peleus till was deposited by an East Antarctic outlet glacier.

Hall et al. (1993) subsequently suggested that the Peleus till correlates with the Asgard till which was deposited in the valleys in the Asgard Range that open into Wright Valley (e.g., Sessrumnir Valley in Fig. 19.22). Marchant et al. (1993b) noted that the Asgard till was deposited by lobes of ice that flowed south

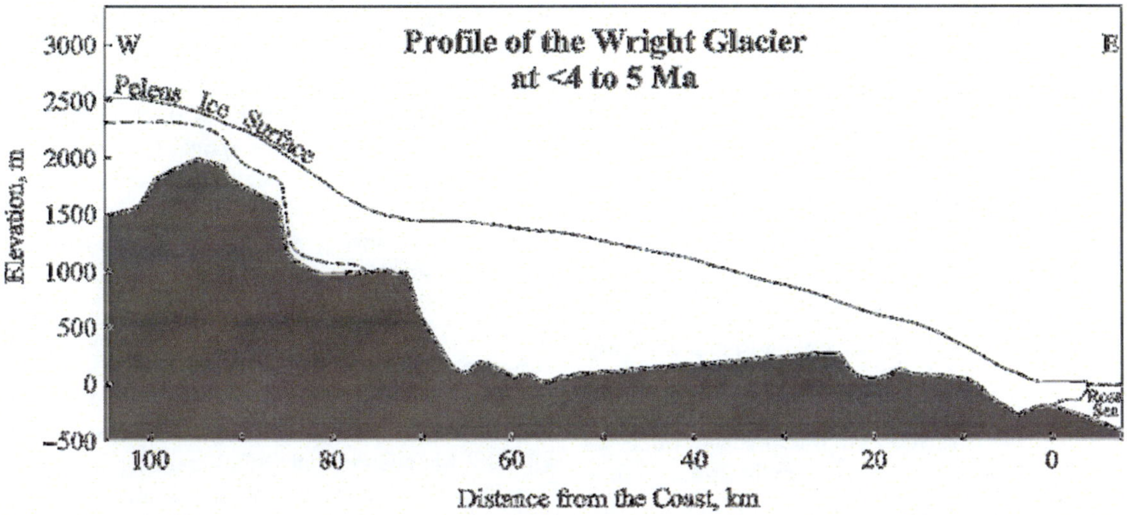

Fig. 19.24 The Peleus till was deposited by a temperate wet-based glacier that flowed across the rocky threshold into the Wright Valley during the early Pliocene less than 4–5 million years ago. The thickening of the East Antarctic ice sheet was caused by an episode of climatic warming. The surface of the present Wright Upper Glacier is indicated by the dashed line. The results of this study indicate that the Peleus till was not deposited by a massive ice sheet that covered the Transantarctic Mountains of southern Victoria Land (Adapted from Fig. 1 of Prentice et al. 1987)

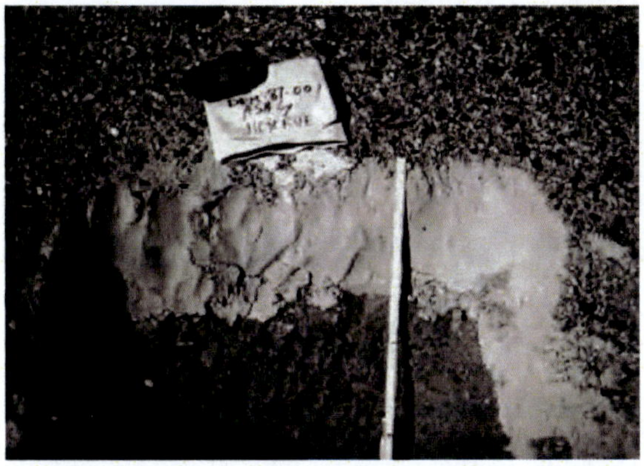

Fig. 19.25 A well preserved layer of white air-fall volcanic ash occurs in the central Wright Valley between the Hart and Goodspeed alpine glaciers. The deposit was preserved in-situ and is covered by a thin layer of lag gravel which Hall et al. (1993) identified as Ross-Sea drift. The K-Ar date (3.9 ± 0.3 Ma) of volcanic glass in the Hart Ash is an important reference point in the glacial history of the Wright Valley (Reproduced by permission from Fig. 11 of Hall et al. 1993)

from an outlet glacier in Wright Valley into the valleys of the Asgard Range analogous to lobes of the ice that flowed from the Taylor Glacier into the valleys of the Quartermain Mountains (e.g., Arena Valley). Marchant et al. (1993b) also concluded that the ice of the Wright outlet glacier originated from the polar plateau, in agreement with the reconstruction of the Wright Glacier by Prentice et al. (1987).

Although the outlet glacier in Wright Valley that deposited the Peleus till may have been wet-based where the ice was thick enough to reach the pressure melting point at its base, this condition does not necessarily mean that the entire glacier was "temperate" and that the climate was subpolar at the time. Hall et al. (1993) pointed out that temperate glaciers form characteristic deposits shaped by the presence of meltwater (e.g., glacial outwash, eskers, and deltas) which are not associated with the Peleus till. Consequently, the Wright outlet glacier was probably cold based at the margins and ablated by sublimation rather than by melting. In other words, the climate of Wright Valley at this time was cold and dry like that in Arena Valley and in the Asgard Range. Hall et al. (1993) also noted that the correlation of the Peleus till with the Asgard till constrains the age of the Peleus till to be more than 13.6 Ma (i.e., Miocene) based on age determinations of volcanic ash by Marchant et al. (1993b). A Miocene age of the Peleus till is consistent with the middle-to-late Miocene age of sediment at the bottom of Lake Vanda (Brady 1979) and with the early-to-middle Pliocene age of the Pecten gravel that overlies the Peleus till at Prospect Mesa at the mouth of Bull Pass in Wright Valley.

19.6.3 Wright Fjord

At Prospect Mesa located south of Bull Pass in the central part of Wright Valley, the Peleus till is overlain by the so-called Pecten gravels. The type section is exposed in a meltwater channel a few hundred meters north of its confluence with the Onyx River. The Pecten gravels which are about 2 m thick contain two horizons with abundant and well preserved pecten shells (*Chlamys tuftsensis*). This marine pelecypod in Fig. 19.26 was named by Turner (1967) from a collection made in 1960/61 by Robert L. Nichols. A closely related species, *Chlamys anderssoni*, occurs in the Scallop Hill Formation on Black Island while the pecten *Adamussium colbecki* (Smith) still occurs in McMurdo Sound at the present time. Turner (1967) dated *Chlamys tuftsensis* as early-to-middle Pliocene. The Pecten gravels also contain marine foraminifera and sponge spicules (Webb 1974).

The lack of abrasion of the pecten shells, the presence of articulated pairs of shells, the lack of sorting, and other criteria cited by Webb (1972) indicate that the shells were deposited in situ in a fjord and were not transported from McMurdo Sound as previously proposed by Nichols (1965). The evidence pertaining to the pecten shells in Wright Valley was also evaluated by McSaveney and McSaveney (1972) and McSaveney (1973) who confirmed the marine invasion of Wright Valley (i.e., Wright Valley was once a fjord). In view of the cold climatic conditions in southern Victoria Land during the Pliocene, the Wright fjord was probably covered by ice which floated on water.

Fig. 19.26 Pecten shells of *Chlamys tuftensis* occur in a gravel bed that overlies the Peleus till at Prospect Mesa which is located at the mouth of Bull Pass in central Wright Valley (Photo by R.D. Turner reproduced with permission)

Webb (1972) proposed that the marine incursion of Wright Valley ended at about 1.9 Ma when the bedrock threshold that is now located under the Wilson Piedmont Glacier at the eastern end of the Wright Valley was uplifted sufficiently to restrict the flow of seawater into Wright Valley. After the marine incursion had ended, ice from the grounded Ross Ice Shelf entered the eastern end of Wright Valley sometime after 1.2 Ma and deposited the Loop Moraine in the vicinity of the present Goodspeed alpine glacier (Denton et al. 1970).

A second deposit of glaciomarine sediment, named Jason diamicton, was discovered in the sediment core of Lake Vanda recovered by the Dry Valley Drilling Project at site 4A (Brady 1979). The Jason diamicton in Lake Vanda is composed of silt and sand containing scattered rock clasts. The thickness of this unit is 7.36 m (Hall et al. 1993). The Jason diamicton also contains shell fragments and marine diatoms of middle-to-late Miocene age. Therefore, Wright Valley was a fjord at least twice during the Neogene: First during the middle Miocene (Jason diamicton in Lake Vanda) and later during the Pliocene (Pecten gravel at Prospect Mesa).

During the time of the two recorded marine incursion into Wright Valley, a deep fjord existed in Taylor Valley from before the late Miocene to the middle Pliocene (3.4 Ma) (Ishman and Rieck 1992). Since Taylor Valley is part of the same structural block, the marine incursion into Wright Valley may also have been *continuous* from middle-to-late Miocene to early-to-middle Pliocene.

19.6.4 Ross-Sea Glaciation

During the Pleistocene, McMurdo Sound was filled with the ice of the grounded Ross Ice Shelf. The Ross-Sea drift that was deposited by this ice is largely unweathered, is underlain by residual bodies of ice, and is bounded by terminal moraines (Stuiver et al. 1981). These deposits occur primarily along the western coast of McMurdo Sound including the Brown

Peninsula, the ice-free valleys along the north side of the Koettlitz Glacier (Garwood, Marshall, and Miers valleys), at the mouth of Taylor Valley, and in the eastern part of Wright Valley. In addition, Ross-Sea drift occurs on Minna Bluff, Black Island, and along the coast of Ross Island (Clapperton and Sugden 1990).

The Ross-Sea drift in Wright Valley is ice-cored, unweathered, and lacks soil. Nichols (1971) originally attributed these deposits to the Trilogy glacial episode. The Ross-Sea drift also forms the Loop Moraine which is convex to the west indicating that the ice flowed from east to west. The alpine glaciers in Wright and Taylor valleys did not contribute to the Ross-Sea ice that invaded these valleys and may actually have retreated because of a decrease in precipitation caused by the fact that the Ross Sea was largely covered by grounded ice at this time.

The Ross-Sea ice that entered Taylor Valley dammed up a proglacial lake called Lake Washburn which flooded the entire valley including the present site of Lake Bonney (Chinn 1993). Lake Washburn was inhabited by freshwater diatoms and by green as well as by blue-green algae. The lacustrine deposits in Taylor Valley contain dried algal mats that occur at different elevations. Carbon-14 dates measured by Stuiver et al. (1981) of these algae in Fig. 19.27 range from 21,200 ± 200 (Lake Bonney drainage) to 8,340 ± 120 years BP. (Explorers Cove drainage at he mouth of Taylor Valley). The range of dates identifies the period of time during which meltwater pooled in Taylor Valley because the mouth of the valley was blocked by the grounded Ross Ice Shelf. The ^{14}C dates of the algae may also indicate that conditions for the growth of algae in Taylor valley were most favorable between about 18,000 and 10,000 years BP.

The ^{14}C-dates of algal mats in Taylor Valley in Fig. 19.28 increase with increasing elevations of the collecting sites such that the oldest dates between 16,000 and 21,200 years BP. apply to algal mats collected at high elevation between 80 and 230 m a.s.l. This pattern of variation, though inexact, indicates that high levels of water persisted in Taylor Valley for about 5,200 years from 21,200 to about 16,600 years BP. when the level LakeWashburn began to decline until the end of the Pleistocene at about 8,300 years BP (= before present).

The lakes that exist in Wright and Taylor valleys at present still contain algae that grow in the warm brines at the bottom of the largest lakes (e.g., Lake Vanda in Wright Valley and Lake Bonney in Taylor Valley).

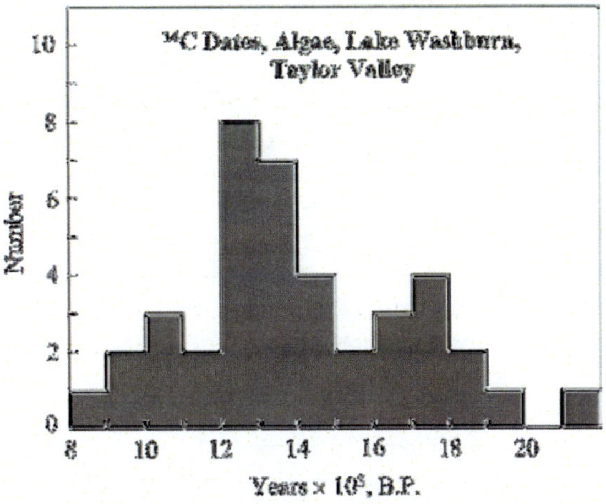

Fig. 19.27 The ^{14}C-dates of algal mats interbedded with lacustrine sediment of Lake Washburn in Taylor Valley range from 21,200 ± 200 (Lake Bonney drainage) to 8340 ± 120 years BP (Explorers Cove drainage at the mouth of Taylor Valley). These dates define the period when the mouth of Taylor Valley was blocked by ice which caused meltwater to collect within the valley. These dates may also indicate that conditions for growth of algae were most favorable between 10,000 and 18,000 years BP (Data from Stuiver et al. 1981)

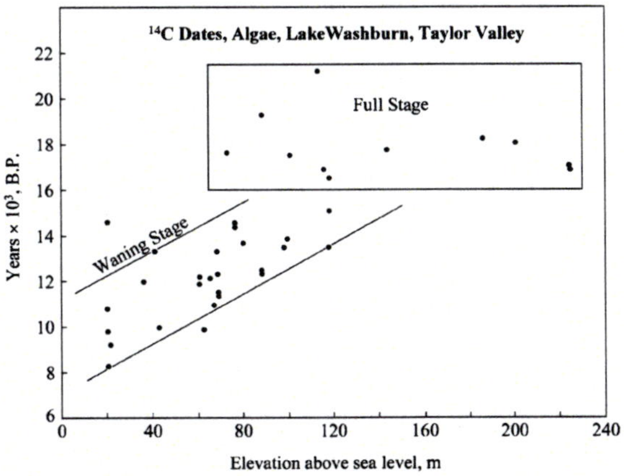

Fig. 19.28 The carbon-14 dates of dried algal mats that are interbedded with lacustrine sediment in Taylor Valley that formed when the mouth of the valley was blocked by the grounded Ross Ice Shelf. The pattern of variation of ^{14}C-dates and corresponding elevations suggests that the lakes reached their highest levels for about 5,300 years from 21,200 to about 16,000 years BP. Subsequently, the lake levels began to decline until about 8,300 years BP. Algal mats are still growing in the warm brine at the bottom of Lake Bonney and in the other deep lakes in the ice-free valleys (Data from Stuiver et al. 1981)

19.7 Lakes of the Ice-Free Valleys

The meltwater lakes in the ice-free valleys of southern Victoria Land have existed for thousands of years and continue to evolve at the present time. The water in some of the lakes is highly stratified in terms of density, salinity, and temperature which has raised questions about the source of the dissolved salts, the elevated temperature of the bottom brines, and about the organisms that inhabit the lakes. These features are being studied by limnologist, geochemists, biologists, and geophysicist all of whom try to uncover the secrets that are proverbially well guarded. The record of these investigations is contained in the relevant scientific literature and in several volumes of the Antarctic Research Series edited by McGinnis (1981), Green and Friedmann (1993), Lyons et al. (1997a, b), and Priscu (1998). In addition, Burton (1981) and Wilson (1981) reviewed the chemical composition, physics, and evolution of saline lakes in Antarctica. More recently, Vincent and Laybourn-Parry (2008) edited a book on Polar lakes and rivers, including those of Antarctica, and Green (2008) published a personal account of his encounter with the lakes of the ice-free valleys of Antarctica.

The physical and biochemical properties of the lakes of southern Victoria Land were reviewed and interpreted by Chinn (1993) who identified the lakes by name and indicated their locations in Appendix 19.9.2. Many of these lakes and ponds are small, most are ice covered, and some are frozen right down to the bottom. Some of the lakes contain meltwater of low salinity while others are density stratified and contain brines of varying salinity. The chemical compositions of the water in the different brine lakes reflect the local environmental conditions that govern their on-going evolution. The water levels of the saline lakes and ponds in closed basins have been rising since the summer of 1968/69 when systematic observations commenced. Information recorded in 1903 indicates that the level Lake Bonney in Taylor Valley has been rising since that year (Chinn 1993).

The principal lakes of southern Victoria Land that attract attention by their size and by the salinities and chemical compositions of the brines they contain are Lake Vida in Victoria Valley (Calkin and Bull 1967), Lake Vanda and Don Juan Pond in Wright Valley, Lake Bonney, Lake Hoare, and Lake Fryxell in Taylor Valley, and Lake Miers in the ice-free valleys along the Koettlitz Glacier (Matsubaya et al. 1979; Matsumoto et al. 1993; Lyons and Finlay 2008).

19.7.1 Lake Vanda, Wright Valley

Wright Valley is a closed basin because its mouth is blocked by the Wilson Piedmont Glacier. Therefore, the water of Lake Vanda can escape only by evaporation of liquid water and by sublimation of ice. Consequently, the salts that are dissolved in the glacial meltwater have accumulated in the soil of Wright Valley (Keys and Williams 1981), in the water of Lake Vanda, and in the saline ponds listed in Appendix 19.9.2.

Lake Vanda in Fig. 19.29 is the largest and deepest lake in southern Victoria Land. It receives meltwater from the Onyx River and from small meltwater streams that enter the lake during the summer (December and January). The Onyx River flows west for about 28 km from Lake Brownworth, which is a proglacial meltwater lake at the terminus of the Wright Lower Glacier. Along the way, the Onyx River receives meltwater discharged by the alpine glaciers along the walls of Wright Valley. In addition, the Onyx River dissolves salts in the glacial drift that covers the floor of Wright Valley.

Lake Vanda in Fig. 19.30 is 5.6 km long, 1.5 km wide, and 69 m deep (Chinn 1993). It was named by a group of geologists from Victoria University in Wellington, New Zealand, in memory of a sled dog (Stonehouse 2002). The lake may have formed during the Pliocene when the Wright fjord lost its water except for a residual lake in the deepest part of the valley. The present lake is surrounded by a set of strand lines (Jones et al. 1971) that record the decreasing water levels from about 145 m a.s.l. to the present level of the lake at about 93 m a.s.l. The former "Great Lake Vanda" was 52 m deeper and significantly larger than the present lake as suggested in Fig. 19.30. Carbon-14 dates of algae collected from the strand lines above the present Lake Vanda range from 3,000 to 2,000 years BP (Yoshida et al. 1975). At its highest level, the lake occupied an area about four times larger than the present lake. Calculations by Wilson and Wellman (1962) based on diffusion of ions between the brine layers in the lake indicate that Lake Vanda actually dried up around 1200 BP, presumably as a result of a prolonged period of low-flow of the Onyx River. The limnology of Lake Vanda was revisited by Bydder and Holdsworth (1977).

19.7.1.1 Lake Levels

Records of the level of Lake Vanda have been kept since the summer of 1968/69 when New Zealand

Fig. 19.29 Lake Vanda in Wright Valley is fed by meltwater from the Onyx River (Fig. 10.1) which flows only during the summer (late December and January). The ice that covers the lake melts along the shore during the summer to form a "moat" of open water up to 50 m wide (Photo by G. Faure)

constructed a small scientific base at the eastern end of the lake close to mouth of the Onyx River (Fig. 19.30). In addition Bull (1966) reported that the level of Lake Vanda increased by 95 cm in 1957/58 and 1958/59. Additional observations were made in the summers of 1961/62 (Calkin and Bull 1967) and in 1969/70 (Heine 1969; Cutfield 1974). The lake levels reported by Chinn (1993) in Fig. 19.31a increased by 2.35 m in 1971/72 and then remained nearly constant until the summer of 1981/82. In the following years, the lake level started to rise steeply until the summer of 1988/89 when it was about 7.50 m above its 1968/69 level.

The level of Lake Vanda depends on the balance between the volume of water that enters the lake annually and the volume of water that is lost by sublimation of the ice cover. The annual discharge of the Onyx River, which is the principal source of water entering the lake, has varied erratically (Fig. 19.31b) in response to the amount of snow and ice that melt in the Wright Valley. The discharge of the Onyx River peaked during the summer of 1970/71 at about 15×10^6 m^3 which explains the increase of 2.35 m in the level of the lake recorded in the following summer of 1971/72. The most recent increases in the level of Lake Vanda, that started in 1982/83 and continued until the summer of 1988/89, coincide with the rising discharge of the Onyx River in Fig. 19.31b.

The annual ablation rate of the ice on Lake Vanda has also varied from 12.1 cm/year in 1985/86 to 68.8 cm/year in 1970/71 for an average of 32.6 ± 12.5 cm/year. The rise of the level of Lake Vanda implies that the volume of water that enters the lake is larger than the volume lost by sublimation of ice. Data published by Chinn (1993) indicate that the levels of all closed-basin lakes in the ice-free valleys have increased between 1971 and 1990, although the levels of lakes Vanda and Bonney have increased more than those of the others.

The increases in the level of Lake Vanda do not correlated well with the average summer temperatures measured at Vanda Station. The reason is that the Onyx River derives most of its water from melting of the Wright Lower Glacier on the coast. In addition, the

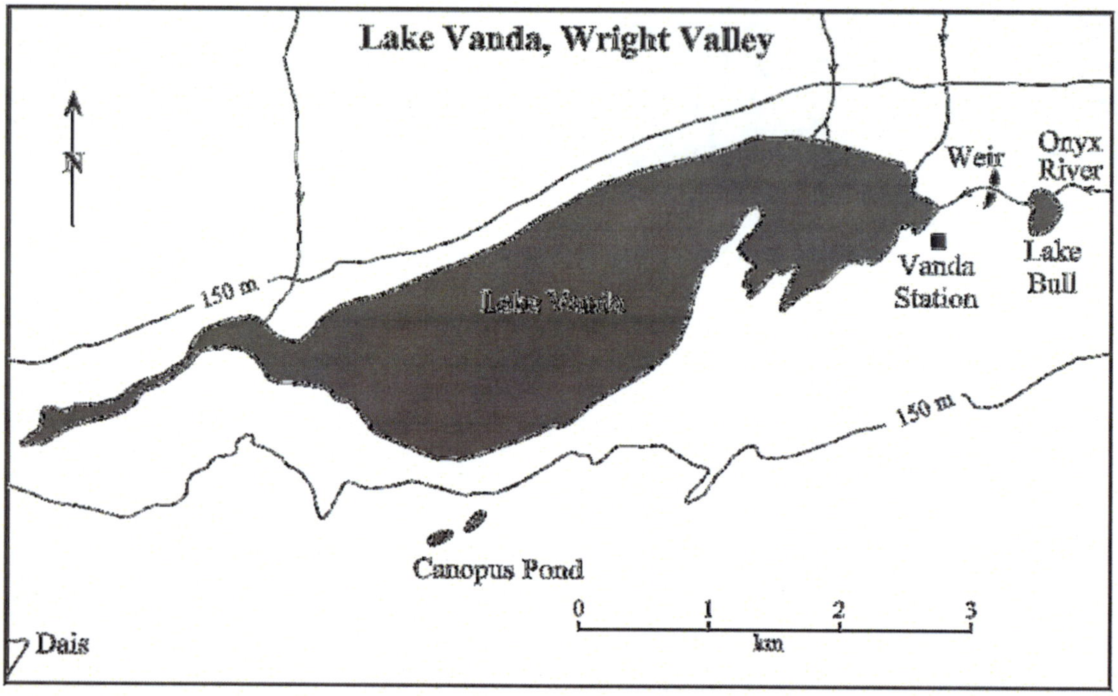

Fig. 19.30 Lake Vanda occupies the lowest area of Wright Valley about 30 km west of the terminus of the Wright Lower Glacier. The lake is 5.6 km long, 1.5 km wide, and 69 m deep. It has a perennial ice cover that is about 4 m thick except for the "moat" of open water that forms along the shore during the summer. Vanda Station was set up in 1968 in order to facilitate the study of the lake. The station was closed in 1991 and the building was removed in 1995. The 150-m contour marks the approximate former shore of "Great Lake Vanda" (Data from Chinn 1993; Stonehouse 2002)

rate of melting of ice depends not only on the air temperature, but is affected also by the extent of the cloud cover, the speed and direction of the wind, and by the elevation above sea level of the ice (Chinn 1993).

19.7.1.2 Temperature Profile

The chemical composition of water in Lake Vanda was first investigated by Armitage and House (1962), Angino and Armitage (1963), Angino et al. (1965), and Boswell et al. (1967), while Nichols (1962) described the geology of the lake. The work of Angino et al. (1965) indicated that the water in Lake Vanda is highly stratified in terms of its chemical composition, electrical conductivity, density, and temperature.

Armitage and House (1962) first reported that a layer of hot brine at the bottom of Lake Vanda has a temperature of 25°C. This observation was subsequently confirmed by Angino et al. (1965) who measured a temperature of 25.5°C at a depth of 67.5 in Lake Vanda. The high temperature of the water at the bottom of Lake Vanda contrasts with the low temperature of the ice that covers the lake. Chinn (1993) reported that a temperature of –25°C was measured in June of 1974 close to the upper surface of the ice on Lake Vanda.

The layer of hot and dense brine at the bottom of Lake Vanda in Fig. 19.32 is a *monimolimnion* and the presence of this non-circulating layer of brine identifies Lake Vanda as a *meromictic* lake. In other words, Lake Vanda is meromictic because its monimolimnion does not mix with the overlying *mixolimnion* which is composed of low-density water that circulates freely.

The potential sources of the heat considered by Angino et al. (1965) were: solar radiation, a high geothermal gradient, and volcanic hot springs at the bottom of the lake. Angino et al. (1965) as well as Ragotzkie and Likens (1964) favored the explanation that Lake Vanda is being heated, at least partly, by a high geothermal gradient and by hot springs.

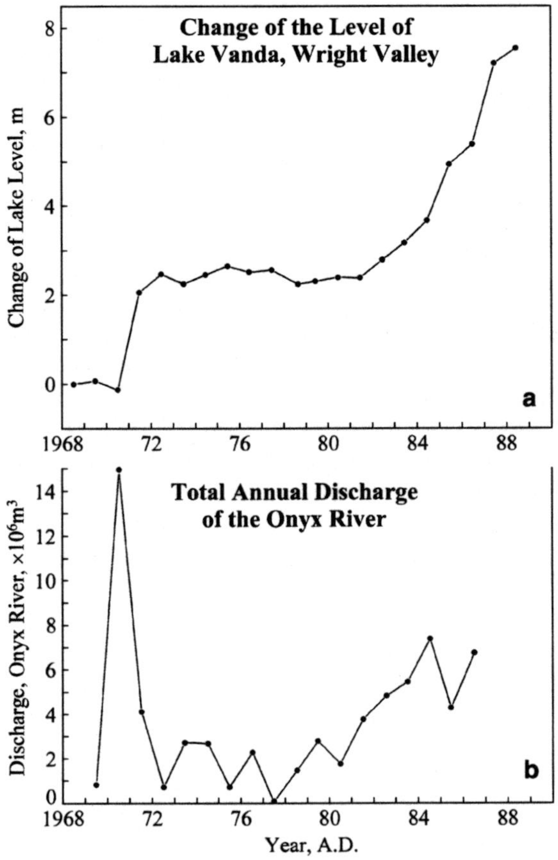

Fig. 19.31 (a) The level of Lake Vanda in Wright Valley has been increasing in steps and has begun to flood the area around it. (b) The total volume of water discharged annually into Lake Vanda by the Onyx River has fluctuated with a major spike in 1970/71 and more regular increases starting in 1981/82 until 1986/87. The water accumulates in Lake Vanda except for losses by sublimation of the ice cover at an average of 32.6 ± 12.5 cm/year (Data from Chinn 1993)

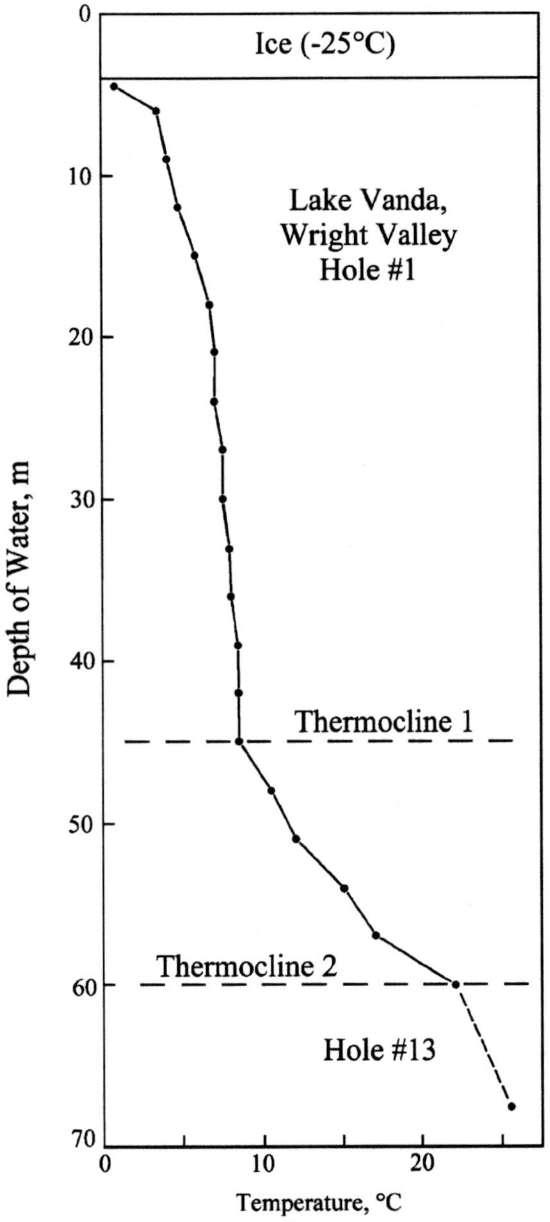

Fig. 19.32 The temperature of the water in Lake Vanda ranges from 0.75°C below the permanent layer of ice to 22°C at a depth of 60 m measured by Angino et al. (1965) on January 1, 1962, in hole #1. The same authors recorded an even higher temperature of 25.5°C at 67.5 m on January 8, 1962, in hole #13 where the lake is deeper. The temperature profile contains two thermoclines at 45 and at 60 m which define three layers of water. The hot brine at the bottom of the lakes (60–67.5 m) is the stagnant and anaerobic monimolimnion of this meromictic lake (Data from Angino et al. 1965) (See also Webster 1994)

Wilson and Wellman (1962) disagreed with this assessment and favored solar heating because the isotherms in the lake are horizontal and extend for several kilometers. In addition, they considered that hot springs are unlikely to occur in the ice-free valleys of Antarctica because of the wide-spread occurrence of permafrost. Wilson and Wellman (1962) reported that the ice is surprisingly transparent to solar radiation and that the water of Lake Vanda is very clear as indicated by its half-thickness of 14 m. Therefore, they concluded that the water of Lake Vanda is being heated by solar radiation rather than by a high geothermal gradient or by hot springs.

A few years later, Wilson et al. (1974a) took advantage of a hole drilled through the sediment at the bottom of Lake Vanda by the Dry Valley Drilling Project to measure the temperature gradient in the rocks beneath the lake. Their results indicate that heat flows from the lake into the underlying bedrock at the rate of 0.5–1.00 cal/cm^2/s. This measurement demonstrated that Lake Vanda is *not* being heated by geothermal heat and it confirmed that the water of the lake is being heated by solar energy.

Solar heating was also demonstrated for Lake Bonney and Lake Fryxell in Taylor Valley by Shirtcliffe (1964), Shirtcliffe and Benseman (1964), and by Hoare et al. (1964, 1965) aided by the clarity of the water in these lakes (Goldman et al. 1967). The temperature of the bottom brines of these lakes is +6.0°C in Lake Bonney and +2.3°C in Lake Fryxell (Chinn 1993).

In general, meromictic lakes in Antarctica and elsewhere are natural collectors of solar energy which accumulates in the stagnant brine layer at the bottom of such lakes. The density of the brine layer is key because it prevents the accumulated solar heat from being dispersed by mixing with the overlying cooler water (e.g., Lyons et al. 1984).

19.7.1.3 Sediment

The sediment in Lake Vanda above a depth of 60 m in the lake is oxidized and is composed of tan-colored sand composed predominantly of quartz and feldspar grains overlain by biological detritus which is up to 13 cm thick (Nelson and Wilson 1972). This organic layer is composed of the tissues of dead algae and dead diatoms mixed with living green algae, diatoms, fungi, bacteria, and terrigenous sand grains. The water below a depth of 60 m is anaerobic and the sediment in the deepest part of the lake consists of gray to gray-green sandy mud with a strong smell of hydrogen sulfide. The sediment contains about 6% of disseminated organic matter and about 10% authigenic calcite (Nelson and Wilson 1972; Lyons et al. 1985).

Additional sediment cores up to 2 m in length were later recovered by Wilson et al. (1974) from the deepest part of Lake Vanda. The sediment is composed of grey to greenish sand that contains several thin layers of evaporite minerals (calcite and gypsum) above a depth of about 100 cm in the core and scattered evaporite nodules at deeper levels down to 200 cm. In addition, the sand contains disseminated organic carbon ranging from 0.3% to 2.2%. A gypsum layer at 55 cm below the sediment-water interface yielded a U-series disequilibrium date of 13,600 ± 1,000 years and less than 2,000 years at 15 cm sub-bottom. Evidently, the sediment ranges in age from late Pleistocene to Holocene.

A much longer core (17.48 m) was recovered from Lake Vanda by the Dry Valley Drilling Project at site 4A in the deepest part of the lake (Cartwright et al. 1974). The core penetrated the entire section of sediment and included 5.06 m of granitic basement rocks at the bottom overlain by 12.42 m of unconsolidated sediment. The stratigraphy of the sediment in Fig. 19.33 includes 6.68 m of marine deposits at the base (M), followed by glacial sand and gravel (2.22 cm) (G), and lacustrine deposits (L) at the top (3.52 cm).

The sediment below 3.84 m sub-bottom in Lake Vanda contains marine microfossils consisting of broken frustules of pennate diatoms. Brady (1979) assigned these diatoms to the lowermost late Miocene and concluded (on p. 382) that:

> "they were deposited by glaciofluvial processes reworking the shallow marine sediments deposited during the final times of the Wright Valley Miocene fjord." (See also Doran et al. 1999).

19.7.1.4 Chemical Composition

The concentrations of ions in the water of Lake Vanda vary as a function of depth by more than three orders of magnitude. For example, the concentrations of sodium ions (Na^+) in Fig. 19.34 increase by a factor of 1,127 from 6 µg/g (ppm) at 4.5 m to 6,761 ppm at 66 m. The concentrations of calcium ions (Ca^{2+}) range by a factor of 1,426 from 17 ppm at the top to 24,254 ppm at the bottom of the lake. The positively charged *cations* detected by Angino et al. (1965) include the alkali metals (Li^+, Na^+, and K^+) and the alkaline earths (Mg^{2+} and Ca^{2+}). The negatively charged *anions* in the water consist of chloride (Cl^-), sulfate (SO_4^{2-}), and bicarbonate (HCO_3^-). In addition, low concentrations of molecular silicic acid (H_4SiO_4) recorded as "SiO_2" are also present.

The chemical analysis of the basal brine in Table 19.9 indicates that Ca^{2+} (24,254 ppm) is the most abundant cation followed by Mg^{2+} (7,684 ppm), Na^+ (6,761 ppm), K^+ (766 ppm), and Li (32 ppm). The concentrations of the anions are Cl^- = 75,870 ppm, SO_4^{2-} = 770 ppm, and HCO_3^- = 126 ppm, whereas the concentration

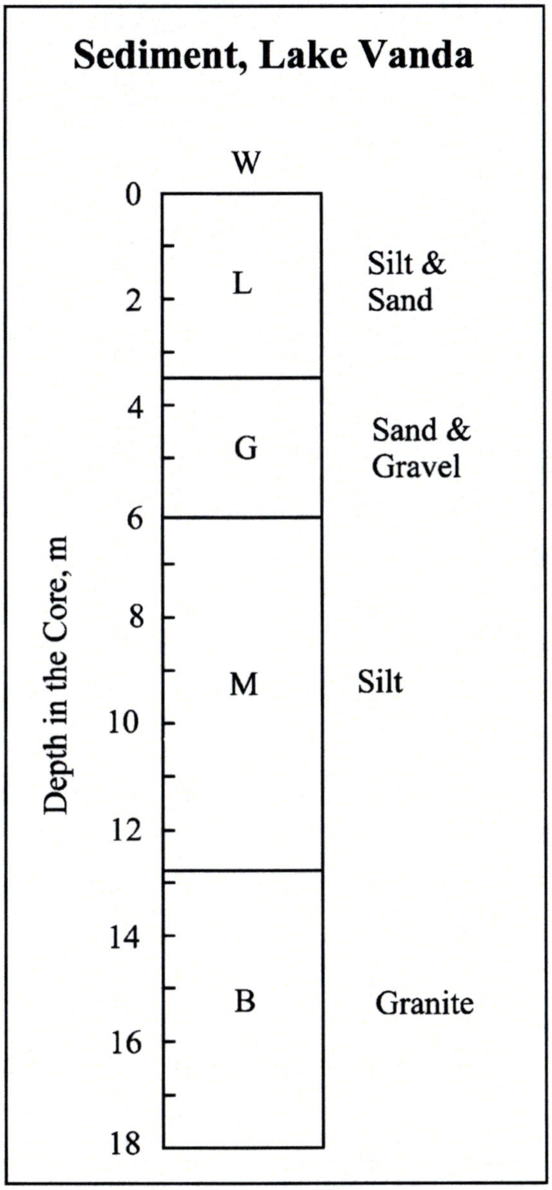

Fig. 19.33 The core recovered from the deepest part of Lake Vanda (Site 4A) by the Dry Valley Drilling Project in November 1973 consists of granitic basement overlain by marine glacial, and lacustrine sediment in about 68.25 m of water. B = basement, M = marine, G = glacial, L = lacustrine, W = water (Data from Cartwright et al. 1974) (See also Gumbley et al. 1974)

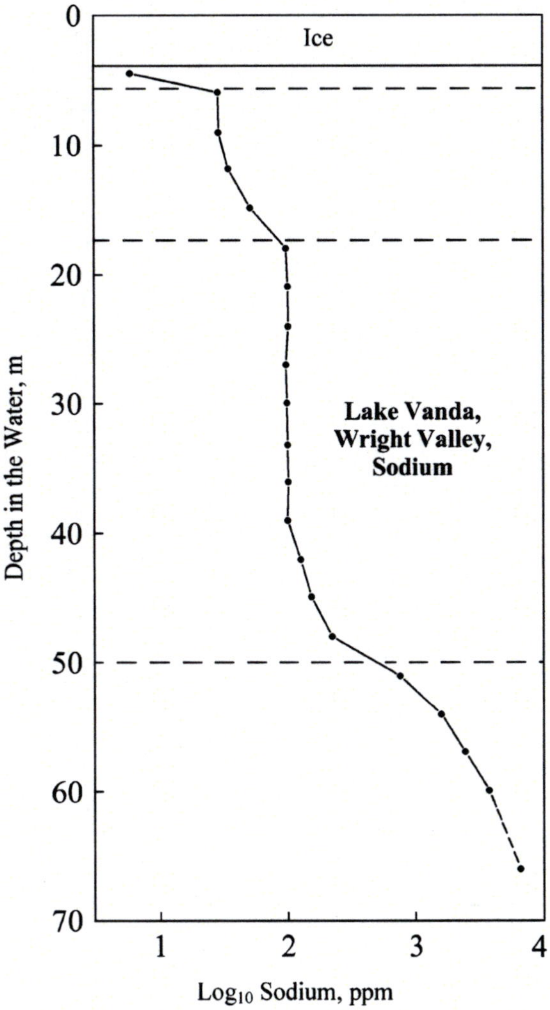

Fig. 19.34 The concentration of sodium (Na^+) and of all other elements in the water of Lake Vanda increase with depth to very high values in the hot brine at the bottom of the lake. The sodium concentrations in this diagram are expressed as the logarithm to the base 10 of the value in parts per million (Data from Angino et al. 1965)

of silicic acid (H_4SiO_4) at 60 m is 110 ppm. The concentrations of these ions were recalculated by use of the atomic masses of the elements to obtain the concentrations of the corresponding *salts* in percent by weight. The results indicate that calcium chloride ($CaCl_2$) is the most abundant salt at 57.8%, followed by magnesium chloride ($MgCl_2$) at 25.2%, sodium chloride (NaCl) at 14.8%, potassium chloride (KCl) at 1.2%, and magnesium sulfate ($MgSO_4$) at 0.8%. Therefore, we conclude that the basal brine of Lake Vanda contains a mixture of chloride salts of Ca, Mg, Na, and K with a trace amount of magnesium sulfate.

The presence of evaporite minerals in the lacustrine sediment (Fig. 19.33) in Lake Vanda suggests that calcite, gypsum, and other minerals may be precipitating

Table 19.9 Chemical composition of the basal brine at a depth of 66 m in Lake Vanda expressed in terms of salts based on data by Angino et al. (1965)

Ions	Concentration	
at 66 m	(ppm)	(μ mole/g)
Na^+	6761	294.1
K^+	766	19.6
Ca^{2+}	24,254	605.1
Mg^{2+}	7,684	316.0
Cl^-	75,870	2140.0
HCO_3^-	126	2.06
SO_4^{2-}	770	8.01
Salt	μg/g	%
NaCl	17,188.0	14.8
$CaCl_2$	67,157.6	57.8
KCl	1,461.2	1.2
$MgCl_2$	29,346.1	25.2
$MgSO_4$	938.9	0.8
Sum	116,091.8	99.2

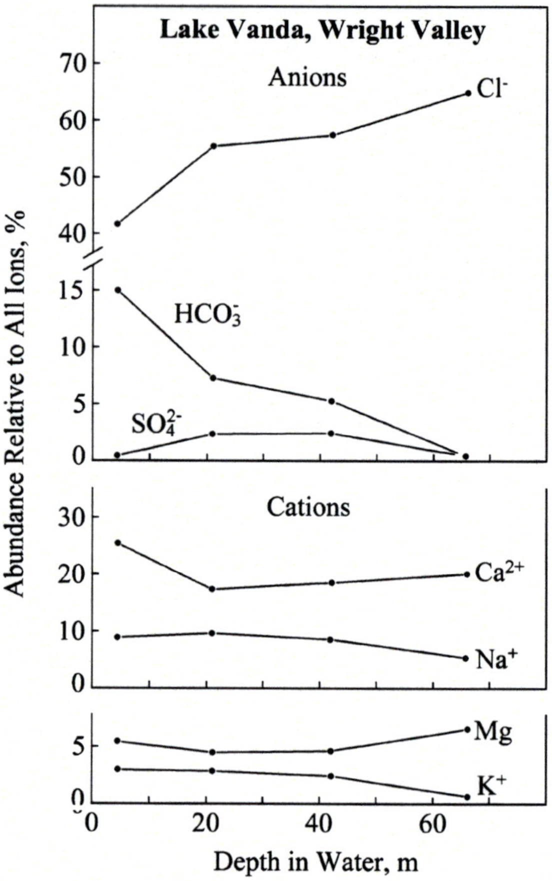

Fig. 19.35 The abundances of the major ions at increasing depths of water in Lake Vanda are expressed as percent values relative to the concentrations of all ions present. The results demonstrate that the abundance of the chloride ions increases while the abundances of the bicarbonate and sulfate ion decrease. These results therefore imply that the carbonates and sulfates of calcium and magnesium are precipitating in the water column, whereas chloride salts are gaining in abundance because they are soluble. In this way, the chemical composition of the water in Lake Vanda is evolving into a solution of chloride salts (Data from Angino et al. 1965)

in the water column above the bottom of the lake. In addition, these minerals may also precipitate from pore water trapped in the sediment which is transported into Lake Vanda by the Onyx River and other meltwater streams. In any case, the ions that form the precipitates are removed from solution in the water which causes the chemical composition of the water in the lake to change. We examine this possibility in Fig. 19.35 by plotting the abundances of the ions expressed as a percent relative to all ions in solution at a given depth. The results reveal that the abundance of Cl^- in the water *rises* from 41.8% at 4.5 m to 65.3% at 66 m, whereas the abundance of the bicarbonate ion (HCO_3^-) *decreases* from 14.9% at 4.5 m to 0.10% at 66 m. The abundance of sulfate rises from 0% at 4.5 m to 2.5% at 42 m and then declines to 0.66% at 66 m. These depth-dependent differences of the abundances of anions may be caused by the precipitation of carbonate and sulfate salts which thereby remove the ions of which they are composed from solution in the water. The chlorides of alkali metals (Li, Na, K) and alkaline earths (Mg, Ca) are soluble in water which explains why the abundance of chloride increase with depth (Webster and Goguel 1988).

The complementary variations of the abundances of Ca^{2+} and Mg^{2+} at increasing depth in Lake Vanda confirm that carbonates and sulfates are precipitating from the water. For example, the abundance of Ca^{2+} decreases from 25.4% at 4.5 m to 17.5% at 21 m, whereas the bicarbonate decreases from 14.9% (4.5 m) to 7.4% (21 m). Similarly, the abundance of Mg^{2+} decreases in the same depth range from 6.0% at 4.5 m to 4.7% at 21 m. Small amounts of sulfate salts may also precipitate at greater depth between 42 and 66 m.

Solids, that precipitate in the upper levels in Lake Vanda and then sink through the water column, encounter increasing temperatures (i.e., Fig. 19.32) and acidity expressed as the pH in Fig. 19.36. As a result, solid particles that form in the upper levels of the lake, such as the oxyhydroxides of iron and manganese, may dissolve at deeper levels. Crystalline as well as amorphous

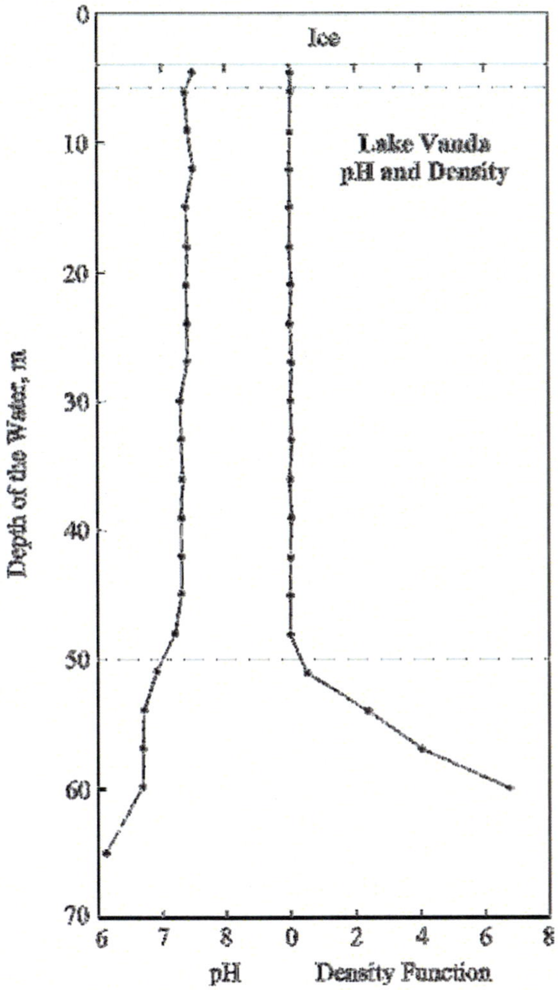

Fig. 19.36 The acidity of the water of Lake Vanda is expressed as the pH (negative of the logarithm to the base 10 of the hydrogen ion activity expressed in moles per liter). The water in the upper 50 m is slightly basic (pH = 7.5 to 7.3). However, the brine at the bottom of the lake is acidic (pH = 6.9 to 6.1). The density of the water is expressed here as the "Density Function" (DF) expressed as DF = (D – 1.000) × 10^2, such that DF = 4.0 corresponds to D = 1.040. The density of the bottom brine increases from 1.000 to greater than 1.060 g/cm^3 (Data from Angino et al. 1965)

particles that are suspended in any body of water have negatively charged sites on their surfaces to which cations are sorbed by electrostatic forces. In cases where the particles dissolve, the cations that are sorbed to their surfaces are released and go back into solution in the water. In this way, the ions of certain metals are transported by sinking particles into deeper water where they may be released into solution when the carrier particles dissolve. This process governs the distribution of trace metals, also called micronutrients, in Lake Vanda and in several other lakes in the ice-free valleys (Masuda et al. 1982; Canfield and Green 1985; Green et al. 1986a, b, 1989, 1993, 1998; Mudrey 1974).

The distribution of micronutrients in the water of Lake Vanda and the penetration of solar radiation during the summer months enable bacteria and colonial algae to inhabit the lake. The biological activity in Lake Vanda causes dissolved organic matter (e.g., fulvic and hydrophilic acids) to occur in the water (Vincent and Vincent 1982; Matsumoto et al. 1984; McKnight et al. 1993). In addition, Green et al. (1993) reported that organic particles contribute to the transport of trace metals especially in the water above the anoxic brine layer of Lake Vanda (De Carlo and Green 2002).

The remarkably abundant flora that exists in Lake Vanda and in the water of other lakes in southern Victoria Land defies the harsh climatic conditions of the surface. For example, the average annual temperature in Wright Valley at Lake Vanda is –19.8°C and the rate of meteoric precipitation is less than 10 mm/year. Although algae thrive in the lakes, fish and other kinds of animals with backbones have not been observed in any of the lakes in southern Victoria Land (Benoit and Hall 1970; Cameron et al. 1970; Horowítz et al. 1972; Parker and Simmons 1978; Parker et al. 1982; Love et al. 1983; Vincent and Howard-Williams 1989; Aiken et al. 1991; Simmons et al. 1993) (See also Lyons et al. 1984).

19.7.1.5 Onyx River

The Onyx River in Fig. 19.37 drains glacial meltwater from Lake Brownworth at the terminus of the Wright Lower Glacier and moves it into Lake Vanda at the opposite (western) end of Wright Valley. The Onyx River also carries glacial meltwater discharged by the alpine glaciers along the south wall of Wright Valley and by snow fields and ice caps on the Asgard and Olympus ranges (Fig. 19.18). The point is that the water that is discharged into Lake Vanda originates as glacial meltwater which has only low concentrations of cations and anions (Jones 1969).

The low solute concentration of glacial meltwater causes problems in the collection and analysis of water samples because of the potential for contamination

Fig. 19.37 The Onyx River flows west (away form the coast) and discharges into Lake Vanda located 28 km west of Lake Brownworth. It carries water only during the summer in December and January. The water originates by melting of ice in the Wright Lower Glacier and of the alpine glaciers along the south wall of Wright Valley (Photo by G. Faure)

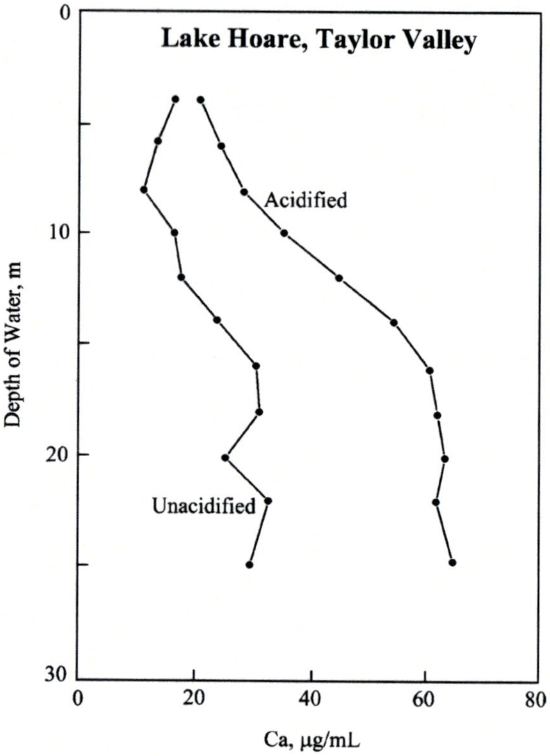

Fig. 19.38 When water samples collected at different depths in Lake Hoare in Taylor Valley are acidified prior to analysis by ion chromatography, the measured calcium concentrations are about two times higher than in unacidified samples. The effect is caused by the displacement of Ca^{2+} by H^+ from negatively charged sites on the surface of suspended colloidal particles (Adapted from Welch et al. 1996)

during collection and storage and because an exceptionally sensitive method is required to analyze them. Jones (1969) used instrumental spectrophotometry to determine cation concentrations of water collected from the Onyx River along its course and from several tributary meltwater streams. Jones (1967) also used isotope dilution to measure the strontium concentrations of the water in Lake Vanda ranging from 0.141 ppm under the ice to 67.1 ppm in the hot bottom brine. The preferred analytical methods nowadays are ion chromatography and inductively coupled plasma spectrometry both of which are capable of detecting cations at very low levels of concentration. For example, Welch et al. (1996) used ion chromatography to analyze glacial meltwaters from Taylor Valley some of which were exceedingly dilute in units of microgram per milliliter:

$$Na^+ = 0.033,\ K^+ = 0.004,\ Ca^{2+} = 0.023$$

$$Mg^{2+} = 0.002;\ Cl^- = 0.017, SO_4^{2-} = 0.025$$

Welch et al. (1996) also demonstrated the effect of acidifying water samples from Lake Hoare in Taylor Valley prior to analysis. Their results in Fig. 19.38 indicate that the measured Ca^{2+} concentrations of acidified water samples from Lake Hoare are higher than those of unacidified samples by a factor of two. The reason is that the hydrogen ions in the acidified water samples displace calcium ions that are sorbed to negatively charged sites on the surfaces of colloidal particles.

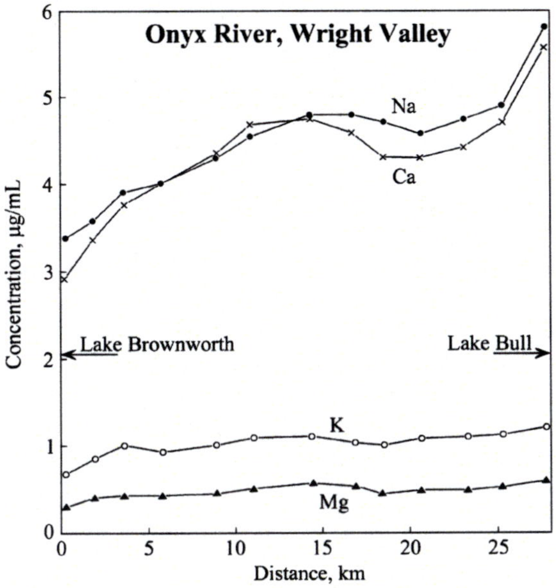

Fig. 19.39 The concentrations of sodium, calcium, potassium, and magnesium in the water of the Onyx River vary along the course of the river and generally double between Lake Brownworth and Lake Bull. Evidently, the water is leaching these ions from soil along its channel (Data from Jones 1973)

The chemical composition of water in the Onyx River was first analyzed by Jones (1973). The samples for this study were collected during the summer of 1969/70 within a 2-h period. The results in Fig. 19.39 indicate that the concentrations of the major cations increase along the course of the river (28 km) between Lake Brownworth and Lake Bull. Green and Canfield (1981, 1984) subsequently measured the composition of the Onyx River at the weir located downstream of Lake Bull and determined that the water of the Onyx River could not have evolved into the Ca-Na-Mg-Cl brine that occurs in Lake Vanda. Instead, they proposed that the water in the lake is a mixture of calcium-chloride brine from the Don Juan basin with water from the Onyx River. According to the proposal of Green and Canfield (1981, 1984), the layer of dense brine at the bottom of Lake Vanda is local groundwater that entered the basin of the lake by subsurface flow, whereas the overlying lake water originated from the Onyx River. The two water masses in the lake have maintained their separate identities except for diffusion of ions from the brine layer into the overlying water derived from glacial meltwater (Wilson 1964, 1979). The meltwater that enters Lake Vanda by discharge of the Onyx River evolves by freezing to the bottom of the permanent ice cover which excludes the ions in solution and thereby increases the total dissolved solids in the water. The evaporative concentration of the water takes place by the subsequent sublimation of the ice on the lake (Thompson and Nelson 1956). Green and Canfield (1981, 1984) also analyzed water collected along the course of the Onyx River and confirmed that the concentrations of ions increase downstream as shown by Jones (1973) in Fig. 19.39.

19.7.2 Sources of Salts: Strontium Isotopes

An important issue in the geochemistry of the ice-free valleys is to identify the sources of the chemical elements that form the salts in the lakes and the soil. Several potential sources have been considered including seawater which is known to have flooded both Wright and Taylor valleys, volcanic activity in the form of hot springs, chemical weathering of minerals, and others mentioned by Torii and Yamagata (1981). All of the lakes contain strontium which has a range of isotopic compositions that identify the provenance of this element by the numerical values of its $^{87}Sr/^{86}Sr$ ratio (Section 3.7.3; Faure and Mensing 2005).

Jones et al. (1967) first reported that the average $^{87}Sr/^{86}Sr$ ratio of 15 water samples from a depth profile of Lake Vanda is 0.7149 ± 0.0001 which is identical within analytical error to the $^{87}Sr/^{86}Sr$ ratio of 0.7146 ± 0.0002 of strontium in the water of the Onyx River. On the other hand, the $^{87}Sr/^{86}Sr$ ratios of strontium in Lake Vanda and the Onyx River differ markedly from the average $^{87}Sr/^{86}Sr$ ratio of the Cenozoic basalt of the McMurdo Volcanics (0.7043) and from the average $^{87}Sr/^{86}Sr$ ratio of two samples of seawater from the Ross Sea (0.7094). These results indicate unequivocally that the strontium at all levels in Lake Vanda and in Onyx River did not originate from marine sources or by weathering of Cenozoic basalt or volcanic ash in Wright Valley.

The profiles of the isotopic composition and concentration of strontium in Fig. 19.40 measured by Jones and Faure (1967) demonstrate that the $^{87}Sr/^{86}Sr$ ratios of the water in Lake Vanda are constant at all levels in the lake including the basal brine at 60 m, but that the *concentrations* vary with depth much like the concentrations of sodium in Fig. 19.34. Soil samples leached with acidified water and even a bulk sample of

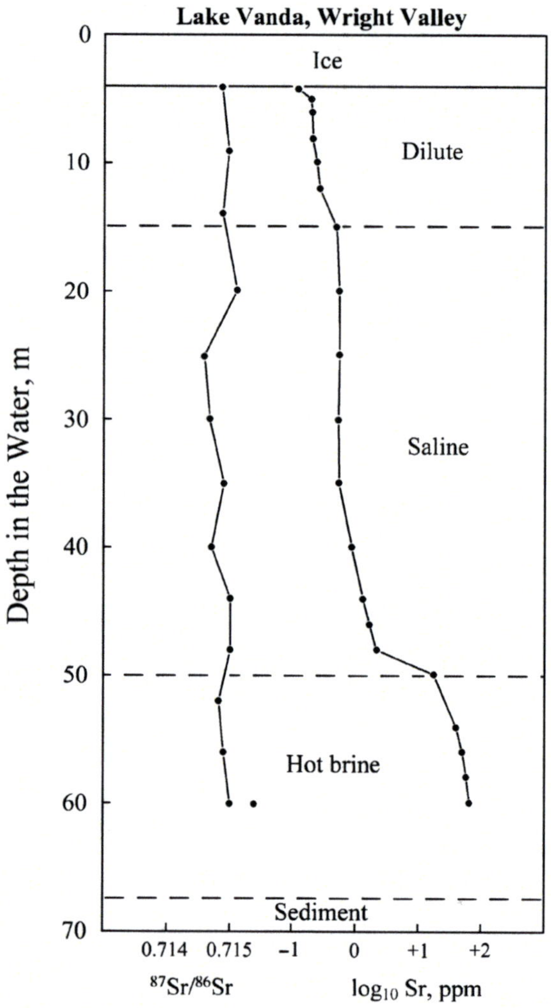

Fig. 19.40 The $^{87}Sr/^{86}Sr$ ratios of strontium in Lake Vanda vary only between 0.7146 and 0.7150 and are essentially independent of depth in the water. These ratios are different from the $^{87}Sr/^{86}Sr$ ratios of basalt (0.7043) and seawater (0.7094) which indicates that the strontium in Lake Vanda did not originate from these sources. The concentrations of strontium range from 0.141 ppm at 4 m to 67.1 ppm at 60 m and were plotted as the logarithms to the base 10 in order to accommodate the wide range of variation. The concentration profile identifies chemoclines at depth of about 15 and 50 m below the surface of the ice (Data from Jones and Faure 1967)

soil analyzed by Jones and Faure (1967) all have $^{87}Sr/^{86}Sr$ ratios (0.7144–0.7148) that are indistinguishable from the $^{87}Sr/^{86}Sr$ ratios of strontium in Lake Vanda and of the Onyx River. Subsequently, Jones and Faure (1978) reported that the $^{87}Sr/^{86}Sr$ ratios of strontium leached from additional soil samples in Wright Valley all range between 0.7119 (Clark Glacier) and 0.7157 (Meserve Glacier) regardless of whether the soil was leached with demineralized water or with dilute hydrochloric acid.

The only strontium of marine origin, found by Jones and Faure (1967), occurred in a sample of ice from the surface of the Meserve Glacier which has an $^{87}Sr/^{86}Sr$ ratio of 0.7090. However, the amount of marine strontium in soil salts of Wright Valley is insignificant. The summary of these results in Fig. 19.41 demonstrates graphically that the principal source of strontium in Lake Vanda and in the Onyx River is the *soil* in Wright Valley. Although this conclusion applies specifically only to the provenance of strontium in Wright Valley, it does demonstrate the importance of chemical weathering of silicate minerals in the soil to the geochemistry of the lakes and streams in the ice-free valleys of southern Victoria Land.

The origin of salts in the ice free valleys has been discussed by many investigators including Gibson (1962), Smith (1965), Bowser et al. (1970), Dort and Dort (1970), Claridge and Campbell (1977), Faure and Felder (1981), Keys and Williams, (1981), Torii and Yamagata (1981) and, more recently, by Sheppard et al. (2002). These studies have contributed to the consensus that the salts in the ice-free valleys have originated by chemical weathering of silicate minerals in the soil and by atmospheric transport from McMurdo Sound. The marine component of the salts is prevalent in the ice of the alpine glaciers of Wright Valley and in the snow that is deposited annually. Although nitrates may also originate from seawater via atmospheric transport (Claridge and Campbell 1968), the abundance of nitrates actually increases with increasing distance from the coast (Keys and Williams 1981) and Wilson and House (1965) first reported that nitrates occur in the snow at the South Pole. More recently, Parker et al. (1978) and Dreschhoff and Zeller (1994) demonstrated that nitrate also forms by the irradiation of the atmosphere by solar protons (Section 17.9.3). The sources of nitrogen, phosphorus, and carbon in Antarctic meltwater streams were discussed by Downes et al. (1986).

19.7.3 Don Juan Pond

Don Juan Pond in Fig. 19.22 is located at 77°34′S and 161°11′E in a topographic basin in the South Fork of Wright Valley about 8 km southwest of Lake Vanda. The pond is about 315 m long, 116 m wide, and less than

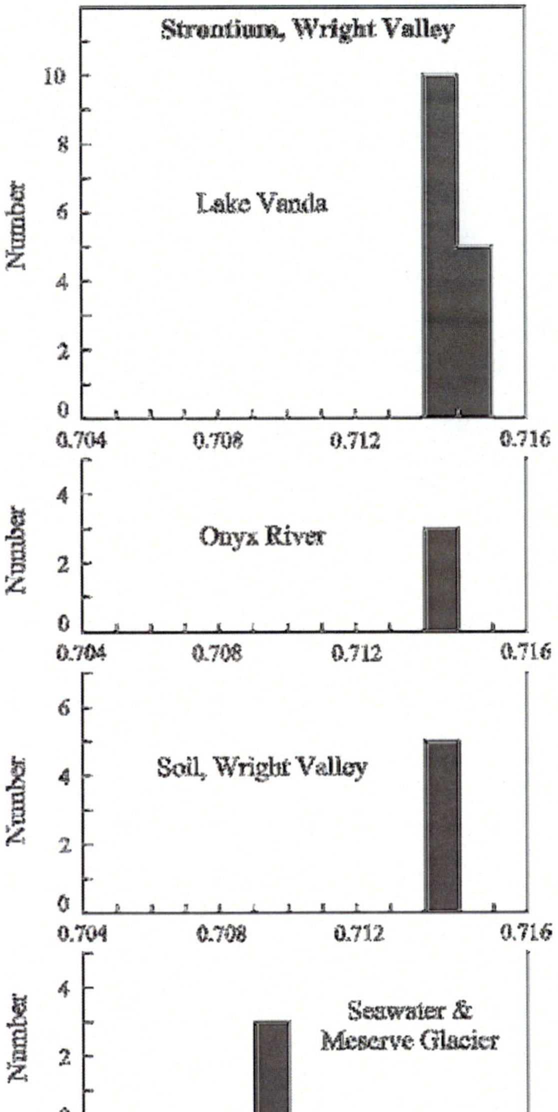

Fig. 19.41 The $^{87}Sr/^{86}Sr$ ratios of strontium in the water of Lake Vanda and in the Onyx River are indistinguishable from the $^{87}Sr/^{86}Sr$ ratios of strontium that is released by chemical weathering of the soil in Wright Valley. Strontium of marine origin is detectable only in the ice of the Meserve Glacier and may also be present in other alpine glaciers in the valley. These results demonstrate that the strontium in the salts of Wright Valley originates primarily by chemical weathering of the soil. Although this conclusion applies only to the origin of strontium, it demonstrates the importance of chemical weathering of silicate minerals to the geochemistry of the ice-free valleys in southern Victoria Land (Data from Jones and Faure 1967)

10 cm deep. It is surrounded by a salt flat and is situated about 200 m east of an active rock glacier (Harris and Cartwright 1982). Strand lines indicate that the water level in Don Juan Pond was up to 21 m higher in the past than it is at present time (Mudrey et al. 1973, 1975). An evocative photograph of Don Juan Pond was published by Porter (1978, p. 101).

Don Juan Pond was discovered on October 11 of 1961 by a group of scientists during a reconnaissance of the ice-free valleys by helicopter. At the time it was discovered, the brine in the pond was not frozen in spite of the fact that the air temperature was –24°C (Meyer et al. 1962). When Japanese scientists visited Don Juan Pond in 1963, the water level and size of the pond had declined markedly (Yamagata et al. 1967). The pond is replenished by two small and intermittent meltwater streams that enter it from the west.

The density of the water in Don Juan Pond is exceptionally high. Meyer et al. (1962) reported a value of 1.2514 g/mL, whereas Yamagata et al. (1967) obtained 1.351 and 1.380 g/mL. Jones (1969) later determined a value of 1.2744 g/mL at 20°C in one sample of brine collected by D.D. Koob during the summer of 1964/65. The shallow depth of the brine (about 10 cm), the lack of an ice cover, and steady wind all prevent the brine from becoming stratified (Harris et al. 1979).

The temperature of the brine in Don Juan Pond varies seasonally depending on the air temperature. The freezing point of the brine ranges from –48°C (Meyer et al. 1962) to –57°C (Tedrow et al. 1963) which means that the brine in Don Juan Pond may not freeze even during the Antarctic winter, although Bull (1966) did record a minimum temperature of –62°C at Lake Vida in Victoria Valley.

The chemical analyses of the brine in Don Juan Pond in Appendix 19.9.3 demonstrate that the concentrations of Ca^{2+} and Na^{+} in samples collected over a span of 4 years from October 1961 to January 1965 vary widely. Nevertheless, all analyses indicate that the salts dissolved in the brine consists of $CaCl_2$ (93.0 ± 5.4%), NaCl (4.82 ± 5.42%), $MgCl_2$ (2.09 ± 0.44%), and KCl (0.09 ± 0.01%) expressed relative to the sum of the total dissolved solids (Mudrey et al. 1973, 1975).

The brine in Don Juan Pond is precipitating a mineral composed of $CaCl_2 \cdot 6H_2O$ called *antarcticite* (Torii and Ossaka 1965). The occurrence of this mineral is unusual because it attracts water and spontaneously dissolved in it (i.e., the mineral is deliquescent). It is stable only in very dry environments such as exist in the vicinity of Don Juan Pond and in Death Valley, California (Torii et al. 1986).

Jones (1969) reported that the strontium in the brine is enriched in radiogenic ^{87}Sr and has an $^{87}Sr/^{86}Sr$ ratio of 0.7183 compared to only 0.7149 in the deep brine of Lake Vanda. The isotopic composition indicates that the strontium was released into solution by chemical weathering of minerals that are enriched in rubidium and have accumulated radiogenic ^{87}Sr by decay of ^{87}Rb during a substantial length of geologic time. Although very small amounts of marine strontium may have originated by melting of ice and snow having an $^{87}Sr/^{86}Sr$ ratio of 0.7092, the marine component is completely overwhelmed by strontium from the local granitic basement rocks.

The $^{87}Sr/^{86}Sr$ ratio of the brine in Don Juan Pond measured by Jones (1969) was confirmed by Friedman et al. (1995) who reported values of 0.7185 for brine collected on January 10, 1975; 0.7186 for crystals of antarcticites collected in November of 1973; and 0.7187 for interstitial brine from sediment at a depth of 10.8 m. They also reported that the average $^{87}Sr/^{86}Sr$ ratio of glacial meltwater that is discharged into Don Juan Pond is 0.7163 ± 0.0009 and that the average strontium concentration of the meltwater is only 0.086 ± 0.011 ppm. Evidently, the strontium in the meltwater that flows into Don Juan Pond is derived by chemical weathering of silicate minerals in the soil west of the pond and marine strontium is not detectable in the water of the tributary streams. The literature containing information relevant to the study of Don Juan Pond is listed in Appendix 19.9.4.

19.7.4 Lake Bonney, Taylor Valley

Although Taylor Valley in Fig. 19.17 is open to McMurdo Sound, it contains a number of lakes that occupy closed basins (Appendix 19.9.2) including Bonney, Hoare, and Fryxell among others. The Taylor Glacier, which is a former outlet glacier, has receded up the valley and now discharges its meltwater directly into Lake Bonney which also receives meltwater from the alpine glaciers at the western end of Taylor Valley. The other lakes in Taylor Valley are also fed by meltwater streams that descend from the Kukri Hills and from the Asgard Range. Most of the meltwater streams flow into the closed-basin lakes of Taylor Valley. The only exception is the Commonwealth stream which discharges meltwater directly into McMurdo Sound.

19.7.4.1 Density and Temperature

Lake Bonney in Fig. 10.3 is located in the western part of Taylor Valley about 30 km from the coast. It is about 6 km long, 1 km wide and is 32 m deep. The lake has two lobes that are connected by a narrow channel which has widened and deepened during the twentieth century because of the increase of the water level in the lake (Shirtcliffe 1964; Jones and Faure 1968, 1978; Chinn 1993). Lake Bonney is perennially covered by ice that is about 4 m thick; but during the summer a moat less than 10 m wide forms around the periphery of the lake, while the eastern end of the lake actually becomes ice-free.

The density of the water in Lake Bonney increases with depth in Fig. 19.42 and identifies the three principal layers of the lake. The dilute water under the ice is underlain by a transitional layer in which the density at 20°C increases from 1.0080 g/mL at 9 m to 1.1509 g/mL at 17 m. The density of the water at the bottom of the lake from 17 to 30 m is nearly constant at an average value of 1.1754 ± 0.0082 g/mL. The density at the base of the brine layer at the bottom of Lake Bonney is 1.1932 g/mL (Jones 1969; Jones and Faure 1978).

The temperature of the water in Fig 19.43 rises with increasing depth to +6°C at about 10 m and then declines to –2°C at the bottom of the lake (Angino et al. 1964; Hoare et al. 1964; Weand et al. 1975). The increase in the temperature was attributed by Shirtcliffe and Benseman (1964) to the entrapment of solar radiation.

19.7.4.2 Chemical Composition

The concentrations of major elements in the lake increase with depth and reach high values in the hypersaline monimolimnion which extends from a depth of about 9 m to the bottom of the lake at 32 m. The presence of this layer of non-circulating brine identifies Lake Bonney as a meromictic lake. The concentration of oxygen in the monimolimnion during the summer of 1972/73 was generally low (<5 mg/L) but increased unexpectedly to about 25 mg/L on January 1, 1973 (Weand et al. 1975). Similar

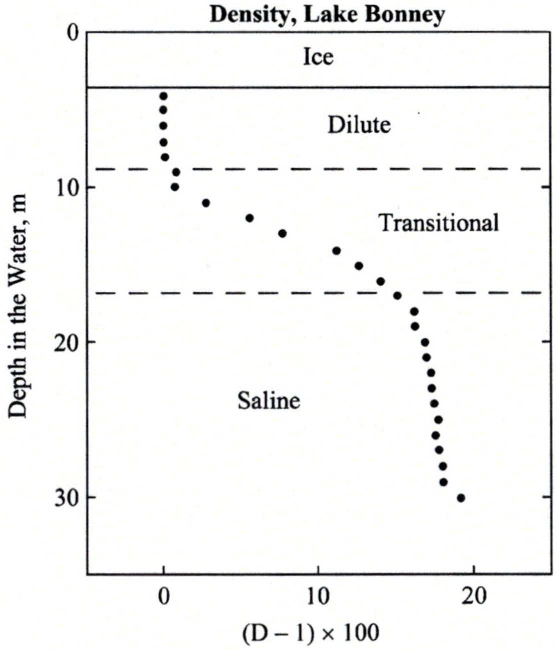

Fig. 19.42 The density of the water in Lake Bonney increases smoothly with increasing depth. A layer of dilute solution extends from the overlying ice to a depth of about 9.0 m. this layer merges into a transitional zone where the density increases from 1.0080 g/mL at 9.0 m to 1.1509 g/mL at 17.0 m. The density of the water at greater depth approaches a constant value at about 1.1750 g/mL. All densities were measure at 20°C. The density is expressed as the function (D – 1) × 100 such that if D = 1.0200, (D - 1) × 100 = 2.00 (Data from Jones (1969) using samples collected by Derry Koob)

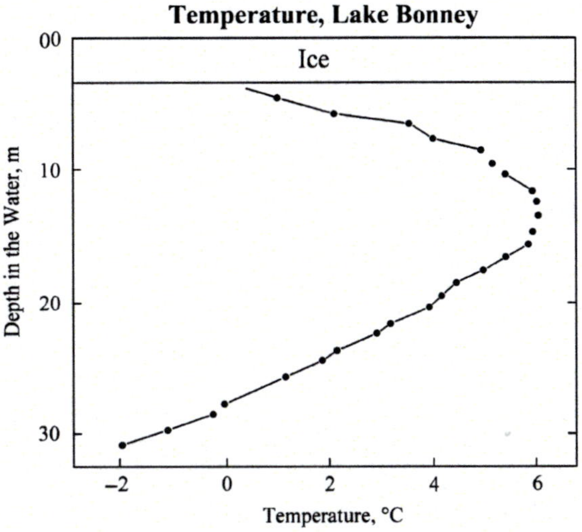

Fig. 19.43 The temperature of the water in Lake Bonney increases from 0°C under the ice to about +6°C at about 12–16 m. The temperature declines at greater depth and reaches –2°C at about 30 m where the density is 1.1932 g/mL. The water at the bottom of the lakes does not freeze because it is too salty. The elevated temperatures were attributed by Shirtcliffe and Benseman (1964) to entrapment of solar radiation. The temperatures were measured on December 16, 1972 at the eastern end of Lake Bonney and were reported by Weand et al. (1975)

erratic changes occurred in the concentrations of the major elements at all levels in the lake. For example, Weand et al. (1975) reported that the concentration of sodium at a depth of 26.0 m at the east end of the lake changed from 43,000 mg/L on 11/25/73, to 48,000 mg/L on 12/12/73, and then to 35,000 mg/L on 1/19/74. The data of Weand et al. (1975) also indicate that the average concentrations of the brine at 26.0 m were: Na = 42,000 mg/L; Mg = 17,933 mg/L; K = 3,250 mg/L; and Ca = 1413 mg/L. Evidently, the salts dissolved in the water of Lake Bonney are primarily composed of the chlorides of Na, Mg, K, and Ca in contrast to Lake Vanda where Ca dominates (Craig et al. 1974).

Weand et al. (1975) also reported a "vigorous effervescence" of the water at their sampling site on Lake Bonney in November of 1973. A similar phenomenon was observed by Koob and Leister (1972), but the odorless gas that was discharged by the lake on these occasions was not identified. Weand et al. (1975) speculated that either carbon dioxide or oxygen may have been released when groundwater entered the hypersaline monimolimnion of Lake Bonney. According to Weand et al. (1975), the erractic fluctuations of the concentrations of chemical elements including oxygen, sulfate, and pH, as well as the temperature of the water in the lake may have been caused by the discharge of groundwater into the lake. Even the abundance of bacteria varied in response to increases in the concentrations of nutrients including inorganic nitrogen (ammonia) and phosphorus (Parker et al. 1973; Hoehn et al. 1974).

Weand et al. (1975, p. 18) concluded that:

> The afore mentioned changes are compatible with the hypothesis that there is a significant, although intermittent, subterranean inflow of fresh meltwater into Lake Bonney at depths below 20 meters.

19.7.4.3 Strontium Isotopes

The $^{87}Sr/^{86}Sr$ ratios of six water samples from different depths in Lake Bonney range only from 0.7129 to 0.7133 even though the concentrations of strontium increase from 0.7345 ppm at 4 m below the surface to 35.88 ppm at a depth of 30 m (Jones and Faure 1968, 1978). A plot of these data in Fig. 19.44 demonstrates that the $^{87}Sr/^{86}Sr$ of the water in the lake is nearly constant, whereas the concentrations of strontium reveal the chemical stratigraphy of the lake. The uppermost layer (4–8 m) of dilute glacial meltwater is underlain by a transitional layer (9–17 m) in which the strontium concentration rises with increasing depth from 3.026 ppm at 9 m to 33.69 ppm at 16 m. The transitional layer grades into a saline brine which extends from about 17 m to the bottom of the lake at 32 m. The strontium concentration of this brine layer ranges only from 35.88 ppm at 30 m to 39.92 ppm at 25 m with an average of 38.4 ± 1.22 ppm for 11 samples. The chemical stratigraphy of Lake Bonney indicated by the strontium concentrations is consistent with the profile of densities in Fig. 19.42 based on data by Jones (1969). The transitional layer is presumably maintained by diffusion of ions from the saline brine at the bottom of the lake into the dilute layer under the ice.

The $^{87}Sr/^{86}Sr$ ratio of strontium leached from the soil in the vicinity of Lake Bonney (0.7136) is similar to the $^{87}Sr/^{86}Sr$ ratio of strontium in the lake (Jones and Faure 1968, 1978). The brine discharged by the Taylor Red Cone at the terminus of the Taylor Glacier also has an $^{87}Sr/^{86}Sr$ ratio of 0.7136. These results suggest that the salts dissolved in Lake Bonney are largely derived by chemical weathering of silicate minerals in the soil in the Bonney watershed. Therefore, the provenance of salts in Lake Bonney closely resembles the derivation of salts in Lake Vanda. The $^{87}Sr/^{86}Sr$ ratios of water in the lakes of Taylor Valley and in soil salts decrease in Fig. 19.45 from Lake Bonney to the coast of McMurdo Sound due to the increasing abundance of marine strontium (Jones and Faure 1978).

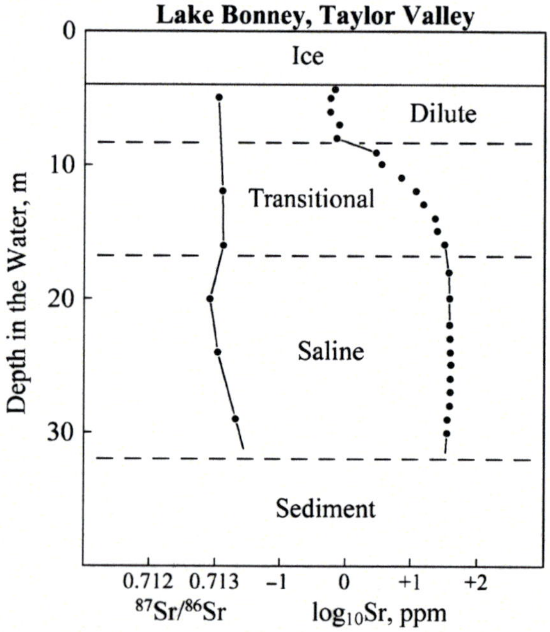

Fig. 19.44 The $^{87}Sr/^{86}Sr$ ratios of the water in Lake Bonney range narrowly with increasing depth and have an average value of 0.71307 ± 0.00006 (1σ). The concentration of strontium defines a layer of dilute water under the ice (about 4 m, Sr = 0.688 ± 0.098 ppm), followed by a transitional layer (about 7 m, Sr = 3.03 to 33.7 ppm), which is underlain by the main body of water (about 13 m, Sr = 38.4 ± 1.2 ppm). The concentration of strontium is expressed as the logarithm to the base 10 of the value in parts per million (micrograms per gram) (Data from Jones and Faure 1978)

19.7.5 Meltwater Streams, Taylor Valley

The principal meltwater streams in Taylor Valley are identified by name in Fig. 19.46 which was prepared by Lyons et al. (1998a). The geochemical evolution of

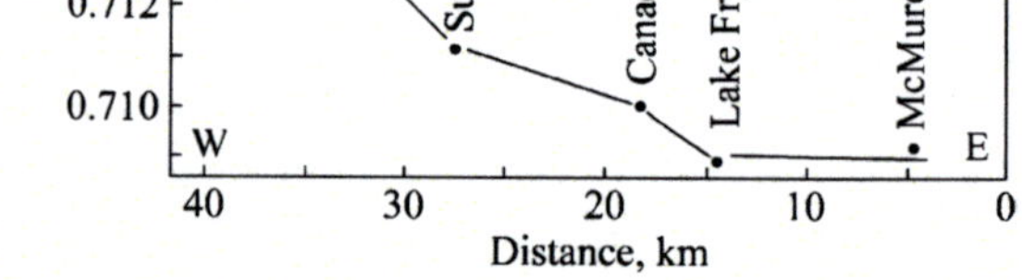

Fig. 19.45 The $^{87}Sr/^{86}Sr$ ratios of water and glacial ice in Taylor Valley decrease down-valley from 0.7136 in the brine discharged at the terminus of the Taylor Glacier (Taylor Red Cone) to 0.7092 in soil at Lake Fryxell located close to the coast. The systematic decrease of the $^{87}Sr/^{86}Sr$ ratios in this profile indicates that strontium of marine origin becomes increasingly abundant and dominates at the eastern end of Taylor Valley. The low $^{87}Sr/^{86}Sr$ ratios of water in Lake Fryxell (0.7090) and soil salts near that lake (0.7089) were first published by Jones and Faure (1968) (This diagram was adapted from Fig. 3 of Jones and Faure 1978)

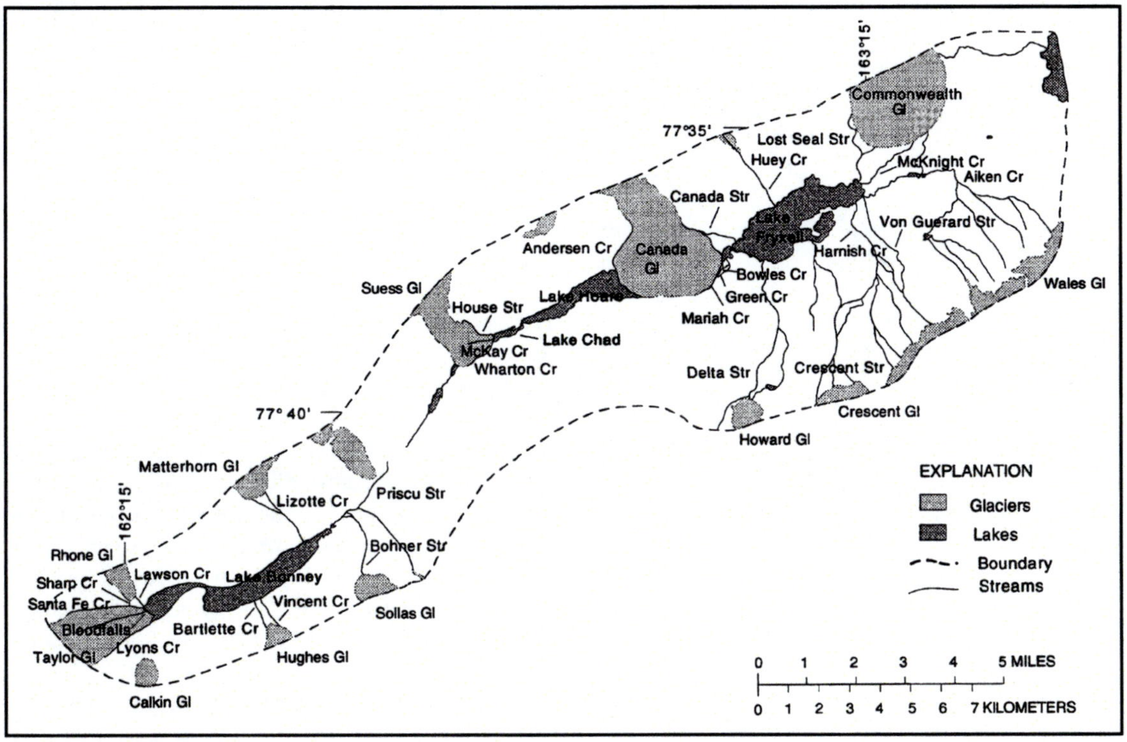

Fig. 19.46 The glacial meltwater in Taylor Valley flows into three closed drainage basins even though the valley itself opens onto McMurdo Sound. The principal lakes in these basins are Lake Bonney (Fig. 3.7), Lake Chad and Lake Hoare, and Lake Fryxell. Lake Bonney receives brine discharged by the Taylor Glacier at Bloodfalls (also known as Taylor Red Cone). Lakes Chad and Hoare collect meltwater directly from the Suess and Canada glaciers. Lake Fryxell, which is located in the widest part of Taylor Valley, receives meltwater from the Commonwealth and Canada glaciers as well as from ice caps on the Kukri Hills. Most of the meltwater streams in Taylor Valley have been given names in order to facilitate their identification (Adapted from Fig. 1 of Lyons et al. (1998) with permission from W.B. Lyons)

lakes Fryxell, Hoare, Joyce, and Miers (north of the Koettlitz Glacier) in relation to their respective tributary streams was investigated by Green et al. (1988) who concluded that:

"[T]hese lakes could have evolved their chemical compositions over a few thousand years of present stream flow. It is not necessary to invoke trapped seawater or other salt sources to explain their evolution".

19.7.5.1 Geochemical Linkages

The relationship between the chemical compositions of the lakes in Taylor Valley and their respective tributaries became even more apparent in the subsequent study of Lyons et al. (1998). These authors demonstrated that a remarkable geochemical continuum exists between concentrations of the major ions in the meltwater streams and the lakes of Taylor Valley, exemplified by the concentrations of Na^+ and Cl^- in Fig. 19.47. This relationship is also displayed by the concentrations of calcium, magnesium, and potassium as well as by the concentrations of sulfate and bicarbonate in each of the three watersheds in Taylor Valley (i.e., Lake Bonney, Lake Hoare, and Lake Fryxell). The geochemical continuum exists even though the concentrations of the ions range over six orders of magnitude, from the very low concentrations of the ions in glacial ice to higher concentrations in the meltwater streams, to the highest concentrations in lakes Hoare, Fryxell, and Bonney. The brine discharged at Blood Falls (Taylor Red Cone) surpasses even the bottom brine of Lake Bonney in terms of its concentrations of major ions. This water may be a cryogenic brine that formed by progressive freezing of glacial meltwater at the base of the ice sheet that is the source of the Taylor Glacier.

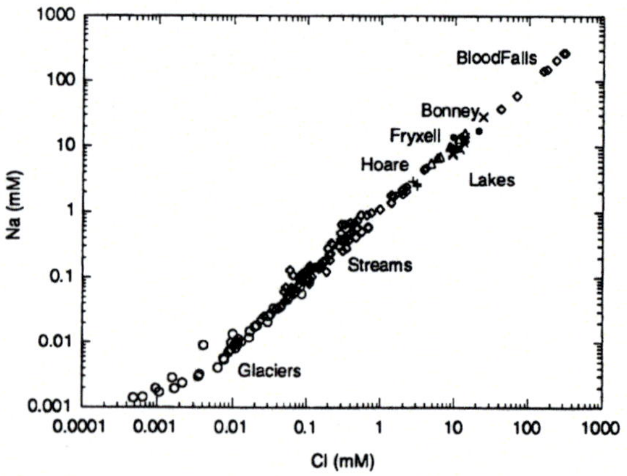

Fig. 19.47 A log-log plot of the concentrations of sodium and chloride ions (in millimoles per liter) of glacial ice, meltwater streams, saline lakes, and the brine discharged by the Blood Falls of the Taylor Glacier form a quasi-linear array. The ions of the other major elements form similar arrays. The coherence of these data is caused by the complex processes that progressively increase the concentrations of the water in the streams and in the lakes of Taylor Valley (Adapted from Lyons et al. 1998)

The chloride concentrations of the meltwater streams analyzed by Lyons et al. (2002) range widely from 0.064 to 2.210 millimoles (mM)/L. The highest chloride concentrations (0.875 ± 0.786 mM) occur in the tributary streams that flow into Lake Bonney which is located about 30 km from the coast. Surprisingly, the streams flowing into Lake Fryxell, which is located only 6.5 km from the coast, have relatively low chloride concentrations (0.278 ± 0.186 mM). Presumably, the soil in the watershed of Lake Bonney contains a residue of sodium chloride that was deposited during previous high stands of the Lake (Craig et al. 1974). In fact, Lyons et al. (1998) reported that the brine discharged by the Taylor Red Cone and the water in Lake Bonney have higher concentrations of sodium and chloride than lakes Fryxell and Hoare (McCraw 1967).

The chloride ion in the streams and lakes of the ice-free valleys is "conservative" which means that it accumulates in the lakes because the chloride salts of most major as well as trace metals are highly soluble. Consequently, the flux of chloride ions that enter a lake annually and the total inventory of chloride in the lake can be used to calculate the "chloride age" of the lake which is defined by:

$$t = \frac{\textit{Total inventory of chloride in kg}}{\textit{Flux in kg/y}}$$

For example, if the inventory of chloride in a lake is 25,000 kg and the annual discharge is 10 kg/year, the chloride age (t) is:

$$t = \frac{25{,}000\,kg}{10\,kg/y} = 2500y$$

19.7.5.2 Chemical Weathering

The isotope ratios of strontium reported by Jones and Faure (1967, 1968, 1978) and by Jones (1969) indicated unequivocally that strontium in water-soluble soil salts and in solution in meltwater streams and closed-basin lakes in Wright and Taylor valleys originated primarily by chemical weathering of silicate minerals in spite of the apparent scarcity of water and the low average annual temperatures in the ice-free valleys. This conclusion has been repeatedly confirmed by Green and Canfield (1984), Green et al. (1986a, b, 1988), by Lyons and Mayewski (1993), Lyons et al. (1997a, b, 1998a, 2002), and by Nezat et al. (2001).

The importance of chemical weathering in the geochemistry of the ice-free valleys of southern Victoria Land is also indicated by the presence of bicarbonate ions and of molecular silicic acid (H_4SiO_4) in the streams and lakes of this area. These constituents are released into solution in surface water by the incongruent dissolution of alumino-silicate minerals such as microcline:

$$\begin{aligned} &2KAlSi_3O_8 + 9H_2O + 2H^+ \rightleftarrows \\ &\underset{\text{kaolinite}}{Al_2Si_2O_5(OH)_4} + 2K^+ + 4H_4SiO_4 \end{aligned} \quad (19.1)$$

Reactions of this kind are by far the most important source of silicic acid in solution in natural waters because the solubility of quartz (SiO_2) is very low. The hydrogen ions (H^+) that are consumed by this reaction originate by dissociation of carbonic acid which forms when carbon dioxide of the atmosphere dissolves and reacts with molecules of water:

$$CO_2 + H_2O \rightleftarrows H_2CO_3 \quad (19.2)$$

Carbonic acid in aqueous solutions dissociates to form the bicarbonate ion (HCO_3^-) and a hydrogen ion (H^+).

$$H_2CO_3 \rightleftarrows HCO_3^- + H^+ \quad (19.3)$$

By combining reactions 19.2 and 19.3 we obtain:

$$CO_2 + H_2O \rightleftarrows HCO_3^- + H^+ \quad (19.4)$$

Solving this equation for H^+ and substituting into Eq. 19.1 yields:

$$2KAlSi_3O_8 + 2CO_2 + 11H_2O \rightleftarrows$$
$$Al_2Si_2O_5(OH)_4 + 2K^+ + 2HCO_3^- + 4H_4SiO_4 \quad (19.5)$$

Equation 19.5 demonstrates why the presence of bicarbonate ions and molecular silicic acid in the streams of Taylor Valley is evidence that chemical weathering of feldspar and other silicate minerals is occurring in spite of the harsh climatic conditions. The only prerequisite to chemical weathering of minerals on the surface of the Earth as well as in the subsurface is the presence of liquid water. The work of Nezat et al. (2001) has demonstrated that silicic acid and bicarbonate ions are present in all meltwater streams in the watersheds of Lake Bonney, Lake Hoare, and Lake Fryxell.

The consumption of atmospheric carbon dioxide by chemical weathering of silicate minerals counteracts global warming caused by increases in the concentration of carbon dioxide in the atmosphere and sets up a feed-back mechanism that causes cyclical variations in the global climate. Nezat et al. (2001) explained it this way:

> high atmospheric concentrations of CO_2 increase global temperatures due to the greenhouse effect. As temperatures rise, the rate of silicate weathering increases, consuming CO_2. As more CO_2 is consumed, atmospheric concentrations of CO_2 decrease, resulting in cooler temperatures. At lower temperatures, the rate of silicate weathering decreases, allowing a buildup of CO_2 in the atmosphere, and the cycle begins again.

The rate of chemical weathering is increased by elevated environmental temperatures and high rates of meteoric precipitation. However, temperature and precipitation are not the only parameters that affect the rate of chemical weathering. Other factors that come into play in certain environments are frost action and glacial grinding, both of which increase the surface areas of rocks and minerals that are exposed to water. In addition, the presence of salt in the soil and on rock surfaces accelerates chemical weathering in the ice-free valleys of Antarctica (Wellman and Wilson 1965; Miotke, 1980, 1981). Spectacular examples of salt weathering of granitic boulders are well known from Bull Pass between the Wright and Victoria valleys (Porter 1978; Green 2008), but occur also elsewhere in the Transantarctic Mountains. Figure 19.48 is an example of cavernous weathering of the Granite Harbor Intrusives in the Wisconsin Range.

Nezat et al. (2001) used the measured concentrations of silicic acid and bicarbonate ion as well as the total annual discharge of the tributary streams in Taylor Valley to calculate denudation rates of the channel of each meltwater stream in units of 10^3 mole/km^2/year. The average concentration of silicic acid in the meltwater streams was 48.0 μmole/L in 1994/95 and 40.9 μmole/L in 1995/96, but the average amount of the discharge increased from 32,322 m^3 in 1994/95 to 41,986 m^3 in 1995/96. The average denudation rate calculated by Nezat et al. (2001) for the stream channels of Taylor Valley is comparable to the silicic acid-based denudation rate of the Amazon and Mekong rivers. Nezat et al. (2001) concluded that:

> These high rates in Taylor Valley suggest that temperatures and precipitation are not the primary controls on chemical denudation. In this case, high stream discharge, high rates of physical weathering (e.g., frost action, salt weathering), and/or the interaction between the stream and hyporheic zone contribute to the high chemical denudation rates in the Taylor Valley stream channels.

The hyporheic zone is the region adjacent to and below the channel of a stream where the interstitial water is exchanging with the water flowing in the main channel of the stream.

19.7.5.3 Strontium Isotopes of Streams

The regional variation of $^{87}Sr/^{86}Sr$ ratios in the lakes and leached from soil of Taylor Valley reported by Jones and Faure (1968, 1978) in Fig. 19.45 was confirmed and elaborated and by Lyons et al. (2002) who measured the $^{87}Sr/^{86}Sr$ ratios of 15 meltwater streams identified in Fig. 19.46. The data are listed in Appendix 19.9.4 and are presented graphically in Fig. 19.49. The $^{87}Sr/^{86}Sr$ ratios of the meltwater streams in each drainage basin of Taylor Valley range both above and below the $^{87}Sr/^{86}Sr$ ratio of the respective lakes and thereby reveal the existence of differences in the ages

Fig. 19.48 Chemical weathering is enhanced by the presence of salt on the surface of large boulders and outcrops of granitic rocks that are exposed to the wind. The example in this picture was observed in a large outcrop of the Granite Harbor Intrusive in the Wisoconsin Range of the Horlick Mountains (Section 7.2) (Photo by J.M. Murtaugh)

and Rb/Sr ratios of the mineral grains in the soil. Nevertheless, the $^{87}Sr/^{86}Sr$ ratios of the tributary streams generally converge to the $^{87}Sr/^{86}Sr$ ratios of the lakes into which they flow.

The strontium in each meltwater stream in Taylor Valley is derived by chemical weathering of the silicate minerals in the soil and contains varying proportions of marine strontium that is transported into Taylor Valley by the wind in the form of snow flakes and atmospheric particulates. The $^{87}Sr/^{86}Sr$ ratios of the lakes in Fig. 19.49 decrease from west to east in Taylor Valley from Lake Bonney (0.71187, west; 0.71204, east) to Lake Hoare (0.71057), and to Lake Fryxell (0.70895). This trend is caused by the increasing abundance of marine strontium with decreasing distance from the coast that is also evident in Fig. 19.45. The elevated $^{87}Sr/^{86}Sr$ ratios in the lakes and tributaries in the central and western parts of Taylor Valley confirm that most of the strontium originated by chemical weathering of silicate minerals in the soil.

19.8 Summary

The geologic history of the Transantarctic Mountains comes to a close with the events of the late Cenozoic. These events are recorded by the glacial deposits in McMurdo Sound and on the high plateaus of the Transantarctic Mountains. The late Cenozoic was a time of change in the geology of Antarctica caused by concurrent uplift of the Transantarctic Mountains, continuing continental glaciation, and vigorous volcanic eruptions caused by the ongoing rift-tectonics that affected East and West Antarctica. As we approach the end of the Pleistocene and the start of the Holocene, the landscape of the Transantarctic Mountains begins to assume the look that we are familiar with.

The evolution of the landscape was especially profound in the valleys of southern Victoria Land which changed from fjords to "dry" valleys in which meltwater pooled in topographic basins and evolved into the ice-covered meromictic lakes and ponds we know

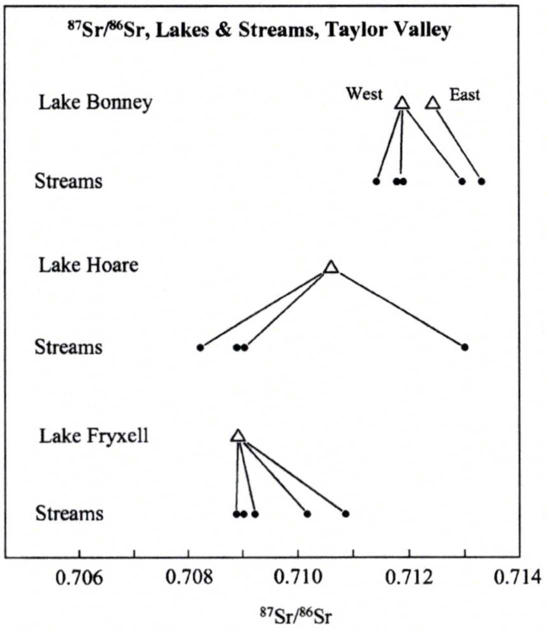

Fig. 19.49 The $^{87}Sr/^{86}Sr$ ratios of the meltwater streams that flow into lakes Bonney, Hoare, and Fryxell in Taylor Valley range widely. However, the $^{87}Sr/^{86}Sr$ ratios of the water in the lakes decrease from west to east because of the increasing abundance of marine strontium having $^{87}Sr/^{86}Sr = 0.70918 \pm 0.0003$ (Faure and Mensing 2005, p. 437) (Data from Lyons et al. 2002)

today. Most of theses changes have occurred in the last 3–4 million years, which is a short time in the context of about 750 million years of geologic history recorded by the Transantarctic Mountains; but it is a long time by our reckoning. Nevertheless, the geological events of the late Pliocene and Pleistocene have made Antarctica the kind of place it is today.

The specific topics that we discussed in this chapter include the study of Neogene sediment in McMurdo Sound by means of drill cores with code names like CIROS, MSSTS, DVDP, CRP, and RISP. We then turned to the lodgment till deposited on Mount Sirius and on many other high places in the Transantarctic Mountains, including the Dominion Range at the head of the Beardmore Glacier. These glacial deposits reveal a history of the waxing and waning of the East Antarctic ice sheet as recorded by layers of massive till and interbedded glacial outwash deposits.

The till of the Sirius Group in the Dominion Range and elsewhere contains reworked marine diatoms of Cenozoic age ranging down to middle Pliocene. In addition, the deposits contain leaves and twigs of Nothofagus (southern beech). The existence of marine diatoms in glacial deposits at high elevations in the Transantarctic Mountains and the presence of unfossilized plant remains in these deposits in the Dominion Range must be integrated into a coherent geologic history of East Antarctica. Part of the challenge is to explain why the climate of the Dominion Range was warmer and more humid during the middle-to-late Pliocene than it is at the present time. The key to understanding this puzzle may be that the Nothofagus grew at a time when the Beardmore valley was a fjord.

The issue that concerns us in the geologic history of southern Victoria Land is to determine the time during the Cenozoic when the valleys were filled with the ice that gave them their present shape. The key to solving this puzzle is that the volcanic rocks that were erupted through small vents in the valleys were deposited on glaciated bedrock surfaces. Therefore, the valley-filling glaciation must have occurred prior to the eruption of lava flows and volcanic ash in the valleys. The insights gained by careful fieldwork and meticulous age determinations indicate that the valley-cutting glaciation occurred more than about 15 million years ago. In addition, Wright and Taylor valleys were invaded by ice from the coast when the Ross Ice Shelf was grounded because of a decrease in sea level.

The last chapter in the geologic history of the ice-free valleys is concerned with the geochemical evolution of the lakes and ponds several of which contain dense brines overlain by layers of dilute water under a permanent cover of ice. These lakes and the associated meltwater streams give the ice-free valleys the appearance of oases on a very cold continent buried under snow and ice. The water in the lakes is warmed by solar radiation that penetrates 4 m of permanent ice and up to 60 m of water in the case of Lake Vanda.

The availability of warm and nutrient-rich water in the lakes of the ice-free valleys enables algal matts and other unicellular organisms to grow in this environment. In addition, lichen composed of a symbiotic arrangement of fungi and algae have colonized porous sandstones in sunny (north-facing) places where solar heat can melt snow and thus provides water to sustain the endolithic lichen. The unexpected presence of living plants in southern Victoria Land and elsewhere along the coast of the Ross Sea adds biological phenomena to the concerns of the scientists who work in the Transantarctic Mountains.

19.9 Appendices

19.9.1 $^{40}Ar/^{39}Ar$ Dates Derived from Volcanic Ash in Arena Valley and from the Western Asgard Range, Southern Victoria Land (Marchant et al. 1993a, b)

Mineral and number ( )	Setting	$^{40}Ar/^{39}Ar$ date (Ma)	Method[a]
Arena Valley			
Sanidine (18) (anorthoclase)	Ash on buried desert pavement	4.33 ± 0.07	Weighted mean[b]
Sanidine (3)	Avalanche deposit, upper Arena Valley	6.37 ± 0.16	Weighted mean
Glass (6)	Same	7.38 ± 0.14	Plateau
Sanidine (5)	Same	7.42 ± 0.04	Weighted mean
Sanidine (3)	Avalanche deposit, lower Arena Valley	11.28 ± 0.05	Weighted mean
Glass (5)	Same	12.90 ± 0.06	Plateau
Sanidine (5)	Same	8.50 ± 0.33	Plateau
Sanidine (5)	Ash disseminated in Quartermain I	11.27 ± 0.12	Weighted mean
Sanidine (9)	Same	11.59 ± 0.07	Weighted mean
Western Asgard Range			
Glass (6)	Ash-filled sand-wedges	10.08 ± 0.17	Plateau
Sanidine (5)	Same	12.07 ± 0.13	Weighted mean
Sanidine (5)	Same	10.12 ± 0.03	Weighted mean
Glass (6)	Same	13.56 ± 0.69	Weighted mean
Sanidine (5)	Same	13.65 ± 0.06	Weighted mean
Glass (11)	Same	14.75 ± 0.03	Weighted mean
Sanidine (10)	Same	15.01 ± 0.02	Weighted mean
Glass (11)	Same	14.24 ± 0.05	Plateau
Sanidine (2)	Ash in rock glaciers	12.56 ± 0.11	Weighted mean
Glass (11)	Same	13.56 ± 0.69	Plateau
Sanidine (3)	Waterlain ash	14.55 ± 0.06	Weighted mean
Glass (3)	Same	15.08 ± 0.03	Weighted mean
Sanidine (4)	Ash overlying desert pavement	14.84 ± 0.28	Weighted mean
Glass (6)	Same	15.24 ± 0.07	Plateau

[a]Sanidines were analyzed by total fusion, whereas volcanic glass was dated by incremental heating.
[b]Samples containing old mineral contaminants were excluded from the weighted mean by the authors.

19.9.2 Names and Locations of 44 Present-Day Lakes and Ponds in southern Victoria Land and on Ross Island (Chinn 1993)

Name	Location
Victoria Valley	
Lake Vida	77°23′S, 161°57′E
Lake Webb	77°20′S, 160°52′E
Hour Glass Lake	77°21′S, 161°04′E
Lake Vashka	77°21′S, 161°11′E
Victoria Upper Lake	77°19′S, 161°35′E
Balham Lake	77°26′S, 160°57′E

Name	Location
Wright Valley	
Lake Vanda	77°32′S, 161°33′E
Lake Bull	77°32′S, 161°42′E
Lake Canopus	77°33′S, 161°31′E
Don Juan Pond	77°4′S, 161°11′E
Don Quixote Pond	
Lake Brownworth	77°26′S, 162°45′E

(continued)

19.9.2 (continued)

Name	Location
Taylor Valley	
Lake Bonney	77°43′S, 162°25′E
Mummy Pond	77°40′S, 162°39′E
Lake Popplewell[a]	
Lake Chad	77°38′S, 162°46′E
Lake Hoare	77°39′S, 165°51′E
Lake Fryxell	77°37′S, 163°11′E
Lake House	77°42′S, 161°24′E
Lake Joyce	77°43′S, 161°37′E
Koettlitz Glacier	
Lake Miers	78°06′S, 163°51′E
Lake Péwé	77°56′S, 164°18′E
Lake Colleen	78°02′S, 163°52′E
Hidden Lake[a]	
Lake Keyhole	78°08′S, 163°31′E
Ward Lake	78°10′S, 163°35′E
Alpha Lake[a]	
Lake Howchin[a]	
Walcott Lake[a]	

Name	Location
Trough Lake[a]	
Lake Penny	78°16′S, 163°12′E
Lake Porkchop	78°16′S, 163°08′E
Lake Morning	
Ross Island (Cape Royds)	
Blue Lake	77°32′S, 166°10′E
Cost Lake[a]	
Clear Lake	77°32′S, 166°09′E
Green Lake	77°33′S, 166°09′E
Ross Island (Cape Evans)	
Skua Lake	77°38′S, 166°25′E
Island Lake	77°38′S, 166°26′E
Ross Island (Cape Barne)	
Deep Lake	77°34′S, 166°13′E
Heart Lake	77°34′S, 166°14′E
Terrace Lake	77°34′S, 166°13′E
Sunk Lake	77°34S′, 166°13′E
Islet Lake[a]	

[a]Informal name.

19.9.3 Chemical Analyses of Brine in Don Juan Pond of Wright Valley, Southern Victoria Land (Compiled by Jones 1969)

Ion (ppm)	1	2	3	4	5
Ions					
Ca	70,000	114,000	108,000	123,900	130,000
Mg	~1,000	1,200	1,320	1,820	2,590
Sr	825.1	–	–	–	–
Na	17,250	11,500	10,900	4,110	2,160
Fe	–	23.7	–	–	–
Cl	–	212,000	190,300	229,400	250,000
SO_4	–	11	160	0	0
HCO_3	–	49	–	–	–
NO_3	–	12.7	–	–	–

1. Jones (1969); 2. Meyer et al. (1962); 3. Tedrow et al. (1963); 4. and 5. Torii and Ossaka (1965).

19.9.4 Reports Concerning the Geochemistry of Don Juan Pond in Wright Valley, Southern Victoria Land

Salts

Salt (%)	1	2	3	4a	4b	5
$CaCl_2$	80.37	90.42	90.23	94.89	95.42	96.32
$MgCl_2$	1.89	1.57	1.80	2.30	3.12	2.02
$SrCl_2$	0.432	–	–	–	–	–
NaCl	17.23	7.95	7.88	2.75	1.38	1.55
KCl	0.087	0.065	0.093	0.058	0.086	0.085

1. Jones (1969); 2. Meyer et al. (1962); 3. Tedrow et al. (1963); 4a, b. Yamagata et al. (1967); 5. Dec. 1964 average of four.

19.9.5 Isotope Compositions of Strontium in the Water of the Principal Tributary Streams in the Three Watersheds of Taylor Valley (Lyons et al. 2002)

Lake	$^{87}Sr/^{86}Sr$	Tributary	$^{87}Sr/^{86}Sr$
Lake Bonney	0.71187	Lawson Cr.	0.71330
west, 6 m[a]		Santa Fe Cr.	0.71182
		Lyons Cr.	0.71189
		Blood Falls	0.71146
Lake Bonney, east, 6 m	0.71204	Priscu St.	0.71333
Lake Hoare	0.71057	Anderson Cr.	0.71305
5 m		McKay Cr.	0.70823
		House Cr.	0.70891
		Wharton Cr.	0.70902
Lake Fryxell	0.70895	Lost Seal Cr.	0.70907
7 m		Green Cr.	0.71017
		Harnish Cr.	0.70894
		Delta Cr.	0.70908
		Canada St.	0.71085
		Von Guerard St.	0.70923

[a]Depth below the surface.

References

Aiken GR, McKnight DM, Wershaw RL, Miller L (1991) Evidence for the diffusion of aquatic fulvic acid from the sediment of Lake Fryxell. Antarctica. In: Baker R (ed) Organic substances and sediments. Lewis, Chelsea, MI, pp 77–88

Anderson DM, Gatto LW, Ugolini FC (1972) An Antarctic analog of Martian permafrost terrain. Antarctic J US 7(4):114

Angino EE, Armitage KB (1963) A geochemical investigation of lakes Bonney and Vanda, Victoria Land, Antarctica. J Geol 71:89–95

Angino EE, Armitage KB, Tash JC (1964) Physicochemical limnology of Lake Bonney, Antarctica. Limnol Oceanogr 9:207–217

Angino EE, Armitage KB, Tash JC (1965) A chemical limnological study of Lake Vanda, Victoria Land, Antarctica. Un Kansas Sci Bull 45(10):1097–1118

Armitage KB, House HB (1962) A limnological reconnaissance in the area of McMurdo Sound, Antarctica. Limnol Oceanogr 7:36–41

Armstrong RL (1978) K-Ar dating: Late Cenozoic McMurdo Volcanic Group and dry valley glacial history, Victoria Land, Antarctica. New Zealand J Geol Geophys 21(6):685–698

Armstrong RL, Hamilton W, Denton GH (1968) Glaciation in Taylor Valley, Antarctica, older than 2.7 million years. Science 159:187–188

Askin RA, Markgraf V (1986) Palynomorphs from the Sirius Formation, Dominion Range, Antarctica. Antarctic J US 21(5) 34–35

Barrera E (1989) Strontium isotope ages. In: Barrett PJ (ed) Antarctic Cenozoic history from the CIROS-1 Drill Hole, McMurdo Sound. DSIR Bull 245:151–152. Wellington, New Zealand

Barrett PJ (ed) (1986) Antarctic Cenozoic history from MSSTS-1 drill hole, McMurdo Sound. DSIR Bull 237. Wellington, New Zealand

Barrett PJ (eds) (1989) Antarctic Cenozoic history from the CIROS-1 drill hole, McMurdo Sound. DSIR Bull 245. Wellington, New Zealand

Barrett PJ (1998a) Studies from the Cape Roberts Project, Ross Sea, Antarctica. Scientific Report of CRP-1, overview. Terra Antartica 5(3):255–258

Barrett PJ (1998b) Chronology of CRP-1. Terra Antartica 5(3):681–682

Barrett PJ, McKelvey BC (1986) Stratigraphy. In: Barrett PJ (ed) Antarctic Cenozoic history from the MSSTS-1 Drill Hole, McMurdo Sound. DSIR Bull 237:9–52. Wellington, New Zealand

Barrett PJ, Powell RD (1981) Mid-Cenozoic glacial beds at Table Mountain, southern Victoria Land. In: Craddock C (ed) Antarctic geosciences. University of Wisconsin Press, Madison, WI, pp 1059–1068

Barrett PJ, Elston DP, Harwood DM, McKelvey BC, Webb PN (1987) Mid-Cenozoic record of glaciation and sea-level change from MSSTS-1 on the margin of the Victoria Land Basin, Antarctica. Geol 15:634–637

Barrett PJ, Hambrey MJ, Harwood DM, Pyne AR, Webb PN (1989) Synthesis. In: Barrett PJ (ed) Antarctic Cenozoic history from the CIROS-1 Drill Hole, McMurdo Sound. DSIR Bull 245:241–251. Wellington, New Zealand

Barrett PJ, Hambrey MJ, Robinson PR (1991) Cenozoic glacial and tectonic history from CIROS-1, McMurdo Sound. In: Thomson MRA, Crame JA, Thomson JW (eds) Geological evolution of Antarctica. Cambridge University Press, Cambridge, UK, pp 651–656

Barrett PJ, Adams CA, McIntosh WC, Swisher CC III, Wilson GS (1992) Geochronological evidence supporting Antarctic deglaciation three million years ago. Nature 359:816–818

Benoit RE, Hall CL Jr (1970) The microbiology of some dry valley soils of Victoria Land, Antarctica. In: Holdgate MW (ed) Antarctic ecology, vol. 2—. Academic, San Diego, CA, pp 697–701

Black RF (1976) Periglacial features indicative of permafrost: ice and soil wedges. Quat Res 6:3–26

Bockheim JG (1978) Soil weathering sequences in Wright Valley, Antarctica. Antarctic J US 13(4):36–39

Boger PD, Faure G (1988) The possible occurrence of volcanic ash in till from Victoria Land, Antarctica. Antarctic J US 23(5):29–30

Bohaty SM, Scherer RP, Harwood DM (1998) Quaternary diatom biostratigraphy and paleoenvironments of the CRP-1 drill core, Ross Sea, Antarctica. Terra Antartica 5(3):431–454

Bonani G, Friedmann EI, Ocampo-Friedmann R, McKay CP, Woelfli W (1988) Preliminary report on radiocarbon dating of cryptoendolithic microorganisms. Polarforschung 58:199

Boswell CR, Brooks RR, Wilson AT (1967) Some trace elements in lakes of McMurdo oasis, Antarctica. Geochim Cosmochim Acta 31:731–736

Bowser CJ, Rafter TA, Black RF (1970) Geochemical evidence for the origin of mirabilite deposits near Hobbs Glacier, Victoria Land, Antarctica. Mineral Soc Amer Spec Paper 3:261–272

Brady HT (1979) A diatom report on DVDP cores 3, 4A, 12, 14, 15, and other related surface sections. In: Nagata T (ed) Proceedings of the Seminar III on the DVDP, 1978. Memoir of National Institute of Polar Research, Tokyo, Japan, pp 165–175

Brady HT, Martin H (1979) Ross Sea region in the middle Miocene: A glimpse into the past. Science 203(4379): 437–438

Broecker WS, Thurber DL, Goddard J, Ku TL, Mathews RK, Mesolella KJ (1968) Milankovitch hypothesis supported by precise dating of coral reefs and deep-sea sediment. Science 159:297–300

Brook EJ, Kurz MD, Ackert R Jr, Denton GH, Brown ET, Raisbeck GM, Yiou F (1993) Chronology of Taylor Glacier advances in Arena Valley, Antarctica, using in-situ cosmogenic ^{3}He and ^{10}Be. Quat Res 39:11–23

Brown ET, Edmund JM, Raisbeck GM, Yiou F, Kurz MD, Brook EJ (1991) Examination of surface exposure ages of Antarctic moraines using in-situ produced ^{10}Be and ^{26}Al. Geochim Cosmochim Acta 55(8):2269–2284

Bull C (1962a) Gravity observations in the Koettlitz Glacier area, southern Victoria Land, Antarctica. New Zealand J Geol Geophys 5:810–819

Bull C (1962b) Quaternary glaciations in southern Victoria Land, Antarctica. J Glaciol 4(32):240–241

Bull C (1966) Climatological observations in ice-free areas of southern Victoria Land, Antarctica. In: Rubin MJ (ed) Studies in Antarctica meteorology. Antarctic Research Series, vol. 9. American Geophysical Union. Washington, DC, pp 177–194

Bull C (1972) Wright Valley: Geoscience workshop of a continent. Antarctic J US 7(6):225

Bull C, McKelvey BC, Webb PN (1962) Quaternary glaciations in southern Victoria Land, Antarctica. J Glaciol 4(31):63–78

Bull C, McKelvey BC, Webb PN (1964) Glacial benches in southern Victoria Land. J Glaciol 5(37):131–134

Burckle LH (1995) Diatoms in igneous and metamorphic rocks from Dronning Maud Land, East Antarctica: Improbable places, improbable surfaces. Antarctic J US 30(5):67–68

Burckle LH, Delaney JS (1999) Terrestrial microfossils in Antarctic ordinary chondrites. Meteorit Planet Sci 34:475–478

Burckle LH, Potter N Jr (1996) Pliocene-Pleistocene diatoms in Paleozoic and Mesozoic sedimentary and igneous rocks from Antarctica: A Sirius problem solved. Geol 24:235–238

Burckle LH, Wasell A (1995) Marine and freshwater diatoms in sediment pockets in igneous rocks from James Ross island, Antarctica. Antarctic J US 30(5):8–9

Burckle LH, Gayley RI, Ram M, Petit J-R (1988) Diatoms in Antarctica ice cores: Some implications for the glacial history of Antarctica. Geol 16(4):326–329

Burckle LH, Kellogg DE, Kellogg TB, Fastook JL (1997) A mechanism for emplacement and concentration of diatoms in glacigenic deposits. Boreas 26:55–60

Burton HR (1981) Chemistry, physics, and evolution of Antarctic saline lakes. Hydrobiologia 82:339–362

Bydder EL, Holdsworth R (1977) Lake Vanda (Antarctica) revisited. New Zealand J Geol Geophys 20(6):1003–1016

Calkin PE, Bull C (1967) Lake Vida, Victoria Valley, Antarctica. J Glaciol 6:833–836

Calkin PE, Bull C (1972) Interaction of the East Antarctic ice sheet, alpine glaciations, and sea level in the Wright Valley area, southern Victoria Land. In: Adie RJ (ed) Antarctic geology and geophysics. Universitetsforlaget, Oslo, Norway, pp 435–440

Calkin PE, Behling RE, Bull C (1970) Glacial history of Wright Valley, southern Victoria Land, Antarctica. Antarctic J US 5(1):22–27

Cameron RE, King J, David CN (1970) Microbiology, ecology, and microclimatology of soil sites in dry valleys of southern Victoria Land, Antarctica. In: Holdgate MW (ed) Antarctic ecology, vol. 2. Academic, San Diego, CA, pp 702–716

Campbell IB, Claridge GGC (1987) Antarctica: Soil, weathering processes, and environment. Developments in Soil Science, vol. 6. Elsevier, New York, pp 1–368

Canfield DE, Green WJ (1985) The cycling of nutrients in a closed-basin Antarctic lake: Lake Vanda. Biogeochemistry 1:233–256

Cape Roberts Science Team (1998) Background to CRP-1, Cape Roberts Project, Antarctica. Terra Antartica 5(1):1–30

Carlquist S (1987) Pliocene Nothofagus wood from the Transantarctic Mountains. ALISO 11(4):571–583

Cartwright K, Treves SB, Torii T (1974) Geology of DVDP 4, Lake Vanda, Wright Valley, Antarctica. Dry Valley Drilling Project (DVDP) Bull 3:49–74

Chinn TJ (1993) Physical hydrology of the Dry Valley lakes. In: Green WJ, Friedmann EI (eds) Physical and biogeochemical processes in Antarctic lakes. Antarctic Research Series, vol. 59. American Geophysical Union, Washington, DC, pp 1–51

Clapperton CM, Sugden DE (1990) Late Cenozoic glacial history of the Ross Embayment, Antarctica. Quat Sci Rev 9:253–272

Claridge GGC, Campbell IB (1968) Origin of nitrate deposits. Nitrate 217:428–430

Claridge GGC, Campbell IB (1977) The salts in Antarctic soils, their distribution and relation to soil processes. Soil Sci 123:337–384

Clough JW, Hansen BL (1979) The Ross Ice Shelf Project. Science 203(4379):433–434

Cooper AK, Davey FJ, Behrendt JC (1987) Seismic stratigraphy and structure of the Victoria Land Basin, western Ross Sea, Antarctica. In: Cooper AK, Davey FJ (eds) The Antarctic continental margin: Geology and geophysics of the Western Ross Sea. Circum-Pacific Council for Energy and Mineral Resources, Earth Science Series, 5B. Houston, TX, pp 27–76

Cooper AK, Davey FJ, Behrendt JC (1991) Structural and depositional controls on Cenozoic and (?) Mesozoic strata beneath the western Ross Sea. In: Thomson MRA, Crame JA, Thomson JW (eds) Geological evolution of Antarctica. Cambridge University Press, Cambridge, UK, pp 279–283

Craig JR, Fortner RD, Weand BL (1974) Halite and hydrohalite from Lake Bonney, Taylor Valley, Antarctica. Geol 2(8):389–390

Cutfield SK (1974) Hydrological aspects of Lake Vanda, Wright Valley, Victoria Land, Antarctica. New Zealand J Geol Geophys 17:645–657

Davey FJ (1987) Geology and structure of the Ross Sea region. In: Cooper AK, Davey JF (eds) The Antarctic continental margin: Geology and geophysics of the Western Ross Sea. Earth Science Series, vol. 5B. Circum-Pacific Council for Energy and Mineral Resources, Houston, TX, pp 1–16

Davey FJ, Christoffel DA (1984) The correlation of MSSTS-1 drillhole results with seismic reflection data from McMurdo Sound, Antarctica. New Zealand J Geol Geophys 27:405–412

David TWE, Priestley RE (1914) Glaciology, physiography, stratigraphy, and tectonic geology of south Victoria Land. Brit. Antarct. Exped. 1907–1909. Report Sci Invest, Geology 1:1–319

Davy BW, Alder G (1989) Seismic reflection surveys. In: Barrett PJ (ed) Antarctic Cenozoic history from the CIROS-1 Drillhole, McMurdo Sound. DSIR Bull 245:15–22

De Carlo EH, Green WJ (2002) Rare earths elements in the water column of Lake Vanda, McMurdo dry valleys, Antarctica. Geochim Cosmochim Acta 66(8):1323–1333

Denton GH, Armstrong RL (1968) Glacial geology and chronology of the McMurdo Sound region. Antarctic J US 3(4):99–101

Denton GH, Hughes TJ (eds) (1981) The last great ice sheets. Wiley, New York

Denton GH, Armstrong RL, Stuiver M (1970) Late Cenozoic glaciation in Antarctica: The record in the McMurdo region. Antarctica J US 5(1):15–21

Denton GH, Prentice ML, Kellogg DE, Kellogg TB (1984) Late Tertiary history of the Antarctic ice sheet: Evidence from the Dry Valleys. Geol 12:263–267

Denton GH, Sugden DE, Marchant DR, Hall BL, Wilch TI (1993) East Antarctica ice sheet sensitivity to Pliocene climate change from a dry valleys perspective. Geogr Ann 75A(4):155–204

Doran PT, Berger GW, Lyons WB, Wharton RA Jr, Davisson ML, Southon J, Dibb JE (1999) Dating Quaternary lacustrine sediments in the McMurdo Dry Valleys, Antarctica. Paleo Paleo Paleo 147:223–240

Dort W Jr, Dort DS (1970) Sodium sulfate deposits in Antarctica. Modern Geol 1:97–117

Downes MT, Howard-Williams C, Vincent WF (1986) Sources of nitrogen, phosphorus, and carbon in Antarctica streams. Hydrobiologia 134:215–225

Dreschhoff GAM, Zeller EJ (1994) 415-year Greenland ice-core record of solar proton events dated by volcanic eruptive episodes. Institute for Tertiary-Quaternary Studies, TERQUA Symp Ser 2:1–24

Eggers AJ (1976) The Scallop Hill Formation, Brown Peninsula, McMurdo Sound, Antarctica. BSc thesis, Victoria University, Wellington, New Zealand

Emiliani C (1992) Planet Earth; cosmology, geology, and the evolution of life and environment. Cambridge University Press, Cambridge, UK

Everett KR (1991) Soils of the Meserve Glacier area, Wright Valley, south Victoria Land, Antarctica. Soil Sci 112(6):425–438

Faure G, Felder RP (1981) Isotopic composition of strontium and sulfur in secondary gypsum crystals, Brown Hills, Transantarctic Mountains. J Geochem Explor 14: 265–270

Faure G, Mensing TM (2005) Isotopes; Principles and applications, Wiley, Hoboken, New Jersey

Faure G, Mensing TM (2007) Introduction to planetary science; the geological perspective. Springer, Dordrecht

Faure G, Taylor KS (1981) Provenance of some glacial deposits in the Transantarctic Mountains based on Rb-Sr dating of feldspars. Chem Geol 32:271–290

Faure G, Taylor KS, Mercer JH (1983) Rb-Sr provenance dates of feldspar in glacial deposits of the Wisconsin Range, Transantarctic Mountains. Geol Soc Amer Bull 94:1275–1280

Faure G, Hagen EH, Johnson KS, Strobel ML, Whiting KS (1995) Geological explorations of East Antarctica: Iron, manganese, and titanium in the heavy-mineral fractions of till in the Transantarctic Mountains. In: Elliot DH, Blaisdell GL (eds) Contributions to Antarctic Research IV. Antarctic Research Series, vol. 67. American Geophysical Union, Washington, DC, pp 19–31

Fleck RJ, Jones LM, Behling RE (1972) K-Ar dates of the McMurdo Volcanics and their relation to the glacial history of Wright Valley. Antarctic J US 7(6):245–246

Forbes RB, Ester DW (1964) Glaciation of Observation Hill, Hut Point Peninsula, Ross Island, Antarctica. J Glaciol 5:87–92

Friedman I, Rafter A, Smith GI (1995) A thermal, isotopic, and chemical study of Lake Vanda and Don Juan Pond, Antarctica. In: Elliot DH, Blaisdell GL (eds) Contributions to Antarctic Research IV. Antarctic Research Series, vol. 67—. American Geophysical Union, Washington, DC, pp 177–194

Friedmann EI (1977) Microorganisms in Antarctic desert rocks from dry valleys and Dufek Massif. Antarctic J US 12(4):26–29

Friedmann EI (1978) Melting snow in the dry valleys is a source of water for endolithic microorganisms. Antarctic J US 13(4):162–163

Friedmann EI (1982) Endolithic microorganisms in the Antarctic cold desert. Science 215:1045–1053

Friedmann EI (ed) (1993) Antarctic microbiology; Examining the microbial habitats and communities that play significant roles in the ecosystem of the Antarctic continent. Wiley, New York

Friedmann EI, Ocampo R (1976) Endolithic blue-green algae in the dry valleys: Primary producers in the Antarctic desert ecosystem. Science 193:1247–1249

Gallikowski CA, Hirsch P (1988) Preliminary characterization and identification of 1984/85 continental Antarctic soil micro-organisms of Linnaeus Terrace (altitude 1600 m, McMurdo Dry Valleys). Polarforschung 58:93

George A (1989) Sand provenance. In: Barrett PJ (eds) Antarctic Cenozoic history from the CIROS-1 drillhole, McMurdo Sound. DSIR Bull 245:159–168. Wellington, New Zealand

Gibson GW (1962) Geological investigations in southern Victoria Land, Antarctica: 8. Evaporite salts in the Victoria Valley region. New Zealand J Geol Geophys 5:361–374

Goldman CR, Mason DT, Hobbie JE (1967) Two Antarctic desert lakes. Limnol Oceanogr 12:295–310

Green WJ (2008) Water, ice, and stone; Science and memory on the Antarctic lakes. Bellevue Literary Press, New York

Green WJ, Canfield DE (1981) A partial geochemical analysis of the Onyx River. Antarctic J US 16(5):42–45

Green WJ, Canfield DE (1984) Geochemistry of the Onyx River (Wright Valley, Antarctica) and its role in the chemical evolution of Lake Vanda. Geochim Cosmochim Acta 48:2457–2467

Green WJ, Friedmann EI (eds) (1993) Physical and biochemical processes in Antarctic Lakes. Antarctic Research Series, vol. 59. American Geophysical Union, Washington, DC

Green WJ, Canfield DE, Lee GF, Jones RA (1986a) Mn, Fe, Cu, and Cd distribution and residence times in closed-basin Lake Vanda (Wright Valley, Antarctica). Hydrobiologia 134:237–248

Green WJ, Ferdelman RG, Gardner TJ, Varner LC, Angle MP (1986b) The residence times of eight trace metals in a closed-basin Antarctic lake: Lake Hoare. Hydrobiologia 134:249–255

Green WJ, Angle MP, Chave KE (1988) The geochemistry of Antarctic streams and their role in the evolution of four lakes in the McMurdo dry valleys. Geochim Cosmochim Acta 52:1265–1274

Green WJ, Ferdelman TG, Canfield DE (1989) Metal dynamics in Lake Vanda (Wright Valley, Antarctica). Chem Geol 76:85–94

Green WJ, Canfield DE, Shengsong Y, Chave KE, Ferdelman TG, Delanois G (1993) Metal transport and release processes in Lake Vanda: The role of oxide phases. In: Green WJ, Friedmann EI (eds) Physical and biogeochemical processes in Antarctic lakes. Antarctic Research Series, vol. 59. American Geophysical Union, Washington, DC, pp 145–163

Green WJ, Canfield DE, Nixon P (1998) Cobalt cycling and fate in Lake Vanda. In: Priscu JC (ed) Econsystem dynamics in a Polar desert: The McMurdo Dry Valleys, Antarctica. Antarctic Research Series, vol. 72. American Geophysical Union, Washington, DC, pp 205–215

Gressitt JL (1965) Ecological survey of land arthropods, Anvers Island, Antarctic Peninsula and entomological research in Antarctica. Bull Antarctic Proj Officer 6(7):23–24

Grew ES, Manton WI (1979) Archean rocks in Antarctica: 2.5 billion year uranium-lead ages of pegmatites in Enderby Land. Science 206:443–445

Grindley GW (1967) The geomorphology of the Miller Range, Transantarctic Mountains. With notes on the glacial history and neotectonics of East Antarctica. New Zealand J Geol Geophys 10(2):557–598

Grindley GW, Laird MG (1969) Geology of the Shackleton Coast. In: Craddock C (ed) Geological map of Antarctica. Folio 12, Sheet 15. American Geographical Society, New York

Gumbley JW, Wilson AT, Hendy CH, Nelson CS (1974) Sedimentology of shallow cores from Lake Vanda. Antarctic J US 9(4):135–137

Hagen EH (1988) Geochemical studies of Neogene till in the Transantarctic Mountains; Evidence for an extraterrestrial component. MSc thesis, The Ohio State University, Columbus, OH

Hagen EH (1995) A geochemical and petrological investigation of meteorite ablation products in till and ice of Antarctica. PhD dissertation, The Ohio State University, Columbus, OH

Hagen EH, Koeberl C, Faure G (1990) Extraterrestrial spherules in glacial sediment, Beardmore Glacier area, Transantarctic Mountains. In: Elliot DH (ed) Contributions to Antarctic Research I. Antarctic Research Series, vol. 50. American Geophysical Union, Washington, DC, pp 19–24

Hagen EH, Faure G, Buchanan D (1992) Recovery of extraterrestrial particles from the Lewis Cliff ice tongue using a passive collection system. Antarctic J US 27(5):31–33

Hall BL, Denton GH, Lux DR, JG Bockheim (1993) Late Tertiary Antarctica paleoclimate and ice-sheet dynamics inferred from surficial deposits in Wright Valley. Geogr Ann 75A(4):239–267

Halpern M (1970) Rubidium-strontium date of possibly 3 billion years for a granitic rock from Antarctica. Science 169:977–978

Hambrey MJ, Barrett PJ, Robinson PH (1989) Stratigraphy. In: Barrett PJ (ed) Antarctic Cenozoic History from the CIROS-1 Drill Hole, McMurdo Sound. DSIR Bull 245:23–48. Wellington, New Zealand

Harris HJ, Cartwright K (1982) Pressure fluctuations in an Antarctic Aquifer: The freight-train response to a moving rock glacier. In: Craddock C (ed) Antarctic earth science. University of Wisconsin Press, Madison, WI, pp 1139–1149

Harris HJ, Cartwright K, Torii T (1979) Dynamic chemical equilibrium in a polar desert pond: A sensitive index of meteorological cycles. Science 204:30–1303 (Erratum, Science 204:909)

Hart CP, Webb PN (1998) Amino acid based geochronology of Unit 3.1 (Quaternary), CRP-1, Victoria Land Basin, Antarctica. Terra Antartica 5(3):697–701

Harwood DM (1983) Diatoms from the Sirius Formation, Transantarctic Mountains. Antarctic J US 18(5):98–100

Harwood DM (1986a) Diatom biostratigraphy and paleoecology with a Cenozoic history of Antarctic ice sheets. PhD dissertation, The Ohio State University, Columbus, OH

Harwood DM (1986b) Recycled siliceous microfossils from the Sirius Formation. Antarctic J US 21(5):101–103

Harwood DM, Webb PN (1998) Glacial transport of diatoms in the Antarctic Sirius Group: Pliocene refrigerator. GSA Today 8(4):1, 4–8

Harwood DM, Barrett PJ, Edwards AR, Rieck HJ, Webb PN (1989) Biostratigraphy and chronology. In: Barrett PJ (ed) Antarctic Cenozoic History from the CIROS-1 Drill Hole, McMurdo Sound. DSIR Bull 245:231–239

Harwood DM, Bohaty SM, Scherer RP (1998) Lower Miocene diatom biostratigraphy of the CRP-1 drill core, McMurdo Sound, Antarctica. Terra Antartica 5(3):499–514

Heine AJ (1971) Seiche observations at Lake Vanda, Victoria Land, Antarctica. New Zealand J Geol Geophys 14:597–599

Heinrich MR (ed) (1976) Extreme environments: Mechanisms of microbial adaptation. Academic, New York

Hill RS, Harwood DM, Webb PM (1991) Last remnant of Antarctica's Cenozoic flora: Nothofagus of the Sirius Group, Transantarctic Mountains. Abstracts, Eighth International Gondwana Symposium, June 24–28, Hobart, Tasmania

Hirsch P, Hoffman B, Gallikowski CA, Mevs U, Siebert J, Sittig M (1988) Diversity and identification of heterotrophs from Antarctic rocks of the McMurdo Dry Valleys (Ross Desert). Polarforschung 58:261

Hoare RA, Popplewell KB, House DA, Henderson RA, Prebble WM, Wilson AT (1964) Lake Bonney, Taylor Valley, Antarctica: A natural solar energy trap. Nature 202:886–888

Hoare RA, Popplewell KB, House DA, Henderson RA, Prebble WM, Wilson AT (1965) Solar heating of Lake Fryxell, a permanently ice covered Antarctica lake. J Geophys Res 70:1555–1558

Hoehn RC, Parker BC, Paterson RA (1974) Toward an ecological model of Lake Bonney. Antarctic J US 9(6):297–300

Holm-Hansen O (1964) Isolation and culture of terrestrial and freshwater algae in Antarctica. Phycologia 4:44–51

Horowítz NH, Cameron RE, Hubbard JS (1972) Microbiology of the dry valleys of Antarctica. Science 176:242–245

Ishman SE, Rieck HJ (1992) A lake Neogene glaclo-enstatic record, victoria Land Basin margin, Antarctica. In, Kennett JP and Warnke DA, eds., The Antarctic Paleoenvironment: A Perspective on Global Change, Part 1. Antarctic Res. Ser. 60:327–348, Washington, DC

IUGS (2002) International stratigraphic chart. International Commission of Stratigraphy, UNESCO

Jones LM (1969) The application of strontium isotopes as natural tracers: The origin of the salts in the lakes and soils of southern Victoria Land, Antarctica. PhD dissertation, The Ohio State University, Columbus, OH

Jones LM (1973) Principal cation concentrations for a length profile of the Onyx River, Wright Valley. Antarctic J US 8(5):274–276

Jones LM, Faure G (1967) Origin of salts in Lake Vanda, Wright Valley, southern Victoria Land, Antarctica. Earth Planet Sci Lett 3:101–106

Jones LM, Faure G (1978) A study of strontium isotopes in lakes and surficial deposits of the ice-free valleys, southern Victoria Land, Antarctica. Chem Geol 22:107–120

Jones LM, Faure G (1968) Origin of the salts in Taylor Valley. Antarctic J US 3(5):177–178

Jones LM, Faure G, Montigny RJE (1967) Geochemical studies in Wright Valley. Antarctic J US 2(4):114

Jones LM, Carver RE, McSaveney ER, Tickhill T (1971) Sediment analyses of the beaches of Lake Vanda, Wright Valley. Antarctic J US 6(5):199–200

Jones LM, Whitney JA, Stormer JC Jr (1973) A volcanic ash deposit, Wright Valley. Antarctic J US 8(5):270–272

Kellogg DE, Kellogg TB (1984) Non-marine diatoms in the Sirius Formation. Antarctic J US 19(5):44–45

Kellogg DE, Kellogg TB (1996a) Glacial/interglacial variations in the flux of atmospherically transported diatoms in Taylor Dome ice core. Antarctic J US 31(2):68–70

Kellogg DE, Kellogg TB (1996b) Diatoms in South Pole ice: Implications for eolian contamination of Sirius Group deposits. Geology 24(2):115–118

Kellogg TB, Kellogg DE, Stuiver M (1990) Late Quaternary history of the southwestern Ross Sea: Evidence from the debris bands on the McMurdo ice shelf, Antarctica. In: Elliot DH (ed) Contributions to Antarctic Research I. Antarctic Research Series, vol. 50. American Geophysical Union, Washington, DC, pp 25–56

Kellogg TB, Burckle LH, Kellogg DE, Fastook JL (1997) A new mechanism for diatom emplacement and concentration in glacigenic deposits. Antarctic J US 32(5):29–30

Keys JR, Williams K (1981) Origin of crystalline, cold desert salts in the McMurdo region, Antarctica. Geochim Cosmochim Acta 45:2299–2309

Kim Y, McGinnis LD, Bowen RH (1986) The Victoria Land Basin: Part of an extended crustal complex between East and West Antarctica. In: Reflection seismology: The continental crust. Geodynamics Series, vol. 14. American Geophysical Union, Washington, DC, pp 323–330

Koeberl C, Hagen E (1989) Extraterrestrial spherules in glacial sediment from the Transantarctic Mountains, Antarctica. Structure, mineralogy, and chemical composition. Geochim Cosmochim Acta 53:937–944

Koeberl C, Hagen EH, Faure G (1988) Chemical composition and morphology of meteorite ablation spherules in Neogene till in the Dominion Range, Transantarctic Mountains. Lunar Planetary Science Conference, vol. 19(2). LPI, Houston, TX, pp 625–626

Koeberl C, Hagen EH, Faure G (1989) Extraterrestrial spherules from glacial sediment in Antarctica. Internal structure, mineralogy, and chemical composition. Lunar Planetary Science Conference, vol. 20(2). LPI, Houston, TX, pp 528–529

Koob DD, Leister GL (1972) Primary productivity and associated physical, chemical, and biological characteristics of Lake Bonney: a perennially ice-covered lake in Antarctica. Antarctic Research Series, vol. 20. American Geophysical Union, Washington, DC, pp 51–68

Kovach J, Faure G (1977) Sources and abundance of volanogenic sediment in piston cores from the Ross Sea, Antarctica. New Zealand J Geol Geophys 20(6):1017–1026

Kurasawa H (1978) Geochemistry of Ferrar dolerite dikes and Cenozoic volcanics in the Dry Valley region. DVDP Bull 8:34–38

Kyle PR (1981) Glacial history of the McMurdo Sound area as indicated by the distribution and nature of McMurdo Volcanic Group rocks. In: McGinnis LD (ed) Dry Valley Drilling Project. Antarctic Research Series, vol. 33. American Geophysical Union, Washington, DC, pp 403–412

Lavelle M (1998) Strontium-isotope stratigraphy of the CRP-1 drill hole, Ross Sea, Antarctica. Terra Antartica 5(3):691–696

Leckie RM, Webb P-N (1979) Scallop Hill Formation and associated Pliocene marine deposits of southern McMurdo Sound. Antarctic J US 14(5):54–56

Leckie RM, Webb P-N (1986) Late Paleogene to early Neogene foraminifera of the Deep Sea Drilling Project Site 270, Ross Sea, Antarctica. In: Kennett JP et al. (eds) Initial reports of the Deep Sea Drilling Project, vol. 90. US Government Printing Office, Washington, DC, pp 1093–1142

Lopez B (1988a) Trouble way down under; Exploiters threaten the eerie beauty of Antarctica. The Washington Post, March 27

Lopez B (1988b) Standing on the South Pole, It's too cold to snow, and any way you look is north. The Washington, Post, Sunday, March 27, C2

Lopez B (1988c) Informed by indifference; A walk in Antarctica. Harper's Magazine, May

Love FG, Simmons GM Jr, Parker BC, Wharton RA Jr, Seaburg KG (1983) Modern conophyton-like microbial mats discovered in Lake Vanda, Antarctica. Geomicrobiol J 3:33–48

Lyons WB, Finlay JC (2008) Biogeochemical processes in high-latitudes lakes and rivers. In: Vincent WF, Laybourn-Parry J (eds) Limnology of Arctic and Antarctic aquatic ecosystems. Oxford University Press, New York, pp 137–156

Lyons WB, Mayewski PA (1993) The geochemical evolution of terrestrial water in the Antarctic: The role of rock-water interactions. In: Green WJ, Friedmann EI (eds) Physical and biochemical processes in Antarctic lakes. Antarctic Research

Series, vol. 59. American Geophysical Union, Washington, DC, pp 135–143

Lyons WB, Long DT, Hines ME, Gaudette HE (1984) The calcification of cyanobacterial mats in solar lake, Sinai. Geol 12:623–626

Lyons WB, Mayewski PA, Chorman FH Jr (1985) Trace metal concentrations in sediments from two closed-basin lakes, Antarctica. Chem Geol 48:265–270

Lyons WB, Welch KA, Nezat CA, McKnight DM, Crick K, Toxey JK, Mastrine JA (1997a) Chemical weathering rates and reactions in the Lake Fryxell basin Taylor Valley: Comparison to temperate river systems. In: Lyons WB, Howard-Williams C, Hawes I (eds) Ecosystem processes in Antarctic ice-free landscapes. Balkema, Rotterman, The Netherlands, pp 147–154

Lyons WB et al. (eds) (1997b) Ecosystem processes in Antarctic ice-free landscapes. Balkema, Rotterdam, The Netherlands

Lyons WB, Welch KA, Neumann K, Toxey JK, McArthur R, Williams C, McKnight DM, Moorhead D (1998a) Geochemical linkages among glaciers, streams, and lakes within Taylor Valley, Antarctica. In: Priscu JC (ed) Ecosystem dynamics in a Polar desert: The McMurdo Dry Valleys, Antarctica. Antarctic Research Series, vol. 72. American Geophysical Union, Washington, DC, pp 77–92

Lyons WB, Tyler SW, Wharton RA Jr, Vaughn B, McKnight DM (1998b) A late Holocene desiccation of Lake Hoare and Lake Fryxell, McMurdo Dry Valleys, Antarctica. Antarctic Sci 10:245–254

Lyons WB, Nezat CA, Benson LV, Bullen TD, Graham EY, Kidd J, Welch KA, Thomas JM (2002) Strontium isotopic signatures of the Taylor Valley lakes, southern Victoria Land, Antarctica: Chemical weathering in a polar climate. Aquat Geochem 8:75–95

Malin MC (1984) Abrasion rate observations in Victoria Valley, Antarctica: 340 day experiment. Antarctic J US 19(5):14–16

Marchant DR, Denton GH, Swisher CC III (1993a) Miocene-Pliocene-Pleistocene glacial history of Arena Valley, Quartermain Mountains, Antarctica. Geogr Ann 75A(4):269–330

Marchant DR, Denton GH, Sugden DE, Swisher CC III (1993b) Miocene glacial stratigraphy and landscape evolution of the western Asgard Range, Antarctica. Geogr Ann 75A(4):303–330

Masuda N, Nishimura M, Torii T (1982) Pathway and distribution of trace elements in Lake Vanda, Antarctica. Nature 298:154–156

Matsubaya D, Sakai H, Torii T, Burton H, Kerry K (1979) Antarctic saline lakes: Stable isotopic ratios, chemical compositions and evolution. Geochim Cosmochim Acta 43:7–25

Matsumoto G, Torii T, Hanya T (1984) Vertical distribution of organic constituents in an Antarctic lake: Lake Vanda. Hydrobiologia 111:119–126

Matsumoto GI (1993) Geochemical features of the McMurdo dry valley lakes, Antarctica. In: Green WJ, Friedmann EI (eds) Physical and biogeochemical processes in Antarctic lakes. Antarctic Research Series, vol. 59. American Geophysical Union, Washington, DC, pp 95–118

Mayewski PA (1972) Glacial geology near McMurdo Sound and comparison with the central Transantarctic Mountains. Antarctic J US 7(4):103–106

Mayewski PA (1975) Glacial geology and late Cenozoic history of the Transantarctic Mountains, Antarctica. Institute of Polar Studies, Report No. 56. The Ohio State University, Columbus, OH, pp 1–168

Mayewski PA, Goldthwait RP (1985) Glacial events in the Transantarctic Mountains: A record of the East Antarctic ice sheet. In: Turner MD, Splettstoesser JF (eds) Geology of the Central Transantarctic Mountains. Antarctic Research Series, vol. 36(12). American Geophysical Union, Washington, DC

McCraw JD (1962) Volcanic detritus in Taylor Valley, Victoria land, Antarctica. New Zealand J Geol Geophys 5(5):740–745

McCraw JD (1967) Soils of Taylor Dry Valley, Victoria Land, Antarctica, with notes on soils from other localities in Victoria Land. New Zealand J Geol Geophys 10:498–539

McDougall I (1977) Potassium-argon dating of glauconite from greensand drilled at Site 270 in the Ross Sea, DSDP Leg 28. In: Barker PF, Dalziel IWD, et al. (eds) Initial reports of the Deep Sea Drilling Project, vol. 36. Government Printing Office, Washington, DC, pp 1071–1072

McGinnis LD (ed) (1981) Dry Valley Drilling Project. Antarctic Research Series, vol. 33. American Geophysical Union, Washington, DC

McGinnis LD, Wilson DD, Burdelik WJ, Larson TH (1983) Crust and upper mantle study of McMurdo Sound. In: Oliver RL, James PR, Jago JB (eds) Antarctic earth science. Australian Academy of Science, Canberra, A.C.T., pp 204–208

McGinnis LD, Bowen RH, Erickson JM, Allred BJ, Kreamer JL (1985) East-West boundary in McMurdo Sound. Tectonophys 114:341–346

McGregor VR (1965) Notes on the geology of the area between the heads of the Beardmore and Shackleton glaciers, Antarctica. New Zealand J Geol Geophys 8(2):278–291

McIntosh WC (1998) $^{40}Ar/^{39}Ar$ geochronology of volcanic clasts and pumice in CRP-1 core, Cape Roberts, Antarctica. Terra Antartica 5(3):683–690

McIntyre DJ, Wilson GJ (1966) Preliminary palynology of some Antarctic Tertiary erratics. New Zealand J Bot 4(3):315–321

McKelvey BC, Mercer JH, Harwood DM, Stott LD (1984) The Sirius Formation: Further considerations. Antarctic J US 19(5):42–43

McKelvey BC, Webb P-N, Harwood DM, Mabin MCG (1991) The Dominion Range Sirius Group: A record of the late Pliocene-early Pleistocene Beardmore Glacier. In: Thomson MRA, Crame JA, Thomson JW (eds) Geological evolution of Antarctica. Cambridge University Press, Cambridge UK, pp 675–682

McKnight DM, Aiken GR, Andrews ED, Bowles ED, Harnish RA (1993) Dissolved organic material in dry valley lakes: A comparison of Lake Fryxell, Lake Hoare, and Lake Vanda. In: Green WJ, Friedmann EI (eds) Physical and biogeochemical processes in Antarctic lakes. Antarctic Research Series, vol. 59. American Geophysical Union, Washington, DC, pp 110–133

McSaveney MJ (1973) New data for a Cenozoic history of Wright Valley, Southern Victoria Land. Antarctic J US 8(5):266–268

McSaveney MJ, McSaveney ER (1972) A reappraisal of the Pecten glacial episode, Wright Valley, Antarctica. Antarctic J US 7(6):235–240

Mercer JH (1963) Glacial geology of Ohio Range, Central Horlick Mountains, Antarctica. Institute of Polar Studies, Report 8. The Ohio State University, Columbus, OH, pp 1–13

Mercer JH (1968) Glacial geology of the Reedy Glacier area, Antarctica. Geol Soc Amer Bull 79(4):471–485

Mercer JH (1972) Some observations on the glacial geology of the Beardmore Glacier area. In: Adie RJ (ed) Antarctic geology and geophysics. Universitetsforlaget, Oslo, Norway, pp 427–433

Mercer JH (1981) Tertiary terrestrial deposits of the Ross Ice Shelf area, Antarctica. In: Hambrey MJ, Harland WB (eds) Earth's Pre-Pleistocene glacial record. Cambridge University Press, Cambridge, UK

Meyer GH, Morrow MB, Wyss O, Berg TE, Littlepage JL (1962) Antarctica: The microbiology of an unfrozen saline pond. Science 138(3545):1103–1104

Meyer MA, Huang GH, Morris GJ, Friedmann EI (1988) The effect of low temperatures on Antarctic endolithic green algae. Polarforschung 58:113

Miller MF, Mabin MCG (1998a) Antarctic Neogene landscapes: In the refrigerator or in the deep freeze? GSA Today 8(4):1–3

Miller MF, Mabin MCG (1998b) Summary. GSA Today 8(4):8

Miotke F-D (1979a) Zur physikalischen Verwitterung im Taylor Valley, Victoria Land, Antarktis. Polarforschung 49(2):117–142

Miotke F-D (1979b) Die Formung und Formungsgeschwindigkeit von Windkantern in Victoria Land, Antarktis. Polarforschung 49(1):30–43

Miotke F-D (1980) Zur Salzsprengung und chemischen Verwitterung in den Darwin Mountains und den Dry Valleys, Victoria Land, Antarctis. Polarforschung 50(1):45–80

Miotke F-D (1981) Geomorphic processes in Victoria Valley. Antarctic J US 16(5):50–52

Montañez IP, Banner JL, Osleger DA, Borg LE, Bosserman PJ (1996) Integrated Sr isotope variations and sea-level history of Middle to Upper Cambrian platform carbonates; Implications for the evolution of Cambrian seawater $^{87}Sr/^{86}Sr$. Geology 24:917–920

Morris EC, Holt HE, Mutch TA, Lindsay JF (1972) Mars analog studies in Wright and Victoria Valleys, Antarctica. Antarctic J US 7(4):113–114

Mudrey MG Jr (1974) A model for chemical evolution of surface water in Wright Valley. Dry Valley Drilling Project Bull 4:41–42

Mudrey MG, Shimp NF, Keighin CW, Oberts GL, McGinnis LD (1973) Chemical evolution of water in Don Juan Pond, Antarctica. Antarctic J US 8(4):164–166

Mudrey MG, Torii T, Harris H (1975) Geology of DVDP No. 13, Don Juan Pond, Wright Valley, Antarctica. DVDP 5:78–99

Neider C (1972) A walk in Taylor Valley. Antarctic J US 7(6):249–253

Neider C (1974) Edge of the world; Ross Island, Antarctica. A personal and historical narrative. Doubleday, Garden City, NY

Nelson CS, Wilson AT (1972) Bathymetry and bottom sediments of Lake Vanda, Antarctica. Antarctic J US 7(4):97–99

Nezat CA, Lyons WB, Welch KA (2001) Chemical weathering in streams of a polar desert Taylor Valley, Antarctica. Geol Soc, Amer Bull 113(11):1401–1408

Nichols RL (1962) Geology of Lake Vanda, Wright Valley, South Victoria Land, Antarctica. In: Wexler H, Rubin MJ, Caskey JE Jr (eds) Geophysical Monograph, vol. 7. American Geophysical Union, Washington, DC, pp 47–52

Nichols RL (1964) Present status of Antarctic glacial geology. In: Adie RJ (ed) Antarctic geology. North-Holland, Amsterdam, The Netherlands, pp 123–137

Nichols RL (1965) Antarctic interglacial features. J Glaciol 5(40):433–443

Nichols RL (1966) Geomorphology of Antarctica. In: Tedrow JCT (ed) Antarctic soils and soil-forming processes. Antarctic Research Series, vol. 8. American Geophysical Union, Washington, DC, pp 1–46

Nichols RL (1971) Glacial geology of the Wright Valley. In: Quam LO (ed) Research in the Antarctic. Amer Assoc Adv Sci, Pub 93:293–340. Washington, DC

Nichols RL (1982) Captain Scott and his last expedition (With an Addendum). In: Craddock C (ed) Antarctic geoscience(in Roman numbers). University of Wisconsin Press, Madison, WI, pp 23–27 (XXIII-XXVII)

Ocampo-Friedmann R, Meyer MA, Chen M, Friedmann EI (1988) Temperature response of Antarctic cryptoendolithic photosynthetic microorganisms. Polarforschung 58:121

Oliver RL (1963) The level of former glaciation near the mouth of Beardmore Glacier. In: Adie RJ (ed) Antarctic geology. North Holland, Amsterdam, The Netherlands, pp 138–142

Palmer RJ Jr, Friedmann EI (1988) Incorporation of inorganic carbon by Antarctic cryptoendolithic fungi. Polarforschung 58:189

Parker BC, Simmons GM Jr (1978) Ecosystem comparisons of oasis lakes and soils. Antarctic J US 13(4):168–169

Parker BC, Hoehn RC, Paterson RA (1973) Ecological model for Lake Bonney, southern Victoria Land, Antarctica. Antarctic J US 8(4):214–216

Parker BC, Heiskill LE, Thompson WJ, Zeller WJ, Zeller EJ (1978) Non-biogenic fixes of nitrogen in Antarctica and some ecological implications. Nature 271:651–652

Parker BC, Simmons GM, Seaburg KG, Cathey D, Allnut FCT (1982) Comparative ecology of plankton communities in seven Antarctic oasis lakes. J Plankton Res 4:271–286

Passchier S (2002) Provenance and depositional environments of Neogene Sirius Group deposits from the Shackleton and Beardmore Glacier areas, central Transantarctic Mountains. In: Gamble JA et al. (eds) Antarctica at the close of the Millennium. Royal Society New Zealand Bulletin, vol. 35. The Royal Society of New Zealand, Wellington, New Zealand, pp 309–318

Pederson DR, Montgomery GE, McGinnis LD, Ervin CP, Wong HK (1981) Aeromagnetic survey of Ross Island, McMurdo Sound and the Dry Valleys. In: McGinnis LD (eds) Dry Valley Drilling Project. Antarctic Retearch Series, vol. 33. American Geophysical Union, Washington, DC, pp 7–27

Péwé TL (1959) Sand-wedge polygons (tessellations) in the McMurdo Sound region, Antarctica; A progress report. Amer J Sci 257:545–552

Porter E (1978) Antarctica. Dutton, New York

Prentice ML (1982) Surficial geology and stratigraphy in central Wright Valley, Antarctica: Implications for Antarctic Tertiary glacial history. MSc thesis, University of Maine, Orono, ME

Prentice ML, Denton GH, Burckle LH, Hodell DA (1987) Evidence from Wright Valley for the response of the Antarctic ice sheet to climate warming. Antarctic J US 22(4):56–58

Priscu JC (1998) (ed) Ecosystem dynamics in a polar desert: The McMurdo dry valleys, Antarctica. Antarctic Research Series, vol. 72. American Geophysical Union, Washington, DC

Ragotzkie RA, Likens GE (1964) The heat balance of two Antarctic lakes. Limnol Oceanogr 9:412–425

Ram M, Gayley RI, Petit J-R (1988) Insoluble particles in Antarctic ice: Background aerosol size distribution and diatom concentration. J Geophys Res 93(D7):8378–8382

Robinson ES (1962) Results of geophysical studies on the McMurdo to South Pole traverse. Research Report, vol. 62(6). Geophysical Polar Research Center., University of Wisconsin, Madison, WI, pp 1–49

Robinson ES (1963) Geophysical investigations in McMurdo Sound, Antarctica. J Geophys Res 68(1):257–262

Robinson ES (1964) Correlation of magnetic anomalies with bedrock geology in the McMurdo Sound area, Antarctica. J Geophys Res 62(2):4319–4326

Schäfer JM, Ivy-Ochs S, Wieler R, Leya I, Baur H, Denton GH, Schlüchter C (1999) Cosmogenic noble gas studies in the oldest landscape on Earth: surface exposure ages of the Dry Valleys, Antarctica. Earth Planet Sci Lett 167:215–226

Scherer RP (1991) Quaternary and Tertiary microfossils from beneath Ice Stream B: Evidence for a dynamic West Antarctic ice-sheet history. Paleo Paleo Paleo 90(4):395–412

Shaffer NR, Faure G (1976) Regional variation of $^{87}S/^{86}Sr$ ratios and mineral compositions of sediment from the Ross Sea, Antarctica. Geol Soc Amer Bull 87:1491–1500

Shaw J, Healy TR (1977 The formation of the Labyrinth, Wright Valley, Antarctica. New Zealand J Geol Geophys 29(5):933–948

Sheppard DS, Claridge GGC, Campbell IB, Lyon GL, Rogers KM (2002) The origin of nitrate and sulphate in Antarctic soils. In: Gamble JA et al. (eds) Antarctica at the close of the Millennium. Roy Soc New Zealand Bull 35:437–441

Shirtcliffe TGL (1964) Lake Bonney, Antarctica: Cause of the elevated temperatures. J Geophys Res 69:5257–5268

Shirtcliffe TGL, Benseman RF (1964) A sun-heated Antarctic lake. J Geophys Res 69:3355–3359

Simmons GM Jr, Parker BC, Allnutt FCT, Brown DP, Cathey DD, Seaburg KG (1979) Ecosystem comparisons of oasis lakes and soils. Antarctic J US 14(5):181–183

Simmons GM Jr, Vestal JR, Wharton RA Jr (1993) Environmental regulators of microbial activity in continental Antarctic lakes. In: Green WJ, Friedmann EI (eds) Physical and biochemical processes in Antarctic lakes. Antarctic Research Series, vol. 59. American Geophysical Union, Washington, DC, pp 165–196

Smith GJ (1965) Evaporite salts from the dry valleys of Victoria Land, Antarctica. New Zealand J Geol Geophys 8:381–382

Speden IG (1962) Fossiliferous Quaternary marine deposits in the McMurdo Sound region, Antarctica. New Zealand J Geol Geophys 5:746–777

Stern TA (1984) A seismic refraction survey near the Dailey Islands, southwestern McMurdo Sound, Antarctica. New Zealand Dept Scient Indust Res, Geophys Div, Rept 198:1–32

Stonehouse B (ed) (2002) Encyclopedia of Antarctica and the southern oceans. Wiley, Chichester, UK

Stott LD, McKelvey BC, Harwood DM, Webb P-N (1983) A revision of the ages of Cenozoic erratics at Mount Discovery and Minna Bluff, McMurdo Sound. Antarctic J US 18(5):36–38

Stroeven AP (1997) The Sirius Group of Antarctica: Age and environment. In: Ricci CA (ed) The Antarctic region; Geological evolution and processes. Terra Antartica, Siena, Italy, pp 747–762

Stroeven AP, Burckle LH, Kleman J, Prentice ML (1998) Atmospheric transport of diatoms in the Antarctic Sirius Group: Pliocene deep freeze. GSA Today 8(4):1, 4–5

Stuiver M, Denton GH, Hughes TJ, Fastook JL (1981) History of the marine ice sheet in West Antarctica during the last glaciation: A working hypothesis. In: Denton GH, Hughes TJ (eds) The last great ice sheets. Wiley, New York, pp 319–436

Sugden DE, Marchant DR, Denton GH (eds) (1993) The case for a stable East Antarctic ice sheet. Geogr Ann 75A(4):151–351

Summerfield MA, Stuart FM, Cockburn HAP, Sugden DE, Denton GH, Dunai T, Marchant DR (1999) Long-term rates of denudation in the Dry Valleys, Transantarctic Mountains, Southern Victoria Land, Antarctica, based on in-situ-produced cosmogenic ^{21}Ne. Geomorphology 27:113–130

Swithinbank CW (1963) Ice movement in valley glaciers flowing into the Ross Ice Shelf, Antarctica. Science 141:523–524

Taylor KS, Faure G (1980) Rb-Sr dating of detrital feldspar: A new method to study till. J Geol 89:97–107

Tedrow JCF, Ugolini FC (1963) An Antarctic saline lake. New Zealand J Sci 6:150–156

Thompson TG, Nelson KH (1956) Concentration of brines and deposition of salts from sea water under frigid conditions. Amer J Sci 254:227–238

Torii T, Ossaka J (1965) Antarcticite: An new mineral, calcium chloride hexahydrate, discovered in Antarctica. Science 145(3687):975–977

Torii T, Yamagata N (1981) Limnological studies of saline lakes in the Dry Valleys. In: McGinnis LD (ed) Dry Valley Drilling Project. Antarctic Research Series, vol. 33. American Geophysical Union, Washington, DC, pp 141–159

Torii T, Yamagata N, Ossaka J (1976) Salt balance in the Don Juan basin. DVDP Bull 6:33–40

Turner RD (1967) A new species of fossil Chlamys from Wright Valley, McMurdo Sound, Antarctica. New Zealand J Geol Geophys 10(2):446–454

Vail P, Hardenbol J (1979) Sea-level changes during the Tertiary. Oceanus 22:71–79

Veizer J, Hoefs J (1976) The nature of $^{18}O/^{16}O$ and $^{13}C/^{12}C$ secular trends in sedimentary carbonate rocks. Geochim Cosmochim Acta 40:1387–1395

Vella PO (1969) Surficial geological sequence, Black Island and Brown Peninsula, McMurdo Sound, Antarctica. New Zealand J Geol Geophys 12:761–770

Vincent WF, Howard-Williams C (1989) Microbial communities in southern Victoria Land streams (Antarctica), II. The effects of low temperature. In: Vincent WF, Ellis-Evans JC (eds) High latitude limnology. Kluwer Academic, Boston, MA, pp 39–50

Vincent WF, Laybourn-Parry J (2008) Polar lakes and rivers; Limnology of Arctic and Antarctic aquatic ecosystems. Oxford University Press, New York

Vincent WF, Vincent CL (1982) Factors controlling phytoplankton production in Lake Vanda (77?S). Canadian J Fish Aquat Sci 39:1602–1609

Weand BL, Fortner RD, Hoehn RC, Parker BC (1975) Subterranean flow into Lake Bonney. Antarctic J US 10:15–19

Webb P-N (1972) Wright Fjord, Pliocene marine transgression of an Antarctic dry valley. Antarctic J US 7(6):225–232

Webb P-N (1974) Micropaleoecology, paleoecology, and correlation of the Pecten Gravels, Wright Valley, Antarctica, and descriptions of Trochoelphidiella onyxi n gen., n.sp. J Foram Res 4:184–199

Webb P-N (1981) Late Mesozoic-Cenozoic geology of the Ross Sector, Antarctica. J Roy Soc New Zealand 11(4):439–446

Webb P-N (1989) Benthic Foraminifera. In, Barrett PJ, ed. Antarctic Cenozoic History of the CIROS-1 Drillhole. New Zealand Dept. Sci. Indust. Res., Miscellaneous Bull. 245:99–118

Webb P-N, Harwood DM (1987) Terrestrial flora of the Sirius Formation: Its significance for late Cenozoic glacial history. Antarctic J US 22(4):7–11

Webb P-N, Harwood DM (1991) Late Cenozoic glacial history of the Ross Embayment, Antarctica. Quat Sci Rev 10:215–223

Webb P-N, Harwood DM (1993) Pliocene fossil *Nothofagus* (southern beech) from Antarctica: Phytogeography, dispersal strategies, and survival in high latitude glacial-deglacial environments. In: Alden J et al. (eds) Forest development in cold climates. Plenum, New York, pp 135–165

Webb P-N, Ronan TE Jr, Lipps JH, DeLaca TE (1979) Miocene glaciomarine sediments from beneath the southern Ross Ice Shelf, Antarctica. Science 203(4379):435–437

Webb P-N, Harwood DM, McKelvey BC, Mercer JH, Stott LD (1983). Late Neogene and older Cenozoic microfossils in high elevation deposits of the Transantarctic Mountains: Evidence for marine sedimentation and ice volume variation on the East Antarctic craton. Antarctic J US 18(5): 96–97

Webb P-N, Harwood DM, McKelvey BC, Mercer JH, Stott LD (1984) Cenozoic marine sedimentation and ice volume variation on the East Antarctic craton. Geol 12:287–291

Webb P-N, McKelvey BC, Harwood DH, Mabin MCG, Mercer JH (1987) Sirius Formation of the Beardmore region. Antarctic J US 22(1/2):8–13

Webb P-N, Harwood DM, Mabin MGC, McKelvey BC (1994) Late Neogene uplift of the Transantarctic Mountains in the Beardmore Glacier region. Terra Antartica 1(2):463–467

Webster JG (1994) Trace metal behavior in oxic and anoxic Ca-Cl brine of the Wright Valley, Antarctica. Chem Geol 112:255–274

Webster JG, Goguel RL (1988) Anions and alkali metals in Lake Vanda, Don Juan Pond, and the Onyx River: Further indications of brine origin. Antarctic J US 23:154–156

Welch KA, Lyons WB, Graham E, Neumann K, Thomas JM, Mikesell D (1996) Determination of major element chemistry in terrestrial waters from Antarctica by ion chromatography. J Chromatography A 73:257–263

Wellman HW, Wilson AT (1965) Salt weathering, a neglected geological erosive agent in coastal and arid environments. Nature 205:1097–1098

White P (1989) Downhole logging. In: Barrett PJ (ed) Antarctic Cenozoic history from the CIROS-1 Drill Hole, McMurdo Sound. DSIR Bull 245:7–14. Wellington, New Zealand

Wight S (1995) Description, chemical composition, and origin of six microscopic spherules collected from the "Meteorite Moraine" in Antarctica; An SEM study. BSc thesis, Department of Geological Sciences, The Ohio State University, Columbus, OH

Wilch TI, Lux DR, McIntosh WC, Denton GH (1989) Plio-Pleistocene uplift of the McMurdo dry valley sector of the Transantarctic Mountains. Ant J US 24(5):30–33

Wilch TI, Lux DR, Denton GH, McIntosh WC (1993a) Minimal Pliocene-Pleistocene uplift of the dry valleys sector of the Transantarctic Mountains: A key parameter in ice-sheet reconstructions. Geol 21:841–844

Wilch TI, Denton GH, Lux DR, McIntosh WC (1993b) Limited Pliocene glacier extent and surface uplift in middle Taylor Valley, Antarctica. Geogr Ann 75A(4):331–350

Wilson AT (1964) Evidence from chemical diffusion of a climate change in the McMurdo dry valleys 1200 years ago. Nature 201:176–177

Wilson AT (1979) Geochemical problems of the Antarctic dry areas. Nature 280:205–208

Wilson AT (1981) A review of the geochemistry and lake physics of the Antarctic dry areas. In: McGinnis LD (ed) Dry Valley Drilling Project. Antarctic Research Series, vol. 33. American Geophysical Union, Washington, DC, pp 185–192

Wilson AT, House DA (1965) Chemical composition of South Polar snow. J Geophys Res 70:5515–5518

Wilson AT, Wellman HW (1962) Lake Vanda: An Antarctic lake. Nature 196:1171–1173

Wilson AT, Holdsworth R, Hendy CH (1974) Lake Vanda: Source of heating. Antarctic J US 9:137–138

Wilson GL (1967) Some new species of lower Tertiary dinoflagellates from McMurdo Sound, Antarctica. New Zealand J Botany 5(1):57–83

Wilson GS, Harwood DM, Askin RA, Levy RH (1998) Late Neogene Sirius Group strata in Reedy Valley, Antarctica: a multiple-resolution record of climate, ice-sheet, and sea-level events. J Glaciol 44:437–447

Wong HK, Christoffel DA (1981) A reconnaissance seismic survey of McMurdo Sound and Terra Nova Bay, Ross Sea. In: McGinnis LD (ed) Dry Valley Project. Antarctica Research Series, vol. 33. American Geophysical Union, Washington, DC, pp 37–62

Yamagata N, Torii T, Murata S (1967) Report of the Japanese summer parties in the dry valleys, Victoria Land, 1963–1965. V. Chemical composition of lake waters. Antarctic Rec (Japan) 29:2339–2361

Yoshida Y, Torii T, Yusa Y, Nakaya S, Moriwaki K (1975) A limnological study of some lakes in the Antarctic. Bull Roy Soc New Zealand 13:311–320

Chapter 20
Antarctica in Retrospect

Antarctica is a place that does not give up her secrets easily. A casual observer might be impressed by the natural beauty of the landscape or by the brutally cold climate, depending on the weather at the time. Geologists who come to study the Transantarctic Mountains are initially overwhelmed by the Niagara of information that confronts them. This is true of mountainsides of exposed granitic gneisses of the Granite Harbor Intrusives, of whole mountain ranges composed of flat-lying sedimentary rocks of the Beacon Supergroup which attains a thickness of 3 km, not to mention the profusion of massive sills of Ferrar Dolerite that intruded the sedimentary rocks, and the thick plateaus of Kirkpatrick Basalt that was poured out onto the eroding surface of the Beacon rocks. How did all these rocks form on the edge of the East Antarctic craton? Whatever geological activity is recorded by the rocks of the Transantarctic Mountains took place in a tectonic setting of continental rifting, compressive subduction, and amalgamation of continental fragments, all of which occurred during 585 million years of geologic time between 750 and 165 Ma.

Starting during the Jurassic Period, the supercontinent of Gondwana which had sheltered the East Antarctic craton, began to break up into large crustal fragments that slowly moved away and became the continents of the southern hemisphere. East Antarctica and smaller crustal fragments were set free and began a slow journey south, which in time caused ice caps to accumulate on the highest mountains. These ice caps gradually expanded into the continental ice sheets of East and West Antarctica. In addition, volcanic activity broke out in mid-Cenozoic time causing Antarctica to become a continent of "fire and ice". At about the same time during the late Oligocene-to-early Miocene, the East Antarctic ice sheet sculpted the peaks and valleys of the Transantarctic Mountains and prepared the stage for the geologic, geochemical, and biological activity of present-day Antarctica.

It took hundreds of scientists to figure all this out in the decades since the IGY which, among other things, led to the creation of an international consortium of nations that have agreed to preserve Antarctica for peaceful purposes. The Antarctic Treaty has created a legal environment that continues to permit scientific study of Antarctica and of the southern ocean around it. The Antarctic Treaty has also demonstrated that the nations that have joined the Treaty have abided by its terms because it is in their own best interest to do so. For this reason, the beneficial influence of the Antarctic Treaty extends far beyond the Antarctic Circle to the governance of the world's seafloor and to the rules that apply to the exploration and uses of planets and asteroids of the solar system.

G. Faure and T.M. Mensing, *The Transantarctic Mountains: Rocks, Ice, Meteorites and Water*,
DOI 10.1007/978-90-481-9390-5_20,

Author Index

G

H

L

M

Y

Z

Subject Index

The subject matter in this index is divided into several categories in order to separate the geological information from the geographic features and to provide space for inclusion of human activities in both arenas. The items in the geographic categories of the Subject Index are listed in alphabetical sequence, whereas the Summaries of Geology are arranged in stratigraphic order. We believe that this arrangement will facilitate the retrieval of information by our readers. The categories are identified by the subheadings which mirror the organization of this book:

1. Abbreviations
2. Topographic features
3A. Human activities
3B. Special topics
4. Glaciers and ice shelves
5. Mountain ranges
6. High mountain peaks

Summary of Geology
7. Ross orogen (The Basement Complex)
8. Beacon Supergroup
9. Special Topics (Beacon Supergroup)
10. Ferrar Group: Kirkpatrick Basalt
11. Ferrar Group: Dolerite Sills
12. Kirwan Volcanic Group
13. Continental tectonics
14. McMurdo Volcanic Group
15. East Antarctic ice sheet
16. Meteorites
17. The Sirius Group of Neogene till
18. Lakes and streams in ice-free valleys
19. Science briefs

1. ABBREVIATIONS

AGU = American Geophysical Union
ANSMET = Antarctic search for meteorites
AP = Antarctic Peninsula
BFC = Berg Field Center, Mc Murdo Station
BPRS = Byrd Polar Research Center, The Ohio State University
CDC = Clothing Distribution Center, Christchurch, New Zealand
CFC = chloro-fluoro-carbon compounds
CTAM = Central Transantarctic Mountains
DJP = Don Juan Pond, Wright Valley
EA = East Antarctic
EAIS = East Antarctic ice sheet
EPICA = European Project of Ice Coring in Antarctica
GANOVEX = German Antarctic northern Victoria Land Expedition
HIMU = high "mu" ($^{238}U/^{204}Pb$) atomic ratio
ICSU = International Council of Scientific Unions
IDP = Interplanetary dust particles
IGY = International Geophysical Year, 1957-1958
IPS = Institute of Polar Studies, The Ohio State University
IUGS = International Union of Geological Sciences
FEM = Far Eastern Mountains
HBM = Harold Byrd Mountains, Queen Maud Mountains
HPP = Hut Point Peninsula, Ross Island
McMS = McMurdo Sound
MMI = Mammoth Mountain Inn, McMurdo Station
MMs = micrometeorites
MORB = mid-ocean ridge basalt
NSF = National Science Foundation
NVL = northern Victoria Land, TAM
NZ = New Zealand
OAE = Old Antarctic Explorer
OPP = Office of Polar Programs
QAR = Queen Alexandra Range, CTAM
QML = Queen Maud Land, EA
QMM = Queen Maud Mountains, TAM
RIS = Ross Ice Shelf
SCAR = Scientific Committee on Antarctic Research, ICSU
SNC = Shergotty, Nakhla, Chassigny (martian meteorites)
SSSI = Sites of Special Scientific Interest
SVL = southern Victoria Land, TAM
SWEAT = Southwest U.S. - East Antarctic Connection
TAM = Transantarctic Mountains
TNB = Terra Nova Bay
USAP = US Antarctic Program
WA = West Antarctica
WAIS = West Antarctic Ice Sheet

2. TOPOGRAPHIC FEATURES (excluding glaciers, mountain ranges, peaks, and volcanoes)

3A HUMAN ACTIVITIES

3B. Special Topics

4. GLACIERS AND ICE SHELVES

Unless otherwise indicated, the name of each feature is followed by the word "Glacier".

5. MOUNTAIN RANGES

Unless otherwise indicated, the name of each feature is followed by the word "Range".

6. HIGH MOUNTAIN PEAKS

Unless otherwise indicated, the name of each mountain is preceded by "Mt." or "Mount"

Geologic Time Scale

The geologic time scale is based on the International Stratigraphic Chart issued by the International Union of Geological Sciences (IUGS) in 2002

Period	Epoch	Age, Ma
	Cenozoic Era	
Quaternary	Holocene	0 –0.01
	Pleistocene	0.01– 1.81
Neogene	Pliocene	1.81 –5.32
	Miocene	5.32 –23.8
Paleocene	Oligocene	23.8 –33.7
	Eocene	33.7– 55.0
	Paleocene	55.0– 65.5
	Mesozoic Era	
Cretaceous	Late Cretaceous	65.6– 98.9
	Early Cretaceous	98.9–142.0
Jurassic	Late Jurassic	142.0–159.4
	Middle Jurassic	159.4–180.1
	Early Jurassic	180.1– 205.1
Triassic	Late Triassic	205.1– 227.4
	Middle Triassic	227.4– 241.7
	Early Triassic	241.7– 251.4
	Paleozoic Era	
Permian	Lopingian	(253.4)[a]
	Guadelupian	(265)[a]
	Cisuralian	(283)[a]
Carboniferous	Pennsylvanian	292– 320
	Mississippian	320– 354
Devonian	Late Devonian	354– 370
	Middle Devonian	370– 391
	Early Devonian	391– 417
Silurian	Peridoli	417– 419
	Ludlow	419– 423
	Wenlock	423–428
	Llandovery	428– 440
Ordovician	Late Ordovician	440– ?
	Middle Ordovician	(467.5)[a]
	Early Ordovician	?– 495
Cambrian	Late Cambrian	495– 500
	Middle Cambrian	500– 520
	Early Cambrian	520– 545

(continued)

(continued)

Eras	Precambrian Eons Periods	Age, Ma
	Proterozoic Eon	
Neoproterozoic	Neoproterozic III	540– 650
	Cryogenian	650–850
	Tonian	850– 1000
Mesoproterozoic	Stenian	1,000–1,200
	Ectasian	1,200–1,400
	Calymmian	1,400–1,600
Paleoproterozoic	Statherian	1,600 –1,800
	Orosirian	1,800 –2,050
	Rhyacian	2,050 – 2,300
	Siderian	2,300– 2,500
	Archean Eon	
Neoarchean	Not defined	2,500 –2,800
Mesoarchean	Not defined	2,800–3200
Paleoarchean	Not defined	3,200– 3600
Eoarchean	Not defined	3,600– ?

[a]The range of dates has not yet been determined.